Cancer Metastasis Through the Lymphovascular System

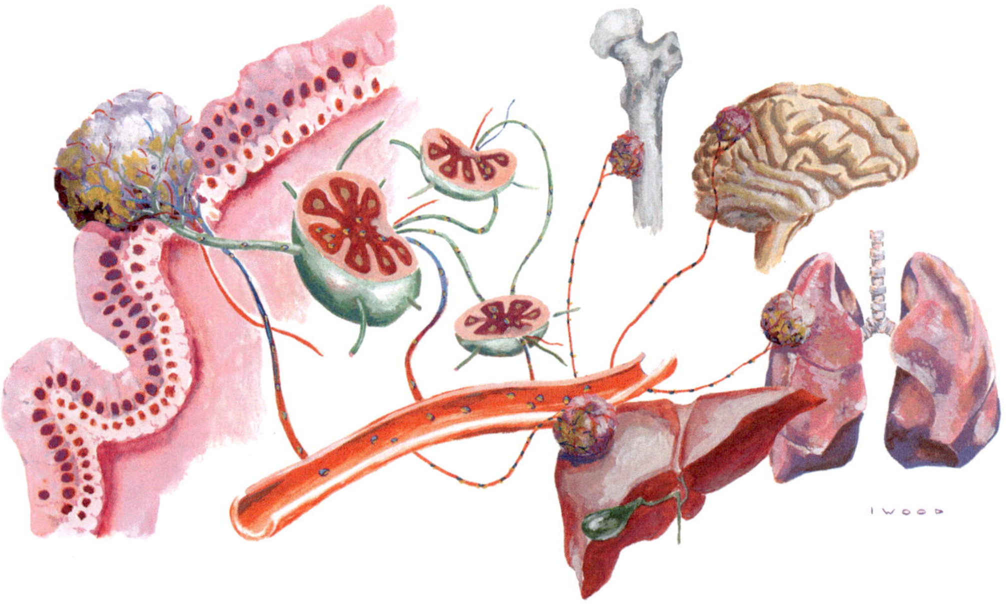

Cover illustration by Ian Wood, MD, resident in the general surgery program at the Henry Ford Health System (HFHS). The artwork for the book cover is supported by the Nathanson/Rands Chair in Breast Cancer Research of the Henry Ford Cancer Institute (HFCI).

Stanley P. Leong • S. David Nathanson
Jonathan S. Zager
Editors

Cancer Metastasis Through the Lymphovascular System

Springer

Editors
Stanley P. Leong
Department of Surgery and Melanoma Center
California Pacific Medical Center and Research Institute
San Francisco, CA, USA

S. David Nathanson
Henry Ford Cancer Institute
Detroit, MI, USA

Jonathan S. Zager
Moffitt Cancer Center
Tampa, FL, USA

ISBN 978-3-030-93083-7 ISBN 978-3-030-93084-4 (eBook)
https://doi.org/10.1007/978-3-030-93084-4

This Springer imprint is published by the registered company Springer Nature Switzerland AG
The registered company address is: Gewerbestrasse 11, 6330 Cham, Switzerland

This book is dedicated to our patients, who are the ones who bravely endure the fight with cancer, and to our colleagues and team members, who help treat cancer every day forging the relationships and fighting the good fight, although sometimes losing the battle. However, when the battle is won, the feeling of accomplishment, of helping someone in need, and victory over this vicious disease could be overwhelming. We want to thank all of our patients, all of our families for their continuous support and encouragement as we help in the fight against cancer, all of the researchers for laying down the groundwork to advance the field, and all of the clinical team members for helping to provide the treatments. Working all together, there is no limit to what can be achieved.

Stanley P. Leong, MD
S. David Nathanson, MD
Jonathan S. Zager, MD

We also dedicate this book to our colleague Dale Han, MD, who served tirelessly as a section editor for this book. Dr. Han graduated with a medical degree from the State University of New York at Stony Brook School of Medicine in 2003. Dr. Han did his general surgery residency at the University of Pennsylvania School of Medicine and his fellowship at Moffitt Cancer Center in Tampa. Dr. Han dedicated his life to treating melanoma and other skin and soft-tissue malignancies. He worked with trainees and other faculty members to publish outcome-based studies, some of which changed the way we look at and treat these malignancies. Dr. Han was devoted to teaching, working with trainees, mentoring them clinically, enabling them to do research, and enriching their lives. He was the recipient of numerous awards. Most importantly to the authors in this book, and to ourselves, he was a beloved and loyal friend. When Dr. Han passed away, we all lost something in the cancer world—a friend, a colleague, an exemplary researcher, and a clinician who excelled on all fronts. We will truly miss him. This book is dedicated to his memory and to his relentless commitment to excellence in clinical care, research, and education.

Giorgos Karakousis, MD
Jonathan S. Zager, MD

Preface

Advances in molecular biology over the past several decades have firmly established the observations that cancer is a heterogeneous disease arising from genomic alterations or mutations. Thus, within a cancer cell population, cellular subclones may develop resulting in heterogeneity within a tumor and between different metastatic deposits. The cancer microenvironment, including immune responses, exerts a selective force akin to Darwinian "natural selection," which promotes the development of clones that invade and metastasize from the primary site to the distant sites, according to the principles of Paget's seed and soil hypothesis for cancer spread. One critical question is whether the initial pathway for cancer metastasis is through tumor-associated lymphatic vessels to the sentinel lymph nodes or through tumor-associated blood vessels, or via both pathways. This important background drove the style and format of this textbook on cancer metastasis.

Molecular and genetic mechanisms of cancer proliferation and metastasis are important in the further definition and development of more targeted and successful treatment modalities. Imaging techniques for the detection of early-stage cancer and cancer metastases and application of liquid biopsies in sarcoma, prostate, gastrointestinal, and lung cancers are highlighted. Several specialized topics, such as the latest findings on cancer biomarkers and personalized cancer therapy, role of stem cells in cancer metastasis, HPV-induced cancer, impact of microbiome on cancer metastasis, and application of artificial intelligence in cancer metastasis research, are discussed in separate chapters.

While many cancer textbooks concentrate on the systemic management of measurable and microscopic metastases, this textbook focuses on molecular, genetic, diagnostic, and surgical aspects of cancer metastasis. Several cancer types are selected to illustrate the common patterns of progression and spread of metastases. The textbook is richly infused with discussions on how cancer should be evaluated in the multidisciplinary setting and ultimately treated with multimodality therapies including surgery, radiotherapy, and systemic therapies emphasized, such as immunotherapy and targeted therapy.

San Francisco, CA, USA — Stanley P. Leong
Detroit, MI, USA — S. David Nathanson
Tampa, FL, USA — Jonathan S. Zager

Contents

Editors and Contributors

About the Editors

Stanley P. Leong is the Chief of the Cutaneous Oncology Department at the Melanoma Center of the California Pacific Medical Center and Senior Investigator of the California Pacific Medical Center Research Institute in San Francisco, California. He is also Emeritus Professor of Surgery at the University of California, San Francisco. He has published several books on cancer metastasis with Springer and over 210 peer-reviewed publications. He is the current President of the Sentinel Node Oncology Foundation. Dr. Leong has been a co-organizer and co-chair of the International Congress on Cancer Metastasis taking place in San Francisco biannually since 2005.

S. David Nathanson is the Professor of Surgical Oncology, Wayne State University, and the Chair of Breast Cancer Research at the Henry Ford Cancer Institute, Detroit, Michigan. Besides his activities as a surgical oncologist, Dr. Nathanson teaches medical students, residents, fellows, nurses, physician assistants, and laboratory scientists at his institute, and he is actively involved in translational research into the mechanisms of lymph node metastasis. He is the author of numerous research articles and book chapters on metastasis and three books.

Jonathan S. Zager is the Chief Academic Officer at Moffitt Cancer Center and Senior Member and Director of Regional Therapies in the Department of Cutaneous Oncology at Moffitt. Dr. Zager is also the Chair of the Department of Oncologic Sciences and Professor of Surgery at the University of South Florida Morsani College of Medicine. Dr. Zager has published over 300 peer-reviewed articles, invited reviews, and book chapters. He is the editor of three textbooks. Dr. Zager has given over 400 podium presentations, invited talks, keynote lectures, and poster presentations. He is a co-organizer and co-chair of the Biannual International Congress on Cancer Metastasis.

Contributors

Tadashi Abe Kyushu University Beppu Hospital, Beppu, Japan

A. Ahmed H. Lee Moffitt Cancer Center, Tampa, FL, USA

Altan Ahmed Department of Diagnostic Imaging and Interventional Radiology, H. Lee Moffitt Cancer Center and Research Institute, Tampa, FL, USA

S. Ali Temple University Hospital, Philadelphia, PA, USA

James V. Alvarez Department of Pharmacology and Cancer Biology, Duke University School of Medicine, Durham, NC, USA

Stephan W. Anderson Department of Radiology, Boston University Medical Center, Boston University School of Medicine, Boston, MA, USA

Vanesa Carlota Andreu-Arasa Department of Radiology, Boston University Medical Center, Boston University School of Medicine, Boston, MA, USA

Mark Ashamalla Department of Radiation Oncology, Stony Brook University Hospital, Stony Brook, NY, USA

Daniel C. Austin Dartmouth Hitchcock Medical Center, Lebanon, NH, USA

Omar Awan Inova Neurosciences and Spine Institute, University of Virginia Medical School, Falls Church, VA, USA

Kirsten Baecher Department of Surgery, Emory University, Atlanta, GA, USA

Aida Bao-Caamano Cancer Epigenomics, Translational Medical Oncology Group (Oncomet), Health Research Institute of Santiago (IDIS), University Clinical Hospital of Santiago (CHUS/SERGAS), Santiago de Compostela, Spain

Raymond Barnhill Department of Translational Research, Institut Curie, Paris, France

University of Paris Réné Descartes Faculty of Medicine, Paris, France

Georgia M. Beasley Department of Surgery, Duke University, Durham, NC, USA

John-Erik Bell Dartmouth Hitchcock Medical Center, Lebanon, NH, USA

Michael Bernas Department of Medical Education, TCU School of Medicine, Fort Worth, TX, USA

William A. Biché Department of Radiology, Boston University Medical Center, Boston University School of Medicine, Boston, MA, USA

Odion Binitie H. Lee Moffitt Cancer Center, Tampa, FL, USA

Jennifer M. Binning Moffitt Cancer Center, Tampa, USA

Francesco Boccardo San Martino Hospital, University of Genoa, Genoa, Italy

Genevieve Boland Department of Surgery, Massachusetts General Hospital, Boston, MA, USA

Amanda Brisco H. Lee Moffitt Cancer Center & Research Institute, Tampa, FL, USA

Zachary Brown Division of Surgical Oncology, Department of Surgery, The James Cancer Hospital and Solove Research Institute, The Ohio State University Wexner Medical Center and James Cancer Hospital, Columbus, OH, USA

Azadeh Carr University of Southern California, Keck School of Medicine, Los Angeles, CA, USA

Peter R. Carroll University of California, San Francisco, CA, USA

Angelika Chachaj Department of Angiology, Wroclaw Medical University, Wroclaw, Poland

Bindupriya Chandrasekaran Indiana University School of Medicine, Indianapolis, IN, USA

Connie Y. Chang Division of Musculoskeletal Imaging and Intervention, Department of Radiology, Massachusetts General Hospital, Boston, MA, USA

Alexandra Charruyer-Reinwald Department of Dermatology, Eli and Edythe Broad Center of Regeneration Medicine and Stem Cell Research, University of California San Francisco, San Francisco, CA, USA

Tom Wei-Wu Chen Graduate Institute of Oncology, College of Medicine, National Taiwan University, Taipei, Taiwan

Department of Oncology, National Taiwan University Hospital, Taipei, Taiwan

Dhananjay A. Chitale Department of Pathology and Laboratory Medicine, Henry Ford Health System, Detroit, MI, USA

Christine Chow Department of Medicine, Division of Hematology Oncology, University of California, San Francisco, San Francisco, CA, USA

Melissa Chow Department of Medicine, Division of Hematology Oncology, University of California, San Francisco, San Francisco, CA, USA

Hubert H. Chuang Department of Nuclear Medicine, The University of Texas MD Anderson Cancer Center, Houston, TX, USA

Denis Cochonneau Nantes Université, CNRS, US2B, UMR 6286, Institut de Cancérologie de l'Ouest, Blvd Jacques Monod, Saint-Herblain, France

Nicolás Costa-Fraga Cancer Epigenomics, Translational Medical Oncology Group (Oncomet), Health Research Institute of Santiago (IDIS), University Clinical Hospital of Santiago (CHUS/SERGAS), Santiago de Compostela, Spain

Colleen M. Costelloe Department of Musculoskeletal Imaging, The University of Texas MD Anderson Cancer Center, Houston, TX, USA

Anne E. Cress Department of Cellular and Molecular Medicine, The University of Arizona Cancer Center, Tucson, AZ, USA

Brian Czerniecki Breast Department, H. Lee Moffitt Cancer Center, Tampa, FL, USA

Sarah K. Daley Advanced Pathology Research Fellow in Lymphology, Department of Surgery, University of Arizona College of Medicine, Tucson, AZ, USA

Adil I. Daud Department of Medicine, Division of Hematology Oncology, University of California, San Francisco, San Francisco, CA, USA

Michael T. Dellinger Division of Surgical Oncology, Department of Surgery and the Hamon Center for Therapeutic Oncology Research, UT Southwestern Medical Center, Dallas, TX, USA

Benjamin Deschner Department of Surgery, Mountain Area Health Education Center, Asheville, NC, USA

Angel Diaz-Lagares Cancer Epigenomics, Translational Medical Oncology Group (Oncomet), Health Research Institute of Santiago (IDIS), University Clinical Hospital of Santiago (CHUS/SERGAS), Santiago de Compostela, Spain

Centro de Investigación Biomédica en Red Cáncer (CIBERONC), Madrid, Spain

Brittney L. Dickey Moffitt Cancer Center, Tampa, USA

Ashley V. DiMarco Department of Pharmacology and Cancer Biology, Duke University School of Medicine, Durham, NC, USA

Robert P. Erickson Department of Pediatrics, University of Arizona College of Medicine, Tucson, AZ, USA

Timothy J. Evans H. Lee Moffitt Cancer Center, Tampa, FL, USA

Mark B. Faries The Angeles Clinic and Research Institute and Cedars-Sinai Medical Center, Los Angeles, CA, USA

Umme Farwa University of California San Diego, La Jolla, CA, USA

Laura Favazza DO Pathology and Laboratory Medicine, Henry Ford Hospital, Detroit, MI, USA

S. Fenerty Temple University Hospital, Philadelphia, PA, USA

Carla S. Fisher Indiana University School of Medicine, Indianapolis, IN, USA

Samuel Ford Department of Sarcoma Surgery, Queen Elizabeth Hospital, Birmingham, UK

Benjamin Franc Department of Radiology, Division of Nuclear Medicine and Molecular Imaging, Stanford University, Stanford, CA, USA

Timothy L. Frankel Department of Surgery, Michigan Medicine, University of Michigan, Ann Arbor, MI, USA

Jeremy S. Frieling Department of Tumor Biology, H. Lee Moffitt Cancer Center & Research Institute, Tampa, FL, USA

Alexander N. Garcia University of Radiation Oncology, Tucson, USA

Nina Marie G. Garcia Department of Pharmacology and Cancer Biology, Duke University School of Medicine, Durham, NC, USA

Xin Geng Cardiovascular Biology Research Program, Oklahoma Medical Research Foundation, Oklahoma City, OK, USA

Department of Cell Biology, University of Oklahoma Health Sciences Center, Oklahoma City, OK, USA

Ruby Ghadially Department of Dermatology, Eli and Edythe Broad Center of Regeneration Medicine and Stem Cell Research, University of California San Francisco, San Francisco, CA, USA

Armando Giuliano Department of Surgery, Cedars-Sinai Medical Center, Los Angeles, CA, USA

Juan C. Gomez-Gelvez Department of Pathology and Laboratory Medicine, Henry Ford Health System, Detroit, MI, USA

Ricardo J. Gonzalez Moffitt Cancer Center, Tampa, FL, USA

Allison C. Greene Albany Medical Center, Albany, NY, USA

Brian D. Griffith Department of Surgery, Michigan Medicine, University of Michigan, Ann Arbor, MI, USA

Valerie Grignol Division of Surgical Oncology, Department of Surgery, The Ohio State University Wexner Medical Center, Columbus, OH, USA

Alessandro Gronchi Department of Surgery, Fondazione IRCCS Istituto Nazionale dei Tumori, Milan, Italy

Sina Habibollahi Division of Musculoskeletal Imaging and Intervention, Department of Radiology, Massachusetts General Hospital, Boston, MA, USA

S. Haider Henry Ford Health System, Detroit, USA

Russell J. Hamilton University of Arizona, Tucson, AZ, USA

Z. Hammoud Department of thoracic surgery Ascension Providence Hospital, Southfield, MI, USA

Department of surgery, Wayne State University School of Medicine, Detroit, MI, USA

Dale Han Division of Surgical Oncology, Oregon Health and Science University, Portland, OR, USA

William L. Harryman The University of Arizona Cancer Center, Tucson, AZ, USA

Eric R. Henderson Dartmouth Hitchcock Medical Center, Lebanon, NH, USA

Dominique Heymann Nantes Université, CNRS, US2B, UMR 6286, Institut de Cancérologie de l'Ouest, Blvd Jacques Monod, Saint-Herblain, France

Department of Oncology and Metabolism, University of Sheffield, Sheffield, UK

Tina J. Hieken Mayo Clinic, Rochester, MN, USA

Paul Hofman Université Côte d'Azur, CNRS, INSERM, IRCAN, Centre Antoine Lacassagne, FHU OncoAge, Team 4, Nice, France

Université Côte d'Azur, CHU Nice, FHU OncoAge, Laboratory of Clinical and Experimental Pathology, Pasteur Hospital, Nice, France

Université Côte d'Azur, CHU Nice, FHU OncoAge, Hospital-Integrated Biobank BB-0033-00025, Nice, France

Q. A. Ho Department of Radiation Oncology, The University of Arizona, Tucson, AZ, USA

Francis J. Hornicek Department of Orthopaedics, Sarcoma Multidisciplinary Care Group, Sylvester Comprehensive Cancer Center, University of Miami Miller School of Medicine, Los Angeles, CA, USA

Department Orthopaedics, University Miami Miller School of Medicine, Miami, FL, USA

Erika Santos Horta University of Arkansas for Medical Sciences, Little Rock, AR, USA

C. C. Hsu Department of Radiation Oncology, The University of Arizona, Tucson, AZ, USA

Ambrose J. Huang Division of Musculoskeletal Imaging and Intervention, Department of Radiology, Massachusetts General Hospital, Boston, MA, USA

Paul Huang Division of Molecular Pathology, The Institute of Cancer Research, London, UK

Rebecca E. Huber University of Massachusetts Amherst, Amherst, MA, USA

Hatim Husain University of California San Diego, La Jolla, CA, USA

James Isom Department of Dermatology and Cutaneous Surgery, University of South Florida Morsani College of Medicine, Tampa, FL, USA

Sivan Izraely The Shmunis School of Biomedicine and Cancer Research, George S. Wise Faculty of Life Science, Tel Aviv University, Tel Aviv, Israel

David G. Jackson MRC Human Immunology Unit, MRC Weatherall Institute of Molecular Medicine, University of Oxford, Oxford, UK

Christopher Javadi Stanford University, Stanford, CA, USA

P. Jonnalagadda Temple University Hospital, Philadelphia, PA, USA

David M. Joyce H. Lee Moffitt Cancer Center, Tampa, FL, USA

Camille Jubelin Nantes Université, CNRS, US2B, UMR 6286, Institut de Cancérologie de l'Ouest, Blvd Jacques Monod, Saint-Herblain, France

Stanley Kalata Section of General Surgery, Department of Surgery, University of Michigan Medical School, Ann Arbor, MI, USA

Giorgos C. Karakousis Division of Endocrine and Oncologic Surgery, Department of Surgery, University of Pennsylvania, Philadelphia, PA, USA

Mohammed Kashani-Sabet California Pacific Medical Center Research Institute, San Francisco, CA, USA

Igor Katsyv Department of Pathology and Cell Biology, Columbia University, New York, NY, USA

Joshua K. Kays Cedars-Sinai Medical Center, Los Angeles, CA, USA

Jessica M. Keilson Emory University, Atlanta, GA, USA

S. Ray Kenney RareCyte, Inc., Seattle, WA, USA

Faisal Khurshid University of California San Diego, La Jolla, CA, USA

John Kiluk Breast Department, H. Lee Moffitt Cancer Center, Tampa, FL, USA

Alex C. Kim Division of Surgical Oncology, Department of Surgery, The James Cancer Hospital and Solove Research Institute, The Ohio State University Wexner Medical Center, Columbus, OH, USA

Hyuna Kim University of Massachusetts Amherst, Amherst, MA, USA

Kevin B. Kim California Pacific Medical Center Research Institute, San Francisco, CA, USA

Richard W. Kim School of Medicine, University of California San Francisco, San Francisco, CA, USA

Olena Kis Department of Pathology and Laboratory Medicine, Henry Ford Health System, Detroit, MI, USA

Yuko Kitagawa Department of Surgery, Keio University School of Medicine, Tokyo, Japan

Scott Kizy Moffitt Cancer Center, Tampa, FL, USA

Lyonell B. Kone University of Illinois at Chicago—Metropolitan Group Hospitals, Chicago, IL, USA

Christina A. LeBedis Department of Radiology, Boston University Medical Center, Boston University School of Medicine, Boston, MA, USA

I. Lee Henry Ford Health System, Detroit, USA

Stanley P. Leong Department of Surgery and Melanoma Center, California Pacific Medical Center and Research Institute, San Francisco, CA, USA

James K. C. Liu H. Lee Moffitt Cancer Center & Research Institute, Tampa, FL, USA

Mike Lowe Department of Surgery, Emory University, Atlanta, GA, USA

Claire Lugassy Department of Translational Research, Institut Curie, Paris, France

Conor C. Lynch Department of Tumor Biology, H. Lee Moffitt Cancer Center & Research Institute, Tampa, FL, USA

John E. Madewell Department of Musculoskeletal Imaging, The University of Texas MD Anderson Cancer Center, Houston, TX, USA

Rita Das Mahapatra University of Massachusetts Amherst, Amherst, MA, USA

Najjia N. Mahmoud Division of Colon and Rectal Surgery, Department of Surgery, University of Pennsylvania, Philadelphia, PA, USA

Shishir K. Maithel Emory University, Atlanta, GA, USA

Ajay V. Maker Department of Surgery, University of California San Francisco, San Francisco, CA, USA

Sindhu J. Malapati Department of Hematology/Oncology, Ascension St. John Hospital, Detroit, MI, USA

Kartik Mani Department of Radiation Oncology, Stony Brook University Hospital, Stony Brook, NY, USA

Dario Marchetti Department of Internal Medicine, Division of Molecular Medicine, The University of New Mexico Health Sciences Center, Albuquerque, NM, USA

Takaaki Masuda Kyushu University Beppu Hospital, Beppu, Japan

Ahmad Mattour Department of Hematology/Oncology, Henry Ford Cancer Institute, Detroit, MI, USA

Shuhei Mayanagi Division of Esophageal Surgery, Shizuoka Cancer Center Hospital, Shizuoka, Japan

James F. McLoughlin Department of Radiology, Boston University Medical Center, Boston University School of Medicine, Boston, MA, USA

Jennifer L. McQuade Department of Melanoma Medical Oncology, The University of Texas MD Anderson Cancer Center, Houston, TX, USA

Winta T. Mehtsun Department of Surgery, Massachusetts General Hospital, Boston, MA, USA

Department of Surgery, Brigham and Women's Hospital, Boston, MA, USA

Dana Farber Cancer Institute, Boston, MA, USA

Nathan W. Mesko Cleveland Clinic, Cleveland, OH, USA

Jane L. Messina Department of Dermatology and Cutaneous Surgery, University of South Florida Morsani College of Medicine, Tampa, FL, USA

Departments of Pathology and Cutaneous Oncology, Moffitt Cancer Center, Tampa, FL, USA

Departments of Pathology and Cell Biology and Oncologic Sciences, University of South Florida Morsani College of Medicine, Tampa, FL, USA

Koshi Mimori Kyushu University Beppu Hospital, Beppu, Japan

Marc Moncrieff Norfolk & Norwich University Hospital NHS Trust, Norwich, UK

Emilie Moranton Nantes Université, CNRS, US2B, UMR 6286, Institut de Cancérologie de l'Ouest, Blvd Jacques Monod, Saint-Herblain, France

B. Morse H. Lee Moffitt Cancer Center, Tampa, FL, USA

Laura Muinelo-Romay Liquid Biopsy Analysis Unit, Translational Medical Oncology Group (Oncomet), Health Research Institute of Santiago (IDIS), University Clinical Hospital of Santiago (CHUS/SERGAS), Santiago de Compostela, Spain

Centro de Investigación Biomédica en Red Cáncer (CIBERONC), Madrid, Spain

Javier Muñoz-Garcia Nantes Université, CNRS, US2B, UMR 6286, Institut de Cancérologie de l'Ouest, Blvd Jacques Monod, Saint-Herblain, France

Takafumi Nakano Kyushu University Beppu Hospital, Beppu, Japan

H. Nasser Henry Ford Hospital, Detroit, MI, USA

Rui P. L. Neves Experimental Surgical Oncology, Department of General, Visceral and Pediatric Surgery, University Hospital and Medical Faculty of the Heinrich-Heine University Düsseldorf, Düsseldorf, Germany

Lukas M. Nystrom Cleveland Clinic, Cleveland, OH, USA

Valeriya Pankova Division of Molecular Pathology, The Institute of Cancer Research, London, UK

Nainesh Parikh Department of Diagnostic Imaging and Interventional Radiology, H. Lee Moffitt Cancer Center and Research Institute, Tampa, FL, USA

Howard Y. Park Department of Orthopaedic Surgery, David Geffen School of Medicine at UCLA, Los Angeles, CA, USA

Jacob A. Pawloski Department of Neurosurgery, Hermelin Brain Tumor Center, Henry Ford Hospital, Detroit, MI, USA

Howard Peach Leeds Teaching Hospitals NHS Trust, Leeds, UK

Shelly R. Peyton University of Massachusetts Amherst, Amherst, MA, USA

Neil Piller Flinders Medical Centre, Department of Surgery, School of Medicine, Bedford Park, SA, Australia

Mara A. Piltin Mayo Clinic, Rochester, MN, USA

Jose M. Pimiento Moffitt Cancer Center, Tampa, FL, USA

Raphael Pollock Division of Surgical Oncology, Department of Surgery, The Ohio State University Wexner Medical Center and James Cancer Hospita, Columbus, OH, USA

A. Popoff Henry Ford Hospital, Detroit, MI, USA

Sima P. Porten University of California, San Francisco, CA, USA

C. Potti Henry Ford Hospital, Detroit, MI, USA

George Poultsides Stanford University, Stanford, CA, USA

Julie Rathwell Moffitt Cancer Center, Tampa, USA

Kristen E. Rhodin Department of Surgery, Duke University, Durham, NC, USA

Adam M. Robin Department of Neurosurgery, Hermelin Brain Tumor Center, Henry Ford Hospital, Detroit, MI, USA

Aitor Rodriguez-Casanova Cancer Epigenomics, Translational Medical Oncology Group (Oncomet), Health Research Institute of Santiago (IDIS), University Clinical Hospital of Santiago (CHUS/SERGAS), Santiago de Compostela, Spain

Roche-Chus Joint Unit, Translational Medical Oncology Group (Oncomet), Health Research Institute of Santiago (IDIS), Santiago de Compostela, Spain

Robert E. Roses University of Pennsylvania, Philadelphia, PA, USA

Samuel Ryu Department of Radiation Oncology, Stony Brook University, Stony Brook, NY, USA

Hideyuki Saito Kyushu University Beppu Hospital, Beppu, Japan

Elizabeth A. Sangalang Department of Medicine, Division of Hematology Oncology, University of California, San Francisco, San Francisco, CA, USA

Nirali Sanghavi University of California San Diego, La Jolla, CA, USA

Luke V. Selby The Ohio State University James Cancer Center, Columbus, OH, USA

Akash A. Shah Department of Orthopaedic Surgery, David Geffen School of Medicine at UCLA, Los Angeles, CA, USA

Brandon M. Shaw Department of Pathology and Laboratory Medicine, Henry Ford Health System, Detroit, MI, USA

David Shibata Department of Surgery, University of Tennessee Health Science Center, Memphis, TN, USA

Qihui Shi Institutes of Biomedical Sciences and Minhang Hospital, Fudan University, Shanghai, China

Emily Siegel Breast Department, H. Lee Moffitt Cancer Center, Tampa, FL, USA

Sunny R. K. Singh Department of Hematology/Oncology, Henry Ford Cancer Institute, Detroit, MI, USA

Andrew J. Sinnamon Moffitt Cancer Center, Tampa, FL, USA

Alexander N. Slade Department of Radiation Oncology, Stony Brook University Hospital, Stony Brook, NY, USA

J. Snyder Henry Ford Health System, Detroit, USA

Gaya Spolverato Department of Surgical, Oncological and Gastroenterological Sciences, University of Padova, Padova, Italy

R. Sathish Srinivasan Cardiovascular Biology Research Program, Oklahoma Medical Research Foundation, Oklahoma City, OK, USA

Department of Cell Biology, University of Oklahoma Health Sciences Center, Oklahoma City, OK, USA

Hayley Standage Division of Surgical Oncology, Oregon Health and Science University, Portland, OR, USA

Baldassarre Stea University of Arizona, Tucson, USA

Nikolas H. Stoecklein Experimental Surgical Oncology, Department of General, Visceral and Pediatric Surgery, University Hospital and Medical Faculty of the Heinrich-Heine University Düsseldorf, Düsseldorf, Germany

Richard J. Straker III Division of Endocrine and Oncologic Surgery, Department of Surgery, University of Pennsylvania, Philadelphia, PA, USA

Division of Colon and Rectal Surgery, Department of Surgery, University of Pennsylvania, Philadelphia, PA, USA

Sheng-Fang Su Graduate Institute of Oncology, College of Medicine, National Taiwan University, Taipei, Taiwan

YongLin Institute of Health, National Taiwan University (National Taiwan University YongLin Scholar), Taipei, Taiwan

Andrzej Szuba Department of Angiology, Wroclaw Medical University, Wroclaw, Poland

Seiichiro Takao Kyushu University Beppu Hospital, Beppu, Japan

Alicia M. Terando University of Southern California, Keck School of Medicine, Los Angeles, CA, USA

Kerry L. Thomas Clinical Associate Professor, Department of Radiology, University of North Carolina at Chapel Hill, School of Medicine, Tampa, FL, USA

Fabio Tirotta Department of Sarcoma Surgery, Queen Elizabeth Hospital, Birmingham, UK

Michael T. Torchia Dartmouth Hitchcock Medical Center, Lebanon, NH, USA

Ning-Hsuan Tseng University of Massachusetts Amherst, Amherst, MA, USA

William W. Tseng Department of Surgery, Division of Breast, Endocrine and Soft Tissue Surgery, Keck School of Medicine of University of Southern California, Los Angeles, CA, USA

Kiran K. Turaga Division of Surgical Oncology, Department of Surgery, University of Chicago, Chicago, IL, USA

Andrew Turk Department of Pathology and Cell Biology, Columbia University, New York, NY, USA

Przemyslaw Twardowski Medical Oncology and Urologic Oncology, Department of Urologic Oncology, John Wayne Cancer Institute, Santa Monica, CA, USA

Raul Fernando Valenzuela Department of Musculoskeletal Imaging, The University of Texas MD Anderson Cancer Center, Houston, TX, USA

P. van der Laan Department of Surgical Oncology, Sarcoma and Melanoma Unit, The Netherlands Cancer Institute – Antoni van Leeuwenhoek, Amsterdam, The Netherlands

Winan J. van Houdt Department of Surgical Oncology, Sarcoma and Melanoma Unit, The Netherlands Cancer Institute – Antoni van Leeuwenhoek, Amsterdam, The Netherlands

Joao R. T. Vicentini Division of Musculoskeletal Imaging and Intervention, Department of Radiology, Massachusetts General Hospital, Boston, MA, USA

Tobias Walbert Hermelin Brain Tumor Center, Henry Ford Health System, Detroit, MI, USA

Christopher P. Wang H. Lee Moffitt Cancer Center & Research Institute, Tampa, FL, USA

Zhiqiang Wang Department of Pathology and Laboratory Medicine, Henry Ford Hospital, Detroit, MI, USA

Jennifer A. Wargo Department of Surgical Oncology, The University of Texas MD Anderson Cancer Center, Houston, TX, USA

Samuel L. Washington III University of California, San Francisco, CA, USA

Michael G. White Department of Surgical Oncology, The University of Texas MD Anderson Cancer Center, Houston, TX, USA

Hunter D. D. Witmer Division of Surgical Oncology, Department of Surgery, University of Chicago, Chicago, IL, USA

Marlys H. Witte Department of Surgery, Neurosurgery, and Pediatrics, University of Arizona College of Medicine, Tucson, AZ, USA

Russell S. Witte Department of Medical Imaging, University of Arizona, Tucson, AZ, USA

Isaac P. Witz The Shmunis School of Biomedicine and Cancer Research, George S. Wise Faculty of Life Science, Tel Aviv University, Tel Aviv, Israel

Ziyuan Zhang Institutes of Biomedical Sciences and Minhang Hospital, Fudan University, Shanghai, China

Mateo Ziu Inova Neurosciences and Spine Institute, University of Virginia Medical School, Falls Church, VA, USA

Part I

Genetic and Molecular Pathways of Cancer Growth and Metastasis

Dhananjay Chitale

In the United States, cancer is the second most leading cause of death after cardiovascular diseases [1]. In addition to high mortality, there is a lot of human emotional and physical suffering inflicted by cancer. The essence of finding new targeted therapies to achieve complete remission is to complement the present mostly anatomic-based treatment protocols with a full understanding of the neoplasms at the molecular genetic level.

It is well known that some primary malignant tumors have a higher propensity to aggressively invade local sites while others metastasize at an early stage. Pathologists have been classifying tumors using morphological correlates such as differentiation or grade based on the closeness of the tumor to its normal counterparts. Nuclear pleomorphism, mitosis, necrosis, infiltration, and lymphatic and vascular invasion are microscopic features that, when correlated with clinical outcomes, predict the aggressiveness of the tumor. The anatomic pathology staging is an integral part of AJCC and TNM staging classifications that consider the tumor size, local lymph node status, and metastasis outside the local, regional anatomic lymphatic drainage. Overall, a larger tumor size is typically associated with a higher chance of metastasis to locoregional lymph nodes or distant sites. These parameters are vital for surgeons and oncologists to tailor therapies based on grade and stage. However, we know that tumor behaviors vary significantly, even within the same AJCC stages. Especially in the early-stage tumors where there is no evidence of local or distant spread of the tumor, it is often difficult to predict the biological potential of the malignant tumor based purely on histologic and radiologic parameters. Therefore, the knowledge of molecular programs of these tumors is fundamental to guide therapies. The current testament to that concept is reflected in the evolution of the staging checklists that are slowly getting transformed from purely anatomic pathology-driven items to a combination of anatomic and molecular biomarkers that have been introduced in the recent 8th AJCC staging classifications to fine-tune tumor stages further to direct targeted therapeutics. These stratification strategies will help separate more aggressive tumors within the same anatomic stage for optimal treatments and outcomes. Breast cancer AJCC staging [2, 3] is one such example where biomarker status such as estrogen receptor (ER), progesterone receptor (PgR), and human epidermal growth factor receptor 2 (Her2) expression status are included along with multigene signature panels such as 21-gene Oncotype Dx Recurrence Score, MammaPrint, the 12-gene EndoPredict, or the PAM50 test making its way into clinical decision-making.

According to Willis, a neoplasm is defined as an abnormal mass of tissue, the growth of which exceeds and is uncoordinated with that of the normal tissues and persists in the same excessive manner after the cessation of the stimuli which evoked the change [4]. From the original descriptions of neoplasm by Willis, after decades of clinical and experimental observations, we know that neoplasms are a mixture of heterogeneous cells that start from "clonal" progeny of a single cell that acquires or less commonly inherits mutations, thus gaining survival advantage through natural selection. The tumor cells sequentially acquire the hallmarks of cancer from a premalignant transformed state to a widely metastatic disease through a multistep process.

The past 40+ years of cancer research has given deep insights into molecular characteristics and underpinnings that drive tumor biology and propensity towards metastasis. There is a large body of knowledge that has further underscored the complexity of cancer as a disease driven by dynamic changes in the genome through a multistep acquisition of large numerical or structural changes to the chromosomal complement or subtle point mutations in oncogenes leading to gain of function and tumor-suppressor genes with loss of function [5]. With the advent of molecular profiling and understanding of tumors at the genomic level, we now know that tumor is not a uniform collection of cancer cells but rather a mixture of different types of cells. It includes tumor cells, immune cells, supporting stromal cells, blood vessels, stem cells, and stromal matrix, all working together as an autonomous organ. Consolidation of the current knowledge of different pathways in human tumorigenesis led to the first conceptual seminal paper by Hanahan and Weinberg in the year 2000 that described the hallmarks of cancer as an organizing principle of oncology [6]. The hallmarks of cancer comprised six biological capabilities acquired by cancer cells through a multistep process. These were updated in 2011 where two additional key properties to a total of 8 and 2 additional functional roles attributed to the tumor microenvironment were added [7]. We have tried to deconstruct these complex hallmarks of cancers described in the following chapters in detail.

In this section of the book, different chapters will describe the hallmarks of cancer with molecular underpinnings of primary tumors as a starting point to set the stage for subsequent chapters, followed by a unifying concept of genomic changes that will elucidate the mutational landscape of cancers. The next chapter will be on changes in the tumor microenvironment as a co-conspirator in tumorigenesis. Then we will deep dive into the hallmarks of metastasis with its molecular underpinnings. And finally, we will summarize hereditary cancer syndromes and their role in cancer metastasis. Through these systematic descriptions of cancer genomics and microenvironments, we hope to give the readers insights and understanding of cancer as a "system biology" where treating the entire system rather than individual components is necessary to defeat this deadly disease effectively, hopefully soon.

References

1. Xu J, Murphy SL, Kockanek KD, Arias E. Mortality in the United States, 2018. NCHS Data Brief. 2020;355:1–8. Retrieved from https://www.ncbi.nlm.nih.gov/pubmed/32487294
2. Giuliano AE, Edge SB, Hortobagyi GN. Eighth edition of the AJCC cancer staging manual: breast cancer. Ann Surg Oncol. 2018;25(7):1783–5. https://doi.org/10.1245/s10434-018-6486-6.
3. Jang N, Choi JE, Kang SH, Bae YK. Validation of the pathological prognostic staging system proposed in the revised eighth edition of the AJCC staging manual in different molecular subtypes of breast cancer. Virchows Arch. 2019;474(2):193–200. https://doi.org/10.1007/s00428-018-2495-x.
4. Willis RA. The spread of tumours in the human body. 2nd ed. London: Butterworth & Co.; 1952. p. ix. 447–28.
5. Kinzler KW, Vogelstein B. Lessons from hereditary colorectal cancer. Cell. 1996;87(2):159–70. https://doi.org/10.1016/s0092-8674(00)81333-1.
6. Hanahan D, Weinberg RA. The hallmarks of cancer. Cell. 2000;100(1):57–70. https://doi.org/10.1016/s0092-8674(00)81683-9.
7. Hanahan D, Weinberg RA. Hallmarks of cancer: the next generation. Cell. 2011;144(5):646–74. https://doi.org/10.1016/j.cell.2011.02.013.

Hallmarks of Cancer: Molecular Underpinnings

1

Dhananjay A. Chitale

Abstract

Cancer, a genetic disease, is an autonomous organ composed of heterogeneous cancer cell clones supported by its microenvironment. The cancer cells gain a set of properties and characteristics called hallmarks of cancer and enabling characteristics, respectively, that highjack the normal cellular mechanisms for their purpose. These are thought to be acquired in a stepwise fashion, in no particular order, giving cancer cells a survival advantage over normal cells, making them self-sustainable, invade and metastasize to distant locations. This chapter provides a brief overview and key concepts on the molecular underpinnings of cancer cells composed of eight hallmarks of cancer and two enabling characteristics.

Learning Objectives

- To understand the molecular genomic alterations in primary cancers conceptually as Hallmarks of Cancer as the basis for cancer metastasis
- Based on the understanding of genomic underpinnings of cancers, learn diagnostic, prognostic, and predictive therapeutic strategies

1.1 Introduction

Cancer is a genetic disease underpinned by mutations, either somatic or rarely genetically inherited, layered with epigenetic alterations that lead to an autonomous organ that is under its own control, defying normal cellular signaling processes. The molecular genetic characteristics of cancer impart tumor cells a set of properties called *hallmarks of cancer* that Hanahan and Weinberg initially published in their seminal paper published in 2000 [1]. They initially described six such properties, viz. self-sufficiency in growth signals, insensitivity to growth-inhibitory (antigrowth) signals, evasion of programmed cell death (apoptosis), limitless replicative potential, sustained angiogenesis, and tissue invasion and metastasis. These properties are acquired by the cancer cell with an inherent assumption that normal cells progressively transform into a neoplastic state through stepwise gain of mutations that confer survival advantage and disseminate at distant sites. The stages or steps of these changes are necessarily not linear, and different tumors may have different sequential gains in these properties. After a decade or so, with additional accumulation of molecular literature and understanding of cancer, in 2011, Hanahan and Weinberg revised the original hallmarks and added two new ones to the original six, and expanded on the functional roles and contributions made by the tumor microenvironment [2].

In this chapter, using Hanahan and Weinberg's publications as the backbone [1, 2], we focus on detailed descriptions of these properties that produce the cellular phenotype and briefly elucidate the molecular underpinnings to gain such properties in the primary tumors with some examples. The details of underlying molecular mechanisms such as point mutations, gene translocations/fusions, deletions, insertions, amplification, chromosomal aneuploidy, transcriptional changes (mRNA profiles), epigenetic changes, MicroRNA profiles, and regulating DNA structural regulatory changes are described in the next chapter titled "Unifying Concept of Genomic Changes: The Mutational Landscape of Cancers."

1.2 Sustaining Proliferative Signaling

Normal tissues regulate the proliferation of cells tightly controlled through growth factor-induced cell signaling, thus controlling the tissue architecture and function. This process is orchestrated through a growth factor binding to its specific

D. A. Chitale (✉)
Department of Pathology and Laboratory Medicine, Henry Ford Health System, Detroit, MI, USA
e-mail: dchital1@hfhs.org

S. P. Leong et al. (eds.), *Cancer Metastasis Through the Lymphovascular System*,
https://doi.org/10.1007/978-3-030-93084-4_1

receptor transiently, activating a cascade of several downstream cytoplasmic signal-transduction proteins transmitting the signal to the nucleus. This, in turn, induces and activates transcription factors and epigenetic alterations initiating DNA transcription. Genes encoding proteins that promote cell cycle progression ultimately result in cell division. The expression of other genes at the same time also leads to changes in metabolic activity (energy metabolism) and proteins supporting cell survival that are needed for optimal growth. The knowledge about the proliferative growth factor signaling through the paracrine mechanism remains limited [2]. However, in cancer cells, the most fundamental trait is their ability to sustain proliferative signals, many through tyrosine kinase activity, even in the absence of growth factors. This self-sustaining property is achieved through several molecular mechanisms involving one or multiple parts of the cell signaling pathways. Primary mechanisms include **mutations in the oncogenes** that constitutively activate cell signaling pathways and **deregulation of inhibitory feedback loops,** enhancing cell signaling.

1.2.1 Mutation in Oncogenes

Normal cellular genes that promote cell proliferation called proto-oncogenes are mutated or overexpressed autonomously in many cancers, which are then termed as oncogenes. An array of mechanisms alters the function of the oncogenes by constitutive activation and resistance to control by normal external signals free from checkpoints. The molecular underpinnings of this hallmark of cancer can involve growth factors, growth factor receptors, proteins involved in signal transduction, nuclear regulatory proteins, and cell cycle regulator.

One pathway is through **increasing growth factor production**. The cancer cells may produce growth factors themselves (autocrine capability); for example, glioblastomas overexpress platelet-derived growth factor (PDGF) and PDGF receptor (PDGFR) [3]. Tumor cells may send signals to stimulate the normal cells in the tumor microenvironment, which supply the growth factors (paracrine function). Amplification of FGF3 growth factor is observed in many cancers (stomach, bladder, breast) [4]. Another mechanism is producing high levels of receptor proteins, such as *Her2 (ERB1B2)* gene amplification in breast cancers. In this case, receptor signaling can be deregulated and sensitized even in the presence of normal amounts of ligand [5–8]. In lung adenocarcinomas, discovery of **activating hot spot mutations** in the tyrosine kinase domain encoded by exon 19 and 21 of *ERB1B1 (EGFR)* receptor has revolutionized the way lung adenocarcinomas are treated using small molecular targeted therapies (tyrosine kinase inhibitors-TKI). Another well-known means of cancer signaling upregulation is achieved through constitutive activation of components of the signaling pathways downstream of the receptors and independent of growth factor ligand activation. Many recurrent **somatic mutations** have been detected by high-throughput DNA sequencing that predicts the **constitutive activation of signal transduction**. Recurrent activating mutations in the *BRAF* gene, usually due to point mutations, have been noted in many solid tumors [9], including 40% of malignant melanomas [10]. These mutations lead to constitutively active signaling through mitogen-activated protein kinase(MAPK) pathway that regulates a wide variety of cellular processes such as proliferation, differentiation, apoptosis, and stress responses [11]. Another canonical signaling pathway that is frequently upregulated due to mutations in the catalytic subunits of phosphoinositide 3-kinase (PI3K) is involved in cellular functions such as cell growth, proliferation, differentiation, motility, survival, and intracellular trafficking [12, 13]. The tumor cells can also attain self-sufficiency due to the **deregulation of nuclear regulatory proteins**. Transcriptional activators, e.g., MYC, a master transcription factor that regulates genes needed for rapid cell growth, are overexpressed either through translocation (Burkitt's lymphoma) [14] or through amplification (breast carcinoma, lung adenocarcinoma, colon adenocarcinoma, prostate adenocarcinoma) [9] in aggressive tumors. **Dysregulation of cell cycle regulators** such as cyclins often occurs in some tumors. For example, cyclin D1 (CCDN1) is upregulated due to translocation (CCND1/IGH) in mantle cell lymphoma, plasma cell myeloma, making the tumor cells proliferate constitutively.

1.2.2 Deregulation of Negative Feedback Loops

Normal cellular mechanisms of cell signaling include negative feedback loops that inactivate the activated signaling to maintain homeostasis. In tumors, the genes controlling this negative feedback are downregulated in diverse tumor types, thus enhancing the mitogenic and proliferative signals. For example, many carcinomas of solid organs such as the colon, pancreas, lung etc., harbor activating mutations in the *KRAS* gene [9]. The effects on downstream signaling are not due to direct activation. But the constitutive activation of *RAS* protein through these mutations leads to loss of intrinsic GTPas activity of RAS, thus disrupting the autoregulatory negative-feedback mechanism [2]. Another example is heterogeneous groups of enzymes called Dual-specificity phosphatases (DUSP) that regulate diverse cellular signaling cascades, including much-researched MAPK phosphatases. Dysregulation of DUSPs has been implicated as major modulators of critical signaling pathways in various tumors [15].

Another well-known example is phosphatase and tensin homolog (PTEN), which degrades phosphatidylinositol 3,4,5-phosphate (PIP3), thus dampening PI3K signaling. Loss of PTEN function due to either somatic mutations such as deletion or gene silencing by hypermethylation leads to amplification of the PI3K signaling in many solid tumors [12, 13]. Germline mutations in *PTEN* in Cowden syndrome, an autosomal dominant inherited condition, predispose individuals to many benign and malignant tumors of the skin, breast, thyroid, endometrium, colorectum, and kidney [16].

1.3 Insensitivity to Growth Suppressors

In the normal cellular proliferation machinery, set of genes categorized as tumor suppressors control the checkpoints to prevent uncontrolled cell growth. These antiproliferative mechanisms lead to quiescence or permanent cell cycle arrest, depending on the scenario. Several tumor suppressor genes are part of this regulatory network limiting cell proliferation or survival responding to activated signaling (Table 1.1). A similar phenomenon occurs in cancer cells with overactive cell signaling that leads to a nonproliferative state called oncogene-induced senescence [18]. Like mitogenic signaling, growth inhibitory signals also arise outside of the cells (paracrine) orchestrated through receptors, signal transducers, and nuclear transcription regulators. Cancer cells evade these checkpoints, thus gaining survival advantage and cell proliferation. These can be categorized based on the molecular underpinnings, viz. genes encoding inhibitors of mitogenic signaling pathways, inhibitors of cell cycle progression, inhibitors of pro-growth programs of metabolism and angiogenesis, inhibitors of invasion and metastasis, enablers of genomic stability, DNA repair factors, and some with unknown mechanisms [17]. It is thought that tumor suppressors have a broader inhibitory effect on many hallmarks of cancer. The key prototypical tumor suppressor proteins include *RB1* and *TP53,* both playing a central role as checkpoint controls for a cell to proliferate or activate growth arrest, senescence, or activate apoptosis. These are described in detail below.

Much of our understanding and concepts regarding tumor suppressor genes and their function are derived from studies on the first tumor suppression gene discovered, the retinoblastoma (*RB1*) gene, the so-called Governor of The Cell Cycle. Based on observations between hereditary retinoblastomas vs. sporadic, Knudson proposed the "two-hit" hypothesis of oncogenesis that states that both the alleles (one from mother and one from father) need to be inactivated by similar or different molecular mechanisms to cause malignancies. In inherited germline *RB1* mutation, the second normal allele is inactivated by somatic mutation silencing the gene function. While in sporadic retinoblastomas, both the alleles are inactivated by somatic mutations [19]. Large-scale genomic sequencing has identified similar somatic *RB1* gene mutations in subsets of the lung, breast, bladder carcinoma, and glioblastoma [9], and among the top 5 somatically altered genes in metastatic tumors [20]. The current model is that loss of normal cell cycle control occurs due to dysregulation of at least one of the four key regulators: *CDKN2A, CCND1, CDK4*, and *RB1* in many human cancers [17]. The loss of tumor suppressor function of *RB1* occurs directly due to mutation in the *RB1* gene or indirectly by the gain of function CDK/Cyclin D activity keeping *RB1* protein in the hyperphosphorylated state, thus inhibiting the binding and sequestering of E2F transcription factors. Release of E2F leads to transcriptional activation and cell cycle progression.

TP53, also known as Guardian of The Genome, is also a tumor suppressor gene regulating cell cycle progression, DNA repair, cellular senescence, and apoptosis. *TP53* is mutated in 50% of all cancers, including lung, colon, breast, ovary, and glioblastoma. Unlike RB1, the loss of function is primarily somatic and rarely seen as germline inheritable genetic mutation seen in Li-Fraumeni syndrome. Like RB1, the loss of function occurs by somatic mutations in both the alleles or in cases of germline mutation, and other allele is inactivated by somatic mutation. Another analogy to *RB1* loss of function is that the functional loss of p53 protein can occur in the absence of TP53 mutations. *MDM2* gene amplification observed in 33% of human sarcomas leads to p53 deficiency due to accelerated degradation. Transforming proteins of several DNA viruses bind to p53 and promote degradation. With the loss of p53 protein function, DNA damage and, therefore, driver mutations accumulate in oncogenes, ultimately leading to the malignant transformation of cells.

RB protein transduces growth-inhibitory signals originating externally and internally and decides whether a cell should proceed through its growth-and-division cycle. In comparison, *TP53* receives inputs from stress and abnormality sensors intracellular signaling. Depending on the degree of genomic damage, the levels of nucleotide pools, growth-promoting signals, glucose, or oxygenation, TP53 can halt cell cycle progression until normalization. Alternatively, if there is irreparable damage to cellular subsystems, TP53 can trigger apoptosis [2]. In the absence of both these proteins, the cell proliferating characteristic of cancer cells remains unimpeded.

1.3.1 Evasion of Contact Inhibition

Normal cells stop proliferating upon contact with other cells. This contact inhibition that is part of tissue homeostasis is lost in tumors. E-cadherin encoded by the *CDH1* gene is a cell adhesion molecule that plays a vital role in

Table 1.1 Selected tumor suppressor genes and associated familial syndrome and cancers grouped by cancer hallmarks are listed in this table

Gene	Protein	Function	Familial syndromes	Sporadic cancers
Inhibitors of mitogenic signaling pathways				
APC	Adenomatous polyposis coli protein	Inhibitor of WNT signaling	Familial colonic polyps and carcinomas	Carcinomas of stomach, colon, pancreas; melanoma
NF1	Neurofibromin-1	Inhibitor of RAS/MAPK signaling	Neurofibromatosis type 1 (neurofibromas and malignant peripheral nerve sheath tumors)	Neuroblastoma, juvenile myeloid leukemia
NF2	Merlin	Cytoskeletal stability, hippo pathway signaling	Neurofibromatosis type 2 (acoustic schwannoma and meningioma)	Schwannoma, meningioma
PTCH	Patched	Inhibitor of hedgehog signaling	Gorlin syndrome (basal cell carcinoma, medulloblastoma, several benign tumors)	Basal cell carcinoma, medulloblastoma
PTEN	Phosphatase and tensin homologue	Inhibitor of PI3K/AKT signaling	Cowden syndrome (variety of benign skin, GI, and CNS growths; breast, endometrial, and thyroid carcinoma)	Diverse cancers, particularly carcinomas and lymphoid tumors
SMAD2, SMAD4	SMAD2, SMAD4	Component of the TGF-β signaling pathway, repressors of MYC and CDK4 expression, inducers of CDK inhibitor expression	Juvenile polyposis	Frequently mutated (along with other components of the TGF-β signaling pathway) in colonic and pancreatic carcinoma
Inhibitors of cell cycle progression				
RB	Retinoblastoma (RB) protein	Inhibitor of G_1/S transition during cell cycle progression	Familial retinoblastoma syndrome (retinoblastoma, osteosarcoma, other sarcomas)	Retinoblastoma; osteosarcoma; carcinomas of breast, colon, lung
CDKN2A	p16/INK4a and p14/ARF	p16: Negative regulator of cyclin-dependent kinases; p14, an indirect activator of p53	Familial melanoma	Pancreatic, breast, and esophageal carcinoma; melanoma; certain leukemias
Inhibitors of pro-growth programs of metabolism and angiogenesis				
VHL	von Hippel–Lindau (VHL) protein	Inhibitor of hypoxia-induced transcription factors (e.g., HIF1α)	von Hippel–Lindau syndrome (cerebellar hemangioblastoma, retinal angioma, renal cell carcinoma)	Renal cell carcinoma
STK11	Liver kinase B1 (LKB1) or STK11	Activator of AMPK family of kinases; suppresses cell growth when cell nutrient and energy levels are low	Peutz-Jeghers syndrome (GI polyps, GI cancers, pancreatic carcinoma, and other carcinomas)	Diverse carcinomas (5%–20% of cases, depending on the type)
SDHB, SDHD	Succinate dehydrogenase complex subunits B and D	TCA cycle, oxidative phosphorylation	Familial paraganglioma, familial pheochromocytoma	Paraganglioma
Inhibitors of invasion and metastasis				
CDH1	E-cadherin	Cell adhesion, inhibition of cell motility	Familial gastric cancer	Gastric carcinoma, lobular breast carcinoma
Enablers of genomic stability				
TP53	p53 protein	Cell cycle arrest and apoptosis in response to DNA damage	Li-Fraumeni syndrome (diverse cancers)	Most human cancers
DNA repair factors				
BRCA1, BRCA2	Breast cancer-1 and breast cancer-2 (BRCA1 and BRCA2)	Repair of double-stranded breaks in DNA	Familial breast and ovarian carcinoma; carcinomas of the male breast; chronic lymphocytic leukemia (BRCA2)	Rare
MSH2, MLH1, MSH6	MSH1, MLH1, MSH6	DNA mismatch repair	Hereditary nonpolyposis colon carcinoma	Colonic and endometrial carcinoma
Unknown mechanisms				
WT1	Wilms tumor-1 (WT1)	Transcription factor	Familial Wilms tumor	Wilms tumor, certain leukemias
MEN1	Menin	Transcription factor	Multiple endocrine neoplasia-1 (MEN1) (pituitary, parathyroid, and pancreatic endocrine tumors)	Pituitary, parathyroid, and pancreatic endocrine tumors

CNS Central nervous system, *GI* Gastrointestinal, *TCA* Tricarboxylic acid
Adapted from Robbins and Cotran Pathologic Basis of Diseases, tenth Edition, Kumar, V., Abbas, A. K., Aster, J. C., Turner, J. R., Robbins, S. L., & Cotran, R. S., Page 292, Table 7.7, Copyright Elsevier, 2021 [17]

contact-mediated growth inhibition of epithelial cells. Disruption of E-cadherin is noted in many solid tumors such as gastric signet ring cell carcinomas and invasive lobular carcinoma of the breast. This allows cells to easily disaggregate due to loss of "stickiness" and promotes distant metastasis. Another function of E-cadherin is that it binds and sequesters β-catenin. Disruption of this complex leads to the translocation of β-catenin to the nucleus, like the WNT pathway that promotes proliferation.

Another mechanism by which cells regulate proliferation is through the NF2 gene. Merlin, the NF2 gene product, promotes contact inhibition by coupling cell-surface adhesion molecules such as E-cadherin to transmembrane receptor tyrosine kinase such as EGFR. This adhesion strengthening also limits the emission of mitogenic signals [21, 22].

One of the complex underpinnings enhancing the tumor suppressor pathway in tumors (e.g., colon, pancreas) is dysregulation of the WNT signaling pathway. WNT pathway activation facilitates β-catenin activity by blocking the formation of a "destruction complex" composed of APC and other proteins. β-catenin, in turn, regulates cell proliferation by modulating transcription factors such as MYC and cyclin D1. β-Catenin activity is kept in check by this destruction complex composed of APC and other proteins. Mutations in the *APC* gene occur in 70–80% of early colonic adenomas, and carcinomas lead to unchecked activity of β-catenin.

1.3.2 Downregulation of Cell Proliferation Inhibitor—Transforming Growth Factor β (TGF-β) Pathway

In many cancers such as colon, stomach, and endometrium, there are loss-of-function mutations in TGF-β receptors leading to loss of inhibitory control on cellular proliferation. There is downregulation of SMAD genes in pancreatic tumors, downstream signal transducers in TGF-β, thus corrupting the TGF-β pathway. In addition to promoting tumor growth, depending on the cellular context, altered TGF-β signaling is found to activate epithelial-mesenchymal transition (EMT) that makes the cancer cell more aggressive [23, 24].

1.4 Resisting Cell Death

Apoptosis, programmed cell death dismantling cells into components, is activated in various cellular conditions orchestrated by many signaling circuitry, including TP53 and RB1 checkpoints. The signaling circuitry governing this program is triggered by physiologic stresses such as exaggerated oncogenic signaling, DNA damage due to chemotherapy, radiation, or hyperproliferative state of the cancer cell. The activation of the apoptotic program is thus achieved through two significant circuits—extrinsic to cell (death receptor—e.g., Fas ligand / Fas receptor) and intrinsic (mitochondrial) cellular programs, and also perforin/granzyme pathway. Intrinsic and extrinsic pathways activate a proteolytic cascade of caspases (caspase 8, 9, and 3) that destroy the cell in an organized manner to form apoptotic bodies observed and reported in pathology literature many decades. Perforin / Granzyme pathway is activated by cytotoxic T cells either by activating caspase 10 or directly lead to DNA cleavage. In the intrinsic pathway, there is disruption of mitochondrial membrane activating BAX/BAK pro-apoptotic proteins that increase the permeability of mitochondrial membrane that allows cytochrome c to leak into the cytoplasm where it activates caspase-9 through APAF-1 binding [25].

Abrogation of the apoptotic program in cancer cells is possible by two central mechanisms and is associated with high-grade tumors and developing resistance to therapies. The tumor cells counterbalance pro-apoptotic signaling by upregulating anti-apoptotic proteins such as Bcl-2 family of proteins [26, 27]. For example, in follicular lymphoma, the molecular underpinning is the t(14;18)(q32;q21) translocation leading to constitutive activation of Bcl2 by transcriptionally active IGH promotor region. The IGH-Bcl2 fusion help escape apoptosis. Another mechanism of escape from apoptosis is achieved through the loss of TP53 function. Loss of TP53 prevents upregulation of a BH3-only protein called PUMA (BH3 proteins neutralize actions of Bcl2 related anti-apoptotic proteins). The evasion of apoptotic circuitry is complex in different tumors and different mechanisms are derailed to evade apoptosis.

1.4.1 Autophagy

In a state of severe nutrient deficiency, cells activate signaling circuits that either arrest cell growth or even cannibalize their own organelles, proteins, and membranes for energy production. In a rapidly growing tumor, the cells may become dormant activating autophagy due to scarcity of nutrients. A similar phenomenon may be used by cancer cells under therapeutic pressure, thus making tumor resistant to therapies and leading to therapeutic failures. The role of autophagy in cancer is still highly debatable and is based on the state of the tumor cell; it can be viewed as tumor protective in certain circumstances and adverse in others [28].

1.5 Enabling Replicative Immortality: The Stem-Cell-Like Properties

Over the last two decades, literature on cancer stem cells has significantly evolved through intense cancer research, albeit still under debate. The first descriptions of cancer stem cells were published on acute myeloid leukemia by Bonnet and Dick in 1997 [29]. The basic property of a cancer as having replicative immortality is conceptualized through its ability to self-replicate and differentiate through cancer stem-cell, a cell that is thought to be like a normal stem-cell counterpart. The debate continues whether the "stemness" arises through the transformation of a normal stem cell or through the ability of the cancer cells to transform into stem-like state through the acquisition of mutation through the selection process. In either pathway, the molecular underpinnings are thought to be by evading senescence and mitotic crisis and self-renewal ability.

Normal cells typically divide 50 to 70 times ("Hayflick limit") [30, 31], after which the cells become senescent, i.e., lose their ability to divide again. But cancer cells gain ability to evade senescence in part by abnormalities in RB and TP53 proteins. Maintaining RB protein in hypophosphorylated (active) state and downregulation of TP53 leads to disruption of G1/S cell cycle checkpoint, thus keeping the cell in a proliferative mode.

Another mechanism by which cancer cells achieve senescence evasion is through the maintenance of telomerase, which restores/"replenishes" the telomere sequences. Normally replicative capacity of a cell is lost after each cell division due to the shortening of the telomere, a repetitive nucleotide sequence at each end of the chromosome protecting the chromosomal deterioration and truncation. Eventually, the cell loses its replicative capacity completely and becomes senescent. But in many of tumors, human telomerase reverse transcriptase (TERT) upregulation leads to overexpression of telomerase enzyme, thus defying mitotic crisis and senescence [32, 33]. The gain of telomerase function may be induced due to cellular mitotic crisis from oncogenic program activation in cancer cells, thus maintaining the replicative potential and the damage to oncogenes and tumor suppressor genes. The telomerase activity is also thought to be upregulated in cancer stem cells.

Tissue stem cells and germ cells retain the capability to divide indefinitely through numerous cell divisions. By expressing telomerase, they resist mitotic crises and the accumulation of mutations. As stem cells divide, during embryogenesis or when a cell is under stress, either both (symmetric) or one (asymmetric) remains as daughter cells [34]. Cancer stem cells are thought to retain their limitless proliferative capacity and immortality, although debate continues as to the mechanism and number of cancer stem cells in different types of cancers. In summary, a small proportion of cancer cells retain or acquire stem-like properties that inactivate senescence signals, reactivate telomerase, and enable replicative immortality.

1.6 Angiogenesis

For tumors to grow and sustain proliferative advantage, an adequate supply of oxygen, nutrients, and metabolic waste management needs an adequate blood supply. Cancers stimulate neoangiogenesis to sustain its growth. The angiogenesis is observed under the microscope in some tumors extensive vascular network, for example, primary and metastatic renal cell carcinoma [35], while in some other tumors, few blood vessels such as scirrhous invasive ductal carcinoma of the breast (Fig. 1.1). The current understanding of angiogenesis is controlled by a balance between promoters and inhibitors of vascular proliferation [36, 37]. In the physiologic states, such as female endometrial cycling and wound healing, "angiogenic switch" is triggered transiently. However, in cancers as the tumor grows in size, the quiescent stage is terminated, and vascular proliferation is initiated by increased local production of angiogenic factors and/or loss

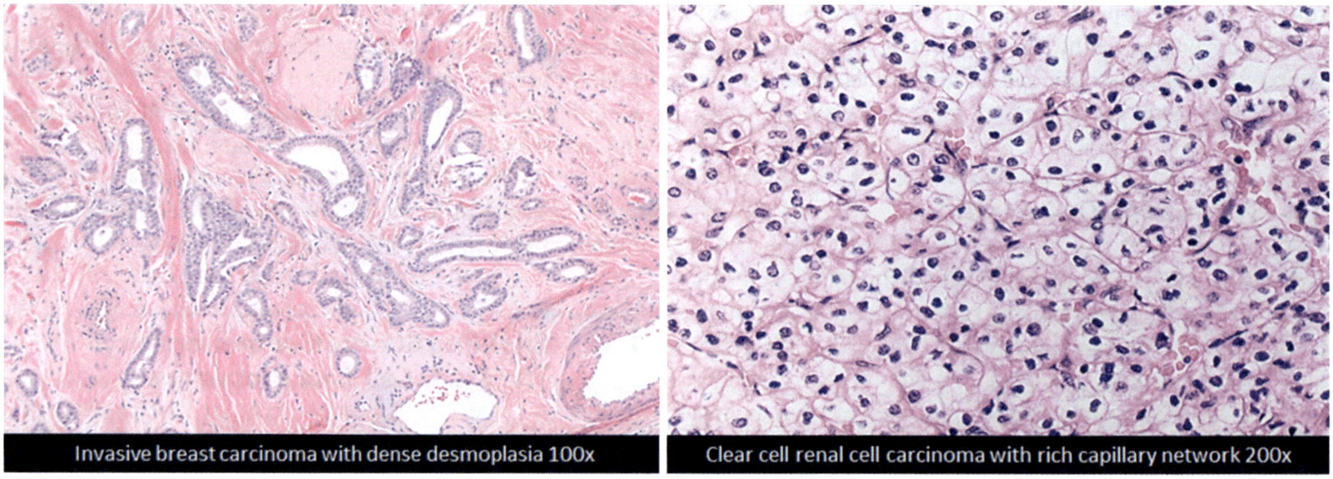

Fig. 1.1 Invasive breast carcinoma with dense desmoplasia and clear cell carcinoma with rich capillary network

of angiogenic inhibitors [17]. The angiogenic factors are locally produced by tumor cells, inflammatory cells, or stromal cells. Some of the inflammatory cells and, in particular, endothelial progenitor cells are thought to be bone marrow-derived in some cases [38]. The molecular underpinnings of angiogenesis in tumor cells include upregulation of many genes, the key genes include hypoxia-inducible transcription factor 1α (HIF1α), vascular endothelial growth factor (VEGF), as angiogenesis inducers, and thrombospondin-1 (TSP-1) as an inhibitor. The pro-angiogenic factors include: Hypoxia led stabilization of HIF1α that upregulates VEGF gene expression. VEGF is also upregulated by oncogene signaling. This stimulates endothelial growth towards the tumor. Normally, TP53 upregulates TSP-1 expression that inhibits angiogenesis by repressing VEGF. TP53 mutations lead to p53 protein loss that permits angiogenesis.

Earlier it was thought that angiogenesis sets in fully when the tumors proliferate. But recent literature has shown that the histologic changes of angiogenesis, such as microvessel density, are noted in the pre-malignant stages of neoplasms in various organs, including squamous cell lesions in the oral mucosa, skin, uterine cervix, vulva, and anal canal [39, 40]. The fundamental properties of pre-malignant conditions for prevention were first described in 1976 by Sporn [41]. Using the knowledge of the molecular basis of hallmarks of cancer, Ryan and Faupel-Badger correlated these concepts with these properties of pre-malignant conditions [42], further supporting the acquisition of many of the hallmarks of cancers in the early stages of development.

1.7 Invasion and Metastasis

One of the most aggressive properties of cancer cells is the invasion and destruction of local tissues and metastasis. The original concept of a pattern of cancer metastasis was described by an English surgeon Stephen Paget in 1889 published as "seed and soil" hypothesis [43] [44]. Paget introduced the concept of the spread of tumor cells through interactions between the cancer cells (seed) and the host organ (soil). Even after a century, this concept is upheld by extensive research with "seed" now identified as the cancer stem cell [45]. It has been shown that this hallmark of cancer occurs through a complex multistep progression called the invasion-metastasis cascade.

The major steps in this process include invasion of the extracellular matrix (ECM), angiolymphatic dissemination, extravasation and tissue homing, and colonization. From primary site to metastatic site, the journey of cancer cells includes breaching the supporting basement membrane, navigating the interstitial connective tissues, and penetrating the vascular wall to gain access to vessels. During homing and colonization, this process is reversed. In the first step, the cells become dis-cohesive and reduce interactions ("Loosening up" of tumor cells). This is followed by degradation of extracellular matrix (ECM), attachment to "remodeled" ECM components, migration, and invasion of tumor cells. In the first step, the well-known alteration in cancer cells is loss of adhesion molecule E-cadherin encoded by the *CDH1* gene through a process called epithelial-mesenchymal transition (EMT). EMT, thought to be an integral part of metastasis, especially in breast and prostate cancers, is controlled by a combination of Snail, Slug, Twist, and Zeb1/2 transcription factors, favoring promigratory properties. The next step is the degradation of the basement membrane and interstitial connective tissues. Through autocrine or paracrine secretions, proteolytic enzymes such as matrix metalloproteinase (MMP), cathepsin D, and urokinase plasminogen activator help cancer cells invade. The final step of invasion is the migration of cancer cells by locomotion through basement membranes and areas of proteolysis by the contractile actin cytoskeleton. Finally, through vascular dissemination, homing, and colonization, a few cancer cells are successful in distant metastasis. Some tumors metastasize to specific sites, thus showing organ tropism. The underlying molecular mechanism is thought to be through the expression of adhesion or chemokine receptors, whose ligands are expressed by endothelial cells at the metastatic site [46].

The details of molecular underpinnings of metastasis are described in a separate chapter in this section, and only high-level key discussion points are described above.

1.8 Emerging Hallmarks and Enabling Characteristics

With mounting literature support, two additional hallmarks of cancer, including **Deregulating Cellular Energetics** and **Avoiding Immune Destruction** and enabling characteristics **Genomic Instability** and **Mutation and Tumor-promoting Inflammation** were introduced [2]. Each of these are described below.

1.9 Growth Promoting Altered Cellular Metabolism (Deregulating Cellular Energetics)

To support the autonomous growth of cancer cells and get oxygen and nutrients through neoangiogenesis, cancer cells reprogram the energy metabolism, which is recently described as a key hallmark of cancer [2]. Cancer metabolism research in cancer biology predates the discovery of oncogenes and tumor suppressor genes [47]. There is overwhelming research support that cancer cells reprogram metabolism to improve cellular sustainability and resilience to gain selective advantage. The cancer cells change their metabolic pathway to the glycolytic pathway as the preferred source of energy where glucose is converted to lactose to generate ATP, even under aerobic conditions. This was first described by Otto Warburg in 1930 and is now known as the Warburg effect or aerobic glycolysis [48]. Although ATP generation is lower than the mitochondrial oxidative metabolic pathway through the tricarboxylic acid (TCA) cycle, this pathway is preferred as its metabolic intermediates are used by the rapidly proliferating cells to synthesize cellular components to fulfill the biosynthetic requirement. Biosynthesis of macromolecules or anabolic pathways is an integral part of cancer metabolism to support growth.

The molecular underpinnings of the altered energy metabolism include upregulation of oncogenes in receptor tyrosine kinase/PI3K-AKT-mTOR signaling that enhances the activity of glucose transporters and multiple glycolytic enzymes promoting lipid, protein, and nucleic acid biosynthesis. Overexpression of transcription factor MYC leads to upregulation of genes involved in anabolic metabolism through glycolytic enzymes and glutaminase. On the tumor suppressor side, STK11, an antagonist of oxidative glycolysis, is downregulated to promote cell growth.

As described above, autophagy is activated in nutrient-deficient conditions where cellular organelles and other components are digested and reused. Depending on the scenario, the tumor inactivates autophagy signaling to promote growth. On the other side, tumor cells may use autophagy to go into a "hibernation" state due to external pressures, such as chemotherapy, to become dormant and prevent cell death.

Recent sequencing data in certain tumors such as gliomas, acute myeloid leukemia, intrahepatic cholangiocarcinoma, and central chondrosarcoma has shown somatic mutations in Krebs cycle (TCA) genes, viz. isocitrate dehydrogenase genes-1 and 2 (*IDH1, IDH2*). Point mutations in *IDH1* or *IDH2* genes lead to mutant *IDH* protein, which loses its normal function and converts α-ketoglutarate to 2-hydrozyglutarate. This protein is considered an oncometabolite which acts as an inhibitor of epigenetic regulatory enzymes leading to cancer gene signaling pathway activation. Similarly, other enzymes from the TCA cycle that are mutated in other cancers include succinate dehydrogenase (SDH), fumarate hydratase (FH), and L-2-hydroxyglutarate dehydrogenase. These oncometabolites present a potential therapeutic target [49].

1.10 Evasion of Host Defense (Avoiding Immune Destruction)

Paul Ehrlich first hypothesized the role of host defense by the immune system in eliminating the tumor cells by recognizing them as "foreign." Lewis Thomas introduced and Macfarlane Burnet then consolidated this hypothesis and termed it as "immune surveillance." This phenomenon is thought to serve as a normal function of the immune system by constantly scanning the body for tumor cells and eliminating them [17, 50]. This concept was tested by several experiments and observations and was initially abundant due to a lack of sufficient evidence. Burnet suggested that the cancer cells' expression of tumor-specific neo-antigens induced an immunological response, thus eliminating such cells before clinical manifestations. We now know that certain tumors are heavily infiltrated by lymphocytes and other immune cells. Such tumors often harbor numerous driver as well as passenger mutations that have been discovered through high-throughput whole-genome sequencing. This "hyper mutator" genotype results in the production of neoantigens that induces a robust host CD8+ cytotoxic T-cell cell response, as reported in many malignant melanomas, many colorectal cancers, lung cancers, among others [51] (Fig. 1.2). Direct demonstration of tumor-specific T cells and antibodies in patients with cancers was noted in a few. Some experiments challenged the hypothesis of immune surveillance, which showed that not all the immunodeficient or immunosuppressed patients develop cancers. However, in immunocompromised patients (acquired or hereditary), viral associated tumors are prevalent (HPV, EBV, HHV8), and so are nonviral-associated solid tumors [52].

Dunn and Schreiber developed a concept of "cancer immunoediting," which includes three stages of "sculpting of tumor" by the immune system. The immune surveillance hypothesis was proposed to be a general part of the cancer editing process and is considered the first stage called the **elimination** of tumor cells. The next step is called the **equilibrium** phase, which appears to be the longest that leads to the formation of new variants in the cancer cells with different mutations that resist the immune pressure. Different molecular mechanisms are triggered when cancer cells go into quiescence or the slow-cycling phase described above. Cancer cells likely activate senescence gene signaling to avoid cell death induced by immune cells. From this stage, some clines emerge and **escape** this immune surveillance and evade immune destruction. The molecular underpinnings include selective outgrowth of antigen-negative variants, loss or reduced expression of major histocompatibility (MHC) class I molecules escaping an attack from CD8+ cytotoxic T cells (CTL). Another mechanism that cancer cells use to evade immune destruction is by inhibiting T cell activation. The tumor cells upregulate inhibitory receptor CTLA-4 on tumor-specific T cells preventing sensitization and extended escape from the tumor-specific T cells. Another mechanism includes overexpression of Programmed death ligand (PD-L)-1 and PD-L2 cell surface proteins that inactivates CD8+ CTL by docking with PD-1 receptor. Thus, tumor cells avoid detection for elimination and escape

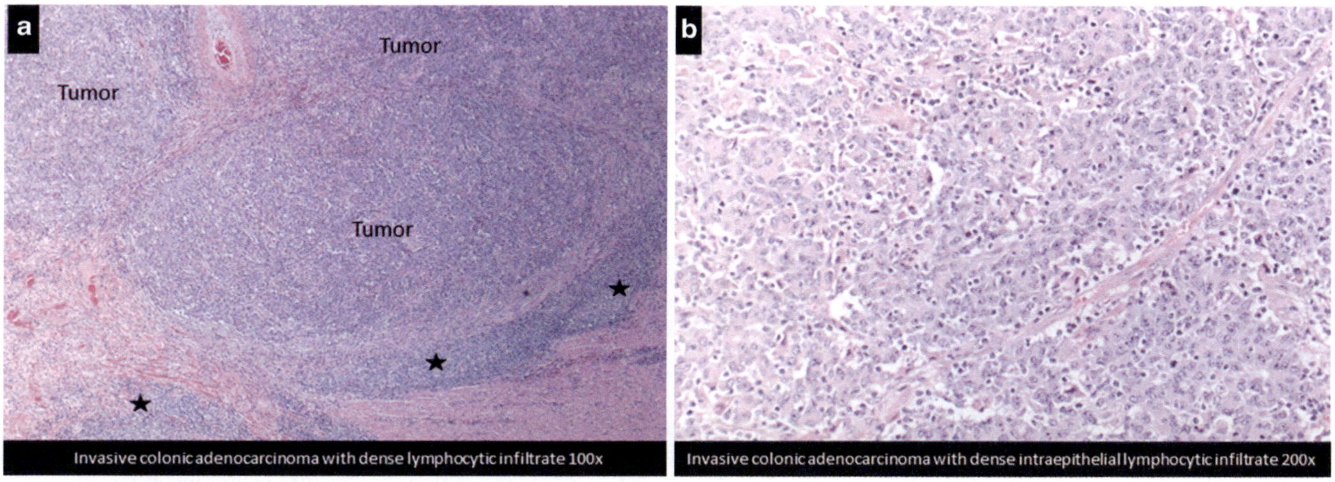

Fig. 1.2 Invasive colonic adenocarcinoma with dense Crohn's-like inflammatory response, (**a**) stars marking peritumoral dense lymphocytic infiltrate (100×), (**b**) depicting intra-epithelial lymphocytes (200×)

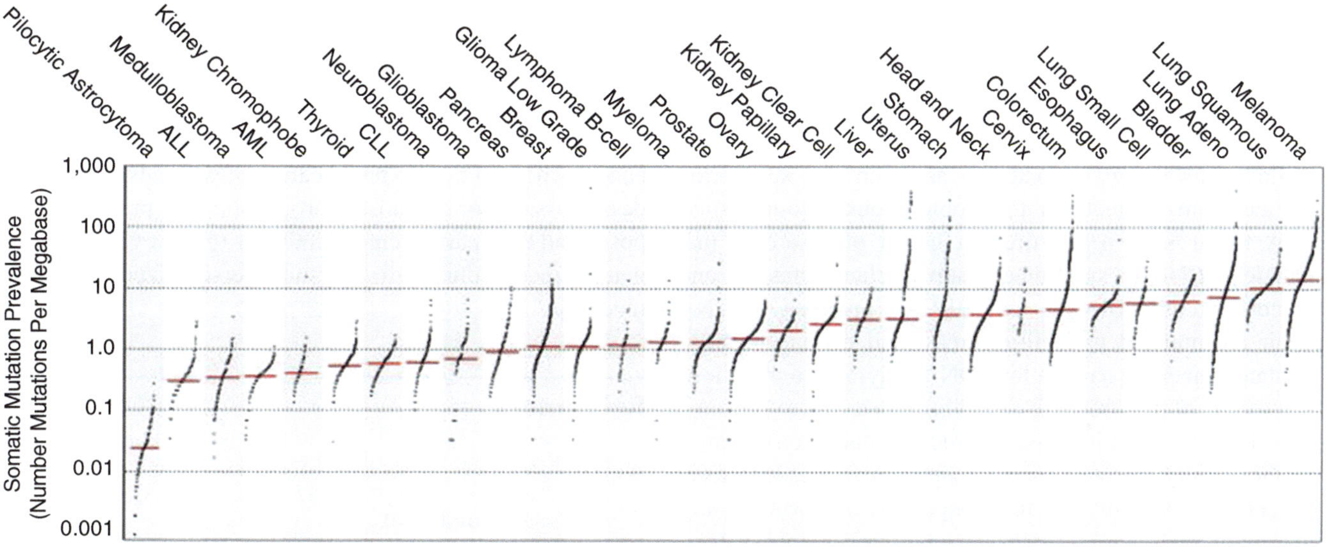

Fig. 1.3 Prevalence of somatic mutations per megabase across human cancer types in ascending order from left to right [53]

immune surveillance. The anti-PD-L1 and anti-CTLA-4 therapies block this inhibition in some solid tumors that overexpress PD-L1 protein (melanoma, lung cancers) and renders them vulnerable to the CD8+ CTL. These therapies (antibodies) that overcome immune evasion by the tumors are called immune checkpoint inhibitors (Fig. 1.3).

Detailed molecular underpinnings of this process is described in the immune microenvironment chapter of this section.

1.11 Genomic Instability

Genomic instability as a distinct enabling characteristics was added to the original six hallmarks in 2011 [2]. The success of all the hallmarks of cancer depends profoundly on the underlying genomic alterations, directly or indirectly. Through multistep gain of mutations, the cancer cell clones get a survival advantage to become more autonomous. This phenomenon is nurtured and amplified by inactivation of genome maintenance system to detect and resolve these defects, either due to somatic alterations (e.g., epigenetic alterations in DNA mutation and histone modifications) or inherited mutations (e.g., TP53 (Li-Fraumeni syndrome) or Lynch (HNPCC) syndrome due to mutations in the mismatch repair (MMR) genes). Broad categories that are part of this genome maintenance system include genes that produce proteins involved in detecting DNA damage and activating the repair machinery, proteins directly repairing damaged DNA, and proteins inactivating or intercepting mutagenic molecules before they have damaged the DNA [54, 55].

Genomic instability is acquired due to the variable loss of these functions. TP53 gene pauses the cell function when there is DNA damage to allow DNA repair to occur. TP53 gene mutations are most frequently observed in many tumors described in the sections above, leading to further accumulation of mutations. DNA MMR deficiency is another mechanism that leads to the accumulation of numerous mutations (hypermutator phenotype) producing many tumors in many organs (e.g., proximal colonic adenocarcinomas, endometrial carcinomas). Loss-of-function mutations in genes encoding nucleoside excision repair system lead to inability to repair UV radiation induced cross-linking of pyrimidine residues thwarting replication, as seen in xeroderma pigmentosum. Defects in the homologous recombination DNA repair system constitute a group of disorders that includes Bloom syndrome, ataxia-telangiectasia, and Fanconi anemia. BRCA1 and BRCA2 involved in DNA repair are mutated in familial breast and ovarian cancers, prostate cancers to name a few. DNA polymerase mutations lead to loss of proofreading capabilities that lead to genomic instability, as noted in some colonic and endometrial carcinomas. Finally, many lymphoid neoplasms show a high level of genomic instability due to errors occurring during immunoglobulin and T cell receptor gene rearrangements (RAG1, RAG2, activation-induced cytosine deaminase).

1.12 Cancer-Enabling Inflammation

In pathology literature, observations of dense inflammatory infiltrate in certain solid tumors have been described for many decades. The density of the infiltrate, from dense to very few noticeable cells, as well as the composition of cells—acute vs. chronic inflammatory cells (innate and adaptive), vary in different morphologic subtypes. Such immune response was thought to be a favorable phenomenon of the immune system attempting to eliminate tumor cells, as described above, shaping the newer clones of tumor cells that eventually escape destruction. But there is enough evidence that the inflammatory response may have a paradoxical effect of enabling cancer progression [2]. Inflammation enables several hallmarks of cancer by having a direct effect on the tumor cells as well as modifying the tumor microenvironment described in the latter part of this section. Few key characteristics include the release of factors that promote proliferation, removal of growth suppressors, enhanced resistance to cell death, inducing angiogenesis, activating invasion and metastasis, and evading immune destruction [17].

1.13 Conclusion

Cancer development is a complex biological process that is underpinned by the multistep acquisition of molecular genetic aberrations imparting the above-described eight hallmarks of cancers and two enabling traits. These are gained either through small changes such as single nucleotide polymorphisms, insertions/deletions, and amplification or through large structural genetic changes such as translocations, aneuploidy, and genomewide methylation, to name a few. These changes confer a survival advantage and ability to invade, recur, and metastasize to distant sites. Several molecularly targeted therapies are currently available that are the standard of practice, while some molecular markers part of AJCC cancer staging. Understanding the molecular basis of cancer will further help healthcare professionals, researchers design assays for diagnostic, prognostic, and predictive purposes and the pharmaceutical industry to discover newer targeted therapeutics to pursue personalized precision medicine.

References

1. Hanahan D, Weinberg RA. The hallmarks of cancer. Cell. 2000;100(1):57–70. https://doi.org/10.1016/s0092-8674(00)81683-9.
2. Hanahan D, Weinberg RA. Hallmarks of cancer: the next generation. Cell. 2011;144(5):646–74. https://doi.org/10.1016/j.cell.2011.02.013.
3. Westermark B. Platelet-derived growth factor in glioblastoma-driver or biomarker? Ups J Med Sci. 2014;119(4):298–305. https://doi.org/10.3109/03009734.2014.970304.
4. Xie Y, Su N, Yang J, Tan Q, Huang S, Jin M, et al. FGF/FGFR signaling in health and disease. Signal Transduct Target Ther. 2020;5(1):181. https://doi.org/10.1038/s41392-020-00222-7.
5. Hynes NE, MacDonald G. ErbB receptors and signaling pathways in cancer. Curr Opin Cell Biol. 2009;21(2):177–84. https://doi.org/10.1016/j.ceb.2008.12.010.
6. Lemmon MA, Schlessinger J. Cell signaling by receptor tyrosine kinases. Cell. 2010;141(7):1117–34. https://doi.org/10.1016/j.cell.2010.06.011.
7. Perona R. Cell signalling: growth factors and tyrosine kinase receptors. Clin Transl Oncol. 2006;8(2):77–82. https://doi.org/10.1007/s12094-006-0162-1.
8. Witsch E, Sela M, Yarden Y. Roles for growth factors in cancer progression. Physiology (Bethesda). 2010;25(2):85–101. https://doi.org/10.1152/physiol.00045.2009.
9. Consortium, A. P. G. AACR project GENIE: powering precision medicine through an international consortium. Cancer

Discov. 2017;7(8):818–31. https://doi.org/10.1158/2159-8290.CD-17-0151.
10. Davies MA, Samuels Y. Analysis of the genome to personalize therapy for melanoma. Oncogene. 2010;29(41):5545–55. https://doi.org/10.1038/onc.2010.323.
11. Guo YJ, Pan WW, Liu SB, Shen ZF, Xu Y, Hu LL. ERK/MAPK signalling pathway and tumorigenesis. Exp Ther Med. 2020;19(3):1997–2007. https://doi.org/10.3892/etm.2020.8454.
12. Osaki M, Oshimura M, Ito H. PI3K-Akt pathway: its functions and alterations in human cancer. Apoptosis. 2004;9(6):667–76. https://doi.org/10.1023/B:APPT.0000045801.15585.dd.
13. Yuan TL, Cantley LC. PI3K pathway alterations in cancer: variations on a theme. Oncogene. 2008;27(41):5497–510. https://doi.org/10.1038/onc.2008.245.
14. Magrath I. Epidemiology: clues to the pathogenesis of Burkitt lymphoma. Br J Haematol. 2012;156(6):744–56. https://doi.org/10.1111/j.1365-2141.2011.09013.x.
15. Patterson KI, Brummer T, O'Brien PM, Daly RJ. Dual-specificity phosphatases: critical regulators with diverse cellular targets. Biochem J. 2009;418(3):475–89. https://doi.org/10.1042/bj20082234.
16. Tan MH, Mester JL, Ngeow J, Rybicki LA, Orloff MS, Eng C. Lifetime cancer risks in individuals with germline PTEN mutations. Clin Cancer Res. 2012;18(2):400–7. https://doi.org/10.1158/1078-0432.CCR-11-2283.
17. Kumar V, Abbas AK, Aster JC, Turner JR, Robbins SL, Cotran RS. Robbins & Cotran pathologic basis of disease; 2021.
18. Collado M, Serrano M. Senescence in tumours: evidence from mice and humans. Nat Rev Cancer. 2010;10(1):51–7. https://doi.org/10.1038/nrc2772.
19. Knudson AG. Mutation and cancer: statistical study of retinoblastoma. Proc Natl Acad Sci. 1971;68(4):820–3. https://doi.org/10.1073/pnas.68.4.820.
20. Robinson DR, Wu YM, Lonigro RJ, Vats P, Cobain E, Everett J, et al. Integrative clinical genomics of metastatic cancer. Nature. 2017;548(7667):297–303. https://doi.org/10.1038/nature23306.
21. Curto M, Cole BK, Lallemand D, Liu CH, McClatchey AI. Contact-dependent inhibition of EGFR signaling by Nf2/Merlin. J Cell Biol. 2007;177(5):893–903. https://doi.org/10.1083/jcb.200703010.
22. Okada T, Lopez-Lago M, Giancotti FG. Merlin/NF-2 mediates contact inhibition of growth by suppressing recruitment of Rac to the plasma membrane. J Cell Biol. 2005;171(2):361–71. https://doi.org/10.1083/jcb.200503165.
23. Ikushima H, Miyazono K. TGFbeta signalling: a complex web in cancer progression. Nat Rev Cancer. 2010;10(6):415–24. https://doi.org/10.1038/nrc2853.
24. Massague J. TGFbeta in cancer. Cell. 2008;134(2):215–30. https://doi.org/10.1016/j.cell.2008.07.001.
25. Elmore S. Apoptosis: a review of programmed cell death. Toxicol Pathol. 2007;35(4):495–516. https://doi.org/10.1080/01926230701320337.
26. Adams JM, Cory S. The Bcl-2 apoptotic switch in cancer development and therapy. Oncogene. 2007;26(9):1324–37. https://doi.org/10.1038/sj.onc.1210220.
27. Lowe SW, Cepero E, Evan G. Intrinsic tumour suppression. Nature. 2004;432(7015):307–15. https://doi.org/10.1038/nature03098.
28. Saha S, Panigrahi DP, Patil S, Bhutia SK. Autophagy in health and disease: a comprehensive review. Biomed Pharmacother. 2018;104:485–95. https://doi.org/10.1016/j.biopha.2018.05.007.
29. Bonnet D, Dick JE. Human acute myeloid leukemia is organized as a hierarchy that originates from a primitive hematopoietic cell. Nat Med. 1997;3(7):730–7. https://doi.org/10.1038/nm0797-730.
30. Burnet Sir M. Intrinsic mutagenesis: a genetic approach to ageing. Dordrecht: Springer Netherlands; 2012.
31. Hayflick L, Moorhead PS. The serial cultivation of human diploid cell strains. Exp Cell Res. 1961;25:585–621. https://doi.org/10.1016/0014-4827(61)90192-6.
32. Heidenreich B, Rachakonda PS, Hemminki K, Kumar R. TERT promoter mutations in cancer development. Curr Opin Genet Dev. 2014;24:30–7. https://doi.org/10.1016/j.gde.2013.11.005.
33. Maciejowski J, de Lange T. Telomeres in cancer: tumour suppression and genome instability. Nat Rev Mol Cell Biol. 2017;18(3):175–86. https://doi.org/10.1038/nrm.2016.171.
34. Batlle E, Clevers H. Cancer stem cells revisited. Nat Med. 2017;23(10):1124–34. https://doi.org/10.1038/nm.4409.
35. Zee YK, O'Connor JP, Parker GJ, Jackson A, Clamp AR, Taylor MB, et al. Imaging angiogenesis of genitourinary tumors. Nat Rev Urol. 2010;7(2):69–82. https://doi.org/10.1038/nrurol.2009.262.
36. Baeriswyl V, Christofori G. The angiogenic switch in carcinogenesis. Semin Cancer Biol. 2009;19(5):329–37. https://doi.org/10.1016/j.semcancer.2009.05.003.
37. Bergers G, Benjamin LE. Tumorigenesis and the angiogenic switch. Nat Rev Cancer. 2003;3(6):401–10. https://doi.org/10.1038/nrc1093.
38. Patenaude A, Parker J, Karsan A. Involvement of endothelial progenitor cells in tumor vascularization. Microvasc Res. 2010;79(3):217–23. https://doi.org/10.1016/j.mvr.2010.01.007.
39. Menakuru SR, Brown NJ, Staton CA, Reed MW. Angiogenesis in pre-malignant conditions. Br J Cancer. 2008;99(12):1961–6. https://doi.org/10.1038/sj.bjc.6604733.
40. Raica M, Cimpean AM, Ribatti D. Angiogenesis in pre-malignant conditions. Eur J Cancer. 2009;45(11):1924–34. https://doi.org/10.1016/j.ejca.2009.04.007.
41. Sporn MB. Approaches to prevention of epithelial cancer during the preneoplastic period. Cancer Res. 1976;36(7 PT 2):2699–702. Retrieved from https://www.ncbi.nlm.nih.gov/pubmed/1277177.
42. Ryan BM, Faupel-Badger JM. The hallmarks of premalignant conditions: a molecular basis for cancer prevention. Semin Oncol. 2016;43(1):22–35. https://doi.org/10.1053/j.seminoncol.2015.09.007.
43. Talmadge JE, Fidler IJ. AACR centennial series: the biology of cancer metastasis: historical perspective. Cancer Res. 2010;70(14):5649–69. https://doi.org/10.1158/0008-5472.CAN-10-1040.
44. Akhtar M, Haider A, Rashid S, Al-Nabet A. Paget's "seed and soil" theory of cancer metastasis: an idea whose time has come. Adv Anat Pathol. 2019;26(1):69–74. https://doi.org/10.1097/PAP.0000000000000219.
45. Langley RR, Fidler IJ. Tumor cell-organ microenvironment interactions in the pathogenesis of cancer metastasis. Endocr Rev. 2007;28(3):297–321. https://doi.org/10.1210/er.2006-0027.
46. Riggi N, Aguet M, Stamenkovic I. Cancer metastasis: a reappraisal of its underlying mechanisms and their relevance to treatment. Annu Rev Pathol. 2018;13:117–40. https://doi.org/10.1146/annurev-pathol-020117-044127.
47. DeBerardinis RJ, Chandel NS. Fundamentals of cancer metabolism. Sci Adv. 2016;2(5):e1600200. https://doi.org/10.1126/sciadv.1600200.
48. Koppenol WH, Bounds PL, Dang CV. Otto Warburg's contributions to current concepts of cancer metabolism. Nat Rev Cancer. 2011;11(5):325–37. https://doi.org/10.1038/nrc3038.
49. Collins RRJ, Patel K, Putnam WC, Kapur P, Rakheja D. Oncometabolites: a new paradigm for oncology, metabolism, and the clinical laboratory. Clin Chem. 2017;63(12):1812–20. https://doi.org/10.1373/clinchem.2016.267666.
50. Ribatti D. The concept of immune surveillance against tumors. The first theories. Oncotarget. 2017;8(4):7175–80. https://doi.org/10.18632/oncotarget.12739.

51. Schumacher TN, Schreiber RD. Neoantigens in cancer immunotherapy. Science. 2015;348(6230):69–74. https://doi.org/10.1126/science.aaa4971.
52. Dunn GP, Bruce AT, Ikeda H, Old LJ, Schreiber RD. Cancer immunoediting: from immunosurveillance to tumor escape. Nat Immunol. 2002;3(11):991–8. https://doi.org/10.1038/ni1102-991.
53. Alexandrov LB, Nik-Zainal S, Wedge DC, Aparicio SA, Behjati S, Biankin AV, et al. Signatures of mutational processes in human cancer. Nature. 2013;500(7463):415–21. https://doi.org/10.1038/nature12477.
54. Ciccia A, Elledge SJ. The DNA damage response: making it safe to play with knives. Mol Cell. 2010;40(2):179–204. https://doi.org/10.1016/j.molcel.2010.09.019.
55. Negrini S, Gorgoulis VG, Halazonetis TD. Genomic instability—an evolving hallmark of cancer. Nat Rev Mol Cell Biol. 2010;11(3):220–8. https://doi.org/10.1038/nrm2858.

Unifying Concept of Genomic Changes: The Mutational Landscape of Cancers

2

Laura Favazza DO

Abstract

The genomic landscape of cancer is complex, can be described in multiple different ways, and is still evolving. As technology evolves and more tumors are sequenced, we discover more each day about patterns of DNA alterations, large- and small-scale chromosomal and copy number changes, epigenetic and methylome modifications, and subsequent RNA alterations among different tumor types. In this chapter we explore the global molecular alterations in different cancer types and how these alterations are ultimately related to one another. The molecular alterations discussed include chromosomal changes, smaller-scale copy number alterations, DNA mutations, epigenetic profiles, mRNA and miRNA variations, and finally we will briefly discuss protein expression in a multitude of different tumor types. These findings may be used to help us understand tumorigenesis and tumor progression. They also can help us further classify tumors based on molecular findings, rather than the traditional, purely histologic findings, which may or may not provide valuable information regarding predictive factors and prognostic information. This chapter goes into detail with examples of several different types of molecular alterations in multiple different tumor types, how these may be used to further classify tumors, and how the various alterations are associated to one another.

Learning Objectives

- Understand the types of chromosomal copy number changes that can occur in cancer.
- Discuss the different types of mutational signatures and their prevalence in different cancer types.
- Learn how DNA hypomethylation and DNA hypermethylation affects genetic transcription.
- Recognize the effects of miRNA on transcription in tumors.

L. Favazza DO (✉)
Pathology and Laboratory Medicine, Henry Ford Hospital, Detroit, MI, USA
e-mail: lfavazz1@hfhs.org

2.1 Introduction

Across cancer there are multiple ways that the human genome, transcriptome, and proteosome may be altered in cancer. The genetic alterations in cancer may range from large chromosomal changes, which may be observed via karyotype, to small single nucleotide DNA substitutions, that can only be seen via DNA sequencing. Regardless of the size of the change, these variations can lead to cell increased cell cycling which promote tumor growth and progression. In this chapter we will explore the general global genomic alterations seen across these areas and across different tumor types.

2.2 Chromosomal Aneuploidy

While traditional chromosomal cytogenetics is less commonly performed in the era of next-generation sequencing, this was one of the first genetic classifications of tumors. Although we use more sensitive and specific techniques presently, classifications and observations have been made through this type of analysis. Chromosomal aneuploidy refers to whole chromosome losses and gains, and in cancer genetics chromosomal arm losses and gains are also included. Chromosomal instability is the process that gives rise to aneuploidy and promotes tumor evolution and intratumoral heterogeneity [1]. Chromosomal instability, for example, is the loss or gain of small genes or clusters of nearby genes and is also frequently associated with loss of p53 tumor suppressive function [2]. Chromosomal instability is associated

S. P. Leong et al. (eds.), *Cancer Metastasis Through the Lymphovascular System*,
https://doi.org/10.1007/978-3-030-93084-4_2

with aneuploidy; however, some tumors can have aneuploidy without chromosomal instability. Aneuploidy may be seen in approximately 80% of solid tumor neoplasms [3].

Tumors can be segregated into those with many alterations (copy number alterations high), moderate number of alterations, and those with few alterations [4]. Approximately one-third of cancers show relatively sparse number of chromosomal alterations. These tumor types include thyroid carcinoma, acute myelocytic leukemia, prostate carcinomas, and thymomas [4].

Patterns of aneuploidy can also be tumor-type specific. This is seen in testicular germ cell tumors which characteristically show chromosome 12p gains and also in IDH wild-type gliomas that have chromosome 7 gain and chromosome 10 loss. An exception to this is seen in endometrial cancers. Endometrial serous-like carcinomas have high copy number alterations and can be separated from other endometrial carcinomas [2]. Tumors from separate but similar origins also show recurring arm losses and gains. Gains of 8q, 13q, and chromosome 20 are seen in gastrointestinal tumors including stomach, pancreas, colorectal, and esophageal adenocarcinomas [2]. Squamous carcinomas, particularly esophageal, lung, and head and neck, characteristically show gain of 3q and loss of 3p [4].

Aneuploidy may be prognostic in multiple tumor types. Hyperdiploid karyotype in multiple myeloma and acute lymphoblastic lymphoma has a favorable overall survival [5]. While the hypodiploid subgroup is an adverse prognostic factor in acute lymphoblastic lymphoma [6]. Losses of 1p19q is disease defining for oligodendrogliomas, which has a favorable prognosis [7]. For many solid tumors, a high degree of aneuploidy is associated with adverse features, including disease progression, decreased overall survival, and decreased progression-free survival [1]. While extreme aneuploidy can sometimes make tumors more sensitive to chemotherapy, as a general rule, increased aneuploidy is usually associated with increased resistance to chemotherapy [8]. Interestingly when comparing the number of mutations to aneuploidy, those that are hypermutated (high microsatellite instability or *POL* ε mutation) show low aneuploidy [9]. However, when hypermutated tumors are excluded, increased number of mutation and increased aneuploidy are frequently seen [2].

There are four major ways to break down smaller structural changes in cancer [10]. The first group is simple cut-and-paste structural variants. These can be further broken down into a deletion of chromosome material, a reciprocal inversion, a fold-back inversion, an unbalanced translocation of chromosomal material from one chromosome to another, and finally a reciprocal translocation [10]. Obviously, in the deletion of chromosomal material there is a copy number change, however, with a balanced translocation there is no such change in copy number since no material is gained or lost. In this case the genetic material has moved to different locations with different promoters. Thus, using only traditional DNA sequencing, this type of translocation can be difficult to detect. The second group is also simple but involves a copy-and-paste structural variant as in a tandem duplication [10]. Depending on the size of the variants listed above, detection can be rather difficult with targeted sequencing panels. Among these structural variants, deletions have been descried as the most common followed by tandem duplications, and finally unbalanced translocations. Remarkably reciprocal translocations and inversions are the least common events [10]. Multiple types of these structural variants are found within specific tumors and tumor types. For example, a study found that esophageal adenocarcinomas harbored several deletions and many complex rearrangements, while ovarian cancers harbored frequent tandem duplications and deletions with fewer complex clustered rearrangements [10].

The last category involves complex cut-and-paste structural variants, which occur by multiple cycles of DNA damage via end-to-end sister chromatid fusions, mitotic bridge fusions, and unstable DNA leading to further breakage [10]. This group contains two types of chromosomal changes. The first, chromoplexy, is defined as many simultaneous double-stranded DNA breaks in several chromosomes which are then reconnected improperly, leading to multiple balanced translocations. This phenomenon is frequently seen in prostate cancers [10]. The second mechanism is called chromothripsis. Chromothripsis is considered a single catastrophic event, where the chromosomes shatter and rearrange, resulting in multiple copy number changes and few to hundreds of chromosomal breakpoints [10].

2.3 Somatic DNA Variants

Molecular variants can occur via multiple mechanisms including single-base substitutions, double-base substitutions, and complex insertions and/or deletions. Those that cause either activation of oncogenes or deleterious effects on tumor suppressor genes are considered pathogenic or likely pathogenic [11].

These patterns may be classified in many ways. One way is via mutational signatures, of which there are now greater than 30 signatures [12]. Most cancers display greater than one mutational signature. These signatures are based on patterns of mutational changes. For example, the single base pair substitutions are characterized by C > A, C > G, C > T, T > A, T > C, and T > G. There are also characterizations for double-base substitutions and insertions/deletions. Signatures specificity of mutation type and sequence context (signature 10), equal amount of all 96 mutations (signature 3), C > T predominant (1A/B, 6, 7, 11, 15, 19), C > A predominant (signatures 4, 8, 18), T > C predominant (signatures 5, 12, 16, 21), and T > G predominant (signatures 9, 17) [13].

Several of these signatures have been well characterized. They include 1A/B also known as the age signature. This is the endogenous mutational process in normal and neoplastic cells that occurs due to aging and involves deamination of 5-methyl-cytosine [13]. Signatures 2 and 13 are also called APOBEC (apolipoprotein B mRNA editing enzyme, catalytic polypeptide-like) signature and are seen in multiple tumor types and lead to cytosine to uracil substitutions from dysfunctional cytidine deaminases [13, 14]. Mutational signature 3 includes those with *BRCA1/2* mutations, whether they are germline or somatic. *BRCA1/2* mutations are most frequently encountered in breast cancers, but are also not uncommonly seen in prostate and pancreatic cancers. This signature is involved in defects in homologous recombination-based DNA double-stranded break repair [13].

Additional signatures are associated with defects in DNA repair. Those include signature 6, associated with DNA mismatch repair deficiency, and signature 10, associated with *Pol E* mutations, which result in tumors harboring numerous mutations. DNA mismatch repair deficient tumors usually have somatic hypermutation and may arise either via germline or somatic mutations of *MLH1*, *PMS2*, *MSH2*, *MSH6*, and *EPCAM*. These are most commonly associated with colorectal carcinoma as well as endometrial carcinomas. Inactivation of *MLH1* can also occur sporadically via promoter hypermethylation, and are also associated with colorectal carcinomas and endometrial carcinomas. Mutations in *POL ε* typically occur via hot spot mutations in endometrial carcinomas, and lead to somatic ultra-mutator phenotype.

Signatures 4 and 7 are associated with exogenous carcinogens. Tobacco use is associated with signature 4, as is frequently seen in lung cancers and head and neck cancers, while exposure to ultraviolet light/radiation is associated with signature 7, most commonly seen in cutaneous melanomas and squamous cell carcinomas. Signatures 8, 12, and 16 show transcription strand bias associated with bulky exogenous carcinogens. A specific signature associated with cancer treatment has also been discovered, signature 11, secondary to treatment with temozolomide. Signature 9 is seen frequently in lymphomas and chronic lymphocytic leukemia/lymphoma as it is associated with immunoglobulin gene hypermutation. This leads to errors in AID-induced cytidine deamination in the error-prone polymerase, polymerase η [13]. Multiple signatures including 5, 8, 10, 12, and 16 show transcription-coupled nucleotide excision repair defects which lead to strand bias. Signature 15 shows a large number of substitutions and small insertions/deletions. Several signatures including 14, 15, and 21 are still not well characterized or understood yet [13].

The number of mutations identified in a tumor type and across tumor types can vary greatly [15]. Typically, a tumor will have a number of driver mutations leading to oncogene activation as well as a number of loss of function mutations in tumor suppressor genes. In addition, multiple passenger mutations may be seen. These passenger mutations may not have a big impact on protein function but can accumulate due to an unstable genome, loss of DNA repair mechanisms, and rapid cell cycling, among others. Lung cancers typically have a high tumor mutation burden due to the accumulation of genetic alterations from DNA damage caused by smoking. This type of phenomenon is also frequently seen in cutaneous malignant melanomas, in which the mutations accumulate over time due to damage from ultraviolet rays from unprotected sun exposure. Conversely, papillary thyroid carcinomas typically have a low tumor mutation burden, with few passenger mutations along with mutations seen in driver oncogenes. Interestingly, tumor mutation burden typically increases in older patients when compared to younger patients, particularly children, who usually have a very low mutational burden [15]. The significance of a high tumor mutation burden may have a therapeutic impact and is discussed in a later chapter. A mechanism in which a high tumor mutation burden is commonly seen is in tumors with loss of function mutations in mismatch repair genes, as mentioned above. These genes, which work in pairs, are responsible for proofreading and fixing errors in DNA particularly in microsatellite regions, or regions in the genome with short one to two base pair repeats. The ultra-mutator phenotype has been described with inactivating hot spot mutations in the gene *POL ε*, which encodes the catalytic part of the protein polymerase epsilon responsible for DNA repair. When this gene is inactivated, the polymerase makes frequent mistakes during replication, resulting in a multitude of mutations throughout the genome.

Noncoding driver mutations are not common but also have been reported [16]. Approximately one-quarter of tumors have at least one noncoding driver point mutation. Of those tumors, about one-third of them affect the *TERT* promoter. Other promoters and enhancers are less frequently seen [16]. But the *TERT* promoter mutation is seen relatively frequently. And usually bears a poor prognosis. This finding has been described in liver, prostate, bladder, and thyroid carcinomas [10, 17].

2.4 DNA Methylation

Epigenetic genes are involved in cancer by regulating and changing the expression of certain genes responsible for cell proliferation and differentiation. These genes can be separated into three groups. Epigenetic modulators send signals to the epigenetic regulators which change the status of the genome. Epigenetic mediators alter their roles based on these signals [18]. Many of these genes are controlled by methylation. DNA methylation occurs when a methyl group is added to the carbon-5 position of the cytosine base, known as 5-methylcytosine. It most frequently occurs at cytosine-guanine dinucleotides, known as CpG, via DNA methyltransferase enzymes [19]. These enzymes are known as the DMNTs. Interestingly, 5-methylcytosine may also be altered by ten-eleven-translocation enzymes (TET), an enzyme which leads to oxidation (DNA hydroxymethylation) and the product 5-hydroxymethylcytosine, which can lead to widespread alterations in the DNA methylation landscape [19].

Global DNA methylation patterns are commonly altered in cancer. DNA hypomethylation events occur at oncogenes (causing the gene to turn on) and at centromeres. DNA hypermethylation typically leads to loss of regulation of enhancers and promoters and silencing of tumor suppressor genes. These global alterations in the promoters of tumor suppressor genes and oncogenes lead to tumorigenesis. Studies have shown that multiple recurrent loci are hypermethylated in different tumor types. About 90% of prostate cancers have hypermethylation of *GSTP*, in 20% of lung cancers hypermethylation occurs at *p16INK4a*, and at *BRCA1* in around 10% of breast and ovarian cancers [19].

As tumors progress, studies have shown there is an increased frequency of *p16INK4a* and *GSTP1* hypermethylation, thus suggesting that DNA hypermethylation changes may be predictive of disease progression [19]. CpG islands are susceptible to DNA methyltransferase, and thus can become hypermethylated. This phenotype, known as the CpG Island Methylator Phenotype (CIMP), was first described in glioblastoma and colon cancer. With this discovery, tumors may be classified by subtypes using the methylation signature [19]. DNMT enzyme expression is also commonly altered in cancer. When this occurs, the DNA methylation patterns also change, which may lead to increased somatic mutations in DNA [19].

Global DNA hypomethylation occurs in the CpG-poor regions. It is seen in multiple tumor types and is predicted to lead to genomic instability and increased chromosomal aneuploidy [20]. When there is a decrease in the amount of DNA methyltransferase 1, mutation rates increase, aneuploidy is more frequent, and tumor development and progression occurs. This suggests that there is a role of DNA hypomethylation possibly leading to chromosomal breakage [19]. Global hypomethylation is associated with abnormal chromosome and chromatin organization, which results in structural variation and genome instability [19]. However, causation has not been definitively shown at this time. Abnormal expression of oncogenes, repeat elements, and transposable elements, such as the *MYC* gene, has been shown to occur with loss of DNA methylation. Amplifications of *MYC* are commonly seen in tumor progression and more aggressive tumors. When this occurs, there is loss of regulation of cell pathways, alteration of chromatin organization, and the overall genome [19].

In addition to the loss of 5-methylcytosine, global loss of 5-hydroxymethylcytosine may be seen in many different types of carcinomas [19]. For example, in breast, colon, prostate, liver, pancreatic, and lung cancers and melanomas, there are significantly decreased levels of 5-hydroxymethylcytosine. This decrease is frequently associated with mutations in the *TET* genes leading to decreased function of TET enzymes. If no *TET* mutations can be found, particularly in leukemia, mutations leading to abnormal activity of isocitrate dehydrogenase 1 and 2 may be seen. This is a cofactor involved in TET enzymatic activity. When patients have *IDH1/2* mutations, TET activity is inhibited, leading to increased CpG island hypermethylation [19].

In one study, tumors arising from the same organ system typically cluster by cancer specific hypermethylation [4]. For example, squamous cell carcinoma from the head and neck, esophagus, lung, and cervix shares similar methylation patterns. As such, adenocarcinomas origination from the esophagus, stomach, and colorectal also shares a distinct set of methylation patterns [4].

2.5 RNA and miRNA Alterations

Alterations of RNA transcriptions in cancer frequently occur due to somatic alterations in genomes. RNA alterations may occur via splice site alterations, gene fusions, and overexpression [21]. On average, tumors have approximately 20% more frequent alternate splicing events when compared to normal tissues [22]. Many of these changes occur as single nucleotide variants occurring at critical areas for binding of splice site machinery [23]. Frameshift mutations in mRNA resulting from intron retention occur most commonly in tumor suppressor genes, while in-frame exonic single nucleotide variants, altering exons, are more frequently seen in oncogenes [23].

Splice site creating mutations may lead to alternate protein production and have been associated with higher expression of T cell associated genes, as well as mRNAs which encode immune-checkpoint blockade molecules (such as PD-1 and PD-L1). These alterations may have a significant therapeutic impact [23].

Driver mutations have been discovered in over 119 splicing factor genes [23]. Hotspot mutations in the RNA-splicing factors *SF3B1*, *SRSF2*, and *U2AF1* have been described to occur more frequently in myeloid leukemias, both acute and chronic, as well as uveal and mucosal melanomas [23].

SF3B1 is the most commonly mutated across all cancers. The effect of this mutation on prognosis varies between tumor types. For example, *SF3B1* mutations in myeloid dysplastic syndrome are associated with a favorable prognosis, while *SF3B1* mutations in CLL, which are typically subclonal, are increasingly seen in patients with aggressive disease [23]. RNA-splicing mutations are relatively uncommon in many solid tumors. However, they are seen much more frequently in several types, including bladder cancer and lung adenocarcinoma [23].

MicroRNA (miRNA) are short RNA sequences, about 22 nucleotides each, which regulate mRNA molecules either by degradation or stopping translation [24]. Each sequence of miRNA may be responsible for controlling hundreds of genes. Changes in miRNA expression play a critical role in cancer development and progression [24]. Depending on the circumstance, miRNAs can function either as promoters of oncogenes or can lead to the silencing of tumor suppressor genes [24]. The first miRNA discovered to be oncogenic is called miR-17-92. It is found to be amplified in B cell lymphomas and has oncogenic activity [24].

Different tumor types have unique miRNA expression profiles. These differences may be used to help distinguish one tumor type from another. These signatures can help further subclassify subtypes of cancer within the same tumor type, and even give important prognostic information [24]. The use of miRNA expression can help distinguish precursor lesions, like tubular adenomas in the colon, from invasive colon cancers [24]. Some of these markers, for example, miR-25 and miR-223, which are elevated in non-small cell lung carcinoma, can be detected in serum [24]. miRNA expression has been used to help classify the aggressiveness of acute leukemias, lung adenocarcinomas, and outcomes of colorectal carcinoma. These types of assays can help classify types of lung, prostate, and hepatocellular cancers [24]. High levels of miR-155 and/or low levels of let-7a-2 miRNAs have been correlated with poor survival in lung adenocarcinomas [24]. While, loss of miR-122 expression in hepatocellular carcinomas have been associated with tumor progression and the ability to metastasize [24]. In colorectal adenocarcinoma, high miR-21 expression has been linked to poor survival [24]. Thus, miRNA profiles may help provide prognostic information in cancer [24].

Finally, the profiling of miRNA signatures may also aid as predictive biomarkers in cancer. miR-21 expression in non-small cell lung cancer has been shotwn to be highly expressed in patients with a shorter disease-free survival, as well as those who have platinum-based chemotherapy resistance. Additional markers have been found to predict chemotherapy responses in advanced colorectal carcinomas. And different miRNA panels have been used to help predict a tumor's response to different types of chemotherapy [24].

2.6 Analysis of Proteins

Protein analysis is not used routinely in clinical practice. However, in a study using reverse phase protein array, tumors were segregated into ten groups [4]. The first group (P1) included gliomas, both low and high grade [4]. The second group (P2) included multiple tumors types and a multiple array of tumors including diffuse large B cell lymphoma, multiple types of soft tissue sarcomas thymomas, and cutaneous melanomas [4]. Another set including P3 and P4 accounted for tumors driven by estrogen, including luminal A and B breast cancers, gynecological carcinomas, and some liver cancers with increased estrogen receptor expression, androgen receptor expression, and increased IGFBP2 expression [4]. Then, there was a pan-kidney tumor group, called P6, as well as a pan-gastrointestinal tumor group known as P8 [4]. This shows that analysis of proteins may also be used to further classify and define tumors.

2.7 Conclusion

Complex computer algorithms can take all of the information listed above and further compile it into more nuanced classifications to give more information on tumor prognosis and prediction. While it may appear as though many of these alterations happen separately from one another, that is not the case, as several of the alterations are associated with one another. For instance, global DNA hypomethylation is associated with genomic instability and chromosomal aneuploidy, while hypermutated tumors tend to have low aneuploidy. However, at this time, comprehensive molecular profiling including microarray, whole genome and/or whole exome sequencing, methylation array, RNA sequencing, and proteomics is not routinely performed as it is cost prohibited, takes a lot of time to analyze this information, and is not necessary for current clinical practice.

This opens up many questions for the future of somatic molecular testing in cancer.

Open Questions

- What types of molecular studies will be recommended for a routine comprehensive cancer work up?
- Will a biopsy sample suffice for the amount of testing required (how much neoplastic material will be necessary for testing)?
- If some of these findings are interrelated, is it necessary to run all of the studies to confirm the findings, or will we proceed in more of a stepwise fashion?
- Will any of this apply to preneoplastic tumors to help predict which ones will progress? How much do these features change in the metastatic setting, and does it matter?

References

1. Ben-David U, Amon A. Context is everything: aneuploidy in cancer. Nat Rev Genet. 2020;21:44–62.
2. Taylor AM, Shih J, Ha G, Gao GF, Zhang X, Berger AC, Schumacher SE, Wang C, Hu H, Liu J, et al. Genomic and functional approaches to understanding cancer aneuploidy. Cancer Cell. 2018;33:676–689.e673.
3. Duesberg P, Li R, Fabarius A, Hehlmann R. The chromosomal basis of cancer. Cell Oncol. 2005;27:293–318.
4. Hoadley KA, Yau C, Hinoue T, Wolf DM, Lazar AJ, Drill E, Shen R, Taylor AM, Cherniack AD, Thorsson V, et al. Cell-of-origin patterns dominate the molecular classification of 10,000 tumors from 33 types of cancer. Cell. 2018;173:291–304.e296.
5. Manier S, Salem KZ, Park J, Landau DA, Getz G, Ghobrial IM. Genomic complexity of multiple myeloma and its clinical implications. Nat Rev Clin Oncol. 2017;14:100–13.
6. Shago M. Recurrent cytogenetic abnormalities in acute lymphoblastic leukemia. Methods Mol Biol. 2017;1541:257–78.
7. Halani SH, Brat DJ. Defining neoplastic diseases differently: an emerging paradigm from the cancer genome atlas lower-grade gliomas project. Mol Cell Oncol. 2016;3:e1074333.
8. Hieronymus H, Murali R, Tin A, Yadav K, Abida W, Moller H, Berney D, Scher H, Carver B, Scardino P, et al. Tumor copy number alteration burden is a pan-cancer prognostic factor associated with recurrence and death. Elife. 2018:7.
9. Schukken KM, Foijer F. CIN and aneuploidy: different concepts, different consequences. Bioessays. 2018;40 https://doi.org/10.1002/bies.201700147.
10. Li Y, Roberts ND, Wala JA, Shapira O, Schumacher SE, Kumar K, Khurana E, Waszak S, Korbel JO, Haber JE, et al. Patterns of somatic structural variation in human cancer genomes. Nature. 2020;578:112–21.
11. Li MM, Datto M, Duncavage EJ, Kulkarni S, Lindeman NI, Roy S, Tsimberidou AM, Vnencak-Jones CL, Wolff DJ, Younes A, et al. Standards and guidelines for the interpretation and reporting of sequence variants in cancer: a joint consensus recommendation of the Association for Molecular Pathology, American Society of Clinical Oncology, and College of American Pathologists. J Mol Diagn. 2017;19:4–23.
12. Alexandrov LB, Kim J, Haradhvala NJ, Huang MN, Tian Ng AW, Wu Y, Boot A, Covington KR, Gordenin DA, Bergstrom EN, et al. The repertoire of mutational signatures in human cancer. Nature. 2020;578:94–101.
13. Alexandrov LB, Nik-Zainal S, Wedge DC, Aparicio SA, Behjati S, Biankin AV, Bignell GR, Bolli N, Borg A, Børresen-Dale AL, et al. Signatures of mutational processes in human cancer. Nature. 2013;500:415–21.
14. Cho RJ, Alexandrov LB, den Breems NY, Atanasova VS, Farshchian M, Purdom E, Nguyen TN, Coarfa C, Rajapakshe K, Prisco M, et al. APOBEC mutation drives early-onset squamous cell carcinomas in recessive dystrophic epidermolysis bullosa. Sci Transl Med. 2018;10:eaas9668.
15. Chalmers ZR, Connelly CF, Fabrizio D, Gay L, Ali SM, Ennis R, Schrock A, Campbell B, Shlien A, Chmielecki J, et al. Analysis of 100,000 human cancer genomes reveals the landscape of tumor mutational burden. Genome Med. 2017;9:34.
16. Rheinbay E, Nielsen MM, Abascal F, Wala JA, Shapira O, Tiao G, Hornshøj H, Hess JM, Juul RI, Lin Z, et al. Analyses of noncoding somatic drivers in 2,658 cancer whole genomes. Nature. 2020;578:102–11.
17. Consortium, I.T.P.-C.A.o.W.G. Pan-cancer analysis of whole genomes. Nature. 2020;578:82–93.
18. Lee CJ, Ahn H, Jeong D, Pak M, Moon JH, Kim S. Impact of mutations in DNA methylation modification genes on genome-wide methylation landscapes and downstream gene activations in pancancer. BMC Med Genet. 2020;13:27.
19. Skvortsova K, Stirzaker C, Taberlay P. The DNA methylation landscape in cancer. Essays Biochem. 2019;63:797–811.
20. Ehrlich M, Lacey M. DNA hypomethylation and hemimethylation in cancer. Adv Exp Med Biol. 2013;754:31–56.
21. Calabrese C, Davidson NR, Demircioğlu D, Fonseca NA, He Y, Kahles A, Lehmann KV, Liu F, Shiraishi Y, Soulette CM, et al. Genomic basis for RNA alterations in cancer. Nature. 2020;578:129–36.
22. Kahles A, Lehmann KV, Toussaint NC, Hüser M, Stark SG, Sachsenberg T, Stegle O, Kohlbacher O, Sander C, Rätsch G, et al. Comprehensive analysis of alternative splicing across tumors from 8,705 patients. Cancer Cell. 2018;34:211–224.e216.
23. Obeng EA, Stewart C, Abdel-Wahab O. Altered RNA processing in cancer pathogenesis and therapy. Cancer Discov. 2019;9:1493–510.
24. Li MH, Fu SB, Xiao HS. Genome-wide analysis of microRNA and mRNA expression signatures in cancer. Acta Pharmacol Sin. 2015;36:1200–11.

Tumor Microenvironment: Coconspirator in Tumorigenesis

3

Zhiqiang Wang

Abstract

The tumor microenvironment is pivotal for tumor establishment, evasion of host immune surveillance, growth, invasion, and metastasis by providing a nutrient meshwork rich in lympho-vasculature, cellular and metabolic coconspirators of the tumor, and a signaling network for cross talks. Of note are a number of players that support or impede tumor growth directly by themselves or indirectly through their products. Given the immensity of the topic and limitation of the scope of this chapter, we choose to discuss the following well-studied players and products that are the focus of research and resource commitment: (1) fibroblasts and cancer-associated fibroblasts, (2) endothelial cells with angiogenesis, (3) immune cells, and a couple of cellular products, i.e., (4) secretomes, and (5) exosomes. We recognize that under each of the five subtitles lies a world of evidence that embodies numerous never-ending topics. Each of them deserves separate in-depth exploration. We hence set out to briefly introduce their basic biological functions, discuss their roles in tumor growth and mechanisms of interaction with tumor cells and among themselves, summarize representative research and milestones of achievements, and finally address their relevant clinical implications and advances.

Learning Objectives

- Understand the notion of the tumor microenvironment (TME) and the basic roles in tumorigenesis
- Learn the names and tumorigenic roles of cancer-associated fibroblasts (CAF), secretome, and exosome
- Comprehend the importance of endothelial cells and angiogenesis in tumorigenesis
- Know different types of immune cells, e.g., tumor-associated macrophages (TAM), myeloid-derived suppressor cells (MDSC), and regulatory T cells (Treg), and their respective roles and functions in tumorigenesis and clinical immunotherapy

3.1 Introduction

Tumor invasion features the departure of tumor cells from the primary normal microanatomic compartment after in situ proliferation to infringe into neighboring territory. During this invasion process, the tumor cells are supported by the surrounding specialized environment to support their growth. Such an environment is frequently referred to as tumor microenvironment (TME). TME is the compartment where tumor cells survive and thrive. By coordinating the interactions and communications among different components via complicated mechanisms, the TME is essential in creating primary niches for the tumor cells and is pivotal for sustained tumor growth, evasion of host immune surveillance, generation of neo-vasculatures, promotion of invasion, and metastasis.

TME consists of physical and chemical components in addition to the tumor cells. TME denotes the stroma with lymph vasculature and immune cells. Stroma is the connective tissue that comprises both noncellular extracellular matrix (ECM) and cellular components. Chemically, the noncellular components of ECM consist of ground substances that form the framework for tumoral and non-tumoral tissue to organize and network with each other. The noncellular components of ECM also consist of functional molecules, including enzymes, metabolic products, and signaling molecules such as chemokines, auto/juxta/paracrine hormones, and growth factors. In aggregate, these proteins are

Z. Wang (✉)
Department of Pathology and Laboratory Medicine, Henry Ford Hospital, Detroit, MI, USA
e-mail: zwang4@hfhs.org

S. P. Leong et al. (eds.), *Cancer Metastasis Through the Lymphovascular System*,
https://doi.org/10.1007/978-3-030-93084-4_3

called secretomes. In contrast, the recently emerging concept of exosomes represents portions of cellular compartments that are shed into the ECM. The cellular components of TME consist of multiple cell types, i.e., fibroblasts, endothelial cells, and immune cells [1].

In this chapter, we limit our review to the fundamental roles of the most relevant components of the TME as coconspirators in tumorigenesis. These coconspirators include fibroblasts and cancer-associated fibroblasts (CAFs), secretomes and exosomes released by CAFs and tumor cells, endothelial cells with angiogenesis, and the immune cells. We also describe the interactions among these different coconspirators and address issues relating to the molecular mechanisms and events. Lastly, we discuss their potential clinical implications.

3.1.1 Fibroblast and CAFs

Fibroblasts are the most cellular component of the stroma. In contrast to fibrocyte, the suffix "-blast" implies immaturity and the potential to function as precursors to differentiate into other cell types. Embryonically, the fibroblast is derived from mesoderm with a subset derived from the neural crest. Functionally, fibroblasts synthesize a variety of proteins, including collagens, fibronectin, and laminin, to constitute the ECM and basement membrane. In addition, they produce various types of ground substances (glycosaminoglycans, proteoglycans, and glycoproteins) and release cytokines, juxta- or paracrine hormones or growth factors, proteases, etc., to maintain and modulate the matrix framework. Fibroblasts can also migrate and proliferate in response to different signals, e.g., leukotrienes and growth factors. Although they lack specific markers, the fibroblasts are morphologically recognized by their spindle shape, oval to elongated nuclei, and their microanatomical location in an organ. When the fibroblasts are activated and/or proliferate, the cytoplasm is plump, bright eosinophilic on H & E stained sections as collagen laying fibroblasts in diseased conditions. The fibroblasts are known for their plasticity. Transformation into myofibroblasts has been well described in tissue repair. Lastly, as it has been well known, in vitro experimental models also showed that human fibroblasts could be reversely transformed into pluripotent stem cells [2].

Since the first descriptions of CAF were brought forth [3], there has been an explosion in knowledge regarding CAFs' roles in ECM remodeling, angiogenesis, signal networking and transduction, harnessing and modulating host immune system, and clinical potential in therapy. While progress continues to be made in this field, challenges remain. Recently, a group of experts in the field of CAF biology summarized and issued a Consensus Statement to elucidate relevant issues of CAFs, including definition, tissue generation, functions, heterogeneity and plasticity, and potential clinical benefits [4].

Although most CAFs are believed to be derived from local fibroblast activation or genetic alterations, other cells of origins have also been described. Several cells are able to "de-differentiate" into CAFs, which include adipocytes [5], endothelial cells [6], and bone marrow-derived mesenchymal stem cells [7]. Epithelial cell as the origin of CAFs has been one of the most extensively studied mechanisms called epithelial-mesenchymal transition (EMT). EMT is associated with tumor progression and metastasis [8]. Both benign and malignant epithelial cells can generate CAFs through EMT [9]. Although the transformation from cancer cells to CAFs is well described, it is interesting to note the nonoverlapping genetic alterations between cancer cells and CAFs [10]. Such discrepancies, however, appear to be due to genetic alterations occurring at early precursor stages with loss or gain of these alterations in subsequent subclonal evolution of tumor cells.

Analogous as they may appear, CAFs are different from normal fibroblasts. Two experimental models are most exemplary that describe the differences: (1) In their mouse model of prostate cancer, Olumi et al. [11] reported that CAFs directed tumor progression by inducing prostatic intraepithelial neoplasia in normal prostate epithelial cells. Under the same experimental conditions, such tumor progression was not observed by normal fibroblasts. These findings were displayed by both in vivo tissue recombination system and in vitro co-culture system. (2) In their breast cancer experimental model systems, Dumont et al. [12] showed that CAFs were protumorigenic and supported tumor dissemination by inducing a mesenchymal-like phenotype of both premalignant and malignant mammary epithelial cells. In comparison, fibroblasts isolated from benign reduction mammaplasty helped maintain the tumor cells epithelial morphology and hindered tumor growth and dissemination.

Once in place, CAFs exert their roles in regulating tumor growth, invasion, and metastasis in various ways through interactions with other ECM components and tumor cells. One school of thought that contributed to the pool of evidence was by implicating the role of physical forces in the tumorigenic process. Using the breast as the model organ, Paszek and Weaver [13] illustrated that the "tumors are rigid because they have a stiff stroma". The stiffness of the stroma was created by increased interstitial tension via infiltrating immune cells and activated resident fibroblasts that will increase the production of macromolecules, e.g., collagens, fibronectin, laminin, and ground substances. The stiffness was further enhanced by cross-linking enzymes [14]. In response, such increased tension would feedback to the tumor cells and increase their invasiveness [15]. One CAF product, i.e., the lysyl oxidase (LOX) family, deserves attention for their roles in contributing to the stiffness of tumor stroma, tumor growth, and progression [16].

CAFs' functions are diverse, which require heterogeneity and plasticity among CAFs to carry out different functions under different conditions. Interestingly, there have been both experimental and clinical model systems that have provided supporting evidence of such specialization. Investigators showed that some CAFs could mutually convert their functional phenotypes under different instruction signals released by different tumor subclones [17]. The molecular events that are implicated behind the functional switch of CAFs phenotypes are still unclear.

Various strategies have been developed to target CAFs at the cellular level for clinical therapeutic purposes based on the knowledge accumulated over the past decades. By targeting fibroblast activation protein (FAP), which CAFs selectively produce, Ostermann et al. [18] demonstrated the antitumor effect of monoclonal antibody (mAb) FAP5-DM1 on xenograft immunodeficient mouse models for lung, pancreas, and head and neck cancers. Sibrotuzumab, a mAb, was tried on patients with advanced colorectal cancer for clinical phase I and early phase II trials, although eventually was not successful [19, 20]. However, the setbacks did not stop investigators from further pursuing alternative means. By conjugating promelittin-containing FAP-cleavable sequences to pegylated phospholipids and anchoring them to reduced graphene oxide (rGO) nanosheets, Kim et al. [21] reported that the resulting nanosheets, PL-rGO, showed tumor inhibitory effect both on in vitro culturing and in vivo mouse models. On the other hand, by taking advantage of CAFs heterogeneity and plasticity, other innovative methods are being devised to reprogram and salvage CAFs back to their "normal" phenotypes to help achieve tumor suppression effects [4].

Progress made in molecular genetics has helped advance our understanding of CAFs and their cross-interactions with cancer cells. The role of Braf, its associated ERK–MAPK signaling pathway, and clinically approved the application of Braf inhibitor for melanoma have been well known [22]. The findings that Hirata et al. [23] published on *BRAF* mutant melanoma cells' response to PLX4720 was most revealing. After the initial response of melanoma cells to PLX4720, a rapid tolerance was developed. Such tolerance was because of PLX4720's effect on TME, leading to "paradoxical" activation of CAFs with more matrix production and remodeling. As a result, an increased integrin β1/FAK/Src signaling was seen in melanoma cells. After adding a FAK inhibitor to the anti-Braf regimen, the investigators showed effective suppression of ERK signaling, and hence more effective tumor treatment.

3.1.2 Secretome

Secretome denotes the collection of proteins released by a cell into the extracellular space as defined by TME. These proteins are of different functional types involved in physiological and pathological processes, including enzymes, metabolic products, chemokines, auto/juxta/paracrine hormones, growth factors, etc. In the context of TME, the Matrix MetalloProteinases (MMPs) family and TGF-beta (TGFβ) are the two major secretory proteins of the secretome that have remained at the center stage of attention over the past decades.

Since its first description in early 1960 by Gross and Lapiere [24], extensive knowledge has accumulated on MMPs over the past decades. These achievements ranged from bioengineering of the first cDNA clone, to protein crystal structure characterization, mouse model establishment, and approval of the first MMP inhibitor for clinical application. MMPs are a family of secreted zinc-dependent endopeptidases, 24 identified so far in humans, comprising collagenases, stromelysins, matrilysins, gelatinases, and others. MMPs are expressed by both tumor cells and CAFs and are known for their roles in tumor invasion as groundbreakers to degrade ECM components at the leading edge of the tumor. Additionally, MMPs also have regulatory functions on cell adhesion molecules processing which eventually impact cell migration and invasion. The roles of MMPs at different stages of cancer survival, growth, invasion, and metastasis were reviewed by Gonzalez-Avila et al. [25] in depth.

Given the roles of MMPs in cancer invasion and metastasis, MMPs have been applied as targets for MMPs inhibitors in clinical trials and as biomarkers for prognostic purposes. Dozens of clinical trials have been conducted, some showing encouraging preclinical data. However, progress on MMP inhibitors has been slow, with none demonstrating antitumor effects in clinical trials. The value of MMPs for their clinical application has been best reflected as biomarkers for prognostic purposes. Two assays exemplify MMPs biomarkers' roles. One of the assays is the 70-gene MammaPrint assay [26]. MammaPrint was approved by the US Food and Drug Administration and regulators in the European Union as an adjunct prognostic assay for patients with stage I/II breast cancer who are below 61 years of age with negative or 1–3 lymph nodes positive disease. MMP9 is listed as 1 of the 5 genes under "altered extracellular matrix adhesion and remodeling" of the "Tissue Invasion and Metastasis" category. In addition, MMP9 is also included as 1 of 6 genes under "altered expression of known angiogenesis effectors" of the "Sustained Angiogenesis" category. The other assay is the Oncotype DX assay [27]. The Oncotype DX assay has been most widely used and well received by clinicians mainly because of the convenience of using formalin-fixed paraffin-embedded tissues and its easy-to-interpret recurrence score-based reporting. MMP11 is included as one of the 16 cancer-related genes.

Of the many growth factors, TGF-beta (TGFβ) has been known as a central player for its complex roles in homeostasis and tumorigenesis by its signaling via canonical SMAD pathway and several non-canonical pathways. On the one hand, TGFβ promotes the transdifferentiation of fibroblasts to CAF and recruits fibroblasts to the tumor. On the other hand, CAFs secrete large amount of TGFβ to affect tumor growth by enhancing cross talk between cancer cells and CAF, and by mediating metabolic activities of tumor. Similarly, another paracrine growth factor, hepatocyte growth factor (HGF), also deserves attention for its contributory roles to promote cancer progression. By overexpression of HGF and TGFβ, Kuperwasser et al. [28] showed that normal and malignant human breast tissues could be reconstructed in mice. Therapeutically, it is not difficult to find clinical trials targeting the CAF-induced signaling molecules, especially TGFβ and HGF inhibitors and mAb. The results are promising but mixed [29].

In contrast, there has been evidence showing biphasic roles of TGFβ in both promoting and hindering tumor growth and progress. In their study on skin tumors using transgenic mouse models, Cui et al. [30] found that TGFβ inhibited benign tumor formation and enhanced invasive carcinomas' progression. Other CAF secretory proteins also demonstrated such inhibitory roles on cancer growth. In the study on colon cancer secretomes using in vitro cell culture, xenograft mouse models, and clinically resected colon cancers, Chen et al. [31] revealed novel molecular signature secretory proteins that exhibited negative modulatory effects on colon cancer cell growth.

In addition to proteinase enzymes and signaling molecules, other aspects of secretomes impacting cancer biology include senescence-associated secretome, inflammatory secretome, angiogenesis secretome. The functions of these diverse secretomes are intertwined. For instance, senescence is a mechanism that prevents and blocks cancer cell proliferation and, therefore, is regarded as a tumor suppressor mechanism. As reviewed in depth by da Cunha [32], driven by NF-κB, the secretome of cells in a senescent state can contain as many as 80 bioactive molecules. These molecules include pro-senescent, pro-apoptotic, and antiangiogenic protein molecules. In association with p16 and SASP (senescence-associated secretory phenotype) that incite immune cells, these secretomes facilitate tissue regeneration and senescent cell removal. On the other hand, however, these molecules also include aforementioned MMPs and growth factors, plus pro-inflammatory leukotrienes and angiogenic factors. As stated by Rodier and Campisi [33], cellular senescence might take part in complex biological processes that include tumor promotion.

3.1.3 Exosome

Other mechanisms involving CAF interactions and cross talk among themselves and cancer cells include effects of metabolites and secretion of exosomes. Exosomes and transfer of information between cancer cells and other cellular components of TME have emerged as a novel and important mechanism that has fueled many research opportunities in recent years. It is worthwhile to mention that given the key roles exosomes play, the term "Exosomics" was used to help carry the notion and its associated importance.

Exosomes are known as a type of small extracellular vesicles (30–150 nm) that form via the cellular process of endocytosis. The endosomes thus formed are shed by the parent cells into ECM with a double-layered lipid membrane. As part of the parent cell, exosome represents a portion of the cellular cytoplasm and membrane containing parent proteins, lipids, and nucleic acids. The proteins in exosomes consist of endocytic proteins and protein molecules that are both membranous (receptors, adhesion molecules, channels) and cytoplasmic (cytoskeletons, enzymes, players of different pathways). Interestingly, on examination of the lipid contents, Llorente et al. [34] found significant differences between exosomes and parent cells. The exosomes lipids were highly enriched in a subset of lipids, e.g., glycosphingolipids, sphingomyelin, cholesterol, and phosphatidylserine. Other non-enriched lipid species were selectively included. Such a pattern of lipid composition helped raise the possibility of a yet-to-be elucidated sorting mechanism and potential tumor biomarkers.

Although largely cytoplasmic, it was demonstrated that exosomes contain different types of DNA species, i.e., single-stranded DNA, mitochondrial DNA, and double-stranded nuclear DNA. Thakur et al. [35] convincingly demonstrated the presence of double-stranded genomic DNA in exosomes and provided further experimental evidence showing that exosomal DNA represents the entire genome and carries detectable mutations that reflect parent tumor cells. Such findings were received with enthusiasm and opened doors for many research opportunities, notably in non-small cell lung carcinomas. Taking advantage of commercial DNA isolation products that have been made available, researchers were able to detect major EGFR mutations with improved sensitivity on liquid biopsy and pleural effusions using exosomal-derived DNA.

In the recent years, exosomal RNA has been extensively investigated due to technical advances and the availability of next-generation sequencing-based RNA sequencing. Various RNA species have been characterized, including coding mRNAs and noncoding RNAs. Through deep RNA sequencing, Nolte-'t Hoen et al. [36] uncovered a variety of noncoding RNAs, including microRNAs (miRNA), small nuclear RNAs (snRNAs), circular RNAs (circRNAs), etc. Interestingly, like lipids, a perplexed sorting mechanism also exists by selectively including certain RNA species but excluding others in the exosomes [37]. There have been different schools of thought trying to address this issue, including the presence of recognizable RNA signal sequences and regulatory pathways.

Research activities exploring the roles of exosomal RNA have been advancing on different fronts from neurodegenerative diseases to different types of cancer. In one translational study aiming for therapeutic purposes against pancreatic ductal adenocarcinoma (PDAC), Kamerkar et al. [38] demonstrated state-of-the-art design of their experiment by taking advantage of the unique features of exosome as a delivery system and molecular features of PDAC. As mentioned above, exosomes are a portion of cellular cytoplasm and membrane that contains transmembrane proteins. CD47 is an integrin-associated transmembrane protein that functions to protect the cells from being phagocytosed by monocytes and macrophages. It, therefore, enhances endocytosis and delivery of the cytoplasmic content to the recipient cells [39]. As PDAC frequently harbor mutations in KRAS gene, investigators developed exosomes derived from normal fibroblast-like mesenchymal cells to carry short interfering RNA to functionally knock down this oncogenic KRAS G12D mutation. This exosome-based treatment by combined use of these features successfully achieved tumor suppression in mouse models of PDAC with a significant increase in overall survival.

The creation of a dedicated program met the enthusiasm in extracellular RNA research under the US National Institutes of Health, the Extracellular RNA Communication Consortium (ERCC). Another program is the International Society for Extracellular Vesicles (ISEV), which surveyed extracellular vesicles and RNA. The ISEV survey showed that although relatively new, the field has generated great interest and commitment both in resource and funding. However challenges remain in elucidating the mechanisms behind RNA production, transfer, and function in recipient cells.

3.1.4 Endothelial Cells and Angiogenesis

Angiogenesis and lymphangiogenesis is an integral part of TME. The clinical implications of TME can be dated back about half a century ago when Dr. Judah Folkman published his article " Tumor angiogenesis: therapeutic implications" [40]. Briefly, tumor growth is accompanied by neovascularization, which is defined by angiogenesis and lymphangiogenesis essential to supply the necessary nutrients and oxygen to support tumor growth, proliferation, and metastasis. Angiogenesis is a dynamic process in response to angiogenic stimulation involving the participation of multiple cell types and factors in the TME. Central to the angiogenic process is the endothelial cells that can come from existing vessels from which new vessels sprout or originate from bone marrow-derived mesenchymal stem cells or even tumor stem cells. In turn, endothelial cells can exhibit different phenotypes, including "dedifferentiating" into CAFs as aforementioned under the CAFs section [6]. Behind the cellular components are multiple factors that impact the angiogenic process, including hypoxia, tumor cell burden, inflammation, genetic alterations of tumor cells, and signaling molecules. Among the many known factors is vascular endothelial growth factor (VEGF) and integrin.

Of the prolific angiogenic growth factors (VEGFs, FGF, PDGFs), VEGF has been the focus of attention because of its pivotal roles in angiogenesis. Bevacizumab (Avastin), the mAb against VEGF, is the most widely used therapeutics in oncology and other FDA-approved anti-VEGF drugs. The family of VEGF have several members, i.e., VEGF A-E and PIGF (Placenta Growth Factor), and a number of isoforms due to alternative splicing. VEGF-A is the major player in regulation of angiogenesis and thus is frequently referred to as VEGF. VEGF-C and VEGF-D are primarily implicated in regulation of lymphangiogenesis, which is recognized as another mode of vascularization in tumors and as alternate route for cancer cell dissemination [41]. Overexpression of the VEGF genes was detected in both tumor and CAFs and the proteins are secreted as soluble growth factors in the ECM that function via binding with VEGF receptors on the surface of endothelial cells or other target cells. The binding triggers a cascade of signaling events in the VEGF regulated pathways that eventually lead to proliferation of endothelial cells and formation of new blood vessels. It is worth mentioning, however, that the tumor neovasculature thus formed are irregular and defective. As a result, the defective vasculature will expose multiple cell types to the TME with various stimulating signals and eventually trigger a cascade of events leading to more aggressive tumor behavior [42].

Knowledge of VEGF led to an investigation using mAB to block its function, which was successfully tested on xenograft mouse models by Kim et al. [43]. The preclinical success led to clinical trials using a humanized murine antibody, i.e., A.4.6.1 and eventual FDA approval of Bevacizumab in 2004 for previously untreated metastatic colorectal cancer and subsequently for patients with non-squamous non-small cell lung carcinoma among other malignancies. The success also provided opportunities for FDA approval of other VEGF inhibitors, including multiple tyrosine kinase inhibitors (TKI); Ramucirumab, which targets the VEGF receptor (VEGFR2); Ziv-aflibercept, which targets a chimeric VEGF (VEGF-A, VEGF-B, PIGF)

Integrin is another TME molecule that deserves attention because of its roles as hub for crosstalk between cells and ECM as well as roles in modulating tumor angiogenesis. Integrin comprises two subunits, i.e., α and β subunits, that form a heterodimer to combine with target receptors found in substances of ECM, which include fibronectin, vitronectin, tenascin, fibrillin, etc. Detection of integrin $\alpha v\beta 3$ in endothelial cells helped link its role to angiogenesis and offered research opportunities for targeted therapy in several different tumors, including glioblastoma (GBM) [44]. GBM has rich vasculature and high-level integrin $\alpha v\beta 3$ expression and matrix protein vitronectin. Taking advantage of these features, MacDonald et al. [45] demonstrated antiangiogenic efficacy of αv integrin antagonist in mouse models of GBM. The successful story helped devise Cilengitide, the first integrin antagonist, and clinical trials. Although the Phase III clinical trial failed to show improved therapeutic outcome in newly diagnosed GBM with methylated MGMT promoter, findings accumulated highlighted the potential of targeting integrins for the treatment of GBM [46]

3.1.5 Immune Cells

Dense inflammatory infiltration, tumor infiltrating lymphocytes (TILs) in many solid tumors have been observed and studied for a long time. The TILs are thought to be an immune response against tumor cells as part of immune surveillance and elimination of tumor cells that are foreign to the immune system. This hypothesis is supported by observations such as patients with immunosuppression (AIDS; organ transplant) developing certain types of cancers. On the other side, inflammation has been known to be tumorigenic, as exemplified by association of inflammatory bowel disease and colorectal adenocarcinoma, viral hepatitis and hepatocellular carcinoma, H. pylori gastritis and lymphoma, etc. Such polarity highlights the intricacy between the host's immune system and cancer.

Along with tumor cells, CAFs, endothelial cells, and other mesenchymal cells, immune cells represent one of the significant cellular types in the TME. They include neutrophils, tumor-associated macrophages (TAM), myeloid-derived suppressor cells (MDSC), dendritic cells (DC), natural killer (NK) cells, or lymphocytes as adaptive immune response. The immune cells can be resident to the TME or recruited into TME by migration via different signaling pathways and mechanisms. Functionally, immune cells can be cytotoxic to tumor cells, e.g., CD8+ cytotoxic T lymphocytes (CTL) and NK cells, or they can be immunosuppressive and protumorigenic. Since tumor-suppressive or cytotoxic functions of immune cells including neutrophils, DCs, CTL, and NK cells have been well documented in the literature, in the following paragraphs, we focus our discussion on three types of immune cells that possess immunomodulatory functions: (1) TAMs, (2) MDSCs, and (3) Regulatory T (Treg) cells.

TAMs are believed to differentiate from monocytes and are recruited from blood to TME by tumor emitted or CAF-derived signals. TAMs are known for their polarizing phenotypes in the tumorigenic process; i.e., M1 macrophages are pro-inflammatory and cytotoxic to tumor cells, M2 macrophages are anti-inflammatory and are friendly to tumor cells. It is worth mentioning that M1 and M2 macrophages should not be simply viewed as physically distinct and separate subclonal cell populations, but instead, they represent functional plasticity that can switch from one phenotype to another upon different signaling they receive from TME at different tumorigenic stages. The concept that TAMs phenotype can mutually switch was supported by DeNardo et al. [47] using their mouse models in which suppression of M2 function of TAM helped improve the tumor toxicity role of M1 type.

The unique polarizing phenotypes of TAMs also caught investigators' attention to extrapolate their inherent clinical significance and implications. Progress is being made on several fronts including prognosticating different types of solid tumors by quantifying TAMs' density and M1/M2 ratio using established immunohistochemical markers or by determining TAMs' locations to correlate with the tumor's clinical behavior [48]. Advances in therapeutics are also noticeable, including suppressing tumor growth by inhibiting M2 TAMs or promoting switching of M2 to M1 TAMs; immunotherapies by PD1/PD-L1 signaling blockade; targeted therapies by mAb and inhibitors; genetic modifications, etc.

MDSCs have been increasingly recognized as a major regulator of host immune response to evade host immunosurveillance. They are mobilized and infiltrate tumors to disrupt or hinder the processes of TAMs polarization, CTL and NK cells cytotoxicity, and dendritic cells antigen presentation. MDSCs were discovered from an early study of inflammation on tumor-bearing mouse models in which a subset of myeloid cells expressing CD11b and Gr-1 were revealed [49]. Since there is no mouse Gr-1 ortholog in humans, MDSC is functionally and phenotypically classified into two major subsets, polymorphonuclear (PMN) and monocytic (M)-MDSC. Criteria to define different types of MDSCs were suggested based on a combination of a set of differentially expressed surface markers, including CD11b, CD14, CD15, etc. Functionally, MDSCs' capabilities to suppress NK cell, lymphocytes, especially CTLs, were demonstrated by a number of experimental models. MDSCs also produce several molecules to help suppress immune responses, e.g., arginases, NO, and ROS. Although knowledge on MDSCs has been improving in recent years, clinical advances in utilizing MDSCs as a prognostic marker and as a therapeutic target by immunomodulatory therapies have not been significant. In addition, the involvement of MDSCs is also being studied in infectious disease, autoimmunity, aging, obesity, pregnancy, and transplantation.

Treg cells are another immune cell type in the TME that possess immune-modulatory functions and functionally demonstrate some overlap with MDSCs in regard to the suppression of tumor-associated antigen presentation and cytotoxic T cell function. The discovery of Treg cells represented efforts of a generation of immunologists that started with the hypothesis in the 1970s that a type of suppressive T cells needs to be present to regulate and maintain homeostasis of the host immune system. The breakthrough came with the finding of CD25 as the surface marker for Treg in 1995. With the subsequent discovery of Treg cells specific forkhead helix transcription factor (Foxp3) as necessary and sufficient for the immune suppressive function, the CD4+CD25+Foxp3+ lymphocytes were therein established as Treg cells [50].

Treg cells are a heterogeneous population and can be either naturally occurring/thymus-derived or induced/peripherally derived. Several mechanisms have been identified on how Treg cells exert their suppressive functions on effector T cells by intercellular interactions, secretome, metabolic interruptions, or by regulating FOXP3 expression. The improved understanding of Treg cells led to many clinical trials. In organ transplantation medicine, researchers took advantage of Treg cells' immune-modulatory functions to infuse patients with ex vivo expanded Treg cells to prevent graft-versus-host disease [51]. On the other hand, investigators also utilized the essence of Treg cells' protumorigenic effects to devise immunotherapeutic strategies by depletion of Treg cells in the hope of augmenting antitumor immune responses by effector T cells. Such depletion strategies can be achieved by direct infusion of Treg-depleted donor lymphocytes to patients with hematopoietic malignancies or by functionally decreasing Treg cells via blocking their migration to TME, targeting immune checkpoint (CTLA-4, PD-1, etc.) and other molecules of Treg cells. With FDA approval of ipilimumab, an anti-CTLA-2 mAb, as the first-line treatment of several solid tumors, including non-small cell lung cancer, therapies targeting immune-modulatory cells hold great promise. However, given the heterogeneity, plasticity, and diverse effects of Treg cells on tumor, more studies are needed on their tumorigenic mechanisms in order to provide new strategies for tumor immunotherapy and develop more effective and less toxic antitumor compounds.

3.2 Conclusion

There have been significant advances in the field of TME with many new discoveries. This chapter set out to cover the most important achievements in both historical perspective and recent progress in TME literature. Discussions were focused on five key coconspirators in TME, namely CAFs, associated secretomes, exosomes, endothelial cells, and immune cells, with emphasis on their fundamental biological roles and functions in tumorigenesis. While each of these coconspirators appears to function as an independent entity, their functions are intertwined and overlap in promoting tumorigenesis. The field is still evolving, but the definitions, basic functions, and roles of the conspirators in TME in tumorigenesis were highlighted. We have tried to review the most current research models, milestones, cutting-edge technologies in targeted therapies, and their roles in providing personalized therapies and potentials for clinical applications.

Study Questions

- With the accumulation of knowledge on CAFs, what directions do investigators need to pursue in the future?
- What can clinical scientists make the best use of CAFs' secretome?
- How does the notion of Exosomics enlighten us?
- What aspects of tumor angiogenesis remain to be further investigated?
- With the advent of immunotherapy, immune cells in TME are getting the attention and resource commitment they deserve. What do we have to do to maintain the momentum?

References

1. Liotta LA, Kohn EC. The microenvironment of the tumour-host interface. Nature. 2001;411:375–9.
2. Takahashi K, Tanabe K, Ohnuki M, et al. Induction of pluripotent stem cells from adult human fibroblasts by defined factors. Cell. 2007;131:861–72.
3. Kalluri R, Zeisberg M. Fibroblasts in cancer. Nat Rev Cancer. 2006;6:392–401.
4. Sahai E, Astsaturov I, Cukierman E, et al. A framework for advancing our understanding of cancer-associated fibroblasts. Nat Rev Cancer. 2020;20:174–86.
5. Bochet L, Lehuédé C, Dauvillier S, et al. Adipocyte-derived fibroblasts promote tumor progression and contribute to the desmoplastic reaction in breast cancer. Cancer Res. 2013;73:5657–68.
6. Piera-Velazquez S, Li Z, Jimenez SA. Role of endothelial-mesenchymal transition (EndoMT) in the pathogenesis of fibrotic disorders. Am J Pathol. 2011;179:1074–80.
7. Quante M, Tu SP, Tomita H, et al. Bone marrow-derived myofibroblasts contribute to the mesenchymal stem cell niche and promote tumor growth. Cancer Cell. 2011;19:257–72.
8. Hay ED. An overview of epithelio-mesenchymal transformation. Acta Anat (Basel). 1995;154:8–20.
9. Petersen OW, Nielsen HL, Gudjonsson T, et al. Epithelial to mesenchymal transition in human breast cancer can provide a nonmalignant stroma. Am J Pathol. 2003;162:391–402.

10. Kurose K, Hoshaw-Woodard S, Adeyinka A, et al. Genetic model of multi-step breast carcinogenesis involving the epithelium and stroma: clues to tumour microenvironment interactions. Hum Mol Genet. 2001;10:1907–13.
11. Olumi AF, Grossfeld GD, Hayward SW, et al. Carcinoma-associated fibroblasts direct tumor progression of initiated human prostatic epithelium. Cancer Res. 1999;59:5002–11.
12. Dumont N, Liu B, Defilippis RA, et al. Breast fibroblasts modulate early dissemination, tumorigenesis, and metastasis through alteration of extracellular matrix characteristics. Neoplasia. 2013;15:249–62.
13. Paszek MJ, Weaver VM. The tension mounts: mechanics meets morphogenesis and malignancy. J Mammary Gland Biol Neoplasia. 2004;9:325–42.
14. Kechagia JZ, Ivaska J, Roca-Cusachs P. Integrins as biomechanical sensors of the microenvironment. Nat Rev Mol Cell Biol. 2019;20:457–73.
15. Mohammadi H, Sahai E. Mechanisms and impact of altered tumour mechanics. Nat Cell Biol. 2018;20:766–74.
16. Wang TH, Hsia SM, Shieh TM. Lysyl oxidase and the tumor microenvironment. Int J Mol Sci. 2016;18:62.
17. Biffi G, Oni TE, Spielman B, et al. IL1-induced JAK/STAT signaling is antagonized by TGFbeta to shape CAF heterogeneity in pancreatic ductal adenocarcinoma. Cancer Discov. 2019;9: 282–301.
18. Ostermann E, Garin-Chesa P, Heider KH, et al. Effective immunoconjugate therapy in cancer models targeting a serine protease of tumor fibroblasts. Clin Cancer Res. 2008;14:4584–92.
19. Hofheinz RD, Al-Batran SE, Hartmann F, et al. Stromal antigen targeting by a humanised monoclonal antibody: an early phase II trial of sibrotuzumab in patients with metastatic colorectal cancer. Onkologie. 2003;26:44–8.
20. Scott AM, Wiseman G, Welt S, et al. A phase I dose-escalation study of sibrotuzumab in patients with advanced or metastatic fibroblast activation protein-positive cancer. Clin Cancer Res. 2003;9:1639–47.
21. Kim MG, Shon Y, Kim J, et al. Selective activation of anticancer chemotherapy by cancerassociated fibroblasts in the tumor microenvironment. J Natl Cancer Inst. 2017;109(1):djw186. https://doi.org/10.1093/jnci/djw186.
22. Dhomen N, Reis-Filho JS, da Rocha DS, et al. Oncogenic Braf induces melanocyte senescence and melanoma in mice. Cancer Cell. 2009;15:294–303.
23. Hirata E, Girotti MR, Amaya Viros A, et al. Intravital imaging reveals how BRAF inhibition generates drug- tolerant microenvironments with high integrin beta1/FAK signaling. Cancer Cell. 2015;27:574–88.
24. Gross J, Lapiere CM. Collagenolytic activity in amphibian tissues: a tissue culture assay. Proc Natl Acad Sci U S A. 1962;47: 1014–22.
25. Gonzalez-Avila G, Sommer B, Mendoza-Posada DA, et al. Matrix metalloproteinases participation in the metastatic process and their diagnostic and therapeutic applications in cancer. Crit Rev Oncol Hematol. 2019;137:57–83.
26. Mook S, Schmidt MK, Viale G, et al. The 70-gene prognosis-signature predicts disease outcome in breast cancer patients with 1-3 positive lymph nodes in an independent validation study. Breast Cancer Res Treat. 2009;116:295–302.
27. Paik S, Shak S, Tang G, et al. A multigene assay to predict recurrence of tamoxifen-treated, node-negative breast cancer. N Engl J Med. 2004;351:2817–26.
28. Kuperwasser C, Chavarria T, Wu M, et al. Reconstruction of functionally normal and malignant human breast tissues in mice. Proc Natl Acad Sci U S A. 2004;101:4966–71.
29. Tarhini AA, Rafique I, Floros T, et al. Phase 1/2 study of rilotumumab (AMG 102), a hepatocyte growth factor inhibitor, and erlotinib in patients with advanced non-small cell lung cancer. Cancer. 2017;123:2936–44.
30. Cui W, Fowlis DJ, Bryson S, et al. TGFbeta1 inhibits the formation of benign skin tumors: but enhances progression to invasive spindle carcinomas in transgenic mice. Cell. 1996;86:531–42.
31. Chen SX, Xu XE, Wang XQ, et al. Identification of colonic fibroblast secretomes reveals secretory factors regulating colon cancer cell proliferation. J Proteome. 2014;110:155–71.
32. da Cunha BR, Domingos C, Stefanini ACB, et al. cellular interactions in the tumor microenvironment: the role of secretome. J Cancer. 2019;10:4574–87.
33. Rodier F, Campisi J. Four faces of cellular senescence. J Cell Biol. 2011;192:547–56.
34. Llorente A, Skotland T, Sylvänne T, et al. Molecular lipidomics of exosomes released by PC-3 prostate cancer cells. Biochim Biophys Acta Mol Cell Biol Lipids. 2013;1831:1302–9.
35. Thakur BK, Zhang H, Becker A, et al. Double stranded DNA in exosomes: a novel biomarker in cancer detection. Cell Res. 2014;24:766–9.
36. Nolte-'t Hoen EN, Buermans HP, Waasdorp M, et al. Deep sequencing of RNA from immune cell-derived vesicles uncovers the selective incorporation of small non-coding RNA biotypes with potential regulatory functions. Nucleic Acids Res. 2012;40:9272–85.
37. Squadrito ML, Baer C, Burdet F, et al. Endogenous RNAs modulate microRNA sorting to exosomes and transfer to acceptor cells. Cell Rep. 2014;8:1432–46.
38. Kamerkar S, Lebleu VS, Sugimoto H, et al. Exosomes facilitate therapeutic targeting of oncogenic KRAS in pancreatic cancer. Nature. 2017;546:498–503.
39. Brown EJ, Frazier WA. Integrin-associated protein (CD47) and its ligands. Trends Cell Biol. 2001;11:130–5.
40. Folkman J. Tumor angiogenesis: therapeutic implications. N Engl J Med. 1971;285:1182–6.
41. Alitalo A, Detmar M. Interaction of tumor cells and lymphatic vessels in cancer progression. Oncogene. 2012;31:4499–508.
42. Apte RS, Chen DS, Ferrara N. VEGF in signaling and disease: beyond discovery and development. Cell. 2019;176:1248–64.
43. Kim KJ, Li B, Winer J, et al. Inhibition of vascular endothelial growth factor-induced angiogenesis suppresses tumour growth in vivo. Nature. 1993;62:841–4.
44. Brooks PC, Clark RA, Cheresh DA. Requirement of vascular integrin $\alpha v\beta 3$ for angiogenesis. Science. 1994;264:569–71.
45. MacDonald TJ, Taga T, Shimada H, et al. Preferential susceptibility of brain tumors to the antiangiogenic effects of an a(v) integrin antagonist. Neurosurgery. 2001;48:151–7.
46. Stupp R, Hegi ME, Gorlia T, et al. Cilengitide combined with standard treatment for patients with newly diagnosed glioblastoma with methylated MGMT promoter (CENTRIC EORTC 26071-22072 study): a multicentre, randomised, open-label, phase 3 trial. Lancet Oncol. 2014;15:1100–8.
47. DeNardo DG, Brennan D, Rexhapaj E, et al. Leukocyte complexity in breast cancer predicts overall survival and functionally regulates response to chemotherapy. Cancer Discov. 2011;1:54–67.
48. Jayasingam SD, Marimuthu Citartan M, Thang TH, et al. Evaluating the polarization of tumor-associated macrophages into M1 and M2 phenotypes in human cancer tissue: technicalities and challenges in routine clinical practice. Front Oncol. 2019;9:1512.
49. Bronte V, Wang M, Overwijk WW, et al. Apoptotic death of CD8 þ T lymphocytes after immunization: induction of a suppressive population of mac-1 þ /gr-1 þ cells. J Immunol. 1998;161:5313–20.
50. Sakaguchi S, Yamaguchi T, Nomura T, et al. Ono M. regulatory T cells and immune tolerance. Cell. 2008;133:775–87.
51. Brunstein CG, Miller JS, Cao Q, et al. Infusion of ex vivo expanded T regulatory cells in adults transplanted with umbilical cord blood: safety profile and detection kinetics. Blood. 2011;117:1061–70.

Hallmarks of Metastasis: Molecular Underpinnings

4

Juan C. Gomez-Gelvez and Dhananjay A. Chitale

Abstract

Metastasis is one of the most important hallmarks in cancer biology responsible for the majority of cancer-related deaths in patients. Cancer cells undergo a series of processes orchestrated by genetic changes in order to acquire invasive properties that allow them to migrate to secondary sites and become a metastatic focus. In this chapter, we will review the molecular bases of the most relevant steps in metastasis, including invasion, intravasation, circulation, extravasation, and metastatic colonization.

Learning Objectives
- To understand the molecular genetic bases of the most important processes contributing to metastasis.
- To recognize the signaling pathways commonly involved in metastasis.
- To identify the interactions of cancer cells with other tumor constituents (e.g., tumor microenvironment and extracellular matrix) that enable metastasis.

4.1 Introduction

Metastasis is defined as the development of secondary tumors in other parts of the body different from the primary tumor site. Metastasis is currently the leading cause of morbidity, mortality, and treatment failure in patients with cancer. Although millions of cancer cells are released into the blood stream daily, metastasis is actually a very inefficient process. As an example, studies performed on animal models have demonstrated that only a vast minority of melanoma cells (approximately 0.01%) have the potential to become a metastatic tumor [1]. Cancer cells must initially leave their primary site to enter and circulate within the bloodstream, which includes acquiring invasion properties as well as the ability of resisting physical stress within the blood vessels. Subsequently, cancer cells must infiltrate distant tissue, a step of the process that is facilitated by alterations that enable formation of a premetastatic niche. Finally, cancer cells must adapt to the new microenvironment of the metastatic site and evade destruction by cellular immunity.

Although advances in molecular biology have contributed to identify key steps of the process, the genetic mechanisms of metastasis remain poorly understood. An improved understanding of the dynamic of this process is crucial to identify possible therapeutic targets that may help stop metastasis and tumor progression. In this chapter, we will review the molecular bases of the most relevant steps in metastasis.

4.2 Dissemination and Invasion

Dissemination and invasion refer to the ability of cancer cells to infiltrate surrounding tissue and constitutes one of the most important steps in malignant progression. The potential of cancer cells to disseminate and invade is influenced by numerous factors capable of modifying their growth, proliferation, motility, and/or migratory ability. These factors are the results of a close interaction between different tumor constituents such as cancer cells, tumor microenvironment, and extracellular matrix (ECM).

Initially, cancer cells are in an epithelial-like state where they are static and bound to other epithelial cells as well as to the ECM. As tumor progresses, cancer cells acquired the ability to detach and escape from the primary tumor through a mechanism that mimics a developmental process known as epithelial-mesenchymal transition (EMT) [2]. This process is crucial for embryogenesis and wound healing; however, in the context of cancer, EMT involves multiple biochemical and morphological changes that grant cancer cells a more

J. C. Gomez-Gelvez (✉) · D. A. Chitale
Department of Pathology and Laboratory Medicine, Henry Ford Health System, Detroit, MI, USA
e-mail: jgomez1@hfhs.org; dchital1@hfhs.org

S. P. Leong et al. (eds.), *Cancer Metastasis Through the Lymphovascular System*,
https://doi.org/10.1007/978-3-030-93084-4_4

mesenchymal-like phenotype with enhanced migratory and invasive capabilities [2]. This process that is crucial for growth and invasion at the primary site as well as at secondary/metastatic sites represents a spectrum of transitional phenotypes between epithelial and mesenchymal characteristics rather than a complete progression in which cancer cell acquires one or the other [3]. Evidence indicates that EMT is operated as a epigenetic process that is not dependent on somatic mutations in normal or neoplastic cells [4]. In fact, EMT is controlled by multiple EMT-associated transcription factors, including the zinc-finger E-box-binding homeobox factors ZEB1 and ZEB2, SNAIL (also known as SNAI1), SLUG (also known as SNAI2), and the basic helix-loop-helix factors TWIST1 and TWIST2 [2]. These factors regulate the expression of each other, repress the expression of genes associated to the epithelial state, and simultaneously induce the expression of genes associated to the mesenchymal state. These changes in gene expression result in disruption of cell-to-cell junctions, loss of cell polarity, reorganization of cytoskeletal proteins, degradation of the underlying basement membrane, and reorganization of ECM. SNAIL represses *CDH1*, the gene encoding for epithelial cadherin (E-cadherin), by binding to E boxes in the *CDH1* promoter and recruiting the polycomb repressive complex [2, 5, 6]. Likewise, ZEB1 represses expression of *CDH1* while inducing expression of genes encoding vimentin and neural cadherin (N-cadherin) through recruitment of chromatin modifiers to their respective promoters [2, 7]. SNAIL and ZEB1 can also repress the expression of proteins associated to tight junctions and apical-basal polarity of epithelial cell such as crumbs (a transmembrane protein), PALS1-associated tight junction protein (PATJ; a scaffolding protein), and lethal giant larvae (LGL; a cytoskeletal protein) [2, 8]. Furthermore, SNAIL and ZEB2 induce expression of matrix metalloproteinases (MMP) which are important for degradation of basement membranes and cell invasion [2, 9, 10]. Other important proteins associated to epithelial phenotype that are downregulated by EMT transcription factors include occludins, claudins, α6β4integrins, cytokeratins, and desmoplakin [2]. Similarly, other important proteins associated to the mesenchymal phenotype that are upregulated by EMT transcription factors include fibronectin, as well as β1 and β3 integrins [2]. Once cancer cells have acquired mesenchymal-like morphology and phenotype, they display spindle shape with front-to-back polarity and reorganized cytoskeleton, which provides motility and invasive capabilities.

EMT is a process activated by tumor microenvironment rather than by cancer cells themselves. The tumor microenvironment, comprised of a mixture of stromal cells and immune cells, secrete a variety of chemokines and growth factors that act on cancer cells to activate EMT and induce the expression of EMT-associated transcription factors through different intracellular signaling pathways (such as TGF-β, Wnt, and Notch pathways) [2]. Cells within the tumor microenvironment that play an important role in activation of EMT include cancer-associated fibroblasts, tumor-associated macrophages, myeloid-derived suppressor cells, and various subtypes of T lymphocytes (i.e., CD4+ helper T cells, CD8+ cytotoxic T cells, CD4+/CD25+ regulatory T cells) [2]. Cancer-associated fibroblasts can secrete chemokines and growth factors such as TGF-β, IL-6, EGF, VEGF, and HGF that have been demonstrated to activate EMT in cancer cells [2]. Tumor-associated macrophages activate EMT via secretion of TGF-β, TNF-α, CSF1, CCL18, and IL-6 [2]. Myeloid-derived suppressor cells are a heterogenous group of immature cells that are recruited by primary tumors and have immunosuppressive properties. These cells accumulate at the invasive front where they promote activation of EMT via secretion of TGF-β, EGF, and HGF [11]. Similarly, CD8+ cytotoxic T cells and CD4+ helper T cells can induce EMT with loss of E-cadherin expression and gain of vimentin expression in pancreatic ductal epithelia cells and pancreatic intraepithelial neoplasia regions, respectively [12]. Although the exact mechanism used by T lymphocytes for induction of EMT has not been elucidated, cytokines such as IL-6, TNF-α, and TGF-β have been suggested to play an important role [13].

EMT is not only significant for tumor dissemination and invasion. This process has also been associated to acquisition of stem cell-like properties and resistance to therapy in cancer cells, which is significant for tumor progression and metastasis. EMT transcription factors can generate cancer stem cells with a combination of epithelial and mesenchymal traits suggesting that they reside at an intermediate state within the EMT spectrum; however, the precise molecular mechanisms have not been established [2]. The mechanism underlying therapy resistance involves regulation of genes associated with cell survival and stem cell maintenance [2]. ZEB1, SNAIL, and SLUG confer resistance to oxaliplatin and cisplatin-based chemotherapies in breast, ovarian colon, and pancreatic cancers [2]. The loss of E-cadherin has been associated with resistance to growth factor and kinase inhibitors in several carcinoma types [2, 14]. The role of EMT in therapy resistance is further supported by the finding that intrinsic molecular mechanisms capable of reversing EMT (such as miR-200 family member which directly repress expression of ZEB1 and ZEB2) have been associated with reversal of resistance to therapy [2]. Interestingly, SNAIL expression has also been associated with resistance to immunotherapy in murine models of melanoma and breast carcinomas [2, 15].

The importance of EMT in malignant progression has recently been a matter of debate since definitive experimental proof of the causal roles of EMT in metastasis formation is limited. Recent studies support the notion that EMT transcription factors such as ZEB1, SNAIL, and SLUG can convert non-metastatic carcinoma cells into cells capable of invading and metastasizing. However, research in lung and pancreas cancer has suggested that EMT might not be essential for metastasis in all types of carcinoma, although it still contributes to chemoresistance [16, 17]. Notably, EMT is currently envisioned as a process that generates a spectrum of cellular states lying between epithelial and mesenchymal poles. Therefore, while some metastasizing carcinomas may exhibit overt mesenchymal properties, other cases may not require the same EMT-associated characteristics [18]. Further evidence is needed to clarify the role of EMT in cancer progression and metastasis.

Multiple genes have been reported to determine invasive potential in cancer cells, suggesting that primary tumor cells may exhibit a metastatic mutational profile [3]. Integrative clinical genomics has shown that the most prevalent genes somatically altered in metastatic cancer include *TP53, CDKN2A, PTEN, PIK3CA,* and *RB1* [19]. Interestingly, putative pathogenic germline variants were present in 12.2% of metastasis specimens, of which 75% were related to defects in DNA repair [19]. Nonetheless, attempts to identify genes that are specifically associated with metastasis have not been successful in spite of the continuous improvement of sequencing technologies. This may suggests that metastasis is driven by largely non-genetic processes such as EMT [2].

Oxygen hemostasis in tumor microenvironment influences expression of genes in cancer cells that are involved in biologic processes related not only to EMT but also to many other aspects of metastasis such as invasion, migration, intravasation, extravasation, establishment of the premetastatic niche, as well as survival and growth at the metastatic site [20]. Hypoxia is a fairly common phenomenon seen in approximately 50% to 60% of solid tumors and is the consequence of an imbalance between oxygen delivery (e.g., abnormalities in tumor vasculature) and oxygen consumption (e.g., increased demand by proliferating cancer cells) [20, 21]. Under hypoxic conditions, tumor microenvironment employs multiple mechanisms to activate hypoxia-inducible factors (HIF). These factors (HIF-1, HIF-2, and HIF-3) are transcription regulators of hundreds of genes that help cancer cells adapt to low oxygen tensions and promote metastasis. HIF are capable of inducing EMT directly by regulating expression of EMT transcription factors such as ZEB1, SNAIL, and TWIST [20]. HIF can also induce EMT indirectly through diverse signaling pathways including Notch, TGF-β, integrin-linked kinase, tyrosine kinase receptors, Wnt, and Hedgehog [20]. AXL, a receptor tyrosine kinase, has been identified as a novel HIF-target driving EMT, invasion and metastasis [22]. Additionally, HIF signaling stimulates secretion of lysyl oxidase (LOX), LOX-like proteins (LOXL2 and 4), and exosomes to create a premetastatic environment in lungs and bones of patients with breast cancer [20]. Clinically, hypoxia is associated to HIF activation, metastasis, resistance to therapy, and poor prognosis [20]. Expression of HIF-1α and HIF-2α is also associated to patient mortality [23]. Similarly, tumors with more extensive hypoxic and anoxic areas have higher rate of metastasis [24].

4.3 Intravasation and Circulation

With malignant progression, cancer cells will eventually invade into the lumen of either normal vasculature in surrounding tissues or neovasculature within tumor microenvironment. This process is denominated intravasation and generates circulating cancer cells (CTC) that have the ability to reach distant organs where they may seed new metastatic colonies. Although tumor vasculature is structurally and functionally abnormal, different factors contributing to intravasation have been documented beyond cancer cell invasiveness. Intravasation requires at least partial degradation of the ECM and basement membrane underlying endothelial cells. Tumor-associated macrophages enhance invasion of cancer cells towards vessels through EGF and CSF1 chemotaxis whereas VEGF causes downregulation of endothelial cell–cell junctions, thus increasing intravasation [25, 26]. Using a micro-fluidic-based approach to investigate cancer cell intravasation, a study showed that the endothelial barrier permeability to cancer cells can be modified by biochemical factors, such as TNF-α, and macrophages to facilitate intravasation and regulate the tumor–endothelial cell interactions [27]. Additionally, live-cell fluorescence microcopy provided evidence for a mitosis-mediated mechanism where tumor cells located along the vessel periphery are able to disrupt the endothelium through cell division and detach into circulation [28]. The mechanical stress imposed by the tumor on invading cancer cells through passive pushing and pulling forces also plays a role in intravasation [3]. Granulocyte macrophage colony-stimulating factor and various types of cytokines, including IL-5, IL-6, IL-8, IL-1β, and IL-17, have also demonstrated a significant role on metastasis [3].

Cancer cells can reach systemic circulation by intravasation into either blood vessels or lymphatics. The latter is having a chief role in lymph node metastasis, which is often the first site of metastasis, and is critical for tumor staging and prognosis. For some cancers, it is the lymphatic vessels rather than blood vessels contributing to the initial dissemination of cancer cells. In fact, carcinomas and melanomas tend to develop lymph node metastases more frequently than sarcomas [29]. Such disparity may be explained by physical or structural differences between lymphatics and blood vessels; however, more active recruitment, including molecular mechanisms, can also play a role. One of the main structural differences is that lymphatic capillaries lack the tight inter-endothelial junctions present in blood vessels. Additionally, lymphatic capillaries lack the surrounding layers of pericytes, smooth muscle cells, and basement membranes [30, 31]. These differences make lymphatic capillaries an easier path of intravasation for cancer cells. It has also been suggested that the flow of interstitial fluid influences the movement of cancer cells towards lymphatics at the periphery of the tumor, which passively increases the chance of intravasation into lymphatics [32, 33].

Molecular mechanisms have also been shown to play a role in lymphatic intravasation of cancer cells. For example, vascular endothelial growth factor receptor-3 (VEGFR3), which is exclusively expressed on lymphatic endothelial cells, and its ligand vascular endothelial growth factor (VEGFC), which is overexpressed by cancer cells, influence the process of lymphangiogenesis and may contribute as an attracting mechanism [31]. However, some studies did not find a significant association between VEGFC and lymph node metastasis, which may be attributed to the probable role of other factors and the heterogeneity of the study samples [34]. Active recruitment of cancer cells towards lymphatics may also occur via chemotaxis of EGF secreted by macrophages, found in the proximity of lymphatics, as well as lymphatic stromal cells. The latter have also been reported to be a source of EGF and IGF-I [35, 36]. Similarly, secretion of chemokines CCL1, CCL19, and CCL21 by lymphatic endothelial cells may attract cancer cells expressing their respective receptors CCR8 (CCL1) and CCR7 (CCL19 and CCL21), mechanism that has been associated to lymph node metastasis [37–39]. Expression of CCR7 on melanoma cells has been shown to increase metastasis to draining lymph nodes, an effect that was blocked by neutralizing anti-CCL21 antibodies [40]. Expression of CCR8 on melanoma cells has also been shown to be crucial for lymph node metastasis following CCL1, which is detected in lymph node lymphatic sinuses but not in peripheral lymphatics [37].

Once in circulation, CTC are found either as single cells or, less frequently, as clusters. As compared to single cells, clusters have an increased metastatic potential since they are more resistant to apoptosis and are more likely to be trapped in small capillaries of distal organs where they may initiate metastatic growth [41]. Additionally, clusters contains stromal cells and immune components from the original tumor microenvironment that contribute to the heterogeneity and enhance their survival as well as metastatic efficiency [3]. Once in circulation, CTC rapidly associate with platelets, an interaction that is triggered by tissue factor displayed on the surface of cancer cells [18]. This interaction facilitates metastasis through formation of a coating shield of platelets around clusters, which prevents detection by immune cells and provides the necessary structure to bear physical stress in circulation [3]. More specifically, the adhered platelets inhibit the activity of natural killer cells by secreting TGF-β and PDGF, as well as by physically shielding cancer cells [18]. Platelets can also alter intracellular signaling pathways of cancer cells that ultimately affect their ability to metastasize. Notably, platelet-derived TGF-β and direct platelet-tumor cell contact synergistically activate the TGF-β/Smad and NF-κB pathways in cancer cells. These pathways induce EMT-like changes in CTC and enhance metastasis [42].

CTC also interact with neutrophils and macrophages to form hetero-aggregates that sustain adhesion to the endothelium and contribute to metastasis. Neutrophils can adopt different functional states and in certain instances they have been found to inhibit metastasis [43]. However, the predominant role of neutrophils appears to be the increase of metastatic potential of CTC [3, 18]. Neutrophils can prolong the retention time of CTC within the vasculature and facilitate adhesive interactions of CTC with sinusoids that contributes to extravasation [18]. Neutrophils can also exert immunosuppressive functions by inhibiting the actions of both cytotoxic CD8+ T cells and natural killer cells [18]. Among patients with advanced breast cancer, further advancement of disease was faster in individuals with CTC-neutrophils clusters than those with them [44]. Additionally, disease-free mice that had been injected with CTC-neutrophils showed and higher number of metastasis as compared to those that were injected with CTC alone [44]. Similarly, in mice with breast cancer, eradication of neutrophils was associated to decreased CTC-neutrophil clusters and delayed metastasis to lungs as compared to mice with breast that did not have their neutrophils depleted [44].

4.4 Extravasation and Colonization

Extravasations consists of CTC leaving circulation and entering the parenchyma of an organ. Initially, CTC arrest in small vessels as a consequence of either physical constrains or adhesion to the intimal layer. Cancer cells must then undergo a process known as transendothelial migration, and finally penetrate the basement membrane in order to form a metastatic colony. In some organs, such as the liver, spleen, and bone, sinusoidal vessels are more permeable allowing for higher rate of metastasis; however, in other organs these vessels are more difficult to trespass, therefore being more dependent on genetic and molecular mediation for metastasis to take place [3]. Clinical observation suggests that most tumors metastasize to specific rather than random organs. This concept, known as metastatic organotropism, is influenced by multiple factors such as genetic alterations in cancer cells different from those that mediate malignant transformation, host microenvironment and the adaptive process of invading cancer cells [3].

Extravasation is a complex process that involves ligand–receptor interactions, chemokines, and circulating non-tumor cells. Although the specific details of this process remain unclear, integrins, a key factor in nearly every step of tumor progression, play an especially important role in extravasation [3]. Endothelial cells in each tissue express unique markers that CTC can recognize and adhere to in a selective manner. This selectivity is thought to be among the mechanisms leading to metastatic organotropism [45]. Both in vivo and in vitro studies show that initial attachment of cancer cells occurs preferentially at endothelial cell junctions, followed by endothelial retraction and adhesion of cancer cells to the underlying basement membrane [45]. Cancer cells have also been demonstrated to adhere preferably to areas of inflammation, which may explain why metastasis often arises at sites of tissue injury [45]. Several proteins secreted by cancer cells, including angiopoietin-like 4 (ANGPTL4), VEGF, MMP, and ADAM12, have been reported to function as disruptors or vascular integrity as well as enhancers of both intravasation and extravasation [18]. Recruitment of CCR2+ inflammatory monocytes in response to CCL2 secretion by cancer or host cells can also facilitate the extravasation and metastatic growth in lung parenchyma [46, 47]. These inflammatory monocytes may differentiate into metastasis-associated macrophages, which contributes to seeding, survival, and growth of cancer cells in the lung parenchyma through the release of VEGF [46]. CCL2 can also act directly on endothelial cells to enhance vascular permeability [47].

Metastatic colonization is the final step of the invasion-metastasis cascade. Although a significant proportion of CTC manage to extravasate, metastatic colonization is a highly inefficient process since the cancer cells are exposed to challenging conditions at the targeted organ which makes survival very difficult. In fact, inefficient colonization is the primary reason for the inefficiency of metastasis seen in animal models [1].

The vast majority of extravasated cancer cells seem almost invariably destined to either be eliminated from the tissue parenchyma or enter into a state of dormancy, in which they persist in an indolent state as single disseminated tumor cells or as small micrometastasis clusters [1, 18]. Of note, dormant cells at secondary sites are thought to be the source of metastasis occurring after successful initial therapy of patients with cancer [48]. Dormancy in metastatic clusters is the result of mechanisms that control the proliferation of cancer cells such as the inability to induce angiogenesis and active suppression by the immune system [18]. Additionally, dormancy programs can be initiated by intrinsic programs in cancer cells, as well as by interactions with the microenvironment that resemble those regulating the self-renewal capacity of adult stem cells [48]. For example, breast cancer cells that have lodged in the bone marrow and possess increased Src activity and expression of CXCR4 are able to activate pro-survival pathways in response to bone-derived CXCL12. On activation of these pathways, cancer cells can resist TNF-related apoptosis-inducing ligand (TRAIL) induced apoptosis, which is a tissue defense mechanism distinctively expressed in bone metastasis microenvironment. Metastatic cells without Src activity would otherwise be eliminated in this microenvironment [49]. Dormant cancer cells can also survive due to their ability to resist anoikis via expression of the tyrosine kinase receptor TrkB or via non-canonical Wnt signaling mediated by WNT2 [18]. Additionally, dormant cell benefit from activation of proliferative signals by interacting with the ECM. This interaction is mediated by the formation of filopodium-like protrusions that are coated with integrin β1, and activates proliferation programs driven by Fak, Src, and Erk signaling [18]. As such, combined inhibition of these pathways induced apoptosis in dormant cells and delayed metastatic colonization [50]. Metabolic homeostasis in dormancy is reached by downregulating two of the most important signaling pathways activated in cancer, RAS/MAPK and PI3K/AKT pathways [51].

Several dormancy-inducing signals have been described in the microenvironment of specific organs. Members of bone morphogenetic proteins (BMP) ligand family, such as BMP7, induce dormancy in cancer cells by activating the metastatic suppressor gene N-myc downstream-regulated gene 1 (*NDGR1*) leading to activation of the p38 MAPK pathway, cell cycle inhibitor p21 expression, and cell cycle arrest [52]. Activation of the p38 MAPK pathway along with absence of mitogenic signals promote an $ERK^{low}/p38^{high}$ state in cancer cells that causes arrest in the G0/G1 phase of the cell cycle and associate quiescence [18, 53]. Interestingly, dormant cells may reside in specific niches such as the endosteal niche during bone metastasis, where they compete with hematopoietic stem cells that normally occupy this location [54]. These specialized niches can support their survival, restrain their proliferation, and possibly provide resistance to therapy [18]. Finally, cancer cells undergoing reactivation after dormancy are nursed by specialized ECM niches that support positive signals, including Wnt and Notch signaling pathways, and attenuated negative signals, such as BMP [48].

Before the arrival of any cancer cell, the metastatic microenvironment, known as premetastatic niche, is primed by a number of systemic signaling tumor-derived factors and bone marrow-derived cells to create supportive conditions for metastatic growth. The formation of this niche includes actions of VEGFR+ bone marrow progenitors, myeloid-derived suppressor cells, or neutrophils [18]. Additionally, cancer cells from primary tumors can secrete exomes containing biologic material (DNA, messenger RNA, microRNA, as well as proteins such as chemokines and growth factors) that can modify the host microenvironment and prepare the premetastatic niche [3, 18]. More recently, a process used to kill microbes in which neutrophils externalize their DNA has been associated to noninfectious inflammatory conditions like cancer and metastasis, and observed to occur at sites where cancer cells colonize more efficiently [55].

Exosomes also contribute to colonization by improving the metastatic properties of cancer cells. For example, malignant tumor cells secrete exosomes containing mRNA transcribed from genes that are involved in cell migration and metastasis. These exomes are then taken up by less malignant tumor cells, located either at the same tumor or at distant tumors, which acquire enhanced migratory behavior and metastatic capacity [56]. Furthermore, exosomes have the ability to remodel the ECM by interacting with fibroblasts, stromal cells, and endothelial cells to degrade protease-associated components such as collagen, laminin, and fibronectin. Exosome-altered ECM shows increased stromal cell proliferation, cancer cell migration and survival, and cancer cell resistance to apoptotic signals [3, 57].

Tumor-derived exosomes play also an important role in metastatic organotropism because they can fuse preferentially with cells at the metastatic site (e.g., lung fibroblasts and epithelial cells, liver Kupffer cells, and brain endothelial cells) and help prepare the premetastatic niche [58]. Exosome proteomics demonstrated distinct integrin expression patterns associated with metastasis at specific organs [58]. Exosome integrins $\alpha6\beta4$ and $\alpha6\beta1$ were associated with lung metastasis, while exosome $\alpha\nu\beta5$ integrin was associated to liver metastasis, suggesting that exosome integrins could be used to predict organ-specific metastasis [58]. Exosome integrin uptake activated Src phosphorylation and pro-inflammatory *S100* gene expression [58]. In line with these findings, targeting integrins $\alpha6\beta4$ and $\alpha\nu\beta5$ decreased exosome uptake, as well as lung and liver metastasis [58].

4.5 Conclusions and Perspectives

- Metastasis is the leading cause of morbidity and mortality patients with cancer.
- Cancer cells modify their phenotype through a mechanism known as epithelial-mesenchymal transition (EMT) in order to acquire motility and invasive capabilities.
- Cancer cells with invasive properties infiltrate into the lumen of vessels and become circulating cancer cells (CTC) that have the ability to reach distant organs where they may seed metastatic colonies.
- Extravasation and transendothelial migration of CTC are complex processes that occur through selective mechanisms contributing to metastatic organotropism.
- The premetastatic niche is primed by tumor-derived factors and bone marrow-derived cells to create supportive conditions for metastatic growth.
- The vast majority of extravasated cancer cells are either eliminated from the tissue parenchyma or enter a state of dormancy, which makes metastatic colonization a highly inefficient process.
- Recognizing and understanding the molecular underpinnings of metastasis will contribute to the development of targeted therapies capable of stopping metastasis and tumor progression; however, the genetic mechanisms of this hallmark of cancer remain poorly understood and require further investigation before its complex nature can be entirely elucidated.

References

1. Luzzi KJ, MacDonald IC, Schmidt EE, Kerkvliet N, Morris VL, Chambers AF, Groom AC. Multistep nature of metastatic inefficiency: dormancy of solitary cells after successful extravasation and limited survival of early micrometastases. Am J Pathol. 1998;153(3):865–73. https://doi.org/10.1016/S0002-9440(10)65628-3.
2. Dongre A, Weinberg RA. New insights into the mechanisms of epithelial-mesenchymal transition and implications for cancer. Nat Rev Mol Cell Biol. 2019;20(2):69–84. https://doi.org/10.1038/s41580-018-0080-4.
3. Fares J, Fares MY, Khachfe HH, Salhab HA, Fares Y. Molecular principles of metastasis: a hallmark of cancer revisited. Signal Transduct Target Ther. 2020;5(1):28. https://doi.org/10.1038/s41392-020-0134-x.
4. Tam WL, Weinberg RA. The epigenetics of epithelial-mesenchymal plasticity in cancer. Nat Med. 2013;19(11):1438–49. https://doi.org/10.1038/nm.3336.
5. Cano A, Perez-Moreno MA, Rodrigo I, Locascio A, Blanco MJ, del Barrio MG, Portillo F, Nieto MA. The transcription factor snail controls epithelial-mesenchymal transitions by repressing E-cadherin expression. Nat Cell Biol. 2000;2(2):76–83. https://doi.org/10.1038/35000025.
6. Herranz N, Pasini D, Diaz VM, Franci C, Gutierrez A, Dave N, Escriva M, Hernandez-Munoz I, Di Croce L, Helin K, Garcia de Herreros A, Peiro S. Polycomb complex 2 is required for E-cadherin repression by the Snail1 transcription factor. Mol Cell Biol. 2008;28(15):4772–81. https://doi.org/10.1128/MCB.00323-08.
7. Sanchez-Tillo E, Lazaro A, Torrent R, Cuatrecasas M, Vaquero EC, Castells A, Engel P, Postigo A. ZEB1 represses E-cadherin and induces an EMT by recruiting the SWI/SNF chromatin-remodeling protein BRG1. Oncogene. 2010;29(24):3490–500. https://doi.org/10.1038/onc.2010.102.
8. Spaderna S, Schmalhofer O, Wahlbuhl M, Dimmler A, Bauer K, Sultan A, Hlubek F, Jung A, Strand D, Eger A, Kirchner T, Behrens J, Brabletz T. The transcriptional repressor ZEB1 promotes metastasis and loss of cell polarity in cancer. Cancer Res. 2008;68(2):537–44. https://doi.org/10.1158/0008-5472.CAN-07-5682.
9. Miyoshi A, Kitajima Y, Kido S, Shimonishi T, Matsuyama S, Kitahara K, Miyazaki K. Snail accelerates cancer invasion by upregulating MMP expression and is associated with poor prognosis of hepatocellular carcinoma. Br J Cancer. 2005;92(2):252–8. https://doi.org/10.1038/sj.bjc.6602266.
10. Miyoshi A, Kitajima Y, Sumi K, Sato K, Hagiwara A, Koga Y, Miyazaki K. Snail and SIP1 increase cancer invasion by upregulat-

ing MMP family in hepatocellular carcinoma cells. Br J Cancer. 2004;90(6):1265–73. https://doi.org/10.1038/sj.bjc.6601685.
11. Toh B, Wang X, Keeble J, Sim WJ, Khoo K, Wong WC, Kato M, Prevost-Blondel A, Thiery JP, Abastado JP. Mesenchymal transition and dissemination of cancer cells is driven by myeloid-derived suppressor cells infiltrating the primary tumor. PLoS Biol. 2011;9(9):e1001162. https://doi.org/10.1371/journal.pbio.1001162.
12. Goebel L, Grage-Griebenow E, Gorys A, Helm O, Genrich G, Lenk L, Wesch D, Ungefroren H, Freitag-Wolf S, Sipos B, Rocken C, Schafer H, Sebens S. CD4(+) T cells potently induce epithelial-mesenchymal-transition in premalignant and malignant pancreatic ductal epithelial cells-novel implications of CD4(+) T cells in pancreatic cancer development. Onco Targets Ther. 2015;4(4):e1000083. https://doi.org/10.1080/2162402X.2014.1000083.
13. Cohen EN, Gao H, Anfossi S, Mego M, Reddy NG, Debeb B, Giordano A, Tin S, Wu Q, Garza RJ, Cristofanilli M, Mani SA, Croix DA, Ueno NT, Woodward WA, Luthra R, Krishnamurthy S, Reuben JM. Inflammation mediated metastasis: immune induced epithelial-to-mesenchymal transition in inflammatory breast cancer cells. PLoS One. 2015;10(7):e0132710. https://doi.org/10.1371/journal.pone.0132710.
14. Singh A, Settleman J. EMT, cancer stem cells and drug resistance: an emerging axis of evil in the war on cancer. Oncogene. 2010;29(34):4741–51. https://doi.org/10.1038/onc.2010.215.
15. Kudo-Saito C, Shirako H, Takeuchi T, Kawakami Y. Cancer metastasis is accelerated through immunosuppression during snail-induced EMT of cancer cells. Cancer Cell. 2009;15(3):195–206. https://doi.org/10.1016/j.ccr.2009.01.023.
16. Fischer KR, Durrans A, Lee S, Sheng J, Li F, Wong ST, Choi H, El Rayes T, Ryu S, Troeger J, Schwabe RF, Vahdat LT, Altorki NK, Mittal V, Gao D. Epithelial-to-mesenchymal transition is not required for lung metastasis but contributes to chemoresistance. Nature. 2015;527(7579):472–6. https://doi.org/10.1038/nature15748.
17. Zheng X, Carstens JL, Kim J, Scheible M, Kaye J, Sugimoto H, Wu CC, LeBleu VS, Kalluri R. Epithelial-to-mesenchymal transition is dispensable for metastasis but induces chemoresistance in pancreatic cancer. Nature. 2015;527(7579):525–30. https://doi.org/10.1038/nature16064.
18. Lambert AW, Pattabiraman DR, Weinberg RA. Emerging biological principles of metastasis. Cell. 2017;168(4):670–91. https://doi.org/10.1016/j.cell.2016.11.037.
19. Freedland SJ, Aronson WJ. Commentary on "Integrative clinical genomics of advanced prostate cancer". Robinson D, Van Allen EM, Wu YM, Schultz N, Lonigro RJ, Mosquera JM, Montgomery B, Taplin ME, Pritchard CC, Attard G, Beltran H, Abida W, Bradley RK, Vinson J, Cao X, Vats P, Kunju LP, Hussain M, Feng FY, Tomlins SA, Cooney KA, Smith DC, Brennan C, Siddiqui J, Mehra R, Chen Y, Rathkopf DE, Morris MJ, Solomon SB, Durack JC, Reuter VE, Gopalan A, Gao J, Loda M, Lis RT, Bowden M, Balk SP, Gaviola G, Sougnez C, Gupta M, Yu EY, Mostaghel EA, Cheng HH, Mulcahy H, True LD, Plymate SR, Dvinge H, Ferraldeschi R, Flohr P, Miranda S, Zafeiriou Z, Tunariu N, Mateo J, Perez-Lopez R, Demichelis F, Robinson BD, Schiffman M, Nanus DM, Tagawa ST, Sigaras A, Eng KW, Elemento O, Sboner A, Heath EI, Scher HI, Pienta KJ, Kantoff P, de Bono JS, Rubin MA, Nelson PS, Garraway LA, Sawyers CL, Chinnaiyan AM. Cell. 21 May 2015;161(5):1215–1228. Urol Oncol. 2017;35(8):535. https://doi.org/10.1016/j.urolonc.2017.05.010.
20. Rankin EB, Giaccia AJ. Hypoxic control of metastasis. Science. 2016;352(6282):175–80. https://doi.org/10.1126/science.aaf4405.
21. Vaupel P, Mayer A. Hypoxia in cancer: significance and impact on clinical outcome. Cancer Metastasis Rev. 2007;26(2):225–39. https://doi.org/10.1007/s10555-007-9055-1.
22. Rankin EB, Fuh KC, Castellini L, Viswanathan K, Finger EC, Diep AN, LaGory EL, Kariolis MS, Chan A, Lindgren D, Axelson H, Miao YR, Krieg AJ, Giaccia AJ. Direct regulation of GAS6/AXL signaling by HIF promotes renal metastasis through SRC and MET. Proc Natl Acad Sci U S A. 2014;111(37):13373–8. https://doi.org/10.1073/pnas.1404848111.
23. Semenza GL. Cancer-stromal cell interactions mediated by hypoxia-inducible factors promote angiogenesis, lymphangiogenesis, and metastasis. Oncogene. 2013;32(35):4057–63. https://doi.org/10.1038/onc.2012.578.
24. Noman MZ, Messai Y, Muret J, Hasmim M, Chouaib S. Crosstalk between CTC, immune system and hypoxic tumor microenvironment. Cancer Microenviron. 2014;7(3):153–60. https://doi.org/10.1007/s12307-014-0157-3.
25. Harney AS, Arwert EN, Entenberg D, Wang Y, Guo P, Qian BZ, Oktay MH, Pollard JW, Jones JG, Condeelis JS. Real-time imaging reveals local, transient vascular permeability, and tumor cell Intravasation stimulated by TIE2hi macrophage-derived VEGFA. Cancer Discov. 2015;5(9):932–43. https://doi.org/10.1158/2159-8290.CD-15-0012.
26. Roussos ET, Condeelis JS, Patsialou A. Chemotaxis in cancer. Nat Rev Cancer. 2011;11(8):573–87. https://doi.org/10.1038/nrc3078.
27. Zervantonakis IK, Hughes-Alford SK, Charest JL, Condeelis JS, Gertler FB, Kamm RD. Three-dimensional microfluidic model for tumor cell intravasation and endothelial barrier function. Proc Natl Acad Sci U S A. 2012;109(34):13515–20. https://doi.org/10.1073/pnas.1210182109.
28. Wong AD, Searson PC. Mitosis-mediated intravasation in a tissue-engineered tumor-microvessel platform. Cancer Res. 2017;77(22):6453–61. https://doi.org/10.1158/0008-5472.CAN-16-3279.
29. Friedl P, Wolf K. Tumour-cell invasion and migration: diversity and escape mechanisms. Nat Rev Cancer. 2003;3(5):362–74. https://doi.org/10.1038/nrc1075.
30. Alitalo K, Carmeliet P. Molecular mechanisms of lymphangiogenesis in health and disease. Cancer Cell. 2002;1(3):219–27. https://doi.org/10.1016/s1535-6108(02)00051-x.
31. Wong SY, Hynes RO. Lymphatic or hematogenous dissemination: how does a metastatic tumor cell decide? Cell Cycle. 2006;5(8):812–7. https://doi.org/10.4161/cc.5.8.2646.
32. Aleskandarany MA, Sonbul SN, Mukherjee A, Rakha EA. Molecular mechanisms underlying lymphovascular invasion in invasive breast cancer. Pathobiology. 2015;82(3–4):113–23. https://doi.org/10.1159/000433583.
33. Dadiani M, Kalchenko V, Yosepovich A, Margalit R, Hassid Y, Degani H, Seger D. Real-time imaging of lymphogenic metastasis in orthotopic human breast cancer. Cancer Res. 2006;66(16):8037–41. https://doi.org/10.1158/0008-5472.CAN-06-0728.
34. Ran S, Volk L, Hall K, Flister MJ. Lymphangiogenesis and lymphatic metastasis in breast cancer. Pathophysiology. 2010;17(4):229–51. https://doi.org/10.1016/j.pathophys.2009.11.003.
35. LeBedis C, Chen K, Fallavollita L, Boutros T, Brodt P. Peripheral lymph node stromal cells can promote growth and tumorigenicity of breast carcinoma cells through the release of IGF-I and EGF. Int J Cancer. 2002;100(1):2–8. https://doi.org/10.1002/ijc.10481.
36. Schoppmann SF, Birner P, Stockl J, Kalt R, Ullrich R, Caucig C, Kriehuber E, Nagy K, Alitalo K, Kerjaschki D. Tumor-associated macrophages express lymphatic endothelial growth factors and are related to peritumoral lymphangiogenesis. Am J Pathol. 2002;161(3):947–56. https://doi.org/10.1016/S0002-9440(10)64255-1.
37. Das S, Sarrou E, Podgrabinska S, Cassella M, Mungamuri SK, Feirt N, Gordon R, Nagi CS, Wang Y, Entenberg D, Condeelis J, Skobe M. Tumor cell entry into the lymph node is controlled by CCL1 chemokine expressed by lymph node lymphatic sinuses. J Exp Med. 2013;210(8):1509–28. https://doi.org/10.1084/jem.20111627.

38. Homey B, Muller A, Zlotnik A. Chemokines: agents for the immunotherapy of cancer? Nat Rev Immunol. 2002;2(3):175–84. https://doi.org/10.1038/nri748.
39. Korbecki J, Grochans S, Gutowska I, Barczak K, Baranowska-Bosiacka I. CC chemokines in a tumor: a review of pro-cancer and anti-cancer properties of receptors CCR5, CCR6, CCR7, CCR8, CCR9, and CCR10 ligands. Int J Mol Sci. 2020;21(20) https://doi.org/10.3390/ijms21207619.
40. Wiley HE, Gonzalez EB, Maki W, Wu MT, Hwang ST. Expression of CC chemokine receptor-7 and regional lymph node metastasis of B16 murine melanoma. J Natl Cancer Inst. 2001;93(21):1638–43. https://doi.org/10.1093/jnci/93.21.1638.
41. Aceto N, Bardia A, Miyamoto DT, Donaldson MC, Wittner BS, Spencer JA, Yu M, Pely A, Engstrom A, Zhu H, Brannigan BW, Kapur R, Stott SL, Shioda T, Ramaswamy S, Ting DT, Lin CP, Toner M, Haber DA, Maheswaran S. Circulating tumor cell clusters are oligoclonal precursors of breast cancer metastasis. Cell. 2014;158(5):1110–22. https://doi.org/10.1016/j.cell.2014.07.013.
42. Labelle M, Begum S, Hynes RO. Direct signaling between platelets and cancer cells induces an epithelial-mesenchymal-like transition and promotes metastasis. Cancer Cell. 2011;20(5):576–90. https://doi.org/10.1016/j.ccr.2011.09.009.
43. Granot Z, Henke E, Comen EA, King TA, Norton L, Benezra R. Tumor entrained neutrophils inhibit seeding in the premetastatic lung. Cancer Cell. 2011;20(3):300–14. https://doi.org/10.1016/j.ccr.2011.08.012.
44. Szczerba BM, Castro-Giner F, Vetter M, Krol I, Gkountela S, Landin J, Scheidmann MC, Donato C, Scherrer R, Singer J, Beisel C, Kurzeder C, Heinzelmann-Schwarz V, Rochlitz C, Weber WP, Beerenwinkel N, Aceto N. Neutrophils escort circulating tumour cells to enable cell cycle progression. Nature. 2019;566(7745):553–7. https://doi.org/10.1038/s41586-019-0915-y.
45. Welch DR, Hurst DR. Defining the hallmarks of metastasis. Cancer Res. 2019;79(12):3011–27. https://doi.org/10.1158/0008-5472.CAN-19-0458.
46. Qian BZ, Li J, Zhang H, Kitamura T, Zhang J, Campion LR, Kaiser EA, Snyder LA, Pollard JW. CCL2 recruits inflammatory monocytes to facilitate breast-tumour metastasis. Nature. 2011;475(7355):222–5. https://doi.org/10.1038/nature10138.
47. Wolf MJ, Hoos A, Bauer J, Boettcher S, Knust M, Weber A, Simonavicius N, Schneider C, Lang M, Sturzl M, Croner RS, Konrad A, Manz MG, Moch H, Aguzzi A, van Loo G, Pasparakis M, Prinz M, Borsig L, Heikenwalder M. Endothelial CCR2 signaling induced by colon carcinoma cells enables extravasation via the JAK2-Stat5 and p38MAPK pathway. Cancer Cell. 2012;22(1):91–105. https://doi.org/10.1016/j.ccr.2012.05.023.
48. Giancotti FG. Mechanisms governing metastatic dormancy and reactivation. Cell. 2013;155(4):750–64. https://doi.org/10.1016/j.cell.2013.10.029.
49. Zhang XH, Wang Q, Gerald W, Hudis CA, Norton L, Smid M, Foekens JA, Massague J. Latent bone metastasis in breast cancer tied to Src-dependent survival signals. Cancer Cell. 2009;16(1):67–78. https://doi.org/10.1016/j.ccr.2009.05.017.
50. El Touny LH, Vieira A, Mendoza A, Khanna C, Hoenerhoff MJ, Green JE. Combined SFK/MEK inhibition prevents metastatic outgrowth of dormant tumor cells. J Clin Invest. 2014;124(1):156–68. https://doi.org/10.1172/JCI70259.
51. Yeh AC, Ramaswamy S. Mechanisms of cancer cell dormancy—another Hallmark of cancer? Cancer Res. 2015;75(23):5014–22. https://doi.org/10.1158/0008-5472.CAN-15-1370.
52. Kobayashi A, Okuda H, Xing F, Pandey PR, Watabe M, Hirota S, Pai SK, Liu W, Fukuda K, Chambers C, Wilber A, Watabe K. Bone morphogenetic protein 7 in dormancy and metastasis of prostate cancer stem-like cells in bone. J Exp Med. 2011;208(13):2641–55. https://doi.org/10.1084/jem.20110840.
53. Sosa MS, Avivar-Valderas A, Bragado P, Wen HC, Aguirre-Ghiso JA. ERK1/2 and p38alpha/beta signaling in tumor cell quiescence: opportunities to control dormant residual disease. Clin Cancer Res. 2011;17(18):5850–7. https://doi.org/10.1158/1078-0432.CCR-10-2574.
54. Shiozawa Y, Pedersen EA, Havens AM, Jung Y, Mishra A, Joseph J, Kim JK, Patel LR, Ying C, Ziegler AM, Pienta MJ, Song J, Wang J, Loberg RD, Krebsbach PH, Pienta KJ, Taichman RS. Human prostate cancer metastases target the hematopoietic stem cell niche to establish footholds in mouse bone marrow. J Clin Invest. 2011;121(4):1298–312. https://doi.org/10.1172/JCI43414.
55. Cedervall J, Zhang Y, Olsson AK. Tumor-induced NETosis as a risk factor for metastasis and organ failure. Cancer Res. 2016;76(15):4311–5. https://doi.org/10.1158/0008-5472.CAN-15-3051.
56. Zomer A, Maynard C, Verweij FJ, Kamermans A, Schafer R, Beerling E, Schiffelers RM, de Wit E, Berenguer J, Ellenbroek SIJ, Wurdinger T, Pegtel DM, van Rheenen J. In vivo imaging reveals extracellular vesicle-mediated phenocopying of metastatic behavior. Cell. 2015;161(5):1046–57. https://doi.org/10.1016/j.cell.2015.04.042.
57. Tarbe N, Losch S, Burtscher H, Jarsch M, Weidle UH. Identification of rat pancreatic carcinoma genes associated with lymphogenous metastasis. Anticancer Res. 2002;22(4):2015–27.
58. Hoshino A, Costa-Silva B, Shen TL, Rodrigues G, Hashimoto A, Tesic Mark M, Molina H, Kohsaka S, Di Giannatale A, Ceder S, Singh S, Williams C, Soplop N, Uryu K, Pharmer L, King T, Bojmar L, Davies AE, Ararso Y, Zhang T, Zhang H, Hernandez J, Weiss JM, Dumont-Cole VD, Kramer K, Wexler LH, Narendran A, Schwartz GK, Healey JH, Sandstrom P, Labori KJ, Kure EH, Grandgenett PM, Hollingsworth MA, de Sousa M, Kaur S, Jain M, Mallya K, Batra SK, Jarnagin WR, Brady MS, Fodstad O, Muller V, Pantel K, Minn AJ, Bissell MJ, Garcia BA, Kang Y, Rajasekhar VK, Ghajar CM, Matei I, Peinado H, Bromberg J, Lyden D. Tumour exosome integrins determine organotropic metastasis. Nature. 2015;527(7578):329–35. https://doi.org/10.1038/nature15756.

5 Hereditary Cancer Syndromes and Cancer Metastasis

Brandon M. Shaw and Olena Kis

Abstract

Hereditary cancer syndromes, also known as familial or inherited cancer syndromes, are inherited genetic disorders that are associated with increased risk for certain types of cancer. Approximately 5–10% of all cancers result from an inherited genetic change (mutation), a defect in a gene that has been passed on from generation to generation within a family. Individuals and families affected with a hereditary cancer syndrome typically present with a specific pattern of cancer within a family, such as the same type of cancer in close relatives, cancer diagnoses at unusually young ages, or multiple different cancers in the same person. This chapter provides a brief overview of hereditary cancer syndromes, their genetic causes, and clinical presentation, with a specific focus on the role of cancer metastasis in hereditary cancer syndromes such as Hereditary Breast and Ovarian Cancer (HBOC) syndrome, Lynch syndrome, and Multiple Endocrine Neoplasia. In addition, this chapter reviews emerging knowledge on the role of the genetic background in the development of metastatic disease and the contribution of specific genes and genetic susceptibility loci to tumor progression and the risk for metastasis.

Learning Objectives

- To understand the differences between hereditary and nonhereditary cancers.
- To learn the main clinical features of several major hereditary cancer syndromes.
- To understand the impact of inherited genetic mutations on cancer metastasis.

B. M. Shaw (✉) · O. Kis
Department of Pathology and Laboratory Medicine, Henry Ford Health System, Detroit, MI, USA
e-mail: bshaw1@hfhs.org; OKis1@hfhs.org

5.1 Introduction

Hereditary cancer syndromes are caused by inherited pathogenic variants that lead to an increased risk for cancer in the individuals carrying these anomalies. Retinoblastoma is the best studied hereditary cancer syndrome that Alfred G. Knudson first described in 1971. Knudson noted a difference in the age of onset in individuals with the inherited form of retinoblastoma versus those with the sporadic form of the disease. This observation led to the development of the two-hit hypothesis of tumorigenesis. The gene responsible for hereditary retinoblastoma, *RB1*, was discovered in 1986.

Inherited genetic pathogenic variants account for 5–10% of all cancer cases. Most hereditary cancers are associated with a germline genetic alteration that was inherited from one of the parents via sperm or egg cell and is consequently present in every cell in the body, making this individual a carrier of a pathogenic (defective) allele in a gene that plays a vital role in protecting against cancer (tumor suppressor gene). Typically, one defective copy of the gene is insufficient to cause cancer, but it makes the individual more susceptible to developing cancer in the future (increased risk for cancer). Suppose any cell in the body loses the second copy of this tumor suppressor gene by acquiring a new genetic alteration (somatic mutation) or loss of the normal allele. In that case, this constitutes the "second hit" resulting in loss of both copies of the tumor suppressor gene in this cell and making it permissive to cancerous growth. The type of cell that acquires this "second hit" somatic mutation will define the type of cancer this individual will develop. Since the inherited genetic variant is present in every cell in the body, carrier individuals are at risk of having multiple types of cancer in their lifetime. They are more likely to develop cancer earlier in life than noncarriers who require two sets of acquired mutations to achieve a similar biallelic loss of this tumor suppressor gene.

In addition to the inherited mutations in high-penetrance cancer predisposition genes that directly increase the risk for

S. P. Leong et al. (eds.), *Cancer Metastasis Through the Lymphovascular System*,
https://doi.org/10.1007/978-3-030-93084-4_5

cancer, recent studies have begun to examine the impact of other germline mutations in low-penetrance genes or metastasis susceptibility loci on tissue susceptibility to metastasis. The emerging data suggest that germline mutations play a much larger role in tumorigenesis than simply acting as predispositions and that inherited germline mutations not only increase cancer risk susceptibility but also act as modulators when shaping tumor evolution. Although the penetrance of other germline genetic defects is very small, collectively, these small changes may explain the apparent heritability in tumor progression and susceptibility to metastasis influenced by the genetic background of the individual. This book chapter will review the impact of inherited genetic variants on tumor microenvironment, transcriptional differences, splice isoforms and expression profiles in various tumors, and tumor immune cell infiltration, and their effect on cancer progression and metastasis.

5.2 Hereditary Cancer Syndromes

Patients who are found to be carriers of a germline pathogenic mutation in a cancer predisposition gene are diagnosed with hereditary cancer syndrome (see Table 5.1 for examples of genes and their associated Hereditary Cancer Syndromes). These individuals are at high risk for developing multiple types of cancer and often at a younger age. They may benefit from more frequent cancer screening to help identify future cancer at an early stage, allow earlier targeted treatment of second cancer, and provide additional risk-reducing surgery options or other clinical management options. In addition, some syndromes have additional clinical manifestations unrelated to cancer (e.g., cardiac defects, immunodeficiency, diabetes, or hearing loss) that may require medical treatment or increased screening. Once the individual is diagnosed with a genetic syndrome, testing close relatives for the identified pathogenic familial variant can also help identify other at-risk family members.

Some hereditary cancer syndromes can be diagnosed without genetic testing, for example, establishing the diagnosis of familial adenomatous polyposis based on the endoscopic findings or presence of bone marrow failure and chromosome breakage in individuals with Fanconi anemia. However, other cancer predisposition syndromes do not have any distinctive clinical manifestations and have a nonspecific pattern of cancer in the family that can be explained by multiple genetic and nongenetic causes. For example, a family history of breast, ovarian, and prostate cancer can be linked to several different hereditary cancer syndromes (see Table 5.1) or may be due to environmental causes or sporadic mutations (acquired). In many cases, diagnosis of a cancer predisposition syndrome requires multigene sequencing using germline DNA from noncancerous cells such as genetic testing of DNA isolated from peripheral blood specimen to identify a deleterious (pathogenic) mutation that can explain the pattern of cancer observed in the family. Since malignancies are prevalent in the population, clinical criteria for genetic testing for hereditary cancer syndromes have been established to help guide the testing and ensure that the indication for molecular genetic analysis is carefully considered.

Many genes associated with hereditary cancer predisposition are involved in DNA repair, cell proliferation, cell death, or signaling pathways. Due to the variation in the roles of these different genes, each cancer syndrome is different in the mechanisms leading to malignant transformation and differs in the types of cancers that are likely to occur (Fig. 5.1). In addition, each hereditary cancer syndrome has a particular characteristic spectrum of tumors. To date, there are ~50 hereditary cancer syndromes described in the literature, and presumably, other genes associated with cancer predisposition are yet to be discovered. Table 5.1 provides an overview of the selected hereditary cancer syndromes with known genetic causes, have a well-characterized spectrum of clinical features, and are associated with moderate to high risk for developing specific types of cancer (high penetrance). Some of the more common syndromes linked to high risk for developing cancer include hereditary breast and ovarian cancer syndrome, Lynch syndrome, also known as hereditary nonpolyposis colorectal cancer (HNPCC) syndrome, and the multiple endocrine neoplasia type 1 and 2. These syndromes will be discussed in more detail, highlighting their genetic underpinnings, mechanisms of cancer development and progression, and the role of cancer metastasis in disease progression and treatment.

5.2.1 Hereditary Breast and Ovarian Cancer Syndrome

The World Health Organization lists breast cancer as the most common cancer in women with 34% of new diagnoses occurring after cancer has disseminated to other locations within the body. Metastasis in breast cancer is responsible for most of the deaths from the disease. Because of this, a better understanding of the individual genetic background and the role it plays in the development of metastatic disease is expected to provide important diagnostic and prognostic information and aid in the management of an individual's treatment regimen. While death rates have been declining due to improved treatments, better screening, and early detection, breast cancer is still one of the leading causes of death in women. Hereditary breast and ovarian cancer syndrome (HBOC) is characterized by an increased risk for female breast cancer (40–80%), male breast cancer (1–10%), ovarian cancer (up to 40%), and to a lesser extent several

Table 5.1 Overview of selected highly penetrant hereditary cancer syndromes

Hereditary cancer syndrome	Gene(s)	Inheritance	Population incidence	Associated cancer risk and other clinical features
Ataxia-Telangiectasia (OMIM# 208900)	ATM	AR	1/40,000–1/100,000	Increased risk for leukemia (ALL), lymphoma (non-Hodgkin lymphoma), and breast cancer. Other clinical features: Progressive cerebellar ataxia from early childhood, telangiectasias, short stature, frequent infections, immunodeficiency
Bloom Syndrome (OMIM# 210900)	BLM	AR	1/48,000	High risk for multiple malignancies (leukemias (AML, ALL), lymphoma, oropharyngeal cancers, colorectal cancer, esophageal and gastric cancer, breast cancer, genitourinary cancers, skin cancer, and Wilms tumor. Other clinical features: Severe pre- and postnatal growth deficiency, short stature, immune abnormalities, sensitivity to sunlight, telangiectases, cutaneous abnormalities (café au lait macules, hyper- or hypopigmentation), insulin resistance
Cowden Syndrome/PTEN Hamartoma Tumor Syndrome (OMIM# 158350)	PTEN	AD	1/200,000	Increased risk for breast, thyroid, endometrial, and renal cancer. Elevated risk for colorectal cancer (up to 9%) and melanoma. Other clinical features: Macrocephaly, hamartomatous overgrowth of tissues, skin hamartomas, uterine fibroids, breast fibrosis, lipomas, ganglioneuromas, trichilemmomas, papillomatous papules, connective tissue, and epidermal nevi.
Familial Adenomatous Polyposis (FAP) and Attenuated FAP (OMIM# 175100)	APC	AD	1/5000–1/33,000	Multiple adenomatous colon polyps and increased risk for colorectal cancer (up to 80%), duodenal polyposis and cancer, increased risk for thyroid cancer (up to 12%), pancreatic cancer, hepatoblastoma, CNS tumors (medulloblastoma), gastric polyps and cancer. Other clinical features: Osteomas, dental abnormalities, benign cutaneous lesions, lipomas, desmoid tumors
Fanconi anaemia (OMIM# 605724, 227650, 615272, 300514, 227645, 227646, 600185, 608111, 603467, 609644, 609054, 610832, 600901, 614083, 613390, 614082, 609053, 61951, 617883, others)	FANCA, FANCB, FANCC, BRCA2 (FANCD1), FANCD2, FANCE, FANCF, FANCG, FANCI, BRIP1 (FANCDJ), FANCL, FANCM, PALB2 (FANCN), RAD51C (FANCO), SLX4 (FANCP), ERCC4 (FANCQ), BRCA1 (FANCS)	AR, XLR, AD	1/100,000–1/160,000	Bone marrow failure and increased risk for malignancy at a young age, most frequently leukemia (AML), MDS, solid tumors of the head and neck, esophagus, liver, and cervix. Other clinical features: Physical abnormalities (short stature, abnormal skin pigmentation, skeletal malformations, microcephaly, others), cutaneous abnormalities (café au lait macules, hyper- or hypopigmentation).
Hereditary breast and ovarian cancer (HBOC) syndrome (OMIM# 113705, 600185)	BRCA1, BRCA2	AD	1/500–1/1000	Increased risk for breast cancer in females (up to 87%) and males (1–9%), ovarian cancer (up to 54%), including fallopian tube and primary peritoneal cancers, prostate cancer (up to 20%), pancreatic cancer (1–7%), and elevated risk for melanoma (cutaneous and ocular), cervical and uterine cancer.
Hereditary Diffuse Gastric Cancer (OMIM# 137215)	CDH1	AD	Very rare	Increased risk for gastric cancer (up to 80%), lobular breast cancer (39–52%), and colorectal cancer.

(continued)

Table 5.1 (continued)

Hereditary cancer syndrome	Gene(s)	Inheritance	Population incidence	Associated cancer risk and other clinical features
Hereditary Paraganglioma-Pheochromocytoma (PGL/PCC) syndromes (OMIM# 115310, 168000, 605373, 601650, 614165, 171300)	SDHB, SDHD, SDHC, SDHAF2, SDHA, MAX, TMEM127	AD	1/1000,000	Characterized by benign or malignant paragangliomas (neuroendocrine tissue tumors) and pheochromocytomas (paragangliomas that are confined to the adrenal medulla). Extra-adrenal sympathetic paragangliomas have an increased likelihood of malignant transformation and metastasis. Risk for developing other cancer types including GIST, pulmonary chondromas, renal clear cell carcinoma, papillary thyroid carcinoma, pituitary adenomas, and neuroendocrine tumors.
Juvenile Polyposis Syndrome, JPS (OMIM# 174900) Juvenile polyposis syndrome/ hereditary hemorrhagic telangiectasia, JPS/HHT (OMIM# 175050)	SMAD4, BMPR1A	AD	1/100,000	Multiple hamartomatous polyps in the GI tract, increased risk for colorectal cancer (up to 70%) and gastric cancer (up to 20%). Other clinical features: Gastrointestinal bleeding, anemia. Individuals with JPS/HHT dues to SMAD4 mutation have additional HHT-related abnormalities such as epistaxis, telangiectases, heart or brain abnormalities, arteriovenous malformations, intestinal malrotation, digital clubbing, and anomalies of the genitalia or urinary tract.
Li-Fraumeni syndrome (OMIM# 151623)	TP53	AD	1/5000–1/20,000	Lifetime risk for cancer >90% in women and >70% in men. Most common cancer types: Breast cancer (up to 80%), soft-tissue sarcomas, osteosarcomas, adrenocortical carcinomas, CNS tumors (glioblastomas, astrocytomas), leukemias (ALL, AML), MDS, lymphoma, and GI tract cancers.
Lynch syndrome (Hereditary non-polyposis colon cancer) (OMIM# 609310, 120,435, 614,337, 614,350)	MLH1, MSH2, MSH6, PMS2, EPCAM deletion	AD	1/400	Increased risk for colorectal cancer (up to 80%), endometrial cancer (up to 55%), ovarian cancer (up to 25%), stomach and small bowel cancer, urinary tract cancers (ureter, renal pelvis, bladder), biliary tract cancer, pancreatic cancer, prostate cancer (up to 30%), CNS tumors (glioblastoma), breast cancer, and sebaceous neoplasms. Other clinical features: Sebaceous adenoma, tumors with microsatellite instability (MSI).
Multiple endocrine neoplasia type 1, MEN1 (OMIM# 131100)	MEN1	AD	1/30,000	Increased risk for tumors of the parathyroid gland (up to 100%), pituitary gland (30–40%), well-differentiated neuroendocrine tumors of the pancreas and the GI tract (up to 70%), elevated risk for carcinoid tumors, adrenocortical tumors, and non-endocrine tumors (facial angiofibromas, collagenomas, lipomas, meningiomas, ependymomas, and leiomyomas)
Multiple endocrine neoplasia type 2, MEN2A/FMTC (OMIM# 171400, 162,300)	RET	AD	1/35,000	Increased risk for medullary thyroid carcinoma (up to 100%), pheochromocytoma (cancer of the adrenal gland), parathyroid adenoma/hyperplasia, hyperparathyroidism
MUTYH-Associated Polyposis (OMIM# 608456)	MUTYH	AR	1/7000–1/22,000	Multiple adenomatous colon polyps and increased risk for colorectal cancer (up to 90%), duodenal adenomas and cancer, ovarian cancer (6–14%), bladder cancer, breast cancer (up to 25%), elevated risk for thyroid, pancreatic, gastric, and skin cancer.

Table 5.1 (continued)

Hereditary cancer syndrome	Gene(s)	Inheritance	Population incidence	Associated cancer risk and other clinical features
Neurofibromatosis type 1 (OMIM# 162200)	NF1	AD	1/3000	Increased risk for optic nerve and other CNS gliomas, malignant peripheral nerve sheath tumors. Other clinical features: Multiple café au lait spots, axillary and inguinal freckling, multiple benign cutaneous neurofibromas, plexiform neurofibromas, iris Lisch nodules, choroidal freckling, musculoskeletal abnormalities (scoliosis, tibial dysplasia), vasculopathy and learning disabilities.
Neurofibromatosis type 2 (OMIM# 101000)	NF2	AD	1/33,000–1/40,000	Characterized by the growth of noncancerous tumors in the nervous system, most frequently bilateral vestibular schwannomas or acoustic neuromas. Risk for schwannomas of other cranial and peripheral nerves, meningiomas, ependymomas, glioma and astrocytomas. Other clinical features: Skin plaques at birth, meningioma, retinal hamartoma, mononeuropathy, neurofibroma, cortical wedge cataracts, unilateral or bilateral hearing loss, balance dysfunction, seizures.
Nevoid Basal Cell Carcinoma or Gorlin syndrome (OMIM# 109400)	PTCH1, SUFU	AD	1/31,000–1/250,000	High risk for developing multiple basal cell carcinomas (the most common form of skin cancer) and several other types of tumors including benign jaw keratocysts, malignant brain tumor (medulloblastoma), and cardiac and ovarian fibromas. Other clinical features: Congenital malformations and bone anomalies, macrocephaly, distinct facial features, pits on the palms of the hands or soles of the feet.
Nijmegen Breakage Syndrome (OMIM# 251260)	NBN	AR	1/100,000	Increased risk for cancer at a young age, most frequently non-Hodgkin's lymphoma. Risk for other cancer types including brain tumors (medulloblastoma, glioma) and muscle tissue tumors (rhabdomyosarcoma). Other clinical features: Microcephaly, growth retardation, distinctive craniofacial features, respiratory tract infections, immunodeficiency, intellectual disability.
Peutz-Jeghers syndrome (PJS) (OMIM# 175200)	STK11	AD	1/25,000–1/300,000	Characterized by the development of noncancerous hamartomatous polyps (Peutz-Jeghers-type) in the gastrointestinal tract, most commonly in the small intestine. Increased risk for colorectal cancer (up to 40%), gastric cancer (up to 30%), pancreatic cancer (11–36%), small intestinal cancer (13%), early-onset breast cancer (32–54%), ovarian sex cord tumors with annular tubules (up to 20%), cervix and uterine cancer (up to 10%), lung cancer (7–17%), and testicular cancer (9%). Other clinical features: Melanocytic macules (dark brown/dark blue).
Retinoblastoma, Familial (OMIM# 180200)	RB1	AD	1/15,00–1/25,000	Characterized by malignant tumor of the developing retina that occurs in children at a young age. Increased risk of developing bilateral retinoblastoma and non-ocular tumors including cancer of the pineal gland in the brain (pineoblastoma), bone cancer (osteosarcomas), soft tissue sarcomas (leiomyosarcomas and rhabdomyosarcomas), and melanomas.

(continued)

Table 5.1 (continued)

Hereditary cancer syndrome	Gene(s)	Inheritance	Population incidence	Associated cancer risk and other clinical features
Tuberous sclerosis (OMIM# 191100, 191,092)	TSC1, TSC2	AD	1/6000	Characterized by benign tumors in many parts of the body including skin, brain, kidneys, and other organs. Increased risk for several types of cancer including CNS tumors (subependymal giant cell astrocytomas) and renal cell carcinomas as well as benign tumors (cardiac rhabdomyomas and renal angiomyolipomas). Other clinical features: Abnormalities of the skin (hypomelanotic macules, confetti skin lesions, facial angiofibromas, shagreen patches, fibrous cephalic plaques, ungual fibromas), renal cysts, brain anomalies (subependymal nodules, cortical dysplasias), heart arrhythmias, eye abnormalities (retinal lesions), seizures, intellectual disability, developmental delay, and psychiatric illness.
Von Hippel-Lindau syndrome (OMIM# 193300)	VHL	AD	1/36,000	Increased risk for a variety of malignant and benign neoplasms, most frequently retinal, cerebellar and spinal hemangioblastoma, renal cell carcinoma (RCC), pancreatic cancer, pheochromocytoma (adrenal gland), and tumors of the reproductive tract. Other clinical features: Cysts in many different parts of the body

Abbreviations: *AD* Autosomal dominant, *ALL* Acute lymphoblastic leukemia, *AML* Acute myeloid leukemia, *AR* Autosomal recessive, *CNS* Central nervous system, *GI* Gastrointestinal, *GIST* Gastrointestinal stromal tumors, *MDS* Myelodysplastic syndromes, *XLR* X-linked recessive
Data Sources: OMIM®, Online Mendelian Inheritance in Man (https://www.omim.org/); GeneReviews (https://www.ncbi.nlm.nih.gov/books/NBK1116/); NIH—Genetic and Rare Diseases Information Center (rarediseases.info.nih.gov); MedlinePlus—Genetics (https://medlineplus.gov/genetics/)

other types of cancer such as prostate cancer, pancreatic cancer, melanoma in BRCA2-mutation carriers, and several other types of cancer. This section will discuss how inherited mutations in BRCA1 and BRCA2 affect the development, progression, and metastasis of breast and ovarian cancers in females and prostate cancer in males.

5.2.1.1 Genetic Causes

Hereditary breast and ovarian cancer syndrome is inherited in an autosomal dominant manner and is caused by germline heterozygous mutations in the BReast CAncer susceptibility genes 1 and 2 (i.e., *BRCA1* and *BRCA2*). *BRCA1* and *BRCA2* are tumor suppressor genes that encode proteins involved in DNA damage response and DNA repair pathways. *BRCA1* is involved in several cellular pathways that maintain genomic stability, including DNA damage repair, cell-cycle checkpoint control, transcriptional regulation of genes, chromatin remodeling, and protein ubiquitination [1]. *BRCA2* plays an important role in DNA double-stranded break repair by homologous recombination and contributes to several other cellular processes that help maintain genome stability, including cell cycle regulation, telomere homeostasis, transcription, and DNA damage response [2]. Proteins encoded by *BRCA1* and *BRCA2* form complexes with other tumor suppressor proteins, encoded by *RAD51* and *PALB2*, to initiate homologous recombination at sites of DNA double-strand breaks [3, 4]. Inherited mutations in *BRCA1* and *BRCA2* genes account for approximately 20–25% of hereditary breast cancers and about 5–10% of all breast cancers as well as approximately 15% of ovarian cancers. Germline mutations in other homologous recombination genes, including *PALB2*, *RAD51C*, *RAD51D*, *BARD1,* and *BRIP1*, have also been linked to increasing the risk for breast and ovarian cancer.

Double-strand breaks simultaneously disrupt both strands of the DNA double helix leading to chromatid breaks. Repair of double-strand breaks can happen in two ways, either through homologous recombination (HR), a preferred, mostly error-free pathway, or by non-homologous end joining (NHEJ), an error-prone process that may result in illegitimate end-joining potentially leading to intrachromosomal or interchromosomal rearrangements, deletion of segments of the chromosome or other genomic abnormalities. Cells with *BRCA1* or *BRCA2* germline mutation acquire additional somatic alterations such as loss of the wild-type *BRCA1* or *BRCA2* allele (i.e., loss of heterozygosity, LOH), leading to a defective HR-mediated repair and forcing the cell to use NHEJ pathway for double-strand breaks repair. This results in the accumulation of genomic alterations that promote malignant progression [5]. Inactivation of other tumor suppressor genes such as *TP53*, *ATM* or *CHEK2* is also commonly seen in BRCA-deficient tumors allowing cells to bypass checkpoint controls and evade apoptosis, leading to malignant transformation. High co-occurrence of *TP53* mutations in

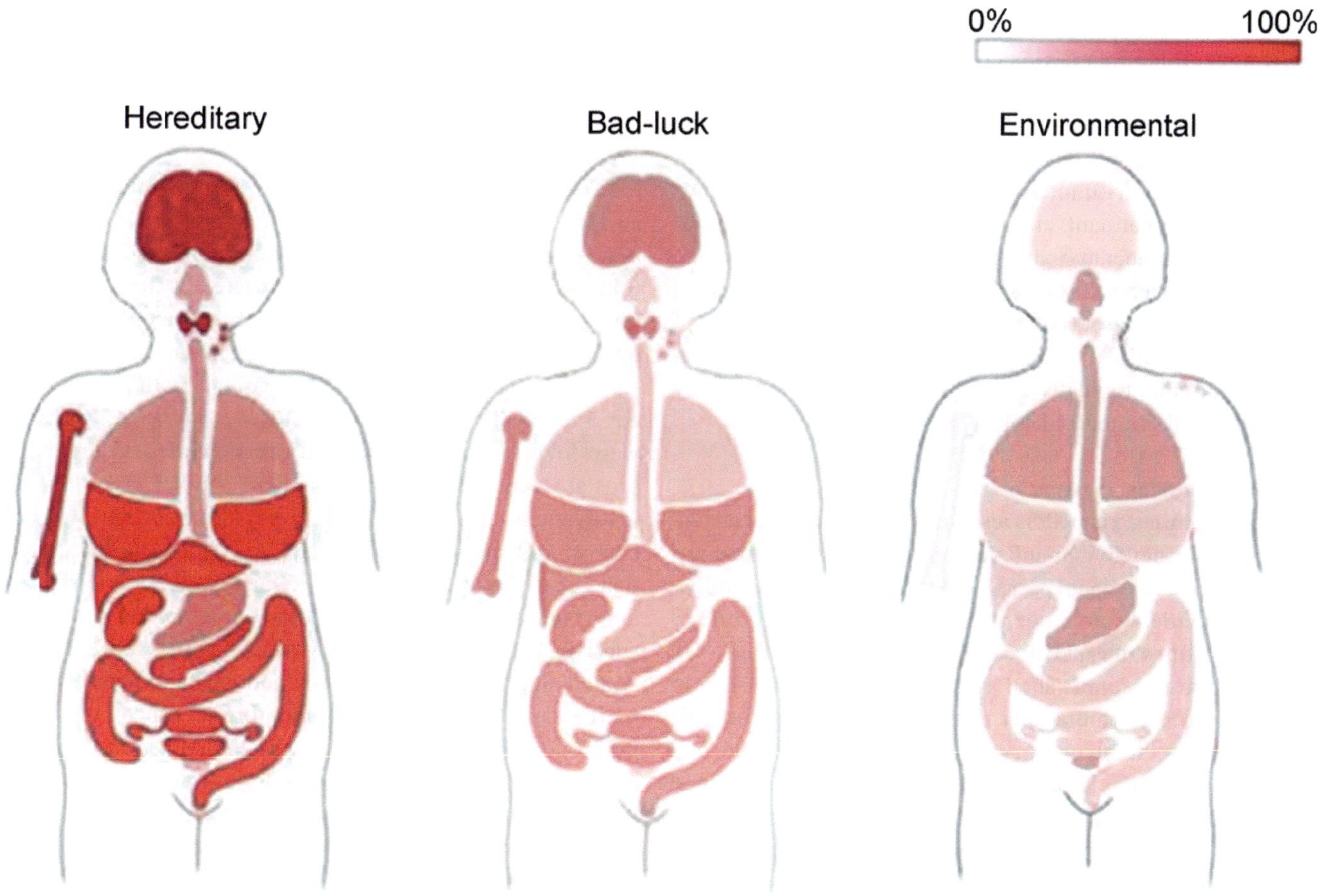

Fig. 5.1 Contributions of germline genomes (i.e., heredity), bad luck (i.e., replicate), and external factors (i.e., environmental) to tumorigenesis and metastasis
The degree of contributions for each factor to tumorigenesis and metastasis in each organ (bone, bone marrows, brain, thymus, respiratory tract, lung, stomach, colon, kidney, liver, ovary, and urinary system) is illustrated by the color depth (from light pink to deep red)
Image was used with permission from Springer Nature and adapted from Jiang X, Asad M, Li L et al. Germline genomes have a dominant-heritable contribution to cancer immune evasion and immunotherapy response. Quant Biol 8, 216–227 (2020). https://doi.org/10.1007/s40484-020-0212-7, Page 223, Table 7.7, Copyright Springer Nature, 2020 (Jiang et al., 2020)

breast tumors with *BRCA1* and *BRCA2* mutations suggests that *TP53* loss of function may be a necessary step in the tumorigenesis of BRCA-associated carcinomas [6].

5.2.1.2 Breast Cancer

Female carriers of *BRCA1* or *BRCA2* mutation have a higher risk for developing breast cancer in their lifetime (46–87% and 38–84%, respectively) compared to females who are noncarriers (12%) based on cumulative data from multiple large cohorts [7–12]. In addition, after the initial breast cancer diagnosis, carriers of inherited *BRCA1* or *BRCA2* mutation have a higher likelihood of developing contralateral breast cancer with estimated cumulative risk by age 70 of 83% in *BRCA1*-mutation carriers and 62% for *BRCA2*-mutation carriers [11]. Breast tumors developing in *BRCA1*- or *BRCA2*-mutation carriers have been shown to have different morphologic and molecular features [13–15]. *BRCA1*-associated tumors are typically high-grade carcinomas that do not express estrogen receptor (ER), progesterone receptor (PR), or human epidermal growth factor receptor 2 (HER2), referred to as triple-negative breast cancer (ER-/PR-/HER2-). *BRCA2*-associated breast cancer tumors typically overexpresses ER (ER+) and in some cases PR (ER+/PR+) but typically do not overexpress the HER2 receptor. Different breast cancer subtypes exhibit distinct gene expression profiles, biological behaviors, and patterns of metastasis [16].

A high likelihood of CNS metastasis (50–53%) has been reported in both *BRCA1*- and *BRCA2*-mutation carriers compared to noncarriers (25%), and the presence of CNS metastasis was associated with increased mortality [16]. CNS metastasis in *BRCA1*-mutation carriers may be due to the higher prevalence of triple-negative breast cancer, which has been documented to increase the risk for CNS involvement [17, 18]. However, *BRCA2* germline mutation was found to be an independent predictor of CNS metastasis [16]. Breast tumors in *BRCA1*-mutation carriers also had frequent lung and distant lymph node metastases (~50%), compared to predominant bone metastasis observed in *BRCA2*-mutation carriers (75%) and noncarriers (53%) [16]. In addition, patients with metastatic cancer who are carriers of a germline *BRCA1* or *BRCA2* mutations are more likely to benefit from treatment with Poly (ADP) ribose polymerase (PARP) inhibitors. Several FDA-approved PARP inhibitors (e.g., olaparib, talazoparib) are included in the NCCN guidelines for systemic treatment in patients with recurrent or metastatic breast cancer harboring *BRCA1* or *BRCA2* germline mutations [19]. Over 65–70% of patients with advanced breast cancer disease develop skeletal metastasis. Carriers of the *BRCA1* or *BRCA2* mutation with metastatic breast cancer have been found to respond better to PARP inhibitor therapy compared to noncarriers [20].

5.2.1.3 Ovarian Cancer

The risk of ovarian cancer is also very high in female carriers of a *BRCA1* (39–63%) or a *BRCA2* (16–27%) mutation compared to noncarriers (1–2%), and presence of a germline *BRCA1* or *BRCA2* mutation is also associated with a higher likelihood of serous adenocarcinoma pathology [7, 8, 10, 21]. High-grade serous adenocarcinomas frequently seen in *BRCA1* or *BRCA2* mutation carriers typically exhibit prominent intraepithelial lymphocytes, marked nuclear atypia, and abundant mitoses [22]. *BRCA1*- and *BRCA2*-associated high-grade serous carcinoma demonstrate characteristic histopathological features and growth patterns in metastatic sites and are associated with either pushing pattern metastases or infiltrative metastases composed only of micropapillae. They differed significantly from cases lacking *BRCA1* or *BRCA2* abnormalities, which have infiltrative metastases with combinations of papillary, glandular, cribriform, and micropapillary architecture [23].

Ovarian cancer in individuals with germline *BRCA1* or *BRCA2* mutation is associated with increased risk for visceral metastasis outside the pelvis as well as a higher risk for CNS metastasis than noncarriers [24–26]. However, CNS metastasis in *BRCA1* or *BRCA2* mutation carriers were more likely to occur without intracranial disease and were found to have a favorable prognosis. In addition, patients with metastatic cancer who are carriers of a germline *BRCA1* or *BRCA2* mutations are more likely to benefit from treatment with Poly (ADP) ribose polymerase (PARP) inhibitors, with several FDA-approved PARP inhibitors available for the treatment of metastatic ovarian and breast cancer in patients with germline *BRCA1* or *BRCA2* mutation. Several studies show improved response to PARP inhibitors in patients with a germline pathogenic *BRCA1* or *BRCA2* mutation [27]. Some PARP inhibitors (e.g., niraparib, velaparib) have been found to cross the blood-brain barrier in animal models and may have efficacy against CNS metastasis in patients with *BRCA1* and *BRCA2* germline mutations [26]. Given the high risk for high-grade ovarian cancer (including fallopian tube and primary peritoneal cancers) in women with *BRCA1* or *BRCA2* germline mutation, prophylactic oophorectomy has shown to result in a significant risk reduction for breast and ovarian cancer.

5.2.1.4 Prostate Cancer

Male carriers of germline *BRCA1* or *BRCA2* mutation are at an increased lifetime risk for prostate cancer. Compared to noncarriers, *BRCA1*- and *BRCA2*-mutation carriers have 3.7-fold and 8.6-fold higher risk, respectively, for developing prostate cancer in their lifetime, with strongest association observed for advanced or metastatic prostate cancer [28–33]. Based on a large cohort of >2000 patients with prostate cancer, individuals with germline *BRCA1* or *BRCA2* mutation were found to have significantly higher rates of aggressive prostate cancer (Gleason score ≥8), higher risk of nodal involvement and distant metastasis, and poor survival outcomes compared to noncarriers [34]. Other studies showed that prostate cancer arising in *BRCA2*-mutation carriers is associated with a higher histologic grade and poorer overall survival [35, 36]. Collectively, these studies have demonstrated that nodal involvement and distant metastasis are more common in patients with prostate cancer who have germline *BRCA1* or *BRCA2* mutations than in noncarriers, and that those carriers with local disease develop metastasis earlier and may require tailored clinical management or earlier treatment intervention. FDA-approved PARP inhibitors (e.g., olaparib, rucaparib) are included in guidelines as second-line and subsequent therapy for patients with metastatic castration-resistant prostate cancer harboring pathogenic mutations in *BRCA1* or *BRCA2*.

5.2.2 Lynch Syndrome

Lynch syndrome, also known as hereditary non-polyposis colorectal cancer (HNPCC), accounts for approximately 1–3% of colon cancers, and 0.8–1.4% of endometrial cancers. With a population prevalence of approximately 1:440, Lynch syndrome is the most common cause of hereditary colorectal cancer and hereditary endometrial cancer in women. It is also associated with increased risk for several other types of cancer, including ovarian, gastrointestinal, renal, hepatobiliary tract, urinary tract, brain, and skin cancer.

5.2.2.1 Genetic Causes

Lynch syndrome is inherited in an autosomal dominant manner and is caused by germline heterozygous mutation in genes involved in the mismatch repair (MMR) pathway, such as *MLH1*, *MSH2*, *MSH6*, or *PMS2*, or in rare cases by deletion within *EPCAM* gene leading to epigenetic silencing of *MSH2* by promoter hypermethylation. The *MLH1*, *MSH2*, *MSH6*, and *PMS2* genes are involved in repairing DNA replication mistakes that occur when DNA is copied during cell division. The MMR pathway functions to correct DNA mismatches through the interaction of MMR proteins as heterodimers. *MSH2* typically heterodimerizes with *MSH6* to form MutSa complex, which recognizes single base-pair mismatches and small insertion/deletion loops and creates a sliding clamp around the DNA. The MutSa complex then binds with the *MLH1-PMS2* (MutLa) complex. Interactions of this four-protein MMR complex with exonuclease-1, PCNA, and DNA polymerase allow excision of the mismatch to occur. In addition, another MMR protein, *MSH3*, plays a role in repair of larger insertion/deletion loops. When *MSH2* heterodimerizes with *MSH3* it forms MutSb complex which can recognize larger insertion/deletion loops. MutSb complements the function of MutSa, which recognizes single pair mismatches and small insertion/deletion loops. Several other MutL heterodimers can also form since *MLH1* can heterodimerize with *PMS2* (most common), *PMS1,* or *MLH3.* Defects in any of the mismatch repair genes can result in the disruption of the MMR complex leading to accumulation of DNA errors as the abnormal cells continue to divide which result in "slippage" errors in areas of repetitive DNA sequences, known as microsatellites. These accumulated replication errors lead to microsatellite instability (MSI) genotype in MMR-deficient tumors such as Lynch syndrome tumors.

While MSI is a defining feature of Lynch syndrome tumors, this molecular finding is not specific for the inherited Lynch syndrome and occurs more frequently in sporadic forms of colorectal and endometrial cancer. MSI is detected in approximately 15–20% of colorectal cancers and 20% of endometrial cancers, most of which are caused by hypermethylation of the *MLH1* gene promoter leading to *MLH1* gene inactivation through transcriptional silencing, or due to biallelic somatic mutations in the MMR genes. *MLH1* promoter hypermethylation occurs sporadically and accounts for 70–80% of MSI-High colorectal tumors. Thus, careful evaluation of patients with MSI-High tumors using multiple testing methodologies is typically needed to differentiate between patients with sporadic cancer and those with inherited Lynch syndrome who may benefit from increased cancer surveillance, earlier and more extensive cancer screening or risk-reducing medical interventions. Presence of germline inherited mutation in *MLH1* and *MSH2* accounts for approximately 90% of pathogenic variants in families with Lynch syndrome, while a smaller proportion is attributed to pathogenic mutations in *MSH6* (7–10%) or *PMS2* (<5%). In addition, germline deletions in *EPCAM* gene located near *MSH2* result in epigenetic silencing of *MSH2* by promoter hypermethylation and are reported in approximately 1% of individuals with Lynch syndrome. Deficiency in the mismatch repair system may also occur sporadically and can be caused by somatic genetic mutation or epigenetic changes that inactivate the mismatch repair genes (e.g., *MLH1* promoter hypermethylation).

5.2.2.2 Colorectal Cancer

Lynch syndrome is the most common hereditary cause of early-onset colorectal cancer accounting for approximately 2–4% of all colorectal cancers. In individuals diagnosed with Lynch syndrome, the relative risk for developing colorectal cancer varies depending on the causative gene (typically *MLH1, MSH2, MSH6,* or *PMS2*). The relative risk of developing colorectal cancer is significantly higher in individuals with germline pathogenic mutation in *MLH1* (46–49%) or *MHS2* (43–52%) compared to carriers of a pathogenic variant in *MSH6* (15–44%) or *PMS2* (12–20%) [37–42]. Overall, all four genes result in an increased risk for colorectal cancer than the general population risk of 4.5%. In addition, carriers of MLH1 or MSH2 mutation typically develop colorectal cancer at a younger age (median age of onset of 43–45 years) compared to carriers of mutations in *MSH6* (51–63 years) or *PMS2* (47–66 years) and imply the need for different cancer surveillance strategies depending on the causative genetic defect [40]. Colorectal cancer tumors that develop in individuals with Lynch syndrome are poorly differentiated, excessively mucinous with a signet ring or cribriform histology, and are characterized by frequent localization in the proximal colon, tendency to develop multiple primary carcinomas, and high levels of tumor infiltrating lymphocytes. Poorly differentiated tumors are typically associated with a poor prognosis. However, Lynch-related colorectal cancer has a better prognosis than sporadic colorectal cancer and is associated with less invasion and metastases than sporadic colorectal cancer. Although the mechanism behind this survival advantage is not fully understood, several possible explanations have been proposed. Immunological response of intraepithelial cytotoxic T lymphocytes against tumor cells in Lynch syndrome or epithelial-mesenchymal transition implicated in the development and progression of Lynch syndrome makes it less susceptible to metastases [43]. Colorectal cancers with high microsatellite instability (MSI-High) tend to have a better prognosis than microsatellite-stable (MSS) colorectal cancers. This improved prognosis may be related to active anti-tumor immune responses. Moreover, MSI-High/MMR deficient status in colorectal cancer is considered a predictive biomarker of response to immunotherapy associated with favorable outcomes compared with microsatellite stable (MSS)/MMR proficient colorectal cancer. The risk of recurrent colorectal cancer, however, is increased in individuals with Lynch syndrome. Specifically, the risk of metachronous colorectal cancer is reported to be highest (up to 43%) in individuals with an *MLH1* or *MSH2* pathogenic variant who have segmental resection but lower in carriers of *MSH6* or *PMS2* germline mutation [44].

5.2.2.3 Endometrial Cancer

Mutations in MMR repair genes are also associated with increased risk for endometrial cancer. The highest risks for endometrial cancer is reported in carriers of pathogenic mutations in *MLH1* (43–57%), *MSH2* (21–57%), or *MSH6* (17–46%), and to a lesser extent PMS2 (0–15%) [37–42]. Furthermore, the mean age at endometrial cancer diagnosis is earlier for *MLH1* and *MSH2*-mutation carriers (47–49 years) compared to individuals with an inherited mutation in *MSH6* (53–55 years) or *PMS2* (49–56 years) and could impact cancer surveillance [40]. Immunotherapy strategies against tumors with high microsatellite instability (MSI-High) have been evaluated in multiple studies including comparison of MSI-High colorectal and endometrial cancers caused by either sporadic *MLH1* promoter methylation or inherited Lynch syndrome. MSI-High/MMR deficient status is thought to be associated with high somatic mutational load, which causes increased production of tumor-specific neoantigens and increased tumor infiltration lymphocytes, and is associated with favorable response to immunotherapy [45, 46].

5.2.3 Multiple Endocrine Neoplasia Syndromes

The multiple endocrine neoplasia (MEN) syndromes are a group of autosomal dominant inherited cancers characterized by the development of tumors in the endocrine system [47]. MEN is separated into multiple endocrine neoplasia type 1 (MEN1), multiple endocrine neoplasia type 2 (MEN2) and MEN4 based on the gene involved and the location of the tumors present.

5.2.3.1 Multiple Endocrine Neoplasia Type 1

Multiple endocrine neoplasia type 1 (MEN1) syndrome occurs both sporadically and as an inherited autosomal dominant disorder. The incidence of MEN1 is around 1 in 30,000 with ~90% of the cases being inherited. MEN1 syndrome results from pathogenic variants in the *MEN1* gene which encodes for the protein menin. Menin acts as a tumor suppressor gene and has a role in controlling the cell cycle. The criteria for diagnosing a patient with MEN1 include a patient with two or more MEN1-associated tumors or one MEN1 associated tumor and a first degree relative with MEN1 [48, 49]. Additionally, a germline pathogenic variant in one of the genes causing a MEN syndrome in an asymptomatic individual is considered diagnostic [49]. The age-related penetrance for clinical features is 50% by age 20 and 95% by age 40.

Individuals with MEN1 have the potential to develop one or a combination of over 20 different endocrine and non-endocrine tumors. Parathyroid tumors occur in 90% of individuals between the ages of 20–25 and are the main MEN1-associated endocrinopathy [47, 49, 50]. This tumor type typically presents with hypercalcemia with symptoms including osteoporosis, lathery, and depression. Pituitary tumors are observed more commonly in females than males with prolactinoma being the most common of the pituitary tumors occurring in 20% of cases. Clinical presentation of pituitary tumors varies based on both the tumor size and the hormone produced. Neuroendocrine tumors occur in 30–70% of cases with gastrinoma and insulinoma being the most common at 40% and 10%, respectively [47, 49, 50]. Gastrinomas present with Zollinger-Ellison syndrome and are usually malignant at the time of diagnosis and tend to metastasize to the lymph nodes or the liver. Carcinoid tumors are non-hormone-secreting tumors and occur in over 10% of cases and are also known to be at increased risk of metastasis. Currently, the risk of metastasis correlates with tumor size with 25–40% of individuals with tumors larger than 4 cm developing liver metastases and 50–70% developing lymph node metastases if the tumor is 2–3 cm in size [50]. Roughly one in four patients develop distant metastases resulting in death. Non-endocrine tumors include facial angiofibromas (can be visible on the lips), collagenomas, lipomas, meningiomas, ependymomas, and leiomyomas.

5.2.3.2 Multiple Endocrine Neoplasia Type 2

Multiple endocrine neoplasia type 2 (MEN2) results from an activating germline mutation in the *RET* gene located at chromosome position 10q11.2. *RET* encodes for a tyrosine kinase receptor located on the surface of cell membranes and functions in cell growth and differentiation. Activating mutations in the *RET* gene are associated with MEN2 and result in the development of medullary thyroid carcinoma (MTC), pheochromocytomas (PCC), and primary hyperparathyroidism (PHPT). MEN2 is separated into Multiple Endocrine Neoplasia 2A, MEN2A (MEN2), Multiple Endocrine Neoplasia 2B, MEN2B (MEN3), and familial medullary thyroid cancer (FMTC) based on the RET mutation present, aggressiveness of the medullary thyroid carcinoma (MTC), and if other endocrine tumors are present. The American Thyroid Association has published guidelines for management of medullary thyroid carcinoma and in those guidelines created subcategories within MEN2A. These subcategories include classical MEN2A (MEN2A), MEN2A and cutaneous lichen amyloidosis (MEN2A/CLA), and MEN2A and Hirschsprung disease (MEN2A/HD) [51]. For this discussion, MEN2A and its subtypes will be discussed as a single entity as the incidence of MTC, PHEO, and PHPT is similar.

MEN2A most commonly results from pathogenic variants in codons 609, 611, 618, and/or 620 of exon 10 or in codon 634 of exon 11 of the *RET* gene [52]. Patients with MEN2A typically present with MTC which develops in ~95% of cases while ~50% of cases develop PCC and around 25%–30% will develop PHPT [53]. While most patients develop MTC, the presence of PCC and PHPT appears to depend on the specific mutations present. Patients with mutations in exon 10 have a lower penetrance of PHEO's than those with other RET mutations. Additionally, the RET mutations in codon 634 have a reportedly higher incidence of HPTH compared to mutations in codons 609, 611, 618, and 620 [54, 55]. Management of MEN2A is dependent on the mutation present and calcitonin levels. Generally, individuals with mutations in 609, 611, 618, 620, 630, and 631 (class B/moderate risk) have a less aggressive phenotype than individuals with a mutation in codon 634 (class C/high risk). Patients with class B mutations can have an age of onset of MTC as early as 5 years of age. Due to this, monitoring of calcitonin levels and physical examination should be performed before the age of 5 with monitoring for PHEO usually starting at age 8. Patients with class C mutations can have MTC develop within the first years of life, so screening is recommended to begin by age three in these individuals [52, 56]. Prophylactic thyroidectomy can be considered before the age of 5.

MEN2B is also characterized by MTC (100% of cases) and the presence of PCC (50% of cases); however, primary hyperparathyroidism (PHPT) is not present. Patients instead exhibit a phenotype of mucosal neuromas of the lips and tongue giving the lips a bumpy appearance, ganglioneuromatosis of the gastrointestinal tract and marfanoid habitus. The majority of MEN2B patients (95%) have a pathogenic missense variant in exon 16 at codon 918 resulting in a Met to Thr change [57, 58]. The next commonly observed pathogenic variant (<5%) is in codon 883 and is also a missense mutation which results in an Ala to Phe change [59, 60]. Individuals with MEN2B have a more aggressive clinical course with the development of MTC within the first year of life and PHEO as early as 8 years of age. Due to the aggressiveness of MTC, patients with inherited MEN2B should have a prophylactic thyroidectomy in the first months of life [52]. Those with de novo MEN2B are usually not diagnosed until a thyroid nodule is detected; however, earlier intervention is possible if the phenotype associated with MEN2B is recognized as it typically precedes MTC in these patients [51]. It is also critical that PHEO are identified prior to before thyroidectomy due to increased morbidity and death due to an undiagnosed PHEO.

5.2.3.3 Metastasis in MEN2A and MEN2B

Individuals with MEN2A and MEN2B are at an increased risk to develop metastasis at an early age. While the mutation present can provide some guidance as to the aggressiveness of the disease, there is a high degree of variability among patients with similar mutations and even within families with the same mutation [61, 62]. Due to this variability, overall tumor size and calcitonin levels are also good predictors of metastasis risk in these individuals and should be used to monitor patients [51]. Individuals with MTC have a similar pattern of metastases as individuals with sporadic MTC. When a thyroid nodule is present, 75% of cases have metastases occurring in the regional lymph nodes and 10–15% have developed distant metastasis [63]. In infants with MEN2B, cases have been reported with metastasis to the lymph nodes at the time of prophylactic thyroidectomy [64] further demonstrating the aggressive nature of this disease. Metastases from MTC tend to develop in the lungs, bone, liver, and less commonly in the brain (1–5% of cases). Metastases to the lungs typically involve multiple sites and can be associated with the mediastinal lymph nodes. Liver metastases occur in 45% of cases and can also involve multiple sites; thus resection is not usually possible. When distant metastases are present, the treatment options are mainly palliative in nature [65].

5.3 Metastasis Susceptibility Loci

In addition to the genes known to be associated with hereditary cancer syndromes, recent studies have begun to examine the impact of an individual's germline mutations and the impact these changes may have on the development and metastasis of tumors (Fig. 5.1). Studies in animal models have shown the role of the genetic background in the development of metastatic disease. On such animal model, mouse mammary tumor virus-driven polyomavirus middle T-antigen (MMTV-PyMT), uses MMTV-LTR to drive expression of a mammary gland specific polyomavirus middle T-antigen which results in development of highly metastatic tumors [66]. These mice were crossed with multiple other strains of mice to generate animal models of breast cancer. The data from these crossbreeding experiments show when MMTV-PyMT is crossed with different strains, the progeny of these mating's show significant variation in the development of pulmonary metastasis from the primary mammary tumor [67]. This difference was as large as 40fold in some cases [68]. The authors attribute this observed variation to inherited polymorphisms which changed the disease process in these animals. Other studies in mice have identified potential modifier genes which include genes known to lead to susceptibility to metastasis in mammary cancer. Furthermore, mouse models used in the study of prostate cancer have led to identification of potential metastasis susceptibility genes suggesting this susceptibility could be due to the genetic background of the individual [68]. Recent studies have been published in the literature which supports the impact of germline variation on tumor progression. This variation is influenced by amino acid changes which alter expression of neighboring genes which then increases the chance a somatic mutation will arise in the same gene or pathway. Germline variants have also been tied to transcriptional differences, splice isoforms, and expression profiles in various tumors. These germline mutations, therefore, lead to genomic instability, potential for more somatic pathogenic variants, increased proliferation, lack of apoptosis, and changes to the tumor microenvironment leading to an increased risk of metastasis.

5.4 Metastasis Genes/Loci

5.4.1 Signal-Induced Proliferation-Associated 1 *(SIPA1)*

SIPA1 encodes for a protein with a C-terminal leucine zipper motif and an N-terminal GTPase activating protein. The protein interacts with member of the Ras family of GTPases and is thought to play a role in regulating cell adhesion. Crawford et al. identified three germline variants, −313G>A (rs931127), 545C>T (rs3741378), and 2760G>A (rs746429) in the SIPA1 gene which were associated with metastasis in breast cancer [69]. The SNP's rs746429 and rs91127 were also shown to be significantly associated with lymph node involvement while rs3741378 was observed more commonly in estrogen receptor negative tumors and with progesterone negative tumors [69]. Functional studies showed SIPA1 is highly expressed in the breast cell cancer line MDA-MB-231 and interacts with the promoter of the ITGB1 gene regulating expression of integrin β1 which has been shown to result in cell invasion and migration thru the FAK/AKT-MMP9 signaling pathway [70].

While most research involving SIPA 1 has focused on Breast cancer, this gene has been shown to potentially lead to susceptibility to metastasis in other cancer types including prostate cancer, oral squamous cell carcinoma, colorectal cancer, cervical cancer, lung cancer, gastric cancer, and melanoma [71–73]). While the available evidence suggests the SIPA1 gene may be a valuable prognostic indicator and potential therapeutic target, there is further work that needs to be done to determine the significance of this gene and how it relates to metastatic disease.

5.4.2 Ribosomal RNA-Processing Protein 1B (*RRP1B*)

RRP1B is predicted to be involved in 60S ribosomal RNA processing and has been shown to interact with *SIPA1* in yeast two hybrid immunoprecipitation assays as well as functional assays. Extracellular matrix (ECM) in vitro assays showed that altered expression of *RRP1B* results in tumor growth and dissemination in metastasis assays [74].

Two separate breast cancer cohorts' constitutional polymorphisms within the RRP1B gene showed a significant association with tumor progression. Specifically, a nonsynonymous SNP g.1421 C>T, p.Pro436Leu (rs9306160) was associated with reduced risk of metastasis risk. This variant has an allele frequency of ~0.42 in the general population. In both cohorts, the T allele was associated with a significant difference in metastasis-free survival and overall survival. The effect was determined to be greater in patients with ER+ tumors. It was also noted when rs9306160 (T) is combined with the rs248490 (A;A) in the *SIPA1* gene the risk to develop distant metastasis was 2.5-fold lower than individual with the rs9306160 (C;C) and rs2448490(G) carriers [74].

5.4.3 Interleukin-6 (*IL-6*)

IL-6 encodes a cytokine that is involved in variety of biological functions including inducing expression of *VEGF* which results in increased angiogenesis. *IL-6* has been implemented in a wide range of disease processes including Kaposi's sarcoma, diabetes, Hodgkin's lymphoma, hypertension, gastric cancer, neuroblastoma, ovarian cancer, breast cancer, and prostate cancer [75–79]. Most of these studies involving cancer link IL6 expression to poor outcomes largely due to treatment resistance.

A SNP within the promotor region of *IL-6* (rs1800795; G;G) has been linked to increased risk of distant metastasis in two separate breast cancer cohorts [80]. This SNP has been previously linked to a poor outcome in several cancer types providing further information regarding its role in cancer. Abana et al. also noted an increased incidence of the rs1800795 G:G allele in black and Asian populations. The higher incidence of G:G individuals in the black population may offer an explanation for the increase in brain metastasis and overall poorer outcome in this population compared to that observed in the white population [80].

5.4.4 Aryl Hydrocarbon Receptor Nuclear Translocator Like 2 *(ARNTL2)*

ARNTL2 encodes a regulator of transcription that is a core component of the circadian clock. The circadian clock is responsible for regulation of metabolism and behavioral rhythms. The circadian clock establishes these rhythms through regulation of gene expression throughout a 24-h cycle.

Two separate studies suggest overexpression of ARNTL2 is strongly associated with metastasis. ER- breast cancer and in lung adenocarcinoma. The first study examined the role of ARNTL2 in ER- breast cancer utilizing mouse models and a patient cohort of individuals with ER-breast cancer. Studies in ARNTL2 knockout mice showed a reduction in pulmonary metastasis while overexpression resulted in increased pulmonary metastasis. Neither of the changes resulted in a change in primary tumor growth [81]. Ha et al. also performed human expression quantitative trait locus (eQTL) analysis in two ER-breast cancer cohorts. This analysis showed polymorphisms in the promoter region of ARNTL2 affect expression levels and in conjunction with the mouse models suggest a significant effect on metastasis in ER-breast cancer [81]. The second study examined the role of ARNTL2 in lung adenocarcinoma and showed this gene is overexpressed in metastases from primary lung tumors compared to non-metastatic lung primary tumors. The authors concluded the data from both mouse and human lung adenocarcinoma is consistent with ARTNL2 overexpression resulting in metastases compared to normal expression levels [82].

5.4.5 BLM RecQ like Helicase (*BLM*)

The BLM gene encodes the DNA helicase RecQl3, an enzyme involved in maintenance of genomic integrity. Homozygous pathogenic variants lead to Bloom syndrome. The associated phenotype includes pre- and postnatal growth retardation, facial sun-sensitive telangiectatic erythema, increased susceptibility to infections, and cancer predisposition with leukemia/lymphoma and adenocarcinoma being the most common.

Carriers of pathogenic variants in the *BLM* gene do not display a phenotype associated with the syndrome yet may have in increased risk for breast cancer in females and colon cancer. A recent study of patients with advanced prostate cancer showed a relative risk of metastasis to be 3.4 in carriers of BLM germline mutations [83]. Based on the data, the possibility exists that carriers of pathogenic loss of function variants in the BLM gene may convey susceptibility to metastasis in prostate cancer. Interestingly, the study also noted the individuals with loss of function germline variants in the *BLM* gene had somatic BRCA2 inactivating mutations in the tumors suggesting a possible interaction between these two genes [83].

5.5 Conclusion

In conclusion, there is a growing body of evidence in the literature which supports the role of germline changes in susceptibility to metastasis. However, the majority of the studies conducted to this point cite small sample sizes and the need to do studies with larger data sets. Additionally, the genes involved in susceptibility to metastasis are likely to have a cumulative effect with each variant/gene contributing to the development of metastases but not being solely responsible. The search for the other involved genes may prove arduous and researchers could find themselves in a "missing heritability" situation not dissimilar to that observed in the search for genes involved in height and other genetic diseases [84].

5.6 Future Directions

Future studies will be tasked with determining the overall effect of individual germline variants on metastasis as well as cumulative effects of multiple variants in a single individual. Currently the field has the capability to do whole genome analysis on both tumor and normal sample from the individual; however, there are limitations of doing these studies such as the ability to consistently detect copy number variants and balanced changes, and rapid data analysis.

Open Questions

- What technologies will be necessary to allow for consistent detection of copy number variants and balanced changes?
- Currently long-read next-generation sequencing and optical genome mapping show promise in this area. Additionally, what new genomic changes may be discovered utilizing these newer technologies?
- Data analysis pipelines have been developed to analyze large amounts of data; however, what adjustments/improvements are necessary to speed up and improve data analysis of large data sets in a clinical setting where answers are critical to patient care?
- Lastly, how can we solve the "missing heritability" issue, where many variants may have a discernable disease effect yet the impact of each is so miniscule as to be undetectable by current data analysis techniques?

References

1. Wu J, Lu LY, Yu X. The role of BRCA1 in DNA damage response. Protein Cell. 2010;1(2):117–23. https://doi.org/10.1007/s13238-010-0010-5.
2. Fradet-Turcotte A, Sitz J, Grapton D, Orthwein A. BRCA2 functions: from DNA repair to replication fork stabilization. Endocr Relat Cancer. 2016;23(10):T1–t17. https://doi.org/10.1530/erc-16-0297.
3. Moynahan ME, Pierce AJ, Jasin M. BRCA2 is required for homology-directed repair of chromosomal breaks. Mol Cell. 2001;7(2):263–72. https://doi.org/10.1016/s1097-2765(01)00174-5.
4. Zhao W, Steinfeld JB, Liang F, Chen X, Maranon DG, Jian Ma C, et al. BRCA1-BARD1 promotes RAD51-mediated homologous DNA pairing. Nature. 2017;550(7676):360–5. https://doi.org/10.1038/nature24060.
5. Roy R, Chun J, Powell SN. BRCA1 and BRCA2: different roles in a common pathway of genome protection. Nat Rev Cancer. 2011;12(1):68–78. https://doi.org/10.1038/nrc3181.

6. Sensi E, Tancredi M, Aretini P, Cipollini G, Naccarato AG, Viacava P, et al. p53 inactivation is a rare event in familial breast tumors negative for BRCA1 and BRCA2 mutations. Breast Cancer Res Treat. 2003;82(1):1–9. https://doi.org/10.1023/B:BREA.0000003836.91844.b5.
7. Chen S, Iversen ES, Friebel T, Finkelstein D, Weber BL, Eisen A, et al. Characterization of BRCA1 and BRCA2 mutations in a large United States sample. J Clin Oncol. 2006;24(6):863–71. https://doi.org/10.1200/jco.2005.03.6772.
8. Easton DF, Ford D, Bishop DT. Breast and ovarian cancer incidence in BRCA1-mutation carriers. Breast cancer linkage consortium. Am J Hum Genet. 1995;56(1):265–71.
9. Ford D, Easton DF, Bishop DT, Narod SA, Goldgar DE. Risks of cancer in BRCA1-mutation carriers. Breast cancer linkage consortium. Lancet. 1994;343(8899):692–5. https://doi.org/10.1016/s0140-6736(94)91578-4.
10. Ford D, Easton DF, Stratton M, Narod S, Goldgar D, Devilee P, et al. Genetic heterogeneity and penetrance analysis of the BRCA1 and BRCA2 genes in breast cancer families. The breast cancer linkage consortium. Am J Hum Genet. 1998;62(3):676–89. https://doi.org/10.1086/301749.
11. Mavaddat N, Peock S, Frost D, Ellis S, Platte R, Fineberg E, et al. Cancer risks for BRCA1 and BRCA2 mutation carriers: results from prospective analysis of EMBRACE. J Natl Cancer Inst. 2013;105(11):812–22. https://doi.org/10.1093/jnci/djt095.
12. Satagopan JM, Offit K, Foulkes W, Robson ME, Wacholder S, Eng CM, et al. The lifetime risks of breast cancer in Ashkenazi Jewish carriers of BRCA1 and BRCA2 mutations. Cancer Epidemiol Biomark Prev. 2001;10(5):467–73.
13. Honrado E, Benítez J, Palacios J. The molecular pathology of hereditary breast cancer: genetic testing and therapeutic implications. Mod Pathol. 2005;18(10):1305–20. https://doi.org/10.1038/modpathol.3800453.
14. Mavaddat N, Barrowdale D, Andrulis IL, Domchek SM, Eccles D, Nevanlinna H, et al. Pathology of breast and ovarian cancers among BRCA1 and BRCA2 mutation carriers: results from the consortium of investigators of modifiers of BRCA1/2 (CIMBA). Cancer Epidemiol Biomark Prev. 2012;21(1):134–47. https://doi.org/10.1158/1055-9965.Epi-11-0775.
15. Musolino A, Bella MA, Bortesi B, Michiara M, Naldi N, Zanelli P, et al. BRCA mutations, molecular markers, and clinical variables in early-onset breast cancer: a population-based study. Breast. 2007;16(3):280–92. https://doi.org/10.1016/j.breast.2006.12.003.
16. Song Y, Barry WT, Seah DS, Tung NM, Garber JE, Lin NU. Patterns of recurrence and metastasis in BRCA1/BRCA2-associated breast cancers. Cancer. 2020;126(2):271–80. https://doi.org/10.1002/cncr.32540.
17. Hung MH, Liu CY, Shiau CY, Hsu CY, Tsai YF, Wang YL, et al. Effect of age and biological subtype on the risk and timing of brain metastasis in breast cancer patients. PLoS One. 2014;9(2):e89389. https://doi.org/10.1371/journal.pone.0089389.
18. Lin NU, Vanderplas A, Hughes ME, Theriault RL, Edge SB, Wong YN, et al. Clinicopathologic features, patterns of recurrence, and survival among women with triple-negative breast cancer in the national comprehensive cancer network. Cancer. 2012;118(22):5463–72. https://doi.org/10.1002/cncr.27581.
19. Kaufman B, Shapira-Frommer R, Schmutzler RK, Audeh MW, Friedlander M, Balmaña J, et al. Olaparib monotherapy in patients with advanced cancer and a germline BRCA1/2 mutation. J Clin Oncol. 2015;33(3):244–50. https://doi.org/10.1200/jco.2014.56.2728.
20. Stickeler E, Aktas B, Behrens A, Belleville E, Ditsch N, Fasching PA, et al. Update breast cancer 2021 part 1 - prevention and early stages. Geburtshilfe Frauenheilkd. 2021;81(5):526–38. https://doi.org/10.1055/a-1464-0953.
21. McLaughlin JR, Rosen B, Moody J, Pal T, Fan I, Shaw PA, et al. Long-term ovarian cancer survival associated with mutation in BRCA1 or BRCA2. J Natl Cancer Inst. 2013;105(2):141–8. https://doi.org/10.1093/jnci/djs494.
22. Fujiwara M, McGuire VA, Felberg A, Sieh W, Whittemore AS, Longacre TA. Prediction of BRCA1 germline mutation status in women with ovarian cancer using morphology-based criteria: identification of a BRCA1 ovarian cancer phenotype. Am J Surg Pathol. 2012;36(8):1170–7. https://doi.org/10.1097/PAS.0b013e31825d9b8d.
23. Reyes MC, Arnold AG, Kauff ND, Levine DA, Soslow RA. Invasion patterns of metastatic high-grade serous carcinoma of ovary or fallopian tube associated with BRCA deficiency. Mod Pathol. 2014;27(10):1405–11. https://doi.org/10.1038/modpathol.2013.237.
24. Jernigan AM, Mahdi H, Rose PG. Epithelial ovarian cancer metastatic to the central nervous system and a family history concerning for hereditary breast and ovarian cancer--a potential relationship. Int J Gynecol Cancer. 2015;25(7):1232–8. https://doi.org/10.1097/igc.0000000000000489.
25. Ratner E, Bala M, Louie-Gao M, Hazard S, Brastianos P. 946P - brain metastases in primary ovarian cancer: real-world data. *Annals of oncology, 29*, viii338. 2018; https://doi.org/10.1093/annonc/mdy285.154.
26. Stasenko M, Cybulska P, Feit N, Makker V, Konner J, O'Cearbhaill RE, et al. Brain metastasis in epithelial ovarian cancer by BRCA1/2 mutation status. Gynecol Oncol. 2019;154(1):144–9. https://doi.org/10.1016/j.ygyno.2019.05.004.
27. Borella F, Bertero L, Morrone A, Gambella A, Bovetti M, Cosma S, et al. Brain metastases from ovarian cancer: current evidence in diagnosis, treatment, and prognosis. Cancers (Basel). 2020;12(8) https://doi.org/10.3390/cancers12082156.
28. Abida W, Armenia J, Gopalan A, Brennan R, Walsh M, Barron D, et al. Prospective genomic profiling of prostate cancer across disease states reveals germline and somatic alterations that may affect clinical decision making. JCO Precis Oncol. 2017;2017 https://doi.org/10.1200/po.17.00029.
29. Agalliu I, Gern R, Leanza S, Burk RD. Associations of high-grade prostate cancer with BRCA1 and BRCA2 founder mutations. Clin Cancer Res. 2009;15(3):1112–20. https://doi.org/10.1158/1078-0432.Ccr-08-1822.
30. Leongamornlert D, Mahmud N, Tymrakiewicz M, Saunders E, Dadaev T, Castro E, et al. Germline BRCA1 mutations increase prostate cancer risk. Br J Cancer. 2012;106(10):1697–701. https://doi.org/10.1038/bjc.2012.146.
31. Na R, Zheng SL, Han M, Yu H, Jiang D, Shah S, et al. Germline mutations in ATM and BRCA1/2 distinguish risk for lethal and indolent prostate cancer and are associated with early age at death. Eur Urol. 2017;71(5):740–7. https://doi.org/10.1016/j.eururo.2016.11.033.
32. Pritchard CC, Mateo J, Walsh MF, De Sarkar N, Abida W, Beltran H, et al. Inherited DNA-repair gene mutations in men with metastatic prostate cancer. N Engl J Med. 2016;375(5):443–53. https://doi.org/10.1056/NEJMoa1603144.
33. van Asperen CJ, Brohet RM, Meijers-Heijboer EJ, Hoogerbrugge N, Verhoef S, Vasen HF, et al. Cancer risks in BRCA2 families: estimates for sites other than breast and ovary. J Med Genet. 2005;42(9):711–9. https://doi.org/10.1136/jmg.2004.028829.
34. Castro E, Goh C, Olmos D, Saunders E, Leongamornlert D, Tymrakiewicz M, et al. Germline BRCA mutations are associated with higher risk of nodal involvement, distant metastasis, and poor survival outcomes in prostate cancer. J Clin Oncol. 2013;31(14):1748–57. https://doi.org/10.1200/jco.2012.43.1882.
35. Gallagher DJ, Gaudet MM, Pal P, Kirchhoff T, Balistreri L, Vora K, et al. Germline BRCA mutations denote a clinicopathologic subset

of prostate cancer. Clin Cancer Res. 2010;16(7):2115–21. https://doi.org/10.1158/1078-0432.Ccr-09-2871.
36. Thorne H, Willems AJ, Niedermayr E, Hoh IM, Li J, Clouston D, et al. Decreased prostate cancer-specific survival of men with BRCA2 mutations from multiple breast cancer families. Cancer Prev Res (Phila). 2011;4(7):1002–10. https://doi.org/10.1158/1940-6207.Capr-10-0397.
37. Baglietto L, Lindor NM, Dowty JG, White DM, Wagner A, Gomez Garcia EB, et al. Risks of lynch syndrome cancers for MSH6 mutation carriers. J Natl Cancer Inst. 2010;102(3):193–201. https://doi.org/10.1093/jnci/djp473.
38. Bonadona V, Bonaïti B, Olschwang S, Grandjouan S, Huiart L, Longy M, et al. Cancer risks associated with germline mutations in MLH1, MSH2, and MSH6 genes in lynch syndrome. JAMA. 2011;305(22):2304–10. https://doi.org/10.1001/jama.2011.743.
39. Møller P, Seppälä TT, Bernstein I, Holinski-Feder E, Sala P, Gareth Evans D, et al. Cancer risk and survival in path_MMR carriers by gene and gender up to 75 years of age: a report from the prospective lynch syndrome database. Gut. 2018;67(7):1306–16. https://doi.org/10.1136/gutjnl-2017-314057.
40. Ryan NAJ, Morris J, Green K, Lalloo F, Woodward ER, Hill J, et al. Association of Mismatch Repair Mutation with age at cancer onset in lynch syndrome: implications for stratified surveillance strategies. JAMA Oncol. 2017;3(12):1702–6. https://doi.org/10.1001/jamaoncol.2017.0619.
41. Senter L, Clendenning M, Sotamaa K, Hampel H, Green J, Potter JD, et al. The clinical phenotype of lynch syndrome due to germline PMS2 mutations. Gastroenterology. 2008;135(2):419–28. https://doi.org/10.1053/j.gastro.2008.04.026.
42. Ten Broeke SW, van der Klift HM, Tops CMJ, Aretz S, Bernstein I, Buchanan DD, et al. Cancer risks for PMS2-associated lynch syndrome. J Clin Oncol. 2018;36(29):2961–8. https://doi.org/10.1200/jco.2018.78.4777.
43. Gu GL, Zhu XQ, Wei XM, Ren L, Li DC, Wang SL. Epithelial-mesenchymal transition in colorectal cancer tissue of patients with lynch syndrome. World J Gastroenterol. 2014;20(1):250–7. https://doi.org/10.3748/wjg.v20.i1.250.
44. Anele CC, Adegbola SO, Askari A, Rajendran A, Clark SK, Latchford A, Faiz OD. Risk of metachronous colorectal cancer following colectomy in lynch syndrome: a systematic review and meta-analysis. Color Dis. 2017;19(6):528–36. https://doi.org/10.1111/codi.13679.
45. Howitt BE, Shukla SA, Sholl LM, Ritterhouse LL, Watkins JC, Rodig S, et al. Association of Polymerase e-mutated and microsatellite-instable endometrial cancers with neoantigen load, number of tumor-infiltrating lymphocytes, and expression of PD-1 and PD-L1. JAMA Oncol. 2015;1(9):1319–23. https://doi.org/10.1001/jamaoncol.2015.2151.
46. Pakish JB, Zhang Q, Chen Z, Liang H, Chisholm GB, Yuan Y, et al. Immune microenvironment in microsatellite-instable endometrial cancers: hereditary or sporadic origin matters. Clin Cancer Res. 2017;23(15):4473–81. https://doi.org/10.1158/1078-0432.Ccr-16-2655.
47. Thakker RV. Multiple endocrine neoplasia type 1 (MEN1) and type 4 (MEN4). Mol Cell Endocrinol. 2014;386(1–2):2–15. https://doi.org/10.1016/j.mce.2013.08.002.
48. Giusti F, Brandi ML. Multiple endocrine neoplasia type 1. GeneReviews 1993–2021; 2005, 2017 Dec 14. https://www.ncbi.nlm.nih.gov/books/NBK1538/
49. Niederle B, Selberherr A, Bartsch D, Brandi ML, Doherty GM, Falconi M, et al. Multiple endocrine neoplasia type 1 (MEN1) and the pancreas - diagnosis and treatment of functioning and non-functioning pancreatic and duodenal neuroendocrine neoplasia within the MEN1 syndrome - an international consensus statement. Neuroendocrinology. 2020; https://doi.org/10.1159/000511791.
50. Thakker RV, Newey PJ, Walls GV, Bilezikian J, Dralle H, Ebeling PR, et al. Clinical practice guidelines for multiple endocrine neoplasia type 1 (MEN1). J Clin Endocrinol Metab. 2012;97(9):2990–3011. https://doi.org/10.1210/jc.2012-1230.
51. Wells SA Jr, Asa SL, Dralle H, Elisei R, Evans DB, Gagel RF, et al. Revised American Thyroid Association guidelines for the management of medullary thyroid carcinoma. Thyroid. 2015;25(6):567–610. https://doi.org/10.1089/thy.2014.0335.
52. Raue F, Frank-Raue K. Genotype-phenotype correlation in multiple endocrine neoplasia type 2. Clinics (Sao Paulo), 67 Suppl. 2012;1(Suppl 1):69–75. https://doi.org/10.6061/clinics/2012(sup01)13.
53. Neumann HPH, Young WF Jr, Eng C. Pheochromocytoma and Paraganglioma. N Engl J Med. 2019;381(6):552–65. https://doi.org/10.1056/NEJMra1806651.
54. Frank-Raue K, Rybicki LA, Erlic Z, Schweizer H, Winter A, Milos I, et al. Risk profiles and penetrance estimations in multiple endocrine neoplasia type 2A caused by germline RET mutations located in exon 10. Hum Mutat. 2011;32(1):51–8. https://doi.org/10.1002/humu.21385.
55. Herfarth KK, Bartsch D, Doherty GM, Wells SA Jr, Lairmore TC. Surgical management of hyperparathyroidism in patients with multiple endocrine neoplasia type 2A. Surgery. 1996;120(6):966–73; discussion 973–964. https://doi.org/10.1016/s0039-6060(96)80042-0.
56. Machens A, Gimm O, Hinze R, Höppner W, Boehm BO, Dralle H. Genotype-phenotype correlations in hereditary medullary thyroid carcinoma: oncological features and biochemical properties. J Clin Endocrinol Metab. 2001;86(3):1104–9. https://doi.org/10.1210/jcem.86.3.7290.
57. Carlson KM, Dou S, Chi D, Scavarda N, Toshima K, Jackson CE, et al. Single missense mutation in the tyrosine kinase catalytic domain of the RET protooncogene is associated with multiple endocrine neoplasia type 2B. Proc Natl Acad Sci U S A. 1994;91(4):1579–83. https://doi.org/10.1073/pnas.91.4.1579.
58. Eng C, Smith DP, Mulligan LM, Nagai MA, Healey CS, Ponder MA, et al. Point mutation within the tyrosine kinase domain of the RET proto-oncogene in multiple endocrine neoplasia type 2B and related sporadic tumours. Hum Mol Genet. 1994;3(2):237–41. https://doi.org/10.1093/hmg/3.2.237.
59. Gimm O, Marsh DJ, Andrew SD, Frilling A, Dahia PL, Mulligan LM, et al. Germline dinucleotide mutation in codon 883 of the RET proto-oncogene in multiple endocrine neoplasia type 2B without codon 918 mutation. J Clin Endocrinol Metab. 1997;82(11):3902–4. https://doi.org/10.1210/jcem.82.11.4508.
60. Smith DP, Houghton C, Ponder BA. Germline mutation of RET codon 883 in two cases of de novo MEN 2B. Oncogene. 1997;15(10):1213–7. https://doi.org/10.1038/sj.onc.1201481.
61. Mian C, Barollo S, Zambonin L, Pennelli G, Bernante P, Pelizzo MR, et al. Characterization of the largest kindred with MEN2A due to a Cys609Ser RET mutation. Familial Cancer. 2009;8(4):379–82. https://doi.org/10.1007/s10689-009-9250-z.
62. Szinnai G, Meier C, Komminoth P, Zumsteg UW. Review of multiple endocrine neoplasia type 2A in children: therapeutic results of early thyroidectomy and prognostic value of codon analysis. Pediatrics. 2003;111(2):E132–9. https://doi.org/10.1542/peds.111.2.e132.
63. Moley J. Medullary thyroid carcinoma: management of lymph node metastases. J Natl Compr Cancer Netw. 2010;8:549–56. https://doi.org/10.6004/jnccn.2010.0042.
64. Zenaty D, Aigrain Y, Peuchmaur M, Philippe-Chomette P, Baumann C, Cornelis F, et al. Medullary thyroid carcinoma identified within the first year of life in children with hereditary multiple endocrine neoplasia type 2A (codon 634) and 2B. Eur J Endocrinol. 2009;160(5):807–13. https://doi.org/10.1530/eje-08-0854.

65. Wells SA. Advances in the management of MEN2: from improved surgical and medical treatment to novel kinase inhibitors. Endocr Relat Cancer. 2018;25(2):T1–T13. https://doi.org/10.1530/erc-17-0325.
66. Guy CT, Cardiff RD, Muller WJ. Induction of mammary tumors by expression of polyomavirus middle T oncogene: a transgenic mouse model for metastatic disease. Mol Cell Biol. 1992;12(3):954–61. https://doi.org/10.1128/mcb.12.3.954-961.1992.
67. Hunter K. The intersection of inheritance and metastasis: the role and implications of germline polymorphism in tumor dissemination. Cell Cycle. 2005;4(12):1719–21. https://doi.org/10.4161/cc.4.12.2258.
68. Hunter KW, Amin R, Deasy S, Ha NH, Wakefield L. Genetic insights into the morass of metastatic heterogeneity. Nat Rev Cancer. 2018;18(4):211–23. https://doi.org/10.1038/nrc.2017.126.
69. Crawford NP, Ziogas A, Peel DJ, Hess J, Anton-Culver H, Hunter KW. Germline polymorphisms in SIPA1 are associated with metastasis and other indicators of poor prognosis in breast cancer. Breast Cancer Res. 2006;8(2):R16. https://doi.org/10.1186/bcr1389.
70. Zhang Y, Gong Y, Hu D, Zhu P, Wang N, Zhang Q, et al. Nuclear SIPA1 activates integrin β1 promoter and promotes invasion of breast cancer cells. Oncogene. 2015;34(11):1451–62. https://doi.org/10.1038/onc.2014.36.
71. Brooks R, Kizer N, Nguyen L, Jaishuen A, Wanat K, Nugent E, et al. Polymorphisms in MMP9 and SIPA1 are associated with increased risk of nodal metastases in early-stage cervical cancer. Gynecol Oncol. 2010;116(3):539–43. https://doi.org/10.1016/j.ygyno.2009.09.037.
72. Gdowicz-Kłosok A, Giglok M, Drosik A, Suwiński R, Butkiewicz D. The SIPA1 -313A>G polymorphism is associated with prognosis in inoperable non-small cell lung cancer. Tumour Biol. 2015;36(2):1273–8. https://doi.org/10.1007/s13277-014-2753-8.
73. Xie C, Yang L, Yang X, Yang R, Li Y, Qiu F, et al. Sipa1 promoter polymorphism predicts risk and metastasis of lung cancer in Chinese. Mol Carcinog. 2013;52(Suppl 1):E110–7. https://doi.org/10.1002/mc.22039.
74. Crawford NP, Qian X, Ziogas A, Papageorge AG, Boersma BJ, Walker RC, et al. Rrp1b, a new candidate susceptibility gene for breast cancer progression and metastasis. PLoS Genet. 2007;3(11):e214. https://doi.org/10.1371/journal.pgen.0030214.
75. DeMichele A, Gray R, Horn M, Chen J, Aplenc R, Vaughan WP, Tallman MS. Host genetic variants in the interleukin-6 promoter predict poor outcome in patients with estrogen receptor-positive, node-positive breast cancer. Cancer Res. 2009;69(10):4184–91. https://doi.org/10.1158/0008-5472.Can-08-2989.
76. Hefler LA, Grimm C, Ackermann S, Malur S, Radjabi-Rahat AR, Leodolter S, et al. An interleukin-6 gene promoter polymorphism influences the biological phenotype of ovarian cancer. Cancer Res. 2003;63(12):3066–8.
77. Hohaus S, Giachelia M, Di Febo A, Martini M, Massini G, Vannata B, et al. Polymorphism in cytokine genes as prognostic markers in Hodgkin's lymphoma. Ann Oncol. 2007;18(8):1376–81. https://doi.org/10.1093/annonc/mdm132.
78. Lagmay JP, London WB, Gross TG, Termuhlen A, Sullivan N, Axel A, et al. Prognostic significance of interleukin-6 single nucleotide polymorphism genotypes in neuroblastoma: rs1800795 (promoter) and rs8192284 (receptor). Clin Cancer Res. 2009;15(16):5234–9. https://doi.org/10.1158/1078-0432.Ccr-08-2953.
79. Ruzzo A, Catalano V, Canestrari E, Giacomini E, Santini D, Tonini G, et al. Genetic modulation of the interleukin 6 (IL-6) system in patients with advanced gastric cancer: a background for an alternative target therapy. BMC Cancer. 2014;14:357. https://doi.org/10.1186/1471-2407-14-357.
80. Abana CO, Bingham BS, Cho JH, Graves AJ, Koyama T, Pilarski RT, et al. IL-6 variant is associated with metastasis in breast cancer patients. PLoS One. 2017;12(7):e0181725. https://doi.org/10.1371/journal.pone.0181725.
81. Ha NH, Long J, Cai Q, Shu XO, Hunter KW. The circadian rhythm gene Arntl2 is a metastasis susceptibility gene for estrogen receptor-negative breast cancer. PLoS Genet. 2016;12(9):e1006267. https://doi.org/10.1371/journal.pgen.1006267.
82. Brady JJ, Chuang CH, Greenside PG, Rogers ZN, Murray CW, Caswell DR, et al. An Arntl2-driven Secretome enables lung adenocarcinoma metastatic self-sufficiency. Cancer Cell. 2016;29(5):697–710. https://doi.org/10.1016/j.ccell.2016.03.003.
83. Ledet EM, Antonarakis ES, Isaacs WB, Lotan TL, Pritchard C, Sartor AO. Germline BLM mutations and metastatic prostate cancer. Prostate. 2020;80(2):235–7. https://doi.org/10.1002/pros.23924.
84. Maher B. Personal genomes: the case of the missing heritability. Nature. 2008;456(7218):18–21. https://doi.org/10.1038/456018a.

Part II

Pathology of Cancer Growth and Metastasis

Jane L. Messina

The early medical literature is rife with clinical descriptions of patients with advanced cancer, from the observation in the Edwin Smith Papyrus that a bulging tumor of the breast was a grave disease for which there is no treatment to the observations by Aulus Celsus that advanced breast cancers have a propensity to spread to the armpit, and that spreading to distant organs may cause death [1]. In 1842, William Budd was the first to note that not only tumor could spread to lymph nodes in the region of the tumor, but also cancer was possessed of such powers that it could be preserved "unimpaired when detached from the parent growth, and carried to distant parts of the organism" [2]. 150 years later, Moreno et al. described a novel pattern of spread of melanoma, described as demonstrating a "striking tropism for small and medium-sized vessels, primarily veins … proliferation within the vessel walls, expanding and replacing the muscle layer but sparing the endothelia," for which they proposed the term angiotropic malignant melanoma [3].

In the ensuing years, knowledge about the mechanisms of the three mainstays of cancer dissemination—lymphatic metastasis with nodal involvement, systemic metastasis through vasculature, and angiotropic metastasis along the perivascular spaces—exploded. The first chapters of this textbook describe the genetics, molecular pathways, and biology that govern each type of metastasis.

This section will focus on the pathologic underpinnings of each of these three types of cancer growth and spread. Beginning with the clinical presentation and significance of the respective types of spread, their role in the staging and prognostication of cancer will be described. The histologic hallmarks of tumor in lymph nodes, systemic/visceral spread, and perivascular spaces will be described. Since many tumors present with metastatic disease in lymph nodes and/or distant organs, the pathologic workup of these tumors will be described. Specific cancers have specific propensity to metastasize by certain routes, and knowledge of these can help navigate diagnostic dilemmas. The pathologic workup and reporting of nodal disease identified at the time of surgical resection of a primary tumor will be addressed, including details of sentinel node reporting for both melanoma and breast cancer. The workup of a tumor presenting as an unknown primary tumor, both as nodal disease and systemically, will be discussed. The chapter on systemic metastasis contains an exhaustive description of methods of histologic and immunohistochemical presentation that can be extrapolated to all cancer types. More recently developed molecular methods of lymph node and systemic metastases will be addressed.

Finally, the mechanisms of extravascular migratory metastasis ("angiotropic metastasis") will be dissected. Although it is the least common form of metastatic spread, it represents a unique method in which tumor cells, particularly melanoma, can spread along low-resistance vascular channels, similar to migration of melanoblasts along the neural crest. Identification of the distinctive pathologic pattern of this type of spread has been shown to be as prognostically significant as nodal and systemic spread in cutaneous and uveal melanoma. Notably, it is the most common pattern of melanoma involvement of the liver. Furthermore, this mechanism, which is slower than intravascular metastasis, may explain a subset of tumors with delayed metastatic presentation. It is our hope that this section provides the clinician with a glimpse of the role of a pathologist in diagnosis and reporting of metastatic disease in its myriad presentation.

References

1. Hajdu SI. A note from history: landmarks in history of cancer, part 1. Cancer. 2011;117(5):1097–102.
2. Budd W. Remarks on the pathology and causes of cancer. Lancet. 1842;2:266–70.
3. Moreno A, Español I, Romagosa V. Angiotropic malignant melanoma. Report of two cases. J Cutan Pathol. 1992;19(4):325–9.

Pathologic Assessment of Lymph Node Metastasis

6

James Isom and Jane L. Messina

Abstract

The pathologic assessment and evaluation of lymph node status plays a central role in the diagnosis, staging, and management of most malignancies. The pathologic status of the lymph node plays a central role in the staging, and thus subsequent treatment of diseases, which is codified in the AJCC/UICC staging system. Lymph node metastasis may be subclinical and detected microscopically during surgical resection of a sentinel node or regional lymphadenectomy for initial therapy, or it may present clinically as the first manifestation of a metastatic malignancy. Different protocols exist for the pathologic evaluation of lymph nodes in these distinct settings. These involve systematic gross assessment, careful light microscopic evaluation, judicious and sometimes stepwise application of immunohistochemical methods, and accurate pathologic reporting. The evaluation of a lymph node metastasis in a patient with unknown primary malignancy requires knowledge of the relative frequencies of malignancies, their most common sites of presentation, and an algorithmic approach to immunohistochemical testing. Molecular analysis by gene expression profiling offers modest improvement in diagnostic accuracy, but these improvements have not yet been translated into clinical improvements.

J. Isom
Department of Dermatology and Cutaneous Surgery, University of South Florida Morsani College of Medicine, Tampa, FL, USA

J. L. Messina (✉)
Department of Dermatology and Cutaneous Surgery, University of South Florida Morsani College of Medicine, Tampa, FL, USA

Departments of Pathology and Cutaneous Oncology, Moffitt Cancer Center, Tampa, FL, USA

Departments of Pathology and Cell Biology and Oncologic Sciences, University of South Florida Morsani College of Medicine, Tampa, FL, USA
e-mail: jane.messina@moffitt.org

Learning Objectives

- Learn the frequency of nodal involvement in patients with malignancy, both at presentation and during the course of their disease.
- Learn the clinical relevance of detection of nodal disease at the time of diagnosis as well as during the course of disease.
- Describe the most common tumors to metastasize to lymph nodes and the pathologic workup of these tumors.
- Explore newer methods of evaluation of lymph nodes.

6.1 Introduction

The clinical importance of metastatic spread of cancer has been recognized for centuries, and the importance of nodal involvement looms large in many historical descriptions of cancer. For example, the "fatal black tumor" described by Hippocrates in 5000 BC [1] that was later termed "melanose" by Rene Laennec in 1804 was recognized to have the propensity to metastasize by William Norris in 1820 [2]. And while the prognosis of melanoma was uniformly acknowledged to be dire, Samuel Cooper described surgical removal as having the potential to improve prognosis [3]. After this, in 1898 Herbert Snow was the first to recognize the potential clinical benefit of removing clinically normal lymph nodes at the time of initial cancer surgery. In describing "anticipatory gland excision," he noted that "it is essential to remove, whenever possible, those lymph glands which first receive the infective protoplasm, and bar its entrance into the blood, before they have undergone increase in bulk" [4]. These early studies solidified the central role of lymph node involvement in tumor staging and prognostication. They also set the stage for the debate that continues today, with persistent questions. Are regional lymph nodes the first stop for

S. P. Leong et al. (eds.), *Cancer Metastasis Through the Lymphovascular System*,
https://doi.org/10.1007/978-3-030-93084-4_6

metastases ("incubator" hypothesis) or does their involvement serve as an indicator of aggressive disease with inherent metastatic potential ("marker" hypothesis) [5]? Is there a better way than nodal staging to improve prediction of disease outcome? Does lymph node removal improve disease outcome?

This chapter will address clinically important questions related to cancer metastasis to lymph nodes, focusing on the clinical relevance of metastatic nodal disease, and the clinicopathologic workup of patients who present with disease in their lymph nodes, either clinically or microscopically detected via sentinel node biopsy. The detection and significance of nodal metastasis discovered through sentinel lymph node (SLN) evaluation will also be covered. A diagnostic algorithm for evaluation of patients presenting with nodal metastasis is presented. Recently developed ancillary techniques for evaluation of lymph nodes will be covered, and pathologic changes in nodes harboring metastasis will be considered.

6.2 Importance of the Lymph Node in Tumor Staging

The centrality of lymph node status in accurate tumor staging was solidified in the development of the tumor-node-metastasis (TNM) staging system by Pierre Denoix in 1952 [6]. This staging system was first adopted by the International Union Against Cancer (UICC), which then unified with the American Joint Committee on Cancer (AJCC) in 1987 [7]. Current TNM classification schemes for all tumors in which nodal resection is part of the initial surgical therapy require quantification of the number of regional lymph nodes with tumor involvement. Most staging schema contain pathologic subclassifications describing the extent of lymph node involvement after surgical resection or biopsy sampling, such as tumor size, and whether tumor was identified by immunohistochemical or molecular methods [8]. Considering the importance of lymph status for prognosis, therefore, almost all attempts at curative surgical resection of malignancy include removal of both clinically negative and positive regional lymph node basins [9]. However, the morbidity of these procedures, and the ability of tumors to metastasize to non-predicted sites, led to attempts to refine surgery for regional lymph node involvement. In 1977, Ramon Cabanas was the first to utilize the term "sentinel lymph node" (SLN) in his seminal work demonstrating that lymphangiography could be used to identify patterns of nodal drainage from the penis, and hypothesizing that this "lymph node center" appears to be the primary site of metastasis from penile carcinoma. A tumor-involved lymph node was found in the groin in 35% of the 80 patients with penile squamous cell carcinoma in this study; patients with involved SLN had a worse five-year overall survival compared to those with negative nodes (70% v 90%) [10]. 15 years later, Donald Morton provided the first description of the operative technique in early-stage melanoma patients [11]. Over the ensuing 28 years, the technique of preoperative lymphoscintigraphy using a radiolabeled tracer (typically technetium-99 labeled sulfur colloid) followed by perioperative injection of blue dye and intraoperative gamma probe detection has become the standard of care in the initial treatment of patients with clinically localized melanoma stage 1b (>0.8 mm in thickness) or greater [12]. According to this procedure, all lymph nodes that exceed institutionally set thresholds for radioactivity and/or blue coloration are surgically removed and labeled as a SLN.

SLN examination in patients with breast cancer became standard of care in parallel with its use in melanoma patients, and the technical details are similar and have been exhaustively described elsewhere [13]. SLN biopsy has also been applied to a variety of other tumors, chiefly those of skin, lung, and upper and lower gastrointestinal tract [14]. In these instances, patients can also derive the benefits of improved staging and identification of non-classic draining nodal basins. Further, the refinement of nodal identification methods using novel tracers such as fluorescent dyes continues to drive knowledge in the field and expansion of the indications for this procedure [15, 16].

Currently, tumor staging using the results of SLN biopsy is employed in breast carcinoma, melanoma, and Merkel cell carcinoma [8]. For other tumors undergoing surgical resection, such as head and neck, lung, gastrointestinal, genitourinary, and gynecologic tumors, lymph node staging continues to be performed in the form of regional node dissection at the time of initial surgery. In all of these cases, the evaluation of lymph nodes is a crucial component of cancer staging using the TNM classifications. In addition, the College of American Pathologists cancer reporting protocols, which include the American Joint Committee on Cancer's most recent recommendations, require evaluation and reporting of lymph node status [17].

6.3 Tumors Most Frequently Involving Lymph Nodes

The incidence of lymph node involvement at the time of definitive primary surgical excision for the most common cancers are: prostate (5%) [18], breast (25%) [19], lung and bronchus (15%) [20], colorectal (33%) [21], urinary bladder (25%) [22], corpus and uterus (21%) [23], thyroid (35%)

Table 6.1 Location at presentation of lymph node metastases and most common anatomic site of primary tumor, adapted from [50, 67]

Site of lymph node	Most common primary site
Cervical-upper	Head and neck Lymphoma
Cervical-lower	Lymphoma Lung Stomach Pancreas
Axillary	Breast Melanoma Lung
Thorax	Lung Breast
Abdomen	Pancreas Lower gastrointestinal Upper gastrointestinal Kidney
Inguinal	Gynecologic Genitourinary Melanoma

[24], and melanoma (18%) [25]. It is worth noting that carcinoma is the dominant cancer morphology in each of these sites, in keeping with the theory that carcinoma favors lymphatic over hematogenous spread. By contrast, hematogenous spread is believed to be the preferred pathway of sarcoma, and lymph node metastases are exceedingly rare in sarcoma cases [26].

Certain primary tumors are more likely to be found in distinct anatomic locations (Table 6.1). Among tumors presenting as lymph node metastases, the most common causes of death are lung cancer (60%), followed by pancreatic, upper aerodigestive tract, liver, and breast cancer [27].

6.4 Pathologic Analysis of Lymph Nodes in the Setting of a Known Primary Tumor

This section will address the pathologic evaluation of lymph nodes containing metastasis in two situations: (1) evaluation of lymph node(s) removed during definitive resection of the primary tumor and (2) evaluation of the SLN. Histologic analysis of lymph nodes harboring metastatic disease has informed our knowledge of how disease spreads via lymphatics. The earliest pattern of nodal metastasis occurs when tumor enters the lymph node via lymphatics, breaches the sinusoidal walls, and appears in the subcapsular sinus. Once inside the lymph node, tumors that escape immune surveillance set up an immunosuppressive niche, adapt metabolically, and can continue to expand [28, 29].

Evaluation of lymph nodes resected in the course of surgical treatment of a malignancy follows a set of general rules for all types of cancers. The pathologic evaluation of these lymph nodes may include the use of immunohistochemical evaluation, which has been shown to increase detection of tumor in a variety of settings, including breast [30], melanoma [31], uterine cervix [32], Merkel cell carcinoma [33], and many others. However, current AJCC guidelines do not require specialized immunohistochemical or molecular techniques for routine examination of lymph node resections [8]. Accurate staging of lymph node status requires reporting not only nodal positivity or negativity, but the total number of nodes identified as positive and negative. For each tumor-involved lymph node, the size of the largest metastasis should be assessed and further subcategorizes the N status of the tumor. For example, nodal metastases of breast, colorectal, and gynecologic cancers use a tumor size of 0.2 cm for different subcategories of nodal involvement. Head and neck cancers use a 3 cm cutoff, and testicular tumors have 2 and 5 cm nodal metastasis size categories [8]. The presence of extracapsular tumor extension must always be noted. Of note, the definition of extracapsular extension is a subject of debate. Most definitions include the requirement that the tumor extends through the lymph node capsule into the surrounding connective tissue, with or without a stromal reaction. However, in cases where incomplete lymph node sections do not demonstrate the capsule, or there is a fibrous reaction to extranodal tumor simulating a lymph node capsule, this can be difficult [34].

Since pathologic evaluation of SLN in melanoma and breast cancer is so pivotal to prognostication and refinement of treatment, immunohistochemical examination is considered a routine part of the workup. However, in the United States there is wide variation among institutions with respect to protocols for examination of SLN, including gross examination, number of sections examined, and number and type of immunohistochemical stain performed. In melanoma, S-100 is most commonly used, followed by Melan-A/Mart-1 [35, 36]. Current AJCC/UICC guidelines recognize two categories of SLN involvement in melanoma: clinically occult micrometastasis and clinically apparent macrometastasis [17]. Micrometastatic melanoma is defined as the presence of morphologically malignant cells positive for at least one immunohistochemical marker (S-100, HMB-45, Melan-A/Mart1, Sox10) and/or malignant melanoma detectable on H&E staining alone [17]. One important pitfall in the diagnosis of SLN in patients with melanoma is the presence of benign melanocytes in the connective tissue capsule or trabeculae of the node. Known as nodal nevi or nevus cell aggregates, these may be found in up to 6.3% of all regional lymph nodes evaluated by melanocytic immunohistochemical stains [37] and up to 22% of melanoma cases [38]. They are characteristically located in the fibrous capsule or intranodal fibrous trabeculae, although they may occasionally be found in the nodal parenchyma [39]. The origin of benign melanocytes in lymph nodes has been debated, and is thought to either represent rests of embryonic origin or "benign metastases." While they may be difficult to distinguish from metastatic melanoma, differential immunohistochemical expression of antibodies typically negative in melanoma (p16 [40]) or positive in melanoma but not nevi (HMB-45 [38], PRAME [41]) is useful. Interestingly, some have hypothesized that nodal nevi may be a source of primary nodal metastatic melanoma presenting as metastasis of unknown primary, but this has never been proven [42].

In contrast, breast cancer SLN involvement is categorized as either micrometastatic or macrometastatic based on size of metastasis and number of cells. There are three pathologic categories of SLN metastases in breast carcinoma. Macrometastases are defined as tumor deposits >2 mm, while micrometastases detected by H&E and/or IHC are >0.2–2 mm and/or >200 cells; both are staged as N1. Lymph nodes are staged as N0(i-) when no metastases are seen histologically or with IHC, and N0(i+) when seen on IHC alone. Per AJCC/UICC guidelines, tumor cell must be enumerated in a single histologic section, regardless of whether multiple sections were performed, rather than added together. The pathology report must also state whether or not IHC was performed [43]. Numerous studies have documented the increased yield of metastasis when serial sectioning and immunohistochemical staining, predominantly with broad spectrum anti-cytokeratin antibodies, are employed [44]. The current controversies concerning the clinical relevance of micrometastases as well as the utility of ALND in this scenario have led to confusion concerning the optimum pathologic evaluation of the SLN. Since the presence of micrometastases/ITC may neither influence decision for further therapy nor impact survival, it has been argued that serial sections and IHC have no use in routine evaluation of breast SLN. Since it has been proven that a single H&E-stained full-face section from each submitted SLN block can identify macrometastases and a high proportion of micrometastases, current CAP and ASCO guidelines do not recommend of the SLN by multilevel node sectioning or require specialized techniques for evaluation of the SLN at this time [45].

6.5 Evaluation of the Lymph Node Metastasis in Patients with Unknown Primary Tumors

According to the American Cancer Society, 1.7% of newly diagnosed cancer cases in the United States in 2021 presented as a cancer of unknown primary origin (CUP) [46]. Greater than 50% of patients present with multiple sites of involvement, while in some a single focus of disease is found. In these latter cases, the most common sites of presentation are lymph nodes, lung, bone, and liver [47]. Approximately 10–40% of patients with CUP have lymph node involvement at presentation [48]. The etiology of this phenomenon has been debated. The most common hypothesis is that primary tumors contain a mixture of cells with varying degrees of motility, and that in patients who present as CUP, the more stationary cells regressed through immunologic surveillance at the primary site, while the more motile ones present at the metastatic site [42]. The prognosis of patients with CUP is poor, with a median overall survival of 3–10 months [49]. It is noteworthy that patients with lymph node involvement alone have a better prognosis than patients with multiple sites of involvement [50]. In addition to patients who present with nodal metastasis, there are several well-defined subsets of CUP, constituting about 20% of patients overall, that have a more favorable prognosis. Nodal presentations with a more favorable outcome include poorly differentiated carcinoma involving midline lymph nodes, and squamous cell carcinoma in cervical nodes or as isolated inguinal nodes [51, 52].

There is a well-defined clinical and diagnostic algorithmic approach to patients who present with CUP, including nodal metastasis. An understanding of the relative frequency of malignancies overall and their most common patterns of spread is essential to the evaluation of these patients and effective deployment of this algorithm. Improvements in diagnostic testing now allow a definitive identification of the primary tumor in up to 75% of patients [53]. In one systematic review of 884 CUP patients evaluated at autopsy or with gene expression profiling, a site of primary origin could be identified in 73% of cases. The most common sites of origin were lung (24%) and pancreas (24%), followed by hepatobiliary tree (8%), kidneys (8%), bowel, genital system, and stomach [54]. Interestingly, the incidence of a CUP presentation has decreased in recent years. Improvements in diagnostic testing are the most likely cause for this recent decrease; for example, age-standardized incidence rates in a Switzerland cancer registry rose from 10.3 to 17.6 per 100,000 from 1981 to 1997, but thereafter decreased to 5.8 in 2014 [55].

Lymph node metastasis of unknown primary origin are identified because of clinical and imaging findings suspicious for malignancy with a resulting lymph node biopsy, or as an unexpected finding during the histologic evaluation of a specimen for an unrelated pathology. The preferred method is core needle biopsy or fine needle aspiration with a cell block [56] [57]. The most common tumor morphologies are carcinoma, sarcoma, lymphoma, and melanoma. Beyond these, less common morphologies include germ cell, mesothelioma, and pheochromocytoma/paraganglioma. In the setting of an undifferentiated malignant neoplasm, the tumor is overwhelmingly likely to be carcinoma (80%)—according to a recent study [58]. Discerning the location of the primary lesion can be extremely challenging. In such cases, a definitive diagnosis is not always possible, but a limited differential can help clinical colleagues with further workup.

When assessing a lymph node with metastasis from unknown primary, the growth pattern and morphology of the tumor cells is the first feature to be examined, and in well-differentiated lesions, it can be sufficient. Broadly, there are three morphologies: epithelioid (cohesive, large) cell, spindle cell, and small round cell. Adenocarcinoma is the prototype of epithelioid cell, and the most common subtype of carcinoma to metastasize, followed by squamous cell carcinoma [59]. Well-differentiated adenocarcinoma may form glands, tubules, and/or papillae. The presence of mucin (cytoplasmic or extracellular) can aid in the diagnosis of adenocarcinoma [58]. The accumulation of keratin within the cytoplasm of squamous cell carcinoma may confer an opaque eosinophilic "glassy" appearance. Additional features such as the presence of intercellular bridges and keratin pearls can further help identify squamous cell carcinoma. Histologic clues to the diagnosis of melanoma include cytoplasmic melanin, nested growth pattern, eosinophilic nucleoli and nuclear pseudoinclusions, but melanin production can appear in other tumors such as psammomatous melanotic schwannoma. The many histomorphologies of melanoma indicate that it should be included in nearly every differential diagnosis when a tumor of unknown primary origin is being assessed [60]. Typically, however, pure morphologic assessment is insufficient to establish a diagnosis. In these cases, immunohistochemical studies are employed. Recent years have seen an explosion of markers touted to be lineage specific, and the repertoire of stains available has greatly expanded. A systematic approach to malignancy of unknown primary origin requires a stepwise approach to performing these stains, combined with knowledge of the most common tumors arising in that region.

It has been estimated that well-performed and adequately interpreted IHC panels can aid in identification of the primary site of origin in 75% of cases [53]. Two general schools of practice exist: a stepwise, directed approach that by definition may take several rounds of staining, or a "shotgun" approach with large numbers of stains ordered simultaneously. The choice of either method may be influenced by the patient's condition and urgency of a diagnosis, as well as pathologist training. The tiered approach is generally recommended. The first round is employed to determine the general lineage of the tumor (carcinoma, sarcoma, lymphoma, melanoma, germ cell tumor, and mesothelioma [61]. If necessary, the second round is performed to refine the primary site or subtype of sarcoma or lymphoma. Once the diagnosis is established, additional biomarker staining may also be useful, such as hormone receptor assays and HER2 evaluation in breast carcinoma. Table 6.2 lists the most commonly used Tier 1 and Tier 2 markers.

In summary, the pathologist has two primary functions with regard to lymph node evaluation: identifying lymph node involvement and determining the primary lesion if it is unknown. The former requires a meticulous work at all stages, including gross examination, dissection, embedding, staining, and histologic examination—and immunohistochemical studies in some cases. The latter requires no less but also demands a deep understanding of tumor biology and immunoprofiles.

6.6 Alternative Methods of Lymph Node Evaluation

Recent advances in molecular analysis have provided additional diagnostic and therapeutic tools for use in the setting of patients with lymph node metastasis of unknown origin. There are several commercially available diagnostic tests that employ gene expression profiling to examine similarity of the mRNA of a patient's tumor to a panel of tumors with known sites of origin. Commonly used tests include CancerTypeID (Biotheranostics) and Tissue of Origin (Cancer Genetics). A recent review of validation analyses for these tests shows sensitivity in the range of ~80–90% and sensitivity of >98% for this test using cases with known diagnoses [62]. The accuracy of CancerTypeID was significantly better than immunohistochemistry using a blinded set of 122 tumors of known origin, with an accuracy of 79% compared to 60% ($p = 0.019$) [63].The Tissue of Origin test accuracy was more similar to immunohistochemistry in a set of well-differentiated tumors (85.3 v 86.8%) but fared better in poorly or undifferentiated tumors (94% vs. 79.1%) [64]. Despite these improvements in diagnostic accuracy, however, evidence that these tests result in improved clinical outcomes is slim [65, 66].

Table 6.2 Algorithmic immunohistochemical evaluation for evaluation of undifferentiated cancer of unknown primary origin presenting as lymph node metastasis

Step 1-Determination of major lineage of tumor		
Immunohistochemical stains	**Histology/diagnosis**	
Step 1-evaluation of lineage of tumor		
Sox10	Melanoma	
Pankeratin (AE1/AE3/ CAM5.2)	Carcinoma	
CK5/6, p63, or p40	Squamous cell carcinoma	
LCA	Lymphoma	
OCT 3/4 or SALL4	Germ cell tumor	
Step 2-evaluation of carcinoma of unknown origin with CK7 and CK20 (80% of cases)		
CK7+/CK20-	Breast. lung, thyroid, Mullerian (endometrial, endocervical, ovarian serous), bladder, upper GI, pancreatobiliary, mucinous ovarian	
CK7+/CK20+	Bladder, upper GI, pancreatobiliary, mucinous ovarian, occ. rectum and lung	
CK7-/CK20+	Colon, Merkel cell, occ. upper GI	
CK7-/CK20-	Prostate, SCC, renal, adrenocortical, hepatocellular	
Step 3-additional markers to clarify origin of most common tumors		
CK7+/CK20-	Breast carcinoma	GATA3, GCDFP-15, mammaglobin, ER/ PR
	Lung adenocarcinoma	TTF1, Napsin 1
		GATA3, GCDFP-15, mammaglobin, ER, PR
	Thyroid carcinoma	TTF-1, PAX8
	Endocervical adenocarcinoma	P16, PAX8
	Endometrial adenocarcinoma	Vimentin, PAX8
	Urothelial carcinoma (subset)	GATA3, p63/p40, CK5/6, uroplakin II
	Upper GI	CDX2, villin
	Pancreatobiliary	
	Serous ovarian	PAX8/WT1
	Salivary gland carcinoma	CK5/6, p63, GATA3, AR
CK7+/CK20 variable	Mucinous ovarian	PAX8+/–, CDX2+/–
	Pancreatic adenocarcinoma (~2/3)	CDX2, CK19, SMAD4 loss CEA, CA19–9, MUC5, CDX2 (variable)
CK7-/CK20+	Lower GI tract adenocarcinoma (small intestine, appendix, colon, rectum)	CDX2, villin, SATB2
	Merkel cell carcinoma (CK20 perinuclear dots)	Chromogranin, synaptophysin, CM2B4, TTF1-negative
CK7+/CK20+	Urothelial	P63, CK4/5, uroplakin, GATA3
	Mucinous ovarian	PAX8/CDK2
CK7-/CK20-	Prostatic adenocarcinoma	NKX3.-1, PSAP, PSA
	Squamous cell carcinoma	CK5/6 or p63 or p40
	Renal cell carcinoma	PAX2, PAX8, CAIX, EMA, vimentin, CD10
	Adrenocortical carcinoma	SF-1, Melan-A, inhibin
	Hepatocellular carcinoma (may be CK7 or CDK20+)	Arginase-1, HepPar-1, Glypican-3, CD30, pericanalicular CEA

Conclusions and Future Perspectives

- Accurate assessment and characterization of lymph node involvement by metastatic disease is an integral part of patient staging and a key driver of patient outcome.
- Focused examination of the sentinel lymph node provides a precise and accurate method of lymph node staging in patients with breast, melanoma, and Merkel cell carcinoma.
- The lymph node is a common site of metastasis in patients presenting with a malignancy of unknown primary origin.
- An algorithmic approach utilizing knowledge of frequent sites of clinical presentation and a tiered approach to immunohistochemical analysis can aid in accurate diagnosis.
- The role of molecular testing in evaluation of patients with lymph node metastasis is an area for future investigation.

References

1. Urteaga O, Pack GT. On the antiquity of melanoma. Cancer. 1966;19(5):607–10.
2. Gorantla VC, Kirkwood JM. State of melanoma: an historic overview of a field in transition. Hematol Oncol Clin North Am. 2014;28(3):415–35.
3. Cooper S. The first lines of the theory and practice of surgery. London: Longman; 1840.
4. Snow HM. Melanoma cancerous disease. Lancet. 1892;140:869–922.
5. Faries MB, et al. Lymph node metastasis in melanoma: a debate on the significance of nodal metastases, conditional survival analysis and clinical trials. Clin Exp Metastasis. 2018;35(5–6):431–42.
6. Denoix P. Nomenclature classification des cancers. Bull Inst Nat Hyg (Paris). 1952;7:743–8.
7. Hutter RV. At last—worldwide agreement on the staging of cancer. Arch Surg. 1987;122(11):1235–9.
8. Amin MB. AJCC cancer staging manual. 8th ed. Switzerland: Springer Nature; 2017.
9. Surgery basic science and clinical evidence. 2nd ed. Springer Nature; 2008.
10. Cabanas RM. An approach for the treatment of penile carcinoma. Cancer. 1977;39(2):456–66.
11. Morton DL, et al. Technical details of intraoperative lymphatic mapping for early stage melanoma. Arch Surg. 1992;127(4):392–9.
12. Fayne RA, et al. Evolving management of positive regional lymph nodes in melanoma: past, present and future directions. Oncol Rev. 2019;13(2):433.
13. Zeitoun J, Babin G, Lebrun JF. Sentinel node and breast cancer: a state-of-the-art in 2019. Gynecol Obstet Fertil Senol. 2019;47(6):522–6.
14. Faries MB. Application of senitnel lymph node surgery outside of melanoma and breast cancer. Clin Exp Metastasis. 2021;
15. Burghgraef TA, et al. In vivo sentinel lymph node identification using fluorescent tracer imaging in colon cancer: a systematic review and meta-analysis. Crit Rev Oncol Hematol. 2021;158:103,149.
16. Vuijk FA, et al. Fluorescent-guided surgery for sentinel lymph node detection in gastric cancer and carcinoembryonic antigen targeted fluorescent-guided surgery in colorectal and pancreatic cancer. J Surg Oncol. 2018;118(2):315–23.
17. Cancer Protocol Templates. 2021 [cited 2021; Available from: https://www.cap.org/protocols-and-guidelines/cancer-reporting-tools/cancer-protocol-templates.
18. Narayanan R, Wilson TG. Sentinel node evaluation in prostate cancer. Clin Exp Metastasis. 2018;35(5–6):471–85.
19. Cotarelo CL, et al. Improved detection of sentinel lymph node metastases allows reliable intraoperative identification of patients with extended axillary lymph node involvement in early breast cancer. Clin Exp Metastasis. 2021;38(1):61–72.
20. Osarogiagbon RU, et al. Survival implications of variation in the thoroughness of pathologic lymph node examination in American College of Surgeons Oncology Group Z0030 (Alliance). Ann Thorac Surg. 2016;102(2):363–9.
21. Gleisner AL, et al. Nodal status, number of lymph nodes examined, and lymph node ratio: what defines prognosis after resection of colon adenocarcinoma? J Am Coll Surg. 2013;217(6):1090–100.
22. Wright JL, Lin DW, Porter MP. The association between extent of lymphadenectomy and survival among patients with lymph node metastases undergoing radical cystectomy. Cancer. 2008;112(11):2401–8.
23. Chan JK, et al. Metastatic gynecologic malignancies: advances in treatment and management. Clin Exp Metastasis. 2018;35(5–6):521–33.
24. Mansour J, et al. Prognostic value of lymph node ratio in metastatic papillary thyroid carcinoma. J Laryngol Otol. 2018;132(1):8–13.
25. Mozzillo N, et al. Sentinel node biopsy in thin and thick melanoma. Ann Surg Oncol. 2013;20(8):2780–6.
26. Leong SP, et al. Clinical patterns of metastasis. Cancer Metastasis Rev. 2006;25(2):221–32.
27. Hemminki K, et al. Site-specific cancer deaths in cancer of unknown primary diagnosed with lymph node metastasis may reveal hidden primaries. Int J Cancer. 2013;132(4):944–50.
28. Kawaguchi T. Pathological features of lymph node metastasis. 2 from morphological aspects. Nihon Geka Gakkai Zasshi. 2001;102(6):440–4.
29. Zhou H, Lei PJ, Padera TP. Progression of metastasis through lymphatic system. Cell. 2021;10(3)
30. Cote RJ, et al. Role of immunohistochemical detection of lymph-node metastases in management of breast cancer. International Breast Cancer Study Group. Lancet. 1999;354(9182):896–900.
31. Messina JL, et al. Pathologic examination of the sentinel lymph node in malignant melanoma. Am J Surg Pathol. 1999;23(6):686–90.
32. Euscher ED, et al. Ultrastaging improves detection of metastases in sentinel lymph nodes of uterine cervix squamous cell carcinoma. Am J Surg Pathol. 2008;32(9):1336–43.
33. Su LD, et al. Immunostaining for cytokeratin 20 improves detection of micrometastatic Merkel cell carcinoma in sentinel lymph nodes. J Am Acad Dermatol. 2002;46(5):661–6.
34. Abdel-Halim CN, et al. Histopathological definitions of extranodal extension: a systematic review. Head Neck Pathol. 2021;15(2):599–607.
35. Dekker J, Duncan LM. Lack of standards for the detection of melanoma in sentinel lymph nodes: a survey and recommendations. Arch Pathol Lab Med. 2013;137(11):1603–9.
36. Cole CM, Ferringer T. Histopathologic evaluation of the sentinel lymph node for malignant melanoma: the unstandardized process. Am J Dermatopathol. 2014;36(1):80–7.
37. Bautista NC, Cohen S, Anders KH. Benign melanocytic nevus cells in axillary lymph nodes. A prospective incidence and immunohistochemical study with literature review. Am J Clin Pathol. 1994;102(1):102–8.

38. Carson KF, et al. Nodal nevi and cutaneous melanomas. Am J Surg Pathol. 1996;20(7):834–40.
39. Biddle DA, et al. Intraparenchymal nevus cell aggregates in lymph nodes: a possible diagnostic pitfall with malignant melanoma and carcinoma. Am J Surg Pathol. 2003;27(5):673–81.
40. Koh SS, Cassarino DS. Immunohistochemical expression of p16 in melanocytic lesions: an updated review and meta-analysis. Arch Pathol Lab Med. 2018;142(7):815–28.
41. Lezcano C, et al. Immunohistochemistry for PRAME in the distinction of nodal nevi from metastatic melanoma. Am J Surg Pathol. 2020;44(4):503–8.
42. Kamposioras K, et al. Malignant melanoma of unknown primary site. To make the long story short. A systematic review of the literature. Crit Rev Oncol Hematol. 2011;78(2):112–26.
43. Lester S et al. Protocol for the examination of specimens from patients with invasive carcinoma of the breast; December 2013.
44. Lyman GH, et al. Sentinel lymph node biopsy for patients with early-stage breast cancer: American Society of Clinical Oncology clinical practice guideline update. J Clin Oncol. 2014;32(13):1365–83.
45. Wiatrek R, Kruper L. Sentinel lymph node biopsy indications and controversies in breast cancer. Maturitas. 2011;69(1):7–10.
46. Siegel RL, et al. Cancer statistics, 2021. CA Cancer J Clin. 2021;71(1):7–33.
47. Buttar A, et al. Cancers of unknown primary. In: Pieters R, editor. Cancer concepts: a guidebook for the non-oncologist. University of Massachusetts Medical School; 2015.
48. Hemminki K, et al. Survival in cancer of unknown primary site: population-based analysis by site and histology. Ann Oncol. 2012;23(7):1854–63.
49. Losa F, et al. 2018 consensus statement by the Spanish Society of Pathology and the Spanish Society of Medical Oncology on the diagnosis and treatment of cancer of unknown primary. Clin Transl Oncol. 2018;20(11):1361–72.
50. Chorost MI, et al. Unknown primary. J Surg Oncol. 2004;87(4):191–203.
51. Hainsworth JD, Fizazi K. Treatment for patients with unknown primary cancer and favorable prognostic factors. Semin Oncol. 2009;36(1):44–51.
52. Jereczek-Fossa BA, Jassem J, Orecchia R. Cervical lymph node metastases of squamous cell carcinoma from an unknown primary. Cancer Treat Rev. 2004;30(2):153–64.
53. Kandalaft PL, Gown AM. Practical applications in immunohistochemistry: carcinomas of unknown primary site. Arch Pathol Lab Med. 2016;140(6):508–23.
54. Pentheroudakis G, Golfinopoulos V, Pavlidis N. Switching benchmarks in cancer of unknown primary: from autopsy to microarray. Eur J Cancer. 2007;43(14):2026–36.
55. Binder C, et al. Cancer of unknown primary-epidemiological trends and relevance of comprehensive genomic profiling. Cancer Med. 2018;7(9):4814–24.
56. Ettinger DS, et al. NCCN Clinical Practice Guidelines Occult primary. J Natl Compr Cancer Netw. 2011;9(12):1358–95.
57. Bochtler T, Löffler H, Krämer A. Diagnosis and management of metastatic neoplasms with unknown primary. Semin Diagn Pathol. 2018;35(3):199–206.
58. Bellizzi AM. An algorithmic Immunohistochemical approach to define tumor type and assign site of origin. Adv Anat Pathol. 2020;27(3):114–63.
59. Houston KA, et al. Patterns in lung cancer incidence rates and trends by histologic type in the United States, 2004-2009. Lung Cancer. 2014;86(1):22–8.
60. Banerjee SS, Harris M. Morphological and immunophenotypic variations in malignant melanoma. Histopathology. 2000;36(5):387–402.
61. Pavlidis N. Cancer of unknown primary: biological and clinical characteristics. Ann Oncol. 2003;14(Suppl 3)):11–8.
62. Lee MS, Sanoff HK. Cancer of unknown primary. BMJ. 2020;371:m4050.
63. Weiss LM, et al. Blinded comparator study of immunohistochemical analysis versus a 92-gene cancer classifier in the diagnosis of the primary site in metastatic tumors. J Mol Diagn. 2013;15(2):263–9.
64. Handorf CR, et al. A multicenter study directly comparing the diagnostic accuracy of gene expression profiling and immunohistochemistry for primary site identification in metastatic tumors. Am J Surg Pathol. 2013;37(7):1067–75.
65. Hainsworth JD, et al. Molecular gene expression profiling to predict the tissue of origin and direct site-specific therapy in patients with carcinoma of unknown primary site: a prospective trial of the Sarah Cannon research institute. J Clin Oncol. 2013;31(2):217–23.
66. Hayashi H, et al. Randomized phase II trial comparing site-specific treatment based on gene expression profiling with carboplatin and paclitaxel for patients with cancer of unknown primary site. J Clin Oncol. 2019;37(7):570–9.
67. Pavlidis N, Pentheroudakis G. Cancer of unknown primary site. Lancet. 2012;379(9824):1428–35.

7 Pathologic Assessment of Systemic or Distant Metastasis

Igor Katsyv and Andrew Turk

Abstract

Pathologists' microscopic examination of tissue is the gold standard for cancer diagnosis. It begins with lesion biopsy or resection and subsequently involves a time- and labor-intensive process of tissue processing, slide preparation, and ancillary testing before a final diagnosis can be rendered. This ancillary testing includes characterization of cancer- and organ-specific protein expression and molecular alterations using immunohistochemical (IHC) stains, and, increasingly, massively parallel sequencing to probe a tumor's genomic, transcriptomic, and epigenomic landscape. Pathologists integrate these modalities into unified diagnoses that form the basis for patients' oncologic care.

Learning Objectives

- Understand processing of tissue that occurs between specimen procurement and histologic evaluation.
- Understand role of ancillary testing, such as immunohistochemistry and molecular assays, in tumor characterization and final diagnosis.
- Appreciate the emergence of novel specimen types.

7.1 Introduction

Despite advances in early detection of cancer, patients often present with metastatic disease at the time of initial diagnosis. The proportion of stage IV disease at diagnosis varies with malignancy, from less than 10% in prostate and female breast cancers to more than 50% in cancers of the lung and bronchus, pancreas, and ovary [1]. With few exceptions, the prognosis for patients with disseminated disease remains poor, particularly for those with cancers of unknown primary (CUP) [2]. Rapid identification of tumor origin and selection of optimal therapy is therefore critical. This chapter will focus on the histologic and molecular methods that, together with clinical and radiologic data, factor into the pathological workup for systemic metastases.

7.2 Clinicopathologic Considerations and Correlation

For every case, consideration of factors related to pathogenesis occurs simultaneously with, and is inseparable from, morphologic review of a specimen's gross and microscopic properties. Clinicopathologic features of specific cancers affect the likelihood of various scenarios and inform a pathologist's differential diagnosis. Pertinent considerations include the routes along which certain cancers metastasize and the manner in which some metastases demonstrate a predilection for stereotypical tissue types and/or anatomic sites.

In terms of available mechanisms of metastasis, cancers travel via lymphatic channels, vasculature (veins and arteries), and/or along nerves. Among these methods, some cancers spread preferentially by predictable avenues rather than others. For instance, follicular carcinoma of the thyroid migrates via hematogenous passageways only, and virtually never involves lymph nodes [3, 4]. By contrast, Hurthle cell carcinoma—another malignancy derived from the same (follicular epithelial) cell—travels via lymphatic as well as hematogenous spread [3, 4]. Other cancers, such as prostatic adenocarcinoma and certain salivary gland tumors, exhibit high propensity for perineural invasion.

In addition to differences between canonical migratory pathways, several cancers also demonstrate recurrent and predictable metastatic sites. For example, prostate cancers tend to metastasize to bones, whereas breast cancers metastasize to liver, bones, lung, pleura, and lymph nodes;

I. Katsyv · A. Turk (✉)
Department of Pathology and Cell Biology, Columbia University, New York, NY, USA
e-mail: ik2425@cumc.columbia.edu; att2101@cumc.columbia.edu

S. P. Leong et al. (eds.), *Cancer Metastasis Through the Lymphovascular System*,
https://doi.org/10.1007/978-3-030-93084-4_7

melanomas tend to have increased risk for metastasis to the central nervous system, spleen, and heart compared to other malignancies [5]. Lung cancers represent the most common metastases to central nervous system, and colorectal cancers are the most frequently observed liver and, in men, peritoneal metastases [2, 5, 6]. The liver is also a common site of uveal melanoma metastases [7, 8]. Among all solid tumors, hepatic and lymph node metastases are most common, followed by lung, bone, and pleura [5].

Knowledge of these generalities (as well as the specific clinical scenario of each particular case) is important prior to and during review of tissue specimens, as these circumstances guide pathologists' assessment, workup, and interpretation.

7.3 Histopathology

7.3.1 Specimen and Slide Preparation

Morphologic examination of biopsy specimens for evaluation of suspected metastasis first requires preparation of hematoxylin- and eosin-stained slides. This process comprises several steps, beginning with placement of sampled tissue in a fixative, generally 10% formalin (which contains 4% formaldehyde, a cross-linking agent). Other preservatives include oxidizing agents, such as $KMnO_4$ and OsO_4. Formalin-fixed tissue is then embedded in paraffin. This phase entails removal of water from tissues via sequential immersion in successive ethyl alcohol solutions of increasing concentration (70% then 95% then 100%), effectively dehydrating tissues and replacing water with alcohol. The specimen is then cleansed of alcohol by immersion in xylene (or a xylene substitute such as toluene or benzene), followed by placement into paraffin. Once embedded in paraffin, tissues are sectioned via microtomy and applied to glass slides. Staining of paraffin-embedded tissues essentially involves reversal of the embedding process. Immersion of tissue affixed to glass slides in xylene first eliminates paraffin, then ethanol solutions of decreasing concentration (100% then 95% then 70%) remove xylene, and tissues are ultimately rehydrated in water. Staining is then performed. Prior to microscopy, however, water must again be removed, due to differences between the refractive indices of water and glass. Accordingly, stained tissues are washed in the same sequence of ethanol concentrations, then xylene. Finally, the stained glass slide is prepared for examination by application of a coverslip using mounting media.

7.3.2 Microscopic Evaluation

Histologic review entails a structured visual algorithm, with successive phases occurring at sequentially higher microscopic resolution. The process begins at low-power magnification, with evaluation of tissue architecture. The pathologist then advances to medium- and high-power views, with examination of cellular and nuclear morphology in greater detail. This procedure comprises a series of interrogations, beginning with the most global and incrementally becoming more granular. The first assessment distinguishes normal from abnormal tissue. With abnormal specimens, the pathologist then determines what category of disease accounts for the irregularity, i.e., whether the lesion represents a neoplastic, inflammatory, or traumatic process, et cetera. For neoplasia, subsequent distinctions parse primary versus metastatic, and (for the former) benign versus malignant tumors. Finally, before proceeding to ancillary testing, the pathologist gleans any apparent indication of tumor type and/or primary site. In general terms, broad categories of malignancy include carcinoma (of epithelial origin), sarcoma (of mesenchymal derivation), melanoma, and lymphoma, among others. These lineages exhibit stereotypical architectural and cytologic features, albeit to variable degrees. Carcinoma, particularly adenocarcinoma, may form glandular structures and/or produce mucin. Sarcoma characteristically displays spindle-shaped cytology. Melanoma demonstrates diverse architectural patterns and cytologic morphologies, but may produce pigment. Lymphoma usually appears discohesive, relative to these other tumor types. For metastatic carcinoma, other properties may permit preliminary assessment of primary site, because carcinoma arising in certain organs possesses various histologic hallmarks. Colorectal carcinoma, for instance, typically shows elongated, "pencillate" nuclei and abundant grungy necrosis; pancreatic ductal adenocarcinoma frequently contains neoplastic cells with clear cytoplasm; et cetera. Familiarity with these largely reproducible properties facilitates generation of a differential diagnosis, refinement of which then depends upon ancillary testing.

7.4 Immunohistochemistry

IHC is the primary ancillary method used by pathologists to characterize cancers beyond their tissue architectural and cytologic features. IHC enables the spatial staining and qualitative microscopic evaluation of protein expression in cells and tissues while preserving tissue architecture and tumor microenvironment. The spatial resolution of IHC is particularly critical when tumors contain extensive stromal and immune cell infiltration. This is exemplified by highly infiltrative tumors, where rare single cells, often detectable only with cell-specific immunostains, are hidden among a sea of intratumoral lymphocytes. Expression of tissue- and cancer-specific proteins enables the diagnostic classification of malignancies, identification of tissue of origin, evaluation of expression of therapeutic targets and their modifiers, and prognostication. IHC staining patterns can also serve as surrogates of molecular alterations that impact protein expression or stability, such as in *TP53*, *ATRX*, *IDH1/2*, and *MDM2*, among many others.

7.4.1 Immunohistochemistry Methodology

Frozen or formalin-fixed, paraffin-embedded (FFPE) tissue is sectioned using a cryostat or microtome, respectively, to sections measuring approximately 4–5 μm in thickness and subsequently mounted on glass slides. In the case of FFPE, mounted tissue is the subject to physical, enzymatic, or chemical antigen retrieval (AR) processes that disrupt protein cross-links and expose epitopes recognized by target-specific primary antibodies. The primary antibody is subsequently applied at an optimized and clinically validated concentration to maximize target epitope binding while minimizing nonspecific background binding. An enzyme-labeled secondary antibody raised against the immunoglobulin used in the primary antibody is then used for signal detection and amplification. Horseradish peroxidase (HRP) and alkaline phosphatase (AP), the most commonly used enzymatic labels, produce a visible color (e.g., brown or red) when a chromogen is applied. A counterstain, such as hematoxylin, labels nuclei of all cells in the section. For quality control, IHC interpretation includes positive control tissue that is known to express the target protein of interest and a negative control (e.g., tissue minus the primary antibody) [9]. The IHC-stained tissue is then interpreted for the presence and subcellular localization, or absence, of target expression in the cells of interest. The intensity of staining can serve as a surrogate for the degree of expression. In many cases, IHC-based detection and interpretation of several protein markers, such as estrogen receptor (ERα), human epidermal growth factor receptor 2 (HER2), Ki-67, and phosphoproteins, can be strongly influenced by tissue ischemia and fixation [10].

7.4.2 Utility of Specific Markers

The utility of immunohistochemistry in surgical pathology derives from the association between various antigens (proteins) and certain tissues. These associations range from very broad to highly specific. Broad associations include the relationships between some cell populations and their characteristic intermediate filaments, such as keratins (the predominant intermediate filament of epithelium), vimentin (an intermediate filament found in fibroblasts and other mesenchymal cells), and desmin (in muscle). The abundance of these proteins within their respective cell types makes them valuable immunohistochemical markers of the corresponding tissues. Among cytokeratins, some members of this intermediate filament class demonstrate greater organ and cellular specificity. Examples include cytokeratins 7 and 20 (denoted as CK7 and CK20), which accordingly represent useful targets during evaluation of carcinoma arising from an unknown primary location. The combinatorial result of stains for CK7 and CK20 facilitates distinction between possible origin sites, as summarized by Wang et al. [11]. For instance, urothelial carcinoma generally stains positively for both markers (CK7+ CK20+). Lung adenocarcinoma and mammary carcinoma usually show positive staining for CK7, whereas staining for CK20 is negative (CK7+ CK20−); the opposite result (CK7− CK20+) is characteristic of colorectal carcinoma. Both stains are typically negative (CK7− CK20−) in hepatocellular carcinoma, renal cell carcinoma, and prostatic adenocarcinoma. In addition to intermediate filaments, other antigens demonstrate tropism for certain lineages and, consequently, value as immunohistochemical targets. Examples are listed in Table 7.1.

Table 7.1 Select immunohistochemical markers

Cell lineage	
Epithelial	Cytokeratins, EMA
Endothelial	CD31, CD34, D2-40, ERG
Smooth Myocytic	SMA, Caldesmon
Skeletal Myocytic	Myogenin, MyoD1, Desmin
Neuroendocrine	Synaptophysin, Chromogranin A, CD56
Hematopoietic	CD45, CD3, CD20, CD1a, CD68, MPO
Melanocytic	S100, SOX10, Melan-A, HMB-45, MITF
Malignancies	
Lung Adenocarcinoma	CK7, TTF-1, NAPSIN
Gastric Adenocarcinoma	CDX2, CK7, CK20
Colon Adenocarcinoma	CDX2, CK20, SATB2
Breast Invasive Ductal Carcinoma	ER, PR, GATA-3, GCDFP-15, MAM
Prostate Adenocarcinoma	PSA, NKX3.1, AR
Endometrial Adenocarcinoma	CK7, CA-125, ER, PR
Clear Cell Renal Cell Carcinoma	CA-IX, PAX8
Squamous Cell Carcinoma	p40/p63, CK5/6, HMWK
Small Cell Carcinoma	Synaptophysin, Chromogranin A, CD56

7.5 Molecular Approaches

7.5.1 Tissue-Based Methods

Detection of molecular alterations within a cancer specimen provides insights beyond the information garnered from morphologic and immunohistochemical evaluation. In some tumor types, results of molecular analysis may have diagnostic, prognostic, and/or predictive implications. Notably, PCR-based methods are not feasible following tissue decalcification in acid. For biopsies of suspected bone metastasis upon which molecular testing may be indicated, alternative decalcifying agents such as EDTA are therefore recommended for this step.

In terms of diagnostic utility, identification of oncogenic mutations may facilitate more precise characterization of some cases that do not permit definitive diagnosis by morphology alone. This prospect is especially pertinent in cases containing scant cellular material, and/or specimens obtained by invasive procedures after which repeat sampling would be inopportune. Three instances illustrate the potential diagnostic value of molecular pathology. First, the presence of unequivocally oncogenic mutations may implicate malignancy in a biopsy specimen that otherwise does not satisfy qualitative morphologic criteria for designation as neoplastic. Second, in the setting of metastatic cancer arising from an unknown source, detection of certain mutations may implicate a specific diagnosis, despite ambiguous morphologic features. For instance, detection of an activating mutation affecting exons 18–21 of *EGFR* within malignant but otherwise unclassifiable cells suggests a diagnosis of metastatic lung adenocarcinoma, in the proper clinical context. Whereas many cancers show recurrent alterations of oncogenes such as *KRAS* and *PIK3CA*, and/or tumor suppressors like *TP53*, *EGFR* mutations are more characteristic of lung adenocarcinoma than many other tumor types. A third potential diagnostic application of molecular pathology involves the scenario of suspected metastasis in a patient whose primary tumor contains a known mutation. The case depicted in Fig. 7.1 demonstrates this situation. A patient with history of melanoma harboring a mutation of *NRAS* (c.37G>C, p. G13R) sustained a humeral fracture, suspected to represent a pathologic fracture. Due to limited and suboptimal tissue sampling, however, histologic review did not yield an unequivocal diagnosis of melanoma. Subsequent molecular testing (which was feasible because the specimen had been

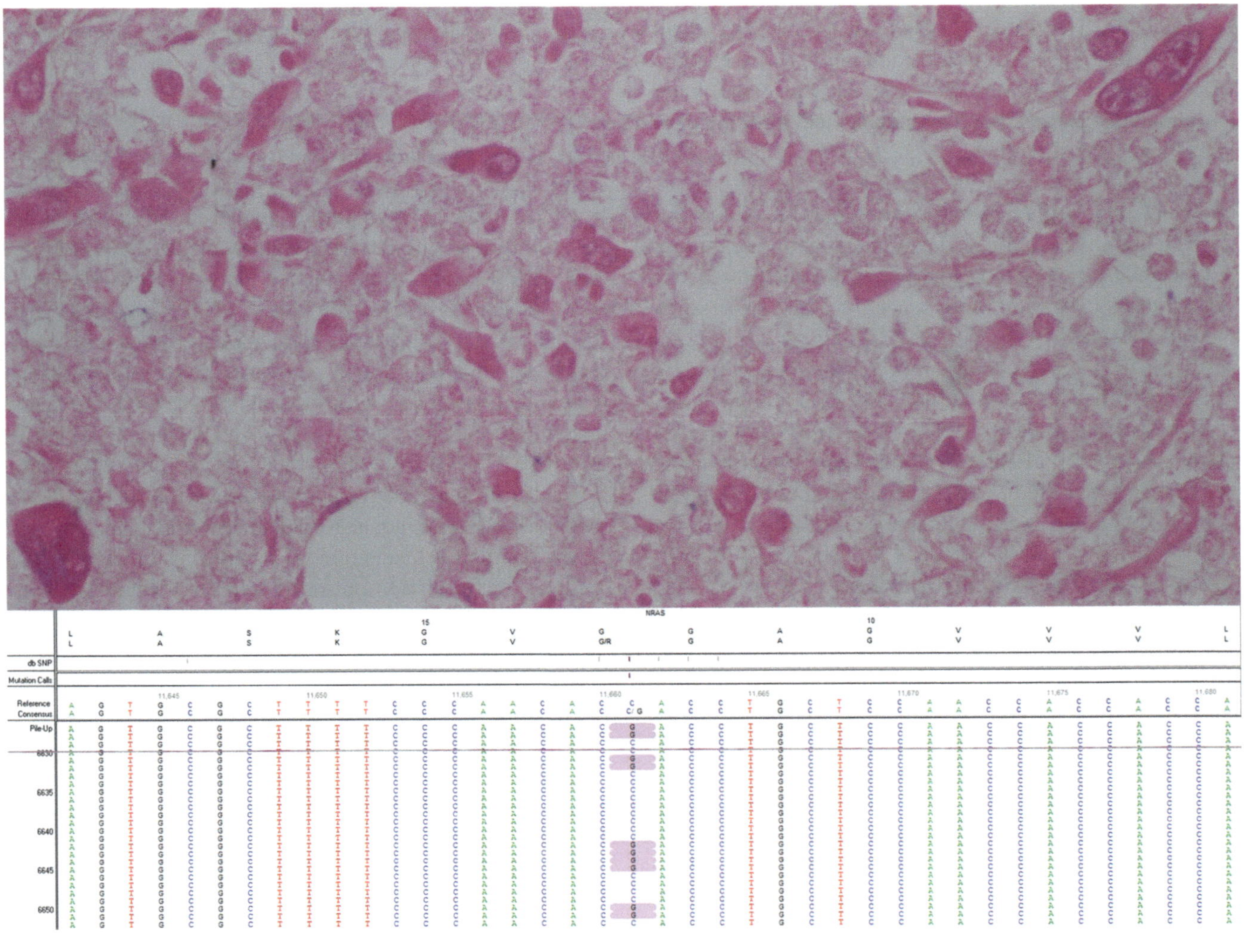

Fig. 7.1 Biopsy (upper panel) and molecular testing (lower panel) of a suspected humeral melanoma metastasis. Histologic review did not yield an unequivocal diagnosis of melanoma, despite presence of scattered atypical cellular forms. Next-generation sequencing showed an NRAS mutation (c.37G>C, p.G13R), identical to that within the antecedent resection specimen, facilitating definitive diagnosis

decalcified in EDTA) exhibited mutation of *NRAS* identical to that within the antecedent resection specimen. Interpreted together with the morphologic findings and clinical history, molecular analysis of this case permitted definitive diagnosis, which would not have been possible if based solely on morphology.

Beyond these diagnostic considerations, molecular pathology also offers prognostic and predictive information in many tumor types. Examples of prognostically significant findings include *KRAS* mutations in lung adenocarcinoma, as these alterations portend shorter survival in early-stage and locally advanced disease (reviewed by Jänne et al. [12]). Predictive contributions relate to mutations associated with targeted therapies, as well as alterations that confer resistance to certain agents. The former category currently includes targetable *EGFR*, *BRAF*, and *MET* mutations (single nucleotide variants and in-frame insertions/deletions), as well as rearrangements involving *ALK*, *ROS1*, and *RET* in lung adenocarcinoma; *BRAF* mutations in melanoma; *FGFR2* and *FGFR3* fusions, and *FGFR3* mutations in urothelial carcinoma; and many other examples. Other alterations may predict drug insensitivity and/or resistance, such as *KRAS* and *BRAF* mutations in lung adenocarcinoma, and *KRAS*/*NRAS*/*BRAF*/PIK3CA mutations in colorectal carcinoma, among many other examples. More information is available via the FDA website (https://www.cancer.gov/about-cancer/treatment/drugs). Overall, the diverse applications of molecular methods in terms of diagnostic, prognostic, and predictive impact demonstrate the tremendous utility of these ancillary tools.

As with protein expression, IHC is typically the first such ancillary tool used to investigate a cancer's molecular profile. Mutation- or fusion-specific antibodies can be used to detect the expression and/or aberrant localization of mutant proteins. In situ hybridization (ISH) uses labeled DNA probes complementary to RNA or DNA sequences to detect gene amplifications, deletions, fusions, and translocations. PCR uses a DNA polymerase and target-specific DNA primers to amplify a DNA region of interest. When monitored in real time, such as in quantitative real-time PCR (qPCR), the abundance or expression of the DNA or mRNA of interest can be quantified. PCR primers spanning specific exon junctions can be used to detect pathogenic transcript splice variants. Similarly, PCR primers spanning known fusion genes, such as *FGFR3-TACC3* in bladder cancers, can be used to amplify and detect tumor-specific gene fusions [13].

Next-generation sequencing (NGS) enables high-throughput, ultrasensitive, massively parallel, and precise detection of genetic alterations. As it provides single nucleotide resolution, NGS has proven indispensable for detecting single-nucleotide variations, insertions and deletions, fusions, translocations, copy number alterations, and transcript-level alterations in cancer driver genes. A significant advantage of NGS over more traditional methods is that it does not require a priori knowledge of genomic alterations. Furthermore, the high-throughput nature of NGS enables comprehensive, whole-genome assessment and identification of novel and/or rare genomic variants, although often such variants are of limited or unknown clinical significance. Tissue foci of interest, typically areas of viable tumor cells, are identified on tissue mounted on glass slides and microdissected. DNA or RNA is then extracted from this microdissected tissue, and cDNA (complementary DNA) libraries are generated for sequencing. Identified genomic variants are subsequently filtered for those with known or predicted impact on tumor suppressor or oncogene function, or those predictive of therapeutic response, such as to receptor tyrosine kinase inhibitors [14].

Mutations in common cancer drivers such as *TP53*, *BRAF*, *PIK3CA*, and *APC* are shared by many cancers [15]. Mutational profiling alone may therefore not offer the necessary resolution to fully characterize a tumor, particularly in cases in which the primary site is not known. Here, analysis of the tumor DNA methylome, though not yet widely used, may be an attractive alternative. DNA methylation occurs early in carcinogenesis, when CpG dinucleotides are methylated by DNA methyltransferases, modulating gene transcription [16, 17]. DNA methylation is also tissue- and disease-specific, and can serve as a molecular fingerprint of cell state [18, 19]. These modifications can be detected by bisulfite treatment, in which unmethylated cytosines are converted to uracils while methylated cytosines remain unchanged, followed by NGS.

7.5.2 Cell-Free DNA

7.5.2.1 Introduction

Although tissue remains the gold standard for cancer diagnosis, cell-free DNA (cfDNA), specifically circulating tumor DNA (ctDNA), in body fluids is emerging as a reliable and minimally invasive method of tumor diagnosis and monitoring. Fragments of double-stranded, nucleosome-associated DNA are released by secretion or following cell death (Fig. 7.2). This process occurs as part of normal cell turnover, as well as in pathological conditions such as trauma, infections, and malignancy [15, 20]. cfDNA profiling is currently used in prenatal care, for example, to detect fetal trisomies in maternal plasma [17]. In the case of malignancies, these DNA fragments can be analyzed to monitor a patient's tumor burden and response to therapy, detect molecular alterations that may be directly therapeutically targetable, and monitor evolution of distinct tumor clones [15].

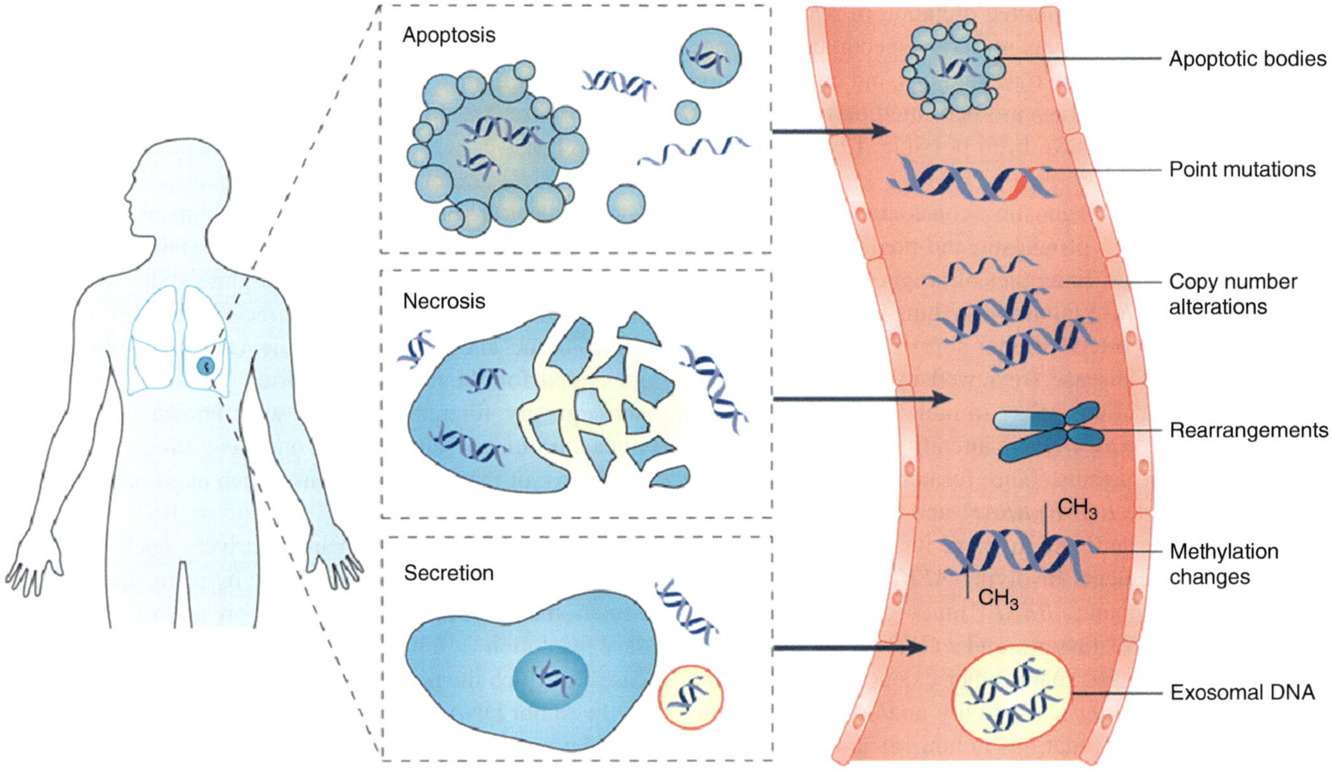

Fig. 7.2 Cellular sources of cfDNA and detectable alterations. Figure reproduced with permission from Wan et al. [15]

7.5.2.2 Isolation and Detection

cfDNA has been isolated from a variety of body fluids, including blood plasma, cerebrospinal fluid (CSF), saliva, urine, stool, pleural and ascitic fluid, luminal gastrointestinal tract brushings, cervical smears, and uterine lavage [15]. Analyte fluid is collected in tubes containing a PCR-compatible anticoagulant, such as EDTA, and centrifuged to remove cells prior to cfDNA extraction. The isolated cfDNA is then profiled using polymerase chain-reaction (PCR)-based and/or next-generation sequencing (NGS)-based methods to identify tumor-specific DNA alterations such as point mutations, structural rearrangements, and altered DNA methylation [15].

7.5.2.3 Clinical Utility of cfDNA

In healthy individuals, most cfDNA in blood comes primarily from normal hematopoietic cells; each milliliter of plasma contains thousands of genome equivalents of cfDNA [20–22]. Patients with malignancies have altered profiles and increased concentration of plasma cfDNA, and those with metastatic disease have as much as 100-fold higher levels of ctDNA compared to patients with early-stage tumors [15, 20, 21, 23, 24]. ctDNA levels may initially increase in patients whose tumors respond to therapy, but subsequently fall, whereas ctDNA levels remained unchanged or increase in those with poor therapeutic response [24]. An added advantage of ctDNA over standard methods of tumor monitoring is its relatively short half-life on the order of minutes to hours. In contrast, serological tumor makers such as cancer antigen 15–3 (CA15–3/MUC1), cancer antigen 125 (CA-125/MUC16), and carcinoembryonic antigen (CEA) have half-lives of weeks to days. ctDNA analysis may thus allow for significantly higher temporal resolution, sensitivity, and specificity in monitoring disease status and therapeutic response [15].

Furthermore, cfDNA profiling may enable detection of minimal residual disease and risk stratification of patients following surgical resection [15]. In 55 patients who underwent neoadjuvant chemotherapy and surgical resection for early-stage breast cancer, detection of ctDNA at postsurgical follow-up was associated with increased risk of metastatic relapse [hazard ratio 25.1 (95% CI 4.08–130.5, logrank $p < 0.0001$)] [25]. Among 230 patients with early-stage colon cancer, detection of ctDNA following surgical resection was similarly associated with increased risk of recurrence [hazard ratio 18 (95% CI 7.9–40, logrank $p < 0.001$)] [26]. In multiple studies, relapse detected by ctDNA preceded clinical relapse by as much as 11 months [15]. Elevated ctDNA may even precede radiographic progression, and thus may serve as an early marker of treatment response [20, 24].

In a given patient, different tumors, different regions within a tumor, or even different cells in a single region of a tumor may show significant molecular heterogeneity corresponding to clones with divergent biological behavior and treatment responsiveness. Compared to biopsy of a single mass, ctDNA may provide a more comprehensive and global portrait of this intra- and inter-tumor heterogeneity and potentially enable earlier identification of tumor clones with resistance to particular treatments. Longitudinal monitoring of ctDNA can help detect the emergence of therapy-resistant clones and obviate the need for repeated invasive tissue biopsies. For example, serial monitoring of colorectal cancer patients undergoing treatment with the anti-EGFR antibody cetixumab showed emergence and positive selection of mutant *KRAS* clones, with subsequent decay following antibody discontinuation [27]. Similarly, ctDNA monitoring of lung cancer patients receiving the EGFR inhibitor erlotinib enabled detection of a resistance conferring T790M mutation clone nearly a full year before clinical progression of disease [28]. The U.S. Food and Drug Administration (FDA) and the European Medicines Agency (EMA) have approved the use of PCR-based ctDNA assays for EGFR mutations prior to initiating therapy with anti-EGFR agents, and additional ctDNA-based assays are being evaluated in clinical trials [15, 20, 29].

7.5.2.4 cfDNA Fragmentation Profile

Cancer and non-cancer cells show significant differences in chromatin structure, some of which underlie transcriptional programs necessary for carcinogenesis. These differences in chromatin structure may be detected as altered cfDNA fragmentation profiles in patients with cancer compared to healthy controls. CfDNA profiles reflect both structural alterations, such as DNA copy number changes, as well as epigenetic modifications that regulate chromatin and nucleosome structure. Differences in sequencing coverage of specific genomic positions can thus distinguish ctDNA from cfDNA derived from normal cells and outperforms DNA copy number and mitochondrial DNA-based methods in distinguishing cfDNA from healthy or cancer patients [21]. In a cohort of 126 patients with carcinomas of the breast, bile duct, colon, stomach, lung, or ovary, a classifier based on cfDNA fragmentation identified 82% of patients with cancer; when combined with mutational analysis, that number rate increased to 91%, and to as high as 100% for patients with metastatic disease. In patients undergoing treatment with EGFR or ERBB2 inhibitors, changes in cfDNA fragmentation profiles paralleled changes in *EGFR* or *ERBB2* mutant allele frequencies, suggesting ctDNA fragmentation may serve as a potential surrogate marker of treatment response [21]. Furthermore, differences in ctDNA fragmentation and the underlying differences in nucleosome occupancy among different cancers may aid in accurate diagnosis in cancers of unknown primary [15].

7.5.2.5 DNA Methylation

cfDNA is chemically stable, and cancer-specific DNA methylation patterns can be detected in ctDNA with improved specificity compared to genetic methods [18, 19]. DNA methylation profiling distinguished benign pelvic masses from high-grade serous ovarian cancers with a specificity of greater than 90%, and identified with a specificity of 88% women who within 2 years developed ovarian cancer [17]. ctDNA methylation markers specific to hepatocellular carcinoma (HCC) were correlated with tumor burden, stage, and treatment response [30]. Methylation profiling of ctDNA has also shown promise in other malignancies, such as papillary thyroid carcinoma and lung cancers [31, 32].

7.5.2.6 Concordance with Tissue-Based Methods

A potential current limitation of cfDNA in tumor monitoring is the high variability in ctDNA concentration among patients, even after matching for malignancy type and stage. Much of this variability is likely due to differences in disease burden between patients. In patients with low tumor burden or with increased release of normal DNA from blood cells, tumor-associated alterations in overall cfDNA can be below the limit of detection and result in low concordance between molecular alterations identified in ctDNA and those identified in tissue biopsies from the same patients. Studies with matched tissue and ctDNA samples show a sensitivity of ctDNA analysis of greater than 60% and up to 98% for detecting tumor-associated mutations [15]. Use of ultrasensitive detection methods, such as droplet digital PCR; beads, emulsion, amplification, magnetics (BEAM) PCR; and targeted and whole-genome NGS can substantially improve sensitivity of ctDNA-based methods and improve their concordance with tissue-based assays [15, 20].

Conclusion and Future Perspective

Pathologists make use of multiple modalities when evaluating a metastatic tumor. When the primary lesion is known, histologic, immunohistochemical, and molecular profiling of metastatic lesions can identify prognostic and treatment-altering genomic, epigenomic, and phenotypic changes. When a primary lesion is not known, this workup can often identify a primary site. Although these assays are currently done on tissue specimens, advances in ctDNA isolation and analysis may enable minimally invasive and global profiling of a patient's systemic disease burden. As pathological diagnosis of cancer evolves, the following questions will drive significant changes in the field:

- To what extent will DNA methylation studies improve diagnostic resolution and classification of cancers?
- Will more cfDNA-based methods supplant with tissue and serological tumor markers?
- Will integration of molecular findings from population-based gene interaction networks, such as those based on the ARACNE algorithm [33], further improve disease classification based on transcriptional dependencies in cancers?
- Relatedly, will artificial intelligence-based tools augment the pathologist's ability to integrate clinical, histologic, and molecular data with improved diagnostic accuracy?
- Will in vivo imaging methods advance sufficiently to provide histologic and cytologic resolution?

References

1. Siegel RL, Miller KD, Fuchs HE, Jemal A. Cancer statistics, 2021. CA Cancer J Clin. 2021;71(1) https://doi.org/10.3322/caac.21654.
2. Riihimäki M, Thomsen H, Hemminki A, Sundquist K, Hemminki K. Comparison of survival of patients with metastases from known versus unknown primaries: survival in metastatic cancer. BMC Cancer. 2013;13(1) https://doi.org/10.1186/1471-2407-13-36.
3. LiVolsi V, Nallib A, Baloch Z, Bartolazzi A, Chan J, DeLellis R, El-Naggar A, Eloy C, Eng C, Fagin J, Ghossein R, Giordano T, Kondo T, Lloyd R, Mete O, Nikiforov Y, Nonaka D, Paschke R, Perren A, Rosai J, Sadow P, Schneider A, Sobrinho S, Tallini G, Williams M (2017b) Follicular thyroid carcinoma. In: Lloyd RV, Osamura R, Kloppel G, Rosai J, editors. WHO classification of tumours of endocrine organs, vol. 10, 4th ed, World Health Organization (IARC), Lyon.
4. LiVolsi V, Baloch Z, Sobrinho S, Tallini G (2017a) Hurthle (oncocytic) cell tumours. In: Lloyd R, Osamura R, Kloppel G, Rosai J, editors. WHO classification of tumours of endocrine organs, vol 10, 4th ed, World Health Organization (IARC), Lyon.
5. Budczies J, von Winterfeld M, Klauschen F, Bockmayr M, Lennerz JK, Denkert C, Wolf T, Warth A, Dietel M, Anagnostopoulos I, Weichert W, Wittschieber D, Stenzinger A. The landscape of metastatic progression patterns across major human cancers. Oncotarget. 2015;6(1) https://doi.org/10.18632/oncotarget.2677.
6. Riihimäki M, Thomsen H, Sundquist K, Sundquist J, Hemminki K. Clinical landscape of cancer metastases. Cancer Med. 2018;7(11) https://doi.org/10.1002/cam4.1697.
7. Lorigan JG, Wallace S, Mavligit GM. The prevalence and location of metastases from ocular melanoma: imaging study in 110 patients. Am J Roentgenol. 1991;157(6):1279–81.
8. Rajpal S, Moore R, Karakousis CP. Survival in metastatic ocular melanoma. Cancer. 1983;52(2)
9. Magaki S, Hojat SA, Wei B, So A, Yong WH. An introduction to the performance of immunohistochemistry. Methods Mol Biol. 2019;1897:289–98.
10. Kim SW, Roh J, Park CS. Immunohistochemistry for pathologists: protocols, pitfalls, and tips. J Pathol Transl Med. 2016;50(6):411–8.
11. Wang N, Zee S, Zarbo R, Bacchi C, Gown A. Coordinate expression of Cytokeratins 7 and 20 defines unique subsets of carcinomas. Appl Immunohistochem. 1995;3(2):99–107.
12. Jänne PA, Engelman JA, Johnson BE. Epidermal growth factor receptor mutations in non–small-cell lung cancer: implications for treatment and tumor biology. J Clin Oncol. 2005;23(14)
13. El-Deiry WS, Goldberg RM, Lenz H, Shields AF, Gibney GT, Tan AR, Brown J, Eisenberg B, Heath EI, Phuphanich S, Kim E, Brenner AJ, Marshall JL. The current state of molecular testing in the treatment of patients with solid tumors, 2019. Cancer J Clin. 2019:caac.21560.
14. Goodwin S, McPherson JD, McCombie WR. Coming of age: ten years of next-generation sequencing technologies. Nat Rev Genet. 2016;17(6)
15. Wan JC, Massie C, Garcia-Corbacho J, Mouliere F, Brenton JD, Caldas C, Pacey S, Baird R, Rosenfeld N. Liquid biopsies come of age: towards implementation of circulating tumour DNA. Nat Rev Cancer. 2017;17(4):223–38.
16. Locke WJ, Guanzon D, Ma C, Liew YJ, Duesing KR, Fung KY, Ross JP. DNA methylation cancer biomarkers: translation to the clinic. Front Genet. 2019;10
17. Widschwendter M, Zikan M, Wahl B, Lempiäinen H, Paprotka T, Evans I, Jones A, Ghazali S, Reisel D, Eichner J, Rujan T, Yang Z, Teschendorff AE, Ryan A, Cibula D, Menon U, Wittenberger T. The potential of circulating tumor DNA methylation analysis for the early detection and management of ovarian cancer. Genome Med. 2017;9(1):116.
18. Fernandez AF, Assenov Y, Martin-Subero JI, Balint B, Siebert R, Taniguchi H, Yamamoto H, Hidalgo M, Tan AC, Galm O, Ferrer I, Sanchez-Cespedes M, Villanueva A, Carmona J, Sanchez-Mut JV, Berdasco M, Moreno V, Capella G, Monk D, Ballestar E, Ropero S, Martinez R, Sanchez-Carbayo M, Prosper F, Agirre X, Fraga MF, Grana O, Perez-Jurado L, Mora J, Puig S, Prat J, Badimon L, Puca AA, Meltzer SJ, Lengauer T, Bridgewater J, Bock C, Esteller M. A DNA methylation fingerprint of 1628 human samples. Genome Res. 2012;22(2):407–19.
19. Lokk K, Modhukur V, Rajashekar B, Märtens K, Mägi R, Kolde R, Koltšina M, Nilsson TK, Vilo J, Salumets A, Tõnisson N. DNA methylome profiling of human tissues identifies global and tissue-specific methylation patterns. Genome Biol. 2014;15(4):3248.
20. Corcoran RB, Chabner BA. Application of cell-free DNA analysis to cancer treatment. N Engl J Med. 2018;379(18):1754–65.
21. Cristiano S, Leal A, Phallen J, Fiksel J, Adleff V, Bruhm DC, Jensen S, Medina JE, Hruban C, White JR, Palsgrove DN, Niknafs N, Anagnostou V, Forde P, Naidoo J, Marrone K, Brahmer J, Woodward BD, Husain H, van Rooijen KL, Ørntoft MBW, Madsen AH, van de Velde CJ, Verheij M, Cats A, Punt CJ, Vink GR, van Grieken NC, Koopman M, Fijneman RJ, Johansen JS, Nielsen HJ, Meijer GA, Andersen CL, Scharpf RB, Velculescu VE. Genome-wide cell-free DNA fragmentation in patients with cancer. Nature. 2019;570(7761):385–9.
22. Phallen J, Sausen M, Adleff V, Leal A, Hruban C, White J, Anagnostou V, Fiksel J, Cristiano S, Papp E, Speir S, Reinert T, Orntoft MBW, Woodward BD, Murphy D, Parpart-Li S, Riley D, Nesselbush M, Sengamalay N, Georgiadis A, Li QK, Madsen MR, Mortensen FV, Huiskens J, Punt C, van Grieken N, Fijneman R, Meijer G, Husain H, Scharpf RB, Diaz LA, Jones S, Angiuoli S, Ørntoft T, Nielsen HJ, Andersen CL, Velculescu VE. Direct detection of early-stage cancers using circulating tumor DNA. Sci Transl Med. 2017;9(403):eaan2415.
23. Bettegowda C, Sausen M, Leary RJ, Kinde I, Wang Y, Agrawal N, Bartlett BR, Wang H, Luber B, Alani RM, Antonarakis ES, Azad NS, Bardelli A, Brem H, Cameron JL, Lee CC, Fecher LA, Gallia GL, Gibbs P, Le D, Giuntoli RL, Goggins M, Hogarty MD, Holdhoff M, Hong SM, Jiao Y, Juhl HH, Kim JJ, Siravegna G,

Laheru DA, Lauricella C, Lim M, Lipson EJ, Marie SKN, Netto GJ, Oliner KS, Olivi A, Olsson L, Riggins GJ, Sartore-Bianchi A, Schmidt K, Shih LM, Oba-Shinjo SM, Siena S, Theodorescu D, Tie J, Harkins TT, Veronese S, Wang TL, Weingart JD, Wolfgang CL, Wood LD, Xing D, Hruban RH, Wu J, Allen PJ, Schmidt CM, Choti MA, Velculescu VE, Kinzler KW, Vogelstein B, Papadopoulos N, Diaz LA. Detection of circulating tumor DNA in early- and late stage human malignancies. Sci Transl Med. 2014;6(224):224ra24.

24. Leon SA, Shapiro B, Sklaroff DM, Yaros MJ. Free DNA in the serum of cancer patients and the effect of therapy. Cancer Res. 1977;37(3):646–50.
25. Garcia-Murillas I, Schiavon G, Weigelt B, Ng C, Hrebien S, Cutts RJ, Cheang M, Osin P, Nerurkar A, Kozarewa I, Garrido JA, Dowsett M, Reis-Filho JS, Smith IE, Turner NC. Mutation tracking in circulating tumor DNA predicts relapse in early breast cancer. Sci Transl Med. 2015;7(302):133–302.
26. Tie J, Wang Y, Tomasetti C, Li L, Springer S, Kinde I, Silliman N, Tacey M, Wong HL, Christie M, Kosmider S, Skinner I, Wong R, Steel M, Tran B, Desai J, Jones I, Haydon A, Hayes T, Price TJ, Strausberg RL, Diaz LA, Papadopoulos N, Kinzler KW, Vogelstein B, Gibbs P. Circulating tumor DNA analysis detects minimal residual disease and predicts recurrence in patients with stage II colon cancer. Sci Transl Med. 2016;8(346):92–346.
27. Siravegna G, Mussolin B, Buscarino M, Corti G, Cassingena A, Crisafulli G, Ponzetti A, Cremolini C, Amatu A, Lauricella C, Lamba S, Hobor S, Avallone A, Valtorta E, Rospo G, Medico E, Motta V, Antoniotti C, Tatangelo F, Bellosillo B, Veronese S, Budillon A, Montagut C, Racca P, Marsoni S, Falcone A, Corcoran RB, Di Nicolantonio F, Loupakis F, Siena S, Sartore-Bianchi A, Bardelli A. Clonal evolution and resistance to EGFR blockade in the blood of colorectal cancer patients. Nat Med. 2015;21(7):795–801.
28. Sorensen BS, Wu L, Wei W, Tsai J, Weber B, Nexo E, Meldgaard P. Monitoring of epidermal growth factor receptor tyrosine kinase inhibitor-sensitizing and resistance mutations in the plasma DNA of patients with advanced non–small cell lung cancer during treatment with erlotinib. Cancer. 2014;120(24):3896–901.
29. Dawson SJ, Tsui DW, Murtaza M, Biggs H, Rueda OM, Chin SF, Dunning MJ, Gale D, Forshew T, Mahler-Araujo B, Rajan S, Humphray S, Becq J, Halsall D, Wallis M, Bentley D, Caldas C, Rosenfeld N. Analysis of circulating tumor DNA to monitor metastatic breast cancer. N Engl J Med. 2013;368(13):1199–209.
30. Xu R, Wei W, Krawczyk M, Wang W, Luo H, Flagg K, Yi S, Shi W, Quan Q, Li K, Zheng L, Zhang H, Caughey BA, Zhao Q, Hou J, Zhang R, Xu Y, Cai H, Li G, Hou R, Zhong Z, Lin D, Fu X, Zhu J, Duan Y, Yu M, Ying B, Zhang W, Wang J, Zhang E, Zhang C, Li O, Guo R, Carter H, Zhu J, Hao X, Zhang K. Circulating tumour DNA methylation markers for diagnosis and prognosis of hepatocellular carcinoma. Nat Mater. 2017;16(11):1155–61.
31. Khatami F, Teimoori-Toolabi L, Heshmat R, Nasiri S, Saffar H, Mohammadamoli M, Aghdam MH, Larijani B, Tavangar SM. Circulating ctDNA methylation quantification of two DNA methyl transferases in papillary thyroid carcinoma. J Cell Biochem. 2019;120(10):17,422–37.
32. Liang W, Zhao Y, Huang W, Gao Y, Xu W, Tao J, Yang M, Li L, Ping W, Shen H, Fu X, Chen Z, Laird PW, Cai X, Fan JB, He J. Non-invasive diagnosis of earlystage lung cancer using high-throughput targeted DNA methylation sequencing of circulating tumor DNA (ctDNA). Theranostics. 2019;9(7):2056–70.
33. Alvarez MJ, Subramaniam PS, Tang LH, Grunn A, Aburi M, Rieckhof G, Komissarova EV, Hagan EA, Bodei L, Clemons PA, Dela Cruz FS, Dhall D, Diolaiti D, Fraker DA, Ghavami A, Kaemmerer D, Karan C, Kidd M, Kim KM, Kim HC, Kunju LP, Langel Li Z, Lee J, Li H, LiVolsi V, Pfragner R, Rainey AR, Realubit RB, Remotti H, Regberg J, Roses R, Rustgi A, Sepulveda AR, Serra S, Shi C, Yuan X, Barberis M, Bergamaschi R, Chinnaiyan AM, Detre T, Ezzat S, Frilling A, Hommann M, Jaeger D, Kim MK, Knudsen BS, Kung AL, Leahy E, Metz DC, Milsom JW, Park YS, Reidy-Lagunes D, Schreiber S, Washington K, Wiedenmann B, Modlin I, Califano A. A precision oncology approach to the pharmacological targeting of mechanistic dependencies in neuroendocrine tumors. Nat Genet. 2018;50(7):979–89.

8 The Role of Angiotropic Extravascular Migratory Metastasis in Metastases

Raymond Barnhill and Claire Lugassy

Abstract

The phenomenon of angiotropism, pericyte mimicry (PM), and extravascular migratory metastasis (EVMM) has questioned the concept that melanoma cells metastasize exclusively within vascular channels. During this phenomenon of cancer spread and metastasis, melanoma cells attach to the abluminal vascular surfaces i.e. angiotropism and spread without intravasation via continuous migration, replacing pericytes, i.e. pericyte mimicry (PM). Angiotropism, PM and EVMM have mainly been studied in melanoma but also occur in other cancer types. PM and EVMM appear to be a reversion to an embryogenesis-derived program. There are many analogies between the neural crest development and cancer progression, including the important role of laminins, epithelial-mesenchymal transition, and the re-activation of embryonic signals by cancer cells. Furthermore, there is no circulation of blood during the first trimester of embryogenesis, despite the fact that there is extensive migration of cells to distant sites and formation of organs and tissues during this period. Embryonic migration therefore is a continuous extravascular migration as are PM and EVMM, supporting the concept that these embryonic migratory events appear to recur abnormally during the metastatic process. This new paradigm may greatly influence the development of new effective therapeutics for metastasis.

R. Barnhill
Department of Translational Research, Institut Curie, Paris, France

University of Paris Réné Descartes Faculty of Medicine, Paris, France

C. Lugassy (✉)
Department of Translational Research, Institut Curie, Paris, France
e-mail: claire.lugassy@curie.fr

Learning Objectives

- The phenomenon of angiotropism, pericyte mimicry (PM), and extravascular migratory metastasis (EVMM) has questioned the concept that melanoma cells metastasize exclusively within vascular channels.
- The long interval to recurrence after resection and other therapies and the development of resistant metastases have proved very problematic. Dormancy remains poorly understood and, as a result, effective therapeutic interventions have not been formulated.
- During EVMM, melanoma cells attach to the abluminal vascular surfaces, i.e., angiotropism and spread without intravasation via continuous migration, replacing pericytes, i.e., pericyte mimicry (PM) toward regional and distant secondary sites.
- There are many analogies between the neural crest development and cancer progression, including the important role of laminins, epithelial-mesenchymal transition, and the re-activation of embryonic signals by cancer cells.
- Importantly, there is no circulation of blood during the first trimester of embryogenesis, despite the fact that there is extensive migration of cells to distant sites and the formation of organs and tissues during this period. Embryonic migration therefore is a continuous extravascular migration as are PM and EVMM.
- Via PM/EVMM, angiotropic cancer cells will follow indefinitely, in a dysregulated way, "embryonic seed and soil" mechanisms in the invaded organ.
- A more comprehensive understanding of the mechanism(s) of tumor spread is greatly needed, and angiotropism and PM/EVMM may profoundly influence and stimulate the development of more effective therapeutics.

S. P. Leong et al. (eds.), *Cancer Metastasis Through the Lymphovascular System*,
https://doi.org/10.1007/978-3-030-93084-4_8

8.1 Introduction

Intravascular dissemination of cancer is still considered the principal if not exclusive route of metastasis. For many years, the relationships between a tumor and its vasculature have been thought to principally involve (i) angiogenesis required for tumor growth at both primary and secondary sites and (ii) intravascular circulation of tumor cells to metastatic sites. However, the precise mechanisms of tumor growth and tumor cell migration and in particular the role of the tumor vasculature in these processes remain unclear [1].

Although it had been assumed that tumor cells in contact with vessel abluminal surfaces were in the process of intravasation, it was shown in the late 1990s that such vessel–tumor interactions could have a completely different biological interpretation. In particular, angiotropism, pericyte mimicry (PM), and extravascular migratory metastasis (EVMM). This new field has raised important questions about cancer biology.

Research on angiotropism, pericyte mimicry, and extravascular migratory.

metastasis or EVMM, opened by Lugassy and Barnhill and reviewed in [2], has questioned the assumption that all tumor cells metastasize exclusively *within* vascular lumina. Angiotropism, PM, and EVMM are three complementary terms that refer to an embryologically derived mechanism of continuous extravascular tumor cell migration without entrance into vascular channels. In addition to migration along vessels as angiotropism and PM, EVMM can also occur along other anatomical tracks such as nerves in neurotropism [2].

Angiotropic tumor cells localize along the abluminal vascular surfaces by competition with and mimicking of pericytes (or other perivascular mesenchymal stem cells): (i) tumor cells are recruited along the abluminal vascular surface instead of pericytes and replace pericytes; (ii) tumor cells migrate along the abluminal surface (Fig. 8.1). However, tumor cells never stabilize neovessels nor the vascular basement membrane matrix as do pericytes. PM allows tumor cells to progress along vessels with all of the needed nutrients and oxygen, and such a process avoids the destruction of tumor cells observed in hematogenous dissemination [3].

8.2 Angiotropism, Pericytic Mimicry (PM), and Extravascular Migratory Metastasis (EVMM) : Terminology and Definitions

In the work initiated by Lugassy, direct contact between tumor cells and endothelium was observed via a basal lamina rich in laminin "which may promote the migration of melanoma cells in contact with vessels" [4, 5]. Electron microscopy was necessary to confirm this direct contact of tumor cells with the basal lamina of endothelial cells. This ultrastructural relationship was termed "the angio-tumoral complex," which in turn corresponds to angiotropism of tumor cells at the light microscopic level.

8.2.1 Angiotropism

Thus, specifically in the cancer literature, angiotropism refers to tumor cells exclusively localized to the external surfaces of the endothelial cells of blood vessels in a pericytic location in the absence of intravasation (Figs. 8.1 and 8.2) [6]. The term angiotropism has also been employed for an uncommon lymphoma—"angiotropic lymphoma," which presents almost exclusively with intravascular rather extravascular tumor. This lymphoma is not recognized as a distinct entity in the Revised European-American Classification of Lymphoid Neoplasms.

8.2.2 Pericytic Mimicry

Since 1999, Lugassy and Barnhill have shown that angiotropism is a prognostic factor and a microscopic marker of tumor cell migration along the abluminal vascular surface in a pericytic location [2], described as "pericytic-like" migration. In 2013, they have replaced "pericytic-like migration" with the term "pericytic mimicry" (PM) [7, 8]. PM describes well this biological process of tumor spread along the periendothelial, abluminal vascular surface, competition with, and mimicking of pericytes [2]. This mode of tumor spread along vessels of all types (including larger vessels, the sinusoids of the liver, etc.) seems to connote tumor cells replacing and mimicking not only pericytes but also other perivascular mesenchymal stem cells [6].

Angiotropism/PM has been shown to occurs along preexisting vessels; however, several data indicate that it could also occur along neovessels (see below).

8.2.3 Extravascular Migratory Metastasis

Extravascular migratory metastasis (EVMM) was originally coined by Lugassy and Barnhill to describe tumor cell spread by way of pericytic mimicry from a primary tumor to secondary metastatic sites [9]. This additional mechanism of metastasis specifically involves tumor cell migration which is a slower, progressive, mode of tumor dissemination (Fig. 8.1) versus the ultra-rapid circulation of tumor cells in lymphatic or blood vascular channels [3, 10]. Angiotropism, PM, and EVMM may be considered three

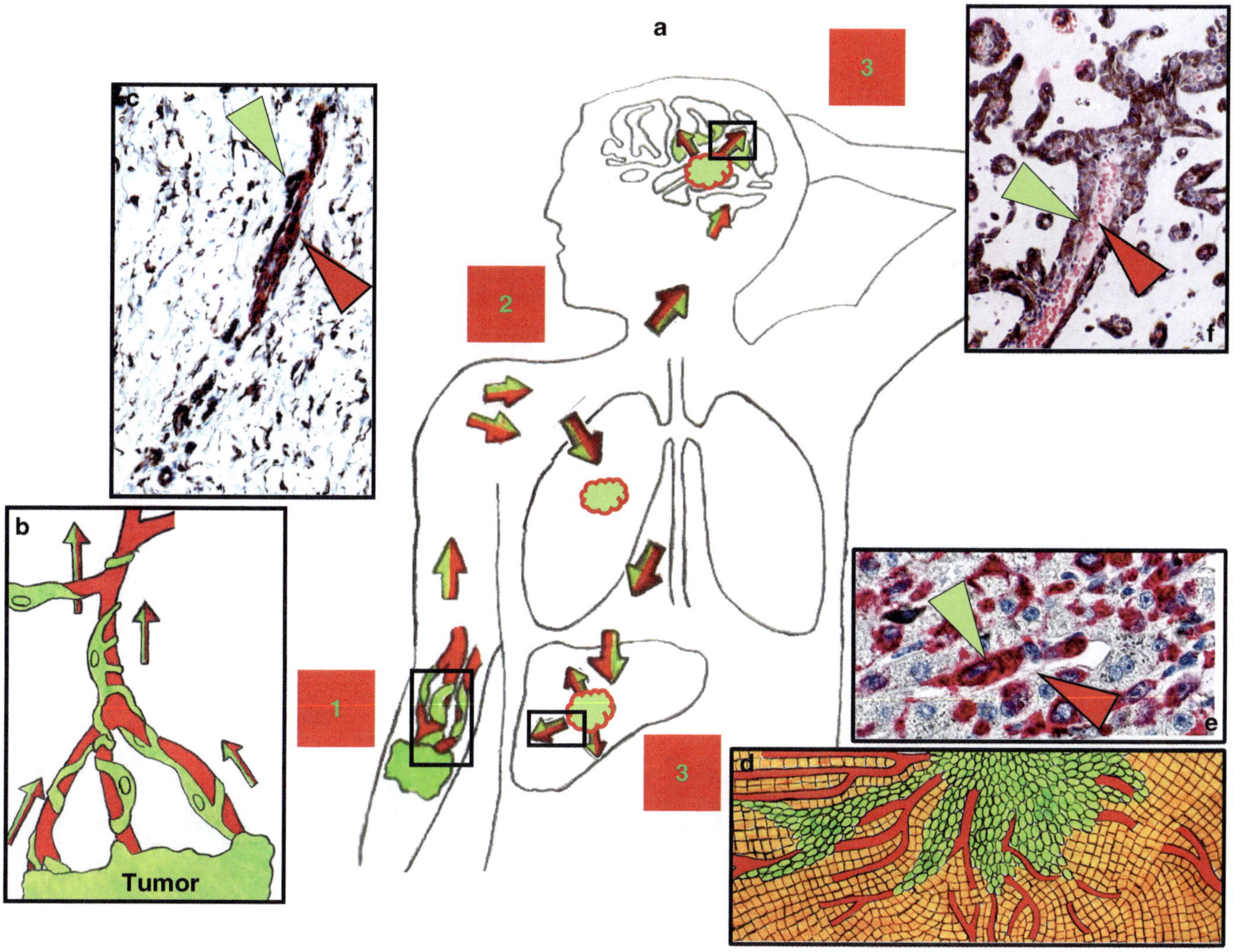

Fig. 8.1 Diagram of PM and EVMM
(**a**) Overview diagram of human body. (**b**) Diagram of PM: tumor cells migrate in a pericytic location along the abluminal vascular surface at the advancing front of the tumor. (**c**) Corresponding human histopathological image of angiotropism/PM at the advancing front of the primary melanoma. Double immunostaining of melanoma cells by melan-A (red chromogen) and vascular channels by CD31 (brown-black chromogen). (**d**) Diagram of the replacement histopathological growth pattern in liver metastasis. Note at the advancing front of the metastasis the progression of green tumor cells replacing hepatocytes (orange) along red vessels. (**e**) Double immunostaining of a corresponding human histopathological image at the advancing front of a liver metastasis. Melanoma cells express HMB45 along endothelial cell expressing CD31 (DAB brown chromogen). F. Human melanoma brain metastasis. Striking angiotropism of pigmented melanoma cells in a pericytic location along microvessels containing erythrocytes
Green arrowheads indicate tumor cells, red arrowheads indicated vessels. Bicolor arrows indicate a schematic view of PC. (1) indicates PM at the advancing front of the primary tumor. (2) indicates EVMM via PM throughout the body. (3) indicates metastases sites [2]

complementary terms describing this mechanism of continuous tumor cell migration toward distant sites (Fig. 8.1). Tumors cells are nourished during their migration along vessel surfaces by obtaining their requisite nutrients and oxygen. Such tumor cell migration also affords the advantage of avoiding destruction upon entry into the blood vascular circulation [11, 12].

EVMM can involve other anatomic structures as tracks for tumor cell migration, as, e.g., nerves, i.e., neurotropism in many tumors and white matter tracks in the brain with glioma invasion [13–15].

8.3 The Biological and Prognostic Significance of Angiotropism in Melanoma

8.3.1 Biological Significance

Lugassy and Barnhill first described the true biological significance of angiotropism as the specific histopathological marker of PM and EVMM [4, 5]. Angiotropism is defined as tumor cells aligned along the external surfaces of vascular

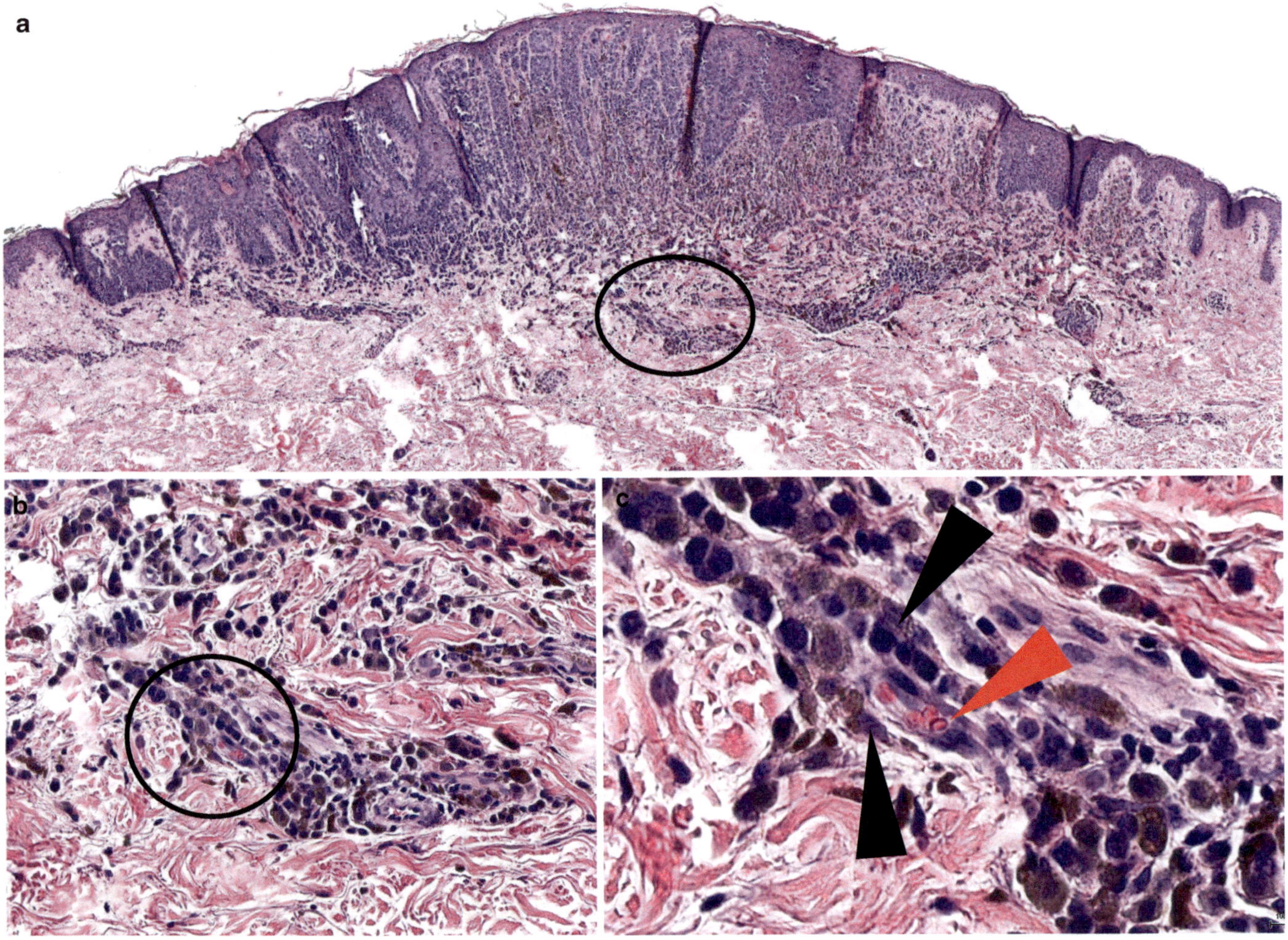

Fig. 8.2 Angiotropism in primary cutaneous melanoma (**a**) Low magnification of the tumor. Encircled area: angiotropism at the advancing front (the base) of the lesion. (**b**, **c**) Higher magnification of the advancing front. Note the angiotropic melanoma cells (black arrows) associated with vascular channels (red arrow) [81]

channels in a pericytic location (PM) (Fig. 8.1a–c). Micro- and macrovascular channels may show comparable angiotropism. Larger vessels are characterized by mimicry of adventitial mesenchymal stem cells. Angiotropism may consist of both single tumor cells or "collective" aggregates of tumor cells and is most frequently detected at the advancing front of the tumor or in nearby tissue and cannot be reliably recognized *within* a tumoral mass per se [6]. Angiotropism occurs in both primary and metastatic tumors (Fig. 8.1). Although studied thus far mainly in melanoma, angiotropism is present not only in other tumors of neural crest origin such as glioma but also in many other neoplasms (see below).

Angiotropism in melanomas and in other neoplasms is recognized at the advancing front of a tumor or in the nearby tissue (usually at least 0.5 to 2 mm away from the tumor) (Fig. 8.2), and in loco-regional microsatellite, satellite, and in transit metastases in nearby skin [2, 6]. In uveal melanoma, melanoma cells from the primary intraocular uveal site traverse and extend beyond the sclera along the abluminal surfaces of vascular channels (including the endothelial-lined channels of Schlemm) [16]. No evidence of intravascular or intra-lymphatic tumor cells are observed.

8.3.1.1 Melanoma liver Metastases

Liver metastases from colorectal and breast cancer [17] and melanoma, both uveal [18] and cutaneous [19], show distinctive "histopathological growth patterns" (HGPs) as described above in the co-option section: "replacement," "desmoplastic," "pushing," and two rare variants—the "sinusoidal" and "portal" HGPs. In particular, the replacement HGP is defined by tumor cells progressively spreading along the abluminal surfaces of sinusoidal vessels (in the space of Disse) or venules of the portal tracts [17–19]. Since hepatic stellate cells, which are considered to be pericytes, are located in this space of Disse [20, 21], tumor cells appear to engage in pericytic mimicry, i.e., competing with and replacing these pericytes. The replacement HGP represents approximately 75% of the liver metastases derived from uveal melanoma [18] and 50% of metastases associated with cutaneous melanoma [19] (Fig. 8.1d,e).

8.3.2 Prognostic Significance

Angiotropism is an independent prognostic factor for the development of metastases and diminished survival in both cutaneous and uveal melanoma, and this includes micro- and macroscopic satellites, in transit metastases, lymph node metastases, and distant metastases [2]. In addition, the replacement HGP in uveal and cutaneous melanoma liver metastases, which exhibits angiotropism, is predictive of poor survival with both univariate and multivariate analysis [18, 19]. As with liver metastases from other neoplasms, the replacement HGP in liver metastases from cutaneous and uveal melanoma is associated with a poorer prognosis than the desmoplastic HGP [17–19]. In addition, angiotropism appears to be a factor in resistance to BRAF inhibitor therapy [22].

8.4 Experimental Models and Molecular Findings

8.4.1 Melanoma Experimental Models

In vitro, ex vivo, and in vivo models have been developed for the study of angiotropism and PM of melanoma cells [2]. These include a 3D in vitro co-culture of tumor cells with endothelial tubules [23]; an ex vivo model of co-culture of melanoma cells with neovessels in rat aortic rings [24, 25], an ex vivo ear tissue invasion model [26]; a modified shell-less chick chorioallantoic membrane (CAM) model [27]; and a genetically engineered murine melanoma model. In the latter, UV irradiation elicited neutrophil-rich inflammation and increased angiotropic melanoma metastases in these murine melanomas [25] (Fig. 8.3a–c). In another murine brain melanoma model, fluorescence-labeled melanoma cells showed progressive outgrowth into the surrounding brain along the abluminal surfaces of preexisting vessels [28] (Fig. 8.3d–f). A new mouse melanoma ear skin model has demonstrated angiotropism and PM by intravital imaging of human melanoma cells [26]. Finally, PM and perivascular migration of human cutaneous and uveal melanoma cells have been visualized in a zebrafish melanoma [29] (Fig. 8.3g,h).

Gene expression studies of human angiotropic primary cutaneous melanoma have identified 15 genes potentially relevant to angiotropism and PM [6]. Utilization of our in vitro model of PM for gene expression microarray analysis has discovered 20 genes related to (i) cell migration (ICAM1, CCL2, PDGFB, RGNEF, RANBP9, IL6), (ii) cancer progression (ICAM1, SELE, CCL2, CXCL6, IL6, SERPINB2, TRAF1, BLID, UPF1, PLAA, MALT1, ZXDC, RGNEF), (iii) EMT (ICAM1, CCL2, PDGFB IL6,), (iv) inflammation (ICAM1, SERPINB2, CCL2, CXCL6, SELE, SLC7A2, C2CD4B, PDGFB, IL6, TRAF1), (v) cancer stem/embryonic cell properties (CCL2, PDGFB, EVX1, CFDP1 and RANBP9, (v) pericyte recruitment (PDGFB). In addition, pericyte-marker expression by C8161 melanoma cells included CD146, NG2, CD44, CD73, CD105, and CD144 [6].

8.4.2 Non-melanoma Tumor Models

Glioma propagation in the brain has been studied with the in vitro co-culture model of PM (see angiotropism/PM and co-option above; for molecular data, see the Co-option section above) [30]. In a similar fashion, the in vitro co-culture and an organotypic brain slice culture models have demonstrated PM by metastatic cells from several different tumor types [31]. Metastatic colonization in various organs via PM occurs after tumor cells are injected into mice. In the latter studies, L1CAM (cell adhesion molecule L1) increases β1 integrin-ILK signaling and in turn YAP nuclear localization [31]. It thus appears that L1CAM and other adhesion molecules are utilized by tumor cells of various origins to facilitate attachment to the abluminal vascular surface during PM [31]. A recent study employing a murine leukemia model suggests that leukemia cells migrate directly along microvascular channels into the central nervous system (CNS) via PM [32] circumventing the need to traverse the blood-brain barrier. PM has also been implicated in prostate cancer progression [33]: CD44, which is a marker of both pericytes and cancer stem cells, appears to promote migration and invasion of docetaxel-resistant prostate cancer cells via induction of Hippo-YAP signaling [34].

8.5 Laminin in Angiotropism and PM

Laminins constitute a large family of αβγ heterotrimeric glycoproteins which serve as the principal component of the BM. Laminins seem to have first appeared about 500 million years ago with the development of multicellular organisms and are considered to be the first extracellular proteins [35]. BM laminins regulate cellular phenotypes and differentiation during organogenesis and cancer progression [6, 35–37]. By remodeling BM and adhesion proteins, embryonic and cancer stem cells are able to develop, grow, and migrate [38]. This will modify cell signaling pathways and in turn the

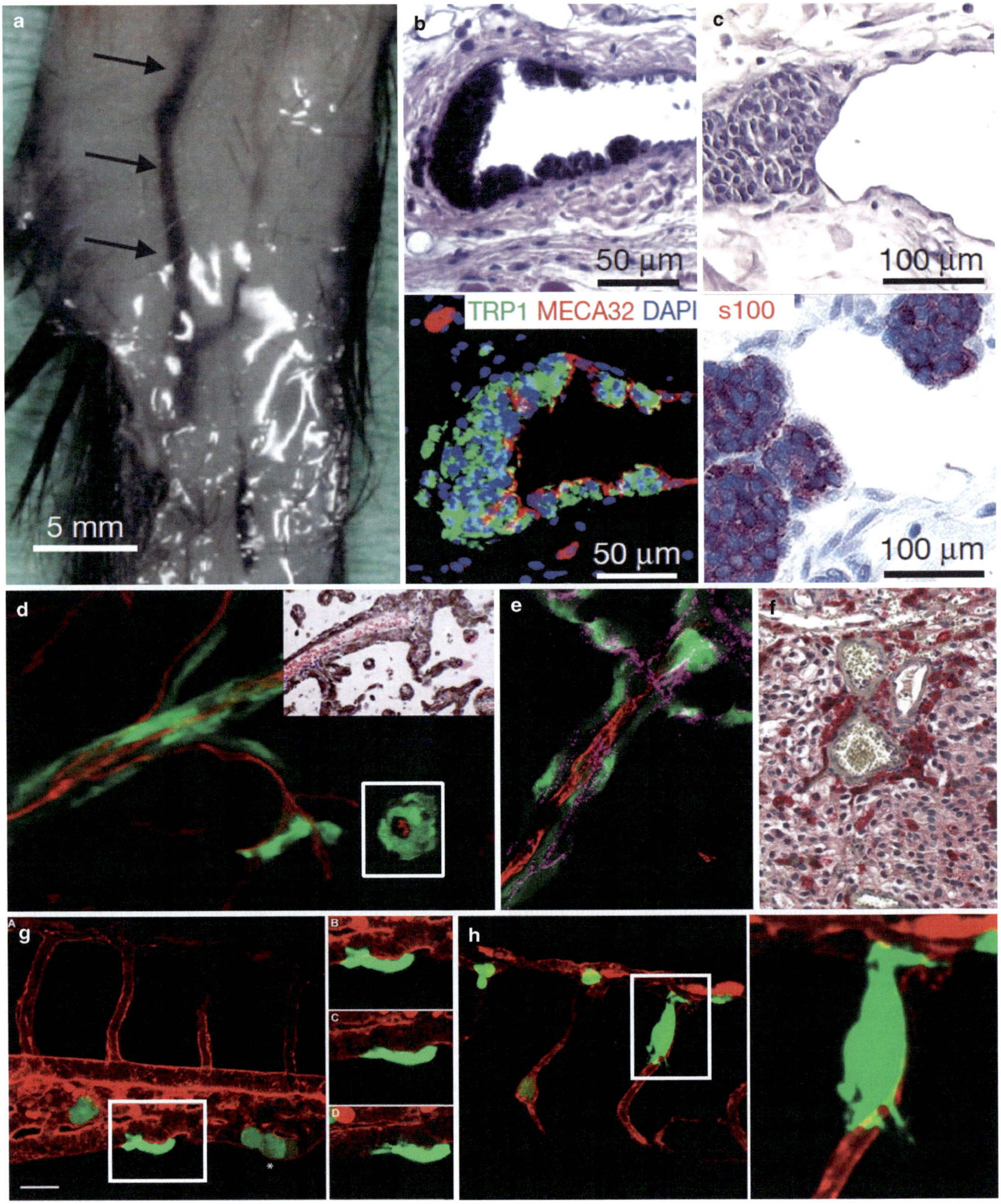
a
5 mm
b
50 µm
c
100 µm
TRP1 MECA32 DAPI
50 µm
s100
100 µm
d
e
f
g
A
B
C
D
h

cytoskeleton by modulating molecules in cryptic sites of the extracellular matrix (ECM). Therefore there is a "remodeling" and not a "degradation" of the ECM and BM during embryogenesis and cancer progression. Indeed, at the ultrastructural level, conventional, stable BM is organized in a supramolecular structure, while during embryogenesis and cancer progression, remodeled BM loses its organizing structure [39].

In PM, tumor cells are present along the vascular abluminal surface, i.e., along a vascular BM which contains laminin. Ultrastructural studies of angiotropism/PM in melanoma have shown an BM material lacking the classical supramolecular structure between tumor and endothelial cells [4, 39]. This material contains mainly laminin, and could expose laminin cryptic sites increasing cell migration [40]. The detection of such alterations may require ultrastructural examination or other more refined analysis. Additional studies are needed to clarify the interactions of tumor cells with the BM, in particular, potential structural and functional alterations of laminins and other associated molecules in PM.

The role for laminins in cancer progression is well known. Notably, experimental data show that PM in melanoma overexpressed laminin chains and receptors [41, 42]. A laminin $\gamma 1$ chain synthetic peptide (peptide C16) significantly promotes cancer progression via adhesion, metastasis, and migration [43]. C16 also enhances PM of human melanoma cells [27]. In addition, melanoma cells produce several laminin isoforms and migrate on the $\alpha 5$ laminin chain [44]. Laminin-332 is implicated in tumor migration and metastasis. Laminin-332 in lungs may stimulate pulmonary metastases of breast carcinoma [45]. Furthermore, the $\gamma 2$ laminin chain, which plays a role in cancer cell motility, is strongly expressed at the invasion front, and this correlates with bad prognosis [46, 47]. $\alpha 4$-Laminins, including laminins 411 and 421, can be expressed by vessels and tumor cells [48]. In co-option and PM progression of glial brain tumors, laminin-421 is switched to laminin-411 ($\alpha 4\beta 1\gamma 1$) [48], and this is associated with higher tumor grade and with expression CSC markers including CD133, Nestin, c-Myc, and Notch pathway members [43, 48].

The interaction of leukemia cells and the abluminal vascular surfaces of bridging vessels triggers the expression of $\alpha 6$ integrin by leukemia cells with vascular laminin, promoting the invasion of the central nervous system via PM [32].

In addition, laminins containing the $\alpha 5$ chain play a role in cell-cell adhesion and migration in several cancers, notably via Lutheran (Lu) (also named basal cell adhesion molecule [49]. Lu/BCAM is a laminin $\alpha 5$ receptor, a subunit of LM-511. Importantly, laminin $\alpha 5$ is essential during cranial NC migration [50]. Recently, it has been also shown that during PM, L1CAM enables the adhesion of metastasis-initiating cells to laminin-rich BM [51].

Concerning the replacement pattern in liver metastases, it is important to emphasize that in normal liver, there is a non-electron dense BM-like structure with reticular collagen 4 and laminin between the endothelium and the hepatocytes [21] and thus the two cell types are in discontinuous close contact. Importantly, stellate cells, corresponding to pericytes, are directly aligned along endothelial cells.

Fig. 8.3 In vivo studies of PM
(**a–c**) PM and angiotropism in a genetically engineered mouse model. A. Macroscopically visible melanoma cell expansion along 25 mm of a dermal blood vessels (arrows). B-C. Histologic(above) and immunohistochemical (below) analysis of angiotropism in the murine model (**b**), in a human primary melanoma (**c**). Note the similar angiotropic images in **b** and **c**. (**d–f**) Human melanoma cells in a murine melanoma model brain model. (**d**) Montage juxtaposing PM in the murine brain melanoma model (main picture) and a human sample of a melanoma metastatic to the brain. On the left part of the image, in the murine model at 4 weeks, PM of green GFP-labeled melanoma cells are visible along red vessels (red tomato lectin). Juxtaposed to the right of the same image: human sample of a melanoma metastatic to the brain. Microvessels (filled by red blood cells) are extensively coated by a layer of melanin-containing melanoma cells, without intravasation, demonstrating significant angiotropism and PM. Note the correspondence of the two juxtaposed images. Scale bars: 100 μm. Insert: XZ cross-section of the murine brain vessel confirming that melanoma cells are external to the vessels, without intravasation. (**e**) Expression of Serpin B2 in the murine brain melanoma model. Strong co-localization of Serpin B2 (purple) to the angiotropic GFP melanoma cells (green) along vessels (red tomato lectin). Scale bars, 50 μm. (**f**) Expression of Serpin B2 by angiotropic melanoma cells in a human brain melanoma metastasis. Striking expression of SERPIN B2 (dark red) by melanoma cells is observed predominately along the abluminal surfaces of microvascular channels. (**g, h**) Intravital observation of PM of melanomas in a zebrafish xenograft. (**g**) Xenograft with GFP green human cutaneous melanoma cells. Angiotropism/PM of green GFP tumor cells along the external surface of the caudal vein (red tomato lectin). On the right: time-lapse images of the angiotropic cell in white square taken at 0, 4, and 8 h after the beginning of the imaging. (**h**) Xenograft with GFP human uveal melanoma cells. Similar image along the external surface of red intersegmental vessels. Note single (**g and h**) and collective (**h**, white square) tumor cell migration along vessels [2]

8.6 Reversion to Embryogenesis-Derived Programs: Implications in Angiotropic EVMM

In the mid-nineteenth Rudolf Virchow observed that "*neoplasms arise in accordance with the same laws that regulate embryonic development*" [52] emphasizing for the first time this important relationship between tumor development and embryonic development. Embryogenesis involves major morphological changes and cellular migrations that are imitated and repeated in cancer development [2, 53]. Furthermore, re-activation of embryonic cellular signaling in adult cells, as a consequence of mutational and epigenetic programming, is prototypic of cancer progression [54].

8.6.1 Epithelial-to-Mesenchymal Transition in the Embryo and in Cancer

The epithelial-to-mesenchymal transition (EMT) is a critical process during embryonic development and organogenesis [55]. In EMT, epithelial cells lose their epithelial properties, including both apical-basal polarity and cell-to-cell adhesion, and acquire migratory and invasive mesenchymal cell characteristics in order to spread from their tissue site of origin. As described below, PM and EVMM share identical properties. In a similar fashion, a switch between EMT and mesenchymal-epithelial-transition (MET) is fundamental during both embryonic and cancer progression. MET facilitates metastatic outgrowth and probably vascular co-option within distant organs [55].

EMT/MET in metastasis, as in embryogenesis, may result in a spectrum of phenotypes between epithelial and mesenchymal cell opposite poles of the spectrum [56]. This reversible "phenotype switch" model of metastasis may be driven by BRN2 in melanoma metastasis [57].

8.6.1.1 Gene Expression in the Embryo and Cancer

It is well known that human tumors express the early gene expression of development, for example, Wnt, Hedgehog, and Notch pathways [26]. YAP is linked to PM [58], and the Hippo pathway is essential in cell migration during organogenesis [59].

As already mentioned, during melanoma PM in vitro there is an expression of genes associated with EMT, embryonic, and cancer stem cells (CSC) properties [7, 8]. Furthermore, development, cell movement, cancer, and embryonic development are among the most significantly enriched functional groups analyzed by Ingenuity-Pathway-Analysis [7].

8.6.1.2 Stem Cells and Competition in Embryogenesis and Cancer Progression

CSCs are a subpopulation of highly plastic tumor cells via EMT to MET switch [55, 56]. CSC properties include self-renewal, migratory, metastasizing, and drug resistance [26]. In the in vitro model of PM, the interaction of tumor cells with endothelial tubules induces the stem-embryonic-like characteristics of tumor cells [7], like tumor cells in vascular niches [26].

Cell competition in development is described as a "fitness-sensing embryonic mechanism" resulting in the replacement of cells less fit by those with greater fitness. Cell competition can be hijacked by cancer cells tumor progression [60], including competition for a niche [60], which is relevant to co-option and PM. In particular, angiotropic cancer cells may compete with pericytes (or other perivascular mesenchymal stem in larger or specialized vessels) during co-option and PM in order to reach the external vascular surfaces of preexisting vessels. The mechanism may also involve nascent vascular structures during neoangiogenesis, as suggested by histological and experimental observations [4, 24].

8.6.1.3 The Seed and the Soil: Relevance to the Embryo and Cancer Development

The "seed and soil" theory as originally described for cancer cells by Paget [58, 61, 62] is unquestionably relevant to organogenesis. Embryonic stem cells are the "seeds" migrating throughout the embryo to reach their ultimate phenotype-specific "soils" in order to grow and develop organs (organogenesis) [2]. This is particularly relevant to the development of the neural crest. The neural crest is a highly migratory embryonic cell population corresponding to the "seeds" giving rise to numerous lineages developing in their specific "soils" [57, 63].

Notably, neural crest cells migrate along developing intersomitic vessels at multiple locations along their travel [64]. This mechanism of neural crest cell migration is very similar to that of tumor cells during angiotropic EVMM. Furthermore, human metastatic melanoma cells, when transplanted into the chick embryonic neural crest, respond to signals from the surrounding microenvironment and migrate along stereotypical migratory routes of neural crest cells [65]. Thus it appears that "*melanoma revives an embryonic migration program to promote plasticity and invasion*" and reversion to an aberrant neural crest genes may promote melanoma plasticity and invasiveness [66]. Likewise, the human gliomas also appear to recapitulate the migratory behavior of glial progenitors, as oligodendrocyte precursors migrating along the vascular system in the developing nervous system [67].

Other stages of development have also demonstrated the same analogies, for example, gastrulation [68]. Importantly, cancer cells in an embryo form a "normal mosaic" organism [69].

8.6.1.4 Absence of a Hematogenous Pathway during Organogenesis

In non-hematogenous cancer migration, or PM/EVMM, the "seed" tumor cells migrate toward secondary sites ("soil"), while in non-angiogenic tumor growth, the "seed" tumor cells grow in primary and secondary organs ("soils"). Intravascular cell migration to reach their soil ("inefficient" hematogenous metastasis) [3]) doesn't exist during organogenesis. Indeed, the maternal circulation to the placenta is not present before the end of the first trimester [70, 71]. Therefore neural crest cell "*migrate as "seeds" during an extensive continuous migration through the embryo to reach their "soils" without entering the vascular lumen,*" i.e., using among other mechanisms PM/EVMM [2].

Embryonic signals activated in cancer are well described [53, 54, 68]. It will be important to consider the similarities between PM/EVMM in cancers with different embryonic origins and the respective pathways of their progenitors during embryonic development. For example, the replacement growth pattern in liver metastases has strong analogies with liver organogenesis [72] and embryonic cell competition [60]. Indeed, early endothelial cells and nascent vascular structures interact with newly derived hepatic endoderm, prior to liver bud emergence, and promote hepatic morphogenesis. Notably, hepatic sinusoids normally lack a continuous BM, although membrane-like structure with reticular collagen 4 and laminin is present [73]. In liver diseases, laminin deposits in BM membrane. Concomitantly, the sinusoidal endothelium loses its fenestrae and is transformed into vascular type endothelium [73]. Such changes in liver metastases may represent a reversion toward embryonic sinusoids, used as tracks for cancer cells [73].

Taken together, these data strongly suggest that angiotropism during PM/EVMM represents a reversion to embryonic migratory pathways.

8.7 Chronology and Latency

Cancer dormancy is a term used to explain the extensive periods of time in which patients remain asymptomatic prior to relapse [74]. Dormancy, which refers to the presence in metastatic niches of dormant CSC/circulating tumor cells after extravasation, is a loosely defined phenomenon that remains poorly understood [74]. In addition, it is possible that the dormancy period corresponds to angiotropic cancer cells migrating toward the metastatic sites [75]. PM/EVMM represents a mesenchymal-type slow movement process [76] and provides an alternative explanation for the delayed formation of metastases in many cancer cases. How long would it take for the tumor cells to reach secondary sites without intravascular dissemination? Migration modes and dynamics can be observed by intravital imaging of invasive human tumor cells [77]. For example, the in vivo velocity (in cm per year) of human melanoma varies from 22 to 525 cm/year, while human breast cancer cells migrate from 13 to 220 cm/year [77]. These speeds are theoretically compatible with the long-time intervals between the recognition of the primary cancer and the formation of metastases and could be applicable to PM/EVMM [2]. These long periods of time for PM/EVMM may explain in part the "dormancy" period observed with many tumors. Interestingly, neural crest cell migration rates are comparable to migrating tumor cells [78].

Taken together, these data may indicate that PM/EVMM of angiotropic melanoma cells represent a reversion to embryonic forms of migration.

8.8 Conclusion and Perspectives

Given (i) the well-recognized analogies between neural crest and melanoma cell migration, (ii) the similarities between neural crest and cancer cell migratory pathways, (iii) the absence of circulation during embryonic organogenesis, and therefore migration of stem cells via PM/extravascular migration, all in concert support the concept that these extravascular embryonic migratory mechanisms recur abnormally during melanoma metastasis. Therefore, embryonic factors linked to migration and cancer metastasis can become relevant targets for future therapies.

Importantly, for several decades, "reductionist" approaches to cancer chemotherapies including recent immunotherapy for solid tumors have resulted in outcome failure rates of 90% (±5) according to governmental agencies and industry" [79]. A more comprehensive understanding of the mechanism(s) of tumor spread is greatly needed. Since angiotropic EVMM is now recognized as a migratory mechanism of cancer spread [80, 81], this may profoundly influence and stimulate the development of more powerful and effective therapeutics.

Future Directions in the Field of Angiotropism, PM, and EVMM

- What is the fundamental biological and molecular mechanism of angiotropic extravascular tumor cell migration?
- What is the biological basis of resistance associated with angiotropic melanoma (cancer) cells?
- What governs the "switch" between extravascular tumor cell migration and tumor cell proliferation?
- What strategies may be used for the inhibition of vascular adhesion and angiotropic tumor migration?
- What other therapeutic initiatives may be developed for targeting PM and EVMM?

References

1. Chitty JL, Filipe EC, Lucas MC, Herrmann D, Cox TR, Timpson P. Recent advances in understanding the complexities of metastasis. F1000Res. 2018;7:F1000 Faculty Rev-1169. https://doi.org/10.12688/f1000research.15064.2.
2. Lugassy C, Kleinman HK, Vermeulen PB, Barnhill RL. Angiotropism, pericytic mimicry and extravascular migratory metastasis: an embryogenesis-derived program of tumor spread. Angiogenesis. 2020;23(1):27–41. https://doi.org/10.1007/s10456-019-09695-9. Epub 2019 Nov 12.
3. Talmadge JE, Fidler IJ. AACR centennial series: the biology of cancer metastasis: historical perspective. Cancer Res. 2010;70(14):5649–69. https://doi.org/10.1158/0008-5472.CAN-10-1040. Epub 2010 Jul 7.
4. Lugassy C, Eyden BP, Christensen L, Escande JP. Angio-tumoral complex in human malignant melanoma characterised by free laminin: ultrastructural and immunohistochemical observations. J Submicrosc Cytol Pathol. 1997;29(1):19–28.
5. Lugassy C, Dickersin GR, Christensen L, Karaoli T, LeCharpentier M, Escande JP, Barnhill RL. Ultrastructural and immunohistochemical studies of the periendothelial matrix in human melanoma: evidence for an amorphous matrix containing laminin. J Cutan Pathol. 1999;26(2):78–83. https://doi.org/10.1111/j.1600-0560.1999.tb01806.x.
6. Barnhill RL, Lugassy C. Angiotropic malignant melanoma and extravascular migratory metastasis: description of 36 cases with emphasis on a new mechanism of tumour spread. Pathology. 2004;36(5):485–90.
7. Lugassy C, Wadehra M, Li X, Corselli M, Akhavan D, Binder SW, Péault B, Cochran AJ, Mischel PS, Kleinman HK, Barnhill RL. Pilot study on "pericytic mimicry" and potential embryonic/stem cell properties of angiotropic melanoma cells interacting with the abluminal vascular surface. Cancer Microenviron. 2013;6(1):19–29.
8. Lugassy C, Péault B, Wadehra M, Kleinman HK, Barnhill RL. Could pericytic mimicry represent another type of melanoma cell plasticity with embryonic properties? Pigment Cell Melanoma Res. 2013;26(5):746–54.
9. Lugassy C, Barnhill RL, Christensen L. Melanoma and extravascular migratory metastasis. J Cutan Pathol. 2000;27(9):481.
10. Celià-Terrassa T, Kang Y. Metastatic niche functions and therapeutic opportunities. Nat Cell Biol. 2018;20(8):868–77.
11. Mitchell MJ, Denais C, Chan MF, Wang Z, Lammerding J, King MR. Lamin a/C deficiency reduces circulating tumor cell resistance to fluid shear stress. Am J Physiol Cell Physiol. 2015;309(11):C736–46.
12. Moose DL, Krog BL, Kim TH, Zhao L, Williams-Perez S, Burke G, et al. Cancer cells resist mechanical destruction in circulation via RhoA/actomyosin-dependent mechano-adaptation. Cell Rep. 2020;30(11):3864–3874.e6.
13. Friedl P, Alexander S. Cancer invasion and the microenvironment: plasticity and reciprocity. Cell. 2011;147(5):992–1009.
14. Gritsenko PG, Ilina O, Friedl P. Interstitial guidance of cancer invasion. J Pathol. 2012;226(2):185–99.
15. Beunk L, Brown K, Nagtegaal I, Friedl P, Wolf K. Cancer invasion into musculature: mechanics, molecules and implications. Semin Cell Dev Biol. 2019;93:36–45.
16. Barnhill RL, Ye M, Batistella A, Stern MH, Roman-Roman S, Dendale R, et al. The biological and prognostic significance of angiotropism in uveal melanoma. Lab Investig. 2017; https://doi.org/10.1038/labinvest.2017.16.
17. van Dam PJ, van der Stok EP, Teuwen LA, Van den Eynden GG, Illemann M, Frentzas S, et al. International consensus guidelines for scoring the histopathological growth patterns of liver metastasis. Br J Cancer. 2017;117(10):1427–41.
18. Barnhill R, Vermeulen P, Daelemans S, van Dam PJ, Roman-Roman S, Servois V, et al. Replacement and desmoplastic histopathological growth patterns: a pilot study of prediction of outcome in patients with uveal melanoma liver metastases. J Pathol Clin Res. 2018;4(4):227–40.
19. Barnhill R, van Dam PJ, Vermeulen P, Champenois G, Nicolas A, Rawson RV, et al. Replacement and desmoplastic histopathological growth patterns in cutaneous melanoma liver metastases: frequency, characteristics, and robust prognostic value. J Pathol Clin Res. 2020;6(3):195–206.
20. Kordes C, Sawitza I, Götze S, Herebian D, Häussinger D. Stellate cells are mesenchymal stem cells. Eur J Med Res. 2014;19(Suppl 1):S6.
21. Häussinger D, Kordes C. Space of Disse: a stem cell niche in the liver. Biol Chem. 2019;401(1):81–95.
22. Lugassy C, Scolyer R, Long G, Menzies A, Mischel P. Barnhill RL PDGFBR expression in anti-BRAF resistant melanoma: are angiotropic melanoma cells a source of BRAF resistance and disease progression? J Cutan Pathol. 2014;41:159–60.
23. Lugassy C, Kleinman HK, Fernandez PM, Patierno SR, Webber MM, Ghanem G, et al. Human melanoma cell migration along capillary-like structures in vitro: a new dynamic model for studying extravascular migratory metastasis. J Invest Dermatol. 2002;119(3):703–4.
24. Lugassy C, Kleinman HK, Engbring JA, Welch DR, Harms JF, Rufner R, et al. Pericyte-like location of GFP-tagged melanoma cells: ex vivo and in vivo studies of extravascular migratory metastasis. Am J Pathol. 2004;164(4):1191–8.
25. Bald T, Quast T, Landsberg J, Rogava M, Glodde N, Lopez-Ramos D, et al. Ultraviolet-radiation-induced inflammation promotes angiotropism and metastasis in melanoma. Nature. 2014;507(7490):109–13.
26. Lee G, Hall RR 3rd, Ahmed AU. Cancer stem cells: cellular plasticity, niche, and its clinical relevance. J Stem Cell Res Ther. 2016;6(10):pii: 363.
27. Lugassy C, Kleinman HK, Vernon SE, Welch DR, Barnhill RL. C16 laminin peptide increases angiotropic extravascular migration of human melanoma cells in a shell-less chick chorioallantoic membrane assay. Br J Dermatol. 2007;157(4):780–2.
28. Bentolila LA, Prakash R, Mihic-Probst D, Wadehra M, Kleinman HK, Carmichael TS, et al. Imaging of angiotropism/vascular cooption in a murine model of brain melanoma: implications for melanoma progression along extravascular pathways. Sci Rep. 2016;6:23,834.
29. Fornabaio G, Barnhill RL, Lugassy C, Bentolila LA, Cassoux N, Roman-Roman S, et al. Angiotropism and extravascular migratory metastasis in cutaneous and uveal melanoma progression in a zebrafish model. Sci Rep. 2018;8(1):10,448.
30. Cheng L, Huang Z, Zhou W, Wu Q, Donnola S, Liu JK, et al. Glioblastoma stem cells generate vascular pericytes to support vessel function and tumor growth. Cell. 2013;153(1):139–52.
31. Er EE, Valiente M, Ganesh K, Zou Y, Agrawal S, Hu J, et al. Pericyte-like spreading by disseminated cancer cells activates YAP and MRTF for metastatic colonization. Nat Cell Biol. 2018;20(8):966–78.
32. Yao H, Price TT, Cantelli G, Ngo B, Warner MJ, Olivere L, et al. Leukaemia hijacks a neural mechanism to invade the central nervous system. Nature. 2018;560(7716):55–60.
33. Lugassy C, Vernon SE, Warner JW, Le CQ, Manyak M, Patierno SR, et al. Angiotropism of human prostate cancer cells: implications for extravascular migratory metastasis. BJU Int. 2005;95(7):1099–103.
34. Lai CJ, Lin CY, Liao WY, Hour TC, Wang HD, Chuu CP. CD44 promotes migration and invasion of docetaxel-resistant prostate

cancer cells likely via induction of hippo-Yap signaling. Cell. 2019;8(4)
35. Ober EA, Lemaigre FP. Development of the liver: insights into organ and tissue morphogenesis. J Hepatol. 2018;68(5):1049–62.
36. Yap L, Tay HG, Nguyen MTX, Tjin MS, Tryggvason K. Laminins in cellular differentiation. Trends Cell Biol. 2019;29(12):987–1000.
37. Engbring JA, Kleinman HK. The basement membrane matrix in malignancy. J Pathol. 2003;200(4):465–70.
38. Qin Y, Rodin S, Simonson OE. Hollande F laminins and cancer stem cells: partners in crime? Semin Cancer Biol. 2017;45:3–12.
39. Rousselle P, Scoazec JY. Laminin 332 in cancer: when the extracellular matrix turns signals from cell anchorage to cell movement. Semin Cancer Biol. 2020;62:149–65.
40. Sekiguchi R, Yamada KM. Basement membranes in development and disease. Curr Top Dev Biol. 2018;130:143–91.
41. Giannelli G, Falk-Marzillier J, Schiraldi O, Stetler-Stevenson WG. Quaranta V induction of cell migration by matrix metalloprotease-2 cleavage of laminin-5. Science. 1997;277(5323):225–8.
42. Lugassy C, Torres-Muñoz JE, Kleinman HK, Ghanem G, Vernon SE, Barnhill RL. Over-expression of malignancy-associated laminins and laminin receptors by angiotropic human melanoma cells in a chick chorioallantoic membrane model. J Cutan Pathol. 2009;36(12):1237–43.
43. Kuratomi Y, Nomizu M, Tanaka K, Ponce ML, Komiyama S, Kleinman HK, et al. Laminin gamma 1 chain peptide, C-16 (KAFDITYVRLKF), promotes migration, MMP-9 secretion, and pulmonary metastasis of B16-F10 mouse melanoma cells. Br J Cancer. 2002;86(7):1169–73.
44. Oikawa Y, Hansson J, Sasaki T, Rousselle P, Domogatskaya A, Rodin S, et al. Melanoma cells produce multiple laminin isoforms and strongly migrate on α5 laminin(s) via several integrin receptors. Exp Cell Res. 2011;317(8):1119–33.
45. Carpenter PM, Sivadas P, Hua SS, Xiao C, Gutierrez AB, Ngo T, et al. Migration of breast cancer cell lines in response to pulmonary laminin 332. Cancer Med. 2017;6(1):220–34.
46. Tsubota Y, Ogawa T, Oyanagi J, Nagashima Y, Miyazaki K. Expression of laminin gamma2 chain monomer enhances invasive growth of human carcinoma cells in vivo. Int J Cancer. 2010;127(9):2031–41.
47. Teng Y, Wang Z, Ma L, Zhang L, Guo Y, Gu M, et al. Prognostic significance of circulating laminin gamma2 for early-stage non-small-cell lung cancer. Onco Targets Ther. 2016;9:4151–62.
48. Sun T, Patil R, Galstyan A, Klymyshyn D, Ding H, Chesnokova A, et al. Blockade of a Laminin-411-notch Axis with CRISPR/Cas9 or a nanobioconjugate inhibits glioblastoma growth through tumor-microenvironment cross-talk. Cancer Res. 2019;79(6):1239–51.
49. Kikkawa Y, Ogawa T, Sudo R, Yamada Y, Katagiri F, Hozumi K, et al. The lutheran/basal cell adhesion molecule promotes tumor cell migration by modulating integrin-mediated cell attachment to laminin-511 protein. J Biol Chem. 2013;288(43):30990–1001.
50. Hutchins EJ, Bronner ME. Draxin alters laminin organization during basement membrane remodeling to control cranial neural crest EMT. Dev Biol. 2019;446(2):151–8.
51. Ganesh K, Basnet H, Kaygusuz Y, Laughney AM, He L, Sharma R, et al. L1CAM defines the regenerative origin of metastasis-initiating cells in colorectal cancer. Nat Cancer. 2020;1(1):28–45.
52. Virchow RLK. Cellular pathology, 1859 special ed. London: John Churchill; 1978. p. 204–7.
53. Cofre J, Abdelhay E. Cancer is to embryology as mutation is to genetics: hypothesis of the cancer as embryological phenomenon. ScientificWorldJournal. 2017;2017:3578090.
54. Aiello NM, Stanger BZ. Echoes of the embryo: using the developmental biology toolkit to study cancer. Dis Model Mech. 2016;9(2):105–14.
55. Campbell K. Contribution of epithelial-mesenchymal transitions to organogenesis and cancer metastasis. Curr Opin Cell Biol. 2018;55:30–5.
56. Saitoh M. Involvement of partial EMT in cancer progression. J Biochem. 2018;164(4):257–64.
57. Fane ME, Chhabra Y, Smith AG, Sturm RA. BRN2, a POUerful driver of melanoma phenotype switching and metastasis. Pigment Cell Melanoma Res. 2019;32(1):9–24.
58. Ribatti D, Mangialardi G, Vacca A. Stephen Paget and the 'seed and soil' theory of metastatic dissemination. Clin Exp Med. 2006;6(4):145–9.
59. Wu Z, Guan KL. Hippo signaling in embryogenesis and development. Trends Biochem Sci. 2021;46(1):51–63.
60. Di Gregorio A, Bowling S, Rodriguez TA. Cell competition and its role in the regulation of cell fitness from development to cancer. Dev Cell. 2016;38(6):621–34.
61. Paget S. The distribution of secondary growths in cancer of the breast. Lancet. 1889;133(3421):571–3.
62. Lugassy C, Barnhill RL. Concepts of cancer and metastasis: historical perspective. from biology to target. Cancer Res. 2020;80(19 Suppl):IA06.
63. Dupin E, Calloni GW, Coelho-Aguiar JM, Le Douarin NM. The issue of the multipotency of the neural crest cells. Dev Biol. 2018;444(Suppl 1):S47–59.
64. George L, Dunkel H, Hunnicutt BJ, Filla M, Little C, Lansford R, et al. In vivo time-lapse imaging reveals extensive neural crest and endothelial cell interactions during neural crest migration and formation of the dorsal root andsympathetic ganglia. Dev Biol. 2016;413(1):70–85.
65. Kulesa PM, Kasemeier-Kulesa JC, Teddy JM, Margaryan NV, Seftor EA, Seftor RE, et al. Reprogramming metastatic melanoma cells to assume a neural crest cell-like phenotype in an embryonic microenvironment. Proc Natl Acad Sci U S A. 2006;103(10):3752–7.
66. Bailey CM, Morrison JA, Kulesa PM. Melanoma revives an embryonic migration program to promote plasticity and invasion. Pigment Cell Melanoma Res. 2012;25(5):573–83.
67. Tsai HH, Niu J, Munji R, Davalos D, Chang J, Zhang H, et al. Oligodendrocyte precursors migrate along vasculature in the developing nervous system. Science. 2016;351(6271):379–84.
68. Cofre J, Saalfeld K, Abdelhay E. Cancer as an embryological phenomenon and its developmental pathways: a hypothesis regarding the contribution of the noncanonical Wnt pathway. ScientificWorldJournal. 2019;2019:4714781.
69. Mintz B, Illmensee K. Normal genetically mosaic mice produced from malignant teratocarcinoma cells. Proc Natl Acad Sci U S A. 1975;72(9):3585–9.
70. Burton GJ, Watson AL, Hempstock J, Skepper JN, Jauniaux E. Uterine glands provide histiotrophic nutrition for the human fetus during the first trimester of pregnancy. J Clin Endocrinol Metab. 2002;87(6):2954–9.
71. Burton GJ, Jauniaux E. Development of the human placenta and fetal heart: synergic or independent? Front Physiol. 2018;9:373.
72. Matsumoto K, Yoshitomi H, Rossant J, Zaret KS. Liver organogenesis promoted by endothelial cells prior to vascular function. Science. 2001;294(5542):559–63.
73. Mak KM, Mei R. Basement membrane type IV collagen and laminin: an overview of their biology and value as fibrosis biomarkers of liver disease. Anat Rec (Hoboken). 2017;300(8):1371–90.
74. Yeh AC, Ramaswamy S. Mechanisms of cancer cell dormancy—another hallmark of cancer? Cancer Res. 2015;75(23):5014–22. https://doi.org/10.1158/0008-5472.CAN-15-1370. Epub 2015 Sep 9.
75. Kienast Y, von Baumgarten L, Fuhrmann M, Klinkert WE, Goldbrunner R, Herms J, Winkler F. Real-time imaging reveals the single steps of brain metastasis formation. Nat Med.

2010;16(1):116–22. https://doi.org/10.1038/nm.2072. Epub 2009 Dec 20.

76. Friedl P, Wolf K. Proteolytic and non-proteolytic migration of tumour cells and leucocytes. Biochem Soc Symp. 2003;70:277–85. https://doi.org/10.1042/bss0700277.
77. Clark AG, Vignjevic DM. Modes of cancer cell invasion and the role of the microenvironment. Curr Opin Cell Biol. 2015;36:13–22. https://doi.org/10.1016/j.ceb.2015.06.004. Epub 2015 Jul 14.
78. Li Y, Vieceli FM, Gonzalez WG, Li A, Tang W, Lois C, Bronner ME. In vivo quantitative imaging provides insights into trunk neural crest migration. Cell Rep. 2019;26(6):1489–1500.e3. https://doi.org/10.1016/j.celrep.2019.01.039.
79. Maeda H, Khatami M. Analyses of repeated failures in cancer therapy for solid tumors: poor tumor-selective drug delivery, low therapeutic efficacy and unsustainable costs. Clin Transl Med. 2018;7(1):11. https://doi.org/10.1186/s40169-018-0185-6.
80. Bald T, Quast T, Landsberg J, Rogava M, Glodde N, Lopez-Ramos D, Kohlmeyer J, Riesenberg S, van den Boorn-Konijnenberg D, Hömig-Hölzel C, Reuten R, Schadow B, Weighardt H, Wenzel D, Helfrich I, Schadendorf D, Bloch W, Bianchi ME, Lugassy C, Barnhill RL, Koch M, Fleischmann BK, Förster I, Kastenmüller W, Kolanus W, Hölzel M, Gaffal E, Tüting T. Ultraviolet-radiation-induced inflammation promotes angiotropism and metastasis in melanoma. Nature. 2014;507(7490):109–13.
81. Massagué J, Ganesh K. Metastasis-initiating cells and ecosystems. Cancer Discov. 2021;11(4):971–94.

Part III

Dormant Cancer Cells and Their Reactivation in the Host Cancer Microenvironment

Maria Soledad Sosa

Metastasis and its effects on the patients' homeostatic functions are the main causes of cancer-related deaths. Clinical and experimental evidence supports that disseminated cancer cells (DCCs) are the source of metastasis in many cancer types [1–5]. The data also support that DCCs remain in a reversible dormant state for prolonged periods of time resisting therapies until, eventually, they reactivate and form overt metastasis.

The program that regulates the cell life cycle of single DCCs is named cellular dormancy. This program involves the ability of DCCs to persist most of the time in a cell cycle G0/G1 arrest which is initiated and/or maintained by the interactions of DCCs with the new niches at secondary organs [6–8]. In addition, several studies suggest that the cellular dormancy programs may also be initiated within the primary tumor itself [9, 10] and then maintained or augmented by secondary organ niche cues. Once at secondary organs, through epigenetic, metabolic, and adaptive stress responses DCCs are reprogramed to evade the immune system and survive the new niche-derived stress signals as well as therapies. Because dormancy is reversible the inherent outcome of this life cycle is the emergence from the G0/G1 arrest and the formation of metastatic outgrowth. Two excellent chapters in this book touch on important aspects on how the pressure of treatment impacts dormancy and reactivation phenotypes and the niches that control DCC dormancy and underlying mechanisms.

The cellular dormancy program can be activated spontaneously in several types of tumors, even in the absence of a detectable primary tumor [11–14] and before any therapeutic interventions. However, experimental and clinical data [15–24] support that therapy-induced dormancy might occur in response to chemotherapy or targeted therapies used in the adjuvant setting, albeit further investigation is required. In this regard, Garcia et al.'s review, among other features, adapted metabolic reprogramming and pro-survival pathways acquired by residual dormant cancer cells after therapies. This chapter provides evidence that many pathways are differentially regulated in residual dormant cancer cells in response to therapies to allow for the survival and further expansion of these tumor cells into metastasis.

As mentioned before, the dormant DCC life cycle has different stages including adaptation to a new microenvironment by interactions with the new extracellular matrix (ECM) [1, 25–28]. Along these lines, Huber et al. review the cellular fates of dormant cancer cell in the presence of fibronectin, collagens, growth factors associated to ECM, proteolysis, and stiffness of ECM. Some of these ECM-derived signals constitute an appropriate niche for dormancy and survival of DCCs whereas others are important for the reactivation of dormant DCCs into metastasis.

Expanding our knowledge on how DCCs survive in a dormant state in their new microenvironments before and after treatments, how therapies modify DCC niches including ECM, and what are the signals that trigger their reactivation programs may likely improve the prediction of patient survival rates and increase the opportunities to design therapies that will extend the remission phase.

References

1. Ghajar CM, Peinado H, Mori H, Matei IR, Evason KJ, Brazier H, et al. The perivascular niche regulates breast tumour dormancy. Nat Cell Biol. 2013;15:807–17. https://doi.org/10.1038/ncb2767.
2. Malladi S, Macalinao DG, Jin X, He L, Basnet H, Zou Y, et al. Metastatic latency and immune evasion through autocrine inhibition of WNT. Cell. 2016;165:45–60. https://doi.org/10.1016/j.cell.2016.02.025.
3. Gao H, Chakraborty G, Lee-Lim AP, Mo Q, Decker M, Vonica A, et al. The BMP inhibitor coco reactivates breast cancer cells at lung metastatic sites. Cell. 2012;150:764–79. https://doi.org/10.1016/j.cell.2012.06.035.
4. Albrengues J, Shields MA, Ng D, Park CG, Ambrico A, Poindexter ME, et al. Neutrophil extracellular traps produced during inflammation awaken dormant cancer cells in mice. Science. 2018;361 https://doi.org/10.1126/science.aao4227.
5. Sosa MS, Parikh F, Maia AG, Estrada Y, Bosch A, Bragado P, et al. NR2F1 controls tumour cell dormancy via SOX9- and RARbeta-driven quiescence programmes. Nat Commun. 2015;6:6170. https://doi.org/10.1038/ncomms7170.
6. Phan TG, Croucher PI. The dormant cancer cell life cycle. Nat Rev Cancer. 2020;20:398–411. https://doi.org/10.1038/s41568-020-0263-0.
7. Aguirre-Ghiso JA. Models, mechanisms and clinical evidence for cancer dormancy. Nat Rev Cancer. 2007;7:834–46.
8. Sosa MS, Bragado P, Aguirre-Ghiso JA. Mechanisms of disseminated cancer cell dormancy: an awakening field. Nat Rev Cancer. 2014;14:611–22. https://doi.org/10.1038/nrc3793.
9. Fluegen G, Avivar-Valderas A, Wang Y, Padgen MR, Williams JK, Nobre AR, et al. Phenotypic heterogeneity of disseminated tumour cells is preset by primary tumour hypoxic microenvironments. Nat Cell Biol. 2017;19:120–32. https://doi.org/10.1038/ncb3465.
10. Kim RS, Avivar-Valderas A, Estrada Y, Bragado P, Sosa MS, Aguirre-Ghiso JA, et al. Dormancy signatures and metastasis in estrogen receptor positive and negative breast cancer. PLoS One. 2012;7:e35569. https://doi.org/10.1371/journal.pone.0035569.
11. Harper KL, Sosa MS, Entenberg D, Hosseini H, Cheung JF, Nobre R, et al. Mechanism of early dissemination and metastasis in Her2(+) mammary cancer. Nature. 2016; https://doi.org/10.1038/nature20609.
12. Hosseini H, Obradovic MM, Hoffmann M, Harper KL, Sosa MS, Werner-Klein M, et al. Early dissemination seeds metastasis in breast cancer. Nature. 2016; https://doi.org/10.1038/nature20785.
13. Linde N, Casanova-Acebes M, Sosa MS, Mortha A, Rahman A, Farias E, et al. Macrophages orchestrate breast cancer early dissemination and metastasis. Nat Commun. 2018;9:21. https://doi.org/10.1038/s41467-017-02481-5.
14. Wang J, Wang G, Cheng D, Huang S, Chang A, Tan X, et al. Her2 promotes early dissemination of breast cancer by suppressing the p38-MK2-Hsp27 pathway that is targetable by Wip1 inhibition. Oncogene. 2020;39:6313–26. https://doi.org/10.1038/s41388-020-01437-2.
15. Quesnel B. Tumor dormancy: long-term survival in a hostile environment. Adv Exp Med Biol. 2013;734:181–200. https://doi.org/10.1007/978-1-4614-1445-2_9.
16. Sharma SV, Lee DY, Li B, Quinlan MP, Takahashi F, Maheswaran S, et al. A chromatin-mediated reversible drug-tolerant state in cancer cell subpopulations. Cell. 2010;141:69–80. https://doi.org/10.1016/j.cell.2010.02.027.
17. Schewe DM, Aguirre-Ghiso JA. Inhibition of eIF2alpha dephosphorylation maximizes bortezomib efficiency and eliminates quiescent multiple myeloma cells surviving proteasome inhibitor therapy. Cancer Res. 2009;69:1545–52.
18. Fox DB, Garcia NMG, McKinney BJ, Lupo R, Noteware LC, Newcomb R, et al. NRF2 activation promotes the recurrence of dormant tumour cells through regulation of redox and nucleotide metabolism. Nat Metab. 2020;2:318–34. https://doi.org/10.1038/s42255-020-0191-z.
19. El Touny LH, Vieira A, Mendoza A, Khanna C, Hoenerhoff MJ, Green JE. Combined SFK/MEK inhibition prevents metastatic outgrowth of dormant tumor cells. J Clin Invest. 2014;124:156–68. https://doi.org/10.1172/JCI70259.

20. Saleh T, Bloukh S, Carpenter VJ, Alwohoush E, Bakeer J, Darwish S, et al. Therapy-induced senescence: an "old" friend becomes the enemy. Cancers (Basel). 2020;12 https://doi.org/10.3390/cancers12040822.
21. Alvarez JV, Pan TC, Ruth J, Feng Y, Zhou A, Pant D, et al. Par-4 downregulation promotes breast cancer recurrence by preventing multinucleation following targeted therapy. Cancer Cell. 2013;24:30–44. https://doi.org/10.1016/j.ccr.2013.05.007.
22. Kurppa KJ, Liu Y, To C, Zhang T, Fan M, Vajdi A, et al. Treatment-induced tumor dormancy through YAP-mediated transcriptional reprogramming of the apoptotic pathway. Cancer Cell. 2020;37:104–122 e112. https://doi.org/10.1016/j.ccell.2019.12.006.
23. Thompson VC, Day TK, Bianco-Miotto T, Selth LA, Han G, Thomas M, et al. A gene signature identified using a mouse model of androgen receptor-dependent prostate cancer predicts biochemical relapse in human disease. Int J Cancer. 2012;131:662–72. https://doi.org/10.1002/ijc.26414.
24. Demicheli R, Desmedt C, Retsky M, Sotiriou C, Piccart M, Biganzoli E. Late effects of adjuvant chemotherapy adumbrate dormancy complexity in breast cancer. Breast. 2020;52:64–70. https://doi.org/10.1016/j.breast.2020.05.002.
25. Aguirre-Ghiso JA, Liu D, Mignatti A, Kovalski K, Ossowski L. Urokinase receptor and fibronectin regulate the ERK(MAPK) to p38(MAPK) activity ratios that determine carcinoma cell proliferation or dormancy in vivo. Mol Biol Cell. 2001;12:863–79.
26. Barkan D, El Touny LH, Michalowski AM, Smith JA, Chu I, Davis AS, et al. Metastatic growth from dormant cells induced by a col-I-enriched fibrotic environment. Cancer Res. 2010;70:5706–16.
27. Boyerinas B, Zafrir M, Yesilkanal AE, Price TT, Hyjek EM, Sipkins DA. Adhesion to osteopontin in the bone marrow niche regulates lymphoblastic leukemia cell dormancy. Blood. 2013;121:4821–31. https://doi.org/10.1182/blood-2012-12-475483.
28. Bragado P, Estrada Y, Parikh F, Krause S, Capobianco C, Farina HG, et al. TGF-beta2 dictates disseminated tumour cell fate in target organs through TGF-beta-RIII and p38alpha/beta signalling. Nat Cell Biol. 2013;15:1351–61. https://doi.org/10.1038/ncb2861.

Tumor Dormancy and Relapse Regulated by the Extracellular Matrix

9

Hyuna Kim, Rebecca E. Huber, Rita Das Mahapatra, Ning-Hsuan Tseng, and Shelly R. Peyton

Abstract

Metastases account for 90% of cancer-related deaths, and many patients experience relapse in distant organs, even after successful initial therapy. This is due to disseminated cells that remain dormant for many years until they regrow in a distant site. There is increasing evidence showing that extracellular matrix (ECM) plays multiple roles in promoting and maintaining dormancy, cell proliferation, and tumor outgrowth. For example, collagen can induce a transition from dormancy to outgrowth in breast cancer, fibronectin facilitates dormant cell survival, and other ECM proteins including TGFB-1 and periostin can accelerate dormant tumor outgrowth. This chapter will give an overview of such studies to elucidate the roles of these ECM proteins, other factors found in ECM, and the mechanical properties of ECM in regulating dormancy and reactivation of disseminated tumor cells.

Learning Objectives

- To understand the known links between extracellular matrix and dormancy/relapse of epithelial cancers.
- Understand the connection between collagen-mediated matrix stiffening and cancer cell dormancy.
- Understand the dual role fibronectin plays in promoting both dormancy and reactivation.

H. Kim · R. E. Huber · R. D. Mahapatra · N.-H. Tseng
S. R. Peyton (✉)
University of Massachusetts Amherst, Amherst, MA, USA
e-mail: hyunakim@umass.edu; rhuber@umass.edu; rdasmahapatr@umass.edu; ntseng@umass.edu; speyton@umass.edu

9.1 Introduction

Even after apparently successful therapy, disseminated tumor cells can remain dormant for many years before outgrowth. Breast cancer is a prominent example, where the presence of disseminated, quiescent tumor cells in the bone marrow is a marker of poor prognosis [1, 2], indicating that dormant disseminated tumor cells contribute to and predict poor outcomes. Dormant metastases are kept quiescent either because individual cells are not proliferative (cellular dormancy) or because an equilibrium is established between proliferation and death across the bulk of the tumor mass (tumor dormancy) [3]. These dormant cancer cells are notoriously difficult to treat, largely because they are not actively cycling, and so cannot be killed by the traditional chemotherapies typically given to patients with metastatic disease. Much of the literature on this topic has focused on the transition of indolent micrometastases to overt metastases following alterations to the tumor microenvironment [4–6]. Immunohistochemistry of fixed clinical and in vivo specimens can provide insight into localization of dormant or proliferating disseminated tumor cells (DTCs) within the extracellular matrix (ECM) via Ki-67 (a commonly used proliferation marker) [7, 8], and studies have begun to identify the factors from the ECM that coincide with tumor outgrowth. Intravital imaging can observe cell dynamics from pre-implanted locations [9–11], biomaterials and chip technologies have started to address these questions in vitro [12].

These studies and others provide overwhelming evidence that cell–ECM interactions regulate metastasis in vitro, in vivo, and in humans [13, 14], and of the role of the ECM in maintaining cell or tumor dormancy. There have been hints of certain extracellular factors regulating dormancy, and the "dormancy program" includes signals downstream of integrin engagement to ECM proteins (for reviews, see [15, 16]). Genetic screens have identified factors that permit dormant cells to outgrow [17–19]. In related studies, cytokines [20], matrix proteins [7], and immune evasion [8] can promote cel-

S. P. Leong et al. (eds.), *Cancer Metastasis Through the Lymphovascular System*,
https://doi.org/10.1007/978-3-030-93084-4_9

lular quiescence, and current work is clarifying how disseminated cells eventually overcome these dormancy-promoting signals and transition to overt metastatic outgrowth.

Although not yet directly linked to dormancy, altered expression of other matrix proteins has been shown to facilitate metastatic survival and outgrowth. As three examples, fibronectin is produced by the stroma to support survival of incoming disseminated tumor cells [21], tumor cells trigger stromal periostin secretion to metastasize to the lung [22], and breast cancer cell-derived tenascin C promotes lung metastasis [23]. Understanding this shift from quiescence to proliferation, alongside concurrent changes in the extracellular microenvironment, could provide oncologists with new avenues for therapy. This chapter will give an overview of the seminal contributions of researchers that have begun to uncover the role of these ECM proteins and other ECM-associated factors in regulating dormancy and relapse of disseminated tumor cells, as well as introduce emerging technologies in the field.

▶ **Extracellular matrix (ECM):** The scaffold of sugars, proteins, and proteoglycans that surrounds cells in tissues. The ECM provides structure, is a depot for growth factors and chemokines, provides cell binding sites, and imparts biophysical signaling to the cells within.

▶ **Integrins:** The transmembrane heterodimeric proteins that allow cells to bind to specific ECM proteins at different tissue sites and initiate biochemical and biophysical signaling within cells upon ECM binding.

▶ **Cellular dormancy:** When a single cancer cell is quiescent for several years.

▶ **Tumor dormancy:** When the proliferation and death rates of a tumor cell population roughly equal, such that a tumor is neither growing nor shrinking.

▶ **Relapse:** When a tumor initiates proliferation after a period of dormancy, resulting in a detectable, growing tumor mass.

9.2 Collagens

Collagens are the predominant structural proteins found in connective tissues. It has a triple helix structure, derived from its primary structure: every third amino acid is a glycine residue [24]. There are multiple isoforms of collagen, and many of them are known to influence cancer cell behaviors, such as invasion, proliferation, and metastasis [25]. Several studies have found that high densities of collagen I in the epidermis [26] and collagen IV in the pancreas [27] stimulate cancer cell proliferation and migration. Deposition of collagen I in breast tissue induces a transition from dormancy to metastatic outgrowth [4]. In this study, they used two mouse mammary cancer cell lines: D2.0R (dormant) and D2A1 (metastatic). These cell lines were derived in the Miller lab (D2A1: [28] and D2.0R: [29]). These are commonly used cell lines for dormancy studies, because both were derived from tumors arising from implants of the same D2 hyperplastic alveolar nodule line. D2A1 cells form metastatic lesions after a few weeks in mice, but D2.0R cells remain dormant for months with rare formation of metastatic lesions. In a three-dimensional culture system, Type I Collagen induces proliferation in otherwise quiescent D2.0R cells via integrin β_1 activation of Src and focal adhesion kinase (FAK), leading to extracellular signal-regulated kinase (ERK)-dependent myosin light chain (MLC) phosphorylation by myosin light chain kinase and actin stress fiber formation (Fig. 9.1a). They also showed that the induction of fibrosis in lung, associated with deposition of Type I Collagen in the in vivo metastatic microenvironment, induces dormant D2.0R cells to form proliferative metastatic lesions through integrin β_1.

Although these and other studies demonstrate that collagen deposition supports cancer cell outgrowth and proliferation, enriched collagen does not always support cancer cell outgrowth. For example, in melanoma, an environment rich in fibrillar collagen I induces quiescence of melanoma cells [30]. Multiple other studies show that tumor cell interaction with fibrillar collagen suppresses cell proliferation and keeps cells dormant [31, 32]. In this study, disorganized (from monomer) collagen gels and organized fibrillar collagen gels were used to study human melanoma cell growth. When cells were on disorganized collagen surfaces, cells proliferated. However, organized collagen gels led to cell growth arrest in the G0/G1 phase of the cell cycle. When they treated the cells on organized collagen gels with the tyrosine kinase inhibitor genistein, discoidin domain receptor 2 (DDR2) was downregulated and stimulated cell proliferation. Their study provides evidence that fibrillar collagen induces G0/G1 cell cycle arrest and restricts proliferation of tumor cells by a direct receptor-mediated mechanism that involves DDR2 (Fig. 9.1b).

9.3 Fibronectin

Fibronectin (FN) is a glycoprotein found in the ECM of all tissues [33]. It forms both linear and branched structures and exists as individual fibrils, complex matrices, or clustered bundles. Converting cell-secreted FN molecules into more complex structures requires both FN–cell and FN–FN interactions. Secreted FN molecules first bind to each other, dimerizing via disulfide bonds on the C terminus end of each molecule. Cells bind to FN via $\alpha_V\beta_3$ and $\alpha_5\beta_1$ integrins on cell surfaces via its RGD and synergy domains, and both of these

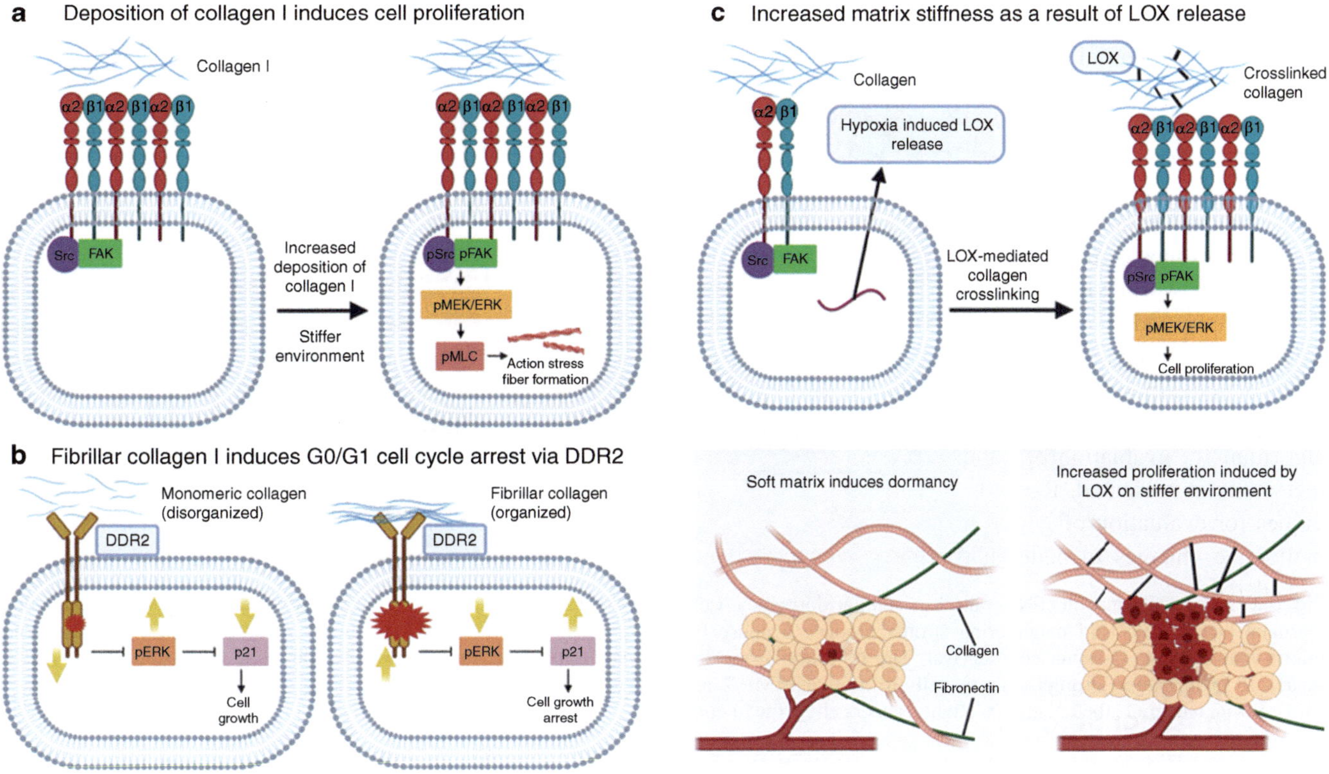

Fig. 9.1 **Collagen in ECM can alter cell proliferation or cell cycle arrest.** (**a**) Increased deposition of collagen I in ECM of breast tissue, the collagen allows cells to metastatically outgrow, escaping from a dormant state. Collagen I induces cell proliferation via integrin β_1 activation, which activates Src and focal adhesion kinase (FAK) signaling pathway in its downstream, leading to extracellular signal-regulated kinase (ERK)-dependent myosin light chain (MLC) phosphorylation. MLC phosphorylation induces actin stress fiber formation, which then promotes cell stiffening and proliferation of preinvasive breast cancer cells. (**b**). When melanoma cells are in fibrillar collagen-rich environment, it activates discoidin domain receptor 2 (DDR2). When DDR2 is activated, it downregulates ERK and thus p21, which triggers cell cycle arrest, is increased. When cells are in disorganized monomeric collagen, DDR2 is downregulated and thus cells can grow. Also, when cells are in fibrillar collagen but are treated with genistein, which inhibits DDR2, cells enter the cell cycle even in the presence of fibrillar collagen. Thus, fibrillar collagen in ECM environment induces cell cycle arrest of melanoma cells by DDR2-mediated mechanism. (**c**) ECM stiffness promotes outgrowth from dormant state. On the left, $\alpha_2\beta_1$ integrin interacts to Collagen type I. Hypoxia induces expression of LOX enzymes. On the right, LOX mediates collagen cross-linking, resulting in increase in ECM stiffness. The increase in ECM stiffness allows dormant cells to proliferate via β_1 activation which activates Src and FAK signaling pathway. Created with BioRender.com

domains must be bound for fibril formation [34]. Once bound, integrin-mediated signaling initiates cell contraction via RhoA-ROCK, exposing cryptic FN-FN binding sites. Integrin clustering also brings FN molecules closer together to promote these FN–FN interactions, which ultimately results in a stiffer, insoluble, fibrillar matrix [35].

Apart from integrins, FN also binds to urokinase plasminogen activator receptor (uPAR) which plays a major role in regulating the mitogen-activated protein kinase (MAPK) pathways. Several studies have shown that a careful balance of MAPK signaling through ERK and $p38^{MAPK}$ (p38) regulates the shift between dormancy and proliferation in cancer cells [35–37]. Studies have shown that persistent activation of ERK causes tumor proliferation, and activation of p38 induces dormancy [35, 37]. FN–uPAR interactions are vital to the orchestration of this balance. In head and neck cancer (HEP3 cells), authors found cells bind to FN via $\alpha_5\beta_1$ and uPAR (a non-integrin vitronectin cell adhesion receptor) simultaneously, which drives ERK activity [35]. With this same cell line, they found that when uPAR is upregulated, it interacts with $\alpha_5\beta_1$ to produce more FN fibrils to amplify the FN signal, increasing ERK activation and suppressing p38, further shifting the balance toward proliferation [35] (Fig. 9.2). The authors then expanded that study to include multiple cancer cell lines, representing several cancer types, including breast, prostate, melanoma, and fibrosarcoma. They found that with the exception of triple negative breast cancer cell line MDA-MB-231, the uPAR/integrin/FN complex regulates ERK activity and the formation of this complex is a "perfect" predictor of proliferation vs. dormancy [37].

In addition to binding preexisting FN in the ECM, stromal and cancer cells can also produce FN, impacting dormancy and relapse. In one study using an in vitro culture system to study an array of cancer cell types, authors found that cells able to produce a FN matrix had better overall survival dur-

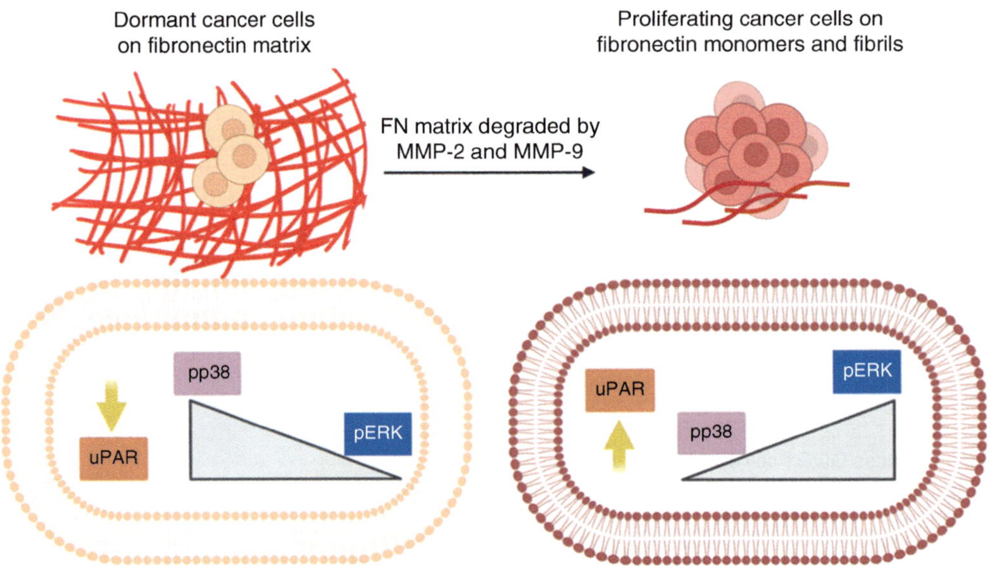

Fig. 9.2 Known fibronectin (FN) control of cellular dormancy. Left: Dormant cancer cells and neighboring stroma produce a dense FN matrix that supports dormant cell survival. The transition from dormancy to proliferation prompts cancer cells to secrete MMP-2 and MMP-9 which degrade the FN matrix. Right: cancer cells grow in number and secrete FN fibrils as they proliferate. Illustrations of cell below highlight major molecular pathways resulting from cell–FN interactions implicated in regulating a dormancy-to-proliferation switch. Created with BioRender.com

ing dormancy than those that did not produce FN. The effect of the FN matrix persevered until the cells secreted matrix metalloprotease 2 (MMP-2) and MMP-9 to degrade the matrix, resulting in tumor proliferation [38]. This mirrors a similar study in the lung, which found that cancer cell interaction with stroma spurs production of FN by the dormant cancer cells and supports survival [39]. However, there are some reports of FN regulating cancer dormancy in a contradictory manner. For example, a study compared the production of FN between dormant and proliferative breast cell lines from the D2A series, and they found the appearance of FN stimulated proliferation [16].

9.4 Other ECM and ECM-Associated Proteins

Several studies have identified other individual matrix proteins that support dormancy and suppress overt metastasis, including thrombospondin 1 [7] and e-selectin [40] on the microvascular endothelium, and osteopontin in the bone marrow [41]. Integrin engagement with these and other matrix-associated proteins, such as VWF and VCAM1, can protect these dormant, disseminated tumor cells from chemotherapy [42]. Tissue factor, a cancer cell associated initiator of coagulation, can regulate dormancy in glioma [43]. They observed that a glioma cell line (U373, which normally do not produce tumors in mice) deficient for TF remained viable but permanently dormant at an injection site for nearly a year, whereas inducing expression of TF led to tumor growth via recruitment of vascular (CD105+) and myeloid (CD11b+ and F4/80+) cells. When they first injected TF-lacking U373 cells in vivo, they were not immediately eliminated from immune-deficient mice. Rather, they remained viable and metabolically active at the injection site, but without evidence of palpable tumor growth. The role of TF was not to initiate cancer cell proliferation directly, but rather to recruit vascular and myeloid cells.

9.4.1 ECM Proteolysis

The ECM is subject to continuous turnover, including breakdown of existing protein fibers and scaffolds by ECM proteases and rebuilding of these same scaffolds via ECM protein deposition. It remains unclear how protease activity and ECM turnover regulate the balance between dormancy and relapse of cancer cells. Several studies have linked activation and proliferation of dormant lung cancer cells to chronic inflammation in the lung [36, 44]. Albrengues et al. demonstrated that chronic inflammation in lungs caused by infection or cigarette smoke results in the formation of neutrophil extracellular traps (NETs, Fig. 9.3) [45]. Two NET-associated proteases, neutrophil elastase (NE) and MMP9, cleave laminin in the local microenvironment, which further induces proliferation of dormant cells through activating $\alpha_3\beta_1$ integrin signaling, resulting in aggressive metastatic tumors. Experiments with breast cancer cell lines (murine D2.0R and human MCF-7) injected intravenously into syngeneic or

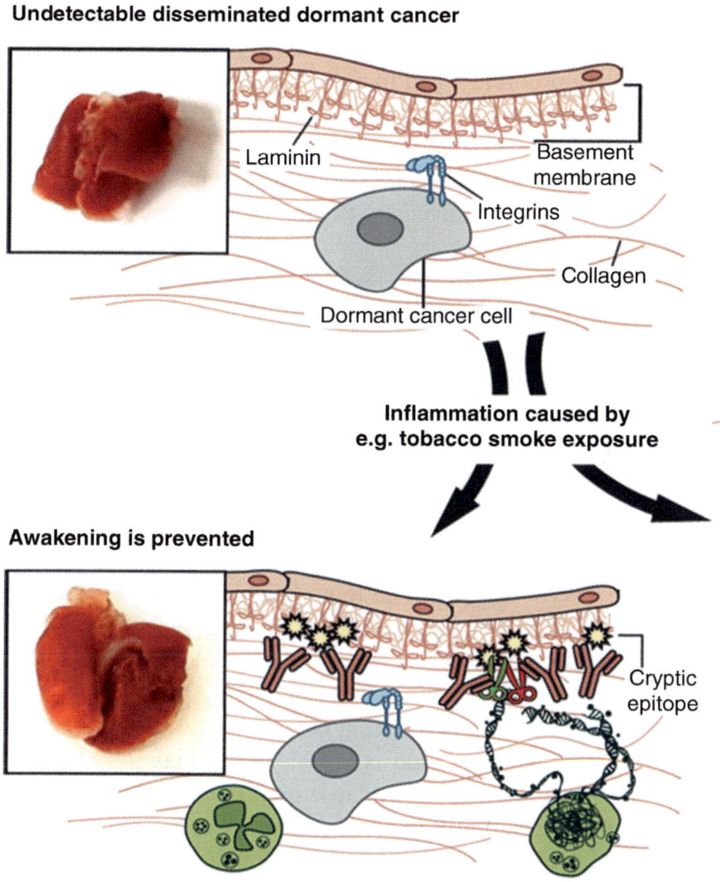

Fig. 9.3 **NETs formed during lung inflammation awaken dormant cancer cells.** Lung inflammation—caused, for instance, by tobacco smoke exposure—leads to NET formation. Two proteases on the NETs sequentially cleave the extracellular matrix protein laminin. This cleavage generates an epitope that activates integrin signaling and proliferation of the previously dormant cancer cells. Antibodies blocking the NET-generated laminin epitope prevent inflammation-induced awakening. Reproduced with permission from The American Association of Science [45]

immunodeficient nude mice, respectively, established a model of cancer cells dormancy-to-proliferation in the lungs, regulated by NETs.

9.4.2 Growth Factors in the ECM

Apart from direct binding to cells, the ECM also provides a binding depot for otherwise soluble growth factors in the local microenvironment. DTCs dormant in these environments could be kept quiescent in this local growth factor milieu or awakened through the acute release of growth factors during ECM remodeling. To understand this, Ghajar et al. [7] created organotypic models of lung- and bone marrow-microvascular niches where dormant DTCs reside upon the microvasculature of lung, bone marrow, and brain. They show that endothelial cells induce and maintain breast cancer DTC quiescence. They identified thrombospondin-1 as an endothelium-derived tumor suppressor, which diminishes near sprouting neovasculature, suggesting that tumors may escape growth regulation in this "sub-niche." Tumor cells grow faster around neovascular tips, which are rich in tumor-promoting factors such as active transforming growth factor beta 1 (TGF-β1) and periostin.

Studies by Chen et al. revealed that abnormal expression of VCAM-1 in breast cancer cells is associated with relapse in the lung [46]. To investigate whether VCAM-1 functions as a mediator of metastasis, they used short hairpin RNA (shRNA) interference to reduce its expression in a VCAM-1-overexpressing lung metastatic cell line (MDA231-LM2-4175) and implanted these cells into mammary glands of immunodeficient mice. They found VCAM-1 tethers metastasis-associated macrophages to cancer cells via α_4-containing integrins. Clustering of cell surface VCAM-1 triggers Akt activation via Ezrin and protects cancer cells from the pro-apoptotic cytokine TRAIL.

The lung has a uniquely high expression of the TGF-β superfamily member BMP4, which promotes the dormancy of DTCs within this tissue. Gao et al. showed that in breast cancer mouse models, DTCs can overcome the suppressive microenvironment of the lung by expressing Coco, a secreted antagonist of TGF-β ligands [47]. Coco binds to BMP4, preventing the activation of cognizant receptors, which triggers relapse of dormant cells.

9.4.3 Lysyl Oxidase and ECM Stiffening

Lysyl oxidase (LOX) is a copper-dependent amine oxidase, which catalyzes the covalent cross-linking of collagen and elastin, thereby increasing matrix stiffness [48]. Previous reports show that a fibrotic microenvironment and remodeling of ECM accounts for the switch of dormant tumor cells to metastatic growth [4, 16] (Fig. 9.1c). LOX family members enable a switch from dormancy to proliferation by establishing a fibrotic environment [4, 49]. Increased expression of LOX has been found in several cancer types (breast, prostate, colorectal, lung, bladder) [4, 48, 50], and hypoxia-inducible factor-1 causes the expression of several members of the LOX family in breast cancer cells, including LOX and Lysyl oxidase Like 2 (LOXL2).

LOXL2 facilitates an epithelial to mesenchymal transition (EMT) and induces the outgrowth of dormant breast cancer cells [51]. In this study, dormant MCF-7 cells expressing LOXL2 (MCF-7-LOXL2#12) underwent EMT, and downregulation of LOXL2 expression in MCF-7-LOXL2#12 cells restored their epithelial phenotype. In vitro, a mammosphere assay confirmed that EMT induced by LOXL2 expression is associated with the cancer stem-like phenotype. Here, MCF-7-LOXL2#12 cells exhibited a significant increase in the sphere-forming capacity whereas control cells remained as single cells, suggesting MCF-7-LOXL2#12 cells overcome the proliferation barrier. In vivo, tail vein injection of nude mice showed MCF-7-LOXL2#12 cells colonizing the lung transitioned from dormancy to metastatic outgrowth, whereas control MCF-7 cells remained dormant. MCF-7-LOXL2#12 cells on the surface of the lungs were imaged after 40 days, which revealed only 20% of the lung lesions remained as single tumor cells and 80% formed multi-cellular lesions. All metastatic lesions from control cells persisted as single quiescent cells (100%).

A stiff mechanical environment induces B16 melanoma tumor-repopulating cells (TRC) to enter dormancy in vivo and in vitro [52]. In vitro experiments on fibrin gels of three different stiffness values (90, 450, and 1050 Pa) have shown that growth of TRCs into spheroids is fastest in more compliant environments, and the most compliant fibrin environments also maximized proliferation in vivo in subcutaneous tumors.

Currently available in vivo dormancy models still have some limitations, which if overcome could lead to quicker advancement in the field. Bioengineers are developing 3D in vitro systems to study dormancy, often focusing on dormancy regulated by matrix stiffness [53–55]. These models have proven successful in allowing dormancy-incapable cells to proliferate and maintaining dormancy-capable cells quiescent. However, still much in vivo research needs to be done to validate these in vitro findings, and to better understand how ECM stiffness is correlated to maintain dormancy and their proliferative growth after several years.

9.5 Conclusions

The role of the ECM in regulating tumor growth and metastasis has been an active field for decades, but only recently have groups begun to establish a link between tissue ECM and dormancy/relapse. Because this field is so new, several questions remain:

- What triggers ECM remodeling that initiates tumor cell growth and relapse at a distant site?
- Does the ECM at certain tissues help maintain dormancy while other ECMs encourage relapse?
- Do certain preexisting conditions that affect ECM composition such as fibrosis and osteoporosis facilitate dormancy or relapse in different patients?
- How can cell–ECM interactions be targeted to prevent escape from dormancy in patients?
- Can ECM-targeting drug regimens be combined with chemotherapy or hormone receptor-targeting drugs to be used as long-term maintenance therapies for patients to prevent relapse?

References

1. Rosen LS, Gordon D, Tchekmedyian NS, Yanagihara R, Hirsh V, Krzakowski M, Pawlicki M, De Souza P, Zheng M, Urbanowitz G, Reitsma D, Seaman J. Long-term efficacy and safety of zoledronic acid in the treatment of skeletal metastases in patients with nonsmall cell lung carcinoma and other solid tumors: a randomized, phase III, double-blind, placebo-controlled trial. Cancer. 2004;100(12):2613–21. Epub 2004/06/16. https://doi.org/10.1002/cncr.20308.
2. Saad F, Gleason DM, Murray R, Tchekmedyian S, Venner P, Lacombe L, Chin JL, Vinholes JJ, Goas JA, Zheng M. Zoledronic acid prostate cancer study G. long-term efficacy of zoledronic acid for the prevention of skeletal complications in patients with metastatic hormone-refractory prostate cancer. J Natl Cancer Inst. 2004;96(11):879–82. Epub 2004/06/03. https://doi.org/10.1093/jnci/djh141.
3. Dittmer J. Mechanisms governing metastatic dormancy in breast cancer. Semin Cancer Biol. 2017;44:72–82. https://doi.org/10.1016/j.semcancer.2017.03.006.
4. Barkan D, El Touny LH, Michalowski AM, Smith JA, Chu I, Davis AS, Webster JD, Hoover S, Simpson RM, Gauldie J, Green JE. Metastatic growth from dormant cells induced by a col-I-

enriched fibrotic environment. Cancer Res. 2010;70(14):5706–16. https://doi.org/10.1158/0008-5472.CAN-09-2356.

5. Lu X, Mu E, Wei Y, Riethdorf S, Yang Q, Yuan M, Yan J, Hua Y, Tiede BJ, Haffty BG, Pantel K, Massague J, Kang Y. VCAM-1 promotes osteolytic expansion of indolent bone micrometastasis of breast cancer by engaging alpha4beta1-positive osteoclast progenitors. Cancer Cell. 2011;20(6):701–14. https://doi.org/10.1016/j.ccr.2011.11.002.
6. Shiozawa Y, Pedersen EA, Patel LR, Ziegler AM, Havens AM, Jung Y, Wang J, Zalucha S, Loberg RD, Pienta KJ, Taichman RS. GAS6/AXL axis regulates prostate cancer invasion, proliferation, and survival in the bone marrow niche. Neoplasia (New York, NY). 2010;12(2):116–27.
7. Ghajar CM, Peinado H, Mori H, Matei IR, Evason KJ, Brazier H, Almeida D, Koller A, Hajjar KA, Stainier DY, Chen EI, Lyden D, Bissell MJ. The perivascular niche regulates breast tumour dormancy. Nat Cell Biol. 2013;15(7):807–17. https://doi.org/10.1038/ncb2767.
8. Malladi S, Macalinao DG, Jin X, He L, Basnet H, Zou Y, de Stanchina E, Massague J. Metastatic latency and immune evasion through autocrine inhibition of WNT. Cell. 2016;165(1):45–60. https://doi.org/10.1016/j.cell.2016.02.025.
9. Lawson MA, McDonald MM, Kovacic N, Hua Khoo W, Terry RL, Down J, Kaplan W, Paton-Hough J, Fellows C, Pettitt JA, Neil Dear T, Van Valckenborgh E, Baldock PA, Rogers MJ, Eaton CL, Vanderkerken K, Pettit AR, Quinn JM, Zannettino AC, Phan TG, Croucher PI. Osteoclasts control reactivation of dormant myeloma cells by remodelling the endosteal niche. Nat Commun. 2015;6:8983. https://doi.org/10.1038/ncomms9983.
10. Boyerinas BZM, Yesilkanal AL, Price TT, Hyjek EM, Sipkins DA. Adhesion to osteopontin in the bone marrow niche regulates lymphoblastic leukemia cell dormancy. Blood. 2013;121(24):4821–31. https://doi.org/10.1182/blood-2012-12-.
11. Price TT, Sivan A, Warner MJ, Cheng R, Lee CH, Olivere L, Comatas K, Magnani J, Lyerly HK, Cheng Q, McCall CM, Sipkins DA. Dormant breast cancer micrometastases reside in specific bone marrow niches that regulate their transit to and from bone. Sci Transl Med. 2016;8(340):340ra73.
12. Hassell BA, Goyal G, Lee E, Sontheimer-Phelps A, Levy O, Chen CS, Ingber DE. Human organ chip models recapitulate orthotopic lung cancer growth, therapeutic responses, and tumor dormancy in vitro. Cell Rep. 2017;21(2):508–16. https://doi.org/10.1016/j.celrep.2017.09.043.
13. Pickup MW, Mouw JK, Weaver VM. The extracellular matrix modulates the hallmarks of cancer. EMBO Rep. 2014;15(12):1243–53. https://doi.org/10.15252/embr.201439246.
14. Jinka R, Kapoor R, Sistla PG, Raj TA, Pande G. Alterations in cell-extracellular matrix interactions during progression of cancers. y3er. 2012;2012:219196. https://doi.org/10.1155/2012/219196.
15. Sosa MS, Bragado P, Aguirre-Ghiso JA. Mechanisms of disseminated cancer cell dormancy: an awakening field. Nat Rev Cancer. 2014;14(9):611–22. https://doi.org/10.1038/nrc3793.
16. Barkan D, Green JE, Chambers AF. Extracellular matrix: a gatekeeper in the transition from dormancy to metastatic growth. Eur J Cancer. 2010;46(7):1181–8. Epub 2010/03/23. https://doi.org/10.1016/j.ejca.2010.02.027.
17. Gao H, Chakraborty G, Lee-Lim AP, Mavrakis KJ, Wendel HG, Giancotti FG. Forward genetic screens in mice uncover mediators and suppressors of metastatic reactivation. Proc Natl Acad Sci U S A. 2014;111(46):16,532–7. https://doi.org/10.1073/pnas.1403234111.
18. Gao H, Chakraborty G, Zhang Z, Akalay I, Gadiya M, Gao Y, Sinha S, Hu J, Jiang C, Akram M, Brogi E, Leitinger B, Giancotti FG. Multi-organ site metastatic reactivation mediated by non-canonical discoidin domain receptor 1 signaling. Cell. 2016;166(1):47–62. https://doi.org/10.1016/j.cell.2016.06.009.
19. Yates LR, Knappskog S, Wedge D, Farmery JHR, Gonzalez S, Martincorena I, Alexandrov LB, Van Loo P, Haugland HK, Lilleng PK, Gundem G, Gerstung M, Pappaemmanuil E, Gazinska P, Bhosle SG, Jones D, Raine K, Mudie L, Latimer C, Sawyer E, Desmedt C, Sotiriou C, Stratton MR, Sieuwerts AM, Lynch AG, Martens JW, Richardson AL, Tutt A, Lonning PE, Campbell PJ. Genomic evolution of breast cancer metastasis and relapse. Cancer Cell. 2017;32(2):169–84 e7. https://doi.org/10.1016/j.ccell.2017.07.005.
20. Bragado P, Estrada Y, Parikh F, Krause S, Capobianco C, Farina HG, Schewe DM, Aguirre-Ghiso JA. TGF-beta2 dictates disseminated tumour cell fate in target organs through TGF-beta-RIII and p38alpha/beta signalling. Nat Cell Biol. 2013;15(11):1351–61. https://doi.org/10.1038/ncb2861.
21. Kaplan RN, Riba RD, Zacharoulis S, Bramley AH, Vincent L, Costa C, MacDonald DD, Jin DK, Shido K, Kerns SA, Zhu Z, Hicklin D, Wu Y, Port JL, Altorki N, Port ER, Ruggero D, Shmelkov SV, Jensen KK, Rafii S, Lyden D. VEGFR1-positive haematopoietic bone marrow progenitors initiate the pre-metastatic niche. Nature. 2005;438(7069):820–7. https://doi.org/10.1038/nature04186.
22. Malanchi I, Santamaria-Martinez A, Susanto E, Peng H, Lehr HA, Delaloye JF, Huelsken J. Interactions between cancer stem cells and their niche govern metastatic colonization. Nature. 2012;481(7379):85–9. https://doi.org/10.1038/nature10694.
23. Oskarsson T, Acharyya S, Zhang XH, Vanharanta S, Tavazoie SF, Morris PG, Downey RJ, Manova-Todorova K, Brogi E, Massague J. Breast cancer cells produce tenascin C as a metastatic niche component to colonize the lungs. Nat Med. 2011;17(7):867–74. https://doi.org/10.1038/nm.2379.
24. Shoulders MD, Raines RT. Collagen structure and stability. Annu Rev Biochem. 2009;78:929–58. https://doi.org/10.1146/annurev.biochem.77.032207.120833.
25. Xu S, Xu H, Wang W, Li S, Li H, Li T, Zhang W, Yu X, Liu L. The role of collagen in cancer: from bench to bedside. J Transl Med. 2019;17(1):309. Epub 2019/09/16. https://doi.org/10.1186/s12967-019-2058-1.
26. Samuel MS, Lopez JI, McGhee EJ, Croft DR, Strachan D, Timpson P, Munro J, Schroder E, Zhou J, Brunton VG, Barker N, Clevers H, Sansom OJ, Anderson KI, Weaver VM, Olson MF. Actomyosin-mediated cellular tension drives increased tissue stiffness and beta-catenin activation to induce epidermal hyperplasia and tumor growth. Cancer Cell. 2011;19(6):776–91. Epub 2011/06/15. https://doi.org/10.1016/j.ccr.2011.05.008.
27. Ohlund D, Franklin O, Lundberg E, Lundin C, Sund M. Type IV collagen stimulates pancreatic cancer cell proliferation, migration, and inhibits apoptosis through an autocrine loop. BMC Cancer. 2013;13:154. https://doi.org/10.1186/1471-2407-13-154.
28. Miller FR, McEachern D, Miller BE. Growth regulation of mouse mammary tumor cells in collagen gel cultures by diffusible factors produced by normal mammary gland epithelium and stromal fibroblasts. Cancer Res. 1989;49(21):6091–7. Epub 1989/11/01
29. Rak JW, McEachern D, Miller FR. Sequential alteration of peanut agglutinin binding-glycoprotein expression during progression of murine mammary neoplasia. Br J Cancer. 1992;65(5):641–8. Epub 1992/05/01. https://doi.org/10.1038/bjc.1992.138.
30. Ossowski L, Aguirre-Ghiso JA. Dormancy of metastatic melanoma. Pigment Cell Melanoma Res. 2010;23(1):41–56. Epub 2009/10/22. https://doi.org/10.1111/j.1755-148X.2009.00647.x.
31. Wall SJ, Zhong ZD, DeClerck YA. The cyclin-dependent kinase inhibitors p15INK4B and p21CIP1 are critical regulators of fibrillar collagen-induced tumor cell cycle arrest. J Biol Chem. 2007;282(33):24471–6. Epub 2007/06/08. https://doi.org/10.1074/jbc.M702697200.
32. Wall SJ, Werner E, Werb Z, DeClerck YA. Discoidin domain receptor 2 mediates tumor cell cycle arrest induced by fibrillar collagen. J

Biol Chem. 2005;280(48):40187–94. Epub 2005/09/28. https://doi.org/10.1074/jbc.M508226200.
33. Singh P, Carraher C, Schwarzbauer JE. Assembly of fibronectin extracellular matrix. Annu Rev Cell Dev Biol. 2010;26:397–419. Epub 2010/08/10. https://doi.org/10.1146/annurev-cellbio-100109-104020.
34. Sottile J, Hocking DC, Langenbach KJ. Fibronectin polymerization stimulates cell growth by RGD-dependent and -independent mechanisms. J Cell Sci. 2000;113(Pt 23):4287–99.
35. Aguirre-Ghiso JA, Liu D, Mignatti A, Kovalski K, Ossowski L. Urokinase receptor and fibronectin regulate the ERK(MAPK) to p38(MAPK) activity ratios that determine carcinoma cell proliferation or dormancy in vivo. Mol Biol Cell. 2001;12(4):863–79. Epub 2001/04/11. https://doi.org/10.1091/mbc.12.4.863.
36. Park SY, Nam JS. The force awakens: metastatic dormant cancer cells. Exp Mol Med. 2020;52(4):569–81. Epub 2020/04/18. https://doi.org/10.1038/s12276-020-0423-z.
37. Aguirre-Ghiso JA, Estrada Y, Liu D, Ossowski L. ERK(MAPK) activity as a determinant of tumor growth and dormancy; regulation by p38(SAPK). Cancer Res. 2003;63(7):1684–95. Epub 2003/04/03
38. Barney LE, Hall CL, Schwartz AD, Parks AN, Sparages C, Galarza S, Platt MO, Mercurio AM, Peyton SR. Tumor cell-organized fibronectin maintenance of a dormant breast cancer population. Sci Adv. 2020;6(11):eaaz4157. Epub 2020/03/21. https://doi.org/10.1126/sciadv.aaz4157.
39. Montagner M, Bhome R, Hooper S, Chakravarty P, Qin X, Sufi J, Bhargava A, Ratcliffe CDH, Naito Y, Pocaterra A, Tape CJ, Sahai E. Crosstalk with lung epithelial cells regulates Sfrp2-mediated latency in breast cancer dissemination. Nat Cell Biol. 2020;22(3):289–96. Epub 2020/02/26. https://doi.org/10.1038/s41556-020-0474-3.
40. Price TT, Burness ML, Sivan A, Warner MJ, Cheng R, Lee CH, Olivere L, Comatas K, Magnani J, Kim Lyerly H, Cheng Q, McCall CM, Sipkins DA. Dormant breast cancer micrometastases reside in specific bone marrow niches that regulate their transit to and from bone. Sci Transl Med. 2016;8(340):340ra73. Epub 2016/05/27. https://doi.org/10.1126/scitranslmed.aad4059.
41. Boyerinas B, Zafrir M, Yesilkanal AE, Price TT, Hyjek EM, Sipkins DA. Adhesion to osteopontin in the bone marrow niche regulates lymphoblastic leukemia cell dormancy. Blood. 2013;121(24):4821–31. Epub 2013/04/17. https://doi.org/10.1182/blood-2012-12-475483.
42. Carlson P, Dasgupta A, Grzelak CA, Kim J, Barrett A, Coleman IM, Shor RE, Goddard ET, Dai J, Schweitzer EM, Lim AR, Crist SB, Cheresh DA, Nelson PS, Hansen KC, Ghajar CM. Targeting the perivascular niche sensitizes disseminated tumour cells to chemotherapy. Nat Cell Biol. 2019;21(2):238–50. Epub 2019/01/22. https://doi.org/10.1038/s41556-018-0267-0.
43. Magnus N, Garnier D, Meehan B, McGraw S, Lee TH, Caron M, Bourque G, Milsom C, Jabado N, Trasler J, Pawlinski R, Mackman N, Rak J. Tissue factor expression provokes escape from tumor dormancy and leads to genomic alterations. Proc Natl Acad Sci U S A. 2014;111(9):3544–9. Epub 2014/02/13. https://doi.org/10.1073/pnas.1314118111.
44. Manjili MH. Tumor dormancy and relapse: from a natural byproduct of evolution to a disease state. Cancer Res. 2017;77(10):2564–9. Epub 2017/05/17. https://doi.org/10.1158/0008-5472.CAN-17-0068.
45. Albrengues J, Shields MA, Ng D, Park CG, Ambrico A, Poindexter ME, Upadhyay P, Uyeminami DL, Pommier A, Kuttner V, Bruzas E, Maiorino L, Bautista C, Carmona EM, Gimotty PA, Fearon DT, Chang K, Lyons SK, Pinkerton KE, Trotman LC, Goldberg MS, Yeh JT, Egeblad M. Neutrophil extracellular traps produced during inflammation awaken dormant cancer cells in mice. Science. 2018;361(6409) Epub 2018/09/29 https://doi.org/10.1126/science.aao4227.
46. Chen Q, Zhang XH, Massague J. Macrophage binding to receptor VCAM-1 transmits survival signals in breast cancer cells that invade the lungs. Cancer Cell. 2011;20(4):538–49. Epub 2011/10/22. https://doi.org/10.1016/j.ccr.2011.08.025.
47. Gao H, Chakraborty G, Lee-Lim AP, Mo Q, Decker M, Vonica A, Shen R, Brogi E, Brivanlou AH, Giancotti FG. The BMP inhibitor Coco reactivates breast cancer cells at lung metastatic sites. Cell. 2012;150(4):764–79. https://doi.org/10.1016/j.cell.2012.06.035.
48. Barker HE, Cox TR, Erler JT. The rationale for targeting the LOX family in cancer. Nat Rev Cancer. 2012;12(8):540–52. Epub 2012/07/20. https://doi.org/10.1038/nrc3319.
49. Barry-Hamilton V, Spangler R, Marshall D, McCauley S, Rodriguez HM, Oyasu M, Mikels A, Vaysberg M, Ghermazien H, Wai C, Garcia CA, Velayo AC, Jorgensen B, Biermann D, Tsai D, Green J, Zaffryar-Eilot S, Holzer A, Ogg S, Thai D, Neufeld G, Van Vlasselaer P, Smith V. Allosteric inhibition of lysyl oxidase-like-2 impedes the development of a pathologic microenvironment. Nat Med. 2010;16(9):1009–17. Epub 2010/09/08. https://doi.org/10.1038/nm.2208.
50. Baker AM, Bird D, Lang G, Cox TR, Erler JT. Lysyl oxidase enzymatic function increases stiffness to drive colorectal cancer progression through FAK. Oncogene. 2013;32(14):1863–8. Epub 2012/05/30. https://doi.org/10.1038/onc.2012.202.
51. Weidenfeld K, Schif-Zuck S, Abu-Tayeh H, Kang K, Kessler O, Weissmann M, Neufeld G, Barkan D. Dormant tumor cells expressing LOXL2 acquire a stem-like phenotype mediating their transition to proliferative growth. Oncotarget. 2016;7(44):71,362–77. Epub 2016/09/23. https://doi.org/10.18632/oncotarget.12109.
52. Liu Y, Lv J, Liang X, Yin X, Zhang L, Chen D, Jin X, Fiskesund R, Tang K, Ma J, Zhang H, Dong W, Mo S, Zhang T, Cheng F, Zhou Y, Xie J, Wang N, Huang B. Fibrin stiffness mediates dormancy of tumor-repopulating cells via a Cdc42-driven Tet2 epigenetic program. Cancer Res. 2018;78(14):3926–37. Epub 2018/05/17. https://doi.org/10.1158/0008-5472.CAN-17-3719.
53. Fang JY, Tan SJ, Wu YC, Yang Z, Hoang BX, Han B. From competency to dormancy: a 3D model to study cancer cells and drug responsiveness. J Transl Med. 2016;14:38. Epub 2016/02/06. https://doi.org/10.1186/s12967-016-0798-8.
54. Preciado JA, Reátegui E, Azarin SM, Lou E, Aksan A. Immobilization platform to induce quiescence in dormancy-capable cancer cells. Technology. 2017;05(03):129–38. https://doi.org/10.1142/S2339547817500078.
55. Pradhan S, Hassani I, Seeto WJ, Lipke EA. PEG-fibrinogen hydrogels for three-dimensional breast cancer cell culture. J Biomed Mater Res A. 2017;105(1):236–52. https://doi.org/10.1002/jbm.a.35899.

Therapy-Induced Dormancy and Residual Disease

10

Ashley V. DiMarco, Nina Marie G. Garcia, and James V. Alvarez

Abstract

In many common epithelial cancers, such as breast and prostate cancer, the majority of deaths are caused by the survival and eventual recurrence of residual tumor cells after therapy. In many cases, residual tumors can survive for years or even decades prior to recurrence, suggesting that these tumors are in a dormant state. Because recurrent tumors are often incurable, it is critical to understand how cancer cells survive treatment, persist as residual cells during a prolonged dormant period, and eventually recur. In this chapter, we review the clinical evidence suggesting that dormant residual tumors are an important reservoir for tumor recurrence, and then discuss some of the molecular and cellular features of dormant residual tumors.

10.1 Introduction

Tumor recurrence following potentially curative therapy is the leading cause of death in many types of cancer, including some of the most common epithelial malignancies like breast cancer, prostate cancer, and melanoma [1]. In many of these cancer types, tumors can recur after a prolonged period. For instance, most breast cancer deaths result from tumor recurrence following therapy, and tumors can recur as long as 20 years after treatment [1]. This suggests that residual tumor cells can persist in a dormant state for decades prior to reemerging as a recurrent tumor [2] [3]. Consistent with this, disseminated tumor cells (DTCs) are present in the bone marrow of up to 40% of breast cancer patients at diagnosis, and the number of DTCs at diagnosis and their persistence several years following adjuvant therapy are independent markers for recurrence risk [4–6]. Furthermore, the extent of residual cancer present locally in the breast following neoadjuvant therapy is strongly predictive of recurrence [7]. Together, these clinical data suggest that the survival and persistence of residual cancer cells following therapy is a major obstacle to curing breast cancer. Here, we review the clinical significance of residual dormant cells and then discuss some of the properties of dormant tumors, with a focus on dormant cell metabolism and the dormant tumor microenvironment.

10.2 Disseminated Tumor Cells (DTCs) Predict Risk of Recurrence

In a seminal paper published over 20 years ago, Klaus Pantel and colleagues examined the bone marrow of breast cancer patients for the presence of cancer cells [8]. Importantly, these patients did not have gross evidence of metastatic disease, and so their cancers were classified as Stages I–III. They found that, despite the absence of clinically evident metastasis, nearly half of these patients had detectable cancer cells in their bone marrow. These cells were referred to as disseminated tumor cells, or DTCs. These DTCs were present both as isolated cancer cells, as well as in DTC clusters. While the frequency of DTCs was highest in Stage III cancers, even a fraction of patients with early-stage, Stage I breast cancer, had detectable DTCs in their bone marrow. To assess the clinical significance of these DTCs, the authors examined their association with prognosis. As described above, metastatic recurrence is the main cause of death from breast cancer. This group found that the presence of DTCs in the bone marrow greatly increased the risk of metastasis and death from breast cancer. This indicates that DTCs are clinically significant, and suggests that these cells may serve as a reservoir for metastatic recurrence.

A. V. DiMarco · N. M. G. Garcia · J. V. Alvarez (✉)
Department of Pharmacology and Cancer Biology, Duke University School of Medicine, Durham, NC, USA
e-mail: ashley.dimarco@duke.edu; ninamarie.garcia@duke.edu; James.Alvarez@duke.edu

S. P. Leong et al. (eds.), *Cancer Metastasis Through the Lymphovascular System*, https://doi.org/10.1007/978-3-030-93084-4_10

These findings have been followed up on and validated by a number of other groups [9–13]. For instance, a pooled analysis of patients from several academic hospitals confirmed these results, and extended them in an important way [9]. This study found that, in addition to the presence of DTCs at diagnosis, the persistence of DTCs for several years after therapy was also predictive of eventual recurrence and survival [9]. This provides further evidence that DTCs serve as a functionally important reservoir for disease relapse.

Evidence that DTCs are dormant has come from several routes, both direct and indirect. Indirect evidence that DTCs are dormant include the clinical observations that the number of DTCs can be quite stable, even in patients who do not develop overt metastases [14–16]. Additionally, there are several case studies where dormant DTCs have reactivated upon organ transplant, giving rise to metastases in the organ recipient [14]. Direct analysis of the proliferation marker Ki67 revealed that the majority of DTCs in breast cancer are negative for Ki67 [17, 18]. Additionally, the dormancy marker NR2F1 was shown to mark the majority of DTCs, and NR2F1-positive DTCs were associated with good outcome in breast cancer patients with DTCs [17]. These data indicate that DTCs frequently exist in a dormant or slowly proliferative state.

Taken together, these studies demonstrate that DTCs represent a clinically relevant population of minimal residual disease, and provide strong clinical evidence that the survival and persistence of residual tumor cells predicts relapse risk.

10.3 Residual Disease Following Neoadjuvant Therapy Predicts Risk of Recurrence

A separate line of clinical investigation has provided evidence that the survival of cancer cells in response to therapy predicts eventual relapse. These studies focused on patients receiving neoadjuvant therapy—that is, adjuvant therapy prior to surgical removal of the tumor. The specific therapy given to patients in the neoadjuvant setting varies depending upon the tumor characteristics, but most frequently includes chemotherapy, such as Adriamycin, cyclophosphamide, and paclitaxel. For Her2-amplified tumors, neoadjuvant therapy might include the anti-Her2 therapies such as Trastuzumab. Notwithstanding the specific therapies given, the benefit of neoadjuvant therapy is that it allows for an assessment of how the tumor responds to therapy, and understanding how this response is related to disease outcome.

A number of studies and meta-analyses have examined the relationship between response to neoadjuvant therapy and various clinical endpoints, including local recurrence, metastatic recurrence, and survival. For instance, pathologic complete response (pCR) in the breast has repeatedly been shown to be associated with a reduced risk of recurrence [19–22]. Similarly, using a different measure, Residual Cancer Burden, it was shown that the presence of residual tumor cells following neoadjuvant chemotherapy is associated with increased risk of relapse [7]. It is important to keep in mind that the response of the primary tumor to therapy is actually a surrogate for what is more important, namely the response of any disseminated tumor cells or micrometastases to therapy, since the primary tumor will be surgically excised following neoadjuvant therapy. These studies make clear that the survival of residual cancer cells following therapy is strongly associated with eventual recurrence.

10.4 Therapy-Induced Dormancy

10.4.1 Chemotherapy

A number of anticancer therapies can induce tumor cells to stop proliferating and enter into a dormant or quiescent state. Pioneering work from Julio Aguirre-Ghiso and colleagues found that cancer cell dormancy is dictated by the ratio between activation of the ERK and p38 pathways. A high ERK:p38 ratio promotes proliferation, while a low ERK:p38 ratio induces dormancy [23]. Because many cellular stresses can activate p38, this provides one mechanism of stress-induced dormancy. For instance, chemotherapy, by inducing p38 activation, can induce cancer cell dormancy [24]. p38 activation, in turn, can lead to induction of the unfolded protein response (UPR) and activation of the stress-responsive transcription factor ATF6 [25]. This stress-response network therefore links therapy response to cell cycle arrest, dormancy, and survival.

10.4.2 Targeted Therapy

Targeted therapy against oncogenes can also induce dormancy. Many oncogenes induce cell proliferation. For instance, receptor tyrosine kinases and growth factor receptors signal through the Ras-MAPK and PI3K-Akt pathways to induce both proliferation and survival. Many of the signaling proteins in these pathways are protooncogenes, and when mutated or overexpressed can induce hyperactivation of these signaling pathways. Consequently, it makes sense that drugs targeting these oncogenes will induce both cell death and a loss of proliferation. In fact, the efficacy of many targeted therapies, including those against receptor tyrosine kinases such as EGFR and Her2, or targeting intracellular kinases like Raf, PI3K, or Akt, is likely due to the combination of cell death and proliferation arrest. However, in many cases some cells can survive these therapies, but remain in a non-proliferative state. For instance, non-small cell lung cancer (NSCLC) cells with mutations in the epidermal growth factor receptor (EGFR) respond to EGFR inhibitors like gefitinib and erlotinib. However, some cells can survive inhibitor treatment and persist in a slowly proliferative state resembling dormancy [26–28]. As described in more detail below, these surviving cells have been referred to as drug-tolerant persister cells. A similar scenario was observed in basal cell carcinoma cells treated with a Smoothened inhibitor, where tumors regress but a population of dormant cells seeds relapse [29].

A number of insights on the mechanism of therapy-induced dormancy have come from conditional genetically engineered mouse models. In these models, doxycycline (dox) administration induces expression of an oncogene—such as Her2/neu, Wnt1, or Myc—leading to mammary tumor formation. Subsequent withdrawal of dox induces oncogene downregulation and complete tumor regression, mimicking response to targeted therapies. However, a population of residual cells survives oncogene downregulation and resides in a dormant, non-proliferative state in the mammary gland, and these cells eventually reinitiate proliferation to form a recurrent tumor. Importantly, the residual dormant cells that survive oncogene downregulation can be isolated and characterized [30]. These models recapitulate residual disease and recurrence as it occurs in breast cancer patients in several key respects. First, analogous to residual DTCs in human patients, the residual cells that survive in these models are non-proliferative but remain viable [31]. Second, the dormant period in these models is long (6–9 months) relative to the lifetime of the mouse, similar to what is observed in human breast cancer. Third, recurrent tumors arising in these models do not express the initiating oncogene, reminiscent of the finding that HER2-amplified tumors are often HER2-negative upon relapse [32].

Using these models, it was found that loss of Her2 [33], Myc [34], Wnt1 [35], and Ras [36] can induce apoptosis and tumor regression. The residual cells that survive oncogene downregulation are non-proliferative, and are thought to reside in the G0 phase of the cell cycle. Thus, targeted therapies directed against oncogenes can induce dormancy in surviving residual tumor cells, and these cells serve as a reservoir for eventual relapse.

10.4.3 Drug-Tolerant Persister Cells

Cells that can survive chemotherapy or targeted therapy are analogous in some ways to so-called "drug-tolerant persister cells." This concept, originally developed in relation to bacteria that survive antibiotic treatment, posits that some cells can survive therapy and persist in a quasi-resistant state. These cells (or bacteria) are not fully resistant to anticancer therapy (or antibiotics), but are viable in the presence of drugs. They can persist in this intermediate resistance state until they acquire additional changes, most frequently genetic mutations, that render them fully resistant [26]. A number of studies have shown that the drug-tolerant state is mediated by epigenetic changes in tumor cells [27]. For instance, in drug-tolerant persistent NSCLC cells following EGFR inhibition, the histone demethylase KDM5A leads to loss of H3K4 methylation at cell-cycle-related genes to prevent their expression. Similarly, the DNA demethylase TET2 promotes dormancy in colon cancer by regulating the expression of proliferation genes [37].

10.5 Dormant Tumor Cell Metabolism

The unique metabolic features of cancer cells were recognized nearly 100 years ago, when Warburg noted that cancer cells exhibit elevated levels of glycolysis, producing lactate even under aerobic conditions. The past decade has seen a renewed interest in understanding the metabolic requirements of tumor cell proliferation. A current model suggests that tumor cells must adopt a glycolytic phenotype to produce the necessary precursors for anabolic growth to support their rapid proliferation [38]. Indeed, the requirements of metabolic pathways to support proliferation, and the regulation of these pathways by tumor suppressors and oncogenes, have been well established. The metabolic demands of dormant cells are likely to be quite different from those of proliferating cells: they do not need to provide precursors for anabolic growth or proliferation, such as nucleotides to synthesize DNA or lipids to synthesize new cell membranes. In contrast to the extensive literature on the metabolism of rapidly proliferating tumor cells, virtually nothing is known about the metabolic requirements of dormant tumor cells, or how metabolic pathways are regulated during dormancy. Similarly, very little is known about the metabolism of recurrent tumors that form when dormant tumor cells reenter the cell cycle or reinitiate rapid proliferation. Considering the intimate relationship between cancer cell proliferation and metabolism, it is likely that the metabolism of cancer cells may change dramatically during these transitions between proliferation and dormancy. However, it is not known whether sustained activity of particular metabolic pathways is functionally important in regulating tumor dormancy and recurrence. Given the clinical importance of these cells, and the fact that they can persist in cancer patients for decades, understanding what nutrient sources they depend upon and how their metabolic pathways are wired may be critical in therapeutically targeting them.

Several recent studies have begun to address these issues. For instance, several studies showed that inhibition of diverse oncogenes—including Her2, Ras, Myc, and Wnt—causes an immediate drop in glucose uptake [36, 39, 40]. In response to decreased glucose metabolism, cells that survive the loss of oncogenic Her2 [40] and KRAS [36] signaling upregulate fatty acid oxidation (FAO) pathways ([36, 40, 41]. Additionally, both decreased glucose metabolism and increased FAO following oncogene inhibition results in increased levels of reactive oxygen species (ROS) [36, 40–42], which can impair tumor cell viability [41, 43, 44]. Similarly, inhibitors targeting the critical cell cycle regulator cyclin D3-CDK6 leads to altered cell metabolism and induction of ROS [45].

These results suggest that tumor cells that survive therapy and eventually develop resistance must adapt to overcome therapy-induced metabolic stress. A number of pathways have been identified that promote metabolic adaptations during tumor dormancy and recurrence remain. In breast tumors following loss of Her2 signaling, activation of the antioxidant transcription factor NRF2 is an adaptive response to elevated ROS [41]. Interestingly, NRF2 remains activated in dormant tumors following Her2 inhibition, and NRF2 activation is both necessary and sufficient for these dormant tumor cells to reinitiate proliferation to form recurrent tumors [41]. Mechanistically, NRF2 functions in dormant tumors to promote nucleotide synthesis and maintain redox homeostasis.

Cells that survive therapy have metabolic dependencies that are unique from the untreated tumor cells. For example, AMPK is activated by diverse therapies [41, 46] consistent with the notion that tumor cells experience bioenergetic stress resulting from therapy. AMPK activation suppresses anabolic metabolism and promotes catabolic metabolism. Pharmacologic activation of AMPK with metformin promotes the survival of dormant breast cancer cells [46]. Consistent with this, lysosomal degradation of proteins and organelles via autophagy is induced by chemotherapy [47] and mTOR inhibition [48–51], and inhibiting autophagy synergizes with these therapies [48, 49]. Autophagy is a central cellular response to nutrient deprivation that serves, in part, to salvage high energy biomolecules for catabolic processes to maintain bioenergetic requirements. Autophagy also promotes dormant cell survival in response to unfavorable cues from the extracellular matrix [52]. These results provide further support that therapies cause metabolic stress and highlight the importance of metabolic adaptation for dormant cell survival.

10.6 Survival Pathways in Dormant Tumor Cells

The fact that dormant tumor cells are able to survive therapy and persist for long time periods suggests that these cells have activated pro-survival pathways or disabled apoptotic pathways. The pro-apoptotic protein Par-4 (encoded by the *Pawr* gene) is upregulated in response to diverse therapies, both targeted therapy and chemotherapy [30]. Loss of Par-4 expression through epigenetic silencing promotes the long-term survival of dormant cells following Her2 inhibition in breast cancer cells [30, 53, 54]. Elevated Wnt signaling is responsible for dormant cell survival in basal cell carcinoma cells treated with a Hedgehog inhibitor [29]. In EGFR-mutant NSCLC, the transcription factor YAP promotes cell survival by repressing the pro-apoptotic protein BMF [28]. In all of these instances, targeting the pro-survival pathway relied on by dormant cells is effective at eliminating these dormant cells. This illustrates the principle that preventing recurrence following treatment of the primary tumor may require a two-pronged strategy. In the first arm, the primary tumor is treated by inhibiting the driving oncogene, for instance Her2 in breast cancer or EGFR in lung cancer. However, a second therapy is needed to eliminate the dormant cells that survive and persist following the primary treatment.

10.7 Epithelial-to-Mesenchymal (EMT) Transition in Dormancy

Several lines of evidence have shown that dormant residual tumor cells frequently undergo an epithelial-to-mesenchymal transition (EMT), and that EMT contributes to tumor recurrence following therapy. A clinical study that tracked the evolution of EGFR-mutant lung tumors as they acquired resistance to EGFR inhibitors found that a fraction of tumors underwent EMT [55]. In conditional GEM models of breast cancer recurrence, recurrent tumors almost universally undergo EMT, and experimental induction of EMT accelerates recurrence [33]. The dormant residual tumors that survive Her2 downregulation also exhibit features of EMT, and barcoding analysis suggests that the induction of EMT is an adaptive response to Her2 inhibition [56]. Several studies have found that many cancer cells exhibit a drug-tolerant state characterized by high expression of mesenchymal gene signatures [57, 58]. Interestingly, this drug-tolerant mesenchymal state can be targeted by inhibitors of the lipid peroxidase GPX4 [57, 58]. Taken together, these studies indicate that EMT promotes dormant tumor survival and suggest that drugs targeting the mesenchymal state may prevent recurrence.

10.8 The Dormant Tumor Microenvironment

The tumor microenvironment is a critical determinant of tumor growth, progression, and response and resistance to therapy. Recent work has focused on characterizing the dormant tumor microenvironment and understanding how the microenvironment contributes to tumor recurrence. A study of NSCLC patients used single-cell RNA-sequencing to study how tumors change in response to targeted therapy and in residual disease [59]. This study revealed both tumor cell-intrinsic changes in gene expression and changes to the tumor immune microenvironment in residual tumors. Residual tumor cells exhibited an alveolar-regenerative cell signature, suggesting that therapy induces a cell-state transition to a more de-differentiated state. The immune composition of residual tumors was markedly different from primary lung tumors, with increased T cell and decreased macrophage infiltration as compared to primary tumors [59].

In a mouse model of Burkitt's lymphoma, chemotherapy leads to secretion of IL-6 and Timp-1 as a response to DNA damage, creating a chemoresistant niche [60]. This pro-survival niche permits a population of minimal residual tumor cells to persist and serve as a reservoir for tumor relapse. Interestingly, thymic endothelial cells serve as one source of the secreted IL-6, highlighting the contribution of the tumor microenvironment to dormant cell survival.

In a conditional mouse model of breast cancer, Her2 downregulation leads to secretion of TNF, which acts in an autocrine manner to stimulate the NFkB pathway [61]. Interestingly, an autocrine TNF pathway was also induced by EGFR inhibition in lung cancer [62]. In both breast cancer and lung cancer, TNF acts to promote the survival of residual cells in response to therapy. Interestingly, in breast cancer this was mediated through increased expression of CCL5 and consequent recruitment of macrophages to dormant tumors [61]. Dormant tumors had abundant macrophages, and these macrophages led to collagen deposition in dormant tumors to promote tumor relapse. These studies illustrate how therapy can induce adaptations in tumor cells that promote dormant cell survival through both cell autonomous mechanisms and via crosstalk with the microenvironment.

10.9 Conclusions and Perspective

- The presence of disseminated tumor cells (DTCs) in patients and their persistence after treatment are associated with increased risk of tumor recurrence
- The survival of residual cancer cells following neoadjuvant therapy predicts poor prognosis
- Dormant residual cancer cells are the reservoir for eventual tumor recurrence
- Chemotherapy and oncogene-targeted therapy can induce cancer cells into a dormant or quiescent state
- Dormant residual tumors have altered metabolism compared to primary tumors
- The dormant tumor microenvironment contributes to tumor recurrence

Future Directions

- Can therapeutic approaches be devised to target residual tumor cells?
- Is it preferable to awaken dormant cells to sensitize these cells to chemotherapy, or attempt to keep these cells dormant to prevent recurrence?
- Are therapy-induced dormancy and metastatic dormancy regulated by similar cellular and molecular pathways?
- How do chemotherapy and targeted therapy differentially induce dormancy?
- How does therapy-induced dormancy differ between different metastatic sites with distinct tumor microenvironments?

References

1. Saphner T, Tormey DC, Gray R. Annual hazard rates of recurrence for breast cancer after primary therapy. J Clin Oncol. 1996;14:2738–46.
2. Aguirre-Ghiso JA. Models, mechanisms and clinical evidence for cancer dormancy. Nat Rev Cancer. 2007;7:834–46.
3. Sosa MS, Bragado P, Aguirre-Ghiso JA. Mechanisms of disseminated cancer cell dormancy: an awakening field. Nat Rev Cancer. 2014;14:611–22.
4. Braun S, Vogl FD, Naume B, Janni W, Osborne MP, Coombes RC, Schlimok G, Diel IJ, Gerber B, Gebauer G, Pierga JY, Marth C, Oruzio D, Wiedswang G, Solomayer EF, Kundt G, Strobl B, Fehm T, Wong GY, Bliss J, Vincent-Salomon A, Pantel K. A pooled analysis of bone marrow micrometastasis in breast cancer. N Engl J Med. 2005;353:793–802.
5. Janni W, Vogl FD, Wiedswang G, Synnestvedt M, Fehm T, Juckstock J, Borgen E, Rack B, Braun S, Sommer H, Solomayer E, Pantel K, Nesland J, Friese K, Naume B. Persistence of disseminated tumor cells in the bone marrow of breast cancer patients predicts increased risk for relapse--a European pooled analysis. Clin Cancer Res. 2011;17:2967–76.
6. Pantel K, Muller V, Auer M, Nusser N, Harbeck N, Braun S. Detection and clinical implications of early systemic tumor cell dissemination in breast cancer. Clin Cancer Res. 2003;9:6326–34.
7. Symmans WF, Peintinger F, Hatzis C, Rajan R, Kuerer H, Valero V, Assad L, Poniecka A, Hennessy B, Green M, Buzdar AU, Singletary SE, Hortobagyi GN, Pusztai L. Measurement of residual breast cancer burden to predict survival after neoadjuvant chemotherapy. J Clin Oncol. 2007;25:4414–22.
8. Braun S, Pantel K, Muller P, Janni W, Hepp F, Kentenich CR, Gastroph S, Wischnik A, Dimpfl T, Kindermann G, Riethmuller G, Schlimok G. Cytokeratin-positive cells in the bone marrow and survival of patients with stage I, II, or III breast cancer. N Engl J Med. 2000;342:525–33.
9. Janni W, Rack B, Schindlbeck C, Strobl B, Rjosk D, Braun S, Sommer H, Pantel K, Gerber B, Friese K. The persistence of isolated tumor cells in bone marrow from patients with breast carcinoma predicts an increased risk for recurrence. Cancer. 2005;103:884–91.
10. Mathiesen RR, Borgen E, Renolen A, Lokkevik E, Nesland JM, Anker G, Ostenstad B, Lundgren S, Risberg T, Mjaaland I, Kvalheim G, Lonning PE, Naume B. Persistence of disseminated

tumor cells after neoadjuvant treatment for locally advanced breast cancer predicts poor survival. Breast Cancer Res. 2012;14:R117.
11. Naume B, Synnestvedt M, Falk RS, Wiedswang G, Weyde K, Risberg T, Kersten C, Mjaaland I, Vindi L, Sommer HH, Saetersdal AB, Rypdal MC, Bendigtsen Schirmer C, Wist EA, Borgen E. Clinical outcome with correlation to disseminated tumor cell (DTC) status after DTC-guided secondary adjuvant treatment with docetaxel in early breast cancer. J Clin Oncol. 2014;32:3848–57.
12. Wiedswang G, Borgen E, Karesen R, Kvalheim G, Nesland JM, Qvist H, Schlichting E, Sauer T, Janbu J, Harbitz T, Naume B. Detection of isolated tumor cells in bone marrow is an independent prognostic factor in breast cancer. J Clin Oncol. 2003;21:3469–78.
13. Wiedswang G, Borgen E, Karesen R, Qvist H, Janbu J, Kvalheim G, Nesland JM, Naume B. Isolated tumor cells in bone marrow three years after diagnosis in disease-free breast cancer patients predict unfavorable clinical outcome. Clin Cancer Res. 2004;10:5342–8.
14. Klein CA. Framework models of tumor dormancy from patient-derived observations. Curr Opin Genet Dev. 2011;21:42–9.
15. Meng S, Tripathy D, Frenkel EP, Shete S, Naftalis EZ, Huth JF, Beitsch PD, Leitch M, Hoover S, Euhus D, Haley B, Morrison L, Fleming TP, Herlyn D, Terstappen LW, Fehm T, Tucker TF, Lane N, Wang J, Uhr JW. Circulating tumor cells in patients with breast cancer dormancy. Clin Cancer Res. 2004;10:8152–62.
16. Weckermann D, Polzer B, Ragg T, Blana A, Schlimok G, Arnholdt H, Bertz S, Harzmann R, Klein CA. Perioperative activation of disseminated tumor cells in bone marrow of patients with prostate cancer. J Clin Oncol. 2009;27:1549–56.
17. Borgen E, Rypdal MC, Sosa MS, Renolen A, Schlichting E, Lonning PE, Synnestvedt M, Aguirre-Ghiso JA, Naume B. NR2F1 stratifies dormant disseminated tumor cells in breast cancer patients. Breast Cancer Res. 2018;20:120.
18. Pantel K, Schlimok G, Braun S, Kutter D, Lindemann F, Schaller G, Funke I, Izbicki JR, Riethmuller G. Differential expression of proliferation-associated molecules in individual micrometastatic carcinoma cells. J Natl Cancer Inst. 1993;85:1419–24.
19. Esserman LJ, Berry DA, DeMichele A, Carey L, Davis SE, Buxton M, Hudis C, Gray JW, Perou C, Yau C, Livasy C, Krontiras H, Montgomery L, Tripathy D, Lehman C, Liu MC, Olopade OI, Rugo HS, Carpenter JT, Dressler L, Chhieng D, Singh B, Mies C, Rabban J, Chen YY, Giri D, van't Veer L, Hylton N. Pathologic complete response predicts recurrence-free survival more effectively by cancer subset: results from the I-SPY 1 TRIAL--CALGB 150007/150012, ACRIN 6657. J Clin Oncol. 2012;30:3242–9.
20. Kuerer HM, Newman LA, Smith TL, Ames FC, Hunt KK, Dhingra K, Theriault RL, Singh G, Binkley SM, Sneige N, Buchholz TA, Ross MI, McNeese MD, Buzdar AU, Hortobagyi GN, Singletary SE. Clinical course of breast cancer patients with complete pathologic primary tumor and axillary lymph node response to doxorubicin-based neoadjuvant chemotherapy. J Clin Oncol. 1999;17:460–9.
21. Ogston KN, Miller ID, Payne S, Hutcheon AW, Sarkar TK, Smith I, Schofield A, Heys SD. A new histological grading system to assess response of breast cancers to primary chemotherapy: prognostic significance and survival. Breast. 2003;12:320–7.
22. Sataloff DM, Mason BA, Prestipino AJ, Seinige UL, Lieber CP, Baloch Z. Pathologic response to induction chemotherapy in locally advanced carcinoma of the breast: a determinant of outcome. J Am Coll Surg. 1995;180:297–306.
23. Sosa MS, Avivar-Valderas A, Bragado P, Wen HC, Aguirre-Ghiso JA. ERK1/2 and p38alpha/beta signaling in tumor cell quiescence: opportunities to control dormant residual disease. Clin Cancer Res. 2011;17:5850–7.
24. Ranganathan AC, Zhang L, Adam AP, Aguirre-Ghiso JA. Functional coupling of p38-induced up-regulation of BiP and activation of RNA-dependent protein kinase-like endoplasmic reticulum kinase to drug resistance of dormant carcinoma cells. Cancer Res. 2006;66:1702–11.
25. Schewe DM, Aguirre-Ghiso JA. ATF6alpha-Rheb-mTOR signaling promotes survival of dormant tumor cells in vivo. Proc Natl Acad Sci U S A. 2008;105:10519–24.
26. Hata AN, Niederst MJ, Archibald HL, Gomez-Caraballo M, Siddiqui FM, Mulvey HE, Maruvka YE, Ji F, Bhang HE, Krishnamurthy Radhakrishna V, Siravegna G, Hu H, Raoof S, Lockerman E, Kalsy A, Lee D, Keating CL, Ruddy DA, Damon LJ, Crystal AS, Costa C, Piotrowska Z, Bardelli A, Iafrate AJ, Sadreyev RI, Stegmeier F, Getz G, Sequist LV, Faber AC, Engelman JA. Tumor cells can follow distinct evolutionary paths to become resistant to epidermal growth factor receptor inhibition. Nat Med. 2016;22:262–9.
27. Sharma SV, Lee DY, Li B, Quinlan MP, Takahashi F, Maheswaran S, McDermott U, Azizian N, Zou L, Fischbach MA, Wong KK, Brandstetter K, Wittner B, Ramaswamy S, Classon M, Settleman J. A chromatin-mediated reversible drug-tolerant state in cancer cell subpopulations. Cell. 2010;141:69–80.
28. Kurppa KJ, Liu Y, To C, Zhang T, Fan M, Vajdi A, Knelson EH, Xie Y, Lim K, Cejas P, Portell A, Lizotte PH, Ficarro SB, Li S, Chen T, Haikala HM, Wang H, Bahcall M, Gao Y, Shalhout S, Boettcher S, Shin BH, Thai T, Wilkens MK, Tillgren ML, Mushajiang M, Xu M, Choi J, Bertram AA, Ebert BL, Beroukhim R, Bandopadhayay P, Awad MM, Gokhale PC, Kirschmeier PT, Marto JA, Camargo FD, Haq R, Paweletz CP, Wong KK, Barbie DA, Long HW, Gray NS, Janne PA. Treatment-induced tumor dormancy through YAP-mediated transcriptional reprogramming of the apoptotic pathway. Cancer Cell. 2020;37:104–122.e112.
29. Sanchez-Danes A, Larsimont JC, Liagre M, Munoz-Couselo E, Lapouge G, Brisebarre A, Dubois C, Suppa M, Sukumaran V, Del Marmol V, Tabernero J, Blanpain C. A slow-cycling LGR5 tumour population mediates basal cell carcinoma relapse after therapy. Nature. 2018;562:434–8.
30. Alvarez JV, Pan TC, Ruth J, Feng Y, Zhou A, Pant D, Grimley JS, Wandless TJ, Demichele A, Investigators IST, Chodosh LA. Par-4 downregulation promotes breast cancer recurrence by preventing multinucleation following targeted therapy. Cancer Cell. 2013;24:30–44.
31. Feng Y, Pan TC, Pant DK, Chakrabarti KR, Alvarez JV, Ruth JR, Chodosh LA. SPSB1 promotes breast cancer recurrence by potentiating c-MET signaling. Cancer Discov. 2014;4:790–803.
32. Lindstrom LS, Karlsson E, Wilking UM, Johansson U, Hartman J, Lidbrink EK, Hatschek T, Skoog L, Bergh J. Clinically used breast cancer markers such as estrogen receptor, progesterone receptor, and human epidermal growth factor receptor 2 are unstable throughout tumor progression. J Clin Oncol. 2012;30:2601–8.
33. Moody SE, Perez D, Pan TC, Sarkisian CJ, Portocarrero CP, Sterner CJ, Notorfrancesco KL, Cardiff RD, Chodosh LA. The transcriptional repressor snail promotes mammary tumor recurrence. Cancer Cell. 2005;8:197–209.
34. Boxer RB, Jang JW, Sintasath L, Chodosh LA. Lack of sustained regression of c-MYC-induced mammary adenocarcinomas following brief or prolonged MYC inactivation. Cancer Cell. 2004;6:577–86.
35. Gunther EJ, Moody SE, Belka GK, Hahn KT, Innocent N, Dugan KD, Cardiff RD, Chodosh LA. Impact of p53 loss on reversal and recurrence of conditional Wnt- induced tumorigenesis. Genes Dev. 2003;17:488–501.
36. Viale A, Pettazzoni P, Lyssiotis CA, Ying H, Sanchez N, Marchesini M, Carugo A, Green T, Seth S, Giuliani V, Kost-Alimova M, Muller F, Colla S, Nezi L, Genovese G, Deem AK, Kapoor A, Yao W, Brunetto E, Kang Y, Yuan M, Asara JM, Wang YA, Heffernan TP, Kimmelman AC, Wang H, Fleming JB, Cantley LC, DePinho RA, Draetta GF. Oncogene ablation-resistant pancreatic cancer cells depend on mitochondrial function. Nature. 2014;514:628–32.

37. Puig I, Tenbaum SP, Chicote I, Arques O, Martinez-Quintanilla J, Cuesta-Borras E, Ramirez L, Gonzalo P, Soto A, Aguilar S, Eguizabal C, Caratu G, Prat A, Argiles G, Landolfi S, Casanovas O, Serra V, Villanueva A, Arroyo AG, Terracciano L, Nuciforo P, Seoane J, Recio JA, Vivancos A, Dienstmann R, Tabernero J, Palmer HG. TET2 controls chemoresistant slow-cycling cancer cell survival and tumor recurrence. J Clin Invest. 2018;128: 3887–905.
38. Vander Heiden MG, Cantley LC, Thompson CB. Understanding the Warburg effect: the metabolic requirements of cell proliferation. Science. 2009;324:1029–33.
39. Alvarez JV, Belka GK, Pan TC, Chen CC, Blankemeyer E, Alavi A, Karp J, Chodosh LA. Oncogene pathway activation in mammary tumors dictates [18F]-FDG-PET uptake. Cancer Res. 2014;74:7583–98.
40. Havas KM, Milchevskaya V, Radic K, Alladin A, Kafkia E, Garcia M, Stolte J, Klaus B, Rotmensz N, Gibson TJ, Burwinkel B, Schneeweiss A, Pruneri G, Patil KR, Sotillo R, Jechlinger M. Metabolic shifts in residual breast cancer drive tumor recurrence. J Clin Invest. 2017;127:2091–105.
41. Fox DB, Garcia NMG, McKinney BJ, Lupo R, Noteware LC, Newcomb R, Liu J, Locasale JW, Hirschey MD, Alvarez JV. NRF2 activation promotes the recurrence of dormant tumour cells through regulation of redox and nucleotide metabolism. Nat Metab. 2020;2:318–34.
42. Krall EB, Wang B, Munoz DM, Ilic N, Raghavan S, Niederst MJ, Yu K, Ruddy DA, Aguirre AJ, Kim JW, Redig AJ, Gainor JF, Williams JA, Asara JM, Doench JG, Janne PA, Shaw AT, McDonald Iii RE, Engelman JA, Stegmeier F, Schlabach MR, Hahn WC. KEAP1 loss modulates sensitivity to kinase targeted therapy in lung cancer. Elife. 2017;6:e18970.
43. Chio IIC, Tuveson DA. ROS in cancer: the burning question. Trends Mol Med. 2017;23:411–29.
44. Schafer ZT, Grassian AR, Song L, Jiang Z, Gerhart-Hines Z, Irie HY, Gao S, Puigserver P, Brugge JS. Antioxidant and oncogene rescue of metabolic defects caused by loss of matrix attachment. Nature. 2009;461:109–13.
45. Wang H, Nicolay BN, Chick JM, Gao X, Geng Y, Ren H, Gao H, Yang G, Williams JA, Suski JM, Keibler MA, Sicinska E, Gerdemann U, Haining WN, Roberts TM, Polyak K, Gygi SP, Dyson NJ, Sicinski P. The metabolic function of cyclin D3-CDK6 kinase in cancer cell survival. Nature. 2017;546:426–30.
46. Hampsch RA, Wells JD, Traphagen NA, McCleery CF, Fields JL, Shee K, Dillon LM, Pooler DB, Lewis LD, Demidenko E, Huang YH, Marotti JD, Goen AE, Kinlaw WB, Miller TW. AMPK activation by metformin promotes survival of dormant ER(+) breast cancer cells. Clin Cancer Res. 2020;26:3707–19.
47. Yun CW, Lee SH. The roles of autophagy in cancer. Int J Mol Sci. 2018;19:3466.
48. Aveic S, Pantile M, Polo P, Sidarovich V, De Mariano M, Quattrone A, Longo L, Tonini GP. Autophagy inhibition improves the cytotoxic effects of receptor tyrosine kinase inhibitors. Cancer Cell Int. 2018;18:63.
49. Egan DF, Chun MG, Vamos M, Zou H, Rong J, Miller CJ, Lou HJ, Raveendra-Panickar D, Yang CC, Sheffler DJ, Teriete P, Asara JM, Turk BE, Cosford ND, Shaw RJ. Small molecule inhibition of the autophagy kinase ULK1 and identification of ULK1 substrates. Mol Cell. 2015;59:285–97.
50. Egan DF, Shackelford DB, Mihaylova MM, Gelino S, Kohnz RA, Mair W, Vasquez DS, Joshi A, Gwinn DM, Taylor R, Asara JM, Fitzpatrick J, Dillin A, Viollet B, Kundu M, Hansen M, Shaw RJ. Phosphorylation of ULK1 (hATG1) by AMP-activated protein kinase connects energy sensing to mitophagy. Science. 2011;331:456–61.
51. Kim J, Kundu M, Viollet B, Guan KL. AMPK and mTOR regulate autophagy through direct phosphorylation of Ulk1. Nat Cell Biol. 2011;13:132–41.
52. Vera-Ramirez L, Vodnala SK, Nini R, Hunter KW, Green JE. Autophagy promotes the survival of dormant breast cancer cells and metastatic tumour recurrence. Nat Commun. 2018;9:1944.
53. Damrauer JS, Phelps SN, Amuchastegui K, Lupo R, Mabe NW, Walens A, Kroger BR, Alvarez JV. Foxo-dependent Par-4 upregulation prevents long-term survival of residual cells following PI3K-Akt inhibition. Mol Cancer Res. 2018;16:599–609.
54. Mabe NW, Fox DB, Lupo R, Decker AE, Phelps SN, Thompson JW, Alvarez JV. Epigenetic silencing of tumor suppressor Par-4 promotes chemoresistance in recurrent breast cancer. J Clin Invest. 2018;128:4413–28.
55. Sequist LV, Waltman BA, Dias-Santagata D, Digumarthy S, Turke AB, Fidias P, Bergethon K, Shaw AT, Gettinger S, Cosper AK, Akhavanfard S, Heist RS, Temel J, Christensen JG, Wain JC, Lynch TJ, Vernovsky K, Mark EJ, Lanuti M, Iafrate AJ, Mino-Kenudson M, Engelman JA. Genotypic and histological evolution of lung cancers acquiring resistance to EGFR inhibitors. Sci Transl Med. 2011;3:75ra26.
56. Walens A, Lin J, Damrauer JS, McKinney B, Lupo R, Newcomb R, Fox DB, Mabe NW, Gresham J, Sheng Z, Sibley AB, De Buysscher T, Kelkar H, Mieczkowski PA, Owzar K, Alvarez JV. Adaptation and selection shape clonal evolution of tumors during residual disease and recurrence. Nat Commun. 2020;11:5017.
57. Hangauer MJ, Viswanathan VS, Ryan MJ, Bole D, Eaton JK, Matov A, Galeas J, Dhruv HD, Berens ME, Schreiber SL, McCormick F, McManus MT. Drug-tolerant persister cancer cells are vulnerable to GPX4 inhibition. Nature. 2017;551:247–50.
58. Viswanathan VS, Ryan MJ, Dhruv HD, Gill S, Eichhoff OM, Seashore-Ludlow B, Kaffenberger SD, Eaton JK, Shimada K, Aguirre AJ, Viswanathan SR, Chattopadhyay S, Tamayo P, Yang WS, Rees MG, Chen S, Boskovic ZV, Javaid S, Huang C, Wu X, Tseng YY, Roider EM, Gao D, Cleary JM, Wolpin BM, Mesirov JP, Haber DA, Engelman JA, Boehm JS, Kotz JD, Hon CS, Chen Y, Hahn WC, Levesque MP, Doench JG, Berens ME, Shamji AF, Clemons PA, Stockwell BR, Schreiber SL. Dependency of a therapy-resistant state of cancer cells on a lipid peroxidase pathway. Nature. 2017;547:453–7.
59. Maynard A, McCoach CE, Rotow JK, Harris L, Haderk F, Kerr DL, Yu EA, Schenk EL, Tan W, Zee A, Tan M, Gui P, Lea T, Wu W, Urisman A, Jones K, Sit R, Kolli PK, Seeley E, Gesthalter Y, Le DD, Yamauchi KA, Naeger DM, Bandyopadhyay S, Shah K, Cech L, Thomas NJ, Gupta A, Gonzalez M, Do H, Tan L, Bacaltos B, Gomez-Sjoberg R, Gubens M, Jahan T, Kratz JR, Jablons D, Neff N, Doebele RC, Weissman J, Blakely CM, Darmanis S, Bivona TG. Therapy-induced evolution of human lung cancer revealed by single-cell RNA sequencing. Cell. 2020;182:1232–1251.e1222.
60. Gilbert LA, Hemann MT. DNA damage-mediated induction of a chemoresistant niche. Cell. 2010;143:355–66.
61. Walens A, DiMarco AV, Lupo R, Kroger BR, Damrauer JS, Alvarez JV. CCL5 promotes breast cancer recurrence through macrophage recruitment in residual tumors. elife. 2019;8
62. Gong K, Guo G, Gerber DE, Gao B, Peyton M, Huang C, Minna JD, Hatanpaa KJ, Kernstine K, Cai L, Xie Y, Zhu H, Fattah FJ, Zhang S, Takahashi M, Mukherjee B, Burma S, Dowell J, Dao K, Papadimitrakopoulou VA, Olivas V, Bivona TG, Zhao D, Habib AA. TNF-driven adaptive response mediates resistance to EGFR inhibition in lung cancer. J Clin Invest. 2018;128:2500–18.

Part IV

The Microenvironment of Site-Specific Metastasis

Isaac P. Witz

Cancer research in the seventies and eighties of the last century was dominated by the "cancer-centric" view postulating that genomic alterations are the only and exclusive determinants of carcinogenesis and progression.

At the same time it became increasingly clear to a minority group of cancer researchers that the cancer-centric view is incapable to solve the many challenges of the pathobiology of cancer including providing a satisfactory mechanism for the progression towards metastasis. It was also realized that cancer cells are part of an ecosystem, that they are "surrounded" by the host (cancer patients or tumor-bearing animals), and that host components take part in tumor progression.

The cancer ecosystem incorporating host-derived cellular and molecular components as well as the landscape they create is defined as the tumor microenvironment [1].

Metastasis is the major killer of cancer patients.

One of the key elements of metastasis is its organ specificity. Tumor cells spread to the vasculature of many organs but they form overt metastases in only a select few to which they have a marked predilection.

Stephen Paget about 130 years ago was the first scientist to recognize that metastasis is site specific. Describing this phenomenon in botanical terms he wrote:

"When a plant goes to seed, its seeds are carried in all directions; but they can only live and grow if they fall on congenial soil." In contemporary terms, the seeds are the cancer cells and the soil is the organ that preferentially and specifically hosts future metastasis [1].

A significant step towards the comprehension of the mechanism underlying the formation of metastasis in specific sites was the discovery of the pre-metastatic niche (PMN) by David Lyden's group [2].

Soluble factors and extracellular vesicles released from the primary tumor induce, in distant organs, the formation of microenvironments that are hospitable for circulating metastasizing cells before their arrival at these organs. These microenvironments are termed pre-metastatic niches [3].

The chapter, The Microenvironment of Site-Specific Metastasis, deals with three main subjects: principles underlying microenvironmental interactions of cancer cells with non-cancer cells, metastatic cascade, and targeting the TME for cancer therapy.

References

1. Maman S, Witz IP. A history of exploring cancer in context. Nat Rev Cancer. 2018;18:359–76.
2. Kaplan RN, Riba RD, Zacharoulis S, Bramley AH, Vincent L, Costa C, MacDonald DD, Jin DK, Shido K, Kerns SA, Zhu Z, Hicklin D, Wu Y, Port JL, Altorki N, Port ER, Ruggero D, Shmelkov SV, Jensen KK, Rafii S, Lyden D. VEGFR1-positive haematopoietic bone marrow progenitors initiate the pre-metastatic niche. Nature. 2005;438(7069):820–7.
3. Wortzel I, Dror S, Kenific CM, Lyden D. Exosome-mediated metastasis: communication from a distance. Dev Cell. 2019;49:347–60.

11 The Microenvironment of Site-Specific Metastasis

Isaac P. Witz and Sivan Izraely

Abstract

Solid tumors can be visualized as an ecosystem composed of cancer and non-cancerous cells and their products, extra cellular matrix, metabolites, and numerous other factors. The cellular components of this ecosystem (termed tumor microenvironment) form an evolving signaling network. The cross talk between cancer and non-cancerous cells brings about a progressing phenotype reprograming of the interacting cells culminating either in tumor demise, dormancy, or progression. The progression of solid tumors on the road to metastasis is a stepwise process comprising well-defined steps. An early step is the formation of a pre-metastatic niche in selected organs that will ultimately host metastasis. Other progression steps are: acquisition of a migratory capacity by a subpopulation of tumor cells resulting in their detachment from the tumor mass, penetration into the circulation, extravasation into organs hosting a pre-metastatic niche, and finally the establishment of a metastatic lesion.

Particular cancers usually metastasize to multiple but favored organ sites. The factors that determine organ-specific metastasis are: reciprocal recognition elements, such as integrins expressed by tumor cells as well as by the endothelium of the target organ and survival and growth factors expressed by this organ.

The realization that interactions between tumors and their microenvironments play pivotal roles in driving or inhibiting tumor progression is being translated to the clinical setting in efforts to boost tumor-microenvironment interactions that inhibit tumor progression and/or oppose interactions that promote tumor progression.

I. P. Witz (✉) · S. Izraely
The Shmunis School of Biomedicine and Cancer Research, George S. Wise Faculty of Life Science, Tel Aviv University, Tel Aviv, Israel
e-mail: IsaacW@tauex.tau.ac.il; sivanisr@taux.tau.ac.il

Learning Objectives

- Define "Tumor Microenvironment" (TME) and review principles of tumor-TME cross talk.
- Describe the progression pathway from primary tumor to metastasis.
- Identify TME-associated factors that determine site-specificity of metastasis.
- Review strategies to modulate the cross talk between the tumor and its microenvironment for the development of cancer therapy.

11.1 Introduction

Stephen Paget, over a century ago, was the first to conceive the idea that tumor formation and especially tumor progression are largely determined by interactions of cancer cells with their microenvironment. This concept, lying dormant until the early seventies of last century, was awakened by studies on angiogenesis and on the immune landscape of tumors [1] conducted by a handful number of researchers.

The comprehension of principles underlying signal transduction and of mechanisms by which signaling networks control gene expression [2, 3] gradually led to the realization that tumor cells communicate with cells in their immediate and distant environments and that the genome of cancer cells and that of non-tumor cells present in the tumor microenvironment (TME) is reciprocally regulated by cues derived from these interaction partners. It became clear that this cross talk is a major driver of tumor progression on the road to metastasis [4].

This concept is now well accepted by the cancer-research community [5].

In addition to tumor cells the TME is composed of host-derived resident and recruited cells such as macrophages, lymphocytes or fibroblasts, secreted products of these cells such as extracellular vesicles and soluble factors such as growth factors, antibodies, components of the extracellular

S. P. Leong et al. (eds.), *Cancer Metastasis Through the Lymphovascular System*, https://doi.org/10.1007/978-3-030-93084-4_11

matrix (ECM) and metabolites. Moreover the TME of solid tumors is characterized by hypoxia and low pH.

All the components of the TME create a dynamic interaction arena in which a web of signaling networks leads to 3 alternative pathways: tumor progression, tumor regression, or tumor dormancy.

The current trends in TME research are mainly focused on exploring pathways leading to metastasis. Targeting interactive tumor-TME pathways such as angiogenesis and immune responses as well as TME constituents such as inflammation has developed into a critical strategy aimed at the development of novel modalities of cancer therapy.

One of the major characteristics of metastasis is their organ specificity [6, 7].

The fact that the tumor-TME cross talk shapes the phenotype of both tumor and non-tumor cells infers that metastases originating in a single primary tumor but metastasizing to different organs (i.e., different microenvironments) will express different phenotypes. Studies on the functions of the metastatic microenvironment in resisting or promoting site-specific metastasis are now in the forefront of TME research.

This chapter will review the principles of tumor-TME interactions, the steps of the metastatic cascade, and the principles of site-specific metastasis. Cancer therapy modalities involving targeting the TME will also be reviewed (Fig. 11.1).

11.2 The Tumor and Its Microenvironment: Principles

Listed below are general features of tumor cells in the context of their immediate milieu.

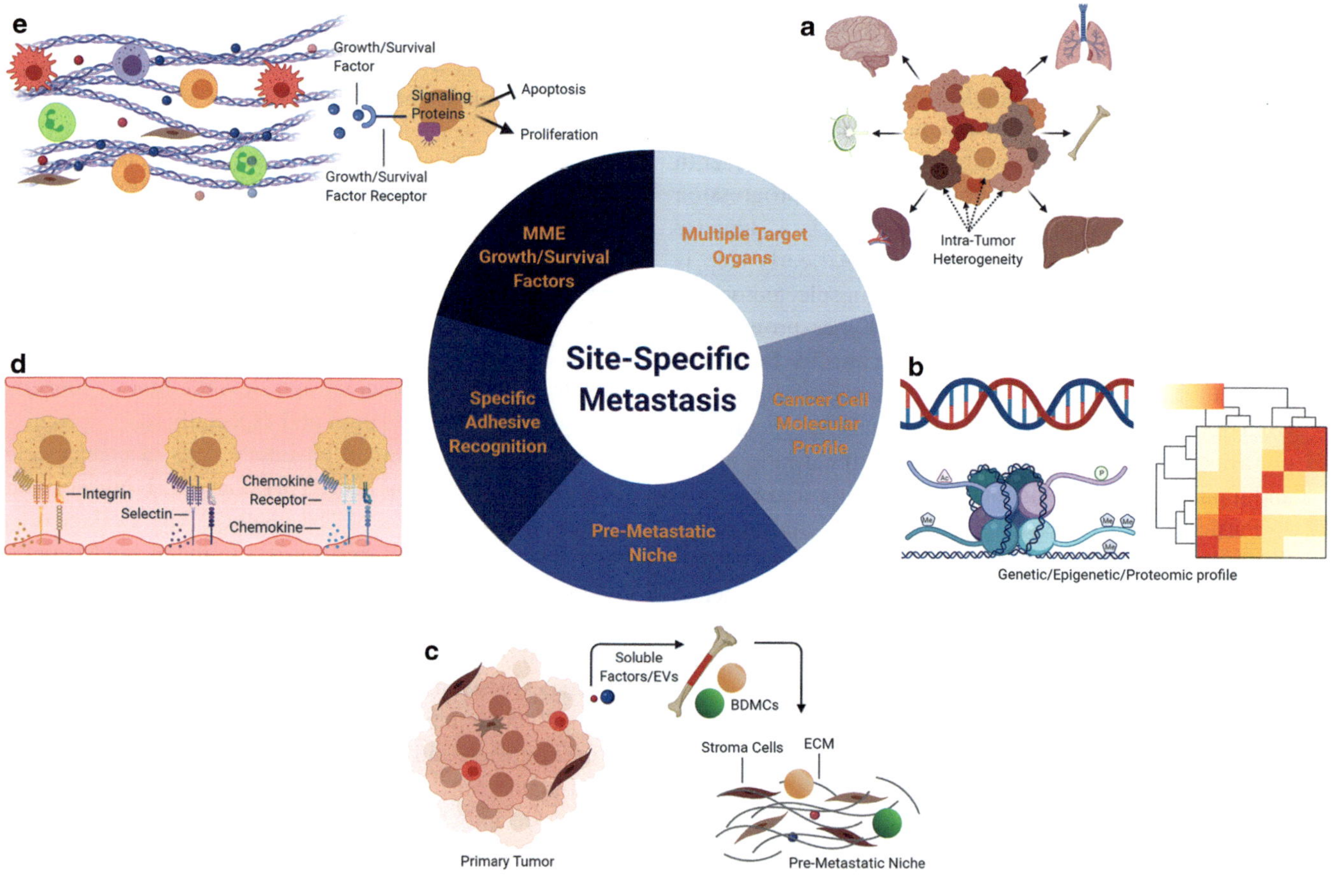

Fig. 11.1 The principles of site-specific metastasis. Site-specific metastasis is manifested by several key principals: (**a**) Tumor cells derived from a single tumor, due to their phenotypic diversity, metastasize to multiple organs, that are characteristic targets to each cancer type. (**b**) Alterations in the genetic, epigenetic, and proteomic profiles of metastasizing cancer cells largely determine the specificity of target organs. (**c**) Extracellular vesicles (EVs) and soluble factors released from primary tumor cells induce the release of progenitor cells (bone marrow-derived cells (BDMCs)) from bone marrow and their targeted migration to a specific future metastatic site, conditioning the pre-metastatic niche for future metastatic colonization. (**d**) Specific adhesive recognition between tumor cells and the endothelium of the target organ is mediated by adhesion molecules such as selectins and integrins and their ligands, as well as by chemokines and their receptors, that are tumor-type/microenvironment-specific. (**e**) Binding of survival and growth factors derived from the metastatic microenvironment to their cognate receptors expressed by tumor cells enable the survival and propagation of tumor cells in the target organ site, and inhibit cell death, through various signaling cascades

1. *The TME is jointly created by the tumor and by the host.* This point was referred to in Sect. 11.1.
2. *Tumor heterogeneity creates diverse evolutionary pathways of tumor cells in a single TME.*
   Intratumor heterogeneity involving genetic, epigenetic, and transcriptomic modifications of tumor cells residing in a specific TME is a pivotal factor that determines tumor evolution in this microenvironment [8].
   Numerous and diverse TME-originating signals are delivered simultaneously to the heterogeneous population of tumor cells present in a given TME. The outcome of this reality is that distinct tumor evolution pathways take place at the same time in a single tumor ecosystem [9, 10].
3. *Tumors having a similar histology but developing in different organs express diverse phenotypes.* This point is very obvious: For example: epithelial tumors (carcinomas) may develop in different organs expressing features specific to that organ. Since, as mentioned above, the microenvironment shapes the phenotype of tumor cells and vice versa, carcinomas in different organ microenvironments are different. This point will be highlighted when discussing site-specific metastasis (see below).
4. *Reciprocal phenotype reprograming between tumor cells and non-tumor microenvironmental cells.* The interactions between tumor cells and non-tumor cells in the TME are reciprocal and bidirectional. Each interaction partner shapes the transcriptomic, proteomic, metabolomic, secretomic, etc. phenotype of the other partner. This mutual reprograming may drive or block tumor progression. Several examples illustrate this point.
   (a) Tumor cells and the extracellular matrix modulate each other's phenotype [11], thereby driving tumor progression.
   (b) Cancer cells activate fibroblasts to become tumor-associated fibroblasts which, in turn, promote (or inhibit) tumor progression and metastasis by a variety of mechanisms [12].
   (c) Tumor-derived chemokines promote tumor progression by recruiting suppressor cells to the TME, thereby inhibiting immune responses. Such tumor-derived chemokines may also function as autocrine growth factors [13]. TME-derived chemokines may direct cancer cells expressing the corresponding receptors to specific organs from which such chemokines are released [14].

   The balance between opposing effects of tumor-TME cross talk determines the extent of malignancy. If the impact of tumor drivers is stronger than that of inhibitors, the tumor will progress. If the opposite occurs, the tumor will regress. If these factors cancel each other, the tumor will enter into a state of dormancy.
5. *Tumor-TME interactions are dynamic*
   The tumor-TME cross talk reprograms the phenotype of these interaction partners and leads to a continuous emergence of novel tumor variants.
6. *The tumor microenvironment may operate as a double agent*
   Single microenvironmental components may exert pro- or anti-malignancy functions [15].
   The following 2 examples, illustrating the Yin Yang functions of TME components on tumor progression, are appropriate illustrations of this reality.
   (a) In breast cancer tumor infiltrating B cells enhance anti-tumor immune responses but may also create an immune-suppressive microenvironment [16].
   (b) Interferon gamma (IFNγ) stimulates both anti-tumor immune activities and immune evasion of tumors. In addition, IFNγ may act directly upon tumor cells and reprograms the expression of molecules that either drive or inhibit the malignant phenotype resulting in opposing effects on tumor progression [17].
7. Tumor cells co-opt (hijack) normal biologic processes such as mechanisms that regulate immune responses, thereby defending themselves from immune attacks [18, 19] or chemokine-chemokine axes to direct their trafficking to secondary sites [20].
   Another physiological process hijacked by tumors is epithelial to mesenchymal transition (EMT) being of prime significance in embryonic development [21].
   This process confers upon epithelial tumor cells plasticity, migratory and invasive and metastasizing capacities [22].

11.3 The Metastatic Cascade

The metastatic cascade is a sequential process in which a subpopulation of primary tumor cells acquires the capacity to migrate, reprogram the extracellular matrix, and invade. These functions endow the cells with the ability to reach lymph and/or blood vessels and invade them. Circulating tumor cells (CTC) in the lymph may reach lymph nodes while CTC circulating in the blood may reach specific target organs (see below—section on site-specific metastasis). The CTC may then form micrometastases undetectable by regular imaging techniques, and/or clinically detectable metastasis.

Each step of the metastatic cascade is jointly controlled by tumor-intrinsic factors and by those originating in the TME [23].

While reviewing the steps of the metastatic cascade we should keep in mind several realities:

- The first steps of the metastatic cascade are common for different types of tumors while other steps are organ specific.
- Metastasis although being an inefficient process is the leading cause of cancer patient death [24].
- Most cancer patients harbor metastasis by the time of diagnosis [25].
- The cancer lesion is heterogeneous, comprising a diverse assortment of cellular variants with distinct genetic and transcriptomic characteristics. Such variants may progress at different speeds so that at a given time point an individual tumor would contain cells at different stages of the metastatic cascade [26].

11.3.1 Site-Specific Metastasis: Basic Principles

The predilection of different cancer types to metastasize to particular organs and the organ specificity of metastasis was first described by Paget in 1889 [27].

The concept of site-specific metastasis lay dormant for about 100 years when the basic principles of site-specific metastasis started to be appreciated [6, 28].

- Different cancer types usually metastasize to multiple but favored organ sites [29] implying that metastases from a single tumor but developing in different organs would express diverse phenotypes.
- The genetic, epigenetic, and proteomic profile of the metastasizing cancer cells as well as factors derived from the metastatic microenvironment govern site-specific metastasis [30, 31].
- Tumor cells reach the vasculature of all organs but metastasis develops only in selected organs that host a pre-metastatic niche. Specific adhesive recognition between cancer cells and the endothelium of the target organ, as well as nonspecific anatomical or mechanical factors such as circulation mechanics are involved in the arrest of circulating cancer cells at the target organ [32, 33].
- The ability of metastasizing tumor cells to successfully establish metastasis depends to a large extent by survival and growth factors from the metastatic microenvironment [34].

11.3.2 The Acquisition of a Migratory and Invasive Phenotype by Epithelial Cancer Cells

The metastatic cascade is initiated when epithelial cancer cells acquire invasive, migratory, and stem cell properties. The driver of this phase is a process termed Epithelial to Mesenchymal Transition (EMT) [35].

The EMT empowers the tumor cells to invade neighboring tissues, reach the circulation, spread throughout the body, and subsequently metastasize to specific organs. EMT is driven by genetic and epigenetic alterations and by TME-derived paracrine cues delivered from resident or infiltrating non-tumor cells such as fibroblasts, macrophages, or immunocytes [36].

The molecular programs that drive EMT function via microenvironmental multi-signaling pathways upregulate several transcription factors such as TWIST, ZEB1, and SNAIL which are identified as the EMT core signature [37, 38].

The EMT process follows distinct intermediate stages. Multiple tumor subpopulations expressing phenotypes ranging from a complete epithelial to a complete mesenchymal one may co-inhabit single solid tumors [39].

In addition to effects on the tumor cells as specified above, EMT also remodels the extracellular matrix and affects various phenotypic traits of microenvironmental cells. These effects could impact tumor progression [38, 40].

11.3.3 Intravasation and the Generation of Circulating Tumor Cells

Cancer cell intravasation is the process by which cancer cells penetrate the vessel wall and enter the circulation. While tumor cells may passively enter the circulation into damaged vessels, they may also actively intravasate into intact vessels. Microenvironmental cells such as macrophages may participate in the intravasation process by secreting cytokines and proteases that increase the permeability of blood and lymphatic vessels. The density and diameter of vessels play also a role in intravasation [41].

Cancer cells may enter lymphatic vessels, migrate to the tumor-draining lymph node (referred to as the Sentinel Lymph Node—SLN), spread to other lymph nodes, develop into lesions in these nodes, and then enter the hematogenous vasculature to spread to secondary organs [42, 43].

The SLN, described for the first time in 1960 [44], is the first "foreign" microenvironment encountered by tumor cells following their emigration from the primary tumor [45].

The primary tumor plays important functions in the establishment and maintenance of metastasis in the SLN by inducing immunosuppression [46] and lymphangiogenesis in this node, thereby generating a pre-metastatic niche [47–49].

Similar to other TMEs, the microenvironment of the SLN may contribute to distant metastasis by shaping and intensifying the malignant phenotype of tumor cells that migrated to this node as well as by serving as a resource for such cells. This issue is however still under debate and investigation [43].

Another unsettled issue is whether tumor cells invade the systemic circulation directly from the primary tumor or indirectly through the intermediate residence in the SLN [50].

Once in the blood stream, CTC have to survive coagulation as well as physical and oxidative stress and to escape damage inflicted by circulating immunocytes such as NK cells. Indeed, most CTCs die in the circulation. Two major mechanisms defend CTCs from these impediments. The first is CTC coating with platelets which provides a physical shield from the destructive effects of shear stress and from being killed by NK cells. The platelets also transduce protective signals and may enhance adhesion to blood vessels of secondary organs, some of which will host metastasis [51].

The second major mechanism that contributes to the survival of CTCs in the circulation is cell clustering. Clustered CTCs are protected from apoptosis caused by lack of contact with other cells (anoikis) and thus have a survival advantage over solitary CTCs. Clustered CTCs also adhere more efficiently to blood vessels of secondary organs [52].

11.3.4 Extravasation of CTCs to Future Metastatic Sites

CTCs may be trapped in small capillaries and invade from there to secondary organs [33].

However the specific interactions between CTCs and the endothelium of the secondary organ are the major facilitating element of site-specific extravasation into organs that are homes for a pre-metastatic niche [53].

Endothelial cells lining a particular organ often express a different profile of cell surface proteins compared to endothelial cells of other organs [54].

This difference accounts for the selective binding of CTCs to the vasculature of particular organs and for their extravasation into these organ sites [55].

Extravasation is a multistep process employed primarily by leukocytes to migrate through the endothelium and home to inflammatory sites. This process is mediated by interactions of selectins, integrins, cytokines, and chemokines with their ligands [56].

Endothelial selectins, by interacting with corresponding ligands expressed by circulating leukocytes, decelerate their flow in the blood causing their rolling on the endothelial layer. Leukocyte-integrins activated by endothelial chemokines interact with their endothelial ligands. This interaction triggers a firm adhesion between the leukocytes and the endothelial cells. Transendothelial migration (diapedesis) is the final step in the migration of leukocytes into the target organ.

Tumor cells hijacked the extravasation strategy of leukocytes and use it to invade specific organs and generate metastasis [57, 58].

11.3.5 Colonization

CTCs that transmigrated the endothelium of specific target organs may succumb to hostile factors (e.g., immune effectors); enter a state of dormancy and form micrometastasis or proliferate and progress to form clinically detectable metastasis. Factors originating in the metastatic microenvironment (MME) mediate these scenarios.

As mentioned above, TME-derived factors may exert anti- and pro-malignancy functions. The same holds true for the MME. A case in point are brain metastases. Microglia (the brain-residing macrophages) in a tumor cell-infiltrated brain may secrete both factors that trigger death of invading tumor cells and factors that promote propagation of such cells [59].

Cell death. Local immunity such as tumor infiltrating T cells is the major mechanism restricting the propagation of CTCs which invaded a secondary organ [60].

Other MME factors may also be involved in killing infiltrating tumor cells. For example, microenvironmental Apo2L/TRAIL, a member of the TNF superfamily, induces apoptosis of metastatic cells of different tumor types [61].

11.3.6 Dormancy and micrometastasis Formation

Tumor dormancy is expressed either as cellular dormancy (quiescent single cells or small micrometastases or as mass dormancy (an equal rate of proliferation and cell death) [62].

Induction and maintenance of dormancy of cancer cells in primary tumors and in metastatic sites are controlled by the TME and by the MME, respectively [63].

Below are a few examples:

Bone-derived factors such as TGFβ2 induce dormancy of bone-metastasizing prostate cancer cells [64].

The beta subunit of human hemoglobin (HBB) produced by alveolar epithelial and endothelial cells mediated growth arrest of lung-metastasizing neuroblastoma metastases [65].

The underlying mechanism of these cases of MME-driven dormancy is a creation of a high p38 and a low ERK signaling activity in cancer cells, thereby inducing dormancy in these cells [66].

MME-derived factors also awaken dormant tumor cells [60].

In addition to their capacity to induce dormancy, MME-derived signals are also able to awaken dormant tumor cells. For example, dormant human breast cancer cells are awakened by IL-6,

IL-8, TNFα, and TGFβ-mediated stromal inflammation [67].

11.3.7 Proliferation and Formation of Metastasis

Newly metastasizing as well as awakened dormant tumor cells need the support of the MME to proliferate and generate metastasis. This support is mediated by factors sustaining survival, fostering proliferation, and neutralizing unfavorable and hostile microenvironmental factors [24, 68].

Below are several examples of mechanisms by which MME-derived factors support, facilitate, and drive metastasis.

VCAM-1 expressing breast cancer cells bind to α4-integrin expressing MME-associated macrophages. This binding protects the cancer cells from apoptosis [69].

The chemokine CXCL12 and tumor necrosis factor-related apoptosis-inducing ligand (TRAIL) expressed in the MME of the bone promote the survival of breast cancer cells in bones [70] and the CX3CL1-CX3CR1 axis plays an important role in organ-specific peritoneal colonization of ovarian cancer. Reducing the interaction of the chemokine CX3CL1 with its receptor reduces metastasis [71].

Neutrophils in the lung MME support the colonization of metastasizing breast cancer cells [72], and IL-1β-expressing cells in the bone MME promoted bone metastasis formation by prostate cancer cells [73].

The local immune system is the major resistance mechanism against tumor formation and its progression towards metastasis [60].

However, tumors have developed several mechanisms to escape from anti-tumor immunity. The major immunity escape mechanisms are: creation of immunosuppressive tumor and metastatic microenvironments; downregulation of MHC class 1 and tumor-associated antigens; recruiting various types of suppressor cells into the TME or MME; and hijacking immune checkpoint strategies [74, 75].

11.4 Targeting Tumor-TME Cross Talk

The majority of cancer therapy modalities target only cancer cells [76].

However both tumors and their microenvironments play pivotal roles in shaping the malignant phenotype of tumors and in metastasis formation (see above).

The huge database on the TME (up to 2020, about 48,000 publications) lead the scientific community to employ discoveries in this area for the benefit of cancer patients. The two main avenues in such endeavors are to boost tumor-microenvironment interactions that antagonize tumor progression and/or block interactions that promote tumor progression [77]. Enhancing local anti-tumor immunity is the most pursued contemporary avenue.

Starting in the 1980s, the notion that the immune microenvironment can be enlisted to curb malignant processes gained increasing attention. Adoptive T cell therapy employing tumor-infiltrating lymphocytes (TILs) that kill cancer cells (mostly CD8-positive T cells) is being successfully used to treat certain cancer patients [78, 79].

Immune checkpoint blockade is a very promising and successful modality of immunotherapy. The 2018 Nobel Prize for Medicine was awarded for the identification of Programmed Cell Death Protein 1 (PD-1) and for conceiving the inhibition of the immune checkpoint strategy hijacked by tumor cells.

The physiological function of immune checkpoints is to regulate immune responses, especially autoimmune ones. This regulation is accomplished by interactions between PD-1 expressed by T cells and its PD-L1 ligand expressed by antigen-presenting cells. Tumor cells acquired the ability to express PD-L1 and thereby to interact with the corresponding receptor on T cells. This interaction blocks the ability of cytotoxic T cells that infiltrated the tumor microenvironment to kill the tumor cells. Inhibiting the PD-L1-PD-1 interaction by antibodies directed either to the receptor or to its ligand blocks this checkpoint and permits the killing of the tumor cells [80].

Targeting interactions between tumor cells and TME cells that drive tumor progression such as tumor-associated macrophages (TAMs), cancer-associated fibroblasts (CAFs), and various immune regulatory cells (T-regs, myeloid-derived suppressor cells-MDSC) is also studied and applied in the clinic [1, 81].

Targeting tumor-TME interactions that promote angiogenesis, a crucial event in tumor survival and progression, was one of the first efforts to target the TME [82]. Several anti-angiogenesis drugs have been approved by the FDA and are being used in the clinic. Inhibiting interactions between receptor tyrosine kinases (RTKs) expressed by tumor cells and TME ligands that stimulate tumor proliferation by RTK inhibitors or antibodies turned out to be an effective means to halt tumor progression [83]. Other modalities are in different stages of clinical development [84].

An exciting possibility to shift the balance of tumor-TME interactions to tumor inhibition is the possibility to "normalize" the TME, for example, by restoring the abnormal and leaky cancer vasculature or the metabolic defects in the TME [85, 86]. Two caveats have to be taken into account when considering the targeting of the tumor-TME cross talk. The first is that the diverse targeting approaches are mostly generally applicable and not restricted by organ specificity. The second is that there are quite many challenges in applying TME targeting. The following two examples illustrate such challenges. In view of the yin-yang functions of many microenvironmental factors (see above), targeting these factors may result in opposite results from those expected. Furthermore, due to the complexity of tumor-TME cross talk, TME-targeted treatments may instigate unintended adverse side effects [1].

11.5 Conclusions and Perspective

- Tumor cells and non-tumor cells in the tumor microenvironment maintain ongoing and dynamic interactions, thereby reprograming each other's phenotype.
- Tumors having a similar histology (e.g., carcinomas) but developing in different organs express diverse phenotypes.
- Microenvironmental components may under certain circumstances promote tumor progression and under different circumstances block it.
- The sequence of events starting with the detachment of tumor cells from the tumor mass and culminating in the formation of frank metastasis in a distant organ is called the metastatic cascade.
- Metastasis is site-specific. Certain cancer types have a marked predilection for metastasis to specific organs. Tumor cells reach the endothelium of most organs; however they can form metastasis in only select ones.
- Metastasis is formed in organs hosting a pre-metastatic niche.
- As discussed above, the tumor and the TME shape each other's phenotype. In view of the fact that the microenvironment of diverse organs is different, metastases originating in a single tumor but residing in diverse organ sites are different.
- In view of the fact that the tumor microenvironment takes part in the formation of metastasis efforts to target the microenvironment are increasingly employed in the clinic.

Questions

- What phenotypic changes may occur in tumor cells following their interaction with microenvironmental components?
- Why do epithelial tumors of the ovaries and of the colon express different phenotypic characteristics?
- What determines the organ specificity of metastasis originating in a single tumor?
- Please describe the process of tumor cell extravasation. Is this process restricted exclusively to tumor cells?
- Can tumor-TME interactions be targeted? If yes, how? Please provide 2 examples for clinical studies employing the strategy of targeting tumor-TME interactions.

Acknowledgments The authors thank their colleagues Dr. Orit Sagi-Assif, Tsipi Meshel and Shlomit Ben-Menachem, the Laboratory of Tumor Microenvironment & Metastasis Research for their valuable help and support. The research conducted by the authors is supported by the Dr. Miriam and Sheldon G. Adelson Medical Research Foundation (Needham, MA, USA), the Sara and Natan Blutinger Foundation (West Orange, NJ, USA), the Fred August and Adele Wolpers Charitable Fund (Clifton, NJ, USA) and the James and Rita Leibman Endowment Fund for Cancer Research (New York, NY, USA).

References

1. Maman S, Witz IP. A history of exploring cancer in context. Nat Rev Cancer. 2018;18(6):359–76.
2. Schlessinger J. Signal transduction by allosteric receptor oligomerization. Trends Biochem Sci. 1988;13(11):443–7.
3. Manning G, Whyte DB, Martinez R, Hunter T, Sudarsanam S. The protein kinase complement of the human genome. Science. 2002;298(5600):1912–34.
4. Witz IP. Tumor-microenvironment interactions: dangerous liaisons. Adv Cancer Res. 2008;100:203–29.
5. Hanahan D, Weinberg RA. Hallmarks of cancer: the next generation. Cell. 2011;144(5):646–74.
6. Fidler IJ. Cancer metastasis. Br Med Bull. 1991;47(1):157–77.
7. Obenauf AC, Massague J. Surviving at a distance: organ-specific metastasis. Trends Cancer. 2015;1(1):76–91.
8. McGranahan N, Swanton C. Clonal heterogeneity and tumor evolution: past, present, and the future. Cell. 2017;168(4):613–28.
9. Zhang M, Rosen JM. Developmental insights into breast cancer Intratumoral heterogeneity. Trends Cancer. 2015;1(4):242–51.
10. de Looff M, de Jong S, Kruyt FAE. Multiple interactions between cancer cells and the tumor microenvironment modulate TRAIL signaling: implications for TRAIL receptor targeted therapy. Front Immunol. 2019;10:1530.
11. Malik R, Lelkes PI, Cukierman E. Biomechanical and biochemical remodeling of stromal extracellular matrix in cancer. Trends Biotechnol. 2015;33(4):230–6.
12. Kalluri R. The biology and function of fibroblasts in cancer. Nat Rev Cancer. 2016;16(9):582–98.
13. Schneider GP, Salcedo R, Welniak LA, Howard OM, Murphy WJ. The diverse role of chemokines in tumor progression: prospects for intervention (review). Int J Mol Med. 2001;8(3):235–44.
14. Klein A, Sagi-Assif O, Meshel T, Telerman A, Izraely S, Ben-Menachem S, Bayry J, Marzese DM, Ohe S, Hoon DSB, Erez N, Witz IP. CCR4 is a determinant of melanoma brain metastasis. Oncotarget. 2017;8(19):31079–91.
15. Witz IP. Yin-yang activities and vicious cycles in the tumor microenvironment. Cancer Res. 2008;68(1):9–13.
16. Zhang Z, Zhu Y, Wang Z, Zhang T, Wu P, Huang J. Yin-yang effect of tumor infiltrating B cells in breast cancer: from mechanism to immunotherapy. Cancer Lett. 2017;393:1–7.
17. Ni L, Lu J. Interferon gamma in cancer immunotherapy. Cancer Med. 2018;7(9):4509–16.
18. Ribas A. Adaptive immune resistance: how cancer protects from immune attack. Cancer Discov. 2015;5(9):915–9.
19. Strazza M, Mor A. The complexity of targeting chemokines to promote a tumor immune response. Inflammation. 2020;43(4):1201–8.
20. Muller A, Homey B, Soto H, Ge N, Catron D, Buchanan ME, McClanahan T, Murphy E, Yuan W, Wagner SN, Barrera JL, Mohar A, Verastegui E, Zlotnik A. Involvement of chemokine receptors in breast cancer metastasis. Nature. 2001;410(6824):50–6.
21. Nieto MA, Huang RY, Jackson RA, Thiery JP. Emt: 2016. Cell. 2016;166(1):21–45.

22. Derynck R, Weinberg RA. EMT and cancer: more than meets the eye. Dev Cell. 2019;49(3):313–6.
23. Valastyan S, Weinberg RA. Tumor metastasis: molecular insights and evolving paradigms. Cell. 2011;147(2):275–92.
24. Massague J, Obenauf AC. Metastatic colonization by circulating tumour cells. Nature. 2016;529(7586):298–306.
25. Fidler IJ, Kripke ML. The challenge of targeting metastasis. Cancer Metastasis Rev. 2015;34(4):635–41.
26. Dagogo-Jack I, Shaw AT. Tumour heterogeneity and resistance to cancer therapies. Nat Rev Clin Oncol. 2018;15(2):81–94.
27. Paget S. The distribution of secondary growths in cancer of the breast. 1889. Cancer Metastasis Rev. 1989;8(2):98–101.
28. Hart IR. 'Seed and soil' revisited: mechanisms of site-specific metastasis. Cancer Metastasis Rev. 1982;1(1):5–16.
29. Nguyen DX, Bos PD, Massague J. Metastasis: from dissemination to organ-specific colonization. Nat Rev Cancer. 2009;9(4):274–84.
30. Horak E, Darling DL, Tarin D. Analysis of organ-specific effects on metastatic tumor formation by studies in vitro. J Natl Cancer Inst. 1986;76(5):913–22.
31. Maman S, Witz IP. The metastatic microenvironment. In: Shurin MR, Umansky V, Malyguine A, editors. The tumor immunoenvironment. New York, NY: Springer; 2013. p. 15–38.
32. Tarin D, Price JE. Influence of microenvironment and vascular anatomy on "metastatic" colonization potential of mammary tumors. Cancer Res. 1981;41(9 Pt 1):3604–9.
33. Wirtz D, Konstantopoulos K, Searson PC. The physics of cancer: the role of physical interactions and mechanical forces in metastasis. Nat Rev Cancer. 2011;11(7):512–22.
34. Radinsky R. Molecular mechanisms for organ-specific colon carcinoma metastasis. Eur J Cancer. 1995;31A(7–8):1091–5.
35. Yang J, Antin P, Berx G, Blanpain C, Brabletz T, Bronner M, Campbell K, Cano A, Casanova J, Christofori G, Dedhar S, Derynck R, Ford HL, Fuxe J, Garcia de Herreros A, Goodall GJ, Hadjantonakis AK, RJY H, Kalcheim C, Kalluri R, Kang Y, Khew-Goodall Y, Levine H, Liu J, Longmore GD, Mani SA, Massague J, Mayor R, McClay D, Mostov KE, Newgreen DF, Nieto MA, Puisieux A, Runyan R, Savagner P, Stanger B, Stemmler MP, Takahashi Y, Takeichi M, Theveneau E, Thiery JP, Thompson EW, Weinberg RA, Williams ED, Xing J, Zhou BP, Sheng G, Association EMTI. Guidelines and definitions for research on epithelial-mesenchymal transition. Nat Rev Mol Cell Biol. 2020;21(6):341–52.
36. Hass R, von der Ohe J, Ungefroren H. The intimate relationship among EMT, MET and TME: A T(ransdifferentiation) E(nhancing) M(ix) to be exploited for therapeutic purposes. Cancers (Basel). 2020;12(12)
37. Peinado H, Olmeda D, Cano A. Snail, Zeb and bHLH factors in tumour progression: an alliance against the epithelial phenotype? Nat Rev Cancer. 2007;7(6):415–28.
38. Dongre A, Weinberg RA. New insights into the mechanisms of epithelial-mesenchymal transition and implications for cancer. Nat Rev Mol Cell Biol. 2019;20(2):69–84.
39. Pastushenko I, Brisebarre A, Sifrim A, Fioramonti M, Revenco T, Boumahdi S, Van Keymeulen A, Brown D, Moers V, Lemaire S, De Clercq S, Minguijon E, Balsat C, Sokolow Y, Dubois C, De Cock F, Scozzaro S, Sopena F, Lanas A, D'Haene N, Salmon I, Marine JC, Voet T, Sotiropoulou PA, Blanpain C. Identification of the tumour transition states occurring during EMT. Nature. 2018;556(7702):463–8.
40. Paolillo M, Schinelli S. Extracellular matrix alterations in metastatic processes. Int J Mol Sci. 2019:20(19).
41. Chiang SP, Cabrera RM, Segall JE. Tumor cell intravasation. Am J Physiol Cell Physiol. 2016;311(1):C1–C14.
42. Hoshida T, Isaka N, Hagendoorn J, di Tomaso E, Chen YL, Pytowski B, Fukumura D, Padera TP, Jain RK. Imaging steps of lymphatic metastasis reveals that vascular endothelial growth factor-C increases metastasis by increasing delivery of cancer cells to lymph nodes: therapeutic implications. Cancer Res. 2006;66(16):8065–75.
43. Zhou H, Lei PJ, Padera TP. Progression of metastasis through lymphatic system. Cell. 2021:10(3).
44. Gould EA, Winship T, Philbin PH, Kerr HH. Observations on a "sentinel node" in cancer of the parotid. Cancer. 1960;13:77–8.
45. Sleeman JP. The lymph node as a bridgehead in the metastatic dissemination of tumors. Recent Results Cancer Res. 2000;157:55–81.
46. Jones D, Pereira ER, Padera TP. Growth and immune evasion of lymph node metastasis. Front Oncol. 2018;8:36.
47. Qian CN, Berghuis B, Tsarfaty G, Bruch M, Kort EJ, Ditlev J, Tsarfaty I, Hudson E, Jackson DG, Petillo D, Chen J, Resau JH, Teh BT. Preparing the "soil": the primary tumor induces vasculature reorganization in the sentinel lymph node before the arrival of metastatic cancer cells. Cancer Res. 2006;66(21):10365–76.
48. Hood JL, San RS, Wickline SA. Exosomes released by melanoma cells prepare sentinel lymph nodes for tumor metastasis. Cancer Res. 2011;71(11):3792–801.
49. Olmeda D, Cerezo-Wallis D, Riveiro-Falkenbach E, Pennacchi PC, Contreras-Alcalde M, Ibarz N, Cifdaloz M, Catena X, Calvo TG, Canon E, Alonso-Curbelo D, Suarez J, Osterloh L, Grana O, Mulero F, Megias D, Canamero M, Martinez-Torrecuadrada JL, Mondal C, Di Martino J, Lora D, Martinez-Corral I, Bravo-Cordero JJ, Munoz J, Puig S, Ortiz-Romero P, Rodriguez-Peralto JL, Ortega S, Soengas MS. Whole-body imaging of lymphovascular niches identifies pre-metastatic roles of midkine. Nature. 2017;546(7660): 676–80.
50. Nathanson SD. ASO author reflections: is breast cancer dissemination lymphatic, Hematogenous, or both; and does it matter? Ann Surg Oncol. 2020;27(Suppl 3):711–2.
51. Micalizzi DS, Maheswaran S, Haber DA. A conduit to metastasis: circulating tumor cell biology. Genes Dev. 2017;31(18):1827–40.
52. Fabisiewicz A, Grzybowska E. CTC clusters in cancer progression and metastasis. Med Oncol. 2017;34(1):12.
53. Peinado H, Zhang H, Matei IR, Costa-Silva B, Hoshino A, Rodrigues G, Psaila B, Kaplan RN, Bromberg JF, Kang Y, Bissell MJ, Cox TR, Giaccia AJ, Erler JT, Hiratsuka S, Ghajar CM, Lyden D. Pre-metastatic niches: organ-specific homes for metastases. Nat Rev Cancer. 2017;17(5):302–17.
54. Augustin HG, Koh GY. Organotypic vasculature: From descriptive heterogeneity to functional pathophysiology. Science. 2017;357(6353)
55. Orr FW, Wang HH, Lafrenie RM, Scherbarth S, Nance DM. Interactions between cancer cells and the endothelium in metastasis. J Pathol. 2000;190(3):310–29.
56. Vestweber D. How leukocytes cross the vascular endothelium. Nat Rev Immunol. 2015;15(11):692–704.
57. Miles FL, Pruitt FL, van Golen KL, Cooper CR. Stepping out of the flow: capillary extravasation in cancer metastasis. Clin Exp Metastasis. 2008;25(4):305–24.
58. Strilic B, Offermanns S. Intravascular survival and extravasation of tumor cells. Cancer Cell. 2017;32(3):282–93.
59. Wu SY, Watabe K. The roles of microglia/macrophages in tumor progression of brain cancer and metastatic disease. Front Biosci (Landmark Ed). 2017;22:1805–29.
60. Angelova M, Mlecnik B, Vasaturo A, Bindea G, Fredriksen T, Lafontaine L, Buttard B, Morgand E, Bruni D, Jouret-Mourin A, Hubert C, Kartheuser A, Humblet Y, Ceccarelli M, Syed N, Marincola FM, Bedognetti D, Van den Eynde M, Galon J. Evolution of metastases in space and time under immune selection. Cell. 2018;175(3):751–65. e16
61. Mace TA, Yamane N, Cheng J, Hylander BL, Repasky EA. The potential of the tumor microenvironment to influence Apo2L/TRAIL induced apoptosis. Immunol Investig. 2006;35(3–4):279–96.

62. Sosa MS, Bragado P, Aguirre-Ghiso JA. Mechanisms of disseminated cancer cell dormancy: an awakening field. Nat Rev Cancer. 2014;14(9):611–22.
63. Linde N, Fluegen G, Aguirre-Ghiso JA. The relationship between dormant cancer cells and their microenvironment. Adv Cancer Res. 2016;132:45–71.
64. Yu-Lee LY, Yu G, Lee YC, Lin SC, Pan J, Pan T, Yu KJ, Liu B, Creighton CJ, Rodriguez-Canales J, Villalobos PA, Wistuba II, de Nadal E, Posas F, Gallick GE, Lin SH. Osteoblast-secreted factors mediate dormancy of metastatic prostate cancer in the bone via activation of the TGFbetaRIII-p38MAPK-pS249/T252RB pathway. Cancer Res. 2018;78(11):2911–24.
65. Maman S, Sagi-Assif O, Yuan W, Ginat R, Meshel T, Zubrilov I, Keisari Y, Lu W, Lu W, Witz IP. The Beta subunit of hemoglobin (HBB2/HBB) suppresses neuroblastoma growth and metastasis. Cancer Res. 2017;77(1):14–26.
66. Aguirre-Ghiso JA, Liu D, Mignatti A, Kovalski K, Ossowski L. Urokinase receptor and fibronectin regulate the ERK(MAPK) to p38(MAPK) activity ratios that determine carcinoma cell proliferation or dormancy in vivo. Mol Biol Cell. 2001;12(4):863–79.
67. Tivari S, Lu H, Dasgupta T, De Lorenzo MS, Wieder R. Reawakening of dormant estrogen-dependent human breast cancer cells by bone marrow stroma secretory senescence. Cell Commun Signal. 2018;16(1):48.
68. Shibue T, Weinberg RA. Metastatic colonization: settlement, adaptation and propagation of tumor cells in a foreign tissue environment. Semin Cancer Biol. 2011;21(2):99–106.
69. Chen Q, Zhang XH, Massague J. Macrophage binding to receptor VCAM-1 transmits survival signals in breast cancer cells that invade the lungs. Cancer Cell. 2011;20(4):538–49.
70. Zhang XH, Wang Q, Gerald W, Hudis CA, Norton L, Smid M, Foekens JA, Massague J. Latent bone metastasis in breast cancer tied to Src-dependent survival signals. Cancer Cell. 2009;16(1):67–78.
71. Gurler Main H, Xie J, Muralidhar GG, Elfituri O, Xu H, Kajdacsy-Balla AA, Barbolina MV. Emergent role of the fractalkine axis in dissemination of peritoneal metastasis from epithelial ovarian carcinoma. Oncogene. 2017;36(21):3025–36.
72. Wculek SK, Malanchi I. Neutrophils support lung colonization of metastasis-initiating breast cancer cells. Nature. 2015;528(7582):413–7.
73. Shahriari K, Shen F, Worrede-Mahdi A, Liu Q, Gong Y, Garcia FU, Fatatis A. Cooperation among heterogeneous prostate cancer cells in the bone metastatic niche. Oncogene. 2017;36(20):2846–56.
74. Drake CG, Jaffee E, Pardoll DM. Mechanisms of immune evasion by tumors. Adv Immunol. 2006;90:51–81.
75. Schaller J, Agudo J. Metastatic colonization: escaping immune surveillance. Cancers (Basel). 2020;12(11)
76. Bashraheel SS, Domling A, Goda SK. Update on targeted cancer therapies, single or in combination, and their fine tuning for precision medicine. Biomed Pharmacother. 2020;125:110,009.
77. Sung SY, Chung LW. Prostate tumor-stroma interaction: molecular mechanisms and opportunities for therapeutic targeting. Differentiation. 2002;70(9–10):506–21.
78. Hinrichs CS, Rosenberg SA. Exploiting the curative potential of adoptive T-cell therapy for cancer. Immunol Rev. 2014;257(1):56–71.
79. Paijens ST, Vledder A, de Bruyn M, Nijman HW. Tumor-infiltrating lymphocytes in the immunotherapy era. Cell Mol Immunol. 2020;
80. Topalian SL, Drake CG, Pardoll DM. Immune checkpoint blockade: a common denominator approach to cancer therapy. Cancer Cell. 2015;27(4):450–61.
81. Laplagne C, Domagala M, Le Naour A, Quemerais C, Hamel D, Fournie JJ, Couderc B, Bousquet C, Ferrand A, Poupot M. Latest advances in targeting the tumor microenvironment for tumor suppression. Int J Mol Sci. 2019;20(19)
82. Ribatti D. History of research on angiogenesis. Chem Immunol Allergy. 2014;99:1–14.
83. Ghosh S, Marrocco I, Yarden Y. Roles for receptor tyrosine kinases in tumor progression and implications for cancer treatment. Adv Cancer Res. 2020;147:1–57.
84. Rajabi M, Mousa SA. The role of angiogenesis in cancer treatment. Biomedicine. 2017;5(2)
85. Jain RK. Normalizing tumor microenvironment to treat cancer: bench to bedside to biomarkers. J Clin Oncol. 2013;31(17):2205–18.
86. Quail DF, Joyce JA. Microenvironmental regulation of tumor progression and metastasis. Nat Med. 2013;19(11):1423–37.

Part V

Circulating Cancer Cells and Liquid Biopsy

Catherine Alix-Panabières and Klaus Pantel

Research on disseminating tumor cells in cancer patients and animal models has opened new avenues for better understanding of the biology of metastasis as the leading cause of cancer-related death. This fundamental work has led to a novel diagnostic concept designated "liquid biopsy" [1] which included first the analysis of circulating tumor cells (CTCs) and has been now extended to cell-free tumor-derived material (in particular circulating tumor DNA, ctDNA) released by tumor cells in the peripheral blood of cancer patients [2].

Over the past decade, various methods have been developed to detect CTCs and ctDNA in the peripheral blood of cancer patients. While reliable information can be easily obtained in patients with advanced disease, early-stage cancer patients usually present with very low concentrations of CTCs and ctDNA [3]. At present, most CTC assays rely on epithelial markers and the majority of CTCs detected are single isolated cells. The clinical relevance of "mesenchymal" CTCs lacking any epithelial markers as well as CTC clusters are still under investigation [4–6]. Although most published studies have been performed on patients with carcinomas and melanomas, CTCs have also been detected in the peripheral blood of patients with primary brain tumors (glioblastomas) despite the blood-brain barrier [7].

Liquid biopsy assays are currently being validated for early detection of cancer, which is supposed to reduce cancer-related mortality [8]. Despite remarkable progress, liquid biopsy-based detection of early stages of cancer remains a challenge, in particular in breast cancer. New blood-based biomarkers for early detection currently validated in clinical trials include miRNAs, exosomes, tumor-associated proteins, and tumor-educated platelets.

In patients with diagnosed cancer, CTCs and ctDNA analyses can obtain independent information on prognosis in early and advanced stages of disease. In particular, CTC counts at initial diagnosis are able to refine the current risk stratification by TNM staging in early-stage breast cancer [9]. Moreover, early detection of relapse by sequential ctDNA (or CTCs) analysis of blood samples obtained post-surgery during the follow-up [10] is possible and may be used in future trials to stratify patients to "post-adjuvant" therapies [11].

Another key application of liquid biopsy is to identify therapeutic targets or mechanisms of resistance of metastatic cells in individual patients [2]. While the analysis of ctDNA focuses on mutations relevant for cancer therapy (e.g., EGFR, KRAS, or ESR1 mutations), CTCs offer a wide spectrum of analyses at the DNA, RNA, and protein levels. Metastatic cells might have unique characteristics that can differ from the bulk of cancer cells in the primary tumor currently used for stratification of patients to systemic therapy. Moreover, monitoring of CTCs and ctDNA before, during, and after systemic therapy (e.g., chemotherapy, hormonal therapy, antibody therapy) might provide unique information for the future clinical management of the individual cancer patients and might serve as a surrogate marker for response to therapy. In the context of recent success in antibody-mediated blockade of immune checkpoint control molecules, expression of the PD-L1 on CTCs might be of interest as a potential predictive marker [12, 13]. Moreover, the expression of androgen receptor variant 7 in CTCs may predict resistance to anti-androgen therapy in prostate cancer [14], while mutations in the estrogen receptor gene (ESR1) provide information on resistance to hormone therapy in breast cancer [15].

Additional therapeutic targets detected on CTCs in cancer patients include the estrogen receptor and HER-2 oncogene [2]. Single-cell RNAseq analysis of CTCs may provide more comprehensive information on relevant pathways.

Considering to introduce liquid biopsy into clinical practice, its clinical utility must be proven [16]. For the first time in this field, clinical utility of CTCs has been proven based on the CTC METABREAST study (NCT01710605) [17]. Indeed, the CTC count is a reliable biomarker method for choosing between chemotherapy and endocrine therapy as the first-line treatment in patients with a metastatic cancer HR^{+} $HER2^{-}$.

For functional analysis of CTCs, the development of in vitro and in vivo test systems has started, which might also serve as models for drug testing. In particular, the development of cell lines and xenografts derived from CTCs can provide novel insights into the biology of tumor cell dissemination and may be used to discover new pathways to target specifically metastatic cells [18, 19].

Besides CTCs and ctDNA the analysis of circulating microRNAs, exosomes, tumor-associated proteins, or tumor-educated platelets may provide complementary information as "liquid biopsy." For example, the integrin composition of exosomes seems to determine the organ site of metastatic niches and the RNA expression pattern of blood platelets reveals information on tumors in cancer patients.

Sensitive methods have also been developed to capture disseminated tumor cells (DTCs) in the bone marrow in cancer patients [11], which provide new insights into the process of "cancer dormancy." The nature of dormant breast cancer cells and the mechanisms leading to their outgrowth are poorly understood. Efforts to unravel the nature of cancer dormancy have been hampered by the lack of sensitive methods to detect dormant cells in cancer patients. The development of novel therapies designed to kill dormant residual tumor cells, or maintain them in a quiescent state, represents a highly attractive approach to prevent late recurrence. Such an approach, however, would require a far more detailed understanding of tumor dormancy and recurrence than that which exists today, as well as biomarkers to enable monitoring of this process and predict recurrence. Analysis of DTCs leads to the discovery of new molecules relevant to the biology of metastasis such as the putative metastasis-suppressor RAI2 [20].

In conclusion, liquid biopsy analysis can be used to obtain new insights into metastasis biology, and as companion diagnostics to improve the stratification of therapies and to obtain insights into therapy-induced selection of cancer cells. Different approaches such as CTC or ctDNA analysis will provide complementary information [16]. Technical and clinical assay validation is very important and can be achieved in international consortia such as the European Liquid Biopsy Society (www.elbs.eu) or the US-based BloodPAC [21].

This book provides a comprehensive overview of current research on cancer metastasis including experimental and translational studies covering the entire spectrum of investigations from discovery of novel genes and pathways to implementation of clinical biomarkers in personalized medicine.

References

1. Pantel K, Alix-Panabieres C. Circulating tumour cells in cancer patients: challenges and perspectives. Trends Mol Med. 2010;16:398–406. https://doi.org/10.1016/j.molmed.2010.07.001.
2. Alix-Panabieres C, Pantel K. Liquid biopsy: from discovery to clinical application. Cancer Discov. 2021;11:858–73. https://doi.org/10.1158/2159-8290.CD-20-1311.
3. Pantel K, Alix-Panabieres C. Liquid biopsy and minimal residual disease - latest advances and implications for cure. Nat Rev Clin Oncol. 2019;16:409–24. https://doi.org/10.1038/s41571-019-0187-3.
4. Szczerba BM, et al. Neutrophils escort circulating tumour cells to enable cell cycle progression. Nature. 2019;566:553–7. https://doi.org/10.1038/s41586-019-0915-y.
5. Gkountela S, et al. Circulating tumor cell clustering shapes DNA methylation to enable metastasis seeding. Cell. 2019;176:98–112 e114. https://doi.org/10.1016/j.cell.2018.11.046.
6. Krol I, et al. Detection of clustered circulating tumour cells in early breast cancer. Br J Cancer. 2021; https://doi.org/10.1038/s41416-021-01327-8.
7. Muller C, et al. Hematogenous dissemination of glioblastoma multiforme. Sci Transl Med. 2014;6:247ra101. https://doi.org/10.1126/scitranslmed.3009095.
8. Pantel K. Blood tests for early detection of lung cancer: challenges and promises. Lancet Respir Med. 2020;8:654–6. https://doi.org/10.1016/S2213-2600(20)30115-6.

9. Bidard FC, et al. Circulating tumor cells in breast cancer patients treated by neoadjuvant chemotherapy: a meta-analysis. J Natl Cancer Inst. 2018;110:560–7. https://doi.org/10.1093/jnci/djy018.
10. Trapp E, et al. Presence of circulating tumor cells in high-risk early breast cancer during follow-up and prognosis. J Natl Cancer Inst. 2019;111:380–7. https://doi.org/10.1093/jnci/djy152.
11. Pantel K, Hayes DF. Disseminated breast tumour cells: biological and clinical meaning. Nat Rev Clin Oncol. 2017; https://doi.org/10.1038/nrclinonc.2017.174.
12. Jacot W, et al. Clinical correlations of programmed cell death ligand 1 status in liquid and standard biopsies in breast cancer. Clin Chem. 2020;66:1093–101. https://doi.org/10.1093/clinchem/hvaa121.
13. Sinoquet, L. *et al.* Programmed cell death ligand 1-expressing circulating tumor cells: a new prognostic biomarker in non-small cell lung cancer. *Clin Chem* In press 2021.
14. El-Heliebi A, et al. In situ detection and quantification of AR-V7, AR-FL, PSA, and KRAS point mutations in circulating tumor cells. Clin Chem. 2018;64:536–46. https://doi.org/10.1373/clinchem.2017.281295.
15. Turner NC, et al. ESR1 mutations and overall survival on Fulvestrant versus Exemestane in advanced hormone receptor-positive breast cancer: a combined analysis of the phase III SoFEA and EFECT trials. Clin Cancer Res. 2020;26:5172–7. https://doi.org/10.1158/1078-0432.CCR-20-0224.
16. Alix-Panabieres C. The future of liquid biopsy. Nature. 2020;579:S9. https://doi.org/10.1038/d41586-020-00844-5.
17. Bidard FC, et al. Efficacy of circulating tumor cell count-driven vs clinician-driven first-line therapy choice in hormone receptor-positive, ERBB2-negative metastatic breast cancer: the STIC CTC randomized clinical trial. JAMA Oncol. 2021;7:34–41. https://doi.org/10.1001/jamaoncol.2020.5660.
18. Koch C, et al. Characterization of circulating breast cancer cells with tumorigenic and metastatic capacity. EMBO Mol Med. 2020;12:e11908. https://doi.org/10.15252/emmm.201911908.
19. Cayrefourcq L, et al. Selective treatment pressure in colon cancer drives the molecular profile of resistant circulating tumor cell clones. Mol Cancer. 2021;20:30. https://doi.org/10.1186/s12943-021-01326-6.
20. Werner S, et al. Suppression of early hematogenous dissemination of human breast cancer cells to bone marrow by retinoic acid-induced 2. Cancer Discov. 2015;5:506–19. https://doi.org/10.1158/2159-8290.CD-14-1042.
21. Connors D, et al. International liquid biopsy standardization alliance white paper. Crit Rev Oncol Hematol. 2020;156:103112. https://doi.org/10.1016/j.critrevonc.2020.103112.

Circulating Tumor Cells and ctDNA in Sarcomas

12

Camille Jubelin, Denis Cochonneau, Emilie Moranton, Javier Muñoz-Garcia, and Dominique Heymann

Abstract

Sarcomas are clustered in two oncological entities named bone and soft tissue sarcomas. Both are rare cancers originating from the mesenchyme, characterized by their propensity to induce the development of lung metastases. Sarcoma cells escaping from the primary tumor site spread to the pulmonary tissue through the bloodstream where they found a favorable microenvironment to establish metastatic foci. The low number of patients, the high histological, genetic, and molecular heterogeneity of sarcomas combined with the absence of specific markers expressed by cancer cells make the detection and follow-up of the minimal residual disease challenging. Over the last decade, tremendous technological progress has been made towards the detection of recurrent diseases. The literature is now enriched of information describing the use of liquid biopsies in clinical care of sarcoma patients. This chapter aims to give a brief overview of the most recent data available on the detection of circulating tumor cells and circulating tumor DNA in sarcomas.

C. Jubelin · D. Cochonneau · E. Moranton · J. Muñoz-Garcia
Nantes Université, CNRS, US2B, UMR 6286, Institut de Cancérologie de l'Ouest, Blvd Jacques Monod, Saint-Herblain, France
e-mail: camille.jubelin@univ-nantes.fr; denis.cochonneau@ico.unicancer.fr; emilie.moranton@ico.unicancer.fr; javier.munoz@ico.unicancer.fr

D. Heymann (✉)
Nantes Université, CNRS, US2B, UMR 6286, Institut de Cancérologie de l'Ouest, Blvd Jacques Monod, Saint-Herblain, France

Department of Oncology and Metabolism, University of Sheffield, Sheffield, UK
e-mail: dominique.heymann@univ-nantes.fr

Learning Objectives
- To get a global overview of the principal methods available for isolating circulating tumor cells in sarcomas.
- To gain insight into the biological value of circulating tumor DNA in sarcoma patients.
- To benefit of a brief summary of the mimimal residual disease in sarcomas.

12.1 Introduction

Sarcomas belong to a large family of malignant entities of mesenchymal origin [1]. These rare cancers can be classified in two large categories, bone sarcomas and soft tissues sarcomas (STS), according to their anatomical site of occurrence. More than 60 subtypes have been defined based on their molecular characteristics. Indeed, in addition to their common origin, sarcomas share a high heterogeneity in terms of molecular/genetic and histological profile, leading to a constant increase of the number of entities increase in parallel with technological advances [2]. The three main subtypes of bone sarcomas are osteosarcoma, Ewing sarcoma, and chondrosarcoma [3]. The two first are mostly observed in adolescent and young adults whereas the peak of incidence of chondrosarcoma is principally seen in the 4 decades of age. These diseases are generally treated using surgery and polychemotherapeutic approaches; however, despite the improvement of multimodal therapies, around one-third of patients still succumb to sarcomas. Nowadays, the overall survival of osteosarcoma and Ewing sarcoma patients is around 70% at 5 years with localized disease. This survival rate drops to 30% when metastases are detected at the time of diagnosis. Among STS, leiomyosarcoma, liposarcoma, and undifferentiated pleomorphic sarcoma are the most common malignant entities. Similar to bone sarcomas, the STS treatment approach mixes surgery and drug administration (e.g., che-

S. P. Leong et al. (eds.), *Cancer Metastasis Through the Lymphovascular System*,
https://doi.org/10.1007/978-3-030-93084-4_12

motherapy, tyrosine kinase inhibitor). The 5-year overall survival for STS is estimated around 60%, depending on the disease stage, the anatomical site, and histological/molecular subtype [2].

The natural history of sarcomas is frequently associated with the development of metastases, which differ significantly from those observed in carcinoma. Indeed, while carcinomas use the lymphatic vasculature for spreading and invading the lymph node chain located at proximity of the tumor site, sarcoma cells predominantly spread through the blood vasculature [4]. Minimal residual disease (MDR) is defined as malignant cells that are resistant to treatment, and that remain in patients after remission, leading to relapse and metastasis. MDR is composed of drug-resistant tumor cells, dynamically presented as persister/dormant/quiescent/cancer cells in residual tumors [5, 6]. Circulating tumor cells (CTCs) in peripheral blood as well as disseminated tumor cells in the bone marrow and other metastatic sites are common examples of MDR. MDR behavior is consequently impacted by the respective microenvironment of the cancer cells concerned (e.g., cytokines/hormones, cells, epigenetic events, therapies). The current biological methods for detecting MDR (including local tumor recurrence and developing metastatic foci) are based on the detection of CTCs and circulating tumor DNA (ctDNA). Whether such approaches have overtaken the proof-of-concept in carcinomas, the clinical significance of liquid biopsies as a prognostic or predictive tool in sarcoma is currently uncertain and under investigation [7–10]. Liquid biopsies are non- or weakly invasive methods based on blood collection that might be useful for the diagnosis and clinical follow-up of cancer. This chapter gives a brief overview of the most recent data available in the literature about the isolation of CTCs and detection of ctDNA and discusses the clinical relevance of such approaches in the context of sarcomas.

12.2 Circulating Tumor Cells in Sarcoma: Methods of Detection and Biological Value

Cancer cells can escape from tumor foci and intravasate for migrating to distant sites before to extravasate and establish metastases. CTCs are then cells released from primary and metastatic tumor foci, which migrate to secondary location through the peripheral blood. In contrast to epithelial tumors, publications describing the isolation of CTCs in sarcomas are currently limited [10]. Over the last decade, accumulating data has described the isolation of CTCs in sarcoma patients. However, researchers have been confronted to several obstacles during the diagnostics of CTC, including the restricted number of patients, the high histological/genetic/molecular heterogeneity of the disease, and the lack of specific markers expressed at the surface of sarcoma cells. This restriction is mainly related to the low number of patients, the high intratumoral heterogeneity of sarcomas, their high diversity, and the absence of universal cell markers expressed by most cancer cells.

In the absence of specific cell markers, researchers have proposed several methods for isolating CTCs from sarcomas. Defined as rare cell events in blood (1 CTC for 10^5–10^8 leukocytes) the isolation of CTCs requires two main technical steps: (i) a first step of pre-enrichment leading to the enrichment of blood samples, followed by (ii) a second step, corresponding to the final cell isolation (Fig. 12.1a). Enrichment methods (Step 1) include physical based approaches, as well as biological based techniques. The high size and low deformability of cancer cells are two important physical features used for pre-enriching CTCs. Indeed, considering these two characteristics, sarcoma cells can easily be trapped in filter and in microcapillaries (microfluidics) [11]. In contrast, biological approaches are mainly based on immunomagnetic isolation techniques (e.g., negative depletion, nonspecific immunostaining). Isolation steps (Step 2) use various technical devises including flow cytometry, dielectrophoretic array, and microfluidics with antibody-based precoated systems (Fig. 12.1b–d). By using these two-step-based enrichment/selection techniques, CTCs have been successfully isolated in sarcomas [10, 12, 13]. For instance, CTCs were previously detected in bone [14, 15] and STS patients [16–18] (Table 12.1).

The high plasticity of cancer cells makes the identification of specific and universal CTCs markers difficult to diagnose. Epithelial cells have been frequently isolated by using epithelial biomarkers (e.g., EpCAM, cytokeratin). However, there is strong evidence that cancer cells isolated only represent a low proportion of CTCs and constitute the emerged part of the iceberg. Indeed, epithelial mesenchymal transition and cell plasticity contribute to the loss or acquisition of epithelial and mesenchymal markers according to the location of cancer cells. Sarcoma cells originate from the mesenchyme but exhibit similar plasticity to carcinoma cells complicating the identification of specific markers. CD99 is a heavily O-glycosylated transmembrane protein expressed by Ewing sarcoma cells and spindle cell tumors such as synovial sarcomas. Ewing sarcoma is one of the first sarcomas from which CTCs were initially isolated thanks to their CD99 expression (Table 12.1). In this context, research groups have been looking for universal sarcoma markers in order to improve the sensitivity and specificity of CTC isolation. Recently, the cell-surface vimentin has emerged as a potential universal biomarker of human sarcoma CTCs [25, 33]. Cell-surface vimentin was expressed by a high diversity of sarcoma subtypes (bone sarcomas: osteosarcoma, Ewing sarcoma; soft tissue sarcomas: angiosarcoma, leiomyosarcoma, undifferentiated pleomorphic sarcoma). Indeed, the majority of the sar-

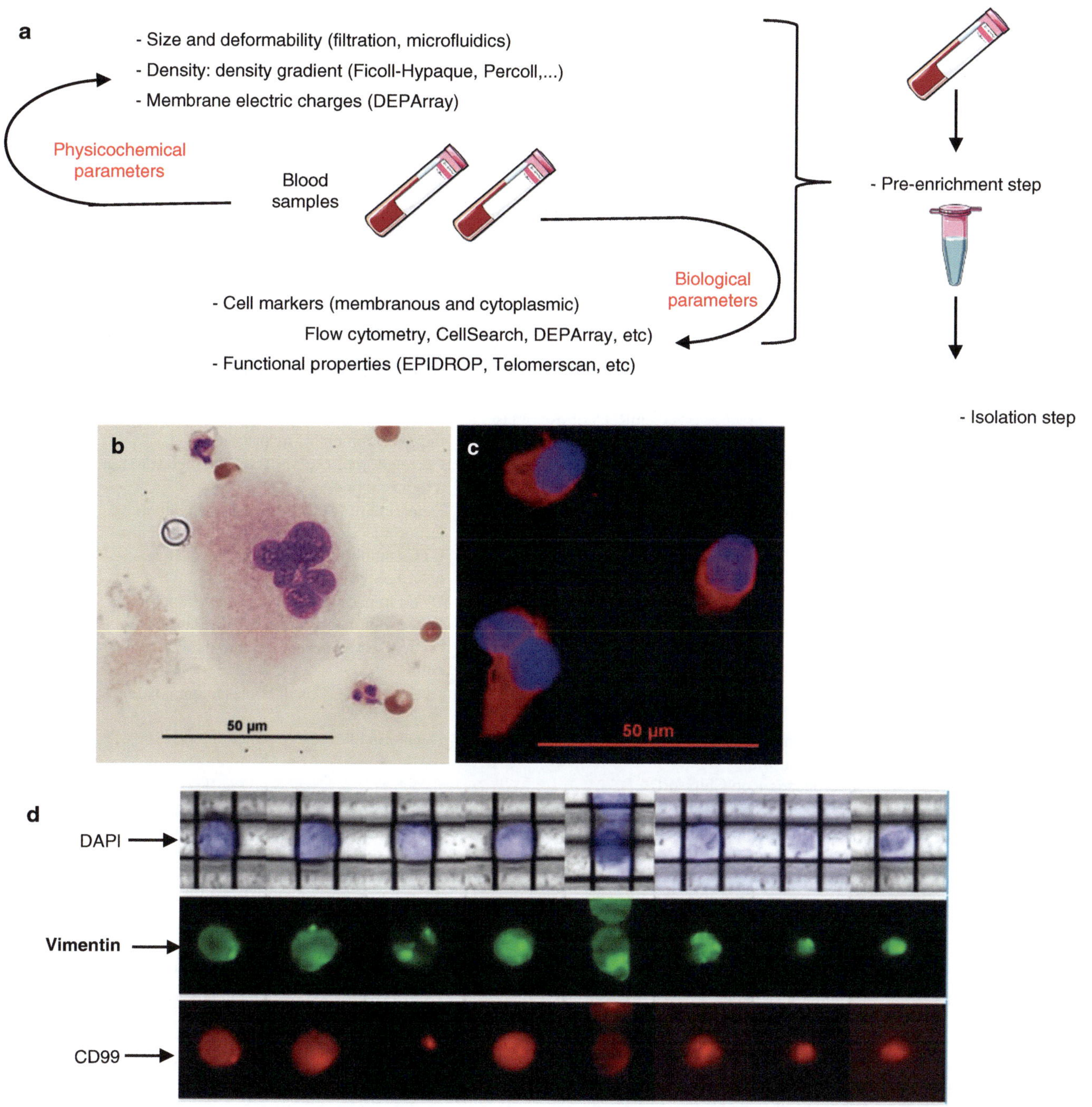

Fig. 12.1 **Detection of circulating tumor cells in sarcomas.** (**a**) Isolation methods of CTCs combine a first pre-enrichment step followed by an isolation step. These steps are based on physical or/and biological properties of CTCs. (**b**) CTCs isolated from blood patient suffering from a pleomorphic liposarcoma. CTC pre-enriched fraction was prepared by PARSORTIX, followed by a cysospin and HE staining. (**c**) CTCs from an undifferentiated epithelioid sarcoma. Vimentin: red; DAPI: blue; (**d**) CTCs isolated for an Ewing sarcoma. Ten mL of blood were pre-enriched by Parsortix followed by CTCs isolation using DEPArray after vimentin/CD99 immunostaining and DAPI staining

coma cell lines studied show high expression levels of cell-surface vimentin. This characteristic was positively used by Satelli et al., who designed a 84-1 monoclonal antibody targeting cell-surface vimentin for isolating CTCs in a short series of blood sarcoma patients [25]. In a series of 77 gastrointestinal stromal tumors, Li et al. identified CTCs, expressing specific macrophage biomarkers (CD14 and CD68) [33]. This new class of CTCs were CD45 negative but showed cell-

Table 12.1 Detection of circulating tumor cells in sarcomas: methods and main results

Bone sarcomas				
Sarcoma type	Detection/isolation method	Biomarker	Main conclusions	References
Ewing sarcoma	RT-PCR	EWS-FLI-1/ERG	CTCs detection in patients with localized disease and association with poor outcome.	[19, 20]
	RT-PCR	EWS-FLI-1/ERG	Correlation between CTC detection and disease progression	[21]
	RT-PCR	EWS-FLI-1/ERG	CTC detection related to reduced survival.	[22]
	RT-PCR	EWS-FLI-1/ERG	No prognostic data	[23, 24]
	Immunomagnetic separation with microbeads	Isolation of $CD99^+$ cells EWSR1/FLI1 followed by RT-PCR	Feasibility of finding CTCs in PB of ES patients by immunoseparation with CD99 antibody and magnetic microbead	[14]
	Immunomagnetic separation with microbeads	Cell-surface vimentin	Identification of the first universal and specific CTC marker described for enumerating CTCs from different types of sarcoma	[25]
Osteosarcoma	RT-PCR	mRNA of osteoblast related genes	Correlation of collagen type I expression and the development of metastases.	[26]
	PCR-ELISA	Osf2 mRNA	Correlation of *Osf2* mRNA and metastasis in a mouse model	[27]
	DEPArray	GFP-cells	Detection of CTCs before any detectable tumor mass in a mouse model. Modulation of CTCs number by chemotherapy? High heterogenetity of CTCs	[28]
	ISET	Cell size	CTCs in patients with high-grade sarcoma	[15]
	Immunomagnetic separation with microbeads	Cell-surface vimentin	Detection of cell surface vimentin+ CTCs in cryopreserved peripheral blood mononuclear cells.	[13]
Ewing sarcoma and osteosarcoma	CellSieve	Cell size	Detection of CTCs in bone sarcoma patients	[12]
Giant cell tumor of bone	Immunomagnetic separation with microbeads + on Chip sort	Cell surface vimentin	Detection of CTCs in patient blood	[29]
Soft tissue sarcomas				
Sarcoma type	Detection method	Marker	Main conclusions	References
Alveolar soft part sarcoma	RT-PCR	ASPSCR1-TFE3	Detection of CTCs in patient blood	[30]
	ApoStream DEP field	$Vimentin^+$, $cytokeratin^{+/-}$ and β-$catenin^+$ + ASPSCR1-TFE3 by FISH	Detection of CTCs in patient blood	[31]
Rhabdomyosarcoma	RT-PCR	PAX3-FKHR	Detection of CTCs in patient blood	[32]
Myxofibrosarcoma	Immunomagnetic separation with microbeads + on Chip sort	Cell surface vimentin	Detection of CTCs in patient blood	[12]
Leiomyosarcoma, angiosarcoma, pleomorphic sarcoma	Immunomagnetic separation with microbeads	Cell-surface vimentin	Cell-surface vimentin as a universal sarcoma CTC marker by using a monoclonal antibody	[25]
Synovial sarcoma, Dedifferentiated liposarcoma	ISET	Cell size	CTCs in patients with high-grade sarcoma	[15]
Gastrointestinal stromal tumors	Immunomagnetic separation with microbeads	Cell-surface vimentin/CD68	Detection of cell surface vimentin+/CD68+ macrophage-like CTCs	[33]
Gastrointestinal stromal tumors, liposarcoma	Immunomagnetic separation with microbeads	Cell-surface vimentin	Detection of cell surface vimentin+ CTCs in cryopreserved peripheral blood mononuclear cells.	[13]
Soft Tissue sarcoma	Aptamer	Cell-surface vimentin	Feasibility in spiking experiments	[34]
Soft tissue sarcoma (clinical trial)	Epic Sciences	$Cytokeratin^-$, $vimentin^+$ expression and image analysis algorithm	No association was observed between the CTC enumeration and clinical outcome.	[35]

CTC Circulating tumor cells, *DEP* Dielectophoresis, *ISET* Isolation by size surface vimentin, CD14 and CD68 positivity in addition to DOG1 and cKIT specific gastrointestinal stromal tumor markers [33]. Interestingly, the number of these vimentin/macrophage CTCs was correlated with metastatic status of patients in contrast to CTCs lacking macrophage markers that failed to predict metastasis formation [33]. Cell-surface vimentin could be extended to carcinoma cells that underwent epithelial mesenchymal transition [36]. In a preclinical model of osteosarcoma, CTCs appeared to be an early biomarker, detectable before any palpable/detectable tumors [28]. The number of CTCs fluctuated along the chemotherapy and was increased after ifosfamide administration. Surprisingly, this study showed an inverse correlation between CTCs enumeration and tumor growth, as well as between CTC number and metastatic disease [28]. These results demonstrated that all CTCs did not possess intrinsic properties of metastases as revealed by the higher number of CTCs in ifosfamide-treated mice. CTCs enumeration should indeed not be the single criteria of their biological value but their molecular profiling appeared mandatory. Overall, these recent studies highlighted the clinical potential of CTCs as prognostic or predictive markers in sarcomas. Recently, multiple center phase 1b clinical trials investigated the feasibility to enumerate CTCs in 51 patients with potentially resectable soft tissue sarcoma treated with olaratumab. This monoclonal antibody was used to target platelet-derived growth factor receptor (PDGFR)-α for one cycle. The first cycle was followed then by up to 6 cycles of olaratumab combined with doxorubicin [35]. The isolation methods of CTCs relied on multiple staining, especially of cellular intracytoplasmic elements like cytokeratins and intermediate filaments. The expression profiles were analyzed using EPIC Sciences algorithm that measured around 90 cell parameters (e.g., immunostaining, morphology, size). CTCs were detectable in collected blood samples; their number increased at cycle 1 and decreased by cycle 3. Interestingly, the decrease of CTCs in patients treated with olaratumab monotherapy was higher in controlled disease. These results showed the potential clinical value of CTCs in sarcoma. Longitudinal clinical trials with large series of patients are now mandatory (Table 12.1).

12.3 Detection of Circulating Tumor DNA in Sarcoma

In healthy and cancer tissues, apoptotic- and necrosis-related cell death results in the release of DNA fragments into the bloodstream, which can be easily detected by PCR. Tissue homeostasis is often associated with the physiological release of cell fragments and contributes to the circulating DNA [37, 38]. When this circulating free DNA fraction is released from a tumor mass, it is entitled circulating tumor DNA (ctDNA). These ctDNAs reflect the genetic heritage of cancer cells composing the tumor tissue and fluctuate along tumor growth, as well as during tumor treatment. In this context, ctDNAs have been used as diagnostic, prognostic, and predictive biomarkers of cancer. More specifically, based on their clinical interest, ctDNA has, over the last decade, been assessed in various sarcomas [39, 40].

The presence of ctDNA was previously detected in plasma patients suffering from bone sarcoma [41–43] and STS [44–48]. Although ctDNA can be released from both healthy and cancer cells, the ctDNA levels were higher in sarcoma patients than in the cancer-free control group. Interestingly, mutations related to cancer initiation and/or development were identified from ctDNA analysis. Mutated genes included *TP53, PIK3CA IDH1*, fusion oncogenes (e.g., *SS18-SSX1/2*), or cytokine receptors. *cKIT* and *PDGFR* are indeed frequently mutated in gastrointestinal of stromal tumors (GIST) and these mutations are not only detectable in ctDNA but mutated ctDNA is quantitatively correlated with clinical progression [49]. The identification of the mutational profile of ctDNA represents an interesting tool for predicting the functional resistance of sarcomas to tyrosine-kinase inhibitors in clinic as observed in GIST patients [50]. Mutational profile can be also predicted from ctDNA in plasmas of bone sarcoma patients. For instance, mutated *IDH1* detected in chondrosarcoma were correlated with tumor grade and poor prognosis [42]. Next-generation sequencing was used to look for somatic single nucleotide variants, insertions and deletions, as well as structural variants in osteosarcoma patients [41, 48]. In Ewing sarcoma, somatic mutations observed from ctDNA were correlated with inferior outcomes [48]. Ewing sarcomas are often characterized by their fusion gene expression in cancer cells, which can be quantified from plasma ctDNA. Mutated ctDNAs represent valuable biomarkers for the diagnostic assessment and therapeutic follow-up of cancer [51]. Cell-free *EWSR1-ETS* fusion sequences have been assessed in Ewing sarcoma patients. Their copy-numbers correlated with patient's risk factors including tumor volume and metastatic status. In addition, a rapid drop of their plasma levels was observed during treatment phases. In contrast, an increased level of their copy-number was a clinical mark of tumor relapse [51].

More recently, studies have detected circulating nucleosomes in blood, revealing that ctDNA retains at least some characteristics of the nuclear chromatin, a useful information for determining the origin of cell-free ctDNA. Whole genome sequencing of ctDNA can give a genome-wide map of nucleosome occupancy, allowing the identification of the cell types that contribute to circulating ctDNA [52, 53]. Such observation highly relevant to Ewing sarcoma led to follow the chromatin state for monitoring therapeutic response in real time [54–56]. Sheffield et al. demonstrated that DNA methylation heterogeneity defined a disease spectrum in Ewing sarcoma [55]. Complementary investigations are mandatory to determine the clinical value of ctDNA methylation heterogeneity.

Overall, the detection and characterization of ctDNA in sarcomas similarly to epithelial cancers represent a promising prognostic/diagnostic tool. Investigation of larger biological sarcoma cohorts with complete clinical annotations is currently needed to validate their clinical values.

12.4 Conclusion

Based on several emerging evidence, it is now conventionally approved that liquid biopsies, including CTCs and ctDNA, represent very efficient tools for the clinical follow-up in oncology. The biological value of CTCs was initially assessed in carcinoma patients and used to compare the genetic/epigenetic profiles of primary tumors and CTCs [57]. More recently, such approaches have been extended to sarcoma. Overall, these revealed that CTCs partly reflect the mutational landscape of the primary and metastatic tumor foci [57–60] and that CTCs should be considered as a snapshot of the tumor tissue heterogeneity at a given time [58, 61]. Recent methods allow to better identify the origin of detected ctDNA. Both ctDNA and CTCs could be an added value for longitudinal patient monitoring. Although the detection of ctDNA appears easier than CTCs, CTCs are advantageous for characterizing the heterogeneity of the mutational spectrum and orientating the future therapeutic options towards the development of new platform for drug screening [62].

Open Questions

- Which is the best method for isolating CTCs in sarcomas?
- What is the prognostic value of CTCs and ctDNA in sarcomas?
- Do CTCs reflect the tumor heterogeneity in sarcomas?
- What is the clonal evolution of CTCs along treatments?
- Do CTCs have the same capacity to induce metastatic foci?
- Do CTCs have dormant/quiescent/stemness properties?

Patient Information and Guidelines

- EuSARC (European network for SARComa): https://eusarc.com/
- NIH website for information to health professional, related to bone cancer: https://www.cancer.gov/types/bone/hp
- World sarcoma network: http://www.worldsarcomanetwork.com/

Acknowledgments The authors are also grateful to M. Clément Heymann for reviewing the English style.

References

1. WHO Classification of Tumours Editorial Board. Soft tissue and bone tumours, WHO classification of tumours series, vol. 3. 5th ed. Lyon: International Agency for Research on Cancer; 2020.
2. Grünewald TG, Alonso M, Avnet S, Banito A, Burdach S, Cidre-Aranaz F, Di Pompo G, Distel M, Dorado-Garcia H, Garcia-Castro J, González-González L, Grigoriadis AE, Kasan M, Koelsche C, Krumbholz M, Lecanda F, Lemma S, Longo DL, Madrigal-Esquivel C, Morales-Molina Á, Musa J, Ohmura S, Ory B, Pereira-Silva M, Perut F, Rodriguez R, Seeling C, Al Shaaili N, Shaabani S, Shiavone K, Sinha S, Tomazou EM, Trautmann M, Vela M, Versleijen-Jonkers YM, Visgauss J, Zalacain M, Schober SJ, Lissat A, English WR, Baldini N, Heymann D. Sarcoma treatment in the era of molecular medicine. EMBO Mol Med. 2020;12:e11131.
3. Brown HK, Schiavone K, Gouin F, Heymann MF, Heymann D. Biology of bone sarcomas and new therapeutic developments. Calcif Tissue Int. 2018;102(2):174–95.
4. Pennacchioli E, Tosti G, Barberis M, De Pas TM, Verrecchia F, Menicanti C, Testori A, Mazzarol G. Sarcoma spreads primarily through the vascular system: are there biomarkers associated with vascular spread? Clin Exp Metastasis. 2012;29(7):757–73.
5. Nicolini A, Rossi G, Ferrari P, Carpi A. Minimal residual disease in advanced or metastatic solid cancers: The G0-G1 state and immunotherapy are key to unwinding cancer complexity. Semin Cancer Biol. 2020;21:1054.
6. Vallette FM, Olivier C, Lézot F, Oliver L, Cochonneau D, Lalier L, Cartron PF, Heymann D. Dormant, quiescent, tolerant and persister cells: four synonyms for the same target in cancer. Biochem Pharmacol. 2019;162:169–76.
7. Alix-Panabières C, Schwarzenbach H, Pantel K. Circulating tumor cells and circulating tumor DNA. Annu Rev Med. 2012;63:199–215.
8. Cortés-Hernández LE, Eslami-S Z, Alix-Panabières C. Circulating tumor cells as the functional aspect of liquid biopsy to understand the metastatic cascade in solide cancer. Mol Asp Med. 2020;72:100816.
9. Coombs CC, Dickherber T, Crompton BD. Chasing ctDNA in patients with sarcoma. Am Soc Clin Oncol Educ Book. 2020;40:e351–60.
10. Tellez-Gabriel M, Brown HK, Young R, Heymann MF, Heymann D. The challenges of detecting circulating tumor cells in sarcoma. Front Oncol. 2016;6:202.

11. Gabriel MT, Calleja LR, Chalopin A, Ory B, Heymann D. Circulating tumor cells: a review of non-EpCam-based approaches for cell enrichment and isolation. Clin Chem. 2016;62(4):571–81.
12. Hayashi M, Zhu P, McCarty G, Meyer CF, Pratilas CA, Levin A, Morris CD, Albert CM, Jackson KW, Tang CM, Loeb DM. Size-based detection of sarcoma circulating tumor cells and cell clusters. Oncotarget. 2017;8(45):78,965–77.
13. Li H, Meng QH, Noh H, Batth IS, Somaiah N, Torres KE, Xia X, Wang R, Li S. Detection of circulating tumor cells from cryopreserved human sarcoma peripheral blood mononuclear cells. Cancer Lett. 2017;403:216–23.
14. Benini S, Gamberi G, Cocchi S, Garbetta J, Alberti L, Righi A, Gambarotti M, Picci P, Ferrari S. Detection of circulating tumor cells in liquid biopsy from Ewing sarcoma patients. Cancer Manag Res. 2018;10:49–60.
15. Chinen LTD, Mello CAL, Abdallah EA, Ocea LM, Buuim ME, Breve NM, Junior JLG, Fanelli MF, Paterlini-Brechot P. Isolation, detection, and immunomorphological characterization of circulating tumor cells (CTCs) from patients with different types of sarcoma using isolation by size of tumor cells: a window on sarcoma-cell invasion. Onco Targets Ther. 2014;7: 1609–17.
16. Braun AC, de Mello CAL, Corassa M, Abdallah EA, Urvanegia AC, Alves VS, Flores BCTCP, Díaz M, Nicolau UR, Silva VSE, Calsavara V, Paterlini-Brechót P, Chinen LTD. EGFR expression in circulating tumor cells from high-grade metastatic soft tissue sarcomas. Cancer Biol Ther. 2018;19(6):454–60.
17. Mihály D, Nagy N, Papp G, Pápai Z, Sápi Z. Release of circulating tumor cells and cell-free nucleic acids is an infrequent event in synovial sarcoma: liquid biopsy analysis of 15 patients diagnosed with synovial sarcoma. Diagn Pathol. 2018;13(1):81.
18. Przybyl, van de Rijn M, Rutkowski P. Detection of SS18-SSX1/2 fusion transcripts in circulating tumor cells of patients with synovial sarcoma. Diagn Pathol. 2019;14(1):24.
19. West DC, Grier HE, Swallow MM, Demetri GD, Granowetter L, Sklar J. Detection of circulating tumor cells in patients with Ewing's sarcoma and peripheral primitive neuroectodermal tumor. J Clin Oncol. 1997;15(2):583–8.
20. Schleiermacher G, Peter M, Oberlin O, Philip T, Rubie H, Mechinaud F, Sommelet-Olive D, Ladman-Parker J, Bours D, Michon J, Delattre O. Société Française de Pédiatrie. Increased risk of systemic relapses associated with bone marrow micrometastasis and circulating tumor cells in localized Ewing tumor. J Clin Oncol. 2003;21(1):85–91.
21. Avigad S, Cohen IJ, Ziberstein J, Liberzon E, Goshen Y, Ash S, Meller I, Kollender Y, Issakov J, Zaizov YI. The predictive potential of molecular detection in the nonmetastatic Ewing family of tumors. Cancer. 2004;100(5):1053–8.
22. Fagnou C, Michon J, Peter M, Bernoux A, Oberlin O, Zucker JM, Magdelenat H, Delattre O. Presence of tumor cells in bone marrow but not in blood is associated with adverse prognosis in patients with Ewing's tumor. Société Française d'Oncologie Pédiatrique. J Clin Oncol. 1998;16(5):1707–11.
23. Peter M, Madgelenat H, Michon J, Melot T, Oberlin O, Zucker JM, Thomas G, Delattre O. Sensitive detection of occult Ewing's cells by the reverse transcriptase-polymerase chain reaction. Br J Cancer. 1995;72(1):96–100.
24. Zoubek A, Ladenstein R, Windhager R, Amann G, Fischmeister G, Kager L, Jugovic D, Ambros PF, Gadner H, Kovar H. Predictive potential of testing for bone marrow involvement in Ewing tumor patients by RT-PCR: a preliminary evaluation. Int J Cancer. 1998;79(1):56–60.
25. Satelli A, Mitra A, Cutrera JJ, Devarie M, Xia X, Ingram DR, Dibra D, Somaiah N, Torres KE, Ravi V, Ludwig JA, Kleinerman ES, Li S. Universal marker and detection tool for human sarcoma circulating tumor cells. Cancer Res. 2014;74(6):1645–50.
26. Wong IH, Chan AT, Johnson PJ. Quantitative analysis of circulating tumor cells in peripheral blood of osteosarcoma patients using osteoblast-specific messenger RNA markers: a pilot study. Clin Cancer Res. 2000;6(6):2183–8.
27. Hatano H, Kawashima H, Ogose A, Hotta T, endo N. A PCR-ELISA assay for the detection of disseminated osteosarcoma cells in a mouse metastatic model. J Orthop Sci. 2001;6(3):269–75.
28. Chalopin A, Tellez-Gabriel M, Brown HK, Vallette F, Heymann MF, Gouin F, Heymann D. Lsolation of circulating tumor cells in a preclinical model of osteosarcoma: effect of chemotherapy. J Bone Oncol. 2018;12:83–90.
29. Hasegawa N, Nakamura IT, Ueno T, Kojima S, Kawazu M, Akaike K, Okubo T, Takagi T, Suehara Y, Hayashi T, Saito T, Kaneko K, Kohsaka S. Detection of circulating sarcoma tumor cells using a microfluidic chip-type cell sorter. Sci Rep. 2019;9(1):20047.
30. Hoshino M, Ogose A, Kawasima H, Izumi T, Hotta T, Hatano H, Morita T, Otsuka H, Yanoma S, Tsukuda M, Endo N. Molecular analyses of cell origin and detection of circulating tumor cells in the peripheral blood in alveolar soft part sarcoma. Cancer Genet Cytogenet. 2009;190(2):75–80.
31. Balasubramanian P, Kinders RJ, Kummar S, Gupta V, Hasegawa D, Menachery A, Lawrence SM, Wang L, Ferry-Galow K, Davis D, Parchment RE, Tomaszewski JE, Doroshow JH. Antibody-independent capture of circulating tumor cells of non-epithelial origin with the ApoStream® system. PLoS One. 2017;12(4):e0175414.
32. Kelly KM, Womer RB, Barr FG. Minimal disease detection in patients with alveolar rhabdomyosarcoma using a reverse transcriptase-polymerase chain reaction method. Cancer. 1996;78(6):1320–7.
33. Li H, Meng QH, Noh H, Somaiah N, Torres KE, Xia X, Batth IS, Joseph CP, Liu M, Wang R, Li S. Cell-surface vimentin-positive macrophage-like circulating tumor cells as a novel biomarker of metastatic gastrointestinal stromal tumors. Onco Targets Ther. 2018;7(5):e1420450.
34. Zheng Y, Zhang J, Huang M, Wang T, Qu X, Wu L, Song J, Wang W, Song Y, Yang C. selection of aptamers against vimentin for isolation and release of circulating tumor cells undergoing epithelial mesenchymal transition. Anal Chem. 2020;92(7):5178–84.
35. Martín-Broto J, Pousa AL, Brohl AS, Van Tine BA, Powers B, Stacchiotti S, Blay JY, Hu JS, Oakley GJ 3rd, Wang H, Szpurka AM, Levy DE, Mo G, Ceccarelli M, Jones RL. Circulating tumor cells and biomarker modulation with Olaratumab monotherapy followed by Olaratumab plus doxorubicin : phase Ib study in patients with soft-tissue sarcoma. Mol Cancer Ther. 2021;20:132–41.
36. Satelli A, Brownlee Z, Mitra A, Meng QH, Li S. Circulating tumor cell enumeration with a combination of epithelial cell adhesion molecule- and cell-surface vimentin-based methods for monitoring breast cancer therapeutic response. Clin Chem. 2015;61(1):259–66.
37. Chen M, Zhao H. Next-generation sequencing in liquid biopsy: cancer screening and early detection. Hum Genomics. 2019;13:34.
38. Volckmar AL, Sültmann H, Riediger A, Fioretos T, Schirmacher P, Endris V, Stenzinger A, Dietz S. A field guide for cancer diagnostics using cell-free DNA: from principles to practice and clinical applications. Genes Chromosomes Cancer. 2018;57:123–39.
39. Coombs CC, Dicherber T, Crompton BD. Chasing ctDNA in patients with sarcoma. Am Soc Clin Oncol Educ Book. 2020;40:e351–60.
40. Namlos HM, Boye K, Meza-Zepeda LA. Cell-free DNA in blood as a noninvasive insight into the sarcoma genome. Mol Aspect Med. 2020;72:100827.
41. Barris DM, Weiner SB, Dubin RA, Fremed M, Zhang X, Piperdi S, Zhang W, Maqbool S, Gill J, Roth M, Hoang B, Geller D, Gorlick R, Weiser DA. Detection of circulating tumor DNA in patients with osteosarcoma. Oncotarget. 2018;9:12695–704.
42. Gutteridge A, Rathbone VM, Gibbons R, Bi M, Archard N, Davies KEJ, Brown J, Plagnol V, Pillay N, Amary F, O'Donnell GM, Tirabisco R, Flanagan AM, Forshew T. Digital PCR analysis of circulating tumor DNA: a biomarker for chondrosarcoma diagno-

sis, prognostication, and residual disease detection. Cancer Med. 2017;6:2194–202.
43. Shukla NN, Patel JA, Magnan H, Zehir A, You D, Tang J, Meng F, Samoila A, Slotkin EK, Ambati SR, Chou AJ, Wexler LH, Meyers PA, Peerschke EI, Viale A, Berger MF, Ladanyi M. Plasma DNA-based molecular diagnosis, prognostication, and monitoring of patients with EWSR1 fusion-positive sarcomas. JCO Precis Oncologia. 2017;2017:PO.16.00028.
44. Boonstra PA, Ter Elst A, Tibbesma M, Bosman LJ, Mathijssen R, Atrafi F, van Coevorden F, Steeghs N, Farag S, Gelderblom H, van der Graaf WTA, Desar OME, Maier J, Overbosch J, Suurmeijer AJH, Gitema J, Schuuring E, Reuners AKL. A single digital droplet PCR assay to detect multiple KIT exon 11 mutations in tumor and plasma from patients with gastrointestinal stromal tumors. Oncotarget. 2018;9:13870–83.
45. Eastley NC, Ottolini B, Neumann R, Luo JL, Hastings RK, Khan I, Moore DA, Esler CP, Shaw JA, Royle NJ, Ashford RU. Circulating tumour-derived DNA in metastatic soft tissue sarcoma. Oncotarget. 2018;9:10549–60.
46. Namløs HM, Boye K, Mishkin SJ, Barøy T, Lorenz S, Bjerkehagen B, Stratford EW, Munthe E, Kudlow BA, Myklebost O, Meza-Zepeda LA. Noninvasive detection of ctDNA reveals intratumor heterogeneity and is associated with tumor burden in gastrointestinal stromal tumor. Mol Cancer Ther. 2018;17:2473–80.
47. Ogino S, Konishi H, Ichikawa D, Hamada J, Shoda K, Arita T, Komatsu S, Shiozaki A, Okamoto K, Yamazaki S, Yaskuwa S, Konishi E, Otsuji E. Detection of fusion gene in cell-free DNA of a gastric synovial sarcoma. World J Gastroenterol. 2018;24: 949–56.
48. Shulman DS, Klega K, Imamovic-Tuco A, Clapp A, Nag A, Thorner AR, Van Allen E, Ha G, Lessnick SL, Gorlick R, et al. Detection of circulating tumour DNA is associated with inferior outcomes in Ewing sarcoma and osteosarcoma: a report from the Children's Oncology Group. Br J Cancer. 2018;2018(119): 615–21.
49. Maier J, Lange T, Kerle I, Specht K, Bruegel M, Wickenhauser C, Jost P, Niederwieser D, Peschel C, Duyster J, von Bubnoff N. Detection of mutant free circulating tumor DNA in the plasma of patients with gastrointestinal stromal tumor harboring activating mutations of CKIT or PDGFRA. Clin Cancer Res. 2013;19:4854–67.
50. Yoo C, Ryu MH, Na YS, Ryoo BY, Park SR, Kang YK. Analysis of serum protein biomarkers, circulating tumor DNA, and dovitinib activity in patients with tyrosine kinase inhibitor-refractory gastrointestinal stromal tumors. Ann Oncol. 2014;25:2272–7.
51. Krumbholz M, Hellberg J, Steif B, Bäuerle T, Gillmann C, Fritscher T, Agaimy A, Frey B, Juengert J, Wardelmann E, Hartmann W, Juergens H, Dirksen U, Metzler M. Genomic EWSR1 fusion sequence as highly sensitive and dynamic plasma tumor marker in Ewing sarcoma. Clin Cancer Res. 2016;22:4356–65.
52. Snyder MW, Kircher M, Hill AJ, Daza RM, Shendure J. Cell-free DNA comprises an in vivo nucleosome footprint that informs its tissues-of-origin. Cell. 2016;164:57–68.
53. Ulz P, Auer M, Heitzer E. Detection of circulating tumor DNA in the blood of cancer patients: an important tool in cancer chemoprevention. Methods Mol Biol. 2016;1379:45–68.
54. Riggi N, Knoechel B, Gillespie SM, Rheinbay E, Boulay G, Suvà ML, Rossetti NE, Boonseng WE, Oksuz O, Cook EB, Formey A, Patel A, Gymrek M, Thapar V, Deshpande V, Ting DT, Hornicek FJ, Nielsen GP, Stamenkovic I, Aryee MJ, Bernstein BE, Rivera MN. EWS-FLI1 utilizes divergent chromatin remodeling mechanisms to directly activate or repress enhancer elements in Ewing sarcoma. Cancer Cell. 2014;26(5):668–81.
55. Sheffield NC, Pierron G, Klughammer J, Datlinger P, Schönegger A, Schuster M, Hadler J, Surdez D, Guillemot D, Lapouble E, Freneaux P, Champigneulle J, Bouvier R, Walder D, Ambros IM, Hutter C, Sorz E, Amaral AT, de Alava E, Schallmoser K, Strunk D, Rinner B, Liegl-Atzwanger B, Huppertz B, Leithner A, de Pinieux G, Terrier P, Laurence V, Michon J, Ladenstein R, Holter W, Windhager R, Dirksen U, Ambros PF, Delattre O, Kover H, Bock C, Tomazou EM. DNA methylation heterogeneity defines a disease spectrum in Ewing sarcoma. Nat Med. 2017;23:386–95.
56. Tomazou EM, Sheffield NC, Schmidl C, Schuster M, Schönegger A, Datlinger P, Kubicek S, Bock C, Kovar H. Epigenome mapping reveals distinct modes of gene regulation and widespread enhancer reprogramming by the oncogenic fusion protein EWS-FLI1. Cell Rep. 2015;10:1082–95.
57. Keller L, Pantel K. Unravelling tumour heterogeneity by single-cell profiling of circulating tumour cells. Nat Rev Cancer. 2019;19(10):553–67.
58. Brown HK, Tellez-Gabriel M, Cartron PF, Vallette FM, Heymann MF, Heymann D. Characterization of circulating tumor cells as a reflection of the tumor heterogeneity: myth or reality? Drug Discov Today. 2019;24(3):763–72.
59. Paoletti C, Cani AK, Larios JM, Hovelson DH, Aung K, Darga EP, Cannell EM, Baratta PJ, Liu CJ, Chu D, Yazdani M, Blevins AR, Sero V, Tokudome N, Thomas DG, Gersch C, Schott AF, Wu YM, Lonigro R, Robinson DR, Chinnaiyan AM, Bischoff FZ, Johnson MD, Park BH, Hayes DF, Rae JM, Tomlins SA. Comprehensive mutation and copy number profiling in archived circulating breast cancer tumor cells. Cancer Res. 2018;78(4):1110–22.
60. Wu ZJ, Tan JC, Qin X, Liu B, Yuan ZC. Significance of circulating tumor cells in osteosarcoma patients treated by neoadjuvant chemotherapy and surgery. Cancer Manag Res. 2018;10(10):3333–9.
61. Tellez-Gabriel M, Heymann MF, Heymann D. Circulating tumor cells as a tool for assessing tumor heterogenetity. Theranostics. 2019;9(16):4580–94.
62. Tellez-Gabriel M, Cochonneau D, Cadé M, Jubellin C, Heymann MF, Heymann D. Circulating tumor cell-derived pre-clinical models for personalized medicine. Cancers (Basel). 2018;11(1):19.

Liquid Biopsy Using Cell-Free Tumor DNA for Gastrointestinal Cancers

13

Takafumi Nakano, Tadashi Abe, Seiichiro Takao, Hideyuki Saito, Takaaki Masuda, and Koshi Mimori

Abstract

Since the advent of precision medicine in 2015, personalized genomic profiles of mutated DNA from tumor tissues have been applied for the selection of compatible drugs for the clinical treatment of genomic aberrations. As a more convenient and less invasive method of precision medicine, liquid biopsy that analyzes mutated cell-free tumor DNA (ctDNA) can be used as a biomarker for postoperative recurrence of genomic aberrations and their appropriate treatment. In this chapter, we explore liquid biopsies for gastrointestinal (GI) cancers in detail. We review the literature on ctDNA-related issues, including the mechanism of ctDNA identification in plasma, the methodology for identifying ctDNA in profiles and targeted sequencing assays, intratumor heterogeneity, recent technological innovations in ctDNA analysis in GI cancers, and the limitations of current approaches. We also describe the clinical aspects of GI cancer, such as actionable targets and clinical trials using mutated ctDNA analysis, the significance of longitudinal observation of ctDNA for clonal evolution after targeted therapy, and the detection of minimal residual disease by ctDNA analysis. Finally, we summarize a few recent studies on ctDNA in colorectal, gastric, and esophageal cancers. (192)

Learning Objectives

- Identification of origin of clones excreting ctDNA.
- We could realize the difference in detectability of mutations in ctDNA among cancers.
- Clinical significance is equivalent between tissue and ctDNA, i.e., actionable targets and prognosis.

T. Nakano · T. Abe · S. Takao · H. Saito · T. Masuda
K. Mimori (✉)
Kyushu University Beppu Hospital, Beppu, Japan
e-mail: mimori.koshi.791@m.kyushu-u.ac.jp

13.1 Introduction

According to global epidemiological data, the mortality rate related to gastrointestinal (GI) tract malignancies is increasing [1]. We believe that early diagnosis of postoperative recurrence, even as minimal residual disease (MRD), will improve clinical outcomes. In current clinical practice, we measure conventional serum tumor markers, such as carcinoembryonic antigen, carbohydrate antigen 19-9, and squamous cell carcinoma antigen. However, due to the limited detectability of these cell surface proteins, we cannot use them as tumor markers for early subclinical diagnosis; thus, there is a need for minimally invasive, convenient, and precise markers and approaches for cancer diagnosis.

A precision medicine initiative was launched and implemented following the 2015 State of the Union Address by the former President of the United States Barack Obama. Liquid biopsy using mutated cell-free tumor DNA (ctDNA) analysis should be the recommended method for precision medicine, as it can identify early markers of postoperative recurrence and actionable targets for appropriate drugs. A recent clinical trial in Japan revealed that the outcomes of metastatic colorectal cancer (CRC) patients with actionable alterations were better than those without them [2].

In this chapter, we review the following aspects of ctDNA as it evolves into the most clinically feasible approach for precision medicine in GI cancers: (1) the basic knowledge of ctDNA and its mechanism in the circulating system, (2) clinical applications of ctDNA, and (3) recent studies on ctDNA in GI cancers.

S. P. Leong et al. (eds.), *Cancer Metastasis Through the Lymphovascular System*,
https://doi.org/10.1007/978-3-030-93084-4_13

13.1.1 Basic Knowledge of ctDNA in GI Cancers

13.1.1.1 General Statement of Liquid Biopsy in GI Cancers

Liquid biopsies have enabled the collection of information regarding malignancies using blood or fluids from cancer patients without using invasive procedures. This methodology was developed gradually to improve its diagnostic accuracy to be at par with conventional biopsy procedures. During the metastasis of cancer cells from primary locations, there are several methods to obtain diverse data reflecting the metastatic processes, including circulating tumor cells, cell-free DNA, cancer-specific transcripts, various noncoding RNAs, and exosomes. These cancer-specific data can evaluate postoperative recurrence and susceptibility to cancer therapy.

The discovery of ctDNA has its origins in CRC. ctDNA in circulating fluid is considered the most informative and robust material for the diagnosis of cancers at the subclinical level. In 1948, Mandel and Métais reported "cell-free DNA" (cfDNA). Sidransky et al. identified KRAS mutations in stool samples from nine CRC patients and proposed early detection of CRC in a noninvasive manner [3]. In 1994, detection of KRAS-mutated DNA in the peripheral blood from CRC patients enabled the identification of cancer in the patients using genomic aberrations in cfDNA from body fluids and excrements (e.g., blood, stool, and urine) [4, 5]. Although cfDNA analysis is unable to detect less than 10 ng of cfDNA/mL of plasma in a healthy patient [6], the levels are increased in cancer patients, thereby enabling easy detection [7]. The use of next-generation sequencing (NGS) can detect low levels of ctDNA; thus, the technique has progressed into a highly feasible method for clinical liquid biopsies.

13.1.1.2 Circulation of ctDNA

In general, increased levels of circulating DNA fragments in the peripheral blood are derived from apoptotic and necrotic cells in tissues by pathological processes, such as malignant and benign lesions, inflammatory diseases, stroke, trauma, and sepsis [8]. In healthy conditions, necrotic or apoptotic cells are eliminated by macrophages; however, the exponential growth of malignant cells simultaneously induces several apoptotic and necrotic cells, which form a major source of ctDNA [9].

Recently, we identified that the detectability of ctDNA was determined not by host immunity in the cancer microenvironment in metastatic CRC. In our study, the mutated genes in ctDNA were derived from malignant clones with an inactivated immune response due to a low binding affinity (50 < IC50) for estimated neoantigen peptides by a calculated pipeline netMHCpan even though high expression of tumor-specific mutated RNA. In addition, the frequency of mutated genes in ctDNA in all samples was significantly associated with the expression of regulatory T cells and exhausted T cell-related genes in metastatic sites. This paradoxical finding was due to increased immunogenicity caused by upregulated immune-resistant molecules in metastatic sites.

13.1.1.3 Two Major Approaches for the Detection of Mutated ctDNA

1. **Profiling of mutated ctDNA by exome sequencing:** Broad coverage assays use NGS-based approaches and have the capability to detect many variants in multiple genes, often over 50 genes, and can be applied to different tumor types. In CRC cases, primary advanced cancer is derived from numerous minor mutations that result in intratumor heterogeneity. However, driver mutations are expressed ubiquitously in primary tumors in a neutral manner [10], thereby reflecting major pathogenic actionable drivers to be diagnosed or cured by molecular targeted therapy.

   In contrast, precancerous lesions or early CRC can contain several driver genes in tumors. A few bona fide drivers must show positive mutations to progress through Darwinian evolution [11]. Therefore, in identifying patients with early cancer using mutated driver ctDNA in plasma, it cannot be determined which altered genes are actionable pathogenic drivers. It is more efficient to survey the whole profile of mutated ctDNA in the early phase of cancer.
2. **Targeted sequencing for detection of cancers:** Targeted sequencing for a single or small number of variants is aimed at greater coverage (1000–10,000×) in comparison to exome sequencing (100–200×). Targeted sequencing uses a "cancer panel" of mutated driver genes from the Cancer Genome Atlas Pan-Cancer Analysis Project, and a small amount of ctDNA (10 ng) is amplified to obtain sequencing results [12]. This assay is appropriate for the longitudinal analysis of a limited number of mutated genes using personalized monitoring targets. In esophageal cancer, recurrence can be identified 3–6 months earlier by ctDNA analysis than by any other conventional method at the subclinical level [13]. In addition, ctDNA analysis has a rapid turnaround time, and results can be obtained earlier than tissue analysis (18 days for tumor tissue analysis and 7 days for ctDNA analysis) [14]. In addition, several solid tumors, such as intrahepatic cholangiocellular carcinoma, harbor heterogeneous mutations; therefore, these cancers might also benefit from a panel with personalized targets instead of using a commercial multipurpose cancer panel.

   For a specific target, a digital PCR or BEAMing assay can yield high sensitivity (variant allele frequency; VAF

<0.1%) and specificity in the detection of altered sequences. Although they are clinically feasible approaches with high performance, there are concerns about contamination from nucleic acids in the air, and only a limited number of mutated loci can be detected simultaneously.

13.1.1.4 Superiority of ctDNA Analysis Compared with Tissue Analysis in Overcoming Intratumor Heterogeneity

Considering intratumor heterogeneity, the analysis of one biopsy specimen might pose a risk for a false-negative diagnosis. Moreover, several studies have reported that ctDNA analysis is better than tissue analysis for GI cancers. Parikh et al. described the therapeutic implications and potential advantages of cfDNA over tissue biopsy in the setting of acquired resistance [15]. They suggested that a single-lesion tumor biopsy alone is inadequate to detect acquired resistance and that ctDNA analysis regularly identifies heterogeneous, clinically relevant resistance alterations critical to the selection of subsequent therapy. Huang et al. reported the clinical benefits of ctDNA assays in the management of hepatocellular carcinoma [16, 17]. Pectasides et al. showed that ctDNA sequencing demonstrated the feasibility of detecting genomic amplifications not detected by tissue sampling of gastroesophageal adenocarcinoma [18]. Multiple genomic changes have been observed during clonal evolution owing to the pressure of therapy in cancer, resulting in acquired secondary resistance [19].

13.1.1.5 New Technologies for Detection of ctDNA in GI Cancers (Table 13.1)

As listed in Table 13.1, Cohen et al. reported a CancerSEEK assay using a combination of ctDNA and protein markers to detect cancer cells in 1005 patients with non-metastatic colorectal, gastric, esophageal, pancreatic, liver, lung, breast, and ovarian cancers [20]. Cristiano et al. reported that the DELFI method detects cancer- and organ-specific fragment patterns using low-coverage sequencing of the whole genome with 91% sensitivity and 98% specificity in multiple cancers [21].

13.1.1.6 Limitations of ctDNA Analysis

The precise method to determine the allele frequency of mutated ctDNA without heterogeneity among examinees and diverse preanalytical variables (e.g., timing of sampling, nucleic acid extraction, transportation system, processing, and storage conditions) is still controversial. In addition, various patient factors might affect the results, such as inflammation, trauma, injuries, metabolites of glucose or fat, immunity, food, and lifestyle.

The ability to detect ctDNA varies among cancers; it is lower in glioma, thyroid cancer, renal cell carcinoma, prostate cancer, and medulloblastoma, while it is higher in bladder cancer, CRC, and gastroesophageal cancer [22]. Additional information from protein assays was added to the mutated ctDNA data in the CancerSEEK study by Cohen et al. According to their findings, the utility of liquid biopsy was different among various malignancies; it was low in breast cancer, lung cancer, and CRC and high in ovarian cancer [20].

Table 13.1 Approaches of liquid biopsy using ctDNA in GI cancers

Cancer	Alteration	Platform	Sensitivity and specificity	Author
Multiple cancers	ctDNA and protein	CancerSEEK assay	1005 cases Sensitivity: 69–98% Specificity: 99%	Cohen et al. [20]
HCC	p16INK4 methylation in ctDNA	qRT-PCR	Tumor burden after surgery	Wong et al. [55]
CRC	Cancer-specific methylation in ctDNA	Methylation-specific dPCR Forms the prediction model	1493 CRC cases Sensitivity: 89.7% Specificity: 86.8%	Luo et al. [56]
Multiple cancers	Evaluate the size of ctDNA	Deep sequence	Focus on 90–150 bp of ctDNA 95% of cases show improved analysis	Mouliere et al. [57]
Multiple cancers	Fragment pattern analysis of ctDNA	DELFI	Sensitivity 91%; specificity 98%	Cristiano et al. [21]
Multiple cancers	Extracellular vesicle in plasma	Cancer panel for proteome assay with AI analysis	Tissue: Sensitivity 90%; specificity 94% Plasma: Sensitivity 95%; specificity 90%	Hoshino et al. [58]
Multiple cancers	ctDNA	Integration of variant reads pipeline	High sensitivity; >one/one million	Wan et al. [59]

Furthermore, somatic mutations accumulate in healthy hematopoietic cells during aging, which can drive the clonal expansion of hematopoietic cells in the absence of dysplasia. These mutations are called clonal hematopoiesis of indeterminate potential (CHIP) [23]. CHIP may cause false-positive ctDNA results in cases where high sensitivity is required, such as MRD and early cancer detection.

13.1.2 Clinical Application of ctDNA Analysis

13.1.2.1 Actionable Targets in ctDNA for the Selection of Anticancer Agents

There are studies comparing PCR-based ctDNA assays to detect RAS mutations and tissue-based RAS testing [24, 25]. The OncoBEAM™ RAS CRC Kit, which detects RAS mutations in ctDNA derived from mCRC by a BEAMing method, is the only ctDNA assay approved in Europe and Japan for ctDNA genomic biomarker testing in GI cancers. Analysis of ctDNA sequences detected genomic alterations in advanced GI cancers, including RAS and BRAF mutations in CRC and HER2 amplification in esophageal cancer [26, 27]. Considering immune checkpoint inhibitor treatment, NGS analysis of ctDNA could determine the microsatellite instability status in GI cancers, with a high concordance with tissue-based testing [28, 29]. Putative germline BRCA1/2 mutations were found in 8 out of 332 (2.4%) patients with pancreatic cancer, suggesting that ctDNA analysis can distinguish germline mutations from somatic mutations by approximately 50% of the mutation allele frequency [30].

13.1.2.2 ctDNA Analyses in Clinical Trials of GI Cancers

Patient selection based on ctDNA has recently been conducted. Metastatic gastroesophageal cancer with FGFR2 amplification was evaluated by ctDNA analysis and FGFR2b overexpression by immunohistochemical analysis (phase I/III NCT03343301; FIGHT). In the HERACLES study, which used trastuzumab plus lapatinib for HER2-positive mCRC, the plasma copy number of HER2 amplification in ctDNA was correlated with the best objective response and progression-free survival [31]. In the VIKTORY study of gastric cancer, ctDNA analysis demonstrated a better correlation between high MET copy number and response to savolitinib than the tissue NGS MET copy number [32]. In the CRICKET study, after longitudinal evaluation of KRAS mutant ctDNA, challenging with anti-EGFR was effective in metastatic CRC cases with KRAS-null mutant clones [33]. TARGET is a molecular profiling program with the primary aim of matching patients with a broad range of advanced cancers to early-phase clinical trials based on analysis of both somatic and KRAS mutations across a 641 cancer-associated gene panel in a single ctDNA assay. Actionable mutations were identified in 41 of 100 patients, and 11 of these patients received matched therapy [34]. In Japan, a large cohort study was conducted in advanced GI tract cancers to evaluate the utility of ctDNA assays compared to tissue assays [2]. In a comparison between the GI-SCREEN study (advanced GI cancers, tumor tissue genotyping using Oncomine Comprehensive Assay v1 (143-gene) or v3 (161-gene)) and the GOZILA study (advanced solid tumors, including GI cancers, plasma ctDNA genotyping using Guardant360, a 74-gene ctDNA assay, Central lab: Guardant Health (Redwood City, CA)), the trial enrollment rate in patients with actionable alterations was significantly higher in GOZILA (41/340, 12.1%) than in GI-SCREEN (85/11714, 5.0%) ($p < 0.0001$) (Nakamura et al., ASCO Annual Meeting, Virtual, May 29–31, 2020). In addition, the objective response rate in tissue was 18.9% (16/85), compared to 17.1% (7/41) in plasma, suggesting that the efficacy of the matched clinical trials was equal.

13.1.2.3 Longitudinal Observation of ctDNA to Detect Clonal Evolution After Targeted Therapy

Several drugs have been approved to target EGFR mutations, but tumors inevitably develop resistance to them. In half of the patients in one study, resistance was caused by a T790M mutation in EGFR; osimertinib was applied, but people who responded to it tended to relapse within a year [35]. Therapy was decided by monitoring the resistance and by tracking blood ctDNA; if there is a reduction in T790M mutation, treatment was switched to a first-generation EGFR inhibitor; however, if the T790M levels rise, the treatment was switched back to osimertinib [35]. A similar finding was observed in GI tract cancers, in which longitudinal ctDNA surveillance detected clonal evolution with minimal invasiveness. After an increased number of RAS mutations, temporal withdrawal of an anti-EGFR antibody resulted in a decline in KRAS mutant allele levels, and eventually, the sensitivity to anti-EGFR antibody was restored [36, 37].

For HER2/neu-amplified gastric cancer cases, the ToGA study indicated treatment with trastuzumab in combination with chemotherapy. In this study, the proportion of NF1 mutations emerged to dominate ctDNA during disease progression in resistant cases [26].

In clinical trials, new molecule-targeting agents have been identified by ctDNA analysis, including PARP1 inhibitors for BRCA1-mutated pancreatic cancer [38], TRK inhibitors for NTRK-mutated cases of TRK fusion-positive GI cancer [39], and FGFR inhibitors for FGFR2 mutations in patients with FGFR2 fusion-positive cholangiocarcinoma [40].

13.1.2.4 ctDNA in MRD

Analysis of ctDNA for the identification of predictive genomic biomarkers for GI cancers has the potential to detect MRD and early cancers with no clinically evident disease. Most patients in whom ctDNA was detectable at the first follow-up visit (13–56 days after surgery) had recurrence, while patients with undetectable ctDNA had no recurrence [41]. Tie et al. used an NGS-based method (Safe-SeqS) and reported that in 178 stage II CRC patients, ctDNA-mutated cases showed 18-fold higher risk of recurrence than non-mutated cases [42].

13.1.3 Summary of Recent Studies of cfDNA in GI Cancers (Table 13.2)

13.1.3.1 CRC

Bettegowda et al. conducted droplet digital PCR (ddPCR) to detect primary tumors and achieved high sensitivity and specificity [22]. Normanno et al. reported that the early detection of KRAS mutations in ctDNA in chemotherapy-free patients was correlated with primary resistance to anti-EGFR mAbs, while both the objective response rate and the overall survival were improved in patients harboring wild-type RAS in ctDNA [43]. Using the OncoBEAM RAS CRC assay, Vidal et al. reported plasma ctDNA RAS mutation analysis results for the diagnosis and treatment monitoring of metastatic CRC patients [25].

In terms of methylation of septin 9 (SEPT9), there were three intriguing studies using qPCR. Warren et al. reported that the FDA granted premarket approval for the first ctDNA-based liquid biopsy for cancer screening using methylated SEPT9 DNA (mSEPT9) [44]. Song et al. evaluated the performance of the SEPT9 gene methylation assay and compared it with other CRC screening tests in a meta-analysis [45]. In addition, Church et al. reported that a blood-based mSEPT9 test showed that CRC signals in blood could be detected in asymptomatic average-risk individuals undergoing screening. However, the utility of the test for population screening for CRC requires improved sensitivity for the detection of early and advanced cancers [46].

13.1.3.2 Gastric Cancer

Frankell et al. showed that clinical ctDNA-NGS testing holds promise for gastroesophageal adenocarcinoma for both the detection of MRD in early-stage disease and as a serial tumor marker [47]. Wang et al. monitored trastuzumab resistance to determine the mechanisms underlying the limited response rate and rapid emergence of resistance in HER2+ metastatic gastric cancer (mGC) and showed that longitudinal ctDNA sequencing provided novel insights into gene alterations underlying trastuzumab resistance in HER2+ mGC [26]. Kim et al. monitored cancer-specific rearrangements and reported that the presence of ctDNA non-shedders is a limitation of this method [48]. Anderson et al. identified a novel cancer-specific methylated DNA marker that undergoes aberrant methylation during oncogenesis [49].

13.1.3.3 Esophageal Cancer

In esophageal cancer cases, genomic DNA from tumors and ctDNA from plasma were obtained from 13 patients undergoing treatment for newly diagnosed esophageal cancer. NGS of ctDNA in plasma was used to identify mutations in 53 cancer-related genes in which recurrent mutations were previously detected in ESCC. We identified more than one concordant somatic mutation in ctDNA and primary tumor samples from 10 patients (83.3%) and the ctDNA and recurrent tumor samples from one patient (100%). Furthermore, to identify accurate biomarkers for identifying ESCC patients, we assessed the usability of the four genes (*TP53*, *FAT3*, *MLL3*, and *AJUBA*) in which mutations were the most frequently detected. Mutations in the four genes were identified with 78.9% sensitivity, 100% specificity, and 92.3% accuracy in tumor and ctDNA samples from patients [13].

A recent study demonstrated a proof-of-principle for ctDNA detection of biomarkers in head and neck cancers [50]. Studies have also correlated the abundance of ctDNA with tumor size and stage, with as much as 40% in metastatic cancers but as low as 0.11% in premalignant or early-stage disease [51]. Thus, a highly sensitive detection system is required for the detection of ctDNA. With the improvement of NGS technologies, it is possible to employ targeted deep-sequencing analysis to detect ctDNA with high sensitivity. Iwaya et al. reported that the clinical validity of ctDNA monitoring was confirmed in 91% (31/34) of patients with ESCC using 31 mutated genes as a panel [52]. Azad et al. detected ctDNA after chemoradiotherapy, which precedes radiographic evidence of tumor progression, and the combined ctDNA and metabolic imaging analyses accurately predicted tumor progression in 100% of the patients [53]. Maron et al. demonstrated that their approach using ctDNA provided clinical benefits not only in HER2/neu-amplified cases but also in EGFR-amplified cases [54].

13.1.4 Conclusion

In this review, we introduced a liquid biopsy using ctDNA analysis as the recommended powerful method for precision medicine to identify early markers of postoperative recurrence and actionable targets for appropriate drugs. This approach is superior to conventional serum tumor markers.

In this chapter, we review the several intriguing aspects of ctDNA. At first, we summarized the basic knowledge of ctDNA and its mechanism in the circulating system. The increased levels of circulating DNA fragments in the periph-

Table 13.2 Summary of liquid biopsy using ctDNA in GI cancers

	Platform	Target alterations	Cases	Clinical application	Findings	Authors
Colorectal cancer	ddPCR	KRAS mutation	12	Detection of primary tumors	Sensitivity: 87.2% Specificity: 100%	Bettegowda et al. [22]
	Digital PCR	KRAS and NRAS mutations	340	Prediction of therapy	RAS mut. 33/92 (35.9%)	Normanno et al. [43]
	qPCR	SEPT9 methylation	50	Detection of primary tumors	Sensitivity: 90%	Warren et al. [44]
	qPCR	SEPT9 methylation	2613 cases6030 controls	CRC screening tests (meta-analysis)		Song et al. [45]
	qPCR	SEPT9 methylation	7941	Detection of asymptomatic CRC(multicenter study)	Sensitivity: I: 35.0%, II: 63.0%, III: 46.0%, IV: 77.4% Specificity: 91.5%	Church et al. [46]
	OncoBEAM™ RAS CRC assay	RAS mutation	115	Diagnosis and treatment monitoring		Vidal et al. [25]
	HTSeq-count	CNV	24	Diagnosis and treatment monitoring	Sensitivity: 79%specificity: 100%	Molparia et al. [60]
	HiSeq 2500 (Illumina)	CNV	Normal 35, Adenoma 20 CRC 80	Staging, diagnosis	AUC I: 0.53, II: 0.54 III: 0.61, IV: 0.60	Li et al. [61]
	dPCR	TP53, KRAS, PIK3CA, APC, FBXW7, BRAF	22	Early detection of metastasis	Sensitivity: 89%specificity: 100%	Furuki et al. [62]
Gastric Cancer	NGS	Mutation profile	1630			Frankell et al. [47]
	Target-Seq	PIK3CA, ERBB2	78	Monitoring trastuzumab resistance	35.3% 29.4%	Wang et al. [26]
	WGS	Personalized cancer-specific Rearrangements	19	Monitoring cancer-specific rearrangements	Recurrence	Kim et al. [48]
	Target-Seq	Molecular tumor Burden index (mTBI)	173 samples (39 cases)	Resistant to trastuzumab	Sensitivity: 94%	Wang et al. [63]
	Methylation PCR	ELMO1, ZNF569, C13orf18		Novel methylated DNA markers (MDM)	Sensitivity: 86% Specificity: 95%	Anderson et al. [49]
	CORD assay	RUNX3	T:50 N:61	Diagnostic tool	Sensitivity: 50 Specificity: 80.3	Hideura et al. [64]
	ddPCR	HER2 copy number	T:60 N:30	ddPCR-based HER2 copy number	Sensitivity: 73.3 Specificity: 93.3	Shoda et al. [65]
Esophageal cancer	Target-Seq	53 genes	13	Early detection of recurrence	Sensitivity: 10/13 (83%)	Ueda et al. [13]
	dPCR	31 genes	34	Early intervention	Validated for 91% (31/34)	Iwaya et al. [52]
	CTC + ctDNA	CTC/ctDNA	57	Serial monitoring for OS		Ko et al. [66]
	CAPP-Seq	607 genes SNV	45	ctDNA and metabolic imaging		Azad et al. [53]
	Target-Seq	483 genes mutation	17	Disease biomarker	Lower frequency In post-surgery	Meng [67]
	ctDNA-NGS	73 genes	203	Clinical decision-making	Benefit in HER2 or EGFR-amplified patients	Maron et al. [54]
	Target-Seq	96 genes	38	Diagnosis of early-stage EAC	18/38 (47%)	Egyud et al. [68]
	Target-Seq	54–73 genes	55	Genomic landscape	ERBB2 alterations poor OS	Kato et al. [69]
	Nano-hmC-seal Method	5hmC 5mC	150	Biomarkers for MRD	Sensitivity: 3.75% Specificity: 85.71%	Tian et al. [70]

eral blood are derived from apoptotic and necrotic cells in tissues by pathological processes. However, the abundant amount of cell-free DNA was derived from cancer cells detected by exome sequencing and/or target sequencing. In terms of clinical applications of ctDNA, we demonstrated recent intriguing studies on ctDNA in GI cancers.

In CRC cases, as primary tumors were comprised of major driver genes as clonal genomic aberrations, such as APC, KRAS, and TP53, we can realize the earlier detection of recurrence and/or evaluate the MRD precisely and continuously during variant therapies by target sequencing. However, in the respect of the early detection of primary site, susceptibility to detect cancer cells by capturing mutated ctDNA within mucosal layer is not enough clinically, further studies will be required. In GC cases, the amplification of HER2/neu is the main driver gene to cause GC. It is worth describing that some cases were applicable to Trastuzumab, which was determined not by tissue biopsy analysis but by ctDNA sequencing. This discrepancy will be provoked by the heterogeneity of the HER2 expression in tissues, while the liquid biopsy can overcome the genomic diversity in tissues. In esophageal cancer cases, just four genes (*TP53*, *FAT3*, *MLL3*, and *AJUBA*) were the most frequently detected. If we improve the clinical outcome in the treatment of the current intractable disease, the early diagnosis and treatment must be required. However, the predisposition of the affected esophageal mucosa was comprised of NOTCH1, NFE2L2, CDKN2A, and FBXW7, which were different from advanced cancers, and we have to bespoke the appropriate cancer panel for target sequencing.

However, there are inherent limitations to the current methodologies. Therefore, clinical oncologists, including surgeons, should consider the specific favorable and unfavorable traits of a given assay before applying it to clinical studies.

In terms of the future of this field, we would like to address the following queries:

Open Questions

- How can detection sensitivity be improved to enable the diagnosis of early-stage cancer?
- What is the significance of detecting epigenomic aberrations in ctDNA?
- Can we identify copy number aberrations/structural variants in ctDNA?
- Which sample is appropriate: urine or blood?
- What mutated genes should be monitored for longitudinal analysis?

References

1. Ferlay J, Colombet M, Soerjomataram I, et al. Estimating the global cancer incidence and mortality in 2018: GLOBOCAN sources and methods. Int J Cancer. 2019;144(8):1941–53.
2. Nakamura Y, Taniguchi H, Ikeda M, et al. Clinical utility of circulating tumor DNA sequencing in advanced gastrointestinal cancer: SCRUM-Japan GI-SCREEN and GOZILA studies. Nat Med. 2020;26(12):1859–64.
3. Sidransky D, Tokino T, Hamilton SR, et al. Identification of ras oncogene mutations in the stool of patients with curable colorectal tumors. Science. 1992;256(5053):102–5.
4. Sorenson GD, Pribish DM, Valone FH, et al. Soluble normal and mutated DNA sequences from single-copy genes in human blood. Cancer Epidemiol Biomark Prev. 1994;3(1):67–71.
5. Vasioukhin V, Anker P, Maurice P, et al. Point mutations of the N-ras gene in the blood plasma DNA of patients with myelodysplastic syndrome or acute myelogenous leukaemia. Br J Haematol. 1994;86(4):774–9.
6. Mouliere F, El Messaoudi S, Pang D, et al. Multi-marker analysis of circulating cell-free DNA toward personalized medicine for colorectal cancer. Mol Oncol. 2014;8(5):927–41.
7. Spindler KL, Appelt AL, Pallisgaard N, et al. Cell-free DNA in healthy individuals, noncancerous disease and strong prognostic value in colorectal cancer. Int J Cancer. 2014;135(12):2984–91.
8. Schwarzenbach H, Hoon DS, Pantel K. Cell-free nucleic acids as biomarkers in cancer patients. Nat Rev Cancer. 2011;11(6):426–37.
9. Jahr S, Hentze H, Englisch S, et al. DNA fragments in the blood plasma of cancer patients: quantitations and evidence for their origin from apoptotic and necrotic cells. Cancer Res. 2001;61(4):1659–65.
10. Uchi R, Takahashi Y, Niida A, et al. Integrated multiregional analysis proposing a new model of colorectal cancer evolution. PLoS Genet. 2016;12(2):e1005778.
11. Saito T, Niida A, Uchi R, et al. A temporal shift of the evolutionary principle shaping intratumor heterogeneity in colorectal cancer. Nat Commun. 2018;9(1):2884.
12. Reinert T, Scholer LV, Thomsen R, et al. Analysis of circulating tumour DNA to monitor disease burden following colorectal cancer surgery. Gut. 2016;65(4):625–34.
13. Ueda M, Iguchi T, Masuda T, et al. Somatic mutations in plasma cell-free DNA are diagnostic markers for esophageal squamous cell carcinoma recurrence. Oncotarget. 2016;7(38):62,280–91.
14. Bachet JB, Bouche O, Taieb J, et al. RAS mutation analysis in circulating tumor DNA from patients with metastatic colorectal cancer: the AGEO RASANC prospective multicenter study. Ann Oncol. 2018;29(5):1211–9.
15. Parikh AR, Leshchiner I, Elagina L, et al. Liquid versus tissue biopsy for detecting acquired resistance and tumor heterogeneity in gastrointestinal cancers. Nat Med. 2019;25(9):1415–21.
16. Chabon JJ, Simmons AD, Lovejoy AF, et al. Circulating tumour DNA profiling reveals heterogeneity of EGFR inhibitor resistance mechanisms in lung cancer patients. Nat Commun. 2016;7:11815.
17. Huang A, Zhang X, Zhou SL, et al. Detecting circulating tumor DNA in hepatocellular carcinoma patients using droplet digital PCR is feasible and reflects intratumoral heterogeneity. J Cancer. 2016;7(13):1907–14.
18. Pectasides E, Stachler MD, Derks S, et al. Genomic heterogeneity as a barrier to precision medicine in gastroesophageal adenocarcinoma. Cancer Discov. 2018;8(1):37–48.
19. McGranahan N, Swanton C. Biological and therapeutic impact of intratumor heterogeneity in cancer evolution. Cancer Cell. 2015;27(1):15–26.

20. Cohen JD, Li L, Wang YX, et al. Detection and localization of surgically resectable cancers with a multi-analyte blood test. Science. 2018;359(6378):926.
21. Cristiano S, Leal A, Phallen J, et al. Genome-wide cell-free DNA fragmentation in patients with cancer. Nature. 2019;570(7761):385–9.
22. Bettegowda C, Sausen M, Leary RJ, et al. Detection of circulating tumor DNA in early- and late-stage human malignancies. Sci Transl Med. 2014;6(224):224ra24.
23. Steensma DP, Bejar R, Jaiswal S, et al. Clonal hematopoiesis of indeterminate potential and its distinction from myelodysplastic syndromes. Blood. 2015;126(1):9–16.
24. Thierry AR, Mouliere F, El Messaoudi S, et al. Clinical validation of the detection of KRAS and BRAF mutations from circulating tumor DNA. Nat Med. 2014;20(4):430–5.
25. Vidal J, Muinelo L, Dalmases A, et al. Plasma ctDNA RAS mutation analysis for the diagnosis and treatment monitoring of metastatic colorectal cancer patients. Ann Oncol. 2017;28(6):1325–32.
26. Wang DS, Liu ZX, Lu YX, et al. Liquid biopsies to track trastuzumab resistance in metastatic HER2-positive gastric cancer. Gut. 2019;68(7):1152–61.
27. Wang H, Li B, Liu Z, et al. HER2 copy number of circulating tumour DNA functions as a biomarker to predict and monitor trastuzumab efficacy in advanced gastric cancer. Eur J Cancer. 2018;88:92–100.
28. Mayrhofer M, De Laere B, Whitington T, et al. Cell-free DNA profiling of metastatic prostate cancer reveals microsatellite instability, structural rearrangements and clonal hematopoiesis. Genome Med. 2018;10(1):85.
29. Willis J, Lefterova MI, Artyomenko A, et al. Validation of microsatellite instability detection using a comprehensive plasma-based genotyping panel. Clin Cancer Res. 2019;25(23):7035–45.
30. Slavin TP, Banks KC, Chudova D, et al. Identification of incidental germline mutations in patients with advanced solid tumors who underwent cell-free circulating tumor DNA sequencing. J Clin Oncol. 2018;36:JCO1800328.
31. Siravegna G, Sartore-Bianchi A, Nagy RJ, et al. Plasma HER2 (ERBB2) copy number predicts response to HER2-targeted therapy in metastatic colorectal cancer. Clin Cancer Res. 2019;25(10):3046–53.
32. Lee J, Kim ST, Kim K, et al. Tumor genomic profiling guides patients with metastatic gastric cancer to targeted treatment: the VIKTORY umbrella trial. Cancer Discov. 2019;9(10):1388–405.
33. Cremolini C, Antoniotti C, Lonardi S, et al. Activity and safety of Cetuximab plus modified FOLFOXIRI followed by maintenance with Cetuximab or bevacizumab for RAS and BRAF wild-type metastatic colorectal cancer: a randomized phase 2 clinical trial. JAMA Oncol. 2018;4(4):529–36.
34. Rothwell DG, Ayub M, Cook N, et al. Utility of ctDNA to support patient selection for early phase clinical trials: the TARGET study. Nat Med. 2019;25(5):738–43.
35. Willyard C. Cancer therapy: an evolved approach. Nature. 2016;532(7598):166–8.
36. Morelli MP, Overman MJ, Dasari A, et al. Characterizing the patterns of clonal selection in circulating tumor DNA from patients with colorectal cancer refractory to anti-EGFR treatment. Ann Oncol. 2015;26(4):731–6.
37. Parseghian CM, Loree JM, Morris VK, et al. Anti-EGFR-resistant clones decay exponentially after progression: implications for anti-EGFR re-challenge. Ann Oncol. 2019;30(2):243–9.
38. Shroff RT, Hendifar A, McWilliams RR, et al. Rucaparib monotherapy in patients with pancreatic cancer and a known deleterious BRCA mutation. JCO Precis Oncologia. 2018;2018:PO.17.00316.
39. Russo M, Misale S, Wei G, et al. Acquired resistance to the TRK inhibitor Entrectinib in colorectal cancer. Cancer Discov. 2016;6(1):36–44.
40. Goyal L, Saha SK, Liu LY, et al. Polyclonal secondary FGFR2 mutations drive acquired resistance to FGFR inhibition in patients with FGFR2 fusion-positive cholangiocarcinoma. Cancer Discov. 2017;7(3):252–63.
41. Diehl F, Schmidt K, Choti MA, et al. Circulating mutant DNA to assess tumor dynamics. Nat Med. 2008;14(9):985–90.
42. Tie J, Wang Y, Tomasetti C, et al. Circulating tumor DNA analysis detects minimal residual disease and predicts recurrence in patients with stage II colon cancer. Sci Transl Med. 2016;8(346):346ra92.
43. Normanno N, Esposito Abate R, Lambiase M, et al. RAS testing of liquid biopsy correlates with the outcome of metastatic colorectal cancer patients treated with first-line FOLFIRI plus cetuximab in the CAPRI-GOIM trial. Ann Oncol. 2018;29(1):112–8.
44. Warren JD, Xiong W, Bunker AM, et al. Septin 9 methylated DNA is a sensitive and specific blood test for colorectal cancer. BMC Med. 2011;9:133.
45. Song L, Jia J, Peng X, et al. The performance of the SEPT9 gene methylation assay and a comparison with other CRC screening tests: a meta-analysis. Sci Rep. 2017;7(1):3032.
46. Church TR, Wandell M, Lofton-Day C, et al. Prospective evaluation of methylated SEPT9 in plasma for detection of asymptomatic colorectal cancer. Gut. 2014;63(2):317–25.
47. Frankell AM, Smyth EC. ctDNA in gastric and gastroesophageal cancer: prognostic, predictive, or preliminary? Clin Cancer Res. 2019;25(23):6893–5.
48. Kim YW, Kim YH, Song Y, et al. Monitoring circulating tumor DNA by analyzing personalized cancer-specific rearrangements to detect recurrence in gastric cancer. Exp Mol Med. 2019;51(8):1–10.
49. Anderson BW, Suh YS, Choi B, et al. Detection of gastric cancer with novel methylated DNA markers: discovery, tissue validation, and pilot testing in plasma. Clin Cancer Res. 2018;24(22):5724–34.
50. Wang Y, Springer S, Mulvey CL, et al. Detection of somatic mutations and HPV in the saliva and plasma of patients with head and neck squamous cell carcinomas. Sci Transl Med. 2015;7(293):293ra104.
51. Perdomo S, Avogbe PH, Foll M, et al. Circulating tumor DNA detection in head and neck cancer: evaluation of two different detection approaches. Oncotarget. 2017;8(42):72621–32.
52. Iwaya T, Endo F, Takahashi F, et al. Frequent tumor burden monitoring of esophageal squamous cell carcinoma with circulating tumor DNA using individually designed digital polymerase chain reaction. Gastroenterology. 2021;160:463–465.e4.
53. Azad TD, Chaudhuri AA, Fang P, et al. Circulating tumor DNA analysis for detection of minimal residual disease after Chemoradiotherapy for localized esophageal cancer. Gastroenterology. 2020;158(3):494–505 e6.
54. Maron SB, Chase LM, Lomnicki S, et al. Circulating tumor DNA sequencing analysis of gastroesophageal adenocarcinoma. Clin Cancer Res. 2019;25(23):7098–112.
55. Wong IH, Zhang J, Lai PB, et al. Quantitative analysis of tumor-derived methylated p16INK4a sequences in plasma, serum, and blood cells of hepatocellular carcinoma patients. Clin Cancer Res. 2003;9(3):1047–52.
56. Luo H, Zhao Q, Wei W, et al. Circulating tumor DNA methylation profiles enable early diagnosis, prognosis prediction, and screening for colorectal cancer. Sci Transl Med. 2020;12(524):eaax7533.
57. Mouliere F, Chandrananda D, Piskorz AM, et al. Enhanced detection of circulating tumor DNA by fragment size analysis. Sci Transl Med. 2018;10(466):eaat4921.
58. Hoshino A, Kim HS, Bojmar L, et al. Extracellular vesicle and particle biomarkers define multiple human cancers. Cell. 2020;182(4):1044–61 e18.
59. Wan JCM, Heider K, Gale D, et al. ctDNA monitoring using patient-specific sequencing and integration of variant reads. Sci Transl Med. 2020;12(548):eaaz8084.

60. Molparia B, Oliveira G, Wagner JL, et al. A feasibility study of colorectal cancer diagnosis via circulating tumor DNA derived CNV detection. PLoS One. 2018;13(5):e0196826.
61. Li J, Dittmar RL, Xia S, et al. Cell-free DNA copy number variations in plasma from colorectal cancer patients. Mol Oncol. 2017;11(8):1099–111.
62. Furuki H, Yamada T, Takahashi G, et al. Evaluation of liquid biopsies for detection of emerging mutated genes in metastatic colorectal cancer. Eur J Surg Oncol. 2018;44(7):975–82.
63. Wang Y, Zhao C, Chang L, et al. Circulating tumor DNA analyses predict progressive disease and indicate trastuzumab-resistant mechanism in advanced gastric cancer. EBioMedicine. 2019:43261–9.
64. Hideura E, Suehiro Y, Nishikawa J, et al. Blood free-circulating DNA testing of methylated RUNX3 is useful for diagnosing early gastric cancer. Cancers (Basel). 2020;12(4)
65. Shoda K, Ichikawa D, Fujita Y, et al. Monitoring the HER2 copy number status in circulating tumor DNA by droplet digital PCR in patients with gastric cancer. Gastric Cancer. 2017;20(1):126–35.
66. Ko JMY, Ng HY, Lam KO, et al. Liquid biopsy serial monitoring of treatment responses and relapse in advanced esophageal squamous cell carcinoma. Cancers (Basel). 2020;12(6)
67. Meng P, Wei J, Geng Y, et al. Targeted sequencing of circulating cell-free DNA in stage II-III resectable oesophageal squamous cell carcinoma patients. BMC Cancer. 2019;19(1):818.
68. Egyud M, Tejani M, Pennathur A, et al. Detection of circulating tumor DNA in plasma: a potential biomarker for esophageal adenocarcinoma. Ann Thorac Surg. 2019;108(2):343–9.
69. Kato S, Okamura R, Baumgartner JM, et al. Analysis of circulating tumor DNA and clinical correlates in patients with esophageal, gastroesophageal junction, and gastric adenocarcinoma. Clin Cancer Res. 2018;24(24):6248–56.
70. Tian X, Sun B, Chen C, et al. Circulating tumor DNA 5-hydroxymethylcytosine as a novel diagnostic biomarker for esophageal cancer. Cell Res. 2018;28(5):597–600.

14 Circulating Tumor Cells and ctDNA in Prostate Cancer

Nikolas H. Stoecklein and Rui P. L. Neves

Abstract

The rapidly expanding treatment options for patients with castration resistant prostate cancer (mCRCP) are promising. However, there is a need for biomarkers to help clinicians delivering the right treatment at the right time and sequence in the final stage of this common male cancer. As the majority of patients present with bone metastases, high-quality tissue biopsies for molecular diagnostics are often not easy to obtain and are not suitable for longitudinal testing. In this situation, CTC and ctDNA assays can seamlessly complement tissue biopsies and provide clinically relevant information for rational therapeutic decision-making in real time can capture the evolutionary dynamics of the disease. In fact, the first liquid biopsy assays started to enter clinical routine diagnostics and it is conceivable that this development continues. However, in view of the low concentration of the liquid biopsy analytes in blood, their detection and characterization remain challenging and newly developed assays require rigorous technical and clinical validation.

Learning Objectives

- To be able to define the major applications of CTC and ctDNA liquid biopsies in PC.
- To understand the current state of CTC-based and ctDNA-based liquid biopsies for clinical decision-making in prostate cancer.
- To be able to describe in which clinical situations the CTC and ctDNA tests are currently used in PC.

N. H. Stoecklein (✉) · R. P. L. Neves
Experimental Surgical Oncology, Department of General, Visceral and Pediatric Surgery, University Hospital and Medical Faculty of the Heinrich-Heine University Düsseldorf, Düsseldorf, Germany
e-mail: Nikolas.stoecklein@hhu.de; Rui.Neves@med.uni-duesseldorf.de

14.1 Introduction

In men, prostate cancer (PC) is the most common non-cutaneous cancer in Western countries and responsible for around 10% of all cancer-related deaths. PC therapy is adapted to the stage of the disease. PC begins as an organ-confined cancer, and evolves through a state displaying rising levels of prostate-specific antigen (PSA) [1] until reaching a metastatic state. Since androgens are major stimulants of PC cell growth, PCs are hormone-sensitive and androgen-deprivation therapies (ADT) have been the cornerstone therapies during these initial PC phases [2]. Finally, PC enters a state termed metastatic castration-resistant PC (mCRCP), in which the cancer cells become unresponsive to previous ADT, and that is lethal for most men. However, during the last decade, the development of a new generation of drugs further interfering with the androgen signaling axis has led to significantly prolonged survival in PC patients previously unresponsive to ADT [3, 4]. It was reported that together with earlier hormonal and chemotherapy administration, the novel inhibitors of the androgen axis could improve survival up to 5 years [5]. Nevertheless, while the 5-year survival rates of non-metastatic PC are nearly 100%, metastatic PC, and especially mCRCP, remain incurable and lethal with overall 5-year survival rates of less than 30% [6]. However, with rapidly expanding therapeutic options for mCRCP patients there is realistic hope to prolong survival. Besides androgen receptor signaling inhibitors (ARSi; e.g. abiraterone and enzalutamide), drugs targeting mitotic-spindle microtubules (docetaxel, cabazitaxel), radioactive calcium mimetics targeting bone metastases (radium-223), stimulators of the immune system (sipuleucel-T), or inhibitors of the poly ADP-ribose polymerase (PARP; olaparib), an enzyme involved in DNA repair, have become available for mCRPC patients [3, 7]. However, selecting the patients that benefit from these therapies is still challenging [8]. This is further complicated by the heterogenous nature of PC and the evolutionary dynamics of the disease [9–11]. Selective

S. P. Leong et al. (eds.), *Cancer Metastasis Through the Lymphovascular System*, https://doi.org/10.1007/978-3-030-93084-4_14

pressures of systemic therapies further shape genetic diversity of tumor cells and, consequently, the response to subsequent therapies [12]. In fact, recent data indicated that one dominant PC cell lineage emerges, which initiates and drives systemic progression. Nonetheless, subsequent polyclonal seeding between different metastatic sites appears to be common [9, 13]. Importantly, the potential genetic drivers of metastasis distinguish the metastatic subclones from those remaining locally confined in the prostate. In this context, it is interesting to note that the metastatic lineage appears to be represented in liquid biopsy samples [13]. This suggests that important information on molecular characteristics of the cancer can be provided by liquid biopsies in real time, complementary to routine tissue biopsies [14]. This can be particularly important, to capture evolutionary developments by longitudinal sampling, which is barley possible with standard tissue biopsies. Another aspect is that approximately 90% of the metastatic PC patients suffer from bone metastases [15]. Bone biopsies can be uncomfortable and contain frequently (in up to 20% of the cases) material insufficient for molecular testing [16]. Thus, given the increasing treatment options and the associated challenges of selecting the right treatment at the right time and in the right sequence, the less invasive liquid biopsies have the potential to improve the care of metastatic PC patients. The goal of this section is to provide an overview on the status of CTC and ctDNA analysis in advanced PC, with a focus on liquid biopsies closest to routine use.

14.2 CTCs in Prostate Cancer

14.2.1 Enumeration of Circulating Tumor Cells in Prostate Cancer

Metastatic prostate cancer is one of three indications (together with breast and colorectal cancer) for which the CellSearch system was cleared by the US Food and Drug Administration (FDA) for therapeutic monitoring. Since back then in 2008, the CellSearch system has been technically unchanged and the only CTC detection platform with regulatory clearance. The focus of most CellSearch studies in PC Patients was mCRPC, in which the prognostic impact of CellSearch CTCs was well established through multiple large phase II and III trials. In PC, unfavorable counts of CTCs (≥5/7.5 mL blood) before treatment or acquired during treatment correlate with reduced disease-free and overall survival [17–21]. In combination with increased serum lactate dehydrogenase (LDH, a biomarker indicating aggressive malignancy) at week 12 of treatment, CTC counts fulfilled all four Prentice criteria as an individual patient-level surrogate for survival [22]. For patients with ≥5 CTCs in 7.5 mL blood before therapy, a ≥30% decline in the CTC counts as early as week four is associated with longer survival [23]. A recent re-analysis using individual patient data of five prospective randomized phase III trials, which enrolled a total of 6081 patients (COU-AA-301, AFFIRM, ELM-PC-5, ELM-PC-4, and COMET-1), established a negative CTC CellSearch test (CTC0) as easily recognizable and clinically meaningful endpoint. Since this outcome can be determined shortly after the treatment initiation and indicating benefit of the therapy, the use of CTC0 as a response endpoint was recommended for early-phase clinical trials [24]. However, the predictive value (i.e., information on the probability of response to a particular therapy) of CTC enumeration when measured before treatment could not be established yet, which limits currently the routine use in the clinic, despite its proven strong prognostic value.

14.2.2 Molecular Characterization of CTCs in PC

Molecular characterization of enriched or individual CTCs is versatile and the potential to provide information for rational therapeutic decision-making is high [25]. In this regard, further phenotypic characterization of CTCs with the CellSearch platform has been applied in PC to assess potential therapeutic targets on CTCs as pharmacodynamic endpoint in early-phase clinical trials (e.g., [26]) or for prediction (e.g., for PSMA, [27]). In terms of therapeutic prediction, the androgen receptor splice variant-7 (AR-V7) has become a major focus of interest. Resistance to ARSi is often associated with elevated expression of both full-length AR (AR-FL) and AR-V7, encoding a truncated protein lacking the C-terminal ligand-binding domain, which can lead to ligand-independent transcriptional activity [28–30]. Detecting AR-V7 in CTCs might therefore be helpful to predict ARSi resistance and improve therapy selection in mCRCP. While several methods for detection of AR-V7 in CTCs have been established [1], the initial study was based on the Adna-Test platform: a multiplexed reverse-transcription polymerase-chain-reaction (qRT-PCR) with primers to detect the presence of CTCs as well as custom primers designed to detect mRNAs for AR-FL and AR-V7 [31]. This detection assay does not allow CTC enumeration since it is based on mRNA expression of blood cells enriched for CTCs by magnetic particles conjugated with a combination of antibodies recognizing prostate cancer cells. However, the data of the initial study indicated the AR-V7 CTC signal to be associated with resistance to ARSi. This was validated in a larger mCRPC patient cohort of 202 patients. In this study, the mRNA signal was corrected for the presence of CellSearch CTCs and it could be demonstrated that CTC AR-V7 was prognostically independent of the CTC enumeration by CellSearch [32]. Another far-developed method to assess AR-V7 positive CTCs was established on Epic sciences "no cell left behind" platform. This automated image-based assay can identify individual CTCs with nuclear localization of AR-V7. Two studies indicated clinical utility of the Epic Sciences assay in mCRPC patients undergoing treatment change to select patients for therapy with taxanes instead of ARSi [33, 34]. Notably, despite the lack of a randomized controlled trial to establish the predictive value for selection of taxanes vs. ARSi, the federal US health insurance program Medicare covers now the AR-V7 protein assay, commercialized as Oncotype DX AR-V7 Nucleus Detect test, for mCRCP patients [35] and will therefore likely enter clinical practice. Supporting its clinical application, the recent PROPHECY study (NCT02269982) demonstrated that both mRNA- and protein-based CTC AR-V7 assays are valuable to select mCRCP patients to ARSi or taxane treatment in the context of other clinical parameters [36]. The authors stressed that the CTC AR-V7 status can explain some of the resistance to AR therapy, but AR-V7 heterogeneity within patients and over time suggests that additional strategies are needed to overcome resistance in such mCRCP patients. A further notion of care concerning the relevance of AR-V7 for ARSi resistance come from the so far largest sequencing effort (n = >400) in mCRPC tumor tissue [37]. In this investigation, an increased expression of AR-V7 and other AR splice variants was observed in samples from patients receiving taxanes or ARSi, but this was not correlated with therapeutic resistance or survival.

In order to establish new predictive biomarkers with the help of CTCs, broader molecular characterizations of individual CTCs have been performed. The Epic Sciences platform has been used to investigate the relevance of phenotypic CTC heterogeneity to predict therapeutic resistance [38]. For this, 9225 morphologic features of individual CTCs from 179 mCRPC patients were recorded and subjected to heterogeneity analysis applying the Shannon-Index. It was observed that a higher CTC diversity was linked to worse overall survival in mCRCP patients treated with an ARSi, while survival was not affected by the degree of diversity in patients treated with taxanes [38]. The cell-to-cell heterogeneity of mCRCP CTCs could also be observed on the genomic level.

Example

Lambros and colleagues generated genome-wide CNA profiles from 185 individual CTCs isolated from 14 mCRCP patients via FACS after their detection by CellSearch. The authors could confirm the typical genomic landscape of advanced prostate cancer for the CellSearch detected CTCs and revealed complex intra-patient, intercellular genomic heterogeneity missed by bulk biopsy [39]. ◄

Similar comprehensive single-cell profiling studies in PC patients on CTCs isolated with different CTC-methods have been performed with next-generation sequencing (NGS) including whole-exome sequencing (WES), whole-genome sequencing (WGS), and RNA sequencing. Collectively, these studies demonstrated the feasibility of comprehensive CTC sequencing on the single-cell level with implications for potential clinical diagnostics and to identify new biomarkers [40].

14.2.3 Functional Analysis of CTCs in PC

The versatility and potential power of CTC-based liquid biopsies can be best illustrated by the possibility to generate cell culture models that allow functional analysis. This became a major interest in the CTC-field [41], but has turned out as technically extremely challenging with so far only anectodical success in most cancer types, including PC [42, 43]. This is despite the fact that the viability of PC CTCs could be demonstrated already more than 15 years ago via the EPISPOT assay [44]. Here, enriched CTCs are plated and short-term cultured on a membrane coated with antibodies to capture secreted prostate-specific antigen (PSA). With this approach, CTCs could be identified in metastatic as well as non-metastatic PC (83% vs. and 42% positivity rate), apparently, with a higher sensitivity when compared to CellSearch. More advanced functional testing would require CTC-derived cell lines or xenografts. An important lesson learned was that high numbers of non-apoptotic CTCs are required for successful expansion of CTCs [45, 46]. This can be enabled by harvesting higher volumes of peripheral blood. Therefore, diagnostic leukapheresis (DLA) might be a key to achieve this [47]. DLA is based on the clinically safe leukapheresis, a method to collect white blood cells with a certain density by continuous centrifugation from liters of peripheral blood. Since epithelial cancer cells have similar densities, CTCs become co-collected during the process. DLA does not only increase dramatically CTC detection but could indeed facilitate the generation of CTC-derived organoid culture and xenografts in PC [39, 48]. Such DLA-derived PC organoid cultures have the potential to serve as relevant pre-clinical models for metastatic PC or even as a personalized drug screening platform [49] in the future.

14.3 ctDNA in Prostate Cancer

Inferring genomic information from plasma samples is technically less complex compared to CTCs. Since the advent of NGS, cell-free DNA (cfDNA) analysis could have been swiftly optimized for routine diagnostic use [50]. Initial NGS-driven cfDNA studies in PC investigated low-coverage whole-genome sequencing (WGS), also known as shallow or low-pass WGS, and could establish tumor-specific copy number alterations (CNA) from the samples [51]. Along with CNAs typical for PC it was possible to detect high-level copy number gains in the AR locus in mCRPC patients, which might be of diagnostic use in the context of ADT resistance. Since circulating tumor DNA (ctDNA) represents only a fraction of cfDNA and low-pass WGS require high ctDNA fractions, these assays are limited to metastatic PC stages in which high tumor DNA concentrations in the plasma are more common [52]. It is of note that higher total ctDNA levels are correlated with poor prognosis and may have a role as biomarker to indicate aggressive cancers or higher disease burden [53, 54]. On the other hand, dropping levels of detectable ctDNA during therapy are correlated with therapeutic response and good prognosis [55, 56], qualifying ctDNA as potential biomarker for therapy monitoring.

Similar to the CNA profiles, ctDNA point mutations were largely concordant with tumor tissue, indicating ctDNA as surrogate material for tumor biopsies [57].

Detection of somatic ctDNA mutations in *TP53* and *AR* genes, in genes of the PI3K pathway, as well as in DNA damage repair genes might have immediate clinical importance for prognosis and prediction. The *TP53* gene is very frequently mutated in mCRPC (around 50% of the cases) and has been associated with poor prognosis [58]. In ctDNA, somatic *TP53* mutations were an independent prognostic factor outperforming the prognostic value of *AR alterations* [59]. While genetic alterations of *AR* are usually not detected in primary tumors or hormone-naïve metastases, copy number gain of *AR* can be detected in mCRCP in 30% and point mutations in approximately 20% of the tumors cases after ADT [57]. Copy number gains of *AR* were associated with resistance to ARSi [60] and were associated with poor prognosis [56, 61, 62]. In contrast to *AR* copy number gains, specific AR point mutations (T878A or L702H) emerged in 13% of the cases during progression on abiraterone [53, 56]. Although detection of *AR* amplifications did not outperform standard prognostic biomarkers in a recent study, structural rearrangements truncating the ligand-binding domain of *AR* were identified in several patients with primary resistance [63]. While the prognostic data of *AR* alterations in ctDNA remain somewhat inconclusive, their detection via ctDNA might have clinical utility for treatment selection.

Approximately 50% of mCRCPs display alterations in *PIK3CA/B*, *PTEN*, or *AKT*, making the phosphatidylinositol-3 kinase (PI3K)/AKT pathway one of the most frequently deregulated signaling pathways in PC [64]. Mutations in these genes are significantly more prevalent in mCRPC than in primary PC (50% vs. <20%) suggesting an important role of this pathway in PC progression [57]. The somatic mutations in the PI3K pathway genes can be traced by ctDNA assays and have been associated with a poor response to ARSi [63]. Since AKT-inhibition is used to therapeutically target *PI3K*-driven cancers and data indicate that a subset of PC patients might benefit from a combination of AKT inhibitors and anti-androgens, ctDNA analysis for PI3K/AKT pathway genes may serve as predictive assay [57]. Another set of signaling pathways commonly altered in mCRCP are belonging to the DNA damage response (DDR). Around 25% of mCRPCs carry mutations in genes of the DDR and this can be therapeutically exploited. Poly (ADP-ribose) polymerase (*PARP*) inhibitors induce programmed cell death in DDR-impaired tumor cells, and monotherapy with the PARP inhibitor olaparib has antitumor activity in mCRCP, delays significantly disease progression, and improves survival. These could be very recently established in the TOPARP A/B [65, 66] and PROfound trials [67]. Importantly, clinically relevant response to olaparib was associated with mutations in *BRCA2*, *BRCA1*, *ATM*, and *PALB2* genes [65]. The DDR mutations can be well captured by ctDNA analysis and were concordant with contemporaneously taken tumor tissue. Importantly, ctDNA analysis revealed the emergence of new mutations at disease progression. Based on the available data the FDA approved On November 6, 2020, a commercially available assay to identify mutations in *BRCA1, BRCA2,* and *ATM* genes [68] as a companion diagnostic device in patients with mCRPC eligible for treatment with olaparib. Thus, DDR gene testing on ctDNA material became the first liquid biopsy test for therapy selection with regulatory approval in prostate cancer.

14.4 Conclusion

With CTC-based AR-V7 testing and the DDR gene ctDNA assay, liquid biopsy just became clinical reality in mCRCP, the lethal advanced stage of PC. With the rapidly expanding treatment armamentarium it is crucial for the clinicians to select the right treatment at the right time in the correct sequence for the patients. With their less invasiveness and the possibility for longitudinal testing, CTC- and ctDNA-based liquid biopsies appear to seamlessly complement classic tissue biopsies in PC. These are often not easy to obtain in PC, as the majority of patients present with bone metastases. Nonetheless, as technology advances, it is conceivable that CTC-based liquid biopsies in particular will continue to evolve with their versatility and yield more clinically relevant assays. mCRPC is also quite suitable for the implementation of such liquid biopsies, as both CTC number and ctDNA concentration are often high, making the assays well applicable. However, in the early stages of the disease, the low abundance of the cancer-derived analytes limits currently their application. Therefore, more sensitive assays or better enrichment methods for the analytes are required. Most importantly, after rigorous technical and clinical validation of new liquid biopsy assays for PC patients, suitable prospective clinical trials are needed to demonstrate their clinical utility.

Open Questions

- Can molecular CTC profiling on single-cell level improve therapeutic management in the clinical routine?
- Does current technology enable the implementation of CTC-derived organoids as individual drug-testing platform?
- Have CTC and ctDNA assays a role for clinical decision-making in earlier PC stages?
- Can ctDNA-based liquid biopsies capture the clinically relevant genetic heterogeneity of the systemic disease?
- What is the best strategy to allow a fast and independent technical and clinical validation of commercialized CTC- and ctDNA liquid biopsy assays?

References

1. Pantel K, Hille C, Scher HI. Circulating tumor cells in prostate cancer: from discovery to clinical utility. Clin Chem. 2019;65(1):87–99. https://doi.org/10.1373/clinchem.2018.287102.
2. Chang AJ, Autio KA, Roach M 3rd, Scher HI. High-risk prostate cancer-classification and therapy. Nat Rev Clin Oncol. 2014;11(6):308–23. https://doi.org/10.1038/nrclinonc.2014.68.
3. Sartor O, de Bono JS. Metastatic prostate cancer. N Engl J Med. 2018a;378(7):645–57. https://doi.org/10.1056/NEJMra1701695.
4. Sartor O, de Bono JS. Metastatic prostate cancer. N Engl J Med. 2018b;378(17):1653–4. https://doi.org/10.1056/NEJMc1803343.
5. Carioli G, Bertuccio P, Boffetta P, Levi F, La Vecchia C, Negri E, Malvezzi M. European cancer mortality predictions for the year 2020 with a focus on prostate cancer. Ann Oncol. 2020;31(5):650–8. https://doi.org/10.1016/j.annonc.2020.02.009.
6. Damodaran S, Kyriakopoulos CE, Jarrard DF. Newly diagnosed metastatic prostate cancer: has the paradigm changed? Urol Clin North Am. 2017;44(4):611–21. https://doi.org/10.1016/j.ucl.2017.07.008.
7. Swami U, McFarland TR, Nussenzveig R, Agarwal N. Advanced prostate cancer: treatment advances and future directions. Trends Cancer. 2020;6(8):702–15. https://doi.org/10.1016/j.trecan.2020.04.010.
8. Valenca LB, Sweeney CJ, Pomerantz MM. Sequencing current therapies in the treatment of metastatic prostate cancer. Cancer Treat Rev. 2015;41(4):332–40. https://doi.org/10.1016/j.ctrv.2015.02.010.
9. Gundem G, Van Loo P, Kremeyer B, Alexandrov LB, Tubio JMC, Papaemmanuil E, Brewer DS, Kallio HML, Högnäs G, Annala M, Kivinummi K, Goody V, Latimer C, O'Meara S, Dawson KJ, Isaacs W, Emmert-Buck MR, Nykter M, Foster C, Kote-Jarai Z, Easton D, Whitaker HC, Neal DE, Cooper CS, Eeles RA, Visakorpi T, Campbell PJ, McDermott U, Wedge DC, Bova GS. The evolutionary history of lethal metastatic prostate cancer. Nature. 2015;520(7547):353–7. https://doi.org/10.1038/nature14347.
10. Haffner MC, Mosbruger T, Esopi DM, Fedor H, Heaphy CM, Walker DA, Adejola N, Gürel M, Hicks J, Meeker AK, Halushka MK, Simons JW, Isaacs WB, De Marzo AM, Nelson WG, Yegnasubramanian S. Tracking the clonal origin of lethal prostate cancer. J Clin Invest. 2013;123(11):4918–22. https://doi.org/10.1172/jci70354.
11. Kumar A, Coleman I, Morrissey C, Zhang X, True LD, Gulati R, Etzioni R, Bolouri H, Montgomery B, White T, Lucas JM, Brown LG, Dumpit RF, DeSarkar N, Higano C, Yu EY, Coleman R, Schultz N, Fang M, Lange PH, Shendure J, Vessella RL, Nelson PS. Substantial interindividual and limited intraindividual genomic diversity among tumors from men with metastatic prostate cancer. Nat Med. 2016;22(4):369–78. https://doi.org/10.1038/nm.4053.
12. Burrell RA, Swanton C. Tumour heterogeneity and the evolution of polyclonal drug resistance. Mol Oncol. 2014;8(6):1095–111. https://doi.org/10.1016/j.molonc.2014.06.005.
13. Woodcock DJ, Riabchenko E, Taavitsainen S, Kankainen M, Gundem G, Brewer DS, Ellonen P, Lepistö M, Golubeva YA, Warner AC, Tolonen T, Jasu J, Isaacs WB, Emmert-Buck MR, Nykter M, Visakorpi T, Bova GS, Wedge DC. Prostate cancer evolution from multilineage primary to single lineage metastases with implications for liquid biopsy. Nat Commun. 2020;11(1):5070. https://doi.org/10.1038/s41467-020-18843-5.
14. Alix-Panabieres C, Pantel K. Liquid biopsy: from discovery to clinical application. Cancer Discov. 2021;11(4):858–73. https://doi.org/10.1158/2159-8290.CD-20-1311.
15. Bubendorf L, Schöpfer A, Wagner U, Sauter G, Moch H, Willi N, Gasser TC, Mihatsch MJ. Metastatic patterns of prostate cancer: an autopsy study of 1,589 patients. Hum Pathol. 2000;31(5):578–83. https://doi.org/10.1053/hp.2000.6698.
16. Lorente D, Omlin A, Zafeiriou Z, Nava-Rodrigues D, Pérez-López R, Pezaro C, Mehra N, Sheridan E, Figueiredo I, Riisnaes R, Miranda S, Crespo M, Flohr P, Mateo J, Altavilla A, Ferraldeschi R, Bianchini D, Attard G, Tunariu N, de Bono J. Castration-resistant prostate cancer tissue acquisition from bone metastases for molecular analyses. Clin Genitourin Cancer. 2016b;14(6):485–93. https://doi.org/10.1016/j.clgc.2016.04.016.
17. de Bono JS, Scher HI, Montgomery RB, Parker C, Miller MC, Tissing H, Doyle GV, Terstappen LW, Pienta KJ, Raghavan D. Circulating tumor cells predict survival benefit from treatment in metastatic castration-resistant prostate cancer. Clin Cancer Res. 2008;14(19):6302–9. https://doi.org/10.1158/1078-0432.CCR-08-0872.
18. Goldkorn A, Ely B, Quinn DI, Tangen CM, Fink LM, Xu T, Twardowski P, Van Veldhuizen PJ, Agarwal N, Carducci MA, Monk JP 3rd, Datar RH, Garzotto M, Mack PC, Lara P Jr, Higano CS, Hussain M, Thompson IM Jr, Cote RJ, Vogelzang NJ. Circulating tumor cell counts are prognostic of overall survival in SWOG S0421: a phase III trial of docetaxel with or without atrasentan for metastatic castration-resistant prostate cancer. J Clin Oncol. 2014;32(11):1136–42. https://doi.org/10.1200/JCO.2013.51.7417.
19. Goodman OB Jr, Fink LM, Symanowski JT, Wong B, Grobaski B, Pomerantz D, Ma Y, Ward DC, Vogelzang NJ. Circulating tumor cells in patients with castration-resistant prostate cancer baseline values and correlation with prognostic factors. Cancer Epidemiol Biomarkers Prev. 2009;18(6):1904–13. https://doi.org/10.1158/1055-9965.epi-08-1173.
20. Scher HI, Jia X, de Bono JS, Fleisher M, Pienta KJ, Raghavan D, Heller G. Circulating tumour cells as prognostic markers in progressive, castration-resistant prostate cancer: a reanalysis of IMMC38 trial data. Lancet Oncol. 2009;10(3):233–9.
21. Vogelzang NJ, Fizazi K, Burke JM, De Wit R, Bellmunt J, Hutson TE, Crane E, Berry WR, Doner K, Hainsworth JD, Wiechno PJ, Liu K, Waldman MF, Gandhi A, Barton D, Jungnelius U, Fandi A, Sternberg CN, Petrylak DP. Circulating tumor cells in a phase 3 study of docetaxel and prednisone with or without lenalidomide in metastatic castration-resistant prostate cancer. Eur Urol. 2017;71(2):168–71. https://doi.org/10.1016/j.eururo.2016.07.051.
22. Scher HI, Heller G, Molina A, Attard G, Danila DC, Jia X, Peng W, Sandhu SK, Olmos D, Riisnaes R, McCormack R, Burzykowski T, Kheoh T, Fleisher M, Buyse M, de Bono JS. Circulating tumor

cell biomarker panel as an individual-level surrogate for survival in metastatic castration-resistant prostate cancer. J Clin Oncol. 2015;33(12):1348–55. https://doi.org/10.1200/JCO.2014.55.3487.
23. Lorente D, Olmos D, Mateo J, Bianchini D, Seed G, Fleisher M, Danila DC, Flohr P, Crespo M, Figueiredo I, Miranda S, Baeten K, Molina A, Kheoh T, McCormack R, Terstappen LW, Scher HI, de Bono JS. Decline in circulating tumor cell count and treatment outcome in advanced prostate cancer. Eur Urol. 2016a;70(6):985–92. https://doi.org/10.1016/j.eururo.2016.05.023.
24. Heller G, McCormack R, Kheoh T, Molina A, Smith MR, Dreicer R, Saad F, de Wit R, Aftab DT, Hirmand M, Limon A, Fizazi K, Fleisher M, de Bono JS, Scher HI. Circulating tumor cell number as a response measure of prolonged survival for metastatic castration-resistant prostate cancer: a comparison with prostate-specific antigen across five randomized phase III clinical trials. J Clin Oncol. 2018;36(6):572–80. https://doi.org/10.1200/jco.2017.75.2998.
25. Keller L, Pantel K. Unravelling tumour heterogeneity by single-cell profiling of circulating tumour cells. Nat Rev Cancer. 2019;19(10):553–67. https://doi.org/10.1038/s41568-019-0180-2.
26. de Bono JS, Attard G, Adjei A, Pollak MN, Fong PC, Haluska P, Roberts L, Melvin C, Repollet M, Chianese D, Connely M, Terstappen LW, Gualberto A. Potential applications for circulating tumor cells expressing the insulin-like growth factor-I receptor. Clin Cancer Res. 2007;13(12):3611–6. https://doi.org/10.1158/1078-0432.ccr-07-0268.
27. Gorges TM, Riethdorf S, von Ahsen O, Nastał YP, Röck K, Boede M, Peine S, Kuske A, Schmid E, Kneip C, König F, Rudolph M, Pantel K. Heterogeneous PSMA expression on circulating tumor cells: a potential basis for stratification and monitoring of PSMA-directed therapies in prostate cancer. Oncotarget. 2016;7(23):34,930–41. https://doi.org/10.18632/oncotarget.9004.
28. Antonarakis ES, Armstrong AJ, Dehm SM, Luo J. Androgen receptor variant-driven prostate cancer: clinical implications and therapeutic targeting. Prostate Cancer Prostatic Dis. 2016;19(3):231–41. https://doi.org/10.1038/pcan.2016.17.
29. Hu R, Dunn TA, Wei S, Isharwal S, Veltri RW, Humphreys E, Han M, Partin AW, Vessella RL, Isaacs WB, Bova GS, Luo J. Ligand-independent androgen receptor variants derived from splicing of cryptic exons signify hormone-refractory prostate cancer. Cancer Res. 2009;69(1):16–22. https://doi.org/10.1158/0008-5472.can-08-2764.
30. Watson PA, Chen YF, Balbas MD, Wongvipat J, Socci ND, Viale A, Kim K, Sawyers CL. Constitutively active androgen receptor splice variants expressed in castration-resistant prostate cancer require full-length androgen receptor. Proc Natl Acad Sci U S A. 2010;107(39):16759–65. https://doi.org/10.1073/pnas.1012443107.
31. Antonarakis ES, Lu C, Wang H, Luber B, Nakazawa M, Roeser JC, Chen Y, Mohammad TA, Chen Y, Fedor HL, Lotan TL, Zheng Q, De Marzo AM, Isaacs JT, Isaacs WB, Nadal R, Paller CJ, Denmeade SR, Carducci MA, Eisenberger MA, Luo J. AR-V7 and resistance to enzalutamide and abiraterone in prostate cancer. N Engl J Med. 2014;371(11):1028–38. https://doi.org/10.1056/NEJMoa1315815.
32. Antonarakis ES, Lu C, Luber B, Wang H, Chen Y, Zhu Y, Silberstein JL, Taylor MN, Maughan BL, Denmeade SR, Pienta KJ, Paller CJ, Carducci MA, Eisenberger MA, Luo J. Clinical significance of androgen receptor splice Variant-7 mRNA detection in circulating tumor cells of men with metastatic castration-resistant prostate cancer treated with first- and second-line Abiraterone and enzalutamide. J Clin Oncol. 2017;35(19):2149–56. https://doi.org/10.1200/jco.2016.70.1961.
33. Scher HI, Graf RP, Schreiber NA, Jayaram A, Winquist E, McLaughlin B, Lu D, Fleisher M, Orr S, Lowes L, Anderson A, Wang Y, Dittamore R, Allan AL, Attard G, Heller G. Assessment of the validity of nuclear-localized androgen receptor splice variant 7 in circulating tumor cells as a predictive biomarker for castration-resistant prostate cancer. JAMA Oncol. 2018;4(9):1179–86. https://doi.org/10.1001/jamaoncol.2018.1621.
34. Scher HI, Lu D, Schreiber NA, Louw J, Graf RP, Vargas HA, Johnson A, Jendrisak A, Bambury R, Danila D, McLaughlin B, Wahl J, Greene SB, Heller G, Marrinucci D, Fleisher M, Dittamore R. Association of AR-V7 on circulating tumor cells as a treatment-specific biomarker with outcomes and survival in castration-resistant prostate cancer. JAMA Oncol. 2016;2(11):1441–9. https://doi.org/10.1001/jamaoncol.2016.1828.
35. CMS.gov. Local coverage determination (LCD): MolDX: androgen receptor variant (AR-V7) protein test (L37701); 2020. https://www.cms.gov/medicare-coverage-database/details/lcd-details.aspx?LCDId=37701. Accessed 13 Jan 2021.
36. Armstrong AJ, Halabi S, Luo J, Nanus DM, Giannakakou P, Szmulewitz RZ, Danila DC, Healy P, Anand M, Rothwell CJ, Rasmussen J, Thornburg B, Berry WR, Wilder RS, Lu C, Chen Y, Silberstein JL, Kemeny G, Galletti G, Somarelli JA, Gupta S, Gregory SG, Scher HI, Dittamore R, Tagawa ST, Antonarakis ES, George DJ. Prospective multicenter validation of androgen receptor splice variant 7 and hormone therapy resistance in high-risk castration-resistant prostate cancer: the PROPHECY study. J Clin Oncol. 2019;37(13):1120–9. https://doi.org/10.1200/jco.18.01731.
37. Abida W, Cyrta J, Heller G, Prandi D, Armenia J, Coleman I, Cieslik M, Benelli M, Robinson D, Van Allen EM, Sboner A, Fedrizzi T, Mosquera JM, Robinson BD, De Sarkar N, Kunju LP, Tomlins S, Wu YM, Nava Rodrigues D, Loda M, Gopalan A, Reuter VE, Pritchard CC, Mateo J, Bianchini D, Miranda S, Carreira S, Rescigno P, Filipenko J, Vinson J, Montgomery RB, Beltran H, Heath EI, Scher HI, Kantoff PW, Taplin ME, Schultz N, deBono JS, Demichelis F, Nelson PS, Rubin MA, Chinnaiyan AM, Sawyers CL. Genomic correlates of clinical outcome in advanced prostate cancer. Proc Natl Acad Sci U S A. 2019;116(23):11,428–36. https://doi.org/10.1073/pnas.1902651116.
38. Scher HI, Graf RP, Schreiber NA, McLaughlin B, Jendrisak A, Wang Y, Lee J, Greene S, Krupa R, Lu D, Bamford P, Louw JE, Dugan L, Vargas HA, Fleisher M, Landers M, Heller G, Dittamore R. Phenotypic heterogeneity of circulating tumor cells informs clinical decisions between AR signaling inhibitors and Taxanes in metastatic prostate cancer. Cancer Res. 2017;77(20):5687–98. https://doi.org/10.1158/0008-5472.can-17-1353.
39. Lambros MB, Seed G, Sumanasuriya S, Gil V, Crespo M, Fontes MS, Chandler R, Mehra N, Fowler G, Ebbs B, Flohr PR, Miranda S, Yuan W, Mackay A, Ferreira A, Pereira R, Bertan C, Figueiredo I, Riisnaes R, Nava Rodrigues D, Sharp A, Goodall J, Boysen G, Carreira S, Bianchini D, Rescigno P, Zafeiriou Z, Hunt J, Moloney D, Hamilton L, Neves RPL, Swennenhuis JF, Andree KC, Stoecklein NH, Terstappen LW, de Bono JS. Single cell analyses of prostate cancer liquid biopsies acquired by apheresis. Clin Cancer Res. 2018; https://doi.org/10.1158/1078-0432.CCR-18-0862.
40. Lu YT, Delijani K, Mecum A, Goldkorn A. Current status of liquid biopsies for the detection and management of prostate cancer. Cancer Manag Res. 2019;11:5271–91. https://doi.org/10.2147/cmar.s170380.
41. Pantel K, Alix-Panabières C. Functional studies on viable circulating tumor cells. Clin Chem. 2016;62(2):328–34. https://doi.org/10.1373/clinchem.2015.242537.
42. Gao D, Vela I, Sboner A, Iaquinta PJ, Karthaus WR, Gopalan A, Dowling C, Wanjala JN, Undvall EA, Arora VK, Wongvipat J, Kossai M, Ramazanoglu S, Barboza LP, Di W, Cao Z, Zhang QF, Sirota I, Ran L, MacDonald TY, Beltran H, Mosquera JM, Touijer KA, Scardino PT, Laudone VP, Curtis KR, Rathkopf DE, Morris MJ, Danila DC, Slovin SF, Solomon SB, Eastham JA, Chi P, Carver B, Rubin MA, Scher HI, Clevers H, Sawyers CL, Chen Y. Organoid cultures derived from patients with advanced pros-

tate cancer. Cell. 2014;159(1):176–87. https://doi.org/10.1016/j.cell.2014.08.016.

43. Rossi E, Rugge M, Facchinetti A, Pizzi M, Nardo G, Barbieri V, Manicone M, De Faveri S, Chiara Scaini M, Basso U, Amadori A, Zamarchi R. Retaining the long-survive capacity of circulating tumor cells (CTCs) followed by xeno-transplantation: not only from metastatic cancer of the breast but also of prostate cancer patients. Onco Targets Ther. 2014;1(1):49–56. https://doi.org/10.18632/oncoscience.8.
44. Alix-Panabières C, Rebillard X, Brouillet JP, Barbotte E, Iborra F, Segui B, Maudelonde T, Jolivet-Reynaud C, Vendrell JP. Detection of circulating prostate-specific antigen-secreting cells in prostate cancer patients. Clin Chem. 2005;51(8):1538–41. https://doi.org/10.1373/clinchem.2005.049445.
45. Baccelli I, Schneeweiss A, Riethdorf S, Stenzinger A, Schillert A, Vogel V, Klein C, Saini M, Bauerle T, Wallwiener M, Holland-Letz T, Hofner T, Sprick M, Scharpff M, Marme F, Sinn HP, Pantel K, Weichert W, Trumpp A. Identification of a population of blood circulating tumor cells from breast cancer patients that initiates metastasis in a xenograft assay. Nat Biotechnol. 2013;31(6):539–44. https://doi.org/10.1038/nbt.2576.
46. Franken A, Driemel C, Behrens B, Meier-Stiegen F, Endris V, Stenzinger A, Niederacher D, Fischer JC, Stoecklein NH, Ruckhaeberle E, Fehm T, Neubauer H. Label-free enrichment and molecular characterization of viable circulating tumor cells from diagnostic leukapheresis products. Clin Chem. 2019;65(4):549–58. https://doi.org/10.1373/clinchem.2018.296814.
47. Fischer JC, Niederacher D, Topp SA, Honisch E, Schumacher S, Schmitz N, Zacarias Fohrding L, Vay C, Hoffmann I, Kasprowicz NS, Hepp PG, Mohrmann S, Nitz U, Stresemann A, Krahn T, Henze T, Griebsch E, Raba K, Rox JM, Wenzel F, Sproll C, Janni W, Fehm T, Klein CA, Knoefel WT, Stoecklein NH. Diagnostic leukapheresis enables reliable detection of circulating tumor cells of nonmetastatic cancer patients. Proc Natl Acad Sci U S A. 2013;110(41):16,580–5. https://doi.org/10.1073/pnas.1313594110.
48. Faugeroux V, Pailler E, Oulhen M, Deas O, Brulle-Soumare L, Hervieu C, Marty V, Alexandrova K, Andree KC, Stoecklein NH, Tramalloni D, Cairo S, NgoCamus M, Nicotra C, Terstappen L, Manaresi N, Lapierre V, Fizazi K, Scoazec JY, Loriot Y, Judde JG, Farace F. Genetic characterization of a unique neuroendocrine transdifferentiation prostate circulating tumor cell-derived eXplant model. Nat Commun. 2020;11(1):1884. https://doi.org/10.1038/s41467-020-15426-2.
49. Mout L, van Dessel LF, Kraan J, de Jong AC, Neves RPL, Erkens-Schulze S, Beaufort CM, Sieuwerts AM, van Riet J, Woo TLC, de Wit R, Sleijfer S, Hamberg P, Sandberg Y, Te Boekhorst PAW, van de Werken HJG, Martens JWM, Stoecklein NH, van Weerden WM, Lolkema MP. Generating human prostate cancer organoids from leukapheresis enriched circulating tumour cells. Eur J Cancer. 2021;150:179–89. https://doi.org/10.1016/j.ejca.2021.03.023.
50. Neumann MHD, Bender S, Krahn T, Schlange T. ctDNA and CTCs in liquid biopsy—current status and where we need to progress. Comput Struct Biotechnol J. 2018;16:190–5. https://doi.org/10.1016/j.csbj.2018.05.002.
51. Heitzer E, Ulz P, Belic J, Gutschi S, Quehenberger F, Fischereder K, Benezeder T, Auer M, Pischler C, Mannweiler S, Pichler M, Eisner F, Haeusler M, Riethdorf S, Pantel K, Samonigg H, Hoefler G, Augustin H, Geigl JB, Speicher MR. Tumor-associated copy number changes in the circulation of patients with prostate cancer identified through whole-genome sequencing. Genome Med. 2013;5(4):30. https://doi.org/10.1186/gm434.
52. Kolinsky MP, Stoecklein N, Lambros M, Gil V, Rodrigues DN, Carreira S, Zafeiriou Z, de Bono JS. Genetic analysis of circulating tumour cells. Recent Results Cancer Res. 2020;215:57–76. https://doi.org/10.1007/978-3-030-26439-0_3.
53. Carreira S, Romanel A, Goodall J, Grist E, Ferraldeschi R, Miranda S, Prandi D, Lorente D, Frenel JS, Pezaro C, Omlin A, Rodrigues DN, Flohr P, Tunariu N, SdB J, Demichelis F, Attard G. Tumor clone dynamics in lethal prostate cancer. Sci Transl Med. 2014;6(254):254ra125. https://doi.org/10.1126/scitranslmed.3009448.
54. Wyatt AW, Annala M, Aggarwal R, Beja K, Feng F, Youngren J, Foye A, Lloyd P, Nykter M, Beer TM, Alumkal JJ, Thomas GV, Reiter RE, Rettig MB, Evans CP, Gao AC, Chi KN, Small EJ, Gleave ME. Concordance of circulating tumor DNA and matched metastatic tissue biopsy in prostate cancer. J Natl Cancer Inst. 2017;109(12) https://doi.org/10.1093/jnci/djx118.
55. Goodall J, Mateo J, Yuan W, Mossop H, Porta N, Miranda S, Perez-Lopez R, Dolling D, Robinson DR, Sandhu S, Fowler G, Ebbs B, Flohr P, Seed G, Rodrigues DN, Boysen G, Bertan C, Atkin M, Clarke M, Crespo M, Figueiredo I, Riisnaes R, Sumanasuriya S, Rescigno P, Zafeiriou Z, Sharp A, Tunariu N, Bianchini D, Gillman A, Lord CJ, Hall E, Chinnaiyan AM, Carreira S, de Bono JS. Circulating cell-free DNA to guide prostate cancer treatment with PARP inhibition. Cancer Discov. 2017;7(9):1006–17. https://doi.org/10.1158/2159-8290.cd-17-0261.
56. Wyatt AW, Azad AA, Volik SV, Annala M, Beja K, McConeghy B, Haegert A, Warner EW, Mo F, Brahmbhatt S, Shukin R, Le Bihan S, Gleave ME, Nykter M, Collins CC, Chi KN. Genomic alterations in cell-free DNA and enzalutamide resistance in castration-resistant prostate cancer. JAMA Oncol. 2016;2(12):1598–606. https://doi.org/10.1001/jamaoncol.2016.0494.
57. Gasi Tandefelt D, de Bono J. Circulating cell-free DNA: translating prostate cancer genomics into clinical care. Mol Asp Med. 2020;72:100837. https://doi.org/10.1016/j.mam.2019.100837.
58. Boerrigter E, Groen LN, Van Erp NP, Verhaegh GW, Schalken JA. Clinical utility of emerging biomarkers in prostate cancer liquid biopsies. Expert Rev Mol Diagn. 2020;20(2):219–30. https://doi.org/10.1080/14737159.2019.1675515.
59. De Laere B, Oeyen S, Mayrhofer M, Whitington T, van Dam PJ, Van Oyen P, Ghysel C, Ampe J, Ost P, Demey W, Hoekx L, Schrijvers D, Brouwers B, Lybaert W, Everaert EG, De Maeseneer D, Strijbos M, Bols A, Fransis K, Beije N, de Kruijff IE, van Dam V, Brouwer A, Goossens D, Heyrman L, Van den Eynden GG, Rutten A, Del Favero J, Rantalainen M, Rajan P, Sleijfer S, Ullén A, Yachnin J, Grönberg H, Van Laere SJ, Lindberg J, Dirix LY. TP53 outperforms other androgen receptor biomarkers to predict abiraterone or enzalutamide outcome in metastatic castration-resistant prostate cancer. Clin Cancer Res. 2019;25(6):1766–73. https://doi.org/10.1158/1078-0432.ccr-18-1943.
60. Azad AA, Volik SV, Wyatt AW, Haegert A, Le Bihan S, Bell RH, Anderson SA, McConeghy B, Shukin R, Bazov J, Youngren J, Paris P, Thomas G, Small EJ, Wang Y, Gleave ME, Collins CC, Chi KN. Androgen receptor gene aberrations in circulating cell-free DNA: biomarkers of therapeutic resistance in castration-resistant prostate cancer. Clin Cancer Res. 2015;21(10):2315–24. https://doi.org/10.1158/1078-0432.ccr-14-2666.
61. Conteduca V, Wetterskog D, Sharabiani MTA, Grande E, Fernandez-Perez MP, Jayaram A, Salvi S, Castellano D, Romanel A, Lolli C, Casadio V, Gurioli G, Amadori D, Font A, Vazquez-Estevez S, González Del Alba A, Mellado B, Fernandez-Calvo O, Méndez-Vidal MJ, Climent MA, Duran I, Gallardo E, Rodriguez A, Santander C, Sáez MI, Puente J, Gasi Tandefelt D, Wingate A, Dearnaley D, Demichelis F, De Giorgi U, Gonzalez-Billalabeitia E, Attard G. Androgen receptor gene status in plasma DNA associates with worse outcome on enzalutamide or abiraterone for castration-resistant prostate cancer: a multi-institution correlative biomarker study. Ann Oncol. 2017;28(7):1508–16. https://doi.org/10.1093/annonc/mdx155.
62. Romanel A, Gasi Tandefelt D, Conteduca V, Jayaram A, Casiraghi N, Wetterskog D, Salvi S, Amadori D, Zafeiriou Z, Rescigno P,

Bianchini D, Gurioli G, Casadio V, Carreira S, Goodall J, Wingate A, Ferraldeschi R, Tunariu N, Flohr P, De Giorgi U, de Bono JS, Demichelis F, Attard G. Plasma AR and abiraterone-resistant prostate cancer. Sci Transl Med. 2015;7(312):312re310. https://doi.org/10.1126/scitranslmed.aac9511.

63. Annala M, Vandekerkhove G, Khalaf D, Taavitsainen S, Beja K, Warner EW, Sunderland K, Kollmannsberger C, Eigl BJ, Finch D, Oja CD, Vergidis J, Zulfiqar M, Azad AA, Nykter M, Gleave ME, Wyatt AW, Chi KN. Circulating tumor DNA genomics correlate with resistance to Abiraterone and enzalutamide in prostate cancer. Cancer Discov. 2018;8(4):444–57. https://doi.org/10.1158/2159-8290.cd-17-0937.
64. Robinson D, Van Allen EM, Wu YM, Schultz N, Lonigro RJ, Mosquera JM, Montgomery B, Taplin ME, Pritchard CC, Attard G, Beltran H, Abida W, Bradley RK, Vinson J, Cao X, Vats P, Kunju LP, Hussain M, Feng FY, Tomlins SA, Cooney KA, Smith DC, Brennan C, Siddiqui J, Mehra R, Chen Y, Rathkopf DE, Morris MJ, Solomon SB, Durack JC, Reuter VE, Gopalan A, Gao J, Loda M, Lis RT, Bowden M, Balk SP, Gaviola G, Sougnez C, Gupta M, Yu EY, Mostaghel EA, Cheng HH, Mulcahy H, True LD, Plymate SR, Dvinge H, Ferraldeschi R, Flohr P, Miranda S, Zafeiriou Z, Tunariu N, Mateo J, Perez-Lopez R, Demichelis F, Robinson BD, Sboner A, Schiffman M, Nanus DM, Tagawa ST, Sigaras A, Eng KW, Elemento O, Sboner A, Heath EI, Scher HI, Pienta KJ, Kantoff P, de Bono JS, Rubin MA, Nelson PS, Garraway LA, Sawyers CL, Chinnaiyan AM. Integrative clinical genomics of advanced prostate cancer. Cell. 2015;162(2):454. https://doi.org/10.1016/j.cell.2015.06.053.
65. Mateo J, Carreira S, Sandhu S, Miranda S, Mossop H, Perez-Lopez R, Nava Rodrigues D, Robinson D, Omlin A, Tunariu N, Boysen G, Porta N, Flohr P, Gillman A, Figueiredo I, Paulding C, Seed G, Jain S, Ralph C, Protheroe A, Hussain S, Jones R, Elliott T, McGovern U, Bianchini D, Goodall J, Zafeiriou Z, Williamson CT, Ferraldeschi R, Riisnaes R, Ebbs B, Fowler G, Roda D, Yuan W, Wu YM, Cao X, Brough R, Pemberton H, A'Hern R, Swain A, Kunju LP, Eeles R, Attard G, Lord CJ, Ashworth A, Rubin MA, Knudsen KE, Feng FY, Chinnaiyan AM, Hall E, de Bono JS. DNA-repair defects and Olaparib in metastatic prostate cancer. N Engl J Med. 2015;373(18):1697–708. https://doi.org/10.1056/NEJMoa1506859.
66. Mateo J, Porta N, Bianchini D, McGovern U, Elliott T, Jones R, Syndikus I, Ralph C, Jain S, Varughese M, Parikh O, Crabb S, Robinson A, McLaren D, Birtle A, Tanguay J, Miranda S, Figueiredo I, Seed G, Bertan C, Flohr P, Ebbs B, Rescigno P, Fowler G, Ferreira A, Riisnaes R, Pereira R, Curcean A, Chandler R, Clarke M, Gurel B, Crespo M, Nava Rodrigues D, Sandhu S, Espinasse A, Chatfield P, Tunariu N, Yuan W, Hall E, Carreira S, de Bono JS. Olaparib in patients with metastatic castration-resistant prostate cancer with DNA repair gene aberrations (TOPARP-B): a multicentre, open-label, randomised, phase 2 trial. Lancet Oncol. 2020;21(1):162–74. https://doi.org/10.1016/S1470-2045(19)30684-9.
67. de Bono J, Mateo J, Fizazi K, Saad F, Shore N, Sandhu S, Chi KN, Sartor O, Agarwal N, Olmos D, Thiery-Vuillemin A, Twardowski P, Mehra N, Goessl C, Kang J, Burgents J, Wu W, Kohlmann A, Adelman CA, Hussain M. Olaparib for metastatic castration-resistant prostate cancer. N Engl J Med. 2020;382(22):2091–102. https://doi.org/10.1056/NEJMoa1911440.
68. fda.gov. FDA approves liquid biopsy NGS companion diagnostic test for multiple cancers and biomarkers; 2020. https://www.fda.gov/drugs/fda-approves-liquid-biopsy-ngs-companion-diagnostic-test-multiple-cancers-and-biomarkers. Accessed 13 Jan 2021.

Circulating Tumor Cells in Lung Cancer

15

Paul Hofman

Abstract

The detection and identification of circulating tumor cells (CTCs) and the analysis of the information they provide are often the subject of discussion in thoracic oncology, in particular for routine clinical use. However, CTCs are being used for an increasing range of applications in several domains of lung carcinoma. Thus, CTCs can participate in better defining not only the prognosis and diagnosis of lung cancer but also screening for the disease. CTCs can also contribute to the major therapeutic orientations through not only the study of the efficacy of targeted molecules but also of immunotherapy. In addition, the study of CTCs provides a better understanding of the natural history of lung cancer and of the mechanisms of metastasis. This review analyzes the different orientations and therapeutic approaches associated with the study of CTCs in the domain of thoracic oncology.

Learning Objectives

- To cite the main current challenges for the detection and the characterization of CTCs in lung cancer patients.
- To develop the potential interest of looking for the presence of CTCs in lung cancer patients in routine clinical practice.
- To inform on the different interests of using CTCs in lung cancer patients as a prognostic and as a diagnostic biomarker, including for lung cancer screening.
- To show the strong interest of combining the analyses of CTCs and other blood components (including the circulating free DNA) to optimize the value of using liquid biopsies for targeted therapy and immunotherapy in lung cancer patients.
- To highlight the importance of translational research projects based on CTCs assessment for better understanding the pathophysiology of lung cancer analyses.
- To inform on the next challenges of using CTCs in thoracic oncology.

15.1 Introduction

Lung cancer is the first cause of death from cancer in the world [1]. Despite progress into early diagnosis and the emergence of novel therapeutics that improve the overall survival of patients, the incidence of cancer has increased in the majority of countries [1]. Thus, it is urgent to make an effort to slow down a disease that concerns more and more young patients, and so has a major impact on health costs worldwide too. To improve the diagnostic and care effectiveness of this cancer a number of approaches are possible including treating better patients and thereby increasing the chances of remission or of prolonging life, even at a metastatic stage. Among the different approaches, the development of liquid biopsies has revolutionized in recent years several domains of lung cancer patient taking care, including the potential capacity of screening, early diagnosis, identification of factors of prognosis and of the predictive response or resistance to treatment as well as the detection of a residual cancer (or a « minimal residual disease »). To this aim

P. Hofman (✉)
Université Côte d'Azur, CNRS, INSERM, IRCAN, Centre Antoine Lacassagne, FHU OncoAge, Team 4, Nice, France

Université Côte d'Azur, CHU Nice, FHU OncoAge, Laboratory of Clinical and Experimental Pathology, Pasteur Hospital, Nice, France

Université Côte d'Azur, CHU Nice, FHU OncoAge, Hospital-Integrated Biobank BB-0033-00025, Nice, France
e-mail: hofman.p@chu-nice.fr

S. P. Leong et al. (eds.), *Cancer Metastasis Through the Lymphovascular System*,
https://doi.org/10.1007/978-3-030-93084-4_15

several components of blood have been studied but it is the analysis of circulating free nucleic acids in plasma that in recent years has been the main domain of daily life application of liquid biopsies in thoracic oncology. More precisely, the analysis of genomic tumor circulating DNA in plasma has progressively become one of the routine clinical approaches in thoracic oncology. This noninvasive approach also provides more and more rapid results, notably for the choice of targeted therapy [2–4].

Other than nucleic acids in plasma several components and compartments of blood, including circulating tumor cells (CTCs) are being investigated in some very important studies in the area of lung cancer [5, 6]. Paradoxically, CTCs were the first to be studied with liquid biopsies but their use in clinical routine and the interest they hold for oncologist in the area of lung cancer have not followed the same momentum and development as the analysis of circulating free DNA (cfDNA). This can be linked to a number of constraints that will be discussed in this chapter. However, beyond its limits, the analysis of CTCs provides essential information and is indispensable to the area of lung cancer. Moreover, the information obtained can be complementary to that provided by the study of other exploitable components of liquid biopsies.

The issues linked to the detection and characterization of CTCs in lung cancer and the perspectives in this domain will be discussed.

15.2 Circulating Tumor Cells in Lung Cancer: Advantages and Limitations

15.2.1 Current Issue

Despite the many improvements in assays and technologies for CTC detection and characterization, the main challenge certainly lies in the selection of the best method according to the project and the different goals associated with CTC assessment [7–10]. The multiple methods and approaches to the isolation, identification, and characterization of CTCs in lung oncology have made the selection of a technique particularly difficult. This has also brought in constraints linked to the comparison of the different studies into validation and reproducibility, placing these methods in competition, notably for their specificity and sensitivity. Thus, the major challenge today is to use CTCs as a biomarker in routine lung oncology, which has not yet been adopted. The only method that has been approved by the Food and Drug Administration (FDA) in the USA is the CellSearch method, but only for specific targeted clinical situations to define prognostic factors for prostate, colon, and breast metastatic cancers. So, no method of detection of CTCs has been approved by either the FDA or a European agency for management of lung cancer and thus for lung cancer patient care.

15.2.2 The Challenges in Using CTCs in Daily Practice for Lung Cancer Patients

Therefore, the usefulness of CTCs in routine clinical practice for lung cancer patients is still under debate, since most applications, recommendations, and guidelines (such as those provided by the International Association for the Study of Lung Cancer/IASLC), for use of a liquid biopsy for better care of lung cancer patients, are mainly based on evaluation of cfDNA [11]. However, many investigators highlight the fact that CTCs are a major component for assessment of lung cancer patients and have strong added value for many different clinical and research applications [12–15] (Fig. 15.1). The issue of the added value will be dealt with below and should also be considered in comparison with the potential use of cfDNA. In fact, some of the challenges in using CTCs in routine clinical practice include, at least, the cost effectiveness of the assay, the reproducibility of the results, the difficulty in obtaining accreditation for this biological test, and good mastering of the pre-analytical phase. For this latter point, and since the detection of CTCs is a rare event, it requires optimal preservation of the blood for several hours.

15.3 Circulating Tumor Cells as a Prognostic Biomarker for Lung Cancer

Several studies have demonstrated the interest in detecting and quantifying CTCs of solid tumors, in particular of lung cancers, and of correlating the number with a more or less unfavorable prognosis [16–18]. These studies were performed on patients at an early stage (stage I-IIIA) or at an advanced or metastatic stage (stage IIIB/IV) [19–21]. A correlation between a high number of CTCs isolated from blood and poor prognosis was observed irrespective of the histological type of the lung cancer, non-small cell or small cell lung cancer [19–22]. The different analyses were performed either at diagnosis or on tumor progression on treatment [19–21]. Finally, some studies have shown an association between a high number of CTCs isolated with the CellSearch method from blood of pulmonary veins before surgical resection and a high frequency of postoperative recurrence [23, 24].

Different methods have been used to detect and quantify CTCs in lung cancer, in particular the CellSearch method or the ISET which is based on blood filtration on polycarbonate filters [20, 21, 25, 26]. Each method has its advantages and disadvantages [7]. The CellSearch method may lack sensitivity due to the use of an anti-cytokeratin antibody for capture of CTCs since a certain number of these cells do not express, or express a low level of, cytokeratins as a result of epithelial-mesenchymal phenomenon, which reduces the sensitivity of this method [25–27]. Blood filtration can identify the CTCs

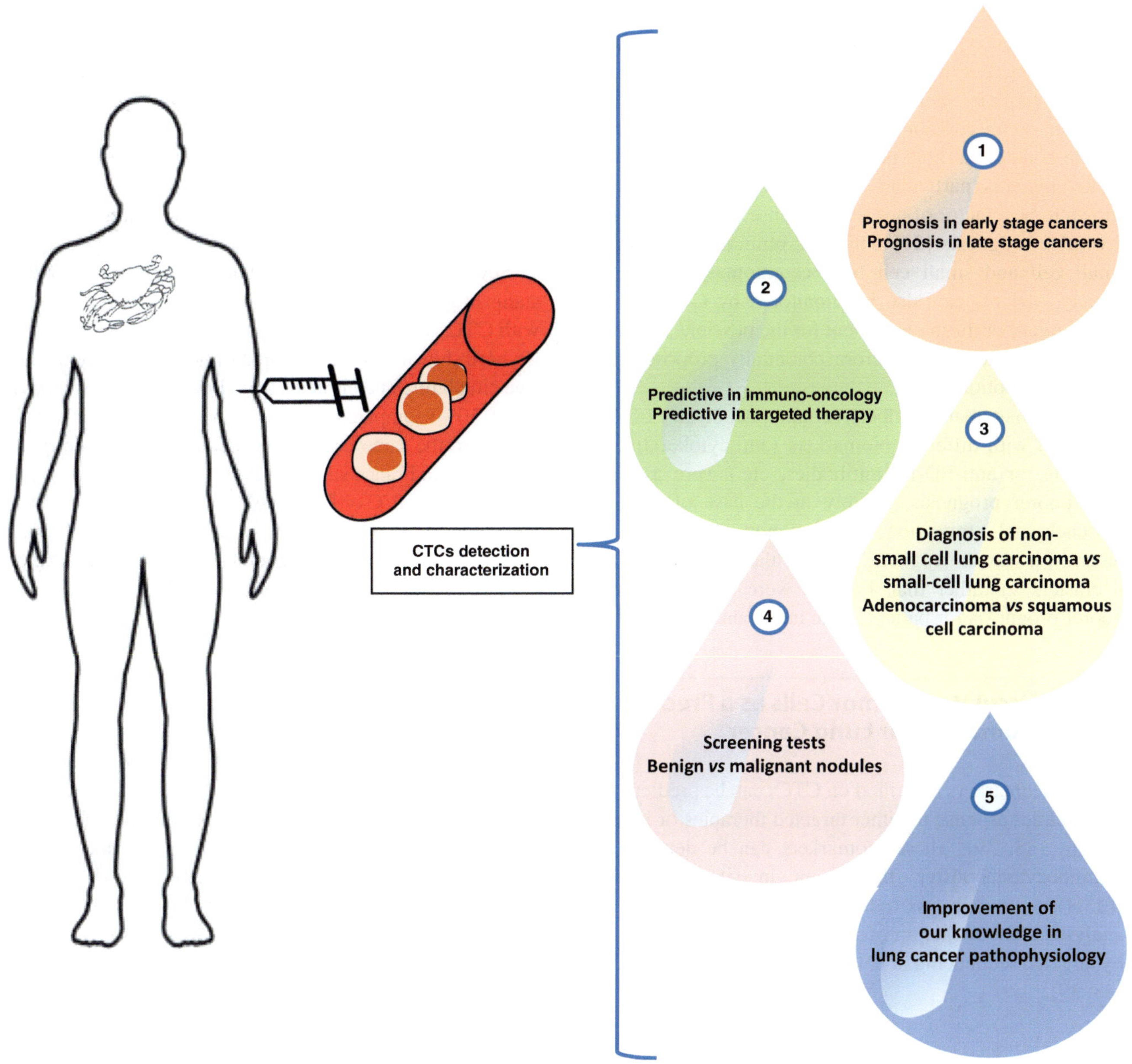

Fig. 15.1 Add value of exploring the circulating tumor cells (CTCs) in lung cancer patients

on the surface of the filters. However, the characterization and identification of these CTCs can be fastidious, since it is very time consuming and requires a strong expertise of a cytopathologist [28]. Some of the cells on the filters have cytopathological aspects that do not allow them to be distinguished with certainty, in the absence at least of immunocytochemistry (ICC), between non-tumor epithelial cells, giant macrophages or large monocytes, activated lymphocytes or even circulating megakaryocytes [28, 29]. Thus, the cytomorphological characterization of CTCs on filters is difficult and requires considerable expertise along with comparative analyses for validation [19]. Studies have compared the sensitivity of the CellSearch and ISET methods using samples from the same lung cancer patients. Overall, the ISET method appears to be more sensitive than the CellSearch method in these populations [25, 26]. Moreover, it cannot be excluded that some small circulating tumor cells may escape detection due to passage through the pores of the filters and so are not counted.

These studies demonstrated that the detection of CTCs before tumor resection can be associated with a poor prognosis and that postoperative recurrence may occur, even after complete tumor resection during surgery [19]. The presence of CTCs in patients with a locally advanced stage of disease and treated with radio-chemotherapy was associated with early recurrence [30]. The presence and high number of CTCs detected in the blood of patients at a metastatic stage before treatment also correlated with poor prognosis, for both non-small cell and small cell lung carcinoma [20, 21, 31]. In contrast, a rapid decrease in the number of CTCs in treated patients, particularly during chemotherapy, has been shown to be a factor of positive response to treatment and of favorable prognosis [21]. This was observed for both non-small cell and small cell lung carcinomas [21, 31, 32]. Finally, patients can also be monitored by CTC detection after surgery or during treatment for the possible presence of a residual tumor and therefore potentially proposed a new therapeutic option [33].

Other than their number, certain phenotypes of CTCs identified with different biomarkers (anti-cytokeratin, anti-vimentin, or anti-PD-L1 antibodies, etc.) were associated with poorer prognosis, notably in the case of epithelial-mesenchymal transition phenomenon [34–36]. More recently, it was shown that CTCs circulating in clumps (or in « clusters ») rather than isolated were associated with a poorer prognosis for patients with lung cancer [37].

15.4 Circulating Tumor Cells as a Predictive Biomarker for Lung Cancer

Certain biomarkers identified on CTCs can be predictive of a therapeutic response to either targeted therapies or immunotherapy [38, 39]. These biomarkers can be detected with immunocytochemistry, fluorescent in situ hybridization (FISH), or, though less frequently, with complex molecular analyses after extraction of nucleic acids and then analysis by targeted sequencing or next-generation sequencing (NGS) [38, 39].

15.4.1 Biomarkers for Targeted Therapies

Thanks to molecular analyses performed with CTCs the interest in detecting mutations in *EGFR* in blood was reported for the first time for patients with non-small cell lung cancer. Consequently, patient treatment could be adjusted with administration of inhibitors of tyrosine kinase targeting these mutations [40]. This approach has also been tested in more recent studies [41]. However, this approach using CTCs was rapidly replaced by the analysis of these mutations using cfDNA, because certain technologies provided more reproducible results, were more easily performed and less costly [2, 3]. It is interesting to note that an integrative approach associating detection of *EGFR* mutations in CTCs and cfDNA from the same patients increased the possibility of detecting these genomic alterations. Nonetheless it is clear that aside from a few recent studies orientated to this type of combination, CTCs are not used in daily practice to evaluate the *EGFR* status of patients with metastatic lung cancer. This is probably due to the difficulty in standardization of the technological approaches, the heterogeneity of the CTCs population, and also the quantitative variations in the number of CTCs from patient to patient. Thus, at present the detection of mutations in *EGFR* is done routinely with cfDNA in plasma using either targeted sequencing technologies or NGS [2–4].

Several studies have shown that it is possible to evaluate the status of gene rearrangements (*ALK*, *ROS1*, and *RET* fusions) with CTCs when using ICC and/or FISH [42, 43]. Most studies have shown that the detection of *ALK* rearrangements with free circulating nucleic acids has a low sensitivity (in the range of 40–50%) compared to analyses performed with tissue samples from the same patient [38]. A recent multi-centric study performed in France validated the possibility of examining the *ALK* status by ICC or FISH with CTCs isolated with filtration techniques [43]. Other studies demonstrated the interest in detecting certain types of profiles of *ALK* amplification in CTCs as a biomarker of efficacy and resistance to therapy targeting *ALK* [42]. The status of *ROS1* has also been evaluated with CTCs isolated by filtration of blood, but a multi-centric validation study and comparison with RT-PCR or NGS from free nucleic acids in plasma are certainly necessary to determine the sensitivity of this approach [42].

Other biomarkers predictive of the response to targeted therapy have been evaluated with CTCs. Thus, the protein expression of MET has been quantified with CTCs isolated and characterized after filtration on polycarbonate filters and the level of expression was compared to that of tissue samples obtained from the same patients [44].

On tumor progression in patients treated with ALK inhibitors the mechanisms of resistance are more often analyzed with circulating free nucleic acids (in particular *ALK* mutations) [45]. However, some studies have also shown that amplifications in *MET*, one of the mechanisms of resistance to ALK inhibitors, can be detected on progression with CTCs [46].

15.4.2 Biomarkers of Immunotherapy

Some biomarkers have been evaluated or used as predictive factors of response to immune checkpoint inhibitors [39]. Both tissue and blood samples are used to detect these biomarkers [36, 39, 47]. The predictive biomarker the most analyzed and validated to date to predict the response to the anti-PD-L1/PDL1 immune checkpoint inhibitors is the protein marker PD-L1 [47]. This biomarker is evaluated by immunohistochemistry with different anti-PD-L1 clones, but has also been evaluated by ICC on different cytological samples [48]. Moreover, several studies have shown good correlation between the expression of PD-L1 on tissue and cytological samples from the same patients [48]. In this context, the expression of PD-L1 has been evaluated by cytology with blood and thus on CTCs obtained with different approaches including methods of filtration [35]. It has been shown that it is possible to identify PD-L1 on CTCs but the prognostic significance and the predictive value of this expression is controversial [35, 49]. A recent study demonstrated that using the CellSearch system, it was possible to detect PD-L1 in a subset of CTCs associated with worse overall survival of late stage NSCLC [36].

Different immunotherapy clinical trials in lung cancer patients have evaluated the tumor mutation burden (TMB) using circulating nucleic acids in plasma [39]. Several clinical trials have shown that a high level of TMB in plasma can be associated with a good response to immunotherapy [39]. Only a few studies have evaluated the prognostic value of the analysis of the TMB with cytological samples but an approach with CTCs may appear to date to have major technological issues and very difficult to perform in clinical practice. The number of CTCs detected before immunotherapy has been associated with a poor response to immune checkpoint inhibitors [39]. However, at present, this parameter is not taken into account when modifying the treatment in clinical practice.

15.5 Circulating Tumor Cells as a Diagnostic Biomarker of Lung Cancer

One of the challenges in thoracic oncology is to diagnose at an early stage a lung cancer using methods that are robust, sensitive and specific, reproducible and noninvasive and with a low cost [6]. Another challenge is to propose to a population of eligible patients an early diagnosis that potentially leads to the detection of a resectable tumor [50]. In this regard, many trials have been developed with the aim of characterizing a molecular biological signature that provides information into the primitive origin of a cancer [50]. These approaches are based mostly on characterization of circulating nucleic acids or microRNAs, or protein signatures [50]. At present, it is more difficult to use CTCs to orientate the diagnosis toward a specific type of solid tumor. Some biomarkers can nonetheless be tested with CTCs (P40 and TTF1 detected with ICC, for example) in providing information concerning the primitive pulmonary origin of these CTCs [8].

15.6 Circulating Tumor Cells as a Screening Tool for Lung Cancer

Recent studies have shown that it is possible to screen for lung cancer, notably in a population at higher risk (subjects aged over 50 years, heavy smokers, developing chronic obstructive pulmonary disease or emphysema), with low-dose computed tomography (LDCT) [51]. Thus, the National Lung Screening Trial (NLST) and Nelson study have shown that screening can reduce the annual mortality from lung cancer for a few thousand individuals [51]. Despite these important findings, several hurdles have slowed down the setup of screening with LDCT in several countries, including France. While minimally invasive this test can appear arduous for patients and patient compliance is sometimes low, as in the USA. The CT images can also give false negative or positive results resulting into needless surgery with a risk of comorbidity and mortality. Knowing this it is of strong interest to develop or associate other tests for early screening of lung cancer including blood tests [50]. The detection of CTCs has aroused hope as a screening approach, in particular for patients at risk. A mono-centric study was thus able, in the absence of CT images of tumors, to predict the emergence of a lung cancer several months before detection by imagery [52]. This promising study has received substantial enthusiasm from the media but needs to be discussed while integrating several parameters. It concerns a retrospective study performed on a relatively small cohort of patients. A multi-centric validation study involving several French hospitals performed did not totally confirm the initial results [53]. This discordance can be explained by several elements, in particular the variation in the pre-analytical and analytical parameters depending on the hospital, but also a slight difference of the design of the two studies [52, 53].

15.7 Studying Circulating Tumor Cells to Better Understand the Pathophysiology of Lung Cancer

The different analyses performed with CTCs, in particular for patients presenting with small cell or non-small cell lung carcinoma, have improved knowledge into the natural history of these cancers and into the molecular anomalies, which have led to envisage novel targeted therapeutics [12, 14]. These studies relied on many technological approaches and innovative methods.

15.7.1 Setting up Different Mice Models Using CTCs

It was under the impetus of the team of Caroline Dive at the Paterson Cancer Institute in Manchester (UK) that models of mice implanted with CTCs (or CDXs) from patients with small cell metastatic lung cancer were developed [31]. These mice models have been used to study tumor progression and response to certain treatments.

15.7.2 Genomic Analysis of a Single CTC

The capture of CTCs makes possible the detection of certain genomic alterations of these cells and also the demonstration of the heterogeneity of the circulating population of cancer cells, in terms of both phenotype and genotype [54]. Thus, a number of applications have been developed with CTCs, in particular for small cell lung cancer [54]. Certain genetic profiles can be distinguished between chemo-sensitive and -resistant tumors [54].

15.7.3 Analysis of Interactions Between CTCs and Circulating Hematological Cells

It is well established that CTCs interact, either directly after cell contact or indirectly through the intermediary of secreted molecules, with circulating hematological cells [55]. Different populations of non-tumor cells circulate in the blood, including myeloid and lymphoid cells (T lymphocytes, natural killer cells), monocytes, and macrophages, but also circulating free endothelial cells and fibroblasts [55]. A strong cellular interaction between CTCs and platelets also exists [55]. The different actors present in the blood circulation can also interact in a complex way with each other through direct physical contact via receptors or cytokines, but also by means of transfer of microvesicles such as exosomes [55]. Thus, it may concern "a three- or four-way cell partnership" involving CTCs, myeloid cells, monocytes/macrophages, and platelets [55]. These cellular interactions may facilitate the progression of CTCs and can protect them from anoikis or from the cytotoxic effect of some circulating lymphoid cells (such as CD8 lymphocytes, natural killer cells, and T reg lymphocytes). Moreover, this allows CTCs to adhere to the wall of vessels and favors their intravasation and extravasation with the development of metastases in different organs [55].

The newly discovered knowledge into interactions between CTCs and different cells present at the parenchymal level have provided a better understanding of the implantation of CTCs in certain target organs and of the phenomenon of "self-seeding" with enrichment of the primitive tumor environment, which leads to renewed growth of the tumor or renewed progression of the tumor at the initial site of surgical resection.

15.7.4 Functional Analysis of CTCs

It is possible to evaluate the function of CTCs using certain approaches, in particular Epispot method [56]. Epispot can detect proteins secreted/released/shed from single epithelial cells, which distinguishes, for example, these viable cells from apoptotic CTCs.

15.8 Perspectives and New Issues

Studies into CTCs have opened up several short- and mid-term prospects for patients with lung cancer. Routine detection and characterization of CTCs has been rapidly replaced for different goals, such as genomic alteration assessment, by the use of circulating nucleic acid in plasma since the tests on cfDNA can be automated, are reproducible, and are in general sensitive and specific as well as less costly [57]. The development of NGS approaches with cfDNA is now possible and can be performed in clinical routine [4]. However, a few perspectives are now open for using CTCs in thoracic oncology.

15.8.1 Setup of Procedures for Accreditation and Harmonization

The International Liquid Biopsy Standardization Alliance (ILSA) is composed of organizations and foundations that recognize the importance of working toward the global use of liquid biopsy in oncology to support clinical decision-making and regulatory considerations and to seek to promote it in the community. An independent liquid biopsy- and standardization-based program has been initiated by ILSA [58]. The procedures of accreditation, notably the ISO 15189 norm, can be adapted to methods of detection of CTCs but the challenge strongly varies according to the technique used. Some of the procedures of the ISO 15189 norm are certainly difficult to apply to clinical validation of the detection of CTCs. These procedures require that an external quality assessment (EQA) be performed. Several constraints can arise when standardizing methods of detection and characterization of CTCs. One of the main challenges certainly concerns mastering the pre-analytical phase. It is also necessary to demonstrate the efficacy and added value of detection of CTCs in the clinic by performing comparative analyses and studies with different tests of detection.

15.8.2 Integrate the Data on CTCs with that Obtained from the Studies of Other Blood Components

Analyses of free circulating nucleic acids and the more difficult to perform detection and characterization of CTCs are often presented in opposition [4, 57]. Integrative biology is certainly a way to optimize and make some approaches more attractive and robust, notably for early detection of lung cancer, by associating radiological images and algorithms based on artificial intelligence [51].

15.8.3 Demonstrate the Contribution of CTCs Analysis to the Clinical Routine

It is essential to demonstrate the added value of routine detection of CTCs in thoracic oncology and in particular for improved care of patients. Notably, the question of integrating the detection of CTCs in clinical trials has been raised [59]. An important consideration concerns the tumor heterogeneity [33]. In fact, there exists a degree of disagreement between the NGS results obtained with circulating nucleic acids and tissue biopsies from the same patients. A complementary CTC analysis can provide additional information concerning the heterogeneity and the discordant results [41]. One of the goals of using CTCs in daily practice could be to detect and monitor the minimal residual disease in early-stage patients or patients after treatment.

15.9 Conclusion

Studies into CTCs have improved the understanding of the pathophysiology of lung cancer and the development of different treatments for patients. It is essential to integrate in a global perspective the analysis of the different components circulating in the blood of patients with lung cancer and to not limit analyses to circulating free nucleic acids in plasma. Improvements need to be made in the standardization of CTCs detection methods, and the characterization and quantification of CTCs, in particular for thoracic oncology. The analysis of CTCs will provide a better understanding of the natural history of lung cancer and lead to the development of innovative treatments. To this aim, it must be kept in mind that the creation of biological collections associating CTCs will allow different analyses and collaborative studies to be performed [60].

Open Questions

- How to optimize the methods for a better detection and quantification of CTCs in thoracic oncology?
- How to determine the metastatic potential of CTCs in thoracic oncology?
- How to be able to better obtain CDXs models from CTCs isolated from NSCLC patients?
- How to develop some tests combining at the same time the analyses of CTCs and other blood components for a better taking care of lung cancer patients?
- How to set up validation methods and how to obtain an accreditation associated with CTCs detection in thoracic oncology?

References

1. Zhang Y, Luo G, Etxeberria J, Hao Y. Global patterns and trends in lung cancer incidence: a population-based study. J Thorac Oncol. 2021;16(6):933–44. https://doi.org/10.1016/j.jtho.2021.01.1626.
2. Heeke S, Benzaquen J, Hofman V, Ilié M, Allegra M, Long-Mira E, Lassalle S, Tanga V, Salacroup C, Bonnetaud C, Fayada J, Gazoppi L, Ribeyre L, Castelnau O, Garnier G, Cattet F, Nanni I, de Fraipont F, Cohen C, Berthet JP, Leroy S, Poudenx M, Marquette CH, Denis MG, Barlesi F, Hofman P. Critical assessment in routine clinical practice of liquid biopsy for EGFR status testing in non-small-cell lung cancer: a single-laboratory experience (LPCE, Nice, France). Clin Lung Cancer. 2020;21(1):56–65.e8. https://doi.org/10.1016/j.cllc.2019.07.010.
3. Heeke S, Hofman V, Benzaquen J, Otto J, Tanga V, Zahaf K, Allegra M, Long-Mira E, Lassalle S, Marquette CH, Ilie M, Hofman P. Detection of EGFR mutations from plasma of NSCLC patients using an automatic cartridge-based PCR system. Front Pharmacol. 2021;12:657743. https://doi.org/10.3389/fphar.2021.657743.
4. Hofman P. Next-generation sequencing with liquid biopsies from treatment-Naïve non-small cell lung carcinoma patients. Cancers (Basel). 2021;13(9):2049. https://doi.org/10.3390/cancers13092049.
5. Alix-Panabières C. The future of liquid biopsy. Nature. 2020;579(7800):S9. https://doi.org/10.1038/d41586-020-00844-5.
6. Pantel K. Blood tests for early detection of lung cancer: challenges and promises. Lancet Respir Med. 2020;8(7):654–6.
7. Hofman V, Ilie M, Long E, Guibert N, Selva E, Washetine K, Mograbi B, Mouroux J, Vénissac N, Reverso-Meinietti J, Milano G, Mazières J, Marquette CH, Paterlini-Bréchot P, Hofman P. Detection of circulating tumor cells from lung cancer patients in the era of targeted therapy: promises, drawbacks and pitfalls. Curr Mol Med. 2014;14(4):440–56.
8. Hofman VJ, Ilie M, Hofman PM. Detection and characterization of circulating tumor cells in lung cancer: why and how? Cancer Cytopathol. 2016;124(6):380–7.
9. Riethdorf S, O'Flaherty L, Hille C, Pantel K. Clinical applications of the CellSearch platform in cancer patients. Adv Drug Deliv Rev. 2018;125:102–21.
10. Xu M, Liu W, Zou K, Wei S, Zhang X, Li E, Wang Q. Design and clinical application of an integrated microfluidic device for circulating tumor cells isolation and single-cell analysis. Micromachines (Basel). 2021;12(1):49.
11. Rolfo C, Mack PC, Scagliotti GV, Baas P, Barlesi F, Bivona TG, Herbst RS, Mok TS, Peled N, Pirker R, Raez LE, Reck M, Riess JW, Sequist LV, Shepherd FA, Sholl LM, Tan DSW, Wakelee HA, Wistuba II, Wynes MW, Carbone DP, Hirsch FR, Gandara DR. Liquid biopsy for advanced non-small cell lung cancer (NSCLC): a statement paper from the IASLC. J Thorac Oncol. 2018;13(9):1248–68.
12. Alix-Panabières C, Pantel K. Liquid biopsy: from discovery to clinical application. Cancer Discov. 2021;11(4):858–73. https://doi.org/10.1158/2159-8290.CD-20-1311.
13. Cortés-Hernández LE, Eslami-S Z, Alix-Panabières C. Liquid biopsy to detect circulating tumor cells: is it ready for a value proposition in laboratory medicine? J Appl Lab Med. 2020;5(5):1027–37.
14. Cortés-Hernández LE, Eslami-S Z, Pantel K, Alix-Panabières C. Molecular and functional characterization of circulating tumor cells: from discovery to clinical application. Clin Chem. 2020;66(1):97–104.
15. Ross K, Pailler E, Faugeroux V, Taylor M, Oulhen M, Auger N, Planchard D, Soria JC, Lindsay CR, Besse B, Vielh P, Farace F. The potential diagnostic power of circulating tumor cell analysis for non-small-cell lung cancer. Expert Rev Mol Diagn. 2015;15(12):1605–29. https://doi.org/10.1586/14737159.2015.1111139.

16. Chemi F, Mohan S, Guevara T, Clipson A, Rothwell DG, Dive C. Early dissemination of circulating tumor cells: biological and clinical insights. Front Oncol. 2021;11:672195. https://doi.org/10.3389/fonc.2021.672195.
17. Lindsay CR, Faugeroux V, Michiels S, Pailler E, Facchinetti F, Ou D, Bluthgen MV, Pannet C, Ngo-Camus M, Bescher G, Caramella C, Billiot F, Remon J, Planchard D, Soria JC, Besse B, Farace F. A prospective examination of circulating tumor cell profiles in non-small-cell lung cancer molecular subgroups. Ann Oncol. 2017;28(7):1523–31.
18. Lindsay CR, Blackhall FH, Carmel A, Fernandez-Gutierrez F, Gazzaniga P, Groen HJM, Hiltermann TJN, Krebs MG, Loges S, López-López R, Muinelo-Romay L, Pantel K, Priest L, Riethdorf S, Rossi E, Terstappen L, Wikman H, Soria JC, Farace F, Renehan A, Dive C, Besse B, Michiels S. EPAC-lung: pooled analysis of circulating tumour cells in advanced non-small cell lung cancer. Eur J Cancer. 2019;117:60–8.
19. Hofman V, Bonnetaud C, Ilie MI, Vielh P, Vignaud JM, Fléjou JF, Lantuejoul S, Piaton E, Mourad N, Butori C, Selva E, Poudenx M, Sibon S, Kelhef S, Vénissac N, Jais JP, Mouroux J, Molina TJ, Hofman P. Preoperative circulating tumor cell detection using the isolation by size of epithelial tumor cell method for patients with lung cancer is a new prognostic biomarker. Clin Cancer Res. 2011;17(4):827–35.
20. Hou JM, Krebs MG, Lancashire L, Sloane R, Backen A, Swain RK, Priest LJ, Greystoke A, Zhou C, Morris K, Ward T, Blackhall FH, Dive C. Clinical significance and molecular characteristics of circulating tumor cells and circulating tumor microemboli in patients with small-cell lung cancer. J Clin Oncol. 2012;30(5):525–32. https://doi.org/10.1200/JCO.2010.33.3716.
21. Krebs MG, Sloane R, Priest L, Lancashire L, Hou JM, Greystoke A, Ward TH, Ferraldeschi R, Hughes A, Clack G, Ranson M, Dive C, Blackhall FH. Evaluation and prognostic significance of circulating tumor cells in patients with non-small-cell lung cancer. J Clin Oncol. 2011;29(12):1556–63. https://doi.org/10.1200/JCO.2010.28.7045.
22. Foy V, Lindsay CR, Carmel A, Fernandez-Gutierrez F, Krebs MG, Priest L, Carter M, Groen HJM, Hiltermann TJN, de Luca A, Farace F, Besse B, Terstappen L, Rossi E, Morabito A, Perrone F, Renehan A, Faivre-Finn C, Normanno N, Dive C, Blackhall F, Michiels S. EPAC-lung: European pooled analysis of the prognostic value of circulating tumour cells in small cell lung cancer. Transl Lung Cancer Res. 2021;10(4):1653–65. https://doi.org/10.21037/tlcr-20-1061.
23. Chemi F, Rothwell DG, McGranahan N, Gulati S, Abbosh C, Pearce SP, Zhou C, Wilson GA, Jamal-Hanjani M, Birkbak N, Pierce J, Kim CS, Ferdous S, Burt DJ, Slane-Tan D, Gomes F, Moore D, Shah R, Al Bakir M, Hiley C, Veeriah S, Summers Y, Crosbie P, Ward S, Mesquita B, Dynowski M, Biswas D, Tugwood J, Blackhall F, Miller C, Hackshaw A, Brady G, Swanton C, Dive C, TRACERx Consortium. Pulmonary venous circulating tumor cell dissemination before tumor resection and disease relapse. Nat Med. 2019;25(10):1534–9.
24. Crosbie PA, Shah R, Krysiak P, Zhou C, Morris K, Tugwood J, Booton R, Blackhall F, Dive C. Circulating tumor cells detected in the tumor-draining pulmonary vein are associated with disease recurrence after surgical resection of NSCLC. J Thorac Oncol. 2016;11(10):1793–7. https://doi.org/10.1016/j.jtho.2016.06.017.
25. Farace F, Massard C, Vimond N, Drusch F, Jacques N, Billiot F, Laplanche A, Chauchereau A, Lacroix L, Planchard D, Le Moulec S, André F, Fizazi K, Soria JC, Vielh P. A direct comparison of CellSearch and ISET for circulating tumour-cell detection in patients with metastatic carcinomas. Br J Cancer. 2011;105(6):847–53.
26. Hofman V, Ilie MI, Long E, Selva E, Bonnetaud C, Molina T, Vénissac N, Mouroux J, Vielh P, Hofman P. Detection of circulating tumor cells as a prognostic factor in patients undergoing radical surgery for non-small-cell lung carcinoma: comparison of the efficacy of the CellSearch Assay™ and the isolation by size of epithelial tumor cell method. Int J Cancer. 2011;129(7):1651–60.
27. Alix-Panabières C, Mader S, Pantel K. Epithelial-mesenchymal plasticity in circulating tumor cells. J Mol Med (Berl). 2017;95:133–42. https://doi.org/10.1007/s00109-016-1500-6.
28. Hofman V, Long E, Ilie M, Bonnetaud C, Vignaud JM, Fléjou JF, Lantuejoul S, Piaton E, Mourad N, Butori C, Selva E, Marquette CH, Poudenx M, Sibon S, Kelhef S, Vénissac N, Jais JP, Mouroux J, Molina TJ, Vielh P, Hofman P. Morphological analysis of circulating tumour cells in patients undergoing surgery for non-small cell lung carcinoma using the isolation by size of epithelial tumour cell (ISET) method. Cytopathology. 2012;23(1):30–8.
29. Manjunath Y, Mitchem JB, Suvilesh KN, Avella DM, Kimchi ET, Staveley-O'Carroll KF, Deroche CB, Pantel K, Li G, Kaifi JT. Circulating Giant tumor-macrophage fusion cells are independent prognosticators in patients with NSCLC. J Thorac Oncol. 2020;15(9):1460–71.
30. Chinniah C, Aguarin L, Cheng P, Decesaris C, Cutillo A, Berman AT, Frick M, Doucette A, Cengel KA, Levin W, Hahn S, Dorsey JF, Simone CB 2nd, Kao GD. Early detection of recurrence in patients with locally advanced non-small-cell lung cancer via circulating tumor cell analysis. Clin Lung Cancer. 2019;20(5):384–90.
31. Blackhall F, Frese KK, Simpson K, Kilgour E, Brady G, Dive C. Will liquid biopsies improve outcomes for patients with small-cell lung cancer? Lancet Oncol. 2018;19(9):e470–e81.
32. Shishido SN, Carlsson A, Nieva J, Bethel K, Hicks JB, Bazhenova L, Kuhn P. Circulating tumor cells as a response monitor in stage IV non-small cell lung cancer. J Transl Med. 2019;17(1):294.
33. Pawlikowska P, Faugeroux V, Oulhen M, Aberlenc A, Tayoun T, Pailler E, Farace F. Circulating tumor cells (CTCs) for the noninvasive monitoring and personalization of non-small cell lung cancer (NSCLC) therapies. J Thorac Dis. 2019;11(Suppl 1):S45–56. https://doi.org/10.21037/jtd.2018.12.80.
34. Ge B, Wang Y, Lei S, Zhang J. Effect of phenotypic detection of circulating tumor cells marked by epithelial-mesenchymal transformation on the prognosis of lung cancer: a protocol for systematic review and meta-analysis. Medicine (Baltimore). 2020;99(44):e22960.
35. Ilié M, Szafer-Glusman E, Hofman V, Chamorey E, Lalvée S, Selva E, Leroy S, Marquette CH, Kowanetz M, Hedge P, Punnoose E, Hofman P. Detection of PD-L1 in circulating tumor cells and white blood cells from patients with advanced non-small-cell lung cancer. Ann Oncol. 2018;29(1):193–9.
36. Sinoquet L, Jacot W, Gauthier L, Pouderoux S, Viala M, Cayrefourcq L, Quantin X, Alix-Panabières C. Programmed cell death ligand 1-expressing circulating tumor cells: a new prognostic biomarker in non-small cell lung cancer. Clin Chem. 2021;6:hvab131. https://doi.org/10.1093/clinchem/hvab131.
37. Murlidhar V, Reddy RM, Fouladdel S, Zhao L, Ishikawa MK, Grabauskiene S, Zhang Z, Lin J, Chang AC, Carrott P, Lynch WR, Orringer MB, Kumar-Sinha C, Palanisamy N, Beer DG, Wicha MS, Ramnath N, Azizi E, Nagrath S. Poor prognosis indicated by venous circulating tumor cell clusters in early-stage lung cancers. Cancer Res. 2017;77(18):5194–206. https://doi.org/10.1158/0008-5472.CAN-16-2072.
38. Hofman P. Liquid biopsy and therapeutic targets: present and future issues in thoracic oncology. Cancers (Basel). 2017;9(11):154.
39. Hofman P, Heeke S, Alix-Panabières C, Pantel K. Liquid biopsy in the era of immuno-oncology: is it ready for prime-time use for cancer patients? Ann Oncol. 2019;30(9):1448–59.
40. Maheswaran S, Sequist LV, Nagrath S, Ulkus L, Brannigan B, Collura CV, Inserra E, Diederichs S, Iafrate AJ, Bell DW, Digumarthy S, Muzikansky A, Irimia D, Settleman J, Tompkins RG, Lynch TJ, Toner M, Haber DA. Detection of mutations in EGFR in

circulating lung-cancer cells. N Engl J Med. 2008;359(4):366–77. https://doi.org/10.1056/NEJMoa0800668.
41. Liu HE, Vuppalapaty M, Wilkerson C, Renier C, Chiu M, Lemaire C, Che J, Matsumoto M, Carroll J, Crouse S, Hanft VR, Jeffrey SS, Di Carlo D, Garon EB, Goldman J, Sollier E. Detection of EGFR mutations in cfDNA and CTCs, and comparison to tumor tissue in non-small-cell-lung-cancer (NSCLC) patients. Front Oncol. 2020;10:572895.
42. Catelain C, Pailler E, Oulhen M, Faugeroux V, Pommier AL, Farace F. Detection of gene rearrangements in circulating tumor cells: examples of ALK-, ROS1-, RET-rearrangements in non-small-cell lung cancer and ERG-rearrangements in prostate cancer. Adv Exp Med Biol. 2017;994:169–79.
43. Ilie M, Mazieres J, Chamorey E, Heeke S, Benzaquen J, Thamphya B, Boutros J, Tiotiu A, Fayada J, Cadranel J, Poudenx M, Moro-Sibilot D, Barlesi F, Thariat J, Clement-Duchene C, Tomasini P, Hofman V, Marquette CH, Hofman P, STALKLUNG01 Study Consortium Investigators. Prospective multi-center validation of the detection of ALK rearrangements of circulating tumor cells for non-invasive longitudinal management of patients with advanced non-small cell lung cancer. J Thorac Oncol. 2021;S1556-0864(21):01664–6. https://doi.org/10.1016/j.jtho.2021.01.1617.
44. Ilie M, Szafer-Glusman E, Hofman V, Long-Mira E, Suttmann R, Darbonne W, Butori C, Lalvée S, Fayada J, Selva E, Yu W, Marquette CH, Shames DS, Punnoose E, Hofman P. Expression of MET in circulating tumor cells correlates with expression in tumor tissue from advanced-stage lung cancer patients. Oncotarget. 2017;8(16):26112–21.
45. Hofman P. Detecting resistance to therapeutic ALK inhibitors in tumor tissue and liquid biopsy markers: an update to a clinical routine practice. Cell. 2021;10(1):168.
46. Berger LA, Janning M, Velthaus JL, Ben-Batalla I, Schatz S, Falk M, Iglauer P, Simon R, Cao R, Forcato C, Manaresi N, Bramlett K, Buson G, Hanssen A, Tiemann M, Sauter G, Bokemeyer C, Riethdorf S, Reck M, Pantel K, Wikman H, Loges S. Identification of a high-level MET amplification in CTCs and cfTNA of an ALK-positive NSCLC patient developing evasive resistance to Crizotinib. J Thorac Oncol. 2018;13(12):e243–e6.
47. Ilie M, Hofman V, Dietel M, Soria JC, Hofman P. Assessment of the PD-L1 status by immunohistochemistry: challenges and perspectives for therapeutic strategies in lung cancer patients. Virchows Arch. 2016;468(5):511–25. https://doi.org/10.1007/s00428-016-1910-4.
48. Gosney JR, Boothman AM, Ratcliffe M, Kerr KM. Cytology for PD-L1 testing: a systematic review. Lung Cancer. 2020;141:101–6. https://doi.org/10.1016/j.lungcan.2020.01.010.
49. Janning M, Kobus F, Babayan A, Wikman H, Velthaus JL, Bergmann S, Schatz S, Falk M, Berger LA, Böttcher LM, Päsler S, Gorges TM, O'Flaherty L, Hille C, Joosse SA, Simon R, Tiemann M, Bokemeyer C, Reck M, Riethdorf S, Pantel K, Loges S. Determination of PD-L1 expression in circulating tumor cells of NSCLC patients and correlation with response to PD-1/PD-L1 inhibitors. Cancers (Basel). 2019;11(6):835.
50. Calabrese F, Lunardi F, Pezzuto F, Fortarezza F, Vuljan SE, Marquette C, Hofman P. Are there new biomarkers in tissue and liquid biopsies for the early detection of non-small cell lung cancer? J Clin Med. 2019;8(3):414. https://doi.org/10.3390/jcm8030414.
51. Benzaquen J, Boutros J, Marquette C, Delingette H, Hofman P. Lung cancer screening, towards a multidimensional approach: why and how? Cancers (Basel). 2019;11(2):212. https://doi.org/10.3390/cancers11020212.
52. Ilie M, Hofman V, Long-Mira E, Selva E, Vignaud JM, Padovani B, Mouroux J, Marquette CH, Hofman P. "sentinel" circulating tumor cells allow early diagnosis of lung cancer in patients with chronic obstructive pulmonary disease. PLoS One. 2014;9(10):e111597.
53. Marquette CH, Boutros J, Benzaquen J, Ferreira M, Pastre J, Pison C, Padovani B, Bettayeb F, Fallet V, Guibert N, Basille D, Ilie M, Hofman V, Hofman P. AIR project study group. Lancet Respir Med. 2020;8(7):709–16.
54. Carter L, Rothwell DG, Mesquita B, Smowton C, Leong HS, Fernandez-Gutierrez F, Li Y, Burt DJ, Antonello J, Morrow CJ, Hodgkinson CL, Morris K, Priest L, Carter M, Miller C, Hughes A, Blackhall F, Dive C, Brady G. Molecular analysis of circulating tumor cells identifies distinct copy-number profiles in patients with chemosensitive and chemorefractory small-cell lung cancer. Nat Med. 2017;23(1):114–9.
55. Heeke S, Mograbi B, Alix-Panabières C, Hofman P. Never travel alone: the crosstalk of circulating tumor cells and the blood microenvironment. Cell. 2019;8(7):714.
56. Alix-Panabières C. EPISPOT assay: detection of viable DTCs/CTCs in solid tumor patients. Recent Results Cancer Res. 2012;195:69–76. https://doi.org/10.1007/978-3-642-28160-0_6.
57. Ilie M, Hofman V, Long E, Bordone O, Selva E, Washetine K, Marquette CH, Hofman P. Current challenges for detection of circulating tumor cells and cell-free circulating nucleic acids, and their characterization in non-small cell lung carcinoma patients. What is the best blood substrate for personalized medicine? Ann Transl Med. 2014;2(11):107. https://doi.org/10.3978/j.issn.2305-5839.2014.08.11.
58. Connors D, Allen J, Alvarez JD, Boyle J, Cristofanilli M, Hiller C, Keating S, Kelloff G, Leiman L, McCormack R, Merino D, Morgan E, Pantel K, Rolfo C, Serrano MJ, Pia Sanzone A, Schlange T, Sigman C, Stewart M. International liquid biopsy standardization alliance white paper. Crit Rev Oncol Hematol. 2020;156:103112.
59. Massard C, Borget I, Farace F, Aspeslagh S, Le Deley MC, Le Tourneau C, Bidard FC, Pierga JY, Dieras V, Hofman P, Spano JP, Ferte C, Lacroix L, Soria JC. RECIST response and variation of circulating tumour cells in phase 1 trials: a prospective multicentric study. Eur J Cancer. 2017;83:185–93.
60. Nejlund S, Smith J, Kraan J, Stender H, Van MN, Langkjer ST, Nielsen MT, Sölétormos G, Hillig T. Cryopreservation of circulating tumor cells for enumeration and characterization. Biopreserv Biobank. 2016;14(4):330–7. https://doi.org/10.1089/bio.2015.007.

CTCs/ctDNA and Brain Metastasis

16

S. Ray Kenney and Dario Marchetti

Abstract

An estimated 20–30% of all patients with cancer will develop brain metastasis, and this percentage increases at autopsy, with advanced stage and aggressive cancers augmenting the risk. However, in spite of the severity of brain metastasis as a daunting clinical problem, advances in systemic therapies have been limited in scope and effectiveness. Brain metastases occur via seeding of circulating tumor cells (CTCs) in the brain microvasculature. This biology guides increasing interest towards interrogating CTCs and circulating tumor DNA (ctDNA) as tumor components of body fluids like blood and cerebral spinal fluid (CSF). Here, we provide a review on research advances implicating CTCs and ctDNAs in relation to the onset of brain metastasis, particularly in breast cancer. Further, we introduce key biomarker discoveries on CTCs associated with breast cancer brain metastasis, and provide overviews of similar roles played by ctDNA as studies reported in the literature. Lastly, we discuss the scientific and technical challenges that remain to be overcome to implement the concept of liquid biopsy in the clinic, notably using liquid biopsy to predict brain metastasis onset as guidance for a specific treatment, or to prevent secondary brain metastasis with more effective therapies.

S. R. Kenney
RareCyte, Inc., Seattle, WA, USA
e-mail: rkenney@rarecyte.com

D. Marchetti (✉)
Department of Internal Medicine, Division of Molecular Medicine, The University of New Mexico Health Sciences Center, Albuquerque, NM, USA
e-mail: Dmarchetti@salud.unm.edu

Learning Objectives
- Introduction of biological and clinical challenges overcoming BM onset.
- Description of latest advances of CTC/ctDNA assays and CTCs/ctDNA implications for clinical monitoring of BM and their therapy.
- Perspectives in applying liquid biopsy concepts by analyses of CTCs and ctDNA for predictive and preventive targets of BM.
- Overview on CTCs/ctDNA profiling to guide novel BM therapeutic strategies.

16.1 Introduction

Brain metastasis (BM) represents a significant clinical and biological challenge with highly devastating and lethal consequences. An estimated 20% of all cancer patients will develop brain metastasis [1, 2]. However, this estimate is likely an underestimate, since autopsy studies have suggested a higher incidence, approximately 40%, of intracranial metastasis in patients with cancer [2, 3]. For example, brain metastasis frequently occurs as an initial site in metastatic melanoma with diagnosis in approximately 60% of patients, and up to 80% of patients at autopsy [3]. Lung, melanoma, and breast cancer are the cancer types that most frequently metastasize to brain, although others include colorectal, pancreatic, renal, and prostate carcinomas [1, 2]. Notably, the overall 2- and 5-year survival rates for patients diagnosed with BM remain dismal at 8.1% and 2.4%, respectively, with a median overall survival of only 4 months from diagnosis [4, 5]. To date, localized BM therapies include neurological resection of BM lesions or stereotactic radiosurgery, while widespread disease includes whole brain radiotherapy, targeted therapy, or immunotherapy. While the latter has shown some significant activity, it is often less than

S. P. Leong et al. (eds.), *Cancer Metastasis Through the Lymphovascular System*,
https://doi.org/10.1007/978-3-030-93084-4_16

what observed for patients with only extracranial metastases [6, 7]. Furthermore, standard treatments result in high neurotoxicity (45% of patients), exacerbated by the fact that the blood-brain barrier (BBB) impairs the ability of chemotherapeutic approaches to achieve therapeutic concentrations in the brain/central nervous system (CNS).

Over the past several years, significant advances have been made in the understanding of the biology of circulating tumor cells (CTCs), the smallest functional units of cancer and "seeds" of fatal metastatic disease [8], and circulating cell-free tumor DNA (ctDNA) [9–12]. Notably, newly discovered biomarkers of CTC subsets and/or ctDNA signatures may be associated with BM. The objective of this review is to outline current CTC/ctDNA-based approaches linked to BM onset and biology. We also describe the advances made that might foster the implementation of these technologies in new clinical trials designed to evaluate the efficacy of CTC/ctDNA biomarkers as novel assays, thus implementing the concept of liquid biopsy (LB) that can be applied either to predict or to prevent BM through targeted therapies.

16.2 CTCs Induce Breast Cancer Brain Metastasis: The BMSM Signature

CTCs are essential components of the metastatic cascade, and crucial not only for understanding mechanisms of cancer metastasis but also for developing new liquid biopsy assays and applying them for initial diagnosis, monitoring of treatment efficacy, and for assessment of therapy efficacy. These concepts are particularly important in BM onset, where the known characteristics of brain as target organ, including the presence of a high microvascular network, circulatory arrest due to decreased blood flow in capillaries, specific CTC-brain endothelial interactions favoring adhesion to/invasion into brain microenvironment, can facilitate an increased exposure to CTCs. However, the comprehensive characterization of CTCs inducing BM remains incomplete, owing to CTC rarity and fragility, high phenotypic heterogeneity, and a lack of universal CTC biomarkers; coupled with CTC technologies which cannot capture and evaluate the entire spectrum of BM-relevant CTCs. For example, the CellSearch™ platform, the only CTC in vitro diagnostic platform approved for use by the U.S. Food and Drug Administration, relies on antibodies targeting the epithelial cell adhesion molecule (EpCAM) and pan-cytokeratin (CK). Therefore, it is only capable of capturing EpCAM+/CK+ CTCs, but not CTCs that are EpCAM/CK-independent, e.g., EpCAM-low or EpCAM-negative. Similarly, alternative approaches to CellSearch based on procedures involving density, immunomagnetic separation, size exclusion, and/or flow manipulation in microfluidic devices are limited by the heterogeneous nature of CTCs, often not capturing EpCAM− CTCs [10]. This is because a significant fraction of CTCs that extravasate and are able to generate distant metastases may have lost EpCAM expression and undergone epithelial-mesenchymal transition (EMT), expressing several EMT and/or cancer stem cell biomarkers [10]. Further, it has been reported that CellSearch is unable to capture CTCs in 30–35% of metastatic breast cancer patients and in more than 60% of patients with breast cancer BM (BCBM) [13, 14]. To this end, we developed new approaches to identify and characterize EpCAM− CTCs in patients clinically diagnosed with or without BCBM, detectable by MRI as standard of care [15].

Our focus has been to investigate biomarkers expressed by those EpCAM− CTCs in their abilities to induce BCBM [16, 17]; moving beyond CTC enumeration to demonstrate that CTCs can be isolated, cause new tumors, and be characterized as brain-homing CTC subsets. We reported the isolation of EpCAM− CTCs from patients with metastatic breast cancer using multiparametric flow cytometry, then assessed these EpCAM− CTCs for metastatic competence by selecting for biomarkers HPSE+/Notch1+/EGFR+/HER2+, which we termed the Brain Metastasis Selected Marker (BMSM) signature. We showed that CTCs selected for this signature are predisposed to metastasize to the brain and lungs once injected in immunodeficient animals [16, 17] (Fig. 16.1). The CTC BMSM signature is significant for two reasons: (1) It is the result of a critical evaluation of CTC biomarkers currently in use; (2) It demonstrates that many biomarkers currently used in the clinic, e.g., EpCAM, may not precisely identify all clinically relevant CTCs. This suggests that CTCs possessing brain specific metastatic competence must be examined to better understand mechanisms of CTC-driven metastasis vs. clinical dormancy.

16.3 The Characterization of CTCs Related to Breast Cancer Dormancy

The molecular switch to differentiate quiescence vs. metastatic competence of CTCs depends on the cross talk between CTCs and the tumor microenvironment. We posited that specific EpCAM-negative CTC subpopulations, shed from the primary or metastatic tumors, avoid organ arrest with extreme efficiency by the concomitant presence of stem cell and quiescence properties. For example, previous studies have established the presence of two neoplastic biomarkers, urokinase plasminogen activator receptor (uPAR) and integrin β1 (ITGβ1), promoting tumor cell growth and proliferation when they interact with the extracellular brain microenvironment, with loss of uPAR and ITGβ1expression, causing a shift from an invasive/metastatic to a dormant state [18–20].

We have reported the isolation of EpCAM-negative breast cancer CTC subsets containing stem-cell properties (CD44+/CD24−) by multiparametric flow cytometry with alternative expression of uPAR and ITGβ1: uPAR+/β1+, uPAR−/β1−, uPAR+/β1−, or uPAR−/β1+ [20, 21]. We investigated these CTC subsets for cell adhesion and proliferation properties, and for their abilities to generate in vitro 3D CTC tumorspheres, and invade the extracellular matrix. We discovered that these properties were dependent upon uPAR/ITGβ1 expression, and propose the idea of uPAR and ITGβ1 as bio-

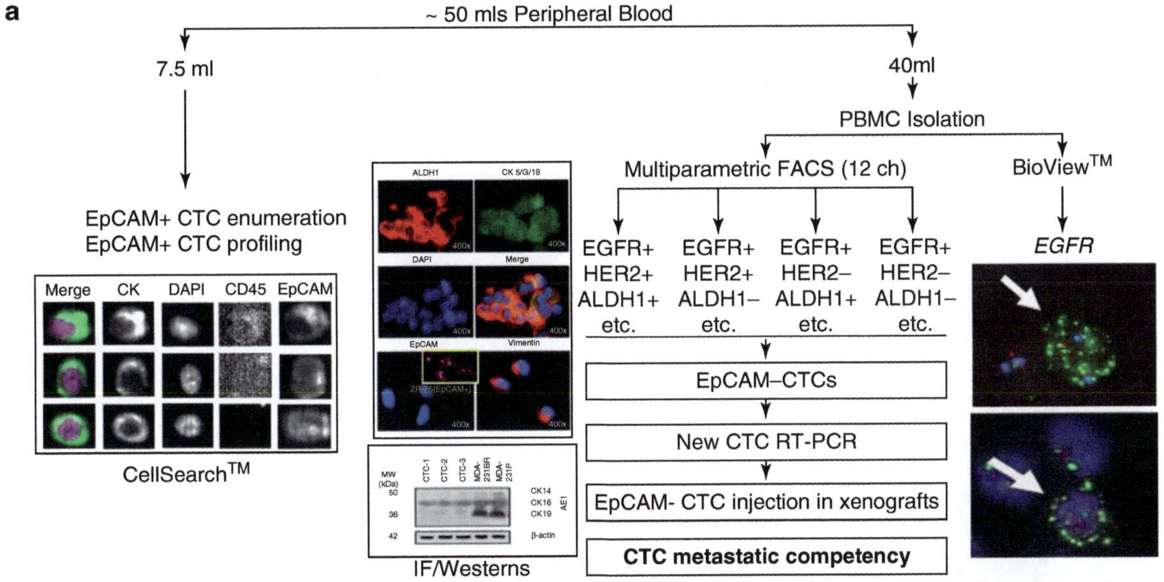

Fig. 16.1 **The isolation of EpCAM-negative CTCs from blood of metastatic breast cancer patients using multiparametric flow cytometry.** (**a**) Panel shows the isolation of peripheral blood mononuclear cells (PBMCs) from patients' blood samples which was followed by multiparametric flow sorting using the negative selection of EpCAM-expressing PBMCs, in synergy with the positive selection of biomarkers (HER2, EGFR), known to relate either with BCBM, or to breast cancer cell stemness (ALDH1), and combinatorial. This strategy resulted in the isolation of EpCAM-negative CTCs which were subsequently shown to be of neoplastic origin by genetic analyses, and for survival/growth abilities as in vitro cultures. CTC cultures (CTC-1, CTC-2, CTC-3) were then interrogated applying a newly developed CTC-PCR assay (Panel **b**). Several positive/negative controls were performed to demonstrate EpCAM-negative CTC selection. They consisted to either use parallel CellSearch profiling to capture EpCAM-positive CTCs within the same blood sample, the visualization of BCBM CTC biomarkers by immunofluorescence microscopy (IF)/Western blotting, or the interrogation of *EGFR* overexpression by BioView™ platform. EpCAM-negative CTC cultures expressing BCBM biomarkers were then injected in immunodeficient mice to evaluate their metastatic competence, notably to brain [16]. (**b**) (Top panel) RT-PCR analyses of isolated EpCAM-negative CTC cultures (CTC-1, CTC-2, CTC-3). Selected genes were classified into four groups: HPSE/Notch1/EGFR/HER2 gene expression profile (termed the BMSM CTC signature), biomarkers for epithelial mesenchymal transition (EMT), cytokeratins (CK), and breast cancer stem cell biomarkers ($CD44^{high}/CD24^{low}$, and CD44 variant v8). PBMCs isolated from patients with (patient) or without (control) breast cancer were used as negative controls for the BMSM CTC signature. In addition, human brain metastatic [MDA-MB-231BRHER2 (brain tropic clone derived from MDA-MB231 parental and HER2 transduced], poorly brain metastatic (MDA-MB-231 parental), and non-metastatic (MCF-7) breast cancer cell lines were used as BMSM biomarkers controls. Glyceraldehyde-3-phosphate dehydrogenase was the loading control. (Bottom panel) Relative gene expression analyses of the BMSM CTC profile in EpCAM-negative CTCs. MDA-MB-231BR, MDA-MB-231P, and SKBR3 cell lines were used as BMSM CTC profile controls. (**c**) (Top left panels) Representative images of BCBM (insets) induced by EpCAM-negative BMSM+ CTCs when injected intracardially into mice. BCBM sections underwent analyses for tumor burden by the CRi Vectra Intelligent system, which highlights and enumerates single tumor cells (in green). (Bottom left panels) Representative immunostaining for BMSM biomarker EGFR in CTC-driven BCBM by IHC of brain macro/micrometastasis. Specific staining for metastatic lesions vs. adjacent brain tissue (negative control) is shown. The same results were found for the other biomarkers of the BMSM CTC profile [16]. Quantification of IHC staining intensity for the BMSM CTC profile in BCBM compared to lung metastasis within the same CTC-injected animal. Data are means ± SEM ($n = 8$ sections per mouse; 10 mice per CTC culture)

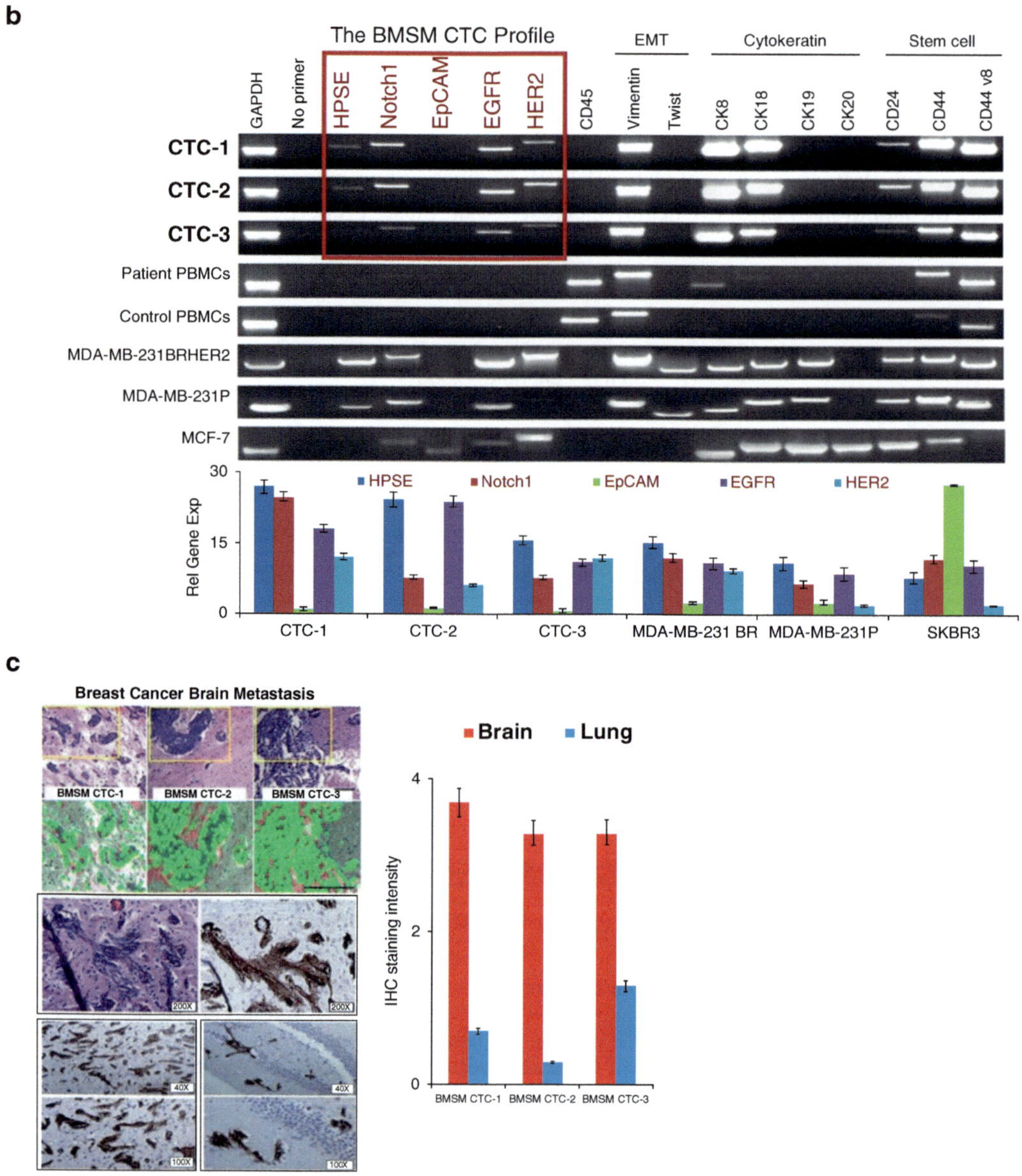

Fig. 16.1 (continued)

markers of the CTC "dormancy axis." Cell adhesion, proliferation, and invasion properties of EpCAM− CTC subsets were found to be relevant to BCBM onset. Of note, because heterogeneous populations of CTCs harbor genetic and epigenetic changes at single-cell level and exhibit distinct breast cancer phenotypes, we used the DEPArray™ single-cell isolation platform to isolate individual CTCs derived from BCBM vs. No BCBM, along with applying MassARRAY™ mutation analyses. We detected the common COSMIC mutation PRKCB G785T in patient-derived CTCs, regardless of the expression markers analyzed (uPAR/ITGβ1/HER2, etc.) or BCBM status [20]. Accordingly, while the variability of genetic mutations at the single CTC level confirmed the high heterogeneity of CTCs, it can provide a better approach to evaluate the biology of CTCs by targeting these mutations and assessing their impact. These findings are significant to clinical dormancy, since CTCs exhibit various phenotypes via mechanisms of expansion which are still unclear. These phenotypes are dependent on the presence of specific biomarkers, such as the expression of uPAR and ITGβ1 axis, in combination with other markers. Multiple CTC phenotypes must exist: they resist apoptosis, undergo evolution and clonal selection via DNA damage and active DNA repair pathways, and avoid arrest and adhesion to target organs with extreme efficiency. Specific CTC clones within the uPAR/ITGβ1 dormancy axis may therefore undergo proliferation and expansion for a long-term niche pool, while CTC clustering can occur leading to secondary tumorsphere formation. Accordingly, the expansion of investigations establishing consistent in vitro CTC cultures for effective drug screening/testing rather than in vivo transplantation will be a clinically useful tool in applications of personalized medicine.

16.4 The Identification of Biomarker Properties and the Proliferative State of CTCs Associated with BCBM

Because of findings which have established the relevance of EpCAM− CTCs to BCBM onset, we expanded upon the characterization of de-differentiated, EpCAM-agnostic CTCs, including both epithelial and stem-like breast cancer CTC subsets. We devised a multiparametric workflow made of three sequential steps—(i) doublet discrimination and dead cell elimination, (ii) depletion of hematopoietic-lineage positive cells normally present in the peripheral blood, and (iii) positive selection of PanCK+ (epithelial) or CD44+/CD24− (stem-like) CTCs. Immunocytochemistry confirmed the selected CTC subsets from this multiparametric flow cytometry approach expressed appropriate biomarkers used for selection. Isolated single CTCs were further interrogated using the DEPArray platform, evaluating the expression of PanCK and CD44, with an absence of CD45 and CD24. The identity of isolated CTCs was validated by genomic and mRNA analyses of these cells [22, 23]. By employing these approaches, we isolated between 101 and 839 CTCs/8 mL of patient blood, whereas CellSearch identified between 0 and 88 CTCs/7.5 mL blood when processed in parallel, implying an ability to capture a larger CTC pool that would have otherwise remained undetected. Of note, transcriptional profiling on CTCs from BC patients with or without BCBM, as confirmed by MRI, was compared to curated data, using the same platform and pathway analysis, from primary patient samples of various BC subtypes. We were able to make three remarkable observations from these analyses.

First, 29,758 genes were downregulated in CTCs compared to 1972 genes upregulated (fold-change <-2 or >2, ANOVA p-value <0.05), indicative of generalized low CTC transcriptional activity, suggesting that most CTCs are quiescent or undergoing apoptosis. Interestingly, none of the CTCs clustered with their corresponding primary breast cancer (pBC) molecular subtype [24, 25]. Conversely, CTCs clustered together as a single group, suggesting they resemble each other more closely than their pBC counterparts [26, 27]. The CTC gene signature was uniformly increased across pBC subtypes. These findings were reflected in cellular functions as CTC upregulated genes were involved in cell death, apoptosis, and cell survival with a concomitant decrease in cellular proliferation and invasive properties, thus substantiating the concept that most CTCs exist in a state of reduced metabolic and/or mitotic activity, enabling them to overcome stress and survive in circulation [22, 23].

Second, considering that Ki67 is routinely used for assessing the proliferative index of cancer tissues, and is the single most important prognostic factor for breast cancer brain metastasis [28], immunostaining of Ki67 in isolated CTCs revealed distinct, nuclear-localized puncta, in both PanCK+ and CD44+/CD24− CTC subtypes. This staining was quantified at the single-cell level using the DEPArray platform. Next, because of our investigations into the dormancy axis using uPAR/ITGβ1 in EpCAM-negative CTCs [20, 21], we evaluated their cell surface expression. Despite the significant concordance between Ki67 and uPAR/ITGβ1 expression in CTCs, there was a low level of discordance between Ki67 and uPAR/ITGβ1 expression in some individual CTCs. Considering that both immunostaining and the DEPArray platform are only semi-quantitative, we designed a single-step flow cytometry method to enumerate Ki67+ and uPAR+/ITGβ1+ CTCs in patient blood. Data analyses demonstrated that BCBM patients had twice as many $Ki67^{High}$ CTCs (ratio of $Ki67^{High}$:$Ki67^{Low}$ ~ 2:1), whereas patients without BCBM had 60% greater $Ki67^{Low}$ CTCs (ratio of $Ki67^{High}$:$Ki67^{Low}$ ~ 1:1.6) [22, 23]. We also compared the ratio of uPAR-/ITGβ1− vs. uPAR+/ITGβ1+ CTCs in these patient samples and found a 2.4-fold increase of uPAR+/ITGβ1+ CTCs in BCBM patients. Consistent with our previous CTC analyses, we found that the Ki67 status of 20–30% CTCs do not correlate with their uPAR/ITGβ1expression and vice versa, emphasizing the complementary roles of these biomarkers defining distinct CTC subsets. Collectively, differences in biomarker expression of these CTC subsets indicate that at least a portion of CTCs in BCBM patient blood were unique from the remaining population in regard to their biology and behavior [22, 23].

Third, we postulated that gene expression profile differences between these two groups would arise specifically from brain metastatic CTCs. Using the presence of MRI identified brain metastases as the clinical discriminator of the study, CTC gene signatures from BCBM vs. No BCBM patients clustered according to respective clinical groups. We discovered a unique 126-gene signature which was significantly altered between these two groups, which we designated the BCBM CTC signature (Fig. 16.2). This signature not only contained the four-gene BMSM CTC profile previously identified [16], but also that individual Ki67+ CTCs

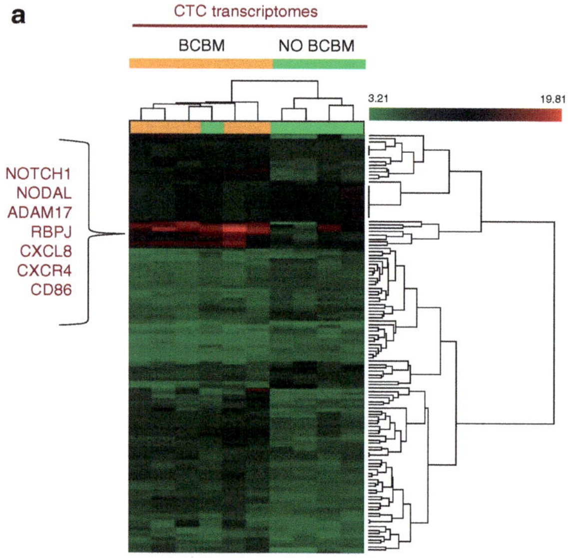

b

Ingenuity Canonical Pathways	-log (p-value)	Ratio	Molecules	Functions Annotation	*p*-Value	Activation *z*-score	Molecules
Communication between Innate and Adaptive Immune Cells	3.09E00	3.37E-02	CXCL8, CD86, IGHA1	cell movement of mononuclear leukocytes	2.09E-04	2.244	ADAM17,AIF1,CD86, CXCL8,CXCR4,IFNGR1, MAP3K2
Notch Signaling	2.55E00	5.26E-02	ADAM17, RBPJ	chemotaxis of leukocytes	2.14E-03	2.129	ADAM17,AIF1, ARHGEF7,CXCL, CXCR4
Hematopoiesis from Pluripotent Stem Cells	2.37E00	4.26E-02	CXCL8, IGHA1	migration of cells	7.69E-04	2.1	ADAM17,AIF1,AREG, ARHGEF,CD86,CXCL8, CXCR4,DDIT4,IFN GR1,MAP3K2, MARCKS,NAMPT, SLC3A2,TNFAIP3
Primary Immunodeficiency Signaling	2.35E00	4.17E-02	IGHA1, TAP2	cellular homeostasis	4.12E-04	2.069	ADAM17,ARHGEF7, CD86,CXCL8,CXCR4, DDIT4,GABARAPL1,IF NGR1,mir-506,NAMPT, RBPJ,SLC3A2,TNFAIP3
Nur77 Signaling in T Lymphocytes	2.21E00	3.51E-02	CD86, MAP3K2	Lymphocyte migration	2.50E-04	2.018	ADAM17,CD86,CXCL8, CXCR4,IFNGR1,MAP3K2

Analyses performed with IPA software (Qiagen) using genes with <>2 fold -change.

Fig. 16.2 Difference in transcriptomic signatures of CTCs isolated from blood of breast cancer patients diagnosed with or without brain metastasis (BCBM/No BCBM). (**a**) Heat map showing CTC transcriptomes in BCBM vs. No BCBM and the discovery of a 126-gene CTC signature found to be upregulated between BCBM and No BCBM, using the Affymetrix HTA2.0 gene panel. Clustering was performed using genes with fold change <−2 or >2, ANOVA *p*-value <0.05. Of note, highest gene expression was detected not only for members of the BMSM profile but also for the Notch pathway/cell stemness biomarkers, along with chemokines involved in communications with immune system cells (displayed in red). (**b**) Table shows activated canonical pathways and cellular functions in BCBM CTCs. Integrated pathway analyses indicate highest activation of pathways involved in communications with cells of the immune system: Notch signaling and pluripotent stem cells [20, 22, 37]

recapitulate this 126-gene signature. Pathway enrichment analyses revealed higher activation of known CTC pathways, e.g., Notch, along with the discovery of novel hematopoietic and immune evasion signaling pathways in CTCs derived from blood of BCBM patients. This was further validated by performing Notch1 immunocytochemistry on patient-derived CTCs isolated by DEPArray. A total of 22 CTCs from 3 patients diagnosed with BCBM and 17 CTCs from 3 patients with No BCBM were analyzed. In the BCBM group, ~72% (16/22) CTCs were positive for Notch staining, whereas 24% (4/17) CTCs were positive for Notch1 staining in the No BCBM group.

These results suggest that upregulation of Notch activity is a feature of BCBM CTCs rather than CD44+/CD24− CTCs. Importantly, cellular functional annotations associated with distant metastases such as cell migration and chemotaxis, which were downregulated in the 10-patient CTC-pBC cohort, were significantly activated in the BCBM vs. No BCBM CTC cohorts; along with an increase of pro-inflammatory chemokines (TNF, IL1β, and NF-κB), immunomodulatory networks (CXCL8, CXCR4, CD86), and mitogenic growth factors (PDGF-BB). Taken together, these findings suggest that growth-arrested CTCs represent a substantial portion of the CTC population, accounting for the anti-proliferative signature of CTC cohorts. However, further segregation of the parent cohort into BCBM vs. No BCBM highlighted the gene signature of transcriptionally and mitotically active CTCs associated with the BCBM phenotype. To this end, important inroads have been recently made of a comprehensive analysis for distinct blood circulatory cell populations from patients diagnosed with metastatic/brain-metastatic BC. This was performed at the single-cell level (10x Genomics Chromium), not only identifying distinct immune cell clusters but also a cluster containing circulatory neoplastic cells, that is, CTCs [29]. These advances extend the definition of what a CTC is, and suggest that the neoplastic cell capacity is distributed across heterogeneous cell profiles in circulation. They will be highly instrumental to replace rigid definitions (eg, presence/absence of EpCAM, etc.) with more solid notions on how a CTC functions in their metastatic/brain-metastatic potency [29].

16.5 The Characterization of ctDNA in Brain Metastasis

Blood plasma cell-free DNA (cfDNA) was first reported in 1948, and the correlation between increased cfDNA and cancer was discovered in 1977 [11, 12]. Circulating tumor DNA (ctDNA), a subset of cfDNA, is derived from tumors as active secretion in extracellular vesicles, leakage after tumor necrosis or apoptosis, lysis of CTC during micrometastasis, and macrophage digestion of tumor DNA fragments and concentration varies due to a number of factors including stage of disease progression, tumor location, and vascularization [12, 30]. While it is known that higher levels of circulating tumor DNA (ctDNA) at diagnosis predict poorer patient outcome [16, 25, 30, 31], there has been significant advancement in the use of ctDNA as prognostic or monitoring tools in several cancer types, including breast, lung, and cancers of the central nervous system (CNS) [32, 33].

The use of plasma ctDNA presents advantages, as well as limitations, as a diagnostic tool. First, plasma can be isolated from the same sample as CTCs, presenting low additional risk to patients, and allowing for a genomic comparison of cfDNA, CTC DNA, and specific ctDNA [33, 34]. These comparisons allow for longitudinal tracking of disease progression and genetic changes at metastatic sites. A second aspect of ctDNA that must be considered is stability and turnover. Estimates of ctDNA clearance from the body range from 1 to 13 h as detected by next-generation sequencing (NGS) [30]. Although ctDNA is rapidly cleared, the high levels of cell death and turnover in tumors contribute a steady supply of ctDNA. Patients with various localized cancer types present with ctDNA approximately 55% of the time [34]. In patients with localized breast, colon, pancreas, or gastroesophageal cancers, detectable ctDNA was present in 49–78% of patients, compared to 86–100% of patients with known metastatic disease. However, in patients with primary brain lesions, detection of plasma ctDNA fell to approximately 20%. In the case of primary brain lesions, plasma ctDNA was only detectable if there were visceral metastases [31]. In patients with localized brain lesions only, ctDNA was only detectable in cerebrospinal fluid (CSF). In individuals with primary or metastatic lesions within the CNS, detection of ctDNA within the CSF is more sensitive and precise than that of plasma [31, 34, 35]. These studies suggest that liquid biopsy of CSF is better suited for the detection and monitoring of primary and metastatic lesions within the CNS.

Quantification of ctDNA from plasma correlates with patient survival [32, 34, 35]. In a study of early-stage BC patients, detection of ctDNA was positive and prognostic in a multivariate model, with ~80% of patients with detectable ctDNA developing progressive disease, and levels of ctDNA correlating with subtype. For example, TNBC patients had highest levels of ctDNA (4.96 copies/mL), HER2+ had intermediate levels (0.81 copies/mL), and ER+/HER2− had lowest (0 copies/mL) [35]. In a study of non-small cell lung cancer, patients had lower ctDNA levels at first follow-up and best radiological response, than at baseline, while patients with stable or progressive disease showed no change in ctDNA levels, indicating ctDNA levels change in response to treatment [32]. Recently, quantification of ctDNA isolated from CSF has been shown to be similarly prognostic of patient outcome [1, 10, 23, 31]. Additionally, quantification of CSF ctDNA is more sensitive and accurate than plasma ctDNA in detection of brain lesions [1, 15, 23, 31, 34]. The use of next-generation sequencing (NGS) detected ctDNA in all patient CSF samples and allowed for clustering of molecular subtypes which correlated with overall survival [10]. In another study, looking at paired detection of CTCs by cytology and ctDNA in patients with primary or secondary brain lesions, ctDNA was present in all 58 patient samples, while CTCs were only present in 76% [23]. The use of NGS to detect mutant allelic frequency (MAF) had two important findings. First, 68 of the 143 cancer associated genes had mutations in at least one BM CSF sample, but all patient samples carried at least one mutation. Second, CSF ctDNA highly correlated with number of CTCs, indicating that CSF ctDNA provides a better marker for disease progression than CTCs. Melanoma patients with progressive BM disease possessed BRAF and MRAS mutations in CSF and 30% of the patients interpreted negative by cytology were positive through ctDNA analysis [15, 34]. Similarly, in patients with BC or melanoma, quantification of CTCs or ctDNA was similarly associated with secondary LM disease and a ctDNA concentration of 0.022 ng/mL was correlated with an increased risk of death [1]. In a recent study, researchers at Toronto Hospital's Neuroscience Center generated cell-free methylated DNA immunoprecipitation (cfMeDIP-seq), a high-throughput sequencing profile of plasma samples from patients presenting with a variety of brain cancers, including BM [36]. They demonstrated not only that this platform can distinguish brain tumors from extracranial cancer types and healthy controls, but also that it can discriminate between different tumors of the CNS that often look undistinguishable by brain scan.

While multiple different techniques have been employed to detect ctDNA, they have all been shown to be similar in sensitivity [34]. However, these techniques are not without drawbacks, as most are highly specific and look at individual cancer-related mutations [10, 15, 35], or the need to have a biopsy or CTCs presence to confirm the source of ctDNA [15, 34]. There are advantages in ctDNA testing. For example, detection of CTCs from brain metastasis can fail [13–15, 34], likely due to segregation within the CSF. Conversely, ctDNA is detectable in plasma at low levels, even in the absence of CTCs. Further, in the presence of brain lesions, CSF ctDNA concentrations are much higher than in plasma [15, 31, 34]. Taken together, these studies provide evidence of clinical validity for the use of ctDNA in progressive disease towards or within the brain. However, similar to the use of CTCs, they have yet to demonstrate clinical utility towards BM detection, monitoring, or therapies.

16.6 Conclusion

Although the application of novel strategies for cancer diagnosis, treatment, and prevention has resulted in many successes, converting some neoplastic conditions from incurable to curable, many challenges still remain. The concept of liquid biopsy to confront the problem of early detection of cancer and its prevention was born on this premise. Liquid

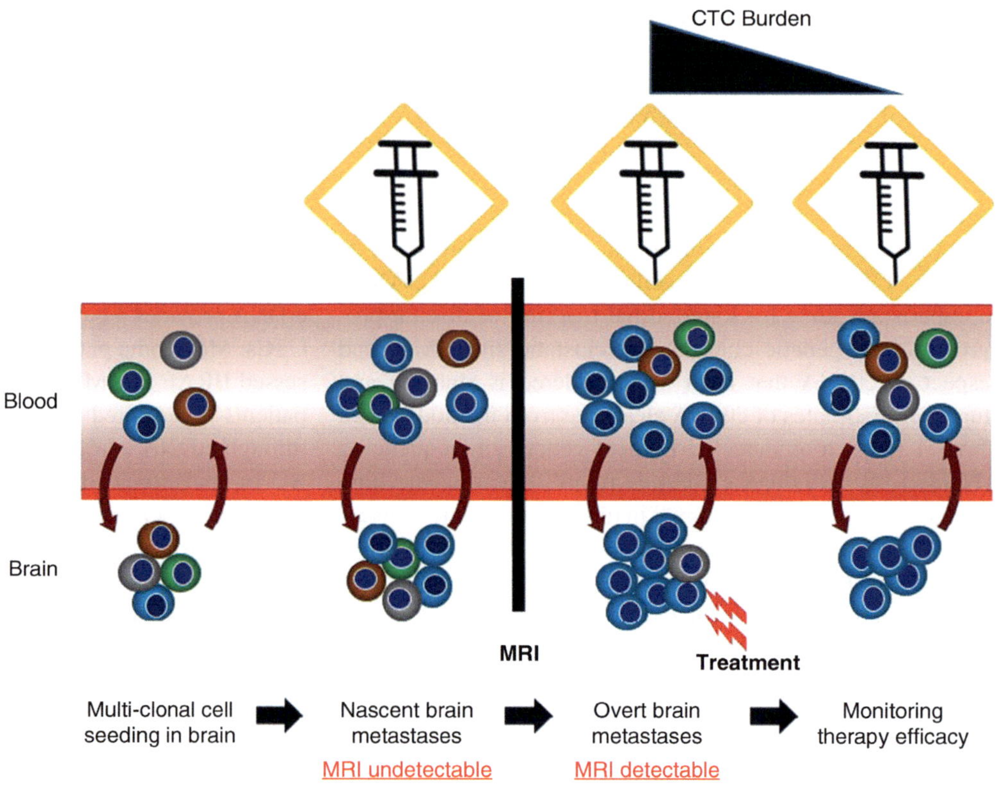

Fig. 16.3 **Translational perspectives for future applications of novel CTC tests to predict or to monitor/prevent BCBM.** The figure shows the clinical implications of BCBM-associated CTC biomarkers Ki67+/uPAR+/ITGβ1+ and/or Notch1+, and their enumeration in relation to BCBM. Enumeration of BCBM CTCs can be used as a sensitive screening method for micrometastasis detection in patients at risk of developing BCBM or who have BCBM prior to detection with MRI. Conversely, serial analyses of BCBM-associated CTC biomarkers can be used as a clinically useful tool to evaluate responses to therapy for patients with overt BCBM disease

biopsy is particularly appealing and has remarkable potential, considering the many platforms and technologies that have evolved allowing its implementation [8, 11, 12, 29]. Complementing our investigations on BCBM-associated CTCs and signaling mechanisms potentiating CTC characteristics which allow colonization of the brain, the studies presented here have important clinical implications, which can be further enhanced and/or complemented by ctDNA analyses relating to BM onset. Building upon these studies, we present two important directions for liquid biopsy: (1) the application of novel liquid biopsy tests in the clinic as a sensitive screening method for detection of micrometastatic brain disease which is undetectable by MRI, and (2) the use of the newly identified CTC/ctDNA signatures as tools for measuring responses to therapy for patients with MRI-detectable BCBM (Fig. 16.3). As outcomes of our CTC studies and after further validation of the BCBM CTC signature, CTC tests could be implemented in the clinic to correlate CTCs/biomarkers of the CTC signature with the detection of brain micrometastasis, thus alerting the oncologist of the potential for BCBM development at the time of pBC treatment or early stages of BCBM. Conversely, in clinical scenarios where BCBM is already detected, a reduction of Ki67+/uPAR+/ITGβ1+ and/or Notch1+ CTC numbers by longitudinal monitoring can correlate with a successful therapy (Fig. 16.3). The latter can be significant in relation to predicting or preventing further secondary metastasis to brain or other organ sites ("metastasis of metastasis"). In conclusion, we foresee that the extension of CTC/ctDNA analyses beyond BCBM and their application will equip clinicians with critical information on the "BM state" in the patient, allowing BM detection in its nascent stages. The information from liquid biopsy tests will aid in formulating rational therapies targeted specifically for brain metastasis and evaluate their efficacy for the benefit of the patient.

Open Questions

- How can CTC/ctDNA testing be improved and complemented for the prediction or prevention of BM?
- Can CTC/ctDNA testing be associated directly with responses to treatment modalities such as focal or whole-brain radiation therapy?
- Is there a common BM CTC/ctDNA profile across all cancer types which increase the risk of BM, brain relapse or minimal residual disease?
- How can CTC/ctDNA testing better define interplays between BM-associated dormancy vs. metastatic competence?
- Are CTC/ctDNA profiles promoting BM consistent across all cancer types? Are there differences in patient populations that may be exploited or limit viable liquid biopsy tests from being developed?

References

1. Nevel KS, DiStefano N, Lin X, Skakodub A, Ogilvie SQ, Reiner AS, Pentsova E, Boire A. A retrospective, quantitative assessment of disease burden in patients with leptomeningeal metastases from non-small-cell lung cancer. Neuro-Oncology. 2019;22:675–83. https://doi.org/10.1093/neuonc/noz208.
2. Simonelli M, Dipasquale A, Orzan F, Lorenzi E, Persico P, Navarria P, Pessina F, Nibali MC, Bello L, Santoro A, et al. Cerebrospinal fluid tumor DNA for liquid biopsy in glioma patients' management: close to the clinic? Crit Rev Oncol Hematol. 2020;146:102.879. https://doi.org/10.1016/j.critrevonc.2020.102879.
3. Baumgartner JM, Riviere P, Lanman RB, Kelly KJ, Veerapong J, Lowy AM, Kurzrock R. Prognostic utility of pre- and postoperative circulating tumor DNA liquid biopsies in patients with peritoneal metastases. Ann Surg Oncol. 2020;27:3259–67. https://doi.org/10.1245/s10434-020-08331-x.
4. Dagogo-Jack I, Carter SL, Brastianos PK. Brain metastasis: clinical implications of branched evolution. Trends Cancer. 2016;2:332–7. https://doi.org/10.1016/j.trecan.2016.06.005.
5. Klein-Scory S, Wahner I, Maslova M, Al-Sewaidi Y, Pohl M, Mika T, Ladigan S, Schroers R, Baraniskin A. Evolution of RAS mutational status in liquid biopsies during first-line chemotherapy for metastatic colorectal cancer. Front Oncol. 2020;10:1115.
6. Ribas A, Hamid O, Daud A, Hodi FS, Wolchok JD, Kefford R, Joshua AM, Patnaik A, Hwu W-J, Weber JS, et al. Association of pembrolizumab with tumor response and survival among patients with advanced melanoma. JAMA. 2016;315:1600–9. https://doi.org/10.1001/jama.2016.4059.
7. Vad-Nielsen J, Meldgaard P, Sorensen BS, Nielsen AL. Cell-free chromatin immunoprecipitation (cfChIP) from blood plasma can determine gene-expression in tumors from non-small-cell lung cancer patients. Lung Cancer. 2020;147:244–51. https://doi.org/10.1016/j.lungcan.2020.07.023.
8. Pantel K, Alix-Panabières C. Liquid biopsy and minimal residual disease—latest advances and implications for cure. Nat Rev Clin Oncol. 2019;16:409–24. https://doi.org/10.1038/s41571-019-0187-3.
9. Alix-Panabieres C, Pantel K. Clinical applications of circulating tumor cells and circulating tumor DNA as liquid biopsy. Cancer Discov. 2016;6:479–91. https://doi.org/10.1158/2159-8290.CD-15-1483.
10. Li Y-S, Zheng M-M, Jiang B-Y, Tu H-Y, Yang J-J, Zhang X-C, Wu Y-L. Association of cerebrospinal fluid tumor DNA genotyping with survival among patients with lung adenocarcinoma and central nervous system metastases. JAMA Netw Open. 2020;3:–e209077. https://doi.org/10.1001/jamanetworkopen.2020.9077.
11. Heitzer E, Haque IS, Roberts CES, Speicher MR. Current and future perspectives of liquid biopsies in genomics-driven oncology. Nat Rev Genet. 2019;20:71–88. https://doi.org/10.1038/s41576-018-0071-5.
12. Diaz LA Jr, Bardelli A. Liquid biopsies: genotyping circulating tumor DNA. J Clin Oncol. 2014;32:579–86. https://doi.org/10.1200/JCO.2012.45.2011.
13. Mego M, De Giorgi U, Dawood S, Wang X, Valero V, Andreopoulou E, Handy B, Ueno NT, Reuben JM, Cristofanilli M. Characterization of metastatic breast cancer patients with nondetectable circulating tumor cells. Int J Cancer. 2011;129:417–23. https://doi.org/10.1002/ijc.25690.
14. Wehrle J, Philipp U, Jolic M, Follo M, Hussung S, Waldeck S, Deuter M, Rassner M, Braune J, Rawluk J, et al. Personalized treatment selection and disease monitoring using circulating tumor DNA profiling in real-world cancer patient management. Diagnostics (Basel). 2020;10 https://doi.org/10.3390/diagnostics10080550.
15. Ballester LY, Glitza Oliva IC, Douse DY, Chen MM, Lan C, Haydu LE, Huse JT, Roy-Chowdhuri S, Luthra R, Wistuba II, et al. Evaluating circulating tumor DNA from the cerebrospinal fluid of patients with melanoma and leptomeningeal disease. J Neuropathol Exp Neurol. 2018;77:628–35. https://doi.org/10.1093/jnen/nly046.
16. Zhang L, Ridgway LD, Wetzel MD, Ngo J, Yin W, Kumar D, Goodman JC, Groves MD, Marchetti D. The identification and characterization of breast cancer CTCs competent for brain metastasis. Sci Transl Med. 2013;5:180ra148. https://doi.org/10.1126/scitranslmed.3005109.
17. Li J, Jiang W, Wei J, Zhang J, Cai L, Luo M, Wang Z, Sun W, Wang S, Wang C, et al. Patient specific circulating tumor DNA fingerprints to monitor treatment response across multiple tumors. J Transl Med. 2020;18:293. https://doi.org/10.1186/s12967-020-02449-y.
18. Aguirre-Ghiso JA. Models, mechanisms and clinical evidence for cancer dormancy. Nat Rev Cancer. 2007;7:834–46. https://doi.org/10.1038/nrc2256.
19. Alcaide M, Cheung M, Hillman J, Rassekh SR, Deyell RJ, Batist G, Karsan A, Wyatt AW, Johnson N, Scott DW, et al. Evaluating the quantity, quality and size distribution of cell-free DNA by multiplex droplet digital PCR. Sci Rep. 2020;10:12,564. https://doi.org/10.1038/s41598-020-69432-x.
20. Vishnoi M, Peddibhotla S, Yin W, Scamardo AT, George GC, Hong DS, Marchetti D. The isolation and characterization of CTC subsets related to breast cancer dormancy. Sci Rep. 2015;5:17,533. https://doi.org/10.1038/srep17533.
21. Lee S-Y, Chae D-K, Lee S-H, Lim Y, An J, Chae CH, Kim BC, Bhak J, Bolser D, Cho D-H. Efficient mutation screening for cervical cancers from circulating tumor DNA in blood. BMC Cancer. 2020;20:694. https://doi.org/10.1186/s12885-020-07161-0.
22. Boral D, Vishnoi M, Liu HN, Yin W, Sprouse ML, Scamardo A, Hong DS, Tan TZ, Thiery JP, Chang JC, et al. Molecular characterization of breast cancer CTCs associated with brain metastasis. Nat Commun. 2017;8:196. https://doi.org/10.1038/s41467-017-00196-1.
23. Zhao Y, He JY, Cui JZ, Meng Z-Q, Zou YL, Guo XS, Chen X, Wang X, Yan L-T, Han WX, et al. Detection of genes mutations in cerebrospinal fluid circulating tumor DNA from neoplastic meningitis patients using next generation sequencing. BMC Cancer. 2020;20:690. https://doi.org/10.1186/s12885-020-07172-x.
24. Sørlie T, Perou CM, Tibshirani R, Aas T, Geisler S, Johnsen H, Hastie T, Eisen MB, van de Rijn M, Jeffrey SS, et al. Gene expression patterns of breast carcinomas distinguish tumor subclasses with clinical implications. Proc Natl Acad Sci. 2001;98:10,869. https://doi.org/10.1073/pnas.191367098.
25. Tang K, Gardner S, Snuderl M. The role of liquid biopsies in pediatric brain tumors. J Neuropathol Exp Neurol. 2020;79:934–40. https://doi.org/10.1093/jnen/nlaa068.
26. Duan H, Hu J-L, Chen Z-H, Li J-H, He Z-Q, Wang Z-N, Zhang G-H, Guo X-Y, Liang L, Mou Y-G. Assessment of circulating tumor DNA in cerebrospinal fluid by whole exome sequencing to detect genomic alterations of glioblastoma. Chin Med J. 2020;133
27. Steinert G, Schölch S, Niemietz T, Iwata N, García SA, Behrens B, Voigt A, Kloor M, Benner A, Bork U, et al. Immune escape and survival mechanisms in circulating tumor cells of colorectal cancer. Cancer Res. 2014;74:1694. https://doi.org/10.1158/0008-5472.CAN-13-1885.
28. Sartori D, Bari M, Pappagallo GL, Rosetti F, Olsen S, Vinante O. Brain metastases in breast cancer: different survival by biological subtype and Ki67 expression. J Clin Oncol. 2009;27:1069. https://doi.org/10.1200/jco.2009.27.15_suppl.1069.
29. Pauken CM, Kenney SR, Brayer KJ, Guo Y, Brown-Glaberman UA, Marchetti D. Heterogeneity of circulating tumor cells neoplastic subpopulations outlined by single-cell transcriptomics. Cancers. 2021;4885:1–21. https://doi.org/10.3390/cancers13194885.
30. Alimirzaie S, Bagherzadeh M, Akbari MR. Liquid biopsy in breast cancer: a comprehensive review. Clin Genet. 2019;95:643–60. https://doi.org/10.1111/cge.13514.

31. De Mattos-Arruda L, Mayor R, Ng CKY, Weigelt B, Martinez-Ricarte F, Torrejon D, Oliveira M, Arias A, Raventos C, Tang J, et al. Cerebrospinal fluid-derived circulating tumour DNA better represents the genomic alterations of brain tumours than plasma. Nat Commun. 2015;6:8839. https://doi.org/10.1038/ncomms9839.
32. Hyun MH, Lee ES, Eo JS, Kim S, Kang EJ, Sung JS, Choi YJ, Park KH, Shin SW, Lee SY, et al. Clinical implications of circulating cell-free DNA quantification and metabolic tumor burden in advanced non-small cell lung cancer. Lung Cancer. 2019;134:158–66. https://doi.org/10.1016/j.lungcan.2019.06.014.
33. Lin C-C, Huang W-L, Wei F, Su W-C, Wong DT. Emerging platforms using liquid biopsy to detect EGFR mutations in lung cancer. Expert Rev Mol Diagn. 2015;15:1427–40. https://doi.org/10.1586/14737159.2015.1094379.
34. Bettegowda C, Sausen M, Leary RJ, Kinde I, Wang Y, Agrawal N, Bartlett BR, Wang H, Luber B, Alani RM, et al. Detection of circulating tumor DNA in early- and late-stage human malignancies. Sci Transl Med. 2014;6:224ra224. https://doi.org/10.1126/scitranslmed.3007094.
35. Garcia-Murillas I, Chopra N, Comino-Méndez I, Beaney M, Tovey H, Cutts RJ, Swift C, Kriplani D, Afentakis M, Hrebien S, et al. Assessment of molecular relapse detection in early-stage breast cancer. JAMA Oncol. 2019;5:1473–8. https://doi.org/10.1001/jamaoncol.2019.1838.
36. Nassiri F, Chakravarthy A, Feng S, Shen SY, Nejad R, Zuccato JA, Voisin MR, Patil V, Horbinski C, Aldape K, et al. Detection and discrimination of intracranial tumors using plasma cell-free DNA methylomes. Nat Med. 2020;26:1044–7. https://doi.org/10.1038/s41591-020-0932-2.
37. Sprouse LM, Welte T, Boral D, Liu NH, Yin W, Vishnoi M, Goswami-Sewell D, Li L, Pei G, Jia P, et al. PMN-MDSCs enhance CTC metastatic properties through reciprocal interactions via ROS/notch/nodal signaling. Int J Mol Sci. 2019;20 https://doi.org/10.3390/ijms20081916.

Epigenetics and Liquid Biopsy in Oncology: Role in Metastasis and Clinical Utility

17

Aitor Rodriguez-Casanova, Aida Bao-Caamano, Nicolás Costa-Fraga, Laura Muinelo-Romay, and Angel Diaz-Lagares

Abstract

The deregulation of epigenetic mechanisms in tumor cells is implicated in multiple steps of cancer development and progression. These epigenetic alterations are tightly associated to the metastatic process, which is one of the hallmarks of cancer and the main cause of cancer-related deaths. By means of lymphatic system or blood vessels, tumors can release different types of tumor material to circulation, including circulating tumor cells (CTCs), nucleic acids (cNAs), and extracellular vesicles (EVs). These circulating tumor elements can present tumor-specific epigenetic alterations with relevant biological information about tumor evolution and dissemination, and with clinical utility as tumor biomarkers. In fact, the analysis of these epigenetic biomarkers by liquid biopsies has shown promising results in different types of tumors for the personalized management of cancer patients. Here, we provide an overview of the main epigenetic alterations in cancer, and describe the implications of these epigenetic marks in cancer dissemination and metastasis, highlighting their clinical utility as circulating tumor biomarkers.

Aitor Rodriguez-Casanova, Aida Bao-Caamano, and Nicolás Costa-Fraga share first authorship.

Learning Objectives

- Describe the concept of epigenetics, the different epigenetic mechanisms, the cross talk between them, and the interest of epigenetic therapy.
- Explain how epigenetic mechanisms are deregulated in cancer and metastasis.
- Define the concept of liquid biopsy and describe the different types of tumor material that can circulate in body fluids.
- Describe the presence of epigenetic alterations in liquid biopsy and their relation to cancer dissemination and metastasis.
- Summarize the role of the epigenetic alterations as tumor biomarkers and their clinical implications in cancer.

17.1 Introduction

Cancer dissemination and metastasis of tumors remain the leading cause of cancer-related deaths [1]. The metastatic process of primary tumors is based on the acquisition of a long series of sequential and interrelated steps, including

A. Rodriguez-Casanova
Cancer Epigenomics, Translational Medical Oncology Group (Oncomet), Health Research Institute of Santiago (IDIS), University Clinical Hospital of Santiago (CHUS/SERGAS), Santiago de Compostela, Spain

Roche-Chus Joint Unit, Translational Medical Oncology Group (Oncomet), Health Research Institute of Santiago (IDIS), Santiago de Compostela, Spain

A. Bao-Caamano · N. Costa-Fraga
Cancer Epigenomics, Translational Medical Oncology Group (Oncomet), Health Research Institute of Santiago (IDIS), University Clinical Hospital of Santiago (CHUS/SERGAS), Santiago de Compostela, Spain

L. Muinelo-Romay
Liquid Biopsy Analysis Unit, Translational Medical Oncology Group (Oncomet), Health Research Institute of Santiago (IDIS), University Clinical Hospital of Santiago (CHUS/SERGAS), Santiago de Compostela, Spain

Centro de Investigación Biomédica en Red Cáncer (CIBERONC), Madrid, Spain

A. Diaz-Lagares (✉)
Cancer Epigenomics, Translational Medical Oncology Group (Oncomet), Health Research Institute of Santiago (IDIS), University Clinical Hospital of Santiago (CHUS/SERGAS), Santiago de Compostela, Spain

Centro de Investigación Biomédica en Red Cáncer (CIBERONC), Madrid, Spain
e-mail: angel.diaz.lagares@sergas.es

S. P. Leong et al. (eds.), *Cancer Metastasis Through the Lymphovascular System*,
https://doi.org/10.1007/978-3-030-93084-4_17

tumor vascularization (angiogenesis), local invasion of the host stroma, the entry in the systemic circulation usually by means of the lymphovascular system, extravasation to the surrounding tissue, and proliferation within the organ parenchyma [2]. Other relevant mechanisms have been identified in relation to the development of metastasis. Thus, tumor cells are able to escape the action of the immune system creating an immunosuppressive microenvironment that facilitates tumor growth and dissemination [3].

All these metastatic steps are tightly regulated by molecular changes, such as deregulation of the epigenetic mechanisms: DNA methylation, posttranslational modifications of histones, and the expression of noncoding RNAs (ncRNAs). These epigenetic marks are implicated in very relevant metastatic features, such as epithelial-mesenchymal transition (EMT), which represents a reversible cellular transition process that allows epithelial cells to adopt characteristics of mesenchymal cells and the properties of tumor-initiating cells (TICs). Of note, this EMT process facilitates the migration and invasion of tumor cells, and colonization at distant sites. In addition, the alteration of epigenetic mechanisms has a relevant role in cancer dissemination from the site of primary tumor growth to distant tissues via the vessels and lymph nodes (LNs) of the lymphatic system. LNs have a fundamental role in metastasis, acting as a "bridgehead" where subpopulations of tumor cells are selected according to their properties to colonize distant sites [4, 5]. In fact, in most of the tumors, the disease progression occurs first through the lymphatic system. However, this is not a general rule, since tumor cells can enter directly the blood bypassing the LNs. Both dissemination paths can lead to venous circulation, as lymphatics drain into blood and lymphatic fluid is filtered by LNs [6]. Therefore, in blood we can find tumor material representative of both dissemination mechanisms, including circulating tumor cells (CTCs), circulating tumor DNA (ctDNA), and other circulating elements, such as tumor-derived extracellular vesicles (EVs) [7] (Fig. 17.1).

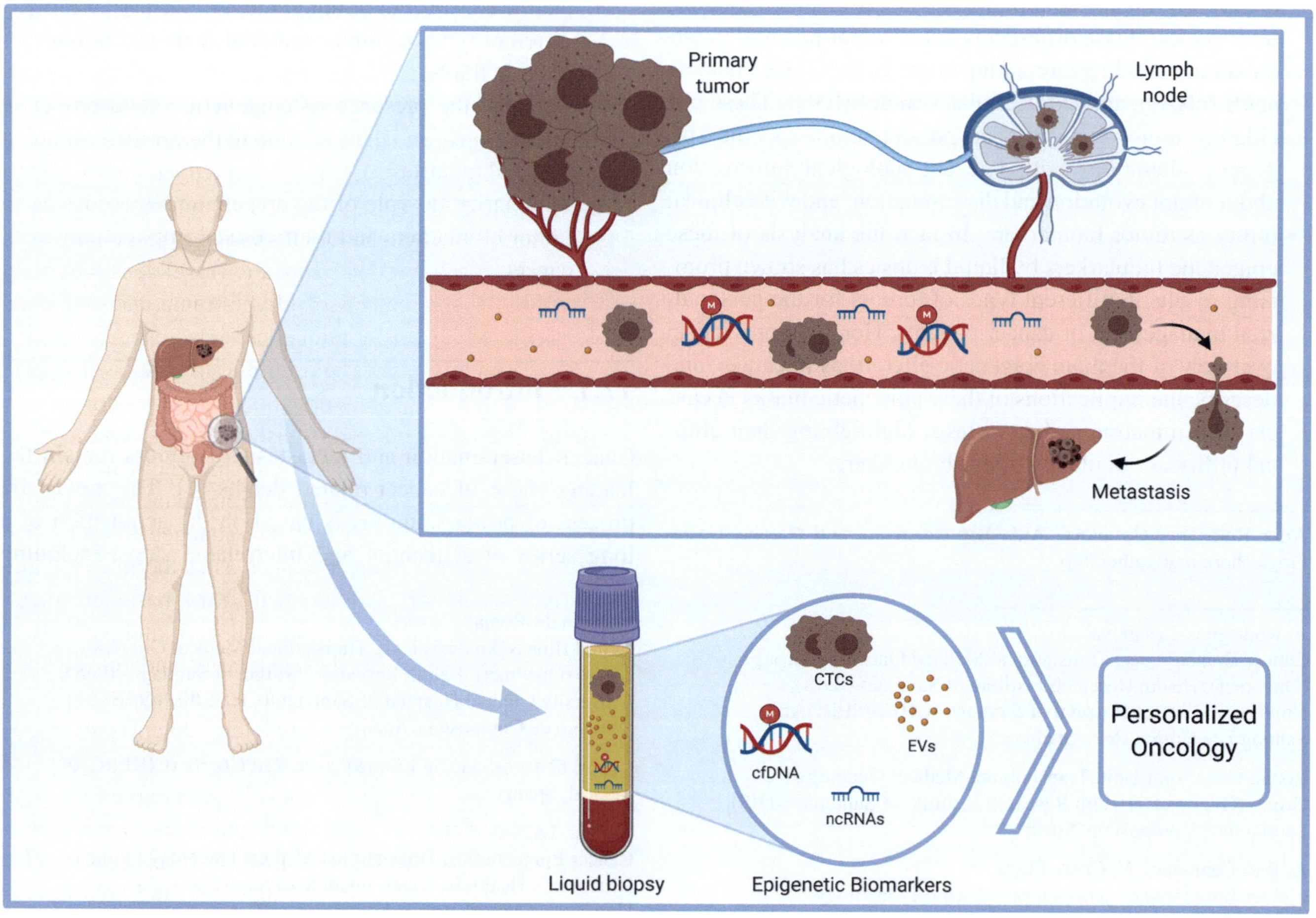

Fig. 17.1 Epigenetic alterations in liquid biopsy and clinical utility in cancer. Primary tumors and metastasis can release different types of tumor material, including circulating tumor cells (CTCs), nucleic acids (cNAs), and extracellular vesicles (EVs), that can reach the circulation via lymphatic system or blood vessels. These circulating tumor elements are able to present tumor-specific epigenetic alterations, showing clinical utility as dynamic tumor biomarkers in different clinical contexts: early detection, prognosis, monitoring, therapy selection, and evaluation of therapy response. The analysis of these circulating epigenetic biomarkers represents a promising tool for personalized oncology. Created with BioRender.com

In this chapter, we provide an overview of the main epigenetic alterations that can be found in cancer, describing their implications in the dissemination of tumor cells and metastasis, and highlighting the clinical utility of these epigenetic modifications as tumor biomarkers for cancer patients in the era of personalized medicine.

17.2 Epigenetic Mechanisms

The term epigenetics refers to the study of heritable modifications in the activity and expression of genes that do not involve alterations of the DNA sequence [8]. The epigenetic machinery includes several mechanisms (Fig. 17.2), such as DNA methylation, histone modifications, and ncRNAs. These mechanisms regulate the normal gene expression in cells and their deregulation has important relevance in cancer development and progression [9].

DNA methylation, which is the best characterized epigenetic mechanism, consists of the incorporation of a methyl group (CH_3) to the 5′ carbon of cytosines in cytosine-phosphate-guanine (CpG) dinucleotides to produce 5-methylcytosine (5mC). This methylation process is mediated by the enzymes DNA methyltransferases (DNMTs) and usually occurs in CpG islands (CGIs), which are genomic regions with a high density of CpGs. These CGIs are often located in the promoter regions of genes but they can also be present in intragenic and intergenic regions. Importantly, DNA methylation can be reversed through the use of epigenetic-based drugs (epidrugs), such as DNMT inhibitors (DNMTi) [9].

Different types of reversible posttranslational modifications (PTM) have been described in histone tails, mainly acetylation and methylation, which are related to gene activation or silencing [10]. These PTMs play an important role in both normal development and pathogenesis. The combina-

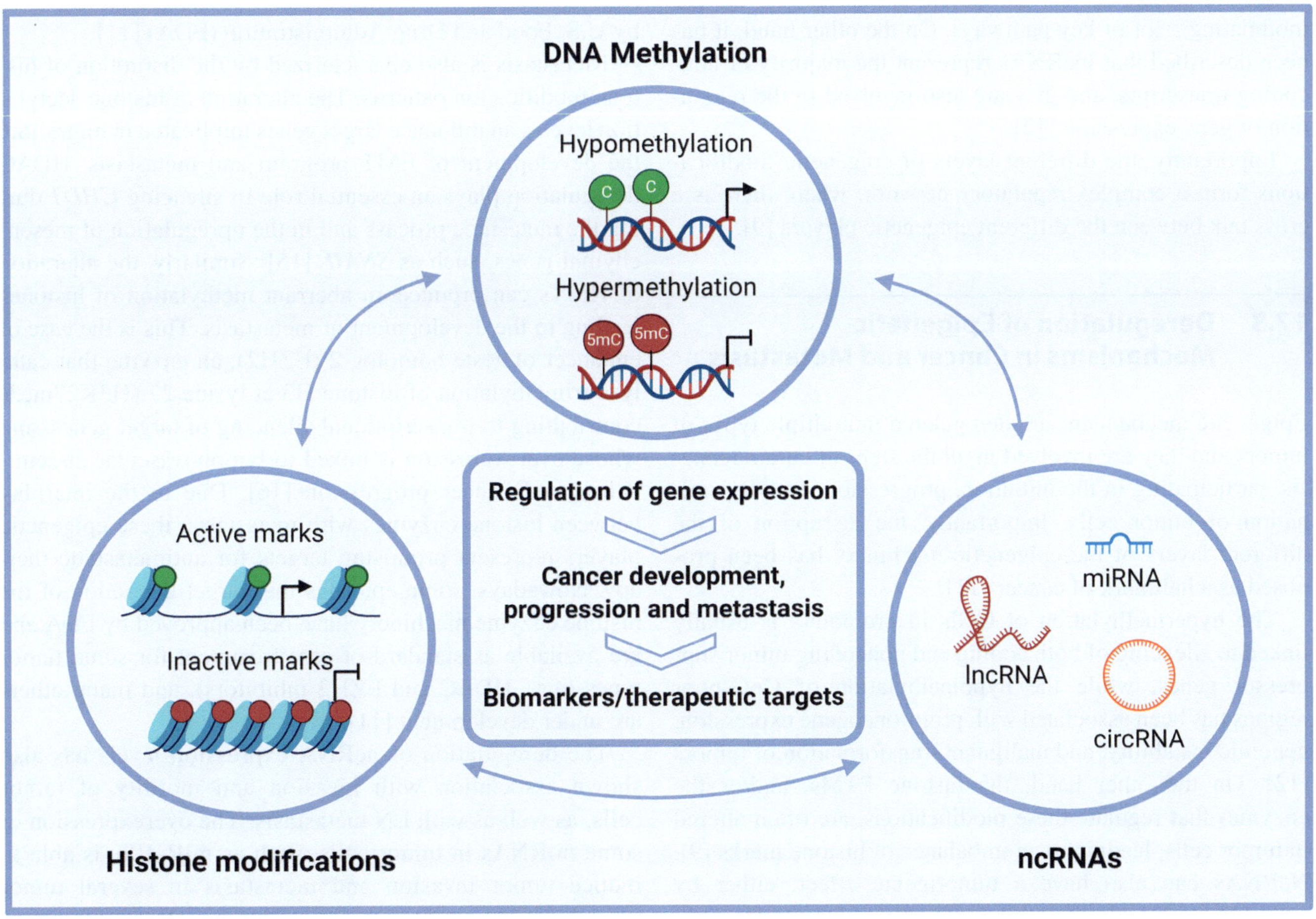

Fig. 17.2 **Epigenetic mechanisms.** There are several interconnected epigenetic layers that normally regulate the gene expression of cells: DNA methylation, posttranslational modifications of histones, and noncoding RNAs. These epigenetic mechanisms can undergo alterations inducing cancer development, progression, and metastasis. This deregulation has shown utility in the clinic as epigenetic biomarkers and therapeutic targets in multiple types of tumors. *5mC* 5-Methylcytosine, *C* Cytosine, *lncRNA* Long noncoding RNA, *miRNA* MicroRNA, *circRNA* Circular RNA. Created with BioRender.com

tion of histone PTMs ("histone code") is regulated by different types of enzymes. In the case of acetylation and methylation, the balance of these histone marks depends on histone acetyltransferases (HATs), histone deacetylases (HDACs), histone methyltransferases (HMTs), and histone demethylases (HDMs). In addition, these marks can be reversed by using epidrugs, including HDAC and HMT inhibitors (HDACi and HMTi) [9, 11].

NcRNAs are also involved in the epigenetic regulation of gene expression. These noncoding transcripts are classified according to their nucleotide length into: (i) small ncRNAs (sncRNAs), with less than 200 nt, which include microRNAs (miRNA), small interfering RNAs (siRNA), and piwi-interacting RNAs (piRNA); and (ii) long ncRNAs (lncRNA), with more than 200 nt, which encompass long intergenic ncRNA (lincRNA) and long intronic ncRNA (intronic lncRNA). MiRNAs are the most widely described ncRNAs, and exert their function by binding to mRNA, blocking its transcription or triggering mRNA degradation. Thus, each miRNA can regulate the expression of hundreds of genes, modulating a lot of key pathways. On the other hand, it has been described that lncRNAs represent the majority of noncoding transcripts, and they are also involved in the regulation of gene expression [12].

Importantly, the different layers of epigenetic modifications form a complex regulatory network, where there is a cross talk between the different epigenetic players [9].

17.3 Deregulation of Epigenetic Mechanisms in Cancer and Metastasis

Epigenetic mechanisms are deregulated in multiple types of tumors and they are involved in all the steps of carcinogenesis, participating in the initiation, progression, and dissemination of tumor cells. Importantly, the disruption of the different layers of the epigenetic machinery has been proposed as a hallmark of cancer [11].

The hypermethylation of CGIs in promoters is usually linked to silencing of both coding and noncoding tumor suppressor genes, while the hypomethylation of CpG-poor regions has been associated with proto-oncogene expression, genomic instability, and malignant transformation of tumors [12]. On the other hand, the histone PTMs, and/or the enzymes that regulate these modifications, are often altered in tumor cells, leading to an imbalance of histone marks [9]. NcRNAs can also have a tumorigenic effect, either by repressing or promoting tumor properties. The normal cross talk among the different epigenetic layers is also disrupted in cancer, where, for example, the alteration of specific methylation and histone modification patterns can modify the expression of different types of ncRNAs [9, 12].

Importantly, several epigenetic events have been described in relation to the metastatic process. Thus, promoter hypermethylation of *CDH1* is able to silence the expression of this gene, leading to the activation of the EMT program and the promotion of invasiveness [13]. Another member of the cadherin family, cell-adherence gene Cadherin-11 *(CDH11),* was found to be unmethylated in primary tumors but hypermethylated and silenced in metastatic tumor cells from LN, suggesting that the epigenetic modulation of this gene could disrupt cell-to-cell contacts, allowing tumor cells intravasation into lymphatic vessels [4]. On the other hand, the methylation status of some genomic regions, such as long interspersed nuclear elements (LINE-1), has been also associated with dissemination of tumor cells. The hypermethylation of LINE-1 contributes to genomic stability, and its hypomethylation is associated with tumor invasion and the presence of positive LN [14]. Importantly, some agents can induce hypomethylation by reversing the tumorigenic effect of hypermethylation. DNMTis, such as Decitabine and Azacitidine, have been described as epidrugs and approved by U.S. Food and Drug Administration (FDA) [11].

Metastasis is also characterized by the disruption of histone modification patterns. The alteration of histone acetylation levels can influence target genes implicated in migration, the development of EMT program and metastasis. HDAC deregulation plays an essential role in silencing *CHD1* during the metastatic process and in the upregulation of mesenchymal genes, such as *SNAIL* [15]. Similarly, the alteration of HMTs can produce an aberrant methylation of histones leading to the development of metastasis. This is the case of enhancer of zeste homolog 2 (EZH2), an enzyme that catalyze trimethylation of histone H3 at lysine 27 (H3K27me3) contributing to transcriptional silencing of target genes, and whose overexpression is linked to lymphovascular dissemination and cancer progression [16]. Due to the interplay between histone enzymes with metastasis, these epigenetic players represent promising targets for antimetastatic therapy. Nowadays, some epidrugs that target the action of the histone enzyme machinery have been approved by FDA and are available as standard-of-care treatment for some tumor types (e.g., HDAC and EZH2 inhibitors), and many others are under development [11].

The deregulation of ncRNA expression levels has also shown association with invasion and motility of tumor cells, as well as with LN metastasis. The overexpression of some miRNAs in tumor cells, such as miR-10b, is able to induce tumor invasion and metastasis in several tumor types [17]. Importantly, the prometastatic effect of this type of ncRNAs can be reversed using targeted therapies, preventing, for example, the metastatic spread from the primary tumor to the LNs and arresting the metastatic growth in the already formed LN metastasis [18]. In addition to miRNAs, the deregulation in the expression levels of other ncRNAs, such as lncRNAs (e.g., HOTAIR), has shown to be implicated in tumor dissemination and LN metastasis of cancer patients [19].

17.4 Liquid Biopsy

All the body fluids that contain tumor material suitable for molecular characterization are encompassed under the term liquid biopsy. Although blood is the most used sample, other fluids, such as urine, pleural effusion, ascitic fluid, cerebrospinal fluid, saliva, and stool are also of great value as liquid biopsies. This tumor material present in body fluids can be released from primary tumors or metastasis and it mainly consists of circulating tumor cells (CTCs) and circulating nucleic acids (cNAs). Both elements constitute a valuable source of information about the molecular mechanisms underlying tumor evolution and dissemination and also of noninvasive biomarkers [20].

The use of cNAs has started to be applied for the therapy selection. These analyses are mainly centered on circulating cell-free DNA (cfDNA), which is released as a consequence of cell death and is present highly fragmentated in body fluids (size, 160–180 bp). This cfDNA generally contains, in a range as low as 0.01–1%, circulating tumor DNA (ctDNA) in patients with advanced tumors [20]. This cfDNA can be analyzed in a noninvasive and comprehensive way to study different genetic alterations such as point mutations, copy number variations, small indels, and translocations, together with epigenetic modifications [20]. In addition, numerous noncoding RNAs (ncRNAs) can be detected in liquid biopsy, and have been described as regulators of specific gene expression signatures in tumor cells and stroma, controlling cell-to-cell communication [21].

On the contrary of cNAs, the study of CTCs has shown limited translation into the routine clinical practice so far. These cells are present in very low numbers in blood and require the implementation of highly sensitive and specific strategies for their enrichment and detection. Although the most common strategy to isolate CTCs was immune enrichment, this strategy has not been widely adopted as a clinical tool, mainly due to the lack of clear benefits for treatment decision-making. Other antigen-independent technologies for the isolation and characterization of CTCs have been developed to isolate a more heterogeneous CTC population [22]. Despite the limitation for their isolation, molecular characterization of CTCs has provided valuable information to understand how tumor cells disseminate and implant at distant locations. Their molecular study, including epigenetic mechanisms, is important to unravel the biological aspects of cancer and identify new clinical and therapeutic biomarkers to manage this disease [23].

17.5 Epigenetic Alterations in Liquid Biopsy and Their Association with Metastasis

During the last decade, different studies have demonstrated the alteration of epigenetic mechanisms in tumor circulating elements, mainly in CTCs and ctDNA, linked to the several steps of tumor generation and dissemination (Fig. 17.1) [24]. Pioneering works characterized the methylation status of tumor-associated genes in CTCs from breast cancer with the goal of identifying new mechanisms that favor this tumor spread. From these studies we learned that *CST6*, *BRMS1*, and *SOX17* promoters were hypermethylated in CTCs, as a mechanism for promoting the tumor dissemination [25]. More recently, a specific methylation program which regulates stemness-associated transcription factors (*OCT4*, *NANOG*, *SOX2*, and *SIN3A*) has been identified in CTCs-clusters and linked to the metastasis formation in breast cancer [26]. In addition to the stem properties, EMT is another important mechanism favoring CTCs dissemination. In fact, in colorectal cancer (CRC) patients, hypermethylation of the *SFRP2* promoter and *VIM* has been described and associated with a more mesenchymal phenotype of this cells and the promotion of metastasis [27]. The epigenetic regulation of EMT in CTCs from prostate cancer has also been described and associated with the blood-borne dissemination [28]. Besides, in patients with advanced prostate cancer CTCs showed hypermethylation patterns in genes with relevant roles for angiogenesis and apoptosis [29].

Different methylation patterns have also been identified in cfDNA from cancer patients and associated with the dissemination process. Thus, hypomethylation of *LINE-1* in plasma cfDNA of CRC patients was correlated with the disease progression, especially, with the presence of LN affectation and the generation of distant metastasis [30]. Contrary, the hypermethylation of other gene promoters such as *RUNX3, GATA5, TFPI2, SFRP2, SHOX2,* and *SEPT9* in cfDNA of CRC patients has been implicated in the lymphatic invasion and tumor recurrence [31–34]. Similarly, in plasma from breast cancer patients the methylation of *SOX17* has been described associated with a more aggressive disease and clearly linked to the presence of lymphovascular dissemination [35]. High levels of methylation of different genes such as *PCDH8* were also associated with high preoperative prostate-specific antigen (PSA) levels and LN metastasis in patients with advanced prostate cancer [36].

In addition to DNA methylation, other epigenetic mechanisms, such as histone modifications, have been described in cfDNA and associated with the process of carcinogenesis. For example, it has been described that H3K27me3 alterations detected in plasma represent an epigenetic mark in prostate carcinogenesis, being associated with the presence of disseminated disease after surgery [37].

On the other hand, different miRNA signatures have shown an impact on cancer progression. The presence of miRNAs related to migration and invasion have been found altered in CTCs in breast tumors [38]. Besides, different miRNAs contained in circulating exosomes of CRC patients, such as miR-25-3p, are increased in patients with metastatic disease [39]. This miRNA regulates the expression of target genes, promoting vascular permeability, angiogenesis, and the formation of liver and lung metastasis in preclinical models [39]. Furthermore, recent results have indicated that decreased levels of serum exosomal miR-638 [40] and miR-548c-5p [41] are associated with liver metastasis in CRC patients. Besides, several circulating miRNAs have shown association with cancer-related thrombosis, which promotes the generation of metastasis [42].

Other circulating ncRNAs have been implicated in the promotion of cancer spread. For example, the lncRNA MALAT1 is well known as an enhancer of the expansion and migration of tumor cells. This lncRNA has been found upregulated in serum and plasma of epithelial ovarian cancer, clearly associated with the presence of lymphovascular affectation [43].

17.6 Clinical Implications of Circulating Epigenetic Biomarkers in Liquid Biopsy

Alterations of epigenetic patterns in liquid biopsies have shown great utility as tumor biomarkers for early detection, prognosis, monitoring, and evaluation of therapeutic response at different stages of cancer disease (Fig. 17.1). In the context of the early detection and screening of CRC, several tests focused on the analysis of methylation signatures have been developed. Epi proColon, based on the detection of methylation in the promoter region of the *SEPT9* gene in cfDNA, was the first test commercialized. The combined analysis of the methylation status of *SEPT9* and *SDC2* led to the development of a new test (ColoDefense), which improved the detection of CRC and adenomas, and represent a valuable tool for CRC screening and early detection [44].

The analysis of epigenetic signatures in cfDNA and CTCs from breast cancer patients has also provided potential clinical biomarkers. Thus, specific methylation profiles of cfDNA have been described as interesting tools for the early diagnosis and the prediction of the therapy response [12, 45]. The hypermethylation of *BRMS1*, *SOX17*, and *CST6* in CTCs has been associated with advanced breast cancer stages and proposed as prognosis biomarkers [12]. Besides, the methylation of *ITIH5* or *RASSF1* in CTCs was also related to poor survival rates [12]. Hypermethylation of other genes such as *ESR1* have shown an impact on the response to targeted therapy in hormone sensitive tumors [46]. In addition, the methylation of several gene panels, such as being *APC*, *RARβ2*, and *RASSF1A*, has demonstrated high accuracy to detect breast cancer [45]. Besides, the analysis of *RASSF1A* and *PITX2* methylation in cfDNA have shown value to predict the overall survival in advanced breast tumors [45].

In bronchial aspirates, pleural effusions, and plasma from patients with lung cancer, *SHOX2* methylation was found clearly increased and associated with lymphovascular affectation, supporting the development of a diagnostic test (Epi proLung BL Reflex Assay) [44]. Beyond the diagnosis, the promoter methylation of *DCLK1*, *SHP1P2,* or *SOX17* has been described as a factor associated with lower survival rates in lung tumors [45]. Besides, in prostate cancer the methylation of *GSTP*1 in cfDNA was described as a specific mark and showed clinical value to predict PSA progression and overall survival in patients with castration resistant tumors [45].

Notably, recently genome-wide methylation approaches on cfDNA have shown great value to detect the presence of different tumor types, alone or in combination with the detection of genomic alterations and specific epigenetic patterns. Moreover, the combination of NGS with machine learning allowed the methylation analysis of cfDNA and the development of a test capable of detecting more than 50 tumor types [47, 48].

Besides, histone modifications have also been proposed as circulating epigenetic biomarkers. Thus, for example, low levels of the circulating histone marks H4K20me3, and H3K27me3, have shown value for CRC detection [49].

NcRNAs are also of great interest as cancer biomarkers. Several circulating miRNA signatures have been proposed as biomarkers for early detection of colon, breast, and lung cancer, among others [21]. Thus, circulating miR-21, miR-155, miR-125b, or miR-19a had shown to be promising diagnostic biomarkers for breast cancer detection. In lung cancer, the most consistent miRNAs biomarkers for the disease diagnosis and prognosis are miR-125b and miR-21. In CRC, the analysis of miR-21 levels has been also proposed for early detection and as a prognostic factor. In addition, increased levels of miR-375 and miR-141 in plasma have been correlated with the presence of metastatic disease in prostate cancer. Finally, several circulating lncRNAs such as HOTAIR, MALAT1, or PCAT18 have an increasing interest as diagnostic and prognostic biomarkers of cancer [21].

Conclusions and Future Perspectives

The epigenetic characterization of circulating tumor material has generated a lot of interest in recent years due to its relevant implication in cancer development and dissemination. The study of epigenetics in liquid biopsy represents an emerging field with great potential to elucidate the main biological and clinical aspects associated with cancer metastasis. Thus, the alteration of epigenetic mechanisms has shown great relevance in the main steps of the metastatic process, such as migration through the EMT process and the lymphovascular-blood dissemination, among others. Epigenetic marks, including DNA methylation, histone modifications, and ncRNAs, can be detected in liquid biopsy and have shown very promising results as tumor biomarkers for early detection, prognosis, monitoring, and evaluation of therapeutic response at different time points of cancer disease. Importantly, epigenetic alterations can be reversed by using epidrugs, some of them approved by FDA, and many others under development.

Overall, the study of epigenetics in liquid biopsy represents an emerging field with enormous perspectives for the future. The characterization of epigenetic marks in circulating tumor material with "omic" and very sensitive technologies will provide a great view of the epigenetic landscape to discover new biomarkers and therapeutic targets, in a noninvasive and dynamic way. Besides, exploring new epigenetic modifications in liquid biopsy, such as DNA hydroxymethylation and different types of ncRNAs, and the analysis of RNA modifications (epitranscriptomic marks), will offer valuable information about cancer biology. In a clinical context, the use of epigenetic modifications as noninvasive tumor biomarkers and therapeutic targets will provide the opportunity to personalize the management of cancer patients. However, some relevant questions must be addressed in the close future to improve the clinical use of circulating epigenetic biomarkers:

- How specific are the epigenetic marks identified in liquid biopsies for being used as diagnostic tools?
- Are all the circulating tumor elements equally useful to characterize the epigenetic marks?
- Can we use the analysis of epigenetic biomarkers to guide the potential treatment of tumors with epidrugs?
- Will we be able to translate the growing knowledge about the epigenetic mechanisms that promote the tumor dissemination into therapeutic targets?
- How are the different epigenetic mechanisms interconnected to promote tumor dissemination?

Acknowledgments This work was co-funded by the ISCIII (PI18/00307) and the European Regional Development Fund (FEDER). AD-L is funded by a contract "Juan Rodés" (JR17/00016) and AB-C by a predoctoral contract PFIS (FI19/00240) co-funded by "Fondo Social Europeo" (FSE) from ISCIII. AR-C is supported by Roche-Chus Joint Unit (IN853B 2018/03) and GAIN. NC-F is funded by a predoctoral contract from "Xunta de Galicia" (IN606A-2020/004). LM-R is supported by AECC.

References

1. Gupta GP, Massague J. Cancer metastasis: building a framework. Cell. 2006;127(4):679–95.
2. Fidler IJ. The pathogenesis of cancer metastasis: the 'seed and soil' hypothesis revisited. Nat Rev Cancer. 2003;3(6):453–8.
3. Soundararajan R, Fradette JJ, Konen JM, et al. Targeting the interplay between epithelial-to-mesenchymal-transition and the immune system for effective immunotherapy. Cancers. 2019;11(5)
4. Carmona FJ, Villanueva A, Vidal A, et al. Epigenetic disruption of cadherin-11 in human cancer metastasis. J Pathol. 2012;228(2):230–40.
5. Dongre A, Weinberg RA. New insights into the mechanisms of epithelial-mesenchymal transition and implications for cancer. Nat Rev Mol Cell Biol. 2019;20(2):69–84.
6. Wong SY, Hynes RO. Lymphatic or hematogenous dissemination: how does a metastatic tumor cell decide? Cell Cycle. 2006;5(8):812–7.
7. Lianidou E, Pantel K. Liquid biopsies. Genes Chromosomes Cancer. 2019;58(4):219–32.
8. Berger SL, Kouzarides T, Shiekhattar R, et al. An operational definition of epigenetics. Genes Dev. 2009;23(7):781–3.
9. Mari-Alexandre J, Diaz-Lagares A, Villalba M, et al. Translating cancer epigenomics into the clinic: focus on lung cancer. Transl Res. 2017;189:76–92.
10. Kouzarides T. Chromatin modifications and their function. Cell. 2007;128(4):693–705.
11. Bates SE. Epigenetic therapies for cancer. N Engl J Med. 2020;383(7):650–63.
12. Bao-Caamano A, Rodriguez-Casanova A, Diaz-Lagares A. Epigenetics of circulating tumor cells in breast cancer. Adv Exp Med Biol. 2020;1220:117–34.
13. Lombaerts M, van Wezel T, Philippo K, et al. E-cadherin transcriptional downregulation by promoter methylation but not mutation is related to epithelial-to-mesenchymal transition in breast cancer cell lines. Br J Cancer. 2006;94(5):661–71.
14. Sunami E, de Maat M, Vu A, et al. LINE-1 hypomethylation during primary colon cancer progression. PLoS One. 2011;6(4):e18884.
15. von Burstin J, Eser S, Paul MC, et al. E-cadherin regulates metastasis of pancreatic cancer in vivo and is suppressed by a SNAIL/HDAC1/HDAC2 repressor complex. Gastroenterology. 2009;137(1):361–71, 371 e361-365.
16. Nienstedt JC, Schroeder C, Clauditz T, et al. EZH2 overexpression in head and neck cancer is related to lymph node metastasis. J Oral Pathol Med. 2018;47(3):240–5.
17. Ma L, Teruya-Feldstein J, Weinberg RA. Tumour invasion and metastasis initiated by microRNA-10b in breast cancer. Nature. 2007;449(7163):682–8.
18. Yigit MV, Ghosh SK, Kumar M, et al. Context-dependent differences in miR-10b breast oncogenesis can be targeted for the prevention and arrest of lymph node metastasis. Oncogene. 2013;32(12):1530–8.
19. Cai B, Wu Z, Liao K, et al. Long noncoding RNA HOTAIR can serve as a common molecular marker for lymph node metastasis: a meta-analysis. Tumour Biol. 2014;35(9):8445–50.

20. Siravegna G, Marsoni S, Siena S, et al. Integrating liquid biopsies into the management of cancer. Nat Rev Clin Oncol. 2017;14(9):531–48.
21. Anfossi S, Babayan A, Pantel K, et al. Clinical utility of circulating non-coding RNAs—an update. Nat Rev Clin Oncol. 2018;15(9):541–63.
22. Barbazan J, Alonso-Alconada L, Elkhatib N, et al. Liver metastasis is facilitated by the adherence of circulating tumor cells to vascular fibronectin deposits. Cancer Res. 2017;77(13):3431–41.
23. Cortes-Hernandez LE, Eslami SZ, Pantel K, et al. Molecular and functional characterization of circulating tumor cells: from discovery to clinical application. Clin Chem. 2020;66(1):97–104.
24. Thomas ML, Marcato P. Epigenetic modifications as biomarkers of tumor development, therapy response, and recurrence across the cancer care continuum. Cancers. 2018;10(4)
25. Chimonidou M, Strati A, Tzitzira A, et al. DNA methylation of tumor suppressor and metastasis suppressor genes in circulating tumor cells. Clin Chem. 2011;57(8):1169–77.
26. Gkountela S, Castro-Giner F, Szczerba BM, et al. Circulating tumor cell clustering shapes DNA methylation to enable metastasis seeding. Cell. 2019;176(1–2):98–112 e114.
27. Lyberopoulou A, Galanopoulos M, Aravantinos G, et al. Identification of methylation profiles of cancer-related genes in circulating tumor cells population. Anticancer Res. 2017;37(3):1105–12.
28. Pixberg CF, Raba K, Muller F, et al. Analysis of DNA methylation in single circulating tumor cells. Oncogene. 2017;36(23):3223–31.
29. Friedlander TW, Ngo VT, Dong H, et al. Detection and characterization of invasive circulating tumor cells derived from men with metastatic castration-resistant prostate cancer. Int J Cancer. 2014;134(10):2284–93.
30. Jung G, Hernandez-Illan E, Moreira L, et al. Epigenetics of colorectal cancer: biomarker and therapeutic potential. Nat Rev Gastroenterol Hepatol. 2020;17(2):111–30.
31. Nishio M, Sakakura C, Nagata T, et al. RUNX3 promoter methylation in colorectal cancer: its relationship with microsatellite instability and its suitability as a novel serum tumor marker. Anticancer Res. 2010;30(7):2673–82.
32. Tang D, Liu J, Wang DR, et al. Diagnostic and prognostic value of the methylation status of secreted frizzled-related protein 2 in colorectal cancer. Clin Invest Med. 2011;34(2):E88–95.
33. Zhang X, Song YF, Lu HN, et al. Combined detection of plasma GATA5 and SFRP2 methylation is a valid noninvasive biomarker for colorectal cancer and adenomas. World J Gastroenterol. 2015;21(9):2629–37.
34. Bergheim J, Semaan A, Gevensleben H, et al. Potential of quantitative SEPT9 and SHOX2 methylation in plasmatic circulating cell-free DNA as auxiliary staging parameter in colorectal cancer: a prospective observational cohort study. Br J Cancer. 2018;118(9):1217–28.
35. Panagopoulou M, Karaglani M, Balgkouranidou I, et al. Circulating cell-free DNA in breast cancer: size profiling, levels, and methylation patterns lead to prognostic and predictive classifiers. Oncogene. 2019;38(18):3387–401.
36. Lin YL, Li YL, Ma JG. Aberrant promoter methylation of Protocadherin8 (PCDH8) in serum is a potential prognostic marker for low Gleason score prostate cancer. Med Sci Monitor. 2017;23:4895–900.
37. Deligezer U, Yaman F, Darendeliler E, et al. Post-treatment circulating plasma BMP6 mRNA and H3K27 methylation levels discriminate metastatic prostate cancer from localized disease. Clin Chim Acta. 2010;411(19–20):1452–6.
38. Markou A, Zavridou M, Sourvinou I, et al. Direct comparison of metastasis-related miRNAs expression levels in circulating tumor cells, corresponding plasma, and primary tumors of breast cancer patients. Clin Chem. 2016;62(7):1002–11.
39. Zeng Z, Li Y, Pan Y, et al. Cancer-derived exosomal miR-25-3p promotes pre-metastatic niche formation by inducing vascular permeability and angiogenesis. Nat Commun. 2018;9(1):5395.
40. Yan S, Han B, Gao S, et al. Exosome-encapsulated microRNAs as circulating biomarkers for colorectal cancer. Oncotarget. 2017;8(36):60,149–58.
41. Peng ZY, Gu RH, Yan B. Downregulation of exosome-encapsulated miR-548c-5p is associated with poor prognosis in colorectal cancer. J Cell Biochem. 2018; https://doi.org/10.1002/jcb.27291.
42. Xue J, Xie VK, Wang P, et al. Interrelationships of circulating tumor cells with metastasis and thrombosis: role of microRNAs. Curr Pharm Des. 2014;20(33):5298–308.
43. Qiu JJ, Lin XJ, Tang XY, et al. Exosomal metastasis associated lung adenocarcinoma transcript 1 promotes angiogenesis and predicts poor prognosis in epithelial ovarian cancer. Int J Biol Sci. 2018;14(14):1960–73.
44. Beltran-Garcia J, Osca-Verdegal R, Mena-Molla S, et al. Epigenetic IVD tests for personalized precision medicine in cancer. Front Genet. 2019;10:621.
45. Constancio V, Nunes SP, Henrique R, et al. DNA methylation-based testing in liquid biopsies as detection and prognostic biomarkers for the four major cancer types. Cell. 2020;9(3)
46. Mastoraki S, Strati A, Tzanikou E, et al. ESR1 methylation: a liquid biopsy-based epigenetic assay for the follow-up of patients with metastatic breast cancer receiving endocrine treatment. Clin Cancer Res. 2018;24(6):1500–10.
47. Liu L, Toung JM, Jassowicz AF, et al. Targeted methylation sequencing of plasma cell-free DNA for cancer detection and classification. Ann Oncol. 2018;29(6):1445–53.
48. Liu MC, Oxnard GR, Klein EA, et al. Sensitive and specific multi-cancer detection and localization using methylation signatures in cell-free DNA. Ann Oncol. 2020;31(6):745–59.
49. Gezer U, Yoruker EE, Keskin M, et al. Histone methylation marks on circulating nucleosomes as novel blood-based biomarker in colorectal cancer. Int J Mol Sci. 2015;16(12):29,654–62.

Metabolic Reprogramming of Circulating Tumor Cells for Metastasis

18

Ziyuan Zhang and Qihui Shi

Abstract

Despite the leading cause of cancer-associated deaths, metastasis is a complex, multi-step, and inefficient biological process for cancer cells, including detachment from the primary tumor site, survival in circulation, and colonization at the distant sites. Metastatic cancer cells thus face a daunting series of challenges, such as anoikis, nutrient deprivation, shear stress in circulation, and metabolic stress at distant organs. Cancer cells rewire metabolic pathways to rapidly generate energy, cellular building blocks and antioxidants for supporting tumor growth and fueling metastasis. Inherent metabolic heterogeneity and additional acquisition of metabolic features during tumor spread promote circulating tumor cell (CTC) survival and seeding metastases by adapting to hostile environments during metastatic cascades. Unlike tumor tissue samples, metabolic assessment of CTCs in blood and other types of body fluids is technologically challenging because these liquid biopsies contain many different types of normal cells, requiring development of efficient and specific markers for accurate CTC identification. In addition to clinical tumor imaging (e.g., positron emission tomography), metabolic reprogramming characteristic of cancer cells represents attractive therapeutic targets for anticancer treatment. Although many metabolism-specific inhibitors are under developing or tested in clinical trials, none of them were approved to enter routine clinical practice due to metabolic plasticity of cancer cells and narrow therapeutic window between normal proliferating cells and cancer cells.

Z. Zhang · Q. Shi (✉)
Institutes of Biomedical Sciences and Minhang Hospital, Fudan University, Shanghai, China
e-mail: qihuishi@fudan.edu.cn

Learning Objectives

- Understand metabolic heterogeneity in primary tumors and how such heterogeneity promotes cancer metastasis.
- Understand metabolic plasticity of CTCs during metastatic spread and how they survive in systematic circulation and successfully seed distant metastases by metabolic adaption to the hostile environment.
- Understand metabolic-specific inhibitors targeting therapeutic vulnerability from altered metabolic circuits in cancer cells and the limitations.

18.1 Introduction

Cancer metastasis accounts for the major cause (~90%) of cancer-associated deaths worldwide [1, 2]. Metastasis is a complex multi-step biological process, including detachment from the primary tumor site, survival in circulation, and successful colonization at the distant sites [3]. Each step poses a significant barrier to the metastatic cascade. Anoikis, shear stress in circulation, and metabolic stress in colonization at the distant organ are extremely challenging for cancer cells to seed distant metastases [4, 5]. Thus, metastasis is an inefficient process for cancer cells [3]. It has been estimated that only less than 0.02% metastatic cancer cells successfully seed metastases from their primary site [3, 4, 6]. To date, how cancer cells detach from primary lesion, how they survive in circulation, and how they select distant sites and cope with hostile microenvironments for full-blown metastasis are largely unknown [1, 7]. In particular, accumulating evidence reveals that cancer cells are able to disseminate to distant sites early during primary tumor growth [8], indicating intrinsic metastatic potential of cancer cells.

For mechanistic understanding of cancer metastasis, many efforts have been devoted to identify metastasis-

S. P. Leong et al. (eds.), *Cancer Metastasis Through the Lymphovascular System*,
https://doi.org/10.1007/978-3-030-93084-4_18

specific mutations but were unsuccessful [9]. Metastatic cancer cells are indistinguishable from their non-metastatic counterparts in primary tumors or in circulation by genetic signatures. Due to the plasticity of cancer cells, reversible mechanisms (e.g., metabolic, epigenetic) rather than irreversible mechanisms (e.g., mutation) are likely used by cancer cells to adapt to hostile environments during cancer metastasis. Metabolic reprogramming is known to be one of the fundamental hallmarks of cancer cells and plays a vital role in the cancer progression and metastasis [10]. Unlike normal cells, cancer cells often utilize aerobic glycolysis, a phenomenon termed Warburg effect, to generate energy, cellular building blocks, and antioxidants to support cell transformation, to fight oxidative stress, and to fuel the metastatic cascade [11]. Complex interplay between the tumor mutational landscape, epigenetic modifications, and microenvironmental influences contributes to rewiring of metabolic pathways within cancer cells [7]. Understanding the tumor metabolic features, especially the acquired metabolic features during metastasis, and metabolic vulnerabilities might open novel diagnostic and therapeutic approaches to improve the efficacy of current anticancer treatments.

In this chapter, we will discuss recent work in understanding the metabolic reprogramming of cancer cells during cancer metastasis. We first discuss the metabolic heterogeneity in primary tumors and how such heterogeneity promotes metastasis. Secondly, we discuss the metabolic plasticity of disseminating tumor cells during metastatic spread and how they survive in systematic circulation and successfully colonizing distant metastases by metabolic adaption. Last, we review potential drugs targeting metabolism-related therapeutic vulnerability of cancer cells, especially those in metastasis.

18.2 Metabolic Heterogeneity in Primary Tumors Relevant to Metastasis

In the 1920s, Otto Warburg introduced the concept of metabolic reprogramming that cancer cells depend on aerobic glycolysis for energy production due to the damaged mitochondrial respiration [12–14]. Given that mitochondria are a main source of reactive oxygen species (ROS) generation, mitochondrial dysfunction may also reduce oxidative stress in cancer cells. The phenomenon that cancer cells are addicted to aerobic glycilysis, termed as Warburg effect, provides a new approach for detecting cancer cells and paves the way for the development of clinical tumor imaging: fluorodeoxyglucose (^{18}F)-positron emission tomography (^{18}F-FDG PET) [15, 16]. However, such concept was debated because recent studies demonstrated fully functional mitochondria activity in cancers [17–21], and equally important roles for mitochondrial metabolism—the tricarboxylic acid (TCA) cycle and oxidative phosphorylation (OXPHOS) in tumor progression and metastasis [22–25]. Compared with lung and colorectal cancer exhibiting elevated glycolytic states, breast cancer and glioblastoma are largely viewed as oxidative tumors [26].LeBieu and coworkers demonstrated invasive breast cancer cells specifically favored mitochondrial respiration for increased ATP production by using the transcription coactivator PGC-1α [27].Cai and coworkers showed crucial role of pyruvate dehydrogenase complex (PDHc) and TCA cycle maintained by AMPK (AMP-activated protein kinase) for promoting cancer metastasis by adapting breast cancer cells to metabolic and oxidative stresses [28].This AMPK-PDHc axis, activated in advanced breast cancer, was found essential for developing full-blown metastatic cancer and associated with poor metastasis-free survival [28]. In addition, Porportato and coworkers reported that both mitochondrial overload and mild respiration dysfunction promoted metastasis by enhanced mitochondrial superoxide production [29].

Interestingly, Tasdogan and coworkers showed that metabolic differences among melanoma cells conferred different metastatic potential as a result of differences in the function of the MCT1 transporter [30]. In patient-derived xenografts, efficiently metastasizing melanomas were found to have higher levels of MCT1, whereas inefficiently metastasizing melanomas exhibited low or negative MCT1 levels. Despite similar tumor-forming capacities between MCT1^{high} and MCT1$^{-/low}$ cells, MCT1^{high} melanoma cells formed more metastases after intravenous injection. Elevated expression of MCT1 enhanced uptake of circulating lactate by cancer cells and this mechanism was used by metastasizing cells to manage oxidative stress and promote metastasis. Thus, metabolic differences among cancer cells conferred differences in metastatic potential.

The diverse metabolic dependencies have been also observed between the primary and metastatic lesions of the same patient, as well as within distinct regions of the same tumor [31, 32].Metabolic heterogeneity arises from genetic and epigenetic diversity, redundant metabolic pathways, and altered microenvironmental conditions [33], and thereby endows cancer cells with enhanced flexibility for rapidly adapting changing environmental conditions during tumor progression and metastasis. Dupuy and coworkers showed extensive metabolic heterogeneity of primary breast cancer cells and that distinct metabolic programs were associated with site-specific metastases [34]. Surprisingly, breast cancer cells with similar genetic backgrounds but distinct metabolic characteristics exhibited preferential metastatic tropism to different sites. Bone- or lung-metastatic breast cancer cells were found engaging in OXPHOS metabolic profile whereas liver-metastatic cells exhibited a glycolytic-dependent metabolic program [34]. Unlike cancer cells, proliferating cells in tumor microenvironment have shown a uniform metabolic signature as revealed by a multi-isotope imaging mass spectrometry with glucose and glutamine tracers [35].

18.3 Metabolic Heterogeneity of Disseminating Tumor Cells During Metastatic Spread

Disseminating tumor cells present in the lymph nodes, blood, and other body fluids offer a direct approach for evaluating metabolic states and heterogeneity of cancer cells during metastatic spread, especially the acquisition of specific metabolic features during the different metastatic cascade. However, these samples (lymph nodes or body fluids) also contain a large number and many different types of normal cells. Thus, disseminating tumor cells should be distinguishable from these normal cells before metabolic evaluation, requiring discovery of markers of disseminating tumor cells in lymph nodes and liquid biopsies.

Pleural effusion is a body fluid that associates with many types of malignant diseases. Pleural effusion containing cancer cells shed from primary tumor is termed malignant pleural effusion (MPE) that denotes an advanced stage of diseased with metastasis (M1a stage). For this reason, MPE offers a direct evaluation of metabolic states of disseminating tumor cells and their metabolic heterogeneity at the single-cell level. Li and coworkers employed two fluorescent metabolic probes (2-NBDG: glucose analog; resazurin: redox indicator) to assay all nucleated cells in MPE from non-small cell lung cancer (NSCLC) patients for identifying metabolically active disseminating tumor cells and resolving their metabolic phenotypes with single-cell resolution [36, 37]. Single-cell sequencing was performed to confirm the malignancy of these metabolically active, CD45 negative cells identified by 2-NBDG and resazurin markers. Three distinct metabolic phenotypes of disseminating tumor cells were identified in MPE. The first subset (2-NBDG^{high}/resazurin^{low}) was ravenous for glucose with high glycolysis but limited mitochondrial oxidation. The second subset (resazurin^{high}/2-NBDG^{low}) had limited glucose uptake and exhibited high mitochondrial oxidation activity. The third subgroup that was a minority phenotype (2-NBDG^{high}/resazurin^{high}) showed enhanced glycolysis and mitochondrial oxidation. This is the first study that reveals the metabolic heterogeneity of disseminating tumor cells at the single-cell level. Despite harboring same driver mutations, the difference in metabolic phenotypes of these cancer cells was found to associate with different therapy response and survival of NSCLC patients [37]. Metabolic signature of disseminating tumor cells in MPE is capable of predicting therapy resistance and patient survival.

Compared with pleural effusion, disseminating tumor cells circulating in blood, termed as circulating tumor cells (CTCs), become subject of intense investigation because CTCs enable seeding metastases in distant organs [38, 39]. Metabolic states and heterogeneity of cancer cells in metastasis can be evaluated by assaying CTCs in blood samples. Traditional CTC detection methods rely on epithelial markers such as EpCAM and keratins (CKs) for patients with epithelial tumors [40]. Yang and coworkers developed a novel metabolic function-associated marker, hexokinase 2 (HK2), for CTC detection and phenotyping in peripheral blood of NSCLC patients [41]. As the first enzymatic step of glycolysis, hexokinase phosphorylates glucose and traps it in the cell [42]. Among hexokinase isoforms, only HK2 is found highly expressed in a wide range of cancers, including epithelial and mesenchymal origins, and exhibits restricted distribution of expression in normal adult tissues [42–45]. As a metabolic function-associated marker, HK2 overcomes the limitation of traditional epithelial CTC markers that is inherently biased toward identifying CTCs with epithelial traits, leading to the capability of detecting CTCs experiencing epithelial-to-mesenchymal transition (EMT) with downregulation of epithelial markers or lineage plasticity [46–49]. For this reason, HK2 enables resolving an underrepresented HK2^{high}/CK^{-} CTC population in peripheral blood of NSCLC patients [41]. Although CK^{-} CTCs were reported in previous studies [49–53], Yang and coworkers surprisingly found that CK^{-}CTCs were a prevalent population in half of NSCLC patients with positive CTC counts [41]. At the molecular level, CK^{-}CTCs exhibited consistent genomic copy number alternations but distinct transcriptome signatures compared to the CK^{+} counterpart. HK2 as a metabolic-function associated marker allows to identify CK^{-}CTCs, leading to classification of CTC based on CK expression. Meanwhile, HK2^{high} and HK2^{low} CTCs represented different CTC subgroups with distinct metabolic phenotype but similar genomic signatures [41].

18.4 Metabolic Adaption of CTCs to Promote Their Survival and Metastasis

To successfully seed distant metastasis, cancer cells detached from primary sites as metastatic precursors face a daunting series of challenges including loss of matrix adhesion, nutrient deprivation, shear stress in circulation, and metabolic stress at distant organs, requiring additional metabolic adaption of cancer cells to these hostile environments. During extracellular matrix detachment, untransformed epithelial cells undergo cell death as a result of reduced glucose uptake, ATP depletion, and oxidative stress. In the suspended condition, epithelial cells were found to undergo metabolic reprogramming by markedly upregulating pyruvate dehydrogenase (PDH) kinase 4 (PDK4) for attenuating mitochondrial respiration and oxidative stress [54]. Decreased glucose oxidation following cell detachment is essential for anoikis resistance. Unlike untransformed cells, most cancer cells exhibit increasing glucose uptake and flux but reduced glucose oxidation due to oncogenic signaling. This inherently possess a survival advantage when suspended [54]. The Warburg effect is thus believed to be a defense mechanism protecting cancer cells from oxidative stress and promoting anoikis resistance and metastasis by increased production of antioxidants (e.g., NADPH, glutathione) [11, 20, 55]. CTCs traveling in the bloodstream accumulate excessive ROS within cells due to decreased NADPH and ATP production and are thus more susceptible to cell death [55]. Metastasis-competent CTCs need to manage the oxidative stress by restoring the skewed redox balance for prolonging cell survival.

Piskounova and coworkers demonstrated that the oxidative stress was a significant hurdle for melanoma dissemination and thereby limited distant metastasis in vivo [56].During metastasis, ROS levels increased in circulating melanoma cells. To combat ROS and to maintain cell survival, successfully metastasizing CTCs underwent reversible metabolic changes that increased NADPH and conferred oxidative stress resistance by promoting the folate pathway in one-carbon metabolism. NADPH-generating enzymes ALDH1L2 and MTHFD1 in the folate pathway were used as suppressors of ROS-mediated toxicity. ALDH1L2 or MTHFD1 knockdown was found to inhibit distant metastasis without significantly affecting the subcutaneous tumors [56]. However, it is also possible that AMPK activation in CTCs reduces NADPH consumption by fatty acid synthesis (FAS) and regenerates NADPH by induction of fatty acid oxidation (FAO), leading to NADPH homeostasis and reduction of elevated ROS.

Zheng and coworkers compared single-cell RNA-seq profiles of CTCs from breast, prostate, and lung cancers, and found consistent induction of β-globin that was overexpressed to suppress intracellular ROS within CTCs [57]. β-Globin expression within CTCs protected them from ROS-induced apoptosis, promoted survival in circulation, and contributed to their ability to initiate distant metastases. Tasdogan and coworkers found that upregulation of cell surface MCT1 transporter in circulating melanoma cells enhanced uptake of circulating lactate for managing oxidative stress of CTCs [30]. MCT1 inhibition reduced lactate uptake and increased ROS levels, resulting in depletion of CTCs and decreased metastatic burden in patient-derived xenografts and in mouse melanomas [30]. Chen and coworkers established a CTC-derived, brain metastatic cell line and these brain metastatic cells showed enhanced mitochondrial respiratory pathways for energy production and upregulation of pentose phosphate pathway and the glutathione system which can minimize production of ROS resulting from an enhanced oxidative metabolism [58]. These metabolic changes were associated with strongly enhanced tumor cell survival and proliferation in the brain microenvironment.

In addition to metabolic adaption for blood-borne metastasis, Lee and coworkers identified a metabolic shift of melanoma cells toward FAO during their lymph node metastasis [59]. Yes-associated protein (YAP), a downstream mediator of the Hippo pathway, was found to be selectively activated in lymph node metastases and thus induce FAO. Interestingly, pharmacological inhibition of FAO or YAP deletion inhibited lymph node metastasis, but not blood-borne metastasis, indicating that the metabolic shift toward FAO was preferentially required for lymph node metastasis.

18.5 Targeting Metabolic Transformation for Improving Cancer Therapeutics

The altered metabolic pathway of cancer cells has been exploited for visualizing tumor in vivo by imaging the uptake of radiolabeled glucose analog ^{18}F-FDG. This metabolic distinction also represents attractive therapeutic targets (e.g., key metabolic enzymes) for anticancer treatment [60–63]. Anti-glycolytic treatment (2-deoxyglucose, 2-DG) was first tested in tumor model [64] and used to treat cancer patients [65] in the 1950s. 2-DG as a glucose analog is phosphorylated by HK to 2-DG-phosphate, which cannot be further metabolized. Accumulation of 2-DG in cells thus inhibits glycolysis causing ATP depletion, cell cycle inhibition, and cell death. Although there are no ongoing clinical trials using 2-DG as a single agent, combining 2-DGwith radiation or chemotherapeutic treatments were found to potentiate the tumor-destroying effects and enhance the clinical efficacy [66].

Besides increasing glucose uptake, cancer cells facilitate the first enzymatic step in elevated glucose metabolism by markedly inducing upregulation of HK2 in addition to HK1, which is already expressed in normal cells. Both HK1 and HK2 bind to mitochondria in a voltage-dependent anion channel (VDAC)-dependent manner, and AKT promotes HK1/HK2 association with mitochondria [67]. HK1 and HK2 possess two tandem catalytic domains, but only one is active in HK1, whereas both are active in HK2 [68]. HK2 induction in cancer cells is driven by multiple oncogenic pathways and is transcriptionally upregulated by MYC [43, 69–71]. However, why cancer cells induce HK2 rather than further increasing HK1 expression is largely unknown. Due to the structural similarities between HK1 and HK2, it could be challenging for developing inhibitors that preferentially inhibit HK2. To date, a few HK2 inhibitors were reported but with limited success [72–74]. Recent studies demonstrated that HK1$^+$/HK2$^+$ cancers could tolerate HK2 silencing or selective inhibition, but the subset of HK1$^-$/HK2$^+$ tumors was found sensitive to therapeutic HK2 inhibition [75–78]. For this reason, stratification of tumors that express HK2, but not the housekeeping HK1 isozyme, should identify tumors treatable with HK2 inhibitors. An interesting example is hepatocellular carcinoma (HCC). Normal hepatocytes typically express HK4 rather than HK1. HCC cells exhibit suppression of HK4 expression and predominant induction of HK2 expression [79]. Thus, systemic delivery of an HK2 inhibitor could selectively target HCC cells but not normal hepatocytes.

Despite accelerated aerobic glycolysis in cancer cells, normal proliferating cells are known to use the same metabolic pathway and glycolytic enzymes as cancer cells. Thus, the inhibition of glycolytic enzymes for treating cancer raises concerns on the risk of adverse and undesirable effects on normal cells. Finding a therapeutic window between normal proliferating cells and cancer cells remains a major challenge in the development of cancer therapy targeting metabolic transformation. Table 18.1 shows metabolic-specific inhibitors that are specifically targeting the altered metabolic circuits of tumors under developing or in clinical trials. Until now, none of the tested metabolic therapies were approved to

Table 18.1 Compounds targeting cancer metabolism in clinical studies

Targeted metabolism	Metabolic inhibitors	Targeted metabolic enzymes	Cancer types	Study phase	Clinical Trials. gov reference
Glycolysis	2-deoxy-D-glucose (2-DG)	HK	Advanced solid tumors	Phase I Phase I/II	NCT00096707[a] NCT00633087[b]
	Lonidamine(LND)	HK	Benign prostatic hyperplasia	Phase III Phase II	NCT00435448[b] NCT00237536[b]
	3-Bromopyruvate (3-BrPA)	HK/GAPDH	Pancreatic cancer, metastatic melanoma	Phase I	N/A
	Silibin-Phytosome	GLUTs	Prostate cancer	Phase II	NCT00487721[a]
	Phloretin	GLUTs	Colon cancer, leukemia	Preclinical	N/A
	WZB117	GLUTs	Lung cancer, breast cancer	Preclinical	N/A
	TLN-232	Pyruvate kinase	Melanoma	Phase II	NCT00735332[b]
	Oxamate	LDHA	Breast cancer	Preclinical	N/A
	FX11	LDHA	Lymphoma	Preclinical	N/A
Citric acid cycle	Dichloroacetate (DCA)	PDK	Many tumor types	Phase II Phase II Phase I	NCT01029925[b] NCT01386632[a] NCT01111097[a]
	CPI-613	PDH/α-ketoglutarate dehydrogenase	Lymphoma, leukemia	Phase II	NCT03793140[c]
	AG-120 (Ivosidenib)	Mutant IDH1	Advanced solid tumors	Phase II	NCT04195555[c]
	IDH305	MutantIDH1	Advanced malignancies	Phase I	NCT02381886[d]
	FT-2102	MutantIDH1	Hepatobiliary tumors	Phase I/II	NCT03684811[d,e]
	AG-221 (Enasidenib)	Mutant IDH2	Hematologic neoplasms	Phase I/II	NCT01915498[d]
	AG-881	Pan-mutant IDH	Advanced hematologic malignancies	Phase I	NCT02492737[a]
	AZD3965	MCTs	Adult solid tumors	Phase I	NCT01791595[d]
Oxidative phosphorylation	Metformin	Mitochondrial AMPK	Colorectal cancer	Phase III	NCT02614339[c]
Arginine pathway	Arginine deiminase (ADI-PEG-20)	ASS1	Melanoma, carcinoma	Phase II Phase I Phase II	NCT00450372[a] NCT00029900[a] NCT00056992[a]
	INCB001158	Arginase	Advanced-stage solid tumors	Phase I/II	NCT02903914[d,e]
Asparagine pathway	Asparaginase	Depletion of circulating asparagine	Acute lymphoblastic leukemia	Phase III Phase III	NCT00165178[a,e] NCT00400946[a,e]
Glutamine pathway	Phenylacetate	Glutamine synthetase and glutaminase	Brain and central nervous system tumors	Phase II Phase II Phase I	NCT00003241[a] NCT00006450[a] NCT00001565[a]
	CB-839	Glutaminase	Solid tumors	Phase I/II Phase I/II	NCT03875313[c,e] NCT02771626[e,f]
	BHV-4157 (Trigriluzole)	EAAT	Metastatic or Unresectable solid tumors or lymphoma	Phase I	NCT03229278[a,e]
Fatty acid synthesis	Omeprazole	Proton pump (H+/K + -ATPase)	Breast cancer	Phase II	NCT02595372[d]
	TVB-2640	FASN	Solid malignant tumors	Phase I/II	NCT03808558[c] NCT02223247[a]
	Orlistat	FASN	N/A	Preclinical	N/A
	C75	FASN	N/A	Preclinical	N/A

(continued)

Table 18.1 (continued)

Targeted metabolism	Metabolic inhibitors	Targeted metabolic enzymes	Cancer types	Study phase	Clinical Trials. gov reference
Nucleotide synthesis	5-fluorouracil(5-FU)	Thymidylate synthase	Many tumor types	US FDA approved	>100
	Cytarabine	DNA polymerase/ ribonucleotide reductase	Many tumor types	US FDA approved	>100
	Methotrexate	DHFR	Advanced colorectal cancer	Phase II	NCT00952016[b]
	Pemetrexed	DHFR/thymidylate synthase /GARFT	Head and neck cancer	Phase II	NCT00293579[a]
	6-mercaptopurine (6-MP) and 6-thioguanine (6-TG)	5-phosphoribosyl pyrophosphate (PRPP) amidotransferase	Acute lymphoblastic leukemia(ALL)	Phase IV	>30
	Leflunomide	DHODH	Brain and central nervous system tumors	Phase III	NCT00003293[a,e]
pH regulation	SLC0111	Carbonic anhydrase	Solid tumors	Phase I Phase I/II	NCT02215850[a] NCT03450018[c,e]
	E7070 (Indisulam)	Carbonic anhydrase	Gastric cancer	Phase I Phase I/II	NCT00060567[a,e] NCT00165594[b]
Akt/mTOR/P13K signaling	Halofuginone hydrobromide	Akt/mTORC1	Kaposi sarcoma	Phase II	NCT00064142[a]
	Rapamycin	mTORC1	Prostate cancer	Phase I/II	NCT00311623[a]
	Temsirolimus and Everlimus	mTORC1	Solid tumors	US FDA approved	>100
	Ridaforolimus and related compounds	mTORC1	Solid tumors	Phase I	>10
	BEZ235	PI3K/mTOR	Malignant solid tumors	Phase I	NCT01343498[a]
	GSK2110183 (Afuresertib)	PI3K/Akt	Many tumor types	Phase II	NCT01531894[a]
HIF signaling	PX-478	HIF1α	Advanced solid tumors	Phase I	NCT00522652[a]
IGF signaling	MK-0646	IGF1R	Non-small cell lung cancer	Phase II	NCT00799240[a,e]
	BIIB022	IGF1R	Solid tumors	Phase I	NCT00555724[a]
	AVE1642	IGF1R	Liver carcinoma	Phase I/II	NCT00791544[b,e]
IDO signaling	INCB024360 (Epacadostat)	IDO1	Urothelial cancer	Phase III	NCT03361865[a,e]
	BMS-986205 (Linrodostat)	IDO1	Advanced cancers	Phase I/II	NCT03459222[c,e]
	Indoximod	IDO1/2	Melanoma	Phase II	NCT03301636[b,e]
Angiogenesis	Bevacizumab and related compounds	Hypoxia/VEGF/VEGFR	Solid tumors	US FDA approved	>100
	Aflibercept	Angiogenesis factors and receptors	Metastatic colorectal cancer	Phase I/II	NCT01661972[a,e]
	Regorafenib	Angiogenesis factors and receptors	Metastatic colorectal cancer	Phase II Phase II	NCT02466009[d] NCT02699073[a]
Metabolic reprogramming	Imprime PGG (β- glucan)	Innate immune cells	Advanced melanoma, triple-negative breast cancer	Phase II	NCT02981303[d,e]
	Tirapazamine	Hypoxia	Cervical cancer, lung cancer	Phase I Phase III	NCT00098995[a,e] NCT00017459[a,e]

HK Hexokinase, *GAPDH* Glyceraldehyde-3-phosphate dehydrogenase, *GLUT* Glucose transporter, *LDHA* Lactate dehydrogenase, *PDK* Pyruvate dehydrogenase kinase, *PDH* Pyruvate dehydrogenase, *IDH* Isocitrate dehydrogenase, *MCT* Monocarboxylate transporter, *AAS1* Argininosuccinate synthase 1, *EAAT*: Excitatory amino acid transporters, *FASN* Fatty acid synthase, *DHFR* Dihydrofolate reductase, *GARFT* Glycinamide ribonucleotide formyltransferase, *DHODH* Dihydroorotate dehydrogenase

[a]Completed

[b]Terminated

[c]Recruiting

[d]Active, not recruiting

[e]In combination with other drugs

[f]Unknown

enter routine clinical practice. Due to the metabolic plasticity of cancer cells, metabolism-inhibiting treatment may require inhibiting multiple metabolic pathways simultaneously or coupled with standard chemo-, targeted or immunotherapy.

18.6 Conclusions

Metabolic reprogramming of cancer cells has been extensively studied over the past century and is now recognized to be a common feature of many cancer cells. The inherent metabolic heterogeneity of cancer cells and acquired metabolic adaptations during dissemination promote CTC survival in circulation and seeding metastases at distant organs. CTC-derived cell lines [59, 80–83] and xenografts [84, 85] have been established from only hundreds of CTCs, as compared to enormous challenges to establish cell lines from millions of primary tumor cells. This might be explained by the advantages of CTCs in survival and growth. Although culturing CTCs in vitro or expanding in vivo using xenografts have significant clinical applications, successful examples are still rare because of low numbers of CTCs in patients. In addition to analyzing CTCs temporally and spatially (at different vascular sites), new analytical approaches such as single-cell, quantitative metabolome profiling would allow us to map the entire metabolic circuits of metastatic cancer cells for exploring the diagnostic and therapeutic potential [86, 87].

Open Questions

- How do cancer cells acquire additional metabolic features to adapt to hostile environment during metastasis?
- How to perform metabolic phenotyping of CTCs for prognosis and prediction of therapy response?
- How to identify and selectively kill CTCs with high metastatic potential by targeting therapeutic vulnerability from altered metabolic circuits?
- How to identify early metastasis in cancer patients by globally detecting CTCs with different phenotypes?

References

1. Gupta GP, Massague J. Cancer metastasis: building a framework. Cell. 2006;127(4):679–95.
2. Steeg PS. Targeting metastasis. Nat Rev Cancer. 2016;16(4):201–18.
3. Vanharanta S, Massagué J. Origins of metastatic traits. Cancer Cell. 2013;24(4):410–21.
4. Chambers AF, Groom AC, MacDonald IC. Dissemination and growth of cancer cells in metastatic sites. Nat Rev Cancer. 2002;2(8):563–72.
5. Senft D, Ronai ZEA. Adaptive stress responses during tumor metastasis and dormancy. Trends Cancer. 2016;2(8):429–42.
6. Luzzi KJ, MacDonald IC, Schmidt EE, Kerkvliet N, Morris VL, Chambers AF, Groom AC. Multistep nature of metastatic inefficiency: dormancy of solitary cells after successful extravasation and limited survival of early micrometastases. Am J Pathol. 1998;153(3):865–73.
7. Elia I, Doglioni G, Fendt SM. Metabolic hallmarks of metastasis formation. Trends Cell Biol. 2018;28(8):673–84.
8. Chemi F, Rothwell DG, McGranahan N, et al. Pulmonary venous circulating tumor cell dissemination before tumor resection and disease relapse. Nat Med. 2019;25(10):1534–9.
9. Valastyan S, Weinberg RA. Tumor metastasis: molecular insights and evolving paradigms. Cell. 2011;147(2):275–92.
10. Hanahan D, Weinberg RA. Hallmarks of cancer: the next generation. Cell. 2011;144(5):646–74.
11. Hamanaka RB, Chandel NS. Cell biology. Warburg effect and redox balance. Science. 2011;334(6060):1219–20.
12. Warburg O, Posener K, Negelein E. Ueber den stoffwechsel der tumoren. Biochem Z. 1924;152(1):319–44.
13. Warburg O, Wind F, Negelein E. The metabolism of tumors in the body. J Gen Physiol. 1927;8(6):519–30.
14. Warburg O. On the origin of cancer cells. Science. 1956;123(3191):309–14.
15. Som P, Atkins HL, Bandoypadhyay D, et al. A fluorinated glucose analog, 2-fluoro-2-deoxy-D-glucose (F-18): nontoxic tracer for rapid tumor detection. J Nucl Med. 1980;21(7):670–5.
16. Hsu PP, Sabatini DM. Cancer cell metabolism: Warburg and beyond. Cell. 2008;134(5):703–7.
17. Weinhouse S. On respiratory impairment in cancer cells. Science. 1956;124(3215):267–9.
18. Warburg O. On respiratory impairment in cancer cells. Science. 1956;124(3215):269–70.
19. Corbet C, Feron O. Cancer cell metabolism and mitochondria: nutrient plasticity for TCA cycle fueling. Biochim Biophys Acta Rev Cancer. 2017;1868(1):7–15.
20. Vander HMG, Cantley LC, Thompson CB. Understanding the Warburg effect: the metabolic requirements of cell proliferation. Science. 2009;324(5930):1029–33.
21. Gatenby R, Gillies R. Why do cancers have high aerobic glycolysis? Nat Rev Cancer. 2004;4(11):891–9.
22. Ahn CS, Metallo CM. Mitochondria as biosynthetic factories for cancer proliferation. Cancer Metab. 2015;3(1):1.
23. Pavlova NN, Thompson CB. The emerging hallmarks of cancer metabolism. Cell Metab. 2016;23(1):27–47.
24. Hensley CT, Wasti AT, DeBerardinis RJ. Glutamine and cancer: cell biology, physiology, and clinical opportunities. J Clin Invest. 2013;123(9):3678–84.
25. Wang MD, Wu H, Fu GB, et al. Acetyl-coenzyme a carboxylase alpha promotion of glucose-mediated fatty acid synthesis enhances survival of hepatocellular carcinoma in mice and patients. Hepatology. 2016;63(4):1272–86.
26. Lehuédé C, Dupuy F, Rabinovitch R, Jones RG, Siegel PM. Metabolic plasticity as a determinant of tumor growth and metastasis. Cancer Res. 2016;76(18):5201–8.
27. LeBleu VS, O'Connell JT, Gonzalez Herrera KN, et al. PGC-1α mediates mitochondrial biogenesis and oxidative phosphorylation in cancer cells to promote metastasis. Nat Cell Biol. 2014;16(10):992–1003.
28. Cai Z, Li CF, Han F, et al. Phosphorylation of PDHA by AMPK drives TCA cycle to promote cancer metastasis. Mol Cell. 2020;80(2):263–78.

29. Porporato PE, Payen VL, Pérez-Escuredo J, et al. A mitochondrial switch promotes tumor metastasis. Cell Rep. 2014;8(3):754–66.
30. Tasdogan A, Faubert B, Ramesh V, et al. Metabolic heterogeneity confers differences in melanoma metastatic potential. Nature. 2020;577(7788):115–20.
31. Hensley CT, Faubert B, Yuan Q, et al. Metabolic heterogeneity in human lung tumors. Cell. 2016;164(4):681–94.
32. Sengupta D, Pratx G. Imaging metabolic heterogeneity in cancer. Mol Cancer. 2016;15(1):1–12.
33. Gerlinger M, Rowan AJ, Horswell S, et al. Intratumor heterogeneity and branched evolution revealed by multiregion sequencing. N Engl J Med. 2012;366(10):883–92.
34. Dupuy F, Tabaries S, Andrzejewski S, et al. PDK1-dependent metabolic reprogramming dictates metastatic potential in breast cancer. Cell Metab. 2015;22(4):577–89.
35. Zhang Y, Guillermier C, De Raedt T, et al. Imaging mass spectrometry reveals tumor metabolic heterogeneity. iScience. 2020;23(8):101355.
36. Tang Y, Wang Z, Li Z, et al. High-throughput screening of rare metabolically active tumor cells in pleural effusion and peripheral blood of lung cancer patients. Proc Natl Acad Sci U S A. 2017;114(10):2544–9.
37. Li Z, Wang Z, Tang Y, et al. Liquid biopsy-based single-cell metabolic phenotyping of lung cancer patients for informative diagnostics. Nat Commun. 2019;10(1):3856.
38. Keller L, Pantel K. Unravelling tumour heterogeneity by single-cell profiling of circulating tumour cells. Nat Rev Cancer. 2019;19(10):553–67.
39. Pantel K, Alix-Panabières C. Liquid biopsy and minimal residual disease—latest advances and implications for cure. Nat Rev Clin Oncol. 2019;16(7):409–24.
40. Werner S, Keller L, Pantel K. Epithelial keratins: biology and implications as diagnostic markers for liquid biopsies. Mol Asp Med. 2020;72(5):100817.
41. Yang L, Yan X, Chen J, et al. Hexokinase 2 discerns a novel circulating tumor cell population associated with poor prognosis in lung cancer patients. Proc Natl Acad Sci U S A. 2021;118(11):e2012228118.
42. Mathupala SP, Ko YH, Pedersen PL. Hexokinase II: cancer's double-edged sword acting as both facilitator and gatekeeper of malignancy when bound to mitochondria. Oncogene. 2006;25(34):4777–86.
43. Patra KC, Wang Q, Bhaskar PT, et al. Hexokinase 2 is required for tumor initiation and maintenance and its systemic deletion is therapeutic in mouse models of cancer. Cancer Cell. 2013;24(2):213–28.
44. Wang L, Xiong H, Wu F, et al. Hexokinase 2-mediated Warburg effect is required for PTEN- and p53-deficiency-driven prostate cancer growth. Cell Rep. 2014;8(5):1461–74.
45. DeWaal D, Nogueira V, Terry AR, et al. Hexokinase-2 depletion inhibits glycolysis and induces oxidative phosphorylation in hepatocellular carcinoma and sensitizes to metformin. Nat Commun. 2018;9(1):446.
46. Keller L, Werner S, Pantel K. Biology and clinical relevance of EpCAM. Cell Stress. 2019;3(6):165–80.
47. Gires O, Pan M, Schinke H, Canis M, Baeuerle PA. Expression and function of epithelial cell adhesion molecule EpCAM: where are we after 40 years? Cancer Metastasis Rev. 2020;39(3):969–87.
48. Quintanal-Villalonga Á, Chan JM, Yu HA, Pe'er D, Sawyers CL, Sen T, Rudin CM. Lineage plasticity in cancer: a shared pathway of therapeutic resistance. Nat Rev Clin Oncol. 2020;17(6):360–71.
49. Stoecklein NH, Fischer JC, Niederacher D, Terstappen LW. Challenges for CTC-based liquid biopsies: low CTC frequency and diagnostic leukapheresis as a potential solution. Expert Rev Mol Diagn. 2016;16(2):147–64.
50. Mohme M, Riethdorf S, Pantel K. Circulating and disseminated tumour cells - mechanisms of immune surveillance and escape. Nat Rev Clin Oncol. 2017;14(3):155–67.
51. Agerbæk MØ, Bang-Christensen SR, Yang MH, et al. The VAR2CSA malaria protein efficiently retrieves circulating tumor cells in an EpCAM-independent manner. Nat Commun. 2018;9(1):3279.
52. Yu M, Bardia A, Wittner BS, et al. Circulating breast tumor cells exhibit dynamic changes in epithelial and mesenchymal composition. Science. 2013;339(6119):580–4.
53. Pecot CV, Bischoff FZ, Mayer JA, et al. A novel platform for detection of CK+ and CK- CTCs. Cancer Discov. 2011;1(7):580–6.
54. Kamarajugadda S, Stemboroski L, Cai Q, Simpson NE, Nayak S, Tan M, Lu J. Glucose oxidation modulates anoikis and tumor metastasis. Mol Cell Biol. 2012;32(10):1893–907.
55. Schafer ZT, Grassian AR, Song L, et al. Antioxidant and oncogene rescue of metabolic defects caused by loss of matrix attachment. Nature. 2009;461(7260):109–13.
56. Piskounova E, Agathocleous M, Murphy MM, et al. Oxidative stress inhibits distant metastasis by human melanoma cells. Nature. 2015;527(7577):186–91.
57. Zheng Y, Miyamoto DT, Wittner BS, et al. Expression of β-globin by cancer cells promotes cell survival during blood-borne dissemination. Nat Commun. 2017;8(1):14344.
58. Chen EI, Hewel J, Krueger JS, et al. Adaptation of energy metabolism in breast cancer brain metastases. Cancer Res. 2007;67(4):1472–86.
59. Lee CK, Jeong SH, Jang C, et al. Tumor metastasis to lymph nodes requires YAP-dependent metabolic adaptation. Science. 2019;363(6427):644–9.
60. Zhao Y, Butler EB, Tan M. Targeting cellular metabolism to improve cancer therapeutics. Cell Death Dis. 2013;4(3):e532.
61. Vander Heiden MG. Targeting cancer metabolism: a therapeutic window opens. Nat Rev Drug Discov. 2011;10(9):671–84.
62. Tennant DA, Duran RV, Gottlieb E. Targeting metabolic transformation for cancer therapy. Nat Rev Cancer. 2010;10(4):267–77.
63. Hay N. Reprogramming glucose metabolism in cancer: can it be exploited for cancer therapy. Nat Rev Cancer. 2016;16(10):635–49.
64. Ely JO. 2-deoxy-D-glucose as an inhibitor of cancerous growth in animals. J Frankl Inst. 1954;258(2):157–60.
65. Landau BR, Laszlo J, Stengle J, Burk D. Certain metabolic and pharmacologic effects in cancer patients given infusions of 2-deoxy-D-glucose. J Natl Cancer Inst. 1958;21(3):485–94.
66. Dwarakanath B, Jain V. Targeting glucose metabolism with 2-deoxy-D-glucose for improving cancer therapy. Fut Oncol. 2009;5(5):581–5.
67. Gottlob K, Majewski N, Kennedy S, Kandel E, Robey RB, Hay N. Inhibition of early apoptotic events by Akt/PKB is dependent on the first committed step of glycolysis and mitochondrial hexokinase. Genes Dev. 2001;15(11):1406–18.
68. Tsai HJ, Wilson JE. Functional organization of mammalian hexokinases: both N- and C-terminal halves of the rat type II isozyme possess catalytic sites. Arch Biochem Biophys. 1996;329(1):17–23.
69. Katabi MM, Chan HL, Karp SE, Batist G. Hexokinase type II: a novel tumor-specific promoter for gene-targeted therapy differentially expressed and regulated in human cancer cells. Hum Gene Ther. 1999;10(2):155–64.
70. Pedersen PL, Mathupala S, Rempel A, Geschwind JF, Ko YH. Mitochondrial bound type II hexokinase: a key player in the growth and survival of many cancers and an ideal prospect for therapeutic intervention. Biochim Biophys Acta. 2002;1555(1–3):14–20.
71. Kim JW, Gao P, Liu YC, Semenza GL, Dang CV. Hypoxia-inducible factor 1 and dysregulated c-Myc cooperatively induce vascular endothelial growth factor and metabolic switches hexo-

kinase 2 and pyruvate dehydrogenase kinase 1. Mol Cell Biol. 2007;27(21):7381–93.
72. Lin H, Zeng J, Xie R, et al. Discovery of a novel 2,6-disubstituted glucosamine series of potent and selective hexokinase 2 inhibitors. ACS Med Chem Lett. 2016;7(3):217–22.
73. Behar V, Pahima H, Kozminsky-Atias A, et al. A hexokinase 2 modulator for field-directed treatment of experimental actinic keratoses. J Invest Dermatol. 2018;138(12):2635–43.
74. Li W, Zheng M, Wu S, et al. Benserazide, a dopadecarboxylase inhibitor, suppresses tumor growth by targeting hexokinase 2. J Exp Clin Canc Res. 2017;36(1):58.
75. Xu S, Catapang A, Doh HM, et al. Hexokinase 2 is targetable for HK1-negative, HK2-positive tumors from a wide variety of tissues of origin. J Nucl Med. 2018:jnumed.118.212365.
76. Xu S, Catapang A, Braas D, et al. A precision therapeutic strategy for hexokinase 1-null, hexokinase 2-positive cancers. Cancer Metab. 2018;6(1):7.
77. Xu S, Herschman HR. A tumor agnostic therapeutic strategy for hexokinase 1-null/hexokinase 2-positive cancers. Cancer Res. 2019;79(23):5907–14.
78. Xu S, Zhou T, Doh HM, et al. An HK2 antisense oligonucleotide induces synthetic lethality in HK1-HK2+ multiple myeloma. Cancer Res. 2019;79(10):2748–60.
79. Mossenta M, Busato D, Dal Bo M, Toffoli G. Glucose metabolism and oxidative stress in hepatocellular carcinoma: role and possible implications in novel therapeutic strategies. Cancers (Basel). 2020;12(6):1668.
80. Cayrefourcq L, Mazard T, Joosse S, et al. Establishment and characterization of a cell line from human circulating colon cancer cells. Cancer Res. 2015;75(5):892–901.
81. Pantel K, Alix-Panabières C. Cell lines from circulating tumor cells. Onco Targets Ther. 2015;2(10):815–6.
82. Zhang L, Ridgway LD, Wetzel MD, et al. The identification and characterization of breast cancer CTCs competent for brain metastasis. Sci Transl Med. 2013;5(180):180ra48.
83. Yu M, Bardia A, Aceto N, et al. Ex vivo culture of circulating breast tumor cells for individualized testing of drug susceptibility. Science. 2014;345(6193):216–20.
84. Simpson KL, Stoney R, Frese KK, et al. A biobank of small cell lung cancer CDX modelselucidates inter- and intratumoral phenotypicheterogeneity. Nat Cancer. 2020;1(4):437–51.
85. Baccelli I, Schneeweiss A, Riethdorf S, et al. Identification of a populationof blood circulating tumor cells from breast cancer patientsthat initiates metastasis in a xenograft assay. NatBiotechnol. 2013;31(6):539–44.
86. Alix-Panabières C, Pantel K. Challenges in circulatingtumour cell research. Nat Rev Cancer. 2014;14(9):623–31.
87. Sun YF, Wu L, Liu SP, et al. Dissecting spatial heterogeneity and theimmune-evasionmechanism of CTCs by single-cellRNA-seq in hepatocellular carcinoma. Nat Commun. 2021;12(1):4091.

Part VI

Lymphatic System and Cancer Metastasis

Marlys H. Witte

Historically, recognition that cancers (tumors) spread through the lymphatic system—as evidenced by enlarged regional lymph nodes observed in dying cancer patients—long preceded Virchow's explanation of the origin of cancer in aberrant cells. There was then not even the most fundamental understanding of lymphatic system function in general and specifically its relation to the involved nodes. Indeed, the view held at the time was that cancers arose in "lymph"—or tissue fluid, a notion that has regained traction recently in expanded investigation of the role of the "tumor microenvironment" in the initiation and spread of cancer—a topic covered in other sections of this textbook.

Gaspar Asellius of Padua discovered the chylous lymph circulation ("lacteis venis") in a postprandial dog in 1627, remarkably the same year that William Harvey described the blood circulation. In the late 1890s, Ernest Starling's elucidation of the laws governing transcapillary fluid movement, which he termed "lymph formation," ushered in the era of lymphatic "systemomics" (Fig. 1) in the first half of the twentieth century—embryonic development and growth and regrowth of the lymphatics (currently termed "lymphvasculogenesis" and "lymphangiogenesis"), transcapillary leakage and recirculation of plasma proteins, lymphocyte trafficking, dietary fat absorption, and immune network. Specific distinguishing features characteristic of lymphatic vessels—ultrastructure, intrinsic contractility, valve function, and lymphatic-venous communications—were described.

Meanwhile, lymphatic system imaging was advancing beyond visual vital dye lymphangiography to include new modalities of direct conventional oil contrast lymphography followed by radionuclide lymphoscintigraphy and fluorescence microlymphangiography. These first two techniques allowed dynamic visualization of the peripheral and central lymphatic vasculature and lymph nodes in continuity in the living human along with characteristic abnormalities arising in disease states, particularly lymphomas and metastatic process but also lymphedemas and lymphatic malformations. They also laid the groundwork for image-guided access for interventional therapies enhanced by magnetic resonance imaging and computed tomography. Further insights into disturbances of lymph circulation resulted from thoracic duct cannulation, lymph sampling, drainage, and also lymph diversion and lymphatic decompression under conditions of lymph overload and lymphatic obstruction.

In the late 1990s, the era of molecular lymphology dawned with the discovery of specific vascular growth factor ligand-receptor combinations (VEGF and angiopoietin families) and a cascade of lymphatic-specific or predominantly lymphatic-directed molecules incorporated in the signaling pathways for lymphvasculogenesis and lymphangiogenesis. Lymphatic systemomics is now being reenergized as these molecules and pathways find expression in clinical syndromes and particularly hereditary lymphedema syndromes, where genotype-phenotype studies are uncovering previously unrecognized and unsuspected links of the lymphatic system to a wide variety of vital processes and disorders, including cancer and its spread.

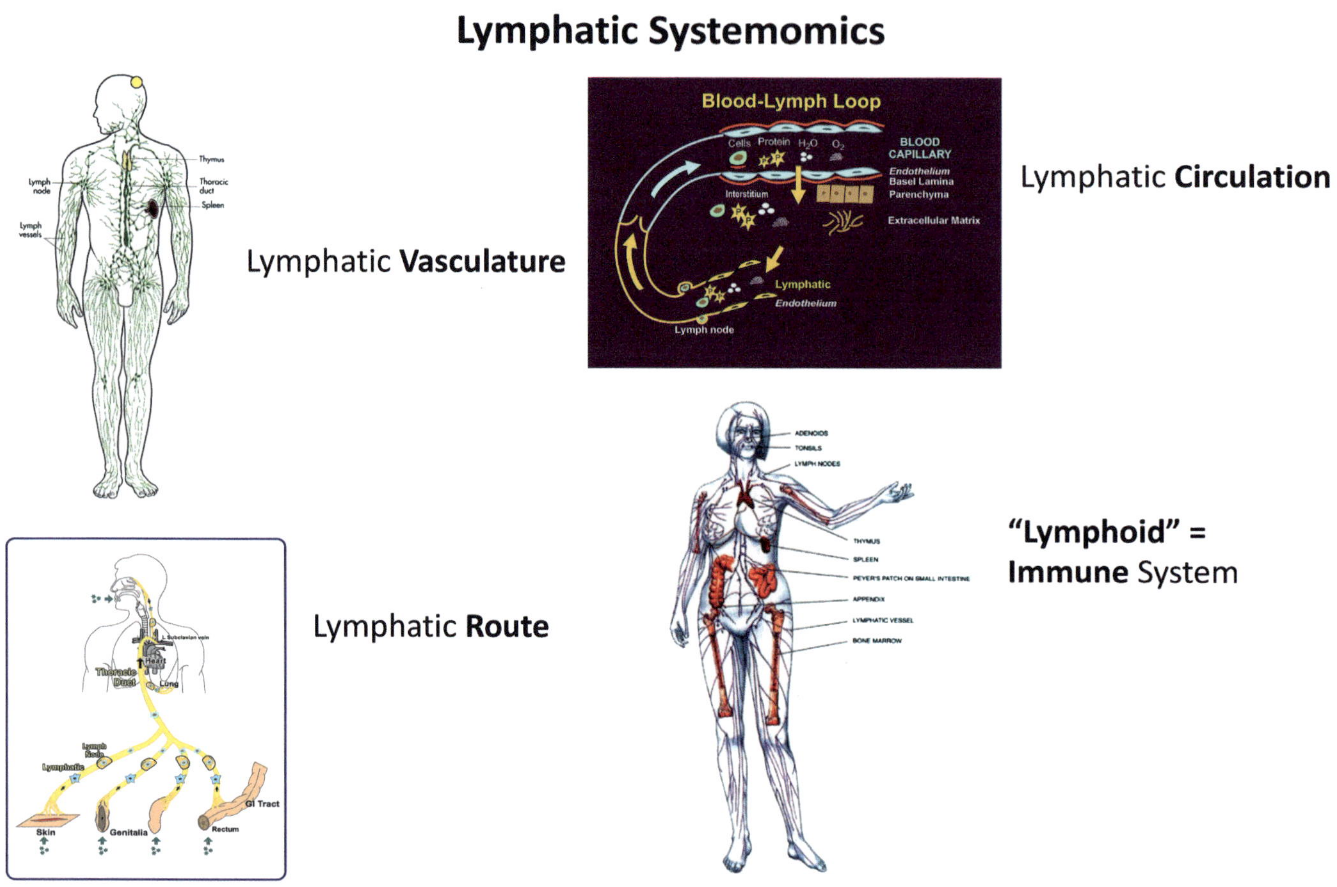

Fig. 1 Lymphatic systemomics encompasses lymphatic system functions as a vasculature, circulation of extravascular extracellular fluid, route of entry into the body, and (image according to Sir Gus Nossal) center of the immune network. Each component can promote or defend against cancer spread

In this section, we will lay the groundwork for understanding the lymphatic system as a vasculature, a circulation of extravascular extracellular fluid, a route of entry and transport of foreign particles and abnormal cells, and the center of the immune system itself. In that light, we will relate the structure and function of the lymphatic system to the events in and manifestations of cancer metastasis and how this system influences both positively and negatively the fate of the cancer cell and ultimately the fate of the host.

19 Lymphatic System Biology, Pathobiology, and Relation to Cancer Metastasis

Marlys H. Witte and Sarah K. Daley

Abstract

Although described since ancient times, the lymphatic system was not fully recognized as an integrated vasculature, circulation, and center of the immune network until the discipline of lymphology was established a little more than half a century ago. This chapter reviews its distinctive macroscopic and microscopic anatomy and the physiologic principles governing lymph formation, absorption, and propulsion. In addition, we discuss the pathobiology and therapeutic implications underlying common lymphatic disorders. Finally, links to cancer, its complications, and spread are highlighted.

Learning Objectives

- Distinguish the vascular, circulatory, and immunologic components of the lymphatic system.
- Define the general macroscopic anatomy and distribution of the peripheral and central lymphatic vasculature.
- Enumerate the distinctive microscopic features of lymphatic vessels and their specific immunostaining properties.
- Define Starling's law of transcapillary fluid exchange (lymph formation) and the principle of lymph formation–lymph absorption balance and the imbalances in edema/lymphedema.
- Name factors/forces involved in lymph flow and propulsion.
- List several ways lymphatic biology and pathobiology relate to cancer theranostics, spread, and complications.

M. H. Witte (✉)
Department of Surgery, Neurosurgery, and Pediatrics, University of Arizona College of Medicine, Tucson, AZ, USA
e-mail: grace@surgery.arizona.edu

S. K. Daley
Advanced Pathology Research Fellow in Lymphology, Department of Surgery, University of Arizona College of Medicine, Tucson, AZ, USA
e-mail: skdaley@arizona.edu

19.1 Introduction

Scarcely visible during life and collapsed after death, lymphatic vessels, in contrast to blood vessels, captured little attention following their discovery in ancient times and even after Gaspar Asellius described the milky, chyliferous lymph circulating in a well-fed dog [1]. On the other hand, William Harvey's discovery of the blood circulation published that same year (1627) excited an explosion of interest and spurred rapidly accumulating knowledge [2]. Nonetheless, the centuries that followed witnessed meticulous postmortem anatomic dissections establishing the nearly ubiquitous presence of lymphatic vessels throughout the body. In 1896 and the decades thereafter, the principles of transcapillary fluid exchange (lymph formation) and transcapillary protein movement and recirculation were elucidated along with the processes of lymphatic propulsion, growth, and development.

Beginning during the last half of the twentieth century, with the ability to dynamically image lymphatic vessels and lymph nodes in the living human, lymphology as a discipline finally emerged. Not merely "lymph nodes held together by strings," the lymphatic system came to be fully recognized as a distinctive vasculature, fluid circulation, route of entry into the body, and center of the immune network ("lymphatic systemomics"), thereby playing a vital role in both health and disease (Figure in Overview, Lymphatic Systemomics).

S. P. Leong et al. (eds.), *Cancer Metastasis Through the Lymphovascular System*,
https://doi.org/10.1007/978-3-030-93084-4_19

The close relationships between the growth of cancer, both its containment and spread, and resulting complications is one such example.

The lymph circulation is a unidirectional vascular system that transports surplus tissue fluid back to the bloodstream. Furthermore, this network stabilizes the mobile intercellular liquid and extracellular matrix microenvironment to ensure parenchymal cellular integrity and function. The lymphatic system is composed of vascular conduits: lymphoid organs including the lymph nodes and cellular elements circulating in the liquid lymph, such as lymphocytes and macrophages. These cells, migrating in the blood vascular system, cross the blood capillary barrier along with immunoglobulins, polypeptides, plasma protein complexes, and cytokines and enter the lymphatics to return, mixed with specific lymph "absorbents" such as chylomicra, to the bloodstream. Whereas body water circulates very rapidly as a plasma suspension of red blood cells within the blood vascular compartment, it percolates slowly outside the bloodstream as a tissue fluid-lymph suspension of lymphocytes through lymph vessels and lymph nodes. As a specialized subcompartment of the extracellular space, therefore, the lymphatic system closes the "blood–lymph loop" of the circulation by returning liquid, macromolecules, trafficking cells, and other blood elements that "escape" or "leak" from blood capillaries.

19.2 Lymphatic Vasculature: Structural–Functional Aspects (Fig. 19.1) [3, 4]

Identification of lymphatic vessels was long hampered by lack of readily identifiable structures. Subsequent anatomic dissections established the presence of lymphatics throughout the body accompanying venous trunks everywhere except in the central nervous system and cortical bony skeleton. The latter have their own special lymph drainage systems.

In general, lymph from the lower torso and viscera enters the bloodstream via the thoracic duct at the left sub-

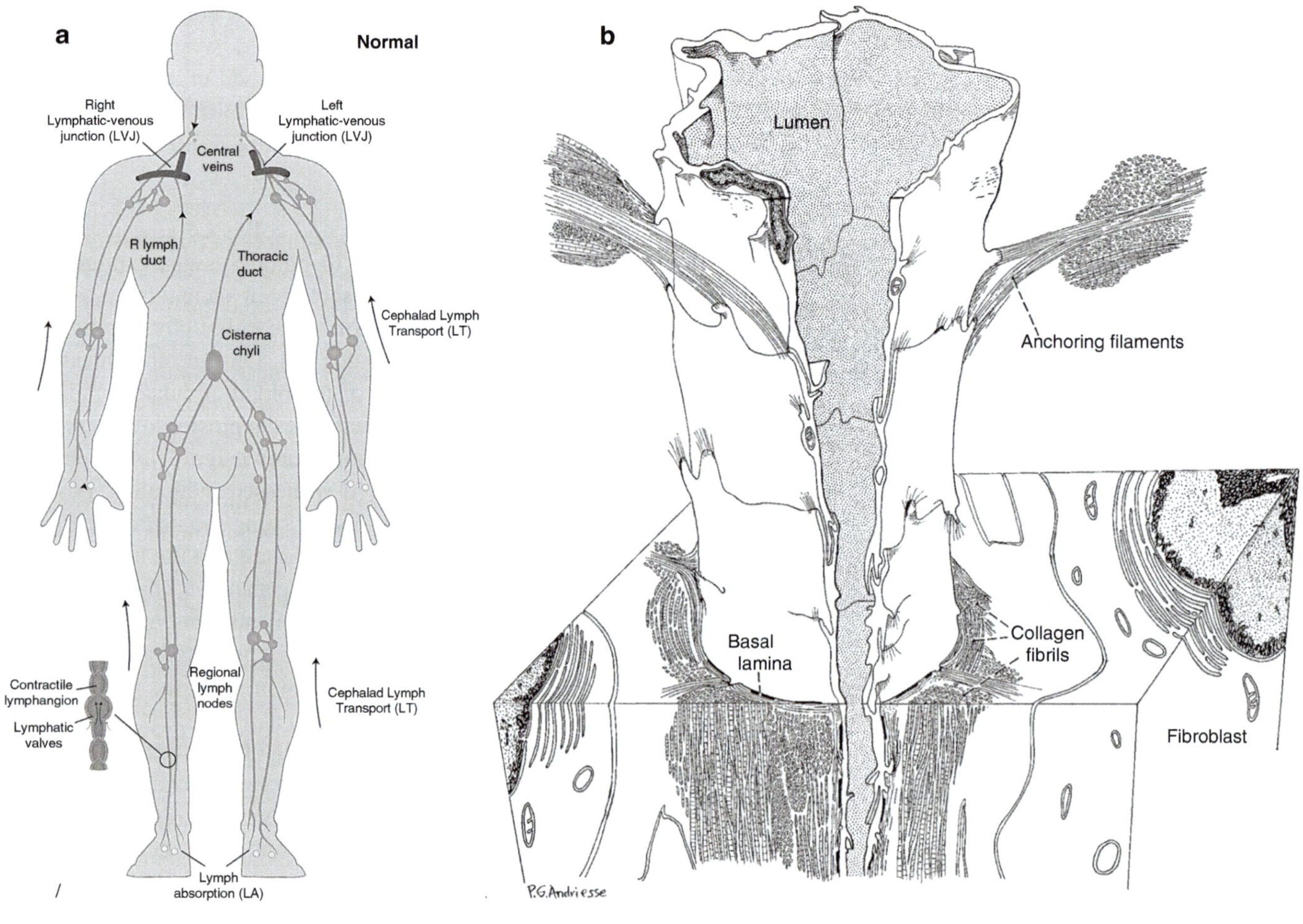

Fig. 19.1 Anatomy of the lymphatic system. (**a**) Schematic depicting the normal macroscopic anatomy and direction of flow of the peripheral and central lymphatic system. (**b**) Lymphatic capillary reconstructed by collated electron micrographs The anchoring filaments originate from the abluminal surface of the lymphatic endothelial cells and extend into the adjacent collagen bundles, thereby forming a firm connection between the capillary wall and the surrounding interstitium. Reproduced with permission of Leak (1970) Microvasc Res 2: 361–391 [49]. See text for further details

clavian–jugular venous junction (Fig. 19.1a). Lymphatics from the head and neck and upper extremities enter central veins either independently or via a common supraclavicular cistern. Numerous interconnections exist within this rich vascular network. The bulk of cardiac and pulmonary lymph empties into the great veins on the right side of the neck whereas intestinal lymph transporting cholesterol, fat-soluble vitamins, and long-chain triglycerides as chylomicra courses retroperitoneally to the aortic hiatus, then joins with other visceral and retroperitoneal lymphatics to form the cisterna chyli and thoracic duct. Most liver lymph flows retrograde or countercurrent to portal blood and joins intestinal lymph collectors just before the origin of the thoracic duct. Recognition of these topographic features and their variants is essential for understanding patterns of cancer spread.

Although the retina and brain do not technically have lymphatic apparatuses, they possess analogous circulations, such as the aqueous humor canal of Schlemm (anterior chamber of the eye) or the cerebrospinal fluid/subarachnoid villus (pacchionian bodies) connections (the brain). Natural vasomotion of cerebral intraparenchymal arteries/arterioles generates a force that moves fluid from the perivascular space into the brain parenchyma crossing the glial perivascular sheet via water pores present in the membrane. Glial elements with their astrocytic foot processes sweeping tissue fluid and nonendothelial-lined intracerebral perivascular (Virchow–Robin) spaces mix with cerebrospinal fluid to form the glymphatic system [5]. The latter empties into the endothelial-lined meningeal lymphatic capillaries and thence to the contractile collecting cervical lymphatics, which pass through the cervical lymph nodes to enter the central lymph and bloodstream.

Regarding development of the lymphatic vasculature, vigorous controversy has persisted since the early 1900s about lymphvasculogenesis and how endothelial precursors or stem cells, such as lymphangioblasts, differentiate and proliferate into a primitive tubular network. This network then expands by subsequent lymphangiogenesis (sprouting from existing vessels) whether derived largely from central veins with growth centrifugally but also arising independently from tissue mesenchyme in peripheral tissues with growth centripetally. This subject is extensively reviewed in the following chapter by Geng and Srinivathan along with its relation to epi(endo)thelial-mesenchymal transition (EMT-MET) and recapitulation in cancer metastasis [6].

Under light microscopy, efferent lymphatics originate within the interstitium as specialized capillaries or less commonly (e.g., liver, brain) from nonendothelialized precapillary channels. Lymphatic capillaries are remarkably porous and readily permit entry of large macromolecules (molecular weight > 1000 kD). In this respect, they resemble the uniquely "leaky" fenestrated hepatic sinusoidal blood capillaries but are in distinct contrast to most blood capillaries, which are relatively impervious to macromolecules even albumin. A variety of cells pass through specialized lymphatic endothelial junctions as well as migrate transcellularly.

It is difficult to distinguish between blood and lymph vessels at the tissue level without special stains, although the latter are usually thin-walled and tortuous, have a wider, more irregular lumen, and are largely devoid of red blood cells. Many staining features have been advocated to differentiate between blood and lymph microvasculature but it was not until recently that histochemical specificity has been refined [7]. One of the most commonly used specific markers currently in use is LYVE-1, with D-240 widely used to stain clinical specimens. LYVE-1 has been applied to tissues ranging from early mouse embryo to adult human and highlights initial collecting vessels and lymphatic capillaries but not larger-caliber lymphatics. The LYVE-1 receptor appears to be a key entry point for dendritic cells to selectively access the lymphatic system [8]. Another strong marker localizing to the nucleus of lymphatic endothelial cells and adaptable to multiple tissues is the transcription factor PROX1 which regulates podoplanin, a transmembrane glycoprotein in lymphatics but not blood vessel endothelial cells. D2-40, the selective antibody toward podoplanin, used routinely in human tissue immunohistochemistry, also sharply distinguishes lymphatic from blood vessel endothelium and larger collecting lymphatics. This feature has been useful in identifying preexisting and new lymphatics in tumors although there is overlap of cell types. These special stains have contributed greatly to delineating lymphatics within cancers and their relationship to cancer metastasis.

Ultrastructurally, lymph capillaries display both open ("button") and "closed" (tight) ("zipper") endothelial junctions and can dramatically adjust their shape and lumen size [9, 10]. Unlike blood capillaries, a basal lamina (basement membrane) is tenuous or altogether lacking. Moreover, complex elastic fibrils, termed anchoring filaments, tether the outer portions of the endothelium to a fibrous gel matrix in the interstitium [11] (Fig. 19.1b). These filaments allow lymph microvessels to open wide causing a sudden increase in tissue fluid load and pressure, in contrast to the simultaneous collapse of adjacent blood capillaries. Just beyond the lymph capillaries, in contrast to more proximal and larger lymph collectors and trunks, the terminal lymphatics with a continuous "zipper" structure are devoid of smooth muscle, although the endothelial lining is rich in the contractile protein actin. Intraluminal bicuspid valves are also prominent features that partition the lymphatic vessels into discrete contractile segments termed lymphangions. These specialized microscopic features support the delicate functions of absorbing and transporting various elements including large protein moieties and cells from the bloodstream but also foreign agents (e.g., viruses, bacteria) that gain access to the interstitial space.

Progressive development of techniques to image the dynamic in vivo structure and function of lymphatic vessels built on centuries of technical advances. After first visualizing small channel "streamers" from intracutaneous vital dyes, direct oil contrast lymphography (LAG) followed in the mid-1950s, opening a window on the lymphatic system and its participation in a variety of disorders [12]. Initial interest focused on primary tumors of the lymphatic system (lymphomas) and cancer primaries and metastases along with staging of these conditions. Modern imaging techniques (discussed at length in a following Chapter by R Witte and Bernas [13]) include the "gold standard" whole-body lymphangioscintigraphy (LAS) after subcutaneous or intradermal injection of protein-bound radiotracers [14]. LAS provides a dynamic image of the peripheral and central lymphatic system noninvasively. Single Photon Emission Computed Tomography (SPECT-CT) greatly improves the sensitivity and spatial localization with 3D higher resolution LAS images. Other agents used for indirect lymphography include various fluorescent or magnetic particles, infrared particles and dyes, immunoglobulin conjugates, and microbubbles for detection with fluorescent microscopes, optical imaging systems, computed tomography (CT), magnetic resonance imaging (MRI; with and without contrast), and ultrasound including in combination with light (photoacoustics). Most important clinically has been the refinement of percutaneous MR imaging with direct puncture of central lymphatics and also inguinal node injection of contrast to open up the central lymphatic system to clear view and with it, the opportunity to intervene therapeutically [15, 16].

19.3 Lymph Circulation: Lymph Formation, Absorption, and Propulsion (Fig. 19.2)

Any protein which leaves these vessels…is lost for the time to the vascular system…it must be collected by lymphatics and restored to the vascular system by way of the thoracic or right lymphatic duct. —Physiologist Ernest Starling, 1909 [17].

As a fine adjuster of the tissue microenvironment, the lymphatic system is neglected in most treatises on vascular diseases. Yet this seemingly delicate system plays a vital role in maintaining the liquid, protein, and osmotic equilibrium around cells and aids in absorption and distribution of nutrients, disposal of wastes, and exchange of oxygen and carbon dioxide in the local milieu intérieur.

Two-thirds of the body is composed of water, and most of this liquid volume is contained within cells. The remainder outside cells continuously circulates. In a series of epochal experiments conducted more than a century ago, Starling outlined the pivotal factors that regulate partitioning of the extracellular fluid. In brief, the distribution of fluid between the blood vascular compartment and tissues and the net flux of fluid escaping from the bloodstream (lymph formation) depends primarily on the transcapillary balance of hydrostatic and protein osmotic (oncotic) pressure gradients as modified by the character (i.e., permeability, surface area, solute reflection coefficient) of the filtering microvascular surface (depicted in Fig. 19.2, top). Normally, a small excess of tissue fluid forms continuously (net capillary filtration), and this surplus enters the lymphatic system and returns to the venous system (here, referred to as "lymph absorption"). In contrast to blood, which flows in a circular pattern at several liters per minute, lymph flows entirely in one direction and at rest at a rate of only 1.5–2.5 L/24 h. This limited volume derives from a slight hydrodynamic imbalance that favors movement of fluid, salt, and macromolecules from plasma into tissue spaces. Although blood capillary beds vary in hydraulic conductance, in general, disturbances in the transcapillary hydrostatic and protein osmotic pressure gradients (Starling forces) tend to promote edema that is low in protein content (<1.0 g/dL [10 g/L]), whereas impedance to lymph flow (lymph stasis) promotes (lymph) edema that is high in protein content (>1.5-3.5 g/dL [15–35 g/L]), reflected in parallel alterations in tissue oncotic pressure.

Although lymphatic vessels, like veins, are thin-walled flexible conduits that return liquid to the heart, the flow–pressure relationships in the venous system and the lymphatic system are different. The energy to drive blood in the venous system derives primarily from the thrust of the heart. The cardiac propulsive boost maintains a pressure head through the arteries and blood capillaries into the veins. In contrast, lymph vessels in tissues are not directly contiguous with the blood vasculature, and the chief source of energy for propulsion of lymph emanates from the intrinsic lymphatic truncal wall contractions (propulsor lymphaticum). To a lesser extent, extrinsic haphazard forces, such as yawning, breathing, muscular squeezing, and transmitted arterial pulsations, contribute to lymph propulsion. A vestige of the powerful amphibian lymph hearts, mammalian lymphatic smooth muscle cells contract in a rhythmically and coordinated fashion. Contractile waves propagate rapidly from one edge of the lymphatic trunk to the other, in the direction of tissue-to-lymph node (afferent lymphatic vessels) or lymph node-to-venous system (efferent lymphatic vessels). These waves, coupled to a well-developed intraluminal valve system, facilitate transport of lymph. In a sense, the lymphatic structures function as micropumps that respond to fluid challenges, and small rises in intraluminal pressure can lead to dramatic increases in contractile frequency. For example, in popliteal afferent lymphatic vessels, raising the intraluminal pressure 1 mmHg triples contractile frequency purportedly improving movement of lymph. Ordinarily, resistance to flow in the lymphatic vessels is relatively high in comparison to the low resistance in the venous system, but the pumping capacity of the lymphatics is able to overcome this imped-

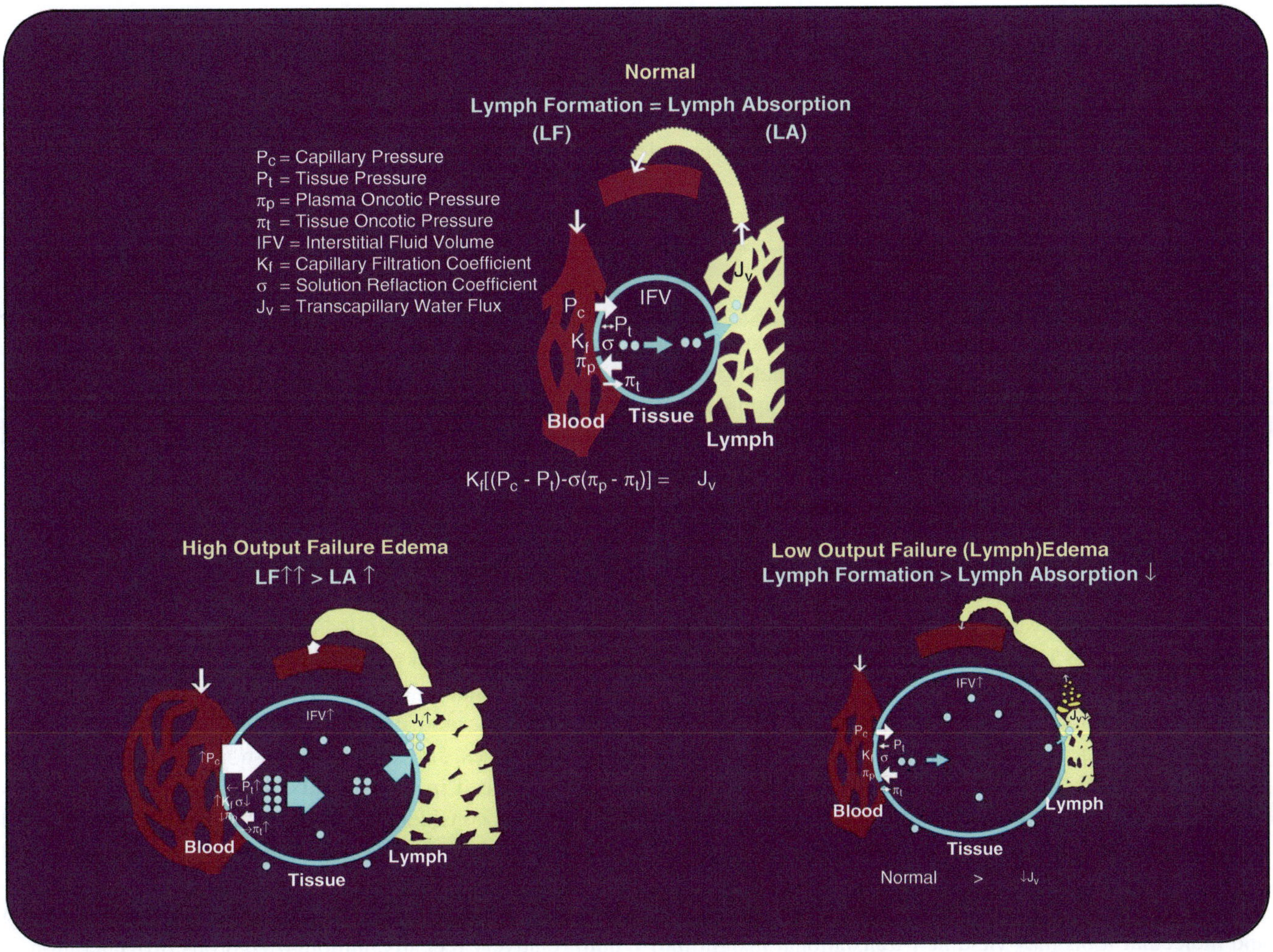

Fig. 19.2 Physiologic factors involved in lymph formation and lymph absorption. Normally (top), these two processes are in balance. Edema appears when lymph formation exceeds lymph absorption either because excessive lymph formation overwhelms the lymphatic absorptive capacity—high output failure of the lymph circulation (bottom, left) or lymph absorption is defective and cannot handle even a normal lymphatic load—low output failure (bottom, right) as in classical lymphedema, or a mixture of both pathomechanisms. Treatment aims to restore the balance between the two processes by either reducing lymph formation or enhancing lymph absorption or both. Examples of high output as well as low output failure of the lymph circulation during the course of cancer spread are common clinical manifestations as illustrated in Fig. 19.4

ance by generating intraluminal pressures of 30–50 mmHg and sometimes even equaling or exceeding arterial pressure. This formidable lymphatic ejection force is modulated not only by filling pressure but also by temperature, sympathomimetic agents, by-products of arachidonic acid metabolism (thromoxanes, prostaglandins), neurogenic stimuli, circulating hormones, and locally released paracrine and autocrine cytokine secretions and other products of endothelial metabolism [18, 19]. The pacemaking activity present in lymphatic vessels is responsible for coordinating contractions and setting their rhythm, consequently affecting lymph flow. Lymphatic contractions are a result of action potentials that originate within the lymphangion, likely due to activity of pressure-sensitive chloride ion channels in lymphatic smooth muscle cells. Theoretically, any smooth muscle cell in the wall of lymphatic vessels can act as a pacemaker, and the rhythm is set by the cell that is faster in firing an action potential, which propagates through the lymphangion wall [19]. Although muscular contraction and external massage clearly accelerate lymph return in the presence of edema, under normal conditions, peripheral lymph flow is regulated primarily by spontaneous contraction of the lymphatics themselves. In peripheral lymphatics, unlike peripheral veins, the column of liquid is incomplete. Accordingly, with normal intralymphatic pressure, external compression is ineffective in propelling lymph onward, although it may increase the frequency and amplitude of lymphatic contractions. During lymphatic obstruction and persistent lymph stasis, the fluid column in the lymphatics becomes continuous, and skeletal muscle or forceful external compression then becomes an effective pumping mechanism that aids lymph transport. Assuming an erect position sharply raises distal venous pressure, but peripheral intralymphatic pressure is unaffected, even though lymphatic truncal pulsation increases in both frequency and amplitude, favoring removal of tissue fluid.

19.4 Lymphatic Transport Route

The lymphatic route of transport of liquid, protein, and macromolecules, particularly chylomicra originating in the small intestine, is well recognized albeit underappreciated. The lymphogenous route of cancer spread forms the conceptual basis for most current treatment approaches to solid organ cancers. The role of lymphatic uptake and clearance of microorganisms in the clinical manifestations (e.g., lymphangitic streaking), pathogenesis, inflammatory and other tissue responses, and immune defense has been underemphasized, particularly in recent years. From the 1930s to the 1960s, a number of carefully performed studies examined the critical lymphatic route in the initial and later phases of viral and bacterial infection and the dynamic physiology of the afferent and efferent arms of the immune response. Lymphologist Joseph Yoffey laid the groundwork for recognition of the importance of the "inert" lymphocyte and the discovery of lymphoid cell trafficking, which are fundamental to the principles of modern immunology [4]. Recasting these older investigations with more modern physiologic techniques and molecular tools may yield new translational insight into disorders ranging from AIDS and tuberculosis to parasitic disorders and disturbances in the external (skin) or internal (gastrointestinal, respiratory, and genitourinary tract) microbiomes while also elucidating the lymphatic spread of cancer [20]. It is likely that SARS-Cov2 (COVID-19) will also turn out to be an important but currently unrecognized lymph-borne infection [21].

19.5 Lymphatic Center of Immune Network: Lymph Nodes, Lymphoid Organs, and Trafficking Cells

In addition to their central immunologic role, lymph nodes are a potential site of impediment to the free flow of lymph. Unlike frogs, which lack lymph nodes but possess four or more strategically placed lymph hearts that propel large quantities of peripheral lymph back to the bloodstream, mammals possess immunoreactive lymph nodes, which, when swollen, fibrotic, or atrophic, may act to initiate or perpetuate lymph stasis. On the other hand, they also perform an important "barrier function" first recognized in the 1860s in regard to the spread of tumors, particulates, and infectious agents. Lymphoid organs also include the spleen, Peyer patches, thymus, and nasopharyngeal tonsils. The importance of immune cell trafficking through the blood–lymph loop is now well recognized for health and in a wide variety of disorders. However, the regulation and intricate mechanisms involved, including in cancer progression and defense, are just beginning to be unraveled and are discussed at length in the following Section Chapter by Jackson [22].

19.6 Molecular Lymphology

Advances in molecular biology and the development of new techniques from the Human Genome Project have ushered in the era of molecular lymphology.

19.6.1 Lymphangiogenesis

In contrast to uncontrolled new growth (lymphangiosarcoma), programmed proliferation of lymphatic endothelium with tube formation (lymphangiogenesis) is critical to a host of physiologic and pathologic processes. During the past several decades, since the phenomenon of angiogenesis was first reproduced in endothelial cell and mixed vascular tissue cultures, considerable attention has been directed toward understanding this process, but largely in the context of blood vessels ('hemangiogenesis") [23]. The lymphatic counterpart (i.e., lymphangiogenesis) had received scant attention until relatively recently, although vigorous (re)generation of lymphatics has been recognized since the early 1900s. In the 1980s, lymphatic endothelial cells were isolated from lymphatic tumors, collecting ducts, and since, from a variety of other tissues and passed in culture with demonstration of "lymphangiogenesis in vitro" [24–27]. Essential to health, lymphatic growth, when uncontrolled, underlies a variety of uncommon, often disfiguring, and sometimes life- and limb-threatening conditions [28]. On the other hand, lymphangiogenesis may contribute to or protect against many other common disorders. This topic is discussed at length in the following Section Chapters and particularly in the Chapter by Detmar [29].

19.6.2 Lymphvascular Genomics–Proteomics

Lymphangiogenesis is clearly under genetic control and multiple genes are continuing to be found specifically associated with human lymphedema syndromes and lymphatic malformations on nearly every chromosome and also in chromosomal deletions, translocations, and aneuploidy as summarized in the Section Chapter by Erickson and Dellinger [30]. Many of these same genes and related proteins have been implicated in cancer growth and metastasis, and many more have been identified putatively in mice and other animal species with uncertain application to human disorders. These include growth factor genes, their products, and transcription factors, most of which were discovered in human studies by reverse and forward genetics and more recently by next-generation sequencing (whole genome and whole exome sequencing) [31–33].

19.6.3 Growth/Inhibitory Factors

Although the underlying mechanisms governing lymphangiogenesis are only beginning to be understood, a variety of stimulatory [most importantly vascular endothelial growth factor C (VEGF-C) discovered by Alitalo [34] and its signaling pathway] and inhibitory factors have been discovered to play a key role. Other factors are context-dependent, such as the angiopoietin-1–angiopoietin-2 yin-yang interaction [35], and these also modulate growth, development, and remodeling of the lymphatic vasculature. Function/dysfunction of the lymphatic vasculature may be influenced in common with or distinct from effects on the blood vasculature [36]. Invaluable insights have been gained from transgenic, spontaneous, and chemical- and/or radiation-induced mutant mouse models. These angiomodulatory factors are crucial to understanding clinical conditions as diverse as cystic hygroma, Klippel–Trénaunay syndrome, (lymph)angiosarcoma, elephantiasis, and a constellation of peripheral and visceral disorders characterized by swelling, scarring, immunodysregulation, and malnutrition.

19.7 Pathobiology of Lymph Formation–Lymph Absorption: The Formula of Edema (Fig. 19.2)

Edema or tissue swelling, a reflection of lymph formation–lymph absorption imbalance, is one of the commonest manifestations of disease, ranging from a cosmetic imperfection as in transient foot swelling after a long flight to the immediate cause of death when surrounding the heart. The common local or generalized edemas ("high-output" lymph flow failures) (Fig. 19.2, bottom, left), where increased lymph formation outpaces enhanced but inadequate lymph absorption, are seen after venous occlusion in congestive heart failure, and in hepatic cirrhosis-associated portal venous hypertension. Here, increased lymph formation results from an imbalance in the Starling forces (elevated blood capillary pressure) along with progressive impediments to increased lymph flow [37–39]. Accordingly, lymphatic overload is approached therapeutically by restoring lymph balance by either reducing lymph formation (e.g., lowering peripheral, central, or portal venous pressure) or accelerating lymph drainage or both [40–42]. Events at the thoracic duct–venous junction in the neck such as central venous hypertension or a pinchcock effect from limited distensibility may be critical in interfering with the free flow of overproduced lymph [42]. In this situation, thoracic duct decompression through shunting or stenting into lower pressure veins may be beneficial.

Because congenital absence and radical excision of regional lymph nodes (and hence, lymphatics) are associated with edema, it has seemed straightforward that lymphedema is simply the direct result of insufficient lymphatic drainage ("low output failure" of the lymphatic circulation) (Fig. 19.2, bottom, right). However, persistent peripheral lymphedema has proved both hard to reproduce experimentally and challenging to treat. Initial experimental attempts to simulate the clinical condition by lymphatic sclerosis and radical excision were unsuccessful and revealed a remarkable capacity of obstructed lymphatics to regenerate and "bridge the gap" or bypass the induced blockage with spontaneous opening of auxiliary lymphatic–venous shunts. It is now nonetheless clear that nonpitting, brawny extremity edema can arise from lymph stasis alone and by the unremitting accumulation of protein-rich fluid in the extracellular matrix. Progressive truncal tortuosity and dilatation give way to marked lymphangiectasia, valvular incompetence, and retrograde flow (dermal backflow) [26]. Serial microscopy discloses mononuclear cell infiltration, intramural destruction of lymphatic collectors, and collagen and adipose deposition throughout the soft tissues. Eventually, the lymph trunks lose their distinctive smooth muscle and endothelial lining, and the boundary lines between lymph collectors and the surrounding matrix progressively blur. Pathophysiology and clinical aspects of lymphedema are discussed at length in the Section Chapter by Szuba, Boccardo, and Piller as well as in referenced textbooks on lymphology [3, 12, 43].

Chylous and non-chylous lymph reflux can also occur including as a complication of cancer and cancer metastasis. Abnormal retrograde transport of lymph is termed reflux. When the lymph derives from outside the intestine (e.g., from the limbs, where it is generally straw-colored to watery clear), it is termed non-chylous reflux, whereas lymph arising from the intestine (generally milky in color) and flowing retrograde is termed chylous reflux. Because cholesterol and long-chain triglycerides in the form of chylomicra are absorbed exclusively by the lymphatic system, dysfunction (in the form of disruption, compression, obstruction, or fistulization) of mesenteric lacteals, the cisterna chyli, and the thoracic duct is directly linked to chylothorax, chylous ascites, chyluria, chylometrorrhagia, and protein-losing enteropathy. In some patients with high-grade blockage of intestinal lymph flow, the peripheral lymphatics gradually dilate, and with progressive valvular incompetence, lactescent lymph refluxes into the soft tissues of the pelvis, scrotum, and lower extremities (chylous vesicles and chyledema) or even into the bronchial tree, manifesting as chyloptysis [44].

The aforementioned pathologic processes can be imaged and intervention undertaken by physical methods, endovascular image-guided techniques, and surgery to decompress or reroute the obstructed or overloaded lymphatic system. To ameliorate edema including lymphedema, either lymph formation must be reduced or lymph drainage enhanced to bring the two processes back into balance. This principle and these therapeutic approaches apply as well to edema, effusions, and chylous reflux during the course of primary cancers, cancer metastasis, and cancer-associated treatment complications.

19.8 Lymphangiotumorigenesis and Tumor Lymphatics

A rare but revealing sequela of long-standing peripheral lymphedema is the occurrence of lymphangiosarcoma and/or angiosarcoma and perhaps also other opportunistic neoplasms. This aggressive vascular malignancy was once thought to arise exclusively after radical mastectomy and irradiation for local control of breast cancer (Stewart–Treves syndrome) (illustrated in Fig. 19.4b, f). However, lymphangiosarcoma has now been documented in other secondary lymphedemas and even in congenital or primary lymphedema. More recently, Kaposi sarcoma (Fig. 19.4c), a vascular tumor akin to Stewart–Treves syndrome and allied closely with AIDS, has been linked to an origin from virally transformed lymphatic endothelium. Perhaps local or generalized immunodeficiency/dysregulation underlies a wide range of common and bizarre vasoproliferative and lymphologic syndromes, including hemolymphangioma (presenting "angiotumorigenesis"), Klippel–Trénaunay syndrome, Gorham–Stout syndrome and multiple lymphangiomatosis (representing as bony lymphangiomatosis or "disappearing bone disease"), angiofollicular hyperplasia (epithelioid hemangioma), lymphangioleiomyomatosis, Kaposiform hemangioendotheliosis, tumor lymphangiogenesis, and lymphangitic metastatic carcinomatosis. Specific genes have been identified in the RAS and MAPK pathway that underlie some of the lymphatic malformations. Moreover, lymphatic-directed angioinhibitory therapy, such as targeting mitogen-activated protein kinase (MEK), has been reported to be effective in controlling and even reversing the condition.

Tumor angiogenesis, which generally refers to hemangiogenesis, has been extensively studied since Folkman's original observations and has also become a major therapeutic target. Tumor lymphangiogenesis, a phenomenon whose very existence until relatively recently was questioned, has now become a "hot topic," discussed at length in the Section Chapter by Detmar [29].

Novel and potential approaches to modulate these processes include gene therapy to promote lymphangiogenesis when inadequate (as in primary lymphedema) or inhibit lymphatic growth when excessive (as in lymphangiomatosis). Stem cell therapy to grow new vessels from undifferentiated embryonic or adult stem cells (e.g., lymphangioblasts) and new biomaterials for artificial lymphatics and scaffolding to drain the tissue also hold promise for the future.

19.9 Lymphatic Systemomics and Cancer Metastasis (Figs. 19.3 and 19.4)

> *When an axillary gland becomes cancerous after disease of the mamma..during a long period remains diseased without the glands next in succession or other organs becoming affected..the gland collects the hurtful ingredients absorbed from the breast and affords protection to the body... but at length proves insufficient, perhaps itself becomes a new independent source of further propagation of the poisonous matter from the diseased parts of the gland.*—Rudolf Virchow, 1868 [46]

This remarkably prescient passage from Virchow, the originator of the "cellular theory" of cancer displacing the "lymph theory" that cancer arises in an altered interstitium [47], elucidates the barrier function of lymph nodes. He had demonstrated this phenomenon previously by subcutaneously injecting particulates. In addition, the passage also suggests the sentinel node concept and points to the further lymphogenous route of dissemination of propagated cancer from the lymph node. We now know that the lymphatic system is intimately involved in all phases of cancer, from its development, progression, and spread to evaluation, prognosis, complications, treatment, treatment-related sequelae, prevention, containment, and opportunities for cure [48]. Specifically, at the biological level (Fig. 19.3), blood capillary hyperpermeability promotes excess lymph formation, lymphangiogenesis is stimulated, and

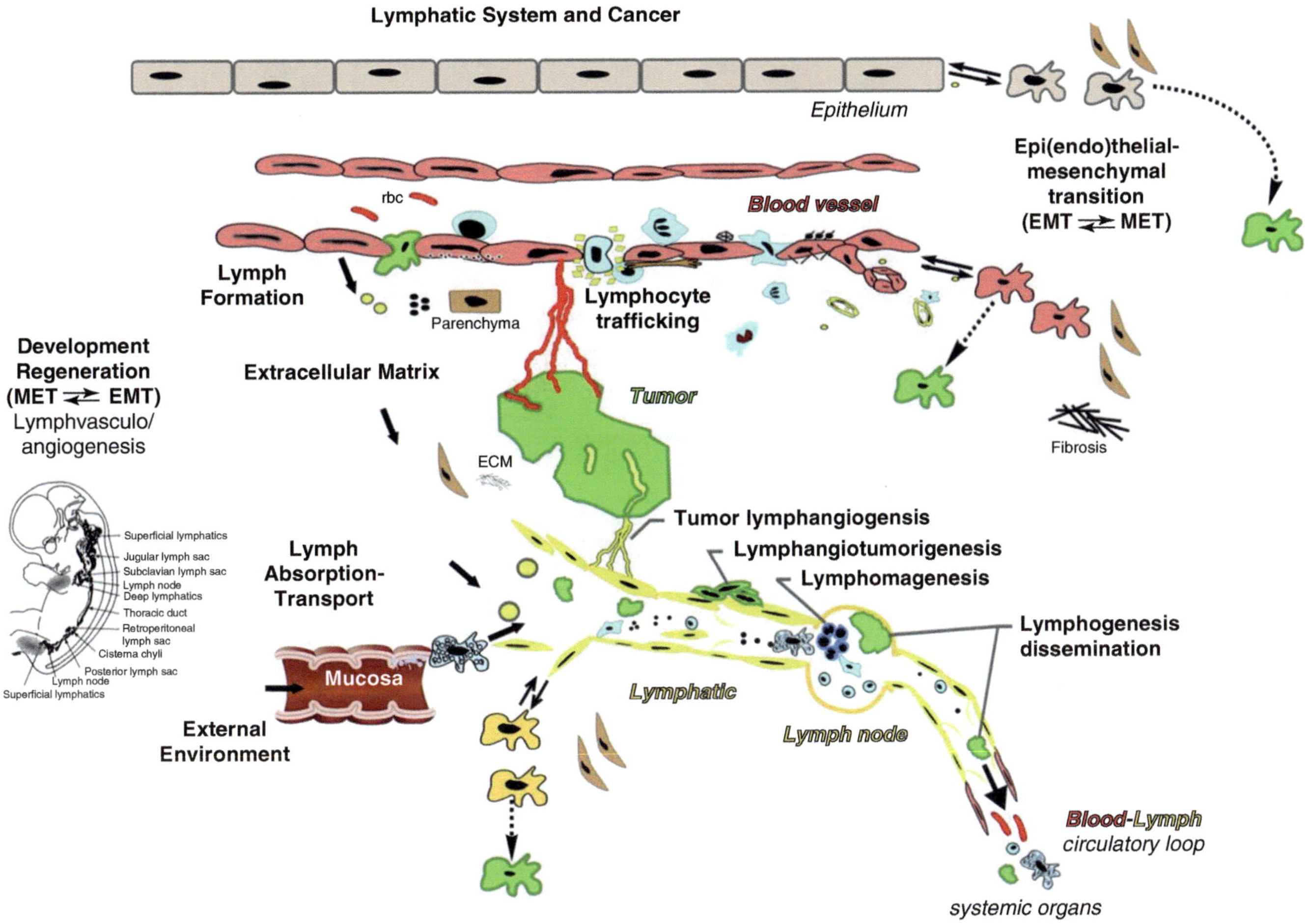

Fig. 19.3 Points of interplay between the developing cancer and ongoing processes within the interstitium and lymphatic system: lymphedema, lymphangiogenesis, lymph absorption-transport, lymphogenous tumor spread, lymphomagenesis, tumor-generated immune response. Potential sites of epi/endothelial-mesenchymal transition (EMT) and the reverse process (MET) in development/regeneration and neoplasia (green) are identified. These complex structural–functional interactions participate in the pathogenesis, clinical manifestations, evaluation, and prognosis as well as the treatment of cancer. Reproduced with permission of Witte, M et al. (2012) Clin Exp Mets 29: 707–712 [45]

lymph drainage may be compromised leading to lymph stasis and resultant edema/effusions, fibrosis, tissue overgrowth, and immunodysregulation. Primary lymphatic malignancies arising in lymphatic endothelial cells (highly aggressive lymphangiosarcomas—Stewart–Treves syndrome and Kaposi sarcoma) and in lymphocytes (lymphomas) also occur. Lymphatic involvement not only alters the local tumor microenvironment affecting crucial physical forces (solid stress, fluid pressure, stiffness, and microarchitecture in the interstitium) [48] but is the main pathway for lymphogenous spread to regional lymph nodes (sentinel nodes and beyond). The lymphatic system in its entirety—vessels, lymph, lymph nodes and lymphocytes—may also function as an immunologic barrier to further spread or alternatively, as a nidus for continued proliferation and further lymphogenous spread through central lymphatic channels to the systemic bloodstream and beyond. It is this dual and contrasting role of the lymphatic system that is a target for both current and future cancer therapy.

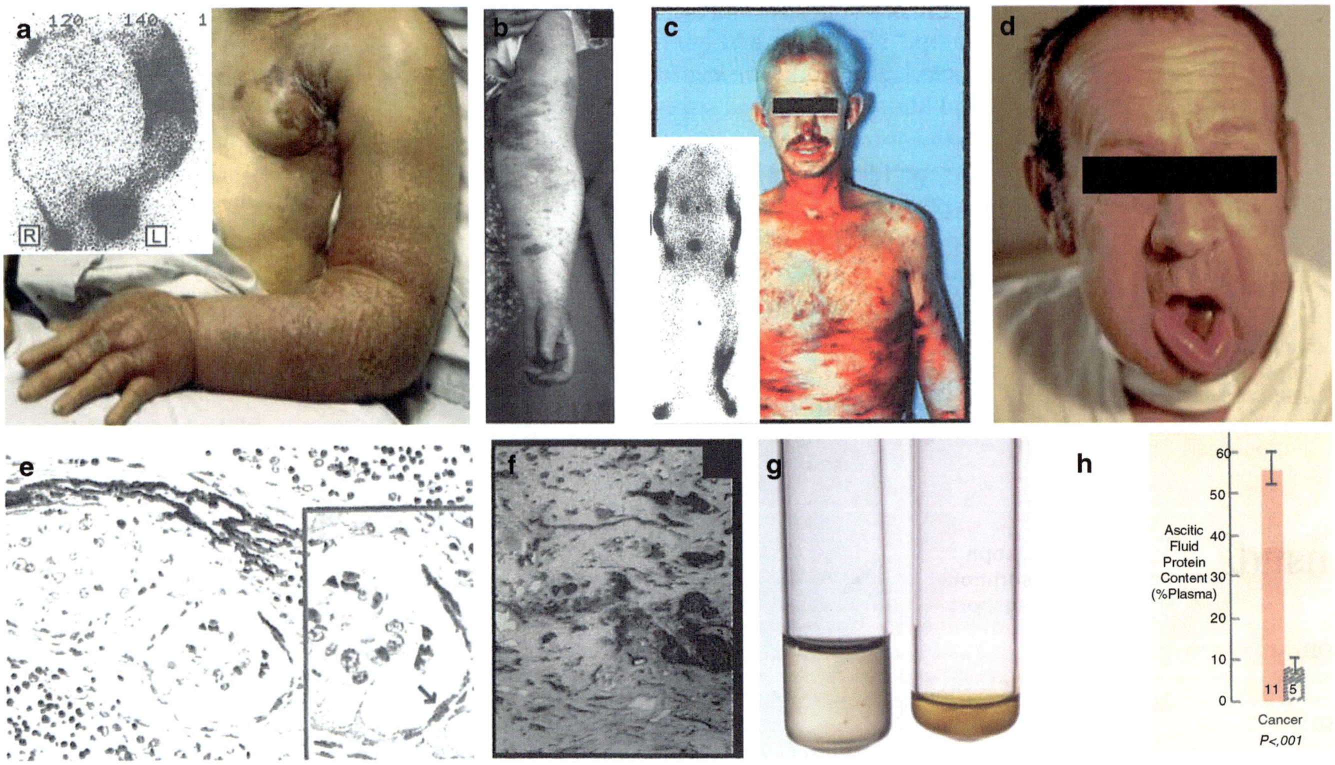

Fig. 19.4 Spectrum of clinical manifestations of lymphatic system disorders in patients with cancer metastasis (**a–d**, top row) and underlying pathology/pathophysiology of these clinical manifestations (**e–h**, lower row). (**a**) Massive lymphedema of left arm and shoulder from untreated invasive left breast cancer with (inset) corresponding left upper extremity lymphangioscintigram displaying complete obstruction of lymphatic channels with extensive dermal reflux of radiotracer. (**b**) Cutaneous spread of aggressive lymphangiosarcoma (Stewart–Treves syndrome) with associated histopathology in E, secondary to long-standing upper extremity lymphedema following initial breast cancer treatment by radical mastectomy. (**c**) Extensive disseminated AIDS-Kaposi sarcoma-associated anasarca (generalized body edema) with inset, upper and lower extremity lymphangioscintigrams displaying bizarre dilated lymphatic collector channels along all four extremities, reflecting the origin of this tumor in lymphatic endothelial cells. (**d**) Severe facial lymphedema acutely developing after radical neck dissection for invasive throat cancer—vital blue dye injected into swollen tongue gradually disappeared subsequently over several weeks as lymphatic vessels regenerated and edema subsided. (**e**) Histopathology showing nest of intralymphatic breast cancer cells in capsule of involved draining lymph node. Reproduced with permission of Nagle et al. (1987) Lymphology 20: 20–24 [50]. (**f**) Corresponding histopathology of lymphangiosarcoma in patient in (**b**). (**g**) One minute flow of thoracic duct lymph in cannulated control patient (~1 ml per minute) compared to markedly reduced minute flow of "jaundiced" (high bilirubin content) lymph in patient with massive ascites (edema fluid in abdominal cavity) secondary to extensive peritoneal lymphatic involvement from biliary tract cancer (example of low flow lymphatic failure). (**h**) illustrates that cancer spread can be associated with a high output failure of the lymph circulation and a resultant low-protein transudative ascites due to portal vein obstruction producing portal hypertension (elevated portal venous pressure) secondary to invasive pancreatic cancer (right bar) compared to more typical high protein exudative ascites associated with low-output failure of the lymph circulation from lymphatic obstruction due to peritoneal carcinomatosis (left bar). Reproduced with permission of Witte, et al. (1972) JAMA 221:1380–1381 [51]. See Fig. 19.2 for pathomechanisms and text for further details

19.10 Conclusions

For lymphatic vessels to function effectively, lymph formed from blood capillary filtrate, cell products, and trafficking cells must be absorbed from the interstitium, carried through non-leaky valveless initial lymphatics and pre-collectors, and then propelled through patent intrinsically contractile valved peripheral collecting channels and interposed lymph nodes to reach the cisterna chyli draining visceral lymph in the abdomen. Lymph finally arrives at the left and right thoracic duct, where it passes through a valved entry into the central bloodstream to complete the "blood–lymph loop" of the extracellular fluid circulation.

Lymphatic dysfunction is associated with disturbances in transcapillary exchange of liquid, macromolecules, and cells across intact and abnormal microvessels. Deranged lymphatic growth patterns and lymph kinetics are individually and together seen when tissue swelling and its sequelae occur. Low output failure of the lymph circulation manifested as peripheral lymphedema is characteristically indolent for many years before lymphatic insufficiency and edema accelerate and become persistent. Chronic lymphedema is characterized by trapping of fluid and extravasated plasma proteins and other macromolecules in the skin and subcutaneous tissues and also in the viscera and body cavities, with an increased propensity for superimposed infection and a progressive pro-growth microenvironment.

On the other hand, still more common is lymphatic overload (high output failure of the lymph circulation) as occurs in hepatic cirrhosis with portal hypertension and in right heart failure from central venous hypertension, leading to increased lymph formation from an imbalance of Starling forces along with progressive resistance to increased lymph drainage, processes that need to be brought back into balance.

Transdifferentiation and transformation of endothelium and other vascular accessory cells associated with lymph stasis may also be pivotal factors in a wide range of dysplastic and neoplastic vascular disorders, including Stewart–Treves syndrome, AIDS-associated Kaposi sarcoma, recurrent lymphangiomatosis, and lymphangitic metastatic carcinomatosis. These phenomena have their origin in controlled and uncontrolled lymphangiogenesis and apparently are regulated at specific sites across the genome. Their modulation may offer new avenues of therapy for both primary and secondary lymphedema and related angiodysplasia syndromes, as well as a variety of benign and malignant neoplasms.

Finally, while pathobiologic features of lymphatic system dysfunction are general ones, they also apply directly to understanding lymphogenous spread of cancer, its complications as well as approaches to intervene and even prevent cancer metastasis.

Unanswered Questions

- Do metastases metastasize, through which route, and under what conditions?
- Which genetic or microenvironmental influences explain why some cancers stay localized, others metastasize widely, and still others metastasize to only single or selected organs?
- What functional changes occur in the local and regional lymphatics during cancer spread and how do these contribute to controlling or favoring spread?
- Could local lymphatic obstruction or deficiency impede cancer spread while promoting primary growth?
- What factors, including lymphatic, explain delayed cancer dormancy, recurrence, and spread?
- How does lymphatic dysfunction in the tumor area contribute to physical forces influencing cancer growth, containment, and spread—i.e., solid stress, fluid pressure, tissue stiffness, and microarchitectural alterations and also to epithelial-mesenchymal transition (EMT)?
- Could "liquid biopsy" sampling central lymph (noninvasive endovascularly now) be more informative about cancer identity and spread than liquid biopsy of peripheral blood?
- What features explain the fine line between benign metastasizing tumors (e.g., lymphangiomatosis) and frank malignancies, including malignant progression/transdifferentiation in some of these benign conditions?

References

1. Asellius G. De lactibus sive lacteis venis quarto vasorum mesaraicorum genere novo inuento. Mediolani: JB Biddellium; 1627.
2. Harvey W. De Motu Cordis. Germany: Frankfurt; 1627.
3. Földi M, Földi E, editors. Földi's textbook of lymphology. 3rd ed. Munich: Urban & Fischer; 2012. 672 pp.
4. Yoffey JM, Courtice FC. Lymphatics, lymph and lymphomyeloid complex. London/New York: Academic Press Inc; 1970. 942 pp
5. Nedergaard M, Goldman SA. Glymphatic failure as a final common pathway to dementia. Science. 2020;370(6512):50–6.
6. Geng X, Srinivasan S. Lymphatic specification and development, EMT-MET, and cancer spread. Section 18, Witte MH (Section ed). In: Leong SPL, Nathanson SD, Zager JS, editors. Cancer metastasis through the lymphovascular system. Switzerland AG: Springer Nature; 2021. (in press).
7. Evangelou E, Kyzas PA, Trikalinos TA. Comparison of the diagnostic accuracy of lymphatic endothelium markers: Bayesian approach. Mod Pathol. 2005;18(11):1490–7.
8. Johnson LA, Banerji S, Lawrance W, et al. Dendritic cells enter lymph vessels by hyaluronan-mediated docking to the endothelial receptor LYVE-1. Nat Immunol. 2017;18(7):762–70.
9. Casley-Smith JR, Florey HW. The structure of normal small lymphatics. Q J Exp Physiol. 1961;46:101–6.
10. Baluk P, Fuxe J, Hashizume H, et al. Functionally specialized junctions between endothelial cells of lymphatic vessels. J Exp Med. 2007;204(10):2349–62.
11. Leak LV. Electron microscopic observation on lymphatic capillaries and the structural components of the connective tissue-lymph interface. Microvasc Res. 1970;2(4):361–91.
12. Kinmonth JB. The lymphatic diseases, lymphography, and surgery. Baltimore: Williams & Wilkins; 1972.
13. Witte RS, Bernas MJ. Lymphatic imaging and intervention: applications to cancer metastasis. Section 18, Witte MH (Section ed). In: Leong SPL, Nathanson SD, Zager JS, editors. Cancer metastasis through the lymphovascular system. Switzerland AG: Springer Nature; 2021. (in press).
14. The Diagnosis and Treatment of Peripheral Lymphedema. 2020 Consensus Document of the International Society of Lymphology. Lymphology. 2020;53(1):3–19.
15. Itkin M, Nadolski GJ. Modern techniques of lymphangiography and interventions: current status and future development. Cardiovasc Intervent Radiol. 2018;41(3):366–76.

16. Schwartz FR, James O, Kuo PH, et al. Lymphatic imaging: current noninvasive and invasive techniques. Semin Intervent Radiol. 2020;37(3):1–13.
17. Starling EH. The fluids of the body. In: The herter lecture. Chicago: WT Keener; 1909. p. 81.
18. McHale NG, Roddie IC, Thornbury KD. Nervous modulation of spontaneous contractions in bovine mesenteric lymphatics. J Physiol. 1980;309:461–72.
19. Scallan JP, Zawieja SD, Castorena-Gonzalez JA, et al. Lymphatic pumping: mechanics, mechanisms and malfunction. J Physiol. 2016;594(20):5749–68.
20. Cole WR, Witte MH, Witte CL. Lymph culture: a new tool for the investigation of human infections. Ann Surg. 1969;170(4): 705–14.
21. Witte MH, Daley SK. SARS-CoV-2/COVID-19, lymphatic vessels, lymph, and lymphology. [Editorial]. Lymphology. 2020;53(3): 97–8.
22. Jackson D. Immune cell trafficking in the lymphatic system, hyaluronan biology, and cancer metastasis. Section 18, Witte MH (Section ed). In: Leong SPL, Nathanson SD, Zager JS, editors. Cancer metastasis through the lymphovascular system. Switzerland AG: Springer Nature; 2021. (in press).
23. Folkman J, Haudenschild C. Angiogenesis in vitro. Nature. 1980;288:551–6.
24. Witte MH, Way DL, Witte CL, et al. Lymphangiogenesis: mechanisms, significance and clinical implications. In: Goldberg IV, Rosen EM, editors. Regulation of angiogenesis. Basel: Birkhäuser; 1997. p. 65–112.
25. Bowman CA, Witte MH, Witte CL, et al. Cystic hygroma reconsidered: hamartoma or neoplasm? Primary culture of an endothelial cell line from a massive cervicomediastinal cystic hygroma with bony lymphangiomatosis. Lymphology. 1984;17(1):15–22.
26. Witte MH, Witte CL. Lymphangiogenesis and lymphologic syndromes. Lymphology. 1986;19(1):21–8.
27. Witte MH, Bernas MJ, Martin CP, et al. Lymphangiogenesis and lymphangiodysplasias: from molecular to clinical lymphology. Microscop Res Tech. 2001;55(2):122–45.
28. Oliver G, Kipnis J, Randolph GJ, et al. The lymphatic vasculature in the 21st century: novel functional roles in homeostasis and disease. Cell. 2020;182(2):270–96.
29. Detmar M. Lymphangiogenesis and tumor lymphangiogenesis. Section 18, Witte MH (Section ed). In: Leong SPL, Nathanson SD, Zager JS, editors. Cancer metastasis through the lymphovascular system. Switzerland AG: Springer Nature; 2021. (in press).
30. Erickson RP, Dellinger MT. Lymphvascular genomics and proteomics, clinical syndromes, and cancer metastasis. Section 18, Witte MH (Section ed). In: Leong SPL, Nathanson SD, Zager JS, editors. Cancer metastasis through the lymphovascular system. Switzerland AG: Springer Nature; 2021. (in press).
31. Brouillard P, Boon L, Vikkula M. Genetics of lymphatic anomalies. J Clin Invest. 2014;124(3):898–904.
32. Michelini S, Paolacci S, Manara E, et al. Genetic tests in lymphatic vascular malformations and lymphedema. J Med Genet. 2018;55(4):222–32.
33. Gordon K, Varney R, Keeley V, et al. Update and audit of the St George's classification algorithm of primary lymphatic anomalies: a clinical and molecular approach to diagnosis. J Med Genet. 2020;57(10):653–9.
34. Karaman S, Leppänen VM, Alitalo K. Vascular endothelial growth factor signaling in development and disease. Development. 2019;145((14):dev151019.
35. Gale NW, Thurston G, Hackett SF, et al. Angiopoietin-2 is required for postnatal angiogenesis and lymphatic patterning, and only the latter role is rescued by angiopoietin-1. Dev Cell. 2002;3(3):411–23.
36. Dellinger MT, Witte MH. Lymphangiogenesis, lymphatic systemomics, and cancer: context, advances and unanswered questions. Clin Exp Metastasis. 2018;35(5-6):419–24.
37. Dumont AE, Witte MH. Clinical usefulness of thoracic duct cannulation. In: Stollerman GH, editor. Advances in internal medicine, vol. XV. Chicago: Year Book Medical Publ., Inc.; 1969. p. 51–72.
38. Witte MH, Dumont AE, Cole WR, et al. Lymph circulation in hepatic cirrhosis: effect of portacaval shunt. Ann Intern Med. 1969;70(2):303–10.
39. Witte MH, Dumont AE, Clauss RH, et al. Lymph circulation in congestive heart failure: effect of external thoracic duct drainage. Circulation. 1969;39(6):723–33.
40. James H, Witte MH, Bernas M, et al. Proposal for prevention or alleviation of protein/lymph-losing enteropathy (PLE/LLE) after Fontan circulation treatment of univentricular hearts: restoration of lymph balance with "lymphatic right-to-left shunt". Lymphology. 2016;49(3):114–27.
41. Kreutzer C, Klinger DA, Chiostri B, et al. Lymphatic decompression concomitant with Fontan/Kreutzer procedure: early experience. World J Pediatr Congenit Heart Surg. 2020;11(3):284–92.
42. Witte M. Thoracic duct decompression: an idea whose time has come – again. Lymphology. 2020;53(2):51–4.
43. Chachaj A, Piller N, Boccardo F, et al. Lymphedema: general pathophysiology, prevention and management in invasive cancer. Section 18, Witte MH (Section ed). In: Leong SPL, Nathanson SD, Zager JS, editors. Cancer metastasis through the lymphovascular system. Switzerland AG: Springer Nature; 2021. (in press).
44. Campisi C, Boccardo F, Witte M, et al. Lymphatic surgery and surgery of lymphatic disorders. In: Dieter RS, Dieter Jr RA, Dieter III RA, editors. Venous and lymphatic diseases, Chapter 42. New York, NY: McGraw Hill; 2011. p. 607–29.
45. Witte MH, Dellinger MT, Papendieck CM, et al. Overlapping biomarkers, pathways, processes and syndromes in lymphatic development, growth and neoplasia. Clin Exp Metastasis. 2012;29(7):707–27.
46. Virchow R (1868) Cellular pathology as based upon physiological and pathological histology. Trans. 2nd ed. of original by Frank Chance. NY: RM De Witt.
47. Triola VA. Nineteenth century foundations of cancer research. Advances in tumor pathology, nomenclature, and theories of oncogenesis. Cancer Res. 1965;25:75–106.
48. Nia H, Munn L, Jain R. Physical traits of cancer. Science. 2020;370:eaaz0868.
49. Leak LV, Burke F. Fine structure of the lymphatic capillary and the adjoining connective tissue area. Am J Anat. 1966;118(3):785–809.
50. Nagle RB, Witte MH, Martinez AP, et al. Factor VIII-associated antigen in human lymphatic endothelium. Lymphology. 1987;20(1):20–4.
51. Witte MH, Witte CL, Davis WM, et al. Peritoneal transudate: a diagnostic clue to portal system obstruction in patients with intraabdominal malignancy or peritonitis. J Am Med Assoc. 1972;221(12):1380–3.

Lymphatic Specification and Development, EMT-MET, and Cancer Spread

20

Xin Geng and R. Sathish Srinivasan

Abstract

Lymphatic vasculature is an integral part of the circulatory system that regulates interstitial fluid homeostasis, inflammation, and lipid absorption. Lymphatic vessels are also a major route for tumor metastasis. Lymphatic vessels were previously considered as passive players that are manipulated by tumors. However, recent evidences suggest that the lymphatics play more complex and conflicting roles in tumor progression. Lymphatic vessels could recruit tumor cells and provide a suitable environment for their survival. Nevertheless, lymphatic vessels could also modulate immune cell behavior and either promote or inhibit immunotherapy. Hence, a better understanding of the mechanisms that regulate the growth and behavior of lymphatic vasculature might reveal innovative opportunities for clinical intervention. In this chapter we will summarize our current understanding of the mechanisms that regulate lymphatic vascular development. We will also briefly describe the epithelial to mesenchymal transition (EMT) and mesenchymal to epithelial transition (MET) that are part of the tumor metastasis process.

Learning Objectives

- Understand the stepwise development of the lymphatic vasculature.
- Know the critical regulators of lymphatic vascular development.
- Learn how defects in lymphatic vascular development could lead to diseases such as lymphedema and chylous ascites.
- Understand the roles of lymphatic vasculature in cancer progression.
- Learn how tumor cells undergo epithelial to mesenchymal transition (EMT) and mesenchymal to epithelial transition (MET) during metastasis.

20.1 Introduction

Lymphatic vasculature regulates fluid homeostasis, lipid absorption, and inflammation [1, 2]. Lymphatic vascular dysfunction leads to lymphedema, metabolic disorders such as obesity and inflammation. Additionally, lymphatic vessels are a major route for tumor metastasis. Tumors promote lymphatic vessel growth (lymphangiogenesis) within or around them. Invasive tumor cells enter these nearby lymphatic vessels and migrate to lymph nodes (LNs) and further upstream to the blood circulation. Subsequently, the tumor cells exit the blood circulation in the lungs, brain, or liver and form secondary tumors. Evidences suggest that the mechanisms that regulate the development of lymphatic vessels are also necessary for tumor lymphangiogenesis. Hence, a better understanding of stepwise development of lymphatic vasculature could reveal novel strategies to treat cancer.

20.2 Lymphatic Endothelial Cell Specification

Imaging analysis and lineage tracing experiments have shown that lymphatic endothelial cells (LECs) originate predominantly from embryonic veins [1, 2]. The homeobox transcription factor PROX1 is the first unique marker that distinguishes LECs from blood vascular endothelial cells

X. Geng · R. S. Srinivasan (✉)
Cardiovascular Biology Research Program, Oklahoma Medical Research Foundation, Oklahoma City, OK, USA

Department of Cell Biology, University of Oklahoma Health Sciences Center, Oklahoma City, OK, USA
e-mail: Xin-Geng@omrf.org; Sathish-Srinivasan@omrf.org

S. P. Leong et al. (eds.), *Cancer Metastasis Through the Lymphovascular System*,
https://doi.org/10.1007/978-3-030-93084-4_20

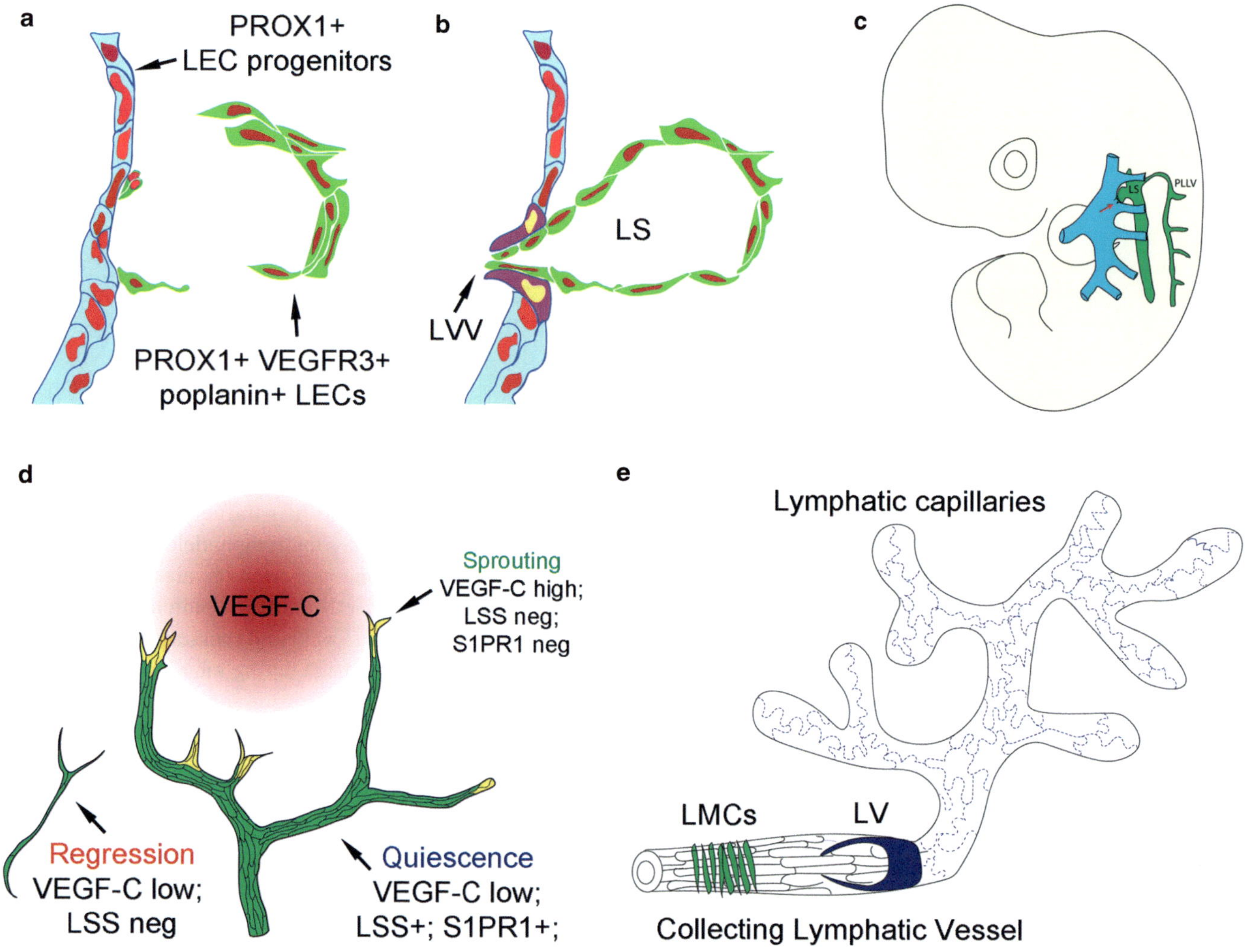

Fig. 20.1 Stepwise development of lymphatic vasculature. (**a**) LEC specification and differentiation: A subset of embryonic venous endothelial cells express PROX1 to become specified into LEC progenitors. These LEC progenitors upregulate the expression of VEGFR3 and podoplanin and migrate out of the veins as clusters of differentiated LECs. (**b**) Lymph sac and LVV formation: The LECs that originate from the anterior portions of the embryos associate with each other to form a pair (left and right) of LS. One of the LS is schematically depicted. Immediately after its formation LS establishes connection with the anterior cardinal vein through a pair of LVVs. One LVV is depicted in the figure for simplicity. LVVs express unique molecules such as GATA2 and FOXC2. (**c**) Primitive lymphatic structures: LS and PLLV are primitive lymphatic structures from which the lymphatic vessels of the anterior portion of the body (skin, head, arms, heart, and lungs) arise. Their overall organization on one side of the embryo is shown. The red arrow indicates the location of LVVs. (**d**) Lymphatic plexus formation: Lymphatic vessels sprout from LS and PLLV to form the lymphatic plexus. The tip cells (yellow) sense and migrate towards the VEGF-C gradient. Tip cells are devoid of S1PR1 activity, lymph flow, and LSS. The more mature lymphatic plexus (green) experience LSS and have S1PR1 activity. LSS enhances VEGF-C signaling, which is buffered by S1PR1. Together they maintain the quiescence of lymphatic vessels in the presence of low levels of VEGF-C. Vessels that have detached from the rest of the vasculature will regress due to the absence of LSS. (**e**) Lymphatic vascular maturation: The lymphatic plexus undergoes maturation to become lymphatic capillaries or collecting lymphatic vessels. Capillaries have button-like junctions to allow fluid entry. Collecting lymphatic vessels have zipper-like junctions to prevent fluid leakage. In addition, collecting lymphatic vessels have LV to regulate the unidirectional flow of fluid and LMCs to regulate contractility

(BECs) (Fig. 20.1a) [3]. Hence, LEC specification is defined as the activation of PROX1 expression in endothelial cells. PROX1 starts to be expressed in a polarized manner in the anterior cardinal veins at Theiler stage 15–16 in mammalian embryos. This stage corresponds to embryonic day (E) 9.75 in mice or ~ E40 in humans. These PROX1$^+$ cells are known as LEC progenitors as they give rise to most of the lymphatic vasculature at subsequent stages [4]. The number of PROX1$^+$ LEC progenitors in veins (anterior cardinal veins, common cardinal vein, intersomitic veins) increase substantially until E12.5 [5, 6]. LECs originate from other embryonic veins such as azygous veins and iliac veins [7]. Therefore, PROX1 is likely expressed in these veins as well.

The expression of PROX1 is directly regulated by the transcription factors COUP-TFII and SOX18 [8, 9]. COUP-TFII is a nuclear hormone receptor that specifies LEC progenitors by initiating PROX1 expression through a highly conserved 16-nucleotide long regulatory element [9]. However, COUP-TFII is expressed in all venous endothelial cells. Hence, COUP-TFII is thought to synergize with additional transcription factors to regulate the expression of PROX1. SOX18 is one such transcription factor that is specifically expressed in a subset of venous endothelial cells that eventually turn on PROX1 [8]. Importantly, SOX18 is necessary for the specification of PROX1 as LEC progenitors are absent in mice lacking SOX18.

Prox1$^{+/-}$ embryos lacking one allele of *Prox1* have approximately half the number of LEC progenitors that are observed in control littermates [10]. This observation suggested that PROX1 regulates itself in LEC progenitors. Indeed, the interaction between PROX1 and COUP-TFII is necessary to maintain *Prox1* expression in LEC progenitors [9]. Additionally, an exquisite feedback loop operates between PROX1 and vascular endothelial growth factor-C (VEGF-C) signaling [11, 12]. Expression of vascular endothelial growth factor receptor-3 (VEGFR3) is regulated by PROX1. In turn, PROX1 expression is positively regulated by VEGF-C signaling. Consequently, LEC progenitors disappear rapidly in *Vegfc*$^{-/-}$ embryos and are dramatically reduced in *Vegfc*$^{+/-}$ embryos [12].

20.3 Lymphatic Endothelial Cell Migration from Veins and Lymph Sac Formation

Starting at E10.5 in mouse embryos the LEC progenitors upregulate the expressions of VEGFR3, which is the cognate receptor of VEGF-C, and the glycoprotein podoplanin and migrate out of the veins as strings of interconnected cells (Fig. 20.1a) [5, 6]. These migrating cells are considered as fully differentiated LECs.

At E12.0 the LECs migrating from the anterior locations of embryos aggregate to form four bilaterally located primitive lymphatic structures (Fig. 20.1b, c). Jugular lymph sac (LS) is located close to the anterior cardinal vein and peripheral longitudinal lymphatic vessel (PLLV) is located dorsally near the dermis [3, 5]. LS and PLLV are connected to each other through an anterior isthmus. LECs that originate from posterior veins have not received the attention of recent studies. However, they are known to assemble and form retroperitoneal lymph sacs and cisterna chyli.

VEGF-C signaling is essential for the migration of LECs from veins [13]. VEGF-C is produced in an inactive form in the mesenchymal cells and it is proteolytically processed into a fully active form by CCBE1 and ADAMTS3 [14, 15]. Mice lacking VEGF-C, CCBE1, or ADAMTS3 also do not have LECs, LS, or PLLV. Although VEGF-C could signal through both VEGFR3 and VEGFR2, VEGFR3 is the primary mediator of VEGF-C signaling during LEC migration and lymph sac formation. Mice lacking the kinase activity of VEGFR3 (*Vegfr3*$^{KD/KD}$) are completely devoid of lymph sacs [16]. Intriguingly, deletion of the VEGF-C binding domain from VEGFR3 does not affect lymph sac formation [16]. These observations suggest that VEGF-C could indirectly activate VEGFR3. The VEGFR3 co-receptor neuropilin-2 is one such molecule that could associate with VEGF-C [17]. Integrin-β1 is another molecule that could enhance VEGFR3 signaling through a cross talk with mechanical signals [18]. Interstitial fluid pressure increases around the embryonic veins prior to the formation of LS. This interstitial fluid pressure activates integrin-β1, which then associates with VEGFR3 to promote the downstream signaling cascade and LEC proliferation.

Additional molecules also regulate PROX1 and VEGF-C signaling. The transcription factor GATA2 is necessary for upregulating the expression of VEGFR3 in migrating LECs [19]. GATA2 could also regulate the expression of PROX1 [20]. Consequently, mice lacking GATA2 from all endothelial cells have hypoplastic LS [19, 21]. Adrenomedullin is a peptide hormone that signals through the G protein-coupled receptor (GPCR) calcitonin receptor-like receptor (CALCRL) and its receptor activity modifying protein RAMP2. Adrenomedullin signaling regulates the expression of PROX1 in LECs. Consequently, deletion of adrenomedullin, CALCRL or RAMP2 from mice results in hypoplastic LS [22].

20.4 Development of Lymphovenous Valves and Establishment of Blood-Lymph Separation

At E12.0 the developing LS establish connections with blood circulation at the junction of jugular and subclavian veins [10, 23]. Lymphovenous valves (LVVs) rapidly develop at these sites of blood/lymph interaction (Fig. 20.1b, c). The primary role of LVVs is to permit lymph return to blood circulation while preventing the entry of blood into the lymphatic vasculature. LVV-forming endothelial cells (LVV-ECs) strongly express the transcription factors PROX1, FOXC2, and GATA2. Integrin- α9 and gap junction molecule connexin-37 are other molecules that are expressed in LVV-ECs. PROX1, FOXC2, and Wnt/β-catenin signaling are necessary for the differentiation of LVV-forming endothelial cells (LVV-ECs) [23, 24]. GATA2, VEGF-C signaling, the transcriptional co-factors YAP/TAZ, and the Ras GTPase activating protein RASA1 are necessary for the maintenance of LVV-ECs [21, 25, 26]. The receptor tyrosine kinase EPHB4 and the gap junction molecule connexin-37 regulate the morphogenesis of LVV-ECs into LVVs [27, 28]. Mice lacking most of the abovementioned molecules have dilated LS or edema due to defective lymph drainage caused by the loss of LVVs.

Although LECs originate from veins, lymphatic vessels are devoid of blood cells except for leukocytes that migrate in and out of lymph nodes (LNs) to regulate immune response. Platelets are important for generating this blood/lymph separation.

Platelet-expressed CLEC2 interacts with LEC-expressed podoplanin to generate transient clots at the sites of LEC migration from veins and LVVs to prevent blood from entering LS without blocking lymphatic drainage [29–31].

In summary, LVVs are the first valves to develop outside the heart and numerous molecules regulate their development. LVVs are the gatekeepers of lymphatic vascular functioning and integrity.

20.5 Lymphatic Vascular Growth and Patterning

The purpose of vascular patterning is to minimize the total energy that is required to efficiently transport fluid. This goal is achieved by minimizing the number of branch points (and therefore the transportation distance) that are necessary to provide nutrients or collect interstitial fluid from the body. This outcome is achieved during lymphatic vascular development in steps.

Lymphatic vascular growth and plexus formation: The first step of lymphatic vascular patterning is the formation of lymphatic plexus. Lymphatic vessels sprout from LS and PLLV to populate the tissues and organs. The sprouting lymphatic vessels branch in a stereotypic manner at regular intervals to form the lymphatic plexus (Fig. 20.1d). VEGF-C signaling is the most important pro-lymphangiogenic molecule. *Vegfc*$^{+/-}$ and *Vegfr3*$^{+/KD}$ mice are almost completely devoid of dermal lymphatic plexus [13, 32]. Dermal lymphatic vascular hypoplasia is also observed in *Nrp2*$^{-/-}$ mice lacking the VEGFR3 co-receptor neuropilin-2 [17]. Delta like 4 (DLL4) and angiopoietin-2 enhance VEGF-C signaling through unidentified mechanisms [33, 34]. Consistent with these findings *Dll4*$^{+/-}$ and *Angpt2*$^{-/-}$ embryos also have hypoplastic lymphatic vessels [33, 35].

Molecules that regulate VEGF-C signaling such as TGF-β, GATA2, YAP/TAZ, and PKD1 are also necessary for the patterning of lymphatic plexus. TGF-β and GATA2 positively regulate the expression of VEGFR3 [19, 36]. YAP and TAZ are able to both positively and negatively regulate the expression of PROX1 in response to VEGF-C signaling [25, 37]. While PKD1 is necessary for the sprouting of LECs in response to VEGF-C, the mechanisms are not fully defined [38, 39].

Additional regulators of lymphatic vascular patterning include the transcription factor FOXC2, and β-catenin [24, 40]. Mice lacking these molecules have dilated lymphatic vessels with abnormal number of branches. FOXC2 is a target of Wnt/β-catenin signaling [24].

Lymphatic vascular pruning and quiescence: Lymphatic vessels of the plexus that do not efficiently transport fluid are removed. Fluid flow generates shear stress on endothelial cells. Therefore, shear stress is likely the signal that dictates the decision between pruning and survival. Indeed, we and others have identified shear stress as an enhancer of cell proliferation and VEGF-C signaling in LECs (Fig. 20.1d) [33, 41, 42]. Laminar shear stress (LSS) inhibits Notch signaling in LECs to induce proliferation. In addition, LSS induces the expression of VEGF-C from LECs and also enhances the sensitivity of LECs to VEGF-C [33, 41]. Statically cultured LECs were unable to activate MAPK pathway in response to low (for example, 1 ng/ml) concentrations of VEGF-C [33]. In contrast, LECs that were previously exposed to LSS for 24 hrs could activate MAPK pathway in response to the same low concentration of VEGF-C. Thus, LSS enhances the response of LECs to VEGF-C.

Mechanisms are also in place to specifically antagonize LSS-enhanced VEGF-C signaling. The GPCR sphingosine 1-phosphate receptor 1 (S1PR1) maintains the quiescence of lymphatic vessels by specifically antagonizing LSS-enhanced VEGF-C signaling in LECs [33]. We propose that S1PR1 buffers the shear stress-induced VEGF-C signaling to keep it within a range that is necessary for the survival of LECs, but not for growth.

Lymphatic vessel maturation: The final step of lymphatic vessel patterning is maturation. During this step some vessels will specialize into lymphatic capillaries that absorb interstitial fluid while others become collecting lymphatic vessels that transport lymph (Fig. 20.1e).

Lymphatic capillaries are thinner, shorter, and more prevalent than collecting lymphatic vessels. In addition, lymphatic capillaries strongly express molecules such as LYVE1, PROX1, and VEGFR3 and are devoid of continuous basement membrane. Furthermore, lymphatic capillaries lack mural cell coverage and acquire button-like junctions that are characterized by discontinuous localization of junctional molecules such as VE-Cadherin, PECAM-1, and claudin-5 [43]. The LEC boundaries overlap to prevent lymph leakage. LECs of capillaries are attached to the surrounding connective tissues by anchoring filaments. The overlapping LEC boundaries are stretched open by the action of anchoring filaments to allow the interstitial fluid to percolate between LECs and into the lymphatic capillaries. Relaxation of anchoring filaments results in the closure of LEC junctions. Subsequently, lymph from the capillaries is transported upstream by yet unknown mechanisms.

Lymph from numerous capillaries drains into bigger collecting lymphatic vessels. Collecting lymphatic vessels express lower levels of LYVE1, PROX1, and VEGFR3 and have zipper-like LEC junctions that are characterized by continuous localization of VE-Cadherin, PECAM1, and claudin-5 between LECs that prevent lymph leakage [43]. FOXC2 and ephrin-B2/EPHB4 signaling are necessary for the maintenance of LEC junctions in the collecting lymphatic vessels [44–46]. FOXC2 maintains junctional integrity at locations that are exposed to oscillatory shear stress (OSS) by preventing cell proliferation. In addition, FOXC2 inhibits MAPK signaling and regulates the cytoskeleton [45, 47]. Ephrin-B2/EPHB4 signaling activates Rac1 and inhibits RhoA activities [44].

Collecting lymphatic vessels recruit specialized mural cells known as lymphatic muscle cells (LMCs) for support. Unlike the mural cells that surround the arteries and veins, LMCs have intrinsic contractile activity to propel lymph [48]. Ephrin-B2, reelin, and PDGF-B regulate the recruitment of LMCs to collecting lymphatic vessels [49–51].

Lymphatic valves (LVs) develop within collecting lymphatic vessels to regulate the unidirectional flow of lymph [52]. LVs and LVVs share a similar molecular profile and require almost an identical set of genes for their development. Shear stress plays critical roles in the morphogenesis of LVs [52–54]. LVs develop at sites such as branch point of vessels or in the middle of long vessels that are exposed to OSS. Importantly, OSS enhances the expression of FOXC2, GATA2, connexin-37, ephrin-B2, and integrin-α9 in LECs. The expression of FOXC2 and GATA2 is dependent on PROX1 and OSS-induced secretion of Wnt ligands from LECs. Furthermore, OSS-induced Ca^{2+} flow between LECs through gap junctions is necessary for the nuclear localization of NFATC1 and FOXC2/NFATC1 complex formation. In addition, syndecan-4-induced inhibition of the planar cell polarity molecule VANGL2 in response to LSS is necessary for the morphogenesis of LVs. Thus, complex cross talk between mechanical and biochemical signals regulates LV formation within collecting lymphatic vessels.

In summary, we described the stepwise development of lymphatic vasculature. In Table 20.1 we have listed some of the genes that are frequently mutated in human patients with congenital lymphedema or lymphatic disorders.

Table 20.1 Genes commonly associated with lymphatic disorders and their mechanisms of action

Gene	Disease	Dev steps in mice
SOX18	Hypotrichosis-lymphedema-telengiectasia	1
VEGFR3	Milroy disease	1, 2, 3, 4, 5
VEGFC	Milroy-like disease	1, 2, 3, 4, 5
ADAMTS3	Hennekam syndrome	2, 3
CCBE1	Hennekam syndrome	2, 3
FAT4	Hennekam syndrome	3, 4, 5
FOXC2	Lymphedema-distichiasis syndrome	3, 4, 5
GATA2	Emberger syndrome	2, 3, 4, 5
RASA1	Capillary malformation; arteriovenous malformation	3, 4, 5
PIEZO1	Generalized lymphatic dysplasia of Fotiou	5
EPHB4	Lymphatic-related hydrops fetalis	5
ITGA9	Fetal chylothorax	5
CALCRL	Nonimmune hydrops fetalis	2

Step 1: LEC specification
Step 2: LEC differentiation and migration
Step 3: Lymphatic vascular patterning and/or integrity
Step 4: Lymphatic vascular maturation
Step 5: LV and LVV morphogenesis

20.6 Tumor Lymphangiogenesis

Tumors secrete VEGF-C directly or indirectly through macrophages to promote the sprouting of preexisting lymphatic vessels. Lymphatic vessels are observed mostly around the periphery of tumors although intratumoral lymphatic vessels are observed in melanoma and head and neck cancer [55]. Tumors could also promote lymphangiogenesis in the draining lymph nodes and in distant organs, which later become sites of secondary tumor formation. Additionally, tumors could transform the identity of LN LECs by downregulating the expression of molecules such as JAM-C and upregulating ITGB2A [56]. Such changes are thought to create pre-metastatic lymphovascular niche to support metastatic cells when they arrive. The heparin-binding chemokine midkine is at least one of the factors that are involved in pre-metastatic niche formation [57].

Lymphatic vessels play dual roles in regulating tumor progression. The lipid-rich environment of lymphatic vessels protects melanoma cells from oxidative stress and helps them proliferate and metastasize [58]. Furthermore, LN LECs could inactivate tumor-specific T cells and provide an immune-compromised environment for the tumors to grow [59]. However, lymphatic vessels could also reduce intratumoral fluid pressure and permit chemotherapeutic agents to enter the tumor microenvironment and reduce tumor burden [60]. Additionally, lymphatic vessels drain tumor antigens to lymph nodes where anti-tumor inflammatory response is activated. Hence, an increase in lymphatic vessel density is associated with better prognosis of glioblastoma and colorectal cancer in mouse models or human patients, respectively [61, 62]. These discrepancies could be due to tissue- or tumor-specific differences in the lymphatic vascular function.

Tumor cells undergo two-transition states, epithelial to mesenchymal transition (EMT) and mesenchymal to epithelial transition (MET), in order to metastasize to distant organs [63]. Primary tumor cells are attached to each other through cell junction molecules such as cadherins and claudins. Tumor cells are also attached to the basement membrane through integrins. These cell–cell and cell–ECM interactions provide apical-basal polarity and reduce their migration ability. Frequently (but not always) malignant cells need to partially or completely downregulate cell junction molecules such as E-cadherin in order to become metastatic. In addition, malignant cells start expressing molecules such as FSP1 and vimentin, which are characteristic of migratory mesenchymal cells. EMT is the process through which tumor cells lose their cell junctions and polarity and acquire mesenchymal characteristics. Signaling molecules such as TGF-β1, transcription factors such as Snail, Slug, ZEB1, ZEB2, and FOXC2 are necessary for tumor cells to successfully undergo EMT (Fig. 20.2a). Detachment of cells from the basement membrane could trigger a form of programmed cell death known as anoikis. Therefore, tumor cells

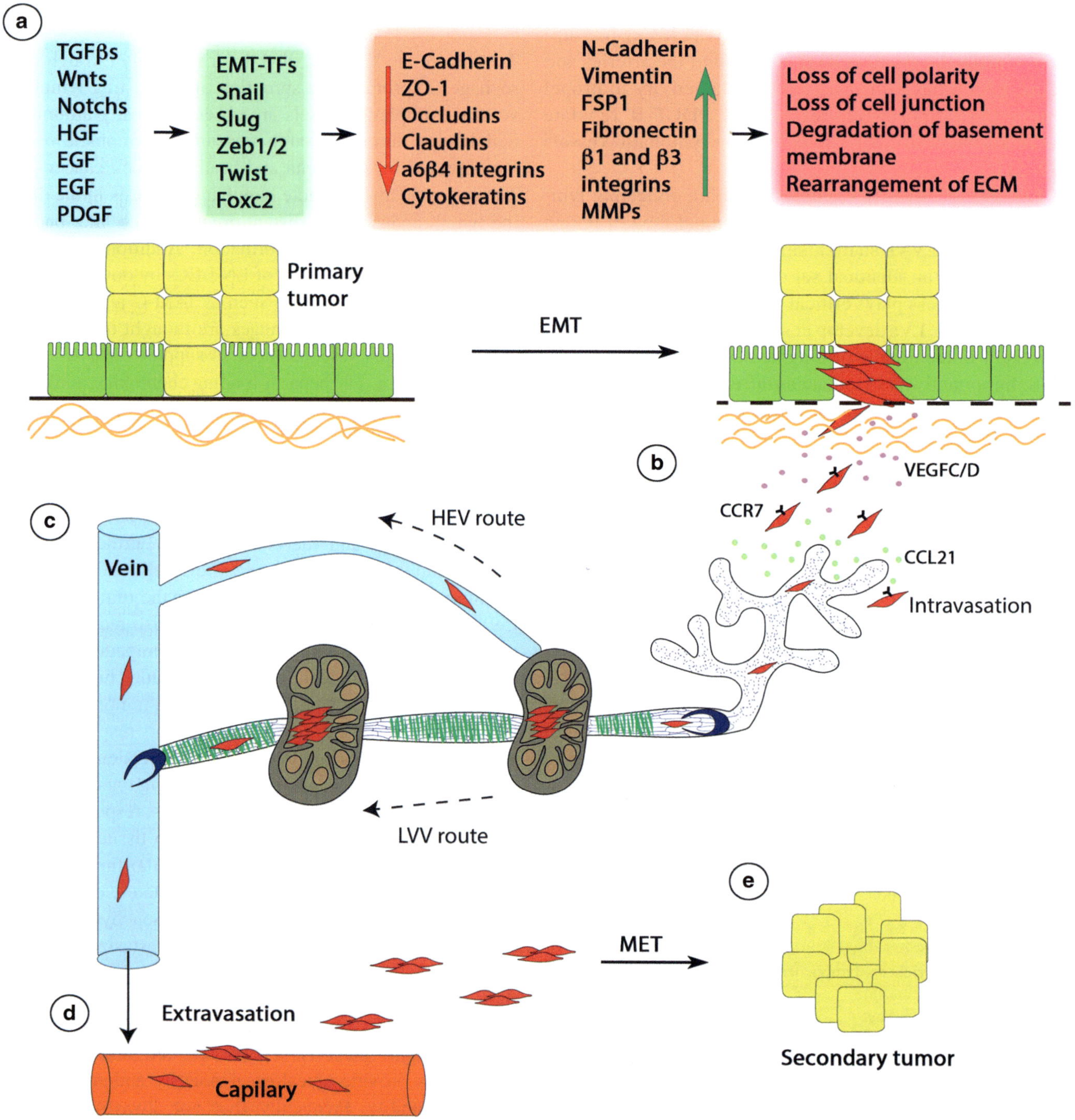

Fig. 20.2 Mechanisms of tumor metastasis through the lymphatic vasculature. (**a**) **EMT:** Primary tumors undergo EMT to become invasive. The molecular events associated with EMT are presented. (**b**) **Tumor intravasation:** Tumors secrete VEGF-C or VEGF-D to attract the lymphatic vessels. LECs secrete factors such as CCL21 that attract tumor cells that express CCR7. (**c**) **Migration of tumor cells:** Tumor cells could exit from LNs via HEVs or continue upstream and enter the blood circulation through LVVs. (**d**) **Tumor extravasation:** Tumor cells exit the blood circulation through capillaries at pre-metastatic niches. (**e**) **Secondary tumor formation:** Tumor cells undergo MET, which is the reversal of EMT, to gain cell junctions and become secondary tumors

should develop additional adaptations to survive the subsequent stages of metastasis.

Lymphatic capillaries with their button-like junctions are conducive for the intravasation (entry) of metastatic tumor cells (Fig. 20.2b). LEC-secreted CCL21 could play an inadvertent role in recruiting and maintaining $CCR7^+$ tumor stem cells in the lymphatic vessels.

Tumor cells exit the lymphatic vasculature and enter the blood circulation through the high endothelial venules (HEVs) of LNs (Fig. 20.2c) [64, 65]. Tumor cells could also circumnavigate the entire lymphatic vasculature and enter the bloodstream through LVVs. Finally, tumor cells extravasate from the vasculature at pre-metastatic niche, downregulate mesenchymal genes, and regain epithelial cell markers through MET (Fig. 20.2d, e). MET has been found to be necessary for the survival of some tumors and the proper localization of some others. Nevertheless, the mechanisms of MET remain to be fully defined.

Summary and Unanswered Questions

- LECs originate primarily from embryonic veins as a homogeneous population of cells.
- LECs aggregate to form LS, which sprout to form the lymphatic plexus.
- Lymphatic plexus undergoes maturation to form lymphatic capillaries, collecting lymphatic vessels and lymphatic valves.
- Lymph is returned to blood through LVVs.
- Tumor cells undergo EMT and enter the lymphatic capillaries.
- Tumor cells exit the lymphatic vasculature and enter blood circulation either from LNs or through LVVs.
- Lymphatic vasculature plays a dual role in tumor progression.
- The signals that regulate LEC specification remain to be identified.
- The signals that coordinate with VEGF-C to regulate lymphatic plexus formation remain to be thoroughly identified.
- The mechanisms of lymphatic vessel maturation are not fully understood.
- The role of shear stress remains to be dissected.
- Whether there are any tissue-specific differences in the identity of LECs remains to be determined.
- Whether the cross talk between tumors and lymphatic vessels depends on tissue-specific differences in LECs needs to be investigated.

Acknowledgments This work is supported by NIH/NHLBI (R01HL131652 and R01HL133216 to RSS), NIH/NIGMS COBRE (P20 GM103441 to XG; PI: Dr. Rodger McEver), and Oklahoma Center for Adult Stem Cell Research (4340 to RSS). We apologize to the authors of numerous articles that we are unable to cite due to space restrictions.

Disclosures: None.

References

1. Aspelund A, Robciuc MR, Karaman S, Makinen T, Alitalo K. Lymphatic system in cardiovascular medicine. Circ Res. 2016;118:515–30.
2. Oliver G, Kipnis J, Randolph GJ, Harvey NL. The lymphatic vasculature in the 21(st) century: Novel functional roles in homeostasis and disease. Cell. 2020;182:270–96.
3. Wigle JT, Oliver G. Prox1 function is required for the development of the murine lymphatic system. Cell. 1999;98:769–78.
4. Srinivasan RS, Dillard ME, Lagutin OV, Lin FJ, Tsai S, Tsai MJ, Samokhvalov IM, Oliver G. Lineage tracing demonstrates the venous origin of the mammalian lymphatic vasculature. Genes Dev. 2007;21:2422–32.
5. Hagerling R, Pollmann C, Andreas M, Schmidt C, Nurmi H, Adams RH, Alitalo K, Andresen V, Schulte-Merker S, Kiefer F. A novel multistep mechanism for initial lymphangiogenesis in mouse embryos based on ultramicroscopy. EMBO J. 2013;32:629–44.
6. Yang Y, Garcia-Verdugo JM, Soriano-Navarro M, Srinivasan RS, Scallan JP, Singh MK, Epstein JA, Oliver G. Lymphatic endothelial progenitors bud from the cardinal vein and intersomitic vessels in mammalian embryos. Blood. 2012;120:2340–8.
7. van der Putte SC. The early development of the lymphatic system in mouse embryos. Acta Morphol Neerl Scand. 1975;13:245–86.
8. Francois M, Caprini A, Hosking B, Orsenigo F, Wilhelm D, Browne C, Paavonen K, Karnezis T, Shayan R, Downes M, Davidson T, Tutt D, Cheah KS, Stacker SA, Muscat GE, Achen MG, Dejana E, Koopman P. Sox18 induces development of the lymphatic vasculature in mice. Nature. 2008;456:643–7.
9. Srinivasan RS, Geng X, Yang Y, Wang Y, Mukatira S, Studer M, Porto MP, Lagutin O, Oliver G. The nuclear hormone receptor coup-tfii is required for the initiation and early maintenance of prox1 expression in lymphatic endothelial cells. Genes Dev. 2010;24:696–707.
10. Srinivasan RS, Oliver G. Prox1 dosage controls the number of lymphatic endothelial cell progenitors and the formation of the lymphovenous valves. Genes Dev. 2011;25:2187–97.
11. Koltowska K, Lagendijk AK, Pichol-Thievend C, Fischer JC, Francois M, Ober EA, Yap AS, Hogan BM. Vegfc regulates bipotential precursor division and prox1 expression to promote lymphatic identity in zebrafish. Cell Rep. 2015;13:1828–41.
12. Srinivasan RS, Escobedo N, Yang Y, Interiano A, Dillard ME, Finkelstein D, Mukatira S, Gil HJ, Nurmi H, Alitalo K, Oliver G. The prox1-vegfr3 feedback loop maintains the identity and the number of lymphatic endothelial cell progenitors. Genes Dev. 2014;28:2175–87.
13. Karkkainen MJ, Haiko P, Sainio K, Partanen J, Taipale J, Petrova TV, Jeltsch M, Jackson DG, Talikka M, Rauvala H, Betsholtz C, Alitalo K. Vascular endothelial growth factor c is required for sprouting of the first lymphatic vessels from embryonic veins. Nat Immunol. 2004;5:74–80.

14. Bos FL, Caunt M, Peterson-Maduro J, Planas-Paz L, Kowalski J, Karpanen T, van Impel A, Tong R, Ernst JA, Korving J, van Es JH, Lammert E, Duckers HJ, Schulte-Merker S. Ccbe1 is essential for mammalian lymphatic vascular development and enhances the lymphangiogenic effect of vascular endothelial growth factor-c in vivo. Circ Res. 2011;109:486–91.
15. Bui HM, Enis D, Robciuc MR, Nurmi HJ, Cohen J, Chen M, Yang Y, Dhillon V, Johnson K, Zhang H, Kirkpatrick R, Traxler E, Anisimov A, Alitalo K, Kahn ML. Proteolytic activation defines distinct lymphangiogenic mechanisms for vegfc and vegfd. J Clin Invest. 2016;126:2167–80.
16. Zhang L, Zhou F, Han W, Shen B, Luo J, Shibuya M, He Y. Vegfr-3 ligand-binding and kinase activity are required for lymphangiogenesis but not for angiogenesis. Cell Res. 2010;20:1319–31.
17. Yuan L, Moyon D, Pardanaud L, Breant C, Karkkainen MJ, Alitalo K, Eichmann A. Abnormal lymphatic vessel development in neuropilin 2 mutant mice. Development. 2002;129:4797–806.
18. Planas-Paz L, Strilic B, Goedecke A, Breier G, Fassler R, Lammert E. Mechanoinduction of lymph vessel expansion. EMBO J. 2012;31:788–804.
19. Frye M, Taddei A, Dierkes C, Martinez-Corral I, Fielden M, Ortsater H, Kazenwadel J, Calado DP, Ostergaard P, Salminen M, He L, Harvey NL, Kiefer F, Makinen T. Matrix stiffness controls lymphatic vessel formation through regulation of a gata2-dependent transcriptional program. Nat Commun. 2018;9:1511.
20. Kazenwadel J, Betterman KL, Chong CE, Stokes PH, Lee YK, Secker GA, Agalarov Y, Demir CS, Lawrence DM, Sutton DL, Tabruyn SP, Miura N, Salminen M, Petrova TV, Matthews JM, Hahn CN, Scott HS, Harvey NL. Gata2 is required for lymphatic vessel valve development and maintenance. J Clin Invest. 2015;125:2979–94.
21. Mahamud MR, Geng X, Ho YC, Cha B, Kim Y, Ma J, Chen L, Myers G, Camper S, Mustacich D, Witte M, Choi D, Hong YK, Chen H, Varshney G, Engel JD, Wang S, Kim TH, Lim KC, Srinivasan RS. Gata2 controls lymphatic endothelial cell junctional integrity and lymphovenous valve morphogenesis through mir-126. Development. 2019;146
22. Fritz-Six KL, Dunworth WP, Li M, Caron KM. Adrenomedullin signaling is necessary for murine lymphatic vascular development. J Clin Invest. 2008;118:40–50.
23. Geng X, Cha B, Mahamud MR, Lim KC, Silasi-Mansat R, Uddin MK, Miura N, Xia L, Simon AM, Engel JD, Chen H, Lupu F, Srinivasan RS. Multiple mouse models of primary lymphedema exhibit distinct defects in lymphovenous valve development. Dev Biol. 2016;409:218–33.
24. Cha B, Geng X, Mahamud MR, Fu J, Mukherjee A, Kim Y, Jho EH, Kim TH, Kahn ML, Xia L, Dixon JB, Chen H, Srinivasan RS. Mechanotransduction activates canonical wnt/beta-catenin signaling to promote lymphatic vascular patterning and the development of lymphatic and lymphovenous valves. Genes Dev. 2016;30:1454–69.
25. Cha B, Ho YC, Geng X, Mahamud MR, Chen L, Kim Y, Choi D, Kim TH, Randolph GJ, Cao X, Chen H, Srinivasan RS. Yap and taz maintain prox1 expression in the developing lymphatic and lymphovenous valves in response to vegf-c signaling. Development. 2020; https://doi.org/10.1242/dev.195453.
26. Chen D, Geng X, Lapinski PE, Davis MJ, Srinivasan RS, King PD. Rasa1-driven cellular export of collagen iv is required for the development of lymphovenous and venous valves in mice. Development. 2020;147
27. Martin-Almedina S, Martinez-Corral I, Holdhus R, Vicente A, Fotiou E, Lin S, Petersen K, Simpson MA, Hoischen A, Gilissen C, Jeffery H, Atton G, Karapouliou C, Brice G, Gordon K, Wiseman JW, Wedin M, Rockson SG, Jeffery S, Mortimer PS, Snyder MP, Berland S, Mansour S, Makinen T, Ostergaard P. Ephb4 kinase-inactivating mutations cause autosomal dominant lymphatic-related hydrops fetalis. J Clin Invest. 2016;126:3080–8.
28. Munger SJ, Geng X, Srinivasan RS, Witte MH, Paul DL, Simon AM. Segregated foxc2, nfatc1 and connexin expression at normal developing venous valves, and connexin-specific differences in the valve phenotypes of cx37, cx43, and cx47 knockout mice. Dev Biol. 2016;412:173–90.
29. Crosswhite PL, Podsiadlowska JJ, Curtis CD, Gao S, Xia L, Srinivasan RS, Griffin CT. Chd4-regulated plasmin activation impacts lymphovenous hemostasis and hepatic vascular integrity. J Clin Invest. 2016;126:2254–66.
30. Hess PR, Rawnsley DR, Jakus Z, Yang Y, Sweet DT, Fu J, Herzog B, Lu M, Nieswandt B, Oliver G, Makinen T, Xia L, Kahn ML. Platelets mediate lymphovenous hemostasis to maintain blood-lymphatic separation throughout life. J Clin Invest. 2014;124:273–84.
31. Uhrin P, Zaujec J, Breuss JM, Olcaydu D, Chrenek P, Stockinger H, Fuertbauer E, Moser M, Haiko P, Fassler R, Alitalo K, Binder BR, Kerjaschki D. Novel function for blood platelets and podoplanin in developmental separation of blood and lymphatic circulation. Blood. 2010;115:3997–4005.
32. Karkkainen MJ, Saaristo A, Jussila L, Karila KA, Lawrence EC, Pajusola K, Bueler H, Eichmann A, Kauppinen R, Kettunen MI, Yla-Herttuala S, Finegold DN, Ferrell RE, Alitalo K. A model for gene therapy of human hereditary lymphedema. Proc Natl Acad Sci U S A. 2001;98:12677–82.
33. Geng X, Yanagida K, Akwii RG, Choi D, Chen L, Ho Y, Cha B, Mahamud MR, Berman de Ruiz K, Ichise H, Chen H, Wythe J, Mikelis CM, Hla T, Srinivasan RS. S1pr1 regulates the quiescence of lymphatic vessels by inhibiting laminar shear stress-dependent vegf-c signaling. JCI. Insight. 2020;5:e137652.
34. Zheng W, Nurmi H, Appak S, Sabine A, Bovay E, Korhonen EA, Orsenigo F, Lohela M, D'Amico G, Holopainen T, Leow CC, Dejana E, Petrova TV, Augustin HG, Alitalo K. Angiopoietin 2 regulates the transformation and integrity of lymphatic endothelial cell junctions. Genes Dev. 2014;28:1592–603.
35. Dellinger M, Hunter R, Bernas M, Gale N, Yancopoulos G, Erickson R, Witte M. Defective remodeling and maturation of the lymphatic vasculature in angiopoietin-2 deficient mice. Dev Biol. 2008;319:309–20.
36. James JM, Nalbandian A, Mukouyama YS. Tgfbeta signaling is required for sprouting lymphangiogenesis during lymphatic network development in the skin. Development. 2013;140:3903–14.
37. Cho H, Kim J, Ahn JH, Hong YK, Makinen T, Lim DS, Koh GY. Yap and taz negatively regulate prox1 during developmental and pathologic lymphangiogenesis. Circ Res. 2019;124:225–42.
38. Coxam B, Sabine A, Bower NI, Smith KA, Pichol-Thievend C, Skoczylas R, Astin JW, Frampton E, Jaquet M, Crosier PS, Parton RG, Harvey NL, Petrova TV, Schulte-Merker S, Francois M, Hogan BM. Pkd1 regulates lymphatic vascular morphogenesis during development. Cell Rep. 2014;7:623–33.
39. Outeda P, Huso DL, Fisher SA, Halushka MK, Kim H, Qian F, Germino GG, Watnick T. Polycystin signaling is required for directed endothelial cell migration and lymphatic development. Cell Rep. 2014;7:634–44.
40. Norrmen C, Ivanov KI, Cheng J, Zangger N, Delorenzi M, Jaquet M, Miura N, Puolakkainen P, Horsley V, Hu J, Augustin HG, Yla-Herttuala S, Alitalo K, Petrova TV. Foxc2 controls formation and maturation of lymphatic collecting vessels through cooperation with nfatc1. J Cell Biol. 2009;185:439–57.
41. Choi D, Park E, Jung E, Seong YJ, Hong M, Lee S, Burford J, Gyarmati G, Peti-Peterdi J, Srikanth S, Gwack Y, Koh CJ, Boriushkin E, Hamik A, Wong AK, Hong YK. Orai1 activates proliferation of lymphatic endothelial cells in response to laminar flow through kruppel-like factors 2 and 4. Circ Res. 2017;120:1426–39.

42. Choi D, Park E, Jung E, Seong YJ, Yoo J, Lee E, Hong M, Lee S, Ishida H, Burford J, Peti-Peterdi J, Adams RH, Srikanth S, Gwack Y, Chen CS, Vogel HJ, Koh CJ, Wong AK, Hong YK. Laminar flow downregulates notch activity to promote lymphatic sprouting. J Clin Invest. 2017;127:1225–40.
43. Baluk P, Fuxe J, Hashizume H, Romano T, Lashnits E, Butz S, Vestweber D, Corada M, Molendini C, Dejana E, McDonald DM. Functionally specialized junctions between endothelial cells of lymphatic vessels. J Exp Med. 2007;204:2349–62.
44. Frye M, Stritt S, Ortsater H, Hernandez Vasquez M, Kaakinen M, Vicente A, Wiseman J, Eklund L, Martinez-Torrecuadrada JL, Vestweber D, Makinen T. Ephrinb2-ephb4 signalling provides rho-mediated homeostatic control of lymphatic endothelial cell junction integrity. elife. 2020;9
45. Norden PR, Sabine A, Wang Y, Demir CS, Liu T, Petrova TV, Kume T. Shear stimulation of foxc1 and foxc2 differentially regulates cytoskeletal activity during lymphatic valve maturation. elife. 2020;9
46. Sabine A, Bovay E, Demir CS, Kimura W, Jaquet M, Agalarov Y, Zangger N, Scallan JP, Graber W, Gulpinar E, Kwak BR, Makinen T, Martinez-Corral I, Ortega S, Delorenzi M, Kiefer F, Davis MJ, Djonov V, Miura N, Petrova TV. Foxc2 and fluid shear stress stabilize postnatal lymphatic vasculature. J Clin Invest. 2015;125:3861–77.
47. Fatima A, Wang Y, Uchida Y, Norden P, Liu T, Culver A, Dietz WH, Culver F, Millay M, Mukouyama YS, Kume T. Foxc1 and foxc2 deletion causes abnormal lymphangiogenesis and correlates with erk hyperactivation. J Clin Invest. 2016;126:2437–51.
48. Muthuchamy M, Zawieja D. Molecular regulation of lymphatic contractility. Ann N Y Acad Sci. 2008;1131:89–99.
49. Lutter S, Xie S, Tatin F, Makinen T. Smooth muscle-endothelial cell communication activates reelin signaling and regulates lymphatic vessel formation. J Cell Biol. 2012;197:837–49.
50. Makinen T, Adams RH, Bailey J, Lu Q, Ziemiecki A, Alitalo K, Klein R, Wilkinson GA. Pdz interaction site in ephrinb2 is required for the remodeling of lymphatic vasculature. Genes Dev. 2005;19:397–410.
51. Wang Y, Jin Y, Mae MA, Zhang Y, Ortsater H, Betsholtz C, Makinen T, Jakobsson L. Smooth muscle cell recruitment to lymphatic vessels requires PDGFB and impacts vessel size but not identity. Development. 2017;144:3590–601.
52. Geng X, Cha B, Mahamud MR, Srinivasan RS. Intraluminal valves: development, function and disease. Dis Models Mech. 2017;10:1273–87.
53. Sabine A, Agalarov Y, Maby-El Hajjami H, Jaquet M, Hagerling R, Pollmann C, Bebber D, Pfenniger A, Miura N, Dormond O, Calmes JM, Adams RH, Makinen T, Kiefer F, Kwak BR, Petrova TV. Mechanotransduction, prox1, and foxc2 cooperate to control connexin37 and calcineurin during lymphatic-valve formation. Dev Cell. 2012;22:430–45.
54. Sabine A, Saygili Demir C, Petrova TV. Endothelial cell responses to biomechanical forces in lymphatic vessels. Antioxid Redox Signal. 2016;25:451–65.
55. Karaman S, Detmar M. Mechanisms of lymphatic metastasis. J Clin Invest. 2014;124:922–8.
56. Commerford CD, Dieterich LC, He Y, Hell T, Montoya-Zegarra JA, Noerrelykke SF, Russo E, Rocken M, Detmar M. Mechanisms of tumor-induced lymphovascular niche formation in draining lymph nodes. Cell Rep. 2018;25:3554–63. e4
57. Olmeda D, Cerezo-Wallis D, Riveiro-Falkenbach E, Pennacchi PC, Contreras-Alcalde M, Ibarz N, Cifdaloz M, Catena X, Calvo TG, Canon E, Alonso-Curbelo D, Suarez J, Osterloh L, Grana O, Mulero F, Megias D, Canamero M, Martinez-Torrecuadrada JL, Mondal C, Di Martino J, Lora D, Martinez-Corral I, Bravo-Cordero JJ, Munoz J, Puig S, Ortiz-Romero P, Rodriguez-Peralto JL, Ortega S, Soengas MS. Whole-body imaging of lymphovascular niches identifies pre-metastatic roles of midkine. Nature. 2017;546:676–80.
58. Ubellacker JM, Tasdogan A, Ramesh V, Shen B, Mitchell EC, Martin-Sandoval MS, Gu Z, McCormick ML, Durham AB, Spitz DR, Zhao Z, Mathews TP, Morrison SJ. Lymph protects metastasizing melanoma cells from ferroptosis. Nature. 2020;585:113–8.
59. Lund AW, Wagner M, Fankhauser M, Steinskog ES, Broggi MA, Spranger S, Gajewski TF, Alitalo K, Eikesdal HP, Wiig H, Swartz MA. Lymphatic vessels regulate immune microenvironments in human and murine melanoma. J Clin Invest. 2016;126:3389–402.
60. Padera TP, Kadambi A, di Tomaso E, Carreira CM, Brown EB, Boucher Y, Choi NC, Mathisen D, Wain J, Mark EJ, Munn LL, Jain RK. Lymphatic metastasis in the absence of functional intratumor lymphatics. Science. 2002;296:1883–6.
61. Mlecnik B, Bindea G, Kirilovsky A, Angell HK, Obenauf AC, Tosolini M, Church SE, Maby P, Vasaturo A, Angelova M, Fredriksen T, Mauger S, Waldner M, Berger A, Speicher MR, Pages F, Valge-Archer V, Galon J. The tumor microenvironment and immunoscore are critical determinants of dissemination to distant metastasis. Sci Transl Med. 2016;8:327ra26.
62. Song E, Mao T, Dong H, Boisserand LSB, Antila S, Bosenberg M, Alitalo K, Thomas JL, Iwasaki A. Vegf-c-driven lymphatic drainage enables immunosurveillance of brain tumours. Nature. 2020;577:689–94.
63. Bakir B, Chiarella AM, Pitarresi JR, Rustgi AK. Emt, met, plasticity, and tumor metastasis. Trends Cell Biol. 2020;30:764–76.
64. Pereira ER, Kedrin D, Seano G, Gautier O, Meijer EFJ, Jones D, Chin SM, Kitahara S, Bouta EM, Chang J, Beech E, Jeong HS, Carroll MC, Taghian AG, Padera TP. Lymph node metastases can invade local blood vessels, exit the node, and colonize distant organs in mice. Science. 2018;359:1403–7.
65. Brown M, Assen FP, Leithner A, Abe J, Schachner H, Asfour G, Bago-Horvath Z, Stein JV, Uhrin P, Sixt M, Kerjaschki D. Lymph node blood vessels provide exit routes for metastatic tumor cell dissemination in mice. Science. 2018;359:1408–11.

Lymphangiogenesis: Lymphatic System and Lymph Nodes; Cancer Lymphangiogenesis and Metastasis

21

Stanley P. Leong and Marlys H. Witte

Abstract

The lymphatic system is considered the major conduit for cancer metastasis. Two key recent developments have made cancer lymphangiogenesis and cancer metastasis more understandable on a molecular and translational level. These include the discovery of lymphatic markers such as VEGF-C, LYVE-1, podoplanin, and Prox-1 and the recognition of the clinical significance of sentinel lymph nodes (SLNs) in melanoma and breast cancer. The lymphatic system develops during the fifth week of embryological life based on Sabin's centrifugal theory and on a minor extent relating to contrasting centripetal theory. The extracellular fluid that escapes at the blood capillaries is absorbed through the lymphatic vessels and passes through multiple levels of lymph nodes. The lymph from most of the body subsequently drains in a large part into the thoracic duct and then to the left subclavian vein. On the upper right side of the body, drainage is largely into the right subclavian vein. Thus, the excess extracellular fluid transported as lymph through the lymphatic vessels and filtered through the lymph nodes becomes sterile from bacteria or other microorganisms prior to reentry into the circulatory system. Although cancer angiogenesis involving blood vessels (hemangiogenesis) occurs in the tumor microenvironment, its prognostic value is unclear. On the other hand, cancer lymphangiogenesis has been extensively studied and it is associated significantly with poor prognosis. Cancer cells in the primary tumor site have the propensity to migrate to the SLN of the tumor draining nodal basin, then, to distant sites. Occasionally, cancer cells may escape into the blood vessels for distant metastasis. In general, the SLN may be considered as the major gateway for cancer metastasis. Micrometastasis in the SLN is an excellent biomarker for a poor clinical outcome. Based on experimental and clinical evidence, the SLN may be preconditioned by factors produced by cancer cells in the primary site to form a pre-metastatic niche that facilitates the arrival of cancer cells to take residence in the SLN. Molecular mechanisms of how cancer cells dissociate from the primary site, follow the lymphatic vessels to reach the SLN, and then, spread beyond to the distant sites remain under investigation.

Learning Objectives

- Describe steps in the development of and growth of lymphatic vessels (lymphangiogenesis) and lymph nodes.
- Profile the various cell populations in lymph nodes, their location, and function.
- Define a SLN.
- Identify the locations relative to the primary tumor and metastatic site where tumor lymphangiogenesis has been documented.
- Describe the postulated relationships between tumor lymphangiogenesis and cancer metastasis.

21.1 Introduction

Since the lymphatic system is the major conduit for cancer metastasis, it is important to understand the development of the lymphatic system including the lymph nodes and associated cell populations. The relationship between the lymphatic system and cancer invasion and metastasis will be

S. P. Leong (✉)
Department of Surgery and Melanoma Center, California Pacific Medical Center and Research Institute, San Francisco, CA, USA
e-mail: leongsx@cpmcri.org

M. H. Witte
Department of Surgery, Neurosurgery and Pediatrics, University of Arizona College of Medicine-Tucson, Tucson, AZ, USA
e-mail: lymph@surgery.arizona.edu

S. P. Leong et al. (eds.), *Cancer Metastasis Through the Lymphovascular System*,
https://doi.org/10.1007/978-3-030-93084-4_21

discussed in detail in this chapter and by others in this section. Recent discovery of several lymphatic-specific biomarkers has opened new vistas in the understanding of the lymphatic system and its pervasiveness in health and multisystem disease. The development of the SLN concept and its clinical utility in the staging of cancer such as melanoma and breast cancer have underscored the significance of the lymphatic system in cancer metastasis. This chapter focuses on the development of the lymphatic system, the process of lymphangiogenesis, and its relationship to cancer invasion and metastasis. Maldevelopment and malformations of the lymphatic system will not be addressed here but are covered in Chap. 23, which goes into more detail on lymphovascular genomics and molecular mechanisms in lymphatic development.

21.2 Development of the Lymphatic System

Although lymphatics were rediscovered by Gaspar Asellius in 1627 [1, 2], study on the lymphatics did not advance much beyond postmortem dissections compared to the immediate interest and rapid progress in exploring the blood circulation being associated with William Harvey's discovery in 1628 [3]. In 1902, Florence Sabin was the first to propose that the lymphatic system was formed by the sprouting of endothelial cells from embryonic veins. This process resulted in the formation of primordial lymph sacs with lymphatic endothelial cells then branching by sprouting into surrounding tissues thus forming mature lymphatic networks [4].

By the end of the fifth week of embryonic development, lymphatic vessels begin to develop from lymph sacs arising from developing veins (Fig. 21.1), which are derived from the mesoderm, according to Sabin [4]. The lymph sacs first appear as paired jugular lymph sacs at the junction of the internal jugular and subclavian veins. Thus, the centrifugal theory of Sabin maintains that the lymphatic plexuses spread from the jugular lymph sacs to the thorax, upper limbs, neck, and head. At least one connection is established between each jugular lymph sac and its jugular vein. The left jugular sac develops into the superior portion of the thoracic duct. Each jugular lymph sac retains at least one connection with its jugular vein, the left one developing into the superior portion of the thoracic duct. The unpaired retroperitoneal lymph sac at the root of the mesentery of the intestine, developing from the primitive vena cava and mesonephric veins sprout and spread from the retroperitoneal lymph sac to form lymphatic vessels and drain the abdominal viscera and diaphragm. The sac establishes connections with the cisterna chyli but loses connections with neighboring veins. The paired lower posterior lymph sacs arise from the iliac veins and form the lymphatic vessels of the abdominal wall, pelvic region, and lower limbs. They then join the cisterna chyli and cut off their connections with adjacent veins. All lymph sacs except the anterior portion from which the cisterna chyli develops become invaded by mesenchymal cells and are converted into groups of lymph nodes. The other lymphoid structures include the spleen, which develops from mesenchymal cells from the dorsal mesentery of the stomach and the thymus, which arises as an outgrowth of the third pharyngeal pouch [5].

In contrast, the centripetal theory as suggested by Kampmeier [6] and others [7] states that the lymphatic vessels arise from the mesenchymal tissues in the periphery and develop toward the central portion of the body to join the central venous system to complete the circulatory loop by delivering the lymph fluid from the final thoracic duct into the left internal jugular vein. This centripetal origin contributes to the final network of lymphatic vessels arising in discrete organs.

21.3 Major Breakthroughs in the Advancement of Research in the Lymphatic System

The vascular endothelial growth factor (VEGF) family members and their receptors were discovered in 1983 [8] and the gene was subsequently cloned in 1989 [9, 10]. VEGF-A has been found from fish to mammals. It plays a major role in the formation of blood vessels during embryonic vasculogenesis and angiogenesis. It is also involved for neovascularization in cancer and other diseases [11]. VEGF-A is a member of a family of related growth factors including placental growth factor (PLGF), VEGF-B, VEGF C, and VEGF-D, and the viral VEGF-Es [12–14]. VEGF-C, discovered in 1996 as a ligand for the orphan receptor VEGFR-3, is crucial in lymphangiogenesis for the growth and proliferation of lymphatic vessels [15].

In 1999, lymphatic vessel endothelial hyaluronan receptor 1 (LYVE1) was discovered. It is a type I intergral membrane glycoprotein and encoded by the LYVE1 gene [16]. LYVE1 binds to hyaluronic acid (HA), homologous to CD44, the main HA receptor. The LYVE-1 molecule binds both soluble and immobilized HA, like CD44. However, it is different from CD44, it binds with HA on the luminal face of the lymphatic vessel wall. Further, it is completely absent from blood vessels, although it has been found also in other tissues [17–19]. Thus, LYVE-1 is a very useful specific biomarker for lymphatic vessels of smaller caliber [16].

Another molecule that is expressed by lymphatic endothelial cells (LECs) and not by blood vascular endothelial cells is podoplanin, which is a small, extensively O-glycosylated, type I transmembrane glycoprotein, first discovered in rats 1997 [20]. Like podoplanin, Prox-1 is also a mucin-type transmembrane protein being expressed in

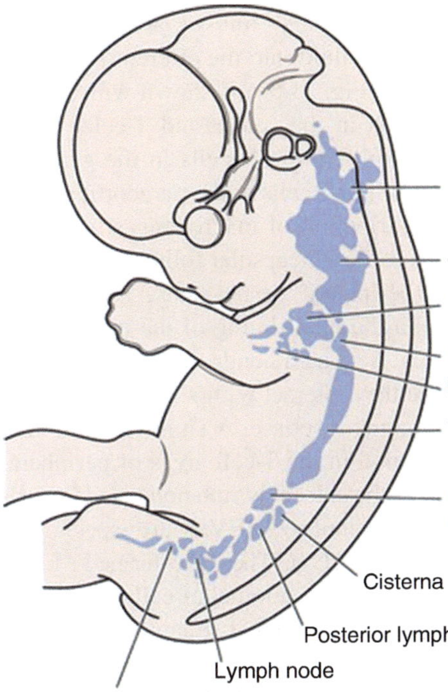

Fig. 21.1 The developing lymphatic system showing the primary lymph sacs and the thoracic duct primordium, as well as several lymph nodes of a 2-month-old human embryo. Figure and legend reproduced with permission from https://thoracickey.com/lymphatic-pathophysiology/ [31]

LECs. The Prox-1 gene was first cloned by homology to the Drosophila melanogaster gene, prospero, and associated with the development of the lymphatic system [21, 22]. It has been found that podoplanin expression is regulated by Prox-1 in the LECs at the transcriptional level [23]. Both podoplanin and PROX-1 are excellent markers for the identification of lymphatic vessels.

Thus, there are now several very useful lymphatic vessel biomarkers available to date including VEGFR-3, LYVE-1, podoplanin, and Prox-1, as described above. These have been used extensively to study the steps in the development of the lymphatic system. Prox-1 has been found to induce the formation of lymphatic vessels from blood vascular endothelium [24, 25].

Using the receptor to the VEGFR-3, Kaipainen et al. [26] have demonstrated the sprouting of lymphatic sacs from the venous endothelium in the mouse embryo. These studies using the lymphatic biomarkers have supported the centrifugal theory of lymphatic development. However, Wilting et al. [27, 28], using the chimeric quail-chick embryos in the engrafted wing lymphatic system, have shown the centripetal mechanism from the lymphangioblasts of the mesenchymal system.

Perhaps, both the centrifugal and centripetal processes contribute to the formation of the lymphatic system and subsequent remodeling. The final outcome is that the lymphatic system while independent of the vascular system ultimately joins the blood circulatory system at the thoracic duct/subclavian vein junction [29] and prompted Witte and Witte to ask the question: How and when are lymphatics different from the blood vessels and under what circumstances do they merge [30]. As noted above, Fig. 21.1 shows the embryological development of the lymphatic system [31].

21.4 Development of the Lymphatic Vessels and Lymph Nodes

As mentioned previously, lymph sacs formed by the mesenchymal cells become elongated to form lymphatic vessels. The lymph sac is surrounded by mesenchymal cells, which then invade the sac and give rise to the lymph node stroma. Along the track of the lymphatic vessels, lymph nodes formed from the lymph sacs are situated as they are clustered in the nodal basin in the axilla, groin, neck, chest, and abdomen.

According to the centrifugal theory, the first lymph sacs arise from the paired jugular lymph sacs at the junction of the internal jugular and subclavian veins. At the root of the mesentery of the intestine from the primitive vena cava and mesonephric veins, the unpaired retroperitoneal lymph sac appears next. The posterior lymph sacs being paired from the iliac veins are the last to develop. They give rise to capillary plexuses and lymphatic vessels of the abdominal wall, pelvic region, and lower limbs. All lymph sacs become invaded by mesenchymal cells and converted into groups of lymph nodes, except for the anterior part of the sac from which the cisterna chyli develops, which is a dilated sac just distal to the thoracic duct into which lymph from the upper body, the intestinal trunk, and two lumbar lymphatic trunk drains. The spleen develops from mesenchymal cells within the layers of the stomach on the dorsal mesentery [5].

The lymphatic system consisting of the lymphatic vessels and lymph nodes is complex. The lymphatic drainage systems of the stomach, pancreas, viscera, pelvis, and lower extremities drain into the final pathway to the cisterna chyli and then to the thoracic duct [32]. In the head and neck, the lymphatic system drains into the left and right jugular lymphatic trunks with the left emptying into the thoracic duct and the right into the right subclavian vein. In the right upper extremity, the lymphatic system drains into the right subclavian trunk, which returns the lymph at the right internal jugular and subclavian vein junction, while the left upper extremity drains into the left subclavian trunk via the thoracic duct. In the upper right portion of the body, the lymph drains into the right lymphatic duct, which returns the fluid through the right subclavian vein. In the upper left portion of the body, the lymph drains into the thoracic duct, which returns the fluid into the left subclavian vein. Furthermore, the lymphatic system includes the mucosa-associated lymphoid tissues like tonsils, adenoids, Peyer's patches in the small bowel, and even the appendix [32]. There are about 600 lymph nodes in the human body within the lymphatic system. About 17 L of the 20 L of the blood at the arteriovenous capillary junctions return through the venous capillaries daily. About 2–3 L of fluid (15%) without cellular components of the blood enter into the extracellular space and drain into the lymphatic vessels with the lymph fluid carrying cellular debris, macromolecules of proteins, excess water, and toxins [33].

The lymphatic vessel has unidirectional semilunar valves with an average diameter of 100 μm. The lymphatic vessel wall is thin and lined by endothelial cells surrounded by smooth muscle with the outermost layer being the adventitia anchoring the lymphatic vessels to the surrounding tissue, mainly the adipose tissue [34]. Fig. 21.2a, b shows the detailed drawing to illustrate the microanatomy of the structure of the lymphatic vessel [35].

Lymphedema manifests when the extracellular fluid is not adequately drained. It is discussed in detail in Chap. 25. The lymph fluid is filtered through multiple lymph nodes and finally drains predominantly into the thoracic duct and subclavian veins as normally nearly sterile lymph fluid to rejoin the blood circulation. For a more detailed review of the anatomy and physiology of the lymphatic system, please see separate reviews [35–37]. The structure of the lymphatics and lymphatic vessels is shown in Fig. 21.2.

Figure 21.3 shows the structure of a mature lymph node. Fig. 21.4 shows the hematoxylin and eosin staining of a lymph node showing the capsule, subcapsular sinus, cortex, lymph follicle, medulla, and medullary sinus. B cells are primarily located in the outer cortex where they are grouped together as follicular B cells in lymphoid follicles. Mature B cells become plasma cells to secrete antibodies against foreign antigens by means of humoral immunity. The medulla is the innermost of the lymph node consisting of large blood vessels, sinuses, and medullary cordis, where antibody-secreting plasma cells, B-cells, and macrophages are found. Macrophages are phagocytic with the ability to engulf and destroy bacteria and other harmful organisms. They may also present antigens to T cells and release inflammatory cytokines. Macrophages take up the carbon such as from tattoos and appear black. Fig. 21.5a shows the germinal center of a lymph node and lymphoid follicles by hematoxylin and eosin staining. Fig. 21.5b shows a close-up of a germinal center and paracortical T cell zone. B cells are primarily clustered in structures called lymphoid follicles, whereas T cells are found mainly in the paracortex. The T cells contribute to the cellular immunity by recognizing the foreign antigen on the infected cell and destroying it. The medullary sinuses or sinusoids are vessel-like spaces separating the medullary cords. The medullary sinuses are vessel-like spaces that drain the lymph fluid into the efferent lymphatic vessels. A lymph follicle (Fig. 21.5b) is shown with its germinal center with B-cells in the center and T-cells in the paracortical area. The proliferating B cells in the germinal center are activated by helper T cells in the paracortical area. Fig. 21.6 shows the distribution of macrophages, dendritic cells, and lymphocytes in the subcapsular follicles. Dendritic cells are present in peripheral tissues such as the skin Langerhans cells and in the inner lining of the nose, lungs, stomach, and intestines. Activated dendritic cells migrate to the *lymph nodes* where they interact with *T cells* and *B cells* to initiate adaptive immune response. A significant number of dendritic cells are found in the T-cell areas of peripheral lymphoid organs such as the spleen, lymph node, and Peyer's patch. High-endothelial venules (HEVs) are specialized postcapillary venules (Fig. 21.7). They are termed "high" because of the thickness of the endothelial cells of the low cuboidal type in contrast to the typical squamous epithelial type of the endothelial cells. HEVs are vascular sites for the entry and exit of lymphocytes from the secondary lymphoid organs. HEV endothelial cells have ligands that bind lymphocytes, facilitating their adhesion and subsequent extravasation into the lymph node. Thus, an adult lymph node consists of primarily the capsule and trabeculae, both are of the dense connective tissue, reticular tissue with reticular cells, and fibers forming the supporting meshwork and macrophages, which process antigen and mingle in with the reticular cells [38–41].

Lymphoid stromal cells are best known as the architectural cells of lymphoid organs [42]. It is now known that the stromal cells are primarily responsible for the adaptive immune response. Fibroblastic reticular cells are derived from mesoderm to form the stroma consisting mainly of the extracellular matrix (type III collagen) with supporting network of cells and extracellular matrix in secondary lymphoid organs namely lymph nodes, spleen, and thymus. The fibers crisscross within the sinuses. They produce reticular fibers to form a network between the capsule and trabeculae allowing the interaction and movement of T cells, B cells, dendritic cells (DCs), plasma cells, and macrophages [43]. These myofibroblasts differentiate into fibroblastic reticular cells in the T cell zone of the cortical and medullar regions. They bind dendritic cells which present antigens to T cells as well as induce remodeling reticular cells in lymph nodes. Further, the reticular cells may regulate the traffic of lymphocytes and activation of specific lymphocyte populations [44]. The myofibroblast markers consist of desmin, vimentin, CD90, CD73, CD103, smooth muscle actin (SMA) and ERTR7 antigen, podoplanin, platelet-derived growth factor receptor, and genes from antigen presentation and cytokine response pathways but lacking CD45 and CD31.

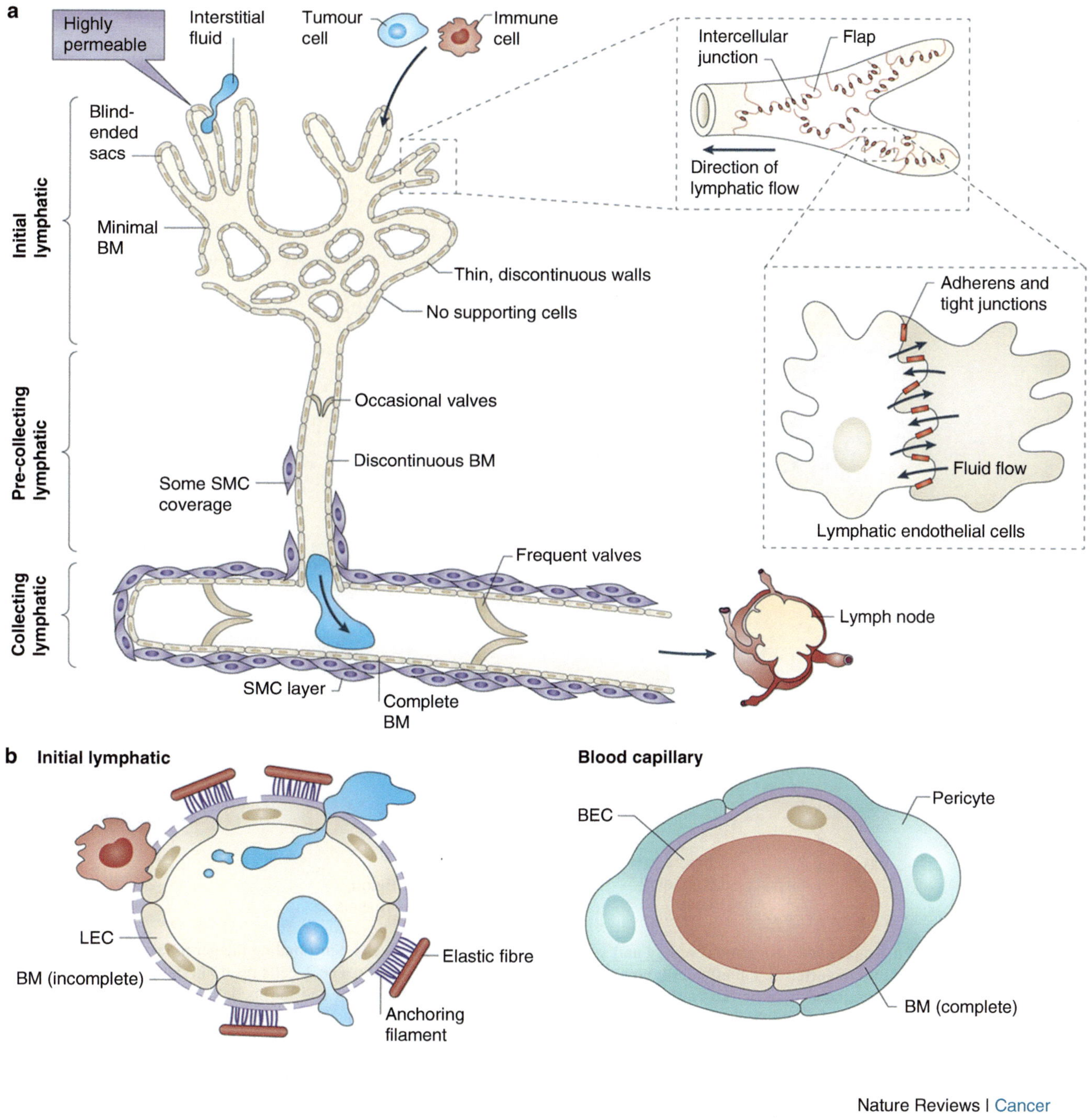

Fig. 21.2 (**a** and **b**) Structure of the lymphatic vasculature and lymphatic endothelial cells from initial lymphatics. The hierarchy of lymphatic vessel subtypes is shown in part **a**, with the characteristics of the subtypes indicated. The association between lymphatic endothelial cells (LECs) in initial lymphatics is shown in the two pull-out boxes, with the contact regions between LECs in the vessels being indicated by the brown lines in the first pullout box. Fluid is thought to enter (as shown by the arrows in the second pullout box) via the tips of the flaps of LECs, without compromising the integrity of the intercellular junctions on the sides of the flaps. The structure of LECs from initial lymphatics is shown in part **b**, illustrating the entry of fluid and cells, which is favored when the interstitial fluid pressure is high, causing stretching of the extracellular matrix. The structure of a blood vascular capillary is shown for comparison. The anchoring filaments that are associated with initial lymphatic vessels connect the abluminal membrane of LECs to surrounding elastic fibers—these filaments are short, but for visual clarity, they are shown larger than to scale. Anchoring filaments are involved in controlling the entry of cells and interstitial fluid into the initial lymphatics. It should be noted that cells can cross lymphatic endothelial layers via a transcellular route, as well as via the paracellular route that is shown here. BEC, blood vascular endothelial cell; BM, basement membrane; SMC, smooth muscle cell. Figure and legend reprinted by permission from Springer Nature: Nature Reviews Cancer, Lymphangiogenesis and lymphatic vessel remodeling in cancer [35]

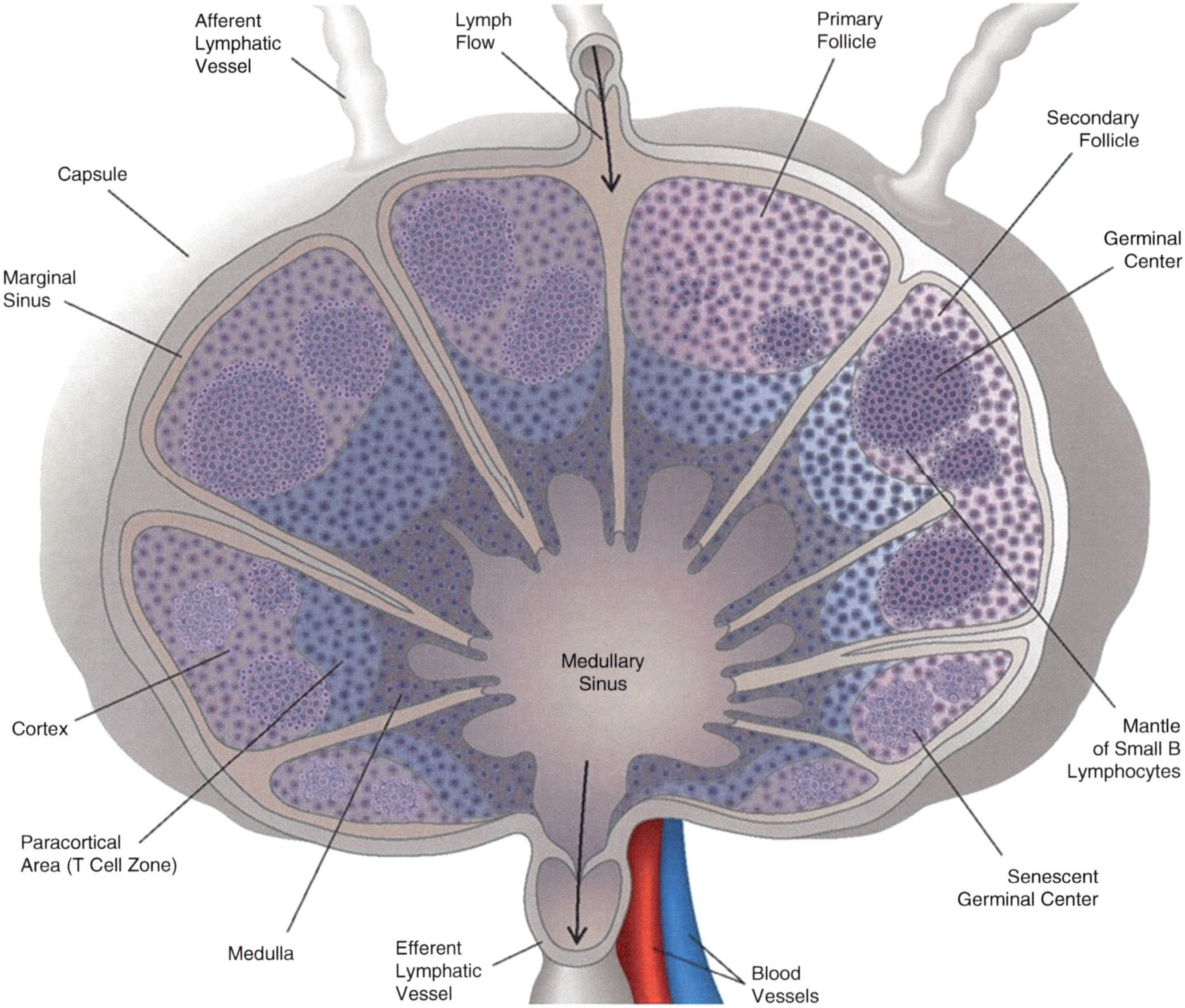

Fig. 21.3 Structure of a lymph node with arterial supply and venous drainage. The diagram shows the afferent and efferent lymphatic vessels. The lymph node is surrounded by a fibrous capsule. Medulla is toward the center of the lymph node. Multiple follicles or lymphatic nodules are noted. Within the lymphatic follicle, there is the germinal center (Fig. 21.4) with lymphoblasts, macrophages, follicular dendritic cells, and plasmas cells. Surrounding the germinal center is the mature B cells with the cortex. The T cells reside in the paracortical area, the space between the cortex and the medulla. Figure and legend reproduced with permission from https://www.iheartpathology.net/post/lymph-nodes-101 [39]

Dendritic cells or antigen-presenting cells originate from the same common progenitor as monocytes. Their role in adaptive immunity consists of presenting antigens and inducing a primary immune response in resting naïve T cells [45]. Recently it has been shown that recirculation of antigen-bearing dendritic cells in remote lymphoid organs can prime T cells in other locations [46].

During the development of the lymphatic system, so-called lymphangiogenesis, as described in Section 21.9, maldevelopment and malfunction may be encountered. Thus, primary lymphedema and lymphangiodysplasia may occur [29, 36, 47]. During cancer treatment, secondary lymphedema may arise [48]. These topics are beyond the scope of this chapter but are discussed in Chap. 25.

21.5 Trafficking of the Lymphocytes

Lymphocytes circulate continuously between the blood and secondary lymphoid tissues such as the lymph nodes, Peyer's patches, and the spleen, where antigens and antigen-presenting dendritic cells are selectively present.

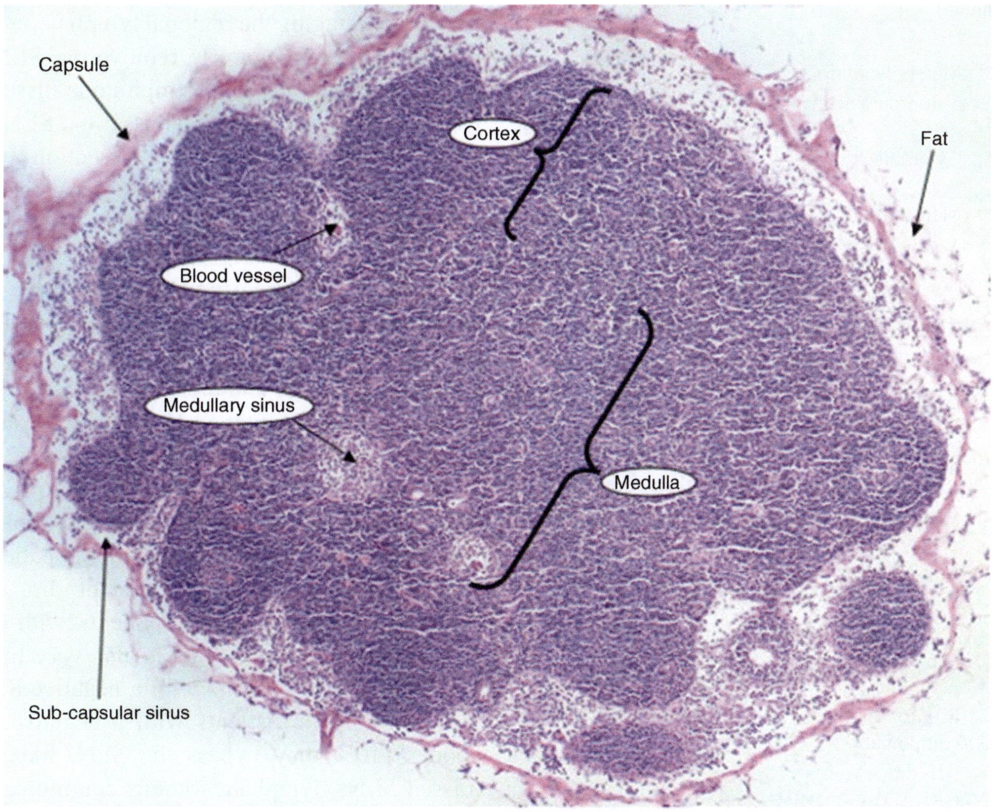

Fig. 21.4 Histologic hematoxylin and eosin staining of part of a lymph node (low power magnification) showing the cortex (B cells) and paracortex space of T cells between the cortex and medulla. Lymph fluid enters the subcapsular sinus gap between the clusters of lymphocytes and the capsule wall. The lymph is filtered through the node. The filtering process ensures anything nonself that is found in the lymph is selected out and processed by the lymphocytes. The cortex of the lymph node is mostly populated by B-cells, and in the paracortex area (slightly deeper region that is sandwiched between the cortex and the medulla) contains mostly T-cell lymphocytes. B-cell lymphocytes respond to antigens that enter the lymph node and produce antibodies once they are activated. T-cell lymphocytes respond to cell-bound antigens as cytotoxic T cells against infected cells with viral particles or even cancer cells. On the other hand, helper T cells may assist B-cells in antibody production. The paracortical region may become expanded in a T cell-dominant immunological response. Courtesy of Ed Uthman MD, Flickr, originally published under CC BY [38]

The naive lymphocytes first bind to and migrate across HEVs. The HEVs are distinctly different from the normal venules in, at least, two ways. First, HEVs express unique adhesion molecules, so-called vascular addressins that act as ligands for homing receptors of the lymphocytes. Second, chemokines and chemokine-binding molecules are produced within the extracellular matrix in and around the HEVs. These chemokines activate integrins on circulating lymphocytes and attract them into the lymph nodes and Peyer's patches. Thus, the lymphocytes can navigate themselves into the lymph nodes or Peyer's patches to launch effective immune responses with the help from the antigen-presenting dendritic cells [49]. The trafficking of lymphocytes is schematically shown in Fig. 21.8. Detailed molecular mechanisms of lymphocyte trafficking and immune responses are areas under extensive investigation.

21.6 Development of the Sentinel Lymph Node Concept

When Halstad developed radical mastectomy including a complete axillary lymph node dissection for breast cancer [50] and Snow championed en bloc resection of the primary melanoma along with a radical regional lymph node dissection [51, 52], they assumed that cancer spread from the primary site then to the regional nodal basin before metastasizing to the distant sites. Halsted's influence was very profound so that radical lymph node dissections became standard for different types of cancer such as the esophagus, gastric, colon, gynecological cancers, and others. Haagensen et al. described the lymphatics in cancer extensively [53]. Although adequate resection of the primary site and simultaneous radical regional lymph node dissection was performed, systemic

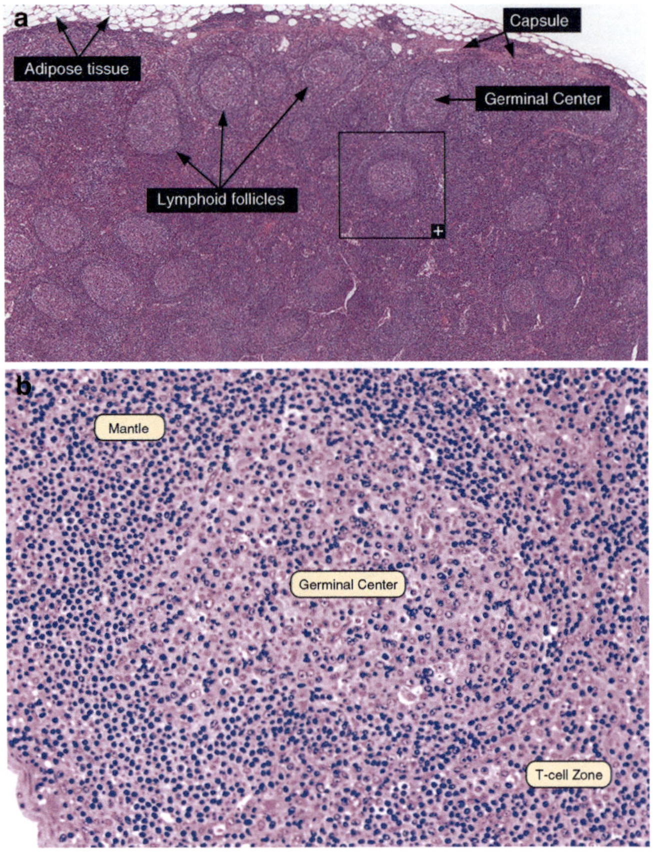

Fig. 21.5 (**a**) Subcapsular location of germinal centers and lymphoid follicles by hematoxylin and eosin staining (low power magnification). Figure and legend reproduced with permission from https://www.iheartpathology.net/post/lymph-nodes-101 [39]. (**b**) Histologic hematoxylin and eosin staining (high power magnification) of the germinal center within a lymph follicle. The germinal center consists of lymphoblasts, macrophages, follicular dendritic cells, and plasmas cells. Surrounding the germinal center is the mature B cells with the cortex. The T cells reside in the paracortical area, the space between the cortex and the medulla. Figure reproduced with permission from http://medcell.med.yale.edu/systems_cell_biology/lymphatics_lab.php [40]

recurrence still occurred. Using breast cancer as a model, Fisher proposed the systemic theory stating that clinically apparent cancer is a systemic disease [54] in that cancer dissemination occurs even when clinically the cancer burden is small. Accordingly, treatment of local or regional disease should not affect overall survival. Thus, Cady concluded that the lymph node was the marker but not the governor of cancer metastasis [55]. An alternate spectrum theory by Hellman was developed to indicate that cancer development was progressive. Cancer may be compartmentalized initially before becoming widespread. Thus, oligometastatic cancer is localized. Appropriate treatment such as surgery or radiation therapy may render extermination of the metastasis either alone or combined with systemic therapy [56].

The pioneering work on the development of the SLN concept by Cabanas [57] and Morton [58] has resolved the dilemma of staging the regional lymph node by a simpler and more reliable procedure to remove the SLN for evaluation and yet to avoid a radical lymph node dissection associated with significant morbidity. Thus, if the SLN biopsy is negative, a radical lymph node dissection may be spared. Furthermore, melanoma [59] and breast cancer [60] patients with a negative SLN biopsy enjoy a significantly improved survival.

In the post-sentinel node era, we have learned that, in most cases such as melanoma and breast cancer, cancer spreads in an orderly process [61] from the primary site to the SLNs primarily and then beyond to the distant sites.

A recent multicenter SLN Working Group study showed that the distant metastatic rate was 4.7% [62] for 4059 melanoma patients with a negative SLN biopsy with a median follow-up of 31.2 months. In another retrospective study, it was found that 126 of 1277 melanoma patients with a negative SLN biopsy developed distant disease (9.9%) [59]. Similarly, for breast cancer patients with a negative SLN biopsy, their recurrence rate is also very low, less than 5% [60]. Thus, most patients with a negative SLN biopsy were disease-free. The patients with a negative SLN biopsy of about 5–10% may bypass the SLN, likely through other missed SLNs, lymphatic-venous communications, or blood vessels, to spread to the distant sites.

Overall, melanoma and breast cancer patients with a negative SLN biopsy manifest much better disease-free and overall survival than those with a positive SLN biopsy [63, 64]. Thus, the spectrum theory of Hellman seems to be most compatible with the clinical outcome of patients undergoing SLN staging procedure. The spectrum theory is applicable for melanoma patients with minimal tumor burden in the SLN. When the micrometastasis has been found to be less than 0.1 mm, their clinical outcome is the same as those patients with a negative SLN biopsy, according to van Akkooi et al. from the Rotterdam criteria [65, 66]. In breast cancer patients, similar findings have been found in that if the breast cancer SLN is found to harbor isolated tumor cells (ITC; ≤0.2 mm), the clinical outcome of these breast cancer patients is the same as those with a negative SLN biopsy [67].

When an SLN is positive with micrometastasis, is it an incubator for cancer cells to proliferate before spreading to the distant sites or is it a marker for systemic disease [59] Perhaps, there is a spectrum of the SLN from an incubator then to a marker, when the tumor burden reaches a certain volume with associated distant metastasis. The cutoff level of tumor burden for this transition has not been determined. In the post-sentinel node era, we have learned that, in most cases such as melanoma and breast cancer, cancer in the early stage spreads in an orderly process [61] from the primary site to the SLNs primarily and then beyond to the distant sites.

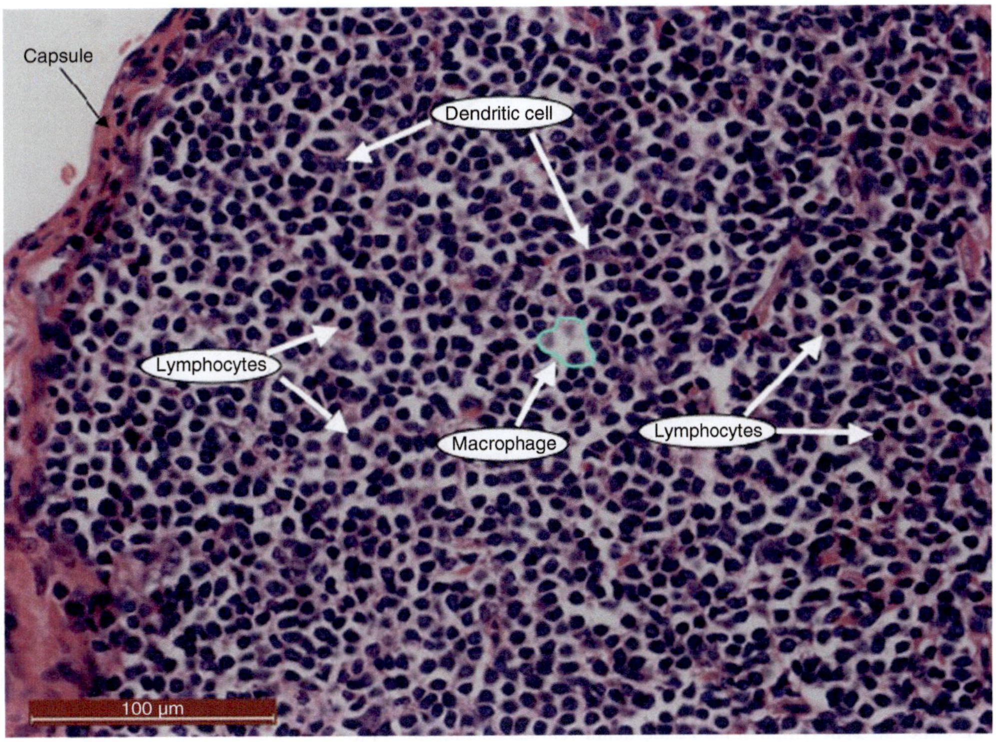

Fig. 21.6 High power view of the subcapsular follicles and sinus of a lymph node. The mature lymphocytes appear to be more in comparison to other cells, such as phagocytic cells like macrophages and dendritic cells, which are in the subcapsular sinus. Most of the other irregular looking cells are lymphocytes in their different developmental stages. Figure reproduced with permission from https://histologyblog.com/2012/04/30/histoquarterly-lymph-node/ [41]

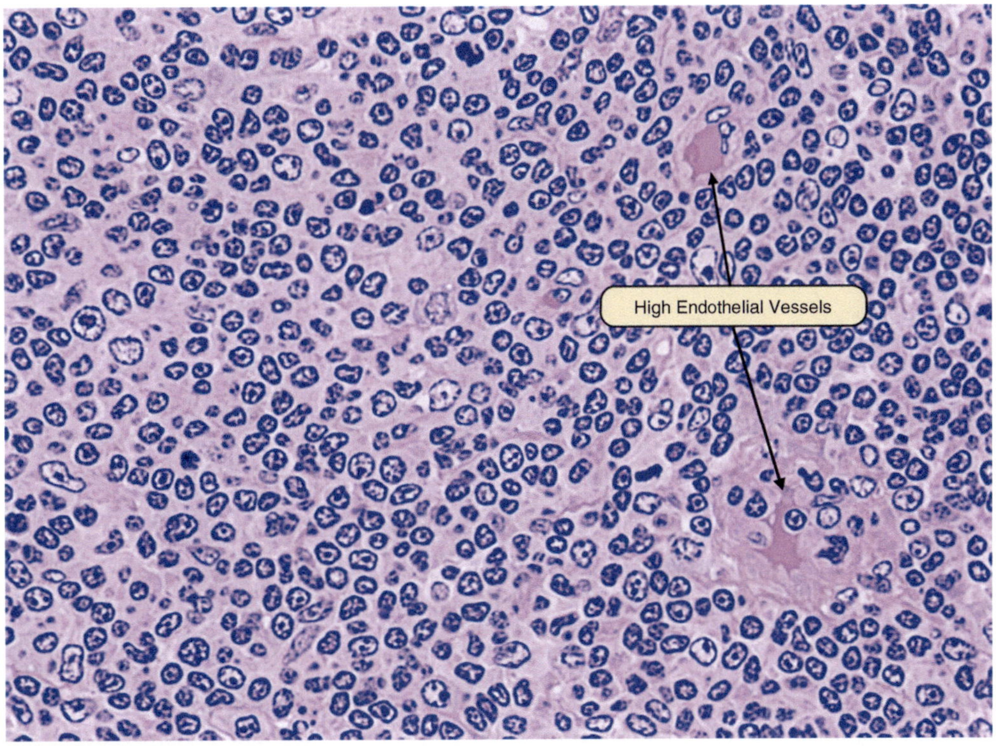

Fig. 21.7 Lumens of the high endothelial venule (HEV), high power view. One significant feature of the paracortex T-cell zone of the lymph node is the HEV, where circulating lymphocytes leave the bloodstream to enter the node. These postcapillary HEVs can be distinguished by their cuboidal endothelial cells. Adhesion molecules called selectins and integrins on the surfaces of HEVs and lymphocytes mediate the attachment of lymphocytes to endothelial cells. Figure reproduced with permission from http://medcell.med.yale.edu/systems_cell_biology/lymphatics_lab.php [40]

21.7 Trafficking of Cancer Cells

The cancer cells utilize the lymphatic system to spread from the primary site as they dissociate themselves from their microenvironment and enter the lymphatic vessels toward the lymph nodes and beyond to the distant sites. The size of a single cancer cell is slightly larger than that of a lymphocyte with an average diameter of 15 μm, thus, the diameter of a cancer cell may be up to 20 μm or more, and it can comfortably pass through the 100 μm lymphatic vessel. Figure 21.9 shows that cancer may spread through the lymphatic and blood vessels.

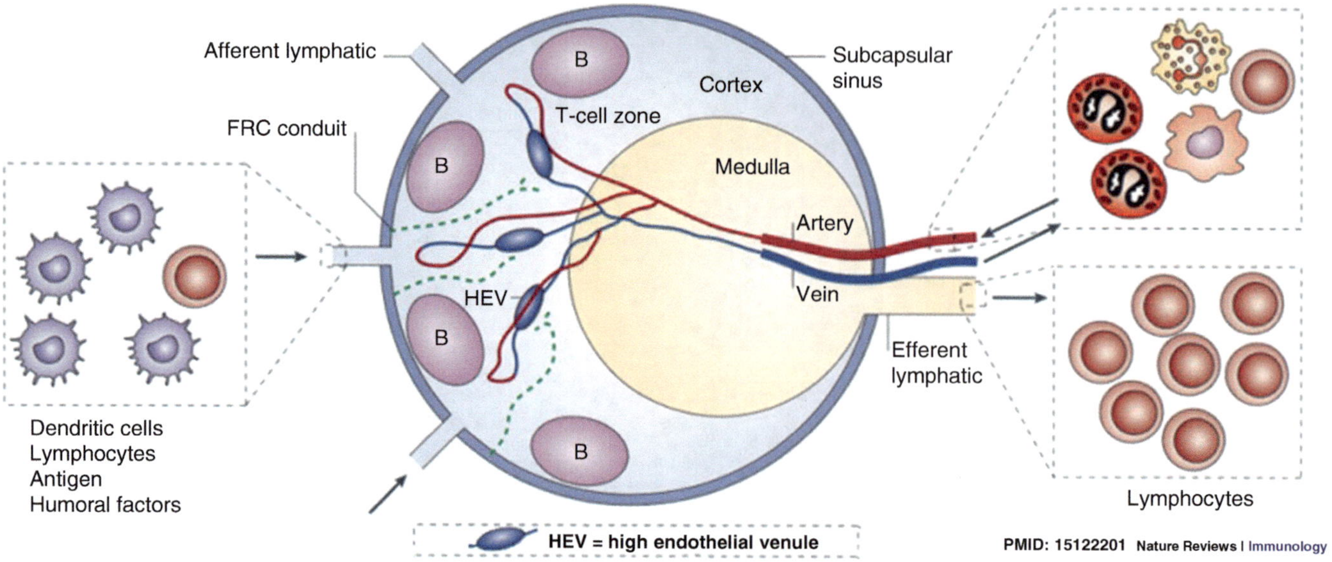

Fig. 21.8 Trafficking of lymphocytes in and out of the lymph node. Unfiltered lymph fluid with cells as lymphoctyes and dendritic cells flow into the lymph node through the afferent (towards lymph node) lymphatic vessels. Efferent (away from lymph node) lymphatic vessels flow out of a lymph node and carry filtered lymph fluid. Lymphocytes use either afferent lymph or high endothelial venules for migration into the lymph node. **B**—B-cell zones. **DC**—dendritic cells. **FRC**—fibroblastic reticular cells. **HEV**—high endothelial venues. Figure and legend reprinted by permission from Springer Nature: Nature Reviews Immunology, Lymphocyte trafficking across high endothelial venules: Dogmas and enigmas [49]

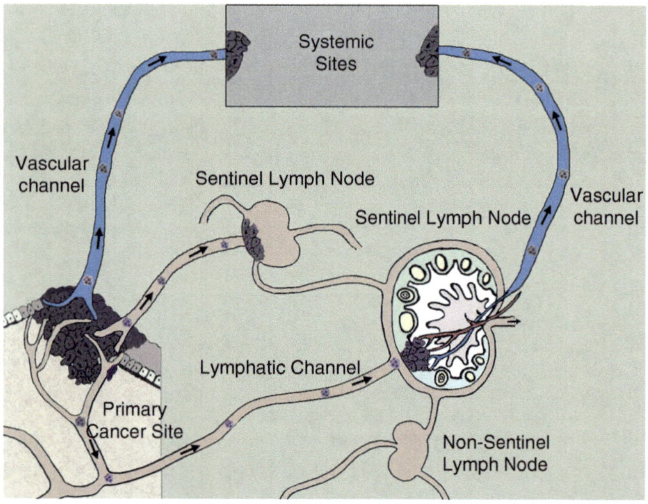

Fig. 21.9 Two major routes of cancer metastasis: one through the lymphatic vessels to the SLNs as the primary gateway and the other through the vascular channels directly to the distant sites bypassing the SLNs. Figure reprinted by permission from Springer: Clinical and Experimental Metastasis, Front page (Special Issue on: The 2017 seventh International Cancer Metastasis Symposium; Cancer Metastasis through the Lymphovascular System: Biology and Treatment [68])

As described previously based on the clinical outcome of melanoma and breast cancer patients with a negative SLN biopsy being very favorable with low rate of distant metastasis, it can be induced that the SLN may be the major gateway in early cancer development for metastasis to the distant sites. Nathanson et al. suggest that systemic metastasis in breast cancer is most often owing to SLN metastasis and only occasionally directly into the bloodstream [69]. Once the breast cancer cells reach the SLN, they may gain direct access to the systemic circulation to the distant sites by invasion into the veins in the lymph node [69]. Figure 21.10 shows the possible mechanism of cancer metastasis to the SLN and then to the distant sites through the venous channel within the lymph node [70].

Thus, a critical question is whether cancer clones that spread through the blood vessels are the same clones that spread through the lymphatic vessels. In other words, are the cancer cells dependent on lymphatic or blood vessels for metastasis? Induction tumor angiogenesis provides a possible explanation for how cancer cells escape their original site by invading the newly formed blood vascular bed [71]. Tumor vascularization has been observed in human melanoma both experimentally [72, 73] and clinically [74]. The importance of tumor angiogenesis for the prognosis of primary malignant melanomas of the skin, however, has remained controversial [75]. Whereas several studies found that tumor microvessel density was associated with decreased disease-free and overall survival [76, 77], other reports did not detect any significant differences of tumor blood microvessel density between metastatic and nonmetastatic melanomas [78]. Therefore, the potential prognostic value of tumor vascularization in malignant melanoma remains at present unclear [79].

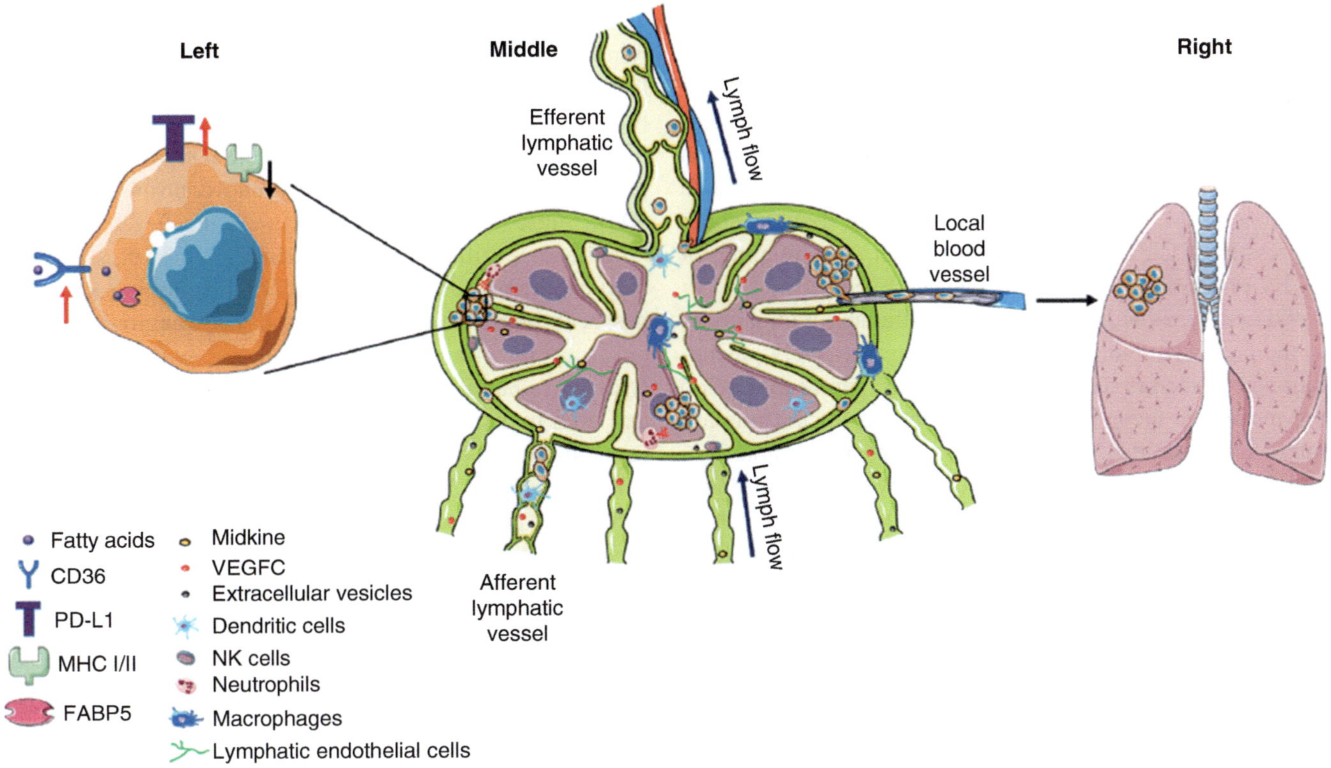

Fig. 21.10 Cancer cells survive in lymph nodes and spread to other organs. (**Left**) To adapt to the lipid-rich environment in lymph nodes, cancer cells enhance their reliance on fatty acid metabolism by upregulating CD36 and FABP5. In addition, to avoid immune destruction, cancer cells increase expression of the immunosuppressive checkpoint molecule PD-L1, while decreasing MHC I/II expression. (**Middle**) Midkine and VEGF-C secreted by cancer cells induce lymphangiogenesis in draining lymph nodes prior to seeding, generating a pre-metastatic niche. Similarly, cancer cells also release extracellular vesicles to interact with B cells in the cortex, promoting lymph node metastasis. Although the lymph node is filled with various immune cell populations, the microenvironment is immunosuppressive. Tolerogenic DCs, M2-like TAMs, and CD15+ TANs were enriched in metastatic lymph nodes. These cells contribute to the inhibition of cytotoxic T cells. The lytic activity of NK cells is also reduced in metastatic lymph nodes. (**Right**) After surviving in the lymph node, cancer cells can invade the lymph node blood vasculature and further spread to distant organs. Cancer cells can also exit through the efferent lymphatic vessel. Figure and legend originally published under CC BY 4.0 [70] with permission to be reprinted here

It is apparent that the lymphatic system draining various body organs is different, some with more complex networks than the others. The SLN drainage of melanoma and breast cancer follows more reliable routes than the internal organs such as the lungs, stomach, pancreas, and other organs, which show complicated lymphatic drainage patterns. Certainly, melanoma and breast cancer allow us to appreciate the biology of early cancer arrival in the SLNs and may form a prototype for cancer spread to the SLNs and beyond to the non-sentinel lymph node compartments [64] and beyond to the distant sites.

21.8 Evidence of Metastatic Niche in Sentinel Lymph Nodes

A microenvironment that facilitates the survival and progression of disseminated cancer cells is termed a pre-metastatic niche. Cancer cells may secrete factors to precondition the microenvironment of the site to allow it to support the cancer cells which may spread to such a site [80]. Other processes may include recruitment of certain cell types that support metastatic implant, induction of growth factors, and cytokines to prepare for the pre-metastatic niche and restructuring of the extracellular matrix [81, 82]. Perhaps, the seed and soil concept of cancer metastasis of Paget [83] may be explained by the fact that the soil is "pre-conditioned" as the pre-metastatic niche for the seed to survive and grow. Evidence for the pre-metastatic niche is extensive including pre-metastatic morphological and histological changes, immunological alterations, tumor-induced immunosuppression in lymph nodes, vascular remodeling, tumor-secreted conditional factors, and remodeling of extracellular matrix [82].

Primary tumors may secrete soluble factors to induce the pre-metastatic niche in the tumor draining or SLNs [82]. In melanoma, immunosuppressive microenvironment suppressor cell activity has been shown in tumor-draining lymph nodes [84–86]. Similarly, in breast cancer SLNs, immunosuppressive activity was noted [87–89]. Further, immunomodulatory cytokines have been found in melanoma SLNs [90, 91].

Thus, Cochran et al. provocatively asked Is SLN susceptibility to metastases related to nodal immune modulation [92]. On a molecular basis, Takeuchi et al. have demonstrated a potential mechanism by tumor-draining lymph nodes expressing CCL21/SLC for recruitment and homing of CCR7-positive metastatic melanoma cells [93].

Of course, as SLNs are the closest draining lymph nodes to the primary tumor, they may be more effectively preconditioned by factors secreted by the primary tumor to develop pre-metastatic niche resulting in preference of the primary tumor to invade the SLNs. To a lesser extent, the non-SLN and distant sites are much less affected by their lower gradient of receiving the "pre-metastatic niche" inducing factors. Perhaps, the orderly progression of cancer metastasis from the primary site to the SLN [61], then to the non-SLNs and distant sites follow this gradient phenomenon. This "gradient" may explain and reflect the rationale of the tumor nodal metastasis (TNM) staging of cancer.

Since the tumor-draining node or SLN is the critical microenvironment for the cancer cells to invade and proliferate, it should be the target of therapeutic manipulation. Granulocyte/macrophage colony-stimulating factor (GM-CSF) has been found to reverse immunosuppression in melanoma SLNs [84]. A recent study showed that nanoparticle targeting of dendritic cells in tumor-draining lymph nodes stimulated maturation of dendritic cells, which then induced a TH1 response. Such a response resulted in increased numbers of tumor antigen-specific CD8+ T cells with resultant slower tumor growth [94]. Thus, targeting the SLNs to reverse the immunosuppression and increase antitumor response may be potentially a viable strategic option. A more extensive discussion of pre-metastatic niche for disseminated cancer cells to establish it as the soil to grow is addressed in a review by Sleeman [82].

21.9 Cancer Lymphangiogenesis: A Marker for Cancer Metastasis

Understanding the growth and development of the lymphatic system has increased dramatically since it was first described in the developing embryo by Sabin in the early 1900s. Lymphangiogenesis was subsequently investigated in vivo during regeneration after severing, inflammation, and ligation [95]. However, the term lymphangiogenesis was not introduced until the early 1980s when lymphatic endothelial cells (LECs) were first isolated in vitro [96–98], and branching structures were seen in vitro (lymphangiogenesis in vitro) [96, 99, 100]. The recognition of tumor lymphangiogenesis did not occur until more than a decade later. Indeed, following the elucidation and extensive investigation of blood vessel angiogenesis and its relation to tumor growth, the presence and potential significance of the phenomenon was still questioned and controversial [101–104]. Subsequently, particularly with the development of lymphatic-specific markers to identify lymphatics and distinguish them from blood vessels on histologic sections, it has been demonstrated that tumor lymphangiogenesis not only occurs within the primary tumor with intratumoral lymphatics but also peritumorally, in draining lymph nodes and even in distant metastatic niches as cancer cells spread through the lymphatic system. Indeed, tumor lymphangiogenesis seems to be a superior predictor of poor prognosis and cancer progression [79] than blood vessel angiogenesis (hemangiogenesis).

Thus, it is now well-recognized that in the cancer microenvironment, cancer cells release VEGF A/C, which promotes local lymphangiogenesis, a process by which new lymphatic vessels are formed. Cancer cells may also undergo epithelial–mesenchymal transition to boost their invasiveness. Further, they can upregulate the expression of CCR7/8 and CXCR4/5, which respond to the corresponding CCL1/21 CXCL10/12 from the lymph node, respectively.

Cancer cells respond to these stimuli by migrating to lymphatic vessels through chemotaxis. Some cancer cells produce ALOX15, which functions as a catalyst to convert arachidonic acid to 12[S]-HETE and 15[S]-HETE. These metabolites cause circular defects on lymphatic endothelial cells (LECs), allowing cancer cells to enter the lymphatic vessels. In addition, interstitial fluid flow forces the cancer cells mechanically into the lymphatic vessels. TWIST1 plays an important role in metastasis, like angiogenesis, invadopodia, and extravasation. SNAIL1/2 downregulates E-cadherin and induces epithelial to mesenchymal transition in the process of cancer metastasis. The flow also assists cancer cells to produce and respond to the autocrine chemokine gradients to lymphatic vessels [70]. Figure 21.11 shows the migration and invasion of lymphatic vessels by cancer cells following lymphangiogenesis.

Cancer-induced lymphatics by lymphangiogenesis may become dilated [105]. It may be speculated that a cluster of cancer cells may accumulate at the valve of the lymphatic vessel [106] and eventually a colony of cancer cells may grow on the inner wall of the lymphatic vessel with subsequent invasion into the surrounding fat tissue as in-transit metastasis, such as the case in melanoma as shown in Fig. 21.12. The other possibility is that a specific receptor on the cancer cell may attach to the inner wall of the lymphatic vessel to establish a nidus to grow. Although the clinical sig-

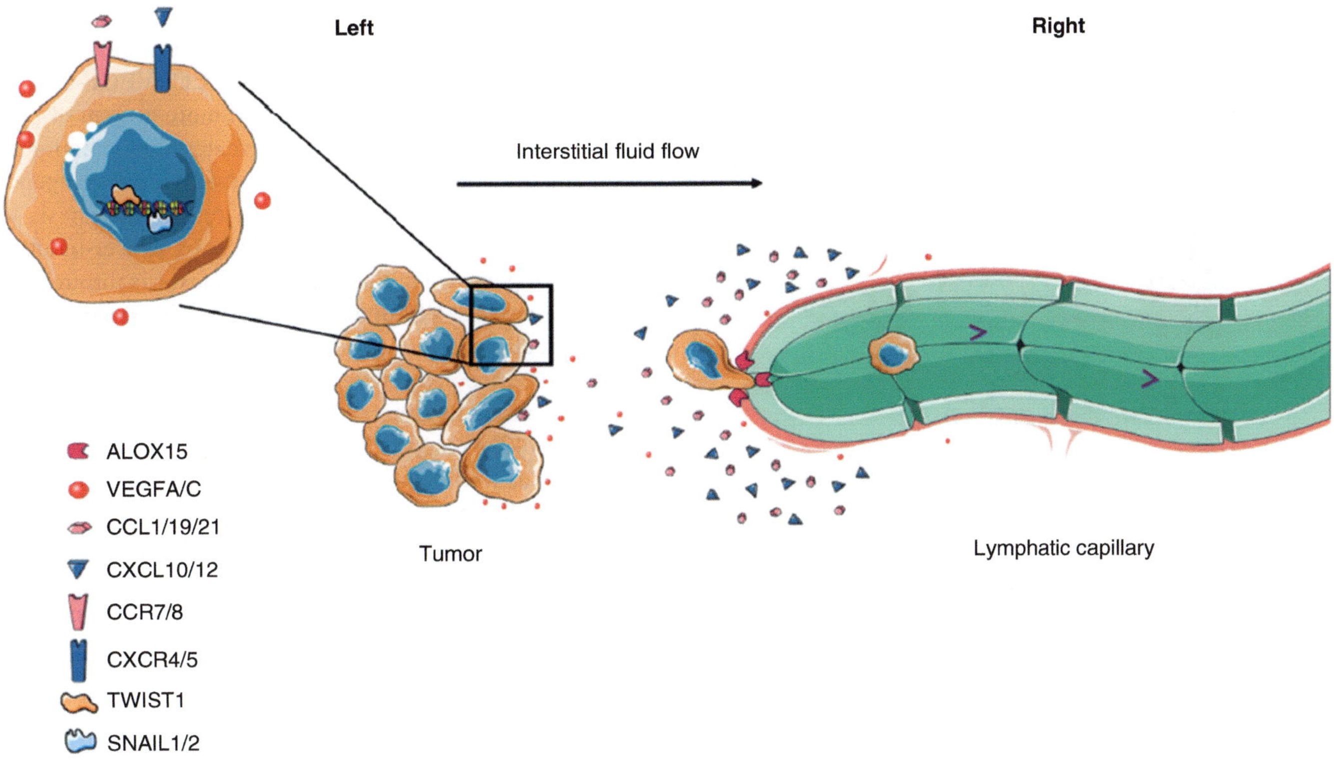

Fig. 21.11 Migration and invasion of cancer cells into the lymphatic vessels following lymphangiogenesis. Figure and legend originally published under CC BY 4.0 [70] with permission to be reprinted here

nificance of SLNs in melanoma [107] and breast cancer [108] has been well established, the molecular mechanism as to how the cancer cells migrate to the SLNs has not been well characterized.

Skobe and Detmar have found that VEGF-C induces cancer lymphangiogenesis associated with breast cancer metastasis [109]. Following orthotopic transplantation of breast cancer in nude mice, they have found intratumoral lymphangiogenesis within human breast cancers using LYVE-1. Since VEGF-C overexpression in breast cancer cells was associated with increased intratumoral lymphangiogenesis associated with the lymph node and lung metastases, they concluded that VEGF-C could be considered as a molecular link between cancer lymphangiogenesis and metastasis.

Dual staining with LYVE-1 and the panvascular marker CD31 was used to study lymphangiogenesis in cutaneous melanoma by Dadras and Detmar [79]. Nineteen melanoma patients with no evidence of metastatic disease and 18 melanoma patients with early nodal metastasis were closely matched for histological and clinical features. Their samples were examined by dual staining as mentioned previously. It was found that the incidence of intratumoral lymphatics was significantly higher in patients with metastatic melanoma. Poor disease-free survival was associated with increased intratumoral lymphatics. A relative lymphatic vessel area of >1.5% was significantly correlated with poor disease-free and overall survival, but there were no differences in the density of tumor-associated blood vessels. They proposed that intratumoral lymphangiogenesis was a novel prognostic marker for the risk of lymph node metastasis in melanoma [79]. Numerous studies have shown that increased lymphangiogenesis associated with increased intratumoral lymphatics and lymphatic invasion is correlated significantly with reduced overall survival in melanoma, breast, colorectal [110, 111] and lung cancer [112, 113]. It has been suggested that lymphangiogenesis and the remodeling of existing lymphatics may play an important role in cancer metastasis [35].

LECs in the lymph node maintain peripheral tolerance by directly inhibiting T-cells against autologous antigens under normal conditions. Using two distinct experimental tumor models, Dieterich et al. have found that PD-L1, a T-cell inhibitory molecule, is expressed by cancer-associated LECs similar to the LECs in the lymph node. However, PDL-1 is downregulated in cancer-associated blood vessels. It was speculated that the upregulation of PDL-1 in the LECs was probably a result of IFN-g released by tumor microenvironment stromal cells. Further, the authors have found that when

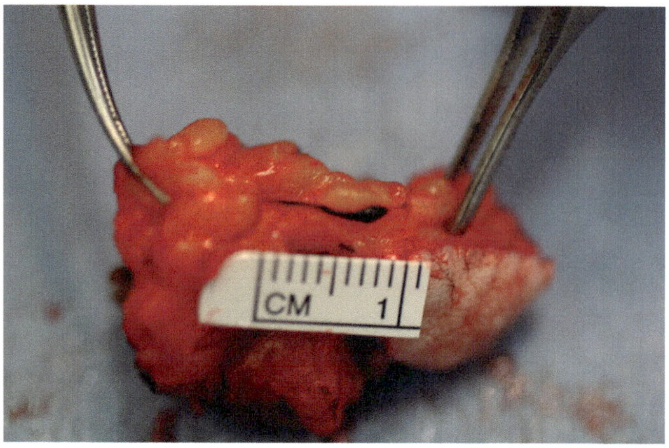

Fig. 21.12 Metastatic melanoma is trapped in the lymphatic vessel within the subcutaneous tissue. Figure and legend reprinted by permission from Springer: Clinical and Experimental Metastasis, Molecular mechanisms of cancer metastasis via the lymphatic versus the blood vessels [114]

PDL-1 is blocked, T-cell stimulation by antigen-presenting LECs was increased in vitro. Based on these findings, the authors assert that tumor-induced lymphatic vessels are critical in the initiation of metastasis and modulation of immune responses against autologous tumor by transporting antigen from the tumor microenvironment to tumor-draining lymph nodes. The T-cell inhibition by LECs is like that of peripheral tolerance by LECs in lymph nodes. Since PDL-1 may be involved in tumor tolerance, checkpoint inhibitors such as anti-PD-1 or anti-PDL-1 may be useful to reverse such a tumor tolerance [115].

Further study on the pre-metastatic niche in draining lymph nodes was carried out in two murine models with the 4 T1 breast cancer and B16F10 melanoma by Commerford et al. [116]. They have found that LEC expansion in tumor-draining lymph nodes representing lymphangiogenesis was noted as early as 4 days after tumor injection. Altered transcriptional profile of LECs from tumor-draining lymph nodes as compared to naive lymph nodes was observed with RNA sequencing in both tumor models. Integrin aIIb upregulated in LECs of tumor-draining lymph nodes. Since integrin aIIb was shown to mediate LEC adhesion to fibrinogen in vitro and LEC-associated fibrinogen was also identified in tumor-draining lymph nodes in vivo, the authors suggested that aIIb could be involved in lymphatic remodeling with possible formation of pre-metastatic niche ahead of the arrival of cancer cells [116].

In a separate mouse metastasis model, Ma et al. found that lymphangiogenesis occurred in distant lung metastases and that some metastatic tumor cells were located in the lymphatic vessels and draining lymph nodes of the distant metastatic sites [117]. In patients with metastatic melanoma, increased lymphatic density was noted within and around the pulmonary metastases. Invasion of lymphatic vessels by melanoma was correlated with poor outcomes (unpublished data from the authors). Using a transgenic murine model in which vascular endothelial growth factor C (VEGF-C) was induced in the lung, the authors found more growth of pulmonary metastases associated with greater dissemination to other organs. Thus, they have shown that lymphangiogenesis is present in metastatic sites with further metastasis to other sites [117]. As described previously, cancer-induced lymphangiogenesis facilitates cancer metastasis through the lymphatic vessels. VEGF receptor-3 (VEGFR-3) expression on tumor cells with autocrine signaling by VEGF-3 has been found to be associated with increased metastatic potential. Also, lymphatic-homing chemokine receptors, particularly C–C chemokine receptor 7 (CCR7), which is expressed in B and T cells as well as mature dendritic cells, may be involved in lymph node metastasis. Issa et al. have found that VEGF-C acts to increase secretion of CCL21 by lymphatic tissue, another lymphoid homing chemokine, which interacts with CCR7 to initiate cancer invasion toward lymphatics. The CCL21/CCR7 axis leads the CCR7 lymphocytes to the tumor-draining lymph node [93] and renders it immunosuppressive [118]. VEGF-C also enhances tumor invasiveness through autocrine signaling by increasing the proteolytic activity and motility of tumor cells in a three-dimensional matrix. Further, VEGF-C induces production of lymphatic CCL21 in vivo when VEGF-C protein was injected intradermally in the mouse. Thus, the authors conclude that their data support the fact that VEGF-C promotes tumor invasion toward lymphatics by both autocrine and CCR7-associated paracrine signaling mechanisms in addition to lymphangiogenesis initiated by VEGF-C. Both of these effects are VEGFR-3 dependent based on the 3-D matrix model [119]. Using the 3D matrix model, Shields et al. have further shown that CCR7 creates a gradient across the cells of increased interstitial flow to enhance tumor cell migration into the lymphatic vessels like that of dendritic cell trafficking [120].

Although lymphangiogenesis by VEGF-C is associated with poor prognosis, Fankhauser and Swartz have found that VEGF receptor-3 (VEGFR-3) blocking antibody was able to reverse immunosuppression in a murine model as VEGF-C was found to enhance the effect of antibody immunotherapy [121]. VEGF-C induced CCL21, which caused tumor infiltration of naive T cells. When the mice were treated with checkpoint inhibition immunotherapy, the antitumor response was significantly increased. Thus, even though VEGF-C causes lymphangiogenesis being associated with poor prognosis, its induction of CCL21 may be exploited for a positive antitumor effect by checkpoint inhibition immunotherapy. The authors further completed a clinical trial to show that gene expression of VEGF-C was strongly associated with T cell inflammation. Further, serum VEGF-C concentrations were correlated with T cell activation and expansion as well as clinical response to peptide tumor vaccination and checkpoint inhibition immunotherapy. The authors propose that serum VEGF-C may serve as a predictive biomarker for immunotherapy response [121].

Interstitial flow and lymphatic drainage within the tumor microenvironment are increased in the edge of the tumor [122, 123]. The increased interstitial flow may facilitate cancer cells to migrate in the direction by chemotaxis from chemokines secreted by cancer cells [120]. Interstitial flow may also affect stromal cells, resulting in cell and matrix alignment [124, 125], increase fibroblast motility via matrix metalloproteinase-1 [126], and induce myofibroblast differentiation via transforming growth factor (TGF)-b1 [124].

Stromal cells in the cancer microenvironment including fibroblasts produce TGF-beta 1. It transforms fibroblast to myofibroblast (cancer-associated fibroblast; [127]), which causes remodeling of fibroblasts. When cancer cell lines were cocultured with dermal fibroblasts in a 3D collagen matrix, Shieh et al. were able to show that cancer invasion was enhanced by fibroblasts when interstitial flow was present. Thus, the authors describe a novel mechanism in which increased interstitial flow in the tumor microenvironment may induce fibroblast remodeling and migration by TGF-beta1 activation and collagen degradation, thus, facilitating tumor invasion into the lymphatic vessels [128]. Swartz and Lund have linked lymphangiogenesis, interstitial flow, mechanobiology, and immunity within the cancer microenvironment [129]. Thus, according to the authors, while lymphangiogenesis is associated with cancer invasion and poor prognosis, it may be of therapeutic benefit with checkpoint inhibition immunotherapy. Figure 21.13a, b summarizes the linkage between mechanobiology, lymphangiogenesis, and antitumor immunity.

21.10 Conclusion: The Lymphatic System as a Conduit for Cancer Metastasis

The cancer cells exploit the lymphatic system to "hitchhike" it to travel from the primary site as they dissociate themselves from their microenvironment and enter the lymphatic vessels toward the lymph nodes and beyond to the distant sites. As a cancer cell is slightly larger than a lymphocyte with an average diameter of 15 μm, the average diameter of a cancer cell may be up to 20 μm or more. The size of a single cancer cell can easily travel within the lymphatic vessel of 100 μm. Lymphatic vessels induced by cancer by means of lymphangiogenesis may become dilated [105].

It is speculative that a cluster of cancer cells may become trapped in the lymphatic vessel at the valve of the lymphatic vessel [106] and eventually grow on the lymphatic vessel inner wall as a colony to invade the surrounding fat tissue as in-transit metastasis, as in melanoma. The other speculation is that cancer cells may have a specific receptor to attach to the lymphatic vessel inner wall and set up a residence. The lymphatic system draining various body organs is different, some are more complex than other. Spread through the SLN can be best illustrated by melanoma and breast cancer, which follow more reliable routes in an orderly fashion [61] than the internal organs such as the lungs, stomach, pancreas, and others, which show complicated lymphatic drainage patterns. Certainly, melanoma and breast cancer allow us to appreciate the biology of early cancer arrival in the SLNs with minimal tumor burden. It has been shown that melanoma patients with a negative SLN biopsy have low recurrence rate of 5–10% [59, 62].

For breast cancer patients with a negative SLN biopsy, their recurrence rate is also very low, less than 5% [60]. Overall, melanoma or breast cancer patients with a negative SLN biopsy fare much better than those with a positive SLN biopsy [63, 64]. Studying the biology of cancer trafficking and lymph node metastasis in early melanoma and breast cancer may allow us to understand the mechanisms of cancer spread from the primary site to lymph nodes and beyond to the distant sites. The routes of spread by the lymphatic versus the blood vessels will be addressed in other review articles [69, 114]. The relative orderly progression of melanoma and breast cancer metastasis through the SLNs, in contrast to the complicated lymphatic system of the stomach, pancreas, and other organs, is like Mendel's discovery of the independent segregation of the dominant and recessive genetic alleles in the pea model with respect to seven features of the peas [130]. If Mendel would have used the human hair color variations to study heredity, he would not have reached such clear and discrete conclusions as there are multiple crossover genes involving the heredity of hair color. Thus, the SLN model for melanoma and breast cancer, being relatively simple, will help us to understand the cancer metastasis to the SLN and its interaction within the lymphatic vessels.

Angiogenesis involving blood vessels induced by cancer has been well established as a means for cancer to support its growth and perhaps for it to gain into the circulation to spread to distant sites [71]. When melanoma transplants were placed in the hamster cheek pouch, new blood vessels were formed [72, 73]. Clinically angiogenesis has been found to be associated with melanoma progression [74]. Tumor microvessel density has been correlated with decreased disease free and overall survival in several studies [76, 77]. On the other hand, other reports found no differences in tumor microvessel density in metastasizing and nonmetastasizing melanomas [78]. Thus, the prognostic value of tumor hemangiogenesis in melanoma has remained controversial [75] and the potential prognostic value of such tumor angiogenesis in melanoma is not clear at this time [79]. On the other hand, based on experimental and clinical data, cancer lymphangiogenesis has been found to be signifi-

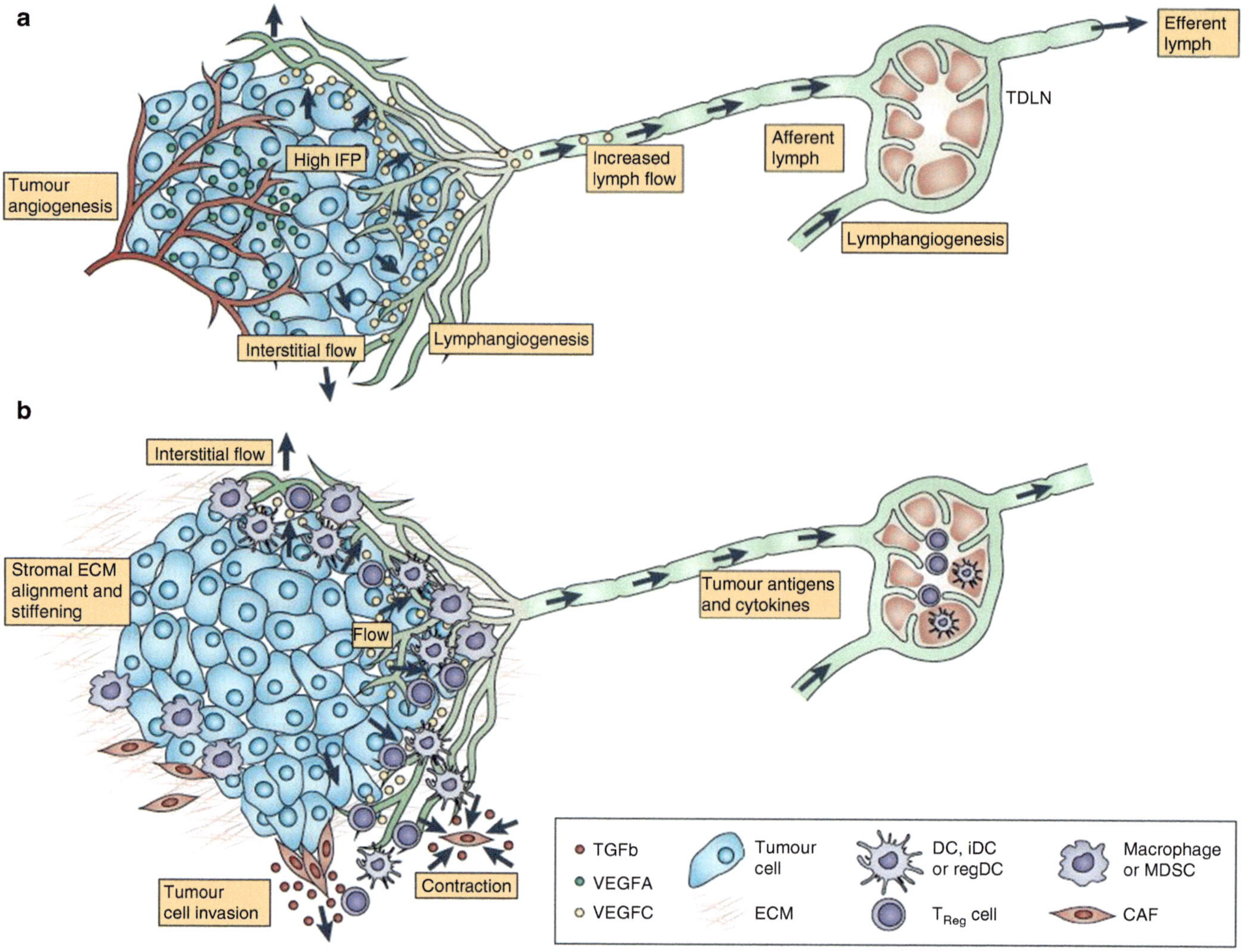

Fig. 21.13 (**a** and **b**) Linking mechanobiology, lymphangiogenesis, and antitumor immunity. Lymphangiogenesis, mechanobiology, and immunity in the tumor microenvironment contain many interdependent features. (**a**) Fluid pathways in the tumor microenvironment are shown. Angiogenic, immature tumor vessels (shown in red) are hyperpermeable, driving heightened interstitial fluid pressure (IFP) in the tumor. This creates steep pressure gradients at the tumor margin that drive interstitial flow through the stroma and into peritumoral lymphatic vessels (green), which are often expanded and hyperplastic. These lymphatic vessels carry tumor interstitial fluid to the sentinel or tumor-draining lymph node (TDLN), where lymphangiogenesis is also seen. (**b**) The heightened interstitial and lymphatic flows (green) in the tumor microenvironment coincide with biomechanical (red) and immunological (purple) changes in the tumor stroma. Biomechanical changes include stiffening and alignment of the extracellular matrix (ECM) of the tumor stroma, owing to cancer-associated fibroblast (CAF) contraction and remodeling that is dependent on transforming growth factor-β (TGFβ). Matrix stiffening promotes tumor cell invasion through CAF-led collective migration and activation of mechanically sensitive stromal components. Interestingly, stromal features of the tumor can mimic those of the TDLN. Immunological changes include biasing of the tumor-infiltrating lymphocyte populations by local factors. Tumor-promoting cytokines promote alternatively activated (M2) macrophages and myeloid-derived suppressor cells (MDSCs) hinder the maturation of immature dendritic cells (iDCs) and stimulate regulatory DCs (regDCs), which in turn promote regulatory T (TReg) cells and suppress local cytotoxic T cell activity. Similar cell populations are biased in the TDLN. VEGF, vascular endothelial growth factor. Figure and legend reprinted by permission from Springer Nature: Nature Reviews Cancer, Lymphatic and interstitial flow in the tumor microenvironment: Linking mechanobiology with immunity [129]

cantly correlated with poor prognosis as presented in this chapter. Such strong evidence that poor prognosis is associated with cancer lymphangiogenesis indicates that the lymphatic system may be the major conduit for cancer metastasis. Furthermore, the strong clinical data that SLN from melanoma and breast cancer are a significant prognostic biomarker, supports the view that the lymphatic system may serve as the major gateway for cancer to spread.

Future Perspectives and Unanswered Questions

The concept and utility of SLNs have been well established in the clinical arena, particularly in the management of melanoma [107] and breast cancer [108] with SLN surgery as a reliable staging modality for the regional nodal basin [32, 114]. The microenvironment with the SLN in relationship to the cancer growth and spread awaits further study [131].

Cancer lymphangiogenesis in the primary site and the tumor-draining node or the SLN has been well established as a poor prognosis marker [132] with detailed discussion in this chapter. Molecular mechanisms of cancer spread to the SLN and then to the distant sites are being pursued in many cancer centers. The release of factors by the primary tumor to the SLN to render it a pre-metastatic niche is gaining evidence. A recent review article on progressive role of the breast cancer SLN with anticancer immunity to develop into a lymph node with immunosuppressive microenvironment allowing the cancer cells to migrate to the SLN and grow in this immunosuppressive state is intriguing [133].

- When does the SLN has the capacity to kill cancer cells with anticancer immunity?
- When does the SLN become immunosuppressive?
- What causes such a change?
- Does it mean that when a SLN harbors cancer cells, the lymph node microenvironment has already become immunosuppressive?
- What is the contribution of T cell repertoire of each patient in such changes in the SLN microenvironment?
- What is the spectrum of heterogeneity with respect to cancer both within the cancer population of each patient and between patients?
- What is the spectrum of heterogeneity in the T cell repertoire of each patient?
- What is the dynamic interaction between cancer cells and T cells in the background of such heterogeneity?

Perhaps, elucidation of molecular mechanisms of these changes can explain the conundrum that a SLN while it should be protective against the invasion of cancer cells to the lymph node, but then, it may become the gateway for distant metastasis with the development of an immunosuppressive microenvironment to allow the arrival and subsequent proliferation of cancer cells in the lymph node to spread. Understanding these important molecular mechanisms of lymph node metastasis may allow us to develop new strategy and novel therapy in the treatment of cancer metastasis.

References

1. Asellius G. De lactibus sive lacteis venis. Milan: Mediolani; 1627.
2. Irschick R, Siemon C, Brenner E. The history of anatomical research of lymphatics — from the ancient times to the end of the European renaissance. Ann Anat Anatomischer Anzeiger. 2019;223:49–69. https://doi.org/10.1016/j.aanat.2019.01.010.
3. Harvey W. Exercitation Anatomica de Motu Cordis et Sanguinis in Animalibus. Frankfurt; 1628.
4. Sabin FR. On the origin of the lymphatic system from the veins and the development of the lymph hearts and thoracic duct in the pig. Am J Anat. 1902;1(3):367–89.
5. Hill MA (2019) Embryology lymph node development. https://embryology.med.unsw.edu.au/embryology/index.php/Lymph_Node_Development. Accessed June 15, 2021.
6. Kampmeier OF. Evolution and comparative morphology of the lymphatic system. Springfield: Charles C. Thomas; 1969.
7. Huntington GS. The development of the mammalian jugular lymphsac, of the tributary primitive ulnar lymphatic and of the thoracic ducts from the view point of recent investigations of vertebrate lymphatic ontogeny, together with a consideration of the genetic relations of lymphatic and haemal vascular channels in the embryos of amniotes. Am J Anat. 1914;16(3):259–316.
8. Senger DR, Galli SJ, Dvorak AM, Perruzzi CA, Harvey VS, Dvorak HF. Tumor cells secrete a vascular permeability factor that promotes accumulation of ascites fluid. Science. 1983;219(4587):983–5. https://doi.org/10.1126/science.6823562.
9. Keck PJ, Hauser SD, Krivi G, Sanzo K, Warren T, Feder J, Connolly DT. Vascular permeability factor, an endothelial cell mitogen related to PDGF. Science. 1989;246(4935):1309–12. https://doi.org/10.1126/science.2479987.
10. Leung DW, Cachianes G, Kuang WJ, Goeddel DV, Ferrara N. Vascular endothelial growth factor is a secreted angiogenic mitogen. Science. 1989;246(4935):1306–9. https://doi.org/10.1126/science.2479986.
11. Ferrara N, Gerber HP, LeCouter J. The biology of VEGF and its receptors. Nat Med. 2003;9(6):669–76. https://doi.org/10.1038/nm0603-669.
12. Holmes DI, Zachary I. The vascular endothelial growth factor (VEGF) family: angiogenic factors in health and disease. Genome Biol. 2005;6(2):209. https://doi.org/10.1186/gb-2005-6-2-209.
13. Li X, Eriksson U. Novel VEGF family members: VEGF-B, VEGF-C and VEGF-D. Int J Biochem Cell Biol. 2001;33(4):421–6. https://doi.org/10.1016/s1357-2725(01)00027-9.
14. Shibuya M. Vascular endothelial growth factor receptor-2: its unique signaling and specific ligand, VEGF-E. Cancer Sci. 2003;94(9):751–6. https://doi.org/10.1111/j.1349-7006.2003.tb01514.x.
15. Rauniyar K, Jha SK, Jeltsch M. Biology of vascular endothelial growth factor C in the morphogenesis of lymphatic vessels. Front Bioeng Biotechnol. 2018;6:7. https://doi.org/10.3389/fbioe.2018.00007.
16. Banerji S, Ni J, Wang S-X, Clasper S, Su J, Tammi R, Jones M, Jackson DG. LYVE-1, a new homologue of the CD44 glycoprotein, is a lymph-specific receptor for hyaluronan. J Cell Biol. 1999;144(4):789–801.
17. Carreira CM, Nasser SM, di Tomaso E, Padera TP, Boucher Y, Tomarev SI, Jain RK. LYVE-1 is not restricted to the lymph vessels: expression in normal liver blood sinusoids and down-regulation in human liver cancer and cirrhosis. Cancer Res. 2001;61(22):8079–84.
18. Gordon EJ, Gale NW, Harvey NL. Expression of the hyaluronan receptor LYVE-1 is not restricted to the lymphatic vasculature; LYVE-1 is also expressed on embryonic blood vessels. Dev Dyn. 2008;237(7):1901–9. https://doi.org/10.1002/dvdy.21605.
19. Maruyama K, Asai J, Ii M, Thorne T, Losordo DW, D'Amore PA. Decreased macrophage number and activation lead to reduced lymphatic vessel formation and contribute to impaired diabetic

wound healing. Am J Pathol. 2007;170(4):1178–91. https://doi.org/10.2353/ajpath.2007.060018.
20. Breiteneder-Geleff S, Matsui K, Soleiman A, Meraner P, Poczewski H, Kalt R, Schaffner G, Kerjaschki D. Podoplanin, novel 43-kd membrane protein of glomerular epithelial cells, is down-regulated in puromycin nephrosis. Am J Pathol. 1997;151(4):1141–52.
21. Oliver G, Sosa-Pineda B, Geisendorf S, Spana EP, Doe CQ, Gruss P. Prox 1, a prospero-related homeobox gene expressed during mouse development. Mech Dev. 1993;44(1):3–16. https://doi.org/10.1016/0925-4773(93)90012-m.
22. Wigle JT, Oliver G. Prox1 function is required for the development of the murine lymphatic system. Cell. 1999;98(6):769–78. https://doi.org/10.1016/s0092-8674(00)81511-1.
23. Pan Y, Wang WD, Yago T. Transcriptional regulation of podoplanin expression by Prox1 in lymphatic endothelial cells. Microvasc Res. 2014;94:96–102. https://doi.org/10.1016/j.mvr.2014.05.006.
24. Hong YK, Detmar M. Prox1, master regulator of the lymphatic vasculature phenotype. Cell Tissue Res. 2003;314(1):85–92. https://doi.org/10.1007/s00441-003-0747-8.
25. Hong YK, Harvey N, Noh YH, Schacht V, Hirakawa S, Detmar M, Oliver G. Prox1 is a master control gene in the program specifying lymphatic endothelial cell fate. Dev Dyn. 2002;225(3):351–7. https://doi.org/10.1002/dvdy.10163.
26. Kaipainen A, Korhonen J, Mustonen T, van Hinsbergh VW, Fang GH, Dumont D, Breitman M, Alitalo K. Expression of the fms-like tyrosine kinase 4 gene becomes restricted to lymphatic endothelium during development. Proc Natl Acad Sci U S A. 1995;92(8):3566–70. https://doi.org/10.1073/pnas.92.8.3566.
27. Wilting J, Papoutsi M, Schneider M, Christ B. The lymphatic endothelium of the avian wing is of somitic origin. Dev Dyn. 2000a;217(3):271–8. https://doi.org/10.1002/(SICI)1097-0177(200003)217:3<271::AID-DVDY5>3.0.CO;2-2.
28. Wilting J, Schneider M, Papoutski M, Alitalo K, Christ B. An avian model for studies of embryonic lymphangiogenesis. Lymphology. 2000b;33(3):81–94.
29. Witte MH, Bernas MJ, Martin CP, Witte CL. Lymphangiogenesis and lymphangiodysplasia: from molecular to clinical lymphology. Microsc Res Tech. 2001;55(2):122–45. https://doi.org/10.1002/jemt.1163.
30. Witte M, Witte C, Ohhashi T. Are lymphatics different from blood vessels. Lymphology. 1987;20(4):169.
31. Witte MH, Bernas MJ (2016) Lymphatic pathophysiology. https://thoracickey.com/lymphatic-pathophysiology/. Accessed July 27 2021.
32. Leong SP, Pissas A, Scarato M, Gallon F, Pissas MH, Amore M, Wu M, Faries M, Lund A. The lymphatic system and sentinel lymph nodes: conduit for cancer metastasis. Clin Exp Metastasis. 2021. https://doi.org/10.1007/s10585-021-10123-w. PMID: 34651243.
33. FuseSchool (2019) The lymphatic system. YouTube.com, https://www.youtube.com/watch?v=o0-1OknbO3M
34. Breslin JW, Yang Y, Scallan JP, Sweat RS, Adderley SP, Murfee WL. Lymphatic vessel network structure and physiology. Compr Physiol. 2018;9(1):207–99. https://doi.org/10.1002/cphy.c180015.
35. Stacker SA, Williams SP, Karnezis T, Shayan R, Fox SB, Achen MG. Lymphangiogenesis and lymphatic vessel remodelling in cancer. Nat Rev Cancer. 2014;14(3):159–72. https://doi.org/10.1038/nrc3677.
36. Dellinger MT, Witte MH. Lymphangiogenesis, lymphatic systemomics, and cancer: context, advances and unanswered questions. Clin Exp Metastasis. 2018;35(5–6):419–24. https://doi.org/10.1007/s10585-018-9907-9.
37. Nieweg OE, Tanis PJ, Leong SP. Anatomy and physiology of lymphatic circulation: application to lymphatic mapping. In: Mariani G, Vidal-Sicart S, Valdes Olmos RA, editors. Atlas of lymphoscintigraphy and sentinel node mapping: a pictorial case-based approach. 2nd ed. Cham, Switzerland: Springer Nature Switzerland AG; 2020. p. 1–6. https://doi.org/10.1007/978-3-030-45296-4.
38. Uthman E (2008) Normal lymph node. https://www.flickr.com/photos/euthman/3113417731/. Accessed August 20 2021.
39. Graham T (2021) Lymph nodes 101. https://www.iheartpathology.net/post/lymph-nodes-101. Accessed August 20 2021.
40. Lymphatics Lab. (2021). http://medcell.med.yale.edu/systems_cell_biology/lymphatics_lab.php. Accessed August 20 2021.
41. Thomas D (2012) HistoQuarterly: lymph node. https://histologyblog.com/2012/04/30/histoquarterly-lymph-node/. Accessed August 20 2021.
42. Gentek R, Bajenoff M. Lymph node stroma dynamics and approaches for their visualization. Trends Immunol. 2017;38(4):236–47. https://doi.org/10.1016/j.it.2017.01.005.
43. Fletcher AL, Acton SE, Knoblich K. Lymph node fibroblastic reticular cells in health and disease. Nat Rev Immunol. 2015;15(6):350–61. https://doi.org/10.1038/nri3846.
44. Alvarenga HG, Marti L. Multifunctional roles of reticular fibroblastic cells: more than meets the eye? J Immunol Res. 2014;2014:402038. https://doi.org/10.1155/2014/402038.
45. Liu K, Nussenzweig MC. Origin and development of dendritic cells. Immunol Rev. 2010;234(1):45–54. https://doi.org/10.1111/j.0105-2896.2009.00879.x.
46. Hervouet C, Luci C, Bekri S, Juhel T, Bihl F, Braud VM, Czerkinsky C, Anjuere F. Antigen-bearing dendritic cells from the sublingual mucosa recirculate to distant systemic lymphoid organs to prime mucosal CD8 T cells. Mucosal Immunol. 2014;7(2):280–91. https://doi.org/10.1038/mi.2013.45.
47. Witte MH, Jones K, Wilting J, Dictor M, Selg M, McHale N, Gershenwald JE, Jackson DG. Structure function relationships in the lymphatic system and implications for cancer biology. Cancer Metastasis Rev. 2006;25(2):159–84.
48. Bernas M, Thiadens SRJ, Stewart P, Granzow J. Secondary lymphedema from cancer therapy. Clin Exp Metastasis. 2021; https://doi.org/10.1007/s10585-021-10096-w.
49. Miyasaka M, Tanaka T. Lymphocyte trafficking across high endothelial venules: dogmas and enigmas. Nat Rev Immunol. 2004;4(5):360–70.
50. Halsted WS. I. the results of operations for the cure of cancer of the breast performed at the Johns Hopkins Hospital from June, 1889, to January, 1894. Ann Surg. 1894;20(5):497–555. https://doi.org/10.1097/00000658-189407000-00075.
51. Neuhaus SJ, Clark MA, Thomas JM. Dr. Herbert Lumley Snow (1847–1930): the original champion of elective lymph node dissection in melanoma. Ann Surg Oncol. 2004;11(9):875–8. https://doi.org/10.1245/ASO.2004.02.031.
52. Snow H. Melanotic cancerous disease. Lancet. 1892;2:872–4.
53. Haagensen CD, Feind CR, Herter FP. The lymphatics in cancer. Philadelphia: WB Saunders Co; 1972.
54. Fisher B, Ravdin RG, Ausman RK, Slack NH, Moore GE, Noer RJ. Surgical adjuvant chemotherapy in cancer of the breast: results of a decade of cooperative investigation. Ann Surg. 1968;168(3):337–56. https://doi.org/10.1097/00000658-196809000-00004.
55. Cady B. Lymph node metastases. Indicators, but not governors of survival. Arch Surg. 1984;119(9):1067–72. https://doi.org/10.1001/archsurg.1984.01390210063014.
56. Hellman S, Weichselbaum RR. Oligometastases. J Clin Oncol. 1995;13(1):8–10. https://doi.org/10.1200/JCO.1995.13.1.8.
57. Cabanas RM. An approach for the treatment of penile carcinoma. Cancer. 1977;39(2):456–66. https://doi.org/10.1002/1097-0142(197702)39:2<456::aid-cncr2820390214>3.0.co;2-i.
58. Morton DL, Wen DR, Wong JH, Economou JS, Cagle LA, Storm FK, Foshag LJ, Cochran AJ. Technical details of intra-

operative lymphatic mapping for early stage melanoma. Arch Surg. 1992;127(4):392–9. https://doi.org/10.1001/archsurg.1992.01420040034005.
59. Morton DL, Hoon DS, Cochran AJ, Turner RR, Essner R, Takeuchi H, Wanek LA, Glass E, Foshag LJ, Hsueh EC, Bilchik AJ, Elashoff D, Elashoff R. Lymphatic mapping and sentinel lymphadenectomy for early-stage melanoma: therapeutic utility and implications of nodal microanatomy and molecular staging for improving the accuracy of detection of nodal micrometastases. Ann Surg. 2003;238(4):538–49.; discussion 549–550. https://doi.org/10.1097/01.sla.0000086543.45557.cb.
60. Kapoor NS, Sim MS, Lin J, Giuliano AE. Long-term outcome of patients managed with sentinel lymph node biopsy alone for node-negative invasive breast cancer. Arch Surg. 2012;147(11):1047–52. https://doi.org/10.1001/archsurg.2012.1563.
61. Reintgen D, Cruse CW, Wells K, Berman C, Fenske N, Glass F, Schroer K, Heller R, Ross M, Lyman G. The orderly progression of melanoma nodal metastases. Ann Surg. 1994;220(6):759–67. https://doi.org/10.1097/00000658-199412000-00009.
62. Thomas DC, Han G, Leong SP, Kashani-Sabet M, Vetto J, Pockaj B, White RL, Faries MB, Schneebaum S, Mozzillo N, Charney KJ, Sondak VK, Messina JL, Zager JS, Han D. Recurrence of melanoma after a negative sentinel node biopsy: predictors and impact of recurrence site on survival. Ann Surg Oncol. 2019;26(7):2254–62. https://doi.org/10.1245/s10434-019-07369-w.
63. Andersson Y, Frisell J, Sylvan M, de Boniface J, Bergkvist L. Breast cancer survival in relation to the metastatic tumor burden in axillary lymph nodes. J Clin Oncol. 2010;28(17):2868–73. https://doi.org/10.1200/JCO.2009.24.5001.
64. Rios-Cantu A, Lu Y, Melendez-Elizondo V, Chen M, Gutierrez-Range A, Fadaki N, Thummala S, West-Coffee C, Cleaver J, Kashani-Sabet M, Leong SP. Is the non-sentinel lymph node compartment the next site for melanoma progression from the sentinel lymph node compartment in the regional nodal basin? Clin Exp Metastasis. 2017;34(5):345–50. https://doi.org/10.1007/s10585-017-9854-x.
65. van Akkooi AC, de Wilt JH, Verhoef C, Schmitz PI, van Geel AN, Eggermont AM, Kliffen M. Clinical relevance of melanoma micrometastases (<0.1 mm) in sentinel nodes : Are these nodes to be considered negative? Ann Oncol. 2006;17(10):1578–85. https://doi.org/10.1093/annonc/mdl176.
66. van Akkooi AC, Nowecki ZI, Voit C, Schäfer-Hesterberg G, Michej W, de Wilt JH, Rutkowski P, Verhoef C, Eggermont AM. Sentinel node tumor burden according to the Rotterdam criteria is the most important prognostic factor for survival in melanoma patients: a multicenter study in 388 patients with positive sentinel nodes. Ann Surg. 2008;248(6):949–55.
67. Reed J, Rosman M, Verbanac KM, Mannie A, Cheng Z, Tafra L. Prognostic implications of isolated tumor cells and micrometastases in sentinel nodes of patients with invasive breast cancer: 10-year analysis of patients enrolled in the prospective East Carolina University/Anne Arundel medical center sentinel node multicenter study. J Am Coll Surg. 2009;208(3):333–40. https://doi.org/10.1016/j.jamcollsurg.2008.10.036.
68. Special Issue on: The 2017 7th International Cancer Metastasis Symposium; Cancer Metastasis through the Lymphovascular System: Biology and Treatment (2018). Volume 35; Number 5–6; June–August 2018 edn. Springer, Clinical and Experimental Metastasis.
69. Nathanson SD, Detmar M, Padera TP, Yates LR, Welch DR, Beadnell TC, Scheid AD, Wrenn ED, Cheung K. Mechanisms of breast cancer metastasis. Clin Exp Metastasis. 2021; https://doi.org/10.1007/s10585-021-10090-2.
70. Zhou H, P-j L, Padera TP. Progression of metastasis through lymphatic system. Cell. 2021;10(3):627.
71. Weidner N, Semple JP, Welch WR, Folkman J. Tumor angiogenesis and metastasis--correlation in invasive breast carcinoma. N Engl J Med. 1991;324(1):1–8. https://doi.org/10.1056/NEJM199101033240101.
72. Hubler WR Jr, Wolf JE Jr. Melanoma. Tumor angiogenesis and human neoplasia. Cancer. 1976;38(1):187–92. https://doi.org/10.1002/1097-0142(197607)38:1<187::aid-cncr2820380129>3.0.co;2-d.
73. Warren B, Shubik P. The growth of the blood supply to melanoma transplants in the hamster cheek pouch. Transplantation. 1966;4(5):652.
74. Barnhill RL, Fandrey K, Levy MA, Mihm MC Jr, Hyman B. Angiogenesis and tumor progression of melanoma. Quantification of vascularity in melanocytic nevi and cutaneous malignant melanoma. Lab Investig. 1992;67(3):331–7.
75. Streit M, Detmar M. Angiogenesis, lymphangiogenesis, and melanoma metastasis. Oncogene. 2003;22(20):3172–9. https://doi.org/10.1038/sj.onc.1206457.
76. Srivastava A, Laidler P, Davies RP, Horgan K, Hughes LE. The prognostic significance of tumor vascularity in intermediate-thickness (0.76–4.0 mm thick) skin melanoma. A quantitative histologic study. Am J Pathol. 1988;133(2):419.
77. Straume O, Salvesen HB, Akslen LA. Angiogenesis is prognostically important in vertical growth phase melanomas. Int J Oncol. 1999;15(3):595–9. https://doi.org/10.3892/ijo.15.3.595.
78. Busam KJ, Berwick M, Blessing K, Fandrey K, Kang S, Karaoli T, Fine J, Cochran AJ, White WL, Rivers J. Tumor vascularity is not a prognostic factor for malignant melanoma of the skin. Am J Pathol. 1995;147(4):1049.
79. Dadras SS, Paul T, Bertoncini J, Brown LF, Muzikansky A, Jackson DG, Ellwanger U, Garbe C, Mihm MC, Detmar M. Tumor lymphangiogenesis: a novel prognostic indicator for cutaneous melanoma metastasis and survival. Am J Pathol. 2003;162(6):1951–60. https://doi.org/10.1016/S0002-9440(10)64328-3.
80. Kaplan RN, Riba RD, Zacharoulis S, Bramley AH, Vincent L, Costa C, MacDonald DD, Jin DK, Shido K, Kerns SA, Zhu Z, Hicklin D, Wu Y, Port JL, Altorki N, Port ER, Ruggero D, Shmelkov SV, Jensen KK, Rafii S, Lyden D. VEGFR1-positive haematopoietic bone marrow progenitors initiate the pre-metastatic niche. Nature. 2005;438(7069):820–7. https://doi.org/10.1038/nature04186.
81. Sleeman JP. The metastatic niche and stromal progression. Cancer Metastasis Rev. 2012;31(3–4):429–40. https://doi.org/10.1007/s10555-012-9373-9.
82. Sleeman JP. The lymph node pre-metastatic niche. J Mol Med (Berl). 2015;93(11):1173–84. https://doi.org/10.1007/s00109-015-1351-6.
83. Paget S. The distribution of secondary growths in cancer of the breast. Lancet. 1889;133(3421):571–3.
84. Cochran AJ, Huang RR, Lee J, Itakura E, Leong SP, Essner R. Tumour-induced immune modulation of sentinel lymph nodes. Nat Rev Immunol. 2006;6(9):659–70. https://doi.org/10.1038/nri1919.
85. Cochran AJ, Wen DR, Farzad Z, Stene MA, McBride W, Lana AM, Hoon DS, Morton DL. Immunosuppression by melanoma cells as a factor in the generation of metastatic disease. Anticancer Res. 1989;9(4):859–64.
86. Hoon DS, Bowker RJ, Cochran AJ. Suppressor cell activity in melanoma-draining lymph nodes. Cancer Res. 1987;47(6):1529–33.
87. Huang RR, Wen DR, Guo J, Giuliano AE, Nguyen M, Offodile R, Stern S, Turner R, Cochran AJ. Selective modulation of paracortical dendritic cells and T-lymphocytes in breast cancer sentinel lymph nodes. Breast J. 2000;6(4):225–32. https://doi.org/10.1046/j.1524-4741.2000.98114.x.
88. Matsuura K, Yamaguchi Y, Ueno H, Osaki A, Arihiro K, Toge T. Maturation of dendritic cells and T-cell responses in sentinel lymph nodes from patients with breast carcinoma. Cancer. 2006;106(6):1227–36. https://doi.org/10.1002/cncr.21729.
89. Mohos A, Sebestyén T, Liszkay G, Plótár V, Horváth S, Gaudi I, Ladányi A. Immune cell profile of sentinel lymph nodes in

patients with malignant melanoma – FOXP3+ cell density in cases with positive sentinel node status is associated with unfavorable clinical outcome. J Transl Med. 2013;11(1):43. https://doi.org/10.1186/1479-5876-11-43.
90. Botella-Estrada R, Dasi F, Ramos D, Nagore E, Herrero MJ, Gimenez J, Fuster C, Sanmartin O, Guillen C, Alino S. Cytokine expression and dendritic cell density in melanoma sentinel nodes. Melanoma Res. 2005;15(2):99–106. https://doi.org/10.1097/00008390-200504000-00003.
91. Leong SP, Peng M, Zhou YM, Vaquerano JE, Chang JW. Cytokine profiles of sentinel lymph nodes draining the primary melanoma. Ann Surg Oncol. 2002;9(1):82–7. https://doi.org/10.1245/aso.2002.9.1.82.
92. Cochran AJ, Huang RR, Su A, Itakura E, Wen DR. Is sentinel node susceptibility to metastases related to nodal immune modulation? Cancer J. 2015;21(1):39–46. https://doi.org/10.1097/PPO.0000000000000094.
93. Takeuchi H, Fujimoto A, Tanaka M, Yamano T, Hsueh E, Hoon DS. CCL21 chemokine regulates chemokine receptor CCR7 bearing malignant melanoma cells. Clin Cancer Res. 2004;10(7):2351–8. https://doi.org/10.1158/1078-0432.ccr-03-0195.
94. Thomas SN, Vokali E, Lund AW, Hubbell JA, Swartz MA. Targeting the tumor-draining lymph node with adjuvanted nanoparticles reshapes the anti-tumor immune response. Biomaterials. 2014;35(2):814–24. https://doi.org/10.1016/j.biomaterials.2013.10.003.
95. Witte MH, Way DL, Witte CL, Bernas M. Lymphangiogenesis: mechanisms, significance and clinical implications. EXS. 1997;79:65–112. https://doi.org/10.1007/978-3-0348-9006-9_5.
96. Bowman CA, Witte MH, Witte CL, Way DL, Nagle RB, Copeland JG, Daschbach CC. Cystic hygroma reconsidered: hamartoma or neoplasm? Primary culture of an endothelial cell line from a massive cervicomediastinal hygroma with bony lymphangiomatosis. Lymphology. 1984;17(1):15–22.
97. Gnepp DR, Chandler W. Tissue culture of human and canine thoracic duct endothelium. In Vitro Cell Dev Biol. 1985;21(4):200–6. https://doi.org/10.1007/BF02620930.
98. Johnston MG, Walker MA. Lymphatic endothelial and smooth-muscle cells in tissue culture. In Vitro. 1984;20(7):566–72. https://doi.org/10.1007/BF02639772.
99. Leak LV, Jones M. Lymphangiogenesis in vitro: formation of lymphatic capillary-like channels from confluent monolayers of lymphatic endothelial cells. In Vitro Cell Dev Biol Anim. 1994;30A(8):512–8.
100. Way D, Hendrix M, Witte M, Witte C, Nagle R, Davis J. Lymphatic endothelial cell line (CH3) from a recurrent retroperitoneal lymphangioma. In Vitro Cell Dev Biol. 1987;23(9):647–52. https://doi.org/10.1007/BF02621074.
101. Folkman J. Clinical applications of research on angiogenesis. N Engl J Med. 1995;333(26):1757–63.
102. Folkman J. Tumor angiogenesis and tissue factor. Nat Med. 1996;2(2):167–8. https://doi.org/10.1038/nm0296-167.
103. van Netten JP, Cann SA, van der Westhuizen NG. Angiogenesis and tumor growth. N Engl J Med. 1996;334(14):920–1. author reply 921
104. Witte MH, Witte CL. On tumor (and other) lymphangiogenesis. Lymphology. 1997;30(1):1–2.
105. He Y, Rajantie I, Pajusola K, Jeltsch M, Holopainen T, Yla-Herttuala S, Harding T, Jooss K, Takahashi T, Alitalo K. Vascular endothelial cell growth factor receptor 3-mediated activation of lymphatic endothelium is crucial for tumor cell entry and spread via lymphatic vessels. Cancer Res. 2005;65(11):4739–46. https://doi.org/10.1158/0008-5472.CAN-04-4576.
106. Karpanen T, Alitalo K. Lymphatic vessels as targets of tumor therapy? J Exp Med. 2001;194(6):F37–42.
107. Faries MB, Thompson JF, Cochran AJ, Andtbacka RH, Mozzillo N, Zager JS, Jahkola T, Bowles TL, Testori A, Beitsch PD, Hoekstra HJ, Moncrieff M, Ingvar C, Wouters M, Sabel MS, Levine EA, Agnese D, Henderson M, Dummer R, Rossi CR, Neves RI, Trocha SD, Wright F, Byrd DR, Matter M, Hsueh E, MacKenzie-Ross A, Johnson DB, Terheyden P, Berger AC, Huston TL, Wayne JD, Smithers BM, Neuman HB, Schneebaum S, Gershenwald JE, Ariyan CE, Desai DC, Jacobs L, McMasters KM, Gesierich A, Hersey P, Bines SD, Kane JM, Barth RJ, McKinnon G, Farma JM, Schultz E, Vidal-Sicart S, Hoefer RA, Lewis JM, Scheri R, Kelley MC, Nieweg OE, Noyes RD, Hoon DSB, Wang HJ, Elashoff DA, Elashoff RM. Completion dissection or observation for sentinel-node metastasis in melanoma. N Engl J Med. 2017;376(23):2211–22. https://doi.org/10.1056/NEJMoa1613210.
108. Giuliano AE. The evolution of sentinel node biopsy for breast cancer: personal experience. Breast J. 2020;26(1):17–21. https://doi.org/10.1111/tbj.13729.
109. Skobe M, Hawighorst T, Jackson D, Prevo R, Janes L, Velasco P, Riccardi L, Alitalo K, Claffey K, Detmar M. Induction of tumor lymphangiogenesis by VEGF-C promotes breast cancer metastasis. Nat Med. 2001;7(2):192–8.
110. Huang C, Chen Y. Lymphangiogenesis and colorectal cancer. Saudi Med J. 2017;38(3):237–44. https://doi.org/10.15537/smj.2017.3.16245.
111. Sundlisaeter E, Dicko A, Sakariassen PO, Sondenaa K, Enger PO, Bjerkvig R. Lymphangiogenesis in colorectal cancer--prognostic and therapeutic aspects. Int J Cancer. 2007;121(7):1401–9. https://doi.org/10.1002/ijc.22996.
112. Sun JG, Wang Y, Chen ZT, Zhuo WL, Zhu B, Liao RX, Zhang SX. Detection of lymphangiogenesis in non-small cell lung cancer and its prognostic value. J Exp Clin Cancer Res. 2009;28:21. https://doi.org/10.1186/1756-9966-28-21.
113. Werynska B, Dziegiel P, Jankowska R. Role of lymphangiogenesis in lung cancer. Folia Histochem Cytobiol. 2009;47(3):333–42. https://doi.org/10.2478/v10042-009-0090-3.
114. Leong SP, Naxerova K, Keller L, Pantel K, Witte M. Molecular mechanisms of cancer metastasis via the lymphatic versus the vascular channels. Clin Exp Metastasis. in press-a;
115. Dieterich LC, Ikenberg K, Cetintas T, Kapaklikaya K, Hutmacher C, Detmar M. Tumor-associated lymphatic vessels upregulate PDL1 to inhibit T-cell activation. Front Immunol. 2017;8:66. https://doi.org/10.3389/fimmu.2017.00066.
116. Commerford CD, Dieterich LC, He Y, Hell T, Montoya-Zegarra JA, Noerrelykke SF, Russo E, Rocken M, Detmar M. Mechanisms of tumor-induced Lymphovascular niche formation in draining lymph nodes. Cell Rep. 2018;25(13):3554–63. e3554. https://doi.org/10.1016/j.celrep.2018.12.002.
117. Ma Q, Dieterich LC, Ikenberg K, Bachmann SB, Mangana J, Proulx ST, Amann VC, Levesque MP, Dummer R, Baluk P, McDonald DM, Detmar M. Unexpected contribution of lymphatic vessels to promotion of distant metastatic tumor spread. Sci Adv. 2018;4(8):eaat4758. https://doi.org/10.1126/sciadv.aat4758.
118. Shields JD, Kourtis IC, Tomei AA, Roberts JM, Swartz MA. Induction of lymphoidlike stroma and immune escape by tumors that express the chemokine CCL21. Science. 2010;328(5979):749–52. https://doi.org/10.1126/science.1185837.
119. Issa A, Le TX, Shoushtari AN, Shields JD, Swartz MA. Vascular endothelial growth factor-C and C-C chemokine receptor 7 in tumor cell-lymphatic cross-talk promote invasive phenotype. Cancer Res. 2009;69(1):349–57. https://doi.org/10.1158/0008-5472.CAN-08-1875.
120. Shields JD, Fleury ME, Yong C, Tomei AA, Randolph GJ, Swartz MA. Autologous chemotaxis as a mechanism of tumor cell homing to lymphatics via interstitial flow and autocrine CCR7 signaling. Cancer Cell. 2007;11(6):526–38.

121. Fankhauser M, Broggi MAS, Potin L, Bordry N, Jeanbart L, Lund AW, Da Costa E, Hauert S, Rincon-Restrepo M, Tremblay C, Cabello E, Homicsko K, Michielin O, Hanahan D, Speiser DE, Swartz MA. Tumor lymphangiogenesis promotes T cell infiltration and potentiates immunotherapy in melanoma. Sci Transl Med. 2017;9(407) https://doi.org/10.1126/scitranslmed.aal4712.
122. Dafni H, Israely T, Bhujwalla ZM, Benjamin LE, Neeman M. Overexpression of vascular endothelial growth factor 165 drives peritumor interstitial convection and induces lymphatic drain: magnetic resonance imaging, confocal microscopy, and histological tracking of triple-labeled albumin. Cancer Res. 2002;62(22):6731–9.
123. Harrell MI, Iritani BM, Ruddell A. Tumor-induced sentinel lymph node lymphangiogenesis and increased lymph flow precede melanoma metastasis. Am J Pathol. 2007;170(2):774–86. https://doi.org/10.2353/ajpath.2007.060761.
124. Ng CP, Hinz B, Swartz MA. Interstitial fluid flow induces myofibroblast differentiation and collagen alignment in vitro. J Cell Sci. 2005;118(Pt 20):4731–9. https://doi.org/10.1242/jcs.02605.
125. Ng CP, Swartz MA. Fibroblast alignment under interstitial fluid flow using a novel 3-D tissue culture model. Am J Physiol Heart Circ Physiol. 2003;284(5):H1771–7. https://doi.org/10.1152/ajpheart.01008.2002.
126. Shi ZD, Ji XY, Qazi H, Tarbell JM. Interstitial flow promotes vascular fibroblast, myofibroblast, and smooth muscle cell motility in 3-D collagen I via upregulation of MMP-1. Am J Physiol Heart Circ Physiol. 2009;297(4):H1225–34. https://doi.org/10.1152/ajpheart.00369.2009.
127. Shi X, Young CD, Zhou H, Wang X. Transforming growth factor-beta signaling in fibrotic diseases and cancer-associated fibroblasts. Biomol Ther. 2020;10(12) https://doi.org/10.3390/biom10121666.
128. Shieh AC, Rozansky HA, Hinz B, Swartz MA. Tumor cell invasion is promoted by interstitial flow-induced matrix priming by stromal fibroblasts. Cancer Res. 2011;71(3):790–800.
129. Swartz MA, Lund AW. Lymphatic and interstitial flow in the tumour microenvironment: linking mechanobiology with immunity. Nat Rev Cancer. 2012;12(3):210–9. https://doi.org/10.1038/nrc3186.
130. Mendel G (2016) Experiments on plant hybrids (1866). http://www.bshs.org.uk/bshs-translations/mendel. Accessed 6 May 2021.
131. Kim R, Emi M, Tanabe K, Arihiro K. Immunobiology of the sentinel lymph node and its potential role for antitumour immunity. Lancet Oncol. 2006;7(12):1006–16. https://doi.org/10.1016/S1470-2045(06)70975-5.
132. Hirakawa S. From tumor lymphangiogenesis to lymphvascular niche. Cancer Sci. 2009;100(6):983–9. https://doi.org/10.1111/j.1349-7006.2009.01142.x.
133. Jana S, Muscarella R, Jones D. The multifaceted effects of breast cancer on tumor-draining lymph nodes. Am J Pathol. 2021;

Immune Cell Trafficking in the Lymphatics, Hyaluronan Biology and Tumour Metastasis

22

David G. Jackson

Abstract

As an integral part of the immune system, the lymphatics enable immune surveillance by shepherding the migration of antigen-laden immune cells to draining lymph nodes (dLNs), as well as handling the clearance of macrophages and neutrophils from sites of injury and infection. In addition, the lymphatics play key roles as conduits for tumour metastasis. The molecular mechanisms controlling entry to the lymphatics in these different contexts have yet to be fully elucidated. However, it is clear that in most cases they involve coordination between the processes of chemotaxis and cell adhesion within the specialized endothelial junctions of initial lymphatic capillaries for transit to the vessel lumen. A fundamental mediator of such adhesion is the large matrix glycosaminoglycan hyaluronan, and growing evidence indicates anchorage of hyaluronan to the surface of migrating cells and appropriate engagement with LYVE-1, its' primary receptor in lymphatic endothelium are critical to the entry process. The present chapter outlines the main properties of hyaluronan and how they support the trafficking of leucocytes for immunity and inflammation. In addition, it discusses the evidence that hyaluronan may also aid the invasion and systemic dissemination of metastasizing tumour cells.

Learning Objectives

- The main immune cell populations that traffic in lymph and their functions in innate and adaptive immunity.
- The biological features of hyaluronan and how immune cells employ hyaluronan and its receptors to enter lymphatic vessels and migrate to draining lymph nodes.
- The distinctive endothelial junctions through which immune cells enter lymphatics.
- How tumours exploit hyaluronan and its receptors to metastasize via lymph.

D. G. Jackson (✉)
MRC Human Immunology Unit, MRC Weatherall Institute of Molecular Medicine, University of Oxford, Oxford, UK
e-mail: David.jackson@imm.ox.ac.uk

22.1 Introduction

The lymphatics play multiple roles in maintaining tissue homeostasis through the uptake and drainage of interstitial fluids and the transport of electrolytes, lipids, cholesterol, extracellular matrix and other dissolved macromolecules for appropriate degradation or delivery to the systemic blood circulation [1]. Of equal importance, lymphatics are integral to the immune system, insofar as they also act as conduits for the migration of immune cells from tissues to the draining lymph nodes for the generation, maintenance and modulation of antigen-specific immune responses [2]. In addition, lymphatic vessels are involved in the timely clearance of phagocytes, apoptotic cell bodies and tissue debris from injured and inflamed tissues after the processes of repair and re-modelling. In the context of cancer, however, these otherwise beneficial properties of lymphatics are co-opted by tumours that exploit them to support their own uncontrolled growth and metastasis, first to draining lymph nodes and then to distant organs with inevitable consequences for patient morbidity and mortality (Pathology of Lymph Node Metastases, Messina, J.).

Despite their undisputed importance, the lymphatics have until recently been relegated to the role of Cinderellas, with most of the limelight being given to the blood vasculature, and how immune cells traffic and tumour cells disseminate

S. P. Leong et al. (eds.), *Cancer Metastasis Through the Lymphovascular System*,
https://doi.org/10.1007/978-3-030-93084-4_22

within it (Hallmarks of Metastasis: Molecular Underpinnings, Gomez-Gelvez, J.). This situation is now changing and in the case of leucocytes, much has been learned about the mechanisms by which the various different populations of these cells enter and migrate within lymphatics and how they are often distinct from the mechanisms used to extravasate from blood vessels. In particular, evidence is emerging for the fundamental involvement of the pro-adhesive extracellular matrix glycosaminoglycan hyaluronan [3]. This ubiquitous molecule, which plays a well-documented role in both immune cell adhesion/migration and tumour metastasis in the blood vasculature via its primary receptor CD44, is now also recognized to play a key role in immune cell entry to the lymphatics, by mediating adhesion to endothelium via the related receptor LYVE-1 [2, 4, 5]. Indeed, preliminary evidence suggests HA and its receptors may also support tumour metastasis via the lymphatics. The present chapter presents the current state of knowledge in this relatively new area of lymphatic biology. Starting with an overview of the main leucocyte populations that migrate in lymph and the novel architecture of the lymphatic capillaries through which they enter, I go on to describe the particular molecular mechanisms by which HA and its receptors mediate immune cell trafficking and potentially tumour metastasis via the lymphatics.

22.2 Immune Cell Populations that Traffic Via Lymph

Within tissues, the continuous flow of lymph and the steady migration of immune cells through afferent capillaries to immediate downstream lymph nodes creates a local neighbourhood watch system that maintains uninterrupted surveillance for injury and the presence of microbial or viral infection. In normal resting tissue, the numbers of such cells are small, comprising almost exclusively T-cells and dendritic cells (DCs). Few if any are naive T lymphocytes, as these enter lymph nodes directly from the circulation via high endothelial venules, and the majority are instead antigen-experienced cells of the so-called effector memory T_{EM} subset. Having originally entered the tissues in search of cognate antigen, the majority of these previously primed and activated T_{EM} cells then migrate via afferent lymphatics to the draining lymph nodes, licensed by their selective expression of CCR7, the receptor for the critical lymph-tropic chemokine CCL21 [6, 7]. They constitute predominantly CD4 cells that provide help for B cell antibody responses as opposed to CD8 cells that engage in target killing and which largely remain within the tissues as immotile CCR7^{-ve} resident memory (T_{RM}) cells. The passage of T_{EM} cells through afferent lymph to downstream lymph nodes and subsequent exit in efferent lymph allows them eventually return to the circulation for continued surveillance and maintenance of the memory T-cell pool [8]. In addition, such migration may allow them to amplify and/or polarize immune responses in draining lymph nodes through their capacity for cytokine release, or even enable their antigen responsiveness to be downmodulated or tolerized by T-cell immune checkpoint receptor ligands such as PDL-1 expressed in the endothelium of lymph node sinuses and tumour-associated lymphatics (Immune microenvironment: Chen, T.). In addition, a significant proportion ($\leq$25%) of the CD4 population in normal afferent lymph are dedicated FOXP3^{+} T regulatory cells (T_{REGs}), an important lymphocyte subset that can suppress other T-cell responses, and one that plays a key role in maintaining tolerance to self-antigens as well as the suppression of anti-tumour responses. Following inflammation, and the resulting influx of T-cells to the tissues from the circulation, the numbers of T-cells in afferent lymph increase, along with those of certain innate immune cell populations as described subsequently.

After T-cells, the next most numerous leucocyte population in normal afferent lymph are DCs, the key mediators of immune surveillance that patrol tissues and endocytose, process and present antigens as bound MHC complexes to stimulate T-cell immune responses [9]. Historically, the migration of these cells has been tracked following their in situ labelling by FITC skin painting, ex vivo labelling and adoptive transfer, or in specialized cases after microneedle injection into lymphatic capillaries. Again, only small numbers of DCs migrate in afferent lymph under normal steady-state conditions in non-inflamed tissues, partly to maintain peripheral tolerance to self-antigens. However, in common with T_{EM} cells, the numbers of these DCs increase dramatically in inflammation due to increased lymph flow and the release of cytokines such as IL-1, TNFα and INFγ within the tissues that induce DCs to mobilize, differentiate and upregulate expression of the chemokine receptor CCR7 for vessel entry (see below).

Other leucocyte populations that may be present in significant numbers in afferent lymph are macrophages and neutrophils. Recruited to tissues from the blood circulation in response to injury or infection, these innate immune cells phagocytose tissue debris and repair and re-model the extracellular matrix (ECM). The short-lived neutrophil population subsequently undergoes programmed cell death, and the apoptotic corpses are taken up by macrophages through efferocytosis and cleared from the resolving tissue sites via afferent lymph. In addition, some neutrophils that phagocytose protozoal parasites and bacteria including *Mycobacterium bovis* BCG have extended tissue survival, and can migrate to dLNs more rapidly than either T-cells or macrophages where they activate and polarize T-cell responses through cytokine release and crosstalk with DCs [10, 11].

In contrast to afferent lymph, the efferent lymph that flows from lymph nodes back to the circulation via the thoracic duct contains predominantly naive T and B cells. Having entered the nodes from the circulation, these migrate along fibroblast reticular cell conduits in the cortex to probe resident DCs for antigen, before exiting through medullary sinuses to the efferent lymphatics, in response to rising gradients of the sinus-derived chemotactic lipid sphingosine-1-phosphate (S1P [12]). Notably, during the onset of infection or inflammation, the efflux of this recirculating population from the lymph node is halted transiently (3–4 days) so as to prolong their residence time and increase the efficiency of immune recognition.

22.3 The Distinctive Features of Lymphatic Vessels, Inter-Endothelial Junctions and Sites of Immune Cell Entry

The lymphatic network differs from the blood vasculature in terms of both architecture and physiology, reflecting the distinct functional requirements of the two vasculatures, and in particular the necessity for lymphatics to handle the passive uptake of fluid and macromolecules as well as selective entry of appropriate immune cell traffic. In addition to being a one-way rather than a circulatory system, the lymphatics fulfill these dual requirements through local variations in vessel structure and function. Accordingly, the initial blind-ended capillaries possess characteristically loose, overlapping endothelial junctions and a rudimentary basement membrane, conducive to fluid absorption and cell entry [13]. Just downstream of these capillaries, the pre-collectors and valved collectors have more conventional tight endothelial junctions, akin to those of blood vessels, and a dense basement membrane more appropriate to the containment of lymph and its transport by oscillatory smooth muscle contraction. As shown by seminal studies of DC migration in acutely inflamed mouse trachea, the majority of immune cells entering the lymphatics target the first few millimetres of the initial capillaries [13]. Here, the constituent endothelial cells have a distinctive oakleaf shape, and their interdigitation creates scalloped borders in which the overlapping membrane flaps are "buttoned" together at their sides by the adherens-junction protein VE–cadherin and tight junction proteins including Claudin-5, ZO-1 (zonula occludens-1), ESAM (endothelial selective adhesion molecule) and JAM-A [13, 14]. In contrast, between these buttons, the flaps are coated exclusively with PECAM-1 (CD31), and most notably, the lymphatic endothelial HA receptor LYVE-1 [5, 15], which gives them the flexibility to remain free and untethered, forming small apertures or "portals" of approximately 0.5–1 μm diameter in the endothelium through which immune cells can squeeze (Fig. 22.1; [13, 14]). As detailed in the following section, the location of LYVE-1 above such portals and its ligand binding characteristics endow this receptor with the ability to mediate vessel entry through engagement with HA in the incoming leucocyte glycocalyx. In addition to such adhesion, the process of immune cell intravasation is driven by actomyosin-mediated motility, directed by the chemokine CCL21, which is constitutively synthesized and secreted by LECs to form long-range haptotactic gradients which are detected by migrating DCs, T_{EM} cells, macrophages and some neutrophils through the receptor CCR7. Intriguingly, contact between transiting DCs and vessel endothelium can also trigger rapid "on demand" discharge of CCL21 from pre-stored depots in trans Golgi vesicles to sites of entry on the basolateral surface of the endothelium by a Ca^{2+} triggered exocytic mechanism that involves microtubules and fibrillar actin [16].

While the foregoing studies indicate immune cells enter lymphatics via overlapping "buttoned" junctions under resting conditions or during the initial response to inflammatory stimuli (i.e. brief exposure to LPS), it is highly likely that additional modes of entry are used during conditions of chronic inflammation, given that newly sprouting lymphatic vessels generated by inflammation-induced lymphangiogenesis have zippered junctions (which lack LYVE-1) rather than buttoned junctions [17]. Furthermore, exposure of existing lymphatic capillaries to inflammatory cytokines leads to the upregulated expression of ICAM-1, VCAM-1 and E-selectin—receptors that are known to mediate leucocyte transit across zippered endothelial junctions in inflamed blood vessels [18]. It may be that the number of default entry sites through button junctions becomes limiting in inflammation, and that induction of ICAM-1 and VCAM-1 is necessary to provide for additional entry sites in downstream zippered junctions.

22.4 Hyaluronan, its Key Biological Functions and its Involvement in Immune Cell Migration Via Lymph

HA, a large (≤ 5MDa) unbranched glycosaminoglycan, is a co-polymer made up from the repeating di-saccharide [N-acetyl glucosamine (β-1,4) glucuronic acid β-1,3)]$_n$, and constitutes a ubiquitous component of the interstitial matrix [19]. Among the many functions of HA in soft tissues is the provision of structural support and a low resistance environment permissive for cell migration, roles that are enabled by its capacity to absorb water equivalent to many times its mass, fill tissue spaces and form hydrated gel-like meshworks. Synthesized by one of three HA synthases (HAS1-3) located at the inner surface of the plasma membrane, these enzymes extrude the nascent HA chains to the cell surface for secretion as free polymers that are subsequently sequestered by

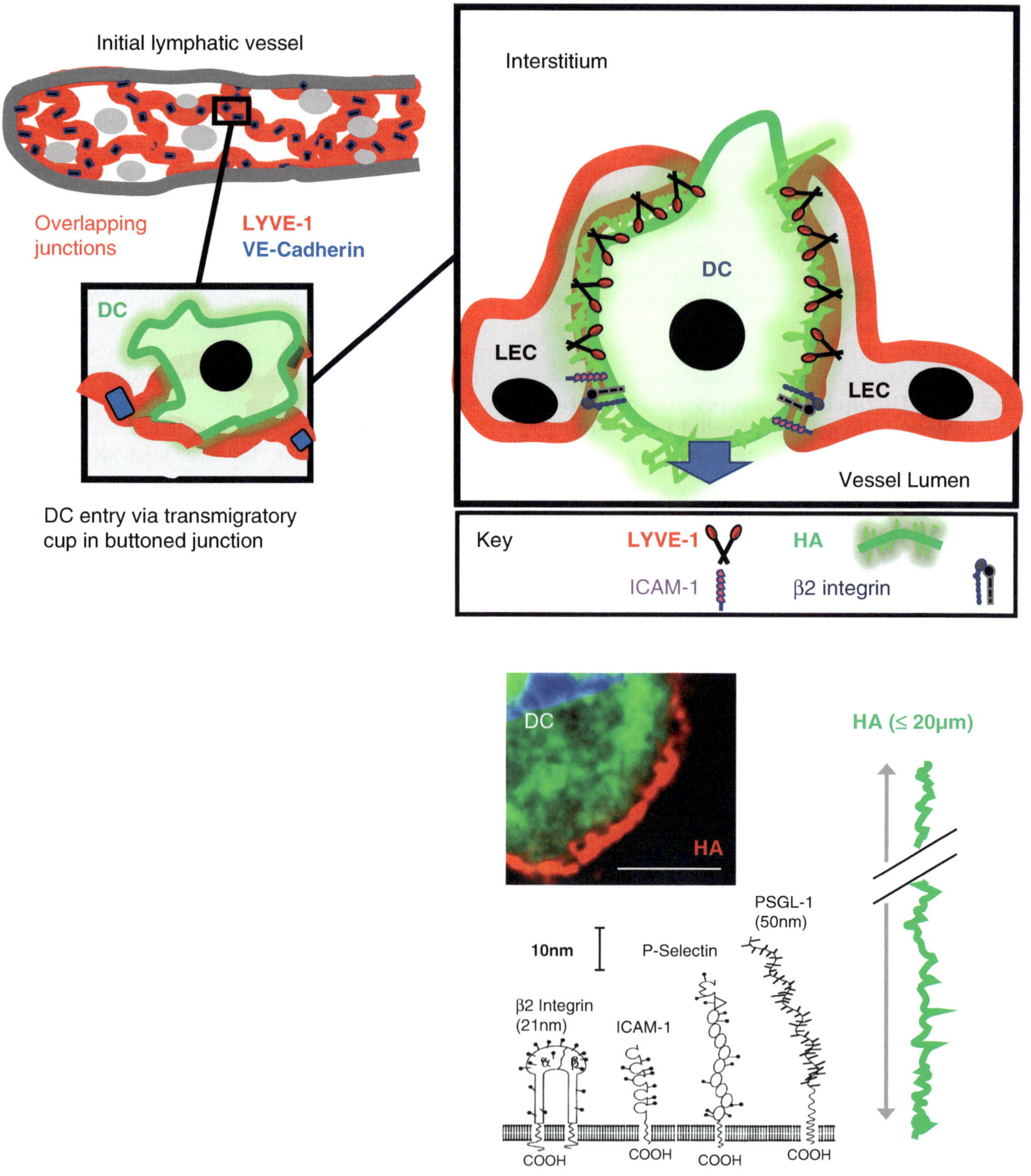

Fig. 22.1 Diagram depicting a dendritic cell entering an initial lymphatic capillary (upper panels) showing the distinctive overlapping junctions between individual oakleaf shaped endothelial cells and the formation of an LYVE-1 transmigratory cup (modified from [3]). The lower panel shows a confocal microscopic image of the DC HA glycocalyx and the dimensions of a typical HA chain relative to underlying adhesion receptors on the DC surface

proteoglycans of the Link superfamily including Aggrecan, Versican and Link protein, which together with type I collagen and fibronectin form cross-linked matrices in most organs [20]. Notably, the metabolism of HA within tissues requires the participation of the lymphatics [21]. Under normal resting conditions, HA in tissues such as skin undergoes regular turnover ($T_{1/2} \geq 48$ h) by cell-associated hyaluronidases, and the intermediate cleavage products are rapidly removed and transported in afferent lymph to draining lymph nodes where they are terminally degraded through uptake by the endocytic HA receptor HARE (HA Receptor for Endocytosis) in sinusoidal endothelium [22]. However, in inflammation or injury, the local turnover of HA is increased and this generates fragments that are detected by the innate immune system as danger signals through binding to toll-like receptors TLR2 and TLR4, and the leucocyte HA receptor CD44 [23, 24]. Unlike intact high molecular weight (0.5–5 MDa) HA which is generally anti-inflammatory, these smaller (< 500 KDa) degradation products are pro-inflammatory and induce cytokine and chemokine release as well as promoting both angiogenesis and lymphangiogenesis and increased lymphatic vessel permeability [25]. The molecular basis for discrimination between high and low MW HA is a hotly debated issue within the field, but is thought by many to reflect the influence of chain length on receptor clustering and downstream signaling events.

Arguably the most important function of HA is the support of cell:cell and cell:matrix adhesion during migration, achieved by means of stable and/or transient interactions with receptors on neighbouring static and migratory cells respectively. Integral to this is the capacity of HA to form a surface glycocalyx in appropriate cell types through retention of the polymer in the plasma membrane by its primary receptor CD44. Indeed, in post-capillary blood venules, such an HA glycocalyx on the luminal surface of the vascular endothelium serves to capture inflammatory cells including T lymphocytes and neutrophils from flow via their own expression of CD44, thus facilitating their extravasation to the underlying tissue through formation of a CD44 • HA • CD44 sandwich (reviewed in [26]). As intimated in the previous section, the discovery that HA plays an analogous role in the intravasation of immune cells from tissue to lymphatic vessels came from the identification of the closely related HA receptor LYVE-1 in the button-like junctions of lymphatic capillaries that serve as leucocyte entry portals [13, 15]. Prior to this, some earlier reports had suggested DCs and activated T-cells can synthesize and sequester cell surface HA, but its role in lymphatic entry and migration was overlooked [27]. HA is synthesized in DCs and monocyte/macrophages primarily by HA synthase-2 (HAS2) which generates long-chain high molecular weight polymers, and a proportion of these is retained by CD44 in the surface glycocalyx as a dense corona of some 200–300 nm thickness (Fig. 22.1; [28]). Curiously, assembly of the glycocalyx in DCs appears to proceed coordinately with the biosynthesis and processing of CD44, insofar as the nascent HA chains associate with the receptor intracellularly and the bound complexes are then exported to the cell surface. As a result, CD44 gene-deleted DCs fail to elaborate a glycocalyx and instead accumulate the glycosaminoglycan within cytoplasmic vesicles [28]. Crucial evidence that the engagement of HA with LYVE-1 is important for vessel entry in vivo came from studies tracking the migration of fluorescently labelled DCs in oxazolone hypersensitized skin. These demonstrated that *Lyve1* gene deletion and mAb blockade led to a perilymphatic accumulation of DCs in the dermis and a consequent delay or inhibition of their trafficking to downstream axial LNs, respectively [4]. Likewise, deletion of CD44 or depletion of the HA glycocalyx using hyaluronidase and the HA biosynthetic inhibitor 4-MU (Hymecromone®) resulted in a dramatic reduction in DC nodal migration compared to controls. Moreover, the significance of LYVE-1 • HA mediated DC migration for in vivo immune function was underscored by experiments with mice immunized intradermally with ovalbumin as a model antigen. These showed that both *Lyve1* gene deletion and mAb blockade disrupted the generation of ova-specific CD4 and CD8 T-cell responses in downstream lymph nodes, confirming that in vivo, the process of LYVE-1 • HA mediated lymphatic entry can be rate-limiting for protective immunity [4].

How engagement of the HA glycocalyx with LYVE-1 mediates vessel entry has been elucidated from rigorous in vitro studies with cultured monolayers of primary dermal lymphatic endothelial cells. These show that contact with the DC glycocalyx triggers redistribution of LYVE-1 into ring-like assemblages termed transmigratory cups (Fig. 22.1) beneath the adherent leukocytes, similar to the structures that form around T-cells during their transit across blood vascular endothelium ([4] and reviewed in [5]). As observed by video microscopy, the DCs attach to these LYVE-1 cups via their posterior pole or uropod, while extending a lamellipodium at the leading edge to explore the endothelial surface for potential sites of transit. Whether such transit occurs through the endothelial cell body (transcellular migration) or at nearby inter-endothelial junctions is not yet clear, although both modes of transmigration are clearly possible [28]. Moreover, the density of the HA glycocalyx can influence the strength of adhesion to endothelium, and hence in vivo it is likely that changes in glycocalyx organization, e.g. through enhanced

HA or CD44 biosynthesis or surface clustering in response to inflammatory cytokines, may tune the rate of leucocyte trafficking as appropriate to local tissue requirements [28].

In addition to its involvement in DC trafficking, the LYVE-1 • HA axis also mediates the trafficking of macrophages in lymph, a process that enables their clearance from inflamed tissues, and one that is vital for resolution of inflammation and the restoration of normal homeostasis. Like DCs, monocyte-derived macrophages assemble a surface HA glycocalyx, and their capacity to synthesize HA increases markedly following their differentiation to an inflammatory M1-like phenotype that can migrate via lymph [29]. Indeed, such macrophages were shown to adhere and transmigrate inflamed LEC monolayers in vitro, in an LYVE-1 and HA-dependent manner [29]. The in vivo importance of the LYVE-1 • HA axis for macrophage migration is underlined by recent studies in mouse models of tissue injury [30]. Firstly, in a ligation-induced model of myocardial infarction, a disease in which ischaemia leads to infiltration of the myocardium by neutrophils and macrophages, it was shown that the latter populations are cleared via cardiac lymphatics to mediastinal lymph nodes, and furthermore, that stimulation of cardiac lymphangiogenesis through local administration of recombinant VEGF-C not only enhanced clearance, but also reduced subsequent fibrotic scarring. Significantly, macrophage exit was attenuated in *Lyve1*$^{-/-}$ mice, which exhibited aggravated inflammation and delayed cardiac repair [30].

Whether an HA glycocalyx is utilized for vessel entry by the T_{EM} cell population that migrates in afferent lymph has yet to be resolved. Although T-cells have been reported to have HA biosynthetic capacity [31], the levels of the glycosaminoglycan detectable on the cell surface appear to be an order of magnitude below those of DCs and macrophages. Moreover, T_{EFF} cells and T_{REGs} are known to enter lymphatics in response to S1P and lymphotoxin (LTα1β3) and its receptor LTβR, so the likelihood remains that these rather than the LYVE-1 • HA axis are used for their transit [32].

Lastly, it is now evident that neutrophils lack a HA glycocalyx and employ a mechanism that is distinct from that of other immune cells to enter lymphatic capillaries. Recruited exclusively to inflamed tissues, these professional phagocytes can in appropriate circumstances exit through the lymphatics and migrate to draining lymph nodes by adhering to endothelium via β2 integrins and then transmigrating through the combined actions of the matrix metalloproteinases MMP8 and MMP9, the serine protease neutrophil elastase and the arachidonate-derived chemorepellent, 12,hydroxyeicosatetraenoate (12(S)HETE) that together promote junctional retraction [33]. The rationale behind this unconventional mechanism may be that it allows a more rapid mode of lymphatic entry than transit through button junctions, in keeping with the strict requirement for these cells to mount a rapid response to tissue injury. Curiously, it has also been reported that this lipoxin-mediated process for endothelial junctional retraction is exploited by certain mammary and colorectal tumours for lymphatic metastasis [34].

Hence, transmigration by means of an HA glycocalyx is not a mechanism adopted by all leucocytes for lymphatic entry. In addition, though engagement of the HA glycocalyx with LYVE-1 and transmigratory cup formation are critical steps in the entry process, these are only the first in a series of adhesion events required for diapedesis and it is evident that further receptor ligand pairs are also involved. Although a full description is beyond the scope of the present discourse, these include ICAM-1, VCAM-1 and E-selectin as mentioned previously, as well as ALCAM/CD166 (activated leucocyte cell adhesion molecule), L1CAM, CD99, macrophage mannose receptor (MMR), 4-1BB/CD137 and CLEVER-1 (combined lymphatic endothelial and vascular endothelial receptor), several of which are selectively upregulated by inflammatory cytokines, and all of which been reported to participate in immune cell adhesion and/or transmigration (see e.g. [2]. for review). Significantly, both ICAM-1 and VCAM-1 are present and appear to co-distribute with LYVE-1 in endothelial transmigratory cups in vitro, and blocking antibodies to ICAM-1 and its β2 integrin ligand LFA-1 inhibit the entry of both DCs and T-cells to lymphatics in oxazolone-hypersensitized and TNFα treated mouse dermis, respectively [4, 35]. One possibility is that these come into play after appropriate disengagement or disruption of HA • LYVE-1 interactions, so as to provide extra traction for the transit of immune cells or their subsequent intraluminal crawling towards downstream collectors. In DCs, the HA glycocalyx appears to remain attached to the cell surface following transendothelial migration [4]. Nevertheless, it could conceivably undergo either discrete re-modelling by hyaluronidases, or regulated redistribution to uncover access to integrins or other underlying adhesion receptor ligands during diapedesis. Determining the precise choreography of such events is a topic of ongoing research in the field of lymphatic trafficking, and one that will require much further research.

22.5 What Is So Special About Hyaluronan?

Why immune cells should have evolved to utilize an HA glycocalyx for adhesion and transit across lymphatic endothelium, rather than more conventional adhesion receptors, may be explained by the unique physical characteristics of HA polymers and the novel binding properties of its primary lymphatic receptor LYVE-1. First, the extreme size of HA molecules (comprising chains of up to 20,000 di-saccharides with contour lengths of up to 1 μm) and their composition from charged sugar residues bestows them with the capacity to form an extended polyanionic coat that reaches far beyond underlying adhesion receptors on the leucocyte surface such as β2 integrins (approx. 20 nm, see Fig. 22.1). This poises the glycocalyx as the first key component on the immune cell surface to make adhesive contact with endothelium and may necessitate its displacement to enable subsequent interactions with other adhesion receptors and their ligands as noted above. Second, because the receptor binding footprint is relatively short (8–20 sugar residues), each polymer chain can accommodate multiple LYVE-1 molecules, thus generating avidity-dependent binding, the magnitude of which can be controlled by changes in either LYVE-1 or CD44 density through surface clustering [5, 29]. Third, the mechanics of the binding interaction between LYVE-1 and HA appear to be optimally tuned for a role in facilitating immune cell entry. As demonstrated by atomic force spectroscopy, LYVE-1 binds HA through a novel interaction that allows the receptor to slide along the polymer chain, as distinct from the conventional "sticking" bonds mediated by CD44, potentially enabling a leucocyte to crawl over the surface of the lymphatic capillary via its glycocalyx prior to transit [5, 36]. Clearly, such characteristics are ideally suited to the low shear environment within the tissues from which immune cells enter lymphatics in response to chemokine-driven haptotaxis as opposed to being captured from rapid laminar flow, as is the case for leucocytes exiting the blood circulation.

22.6 Hyaluronan and Tumour Metastasis

The majority of cancers, and in particular carcinomas, metastasize via the lymphatic vasculature, and lymph node involvement is one of the first clinical signs of tumour progression (Overview of Principles of Basic Lymphology and Relation to Cancer Metastasis: Witte, M.), (Primary Tumor Staging and Detection of Common Sites of Distant Metastatic Disease: Thomas, K., Ahmed, A., Morse, B.). Nevertheless, the impact of lymphatic metastasis remains a topic for debate, and the common assumption is that dissemination by the lymphatic route enables seeding of distant metastases only indirectly, after sequential spread through sentinel and post-sentinel nodes followed by transport in efferent lymph to the systemic circulation via the thoracic duct. Recent studies, however, have presented persuasive evidence that tumours can metastasize directly from the first draining node by entering the circulation through the invasion of high endothelial venules (HEVs) in the lymph node medulla [37, 38]. Notably, ductal breast carcinomas microinjected into pre-nodal lymphatics have been shown to enter the systemic vasculature by such a route without accessing successive downstream nodes, seeding distant lung metastases at a much faster rate than would be achieved by invasion of local blood vessels at the primary tumour site [38]. Hence, it is clear that lymphatic vessels and not blood vessels are preferentially targeted by such tumours for early systemic spread (Pathology of Distant/Systemic Metatases: Turk, A.). Yet in spite of these insights, the molecular mechanisms that are employed by tumours to enter the lymphatics and exit across subcapsular sinuses to access lymph node HEVs are poorly understood, and the degree to which they overlap with the transit mechanisms used by immune cells remains uncertain. Within the primary tumour mass, cancer cells are initially attracted towards lymphatic vessels by chemotaxis, promoted by the same endothelial-derived chemokines CCL21 and CXCL12 that direct DC and T_{MEM} cell entry, while in draining lymph nodes their transit across the subcapsular sinus to the parenchyma is directed by CCL1, a chemokine that mediates similar transit of T-helper cells and DCs via its receptor CCR8 during inflammation [39]. Individual tumour cells or emboli then initiate metastasis by invading either pre-existing or newly sprouting vessels (Lymphangiogenesis and Tumor Lymphangiogenesis: Detmar, M.) generated by secretion of the lymphangiogenic growth factor VEGF-C and to a lesser extent VEGF-D, both by the tumour and by macrophages in its' local microenvironment (Tumor Microenvironment: Co-conspirator in Tumorigenesis: Wang, Z.). This expansion of tumour-associated lymphatics increases the number of potential sites for tumour entry, while their greater leakiness makes them more amenable for infiltration, either through button junctions in the manner of immune cells, or through zippered pre-collectors/collectors rendered more permeable by VEGF-C induced vessel dilation. One might therefore anticipate that tumours with the capacity to assemble a HA glycocalyx could co-opt LYVE-1 for entry to initial lymphatic capillaries, and as discussed subsequently, there is now evidence suggesting this is indeed the case.

The association of HA with tumour growth and invasion has been recognized for many years. A majority of different tumour cell lines have elevated levels of both HA and CD44, and upregulation of HAS gene expression and HA deposition correlates with aggressive and metastatic cancers [40]. Moreover, upregulated synthesis of HA and CD44 is hallmark of epithelial to mesenchymal transition, a key process in carcinoma progression during which tumour cells acquire enhanced migratory capacity and in some cases development of cancer stem cells (Hallmarks of Cancer: Molecular Underpinnings: Chitale, D.).

While studies have linked high levels of CD44 expression with metastasis and disease progression in multiple different cancers, their focus on lymphatic metastasis has been mostly indirect. For example, in a histological analysis of gastric carcinoma, most tumour cells showed histological positivity for HA, and a high proportion of such cells was significantly associated with lymphatic invasion, nodal metastasis, poor differentiation grade and adverse prognosis [41]. Similarly, studies of colorectal carcinoma revealed a strong correlation between the proportion of tumour cells with HA positivity and poor survival, and between HA staining intensity and advanced metastatic dissemination [42]. Indeed, in breast cancer metastasis, the appearance of a CD44-associated pericellular HA matrix was observed around tumour emboli present within newly invaded lymphatic vessels [43]. This has also been substantiated by in vitro studies demonstrating that breast tumour HS578T and MCF7 cells synthesize an HA pericellular coat and can adhere via HA to LYVE-1 in transfected fibroblasts [44]. Furthermore, preliminary work in the authors laboratory has revealed that the 4 T1 breast carcinoma line which can metastasize directly to the blood circulation from sentinel lymph nodes also assemble an HA glycocalyx, and can adhere and transit across primary lymphatic endothelial cells via LYVE-1 transmigratory cups. Moreover, when injected orthotopically into syngeneic mice, their capacity to invade local lymphatics and metastasize to draining inguinal lymph nodes is disrupted in *Lyve1*$^{-/-}$ mice (Johnson LA and Jackson DG unpublished). Whether this truly represents the same mechanism used by DCs and macrophages to enter initial lymphatics will require careful validation. However, independent observations using EM, that transgenic mouse prostate adenocarcinoma tumours, colon carcinoma cells and B16 melanomas extend filopodia-like protrusions towards the endothelium during lymphatic invasion, and that endothelial cells simultaneously extend transmigratory cup-like projections around the transiting tumour cells, add further credence to the notion [45]. It is also likely that the tumour cell HA glycocalyx plays a role in downstream invasion of lymph node subcapsular or medullary sinuses as a prelude to systemic metastasis (Pathology of Distant/Systemic Metastases: Turk, A.). Notably, both LYVE-1 and the endocytic HA receptor HARE are expressed in such sinuses, and administration of HARE blocking mAbs in an orthotopic mouse model of prostate cancer has been shown to dramatically reduce the metastatic burden in tumour draining para-aortic lymph nodes [46].

Finally, the secretion of small HA oligosaccharides by tumour cells as a consequence of the rapid HA turnover in neoplastic tissue may also aid lymphatic entry through effects on vessel permeability and growth. Specifically, it has been shown that binding of low molecular weight HA to LYVE-1 not only decreases endothelial barrier function in vitro, by transducing signals for the junctional disassembly of VE–cadherin, but also promotes lymphangiogenesis in vitro as well as in vivo, most likely through collaboration with tyrosine kinase linked growth factor receptors [47–49]. Indeed, in studies on a cohort of human colorectal carcinoma patients, an increase in the concentration of 12-50mer HA oligosaccharides in the tumour interstitium was associated with both peritumoral lymphatic invasion and lymph node metastasis [50].

22.7 Conclusion and Perspective

The present chapter has detailed how the extracellular matrix polysaccharide HA forms a glycocalyx on the surface of many immune cells, and how this structure mediates their entry to lymphatic capillaries by engaging the key HA receptor LYVE-1 in lymphatic endothelium for subsequent trafficking to downstream lymph nodes during immune surveillance and the resolution of tissue injury. In addition, it outlines how an equivalent glycocalyx on certain cancer cells may facilitate their invasion and metastasis via lymphatics, and how HA oligosaccharides generated by high turnover likely enhance such invasion by promoting peri-tumoral lymphangiogenesis and increased lymphatic permeability. These properties of HA and its novel binding interaction with LYVE-1 appear specifically adapted to the trafficking of both cell populations in the low shear environment around lymphatic vessels, and may even offer new targets for appropriate adjunct therapy of metastatic cancers.

Key Questions for Future Research

- What are the events subsequent to engagement of the HA glycocalyx that enable immune cells to enter lymphatic vessels, and which additional receptor and ligands are involved?
- Does the HA glycocalyx also mediate the transit of immune cells and tumour cells across lymph node sinuses?
- How widespread is HA glycocalyx assembly among different cancer types, and how closely does it correlate with metastatic proclivity?
- Could the HA glycocalyx have other functions during tumour growth and survival, e.g. by providing camouflage to protect against immune recognition?
- Could targeted inhibition of HA synthesis or LYVE-1 antibody blockade be exploited as an adjunct for metastatic cancer therapy?

References

1. Randolph GJ, Ivanov S, Zinselmeyer BH, Scallan JP. The lymphatic system: integral roles in immunity. Annu Rev Immunol. 2017;35:31–52.
2. Jackson DG. Leucocyte trafficking via the lymphatic vasculature—mechanisms and consequences. Front Immunol. 2019;10:471.
3. Jackson DG. Immunological functions of hyaluronan and its receptors in the lymphatics. Immunol Rev. 2009;230(1):216–31.

4. Johnson LA, Lawrance W, Roshorn YM, Hanke T, Banerji S, Jackson DG. The lymphatic vessel endothelial receptor LYVE-1 mediates dendritic cell entry to the afferent lymphatics via transmigratory cups that engage the leukocyte hyaluronan glycocalyx. Nat Immunol. 2017;18:762–70.
5. Jackson DG. Hyaluronan in the lymphatics: The key role of the hyaluronan receptor LYVE-1 in leucocyte trafficking. Matrix Biol. 2019;78–79:219–235.
6. Bromley SK, Thomas SY, Luster AD. Chemokine receptor CCR7 guides T cell exit from peripheral tissues and entry into afferent lymphatics. Nat Immunol. 2005;6(9):895–901.
7. Bromley SK, Yan S, Tomura M, Kanagawa O, Luster AD. Recirculating memory T cells are a unique subset of CD4+ T cells with a distinct phenotype and migratory pattern. J Immunol. 2013;190(3):970–76.
8. Forster R, Braun A, Worbs T. Lymph node homing of T cells and dendritic cells via afferent lymphatics. Trends Immunol. 2012;33(6):271–80.
9. Randolph GJ, Angeli V, Swartz MA. Dendritic-cell trafficking to lymph nodes through lymphatic vessels. Nat Rev Immunol. 2005;5(8):617–28.
10. Abadie V, Badell E, Douillard P, Ensergueix D, Leenen PJ, Tanguy M, et al. Neutrophils rapidly migrate via lymphatics after Mycobacterium bovis BCG intradermal vaccination and shuttle live bacilli to the draining lymph nodes. Blood. 2005;106(5):1843–50.
11. Hampton HR, Bailey J, Tomura M, Brink R, Chtanova T. Microbe-dependent lymphatic migration of neutrophils modulates lymphocyte proliferation in lymph nodes. Nat Commun. 2015;6:7139.
12. Pham TH, Baluk P, Xu Y, Grigorova I, Bankovich AJ, Pappu R, et al. Lymphatic endothelial cell sphingosine kinase activity is required for lymphocyte egress and lymphatic patterning. J Exp Med. 2010;207(1):17–27.
13. Baluk P, Fuxe J, Hashizume H, Romano T, Lashnits E, Butz S, et al. Functionally specialized junctions between endothelial cells of lymphatic vessels. J Exp Med. 2007;204(10):2349–62.
14. Zhang F, Zarkada G, Yi S, Eichmann A. Lymphatic endothelial cell junctions: molecular regulation in physiology and diseases. Front Physiol. 2020;11:509.
15. Banerji S, Ni J, Wang SX, Clasper S, Su J, Tammi R, et al. LYVE-1, a new homologue of the CD44 glycoprotein, is a lymph-specific receptor for hyaluronan. J Cell Biol. 1999;144(4):789–801.
16. Vaahtomeri K, Brown M, Hauschild R, De Vries I, Leithner AF, Mehling M, et al. Locally triggered release of the chemokine CCL21 promotes dendritic cell transmigration across lymphatic endothelia. Cell Rep. 2017;19(5):902–9.
17. Yao LC, Baluk P, Srinivasan RS, Oliver G, McDonald DM. Plasticity of button-like junctions in the endothelium of airway lymphatics in development and inflammation. Am J Pathol. 2012;180(6):2561–75.
18. Johnson LA, Clasper S, Holt AP, Lalor PF, Baban D, Jackson DG. An inflammation-induced mechanism for leukocyte transmigration across lymphatic vessel endothelium. J Exp Med. 2006;203(12):2763–77.
19. Monslow J, Govindaraju P, Pure E. Hyaluronan—a functional and structural sweet spot in the tissue microenvironment. Front Immunol. 2015;6:231.
20. Weigel PH, DeAngelis PL. Hyaluronan synthases: a decade-plus of novel glycosyltransferases. J Biol Chem. 2007;282(51):36777–81.
21. Armstrong SE, Bell DR. Relationship between lymph and tissue hyaluronan in skin and skeletal muscle. Am J Physiol Heart Circ Physiol. 2002;283(6):H2485–94.
22. Harris EN, Kyosseva SV, Weigel JA, Weigel PH. Expression, processing, and glycosaminoglycan binding activity of the recombinant human 315-kDa hyaluronic acid receptor for endocytosis (HARE). J Biol Chem. 2007;282(5):2785–97.
23. Jiang D, Liang J, Noble PW. Hyaluronan as an immune regulator in human diseases. Physiol Rev. 2011;91(1):221–64.
24. Termeer C, Benedix F, Sleeman J, Fieber C, Voith U, Ahrens T, et al. Oligosaccharides of Hyaluronan activate dendritic cells via toll-like receptor 4. J Exp Med. 2002;195(1):99–111.
25. Petrey AC, de la Motte CA. Hyaluronan, a crucial regulator of inflammation. Front Immunol. 2014;5:101.
26. Pure E, Cuff CA. A crucial role for CD44 in inflammation. Trends Mol Med. 2001;7(5):213–21.
27. Mummert ME, Mummert D, Edelbaum D, Hui F, Matsue H, Takashima A. Synthesis and surface expression of hyaluronan by dendritic cells and its potential role in antigen presentation. J Immunol. 2002;169(8):4322–31.
28. Johnson L, Banerji S, Lagerholm BC, Jackson DG. Dendritic cell entry to lymphatic capillaries is regulated by CD44 and the hyaluronan glycocalyx. Life Sci Alliance. 2021;4(5):e202000908.
29. Lawrance W, Banerji S, Day AJ, Bhattacharjee S, Jackson DG. Binding of Hyaluronan to the native lymphatic vessel endothelial receptor LYVE-1 is critically dependent on receptor clustering and Hyaluronan organization. J Biol Chem. 2016;291(15):8014–30.
30. Vieira JM, Norman S, Villa Del Campo C, Cahill TJ, Barnette DN, Gunadasa-Rohling M, et al. The cardiac lymphatic system stimulates resolution of inflammation following myocardial infarction. J Clin Invest. 2018;128(8):3402–12.
31. Mahaffey CL, Mummert ME. Hyaluronan synthesis is required for IL-2-mediated T cell proliferation. J Immunol. 2007;179(12):8191–99.
32. Brinkman CC, Iwami D, Hritzo MK, Xiong Y, Ahmad S, Simon T, et al. Treg engage lymphotoxin beta receptor for afferent lymphatic transendothelial migration. Nat Commun. 2016;7:12021.
33. Rigby DA, Ferguson DJ, Johnson LA, Jackson DG. Neutrophils rapidly transit inflamed lymphatic vessel endothelium via integrin-dependent proteolysis and lipoxin-induced junctional retraction. J Leukoc Biol. 2015;98:897–912.
34. Kerjaschki D, Bago-Horvath Z, Rudas M, Sexl V, Schneckenleithner C, Wolbank S, et al. Lipoxygenase mediates invasion of intrametastatic lymphatic vessels and propagates lymph node metastasis of human mammary carcinoma xenografts in mouse. J Clin Invest. 2011;121(5):2000–12.
35. Johnson LA, Prevo R, Clasper S, Jackson DG. Inflammation-induced uptake and degradation of the lymphatic endothelial hyaluronan receptor LYVE-1. J Biol Chem. 2007;282(46):33671–80.
36. Bano F, Banerji S, Howarth M, Jackson D, Richter R. A Single molecule assay to probe monovalent and multivalent bonds between hyaluronan and its key leukocyte receptor CD44 under force. Scientific Rep. 2016;6:34176.
37. Padera TP, Kadambi A, di Tomaso E, Carreira CM, Brown EB, Boucher Y, et al. Lymphatic metastasis in the absence of functional intratumor lymphatics. Science. 2018;296(5574):1883–6.
38. Brown M, Assen FP, Leithner A, Abe J, Schachner H, Asfour G, et al. Lymph node blood vessels provide exit routes for metastatic tumor cell dissemination in mice. Science. 2018;359(6382):1408–11.
39. Das S, Sarrou E, Podgrabinska S, Cassella M, Mungamuri SK, Feirt N, et al. Tumor cell entry into the lymph node is controlled by CCL1 chemokine expressed by lymph node lymphatic sinuses. J Exp Med. 2013;210(8):1509–28.
40. Liu M, Tolg C, Turley E. Dissecting the dual nature of Hyaluronan in the tumor microenvironment. Front Immunol. 2019;10:947.
41. Setala LP, Tammi MI, Tammi RH, Eskelinen MJ, Lipponen PK, Agren UM, et al. Hyaluronan expression in gastric cancer cells is associated with local and nodal spread and reduced survival rate. Br J Cancer. 1999;79(7–8):1133–38.
42. Ropponen K, Tammi M, Parkkinen J, Eskelinen M, Tammi R, Lipponen P, et al. Tumor cell-associated hyaluronan as an unfa-

vorable prognostic factor in colorectal cancer. Cancer Res. 1998;58(2):342–7.
43. Williams CS, Leek RD, Robson AM, Banerji S, Prevo R, Harris AL, et al. Absence of lymphangiogenesis and intratumoural lymph vessels in human metastatic breast cancer. J Pathol. 2003;200(2):195–206.
44. Du Y, Liu Y, Wang Y, He Y, Yang C, Gao F. LYVE-1 enhances the adhesion of HS-578T cells to COS-7 cells via hyaluronan. Clin Invest Med. 2011;34(1):E45–54.
45. Azzali G. On the transendothelial passage of tumor cell from extravasal matrix into the lumen of absorbing lymphatic vessel. Microvasc Res. 2006;72(1–2):74–85.
46. Simpson MA, Weigel JA, Weigel PH. Systemic blockade of the hyaluronan receptor for endocytosis prevents lymph node metastasis of prostate cancer. Int J Cancer. 2012;131(5):E836–40.
47. Wu M, Du Y, Liu Y, He Y, Yang C, Wang W, et al. Low molecular weight hyaluronan induces lymphangiogenesis through LYVE-1-mediated signaling pathways. PLoS One. 2014;9(3):e92857.
48. Bauer J, Rothley M, Schmaus A, Quagliata L, Ehret M, Biskup M, et al. TGFbeta counteracts LYVE-1-mediated induction of lymphangiogenesis by small hyaluronan oligosaccharides. J Mol Med. 2018;96(2):199–209.
49. Hou WH, Liu IH, Tsai CC, Johnson FE, Huang SS, Huang JS. CRSBP-1/LYVE-1 ligands disrupt lymphatic intercellular adhesion by inducing tyrosine phosphorylation and internalization of VE-cadherin. J Cell Sci. 2011;124(Pt 8):1231–44.
50. Schmaus A, Klusmeier S, Rothley M, Dimmler A, Sipos B, Faller G, et al. Accumulation of small hyaluronan oligosaccharides in tumour interstitial fluid correlates with lymphatic invasion and lymph node metastasis. Br J Cancer. 2014;111(3):559–67.

23 Lymphovascular Genomics and Proteomics, Clinical Syndromes, and Cancer Metastasis

Robert P. Erickson and Michael T. Dellinger

Abstract

The number of genetic mutations that cause lymphedema and other diseases of the lymphatic system has significantly grown in recent years. This information has shed light on the molecular mechanisms controlling the development of the lymphatic system and has increased our understanding of the etiology of inherited and non-inherited diseases of the lymphatic system. Here, we briefly describe the different genetic mutations that cause diseases of the lymphatic system, some of which are also mutated in some cancers. Additionally, we discuss the similarities between lymphatic diseases and cancer.

Learning Objectives

By the end of this chapter, readers will be able to:

- Name pathways that serve a critical role in the development of lymphatic vessels in health and disease.
- Name genes that control the development of lymphatic vessels.
- List several characteristics of different lymphatic diseases.
- Give examples of treatments for lymphatic diseases.

R. P. Erickson (✉)
Department of Pediatrics, University of Arizona College of Medicine, Tucson, AZ, USA
e-mail: erickson@peds.arizona.edu

M. T. Dellinger
Division of Surgical Oncology, Department of Surgery and the Hamon Center for Therapeutic Oncology Research, UT Southwestern Medical Center, Dallas, TX, USA
e-mail: michael.dellinger@utsouthwestern.edu

23.1 Introduction

Over the past 30 years, substantial progress has been made in the field of molecular lymphology. Molecular markers of lymphatics have been identified, genes that control lymphatic development have been uncovered, and treatments for lymphatic diseases have been discovered. Additionally, lymphatics have been shown to serve a critical role in the tumor microenvironment. Cancer cells can stimulate the growth of lymphatics and use these newly formed lymphatics to escape the primary tumor. The growth of lymphatics involves cell proliferation and migration, two tightly regulated processes that are dysregulated in cancer cells. Interestingly, the molecular pathways controlling these behaviors in cancer cells serve overlapping functions in lymphatic endothelial cells (LECs). There is also growing evidence that the genetic mutations that cause cancer can also cause rare lymphatic malformations. Thus, a better understanding of the molecular mechanisms underlying rare diseases of the lymphatic system could yield novel therapies to block tumor lymphangiogenesis and/or halt the progression of cancer. In this chapter, we briefly review the genetics of human lymphatic diseases, discuss the surprising similarities between some lymphatic diseases and cancer, and point to directions for studies of lymphatics in metastasis.

23.2 Early Genetic Discoveries

Lymphedema is abnormal swelling due to accumulated interstitial fluid that generally occurs in the arms or legs. The most common form of lymphedema is secondary lymphedema. In the Western world, this form of lymphedema is frequently observed in cancer patients that have had lymph nodes removed for the staging of their disease. Primary lymphedema occurs less frequently and is caused by inherited genetic mutations. Our evolving knowledge of genes in which mutations cause primary lymphedema has depended on syn-

S. P. Leong et al. (eds.), *Cancer Metastasis Through the Lymphovascular System*,
https://doi.org/10.1007/978-3-030-93084-4_23

ergistic findings in both man and mice. The first major discoveries were based on classical, human genetic approaches using linkage studies in large families to locate, and then clone, genes involved in lymphatic development and function. Mutations in *VEGFR3* (*FLT4*, the receptor for the major lymphatic growth factor, VEGFC) in Milroy's, *FOXC2* (a transcription factor important in the development of lymphatics) in lymphedema/distichiasis syndrome, and *SOX18* (another transcription factor) in hypotrichosis–lymphedema–telangiectasia syndrome are early examples [1–3].

23.3 Mutations Affecting Early Stages of Lymphatic Development

Multiple genes are involved in the early development of lymphatics. The discovery of prospero-related homeobox-1 (Prox1) provided the first marker for early lymphatic development [4]. It is required for the development and maintenance of LECs [5]. *CCBE1* (collagen and calcium-binding EGF domain-containing protein 1), *FAT4* (encoding an atypical cadherin), and *ADAMTS3* (encoding a metalloproteinase) also affect early lymphatic development and mutations in these genes cause Hennekam's syndrome, an autosomal recessive disorder characterized by lymphangiectasia/lymphedema [6]. Rare mutations in *VEGFC*, the major stimulator of lymphatic growth and development, can also cause lymphedema [7]. VEGF-C also stimulates tumor lymphangiogenesis and metastasis and it was recently reported that VEGF-C-mediated lymphatic drainage enables immunosurveillance of brain tumors [8]. VEGF-C activates VEGFR3, a receptor tyrosine kinase that promotes PI3K/AKT/mTOR and RAS/MAPK signaling in LECs. Importantly, these pathways have many components in which mutations are associated with disorders that exhibit lymphedema. They include Noonan's syndrome (*PTPN11, SOS1, RAF1, KRAS, BRAF, SHOC2,* and *CBL*), Costello syndrome (*HRAS*), cardio-facial-cutaneous syndrome (*BRAF*), lymphedema–choanal atresia syndrome (*PTPN14*), and rare cases of chylothorax (*RASA1*) [9].

23.4 Mutations Affecting Lymphatic Valve Formation

Lymphatic valves are bicuspid structures that prevent the retrograde flow of lymph. In general, mutations in genes that regulate lymphatic valve development cause lymphangiectasia and lymph reflux. Lymphatic valve development begins when clusters of LECs upregulate the transcription factors PROX1, FOXC2, and GATA2 [10]. These valve-forming cells elongate, reorient with respect to the longitudinal axis of the vessel, and then collectively invade the vessel to form a valve comprised of two layers of LECs. Importantly, mutations in *FOXC2* and *GATA2* have been shown to affect lymphatic valve development in mice and cause lymphedema in humans [9, 11, 12]. The planar cell polarity (PCP) gene *CELSR1* is also essential for normal lymphatic valve development. PCP refers to the coordinated orientation of cells in epithelia in the direction perpendicular to the apical-basal orientation. Loss of *Celsr1* in LECs in mice impairs lymphatic valve development and mutations in *CELSR1* have been identified in families with lymphedema [9, 13].

Connexins are gap junction proteins that are important for lymphatic development and three have been found to be expressed by lymphatic valves: Cx37, Cx43, and Cx47 [14]. In mice, knockouts of *Cx37* and *Cx43* disrupt lymphatic valve development and result in embryonic lymphedema and chylothorax with markedly reduced postnatal survival [15]. Ferrell and colleagues first reported *GJC2* (Cx47) mutations in two families with dominantly inherited lymphedema but "uncomplicated lymphedema of the leg or hand" was the only nearly constant feature and non-penetrance occurred in one individual [16]. Other families followed and one family with a mutation in *GJA1* (Cx43) was found [17]. Recently, the CRISPR/Cas9 system was used to create mice that harbor a Cx47 mutation found in humans. These mice showed a lymphatic phenotype of increased numbers of lymph nodes and lymphatic channels and decreased lymphatic contractility [18]. Genetic variation in *CX47* [19] and *VEGFR3* [20] has been implicated in susceptibility to post-breast cancer surgery lymphedema.

23.5 Mutations Affecting Flow Sensing and Signaling in Lymphatics

After the initiation of lymphangiogenesis, laminar fluid flow serves a function in expanding the growing lymphatic network. ORAI1 is a pore subunit of the calcium release–activated calcium channel that promotes LEC proliferation in response to laminar fluid flow [21]. Mechanistically, ORAI1 induces LEC proliferation by upregulating Krüppel-like factor (KLF)-2 and KLF4 in flow-activated LECs. These transcription factors upregulate *VEGF-A*, *VEGF-C*, and *FGFR3* and downregulate the cell cycle inhibitor, p57 [21]. PIEZO1 is also an ion channel whose activity is regulated by fluid flow. Interestingly, loss of *Piezo1* in mice impairs the development of lymphatic valves, and mutations in *PIEZO1* can cause lymphedema in humans [22]. Fluid shear stress signaling in LECs is also mediated by a mechanosensory complex comprised of PECAM-1 and VE-cadherin and loss of either gene impairs the development of lymphatic valves in mice [23, 24].

23.6 Polygenic Causes of Lymphedema

Several chromosomal disorders are associated with lymphedema. Turner's syndrome (45,XO) frequently has infantile generalized lymphedema (so much so that, before the chromosomal cause was known, it was considered a separate syndrome, Bonnevie-Ullrich) and in children/adults it can reoccur. While an X-chromosomal location has been found, a specific gene has not been pinpointed and it is likely that several loci's gene products contribute to the phenotype [25]. Similarly, trisomy 21 (and the rarer trisomies 13 and 18) is frequently associated with lymphedema and these chromosomal mutations are sometimes diagnosed in utero because of ultrasound-detected nuchal edema (posterior lateral neck swellings thought to be related to enlargement of the cervical lymphatic sacs). Again, multiple genes are likely to be involved. The not-infrequently diagnosed Prader–Willi syndrome, which involves abnormal imprinting of a portion of chromosome 15 caused by gene or chromosomal mutations, often shows lymphedema along with other syndromic features [26].

23.7 Additional Pathways Regulating Lymphangiogenesis

While the emphasis in the preceding section was on the VEGFC/VEGFR3 pathway, there are many other signaling pathways involved in lymphatic development and function. These pathways control LEC growth, proliferation, migration, adhesion, differentiation, and apoptosis. EphB4 is a receptor tyrosine kinase that is activated by the transmembrane ligand EphrinB2 [27]. Interestingly, engagement of EphB4 by EphrinB2 causes "reverse signaling" from the cytoplasmic tail of EphrinB2. This unusual form of signaling serves a critical role in the maturation of lymphatics in mice [27]. Additionally, EphB4 "forward signaling" has been shown to function in lymphatic valve development in mouse and man [28, 29]. Another pathway highly important for lymphatic development is the Angiopoietin-Tie pathway. Angiopoietin-2 (Angpt2) activates Tie2 on LECs and loss of either of the factor impairs lymphatic development in mice [30, 31]. Underscoring the importance of this pathway in lymphatic development is the recent observation that mutations in *ANGPT2* may cause primary lymphedema in humans [32].

23.8 Overview of Sporadic Lymphatic Anomalies

The previous sections focused on germline mutations and inherited diseases of the lymphatic system. The following sections focus on somatic mutations and sporadic lymphatic anomalies. Sporadic lymphatic anomalies are non-inherited diseases caused by errors in the development of the lymphatic system. According to the International Society for the Study of Vascular Anomalies (ISSVA), there are several different types of sporadic lymphatic anomalies [33]. These include common cystic lymphatic malformations (CCLMs), generalized lymphatic anomaly (GLA), central conducting lymphatic anomaly (CCLA), kaposiform lymphangiomatosis (KLA), and Gorham–Stout disease (GSD) [33]. Patients with these diseases can have tortuous dilated lymphatics, large disfiguring lymphatic malformations (LMs), multifocal LMs that infiltrate surrounding tissue, and ectopic lymphatics in bone [34]. The overlapping signs, symptoms, and complications of these diseases make them difficult to diagnose and the etiology of these diseases is poorly understood [34]. However, recent genetic studies have begun to shed light on the genetic basis of sporadic lymphatic anomalies. In the following sections, we briefly review the latest advances in research on the genetic underpinnings and treatment of sporadic lymphatic anomalies.

23.9 Common Cystic Lymphatic Malformations (CCLMs) and Generalized Lymphatic Anomaly (GLA)

CCLMs are isolated lesions comprised of irregular lymphatic vessels [35, 36]. Based on radiographic imaging, these malformations are classified as being either macrocystic, microcystic, or mixed [35, 36]. A majority of CCLMs are apparent at birth and affect the head and neck [35, 36]. Unfortunately, CCLMs can severely disfigure affected areas and impair normal organ function by obstructing or compressing structures [35, 36]. GLA (formerly called lymphangiomatosis) is a related disorder that is characterized by multiple diffuse or multifocal LMs [37]. Patients with GLA frequently display lymphatic abnormalities in their soft tissues and abdominal and thoracic viscera. Interestingly, they can also develop lymphatics in multiple noncontiguous bones [37]. Recently, somatic activating mutations in *PIK3CA* have been shown to cause both CCLM and GLA [35, 36, 38, 39]. *PIK3CA* encodes for p100α, which is the catalytic subunit of phosphatidylinositol 3-kinase (PI3K). The *PIK3CA* mutations in CCLM and GLA patients are "hotspot" mutations that occur in cancer and cause hyperactive PI3K/AKT/mTOR signaling [39]. Importantly, the FDA-approved mTOR inhibitor, sirolimus, has been reported to reduce lesion size in patients with CCLM and to reduce signs and symptoms of disease in patients with GLA [39–42].

23.10 Central Conducting Lymphatic Anomaly (CCLA)

CCLA (formerly called lymphangiectasia) is characterized by the pathologic dilation of lymphatic vessels [34]. Common complications of CCLA include chylothorax, chylous ascites, and protein-losing enteropathy [34]. Despite intense research efforts to identify effective therapies for CCLA, it remains an extremely challenging condition to treat. To gain a better understanding of the etiology of CCLA, affected tissues from two patients were analyzed by next-generation sequencing [43]. This revealed that both patients had a somatic mutation in *ARAF* (p.S214P) [43]. ARAF is a serine/threonine kinase and the p.S214P mutation increased MAPK signaling in LECs and caused aberrant lymphatic development in zebrafish [43]. Importantly, one patient with an *ARAF* (p.S214P) mutation was treated with trametinib (an FDA-approved MEK1/2 inhibitor) and their lung function and lymphedema significantly improved while on drug [43]. Whether trametinib is effective in other patients with CCLA remains to be determined.

23.11 Kaposiform Lymphangiomatosis (KLA)

KLA is an aggressive lymphatic anomaly that is typically diagnosed in children and has a 5-year survival rate of 51% [44]. Patients with KLA exhibit diffuse lymphatic abnormalities in multiple tissues with the mediastinum, lung, pleura, spleen, and skeleton being the most frequently affected sites [44]. Similar to GLA, bone involvement in KLA is associated with the loss of medullary bone, not cortical bone [44]. However, hemorrhagic pleural or pericardial effusions are more common in KLA than GLA [44]. Thrombocytopenia is also more common in KLA than GLA [44]. KLA is diagnosed based on clinical and histological observations. KLA has a unique histology with spindle-shaped LECs residing near abnormal lymphatics [44]. Several therapeutics have been used to treat KLA with the most common being vincristine, interferon-α, and sirolimus [44, 45]. However, these treatments are inadequate for some patients. A better understanding of the genetic causes of KLA could lead to new treatments for patients. To identify a disease-causing mutation for KLA, LECs were isolated from a KLA patient and analyzed by whole-exome sequencing [46]. This revealed that the cells harbored a "hotspot" mutation in *NRAS* (p.Q61R) [46]. The p.Q61R mutation occurs in cancer and causes hyperactive RAS/MAPK signaling. Interestingly, this same *NRAS* variant has been observed in affected tissues from 10 other KLA patients and in cell-free DNA from patients [47, 48]. However, not all patients with KLA have a mutation in *NRAS*. Recently, a causative mutation in *CBL* was identified in a patient with KLA [49]. Mutations in *CBL* also increase RAS/MAPK signaling and the KLA patient with the *CBL* mutation had a dramatic improvement in their disease following treatment with trametinib [49].

23.12 Gorham–Stout Disease (GSD)

GSD (also known as vanishing bone disease) is a rare disease that is characterized by the development of lymphatics in bone and progressive osteolysis. J.B.S. Jackson is credited with publishing the first case of GSD in 1838 [50]. However, the clinical and histologic features of the disease were not defined until Gorham and Stout published their seminal paper in 1955 [51]. Since 1955, more than 300 cases of GSD have been described in the literature [50]. These reports have revealed that GSD does not display clear sex predilection or inheritance pattern. Although GSD can involve any bone in the body, it typically affects the ribs and vertebrates [50]. Areas of bone resorption can arise in a single bone, but usually occur in multiple contiguous bones [50]. In severe cases of the disease, osteolysis continues until entire bones are resorbed and replaced by fibrous tissue [50]. A somatic activating *KRAS* mutation (p.Q61R) was recently identified in one patient with GSD [52]. The *KRAS* (p.Q61R) variant is a known "hotspot" mutation that frequently occurs in cancer and causes excessive RAS/MAPK signaling [52]. The identification of this mutation raises the possibility that inhibitors of the RAS/MAPK pathway could be novel therapies for GSD.

23.13 Precision Medicine for Sporadic Lymphatic Anomalies

Sporadic lymphatic anomalies are confusing and devastating diseases. If left untreated, patients suffer from a poor quality of life and can develop life-threatening complications. The reports described above show that sporadic lymphatic anomalies are caused by actionable mutations in oncogenes and respond to FDA-approved cancer therapies [39, 43, 49]. Therefore, supportive care should no longer be the standard of care for patients. Clinicians who encounter patients should attempt to obtain a molecular diagnosis because targeted therapies can produce miraculous results when administered to patients. Future clinical trials with molecularly diagnosed patients and targeted therapies could ultimately lead to the repurposing of FDA-approved cancer therapies for sporadic lymphatic anomalies.

23.14 Targeted Therapies to Block Tumor Lymphangiogenesis

Cancer patients are frequently treated with drugs that block either PI3K/AKT/mTOR or RAF/MEK/MAPK signaling because these pathways serve a critical role in tumorigenesis. These pathways, when active in LECs, also stimulate lymphangiogenesis (discussed in previous sections). Therefore, PI3K/AKT/mTOR and RAF/MEK/MAPK inhibitors could potentially be used to inhibit tumor lymphangiogenesis and metastasis to lymph nodes. Future investigations could reveal the effect of targeted therapies on tumor lymphangiogenesis and metastasis.

23.15 Additional Links between Lymphatics and Cancer

There are several additional links between lymphatics, lymphatic diseases, and cancer. As mentioned previously, cancer cells metastasize to lymph nodes by traveling through lymphatic vessels. Sporadic lymphatic anomaly patients can also have multifocal disease. Whether this is a result of LEC "metastasis" remains to be determined. Additionally, some secondary lymphedema patients go on to develop an aggressive cancer called lymphangiosarcoma. The mechanisms by which the altered microenvironment in lymphedematous limbs facilitate this transformation is poorly understood. Lastly, many of the genes that serve a role in the development of the lymphatic system function in processes that form the biological hallmarks of cancer. These processes include proliferative signaling, cell invasion, regulation of cellular energetics, and angiogenesis. Thus, advances in the study of lymphatic diseases could impact our understanding of the biological hallmarks of cancer and the treatment of cancer patients.

Conclusion and Perspective

Traditional genetic approaches and next-generation sequencing studies have led to the identification of germline and somatic mutations that cause diseases of the lymphatic system. This information has shed light on the molecular mechanisms controlling the development of the lymphatic system and has increased our understanding of the etiology of these disabling, disfiguring, and life-threatening diseases. Mutations occurring in sporadic lymphatic abnormalities also occur in some cancers and the lymphatic system is highly implicated in tumor spread. Although substantial progress has been made in the field, many questions still exist surrounding the development of lymphatic vessels in health and disease.

- What is the full complement of mutations that cause lymphatic disease in humans?
- Are there tissue-specific differences in the development and function of lymphatics?
- Can molecular-based therapies be used to treat primary lymphedema?
- Why do mutations in oncogenes in LECs cause sporadic lymphatic anomalies instead of cancer?
- Can synthetically lethal genetic interactions be exploited to develop novel therapies for sporadic lymphatic anomalies?
- Do targeted therapies inhibit lymphangiogenesis in pathological settings (e.g., tumor lymphangiogenesis in cancer)?

Importantly, sustained research efforts will allow scientists to fill these gaps in knowledge and increase the ability of clinicians to treat patients suffering from primary lymphedema or a sporadic lymphatic anomaly. They could also result in new treatments to block the spread and growth of cancer.

References

1. Fang J, Dagenais SL, Erickson RP, Arlt MF, Glynn MW, Gorski JL, Seaver LH, Glover TW. Mutations in FOXC2 (MFH-1), a forkhead family transcription factor, are responsible for the hereditary lymphedema-distichiasis syndrome. Am J Hum Genet. 2000;67(6):1382–8. https://doi.org/10.1086/316915.
2. Ferrell RE, Levinson KL, Esman JH, Kimak MA, Lawrence EC, Barmada MM, Finegold DN. Hereditary lymphedema: evidence for linkage and genetic heterogeneity. Hum Mol Genet. 1998;7(13):2073–8.
3. Irrthum A, Devriendt K, Chitayat D, Matthijs G, Glade C, Steijlen PM, Fryns JP, Van Steensel MA, Vikkula M. Mutations in the transcription factor gene SOX18 underlie recessive and dominant forms of hypotrichosis-lymphedema-telangiectasia. Am J Hum Genet. 2003;72(6):1470–8. https://doi.org/10.1086/375614.
4. Wigle JT, Oliver G. Prox1 function is required for the development of the murine lymphatic system. Cell. 1999;98(6):769–78. doi:S0092-8674(00)81511-1 [pii]
5. Johnson NC, Dillard ME, Baluk P, McDonald DM, Harvey NL, Frase SL, Oliver G. Lymphatic endothelial cell identity is reversible and its maintenance requires Prox1 activity. Genes Dev. 2008;22(23):3282–91. https://doi.org/10.1101/gad.1727208.
6. Scheuerle AE, Sweed NT, Timmons CF, Smith ED, Alcaraz WA, Shinde DN. An additional case of Hennekam lymphangiectasia-lymphedema syndrome caused by loss-of-function mutation in ADAMTS3. Am J Med Genet A. 2018;176(12):2858–61. https://doi.org/10.1002/ajmg.a.40633.
7. Gordon K, Schulte D, Brice G, Simpson MA, Roukens MG, van Impel A, Connell F, Kalidas K, Jeffery S, Mortimer PS, Mansour S, Schulte-Merker S, Ostergaard P. Mutation in vascular endothelial growth factor-C, a ligand for vascular endothelial growth factor receptor-3, is associated with autosomal dominant Milroy-like primary lymphedema. Circ Res. 2013;112(6):956–60. https://doi.org/10.1161/CIRCRESAHA.113.300350.

8. Song E, Mao T, Dong H, Boisserand LSB, Antila S, Bosenberg M, Alitalo K, Thomas JL, Iwasaki A. VEGF-C-driven lymphatic drainage enables immunosurveillance of brain tumours. Nature. 2020;577(7792):689–94. https://doi.org/10.1038/s41586-019-1912-x.
9. Michelini S, Paolacci S, Manara E, Eretta C, Mattassi R, Lee BB, Bertelli M. Genetic tests in lymphatic vascular malformations and lymphedema. J Med Genet. 2018;55(4):222–32. https://doi.org/10.1136/jmedgenet-2017-105064.
10. Sabine A, Saygili Demir C, Petrova TV. Endothelial cell responses to biomechanical forces in lymphatic vessels. Antioxid Redox Signal. 2016;25(7):451–65. https://doi.org/10.1089/ars.2016.6685.
11. Kazenwadel J, Betterman KL, Chong CE, Stokes PH, Lee YK, Secker GA, Agalarov Y, Demir CS, Lawrence DM, Sutton DL, Tabruyn SP, Miura N, Salminen M, Petrova TV, Matthews JM, Hahn CN, Scott HS, Harvey NL. GATA2 is required for lymphatic vessel valve development and maintenance. J Clin Invest. 2015;125(8):2979–94. https://doi.org/10.1172/JCI78888.
12. Petrova TV, Karpanen T, Norrmen C, Mellor R, Tamakoshi T, Finegold D, Ferrell R, Kerjaschki D, Mortimer P, Yla-Herttuala S, Miura N, Alitalo K. Defective valves and abnormal mural cell recruitment underlie lymphatic vascular failure in lymphedema distichiasis. Nat Med. 2004;10(9):974–81. https://doi.org/10.1038/nm1094nm1094.
13. Erickson RP, Lai LW, Mustacich DJ, Bernas MJ, Kuo PH, Witte MH. Sex-limited penetrance of lymphedema to females with CELSR1 haploinsufficiency: a second family. Clin Genet. 2019;96(5):478–82. https://doi.org/10.1111/cge.13622.
14. Kanady JD, Simon AM. Lymphatic communication: connexin junction, what's your function? Lymphology. 2011;44(3):95–102.
15. Kanady JD, Dellinger MT, Munger SJ, Witte MH, Simon AM. Connexin37 and Connexin43 deficiencies in mice disrupt lymphatic valve development and result in lymphatic disorders including lymphedema and chylothorax. Dev Biol. 2011;354(2):253–66. https://doi.org/10.1016/j.ydbio.2011.04.004.
16. Ferrell RE, Baty CJ, Kimak MA, Karlsson JM, Lawrence EC, Franke-Snyder M, Meriney SD, Feingold E, Finegold DN. GJC2 missense mutations cause human lymphedema. Am J Hum Genet. 2010;86(6):943–8. https://doi.org/10.1016/j.ajhg.2010.04.010.
17. Brice G, Ostergaard P, Jeffery S, Gordon K, Mortimer PS, Mansour S. A novel mutation in GJA1 causing oculodentodigital syndrome and primary lymphoedema in a three generation family. Clin Genet. 2013;84(4):378–81. https://doi.org/10.1111/cge.12158.
18. Mustacich DJ, Kylat RI, Bernas MJ, Myles RJ, Jones JA, Kanady JD, Simon AM, Georgieva TG, Witte MH, Erickson RP, Pires PW. Abnormal lymphatic phenotype in a crispr mouse model of the human lymphedema-causing connexin47 R260c point mutation. Lymphology. 2021;54:78–91.
19. Finegold DN, Baty CJ, Knickelbein KZ, Perschke S, Noon SE, Campbell D, Karlsson JM, Huang D, Kimak MA, Lawrence EC, Feingold E, Meriney SD, Brufsky AM, Ferrell RE. Connexin 47 mutations increase risk for secondary lymphedema following breast cancer treatment. Clin Cancer Res. 2012;18(8):2382–90. https://doi.org/10.1158/1078-0432.CCR-11-2303.
20. Newman B, Lose F, Kedda MA, Francois M, Ferguson K, Janda M, Yates P, Spurdle AB, Hayes SC. Possible genetic predisposition to lymphedema after breast cancer. Lymphat Res Biol. 2012;10(1):2–13. https://doi.org/10.1089/lrb.2011.0024.
21. Choi D, Park E, Jung E, Seong YJ, Hong M, Lee S, Burford J, Gyarmati G, Peti-Peterdi J, Srikanth S, Gwack Y, Koh CJ, Boriushkin E, Hamik A, Wong AK, Hong YK. ORAI1 activates proliferation of lymphatic endothelial cells in response to laminar flow through Kruppel-like factors 2 and 4. Circ Res. 2017;120(9):1426–39. https://doi.org/10.1161/CIRCRESAHA.116.309548.
22. Nonomura K, Lukacs V, Sweet DT, Goddard LM, Kanie A, Whitwam T, Ranade SS, Fujimori T, Kahn ML, Patapoutian A. Mechanically activated ion channel PIEZO1 is required for lymphatic valve formation. Proc Natl Acad Sci U S A. 2018;115(50):12817–22. https://doi.org/10.1073/pnas.1817070115.
23. Wang Y, Baeyens N, Corti F, Tanaka K, Fang JS, Zhang J, Jin Y, Coon B, Hirschi KK, Schwartz MA, Simons M. Syndecan 4 controls lymphatic vasculature remodeling during mouse embryonic development. Development. 2016;143(23):4441–51. https://doi.org/10.1242/dev.140129.
24. Yang Y, Cha B, Motawe ZY, Srinivasan RS, Scallan JP. VE-cadherin is required for lymphatic valve formation and maintenance. Cell Rep. 2019;28(9):2397–412. e2394. https://doi.org/10.1016/j.celrep.2019.07.072.
25. Boucher CA, Sargent CA, Ogata T, Affara NA. Breakpoint analysis of turner patients with partial Xp deletions: implications for the lymphoedema gene location. J Med Genet. 2001;38(9):591–8. https://doi.org/10.1136/jmg.38.9.591.
26. Lee BB, Andrade M, Antignani PL, Boccardo F, Bunke N, Campisi C, Damstra R, Flour M, Forner-Cordero I, Gloviczki P, Laredo J, Partsch H, Piller N, Michelini S, Mortimer P, Rabe E, Rockson S, Scuderi A, Szolnoky G, Villavicencio JL, International Union of P (2013) Diagnosis and treatment of primary lymphedema. Consensus document of the International Union of Phlebology (IUP)-2013. Int Angiol 32 (6):541–574.
27. Makinen T, Adams RH, Bailey J, Lu Q, Ziemiecki A, Alitalo K, Klein R, Wilkinson GA. PDZ interaction site in ephrinB2 is required for the remodeling of lymphatic vasculature. Genes Dev. 2005;19(3):397–410. https://doi.org/10.1101/gad.330105.
28. Li D, Wenger TL, Seiler C, March ME, Gutierrez-Uzquiza A, Kao C, Bhoj E, Tian L, Rosenbach M, Liu Y, Robinson N, Behr M, Chiavacci R, Hou C, Wang T, Bakay M, Pellegrino da Silva R, Perkins JA, Sleiman P, Levine MA, Hicks PJ, Itkin M, Dori Y, Hakonarson H. Pathogenic variant in EPHB4 results in central conducting lymphatic anomaly. Hum Mol Genet. 2018;27(18):3233–45. https://doi.org/10.1093/hmg/ddy218.
29. Martin-Almedina S, Martinez-Corral I, Holdhus R, Vicente A, Fotiou E, Lin S, Petersen K, Simpson MA, Hoischen A, Gilissen C, Jeffery H, Atton G, Karapouliou C, Brice G, Gordon K, Wiseman JW, Wedin M, Rockson SG, Jeffery S, Mortimer PS, Snyder MP, Berland S, Mansour S, Makinen T, Ostergaard P. EPHB4 kinase-inactivating mutations cause autosomal dominant lymphatic-related hydrops fetalis. J Clin Invest. 2016;126(8):3080–8. https://doi.org/10.1172/JCI85794.
30. Dellinger M, Hunter R, Bernas M, Gale N, Yancopoulos G, Erickson R, Witte M. Defective remodeling and maturation of the lymphatic vasculature in Angiopoietin-2 deficient mice. Dev Biol. 2008;319(2):309–20. https://doi.org/10.1016/j.ydbio.2008.04.024.
31. Souma T, Thomson BR, Heinen S, Carota IA, Yamaguchi S, Onay T, Liu P, Ghosh AK, Li C, Eremina V, Hong YK, Economides AN, Vestweber D, Peters KG, Jin J, Quaggin SE. Context-dependent functions of angiopoietin 2 are determined by the endothelial phosphatase VEPTP. Proc Natl Acad Sci U S A. 2018;115(6):1298–303. https://doi.org/10.1073/pnas.1714446115.
32. Leppanen VM, Brouillard P, Korhonen EA, Sipila T, Jha SK, Revencu N, Labarque V, Fastre E, Schlogel M, Ravoet M, Singer A, Luzzatto C, Angelone D, Crichiutti G, D'Elia A, Kuurne J, Elamaa H, Koh GY, Saharinen P, Vikkula M, Alitalo K. Characterization of ANGPT2 mutations associated with primary lymphedema. Sci Transl Med. 2020;12(560) https://doi.org/10.1126/scitranslmed.aax8013.
33. Wassef M, Blei F, Adams D, Alomari A, Baselga E, Berenstein A, Burrows P, Frieden IJ, Garzon MC, Lopez-Gutierrez JC, Lord DJ, Mitchel S, Powell J, Prendiville J, Vikkula M, Committee. IBaS. Vascular anomalies classification: recommendations from the International Society for the Study of vascular anomalies. Pediatrics. 2015;136(1):e203–14. https://doi.org/10.1542/peds.2014-3673.

34. Trenor CC 3rd, Chaudry G. Complex lymphatic anomalies. Semin Pediatr Surg. 2014;23(4):186–90. https://doi.org/10.1053/j.sempedsurg.2014.07.006.
35. Boscolo E, Coma S, Luks VL, Greene AK, Klagsbrun M, Warman ML, Bischoff J. AKT hyper-phosphorylation associated with PI3K mutations in lymphatic endothelial cells from a patient with lymphatic malformation. Angiogenesis. 2015;18(2):151–62. https://doi.org/10.1007/s10456-014-9453-2.
36. Osborn AJ, Dickie P, Neilson DE, Glaser K, Lynch KA, Gupta A, Dickie BH. Activating PIK3CA alleles and lymphangiogenic phenotype of lymphatic endothelial cells isolated from lymphatic malformations. Hum Mol Genet. 2015;24(4):926–38. https://doi.org/10.1093/hmg/ddu505.
37. Lala S, Mulliken JB, Alomari AI, Fishman SJ, Kozakewich HP, Chaudry G. Gorham-Stout disease and generalized lymphatic anomaly-clinical, radiologic, and histologic differentiation. Skelet Radiol. 2013;42:917–24. https://doi.org/10.1007/s00256-012-1565-4.
38. Luks VL, Kamitaki N, Vivero MP, Uller W, Rab R, Bovee JV, Rialon KL, Guevara CJ, Alomari AI, Greene AK, Fishman SJ, Kozakewich HP, Maclellan RA, Mulliken JB, Rahbar R, Spencer SA, Trenor CC 3rd, Upton J, Zurakowski D, Perkins JA, Kirsh A, Bennett JT, Dobyns WB, Kurek KC, Warman ML, SA MC, Murillo R. Lymphatic and other vascular malformative/overgrowth disorders are caused by somatic mutations in PIK3CA. J Pediatrics. 2015;166(4):1048–54. e1041–1045. https://doi.org/10.1016/j.jpeds.2014.12.069.
39. Rodriguez-Laguna L, Agra N, Ibanez K, Oliva-Molina G, Gordo G, Khurana N, Hominick D, Beato M, Colmenero I, Herranz G, Torres Canizalez JM, Rodriguez Pena R, Vallespin E, Martin-Arenas R, Del Pozo A, Villaverde C, Bustamante A, Ayuso C, Lapunzina P, Lopez-Gutierrez JC, Dellinger MT, Martinez-Glez V. Somatic activating mutations in PIK3CA cause generalized lymphatic anomaly. J Exp Med. 2019;216(2):407–18. https://doi.org/10.1084/jem.20181353.
40. Adams DM, Trenor CC 3rd, Hammill AM, Vinks AA, Patel MN, Chaudry G, Wentzel MS, Mobberley-Schuman PS, Campbell LM, Brookbank C, Gupta A, Chute C, Eile J, McKenna J, Merrow AC, Fei L, Hornung L, Seid M, Dasgupta AR, Dickie BH, Elluru RG, Lucky AW, Weiss B, Azizkhan RG. Efficacy and safety of Sirolimus in the treatment of complicated vascular anomalies. Pediatrics. 2016;137(2):1–10. https://doi.org/10.1542/peds.2015-3257.
41. Ricci KW, Hammill AM, Mobberley-Schuman P, Nelson SC, Blatt J, Bender JLG, McCuaig CC, Synakiewicz A, Frieden IJ, Adams DM. Efficacy of systemic sirolimus in the treatment of generalized lymphatic anomaly and Gorham-Stout disease. Pediatr Blood Cancer. 2019;66(5):e27614. https://doi.org/10.1002/pbc.27614.
42. Strychowsky JE, Rahbar R, O'Hare MJ, Irace AL, Padua H, Trenor CC 3rd. Sirolimus as treatment for 19 patients with refractory cervicofacial lymphatic malformation. Laryngoscope. 2018;128(1):269–76. https://doi.org/10.1002/lary.26780.
43. Li D, March ME, Gutierrez-Uzquiza A, Kao C, Seiler C, Pinto E, Matsuoka LS, Battig MR, Bhoj EJ, Wenger TL, Tian L, Robinson N, Wang T, Liu Y, Weinstein BM, Swift M, Jung HM, Kaminski CN, Chiavacci R, Perkins JA, Levine MA, Sleiman PMA, Hicks PJ, Strausbaugh JT, Belasco JB, Dori Y, Hakonarson H. ARAF recurrent mutation causes central conducting lymphatic anomaly treatable with a MEK inhibitor. Nat Med. 2019;25(7):1116–22. https://doi.org/10.1038/s41591-019-0479-2.
44. Croteau SE, Kozakewich HP, Perez-Atayde AR, Fishman SJ, Alomari AI, Chaudry G, Mulliken JB, Trenor CC 3rd. Kaposiform lymphangiomatosis: a distinct aggressive lymphatic anomaly. J Pediatr. 2014;164(2):383–8. https://doi.org/10.1016/j.jpeds.2013.10.013.
45. Ozeki M, Fujino A, Matsuoka K, Nosaka S, Kuroda T, Fukao T. Clinical features and prognosis of generalized lymphatic anomaly, Kaposiform Lymphangiomatosis, and Gorham-Stout disease. Pediatr Blood Cancer. 2016;63(5):832–8. https://doi.org/10.1002/pbc.25914.
46. Manevitz-Mendelson E, Leichner GS, Barel O, Davidi-Avrahami I, Ziv-Strasser L, Eyal E, Pessach I, Rimon U, Barzilai A, Hirshberg A, Chechekes K, Amariglio N, Rechavi G, Yaniv K, Greenberger S. Somatic NRAS mutation in patient with generalized lymphatic anomaly. Angiogenesis. 2018;21(2):287–98. https://doi.org/10.1007/s10456-018-9595-8.
47. Barclay SF, Inman KW, Luks VL, McIntyre JB, Al-Ibraheemi A, Church AJ, Perez-Atayde AR, Mangray S, Jeng M, Kreimer SR, Walker L, Fishman SJ, Alomari AI, Chaudry G, Trenor Iii CC, Adams D, Kozakewich HPW, Kurek KC. A somatic activating NRAS variant associated with kaposiform lymphangiomatosis. Genet Med. 2019;21(7):1517–24. https://doi.org/10.1038/s41436-018-0390-0.
48. Ozeki M, Aoki Y, Nozawa A, Yasue S, Endo S, Hori Y, Matsuoka K, Niihori T, Funayama R, Shirota M, Nakayama K, Fukao T. Detection of NRAS mutation in cell-free DNA biological fluids from patients with kaposiform lymphangiomatosis. Orphanet J Rare Dis. 2019;14(1):215. https://doi.org/10.1186/s13023-019-1191-5.
49. Foster JB, Li D, March ME, Sheppard SE, Adams DM, Hakonarson H, Dori Y. Kaposiform lymphangiomatosis effectively treated with MEK inhibition. EMBO Mol Med. 2020:e12324. https://doi.org/10.15252/emmm.202012324.
50. Dellinger MT, Garg N, Olsen BR. Viewpoints on vessels and vanishing bones in Gorham-Stout disease. Bone. 2014;63:47–52. https://doi.org/10.1016/j.bone.2014.02.011.
51. Gorham LW, Stout AP. Massive osteolysis (acute spontaneous absorption of bone, phantom bone, disappearing bone); its relation to hemangiomatosis. J Bone Joint Surg Am. 1955;37-A(5):985–1004.
52. Nozawa A, Ozeki M, Niihori T, Suzui N, Miyazaki T, Aoki Y. A somatic activating KRAS variant identified in an affected lesion of a patient with Gorham-Stout disease. J Hum Genet. 2020; https://doi.org/10.1038/s10038-020-0794-y.

Imaging of the Lymphatic System with Relevance to Cancer and Cancer Metastasis

24

Russell S. Witte and Michael Bernas

Abstract

Imaging plays a crucial role in the fundamental understanding of lymphatic system structure and function throughout the body both in health and a wide variety of clinical disorders. It also has application to identify primary and metastatic cancer spread to regional nodes/organs as well as its importance in cancer assessment/treatment planning using the sentinel node protocol. There are multiple different techniques utilizing associated tracers for in vivo dynamic imaging in both humans and experimental models, and the future continues to offer advancements including theranostics. Each technique has strengths and weaknesses for specific applications. Individual techniques (and sometimes in combination) produce images for the diagnosis, treatment monitoring, and even to develop new treatment strategies for cancer and cancer-related sequelae.

Learning Objectives

- Describe more than five techniques currently used to image the lymphatic system.
- List at least one advantage and one disadvantage to each of the techniques.
- Delineate at least four applications of lymphatic imaging in the management of patients with metastatic cancer.

R. S. Witte (✉)
Department of Medical Imaging, University of Arizona, Tucson, AZ, USA
e-mail: rwitte@arizona.edu

M. Bernas
Department of Medical Education, TCU School of Medicine, Fort Worth, TX, USA
e-mail: m.bernas@tcu.edu

24.1 Introduction

There are many techniques currently in use for imaging the lymphatic system, and they all focus on producing images to help understand the structure and function of the lymphatic system in health and disease. These various imaging techniques and technologies have unique strengths and weaknesses, including their application in the setting of cancer metastasis.

Although in some cases lymphatic vessels and nodes can be visualized without contrast, most techniques require contrast agents of relatively large molecules (the size of albumin and greater), which are selectively taken up by lymphatic capillaries rather than blood vessels when deposited intradermally, subcutaneously, interstitially, directly into lymph nodes, or by image-guided direct puncture of large channels. Tracers have different and unique characteristics depending on the modality utilized and the individual energy sources (X-ray, gamma ray, near infrared, magnetic, and ultrasound) as they produce images of the lymphatic system at different spatial resolutions, fields-of-view, and penetration depths. Images depict both functional and dynamic flow through the lymphatics as well as the anatomical structure of nodes and channels. The most common imaging techniques and applications are described below.

24.2 Modalities

24.2.1 X-Ray or CT Lymphography (LAG)

24.2.1.1 Technique Summary

Classical LAG is performed by first identifying a pedal lymphatic through the use of blue dye and then cannulation for instilling the contrast agent [1, 2]. The agent (ethiodized oil or injecting iodized water-soluble solution [300 mg iodine/mL]) is injected at a very slow rate (6 mL/h) in order to not cause rupture of the fine peripheral lymphatics. Patients must

S. P. Leong et al. (eds.), *Cancer Metastasis Through the Lymphovascular System*,
https://doi.org/10.1007/978-3-030-93084-4_24

remain relatively immobile for 60–90 min. A drawback of this procedure is the invasive cut down needed to cannulate the lymphatic vessels as well as the time for patient immobilization. Whereas early studies required X-ray systems for imaging, CT has become more widely used due to its 3D capability particularly for those cases involving chylous leakage. An obvious advantage of this technique is that both X-ray and CT systems are widely available throughout the world. However, there require exposure to ionizing radiation. LAG can also be performed by an intranodal injection of the contrast agent for access to the deeper lymphatic structures of the cisterna chyli and thoracic duct [3, 4].

24.2.1.2 Imaging Summary

From its discovery in the 1950s conventional bipedal oily contrast direct lymphography (LAG) was widely used and considered safe for staging cancers particularly lymphomas to determine lymph nodal involvement prior to surgery or chemotherapy, to deliver endolymphatic chemotherapy, and also to delineate other nonneoplastic lymphadenopathic disorders [2]. LAG produces images of fine anatomical detail (Fig. 24.1). Although these images are static, real-time fluoroscopy can demonstrate the function of the lymphatic system. Intranodal LAG is particularly invaluable for precise anatomic depiction of the lymphatic vessels and nodes in cases of chylous or non-chylous reflux where a surgeon or interventional radiologist can subsequently perform the treatment to inhibit leakage. Recent work using intranodal injection has clearly demonstrated the thoracic duct including identification of areas of leakage [3] (Fig. 24.1) and has greatly impacted the field of lymphatic imaging and interventions [5]. Using US guidance, the placement of a needle directly into a node allows injection of the oil-based iodinated contrast. These advances have been adopted worldwide and helped bring lymphatic interventional treatment procedures to the forefront [6]. A drawback is that the oil-based contrast agent has increased viscosity which can inhibit clear visualization of a leakage.

24.2.2 Lymphoscintigraphy (LAS)

24.2.2.1 Technique Summary

LAS is a primary imaging modality to study and quantify lymphatic drainage and flow to help diagnose and stage lymphatic disorders, as well as to assess and monitor treatment. This minimally invasive technique uses a single small (0.05 mL) injection into each hand or foot. The standard radiotracer (in the United States) is Tc99m filtered sulfur colloid with particle sizes of approximately 50 nm.

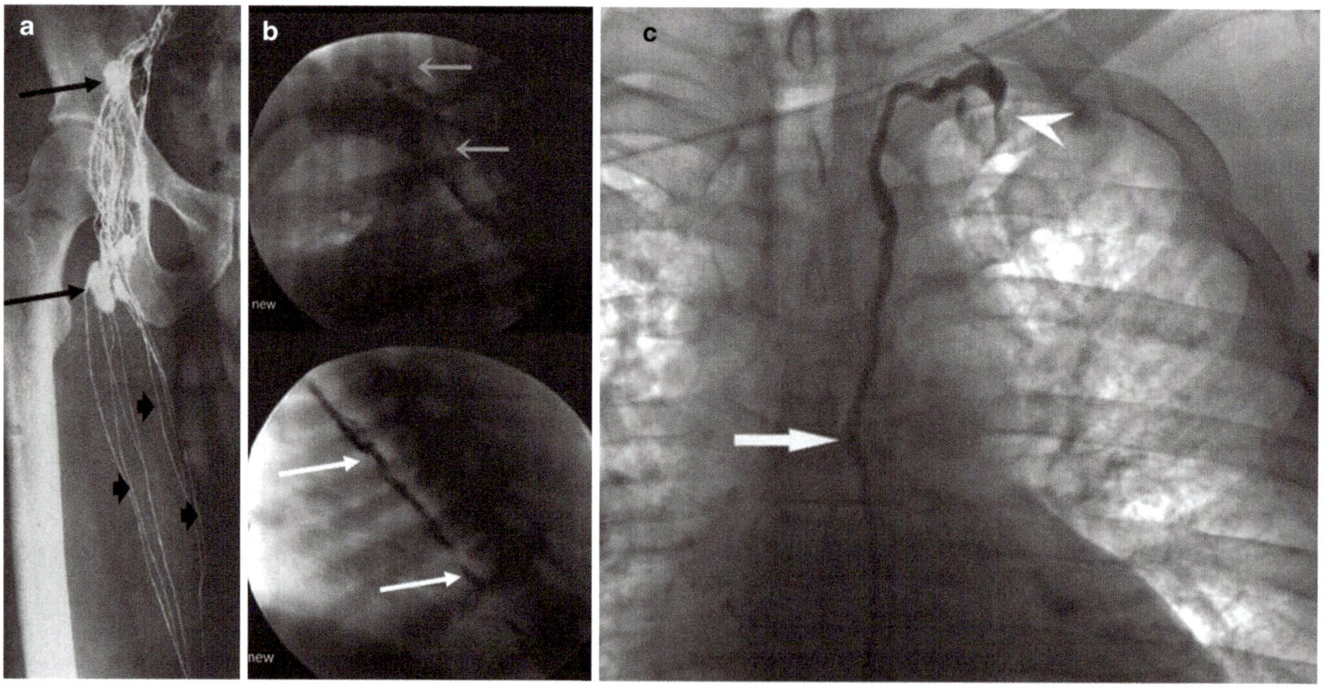

Fig. 24.1 Lymphography. Classical pedal lymphography (**a** and **b**) in a man (**a**) clearly displaying normal lymphatic imaging in the upper thigh with discrete lymphatic vessels (arrowheads) and lymph nodes (arrows). Lymphography (captured by fluoroscopy) in experimental sheep imaging demonstrating thoracic duct in the upper chest (upper image arrows) as well as in the abdomen (lower image arrows). In the lower image, "packets" of lymph are clearly demonstrated depicting the noncontinuous column of lymphatic fluid in normal flow. (**c**) Fluoroscopic image of the thoracic duct (arrow) following injection intranodal injection of the contrast. The injected contrast is seen being transported into the subclavian vein at thoracic duct to subclavian vein junction (arrowhead). (**a** modified from [2], **c** modified from [3])

Technetium has a half-life of only 6 h, making it available to almost any patient with few restrictions. It also has an excellent safety profile (total radiation dose is approximately 1/20th of a chest X-ray), readily available, and relatively inexpensive. A drawback concerning LAS is that it delivers poor anatomical resolution that can limit its role in both diagnostics and pre-treatment procedural planning. A gamma camera is utilized for imaging and it can visualize the functional lymphatic system from the distal extremities to the deep thoracic duct. Most systems can now produce a single "whole body" image following bilateral simultaneous injections (Fig. 24.2). A dual-head camera is typically used to acquire early static and delayed (~3 h) dynamic images.

24.2.2.2 Imaging Summary

LAS is the current gold standard for imaging of the lymphatic system as recommended by the Consensus Document [10]. The technique produces images for visualization of site(s) of obstruction (no or reduced flow), backflow, and leakage which are valuable for determining diagnosis and staging as well as helping to delineate potential approaches for future treatment in patients with complications such as chylous ascites or chylothorax (either due to situations involving cancer or not) [7]. See Ref. [11] for an extensive review of LAS imaging depicting multiple patients with various lymphatic abnormalities and clinical presentations as well as their corresponding LAS images. The main limitation of standard LAS is poor spatial resolution and lack of anatomical landmarks. SPECT/CT (below) can be added to provide such landmarks and regional localization of lymphatic transport of the tracer. There have been investigators who have proposed quantitative evaluation algorithms for LAS images [12, 13] as well as classification systems for lymphedema based on LAS imaging [14].

SPECT/CT addresses limitations with lymphoscintigraphy, which produces only planar images without clear anatomical landmarks. SPECT/CT overcomes this limitation by taking advantage of state-of-the-art gamma cameras combined with CT tomography [8, 15]. Cameras are placed at different angles to produce multiple 2D images and a tomo-

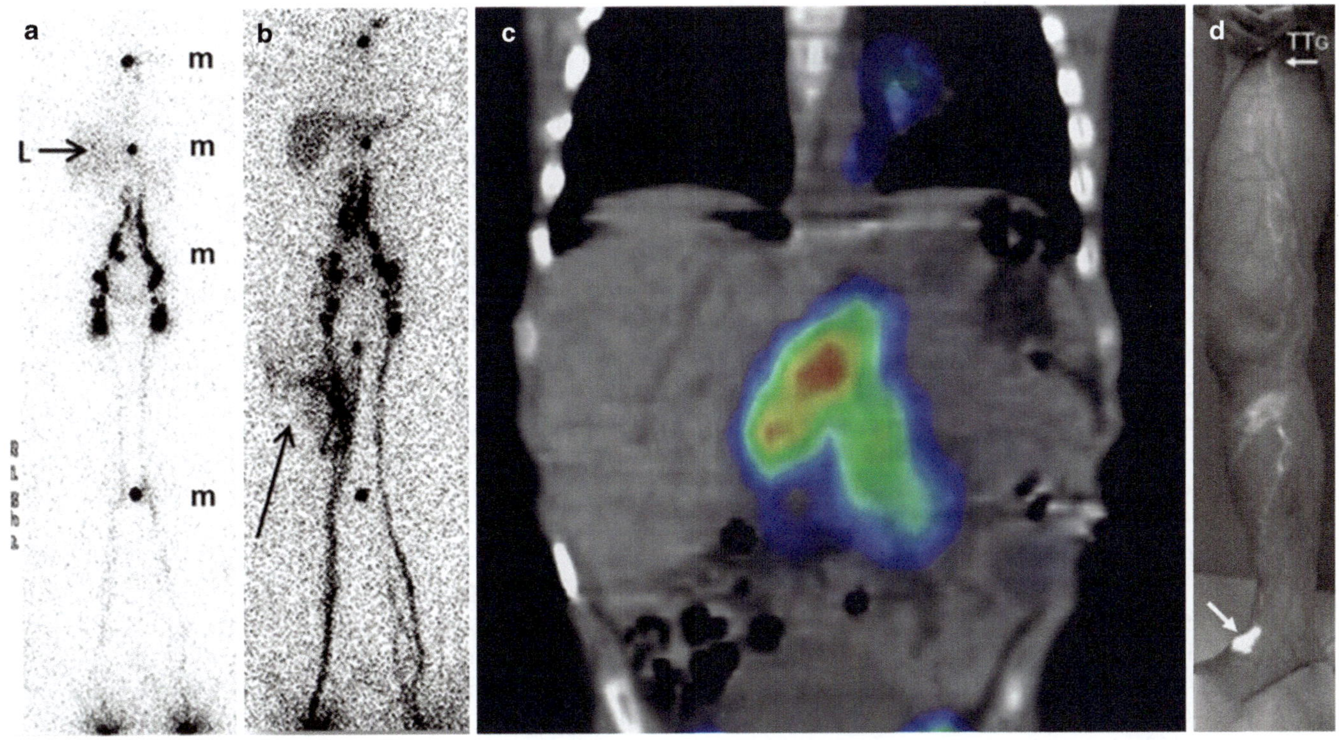

Fig. 24.2 (**a**, **b**) Lymphscintigraphy. (**a**) 2-D whole body image of normal late (3.5 h) image of the legs depicting almost complete tracer clearance through the lymphatic vessels following injection into both feet. Symmetric, bilateral inguinal nodes, and retroperitoneal nodes are clearly seen. Markers (m) are at the knees, pubis, xiphoid, and sterna notch. Liver (L) can be seen in this later image as the tracer has been successfully transported through the lymphatic system into the blood pool. (**b**) Late image of a patient with swelling of the right upper thigh following treatment for sarcoma. The patient developed a seroma which was being treated with a drain (arrow) and both collected tracer as it was transported through the lymphatic system. (**c**) SPECT-CT image in a child demonstrating advantageous 3D localization compared to 2D planar LAS and also co-registration of SPECT and CT for anatomic correlation. Low radioactivity appears as blue and highest radioactivity as orange with intermediate in yellow. Patient has severe protein-losing enteropathy. Note radiotracer highlighting the duodenal loop location of leakage. (**d**) Panoramic views of ICG fluorescence lymphography in a healthy volunteer demonstrate injection site (arrow) and transport of the tracer up to the groin. (**a** and **b** modified from Bernas [7], **c** modified from Kuo [8], **d** modified from Unno [9])

graphic reconstruction produces a 3D image with the localized radioactive signal (typically produced by 99 m-Tc) displaying lymphatic structure and function, including leakage (Fig. 24.2). This method is quite useful for mapping of SLNs before resection and surgical treatment for several types of cancer.

24.2.3 NIR (ICG) Imaging

24.2.3.1 Technique Summary

Fluorescent contrast agents have been utilized to image the anatomy and function of the lymphatic system. Typically, they are injected intradermally to access the initial lymphatics beneath the epidermis and then are transported through the lymphatic system. Some of the earliest work in humans was performed by Bollinger and his colleagues [16] using FITC-labeled dextran. They were able to show lymphatic microvessels since the techniques did not allow deeper imaging. The fluorescein sodium is excited by visible light and the fluorescent photons are absorbed by many common tissues (blood, water, etc.) which also inhibits its usefulness for deeper structures [17].

A newer technique using indocyanine green (ICG) has become much more widely used. ICG is excited by near-infrared light (700–900 nm) and that is why these techniques are termed near-infrared (NIR) imaging (also sometimes referred to as ICG imaging) [18]. Since near-infrared light can penetrate 3–4 cm beneath the tissue surfaces, there is minimal autofluorescence and, as a result, NIR imaging with ICG has been more recently employed for imaging the initial lymphatic system as well as imaging of deeper lymphatic structures. There are also multiple commercial cameras and imaging systems, which are relatively low-cost and in use in a variety of settings. A drawback of ICG imaging is its lack of whole-body imaging due to relatively small fields of view from the cameras and limited depth resolution.

24.2.3.2 Imaging Summary

NIR imaging using ICG has become more common to assist in the diagnosis of lymphedema [9, 19], and imaging patterns suggesting different levels of lymphatic dysfunction in lymphedema have been defined [20]. There have also been publications showing lymphatic system changes in the setting of cancer [21, 22] as well as used to visualize and quantitate in vivo lymphatic function [18]. Additionally, ICG imaging has moved into the operating room as a planning tool for microsurgical anastomosis of lymphatic to venous vessels for the treatment of lymphedema and surgical resection of sentinel lymph nodes [23, 24].

24.2.4 Magnetic Resonance Lymphangiography (MRL)

24.2.4.1 Technique Summary

MRL is increasingly used to visualize lymphatic vessels, cisterna chyli, and the thoracic duct in humans instead of invasive lymphangiography. This is due to its increased spatial resolution and the availability of improved equipment and newer imaging sequences. MR lymphagiography can largely be divided into two types of imaging contrast-enhanced and non-contrast images. They each have their strengths and weaknesses and can produce images of high quality (Fig. 24.3).

MR lymphangiography without contrast can be performed at most facilities with a 1.5-T or 3-T clinical MR machine. Although specific sequences vary, in general a coronal or sagittal heavy T2 three-dimensional (3D) volume sequence [15] or turbo-spin-echo (TSE) pulse train with long echo time is utilized [25]. There has also been work published using a 3-T system and a noninvasive arterial spin labeling (ASL) MR method for measuring blood flow to evaluate lymphatic flow [26]. The advantage of not using contrast agents is balanced by potential interference of large pleural effusions that may be related to a lymphatic disorder on image quality and respiratory artifact. In addition, visualization of smaller lymphatic vessels in the periphery is not as successful without contrast [15].

Dynamic contrast-enhanced MR lymphangiography was initially described in 2015 [27]. It is performed by direct intranodal injection of diluted gadolinium into the inguinal lymph nodes using ultrasound guidance [3, 28]. It provides superior information on lymphatic flow dynamics and is best for suspected lymphatic flow pathologies and lymphatic malformations that may be seen in the setting of cancer. Opposed to non-contrast imaging, dynamic contrast-enhanced MR lymphangiography of the extremities can be obtained using a subdermal injection of gadolinium contrast media [15, 29].

24.2.4.2 Imaging Summary

Clear images of central lymphatic structures can be obtained using either contrasted or non-contrast MR lymphangiography (Fig. 24.3). Imaging of the central structures and their pathology (leaks, obstructions, lymphatic aplasia, hypo- and hyperplasia, etc.) has opened the window into designing and performing various treatments. The first percutaneous lymphatic intervention in the setting of the chylous leak was performed by Cope in 2002 [30]. Since that time and with the newer imaging, multiple interventional treatments have been carried out for a variety of lymphatic abnormalities [3, 4, 31].

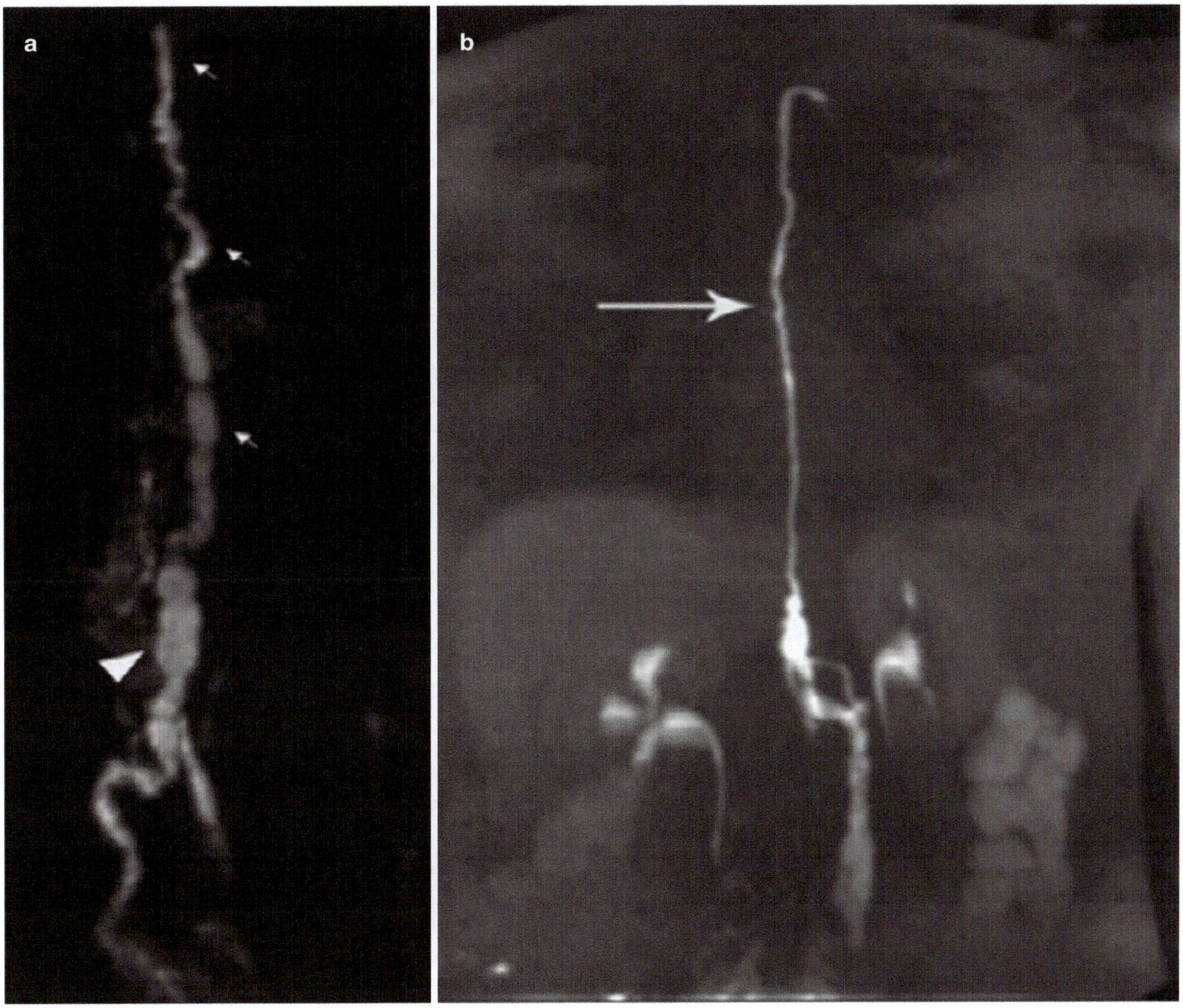

Fig. 24.3 Contrasted and non-contrasted MR lymphangiography. (**a**) Non-contrasted MR imaging using heavily T2-weighted sequence clearly displays the lower portion of the thoracic duct (small arrows) and the cisterna chyli (arrowhead). Contrasted MR lymphangiography with gadolinium-based contrast agent in a dynamic contrast-enhanced MR lymphangiography (DCRML) displays visualization of a normal single thoracic duct (white arrow). (**a** modified from Schwartz et al. [15], **b** modified from Itkin et al. [3])

24.2.5 Ultrahigh Frequency Ultrasound (UHFUS) Imaging

24.2.5.1 Technique Summary

For decades, pulse-echo ultrasound has been used as a non-invasive screening tool to assess lymph node status at the submillimeter scale. Conventional ultrasound detects structural changes suggestive of lymph node metastases, such as enlargement, irregular borders, and hypoechogenicity [32–35]. While standard B mode ultrasound is sensitive to subtle differences in acoustic backscatter, it is not commonly used for defining lymph node status because it typically has poor specificity for discriminating normal, malignant, and metastatic tissue. For example, a second Multicenter Selective Lymphadenectomy Trial (MSLT-II) revealed a sensitivity of 6.6%, increasing only slightly with higher risk primaries [36]. However, as the ultrasound frequency increases, so does the spatial resolution. New commercial systems like the Vevo MD (Visualsonics by Fujifilm) offer UFUS probes (>50 MHz) for high-resolution imaging of lymphatic channels approaching the scale of single cells (<30 μm). The primary disadvantage of UFUS is that the attenuation increases dramatically at frequencies >50 MHz, limiting the depth of penetration to ~10 mm.

24.2.5.2 Imaging Summary

Resolution improves dramatically as the ultrasound frequency increases. A recent paper evaluated ultrahigh frequency ultrasound (UHFUS, 48 and 70 MHz) for assessing the structure and degeneration status of human peripheral lymphatic vessels prior to surgical resection [37]. There was good agreement between UHFUS captured on the Vevo MD system (Visualsonics, Fujifilm) and subsequent histological examination of the lumen inner and outer diameters and wall thickness. The data support UHFUS as a noninvasive technique for planning and guiding lymphatic surgery by helping to identify lymphatic channels suitable for anastomosis. Although UHFUS is suitable for mapping superficial structures, penetration depth is limited to <10 mm.

24.2.6 Contrast-Enhanced Ultrasound (CEUS) Imaging

24.2.6.1 Technique Summary

Contrast-enhanced ultrasound (CEUS), on the other hand, employs a subcutaneous injection of microbubbles (2–6 μm) to enhance the contrast of the lymphatic system. CEUS provides a real-time functional assessment of transport into and out of peripheral lymphatic vessels and nodes [38]. Although microbubble contrast agents are FDA-approved for cardiac imaging and other clinical applications, CEUS is not a standard for imaging the human lymphatic system, and the method remains primarily a preclinical technique for investigating function and dysfunction of the lymphatic system and developing novel therapeutics. An alternative approach for CEUS uses an intravenous (IV) injection of microbubbles combined with color Doppler to enhance the blood vascular network and help classify lymph nodes as metastatic, tuberculous, cystic, or enlarged secondary to lymphoma [39–41].

24.2.6.2 Imaging Summary

CEUS with microbubbles has great potential to identify sentinel lymph nodes and aid with axillary staging in breast cancer patients. The method can be performed noninvasively with a standard clinical scanner without the need for radioactive contrast agents. A recent study in rabbits improved the feasibility of CEUS for mapping peripheral lymphatic vessels and nodes by implementing high frame rates to reduce tissue artifacts, limit bubble disruption, and improve SNR for detecting slow flow with Doppler-based imaging techniques [42]. A similar approach has recently been applied to diagnose central lymph node metastasis in papillary thyroid cancer [42]. Another group demonstrated deep contrast-enhanced ultrasound imaging of lymphatic channels and sentinel nodes with Sonzazoid in a porcine model [43]. In recent years, there has been an increased interest in applying CEUS to evaluate lymphatic channels and sentinel lymph nodes in breast cancer patients for presurgical planning and needle biopsy guidance [44–46]. Other investigators have combined CEUS, B mode, and Doppler flow imaging to improve sensitivity and specificity for detecting lymph node metastasis. Figure 24.4 provides an example of an intravenous injection of SonoVue microbubbles for comparing metastatic and nonmetastatic cervical lymph nodes using CEUS in patients with papillary thyroid carcinoma [47].

24.2.7 Multispectral Photoacoustics

24.2.7.1 Technical Summary

Photoacoustic imaging (PAI) is a hybrid technique that uses a nanosecond laser pulse to transiently heat the tissue and, due to thermoelastic expansion, generate ultrasonic waves that are detected to produce images proportional to the optical absorption of the underlying structures. PAI is a scalable technology that can provide real-time, high resolution

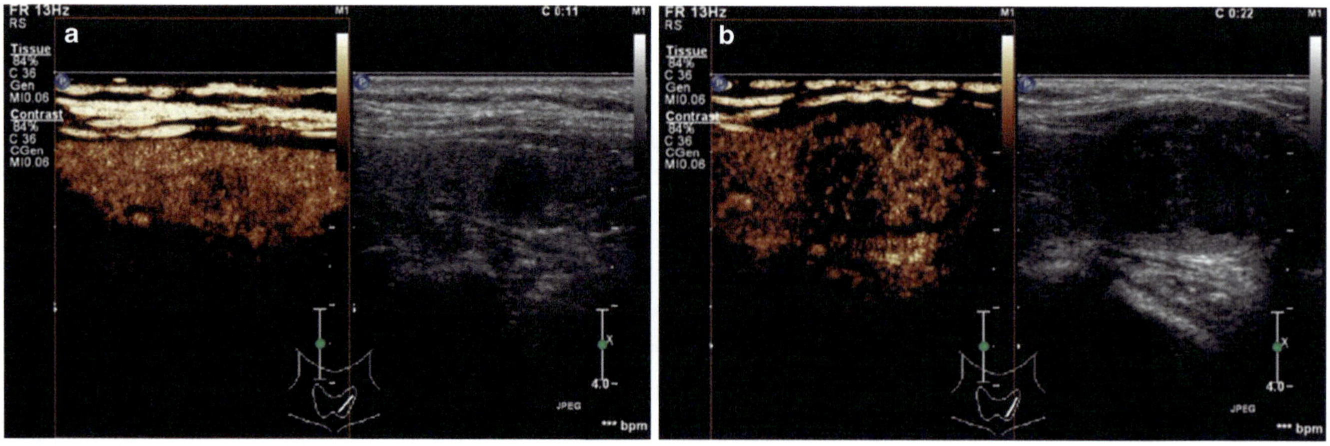

Fig. 24.4 CEUS and B mode US images of the thyroid and cervical lymph node from a patient with nonmetastatic (**a**) and another with metastatic (**b**) papillary thyroid carcinoma confirmed by histology. The authors of this study combined features from CEUS, B mode, and Doppler to identify nodal metastasis and develop predictive models of metastatic disease based on preoperative clinical and ultrasonographic features [47]

(<200 μm) images deep into tissue (>20 mm), including volumetric images of the human breast within a single breath hold [48, 49]. Without contrast agents, PAI is able to map blood vessels and blood oxygen saturation (SO_2) based on the optical absorption properties of hemoglobin. Melanoma cells also have strong near-infrared absorption and can be identified and tracked with PAI. Near-infrared absorbing tracers, such as FDA-approved dyes ICG and Methylene Blue, are required to visualize lymphatic vessels and nodes with PAI [50]. By sweeping the optical wavelength, multi-spectrum PAI with spectral unmixing is able to display both blood and lymphatic vessels in the same image. Because it is inherently a 3D modality, PAI has better depth resolution than fluorescent imaging with better depth penetration. However, PAI requires custom or commercial hardware that integrates a tunable pulsed laser with ultrasound technology, making it cost-prohibitive at the present time compared to standard NIR fluorescent imaging with ICG. PAI also typically has worse sensitivity than fluorescent and nuclear imaging techniques for detecting contrast agents.

24.2.7.2 Imaging Summary

Preclinical studies have demonstrated the capability of real-time PAI to quantify lymphatic flow, pumping, and pharmacokinetics using ICG tracers and encapsulated nanoparticles [51]. Another study in mice developed a targeted contrast agent for immunofunctional photoacoustic imaging of lymph node metastasis [52]. Glycol-chitosan-coated gold nanoparticles were injected into the footpad and within 24 h filled the sentinel lymph node that was not tumor-bearing metastatic. However, the nanoparticles were not able to accumulate inside nodes that were expressed metastatic disease (Fig. 24.5). The high specificity of the results in mice suggests that PAI could be a valuable tool for both diagnosing metastatic disease and monitoring effects of treatment (i.e., theranostics).

A large-field-of-view photoacoustic platform has recently been developed for mapping lymphatic vessels in the human hand, foot, or limb [54, 55]. ICG was injected intradermally or subdermally into the webspace of the foot in both healthy subjects and patients with secondary lymphedema (Fig. 24.6). Photoacoustic lymphangiography revealed 3D complex lymphatic structures near regions of dermal backflow (DBF), which was confirmed on lymphoscintigraphy and NIR fluorescent imaging. Compared to other modalities, the system provided 3D high-resolution images (<200 μm) with a penetration depth > 20 mm. In addition, the platform was able to identify small blood vessels near the lymphatic structures and distinguish arteries from veins based on the intrinsic NIR absorption of hemoglobin and the dependence of oxygen on the absorption spectrum. One of the clinical applications of the imaging system is for postoperative assessment of anastomoses.

Clinical studies have also been performed using multi-spectral photoacoustic imaging to evaluate metastatic status in sentinel lymph nodes (SLNs) based on the presence of melanoma cells [56, 57]. However, questions remain regarding the sensitivity of the system, and large clinical trials are

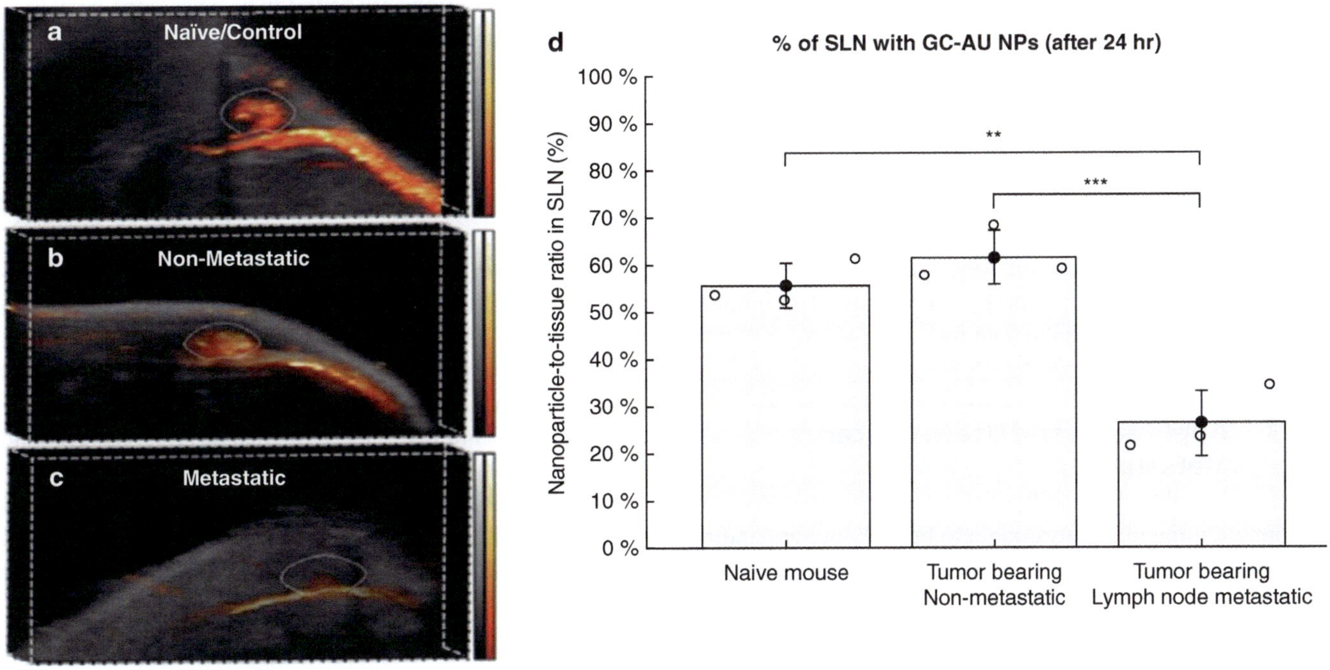

Fig. 24.5 Detection of lymph node metastasis with targeted gold nanoparticles and photoacoustic imaging in mice. (**a**) PAI (hot colors) of sentinel lymph node and afferent lymphatic vessel 24 h after subcutaneous injection of GC-AuNPs in the footpad in (**a**) control; (**b**) non-metastatic; and (**c**) metastatic mice. PAI is superimposed with grayscale pulse-echo image displaying anatomical structure; (**d**) Nanoparticle-to-tissue ratio for each type of mouse, indicating that mice with metastatic disease had poorly functioning SNLs compared to the other categories (modified from [52, 53])

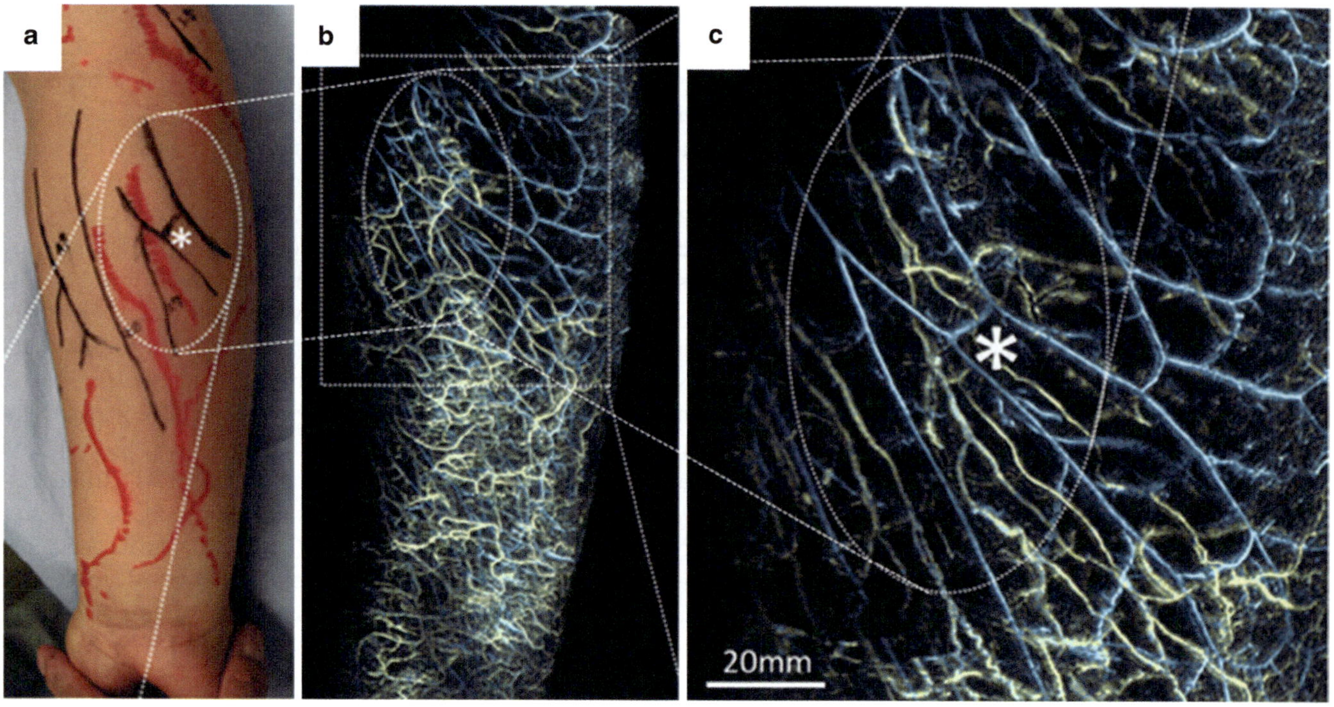

Fig. 24.6 Wide-field photoacoustic lymphangiography (PAL) in male patient with secondary lymphedema. (**a**) Top view photograph of upper limb before lymphaticovenular anastomosis surgery with red and black marks denoting approximate location of lymphatic vessels identified with NIR and small veins identified with B-mode US, respectively. (**b**) Corresponding 3D representation of the forearm captured with PAL indicating complex network of lymphatic channels (enhanced following a subdermal injection of ICG) and veins identified by the photoacoustic signal corresponding to deoxygenated hemoglobin. (**c**) Zoomed in representation of dotted oval (in **a** and **b**) clearly illustrating the high resolution (<200 μm) and deep imaging capability (>20 mm) of PAL. Compared to control images, patients with secondary lymphedema manifested a much higher quantity of lymphatic vessels, which also appeared twisted and tortuous [55]

necessary to determine the efficacy of noninvasive PAI for detecting SLN metastases compared to the gold standard. Another approach applies real-time, label-free photoacoustic detection and tracking of circulating tumor cells (CTCs) and distant metastasis [58]. The photoacoustic device is called the Cytophone and provides a liquid biopsy based on the intrinsic contrast of circulating melanoma cells, red blood cells, and other intrinsic optical absorbers. The system was also combined as a theranostic tool by counting CTCs before and after photodynamic therapy consisting of continuous laser exposure in the cubital vein for 1 h.

24.3 Application and Use in Cancer Metastasis

Lymphatic imaging is invaluable to broadly understanding, evaluating, and treating cancer metastasis as illustrated throughout this chapter as well as in most other chapters of the book. Applications are found in basic science investigations of in vitro as well as in vivo phantom and animal models to delineate mechanisms, discover and improve imaging techniques, and test potential therapeutic targets. These findings as well as clinical and translational research can then be utilized in clinical settings to determine (singularly or in multimodal combinations) whether, when, and where: specific cancers metastasize; the impact and alterations of the lymphatic system and flow; and potential avenues and points for the treatment and monitoring of both the lymphatic and cancer alterations [59]. These treatments could entail image-guided therapy (and monitoring) as well as in select instances (i.e., high-frequency ultrasound) use of imaging technology itself to treat primary cancer and metastases.

Table 24.1 summarizes the most common and emerging imaging modalities and their attributes for studying the lymphatic system with relevance to cancer and cancer metastasis.

Table 24.1 Common and emerging imaging modalities and their attributes for studying the lymphatic system with relevance to cancer and cancer metastasis

Modality	Clinical tracer	Basic	Clinical	Spatial resolution	Contrast resolution	Penetration	Sensitivity
Lymphography-classical	Lipidol		X	+	+++	+++	+++
Lymphography- intra-nodal	Lipidol		X	+	+++	+++	+++
Lymphscintigraphy	99mTc	X	X	+	+++	+++	+++
Lymphscintigraphy- CT-SPECT	Lipidol, 99mTc		X	++	+++	+++	+++
MRI	Gd (optional)	X	X	++	++	+++	++
NIR	ICG	X	X	+++	+++	+	+++
US	Microbubbles	X	X	+++	++	+++	++
PAI	ICG, meth. Blue	X	X	+++	+++	++	+

24.4 Conclusions

Imaging of the lymphatic system is crucial for depicting both lymphatic structure and function in health and a wide variety of clinical disorders. The multiple different techniques described utilize various tracers and techniques for humans and experimental models and each have their own strengths, weaknesses, and specific applications (summarized in Table 24.1). Beginning with a history of postmortem dissections and contrast injections centuries ago, these techniques have evolved from simple intradermal visualization of blue dye-colored lymphatic "streamers" and invasive oil-contrast lymphography to radionuclide lymphoscintigraphy, single photon-emission computed tomography, magnetic resonance imaging with and without contrast, a variety of fluorescent imaging, and most-recently, photoacoustic applications. These techniques are now used in combinations and particularly for image-guided procedures and to directly intervene in an expanding number of lymphatic disorders. Applications to cancer and cancer metastasis are extensive, ranging from detection and monitoring of cancer spread using molecular tools to endolymphatic administration of anticancer drugs and decompression of central lymphatic obstruction. The future is encouraging as the field continues to offer advancements and additional applications.

Unanswered Questions

- There are many unanswered questions as well as points to consider for advancing the field in the future. Some of these include:
- How can devices, e.g., gamma camera, MRI be made more portable and less expensive to facilitate point-of-care use?
- How can spatial resolution and depth penetration be improved for each imaging modality?
- What multimodal combinations are optimal for imaging specific lymphatic disorders and for intervention in reversing pathologic processes?
- Which techniques (or combinations) are optimal for imaging primary and metastatic cancers?
- Can central (thoracic duct) access be perfected to facilitate stenting into venous system to reverse thoracic duct obstruction, i.e., for decompression after tumor invasion and lymph leakage? For sampling lymph noninvasively as "liquid biopsy" to detect and characterize cancers?
- Would endolymphatic therapy provide a better approach to regional chemotherapy in certain cancers?
- Can the lymphatic system itself (absorption and transport of various agents) be utilized for cancer treatment?
- What other techniques are being developed that use various energies (nuclear, light, sound, or magnetism) and tracers (e.g., dyes, nanoparticles) for imaging the lymphatic system?
- Does multispectral photoacoustic imaging have sufficient sensitivity and specificity to assess lymph node status and reliably detect metastasis?

References

1. Kinmonth JB, Taylor GW, Harper RK. Lymphangiography; a technique for its clinical use in the lower limb. BMJ. 1955;1(4919):940–2.
2. Kinmonth JB. The lymphatics: diseases, lymphography and surgery. Baltimore: Williams & Wilkins; 1972.
3. Itkin M, Nadolski GJ. Modern techniques of lymphangiography and interventions: current status and future development. Cardiovasc Intervent Radiol. 2018;41(3):366–76. https://doi.org/10.1007/s00270-017-1863-2. Epub 2017 Dec 18
4. Dori Y. Novel lymphatic imaging techniques. Tech Vasc Interv Radiol. 2016;19(4):255–61. https://doi.org/10.1053/j.tvir.2016.10.002.
5. Guermazi A, Brice P, Hennequin C, Sarfati E. Lymphography: an old technique retains its usefulness. Radiographics. 2003;23:1541–60.
6. Peterss S, Gratz KF, Khaladj N, Hagl C, Warnecke G, Fischer S, et al. Lymphoscintigraphic localization of a high-output chylous leak after bilateral lung transplantation. J Heart Lung Transpl. 2011;30:481–3.

7. Bernas M. Imaging of the lymphatic system. In: Lymphedema management: the comprehensive guide for practitioners. 3rd ed. New York: Thieme; 2013.
8. Kuo PH, Barber BJ, Kylat RI, Klewer SE, Behan S, Lau-Braunhut S, Bernas MJ, Moedano L, Bedrick AD, Mustacich DJ, Witte MH. Whole-body lymphangioscintigraphy and SPECT/CT in children with lymphatic complications after surgery for complex congenital heart disease. Lymphology. 2019;52(4):157–65.
9. Unno N, Nishiyama M, Suzuki M, Yamamoto N, Inuzuka K, Sagara D, Tanaka H, Konno H. Quantitative lymph imaging for assessment of lymph function using indocyanine green fluorescence lymphography. Eur J Vasc Endovasc Surg. 2008;36(2):230–6. https://doi.org/10.1016/j.ejvs.2008.04.013. Epub 2008 Jun 4
10. International Society of Lymphology. The diagnosis and treatment of peripheral lymphedema: 2020 consensus document of the international society of lymphology. Lymphology. 2020;53:3–19.
11. Witte CL, Witte MH, Unger EC, Williams WH, Bernas MJ, McNeill GC, Stazzone AM. Advances in imaging of lymph flow disorders. Radiographics. 2000;20(6):1697–719.
12. Kleinhans E, Baumeister RG, Hahn D, Siuda S, Büll U, Moser E. Evaluation of transport kinetics in lymphoscintigraphy: follow-up study in patients with transplanted lymphatic vessels. Eur J Nucl Med. 1985;10(7–8):349–52. https://doi.org/10.1007/BF00251310.
13. Cambria RA, Gloviczki P, Naessens JM, Wahner HW. Noninvasive evaluation of the lymphatic system with lymphoscintigraphy: a prospective, semiquantitative analysis in 386 extremities. J Vasc Surg. 1993;18(5):773–82. https://doi.org/10.1067/mva.1993.50510.
14. Pecking AP, Alberini JL, Wartski M, Edeline V, Cluzan RV. Relationship between lymphoscintigraphy and clinical findings in lower limb lymphedema (LO): toward a comprehensive staging. Lymphology. 2008;41(1):1–10. Epub 2008/06/28
15. Schwartz FR, James O, Kuo PH, Witte MH, Koweek LM, Pabon-Ramos WM. Lymphatic imaging: current noninvasive and invasive techniques. Semin Intervent Radiol. 2020;37(3):237–49. https://doi.org/10.1055/s-0040-1713441. Epub 2020 Jul 31
16. Bollinger A, Jäger K, Sgier F, Seglias J. Fluorescence microlymphography. Circulation. 1981;64(6):1195–200. https://doi.org/10.1161/01.cir.64.6.1195.
17. O'Donnell TF Jr, Rasmussen JC, Sevick-Muraca EM. New diagnostic modalities in the evaluation of lymphedema. J Vasc Surg Venous Lymphat Disord. 2017;5(2):261–73. https://doi.org/10.1016/j.jvsv.2016.10.083. Epub 2017 Jan 16. Epub 2017 Jan 16. PMID: 28214496; PMCID: PMC5325714
18. Rasmussen JC, Tan IC, Marshall MV, Fife CE, Sevick-Muraca EM. Lymphatic imaging in humans with near-infrared fluorescence. Curr Opin Biotechnol. 2009;20(1):74–82. https://doi.org/10.1016/j.copbio.2009.01.009.
19. Yamamoto T, Matsuda N, Doi K, Oshima A, Yoshimatsu H, Todokoro T, Ogata F, Mihara M, Narushima M, Iida T, Koshima I. The earliest finding of indocyanine green lymphography in asymptomatic limbs of lower extremity lymphedema patients secondary to cancer treatment: the modified dermal backflow stage and concept of subclinical lymphedema. Plast Reconstr Surg. 2011a;128(4):314e–21e.
20. Yamamoto T, Yamamoto N, Doi K, Oshima A, Yoshimatsu H, Todokoro T, Ogata F, Mihara M, Narushima M, Iida T, Koshima I. Indocyanine green-enhanced lymphography for upper extremity lymphedema: a novel severity staging system using dermal backflow patterns. Plast Reconstr Surg. 2011b;128(4):941–7. https://doi.org/10.1097/PRS.0b013e3182268cd9.
21. Rasmussen JC, Kwon S, Sevick-Muraca EM, Cormier JN. The role of lymphatics in cancer as assessed by near-infrared fluorescence imaging. Ann Biomed Eng. 2012;40:408–21. https://doi.org/10.1007/s10439-011-0476-1.
22. Aldrich MB, Rasmussen JC, Fife CE, Shaitelman SF, Sevick-Muraca EM. The development and treatment of lymphatic dysfunction in cancer patients and survivors. Cancers (Basel). 2020;12(8):2280. https://doi.org/10.3390/cancers12082280.
23. Chang EI, Skoracki RJ, Chang DW. Lymphovenous anastomosis bypass surgery. Semin Plast Surg. 2018;32(1):22–7. https://doi.org/10.1055/s-0038-1636510.
24. Hackethal A, Hirschburger M, Eicker SO, Mücke T, Lindner C, Buchweitz O. Role of Indocyanine green in fluorescence imaging with near-infrared light to identify sentinel lymph nodes, lymphatic vessels and pathways prior to surgery: a critical evaluation of options. Geburtshilfe Frauenheilkd. 2018;78(1):54–62. https://doi.org/10.1055/s-0043-123937.
25. Crescenzi R, Donahue PMC, Hartley KG, Desai AA, Scott AO, Braxton V, Mahany H, Lants SK, Donahue MJ. Lymphedema evaluation using noninvasive 3T MR lymphangiography. J Magn Reson Imaging. 2017;46(5):1349–60. https://doi.org/10.1002/jmri.25670. Epub 2017 Feb 28
26. Rane S, Donahue PMC, Towse T, Ridner S, Chappell M, Jordi J, Gore J, Donahue MJ. Clinical feasibility of noninvasive visualization of lymphatic flow with principles of spin labeling MR imaging: implications for lymphedema assessment. Radiology. 2013;269:893–902.
27. Krishnamurthy R, Hernandez A, Kavuk S, Annam A, Pimpalwar S. Imaging the central conducting lymphatics: initial experience with dynamic MR lymphangiography. Radiology. 2015;274(03):871–8.
28. Chavhan GB, Amaral JG, Temple M, Itkin M. MR lymphangiography in children: technique and potential applications. Radiogr Radiol Soc N Am. 2017:1775–90.
29. Pieper CC. Nodal and Pedal MR lymphangiography of the central lymphatic system: techniques and applications. Semin Intervent Radiol. 2020;37(3):250–62. https://doi.org/10.1055/s--0040-1713442. Epub 2020 Jul 31
30. Cope C, Kaiser LR. Management of unremitting chylothorax by percutaneous embolization and blockage of retroperitoneal lymphatic vessels in 42 patients. J Vasc Interv Radiol. 2002;13:1139–48. 374
31. Dori Y, Keller MS, Fogel MA, Rome JJ, Whitehead KK, Harris MA, Itkin M. MRI of lymphatic abnormalities after functional single-ventricle palliation surgery. AJR. 2014;203:426–31.
32. Vassallo P, Wernecke K, Roos N, Peters PE. Differentiation of benign from malignant superficial lymphadenopathy: the role of high-resolution US. Radiology. 1992;183(1):215–20. (ISSN: 0033–8419)
33. Tregnaghi A, De Candia A, Calderone M, Cellini L, Rossi CR, Talenti E, Blandamura S, Borsato S, Muzzio PC, Rubaltelli L. Ultrasonographic evaluation of superficial lymph node metastases in melanoma. Eur J Radiol. 1997;24(3):216–21. ISSN 0720-048X. https://doi.org/10.1016/S0720-048X(96)01102-3.
34. Sureshkannan P, Vijayprabhu JR. Role of ultrasound in detection of metastatic neck nodes in patients with oral cancer. Indian J Dent Res. 2011;22(3):419–23. https://doi.org/10.4103/0970-9290.87064.
35. Maxwell FM, de Margerie Mellon C, Bricout M, Cauderlier E, Chapelier M, Albiter M, Bourrier P, Espié M, de Kerviler E, de Bazelaire C. Diagnostic strategy for the assessment of axillary lymph node status in breast cancer. Diagn Interv Imaging. 2015;96(10):1089–101. ISSN 2211-5684
36. Thompson JF, Haydu LE, Uren RF, et al. Preoperative ultrasound assessment of regional lymph nodes in melanoma patients does not provide reliable nodal staging: results from a large multicenter trial. Ann Surg. 2019; Epub 7 Jun 2019
37. Bianchia A, Viscontib G, Hayashic A, Santorod A, Longoe V, Salgarellof M. Ultra-high frequency ultrasound imaging of lymphatic channels correlates with their histological features: a step forward in lymphatic surgery. J Plast Reconstr Aesthet Surg. 2020;73:1622–9.

38. Choi SH, Kono Y, Corbeil J, Lucidarme O, Mattrey RF. Model to quantify lymph node enhancement on indirect sonographic lymphography. AJR Am J Roentgenol. 2004;183(2):513–7. https://doi.org/10.2214/ajr.183.2.1830513. PMID: 15269049.
39. Steinkamp HJ, Wissgott C, Rademaker J, Felix R. Current status of power Doppler and color Doppler sonography in the differential diagnosis of lymph node lesions. Eur Radiol. 2002;12(7):1785–93. https://doi.org/10.1007/s003300101111. Epub 2001 Sep 7. PMID: 12111070.
40. Moritz JD, Ludwig A, Oestmann JW. Contrast-enhanced color Doppler sonography for evaluation of enlarged cervical lymph nodes in head and neck tumors. Am J Roentgenol. 2000;174:1279.
41. Yang L, Zhao H, He Y, Zhu X, Yue C, Luo Y, Ma B. Contrast-Enhanced Ultrasound in the Differential Diagnosis of Primary Thyroid Lymphoma and Nodular Hashimoto's Thyroiditis in a Background of Heterogeneous Parenchyma. Front Oncol. 2021;10:597975. https://doi.org/10.3389/fonc.2020.597975. PMID: 33489895; PMCID: PMC7817885.
42. Zhu J, Lin S, Leow CH, Rowland EM, Riemer K, Harput S, Weinberg PD, Tang MX. High frame rate contrast-enhanced ultrasound imaging for slow lymphatic flow: influence of ultrasound pressure and flow rate on bubble disruption and image persistence. Ultrasound Med Biol. 2019;45(9):2456–70. https://doi.org/10.1016/j.ultrasmedbio.2019.05.016. Epub 2019 Jul 3
43. Kawai Y, Ajima K, Nagai T, Kaidoh M, Ohhashi T. Real-time imaging of the lymphatic channels and sentinel lymph nodes of the stomach using contrast-enhanced ultrasonography with Sonazoid in a porcine model. Cancer Sci. 2011;102(11):2073–81. https://doi.org/10.1111/j.1349-7006.2011.02048.x. Epub 2011 Aug 24
44. Wang Y, Zhou W, Li C, Gong H, Li C, Yang N, Zha X, Chen L, Xia T, Liu X, Wang M, Ding Q. Variation of sentinel lymphatic channels (SLCs) and sentinel lymph nodes (SLNs) assessed by contrast-enhanced ultrasound (CEUS) in breast cancer patients. World J Surg Oncol. 2017;15(1):127. https://doi.org/10.1186/s12957-017-1195-3.
45. Cui Q, Dai L, Li J, Xue J. Accuracy of CEUS-guided sentinel lymph node biopsy in early-stage breast cancer: a study review and meta-analysis. World J Surg Oncol. 2020;18(1):112. https://doi.org/10.1186/s12957-020-01890-z.
46. Li J, Lu M, Cheng X, Hu Z, Li H, Wang H, Jiang J, Li T, Zhang Z, Zhao C, Ma Y, Tan B, Liu J, Yu Y. How Pre-operative Sentinel Lymph Node Contrast-Enhanced Ultrasound Helps Intra-operative Sentinel Lymph Node Biopsy in Breast Cancer: Initial Experience. Ultrasound Med Biol. 2019;45(8):1865–73. https://doi.org/10.1016/j.ultrasmedbio.2019.04.006. Epub 2019 May 14. PMID: 31101445.
47. Wang Y, Nie F, Wang G, Liu T, Dong T, Sun Y. Value of combining clinical factors, conventional ultrasound, and contrast-enhanced ultrasound features in preoperative prediction of central lymph node metastases of different sized papillary thyroid carcinomas. Cancer Manag Res. 2021;13:3403–15. Published 2021 Apr 19. https://doi.org/10.2147/CMAR.S299157.
48. Lin L, Hu P, Shi J, Appleton CM, Maslov K, Li L, Zhang R, Wang LV. Single-breath-hold photoacoustic computed tomography of the breast. Nat Commun. 2018;9(1):2352.
49. Witte RS, Karunakaran C, Zuniga AN, Schmitz H, Arif H. Frontiers of cancer imaging and guided therapy using ultrasound, light, and microwaves. Clin Exp Metastasis. 2018;
50. Garcia-Uribe A, Erpelding TN, Krumholz A, Ke H, Maslov K, Appleton C, Margenthaler JA, Wang LV. Dual-modality photoacoustic and ultrasound imaging system for noninvasive sentinel lymph node detection in patients with breast cancer. Sci Rep. 2015;5:15748.
51. Forbrich A, Heinmiller A, Zemp RJ. Photoacoustic imaging of lymphatic pumping. J Biomed Opt. 2017;22(10):106003.
52. Dumani DS, Sun IC, Emelianov SY. Ultrasound-guided immunofunctional photoacoustic imaging for diagnosis of lymph node metastases. Nanoscale. 2019;11(24):11649–59.
53. Witte RS, Tamimi EA. Emerging photoacoustic and thermoacoustic imaging technologies for detecting primary and metastatic cancer and guiding therapy. Clin Exp Metastasis. 2021; https://doi.org/10.1007/s10585-021-10095-x.
54. Kajita H, Suzuki Y, Sakuma H, Imanishi N, Tsuji T, Jinzaki M, Aiso S, Kishi K. Visualization of lymphatic vessels using photoacoustic imaging. Keio J Med. 2021; https://doi.org/10.2302/kjm.2020-0010-OA. Epub ahead of print
55. Nagae K, Asao Y, Sudo Y, Murayama N, Tanaka Y, Ohira K, Ishida Y, Otsuka A, Matsumoto Y, Saito S, Furu M, Murata K, Sekiguchi H, Kataoka M, Yoshikawa A, Ishii T, Togashi K, Shiina T, Kabashima K, Toi M, Yagi T. Real-time 3D photoacoustic visualization system with a wide field of view for imaging human limbs. F1000Res. 2018;7:1813. https://doi.org/10.12688/f1000research.16743.2.
56. Sinnamon AJ, Neuwirth MG, Song Y, Schultz SM, Liu S, Xu X, et al. Multispectral photoacoustic imaging for the detection of subclinical melanoma. J Surg Oncol. 2019;119(8):1070–6.
57. Stoffels I, Morscher S, Helfrich I, et al. Metastatic status of sentinel lymph nodes in melanoma determined noninvasively with multispectral optoacoustic imaging. Sci Transl Med. 2015; 7(317):317.
58. Galanzha EI, et al. In vivo liquid biopsy using Cytophone platform for photoacoustic detection of circulating tumor cells in patients with melanoma. Sci Transl Med. 2019;11(496):eaat5857.
59. Bernas M, Thiadens SRJ, Stewart P, Granzow J. Secondary lymphedema from cancer therapy. Clin Exp Metastasis. 2021; https://doi.org/10.1007/s10585-021-10096-w. Published online May 05

Lymphedema: General Pathophysiology, Prevention, and Management in Invasive Cancer

25

Angelika Chachaj, Neil Piller, Francesco Boccardo, and Andrzej Szuba

Abstract

Lymphedema is a common complication in cancer survivors. It results from lymphatic system injury due to lymphadenectomy, radio-, and chemotherapy. It also can result from lymphatic obstruction by malignant cells in advanced cancer. Reported incidence of CRL varies widely depending on the type of malignancy, lymphedema diagnostic criteria, and follow-up period. CRL is a debilitating condition and significantly impairs the quality of life in cancer survivors in all, physical, emotional, social, and economic aspects of daily life. Prospective surveillance of persons after cancer surgery helps early recognition and treatment. The *decongestive lymphatic therapy* is a gold standard in conservative treatment of lymphedema in the acute phase and custom compression garments are used in the maintenance therapy. Microsurgical methods are gaining more popularity and have the potential to restore lymphatic flow in affected extremities both to prevent lymphedema and improve existing lymphedema. Excess of subcutaneous fat in lymphedematous limbs can be effectively removed using liposuction; however, continuous compression with garments is necessary. No specific, FDA-approved drug exists for lymphedema. A range of drugs, including anti-T cell, anti-inflammatory, and antibiotics add to the potential of pharmacological treatment/management of lymphedema. Coumarin, studied years ago, shows it has a potential worth further exploration.

Learning Objectives

- To identify patients at the highest risk for developing cancer-related lymphedema (CRL).
- To recognize the main clinical signs and symptoms of CRL.
- To be aware of significant physical and psychosocial consequences of CRL.
- To know the prevention and treatment methods in CRL.
- To become more aware of drugs that may help lymphedema treatment.
- To know how to prevent lymphedema following nodal dissection by microsurgical lymphatic venous bypass technique (LYMPHA procedure), indications, tips, and tricks.
- To know how to treat chylous leakage following lymph nodal dissection for the treatment of abdominal organ tumors, diagnostics, and management.

Sections 7.1–7.5.1 have been prepared by Andrzej Szuba and Angelika Chachaj, Section 7.52 has been prepared by Neil Piller and Sections 7.53 and 7.54 have been prepared by Francesco Boccardo.

A. Chachaj · A. Szuba (✉)
Department of Angiology, Wroclaw Medical University, Wroclaw, Poland
e-mail: angelika.chachaj@umed.wroc.pl

N. Piller
Flinders Medical Centre, Department of Surgery, School of Medicine, Bedford Park, SA, Australia
e-mail: neil.piller@flinders.edu.au

F. Boccardo
San Martino Hospital, University of Genoa, Genoa, Italy
e-mail: francesco.boccardo@unige.it

25.1 Introduction

Lymphedema is a chronic and progressive condition caused by lymph stasis due to lymphatic vessels dysfunction [1].

Primary lymphedema results from abnormal development of lymphatic vasculature due to germinal or somatic genetic mutations. *Secondary or acquired lymphedema* is a result of lymphatic injury, which can be iatrogenic, post-traumatic, or caused by infection. Filariasis is the most prevalent cause of

S. P. Leong et al. (eds.), *Cancer Metastasis Through the Lymphovascular System*,
https://doi.org/10.1007/978-3-030-93084-4_25

secondary lymphedema worldwide. Lymph node metastases and/or lymphatic vessels' obliteration by neoplastic cells in advanced stage of cancer may also cause lymph stasis and lymphedema (malignant lymphedema). The most common cause of secondary lymphedema in countries not affected by filariasis is oncological therapy [2]. With improved prognosis for cancer survivors, the number of patients who suffer from cancer-related lymphedema (CRL) may increase in the coming years [3]. The incidence of CRL varies with the type of cancer, the treatment used, measurement methods, and the duration of follow-up [4, 5]. CRL most frequently develops within 2 years after oncological treatment, but it can also appear years or even decades later [4].

CRL affects areas drained by dysfunctional/dissected lymph nodes: upper and lower limbs, head, and neck, but also breast, trunk, pubic area, groin, and scrotum. CRL is usually limited to the skin and subcutaneous tissue; however, head and neck lymphedema may affect deeper tissues and organs, e.g., tongue, larynx, and throat.

Breast cancer survivors are the major group of patients affected with CRL. According to large meta-analysis of 72 studies, the overall incidence of arm lymphedema in this group of patients was estimated to be approximately 17%. Data from prospective cohort studies suggest higher incidence of arm lymphedema in breast cancer survivors—21%. Arm lymphedema is nearly four times more prevalent in women after axillary lymph node dissection (ALND) than in those who had sentinel lymph node dissection (SLND) [4].

Lymphedema is a common complication of oncological treatment in all solid malignancies including melanoma, gynecological and genitourinary tumors, head and neck cancers, and sarcoma (Fig. 25.1). The overall prevalence of CRL excluding breast cancer was estimated as 15% and varied by malignancy (Table 25.1) [5].

Definition

Cancer-related lymphedema (CRL)—edema caused by lymphatic system injury and lymph stasis due to oncological treatment (lymph nodes dissection, radio, and chemotherapy) or by tumor spread into lymphatics. Lymphedema is characterized by progressive swelling, fibrosis, and abnormal fat deposition of the affected region.

25.2 General Pathophysiology of Cancer-Related Lymphedema*

The cardinal function of the lymphatic system is absorption and transport of interstitial fluid from extracellular space. Under physiological conditions, the extracellular fluid is drained and transported solely by lymphatic vessels. Reabsorption into capillary venules at the level of microcirculation is transient and minimal [8]. Impairment of lymphatic drainage results in lymph stasis and lymphedema [1].

Regional lymph node dissection (axillary, inguinofemoral, pelvic, neck dissection) and/or radiotherapy results in disruption of lymphatic vessels and failure in drainage of appropriate lymphatic territory. *Malignant lymphedema* develops when neoplastic cells cause an obstruction of lymph nodes and/or lymphatic vessels [9].

Lymphatic vessels' disruption leads to a rise in lymph pressure inside collecting lymphatic vessels and progressive structural changes—from initial dilation to fibrosis and complete obliteration [10]. *Lymph stasis* is accompanied by the accumulation of immune cells, glycosaminoglycans, metabolic products, and inflammatory mediators leading to chronic inflammation. The subsequent steps in lymphedema development are proliferation of keratinocytes, fibroblasts and adipocytes, destruction of elastic fibers, and collagen deposition [9].

Lymphedema, initially reversible, becomes irreversible at the later stages. Skin thickens and hardens and it is impossible to pinch a fold of skin at the base of the second toe. This symptom is known as positive Stemmer's sign, considered pathognomonic for lymphedema. Lymph stasis impairs local immune response leading to recurrent infections of the skin and subcutaneous tissue (*dermatolymphangioadenitis*). Recurrent infections cause further damage of skin lymphatics and aggravation of lymphedema [9].

International Society of Lymphology distinguishes four stages of lymphedema progression: stage 0–subclinical lymph stasis, stage 1—reversible edema, stage 2—irreversible edema, stage 3—irreversible edema with limb deformation and skin changes (elephantiasis) (ISL Consensus 2020) [11].

*The section on "General pathophysiology of cancer-related lymphedema" was solely prepared by Angelika Chachaj and Andrzej Szuba

Chronic lymphedema is a risk factor of secondary *malignancy–lymphedema-associated angiosarcoma* (earlier called lymphangiosarcoma). Initially described in women with breast cancer–related lymphedema (Stewart–Treves syndrome), this is an extremely aggressive tumor that originated from the lymphatic endothelium, with extremely poor prognosis. Five-year survival is less than 10% [12]. Other malignancies with increased prevalence in lymphedema include Kaposi's sarcoma, squamous cell carcinoma, malignant lymphoma, and melanoma [13].

Extremity lymphedema leads to remarkable functional impairment depending on its severity. Asymmetric limb

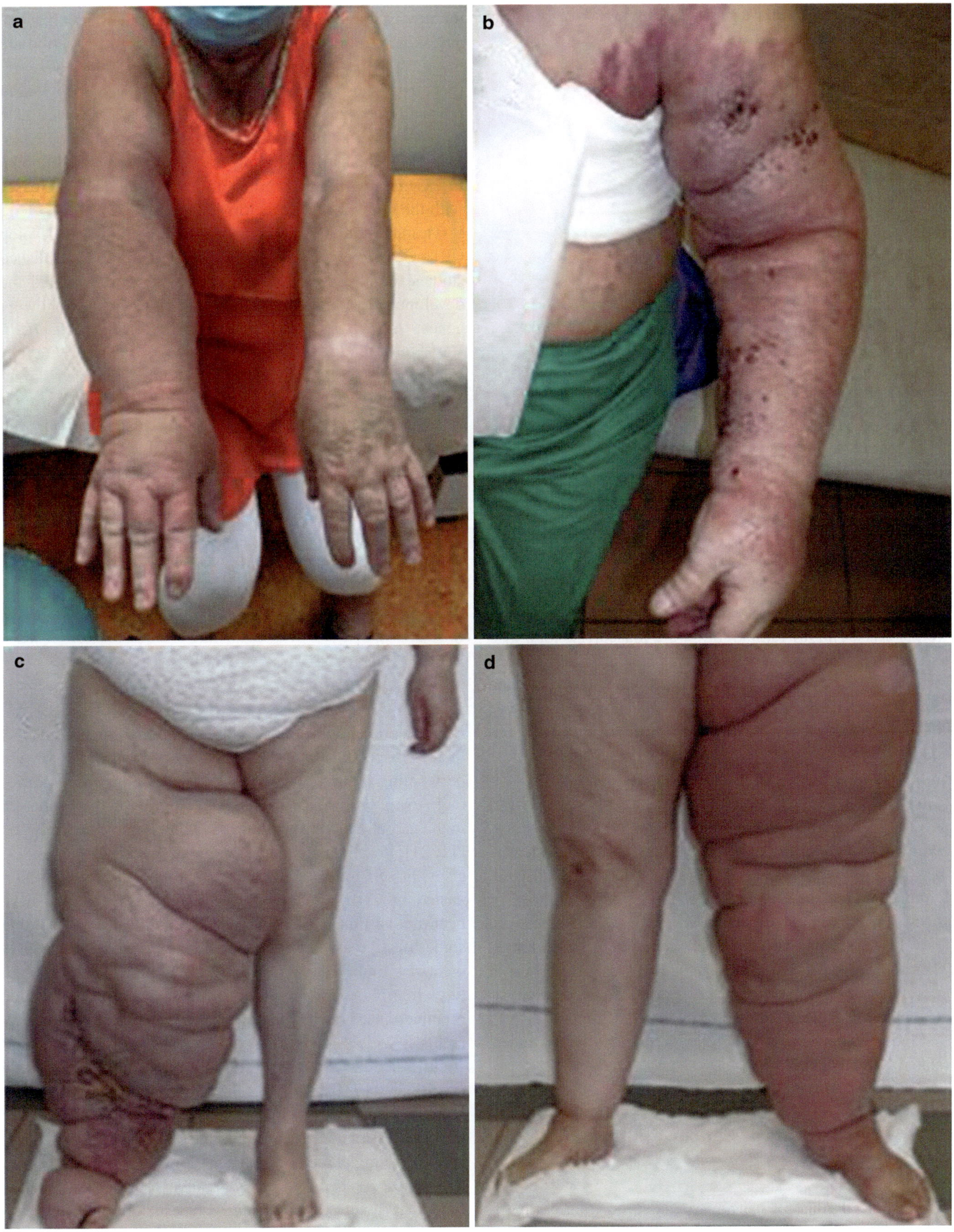

Fig. 25.1 (**a–d**) Examples of cancer-related lymphedema: Right-arm lymphedema secondary to breast cancer treatment (**a**), left arm "malignant" lymphedema (**b**), right leg lymphedema after treatment of cervical cancer (surgery + radiotherapy) (**c**), left leg lymphedema secondary to treatment of vulvar cancer (**d**)

Table 25.1 Cancer-related lymphedema (CRL) prevalence according to the literature data

Type of cancer	Average (range) prevalence of CRL [%]	References
Breast cancer	17 (8–21)	[4]
– with axillary lymph node dissection (ALND)	20	
– with sentinel lymph node dissection (SLND)	6	
Melanoma		[5]
– with axillary lymph node dissection (ALND)	3 (1–39)	
– with inguinofemoral lymph node dissection	18 (6–66)	
Gynecologic malignancies	25 (0–73)	[5]
– cervical	27 (2–49)	
– vulvar	30 (0–73)	
– endometrial	1	
Genitourinary malignancies	11 (1–23)	[5]
– penile cancer	21 (20–21)	
– bladder cancer	4 (15–23)	
– prostate cancer	16 (1–18)	
Head/neck cancer		[6]
– external lymphedema	90	
– internal lymphedema	89	
– internal and external lymphedema	81	
Sarcoma	30	[7]

weight causes significant strain on the musculoskeletal system and peripheral nerves. Lymphedema complications include among others: brachial plexopathy, cervical and lumbar radiculopathy, carpal tunnel syndrome, degenerative spine, and joint changes [1]. People with advanced lymphedema have difficulties to move around and have limited ability to work and to carry out daily activities [14].

CRL is also one of the most important conditions leading to psychosocial dysfunction. Lymphedema is a chronic, lifelong, and incurable condition. It is visible, hence can be stigmatizing. It constantly reminds the patient of cancer disease. Lymphedema makes it difficult to find suitable clothes or shoes. It often causes withdrawal from social life, or leads to loss of job. Anxiety and depressive symptoms are common. Therefore, lymphedema significantly contributes to an overall poor quality of life. It affects every aspect of daily life: physical, emotional, social, and financial [14].

Figure 25.1 presents examples of patients with cancer-related lymphedema.

25.3 Risk Factors of Cancer-Related Lymphedema

The advanced stage of cancer, extensive oncological treatment, including more extensive surgical procedures, removal of significant number of lymph nodes, receipt of adjuvant therapy (radiation therapy, but also chemotherapy taxanes), and skin infections in the peri- and postoperative period are the most important risk factors for lymphedema development in cancer patients [4, 5, 11, 15]. Increased body weight (particularly >25) and being insufficiently active are also associated with higher risk, however, the exact pathophysiological mechanism mediating this association is not known [4]. Generally, the possibility of developing CRL increases with the number of risk factors [11].

Despite similar risk factors, only some patients develop CRL. Genetic predisposition may also play a role. Published studies found associations between genetic factors and the development of breast CRL for variations in the following genes: *HGF, MET, GJC2, IL1A, IL4, IL6, IL10, IL13, VEGF-C, NFKB2, LCP-2, NRP-2, SYK, VCAM1, FOXC2, VEGFR2, VEGFR3*, and *RORC* [16]. Other possible risk factors are lower efficiency of lymphatic drainage and transport and concomitant chronic venous insufficiency. Women who developed breast CRL had higher pumping pressures and lymph transport before cancer treatment, what could indicate initial lymphatic overload [17]. Individual anatomical arrangement of lymph nodes, specific lymphatic and venous network, higher peripheral filtration rates within microcirculation, and individual weakness or underdevelopment of lymphatics and veins could constitute additional risk factors [18–20].

25.4 Prevention of Cancer-Related Lymphedema

Primary prevention of CRL is based on modification of risk factors, especially maintaining a proper body weight and performing regular physical activity. Avoiding skin injuries (e.g., using protective gloves when working in the kitchen or in the garden) and insect bites using repellants is also advised [11, 21]. Prophylactic *compression garments*, physical exercises, or self-massage after cancer treatment are often suggested in CRL prevention, however, these methods have been not sufficiently evaluated in clinical studies. Avoiding some lifestyle-associated behaviors and medical procedures on at-risk limb, such as air travel, effect of extreme temperatures (e.g., hot tubs or sauna), intravenous blood collection or injections, and ipsilateral arm blood pressure measurements is widely advised to cancer survivors; however, their significance in CRL prevention has also been insufficiently investigated [9, 11, 15].

Early diagnosis of CRL gives the opportunity for greater effectiveness of conservative treatment and in some patients of applying surgical methods, e.g., creation of lymphatic-venous anastomoses. Therefore, postoperative follow-up evaluation of patients assures early diagnosis (e.g., in 3 months intervals during the first post-cancer treatment and then less frequently) and is highly recommended [11].

25.5 Management of Cancer-Related Lymphedema

Treatment of lymphedema should be complex and multimodal. Optimally, it should involve cooperation of medical team consisting of lymphologist and physiotherapist and in some patients also of nurse, surgeon, dietician, and psychologist. Therapy of peripheral lymphedema is divided into conservative (nonoperative) and surgical methods [11].

25.5.1 Conservative Treatment

Complete decongestive therapy (CDT) is widely recognized as primary and effective conservative treatment. It can be used in both children and adults. CDT may be also applied as palliative method in cases of tumor-blocked lymphatics to control volume of malignant lymphedema [11, 22].

CDT includes skincare (protection from injuries, moisturizing, active treatment of possible ulcerations), use of manual lymphatic drainage (MLD) to augment lymphatic flow, compression therapy for volume reduction, and decongestive exercises, both aerobic and resistive. The first phase of CDT is applied to improve the condition of the skin and to reduce limb volume. The second phase of CDT aims to maintain and optimize the effect achieved during the first phase [9, 11, 22].

The most important component of CDT is *compression therapy*. The mechanism of its effectiveness in reducing edema volume depends on producing gradual pressure—the highest at the distal part of the limb and the lowest proximally. Major medical contraindications to compression therapy include heart failure, peripheral arterial disease, acute infections, and some skin disorders. Compression therapy during the first phase of CDT is provided using multilayered bandage wrapping with the low-stretch bandages worn for 23 h a day. After achieving the maximum reduction in volume, compression therapy in the second stage of CDT may be replaced by compression garments: preferably custom-made flat-knitted compression stockings or sleeves. The compression garments should be worn every day [11]. Compression systems with Velcro were recently reported to as effective as low-stretch bandages in the treatment of lymphedema [23]. This type of compression garment is easier to put on by patients and therapists and can effectively reduce limb volume in ambulatory patients who have no access to CDT.

Intermittent pneumatic compression, lymphotaping, low-level laser therapy and far infrared radiation therapy, play a supporting role in the management of lymphedema [11].

There is limited evidence to support weight loss in lymphedema treatment. However, higher body mass index is a risk factor of lymphedema development and weight loss contributes to relief in lymphedema-associated symptoms, improve body image, quality of life, and insulin control. Thus, weight loss should be recommended for patients with CRL [11, 21].

Psychological support and patient education in self-compression therapy, MLD, infection prevention, exercises, and diet for weight control are also crucial in successful CRL therapy [11].

25.5.2 The Pharmacological Management of Lymphedema and its Risk Factors

There are no FDA-approved drugs for lymphedema. Lymphedema has an inflammatory basis. The factors leading to its precipitation/worsening are infection and increasing fibrosis. By exploring/using drugs affecting these processes, we could expect increased benefits from the traditional core treatment modalities, reducing health system/patient costs, and improving functional outcomes.

> The Section "The Pharmacological Management of Lymphoedema and its risk factors" was solely prepared by Neil Piller

25.5.2.1 Anti T-Cell drugs

Regulatory T cells ensure that our immune responses do not overshoot the target. Lymphedema can be kept under control by suppressing inflammation, but the danger in this is that tumor growth may be promoted.

In a mouse model, increasing the number of T-reg cells reduced lymphedema development and reduced edema, inflammation, and fibrosis, and increased lymphatic drainage in established lymphedema [23].

Ketoprofen (an anti-inflammatory drug) was found to improve lymphatic function in a small trial [24].

Ubenimex (an inhibitor of Leukotriene LT (LTB4) synthesis) was investigated in a double-blind, placebo-controlled study with a 6-month treatment period, in adults with primary/secondary lymphedema, but there was no improvement over placebo in some primary and secondary endpoints [25].

Tacrolimus (an anti-T-cell immunosuppressive drug), applied externally prevented lymphedema development and improved established lymphedema. This study showed reduced swelling, T-cell infiltration and tissue fibrosis, and increased formation of lymphatic collaterals. There was minimal systemic absorption of the drug [26]. These studies raise the possibility of the use of this range of drugs in the treatment of secondary lymphedema but further human testing is essential.

25.5.2.2 Other Drugs

A new and relatively untrialed drug with potential anti-inflammatory/anti-fibrotic activities is deupirfenidone. It is a deuterated form of the synthetic antifibrotic agent pirfenidone. Animal studies have shown it inhibits a variety of pro-inflammatory mediators, such as interleukin-6 (IL-6), tumor necrosis factor-alpha (TNF-α), and transforming growth factor-beta (TGF-β). This means it has the potential to reduce fibrosis and inflammation and may facilitate improvements in impaired lymphatic flow.

25.5.2.3 Coumarin and the Benzopyrones

However, the focus of this section of this chapter revolves around changing the pathophysiology of the progression of lymphedema. This has led to a focus on the *Benzopyrones*. All are derived from an alpha or gamma pyrone united with a benzene ring. There are two main groups, *Coumarin* and its derivatives and *Flavones* and derivatives.

They have been shown to have a degree of clinical efficacy in lymphedema management. Some cases of idiosyncratic hepatotoxicity, however, were documented leading to a reluctance to use them. However, by considering their metabolic pathways and identifying relevant alleles needed to take a targeted pharmacogenetic approach in their future use this risk can be minimized.

Benzopyrones have reasonably high levels of evidence to indicate a benefit in reducing primary/secondary lymphedema [27, 28]. They reduce high protein edema by stimulating proteolysis by tissue macrophages [29]. Removing/reducing protein means a reduced osmotic force so they are freed to leave the accumulation sites. Since these often contain a range of inflammatory mediators, its removal reduces chronic inflammation, fibrosis formation, and the risk of secondary infection [27].

However, reports of hepatotoxic effects led to their use for lymphedema management being curtailed. The cytotoxic outcomes relate to its metabolism. Coumarin has two main hepato-detoxification pathways. First, the nontoxic 7-hydroxylation catalyzed by CYP2A6 and the toxic heterocyclic "ring-splitting" epoxidation pathways catalyzed likely by CYP1A and CYP2E. Acetaldehyde dehydrogenase (ALDH) clears toxic aldehyde intermediates.

Dysfunctional polymorphisms of the above genes are significantly more prevalent in Eastern Asian populations and relatively uncommon in Caucasian populations.

The problem of coumarin toxicity can relatively easily be solved by considering and knowing its metabolic pathways and identifying relevant alleles. Knowing these precautions, a targeted pharmaco-genetic approach can be taken in its use [30, 31]. CYP2A6 polymorphism screening methods, including genotyping, by real-time polymerase chain reaction, and chromatography–mass spectroscopy functional metabolite assays are available and increasingly reliable and clinically feasible, but some uncertainty remains about the link between the genotype, metabolic phenotype, and the exact gene products involved, these are being elucidated and once clarified, a targeted pharmaco-genomic approach may be appropriate.

25.5.2.4 Conclusion

A range of drugs, including anti-T cell, anti-inflammatory, and antibiotics, adds to the potential of pharmacological treatment/management of lymphedemas becoming mainstream. But refocusing on Coumarin shows it has a potential as a pharmacological means of targeting edemas where protein levels are elevated. Its core action is the stimulation of proteolysis by tissue macrophages. By removing protein, there is less osmotic force holding the fluids, freeing them from the accumulation sites, reducing associated chronic inflammation, and fibrosis formation.

A barrier to its common use (hepatotoxicity) can probably be removed through improved screening.

In combination with (or pre and post use) with other treatments such as compression and massage synergistic improved outcomes might be expected, but need to be further explored.

25.5.3 Surgical Approaches to Lymphedema Prevention and Treatment in Oncological Patients*

25.5.3.1 Background

A side effect of axillary lymph node excision and radiation therapy for breast cancer is arm lymphedema, which occurs in approximately 25% of patients (13% to 52%). *Sentinel lymph node (SLN) biopsy* reduced the severity of swelling in nearly 6% of patients (2% to 7%) but, in the case of positive SLN, complete axillary dissection (AD) is still required. This is why the *Axillary Reverse Mapping (ARM)* method was developed, with the aim of identifying and preserving the lymphatic vessels that drain the arm [32–34]. It consists of injecting a small amount (1–2 mL) of blue dye intradermally and subcutaneously on the medial surface of the arm, which helps to localize the lymphatic tracts of the draining arm. Leaving the lymph nodes related to lymphatic drainage in the arm in place would reduce the risk of arm lymphedema, but the main risk is to leave the metastatic disease in the axilla. On the contrary, the lymphatic pathways of the arm that enter the armpit are not considered as a usual site of breast cancer and their preservation would certainly lead to a significant decrease in the onset rate of lymphedema.

*The section on "Surgical Approaches to Lymphedema Prevention and Treatment in Oncological Patients" was solely prepared by Francesco Boccardo

25.5.3.2 Clinical Applications

Based on long-term experience in *lymphatic-venous anastomoses (LVA)* for the treatment of lymphedema [35, 36], we conceived and carried out a preventive LVA during nodal dissection, which consists of anastomosing the lymphatics of the arm to a collateral branch of the axillary vein (Lymphatic Microsurgical Preventing Healing Approach—*LYMPHA technique*) [37]. Indications for the LYMPHA technique are based on clinical and lymphoscintigraphic parameters. Obesity is known to predispose to lymphedema and therefore patients with BMI greater than 30 are considered to be at high risk of lymphedema and are candidates for LYMPHA. Patients with normal BMI are studied with lymphoscintigraphy which is to detect latent lymphatic impairment, not yet clinically evident. Radiotherapy causes temporary arm edema but do not jeopardize the LVA.

25.5.3.3 Discussion

Recent advances in breast cancer treatment, particularly regarding the prevention of lymphatic complications following sentinel lymph node biopsy and axillary dissection, have led to the proposal of a new technique to primarily prevent lymphedema by microsurgical lymphatic-venous anastomosis (LYMPHA). LVA has been shown not only to prevent lymphedema but also to reduce early lymphatic complications (e.g., lymphorrhea, lymphocele) thanks to the reduced regional intralymphatic pressure. The proposed surgical technique for patients with operable breast cancer requiring axillary dissection consisted of performing LVA [38, 39] between the identified arm lymphatic vessels by simultaneously injecting the blue dye into the arm and a branch of the axillary vein (Fig. 25.2). The incidence of secondary arm lymphedema after the LYMPHA technique is about 4.05%,

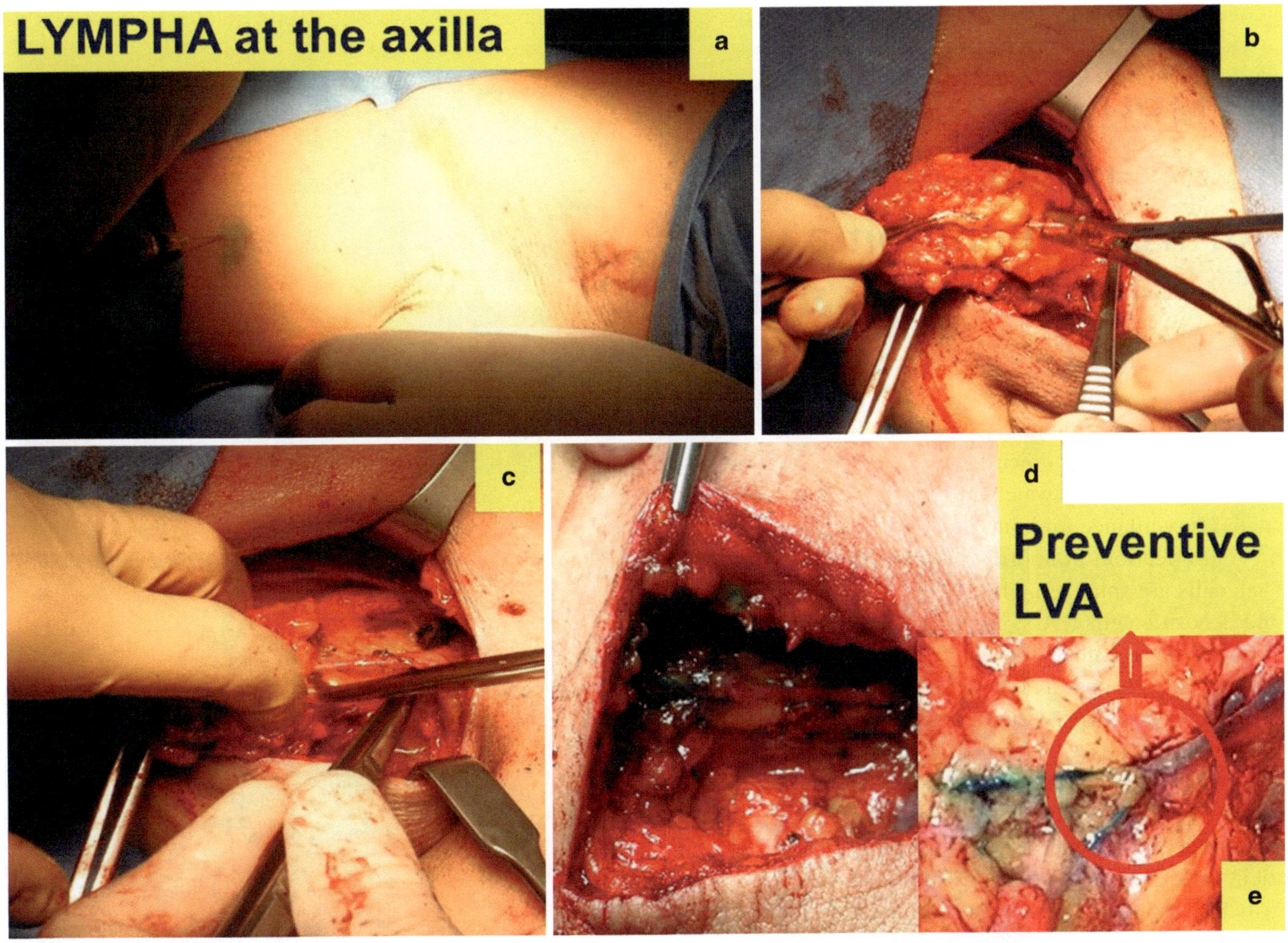

Fig. 25.2 (**a–e**) Injection of BPV dye to the arm (**a**), isolation of the brachial lymphatics during lymph node dissection and identification of the same using metal clips (**b**), isolation and preparation of a collateral of the axillary vein (**c**), lymphatic-venous anastomosis at the end (**d**), LYMPHA technique at higher magnification (**e**)

compared to the incidence rate after complete axillary dissection alone reported in the literature, which varies from 13% to 65%, depending on the criteria used and the evaluation methods.

25.5.3.4 LyMPHA Technique Beyond Breast Cancer

The LYMPHA technique is also useful in patients with trunk melanoma and vulvar cancer, in whom preventive lymphatic-venous anastomosis (LVA) can be performed simultaneously with inguinal lymphadenectomy [40] (Fig. 25.3). Patients at risk can be assessed peroperitively with lymphoscintigraphy and be referred to preventive measures.

25.5.3.5 Lymphedema Surgical Options

Different surgical techniques are nowadays proposed for the treatment of lymphedema. Most of them are reconstructive microsurgical techniques and some of them are debulking, modernly performed through mini-invasive procedures. There is still not a consensus on which are the specific indications for each technique, how to evaluate results, pros, and cons. The aim is to give an outline of the state of the art of surgical treatment of lymphedema. Various review articles and published clinical experiences from different groups of surgeons devoted to lymphatic surgery report about the usefulness of lymphatic reconstructive microsurgery in early stages, where there are less fibrotic tissue and lymphatic changes (8). It is much more difficult to find good lymphatics in more advanced stages, because of the severe sclerotic changes of the tissues and the aponeurotic fasciae, even though searching for both superficial and deep lymphatic collectors, actually a few numbers of lymphatic vessels may be found. It is mandatory to perform a proper preoperative diagnostic assessment through superficial and deep lymphoscintigraphy and echo-duplex venous scan. *Lymphangio-MR* is used in the most advanced stages to study the morphology of lymphatic drainage. *ICG lymphangiogra-*

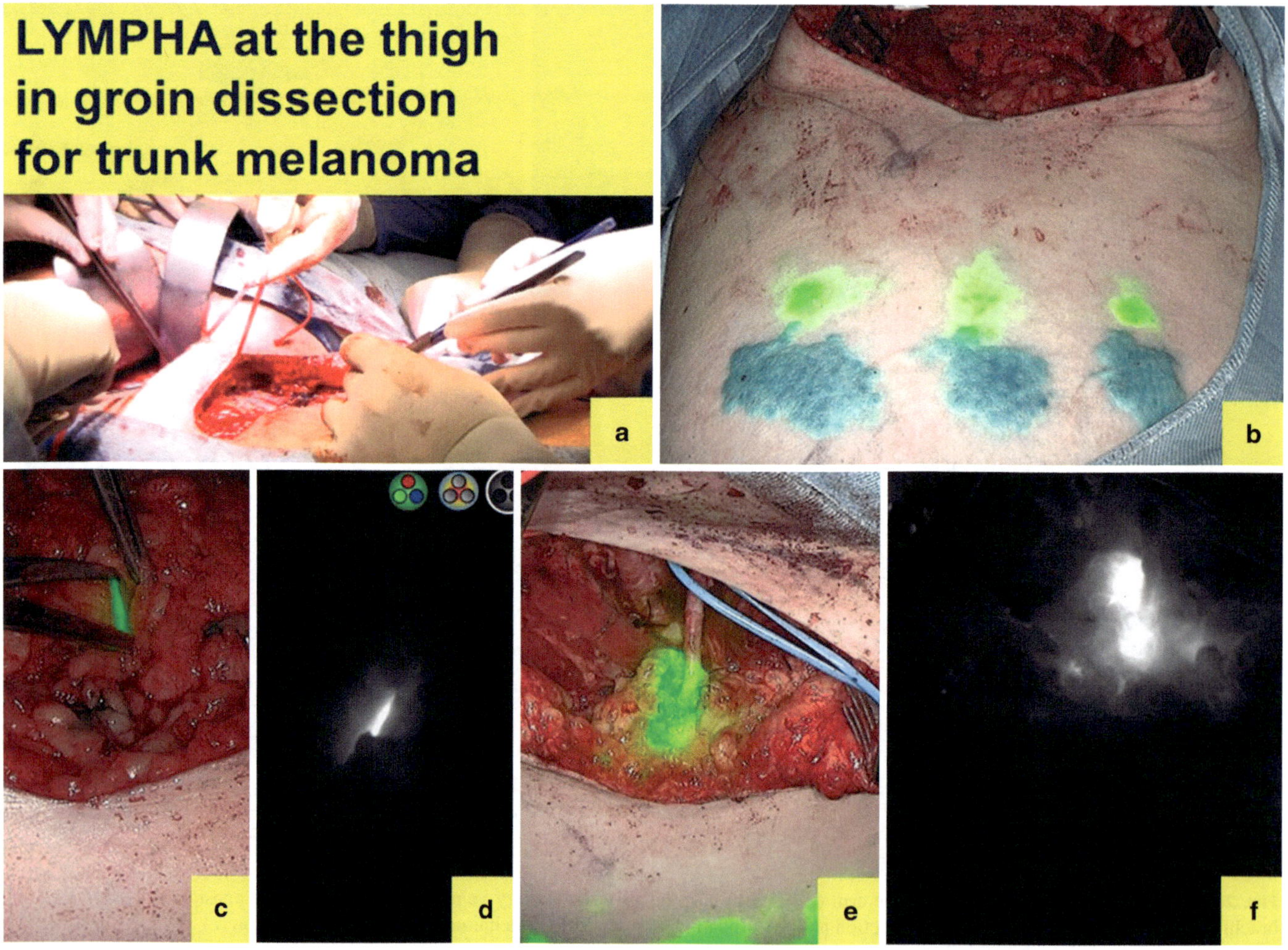

Fig. 25.3 (**a**–**f**) Iliac-obturator and femoro-inguinal lymph node dissection for trunk melanoma (**a**), injection of BPV dye and indocyanine green (ICG) distal to the surgical area (**b**), identification and isolation of the limb lymphatic tracts by fluorescence (**c**, **d**), lymphatic-venous anastomosis at the end, with control of patency by fluorescence (**e**, **f**)

phy is used peroperitively together with the blue dye (BPV) to find sound lymphatic collectors during surgery. Results are evaluated by clinical assessment and lymphoscintigraphy. Minimally performed debulking techniques are used in the most advanced cases to reduce the fibrotic tissue after CDP plus microsurgery.

The best results are obtained by microsurgery in the earlier stages and in secondary lymphedemas in terms of reduction of lymphangitis and less necessity of postoperative physical treatment and elastic garments. Microsurgery may be of great help also in late stages, which may however need proper postoperative treatment to improve the outcome.

The role of surgery in lymphedema treatment is of utmost importance mainly for secondary lymphedemas and obstructive primary ones. The meaning of microsurgical techniques is to open the tap and allow the lymph to pass through a lymphatic-venous bypass as it happens in arterial surgery and in both cases, the vein is the queen.

25.5.3.6 Final Considerations

Lymphedema is a consequence of cancer treatment and patients should be informed of the prevention options of lymphedema. The use of blue dye, Indocyanine Green (ICG), and LVA helps to solve the problem of the prevention of secondary lymphedema of the arms and legs while maintaining the oncological radicality. LYMPHA, therefore, could represent a rational approach to the prevention of lymphedema and reduce other lymphatic complications after axillary and inguinal surgery in the treatment of breast cancer, melanoma, vulvar cancer, and other cancers.

25.5.4 Treatment of Chylous Effusions from Metastatic Disease*

25.5.4.1 Background

Chylous disorders can manifest themselves with chyloperitoneum, chyledema of the lower limbs and/or of the external genitalia, chyluria, chylothorax, chylometrorrea, chylarthrosis, variously associated with one another and involve a demanding diagnostic and therapeutic process [41].

25.5.4.2 Diagnostics

The main diagnostic investigations include an accurate hematochemical study (including protein electrophoresis, lipidogram, calcemia), echotomography, lymphoscintigraphy, lymphangio-CT, lymphangio-MR, and laparoscopy, which is able to provide precise data on the clinical situation, severity, and extent of the disease [42].

25.5.4.3 Treatment

Chylous pathologies are initially treated by an adequate compensation of the metabolic conditions, with an appropriate dietary regime, based on protein reintegration and limiting the lipid intake exclusively to fats based on medium-chain triglycerides (MCT). The surgical timing will start articulating itself on the basis of the result obtained with the conservative therapy. The surgical treatment will be modulated according to the case and the different possible surgical procedures will involve:

- drainage of the chyloperitoneum;
- identification and closure of the sites of the chylorrhagia;
- removal of chylomas;
- ligatures of dysplastic chyliferous vessels;
- derivative microsurgical techniques (chyle-venous anastomosis);
- application of fibrin glue or platelet gel.

> **The section on "Treatment of chylous effusions from metastatic disease" was solely prepared by Francesco Boccardo

Finally, alternative therapy to the above-described techniques is represented by the peritoneo-jugular shunt (Denver shunt), in nonresponsive cases.

25.6 Conclusion and Perspective

- Cancer-related lymphedema (CRL) is a common consequence of malignancy and cancer treatment.
- The frequency of this complication will probably increase as a result of improvement in effectiveness of cancer treatment.
- CRL is a debilitating condition which significantly impairs the quality of life in cancer-survivors and demands lifelong therapy.
- Screening of cancer survivors toward lymphedema is essential in early diagnosis and successful treatment of CRL.
- Understanding the pathophysiology, recognition of risk factors and predisposition, and knowledge about prevalent methods of CRL treatment will provide the basis for the development of new preventive, diagnostic, and therapeutic methods in the management of CRL.

Open Questions

- Can we improve cancer treatment to avoid/minimize the risk of treatment-related lymphedema?
- What prophylactic measures should be mandatory in cancer patients at risk of lymphedema?
- Should early microsurgical restoration of lymphatic flow be recommended for all cancer patients with early lymphedema?
- Is therapeutic lymphangiogenesis a valid and safe option for lymphedema treatment?
- What exactly are protective mechanisms in cancer survivors, who do not develop limb lymphedema after treatment?

Patient Information and Guidelines

Support groups and resources can be found at the National Lymphedema Network.

References

1. Szuba A, Rockson SG. Lymphedema: anatomy, physiology and pathogenesis. Vasc Med. 1997;2(4):321–6.
2. Szuba A, Rockson SG. Lymphedema: classification, diagnosis and therapy. Vasc Med. 1998;3(2):145–56.
3. Brayton KM, Hirsch AT, O Brien PJ, Cheville A, Karaca-Mandic P, Rockson SG. Lymphedema prevalence and treatment benefits in cancer: impact of a therapeutic intervention on health outcomes and costs. PLoS One 2014;9(12):e114597.
4. DiSipio T, Rye S, Newman B, Hayes S. Incidence of unilateral arm lymphoedema after breast cancer: a systematic review and meta-analysis. Lancet Oncol. 2013;14(6):500–15.
5. Cormier JN, Askew RL, Mungovan KS, Xing Y, Ross MI, Armer JM. Lymphedema beyond breast cancer: a systematic review and meta-analysis of cancer-related secondary lymphedema. Cancer. 2010;116(22):5138–49.
6. Ridner SH, Dietrich MS, Niermann K, Cmelak A, Mannion K, Murphy B. A prospective study of the lymphedema and fibrosis continuum in patients with head and neck cancer. Lymphat Res Biol. 2016;14(4):198–205.
7. Robinson MH, Spruce L, Eeles R, Fryatt I, Harmer CL, Thomas JM, et al. Limb function following conservation treatment of adult soft tissue sarcoma. Eur J Cancer. 1991;27(12):1567–74.
8. Levick JR, Michel CC. Microvascular fluid exchange and the revised Starling principle. Cardiovasc Res. 2010;87(2):198–210.
9. Rockson SG, Keeley V, Kilbreath S, Szuba A, Towers A. Cancer-associated secondary lymphoedema. Nat Rev Dis Primers. 2019;5(1):22.
10. Mihara M, Hara H, Hayashi Y, Narushima M, Yamamoto T, Todokoro T, et al. Pathological steps of cancer-related lymphedema: histological changes in the collecting lymphatic vessels after lymphadenectomy. PLoS One. 2012;7(7):e41126.
11. The diagnosis and treatment of peripheral lymphedema. 2020 Consensus Document of the International Society of Lymphology. Lymphology. 2020;53(1):3–19.
12. Sharma A, Schwartz RA. Stewart-Treves syndrome: pathogenesis and management. J Am Acad Dermatol. 2012;67(6):1342–8.
13. Peyron N, Dandurand M, Guillot B. Malignant tumors as complications of lymphedema. J Mal Vasc. 1993;18(4):293–8.
14. Chachaj A, Malyszczak K, Pyszel K, Lukas J, Tarkowski R, Pudelko M, et al. Physical and psychological impairments of women with upper limb lymphedema following breast cancer treatment. Psychooncology. 2010;19(3):299–305.
15. Asdourian MS, Skolny MN, Brunelle C, Seward CE, Salama L, Taghian AG. Precautions for breast cancer-related lymphoedema: risk from air travel, ipsilateral arm blood pressure measurements, skin puncture, extreme temperatures, and cellulitis. Lancet Oncol. 2016;17(9):e392–405.
16. Visser J, van Geel M, Cornelissen AJM, van der Hulst R, Qiu SS. Breast cancer-related lymphedema and genetic predisposition: a systematic review of the literature. Lymphat Res Biol. 2019;17(3):288–93.
17. Cintolesi V, Stanton AW, Bains SK, Cousins E, Peters AM, Purushotham AD, et al. Constitutively enhanced lymphatic pumping in the upper limbs of women who later develop breast cancer-related lymphedema. Lymphat Res Biol. 2016;14(2):50–61.
18. Szuba A, Chachaj Z, Koba-Wszedybylb M, Hawro R, Jasinski R, Tarkowski R, et al. Axillary lymph nodes and arm lymphatic drainage pathways are spared during routine complete axillary clearance in majority of women undergoing breast cancer surgery. Lymphology. 2011;44(3):103–12.
19. Pain SJ, Barber RW, Ballinger JR, Solanki CK, Mortimer PS, Purushotham AD, et al. Tissue-to-blood transport of radiolabelled immunoglobulin injected into the web spaces of the hands of normal subjects and patients with breast cancer-related lymphoedema. J Vasc Res. 2004;41(2):183–92.
20. Britton TB, Solanki CK, Pinder SE, Mortimer PS, Peters AM, Purushotham AD. Lymphatic drainage pathways of the breast and the upper limb. Nucl Med Commun. 2009;30(6):427–30.
21. Shaw C, Mortimer P, Judd PA. Randomized controlled trial comparing a low-fat diet with a weight-reduction diet in breast cancer-related lymphedema. Cancer. 2007;109(10):1949–56.
22. Lasinski BB, McKillip Thrift K, Squire D, Austin MK, Smith KM, Wanchai A, et al. A systematic review of the evidence for complete decongestive therapy in the treatment of lymphedema from 2004 to 2011. PM R. 2012;4(8):580–601.
23. Gousopoulos E, Proulx ST, Bachmann SB, Scholl J, Dionyssiou D, Demiri E, et al. Regulatory T cell transfer ameliorates lymphedema and promotes lymphatic vessel function. JCI Insight. 2016;1(16):e89081.
24. Rockson SG, Tian W, Jiang X, Kuznetsova T, Haddad F, Zampell J, et al. Pilot studies demonstrate the potential benefits of antiinflammatory therapy in human lymphedema. JCI. Insight. 2018;3(20)
25. ULTRA Study EP. www.eigerbio.com/eiger-biopharmaceuticals-announces-phase-2-ultra-results-of-ubenimex-in-lower-leg-lymphedema-study-did-not-meet-primary-or-secondary-endpoint http://www.eigerbio.com/eiger-biopharmaceuticals-announces-phase-2-ultra-results-of-ubenimex-in-lower-leg-lymphedema-study-did-not-meet-primary-or-secondary-endpoint. 2017.
26. Gardenier JC, Kataru RP, Hespe GE, Savetsky IL, Torrisi JS, Nores GD, et al. Topical tacrolimus for the treatment of secondary lymphedema. Nat Commun. 2017;8:14345.
27. Casley-Smith JR, Morgan RG, Piller NB. Treatment of lymphedema of the arms and legs with 5,6-benzo-[alpha]-pyrone. N Engl J Med. 1993;329(16):1158–63.
28. Badger C, Preston N, Seers K, Mortimer P. Benzo-pyrones for reducing and controlling lymphoedema of the limbs. Cochrane Database Syst Rev. 2004;(2):CD003140.
29. Piller NB. The induction of controlled proteolysis in high protein oedemas by coumarin. Z Lymphol. 1979;3(2):110–3.
30. Farinola N, Piller NB. CYP2A6 polymorphisms: is there a role for pharmacogenomics in preventing coumarin-induced hepatotoxicity in lymphedema patients? Pharmacogenomics. 2007;8(2):151–8.
31. Hu M, Piller NB. Strategies for avoiding benzopyrone hepatotoxicity in lymphedema management: the role of pharmacogenet-

ics, metabolic enzyme gene identification, and patient selection. Lymphat Res Biol. 2017;15(4):317–23.
32. Thompson M, Korourian S, Henry-Tillman R, Adkins L, Mumford S, Westbrook KC, et al. Axillary reverse mapping (ARM): a new concept to identify and enhance lymphatic preservation. Ann Surg Oncol. 2007;14(6):1890–5.
33. Nos C, Lesieur B, Clough KB, Lecuru F. Blue dye injection in the arm in order to conserve the lymphatic drainage of the arm in breast cancer patients requiring an axillary dissection. Ann Surg Oncol. 2007;14(9):2490–6.
34. Ponzone R, Mininanni P, Cassina E, Sismondi P. Axillary reverse mapping in breast cancer: can we spare what we find? Ann Surg Oncol. 2008;15(1):390–1. author reply 92-3
35. Boccardo F, Casabona F, De Cian F, Friedman D, Villa G, Bogliolo S, et al. Lymphedema microsurgical preventive healing approach: a new technique for primary prevention of arm lymphedema after mastectomy. Ann Surg Oncol. 2009;16(3):703–8.
36. Francesco B, Corrado C, Giuseppe M, Emanuela B, Chiara B, Francesco P, et al. Prevention of lymphatic injuries in surgery. Microsurgery. 2010;30(4):261–5.
37. Boccardo FM, Casabona F, Friedman D, Puglisi M, De Cian F, Ansaldi F, et al. Surgical prevention of arm lymphedema after breast cancer treatment. Ann Surg Oncol. 2011;18(9):2500–5.
38. Boccardo F, Fulcheri E, Villa G, Molinari L, Campisi C, Dessalvi S, et al. Lymphatic microsurgery to treat lymphedema: techniques and indications for better results. Ann Plast Surg. 2013;71(2):191–5.
39. Witte MH, Dellinger MT, Papendieck CM, Boccardo F. Overlapping biomarkers, pathways, processes and syndromes in lymphatic development, growth and neoplasia. Clin Exp Metastasis. 2012;29(7):707–27.
40. Boccardo F, Valenzano M, Costantini S, Casabona F, Morotti M, Sala P, et al. LYMPHA Technique to prevent secondary lower limb lymphedema. Ann Surg Oncol. 2016;23(11):3558–63.
41. Boccardo F, Bellini C, Eretta C, Pertile D, Da Rin E, Benatti E, et al. The lymphatics in the pathophysiology of thoracic and abdominal surgical pathology: immunological consequences and the unexpected role of microsurgery. Microsurgery. 2007;27(4):339–45.
42. Dessalvi S, Boccardo F, Molinari L, Spinaci S, Campisi C, Ferrari GM. Chyloperitoneum: diagnostic and therapeutic options. Lymphology. 2016;49(1):1–7.

Part VII

Radiological Imaging of Early Cancer and Cancer Metastases

Jamie T. Caracciolo

Diagnostic imaging is a critical component in the ongoing fight against cancer and is widely utilized throughout all phases of patient management and cancer care from early tumor detection through long-term patient surveillance. A multitude of diagnostic imaging examinations and image-guided interventions are available to primary care providers and specialized cancer care physicians alike in the detection and diagnosis of various types and stages of malignant diseases, surgical and medical treatment planning and execution, assessment of treatment response, and post-treatment follow-up of patients living with cancer. Diagnostic and interventional radiologists working closely with primary care physicians, surgical specialists, medical and radiation oncologists, and pathologists in a multidisciplinary approach to cancer diagnosis and treatment provide for optimal patient management and clinical care. Ideally performed in close consultation with referring physicians, diagnostic imaging provides clinicians with critical data to inform many aspects of the clinical decision-making process.

Diagnostic imaging examinations including radiography, mammography, ultrasound, computed tomography (CT), magnetic resonance imaging (MRI), and nuclear medicine examinations, such as positron-emission tomography (PET), allow for detection and characterization of primary malignant tumors, initial tumor staging, identification of distant metastatic disease, and assessment of treatment response. Screening examinations of small high-risk patient groups or large patient populations are performed in an effort to detect earlier stage disease allowing for earlier initiation of treatment which often leads to improved patient outcomes such as disease-free survival and overall survival. Following detection of a suspicious abnormality, image-guided biopsies are available for tissue sampling to determine tumor histology and, in many cases, confirm the presence or absence of distant metastatic disease. Diagnostic imaging also informs many important decisions throughout the course of therapy. Cross-sectional imaging is widely available to assess the local extent of primary diseases helping to determine surgical resectability or the need for neoadjuvant therapy. CT, MRI, and PET/CT are often used to assess tumor response to chemotherapy in patients with non-operable or metastatic disease. Numerous targeted therapies delivered under fluoroscopic guidance, such as hepatic chemo- or radio-embolization, are currently available in the treatment of primary and metastatic disease. Cross-sectional imaging is also typically performed throughout the course of long-term surveillance of previously treated patients to detect recurrent disease. These are several of the many important and diverse roles diagnostic and interventional radiologists play in the management of patients with malignant diseases.

The following chapters focus on several important topics in diagnostic and interventional radiology relevant to the detection, diagnosis, and management of the most common primary malignancies in the United States and associated patterns of metastatic disease. First, we review several currently available and commonly used imaging modalities and examinations to detect, characterize, and stage malignant tumors, and present emerging techniques such as radiomics and artificial intelligence which promise to become part of the routine evaluation of malignant diseases in the future. We discuss typical imaging findings of the most common primary malignancies in the US emphasizing primary tumor characterization, important findings defining the local extent of disease which direct surgical or medical management, and common sites and patterns of distant metastatic disease for each primary tumor type. We expand our discussion of distant metastatic disease with dedicated chapters highlighting pertinent imaging findings in pulmonary metastatic disease, visceral and lymph node metastatic disease, and osseous metastatic disease. We also discuss relevant site-specific risks associated with metastatic disease, some of the available treatment options, and imaging findings reflective of response or resistance to therapy. Finally, we discuss the role of interventional radiology in the initial diagnosis and treatment of metastatic disease including image-guided biopsy and targeted interventions for metastatic disease.

Primary Tumor Staging and Detection of Common Sites of Distant Metastatic Disease

26

Kerry L. Thomas, A. Ahmed, and B. Morse

Abstract

Diagnostic imaging is a cornerstone of cancer diagnosis, staging, assessment of treatment response, and long-term surveillance. The most appropriate choice of imaging modality for primary tumor staging and detection of the metastatic disease depends on the tumor type and common sites of distant disease. This chapter reviews several of the most encountered cancers in the United States, discusses typical patterns of metastatic disease, and suggests appropriate imaging examinations to evaluate primary and metastatic disease.

Learning Objectives

After reviewing this chapter, the reader with better understand:

- Most common malignancies in men and women.
- Staging and prognosis of the malignancies reviewed.
- Common sites of metastatic disease.

26.1 Introduction

Cancer is a leading cause of death in the United States, second only to heart disease, according to the Centers for Disease Control, accounting for nearly 600,000 deaths in 2017 [1]. Diagnostic imaging plays a crucial role in initial detection and characterization of primary tumors, subsequent tumor staging, and surveillance of cancer. Depending on the primary site of disease, various imaging studies and modalities are available to characterize the primary tumor and to assess for locoregional and distant metastatic disease. This chapter discusses several of the most common primary malignancies affecting patients in the United States and how diagnostic imaging can be used in primary tumor assessment and detection of metastatic disease.

26.2 Breast Cancer

Breast cancer is expected to be the leading site of all new cancer cases and the second most common cause of cancer death in women in the United States in 2020. There will be an estimated 276,480 new cases and 42,170 new deaths related to breast cancer this year. These estimates account for 30% of all new cancer cases in women and 15% of cancer deaths in women for 2020 [2, 3]. The 5-year relative survival rate for breast cancer is 90%, and the lifetime risk of developing breast cancer is about 13% [2].

Staging of breast cancer follows the tumor, node, metastasis (TNM) staging system set forth by the American Joint Committee on Cancer (AJCC) [4]. Historically, the TNM staging system stages cancers by anatomic extent of disease, including tumor size, presence or absence of locoregional nodal metastases, and the presence or absence of distant metastases. However, for breast cancer staging, the eighth edition of the AJCC Cancer Staging Manual also includes a prognostic stage, which incorporates biologic factors, including tumor grade and hormone receptor status [4].

Diagnostic imaging used in initial staging of invasive breast cancer includes diagnostic mammogram or breast tomosynthesis and ultrasound as necessary for primary tumor assessment [5, 6]. Breast magnetic resonance imaging (MRI) is considered optional and may be helpful for defining tumor extent in patients with multifocal, multicentric, and

K. L. Thomas (✉)
Clinical Associate Professor, Department of Radiology, University of North Carolina at Chapel Hill, School of Medicine, Tampa, FL, USA
e-mail: Kerry_Thomas@med.unc.edu

A. Ahmed · B. Morse
H. Lee Moffitt Cancer Center, Tampa, FL, USA
e-mail: abraham.ahmed@moffitt.org; brian.morse@moffitt.org

S. P. Leong et al. (eds.), *Cancer Metastasis Through the Lymphovascular System*,
https://doi.org/10.1007/978-3-030-93084-4_26

occult or poorly defined tumors by other imaging modalities [5, 6].

The presence or absence of lymph node metastasis is an important predictor of overall survival and recurrence. Regional lymph nodes for the breast include axillary, ipsilateral internal mammary, and ipsilateral supraclavicular lymph nodes [4]. Axillary lymph nodes levels are determined by their relationship to the pectoralis minor muscle and include three levels: level I: lateral to pectoralis minor; level II: between the muscular borders of the pectoralis minor; and level III: medial to the pectoralis minor. Imaging assessment of regional axillary lymph nodes can be performed with diagnostic ultrasound and, if indicated, ultrasound-guided biopsy. However, controversy remains as to whether or not the routine use of axillary ultrasound should be employed preoperatively [4, 5]. MRI, when performed for evaluating primary tumor extent, is also helpful in evaluating the axilla for lymph node involvement as it is included in the field of view. MRI also provides information about internal mammary and supraclavicular lymph nodes (level III), which should be assessed if there is evidence of level I or level II adenopathy. If MRI is not performed, internal mammary and supraclavicular nodes may be assessed with contrast-enhanced chest computed tomography (CT).

National Comprehensive Cancer Network (NCCN) recommendations suggest that imaging evaluation for distant metastatic disease only be performed if there are other clinical signs, symptoms, or laboratory values that indicate metastasis [6].

The most common sites of distant metastatic disease for breast cancer include bone, liver, lung, and brain, and patients may have multiple sites of metastatic disease [7, 8]. Imaging studies recommended by the NCCN for metastatic evaluation, when indicated, includebone scan or sodium fluoride PET/CT to assess for bone metastases, chest CT for pulmonary metastases, and either CT or MRI of the abdomen and pelvis for visceral metastases. FDG PET/CT is optional [6].

26.3 Prostate Cancer

Prostate cancer is expected to be the leading site of all new cancer cases and the second most common cause of cancer death in men in the United States in 2020. It is estimated that there will be 191,930 new cases and 33,330 deaths from prostate cancer this year. This accounts for 21% of all new cancer cases and 10% of cancer related death in men [3, 9]. The 5-year relative survival rate for prostate cancer is 97.8% and the lifetime risk of developing prostate cancer is approximately 12% [9, 10].

Like other tumors, prostate cancer is staged utilizing the AJCC TNM staging system, but T stage is assessed both clinically and pathologically with Gleason score and group score, as well as serum prostate-specific antigen (PSA) [11].

Transrectal ultrasound (TRUS) is used for guiding prostate biopsy in clinically suspected prostate cancer. Multiparametric MRI of the prostate gland is useful in local T staging of prostate cancer and MRI may also be used for targeted biopsy [11, 12]. The T stage of the tumor reflects local extent of disease within the prostate gland and any extra-prostatic extension. While CT may demonstrate gross extracapsular extension, in general, CT is not considered sufficient to assess the prostate gland, but it is helpful in evaluating for nodal metastasis [13]. Nodal (N) staging for prostate cancer is less extensive than other diseases, whereby nodal staging consists of only three categories: not assessed, no positive nodes, or presence of positive nodes. Nodes refer to regional lymph nodes in the true pelvis. However, it should be mentioned, that the NCCN does not recommend nodal staging assessment in very low and low-risk patients. Nodal assessment for disease is reserved for patients in intermediate or higher risk categories where there is a greater probability (>10% risk) of nodal disease [13].

Distant sites of metastatic disease in prostate cancer most commonly occur in bones and lymph nodes, followed by liver, lung and pleura, and adrenal glands. Other sites are reported but are less common [14, 15]. Because bones are the most common site of distant metastatic disease, skeletal imaging is the mainstay of distant tumor detection. However, not all patients with prostate cancer should undergo bone imaging at initial presentation. Guidelines from the NCCN, American Urologic Association (AUA), and American College of Radiology Appropriateness Criteria recommend bone imaging in cases of clinical stage T2 and PSA >10 ng/mL, baseline PSA >20 ng/mL, clinical stage T3 or higher, or Gleason score of 8 or greater [12, 13]. Osseous metastases are usually osteoblastic, and Tc-99 whole-body bone scan has been standard to assess for osseous metastases. However, there are several novel radiotracers with promising results for improved accuracy compared with conventional bone scan. These tracers, which may be performed with both PET/CT and PET/MRI include 18-F sodium fluoride, 11-C Choline, and Fluciclovine [16]. Other imaging studies under investigation, including 68-Ga PMSA (prostatic membrane specific antigen), may prove useful in initial staging and detection of recurrent disease [16].

26.4 Bronchogenic/Lung Cancer

Lung cancer remains the leading cause of cancer deaths in the United States. Nearly 25% of all oncologic deaths occur as a result of primary lung carcinoma, and it accounts for more mortality than colon, breast, and prostate cancers combined [17]. Lung carcinomas are generally divided into two major categories: small cell and non-small cell carcinomas (NSCLC). Most lung malignancies, nearly 80–85%, are of the non-small cell type. Adenocarcinomas are the most common non-small cell cancers and the prevalent histologic type found in smokers. They are found predominantly in the periphery of the lung, owing to their origin in mucus-secreting cells lining the alveolar sacs.

Lymphogenic spread is the most common route of spread for lung malignancies. This is reflected in the TNM staging criteria which are the prevalent means of classifying the anatomic extent of tumor, where the N designates regional nodal involvement. N1 implies ipsilateral peribronchial or hilar lymph nodal spread, i.e., a right lung cancer with spread to a right hilar node. N2 implies spread to ipsilateral mediastinal and/or subcarinal lymph nodes. N3 indicates tumor spread to contralateral mediastinal, contralateral hilar, or supraclavicular nodal stations or any combination thereof [18].

Most lung cancers spread to ipsilateral hilar nodes before eventually reaching mediastinal nodes. Lung cancers in the lower lobes and middle lobe usually drain via the posterior mediastinum to subcarinal and subaortic nodal stations. Tumors in right upper lobe mostly drain to the superior mediastinum, particularly right paratracheal nodes. Tumors in the left upper lobe drain typically to the anterior mediastinum and involve the subaortic station, as well as to the upper mediastinum in nearly 1/3 of all cases. Retrograde lymphatic drainage and subsequently tumor spread to the pleura can occur, especially from peripheral tumors. Hematogenous spread to other more distant sites can also occur, with spread to other sites in the lungs, bone, brain, and liver being the most commonly observed sites of metastatic spread in NSCLC [19, 20].

NCCN guidelines recommend initial staging with contrast-enhanced chest CT. Pretreatment evaluation for resectability involves nodal measurement at different stations, with suspected nodal involvement warranting a multi-modality treatment approach rather than a primary surgical approach. A relatively low sensitivity of 57% has been reported when using a 1 cm short axis measurement threshold for determining metastatic involvement [21–24]. Hence, CT is useful only as an initial tool to guide further investigation into nodal involvement, potentially providing targets for guiding endobronchial ultrasound (EBUS) sampling which is ultimately necessary in most cases for accurate staging. PET/CT may be performed as an initial staging tool or in surveillance after treatment. It has a higher reported sensitivity for detection of nodal metastases [25] and has shown the ability to alter management in a significant number of patients, potentially avoiding mediastinoscopy and nodal sampling in a smaller subset [26, 27]; however, this last assertion remains controversial [28]. Contrast-enhanced abdominal CT is also typically performed to evaluate common sites of distant disease such as the liver and adrenal glands. MRI is not typically used in staging of lung cancers. However, it may be useful in establishing the relationship of the tumor and chest wall, confirm invasion of mediastinal structures, neural axis involvement, or determine the involvement of the brachial plexus in cases of Pancoast tumors [29, 30].

26.5 Colorectal Cancer

Colorectal cancer (CRC) is one of the most common causes of cancer-related death in the United States (third leading for both men and women) [31]. Worldwide CRC accounts for approximately 10% of newly diagnosed cancers and cancer-related deaths [32]. The rate of newly diagnosed CRC cases has dropped in the United States since the 1980s, a trend thought to be secondary to improved screening and population changes of modifiable risk factors [31]. Unfortunately, a new trend is increasing incidence in those younger than age 55 (increased 2% per year from 2007 to 2016) [31]. CRC is more common in developed countries and worldwide cases are expected to increase with improved living standards in less developed countries over the next several decades [32].

CRC is staged using the TNM system. A key component of primary tumor (T) staging is determination of local extent of disease and depth of mural invasion of the colon or rectum. This distinction is less important with colon cancer than rectal cancer. In general, colon cancer will be resected unless there is bulky nodal disease or invasion of adjacent organs [33]. Rectal cancer has a more nuanced approach. With rectal cancer, contrast-enhanced pelvic MRI is commonly used to determine whether the tumor is confined to the rectal wall or extends into the surrounding soft tissues. Imaging findings at pelvic MRI will inform timing and type of surgery, assist in the selection of patients for neoadjuvant therapy, and identify tumors with high-risk features [34]. Tumors confined to the rectum can proceed to upfront surgery while those extending beyond the wall may be downsized prior to surgery with chemotherapy and radiation therapy.

N staging is based on the number and location of lymph nodes involved by disease. Nodes outside the regional colon and/or rectal lymph node basin are considered metastatic. The M component of staging refers to distant metastatic disease. Common sites of metastatic disease for CRC are the liver, lung, and peritoneum [35]. Treatment varies based on the degree and location of metastatic disease, but in general there is a role for surgery in cases of limited metastatic disease. In most cases, initial assessment for metastatic disease in colorectal carcinoma is accomplished with contrast-enhanced CT of the chest, abdomen, and pelvis. PET/CT is not routinely used for initial staging, but could be considered when evaluating cases with limited metastatic disease to assist with therapy planning [33]. Contrast-enhanced abdominal MRI may be used to evaluate indeterminate liver lesions detected by other imaging modalities or surgical planning when CT is deemed inadequate.

26.6 Anal Cancer

Anal cancer is relatively uncommon although the worldwide incidence has been rising for many years [36]. In the United States, the lifetime risk of anal cancer is approximately 1 in 500 [36]. Most cases of anal cancer are now thought to be secondary to infection with oncogenic subtypes of the human papillomavirus (HPV). Worldwide it is estimated that 90% of anal cancer cases are secondary to HPV infection [37]. Risk factors for men include anal receptive sexual intercourse and in women risk increases with number of sexual partners and early age at first intercourse [36].

Anal cancer can be staged using the TNM system. The primary tumor stage (T) is determined by the size of the tumor and whether it has invaded adjacent structures. N staging is based on the number and location of involved lymph nodes. Non-regional lymph nodes qualify as metastatic disease. Common sites of metastatic disease (M) include the liver and lungs [38]. Therapy is generally chemotherapy and radiation therapy with surgery reserved for very small tumors at initial detection or in a salvage setting.

The main imaging modality used for staging of anal cancer is contrast-enhanced CT. Pelvic MRI can be useful to determine if there is invasion of adjacent structures with large or aggressive tumors. PET/CT can be considered at baseline or to evaluate for viable tumor after therapy. Recent work has found PET/CT can be useful in initial staging for characterization of metastatic lymph nodes compared to CT alone [38].

26.7 Kidney and Renal Pelvis Cancer

It is estimated that in 2020, there will be 73,750 new cases of kidney and renal pelvis cancer diagnosed in the United States [3]. Kidney and renal pelvis cancers represent the sixth most common cancer in men and eighth most common cancer in women in 2020, accounting for 5% and 3% of all new cancers cases, respectively [1, 3]. The rate of new cases of kidney and renal pelvis cancer has been on the rise, thought at least in part due to increased incidental detection with newer imaging techniques [39, 40]. According to the American Cancer Society, the overall 5-year relative survival rate of kidney and renal pelvis cancer is 75% for all stages combined, although the 5-year relative survival rate in those who present with localized disease is 93% [41].

While several imaging modalities are available to characterize indeterminate renal masses, it is recommended by both the American College of Radiology Appropriateness Criteria and the NCCN to perform multiphasic unenhanced and contrast-enhanced CT or MRI (CT urography or MR urography) for lesion characterization including unenhanced, nephrographic, and excretory phase imaging. Abdominopelvic CT or MRI is useful for assessing the primary tumor as well as tumor staging. Kidney cancers are staged according to the TNM classification from the AJCC. T staging is determined by tumor size as well as local invasion of the adjacent adrenal gland, perirenal soft tissues, and presence of renal vein and IVC tumor thrombus. Nodal staging is determined by the number of regional lymph nodes: none (N0); 1 node (N1); or multiple nodes (N2) [40].

Renal cell carcinoma (RCC) is more likely to spread hematogenously than lymphatically. Common sites of metastatic disease include the lungs and bones [39, 40]. Brain metastases can be seen in up to 17% of renal cancer patients [42]. As such, imaging for staging of RCC is aimed at these more common sites of disease. Guidelines recommend chest radiography as part of initial staging with consideration of chest CT when clinical presentation suggests distant disease. Similarly, bone scan may also be considered if there is clinical concern of osseous metastasis, whereby CT or MRI of the head/brain is useful to screen for brain metastasis. At present, the role of PET/CT in the staging of renal cancer is unclear and often limited by excreted radiotracer, and thus, not usually recommended [39].

26.8 Urinary Bladder Cancer

Bladder cancer is the sixth most common cancer in the United States [43]. It is most commonly diagnosed later in life (mean age of 73 years) [43]. Risk factors for bladder cancer include male sex, tobacco exposure, exposure to some industrial chemicals and medications, certain birth defects and genetic syndromes, and chronic bladder irritation [44].

Bladder cancer is staged using the TNM staging system. Primary tumor (T) staging is defined by depth of mural invasion and whether the tumor extends to the muscular layer of the bladder wall. Non-muscle invasive bladder cancer can be treated with local transurethral resection and intravesical instillation of chemotherapy. When tumor invades the muscular layer of the bladder wall, cystectomy is necessary for complete tumor removal and local tumor control. T staging is often performed using transurethral resection and pathological evaluation of the depth of invasion, although recent advances may increase the role of MRI in the future. Contrast-enhanced pelvic MRI including high-resolution T2-weighted imaging is gaining acceptance in the initial evaluation of bladder cancer including muscular invasion [45]. Several studies have also shown the ability of MRI to detect residual or recurrent tumor after therapy [45].

Complete staging of bladder cancer typically includes contrast-enhanced abdominopelvic CT to assess the size, location, and extent of the primary bladder tumor; assessment of locoregional lymph nodes in the pelvis; evaluate the kidneys and upper urinary tracts for synchronous genitourinary tumors; and identify distant metastatic disease to abdominal viscera or retroperitoneum. N staging is based on number and location of locoregional lymph nodes with non-regional nodes classified as metastatic disease. Common sites of metastatic disease include liver, lungs, and bones [35]. Chest imaging can be performed with either radiographs or CT based on clinical suspicion of pulmonary metastatic disease (whereby higher suspicion warrants chest CT) [46]. Whole-body bone scan may also be performed in cases with increased clinical concern for bony metastases [46]. PET/CT is generally not indicated but may be used as a problem-solving tool.

26.9 Melanoma

According to the American Cancer Society, over 100,000 new cases of melanoma will be diagnosed in the United States in 2020, resulting in nearly 7000 deaths. The risk for developing melanoma increases with age, however, it is not an uncommon cancer in young adults, particularly females [3].

Five-year survival rates for melanoma vary greatly based on classification into localized, regional, and distant stages. A 99% 5-year survival rate is reported for localized cancers, while that percentage drops to 66% for regional spread and only 27% for distant metastatic disease [47]. Accurate initial staging is important not only for risk stratification and prognosis but also for guiding treatment.

For detection of melanoma nodal spread, nuclear lymphoscintigraphy has become standard of care for identification of the sentinel draining node as well as in-transit nodes. These represent embryonic lymphatic rests and can also harbor tumor along lymphatic channels. ^{99m}Tc sulfur colloid is the FDA-approved radiopharmaceutical injected and imaged both dynamically and intraoperatively to identify these sites, with performance superior to vital dyes [48, 49].

NCCN guidelines do not recommend baseline imaging for stage I or stage II melanoma, defined as local disease. Baseline imaging including contrast-enhanced CT of the chest, abdomen, and pelvis should be performed for stage III or regional melanoma, defined as spread either to adjacent skin, soft tissues or draining nodal basin. In more advanced stage III disease, a brain MRI should also be considered. Baseline ultrasound is recommended in cases where nodal disease is equivocal by physical examination; otherwise, sonography is typically utilized in nodal basin surveillance after resection. Stage IV metastatic disease should also be staged by similar CT and MRI exams, with consideration to whole-body PET/CT scanning [50, 51]. PET/CT has a relatively high sensitivity for the detection of metastatic disease, particularly in more advanced disease and may result in disease management changes in nearly a third of patients [52].

26.10 Pancreatic Cancer

While not among the most common cancers, pancreatic cancer still accounts for nearly 8% of all cancer deaths and only has a 10% combined 5-year survival rate [3]. Some of the challenges in effective management of this deadly cancer include a relatively large median tumor size of 30 mm at the time of presentation and metastatic disease affecting a majority of patients [53]. While resectability numbers are dismal for tumors 30 mm or greater at detection, these have been reported as high as 83% for tumors that are smaller than 2 cm [54]. Larger tumors also have a higher propensity to grow outside the confines of the pancreas via perineural spread, precluding curative intent via surgery, and resulting in a high recurrence rate [55]. Hence, early detection is the largest factor in improving survival from pancreatic cancer.

Staging of pancreatic cancer is via the AJCC TNM classification. NCCN guidelines have been established for stratifying nonmetastatic tumors into primary resectable, borderline resectable, or nonresectable categories based on the degree of vascular encasement [56]. Tumors without any vascular involvement are deemed primary resectable, while there is some debate about the borderline resectable and nonresectable categories [57, 58]. Despite differences between institutions on management, an accurate assessment of the anatomic relation of the tumor to critical vessels remains key in determining the surgical outcome.

Contrast-enhanced multi-detector CT utilizing a pancreatic protocol and multiplanar reconstructions is the primary means of staging pancreatic cancer and guiding treatment. CT has a relatively high sensitivity for detection of tumor vascular encasement [59] and may also be able to discern extra-pancreatic and metastatic disease which would preclude an R0 or microscopically negative surgical resection, and patients typically may receive neoadjuvant therapy followed by reassessment of resectablility in this scenario. Limitations for CT include an inability to diagnose or accurately assess the size of isoattenuating tumors, limited detection of nodal metastases with enlarged nodes being frequently reactive and small nodes harboring micrometastases, and poor sensitivity for small liver metastases [60]. MRI is a helpful adjunct to CT with better detection of isoattenuating tumors and small liver metastases and should be performed preoperatively in surgical candidates [61] as well in patients undergoing neoadjuvant therapy. PET/CT may also serve as an adjunct to CT staging, with increased sensitivity for distant metastatic disease with the caveat that nearly 30% of all tumors are not PET avid [62].

There is ongoing research in increasing tumor conspicuity with spectral CT utilizing low energy acquisition and split contrast bolus protocols [63, 64]. Newer MR-guided radiation therapy systems have been developed with superior tissue contrast to guide real-time tumor demarcation and track normal tissue motion with higher radiation doses [65].

Conclusion and Perspective

- Imaging plays a key role in the diagnosis and staging of cancer.
- Will improvements in imaging technology result in earlier cancer detection?
- Will advances in medical imaging help in the development of more effective anticancer agents?
- Will functional imaging (e.g., PET/CT) play a larger role in the detection and staging of cancer in the future?
- How will the growing field of artificial intelligence affect cancer imaging?
- Can improvements in medical imaging decrease the financial cost of cancer care?

References

1. FastStats—Leading Causes of Death [Internet]. [cited 2020 Oct 30]. Available from: https://www.cdc.gov/nchs/fastats/leading-causes-of-death.htm
2. Female Breast Cancer—Cancer Stat Facts [Internet]. [cited 2020 Oct 30]. Available from: https://seer.cancer.gov/statfacts/html/breast.html
3. American Cancer Society. Annual Cancer Facts and Figures 2020. [cited 2020 Oct 30]. Available from: https://www.cancer.org/content/dam/cancer-org/research/cancer-facts-and-statistics/annual-cancer-facts-and-figures/2020/cancer-facts-and-figures-2020.pdf
4. Kalli S, Semine A, Cohen S, Naber SP, Makim SS, Bahl M. American joint committee on cancer's staging system for breast cancer, eighth edition: what the radiologist needs to know. Radiographics. 2018;38:1921–33. https://doi.org/10.1148/rg.2018180056.
5. Lee SC, Jain PA, Jethwa SC, Tripathy D, Amashita MWY. Radiologists' role in breast cancer staging: providing key information for clinicians. Radiographics. 2014;34:330–42. Available from: www.rsna.org/rsnarights
6. Lurie RH, Anderson BO, Abraham J, Aft R, Agnese D, Allison KH, et al. NCCN Guidelines Version 6. 2020 Breast Cancer 2020.
7. DiSibio G, French SW. Metastatic patterns of cancers : results from a large autopsy study. Arch Pathol Lab Med. 2008;132:931–9. Available from: https://pubmed.ncbi.nlm.nih.gov/18517275/
8. Chen MT, Sun HF, Zhao Y, Fu WY, Yang LP, Gao SP, et al. Comparison of patterns and prognosis among distant metastatic breast cancer patients by age groups: a SEER population-based analysis. Sci Rep. 2017;7. Available from: https://pubmed.ncbi.nlm.nih.gov/28835702/
9. Prostate Cancer—Cancer Stat Facts [Internet]. [cited 2020 Oct 31]. Available from: https://seer.cancer.gov/statfacts/html/prost.html
10. Siegel RL, Miller KD, Jemal A. Cancer statistics, 2020. CA Cancer J Clin. 2020;70:7–30. https://doi.org/10.3322/caac.21590. Wiley
11. Buyyounouski MK, Choyke PL, McKenney JK, Sartor O, Sandler HM, Amin MB, et al. Prostate cancer—major changes in the American Joint Committee on Cancer eighth edition cancer staging manual. CA Cancer J Clin. 2017;67:245–53. Available from: https://pubmed.ncbi.nlm.nih.gov/28222223/
12. Coakley FV, Oto A, Alexander LF, Allen BC, Davis BJ, Froemming AT, et al. ACR appropriateness criteria® prostate cancer—pretreatment Detection, surveillance, and staging. J Am Coll Radiol. 2017;14:S245–57. Available from: https://pubmed.ncbi.nlm.nih.gov/28473080/
13. Freedman-Cass D, Shead DA, Schaeffer E, Antonarakis ES, Armstrong AJ, et al. NCCN Guidelines Version 2. 2020 Prostate Cancer 2020.
14. Bubendorf L, Schöpfer A, Wagner U, Sauter G, Moch H, Willi N, et al. Metastatic patterns of prostate cancer: an autopsy study of 1,589 patients. Hum Pathol. 2000;31:578–83. WB Saunders
15. Vinjamoori AH, Jagannathan JP, Shinagare AB, Taplin M-E, Oh WK, Van den Abbeele AD, et al. Atypical metastases from prostate cancer: 10-year experience at a single institution. Am J Roentgenol. 2012;199:367–72. https://doi.org/10.2214/AJR.11.7533.
16. Tanaka T, Yang M, Froemming AT, Bryce AH, Inai R, Kanazawa S, et al. Current imaging techniques for and imaging spectrum of prostate cancer recurrence and metastasis: a pictorial review. Radiographics. 2020;40:709–26. Radiological Society of North America Inc. Available from: https://pubmed.ncbi.nlm.nih.gov/32196428/
17. Siegel RL, Miller KD, Jemal A. Cancer statistics, 2020. CA A Cancer J Clin. 2020;70:7–30. https://doi.org/10.3322/caac.21590.
18. Rami-Porta R, Asamura H, Travis WD, et al. Lung cancer - major changes in the American joint committee on cancer eighth edition cancer staging manual. CA A Cancer J Clin. 2017;67:138–55. https://doi.org/10.3322/caac.21390.
19. Hess KR, Varadhachary GR, Taylor SH, et al. Metastatic patterns in adenocarcinoma. Cancer. 2006;106:1624–33. https://doi.org/10.1002/cncr.21778.
20. Viadana E, Au K-L. Patterns of metastases in adenocarcinomas of man. An autopsy study of 4728 cases. J Med. 1975;6:1–14.
21. Toloza EM, Harpole L, McCrory DC. Noninvasive staging of non-small cell lung cancer*: a review of the current evidence. Chest. 2003;123:137–46. https://doi.org/10.1378/chest.123.1_suppl.137S.
22. Fritscher-Ravens A, Bohuslavizki KH, Brandt L, et al. Mediastinal lymph node involvement in potentially resectable lung cancer: comparison of CT, positron emission tomography, and endoscopic ultrasonography with and without fine-needle aspiration. Chest. 2003;123:442–51. https://doi.org/10.1378/chest.123.2.442.
23. Arita T, Matsumoto T, Kuramitsu T, et al. Is it possible to differentiate malignant mediastinal nodes from benign nodes by size? Reevaluation by CT, transesophageal echocardiography, and nodal specimen. Chest. 1996;110:1004–8. https://doi.org/10.1378/chest.110.4.1004.
24. Prenzel KL, Mönig SP, Sinning JM, et al. Lymph node size and metastatic infiltration in non-small cell lung cancer. Chest. 2003;123:463–7. https://doi.org/10.1378/chest.123.2.463.
25. Steinert HC, Hauser M, Allemann F, et al. Non-small cell lung cancer: nodal staging with FDG PET versus CT with correlative lymph node mapping and sampling. Radiology. 1997;202:441–6. https://doi.org/10.1148/radiology.202.2.9015071.
26. Stamatis G. Staging of lung cancer: the role of noninvasive, minimally invasive and invasive techniques. Eur Respir J. 2015;46:521–31. https://doi.org/10.1183/09031936.00126714.
27. Schimmer C, Neukam K, Elert O. Staging of non-small cell lung cancer: clinical value of positron emission tomography and mediastinoscopy. Interact Cardiovasc Thorac Surg. 2006;5:418–23. https://doi.org/10.1510/icvts.2006.129478.

28. Xia Y, Zhang B, Zhang H, et al. Evaluation of lymph node metastasis in lung cancer: who is the chief justice? J Thorac Dis. 2015;7:231–7. https://doi.org/10.3978/j.issn.2072-1439.2015.11.63.
29. Komaki R, Putnam JB Jr, Walsh G, et al. The management of superior sulcus tumors. Semin Surg Oncol. 2000;18:152–64. https://doi.org/10.1002/(sici)1098-2388(200003)18:2<152:aid-ssu9>3.0.co;2-h.
30. MacDonald SLS, Hansell DM. Staging of non-small cell lung cancer: imaging of intrathoracic disease. Eur J Radiol. 2003;45:18–30. https://doi.org/10.1016/s0720-048x(02)00300-5.
31. Colorectal Cancer Statistics | How Common Is Colorectal Cancer? [Internet]. [cited 2020 Oct 31]. Available from: https://www.cancer.org/cancer/colon-rectal-cancer/about/key-statistics.html
32. Dekker E, Tanis PJ, Vleugels JLA, Kasi PM, Wallace MB. Colorectal cancer [internet]. Lancet. 2019:1467–80. Lancet Publishing Group. Available from: https://pubmed.ncbi.nlm.nih.gov/31631858/
33. Benson AB, Al-Hawary MM, Arain MA, Chen Y-J, Ciombor KK, Cohen S, et al. NCCN Guidelines Version 42020 Colon Cancer Continue NCCN Guidelines Panel Disclosures; 2020.
34. Horvat N, Rocha CCT, Oliveira BC, Petkovska I, Gollub MJ. MRI of rectal cancer: tumor staging, imaging techniques, and management. Radiographics. 2019;39:367–87. https://doi.org/10.1148/rg.2019180114.
35. Metastatic Cancer—National Cancer Institute [Internet]. [cited 2020 Oct 31]. Available from: https://www.cancer.gov/types/metastatic-cancer
36. Key Statistics for Anal Cancer [Internet]. [cited 2020 Oct 31]. Available from: https://www.cancer.org/cancer/anal-cancer/about/what-is-key-statistics.html
37. Valvo F, Ciurlia E, Avuzzi B, Doci R, Ducreux M, Roelofsen F, et al. Cancer of the anal region [internet]. Crit Rev Oncol Hematol. 2019:115–27. Elsevier Ireland Ltd.; Available from: https://pubmed.ncbi.nlm.nih.gov/30819440/
38. Durot C, Dohan A, Boudiaf M, Servois V, Soyer P, Hoeffel C. Cancer of the anal canal: diagnosis, staging and follow-up with MRI. Korean J Radiol. 2017;18:946–56. Korean Radiological Society; Available from: https://pubmed.ncbi.nlm.nih.gov/29089827/
39. Vikram R, Beland MD, Blaufox MD, Moreno CC, Gore JL, Harvin HJ, et al. ACR Appropriateness criteria renal cell carcinoma staging. J Am Coll Radiol. 2016;13:518–25. Available from: https://pubmed.ncbi.nlm.nih.gov/27016804/
40. Ng CS, Wood CG, Silverman PM, Tannir NM, Tamboli P, Sandler CM. Renal cell carcinoma: diagnosis, staging, and surveillance. AJR Am J Roentgenol. 2008:1220–32. Available from: https://pubmed.ncbi.nlm.nih.gov/18806169/
41. Kidney Cancer Early Detection, Diagnosis, and Staging. [cited 2020 Oct 31]; Available from: www.cancer.org/treatment/understanding-your-diagnosis/tests/imaging-radiology-
42. Shuch B, La Rochelle JC, Klatte T, Riggs SB, Liu W, Kabbinavar FF, et al. Brain metastasis from renal cell carcinoma. Cancer. 2008;113:1641–8. https://doi.org/10.1002/cncr.23769. John Wiley & Sons, Ltd
43. Degeorge KC, Holt HR, Hodges SC. Bladder cancer: diagnosis and treatment. Am Fam Physician. 2017; Available from: www.aafp.org/afp
44. Bladder Cancer Risk Factors [Internet]. [cited 2020 Oct 31]. Available from: https://www.cancer.org/cancer/bladder-cancer/causes-risks-prevention/risk-factors.html
45. Gurram S, Muthigi A, Egan J, Stamatakis L. Imaging in localized bladder cancer: can current diagnostic modalities provide accurate local tumor staging? Curr Urol Rep. 2019;20(12):1–7. Springer; Available from: https://pubmed.ncbi.nlm.nih.gov/31781871/
46. Flaig TW, Spiess PE, Chair V, Agarwal N, Bangs R, Patient Advocate M, et al. NCCN Guidelines Version 6.2020 Bladder Cancer 2020.
47. National Cancer Institute Surveillance, Epidemiology, and End Results Database. Cancer Stat Facts: Melanoma of the Skin.
48. Cascinelli N, Belli F, Santinami M, et al. Sentinel lymph node biopsy in cutaneous melanoma: the WHO Melanoma Program Experience. Ann Surg Oncol. 2000;7:469–74. https://doi.org/10.1007/s10434-000-0469-z.
49. Intenzo CM, Truluck CA, Kushen MC, et al. Lymphoscintigraphy in cutaneous melanoma: an updated Total body atlas of sentinel node mapping. Radiographics. 2009;29:1125–35. https://doi.org/10.1148/rg.294085745.
50. Freeman M, Laks S. Surveillance imaging for metastasis in high-risk melanoma: importance in individualized patient care and survivorship. Melanoma Manag. 2019;6:12. https://doi.org/10.2217/mmt-2019-0003.
51. Coit DG, Thompson JA, Albertini MR, et al. Cutaneous melanoma, Version 2.2019, NCCN clinical practice Guidelines in oncology. J Natl Compr Cancer Netw. 2019;17:367–402. https://doi.org/10.6004/jnccn.2019.0018.
52. Krug B, Crott R, Lonneux M, et al. Role of PET in the initial staging of cutaneous malignant melanoma: systematic review. Radiology. 2008;249:836–44. https://doi.org/10.1148/radiol.2493080240.
53. Hidalgo M. Pancreatic cancer. N Engl J Med. 2010;362:1605–17. https://doi.org/10.1056/NEJMra0901557.
54. Agarwal B, Correa AM, Ho L. Survival in pancreatic carcinoma based on tumor size. Pancreas. 2008;36:15–20. https://doi.org/10.1097/mpa.0b013e31814de421.
55. Chang ST, Jeffrey RB, Patel BN, et al. Preoperative multidetector CT diagnosis of extrapancreatic perineural or duodenal invasion is associated with reduced postoperative survival after pancreaticoduodenectomy for pancreatic adenocarcinoma: preliminary experience and implications for patient care. Radiology. 2016;281:816–25. https://doi.org/10.1148/radiol.2016152790.
56. Tempero MA, Malafa MP, Chiorean EG, et al. Pancreatic adenocarcinoma, version 1.2019. J Natl Compr Cancer Netw. 2019;17:202–10. https://doi.org/10.6004/jnccn.2019.0014.
57. Varadhachary GR, Tamm EP, Abbruzzese JL, et al. Borderline resectable pancreatic cancer: definitions, management, and role of preoperative therapy. Ann Surg Oncol. 2006;13:1035–46. https://doi.org/10.1245/ASO.2006.08.011.
58. Khorana AA, Mangu PB, Berlin J, et al. Potentially curable pancreatic cancer: American Society of Clinical Oncology clinical practice guideline update. J Clin Oncol. 2017;35:2324–8. https://doi.org/10.1200/JCO.2017.72.4948.
59. Zamboni GA, Kruskal JB, Vollmer CM, et al. Pancreatic adenocarcinoma: value of multidetector CT angiography in preoperative evaluation. Radiology. 2007;245:770–8. https://doi.org/10.1148/radiol.2453061795.
60. Zins M, Matos C, Cassinotto C. Pancreatic adenocarcinoma staging in the era of preoperative chemotherapy and radiation therapy. Radiology. 2018;287:374–90. https://doi.org/10.1148/radiol.2018171670.
61. Kim HW, Lee J-C, Paik K-H, et al. Adjunctive role of preoperative liver magnetic resonance imaging for potentially resectable pancreatic cancer. Surgery. 2017;161:1579–87. https://doi.org/10.1016/j.surg.2016.12.038.
62. Rijkers AP, Valkema R, Duivenvoorden HJ, et al. Usefulness of F-18-fluorodeoxyglucose positron emission tomography to confirm suspected pancreatic cancer: a meta-analysis. Eur J Surg Oncol. 2014;40:794–804. https://doi.org/10.1016/j.ejso.2014.03.016.
63. Macari M, Spieler B, Kim D, et al. Dual-source dual-energy MDCT of pancreatic adenocarcinoma: initial observations with data generated at 80 kVp and at simulated weighted-average 120 kVp. Am J Roentgenol. 2010;194:27–32. https://doi.org/10.2214/AJR.09.2737.
64. Brook OR, Gourtsoyianni S, Brook A, et al. Split-bolus spectral multidetector CT of the pancreas: assessment of radiation dose and tumor conspicuity. Radiology. 2013;269:139–48. https://doi.org/10.1148/radiol.13121409.
65. Kupelian P, Sonke J-J. Magnetic resonance-guided adaptive radiotherapy: a solution to the future. Semin Radiat Oncol. 2014;24:227–32. https://doi.org/10.1016/j.semradonc.2014.02.013.

Imaging Modalities in the Detection and Diagnosis of Metastatic Disease

27

S. Ali, S. Fenerty, and P. Jonnalagadda

Abstract

Imaging is a critical tool in the evaluation and workup of patients with known or suspected metastatic disease and in the follow-up of disease progression after surgery, chemotherapy, and radiation therapy. The imaging modality utilized is dependent on the primary tumor and location of the metastatic deposit(s). Traditional imaging modalities include plain film radiography, computed tomography (CT), ultrasound, magnetic resonance imaging (MRI), nuclear medicine, and more recently fused imaging such as PET-CT and PET-MRI. A single modality may be used in isolation; however, it is more common that various combinations are used depending on the specific tumor. Interventional radiology complements these diagnostic imaging modalities for diagnosis (biopsy and tissue sampling) as well as therapy (ablation, intravascular chemotherapy) and preoperative or palliative tumor embolization. In this chapter, we outline the basics of physics and the utility of each modality and indicate future directions.

Learning Objectives
- Understand the basic physics of the respective imaging modalities.
- Understand the benefits and limitations of each modality.
- Understand radiation dose for different modalities.
- Apply knowledge of the various modalities to maximize benefit.

S. Ali (✉) · S. Fenerty · P. Jonnalagadda
Temple University Hospital, Philadelphia, PA, USA
e-mail: sayed.ali@tuhs.temple.edu; Sarah.Fenerty@tuhs.temple.edu; Padmaja.Jonnalagadda@tuhs.temple.edu

27.1 Introduction

Metastatic disease is an indication of advanced tumor and generally indicates a poor prognosis. Historically, before the advent of medical imaging, patients frequently presented with advanced disease at the time of diagnosis. With the advent of plain film radiography little changed, as the sensitivity was such that only advanced disease was detectable, and generally only advanced bone or pulmonary metastases were diagnosed. CT imaging revolutionized the management of cancer patients as the spatial and contrast resolution far exceeds that of plain films. For the first time, abdominal, pelvic, and central nervous system structures could be imaged, tumors staged, and more appropriate therapy started. MRI, with its superior contrast resolution, was another step in earlier disease diagnosis, particularly in the central nervous system, abdomen, and pelvis. Fusion of cross-sectional imaging with nuclear medicine (PET-CT and recently PET-MRI) subsequently enabled a functional imaging overlay of the metabolic information provided by PET with the anatomic detail of CT and MRI. This has reduced the amount of unnecessary biopsies and allows for follow-up after treatment. Historically, nuclear medicine proved most useful in metastatic bone tumor assessment, but more recently has afforded significant advances in the early detection and diagnosis of other tumors.

27.2 Plain Film Radiography

Radiography relies on X-ray radiation to produce a two-dimensional image. X-rays are a form of ionizing radiation, which means they possess sufficient energy to potentially remove electrons from an atom, thus giving it a charge and creating an ion. X-rays were discovered by German physicist Wilhelm Roentgen in 1895 when he attempted to see if cathode rays could pass through glass and realized that the rays could pass through the heavy black paper covering the cathode tube. He named these mysterious rays "X" rays and soon

S. P. Leong et al. (eds.), *Cancer Metastasis Through the Lymphovascular System*,
https://doi.org/10.1007/978-3-030-93084-4_27

ascertained that the rays could also pass through human tissue, hence the term X-rays, which has become widely known in the medical literature.

A typical X-ray unit consists of a source, patient, image receptor, processor, and viewbox or PACS (Picture Archiving and Communication System) [1]. The source is the X-ray tube, which emits a beam containing a spectrum of photon energy (the X-ray beam). When the beam enters the body, a portion of the radiation is absorbed in a process known as attenuation. The remnant beam passes through the patient and is detected by radiographic film or, as is much more common in current practice, a digital plate detector. Due to differential absorption of the X-ray photons, areas of image receptor receive different amounts of radiation, resulting in an image with areas of contrasting density [2]. Those portions of the image receptor that receive the most radiation (i.e., portions of the beam experiencing the least attenuation) will be more exposed and are processed to appear darker. Conversely, those receiving the least radiation will appear lighter. The denser the structure encountered by the beam, the greater the photon attenuation. This is why bone, which is high density, appears white, while lungs, which contain mostly air, appear black, with soft tissues appearing varying shades of gray. Therefore, if the tumor destroys bone then the affected area will attenuate fewer photons and appear darker (lytic) on a background of whiter normal bone. Conversely, osteoblastic metastases will attenuate more photons and appear whiter than normal bone. Similarly, lung metastases in a background of very low-density air will appear as denser white lesions.

In the past, image intensifying screens were used to decrease the radiation dose. With the introduction of digital plates and computerized radiography systems, intensifying screens are no longer necessary [3]. Processing of films is also of historical interest, as the images generated by the digital plate are now transmitted electronically to the PACS system and viewed and stored electronically. The format used for storage and transfer universally is called DICOM (Digital Imaging and Communications in Medicine). Image storage can be done locally, however, with the vast amount of data not just from radiographs but also other imaging that is currently generated by nationwide radiology departments, data is also usually backed up by large third-party vendors [4].

27.3 Computed Tomography (CT)

Computed tomography (CT), formerly known as computed axial tomography (CAT), has probably been the greatest advancement in the detection and staging of metastatic disease. Its exquisite spatial and contrast resolution has revolutionized the field of oncologic imaging (Fig. 27.1). These scans are quick, fairly cost-effective, and well tolerated by patients. Multiplanar reconstructions can aid in visualizing pathology and in surgical planning. The administration of intravenous iodinated contrast media increases contrast resolution and therefore diagnostic accuracy. More recent formulations of low osmolar contrast media have resulted in significantly fewer side effects than prior formulations and are generally well tolerated [5]. Oral contrast media in abdominal/pelvic imaging also serves the same purpose, although this may not be routinely required in oncology follow-up imaging [6]. One caveat is that there is an increased radiation dose compared to plain films, especially when patients undergo multiple scans as is typical for oncologic imaging. However, the benefits usually greatly outweigh the risks in these patients.

The fundamental principle of CT imaging is the use of a rotating X-ray tube within a gantry. Detectors located 180° from the tube perform the same function as a conventional radiograph. The rotation of the tube allows multiple X-ray measurements of the same point from different angles which can be reconstructed using computer algorithms to generate an image or "slice" of tissue. The density of the tissue is detected and scaled using Hounsfield Units (HU), ranging from −1000 HU for air to +2000 HU for very dense bone and greater than 3000 for metal. Over the past three decades, the use of a two-dimensional array of detector elements (multidetector CT) compared to the linear array seen in conventional or helical scanners has dramatically increased the speed and resolution of these scans. Multidetector CT reduces the effect of patient motion and increases patient throughput without significantly increasing radiation dose if key imaging parameters are observed, including the use of automatic exposure control (AEC) [7]. When using AEC, care must be taken in obese patients or when imaging areas with metallic hardware, which can result in paradoxical dose increases [8]. More recent advances include dual-energy CT, where source techniques include the use of two X-ray tubes or fast Kev switching (among other techniques), or retrospective dual-layer technology where the energy separation is detector based with the outer layer absorbing higher energy photons and the inner layer lower energy photons [9]. This technology provides additional functional information including detection of bone marrow edema, which can be extremely useful in suspected metastatic disease where the morphologic CT appearance is normal although bone marrow edema is present, and for metal artifact reduction which can be especially useful in postoperative cancer imaging [10].

27.4 X-Ray Tomography

The most relevant application of tomosynthesis is digital breast tomosynthesis (DBT), a technique that generates a volumetric reconstruction of the whole breast from a set

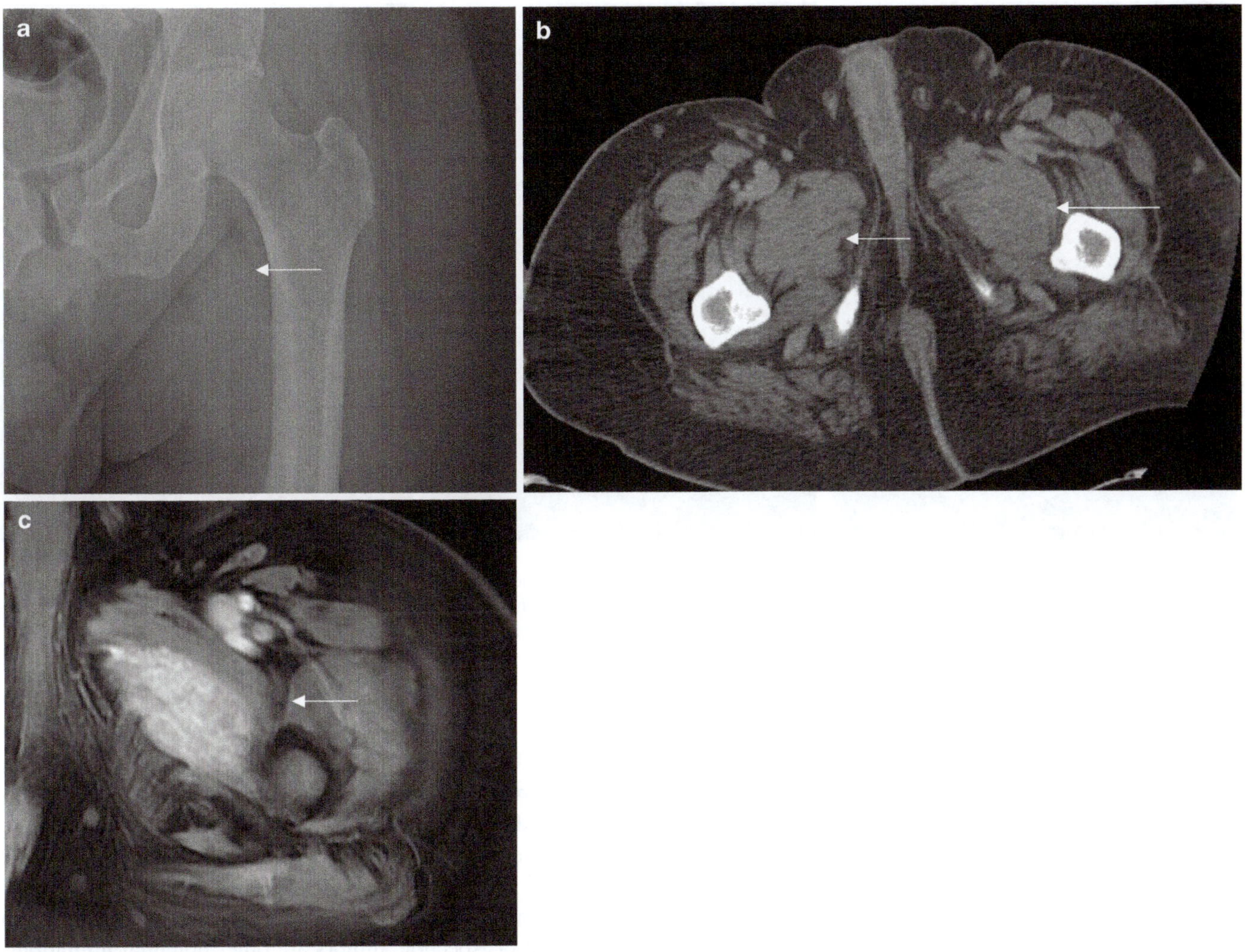

Fig. 27.1 56-year-old man with melanoma. (**a**) Plain films are normal. (**b**) Noncontrast CT shows subtle effacement of the left adductor muscle fat relative to the right. (**c**) Axial T2W MRI shows a large soft tissue mass consistent with metastasis. This illustrates the increasing contrast resolution from plain films through CT and MRI

number of two-dimensional projections obtained at different X-ray tube angles, which is emerging as the standard of care for breast imaging in both screening and diagnostic settings. Conventional mammography is two dimensional; thus, overlap of breast tissue, particularly in women with denser breasts, may both obscure cancers and simulate abnormalities where none are present. In DBT, the X-ray tube pivots in an arc (ranging from 15° to 60°) over the breast while acquiring multiple low-dose images, with each image delivering a fraction of the radiation dose provided in a standard mammogram and obtaining structural information for each layer. Through a computer algorithm, these are reconstructed into a series of stacked images, not dissimilar to CT "slices," providing localization information and improved lesion characterization [11] (Fig. 27.2), which may decrease or eliminate the need for additional diagnostic workup, particularly in those with dense breasts.

27.5 Ultrasound

Ultrasound (US) imaging uses a probe containing multiple acoustic transducers to send pulses of high-frequency sound ("ultrasound" waves, generally ranging from 2 to 20 MH), into the body [12]. When the sound waves encounter material of a different density (acoustic impedance), a portion of the wave is reflected back to the transducer and detected as an echo, which is then converted back into electrical impulses by the transducer crystals and processed to form the ultrasound image displayed on the screen. A greater difference in acoustic impedance (the resistance an ultrasound beam encounters as it passes through tissue) results in a higher proportion of reflected sound and proportionate decrease in the amount of transmitted sound. With significant differences in tissue density, sound is entirely reflected, resulting in com-

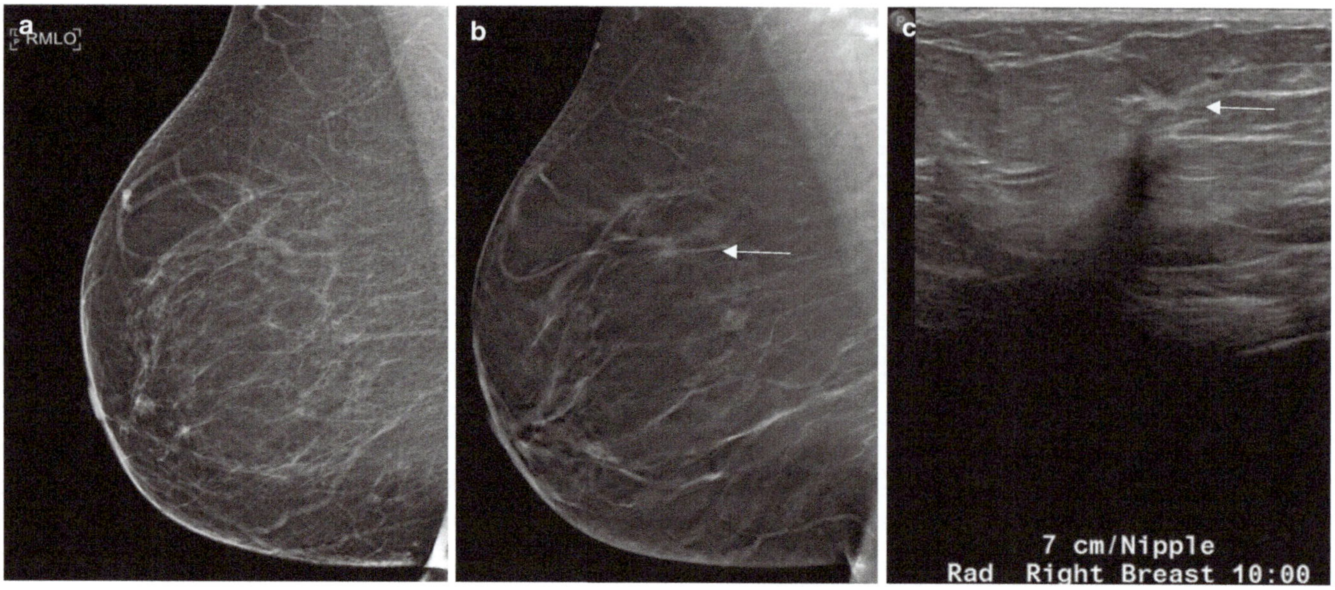

Fig. 27.2 55 year old woman for screening mammogram. (**a**) 2D Right MLO shows scattered fibroglandular tissue with no discrete mammographic abnormality (**b**) Corresponding tomosynthesis view shows architectural distortion in the upper outer right breast (arrow) (**c**) Ultrasound shows area of shadowing and distortion corresponding to tomosynthesis finding, biopsied to reveal invasive mammary carcinoma. Images courtesy of Mindy Yang, MD

plete acoustical shadowing and the inability to see deeper, as occurs behind bones, dense calcifications, and air.

Four different modes of ultrasound are used in medical imaging:

- A-mode (amplitude mode), the simplest type of ultrasound, in which a single transducer scans a line through the body with the echoes plotted on screen as a function of depth. This is used for echoencephalography and echo-ophthalmoscopy.
- B-mode (brightness mode, 2D mode, or gray-scale imaging), in which a linear array of transducers simultaneously scan a plane through the body which is displayed as a two-dimensional image on screen.
- M-mode (motion mode), a time motion display of the ultrasound wave along a chosen ultrasound line, used for analyzing moving body parts, typically used in cardiac, fetal cardiac, and lung US.
- Doppler mode, typically used to assess blood flow, allowing visualization of flow direction and velocity (color Doppler) and yielding graphical representations of flow velocity over time (spectral Doppler).

US is a convenient, cost-effective, and widely available imaging tool, free of ionizing radiation and capable of being performed bedside or brought into the operating room through the use of portable and point-of-care ultrasound machines. Most importantly, it can provide accurate diagnostic information comparable to CT and MRI. The main limitations of US include its greater dependence on operator skill, limited reproducibility, and potential for inadequate acoustical windows (often due to bowel gas or patient obesity). Incomplete visualization of lesions due to large lesion size or acoustic shadowing from calcifications, variable sonographic appearance of malignancies, and poor sensitivity in detection of small or isoechoic tumors are other potential limitations. Furthermore, US examinations are typically targeted to an area of interest, and hence are seldom the primary modality for comprehensive staging, but may be a useful adjunct in the focused assessment of certain organs (e.g., detection of liver or ovarian metastases or residual axillary nodal disease in breast cancer), can assess vascular tumor involvement, and demonstrate malignant ascites when present. In some instances, US examination performed for underlying symptoms (e.g., abdominal pain) may reveal a previously undiagnosed malignancy (Fig. 27.3).

Doppler imaging can provide important directional blood flow information, but it is most effective for evaluating large vessels such as carotid arteries, leg veins, and major visceral vessels of the abdomen, including hepatic artery, central portal, and hepatic veins. The ability of Doppler imaging to detect blood flow at the perfusional level is limited. Contrast-enhanced US (CEUS), which involves the intravenous administration of microbubble contrast agents (microbubbles of perfluorocarbon, nitrogen gas, or sulfur hexafluoride stabilized in a phospholipid membrane) and specialized imaging techniques, can provide dynamic real-time imaging blood flow and tissue perfusion information with high spatial

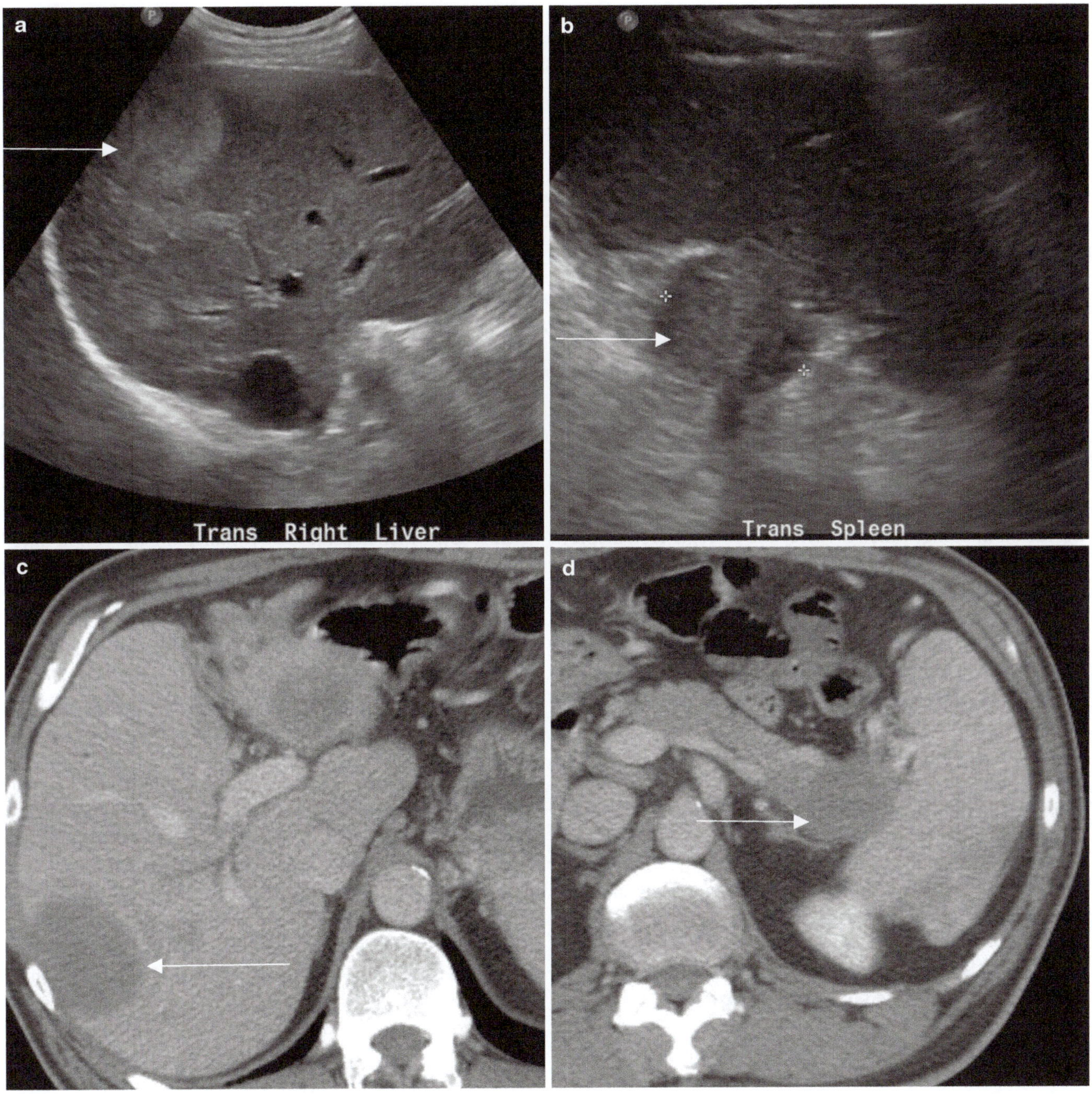

Fig. 27.3 67 year old man underwent abdominal ultrasound for evaluation of abdominal pain, which showed multiple hyperechoic liver masses, one of which is shown in (**a**); and a mass abutting the splenic hilum (**b**). Correlative portal venous phase CT images show hypoenhancing masses in the liver (**c**) and large mass involving pancreatic tail/splenic hilum (**d**). Subsequent biopsy of the depicted right hepatic lobe mass showed metastatic pancreatic adenocarcinoma

and temporal capability, thus allowing ultrasound to play a competitive role relative to CT and MRI in the evaluation of both solid and hollow organ pathology. Vascular CEUS contrast agents show distinct enhancement phases much like CT and MRI contrast media, permitting dynamic evaluation of the vascularity of a target lesion, most commonly in the liver or kidney, which may aid in diagnosis [13]. Supporting evidence for use of microbubble contrast agents includes their ease of use, positive contribution to patient care, lack of nephrotoxicity, and favorable safety profile [14].

Variations of CEUS such as dynamic contrast-enhanced ultrasound (DCE-US) and ultrasound molecular imaging (targeted CEUS), which utilizes molecularly targeted ultrasound contrast agents, are emerging techniques for quantitative monitoring of treatment effects and earlier cancer detection. CEUS has an advantage over contrast-enhanced MRI and CT in patients with contraindications such as renal failure or iodinated contrast allergy and also allows for dynamic and repeat examinations. In the United States, a greater availability of CT and MRI and limited availability of US contrast agents have contributed to its lesser role in oncologic imaging, although in Europe and Canada, where it is widely used, CEUS has shown high (87–91%) accuracy in detection and characterization of liver lesions [15].

US guidance is also a useful tool for biopsy of suspected malignancies, while intraoperative US (IOUS) can be important in localization of tumors for ablative techniques or to guide intraoperative biopsy or surgical resection [16]. In many centers with equivalent expertise in US and CT, US guidance is the method of choice for percutaneous interventional procedures if the lesion is easily visible and accessible sonographically, as it provides real-time imaging, does not use ionizing radiation, is readily available, and inexpensive. US fusion imaging, which involves the co-registered display of real-time US with a reference series from another modality, such as CT, MRI, or PET, arranged side-by-side or overlaid on the US monitor, can further aid in the detection and localization of lesions with low conspicuity on standard B-mode US [17]. It is also possible to indicate the desired needle trajectory for guidance during the procedure. US fusion imaging can be associated with the use of different ultrasound techniques such as color Doppler US, elastography, and CEUS, for better localization and characterization of lesions. During biopsy, the use of color Doppler US allows for identification of major vascular structures which should preferably be avoided during needle insertion, while CEUS can increase the conspicuity of lesions to be treated and vascular structures to be avoided.

27.6 Magnetic Resonance Imaging

Unlike CT and radiographs, magnetic resonance imaging (MRI) uses nonionizing radiation to yield diagnostic images. MRI is based on the magnetization properties of atomic nuclei, taking advantage of the high prevalence of hydrogen in the body and the magnetic properties of protons in a hydrogen atom. Each hydrogen proton acts like a tiny magnet which rotates along its long axis and generates a small magnetic field. These are randomly oriented so that their magnetic fields cancel out. When subject to powerful, uniform, external magnetic field (referred to as B0), protons will either align either parallel (spin up) or antiparallel (spin down) to the direction of the magnetic field, with a slight excess in the lower energy parallel direction, producing a "net magnetization" parallel to the main magnetic field. This net magnetization becomes the source of our MR signal and is used to produce MR images. This alignment (or magnetization) is subsequently perturbed by introducing an external radio frequency (RF) energy to the body. The nuclei return to their resting alignment through various relaxation processes, in the process emitting RF energy, which is detected by a radiofrequency receiver. Different tissues realign at different rates and image acquisition relies on subtle differences in the intrinsic behavior of hydrogen protons bound to different soft tissues and fluids to generate image contrast [18]. Using Fourier transformation, the frequency information from each location in the imaged plane is converted to corresponding intensity levels, which are displayed as shades of gray in a matrix arrangement of pixels. By varying the sequence of RF pulses, different types of images are created.

The main magnetic field of an MRI scanner is generated by a large electric current flowing through wires formed into a loop in the magnet of the imaging system in which the patient lies. These wires have zero resistance to electrical current when maintained at superconducting temperatures (−269 °C), which is achieved by immersing the coil in liquid helium. A power supply is used to inject an electric current into the coils until the desired magnetic field is reached. The power supply can then be removed. Once established, the current continues in the closed-loop coil for years without significant decline, provided that liquid helium levels are maintained. Consequently, the magnetic field is always present. The majority of MRI systems in clinical use have a magnetic field strength of 1.5 T (tesla) or 3 T, although field strengths of 7 T are now commercially available [19]. One Tesla is equal to 10,000 gausses (G); in comparison, the earth's magnetic field is only about 0.5 Gauss. There are unique potential risks associated with MR imaging such as projectile injury, translational force and torque from the main magnet, burns and conductive loops from radiofrequency coils, and acoustic injury from the gradient coils. In order to avoid MRI-related accidents, there is an MRI safety zoning system (zones 1–4) with progressive restriction of entry and increased supervision for higher zones closer to the magnet.

The basic MRI sequences are T1-weighted (T1W) and T2-weighted (T2W) sequences, which are predominately determined by intrinsic T1 and T2 properties of tissue, respectively. A detailed discussion of MRI physics is beyond the scope of this section, however, in general, fluid is bright (or white) on T2W images and dark on T1W images. Fat is bright on both but can be suppressed (darkened) using fat suppression techniques to create appropriate contrast. T1W imaging can also be performed with intravenous Gadolinium-based contrast agents. Gadolinium is a paramagnetic contrast enhancement agent which shortens T1 relaxation time, thus demonstrating increased signal on T1W sequences. Fluid-attenuated inversion recovery (FLAIR) is a sequence commonly used in CNS imaging, which removes or nulls the signal of cerebrospinal fluid, making it easier to differentiate

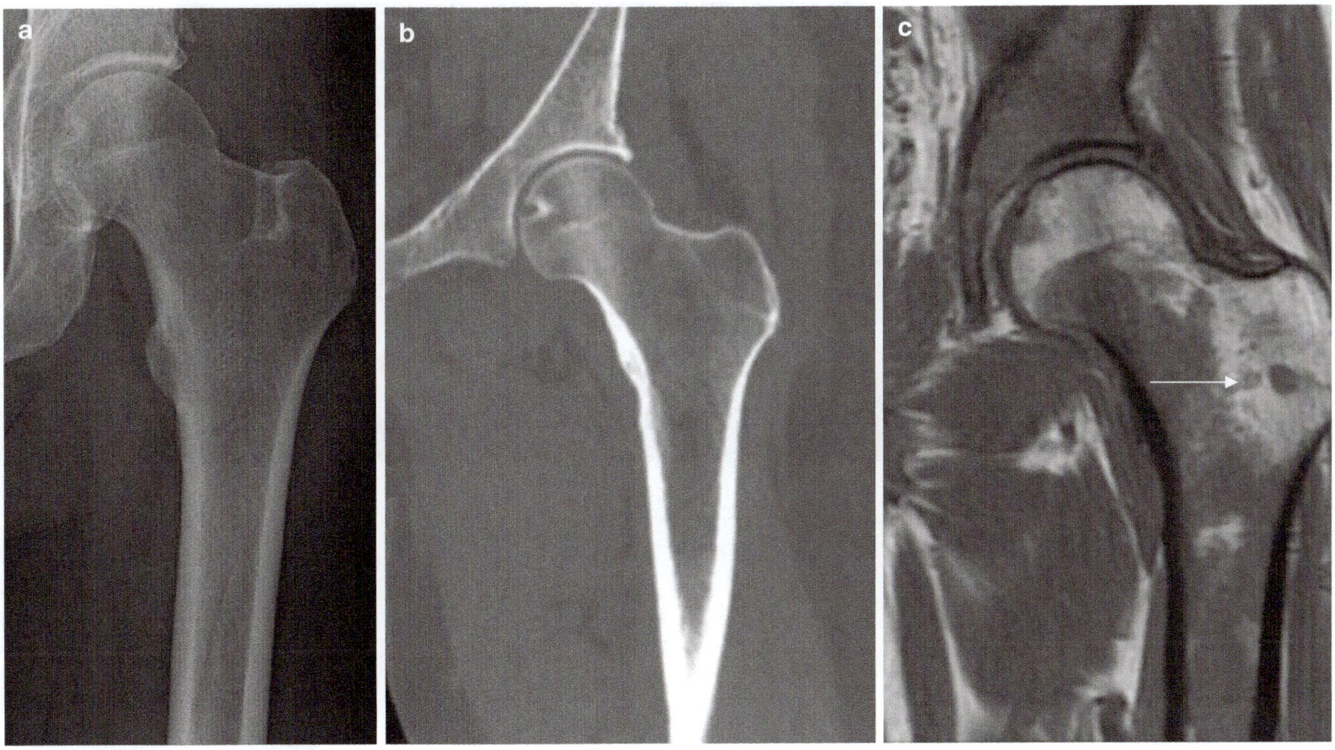

Fig. 27.4 65 year old man with prostate cancer and increasing left hip pain. (**a**) AP radiograph is negative. (**b**) Coronal CT image shows no focal lesion. (**c**) Coronal T1W MRI shows diffuse and focal tumor infiltration consistent with metastases

abnormal hyperintense lesions. Diffusion-weighted imaging (DWI) is another commonly utilized sequence based on measuring the random Brownian motion of water molecules and is particularly useful in tumor characterization and cerebral ischemia. In simplified terms, highly cellular tissues, or those with cellular swelling, will exhibit lower diffusion coefficients.

While CT has better spatial resolution, the soft-tissue contrast resolution of MRI (or ability to distinguish between differences in image intensity) is much better than CT. MRI is therefore often better able to delineate the extent of tumor invasion and assess for residual disease after resection or treatment. MRI is highly sensitive in the detection of many osseous and abdominopelvic metastases and is often used as a problem-solving tool for equivocal imaging findings on CT or US. Although the MRI appearance of metastatic disease is variable and typically dependent on the primary tumor, certain MR enhancement characteristics, such as an early ring enhancement and peripheral washout of liver lesions, are considered specific to metastases and allow differentiation of metastases from benign lesions. Metastatic disease may be visible on MRI prior to other imaging modalities. Osseous metastases, for example, are detectable on MRI before radionuclide bone scan [20]. Normal bone marrow demonstrates high signal on T1W sequence due to a high percentage of fat. On T1W sequences most osseous metastases demonstrate discrete foci or diffuse low signal due to the replacement of normal fatty marrow by malignant cells (Fig. 27.4). On a T2W fat-saturated sequence, osteolytic metastases usually demonstrate hyperintense signal due to elevated water content and enhancement on post-contrast study due to increased vascularity [21]. Osteoblastic metastases are typically low signal (dark) on both sequences.

27.7 Nuclear Medicine

Nuclear medicine techniques quantitatively assess the function of tissue or tumor cells as compared to morphologic imaging techniques such as plain films, CT, and MRI where only the anatomy of a lesion is demonstrated. Two radionuclide functional techniques, planar scintigraphy and 18F-FDG positron emission tomography (PET), with or without CT/MRI integration, play a major role in the diagnosis and evaluation of tumors including skeletal metastasis [22].

27.7.1 Bone Scan

To illustrate the concept of planar scintigraphy, we will use osseous metastases as an example. Bone scan is a highly sensitive, frequently performed, relatively inexpensive, and widely available radionuclide technique for the detection of these lesions. A major advantage of a bone scan is that imag-

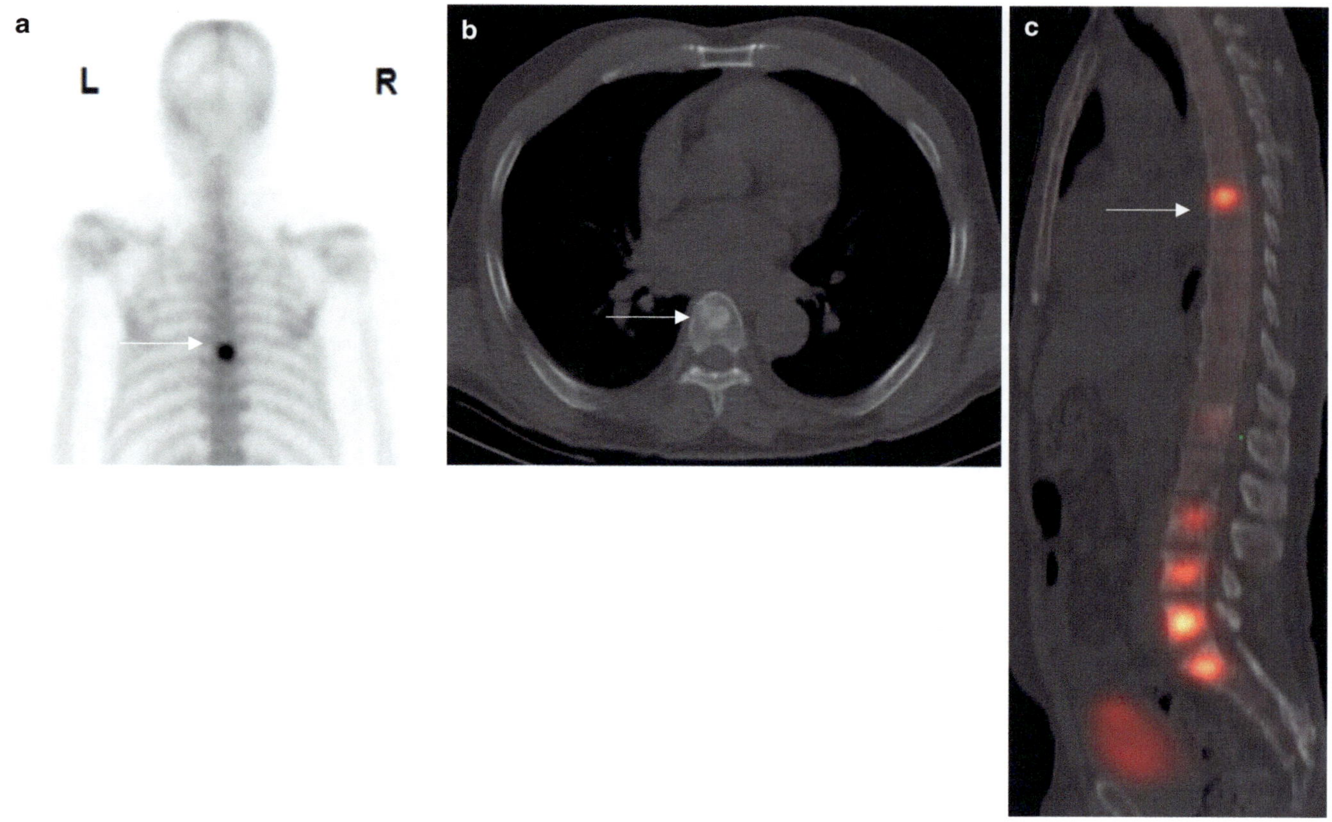

Fig. 27.5 73-year-old man with prostate cancer. (**a**) Planar bone scan demonstrates focal intense tracer uptake in midthoracic spine. (**b**, **c**) SPECT CT aids in anatomic localization of the lesion allowing for targeted treatment. Images courtesy of Bhishak Kamat, MD

ing of the entire skeleton can be performed. The procedure is performed after intravenous administration of technetium-99 m labeled methylene diphosphonate. Images are obtained 2–6 hours later. The tracer rapidly accumulates in bone and emits radiation (gamma rays), which are detected by a gamma camera from which planar anterior and posterior images are generated. The labeled diphosphonates enable visualization of metastases in metabolically active tumors such as prostate and breast cancers. A two or three-phase protocol at earlier time points can provide additional information regarding blood flow and soft tissue uptake.

Metastases that are not intensely hypermetabolic may not be detected in planar imaging and the detection of these metastases is markedly improved by the addition of single-photon emission computed tomography (SPECT) or fused SPECT-CT (Fig. 27.5). The sensitivity and specificity for detection of metastasis increase over 90% when SPECT-CT is used, especially for prostate metastases [23].

27.8 PET-CT and PET-MRI

PET-CT is a fusion of functional planar radionuclide imaging and cross-sectional anatomic imaging (CT). Fluorodeoxyglucose F18 (^{18}F-FDG) is most commonly used in oncology because it is a radiopharmaceutical analog of glucose and there is high glucose metabolism of malignant cells. This radiotracer is produced in a cyclotron. The patient is injected intravenously and is imaged approximately 60 min following the injection of ^{18}F-FDG. The tracer is taken up by metabolically active tumor cells, and unlike glucose, the breakdown product does not undergo further metabolism and is trapped in metabolically active cells. Positrons emitted from the F18 annihilate electrons releasing two annihilation photons at 511 keV traveling in opposite directions that are detected by PET detectors using a technique called coincidence detection.

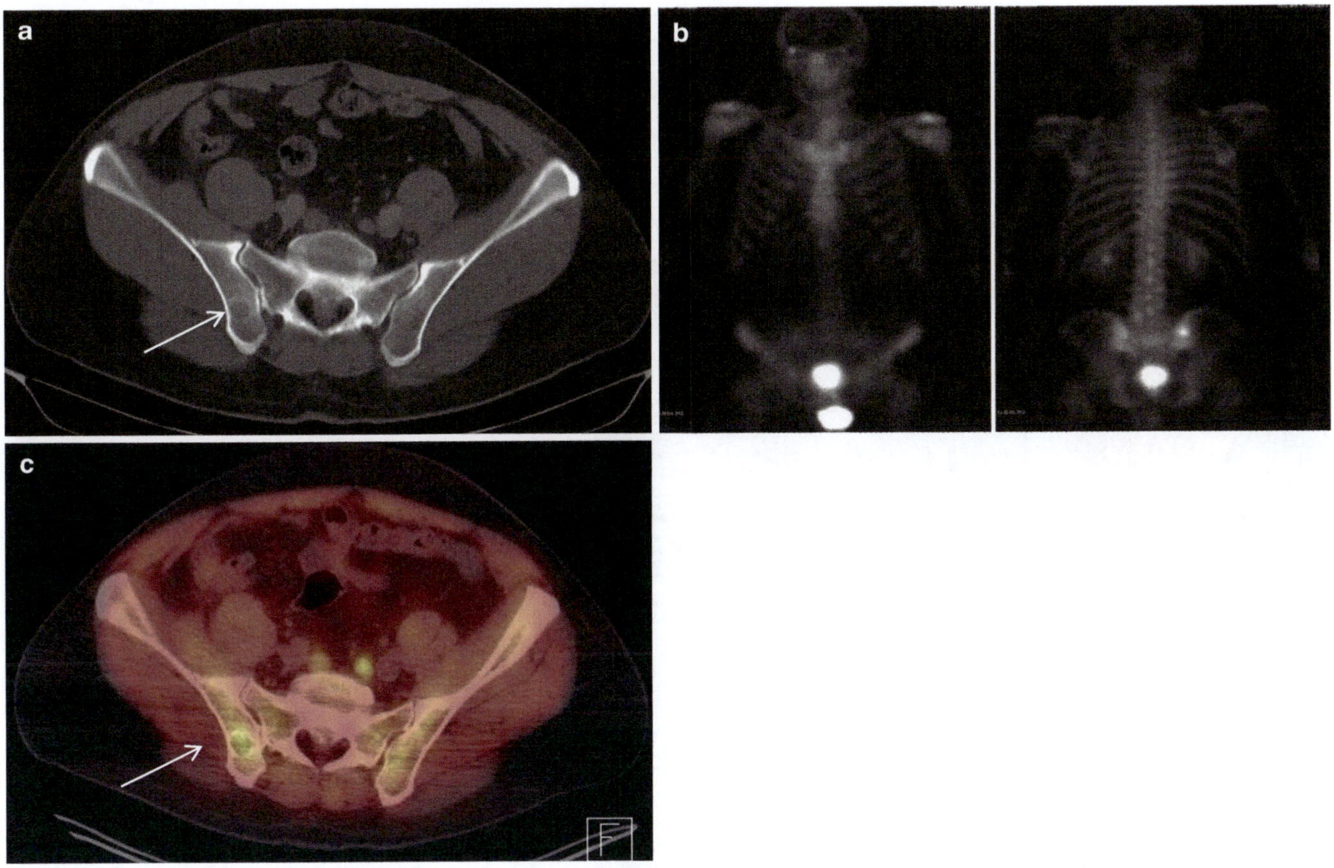

Fig. 27.6 50-year-old man with prostate cancer and rising PSA. (**a**) CT shows predominantly lytic lesion in right iliac bone. (**b**) Bone scan shows no tracer uptake. (**c**) PET-CT F-18 Axumin shows focal increased uptake in the lesion consistent with metastasis

The PET scanner is located behind the CT scanner in the same gantry. ^{18}F-FDG PET-CT acquires PET and CT data in the same imaging session, therefore allow accurate anatomical localization of lesions detected on ^{18}F-FDG PET scan. The assessment of radiotracer uptake by normal physiologic and pathologic tissues is performed by visual inspection and the standardized uptake value (SUV) which is a semiquantitative method. Normal physiologic uptake of the agent is seen in the brain, thymus, bone marrow, intestines, and brown fat. Increased SUV values are seen in tumors. A major advantage of ^{18}F-FDG PET is the ability to compare the maximum SUV value of metastases on sequential studies which can provide an objective measure of the response to treatment. It is more sensitive and specific in certain cancers and primarily aids in staging, restaging, and recurrence [24]. ^{18}F-FDG PET-CT Axumin (Fluciclovine) demonstrates intense focal uptake in lytic prostate cancer metastases and is a problem solver in those patients with a negative conventional bone scan due to lack of osteoblastic activity (Fig. 27.6).

PET-MRI is a more recent imaging modality where the exquisite contrast resolution of MRI is combined with functional data of ^{18}F-FDG, similar to PET-CT, however, with better anatomic detail. Its clinical efficacy is currently being evaluated [25] (Fig. 27.7).

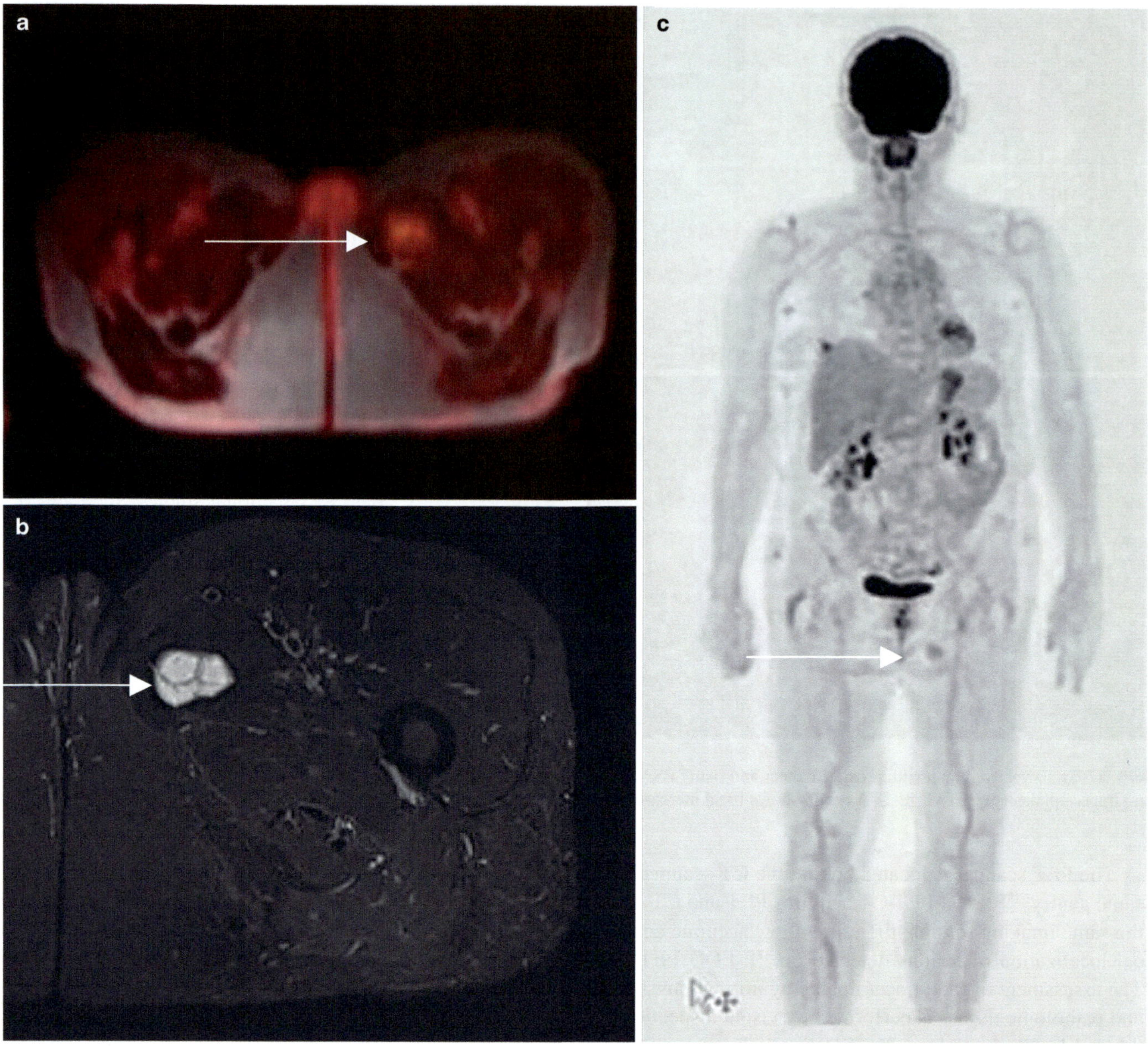

Fig. 27.7 70-year-old woman with history of breast cancer and DVT. (**a**) Fused PET- MRI image shows hypermetabolic mass (SUV 2.69) in left thigh corresponding to T2 hyperintense mass on axial T2W MRI image (**b**) and hypermetabolic focus on planar PET image (**c**). The fused image increases diagnostic accuracy. Biospy confirmed intramuscular myxoma. Images courtesy of Andrew Newberg, MD and Anthony Bazzan, MD

Conclusion

Early diagnosis of metastatic disease allows for appropriate and targeted treatments and can reduce morbidity and mortality. Current treatment planning is heavily dependent on imaging as is follow-up. Recent advances in diagnostic imaging have progressed at an exponential pace with faster and more accurate imaging. Functional imaging and functional overlay of diagnostic scans (PET-CT and PET-MRI) have revolutionized the field and continue to grow. Although beyond the scope of this chapter, the field of radiogenomics is rapidly advancing and is likely to play an integral role in tumor imaging.

Ongoing and future research will strive to answer:

- Can radiation dose be decreased by increasing the use of modalities utilizing nonionizing radiation (such as ultrasound and MRI) particularly in cases where traditionally radiography and CT have been utilized?
- Plain film radiography is the most commonly performed imaging study. Will further refinements in computerized radiography further decrease radiation dose?
- Will functional imaging play a greater role in tumor diagnosis and follow-up?
- What is the future role of contrast-enhanced ultrasound in tumor follow-up?
- What is the future potential of radiogenomics?

References

1. Ritenour ER. Physics overview of screen-film radiography. Radiographics. 1996;16(4):903–16.
2. Bushberg JT. The essential physics of medical imaging. 2nd ed. Philadelphia: Lippincott Williams & Wilkins; 2002. p. 210.
3. Bansal GJ. Digital radiography. A comparison with modern conventional imaging. Postgraduate Med J. 2006;82(969):425428.
4. Ratib O. From PACS to the clouds. Eur J Radiol. 2011;78(2):161–2.
5. Hunt CH, Hartman RP, Hesley GK. Frequency and severity of adverse effects of iodinated and gadolinium contrast materials: retrospective review of 456,930 doses. AJR Am J Roentgenol. 2009;193:1124–12.
6. Harieaswar S, Rajesh A, Griffin Y, Tyagi R, Morgan B. Routine use of positive oral contrast material is not required for oncology patients undergoing follow-up multidetector CT. Radiology. 2009;250(1):246–53.
7. Mahesh M. MDCT physics: the basics—technology, image quality and radiation dose. Philadelphia, PA: Lippincott Williams & Wilkins; 2009.
8. Raman SP, Mahesh M, Blasko RV, Fishman EK. CT scan parameters and radiation dose: practical advice for radiologists. J Am Coll Radiol. 2013;10(11):840–6.
9. Johnson TR. Dual-energy CT: general principles. AJR Am J Roentgenol. 2012;1999(Suppl. 5):S3–8.
10. Rajiah P, Sundaram M, Subhas N. Dual-energy CT in musculoskeletal imaging: what is the role beyond gout? AJR Am J Roentgenol. 2019;213(3):493–505.
11. Sujlana PS, Mahesh M, Vedantham S, Harvey SC, Mullen LA, Woods RW. Digital breast tomosynthesis: image acquisition principles and artifacts. Clin Imaging. 2019;55:188–95.
12. Hoskins PR, Martin K, Thrush A. Diagnostic ultrasound. 2nd ed. Cambridge University Press; 2010.
13. Kloth C, Kratzer W, Schmidberger J, Beer M, Clevert DA, Graeter T. Ultrasound diagnostics & therapy: on the way to multimodal ultrasound: contrast-enhanced ultrasound (CEUS), microvascular Doppler techniques, fusion imaging, Sonoelastography, Interventional Sonography. In: RöFo-Fortschritte auf dem Gebiet der Röntgenstrahlen und der bildgebenden Verfahren, vol. 2020. Georg Thieme Verlag; 2020. https://doi.org/10.1055/a-1217-7400.
14. Wilson SR, Greenbaum LD, Goldberg BB. Contrast-enhanced ultrasound: what is the evidence and what are the obstacles? Am J Roentgenol. 2009;193(1):55–60.
15. Kaur H, Hindman NM, Al-Refaie WB, et al. ACR appropriateness criteria® suspected liver metastases. Available at https://www.jacr.org/article/S1546-1440(17)30144-8/pdf. American College of Radiology.
16. Pomfret EA, Washburn K, Wald C, et al. Report of a national conference on liver allocation in patients with hepatocellular carcinoma in the United States. Liver Transpl. 2010;16:262–78.
17. European Society of Radiology (ESR). Abdominal applications of ultrasound fusion imaging technique: liver, kidney, and pancreas. Insights. Imaging. 2019;10(1):6.
18. Currie S, Hoggard N, Craven IJ, et al. Understanding MRI: basic MR physics for physicians. Postgrad Med J. 2013;89:209–23.
19. Tsai LL, Grant AK, Mortele KJ, Kung JW, Smith MP. A practical guide to MR imaging safety: what radiologists need to know. Radiographics. 2015;35(6):1722–37.
20. O'Sullivan GJ, Carty FL, Cronin CG. Imaging of bone metastasis: an update. World J Radiol. 2015;7(8):202–11.
21. Yu HH, Tsai YY, Hoffe SE. Overview of diagnosis and management of metastatic disease to bone. Cancer Control. 2012;19(2):84–91.
22. Cuccurullo V, Cascini GL, Tamburrini O, Rotondo A, Mansi L. Bone metastases radiopharmaceuticals: an overview. Curr Radiopharm. 2013;6(1):41–7.
23. Heindel W, Gübitz R, Vieth V, Weckesser M, Schober O, Schäfers M. The diagnostic imaging of bone metastases. Dtsch Arztebl Int. 2014;111(44):741–7.
24. Vassiliou V, Andreopoulos D, Frangos S, Tselis N, Giannopoulou E, Lutz S. Bone metastases: assessment of therapeutic response through radiological and nuclear medicine imaging modalities. Clin Oncol (R Coll Radiol). 2011;23(9):632–45.
25. Çelebi F. What is the diagnostic performance of 18F-FDG-PET/MRI in the detection of bone metastasis in patients with breast cancer? Eur J Breast Health. 2019;15(4):213–6. Published 2019 Oct 1

Imaging of Bone Metastases

28

Colleen M. Costelloe, Raul Fernando Valenzuela,
Hubert H. Chuang, and John E. Madewell

Abstract

Bone metastases have become more common due to the successful prolongation of the lives of many cancer patients. Imaging plays an integral role in the staging of newly diagnosed malignancies and is essential for the follow-up of these patients throughout their therapeutic course. The roles of standard imaging such as radiography, CT, conventional MRI; functional imaging such as skeletal scintigraphy and positron emission tomography (PET/CT and PET/MRI); and advanced imaging such as whole-body MRI and diffusion-weighted imaging will be discussed.

Learning Objectives

- Learn how conventional imaging can be used for the staging of bone metastases.
- Understand the advantages of indirect and direct functional imaging.
- Learn that radiopharmaceuticals can be used to treat bone metastases.
- Understand diffusion-weighted imaging with respect to bone metastases.

C. M. Costelloe (✉) · R. F. Valenzuela · J. E. Madewell
Department of Musculoskeletal Imaging, The University of Texas MD Anderson Cancer Center, Houston, TX, USA
e-mail: ccostelloe@mdanderson.org; rfvalenzuela@mdanderson.org; jmadewell@mdanderson.org

H. H. Chuang
Department of Nuclear Medicine, The University of Texas MD Anderson Cancer Center, Houston, TX, USA
e-mail: hhchuang@mdanderson.org

28.1 Introduction

Patients are living longer due to the successful treatment of their malignancies. This encouraging phenomenon has, nevertheless, led to an increase in the incidence of bone metastases [1]. Standard imaging modalities such as radiography, computed tomography (CT), conventional magnetic resonance imaging (MRI); functional imaging such as skeletal scintigraphy/bone scan (BS), and positron-emission tomography (PET) with fused CT (PET/CT) or MRI (PET/MRI), and advanced imaging such as whole-body MRI (WBMRI) and diffusion-weighted imaging (DWI) play important roles in the initial and on-going assessment of patients with malignancies.

28.2 Conventional Imaging

Radiographs are highly specific for the detection and identification of bone metastases [2], but a minimum of 50% of bone loss must occur for many lytic metastases to be detectable on radiographs [3] and lesions in the axial skeleton can be obscured by overlapping structures. When the clinical question is answered on radiography, it is a highly cost-effective method of diagnosis, making it a first-line imaging modality [4] (Fig. 28.1a).

Radiography is also a cost-effective method of assessing pathologic fracture risk in the long bones and Mirel's scoring system [5] can be used to quantify the risk. Points are given based on four critical factors including the site of metastatic disease (upper limb = 1, lower limb = 2, peritrochanteric femur = 3), pain (mild = 1, moderate = 2, aggravated by function = 3), lesion density (blastic = 1, mixed = 2, lytic = 3), and size with respect to the diameter of the bone (<1/3 = 1, 1/3–2/3 = 2, >2/3 = 3). Based on a study of 61 lesions, Mirel et al. found that a score of 8 indicated a 15% probability of fracture, while a score of 9 indicated a 33% probability. A high score, such as 8, can

S. P. Leong et al. (eds.), *Cancer Metastasis Through the Lymphovascular System*,
https://doi.org/10.1007/978-3-030-93084-4_28

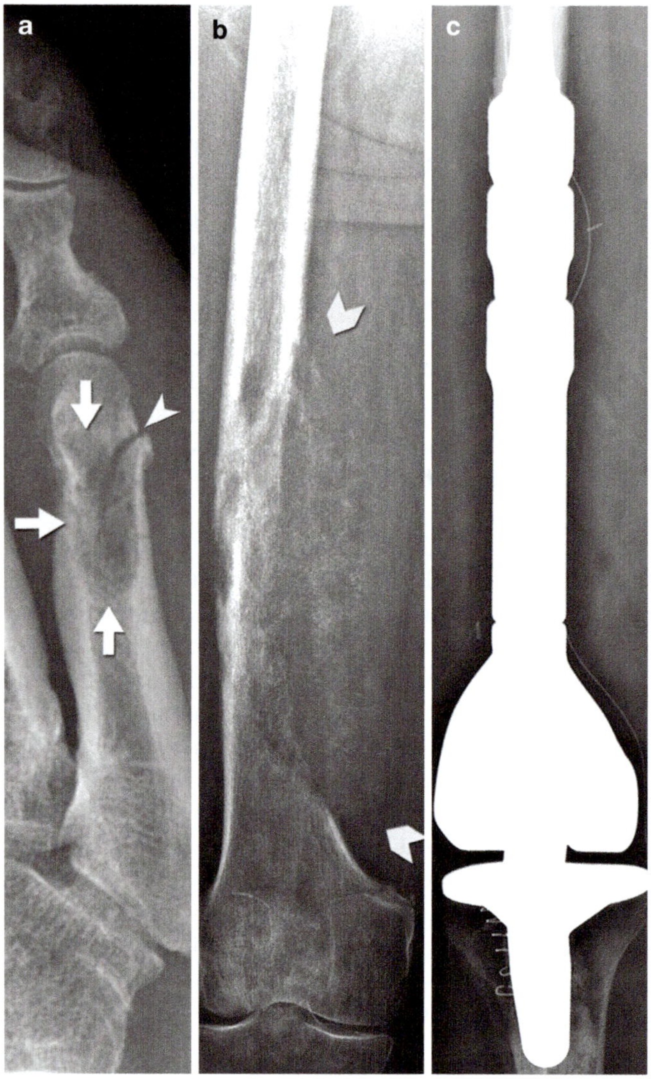

Fig. 28.1 (**a**) Anteroposterior (AP) radiograph of the fifth metatarsal demonstrates a lytic, permeative lesion (arrows) with pathologic fracture (arrowhead). It is likely a bone metastasis in this patient with renal cell carcinoma. Biopsy is recommended to confirm the diagnosis in the absence of other bone metastases or if infection or other etiologies are suspected. No advanced imaging is necessary unless MRI is needed to further assess the extent of the disease. The lesion and fracture can be followed with radiography. (**b**) AP radiograph of the femur in a patient with metastatic urothelial carcinoma demonstrates a lytic metastasis with soft-tissue extension (chevrons). According to Mirel's system, this lesion scores 11 (lower limb = 2, pain aggravated by function = 3, lytic = 3, size >2/3 of the diameter of the bone = 3), indicating a very high fracture risk. (**c**) Treatment with resection arthroplasty

be used as an indication for intervention, such as prophylactic internal fixation, or resection with reconstruction (Fig. 28.1b).

CT is an excellent modality for evaluating mineralized structures. The cross-sectional nature of CT with multiplanar reformations (sagittal, coronal, and oblique) makes it more effective than radiography for assessing the axial skeleton and irregular bones. MRI provides superior soft-tissue contrast resolution in comparison to CT. MRI is commonly used to delineate the extent of disease inside of the medullary cavity and adjacent soft tissues [4], and as a first-line modality for assessing commonly involved anatomy in high-risk patients (spine and pelvis). Conventional pulse sequences for evaluation of bone metastases include fast spin-echo (FSE) T1, fat-saturated (FS) T2, and FS FSE T1-weighted images with intravenous gadolinium contrast [6].

The behavior of bone metastases following therapy can be assessed using the MD Anderson bone metastases response criteria (MDA criteria) [7] which provide for quantitative and nonquantitative evaluation. The nonquantitative portion of the criteria allows the use of factors other than size to influence the response evaluation. For example, complete response (CR) is defined as a return to normal bone density or complete osteoblastic (sclerotic) fill-in of previously lytic or mixed lesions on radiography or CT, normalization of tracer uptake on bone scan (BS), and normalization of signal intensity on MRI. Partial response (PR) is defined as the development of a sclerotic rim and/or partial sclerotic fill-in of lytic metastases on radiography (Fig. 28.2) or CT (Fig. 28.3), a decrease of 50% in unmeasurable lesions or of tracer uptake on BS. Progressive disease (PD) is defined as an increase of 25% or the development of new lesions.

Quantitatively, the perpendicular sum of the greatest diameters of well-defined intramedullary lesions can be used. PR requires a 50% or greater decrease on radiography, CT, or MRI, PD a 25% or greater increase, and stable disease (SD) does not meet these requirements.

The MDA criteria have been shown to be clinically efficacious [8–10]. A study of 55 patients that assessed the effect of radiation therapy on vertebral bone metastases found that sclerotic fill-in demonstrated a significant trend with a positive outcome (reduced pain, $p = 0.021$), [8]. In a study that compared the World Health Organization and International Union Against Cancer criteria to the MDA criteria in 41 breast cancer patients with bone-only metastases found a significant correlation to progression-free survival (PFS) ($p = 0.025$), [9]. In another study of 29 breast cancer patients with bone and measurable soft tissue metastases [10], the MDA criteria indicated PFS at 6 months ($p = 0.002$). The criteria have also been incorporated into clinical protocols [11, 12]. The MDA criteria can be used as a formal or an

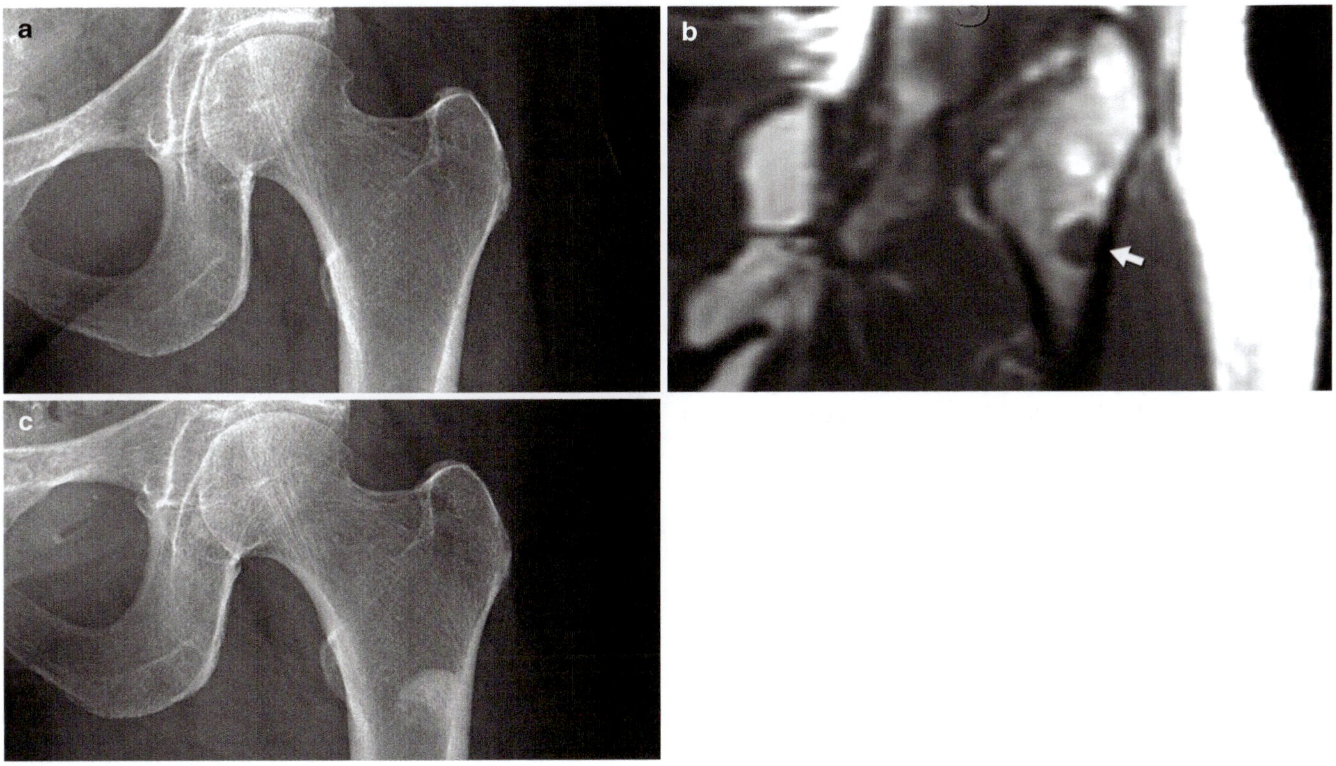

Fig. 28.2 (**a**) Unremarkable pre-therapeutic radiograph of the proximal femur. (**b**) Simultaneous T1 coronal MRI demonstrates a radiographically occult lung cancer metastasis (arrow). (**c**) Radiography 1 year later shows an osteoblastic metastasis, indicating interval healing sclerosis. Without the benefit of the prior MRI, the lesion would appear to represent new disease (PD rather than PR)

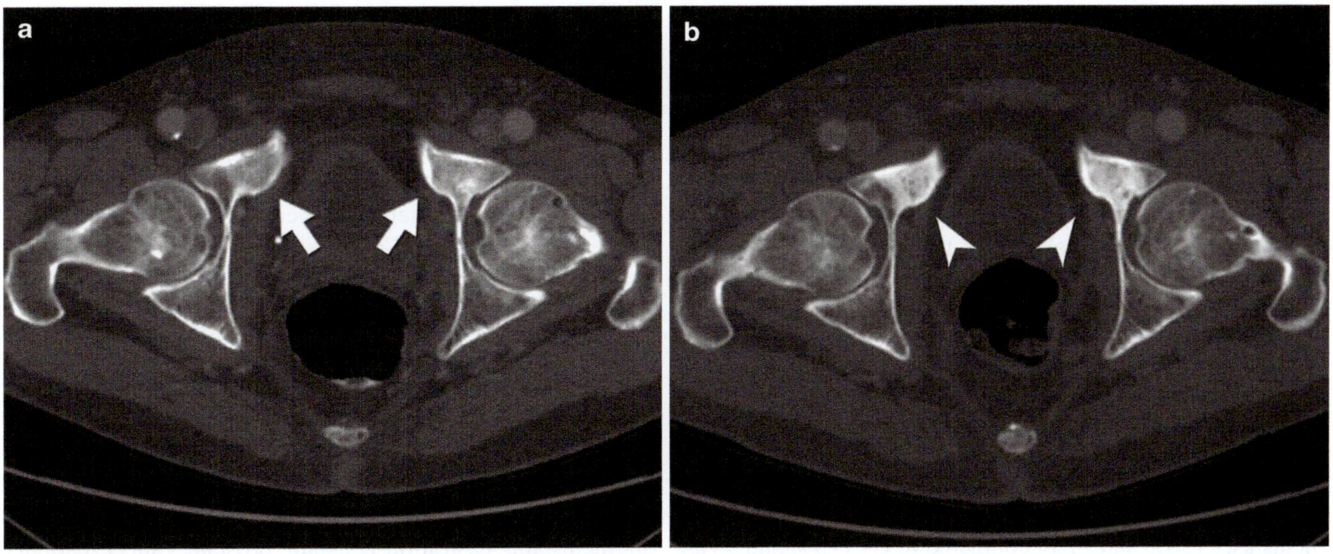

Fig. 28.3 (**a**) Initial CT of faintly sclerotic prostate cancer metastases in bilateral superior pubic rami (arrows). (**b**) Increased density following therapy indicates PR (arrowheads). Knowledge of the therapeutic history is essential for proper image interpretation

informal guide to the interpretation of the imaging of bone metastases following therapy.

28.3 Functional Imaging

Functional imaging of bone metastases can be divided into direct and indirect mechanisms. Direct detection relies upon direct uptake by tumor, such as with ^{18}F-FDG, ^{123}I-mIBG, somatostatin analogs (e.g., ^{68}Ga-DOTA-TATE), ^{18}F-Fluciclovine, and PSMA agents. BS, using ^{18}F-NaF or skeletal scintigraphy, indirectly detect lesions through osteoblastic reaction. Each method has consequences for lesion detection and response assessment.

BS with diphosphonates like ^{99m}Tc-MDP is widely available, images the whole body, and is relatively inexpensive. However, uptake can be nonspecific and related to tumors, inflammation, or trauma. Tomographic imaging with single-photon emission tomography (SPECT), combined with anatomic imaging into SPECT/CT scans, increases lesion contrast and improves localization and characterization, resulting in higher diagnostic accuracy [13].

^{18}F-NaF PET/CT has higher spatial resolution and generates whole-body tomographic images, resulting in higher lesion detection [14], making this the ideal indirect detection method. However, it is not as widely available, is more expensive, and more complex to interpret as compared to BS.

Bone-seeking agents are better for detecting sclerotic/osteoblastic metastases as compared to lytic metastases [15]. However, bone metastases can undergo sclerotic change with therapy [7, 16], resulting in paradoxical "scintigraphic flare" on BS (Fig. 28.4), similar to the "sclerotic flare" seen on CT scans and paralleling clinical flare (with increased bone markers and bone pain), which can be misinterpreted as progressive disease [17]. Moreover, after therapy, activity with bone remodeling may take years to normalize [18]. Thus, indirect detection of bone metastases may result in a signal that persists after the eradication of tumor.

Direct detection of tumor is better at identifying lytic or exclusively marrow-based metastases as compared to sclerotic lesions. The most experience is with ^{18}F-FDG PET/CT, and the increased detection often results in treatment changes, highlighting its utility. A meta-analysis reported higher sensitivity and specificity of 18F-FDG PET for bone metastases (90% and 97%) compared to CT (73% and 95%) or BS (86% and 81%), but similar to MRI (91% and 95%) [19]. However, lesion detection can be radiopharmaceutical specific (Fig. 28.5), so the choice of the imaging agent is important.

Responses with direct tumor imaging agents can be seen much more rapidly, and flare phenomenon has not been reported. Thus, direct tumor imaging agents may be ideal for detecting and assessing response in bone metastases (Fig. 28.4).

28.4 PET/MRI

Hybrid imaging with PET/MRI would seem to be the ideal method for identifying osseous metastases, by taking advantage of the specificity of PET tracers and combining it with the anatomic sensitivity of MRI. Catalano et al. reported higher detection of osseous lesions in breast cancer patients (96% vs. 85%) [20]. Botsikas et al. found similar sensitivities (92% vs. 69%) during the staging of breast cancer, but also reported more false-positive PET/MRI findings (7 vs. 0), [21]. Beiderwellen et al. also reported better detection of benign and metastatic bone lesions in a mix of patients with PET/MRI (48/48) compared to PET/CT (45/48 or 94%) [22]. However, not all studies have found greater detection of osseous lesions by PET/MRI. Guberina et al. reported equivalent detection of 23 bone metastases in prostate cancer patients [23], and Eiber et al. reported no significant difference in detection of bone metastases in a cohort of 119 patients with different cancers [24]. Moreover, Ishii et al. reported the identification of 34/36 bone lesions by PET/MRI but 35/36 lesions on PET/CT [25]. PET/MRI is a relatively new modality, and although there appears to be a small improvement in detection of bone metastases in certain groups, it is unclear if this improved lesion detection is enough to make significant changes in overall patient management.

28.5 Radiopharmaceutical Therapy

Radiopharmaceuticals, Strontium-89, and Samarium-153 can be used for the palliative treatment of painful osseous metastases. About 25–80% of patients report reduced bone pain that lasts about 6 months, with minimal toxicity [26]. Radium-223 also improves survival (14.9 vs. 11.3 months) in metastatic castration-resistant prostate cancer patients, despite potential lack of PSA response [27]. However,

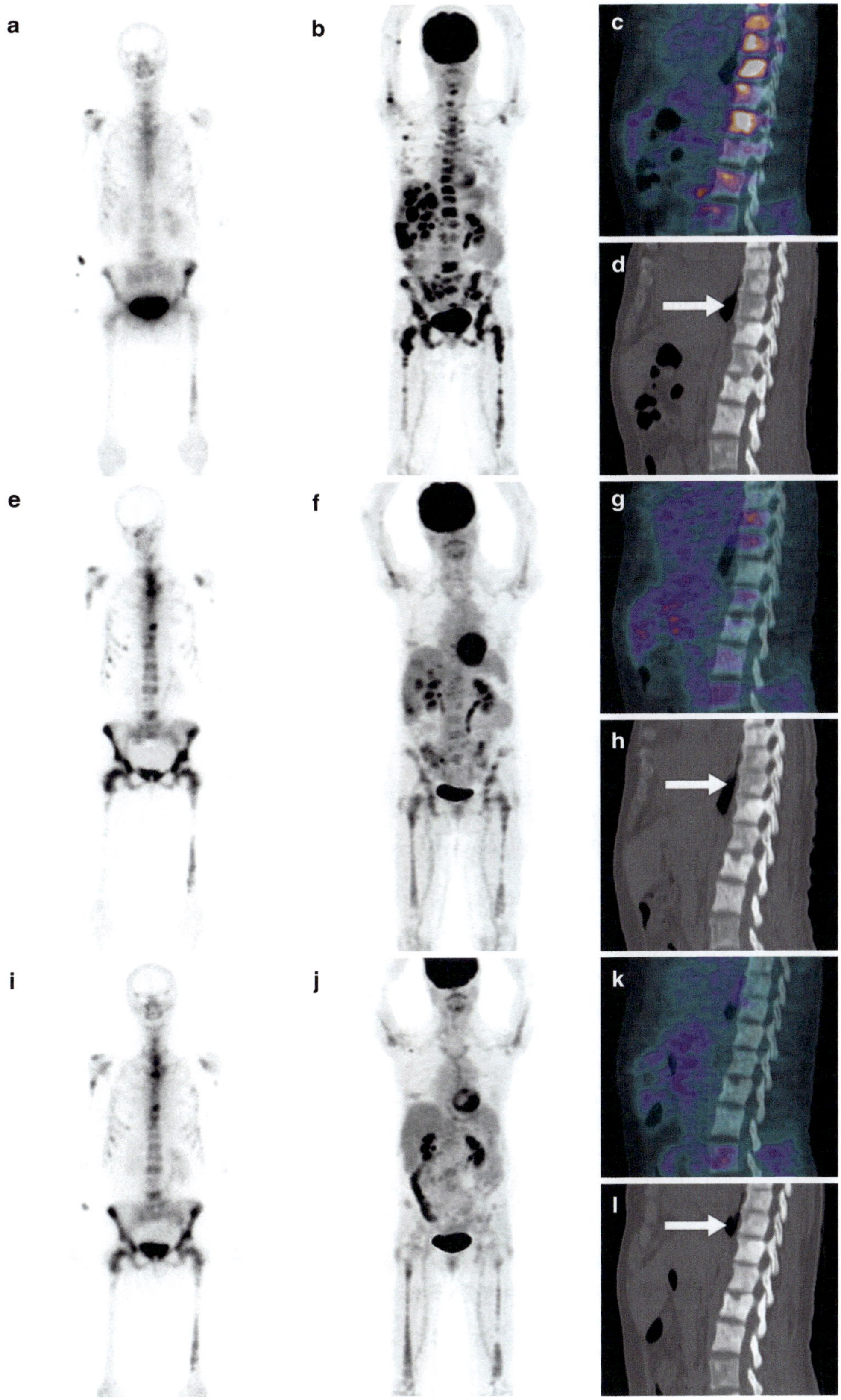

Fig. 28.4 (**a**) BS shows osseous metastases from breast cancer. (**b**–**d**) PET/CT the same day shows metastatic disease involving bones and liver, reported as progressive. Two months following new therapy, (**e**) BS shows increased uptake, indicative of progressive disease versus scintigraphic flare. (**f**–**h**) PET/CT shows decreased uptake in hepatic and osseous metastases, but little anatomic change (arrows). After 2 more months of therapy (**i**) BS shows persistently increased uptake, consistent with scintigraphic flare because (**j**–**l**) PET/CT shows near-complete metabolic response and healing sclerotic change (arrow)

combining radium-223 with other therapies may result in increased complications or toxicity without survival benefit, so caution is advised [28].

28.6 Advanced Imaging

Magnetic resonance imaging (MRI) with functional sequences including diffusion-weighted imaging (DWI) increases diagnostic accuracy by permitting the assessment of metastatic tumor burden, treatment response, and can help differentiate metastases from other lesions, when conventional MRI may be inconclusive. DWI is a nonenhanced functional MRI technique offering useful information regarding the cellularity of musculoskeletal lesions. Tumor foci demonstrate increased signal intensity on DWI [29]. Also, the amount of signal loss from restricted free diffusion across the intact cell membranes of viable malignant cells is quantified as correspondingly low measured apparent diffusion coefficient (ADC), [30]. The ADC value increases as cell membranes break down, allowing free water diffusion following tumor kill from effective therapy.

Whole-body DWI is often utilized for tumor detection (Fig. 28.6), and localized DWI for characterizing lesions and assessing treatment response. The typical spinal marrow ADC value in children and young adults is approximately 0.5×10^{-3} mm²/s. In the elderly population, due to normal

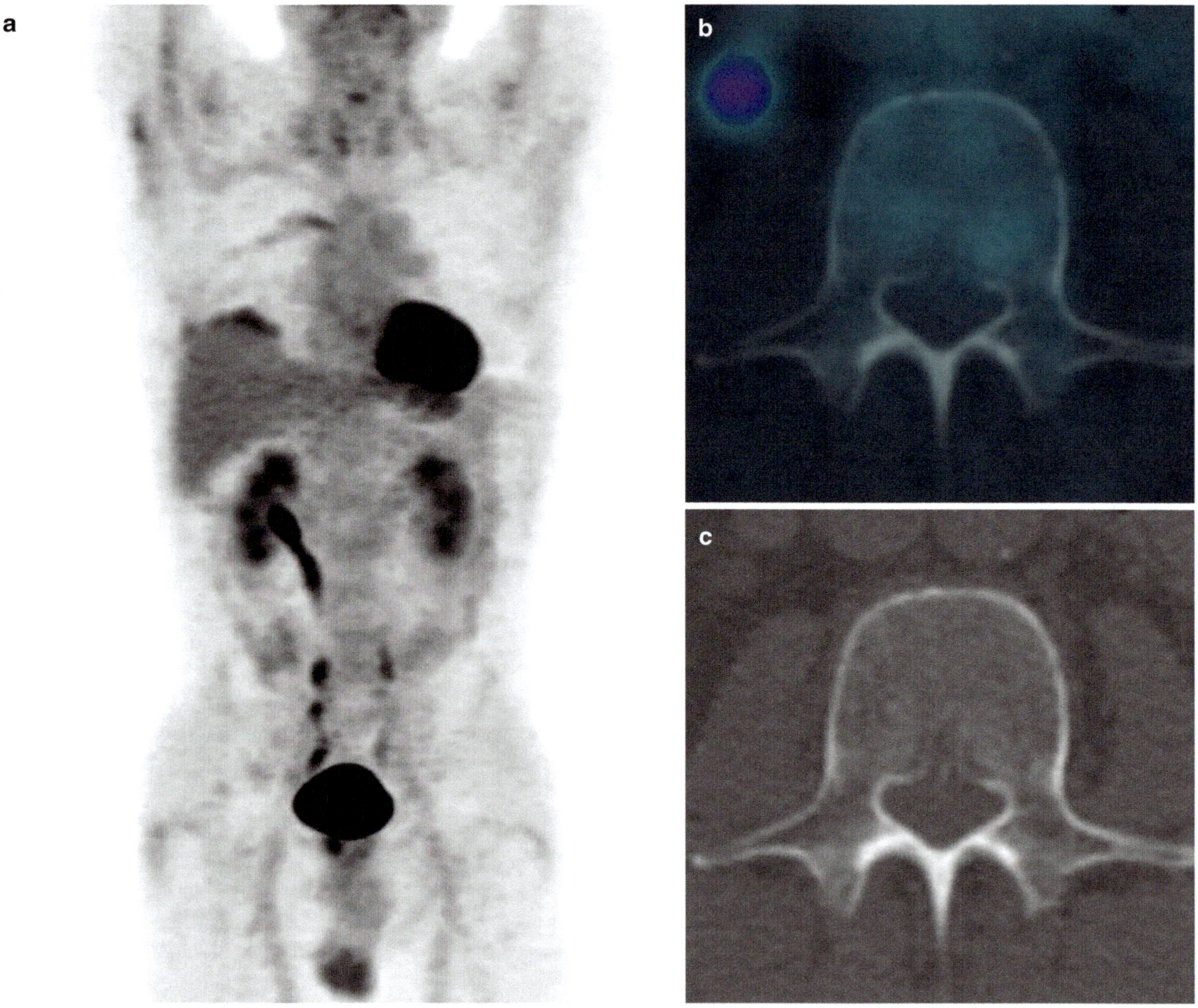

Fig. 28.5 (**a–c**) ^{18}F-FDG PET/CT negative for active tumor in this patient with prostate cancer. (**d**) ^{18}F-Fluciclovine PET/CT reveals pelvic nodal metastasis (arrow) in addition to bone metastases (**e–f**)

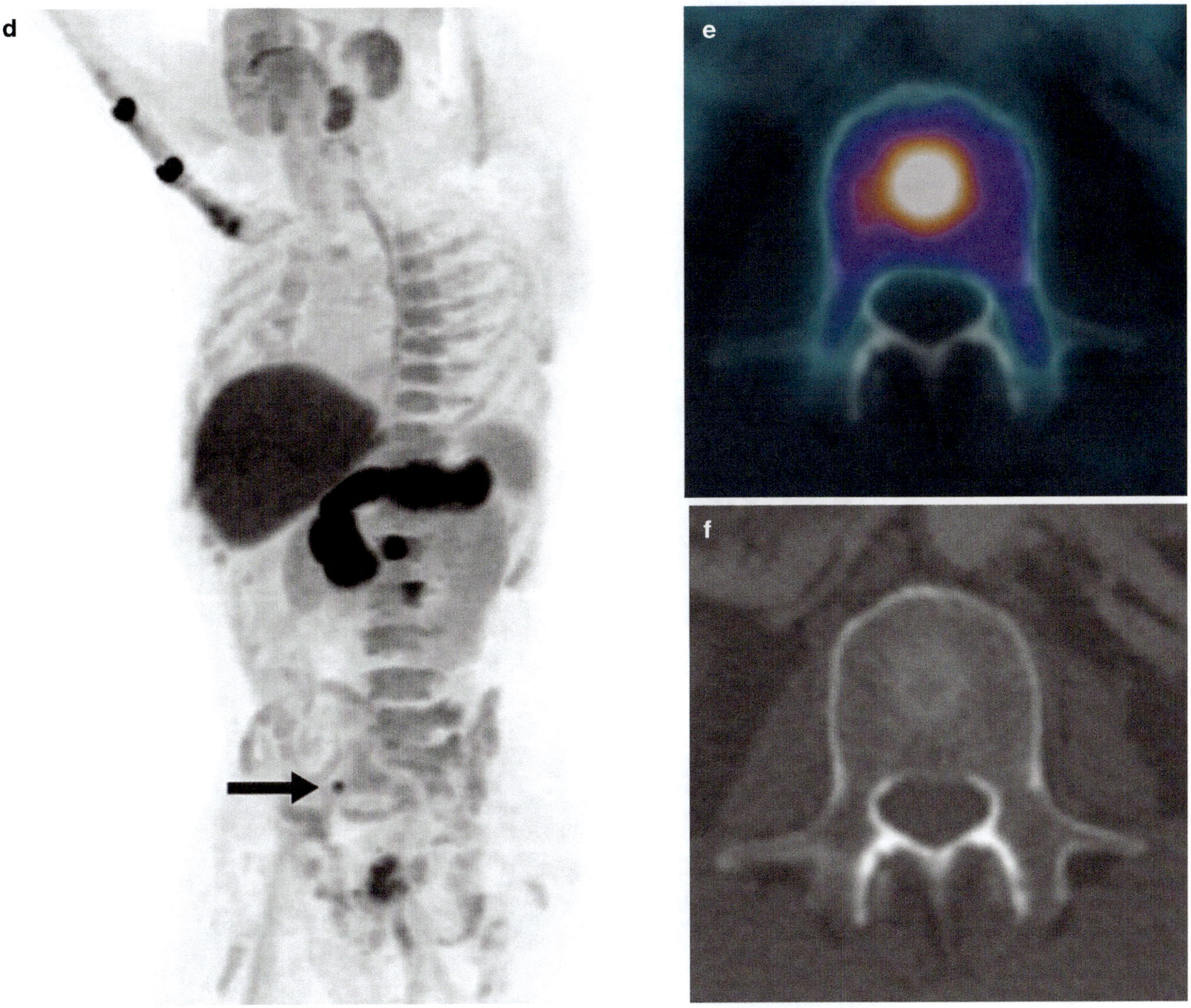

Fig. 28.5 (continued)

increased marrow fat content, ADC value drops to approximately 0.4×10^{-3} mm^2/s. Most cellular malignant tumors demonstrate ADC values ranging between 0.7 and 1.0×10^{-3} mm^2/s [29].

ADC can be useful in assessing response to systemic therapy. In metastatic prostate carcinoma, resistance to therapy and shorter time to progression is indicated when a significant percentage of tumor voxels continues to demonstrate low ADC (below 1.0×10^{-3} mm^2/s) after 1 month of antiandrogenic treatment [31]. Findings on conventional imaging can be nonspecific while ADC increases only after effective therapy (Fig. 28.7).

Normal marrow fat demonstrates low ADC values. During treatment, bone metastases can undergo yellow marrow reconversion, leading to post-therapeutic ADC measurements below the pre-therapeutic values. This counterintuitive lowering of ADC values represents a positive treatment response [29]. Therefore, when assessing response after 12 weeks of systemic treatment, the measurement of the mean ADC from a whole-lesion ROI (region of interest) may not be the most appropriate approach, because the changing proportions of bone marrow and tumor within the same ROI may have opposing effects. Instead, whole-lesion ADC histograms can accurately demonstrate the proportions of both normal and diseased marrow, helping to discriminate between responding and progressing patients [32].

Most viable malignant tumor components tend to demonstrate a concentration of pixels with ADC values of 1.0×10^{-3} mm^2/s or less on the left side of the histogram curve. Bharwani et al. demonstrated that patients in whom the tumor ADC histogram showed a higher number of pixels within the lowest 25th percentile values at baseline have a reduced overall survival [33]. Typically, a posttreatment displacement of the histogram to the right indicates decreased cellularity and positive treatment response [29] (Fig. 28.8).

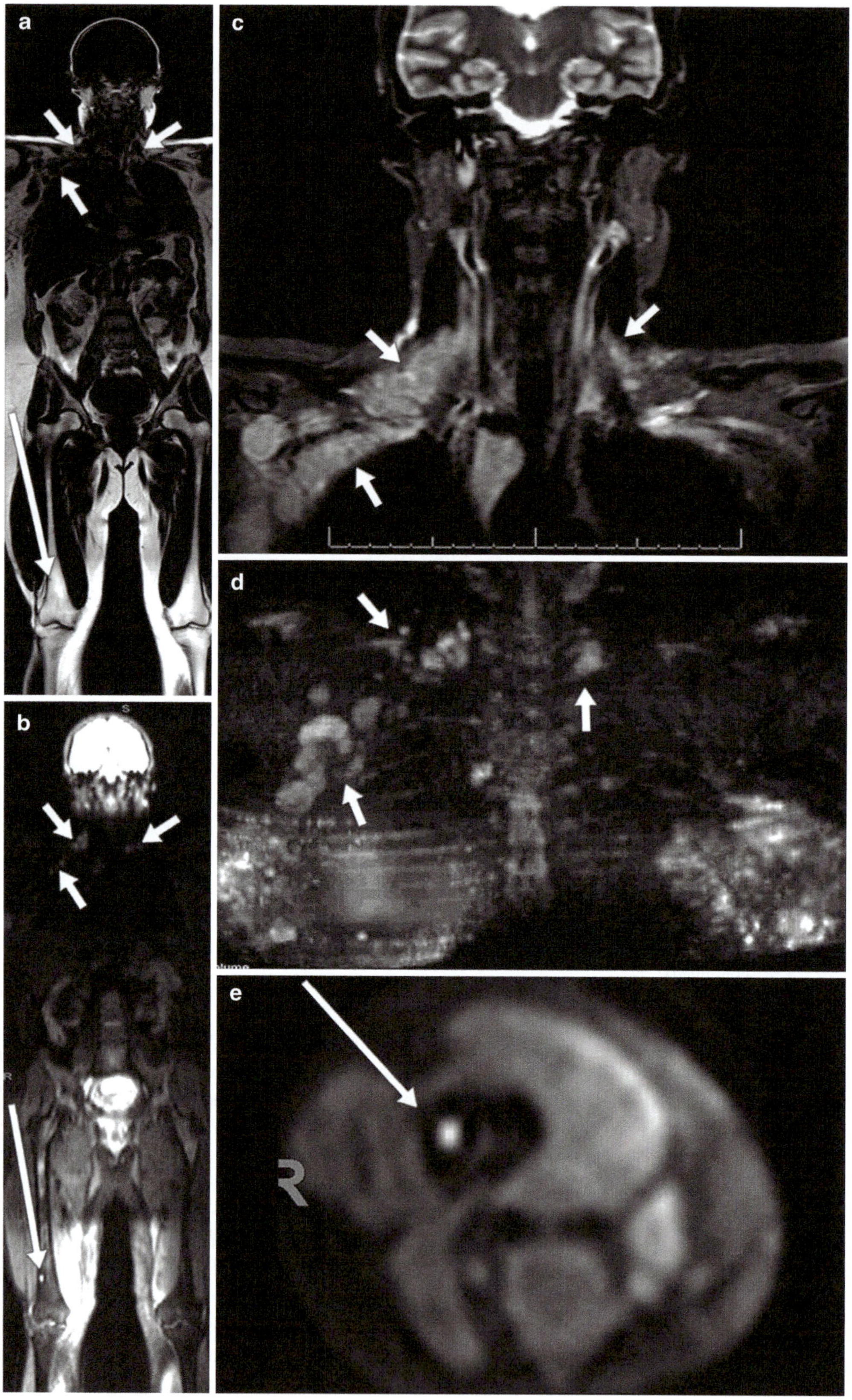

Fig. 28.6 WB-MRI-DWI demonstrates one right femoral (long arrows) and multiple nodal (short arrows) breast cancer metastases in the neck and chest on a single whole-body study. (**a**) WB-MRI Dixon fat-only coronal, (**b**) WB-DWI b800, (**c**) WB-STIR coronal, (**d**) WB-DWI b800 coronal MIP, and (**e**) DWI b800 axial images

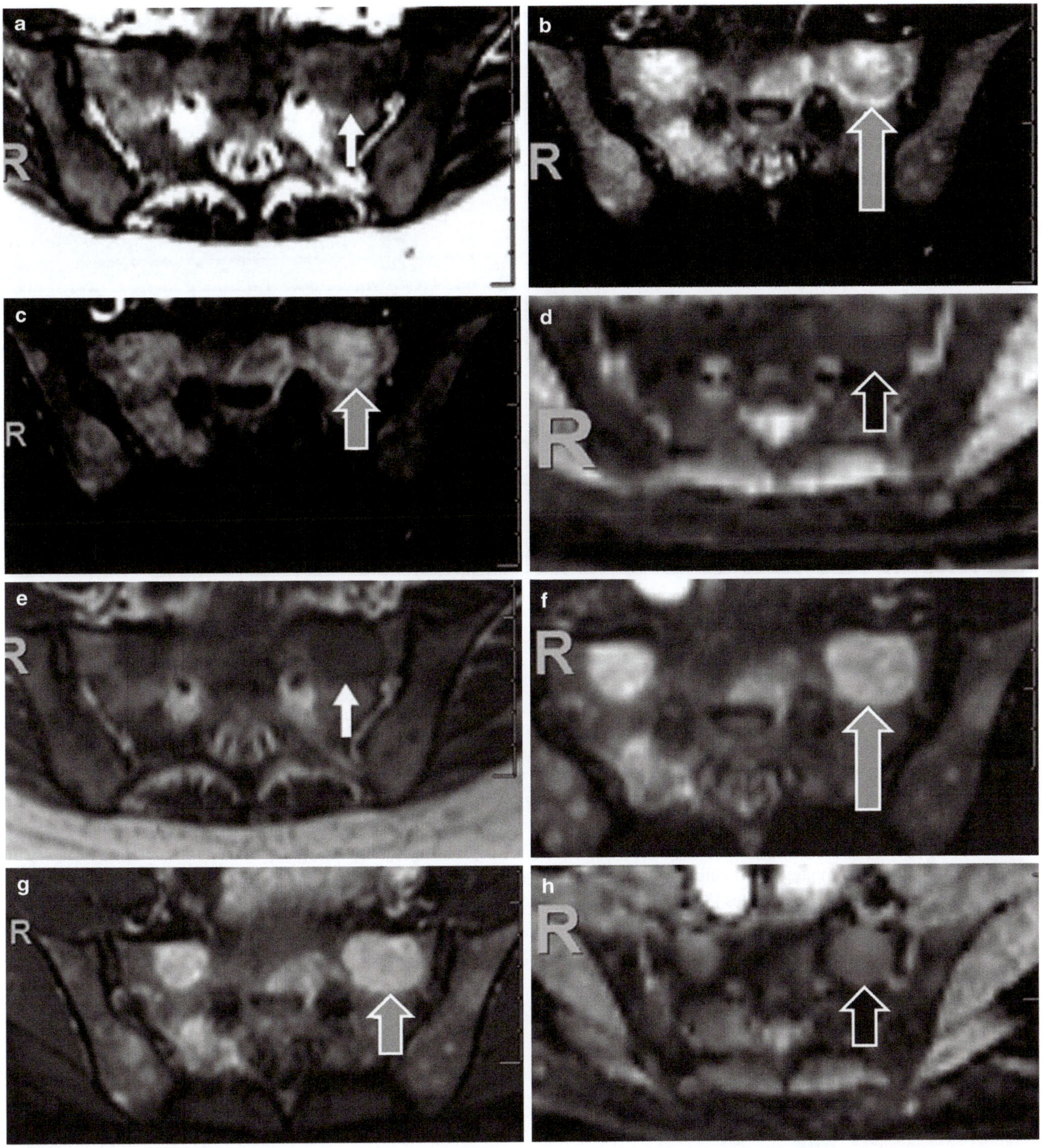

Fig. 28.7 Medullary thyroid carcinoma patient with multiple pelvic bone lesions, without size or appearance change during treatment on conventional MRI. A metastasis is demonstrated in the left sacrum. From left to right in three sets of four images: T1 axials (white arrows), FST2 axials (long gray arrows), FST post-contrast axials (short gray arrows), and axial ADC maps (black arrows). (**a–d**) Pretreatment examination. (**e–h**) 2 months after the beginning of therapy, and (**i–l**) at 4 months. ADC values increased from 0.8×10^{-3} mm^2/s at baseline (**d**) to 1.0×10^{-3} mm^2/s at 2 months (**h**) and 1.5×10^{-3} mm^2/s at 4 months (**l**), indicative of treatment response

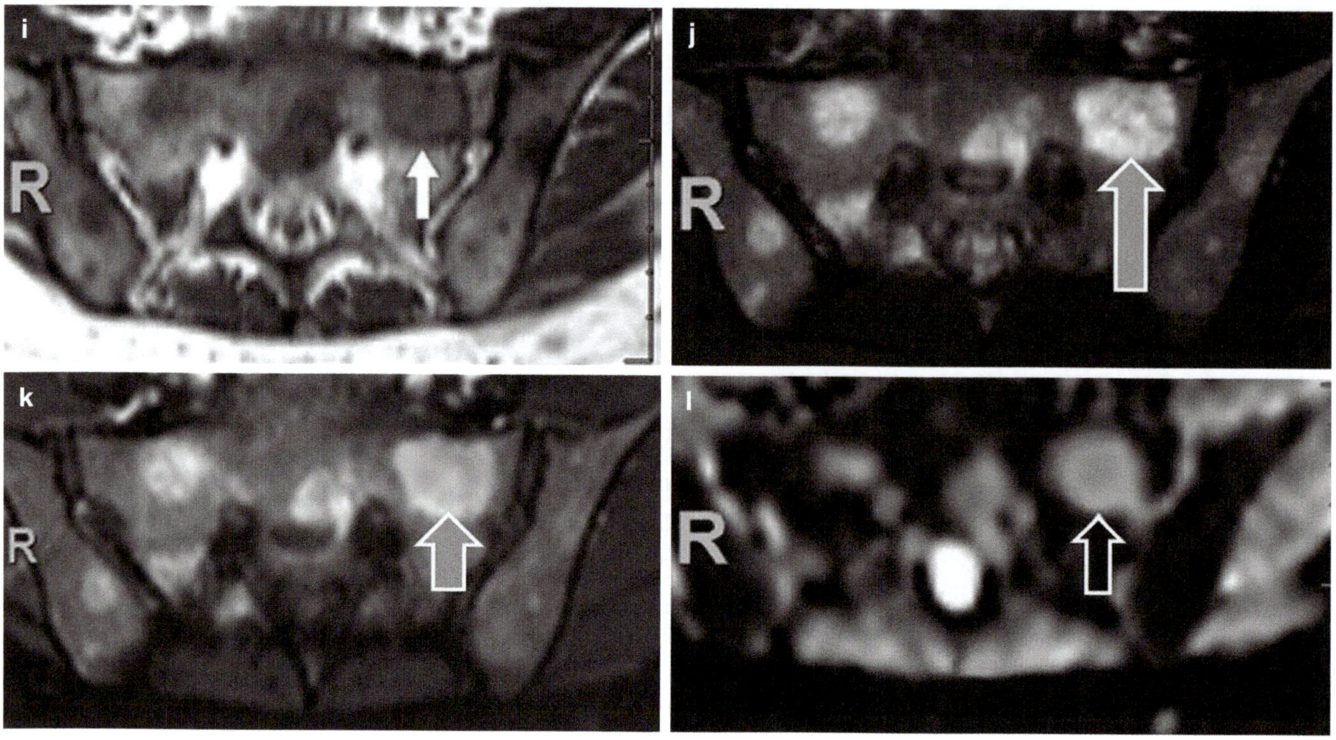

Fig. 28.7 (continued)

With respect to radiation therapy, DWI/ADC can assess cancer cell kill, allowing an objective assessment of radiation effect [34]. Decreased ADC values, indicative of increased lesion cellularity and progression of disease, are seen only in 1.72% of radiated bone metastases, while the vast majority demonstrate increased postradiation ADC values [35]. Multifractionated radiation therapy shows a higher increase in ADC values and a lower retreatment rate compared with single-fraction radiation therapy, despite equal pain control rates [34].

28.7 WB-MRI-DWI

Unlike BS and CT, which rely primarily on an osteoblastic response for the detection of bone metastases, whole-body magnetic resonance with diffusion-weighted imaging (WB-MRI-DWI) provides an evaluation of tumor viability. WB-MRI-DWI plays an increasing role in assessing the response to therapy of tumors with a high prevalence of bone disease, such as prostate and breast cancer, allowing the early detection of nonresponders, with a significant impact on therapeutic assessment [36] (Fig. 28.9). WB-MRI-DWI benefits from improved tumor to background contrast on high b-value images, permitting the creation of multiple metrics for tumor analysis such as the global apparent diffusion coefficient (gADC) and total diffusion volume (tDV) which reflect tissue cellularity and total tumor burden, respectively. Responding patients show a more considerable increase in gADC and nonresponding patients demonstrate an increase in tDV [37].

In thyroid tumors and melanoma bone metastases, the WB-MRI-DWI shows higher sensitivity than ^{18}F-FDG PET/CT and BS for lesion detection. In lung cancer bone metastases, WB-MR-DWI demonstrates similar sensitivity to ^{18}F-FDG PET/CT [38]. For the detection of prostate cancer bone metastases, WB-MRI shows a higher specificity but lower sensitivity than ^{18}F-NaF PET/CT, as the latter is considered a highly sensitive modality [39]. As WB-MRI-DWI becomes more widely available, it may replace BS as the most common method of evaluating the entire skeleton for metastases on a single imaging study. BS will likely remain a cost-effective study.

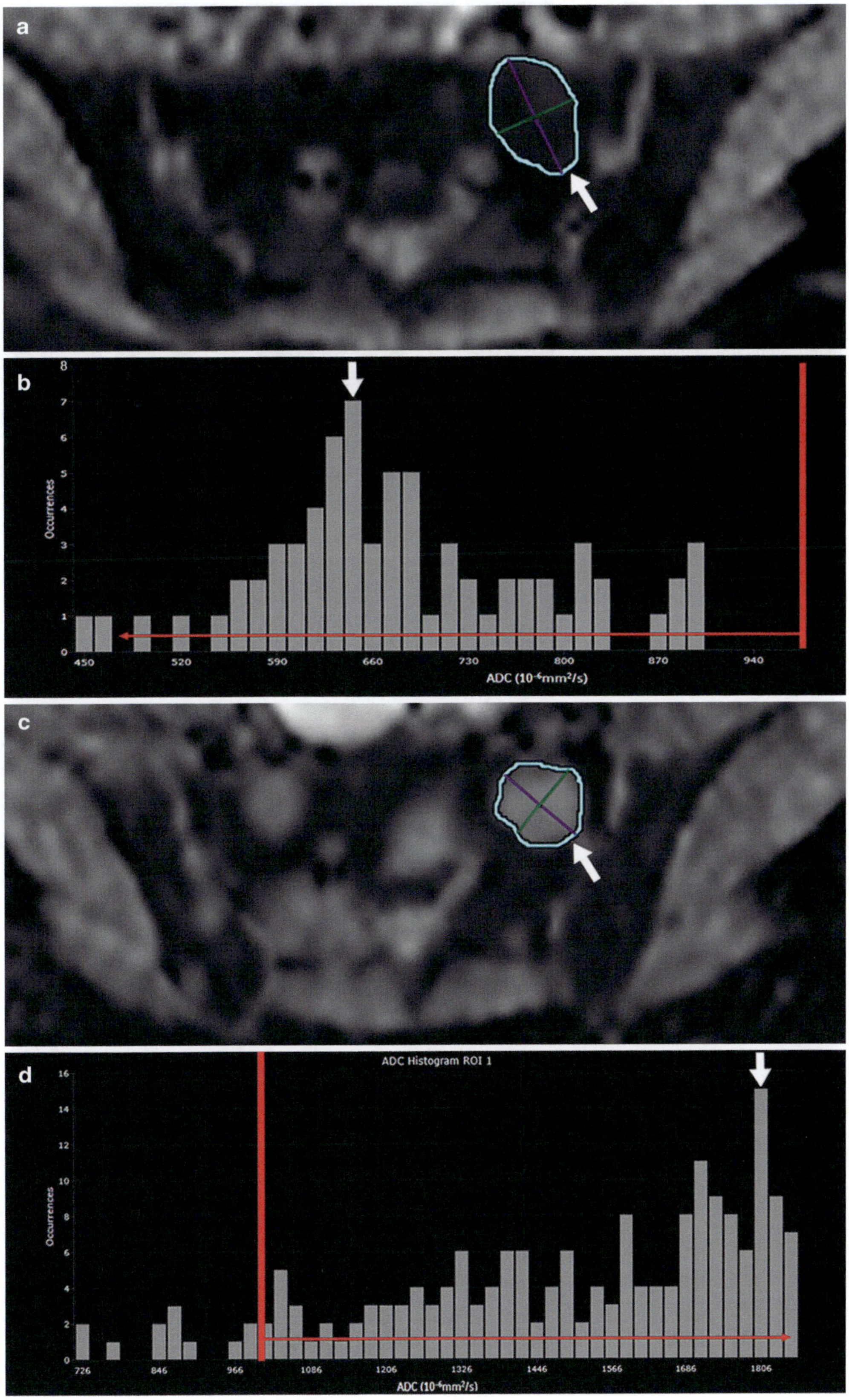

Fig. 28.8 The same patient as Fig. 28.7. Sacral bone lesion (arrows) with interval increased ADC values from (**a**) 0.7×10^{-3} mm^2/s at baseline to (**c**) 1.5×10^{-3} mm^2/s at 4 months, indicative of positive treatment response. (**b**) Pretreatment histogram demonstrates most pixels under 1.0×10^{-3} mm^2/s, with curve centered at 0.9×10^{-3} mm^2/s (arrow), (**d**) posttreatment histogram demonstrates a right-sided curve shift, with most pixels above 1.0×10^{-3} mm^2/s, centered at 1.8×10^{-3} mm^2/s (arrow)

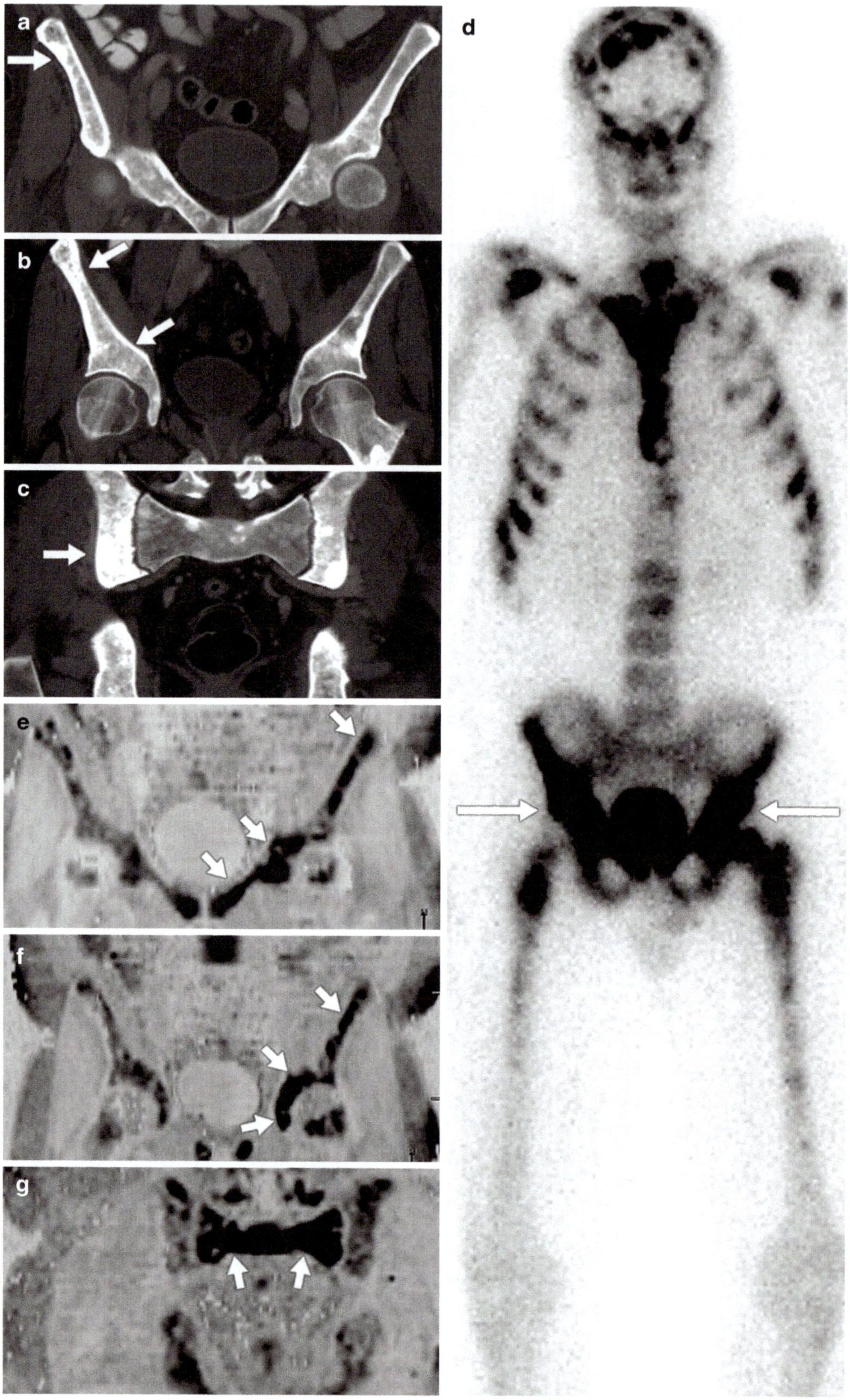

Fig. 28.9 (**a**–**c**) Coronal CT reformations, displaying widespread sclerotic prostate cancer metastases, such as in the right iliac bone and acetabulum (arrows). (**d**) ^{99m}Tc MDP BS demonstrates increase radiotracer uptake in the bilateral pelvis (long arrows). These findings can remain unchanged for a prolonged period despite therapeutic response or lack thereof. (**e**–**g**) Coronal pelvic DWI reformations demonstrate low signal, indicating that the disease is active in the left hemipelvis and sacrum (small arrows) but not in portions of the right hemipelvis despite persistent MDP uptake on BS

Conclusion

- The MDA criteria can be used as a guide to the interpretation of the conventional imaging of bone metastases following therapy. Future research endeavors can investigate the addition of functional imaging and advanced imaging to increase the efficacy of the criteria.
- Bone scans are indirect tumor imaging agents that are more sensitive for sclerotic than lytic metastases, but may show paradoxical flare after initiation of successful therapy. In contrast, direct tumor imaging agents are better at identifying lytic or marrow-based metastases and response to therapy, but detection is dependent on the imaging agent. Development of an imaging agent that both accurately identifies and reflects response in all types of bone metastases would be ideal.
- Radium-223 is an alpha-emitting bone-seeking radionuclide that can increase the survival of castration-resistant prostate cancer patients with osseous metastases. Its utility in treating osseous metastases from other cancer types is under investigation.
- While conventional MRI can be inconclusive in assessing bone metastases for treatment response, DWI has the potential to evaluate more reliably the effect of systemic cytotoxic and radiation therapy.
- Future refinements in post-processing may grant WB-MRI-DWI a central role in oncologic imaging for total tumor burden and therapy response assessments.

References

1. Langer C, Hirsh V. Skeletal morbidity in lung cancer patients with bone metastases: demonstrating the need for early diagnosis and treatment with bisphosphonates. Lung Cancer. 2010;67(1):4–11.
2. Hortobagyi GN, Libshitz HI, Seabold JE. Osseous metastases of breast cancer. Clinical, biochemical, radiographic, and scintigraphic evaluation of response to therapy. Cancer. 1984;53(3):577–82.
3. Galasko CS. Skeletal metastases and mammary cancer. Ann R Coll Surg Engl. 1972;50(1):3–28.
4. Costelloe CM, Rohren EM, Madewell JE, Hamaoka T, Theriault RL, Yu TK, et al. Imaging bone metastases in breast cancer: techniques and recommendations for diagnosis. Lancet Oncol. 2009;10(6):606–14.
5. Mirels H. Metastatic disease in long bones: a proposed scoring system for diagnosing impending pathologic fractures. Clin Orthop Relat Res. 1989;2003(Suppl. 415):S4–13.
6. Costelloe CM, Amini B, Madewell JE. Risks and benefits of gadolinium-based contrast-enhanced MRI. Semin Ultrasound CT MR. 2020;41(2):170–82.
7. Hamaoka T, Madewell JE, Podoloff DA, Hortobagyi GN, Ueno NT. Bone imaging in metastatic breast cancer. J Clin Oncol. 2004;22(14):2942–53.
8. Nakata E, Sugihara S, Kataoka M, Yamashita N, Furumatsu T, Takigawa T, et al. Early response assessment of re-ossification after palliative conventional radiotherapy for vertebral bone metastases. J Orthop Sci. 2019;24(2):332–6.
9. Hamaoka T, Costelloe CM, Madewell JE, Liu P, Berry DA, Islam R, et al. Tumour response interpretation with new tumour response criteria vs the World Health Organisation criteria in patients with bone-only metastatic breast cancer. Br J Cancer. 2010;102(4):651–7.
10. Hayashi N, Costelloe CM, Hamaoka T, Wei C, Niikura N, Theriault RL, et al. A prospective study of bone tumor response assessment in metastatic breast cancer. Clin Breast Cancer. 2013;13(1):24–30.
11. Tsai YC, Lee HL, Kuo CC, Chen CY, Hsieh KL, Wu MH, et al. Prognostic and predictive factors for clinical and radiographic responses in patients with painful bone metastasis treated with magnetic resonance-guided focused ultrasound surgery. Int J Hyperth. 2019;36(1):932–7.
12. Yu T, Choi CW, Kim KS. Treatment outcomes of stereotactic ablative radiation therapy for non-spinal bone metastases: focus on response assessment and treatment indication. Br J Radiol. 2019;92(1099):20181048.
13. Donohoe KJ, Cohen EJ, Giammarile F, Grady E, Greenspan BS, Henkin RE, et al. Appropriate use criteria for bone scintigraphy in prostate and breast cancer: summary and excerpts. J Nucl Med. 2017;58(4):14N–7N.
14. Even-Sapir E, Metser U, Mishani E, Lievshitz G, Lerman H, Leibovitch I. The detection of bone metastases in patients with high-risk prostate cancer: 99mTc-MDP Planar bone scintigraphy, single- and multi-field-of-view SPECT, 18F-fluoride PET, and 18F-fluoride PET/CT. J Nucl Med. 2006;47(2):287–97.
15. Cook GJ, Houston S, Rubens R, Maisey MN, Fogelman I. Detection of bone metastases in breast cancer by 18FDG PET: differing metabolic activity in osteoblastic and osteolytic lesions. J Clin Oncol. 1998;16(10):3375–9.
16. Costelloe CM, Chuang HH, Madewell JE. FDG PET for the detection of bone metastases: sensitivity, specificity and comparison with other imaging modalities. PET Clin. 2010;5(3):281–95.
17. Coleman RE, Mashiter G, Whitaker KB, Moss DW, Rubens RD, Fogelman I. Bone scan flare predicts successful systemic therapy for bone metastases. J Nucl Med. 1988;29(8):1354–9.
18. Matin P. The appearance of bone scans following fractures, including immediate and long-term studies. J Nucl Med. 1979;20(12):1227–31.
19. Yang HL, Liu T, Wang XM, Xu Y, Deng SM. Diagnosis of bone metastases: a meta-analysis comparing (1)(8)FDG PET, CT, MRI and bone scintigraphy. Eur Radiol. 2011;21(12):2604–17.
20. Catalano OA, Nicolai E, Rosen BR, Luongo A, Catalano M, Iannace C, et al. Comparison of CE-FDG-PET/CT with CE-FDG-PET/MR in the evaluation of osseous metastases in breast cancer patients. Br J Cancer. 2015;112(9):1452–60.
21. Botsikas D, Bagetakos I, Picarra M, Da Cunha Afonso Barisits AC, Boudabbous S, Montet X, et al. What is the diagnostic performance of 18-FDG-PET/MR compared to PET/CT for the N-and M- staging of breast cancer? Eur Radiol. 2019;29(4):1787–98.
22. Beiderwellen K, Huebner M, Heusch P, Grueneisen J, Ruhlmann V, Nensa F, et al. Whole-body [(1)(8)F]FDG PET/MRI vs. PET/CT in the assessment of bone lesions in oncological patients: initial results. Eur Radiol. 2014;24(8):2023–30.
23. Guberina N, Hetkamp P, Ruebben H, Fendler W, Grueneisen J, Suntharalingam S, et al. Whole-body integrated [(68)Ga]PSMA-11-PET/MR imaging in patients with recurrent prostate cancer: comparison with whole-body PET/CT as the standard of reference. Mol Imaging Biol. 2020;22(3):788–96.

24. Eiber M, Takei T, Souvatzoglou M, Mayerhoefer ME, Furst S, Gaertner FC, et al. Performance of whole-body integrated 18F-FDG PET/MR in comparison to PET/CT for evaluation of malignant bone lesions. J Nucl Med. 2014;55(2):191–7.
25. Ishii S, Shimao D, Hara T, Miyajima M, Kikuchi K, Takawa M, et al. Comparison of integrated whole-body PET/MR and PET/CT: Is PET/MR alternative to PET/CT in routine clinical oncology? Ann Nucl Med. 2016;30(3):225–33.
26. Pandit-Taskar N, Batraki M, Divgi CR. Radiopharmaceutical therapy for palliation of bone pain from osseous metastases. J Nucl Med. 2004;45(8):1358–65.
27. Parker C, Nilsson S, Heinrich D, Helle SI, O'Sullivan JM, Fossa SD, et al. Alpha emitter radium-223 and survival in metastatic prostate cancer. N Engl J Med. 2013;369(3):213–23.
28. Smith M, Parker C, Saad F, Miller K, Tombal B, Ng QS, et al. Addition of radium-223 to abiraterone acetate and prednisone or prednisolone in patients with castration-resistant prostate cancer and bone metastases (ERA 223): a randomised, double-blind, placebo-controlled, phase 3 trial. Lancet Oncol. 2019;20(3):408–19.
29. Valenzuela RF, Madewell JE, Kundra V. Advanced imaging in musculoskeletal oncology: moving away from RECIST and embracing advanced bone and soft tissue tumor imaging (ABASTI)-Part I: Novel functional imaging techniques. Semin Ultrasound CT MRI. 2020:1–14.
30. Lecouvet FE, Talbot JN, Messiou C, Bourguet P, Liu Y, de Souza NM, et al. Monitoring the response of bone metastases to treatment with magnetic resonance imaging and nuclear medicine techniques: a review and position statement by the European Organisation for Research and Treatment of Cancer imaging group. Eur J Cancer. 2014;50(15):2519–31.
31. Reischauer C, Koh DM, Froehlich JM, Patzwahl R, Binkert CA, Gutzeit A. Pilot study on the detection of antiandrogen resistance using serial diffusion-weighted imaging of bone metastases in prostate cancer. J Magn Reson Imaging. 2016;43(6):1407–16.
32. Messiou C, Collins DJ, Giles S, de Bono JS, Bianchini D, de Souza NM. Assessing response in bone metastases in prostate cancer with diffusion weighted MRI. Eur Radiol. 2011;21(10):2169–77.
33. Bharwani N, Miquel ME, Powles T, Dilks P, Shawyer A, Sahdev A, et al. Diffusion-weighted and multiphase contrast-enhanced MRI as surrogate markers of response to neoadjuvant sunitinib in metastatic renal cell carcinoma. Br J Cancer. 2014;110(3):616–24.
34. Musio D, De Francesco I, Galdieri A, Marsecano C, Piciocchi A, Napoli A, et al. Diffusion-weighted magnetic resonance imaging in painful bone metastases: using quantitative apparent diffusion coefficient as an indicator of effectiveness of single fraction versus multiple fraction radiotherapy. Eur J Radiol. 2018;98:1–6.
35. Reginelli A, Silvestro G, Fontanella G, Sangiovanni A, Conte M, Nuzzo I, et al. Validation of DWI in assessment of radiotreated bone metastases in elderly patients. Int J Surg. 2016;33(Suppl 1):S148–53.
36. Padhani AR, Makris A, Gall P, Collins DJ, Tunariu N, de Bono JS. Therapy monitoring of skeletal metastases with whole-body diffusion MRI. J Magn Reson Imaging. 2014;39(5):1049–78.
37. Blackledge MD, Collins DJ, Tunariu N, Orton MR, Padhani AR, Leach MO, et al. Assessment of treatment response by total tumor volume and global apparent diffusion coefficient using diffusion-weighted MRI in patients with metastatic bone disease: a feasibility study. PLoS One. 2014;9(4):e91779.
38. Stecco A, Trisoglio A, Soligo E, Berardo S, Sukhovei L, Carriero A. Whole-body MRI with diffusion-weighted imaging in bone metastases: a narrative review. Diagnostics (Basel). 2018;8(3)
39. Mosavi F, Johansson S, Sandberg DT, Turesson I, Sorensen J, Ahlstrom H. Whole-body diffusion-weighted MRI compared with (18)F-NaF PET/CT for detection of bone metastases in patients with high-risk prostate carcinoma. AJR Am J Roentgenol. 2012;199(5):1114–20.

Imaging of Lung and Soft Tissue Metastases

29

William A. Biché, James F. McLoughlin,
Vanesa Carlota Andreu-Arasa, Stephan W. Anderson,
and Christina A. LeBedis

Abstract

Excepting our surgery and pathology colleagues, most physicians only "see" cancer via medical imaging. This evaluation is key to the diagnosis and staging of nearly every cancer, driving the choice of therapeutic approach and prognosis. While radiologists should be available for consultation, the ability to recognize potential metastatic disease on imaging is a valuable skillset for any practitioner, particularly those most involved in oncology. This chapter focuses on non-osseous metastases, introducing a framework and lexicon aimed at non-radiologists to assist in accurate diagnosis and appropriate therapy selection. We cover numerous modalities, including MR, CT, and ultrasound, and move organ by organ, describing the imaging features of metastases to each.

Learning Objectives

- Describe several imaging findings concerning metastatic disease in the central nervous system, chest, abdomen, and pelvis.
- Know the most common primary cancers for metastases found in various organs.
- Correctly distinguish between normal lymph nodes and those concerning metastasis based on imaging characteristics.

While pathology remains the gold standard for the diagnosis of malignancy, imaging plays a critical role in screening, diagnosis, and monitoring. In the latter case, the role of imaging has continued to expand, with studies like RECIST [1] formalizing imaging as a primary method of monitoring cancer response to therapy. In this chapter, we will review the myriad appearances of metastatic disease in the soft tissues, proceeding by organ.

It should be noted as a warning to the reader that these descriptions are not comprehensive and serve only as a guide to stereotypical and common appearances. For one, the metastatic disease can present in more ways than any book could describe. For two, metastasis should be on the differential for any unusual lesion in a patient with known malignancy. While this chapter should serve as a guide toward familiarity with the most typical presentations, it should not dissuade practitioners from consulting their colleagues in radiology for further insight into a given lesion.

29.1 Intracranial Metastases

29.1.1 Introduction

Approximately 15–20% of patients with cancer will develop intracranial metastases [2]. The true incidence may be even higher, as screening guidelines for the majority of solid tumors do not recommend routine brain scans. Intracranial metastatic lesions may involve the parenchyma—the cerebral hemispheres, cerebellum, and the brain stem—and/or the meninges.

Although any primary tumor can metastasize to the brain parenchyma or the meninges, several large series have identified lung cancer as universally the most common solid primary tumor resulting in intracranial metastasis, seen in 18–64% of cases, followed by breast cancer (2–21%), melanoma (4–16%), and colorectal cancer (2–11%). When included, hematologic cancers cause approximately 10% of intracranial metastases, primarily to the meninges [3].

Within the brain parenchyma, metastases occur in approximately 85% of cases to the cerebral hemispheres, 15% to the

William Biche and James McLoughlin are co-first authors, contributed equally to this chapter.

W. A. Biché · J. F. McLoughlin · V. C. Andreu-Arasa
S. W. Anderson · C. A. LeBedis (✉)
Department of Radiology, Boston University Medical Center, Boston University School of Medicine, Boston, MA, USA
e-mail: christina.lebedis@bmc.org

S. P. Leong et al. (eds.), *Cancer Metastasis Through the Lymphovascular System*,
https://doi.org/10.1007/978-3-030-93084-4_29

cerebellum, and 3% to the brainstem, in proportion to blood flow to these structures [4, 5]. Less commonly, tumors may metastasize to the choroid plexus, ventricles, or pituitary gland. The most common location for parenchymal metastasis is the gray–white matter junction due to the significant change in the size of arterioles at this location. Retroperitoneal tumors such as colorectal, genitourinary, and uterine cancer metastasize disproportionately to the infratentorial region (posterior fossa and spine), which may be due to retrograde dissemination to this location via Batson's venous plexus [4].

Leptomeningeal carcinomatosis represents metastases, which grow within the leptomeninges. Metastases may spread to leptomeninges via different mechanisms, including direct seeding from the brain parenchyma, seeding from the dura and skull, endoneural/perineural invasion, and hematogenous seeding. It is an uncommon and late complication seen in 5–8% of cases of solid tumors and 5–15% of cases of hematological cancers [6]. The incidence of leptomeningeal carcinomatosis varies among different cancer types, but is diagnosed most frequently in patients with breast or lung cancer (10–35%), as well as melanoma (5–25%) [7, 8].

Dural metastases are less common than leptomeningeal metastases and are most frequently associated with contiguous extension from calvarial metastases in approximately 61% of cases. Isolated dural metastases can also be seen, but are much less common, occurring in 4% of cases. Particular cancers that are associated with dural-based metastases include breast (34%), prostate (17%), and lung (13%). Mechanisms of spread to the dura are similar to the leptomeninges and include hematogenous spread, lymphatic spread, or retrograde spread via the vertebral venous plexus [9].

29.1.2 Imaging Findings

29.1.2.1 Computed Tomography (CT)

Non-enhanced computed tomography (NECT) may be the first imaging modality a patient with brain metastasis undergoes, either in detecting a previously unknown lesion or with the development of new neurologic findings. NECT alone is not sensitive enough to accurately characterize cerebral metastases, but findings can suggest the presence of a lesion or signs of mass effect.

Brain parenchymal metastases on CT appear as solitary or multiple well-circumscribed round mass lesions, which are present a majority of the time at the gray–white matter interface. They are typically small and surrounded by digitiform hypodensity representing vasogenic edema. Metastases may be hypodense, isodense, or hyperdense to the brain parenchyma. Approximately 20% of metastases are hemorrhagic, and a hemorrhagic parenchymal neoplasm is more likely to be metastatic than primary. Melanoma metastases tend to be hyperdense even in the absence of hemorrhage [10]. Almost all parenchymal metastases demonstrate avid enhancement on post-contrast imaging. Enhancement patterns may be solid, nodular, or ring-enhancing, although specific enhancement pattern offers little insight into the primary tissue of origin. Contrast-enhanced computed tomography (CECT) may be used to screen for metastases if magnetic resonance imaging (MRI) is contraindicated or unavailable, however is less sensitive than contrast-enhanced MRI [11, 12].

Metastases involving the meninges may not have any visible signs on NECT, especially when not large enough to cause mass effect of the brain parenchyma. When large enough to be detectable, they present as extra-axial biconvex hyperdensities. On CECT, metastases to the meninges may appear as enhancing biconvex masses displacing the brain parenchyma. Involvement of the calvarium would be more suggestive of dural-bases metastasis than leptomeningeal involvement. MRI is the preferable modality for the assessment of metastasis involving the meninges.

29.1.2.2 Magnetic Resonance (MR)

MR with contrast is the most sensitive screening test available for detecting intracranial metastasis. It is also useful to further characterize mass lesions found on CT in order to refine the differential diagnosis. In fact, CT may underestimate the number of parenchymal metastatic lesions in up to 20% of cases, even when contrast medium is used [13].

On MR, intraparenchymal brain metastases are usually iso- or hypointense on T1, hyperintense on T2, and exhibit avid contrast enhancement. Commonly there will be surrounding T2/FLAIR hyperintense vasogenic edema (Fig. 29.1). Melanoma metastases can be T1 hyperintense due to the free radical content of melanin. Hemorrhagic metastases can also be T1 hyperintense, depending on the age of hemorrhage. Susceptibility-weighted images (SWI) are highly sensitive to the deposition of hemosiderin and other blood breakdown products, and the presence of blooming artifact on these sequences aids in differentiating melanin from hemorrhage. Low T2 signal is seen in some metastatic neoplasms and may offer a clue to tissue of origin. Mucin-secreting metastases, which can be seen in GI and GU adenocarcinomas, and at times lung, may demonstrate low T2 signal. Restricted diffusion (high signal on diffusion-weighted imaging with reduced apparent diffusion coefficient maps) may also occasionally be seen in mucinous-producing lesions and metastasis from other highly cellular primary neoplasms such as breast, colon, testicular, and small and non-small cell lung carcinoma [5, 14].

Nearly all parenchymal metastases demonstrate enhancement on contrast-enhanced MR due to lack of the normal blood-brain barrier architecture. Like CT, enhancement patterns may demonstrate solid, nodular, or ring-like enhancement, however provide little insight into the primary site of tumor. Although the vast majority of contrast-enhancing lesions

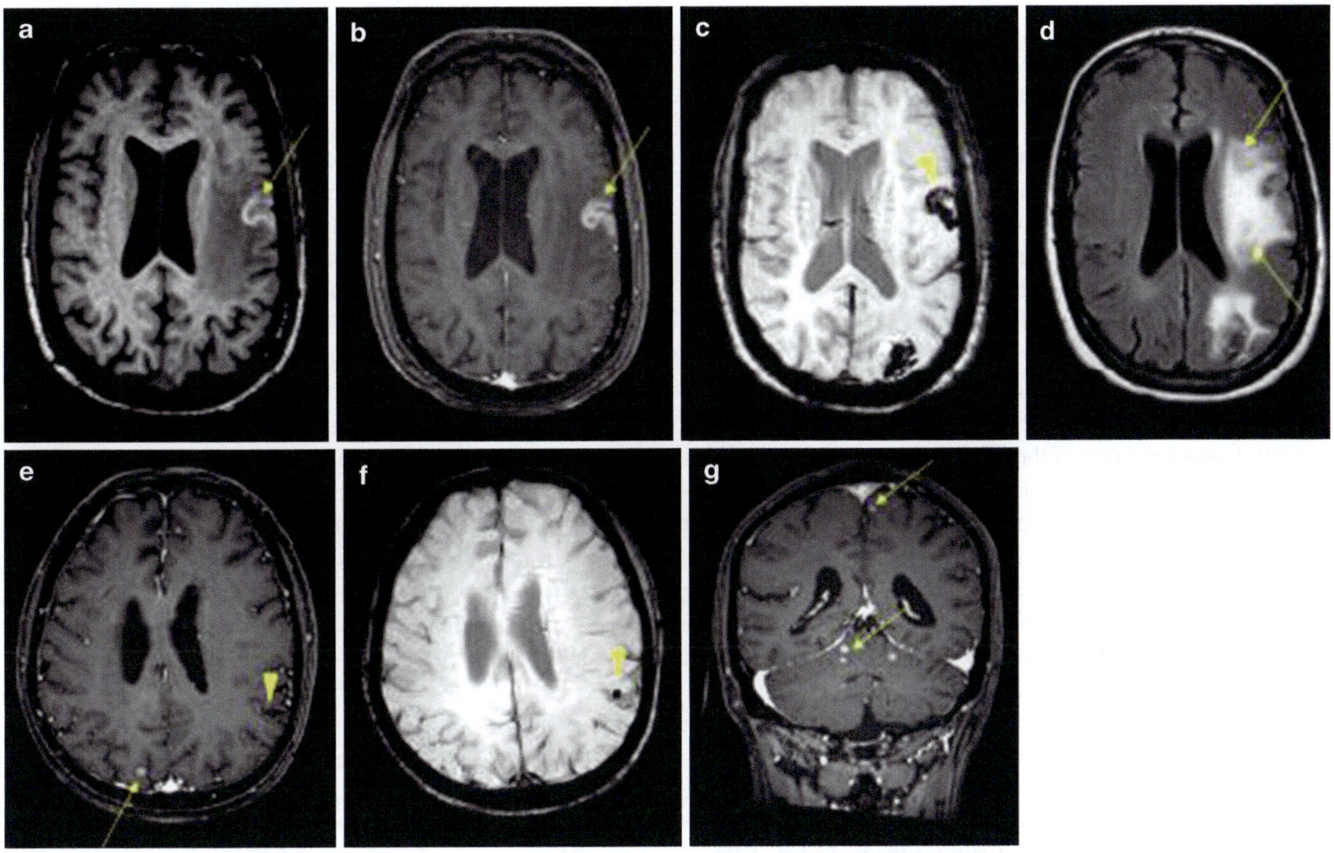

Fig. 29.1 (**a–d**) A 63-year-old male with melanoma. (**a** and **b**) Axial pre and postcontrast T1-WI demonstrate cortical thickening and hyperintensity on pre-contrast images (arrows). (**c**) Axial SWI shows foci of susceptibility artifact (arrowheads) from hemosiderin deposition due to hemorrhage. (**d**) Axial FLAIR shows significant FLAIR hyperintensity representing vasogenic edema (arrows). (**e–g**) A 67-year-old male with stage IV small cell lung carcinoma. (**e**) Axial postcontrast T1-WI demonstrates multiple small nodular foci of enhancement consistent with metastasis (arrow and arrowhead). (**f**) Axial SWI show susceptibility artifact (arrowhead) in some of the lesion representing blood products. (**g**) Coronal post-contrast T1-WI again shows multiple scattered metastatic lesions (arrows)

in a patient with a primary tumor are metastases, the differential diagnosis includes primary brain tumor, abscess, infarction, or radiation necrosis in a previously treated patient [5].

MR is more sensitive for detecting intracranial metastasis to the leptomeninges compared to CT and has a specificity of approximately 77–100% [6]. Leptomeningeal metastases present as either smooth or nodular thickening of the leptomeninges. Lesions are typically T1 iso- or hypointense relative to the gray matter, and T2/FLAIR sequences demonstrate hyperintensity following the sulci and CSF spaces. Leptomeninges demonstrate avid enhancement, which is usually diffuse but may include areas of focal nodularity. Common locations to be affected by leptomeningeal metastasis include the basilar cisterns and cerebellar folia [14, 15].

Dural metastases appear as thickening of the dura, which may be smooth or nodular, and following the contour of the calvarium. Like leptomeningeal metastases, lesions may appear T1 iso- or hypointense and variable on T2-weighted images. They may cause mass effect on the underlying parenchyma with associated vasogenic edema, which presents as parenchymal hyperintensity on FLAIR sequences. Dural metastases enhance avidly, and like meningiomas, an enhancing dural tail may be present in nearly half of cases. There may be calvarial involvement of an enhancing lesion, suggestive of contiguous extension into the skull [9, 16].

29.2 Extra-Osseous Spine Metastases

29.2.1 Introduction

Metastases to the spine can involve the bone, epidural space, leptomeninges, and/or spinal cord. Extradural lesions account for up to 95% of spinal lesions, and can be divided into pure epidural lesions, and those originating from the vertebrae and subsequently impinging on the thecal sac. Intradural extramedullary and intramedullary metastases are less common than extradural metastases, accounting for 5–6% and 0.5–1% of spinal metastases, respectively [17]. Routes of tumor dissemination to the spinal cord and meninges include hematogenous spread via Batson's venous plexus, arterial spread, perineural and perivascular migration from systemic tumors, contiguous extension from the vertebrae, and spread through the CSF as drop metastases.

The vast majority of epidural metastases originate in either the vertebrae (85%) or the paravertebral tissues (10–15%) with extension into the epidural space, with the remainder originating in the epidural space itself [18]. As tumors grow in the epidural space, they can encroach on the thecal sac and compresses the spinal cord, causing neurological disturbance.

In adults, intradural extramedullary metastases of the spine have an incidence as high as 4–15% in patients with solitary tumors not part of the CNS [19]. They are most commonly found in primary tumors of the breast, lung, and melanoma; melanoma and lung cancer have the highest rates of metastasis to this location at 20% and 11%, respectively [20].

Intramedullary spinal metastases are a rare but devastating manifestation of systemic malignancy. Lung cancer is the most common primary sources for intramedullary metastasis (51.4%), followed by breast cancer (18.6%), renal cell carcinoma (8.6%), and melanoma (3.9%) [21]. The presence of an intramedullary spinal metastatic lesion suggests a biologically aggressive form of cancer. The mechanism of metastatic spread from the primary tumor to the spinal cord is thought to be hematogenous via an arterial route in the majority of cases.

29.2.2 Imaging Findings

Although CT can be useful for assessing the osseous structures of the spine, it is inadequate for the assessment of possible meningeal and intramedullary metastatic disease. MR is used for this purpose and can aid in the diagnosis of metastasis, characterization of levels of involvement, and the diagnosis of potential associated cord compression.

Extradural metastases are located external to the dura mater as the epidural space is extradural. Typically, epidural metastases appear as a region of iso- or hypointense T1 signal, increased T2 signal, and contrast enhancement within the epidural space. Involvement may be focal or multilevel depending on the extent of disease. Increased T2 signal intensity in the spinal cord in this setting is suggestive compressive edema or infarction from vascular compression.

Intradural extramedullary metastases are located inside the pachymeninges but outside the spinal cord. Most lesions are located within the subarachnoid space and involve the leptomeninges. Lesions demonstrate enhancing smooth or nodular thickening along the surface of the spinal cord and/or the cord roots, which can be focal or diffuse. Involvement can be isolated to just the spine or can involve both the intracranial structures and the spine (Fig. 29.2). Metastatic lesions may be seen as small T2 isointense nodules, but are generally poorly seen on T1-weighted images. On post-contrast imaging, lesions typically demonstrate avid enhancement. In cases where there is diffuse enhancement coating the spinal cord, it is referred as "sugar coating" due to the appearance of extensive bright enhancement along the spinal cord, brainstem, and sulci [17].

Intradural intramedullary metastases are located deep to the pia mater, typically within the substance of the cord itself. These are typically small, well-circumscribed spherical, or ovoid lesions with surrounding edema seen as hyperintensity on T2-weighted images that may be out of proportion to the size of the lesion. Post-contrast images demonstrate avid enhancement of the lesion within the spinal cord. Intramedullary metastases can be hemorrhagic, and in these cases, demonstrate similar MR findings to hemorrhagic brain lesions. A thin line of increased enhancement surrounding the margins of the lesion ("rim sign") with ill-defined triangular extension along the cranial and caudal aspect of the lesion ("flame sign") on sagittal post-contrast images is considered highly specific for intramedullary metastasis [16, 17, 21].

29.3 Lung Metastases

29.3.1 Introduction

Lung metastases are seen in 20–54% of extrathoracic malignancies. The lungs are the second most frequent site of metastasis, after the liver. Approximately 20% of metastatic disease is isolated to just the lungs. For these reasons, assessment of lung imaging is of paramount importance in the management and staging of cancer patients [22, 23].

A retrospective study conducted in 288 patients with confirmed metastatic lung nodules revealed the primary sites of tumor were as follows: colorectal 25.8%, head and neck 19.4%, urologic (kidneys, ureter, bladder, prostate, testes) 14.7%, gastrointestinal non-colorectal 10.9%, breast 10.5%, melanoma 6.5%, gynecologic (ovarian, endometrial, cervical) 6.1%, and other sites (e.g., sarcoma, thyroid) in 6.1% of cases [24].

Malignancies can reach the lung parenchyma by hematogenous spread through the pulmonary or bronchial arteries, lymphatic spread, or direct invasion. The most common path by far is hematogenous spread—metastases by this route tend to be located in the more highly vascularized basal and peripheral regions of the lungs. In extensive metastatic disease, the distribution becomes more random and multifocal [25].

29.3.2 Imaging Findings

29.3.2.1 Radiography

Chest radiographs are known to have low sensitivity compared to computed tomography (CT) imaging in the detection

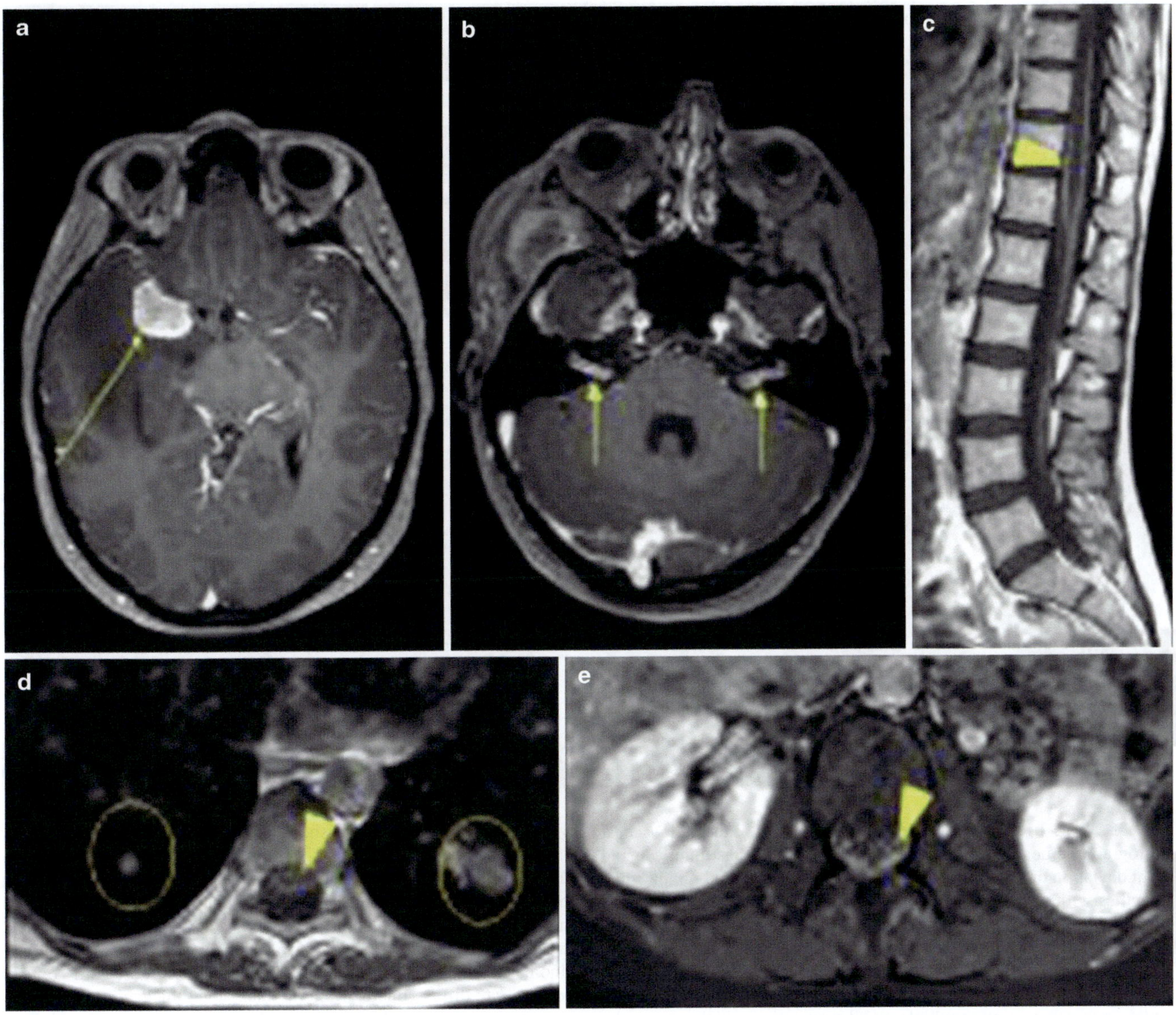

Fig. 29.2 A 31-year-old female with breast carcinoma. (**a** and **b**) Axial post-contrast T1-WI shows a large enhancing mass in the right temporal fossa (**a**) that was thought to be extra-axial, surgically removed, and demonstrated to be a parenchymal lesion (arrow). In addition, there is diffuse enhancement along the bilateral internal auditory canals (arrows) representing leptomeningeal disease. (**c**) Sagittal T1-WI after contrast administration shows extensive enhancement coating the spinal cord (arrowhead) and cauda equina nerve roots (arrowheads), which is also demonstrated on axial post-contrast T1-WI at the lower thoracic and lumbar spine levels (**d** and **e**, respectively). Also, lung nodules are noted (**d**) consistent with lung metastasis (circles)

of metastatic nodules. Radiographs fail to depict metastatic lesions smaller than 7 mm and are limited by overlapping structures and relatively low contrast. In addition, radiographs are known to underestimate metastases compared to CT; a solitary lesion seen with radiograph may be associated with multiple nodules on CT [26, 27].

29.3.2.2 Computed Tomography (CT)

CT is the modality of choice for the detection and follow-up of pulmonary metastases because of its high spatial resolution. No contrast is required for the detection of pulmonary metastases. Typical findings include multiple nodules of different sizes, diffuse thickening of the interstitium (lymphangitic carcinomatosis), and involvement of the pleura. Atypical patterns of lung metastasis can confound the differential and include diffuse miliary seeding, cavitation, and calcified metastases [28] (Fig. 29.3).

The most common CT pattern of pulmonary metastases is the presence of multiple solid round pulmonary nodules. Nodules have soft-tissue attenuation and can be smooth or irregular. When nodules are numerous, they are distributed diffusely throughout the lungs without any anatomic

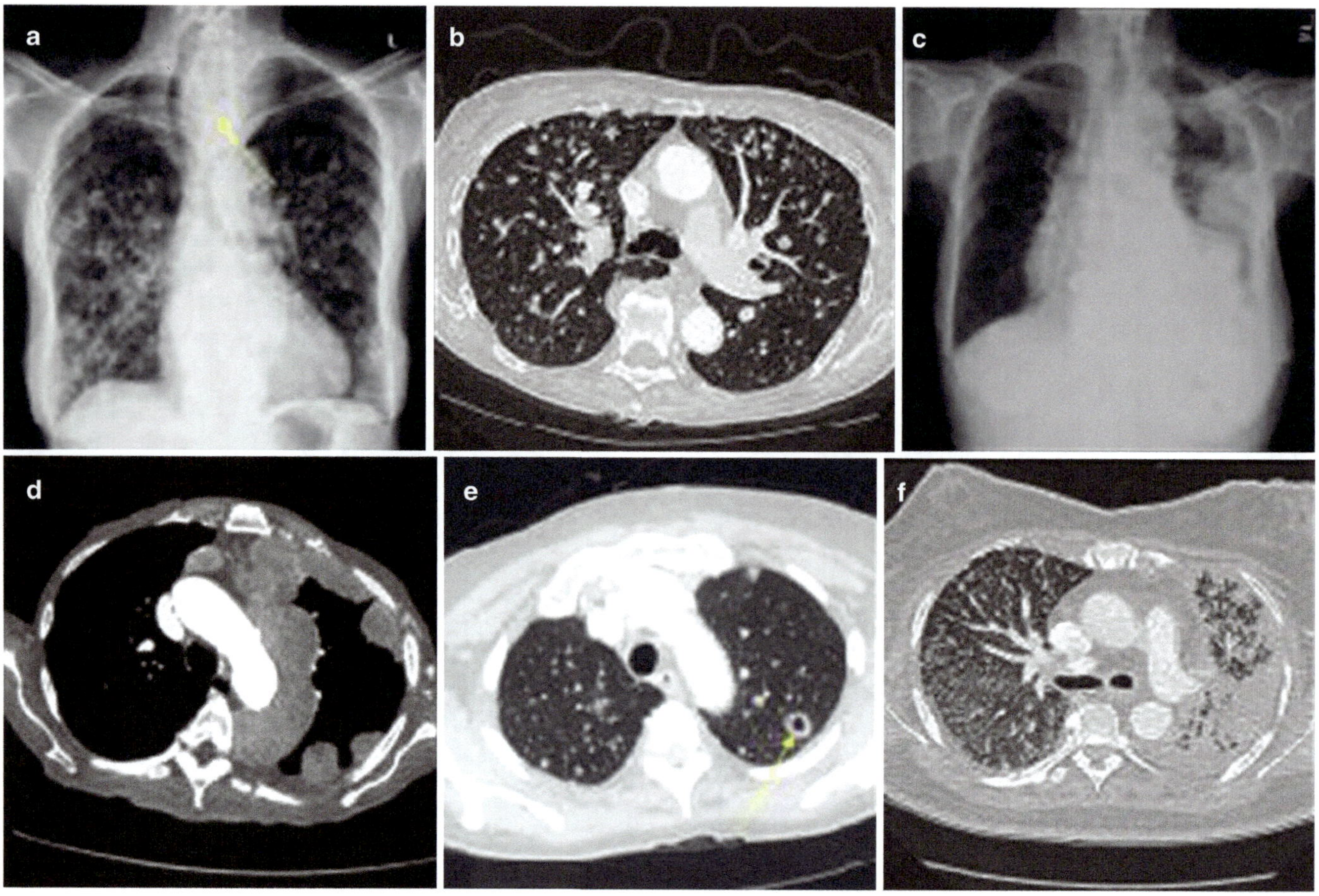

Fig. 29.3 (**a**, **b**) 82-year-old female with thyroid cancer. PA chest radiograph demonstrates opacity of the superior mediastinum causing right deviation of the trachea (arrow). There are multifocal nodular-like opacities within the bilateral lungs. A follow-up axial CT of the chest with contrast demonstrates innumerable solid pulmonary nodules in a random distribution. The majority of the nodules are smooth but few have irregular margins, highly suspicious for metastatic disease. (**c**, **d**) 54-year-old female with history of metastatic melanoma. PA chest radiograph demonstrates multiple nodular opacities, which occupy the majority of the left hemithorax, favoring the periphery. Axial CT images of the chest with contrast demonstrate multiple pleural-based masses, which also involve the lateral aspect of the mediastinum and encase part of the aortic arch. Such findings are most suspicious for pleural metastases. (**e**) 68-year-old female with uterine sarcoma. Axial CT image of the chest with contrast demonstrates multiple smooth-walled nodules throughout the lungs. In addition, there is a cavitary lesion with thick, irregular walls (arrow). (**f**) 60-year-old female with stage IV adenocarcinoma of the left lung. Axial CT images of the chest with contrast demonstrate diffuse bilateral interlobular septal thickening with nodularity and irregularity along the interlobular septa, consistent with lymphangitic carcinomatosis. Loss of lung volume and pleural effusion is noted in the left lung

distribution; when nodules are few or solitary, they are predominantly peripheral and within the lung bases. It is also not unusual for nodules to be of different sizes due to multiple different episodes of tumor embolization [25, 29].

Lymphangitic carcinomatosis refers to the spread of a neoplasm through the lymphatics. It is most commonly seen in adenocarcinomas, particularly primary tumors of the lung, stomach, breast, pancreas, uterus, rectum, and prostate. The mechanisms of metastatic involvement of the lymphatics include antegrade lymphatic invasion through the diaphragm or pleural surfaces or retrograde spread from hilar metastases. The typical appearance is characterized by interlobular septal thickening, with areas of nodularity or irregularity along the interlobular septa extending toward the hilum. In most cases, there is the preservation of the underlying lung architecture [30].

While the classic imaging appearances of lung metastases are extremely helpful in narrowing the differential down to metastasis, especially with a known primary, atypical features can make it difficult to distinguish lung metastases from other lung pathologies.

Miliary nodules refer to innumerable small randomly distributed nodular opacities approximately <5 mm in size that resembles millet seeds. Miliary metastases can be seen in cancers of the thyroid, kidney, breast, and pancreas, as well as more rarely in malignant melanoma and osteosarcoma. The differential for miliary patterns on chest imaging includes tuberculosis, fungal infections, sarcoidosis, pneumoconiosis [31].

Cavitation is seen in approximately 5% of lung metastasis. Tumor necrosis of the central portion of the lesion and subsequent discharge of the necrotic material is the mechanism which causes the cavitating appearance. Cavitating lesions generally feature thick and irregular walls. Squamous cell carcinomas are the most common primary tumor to cause cavitation, although adenocarcinomas (GI tract, breast), transitional cell tumors, and sarcomas are also known to cavitate. In addition, primary lung cancers can also cavitate, which confounds the differential when there is a solitary cavitating pulmonary nodule present instead of multiple lesions [32].

Calcification of metastatic nodules is often seen only on CT scans. Calcification is seen in metastases from osteosarcoma, chondrosarcoma, giant cell tumor of the bone, and mucinous adenocarcinoma. Reasons for calcification vary but can include bone formation in primary bone tumors, dystrophic calcification, and mucoid calcifications [28].

In addition to tumor disseminating to the lung parenchyma, the pleura can also be involved. Pleural metastases originate from hematogenous spread to the pleura, but can also be caused by lymphangitic spread, or contiguous spread of tumor to the chest wall or mediastinum from established metastases. Pleural metastases are present on imaging as nodularities along the pleura. There may or may not be an associated malignant pleural effusion. Typically pleural metastases are visible on non-contrast CT as irregularities of the pleura, which do not typically require contrast, however without contrast may be obscured by effusions or other airspace disease [33].

29.3.2.3 Magnetic Resonance Imaging (MR)

MR has lower sensitivity than CT in the detection of metastatic pulmonary nodules and for that reason is not routinely used in the evaluation of lung metastases. Metastatic nodules have low or intermediate intensities on T1-weighted images and higher intensity on T2-weighted images. T1 weighted, T2 weighted, and STIR sequences can help distinguish metastatic lesions from other lung lesions which may confound the diagnosis such as tuberculoma, bronchocele, mucin-containing tumors, hamartoma, and aspergilloma [34].

29.3.2.4 Nuclear Imaging

Fluorodeoxyglucose positron emission tomography (FDG-PET) is the primary scintigraphic modality used in the evaluation of pulmonary metastases. FDG-PET is important in differentiating between benign and malignant nodules, especially in the setting of a solitary pulmonary nodule. It can aid in raising suspicion of malignancy without proceeding to more invasive diagnostic tests such as biopsy.

FDG-PET has been shown to have a sensitivity of 96% and specificity of 88% for diagnosing a pulmonary nodule as malignant [35]. The positive predictive value is lower because false positives caused by infectious or inflammatory etiologies also demonstrate avidity on FDG-PET. False-negative FDG-PET pulmonary nodules are usually caused by lesions smaller than 1 cm. False negatives can also occur due to malignant pulmonary lesions which are not FDG avid, which include renal and testicular cancers. For these reasons, a negative FDG-PET scan in the presence of suspicious findings on chest CT cannot reliably exclude metastatic disease [36].

29.4 Abdomen Metastases

29.4.1 Liver, Biliary Tree, and Pancreas

The liver is one of the most common sites of metastatic disease in the body, second only to lymph nodes [37]. Almost any tumor is capable of finding its way to the hepatic parenchyma, either via the organ's prodigious blood supply or via its complex lymphatic network. The most common primary malignancy is, of course, adenocarcinoma of the colon, but most GI malignancies are likely to metastasize to the liver. This is facilitated by the portal venous system and thus is less commonly seen in cirrhotic patients where blood is shunted away from the liver.

Metastatic lesions in the liver are diverse in appearance. Most classically, they appear as well-defined hypoattenuating lesions, which demonstrate mild peripheral enhancement on portal venous imaging [38] (Fig. 29.4a, h). This appearance is most associated with hypovascular colon cancer metastases. Hypervascular metastases generally enhance avidly on arterial phase imaging (Fig. 29.4e), and may even be undetectable on portal venous imaging. Neuroendocrine tumors, choriocarcinomas, and melanomas are common, but thyroid, RCC, and a small percentage of breast cancers may also develop hypervascular liver metastases. Liver metastases may also be cystic or necrotic, with nonenhancing, low, central attenuation. This appearance is more commonly seen in lung cancer, melanoma, mucinous colon cancer, or occasionally carcinoid.

On MRI, metastatic lesions are most often T1 hypointense and T2 hyperintense [39] (Fig. 29.4g), although this can flip in metastases with significant hemorrhage or melanin. Metastatic lesions also tend to restrict diffusion. One particular advantage to MR imaging of the liver is the advent of the use of contrast agents with hepatobiliary secretion, adding another tool in the characterization of hepatic lesions. As one would predict, metastatic lesions in the liver do not contain functional hepatocytes, and thus do not enhance with these agents.

Hepatic metastases most typically have a targetoid appearance on ultrasound (Fig. 29.4f), with a hyperechoic or isoechoic center and a hypoechoic periphery [40]. This appearance is most typical in colon cancer, but HCC and

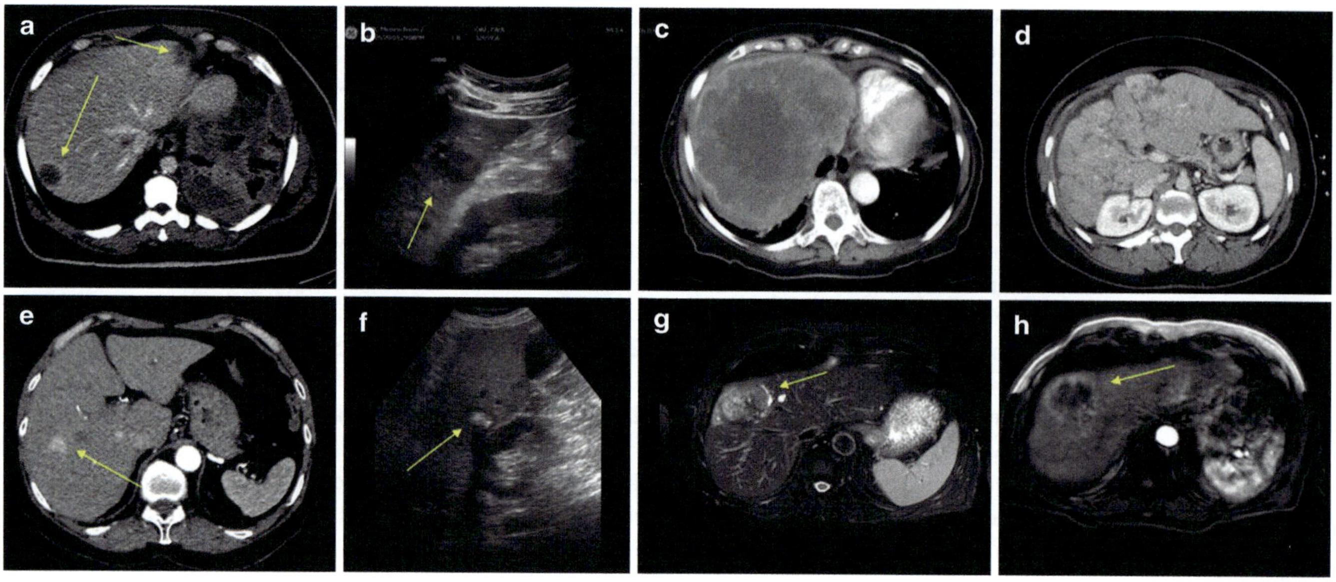

Fig. 29.4 Hepatic metastases. (**a** and **b**) A 51-year-old male with pancreatic cancer metastatic to the liver. (**a**) Axial CT image with IV contrast showing multiple rim enhancing hypoattenuating masses in the liver (arrows). (**b**) Ultrasound of hypoechoic metastasis in right hepatic lobe (arrow) (**c**) 76-year-old female with colon cancer. Axial CT image with IV contrast shows a large hypoattenuating mass in the liver with psammomatous calcifications (**d**) 57-year-old female with breast cancer metastatic to liver. Axial CT image with IV contrast with liver pseudocirrhosis, caused by innumerable small treated metastases. (**e**) A 63-year-old female with small bowel neuroendocrine tumor. Axial CT with IV contrast in the arterial phase shows an enhancing hypervascular metastasis (arrow). (**f**) A 71-year-old male with mucinous colon adenocarcinoma. US of the liver shows a hyperechoic lesion with a hypoechoic halo, a classic "target" metastasis. (**g** and **h**) A 65-year-old man with rectal cancer. Right hepatic lobe metastasis manifesting as a hyperintensity on T2 fat-suppressed imaging (arrow, **g**). This lesion demonstrates continuous peripheral contrast enhancement which was persistent on T1-delayed phase imaging (arrow, **h**)

lymphoma can also present in this manner. Ultrasound can also appreciate the hypervascularity and cystic nature of certain cancers (as above). Another feature, appreciable on ultrasound, is calcification. While this can be seen in the setting of chronic granulomatous changes, metastases from mucinous tumors of the GI tract, breast, and ovaries may also show calcifications, which may also be appreciated on CT and MRI (Fig. 29.4c). Lastly, in patients with diffuse small metastases throughout the liver, the parenchyma may appear diffusely heterogeneous on ultrasound; this is notably seen in some metastatic breast cancers. Once these are treated, they may involute, causing nodularity of the liver known as pseudocirrhosis (Fig. 29.4d).

Metastases to the gallbladder and biliary tree are quite rare, although are occasionally seen in aggressive melanoma, gastric, or breast cancers. The route is typically hematogenous. These lesions generally enhance, and demonstrate either infiltrative or polypoid morphology, sometimes with internal foci of hemorrhage [41]. These lesions are hyperechoic on ultrasound with minimal acoustic shadowing and may show internal vascularity.

Pancreatic metastases are uncommon. Cadaver studies report approximately 3–12% of patients with advanced malignancy have metastatic disease to the pancreas, and metastases are outnumbered by primary lesions 30:1. Unsurprisingly, by the time metastatic disease reaches the pancreas, it has typically seeded other organs first. The most common primaries are melanoma, RCC, lung carcinoma, and breast, and less commonly, colon [42]. These lesions are typically round, well marginated, and heterogeneously enhancing on cross-sectional imaging (Fig. 29.7a, b). While multiple lesions are always suggestive of metastases, solitary lesions are more common. On MRI, these metastases are typically T1 hypointense and T2 hyperintense (Fig. 29.7c) compared to healthy pancreatic parenchyma and they appear solid and hypoechoic on ultrasound.

29.4.2 Spleen, Kidneys, and Ureters

Despite its considerable blood flow, the spleen is an uncommon site for metastatic disease. Only 2–7% of cadavers with advanced metastatic disease have splenic lesions, and the metastatic disease usually presents elsewhere first [43]. Most commonly they are from melanoma (>50%), breast, lung, ovary, gastric, or prostate cancer. The lesions are classically ill-defined, soft tissue attenuating masses with peripheral enhancement (Fig. 29.7d). They may be cystic, particularly in melanoma. MRI shows T1 hypointensity and T2 hyperintensity, with variable enhancement [44]. Of note, implantations to the serosa from intraperitoneal seeding (most commonly from gastric, ovarian, or pancreatic cancers) may

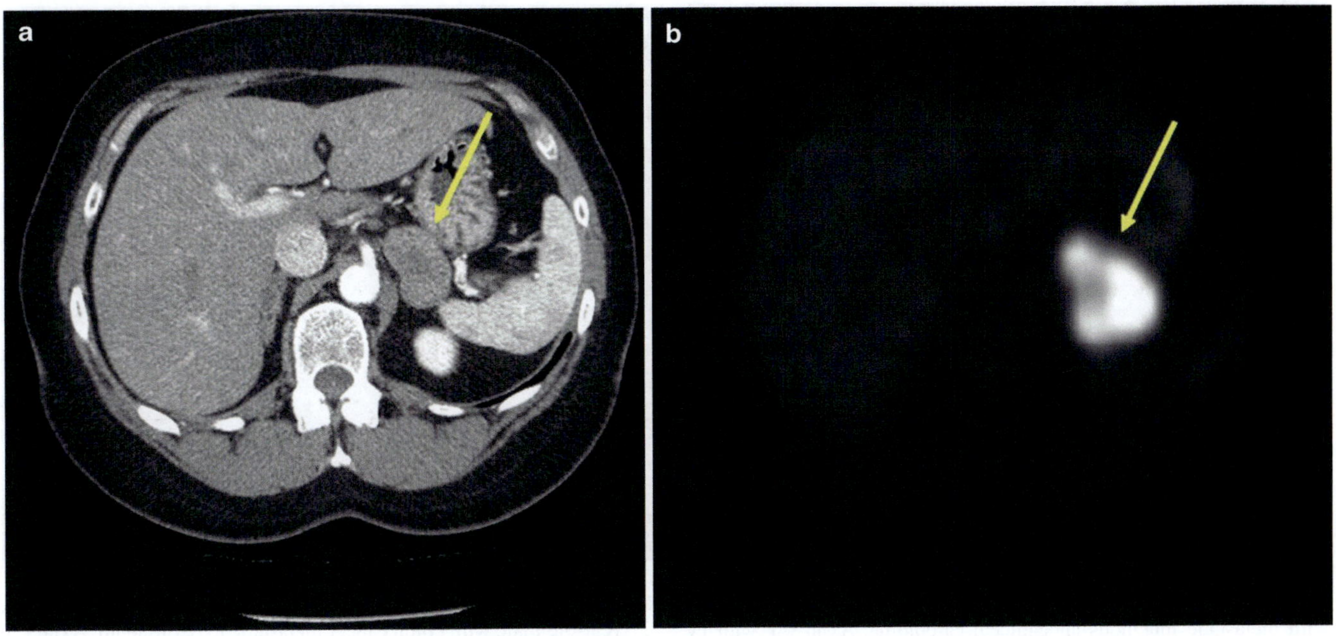

Fig. 29.5 Adrenal metastases. A 60-year-old female with lung cancer metastatic to left adrenal gland. (**a**) Axial CT with IV contrast showing large enhancing left adrenal nodule (arrow). (**b**) Nodule demonstrates hypermetabolic activity (arrow) on PET

cause scalloping (Fig. 29.6a) as the masses tug at the splenic capsule.

Metastases to the kidneys are relatively common, being seen in ~7% of all patients with extrarenal cancers on autopsy. Despite this, they tend to be involved later in the course of disease relative to organs like the lungs or liver. The primary cancers are most commonly lung, breast, or GI adenocarcinomas. When seen on imaging, they are typically multiple and bilateral, with low attenuation on CT. Metastases tend to be solid and infiltrative, with cystic appearance being uncommon. Relative to the kidney parenchyma, they tend to be lower attenuation, with variable enhancement.

29.4.3 Adrenal Glands

The adrenal glands are a very common location for metastatic disease, representing the fifth most common site in the body (after the liver, lungs, bones, and lymph nodes). They are most frequently involved in the lung (especially bronchogenic carcinoma), breast, colon adenocarcinoma, and melanoma, but RCC, thyroid, and lymphoma are also common [45].

Adrenal metastases are usually bilateral and mostly attenuate higher than the normal adrenal parenchyma on CT (Fig. 29.5a, b), with lesions >43 Hounsfield units being very likely metastatic [46]. Exceptions are RCC and rare HCC metastases, which may be hypoattenuating. When large (>3 cm), this attenuation may be more heterogeneous, while the lesion itself may have a poorly defined, lobulated morphology, and may bleed or calcify. Smaller lesions, in contrast, tend to be more homogenous and well defined, making them harder to differentiate from lipid-poor benign adenomas. This diagnostic difficulty is compounded by several similar imaging characteristics including diffusion restriction on MRI and enhancement. Metastases do washout more slowly compared to benign lesions, with absolute washout <60% and relative washout <40% [47]. Both lipid-poor adenomas and metastases are also usually T1 hypointense and T2 hyperintense [48], although the latter moderately favors metastases. In uncertain scenarios, the best imaging differentiator may simply be growth over time.

29.4.4 Gastrointestinal Tract

The GI tract is most commonly involved in metastatic disease through the peritoneal spread of malignancy, from ovarian, gastric, or colonic primaries. In this setting, small, soft tissue attenuation lesions may be identified in the serosa, which frequently causes adhesions to nearby loops of the bowel, predisposing to obstruction or encasement. They are very rarely solitary and frequently innumerable.

Less commonly, it may be seeded hematogenously, most commonly by melanoma, breast, or lung cancers. In contrast to peritoneal spread, these lesions tend to be submucosal or intramural, with contrast enhancement. In these locations, they may cause intussusception, in addition to obstructions.

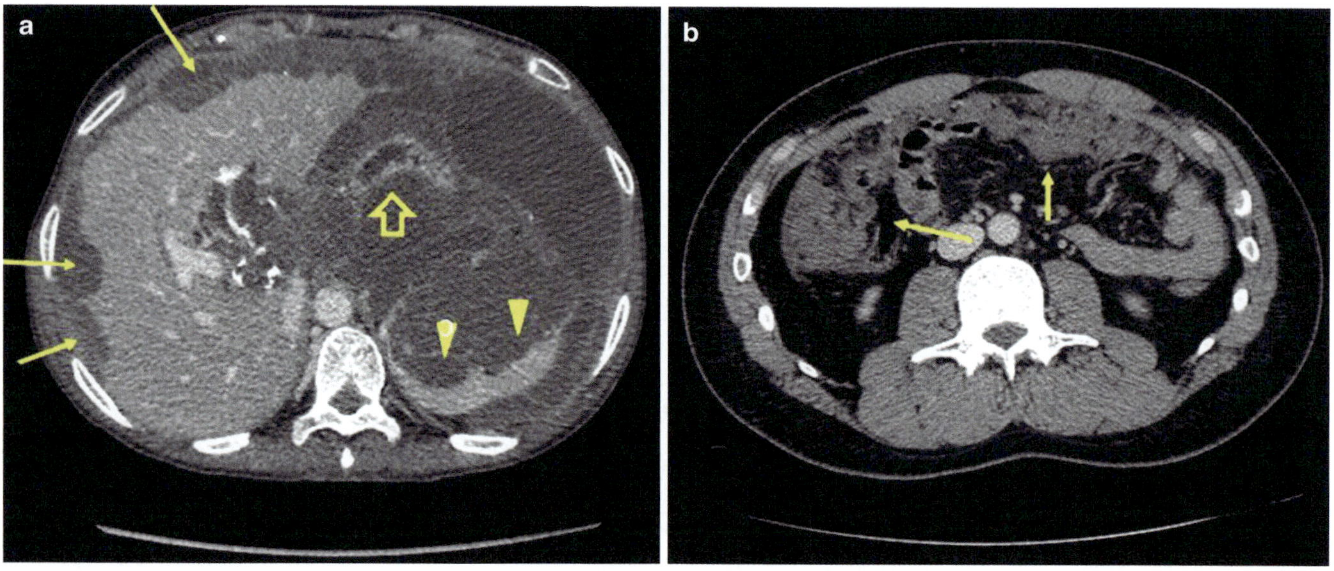

Fig. 29.6 Peritoneal/omental metastases. (**a**) A 38-year-old female with metastatic mucinous appendiceal carcinoma. Axial CT with IV contrast through the upper abdomen shows mucinous metastases throughout the peritoneum consistent with pseudomyxoma peritonei. Note the "scalloping" of the stomach (hollow arrow), liver (arrows), and spleen (arrowheads), consistent with serosal deposits. (**b**) A 27-year-old male with colon cancer metastatic to peritoneum. Axial CT with IV contrast shows omental thickening and soft tissue deposits (arrows) consistent with "omental caking"

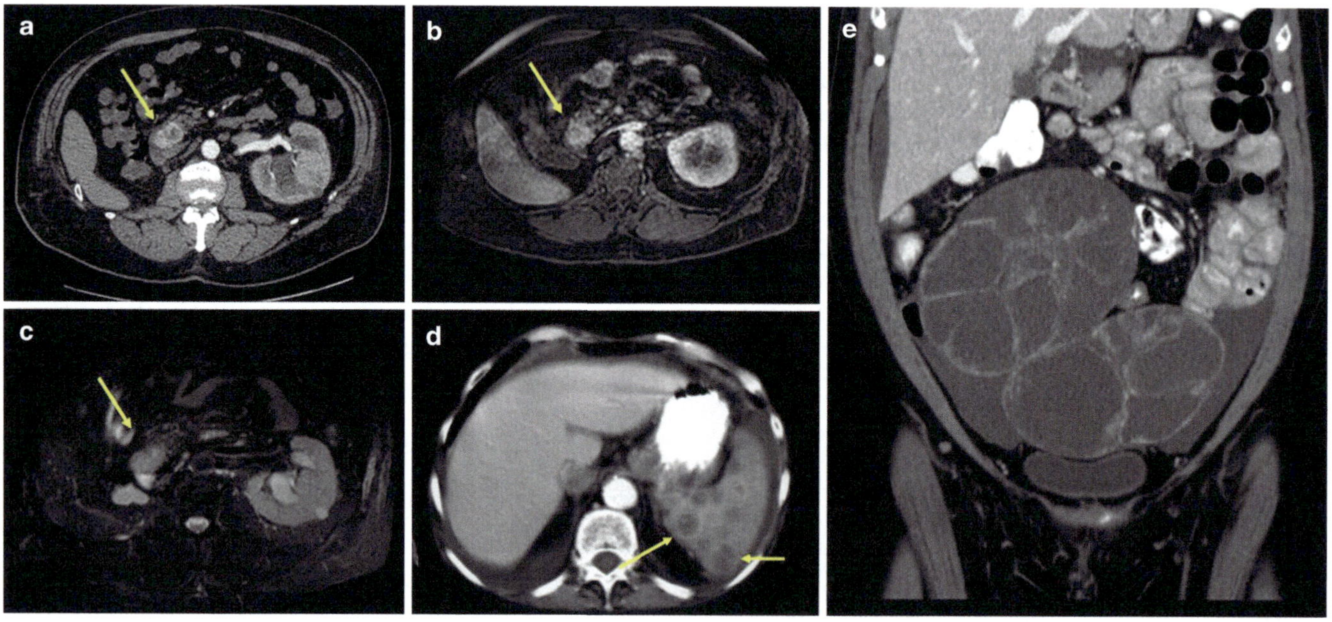

Fig. 29.7 Pancreatic and ovarian metastases. (**a**–**d**) A 54-year-old male with renal cell carcinoma metastatic to the pancreatic head. The metastasis demonstrated arterial enhancement on axial CT (**a**, arrow) and T1 sequence MRI (**b**, arrow). Note internal hypoattenuation suggestive of necrosis. This lesion was hyperintense on T2 fat-suppressed MR imaging (**c**, arrow). (**d**) A 65-year-old female with small cell lung cancer. Axial CT with IV contrast and numerous hypoattenuating cystic metastases in the spleen (arrows). Image credit Ashley Davidoff, MD. (**e**) 51-year-old female with gastric cancer metastatic to ovaries (Krukenberg tumor). Coronal CT with IV contrast shows large hypoattenuating polycystic masses in the bilateral ovaries with mild ascites

29.4.5 Peritoneum

Peritoneal metastases are soft tissue typically small, soft-tissue attenuating, and demonstrate contrast enhancement. Similar to most metastatic lesions, they show T1 hypointensity and T2 hyperintensity on MRI. They are often accompanied by ascites, and where they involve the omentum they may be mass-like, or broad and flat (Fig. 29.6b).

29.5 Pelvic Metastases

The bladder is infrequently involved in metastatic disease, and, where present, it tends to be direct invasion from other primary pelvic malignancies like prostate or ovarian cancers [49]. These cases tend to be less equivocal, with irregular enhancing masses infiltrating into and distorting normal bladder wall contours. Rarely, hematogenous spread may bring gastric, breast, or lung cancers to intramural locations within the bladder wall, where they form small, contrast-enhancing, soft tissue–attenuating lesions [50]. Bladder involvement is typically late, with metastatic disease presenting elsewhere first. Metastatic nodules may be indistinguishable from primary bladder cancer on imaging, requiring tissue sampling for diagnosis. Notably, the bladder may be involved in lymphoma, where imaging may identify a solid solitary intramural mass.

Metastatic disease to the female reproductive tract favors the ovaries, with uterine metastases being extremely rare. Ovarian metastases are often bilateral on presentation, with T1/T2 hypointensity on MRI, and usually are at least partially cystic [51] (Fig. 29.7e). Primaries include GI tract tumors, lungs, and endometrium [52]. On ultrasound, they are heterogeneous, generally with hypervascularity of their solid components. Metastases to the ovary may demonstrate a desmoplastic reaction, with fibrous connective tissue bands pulling on the surrounding tissues. They also commonly present with ascites.

The male reproductive tract is also a rare target of metastatic disease. Case reports of metastases from colon, renal, and bladder cancers exist in the literature, but what few case series exist point toward the prostate as the most common primary, representing approximately 15% of metastases [53]. In these rare cases, ultrasound examination tended to show heterogeneous testicles with multiple distinct masses and occasional calcifications [54]. The prostate is likewise a rare target, with occasional metastases from colon cancer [55]. In this setting, case reports have described a hypoattenuating mass with minimal contrast enhancement.

29.6 Lymph Node Metastases

Irregularities of the regional lymph nodes are frequently the first radiologic sign of metastatic disease, particularly in carcinomas and melanomas [56]. Tumors nearly always spread from their most proximal, or "sentinel," node more centrally through the lymphatic tree [57]. While the involvement of any specific set of nodes is thus widely variable based on the site and lymphatic drainage of the primary tumor, numerous suspicious features remain common between them. It also should be noted that while imaging is a necessary step in evaluating lymph nodes, conventional imaging such as CT and MR remain moderately limited in both sensitivity and specificity for metastatic involvement, and that here, as always, tissue diagnosis remains king.

The most commonly noted finding for lymphadenopathy is size (Fig. 29.8a) usually reported as maximum short-axis axial diameter. Specific size cutoffs are both controversial and vary by lymph node site, ranging from 11 mm for subcarinal nodes to 5 mm in the lateral retropharyngeal group [58]. Unfortunately, limits of imaging dictate that any improvement in sensitivity will sacrifice specificity and vice versa. That said, a commonly used rule of thumb is 10 mm or larger, at which point most nodes should be looked at with suspicion.

Location of irregular lymph nodes should also be considered when gauging the likelihood of their involvement in metastatic disease. As tumor spread through lymphatics is circumscribed by the route those lymphatics travel, the likelihood of remote lymphatic involvement with normal proximal nodes is very unlikely [59]. Further, diffuse lymphadenopathy is an uncommon sign of metastatic (non-lymphoma) cancer. In these instances, alternative etiologies for lymphadenopathy should be considered.

Lastly, and perhaps most importantly, is consideration of the internal nodal architecture. Typically, lymph nodes are soft tissue attenuation on CT and are isointense to muscle on T1-weighted MR sequences, with isointensity to mild hyperintensity on T2 [60]. Normal lymph nodes demonstrate a reniform morphology with a fatty hilum. As metastatic disease distorts physiologic nodal architecture, this hilum may be lost, a sign of metastatic involvement. Lymph nodes do show mild homogeneous enhancement physiologically, but heterogenous or avid enhancement (usually corresponding to that of the primary tumor) may be a sign of metastasis. Lymph node necrosis, demonstrated by heterogeneous low attenuation and increased central T2 signal (Fig. 29.8c), or heterogenous calcification (Fig. 29.8b), should also raise suspicion [61]. It should be noted, however, that necrosis may be seen as a sign of tumor response to therapy, and granulomatous infection may cause calcification.

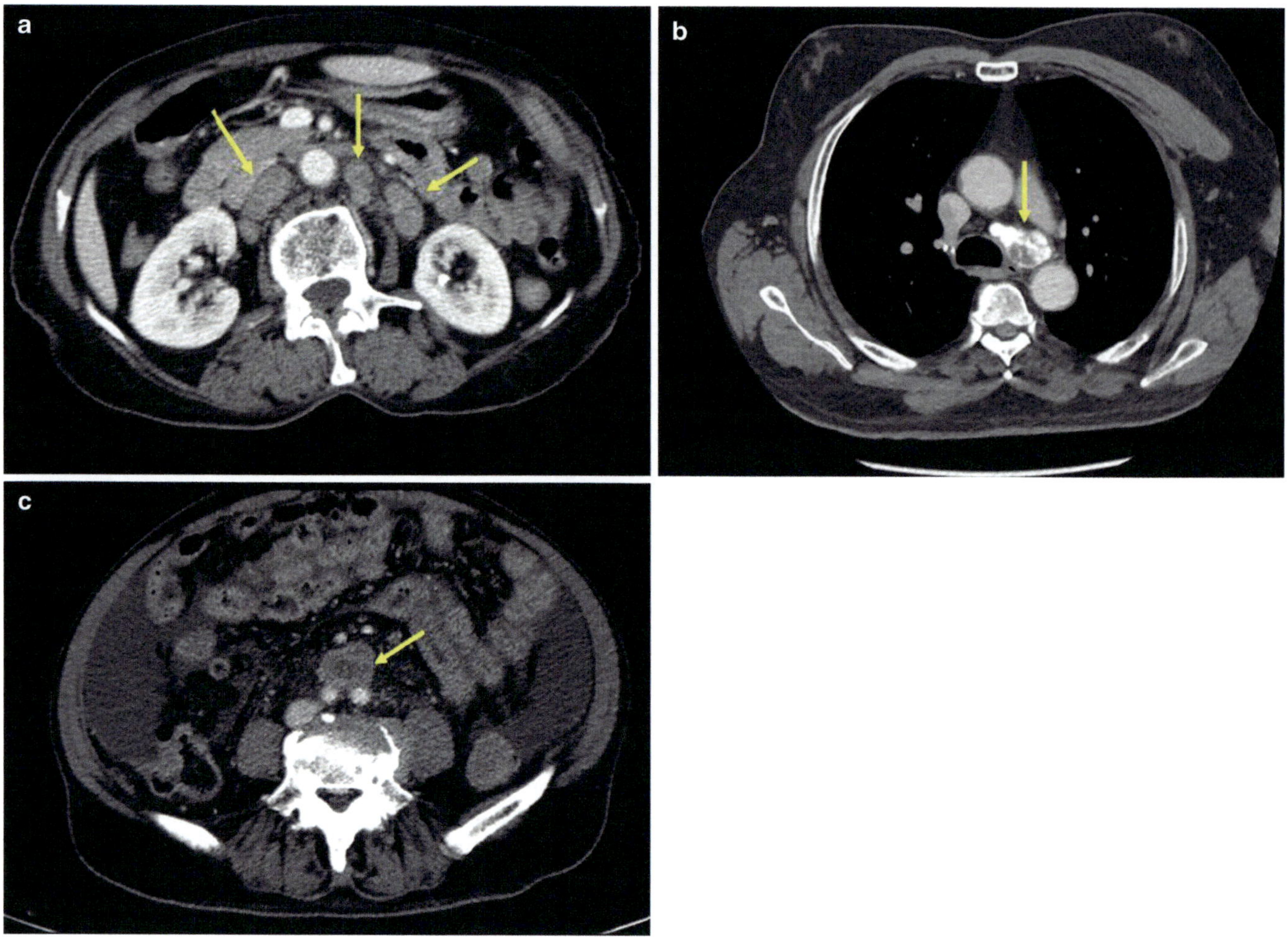

Fig. 29.8 Lymph node metastases. (**a**) A 76-year-old female with colon cancer. Axial CT images with IV contrast demonstrating bulky retroperitoneal lymph nodes. Note rounded shape and loss of fatty hila. (**b**) 67-year-old male with metastatic recurrence of thyroid cancer. Axial CT images with IV contrast demonstrate enlarged heterogeneously calcified mediastinal lymph node. (**c**) A 63-year-old male with rectal cancer metastatic to retroperitoneal lymph nodes. Axial CT images with IV contrast show an enlarged, irregularly shaped lymph node with internal hypoattenuation consistent with necrosis

Unfortunately, the reality is that these single features are rarely both sensitive and specific for metastatic involvement per se, and so acceptable diagnostic accuracy is only achieved by considering them together.

29.7 Conclusion

- Knowledge of typical metastatic patterns of specific cancers can aid in their detection.
- CT, US MRI, and PET play complementary roles with respect to lesion detection, characterization, and targeting for tissue sampling).
- Radiology will continue to provide critical diagnostic and prognostic indicators for the foreseeable future as the premier noninvasive tool.
- Imaging remains the primary metric to gauge response to therapy.
- As novel immunologics and "theranostics" enter clinical practice, both traditional imaging and novel imaging modalities (including specialized PETs) will require further development to monitor treatment response.

Disclosure The authors have no relevant financial disclosures related to this topic.

References

1. Schwartz LH, Litière S, de Vries E, et al. RECIST 1.1-Update and clarification: from the RECIST committee. Eur J Cancer Oxf Engl. 1990;2016(62):132–7. https://doi.org/10.1016/j.ejca.2016.03.081.

2. Johnson JD, Young B. Demographics of brain metastasis. Neurosurg Clin N Am. 1996;7(3):337–44.
3. Posner JB, Chernik NL. Intracranial metastases from systemic cancer. Adv Neurol. 1978;19:579–92.
4. Delattre JY, Krol G, Thaler HT, Posner JB. Distribution of brain metastases. Arch Neurol. 1988;45(7):741–4. https://doi.org/10.1001/archneur.1988.00520310047016.
5. Lignelli A, Khandji AG. Review of imaging techniques in the diagnosis and management of brain metastases. Neurosurg Clin N Am. 2011;22(1):15–25. https://doi.org/10.1016/j.nec.2010.09.002.
6. Batool A, Kasi A. Leptomeningeal carcinomatosis. In: StatPearls. StatPearls Publishing; 2020. Accessed December 17, 2020. http://www.ncbi.nlm.nih.gov/books/NBK499862/.
7. Clarke JL, Perez HR, Jacks LM, Panageas KS, Deangelis LM. Leptomeningeal metastases in the MRI era. Neurology. 2010;74(18):1449–54. https://doi.org/10.1212/WNL.0b013e3181dc1a69.
8. El Shafie RA, Böhm K, Weber D, et al. Outcome and prognostic factors following palliative craniospinal irradiation for leptomeningeal carcinomatosis. Cancer Manag Res. 2019;11:789–801. https://doi.org/10.2147/CMAR.S182154.
9. Lyndon D, Lansley JA, Evanson J, Krishnan AS. Dural masses: meningiomas and their mimics. Insights Imaging. 2019;10 https://doi.org/10.1186/s13244-019-0697-7.
10. McGann GM, Platts A. Computed tomography of cranial metastatic malignant melanoma: features, early detection and unusual cases. Br J Radiol. 1991;64(760):310–3. https://doi.org/10.1259/0007-1285-64-760-310.
11. Schellinger PD, Meinck HM, Thron A. Diagnostic accuracy of MRI compared to CCT in patients with brain metastases. J Neuro-Oncol. 1999;44(3):275–81. https://doi.org/10.1023/a:1006308808769.
12. Yokoi K, Kamiya N, Matsuguma H, et al. Detection of brain metastasis in potentially operable non-small cell lung cancer: a comparison of CT and MRI. Chest. 1999;115(3):714–9. https://doi.org/10.1378/chest.115.3.714.
13. Hanssens P, Karlsson B, Yeo TT, Chou N, Beute G. Detection of brain micrometastases by high-resolution stereotactic magnetic resonance imaging and its impact on the timing of and risk for distant recurrences. J Neurosurg. 2011;115(3):499–504. https://doi.org/10.3171/2011.4.JNS101832.
14. Fink KR, Fink JR. Imaging of brain metastases. Surg Neurol Int. 2013;4(Suppl. 4):S209–19. https://doi.org/10.4103/2152-7806.111298.
15. Wang N, Bertalan MS, Brastianos PK. Leptomeningeal metastasis from systemic cancer: review and update on management. Cancer. 2018;124(1):21–35. https://doi.org/10.1002/cncr.30911.
16. Rykken JB, Diehn FE, Hunt CH, et al. Rim and flame signs: postgadolinium MRI findings specific for non-CNS intramedullary spinal cord metastases. AJNR Am J Neuroradiol. 2013;34(4):908–15. https://doi.org/10.3174/ajnr.A3292.
17. Shah LM, Salzman KL. Imaging of spinal metastatic disease. Int J Surg Oncol. 2011;2011:769753. https://doi.org/10.1155/2011/769753.
18. Yáñez ML, Miller JJ, Batchelor TT. Diagnosis and treatment of epidural metastases. Cancer. 2017;123(7):1106–14. https://doi.org/10.1002/cncr.30521.
19. Kaplan JG, DeSouza TG, Farkash A, et al. Leptomeningeal metastases: comparison of clinical features and laboratory data of solid tumors, lymphomas and leukemias. J Neuro-Oncol. 1990;9(3):225–9.
20. Amer MH, Al-Sarraf M, Baker LH, Vaitkevicius VK. Malignant melanoma and central nervous system metastases: incidence, diagnosis, treatment and survival. Cancer. 1978;42(2):660–8. https://doi.org/10.1002/1097-0142(197808)42:2<660::aid-cncr2820420237>3.0.co;2-e.
21. Goyal A, Yolcu Y, Kerezoudis P, Alvi MA, Krauss WE, Bydon M. Intramedullary spinal cord metastases: an institutional review of survival and outcomes. J Neuro-Oncol. 2019;142(2):347–54. https://doi.org/10.1007/s11060-019-03105-2.
22. Mohammed T-LH, Chowdhry A, Reddy GP, et al. ACR Appropriateness criteria® screening for pulmonary metastases. J Thorac Imaging. 2011;26(1):W1–3. https://doi.org/10.1097/RTI.0b013e3182010bf9.
23. Stella GM, Kolling S, Benvenuti S, Bortolotto C. Lung-seeking metastases. Cancers. 2019;11:7. https://doi.org/10.3390/cancers11071010.
24. Caparica R, Mak MP, Rocha CH, et al. Pulmonary nodules in patients with nonpulmonary cancer: not always metastases. J Glob Oncol. 2016;2(3):138–44. https://doi.org/10.1200/JGO.2015.002089.
25. Hirakata K, Nakata H, Nakagawa T. CT of pulmonary metastases with pathological correlation. Semin Ultrasound CT MR. 1995;16(5):379–94. https://doi.org/10.1016/0887-2171(95)90027-6.
26. Girvin F, Ko JP. Pulmonary nodules: detection, assessment, and CAD. AJR Am J Roentgenol. 2008;191(4):1057–69. https://doi.org/10.2214/AJR.07.3472.
27. Ketai L, Malby M, Jordan K, Meholic A, Locken J. Small nodules detected on chest radiography: does size predict calcification? Chest. 2000;118(3):610–4. https://doi.org/10.1378/chest.118.3.610.
28. Seo JB, Im JG, Goo JM, Chung MJ, Kim MY. Atypical pulmonary metastases: spectrum of radiologic findings. Radiogr Rev Publ Radiol Soc N Am Inc. 2001;21(2):403–17. https://doi.org/10.1148/radiographics.21.2.g01mr17403.
29. Gross BH, Glazer GM, Bookstein FL. Multiple pulmonary nodules detected by computed tomography: diagnostic implications. J Comput Assist Tomogr. 1985;9(5):880–5. https://doi.org/10.1097/00004728-198509000-00007.
30. Bruce DM, Heys SD, Eremin O. Lymphangitis carcinomatosa: a literature review. J R Coll Surg Edinb. 1996;41(1):7–13.
31. Pillai S, Khan A, Khan S. Adenocarcinoma of the lung presenting with intrapulmonary miliary metastasis. Cureus. 2019;11(8):e5430. https://doi.org/10.7759/cureus.5430.
32. Chaudhuri MR. Cavitary pulmonary metastases. Thorax. 1970;25(3):375–81. https://doi.org/10.1136/thx.25.3.375.
33. Purandare NC, Rangarajan V. Imaging of lung cancer: implications on staging and management. Indian J Radiol Imaging. 2015;25(2):109–20. https://doi.org/10.4103/0971-3026.155831.
34. Biederer J, Hintze C, Fabel M. MRI of pulmonary nodules: technique and diagnostic value. Cancer Imaging Off Publ Int Cancer Imaging Soc. 2008;8:125–30. https://doi.org/10.1102/1470-7330.2008.0018.
35. Patz EF, Lowe VJ, Hoffman JM, et al. Focal pulmonary abnormalities: evaluation with F-18 fluorodeoxyglucose PET scanning. Radiology. 1993;188(2):487–90. https://doi.org/10.1148/radiology.188.2.8327702.
36. Roh J-L, Ryu CH, Kim JS, et al. Clinical significance of intrathoracic lesions detected by 18F-fluorodeoxyglucose positron emission tomography in the management of patients with head and neck cancer. Oral Oncol. 2007;43(8):757–63. https://doi.org/10.1016/j.oraloncology.2006.09.004.
37. Roth CG, Deshmukh S. Fundamentals of body MRI.; 2017. Accessed December 13, 2020. http://search.ebscohost.com/login.aspx?direct=true&scope=site&db=nlebk&db=nlabk&AN=1288236
38. Sica GT, Ji H, Ros PR. CT and MR imaging of hepatic metastases. AJR Am J Roentgenol. 2000;174(3):691–8. https://doi.org/10.2214/ajr.174.3.1740691.
39. Namasivayam S, Martin DR, Saini S. Imaging of liver metastases: MRI. Cancer Imaging Off Publ Int Cancer Imaging Soc. 2007;7:2–9. https://doi.org/10.1102/1470-7330.2007.0002.
40. Wernecke K, Vassallo P, Bick U, Diederich S, Peters PE. The distinction between benign and malignant liver tumors on sonog-

raphy: value of a hypoechoic halo. AJR Am J Roentgenol. 1992;159(5):1005–9. https://doi.org/10.2214/ajr.159.5.1329454.
41. Choi WS, Kim SH, Lee ES, et al. CT findings of gallbladder metastases: emphasis on differences according to primary tumors. Korean J Radiol. 2014;15(3):334–45. https://doi.org/10.3348/kjr.2014.15.3.334.
42. Klein KA, Stephens DH, Welch TJ. CT characteristics of metastatic disease of the pancreas. Radiogr Rev Publ Radiol Soc N Am Inc. 1998;18(2):369–78. https://doi.org/10.1148/radiographics.18.2.9536484.
43. Compérat E, Bardier-Dupas A, Camparo P, Capron F, Charlotte F. Splenic metastases: clinicopathologic presentation, differential diagnosis, and pathogenesis. Arch Pathol Lab Med. 2007;131(6):965–9. https://doi.org/10.1043/1543-2165(2007)131[965:SMCPDD]2.0.CO;2.
44. Elsayes KM, Narra VR, Mukundan G, Lewis JS, Menias CO, Heiken JP. MR imaging of the spleen: spectrum of abnormalities. Radiogr Rev Publ Radiol Soc N Am Inc. 2005;25(4):967–82. https://doi.org/10.1148/rg.254045154.
45. Lam K-Y, Lo C-Y. Metastatic tumours of the adrenal glands: a 30-year experience in a teaching hospital. Clin Endocrinol. 2002;56(1):95–101. https://doi.org/10.1046/j.0300-0664.2001.01435.x.
46. Richard WW. Fundamentals of body Ct. 5th ed. Elsevier; 2019.
47. Johnson PT, Horton KM, Fishman EK. Adrenal mass imaging with multidetector CT: pathologic conditions, pearls, and pitfalls. Radiogr Rev Publ Radiol Soc N Am Inc. 2009;29(5):1333–51. https://doi.org/10.1148/rg.295095027.
48. Elsayes KM, Mukundan G, Narra VR, et al. Adrenal masses: MR imaging features with pathologic correlation. Radiogr Rev Publ Radiol Soc N Am Inc. 2004;24(Suppl. 1):S73–86. https://doi.org/10.1148/rg.24si045514.
49. Modh RA, Corbyons KA, Yeung LL. A rare metastasis to the bladder. Case Rep Urol. 2013;2013:789039. https://doi.org/10.1155/2013/789039.
50. Karjol U, Jonnada P, Cherukuru S, Chandranath A. Bladder metastasis from breast cancer: a systematic review. Cureus. 2020;12(3):e7408. https://doi.org/10.7759/cureus.7408.
51. Yada-Hashimoto N, Yamamoto T, Kamiura S, et al. Metastatic ovarian tumors: a review of 64 cases. Gynecol Oncol. 2003;89(2):314–7. https://doi.org/10.1016/s0090-8258(03)00075-1.
52. Lee S-J, Bae J-H, Lee A-W, Tong S-Y, Park Y-G, Park J-S. Clinical characteristics of metastatic tumors to the ovaries. J Korean Med Sci. 2009;24(1):114–9. https://doi.org/10.3346/jkms.2009.24.1.114.
53. Turk A, Graff JM, Memo M. Case report of metastatic prostate cancer to testicles: an ominous sign of advanced disease. Urol Case Rep. 2019;26:100935. https://doi.org/10.1016/j.eucr.2019.100935.
54. Ulbright TM, Young RH. Metastatic carcinoma to the testis: a clinicopathologic analysis of 26 nonincidental cases with emphasis on deceptive features. Am J Surg Pathol. 2008;32(11):1683–93. https://doi.org/10.1097/PAS.0b013e3181788516.
55. Youssef FR, Hunt L, Meiring PD, Taraporewalla DR, Gupta R, James MJ. Metastasis of a cecal adenocarcinoma to the prostate five years after a right hemicolectomy: a case report. J Med Case Rep. 2011;5:223. https://doi.org/10.1186/1752-1947-5-223.
56. Wong SY, Hynes RO. Lymphatic or hematogenous dissemination: how does a metastatic tumor cell decide? Cell Cycle Georget Tex. 2006;5(8):812–7. https://doi.org/10.4161/cc.5.8.2646.
57. Nathanson SD. Insights into the mechanisms of lymph node metastasis. Cancer. 2003;98(2):413–23. https://doi.org/10.1002/cncr.11464.
58. Lan M, Huang Y, Chen C-Y, et al. Prognostic value of cervical nodal necrosis in nasopharyngeal carcinoma: analysis of 1800 patients with positive cervical nodal metastasis at MR imaging. Radiology. 2015;276(2):536–44. https://doi.org/10.1148/radiol.15141251.
59. Paño B, Sebastià C, Buñesch L, et al. Pathways of lymphatic spread in male urogenital pelvic malignancies. Radiogr Rev Publ Radiol Soc N Am Inc. 2011;31(1):135–60. https://doi.org/10.1148/rg.311105072.
60. Koh DM, Hughes M, Husband JE. Cross-sectional imaging of nodal metastases in the abdomen and pelvis. Abdom Imaging. 2006;31(6):632–43. https://doi.org/10.1007/s00261-006-9022-2.
61. Yang WT, Lam WW, Yu MY, Cheung TH, Metreweli C. Comparison of dynamic helical CT and dynamic MR imaging in the evaluation of pelvic lymph nodes in cervical carcinoma. AJR Am J Roentgenol. 2000;175(3):759–66. https://doi.org/10.2214/ajr.175.3.1750759.

Image-Guided Biopsy and Intervention for Metastatic Disease

30

Joao R. T. Vicentini, Sina Habibollahi, Ambrose J. Huang, and Connie Y. Chang

Abstract

The field of image-guided musculoskeletal intervention includes a wide variety of diagnostic and therapeutic procedures in the management of metastatic disease. This chapter provides an overview of percutaneous musculoskeletal biopsies, including information on preprocedural planning, indications, technique, strategies to maximize diagnostic yield, and potential risks and complications. In recent years, new image-guided treatment options have been developed for the management of metastases-related pain and oligometastatic disease. These techniques are also reviewed, with a focus on radiofrequency ablation and cryoablation, and additional information on microwave ablation, laser ablation, and MR imaging-guided extracorporeal focused ultrasound.

Learning Objectives

- Preprocedural planning is the most important step of every image-guided musculoskeletal intervention and should include a review of the indications, a plan for the safest approach, and anticipation of potential issues, with the help of a multidisciplinary team whenever possible.
- Biopsy technique, including the choice of imaging modality, will vary depending on the type and location of the lesion, but the main goals are similar for each case: (1) secure a safe access to the target; (2) obtain a satisfactory number of viable specimens; and (3) place the correct orders so adequate testing is performed.
- Most image-guided ablation techniques involve thermal energy, and the main applications for osseous metastases are pain palliation and treatment of oligometastatic disease. The choice of technique will depend mostly on location of the lesion and resource availability.

30.1 Introduction

Image-guided percutaneous biopsy has become the first-line option for pathologic confirmation of the metastatic disease in the musculoskeletal system. When compared to surgical biopsy, image-guided procedures are less invasive, safer, and lower cost, while maintaining similar diagnostic yield [1]. Image-guided intervention can also be used for therapeutic purposes, and percutaneous ablation has been expanding as an alternative option for treatment and pain palliation in the setting of metastatic disease. This chapter offers an overview of the main image-guided diagnostic and therapeutic procedures currently available for musculoskeletal metastases.

30.2 Percutaneous Musculoskeletal Biopsy

30.2.1 Preprocedural Planning

30.2.1.1 Imaging Review and Biopsy Indication

The first step when planning a biopsy is reviewing all the available imaging to confirm that the procedure is necessary. Many incidental lesions found during staging turn out to be benign or non-related to the primary malignancy. The radiologist can make that determination when classic benign

J. R. T. Vicentini · S. Habibollahi · A. J. Huang · C. Y. Chang (✉)
Division of Musculoskeletal Imaging and Intervention, Department of Radiology, Massachusetts General Hospital, Boston, MA, USA
e-mail: cychang@mgh.harvard.edu

S. P. Leong et al. (eds.), *Cancer Metastasis Through the Lymphovascular System*,
https://doi.org/10.1007/978-3-030-93084-4_30

features are present, or when long-term stability can be ensured by comparison with older studies.

A common source of uncertainty during imaging workup is the differentiation of osteoblastic metastases and enostoses or bone islands. Enostoses tend to be smaller (1 mm–2 cm) and usually remain stable in size over time, although slow growth has been described [2, 3]. Irregular or spiculated margins are a common feature, and some can show scintigraphic activity on 99 m-Tc-methylene diphosphate (MDP) bone scan [3]. Since enostoses are composed of compact cortical bone, these lesions are usually very dense, and that characteristic can be used to differentiate them from metastatic lesions. On CT, threshold values of 885 HU for mean attenuation and 1060 HU for maximal attenuation have been associated with 95% sensitivity and 96% specificity for differentiating benign enostoses from untreated osteoblastic metastases [2]. For borderline values or variable appearance, correlation with other imaging modalities, such as MRI or bone scan, may be helpful. Degenerative changes may also present with sclerosis and increased uptake on radionuclide bone scans, but the typical pattern of distribution makes them less likely to be mistaken for a neoplastic process.

Infection is another mimicker of bone malignancy (Fig. 30.1). Osteomyelitis can present with an indolent course and cause progressive bone destructive changes and periosteal reaction, making it difficult to rule out an underlying marrow replacing lesion [4]. In many cases, clinical data may point toward the correct diagnosis; however, classic signs of infection may not be present in chronic osteomyelitis, and sampling of the affected area may be necessary for confirmation.

Incidental lucent lesions need to be differentiated from osteolytic metastases. The broad differential diagnosis of osseous lucent lesions is beyond the scope of this chapter, but as a general rule, nonaggressive imaging features suggest benignity. These features include well-defined margins, narrow transition zone with the normal bone, presence of intralesional fat, and preservation of adjacent trabecular and cortical bone [3]. Common incidental benign lucent lesions include intraosseous lipomas and ganglion cysts. Non-ossifying fibromas are benign lesions that can show hypermetabolism on ^{18}F-fluorodeoxyglucose positron emission tomography (^{18}F-FDG PET) [5]. MRI can be helpful to help identify the different components of each lesion.

Metastases to subcutaneous tissues and skeletal muscle are uncommon, with reported rates below 2% among patients investigated for a soft tissue mass [6]. Most authors believe this is likely related to variable blood flow and decreased capillary density when compared to other common sites of metastatic disease such as the liver and lungs. The most common primary tumor associated with soft tissue metastasis is lung cancer [6, 7]. Imaging workup of soft tissue masses can be performed with ultrasound, MRI, or a combination of both methods. Benign lesions such as lipomas and ganglion cysts can be easily identified due to their typical imaging features. Hibernomas are a distinct type of adipocytic lesion composed primarily of brown fat, which can be found in typical locations for brown fat deposits, such as the neck, axilla, shoulder, chest wall, retroperitoneum, and thighs. Hibernomas have characteristics similar to lipomas but tend to be more heterogeneous, with low-signal serpentine or branching vascular structures, and can also show increased ^{18}F-FDG avidity on PET [8].

Radiographs can be helpful during workup of soft tissue masses. For example, hemangiomas and other types of vascular lesions can show radiopaque phleboliths [9]. Radiographs can also show the typical peripheral ossification pattern associated with myositis ossificans, which may mimic an aggressive lesion on MRI. Although CT scans are not typically part of the diagnostic workup of soft tissue masses, CT can help with pre-procedure planning.

When the imaging review of an osseous or soft tissue lesion cannot definitively rule out malignancy, tissue sampling is indicated. Then, the next step is to determine the best way to obtain an adequate sample.

30.2.1.2 Planning the Approach

The goal of preprocedural planning is to determine the safest approach to the target and increase the chances of obtaining a diagnostic sample. This process includes the choice of imaging modality for guidance, which is mainly influenced by the type of targeted lesion (osseous vs. soft tissue) and its location. If there are multiple lesions, which is the case for many metastatic tumors, the choice of a biopsy target usually involves a combined evaluation of safety profile, technical feasibility, and size of the lesion. For example, pelvic or long bones are usually preferred over vertebrae given safer access through the pelvic soft tissues. Larger lesions offer the possibility of obtaining more samples but may also contain areas of hemorrhage or necrosis.

In general, the areas with more aggressive features are selected for biopsy to avoid underestimating tumor grade [10]. Some lesions will demonstrate non-enhancing areas, which can reflect nonviable, necrotic tissue; therefore, the enhancing solid components are usually chosen for sampling to increase the likelihood of obtaining diagnostic tissue (Fig. 30.2) [1, 11, 12]. Although necrotic tissue does not contribute to the final histologic diagnosis, documenting the presence of necrosis may be helpful given its association with worse prognosis in some types of sarcoma [13, 14]. Pathologic fractures should be avoided whenever possible. Sampling the fractured area makes histopathologic evaluation more challenging given the presence of granulation tissue and immature osteoid as part of the healing process, which can mimic bone-forming tumors [10].

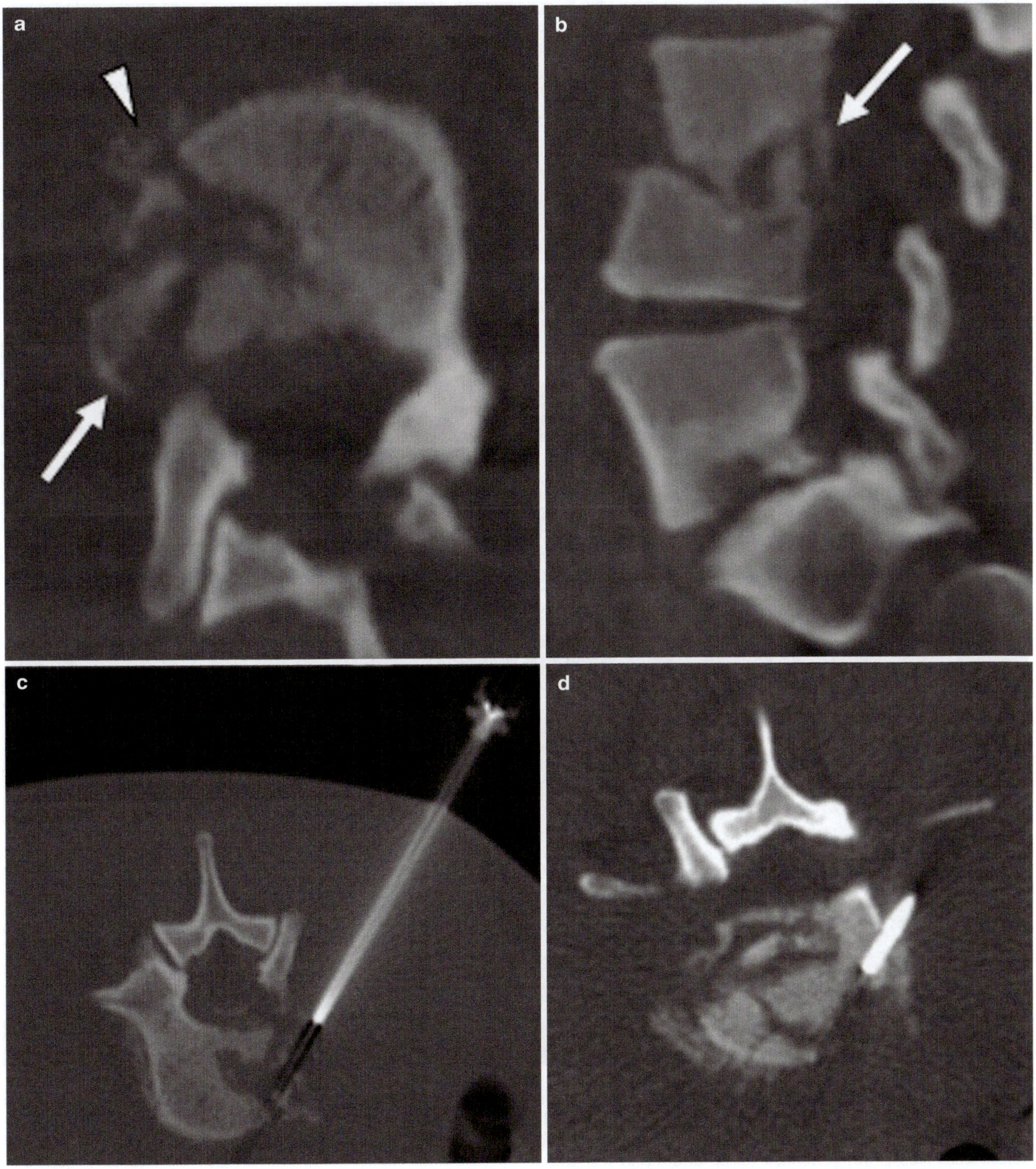

Fig. 30.1 A 33-year-old male with history of HCV admitted after low-grade trauma. (**a**) Axial and (**b**) sagittal CT images of the lumbar spine show fractures with fragmentation of the vertebral endplates at L3–L4 (arrows). Mineralization at the right anterolateral aspect of L3 raised concern for associated matrix-forming lesion (arrowhead). (**c**) Sampling of disc material and (**d**) bone was performed and sent for surgical pathology and microbiology. Results were consistent with discitis–osteomyelitis with cultures positive for *Pseudomonas aeruginosa*

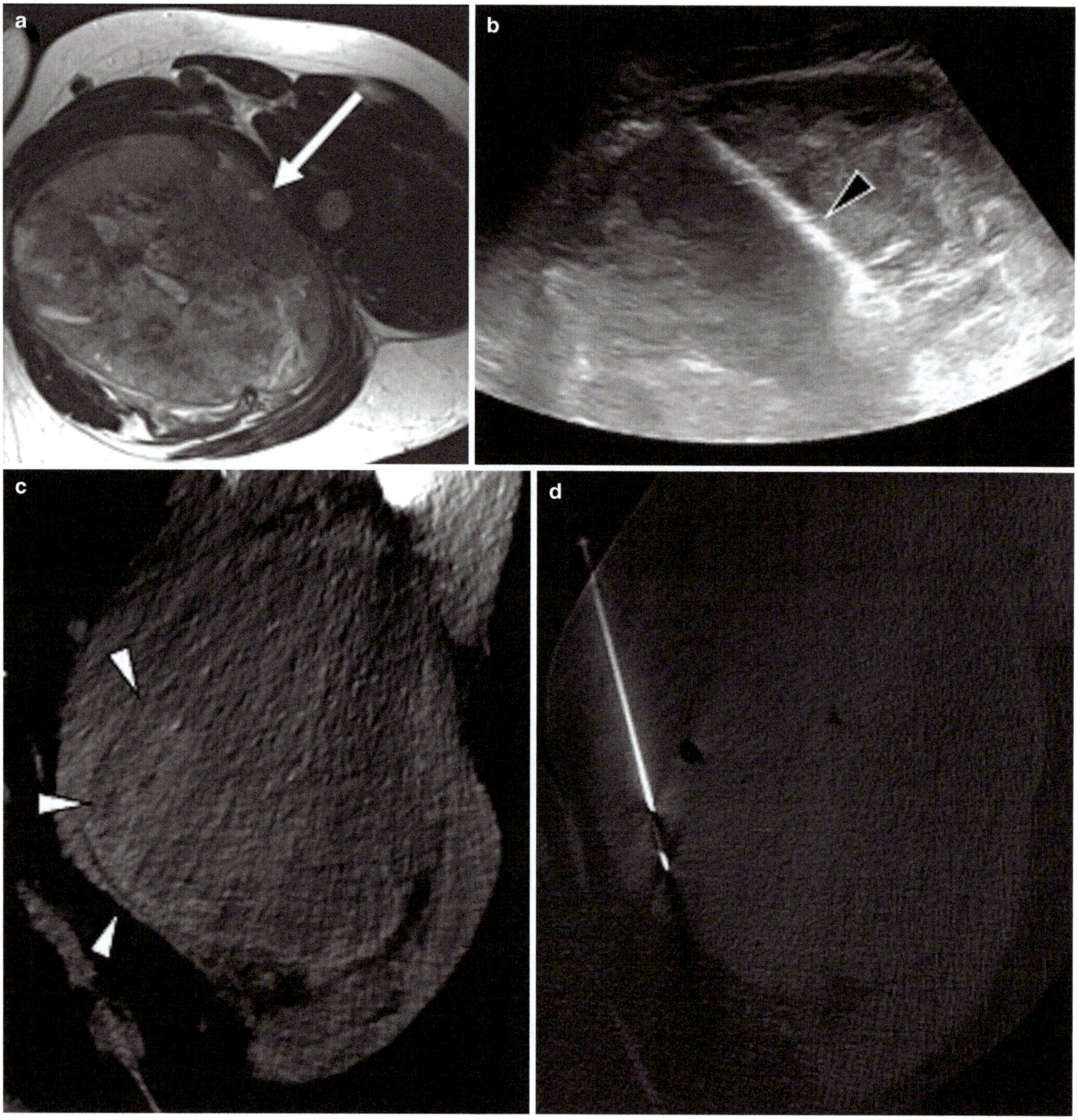

Fig. 30.2 A 65-year-old male presented with chronic left hip pain. (**a**) Axial T2-weighted MR image shows a large mass centered within the adductor musculature in the left proximal thigh (arrow). (**b**) Ultrasound-guided core biopsy was performed (black arrowhead) and frozen pathology showed predominantly nondiagnostic necrotic tissue. Additional samples were obtained; however, these were also nondiagnostic due to extensive necrosis. Necrotic and viable tissue could not be distinguished through ultrasound. (**c**, **d**) Patient returned for a CT-guided biopsy with the use of intravenous iodinated contrast. (**c**) Contrast-enhanced CT image showed subtle enhancement along the peripheral portion of the mass (white arrowheads). (**d**) This area was targeted during the biopsy, and pathology was consistent with undifferentiated pleomorphic sarcoma

Another important factor when considering the approach in musculoskeletal biopsies is the risk of seeding the biopsy track, particularly in primary sarcomas. Although this is a controversial topic in the literature, most studies suggest that the risk of needle tract seeding is likely below 1% [15–17]. For metastatic disease, this is a less important issue, since treatment will rarely involve surgical intervention, so lesions with the highest yield potential and safest approach are prioritized. It is still good practice to communicate with the referring provider to discuss any potential options and preferences for each patient, especially for solitary lesions that may turn out to be a synchronous primary malignancy. As a general rule, when performing a percutaneous needle biopsy, one should avoid traversing uninvolved compartments, different joint spaces, and neurovascular bundles [1, 10, 11, 18]. Another strategy to minimize the risk of needle tract seeding is the use of coaxial systems, further discussed in the technique section.

30.2.1.3 Orders

The final step in biopsy planning is placing adequate lab orders and anticipating the need for specific tests or analyses besides standard surgical pathology. If a lesion shows a cystic component from which fluid can be aspirated, the obtained material can also be sent to cytology. When infection is among the differential diagnoses, additional samples should be obtained and sent for microbiology analysis. Other specific tests or requests may be included when they can influence the treatment plan, such as genetic testing and dosage of hormone receptors. Research biopsies are also common for patients with metastatic disease and can be used to guide treatment and to determine eligibility for a specific therapy or clinical trial.

30.2.1.4 Plan for Sedation

Most musculoskeletal biopsies can be performed with local anesthesia. This is a common choice for soft tissue lesions and lytic osseous lesions with cortical breakthrough. In some cases, however, conscious sedation can be helpful, especially for purely intraosseous lesions, when there is surrounding inflamed tissue, for lesions of difficult access or in proximity to vital structures, or if the patient experiences extreme anxiety. General anesthesia is rarely needed for most musculoskeletal procedures and is reserved for special cases, such as for pediatric patients or image-guided ablation.

30.2.1.5 Multidisciplinary Approach

Multidisciplinary review is key whenever planning any kind of intervention for musculoskeletal tumors, primary or metastatic. Some factors that can be discussed ahead of time include clinical findings, need for further imaging before the procedure, biopsy approach, plan for frozen pathology, potential complications, specific tests to be ordered, and treatment plans. Multidisciplinary teams should include all the specialties involved in patient care, including orthopedic surgeons or surgical oncologists, radiologists, medical oncologists, pathologists, and radiation oncologists.

30.2.2 Technique

30.2.2.1 Ultrasound Guidance

When the target is mostly composed of soft tissue in a superficial location, ultrasound is the preferred modality for image guidance. Real-time visualization allows the operator to find the most promising area based on appearance and size and also helps to avoid damage to adjacent structures, including blood vessels, which can be evaluated with color Doppler, and tendons and nerves. Differentiation between solid and cystic components can also be helpful, since in most cases the solid component is targeted for sampling [11]. Superficial, small lesions are also better assessed under ultrasound when compared to CT. Another advantage is the lack of ionizing radiation, which is especially important for younger patients. Limitations of ultrasound guidance include limited access to deep lesions, poor evaluation of osseous lesions, and limited correlation with other cross-sectional studies, such as MRI and PET-CT.

30.2.2.2 CT Guidance

CT is the modality of choice for purely intraosseous lesions or those located within deep planes. Advantages of CT include the larger field of view and the possibility to access difficult locations. Disadvantages include the use of ionizing radiation and technical difficulties for small, superficial lesions [11]. Gantry size may also influence the final positioning and choice of imaging modality. The total length of the external portions of the biopsy tools should be added to the diameter of the selected body region and then compared with the CT gantry size [10]. This is especially important for obese patients and procedures in the thoracoabdominal region.

30.2.2.3 Other Imaging Modalities

Fluoroscopy can be used for biopsy guidance, and common uses include bone marrow aspiration and biopsy of large osseous lesions in the pelvis or appendicular skeleton. Advantages include higher flexibility in terms of positioning, potentially shorter procedure times, and less radiation exposure when compared to CT [1]. MRI and PET-CT guidance are potential options for CT occult bone lesions; however, these modalities are usually associated with higher cost and longer procedure times and may not be readily available in many institutions [19, 20]. A more common use of MRI and PET-CT is to help establish landmarks for sampling lesions that are not well seen on CT. Although this approach may be difficult for smaller findings, biopsy of CT occult lesions using anatomical landmarks has been shown to result in similar diagnostic yield when compared to CT evident lesions [1, 21].

30.2.2.4 Coaxial Technique

For many osseous and deep soft tissue lesions, it is helpful to secure direct access to the biopsy target. This can be achieved with coaxial technique, which involves the use of an introducer device that is positioned next to the area to be biopsied. The biopsy needle can then be advanced through the introducer to obtain the samples without creating new biopsy tracks. This practice avoids the need for multiple needle passes from the skin through the soft tissues into the lesion, minimizing the risk of injury to adjacent structures [1]. Larger diameter introducers can accommodate different types of biopsy needles. They can also be used for aspiration and placement of probes for ablation therapy.

30.2.2.5 Biopsy Device

The choice of which biopsy device to use is usually influenced by the composition of the targeted lesion (bone or soft tissue), location, operator preference, and resource availability. Sampling of lesions arising in the soft tissues or of the extraosseous soft tissue component of bone lesions can be done with a spring-loaded side-cutting biopsy needle. For osseous lesions with an intact cortex, a manual or battery-powered drill can be used to insert an introducer or cannula through the cortex. Then, a biopsy needle can be inserted coaxially through the introducer to obtain the samples. Fine needle aspiration (FNA) can also be performed through the introducer. FNA can be especially helpful for lesions in challenging locations, such as in the craniofacial and cervical regions.

30.2.3 Accuracy

Reported accuracy of image-guided biopsy of musculoskeletal lesions ranges from 74% to 97% [11, 22–27]. Even when a biopsy is not necessarily accurate in terms of identifying the histopathologic features and tumor grade, the procedure can still be considered successful when it allows the referring provider to make a decision about treating or monitoring a lesion, or at least to rule out a more urgent concern, such as infection. Accuracy is higher for malignant and high-grade lesions, large tumors, osteolytic lesions, and metastases [1, 11, 22, 23]. Factors that may decrease biopsy success are usually related to difficulty in obtaining adequate samples, including densely sclerotic lesions or presence of necrosis. Lower success rate has also been reported for benign or low-grade tumors, spine lesions, lack of soft tissue component, myxoid lesions, and infection [11, 22, 23, 28].

The size and number of samples may also influence the biopsy success rate. Wu et al. reported higher diagnostic yield with samples measuring at least 10 mm in length and noticed increased yield with additional specimens, reaching a plateau at four samples for soft tissue lesions and three for bone lesions [29].

Fine needle aspiration (FNA) alone has lower accuracy when compared to core needle biopsy and may underestimate tumor grading due to lack of information on tissue architecture and appearance of adjacent structures [23, 27, 30]. However, it can be used as an adjunct during a core needle biopsy, especially in the investigation of a potentially metastatic lesion, when the visualization of cells with malignant features is usually enough to confirm the diagnosis.

30.2.4 Complications

The complication rate of percutaneous needle biopsies is considered low, and most studies report complication rates below 10%, and some as low as 1.1%, versus up to 16% for open surgical biopsies [9, 11, 18]. Medication-related complications include allergic reaction to local anesthetic or medications used for conscious sedation [11]. Minor technical complications include increased pain, swelling, and bruising. Major complications include bleeding, infection, and damage to adjacent structures. In spine biopsies, there is a potential risk of hematoma formation and compression of nerve roots or the spinal cord. For biopsies of the thoracic spine, ribs, or sternum, pneumothorax is also a potential complication. Given the low incidence of hemorrhagic complications, in some scenarios, anticoagulation does not need to be suspended. For patients on warfarin, international normalized ratio of 1.5 or less is considered safe, and platelet count equal or above 50.000/μL is usually recommended [1].

30.3 Therapeutic Options: Percutaneous Ablation

Since the first report describing the use of radiofrequency ablation (RFA) for treatment of osteoid osteoma in 1992, new techniques have been developed, and the indications for ablation have expanded to include treatment of metastatic malignancy [31, 32]. The main principle of ablative therapy is to eradicate tumor cells in situ without the need for surgical resection. When compared to surgery, percutaneous ablation is less invasive and usually associated with lower cost, shorter hospitalization times, and similar or better safety profile [32].

30.3.1 Indications and Patient Selection

Percutaneous ablation has shown many benefits in the treatment of metastatic disease in the musculoskeletal system [32, 33]. Indications include chronic pain that failed to

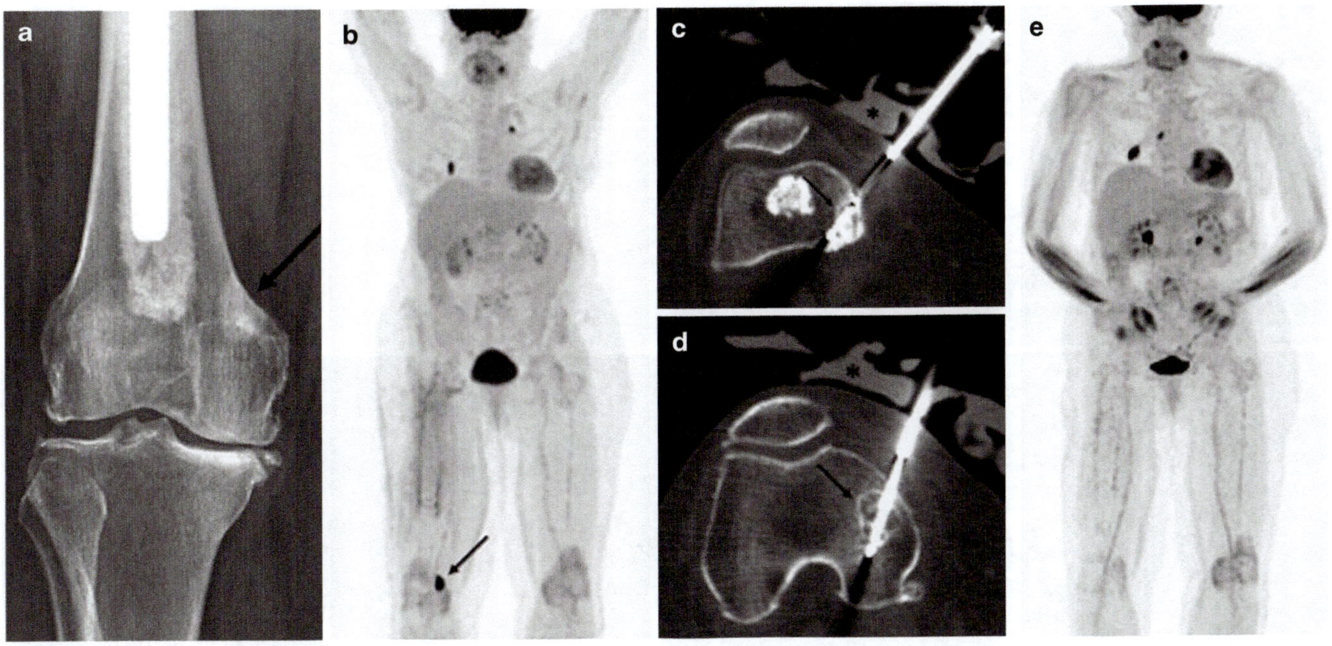

Fig. 30.3 A 64-year-old female with a history of metastatic breast cancer presents with moderate right knee pain. (**a**) Anteroposterior radiograph of the right distal femur shows a radiodense metastasis in the medial femoral condyle that had been previously irradiated (arrow). (**b**) Coronal maximum intensity projection (MIP) image from a PET shows increased FDG activity within the right medial femoral condyle lesion, consistent with recurrence (arrow). (**c, d**) Axial CT images of the right medial femoral condyle show two cryoablation probes within the mixed lytic and blastic lesion (arrows). Heat packs were placed on the skin surface during the cryoablation to protect the skin from freezing (*). Biopsy performed during the cryoablation confirmed metastatic breast cancer. (**e**) Coronal MIP image from a PET performed 7 months after the cryoablation shows resolution of the FDG activity within the right medial femoral condyle lesion. The patient also reported decreased knee pain after the cryoablation

improve with standard therapy (Fig. 30.3), risk of increased morbidity with progression of disease (e.g., higher risk for a pathologic fracture), adjuvant therapy before surgical resection, and oligometastatic disease in which surgery is not indicated (Fig. 30.4).

Pain palliation is the most common indication for percutaneous image-guided ablation in the setting of metastatic disease. Up to 50–79% of cancer patients experience metastasis-related pain, but only 20–30% achieve long-lasting pain relief with conventional therapies, which usually involve external beam radiation therapy or chemotherapy [34, 35]. Ablation may help decrease pain by destroying nerve endings, decreasing tumor volume, reducing neurostimulating cytokines from tumor cells, and inhibiting osteoclastic activity at the interface between normal bone and tumor [34]. In a multicenter trial with 61 patients undergoing cryoablation for treatment of painful musculoskeletal metastases, 75% of them achieved 90% or higher pain relief at some point during follow-up, and pain recurrence was observed in only 14% of the cases [36]. RFA has also been proven to be an effective tool for the treatment of painful metastases on prior reports and at least two multicenter trials [37, 38]. To be an adequate candidate for treatment, the patient should be experiencing at least moderate pain, usually a visual analog scale score of 4 out of 10, according to most authors [32, 39]. Ideally, symptoms should also be localized in an area that correlates to the presence of a metastatic lesion on cross-sectional imaging. Lesions with osteolytic or soft tissue components are associated with better treatment response [39, 40]. The location of the targeted area also plays a role in patient selection, since proximity to other important structures, such as peripheral nerves or the spinal cord, may preclude the use of certain ablation techniques due to the risk of thermal injury. Patients should also be able to tolerate conscious sedation or anesthesia, which is required in many cases.

Oligometastatic disease is another indication for percutaneous image-guided ablation and is defined as the presence of 5 or fewer metastatic lesions. Studies have demonstrated rates of local control varying from 36% to 97% with the use of different ablation techniques [41, 42]. The most common primary malignancies in these studies were renal, breast, and prostate cancer.

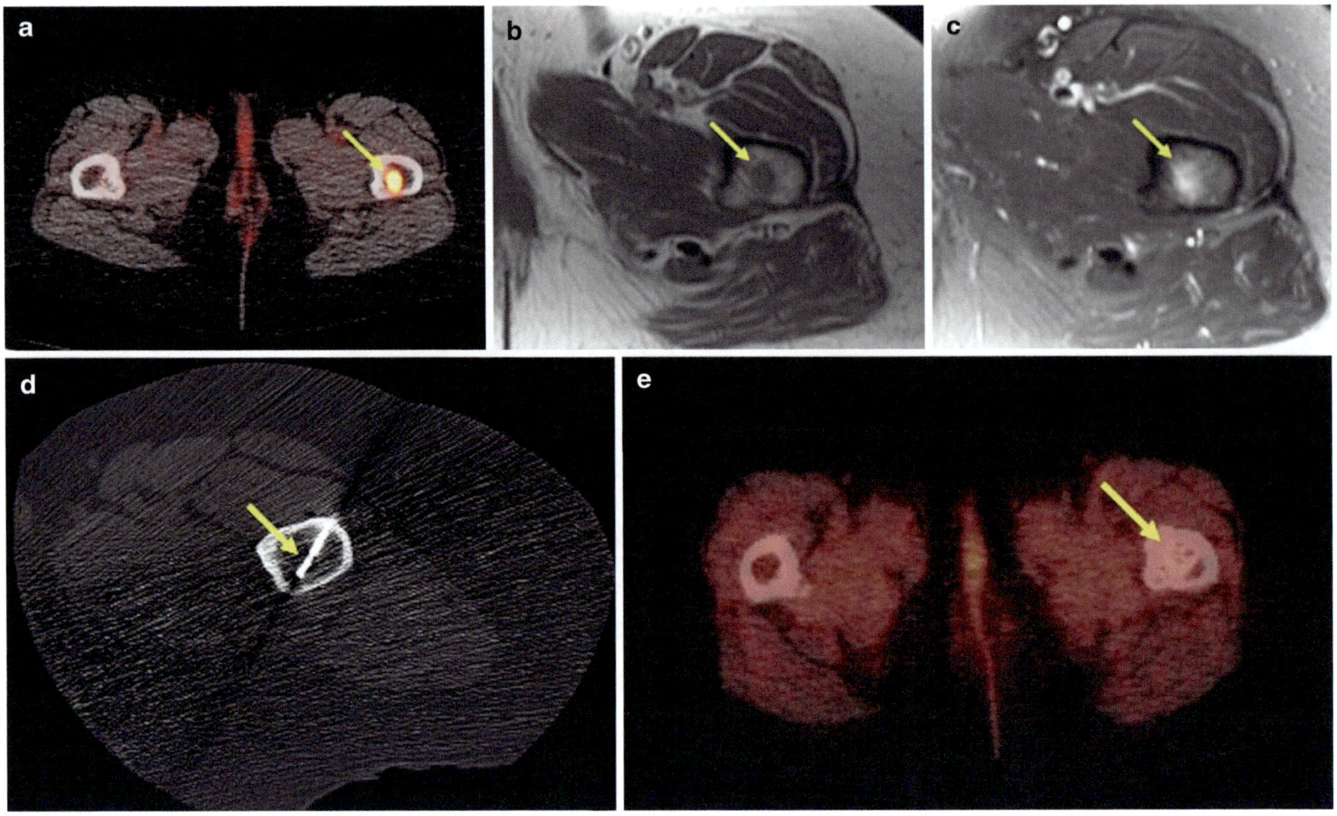

Fig. 30.4 A 73-year-old female with a history of thyroid cancer, status post 100 mCi I-131 treatment, presents with rising thyroglobulin. (**a**) Fused axial image from a PET-CT shows an FDG-avid lesion in the left proximal femur (arrow), concerning metastatic disease. Axial (**b**) T1-weighted and (**c**) T2-weighted fat-suppressed MR images show the low T1 signal intensity, high T2 signal intensity lesion within the left proximal femur (arrows), concerning metastatic disease. (**d**) Axial CT image during radiofrequency ablation shows the radiofrequency probe within the left proximal femur lesion (arrow). Biopsy at the time of ablation confirmed metastatic thyroid cancer. (**e**) Fused axial image from a PET-CT performed 2.5 years after the radiofrequency ablation shows increased attenuation of the left proximal femur lesion, representing post-ablation change, and resolution of the FDG activity

30.3.2 Radiofrequency Ablation (RFA)

(a) Background

RFA uses an alternating electric current to generate heat within the targeted tissue. Most systems include a generator, dispersive grounding pads, and a probe with an exposed tip, which is placed at the area of treatment. The generator creates a high-frequency alternating current that is transmitted to the probe and dispersed by the grounding pads. Ions in the vicinity of the probe tip will oscillate in an attempt to align with the alternating current, which creates an elliptical-shaped area of frictional heat centered at the tip of the probe [34]. The high temperatures lead to protein denaturation and cell membrane damage, eventually leading to cell death. Irreversible cell damage occurs in 4–6 min for temperatures around 50 °C, and it is instantaneous for temperatures in the range of 60–100 °C [34].

(b) Technical details

As with biopsies, all the available imaging is reviewed prior to the procedure in order to identify the ablation target and surrounding structures. Proximity to nerves or vital organs may influence the choice of probe size and planned ablation area. For pain palliation, a successful ablation may prioritize the bone–tumor interface without necessarily covering the entire lesion, especially if safety to surrounding structures is a concern. On the other hand, for oligometastatic disease, it is important to be able to ablate the lesion plus a margin in order to achieve local control [32]. Large vessels should be avoided because they act as heat sinks and dissipate the heat away from the ablation zone.

The choice of access device, as well as size and number of probes, will depend on size and location of the lesion. Imaging guidance is usually performed with CT. However, ultrasound and MRI can also be used when compatible equipment is available. For larger lesions, an array of probes may be used to increase the size of the ablation zone. Once the probe tip is positioned, the system is monitored until the target temperature is achieved. This should be maintained for at least 5 min and up to 15 min. Larger tumors may require additional ablation cycles.

(c) Complications

Post-ablation syndrome presenting predominantly with malaise has been described, likely due to the release of cytokines following tumor cell death [34]. Paradoxical increase in pain can also occur in the first few days after the ablation. More important complications such as adjacent nerve damage, skin burns, and subsequent fractures have been reported in a minority of patients [38].

30.3.3 Cryoablation

(a) Background

Cryoablation promotes cell death by freezing the affected area. Different cellular death mechanisms happen simultaneously during cryoablation. Differences in solute concentration resulting from extracellular ice formation lead to cell dehydration and subsequent damage to enzymatic machinery and the cell membrane. Intracellular ice crystals formed during this process also damage the cell components through mechanical shear injury. Lastly, the lower temperatures promote ischemia and endothelial damage in the affected tissues [34].

(b) Technical details

Most cryoablation systems use pressurized argon gas as a coolant and pressurized helium for thawing. Cell death usually occurs at sustained temperatures of −40 °C or lower. Studies have demonstrated that this process is more effective using cycles of freeze-thaw–freeze rather than continuous freezing only [34]. Most protocols involve freeze-thaw–freeze cycles lasting 10, 5, and 10 min, respectively [32]. Each cryoprobe produces an ice ball whose size depends on the length of the non-insulated probe tip, and multiple cryoprobes can be arranged so that the ice balls they generate cover a larger area. For pain palliation, the bone–tumor interface is prioritized when the entire lesion cannot be safely treated because of its proximity to important structures. To achieve local control of a tumor, the ice ball should ideally extend at least 3 mm beyond its margins [40]. This is necessary because at the margins of the ice ball, the temperature remains around 0 °C, which would still allow some cells to be viable after the procedure. One of the main advantages of cryoablation when compared to RFA is the possibility of monitoring the size of the ice ball with CT or MR imaging throughout the procedure. Larger tumors may require more than one session to complete the treatment.

(c) Complications

Complications are rare in cryoablation, in part due to adequate patient selection. A multicenter trial reported a 2% complication rate (1 out of 61 patients) with the development of osteomyelitis of the treated area [36]. In a study with 40 patients who underwent cryoablation for treatment of oligometastatic disease, 2 patients (5%) developed major complications including avascular osteonecrosis of the femoral head and ureteral stricture [42]. Permanent skin damage is a potential complication that should be addressed for superficial lesions.

30.3.4 Other Techniques

30.3.4.1 Microwave Ablation

Microwave ablation (MWA) is a relatively new, yet promising technique for the treatment of musculoskeletal tumors. Similar to RFA, MWA utilizes thermal energy to cause cell death. However, instead of using an alternating electric current, MWA devices use an electromagnetic field. Malignant tumors are well perfused and usually show high water content, which makes them especially susceptible to the effects of the electromagnetic field. Because microwave propagation does not depend on electric conductivity, heat distribution is more uniform, and larger ablation zones (up to 5–7 cm) can be achieved more rapidly than with RFA, even in tissues with high impedance, such as bone and lung [43]. Another advantage is that microwave systems do not require grounding pads or other additional equipment. The microwave antenna design, microwave frequency used, and power applied are the main determinants of the final size and shape of the ablation zone. Disadvantages include the lack of smaller gauge antennas in some MWA systems, which can make the treatment of smaller lesions more challenging [44]. Potential complications include posttreatment instability of the treated area with a higher risk of pathologic fracture, which has led some practices to combine cement osteoplasty following MWA therapy [45].

30.3.4.2 MR Imaging-Guided Extracorporeal Focused Ultrasound (MRgFUS)

MRgFUS combines MRI guidance and extracorporeal focused ultrasound energy to generate heat and promote the destruction of periosteal innervation. It has shown promising results in pain palliation for osseous metastatic disease without significant adverse effects [46]. This indication was approved by the Food and Drug Administration (FDA) in the United States in 2012 and is now reimbursed by many insurance companies [47]. Because of the high acoustic energy absorption by osseous tissue, energy levels as low as 30% of that used for soft tissue lesions can be employed on bone lesions with similar effects [48]. Additionally, specific MR techniques can be used for real-time thermometry and noninvasive temperature monitoring during the procedure. MRgFUS is truly noninvasive and can be used in areas previously exposed to radiation therapy. Potential complications include skin burns and thermal damage to adjacent structures. Sonication-related pain has also been reported as a limiting factor for some patients.

30.3.4.3 Laser Ablation

Another alternative to thermal ablation is the use of infrared light energy or laser ablation. The laser energy is administered using optic fibers, and the equipment consists of a commercially available laser, such as diode or Nd:YAG laser, a disposable applicator, and a built-in cooling system. Advantages of laser ablation include MRI compatibility, rapid rate of heating, which makes the procedure time-efficient, and the possibility of creating temperature maps to monitor the treatment zones [39, 49]. To achieve a sizable ablation zone, more than one optic fiber is usually needed. Potential complications are similar to other types of thermal ablation and usually involve damage to adjacent structures.

Conclusion

Review of the indications and careful preprocedural planning are vital to increase the chances of success of diagnostic and therapeutic image-guided interventions. A multidisciplinary approach is the best option to ensure an effective and safe procedure. The wide variety of biopsy tools and ablation techniques are being continually improved, and new innovations are expected.

- Can artificial intelligence improve our accuracy by helping select the most promising targets for biopsy?
- Will there be new tools or testing for intraprocedural specimen analysis, decreasing the need for frozen pathology?
- Can new biopsy devices increase safety to allow sampling in even more challenging locations?
- With the development of improved systemic treatment regimens to increase survival, will percutaneous ablations become a standard option for the treatment of oligometastatic disease?
- Will new artificial intelligence algorithms be able to predict the ablation success rate based on pre- or posttreatment appearance of the lesions?

The possibilities are many and are likely to continue expanding in future years.

References

1. Tomasian A, Hillen TJ, Jennings JW. Bone biopsies: what radiologists need to know. Am J Roentgenol. 2020;215:523–33.
2. Ulano A, Bredella MA, Burke P, Chebib I, Simeone FJ, Huang AJ, et al. Distinguishing untreated osteoblastic metastases from enostoses using CT attenuation measurements. Am J Roentgenol. 2016;207:362–8.
3. Bernard S, Walker E, Raghavan M. An approach to the evaluation of incidentally identified bone lesions encountered on imaging studies. AJR Am J Roentgenol. 2017;208:960–70.
4. Moser T, Ehlinger M, Chelli Bouaziz M, Fethi Ladeb M, Durckel J, Dosch J-C. Pitfalls in osteoarticular imaging: how to distinguish bone infection from tumour? Diagn Intervent Imaging. 2012;93:351–9.
5. Goodin GS, Shulkin BL, Kaufman RA, McCarville MB. PET/CT Characterization of fibroosseous defects in children: ^{18}F-FDG uptake can mimic metastatic disease. Am J Roentgenol. 2006;187:1124–8.
6. Plaza JA, Perez-Montiel D, Mayerson J, Morrison C, Suster S. Metastases to soft tissue: a review of 118 cases over a 30-year period. Cancer. 2008;112:193–203.
7. Sammon J, Jain A, Bleakney R, Mohankumar R. Magnetic resonance imaging appearance of soft-tissue metastases: our experience at an orthopedic oncology center. Skelet Radiol. 2017;46:513–21.
8. Murphey MD, Carroll JF, Flemming DJ, Pope TL, Gannon FH, Kransdorf MJ. From the archives of the AFIP: benign musculoskeletal lipomatous lesions. Radiographics. 2004;24:1433–66.
9. Wu JS, Hochman MG. Soft-tissue tumors and tumorlike lesions: a systematic imaging approach. Radiology. 2009;253:297–316.
10. Meek RD, Mills MK, Hanrahan CJ, Beckett BR, Leake RL, Allen H, et al. Pearls and pitfalls for soft-tissue and bone biopsies: a cross-institutional review. Radiographics. 2020;40:266–90.
11. Huang AJ, Kattapuram SV. Musculoskeletal neoplasms: biopsy and intervention. Radiol Clin N Am. 2011;49:1287–305. vii
12. Filippiadis DK, Charalampopoulos G, Mazioti A, Keramida K, Kelekis A. Bone and soft-tissue biopsies: what you need to know. Semin Intervent Radiol. 2018;35:215–20.
13. Kim MS, Lee S-Y, Cho WH, Song WS, Koh J-S, Lee JA, et al. Tumor necrosis rate adjusted by tumor volume change is a better predictor of survival of localized osteosarcoma patients. Ann Surg Oncol. 2008;15:906–14.
14. Rakheja R, Makis W, Tulbah R, Skamene S, Holcroft C, Nahal A, et al. Necrosis on FDG PET/CT correlates with prognosis and mortality in sarcomas. Am J Roentgenol. 2013;201:170–7.
15. Oliveira MP, Lima PM de A, Silva HJ da, Mello RJV de. Neoplasm seeding in biopsy tract of the musculoskeletal system. A systematic review. Acta Ortop Bras 2014;22:106–110.
16. UyBico SJ, Motamedi K, Omura MC, Nelson SD, Eilber FC, Eckardt J, et al. Relevance of compartmental anatomic guidelines for biopsy of musculoskeletal tumors: retrospective review of 363 biopsies over a 6-year period. J Vasc Intervent Radiol. 2012;23:511–518.e2.
17. Siddiqi MA, Kim H-S, Jede F, Han I. Association of core needle biopsy tract resection with local recurrence in extremity soft tissue sarcoma. Skelet Radiol. 2017;46:507–12.
18. Espinosa LA, Jamadar DA, Jacobson JA, DeMaeseneer MO, Ebrahim FS, Sabb BJ, et al. CT-guided biopsy of bone: a radiologist's perspective. AJR Am J Roentgenol. 2008;190:W283–9.
19. Kumar R, Mittal BR, Bhattacharya A, Singh H, Bal A, Prakash G, et al. ^{18}F-FDG PET/CT-Guided real-time automated robotic arm–assisted needle navigation for percutaneous biopsy of hypermetabolic bone lesions: diagnostic performance and clinical Impact. Am J Roentgenol 2019;212:W10–W18.
20. Carrino JA, Khurana B, Ready JE, Silverman SG, Winalski CS. Magnetic resonance imaging-guided percutaneous biopsy of musculoskeletal lesions. J Bone Joint Surg Am. 2007;89:2179–87.

21. Hillen TJ, Talbert RJ, Friedman MV, Long JR, Jennings JW, Wessell DE, et al. Biopsy of CT-occult bone lesions using anatomic landmarks for CT guidance. AJR Am J Roentgenol. 2017;209:214–21.
22. Omura MC, Motamedi K, UyBico S, Nelson SD, Seeger LL. Revisiting CT-guided percutaneous core needle biopsy of musculoskeletal lesions: contributors to biopsy success. AJR Am J Roentgenol. 2011;197:457–61.
23. Hau A, Kim I, Kattapuram S, Hornicek FJ, Rosenberg AE, Gebhardt MC, et al. Accuracy of CT-guided biopsies in 359 patients with musculoskeletal lesions. Skelet Radiol. 2002;31:349–53.
24. Stoker DJ, Cobb JP, Pringle JA. Needle biopsy of musculoskeletal lesions. A review of 208 procedures. J Bone Joint Surg Br. 1991;73:498–500.
25. Hryhorczuk AL, Strouse PJ, Biermann JS. Accuracy of CT-guided percutaneous core needle biopsy for assessment of pediatric musculoskeletal lesions. Pediatr Radiol. 2011;41:848–57.
26. Jelinek JS, Murphey MD, Welker JA, Henshaw RM, Kransdorf MJ, Shmookler BM, et al. Diagnosis of primary bone tumors with image-guided percutaneous biopsy: experience with 110 tumors. Radiology. 2002;223:731–7.
27. Rimondi E, Rossi G, Bartalena T, Ciminari R, Alberghini M, Ruggieri P, et al. Percutaneous CT-guided biopsy of the musculoskeletal system: results of 2027 cases. Eur J Radiol. 2011;77:34–42.
28. Datir A, Pechon P, Saifuddin A. Imaging-guided percutaneous biopsy of pathologic fractures: a retrospective analysis of 129 cases. AJR Am J Roentgenol. 2009;193:504–8.
29. Wu JS, Goldsmith JD, Horwich PJ, Shetty SK, Hochman MG. Bone and soft-tissue lesions: what factors affect diagnostic yield of image-guided core-needle biopsy? Radiology. 2008;248:962–70.
30. Yang YJ, Damron TA. Comparison of needle core biopsy and fine-needle aspiration for diagnostic accuracy in musculoskeletal lesions. Arch Pathol Lab Med. 2004;128:759–64.
31. Rosenthal DI, Alexander A, Rosenberg AE, Springfield D. Ablation of osteoid osteomas with a percutaneously placed electrode: a new procedure. Radiology. 1992;183:29–33.
32. Rosenthal D, Callstrom MR. Critical review and state of the art in interventional oncology: benign and metastatic disease involving bone. Radiology. 2012;262:765–80.
33. Callstrom MR, Charboneau JW, Goetz MP, Rubin J, Atwell TD, Farrell MA, et al. Image-guided ablation of painful metastatic bone tumors: a new and effective approach to a difficult problem. Skelet Radiol. 2006;35:1–15.
34. Rybak LD. Fire and ice: thermal ablation of musculoskeletal tumors. Radiol Clin N Am. 2009;47:455–69.
35. Simon CJ, Dupuy DE. Percutaneous minimally invasive therapies in the treatment of bone tumors: thermal ablation. Sem Musculoskel Radiol. 2006;10:137–44.
36. Callstrom MR, Dupuy DE, Solomon SB, Beres RA, Littrup PJ, Davis KW, et al. Percutaneous image-guided cryoablation of painful metastases involving bone: multicenter trial. Cancer. 2013;119:1033–41.
37. Dupuy DE, Liu D, Hartfeil D, Hanna L, Blume JD, Ahrar K, et al. Percutaneous radiofrequency ablation of painful osseous metastases: a multicenter American College of Radiology Imaging Network trial. Cancer. 2010;116:989–97.
38. Goetz MP, Callstrom MR, Charboneau JW, Farrell MA, Maus TP, Welch TJ, et al. Percutaneous image-guided radiofrequency ablation of painful metastases involving bone: a multicenter study. J Clin Oncol. 2004;22:300–6.
39. Kurup AN, Callstrom MR. Ablation of skeletal metastases: current status. J Vasc Intervent Radiol. 2010;21:S242–50.
40. Callstrom MR, Kurup AN. Percutaneous ablation for bone and soft tissue metastases—why cryoablation? Skelet Radiol. 2009;38:835–9.
41. Kurup AN, Morris JM, Callstrom MR. Ablation of musculoskeletal metastases. American J Roentgenol. 2017;209:713–21.
42. McMenomy BP, Kurup AN, Johnson GB, Carter RE, McWilliams RR, Markovic SN, et al. Percutaneous cryoablation of musculoskeletal oligometastatic disease for complete remission. J Vasc Intervent Radiol. 2013;24:207–13.
43. Hinshaw JL, Lubner MG, Ziemlewicz TJ, Lee FT, Brace CL. Percutaneous tumor ablation tools: microwave, radiofrequency, or cryoablation—what should you use and why? Radiographics. 2014;34:1344–62.
44. Lubner MG, Brace CL, Hinshaw JL, Lee FT. Microwave tumor ablation: mechanism of action, clinical results, and devices. J Vasc Intervent Radiol. 2010;21:S192–203.
45. Deib G, Deldar B, Hui F, Barr JS, Khan MA. Percutaneous microwave ablation and cementoplasty: clinical utility in the treatment of painful extraspinal osseous metastatic disease and myeloma. Am J Roentgenol. 2019;212:1377–84.
46. Liberman B, Gianfelice D, Inbar Y, Beck A, Rabin T, Shabshin N, et al. Pain palliation in patients with bone metastases using MR-guided focused ultrasound surgery: a multicenter study. Ann Surg Oncol. 2009;16:140–6.
47. Scipione R, Anzidei M, Bazzocchi A, Gagliardo C, Catalano C, Napoli A. HIFU for Bone metastases and other musculoskeletal applications. Semin Intervent Radiol. 2018;35:261–7.
48. Tempany CMC, McDannold NJ, Hynynen K, Jolesz FA. Focused ultrasound surgery in oncology: overview and principles. Radiology. 2011;259:39–56.
49. Ahrar K, Stafford RJ. Magnetic resonance imaging-guided laser ablation of bone tumors. Tech Vasc Interv Radiol. 2011;14:177–82.

Part VIII

Sentinel Lymph Node Surgery for Solid Cancer

Mark B. Faries

> "The danger lies in the diffusion of malignant cell particles from this primary focus; these always implicate the nearest lymph glands, which intercept them for a time. Eventually they pass beyond such "traps" into the blood current; then death, with multiple visceral metastases, ensues. Palpable enlargement of these glands is unfortunately but a late symptom of deposit therein; by the time it occurs there is almost always implication of deeper organs or tissues ... We further see the paramount importance of securing, whenever possible, the perfect eradication of those lymph glands which will necessarily be first infected; before enlargement takes place radical removal of such organs in the axilla, groin, surface regions of the neck &c, before they have undergone appreciable increase in bulk, is a safe and easy measure, which, under the conditions indicated, should never be neglected."
> -Herbert Snow MD, Lancet, 1892 [1]

For many solid tumors, the importance of lymphatic spread and lymph node metastases has been abundantly clear from the time of their earliest descriptions. Regional lymph nodes represent the initial site of metastasis for most solid tumors and the presence of disease in those nodes is generally associated with a marked decrease in prognosis. However, although these central observations are self-evident, their implications for clinical medicine and the treatment of patients with cancer have long been the source of great controversy and remain the subject of extensive research. Initial proposals, such as that of Herbert Snow above, were for early and radical excision of draining regional lymph nodes automatically, even before any evidence of tumor spread to those locations.

This reaction was intuitive, particularly in an era when most primary tumors were relatively advanced at diagnosis. However, the price of such a radical approach was also apparent in the morbidity associated with complete dissections. Patients undergoing groin or axillary dissections are at risk for pain, nerve dysfunction, and, most concerningly, lymphedema of the adjacent extremity. This last toxicity, though treatable, is often permanent at least to some degree. As primary tumors were diagnosed earlier, the frequency of nodal involvement decreased and the wisdom of complete nodal dissection for all patients was increasingly questioned. In breast cancer and melanoma, randomized trials were conducted to evaluate the value of immediate or elective lymph node dissection in the absence of clinically apparent metastases [2, 3]. In gastric, lung, and esophageal cancers, more extensive nodal dissections were compared to less extensive ones, and debates raged in surgical and oncologic meetings around the world [4–6].

Into this swirl of controversy Donald Morton, Alistair Cochran, and their colleagues at the UCLA introduced a novel technique for intraoperative mapping of the lymphatic drainage from the primary tumor site to the initial node(s) along the pathway [7]. The pathologic status of these sentinel lymph nodes (SLN) was found to be representative of that of the entire basin. This technique has gone on to revolutionize the care of not only melanoma but also breast cancer, and it is being productively explored in multiple other malignancies as outlined in this section.

The concept of a "sentinel" node was not new. The idea of a node that was the specific site of drainage from a tumor had arisen several times, likely independently, in the past [8]. Cabanas proposed the idea for penile cancers as did Gould for the parotid and Kett in breast cancer

[9–11]. Similar ideas regarding specific patterns of lymphatic drainage were put forward as early as 1923 by Braithwaite, while Virchow described the drainage of tattoo ink in the skin to regional nodes as early as the nineteenth century. However, a practical, reproducible, and reliable surgical technique for finding the critical node eluded the medical community until Morton described it.

With SLN biopsy it became possible to ascertain the pathologic status of the regional nodal basin with a minimally invasive, outpatient procedure. A revolution was launched that has had far-reaching implications that have benefitted countless patients around the world.

Why was SLN biopsy such an advance? It offers substantial advantages over complete dissection of an entire nodal basin. First it is less morbid than full dissection. It can be completed through much smaller incisions with much less dissection. Nerves, blood vessels, and lymphatic channels that are transected in dissections are largely preserved in SLN biopsy. The risk of long-term toxicity, including lymphedema, is also markedly reduced. Second, this decrease in the extent of surgery is not accompanied by any reduction in the accuracy of regional staging. In fact, the opposite is true. Since the number of nodes that must be examined by the pathologist is greatly reduced, the intensity of evaluation of the SLN is able to be increased. Routine use of immunohistochemical markers is possible in SLN, making small metastases more readily detected. In melanoma, as an example, this has resulted in greatly improved prognostic discrimination for node-negative and node-positive patients, which remarkably improved staging in the current 8th Edition of the American Joint Commission on Cancer staging system [12].

Finally, it appears that SLN biopsy may be therapeutic in some respects, and not merely a staging procedure on the way to a therapeutic nodal dissection. The concept appears to be most clearly evaluated for melanoma, where in the second Multicenter Selective Lymphadenectomy Trial (MSLT-II) patients with SLN metastases were randomized to either completion dissection or nodal observation [13]. One of the less appreciated results of the trial was that among observation arm patients, all of whom had nodal disease, only a minority ever developed regional nodal recurrence and never required full dissection. So for about three-quarters of the patients, all nodal disease was eradicated with the SLN biopsy alone. From MSLT-I we also know that early removal of nodal disease likely results in improved long-term melanoma-specific survival [14]. The degree to which these observations apply in other diseases will likely vary based on the biology of those malignancies. In breast cancer, as a different example, regional nodal disease may be more likely to be controlled with nonsurgical measures, making a therapeutic impact less likely [15]. Additional study will be required to fully elucidate the relative therapeutic value of SLN biopsy in different malignancies.

SLN biopsy continues to be developed in other malignancies, including upper gastrointestinal cancer, colon cancer, and lung cancer. These diseases have proved more challenging both technically and in terms of evaluation of efficacy. Advances in technology have increased the likelihood of accurate and practical application of lymphatic mapping beyond breast cancer and melanoma. These advances include the use of fluorescent dyes that may be visualized during laparoscopic or robotic surgical approaches [16]. Ferromagnetic mapping agents may also come into increased use as research continues. SPECT/CT may enhance the accuracy of mapping in anatomically difficult regions such as the head and neck [17]. These developments are likely to continue to open avenues of investigation to attempt to improve care for patients in the future.

A final area of active investigation relates to the increasingly multidisciplinary nature of oncology in the modern era. In the past, the pathway between the surgeon and the medical oncologist was a one-way street. With the advent of neoadjuvant approaches to many cancers, treatments have become better and more sophisticated, albeit more complicated. In breast cancer, where SLN biopsy and neoadjuvant systemic treatment have had perhaps the most interaction, it appears that lymphatic mapping is more challenging than it is in the de novo setting. Work continues to optimize the interaction of the two techniques to render treatment that is both accurate and minimally morbid.

References

1. Snow HM. Melanotic cancerous disease. Lancet. 1892;140:869–922.
2. Fisher B, Jeong JH, Anderson S, et al. Twenty-five-year follow-up of a randomized trial comparing radical mastectomy, total mastectomy, and total mastectomy followed by irradiation. N Engl J Med. 2002;347:567–75.

3. Balch CM, Cascinelli N, Sim FH. Elective lymph node dissection: results of prospective randomized surgical trials. In: Balch CM, Houghton AN, Sober AJ, et al., editors. Cutaneous melanoma. Saint Louis, MO: Quality Medical Publishing; 2003. p. 379–95.
4. Songun I, Putter H, Kranenbarg EM, et al. Surgical treatment of gastric cancer: 15-year follow-up results of the randomised nationwide Dutch D1D2 trial. Lancet Oncol. 2010;11:439–49.
5. Fok M, Siu KF, Wong J. A comparison of transhiatal and transthoracic resection for carcinoma of the thoracic esophagus. Am J Surg. 1989;158:414–9.
6. Izbicki JR, Thetter O, Habekost M, et al. Radical systematic mediastinal lymphadenectomy in non-small cell lung cancer: a randomized controlled trial. Br J Surg. 1994;81:229–35.
7. Morton D, Wen D, Wong J, et al. Technical details of intraoperative lymphatic mapping for early stage melanoma. Arch Surg. 1992;127:392–9.
8. Nieweg OE, Uren RF, Thompson JF. The history of sentinel lymph node biopsy. Cancer J. 2015;21:3–6.
9. Cabanas R. An approach for the treatment of penile carcinoma. Cancer. 1977;39:456–66.
10. Gould EA, Winship T, Philbin PH, et al. Observations on a "sentinel node" in cancer of the parotid. Cancer. 1960;13:77–8.
11. Kett K, Lukacs L. Diagnosis of regional lymph node metastases in cancer of the breast. A study of the lymphatic system of the diseased breast. Am J Surg. 1974;128:398–401.
12. Gershenwald JE, Scolyer RA, Hess KR, et al. Melanoma staging: evidence-based changes in the American joint committee on cancer eighth edition cancer staging manual. CA Cancer J Clin. 2017;67:472–92.
13. Faries MB, Thompson JF, Cochran AJ, et al. Completion dissection or observation for sentinel-node metastasis in melanoma. N Engl J Med. 2017;376:2211–22.
14. Morton DL, Thompson JF, Cochran AJ, et al. Final trial report of sentinel-node biopsy versus nodal observation in melanoma. N Engl J Med. 2014;370:599–609.
15. Giuliano AE, Ballman K, McCall L, et al. Locoregional recurrence after sentinel lymph node dissection with or without axillary dissection in patients with sentinel lymph node metastases: long-term follow-up from the American College of Surgeons oncology group (Alliance) ACOSOG Z0011 randomized trial. Ann Surg. 2016;264:413–20.
16. Ito N, Fukuta M, Tokushima T, et al. Sentinel node navigation surgery using indocyanine green in patients with lung cancer. Surg Today. 2004;34:581–5.
17. Chapman BC, Gleisner A, Kwak JJ, et al. SPECT/CT improves detection of metastatic sentinel lymph nodes in patients with head and neck melanoma. Ann Surg Oncol. 2016;23:2652–7.

Sentinel Lymph Node Biopsy for Primary Cutaneous Malignancy

31

Marc Moncrieff and Howard Peach

Abstract

Sentinel lymph node biopsy (SLNB) is the international standard of care for providing prognostic information in cutaneous melanoma. This chapter describes the rationale behind development of sentinel node biopsy in this role, as well as the important practical considerations for lymphoscintigraphic imaging. It chronicles the scientific evidence gleaned from large international clinical trials and how this evidence was used to shape current clinical practice. This includes detailed analysis of the MSLT-1, MSLT-2 and DeCOG trials, which has permitted the safe reduction in recommended surgical procedures. There is a distillation of the evidence on the relevance of microscopic deposits within the node and other pathological findings and how the extent of this disease impacts on prognosis. It concludes with a summary of current recommended practice.

Finally, the role of SLNB in the context of other skin cancers is less transparent. There is an explanation why the natural biology and the clinical presentation of other skin cancers questions the validity of SLNB in these pathologically different cancers.

Learning Objectives

This chapter will inform the reader about:

- Why elective lymph node dissection was initially advocated, what the clinical concerns about it were and how it helped develop the role of SLNB.
- How the new procedure of SLNB was developed, introduced into clinical practice and how it impacts on prognosis.
- The logic of using the current radio-labelled tracers and how the use of SPECT-CT improves localisation accuracy.
- The results of the MSLT-1, -2 and DeCOG studies. It explains the controversy around particularly the MSLT-1 study and how each trial has helped define the role of SLNB and the cohort of patients who are most likely to benefit.
- The importance of micrometastatic disease pathology and how it impacts on predicted future melanoma recurrence.
- Which cohort of melanoma patients and patients with other skin malignancies should be offered SLNB.

31.1 Introduction

Primary cutaneous malignancies, particularly melanoma, readily metastasise to the regional lymph nodes by accessing the rich lymphovascular network afforded to the skin. Unlike other cancers, the lymphatic drainage pattern of the primary tumour cannot be easily predicted based on its anatomical location, making elective lymph node dissection an unreli-

M. Moncrieff (✉)
Norfolk & Norwich University Hospital NHS Trust, Norwich, United Kingdom
e-mail: marc.moncrieff@nnuh.nhs.uk

H. Peach
Leeds Teaching Hospitals NHS Trust, Leeds, United Kingdom
e-mail: h.peach@nhs.net

S. P. Leong et al. (eds.), *Cancer Metastasis Through the Lymphovascular System*,
https://doi.org/10.1007/978-3-030-93084-4_31

able and ineffective treatment strategy. Sentinel lymph node biopsy (SLNB) is a technique developed to identify the lymphatic drainage of the primary tumour and to stage the regional lymph nodes to identify patients in need of further treatment. Since its development, SLNB has evolved to become the standard of care for primary cutaneous melanoma and is incorporated in the current AJCC classification system. [1] This chapter describes the use of SLNB for the staging and treatment of melanoma, and compares its application to other cutaneous malignancies, such as Merkel cell and squamous cell carcinomas.

31.2 Evolution of Sentinel Lymph Node Biopsy for Cutaneous Malignancy

Primary cutaneous melanoma readily metastasises through the draining lymphatic system to the regional lymph nodes. The risk of this event is directly proportionate to the depth of invasion of the primary into the skin, termed Breslow thickness. Whilst the probability of nodal metastasis given a particular Breslow thickness is consistent across the literature, [2] the process is almost certainly stochastic, meaning that a nodal metastasis may not be predicted precisely in any given patient with the current biomarkers available to clinicians at present.

31.3 Elective Lymph Node Dissection

Clinicians have been aware for decades that the most important predictor of patient outcome or survival is the nodal status at the time of presentation of the primary melanoma. Accordingly, once a nodal metastasis has been identified, then the contribution of the prognostic factors of the primary melanoma seemingly have little or no further influence on the patient's prognosis. Prior to the 1990s, elective lymph node dissection (ELND) was routinely practiced using the rationale that removing microscopic disease may reduce the risk of distant spread and accurately stage the patient. Unfortunately, the majority of the patients did not harbour microscopic disease and underwent a morbid procedure for which they received no clinical benefit. Clinicians were aware that the risk of regional nodal metastasis increased according to the Breslow thickness, and that a dual-track risk of nodal and distant metastasis appeared to operate whereby, the risk of distant metastasis outweighed the risk of nodal metastasis for tumours >4 mm in thickness but the situation was reversed in the early or intermediate thickness tumours between 0.75 and 4 mm thick. [3]

Several international randomised controlled trials failed to show any survival benefit for melanoma patients who underwent ELND at the time of surgical management of the primary tumour. [4–7] However, analyses from these trials and others [8] suggested that a subgroup of patients with intermediate thickness melanoma may benefit from early nodal intervention. The Intergroup Melanoma Surgical Trial [9] randomised patients with melanomas 1–4 mm thick to ELND or observation. This was the first trial to mandate lymphatic mapping using preoperative lymphoscintigraphy to direct the nodal dissection and found that, without this information, ELND would have been misdirected in more than 50% of patients. Whilst the trial failed to show an overall survival benefit for ELND, a planned subgroup analysis found an increased survival benefit for patients with thinner melanomas (1.0–2.0 mm thick) and patients younger than 60 years old. This was the first trial to show a survival benefit for early removal of microscopic nodal disease and, with preoperative lymphoscintigraphy, became the foundation of the sentinel node biopsy paradigm.

31.4 Sentinel Lymph Node Biopsy

Donald Morton developed the technique for identifying the draining nodal basin and identifying the sentinel node. In the landmark paper, [10] the initial procedure termed, 'intraoperative lymphatic mapping and selective lymphadenectomy' incorporated preoperative lymphoscintigraphy to identify the draining regional nodes and intraoperative injection of blue dye to identify the sentinel nodes visually. This was the culmination of work from animal studies demonstrating variable nodal drainage in the same nodal field from different primary sites. [11] Crucially they also demonstrated a predictable, repeatable and orderly nodal drainage pattern via the sentinel node. Using this technique in humans, Morton's team identified sentinel nodes in 82% of patients and had a false-negative rate of 1%. These data led them to conclude that, 'this technique identifies, with a high degree of accuracy, patients with early stage melanoma who have nodal metastases'. These results confirmed that the sentinel node is the initial site of regional lymphatic node metastases.

Shortly afterwards, Morton and others described the use of intraoperative radio-localisation of the sentinel node using a handheld gamma probe. [12] This technique was rapidly validated by other centres internationally. [13–15] Subsequently, Morton published a multicentre experience of the technique which described both a 'learning phase' of 30 cases for a centre to become proficient at the technique and a case volume of at least 25 cases per annum to maintain that proficiency. [16] Furthermore, the paper cemented the current technique of dual-localisation using preoperative lymphoscintigraphy and intraoperative blue dye and radio-localisation as the most accurate method for identifying the sentinel node.

31.5 Technical Details of Lymphoscintigraphy

The technical aspects for a modern sentinel lymph node practice are covered in comprehensive detail by two review articles from North America [17] and Europe [18] and it is unnecessary to replicate those details here. Whilst the surgical technique has not changed substantially over the last decade, there have been significant improvements in the preoperative sentinel node localisation, brought about by changes in the imaging protocols and the choice of tracer agents. [19]

The European Association of Nuclear Medicine (EANM) states that, 'hybrid imaging with SPECT and CT including anatomical information improves the localisation of SLNs and reduces misinterpretation of images. Images obtained by SPECT/CT are three-dimensional and have better contrast and spatial resolution than planar images. For SPECT/CT, a higher overall SLN detection rate and better detection of SLNs located next to the injection site have been reported'. A large prospective multicentre trial found that SPECT/CT is also able to modify clinical decision-making. The International Atomic Energy Agency Sentinel Node Trial demonstrated that SPECT/CT had modified the surgical approach in 37% of patients with melanoma, with greatest improvement in accuracy seen in the head and neck and truncal regions. [20]

Novel tracer agents have entered into mainstream use to improve accuracy of localisation both preoperatively. One of the significant details of the procedure is timing regarding the decay in gamma radiation of ^{99m}Tc which constrains the surgery to within, at most, 24 h of the preoperative localisation. Further concerns arise with temporal migration of the tracer agent to the second echelon nodes, theoretically increasing the risk of harvesting an adjacent non-sentinel node, with the consequent risk of a false-negative biopsy. A recent European multicentre study found a significantly reduced survival (hazard ratio 1.36, 95% CI 1.05–1.74; $p = 0.018$) and an increased nodal relapse rate (4.5% vs. 2.5%; $p = 0.008$) when using ^{99m}Tc-nanocolloid, in patients who underwent SLNB >12 h after their preoperative SLN localisation, compared to those who had their SLNB within 12 h. [21] Novel tracer agents, such as ^{99m}Tc-tilmanocept, have been approved as alternatives to address these issues. ^{99m}Tc-tilmanocept is composed of a dextran backbone, attached to units of mannose and DTPA and avidly binds to a receptor specific to reticuloendothelial cells. [22] Alternative tracer agents, indocyanine green and MRI-visible, magnetic nanoparticles have been investigated to replace the blue dye and radiolabelled ^{99m}TC tracer agents with promising results. [23, 24]

31.6 Outcomes of Sentinel Lymph Node Biopsy for Melanoma

The purpose of developing SLNB was to identify those patients harbouring micrometastatic disease in the draining regional lymph nodes to subsequently offer a lymphadenectomy to prevent distant relapse and death. There were signals from previous ELND studies that this may be a beneficial therapeutic strategy for a specific subgroup of patients, particularly those with intermediate thickness melanomas, 1.2–3.5 mm thick. [8, 9, 25] Accordingly, Morton and colleagues devised the Multicenter Selective Lymphadenectomy Trial-1 (MSLT-1) to investigate the therapeutic utility of SLNB with subsequent completion lymph node dissection (CLND) for micrometastatic nodal disease.

31.6.1 MSLT-1 Study

The MSLT-1 study [26] can rightly be described as a landmark trial in the surgical management of cutaneous melanoma [27]. The study recruited 2001 patients with primary cutaneous melanoma and randomised patients to either SLNB with subsequent CLND for a positive SLNB (SN+) or nodal observation and subsequent therapeutic lymph node dissection (TLND) for patients who developed palpable lymphadenopathy during follow-up. The principal target cohort was intermediate thickness tumours (1.2–3.5 mm) but eligibility criteria allowed for melanomas >3.5 mm thickness, or <1.2 mm thickness if the tumour was Clark's level IV or V. The randomisation cohorts were allocated in a 60:40 manner in favour of the SLNB group to power the study appropriately for nodal events in the SLNB field.

The study results showed a non-significant survival difference in the primary endpoint of melanoma-specific survival (MSS) for the intermediate thickness group. This was unsurprising given that the incidence of SN+ was 16.0% and only 20% developed nodal metastases in the observation group. Accordingly, 80% of patients in either group could not have benefitted from any intervention, and this majority meant that the study was not statistically powered to detect a survival difference, though at 10 years, there was a non-significant 4% higher survival in the SLNB arm.

Some commentators seized upon the negative outcome of the primary endpoint and dismissed the therapeutic utility of SLNB and the trial as a result. However, the subgroup analysis of the data revealed significant findings regarding the biology of the tumour, which merit further discussion. One of the major findings demonstrated the importance of SLNB as a staging tool for primary cutaneous melanoma. For SN+

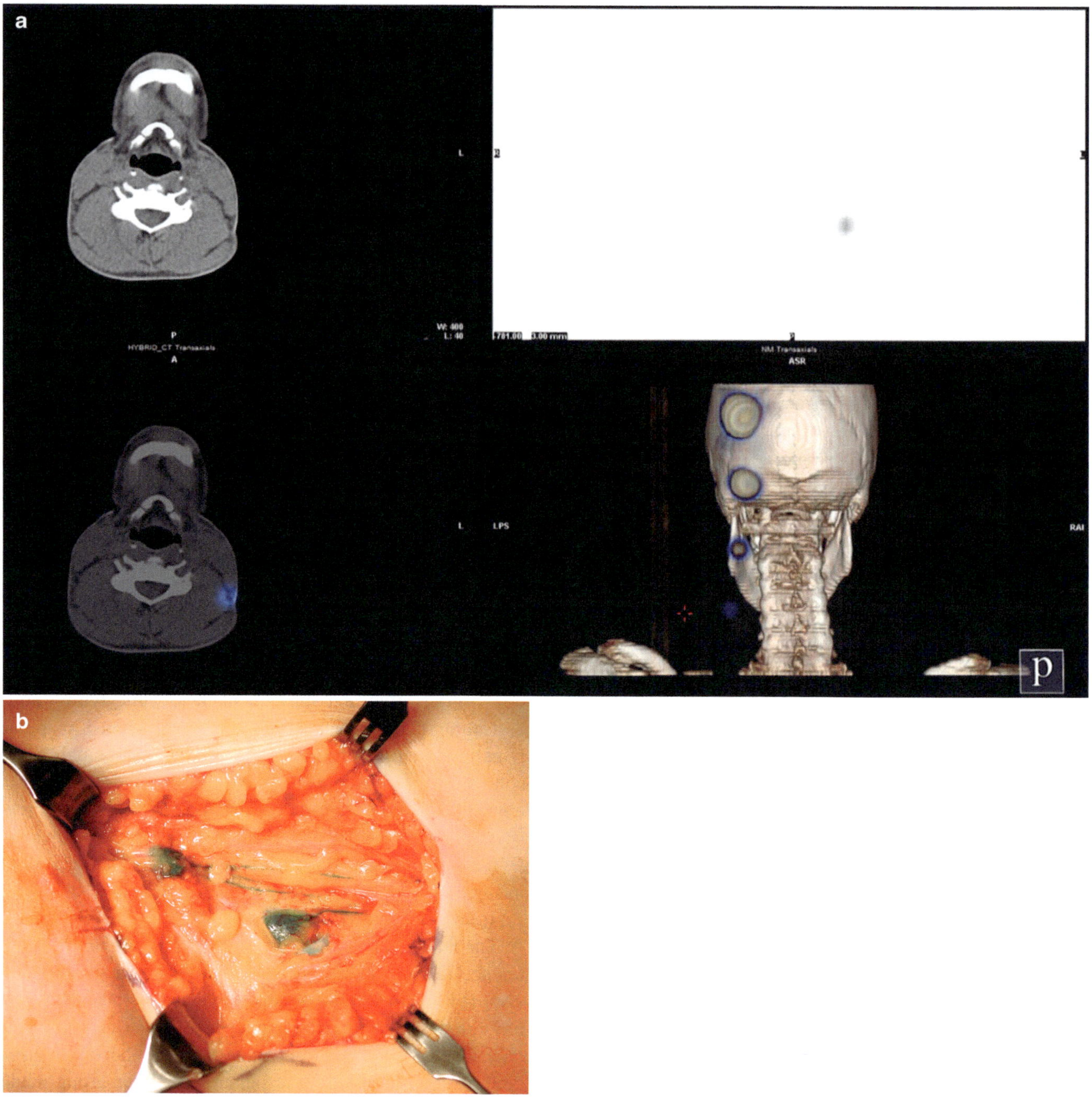

Fig. 31.1 (**a**) Pre-operative SPECT-CT imaging demonstrating lymphatic mapping of a patient with a melanoma on the left posterior scalp. (**b**) Intraoperative photograph of a sentinel nodes in the right axilla demonstrating lymphatic channels and sentinel nodes stained with patent blue dye. (Photographs courtesy of Howard Peach)

patients the 10-year melanoma-specific survival rate was 62.1 ± 4.8% compared with 85.1 ± 1.5% for SN-negative patients (hazard ratio for death from melanoma, 3.09; 95% CI, 2.12–4.49; $p < 0.001$). Multivariate analysis revealed that SLNB was the most important prognostic indicator for MSS (HR: 2.40; 95% CI: 1.61–3.56; $p < 0.001$) (Fig. 31.1).

Another major finding was the highly significant difference in nodal disease burden and behaviour between the two groups. The mean number of nodes involved, for the intermediate thickness cohort, was 1.4 ± 0.1 in the SLNB group versus 3.2 ± 0.4 in the observation group ($p = 0.0001$) for the intermediate thickness cohort. Patients in the SN group who subsequently relapsed in the nodal field (termed 'false negative') had a similar nodal burden to the observation group (4.3 ± 1.3; $p < 0.0001$). The cumulative incidence of nodal metastasis in the SLNB and observation groups respectively were (21.9% vs. 19.5%; p = NS) for the intermediate thickness cohort and identical (42.0% vs. 41.4%; p = NS)—Fig. 31.2a, b in the thick melanoma cohort. The disease-free survival for both the intermediate and thick melanoma cohorts was significantly better in the SLNB groups, which was attributable to the reduced incidence of nodal relapse. Finally, the median time to nodal relapse was the same for the false-negative SLNB group and the observation groups in both the intermediate thickness and thick melanoma cohorts, though the for the thick melanoma cohort overall the time to relapse was approximately half the time of the intermediate thickness cohort overall (9.2 vs. 19.2 months).

The most plausible interpretation of these data is that a nodal micrometastasis will progress to palpable disease without intervention, and that intervention (i.e. SLNB) prevents the nodal disease progressing to second echelon nodes within the nodal field, thereby preventing further nodal relapse. Figure 31.2c, d shows the survival curves for the intermediate-thickness cohort, which was a planned subgroup analysis. The analysis demonstrates that among patients with nodal metastases from intermediate-thickness melanomas, the 10-year melanoma-specific survival rate was 62.1 ± 4.8% in the SLNB group as compared with 41.5 ± 5.6% in the observation group (hazard ratio for death from melanoma, 0.56; 95% CI, 0.37–0.84; p = 0.006). Similarly, there was no difference in MSS between the false-negative group and the observation group (HR, 1.15 (95% CI, 0.68–1.94); $p = 0.60$). Whilst, the comparison is statistically problematic, since the randomisation took place at the time of treatment of the primary, the data make a highly compelling argument for a survival benefit for a small group of patients undergoing SLNB who have their focus of micrometastatic disease excised in the draining node. A further, important therapeutic benefit, and a survivorship issue, was the significant reduction in lymphoedema rates seen in the patients who underwent CLND in the SLNB cohort compared to those who underwent TLND in the observation cohort (20.4% vs. 12.4%, $p = 0.04$). [28]

31.6.2 Completion Lymph Node Dissection for Sentinel Node Positive Patients: MSLT-2 and DeCOG Studies

The original rationale for undertaking SLNB was to identify patients who were harbouring micrometastatic disease in the regional nodal field and who would subsequently benefit from completion lymph node dissection. However, it became rapidly apparent that there was a equipoise regarding the need for the CLND. On the one hand, patients benefitted from improved regional control and the extra prognostic information that the procedure afforded, whilst on the other, the majority of patients had no further disease other than the positive sentinel node (about 90%) and were subject to the additional morbidity of the procedure, from which they would derive no benefit. Moreover, CLND was mandated at that time, for most large phase III adjuvant systemic therapy clinical trials, whilst those patients who demonstrated additional metastases in their non-sentinel nodes (NSN) had an overall poor prognosis, for which there were no effective systemic therapeutic options. [27]

To resolve these concerns, Dr. Morton and colleagues devised the MSLT-2 study which randomised patients to either observation of the metastatic nodal field with four-monthly targeted ultrasound (and TLND on clinical nodal recurrence) or CLND at the time of diagnosis of the positive SLNB. [29] The study also incorporated a screening phase, with histologically negative SLNBs being screened for disease using reverse transcription polymerase chain reaction (RT-PCR) testing. Those patients who were RT-PCR positive were then randomised to CLND or nodal observation. The primary endpoint was melanoma-specific survival (MSS) at 10 years. At approximately the same time, the Dermatologic Cooperative Oncology Group (DeCOG) opened a trial with a very similar trial design, [30] with minor differences that included patients being screened with ultrasound every 3 months with a serum S100B measurement performed at the same time, and with head and neck melanomas being excluded. Both trials excluded metastatic nodes demonstrating extracapsular spread (syn. extracapsular or extranodal extension) and patients who were immunocompromised. The major difference between the two trials was that the DeCOG-SLT had distant disease-free survival (distant-DFS) as the primary endpoint.

For the MSLT-2 study, [29] a total of 3531 patients were enrolled in the screening phase and 1939 patients underwent randomisation, with 1713 pathology positive and 226 RT-PCR positive. The median size of the deposits in the SLNB was 0.61 mm in the dissection group and 0.67 mm in the observation group. The majority of patients had only one positive sentinel node: 72.3% in the CLND group and 69.1% in the observation group. There was no difference in MSS between the two groups (86 ± 1.3% and 86 ± 1.2%; $p = 0.42$). The rate of disease-free survival was slightly higher in the dissection group than in the observation group (68 ± 1.7% and 63 ± 1.7%, respectively; $p = 0.05$) at 3 years. The authors assumed that this was based on the significantly reduced regional nodal relapse rate in the CLND group at 3 years (92 ± 1.0% vs. 77 ± 1.5%; $P < 0.001$). It is noteworthy that the incidence of 'node-only' relapses in the observation group was 7.7% and the overall incidence was 4.7%. The presence of non-SLN metastases in the CLND arm was an independent predictor of MSS (HR 1.78; 95% CI 1.19–2.68, p = 0.005). Subgroup analyses did not identify any groups with significant survival benefit from CLND. However, CLND was associated with a significantly lymphoedema

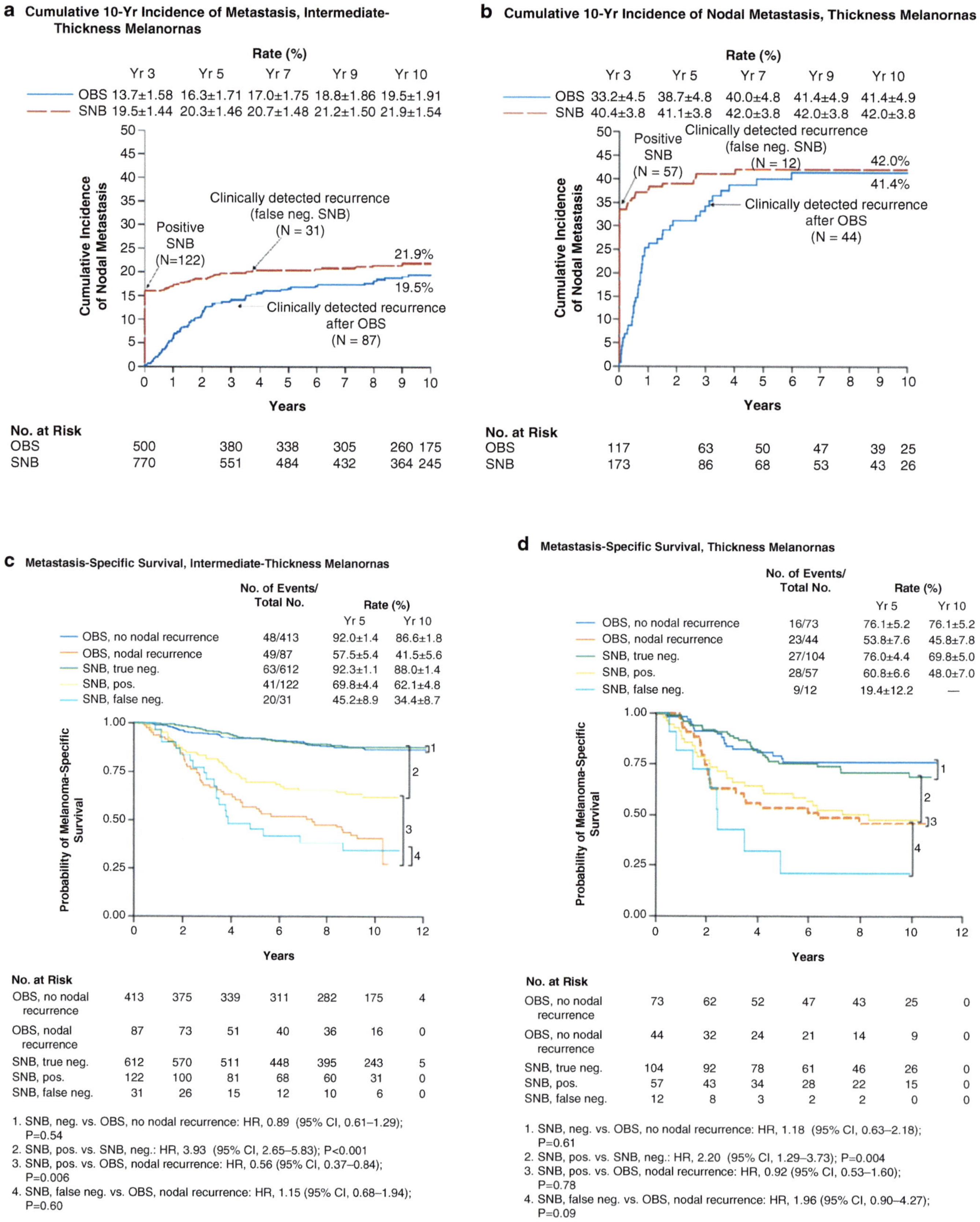
a Cumulative 10-Yr Incidence of Metastasis, Intermediate-Thickness Melanornas
Rate (%)
Yr 3 Yr 5 Yr 7 Yr 9 Yr 10
OBS 13.7±1.58 16.3±1.71 17.0±1.75 18.8±1.86 19.5±1.91
SNB 19.5±1.44 20.3±1.46 20.7±1.48 21.2±1.50 21.9±1.54
Cumulative Incidence of Nodal Metastasis
Positive SNB (N=122)
Clinically detected recurrence (false neg. SNB) (N = 31)
21.9%
19.5%
Clinically detected recurrence after OBS (N = 87)
Years
No. at Risk
OBS 500 380 338 305 260 175
SNB 770 551 484 432 364 245
b Cumulative 10-Yr Incidence of Nodal Metastasis, Thickness Melanornas
Rate (%)
Yr 3 Yr 5 Yr 7 Yr 9 Yr 10
OBS 33.2±4.5 38.7±4.8 40.0±4.8 41.4±4.9 41.4±4.9
SNB 40.4±3.8 41.1±3.8 42.0±3.8 42.0±3.8 42.0±3.8
Cumulative Incidence of Nodal Metastasis
Positive SNB (N = 57)
Clinically detected recurrence (false neg. SNB) (N = 12)
42.0%
41.4%
Clinically detected recurrence after OBS (N = 44)
Years
No. at Risk
OBS 117 63 50 47 39 25
SNB 173 86 68 53 43 26
c Metastasis-Specific Survival, Intermediate-Thickness Melanornas
No. of Events/Total No.
Rate (%)
Yr 5 Yr 10
OBS, no nodal recurrence 48/413 92.0±1.4 86.6±1.8
OBS, nodal recurrence 49/87 57.5±5.4 41.5±5.6
SNB, true neg. 63/612 92.3±1.1 88.0±1.4
SNB, pos. 41/122 69.8±4.4 62.1±4.8
SNB, false neg. 20/31 45.2±8.9 34.4±8.7
Probability of Melanoma-Specific Survival
Years
No. at Risk
OBS, no nodal recurrence 413 375 339 311 282 175 4
OBS, nodal recurrence 87 73 51 40 36 16 0
SNB, true neg. 612 570 511 448 395 243 5
SNB, pos. 122 100 81 68 60 31 0
SNB, false neg. 31 26 15 12 10 6 0
1. SNB, neg. vs. OBS, no nodal recurrence: HR, 0.89 (95% CI, 0.61–1.29); P=0.54
2. SNB, pos. vs. SNB, neg.: HR, 3.93 (95% CI, 2.65–5.83); P<0.001
3. SNB, pos. vs. OBS, nodal recurrence: HR, 0.56 (95% CI, 0.37–0.84); P=0.006
4. SNB, false neg. vs. OBS, nodal recurrence: HR, 1.15 (95% CI, 0.68–1.94); P=0.60
d Metastasis-Specific Survival, Thickness Melanornas
No. of Events/Total No.
Rate (%)
Yr 5 Yr 10
OBS, no nodal recurrence 16/73 76.1±5.2 76.1±5.2
OBS, nodal recurrence 23/44 53.8±7.6 45.8±7.8
SNB, true neg. 27/104 76.0±4.4 69.8±5.0
SNB, pos. 28/57 60.8±6.6 48.0±7.0
SNB, false neg. 9/12 19.4±12.2 —
Probability of Melanoma-Specific Survival
Years
No. at Risk
OBS, no nodal recurrence 73 62 52 47 43 25 0
OBS, no nodal recurrence 44 32 24 21 14 9 0
SNB, true neg. 104 92 78 61 46 26 0
SNB, pos. 57 43 34 28 22 15 0
SNB, false neg. 12 8 3 2 2 0 0
1. SNB, neg. vs. OBS, no nodal recurrence: HR, 1.18 (95% CI, 0.63–2.18); P=0.61
2. SNB, pos. vs. SNB, neg.: HR, 2.20 (95% CI, 1.29–3.73); P=0.004
3. SNB, pos. vs. OBS, nodal recurrence: HR, 0.92 (95% CI, 0.53–1.60); P=0.78
4. SNB, false neg. vs. OBS, nodal recurrence: HR, 1.96 (95% CI, 0.90–4.27); P=0.09

rate was more common in patients in the CLND arm (24% vs. 6%, $p < 0.001$).

Similar results were seen in the DeCOG-SLT study. [30] Although the study closed early due to a reduced accrual rate and unexpectedly low event rate, the distant metastasis-free survival (DMFS) was identical between the two treatment groups at 3 years: 77.0 ± 5.1% in the observation group and 74.9 ± 5.4% in the CLND group. The same outcomes were seen for both overall survival (81.7 ± 4.9% vs. 81.2 ± 5.1%) and recurrence-free survival (67.4 ± 4.8% vs. 66.8 ± 5.9%). The node-only recurrence rate was 4.9% at 3 years with no significant difference seen between the two groups. A total of 65.8% patients had SLNB deposits ≤1 mm and this group demonstrated a significantly improved DMFS in multivariate analysis (hazard ratio 2.33; 90% CI: 1.63–3.34; $p < 0.0001$).

31.7 Classification of the Positive Sentinel Lymph Node Biopsy

The MSLT-1, MSLT-2 and DeCOG-SLT studies utilised the crude binary outcome of SN+ or SN− for analysis or entry criteria, however, centres providing SLNB services rapidly began to appreciate that more refined prognostic information can be obtained from a more granular analysis of the phenotypical appearance of the tumour deposits within the SLNBs. These additional measurements were investigated in the later surgical trials and have become incorporated into recent iterations of the AJCC classification system for melanoma.

31.7.1 TNM (N-Stage)

The simplest classification system for SN+ disease is the absolute number of nodes. This gives a relatively crude measurement of nodal tumour burden and was the first refinement to be incorporated into the AJCC classification system. [31, 32] Ostensibly, the classification system gives a good separation of survival outcomes, with the N-stage being categorised as N1, N2 or N3 for involvement of 1, 2–3 or ≥4 nodes, respectively. The issue is that the distribution of the nodal burden is skewed. Both the MSLT-2 studies demonstrated that the N1 disease comprises ~80% of the SN+ cases, with N3 disease only 1%. Furthermore, the nodal burden of the SN+ and the Breslow thickness of the primary are not independent of each other. [2, 33, 34] These issues are highlighted in the AJCC eighth edition classification system for melanoma. [1] Inspection of the survival curves reveals that there is no clinically relevant survival difference between the N1 and N2 stages (Fig. 31.3a, b), which pertains to 99% of the SN+ cohort. In addition, the classification system for stage III disease takes into account the Breslow thickness and ulceration status of the primary melanoma, highlighting the competing and intertwined risks of primary tumour burden and nodal tumour burden in determining overall patient prognosis.

31.7.2 Rotterdam Criteria

The Rotterdam criteria uses a very simple method of measuring the maximum diameter of the largest deposit in the SN+s and sub-classifying the measurement into three groups; <0.1 mm, 0.1–1 mm and >1 mm. [35, 36] Using this classification, a multicentre multicentre study involving 388 SLN-positive melanoma patients reported that among the 10% of patients with SLN micrometastases of less 0.1 mm, only 3% had additional positive non-SLNs. Furthermore, patients with SN+ <0.1 mm had the same prognosis as the SN-patients. [36] The prognostic relevance of the Rotterdam criteria was also validated in the DeCOG-SLT, with a diameter >1 mm being a significant and independent predictor for worse DMFS (HR: 2.33; 90% CI: 1.63–3.34; $p < 0.0001$), overall survival (HR: 3.46; 90% CI: 1.93–6.21; $p < 0.0001$) and recurrence-free survival (HR: 1.85; 90% CI: 1.34–2.55; $p = 0.002$) on multivariate analysis. [30] However, these observations were not replicated in the analysis undertaken in the John Wayne Cancer Institute, where the incidence of subclinical SN+ detected by RT-PCR was 30%. [37] This group had a significantly worse recurrence-free and for overall survival (hazard ratio: 11.42; 95% CI, 3.17–41.1; $p = 0.0002$) on multivariate analysis. Data from the AJCC

Fig. 31.2 Estimated 10-year incidence of nodal metastasis and melanoma-specific survival, according to study group, melanoma thickness and presence or absence of nodal recurrence. Panels (**a**) and (**b**) show the cumulative incidence of nodal metastasis at 10 years among patients with intermediate-thickness melanomas and those with thick melanomas, respectively. Data are from the per-protocol analysis; patients who left the study or were lost to follow-up were excluded. In the biopsy group, nodal recurrence in patients whose sentinel nodes were negative for tumour (i.e. false-negative biopsy results) was assessed. Plus–minus values are means ±SE for the estimated rate of nodal metastasis. The abbreviation neg. denotes negative, and pos. positive. Panels (**c**) and (**d**) show the probability of melanoma-specific survival (i.e. survival until death from melanoma) among patients with intermediate-thickness melanomas and those with thick melanomas, respectively. Numbers (1–4) to the right of the survival curves refer to the numbered comparisons below each graph. Plus–minus values are means ±SE for the estimated probability of melanoma-specific survival. (From: Morton DL, Thompson JF, Cochran AJ, et al. Final Trial Report of Sentinel-Node Biopsy versus Nodal Observation in Melanoma. N Engl J Med. 2014;370(7):599–609 Copyright © (2014) Massachusetts Medical Society. Reprinted with permission)

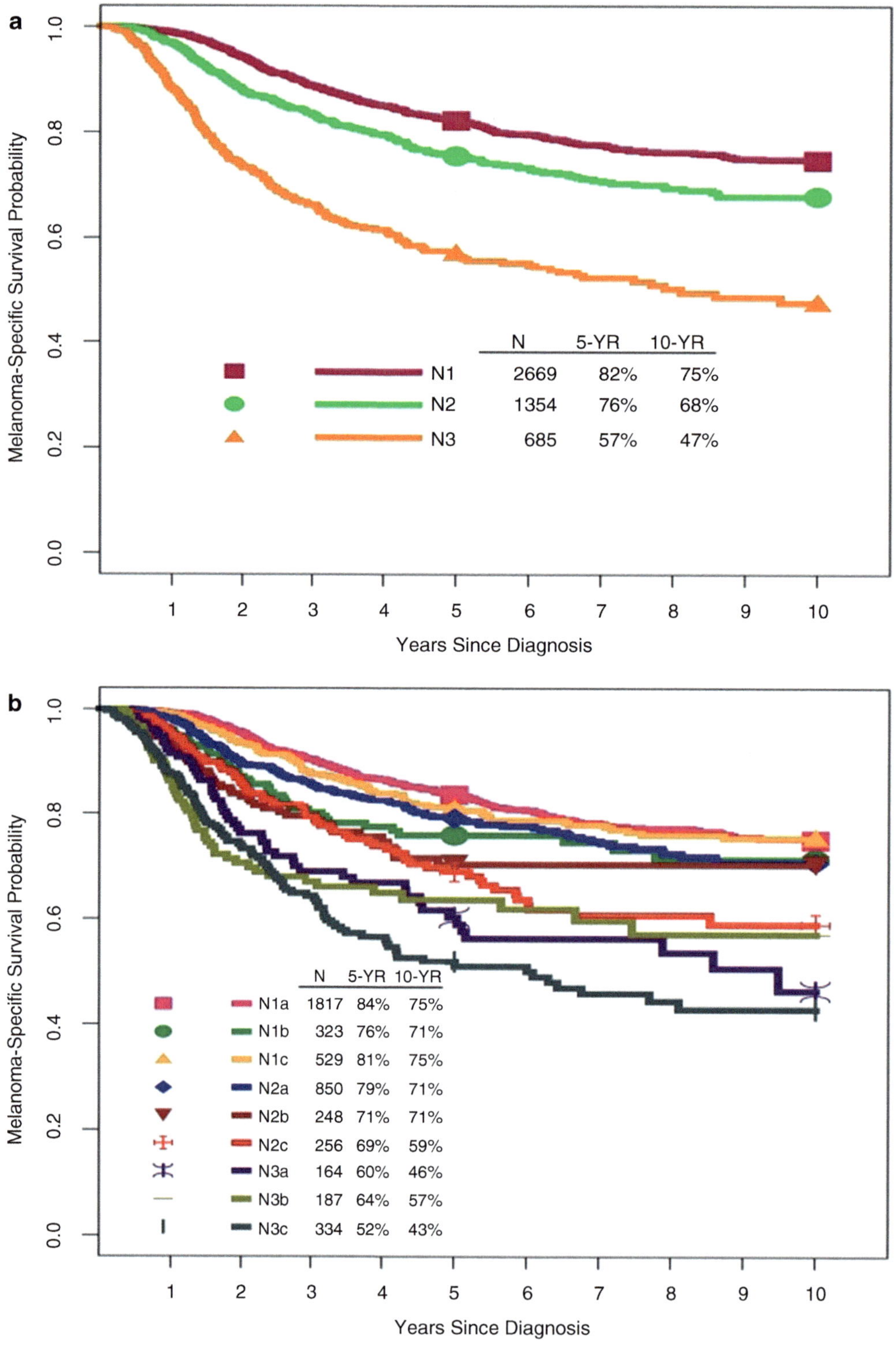

Fig. 31.3 Kaplan–Meier melanoma-specific survival curves according to (**a**) N categories, (**b**) subcategories and (**c**) maximum dimension of sentinel lymph node metastatic focus (mm) from the eighth edition international melanoma database. (Note that there were insufficient data (<10 cases) to estimate 10-year melanoma-specific survival for patients who had a maximum sentinel lymph node metastatic focus of 2–4 mm). From the Eighth Edition International Melanoma Database. (From: Gershenwald JE, Scolyer RA, Hess KR, et al. Melanoma staging: Evidence-based changes in the American Joint Committee on Cancer eighth edition cancer staging manual. CA Cancer J Clin. 2017;67(6):472–492. © 2017 American Cancer Society. Reprinted with permission)

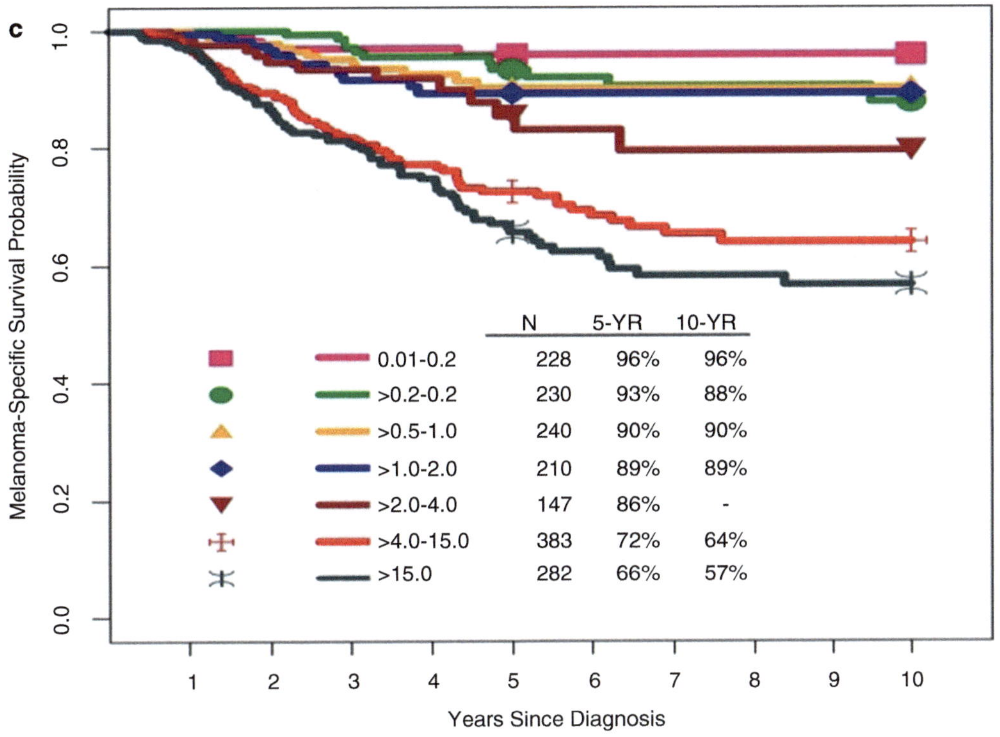

Fig. 31.3 (continued)

eighth edition classification system (Fig. 31.3c) demonstrated that the nodal tumour burden has significant prognostic implications. The 10-year melanoma-specific survival for patients with tumour deposits <0.2 mm and 1.01–2 mm were 96% and 89%, respectively.

31.7.3 Dewar's Microanatomic Classification

Dewar *et al.* developed a separate classification system to determine patients eligible for CLND. [38] The SN+ was assessed according to its microanatomic location within the node (subcapsular, combined subcapsular and parenchymal, parenchymal, multifocal or extensive), and this was correlated with the presence of involved non-sentinel nodes. They identified that the risk of a positive non-SLN increased with changing microanatomical involvement of the node 0% (subcapsular), 11.1% (combined subcapsular and parenchymal), 18.8% (parenchymal), 36.8% (multifocal), 42.1% (extensive—extracapsular or metastasis >5 mm). A separate study combined both the Rotterdam criteria and Dewar's classification which identified low-risk patients in particular with a SN+, where the Kaplan–Meier 5- and 10-year MSS rates were both 95% for patients with Rotterdam criteria <0.1 mm, confined to the subcapsular space. [39] Concerns have been raised about the reproducibility of Dewar's classification. One study identified that, whilst the interobserver agreement for quantifiable nodal measurements, such as the Rotterdam criteria was good to excellent, the agreement for the identification of subcapsular non-capsular deposits was only fair. [40] A separate study [39] showed that the several parameters of SN tumour burden including intranodal location, tumour-penetrating depth and maximum size provided prognostic information, but their prognostic significance varied considerably between the different centres, leading the authors to conclude that further studies are required, 'to establish a classification system for SN tumour burden that consistently and accurately stratifies patients into meaningful prognostic groups'.

31.7.4 Extracapsular Spread

Data from two UK centres showed that SN+ can be alternatively stratified according to the presence of extracapsular spread (ECS). [41] ECS is a well-known risk factor for melanoma recurrence in macroscopic disease, but was an exclusion criterion for both the MSLT-2 and DeCOG-SLT studies. Accordingly, clinical data and treatment guidance were missing for this subgroup. In a sample of 265 SN+ patients, the incidence of ECS was 16.2%. When the cohort was compared with a contemporaneous sample of 250 patients with palpable disease, the researchers found that, where ECS was absent, there was a significantly increased median number of nodes in the palpable disease cohort (1 node vs. 2 nodes; $p < 0.0001$), as with the MSLT-1 study. Whereas, the cohort where ECS was present, the median number of nodes was the same for both SN+ and the palpable disease cohorts (3, both). Furthermore, when the MMS curves were compared between the two groups, the presence of ECS in the microscopic involved nodal group, upstaged these N+ patients to the same outcome as the palpable disease group, in a similar fashion seen in the AJCC classification system curves when analysing the presence of ulceration of the primary tumour. Multivariate analysis demonstrated that the presence of ECS was the most important independent predictor of overall survival (HR: 2.47; 95% CI 1.87–3.26; $p < 0.0001$). These findings led the researchers to hypothesise that the presence of ECS in the sentinel node is a biomarker for a high-risk genotype and led them to coin the term '*nodal ulceration*' to describe the entity and its behaviour. Finally, the authors suggest that substitution of the N-Stage for ECS status in the AJCC eighth edition stage III classification system identifies low-risk, intermediate Breslow thickness melanoma, SN+ patients, which may guide further clinical management.

31.7.5 Molecular Analysis

The treatment of metastatic melanoma with systemic therapy is an established standard of care for AJCC stage III and IV disease. [42] Phase III clinical trials have also demonstrated a relapse-free survival benefit for resected stage III patients treated with adjuvant systemic therapy, either with targeted MAP-Kinase pathway inhibitors for *BRAF*-mutant melanomas [43] or with checkpoint inhibitors. [44, 45] The incidence of *BRAF* mutations in metastatic melanoma is ~45–50% [46] and the incidence of *NRAS* mutations is ~15–20%, [47] though the latter are typically classified as *BRAF* wild type, despite the known true mutation. Limited studies have suggested that the presence of *BRAF* mutation in resected stage III disease is associated with a worse prognosis. [48–50] Current guidelines recommend testing to determine *BRAF* mutation status on the SLNB metastasis, to direct future adjuvant systemic therapy. [51] Rapid and reliable results can be achieved with immunohistochemistry (IHC) markers specific for *BRAF* V600E mutations, with molecular testing reserved for the IHC-negative samples to ensure V600K mutations are not overlooked. [52, 53]

31.8 Current Role of Sentinel Lymph Node Biopsy in the Management of Cutaneous Melanoma

What information does the outcome data from these three 'landmark' surgical trials provide for the surgeon managing primary cutaneous melanoma? The MSLT-1 study confirmed the findings of several studies, including the datasets used to produce previous iterations of the AJCC staging system for melanoma, [31, 54] that the SLNB status is the single most important prognostic biomarker for cutaneous melanoma. Accordingly, the current challenge remains in identifying subgroups of patients who are at significant risk of harbouring a micrometastasis in their sentinel node. The current AJCC staging system has suggested that the threshold for offering SLNB in thinner melanomas is where the risk of SN+ is ≥5%, as indicated by the T1b subcategory. [1]

The prognostic significance of SLNB is particularly true for thin melanoma (0.8–1.0 mm, AJCC IB), and for intermediate thickness melanoma, (1.0–4 mm thick, AJCC IB-IIA). [1] For this group, the outcome of the SLNB has significant management implications, since the risk of distant relapse or death for the sentinel node negative (SN−) patients is extremely low, such that a negative SLNB result in the 0.8–1.0 mm group indicates a downstaging to AJCC IA. [1] An ongoing phase III surgical RCT reported its interim results, [55] which demonstrated that a low-frequency, patient-directed follow-up programme for SN- patients had a very high satisfaction rate (>93%) and no difference in quality of life scores when compared with a nationally directed, higher frequency follow-up regimen. Importantly, there was no difference in recurrence detection rates, which was by the patient in 70% of cases, regardless of which follow-up regimen was employed. This study, and its reciprocal study in the Netherlands, [56] has important implications for service planning and resource utilisation for cancer centres managing melanoma patients, particularly since the MSLT-1 study demonstrated that the great majority of SLNB are negative and the prognosis for this group is excellent.

The current management of SN+ in the thin and intermediate thickness group is still evolving. The MSLT-1 study demonstrated that the risk of recurrence is increased, which would suggest that a regular systemic imaging schedule is indicated, [57] though even this is controversial for some. Whilst the MSLT-II and DeCOG-SLT studies utilised regional ultrasound as the follow-up imaging modality in their protocols, the efficacy of this investigation is questionable given that both studies also found the incidence of 'node-only' relapse was very low. The MSLT-1 study was highly suggestive of a therapeutic benefit from removing the micrometastatic focus in the sentinel node for a subgroup of patients in the intermediate risk group (post hoc analysis 1–4 mm). It is clear from the MSLT-2 and DeCOG-SLT studies, that no further therapeutic benefit can be derived by undertaking a CLND, except in extenuating circumstances. Accordingly, one of the modern challenges for clinicians remains to identify those SN+ patients with thin or intermediate thickness melanomas who have been adequately treated with surgery alone, whilst considering adjuvant systemic therapy for those patients in this group who are at risk of further relapse. Data from the DeCOG-SLT study would suggest that patients with tumour deposit(s) greater than 1 mm are at increased risk of relapse, though data from the AJCC eighth edition analysis would suggest that 10-year survival of patients with tumour deposits 0.2–0.5 mm, 0.5–1 mm and 1—2 mm are identical. [1] An alternative strategy taking into account the ECS status of the SN+ has been suggested instead. [41]

The role of SLNB for thick and high-risk melanomas would appear to be a more straightforward process of identifying patients to be referred for adjuvant systemic therapy. This treatment strategy for this group may also evolve, however, based on the future outcome of data from two phase III adjuvant systemic immunotherapy trials for AJCC IIB-IIC (eighth edition) patients ('KeyNote-716', ClinicalTrials.gov Identifier: NCT03553836 and 'CheckMate-76 K', ClinicalTrials.gov Identifier: NCT04099251). The MSLT-1 study showed no survival benefit from removing the positive sentinel node(s) for patients with primary melanomas thicker than 3.5 mm. The likely explanation is that the SN+ nodal status is a biomarker for systemic disease in this subgroup of patients, particularly if this is associated with ECS [41]. Accordingly, the emphasis is placed on identifying all patients at higher risk of metastatic disease, and the role of SLNB has become one that now influences access to adjuvant systemic therapy. [51, 58] Furthermore, the last two iterations of the AJCC systems classified the presence of microsatellites at the primary site as stage IIIB-C disease, an indication for direct referral for adjuvant systemic therapy. Accordingly, it is hard to justify the use of SLNB for this group of patients.

31.9 Sentinel Node Biopsy for Other Skin Cancers

The overwhelming majority of the literature describing SLNB for cutaneous malignancy is dedicated to melanoma. The main reasons that SLNB is an effective treatment strategy for melanoma are that the primary is discrete measuring <10 mm in diameter, it is a common tumour, nearly 50% of patients diagnosed are <65 years old, and it is very unusual for patients to have more than one melanoma in their lifetime.

31.9.1 Merkel Cell Carcinoma

Merkel cell carcinoma (MCC) is a very rare tumour that usually develops on the head and neck region. It commonly affects the elderly with a mortality rate of 40%. [59] Rapid growth of the primary is common, with the tumour usually >15 mm in diameter at diagnosis. [60] UV-B radiation exposure and polyomavirus virus infection are proposed as the two major aetiologies. [59] Unlike melanoma, MCC is a radiosensitive tumour, which affords the advantage of further treatment options in this often-frail demographic. [61] Very little RCT data exist on this disease; however, retrospective case series from large institutions suggest that either surgery or primary radiotherapy is equally effective in the management of both the primary disease and nodal metastases. [62–64] SLNB for MCC is a controversial subject, [65] but large retrospective case series indicate that the incidence of SN+ disease is common (~20–40%) and the finding of a SN+ helps to guide adjuvant therapy in the context of the patient's underlying health status. On a technical level, one would caution the accuracy of the SLNB in the head and neck region and radiological screening with PET-CT is recommended for surveillance. [66] Furthermore, consideration should be given to omitting intraoperative blue dye injection, which causes highly visible dermal staining, if treatment of the primary site with radiotherapy alone is likely.

31.9.2 Cutaneous SCC

Cutaneous squamous cell carcinoma (cSCC) is the second most common skin malignancy. It routinely affects the elderly and most commonly develops on the head and neck region. Multiple primaries are very common and rapid growth of the primary resulting in large lesions at presentation are frequently seen. Unlike MCC and melanoma, the overall incidence of lymph node metastases is low, though nodal metastases are associated with a poor prognosis. When targeting high-risk head and neck cSCC, the incidence of SN+ disease was ~15% in two single-centre studies. [67, 68] The incidence of false negativity was 22–45% indicating the limitations of SLNB accuracy in the head and neck region. Given the limitations of accuracy of SLNB for cSCC, the lack of high-level evidence for the utility of the procedure, and the lack of consensus regarding the management of a SN+ and the lack of consensus regarding the definition of a high-risk primary, [69] there are no current standardised guidelines for SLNB for cSCC. [70]

31.10 Conclusion

High-level evidence and consistent international experience has firmly established the role of SLNB for the staging and management of primary cutaneous melanoma, though the paradigm has permanently shifted from a role of identifying patients suitable for regional surgery to one of identifying patients eligible for adjuvant systemic therapy. SLNB for Merkel cell carcinoma currently has a role for directing regional therapy to the draining nodal field, though high-level evidence for the efficacy of the procedure is currently lacking. Evidence is lacking for the utility of SLNB for cutaneous SCC, accordingly it is not routinely recommended for this group of patients at present.

Open Questions

- Can methods of patient selection for sentinel lymph node biopsy in melanoma be refined with standard clinicopathologic variables?
- Can methods of patient selection for patients with sentinel node melanoma metastases be refined to identity low risk patients who can be observed and high risk patients who may benefit from adjuvant systemic therapy?
- What is the role of sentinel node lymph node biopsy for patients with primary Merkel cell carcinoma?
- What is the role of sentinel lymph node biopsy in cutaneous squamous cell carcinoma and in which patients?

References

1. Gershenwald JE, Scolyer RA, Hess KR, et al. Melanoma staging: evidence-based changes in the American Joint Committee on Cancer eighth edition cancer staging manual. CA Cancer J Clin. 2017;67(6):472–92.
2. Sondak VK, Taylor JMG, Sabel MS, et al. Mitotic rate and younger age are predictors of sentinel lymph node positivity: lessons learned from the generation of a probabilistic model. Ann Surg Oncol. 2004;11(3):247–58.
3. Balch CM. Surgical management of regional lymph nodes in cutaneous melanoma. J Am Acad Dermatol. 1980;3(5):511–24.
4. Veronesi U, Adamus J, Bandiera DC, et al. Stage I melanoma of the limbs. Immediate versus delayed node dissection. Tumori. 1980;66(3):373–96.
5. Veronesi U, Adamus J, Bandiera DC, et al. Inefficacy of immediate node dissection in stage 1 melanoma of the limbs. N Engl J Med. 1977;297(12):627–30.
6. Sim FH, Taylor WF, Pritchard DJ, Soule EH. Lymphadenectomy in the management of stage I malignant melanoma: a prospective randomized study. Mayo Clin Proc. 1986;61(9):697–705.
7. Cascinelli N, Morabito A, Santinami M, MacKie RM, Belli F. Immediate or delayed dissection of regional nodes in patients with melanoma of the trunk: a randomised trial. WHO Melanoma Programme. Lancet. 1998;351(9105):793–6.
8. Essner R, Conforti A, Kelley MC, et al. Efficacy of lymphatic mapping, sentinel lymphadenectomy, and selective complete lymph node dissection as a therapeutic procedure for early-stage melanoma. Ann Surg Oncol. 1999;6(5):442–9.
9. Balch CM, Soong S-J, Bartolucci AA, et al. Efficacy of an elective regional lymph node dissection of 1 to 4 mm thick melanomas for patients 60 years of age and younger. Ann Surg. 1996;224(3):255.
10. Morton DL, Wen DR, Wong JH, et al. Technical details of intraoperative lymphatic mapping for early stage melanoma. Arch Surg. 1992;127(4):392–9.
11. Wong JH, Cagle LA, Morton DL. Lymphatic drainage of skin to a sentinel lymph node in a feline model. Ann Surg. 1991;214(5):637–41.
12. Alex JC, Weaver DL, Fairbank JT, Rankin BS, Krag DN. Gamma-probe-guided lymph node localization in malignant melanoma. Surg Oncol. 1993;2(5):303–8.
13. Thompson JF, McCarthy WH, Bosch CM, et al. Sentinel lymph node status as an indicator of the presence of metastatic melanoma in regional lymph nodes. Melanoma Res. 1995;5(4):255–60.
14. Reintgen D, Cruse CW, Wells K, et al. The orderly progression of melanoma nodal metastases. Ann Surg. 1994;220(6):759–67.
15. Albertini JJ, Cruse CW, Rapaport D, et al. Intraoperative radio-lympho-scintigraphy improves sentinel lymph node identification for patients with melanoma. Ann Surg. 1996;223(2):217–24.
16. Morton DL, Thompson JF, Essner R, et al. Validation of the accuracy of intraoperative lymphatic mapping and sentinel lymphadenectomy for early-stage melanoma: a multicenter trial. Multicenter Selective Lymphadenectomy Trial Group. Ann Surg. 1999;230(4):453–63; discussion 463–465.
17. Wong SL, Balch CM, Hurley P, et al. Sentinel lymph node biopsy for melanoma: American Society of Clinical Oncology and Society of Surgical Oncology joint clinical practice guideline. J Clin Oncol. 2012;30(23):2912–8.
18. Bluemel C, Herrmann K, Giammarile F, et al. EANM practice guidelines for lymphoscintigraphy and sentinel lymph node biopsy in melanoma. Eur J Nucl Med Mol Imaging. 2015;42(11):1750–66.
19. Quartuccio N, Siracusa M, Pappalardo M, Arnone A, Arnone G. Sentinel node identification in melanoma: current clinical impact, new emerging SPECT radiotracers and technological advancements. An update of the last decade. Curr Radiopharm. 2020;13(1):32–41.
20. Jimenez-Heffernan A, Ellmann A, Sado H, et al. Results of a prospective multicenter International Atomic Energy Agency sentinel node trial on the value of SPECT/CT over planar imaging in various malignancies. J Nucl Med. 2015;56(9):1338–44.
21. Moncrieff MD, O'Leary FM, Beadsmoore CJ, et al. Effect of delay between nuclear medicine scanning and sentinel node biopsy on outcome in patients with cutaneous melanoma. Br J Surg. 2020;107(6):669–76.
22. Sondak VK, King DW, Zager JS, et al. Combined analysis of phase III trials evaluating [^{99m}Tc]tilmanocept and vital blue dye for identification of sentinel lymph nodes in clinically node-negative cutaneous melanoma. Ann Surg Oncol. 2013;20(2):680–8.
23. Knackstedt R, Couto RA, Ko J, Cakmakoglu C, Wu D, Gastman B. Indocyanine green fluorescence imaging with lymphoscintigraphy for sentinel node biopsy in melanoma: increasing the sentinel lymph node-positive rate. Ann Surg Oncol. 2019;26(11):3550–60.
24. Anninga B, White SH, Moncrieff M, et al. Magnetic technique for sentinel lymph node biopsy in melanoma: the MELAMAG trial. Ann Surg Oncol. 2016;23(6):2070–8.
25. Morton DL, Wanek L, Nizze JA, Elashoff RM, Wong JH. Improved long-term survival after lymphadenectomy of melanoma metastatic to regional nodes. Analysis of prognostic factors in 1134 patients from the John Wayne Cancer Clinic. Ann Surg. 1991;214(4):491–9; discussion 499–501.
26. Morton DL, Thompson JF, Cochran AJ, et al. Final trial report of sentinel-node biopsy versus nodal observation in melanoma. N Engl J Med. 2014;370(7):599–609.
27. Bello DM, Faries MB. The landmark series: MSLT-1, MSLT-2 and DeCOG (management of lymph nodes). Ann Surg Oncol. 2020;27(1):15–21.
28. Faries MB, Thompson JF, Cochran A, et al. The impact on morbidity and length of stay of early versus delayed complete lymphadenectomy in melanoma: results of the multicenter selective lymphadenectomy trial (I). Ann Surg Oncol. 2010;17(12):3324–9.
29. Faries MB, Thompson JF, Cochran AJ, et al. Completion dissection or observation for sentinel-node metastasis in melanoma. N Engl J Med. 2017;376(23):2211–22.
30. Leiter U, Stadler R, Mauch C, et al. Complete lymph node dissection versus no dissection in patients with sentinel lymph node

biopsy positive melanoma (DeCOG-SLT): a multicentre, randomised, phase 3 trial. Lancet Oncol. 2016;17(6):757–67.
31. Balch CM, Soong S-J, Gershenwald JE, et al. Prognostic factors analysis of 17,600 melanoma patients: validation of the American Joint Committee on Cancer melanoma staging system. J Clin Oncol. 2001;19(16):3622–34.
32. Balch CM, Buzaid AC, Soong SJ, et al. Final version of the American Joint Committee on cancer staging system for cutaneous melanoma. J Clin Oncol. 2001;19(16):3635–48.
33. Sabel MS, Griffith K, Sondak VK, et al. Predictors of nonsentinel lymph node positivity in patients with a positive sentinel node for melanoma. J Am Coll Surg. 2005;201(1):37–47.
34. Quaglino P, Ribero S, Osella-Abate S, et al. Clinico-pathologic features of primary melanoma and sentinel lymph node predictive for non-sentinel lymph node involvement and overall survival in melanoma patients: a single centre observational cohort study. Surg Oncol. 2011;20(4):259–64.
35. van Akkooi ACJ, Nowecki ZI, Voit C, et al. Sentinel node tumor burden according to the Rotterdam criteria is the most important prognostic factor for survival in melanoma patients: a multicenter study in 388 patients with positive sentinel nodes. Ann Surg. 2008;248(6):949–55.
36. van Akkooi ACJ, de Wilt JHW, Verhoef C, et al. Clinical relevance of melanoma micrometastases (<0.1 mm) in sentinel nodes: are these nodes to be considered negative? Ann Oncol. 2006;17(10):1578–85. https://doi.org/10.1093/annonc/mdl176.
37. Takeuchi H, Morton DL, Kuo C, et al. Prognostic significance of molecular upstaging of paraffin-embedded sentinel lymph nodes in melanoma patients. J Clin Oncol. 2004;22(13):2671–80.
38. Dewar DJ, Newell B, Green MA, Topping AP, Powell BWEM, Cook MG. The microanatomic location of metastatic melanoma in sentinel lymph nodes predicts nonsentinel lymph node involvement. J Clin Oncol. 2004;22(16):3345–9.
39. van der Ploeg APT, van Akkooi ACJ, Rutkowski P, et al. Prognosis in patients with sentinel node-positive melanoma is accurately defined by the combined Rotterdam tumor load and Dewar topography criteria. J Clin Oncol. 2011;29(16):2206–14.
40. Murali R, Cochran AJ, Cook MG, et al. Interobserver reproducibility of histologic parameters of melanoma deposits in sentinel lymph nodes: implications for management of patients with melanoma. Cancer. 2009;115(21):5026–37.
41. Lo M, Robinson A, Wade R, et al. Extracapsular spread in melanoma lymphadenopathy: prognostic implications, classification, and management. Ann Surg Oncol. 2021;28(3):1642–53. https://doi.org/10.1245/s10434-020-09099-w.
42. Baetz TD, Fletcher GG, Knight G, et al. Systemic adjuvant therapy for adult patients at high risk for recurrent melanoma: a systematic review. Cancer Treat Rev. 2020;87:102032.
43. Dummer R, Hauschild A, Santinami M, et al. Five-year analysis of adjuvant Dabrafenib plus Trametinib in stage III melanoma. N Engl J Med. 2020;383(12):1139–48.
44. Weber J, Mandala M, Del Vecchio M, et al. Adjuvant Nivolumab versus Ipilimumab in resected stage III or IV melanoma. N Engl J Med. 2017;377:1824–35. https://doi.org/10.1056/NEJMoa1709030.
45. Eggermont AMM, Blank CU, Mandala M, et al. Adjuvant Pembrolizumab versus placebo in resected stage III melanoma. N Engl J Med. 2018;378(19):1789–801.
46. Greaves WO, Verma S, Patel KP, et al. Frequency and spectrum of BRAF mutations in a retrospective, single-institution study of 1112 cases of melanoma. J Mol Diagn. 2013;15(2):220–6.
47. Adler NR, Wolfe R, Kelly JW, et al. Tumour mutation status and sites of metastasis in patients with cutaneous melanoma. Br J Cancer. 2017;117(7):1026–35.
48. Barbour AP, Tang YH, Armour N, et al. BRAF mutation status is an independent prognostic factor for resected stage IIIB and IIIC melanoma: implications for melanoma staging and adjuvant therapy. Eur J Cancer. 2014;50(15):2668–76.
49. Picard M, Pham Dang N, D'Incan M, et al. Is BRAF a prognostic factor in stage III skin melanoma? A retrospective study of 72 patients after positive sentinel lymph node dissection. Br J Dermatol. 2014;171(1):108–14.
50. Heppt MV, Siepmann T, Engel J, et al. Prognostic significance of BRAF and NRAS mutations in melanoma: a German study from routine care. BMC Cancer. 2017;17(1):536.
51. Coit DG, Thompson JA, Albertini MR, et al. Cutaneous melanoma, version 2.2019, NCCN clinical practice guidelines in oncology. J Natl Compr Cancer Netw. 2019;17(4):367–402.
52. Lo MCI, Paterson A, Maraka J, et al. A UK feasibility and validation study of the VE1 monoclonal antibody immunohistochemistry stain for BRAF-V600E mutations in metastatic melanoma. Br J Cancer. 2016;115(2):223–7.
53. Long GV, Wilmott JS, Capper D, et al. Immunohistochemistry is highly sensitive and specific for the detection of V600E BRAF mutation in melanoma. Am J Surg Pathol. 2013;37(1):61–5.
54. Balch CM, Gershenwald JE, Soong S-J, et al. Final version of 2009 AJCC melanoma staging and classification. J Clin Oncol. 2009;27(36):6199–206.
55. Moncrieff MD, Underwood B, Garioch JJ, et al. The MelFo study UK: effects of a reduced-frequency, stage-adjusted follow-up schedule for cutaneous melanoma 1B to 2C patients after 3-years. Ann Surg Oncol. 2020;27(11):4109–19. https://doi.org/10.1245/s10434-020-08758-2.
56. Deckers EA, Hoekstra-Weebers JEHM, Damude S, et al. The MELFO study: a multicenter, prospective, randomized clinical trial on the effects of a reduced stage-adjusted follow-up schedule on cutaneous melanoma IB-IIC patients-results after 3 years. Ann Surg Oncol. 2020;27(5):1407–17. https://doi.org/10.1245/s10434-019-07825-7.
57. Stodell M, Thompson JF, Emmett L, Uren RF, Kapoor R, Saw RPM. Melanoma patient imaging in the era of effective systemic therapies. Eur J Surg Oncol. 2017;43(8):1517–27.
58. Peach H, Board R, Cook M, et al. Current role of sentinel lymph node biopsy in the management of cutaneous melanoma: a UK consensus statement. J Plast Reconstr Aesthet Surg. 2020;73(1):36–42.
59. Becker JC, Stang A, Hausen AZ, et al. Epidemiology, biology and therapy of Merkel cell carcinoma: conclusions from the EU project IMMOMEC. Cancer Immunol Immunother. 2018;67(3):341–51.
60. Smith FO, Yue B, Marzban SS, et al. Both tumor depth and diameter are predictive of sentinel lymph node status and survival in Merkel cell carcinoma. Cancer. 2015;121(18):3252–60.
61. Strom T, Carr M, Zager JS, et al. Radiation therapy is associated with improved outcomes in Merkel cell carcinoma. Ann Surg Oncol. 2016;23(11):3572–8.
62. Gunaratne DA, Howle JR, Veness MJ. Sentinel lymph node biopsy in Merkel cell carcinoma: a 15-year institutional experience and statistical analysis of 721 reported cases. Br J Dermatol. 2016;174(2):273–81.
63. Perez MC, Oliver DE, Weitman ES, et al. Management of sentinel lymph node metastasis in Merkel cell carcinoma: completion lymphadenectomy, radiation, or both? Ann Surg Oncol. 2019;26(2):379–85.
64. Lee JS, Durham AB, Bichakjian CK, et al. Completion lymph node dissection or radiation therapy for sentinel node metastasis in Merkel cell carcinoma. Ann Surg Oncol. 2019;26(2):386–94.
65. Hruby G, Guminski A, Thompson JF. Management of regional lymph nodes in patients with Merkel cell carcinoma following a positive sentinel node biopsy: less may be more, but is either enough? Ann Surg Oncol. 2019;26(2):315–7.

66. Bichakjian CK, Olencki T, Aasi SZ, et al. Merkel cell carcinoma, version 1.2018, NCCN clinical practice guidelines in oncology. J Natl Compr Cancer Netw. 2018;16(6):742–74.
67. Gore SM, Shaw D, Martin RCW, et al. Prospective study of sentinel node biopsy for high-risk cutaneous squamous cell carcinoma of the head and neck. Head Neck. 2016;38(Suppl 1):E884–9.
68. Durham AB, Lowe L, Malloy KM, et al. Sentinel lymph node biopsy for cutaneous squamous cell carcinoma on the head and neck. JAMA Otolaryngol Head Neck Surg. 2016;142(12):1171–6.
69. Ahadiat O, Higgins S, Sutton A, Ly A, Wysong A. SLNB in cutaneous SCC: a review of the current state of literature and the direction for the future. J Surg Oncol. 2017;116(3):344–50.
70. Allen JE, Stolle LB. Utility of sentinel node biopsy in patients with high-risk cutaneous squamous cell carcinoma. Eur J Surg Oncol. 2015;41(2):197–200.

Sentinel Lymph Node Biopsy in Breast Cancer

32

Emily Siegel, John Kiluk, Armando Giuliano, and Brian Czerniecki

Abstract

Lymphatic metastasis is a strong predictor of poor outcome in breast cancer patients. Most often, lymphatic metastasis occurs through the axillary lymph nodes. While axillary dissection (ALND) was previously routine, sentinel lymph node biopsy (SLNB) has become the standard of care for clinically and pathologically node-negative breast cancer given its reduced morbidity. SLNB has replaced ALND for node-positive disease in patients with small tumors undergoing breast conservation therapy and is rapidly evolving to replace ALND in patients who are converted from node positive to node negative with the use of neoadjuvant chemotherapy. Controversy remains about whether axillary staging, and therefore SLNB may be omitted in a certain subset of patients, such as the elderly and those with exceptional responses to neoadjuvant therapy.

Learning Objectives

- To be able to describe the effectiveness of SLNB to accurately stage the axilla.
- To understand the utility of SLNB in lymph node-positive breast cancer patients.
- To understand the current controversies in the scope of SLNB compared to ALND.

E. Siegel · J. Kiluk · B. Czerniecki (✉)
Breast Department, H. Lee Moffitt Cancer Center, Tampa, FL, USA
e-mail: emily.siegel@moffitt.org; John.Kiluk@moffitt.org; brian.czerniecki@moffitt.org

A. Giuliano
Department of Surgery, Cedars-Sinai Medical Center, Los Angeles, CA, USA
e-mail: armando.giuliano@cshs.org

32.1 Introduction

Lymphatic metastasis is an important prognostic indicator for outcomes in breast cancer. The evaluation of lymph nodes and axillary staging is a vital tool in the treatment of patients with breast cancer and can be accomplished in multiple ways. The surgical assessment of lymphatic metastasis for breast cancer has evolved significantly since first introduced in the 1970s and can now be achieved in many circumstances with axillary sentinel lymph node biopsy.

32.2 History of SLNB and Technical Considerations

Lymphatic metastasis is a strong prognostic indicator of poor outcome in breast cancer. Nodal metastasis occurs through a variety of mechanisms and in a stepwise pattern which begins with tumor lymphangiogenesis. Cancer cells then invade from the tumor to lymphatic vessels at the site of primary disease, from which they travel to the sentinel lymph node. There, the cancer cells survive through immunogenicity and nodal immunosuppression [1].

Given its prognostic indication, axillary nodal status is an important factor for guiding adjuvant treatment decisions. Surgical identification of axillary lymph nodes has long been the gold standard of axillary staging, though the scope of axillary surgery in breast cancer has changed dramatically over the past century. In 1977, Morton et al. used cutaneous lymphoscintigraphy to identify lymphatic drainage to nodal basins of melanomas [2]. They also showed that preoperative lymphoscintigraphy could demonstrate a single lymph node receiving drainage [3]. The ability to safely and accurately perform sentinel lymph node biopsy (SLNB) in breast cancer was then demonstrated in 1994 by Giuliano and Morton [4]. The initial studies defined the technique and demonstrated a significant learning curve, with the final patients enrolled exhibiting a 100% predictive rate of negative senti-

S. P. Leong et al. (eds.), *Cancer Metastasis Through the Lymphovascular System*, https://doi.org/10.1007/978-3-030-93084-4_32

nel nodes. A prospective trial by the same group then demonstrated that in addition to technical feasibility, SLNB was safe and highly accurate [5]. Since that time, multiple large-scale studies have validated the technique and feasibility of SLNB [6, 7]. The low rate of locoregional tumor recurrence in patients with negative SLN made a tremendous impact on axillary surgery, further promoting the use of SLNB [8, 9].

To improve the learning curve and increase identification, Cox et al. demonstrated the technique of combination blue-dye and radiocolloid [14]. In early single-centered trials, the combination technique showed a quick learning curve and low false-negative rate [15]. Studies of particle kinetics and comparison studies of different commercially available tracers including albumin, sulfur colloid, and albumin colloid, were completed to classify the best technique [10, 11]. The mechanism by which these tracers are retained in sentinel lymph nodes was demonstrated by Faries et al. Their group demonstrated that antigen-presenting cells rapidly take up radiotracer [12]. These cells, expressing major histocompatibility class II proteins, then appear in sentinel lymph nodes with increasing amounts over time [12]. Since that time, other receptor-binding radiopharmaceuticals have appeared on the market. *Lymphoseek*, specifically designed for SLN mapping, was compared to sulfur colloid, and showed equal uptake in sentinel nodes with faster injection site clearance, though it is not in widespread use given the need for a 2-day protocol and cost associated [13].

Attention then turned to the location of injection. Initial studies advocated for the use of peritumoral injection sites [17, 18]. However, this technique requires preoperative imaging of the breast and injection under radioguidance. In addition, injections for tumors in the upper outer quadrant create difficulty in determining which activity is in the nodes versus in the tumor bed, called "shine through." Therefore, interest in periareolar injections surfaced. In 2004, the first prospective multicenter trial using periareolar injections was published. In this study, patients underwent SLNB and completed axillary dissection. The SLN was successfully identified in 99.3% of patients who received subareolar injections and 95.6% of patients who underwent periareolar injections, significantly higher than the 91.1% of patients who underwent peritumoral injection [19]. Thus, the current technique to identify axillary SLN became periareolar or subareolar injections.

32.2.1 Increasing Use of SLNB in Clinically Node-Negative Patients

The American College of Surgeons Oncology Groups (ACOSOG) Z0010 trial was then completed, a multi-institutional study evaluating both the accuracy and prognostic value of SLNB. In this study, identification of SLN was performed using a combination of radiopharmaceutical and blue dye in 79.4% of cases, blue dye alone in 14.8% cases, and radiopharmaceutical alone in 5.7% of cases, with an overall success rate of 98.7%. The type of dye or radioactive isotope did not affect failure rates as unsuccessful identification with blue dye only (1.4%), radiopharmaceutical only (2.3%), and combined (1.2%) were not statistically different ($p = 0.28$). Further evaluation of isosulfan blue compared to methylene blue was performed and no significant differences were seen in identification rates and adverse reactions [16]. After proving technical feasibility, sentinel lymph node biopsy became the preferred nodal staging procedure for patients undergoing surgery for breast cancer. SLNB is preferred over complete axillary lymph node dissections (ALND) given decreased rates of morbidity. Initial reports in 2000 on small numbers of patients suggested significantly higher patient-reported rates of lymphedema and pain in those who received ALND [20] compared to those treated with SLNB. Further retrospective data confirmed these findings, suggesting that patients with SLNB had 3.2-fold lower risk of pain, five-fold lower risk of lymphedema, and 7.7-fold lower risk of numbness [21]. Prospective multicenter studies followed. One which included 395 patients showed significantly higher arm volume (and therefore potential for lymphedema) and increased shoulder and arm pain in patients with ALND compared to SLNB [22]. The multicenter randomized National Surgical Adjuvant Breast and Bowel Project (NSABP) B-32 trial randomized 5611 pathologically node-negative patients to either SLNB alone or SLNB and completed axillary dissection. Patients in both groups had similar disease-free and overall survival. Furthermore, there was no difference in axillary or regional-node recurrences between groups [23]. Given the utility in node-negative patients, interest in SLNB for clinically node-negative but pathologically node-positive patients grew.

32.2.2 Adequacy of Sentinel Node Biopsy in Node-Positive Patients

The landmark ACOSOG Z0011 trial aimed to determine if SLNB could replace ALND for node-positive patients undergoing breast conservation therapy (BCT). Inclusion criteria were female patients with T1–T2 invasive breast cancer without palpable adenopathy and 1–2 positive SLN identified during surgery. Patients were randomized to either SLNB alone or ALND with >10 nodes removed at the time of surgery. Importantly, all patients underwent adjuvant radiation treatment. With a median follow-up of 6.3 years at the time of initial publication, there was no statistically significant difference in 5-year overall survival (92.5% vs. 91.8% in ALND and SLNB groups, respectively), with a low axillary recurrence rate of fewer than 1% in both groups. There was also no difference in outcomes between patients which were estrogen receptor (ER) +/Progesterone Receptor (PR)+, and those who were ER−/PR− [24]. Criticism of the ACOSOG z0011 trial focused on its short-term follow-up, generally low-risk patients (given high percentage of ER+/PR+), and lack of blinding of sentinel node status to radiation oncologists. In response, the long-term 10-year follow-up was published in 2017. With a median follow-up of 9.3 years, SLNB was non-inferior to ALND for overall survival (survival rate 86.3% SLNB alone vs. 83.6% ALND, non-inferiority $p = 0.02$). The 10-year disease-free survival was not statistically significantly different between groups (80.2% SLNB alone vs. 78.2% ALND, $p = 0.32$). A subset analysis was performed on those patients with specific radiation field data available. Two hundred and twenty-eight patients had this data available, which showed no difference in disease-free survival, overall survival, or locoregional recurrence for those treated with nodal-field irradiation compared with those who did not receive irradiation [25]. ACOSOG Z0011 allowed SLNB to replace routine ALND in node-positive patients with T1–T2 invasive breast cancer, no palpable axillary lymphadenopathy undergoing breast conservation therapy. The ACOSOG Z0011 trial was further validated with the "Radiotherapy or surgery of the axilla after a positive sentinel node in breast cancer (EORTC 10981-22023 AMAROS)" trial which again evaluated surgery or radiotherapy of the axilla after a positive SLN. Unlike the ACOSOG Z0011 trial, patients undergoing mastectomy were included in the AMAROS trial. There was no difference in axillary recurrence in patients who were randomized to axillary dissection compared to those in the axillary radiotherapy group (0.43% vs. 1.19%, respectively), though overall number of patients undergoing mastectomy was small [26]. Whether or not patients who are node positive and treated with mastectomy can spare full ALND or can undergo radiation therapy remains unstudied in large randomized groups.

32.3 SLNB in Patients Receiving Neoadjuvant Chemotherapy

Further attempts at integrating SLNB into practice were then made. As neoadjuvant chemotherapy (NAC) became more widely adopted in the treatment of breast cancer, timing and technical feasibility of SLNB became widely debated. It was initially suggested that operative SLNB be performed prior to administration of neoadjuvant chemotherapy in order to accurately stage the axilla and determine the need for nodal radiation [27]. However, if axillary metastasis were eradicated by neoadjuvant chemotherapy, performing SLNB after the chemotherapy regimen could eliminate the need for a full ALND. The technical ability to perform SLNB after NAC was initially concerning, with thoughts that lymphatic mapping would be disrupted by excessive fibrosis of tumor lymphatics or blocking of lymphatic channels with cellular material or necrotic material created during chemotherapeutic destruction of the primary tumor and/or axillary metastasis. Indeed, initial studies of SLNB after NAC had high rates of failure to identify the node, up to 20%, with up to 16% false-negative rates [28, 29]. However, these initial studies included patients with initial bulky axillary metastasis and generally had a wide range of inclusion criteria. Patients with clinically node-negative axilla prior to initiating NAC had improved ability to identify SLN. Initial retrospective data showed comparable rates of identification (IR) and false negativity (FNR) in patients undergoing NAC as those who had surgery upfront, with IR of 99% and 97% and FNR of 4% and 6% [30]. Improved identification rates were seen in those patients whose lymphatics were mapped using dual tracer [31]. These improved IR and FNR rates were validated in meta-analyses [32].

The rates of negative SLNB were consistently higher in patients who received NAC in both single institution and multi-institutional studies [30, 33, 34]. Increasing node-negativity, the ability to potentially downstage the axilla, and therefore avoid ALND and/or nodal radiation led to increase use of SLNB in patients who were initially node positive. These attempts were supported by low rates of axillary recurrence in those patients who achieved nodal pCR [35]. Many prospective multi-institutional studies then proceeded to examine the ability to accurately identify SLN in patients who were clinically node positive prior to NAC. The ACOSOG 1071 trial was a phase II clinical trial which occurred in centers across the United States. Women with biopsy confirmed axillary nodal disease undergoing NAC were included. Patients with a previous history of axillary surgery or prior SLNB were excluded. Those enrolled in ACOSOG 1071 underwent lymphatic mapping with dual tracer following the completion of NAC, and all SLN were excised prior to proceeding with ALND. The FNR in patients with two or more nodes removed was 13%, while the FNR with three or more nodes removed fell to 9%, however, only a small percentage (56%) of patients in the trial met this criterion [36]. The SENTInel NeoAdjuvant (SENTINA) trial was another multicenter prospective study investigating patients undergoing NAC. The primary endpoint of the study was the accuracy and FNR of patients converting from positive SLN to negative SLN after NAC. FNR of these patients was over 14% and the detection rate was only 80% [37]. Further attempts at accurately detecting SLN after NAC continued. The Sentinel Node Biopsy Following Neoadjuvant Chemotherapy (SN FNAC) study also enrolled patients with biopsy-proven node-positive patients, attempting to reduce FNR to under 10%. However, this study was stopped early when the ACOSOG Z1071 and other trials failed to meet their criteria of FNR <10%. Prior to ending, the SN FNAC showed an IR of 88% and FNR of 8% [38].

Given these unacceptably high FNR, further attempts to improve SLNB after NAC for patients initially node positive ensued. The addition of placing a clip in the biopsied node and subsequently performing targeted SLNB was examined. Initial retrospective reviews showed IR of 92% when adding a marker and wire-localization of the biopsied axillary node to the operation [39]. A prospective trial performed by MD Anderson Cancer Center enrolled patients with biopsy confirmed axillary metastasis who underwent clip placement in that node. After NAC, these patients underwent localization with placement of an iodine-125 radioactive seed. FNR in these patients was significantly lower, 4.2% for the clipped node. The FNR of the combination of SLNB and targeted dissection of the clipped node reduced FNR to 1.4%. In this study, 23% of patients were unable to have their clipped node identified. The only factor which was associated with this discordance was having >3 abnormal nodes on initial ultrasound [40, 41].

The long-term oncologic safety and axillary recurrence rates in patients who undergo targeted axillary dissection remain to be seen. Recent data from a single institution retrospective study comparing patients who were initially clinically node negative to those who were clinically node positive with negative SLNB treated with ALND, and to those who were node positive with negative SLNB without ALND showed an IR of 98.3%. There was no significant difference in axillary recurrence-free survival ($p = 0.12$), disease-free survival ($p = 0.27$), and overall survival ($p = 0.15$) between groups, suggesting oncologic safety. Though there remains a paucity of prospective data on axillary or distant recurrence in those with negative SLNB (and no subsequent ALND) after NAC, the technique has become widely adopted [42]. Further studies examining the rates of locoregional recurrence in these patients will be necessary long term.

Currently, a large-scale multicenter randomized trial is accruing patients who have received NAC to determine whether ALND is necessary or whether patients can be treated with SLNB. The Alliance A011202 trial is a phase III study comparing patients who are clinically node positive prior to treatment who then undergo NAC. Those patients then undergo surgery with SLNB. Patients with a positive SLN after NAC at the time of surgery are then randomized to either ALND with nodal radiation or radiation to both nodal fields and the axillary basin. The results of this study stand to be groundbreaking and practice changing should nodal radiation have similar outcomes to ALND in patients who are node-positive after NAC. In addition, the NSABP B-51 trial is examining whether SLNB can inform decisions about radiation. In this multicenter randomized trial, patients who are node-positive undergo NAC and are subsequently SLN negative at surgery are randomized to either receive nodal radiation or not. The use of targeted dissection and subsequent results of the NSABP B-51 trial also have the potential for dramatic shifts in the treatment of breast cancer patients after neoadjuvant therapy.

32.4 When to Omit SLNB

Given a still-present 1–3% risk of lymphedema with SLNB, and a potential for axillary nodal status to not affect treatment in certain groups, attempts have been made to omit SLNB certain clinical situations. This is true in patients undergoing NAC, specifically those with triple negative (TN) or HER2-positive tumors with clinical complete response to NAC. A retrospective analysis of 646 T1–4, N0–2 patients undergoing NAC showed that HER2 positive and TN tumors who started as clinically node negative remain node negative if they achieve an in-breast pathologic complete response (pCR) [43]. A prospective study from the group at MD Anderson also examined the role of SLNB in patients who were so-called exemplary responders to NAC. A total of 527 patients were included, 290 of whom were node-negative by ultrasound prior to NAC. Of those, 116 (40%) had a breast pCR, 100% of whom also had no evidence of axillary lymph node metastasis. Those with initial axillary metastasis showed an in-breast pCR rate of 32% ($N = 77$), of whom 89.6% also had negative nodal disease [44]. Given these studies, the authors suggest that patients may be able to be spared SLNB if they are initially node negative and have an in-breast pathologic complete response. However, there is no strong or randomized data to suggest that an in-breast complete pathologic response is associated with a nodal complete pathologic response. In addition, long-term safety data are not yet available.

Additional controversy exists on whether elderly patients benefit from SLNB. Prospective trials on patients aged 65–80 with clinically node-negative axilla have been performed in attempts to clarify its utility. Omission of any axillary staging in patients aged 65–80 years old resulted in very low 15-year axillary recurrence and that breast-cancer-specific mortality is not affected by axillary staging. However, these patients had small, ER+ tumors [45]. Other single-institution retrospective studies have evaluated patients over the age of 70 with small (T1), ER+, ductal tumors. Christian et al. showed that in this group of patients, omission of SLNB increased with time from 2009 to the end of their study period, 2017 [46]. A meta-analysis and systematic review examined the differences in risk of axillary recurrence in elderly patients either undergoing or omitting axillary staging. While the risk of axillary recurrence was lower in patients who underwent axillary staging (RR 0.24, CI 0.06–0.95, $p = 0.04$), there was no significant difference in in-breast recurrence or distant recurrence (RR 1.20, 95% CI: 0.55–2.64, $I^2 = 62\%$, $p = 0.65$, RR 1.17, 95% CI: 0.75–1.82, $I^2 = 0\%$, $p = 0.48$, respectively), as well as no difference in overall or breast-cancer specific mortality (RR 0.99, 95% CI: 0.79–1.24, $I^2 = 0\%$, $p = 0.92$, RR 1.07, 95% CI: 0.72–1.57, $I^2 = 0\%$, $p = 0.75$, respectively), inciting questions on the clinical significant of axillary staging in this group of patients [45].

Conversely, a large-scale retrospective analysis using the National Cancer Database (NCDB) showed a potential role for lymph node surgery in the elderly (>70 years old). Patients with clinical T1–T3 and clinical N0 status were divided into two groups—those with lymph nodes evaluated and those without. Over 133,700 patients were included in the analysis. Seventy-six percent of those underwent nodal surgery, and those were more likely to receive chemotherapy, radiation, and endocrine therapy. Those who did not receive nodal surgery had worse overall survival (hazard ratio 1.66, 95% confidence interval 1.61–1.7) [47]. However, selection bias weighs heavily, as those patients who underwent chemotherapy and radiation may have had overall better baseline status and potential longer term survival from other conditions. Given mixed data, guidelines have yet to explicitly adopt this recommendation. The Society of Surgical Oncology "Choosing Wisely" guidelines state that routine use of sentinel lymph node biopsy can be omitted in women older than 70 with early stage, hormone receptor positive, and HER2 negative disease. The NCCN states that "the performance of axillary staging may be considered optional" for elderly patients "in the absence of definitive data."

32.5 SLNB in Patients with Previous Axillary Surgery

When the use of SLNB continued to expand, concerns arose about whether it was technically feasible in patients who had previously undergone breast or axillary surgery. Initial concern arose from case reports which demonstrated disrupted lymphatics after ipsilateral breast surgery. However, further studies thereafter showed more promising results. Using a dual-tracer technique, patients who underwent repeat SLNB after an ipsilateral SLNB were able to be mapped in over 80% of cases. Oncologic safety was also demonstrated, with no axillary recurrences seen [48]. Although thought to be technically impractical, repeat SLNB was also shown to be possible after ipsilateral ALND. A retrospective analysis of 45 patients who underwent breast conservation therapy and axillary lymph node dissection was performed. Of those, 29% had successful lymphatic mapping and SLNB [49]. Indeed, even studies looking at SLNB after mastectomy showed promise. A small review of 20 patients who had previously undergone mastectomy utilized radioactive tracer and blue dye injected superior to the mastectomy incision in the skin flap. When performed using this technique, SLNB was successful in 65% of cases [50]. Small data sets also examined patients undergoing repeat axillary staging after ALND. Indeed, redo SLNB was more successful after SLNB than after ALND, with success rates of 74% versus 38%, respectively [51]. Subsequently, systematic review and meta-analyses confirmed these initial findings. SLNB was successful in 81% (95% CI 76–85.2%) of patients who had previously undergone SLNB, and 52.2% (95% CI 46.6–57.8%) of patients with previous ALND [52]. A negative redo SLN has an excellent negative predictive value, 96.5%, suggesting a significant benefit to patients in this setting [53].

However, whether to stage the axilla in the context of recurrent breast cancer remains best considered in a multidisciplinary context. Small retrospective studies showed patients who underwent re-operative axillary staging had similar rates of subsequent radiation and systemic therapy. In addition, there was no statistically significant difference in times to recurrence, axillary failure, or non-axillary recurrence, nor was there a difference in rates of distant metastasis or death, though follow-up time was relatively short (4.2 years) [54]. Further multicenter trials are needed to determine the overall utility of axillary staging for recurrent disease.

32.6 Alternative Drainage Sites

While the overwhelming amount of data for SLNB is for those nodes in the ipsilateral axilla, it is important to note that alternative lymphatic drainage sites for primary breast tumors exist. Drainage to these alternative sites is rare but not impossible. In a series of 549 patients undergoing lymphoscintigraphy, drainage to internal mammary nodes (IMN) occurred in 15.7% of patients, and other non-axillary and non-internal mammary nodes in 8% of patients [55]. However, the decision to remove these non-axillary nodes remains controversial. A systematic review and meta-analysis were performed examining the utility of IMN excision. A total of 2427 patients and 18 articles were included. The pooled IMN positivity rate was 15%, and axillary nodal metastasis was a strong, independent predictor of IMN involvement (odds ratio 6.01, 95% CI, 3.49, 10.34). However, little data exist on whether removal of these nodes either improves survival or decreases recurrence, and isotope drainage to the IMN nodes may also depend on the depth of injection site within the breast [56–58].

As the utility of SLNB has grown, so has the interest in the biology of these metastases. While the goal was originally to identify tumor cells within the lymph node, current research is focused on the biology of these cells and whether they match the primary tumor. Recent publications have shed light on the unique characteristics of these metastases, showing a distinct genomic profile in comparison to the primary tumor, with upregulation of chemotaxis genes and downregulation of those which regulate cell–cell interactions including collagen and metallopeptidase genes [59]. Therefore, treatment of axillary or nodal metastases may have different targets than the primary breast lesion. Management of the axilla continues to progress. From formal axillary dissections for every patient to decisions based on SLNB, the spectrum of surgical intervention has changed dramatically in the past 30 years. As such, further interventions will occur and the extent of axillary surgery will continue to evolve.

32.7 Conclusion

The evolution of axillary staging has been dramatic over the past half-century, with a striking shift from ALND to SLNB. Indeed, ALND has been abandoned for DCIS, and SLNB has replaced ALND for node-negative patients undergoing both BCT and mastectomy, in addition to those patients with node-positive, small primary tumors undergoing BCT. SLNB has been shown to be technically feasible with low false-negative rates in patients following neoadjuvant chemotherapy treatment, and its long-term utility is being further evaluated in patients who have been converted from node-positive to node-negative. Further trials are also ongoing to determine whether SLNB with radiation can replace ALND in patients who remain node-positive after neoadjuvant systemic therapy. As the extent of axillary surgery continues to fall, there will likely be further groups of patients who do not receive SLNB, even as its utility continues to expand.

Questions for Future Directions

- Can we rely only on SLNB and omit ALND for patients treated with mastectomy with favorable tumor biology found to be SLN positive?
- When can SLNB be omitted for recurrent breast cancer and does that answer depend not only on length of time from primary disease but also on receptor status?
- What are the long-term outcomes of omitting ALND in patients who are treated with neoadjuvant chemotherapy and converted from node-positive to node-negative on targeted axillary dissection?
- What are the long-term outcomes of omitting SLNB in patients who are "extraordinary" responders to neoadjuvant chemotherapy?
- Given differences in gene expressions between primary tumors and axillary metastases, will targeted therapies to axillary metastases improved locoregional control and potentially overall survival?

References

1. Jones D, Pereira ER, Padera TP. Growth and immune evasion of lymph node metastasis. Front Oncol. 2018;8:36.
2. Robinson DS, Sample WF, Fee HJ, et al. Regional lymphatic drainage in primary malignant melanoma of the trunk determined by colloidal gold scanning. Surg Forum. 1977;28:147–8.
3. Morton DL, Wen DR, Foshag LJ, et al. Intraoperative lymphatic mapping and selective cervical lymphadenectomy for early-stage melanomas of the head and neck. J Clin Oncol. 1993;11:1751–6.
4. Giuliano AE, Kirgan DM, Guenther JM, Morton DL. Lymphatic mapping and sentinel lymphadenectomy for breast cancer. Ann Surg. 1994;220:391–8; discussion 398–401.
5. Giuliano AE, Jones RC, Brennan M, Statman R. Sentinel lymphadenectomy in breast cancer. J Clin Oncol. 1997;15:2345–50.
6. Morrow M, Rademaker AW, Bethke KP, et al. Learning sentinel node biopsy: results of a prospective randomized trial of two techniques. Surgery. 1999;126:714–20; discussion 720–712.
7. Veronesi U, Paganelli G, Galimberti V, et al. Sentinel-node biopsy to avoid axillary dissection in breast cancer with clinically negative lymph-nodes. Lancet. 1997;349:1864–7.
8. Langer I, Marti WR, Guller U, et al. Axillary recurrence rate in breast cancer patients with negative sentinel lymph node (SLN) or SLN micrometastases: prospective analysis of 150 patients after SLN biopsy. Ann Surg. 2005;241:152–8.
9. Smidt ML, Janssen CM, Kuster DM, et al. Axillary recurrence after a negative sentinel node biopsy for breast cancer: incidence and clinical significance. Ann Surg Oncol. 2005;12:29–33.
10. Strand SE, Persson BR. Quantitative lymphoscintigraphy I: basic concepts for optimal uptake of radiocolloids in the parasternal lymph nodes of rabbits. J Nucl Med. 1979;20:1038–46.
11. Wong JH, Terada K, Ko P, Coel MN. Lack of effect of particle size on the identification of the sentinel node in cutaneous malignancies. Ann Surg Oncol. 1998;5:77–80.
12. Faries MB, Bedrosian I, Reynolds C, et al. Active macromolecule uptake by lymph node antigen-presenting cells: a novel mechanism in determining sentinel lymph node status. Ann Surg Oncol. 2000;7:98–105.
13. Wallace AM, Hoh CK, Limmer KK, et al. Sentinel lymph node accumulation of lymphoseek and Tc-99m-sulfur colloid using a "2-day" protocol. Nucl Med Biol. 2009;36:687–92.
14. Cox CE. Lymphatic mapping in breast cancer: combination technique. Ann Surg Oncol. 2001;8:67S–70S.
15. Cox CE, Salud CJ, Cantor A, et al. Learning curves for breast cancer sentinel lymph node mapping based on surgical volume analysis. J Am Coll Surg. 2001;193:593–600.
16. Eldrageely K, Vargas MP, Khalkhali I, et al. Sentinel lymph node mapping of breast cancer: a case-control study of methylene blue tracer compared to isosulfan blue. Am Surg. 2004;70:872–5.
17. Giuliano AE. Sentinel lymphadenectomy in primary breast carcinoma: an alternative to routine axillary dissection. J Surg Oncol. 1996;62:75–7.
18. Krag DN, Weaver DL, Alex JC, Fairbank JT. Surgical resection and radiolocalization of the sentinel lymph node in breast cancer using a gamma probe. Surg Oncol. 1993;2:335–9; discussion 340.
19. Chagpar A, Martin RC 3rd, Chao C, et al. Validation of subareolar and periareolar injection techniques for breast sentinel lymph node biopsy. Arch Surg. 2004;139:614–8; discussion 618–620.
20. Schrenk P, Rieger R, Shamiyeh A, Wayand W. Morbidity following sentinel lymph node biopsy versus axillary lymph node dissection for patients with breast carcinoma. Cancer. 2000;88:608–14.
21. Schijven MP, Vingerhoets AJ, Rutten HJ, et al. Comparison of morbidity between axillary lymph node dissection and sentinel node biopsy. Eur J Surg Oncol. 2003;29:341–50.
22. Husted Madsen A, Haugaard K, Soerensen J, et al. Arm morbidity following sentinel lymph node biopsy or axillary lymph node dissection: a study from the Danish Breast Cancer Cooperative Group. Breast. 2008;17:138–47.

23. Krag DN, Anderson SJ, Julian TB, et al. Sentinel-lymph-node resection compared with conventional axillary-lymph-node dissection in clinically node-negative patients with breast cancer: overall survival findings from the NSABP B-32 randomised phase 3 trial. Lancet Oncol. 2010;11:927–33.
24. Giuliano AE, Hunt KK, Ballman KV, et al. Axillary dissection vs no axillary dissection in women with invasive breast cancer and sentinel node metastasis: a randomized clinical trial. JAMA. 2011;305:569–75.
25. Giuliano AE, Ballman KV, McCall L, et al. Effect of axillary dissection vs no axillary dissection on 10-year overall survival among women with invasive breast cancer and sentinel node metastasis: the ACOSOG Z0011 (Alliance) Randomized Clinical Trial. JAMA. 2017;318:918–26.
26. Donker M, van Tienhoven G, Straver ME, et al. Radiotherapy or surgery of the axilla after a positive sentinel node in breast cancer (EORTC 10981-22023 AMAROS): a randomised, multicentre, open-label, phase 3 non-inferiority trial. Lancet Oncol. 2014;15:1303–10.
27. Sabel MS, Schott AF, Kleer CG, et al. Sentinel node biopsy prior to neoadjuvant chemotherapy. Am J Surg. 2003;186:102–5.
28. van Deurzen CH, Vriens BE, Tjan-Heijnen VC, et al. Accuracy of sentinel node biopsy after neoadjuvant chemotherapy in breast cancer patients: a systematic review. Eur J Cancer. 2009;45:3124–30.
29. Nason KS, Anderson BO, Byrd DR, et al. Increased false negative sentinel node biopsy rates after preoperative chemotherapy for invasive breast carcinoma. Cancer. 2000;89:2187–94.
30. Hunt KK, Yi M, Mittendorf EA, et al. Sentinel lymph node surgery after neoadjuvant chemotherapy is accurate and reduces the need for axillary dissection in breast cancer patients. Ann Surg. 2009;250:558–66.
31. Xing Y, Foy M, Cox DD, et al. Meta-analysis of sentinel lymph node biopsy after preoperative chemotherapy in patients with breast cancer. Br J Surg. 2006;93:539–46.
32. Shirzadi A, Mahmoodzadeh H, Qorbani M. Assessment of sentinel lymph node biopsy after neoadjuvant chemotherapy for breast cancer in two subgroups: initially node negative and node positive converted to node negative—a systemic review and meta-analysis. J Res Med Sci. 2019;24:18.
33. Fisher B, Brown A, Mamounas E, et al. Effect of preoperative chemotherapy on local-regional disease in women with operable breast cancer: findings from National Surgical Adjuvant Breast and Bowel Project B-18. J Clin Oncol. 1997;15:2483–93.
34. Bear HD, Anderson S, Brown A, et al. The effect on tumor response of adding sequential preoperative docetaxel to preoperative doxorubicin and cyclophosphamide: preliminary results from National Surgical Adjuvant Breast and Bowel Project protocol B-27. J Clin Oncol. 2003;21:4165–74.
35. Mamounas EP, Anderson SJ, Dignam JJ, et al. Predictors of locoregional recurrence after neoadjuvant chemotherapy: results from combined analysis of National Surgical Adjuvant Breast and Bowel Project B-18 and B-27. J Clin Oncol. 2012;30:3960–6.
36. Boughey JC, Suman VJ, Mittendorf EA, et al. Sentinel lymph node surgery after neoadjuvant chemotherapy in patients with node-positive breast cancer: the ACOSOG Z1071 (Alliance) Clinical Trial. JAMA. 2013;310:1455–61.
37. Kuehn T, Bauerfeind I, Fehm T, et al. Sentinel-lymph-node biopsy in patients with breast cancer before and after neoadjuvant chemotherapy (SENTINA): a prospective, multicentre cohort study. Lancet Oncol. 2013;14:609–18.
38. Boileau JF, Poirier B, Basik M, et al. Sentinel node biopsy after neoadjuvant chemotherapy in biopsy-proven node-positive breast cancer: the SN FNAC study. J Clin Oncol. 2015;33:258–64.
39. Balasubramanian R, Morgan C, Shaari E, et al. Wire guided localisation for targeted axillary node dissection is accurate in axillary staging in node positive breast cancer following neoadjuvant chemotherapy. Eur J Surg Oncol. 2020;46:1028–33.
40. Caudle AS, Yang WT, Krishnamurthy S, et al. Improved axillary evaluation following neoadjuvant therapy for patients with node-positive breast cancer using selective evaluation of clipped nodes: implementation of targeted axillary dissection. J Clin Oncol. 2016;34:1072–8.
41. Choi HJ, Kim I, Alsharif E, et al. Use of sentinel lymph node biopsy after neoadjuvant chemotherapy in patients with axillary node-positive breast cancer in diagnosis. J Breast Cancer. 2018;21:433–41.
42. Vugts G, Maaskant-Braat AJ, de Roos WK, et al. Management of the axilla after neoadjuvant chemotherapy for clinically node positive breast cancer: a nationwide survey study in the Netherlands. Eur J Surg Oncol. 2016;42:956–64.
43. Esgueva A, Siso C, Espinosa-Bravo M, et al. Leveraging the increased rates of pathologic complete response after neoadjuvant treatment in breast cancer to de-escalate surgical treatments. J Surg Oncol. 2021;123(1):71–9.
44. Tadros AB, Yang WT, Krishnamurthy S, et al. Identification of patients with documented pathologic complete response in the breast after neoadjuvant chemotherapy for omission of axillary surgery. JAMA Surg. 2017;152:665–70.
45. Liang S, Hallet J, Simpson JS, et al. Omission of axillary staging in elderly patients with early stage breast cancer impacts regional control but not survival: a systematic review and meta-analysis. J Geriatr Oncol. 2017;8:140–7.
46. Christian N, Heelan Gladden A, Friedman C, et al. Increasing omission of radiation therapy and sentinel node biopsy in elderly patients with early stage, hormone-positive breast cancer. Breast J. 2020;26:133–8.
47. Tamirisa N, Thomas SM, Fayanju OM, et al. Axillary nodal evaluation in elderly breast cancer patients: potential effects on treatment decisions and survival. Ann Surg Oncol. 2018;25:2890–8.
48. Cox CE, Furman BT, Kiluk JV, et al. Use of reoperative sentinel lymph node biopsy in breast cancer patients. J Am Coll Surg. 2008;207:57–61.
49. Kaur P, Kiluk JV, Meade T, et al. Sentinel lymph node biopsy in patients with previous ipsilateral complete axillary lymph node dissection. Ann Surg Oncol. 2011;18:727–32.
50. Karam A, Stempel M, Cody HS 3rd, Port ER. Reoperative sentinel lymph node biopsy after previous mastectomy. J Am Coll Surg. 2008;207:543–8.
51. Port ER, Garcia-Etienne CA, Park J, et al. Reoperative sentinel lymph node biopsy: a new frontier in the management of ipsilateral breast tumor recurrence. Ann Surg Oncol. 2007;14:2209–14.
52. Maaskant-Braat AJ, Voogd AC, Roumen RM, Nieuwenhuijzen GA. Repeat sentinel node biopsy in patients with locally recurrent breast cancer: a systematic review and meta-analysis of the literature. Breast Cancer Res Treat. 2013;138:13–20.
53. Poodt IGM, Vugts G, Schipper RJ, Nieuwenhuijzen GAP. Repeat sentinel lymph node biopsy for ipsilateral breast tumor recurrence: a systematic review of the results and impact on prognosis. Ann Surg Oncol. 2018;25:1329–39.
54. Ugras S, Matsen C, Eaton A, et al. Reoperative sentinel lymph node biopsy is feasible for locally recurrent breast cancer, but is it worthwhile? Ann Surg Oncol. 2016;23:744–8.
55. Tanis PJ, Nieweg OE, Valdes Olmos RA, et al. Impact of non-axillary sentinel node biopsy on staging and treatment of breast cancer patients. Br J Cancer. 2002;87:705–10.
56. Tan C, Caragata R, Bennett I. Is sentinel node biopsy of the internal mammary lymph nodes relevant in the management of breast cancer? Breast J. 2017;23:410–4.
57. Qiu PF, Zhao RR, Wang W, et al. Internal mammary sentinel lymph node biopsy in clinically axillary lymph node-positive breast can-

cer: diagnosis and implications for patient management. Ann Surg Oncol. 2020;27:375–83.
58. Gong J, Yu Y, Wu G, et al. Should internal mammary lymph node sentinel biopsy be performed in breast cancer: a systematic review and meta-analysis. World J Surg Oncol. 2019;17:135.
59. Srour MK, Gao B, Dadmanesh F, et al. Gene expression comparison between primary triple-negative breast cancer and paired axillary and sentinel lymph node metastasis. Breast J. 2020;26:904–10.

Sentinel Node Navigation Surgery for Upper Gastrointestinal Cancer

33

Shuhei Mayanagi and Yuko Kitagawa

Abstract

Radical resection with systematic lymph node dissection is traditionally performed in patients with upper gastrointestinal cancer because the lymphatic system of these cancers is complex and skip metastases are often present. Although sentinel lymph node mapping for upper gastrointestinal cancer is still in clinical trials, the sentinel node concept could be applied for early-stage cancers. In early gastric cancer, in particular, cancer metastasis can be identified with high diagnostic accuracy using the dual tracer method and sentinel basin dissection. The usefulness of sentinel node navigation surgery for early gastric cancer has been verified in multicenter prospective studies. In the near future, it will likely be possible to safely perform radical surgery with a minimal extent of gastrectomy based on the sentinel node concept.

Learning Objectives

- The sentinel lymph node in upper gastrointestinal cancer is not necessarily the node that is located anatomically closest to the primary lesion and is not necessarily limited to a single lymphatic drainage area.
- The dual tracer method is the gold standard for sentinel lymph node navigation surgery for gastric cancer: this method uses radioisotopes and dye, and involves dissection of the lymphatic basin defined as the lymphatic drainage area along the main gastric arteries.
- For clinical T1 gastric cancer <3–4 cm, large-scale prospective clinical trials for sentinel lymph node navigation surgery are ongoing, with overall survival as the primary endpoint.

33.1 Introduction

The presence and extent of lymph node metastasis are important prognostic factors for upper gastrointestinal cancers, such as gastric cancer, esophagogastric junction cancer, and esophageal cancer, and systematic regional lymph node dissection is the standard surgical procedure for these patients. Even today, with the development of chemotherapy and radiation therapy, surgery still plays a very important role in the treatment of upper gastrointestinal cancers. In particular, early gastric cancer (EGC) in which the invasion of the primary lesion is limited to the submucosa (T1), is curable by gastrectomy with lymph node dissection alone. Several factors contribute to the favorable outcome of surgery for EGC, the main reason being the small number of lymph node metastases. The frequency of lymph node metastasis in EGC is 1–8% in mucosal cancer and 5–19% in submucosal invasive cancer [1]. Even if lymph node metastasis is present, the majority of patients with EGC are cured by appropriate lymph node dissection. However, patients who undergo standard gastrectomy with systematic lymph node dissection often suffer from postgastrectomy syndromes, including heartburn, early satiety, and abnormal bowel movements [2, 3].

In recent years, most surgeries for upper gastrointestinal cancers have been performed endoscopically, laparoscopically, or thoracoscopically, and the quality of life of patients

S. Mayanagi
Division of Esophageal Surgery, Shizuoka Cancer Center Hospital, Shizuoka, Japan
e-mail: s.mayanagi@scchr.jp

Y. Kitagawa (✉)
Department of Surgery, Keio University School of Medicine, Tokyo, Japan
e-mail: kitagawa.a3@keio.jp

S. P. Leong et al. (eds.), *Cancer Metastasis Through the Lymphovascular System*,
https://doi.org/10.1007/978-3-030-93084-4_33

has improved, especially in the early postoperative period. The main difference between laparoscopic surgery and open surgery is the trauma to the abdominal wall, while the extent of gastrectomy and lymph node dissection is the same in both; therefore, there is almost no difference in long-term outcomes, such as postgastrectomy syndromes, between the two types of surgery. Standard gastrectomy and systematic lymph node dissection have produced excellent overall survival for EGC, but with significant risk for postoperative dysfunction; thus the next logical goal is trying to achieve similar cure rates without the requirement for extensive gastrectomy and lymph node dissection.

Historically, it is essential to develop research on the lymph flow around the stomach in order to establish the extent of lymph node dissection for gastric cancer. In line with this, a systematic report on the lymphatic system of the stomach was published in the early 1900s [4]. Since then, the lymphatic system, based on lymph node metastasis and recurrence in clinical practice, has been extensively studied for more than 100 years. According to the Japanese Classification of Gastric Cancer, regional lymph nodes of gastric cancer are defined as the "lymph node group targeted for dissection," and are classified into three groups [5]. In this classification, the frequency of lymph node metastasis and the therapeutic effect of lymph node dissection are considered in addition to the anatomical position of the lymph nodes: The perigastric lymph nodes are classified into Group (Compartment) 1, along the main trunk arteries into Group 2, and periaortic and deep lymph nodes into Group 3. Depending on the clinical stage of the primary tumor, the extent of the lymph node groups that should be removed is determined prophylactically, e.g., D2 dissection for removal of up to Group 2 lymph nodes.

Systematic lymph node dissection, represented by D2 dissection, sacrifices the major arteries that feed the stomach and results in extensive gastrectomy. For instance, a distal gastrectomy with D2 lymph node dissection for gastric cancer located at the antrum of the stomach requires resection of the distal two-thirds of the stomach. Therefore, minimization of lymph node dissection is essential to reduce the extent of gastrectomy. As mentioned previously, the rate of lymph node metastasis in EGC is low, so routine systematic lymph node dissection and extensive gastrectomy are not always required for EGCs. However, since most lymph node metastases in EGC are microscopic (histological lymph node metastasis), there is currently no reliable method of nonsurgical diagnostic for the detection of lymph node metastasis. It is unclear which of the myriad regional lymph nodes should be biopsied in order to make a rapid intraoperative pathological diagnosis of lymph node metastasis; this is in contrast to malignant tumors on the body surface, such as breast cancer. The sentinel lymph node biopsy represents the long-awaited method required for a breakthrough in EGC treatment.

33.2 Gastric Cancer

33.2.1 Sentinel Node Mapping Procedure

Unlike malignancies on the surface of the body, such as melanoma and breast cancer, several points require attention in sentinel lymph node mapping for upper gastrointestinal cancers that are deep in the body cavity.

First, regarding the tracer, as with other cancers, there are two main methods for identifying sentinel lymph nodes in EGC: the radio-guided method and dye method. Technetium-99m colloid is frequently used as a tracer in the radio-guided method, and isosulfan blue, patent blue, and indocyanine green (ICG) are frequently used in the dye method. However, each method has advantages and disadvantages. For instance, the radio-guided method is able to confirm the complete harvest of sentinel nodes by gamma probing, even deep inside the body cavity; however, its use is severely restricted due to radiation exposure. While the dye method allows real-time observation of the visualized lymphatic vessels, it may be difficult to see with dye in deep adipose tissue. The improvement of the detection rate and diagnostic accuracy using a combined method has been reported in lymphatic mapping for both breast cancer and malignant melanoma, and consequently, it has become the gold standard for gastric cancer. The technical errors that are present using the single mapping agent approach can be reduced by adding a different approach for lymphatic mapping. Therefore, a combination of the dye and radio-guided methods is recommended for systematic lymphatic mapping of EGC. In our institute, we have chosen to use technetium-99m tin colloid, which has a relatively large particle size [6]. In our experience, tin colloid with a diameter of 500 nm migrates into the sentinel nodes within 2 h and remains there for more than 20 h following phagocytosis by macrophages in the sentinel nodes. The radio-guided method allows quantitative detection of sentinel nodes due to the stable incorporation of the radioactive tracer, even in blind areas. Moreover, the use of a small gamma probe is also useful for sentinel node identification in laparoscopic surgery. In contrast, ICG, which is frequently used as a dye tracer for sentinel node identification, has an extremely small particle size and is effectively taken up into lymphatic vessels, and is therefore suitable for real-time observation of lymphatic flow during surgery [7]. In addition, a method of irradiating near-infrared light to capture the fluorescence emitted by ICG under a laparoscope has been developed, and good results have been reported that surpass the dye method that relies on visual recognition [8, 9]. In this method, a 2.0 mL solution of technetium-99m tin colloid (150 MBq) is injected into the submucosal layer around the primary tumor in all four abdominal quadrants the day before surgery, and ICG is also initially injected using intraoperative endoscopy via an approach similar to that used for radioisotopes. Stained lymphatic vessels and nodes can be detected visually within 15 min of injecting the dye, and a hand-held gamma probe is simultaneously used to identify the radioactive sentinel nodes.

Second, two types of sentinel lymph node sampling procedures have been described in the treatment of EGC, the pickup method or lymphatic basin dissection (Fig. 33.1). The pickup method is a well-established and simple method that is currently used to assess breast cancer and melanoma; however, this method is not considered to be suitable for EGC given its diverse and complex lymphatic drainage. To address this issue, the concept of lymphatic basin dissection has been developed in EGC. The lymphatic basin is defined as the lymphatic zone, which is divided by the stream of dye mapping [10]. The proximal border of the lymphatic basin is the fatty tissue attached to the stomach wall, and the distal border of the basin is the front of the stained node furthest from the stomach. All sentinel lymph nodes exist within the lymphatic basin, and the lymphatic basins are thought to be the primary lymphatic drainage areas, with one or a few basin(s) normally present in EGC. Miwa and Kinami proposed the basin concept and classified the basins into five directions along the following main arteries: left gastric artery, right gastric artery, left gastroepiploic artery, right gastroepiploic artery, and posterior gastric artery in which sentinel node basins contained true positive nodes [10, 11] (Fig. 33.2). Compared to the pickup method, basin dissection has been reported to be superior in terms of detection rate, sensitivity, and specificity [12] and it was also shown to be valid in a prospective multicenter trial [13]. Sentinel basin dissection is considered a minimally focused lymphadenectomy method for EGC with a reasonable safety net that avoids recurrence after a false-negative sentinel node biopsy.

33.2.2 Indication for Sentinel Node Mapping

One of the most common reasons for a false-negative result of sentinel node mapping in gastric cancer is a lymphatic vessel obstructed by cancer invasion. In these cases, the administered tracer cannot migrate into the real sentinel nodes and its flow instead takes the path of least resistance, leading it to the second-echelon, or false, sentinel nodes. Thus, clinically positive node cases and advanced primary tumor cases are not indications for sentinel node mapping. The purpose of this technique is to identify clinically undetectable lymph node involvement. Although muscular invasive tumor (T2) is the indication in many studies as well as T1 tumors, false-negative cases were reported in patients with T2 gastric cancer in a previous prospective study [13]. Therefore, this technique is now excluded from indications in order to eliminate false positives. Moreover, lateral spread is also an important factor in the indication, and initially, it was also indicated for tumors that were spread widely in the lateral direction. However, a large tumor size >4 cm strongly correlated with skip metastases [14]. In short, clinical T1 tumors with a primary lesion diameter <4 cm are considered as suitable indications for sentinel node mapping for gastric cancer.

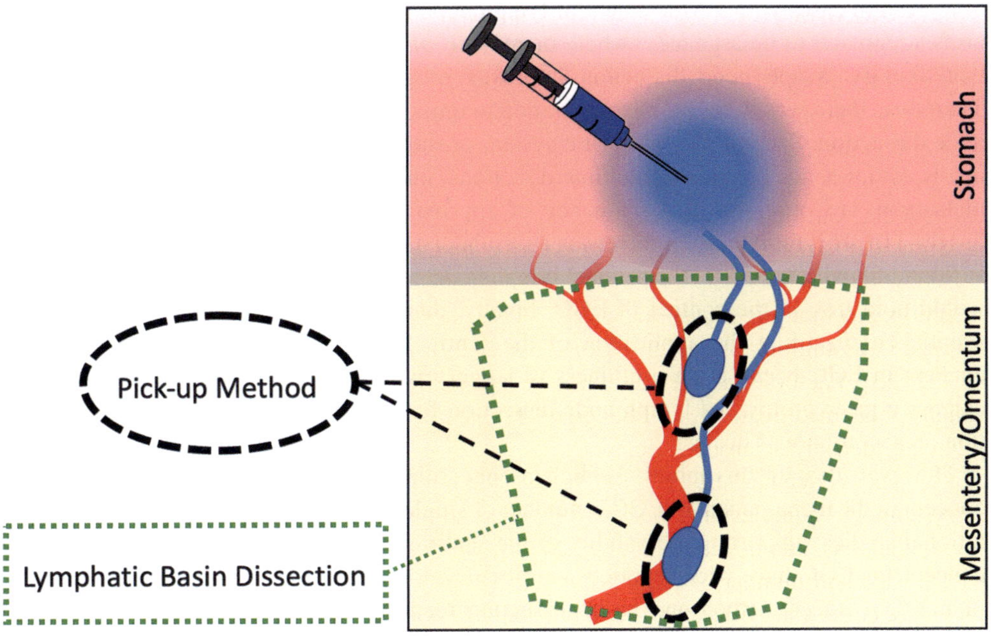

Fig. 33.1 Sentinel lymph node mapping procedures: Pickup method versus lymphatic basin dissection. Pickup method is not considered to be suitable for early gastric cancer given its diverse and complex lymphatic drainage. The lymphatic basins are thought to be the primary lymphatic drainage areas, with one or a few basin(s) normally present in early gastric cancer.

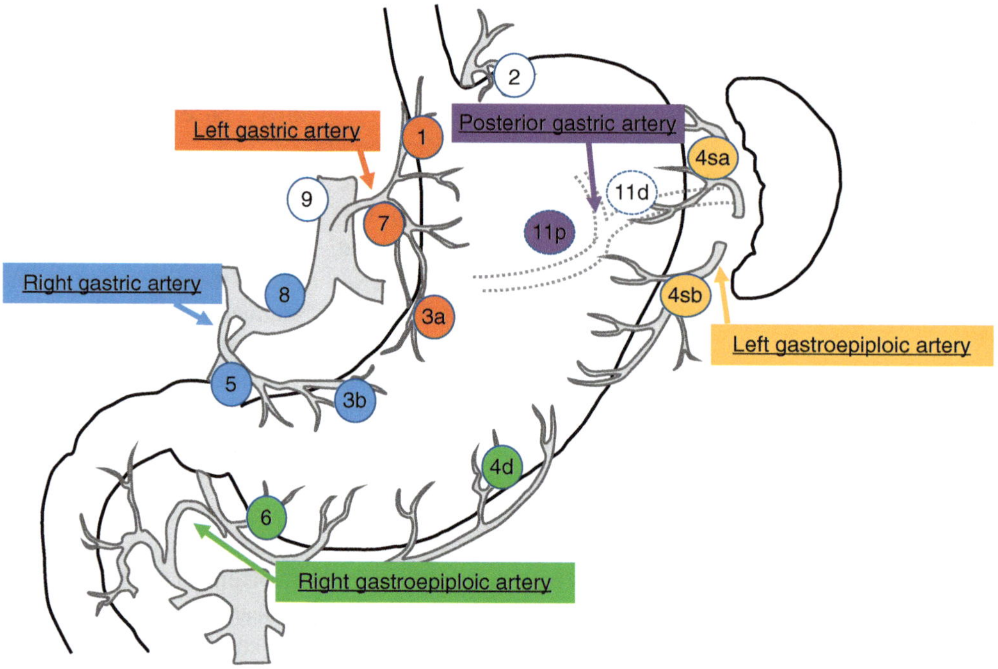

Fig. 33.2 Gastric lymphatic basins. The sentinel basins are classified into five directions along the following main arteries: left gastric artery, right gastric artery, left gastroepiploic artery, right gastroepiploic artery, and posterior gastric artery

33.3 Current Status and Future Prospects

The generalization of the sentinel node concept is remarkable in both malignant melanoma and breast cancer, and lymph node dissection using biopsy is currently omitted in these cancers. In contrast, the development of the sentinel node concept in EGC has been delayed and is still in the clinical trial setting. There are several reasons for this. First, gastric cancer may be cured by thorough lymph node dissection, even in cases with a wide extent of lymph node metastases, compared to breast cancer where there is no evidence that axillary lymph node dissection improves prognosis. Therefore, there was initially little motivation to omit lymph node dissection. Second, the lymphatic system of the stomach is complex and spreads in multiple directions, and skip metastases are observed at a frequency of approximately 5–10% [15, 16]. Furthermore, there was a misunderstanding surrounding whether or not the sentinel lymph node concept would hold true in the context of EGC. Finally, there was skepticism regarding the application of the sentinel node concept in EGC because the usefulness of surgery without extensive gastrectomy and lymph node dissection for function preservation was unclear.

However, despite this initial skepticism concerning this concept in the management of EGC, a number of single institutional studies supporting the validity of the sentinel node concept for EGC have been reported with the development of the dual tracer method and basin dissection mentioned above [6, 17–23]. It is now hypothesized that the sentinel lymph node of gastric cancer is not necessarily the node that is located anatomically closest to the primary lesion and is not necessarily limited in a single lymphatic drainage area. Furthermore, in gastrointestinal cancer, the sentinel node may be both patient- and lesion-specific. Sentinel nodes with multiple distributions are not a major problem for lymphatic mapping. In addition, the development of local gastrectomy performed at the same time as sentinel lymph node biopsy has also progressed. With the development of laparoscopic endoscopic cooperative surgery, such as nonexposed endoscopic wall inversion surgery [24], endoscopic full-thickness resection has become safer and more reliable, and the extent of gastrectomy has become smaller. Furthermore, it has become possible to realize function-preserving minimizing surgery for EGC using sentinel lymph node biopsy.

A large-scale multicenter prospective clinical trial was conducted by the Japanese Society for Sentinel Node Navigation Surgery (SNNS) in order to verify the validity of sentinel node mapping for EGC [13]. Of the 433 cT1 or cT2 gastric cancer patients, 397 were considered eligible based on surgical findings. Sentinel node biopsies were performed in all patients, with a detection rate of 97.5% (387/397). There were no serious adverse effects associated with endoscopic tracer infusion. Of the 57 patients with conventional hematoxylin and eosin-stained lymph node metastases, 93% (53 of 57) were sentinel node positive, and the accuracy of the lymph node evaluation of metastases was 99% (383 of 387). Only four false-negative sentinel node biopsies were observed, and pathological analysis revealed that three of these biopsies had pT2 or tumor >4 cm. It has been suggested that the sentinel node concept is sufficiently clinically applicable to relatively small cT1 gastric cancer of ≤4 cm using the dual tracer method. In addition, the short-term outcomes of another large-scale prospective multicenter randomized controlled trial (the SENORITA trial) were reported in South Korea [25]. This trial was designed to compare laparoscopic standard gastrectomy and laparoscopic SNNS using the dual tracer method with 3-year disease-free survival in EGC tumors <3 cm. A total of 580 patients were randomly enrolled in this trial. In the SNNS group, sentinel node basins were detected in 96.7%, of the 245 patients: 127 (52%) had 1, 89 (36%) had 2, and 21 (9%) had 3 sentinel basins. Stomach-preserving surgery, such as wedge resection and segmental resection, was carried out in 210 of 258 patients, and the rate and severity of postoperative complications are comparable to those after standard gastrectomy.

Meanwhile, an unacceptably high false-negative rate of 46.4% (13/28) was reported in the multicenter prospective study conducted by the Japan Clinical Oncology Group (JCOG0302) [26]. This study was conducted to evaluate the feasibility and accuracy of diagnosis using sentinel node biopsy using the only dye method with intraoperative subserosal direct injection of ICG. The most serious limitations of this study were the inclusion criteria for the participating institutes in terms of prior experience with sentinel node mapping for gastric cancer. In the field of breast cancer surgery, the significance of the technical learning curve for this technique has already been clearly demonstrated. For sentinel node mapping of gastric cancer, the experience of 26 cases has been suggested as the minimum requirement for high detection rates [27]. These studies showed that SNNS for EGC within 3–4 cm can be performed with high diagnostic accuracy using the dual tracer method in a facility proficient in sentinel mapping. A relatively high incidence of skip metastases in EGC is not an obstacle to the clinical application of the sentinel node concept. On the contrary, this technique is essential for the accurate detection of micrometastases in EGC due to the variations in the lymphatic drainage routes.

It has also been suggested that the sentinel concept could be applied to lesions after endoscopic resection [28]. As a next step, it may be possible to perform further minimizing surgery by combining endoscopic resection with sentinel biopsy in order to preserve the entire stomach. At present, the enrollment of patients in clinical trial to verify non-inferiority with overall survival as the primary outcome has been completed in Japan as well as in the SENORITA trial. Based on the results of these large-scale prospective studies, it is expected that SNNS will improve the quality of life for patients with EGC in the future.

33.4 Esophageal Cancer

Sentinel node mapping for thoracic esophageal cancer is technically more difficult than that for gastric cancer. Because the esophagus is deep in the mediastinum, even though surgical mobilization of the esophagus is essential for esophagectomy, the mobilization can interfere with active lymphatic flow from the primary tumor site. For this reason, the radio-guided method is preferred over the dye method to identify sentinel lymph nodes in esophageal cancer. Although several types of radioisotope tracers have been used in sentinel mapping of esophageal cancer, radioactive colloids with relatively large particle sizes have long deposits of tracers in the lymph nodes and are effective in identifying sentinel lymph nodes. In our institute, we use technetium-99m tin colloid that is endoscopically injected around the primary tumor in the same way as is performed for gastric cancer [29]. Preoperative lymphatic scintigraphy is usually performed 3–4 h after injection of the tracer. Lymphatic scintigraphy makes it easy to understand the distribution of the sentinel lymph nodes, while gamma probing can identify the sentinel lymph node intraoperatively.

Although few studies have shown the validity of the sentinel lymph node concept in early esophageal cancer compared to that in EGC, some single-center studies have shown acceptable outcomes not only in thoracic esophageal squamous cell carcinoma but also in esophagogastric junctional adenocarcinoma [30–34]. We have also previously reported a feasibility study on radio-guided sentinel mapping for 75 patients with cT1 or cT2 esophageal cancer [35]. Our study demonstrated acceptable sentinel node detection in 95% and diagnostic accuracy in 94% of nodal metastases based on sentinel node status. Interestingly, the distribution of SN showed a widespread from cervical to abdominal areas in patients with thoracic esophageal cancer. Even in lower thoracic esophageal cancer, the sentinel nodes were mainly located in the lower and middle mediastinum and the abdominal area, although some patients showed sentinel nodes in the upper mediastinum. In more than 85% of patients with thoracic esophageal cancer, at least one sentinel node was found to be located in the Group 2 or 3 regional lymph nodes. A systematic review and meta-analysis reported the diagnostic value of sentinel node mapping in esophageal cancer [36]. A meta-analysis of 23 related studies including 623 patients, the detection rate of sentinel node and diagnostic accuracy was 93% and 88%, respectively. There was no significant difference in histological type between squamous cell carcinoma or adenocarcinoma and it was concluded that SN mapping was technically feasible for patients with esophageal cancer.

Radical esophagectomy with extended three-field lymphadenectomy, including the neck, mediastinum, and abdomen, has been established in Japan as a potential curative surgery for patients with thoracic esophageal squamous cell carcinoma. Esophagectomy with three-field lymphadenectomy may be appropriate due to the wide distribution of sentinel nodes and the presence of unpredictable metastatic lymph nodes. Therefore, the application of the sentinel node concept to thoracic esophageal squamous cell carcinoma may be limited. However, it may be possible to select cases and apply sentinel mapping with the aim to improve the curability of surgery and reduce postoperative complications. For example, if a sentinel node is identified along the recurrent laryngeal nerves in the upper mediastinum and is positive for cancer metastasis in a patient with lower esophageal cancer, extended lymphadenectomy for the upper mediastinum might be considered; this may be the case even in Ivor Lewis esophagectomy, which does not normally require upper mediastinum lymph node dissection. Also, if sentinel nodes are only identified in the abdominal area, and are negative for metastasis in patients with adenocarcinoma of the distal esophagus, it may be possible to treat with limited resection of the distal esophagus without the need for extensive mediastinal lymphadenectomy. In addition, if a combination of endoscopic submucosal dissection and sentinel lymph node biopsy can be performed simultaneously, as in gastric cancer, esophageal-preserving treatment may be possible in patients with early cancer located in the thoracic esophagus or at the esophagogastric junction. However, no multicenter prospective trial has been conducted to evaluate the sentinel node concept of esophageal cancer. Although further accumulation of evidence based on multicenter trials using standard protocols for sentinel node mapping is needed, SNNS has the potential to be an ideal and minimally invasive personalized treatment for esophageal cancer.

33.5 Conclusion

SNNS for early-stage upper gastrointestinal cancer is one of the most promising treatment options, as long as it is adapted by an appropriate method for the indication. With the development of procedures and devices, it is expected that SNNS will become more sophisticated and will be of considerable benefit to patients with EGC.

Open Questions

- To what extent can function-preserving limited gastrectomy be applied to gastric cancer using lymphatic mapping and the sentinel lymph node concept?
- Can training of surgeons in performing lymphatic mapping and sentinel lymph node biopsy be improved and standardized to allow accurate performance of the technique outside of current centers of excellence?
- Will ongoing clinical trials validate acceptable survival with improved morbidity with endoscopic resection and sentinel lymph node mapping in gastric cancer?
- Which patients with esophageal cancer are optimal candidates for lymphatic mapping and sentinel lymph node biopsy?

References

1. Japanese Gastric Cancer Association. Japanese gastric cancer treatment guidelines 2010 (ver. 3). Gastric Cancer. 2011;14(2):113–23.
2. Eagon JC, Miedema BW, Kelly KA. Postgastrectomy syndromes. Surg Clin North Am. 1992;72(2):445–65.
3. Nakada K, Ikeda M, Takahashi M, Kinami S, Yoshida M, Uenosono Y, Kawashima Y, Oshio A, Suzukamo Y, Terashima M, et al. Characteristics and clinical relevance of postgastrectomy syndrome assessment scale (PGSAS)-45: newly developed integrated questionnaires for assessment of living status and quality of life in postgastrectomy patients. Gastric Cancer. 2015;18(1):147–58.
4. Jamieson JK, Dobson JF. Lectures on the lymphatic system of the stomach. Lancet. 1907;169(4364):1061–6.
5. Japanese Gastric Cancer Association. Japanese classification of gastric carcinoma—2nd English edition. Gastric Cancer. 1998;1(1):10–24.
6. Kitagawa Y, Fujii H, Mukai M, Kubota T, Otani Y, Kitajima M. Radio-guided sentinel node detection for gastric cancer. Br J Surg. 2002;89(5):604–8.
7. Takeuchi H, Kitagawa Y. New sentinel node mapping technologies for early gastric cancer. Ann Surg Oncol. 2013;20(2):522–32.
8. Nimura H, Narimiya N, Mitsumori N, Yamazaki Y, Yanaga K, Urashima M. Infrared ray electronic endoscopy combined with indocyanine green injection for detection of sentinel nodes of patients with gastric cancer. Br J Surg. 2004;91(5):575–9.
9. Tajima Y, Yamazaki K, Masuda Y, Kato M, Yasuda D, Aoki T, Kato T, Murakami M, Miwa M, Kusano M. Sentinel node mapping guided by indocyanine green fluorescence imaging in gastric cancer. Ann Surg. 2009;249(1):58–62.

10. Miwa K, Kinami S, Taniguchi K, Fushida S, Fujimura T, Nonomura A. Mapping sentinel nodes in patients with early-stage gastric carcinoma. Br J Surg. 2003;90(2):178–82.
11. Kinami S, Fujimura T, Ojima E, Fushida S, Ojima T, Funaki H, Fujita H, Takamura H, Ninomiya I, Nishimura G, et al. PTD classification: proposal for a new classification of gastric cancer location based on physiological lymphatic flow. Int J Clin Oncol. 2008;13(4):320–9.
12. Lee YJ, Ha WS, Park ST, Choi SK, Hong SC, Park JW. Which biopsy method is more suitable between a basin dissection and pick-up biopsy for sentinel nodes in laparoscopic sentinel-node navigation surgery (LSNNS) for gastric cancer? J Laparoendosc Adv Surg Tech A. 2008;18(3):357–63.
13. Kitagawa Y, Takeuchi H, Takagi Y, Natsugoe S, Terashima M, Murakami N, Fujimura T, Tsujimoto H, Hayashi H, Yoshimizu N, et al. Sentinel node mapping for gastric cancer: a prospective multicenter trial in Japan. J Clin Oncol. 2013;31(29):3704–10.
14. Lee SE, Lee JH, Ryu KW, Cho SJ, Lee JY, Kim CG, Choi IJ, Kook MC, Nam BH, Park SR, et al. Sentinel node mapping and skip metastases in patients with early gastric cancer. Ann Surg Oncol. 2009;16(3):603–8.
15. Maruyama K, Sasako M, Kinoshita T, Sano T, Katai H. Can sentinel node biopsy indicate rational extent of lymphadenectomy in gastric cancer surgery? Fundamental and new information on lymph-node dissection. Langenbecks Arch Surg. 1999;384(2):149–57.
16. Siewert JR, Sendler A. Potential and futility of sentinel node detection for gastric cancer. Recent Results Cancer Res. 2000;157:259–69.
17. Hiratsuka M, Miyashiro I, Ishikawa O, Furukawa H, Motomura K, Ohigashi H, Kameyama M, Sasaki Y, Kabuto T, Ishiguro S, et al. Application of sentinel node biopsy to gastric cancer surgery. Surgery. 2001;129(3):335–40.
18. Hayashi H, Ochiai T, Mori M, Karube T, Suzuki T, Gunji Y, Hori S, Akutsu N, Matsubara H, Shimada H. Sentinel lymph node mapping for gastric cancer using a dual procedure with dye- and gamma probe-guided techniques. J Am Coll Surg. 2003;196(1):68–74.
19. Ichikura T, Morita D, Uchida T, Okura E, Majima T, Ogawa T, Mochizuki H. Sentinel node concept in gastric carcinoma. World J Surg. 2002;26(3):318–22.
20. Ohdaira H, Nimura H, Mitsumori N, Takahashi N, Kashiwagi H, Yanaga K. Validity of modified gastrectomy combined with sentinel node navigation surgery for early gastric cancer. Gastric Cancer. 2007;10(2):117–22.
21. Gretschel S, Bembenek A, Hünerbein M, Dresel S, Schneider W, Schlag PM. Efficacy of different technical procedures for sentinel lymph node biopsy in gastric cancer staging. Ann Surg Oncol. 2007;14(7):2028–35.
22. Lee JH, Ryu KW, Kim CG, Kim SK, Lee JS, Kook MC, Choi IJ, Kim YW, Chang HJ, Bae JM. Sentinel node biopsy using dye and isotope double tracers in early gastric cancer. Ann Surg Oncol. 2006;13(9):1168–74.
23. Zulfikaroglu B, Koc M, Ozmen MM, Kucuk NO, Ozalp N, Aras G. Intraoperative lymphatic mapping and sentinel lymph node biopsy using radioactive tracer in gastric cancer. Surgery. 2005;138(5):899–904.
24. Goto O, Takeuchi H, Kawakubo H, Sasaki M, Matsuda T, Matsuda S, Kigasawa Y, Kadota Y, Fujimoto A, Ochiai Y, et al. First case of non-exposed endoscopic wall-inversion surgery with sentinel node basin dissection for early gastric cancer. Gastric Cancer. 2015;18(2):434–9.
25. Eom BW, Yoon HM, Min JS, Cho I, Park JH, Jung MR, Hur H, Kim YW, Park YK, Nam BH, et al. Prospective multicenter feasibility study of laparoscopic Sentinel Basin dissection after endoscopic submucosal dissection for early gastric cancer: SENORITA 2 trial protocol. J Gastric Cancer. 2019;19(2):157–64.
26. Miyashiro I, Hiratsuka M, Sasako M, Sano T, Mizusawa J, Nakamura K, Nashimoto A, Tsuburaya A, Fukushima N. High false-negative proportion of intraoperative histological examination as a serious problem for clinical application of sentinel node biopsy for early gastric cancer: final results of the Japan Clinical Oncology Group multicenter trial JCOG0302. Gastric Cancer. 2014;17(2):316–23.
27. Lee JH, Ryu KW, Lee SE, Cho SJ, Lee JY, Kim CG, Choi IJ, Kook MC, Kim MJ, Park SR, et al. Learning curve for identification of sentinel lymph node based on a cumulative sum analysis in gastric cancer. Dig Surg. 2009;26(6):465–70.
28. Mayanagi S, Takahashi N, Mitsumori N, Arigami T, Natsugoe S, Yaguchi Y, Suda T, Kinami S, Ohi M, Kawakubo H, et al. Sentinel node mapping for post-endoscopic resection gastric cancer: multicenter retrospective cohort study in Japan. Gastric Cancer. 2020;23(4):716–24.
29. Takeuchi H, Kitagawa Y. Sentinel node navigation surgery for esophageal cancer. Gen Thorac Cardiovasc Surg. 2008;56(8):393–6.
30. Yasuda S, Shimada H, Chino O, Tanaka H, Kenmochi T, Takechi M, Nabeshima K, Okamoto Y, Kato Y, Kijima H, et al. Sentinel lymph node detection with Tc-99m tin colloids in patients with esophagogastric cancer. Jpn J Clin Oncol. 2003;33(2):68–72.
31. Burian M, Stein HJ, Sendler A, Piert M, Nährig J, Feith M, Siewert JR. Sentinel node detection in Barrett's and cardia cancer. Ann Surg Oncol. 2004;11(3 Suppl):255s–8s.
32. Lamb PJ, Griffin SM, Burt AD, Lloyd J, Karat D, Hayes N. Sentinel node biopsy to evaluate the metastatic dissemination of oesophageal adenocarcinoma. Br J Surg. 2005;92(1):60–7.
33. Bhat MA, Naikoo ZA, Dass TA, Lone RA, Dar AM. Role of intraoperative sentinel lymph node mapping in the management of carcinoma of the esophagus. Saudi J Gastroenterol. 2010;16(3):168–73.
34. Kim HK, Kim S, Park JJ, Jeong JM, Mok YJ, Choi YH. Sentinel node identification using technetium-99m neomannosyl human serum albumin in esophageal cancer. Ann Thorac Surg. 2011;91(5):1517–22.
35. Takeuchi H, Fujii H, Ando N, Ozawa S, Saikawa Y, Suda K, Oyama T, Mukai M, Nakahara T, Kubo A, et al. Validation study of radio-guided sentinel lymph node navigation in esophageal cancer. Ann Surg. 2009;249(5):757–63.
36. Nagaraja V, Eslick GD, Cox MR. Sentinel lymph node in oesophageal cancer—a systematic review and meta-analysis. J Gastrointest Oncol. 2014;5(2):127–41.

Sentinel Lymph Node Mapping in Non-small Cell Lung, Colon, and Thyroid Carcinomas

34

Joshua K. Kays and Mark B. Faries

Abstract

Lymphatic mapping and sentinel lymph node (SLN) biopsy is now a standard procedure for melanoma and breast cancer, providing the most accurate staging in those diseases. This technique is now also being applied to other malignancies, including non-small cell lung carcinoma, colon cancer, and thyroid cancer, although without widespread acceptance as yet. This chapter reviews the use of SLN biopsy in those malignancies including technical details, SLN detection rates, frequency of nodal upstaging, and potential clinical applications and limitations.

Learning Objectives

- Understand the importance of accurate lymph node staging for NSCLC, colon, and thyroid cancer prognosis and treatment.
- Appreciate the variability in lymphatic drainage and how it contributes to the limitations of traditional lymph node harvesting and pathologic examination for NSCLC, colon, and thyroid cancer.
- Describe the various techniques and success rates for SLN mapping.
- Understand how SLN mapping improves pathologic staging compared to traditional methods.
- Appreciate SLN mapping's role in the ever-changing care of NSCLC, colon, and thyroid cancer patients.

J. K. Kays
Cedars-Sinai Medical Center, Los Angeles, CA, USA
e-mail: Joshua.Kays@cshs.org

M. B. Faries (✉)
The Angeles Clinic and Research Institute and Cedars-Sinai Medical Center, Los Angeles, CA, USA
e-mail: mfaries@theangelesclinic.org

34.1 Introduction

Lymph node involvement is the single most important locoregional prognostic factor in nearly all solid tumors. Sentinel lymph node (SLN) mapping is a technique that involves injection of a dye or tracer at the primary tumor site in order to map the lymphatic drainage of that site and determine the location of the lymph node(s) receiving direct drainage from the tumor. This technique has proven to be successful in breast cancer and melanoma, revolutionizing the way these malignancies are treated. It has allowed for more precise staging, which has ultimately resulted in improved treatments and outcomes for patients with these cancers. More recently, this technique has been applied to other solid tumors with similar goals.

The concept of an SLN as a site of specific lymphatic drainage from an organ or body site of interest has been speculated and investigated for many years [1]. However, broad applicability remained elusive until Morton and colleagues described a surgical technique involving injection of vital blue dye at the site of a primary melanoma and identification of the blue, SLN in the draining nodal basin [2]. This technique proved accurate and reproducible, allowing the determination of the pathologic status of the nodal basin without the need for a complete nodal dissection. Further, this study showed the likelihood of a non-sentinel node harboring cancer with a negative sentinel lymph node was less than 1%.

Given the advantages of SLN biopsy that were demonstrated in melanoma, it was natural that the technique was investigated in other solid tumors. In breast cancer, it has also become a standard component of initial therapy. In other malignancies, though, several challenges have kept the technique from broad application. However, in non-small cell carcinoma of the lung (NSCLC), colon cancer, and thyroid cancer, the technique remains under active investigation. Its use in other diseases, including cancers of the upper gastrointestinal organs, is discussed elsewhere in this text.

S. P. Leong et al. (eds.), *Cancer Metastasis Through the Lymphovascular System*,
https://doi.org/10.1007/978-3-030-93084-4_34

34.2 Potential Benefits of Sentinel Lymph Node Biopsy in General

Prior to SLN mapping, regional lymph node basins were staged by removal of the entire basin and histopathologic examination. This approach not only engenders the risk of surgical morbidity for a larger procedure, it also burdens the pathologist with an increased amount of material to examine. In breast cancer and melanoma, the risks of dissection include lymphedema, which is probably one of the most feared complications. While that particular risk is not a major concern for nodal dissections for NSCLC, colon cancer or thyroid cancer, dissections for these cancers also present risks for increased morbidity to the patient. As discussed subsequently, the scrutiny a pathologist can practically place on a full lymph node dissection is less than that which can be used for a single, or small number of, SLNs. This enhanced pathologic analysis may also increase the accuracy of staging, in addition to decreasing morbidity.

34.3 Non-small Cell Lung Carcinoma

Lung cancer is the second most common cancer diagnosis in the United States and is the number one cause of cancer mortality. NSCLC represents over 80% of all lung cancers. Early-stage NSCLC can generally be defined as NSCLC without clinically apparent lymph node involvement. The primary treatment for these tumors is surgical resection with mediastinal lymph node sampling to stage the lymph nodes. Patients without lymph node involvement are considered to be free of disease and undergo surveillance, while those with lymph node involvement are generally treated with adjuvant chemotherapy with the goals of decreasing recurrence and improving survival.

The fact that patients with negative lymph node involvement by traditional pathologic methods have a 20–30% recurrence rate contradicts the notion that these patients are completely disease free. [3] It is suspected that occult lymph node metastases lead to this relapse. Furthermore, the presence of locoregional lymph node disease decreases 5-year survival by approximately 50% [4]. SLN biopsy has the potential to improve staging accuracy in NSCLC as it has in other malignancies, but technical issues that are particular to intrathoracic malignancies make mapping for these cancers more challenging. These challenges include the close proximity of many lymph nodes to the primary tumor site, the physically confined nature of the thoracic cavity, and the anthracotic nature of pulmonary lymph nodes making blue dye more difficult to see.

Table 34.1 shows an overview of studies conducted with corresponding SLN identification success rates. The first published study using an SLN technique in NSCLC was performed by Little et al. in 1999 [5]. They performed intraop-

Table 34.1 Sentinel lymph node detection rates

Non-small cell lung carcinoma				
Study	***N***	**Tracer**	**Injection method**	**SLN detection**
Little et al. [5]	36	ISB	PT	47% (17/36)
Liptay et al. [6]	45	^{99m}Tc-SC	PT	82% (37/45)
Norori et al. [7]	47	^{99m}Tc tin colloid	PT	87% (40/47)
Schmidt et al. [8]	31	^{99m}Tc -SC+ ISB	Intratumoral	80% (25/31)
Sugi et al. [9]	48	Indocyanine green	Intra-op PT	6.3% (1/16)
		ISB	Intra-op PT	50% (9/18)
		^{99m}Tc tin colloid	CT-guided PT	64.3% (9/14)
Melfi et al. [10]	26	^{99m}Tc-SC	Intra-op PT or[a] CT-guided PT	96% (25/26)
Lardinois et al. [11]	20	^{99m}Tc-SC	Endoscopic injection	95% 919/20)
Nakagawa et al. [12]	38	Ferumoxides	Intra-op PT	82% (31/38)
Faries et al. [13]	28	ISB + ^{99m}Tc-SC	Intra-op PT	100% (28/28)
Colon cancer				
Study	***N***	**Tracer**	**Injection method**	**SLN detection**
Bilchik et al. [14]	120	ISB and ^{99m}Tc-SC	Intra-op, in vivo + ex vivo, PT	96% (115/120)
Saha et al. [15]	86	ISB	Intra-op PT	99% (85/86)
Patten et al. [16]	57	ISB + ^{99m}Tc sulfur	Intra-op PT	98% (56/57)
Blichik et al. [17]	100	ISB	Intra-op, in vivo + ex vivo, PT	97% (97/100)
Saha et al. [18]	209	ISB/fluorescein + ^{99m}Tc-SC		100% (209/209)
Wong et al. [19]	124	ISB	Ex vivo, PT	98% (121/124)
Wood et al. [20]	75	ISB	Intra-op, in vivo + ex vivo, PT	96% (72/75)
Bertagnolli et al. [21]	72	ISB	Intra-op PT	92% (66/72)
Weisse, et al. [35]	83	ISB	Intra-op PT	99% (82/83)

ISB isosulfan blue, ^{99m}*Tc-SC* technetium-99m sulfur colloid, *PT* peritumoral
a—did not differentiate rates between methods/Tc-99m—Technetium 99m colloid

erative peritumoral injections of isosulfan blue. While they showed a relatively low SLN detection rate of 42%, they had a 100% negative predictive value, as no patient with a negative SLN was found to have disease in any other node after systematic mediastinal lymphadenectomy.

More recent studies have shown significantly improved SLN detection rates using different tracers, most notably technetium-99 colloids. Melfi et al. used pre- and intraoperative peritumoral injections of technetium-99 albumin nanocolloid to achieve 96% SLN detection rate [10], while Lardinois used transbronchial injections technetium-99 just prior to incision to obtain a 95% SLN detection rate with a 90% negative predictive value of the SLN [11]. Several others have shown similar results.

The ipsilateral peribronchial, hilar, and intrapulmonary nodes are the closest lymph node basins to a lung tumor. These are designated as the N1 nodal station and represent the most likely first place for the lymphatic spread of NSCLC. However, studies have shown that between 13 and 21% of SLN in NSCLC were in locations other than the N1 nodal station [11, 13]. This opens up the possibility of having skip metastases, which are metastases in nodes beyond the N1 station despite the most proximal nodes being negative. These skip metastases may be responsible for the high rate of recurrence of node-negative NSCLC. SLN mapping is a systematic way to identify aberrant lymph node drainage and may lead to more reliable identification of skip metastases than the current standard of mediastinal lymph node sampling.

Another significant value of SLN mapping is that it allows for "ultrastaging" of the regional lymph nodes. Traditional histopathologic evaluation involves hematoxylin and eosin (H&E) staining to allow examination of the cell structure. Immunohistochemistry (IHC) uses antibodies targeting cell-specific antigens to identify tissues and has been shown to accurately detect even isolated tumor cells. IHC, however, is more costly and time-consuming than traditional H&E staining, and its application to all lymph nodes harvested from a regional lymph node basin is not practical. SLN mapping allows for targeted use of IHC, which has been shown to discover metastatic disease in lymph nodes previously declared disease negative by H&E staining.

Upstaging rates for NSCLC from N0 to N+ disease using this ultrastaging technique have been reported as high as 30%. Having the nodes upstaged to N+ also appears to have significant implication on patient outcomes as studies have shown those with SLN involvement discovered by IHC have increased risk of recurrence, shorter recurrence-free survival (RFS), disease-free survival (DFS), and overall survival (OS).

Digesu et al. showed that SLN mapping with IHC was able to detect occult metastatic disease in 30% of all SLN specimens in patients with NSCLC [22]. Those with negative SLN determined by IHC had lower rates of recurrence and improved 5-year DFS and OS when compared to those who were deemed to be nodal negative by traditional mediastinal lymph node sampling. While this could be attributed to a sampling error, other studies have shown similar results.

Kuboschok et al. demonstrated a 21.6% upstaging rate of previously negative SLN after IHC [23]. Those discovered to have disseminated tumor cells by IHC had decreased DFS and OS on univariate analysis. Furthermore, multivariate analysis showed IHC positive nodes were associated with a 2.7 times increased risk of recurrence and a 2.5 times increased risk of shorter survival.

To date, SLN biopsy for NSCLC remains confined to certain specialized centers with an interest in the technique. More clinical data will likely be required to demonstrate the utility of the technique for this cancer before the adoption becomes more widespread.

34.4 Colon Cancer

Colon cancer is the third most common cancer diagnosis and the second leading cause of cancer deaths. Similar to NSCLC, early-stage colon cancer is generally defined as colon cancer without lymph node involvement. The primary treatment is colectomy with removal of the adjacent mesentery and the lymph nodes it contains. Examination of at least 12 lymph nodes is required as a minimum for accurate staging. Patients with negative lymph nodes are typically managed with surveillance, while those with positive nodes are treated with adjuvant chemotherapy. However, patients with "high-risk" Stage II colon cancers, including those with inadequate nodal assessment are considered for adjuvant chemotherapy as well.

In the same manner that NSCLC has a high recurrence rate among node-negative patients, a significant number of patients with early-stage colon cancer will also experience recurrence of the disease. Approximately 1/3 of node-negative patients will have disease recurrence and node positivity is associated with a relative decrease in survival by about 50% [24–26].

Table 34.1 shows an overview of studies conducted on SLN mapping in colon cancer and the corresponding rates of SLN identification. While nearly all studies use intraoperative peritumoral injections, mapping agents vary. The overall success rate of SLN identification is between 58 and 100% with most studies reporting >80% success rates. This is in contrast to mapping for rectal cancers, which has not been found to have acceptable accuracy in most studies.

Variability in lymphatic drainage with aberrant or unexpected SLN locations has also been encountered in colon cancer. While most studies show the SLN to be within the nodes in most immediate proximity to the tumor, there are studies that have demonstrated a not insignificant number of patients to have aberrant nodal drainage to locations that might not be removed in standard colectomy procedures.

Saha et al. used peritumoral injections of blue dye at the time of colon resection and found that 22% of patients had aberrant lymphatic drainage that required a change in operation to ensure removal of these nodes [27]. In addition, nodal positivity was higher in patients who required a change in operation, and 10% of the positive SLN were found in aberrant locations.

Ultrastaging sentinel lymph nodes with IHC have been shown to have the potential to upstage colon cancer as well. Weixler et al. showed that SLN mapping with blue dye followed by IHC examination resulted in a 26% upstaging rate to node-positive disease and that micrometastases and ITC were associated with decreased survival [28]. Protic et al. showed significantly lower disease recurrence rates in patients with IHC negative nodes (2.6% vs. 16.7%) and a significantly longer DFS (92.9 months vs. 71.8 months) compared to patients with IHC positive nodes [29]. While Bilchik et al. found a lower rate of upstaging after IHC at only 8%, they did show that 83% of recurrences at 25 months had lymph node metastases that were only detected by IHC [30]. They concluded that detection of SLN metastases by IHC after being deemed node negative by traditional staining methods may improve selection of patients for adjuvant chemotherapy and those with SLN negative by IHC are likely to be cured with surgery alone.

34.5 Thyroid Cancer

While thyroid cancer is not nearly as deadly as NSCLC or colon cancer, nodal staging is a standard component of treatment for many patients and lymph node involvement is a poor prognostic feature, particularly among older patients. Nodal involvement also requires additional treatment, which in the case of thyroid cancer may involve a complete central or lateral lymph node dissection. While complete central and/or lateral neck dissection is generally a safe procedure, there are risks including nerve injury or inadvertent removal of parathyroid glands. SLN mapping has been investigated as a means to allow the identification of patients who should have a neck dissection performed and those in whom it may be possible to omit this procedure.

The data on the utility of SLN mapping in thyroid cancer are not as extensive as it is for other previously mentioned malignancies. Albers et al. found an SLN detection rate that was 67% by preoperative scintigraphy, but dropped to 46% when using intraoperative gamma probe [31]. Cabera et al. showed a much higher rate of detection of an SLN by scintigraphy at 91%. However, the final pathologic examination showed a not insignificant false-negative rate with 13% of patients with a negative SLN having non-SLN with metastases [32]. Keleman et al. showed that 11.8% of patients who underwent SLN mapping had SLN in the internal jugular region in addition to the paratracheal SLN [33]. Finally, Balasubrammania et al. presented an analysis of the theoretical cost and benefit of SLN biopsy suggested the procedure would only save seven patients from central neck lymph node dissection in a 5-year period at their institution [34].

34.6 Clinical Application and Future Direction

The data on SLN mapping for NSCLC and colon cancer clearly demonstrate that it is technically feasible, readily reproducible, and has a high success rate of SLN identification. Adding IHC to SLN that were determined to be negative by traditional pathologic methods appears to identify a significant proportion of patients that may be upstaged to node-positive and may benefit from adjuvant therapy to decrease recurrence and improve survival. Conversely, patients with colon cancer who have negative SLN may have a sufficiently favorable prognosis that adjuvant systemic therapy and its attendant morbidity is not necessary for them. Future trials enrolling patients with micrometastases and ITC discovered by SLN mapping and IHC will need to be performed to determine the true benefit. As of this writing the benefit of SLN mapping in thyroid cancer is less clear and additional studies are needed to establish its utility.

34.7 Limitations of SLN Mapping

One of the major limitations preventing the widespread application of SLN mapping in NSCLC and colon cancer is the lack of standardization with the technique. Among the published studies, various techniques have been used with various results; preoperative versus intraoperative injection, peritumoral versus CT-guided versus endoscopic, blue dye versus technetium-99 colloids versus ICG. A standardized approach may result in more acceptance among clinicians and widespread usage. As technology improves, optimal techniques may emerge in each of these malignancies.

Another factor limiting interest in SLN biopsy for these malignancies is the paucity of data proving clinical benefit. While it has been shown SLN mapping is feasible and that it can upstage disease, studies have yet to show a clinical benefit with regard to decreased recurrence or improved survival. Further studies need to be performed to demonstrate the clinical benefit before SLN mapping will be likely to be widely adopted for these cancers.

34.8 Conclusions

While regional nodal assessment using lymphatic mapping and sentinel lymph node biopsy has proven to be an invaluable technique in several common malignancies, its role in the lung, colon, and thyroid cancer is still being defined. In NSCLC, improvements in technology and continued clinical research evaluations will help improve the applicability of the technique there. In colon cancer, adaptation of lymphatic mapping to the minimally invasive techniques that are more commonly employed for that disease will likely be essential to broadening the utility of sentinel node evaluation in that disease. In particular, if the accuracy of nodal staging can be improved with SLN mapping, more patients may be able to avoid the toxicity of adjuvant chemotherapy. The biology of thyroid cancer may limit the utility of SLN mapping in that disease, but additional research may demonstrate populations for whom the technique is of particular value.

Future Perspective Questions

- Can prognostic evaluation of patients with NSCLC be improved by SLN mapping?
- Will technologies, including fluorescence imaging in minimally invasive thoracic surgery enable more widespread application of SLN mapping in that disease?
- Can ultrastaging of regional lymph nodes be made feasible using SLN mapping and would such staging enable more patients to safely avoid adjuvant chemotherapy?
- Which patients with thyroid cancer may benefit from nodal staging using SLN mapping?

References

1. Kett K, Varga G, Lukacs L. Direct lymphography of the breast. Lymphology. 1970;3(1):3–12.
2. Morton DL, Wen DR, Wong JH, Economou JS, Cagle LA, Storm FK, Foshag LJ, Cochran AJ. Technical details of intraoperative lymphatic mapping for early stage melanoma. Arch Surg. 1992;127(4):392–9.
3. Lou F, Huang J, Sima CS, Dycoco J, Rusch V, Bach PB. Patterns of recurrence and second primary lung cancer in early stage lung cancer survivors followed with routine CT surveillance. J Thorac Cardiovasc Surg. 2013;145(1):75–82.
4. Rusch VW, Hawes D, Decker PA, Martin SE, Abati A, Landreneau RJ, Patterson GA, Inculet RI, Jones DR, Malthaner RA, Cohen RG. Occult metastases in lymph nodes predict survival in resectable non-small cell lung cancer: report of the ACOSOG Z0040 trial. J Clin Oncol. 2011;29(32):4313.
5. Little AG, DeHoyos A, Kirgan DM, Arcomano TR, Murray KD. Intraoperative lymphatic mapping for non-small cell lung cancer: the sentinel node technique. J Thorac Cardiovasc Surg. 1999;117(2):220–4.
6. Liptay MJ, Masters GA, Winchester DJ, Edelman BL, Garrido BJ, Hirschtritt TR, Perlman RM, Fry WA. Intraoperative radioisotope sentinel lymph node mapping in non-small cell lung cancer. Ann Thorac Surg. 2000;70(2):384–9.
7. Normori H, Horio H, Naruke T, Orikasa H, Yamazaki K, Suemasu K. Use of technetium-99m tin colloid for sentinel lymph node identification in non-small cell lung cancer. J Thorac Cardiovasc Surg. 2002;124(3):486–92.
8. Schmidt FE, Woltering EA, Webb WR, Garcia OM, Cohen JE, Rozans MH. Sentinel nodal assessment in patients with carcinoma of the lung. Ann Thorac Surg. 2002;74(3):870–5.
9. Sugi K, Fukuda M, Nakamura H, Kaneda Y. Comparison of three tracers for detecting sentinel lymph nodes in patients with clinical N0 lung cancer. Lung Cancer. 2003;39(1):37–40.
10. Melfi FM, Chella A, Menconi GF, Givigliano F, Boni G, Mariani G, Sbragia P, Angeletti CA. Intraoperative radioguided sentinel lymph node biopsy in non-small cell lung cancer. Eur J Cardiothorac Surg. 2003;23(2):214–20.
11. Lardinois D, Brack T, Gaspert A, Spahr T, Schneiter D, Steinert HC, Weder W. Bronchoscopic radioisotope injection for sentinel lymph node mapping in potentially resectable non-small cell lung cancer. Eur J Cardiothorac Surg. 2003;23(5):824–7.
12. Nakagawa T, Minamiya Y, Katayose Y, Saito H, Taguchi K, Imano H, Watanabe H, Enomoto K, Sageshima M, Ueda T, Ogawa JI. A novel method for sentinel lymph node mapping using magnetite in patients with non-small cell lung cancer. J Thorac Cardiovasc Surg. 2003;126(2):563–7.
13. Faries MB, Bleicher RJ, Ye X, Essner R, Morton DL. Lymphatic mapping and sentinel lymphadenectomy for primary and metastatic pulmonary malignant neoplasms. Arch Surg. 2004;139(8):870–7.
14. Bilchik AJ, Nora DT, Sobin LH, Turner RR, Trocha S, Krasne D, Morton DL. Effect of lymphatic mapping on the new tumor-node-metastasis classification for colorectal cancer. J Clin Oncol. 2003;21(4):668–72.
15. Saha S, Wiese D, Badin J, Beutler T, Nora D, Ganatra BK, Desai D, Kaushal S, Nagaraju M, Arora M, Singh T. Technical details of sentinel lymph node mapping in colorectal cancer and its impact on staging. Ann Surg Oncol. 2000;7:120–4.
16. Patten LC, Berger DH, Rodriguez-Bigas M, Mansfield P, Delpassand E, Cleary KR, Fagan SP, Curley SA, Hunt KK, Feig BW. A prospective evaluation of radiocolloid and immunohistochemical staining in colon carcinoma lymphatic mapping. Cancer. 2004;100(10):2104–9.
17. Blichik AJ, Nora D, Tollenaar RA, Van de Velde CJ, Wood T, Turner R, Morton DL, Hoon DS. Ultrastaging of early colon cancer using lymphatic mapping and molecular analysis. Eur J Cancer. 2002;38(7):977–85.
18. Saha S, Dan AG, Beutler T, Wiese D, Schochet E, Badin J, Branigan T, Ng P, Bassily N, David D. Sentinel lymph node mapping technique in colon cancer. Semin Oncol. 2004;31(3):374–81.
19. Wong JH, Johnson DS, Namiki T, Tauchi-Nishi P. Validation of ex vivo lymphatic mapping in hematoxylin-eosin node-negative carcinoma of the colon and rectum. Ann Surg Oncol. 2004;11(8):772–7.
20. Wood TF, Saha S, Morton DL, Tsioulias GJ, Rangel D, Hutchinson W, Foshag LJ, Bilchik AJ. Validation of lymphatic mapping in colorectal cancer: in vivo, ex vivo and laparoscopic techniques. Ann Surg Oncol. 2001;8(2):150–7.
21. Bertagnolli M, Miedema B, Redston M, Dowell J, Niedzwiecki D, Fleshman J, Bem J, Mayer R, Zinner M, Compton C. Sentinel node staging of resectable colon cancer. Results of a multicenter study. Ann Surg. 2004;240(4):624.

22. Digesu CS, Hachey KJ, Gilmore DM, Khullar OV, Tsukada H, Whang B, Chirieac LR, Padera RF, Jaklitsch MT, Colson YL. Long term outcomes after near-infrared sentinel lymph node mapping in non-small cell lung cancer. J Thorac Cardiovasc Surg. 2018;155(3):1280–91.
23. Kubuschok B, Passlick B, Izbicki JR, Thetter O, Pantel K. Disseminated tumor cells in lymph nodes as a determinant for survival in surgically resected non-small cell lung cancer. J Clin Oncol. 1999;17(1):19–9.
24. Le Voyer TE, Sigurdson ER, Hanlon AL, Mayer RJ, Macdonald JS, Catalano PJ, Haller DG. Colon cancer survival is associated with increasing number of lymph nodes analyzed: a secondary survey of intergroup trial INT-0089. J Clin Oncol. 2003;21(15):2912–9.
25. Cohen AM, Kelsen D, Saltz L, Minsky BD, Nelson H, Farouk R, Gunderson LL, Michelassi F, Arenas RB, Schilsky RL, Willet CG. Adjuvant therapy for colorectal cancer. Curr Probl Surg. 1997;34(8):601–76.
26. Cohen AM, Tremiterra S, Candela F, Thaler HT, Sigurdson ER. Prognosis of nodal positive cancer. Cancer. 1991;67(7):1859–61.
27. Saha S, Elgamal M, Cherry M, Buttar R, Pentapati S, Mukkamala S, Devisetty K, Kaushal S, Alnounou M, Singh T, Grewal S. Challenging the conventional treatment of colon cancer by sentinel lymph node mapping and its role of detecting micrometastases for adjuvant chemotherapy. Clin Exp Metastasis. 2018;36(5–6):463–9.
28. Weixler B, Rickenbacher A, Raptis DA, Viehl CT, Guller U, Rueff J, Zetti A, Zuber M. Sentinel lymph node mapping with isosulfan blue or indocyanine green in colon cancer shows comparable results and identifies patients with decreased survival: a prospective single-center trial. World J Surg. 2017;41(9):2378–86.
29. Protic M, Stojadinovic A, Nissan A, Wainberg Z, Steele SR, Chen DC, Avital I, Bilchik AJ. Prognostic effect of ultra-staging node-negative colon cancer without adjuvant chemotherapy: a prospective national cancer institute-sponsored clinical trial. J Am Coll Surg. 2015;221(3):643–51.
30. Bilchik AJ, Hoon DS, Saha S, Turner RR, Wiese D, DiNome M, Koyanagi K, McCarter M, Shen P, Iddings D, Chen SL. Prognostic impact of micrometastases in colon cancer. Interim results of a prospective multicenter trial. Ann Surg. 2007;246(4):568–77.
31. Albers MB, Nordenstrom E, Wohlfahrt J, Bergenfelz A, Almquist M. Sentinel lymph node biopsy in thyroid cancer. World J Surg. 2020;44(1):142–7.
32. Cabrera RN, Chone CT, Zantut-Wittmann D, Matos P, Perreira DM, Pereira PS, Ferrari RJ, Santos AO, Crespo AN, Etchebehere EC. Value of sentinel lymph node biopsy in papillary thyroid cancer: initial results of a prospective trial. Eur Arch Otorhinolaryngol. 2015;272(4):971–9.
33. Keleman PR, Van Herle AJ, Giuliano AE. Sentinel lymphadenectomy in thyroid malignant neoplasms. Arch Surg. 2008;133(3):288–92.
34. Balasubramanian SP, Brignall J, Lin HY, Stephenson TJ, Wadsley J, Harrison BJ, Craig WL, Smart L, Krukowski. Sentinel node biopsy in papillary thyroid cancer—what is the potential? Langenbecks Arch Surg. 2014;399(2):245–51.
35. Wiese DA, Saha S, Badin J, Ng PS, Gauthier J, Ahsan A, Yu L. Pathologic evaluation of sentinel lymph nodes in colorectal carcinoma. Arch Pathol Lab Med. 2000;124(12):1759–63.

Part IX

Surgical Treatment of Primary and Metastatic Cancer

Giorgos Karakousis and Dale Han

The surgical management of cancer has undergone remarkable changes over the past century based on an improved understanding of tumor biology, the development of effective systemic therapies, and the introduction of technologies to detect patient-specific molecular aberrations and therapeutic targets, thereby allowing for individualized, precision oncologic care. In contrast, prior to the mid-twentieth century, most oncologic surgeries were performed with a rudimentary understanding of cancer biology and a lack of other treatment options. The ability to control cancer and provide a durable cure was relatively limited during this early period of oncologic care and the prognosis for cancer patients was generally poor. However, the introduction of radical surgery by William Stewart Halsted in the late nineteenth century revolutionized the thought process of the treatment of tumors and ultimately helped to pave the way for the development of newer theories on cancer biology, such as the systemic theory and later the spectrum theory, which challenged the Halstedian view of orderly cancer progression. In parallel with an improved understanding of cancer biology, there were significant developments in other modalities for cancer treatment, including radiation therapy and systemic therapies. In particular, systemic therapy has seen an evolution from a relatively unselective approach with uniform usage of cytotoxic chemotherapeutics to more recently the development of targeted therapies taking advantage of specific genetic aberrations in certain cancers, and finally to harnessing the power of the innate host immune system to eradicate cancers and provide durable regression of disease.

The efficacy of surgical therapy for cancers is now determined by several factors, and this section will focus on the development and evolution of surgical treatments in specific cancer types based on cancer biology and the availability of other therapies that can be used as multimodality therapy in conjunction with surgery. Drs. Georgia Beasley, Michael Lowe, and Genevieve Boland will specifically discuss the role of surgery in managing primary and metastatic melanoma, while Drs. Alicia Terando, Mara Piltin, Tina Hieken, and Carla Fisher will evaluate the role of surgery in managing primary and metastatic breast cancer. Our focus will then change to a discussion of the role of surgery in treating several intra-abdominal cancers with Drs. Najjia Mahmoud, Kiran Turaga, and David Shibata discussing the role of surgery in managing primary and metastatic colorectal cancer and Drs. Robert Roses and Jose Pimiento assessing the role of surgery in managing primary and metastatic gastric cancer. Finally, Drs. Ajay Maker, George Poultsides, and Shishir Maithel will discuss the role of surgery in specifically managing primary and metastatic hepatopancreaticobiliary cancers.

In demonstrating how the role of surgery has evolved and developed in these various cancer types, our goals are to demonstrate how our armamentarium of therapy options has increased as our comprehension of cancer biology has evolved, and show that while surgery remains an integral component to oncologic treatment, it is but one of many treatment options we have at hand which should be used in conjunction with, not exclusively from, other available treatment modalities. In light of these developments and in line with the shift from a more radical to conservative approach, the role of surgery in the treatment of cancer remains essential but is evolving, with surgical oncologists now playing an active role in all stages of the oncologic treatment spectrum, including preventive counseling and screening, initial cancer diagnosis, collaboration with multidisciplinary teams to determine the optimal sequence of treatment and interventions, postoperative surveillance monitoring, end-of-life care, and high-quality research.

The Efficacy and Evolution of Surgical Management Based on Cancer Biology

35

Richard J. Straker III, Hayley Standage, Giorgos C. Karakousis, and Dale Han

Abstract

The surgical management of cancer has changed dramatically over the past century. As our understanding of cancer biology has evolved, surgical techniques have also changed to align with the concepts of tumor growth and metastasis. William Stewart Halsted's theory of radical surgery at the end of the nineteenth century represented early methods for curative therapy for cancers. However, the introduction of alternative theories, including the systemic theory and later the spectrum theory, has helped to change our approach to cancer surgery, such that some cancers are treated with more radical surgery, while other tumors may benefit from more conservative approaches. Furthermore, the development of effective systemic therapies and precision molecular medicine has also influenced how certain cancers are managed, and while the role and timing of surgery may be changing, it remains an integral component of modern cancer care. As such, a universal treatment algorithm cannot be mandated, but each cancer must be approached individually and treated with the best combination of available therapies at hand for that specific malignancy.

Richard J. Straker and Hayley Standage contributed equally to the manuscript and share first authorship.

Giorgos C. Karakousis and Dale Han contributed equally to the manuscript and share senior authorship.

R. J. Straker III (✉) · G. C. Karakousis
Division of Endocrine and Oncologic Surgery, Department of Surgery, University of Pennsylvania, Philadelphia, PA, USA
e-mail: Richard.straker@pennmedicine.upenn.edu; giorgos.karakousis@pennmedicine.upenn.edu

H. Standage · D. Han
Division of Surgical Oncology, Oregon Health and Science University, Portland, OR, USA
e-mail: standagh@ohsu.edu; handal@ohsu.edu

Learning Objectives

- To understand the different theories on cancer biology that have developed over time.
- To understand the historical and current role of systemic therapy in the management of cancer.
- To understand the evolution of surgical management of cancers in light of developing theories of cancer biology and the development of effective systemic therapies.
- To understand the modern molecular techniques being used to diagnose, identify, and treat cancer in the setting of evaluating patients for surgical management.

35.1 Introduction

The surgical management of cancer has undergone considerable changes over the past century based on an improved understanding of tumor biology and the development of effective systemic therapies. Prior to the mid-twentieth century, most oncologic surgeries were performed for palliation [1]. Other treatment options, such as radiation and systemic therapies, either were not available or were novel approaches that were not generally efficacious. The ability to control cancer and provide a durable cure was relatively limited, and the prognosis for cancer patients was generally poor [2]. However, the introduction of radical surgery by William Stewart Halsted in the late nineteenth century revolutionized the thought process of the treatment of tumors [3]. This philosophy dominated oncology therapy for most of the next century.

Oncologic care in the latter half of the twentieth century evolved as adjunct treatments such as radiation and systemic therapies demonstrated beneficial outcomes for cancer patients. Alternative theories of cancer biology also influenced the view of cancer development and treatment. For

S. P. Leong et al. (eds.), *Cancer Metastasis Through the Lymphovascular System*,
https://doi.org/10.1007/978-3-030-93084-4_35

instance, Bernard Fisher challenged the Halstedian paradigm with the systemic theory of cancer biology, which postulated that the systemic spread of cancer could occur earlier in the disease process [4]. Radical surgery would therefore be inadequate to halt the spread of cancer. The acceptance of the systemic theory of cancer biology led to fundamental changes in the management of breast cancer. However, it has since been realized that the biology of cancers and even different tumors of the same cancer category can vary tremendously. It is now theorized that cancer biology is a heterogeneous spectrum. Tumors may remain localized throughout the disease course, while other cancers may spread systemically relatively early. Therefore, treatment of a specific tumor should be tailored to the specific biology of that cancer and may include surgery or a multimodal approach with a combination of surgery, radiation therapy, and systemic therapy [4].

The efficacy of surgical therapy for cancers is determined by several factors, such as the biology of a specific tumor and the availability and effectiveness of therapies that can be utilized in conjunction with surgery to optimize outcomes [5]. For example, surgical treatment of a primary tumor may provide more benefit in cancers with a greater tendency to remain localized or in cancer types in which effective systemic therapies are available. In contrast, metastasectomy may have limited utility in cases in which there is more aggressive biology or in which systemic therapies are less effective in controlling the metastatic disease. This section will focus on the development and evolution of surgical treatments in specific cancer types based on cancer biology and the availability of other therapies that can be used as multi-modality therapy in conjunction with surgery.

35.2 Historical Evolution of Theories on Cancer Biology

An understanding of cancer biology was limited until the mid-twentieth century. Early surgical treatment for cancer was based on anatomic principles and an early rudimentary understanding of tumor and disease spread. As such, surgeons simply excised the affected anatomical structures. Additionally, adjunctive options for disease control such as radiation therapy or chemotherapy were not yet available [1]. As a result, operations were often palliative, and the prognosis was generally poor [2]. In the late nineteenth century, Dr. William Stewart Halsted introduced the concept of radical surgery that transformed the approach to cancer treatment. The Halstedian model of cancer progression stated that tumors progressed by spread through contiguous structures in an orderly fashion from the primary site to the level of the regional lymph nodes, and then to the systemic level at distant sites. Appropriate therapy that removed not only the cancer but also adequate surrounding tissue to control that level (e.g., radical surgery) could prevent cancer from spreading to the subsequent level. This idea was first introduced in his discussion of the radical mastectomy [3] and was then applied to other cancer types. The philosophy of radical surgical control of cancer dominated oncology therapy for much of the next century.

The Halstedian paradigm was challenged in the latter half of the twentieth century by Dr. Bernard Fisher, who theorized that breast cancer was a systemic disease. Fisher observed evidence of cancer recurrence at various sites in many patients despite radical surgery. In animal models, he found cancer cells in circulation when the tumor otherwise appeared clinically localized. The Fisher theory of cancer biology, also known as the systemic theory, postulated that cancers did not progress in an orderly fashion and could instead spread both via the lymphatic system and hematogenously. Therefore, the systemic spread of cancer could occur earlier in the disease process. A radical surgical approach would have limited utility to halt the progression of cancer that was already present at distant sites at disease presentation [6]. Fisher's theory led to fundamental changes in breast cancer treatment. For instance, less radical surgeries were completed to treat the primary tumor (e.g., lumpectomy), and additional therapies such as radiation therapy were used to further control the primary site. Systemic therapy was utilized in cases that appeared clinically localized but were at high risk for disease recurrence.

The Fisher model of cancer spread was controversial and met with resistance, but the preponderance of data supporting the doctrines of this theory, through the National Surgical Adjuvant Breast and Bowel Project clinical trials, led to general acceptance of his theory and the development of new management paradigms in breast cancer [6]. Subsequently, however, it was observed that the biology of cancers appeared to vary according to different cancer types and to be far more complex than what could be explained by the models developed by Halsted and Fisher. In 1994, Hellman et al. introduced the "spectrum theory" of metastases, which theorized that cancer biology was a heterogeneous spectrum, with some tumors remaining localized throughout the disease course and other tumors presenting with systemic disease at first detection [7]. The biology of a specific cancer was the function of that specific type of tumor growth and progression. Persistence of disease increased the risk for cancer spread and progression at secondary sites. For instance, lymph node metastasis was a marker of more aggressive biology but could also serve as a nidus for further disease metastases. Therefore, locoregional therapy could potentially have an important role in the prevention of disease progression and decrease opportunities for cancer cells to develop into metastatic lesions. Moreover, the importance of the host, particularly the immune system, in influencing the progression of cancers continues to be an area of active investigation [8].

As the biological theories of cancer continue to evolve, the approach to treatment is also adapting. Early surgeons focused on anatomical resections to provide palliation. These surgeries were subsequently modified to optimize surgical outcomes and survival. However, alternative theories based on biological principles, which seemed radical at the time, challenged widely accepted models, and provided important foundations for the modern understanding of cancer biology. Furthermore, the development of efficacious treatment options like radiation and systemic therapies, the addition of cancer screening tools, and innovations in molecular science have changed how we view cancer biology and treatment.

35.3 Evolution of Systemic Therapy

First introduced in the mid-1900s, initial systemic chemotherapeutic agents were broadly used for many types of cancer. These drugs were nondiscriminatory against which cells they acted upon, resulting in severe toxicities and poor tolerance [9]. As our knowledge of cellular biology progressed in parallel with medical technological advancements, these treatments have evolved into the systemic agents of the modern era which have the ability to precisely target individual cellular components and can be modified to fit patient-specific genetic mutations.

Systemic chemotherapy was first studied during World War I when scientists discovered on autopsy evaluations that mustard gas, at that time being utilized as a chemical warfare agent, led to extensive myelosuppression in those poisoned [10]. Further research identified these nitrogen and sulfur mustard compounds to be DNA-alkylating agents, which formed covalent bonds with purines and induced apoptosis [11]. These and other nitrogen mustards, including cyclophosphamide, became first-line agents for treating lymphomas, leukemias, and sarcomas, and some of these drugs are still used today [12, 13]. Discovery of additional alkylating agents, comprising the well-known classes of alkyl sulfates (busulfan), triazenes (dacarbazine), and platinum compounds, ensued, and many of these agents remain mainstay components of modern systemic treatment regimens [14].

The next major breakthrough came in the form of antimetabolite therapy, which was initially developed in the late 1940s. Antifolate compounds were the first antimetabolites studied and demonstrated the ability to induce remission in patients with acute lymphoblastic leukemia (ALL) [15]. Initially not understood, the mechanism behind antifolate therapy was eventually discovered to be permanent inhibition of dihydrofolate reductase (DHFR), leading to the inability of cells to complete purine production with subsequent apoptosis. Methotrexate is one of the most well-known antifolates and is very prevalent in the modern treatment of several lymphoma subtypes, osteosarcoma, and choriocarcinoma [9]. Expanding on the work with antifolates, numerous additional classes of antimetabolites were created, mostly functioning via the inhibition of purine or pyrimidine production, and having major impacts on both solid and hematologic malignancies. Five-fluorouracil in particular, a pyrimidine antimetabolite, has revolutionized the treatment of gastrointestinal malignancies, especially colorectal cancer [16].

As these aforementioned categories of drugs were being modified and expanded upon, the next few decades witnessed the introduction of several novel cytotoxic chemotherapeutic classes. These included the Vinca alkaloids (vincristine, vinblastine), which function to destabilize microtubules and are used today for treating ALL, lymphomas, metastatic breast cancer and small cell lung cancer, topoisomerase inhibitors (irinotecan), which inhibit DNA replication and are used today for treating colorectal cancer and ovarian cancer, and taxanes (paclitaxel, docetaxel), which impair microtubule dynamics and are frequently used for treating patients with metastatic breast or pancreatic cancer, non-small cell lung cancer, and gastric cancer [17, 18]. Antibiotics with antitumor effects were discovered and include anthracyclines (doxorubicin, daunorubicin), which act on DNA topoisomerase, and bleomycin, which is commonly used today in treating lymphomas, testicular cancer, and cervical cancer [19].

Starting in the 1980s and continuing through the present time, rapid advances in technology led to landmark discoveries in the fields of immunology, biochemistry, and cellular biology. These findings have allowed for the highly sophisticated understanding of the molecular mechanisms driving cancer development that we hold today, and have permitted the identification of numerous molecular compounds that act as therapeutic targets including growth factors, signaling molecules, and cell-cycle proteins. Key to these new drugs is their ability to pinpoint the products of mutated oncogenes and tumor suppressor genes while avoiding the wild-type products, thereby markedly reducing the toxic effects of systemic therapy that are observed with cytotoxic agents [9, 20].

One of the most influential breakthroughs at the beginning of the targeted therapeutic era was the development of the tyrosine kinase inhibitor imatinib mesylate (Gleevec), which targets the BCR-ABL receptor, c-KIT, and platelet-derived growth factor receptor. Used to treat chronic myelogenous leukemia and gastrointestinal stromal tumors, this drug was shown to be slowly metabolized, well-absorbed when taken orally, and incited minimal adverse events [21]. It was the prototypical example for all future targeted therapies to be modeled after and paved the way for the development of agents targeting epidermal growth factor receptor, MAPK/MEK pathway, vascular endothelial growth factor receptor, mTOR, and many more that are used to treat a variety of both solid and hematologic malignancies [18].

The next epochal turning point was the development of immune-mediated therapies in the form of monoclonal antibodies directed at cellular checkpoint pathways, which function to regulate host immune responses. Termed immune checkpoint inhibitors, the most common agents act against cytotoxic T-lymphocyte-associated antigen 4 (ipilimumab) and programmed cell death protein 1 (nivolumab, pembrolizumab), both of which lead to the activation of pathways that downregulate the host immune system. By blocking these inhibitory pathways, these immune checkpoint inhibitors prevent tumors from taking advantage of them and evading immune detection [22]. Ipilimumab, the first approved of these agents, is currently used alone or in combination with nivolumab for unresectable or metastatic melanoma, and is also being evaluated in the management of renal cancer, non-small cell lung cancer, and other types of cancer [23]. Nivolumab and pembrolizumab are frequently used for the treatment of locoregionally advanced and metastatic melanoma and for non-small cell lung cancer, along with urothelial carcinoma [9, 24].

35.4 Evolution of Surgical Management of Cancer

The surgical management of various cancer types has evolved and changed as our understanding of cancer biology has improved and adjunct therapies have become available. For instance, the introduction of the radical mastectomy by Halsted in the late nineteenth century represented an important shift in breast cancer management. Halsted's predecessors performed a variety of anatomical excisions for breast tumors, but these operations had high rates of complications and disease recurrence [25]. As a result, early breast cancer surgeries were generally palliative. Halsted found that many patients developed cancer recurrences near the initial primary tumor site, and he theorized that cancer spread contiguously from local, to regional, and then to distant sites [6, 25]. He subsequently developed the radical, or en bloc, method to remove all vulnerable tissue for locoregional control before the disease spread elsewhere [6, 25]. Halsted's patients had lower rates of local recurrence [6], and as a result, the radical mastectomy was considered the standard of care for much of the next century [7]. However, surgeons found that improved locoregional control did not necessarily improve survival [26].

Fisher's challenging of Halsted's theory of cancer spread with his own systemic theory resulted in the shift of breast cancer surgery from a radical approach to more conservative techniques like lumpectomy in conjunction with radiotherapy and systemic treatments [6]. Subsequently, results of the ACOSOG Z0011 trial demonstrated no survival benefit for performing lymphadenectomy in certain positive sentinel node patients, thereby further reducing the role of surgery in the locoregional management of specific breast cancer patients [6]. In addition, the introduction of screening mammography at the end of the twentieth century enabled earlier detection of breast cancers [10] and identified patients more likely to have disease restricted to the primary site, thus leading to the "spectrum theory" in which breast cancer was considered a heterogeneous disease with a variety of presentations. Tumors may remain locoregional or possess a higher propensity for spread and progression [27].

For melanoma, early surgical management was also palliative with techniques that ranged from cauterization to amputation [27, 28]. Extreme margins of 4–5 cm were used to control the primary site initially, but these margins were subsequently decreased to 1–2 cm based on several important clinical trials that showed no significant differences in local recurrence rates with narrower margins. In addition, in the late nineteenth century, surgeons recognized that melanomas metastasized to regional lymph nodes. Herbert Snow recommended the addition of lymphadenectomy in all melanoma excisions in 1892 [27]. Analogous to Halsted's radical mastectomy, the en bloc approach for melanoma, in which the primary melanoma was removed along with the draining nodal basin, whether therapeutically or electively, remained the standard of care for many decades [29].

The development of sentinel lymph node biopsy (SLNB) was another important advancement in melanoma surgery. Based on the theory of lymphatic spread, Donald Morton demonstrated that determining the status of the SLN provided the status of the entire draining nodal basin using a less extensive technique. Positive SLN patients underwent complete lymph node dissection (CLND), whereas negative SLN patients were spared unnecessary lymphadenectomies. However, the biology of the SLN remained debated with two theories that were developed [30]. The "incubator hypothesis" theorized that tumor cells spread in an orderly fashion from the primary site to the regional lymph nodes and then to distant sites. Therefore, control of disease at certain levels could potentially prevent further disease progression. In contrast, the "marker hypothesis" suggested that tumor cells in an SLN served as an early marker of systemic disease spread. In this theory, control of nodal disease would not affect the ultimate outcome of patients with nodal metastases. DeCOG-SLT and Multicenter Selective Lymphadenectomy Trial-II provided support for the marker hypothesis by demonstrating no differences in survival between positive SLN patients who were treated with CLND versus nodal observation [29]. Despite these results, the biological relevance of the SLN continues to be debated. More recently, metastasectomy is being used more frequently in appropriately selected patients with stage IV melanoma, given the availability of effective systemic therapies, particularly immune checkpoint inhibitors [31]. Even so, surgery for early-stage melanoma, like for breast cancer, has continued to become less extensive both at the primary site and at the level of the regional lymph nodes as we have increased our understanding of cancer biology.

The theory of lymphatic spread also guided the evolution of the radical gastrectomy. From Theodor Billroth's first successful distal gastrectomy in 1881, surgeons described varying degrees of anatomical resection of the stomach. Most based surgical techniques on the idea of disease progression through the stomach wall layers to adjacent lymphatics [31]. Throughout the next several decades, the concept of radical resection would continue to develop as surgeons performed en bloc operations to include adjacent organs, such as the pancreas and spleen. Despite these innovations, survival rates for patients with gastric cancer remained poor. It was not until the mid-1900s that surgeons determined that these risks exceeded the safety and benefit of gastrectomy. For instance, randomized controlled trials in the late twentieth century demonstrated that there was no survival difference in patients who underwent total versus subtotal gastrectomy, which supported a less radical surgical approach. Furthermore, clinical trials evaluated the extent of lymphadenectomy for gastric cancer. Fifteen-year follow-up of the Dutch D1D2 trial showed significant improvements in recurrence and death rates after a D2 resection, consisting of a more extensive lymph node dissection, over a D1 resection [32, 33]. However, the extent of lymphadenectomy for gastric cancer continues to be controversial with the development of D3 lymphadenectomies, which are often performed in Asian countries [34].

Another pioneering en bloc operation by Halsted was the first radical resection for periampullary carcinoma in 1898. Allen Whipple subsequently modeled his approach on Halsted's theories in a two-step operation that was later amended to a more radical one-step procedure [33]. The latter, named the standard Whipple, was successfully completed by Alexander Brunschwig in 1937. In contrast, Traverso and Longmire proposed a similar procedure with reduced margins, known as the pylorus-preserving pancreaticoduodenectomy. Despite concerns that microscopic disease may persist with decreased margins, trials showed no differences in morbidity, mortality, or survival between the standard Whipple and a total pancreatectomy. Advances in technique, such as en bloc vascular resection with reconstruction, as well as the use of neoadjuvant therapies in borderline resectable cases, have led to enhanced locoregional control and identified more patients who are amenable to surgical resection [35, 36].

Similarly, early surgical management of colorectal cancers was initially palliative due to high recurrence rates. However, more radical versions of these operations would transform colorectal surgery into potentially curative procedures. Mikulicz devised a two-stage colon resection that included mesenteric lymphadenectomy based on his theory of lymphatic spread. Subsequently, in 1908, Ernest Miles published a more extended approach of lymphadenectomy with abdominoperineal resection (APR) to include proximal lymph nodes, which he postulated were responsible for local recurrences [35]. However, surgeons challenged that APR was too radical for all cases of colorectal carcinoma. This led to the development of more conservative techniques that included Henri Hartmann's end colostomy to salvage the distal rectum and Claude Dixon's creation of the sphincter-sparing anterior resection [37, 38].

Consistent with the importance of locoregional control in en bloc resections, Bill Heald introduced total mesorectal excision (TME) for treating rectal cancers in 1982. TME preserved the embryological plane in the mesentery to prevent disruption of lymphatics and the release of tumor cells [38]. Although colon and rectal cancers were historically approached in a similar manner, the inclusion of adjunct treatments has led to their acknowledgment as two distinct entities. For instance, further locoregional control for primary rectal carcinoma can be achieved with neoadjuvant radiation therapy [37], and may alter the extensiveness of the planned surgical resection. Finally, hepatic metastasectomy may be appropriate in select cases of limited-stage IV colorectal cancer and can be performed simultaneously or metachronously with primary colon surgery [38]. Adjuvant systemic chemotherapy is also utilized and has been shown to improve survival in colon cancer patients with nodal involvement [39, 40].

Pelvic exenteration is one of the most radical forms of surgery that is still performed in the modern world. It was first introduced by Brunschwig for recurrent cervical carcinoma in 1948. This en bloc resection of multiple pelvic organs, employed for locally advanced tumors, was initially associated with high rates of mortality and was performed for palliation [39, 40]. However, advances in perioperative care and imaging techniques have assisted with operative planning and patient selection, ultimately leading to improved postoperative outcomes. Similarly, the radical hysterectomy was a mainstay in the treatment of cervical cancer for many decades [41]. Although modified forms of this operation continue to be utilized for locoregional treatment of select cases of gynecologic cancers, the addition of or primary use of radiation therapy has allowed for more conservative treatment approaches.

Surgery for genitourinary cancers has also undergone a similar transformation. For instance, radical prostatectomy, first performed in 1904, was the standard surgical treatment for prostate cancer for several decades [42]. Moreover, the retropubic approach, suggested by Terrence Millin in 1945, enabled surgeons to access pelvic lymph nodes. Subsequently, transurethral resection of the prostate was introduced as a less extensive method for treating certain prostate cancers [42]. However, unlike other cancers of the pelvis, many prostate cancers are now treated medically, specifically with androgen-ablation therapy, and surgery has a lesser role in the management of prostate cancer [42].

35.5 Future Perspectives of Surgery in Cancer Management

Monumental innovations in human genetics and biomedical technology have led to the development of cutting-edge techniques for the diagnosis and characterization of cancer. Remarkably accurate, these novel assays can identify patient-to-patient variations for specific cancers at the molecular and genomic level with extreme precision. This allows for individually tailored treatment protocols based on pathologic responses to therapies or dynamic changes in levels of biomarkers. In light of these developments and in line with the shift from a more radical to conservative approach, the role of surgery in the treatment of cancer remains essential but is evolving.

The ability to sequence the human genome led to the development of the DNA microarray. First described in 1995, this groundbreaking diagnostic assay allows for the characterization of genome-wide gene expression at any given point in time. Initial studies with yeast and bacteria demonstrated the wide-reaching potential of this new technology, including its capability to detect induction and repression of genes due to environmental stresses and its ability to bring to light novel cellular regulatory pathways by the detection of transcription and other regulatory factors based on upstream gene expression patterns [43].

Some of the most noteworthy applications of DNA microarrays have been with its use to define cancers by their gene expression profiles (GEPs). Gene profiling allows for accurate identification of tumors of unknown etiology and can detect subtle molecular differences between two otherwise seemingly similar cancers, thereby leading to individualized treatment regimens and significant improvements in outcomes [44]. Changes in GEPs following therapy administration can provide important prognostic information, guide therapy, and identify novel biomarkers and therapeutic drug targets. In modern practice, gene profiling has been instrumental in the management of breast cancer and is commonly utilized in the treatment of lung, prostate, and colon cancers, among many others [44]. Future research with gene profiling and other molecular techniques will likely focus on combining these already established methods with high-throughput technologies, integrating tumor molecular patterns with intraoperative imaging modalities, and detecting new and downstream activating mutations in currently targeted molecular pathways, which may be responsible for treatment resistance [45].

As previously discussed, our understanding of the biology underlying cancer development continues to progress. In contrast to Halsted's theory, it is abundantly clear that simply removing the malignant disease is not sufficient in many cases, as this does not address microscopic metastatic disease which may be present and is ultimately the cause of most cancer-related deaths. Likewise, using toxic chemicals at concentrations necessary to eradicate every malignant cell causes intolerable toxicities to their nonmalignant counterparts. New and emerging treatments will need to focus on exploiting molecular and genetic differences between malignant and nonmalignant cells, such as blocking constitutively activated cellular receptors on cancer cells, preventing cancer cells from utilizing growth factors or cellular machinery, and blocking cancer cells from hijacking regulatory immune mechanisms to evade detection. However, nonimmune systemic therapies to date generally fall short of demonstrating the capacity to completely exterminate all areas of malignant disease [46].

As we enter the modern era of cancer treatment, evidence is emerging that combination therapy, both in the form of multidrug chemotherapy regimens and multimodal treatment approaches combining systemic treatment, radiation therapy, and surgery, are associated with improved outcomes [47]. Neoadjuvant therapy is also becoming increasingly common, with subsequent therapies dependent on the response to the initial treatment [48]. As all aspects of cancer management are evolving, so too is the role of the surgeon in the cancer patient's care. While surgical resection is still the mainstay of treatment for most early-stage malignancies, the scope of the modern surgeon is no longer simply that of a technician tasked with excising a malignancy. He or she now has an active role in all stages of the treatment spectrum, including preventive counseling and screening, initial cancer diagnosis, collaboration with multidisciplinary teams to determine the optimal sequence of treatment and interventions, postoperative surveillance monitoring, end-of-life care, and high-quality research [49]. Trials have already begun studying the extent of surgical resection based on pathologic response to neoadjuvant treatment, and as the efficacy of systemic therapy continues to improve, more patients who initially would have had unresectable disease will now become surgical candidates [50]. Just as with the other disciplines of care for the cancer patient, the role of surgical therapy will continue to evolve alongside the entire field of oncology.

35.6 Conclusions

- As our comprehension of cancer biology continues to evolve, so too does our armamentarium of therapy options.
- We now know that cancer does not only ascribe to the initial Halstedian dogma of strict local to regional to distant spread; yet, the Fisher model of cancer as an inherently systemic disease at presentation falls short of accurately capturing the complexity of cancer biology for all tumor types.
- In accordance with the spectrum theory, it is likely that each unique cancer type, and subtypes within these categories, has specific molecular and biological attributes, allowing it to develop and progress in a distinct fashion; progression of a particular cancer is also likely contingent upon the host (immune) factors.
- Surgery remains an integral component to oncologic treatment, but it is important to remember that it is but one of many treatment options we have at hand, and should be used in conjunction with, not exclusively from, other available treatment modalities.
- The development of technology to detect patient-specific molecular aberrations and identify novel biomarker and therapeutic targets will allow for individualized, precision oncologic care.
- Ongoing and future research will strive to answer:

Open Questions

- Is there a synergistic effect of combining immune-mediated or targeted molecular therapies together or with traditional chemotherapeutic agents?
- How will the use of intraoperative or improved more sensitive preoperative molecular imaging techniques change the landscape of oncologic surgery?
- How will pathologic response to neoadjuvant treatment be used to influence subsequent cancer treatment including extensiveness of surgical resection?
- How will the discovery of novel biomarkers reshape treatment regimens and therapy duration?
- Will the development of more precise and molecular-based systemic therapies allow more patients with advanced-stage disease to become candidates for curative surgical resection? What will the role of surgery be for the removal of cancer sites resistant to systemic therapies?

Disclosures The authors declare no conflicts of interest or external funding for this publication.

References

1. Onuigbo WI. Historical trends in cancer surgery. Med Hist. 1962;6:154–61. https://doi.org/10.1017/s0025727300027125.
2. Halsted WS. I. The results of operations for the cure of cancer of the breast performed at the Johns Hopkins Hospital from June, 1889, to January, 1894. Ann Surg. 1894;20(5):497–555. https://doi.org/10.1097/00000658-189407000-00075.
3. Fisher B. Laboratory and clinical research in breast cancer—a personal adventure: the David A. Karnofsky memorial lecture. Cancer Res. 1980;40(11):3863.
4. Leong SP, Tseng WW. Micrometastatic cancer cells in lymph nodes, bone marrow, and blood: clinical significance and biologic implications. CA Cancer J Clin. 2014;64(3):195–206. https://doi.org/10.3322/caac.21217.
5. Lawrence W, Lopez MJ. Radical surgery for cancer: a historical perspective. Surg Oncol Clin N Am. 2005;14(3):441–6., , v. https://doi.org/10.1016/j.soc.2005.05.003.
6. Özmen V. Paradigm shift from Halstedian radical mastectomy to personalized medicine. J Breast Health. 2017;13(2):50–3. https://doi.org/10.5152/tjbh.2017.312017.
7. Hellman S. Karnofsky memorial lecture. Natural history of small breast cancers. J Clin Oncol. 1994;12(10):2229–34. https://doi.org/10.1200/JCO.1994.12.10.2229.
8. Wilky BA. Immune checkpoint inhibitors: the linchpins of modern immunotherapy. Immunol Rev. 2019;290(1):6–23. https://doi.org/10.1111/imr.12766.
9. Chabner BA, Roberts TG. Timeline: chemotherapy and the war on cancer. Nat Rev Cancer. 2005;5(1):65–72. https://doi.org/10.1038/nrc1529.
10. Gilman A, Philips FS. The biological actions and therapeutic applications of the B-Chloroethyl amines and sulfides. Science. 1946;103(2675):409–36. https://doi.org/10.1126/science.103.2675.409.
11. Brookes P, Lawley PD. The reaction of mustard gas with nucleic acids in vitro and in vivo. Biochem J. 1960;77(3):478–84. https://doi.org/10.1042/bj0770478.
12. Rhoads CP. Nitrogen mustards in the treatment of neoplastic disease; official statement. J Am Med Assoc. 1946;131:656–8. https://doi.org/10.1001/jama.1946.02870250010003.
13. Emadi A, Jones RJ, Brodsky RA. Cyclophosphamide and cancer: golden anniversary. Nat Rev Clin Oncol. 2009;6(11):638–47. https://doi.org/10.1038/nrclinonc.2009.146.
14. Puyo S, Montaudon D, Pourquier P. From old alkylating agents to new minor groove binders. Crit Rev Oncol Hematol. 2014;89(1):43–61. https://doi.org/10.1016/j.critrevonc.2013.07.006.
15. Farber S, Diamond LK. Temporary remissions in acute leukemia in children produced by folic acid antagonist, 4-aminopteroyl-glutamic acid. N Engl J Med. 1948;238(23):787–93. https://doi.org/10.1056/NEJM194806032382301.
16. Carrillo E, Navarro SA, Ramírez A, et al. 5-Fluorouracil derivatives: a patent review (2012–2014). Expert Opin Ther Pat. 2015;25(10):1131–44. https://doi.org/10.1517/13543776.2015.1056736.
17. Rosen LS. Irinotecan in lymphoma, leukemia, and breast, pancreatic, ovarian, and small-cell lung cancers. Oncology (Williston Park). 1998;12(8 Suppl 6):103–9.
18. Falzone L, Salomone S, Libra M. Evolution of cancer pharmacological treatments at the turn of the third millennium. Review. Front Pharmacol. 2018;9:1300. https://doi.org/10.3389/fphar.2018.01300.
19. Bolzán AD, Bianchi MS. DNA and chromosome damage induced by bleomycin in mammalian cells: an update. Mutat Res. 2018;775:51–62. https://doi.org/10.1016/j.mrrev.2018.02.003.

20. Pucci C, Martinelli C, Ciofani G. Innovative approaches for cancer treatment: current perspectives and new challenges. Ecancermedicalscience. 2019;13:961. https://doi.org/10.3332/ecancer.2019.961.
21. Waller CF. Imatinib mesylate. Recent Results Cancer Res. 2018;212:1–27. https://doi.org/10.1007/978-3-319-91439-8_1.
22. Seidel JA, Otsuka A, Kabashima K. Anti-PD-1 and anti-CTLA-4 therapies in cancer: mechanisms of action, efficacy, and limitations. Front Oncol. 2018;8:86. https://doi.org/10.3389/fonc.2018.00086.
23. Sakamuri D, Glitza IC, Betancourt Cuellar SL, et al. Phase I dose-escalation study of anti-CTLA-4 antibody ipilimumab and lenalidomide in patients with advanced cancers. Mol Cancer Ther. 2018;17(3):671–6. https://doi.org/10.1158/1535-7163.MCT-17-0673.
24. Palumbo MO, Kavan P, Miller WH, et al. Systemic cancer therapy: achievements and challenges that lie ahead. Front Pharmacol. 2013;4:57. https://doi.org/10.3389/fphar.2013.00057.
25. Sakorafas GH, Safioleas M. Breast cancer surgery: an historical narrative. Part II. 18th and 19th centuries. Eur J Cancer Care. 2010;19(1):6–29. https://doi.org/10.1111/j.1365-2354.2008.01060.x.
26. Cotlar AM, Dubose JJ, Rose DM. History of surgery for breast cancer: radical to the sublime. Curr Surg. 2003;60(3):329–37. https://doi.org/10.1016/S0149-7944(02)00777-8.
27. Rebecca VW, Sondak VK, Smalley KS. A brief history of melanoma: from mummies to mutations. Melanoma Res. 2012;22(2):114–22. https://doi.org/10.1097/CMR.0b013e328351fa4d.
28. Gonzalez A. Sentinel lymph node biopsy: past and present implications for the management of cutaneous melanoma with nodal metastasis. Am J Clin Dermatol. 2018;19(1):24–30. https://doi.org/10.1007/s40257-018-0379-0.
29. Faries MB, Han D, Reintgen M, Kerivan L, Reintgen D, Caracò C. Lymph node metastasis in melanoma: a debate on the significance of nodal metastases, conditional survival analysis and clinical trials. Clin Exp Metastasis. 2018;35(5–6):431–42. https://doi.org/10.1007/s10585-018-9898-6.
30. Lee DJ, Faries MB. Practical manual for dermatologic and surgical melanoma management. 1st ed. Springer International Publishing; 2020.
31. Zhang CD, Yamashita H, Seto Y. Gastric cancer surgery: historical background and perspective in Western countries versus Japan. Ann Transl Med. 2019;7(18):493. https://doi.org/10.21037/atm.2019.08.48.
32. Are C, Dhir M, Ravipati L. History of pancreaticoduodenectomy: early misconceptions, initial milestones and the pioneers. HPB (Oxford). 2011;13(6):377–84. https://doi.org/10.1111/j.1477-2574.2011.00305.x.
33. Griffin JF, Poruk KE, Wolfgang CL. Pancreatic cancer surgery: past, present, and future. Chin J Cancer Res. 2015;27(4):332–48. https://doi.org/10.3978/j.issn.1000-9604.2015.06.07.
34. Douridas GN, Pierrakakis SK. Is there any role for D3 lymphadenectomy in gastric cancer? Front Surg. 2018;5:27. https://doi.org/10.3389/fsurg.2018.00027.
35. Galler AS, Petrelli NJ, Shakamuri SP. Rectal cancer surgery: a brief history. Surg Oncol. 2011;20(4):223–30. https://doi.org/10.1016/j.suronc.2010.01.001.
36. Lirici MM, Hüscher CGS. Techniques and technology evolution of rectal cancer surgery: a history of more than a hundred years. Minim Invasive Ther Allied Technol. 2016;25(5):226–33. https://doi.org/10.1080/13645706.2016.1198381.
37. Rentsch M, Schiergens T, Khandoga A, Werner J. Surgery for colorectal cancer—trends, developments, and future perspectives. Visc Med. 2016;32(3):184–91. https://doi.org/10.1159/000446490.
38. Tamas K, Walenkamp AM, de Vries EG, et al. Rectal and colon cancer: not just a different anatomic site. Cancer Treat Rev. 2015;41(8):671–9. https://doi.org/10.1016/j.ctrv.2015.06.007.
39. Brown KGM, Solomon MJ, Koh CE. Pelvic Exenteration surgery: the evolution of radical surgical techniques for advanced and recurrent pelvic malignancy. Dis Colon Rectum. 2017;60(7):745–54. https://doi.org/10.1097/DCR.0000000000000839.
40. Sevin BU, Koechli OR. Pelvic exenteration. Surg Clin North Am. 2001;81(4):771–9. https://doi.org/10.1016/s0039-6109(05)70164-x.
41. Dursun P, Gultekin M, Ayhan A. The history of radical hysterectomy. J Low Genit Tract Dis. 2011;15(3):235–45. https://doi.org/10.1097/LGT.0b013e31820eb038.
42. Denmeade SR, Isaacs JT. A history of prostate cancer treatment. Nat Rev Cancer. 2002;2(5):389–96. https://doi.org/10.1038/nrc801.
43. Brewster JL, Beason KB, Eckdahl TT, Evans IM. The microarray revolution: perspectives from educators. Biochem Mol Biol Educ. 2004;32(4):217–27. https://doi.org/10.1002/bmb.2004.494032040362.
44. Pierouli K, Mitsis T, Papakonstantinou E, Vlachakis D. Introductory chapter: gene profiling in cancer in the era of metagenomics and precision medicine; 2019.
45. Szalat R, Avet-Loiseau H, Munshi NC. Gene expression profiles in myeloma: ready for the real world? Clin Cancer Res. 2016;22(22):5434–42. https://doi.org/10.1158/1078-0432.CCR-16-0867.
46. Somarelli JA, Gardner H, Cannataro VL, et al. Molecular biology and evolution of cancer: from discovery to action. Mol Biol Evol. 2020;37(2):320–6. https://doi.org/10.1093/molbev/msz242.
47. Wolmark N, Colangelo L, Wieand S. National surgical adjuvant breast and bowel project trials in colon cancer. Semin Oncol. 2001;28(1 Suppl 1):9–13. https://doi.org/10.1016/s0093-7754(01)90245-3.
48. Imyanitov EN, Yanus GA. Neoadjuvant therapy: theoretical, biological and medical consideration. Chin Clin Oncol. 2018;7(6):2.
49. Elder K, Barber M. The role of the surgeon in cancer care. Surgery (Oxford). 2018;36. https://doi.org/10.1016/j.mpsur.2017.12.005.
50. Reijers ILM, Dimitriadis P, Rozeman EA, et al. Personalized combination of neoadjuvant domatinostat, nivolumab and ipilimumab in macroscopic stage III melanoma patients stratified according to the interferon-gamma signature: the DONIMI study. J Clin Oncol. 2020;38(15_suppl):TPS10087. https://doi.org/10.1200/JCO.2020.38.15_suppl.TPS10087.

The Role of Surgery in Managing Primary and Metastatic Melanoma

36

Kristen E. Rhodin, Kirsten Baecher, Winta T. Mehtsun, Mike Lowe, Genevieve Boland, and Georgia M. Beasley

Abstract

This chapter describes updates in the surgical management of melanoma including primary melanoma, regional melanoma, and distant metastatic disease. Surgical management is discussed in the context of a multidisciplinary approach. Specifically, surgical therapy needs to be considered in the context of new effective systemic therapy agents for melanoma. With the approval of multiple new treatment strategies for melanoma, the surgical approach has changed dramatically in the last 10 years. Here we review the rationale for appropriate contemporary surgical approaches to melanoma including minimizing morbidity from surgery and introduction of neoadjuvant therapy approaches.

K. E. Rhodin · G. M. Beasley (✉)
Department of Surgery, Duke University, Durham, NC, USA
e-mail: kristen.rhodin@duke.edu; Georgia.beasley@duke.edu

K. Baecher · M. Lowe
Department of Surgery, Emory University, Atlanta, GA, USA
e-mail: k.m.baecher@emory.edu; mlowe3@emory.edu

W. T. Mehtsun
Department of Surgery, Massachusetts General Hospital, Boston, MA, USA

Department of Surgery, Brigham and Women's Hospital, Boston, MA, USA

Dana Farber Cancer Institute, Boston, MA, USA
e-mail: wmehtsun@partners.org

G. Boland
Department of Surgery, Massachusetts General Hospital, Boston, MA, USA
e-mail: gmboland@mgh.harvard.edu

Learning Objectives

After reading this chapter, readers should be able to:

- Describe appropriate margins for primary melanoma
- Understand indications for sentinel lymph node biopsy
- List options for management of regionally advanced melanoma
- Describe neoadjuvant approaches for resectable melanoma.

36.1 Introduction

The incidence of cutaneous melanoma has been rapidly increasing in recent years [1]. In the last decade (2010–2020), there have also been dramatic changes in both the surgical and medical management of patients with melanoma. Given the introduction of effective systemic therapy for melanoma in the last decade, the current approach to surgery needs to be viewed as just one component of an effective strategy to minimize disease mortality.

36.2 Background

A number of distinct features of the primary melanoma are well known to be prognostic of outcomes—notably Breslow thickness and the presence of ulceration [2]. Metastatic spread of cutaneous melanoma usually occurs as regional lymph node metastases, locoregional metastases, or distant metastases (in 50%, 20%, and 30% of patients with recurrent melanoma, respectively) [3]. The presence of lymph node metastases either detected clinically or via sentinel lymph node biopsy (SLNB), and the presence of distant metastases

S. P. Leong et al. (eds.), *Cancer Metastasis Through the Lymphovascular System*,
https://doi.org/10.1007/978-3-030-93084-4_36

is also critical components in the staging system with increasing disease burden associated with worse outcomes.

Increasingly, host immunity and molecular profiling are also recognized as key determinants of melanoma biology. Recently, several somatic mutations have been identified causing deregulation of key intracellular signaling pathways important in the proliferation of melanoma, including activating mutations in the BRAF serine/threonine protein kinase in approximately 50% of cases [4], activating mutations in NRAS, and C-KIT mutations. This has led to the use and success of targeted therapy for appropriate melanoma patients, including combination BRAF/MEK inhibitors [5]. Additionally, the degree of host lymphoid infiltration into the primary tumor and into metastatic lesions may be important in disease control and response to modern therapies, which include immune checkpoint inhibitors (ICI) [6]. Since 2011, multiple new immune therapies and targeted therapies have been FDA (United States Food and Drug Administration) approved for melanoma based on encouraging clinical trial results for those therapies [7].

While complete surgical resection of melanoma offered one of the few options of effective disease control in the past, surgeons managing melanoma in the modern era now have updated evidence-based guidelines for the surgical management of the primary melanoma lesion and the regional nodal basin and for the current role of metastasectomy. In this chapter, we will highlight key recent updates, especially with regard to the surgical management of the draining nodal basin.

36.2.1 Management of Primary Melanoma

In 1907, Dr. Handley described extensive surgical resection of skin and subcutaneous tissue for melanoma [8]. For decades, a wide variation existed in surgical practice in part due to wide variability in melanoma biology. Several retrospective studies then found that resection margins of 2 cm or less in melanomas with Breslow depth <1 mm were associated with low rates of local recurrence [8]. The World Health Organization published the first prospective randomized trial supporting adequate disease control with narrow margins (1 cm) compared to 3 cm margins for melanomas less than 2 mm in thickness [9]. Disease-free and overall survival rates were similar between the two groups. Other randomized trials completed around the world also supported more conservative margins [10]. A Cochrane review supported that margins greater than 2 cm could safely be avoided [10].

Based on data from these clinical trials, current National Comprehensive Care Network (NCCN) guidelines recommend wide excision with 1 cm margins for melanomas 1.0 mm or less, wide excision with 1–2 cm margin for melanomas 1.01–2 mm in thickness, and 2 cm margins for melanomas greater than 2 mm in depth. Tissue removed should include skin and subcutaneous tissue to the level of the fascia which contains all local lymphatic tissue. There are ongoing efforts to determine if 1 cm margins may be adequate in patients regardless of Breslow depth. Long-term data from a large randomized trial in the United Kingdom comparing 1–3 cm margins for patients with primary melanomas at least 2 mm in thickness found that melanoma-specific survival was significantly worse for those in the 1 cm group compared to the 3 cm margin group (unadjusted hazard ratio, 1.24; 95% CI 1.01–1.53; $p = 0.041$) [11]. However, surgical complications were nearly double in the 3 cm group (8% vs. 15%) and SLN biopsy (now considered standard of care for this thickness group) was not routinely performed [11]. To further investigate the optimal margins for thicker melanomas, the MelmarT-I study was conducted as an international randomized study comparing 1–2 cm margins in patients with greater than 1 mm thick tumors [12]. Patients in the 2 cm margin group more often required reconstruction (34.9 vs. 13.6%; $p < 0.0001$) and had an increased wound necrosis rate (3.6% vs. 0.5%; $p = 0.036$) compared to the 1 cm margin group. The MelMarT-II study (NCT03860883) is ongoing. Patients with >2 mm thickness or 1- to 2-mm thickness with ulceration are being randomized to either 1 or 2 cm margins. The primary outcome is disease-free survival with secondary outcomes that include local recurrence, melanoma-specific survival, and surgery-related adverse events. This trial should further clarify the application of 1 cm margins to patients with melanoma regardless of Breslow depth.

Certain anatomic sites including but not limited to digits, ears, the face, and the plantar surface of the foot may not be as easily amenable to rigid surgical criteria without significant concerns for poor cosmetic and functional outcomes. For subungual lesions, conservative amputation at the most distal interphalangeal joint is appropriate. Certain melanoma subtypes also may prompt special consideration. Wider margins (2 cm) are generally considered for desmoplastic melanoma, particularly for the pure subtype of desmoplastic melanoma, given that 1 cm margins have been associated with higher rates of local recurrence [13].

Among patients with clinical stage I and II melanoma, metastasis to the sentinel node is one of the strongest prognostic factors for melanoma [14]. The application of SLNB to melanoma by Morton in 1992 proved to be practice changing. In the multicenter selective lymphadenectomy trial 1 (MSLT-1), patients undergoing wide excision for melanoma were randomized to receive either SLNB with immediate completion lymph node dissection (CLND) for a positive SLN or observation with interval therapeutic lymph node dissection (TLND) for a nodal relapse. There was no difference in 10-year melanoma-specific survival between the two arms; however, in secondary analysis, SLNB did improve the 10-year rate of disease-free survival in patients with intermediate thickness (1.2–3.5 mm) melanoma [14]. In patients with intermediate-thickness melanomas, the 10-year melanoma-specific survival rate was 62.1 ± 4.8% among those with SLN metastasis versus 85.1 ± 1.5% for those without SLN metastasis [14]. On multivariate analysis, SLN node status was the strongest predictor of both disease recurrence and death from melanoma. Recently, both the American Society of Clinical Oncology and Society of Surgical Oncology reviewed available data and published guidelines for current SLNB indications [15]. SLNB is recommended for intermediate thickness T2 or T3 melanomas (1.0–4.0 mm thickness) and may be considered in T1b thin melanomas (0.8–1.0 mm or <0.8 mm thick with ulceration) [15]. Furthermore, SLNB may be performed for patients with T4 thick melanomas (>4.0 mm thickness) because SLNB appears to still provide prognostic data and a positive SLNB in the current era renders these patients eligible for effective systemic adjuvant therapy [7]. Although the rate of SLN metastasis is generally less than 5% for patients with lesions less than 0.8 mm in thickness (T1a), there may be select patients at higher risk for whom SLN may be considered. Additional recent efforts have focused on identifying patients with greater than 1 mm thick tumors for whom the risk of SLN metastases is low, thereby potentially allowing these patients to forego SLNB [16].

Approximately 10–20% of patients with tumors >1 mm in depth will have a positive SLN [14]. SLNB-based management of patients resulted in improved melanoma-specific survival among patients with nodal metastases and intermediate depth melanomas in the MSLT-1 trial (10-year melanoma-specific survival: 62.1% ± 4.8% vs. 41.5% ± 5.6% in observation group). In addition, complication rates for SLNB are much lower than what is seen for lymph node dissections [17]. For example, in the multicenter selective lymphadenectomy trial 2 (MSLT-2), 24.1% of patients in the CLND group experienced lymphedema compared to only 6.3% of patients who had SLNB only [18]. In addition to biologic factors of Breslow depth and ulceration, the decision to perform SLNB should also be based on overall patient status and how results can or will impact further management.

36.3 Management of In-Transit Disease

In-transit melanoma (ITM) is a distinct clinical entity whereby melanoma tumor cells recur as dermal or subcutaneous nodules arising between the primary site and the draining regional lymph node basin. This definition encompasses a wide array of presentations, with ITM lesions ranging in size from 0.2 to 3 cm and in quantity from single to hundreds of lesions [19]. Despite optimal primary resection of melanoma, ITM arises in 4–12% of patients within a median of 18 months after surgery and is associated with worse survival compared to those with localized disease [19].

ITM can become symptomatic with patients often reporting discomfort, pruritus, bleeding, and infection; however, its heterogeneous presentation has made its management challenging. Historically, surgical excision has had a limited role in the management of ITM—limited primarily to resection of few and small lesions. Unfortunately, ITM lesions are frequently not as amenable to metastasectomy as isolated distant metastases, with resection often being morbid or impractical [19]. The utility of SLNB for ITM is debated, although SLNB may refine the stage III subgroup staging and thus carry some minor prognostic value. In the modern era, patients with ITM are eligible for newer systemic therapy agents, given the Stage III classification regardless of SLN status.

For the majority of patients who are unable to undergo resection of ITM, its superficial nature lends accessibility for regional and intralesional therapies. Regional chemotherapy remains an option for ITM limited to an extremity. This strategy involves isolating the affected limb's blood supply and administering high-dose chemotherapy. Two variants of this include hyperthermia isolated limb perfusion (HILP) and isolated limb infusion (ILI). While both provide high-dose regional chemotherapy, they differ in the invasiveness of vascular access, limb-threatening toxicity, and operative times [19]. In HILP, the major vessels of the extremity are dissected and cannulated, followed by the application of a tourniquet. The extremity is heated to 41 °C prior to infusion of melphalan, and a bypass circuit is used to maintain oxygenation. Alternatively, ILI utilizes percutaneous catheterization and infusion of melphalan plus actinomycin D at normothermic to slightly hyperthermic temperatures under progressively hypoxic conditions. HILP has superior complete response rates that must be balanced with higher rates of limb-threatening toxicity [20]. Proponents of ILI favor its use as it is less invasive and can be performed multiple times. While no studies have demonstrated superior survival with HILP versus ILI (or prospectively in a randomized fashion directly comparing these two forms of regional therapy), based on the results of Chai and colleagues' multi-institutional findings, it has been proposed that ILI be utilized upfront, except in patients with bulky disease, while saving HILP as a salvage strategy for those with high-volume disease, progression on ILI, or rapid recurrence [20].

Alternatively, intralesional therapies have been an active area of investigation for the management of ITM [19]. These therapies are hypothesized to incite a local inflammatory response that induces antitumor immunity in the local environment and perhaps systemically. The accessibility of superficial lesions makes intralesional therapy a feasible and attractive strategy. Further, local therapies have less systemic adverse effects and lower toxicity risk to patients. Several agents that have been trialed in this setting include, but are not limited to, interleukin-2 (IL-2), interferon-alpha, Bacillus Calmette–Guerin (BCG), and PV-10 (Rose Bengal) [21]. Many of these therapies demonstrated promising responses in injected lesions, and for some agents such as PV-10, responses were also seen in noninjected bystander lesions; however, in trials, significant survival benefits have not been consistently shown.

Oncolytic viruses have become a promising intralesional therapy for ITM. Notably, Talimogene Laherparepvec (T-VEC), a genetically modified herpes simplex virus type 1, has demonstrated efficacy in patients with advanced melanoma and was FDA approved in 2015. T-VEC, like other oncolytic viruses, has a direct effect by lysing tumor cells [22]. T-VEC also acts by producing granulocyte-macrophage colony-stimulating factor (GM-CSF), which recruits immune cells to the tumor microenvironment. While T-VEC has a durable response rate of 16.3% (objective response lasting at least 6 months), it is important to note that it has not demonstrated a survival benefit when utilized as a single therapy [23]. Further exploration into other oncolytic viruses and intratumoral agents, including ONCOS-102, toll-like receptors, and many others, is ongoing [21].

Although patients with isolated ITM are candidates for the aforementioned regional and intralesional therapies, the majority of patients with ITM will develop nodal or distant metastasis. Furthermore, even for patients with isolated ITM in the absence of distant metastases, no regional therapy has demonstrated improvements in overall survival, although select patients who respond to therapy may achieve a survival benefit. Conversely, systemic therapy with ICI or BRAF/MEK inhibitor therapy has been associated with durable response rates and improved survival in patients with unresectable stage III melanoma, which includes patients with ITM [24]. As such, in the current era, in patients presenting with unresectable ITM, systemic therapy should be strongly considered as the first-line treatment. Regional and intralesional therapies can be used for patients who have a low volume of ITM only, have contraindications to systemic therapy, or have failed systemic therapies. One exciting approach is to combine intralesional or regional chemotherapy with systemic therapy, and multiple trials are ongoing exploring these combination therapies [25, 26].

36.4 Evaluation of the Nodal Basin and Management of Node-Positive Disease

The eighth edition of the AJCC guidelines attempted to take into account the recent advancements in both the understanding and the management of cutaneous melanoma when considering treatment for patients with regional disease. The diagnosis of microscopic nodal disease (N1a, N2a, N3a in the eighth edition of the staging system) is made through SLNB, while macroscopic and in transit/satellite disease is diagnosed clinically or radiographically. Management of patients with macroscopic nodal disease has largely remained unchanged, and these patients should undergo TLND, although the extent of lymphadenectomy is evolving and neoadjuvant approaches are being evaluated.

Upon diagnosis of microscopic nodal disease by SLNB or of macroscopic disease clinically or radiographically, management includes some combination of surgery, systemic therapy (ICIs and BRAF/MEK inhibitors), and potentially radiation therapy in select cases, depending on the extent of disease in the nodal basin. The intent of adjuvant therapy in stage III melanoma is to "reduce the recurrence rate of radically operated patients at high risk… and to potentially improve survival." [27] A thorough understanding of the mechanisms underlying the development of nodal metastases is critical to determining which therapy or combination of therapies may be most beneficial to patients with nodal disease.

36.5 Surgical Management of the Nodal Basin

Prior to the development of SLNB, elective lymph node dissection (ELND) was routinely performed to evaluate for nodal metastases and theoretically prevent the spread of melanoma beyond the lymph nodes. However, due to the relatively low prevalence of microscopic nodal metastasis in patients with primary cutaneous melanoma, which is approximately 20% [28], routine ELND subjected a majority of patients with no metastatic disease to the complications of ELND without deriving any measurable benefit. With the advent of lymphoscintigraphy and SLNB, the prognostic benefit of lymph node dissection could be achieved without subjecting the patient to the morbidities associated with ELND. However, the question still remained: in a patient with a positive SLN, should we return to the operating room to perform a CLND?

Following the validation of SLNB, CLND was nearly uniformly adopted as the standard of care for patients with a positive SLN. However, several conceptual questions underlie this issue. Does the result of SLNB correlate with the likelihood of non-sentinel lymph node (NSLN) disease? Do SLNs become positive before NSLNs, and if so, does removal of the SLN protect against disease progression in the nodal basin and beyond? We know that only 10–30% of patients with a positive SLN will have positive NSLNs [29]. Various systems (e.g., Rotterdam–Dewar) have attempted to predict NSLN status based on features (size, location) of the

SLN metastasis [30]. A 2008 study of stage I and II melanomas found that positive SLNs with micrometastases (metastases less than 2 mm in diameter) only were associated with negative NSLNs [31]. However, we have also learned that patients who do have positive NSLNs have a worse prognosis than patients with only positive SLNs, regardless of the number of overall positive nodes [32]. This suggests that positive NLSNs may reflect a different tumor biology and carry prognostic significance beyond just increasing the nodal metastatic burden. For these reasons, some proposed that the prognostic value of the NSLNs merits any invasive means necessary to find this prognostic information, further cementing CLND as a necessary step in the management of SLN-positive patients. Later studies investigated additional noninvasive methods of determining the prognostic information of NSLNs; for example, the N-SNORE (Non sentinel node risk score) was a tool developed in 2010 to predict the likelihood of NSLN disease based on characteristics derived from the SLNs [33]. Since then it has been demonstrated that even in patients with positive NSLNs, the melanoma metastatic burden within those NSLNs are significantly diluted as compared to their SLNs [34]. Overall, data suggest that while a small proportion of patients may have some NSLN disease left behind after SLNB, the disease burden is likely to be lower than in the SLNs. Furthermore, although use of the presence of positive NSLNs may aid in prognosis and risk of recurrence, prognosis and risk can also be ascertained based on known variables without subjecting the patient to additional surgery.

The role of CLND as simply a prognostic tool was foreshadowed when several studies showed no improvement in melanoma-specific survival with CLND [18]. The second Multicenter Selective Lymphadenectomy Trial (MSLT-II) and DeCOG-SLT provided definitive evidence confirming that melanoma-specific survival is not significantly different between patients with positive SLNs who underwent CLND or nodal observation [35, 36]. Since that time, CLND is no longer routinely recommended for patients with microscopic nodal stage III disease. The preferred alternative for patients with SNLB-positive nodes per NCCN guidelines is clinical follow-up and nodal basin ultrasound surveillance, as well as consideration for adjuvant therapy (notably all adjuvant therapy trials were completed mandating CLND).

While CLND has been abandoned for most patients with positive SLNs, some indications may remain for CLND. The main indication for CLND is the inability of the patient to undergo routine surveillance. Additional indications are more controversial, and include head and neck location, presence of more than 1 mm of disease in an SLN node and having more than one positive SLN. For example, the authors of MSLT-II declined to recommend against CLND in patients with positive SLNs with tumor deposits greater than 1 mm. In general, these represent patient populations that were underrepresented in the clinical trials and are thought to be at higher risk for harboring additional regional nodal disease. The decision to perform CLND in these populations should be determined on an individual basis and take into consideration the risks of the specific operation (e.g., higher risk of lymphedema with inguinal vs. axillary or cervical lymphadenectomies).

For those patients with clinically positive nodes, TLND remains the standard of care. With clinically positive nodes, the rates of local and systemic recurrence are significantly increased, and overall survival is significantly lower [37]. These patients should be strongly considered for adjuvant systemic therapy and possibly radiation therapy. In determining adjuvant treatment strategies, the patient-specific prognosis must be assessed. Factors that portend a poorer prognosis, and therefore warrant discussions of adjuvant therapy, include stage IIIA disease with SLN tumor deposit >1 mm and stage IIIB/IIIC/IIID disease. Patients with stage IIIA melanoma with <1 mm tumor deposit in the SLN may be offered surveillance without adjuvant therapy. Stage III melanomas should be evaluated for mutational profile, specifically BRAF mutations status, to assess for eligibility of BRAF/MEK inhibitor targeted therapy.

36.6 Adjuvant Systemic Immunotherapy

Checkpoint inhibitors are a recent development within the past decade and have revolutionized the management of many cancers, including cutaneous melanoma. As a class, these monoclonal antibodies are directed against proteins involved in downregulating the T-cell-dependent immune activity of the patient. Programmed cell death protein 1 (PD-1) and cytotoxic T lymphocyte antigen 4 (CTLA-4) are T-cell membrane proteins that affect immune self-tolerance via T-cell suppression; PD-L1 is a ligand for PD-1. Melanomas are able to evade these T-cell-dependent immune pathways by constitutively expressing proteins that promote self-tolerance, such as PD-L1 [38]. The anti-CTLA-4 drug ipilimumab was the first molecule to gain FDA approval for use in metastatic melanoma; FDA approval for adjuvant use in stage III disease was granted in 2015 [39] based on a phase 3 randomized controlled trial that demonstrated improved survival with ipilimumab versus placebo in patients with resected stage III disease [40]. Since that time, the anti-PD-1 drugs nivolumab and pembrolizumab have also been approved for use in the adjuvant setting [7]. Direct comparison between nivolumab and ipilimumab demonstrated improved recurrence-free survival and fewer adverse events with nivolumab [41]; consequently, in a 2020 clinical practice guideline, the American Society of Clinical Oncology recommended that adjuvant nivolumab or pembrolizumab should be offered to patients with stage III melanoma, and cautioned against routine use of ipilimumab in the adjuvant setting.

36.7 Adjuvant Systemic Targeted Therapy: BRAF, MEK Inhibitors

Melanoma is also able to evade the immune response by manipulating the mitogen-activated protein kinase (MAPK) signaling pathway. Activating mutations in the downstream molecules RAF and MAPK-kinase (MEK) are classic culprits in this deregulation, allowing for the loss of cell cycle regulation and promotion of cell cycle progression. Consequently, agents targeting these mutated proteins have been developed and are now approved adjuvant medications for treating melanoma. Vemurafenib was the first mutant BRAF inhibitor to be approved for use in melanoma. In a randomized controlled trial in 2013, vemurafenib was demonstrated to be superior to dacarbazine (a chemotherapeutic agent) in unresectable stage III and stage IV melanoma with BRAF V600E mutations [5]. Since that time, newer BRAF inhibitors such as dabrafenib have been utilized in combination with MEK inhibitors such as trametinib to overcome melanoma's resistance to single-agent therapy [42]. At present, specifically for patients whose melanoma carries the V600E/K BRAF mutation, a combination of adjuvant dabrafenib and trametinib may be offered as an alternative to adjuvant ICI in patients with BRAF-mutated melanoma based on results of the COMBI-AD trial [42].

36.8 The Role of Neoadjuvant Therapy in Managing Advanced Stage Melanoma

The success of neoadjuvant approaches in other solid tumor malignancies underscores its potential application in melanoma patients. Pathologic response is an early readout of patient outcomes, correlating with subsequent recurrence-free survival and is a marker of patient-specific therapy response, potentially informing the choice of future therapies. Furthermore, neoadjuvant approaches enable a deeper understanding of mechanisms of resistance and response through the identification of biomarkers. There is also evidence that T-cell checkpoint blockade given as neoadjuvant therapy induces a stronger and broader tumor-specific T-cell response than adjuvant therapy [43]. Neoadjuvant trials in melanoma predate the introduction of ICIs and targeted therapy; however, for the purposes of this section, we will focus on the recent targeted therapy and immunotherapy trials (Table 36.1).

36.8.1 Targeted Therapy

Amaria et al. examined patients with bulky stage III or resectable stage IV BRAF V600E/K mutated melanoma and randomized them in a 1:2 ratio to either standard of care (SOC) surgery and adjuvant therapy versus neoadjuvant dabrafenib plus trametinib therapy for 8 weeks followed by surgery and up to 44 weeks of adjuvant dabrafenib plus trametinib therapy [44]. Fifty-eight percent of the neoadjuvant group who underwent surgery achieved a pathologic

Table 36.1 Neoadjuvant targeted therapy and immunotherapy trials for locally advanced melanoma

Study	Population	Regimen	*N*	pCR (%)	Median RFS (mo)	Median FU (mo)
NCT02231775 Amaria (2018) [18]	Stage II, IV BRAF V600E/K	Dab/Tram x 8w-surgery- Dab/Tram x 44w	21	58	19.7	18.6
NCT01972347 Long (2019) [19]	Stage III BRAF V600E/K	Dab/Tram x 12w-surgery Dab/Tram x 40w	35	49	23.0	27.0
NCT02437279 Blank (2018) [20]	Stage III	I3N1 x 2-surgery-I3N1 x 2	10	33	NR	32
NCT02519322 Amaria (2019) [22]	Stage III, IV (resectable)	A: Nivo x 4-surgery-Nivo x 13 B: I3N1 x 3-surgery-Nivo x 13	A: 12 B: 11	25 45	NR NR	20
NCT02519322 Huang (2019) [23]	Stage III, IV (resectable)	Pembro x 1- surgery-Pembro x 17	30	19	NR	18
NCT02977052 Rozeman (2019) [24]	Stage III	A: I3N1 x 2 surgery B: I3N1 x 2 surgery C: Ipi x 2-Nivo x 2- surgery	A: 30 B: 30 C: 26	B: 57	NR	8.3

NR not reported, *Dab* dabrafenib, *Tram* trametinib, *I* ipilimumab, *N* nivolumab, *Nivo* nivolumab, *Pembro* pembrolizumab, *pCR* pathologic complete response, *RFS* recurrence-free survival, *FU* follow up complete response (pCR). Patients in the neoadjuvant group had a 60-fold reduced risk of relapse after surgery in comparison with patients who underwent SOC (median event-free survival was 19.7 months vs. 2.9 months, respectively, hazard ratio 0.016, $p < 0.0001$), and the trial was stopped early because it reached the primary objective of improved event-free survival with neoadjuvant treatment [44]. Long et al. examined patients with bulky stage IIIB/C BRAF V600E/K mutated melanoma in a single-arm phase II trial where they administered dabrafenib and trametinib for 12 weeks prior to surgery followed by 40 weeks of dabrafenib and trametinib [45]. At the time of surgical evaluation, 17 patients achieved a pCR (49%). No progression of disease was observed during the neoadjuvant therapy period. At a median follow-up period of 12.1 months (95% CI 8.8–14.8 months), recurrence of disease was observed in 12 patients (36%), with 4 patients having recurrence while undergoing adjuvant targeted therapy [45]. In summary, both these trials demonstrated that patients with a pCR had significantly longer distant metastasis-free survival.

36.9 Immunotherapy Trials

Blank et al. examined adjuvant versus neoadjuvant combination therapy (ipilimumab 3 mg/kg + nivolumab 1 mg/kg) in patients presenting with clinical stage III melanoma (OpACIN phase 1b trial) [43]. After a median follow-up 32 months, none of the seven patients that achieved a pathologic response in the neoadjuvant arm relapsed. At 30 months, estimated recurrence-free survival rates were 80% for the neoadjuvant arm and 60% for the adjuvant arm, while overall survival rates were 90% and 67%, respectively [43]. Amaria et al. examined a combination immunotherapy approach in a phase II randomized trial of neoadjuvant nivolumab 3 mg/kg versus neoadjuvant ipilimumab 3 mg/kg and nivolumab 1 mg/kg in clinical stage III and oligometastatic stage IV melanoma patients [46]. Improved relapse-free survival was shown in patients achieving a pCR, but the trial was stopped early due to high rates of disease progression in the monotherapy arm and high rates of toxicities in the combination arm. Huang et al. examined monotherapy neoadjuvant pembrolizumab in clinical stage III and oligometastatic stage IV melanoma patients [47]. After a single dose of neoadjuvant pembrolizumab, 5 of 27 patients achieved a pCR, and all patients with a pCR or major pathologic response remained relapse free. This study showed that pathological response and immunological analyses after a single neoadjuvant dose can predict clinical outcome and unveil underlying mechanisms in checkpoint blockade.

The OpACIN-neo study sought to identify the best dosing and scheduling of neoadjuvant immunotherapy in clinical stage III melanoma [48]. Patients were randomized to three arms (ipilimumab 3 mg/kg and nivolumab 1 mg/kg, ipilimumab 1 mg/kg and nivolumab 3 mg/kg, and sequential ipilimumab 3 mg/kg followed by nivolumab 3 mg/kg). Based on response and toxicities, ipilimumab 1 mg/kg and nivolumab 3 mg/kg (called "flip dose") were determined to be the best regimen to study in future trials. The MeMaloc substudy of OpACIN-neo investigated the feasibility of targeted removal of a single pathological (index) node and whether this node was a reliable indicator of pathologic response in the entire lymph node basin [49]. The index node was marked with a magnetic seed, removed surgically, and analyzed separately from the rest of the lymph node basin after two cycles of therapy. Results showed that the index node was congruent with the entire lymph node basin in 12 out of 12 nodes, which prompted further investigation. Personalized Response-driven Adjuvant therapy after Combination of Ipilimumab and Nivolumab (PRADO) is an extension cohort of the OpACIN-neo study which is actively examining three objectives: (1) the pathologic response rate and safety of ipilimumab 1 mg/kg + nivolumab 3 mg/kg combination scheme, (2) whether patients with a pCR or near-pCR in the index lymph node can be safely spared a lymphadenectomy without affecting recurrence-free survival, and (3) whether recurrence-free survival can be prolonged among pathological nonresponders with additional adjuvant therapy. Patients receive two cycles of neoadjuvant ipilimumab 1 mg/kg and nivolumab 3 mg/kg after marker placement in the index lymph node. Patients that achieve a pathologic response in the index lymph node do not undergo lymphadenectomy; patients with partial response (10–50% viable tumor cells) undergo lymphadenectomy; and patients with no pathologic response undergo lymphadenectomy and receive adjuvant nivolumab or targeted therapy for 52 weeks +/− radiotherapy. Preliminary data confirm the high pathologic response rate and safety observed previously in the OpACIN-neo arm B (ipilimumab 1 mg/kg + nivolumab 3 mg/kg), with a pathologic response rate of 71% and grade 3–4 adverse event rate of 22% in the first 12 weeks. Lymphadenectomy was omitted in 59 (60%) patients [46]. Longer follow-up is needed to report safety and recurrence-free survival when lymphadenectomy is omitted in pathologic responders.

The dosing and duration of therapy have been heterogeneous in the above-highlighted trials. The International Neoadjuvant Melanoma Consortium (INMC) recommends that neoadjuvant protocols are 6–8 weeks in the preoperative phase and that trials should complete an adjuvant phase to total a year of systemic therapy [50]. The degree of pathological response may direct the inclusion of an adjuvant phase, as is occurring in the PRADO study, but this approach is still very much under review. Data on surgical resectability, complications, and delays of surgical intervention have not been robustly collected in the above-mentioned trials, but some preliminary data demonstrate that rates of surgical complications in the neoadjuvant setting are no different than upfront surgery. The long-term results of the PRADO study will shed key insights on our approach to a more focused versus more extensive lymph node dissection and on the use of radiographic markers to assess and monitor disease burden as outlined in the PRADO cohort. These results may also further inform treatment strategies for satellite, in-transit, or oligometastatic disease within the neoadjuvant trial setting. The INMC Surgery Working Group is preparing a specific surgical assessment tool that will standardize surgical outcomes collected and implemented in future trials.

36.10 Conclusion

Multiple new systemic therapies have been approved for patients with melanoma based on results showing efficacy. Along with these new agents, landmark surgical trials have been published demonstrating no survival benefit for routine CLND for a positive SLNB. Therefore, the current role for surgery in managing melanoma must take into account these new results. However, surgical therapy for primary melanoma is relatively unchanged, although efforts to define appropriate margins are ongoing. SLNB is still utilized primarily for prognostication, and is now increasingly to direct effective adjuvant treatment, but SLNB may harbor a possible therapeutic benefit in some cases. Surgery will likely continue to play a role in the management of select patients with isolated metastatic disease in conjunction with systemic therapy and will also allow for assessment of pathologic response after neoadjuvant therapy.

Ongoing and future research will strive to answer:

- What are standard margins for a patient with early-stage localized melanoma?
- What is the appropriate surgery for a patient with clinically involved regional lymph nodes and no other distant disease?
- Which patients should be offered a sentinel lymph node biopsy?
- Is neoadjuvant therapy standard of care for patients with advanced melanoma?
- Which intralesional agents or regional therapies should be considered for patients with metastatic dermal and in-transit melanoma and how can they be combined with systemic therapies?
- When should surgery be considered for patients with distant metastases?

Disclosures GMB (Beasley) served on a scientific advisory board for Regeneron (2019) and has currently clinical trial funding paid to institution from Oncosec, Istari oncology, Replimune, and Delcath. GMB (Boland) served on a scientific advisory board and currently serves on a steering committee for Nektar Therapeutics. She served on a scientific advisory board and as a speaker for Novartis. She has sponsored research agreements with Olink proteomics and Palleon Pharmaceuticals. MCL serves on an advisory board for BMS and receives research support from Vaccinex, BMS, Delcath, and Amgen.

References

1. Whiteman DC, Green AC, Olsen CM. The growing burden of invasive melanoma: projections of incidence rates and numbers of new cases in six susceptible populations through 2031. J Invest Dermatol. 2016;136:1161–71.
2. Gardner LJ, Ward M, Andtbacka RHI, Boucher KM, Bowen GM, Bowles TL, Cohen AL, Grossmann K, Hitchcock YJ, Holmen SL, Hyngstrom J, Khong H, McMahon M, Monroe MM, Ross CB, Suneja G, Wada D, Grossman D. Risk factors for development of melanoma brain metastasis and disease progression: a single-center retrospective analysis. Melanoma Res. 2017;27:477–84. PMC5583702.
3. Leiter U, Meier F, Schittek B, Garbe C. The natural course of cutaneous melanoma. J Surg Oncol. 2004;86:172–8.
4. Davies H, Bignell GR, Cox C, Stephens P, Edkins S, Clegg S, Teague J, Woffendin H, Garnett MJ, Bottomley W, Davis N, Dicks E, Ewing R, Floyd Y, Gray K, Hall S, Hawes R, Hughes J, Kosmidou V, Menzies A, Mould C, Parker A, Stevens C, Watt S, Hooper S, Wilson R, Jayatilake H, Gusterson BA, Cooper C, Shipley J, Hargrave D, Pritchard-Jones K, Maitland N, Chenevix-Trench G, Riggins GJ, Bigner DD, Palmieri G, Cossu A, Flanagan A, Nicholson A, Ho JW, Leung SY, Yuen ST, Weber BL, Seigler HF, Darrow TL, Paterson H, Marais R, Marshall CJ, Wooster R, Stratton MR, Futreal PA. Mutations of the BRAF gene in human cancer. Nature. 2002;417:949–54.
5. Chapman PB, Hauschild A, Robert C, Haanen JB, Ascierto P, Larkin J, Dummer R, Garbe C, Testori A, Maio M, Hogg D, Lorigan P, Lebbe C, Jouary T, Schadendorf D, Ribas A, O'Day SJ, Sosman JA, Kirkwood JM, Eggermont AM, Dreno B, Nolop K, Li J, Nelson B, Hou J, Lee RJ, Flaherty KT, McArthur GA. Improved survival with vemurafenib in melanoma with BRAF V600E mutation. N Engl J Med. 2011;364:2507–16. PMC3549296.
6. Ayers M, Lunceford J, Nebozhyn M, Murphy E, Loboda A, Kaufman DR, Albright A, Cheng JD, Kang SP, Shankaran V, Piha-Paul SA, Yearley J, Seiwert TY, Ribas A, McClanahan TK. IFN-gamma-related mRNA profile predicts clinical response to PD-1 blockade. J Clin Invest. 2017;127:2930–40. PMC5531419.
7. Kwak M, Farrow NE, Salama AKS, Mosca PJ, Hanks BA, Slingluff CL, Beasley GM. Updates in adjuvant systemic therapy for melanoma. J Surg Oncol. 2019;119:222–31. PMC6330126.
8. Hewitt BB, Bradley E, Berger AC. Margins of melanoma excision and modifications to standards. Surg Oncol Clin N Am. 2020;29:339–47.
9. Veronesi U, Cascinelli N, Adamus J, Balch C, Bandiera D, Barchuk A, Bufalino R, Craig P, De Marsillac J, Durand JC, et al. Thin stage I primary cutaneous malignant melanoma. Comparison of excision with margins of 1 or 3 cm. N Engl J Med. 1988;318:1159–62.
10. Sladden MJ, Balch C, Barzilai DA, Berg D, Freiman A, Handiside T, Hollis S, Lens MB, Thompson JF. Surgical excision margins for primary cutaneous melanoma. Cochrane Database Syst Rev. 2009:CD004835.

11. Hayes AJ, Maynard L, Coombes G, Newton-Bishop J, Timmons M, Cook M, Theaker J, Bliss JM, Thomas JM. Wide versus narrow excision margins for high-risk, primary cutaneous melanomas: long-term follow-up of survival in a randomised trial. Lancet Oncol. 2016;17:184–92. PMC4737890.
12. Moncrieff MD, Gyorki D, Saw R, Spillane AJ, Thompson JF, Peach H, Oudit D, Geh J, Dziewulski P, Wilson E, Matteucci P, Pritchard-Jones R, Olofsson Bagge R, Wright FC, Crampton N, Cassell O, Jallali N, Berger A, Kelly J, Hamilton S, Durrani A, Lo S, Paton E, Henderson MA. 1 versus 2-cm excision margins for pT2-pT4 primary cutaneous melanoma (MelMarT): a feasibility study. Ann Surg Oncol. 2018;25:2541–9. PMC6097739.
13. Maurichi A, Miceli R, Camerini T, Contiero P, Patuzzo R, Tragni G, Crippa F, Romanidis K, Ruggeri R, Carbone A, Santinami M. Pure desmoplastic melanoma: a melanoma with distinctive clinical behavior. Ann Surg. 2010;252:1052–7.
14. Morton DL, Thompson JF, Cochran AJ, Mozzillo N, Nieweg OE, Roses DF, Hoekstra HJ, Karakousis CP, Puleo CA, Coventry BJ, Kashani-Sabet M, Smithers BM, Paul E, Kraybill WG, McKinnon JG, Wang HJ, Elashoff R, Faries MB, Group M. Final trial report of sentinel-node biopsy versus nodal observation in melanoma. N Engl J Med. 2014;370:599–609. PMC4058881.
15. Wong SL, Faries MB, Kennedy EB, Agarwala SS, Akhurst TJ, Ariyan C, Balch CM, Berman BS, Cochran A, Delman KA, Gorman M, Kirkwood JM, Moncrieff MD, Zager JS, Lyman GH. Sentinel lymph node biopsy and management of regional lymph nodes in melanoma: American Society of Clinical Oncology and Society of Surgical Oncology clinical practice guideline update. Ann Surg Oncol. 2018;25:356–77.
16. Lo SN, Ma J, Scolyer RA, Haydu LE, Stretch JR, Saw RPM, Nieweg OE, Shannon KF, Spillane AJ, Ch'ng S, Mann GJ, Gershenwald JE, Thompson JF, Varey AHR. Improved risk prediction calculator for sentinel node positivity in patients with melanoma: the Melanoma Institute Australia Nomogram. J Clin Oncol. 2020;38(24):2719–27.
17. Morton DL, Thompson JF, Cochran AJ, Mozzillo N, Elashoff R, Essner R, Nieweg OE, Roses DF, Hoekstra HJ, Karakousis CP, Reintgen DS, Coventry BJ, Glass EC, Wang HJ. Sentinel-node biopsy or nodal observation in melanoma. N Engl J Med. 2006;355:1307–17.
18. Faries MB, Thompson JF, Cochran AJ, Andtbacka RH, Mozzillo N, Zager JS, Jahkola T, Bowles TL, Testori A, Beitsch PD, Hoekstra HJ, Moncrieff M, Ingvar C, Wouters M, Sabel MS, Levine EA, Agnese D, Henderson M, Dummer R, Rossi CR, Neves RI, Trocha SD, Wright F, Byrd DR, Matter M, Hsueh E, MacKenzie-Ross A, Johnson DB, Terheyden P, Berger AC, Huston TL, Wayne JD, Smithers BM, Neuman HB, Schneebaum S, Gershenwald JE, Ariyan CE, Desai DC, Jacobs L, McMasters KM, Gesierich A, Hersey P, Bines SD, Kane JM, Barth RJ, McKinnon G, Farma JM, Schultz E, Vidal-Sicart S, Hoefer RA, Lewis JM, Scheri R, Kelley MC, Nieweg OE, Noyes RD, DSB H, Wang HJ, Elashoff DA, Elashoff RM. Completion dissection or observation for sentinel-node metastasis in melanoma. N Engl J Med. 2017;376:2211–22. PMC5548388.
19. Perone JA, Farrow N, Tyler DS, Beasley GM. Contemporary approaches to in-transit melanoma. J Oncol Pract. 2018;14:292–300.
20. Chai CY, Deneve JL, Beasley GM, Marzban SS, Chen YA, Rawal B, Grobmyer SR, Hochwald SN, Tyler DS, Zager JS. A multi-institutional experience of repeat regional chemotherapy for recurrent melanoma of extremities. Ann Surg Oncol. 2012;19:1637–43. PMC4515968.
21. Farrow NE, Leddy M, Landa K, Beasley GM. Injectable therapies for regional melanoma. Surg Oncol Clin N Am. 2020;29:433–44. PMC7274146.
22. Louie RJ, Perez MC, Jajja MR, Sun J, Collichio F, Delman KA, Lowe M, Sarnaik AA, Zager JS, Ollila DW. Real-world outcomes of Talimogene Laherparepvec therapy: a multi-institutional experience. J Am Coll Surg. 2019;228:644–9.
23. Andtbacka RH, Kaufman HL, Collichio F, Amatruda T, Senzer N, Chesney J, Delman KA, Spitler LE, Puzanov I, Agarwala SS, Milhem M, Cranmer L, Curti B, Lewis K, Ross M, Guthrie T, Linette GP, Daniels GA, Harrington K, Middleton MR, Miller WH Jr, Zager JS, Ye Y, Yao B, Li A, Doleman S, VanderWalde A, Gansert J, Coffin RS. Talimogene Laherparepvec improves durable response rate in patients with advanced melanoma. J Clin Oncol. 2015;33:2780–8.
24. Robert C, Long GV, Brady B, Dutriaux C, Giacomo AMD, Mortier L, Rutkowski P, Hassel JC, McNeil CM, Kalinka EA, Lebbé C, Charles J, Hernberg MM, Savage KJ, Chiarion-Sileni V, Mihalcioiu C, Mauch C, Arance A, Cognetti F, Ny L, Schmidt H, Schadendorf D, Gogas H, Zoco J, Re S, Ascierto PA, Atkinson V. Five-year outcomes with nivolumab in patients with wild-type braf advanced melanoma. J Clin Oncol. 2020;38(33):3937–46.
25. Ribas A, Dummer R, Puzanov I, VanderWalde A, Andtbacka RHI, Michielin O, Olszanski AJ, Malvehy J, Cebon J, Fernandez E, Kirkwood JM, Gajewski TF, Chen L, Gorski KS, Anderson AA, Diede SJ, Lassman ME, Gansert J, Hodi FS, Long GV. Oncolytic virotherapy promotes intratumoral T cell infiltration and improves anti-PD-1 immunotherapy. Cell. 2017;170:1109–1119.e10.
26. Ariyan CE, Lefkowitz RA, Panageas K, Callahan MK, Misholy O, Bello D, Fisher C, Kotin A, Siegelbaum R, Coit DG, Wolchok JD, Brady MS. Safety and clinical activity of combining systemic ipilimumab with isolated limb infusion in patients with in-transit melanoma. J Clin Oncol. 2014;32:9078.
27. Testori AAE, Ribero S, Indini A, Mandalà M. Adjuvant treatment of melanoma: recent developments and future perspectives. Am J Clin Dermatol. 2019;20:817–27.
28. Morton DL, Wen DR, Wong JH, Economou JS, Cagle LA, Storm FK, Foshag LJ, Cochran AJ. Technical details of intraoperative lymphatic mapping for early stage melanoma. Arch Surg. 1992;127:392–9.
29. McMasters KM, Wong SL, Edwards MJ, Chao C, Ross MI, Noyes RD, Viar V, Cerrito PB, Reintgen DS. Frequency of non-sentinel lymph node metastasis in melanoma. Ann Surg Oncol. 2002;9:137–41.
30. Dewar DJ, Newell B, Green MA, Topping AP, Powell BW, Cook MG. The microanatomic location of metastatic melanoma in sentinel lymph nodes predicts nonsentinel lymph node involvement. J Clin Oncol. 2004;22:3345–9.
31. Glumac N, Hocevar M, Zadnik V, Snoj M. Sentinel lymph node micrometastasis may predict non-sentinel involvement in cutaneous melanoma patients. J Surg Oncol. 2008;98:46–8.
32. Rossi CR, Mocellin S, Campana LG, Borgognoni L, Sestini S, Giudice G, Caraco C, Cordova A, Solari N, Piazzalunga D, Carcoforo P, Quaglino P, Caliendo V, Ribero S. Prediction of non-sentinel node status in patients with melanoma and positive sentinel node biopsy: an Italian Melanoma Intergroup (IMI) Study. Ann Surg Oncol. 2018;25(1):271–9.
33. Murali R, Desilva C, Thompson JF, Scolyer RA. Non-sentinel node risk score (N-SNORE): a scoring system for accurately stratifying risk of non-sentinel node positivity in patients with cutaneous melanoma with positive sentinel lymph nodes. J Clin Oncol. 2010;28:4441–9.
34. Ulmer A, Dietz K, Werner-Klein M, Häfner H-M, Schulz C, Renner P, Weber F, Breuninger H, Röcken M, Garbe C, Fierlbeck G, Klein CA. The sentinel lymph node spread determines quantitatively melanoma seeding to non-sentinel lymph nodes and survival. Eur J Cancer. 2018;91:1–10.
35. Faries MB, Cochran AJ, Elashoff RM, Thompson JF. Multicenter selective lymphadenectomy trial-I confirms the central role of sentinel node biopsy in contemporary melanoma management: response to 'No survival benefit for patients with melanoma undergoing

sentinel lymph node biopsy: critical appraisal of the multicenter selective lymphadenectomy trial-I final report'. Br J Dermatol. 2015;172:571–3. PMC4437591.
36. Leiter U, Stadler R, Mauch C, Hohenberger W, Brockmeyer N, Berking C, Sunderkötter C, Kaatz M, Schulte K-W, Lehmann P, Vogt T, Ulrich J, Herbst R, Gehring W, Simon J-C, Keim U, Martus P, Garbe C. Complete lymph node dissection versus no dissection in patients with sentinel lymph node biopsy positive melanoma (DeCOG-SLT): a multicentre, randomised, phase 3 trial. Lancet Oncol. 2016;17:757–67.
37. Nessim C, Law C, McConnell Y, Shachar S, McKinnon G, Wright F. How often do level III nodes bear melanoma metastases and does it affect patient outcomes? Ann Surg Oncol. 2013;20:2056–64.
38. Byrne EH, Fisher DE. Correlates of metastatic melanoma response to checkpoint inhibition. Cancer. 2018;123(Suppl 11):2143–53.
39. Weber J, Mandala M, Vecchio MD, Gogas HJ, Arance AM, Cowey CL, Dalle S, Schenker M, Chiarion-Sileni V, Marquez-Rodas I, Grob J-J, Butler MO, et al. ftCC: adjuvant Nivolumab versus Ipilimumab in resected stage III or IV melanoma. NEJM. 2017;377:1824–35.
40. Eggermont AMM, Chiarion-Sileni V, Grob JJ, Dummer R, Wolchok JD, Schmidt H, Hamid O, Robert C, Ascierto PA, Richards JM, Lebbe C, Ferraresi V, Smylie M, Weber JS, Maio M, Hosein F, de Pril V, Kicinski M, Suciu S, Testori A. Adjuvant ipilimumab versus placebo after complete resection of stage III melanoma: long-term follow-up results of the European Organisation for Research and Treatment of Cancer 18071 double-blind phase 3 randomised trial. Eur J Cancer. 2019;119:1–10.
41. Weber J, Mandala M, Del Vecchio M, Gogas HJ, Arance AM, Cowey CL, Dalle S, Schenker M, Chiarion-Sileni V, Marquez-Rodas I, Grob JJ, Butler MO, Middleton MR, Maio M, Atkinson V, Queirolo P, Gonzalez R, Kudchadkar RR, Smylie M, Meyer N, Mortier L, Atkins MB, Long GV, Bhatia S, Lebbe C, Rutkowski P, Yokota K, Yamazaki N, Kim TM, de Pril V, Sabater J, Qureshi A, Larkin J, Ascierto PA. Adjuvant Nivolumab versus Ipilimumab in resected stage III or IV melanoma. N Engl J Med. 2017;377(19):1824–35.
42. Long GV, Hauschild A, Santinami M, Atkinson V, Mandala M, Chiarion-Sileni V, Larkin J, Nyakas M, Dutriaux C, Haydon A, Robert C, Mortier L, Schachter J, Schadendorf D, Lesimple T, Plummer R, Ji R, Zhang P, Mookerjee B, Legos J, Kefford R, Dummer R, Kirkwood JM. Adjuvant Dabrafenib plus Trametinib in stage III BRAF-mutated melanoma. N Engl J Med. 2017;377:1813–23.
43. Blank CU, Rozeman EA, Fanchi LF, Sikorska K, van de Wiel B, Kvistborg P, Krijgsman O, van den Braber M, Philips D, Broeks A, van Thienen JV, Mallo HA, Adriaansz S, Ter Meulen S, Pronk LM, Grijpink-Ongering LG, Bruining A, Gittelman RM, Warren S, van Tinteren H, Peeper DS, Haanen J, van Akkooi ACJ, Schumacher TN. Neoadjuvant versus adjuvant ipilimumab plus nivolumab in macroscopic stage III melanoma. Nat Med. 2018;24:1655–61.
44. Amaria RN, Prieto PA, Tetzlaff MT, Reuben A, Andrews MC, Ross MI, Glitza IC, Cormier J, Hwu WJ, Tawbi HA, Patel SP, Lee JE, Gershenwald JE, Spencer CN, Gopalakrishnan V, Bassett R, Simpson L, Mouton R, Hudgens CW, Zhao L, Zhu H, Cooper ZA, Wani K, Lazar A, Hwu P, Diab A, Wong MK, McQuade JL, Royal R, Lucci A, Burton EM, Reddy S, Sharma P, Allison J, Futreal PA, Woodman SE, Davies MA, Wargo JA. Neoadjuvant plus adjuvant dabrafenib and trametinib versus standard of care in patients with high-risk, surgically resectable melanoma: a single-centre, open-label, randomised, phase 2 trial. Lancet Oncol. 2018;19:181–93.
45. Long GV, Saw RPM, Lo S, Nieweg OE, Shannon KF, Gonzalez M, Guminski A, Lee JH, Lee H, Ferguson PM, Rawson RV, Wilmott JS, Thompson JF, Kefford RF, Ch'ng S, Stretch JR, Emmett L, Kapoor R, Rizos H, Spillane AJ, Scolyer RA, Menzies AM. Neoadjuvant dabrafenib combined with trametinib for resectable, stage IIIB-C, BRAF(V600) mutation-positive melanoma (NeoCombi): a single-arm, open-label, single-centre, phase 2 trial. Lancet Oncol. 2019;20:961–71.
46. Amaria RN, Reddy SM, Tawbi HA, Davies MA, Ross MI, Glitza IC, Cormier JN, Lewis C, Hwu WJ, Hanna E, Diab A, Wong MK, Royal R, Gross N, Weber R, Lai SY, Ehlers R, Blando J, Milton DR, Woodman S, Kageyama R, Wells DK, Hwu P, Patel SP, Lucci A, Hessel A, Lee JE, Gershenwald J, Simpson L, Burton EM, Posada L, Haydu L, Wang L, Zhang S, Lazar AJ, Hudgens CW, Gopalakrishnan V, Reuben A, Andrews MC, Spencer CN, Prieto V, Sharma P, Allison J, Tetzlaff MT, Wargo JA. Neoadjuvant immune checkpoint blockade in high-risk resectable melanoma. Nat Med. 2018;24:1649–54. PMC6481682.
47. Huang AC, Orlowski RJ, Xu X, Mick R, George SM, Yan PK, Manne S, Kraya AA, Wubbenhorst B, Dorfman L, D'Andrea K, Wenz BM, Liu S, Chilukuri L, Kozlov A, Carberry M, Giles L, Kier MW, Quagliarello F, McGettigan S, Kreider K, Annamalai L, Zhao Q, Mogg R, Xu W, Blumenschein WM, Yearley JH, Linette GP, Amaravadi RK, Schuchter LM, Herati RS, Bengsch B, Nathanson KL, Farwell MD, Karakousis GC, Wherry EJ, Mitchell TC. A single dose of neoadjuvant PD-1 blockade predicts clinical outcomes in resectable melanoma. Nat Med. 2019;25:454–61. PMC6699626.
48. Rozeman EA, Blank CU, Akkooi ACJV, Kvistborg P, Fanchi L, Thienen JVV, Stegenga B, Lamon B, Haanen JBAG, Schumacher T. Neoadjuvant ipilimumab + nivolumab (IPI+NIVO) in palpable stage III melanoma: updated data from the OpACIN trial and first immunological analyses. J Clin Oncol. 2017;35:9586.
49. Schermers B, Franke V, Rozeman EA, van de Wiel BA, Bruining A, Wouters MW, van Houdt WJ, Ten Haken B, Muller SH, Bierman C, Ruers TJM, Blank CU, van Akkooi ACJ. Surgical removal of the index node marked using magnetic seed localization to assess response to neoadjuvant immunotherapy in patients with stage III melanoma. Br J Surg. 2019;106:519–22. PMC6593699.
50. Amaria R, Caffrey M, et al. Am J Manag Care. 2019;25:sp43–4.

The Role of Surgery in Managing Primary and Metastatic Breast Cancer

37

Alicia M. Terando, Azadeh Carr, Tina J. Hieken, Mara A. Piltin, Bindupriya Chandrasekaran, and Carla S. Fisher

Abstract

Modern breast cancer surgery began with Dr. William Halsted's hypothesis that breast cancer spreads in a stepwise fashion from local to regional to distant sites and his description of the radical mastectomy in 1894. With the introduction of antineoplastic chemotherapy and radiation therapy, and improved knowledge of the biology and behavior of breast cancer, the treatment of breast cancer has evolved substantially. It is now understood that breast cancer metastasis is driven by tumor biology, hence en bloc resection does not improve outcomes over operations that extirpate cancer while minimizing sacrifice of normal tissue. In turn, efficacious new systemic therapies, including biologically targeted therapies, permit further de-escalation of surgical therapy. For patients presenting with an intact primary breast tumor and synchronous distant metastases, recent data suggest that surgery for the primary site does not improve overall survival, providing further evidence that tumor biology cannot be overcome by surgery.

Learning Objectives

- Understand the biological subtypes of breast cancer.
- Understand the differences between breast-conserving surgery and mastectomy in the surgical management of early-stage breast cancer.
- Define standard resection margins in the treatment of breast cancer.
- Understand appropriate surgical management of T4 breast cancer.
- Describe the surgical management of breast tumors following neoadjuvant chemotherapy.
- Understand the evolution of the surgical management of the axillary lymph nodes.
- Understand the role of surgery in the management of stage IV breast cancer.

37.1 Introduction

Modern breast cancer surgery began with Dr. William Halsted's hypothesis that breast cancer spreads in a stepwise fashion from local to regional to distant sites and his description of the radical mastectomy in 1894. With the introduction of antineoplastic chemotherapy and radiation therapy, and improved knowledge of the biology and behavior of breast cancer, the treatment of breast cancer has evolved substantially. It is now understood that breast cancer metastasis is driven by tumor biology, hence en bloc resection does not improve outcomes over operations that extirpate cancer while minimizing sacrifice of normal tissue. In turn, efficacious new systemic therapies, including biologically targeted therapies, permit further de-escalation of surgical therapy. For patients presenting with an intact primary breast tumor and synchronous distant metastases, recent data suggest that surgery for the primary site does not improve overall survival,

A. M. Terando (✉) · A. Carr
University of Southern California, Keck School of Medicine, Los Angeles, CA, USA
e-mail: Alicia.Terando@med.usc.edu; Azadeh.carr@med.usc.edu

T. J. Hieken · M. A. Piltin
Mayo Clinic, Rochester, MN, USA
e-mail: Hieken.Tina@Mayo.edu; Piltin.Mara@Mayo.edu

B. Chandrasekaran · C. S. Fisher
Indiana University School of Medicine, Indianapolis, IN, USA
e-mail: BiChan@iu.edu; FisherCS@iu.edu

S. P. Leong et al. (eds.), *Cancer Metastasis Through the Lymphovascular System*,
https://doi.org/10.1007/978-3-030-93084-4_37

providing further evidence that tumor biology cannot be overcome by surgery.

37.2 Breast Cancer Tumor Biology

Beyond the key markers used to stratify breast cancer subtypes (estrogen receptor [ER], progesterone receptor [PR], and human epidermal growth factor [Her2/neu]), more sophisticated gene expression profiling is now used to predict tumor behavior and selection of adjuvant therapies even beyond that of the four main breast cancer subtypes: Luminal A, Luminal B, Her2/neu amplified, and basal like. Due to variation in prognosis and treatment sensitivity, knowledge of these subtypes is important in guiding treatment. Tumors may be classified into intrinsic subtypes by multigene expression molecular assays (PAM50—predictor analysis of Microarray 50). Since these tests are not readily available, pathology-based surrogate definitions were proposed and have since been revised to further distinguish Luminal A and Luminal B subtypes [1–3]. Luminal A is defined as ER and PR positive (≥20%), Her2/neu non-amplified, Ki-67 low (<14%), with a low recurrence risk based on multigene expression assay results, if available. Luminal B is defined as ER positive, Her2/neu non-amplified, and at least one of the following: Ki67 high (>14%), PR negative or low positive (<20%), with a high risk of recurrence based on multigene expression assay results. Other studies have recommended higher cutoffs for Ki 67 of ≥20% [4]. Tumors that are Her2/neu-amplified and ER/PR positive are also classified as Luminal B. Basal-like tumors are ER/PR negative, and Her2/neu non-amplified, commonly referred to as triple-negative breast carcinoma (TNBC). However, not all TNBCs are basal like: uncommon types of TNBC such as medullary and adenoid cystic carcinoma are not considered basal like and have a low risk of distant recurrence [1, 5]. Conversely, metaplastic breast cancer, while being triple-negative, has a worse prognosis than classic basal-like TNBC and responds differently to systemic therapies [6].

The choice of systemic treatment for breast cancer is driven by subtype. Endocrine therapy is an integral part of treatment for all patients with hormone receptor-positive tumors. There are a number of prognostic assays available for Luminal A and Luminal B tumors (not Her2/neu amplified) that estimate recurrence risk (OncotypeDx®, EndoPredict, PAM50, Breast Cancer Index, and MammaPrint), however only OncotypeDx® Recurrence Score is validated for predicting adjuvant chemotherapy benefit in reducing the risk of recurrence and thus included in National Comprehensive Cancer Network Treatment Guidelines for Breast Cancer [7–9]. For Luminal A tumors, endocrine therapy is often used alone except in situations of high-risk multigene expression assay results, grade 3 disease, or involvement of multiple lymph nodes. Luminal B tumors have a worse prognosis than Luminal A and are most often recommended cytotoxic systemic therapy along with endocrine therapy [10]. Due to the aggressive nature of Her2/neu-amplified tumors, anti-Her2/neu therapy in combination with cytotoxic chemotherapy is recommended for tumors ≥0.5 cm and/or those with nodal metastases. Basal-like TNBC is generally treated with cytotoxic chemotherapy due to the lack of known targetable proteins or mutations. Interestingly, there is a higher incidence of basal-like tumors in African Americans [11].

Different molecular subtypes of breast cancer also behave differently as far as the risk of early recurrence and metastasis [10, 12, 13]. Luminal A tumors, which are the most common, carry the best prognosis of all breast cancer subtypes [10, 14, 15]. They are associated with a lower overall risk of distant metastasis, although metastases have been known to present much later, on the order of 5–15 years following completion of primary therapy [12, 16, 17]. On the other hand, basal like and Her2/neu amplified/ER-negative tumors are more likely to metastasize early, within 5 years after diagnosis and treatment [18]. Additional studies of Her2/neu amplified tumors have identified significant heterogeneity within this group, with some Her2/neu amplified tumors exhibiting luminal A features or basal-like features, which may affect outcomes and response to treatment [19–21].

Disease-free survival and specific sites of metastasis may also differ based on tumor subtype. A retrospective review of 1951 patients with node-negative, early-stage breast cancer showed that disease-free survival at 10 years was higher for Luminal A breast cancers (86%) compared to Luminal B (76%), Her2/neu amplified (73%), and triple-negative breast cancers (71%, $p < 0.001$) [16]. Luminal B tumors are more likely than other subtypes to have first recurrences in the bones. A retrospective review of 3762 patients with a median follow-up of 14.8 years found, on multivariate analysis, which compared with luminal A tumors, luminal B Her2/neu amplified, and Her2/neu amplified tumors had significantly higher rates of brain, liver, and lung metastases [12]. Basal-like tumors had higher rates of brain, lung, and distant nodal metastases, but lower rates of liver and bone metastases. In a study of breast cancers diagnosed from 2010 to 2014 and reported in the SEER database, the rates of concurrent distant metastases were 3.28% for bone, 1.52% for lung, 1.2% for liver, and 0.35% for the brain [13]. On multivariate analysis, when compared to hormone receptor-positive, Her2/neu non-amplified tumors (luminal A and Her2/neu non-amplified luminal B), Her2/neu amplified subtypes, whether hormone receptor-positive or not, had significantly higher rates of liver and lung metastasis. TNBCs had a higher rate of the brain (OR 1.95), liver (OR, 1.35), and lung (OR, 1.34) metastasis but significantly lower rate of bone metastasis than hormone receptor-positive, Her2/neu non-amplified. Overall prognosis stratified by metastatic site was also worse for triple-negative subtypes. In a cohort study of patients treated with breast-conserving therapy for early-stage breast cancer, the overall rate of brain metastasis was less than 2% at 5 years and less than 5% at 10 years. However, luminal B (12%), luminal B Her2/neu amplified (7%), Her2/neu amplified (12%), and basal-like (7%) tumors were more likely to develop brain metastasis [22]. Even for patients whose tumors fall into the same subtype, there remains considerable heterogeneity as far as clinical outcomes. Further refinement of breast cancer subtypes is an area of active research which may help better determine the prognosis for individual patients in the future [23].

37.3 Surgery for the Primary Site in Breast Cancer

Early breast cancer refers to breast cancer limited to the breast and/or regional lymph nodes that are presumed curable. Notwithstanding notable advances in systemic therapy, surgery remains an essential component of treatment for early breast cancer.

37.3.1 Breast Conservation Versus Mastectomy

There are two main options for surgical management of breast tumors, namely mastectomy or breast-conserving surgery. Mastectomy is recommended for multicentric disease and for large or unfavorably located tumors when resection with negative margins would preclude acceptable cosmesis. Mastectomy can be done with or without immediate reconstruction (first stage or definitive) thus enabling preservation of an envelope of the breast skin and, in selected cases, even the nipple–areolar complex [8, 24, 25]. For the remainder of patients, either approach is acceptable, and decision-making is multifaceted. In 2002, publication of the 20-year follow-up from two large randomized controlled clinical trials comparing lumpectomy with whole breast radiation to mastectomy undisputedly established the oncologic safety of breast preservation showing no difference in overall or disease-free survival [26] [27]. In-breast tumor recurrence (IBTR) is higher among patients with a retained breast, as would be expected, but the risk of local recurrence is not nil with mastectomy. The risk of local recurrence parallels the risk of regional and distant recurrence for a given tumor and is influenced by biologic subtype and adjuvant systemic therapies to a greater extent than the choice for or against breast preservation when appropriate surgical care is provided. Therefore, shared decision-making is a key component of breast surgical care. When an adequate oncologic surgical resection can be performed, the correct choice is one made by an informed patient reflecting their own values and preferences [28].

Technical standards for the conduct of breast oncologic surgery have been published under the auspices of the Cancer Programs of the American College of Surgeons and provide an excellent resource for the surgeon [28].

37.3.2 Pathologic Margins for Breast Cancer Surgery

One critical technical standard for breast cancer operations is resection to achieve microscopically negative margins with care taken to avoid violating the tumor. The surgeon must properly orient the specimen for the pathologist who then reports on both gross and microscopic margins [8, 28]. Based on a meta-analysis of 33 studies published in 2014 [29], in an effort to decrease secondary operations for margin re-excision which averaged 30% at the time [30], the Society of Surgical Oncology and the American Society of Radiation Oncology published consensus guidelines regarding appropriate negative margin width in the setting of breast-conserving surgery with planned whole breast radiation. These guidelines, subsequently endorsed by the American Society for Clinical Oncology, state that no-tumor-on-ink is an acceptable margin for patients with stage I and II breast cancer treated with lumpectomy and adjuvant radiation [31, 32].

No consensus guidelines exist for pathologic margins when mastectomy is performed for breast cancer treatment. A recent systematic review of 34 studies concluded that positive margins were associated with a 2.6-fold increased risk of local recurrence, but there was no clear association with margin width. Acknowledging that margin width was defined differently across the studies, similar hazard ratios were seen for positive/close margins within 2 mm [33]. A study of 1206 mastectomies for cancer with meticulous pathologic margin assessment, including 434 skin- or nipple-sparing mastectomies performed from 2006 to 2010 with a minimum 5-year follow-up, reported a 3.1% local recurrence rate at 5 years, a lower local recurrence rate with negative margins of >2 mm versus closer margins, a linear decrease in recurrence for negative margin width up to 10 mm and no difference between patients treated with total mastectomy versus skin- or nipple-sparing mastectomy [34].

37.3.3 Surgical Treatment of T4 Breast Cancers

T4 tumors of the breast include those with invasion of the chest wall beyond the pectoralis major muscle (T4a), skin ulceration or nodules (T4b), or inflammatory breast cancer (T4d) defined clinically as erythema or edema of less than or equal to one-third of the breast for ≤6 months with biopsy-confirmed underlying invasive breast cancer. Notably, a skin biopsy to assess tumor emboli in dermal lymphatics neither confirms nor excludes the diagnosis [8, 35]. These tumors pose local management challenges for the breast surgical oncologist. The standard-of-care recommendation in these situations is for total mastectomy with en bloc resection of the affected chest wall for T4a tumors, and resection including skin nodules or ulcers for T4b tumors, although breast conservation may be considered for highly selected patients with favorable biology and noninflammatory T4 disease with limited skin and/or chest wall involvement [8, 35, 36]. Neoadjuvant systemic therapy is recommended for inflammatory breast cancer and often for other T4 tumors as well. The often-widespread dermal lymphatic involvement seen in inflammatory breast cancer, along with high rates (~90%) of regional nodal involvement at presentation, has thus far stymied efforts at breast and/or axillary conservation with high false-negative rates reported for sentinel node surgery. De-escalating surgical treatment in exceptional responders to neoadjuvant systemic therapy might be investigated in the future as novel and increasingly efficacious regimens are developed.

37.3.4 Management of the Primary Breast Tumor After Neoadjuvant Therapy

For patients for whom chemotherapy is indicated at the time of diagnosis, administration of chemotherapy in the neoadjuvant setting can allow de-escalation of surgical treatment. The index tumor marked with a clip may substantially decrease in size following neoadjuvant therapy to permit breast-conserving surgery, especially in patients with ER-negative-Her2/neu non-amplified or Her2/neu amplified tumors [37]. With pathologic complete response rates in the breast as high as 50% for some tumor types, clinical trials testing whether or not surgery might be avoided altogether in selected patients have been initiated. The NRG BR005 trial evaluated the use of image-guided core needle biopsies of the primary breast tumor bed following neoadjuvant chemotherapy. Recently reported results from this study showed a low sensitivity (50%) for detection of residual disease in the breast, suggesting further refinements are needed before surgical treatment can be safely omitted for these patients [38].

37.4 Surgery for Nodal Metastases in Breast Cancer

37.4.1 Axillary Node Dissection for All

Until the 1990s, axillary lymph node dissection (ALND) remained an important part of the surgical treatment for all patients with breast cancer because of the pathological status of the lymph nodes contributed to breast cancer staging and prognostication. Additionally, at the time, removal of the axillary lymph nodes was felt to be necessary to achieve locoregional control. The presence of axillary metastasis is known to be associated with reduced disease free and overall survival compared to patients without nodal disease [39]. ALND involves removal of nodal tissue between the anatomic landmarks of the axillary vein superiorly, the chest wall medially, and the thoracodorsal bundle and latissimus dorsi muscle laterally, generally yielding approximately 10–20 nodes. While ALND, with or without mastectomy, has less morbidity than the radical mastectomy, it is still associated with significant morbidity including paresthesia, seroma, lymphedema, axillary web formation, pain, decreased shoulder mobility and rarely, nerve palsy [40–42]. Thus, surgeons sought to develop an alternative procedure to stage breast cancer to avoid the morbidity of ALND.

37.4.2 Introduction of the Sentinel Lymph Node Biopsy

The concept of the sentinel lymph node (SLN) was first developed for melanoma surgery in the late 1980s [43]. The SLN is defined as the first lymph node to receive lymphatic drainage from a tumor and therefore should be the first site of lymphatic metastases. Identification of these nodes was performed via injection of radioactive tracers and/or vital blue dye into the breast tissue. The sentinel lymph node biopsy (SLNB) technique was subsequently adapted for patients with breast cancer by Giuliano et al. who published their experience with the procedure in 114 patients with breast cancer in 1994 [44]. The SLNB was performed in patients with clinically negative axillae, meaning that there was no evidence of metastatic nodal disease identified on physical examination or by imaging. For patients with clinically positive nodes, a failed SLNB, or SLNB that returned showing metastatic breast cancer, a complete ALND was performed. SLNB in patients with early-stage breast cancer was shown to be an oncologically safe procedure to establish selective application of complete axillary node dissection in clinically node-negative patients in two large randomized controlled trials: the NSABP B-32 and the Axillary Lymphatic Mapping Against Nodal Axillary Clearance (ALMANAC) trials [45–

47]. Not surprisingly, patients undergoing SLNB had lower rates of lymphedema, pain, and improved arm mobility compared to ALND [48, 49].

37.4.3 Sentinel Lymph Node Biopsy and Micrometastases

Although the SLNB technique was a great advancement for patients and surgeons, it introduced a new pathologic finding which needed to be addressed. With more thorough examination of the SLNs and advances in pathology, there was enhanced detection of very small amounts of disease in the lymph nodes. Micrometastatic disease in lymph nodes is defined as disease measuring between 0.2 and 2 mm. In a review of eight studies with at least 100 patients each and 5 years follow-up, Tjan-Heijen et al. concluded that there was insufficient evidence to confirm that micrometastases were of prognostic utility [50]. The International Breast Cancer Study Group (IBCSG 23-01) published results comparing T1–T2 patients with isolated tumor cells and micrometastatic nodal disease who underwent ALND with patients who did not undergo ALND and showed no difference in disease-free survival between the two groups [51, 52]. This trial confirmed that ALND can be safely omitted in patients with micrometastatic nodal disease.

37.4.4 American College of Surgeons Oncology Group (ACOSOG) Z0011 Trial

As experience with SLNB grew, the question of the need for ALND in all patients with node positivity was raised. Many patients with a positive sentinel node or nodes (micro- and macro-metastatic disease) had no further axillary nodal involvement at the time of ALND. Moreover, as medical therapy was increasingly dictated by the receptor profile of the tumor as well as molecular assays, the morbidity associated with ALND became hard to justify for the small amount of additional information gleaned from completion axillary dissection.

The Z0011 trial enrolled patients with T1–T2 breast cancers who were clinically node negative and planned to undergo breast-conserving surgery (BCS) and whole breast radiation, and who were found to have one or two positive SLNs. These patients were randomized to completion ALND or no further surgery. All patients received whole breast radiation and adjuvant endocrine and/or chemotherapy. The study showed no differences in overall survival, disease-free survival, or locoregional recurrence between the two groups and concluded that there was no clinical benefit to ALND in patients with low-volume axillary nodal disease [53, 54]. These findings remained true in a 10-year follow-up analysis of the original Z0011 trial [55]. While the conduct and results of the Z0011 trial were initially met with skepticism, many studies since its original publication have confirmed that ALND can be safely omitted for patients meeting "Z0011 criteria." The findings of the Z0011 trial were quickly accepted by surgical oncologists, leading to a profound and sweeping change in clinical practice that was soon reflected in national guidelines supporting the omission of ALND in patients meeting the inclusion criteria of the Z0011 trial [56].

37.4.5 Nodal Evaluation After Neoadjuvant Chemotherapy

The increasing use of neoadjuvant chemotherapy (NAC) for breast cancer has led to an increasing and substantial number of patients who require axillary nodal evaluation. Initially, there were no data to guide how and when axillary nodal evaluation should be performed relative to neoadjuvant treatment. For several years, the management of the axilla in these patients was determined by surgeon preference: some surgeons favored SLNB prior to chemotherapy and others espoused complete ALND after chemotherapy. In the absence of data, both approaches were criticized. On view was that SLNB prior to chemotherapy did not give patients an opportunity to "downstage" following chemotherapy, obligating patients with a positive sentinel lymph node to a complete axillary lymph node dissection. On the other hand, there were concerns related to the feasibility and accuracy of SLNB after chemotherapy, related to fibrosis in lymphatic channels following chemotherapy that may result in altered lymphatic drainage and potentially a nonuniform response to treatment in the axillary nodes that would cause an unacceptably high false-negative rate (FNR) for SLNB after NAC. Several studies, including NSABP B-27, demonstrated that sentinel node identification rates and FNRs after chemotherapy were not significantly different from the pre-neoadjuvant therapy scenario [57, 58]. In clinically node-negative patients, SLNB following NAC was soon considered to be reliable and became the standard of care. However, if the SLNB was positive or unable to be identified, a completion ALND was performed.

While SLNB became routine after neoadjuvant therapy for patients who were clinically node negative at presentation, all patients with nodal involvement at initial presentation continued to undergo complete ALND following NAC, regardless of clinical or radiologic response. However, it began to be appreciated that a substantial number of patients with clinical N1 disease at presentation experienced a complete response to therapy. Specifically, two studies reported pathologic complete responses in the lymph nodes in

12–65% of clinically node-positive patients who received neoadjuvant systemic therapy [59, 60]. The pathologic complete response rates were highest in triple-negative and Her2/neu amplified cancers. These impressive rates of pathologic complete response led to a reevaluation of the need for complete ALND in patients who have a clinical response to neoadjuvant therapy. Several studies were performed to examine the safety and accuracy of SLNB following NAC in patients who were clinically node positive at presentation [61, 62]. The key outcome for these studies was an acceptable FNR. While the reported FNRs in these trials were felt to be prohibitively high, further reviews noted that FNR could be lowered further by removing at least two sentinel nodes, using dual tracers, and by excising the lymph node that was initially biopsy proven to contain metastatic disease, which could be identified via use of a clip placed at the time of initial biopsy [63]. Piltin et al. showed that omission of ALND for patients with nodal pathologic complete response to neoadjuvant systemic therapy did not worsen 2-year oncologic outcomes [64].

Unfortunately, there remains a substantial number of patients with nodal involvement at the time of presentation who do not have a pathologic complete response to chemotherapy in the lymph nodes. At the time of this writing, the standard of care in this scenario is to perform a complete ALND followed by comprehensive nodal radiation. As a continuation of the trend for de-escalation of surgical therapy, it is now being questioned whether complete ALND provides any benefit over comprehensive nodal radiation in these patients. The clinical trial, ALLIANCE A11202 [65], will specifically examine the clinical outcomes of patients with residual nodal disease following neoadjuvant systemic therapy who undergo comprehensive nodal radiation with ALND or comprehensive nodal radiation alone.

37.5 Surgery for Distant Metastases in Breast Cancer

Patients who present with or develop metastatic breast cancer are unlikely to be cured of their disease. The development of newer treatment agents over the past 3 decades has led to significant improvements in overall survival for patients with metastatic breast cancer [66]. Generally, patients who develop metastatic disease are treated with systemic therapy but there may be situations in which surgical treatment of the primary or distant disease may be considered. At this time, there is no clear evidence that local treatment improves survival, although some retrospective trials have suggested a possible benefit. When surgery for metastatic disease is considered, a thorough evaluation for additional sites of disease is imperative.

37.5.1 Treatment of Distant Nodal Metastases

Metastatic involvement of contralateral axillary lymph nodes has traditionally been considered a distant metastatic disease process. However, there is evidence to suggest that in breast cancer this may be more appropriately considered local–regional disease. Contralateral lymph node recurrence may be the result of aberrant lymphatic drainage from the ipsilateral breast to the contralateral nodal basin in light of previous breast and axillary surgery. A prospective study of sentinel node biopsy in recurrent breast cancer found that 58.9% of patients had aberrant drainage pathways—with a significantly higher rate seen in patients who underwent prior ALND (79.3%) compared to previous SLNB (25.0%, $p < 0.001$) [67]. Of 95 patients with successful lymphatic mapping, 14 (14.7%) mapped to the contralateral axilla, all of whom all had undergone previous ALND. A systematic literature review was recently performed of patients with an isolated contralateral nodal recurrence of breast cancer following prior curative treatment, with or without ipsilateral breast cancer recurrence [68]. The majority of these patients received surgical and systemic treatment with curative intent rather than for palliation. Although the numbers were small and the study was retrospective, the overall survival for these patients was higher than would generally be expected for metastatic breast cancer (82.6% after a mean of 50 months). While these results should be interpreted with caution, they suggest that surgery may be considered for contralateral axillary nodal metastases in the absence of other distant metastases.

37.5.2 Treatment of Oligometastatic Disease

Treatment of distant metastases in breast cancer is aimed at prolonging survival and alleviating possible associated symptoms while maintaining an appropriate quality of life. Individual responses to treatment and outcomes are dependent upon a number of factors, including tumor subtype, site of disease, and overall disease burden. Patients with visceral metastases and short disease-free intervals following initial therapy will generally have a poorer outcome compared to patients with soft tissue and bone metastases [69]. Because systemic treatment is directed by tumor subtype, a repeat biopsy to establish the diagnosis and to reevaluate the receptor status should be considered at the time potential metastases are identified.

In specific situations, local intervention for isolated metastases may be considered. Patients are considered to have oligometastatic disease when a limited number of metastatic sites are identified which may be considered amenable to local treatment [70]. Patients may be considered suitable for local treatment based on a limited extent of disease, an overall good performance status, a longer disease-free interval between primary definitive therapy and the identification of metastatic disease, and the likelihood of achieving complete resection. These situations are different from instances when urgent intervention may be performed to alleviate symptoms causing significant pain or loss of function such as cord compression or pathologic fracture. The goal in these settings is palliation of symptoms and not to improve survival. Patients with bone disease are generally considered for local treatment if there is evidence of a fracture or impending fracture, debilitating pain, joint restriction, or spinal cord compression. Local treatment is generally not recommended for asymptomatic patients with bony metastases due to the general slow growth and progression of bone disease in breast cancer [71].

An estimated 10–30% of women with breast cancer will develop brain metastases [72]. As discussed previously, patients with Her2/neu amplified and basal-like breast cancer are at higher risk of developing brain metastasis [12, 73]. Treatment options include surgery, whole-brain radiotherapy, and stereotactic radiosurgery [74]. The Breast Graded Prognostic Assessment (GPA) is a prognostic index for breast cancer patients with brain metastases that have undergone a number of subsequent modifications [75]. This index may be used to determine prognosis and help direct individualized treatment. There has been conflicting evidence on the benefit of surgery followed by whole-brain radiation therapy versus whole brain radiation alone [76–78]. However, guidelines do recommend considering surgical resection in patients with a limited number of brain metastases, especially when other systemic disease is absent and the patient has a good performance score [79].

Patients with lung metastases will rarely become symptomatic until there is significant, unresectable tumor burden [80]. In fact, most lung metastases are asymptomatic and found incidentally [81]. In addition to metastatic disease, lung lesions identified in breast cancer patients may also represent primary lung cancer or benign, inflammatory, or infectious processes. Depending on the size and location of the lesion, image-guided biopsy or surgical excision may be used to obtain a tissue diagnosis [82]. Retrospective studies have demonstrated an improved survival for patients undergoing pulmonary metastasectomy, although selection bias likely plays a role in these findings [83, 84].

While liver metastases occur in half of the patients with metastatic breast cancer, there is limited data on the benefit of liver resection. Only a small number of patients, (5–25%) will have isolated liver metastases [85]. Urgent indications for local therapy include pain, uncontrolled bleeding, or biliary obstruction. Treatment options include hepatic resection and stereotactic body radiation therapy (SBRT), with the goal being to preserve liver function as much as possible and to obtain a complete resection/treatment. A systematic retrospective review of 553 patients in 19 studies who underwent liver resection for metastatic breast cancer was reported in 2011 [86]. The median post-hepatectomy survival for patients in this study was 40 months, with a 5-year survival of 40%, which is longer than would typically be expected for patients with metastatic breast cancer. A cost-utility analysis of liver resection performed for metastatic breast cancer found that liver resection followed by postoperative conventional systemic therapy compared to long-term treatment with systemic therapy was cost-effective [87]. Poor prognostic features include involved surgical margins and hormone receptor-negative disease. Other local therapy options may include SBRT and radiofrequency ablation (RFA) as well as various other less studied liver-directed therapies, which have been used for colorectal cancer liver metastases, but the data are limited on their utility in patients with breast cancer.

37.5.3 Management of the Primary Site in Patients with Distant Metastatic Disease at Presentation

A number of studies have attempted to determine whether surgical resection of the primary breast tumor is of benefit for patients with metastatic breast cancer. Several retrospective studies have suggested a survival advantage for patients with metastatic breast cancer who undergo surgery of the primary site. However, this has not been clearly demonstrated in prospective studies. Retrospective studies include reviews of the National Cancer Database [88] and the Surveillance, Epidemiology, and End Results database [89] as well as institutional reviews [90], each of which found survival advantages for patients who received local extirpative therapy as well as systemic therapy. However, matched-pair analyses have found similar survival outcomes for patients treated with surgery or without [91, 92], suggesting that the survival benefit demonstrated in the forgoing retrospective studies is due to selection bias.

Two international prospective trials have attempted to evaluate the potential benefit of surgery for the intact primary site in patients with metastatic breast cancer. The first trial randomized 350 women with stage IV breast cancer who had partial or complete responses to chemotherapy to surgery for the primary site versus no locoregional treatment, finding no difference in overall survival between the groups [93]. Notably, however, this study did not use standard chemotherapy and endocrine therapy regimens. A second study randomized patients with metastatic breast cancer and intact primary tumors to surgery first followed by systemic therapy versus systemic therapy alone [94]. At 3 years, no significant difference could be seen between the groups, but at 5 years there was improved survival in patients who received local therapy (46.4% vs. 26.4%; HR 0.66, 95% CI 0.49–0.88). It should be noted, however, that the local therapy group had few patients with TNBC and visceral metastases. Furthermore, surgery for the primary site was performed before systemic therapy so patients' response to chemotherapy could not be ascertained. A recent randomized phase III trial presented at the American Society of Clinical Oncology Annual meeting in 2020—Eastern Cooperative Oncology Group E2108—evaluated systemic therapy plus early local therapy versus systemic therapy alone in women with de novo stage IV breast cancer [95]. Enrolled patients received optimal systemic therapy up-front and those who did not progress during 4–8 months of treatment were randomized to local therapy versus none. There were no significant differences observed in overall survival or progression-free survival. The current data suggest that locoregional treatment for metastatic breast cancer does not show a clear survival benefit and should be considered only in patients with good response to systemic therapy, oligometastatic disease, and as part of a multidisciplinary treatment plan.

37.6 Conclusion

Surgery plays a critical role in the multidisciplinary management of breast cancer. Appropriate adherence to best clinical practices, as they evolve, from the standpoint of selection of the appropriate surgical treatment plan, sequencing with other therapies, and the technical conduct of breast cancer operations will ensure optimal patient outcomes.

Open Questions

- How might consideration of tumor biologic subtype further refine the choice of operation for the breast and axillary lymph nodes?
- Will tumor biologic subtype be further refined by enhanced genomic profiling? If so, will this be utilized in clinical practice?
- How will a clinical complete response to neoadjuvant chemotherapy be defined in the future?
- Will surgery always be required for patients who have a complete clinical response to neoadjuvant chemotherapy?
- Can axillary radiation replace axillary dissection in patients with residual nodal disease following neoadjuvant chemotherapy?

Author Disclosures Alicia M. Terando, MD has no disclosures. Azadeh Carr, MD has no disclosures. Tina J. Hieken, MD receives research support from Genentech, Breast Cancer Research Foundation and SkylineDx. Mara A. Piltin, DO has no disclosures. Bindupriya Chandrasekaran, MD has no disclosures. Carla S. Fisher, MD, MBA has no disclosures

References

1. Goldhirsch A, Winer EP, Coates AS, Gelber RD, Piccart-Gebhart M, Thurlimann B, Senn HJ, Panel members. Personalizing the treatment of women with early breast cancer: highlights of the St Gallen international expert consensus on the primary therapy of early breast cancer 2013. Ann Oncol. 2013;24(9):2206–23. https://doi.org/10.1093/annonc/mdt303.
2. Goldhirsch A, Wood WC, Coates AS, Gelber RD, Thurlimann B, Senn HJ, Panel members. Strategies for subtypes—dealing with the diversity of breast cancer: highlights of the St. Gallen international expert consensus on the primary therapy of early breast cancer 2011. Ann Oncol. 2011;22(8):1736–47. https://doi.org/10.1093/annonc/mdr304.
3. Prat A, Pineda E, Adamo B, Galvan P, Fernandez A, Gaba L, Diez M, Viladot M, Arance A, Munoz M. Clinical implications of the intrinsic molecular subtypes of breast cancer. Breast. 2015;24(Suppl 2):S26–35. https://doi.org/10.1016/j.breast.2015.07.008.
4. Maisonneuve P, Disalvatore D, Rotmensz N, Curigliano G, Colleoni M, Dellapasqua S, Pruneri G, Mastropasqua MG, Luini A, Bassi F, Pagani G, Viale G, Goldhirsch A. Proposed new clinicopathological surrogate definitions of luminal A and luminal B (HER2-negative) intrinsic breast cancer subtypes. Breast Cancer Res. 2014;16(3):R65. https://doi.org/10.1186/bcr3679.
5. Lehmann BD, Bauer JA, Chen X, Sanders ME, Chakravarthy AB, Shyr Y, Pietenpol JA. Identification of human triple-negative breast cancer subtypes and preclinical models for selection of targeted therapies. J Clin Invest. 2011;121(7):2750–67. https://doi.org/10.1172/JCI45014.
6. Aydiner A, Sen F, Tambas M, Ciftci R, Eralp Y, Saip P, Karanlik H, Fayda M, Kucucuk S, Onder S, Yavuz E, Muslumanoglu M, Igci A. Metaplastic breast carcinoma versus triple-negative breast cancer: survival and response to treatment. Medicine (Baltimore). 2015;94(52):e2341. https://doi.org/10.1097/MD.0000000000002341.
7. Krop I, Ismaila N, Andre F, Bast RC, Barlow W, Collyar DE, Hammond ME, Kuderer NM, Liu MC, Mennel RG, Van Poznak C, Wolff AC, Stearns V. Use of biomarkers to guide decisions on adjuvant systemic therapy for women with early-stage invasive breast cancer: American Society of Clinical Oncology clinical practice guideline focused update. J Clin Oncol. 2017;35(24):2838–47. https://doi.org/10.1200/JCO.2017.74.0472.
8. NCCN (2020) National Comprehensive Cancer Network Clinical Practice Guidelines in Oncology. Breast Cancer v6.2020.
9. Sparano JA, Gray RJ, Makower DF, Pritchard KI, Albain KS, Hayes DF, Geyer CE Jr, Dees EC, Goetz MP, Olson JA Jr, Lively T, Badve SS, Saphner TJ, Wagner LI, Whelan TJ, Ellis MJ, Paik S, Wood WC, Ravdin PM, Keane MM, Gomez Moreno HL, Reddy PS, Goggins TF, Mayer IA, Brufsky AM, Toppmeyer DL, Kaklamani VG, Berenberg JL, Abrams J, Sledge GW Jr. Adjuvant chemotherapy guided by a 21-gene expression assay in Breast cancer. N Engl J Med. 2018;379(2):111–21. https://doi.org/10.1056/NEJMoa1804710.

10. Voduc KD, Cheang MC, Tyldesley S, Gelmon K, Nielsen TO, Kennecke H. Breast cancer subtypes and the risk of local and regional relapse. J Clin Oncol. 2010;28(10):1684–91. https://doi.org/10.1200/JCO.2009.24.9284.
11. Millikan RC, Newman B, Tse CK, Moorman PG, Conway K, Dressler LG, Smith LV, Labbok MH, Geradts J, Bensen JT, Jackson S, Nyante S, Livasy C, Carey L, Earp HS, Perou CM. Epidemiology of basal-like breast cancer. Breast Cancer Res Treat. 2008;109(1):123–39. https://doi.org/10.1007/s10549-007-9632-6.
12. Kennecke H, Yerushalmi R, Woods R, Cheang MC, Voduc D, Speers CH, Nielsen TO, Gelmon K. Metastatic behavior of breast cancer subtypes. J Clin Oncol. 2010;28(20):3271–7. https://doi.org/10.1200/JCO.2009.25.9820.
13. Xiao W, Zheng S, Yang A, Zhang X, Zou Y, Tang H, Xie X. Breast cancer subtypes and the risk of distant metastasis at initial diagnosis: a population-based study. Cancer Manag Res. 2018;10:5329–38. https://doi.org/10.2147/CMAR.S176763.
14. Sorlie T, Perou CM, Tibshirani R, Aas T, Geisler S, Johnsen H, Hastie T, Eisen MB, van de Rijn M, Jeffrey SS, Thorsen T, Quist H, Matese JC, Brown PO, Botstein D, Lonning PE, Borresen-Dale AL. Gene expression patterns of breast carcinomas distinguish tumor subclasses with clinical implications. Proc Natl Acad Sci U S A. 2001;98(19):10869–74. https://doi.org/10.1073/pnas.191367098.
15. Sotiriou C, Pusztai L. Gene-expression signatures in breast cancer. N Engl J Med. 2009;360(8):790–800. https://doi.org/10.1056/NEJMra0801289.
16. Metzger-Filho O, Sun Z, Viale G, Price KN, Crivellari D, Snyder RD, Gelber RD, Castiglione-Gertsch M, Coates AS, Goldhirsch A, Cardoso F. Patterns of recurrence and outcome according to breast cancer subtypes in lymph node-negative disease: results from international breast cancer study group trials VIII and IX. J Clin Oncol. 2013;31(25):3083–90. https://doi.org/10.1200/JCO.2012.46.1574.
17. Minicozzi P, Bella F, Toss A, Giacomin A, Fusco M, Zarcone M, Tumino R, Falcini F, Cesaraccio R, Candela G, La Rosa F, Federico M, Sant M. Relative and disease-free survival for breast cancer in relation to subtype: a population-based study. J Cancer Res Clin Oncol. 2013;139(9):1569–77. https://doi.org/10.1007/s00432-013-1478-1.
18. Savci-Heijink CD, Halfwerk H, Hooijer GK, Horlings HM, Wesseling J, van de Vijver MJ. Retrospective analysis of metastatic behaviour of breast cancer subtypes. Breast Cancer Res Treat. 2015;150(3):547–57. https://doi.org/10.1007/s10549-015-3352-0.
19. Carey LA, Berry DA, Cirrincione CT, Barry WT, Pitcher BN, Harris LN, Ollila DW, Krop IE, Henry NL, Weckstein DJ, Anders CK, Singh B, Hoadley KA, Iglesia M, Cheang MC, Perou CM, Winer EP, Hudis CA. Molecular heterogeneity and response to neoadjuvant human epidermal growth factor receptor 2 targeting in CALGB 40601, a randomized phase III trial of paclitaxel plus Trastuzumab with or without Lapatinib. J Clin Oncol. 2016;34(6):542–9. https://doi.org/10.1200/JCO.2015.62.1268.
20. Llombart-Cussac A, Cortes J, Pare L, Galvan P, Bermejo B, Martinez N, Vidal M, Pernas S, Lopez R, Munoz M, Nuciforo P, Morales S, Oliveira M, de la Pena L, Pelaez A, Prat A. HER2-enriched subtype as a predictor of pathological complete response following trastuzumab and lapatinib without chemotherapy in early-stage HER2-positive breast cancer (PAMELA): an open-label, single-group, multicentre, phase 2 trial. Lancet Oncol. 2017;18(4):545–54. https://doi.org/10.1016/S1470-2045(17)30021-9.
21. Prat A, Carey LA, Adamo B, Vidal M, Tabernero J, Cortes J, Parker JS, Perou CM, Baselga J. Molecular features and survival outcomes of the intrinsic subtypes within HER2-positive breast cancer. J Natl Cancer Inst. 2014;106(8):dju152. https://doi.org/10.1093/jnci/dju152.
22. Arvold ND, Oh KS, Niemierko A, Taghian AG, Lin NU, Abi-Raad RF, Sreedhara M, Harris JR, Alexander BM. Brain metastases after breastconserving therapy and systemic therapy: incidence and characteristics by biologic subtype. Breast Cancer Res Treat. 2012;136(1):153–60. https://doi.org/10.1007/s10549-012-2243-x.
23. Rueda OM, Sammut SJ, Seoane JA, Chin SF, Caswell-Jin JL, Callari M, Batra R, Pereira B, Bruna A, Ali HR, Provenzano E, Liu B, Parisien M, Gillett C, McKinney S, Green AR, Murphy L, Purushotham A, Ellis IO, Pharoah PD, Rueda C, Aparicio S, Caldas C, Curtis C. Dynamics of breast-cancer relapse reveal late-recurring ER-positive genomic subgroups. Nature. 2019;567(7748):399–404. https://doi.org/10.1038/s41586-019-1007-8.
24. Sada A, Day CN, Hoskin TL, Degnim AC, Habermann EB, Hieken TJ. Mastectomy and immediate breast reconstruction in the elderly: trends and outcomes. Surgery. 2019;166(4):709–14. https://doi.org/10.1016/j.surg.2019.05.055.
25. Young WA, Degnim AC, Hoskin TL, Jakub JW, Nguyen MD, Tran NV, Harless CA, Manrique OJ, Boughey JC, Hieken TJ. Outcomes of >1300 nipple-sparing mastectomies with immediate reconstruction: the impact of expanding indications on complications. Ann Surg Oncol. 2019;26(10):3115–23. https://doi.org/10.1245/s10434-019-07560-z.
26. Fisher B, Anderson S, Bryant J, Margolese RG, Deutsch M, Fisher ER, Jeong JH, Wolmark N. Twenty-year follow-up of a randomized trial comparing total mastectomy, lumpectomy, and lumpectomy plus irradiation for the treatment of invasive breast cancer. N Engl J Med. 2002;347(16):1233–41. https://doi.org/10.1056/NEJMoa022152.
27. Veronesi U, Cascinelli N, Mariani L, Greco M, Saccozzi R, Luini A, Aguilar M, Marubini E. Twenty-year follow-up of a randomized study comparing breast-conserving surgery with radical mastectomy for early breast cancer. N Engl J Med. 2002;347(16):1227–32. https://doi.org/10.1056/NEJMoa020989.
28. Nelson H, Hunt K, Veeramachaneni N, Blair S, Chang G, Halverson A, Katz M, Posner M, editors. Operative standards for cancer surgery, Breast, lung, pancreas, colon, vol. 1. Philadelphia: Lippincott, Williams & Wilkins; 2015.
29. Houssami N, Macaskill P, Marinovich ML, Morrow M. The association of surgical margins and local recurrence in women with early-stage invasive breast cancer treated with breast-conserving therapy: a meta-analysis. Ann Surg Oncol. 2014;21(3):717–30. https://doi.org/10.1245/s10434-014-3480-5.
30. Boughey JC, Hieken TJ, Jakub JW, Degnim AC, Grant CS, Farley DR, Thomsen KM, Osborn JB, Keeney GL, Habermann EB. Impact of analysis of frozen-section margin on reoperation rates in women undergoing lumpectomy for breast cancer: evaluation of the National Surgical Quality Improvement Program data. Surgery. 2014a;156(1):190–7. https://doi.org/10.1016/j.surg.2014.03.025.
31. Buchholz TA, Somerfield MR, Griggs JJ, El-Eid S, Hammond ME, Lyman GH, Mason G, Newman LA. Margins for breast-conserving surgery with whole-breast irradiation in stage I and II invasive breast cancer: American Society of Clinical Oncology endorsement of the Society of Surgical Oncology/American Society for Radiation Oncology consensus guideline. J Clin Oncol. 2014;32(14):1502–6. https://doi.org/10.1200/JCO.2014.55.1572.
32. Moran MS, Schnitt SJ, Giuliano AE, Harris JR, Khan SA, Horton J, Klimberg S, Chavez-MacGregor M, Freedman G, Houssami N, Johnson PL, Morrow M. Society of Surgical Oncology-American Society for Radiation Oncology consensus guideline on margins for breast-conserving surgery with whole-breast irradiation in stages I and II invasive breast cancer. Ann Surg Oncol. 2014;21(3):704–16. https://doi.org/10.1245/s10434-014-3481-4.
33. Bundred J, Michael S, Bowers S, Barnes N, Jauhari Y, Plant D, Maishman T, Cutress R, Holleczek B, Dodwell D, Bundred N. Do surgical margins matter after mastectomy? A systematic review. Eur J Surg Oncol. 2020;46(12):2185–94. https://doi.org/10.1016/j.ejso.2020.08.015.
34. Glorioso JM, Gonzalez Juarrero AB, Rodysill BR, Harmsen WS, Habermann EB, Carter JM, Mutter RW, Degnim AC, Jakub

JW. Margin proximity correlates with local recurrence after mastectomy for patients not receiving adjuvant radiotherapy. Ann Surg Oncol. 2017;24(11):3148–56. https://doi.org/10.1245/s10434-017-5968-2.
35. Murphy BL, Hoskin TL, Boughey JC, Degnim AC, Carter JM, Glazebrook KN, Hieken TJ. Contemporary operative management of T4 breast cancer. Surgery. 2016;160(4):1059–69. https://doi.org/10.1016/j.surg.2016.06.030.
36. Hieken TJ, Murphy BL, Boughey JC, Degnim AC, Glazebrook KN, Hoskin TL. Influence of biologic subtype of inflammatory breast cancer on response to neoadjuvant therapy and cancer outcomes. Clin Breast Cancer. 2018;18(4):e501–6. https://doi.org/10.1016/j.clbc.2017.10.003.
37. Leon Ferre R, Hieken TJ, Boughey JC. The landmark series: neoadjuvant chemotherapy for breast cancer. Ann Surg Oncol. 2020;28(4):2111–9.
38. Basik M, Cecchini RS, De Los Santos JF. Abstract GS5-05: primary analysis of NRG-BR005, a phase II trial assessing accuracy of tumor bed biopsies in predicting pathologic complete response (pCR) in patients with clinical/radiological complete response after neoadjuvant chemotherapy (NCT) to explore the feasibility of breast-conserving treatment without surgery. Cancer Res. 2020;80 (4).
39. Dings PJ, Elferink MA, Strobbe LJ, de Wilt JH. The prognostic value of lymph node ratio in node-positive breast cancer: a Dutch nationwide population-based study. Ann Surg Oncol. 2013;20(8):2607–14. https://doi.org/10.1245/s10434-013-2932-7.
40. Ashikaga T, Krag DN, Land SR, Julian TB, Anderson SJ, Brown AM, Skelly JM, Harlow SP, Weaver DL, Mamounas EP, Costantino JP, Wolmark N, National Surgical Adjuvant Breast, Bowel Project. Morbidity results from the NSABP B-32 trial comparing sentinel lymph node dissection versus axillary dissection. J Surg Oncol. 2010;102(2):111–8. https://doi.org/10.1002/jso.21535.
41. Warmuth MA, Bowen G, Prosnitz LR, Chu L, Broadwater G, Peterson B, Leight G, Winer EP. Complications of axillary lymph node dissection for carcinoma of the breast: a report based on a patient survey. Cancer. 1998;83(7):1362–8. https://doi.org/10.1002/(sici)1097-0142(19981001)83:7<1362::aid-cncr13>3.0.co;2-2.
42. Zintel HA, Nay HR. Postoperative complications of radical mastectomy. Surg Clin North Am. 1964;44:313–23. https://doi.org/10.1016/s0039-6109(16)37231-0.
43. Morton DL, Wen DR, Wong JH, Economou JS, Cagle LA, Storm FK, Foshag LJ, Cochran AJ. Technical details of intraoperative lymphatic mapping for early stage melanoma. Arch Surg. 1992;127(4):392–9. https://doi.org/10.1001/archsurg.1992.01420040034005.
44. Giuliano AE, Kirgan DM, Guenther JM, Morton DL. Lymphatic mapping and sentinel lymphadenectomy for breast cancer. Ann Surg. 1994;220(3):391.
45. Clarke D, Khonji NI, Mansel RE. Sentinel node biopsy in breast cancer: ALMANAC trial. World J Surg. 2001;25(6):819–22. https://doi.org/10.1007/s00268-001-0011-x.
46. Krag DN, Anderson SJ, Julian TB, Brown AM, Harlow SP, Costantino JP, Ashikaga T, Weaver DL, Mamounas EP, Jalovec LM, Frazier TG, Noyes RD, Robidoux A, Scarth HM, Wolmark N. Sentinel-lymph-node resection compared with conventional axillary-lymph-node dissection in clinically node-negative patients with breast cancer: overall survival findings from the NSABP B-32 randomised phase 3 trial. Lancet Oncol. 2010;11(10):927–33. https://doi.org/10.1016/S1470-2045(10)70207-2.
47. Mansel RE, Fallowfield L, Kissin M, Goyal A, Newcombe RG, Dixon JM, Yiangou C, Horgan K, Bundred N, Monypenny I, England D, Sibbering M, Abdullah TI, Barr L, Chetty U, Sinnett DH, Fleissig A, Clarke D, Ell PJ. Randomized multicenter trial of sentinel node biopsy versus standard axillary treatment in operable breast cancer: the ALMANAC trial. J Natl Cancer Inst. 2006;98(9):599–609. https://doi.org/10.1093/jnci/djj158.
48. Purushotham AD, Upponi S, Klevesath MB, Bobrow L, Millar K, Myles JP, Duffy SW. Morbidity after sentinel lymph node biopsy in primary breast cancer: results from a randomized controlled trial. J Clin Oncol. 2005;23(19):4312–21. https://doi.org/10.1200/JCO.2005.03.228.
49. Veronesi U, Paganelli G, Viale G, Luini A, Zurrida S, Galimberti V, Intra M, Veronesi P, Robertson C, Maisonneuve P, Renne G, De Cicco C, De Lucia F, Gennari R. A randomized comparison of sentinel-node biopsy with routine axillary dissection in breast cancer. N Engl J Med. 2003;349(6):546–53. https://doi.org/10.1056/NEJMoa012782.
50. Tjan-Heijnen VC, Buit P, de Widt-Evert LM, Ruers TJ, Beex LV. Micro-metastases in axillary lymph nodes: an increasing classification and treatment dilemma in breast cancer due to the introduction of the sentinel lymph node procedure. Breast Cancer Res Treat. 2001;70(2):81–8. https://doi.org/10.1023/a:1012938825396.
51. Galimberti V, Cole BF, Zurrida S, Viale G, Luini A, Veronesi P, Baratella P, Chifu C, Sargenti M, Intra M, Gentilini O, Mastropasqua MG, Mazzarol G, Massarut S, Garbay JR, Zgajnar J, Galatius H, Recalcati A, Littlejohn D, Bamert M, Colleoni M, Price KN, Regan MM, Goldhirsch A, Coates AS, Gelber RD, Veronesi U, International Breast Cancer Study Group Trial 23-01 Investigators. Axillary dissection versus no axillary dissection in patients with sentinel-node micrometastases (IBCSG 23-01): a phase 3 randomised controlled trial. Lancet Oncol. 2013;14(4):297–305. https://doi.org/10.1016/S1470-2045(13)70035-4.
52. Hieken TJ, Boughey JC. Axillary dissection versus no axillary dissection in patients with sentinel-node micrometastases: commentary on the IBCSG 23-01 trial. Gland Surg. 2013;2(3):128–32. https://doi.org/10.3978/j.issn.2227-684X.2013.07.04.
53. Giuliano AE, Hunt KK, Ballman KV, Beitsch PD, Whitworth PW, Blumencranz PW, Leitch AM, Saha S, McCall LM, Morrow M. Axillary dissection vs no axillary dissection in women with invasive breast cancer and sentinel node metastasis: a randomized clinical trial. JAMA. 2011;305(6):569–75. https://doi.org/10.1001/jama.2011.90.
54. Giuliano AE, McCall L, Beitsch P, Whitworth PW, Blumencranz P, Leitch AM, Saha S, Hunt KK, Morrow M, Ballman K. Locoregional recurrence after sentinel lymph node dissection with or without axillary dissection in patients with sentinel lymph node metastases: the American College of Surgeons Oncology Group Z0011 randomized trial. Ann Surg. 2010;252(3):426–32; discussion 432-423. https://doi.org/10.1097/SLA.0b013e3181f08f32.
55. Giuliano AE, Ballman KV, McCall L, Beitsch PD, Brennan MB, Kelemen PR, Ollila DW, Hansen NM, Whitworth PW, Blumencranz PW, Leitch AM, Saha S, Hunt KK, Morrow M. Effect of axillary dissection vs no axillary dissection on 10-year overall survival among women with invasive breast cancer and sentinel node metastasis: the ACOSOG Z0011 (Alliance) randomized clinical trial. JAMA. 2017;318(10):918–26. https://doi.org/10.1001/jama.2017.11470.
56. Black DM, Mittendorf EA. Landmark trials affecting the surgical management of invasive breast cancer. Surg Clin North Am. 2013;93(2):501–18. https://doi.org/10.1016/j.suc.2012.12.007.
57. Hunt KK, Yi M, Mittendorf EA, Guerrero C, Babiera GV, Bedrosian I, Hwang RF, Kuerer HM, Ross MI, Meric-Bernstam F. Sentinel lymph node surgery after neoadjuvant chemotherapy is accurate and reduces the need for axillary dissection in breast cancer patients. Ann Surg. 2009;250(4):558–66. https://doi.org/10.1097/SLA.0b013e3181b8fd5e.
58. Mamounas EP, Brown A, Anderson S, Smith R, Julian T, Miller B, Bear HD, Caldwell CB, Walker AP, Mikkelson WM, Stauffer JS, Robidoux A, Theoret H, Soran A, Fisher B, Wickerham DL, Wolmark N. Sentinel node biopsy after neoadjuvant chemotherapy in breast cancer: results from National Surgical Adjuvant Breast and Bowel Project Protocol B-27. J Clin Oncol. 2005;23(12):2694–702. https://doi.org/10.1200/JCO.2005.05.188.

59. Boughey JC, McCall LM, Ballman KV, Mittendorf EA, Ahrendt GM, Wilke LG, Taback B, Leitch AM, Flippo-Morton T, Hunt KK. Tumor biology correlates with rates of breast-conserving surgery and pathologic complete response after neoadjuvant chemotherapy for breast cancer: findings from the ACOSOG Z1071 (Alliance) prospective multicenter clinical trial. Ann Surg. 2014b;260(4):608–14; discussion 614–606. https://doi.org/10.1097/SLA.0000000000000924.
60. Hennessy BT, Hortobagyi GN, Rouzier R, Kuerer H, Sneige N, Buzdar AU, Kau SW, Fornage B, Sahin A, Broglio K, Singletary SE, Valero V. Outcome after pathologic complete eradication of cytologically proven breast cancer axillary node metastases following primary chemotherapy. J Clin Oncol. 2005;23(36):9304–11. https://doi.org/10.1200/JCO.2005.02.5023.
61. Boileau JF, Poirier B, Basik M, Holloway CM, Gaboury L, Sideris L, Meterissian S, Arnaout A, Brackstone M, McCready DR, Karp SE, Trop I, Lisbona A, Wright FC, Younan RJ, Provencher L, Patocskai E, Omeroglu A, Robidoux A. Sentinel node biopsy after neoadjuvant chemotherapy in biopsy-proven node-positive breast cancer: the SN FNAC study. J Clin Oncol. 2015;33(3):258–64. https://doi.org/10.1200/JCO.2014.55.7827.
62. Kuehn T, Bauerfeind I, Fehm T, Fleige B, Hausschild M, Helms G, Lebeau A, Liedtke C, von Minckwitz G, Nekljudova V, Schmatloch S, Schrenk P, Staebler A, Untch M. Sentinel-lymph-node biopsy in patients with breast cancer before and after neoadjuvant chemotherapy (SENTINA): a prospective, multicentre cohort study. Lancet Oncol. 2013;14(7):609–18. https://doi.org/10.1016/S1470-2045(13)70166-9.
63. Boughey JC, Ballman KV, Le-Petross HT, McCall LM, Mittendorf EA, Ahrendt GM, Wilke LG, Taback B, Feliberti EC, Hunt KK. Identification and resection of clipped node decreases the false-negative rate of sentinel lymph node surgery in patients presenting with node-positive Breast cancer (T0-T4, N1-N2) who receive neoadjuvant chemotherapy: results from ACOSOG Z1071 (Alliance). Ann Surg. 2016;263(4):802–7. https://doi.org/10.1097/SLA.0000000000001375.
64. Piltin MA, Hoskin TL, Day CN, Davis J Jr, Boughey JC. Oncologic outcomes of sentinel lymph node surgery after neoadjuvant chemotherapy for node-positive breast cancer. Ann Surg Oncol. 2020;27(12):4795–801. https://doi.org/10.1245/s10434-020-08900-0.
65. NCT01901094 (2014) Comparison of axillary lymph node dissection with axillary radiation for patients with node-positive breast cancer treated with chemotherapy. https://ClinicalTrials.gov/show/NCT01901094.
66. Caswell-Jin JL, Plevritis SK, Tian L, Cadham CJ, Xu C, Stout NK, Sledge GW, Mandelblatt JS, Kurian AW. Change in survival in metastatic breast cancer with treatment advances: meta-analysis and systematic review. JNCI Cancer Spectr. 2018;2(4):pky062. https://doi.org/10.1093/jncics/pky062.
67. Maaskant-Braat AJ, Roumen RM, Voogd AC, Pijpers R, Luiten EJ, Rutgers EJ, Nieuwenhuijzen GA. Sentinel node and recurrent Breast cancer (SNARB): results of a nationwide registration study. Ann Surg Oncol. 2013;20(2):620–6. https://doi.org/10.1245/s10434-012-2625-7.
68. Moossdorff M, Vugts G, Maaskant-Braat AJ, Strobbe LJ, Voogd AC, Smidt ML, Nieuwenhuijzen GA. Contralateral lymph node recurrence in breast cancer: regional event rather than distant metastatic disease. A systematic review of the literature. Eur J Surg Oncol. 2015;41(9):1128–36. https://doi.org/10.1016/j.ejso.2015.05.015.
69. Leone BA, Vallejo CT, Romero AO, Machiavelli MR, Perez JE, Leone J, Leone JP. Prognostic impact of metastatic pattern in stage IV breast cancer at initial diagnosis. Breast Cancer Res Treat. 2017;161(3):537–48. https://doi.org/10.1007/s10549-016-4066-7.
70. Al-Shafa F, Arifin AJ, Rodrigues GB, Palma DA, Louie AV. A review of ongoing trials of stereotactic ablative radiotherapy for oligometastatic cancers: where will the evidence lead? Front Oncol. 2019;9:543. https://doi.org/10.3389/fonc.2019.00543.
71. Briasoulis E, Karavasilis V, Kostadima L, Ignatiadis M, Fountzilas G, Pavlidis N. Metastatic breast carcinoma confined to bone: portrait of a clinical entity. Cancer. 2004;101(7):1524–8. https://doi.org/10.1002/cncr.20545.
72. Lin NU, Bellon JR, Winer EP. CNS metastases in breast cancer. J Clin Oncol. 2004;22(17):3608–17. https://doi.org/10.1200/JCO.2004.01.175.
73. Gabos Z, Sinha R, Hanson J, Chauhan N, Hugh J, Mackey JR, Abdulkarim B. Prognostic significance of human epidermal growth factor receptor positivity for the development of brain metastasis after newly diagnosed breast cancer. J Clin Oncol. 2006;24(36):5658–63. https://doi.org/10.1200/JCO.2006.07.0250.
74. Raghunath A, Desai K, Ahluwalia MS. Current treatment options for Breast cancer brain metastases. Curr Treat Options in Oncol. 2019;20(3):19. https://doi.org/10.1007/s11864-019-0618-5.
75. Sperduto PW, Mesko S, Li J, Cagney D, Aizer A, Lin NU, Nesbit E, Kruser TJ, Chan J, Braunstein S, Lee J, Kirkpatrick JP, Breen W, Brown PD, Shi D, Shih HA, Soliman H, Sahgal A, Shanley R, Sperduto W, Lou E, Everett A, Boggs DH, Masucci L, Roberge D, Remick J, Plichta K, Buatti JM, Jain S, Gaspar LE, Wu CC, Wang TJC, Bryant J, Chuong M, Yu J, Chiang V, Nakano T, Aoyama H, Mehta MP. Beyond an updated graded prognostic assessment (Breast GPA): a prognostic index and trends in treatment and survival in Breast cancer brain metastases from 1985 to today. Int J Radiat Oncol Biol Phys. 2020;107(2):334–43. https://doi.org/10.1016/j.ijrobp.2020.01.051.
76. Mintz AH, Kestle J, Rathbone MP, Gaspar L, Hugenholtz H, Fisher B, Duncan G, Skingley P, Foster G, Levine M. A randomized trial to assess the efficacy of surgery in addition to radiotherapy in patients with a single cerebral metastasis. Cancer. 1996;78(7):1470–6. https://doi.org/10.1002/(sici)1097-0142(19961001)78:7<1470::aid-cncr14>3.0.co;2-x.
77. Patchell RA, Tibbs PA, Walsh JW, Dempsey RJ, Maruyama Y, Kryscio RJ, Markesbery WR, Macdonald JS, Young B. A randomized trial of surgery in the treatment of single metastases to the brain. N Engl J Med. 1990;322(8):494–500. https://doi.org/10.1056/NEJM199002223220802.
78. Vecht CJ, Haaxma-Reiche H, Noordijk EM, Padberg GW, Voormolen JH, Hoekstra FH, Tans JT, Lambooij N, Metsaars JA, Wattendorff AR, et al. Treatment of single brain metastasis: radiotherapy alone or combined with neurosurgery? Ann Neurol. 1993;33(6):583–90. https://doi.org/10.1002/ana.410330605.
79. Soffietti R, Abacioglu U, Baumert B, Combs SE, Kinhult S, Kros JM, Marosi C, Metellus P, Radbruch A, Villa Freixa SS, Brada M, Carapella CM, Preusser M, Le Rhun E, Ruda R, Tonn JC, Weber DC, Weller M. Diagnosis and treatment of brain metastases from solid tumors: guidelines from the European Association of Neuro-Oncology (EANO). Neuro Oncol. 2017;19(2):162–74. https://doi.org/10.1093/neuonc/now241.
80. Yousefi M, Nosrati R, Salmaninejad A, Dehghani S, Shahryari A, Saberi A. Organ-specific metastasis of breast cancer: molecular and cellular mechanisms underlying lung metastasis. Cell Oncol (Dordr). 2018;41(2):123–40. https://doi.org/10.1007/s13402-018-0376-6.
81. Rashid OM, Takabe K. The evolution of the role of surgery in the management of breast cancer lung metastasis. J Thorac Dis. 2012;4(4):420–4. https://doi.org/10.3978/j.issn.2072-1439.2012.07.16.
82. Nichols FC. Pulmonary metastasectomy. Thorac Surg Clin. 2012;22(1):91–9, vii. https://doi.org/10.1016/j.thorsurg.2011.08.017.
83. Meimarakis G, Ruttinger D, Stemmler J, Crispin A, Weidenhagen R, Angele M, Fertmann J, Hatz RA, Winter H. Prolonged overall survival after pulmonary metastasectomy in patients with

breast cancer. Ann Thorac Surg. 2013;95(4):1170–80. https://doi.org/10.1016/j.athoracsur.2012.11.043.

84. Yoshimoto M, Tada K, Nishimura S, Makita M, Iwase T, Kasumi F, Okumura S, Sato Y, Nakagawa K. Favourable long-term results after surgical removal of lung metastases of breast cancer. Breast Cancer Res Treat. 2008;110(3):485–91. https://doi.org/10.1007/s10549-007-9747-9.
85. Caralt M, Bilbao I, Cortes J, Escartin A, Lazaro JL, Dopazo C, Olsina JJ, Balsells J, Charco R. Hepatic resection for liver metastases as part of the "oncosurgical" treatment of metastatic breast cancer. Ann Surg Oncol. 2008;15(10):2804–10. https://doi.org/10.1245/s10434-008-0072-2.
86. Chua TC, Saxena A, Liauw W, Chu F, Morris DL. Hepatic resection for metastatic breast cancer: a systematic review. Eur J Cancer. 2011;47(15):2282–90. https://doi.org/10.1016/j.ejca.2011.06.024.
87. Spolverato G, Vitale A, Bagante F, Connolly R, Pawlik TM. Liver resection for breast cancer liver metastases: a cost-utility analysis. Ann Surg. 2017;265(4):792–9. https://doi.org/10.1097/SLA.0000000000001715.
88. Khan SA, Stewart AK, Morrow M. Does aggressive local therapy improve survival in metastatic breast cancer? Surgery. 2002;132(4):620–6; discussion 626–627. https://doi.org/10.1067/msy.2002.127544.
89. Gnerlich J, Jeffe DB, Deshpande AD, Beers C, Zander C, Margenthaler JA. Surgical removal of the primary tumor increases overall survival in patients with metastatic breast cancer: analysis of the 1988-2003 SEER data. Ann Surg Oncol. 2007;14(8):2187–94. https://doi.org/10.1245/s10434-007-9438-0.
90. Lang JE, Tereffe W, Mitchell MP, Rao R, Feng L, Meric-Bernstam F, Bedrosian I, Kuerer HM, Hunt KK, Hortobagyi GN, Babiera GV. Primary tumor extirpation in breast cancer patients who present with stage IV disease is associated with improved survival. Ann Surg Oncol. 2013;20(6):1893–9. https://doi.org/10.1245/s10434-012-2844-y.
91. Cady B, Nathan NR, Michaelson JS, Golshan M, Smith BL. Matched pair analyses of stage IV breast cancer with or without resection of primary breast site. Ann Surg Oncol. 2008;15(12):3384–95. https://doi.org/10.1245/s10434-008-0085-x.
92. Dominici L, Najita J, Hughes M, Niland J, Marcom P, Wong YN, Carter B, Javid S, Edge S, Burstein H, Golshan M. Surgery of the primary tumor does not improve survival in stage IV breast cancer. Breast Cancer Res Treat. 2011;129(2):459–65. https://doi.org/10.1007/s10549-011-1648-2.
93. Badwe RA, Kataria K, Srivastava A. Surgical resection of Phyllodes tumour: a radical approach as a safeguard against local recurrence. Indian J Surg. 2015;77(2):161–3. https://doi.org/10.1007/s12262-013-0935-z.
94. Soran A, Ozmen V, Ozbas S, Karanlik H, Muslumanoglu M, Igci A, Canturk Z, Utkan Z, Ozaslan C, Evrensel T, Uras C, Aksaz E, Soyder A, Ugurlu U, Col C, Cabioglu N, Bozkurt B, Uzunkoy A, Koksal N, Gulluoglu BM, Unal B, Atalay C, Yildirim E, Erdem E, Salimoglu S, Sezer A, Koyuncu A, Gurleyik G, Alagol H, Ulufi N, Berberoglu U, Dulger M, Cengiz O, Sezgin E, Johnson R. Randomized trial comparing resection of primary tumor with no surgery in stage IV Breast cancer at presentation: protocol MF07-01. Ann Surg Oncol. 2018;25(11):3141–9. https://doi.org/10.1245/s10434-018-6494-6.
95. Khan SA, Zhao F, Solin LJ, Goldstein LJ, Cella D, Basik M, Golshan M, Julian TB, Pockaj BA, Lee CA, Razaq W, Sparano JA, Babiera GV, Dy IA, Jain S, Silverman P, Fisher C, Tevaarwerk AJ, Wagner LI, Sledge GW. A randomized phase III trial of systemic therapy plus early local therapy versus systemic therapy alone in women with de novo stage IV breast cancer: a trial of the ECOG-ACRIN Research Group (E2108). J Clin Oncol. 2020;38(18_suppl):LBA2. https://doi.org/10.1200/JCO.2020.38.18_suppl.LBA2.

The Role of Surgery in Managing Primary and Metastatic Colorectal Cancer

38

Richard J. Straker III, Hunter D. D. Witmer, Benjamin Deschner, David Shibata, Kiran K. Turaga, and Najjia N. Mahmoud

Abstract

The surgical management of metastatic colorectal cancer (CRC) has evolved dramatically over the last few decades. Proper lymph node evaluation is essential, as not only does detection of regional lymph node metastases carry important prognostic implications, but it also dictates the need for adjuvant therapy following surgical resection. Attainment of an adequate number of lymph nodes is necessary for complete staging and is used as a quality care metric, but debate continues to exist about the clinical significance of harvesting greater or fewer nodes than this minimum required amount. Data have shown that with careful patient selection, resection of hepatic CRC metastases can provide a survival benefit, but the ideal timing and sequence of resection remain unclear. The use of cytoreductive surgery, with or without hyperthermic intraperitoneal chemotherapy, for the management of peritoneal metastases continues to be an area of active ongoing research, as well. As we enter the era of precision medicine, growing evidence supports the prognostic implications provided by genetic analysis of tumor specimens, and emerging molecular techniques may offer more accurate identification of tumor metastases within harvested lymph nodes.

Learning Objectives

- To understand the current literature describing the importance of lymph node harvest and its association with survival in patients with colorectal cancer.
- To understand the different lymph node dissection techniques currently practiced for patients with colorectal cancer and describe the advantages and disadvantages of a more versus less extensive dissection.
- To understand the advantages and disadvantages of synchronous versus staged resection of colorectal cancer hepatic metastases.
- To understand that current data on cytoreductive surgery/heated intraperitoneal chemotherapy for colorectal cancer peritoneal metastases.
- To understand the potential role of advanced molecular and genetic analyses in the management of metastatic colorectal cancer.

38.1 Introduction

Modern paradigms of tumor metastases continue to evolve as our biological understanding of cancer expands. Newer theories of metastases have emerged out of advances in biotechnology and analytics. The diaspora theory [1] consists of three distinct primary considerations: (a) the suitability of

Richard J. Straker and Hunter D. D. Witmer contributed equally to the manuscript and share first authorship.

Kiran K. Turaga and Najjia N. Mahmoud contributed equally to the manuscript and share senior authorship.

R. J. Straker III (✉) · N. N. Mahmoud
Division of Colon and Rectal Surgery, Department of Surgery, University of Pennsylvania, Philadelphia, PA, USA
e-mail: Richard.straker@pennmedicine.upenn.edu; najjia.mahmoud@pennmedicine.upenn.edu

H. D. D. Witmer · K. K. Turaga
Division of Surgical Oncology, Department of Surgery, University of Chicago, Chicago, IL, USA
e-mail: hunter.witmer@uchospitals.edu; Kturaga@surgery.bsd.uchicago.edu

B. Deschner
Department of Surgery, Mountain Area Health Education Center, Asheville, NC, USA

D. Shibata
Department of Surgery, University of Tennessee Health Science Center, Memphis, TN, USA
e-mail: dshibata@uthsc.edu

S. P. Leong et al. (eds.), *Cancer Metastasis Through the Lymphovascular System*,
https://doi.org/10.1007/978-3-030-93084-4_38

the primary tumor's microenvironment, (b) the adaptability and proficiency of various tumor clonal populations to migrate to other sites, and (c) the suitability of the microenvironment at remote, potential metastatic sites (Fig. 38.1). These variables help to explain some of the driving forces underpinning the mechanisms of metastatic spread, such as the relative degree of selective pressure for epithelial to mesenchymal transition and the inherent propensity for malignant cells to shed into the lymphovascular system. The diaspora theory also provides a conceptual framework for the proclivity of disseminated tumor cell subpopulations for certain tissues irrespective of patterns of anatomic drainage or circulation and for variations in the subsequent aggressiveness at these sites. The related theory of oligometastases holds that there are intermediate steps between initial tumor cell dispersion and fulminant metastatic disease. Moreover, metastatic potential is limited to certain sites, and, by implication, is potentially curable through localized treatment [2, 3]. Given the substantial incidence of metastatic colorectal cancer (CRC), we outline some of the nuances of surgical management of patients with regional and distant metastases in the context of these paradigms.

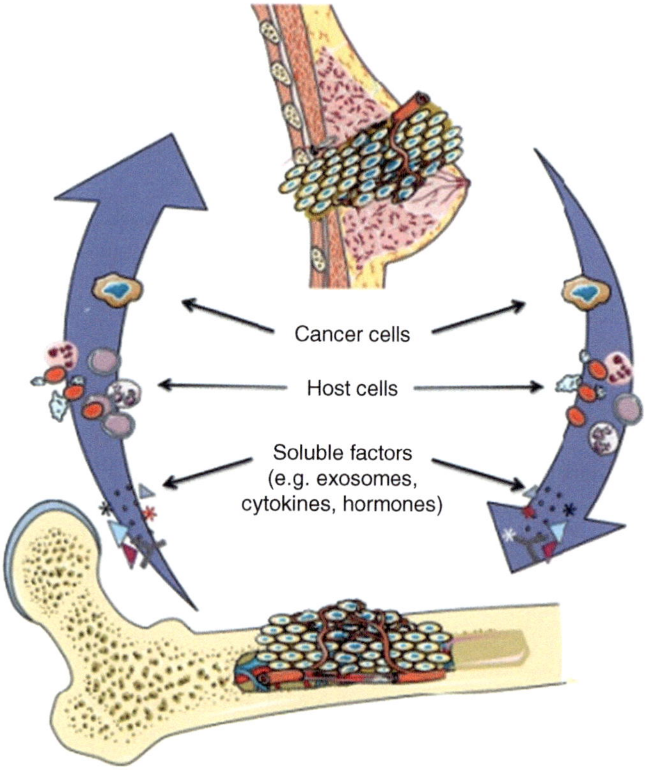

Fig. 38.1 Illustrative representation of the many complex interactions of cancer progression under the diaspora theory. The microenvironment at both the primary tumor site as well as in other tissues, and the interplay between them, may determine where metastases develop and their behavior at these sites

38.2 The Significance of Lymph Node Harvest in Colorectal Cancer

For patients with CRC without distant metastases, histopathologic lymph node status represents the most important prognostic factor for survival [4]. Estimated 5-year survival decreases from 70 to 90% for those without lymph node involvement to 30–60% for those with spread to regional nodes, and survival rates further vary based on the number of nodes involved [5]. Standard of care for patients with lymph node-positive disease entails adjuvant chemotherapy with a fluoropyrimidine-based regimen with or without oxaliplatin, which has been shown to both reduce disease recurrence and improve overall survival (OS) [6, 7]. To ensure adequate staging among node-negative patients, current guidelines recommend a minimum of 12 lymph nodes be evaluated in both colon and rectal cancer [8]. In the absence of adequate lymph node sampling, positive lymph nodes may not be detected, thus leading to patients potentially being under-staged [9].

Given the clinically important correlations between lymph node involvement and outcomes, numerous studies have evaluated the prognostic implications of the number of nodes harvested, with many studies finding a positive association between higher nodal retrieval and improved survival [10]. Specifically among patients with stage II colon cancer, Chen et al. demonstrated that patients with 15 or more lymph nodes evaluated had better survival than those in which fewer nodes were assessed, and similar results have been seen with rectal cancer, as well [11, 12]. While this would lead one to believe that the standard should shift toward harvesting more than the recommended 12 nodes, significant debate exists regarding whether these aforementioned studies are capturing the true reasons behind the association between node number and survival. Higher numbers of harvested nodes likely lead to a stage migration effect, resulting in the appearance of improved survival among stage I/II patients as more stage III patients are placed into the correct staging category [13].

However, critics of this, "Will Rogers phenomenon," theory argues that more extensive lymph node yields don't necessarily correlate to higher rates of node positivity, suggesting that other confounding factors may be present [14]. Baxter et al. showed that among patients with pathologic T3 colon cancers, staging is improved with higher nodal yields when the nodal count is low (<5–7 nodes), but that this effect is mitigated at higher nodal counts [15]. An alternative theory gaining much traction postulates that host immune responses could be the dominant variable responsible for the association between lymph node count and survival, discussed in more detail subsequently [13]. Additionally, as discussed in the following sections, surgical excision is just one of the many factors that ultimately dictates the number of lymph nodes evaluated in a given resection specimen. As such, these uncertainties surrounding the association between lymph node count and survival have led some to propose that the lymph node ratio, calculated as the ratio of metastatic lymph nodes to total lymph nodes harvested, may be a more reliable prognostic factor than the absolute number of lymph nodes retrieved [16].

Finally, the retrieval of at least 12 nodes for pathologic evaluation has emerged as a metric of quality care and a surrogate marker for an adequate surgical resection in colon cancer [17]. Much enthusiasm exists for its use as a quality indicator due to its relative ease of measurement and comparability across institutions [18]. Early studies showed that a majority of institutions failed to achieve compliance with this 12-node metric, but more recent data demonstrate improvements in the proportion of hospitals meeting this benchmark [4, 14, 19]. However, improvements in clinical outcomes have not reliably paralleled this increased compliance, and many investigators have outlined the numerous limitations associated with using lymph node harvest in this capacity, including the fact that surgical resection is only one of many known factors associated with lymph node yield [18]. As we continue to parse out the true associations between lymph node yield and survival, the role of this metric in the assessment of quality surgical care may come to change.

38.2.1 Nonsurgical Factors Affecting Lymph Node Evaluation

While surgical excision plays an obvious role in the quantity of lymph nodes available for analysis, the histopathologic methods by which lymph nodes are defined and examined, as well as immunologic, patient, and tumor-specific characteristics also influence the number of lymph nodes that are ultimately available for evaluation.

Significant alterations to the pathologic classification of a lymph node metastasis have occurred over successive editions of the TNM staging system. Although minimal changes were made from the previously used seventh edition to the currently used eighth edition, the seventh edition introduced the N1c stage as a separate category for tumor deposits present in the subserosa, mesentery, or nonperitonealized pericoloc or perirectal/mesorectal tissues without concomitant lymph node involvement [5, 20]. Prior editions defined tumor deposits differently, and many of these deposits were classified as regional lymph node metastases. However, the seventh edition no longer identified these deposits as lymph nodes with tumor involvement, resulting in a significant rise in the number of cases with less than 12 lymph nodes detectable for evaluation, and thus increasing the proportion of patients not meeting the minimum lymph node requirement for adequate staging [21]. Thorough and complete dissection of the surgical specimen by the pathologist is also critical for the identification of lymph nodes and is highly operator dependent. Studies have shown that marked variation exists across institutions in the methods utilized and time spent dissecting out lymph nodes, with one study demonstrating that an additional 20 min of time spent searching led to the discovery of a mean of six additional lymph nodes [22, 23].

Much data are emerging linking the host immune system with lymph node yield and outcomes in CRC. Enhanced immunologic reactions against the tumor can result in lymph node hyperplasia and enlargement allowing for easier detection, and lymph node hyperplasia has been shown to be associated with improved survival, as well [13]. Larger tumors may induce more profound immune responses in the mesentery, leading to a higher absolute number of lymph nodes being identified for pathologic evaluation from the resected mesenteric specimen. The more advanced bowel wall penetration exhibited by higher T-stage tumors may also induce greater immune responses, again leading to increased numbers of nodes retrieved following resection. Studies have also shown a positive correlation between the presence of tumor-infiltrating lymphocytes and lymph node retrieval, and likewise improved survival when increasing numbers of these immunologic cells are present. Finally, the established association between microsatellite instability and improved survival is thought, at least in part, to be due to immunologic factors [13]. Based on this evidence, the immune system appears to play an increasingly prevalent role in both the absolute number of lymph nodes harvested and the association between lymph node yield and survival in colorectal cancer.

A number of patient-specific factors have been shown to influence lymph node harvest. Numerous reports have identified age as an important factor affecting lymph node yield, with increasing age consistently being associated with lower recovery counts [24, 25]. This may be related to decreased surgical aggressiveness in older patients, or alternatively may be due to a diminished immune response with older age, leading to smaller and less easily detectable lymph nodes [13]. The association between gender and body mass index (BMI) on lymph node recovery is less clear, with some studies finding that females and lower BMI patients have higher lymph node harvests, while others have shown higher BMI leads to lower yields [26, 27].

Several tumor-specific factors have been associated with lymph node recovery, as well, including tumor location, T-stage, and lymph node size. Overwhelming evidence has demonstrated that right-sided colon cancers yield significantly higher lymph node counts compared to left-sided tumors. Many explanations for this observation exist, including anatomic differences between the right and left colon, as well as a higher rate of microsatellite instability among right-sided cancers, a factor that has also been shown to correlate with higher lymph node counts [24, 25, 27–29]. As previously mentioned, higher T-stage tumors and larger lymph node size are also consistently found to be associated with greater lymph node yields, possibly due to more robust immunologic responses against more advanced tumors, leading to easier detection [24, 25, 27, 28, 30].

38.2.2 Extent of Lymph Node Dissection

In addition to removal of the malignant lesion with appropriate bowel margins, surgical resection also includes mesenteric excision which provides the lymph nodes for staging. Much debate exists regarding whether there is a direct therapeutic benefit from this lymph node dissection, what the optimal extent of this excision is, and whether outcomes are significantly improved when more extended dissections are employed.

Indirect evidence for a therapeutic benefit to lymphadenectomy may be derived from the Intergroup Trial INT-0089. While the authors of this trial found that higher lymph node counts in general were associated with improved survival, this improvement was also seen specifically among subgroup analyses of patients with node-positive disease, in which stage N2 patients with >35 nodes removed had a 5-year survival of 71% compared to a 51% survival for N2 patients with ≤35 nodes [31]. Given that all patients within this analysis had N2 disease, stage migration, and more accurate staging incompletely explain these results, suggesting that there may have been a therapeutic component to the higher lymph node yield. However, the data remain conflicting, with other reports finding incomplete evidence to support a direct therapeutic benefit [32].

Whether a therapeutic benefit is derived from a more extensive lymphadenectomy directly relates to the debate on what is the optimal extent of the mesenteric dissection is. Vascular ligation combined with lymphadenectomy up to the root of the main feeding vessel is defined as a D2 resection and results in the removal of the pericolic and intermediate level nodes, leaving behind the main/apical level nodes that are present along the length of the main feeding vessel. The more extended D3 resection, commonly practiced in Japan, includes removal of these main level nodes, requiring dissection along the length of the feeding vessels in order to excise the lymphoadipose tissue they reside in [33]. Similar to the D3 resection, Hohenberger et al. described the complete mesocolic excision (CME) with central vascular ligation (CVL), although this technique has been shown to have significantly longer lateral resection margins than are seen with the D3 method [34]. Proponents of these extended dissections suggest that the higher nodal yields achieved are associated with improved survival as evidenced by studies demonstrating low recurrence and improved survival rates with employment of these techniques [35, 36]. Furthermore, these central dissections allow for removal of the apical nodes, which are involved in an estimated 3–11% of tumors, and may also lead to the removal of skip metastases (node-positive main nodes in the absence of positive intermediate nodes), both of which would have otherwise been left behind had an extended lymphadenectomy been deferred [35].

However, despite the theoretical benefits of these extended lymphadenectomies, data have not indisputably demonstrated improved oncologic outcomes following a more extensive dissection. Clinical trials have failed to show a survival advantage even with high ligation of the inferior mesenteric artery (IMA) in patients with left-sided colon or rectal cancer, and lateral pelvic nodal dissection in these patients has not been shown to improve survival, either [37, 38]. Moreover, extended lymphadenectomy techniques are associated with increased morbidity, with higher rates of direct injury to neighboring structures, including the duodenum, ureters, spleen, and nerve plexuses in some reports [38, 39]. Concern also continues to exist over ischemia-related anastomotic complications due to decreased blood flow following high ligation, although the evidence surrounding this specific topic is controversial [40]. Given the unproven benefit coupled with the potential for increased morbidity, current evidence is insufficient to recommend extended lymphadenectomy, i.e., D3 dissection, as the standard of care [32].

38.2.3 Precision Medicine and Lymph Node Harvest

As our comprehension of cancer biology continues to evolve, we are gaining more insight into the importance of molecular pathogenesis, not just in CRC, but within the entire discipline of oncology. Genetic analyses and precision medicine are revolutionizing the modern treatment of cancer, and it is likely that these techniques will have a pivotal role in the harvest and evaluation of lymph nodes in CRC, too. Oncogenic mutations in *KRAS* and *TP53* have been shown to correlate with higher rates of regional lymph node metastasis, while mutations in *APC* and *PIK3CA* have not. Similarly, certain transcription factors and copy number variations have been demonstrated to confer an increased risk for nodal disease [41]. Zhang et al. correlated the level of lymph node metastases with underlying genetic mutations, allowing for the discovery of mutations which confer a higher risk for development of skip metastases [42]. Identification of these molecular variants may allow for classification of patients as high or low risk for lymph node involvement and allow for a personalized approach to deciding who will benefit from a higher nodal harvest or more extensive lymphadenectomy. Advanced molecular techniques have also been applied to the evaluation of harvested lymph nodes and can more precisely identify the presence of cancer within the lymph node, as well as quantify the amount of tumor present [43]. Further research is needed to determine whether these advanced analyses will become mainstay in the management of colorectal cancer.

38.3 Hepatic Metastases

38.3.1 Criteria for Surgical Resection of Liver Metastases

Current category 2A recommendations from the National Comprehensive Cancer Network (NCCN) [6] state resection as the treatment of choice for liver metastases from CRC when (a) complete resection is feasible, (b) adequate postoperative liver function is expected, (c) the primary tumor has undergone an R0 resection, and (d) no unresectable extrahepatic disease (EHD) is present (Table 38.1). Other guidelines recommend synchronous resection when the hepatic phase of the operation is expected to be less extensive (i.e., when liver metastases are small, peripheral, and/or the expected hepatic specimen is <50% of liver volume) and that patients with peritoneal metastases (PM) can also be considered for curative surgical therapy [44]. Prospective randomized data and meta-analyses also show laparoscopic resection to be safe and oncologically sound, and without any differences in disease-free survival (DFS) or OS [46, 47].

A 2013 consensus statement from the Americas Hepato-Pancreato-Biliary Association (AHPBA), Society of Surgical Oncology (SSO), and Society for Surgery of the Alimentary Tract (SSAT) further specified that patients with limited and resectable EHD or a reasonable expectation for long-term control with adjuvant treatment could be considered for hepatic resection [45].

In terms of prognostication, Fong et al. [48] established a clinically relevant scoring system for predicting survival in patients undergoing resection of CRC liver metastases in which one point is assigned for each of the following risk factors: lymph node-positive primary tumor, preoperative carcinoembryonic antigen (CEA) level > 200, disease-free interval of <12 months, >1 metastasis, and largest liver metastasis >5 cm. In the authors' validation series of 1001 patients, median survival in months was 74, 51, 47, 33, 20, and 22 for scores of 0 through 5, respectively.

38.3.2 Synchronous, Staged and Liver-First Resections

There has been substantial interest in the optimal timing of hepatic metastasis resection given the number of patients who present with synchronous primary and isolated hepatic metastases and the risk-benefit assessment in the context of

Table 38.1 Society and national guidelines for resection of hepatic CRC metastases

	NCCN guidelines [6]	Chinese guidelines [44]	AHPBA/SSO/SSAT guidelines [45]
Technical considerations	– Feasibility of complete resection – Maintenance of adequate hepatic function – Consider PVE or staged resection when insufficient FLR is expected	– R0 resection expected – Adequate FLR volume: • >40% for synchronous resection • >30% for staged resection – Preservation of at least one hepatic vein – Assessment of patient intolerance	– R0 resection expected – Ability to preserve two contiguous segments – Adequate vascular inflow/outflow & biliary drainage – Adequate FLR volume (20–30%) – Appropriate regenerative response after PVE in patients with marginal FLR
Considerations for EHD	– Prior R0 resection of primary tumor – No unresectable EHD	– Prior, or expected, R0 resection of primary tumor – No unresectable EHD or not amenable local ablative treatment	– Resectable EHD or reasonable expectation of long-term control with adjuvant therapy – Defer resection in the setting of metastatic disease progression while on first-line systemic therapy
Synchronous versus staged resection	Synchronous or staged depending on complexity, comorbidities, surgical exposure, and surgeon expertise	Recommend synchronous resection when: – Non-emergent – Small and/or peripherally located hepatic metastases or confined to single liver lobe – Expected hepatectomy volume <50% – Suspicious portal lymph nodes, intra-abdominal or other distant metastases can be completely resected	Not mentioned

EHD Extrahepatic disease, *FLR* Future liver remnant, *PVE* Portal vein embolization multimodal treatment options. While a staged approach may reduce surgical morbidity and avert mechanical complications associated with primary tumors, it may also prolong the time from diagnosis to initiation of systemic chemotherapy [49]. Much of the literature is retrospective, yet supports a synchronous over a staged approach despite a higher associated morbidity and mortality, especially with more substantial hepatectomies [50–53]. Recent prospective data suggest that OS improves with synchronous resection compared to a delayed approach without an increase in complications [54]. There is also an ongoing clinical trial (RESECT, NCT02954913) seeking to further delineate the complication profile associated with synchronous resection.

Another sequencing option, known as the liver-first (or reverse) approach, has also been accepted as a valid treatment strategy in which patients receive systemic chemotherapy and undergo resection of liver metastases first, while the primary tumor is removed in a delayed fashion [55]. The predominant advantages of this approach are performing metastasectomy prior to progression that may preclude resectability and a reduction in time to initiating systemic chemotherapy; however, local complications related to the primary tumor can still occur in the surgical interval, albeit infrequently [49].

38.3.3 Sequential Hepatic Resections

The surgical community has also sought innovative approaches to increasing the number of patients eligible for resection, and therefore cure, without untenable risks of complications such as postoperative liver failure. One such technique is the two-stage hepatectomy, classically for patients with bi-lobar metastases in whom a single step resection would be impossible. In the first operation, disease in one hemi-liver is resected while the remaining liver parenchyma is left to undergo compensatory hypertrophy. The remaining metastatic disease is then resected in the second stage. This can be done alone, with simultaneous resection of the primary, and/or in conjunction with ablative or embolization techniques [56–58]. Although the morbidity associated with two major hepatic resections is high, the majority of patients undergoing two-stage hepatectomy reported in the literature ultimately underwent the second stage and had R0 resections [56]. Five-year OS is reported to be an encouraging 32–64% considering that these patients traditionally would have been deemed to have unresectable stage IV disease and only offered palliative chemotherapy [56].

38.3.4 Associated Liver Partition and Portal Vein Ligation for Staged Hepatectomy

A newer option for patients with marginal or insufficient projected postoperative future liver remnant (FLR) has emerged and is dubbed associated liver partition and portal vein ligation for staged hepatectomy (ALPPS). This approach is divided into two stages. The initial operation involves mobilization of the liver and ligation of the portal vein branch supplying the diseased side at the bifurcation without disrupting the arterial or biliary systems. The liver parenchyma is subsequently divided along the intended resection plane (i.e., the in situ splitting phase) and the de-portalized segments of liver are then left in place and the abdomen closed. The second stage, or completion of resection, occurs after a delay of 1–2 weeks allowing for compensatory hypertrophy of the FLR [59]. Parenchymal division in the first stage is thought to promote more rapid compensatory hypertrophy by preventing any collateral portal blood flow to the de-portalized side [60], an effect not possible through percutaneous approaches to portal exclusion. Although this approach can facilitate curative resection of hepatic metastases in patients with borderline predicted FLR, it is associated with substantial morbidity and mortality and has not been proven to be superior to staged resections in retrospective analysis [61–64].

38.3.5 Administration of Chemotherapy Before and After Liver Resection

The timing of chemotherapy with respect to hepatic metastasectomy is also a germane consideration given that perioperative chemotherapy has been shown to improve 3-year progression-free survival (PFS) in patients with resectable hepatic metastases undergoing surgery, but is also associated with higher rates of postoperative complications [65]. Unfortunately, the survival benefit in this population does not appear to be durable [66]. Several agents used in standard of care chemotherapeutic regimens are associated with specific hepatic complications: the use of irinotecan is associated with higher rates of steatohepatitis and postoperative mortality, whereas oxaliplatin has been linked to vascular lesion development in non-tumor-bearing liver specimens and more significant intraoperative transfusion requirements, but no difference in postoperative mortality [67–69]. The use of neoadjuvant chemotherapy also necessitates a larger FLR volume in order to mitigate the risks of postoperative liver failure due to impaired liver function [70, 71].

Evaluation of the clinical response to neoadjuvant chemotherapy may also aid in the identification of patients who will benefit from subsequent metastasectomy. Those who demonstrate disease progression while on upfront chemotherapy are less likely to benefit from resection, and the decision for surgical intervention can be reconsidered, whereas those with stable disease or who respond well to chemotherapy can then still undergo their planned surgical resection [72, 73]. However, even in patients who appear to have a complete radiographic response to neoadjuvant treatment, the majority are still likely to have residual disease, and therefore, complete clinical response should not be equated with a cure [74].

Neoadjuvant therapy may also be administered through a regional approach via hepatic arterial infusion chemotherapy [75]. Direct administration of chemotherapy to the liver allows delivery and exposure of greater drug concentrations directly to the tumor over an extended period of time while limiting systemic side effects through implantation of a hepatic artery infusion pump. High objective radiologic response rates have been observed with this method of administration, but a definitive survival advantage has not been demonstrated [75]. Hepatic toxicity and technical complications related to the pump or accompanying ports/catheters limit this treatment approach, and its use is mostly confined to specialized centers with extensive experience in this treatment modality [76].

38.3.6 Timing of Radiation Therapy for Rectal Cancer with Metastases

Neoadjuvant radiation therapy (RT) currently plays a significant role in treatment algorithms for locally advanced rectal cancer in order to improve long-term local control; however, appropriate sequencing of treatment is less clear in the setting of synchronous liver metastases [77]. NCCN guidelines for nonmetastatic rectal cancer advocate the use of long course RT (45–50 Gy in 25–28 fractions) in conjunction with radio-sensitizing fluoropyrimidine-based chemotherapy followed by resection 4–6 weeks after completion [7]. These guidelines also stipulate that patients with T3 tumors can be considered for a short course regimen (25 Gy in 5 fractions) followed by surgery within 3–7 days, or after 4–6 weeks, based on current evidence [78]. The concern with long course RT directed at the primary tumor in the setting of hepatic metastases is the prolonged delay in treatment of the metastatic disease, which may no longer be resectable within the recommended surgical window for the primary tumor (4–6 weeks after completing RT). Short course RT has shown equivalency in terms of 5-year local recurrence, OS, and late toxicity for patients with T3 tumors within 12 cm of the anal verge, and T3/T4 tumors without sphincter involvement [78, 79]. Although neither of these studies involved patients with evidence of metastatic disease, short course RT is attractive in reducing this treatment interval without foregoing the benefits of improved local control. Any risks or benefits in delaying liver-directed therapies in favor of long-course RT remain to be delineated, and individualized treatment decisions should be made with a multidisciplinary tumor board discussion.

38.4 Peritoneal Metastases

The estimated incidence of CRC PM is 8.3–26.6% and varies in the literature depending on study type and inclusion criteria [80]. Autopsy series show that in patients who die from CRC cancer, 20% with adenocarcinoma, 48% with mucinous subtypes, and 51% with signet ring cell tumors ultimately develop PM, while nearly 15% have isolated PM regardless of histology [80, 81]. Cytoreductive surgery (CRS), with or without hyperthermic intraperitoneal chemotherapy (HIPEC), therefore continues to be an attractive treatment modality in the setting of synchronous and metachronous PM and is an area of ongoing research [82].

38.4.1 Clinical Evidence

The highest quality evidence for CRS/HIPEC in the setting of PM from CRC comes from a randomized clinical trial of patients with colorectal adenocarcinoma without solid organ metastases who were randomized to either systemic chemotherapy alone or CRS/HIPEC with mitomycin C and adjuvant systemic chemotherapy [83]. Both treatment arms received systemic fluorouracil and leucovorin alone, which is no longer current first-line therapy [84]. The addition of CRS/HIPEC improved both PFS (12.6 vs. 7.7 months, $p = 0.020$) and disease-specific survival (DSS) (22.2 vs. 12.6 months, $p = 0.028$) after long-term follow-up, and represents a substantial improvement compared to the expected survival with chemotherapy alone [85–87]. Furthermore, patients who underwent an R0–1 resection had a median survival of 48 months and a 5-year OS of 45%, demonstrating a durable survival benefit. Notably, this study did not parse out the individual contributions of HIPEC versus CRS (which was evaluated in the PRODIGE 7 trial, discussed subsequently), and also did not include combinations of chemotherapy regimens currently used today. Therefore, it is plausible that the systemic therapy alone arm would have significantly better outcomes with modern chemotherapeutics.

More recently, preliminary data from the unpublished PRODIGE 7 [88] trial (NCT00769405) presented in 2018 included 265 patients with synchronous PM from CRC and a peritoneal cancer index ≤25 who were randomized to either CRS/HIPEC with oxaliplatin or CRS alone. Both groups received standard neoadjuvant and adjuvant systemic chemotherapy for a total of 6 months. Interestingly, the addition of oxaliplatin-HIPEC to CRS did not prolong OS (41.2 vs. 41.7 months, $p = 0.995$) or relapse-free survival (11.1 vs. 13.1 months, $p = 0.486$), which implies that the comparative survival advantage for this population over systemic chemotherapy alone may derive primarily from the resection of their metastatic disease.

38.4.2 Adjuvant HIPEC and Second Look Procedures

Two recent clinical trials investigated the role of adjuvant HIPEC in patients without synchronous PM but at high risk for subsequent metachronous disease due to perforated or T4 primary CRC tumors. In the COLOPEC trial [89] (NCT02231086), patients underwent resection of the primary tumor and received either HIPEC with oxaliplatin (460 mg/m^2) and routine adjuvant systemic chemotherapy or the standard of care with adjuvant systemic chemotherapy alone. Ultimately, no differences in 18-month PM-free survival, DFS, or OS were observed between groups. Criticisms of this trial include that only 87% of patients in the experimental arm actually underwent HIPEC, some of which were performed concurrent with their initial resection, whereas others were delayed 5–8 weeks, and that patients in the control arm were eligible for cross over if they developed PM during surveillance. Ultimately, the results of this trial do not support the adjuvant use of HIPEC in patients with high-risk CRC primary tumors without synchronous PM.

The other trial (PROPHYLOCHIP [90], NTC01226394) included patients with perforated primary tumors and those with isolated PM resected with the primary in order to assess the efficacy of routine delayed CRS/HIPEC in patients at high risk for metachronous peritoneal disease. All patients received adjuvant first-line systemic chemotherapy following their initial resection and had negative CT scan and tumor markers at 6 months. Randomization was then to either continued standard surveillance or a second-look surgery with CRS/HIPEC (with either oxaliplatin and irinotecan, oxaliplatin alone, or mitomycin C alone) and simultaneous intravenous fluorouracil and leucovorin. At a median follow-up of 50.8 months, there were no differences in recurrence rates, 5-year DFS, or OS between groups, therefore, arguing against routine second-look surgery. However, it should be noted that, there was significant heterogeneity in the HIPEC chemotherapeutic agents employed and the trial was not designed to assess the efficacy of prophylactic HIPEC in high-risk patients without PM.

38.4.3 Ongoing Clinical Trials

There are currently several ongoing trials seeking to further assess the role and approach to CRS/HIPEC in patients with CRC. The ICARuS trial (NCT01815359) is a multicenter, phase II study comparing CRS/HIPEC with mitomycin C to CRS with early postoperative intraperitoneal chemotherapy (EPIC) with 5-fluorodeoxyuridine and leucovorin in the treatment of PM for CRC and appendiceal cancers. Two other ongoing phase III trials in Spain (PROMENADE, NCT02974556) and China (APEC, NCT02965248) are also investigating the role of HIPEC at the time of primary resection despite the aforementioned evidence against prophylactic HIPEC in high-risk patients. Additionally, CAIRO6 (NCT02758951) will assess the oncologic efficacy of CRS/HIPEC alone versus in addition to standard systemic chemotherapeutic regimens for patients with PM from non-appendiceal CRC.

38.5 Evidence for Hepatic Resection in the Setting of Extrahepatic Disease

Despite the traditional paradigm regarding the futility of resection in the setting of multisite metastatic disease, the benefits of performing hepatic resection in this setting are increasingly supported by the literature [91]. In patients undergoing resection of EHD, OS appears to depend on the extrahepatic metastatic site with longer median survival associated with metastasis to the lungs compared to portal or para-aortic lymph nodes or the peritoneum [92]. This observation challenges the paradigm of surgical futility in the treatment of oligometastatic disease. Only one prospective randomized trial comparing pulmonary metastasectomy to active surveillance has been published and was terminated early due to poor accrual, thereby precluding significant interpretation [93]. Some retrospective evidence shows that long-term survival for patients undergoing pulmonary and hepatic metastasectomy is comparable to that seen with isolated hepatic resection, although it is unclear to what extent this is a reflection of treatment versus tumor biology [94, 95].

Available data suggest that no medical intervention is as efficacious in increasing survival as surgical resection for both isolated hepatic metastases and hepatic metastases with EHD. Even so, the median survival reported in patients undergoing resection of multisite EHD is comparable to that seen with chemotherapy alone and is rarely curative [96, 97]. Proposed scoring systems for OS and recurrence-free survival (RFS) based on preoperatively known factors include largest liver metastasis >3 cm, >5 liver metastases, and unfavorable EHD site (portal/retroperitoneal lymph nodes or multiple sites) with each criterion associated with worse outcomes [91, 98]. Abstracting any uniform recommendation from the present literature is challenging due to substantial heterogeneity in patient selection, perioperative chemotherapy regimens, adjunctive treatment modalities, and possible reporting bias [62, 92, 96, 98].

38.6 Clinical Prognostication Through Histologic and Genetic Factors

38.6.1 Immunoscores

Growing translational research continues to offer useful prognostic information in patients with metastatic CRC based on histologic factors of the primary and metastatic tumors [99]. For example, cellular expression of COX-2, HLA, and PD-1 in resected metastatic deposits is independently associated with prolonged RFS and OS after systemic chemotherapy [100]. A general indication of the host immune reaction to the tumor can be determined through assessment of T-cell infiltration at the tumor center and invasive margin, which is then used to calculate the immunoscore of a pathologic specimen [101]. Some data suggest that these scores prognostically outperform current clinical staging systems in terms of DSS and OS, with a more robust immune response associated with better outcomes [99, 100, 102]. Furthermore, activation of known oncogenic signaling pathways, such as WNT/β-catenin, dampens T-cell recruitment and reduces the efficacy of T-cell targeted monoclonal antibody therapy in a mouse model [103]. Even so, an increasingly sophisticated understanding of the mechanisms of metastasis, tumor modulation of the cellular microenvironment, and the maladaptive host response is leading to promising new immunotherapies. Indeed, immune checkpoint blockade has already shown efficacy for patients with microsatellite instability–high metastatic colorectal cancer [104, 105].

38.6.2 Patient Selection Through Mutational Analysis

Growing evidence also supports genetic analysis of tumor specimens as an emerging prognostic factor in CRC. As described previously, older clinical scoring systems incorporating clinical markers can be determined preoperatively to infer oncologic benefit from resection of CRC liver metastases [48]. However, these systems are now being modified to include mutational analyses in specific signaling pathways that are associated with OS, such as RAS-RAF and SMAD [106, 107]. Other nuanced associations with potential clinical implications continue to be described such as RAS mutations having a more deleterious association with outcomes in younger compared with older patients undergoing resection of CRC liver metastases compared to older ones [108], and increasing prognostic precision when considering combinations of coexisting mutations [109]. Such factors may ultimately be incorporated into risk-benefit stratification paradigms.

38.6.3 Future Directions

Promising avenues of investigation include the use of cell-free DNA, not only in the earlier detection of disease progression but also prediction of which patients will maximally benefit from resection of oligometastatic disease based on genetic profile [110]. Further research is needed in order to delineate the effects of primary tumor and/or metastatic resection on responsiveness to immunotherapy or other systemic treatments [111].

38.7 Conclusions

- Histopathologic lymph node status remains the most important prognostic factor for patients diagnosed with colorectal cancer without distant metastases.
- Although absolute lymph node count is currently used as a metric of quality surgical care, a complex interplay of multiple factors is likely responsible for the lymph node yield harvested at the time of surgery, bringing into question the utility of nodal count as a quality metric.
- More extensive nodal dissections lead to higher nodal counts, but whether these extended dissections lead to improved survival remains unclear, and must be weighed against the associated increased morbidity observed when they are performed.
- With careful patient selection, resection of colorectal cancer hepatic metastases is associated with improved survival, but the optimal timing and sequence of metastasectomy continue to be debated and must be weighed against surgical morbidity that can lead to delays in systemic therapies.
- While cytoreductive surgery may provide a survival benefit for patients with colorectal cancer peritoneal metastases, studies are ongoing to determine the exact role of cytoreductive surgery/heated intraperitoneal chemotherapy in the management of colorectal cancer.
- Advanced molecular evaluations and genetic analyses may allow for a more targeted and individualized approach to the management of metastatic colorectal cancer.
- Ongoing and future research will strive to answer:

Open Questions

- Is there a better metric of quality surgical care than absolute lymph node harvest for patients undergoing surgical resection of primary colorectal cancer?
- Do more extensive lymph node dissections lead to improved survival?
- Can patients undergo major resections of primary colorectal cancer and hepatic metastases in a synchronous fashion without suffering increased morbidity in comparison to a staged approach?
- What is the optimal combination of systemic therapy and cytoreductive surgery/heated intraperitoneal chemotherapy for patients with colorectal cancer peritoneal metastases?
- Can the surgical management of metastatic colorectal cancer be stratified based on genetic and molecular evaluations?

Disclosures The authors declare no conflicts of interest or external funding for this publication.

References

1. Pienta KJ, Robertson BA, Coffey DS, Taichman RS. The cancer diaspora: metastasis beyond the seed and soil hypothesis. Clin Cancer Res. 2013;19(21):5849–55. https://doi.org/10.1158/1078-0432.CCR-13-2158.
2. Weichselbaum RR, Hellman S. Oligometastases revisited. Nat Rev Clin Oncol. 2011;8(6):378–82. https://doi.org/10.1038/nrclinonc.2011.44.
3. Pitroda SP, Khodarev NN, Huang L, et al. Integrated molecular subtyping defines a curable oligometastatic state in colorectal liver metastasis. Nat Commun. 2018;9(1):1793. https://doi.org/10.1038/s41467-018-04278-6.
4. Baxter NN, Virnig DJ, Rothenberger DA, Morris AM, Jessurun J, Virnig BA. Lymph node evaluation in colorectal cancer patients: a population-based study. J Natl Cancer Inst. 2005;97(3):219–25. https://doi.org/10.1093/jnci/dji020.
5. Edge S, Byrd D, Compton C, Fritz A, Greene F, Trotti A. AJCC cancer staging manual 7th edition. 7th edn. Springer; 2010.
6. National Comprehensive Cancer Network. Colon cancer, version 4.2020, NCCN clinical practice guidelines in oncology. https://www.nccn.org/professionals/physician_gls/pdf/colon_blocks.pdf. Accessed 17 Nov 2020.
7. National Comprehensive Cancer Network. Rectal cancer, version 6.2020, NCCN clinical practice guidelines in oncology. https://www.nccn.org/professionals/physician_gls/pdf/rectal_blocks.pdf. Accessed 17 Nov 2020
8. Nelson H, Petrelli N, Carlin A, et al. Guidelines 2000 for colon and rectal cancer surgery. J Natl Cancer Inst. 2001;93(8):583–96. https://doi.org/10.1093/jnci/93.8.583.
9. Joseph NE, Sigurdson ER, Hanlon AL, et al. Accuracy of determining nodal negativity in colorectal cancer on the basis of the number of nodes retrieved on resection. Ann Surg Oncol. 2003;10(3):213–8. https://doi.org/10.1245/aso.2003.03.059.
10. Chang GJ, Rodriguez-Bigas MA, Skibber JM, Moyer VA. Lymph node evaluation and survival after curative resection of colon cancer: systematic review. J Natl Cancer Inst. 2007;99(6):433–41. https://doi.org/10.1093/jnci/djk092.
11. Chen SL, Bilchik AJ. More extensive nodal dissection improves survival for stages I to III of colon cancer: a population-based study. Ann Surg. 2006;244(4):602–10. https://doi.org/10.1097/01.sla.0000237655.11717.50.
12. Tepper JE, O'Connell MJ, Niedzwiecki D, et al. Impact of number of nodes retrieved on outcome in patients with rectal cancer. J Clin Oncol. 2001;19(1):157–63. https://doi.org/10.1200/JCO.2001.19.1.157.
13. Märkl B. Stage migration vs immunology: the lymph node count story in colon cancer. World J Gastroenterol. 2015;21(43):12218–33. https://doi.org/10.3748/wjg.v21.i43.12218.
14. van Erning FN, Crolla RM, Rutten HJ, Beerepoot LV, van Krieken JH, Lemmens VE. No change in lymph node positivity rate despite increased lymph node yield and improved survival in colon cancer. Eur J Cancer. 2014;50(18):3221–9. https://doi.org/10.1016/j.ejca.2014.10.011.
15. Baxter NN, Ricciardi R, Simunovic M, Urbach DR, Virnig BA. An evaluation of the relationship between lymph node number and staging in pT3 colon cancer using population-based data. Dis Colon Rectum. 2010;53(1):65–70. https://doi.org/10.1007/DCR.0b013e3181c70425.
16. Parnaby CN, Scott NW, Ramsay G, et al. Prognostic value of lymph node ratio and extramural vascular invasion on survival for patients undergoing curative colon cancer resection. Br J Cancer. 2015;113(2):212–9. https://doi.org/10.1038/bjc.2015.211.
17. American College of Surgeons. Commission on cancer quality of care measures. https://www.facs.org/quality-programs/cancer/ncdb/qualitymeasures. Accessed 18 Nov 2020
18. Baxter NN. Is lymph node count an ideal quality indicator for cancer care? J Surg Oncol. 2009;99(4):265–8. https://doi.org/10.1002/jso.21197.
19. Parsons HM, Tuttle TM, Kuntz KM, Begun JW, McGovern PM, Virnig BA. Association between lymph node evaluation for colon cancer and node positivity over the past 20 years. JAMA. 2011;306(10):1089–97. https://doi.org/10.1001/jama.2011.1285.
20. Amin M, Edge S, Greene F et al AJCC cancer staging manual, 8th edn. Springer International Publishing: American Joint Commission on Cancer; 2017.
21. Jin M, Roth R, Rock JB, Washington MK, Lehman A, Frankel WL. The impact of tumor deposits on colonic adenocarcinoma AJCC TNM staging and outcome. Am J Surg Pathol. 2015;39(1):109–15. https://doi.org/10.1097/PAS.0000000000000320.
22. Evans MD, Barton K, Rees A, Stamatakis JD, Karandikar SS. The impact of surgeon and pathologist on lymph node retrieval in colorectal cancer and its impact on survival for patients with Dukes' stage B disease. Color Dis. 2008;10(2):157–64. https://doi.org/10.1111/j.1463-1318.2007.01225.x.
23. de Burlet KJ, van den Hout MF, Putter H, Smit VT, Hartgrink HH. Total number of lymph nodes in oncologic resections, is there more to be found? J Gastrointest Surg. 2015;19(5):943–8. https://doi.org/10.1007/s11605-015-2764-9.
24. Stocchi L, Fazio VW, Lavery I, Hammel J. Individual surgeon, pathologist, and other factors affecting lymph node harvest in stage II colon carcinoma. Is a minimum of 12 examined lymph nodes sufficient? Ann Surg Oncol. 2011;18(2):405–12. https://doi.org/10.1245/s10434-010-1308-5.
25. Gonsalves WI, Kanuri S, Tashi T, et al. Clinicopathologic factors associated with lymph node retrieval in resectable colon cancer: a Veterans' Affairs Central Cancer Registry (VACCR)

database analysis. J Surg Oncol. 2011;104(6):667–71. https://doi.org/10.1002/jso.21886.
26. Kuo YH, Lee KF, Chin CC, Huang WS, Yeh CH, Wang JY. Does body mass index impact the number of LNs harvested and influence long-term survival rate in patients with stage III colon cancer? Int J Color Dis. 2012;27(12):1625–35. https://doi.org/10.1007/s00384-012-1496-5.
27. Mekenkamp LJ, van Krieken JH, Marijnen CA, van de Velde CJ, Nagtegaal ID, Pathology Review Committee and the Co-operative Clinical Investigators. Lymph node retrieval in rectal cancer is dependent on many factors—the role of the tumor, the patient, the surgeon, the radiotherapist, and the pathologist. Am J Surg Pathol. 2009;33(10):1547–53. https://doi.org/10.1097/PAS.0b013e3181b2e01f.
28. Ahmadi O, Stringer MD, Black MA, McCall JL. Influence of age and site of disease on lymph node yield in colorectal cancer. N Z Med J. 2014;127(1395):31–40.
29. Søreide K, Nedrebø BS, Søreide JA, Slewa A, Kørner H. Lymph node harvest in colon cancer: influence of microsatellite instability and proximal tumor location. World J Surg. 2009;33(12):2695–703. https://doi.org/10.1007/s00268-009-0255-4.
30. Märkl B, Rößle J, Arnholdt HM, et al. The clinical significance of lymph node size in colon cancer. Mod Pathol. 2012;25(10):1413–22. https://doi.org/10.1038/modpathol.2012.92.
31. Le Voyer TE, Sigurdson ER, Hanlon AL, et al. Colon cancer survival is associated with increasing number of lymph nodes analyzed: a secondary survey of intergroup trial INT-0089. J Clin Oncol. 2003;21(15):2912–9. https://doi.org/10.1200/JCO.2003.05.062.
32. Ong ML, Schofield JB. Assessment of lymph node involvement in colorectal cancer. World J Gastrointest Surg. 2016;8(3):179–92. https://doi.org/10.4240/wjgs.v8.i3.179.
33. Brouwer NPM, Hugen N, Nagtegaal ID. More extensive lymphadenectomy in colon cancer; how far are we willing to go for a biomarker? Tech Coloproctol. 2020;24(7):761–4. https://doi.org/10.1007/s10151-020-02239-0.
34. West NP, Kobayashi H, Takahashi K, et al. Understanding optimal colonic cancer surgery: comparison of Japanese D3 resection and European complete mesocolic excision with central vascular ligation. J Clin Oncol. 2012;30(15):1763–9. https://doi.org/10.1200/JCO.2011.38.3992.
35. Kanemitsu Y, Komori K, Kimura K, Kato T. D3 lymph node dissection in right hemicolectomy with a no-touch isolation technique in patients with colon cancer. Dis Colon Rectum. 2013;56(7):815–24. https://doi.org/10.1097/DCR.0b013e3182919093.
36. Hohenberger W, Weber K, Matzel K, Papadopoulos T, Merkel S. Standardized surgery for colonic cancer: complete mesocolic excision and central ligation—technical notes and outcome. Color Dis. 2009;11(4):354–64; discussion 364–5. https://doi.org/10.1111/j.1463-1318.2008.01735.x.
37. Rouffet F, Hay JM, Vacher B, et al. Curative resection for left colonic carcinoma: hemicolectomy vs. segmental colectomy. A prospective, controlled, multicenter trial. French Association for Surgical Research. Dis Colon Rectum. 1994;37(7):651–9. https://doi.org/10.1007/BF02054407.
38. Nagawa H, Muto T, Sunouchi K, et al. Randomized, controlled trial of lateral node dissection vs. nerve-preserving resection in patients with rectal cancer after preoperative radiotherapy. Dis Colon Rectum. 2001;44(9):1274–80. https://doi.org/10.1007/BF02234784.
39. Bertelsen CA, Neuenschwander AU, Jansen JE, et al. Short-term outcomes after complete mesocolic excision compared with 'conventional' colonic cancer surgery. Br J Surg. 2016;103(5):581–9. https://doi.org/10.1002/bjs.10083.
40. Tsujinaka S, Kawamura YJ, Tan KY, et al. Proximal bowel necrosis after high ligation of the inferior mesenteric artery in colorectal surgery. Scand J Surg. 2012;101(1):21–5. https://doi.org/10.1177/145749691210100105.
41. Kandagatla P, Maguire LH, Hardiman KM. Biology of nodal spread in colon cancer: insights from molecular and genetic studies. Eur Surg Res. 2018;59(5–6):361–70. https://doi.org/10.1159/000494832.
42. Zhang C, Zhang L, Xu T, et al. Mapping the spreading routes of lymphatic metastases in human colorectal cancer. Nat Commun. 2020;11(1):1993. https://doi.org/10.1038/s41467-020-15886-6.
43. Güller U, Zettl A, Worni M, et al. Molecular investigation of lymph nodes in colon cancer patients using one-step nucleic acid amplification (OSNA): a new road to better staging? Cancer. 2012;118(24):6039–45. https://doi.org/10.1002/cncr.27667.
44. Xu J, Fan J, Qin X, et al. Chinese guidelines for the diagnosis and comprehensive treatment of colorectal liver metastases (version 2018). J Cancer Res Clin Oncol. 2019;145(3):725–36. https://doi.org/10.1007/s00432-018-2795-1.
45. Adams RB, Aloia TA, Loyer E, et al. Selection for hepatic resection of colorectal liver metastases: expert consensus statement. HPB (Oxford). 2013;15(2):91–103. https://doi.org/10.1111/j.1477-2574.2012.00557.x.
46. Fretland Å, Dagenborg VJ, Bjørnelv GMW, et al. Laparoscopic versus open resection for colorectal liver metastases: the OSLO-COMET randomized controlled trial. Ann Surg. 2018;267(2):199–207. https://doi.org/10.1097/SLA.0000000000002353.
47. Pan L, Tong C, Fu S, et al. Laparoscopic procedure is associated with lower morbidity for simultaneous resection of colorectal cancer and liver metastases: an updated meta-analysis. World J Surg Oncol. 2020;18(1):251. https://doi.org/10.1186/s12957-020-02018-z.
48. Fong Y, Fortner J, Sun RL, Brennan MF, Blumgart LH. Clinical score for predicting recurrence after hepatic resection for metastatic colorectal cancer: analysis of 1001 consecutive cases. Ann Surg. 1999;230(3):309–18; discussion 318–21. https://doi.org/10.1097/00000658-199909000-00004.
49. Lillemoe HA, Vauthey JN. Surgical approach to synchronous colorectal liver metastases: staged, combined, or reverse strategy. Hepatobiliary Surg Nutr. 2020;9(1):25–34. https://doi.org/10.21037/hbsn.2019.05.14.
50. Martin RC 2nd, Augenstein V, Reuter NP, Scoggins CR, McMasters KM. Simultaneous versus staged resection for synchronous colorectal cancer liver metastases. J Am Coll Surg. 2009;208(5):842–50; discussion 850–2. https://doi.org/10.1016/j.jamcollsurg.2009.01.031.
51. Thelen A, Jonas S, Benckert C, et al. Simultaneous versus staged liver resection of synchronous liver metastases from colorectal cancer. Int J Color Dis. 2007;22(10):1269–76. https://doi.org/10.1007/s00384-007-0286-y.
52. Reddy SK, Pawlik TM, Zorzi D, et al. Simultaneous resections of colorectal cancer and synchronous liver metastases: a multi-institutional analysis. Ann Surg Oncol. 2007;14(12):3481–91. https://doi.org/10.1245/s10434-007-9522-5.
53. Feng Q, Wei Y, Zhu D, et al. Timing of hepatectomy for resectable synchronous colorectal liver metastases: for whom simultaneous resection is more suitable—a meta-analysis. PLoS One. 2014;9(8):e104348. https://doi.org/10.1371/journal.pone.0104348.
54. Boudjema K, Locher C, Sabbagh C, et al. Simultaneous versus delayed resection for initially resectable synchronous colorectal cancer liver metastases: a prospective, open-label, randomized, controlled trial. Ann Surg. 2021;273(1):49–56. https://doi.org/10.1097/SLA.0000000000003848.
55. Mentha G, Majno PE, Andres A, Rubbia-Brandt L, Morel P, Roth AD. Neoadjuvant chemotherapy and resection of advanced synchronous liver metastases before treatment of the colorectal

primary. Br J Surg. 2006;93(7):872–8. https://doi.org/10.1002/bjs.5346.
56. Lam VW, Laurence JM, Johnston E, Hollands MJ, Pleass HC, Richardson AJ. A systematic review of two-stage hepatectomy in patients with initially unresectable colorectal liver metastases. HPB (Oxford). 2013;15(7):483–91. https://doi.org/10.1111/j.1477-2574.2012.00607.x.
57. Jaeck D, Oussoultzoglou E, Rosso E, Greget M, Weber JC, Bachellier P. A two-stage hepatectomy procedure combined with portal vein embolization to achieve curative resection for initially unresectable multiple and bilobar colorectal liver metastases. Ann Surg. 2004;240(6):1037–49; discussion 1049–51. https://doi.org/10.1097/01.sla.0000145965.86383.89.
58. Karoui M, Vigano L, Goyer P, et al. Combined first-stage hepatectomy and colorectal resection in a two-stage hepatectomy strategy for bilobar synchronous liver metastases. Br J Surg. 2010;97(9):1354–62. https://doi.org/10.1002/bjs.7128.
59. Schnitzbauer AA, Lang SA, Goessmann H, et al. Right portal vein ligation combined with in situ splitting induces rapid left lateral liver lobe hypertrophy enabling 2-staged extended right hepatic resection in small-for-size settings. Ann Surg. 2012;255(3):405–14. https://doi.org/10.1097/SLA.0b013e31824856f5.
60. Selby K, Hernandez-Alejandro R. Two-stage hepatectomy for liver metastasis from colorectal cancer. CMAJ. 2014;186(15):1163–6. https://doi.org/10.1503/cmaj.131022.
61. Bertens KA, Hawel J, Lung K, Buac S, Pineda-Solis K, Hernandez-Alejandro R. ALPPS: challenging the concept of unresectability—a systematic review. Int J Surg. 2015;13:280–7. https://doi.org/10.1016/j.ijsu.2014.12.008.
62. Stewart CL, Warner S, Ito K, et al. Cytoreduction for colorectal metastases: liver, lung, peritoneum, lymph nodes, bone, brain. When does it palliate, prolong survival, and potentially cure? Curr Probl Surg. 2018;55(9):330–79. https://doi.org/10.1067/j.cpsurg.2018.08.004.
63. Ratti F, Schadde E, Masetti M, et al. Strategies to increase the resectability of patients with colorectal liver metastases: a multi-center case-match analysis of ALPPS and conventional two-stage hepatectomy. Ann Surg Oncol. 2015;22(6):1933–42. https://doi.org/10.1245/s10434-014-4291-4.
64. Moris D, Ronnekleiv-Kelly S, Kostakis ID, et al. Operative results and oncologic outcomes of associating liver partition and portal vein ligation for staged hepatectomy (ALPPS) versus two-stage hepatectomy (TSH) in patients with unresectable colorectal liver metastases: a systematic review and meta-analysis. World J Surg. 2018;42(3):806–15. https://doi.org/10.1007/s00268-017-4181-6.
65. Nordlinger B, Sorbye H, Glimelius B, et al. Perioperative chemotherapy with FOLFOX4 and surgery versus surgery alone for resectable liver metastases from colorectal cancer (EORTC Intergroup trial 40983): a randomised controlled trial. Lancet. 2008;371(9617):1007–16. https://doi.org/10.1016/S0140-6736(08)60455-9.
66. Nordlinger B, Sorbye H, Glimelius B, et al. Perioperative FOLFOX4 chemotherapy and surgery versus surgery alone for resectable liver metastases from colorectal cancer (EORTC 40983): long-term results of a randomised, controlled, phase 3 trial. Lancet Oncol. 2013;14(12):1208–15. https://doi.org/10.1016/S1470-2045(13)70447-9.
67. Vauthey JN, Pawlik TM, Ribero D, et al. Chemotherapy regimen predicts steatohepatitis and an increase in 90-day mortality after surgery for hepatic colorectal metastases. J Clin Oncol. 2006;24(13):2065–72. https://doi.org/10.1200/JCO.2005.05.3074.
68. Abdalla EK, Vauthey JN. Steatosis as a risk factor in liver surgery. Ann Surg. 2007;246(2):340–1; author reply 341. https://doi.org/10.1097/SLA.0b013e31811ea9d2.
69. Aloia T, Sebagh M, Plasse M, et al. Liver histology and surgical outcomes after preoperative chemotherapy with fluorouracil plus oxaliplatin in colorectal cancer liver metastases. J Clin Oncol. 2006;24(31):4983–90. https://doi.org/10.1200/JCO.2006.05.8156.
70. Narita M, Oussoultzoglou E, Fuchshuber P, et al. What is a safe future liver remnant size in patients undergoing major hepatectomy for colorectal liver metastases and treated by intensive preoperative chemotherapy? Ann Surg Oncol. 2012;19(8):2526–38. https://doi.org/10.1245/s10434-012-2274-x.
71. Kishi Y, Zorzi D, Contreras CM, et al. Extended preoperative chemotherapy does not improve pathologic response and increases postoperative liver insufficiency after hepatic resection for colorectal liver metastases. Ann Surg Oncol. 2010;17(11):2870–6. https://doi.org/10.1245/s10434-010-1166-1.
72. Kopetz S, Vauthey JN. Perioperative chemotherapy for resectable hepatic metastases. Lancet. 2008;371(9617):963–5. https://doi.org/10.1016/S0140-6736(08)60429-8.
73. Khan K, Wale A, Brown G, Chau I. Colorectal cancer with liver metastases: neoadjuvant chemotherapy, surgical resection first or palliation alone? World J Gastroenterol. 2014;20(35):12391–406. https://doi.org/10.3748/wjg.v20.i35.12391.
74. Benoist S, Brouquet A, Penna C, et al. Complete response of colorectal liver metastases after chemotherapy: does it mean cure? J Clin Oncol. 2006;24(24):3939–45. https://doi.org/10.1200/JCO.2006.05.8727.
75. Bartlett DL, Berlin J, Lauwers GY, Messersmith WA, Petrelli NJ, Venook AP. Chemotherapy and regional therapy of hepatic colorectal metastases: expert consensus statement. Ann Surg Oncol. 2006;13(10):1284–92. https://doi.org/10.1245/s10434-006-9018-8.
76. Chouliaras K, Russell G, Levine E, et al. Hepatic arterial infusion chemotherapy for colorectal liver metastases revisited. HPB (Oxford). 2020;22(9):1265–70. https://doi.org/10.1016/j.hpb.2019.12.003.
77. Sauer R, Liersch T, Merkel S, et al. Preoperative versus postoperative chemoradiotherapy for locally advanced rectal cancer: results of the German CAO/ARO/AIO-94 randomized phase III trial after a median follow-up of 11 years. J Clin Oncol. 2012;30(16):1926–33. https://doi.org/10.1200/JCO.2011.40.1836.
78. Ngan SY, Burmeister B, Fisher RJ, et al. Randomized trial of short-course radiotherapy versus long-course chemoradiation comparing rates of local recurrence in patients with T3 rectal cancer: trans-Tasman radiation oncology group trial 01.04. J Clin Oncol. 2012;30(31):3827–33. https://doi.org/10.1200/JCO.2012.42.9597.
79. Bujko K, Nowacki MP, Nasierowska-Guttmejer A, Michalski W, Bebenek M, Kryj M. Long-term results of a randomized trial comparing preoperative short-course radiotherapy with preoperative conventionally fractionated chemoradiation for rectal cancer. Br J Surg. 2006;93(10):1215–23. https://doi.org/10.1002/bjs.5506.
80. Jacobson R, Sherman SK, Dahdaleh F, Turaga KK. Peritoneal metastases in colorectal cancer. Ann Surg Oncol. 2018;25(8):2145–51. https://doi.org/10.1245/s10434-018-6490-x.
81. Hugen N, van de Velde CJH, de Wilt JHW, Nagtegaal ID. Metastatic pattern in colorectal cancer is strongly influenced by histological subtype. Ann Oncol. 2014;25(3):651–7. https://doi.org/10.1093/annonc/mdt591.
82. Ellis RJ, Schlick CJR, Yang AD, Barber EL, Bilimoria KY, Merkow RP. Utilization and treatment patterns of cytoreduction surgery and intraperitoneal chemotherapy in the United States. Ann Surg Oncol. 2020;27(1):214–21. https://doi.org/10.1245/s10434-019-07492-8.
83. Verwaal VJ, van Ruth S, de Bree E, et al. Randomized trial of cytoreduction and hyperthermic intraperitoneal chemo-

therapy versus systemic chemotherapy and palliative surgery in patients with peritoneal carcinomatosis of colorectal cancer. J Clin Oncol. 2003;21(20):3737–43. https://doi.org/10.1200/JCO.2003.04.187.

84. Chicago Consensus Working Group. The Chicago consensus on peritoneal surface malignancies: management of colorectal metastases. Ann Surg Oncol. 2020;27(6):1761–7. https://doi.org/10.1245/s10434-020-08315-x.
85. Verwaal VJ, Bruin S, Boot H, van Slooten G, van Tinteren H. 8-year follow-up of randomized trial: cytoreduction and hyperthermic intraperitoneal chemotherapy versus systemic chemotherapy in patients with peritoneal carcinomatosis of colorectal cancer. Ann Surg Oncol. 2008;15(9):2426–32. https://doi.org/10.1245/s10434-008-9966-2.
86. Jayne DG, Fook S, Loi C, Seow-Choen F. Peritoneal carcinomatosis from colorectal cancer. Br J Surg. 2002;89(12):1545–50. https://doi.org/10.1046/j.1365-2168.2002.02274.x.
87. Glehen O, Kwiatkowski F, Sugarbaker PH, et al. Cytoreductive surgery combined with perioperative intraperitoneal chemotherapy for the management of peritoneal carcinomatosis from colorectal cancer: a multi-institutional study. J Clin Oncol. 2004;22(16):3284–92. https://doi.org/10.1200/JCO.2004.10.012.
88. Quenet F, Elias D, Roca L, Goere D, Ghouti L, Pocard M, Facy O, Arvieux C, Lorimier G, Pezet D, Marchal F, Loi V, Meeus P, De Forges H, Stanbury T, Paineau L, Glehen O. A UNICANCER phase III trial of hyperthermic intra-peritoneal chemotherapy for colorectal peritoneal carcinomatosis: PRODIGE 7. J Clin Oncol. 2018;36(abstr LBA3503.)
89. Klaver CEL, Wisselink DD, Punt CJA, et al. Adjuvant hyperthermic intraperitoneal chemotherapy in patients with locally advanced colon cancer (COLOPEC): a multicentre, open-label, randomised trial. Lancet Gastroenterol Hepatol. 2019;4(10):761–70. https://doi.org/10.1016/S2468-1253(19)30239-0.
90. Goere D, Glehen O, Quenet F, et al. Second-look surgery plus hyperthermic intraperitoneal chemotherapy versus surveillance in patients at high risk of developing colorectal peritoneal metastases (PROPHYLOCHIP-PRODIGE 15): a randomised, phase 3 study. Lancet Oncol. 2020;21(9):1147–54. https://doi.org/10.1016/S1470-2045(20)30322-3.
91. Elias D, Sideris L, Pocard M, et al. Results of R0 resection for colorectal liver metastases associated with extrahepatic disease. Ann Surg Oncol. 2004;11(3):274–80. https://doi.org/10.1245/aso.2004.03.085.
92. Hwang M, Jayakrishnan TT, Green DE, et al. Systematic review of outcomes of patients undergoing resection for colorectal liver metastases in the setting of extra hepatic disease. Eur J Cancer. 2014;50(10):1747–57. https://doi.org/10.1016/j.ejca.2014.03.277.
93. Treasure T, Farewell V, Macbeth F, et al. Pulmonary metastasectomy versus continued active monitoring in colorectal cancer (PulMiCC): a multicentre randomised clinical trial. Trials. 2019;20(1):718. https://doi.org/10.1186/s13063-019-3837-y.
94. Marudanayagam R, Ramkumar K, Shanmugam V, et al. Long-term outcome after sequential resections of liver and lung metastases from colorectal carcinoma. HPB (Oxford). 2009;11(8):671–6. https://doi.org/10.1111/j.1477-2574.2009.00115.x.
95. Miller G, Biernacki P, Kemeny NE, et al. Outcomes after resection of synchronous or metachronous hepatic and pulmonary colorectal metastases. J Am Coll Surg. 2007;205(2):231–8. https://doi.org/10.1016/j.jamcollsurg.2007.04.039.
96. Hadden WJ, de Reuver PR, Brown K, Mittal A, Samra JS, Hugh TJ. Resection of colorectal liver metastases and extra-hepatic disease: a systematic review and proportional meta-analysis of survival outcomes. HPB (Oxford). 2016;18(3):209–20. https://doi.org/10.1016/j.hpb.2015.12.004.
97. de Jong MC, Pulitano C, Ribero D, et al. Rates and patterns of recurrence following curative intent surgery for colorectal liver metastasis: an international multi-institutional analysis of 1669 patients. Ann Surg. 2009;250(3):440–8. https://doi.org/10.1097/SLA.0b013e3181b4539b.
98. Leung U, Gonen M, Allen PJ, et al. Colorectal cancer liver metastases and concurrent extrahepatic disease treated with resection. Ann Surg. 2017;265(1):158–65. https://doi.org/10.1097/SLA.0000000000001624.
99. Bruni D, Angell HK, Galon J. The immune contexture and immunoscore in cancer prognosis and therapeutic efficacy. Nat Rev Cancer. 2020;20(11):662–80. https://doi.org/10.1038/s41568-020-0285-7.
100. Moretto R, Corallo S, Belfiore A, et al. Prognostic impact of immune-microenvironment in colorectal liver metastases resected after triplets plus a biologic agent: a pooled analysis of five prospective trials. Eur J Cancer. 2020;135:78–88. https://doi.org/10.1016/j.ejca.2020.04.045.
101. Angell HK, Bruni D, Barrett JC, Herbst R, Galon J. The immunoscore: colon cancer and beyond. Clin Cancer Res. 2020;26(2):332–9. https://doi.org/10.1158/1078-0432.CCR-18-1851.
102. Mlecnik B, Tosolini M, Kirilovsky A, et al. Histopathologic-based prognostic factors of colorectal cancers are associated with the state of the local immune reaction. J Clin Oncol. 2011;29(6):610–8. https://doi.org/10.1200/JCO.2010.30.5425.
103. Spranger S, Bao R, Gajewski TF. Melanoma-intrinsic beta-catenin signalling prevents anti-tumour immunity. Nature. 2015;523(7559):231–5. https://doi.org/10.1038/nature14404.
104. Yu X, Zhu L, Liu J, Xie M, Chen J, Li J. Emerging role of immunotherapy for colorectal cancer with liver metastasis. Onco Targets Ther. 2020;13:11645–58. https://doi.org/10.2147/OTT.S271955.
105. Biller LH, Schrag D. Diagnosis and treatment of metastatic colorectal cancer: a review. JAMA. 2021;325(7):669–85. https://doi.org/10.1001/jama.2021.0106.
106. Lang H, Baumgart J, Heinrich S, et al. Extended molecular profiling improves stratification and prediction of survival after resection of colorectal liver metastases. Ann Surg. 2019;270(5):799–805. https://doi.org/10.1097/SLA.0000000000003527.
107. Brudvik KW, Jones RP, Giuliante F, et al. RAS mutation clinical risk score to predict survival after resection of colorectal liver metastases. Ann Surg. 2019;269(1):120–6. https://doi.org/10.1097/SLA.0000000000002319.
108. Jacome AA, Vreeland TJ, Johnson B, et al. The prognostic impact of RAS on overall survival following liver resection in early versus late-onset colorectal cancer patients. Br J Cancer. 2021;124(4):797–804. https://doi.org/10.1038/s41416-020-01169-w.
109. Kawaguchi Y, Kopetz S, Newhook TE, et al. Mutation status of RAS, TP53, and SMAD4 is superior to mutation status of RAS alone for predicting prognosis after resection of colorectal liver metastases. Clin Cancer Res. 2019;25(19):5843–51. https://doi.org/10.1158/1078-0432.CCR-19-0863.
110. Phallen J, Sausen M, Adleff V, et al. Direct detection of early-stage cancers using circulating tumor DNA. Sci Transl Med. 2017;9(403):eaan2415. https://doi.org/10.1126/scitranslmed.aan2415.
111. Nakayama I, Hirota T, Shinozaki E. BRAF mutation in colorectal cancers: from prognostic marker to targetable mutation. Cancers (Basel). 2020;12(11):3236. https://doi.org/10.3390/cancers12113236.

The Role of Surgery in Management of Gastric Cancer

39

Andrew J. Sinnamon, Jose M. Pimiento, and Robert E. Roses

Abstract

Gastric adenocarcinoma is an aggressive tumor with significant global burden. Incidence is decreasing in Western countries but gastric cancer remains a major cause of cancer death in East Asian and developing countries. Gastric adenocarcinomas have been historically subdivided according to dominant histology by the Lauren classification into gland-forming intestinal-type lesions and infiltrative, non-gland forming diffuse-type lesions. Loss of E-cadherin expression encoded by the CDH1 gene plays a major role in carcinogenesis of diffuse-type cancers while *Helicobacter pylori* infection contributes to intestinal-type cancers. Gastric adenocarcinoma has a proclivity for early invasion into lymphatic channels with rates of lymph node involvement in the range of 20–30% with submucosal or deeper involvement by the primary tumor. Adequate lymphadenectomy in potentially resectable cases has therefore been a topic of substantial study. A surgical approach accounting for the typical lymphatic drainage of gastric tumors has been well described by Japanese surgeons and D2 lymphadenectomy, encompassing surgical clearance of the extended regional basin, has increasingly been adopted worldwide. Multimodal therapy plays an important role in clinical stage II–III cancers, with perioperative chemotherapy gaining momentum in recent years following the publication of randomized trial data supporting this approach. The role of adjuvant chemoradiotherapy compared to adjuvant chemotherapy alone remains a subject of controversy. Therapy directed at the HER2 receptor and the programmed death receptor immune checkpoint pathway are under active investigation and show promise for appropriately selected patients.

A. J. Sinnamon · J. M. Pimiento
Moffitt Cancer Center, Tampa, FL, USA
e-mail: Andrew.Sinnamon@moffitt.org; Jose.Pimiento@moffitt.org

R. E. Roses (✉)
University of Pennsylvania, Philadelphia, PA, USA
e-mail: Robert.Roses@pennmedicine.upenn.edu

Learning Objectives

- Understand classic histopathologic classification of gastric adenocarcinoma into intestinal- and diffuse-type and correlation with molecular characteristics.
- Appreciate lymphatic drainage patterns of gastric tumors.
- Be able to define the extent of surgical lymphadenectomy and identify evidence supporting D2 versus D1 and D3 lymphadenectomy.
- Recognize the implications of the anatomic epicenter of a gastric tumor.
- Identify the options for multimodality therapy including perioperative chemotherapy, adjuvant chemoradiotherapy, adjuvant chemotherapy alone, as well as an emerging role for immune therapy.

39.1 Introduction

Gastric cancer—specifically adenocarcinoma—is an aggressive entity with a propensity for early lymphatic spread. It is one of the more common cancers worldwide, but there is marked variation in geographic distribution related to environmental and genetic risk factors. While the incidence continues to decrease in Western countries, including the United States and Europe, it is common relative to other cancers in East Asian countries, most notably Japan, Korea, and China, as well as in other developing countries. While understanding of the underlying biology continues to evolve, gastric cancers are generally classified into two histologic subtypes—diffuse and intestinal. Surgery plays a critical role in potentially curable gastric cancer as well as in select cases for palliative intent. The merits of multimodality therapy are

S. P. Leong et al. (eds.), *Cancer Metastasis Through the Lymphovascular System*,
https://doi.org/10.1007/978-3-030-93084-4_39

well supported by a growing body of evidence, with surgery, chemotherapy, radiotherapy, and immune therapy playing roles depending on the clinical scenario.

39.2 Histology and Pathogenesis

Gastric adenocarcinomas are commonly classified as either of two histologic subtypes according to the Lauren classification [1]. Intestinal-type tumors maintain a gland-forming morphology, while diffuse-type cancers lack cohesion between cancer cells and thus do not exhibit gland-forming morphology. The overarching difference between these subtypes is the tendency for the intercellular adhesion responsible for the formation of recognizable gland structures. This feature is maintained in intestinal-type tumors and lost in diffuse cancers. Lacking cohesion, cells from diffuse carcinomas tend to be more infiltrative into surrounding tissues, which may result in the clinical entity known as linitis plastica. Linitis plastica is characterized by a rigid, leathery consistency of the stomach on a macroscopic level. Moreover, signet cells are often seen within diffuse-type carcinomas.

The subdivision by histologic appearance of gastric cancers is now thought to reflect genetic pathways unknown to Lauren at the time of description in 1965. The E-cadherin ("epithelial cadherin") protein is encoded by the CDH1 gene and contributes to cell-to-cell adhesion and thus maintenance of normal epithelial architecture [2, 3]. Loss of E-cadherin protein function has been shown to result in abnormal tissue architecture with unregulated cell growth and invasion into adjacent tissues. A variety of genetic and epigenetic alterations can lead to a defect in CDH1 genetic function. A frequently observed mechanism of such loss of function is promoter region hypermethylation. Hereditary diffuse gastric cancer (HDGC) is a manifestation of these genetics. Families with HDGC typically carry a pathogenic variant in the CDH1 gene, although a number of genetic and epigenetic variants have been described which result in the same phenotype [4, 5]. Lifetime risk of developing diffuse-type gastric cancer is very high and often occurs at an early age; cumulative incidence is 70% among men and 56% among women, with an average age of onset of 38 years. Individuals with HDGC are also at increased risk of lobular breast cancer, with a cumulative incidence of 42% for women.

Intestinal-type gastric adenocarcinoma, in contrast to diffuse-type, is not thought to be related to a single identifiable genetic signature and is often associated with *Helicobacter pylori* infection. Anatomically, these cancers typically arise from the gastric body, fundus, and antrum. In fact, it is estimated that worldwide, 89% of non-cardia gastric cancers are attributable to *H. pylori*, making this pathogen a major global cause of cancer [6]. A multistep process beginning with infection and ending with gastric adenocarcinoma over the course of decades has been described [7]. This process begins with active superficial gastritis related to *H. pylori* infection. In a subset of individuals, this progresses to chronic atrophic gastritis with loss of normal gastric glandular tissue. The next step in this gastric carcinogenesis pathway is intestinal metaplasia, first resembling small intestinal epithelium and later resembling colonic epithelium. It is possible that during this sequence, the stomach may be colonized by bacteria more likely to produce carcinogenic nitrosamines. It should be noted that while this cascade of events is thought to result from *H. pylori* infection, a very small fraction of infected individuals develops gastric cancer. However, given the worldwide prevalence of *H. pylori* infection, the cumulative burden of disease is substantial.

Efforts have been made to define gastric cancers according to molecular signature as improvements in genomic technologies evolve. Pooling tumors with similar genetic and epigenetic patterns into subtypes may allow for better targeted, more effective treatment approaches. For example, The Cancer Genome Atlas has defined four molecular subtypes of gastric adenocarcinoma [8]. These include chromosomally unstable gastric cancers (~50% of gastric cancers), genomically stable gastric cancers (~20%), microsatellite unstable tumors (~20%), and EBV-associated tumors (10%). Chromosomally unstable cancers are typically intestinal-type by histology. Genetically stable tumors are frequently diffuse-type by histology, and CDH1-associated tumors make up about a third of these. Microsatellite unstable (MSI) tumors have a deficiency in the mismatch repair machinery (dMMR), which commonly results from epigenetic hypermethylation and silencing of gene expression. Inherited dMMR is known as Lynch syndrome, which most commonly includes colorectal and endometrial cancers but can be associated with gastric cancer as well.

Other molecular markers that have drawn particular interest because of their therapeutic implications include HER2 and PD-L1 overexpression. HER2 (human epidermal growth factor receptor 2) is a receptor tyrosine kinase in the EGFR family, and overexpression of HER2 is thought to drive tumor growth. HER2 overexpression is more commonly seen in proximal and intestinal-type adenocarcinomas [9]. Tumors with PD-L1 (ligand of programmed cell death receptor 1, PD-1) overexpression are capable of evading immune clearance [10, 11]. PD-L1 on the tumor cell binds PD1 on T-cells, thus causing T-cell inhibition, also known as immune checkpoint activation. This phenomenon is under active study in a variety of cancers and is of growing interest in gastric cancer as well.

Histologic depth of invasion defines T stage. In the eighth edition of the American Joint Committee on Cancer (AJCC) Staging Manual, T1 tumors are subdivided into T1a lesions which invade the lamina propria or muscularis mucosae and T1b lesions which invade the submucosa [12]. T2 tumors invade the muscularis propria and T3 tumors invade the subserosal connective tissue without invasion of adjacent structures or serosa. T4 tumors that penetrate the serosa only are classified as T4a lesions while invasion through the serosa and into adjacent structures are classified as T4b. T stage is correlated with the rate of lymphatic spread (N stage). This has been demonstrated in multiple surgical series. For example, Maruyama and colleagues meticulously dissected specimens from 1931 gastric resections and identified increasing rates of regional nodal involvement in a stepwise fashion for tumors with invasion into muscularis propria, serosa, and into adjacent structures [13]. Similarly, in an analysis of 920 gastric cancer patients undergoing resection, the rate of N+ disease was shown to increase with higher T stage [14]. Among patients with primary tumors confined to the mucosa (T1a), the rate of lymph node metastases was 15.0% (n = 3/20). This increased to 23.4% (n = 15/64) with T1b disease, 48.2% (n = 52/108) with T2 disease, 69.8% (n = 280/358) with T3 disease, and 75.7% (n = 212/280) with T4a disease. These series may be somewhat limited with respect to early gastric cancers because of underrepresentation of early tumors. As frequency of upper endoscopy has increased over time, earlier cancers with submucosal only invasion have been increasingly detected. Rate of lymph node metastases is approximately 10% in mucosal cancers (T1a) and 20–30% in submucosal (T1b) cancers [15, 16]. Independent risk factors for lymph node metastases include tumor size, depth of submucosal invasion, poor differentiation, and lymphovascular invasion. Depth of submucosal invasion can be further separated into sm1, sm2, and sm3 lesions, which represent invasion into the upper, middle, and lower third of the submucosal layer. Patients with multiple risk factors may have appreciable rates of lymph node metastases despite only submucosal invasion of the primary lesion [15].

39.3 Anatomic Considerations

As described previously, carcinogenesis is often related to the anatomic site of the tumor, with non-cardia cancers commonly resulting from *H. pylori* infection. In contrast, tumors of the gastric cardia and gastroesophageal junction (GEJ) may represent an overlap region, both in terms of anatomy and disease pathogenesis. Adenocarcinomas above this point are thought to arise from the intestinal metaplasia sequence seen in Barrett's esophagus, while adenocarcinomas arising below follow patterns seen in the development of gastric cancers.

The Siewert classification is a commonly used framework when considering GEJ adenocarcinomas [17]. Tumors are divided into three types based on the location of their epicenter. Siewert type I tumors arise from 1 to 5 cm above the anatomic esophagogastric junction. Siewert type II lesions straddle the junction and are located from 1 cm above to 2 cm below the GEJ. Lastly, Siewert type III tumors extend 2–5 cm below the GEJ. In their original studies, Siewert and colleagues suggested that the pattern of lymphatic drainage of type II and III tumors is primarily directed inferiorly, toward the paracardial, lesser curvature, and left gastric artery nodes [18]. This is consistent with analyses performed by Nigro and colleagues, which further demonstrated GEJ tumors with intramuscular invasion had a higher likelihood of spreading to abdominal lymph nodes along the left gastric artery and greater curvature [19].

39.4 Patterns of Lymphatic Drainage and Clinical Assessment

Lymphatic drainage of the stomach includes two different networks [20]. The intrinsic network is located between the muscularis mucosae and the submucosa, and the extrinsic network follows the arterial blood supply to the stomach, which defines the regional nodal basin. Gastric lymphatics are designated by the arteries they follow and often drain to lymph nodes along these arteries. The final lymph nodes collecting stomach lymph are the preaortic (also "celiac") nodes found in front of the aorta around the celiac trunk as it arises from the abdominal aorta. Lymph from the gastric and duodenal regions is collected from gastric, gastro-epiploic, pancreaticosplenic, pyloric, and hepatic nodes that pass lymph into the celiac nodes, intestinal lymph trunks, union of abdominal lymph trunks and finally, to the thoracic duct.

The perigastric lymph nodes are thought to be the first echelon nodes to harbor disease. These include the right and left paracardial nodes superiorly, nodes along the lesser and greater curvature, and the supra- and infrapyloric nodes. Much of the lymphatic drainage, particularly in the proximal stomach, coalesces around the left gastric, common hepatic, and splenic arteries, as well as those nodes within the splenic hilum and hepatoduodenal ligament. This lymphatic drainage paradigm has been described by the Japanese Gastric Cancer Association, which defines the gastric cancer lymph node stations [21]. However, it has been reported that lymph node metastases can "skip" nodal basins.

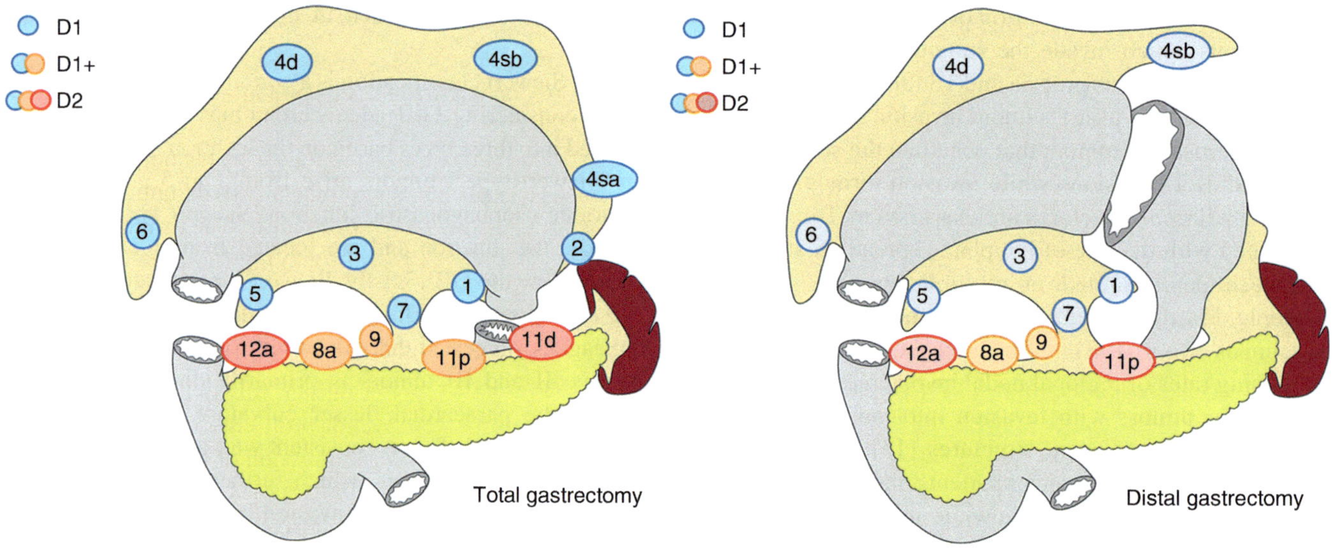

Figure courtesy of Japanese Gastric Cancer Association. Japanese gastric cancer treatment guidelines 2018 (fifth edition). Gastric Cancer 24, 1–21 (2021). https://doi.org/10.1007/s10120-020-01042-y. *Open access material.*

Gastric cancer may spread beyond regional lymph nodes to involve the superior mesenteric, middle colic, retropancreatic, pancreaticoduodenal, peripancreatic, para-aortic, and other retroperitoneal lymph nodes. These nodes are not considered part of the regional nodal basin. Thus, involvement of these lymph nodes constitutes M1 disease according to the Eighth Edition of the AJCC Cancer Staging Manual.

Involvement of the regional lymph nodes has important implications for use of adjuvant and perioperative multimodality therapy (to be discussed subsequently). Accurate lymph node staging is therefore critical in clinical practice and for treatment selection. A variety of diagnostic techniques exist with varying sensitivity and specificity for regional nodal staging. Endoscopic ultrasound (EUS), cross-sectional CT, and FDG-PET each may play a role in nodal staging for newly diagnosed gastric cancer.

Endoscopic ultrasound is an effective diagnostic modality for the assessment of depth of mural invasion of the primary tumor (T stage), more so than cross-sectional imaging, and is therefore recommended for routine staging by the National Comprehensive Cancer Network (NCCN) and European Society for Medical Oncology (ESMO) [22, 23]. Its accuracy for assessing nodal involvement (N stage), however, is less clear. In a large meta-analysis including 44 individual studies reporting diagnostic accuracy assessment of regional nodal disease in over 3500 patients, sensitivity and specificity of EUS were determined to be 83% and 67%, respectively [24]. Importantly, as EUS is a highly user-dependent technique due to both endoscopic capability and radiographic interpretation, an appreciable amount of inter-study heterogeneity was observed in this meta-analysis. Diagnostic accuracy of EUS is enhanced with the addition of fine-needle aspiration (FNA) of suspicious lymph nodes for histopathologic confirmation [25, 26].

Cross-sectional CT of the chest abdomen and pelvis is critical for staging in gastric cancer and is recommended in all patients by the NCCN. Its primary utility is in the detection of distant metastatic disease, as its accuracy in assessing T stage and N stage may be eclipsed by EUS. Small regional lymph nodes harboring metastatic disease may not be detected by CT, while conversely, enlarged nodes due to inflammatory response may be falsely identified as positive. Sensitivity, specificity, and overall accuracy for regional nodal staging are reported to be 86–97%, 65–76%, and 58–87%, respectively [27–29]. The sensitivity for CT for the detection of nodal metastases increases with greater N stage and number of metastatic lymph nodes. The use of FDG-PET may be helpful in individualized cases when CT and EUS are equivocal [30].

39.5 Fundamentals of Surgery for Gastric Cancer

Surgical resection remains a cornerstone of management for clinically localized gastric cancer. The goals of surgery are rooted in the biologic behavior of the disease; namely, adequate assessment for occult metastatic disease, excision of the primary tumor with adequate margins, and adequate lymphadenectomy, given the propensity for early regional spread. In patients with an inherited predisposition for the development of gastric cancer, mitigation of risk for clinically significant cancer is an additional goal.

39.5.1 Role of Staging Laparoscopy

Several modalities are useful in the proper staging of a newly diagnosed gastric cancer, including endoscopy with biopsy for molecular classification, EUS, cross-sectional CT of the chest, abdomen, and pelvis, and PET scan. While these studies are complementary to one another and provide a thorough locoregional and metastatic evaluation, one limitation is the possibility of failing to detect occult peritoneal disease. Laparoscopy has emerged as an effective remedy to this limitation. Peritoneal carcinomatosis that is occult on cross-sectional imaging can be found and biopsied at the time of laparoscopy. Additionally, peritoneal washings may be performed for cytologic examination. Presence of malignant cells on cytology constitutes M1 disease and generally precludes surgical resection for curative intent. For this reason, laparoscopy is best performed as a part of the staging workup and before initiation of systemic therapy. Generally, the yield of diagnostic laparoscopy—in terms of detecting previously unknown peritoneal disease—increases with clinical T stage (typically determined on EUS). The yield of laparoscopy has been reported to be up to 36%, with approximately 20% with macroscopic peritoneal disease seen on laparoscopy and an additional 10–15% with microscopic disease found on cytologic washings [31, 32]. The NCCN currently recommends consideration of laparoscopy with peritoneal cytology for lesions that are thought to be T1b or higher.

39.5.2 Extent of Gastric Resection

Curative intent resection for gastric adenocarcinoma is generally felt to require ample tissue resection to obtain a negative microscopic margin. Classically, a gross margin of approximately 5 cm was required to ensure adequate clearance. This requirement seems to stand for early gastric cancers but not for more advanced disease wherein the risk of systemic disease drives survival. For example, an analysis by the US Gastric Cancer Collaborative, focusing on distal gastric cancers, demonstrated 3.1–5.0 cm margins to be associated with the same favorable overall survival as >5.0 cm margins for stage I disease, whereas T stage and nodal involvement dictated overall survival and not margins >5.0 cm for stage II–III disease [33]. In an analysis from Memorial Sloan-Kettering, focusing on cancers of the GEJ, a margin length of 3.8 cm ex vivo was identified as a target for resection [34]. It should also be noted that the rate of incomplete resection remains appreciable even in the clinical trial setting as illustrated in the recent FLOT4 trial (R0 resection rate of 78% and 85% by arm, to be discussed further subsequently). Distal tumors are generally thought to be adequately resected with distal, or subtotal, gastrectomy. Reconstruction may be performed with gastrojejunostomy in a Billroth II or a Roux-en-Y configuration. More proximal tumors generally require a total gastrectomy with Roux-en-Y esophagojejunostomy. Proximal gastrectomies with esophagogastric reconstructions have been advocated in the East, but in general, poor functional outcomes have limited the adoption of this surgical technique in the West.

Optimal management of GEJ tumors remains controversial given overlapping but distinct recommendations for surgical and multimodality approaches in pure esophageal and gastric cancers. Therefore, when approaching these tumors, careful attention to endoscopy is critical in order to attempt to identify the epicenter of the cancer. Siewert type I and II tumors are more often treated as esophageal cancers, while type III lesions are more often treated as gastric cancers. In the latter, total gastrectomy is often warranted, usually with en bloc resection of the distal esophagus in order to obtain an adequate margin. Frozen section pathology may be obtained on the proximal margin in order to ensure adequate clearance. In cases where proximal clearance cannot be achieved without more extensive esophagectomy, a colonic interposition or jejunal conduit may be considered in order to obtain enteral continuity. However, one should carefully balance the likelihood of a curative operation for such poor disease biology with the higher morbidity of such an approach.

Minimally invasive resections utilizing laparoscopic and robotic techniques are being increasingly performed. Minimally invasive gastrectomy—as compared to open—appears to be associated with decreased length of hospitalization without compromise of the R0 resection rate or immediate or overall survival [35]. In an analysis using the National Cancer Data Base, comparing robotic and laparoscopic gastrectomy, the average number of lymph nodes sampled was higher among patients undergoing robotic resection (19.6 vs. 17.4 lymph nodes, respectively). Furthermore, the proportion of resections in which at least 15 lymph nodes were removed was higher in the robotic group compared with the laparoscopic group (64% vs. 58%, respectively; [36]). This may be related to improved visualization during surgery or maybe an epiphenomenon of care at expert centers.

39.5.3 Extent of Lymphadenectomy

A focus of surgical therapy for gastric cancer is the performance of adequate lymphadenectomy given the propensity for lymphatic spread early in the disease process. The extent of lymphadenectomy required during gastrectomy has been a topic of great study and debate, which is a consequence of an evolution of balancing aggressiveness of lymph node dissection with perioperative morbidity and mortality. Extent of lymphadenectomy is denoted with the prefix letter D and may be defined as D0, D1, or D2. In general, D1 and D2 resections correlate with the removal of N1 and N2 lymph node basins, respectively. These lymphadenectomies vary according to the extent of gastric resection being performed (e.g., total vs. distal gastrectomy). A more aggressive D3 resection has historically been proposed and prospectively studied in Japanese literature. However, it is no longer defined by the NCCN and is generally not currently part of the framework for surgical management.

A D1 lymphadenectomy for a total gastrectomy is defined by the NCCN as the removal of the nodes of the greater and lesser omenta, which consist of the nodes of the right and left cardia, greater and lesser curvature, and supra- and infrapyloric areas. The Japanese classification of a D1 resection additionally includes resection of the nodes of the left gastric artery (station 7 nodes for a complete removal of stations 1–7). A D1 resection during a distal gastrectomy may omit removal of the left paracardial nodes (station 2). A D2 resection is defined by the NCCN as a D1 resection with additional removal of the nodes of the left gastric, common hepatic, celiac, and splenic arteries, which corresponds to the Japanese classification of a D2 resection as removal of stations 1–9, 11, and 12. Again, for a distal gastrectomy, left paracardial nodes (station 2) may be omitted. A D3 lymphadenectomy includes peripancreatic, superior mesenteric, mesocolic, and para-aortic nodes.

The debate regarding the importance of D2 versus D1 lymphadenectomy for gastric cancer must be considered in historical context. Japanese series published in the 1970s and 1980s, which reported outcomes following D2 resections, generally demonstrated favorable oncologic outcomes with acceptable rates of morbidity and mortality. However, at that time, randomized prospective studies comparing this approach to D1 lymphadenectomy had not been performed. Western surgeon investigators sought to perform such trials in the 1980s. Initial results from these endeavors did not support D2 lymphadenectomy. In the UK Medical Research Council (MRC) Gastric Cancer Surgical Trial, patients were randomized to undergo gastric resection with D1 versus D2 lymphadenectomy, with the D2 group having significantly higher and unacceptably high overall postoperative mortality [37]. Importantly, rates of distal pancreatectomy and splenectomy were high among the D2 resection group, and in subsequent analysis, these combined operations were felt to be the primary cause of increased mortality compared to the more limited D1 resection. Contemporaneously, Bonenkamp and colleagues similarly randomized patients to undergo gastrectomy with D1 versus D2 lymphadenectomy in what is known as the Dutch trial [38]. Again, the D2 group had a significantly higher rate of complications and consequent mortality (10% vs. 4%) when compared to the D1 group. Rates of pancreatectomy and splenectomy were again high in the D2 arm (30% and 37%, respectively). Splenectomy was once again found to be associated with a higher rate of mortality. The results from these trials initially shed doubt on the routine performance of D2 lymphadenectomy for gastric cancer.

However, long-term follow-up data from these studies suggested oncologic benefit with extended lymphadenectomy. In the Dutch trial, gastric cancer-related death was significantly higher in the D1 group than the D2 group at 15 years of follow-up (48% vs. 37%, respectively; [39]). Similarly, local recurrence rates were higher with D1 compared to D2 resections (22% vs. 12%, respectively). Importantly, surgical techniques improved during this timeframe allowing for safer spleen-preserving methods of gastrectomy with D2 lymphadenectomy. Increased postoperative morbidity and mortality from the MRC and Dutch trials have been attributed to the performance of splenectomy and distal pancreatectomy in posttrial analyses, and more modern spleen- and pancreas-preserving resections should function to mitigate this increase [40]. To this point, the authors of the Dutch trial ultimately felt their results supported the use of D2 lymphadenectomy despite initial trial results.

These trials suggest a modest advantage of D2 lymphadenectomy given that an extended length of follow-up—up to 15 years—was required to overcome the immediate increased surgical morbidity and mortality. Importantly, the increased surgical morbidity and mortality of D2 resection are significantly lower in modern series than in these early trials. This was demonstrated by the Italian Gastric Cancer Study Group, which randomized patients to D1 or D2 resection in specialized centers and demonstrated no significant differences in immediate postoperative morbidity (12% vs. 18%, respectively) or 30-day mortality (3.0% vs. 2.2%, respectively; [41]).

Extended lymphadenectomy *beyond* D2, including the paraaortic nodes (D3 resection), has also been prospectively studied and shown to be of no benefit. The large Japan Clinical Oncology Group (JCOG) study 9501 randomly assigned patients undergoing curative intent gastrectomy to either D2 or D3 lymphadenectomy [42]. Postoperative morbidity was higher in the D3 group versus D2 group (28% vs. 21%, respectively), with no difference in operative mortality. Importantly, at 5 years, recurrence-free survival and overall survival were no differences between the D2 and D3 groups. This data would suggest D3 lymphadenectomy carries little or no oncologic benefit and is associated with a higher complication rate. A smaller randomized study by Yonemura and colleagues produced similar results [43].

Accounting for the data produced in the aforementioned studies, the NCCN currently recommends D1 lymphadenectomy plus the resection of the associated named vessel lymph nodes or what could be considered a modified D2 resection, with a goal of examining at least 16 lymph nodes. The NCCN emphasizes that D2 lymphadenectomy should be performed by experienced surgeons in high-volume centers.

Furthermore, the NCCN recommends against routine distal pancreatectomy and cautions against splenectomy except in instances with splenic tumor involvement or extensive hilar lymphadenopathy.

Individuals found to be carriers of a pathogenic CDH1 variant should be considered for prophylactic total gastrectomy given their substantial lifetime risk of developing gastric cancer. Goals of surgical therapy in this setting are to remove all gastric mucosa, which may involve intraoperative pathologic assessment of proximal and distal resection margins to ensure presence of esophageal and duodenal tissue, respectively. As gastrectomy in this setting is prophylactic, a D2 lymphadenectomy is not required. Endoscopy should be performed preoperatively to assess for the presence of a clinically significant cancer, as this would change management from a prophylactic approach to a therapeutic one. The NCCN currently recommends that gastrectomy be offered to these patients after the age of 18 years. Effective surveillance in these patients can be challenging but may be considered selectively [44].

39.5.4 Role of Sentinel Lymph Node Biopsy

While a "standard oncologic lymphadenectomy" is currently recommended during gastrectomy for gastric cancer, the potential role of sentinel lymph node biopsy to specifically identify draining lymph nodes from a gastric tumor has been studied. Sentinel lymph node biopsy could play a role for selective lymphadenectomy when performing endoscopic or local surgical resection of early gastric cancer to determine the need for formal gastrectomy with regional lymphadenectomy. A multicenter Japanese trial included 397 patients with early gastric cancers with performance of sentinel lymph node biopsy using both preoperative endoscopic injection of radiolabeled technetium and intraoperative injection of blue dye [45]. The sentinel lymph nodes were positive in 93% (53 of 57) of patients with involved lymph nodes, and the overall accuracy of sentinel lymph node staging was 99%. Mueller and colleagues have shown promising preliminary results in the West, also using a dual tracer technique to identify the sentinel lymph nodes [46].

39.6 Adjuncts to Surgery and Multimodality Therapy for Gastric Cancer

Given the propensity of gastric cancer to metastasize, adjuncts to surgery to treat and prevent local, regional, and distant recurrences play an important role in management. Use of systemic therapy, radiotherapy, and the combination of the two have been studied prospectively in both the neoadjuvant and adjuvant settings; however, the optimal regimen remains unclear. Furthermore, choice of therapy should be tailored to the individual patient after multidisciplinary discussion. Several landmark trials have established effective neoadjuvant, adjuvant, and perioperative regimens.

The use of perioperative chemotherapy was established with the Medical Research Council Adjuvant Gastric Infusional Chemotherapy—or MAGIC—trial [47]. In this landmark trial, patients with potentially curable at least clinical stage II gastric, distal esophageal, or GEJ adenocarcinomas were randomly assigned to undergo surgery alone or surgery with perioperative (both preoperative and postoperative) chemotherapy. The regimen used was epirubicin, cisplatin, and fluorouracil (ECF), with three cycles administered before surgery and three given after. The majority of the patients (~74%) had true gastric cancers. Overall survival was demonstrated to be superior for the patients undergoing perioperative therapy compared with surgery alone (36% vs. 23%, respectively, at 5 years). This was despite the fact that only 42% of the patients in the perioperative chemotherapy arm completed the full regimen, underscoring the frequent difficulties in completing *post*operative therapy in practice.

The chemotherapy regimen in the MAGIC trial (ECF) was subsequently compared to a regimen of docetaxel, oxaliplatin, and leucovorin/fluorouracil (FLOT) given in four cycles preoperatively and four cycles postoperatively ("FLOT4"; [48]). In this trial, the ECF arm also included patients treated with oral capecitabine in place of infusional fluorouracil (ECX). Approximately one-quarter of these patients had Siewert type 1 lesions and were treated with transthoracic esophagectomy. Median overall survival was longer among patients undergoing FLOT perioperative therapy than for ECF/ECX (50 months vs. 35 months, respectively). The FLOT regimen currently is preferred by the NCCN for the use of perioperative therapy in medically fit patients who are felt to be able to tolerate it.

Adjuvant chemotherapy is also an option for gastric cancer in addition to perioperative therapy. Multiple regimens are acceptable. The use of oral capecitabine and oxaliplatin (CAPOX) was established in the CLASSIC trial ("capecitabine and oxaliplatin adjuvant study in stomach cancer"), where patients who underwent curative gastrectomy with D2 lymphadenectomy for stages II–IIIB tumors were randomized to eight cycles of CAPOX versus surgery only. In the initial trial report, receipt of CAPOX was associated with improved progression-free survival [49]. With longer follow-up, improved overall survival was also seen with the adjuvant treatment compared with surgery alone (78% vs. 69%, respectively; [50]). In Japan and Europe, S-1 is an option for adjuvant therapy following gastrectomy with D2 lymphadenectomy for patients with stage II/III disease. This agent, not available in the United States, was shown to be associated with improved overall survival at 5 years when compared to surgery alone in the ACTS-GC trial [51].

The use of adjuvant chemoradiotherapy following resection of gastric and GEJ cancers was established with the Intergroup 0116 trial [52]. Patients were randomized to surgery alone or a course of fluorouracil-based chemoradiation. Overall survival and relapse-free survival were both improved with chemoradiotherapy. However, a major criticism of this trial is that surgical resections included fairly limited lymphadenectomies, with only 10% of patients undergoing a D2 resection, and over half of the patients undergoing a D0 resection. Several subsequent trials sought to answer whether combined chemoradiotherapy in the adjuvant setting provides benefit over chemotherapy alone, including the ARTIST, ARTIST 2, and CRITICS trials [53–55]. These trials included more extensive surgical lymphadenectomy than the Intergroup trial and thus may be more appropriate to assess whether an adjuvant locoregional therapy (radiation) provides benefit following adequate lymphadenectomy. However, none of these trials showed benefit of chemoradiotherapy over chemotherapy alone. It therefore remains an open question of whether chemoradiotherapy provides a significant benefit over chemotherapy alone in the adjuvant setting following an adequate clearance of regional lymph nodes with surgery.

The addition of therapy targeting the HER2 receptor (i.e., trastuzumab, pertuzumab) or inhibiting the PD1–PDL1 immune checkpoint (i.e., pembrolizumab, nivolumab) to systemic therapy regimens is currently under investigation. In a recent randomized phase II trial, the addition of trastuzumab and pertuzumab to perioperative FLOT therapy for HER2 overexpressing gastric cancers significantly improved pathologic complete response rates when compared to perioperative FLOT only (35% vs. 12%, respectively) and pathologic nodal negativity rates (68% vs. 39%, respectively) in the resected surgical specimens [56]. Several other trials are currently underway investigating the addition of HER2-directed therapy to systemic regimens for potentially resectable gastric cancers with HER2 overexpression. Similarly, while PD1–PDL1 inhibitor immunotherapies are approved for use in the metastatic setting, their role in the perioperative and adjuvant setting is currently under investigation.

39.7 Conclusions

Gastric cancer is an aggressive entity with early spread to lymph nodes. Surgical resection with an adequate lymphadenectomy is a cornerstone of therapy. Accounting for long-term follow-up data in the landmark surgical lymphadenectomy trials, a D2 or modified D2 lymphadenectomy is currently recommended. Multimodal therapy for clinical stage II–III disease has gained traction and is supported by several prospective randomized trials, with perioperative chemotherapy, adjuvant chemotherapy, and adjuvant chemoradiotherapy all accepted regimens.

Future Directions

- How will individualized treatment paradigms evolve as the understanding of molecular classification improves?
- What is the role of adjuvant chemoradiotherapy relative to adjuvant chemotherapy alone in the setting of an adequate surgical lymphadenectomy?
- What is the preferred systemic regimen for HER2-overexpressing tumors and what is the best sequence of therapy relative to surgery?
- What is the role of sentinel lymph node biopsy for selective lymphadenectomy in the setting of endoscopic or local resections for early gastric cancers?
- Do immunotherapies using PD1-PDL1 inhibitors provide benefit in the perioperative and adjuvant setting?

Authorship Disclosures None.

References

1. Lauren P. The two histological main types of gastric carcinoma: diffuse and so-called intestinal-type carcinoma. An attempt at a histo-clinical classification. Acta Pathol Microbiol Scand. 1965;64(31)
2. Berx G, Becker KF, Hofler H, van Roy F. Mutations of the human E-cadherin (CDH1) gene. Hum Mutat. 1998;12:226.
3. Graziano F, Humar B, Guilford P. The role of the E-cadherin gene (CDH1) in diffuse gastric cancer susceptibility: from the laboratory to clinical practice. Ann Oncol. 2003;14(12):1705.
4. Hansford S, et al. Hereditary diffuse gastric cancer syndrome: CDH1 mutations and beyond. JAMA Oncol. 2015;1(1):23.
5. van der Post RS, et al. Hereditary diffuse gastric cancer: updated clinical guidelines with an emphasis on germline CDH1 mutation carriers. J Med Genet. 2015;52(6):361–74.
6. Plummer M, et al. Global burden of gastric cancer attributable to Helicobacter pylori. Int J Cancer. 2015;136(2):487.
7. Correa P. Human gastric carcinogenesis: a multistep and multifactorial process—First American Cancer Society Award Lecture on Cancer Epidemiology and Prevention. Cancer Res. 1992;52(24):6735.
8. Cancer Genome Atlas Research Network. Comprehensive molecular characterization of gastric adenocarcinoma. Nature. 2014;513(7517):202–9.
9. Bang YJ, et al. Trastuzumab in combination with chemotherapy versus chemotherapy alone for treatment of HER2-positive advanced gastric or gastro-oesophageal junction cancer (ToGA): a phase 3, open-label, randomised controlled trial. Lancet. 2010;376(9742):687.

10. Dong H, et al. B7-H1, a third member of the B7 family, co-stimulates T-cell proliferation and interleu-kin-10 secretion. Nat Med. 1999;5:1365–9.
11. Topalian SL, Drake CG, Pardoll DM. Targeting the PD-1/B7-H1(PD-L1) pathway to activate anti-tumor immunity. Curr Opin Immunol. 2012;24:207–12.
12. Amin MB et al. (Eds.). AJCC cancer staging manual, 8th edn. Springer International Publishing: American Joint Commission on Cancer; 2017.
13. Maruyama K, et al. Lymph node metastases of gastric cancer. General pattern in 1931 patients. Ann Surg. 1989;210(5):596–602.
14. Zuo CH, et al. Characterization of lymph node metastasis and its clinical significance in the surgical treatment of gastric cancer. Mol Clin Oncol. 2014;2(5):821–6.
15. Tajima Y, et al. Risk factors for lymph node metastasis from gastric cancers with submucosal invasion. Ann Surg Oncol. 2010;17(6):1597–604.
16. Ohashi S, Okamura S, Urano F, Maeda M. Clinicopathological variables associated with lymph node metastasis in submucosal invasive gastric cancer. Gastric Cancer. 2007;10(4):241–50.
17. Siewert JR, Stein HJ. Classification of adenocarcinoma of the oesophagogastric junction. Br J Surg. 1998;85(11):1457–9.
18. Siewert JR, Feith M, Werner M, Stein HJ. Adenocarcinoma of the esophagogastric junction: results of surgical therapy based on anatomical/topographic classification in 1,002 consecutive patients. Ann Surg. 2000;232(3):353–61.
19. Nigro JJ, et al. Prevalence and location of nodal metastases in distal esophageal adenocarcinoma confined to the wall: implications for therapy. J Thorac Cardiovasc Surg. 1999;117(1):16–23.
20. Alexander JS, Ganta VC, Jordan PA, Witte MH. Gastrointestinal lymphatics in health and disease. Pathophysiology. 2010;17(4):315–35.
21. Japanese Gastric Cancer Association. Japanese gastric cancer treatment guidelines 2018, (5th edition). Gastric Cancer. 2020;24: 1–21.
22. National Comprehensive Cancer Network (NCCN). NCCN clinical practice guidelines in oncology; n.d.. https://www.nccn.org/professionals/physician_gls.
23. Smyth EC et al; ESMO Guidelines Committee. Gastric cancer: ESMO clinical practice guidelines for diagnosis, treatment and follow-up. Ann Oncol. 2016;27(suppl 5):v38–v49.
24. Mocellin S, Pasquali S. Diagnostic accuracy of endoscopic ultrasonography (EUS) for the preoperative locoregional staging of primary gastric cancer. Cochrane Database Syst Rev. 2015, 2015;(2):CD009944.
25. Chang KJ, et al. Endoscopic ultrasound-guided fine-needle aspiration. Gastrointest Endosc. 1994;40(6):694.
26. Pollack BJ, Chak A, Sivak MV Jr. Endoscopic ultrasonography. Semin Oncol. 1996;23(3):336.
27. D'Elia F, Zingarelli A, Palli D, Grani M. Hydro-dynamic CT preoperative staging of gastric cancer: correlation with pathological findings. A prospective study of 107 cases. Eur Radiol. 2000;10(12):1877.
28. Sohn KM, et al. Comparing MR imaging and CT in the staging of gastric carcinoma. AJR Am J Roentgenol. 2000;174(6):1551.
29. Yan C, et al. Value of multidetector-row computed tomography in the preoperative T and N staging of gastric carcinoma: a large-scale Chinese study. J Surg Oncol. 2009;100(3):205.
30. Yun M, Lim JS, Noh SH, Hyung WJ, Cheong JH, Bong JK, Cho A, Lee JD. Lymph node staging of gastric cancer using (18) F-FDG PET: a comparison study with CT. J Nucl Med. 2005 Oct;46(10):1582–8.
31. Bentrem D, et al. The value of peritoneal cytology as a preoperative predictor in patients with gastric carcinoma undergoing a curative resection. Ann Surg Oncol. 2005;12(5):347–53.
32. Ikoma N, et al. Yield of staging laparoscopy and lavage cytology for radiologically occult peritoneal carcinomatosis of gastric cancer. Ann Surg Oncol. 2016;23(13):4332–7.
33. Squires MH 3rd, et al. Is it time to abandon the 5-cm margin rule during resection of distal gastric adenocarcinoma? A multi-institution study of the U.S. Gastric Cancer Collaborative. Ann Surg Oncol. 2015;22(4):1243–51.
34. Barbour AP, et al. Adenocarcinoma of the gastroesophageal junction: influence of esophageal resection margin and operative approach on outcome. Ann Surg. 2007;246(1):1–8.
35. Sweigert PJ, et al. Total gastrectomy in patients with gastric adenocarcinoma: is there an advantage to the minimally invasive approach? Surgery. 2019;166(4):623–31.
36. Ryan S, et al. Robotic versus laparoscopic gastrectomy for gastric adenocarcinoma: propensity-matched analysis. Surg Innov. 2020;27(1):26–31.
37. Cuschieri A et al. Postoperative morbidity and mortality after D1 and D2 resections for gastric cancer: preliminary results of the MRC randomised controlled surgical trial. The Surgical Cooperative Group. Lancet 1996;347(9007):995–9.
38. Bonenkamp JJ, et al. Extended lymph-node dissection for gastric cancer. N Engl J Med. 1999;340:908–14.
39. Songun I, et al. Surgical treatment of gastric cancer: 15-year follow-up results of the randomised nationwide Dutch D1D2 trial. Lancet Oncol. 2010;11:439–49.
40. Cuschieri A, et al. Patient survival after D1 and D2 resections for gastric cancer: long-term results of the MRC randomized surgical trial. Br J Cancer. 1999;79(9–10):1522–30.
41. Degiuli M, et al. Morbidity and mortality in the Italian Gastric Cancer Study Group randomized clinical trial of D1 versus D2 resection for gastric cancer. Br J Surg. 2010;97(5):643–9.
42. Sasako M, et al. D2 lymphadenectomy alone or with para-aortic nodal dissection for gastric cancer. N Engl J Med. 2008;359(5):453–62.
43. Yonemura Y, et al. Randomized clinical trial of D2 and extended paraaortic lymphadenectomy in patients with gastric cancer. Int J Clin Oncol. 2008;13(2):132–7.
44. Blair VR, et al. Hereditary diffuse gastric cancer: updated clinical practice guidelines. Lancet Oncol. 2020;21(8):e386–97.
45. Kitagawa Y, et al. Sentinel node mapping for gastric cancer: a prospective multicenter trial in Japan. J Clin Oncol. 2013;31(29):3704–10.
46. Mueller CL, et al. Sentinel lymph node sampling for early gastric cancer-preliminary results of a north American prospective study. J Gastrointest Surg. 2019;23(6):1113–21.
47. Cunningham D, et al. Perioperative chemotherapy versus surgery alone for resectable gastroesophageal cancer. N Engl J Med. 2006;355(1):11–20.
48. Al-Batran SE, et al. Perioperative chemotherapy with fluorouracil plus leucovorin, oxaliplatin, and docetaxel versus fluorouracil or capecitabine plus cisplatin and epirubicin for locally advanced, resectable gastric or gastro-oesophageal junction adenocarcinoma (FLOT4): a randomised, phase 2/3 trial. Lancet. 2019;393(10184):1948–57.
49. Bang YJ, et al. Adjuvant capecitabine and oxaliplatin for gastric cancer after D2 gastrectomy (CLASSIC): a phase 3 open-label, randomised controlled trial. Lancet. 2012;379(9813): 315–21.
50. Noh SH, et al. Adjuvant capecitabine plus oxaliplatin for gastric cancer after D2 gastrectomy (CLASSIC): 5-year follow-up of an open-label, randomised phase 3 trial. Lancet Oncol. 2014;15(12):1389–96.
51. Sakuramoto S, et al. Adjuvant chemotherapy for gastric cancer with S-1, an oral fluoropyrimidine. N Engl J Med. 2007;357(18): 1810–20.

52. Macdonald JS, et al. Chemoradiotherapy after surgery compared with surgery alone for adenocarcinoma of the stomach or gastro-esophageal junction. N Engl J Med. 2001;345(10):725–30.
53. Lee J, et al. Phase III trial comparing capecitabine plus cisplatin versus capecitabine plus cisplatin with concurrent capecitabine radiotherapy in completely resected gastric cancer with D2 lymph node dissection: the ARTIST trial. J Clin Oncol. 2012;30(3):268–73.
54. Park SH, et al. A randomized phase III trial comparing adjuvant single-agent S1, S-1 with oxaliplatin, and postoperative chemoradiation with S-1 and oxaliplatin in patients with node-positive gastric cancer after D2 resection: the ARTIST 2 trial. Ann Oncol. 2021;32(3):368–74.
55. Cats A, et al. Chemotherapy versus chemoradiotherapy after surgery and preoperative chemotherapy for resectable gastric cancer (CRITICS): an international, open-label, randomised phase 3 trial. Lancet Oncol. 2018;19(5):616–28.
56. Hofheinz RD, et al. Perioperative trastuzumab and pertuzumab in combination with FLOT versus FLOT alone for HER2-positive resectable esophagogastric adenocarcinoma: final results of the PETRARCA multicenter randomized phase II trial of the AIO. J Clin Oncol. 2020;38S:ASCO #4502.

The Role of Surgery in Managing Primary and Metastatic Hepatopancreaticobiliary Cancers

40

Lyonell B. Kone, Christopher Javadi, Jessica M. Keilson, Shishir K. Maithel, George Poultsides, and Ajay V. Maker

Abstract

This chapter focuses on the effectiveness of surgery in the management of primary and metastatic hepatopancreaticobiliary (HPB) cancers. Tumor biology is reviewed, as is the role of lymphovascular spread and how surgery is tailored to treat each HPB cancer. The importance of surgical intervention to achieve long-term survival is addressed in the context of its significant limitations to control microscopic disease and metastasis. Recent and upcoming adjunctive therapies hold great promise for the management of HPB patients and are also addressed in the context of increasing eligibility and efficacy of surgical intervention for these complex cancers.

Learning Objectives

At the conclusion of this chapter, readers will be able to:

- describe the complex tumor biology and lymphovascular spread of hepatopancreaticobiliary cancers
- identify the benefits and limitations of surgical therapy for hepatopancreaticobiliary cancers
- recognize the benefits of adjunctive therapies in managing microscopic and distant metastatic disease in the setting of hepatopancreaticobiliary cancers

L. B. Kone
University of Illinois at Chicago—Metropolitan Group Hospitals, Chicago, IL, USA
e-mail: Lyonell.kone@aah.org

C. Javadi · G. Poultsides
Stanford University, Stanford, CA, USA
e-mail: javadi@stanford.edu; gpoultsides@stanford.edu

J. M. Keilson · S. K. Maithel
Emory University, Atlanta, GA, USA
e-mail: jessica.keilson@emory.edu; smaithe@emory.edu

A. V. Maker (✉)
Department of Surgery, University of California San Francisco, San Francisco, CA, USA
e-mail: ajay.maker@ucsf.edu

40.1 Introduction

Hepatopancreaticobiliary (HPB) cancers are defined by their complex and aggressive tumor biology. Our understanding of these inherent properties dictates surgical management as well as the need for adjuvant therapies to control disease progression and limit recurrence. In pancreaticobiliary cancers, the complexities of lymphovascular invasion (LVI) highlight both the need for aggressive surgical management and its limitations. As locoregional spread occurs via lymphatics for several malignancies, LVI may serve as a window into the dissemination of disease beyond its origin. In colorectal liver metastasis and hepatocellular carcinoma, tumor-specific molecular and genetic subclassification, as well as patient-specific factors such as organ dysfunction and functional status, guide our surgical approach. Surgical resection provides the only curative-intent option for these diseases; however, the efficacy of these approaches is limited by tumor-intrinsic factors that drive disease aggressiveness, underscoring the need for adjunctive therapies.

Below we will discuss, in a site-specific manner, tumor biology and disease progression, surgical approaches and their limitations, and the role of adjunctive therapies in managing disease progression and recurrence.

40.2 Gallbladder carcinoma

Gallbladder carcinoma (GBC) is a rare disease most often detected at advanced stages, with an estimated 5-year survival rate of 5–13%. The gallbladder is drained by a plexus of lymphatic vessels within the subserosa that runs along the anterior and posterior surfaces, following a similar distribution to the arterial system and ultimately emptying into lymphatic collecting ducts located on the inferior surface of the gallbladder. Ito and colleagues describe the lymphatic drainage of the gallbladder as following three distinct routes: (1) the cholecystoretropancreatic pathway, running along the

S. P. Leong et al. (eds.), *Cancer Metastasis Through the Lymphovascular System*,
https://doi.org/10.1007/978-3-030-93084-4_40

anterior and posterior surfaces of the common bile duct to converge at the retroportal nodes; (2) the cholecystoceliac pathway, through the hepatoduodenal ligament to the celiac nodes; and (3) the cholecystomesenteric pathway, running anterior to the portal vein toward the posterior pancreaticoduodenal nodes, all of which ultimately converge at the aortocaval and paraaortic nodes [1]. Lymph also drains directly into the lymphatics of the liver through pathways within the gallbladder fossa. As surgical resection confers the only chance of cure in GBC, radical resection is recommended, including regional lymphadenectomy dictated by lymphovascular drainage patterns. Hepatic resection of segments IVB and V is indicated for lesions with subserosal invasion [2], whereas simple cholecystectomy is sufficient for the treatment of T1a disease [3].

T stage has a strong correlation with LN positivity. Up to 20% of patients with T1b disease are found to be lymph node (LN) positive, underscoring the importance of completing regional lymphadenectomy at the time of resection even in the setting of superficial disease. While the value of resection in patients with known LN metastases is debatable, up to 20% 5-year survival rate is possible in patients with LN metastases following resection [4]. Regional and paraaortic lymphadenectomy does not improve outcomes in patients with the paraaortic disease and confers survival equivalent to that of distant metastatic disease [5]. However, among patients with regional N1 or N2 disease who underwent systematic lymphadenectomy, survival was equivalent to patients without nodal disease. Thus, resection remains standard of care for patients with locoregional disease.

In the setting of incidental GBC identified following cholecystectomy, T-stage has been largely associated with the presence of residual disease and thus decreased survival. However, prognostic risk scores incorporating LVI have demonstrated superior prediction of locoregional and distant disease compared to T-stage alone [6]. Therefore, the presence of LVI detected after the index cholecystectomy may potentially guide the use of preoperative therapy prior to re-resection for incidental GBC.

The eighth edition American Joint Committee on Cancer (AJCC) tumor TNM staging system defines N stage based on the number of metastatic LNs, an advancement from the seventh edition which stratified node-positive disease based on the location of involved LNs [7]. Resection and evaluation of at least six LNs, including hilar, celiac, periduodenal, peripancreatic, superior mesenteric, and posterior pancreatic head nodes, improve stage stratification. Distant LN involvement, such as paraaortic LNs, is still regarded as stage IV metastatic disease. A number of studies have demonstrated increased survival following regional LN dissection for node-positive GBC following R0 resection. Liu et al. showed improved stage stratification and disease-specific survival in node-positive patients to 33 months with a total LN count ≥ 6 compared to 15 months for a total LN count <6 [8]. However, the use of absolute LN count introduces a certain degree of bias as it relies on adequate LN retrieval during surgery and accurate histolopathologic examination. To mitigate this bias, some investigators have suggested the utilization of LN ratio to more accurately reflect disease burden and to better prognosticate following curative resection [9]. More recent studies have proposed a ratio of the number of positive LNs to the number of negative LNs, the log odds of positive LNs, as a more reliable marker of outcome than either LN ratio or AJCC N-stage [10].

Indeed, LVI and LN positivity strongly correlate with survival outcomes; however, the role of lymphadenectomy extends beyond prognostication and may provide therapeutic benefits. In patients with early-stage disease, where 24% of T1b and 45% of T2 lesions demonstrate LN involvement, radical liver resection or cholecystectomy alone with LN evaluation improves cancer-specific and overall survival over resection of the primary tumor alone [11]. This same finding is seen in patients with T3N0 disease; however, T3N1 and T4 disease do not demonstrate improved outcomes with radical resection [12]. Thus, while LN evaluation is a requisite factor in disease staging and prognosis, lymphadenectomy may provide therapeutic value for earlier stage disease.

40.3 Cholangiocarcinoma

Cholangiocarcinoma (CCA) is a rare malignancy of the biliary tract that also carries a poor prognosis and high mortality. Complete resection is the standard of care and remains the only potentially curative approach in highly selected patients. As in GBC, LN-positive disease is associated with worse survival outcomes. This is reflected in the AJCC staging system where LN-positive disease was considered stage IV in the seventh edition and is currently stage IIIb in the eighth edition.

Intrahepatic cholangiocarcinoma (ICCA) arises from the bile ducts within the hepatic parenchyma and accounts for approximately 20% of all CCA. LN metastasis is the most significant predictor of poor survival after resection [13], with up to 34% of patients with ICCA having LN-positive disease. The lymphatic system of the liver includes a deep lymphatic system, which is contained within the portal triad and communicates with the hilar and peripancreatic LNs or runs along the hepatic venous system toward the inferior vena cava to drain into the general lymphatic system, and a superficial lymphatic system, which runs within the subserosal connective tissue of the liver and drains into the general lymphatic system or distant LNs [14]. Tumors originating from the right hepatic lobe spread to hilar, periduodenal, and peripancreatic LNs, whereas left hepatic lobe tumors typically spread to hilar, inferior phrenic, and gastrohepatic LNs.

Though the current AJCC eighth edition offers no recommendation for the optimal number of LNs for accurate staging, a major limitation in LN evaluation for ICCA is that portal lymphadenectomy often contains less than three LNs [15]. Further, up to 13% of patients reported to have LN negative disease following resection have occult portal LN metastases on detailed pathologic reexamination [16]. This suggests that the rate of LN positivity is not only influenced by the extent of LN harvest but also by the detail with which LNs are evaluated following resection. As such, several authors have suggested LVI as an independent prognostic factor for reduced survival and a surrogate marker for aggressive tumor biology that can be utilized in the setting of inadequate LN sampling or as an indication for adjuvant chemotherapy [15].

Hilar cholangiocarcinoma (HCCA) accounts for 50–60% of CCA. Lymphatic drainage courses along the common hepatic artery to the celiac LNs, the common bile duct toward the posterior pancreatic head and paraaortic LNs, and the portal vein and superior mesenteric artery to drain into the superior mesenteric LNs [17]. Due to its infiltrative growth pattern, extrahepatic bile duct resection for HCCA is often associated with high positive margin rates, which carries a survival rate similar to that of metastatic disease [15].

Node-positive disease is an independent prognostic factor for both locoregional and distant recurrence among HCCA. The incidence of nodal involvement in patients with HCCA following resection is as high as 31–58% [17], and occurs most commonly around the common bile duct and portal vein. Retrieval of LNs along the hepatoduodenal ligament and posterior pancreatic head is considered the standard of care. LN involvement beyond these nodal basins is considered M1 metastatic disease and is a contraindication to resection.

Recurrence rates in patients with resected HCCA are estimated to be as high as 76% and have been demonstrated up to 8 years following resection [18]. Locoregional recurrence is typically the first site of disease recurrence, resulting from positive margins along the proximal or distal bile duct margins or within the peribiliary tissue. Tumor spread along the length of the bile duct or within the surrounding tissue often occurs through the lymphovascular system. As a result, standard of care includes hemi-hepatectomy dictated by the extent of tumor and Bismuth classification in order to clear any tissue with lymphovascular involvement and to increase R0 resection rates. There remains controversy in balancing postoperative mortality associated with more extensive resections, including vascular en bloc resections or hepaticopancreatoduodenectomy, against a reduction in locoregional recurrence rates [18]. In retrospective series, total hepatectomy with orthotopic liver transplant, the most extreme option to mitigate the incidence of residual disease, has been associated with improved recurrence-free and overall survival compared to resection alone [18, 19]. However, these patients are highly selected and receive 6 months of high-dose radiation and chemotherapy prior to transplantation, suggesting that decreased recurrence rates and improved survival outcomes may be in part a consequence of patient selection and inherent tumor biology, as well as this additional therapy received prior to transplantation.

Distal cholangiocarcinoma (DCCA) involves the distal extrahepatic, or intrapancreatic, portion of the bile duct. Similar to ICCA and HCCA, surgical resection is the only potentially curative option. As with other biliary malignancies, LN metastasis is a prognostic indicator of decreased survival. The frequency of LN-positive disease among patients with DCCA has been reported as high as 68% [20], thus assessment of at least 12 LNs is suggested for accurate staging [7]. As surgical resection of DCCA necessitates pancreatoduodenectomy, adequate lymphadenectomy is often achieved.

LVI has also been shown to be independently associated with decreased overall survival in DCCA. Thus, a new T-stage system that incorporates both tumor size >3 cm and presence of LVI has been proposed [21].

40.4 Adjuvant Therapy for Biliary Tract Disease

Surgical resection is the only potentially curative option for biliary tract cancers; however, these cancers are often detected at advanced stages and carry a high risk of recurrence, underscoring the need for effective adjuvant treatments.

The landmark BILCAP (Adjuvant Capecitabine for Biliary Tract Cancer) phase III randomized trial demonstrated improved overall survival with adjuvant capecitabine compared to observation alone in a prespecified intention-to-treat analysis adjusting for nodal status, disease grade, and sex [22]. Adjuvant capecitabine is therefore recommended for patients with resected biliary tract cancers for a duration of 6 months. However, gemcitabine plus cisplatin remains an effective and acceptable adjuvant treatment regimen.

The SWOG 0809 non-randomized phase II trial of capecitabine plus gemcitabine followed by capecitabine-based chemoradiation demonstrated comparable disease-

Table 40.1 Retrospective studies examining the impact on survival of extending pancreatic ductal adenocarcinoma resections to convert positive margins based on intraoperative frozen section to negative

Author	Institution	Period	Margin status	*N*	Median overall survival (mos)	*P* (R1-to-R0 vs. R1)
Hernandez (2009)	U. of South Florida	1995–2006	R1-to-R0	17	11	0.91
			R1	44	13	
Fatima (2010)	Mayo Clinic	1981–2007	R1-to-R0	57	18	NS
			R1	127	15	
Pang (2014)	Royal North Shore Hospital, Australia	2007–2012	R1-to-R0	7	16	0.31
			R1	67	23	
Mathur (2014)	Florida Hospital Tampa, U. of South Florida	1995–2012	R1-to-R0	40	14	0.19
			R1	110	12	
Kooby (2014)	Emory, U. of Wisconsin, Northwestern, U. of Cincinnati, U. of North Carolina, U. of Louisville, Vanderbilt, Washington U.	2000–2012	R1-to-R0	72	12	0.98
			R1	131	14	

free and median overall survival between R0 and R1 subgroups as well as tolerability in patients with extrahepatic CCA and GBC [23]. Based on the 2020 ASCO Clinical Practice Guidelines for resected biliary tract cancers, patients with microscopically positive margins following resection for extrahepatic CCA and GBC can be offered chemoradiotherapy [24]. While there remains limited evidence on the role of radiotherapy for biliary tract cancers, chemoradiation in this setting may be efficacious in controlling local and distant disease recurrence.

40.5 Pancreatic Adenocarcinoma

In the United States, pancreatic ductal adenocarcinoma (PDAC) is expected to become the second leading cause of cancer-related death by 2030 [25]. The 5-year survival over the last several decades has improved only modestly, from 3% in 1975, to 5% in 2005, and to 10% in 2010. For patients with localized disease, surgery followed by systemic chemotherapy remains the only potentially curative approach. However, even after a successful, curative-intent operation, up to a third of patients with PDAC die within a year from surgery [26]. These early recurrences, largely secondary to microscopic distant disease at the time of surgery, have fueled debates on the extent of surgical resection, particularly in regard to margins and the appropriate extent of lymphadenectomy. The biological issue with these tumors is that the first tumor driving mutations are often present two decades before the radiographic appearance of the tumor [27]. Thus by the time surgery is considered, tumor cells are often already circulating systemically, and the window upon which to select patients with localized disease has often passed. Extensive genomic analyses have revealed the existence of distinct molecular subtypes of pancreatic cancer that may enhance clinical stratification of patients for both personalized treatments and surgical selection in the future, but until this time, surgical treatment consists of pancreatectomy with lymphadenectomy.

40.5.1 Extending the Resection to Obtain Negative Margins

The status of the surgical margin is an important prognostic factor for PDAC and is considered positive if the tumor is within 1 mm of the cut surface. The uncinate margin is at the highest risk for residual disease for pancreatic head tumors and consists of the connective tissue separating the uncinate process from the superior mesenteric artery (SMA) and the neural and lymphatic tissue surrounding the celiac trunk (CA) [28].

Multiple studies have found improved survival in patients with microscopically negative tumor margins. However, multiple retrospective studies have examined the utility of re-resection of the pancreatic neck to achieve an R0 pancreatic margin, revealing no association with survival compared to an R1 resection. Patients with negative initial frozen sections had significantly longer survival than re-resection R0 and R1 patients, suggesting that a positive margin is a reflection of aggressive tumor biology rather than a driver of disease progression (Table 40.1).

Furthermore, the dominant pattern of recurrence, regardless of margin status, is distant disease. A large series from Belgium found that neither an R1 resection nor a tumor-free margin less than 1 mm resulted in a greater number of locoregional recurrences. In their study, only 8% of patients who recurred had an isolated local recurrence [29]. Collectively, these studies suggest that subclinical systemic disease determines the survival for the overwhelming majority of PDAC patients, not local recurrence.

40.5.2 Lymph Node Status Predicts Prognosis

Five-year survival after pancreaticoduodenectomy is approximately 30% for node-negative patients but only 10% for patients with nodal involvement, even if only one node is positive [30, 31]. Moreover, an increasing number of positive lymph nodes correlates with decreased survival. In a study of

811 patients who underwent pancreatoduodenectomy for PDAC between 2001 and 2012, median survival was 26.1 months for patients with two to three positive nodes, 21.9 months for cases with four to seven positive nodes, and 18.3 months for patients with greater than eight positive nodes [32].

An analysis of SEER data found that among pathologically node-negative patients, there is improved survival associated with an increasing number of examined nodes [33]. Examination of 15 lymph nodes appears to be optimal for accurate staging of node-negative adenocarcinoma of the pancreas after pancreaticoduodenectomy. The median survival for the group of patients with pN0 disease who had 15 or more nodes examined was 27 months, whereas the median survival for the group with fewer than 15 nodes examined was 19 months. Thus, adequate surgery includes a lymphadenectomy, the extent of which has been studied by multiple groups, although these studies remain steeped in institutional biases.

40.5.3 Randomized Trials of Extended Lymphadenectomy

Pathologic assessment after extended lymphadenectomy (LAD) revealed that 30–40% of patients with resectable pancreatic head tumors had involved lymph nodes outside the confines of a standard Whipple operation, typically in the region between the CA and the SMA. Early reports suggested that extended LAD of the CA, SMA, middle colic, and para-aortic nodes was associated with decreased local recurrence and increased survival [34–36]. However, since these early reports, extended LAD has been investigated in five prospective, randomized controlled trials (RCTs) (Table 40.2). Despite differences in study design and surgical outcomes, these RCTs showed that extended LAD does not improve R0 resection rates, overall recurrence rates, or long-term survival compared with standard pancreatectomy. Extended LAD, however, is associated with increased morbidity, particularly if the celiac artery and SMA are skeletonized circumferentially, resulting in high rates of postoperative diarrhea with malnutrition.

40.5.4 Systemic Therapy for PDAC

The more widespread use of systemic agents, including neoadjuvant and adjuvant chemotherapy, and radiation has become a major focus of cancer centers treating pancreatic ductal adenocarcinoma. ESPAC-4 found a prolonged survival in patients who received adjuvant gemcitabine and capecitabine, with 5-year survival rates of 30% for all patients, rising to 40% in those with an R0 resection margin, and to nearly 50% in node-negative patients [37]. The PRODIGE 24 trial demonstrated an impressive survival benefit of adjuvant oxaliplatin, irinotecan, fluorouracil, and leucovorin (mFOLFIRINOX), with a median OS reaching 54.4 months [38]. Other trials, ESPAC-5F and SWOG S1505, have shown a benefit of neoadjuvant or perioperative approaches using either mFOLFIRINOX or gemcitabine plus nanoalbumin-bound (nab)-paclitaxel [39, 40].

40.6 Pancreatic Neuroendocrine Tumors

Pancreatic neuroendocrine tumors (PNETs) are a rare group of neoplasms arising from pancreatic islet cells that have a wide spectrum of biologic behavior. Traditionally, PNETs are more indolent than their exocrine counterpart PDACs, but they can exert locoregional effects due to location and size, and can metastasize to regional lymph nodes and distant metastatic sites. The incidence of PNETs is rising due to increased use of cross-sectional imaging, endoscopic ultrasound, and nuclear medicine imaging (Ga-DOTATATE-PET), which has led to the detection of smaller tumors at an earlier stage [41].

The 2019 World Health Organization classification divides PNETs into well-differentiated NETs and poorly

Table 40.2 Randomized trials examining the impact of extended lymphadenectomy on post-resection survival for pancreatic ductal adenocarcinoma

Author	Institution	Period	Lymph-adenectomy	N	Nodes retrieved	Median OS (mos)	*P*	Morbidity
Pedrazzoli (1998)	Italy, Germany, USA	1991–1994	Standard	40	13	11	0.65	No difference
			Extended	41	20	16		
Yeo (2002)	Johns Hopkins	1996–2001	Standard	81	17	21	0.79	Extended LAD had increased LOS, DGE, fistulae, and infection rate
			Extended	82	28	20		
Farnell (2005)	Mayo Clinic	1997–2003	Standard	40	15	26	0.32	Extended LAD had increased transfusion rate and OR time
			Extended	39	34	19		
Nimura (2012)	Multicenter, Japan	2000–2003	Standard	51	13	20	0.12	Extended LAD had higher OR time and intra-operative blood loss
			Extended	50	40	14		
Jang (2014, 2017)	South Korea	2006–2009	Standard	83	17	19	0.39	Two extended patients had SMA aneurysms
			Extended	86	34	16		

LAD lymphadenectomy, *LOS* length of stay, *DGE* delayed gastric emptying, *OS* overall survival, *SMA* superior mesenteric artery differentiated neuroendocrine carcinomas (NECs) based on tumor morphology, with further classification of well-differentiated NETs into grades 1–3 based on the Ki67 proliferation index and mitotic index. Surgical resection remains the mainstay of treatment for functional tumors, larger localized tumors, and even for highly selected patients with metastatic disease.

40.6.1 Observation

The natural history of small, well-differentiated PNETs is commonly one of little to no growth nor metastases; therefore, many advocate for surveillance for tumors <2 cm. A study of 95 patients with small nonfunctioning PNETs (NF-PNETs) at the Mayo Clinic found a median growth rate of 0.03 mm/month. No patients progressed beyond resectability or developed metastatic disease during the observation period [42]. A European study found observation to be a safe strategy for small NF-PNETs in 327 patients, none of whom had any disease-related mortality [43].

40.6.2 Surgical Resection

Upfront surgical resection is often considered with curative intent for localized PNETs that are>2 cm, <grade 3, symptomatic, functional, or with evidence of regional lymph node metastases in patients who are fit for surgery. Population-based studies have demonstrated significantly improved survival for patients with localized tumors who have undergone surgical resection versus no surgery, but these studies have been limited by significant selection bias [44, 45]. A large retrospective review found that the association of survival with resection of NF-PNETs increased as tumor size increased, with a cutoff of >1.5 cm [46]. Most current guidelines recommend observation for asymptomatic NF-PNETs < 1 cm, resection for appropriate surgical candidates with PNETs >2 cm, and to individualize the decision for resection versus observation for PNETs 1–2 cm based on patient age, comorbidities, tumor location (head vs. body/tail), grade, radiographic characteristics (pancreatic duct dilatation, presence of calcifications, solid morphology), and patient preference.

40.6.3 Role of Lymphadenectomy

A recent meta-analysis of 13,374 patients with resected PNETs found a strong association between lymph node metastasis and decreased overall survival, with a hazard ratio of 3.87 for all PNETs and 4.98 for NF-PNETs [47]. The total number of nodes evaluated also impacts adequate staging. Using a multi-institutional database along with SEER data, Zhang et al. found that eight or more lymph nodes were required for optimal N staging [48].

Unlike PDAC, there are no prospective RCTs investigating a survival benefit of regional lymphadenectomy for PNETs. In a retrospective review of 2664 patients in the National Cancer Database, of whom 80.6% received lymphadenectomy with a median of nine nodes resected, overall survival was similar between patients who did and did not receive lymphadenectomy [49]. Furthermore, evidence for pancreas-preserving approaches (enucleation and central pancreatectomy) and spleen preservation, in which fewer lymph nodes are typically harvested, also suggests that a formal lymphadenectomy may not be necessary in all patients. Enucleations generally are performed for PNETs away from the pancreatic duct as a way to avoid transection. Central pancreatectomy may also be considered for tumors at the pancreatic neck or proximal body in order to preserve parenchyma.

Two meta-analyses have compared the outcomes of patients who underwent either standard resection or enucleation. Enucleation was associated with a significantly reduced incidence of exocrine and endocrine insufficiency, but the pancreatic fistula rate was higher [50, 51]. Similarly, a recent meta-analysis concluded that central pancreatectomy was associated with preserved pancreatic function compared with distal pancreatectomy, but at the expense of higher postoperative complications and length of stay [52]. Since neither enucleation nor central pancreatectomy typically removes regional lymph nodes, these pancreas-preserving approaches are recommended for tumors at low risk of lymph node metastasis, such as insulinomas or small NF-PNETs. Among patients who are candidates for pancreas-preserving resections, regional lymph node dissection and resection of suspicious lymph nodes on preoperative imaging should be considered. Radiographic criteria and scoring systems that include size, grade, and location, have been proposed to aid in identifying patients at highest risk for lymph node metastases who may benefit from lymphadenectomy [53].

40.7 Hepatocellular Carcinoma

Hepatocellular carcinoma (HCC) is the fourth leading cause of cancer-related death worldwide, and unlike many other cancers, the overall burden of disease has been rising over time. In the United States, the incidence of HCC has risen two- to threefold over the past three decades, attributed principally to hepatitis C infection in the mid-twentieth century, and the more recent rise in obesity-related non-alcohol steatohepatitis (NASH).

40.7.1 Molecular Subtype

HCC is the most common form of primary liver cancer, and usually occurs in the setting of chronic liver disease (>90%), namely viral hepatitis B and C infections, alcoholic cirrhosis, and non-alcoholic fatty liver disease (NAFLD). Dysplastic nodules, which consist of clonal hepatocytes with premalignant features, arise in patients with chronic liver disease. A series of genetic and epigenetic changes then promote a progression of low-grade to high-grade dysplasia to invasive cancer, also known as the dysplasia-carcinoma sequence. Telomerase reactivation is the principal molecular driver of this sequence. In the non-cirrhotic liver, however, HCC development arises in an adenoma-carcinoma sequence. The initial mutation in this sequence generally occurs in the proto-oncogene CTNNB1, which encodes for Β-catenin. TERT mutation also plays a role, but arises later in the sequence [54]. Given the complex molecular heterogeneity of HCC development, subclassification is challenging, but a popular classification scheme based on genetic signature divides HCC into proliferation and nonproliferation subtypes. The proliferation subtype is associated with TP53 mutation, chromosomal instability, hepatitis B infection, high AFP levels, high tumor grade, high vascular invasion, and worse clinical outcome. The nonproliferation subtype is associated with CTNNB1 mutation, immune exclusion, hepatitis C infection, alcohol use, low AFP levels, low tumor grade, low vascular invasion, and better clinical outcomes [55]. While these current subtypes do not yet dictate the surgical approach, they can help guide prognosis after surgery, and may help guide precision medicine approaches to systemic therapies.

40.7.2 Staging and Management

The surgical and clinical management of HCC is complex and depends on resource availability, local expertise, liver function, and patient condition. The Barcelona Clinic Liver Cancer (BCLC) staging system can help determine appropriate patients for surgery and takes into account tumor burden, liver function, and patient performance status (PS) (Figs. 40.1 and 40.2). Liver function is graded with the Child–Pugh system and PS is determined using the ECOG scale that ranges from 0 (fully active) to 4 (in bed or chair at all times with need for complete assistance with all tasks). In patients with limited and compensated liver disease, the goals of surgical resection are to remove HCC with negative margins, while preserving an adequate functional liver remnant. Therefore, major resections are often avoided in favor of parenchymal preserving techniques and performed either via an open laparotomy or minimally invasive technique. Techniques to increase the size of the functional liver remnant may be considered, e.g., portal vein embolization, though these are carefully considered on a patient-by-patient basis. As opposed to liver resection in non-cirrhotic cases, operations on patients with compensated cirrhosis are potentially complicated by portal hypertension, thrombocytopenia, bleeding, and poor

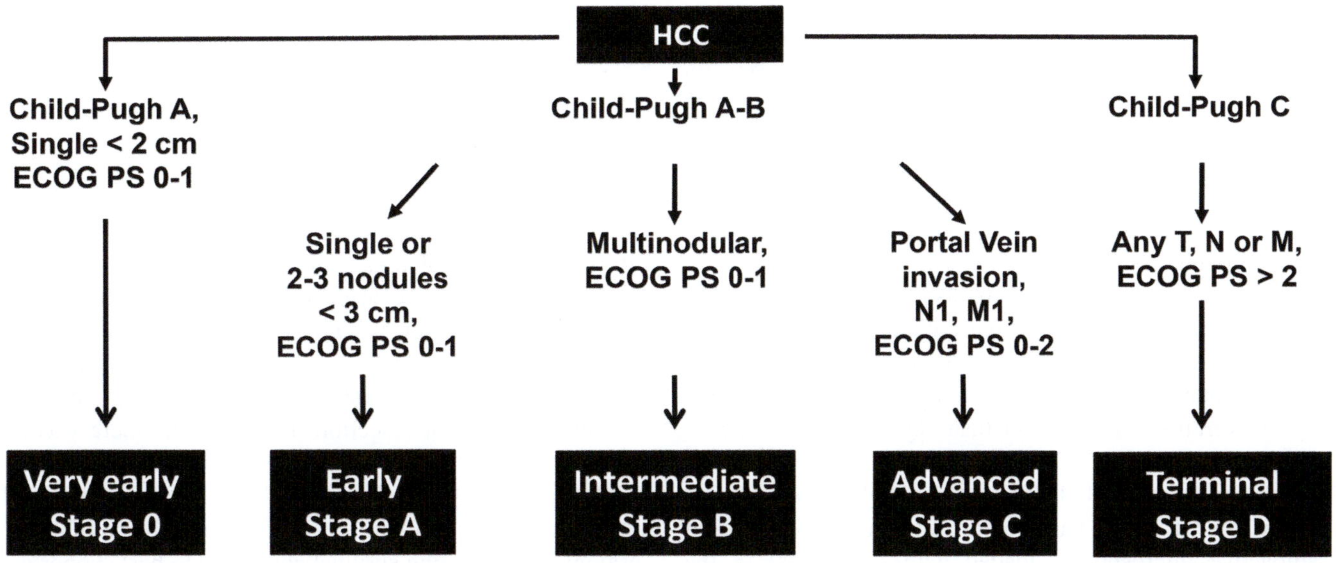

Fig. 40.1 BCLC HCC staging system. (*Adapted from "Diagnosis, Staging, and Management of Hepatocellular Carcinoma: 2018 Practice Guidance by the American Association for the Study of Liver Diseases" by Jorge A Marrero, Laura M Kulik, Claude B Sirlin, Andrew X Zhu, Richard S Finn, Michael M Abecassis, Lewis R Roberts, Julie K Heimbach, 2018, Hepatology, 68(2):723–750.* Reprinted with permission). *BCLC* Barcelona Clinic Liver Cancer, *HCC* Hepatocellular Carcinoma, *ECOG* Eastern Cooperative Oncology Group, *PS* Performance status

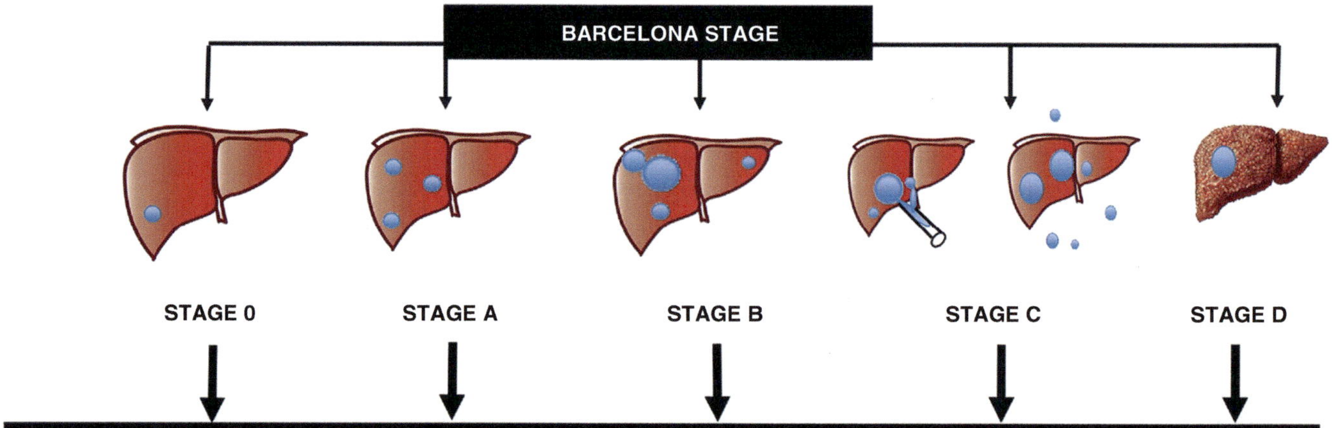

Level of Evidence	Stage 0	Stage A	Stage B	Stage C	Stage D
1	Resection		TACE	Sorafenib (1L) Lenvatinib (1L) Regorafenib (2L) Cabozantinib (2L)	
2	RFA MWA	Resection OLT RFA MWA TARE TACE SBRT	TARE Downsize OLT	Nivolumab (2L)	OLT BSC
3				TARE	

Fig. 40.2 Treatment recommendations according to BCLC Stage. (*Adapted from "Diagnosis, Staging, and Management of Hepatocellular Carcinoma: 2018 Practice Guidance by the American Association for the Study of Liver Diseases" by Jorge A Marrero, Laura M Kulik, Claude B Sirlin, Andrew X Zhu, Richard S Finn, Michael M Abecassis, Lewis R Roberts, Julie K Heimbach, 2018, Hepatology, 68(2):723–750.* Reprinted with permission). *BCLC* Barcelona Clinic Liver Cancer, *MWA* microwave ablation, *RFA* radiofrequency ablation, *OLT* Orthotopic liver transplant, *TARE* Transarterial radioembolization, *TACE* Transarterial chemoembolization, *SBRT* Stereotactic Body Radiation Therapy, *BSC* best supportive care, *1 L* first-line therapy, *2 L* second-line therapy

liver function of the remnant liver, which make technical considerations of parenchymal sparing techniques, hemostatic agents, and the use of ablation or energy devices all the more pertinent.

Patients with early HCC may also be eligible for orthotopic liver transplant. The Milan criteria identifies patients with a single lesion less than 5 cm, or two-to-three lesions each less than 3 cm, as eligible for transplant with an expected 5-year overall survival greater than 70%. Transarterial chemoembolization (TACE) consists of intra-arterial infusion of chemotherapy and embolizing particles of the tumor-feeding artery, while TARE is based on the intra-arterial infusion of radioactive microspheres loaded with Yttrium-90 in the tumor-feeding artery. There are currently no major differences in overall survival between the two modalities and both remain important tools in the armamentarium for control of unresectable or multifocal disease and as a bridge to transplant.

40.7.3 Systemic Therapy

Unfortunately, long-term survival, even after complete surgical resection or ablation, remains elusive, with half of patients experiencing recurrences within 18 months, and up to 80% recurring within 5 years. Thus, including the 30–40% of patients that present with advanced disease, most patients will not be eligible for curative intent surgical options such as liver transplant or resection. Before 2007, there was no effective systemic therapy for patients with advanced-stage HCC. However, since that time there has been regained enthusiasm due to the success of the multikinase inhibitor sorafenib in improving survival compared to best supportive care. In recent years, additional multikinase inhibitors (lenvatinib and regorafenib) have gained approval for treatment of advanced HCC, which have shown marginal advantages over sorafenib but without improving overall survival. Though these agents have redefined the role of systemic therapy, the survival benefit remains measured in weeks. With the great hope inspired by the success of PD-1/PD-L1 checkpoint blockade immunotherapy in other tumor types, immunotherapy for HCC has also been explored. Unfortunately, the PD-1 inhibitor nivolumab demonstrated no survival benefit compared to sorafenib in the CheckMate040 trial, and pembrolizumab was only evaluated in the second-line setting. Since vascular endothelial growth factor (VEGF) expression in the tumor microenvironment both promotes angiogenesis and enhances evasion of tumors from the immune system via increased expression of PD-1, combination therapy of atezolizumab (PD-L1 inhibitor) and bevacizumab (anti-VEGF) was explored and was recently demonstrated to be superior to first-line sorafenib. As of May 2020, this combination is recommended as first-line treatment for unresectable or metastatic HCC in appropriate patients [56].

40.8 Colorectal Liver Metastasis

Colorectal cancer (CRC) is the second leading cause of cancer-related death worldwide, with 20% presenting with synchronous colorectal liver metastases (CRLM) and another 20–30% eventually presenting with metachronous CRLM [57, 58]. In selected patients with liver-limited disease and an appropriate functional liver remnant, surgical resection has been demonstrated to provide a greater than 20% 10-year survival. The clinical risk score (CRS) was one of the first reliable predictors of survival in patients with CRLM that helped select patients for surgical resection. Predictors of poor outcome in this score include positive margins, extrahepatic disease, node-positive primary, disease-free interval <12 months, >1 hepatic tumor, largest tumor >5 cm, and CEA > 200 ng/mL [59]. More recently, the GAME score was created utilizing only preoperative data and was demonstrated to be a superior predictor of outcome compared to the CRS [60].

40.8.1 Molecular Subtype

CRC has been classified into three classical pathways. The most common is the chromosomal instability pathway, which follows the adenoma-carcinoma sequence starting with APC mutation, followed by KRAS mutation, and TP53 inactivation. Second is the microsatellite instability (MSI) pathway resulting from inactivation of mismatch repair genes (MLH1, MSH2). Third is the CpG island methylation phenotype (CIMP) pathway, which often occurs concurrently with the MSI pathway, and results in inactivating methylation of tumor suppressor genes (MINT1, MLH1, TIMP3) [61]. These pathways become important in consideration of CRLM not only in prognosis but also for treatment since recent trials have demonstrated durable antitumor responses to checkpoint blockade immunotherapy in MSI CRC.

Recent advances in high throughput sequencing technology may enable subclassification pertinent to directed treatment of metastatic CRC. Based on 1290 CRCs, Sadanandam et al. subclassified tumors into six groups associated with response to systemic therapies. For instance, stem-like and goblet-like subgroups may benefit from FOLFIRI combination chemotherapy, while the transit-amplifying subgroup may be more responsive to either cetuximab or cMET inhibitors depending on the sensitivity subtype [62]. High throughput analysis also improves the ability to predict long-term outcomes. Indeed, Pitroda et al. combined clinical data with integrated molecular analysis of colorectal liver metastases and were able to create three subgroup risk strata among patients who were eligible for surgical resection, with an estimated 10-year overall survival of 94% (low risk), 45% (intermediate risk), and 19% (high risk) [63]. Recent studies of BRAF, RAS, TP53, and SMAD4 mutations in resected CRLM have also determined pathways associated with early recurrence after surgery [64]. It is possible that these mutational signatures will be incorporated into the surgical decision-making process of liver metastatic disease in the future. Furthermore, it has become evident that the immunologic landscape of CRLM tumor microenvironment is a predictor of outcome. It has been demonstrated that the presence of activated and proliferating tumor-infiltrating lymphocytes is associated with improved survival [65]. Moreover, the presence of regulatory T-cells is associated with worse survival [66], and the ratio of cytotoxic T-cells to regulatory T-cells is predictive of outcomes. These studies will help determine patients most likely to benefit from new immunotherapeutic strategies that enhance T-cell infiltration and suppress immunosuppressive signals.

40.8.2 Surgical Management

Surgical resection of CRLM is currently the only curative-intent treatment and it provides the best chance of long-term survival in highly selected patients. Historically, a 1-cm margin was deemed necessary, however, it is now accepted that the width of the margin is not a predictor of outcome [67]. Thus, similar to HCC, parenchymal sparing hepatectomy is preferred to anatomical hepatectomies where appropriate. In patients with bilateral CRLM, surgical resection may still be pursued with specific strategies. In a two-staged hepatectomy (TSH), one side of the liver can be surgically cleared of tumor through resection and/or ablation, and portal venous ligation/embolization of the contralateral vessel is utilized to increase the functional liver remnant prior to resection of disease in the contralateral lobe. Another less common approach performed in highly selected patients is associating liver partition and portal vein ligation for staged hepatectomy (ALPPS) , which was first described in 2011 and appears to increase liver growth rate and volume in a shorter period of time. The LIGRO trial (NCT02215577) demonstrated a higher resection rate with ALLPS over TSH and equal safety profile, but these procedures remain isolated to a few specialized centers around the world due to their morbidity and mortality. In the setting of synchronous disease, the timing of hepatic resection with respect to the primary tumor resection is also an area of study. While there is no strong data to support liver-first versus colorectal-first versus combined resection strategies, some have advocated that combined resection be reserved for healthy, non-rectal cancer patients requiring minor hepatectomies. Interestingly, the SECA-2 trial (NCT01479608) in Norway reported a 5-year overall survival of 83% after liver transplant in patients with unresectable CRLM and good response to chemotherapy by RECIST criteria (10% decrease). However, these are highly selected patients and liver transplant for CRLM is uncommon in the United States, compounded by the low availability of liver donors.

40.8.3 Perioperative Therapy

Numerous studies have demonstrated that the timing of systemic chemotherapy (neoadjuvant or adjuvant) does not affect long-term survival in resectable CRLM. Furthermore, perioperative chemotherapy has not been shown to impact overall survival when compete extirpation of disease is performed. Regardless, systemic chemotherapy plays a significant role in the surgical treatment of CRLM by reducing the risk of relapse, assessing for chemosensitivity, testing the biology of disease, and by potentially converting an unresectable to a resectable tumor. This is at the expense of a potentially disappearing small tumor and chemotoxicity, specifically steatosis and steatohepatitis.

Patients with unresectable disease that receive chemotherapy to extend survival will benefit from regimens that include combination chemotherapy FOLFOX or FOLFIRI, anti-EGFR antibodies (Panitumumab, Cetuximab), and/or the addition of anti-VEGF (Bevacizumab) therapy. Intra-arterial therapies (TACE and TARE) are also adjuncts for the management of unresectable CRLM, with some reporting conversion to resectable CRLM as high as 52% [68]. This data builds on the long-standing data reporting the efficacy of Floxuridine-based (FUDR) hepatic artery infusion pumps in improving survival, response rates, and conversion to resectable disease in bilateral CRLM [69].

40.9 Conclusion and Perspective

Tumor biology, surgical management, and adjunctive therapy for HPB cancers were reviewed. Surgery alone, or as a part of an adjunctive treatment plan, remains the best and only curative-intent strategy to enable long-term survival. Survival after surgical resection is dependent on tumor biology and lymphovascular invasion/involvement. Advances in systemic, ablative, and radiation therapy are likely to both increase operative eligibility and improve outcomes.

Future Directions in this Field will Likely be Driven by Answers to the Following Questions

- What is the role of neoadjuvant chemotherapy in improving survival in gallbladder and biliary cancers?
- How can pancreatic cancer be diagnosed and treated at an earlier stage?
- Can we biologically determine which pancreatic neuroendocrine tumors, regardless of size, have malignant potential?
- Is it possible to stop or reverse the biologic pathways that lead to progressive dysplasia in cirrhotic livers?
- How can we harness the immune system to recognize and attack colon cancers that have metastasized to the liver?

Authors Disclosures Lyonell Kone, Christopher Javadi, Jessica Keilson, Shishir Maithel, George Poultsides, and Ajay Maker have no disclosures to report.

References

1. Ito M, Mishima Y, Sato T. An anatomical study of the lymphatic drainage of the gallbladder. Surg Radiol Anat. 1991;13(2):89–104. https://doi.org/10.1007/BF01623880.
2. Chijiiwa K, Nakano K, Ueda J, Noshiro H, Nagai E, Yamaguchi K, Tanaka M. Surgical treatment of patients with T2 gallbladder carcinoma invading the subserosal layer. J Am Coll Surg. 2001;192(5):600–7. https://doi.org/10.1016/s1072-7515(01)00814-6.
3. Lee SE, Jang JY, Lim CS, Kang MJ, Kim SW. Systematic review on the surgical treatment for T1 gallbladder cancer. World J Gastroenterol. 2011;17(2):174–80. https://doi.org/10.3748/wjg.v17.i2.174.
4. D'Angelica M, Dalal KM, DeMatteo RP, Fong Y, Blumgart LH, Jarnagin WR. Analysis of the extent of resection for adenocarcinoma of the gallbladder. Ann Surg Oncol. 2009;16(4):806–16. https://doi.org/10.1245/s10434-008-0189-3.

5. Kondo S, Nimura Y, Hayakawa N, Kamiya J, Nagino M, Uesaka K. Regional and para-aortic lymphadenectomy in radical surgery for advanced gallbladder carcinoma. Br J Surg. 2000;87(4):418–22. https://doi.org/10.1046/j.1365-2168.2000.01384.x.
6. Ethun CG, Postlewait LM, Le N, Pawlik TM, Buettner S, Poultsides G, Tran T, Idrees K, Isom CA, Fields RC, Jin LX, Weber SM, Salem A, Martin RC, Scoggins C, Shen P, Mogal HD, Schmidt C, Beal E, Hatzaras I, Shenoy R, Merchant N, Cardona K, Maithel SK. A novel pathology-based preoperative risk score to predict locoregional residual and distant disease and survival for incidental gallbladder cancer: a 10-institution study from the U.S. Extrahepatic Biliary Malignancy Consortium. Ann Surg Oncol. 2017;24(5):1343–50. https://doi.org/10.1245/s10434-016-5637-x.
7. Amin MB, Greene FL, Edge SB, Compton CC, Gershenwald JE, Brookland RK, Meyer L, Gress DM, Byrd DR, Winchester DP. The eighth edition AJCC cancer staging manual: continuing to build a bridge from a population-based to a more "personalized" approach to cancer staging. CA Cancer J Clin. 2017;67(2):93–9. https://doi.org/10.3322/caac.21388.
8. Liu GJ, Li XH, Chen YX, Sun HD, Zhao GM, Hu SY. Radical lymph node dissection and assessment: impact on gallbladder cancer prognosis. World J Gastroenterol. 2013;19(31):5150–8. https://doi.org/10.3748/wjg.v19.i31.5150.
9. Negi SS, Singh A, Chaudhary A. Lymph nodal involvement as prognostic factor in gallbladder cancer: location, count or ratio? J Gastrointest Surg. 2011;15(6):1017–25. https://doi.org/10.1007/s11605-011-1528-4.
10. Xiao Z, Shi Z, Hu L, Gao Y, Zhao J, Liu Y, Xu Q, Huang D. A new nomogram from the SEER database for predicting the prognosis of gallbladder cancer patients after surgery. Ann Transl Med. 2019;7(23):738. https://doi.org/10.21037/atm.2019.11.112.
11. Jensen EH, Abraham A, Habermann EB, Al-Refaie WB, Vickers SM, Virnig BA, Tuttle TM. A critical analysis of the surgical management of early-stage gallbladder cancer in the United States. J Gastrointest Surg. 2009;13(4):722–7. https://doi.org/10.1007/s11605-008-0772-8.
12. Foster JM, Hoshi H, Gibbs JF, Iyer R, Javle M, Chu Q, Kuvshinoff B. Gallbladder cancer: defining the indications for primary radical resection and radical re-resection. Ann Surg Oncol. 2007;14(2):833–40. https://doi.org/10.1245/s10434-006-9097-6.
13. Murakami Y, Uemura K, Sudo T, Hashimoto Y, Nakashima A, Kondo N, Sakabe R, Ohge H, Sueda T. Prognostic factors after surgical resection for intrahepatic, hilar, and distal cholangiocarcinoma. Ann Surg Oncol. 2011;18(3):651–8. https://doi.org/10.1245/s10434-010-1325-4.
14. Morine Y, Shimada M. The value of systematic lymph node dissection for intrahepatic cholangiocarcinoma from the viewpoint of liver lymphatics. J Gastroenterol. 2015;50(9):913–27. https://doi.org/10.1007/s00535-015-1071-2.
15. Patel SH, Kooby DA, Staley CA 3rd, Sarmiento JM, Maithel SK. The prognostic importance of lymphovascular invasion in cholangiocarcinoma above the cystic duct: a new selection criterion for adjuvant therapy? HPB (Oxford). 2011;13(9):605–11. https://doi.org/10.1111/j.1477-2574.2011.00335.x.
16. Ocuin LM, Bagci P, Fisher SB, Patel SH, Kooby DA, Sarmiento JM, Cardona K, Russell MC, Staley CA, Volkan Adsay N, Maithel SK. Discordance between conventional and detailed lymph node analysis in resected biliary carcinoma at or above the cystic duct: are we understaging patients? Ann Surg Oncol. 2013;20(13):4298–304. https://doi.org/10.1245/s10434-013-3143-y.
17. Li J, Zhou MH, Ma WJ, Li FY, Deng YL. Extended lymphadenectomy in hilar cholangiocarcinoma: what it will bring? World J Gastroenterol. 2020;26(24):3318–25. https://doi.org/10.3748/wjg.v26.i24.3318.
18. Groot Koerkamp B, Wiggers JK, Allen PJ, Besselink MG, Blumgart LH, Busch OR, Coelen RJ, D'Angelica MI, DeMatteo RP, Gouma DJ, Kingham TP, Jarnagin WR, van Gulik TM. Recurrence rate and pattern of perihilar cholangiocarcinoma after curative intent resection. J Am Coll Surg. 2015;221(6):1041–9. https://doi.org/10.1016/j.jamcollsurg.2015.09.005.
19. Ethun CG, Lopez-Aguiar AG, Anderson DJ, Adams AB, Fields RC, Doyle MB, Chapman WC, Krasnick BA, Weber SM, Mezrich JD, Salem A, Pawlik TM, Poultsides G, Tran TB, Idrees K, Isom CA, Martin RCG, Scoggins CR, Shen P, Mogal HD, Schmidt C, Beal E, Hatzaras I, Shenoy R, Cardona K, Maithel SK. Transplantation versus resection for hilar cholangiocarcinoma: an argument for shifting treatment paradigms for resectable disease. Ann Surg. 2018;267(5):797–805. https://doi.org/10.1097/SLA.0000000000002574.
20. Ito K, Ito H, Allen PJ, Gonen M, Klimstra D, D'Angelica MI, Fong Y, DeMatteo RP, Brennan MF, Blumgart LH, Jarnagin WR. Adequate lymph node assessment for extrahepatic bile duct adenocarcinoma. Ann Surg. 2010;251(4):675–81. https://doi.org/10.1097/SLA.0b013e3181d3d2b2.
21. Postlewait LM, Ethun CG, Le N, Pawlik TM, Buettner S, Poultsides G, Tran T, Idrees K, Isom CA, Fields RC, Krasnick B, Weber SM, Salem A, Martin RCG, Scoggins C, Shen P, Mogal HD, Schmidt C, Beal E, Hatzaras I, Vitiello G, Cardona K, Maithel SK. Proposal for a new T-stage classification system for distal cholangiocarcinoma: a 10-institution study from the U.S. Extrahepatic Biliary Malignancy Consortium. HPB (Oxford). 2016;18(10):793–9. https://doi.org/10.1016/j.hpb.2016.07.009.
22. Primrose JN, Fox RP, Palmer DH, Malik HZ, Prasad R, Mirza D, Anthony A, Corrie P, Falk S, Finch-Jones M, Wasan H, Ross P, Wall L, Wadsley J, Evans JTR, Stocken D, Praseedom R, Ma YT, Davidson B, Neoptolemos JP, Iveson T, Raftery J, Zhu S, Cunningham D, Garden OJ, Stubbs C, Valle JW, Bridgewater J, BILCAP Study Group. Capecitabine compared with observation in resected biliary tract cancer (BILCAP): a randomised, controlled, multicentre, phase 3 study. Lancet Oncol. 2019;20(5):663–73. https://doi.org/10.1016/S1470-2045(18)30915-X.
23. Ben-Josef E, Guthrie KA, El-Khoueiry AB, Corless CL, Zalupski MM, Lowy AM, Thomas CR Jr, Alberts SR, Dawson LA, Micetich KC, Thomas MB, Siegel AB, Blanke CD. SWOG S0809: a phase II intergroup trial of adjuvant capecitabine and gemcitabine followed by radiotherapy and concurrent capecitabine in extrahepatic cholangiocarcinoma and gallbladder carcinoma. J Clin Oncol. 2015;33(24):2617–22. https://doi.org/10.1200/JCO.2014.60.2219.
24. Shroff RT, Kennedy EB, Bachini M, Bekaii-Saab T, Crane C, Edeline J, El-Khoueiry A, Feng M, Katz MHG, Primrose J, Soares HP, Valle J, Maithel SK. Adjuvant therapy for resected biliary tract cancer: ASCO clinical practice guideline. J Clin Oncol. 2019;37(12):1015–27. https://doi.org/10.1200/JCO.18.02178.
25. Rahib L, Smith BD, Aizenberg R, Rosenzweig AB, Fleshman JM, Matrisian LM. Projecting cancer incidence and deaths to 2030: the unexpected burden of thyroid, liver, and pancreas cancers in the United States. Cancer Res. 2014;74(11):2913–21. https://doi.org/10.1158/0008-5472.CAN-14-0155.
26. Winter JM, Brennan MF, Tang LH, D'Angelica MI, Dematteo RP, Fong Y, Klimstra DS, Jarnagin WR, Allen PJ. Survival after resection of pancreatic adenocarcinoma: results from a single institution over three decades. Ann Surg Oncol. 2012;19(1):169–75. https://doi.org/10.1245/s10434-011-1900-3.
27. Yachida S, Jones S, Bozic I, Antal T, Leary R, Fu B, Kamiyama M, Hruban RH, Eshleman JR, Nowak MA, Velculescu VE, Kinzler KW, Vogelstein B, Iacobuzio-Donahue CA. Distant metastasis occurs late during the genetic evolution of pancreatic cancer. Nature. 2010;467(7319):1114–7. https://doi.org/10.1038/nature09515.
28. Evans DB, Farnell MB, Lillemoe KD, Vollmer C, Strasberg SM, Schulick RD. Surgical treatment of resectable and borderline resectable pancreas cancer: expert consensus statement. Ann Surg Oncol. 2009;16(7):1736–44. https://doi.org/10.1245/s10434-009-0416-6.

29. Van den Broeck A, Sergeant G, Ectors N, Van Steenbergen W, Aerts R, Topal B. Patterns of recurrence after curative resection of pancreatic ductal adenocarcinoma. Eur J Surg Oncol. 2009;35(6):600–4. https://doi.org/10.1016/j.ejso.2008.12.006.
30. Allen PJ, Kuk D, Castillo CF-D, Basturk O, Wolfgang CL, Cameron JL, Lillemoe KD, Ferrone CR, Morales-Oyarvide V, He J, Weiss MJ, Hruban RH, Gönen M, Klimstra DS, Mino-Kenudson M. Multi-institutional validation study of the American Joint Commission on cancer (8th edition) changes for T and N staging in patients with pancreatic adenocarcinoma. Ann Surg. 2017;265(1):185–91. https://doi.org/10.1097/SLA.0000000000001763.
31. Showalter TN, Winter KA, Berger AC, Regine WF, Abrams RA, Safran H, Hoffman JP, Benson AB, MacDonald JS, Willett CG. The influence of total nodes examined, number of positive nodes, and lymph node ratio on survival after surgical resection and adjuvant chemoradiation for pancreatic cancer: a secondary analysis of RTOG 9704. Int J Radiat Oncol Biol Phys. 2011;81(5):1328–35. https://doi.org/10.1016/j.ijrobp.2010.07.1993.
32. Strobel O, Hinz U, Gluth A, Hank T, Hackert T, Bergmann F, Werner J, Büchler MW. Pancreatic adenocarcinoma: number of positive nodes allows to distinguish several N categories. Ann Surg. 2015;261(5):961–9. https://doi.org/10.1097/SLA.0000000000000814.
33. Tomlinson JS, Jain S, Bentrem DJ, Sekeris EG, Maggard MA, Hines OJ, Reber HA, Ko CY. Accuracy of staging node-negative pancreas cancer: a potential quality measure. Arch Surg. 2007;142(8):767–23; discussion 773–764. https://doi.org/10.1001/archsurg.142.8.767.
34. Cubilla AL, Fortner J, Fitzgerald PJ. Lymph node involvement in carcinoma of the head of the pancreas area. Cancer. 1978;41(3):880–7.
35. Fortner JG. Regional resection of cancer of the pancreas: a new surgical approach. Surgery. 1973;73(2):307–20. papers3://publication/uuid/502CAAC8-A199-4028-9419-52EFF08CE121.
36. Ishikawa O, Ohigashi H, Imaoka S, Furukawa H, Sasaki Y, Fujita M, Kuroda C, Iwanaga T. Preoperative indications for extended pancreatectomy for locally advanced pancreas cancer involving the portal vein. Ann Surg. 1992;215(3):231–6. https://doi.org/10.1097/00000658-199203000-00006.
37. Neoptolemos JP, Palmer DH, Ghaneh P, Psarelli EE, Valle JW, Halloran CM, Faluyi O, O'Reilly DA, Cunningham D, Wadsley J, Darby S, Meyer T, Gillmore R, Anthoney A, Lind P, Glimelius B, Falk S, Izbicki JR, Middleton GW, Cummins S, Ross PJ, Wasan H, McDonald A, Crosby T, Ma YT, Patel K, Sherriff D, Soomal R, Borg D, Sothi S, Hammel P, Hackert T, Jackson R, Buchler MW, European Study Group for Pancreatic Cancer. Comparison of adjuvant gemcitabine and capecitabine with gemcitabine monotherapy in patients with resected pancreatic cancer (ESPAC-4): a multicentre, open-label, randomised, phase 3 trial. Lancet. 2017;389(10073):1011–24. https://doi.org/10.1016/S0140-6736(16)32409-6.
38. Conroy T, Hammel P, Hebbar M, Ben Abdelghani M, Wei AC, Raoul JL, Chone L, Francois E, Artru P, Biagi JJ, Lecomte T, Assenat E, Faroux R, Ychou M, Volet J, Sauvanet A, Breysacher G, Di Fiore F, Cripps C, Kavan P, Texereau P, Bouhier-Leporrier K, Khemissa-Akouz F, Legoux JL, Juzyna B, Gourgou S, O'Callaghan CJ, Jouffroy-Zeller C, Rat P, Malka D, Castan F, Bachet JB, Canadian Cancer Trials Group and the Unicancer-GI–PRODIGE Group. FOLFIRINOX or gemcitabine as adjuvant therapy for pancreatic cancer. N Engl J Med. 2018;379(25):2395–406. https://doi.org/10.1056/NEJMoa1809775.
39. Ahmad SA, Duong M, Sohal DPS, Gandhi NS, Beg MS, Wang-Gillam A, Wade JL 3rd, Chiorean EG, Guthrie KA, Lowy AM, Philip PA, Hochster HS. Surgical outcome results from SWOG S1505: a randomized clinical trial of mFOLFIRINOX versus gemcitabine/Nab-paclitaxel for perioperative treatment of resectable pancreatic ductal adenocarcinoma. Ann Surg. 2020;272(3):481–6. https://doi.org/10.1097/SLA.0000000000004155.
40. Ghaneh P, Palmer DH, Cicconi S, Halloran C, Psarelli EE, Rawcliffe CL, Sripadam R, Mukherjee S, Wadsley J, Al-Mukhtar A, Jiao LR, Wasan HS, Carter R, Graham JS, Ammad F, Evans J, Tjaden C, Hackert T, Buchler MW, Neoptolemos JP, European Study Group for Pancreatic Cancer (ESPAC). ESPAC-5F: four-arm, prospective, multicenter, international randomized phase II trial of immediate surgery compared with neoadjuvant gemcitabine plus capecitabine (GEMCAP) or FOLFIRINOX or chemoradiotherapy (CRT) in patients with borderline resectable pancreatic cancer. J Clin Oncol. 2020;38(15_suppl):4505. https://doi.org/10.1200/JCO.2020.38.15_suppl.4505.
41. Dasari A, Shen C, Halperin D, Zhao B, Zhou S, Xu Y, Shih T, Yao JC. Trends in the incidence, prevalence, and survival outcomes in patients with neuroendocrine tumors in the United States. JAMA Oncol. 2017;3(10):1335–42. https://doi.org/10.1001/jamaoncol.2017.0589.
42. Yohanathan L, Dossa F, St Germain AT, Golbafian F, Moulton C-A, McGilvray ID, Greig PD, Serra S, Wei AC, Jhaveri KS, Gallinger S, Cleary SP. Management and surveillance of non-functional pancreatic neuroendocrine tumours: retrospective review. Pancreatology. 2019;19(2):360–6. https://doi.org/10.1016/j.pan.2019.02.008.
43. Partelli S, Cirocchi R, Crippa S, Cardinali L, Fendrich V, Bartsch DK, Falconi M. Systematic review of active surveillance versus surgical management of asymptomatic small non-functioning pancreatic neuroendocrine neoplasms. Br J Surg. 2017;104(1):34–41. https://doi.org/10.1002/bjs.10312.
44. Assi HA, Mukherjee S, Kunz PL, Machiorlatti M, Vesely S, Pareek V, Hatoum H. Surgery versus surveillance for well-differentiated, nonfunctional pancreatic neuroendocrine tumors: an 11-year analysis of the National Cancer Database. Oncologist. 2020;25(2):e276–83. https://doi.org/10.1634/theoncologist.2019-0466.
45. Hill JS, McPhee JT, McDade TP, Zhou Z, Sullivan ME, Whalen GF, Tseng JF. Pancreatic neuroendocrine tumors: the impact of surgical resection on survival. Cancer. 2009;115(4):741–51. https://doi.org/10.1002/cncr.24065.
46. Zhang IY, Zhao J, Fernandez-Del Castillo C, Braun Y, Razmdjou S, Warshaw AL, Lillemoe KD, Ferrone CR. Operative versus non-operative management of nonfunctioning pancreatic neuroendocrine tumors. J Gastrointest Surg. 2016;20(2):277–83. https://doi.org/10.1007/s11605-015-3043-5.
47. Tanaka M, Heckler M, Mihaljevic AL, Probst P, Klaiber U, Heger U, Schimmack S, Büchler MW, Hackert T. Systematic review and metaanalysis of lymph node metastases of resected pancreatic neuroendocrine tumors. Ann Surg Oncol. 2020;21:9512. https://doi.org/10.1245/s10434-020-08850-7.
48. Zhang X-F, Xue F, Dong D-H, Lopez-Aguiar AG, Poultsides G, Makris E, Rocha F, Kanji Z, Weber S, Fisher A, Fields R, Krasnick BA, Idrees K, Smith PM, Cho C, Beems M, Lv Y, Maithel SK, Pawlik TM. New nodal staging for primary pancreatic neuroendocrine tumors: a multi-institutional and national data analysis. Ann Surg. 2019;274(1):e28–35. https://doi.org/10.1097/SLA.0000000000003478.
49. Mao R, Zhao H, Li K, Luo S, Turner M, Cai J-Q, Blazer D. Outcomes of lymph node dissection for non-metastatic pancreatic neuroendocrine tumors: a propensity score-weighted analysis of the National Cancer Database. Ann Surg Oncol. 2019;26(9):2722–9. https://doi.org/10.1245/s10434-019-07506-5.
50. Chua TC, Yang TX, Gill AJ, Samra JS. Systematic review and meta-analysis of enucleation versus standardized resection for small pancreatic lesions. Ann Surg Oncol. 2016;23(2):592–9. https://doi.org/10.1245/s10434-015-4826-3.
51. Hüttner FJ, Koessler-Ebs J, Hackert T, Ulrich A, Büchler MW, Diener MK. Meta-analysis of surgical outcome after enucleation

versus standard resection for pancreatic neoplasms. Br J Surg. 2015;102(9):1026–36. https://doi.org/10.1002/bjs.9819.
52. Dragomir MP, Sabo AA, Petrescu GED, Li Y, Dumitrascu T. Central pancreatectomy: a comprehensive, up-to-date meta-analysis. Langenbeck's Arch Surg. 2019;404(8):945–58. https://doi.org/10.1007/s00423-019-01829-3.
53. Poultsides GA, Huang LC, Chen Y, Visser BC, Pai RK, Jeffrey RB, Park WG, Chen AM, Kunz PL, Fisher GA, Norton JA. Pancreatic neuroendocrine tumors: radiographic calcifications correlate with grade and metastasis. Ann Surg Oncol. 2012;19(7):2295–303. https://doi.org/10.1245/s10434-012-2305-7.
54. Craig AJ, von Felden J, Garcia-Lezana T, Sarcognato S, Villanueva A. Tumour evolution in hepatocellular carcinoma. Nat Rev Gastroenterol Hepatol. 2020;17(3):139–52. https://doi.org/10.1038/s41575-019-0229-4.
55. Villanueva A. Hepatocellular carcinoma. N Engl J Med. 2019;380(15):1450–62. https://doi.org/10.1056/NEJMra1713263.
56. Finn RS, Qin S, Ikeda M, Galle PR, Ducreux M, Kim TY, Kudo M, Breder V, Merle P, Kaseb AO, Li D, Verret W, Xu DZ, Hernandez S, Liu J, Huang C, Mulla S, Wang Y, Lim HY, Zhu AX, Cheng AL, Investigators IM. Atezolizumab plus Bevacizumab in unresectable hepatocellular carcinoma. N Engl J Med. 2020;382(20):1894–905. https://doi.org/10.1056/NEJMoa1915745.
57. Bray F, Ferlay J, Soerjomataram I, Siegel RL, Torre LA, Jemal A. Global cancer statistics 2018: GLOBOCAN estimates of incidence and mortality worldwide for 36 cancers in 185 countries. CA Cancer J Clin. 2018;68(6):394–424. https://doi.org/10.3322/caac.21492.
58. Leporrier J, Maurel J, Chiche L, Bara S, Segol P, Launoy G. A population-based study of the incidence, management and prognosis of hepatic metastases from colorectal cancer. Br J Surg. 2006;93(4):465–74. https://doi.org/10.1002/bjs.5278.
59. Fong Y, Fortner J, Sun RL, Brennan MF, Blumgart LH. Clinical score for predicting recurrence after hepatic resection for metastatic colorectal cancer: analysis of 1001 consecutive cases. Ann Surg. 1999;230(3):309–18; discussion 318–321. https://doi.org/10.1097/00000658-199909000-00004.
60. Margonis GA, Sasaki K, Gholami S, Kim Y, Andreatos N, Rezaee N, Deshwar A, Buettner S, Allen PJ, Kingham TP, Pawlik TM, He J, Cameron JL, Jarnagin WR, Wolfgang CL, D'Angelica MI, Weiss MJ. Genetic And Morphological Evaluation (GAME) score for patients with colorectal liver metastases. Br J Surg. 2018;105(9):1210–20. https://doi.org/10.1002/bjs.10838.
61. Markowitz SD, Bertagnolli MM. Molecular origins of cancer: molecular basis of colorectal cancer. N Engl J Med. 2009;361(25):2449–60. https://doi.org/10.1056/NEJMra0804588.
62. Sadanandam A, Lyssiotis CA, Homicsko K, Collisson EA, Gibb WJ, Wullschleger S, Ostos LC, Lannon WA, Grotzinger C, Del Rio M, Lhermitte B, Olshen AB, Wiedenmann B, Cantley LC, Gray JW, Hanahan D. A colorectal cancer classification system that associates cellular phenotype and responses to therapy. Nat Med. 2013;19(5):619–25. https://doi.org/10.1038/nm.3175.
63. Pitroda SP, Khodarev NN, Huang L, Uppal A, Wightman SC, Ganai S, Joseph N, Pitt J, Brown M, Forde M, Mangold K, Xue L, Weber C, Segal JP, Kadri S, Stack ME, Khan S, Paty P, Kaul K, Andrade J, White KP, Talamonti M, Posner MC, Hellman S, Weichselbaum RR. Integrated molecular subtyping defines a curable oligometastatic state in colorectal liver metastasis. Nat Commun. 2018;9(1):1793. https://doi.org/10.1038/s41467-018-04278-6.
64. Kawaguchi Y, Kopetz S, Newhook TE, De Bellis M, Chun YS, Tzeng CD, Aloia TA, Vauthey JN. Mutation status of RAS, TP53, and SMAD4 is superior to mutation status of RAS alone for predicting prognosis after resection of colorectal liver metastases. Clin Cancer Res. 2019;25(19):5843–51. https://doi.org/10.1158/1078-0432.CCR-19-0863.
65. Maker AV, Ito H, Mo Q, Weisenberg E, Qin LX, Turcotte S, Maithel S, Shia J, Blumgart L, Fong Y, Jarnagin WR, DeMatteo RP, D'Angelica MI. Genetic evidence that intratumoral T-cell proliferation and activation are associated with recurrence and survival in patients with resected colorectal liver metastases. Cancer Immunol Res. 2015;3(4):380–8. https://doi.org/10.1158/2326-6066.CIR-14-0212.
66. Katz SC, Bamboat ZM, Maker AV, Shia J, Pillarisetty VG, Yopp AC, Hedvat CV, Gonen M, Jarnagin WR, Fong Y, D'Angelica MI, DeMatteo RP. Regulatory T cell infiltration predicts outcome following resection of colorectal cancer liver metastases. Ann Surg Oncol. 2013;20(3):946–55. https://doi.org/10.1245/s10434-012-2668-9.
67. Pawlik TM, Scoggins CR, Zorzi D, Abdalla EK, Andres A, Eng C, Curley SA, Loyer EM, Muratore A, Mentha G, Capussotti L, Vauthey JN. Effect of surgical margin status on survival and site of recurrence after hepatic resection for colorectal metastases. Ann Surg. 2005;241(5):715–22; discussion 722-714. https://doi.org/10.1097/01.sla.0000160703.75808.7d.
68. Datta J, Narayan RR, Kemeny NE, D'Angelica MI. Role of hepatic artery infusion chemotherapy in treatment of initially unresectable colorectal liver metastases: a review. JAMA Surg. 2019;154(8):768–76. https://doi.org/10.1001/jamasurg.2019.1694.
69. Thiels CA, D'Angelica MI. Hepatic artery infusion pumps. J Surg Oncol. 2020;122(1):70–7. https://doi.org/10.1002/jso.25913.

Part X

Cancer Metastasis to the Lung

Chinmayee Potti, Hassan Nasser
Andrew Popoff and Zane Hammoud

Lung cancer kills more people than breast, colon, and prostate cancer combined. The American Cancer Society estimates that for 2020, 229,000 Americans will be diagnosed with lung cancer and roughly 136,000 deaths will be attributed to the disease. The majority of lung cancers present at advanced stages or are metastatic (stage IV) at the time of diagnosis. The prognosis in advanced disease is dramatically worse than when it is detected early. Lung cancer screening programs, now well established in Western Europe, the United States, Canada, and elsewhere, aim to shift this survival curve by detecting more early, curable disease. The penetrance of lung cancer screening, however, remains low at present.

Knowledge of the anatomy, pathophysiology, and management of advanced and metastatic disease is crucial for appropriate, guideline-directed clinical care. This chapter provides an overview of anatomy, presentation, and prognostic implications of metastatic disease, and management of various clinical scenarios that cancer clinicians may encounter. Less common clinical situations, such as the management of oligometastatic disease, as well as combined modality and palliative approaches to the care of metastatic disease, are also discussed. Other primary malignancies, such as colorectal, renal cell, and uterine cancers with metastases to the lung, and pulmonary metastasectomy are also briefly addressed.

Scientific advances in the treatment of advanced and metastatic disease, such as targeted therapies directed at specific tumor genetics, as well as immunotherapy, either alone or in concert with conventional chemotherapy, have strengthened the armamentarium of clinicians who routinely care for patients with metastatic disease. It is likely that as more refined therapeutics become available, such as chimeric antigen receptor T-cell (CAR T-cell) therapy for lung cancer and other solid tumors, the arsenal of treatment options will continue to evolve. This, and other novel modalities, remains an area of intense interest in the future to have an even greater impact on the treatment of a disease that carries what has traditionally been a dismal prognosis.

Metastatic Disease of the Lung

41

Z. Hammoud, A. Popoff, Chinmayee Potti, and H. Nasser

Abstract

Lung cancer remains a malignancy in which patients with metastatic disease have a much poorer prognosis compared to those without metastatic disease. With the majority of patients presenting with advanced or metastatic disease, knowledge of the process of metastasis and clinical management is critical. With advancements in therapy, numerous guideline changes have occurred in the past decade. The main aim of this chapter is to shed light on some of the less commonly answered questions about lung cancer metastasis. The chapter discusses the processes behind lung cancer metastasis and the tendency for certain histological types to metastasize to certain organs. The chapter then focuses on diagnosing and managing lung cancer metastasis to lymph nodes and to distant organs, with emphasis on small cell versus non-small cell lung carcinoma. The chapter ends with a brief description of oncological indications for pulmonary metastasectomy.

Learning Objectives

- Relevant clinical anatomy of the lung
- Lymph node mapping
- Understanding lung metastases
  - Incidence and implications
  - Morbidity and mortality
  - Histological tropism of lung metastases to certain organs based
  - Molecular basis of metastasis to and from the lung
    Why does lung cancer metastasize to certain organs, and why do tumors from elsewhere metastasize to the lungs?
  - Sites of lung cancer metastasis
    Lymph nodes
    Hematogenous to other organs
  - Diagnosing and staging metastatic lung cancer
  - Management of lung cancer metastasis
    Non-small cell lung cancer
    - Treatment options
    - Oligometastasis

    Small cell lung cancer
  - Surgical management of metastasis to the lung: Pulmonary metastasectomy
    Oncological indications and recommendations

41.1 Introduction

Lung cancer is a major public health problem throughout the world. Worldwide, lung cancer is the most common cancer, both in incidence and mortality. In the United States, lung cancer kills more people than breast, colon, and prostate cancer combined. Lung cancer is the leading cause of cancer death in both men and women. The American Cancer Society estimates that for 2020, 229,000 Americans will be diagnosed with lung cancer and roughly 136,000 deaths will be attributed to the disease [3]. The majority of lung cancers present at advanced stages or are metastatic (Stage IV) at the time of diagnosis, thereby limiting curative treatment options. Despite notable improvements in detection and advances in treatment, overall 5-year survival remains dismal. Like other cancers, the prognosis in advanced disease is dramatically worse than

Z. Hammoud (✉)
Department of thoracic surgery Ascension Providence Hospital, Southfield, MI, USA

Department of surgery, Wayne State University School of Medicine, Detroit, MI, USA
e-mail: Zane.hammoud@ascension.org

A. Popoff · C. Potti · H. Nasser
Henry Ford Hospital, Detroit, MI, USA
e-mail: apopoff2@hfhs.org; cpotti@hfhs.org; hnasser2@hfhs.org

S. P. Leong et al. (eds.), *Cancer Metastasis Through the Lymphovascular System*,
https://doi.org/10.1007/978-3-030-93084-4_41

when it is detected early. Therefore, efforts aimed at earlier detection have been undertaken. Lung cancer screening programs, now well established in Western Europe, the United States, Canada, and elsewhere aim to shift this survival curve by detecting more early, curable disease. The penetrance of lung cancer screening, however, remains low at present.

Tobacco use is well established as the predominant risk factor for the development of lung cancer, accounting for up to 90% of lung cancers [3]. While not a main topic of this chapter, no discussion of lung cancer can be considered complete without mention of its main risk factor. Both the number of cigarettes and the duration of smoking contribute to the risk of lung cancer. Furthermore, tobacco use is associated with certain histologies of lung cancer. For instance, there is a strong association between tobacco use and small cell lung cancer while nonsmokers tend to be afflicted predominantly by adenocarcinoma. Efforts at curbing tobacco use, such as widespread smoking cessation programs, have been met with limited success. Other causes of lung cancer include secondhand smoke exposure, exposure to radon gas as well as other inhaled chemicals, and asbestos exposure among others.

Knowledge of the anatomy, pathophysiology, and management of advanced and metastatic disease is crucial for appropriate, guideline-directed clinical care. This chapter provides an overview of anatomy, the presentation, prognostic implications of metastatic disease, and management of various clinical scenarios cancer clinicians may encounter. Accurate staging of the patient with lung cancer is crucial to the management and a brief discussion of staging is undertaken in the chapter. Less common clinical situations, such as the management of oligometastatic disease, as well as combined modality and palliative approaches to the care of metastatic disease are also discussed. Other primary malignancies, such as colorectal, renal cell, and uterine cancers with metastases to the lung, and pulmonary metastasectomy are also briefly addressed. Surgical management of various scenarios is discussed. Indications for other therapies are also briefly mentioned.

Scientific advances in the treatment of advanced and metastatic disease, such as targeted therapies directed at specific tumor genetics, as well as immunotherapy, either alone or in concert with conventional chemotherapy, have strengthened the armamentarium of clinicians who routinely care for patients with metastatic disease. Agents directed at the epidermal growth factor receptor (EGFR) and tyrosine kinase inhibitors (TKI) have shown efficacy in the treatment of advanced and/or metastatic disease. It is likely that as more refined therapeutics become available, such as chimeric antigen receptor T-cell (CAR T-cell) therapy for lung cancer and other solid tumors, the arsenal of treatment options will continue to evolve. This, and other novel modalities, remain an area of intense interest in the future to have an even greater impact on the treatment of a disease that carries what has traditionally been a dismal prognosis.

41.2 Section I

41.2.1 Lung and Lymph: Anatomy

41.2.1.1 Clinical Anatomy of the Lung

The functional design of the thorax facilitates the complex process of respiration—the exchange of oxygen and carbon dioxide. Each lung is covered by pleura, a serous membrane arranged as a closed invaginated sac. The parietal pleura is constituted by costovertebral, diaphragmatic, cervical (pleural dome), and mediastinal pleura. The visceral pleura is closely adherent to the lung and follows the interlobar fissures to eventually continue over the hila as the parietal pleura. The right lung has three lobes (upper, middle, and lower) and two fissures (oblique and horizontal). The left lung has two lobes (upper and lower) separated by the oblique fissure. The lobes are further divided into bronchopulmonary segments, with each segment having its own bronchus and blood supply.

The lungs have a dual blood supply via the pulmonary and bronchial vessels. Arising from the right ventricle, the main pulmonary artery divides into left and right branches at the left of midline, lying above the left main bronchus. Further branches tend to lie close to the segmental bronchi and their branches. The bronchial circulation consists of a paired group of left bronchial branches arising directly from the descending thoracic aorta, and a right bronchial branch arising from a right-sided posterior intercostal artery, most commonly the third posterior intercostal artery [1]. Through an extensive capillary plexus, the pulmonary arterial system is responsible for alveolar gas exchange whereas the high-pressure bronchial circulation, which is a part of the systemic circulation, is responsible for the nutrient supply of the parenchyma, visceral pleura, and tracheobronchial tree. While the pulmonary vein tributaries are constituted by draining capillaries of bronchial and pulmonary arteries, the bronchial veins drain the larger bronchi.

Lying beneath the visceral pleura, lymphatics in the interlobular septa drain proximally toward the hilum, containing smooth muscle and valves as they get larger. On occasion, lymphatics can bypass lung lymph nodes to enter the mediastinum, likely the basis of "skip metastases" [1]. Similarly, while bronchial lymphatics usually remain ipsilateral, they can cross to the contralateral mediastinum in front of lower trachea and carina. Located either within the parenchyma or underneath the visceral pleura, intrapulmonary lymph nodes are further connected to bronchopulmonary lymph nodes which are found at the bifurcations of segmental, lobar bronchi, and bifurcations of the pulmonary artery. These lymph nodes are connected to the lobar lymph nodes, which in turn are connected to the mediastinal lymph nodes. Mediastinal lymph nodes consist of anterior (prevascular), tracheobronchial, paratracheal, and posterior nodes in the posterior mediastinum.

41.2.1.2 Lymph Node Mapping

Numerous studies have mapped the lymphatic drainage of various lung segments (see Fig. 41.1). Right upper lobe tumors usually drain to the right paratracheal lymph nodes; left upper lobe tumors drain to the periaortic and subaortic lymph nodes; and middle and lower lobe tumors drain to the subcarinal and right paratracheal lymph nodes. "Skip metastases," where there is direct drainage to the mediastinal lymph nodes without the involvement of the hilar and interlobar lymph nodes, occur in up to 25% of cases [1].

41.2.2 Understanding Lung Metastasis

41.2.2.1 Incidence and Implications

The 5-year survival rate in metastatic lung cancer (5%) is much lower than colon (64.5%), breast (89.6%), and pros-

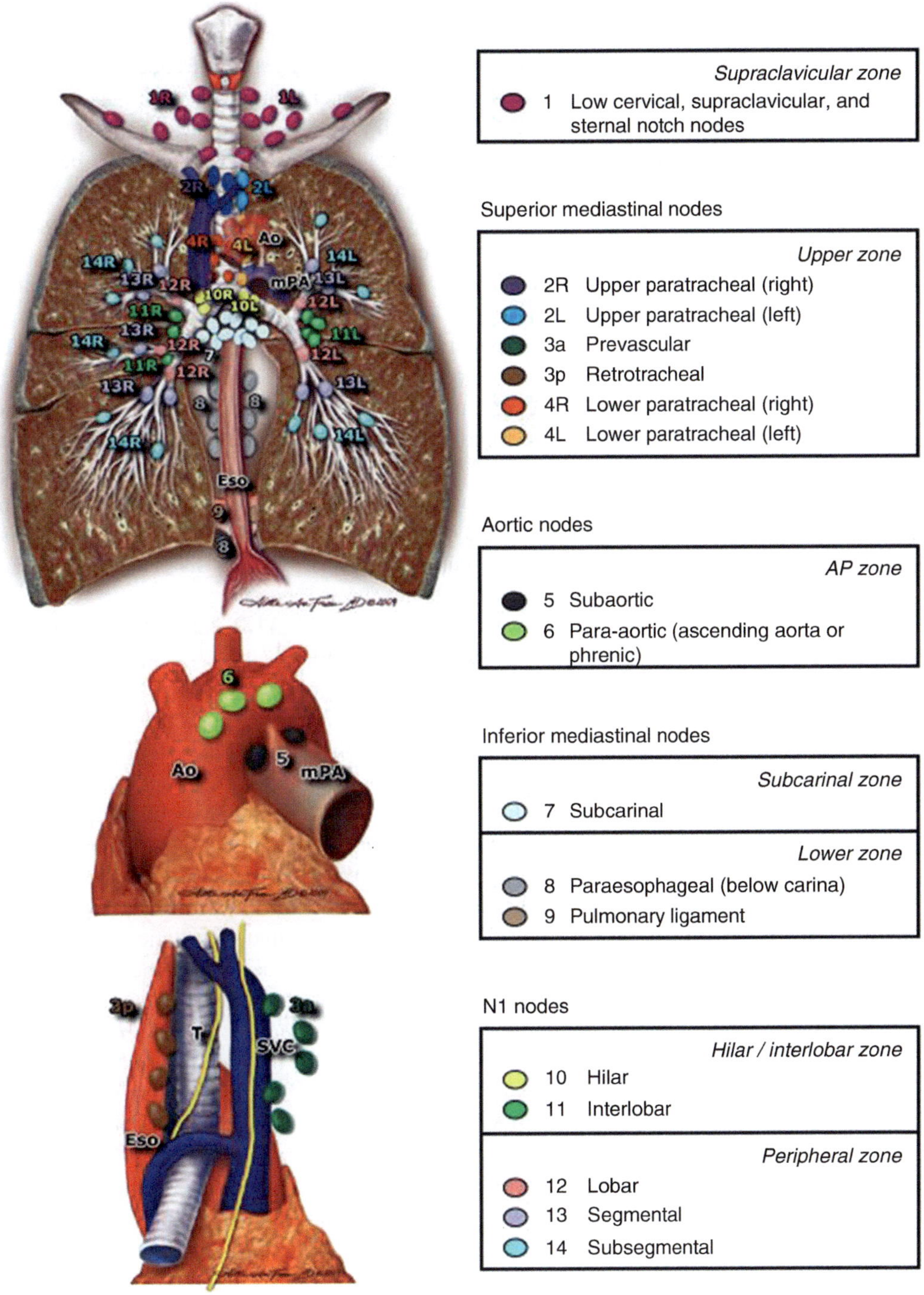

Fig. 41.1 A map of regional lymph nodes for the determination of the N descriptor during tumor, node, and metastasis staging of lung cancer, published by the International Association for the Study of Lung Cancer. Abbreviations: *Ao* aorta; *AP* aortopulmonary; *Eso* esophagus; *mPA* main pulmonary artery; *SVC* superior vena cava; *T* trachea. (Reprinted with permission from the International Association for the Study of Lung Cancer, Staging Manual in Thoracic Oncology, Orange Park, FL: Editorial Rx Press, 2009) [2]

tate (98.2%) cancer. Only 16% of lung cancers are diagnosed in the early stages [3]. Lung cancer represents a typical example for which metastatic patients tend to have extraordinarily poorer prognosis than nonmetastatic cases [4]. Overall, 70–95% of lung cancer patients present with locally advanced or metastatic disease at diagnosis [5]. In the United States, the proportions of early-stage, locally advanced, and advanced/metastatic cases were 15.9%, 22.0%, and 57.0%, respectively [6].

41.2.2.2 Histological Tropism and Sites of Metastasis

Tumor hallmarks, metastatic patterns, and prognostic outcomes differ greatly among different histological types of lung cancer. Histologically, non-small cell lung cancer (NSCLC), including adenocarcinoma, squamous cell carcinoma, large-cell lung cancer (LCLC), and others that are not otherwise specified (NOS), accounts for more than 80% of all lung cancers [7]. Small-cell lung cancer (SCLC), accounting for less than 20% of cases, is the most aggressive form of lung cancer and is characterized by rapid growth and early invasive spread [8].

Although lung cancer metastasizes hematogenously as well as by lymphatics, Popper et al. [9] report that careful evaluation of resected lung carcinomas revealed that vascular invasion is often seen in low-stage tumors, which usually results in increased incidence of recurrence as well as shortened survival of the patient. While metastasis via lymphatics usually takes longer until distant metastases are present, hematogenous spread leads to early distant metastases. Preferentially, lung carcinomas metastasize to bones, brain, liver, and adrenal glands. The frequency of lung cancer metastasis to other organs is as follows: bone 34.3%, lung 32.1%, brain 28.4%, adrenals 16.7%, and liver 13.4% [10]. Clinicopathologically, metastatic patients tended to have poorer tumor differentiation, larger tumor size, and a higher rate of regional lymph node involvement, an indicator of a more aggressive and invasive tumor biology [4].

The tropism of different histological cancer types for certain tissues or organs is now better recognized. According to previously reported data, bone and brain were two leading distant targets for metastasis in lung cancer [11]. While hematogenous metastases were most common to liver, bone, brain, and adrenals, the lymphatic spread was most prevalent in hilar lymph nodes. Understanding metastatic patterns of lung cancer is crucial to clinical management and therapy. In an autopsy study by Milovanovic *et al.* [12], lung adenocarcinoma was noted to be the most frequent histological type of metastasizing primary lung cancer, with metastasis more common in younger patients. Biologically aggressive tumor forms in younger patients, age-dependent metabolic dynamic differences, and survival rate differences are postulated hypotheses to explain this finding [4]. Furthermore, some mechanisms behind reduced aggressiveness of malignancies in elderly patients have also been identified: cell proliferation is slower, the concentration of hormones and the responsiveness of cells to hormones are altered, and angiogenesis is impaired [4].

In a study of 159,241 lung cancer cases by Wang *et al.* [4], the frequency of bone metastasis was the highest for adenocarcinoma, squamous cell carcinoma, LCLC, and NSCLC/NOS, while the liver was the most common metastatic site in SCLC. More than 30% of adenocarcinoma and LCLC patients demonstrated distant metastasis, while squamous cell cancer had the lowest rate of distal metastasis. Moreover, among all histological types of lung cancer, SCLC had the highest frequency of extrathoracic metastasis, especially to the liver, which is consistent with the reported data in previous studies [13, 14]. Although the majority of brain metastasis was noted to be from adenocarcinoma, this has been attributed to the higher incidence of adenocarcinoma, while small cell carcinoma has a higher predilection and risk for brain metastasis [10, 12]. While large and small intestine metastasis was most commonly noted from large cell carcinoma, adrenal metastasis mostly arose from adenocarcinoma [12]. Wang et al. further note that all single-site metastases were independent of prognostic factors [4]. Multisite metastasis studies in lung cancer established that that bone preferentially tended to co-metastasize with liver and distal lymph nodes while liver metastasis was significantly correlated with distant lymph node metastasis. Combined metastases resulted in worse prognosis than the separated single-organ metastasis [4].

41.2.2.3 Molecular Basis of Cancer Metastasis: From and to the Lung

Many theories have been proposed to explain organ-specific metastasis. However, only two have withstood the test of time [15]. Stephen Paget's seed and soil hypothesis, reported in 1889, proposed that metastasis occurs as a result of favorable interactions between the cancer cells themselves (seed) and their specific organ microenvironment (soil). Thus, only when the seed and soil are compatible will metastasis occur. In 1929, James Ewing challenged this theory, postulating that metastatic spread occurs strictly as a result of mechanical factors inherent to the circulatory system and that preferential metastasis can be accounted for by circulatory patterns alone [15]. There is likely an existential interplay between the two theories when applied to lung-specific metastasis.

A comprehensive review by Popper et al. [9] gives insight into the science behind this histological tropism, essentially contingent on tumor microenvironment. Metastasis requires cell migration toward higher oxygen tension, which is based on changing the structure of the cell (epithelial–mesenchymal transition), orientation within the stroma and stromal interaction, and communication with the immune system to avoid attack. Once in the bloodstream, cells must survive trapping by the coagulation system, shear stress in small blood vessels, and must find the right location for extravasation. Once outside in the metastatic locus, tumor cells have to learn the communication with the "foreign" stroma cells to establish vascular supply and again express molecules, which induce immune tolerance. Although detailed description is beyond the scope of this chapter, different expression of genes and proteins including upregulation of MLH 1, downregulation of COX 1, FGFR4, SMAD3, p120, serine protease fibroblast activation protein, fibroblast growth factor receptor 1 in bone metastases, cytokine, and adhesion molecule regulation are involved in metastatic patterns. Macrophage signature seems to play a key role in prognosis in metastatic lung cancer [9, 16]. Prolonged patient survival has been correlated with a high density of M1 macrophages, while poor prognosis correlated with a high density of M2 macrophages in tumor islets [17]. Mechanism of tumor migration differs among histological variants of lung cancer, with single-cell or small cell cluster movement noted in small cell and undifferentiated NSCLC, whereas movement by large clusters of organized cells noted in acinar adenocarcinoma and some cases of squamous cell carcinoma [9].

41.3 Section II

The overall median survival of metastatic lung cancer was noted to be 5 months, ranging from less than 3 months to over 12 months [18]. In one of the largest studies on metastatic locations, 18% of patients had lung metastasis, 16% bone metastasis, 12% brain metastasis, 7% liver metastasis, and 6% adrenal gland metastasis at diagnosis [19]. Adding to previous reports that have associated liver metastasis with a short survival, Riihimaki et al. [20] reported comparable survival between metastasis to the nervous system and to the respiratory system and adrenals, whereas survival was poor in bone and liver metastasis, especially from LCLC. They also noted a better prognosis in younger patients and women. Younger cancer patients are more often diagnosed with advanced metastatic involvement to lymph nodes and other organs, possibly because tumors need to be more aggressive to escape immunosurveillance in the young immunocompetent individual [20].

41.3.1 Sites of Lung Cancer Metastases

41.3.1.1 Lymphatic Spread

Lymph nodes are among the first metastatic foci of lung carcinomas [9]. Generally, the frequency of lymph node metastasis has been shown to increase as the tumor size increases, ranging from 2 to 40% [21, 22]. There are several well-known risk factors for lymph node metastasis in NSCLC. These include degree of differentiation, visceral pleural invasion, amount of ground-glass opacity, lymphovascular invasion, high SUVmax on PET scan [23], and tumor size, especially the solid tumor size [24]. Rather than acting individually, these factors interact together and influence lymph node metastasis. Despite multiple previous publications with differing opinions, Riquet et al. [25] concluded that lymphatic spread of NSCLC cannot be considered lobe-specific. Therefore, systematic ipsilateral mediastinal lymph node dissection is necessary to accurately determine nodal status and to interrupt lymphatic pathways.

41.3.1.2 Hematogenous Spread

Bone

The skeletal system is one of the most common distant metastatic sites in patients with lung cancer, with 30–40% of those with advanced lung cancer developing skeletal metastases, mainly to the spine and ribs [26]. The reported prevalence of bone metastasis at the time of lung cancer diagnosis is 15–40% [27]. Such patients may present with metabolic disorders such as hypercalcemia, pathologic fractures, and spinal cord compression, any of which reduces median survival from 4.4 months to 2.8 months in one study [26]. Sugiura et al. [28] estimated the survival rates in the patients with bone metastases at 9.7 months after diagnosis. Better prognosis was recorded in women, in individuals where adenocarcinoma was the histological type, in cases of solitary bone metastasis, better performance status, lower number of skeletal metastases, use of an epithelial growth factor receptor inhibitor, as well as in the individuals without pathological fractures. Worse prognosis was noted in patients with elevated CA-125 and alkaline phosphatase (ALP) [29]. A possible mechanism for spinal metastasis is access to the vertebral bodies in the thoracic and lumbar spine through the plexus vertebral system and the high bone marrow flow of some skeletal elements [30].

Bone metastasis from lung cancer are treated palliatively not only because of poorer prognosis when compared to other cancers but also because of decreased pulmonary capacity to tolerate surgical procedures in this patient population. Systemic therapies are found to have a modest impact on survival [30]. Receptor activator for nuclear factor-κβ ligand (RANKL) inhibitors have demonstrably superior efficacy over bisphosphonates that block osteoclast activity in reducing the incidence of skeletal-related events in NSCLC in a randomized trial [26]. Radiotherapy is considered standard for small lesions, whereas surgical management is reserved for fractures or impending fractures.

Brain

Brain metastases are thought to occur via seeding of circulating tumor cells into the brain microvasculature; within this unique microenvironment, tumor growth is promoted and the penetration of systemic medical therapies is limited. Development of brain metastases remains a substantial contributor to overall cancer mortality [31]. About 25% of patients with lung carcinoma show brain metastases in advanced stages of the disease. Solitary brain metastasis may clinically present with seizures, symptoms of increased intracranial pressure (e.g., headache), or other nonspecific symptoms even prior to diagnosis of lung cancer [12]. These have a poor prognosis that strongly depends on the number of brain metastases, presence of metastases in other organs, and the stage of the primary lung cancer [32]. Although more cases of brain metastases have been reported previously with adenocarcinoma given its higher incidence, SCLC has the highest risk of brain metastases [32]. In recent years, brain metastases are increasingly seen in adenocarcinomas with epidermal growth factor receptor (EGFR) mutations and EML4ALK1 rearrangement, whereas squamous cell carcinomas have a tendency to locally invade the thoracic wall [9].

Liver

SCLC has been associated with an increased metastatic potential to liver and brain [12, 20, 33]. The incidence of liver metastasis in newly diagnosed SCLC patients is as high as 17% compared to 4% in NSCLC patients [33]. Liver is a frequent metastatic focus in other histological types [12, 20]. However, in autopsy studies, the incidence of liver metastasis in patients with lung cancer has been reported to be 33–51% [12, 33]. Extensive and diffuse metastases of SCLC in the liver parenchyma were reported to potentially cause fulminant hepatic insufficiency (jaundice, coagulopathy, encephalopathy, and organs dysfunction) with poor prognosis and often quick lethal outcome [12]. Lung cancer may also cause biliary tract obstruction by metastasizing to lymph nodes; this should be differentiated from parenchymal liver failure as systemic chemotherapy might relieve this obstruction while parenchymal involvement has a poor prognosis with some chemotherapeutic agents worsening liver failure [33].

Adrenal

Following the initial diagnosis of NSCLC, metastasis to the adrenal glands was reported in 10% of cases [34]. Furthermore, in advanced stages of the disease, NSCLC metastases are usually bilateral and appear in 40% of patients [35]. Survival in patients with adrenal metastasis is comparable to patients with metastasis to the nervous system and respiratory system from lung [20]. In an autopsy study, it was noted that 60% of adrenal metastases originated from lung adenocarcinoma, and more than 50% of those who died of adenocarcinoma in the first year of diagnosis had adrenal metastases. This is attributed to the likelihood of serious complications such as bleeding or adrenal insufficiency [12].

Other Metastasis

Other rare metastases, such as gastric, intestinal, and pancreatic metastases, were found in higher percentage in autopsy studies but are rarely diagnosed in clinical settings [12]. Autopsy studies [12, 36] note that intestinal metastases were most often from LCLC, while adenocarcinoma and squamous cell carcinoma were rarely the sources of intestinal metastases. Patients with intestinal metastases died within the first year after clinical diagnosis of lung cancer.

41.4 Section III

41.4.1 Diagnosing Lung Cancer Metastasis

In patients with NSCLC, determining the extent of disease will strongly impact management and prognosis. While palliation is the objective of management of distant metastases, localized disease is treated with an intention to cure, especially based on lymph node involvement. The American College of Chest Physicians guidelines recommends surgical resection for patients with stage IA, IB, IIA, and IIB while patients stages III and IV rarely meet criteria for surgery [37]. In patients with clinical stage III or IV NSCLC, routine imaging of the brain with head MRI is recommended, even if they have a negative clinical evaluation.

While the potential for surgical resection exists in patients with NSCLC, patients with SCLC are rarely treated with resection and are most often treated with systemic chemotherapy plus/minus radiation therapy.

When there is a strong suspicion of lung cancer, staging workup may be pursued prior to or simultaneously while confirming diagnosis [38]. Extensive clinical evaluation for risk factors, history, and physical examination is the best predictor of distant metastasis. Site-specific symptoms warrant directed evaluation with appropriate studies [37]. Staging modalities are of varying sensitivities and specificities (see Table 41.1). Positron emission tomography–computed tomography (PET-CT) scan has emerged as a standard recommendation for the staging of lung cancer. PET can diagnose unsuspected mediastinal and/or distant metastasis and thus is critical to determining accurate stage. Patients with apparent peripheral early-stage (Ia) disease on computed tomography (CT) or those with clear evidence of advanced disease or metastasis on CT may not require a PET scan or preoperative invasive mediastinal staging [37]. If needle biopsy techniques are negative, surgical biopsy is recommended. Tissue biopsy confirmation is recommended for all abnormal scans, with more evidence pointing toward

Table 41.1 Lymph node staging modalities Gelberg J, Grondin S, Tremblay A. Mediastinal staging for lung cancer. *Can Respir J.* 2014;21(3):159–161. doi:10.1155/2014/890108 (reprinted with permission) [39]

Staging modality	Lymph node access	% Sensitivity[a]	Specificity[a]	NPV[a]
Noninvasive				
CT chest	All	55	81	83
PET	All	80	88	91
PET-CT	All	62	90	90
Invasive				
Mediastinoscopy[b]	2R, 2L, 4R, 4L, 7	78	100[c]	91
EBUS[b]	2R, 2L, 4R, 4L, 7, 10R, 10L, 11R, 11L	89	100[c]	91
EUS	4L, 7, 5, 8, 9	89	100[c]	86
Combined EBUS/EUS	2R, 2L, 4R, 4L, 7, 10R, 10L, 11R, 11L, 5, 8, 9	91	100[c]	96

Source: American College of Chest Physicians lung cancer guidelines, 3rd edn (1). *CT* computed tomography; *EUS* endoscopic ultrasound; *PET* positron emission tomography

[a]Sensitivity, specificity and negative predictive value (NPV) for each staging modality were obtained from studies that may have different patient populations and direct comparison between modalities based on these values alone should be avoided

[b]Studies directly comparing endobrachial ultrasound (EBUS) versus mediastinoscopy are described in the text

[c]Results cannot be verified in studies reporting 100% specificity because no follow-up test was performed to confirm the results

improved patient outcomes with more complete staging. Furthermore, significantly lower risk of death is reported in patients with bi- and trimodality staging versus single-modality staging [40].

41.4.2 Managing Lung Cancer Metastasis

41.4.2.1 Non-small Cell Lung Cancer (NSCLC)

For management of metastatic NSCLC, the National Comprehensive Cancer Network (NCCN) guidelines recommend first establishing a histological subtype, followed by molecular testing for EGFR mutation, ALK, ROS1, BRAF, and PD-L1. Chemotherapy with platinum-based agents has emerged as the first line of management in metastatic NSCLC. However, more focus is currently being placed on upcoming immunotherapeutic agents. In the 2018 NCCN guidelines update: crizotinib was recommended in ROS1 rearrangements; alectinib, ceritinib, and crizotinib in ALK rearrangement; dabrafenib with trametinib with or without cytotoxic agents in BRAF V600E mutations. In the 2020 update, NCCN recommends single-agent pembrolizumab as a first-line therapy option for certain patients with metastatic NSCLC and PD-L1 expression levels of ≥50%. Pembrolizumab is also recommended for patients with metastatic NSCLC and PD-L1 levels of 1–49% who cannot tolerate platinum-based chemotherapy; in patients with negative test results for EGFR, ALK, ROS1, or BRAF genetic alterations, regardless of their PD-L1 expression levels [41].

Multiple reports suggest that a subset of these patients with oligometastatic disease (generally defined as five or fewer metastatic sites) can benefit from definitive management of the disease with improved long-term survival [42]. Furthermore, synchronous versus metachronous presentation (metachronous being more favorable), nodal status, number, and location of oligometastatic sites have all been shown to impact survival [43].

Intracranial Oligometastasis

Patel et al. [42] reported that surgical resection of oligometastatic brain metastases results in long-term survival rates of up to 37%. Solitary brain metastasis should be considered for surgical resection [44]. Additionally, it is recommended that patients with neurological symptoms undergo local therapy for intracranial disease in the form of whole-brain radiation therapy (WBRT) or surgery as appropriate, while considering performance status and other prognostic factors [44].

Extracranial Oligometastasis

Improved long-term survival rates ranging from 23 to 86% with surgical resection of extracranial oligometastatic NSCLC has been demonstrated, with higher survival in metachronous than synchronous disease [42]. Additionally, the efficacy of less invasive surgery in the resection of adrenal and contralateral lung oligometastatic disease may allow for fewer complications from such resection. Radiation therapy has also shown a survival advantage in metachronous oligometastatic disease [42]. However, performance status should be carefully considered when deciding on therapeutic options.

41.4.2.2 Small Cell Lung Cancer (SCLC)

SCLC is generally rapidly growing with a tendency for early metastases and high recurrence rates. It typically presents with bulky symptomatic masses and mediastinal involvement, often associated with a number of neurologic and endocrine paraneoplastic syndromes, with extrathoracic spread in 75–80% patients at presentation [45]. Limited-stage small-cell lung cancer (LS-SCLC) is defined as tumor confined to 1 hemithorax, with or without regional lymph-node involvement, which can be safely encompassed in a tolerable radiation field. Extensive-stage small cell lung cancer (ES-SCLC) is defined as disease that cannot be safely encompassed in a tolerable radiation field [45]. The median survival for LS-SCLC is 18–24 months with a 5-year survival of 20–25%. For ES-SCLC, the median survival is 9–10 months, with 10% 2-year survival [45]. Chemotherapy and radiation form the mainstay of treatment for SCLC, with very rare exception of certain cases of LS-SCLC where surgical management with intention to cure is preferred. Chest radiation and prophylactic cranial radiation (25 Gy over 10 daily fractions) have shown increased survival. Immunotherapy, including nivolumab with or without ipilimumab, is now available for refractory disease.

A staging evaluation consists of medical history and physical examination, complete blood count (CBC), and comprehensive chemistry panel with renal and hepatic function tests, CT of the chest and abdomen with intravenous contrast, MRI or CT of the brain, and bone scan. PET is usually added as well.

LS-SCLC is typically treated with chemoradiation: a combination of etoposide and cisplatin administered concomitantly with thoracic irradiation. The NCCN and American College of Chest Physicians (ACCP) guidelines suggest that there may also be a role for surgery in the management of patients with very early-stage SCLC (stage I) generally without nodal involvement, with some studies showing an improved 5-year survival [46]. Prophylactic cranial irradiation is recommended in patients who had a response to initial treatment [46]. In LS-SCLC, with T3 or node positivity, concurrent chemoradiation with prophylactic cranial irradiation is recommended in patients who had a response to initial treatment.

For ES-SCLC, systemic chemoradiation with a combination of platinum agent and etoposide and organ-specific therapy (radiation or surgery) is recommended. Prophylactic cranial irradiation is controversial in this patient population. Although there is 40–70% objective response rate to chemotherapy, 2-year and 5-year survival in ES-SCLC remains <5% and <2%, respectively [46].

41.4.3 Resectability of Metastases to the Lung

Lung is the second most common site of metastases, making it important to define management and resectability. The present oncological criteria for pulmonary metastasectomy include [47]:

(I) Primary cancer is controllable
(II) No extrathoracic metastasis that is not controlled or controllable
(III) All of the tumor must be resectable, with adequate pulmonary reserve
(IV) No alternative medical treatment options with lower morbidity.

Generally, favorable prognostic features in patients with pulmonary metastases include one or few metastases, long disease-free interval, and normal CEA in colorectal cancers [49]. Radical resection, histology, and disease-free interval seem to be independent prognostic factors identifying a cohort of patients maximally benefitting from lung metastasectomy [48]. Active primary cancer, extrathoracic metastases, incomplete surgical resection, and mediastinal lymphatic spread are negative prognostic features [47].

In urological cancers, metastasectomy for single metastasis was statistically related to longer time to progression especially in patients who previously responded to chemotherapy, with resectable tumor, and documented period of stability [48]. Lung metastasectomy has been shown to prolong survival and should always be considered when feasible in patients with sarcoma, given that up to 20% with soft tissue sarcoma and 40% with osteosarcoma present with isolated pulmonary metastases [48]. Younger age at diagnosis and low-grade tumors with longer disease-free interval offer the greatest survival advantage to radically resected patients [49]. Five-year survival after lung metastasectomy of head and neck cancers is reported to be 50–60% [50] with worse outcomes in older age, squamous cell carcinoma, and short disease-free interval and mediastinal lymph nodal involvement [51]. In gynecological cancers, NCCN recommends surgical resection for resectable regional uterine cancer metastasis, with a 5-year survival rate of 32.9–46.8% [48]. Overall, lung metastasectomy, when feasible, has shown to improve long-term survival and prognosis.

41.5 Conclusions

1. Lung cancer remains the most common cause of cancer-related mortality among both men and women.
2. An understanding of the common sites of metastasis, their detection, and their management is critical to improving outcomes.
3. The lung is a common metastatic site for other types of cancers.
4. The understanding and management of metastases to the lung from other cancers may also lead to improved outcomes in these cancers.

Open Questions

- What diagnostic modalities will improve staging in lung cancer?
- What explains the tropism of certain lung cancers to specific organs?
- Which genes/proteins are responsible for metastasis in lung cancer?
- Which targeted agents will prove efficacious in treating lung cancer?
- What makes the lung a suitable host organ for metastasis from other cancers?

References

1. Susan Standring MBE, PhD, DSc, FKC, Hon FAS, Hon FRCS, Gray's Anatomy, Chapter 54, pp. 953–69.e2.
2. The International Association for the Study of Lung Cancer, Staging Manual in Thoracic Oncology, Orange Park, FL: Editorial Rx Press; 2009.
3. Siegel RL, Miller KD, Jemal A. Cancer statistics, 2019. CA Cancer J Clin. 2019;69:7–34. https://doi.org/10.3322/caac.21551.
4. Wang X, Wang Z, Pan J, et al. Patterns of extrathoracic metastases in different histological types of lung cancer. Front Oncol. 2020;10:715. Published 2020 May 19. https://doi.org/10.3389/fonc.2020.00715.
5. Araujo LH, Baldotto C, Castro G Jr, Katz A, Ferreira CG, Mathias C. Lung cancer in Brazil. J Bras Pneumol. 2018;44(1):55–64. https://doi.org/10.1590/s1806-37562017000000135.
6. Cancer Research UK. Lung cancer statistics. London: Cancer Research UK; 2015.
7. Goldstraw P, Ball D, Jett JR, Le Chevalier T, Lim E, Nicholson AG, et al. Non-small-cell lung cancer. Lancet. 2011;378:1727–40. https://doi.org/10.1016/S0140-6736(10)62101-0.
8. Rudin CM, Ismaila N, Hann CL, Malhotra N, Movsas B, Norris K, et al. Treatment of small-cell lung cancer: American society of clinical oncology endorsement of the American college of chest physicians guideline. J Clin Oncol. 2015;33:4106–11. https://doi.org/10.1200/JCO.2015.63.7918.
9. Popper HH. Progression and metastasis of lung cancer. Cancer Metastasis Rev. 2016;35(1):75–91. https://doi.org/10.1007/s10555-016-9618-0.
10. Tamura T, Kurishima K, Nakazawa K, Kagohashi K, Ishikawa H, Satoh H, Hizawa N. Specific organ metastases and survival in metastatic non-small-cell lung cancer. Mol Clin Oncol. 2015;3:217–21.
11. Hattori A, Takamochi K, Oh S, Suzuki K. New revisions and current issues in the eighth edition of the TNM classification for non-small cell lung cancer. Jpn J Clin Oncol. 2019;49:3–11. https://doi.org/10.1093/jjco/hyy142.
12. Milovanovic IS, Stjepanovic M, Mitrovic D. Distribution patterns of the metastases of the lung carcinoma in relation to histological type of the primary tumor: an autopsy study. Ann Thorac Med. 2017;12(3):191–8. https://doi.org/10.4103/atm.ATM_276_16.
13. Cai H, Wang H, Li Z, Lin J, Yu J. The prognostic analysis of different metastatic patterns in extensive-stage small-cell lung cancer patients: a large population-based study. Future Oncol. 2018;14:1397–407. https://doi.org/10.2217/fon-2017-0706.
14. Nakazawa K, Kurishima K, Tamura T, Kagohashi K, Ishikawa H, Satoh H, et al. Specific organ metastases and survival in small cell lung cancer. Oncol Lett. 2012;4:617–20. https://doi.org/10.3892/ol.2012.792.
15. Piaseczny MM, Allan AL. Why does breast cancer often spread to the lung? Womens Health. 2014;10:561–4. https://doi.org/10.2217/WHE.14.45.
16. Sarode P, Schaefer MB, Grimminger F, Seeger W, Savai R. Macrophage and tumor cell cross-talk is fundamental for lung tumor progression: we need to talk. Front Oncol. 2020;10:324. . Published 2020 Mar 11. https://doi.org/10.3389/fonc.2020.00324.
17. Cao L, Che X, Qiu X, Li Z, Yang B, Wang S, et al. M2 macrophage infiltration into tumor islets leads to poor prognosis in non-small-cell lung cancer. Cancer Manag Res. 2019;11:6125–38. https://doi.org/10.2147/CMAR.S199832.
18. Socinski M, Morris D, Masters G, Lilenbaum R. Chemotherapeutic management of stage IV non-small cell lung cancer. Chest. 2003;123:226S–43S.
19. Oikawa A, Takahashi H, Ishikawa H, Kurishima K, Kagohashi K, Satoh H. Application of conditional probability analysis to distant metastases from lung cancer. Oncol Lett. 2012;3:629–34.
20. Riihimäki M, Hemminki A, Fallah M, Thomsen H, Sundquist K, Sundquist J, et al. Metastatic sites and survival in lung cancer. Lung Cancer. 2014;86:78–84.
21. Oda M, Watanabe Y, Shimizu J, et al. Extent of mediastinal node metastasis in clinical stage I non-small-cell lung cancer: the role of systematic nodal dissection. Lung Cancer. 1998;22:23–30.
22. Ohta Y, Oda M, Wu J, et al. Can tumor size be a guide for limited surgical intervention in patients with peripheral non-small cell lung cancer? Assessment from the point of view of nodal micrometastasis. J Thorac Cardiovasc Surg. 2001;122:900–6.
23. Miyata Y, Tsutani Y, Okada M. Use of high-resolution computed tomography and positron emission tomography/computed tomography in the management of stage IA adenocarcinoma. Semin Thorac Cardiovasc Surg. 2012;24:267–74.
24. Seok Y, Yang HC, Kim TJ, Lee KW, Kim K, Jheon S, Cho S. Frequency of lymph node metastasis according to the size of tumors in resected pulmonary adenocarcinoma with a size of 30 mm or smaller. J Thorac Oncol. 2014 Jun;9(6):818–24. https://doi.org/10.1097/JTO.0000000000000169.
25. Riquet M, Rivera C, Pricopi C, Arame A, Mordant P, Foucault C, Dujon A, Le Pimpec-Barthes F. Is the lymphatic drainage of lung cancer lobe-specific? A surgical appraisal. Eur J Cardiothorac Surg. 2015;47(3):543–9. https://doi.org/10.1093/ejcts/ezu226.
26. Rosen LS, Gordon D, Tchekmedyian NS, Yanagihara R, Hirsh V, Krzakowski M, Pawlicki M, De Souza P, Zheng M, Urbanowitz G, et al. Long-term efficacy and safety of zoledronic acid in the treatment of skeletal metastases in patients with nonsmall cell lung carcinoma and other solid tumors: a randomized, phase III, double-blind, placebo-controlled trial. Cancer. 2004;100(12):2613–21. https://doi.org/10.1002/cncr.20308.
27. Oliveira MB, Mello FC, Paschoal ME. The relationship between lung cancer histology and the clinicopathological characteristics of bone metastases. Lung Cancer. 2016;96:19–24.
28. Sugiura H, Yamada K, Sugiura T, Hida T, Mitsudomi T. Predictors of survival in patients with bone metastasis of lung cancer. Clin Orthop Relat Res. 2008;466(3):729–36. https://doi.org/10.1007/s11999-007-0051-0.
29. Zhou Y, Yu QF, Peng AF, et al. The risk factors of bone metastases in patients with lung cancer. Sci Rep. 2017;7:8970. https://doi.org/10.1038/s41598-017-09650-y.
30. Cho YJ, Cho YM, Kim SH, Shin KH, Jung ST, Kim HS. Clinical analysis of patients with skeletal metastasis of lung cancer. BMC Cancer. 2019;19(1):303. . Published 2019 Apr 3. https://doi.org/10.1186/s12885-019-5534-3.
31. Achrol AS, Rennert RC, Anders C, Soffietti R, Ahluwalia MS, Nayak L, Peters S, Arvold ND, Harsh GR, Steeg PS, Chang SD. Brain metastases. Nat Rev Dis Primers. 2019;5(1):5. https://doi.org/10.1038/s41572-018-0055-y.
32. Arrieta O, Villarreal-Garza C, Zamora J, Blake-Cerda M, de la Mata MD, Zavala DG, et al. Long-term survival in patients with non-small cell lung cancer and synchronous brain metastasis treated with whole-brain radiotherapy and thoracic chemoradiation. Radiat Oncol. 2011;6:166.
33. Kagohashi K, Satoh H, Ishikawa H, Ohtsuka M, Sekizawa K. Liver metastasis at the time of initial diagnosis of lung cancer. Med Oncol. 2003;20:25–8.
34. Allard P, Yankaskas BC, Fletcher RH, Parker LA, Halvorsen RA. Jr Sensitivity and specificity of computed tomography for the detection of adrenal metastatic lesions among 91 autopsied lung cancer patients. Cancer. 1990;66:457–62.
35. Matthews MJ. Problems in morphology and behavior of bronchopulmonary malignant disease. In: Lung cancer: natural history, prognosis and therapy. New York: Academic Press; 1976.
36. McNeill PM, Wagman LD, Neifeld JP. Small bowel metastases from primary carcinoma of the lung. Cancer. 1987;59:1486–9.

37. Silvestri GA, Gonzalez AV, Jantz MA, Margolis ML, Gould MK, Tanoue LT, Harris LJ, Detterbeck FC. Methods for staging non-small cell lung cancer: diagnosis and management of lung cancer, 3rd ed: American College of Chest Physicians evidence-based clinical practice guidelines. Chest. 2013 May;143(5 Suppl):e211S–50S. https://doi.org/10.1378/chest.12-2355.
38. Rivera MP, Mehta AC, Wahidi MM. Establishing the diagnosis of lung cancer: diagnosis and management of lung cancer, 3rd ed: American College of Chest Physicians evidence-based clinical practice guidelines. Chest. 2013;143(5 suppl):e142S–65S.
39. Gelberg J, Grondin S, Tremblay A. Mediastinal staging for lung cancer. Can Respir J. 2014;21(3):159–61. https://doi.org/10.1155/2014/890108.
40. Farjah F, Flum DR, Ramsey SD, Heagerty PJ, Symons RG, Wood DE. Multi-modality mediastinal staging for lung cancer among medicare beneficiaries. J Thorac Oncol. 2009;4(3):355–63.
41. Ettinger DS, Wood DE, Aggarwal C, Aisner DL, Akerley W, Bauman JR, Bharat A, Bruno DS, Chang JY, Chirieac LR, et al. NCCN guidelines insights: non-small cell lung cancer, version 1.2020. J Natl Compr Cancer Netw. 2019;17(12):1464–72.
42. Patel AN, Simone CB 2nd, Jabbour SK. Risk factors and management of oligometastatic non-small cell lung cancer. Ther Adv Respir Dis. 2016;10(4):338–48. https://doi.org/10.1177/1753465816642636.
43. Pirker R, Pereira J, Szczesna A, Von Pawel J, Krzakowski M, Ramlau R, et al. Prognostic factors in patients with advanced non-small cell lung cancer: data from the phase III FLEX study. Lung Cancer. 2012;77:376–82.
44. Tsao M, Rades D, Wirth A, Lo S, Danielson B, Gaspar L, et al. Radiotherapeutic and surgical management for newly diagnosed brain metastasis(es): an American Society for Radiation Oncology evidence-based guideline. Pract Radiat Oncol. 2012;2:210–25.
45. Wang S, Zimmermann S, Parikh K, Mansfield AS, Adjei AA. Current diagnosis and management of small-cell lung cancer. Mayo Clin Proc. 2019;94(8):1599–622. https://doi.org/10.1016/j.mayocp.2019.01.034.
46. Yu JB, Decker RH, Detterbeck FC, Wilson LD. Surveillance epidemiology and end results evaluation of the role of surgery for stage I small cell lung cancer. J Thorac Oncol. 2010;5(2):215–9.
47. Royal Brompton & Harefield NHS Foundation Trust, University of Cambridge, University of Sussex. A randomised Trial of Pulmonary Metastasectomy in Colorectal Cancer (PulMiCC). https://clinicaltrials.gov/ct2/show/NCT01106261.
48. Treasure T, Milošević M, Fiorentino F, et al. Pulmonary metastasectomy: what is the practice and where is the evidence for effectiveness? Thorax. 2014;69:946–9. https://doi.org/10.1136/thoraxjnl-2013-204528.
49. Giuliano K, Sachs T, Montgomery E, et al. Survival following lung metastasectomy in soft tissue sarcomas. Thorac Cardiovasc Surg. 2016;64:150–8. https://doi.org/10.1055/s-0035-1563538.
50. Haro A, Yano T, Yoshida T, et al. Results of a surgical resection of pulmonary metastasis from malignant head and neck tumor. Interact Cardiovasc Thorac Surg. 2010;10:700–3. https://doi.org/10.1510/icvts.2009.219766.
51. Galetta D, Petrella F, Leo F, et al. Treatment of pulmonary metastases from primary intraosseous odontogenic carcinoma. Lancet Oncol. 2006;7:272–3. https://doi.org/10.1016/S1470-2045(06)70620-9.

Part XI

Cancer Metastasis to the Liver

Alex C. Kim and Clifford S. Cho

Cancer metastasis involves a complex pathologic cellular process that allows for malignant cells to travel from the primary tumor site to regional lymph nodes or distant organs. It is the major contributor to host morbidity and is responsible for 90% of cancer-related mortality [1, 2]. The mechanism of metastatic cascade consisting of initiation, migration, colonization, and re-propagation is a very complex process. Detailed description of the cascade is beyond the scope of this chapter. The concepts introduced here are intended to provide a general overview of hypotheses that are viewed as important in hepatic metastasis.

The liver is a unique, multifunctional organ responsible for metabolic homeostasis, bile synthesis, and paracrine and endocrine secretions. Anatomically, the liver derives its blood supply from both the splanchnic (portal) and systemic circulations. The portal system derives its blood volume from intra-abdominal organs and is very rich in nutrients and metabolic by-products. The hepatic artery delivers well-oxygenated blood from the systemic circulation to the liver. The liver is further organized into eight Couinaud segments with each possessing its independent portal and systemic vascular inflow, hepatic outflow, and biliary pedicles. At the level of the parenchyma, hepatocytes and cholangiocytes make up the majority of the liver and accommodate for metabolic detoxification, bile synthesis, and endocrine function. Additional mesenchymal and hematopoietic derived cells make up the rest of the liver, and mediate liver microcirculation and regenerative, inflammatory, and immune functions. The cells are arranged in a complex, dense network of polyhedral shaped hepatic lobules composed of peripheral portal triads and a central vein. The portal triad contains the hepatic arteriole branch, portal venous branch, bile ducts, lymphatics, and nerve branch. While the microcirculation flows through the hepatic arterioles and the portal venules and drains through the central vein, the lymphatic and bile flows are in the opposite direction. This evolutionarily conserved complex microvascular network is a perfectly fertile environment for entry and seeding of malignant cells, and provides a mechanism for tumor growth and proliferation and for immune escape or tolerance. As such, the liver serves as an ideal niche for cancer metastasis, especially for colorectal cancer.

One of the hallmarks of cancer is the ability to invade and to metastasize [2–4]. In order to disseminate to distant tissue and to establish metastatic colonies, neoplastic cells commandeer a complex biologic program known as epithelial-mesenchymal transition (EMT) [5–8]. EMT is a dynamic molecular process that triggers a transient switch in cellular phenotype from a polarized, differentiated epithelial state to a mobile, dedifferentiated mesenchymal state. This process is readily evident during embryogenesis and wound healing, and allows for spatial and temporal cellular distribution and generation of new organs or tissues [4, 6, 9]. During the initiation of invasion, aberrant activation of EMT is hypothesized to be the principal mechanism for the acquisition of metastatic phenotype by epithelial tumors [5, 6]. The transition allows for nonmetastatic cells to suppress their epithelial/differentiated state (loss of cell-cell adhesion and polarity) while activating their mesenchymal/dedifferentiated state (gain of motility, invasion, and stemness) [9]. The activation of EMT is dependent on a convergence of various cellular signals from the tumor microenvironment, including the TGF-β signaling cascade [4, 7, 9, 10]. The resultant induction of intracellular signaling pathways provokes expression and activation of multiple EMT transcription factors, such as TWIST1 and SNAI1, to trigger the cellular transition [4, 6, 9]. In addition, activation of these factors, especially in colorectal cancer, can trigger an undifferentiated state through TGF-β signaling that allows cells to behave as cancer stem cells [11]. These cells gain an ability to self-renew, produce differentiated progeny, and re-metastasize to a new site. The cancer stem cell theory further explains that these cells are thought to be resilient to cytotoxic and hypoxic conditions that allow survival until they arrive at their destination organ [12]. Colorectal cancer cells with tumor-propagating ability are characterized by cell surface molecules including CD44, CD133, and LGR5 [12–16]. CD44 is a cell surface adhesion peptide and consists of various isoforms stemming from alternate splicing products [17]. Particularly, CD44v6 gained wide interest due to its ability to bind to human growth factor (HGF) and vascular endothelial growth factor (VEGF) [12, 18, 19]. These interactions allow for activation of downstream signal transduction pathways that cause tumor proliferation and angiogenesis. Increased CD44v6 expression is often correlated with advanced stages of colon cancer and is associated with poorer prognosis [20]. CD133 is another molecule expressed particularly in colon cancer stem cells [12]. Cells positive for this molecule are able to initiate tumor formation in in vivo models. Only a small percentage of cells are positive for these molecules, supporting their representation of rare populations of cancer stem cells [12, 21]. LGR5 was identified as the defining marker for colonic intestinal adult stem cell populations [16]. This molecule is a downstream target for the WNT pathway and also serves as the receptor for R-spondin [16, 22]. It is hypothesized that the constitutive activation of the β-catenin pathway in LGR5 results in the transformation of adult stem cells into cancer stem cells [23]. Through EMT, these cells gain the ability to migrate and establish distant metastatic colonies in favorable sites like the liver.

In 1889, Stephen Paget postulated that cancer metastasis requires an optimal interaction between "seeds," or the metastatic cells, and the "soil," or the premetastatic niche [24]. The entry of the metastatic cells into the liver and their subsequent colonization are a result of bidirectional interaction between malignant cells and liver microenvironment. This is often referred to as a premetastatic niche, or a microenvironment that is permissive for metastatic growth [25, 26]. The hepatic microcirculation consists of hepatic sinusoids, Kupffer cells, and extracellular matrix (ECM), all representing an example of a premetastatic niche. In murine models of metastatic pancreatic ductal adenocarcinoma, it was recently observed that exosomes and IL-6 released from primary pancreas cancers, respectively, trigger signaling pathways and protein expression changes in distant Kupffer cells and hepatocytes that make the liver more receptive to infiltration and growth of metastatic cancer cells—demonstrating a paracrine cross talk that evolves between primary tumors and liver which favors the development of liver metastases [27, 28]. The initial implantation of cancer cells induces local inflammatory responses. Particularly, neutrophils are hypothesized to play a significant role in the infiltration of cancer cells to the metastatic site. Activation of neutrophils at the site of injury results in the release of highly adhesive neutrophil extracellular traps (NETs) composed of DNA and antimicrobial enzymes. It is hypothesized that these NETs may facilitate early adhesion of cancer cells into the metastatic niche of the liver [29, 30]. The subsequent ECM remodeling through enzymes such as tissue inhibitor of metalloproteninases-1 (TIMP-1) and the increased expression of molecules including E-selectin and vascular cell adhesion molecule (VCAM)-1 under the control of cytokine and signaling pathway activation by Kupffer cells are thought to enable further implantation of metastatic cells [28, 31]. Indeed, blockade of this inflammation-induced cell surface molecule expression has resulted in decreased liver metastasis [32–34].

The next steps in hepatic metastasis involve early proliferation and cellular growth within the hepatic microcirculation. Under normal conditions, hepatic stellate cells (HSC) are quiescent cells that contribute to extracellular matrix turnover and participate in liver homeostasis through the production of stimulating factors [35]. Upon contact with cancer cells, HSCs are activated and undergo differentiation into myofibroblasts, with the expression of a-SMA, vimentin, and PDGFR-a/b [36]. Activated myofibroblasts are in turn responsible for remodeling of the microenvironment, with production of desmoplastic stroma. In addition, they promote metastasis through the secretion of potent mitogens including TGF-β, enabling further recruitment of stromal cells for colony establishment [35, 37]. The activated TGF-β signaling pathway additionally promotes increased expression of metalloproteinases that allow for metastatic cell migration and invasion, promote antitumor responses by downregulating inflammatory cells including NK cells, and activate angiogenesis by increasing production of VEGF [37].

Upon establishment of the initial metastatic colony, the activation of angiogenesis becomes crucial for the survival of these cells [38]. The local injury caused by cancer cell entrapment and disruption of microvasculature results in paracrine stimulation from local cells including myofibroblasts. The resulting hypoxia causes increased expression of hypoxia-inducible factor 1-α (HIF1-α), promoting the secretion of pro-angiogenic factors such as VEGF [39]. Subsequent interaction between HSC, myofibroblasts, and tumor cells results in further secretion of IL-6, prostaglandin E2, VEGF, and MMP2 to promote endothelial cell migration and angiogenesis that establishes new metastatic vascular circuits.

The last phase, or the established hepatic metastasis phase, results in further growth of the established metastatic colony through further alteration of the local environment, tumor proliferation, and differentiation. This growth phase would be impossible without evasion from and suppression of the immune system through two generalized mechanisms [40]. First, downregulation of MHC class I on hepatic antigen-presenting cells and increased expression of immune inhibitory molecules such as PD-1 on hepatic lymphocytes result in immune evasion and suppression [40]. Next, the active suppression of adaptive immune responses by expanding local populations of regulatory T cells within the metastasis enables further immune escape. With these processes contributing to a growth circuit, the result is progression and enlargement of the metastatic deposit.

Stemming from these complex processes of metastasis, it is easy to envision multiple mechanisms of treatment failure resulting in recurrence, morbidity, and mortality. Treatment goals should target, in a multimodal fashion, the disruption of individual processes of cancer metastasis development. The next two chapters examine the role of locoregional therapies including surgical resection and cancer regional therapies to control liver metastasis. However, it is prudent to realize that the application of locoregional therapy alone is insufficient for disease eradication.

References

1. Chaffer CL, Weinberg RA. A perspective on cancer cell metastasis. Science. 2011;331:1559–64.
2. Seyfried TN, Huysentruyt LC. On the origin of cancer metastasis. Crit Rev Oncogenesis. 2012;18:43–73.
3. Thiery JP. Epithelial–mesenchymal transitions in tumour progression. Nat Rev Cancer. 2002;2:442–54.
4. Kalluri R. EMT: when epithelial cells decide to become mesenchymal-like cells. J Clin Invest. 2009;119:1417–9.
5. Hanahan D, Weinberg RA. Hallmarks of cancer: the next generation. Cell. 2011;144:646–74.
6. Tam W, Weinberg RA. The epigenetics of epithelial-mesenchymal plasticity in cancer. Nat Med. 2013;19:1438–49.
7. Fabregat I, Malfettone A, Soukupova J. New insights into the crossroads between EMT and Stemness in the context of cancer. J Clin Medicine. 2016;5:37.
8. Chuang H-YY, et al. Aminopeptidase A initiates tumorigenesis and enhances tumor cell stemness via TWIST1 upregulation in colorectal cancer. Oncotarget. 2017;8:21266–80.
9. Kalluri R, Weinberg RA. The basics of epithelial-mesenchymal transition. J Clin Invest. 2009;119:1420–8.
10. Xu J, Lamouille S, Derynck R. TGF-β-induced epithelial to mesenchymal transition. Cell Res. 2009;19:cr20095.
11. Nakano M, et al. Dedifferentiation process driven by TGF-beta signaling enhances stem cell properties in human colorectal cancer. Oncogene. 2019;38:780–93.
12. Ricci-Vitiani L, et al. Identification and expansion of human colon-cancer-initiating cells. Nature. 2007;445:111–5.

13. Dalerba P, et al. Single-cell dissection of transcriptional heterogeneity in human colon tumors. Nat Biotechnol. 2011;29:1120–7.
14. Vermeulen L, et al. Single-cell cloning of colon cancer stem cells reveals a multi-lineage differentiation capacity. Proc National Acad Sci. 2008;105:13427–32.
15. Dalerba P, et al. Phenotypic characterization of human colorectal cancer stem cells. Proc National Acad Sci. 2007;104:10158–63.
16. Barker N, et al. Identification of stem cells in small intestine and colon by marker gene Lgr5. Nature. 2007;449:1003–7.
17. Wielenga VJ, et al. Expression of CD44 variant proteins in human colorectal cancer is related to tumor progression. Cancer Res. 1993;53:4754–6.
18. Tremmel M, et al. A CD44v6 peptide reveals a role of CD44 in VEGFR-2 signaling and angiogenesis. Blood. 2009;114:5236–44.
19. Orian-Rousseau V, Chen L, Sleeman JP, Herrlich P, Ponta H. CD44 is required for two consecutive steps in HGF/c-Met signaling. Genes Dev. 2002;16:3074–86.
20. Finke, Lothar H. et al. Colorectal cancer prognosis and expression of exon-v6-containing CD44 proteins. Lancet. 1995;345:583–4.
21. Todaro M, et al. Colon cancer stem cells dictate tumor growth and resist cell death by production of interleukin-4. Cell Stem Cell. 2007;1:389–402.
22. Carmon KS, Gong X, Lin Q, Thomas A, Liu Q. R-spondins function as ligands of the orphan receptors LGR4 and LGR5 to regulate Wnt/β-catenin signaling. Proc National Acad Sci. 2011;108:11452–7.
23. Barker N, et al. Crypt stem cells as the cells-of-origin of intestinal cancer. Nature. 2009;457:608–11.
24. Paget S. The distribution of secondary growths in cancer of the breast. Lancet. 1889;133:571–3.
25. Liu Y, Cao X. Characteristics and significance of the pre-metastatic niche. Cancer Cell. 2016;30:668–81.
26. Peinado H, et al. Pre-metastatic niches: organ-specific homes for metastases. Nat Rev Cancer. 2017;17:302–17.
27. Costa-Silva B, et al. Pancreatic cancer exosomes initiate pre-metastatic niche formation in the liver. Nature Cell Biol. 2015;17:816–26.
28. Lee JW, et al. Hepatocytes direct the formation of a pro-metastatic niche in the liver. Nature. 2019;567:249–52.
29. Cools-Lartigue J, et al. Neutrophil extracellular traps sequester circulating tumor cells and promote metastasis. J Clin Invest. 2013;123:3446–58.
30. Erpenbeck L, Schön MP. Neutrophil extracellular traps: protagonists of cancer progression? Oncogene. 2017;36:2483–90.
31. Seubert B, et al. Tissue inhibitor of metalloproteinases (TIMP)-1 creates a premetastatic niche in the liver through SDF-1/CXCR4-dependent neutrophil recruitment in mice. Hepatology. 2015;61:238–48.
32. Ham B, et al. TNF receptor-2 facilitates an immunosuppressive microenvironment in the liver to promote the colonization and growth of hepatic metastases. Cancer Res. 2015;75:5235–47.
33. Jiao S-F, et al. Inhibition of tumor necrosis factor alpha reduces the outgrowth of hepatic micrometastasis of colorectal tumors in a mouse model of liver ischemia-reperfusion injury. J Biomed Sci. 2014;21:1.
34. Khatib A-M, Fallavollita L, Wancewicz EV, Monia BP, Brodt P. Inhibition of hepatic endothelial E-selectin expression by C-raf antisense oligonucleotides blocks colorectal carcinoma liver metastasis. Cancer Res. 2002;62:5393–8.
35. Tsuchida T, Friedman SL. Mechanisms of hepatic stellate cell activation. Nat Rev Gastroentero. 2017;14:397–411.
36. Tadashi T, Kiyoko M, Noboru T, Yasuo S, Yasuni N. Alpha-smooth muscle actin-positive stromal cells in cholangiocarcinomas, hepatocellular carcinomas and metastatic liver carcinomas. J Hepatol. 1996;24:706–12.
37. Marvin DL, Heijboer R, ten Dijke P, Ritsma L. TGF-β signaling in liver metastasis. Clin Transl Medicine. 2020;10:e160.

38. Olaso E, et al. Proangiogenic role of tumor-activated hepatic stellate cells in experimental melanoma metastasis. Hepatology. 2003;37:674–85.
39. Pezzuto A, Carico E. Role of HIF-1 in cancer progression: novel insights. A review. Curr Mol Med. 2019;18:343–51.
40. Butt AQ, Mills KHG. Immunosuppressive networks and checkpoints controlling antitumor immunity and their blockade in the development of cancer immunotherapeutics and vaccines. Oncogene. 2014;33:4623–31.

Tumor Resection and Ablation as a Means of Controlling Hepatic Metastases

42

Brian D. Griffith and Timothy L. Frankel

Abstract

The liver is a common site for metastasis in colorectal cancer. Hepatectomy has emerged as the gold standard for the treatment of colorectal liver metastases and is the only treatment modality that offers a potential for long-term cure. However, only a small proportion of patients are candidates for surgical resection. Adjunctive treatments like portal vein embolization, two-stage hepatectomy, and/or neoadjuvant chemotherapy can be used to downstage disease or convert patients to surgical candidates. Less invasive therapies like ablation, radiotherapy, and systemic chemotherapy have emerged as treatment options for poor surgical candidates.

Learning Objectives

- To identify indications for resection of hepatic metastases and understand the appropriate preoperative imaging
- To understand indications for adjunct therapies like portal vein embolization, two-stage hepatectomy, or neoadjuvant chemotherapy in patients who are not initially surgical candidates
- To understand the role of ablation, radiotherapy, and systemic chemotherapy for the treatment of hepatic metastasis in patients who are not surgical candidates

B. D. Griffith · T. L. Frankel (✉)
Department of Surgery, Michigan Medicine, University of Michigan, Ann Arbor, MI, USA
e-mail: briangr@med.umich.edu; timofran@umich.edu

42.1 Introduction

Colorectal cancer is the third most common malignancy in the United States, with up to 145,000 new cases per year [1]. As many as 20% of patients present with metastatic disease, with liver metastasis present in almost 80% of patients with stage IV disease [2]. As surgical and anesthetic techniques for hepatectomy have improved over the past decades, hepatic resection has emerged as the gold standard for the treatment of colorectal liver metastases (CLM). While operative mortality was initially as high as 30% in early series of hepatic resection for colorectal cancer [3], perioperative mortality has improved and is now as low as 1–3% [4–6]. In appropriately selected patients, hepatic resection provides excellent disease control when compared to systemic chemotherapy and can lead to a long-term cure [7]. A multidisciplinary approach is needed to achieve the best long-term outcomes and should involve medical oncology, radiology, anesthesia, and surgery.

42.2 Hepatic Resection

42.2.1 Indications for Resection and Efficacy

The essential tenets of resection of metastatic disease, first popularized by Alexander and Haigh, include a resectable primary tumor, a patient who can tolerate a major surgery, and the ability to obtain a negative margin [8]. As the effectiveness of systemic chemotherapy and the safety of the operation have improved, indications for resection of CLM have expanded. Multiple predictive models based on large retrospective reviews have been developed to better select patients for resection. Fong et al. [9] reviewed 1001 patients undergoing liver resection over a 13-year period and discovered that positive margin, presence of extrahepatic disease, positive nodal status of the primary,

S. P. Leong et al. (eds.), *Cancer Metastasis Through the Lymphovascular System*,
https://doi.org/10.1007/978-3-030-93084-4_42

disease-free interval from primary to metastatic disease <12 months, number of tumors >1, preoperative carcinoembryonic antigen (CEA) level >200 ng/mL, and size of largest tumor >5 cm were all significant and independent predictors of poor survival. By assigning one point to each of the latter five criteria, a score was created which was highly predictive of cancer-specific outcome; low-score patients had a median survival of 74 months, while high-score patients had a median survival of only 22 months. These scoring systems have been externally validated in independent series [10], and other similar prognostic systems have been developed [4, 5]. A more recent examination of cancer survival by Nordlinger et al. [5] evaluated 1568 patients undergoing hepatic resection for CLM and found similar prognostic factors with the addition of age and stage of the primary tumor. In addition to clinical factors, recent attention has turned to aspects of tumor biology that impact prognosis. Genetic sequencing of tumors has revealed that the presence of both KRAS mutations [11] and BRAF mutations [12] is associated with decreased survival in patients with CLM. As our understating of the biologic factors driving tumorigenesis advances, incorporation into prognostic scoring will likely improve patient selection for surgical resection.

In patients with unresectable metastatic colon cancer, the median survival with systemic chemotherapy is approximately 21 months, with very few alive at 5 years [13]. In comparison, the overall 5-year actuarial survival of patients with CLM after hepatic resection with curative intent ranges from 27 to 39% and 10-year actuarial survival of 17–22% [7, 9, 14]. Tomlinson et al. [7] evaluated 612 patients with CLM treated with hepatic resection and of the 102 patients that were alive at 10 years, 97% were disease free and only one experienced a disease-specific death after 10 years, suggesting that those who survive 10 years were likely cured of the disease.

While survival can be dramatically improved, liver surgery can be associated with significant morbidity, and proper patient education and medical optimization are paramount. As cancer patients tend to be older, prior to any resection, patients should undergo evaluation to ensure appropriate medical fitness. Patients should be screened to ensure adequate cardiac and pulmonary function, performance status, and sufficient future liver remnant (FLR). The FLR is the volume of functional liver parenchyma present after the resection relative to total liver volume excluding cysts, tumors, and ablation cavities and is useful to predict morbidity and mortality after hepatic resection. In general, an FLR of 25–30% for chemotherapy naive patients or 40% for those with parenchymal damage like steatosis or fibrosis is associated with a low risk of postoperative risk of liver failure [15].

42.2.2 Preoperative Imaging

Common poor predictors of survival among patients with CLM include size of the primary tumor, presence of multiple hepatic lesions, and evidence of extrahepatic disease [5, 9]. Appropriate preoperative cross-sectional imaging is essential to best identify these factors as well as to plan the resection and improve the overall safety and efficacy of resection. Except in the case of two-staged hepatectomy, the overall goal of resection is to remove all metastatic disease, preserve unaffected tissue, and maintain the segmental blood supply and biliary and venous drainage to the remaining liver remnant. Thus, when assessing the feasibility of a resection, one must consider the number of involved segments as well as proximity to arteries, veins, and bile ducts.

There are multiple imaging modalities, each with their respective advantages and disadvantages. Multidetector computed tomography (MDCT) is routinely used in workup and surveillance of oncology patients, most often with the addition of iodinated contrast agents. With contrast enhancement, colorectal liver metastases appear hypovascular with variable heterogeneity on portal venous phase [16]. The addition of arterial phase imaging generally does not improve the detection of CLM [16], but is helpful for presurgical planning to identify aberrant anatomy. CT angiography is particularly important for planning hepatic artery pump placement if hepatic artery infusion (HAI) chemotherapy is planned. Positron emission tomography (PET) is an imaging modality often combined with CT that detects the uptake of a glucose analog, ^{18}F-fluorodeoxyglucose (FDG) in hypermetabolic tumors and is most often used for detection of extrahepatic metastases [17]. Metastases demonstrate high uptake on PET, but sensitivity in the liver is limited due to high physiologic background hepatic uptake [18].

Magnetic resonance imaging (MRI), by comparison, offers improved soft-tissue contrast, particularly in the presence of steatosis or fibrosis. In general, MR images are obtained in T1 weighted (T1w), T2 weighted (T2w), and diffusion-weighted images (DWI) before and after gadolinium-binding contrast agent, with CLM hypointense on T1w, hyperintense on T2w, and hyperintense on DWI. After IV contrast, CLM is hypovascular with irregular rim enhancement. Over the past decade, hepatocyte-specific contrast agents have also been developed. These contrast agents, including gadobenate dimeglumine (Gd-BOPTA) and gadoexitic acid (Gd-EOB-DTPA), are taken up by functioning hepatocytes and excreted in bile [19]. This allows improved detection of metastatic disease, as normal parenchyma is hyperintense relative to liver metastatic disease in T1-weighted delayed hepatobiliary phased images. The combination of diffusion-weighted images and hepatobiliary-phased MR images allows the highest sensitivity for small CLM, particularly metastases less than 1 cm in size, relative to other imaging modalities [20, 21].

42.2.3 Surgical Technique

The goal of hepatectomy for CLM is to obtain a complete resection with negative margins, while preserving healthy parenchyma. Historically, resections for CLM were frequently performed as anatomic resections like segmentectomies, hemi-hepatectomy, or tri-sectionectomy, as non-anatomics resection was felt to impose higher risk of a positive margin [22, 23]. However, as further retrospective reviews have suggested no difference in rates of positive margins, overall recurrence, or overall survival [24, 25], nonanatomic, more limited resections are performed with increasing frequency. The specific resection performed is dependent on the location of the tumors in relation to the artery and portal venous inflow pedicles, as well as the outflow veins. Frequently, operative plans are developed at a multidisciplinary conference with the assistance of radiologists to identify tumors and anatomic structures, as well as medical oncologists and interventional radiologists to discuss preoperative strategies to downsize tumors or treat with adjunct therapies.

Open exploration for CLM begins with a midline laparotomy or subcostal incision large enough to allow for inspection of the entire liver and should consider any planned combined procedure for synchronous disease. The abdominal cavity, including peritoneal surfaces and the portal lymph nodes, should be inspected to exclude extrahepatic disease. Hepatic exploration is performed with palpation and intraoperative ultrasound to identify deeper lesions and clearly define the anatomy. Once the degree of disease burden has been established, inflow and outflow control can be established prior to hepatic transection. This is particularly important if hemi-hepatectomy or tri-sectionectomy is planned and/or tumors are close to the major inflow or outflow vessels. Once inflow and outflow are controlled, the plane of transection is marked. Multiple techniques have been described for transection including the crush–clamp method, ultrasonic vibrations, or radiofrequency ablation. Regardless of the technique used, transection should be performed as a careful dissection to avoid injury to inflow and outflow structures of the remnant liver, with frequent reorientation or application of the ultrasound probe to ensure an adequate margin and reconfirm anatomic landmarks. After transection, hemostatic agents or an argon beam coagulator can be used to ensure hemostasis and biliostasis. Although traditionally performed as an open resection, minimally invasive laparoscopic or robot-assisted resection of CLM is increasingly performed, particularly for small peripheral tumors. Studies have confirmed that overall long-term survival for laparoscopic resection does not appear to be significantly different than open resection in propensity-matched patients [26, 27].

As more is learned about the biology of metastatic disease and with improvements in chemotherapy, an increasing pool of patients with initially unresectable disease can often be converted to surgical candidates with curative intent. Two-stage hepatectomy is one such strategy to convert initially unresectable bilobar disease into resectable disease. Initially described by Adam et al. [28], the goal of a staged hepatectomy is to clear tumors from one lobe of the liver during an initial operation followed by a period of convalescence where ipsilateral hypertrophy can occur. At a second operation, the contralateral lobe of the liver is resected or cleared of remaining metastases. In cases where the FLR is predicted to be below threshold, preoperative portal vein embolization (PVE) can be used to induce hypertrophy of the FLR. Occlusion of portal flow to the hepatic lobe has been shown to cause compensatory hypertrophy to the contralateral lobe [29]. Preoperative portal vein embolization can be used to induce up to 30–50% increase in relative size of the contralateral lobe at 6 weeks [30, 31]. Hypertrophy can continue for up to 6 months, but most often plateaus 6–8 weeks later. Two-stage hepatectomy has been shown to be a safe procedure, with perioperative mortality of approximately 0–7% and median overall survival of 37 months [32]. Series show that approximately 70–100% of patients scheduled for two-stage hepatectomy were able to successfully complete the treatment [28, 33].

42.2.4 Postoperative Care and Follow-Up

Postoperatively, patients should be monitored for evidence of bleeding and adequate analgesia provided. Hepatic function can be measured with INR to assess liver synthetic function and serum bilirubin to assess hepatic clearance. A trend of rising INR or bilirubin over the first few postoperative days should raise concern for postoperative liver failure. Supportive measures such as phosphorus repletion and avoidance of hypotension, as well as imaging to rule out potentially correctable biliary and inflow or outflow vascular issues, should be considered.

Common complications after major hepatectomy include pleural effusion, wound breakdown, or ascites. Early identification of ascites, which can lead to wound leakage, protein wasting, and early satiety if severe, and treatment with diuretics or aldosterone antagonists is prudent to minimize sequala. An important complication that can lead to liver failure is unrecognized infection, as the systemic inflammatory response impairs liver regeneration. Biliary leak should be suspected in the setting of rising bilirubin and appropriate cross-sectional imaging is considered to identify a leak. The risk of venous thromboembolism is similar to that of other major abdominal operations, and prophylaxis should be considered [34].

42.2.5 Hepatic Artery Infusion Chemotherapy

Hepatic artery infusion (HAI) chemotherapy is a treatment that can be used either as stand-alone therapy for liver dominant disease or to adjunctively deliver chemotherapy to the functional liver remnant after metastasectomy, as the liver is a common site of recurrence. The infusion catheter is most often placed into the gastroduodenal artery, and chemotherapy, typically Fluorodeoxyuridine based, is delivered intra-arterially starting about 4 weeks after surgery. Hepatic metastases derive blood supply primarily from the hepatic artery, while normal liver cells derive it from the portal vein. That biologic phenomenon, in addition to a high first-pass metabolism, allows for the delivery of concentrated chemotherapy with minimal systemic effects [35]. Studies on the use of adjuvant HAI and systemic therapy versus systemic therapy alone found improved hepatic relapse-free survival and overall survival in patients with CLM treated with regional chemotherapy [36, 37]. Additionally, HAI chemotherapy can be effective in converting patients with unresectable disease into surgical candidates [38].

42.3 Alternatives to Resection for Local Control

42.3.1 Ablation

Tumor ablation is a minimally invasive local therapy that uses hyperthermic treatment modalities to treat metastases. Options for ablation include radiofrequency ablation (RFA), microwave ablation (MWA), and high-intensity focused ultrasound (HIFU). The goal of treatment is to induce complete necrosis of the metastasis and a safety margin similar to what would be achieved with an R0 resection. While surgical resection remains the gold standard treatment for hepatic metastases, a significant population of patients is not surgical candidates due to inadequate FLR, metastases location, or medical comorbidities. A randomized control trial investigating systemic therapy with or without RFA in patients with unresectable CLM found a significantly improved long-term survival in the combined treatment group [39, 40]. Ablation can also be used as an adjunct to surgery particularly for deep, small lesions where removal of a large volume of normal liver would be needed for tumor resection.

In the past, RFA was the most commonly used ablative therapy and relies on high-energy radio waves to heat tissues. It suffers from a relative lack of tissue penetration and is typically reserved for tumors less than 3 cm in diameter [41], as larger tumors have high rates of local recurrence [42]. RFA is also limited near major blood vessels where a heat sink effect causes relatively low temperatures and poor tumor destruction. The intense heat of ablation can cause damage to surrounding structures including bile ducts, the gallbladder as well as colon, stomach, or diaphragm [43]. As such, the size of tumors, number of tumors, and location of metastases are essential to consider prior to ablation. RFA is generally recommended for patients with the following characteristics: tumor size <3 cm, less than 5 lesions, in which the tumor is not located near major bile ducts or other structures at risk for injury [44].

MWA is a second hyperthermic ablation technique that is gaining widespread adoption. Benefits over RFA include a larger volume of necrosis, shorter procedure time, higher temperatures within the target lesion, and less variability within the treatment zone due to heat sink effect [45]. MWA may offer improved local control for perivascular tumors, while the more controlled RFA may be preferred for peribiliary lesions [46, 47]. However, specific indications for RFA and MWA continue to evolve.

When comparing ablation and surgery, a recent meta-analysis found significantly shorter hospital stay and lower complication rate (13% vs. 26%) among the ablation group, driven primarily by lower wound complications, gastrointestinal complications, and biliary complications [48]. However, ablative techniques have been associated with higher rates of local recurrence [48–50] and decreased overall survival [48, 51]. Comparisons between ablation and surgery are primarily retrospective in nature, which may be subject to patient section bias, as ablation is more often used for unresectable disease or as an adjunct to surgery. To better compare ablation and surgery in patients with CLM, the Colorectal Liver Metastases: Surgery versus Thermal Ablation (COLLISION) trial is currently enrolling to evaluate whether thermal ablation with RFA or MWA is non-inferior to hepatic resection in patients with at least one resectable or ablatable CLM [52].

42.3.2 Stereotactic Body Radiotherapy

Stereotactic body radiotherapy (SBRT) is an alternative, noninvasive treatment modality for treatment liver metastases, in which targeted radiation is delivered to a tumor in a hypofractionation schedule while minimizing radiation to adjacent normal tissue. Radiotherapy has historically been limited by poor tolerance from liver parenchyma, but the focused delivery in SBRT limits radiation-induced liver disease and liver failure [53]. SBRT has safely been administered up to 75 gray for treatment of metastases [54]. Similar to the ablative techniques previously described, SBRT offers an adjunct for local treatment of metastases in patients who are not candidates for resection. Early evidence has shown promising results for local control of 55–90%, and actuarial overall survival of 40–70% at 2 years, in patients with unresectable CLM [54, 55], with differences likely dependent on tumor volume and pathology, prior therapy, SBRT doses, and fractionation regimens [56]. The role of SBRT for local therapy and its comparison to other ablative techniques in treatment of metastasis has yet to be defined. However, SBRT may provide local control when considered for treatment of one to five metastases, which are <6 cm in diameter, and in patients with good performance status and adequate pre-SBRT liver function [56].

42.4 Conclusion

Colon cancer is the third most common malignancy in the United States, and up to 20% of patients present with metastatic disease. Hepatectomy is the gold standard for the treatment of colorectal liver metastases and is the most proven treatment option with regards to potential for a long-term cure. Adjunct therapies like portal vein embolization and two-stage hepatectomy have expanded the population of patients in whom safe resection can be performed. For patients who are poor surgical candidates, therapies like ablation and radiotherapy, offer alternative approaches for tumor eradication which may improve survival when compared to systemic therapy alone.

Open Questions

- What tumor biomarkers are predictive of outcome, and what is the role of tumor biology in determining candidacy for resection in patients with CLM?
- Is there a role for ablation in the treatment of CLM that are surgically resectable, particularly in patients with significant comorbidities?
- What is the role of hepatic artery infusion chemotherapy in unresectable CLM?
- What is the role of radiotherapy for both local control and treatment of unresectable CLM?
- What is the role of resection of CLM in the setting of extra-hepatic metastatic disease?

References

1. Siegel RL, Miller KD, Jemal A. Cancer statistics, 2019. CA Cancer J Clin. 2019;69(1):7–34.
2. Siegel RL, Ward EM, Jemal A. Trends in colorectal cancer incidence rates in the United States by tumor location and stage, 1992-2008. Cancer Epidemiol Biomark Prev. 2012;21(3):411–6.
3. Foster JH, Berman MM. Solid liver tumors. Major Probl Clin Surg. 1977;22:1–342.
4. Rees M, Tekkis PP, Welsh FK, O'Rourke T, John TG. Evaluation of long-term survival after hepatic resection for metastatic colorectal cancer: a multifactorial model of 929 patients. Ann Surg. 2008;247(1):125–35.
5. Nordlinger B, Guiguet M, Vaillant JC, et al. Surgical resection of colorectal carcinoma metastases to the liver. A prognostic scoring system to improve case selection, based on 1568 patients. Association Francaise de Chirurgie. Cancer. 1996;77(7):1254–62.
6. Simmonds PC, Primrose JN, Colquitt JL, Garden OJ, Poston GJ, Rees M. Surgical resection of hepatic metastases from colorectal cancer: a systematic review of published studies. Br J Cancer. 2006;94(7):982–99.
7. Tomlinson JS, Jarnagin WR, DeMatteo RP, et al. Actual 10-year survival after resection of colorectal liver metastases defines cure. J Clin Oncol. 2007;25(29):4575–80.
8. Alexander J, Haight C. Pulmonary resection for solitary metastatic sarcomas and carcinomas. Surg Gynecol Obstet. 1947;85(2):129–46.
9. Fong Y, Fortner J, Sun RL, Brennan MF, Blumgart LH. Clinical score for predicting recurrence after hepatic resection for metastatic colorectal cancer: analysis of 1001 consecutive cases. Ann Surg. 1999;230(3):309–18; discussion 318–321.
10. Mala T, Bohler G, Mathisen O, Bergan A, Soreide O. Hepatic resection for colorectal metastases: can preoperative scoring predict patient outcome? World J Surg. 2002;26(11):1348–53.
11. Vauthey JN, Zimmitti G, Kopetz SE, et al. RAS mutation status predicts survival and patterns of recurrence in patients undergoing hepatectomy for colorectal liver metastases. Ann Surg. 2013;258(4):619–26; discussion 626–617.
12. Pikoulis E, Margonis GA, Andreatos N, et al. Prognostic role of BRAF mutations in colorectal cancer liver metastases. Anticancer Res. 2016;36(9):4805–11.
13. Meyerhardt JA, Mayer RJ. Systemic therapy for colorectal cancer. N Engl J Med. 2005;352(5):476–87.
14. Choti MA, Sitzmann JV, Tiburi MF, et al. Trends in long-term survival following liver resection for hepatic colorectal metastases. Ann Surg. 2002;235(6):759–66.
15. Ferrero A, Vigano L, Polastri R, et al. Postoperative liver dysfunction and future remnant liver: where is the limit? Results of a prospective study. World J Surg. 2007;31(8):1643–51.
16. Soyer P, Poccard M, Boudiaf M, et al. Detection of hypovascular hepatic metastases at triple-phase helical CT: sensitivity of phases and comparison with surgical and histopathologic findings. Radiology. 2004;231(2):413–20.
17. Sacks A, Peller PJ, Surasi DS, Chatburn L, Mercier G, Subramaniam RM. Value of PET/CT in the management of liver metastases, part 1. AJR Am J Roentgenol. 2011;197(2):W256–9.
18. Kong G, Jackson C, Koh DM, et al. The use of 18F-FDG PET/CT in colorectal liver metastases—comparison with CT and liver MRI. Eur J Nucl Med Mol Imaging. 2008;35(7):1323–9.
19. Seale MK, Catalano OA, Saini S, Hahn PF, Sahani DV. Hepatobiliary-specific MR contrast agents: role in imaging the liver and biliary tree. Radiographics. 2009;29(6):1725–48.
20. Muhi A, Ichikawa T, Motosugi U, et al. Diagnosis of colorectal hepatic metastases: comparison of contrast-enhanced CT, contrast-enhanced US, superparamagnetic iron oxide-enhanced MRI, and gadoxetic acid-enhanced MRI. J Magn Reson Imaging. 2011;34(2):326–35.
21. Seo HJ, Kim MJ, Lee JD, Chung WS, Kim YE. Gadoxetate disodium-enhanced magnetic resonance imaging versus contrast-enhanced 18F-fluorodeoxyglucose positron emission tomography/computed tomography for the detection of colorectal liver metastases. Investig Radiol. 2011;46(9):548–55.
22. Welsh FK, Tekkis PP, O'Rourke T, John TG, Rees M. Quantification of risk of a positive (R1) resection margin following hepatic resection for metastatic colorectal cancer: an aid to clinical decision-making. Surg Oncol. 2008;17(1):3–13.

23. DeMatteo RP, Palese C, Jarnagin WR, Sun RL, Blumgart LH, Fong Y. Anatomic segmental hepatic resection is superior to wedge resection as an oncologic operation for colorectal liver metastases. J Gastrointest Surg. 2000;4(2):178–84.
24. Kokudo N, Tada K, Seki M, et al. Anatomical major resection versus nonanatomical limited resection for liver metastases from colorectal carcinoma. Am J Surg. 2001;181(2):153–9.
25. Sarpel U, Bonavia AS, Grucela A, Roayaie S, Schwartz ME, Labow DM. Does anatomic versus nonanatomic resection affect recurrence and survival in patients undergoing surgery for colorectal liver metastasis? Ann Surg Oncol. 2009;16(2):379–84.
26. Lewin JW, O'Rourke NA, Chiow AKH, et al. Long-term survival in laparoscopic vs open resection for colorectal liver metastases: inverse probability of treatment weighting using propensity scores. HPB (Oxford). 2016;18(2):183–91.
27. Beppu T, Wakabayashi G, Hasegawa K, et al. Long-term and perioperative outcomes of laparoscopic versus open liver resection for colorectal liver metastases with propensity score matching: a multi-institutional Japanese study. J Hepatobiliary Pancreat Sci. 2015;22(10):711–20.
28. Adam R, Laurent A, Azoulay D, Castaing D, Bismuth H. Two-stage hepatectomy: a planned strategy to treat irresectable liver tumors. Ann Surg. 2000;232(6):777–85.
29. Makuuchi M, Thai BL, Takayasu K, et al. Preoperative portal embolization to increase safety of major hepatectomy for hilar bile duct carcinoma: a preliminary report. Surgery. 1990;107(5):521–7.
30. van Lienden KP, van den Esschert JW, de Graaf W, et al. Portal vein embolization before liver resection: a systematic review. Cardiovasc Intervent Radiol. 2013;36(1):25–34.
31. Abulkhir A, Limongelli P, Healey AJ, et al. Preoperative portal vein embolization for major liver resection: a meta-analysis. Ann Surg. 2008;247(1):49–57.
32. Lam VW, Laurence JM, Johnston E, Hollands MJ, Pleass HC, Richardson AJ. A systematic review of two-stage hepatectomy in patients with initially unresectable colorectal liver metastases. HPB (Oxford). 2013;15(7):483–91.
33. Adam R, Miller R, Pitombo M, et al. Two-stage hepatectomy approach for initially unresectable colorectal hepatic metastases. Surg Oncol Clin N Am. 2007;16(3):525–36, viii.
34. Tzeng CW, Katz MH, Fleming JB, et al. Risk of venous thromboembolism outweighs post-hepatectomy bleeding complications: analysis of 5651 National Surgical Quality Improvement Program patients. HPB (Oxford). 2012;14(8):506–13.
35. Sigurdson ER, Ridge JA, Kemeny N, Daly JM. Tumor and liver drug uptake following hepatic artery and portal vein infusion. J Clin Oncol. 1987;5(11):1836–40.
36. Kemeny N, Jarnagin W, Gonen M, et al. Phase I/II study of hepatic arterial therapy with floxuridine and dexamethasone in combination with intravenous irinotecan as adjuvant treatment after resection of hepatic metastases from colorectal cancer. J Clin Oncol. 2003;21(17):3303–9.
37. Kemeny N, Huang Y, Cohen AM, et al. Hepatic arterial infusion of chemotherapy after resection of hepatic metastases from colorectal cancer. N Engl J Med. 1999;341(27):2039–48.
38. Kemeny NE, Melendez FD, Capanu M, et al. Conversion to resectability using hepatic artery infusion plus systemic chemotherapy for the treatment of unresectable liver metastases from colorectal carcinoma. J Clin Oncol. 2009;27(21):3465–71.
39. Ruers T, Punt C, Van Coevorden F, et al. Radiofrequency ablation combined with systemic treatment versus systemic treatment alone in patients with non-resectable colorectal liver metastases: a randomized EORTC Intergroup phase II study (EORTC 40004). Ann Oncol. 2012;23(10):2619–26.
40. Ruers T, Van Coevorden F, Punt CJ, et al. Local treatment of unresectable colorectal liver metastases: results of a randomized phase II trial. J Natl Cancer Inst. 2017;109(9):djx015.
41. van Duijnhoven FH, Jansen MC, Junggeburt JM, et al. Factors influencing the local failure rate of radiofrequency ablation of colorectal liver metastases. Ann Surg Oncol. 2006;13(5):651–8.
42. Shady W, Petre EN, Gonen M, et al. Percutaneous radiofrequency ablation of colorectal cancer liver metastases: factors affecting outcomes—a 10-year experience at a single center. Radiology. 2016;278(2):601–11.
43. Cirocchi R, Trastulli S, Boselli C, et al. Radiofrequency ablation in the treatment of liver metastases from colorectal cancer. Cochrane Database Syst Rev 2012(6):CD006317.
44. Gillams A, Goldberg N, Ahmed M, et al. Thermal ablation of colorectal liver metastases: a position paper by an international panel of ablation experts, The Interventional Oncology Sans Frontieres meeting 2013. Eur Radiol. 2015;25(12):3438–54.
45. Brace CL. Radiofrequency and microwave ablation of the liver, lung, kidney, and bone: what are the differences? Curr Probl Diagn Radiol. 2009;38(3):135–43.
46. Shady W, Petre EN, Do KG, et al. Percutaneous microwave versus radiofrequency ablation of colorectal liver metastases: ablation with clear margins (A0) provides the best local tumor control. J Vasc Interv Radiol. 2018;29(2):268–275 e261.
47. van Tilborg AA, Scheffer HJ, de Jong MC, et al. MWA versus RFA for perivascular and peribiliary CRLM: a retrospective patient- and lesion-based analysis of two historical cohorts. Cardiovasc Intervent Radiol. 2016;39(10):1438–46.
48. van Amerongen MJ, Jenniskens SFM, van den Boezem PB, Futterer JJ, de Wilt JHW. Radiofrequency ablation compared to surgical resection for curative treatment of patients with colorectal liver metastases—a meta-analysis. HPB (Oxford). 2017;19(9):749–56.
49. Otto G, Duber C, Hoppe-Lotichius M, Konig J, Heise M, Pitton MB. Radiofrequency ablation as first-line treatment in patients with early colorectal liver metastases amenable to surgery. Ann Surg. 2010;251(5):796–803.
50. Aliyev S, Agcaoglu O, Aksoy E, et al. Efficacy of laparoscopic radiofrequency ablation for the treatment of patients with small solitary colorectal liver metastasis. Surgery. 2013;154(3):556–62.
51. Agcaoglu O, Aliyev S, Karabulut K, et al. Complementary use of resection and radiofrequency ablation for the treatment of colorectal liver metastases: an analysis of 395 patients. World J Surg. 2013;37(6):1333–9.
52. Puijk RS, Ruarus AH, Vroomen L, et al. Colorectal liver metastases: surgery versus thermal ablation (COLLISION)—a phase III single-blind prospective randomized controlled trial. BMC Cancer. 2018;18(1):821.
53. Dawson LA, Normolle D, Balter JM, McGinn CJ, Lawrence TS, Ten Haken RK. Analysis of radiation-induced liver disease using the Lyman NTCP model. Int J Radiat Oncol Biol Phys. 2002;53(4):810–21.
54. Scorsetti M, Comito T, Tozzi A, et al. Final results of a phase II trial for stereotactic body radiation therapy for patients with inoperable liver metastases from colorectal cancer. J Cancer Res Clin Oncol. 2015;141(3):543–53.
55. Chang DT, Swaminath A, Kozak M, et al. Stereotactic body radiotherapy for colorectal liver metastases: a pooled analysis. Cancer. 2011;117(17):4060–9.
56. Cacciola A, Parisi S, Tamburella C, et al. Stereotactic body radiation therapy and radiofrequency ablation for the treatment of liver metastases: how and when? Rep Pract Oncol Radiother. 2020;25(3):299–306.

43 Regional Arterial Infusional Therapy as a Means of Controlling Hepatic Metastases

Zachary Brown, Stanley Kalata, and Alex C. Kim

Abstract

Colorectal cancer (CRC) is one of the most prevalent cancers diagnosed in the United States. Many patients are diagnosed with synchronous or metachronous liver metastasis. Liver is the most common site of metastasis and is the leading contributing factor for cancer-related morbidity and mortality for CRC. The prognosis of the patient is highly dependent on the ability to eradicate disease through surgery. In addition, modern systemic chemotherapy has significantly contributed to increased overall survival as well as conversion of initially unresectable disease to resection in patients with colorectal liver metastasis (CRLM). Despite these improvements, there are still significant volume of patients who either experience early recurrence following surgery or remain with unresectable disease. Regional infusional therapy, such as the hepatic artery pump therapy, is gaining wider acceptance with its role established in adjuvant therapy following resection in high-risk patients, definitive treatment of high-volume disease with a goal of conversion to resection, and palliation for unresectable disease.

Zachary Brown and Stanley Kalata contributed equally to this work.

Z. Brown · A. C. Kim (✉)
Division of Surgical Oncology, Department of Surgery, The James Cancer Hospital and Solove Research Institute, The Ohio State University Wexner Medical Center, Columbus, OH, USA
e-mail: Zachary.brown2@osumc.edu; alex.kim@osumc.edu

S. Kalata
Section of General Surgery, Department of Surgery, University of Michigan Medical School, Ann Arbor, MI, USA
e-mail: stkalata@med.umich.edu

Learning Objectives

- History and conception of hepatic artery infusional therapy
- Understanding pharmacology of floxuridine (FUDR) infusion
- Introduction to operative approaches for hepatic artery infusion pump placement
- Recognition of pitfalls and complications from the hepatic artery infusional therapy
- Assessment of critical trials behind hepatic artery infusion pump therapy
- Identification of future application of hepatic artery infusional therapy

43.1 Introduction

Colorectal cancer (CRC) is the fourth most prevalent cancer in the United States with 147,950 estimated cases and with 53,200 estimated cancer deaths in 2020 [1]. Colorectal liver metastasis (CRLM) is the most common site of distant disease, accounting for greater than 60% of all metastatic sites and remains the top cause of cancer-related mortality [2–4]. It is believed that approximately 15–25% of patients present with synchronous metastasis and additional 18–25% eventually develop metachronous disease within 5 years of initial diagnosis [4, 5]. The patient's prognosis is highly dependent on the resectability of the CRLM, with R0 resection being potentially curative [6, 7]. The advent of modern systemic chemotherapy (5-FU, leucovorin, and oxaliplatin (FOLFOX); 5-FU, leucovorin, and irinotecan (FOLFIRI), with or without targeted therapy such as bevacizumab (anti-VEGF) and cetuximab (anti-EGF)) has not only increased survival in patients with any metastatic disease but also allowed for the conversion of unresectable CRLM patients to resection [8, 9]. Despite these advances, a majority of patients are still

S. P. Leong et al. (eds.), *Cancer Metastasis Through the Lymphovascular System*,
https://doi.org/10.1007/978-3-030-93084-4_43

faced with unresectable metastatic disease with limited treatment options with only 15–25% of patients becoming candidates for hepatectomy [7, 10]. Recently, the hepatic artery infusion (HAI) pump therapy has gained more widespread acceptance with its role established in adjuvant therapy following resection in high-risk patients, definitive treatment of high-volume disease with a goal of conversion to resection, and palliation for unresectable disease.

The concept of liver-directed arterial infusion therapy was initially developed in the 1950s with a goal of treating primary and secondary liver cancers. The rationale of direct delivery of chemotherapy to the tumor was to capitalize on the unique hepatic circulation. The liver is supplied by a dual vascular inflow consisting of the portal vein and the hepatic artery. The portal vein is a low-pressure system with a 70–80% oxygen saturation while the hepatic artery maintains systemic arterial pressure and oxygen saturation [11]. Based on this fact, preclinical studies utilizing animal models examined and demonstrated a differential delivery of blood flow to either the native liver parenchyma or hepatic tumors from the portal vein and hepatic artery. The injection of India ink through the portal vein resulted in preferential native parenchymal uptake, while the injection through the hepatic artery demonstrated preferential tumor uptake [12]. Subsequent research established that the malignant tumor vasculature was mostly derived from the arterial system due to portal system invasion and occlusion by the tumor [12, 13]. In 1966, Nilsson et al. performed an in vivo study demonstrating selective hepatic tumor necrosis following hepatic artery ligation and improved survival in the animal models [14]. Furthermore, inadvertent and deliberate hepatic artery interruptions in patients confirmed the preferential effect in both primary and secondary liver cancers [14–18]. As such, the concept of liver regional therapy was conceived by the landmark study from Sullivan et al. In the study, the authors described their experience of placement of a catheter in the common hepatic artery and directly infusing chemotherapy for 1–2 months to treat primary and secondary liver cancer [19]. The results of the study were astounding and found that of 16 patients who received a full course of therapy, 10 patients derived true clinical benefit from the treatment. These patients exhibited a reduction of liver tumor volume and significant improvement in liver function tests with minimal toxicity. The subsequent studies by groups at the University of Wisconsin and Memorial Sloan Kettering Cancer Center revealed similar results revealing safety and efficacy [13, 20].

The early delivery systems for hepatic artery infusion (HAI) were fraught with many technical and mechanical complications. These involved placing an external catheter or port during an exploratory laparotomy or via a percutaneous approach. The delivery catheter itself had risks of clotting, kinking, leaking, dislodgement, with more serious complications including arterial thrombosis, sepsis, and air embolism. The percutaneous catheter required a connection to an extracorporeal pump device that was unwieldy and restrictive for patients. One of the largest series from that era showed an operative mortality of 12% and catheter-related morbidity of 24% [21]. In 1980, Buchwald published the first study on five patients who underwent placement of an implantable infusion pump [22]. Subsequent studies comparing implantable pumps versus ports found that unsalvageable complications occurred in 30% of pump patients and in 47% of port patients, with a complication-free survival of 12.2 months in the pump group compared to 7.3 months in the port group [23]. A metanalysis of over 3000 patients found that the complication rate was 16% with implantable pumps and over 30% for port systems. Furthermore, the median number of cycles delivered in the pump group was 12 versus 9 in the port group which also suggested increased durability and reliability of the implantable pump system [24]. There are now several models commercially available for implantation, with one of the most commonly used pumps including a 20–50 mL titanium disk that uses two chambers to separate the drug infusion from the energy source. The pump design allows the process of filling the drug infusion chamber to recharge the chemical energy power source such that only periodic refills are required for maintenance [25].

43.2 Pharmacology

Once the delivery mechanism was established, extensive studies to identify the optimal chemotherapy regimen ensued. The ideal agent for this type of therapy was required to exhibit a steep dose–response curve in which greater antitumor and toxic effects directly correlate to increased dose and/or concentration. The agent also needed to demonstrate a high first-pass hepatic metabolism and short plasma half-life which limit the systemic toxicity and increase the therapeutic index [26]. Early candidate agents included antimetabolites [5-fluoruouracil (5-FU) and floxuridine (FUDR)] [27–29], mitomycin C, cisplatin [30], and doxorubicin [31]. FUDR was quickly identified as the leading candidate. In vivo studies demonstrated that the infusion of FUDR resulted in a 15-fold higher drug concentration within tumor cells in comparison to normal hepatocytes. Moreover, there was 92% first-pass extraction which limited the systemic exposure. The result was a significant hepatic-to-systemic concentration of 400:1 [32].

Stemming from the unique delivery mechanism, the side effects of FUDR infusion are mainly limited to gastrointestinal (GI) tract inflammation/ulceration and hepatotoxicity manifested by abnormalities in liver function tests.[33]As such, prophylaxis with PPI or H2 blockers and close laboratory monitoring is crucial. Initially, it was hypothesized that the upper GI inflammation may be the result of collateralization of the blood vessels following the operation or possible exposure of the tract to the FUDR metabolites in the bile [33]. While majority of these cases can be managed with GI prophylaxis, some patients warrant endoscopic or interventional radiology procedures to appropriately treat them.

Mild biliary toxicity, manifested by elevation of aspartate transaminase (AST), alkaline phosphatase (ALT), and bilirubin, warrants dose reduction and/or infusion of dexamethasone through the pump. Due to the variability in liver enzyme derangement from metastatic disease burden, hepatic toxicity is generally defined as at least doubling of alkaline phosphatase, tripling of AST, and a 1.5-fold increase in bilirubin levels. Occasionally, treatment cessation is necessary in a small number of patients. Biliary sclerosis is a dreaded complication likely resulting from direct drug toxicity and ischemic injury from decreasing the blood supply to the bile duct via ligation of the gastroduodenal artery collaterals. The addition of dexamethasone was shown to decrease rates of biliary sclerosis (30% FUDR alone vs. 9% FUDR/dexamethasone) [34]. Systemic bevacizumab in conjunction with FUDR therapy has been shown to significantly increase the rates of biliary sclerosis [35]. As such, the addition of bevacizumab to the therapy is not recommended [35]. Maintaining a high index of suspicion is key when hepatic enzymes fail to improve after dose reduction and warrants radiographic investigation of extrinsic compression of the biliary tree from recurrence via CT with subsequent MRCP or ERCP to investigate for biliary stricture. ERCP is particularly helpful in diagnosing extrahepatic or intrahepatic biliary tree strictures not related to tumor progression and therefore stenting/dilation or external biliary drainage can be pursued [36].

43.3 Operation

The operative steps to hepatic artery infusion pump placement remain relatively similar for both open and minimally invasive approaches. Careful preoperative assessment of the hepatic arterial anatomy and its collaterals is essential. High-resolution CT angiography remains the preferred diagnostic modality in most major centers. The presence of aberrant hepatic artery anatomy is usually not a contraindication to pump placement. A retrospective review of 200 angiograms of patients with hepatic artery infusion catheters placed found that 70% demonstrated typical anatomy with the gastroduodenal artery (GDA) proximal to the bifurcation of the hepatic artery [37]. For patients with an aberrant hepatic artery, dual lumens pumps were used to cannulate the GDA and aberrant artery for uniform delivery to the liver [38]. These pumps are no longer manufactured, so the operative strategy has evolved to ligating the anomalous vessel and placing the pump in the GDA. This strategy was discovered after patients with ligated hepatic lobar arteries demonstrated uniform bilobar hepatic perfusion for 100% of patients by postoperative day 5 [39]. In the rare case of a trifurcation of the left hepatic artery/right hepatic artery/GDA off the common hepatic artery (CHA), it is still preferable to ligate the smaller lobar vessel and insert the catheter in the GDA. If the GDA is absent or unable to be cannulated for catheter placement, retrograde cannulation via the splenic artery can be attempted with ligation of the GDA, right gastric artery, left gastric artery, and smaller collateral vessels. Unfortunately, cannulating any vessel other than the GDA results in an increased risk of pump related complications and failure [40].

Staging laparoscopy to rule out extrahepatic disease should be considered for patients with unresectable liver masses to avoid committing patients to open exploration. A laparotomy is performed via an upper midline or subcostal incision followed by an extensive search for extrahepatic disease. If no extrahepatic disease is found, a portal lymphadenectomy is performed by dissecting inferolateral to the confluence of the cystic duct and common hepatic duct before dissecting the medial aspect of the bile duct. A cholecystectomy is then performed to prevent chemical cholecystitis as the cystic artery originates from the hepatic artery branches. The CHA and GDA are palpated in the porta hepatis. The distal CHA, GDA, and proper hepatic artery (PHA) are skeletonized circumferentially and all collateral branches arising from these vessels for at least 2 cm are ligated to prevent extrahepatic perfusion [41]. The right gastric artery is also ligated in a similar fashion. It is also crucial to rule out critical stenosis at the origin of the celiac axis necessitating retrograde flow through the GDA by clamping the GDA and palpating pulses in the PHA. A 2–3 cm stump of the GDA is necessary to safely secure the infusion catheter. After the target arterial anatomy is sufficiently isolated, a pump pocket is created by making an 8 cm transverse incision and forming a pocket superficial to the abdominal wall fascia. The pump pocket should be chosen so that the pump will not touch the anterior superior iliac spine or costal margin, and the pump must easily be filled with a needle through the abdominal wall. If the patient is obese with a great deal of subcutaneous fat placement on the chest may be necessary to ensure subsequent filling.

Prior to cannulating the GDA, the catheter is flushed and pump filled with heparinized saline. An aperture is made at the center of the pump pocket to facilitate catheter placement into the abdominal cavity. The pump is then secured to the pocket with stay sutures ensuring the catheter is positioned behind it. Proximal and distal control of the GDA is obtained by ligating the distal GDA with nonabsorbable sutures and placement of vascular clamps on the CHA and PHA. Another technique that has been performed is to place a clamp directly on the GDA at its takeoff from the CHA. The latter technique prevents the inadvertent advancement of the catheter into the

CHA. An arteriotomy on the GDA is created and the catheter is advanced and secured with two or three nonabsorbable ties. The catheter tip should lie exactly at the junction of the CHA and GDA. Catheter placement in the CHA may result in turbulent flow and thrombosis while placement in the distal GDA will leave this vessel segment directly exposed to cytotoxic drugs with minimal flow leading to sclerosis, pseudoaneurysm formation, or thrombosis [42]. With the catheter secure, methylene blue (using visible light) or fluorescein (using a Wood's lamp) is injected into the pump, and the liver is assessed for uniform dye distribution while extrahepatic perfusion or dye extravasation is evaluated. If any abnormalities are identified, an extensive search for potential accessory hepatic arteries or collateral vessels should be performed and those vessels should be ligated. Once the perfusion test is adequate, the surgical incisions should be closed, and the procedure is complete. Although not universally performed, a radionucleotide pump flow study can be done on postoperative day 3–5 by intravenously injecting radiolabeled sulfur colloid followed by a technetium-labeled macroaggregated albumin injection via the hepatic artery pump to ensure uniformity of hepatic perfusion and confirm the absence of extrahepatic perfusion. Hepatic arterial chemotherapy is typically initiated 7–14 days after the operation and requires biweekly labs with monthly clinical examination.

Percutaneous catheter placement via a transaxillary approach has been attempted to avoid a major abdominal operation, and while the length of stay was shorter, catheter-related complications resulting in a delay or cessation of treatment was seen in 43% of percutaneous catheter patients compared to 7% in the standard laparotomy group [43]. Minimally invasive platforms including both laparoscopic and robotic-assisted operation have been utilized and shown to have a shorter length of stays, less pain, and more immediate commencement of regional chemotherapy [44]. These are becoming increasingly popular as concomitant procedures to address primary and secondary tumors are almost universally attempted laparoscopically. Unfortunately, the complexity of the minimally invasive maneuvers required limits this approach to specialized high-volume centers. Robotic-assisted surgery for HAI pump placement has been gaining popularity with multiple single center experience studies proving them to be safe and feasible [45, 46]. Further studies are needed to determine the durability and generalizability of the results outside of these specific institutions.

43.4 Technical Complications

The most common complications after HAI pump placement are thrombosis, infection, hemorrhage, incomplete or extrahepatic perfusion, and catheter kinking/migration. Early studies as late as the 1990s reported pump complication rates between 40% and 60% [32, 47], with improvement in rates to 16–22% over time [48]. The largest review of a single-center experience was published by Allen et al. in 2005 which demonstrated 22% of patients developed a pump-related complication with a resultant 45% rate of pump salvage to complete hepatic artery infusion therapy [40]. Arterial thrombosis can result acutely intraoperatively during insertion or due to chronic exposure to cytotoxic therapy with resultant sclerosis/thrombosis. A systematic review of 1371 implantable pump placements found a thrombosis rate of 6.6%, while the use of thrombolytic therapy and/or anticoagulation results in salvage rates of approximately 30% [24]. Infection associated with the pump pocket occurs in 2–3% of patients and typically treated with oral antibiotics for simple cellulitis. The failure to improve on antibiotics, presence of infected collections, or pump extrusion necessitates exploration. Operative exploration typically involves a local operation within the pump bed to remove the infected pump and catheter, splicing the old catheter and new catheter together, and creating a new pump pocket. Early postoperative hemorrhage or rapidly expanding hematomas warrant immediate reexploration to control the source of bleeding.

Incomplete hepatic perfusion occurs in about 2% of cases from failure to ligate an aberrant hepatic artery or failure of collateralization after lobar artery ligation. Given the concern for a decrease in treatment efficacy, these patients can be followed with repeat imaging in 2–4 weeks to determine if incomplete hepatic perfusion persists. Nearly 100% of patients have a resolution of this finding, while a persistent patent accessory artery can be managed with angioembolization [49]. A more worrisome finding is extrahepatic perfusion of the stomach, duodenum, or pancreas, which occurs from collateral vessels arising distal to the catheter or catheter malposition. This presents as severe epigastric pain/diarrhea with infusions from ulcers or pancreatitis. After the diagnosis is confirmed with endoscopic or pump flow studies, embolizing the culprit vessel allows for pump salvage. A study by Sofocleous found 43 (9%) of patients had signs of extrahepatic perfusion and 19 of 24 (79%) patients with a vessel to target were able to undergo embolization with clinical success [50]. Pump malfunction is typically discovered early in the postoperative setting and can universally be salvaged with replacing the pump. Catheter occlusion/dislodgement and pump migration have also been described with an 11–36% rate of salvage, while catheter erosion necessitates pump removal.

43.5 Clinical Outcomes

43.5.1 Unresectable Colorectal Cancer Liver Metastasis

In patients with CRLM, long-term survival is achievable in properly selected patients after resection of isolated, low-volume disease with a 5-year survival up to 70% [51–53]. Unfortunately, 80–90% of patients initially present with unresectable liver disease [3, 54–56]. HAI pump therapy for patients with unresectable colorectal liver metastasis was originally adopted in the 1980s where Kemeny et al. utilized FUDR with promising results [33] (Table 43.1). The study demonstrated a partial response in 12 of 23 patients (52%) with a mean duration of response of 8 months. Interestingly, patients previously treated with systemic chemotherapy exhibited a smaller response rate of 3 of 18 (17%) patients with a mean duration of 4 months. Although the study cohort size was small, the therapeutic responses were promising especially in previously untreated patients. It was hypothesized that the degree of tumor burden in the liver had an impact of survival, especially in patients with more than 60% hepatic involvement, resulting in median survival of 13 months. This is in comparison to patients with less than 20% tumor burden where a median survival was not reached at the end of the study. Importantly, this investigation defined the adverse events stemming from the HAI pump therapy. Severe GI side effects, such as gastrointestinal ulcers, diffuse gastritis or duodenitis, were present in 46% of patients as well as 71% of patients having hepatic-enzyme abnormalities [33]. Nonetheless, this monumental study ushered the era for hepatic artery infusion pump therapy.

Building on this trial, Chang et al. hypothesized that regional delivery of FUDR was more effective compared to systemic delivery [57]. The authors randomized 64 patients, 32 for regional intra-arterial infusion and 32 for intravenous systemic infusion of FUDR. It was noted that there was a significantly improved response for patients who received intra-arterial infusion of FUDR versus systemic FUDR (62% vs. 17%, respectively) ($p < 0.0003$). The trial did not demon-

Table 43.1 Studies of hepatic artery infusion therapy and unresectable colorectal liver metastasis

Study	Treatment regimen	Control arm	Sample size	Results
Kemeny et al. 1984	HAI FUDR	None	41	PR: 12 of 23 (52%) previously treated patients Mean duration of response: 8 months PR: 3 of 18 (17%) previously treated patients Mean duration of response: 4 months
Chang et al. 1987	HAI FUDR	Systemic FUDR	64	Significantly improved response for HAI FUDR (62%) compared to systemic (17%) $p < 0.003$ No difference is OS, HAI (22%) and systemic (15%), $p = 0.27$
Kemeny et al. 2006 (CALGB 9481)	HAI FUDR + dex	Systemic 5-FU and leucovorin	135	HAI with longer OS (24.4 vs. 20 months) $p = 0.0034$, greater response rates (47% vs. 24%) $p = 0.012$, longer time to progression in the liver (9.8 vs. 7.3 months) $p = 0.034$ and a shorter time to extrahepatic progression (7 vs. 14.8) $p = 0.29$
Kemeny et al. 2009	Systemic oxaliplatin and irinotecan with HAI FUDR + dex	None	49	8% CR, 84% PR 47% of patients were able to undergo resection Median OS chemo-naive 50.8, previously treated 35 months
Ammori et al. 2012	All patients treated with HAI and systemic chemotherapy at a single institution from 2000 to 2009	None	373	Patients without extrahepatic disease had a median survival of 32 months compared to 16 months in patients with extrahepatic disease $p < 0.001$
Ammori et al. 2013	All patients treated with HAI and systemic chemotherapy at a single institution from 2000 to 2009	None	373	25% of patient underwent subsequent resection or ablation Median 5-year survival from time of HAI pump placement was 59 months in patients who underwent resection or ablation and 16 months in patients who did not ($p < 0.001$)
D'Angelica et al. 2015	Patients who received <2 cycles of oxaliplatin with no symptoms received oxaliplatin, irinotecan, and bev Patients who received >2 cycles of oxaliplatin received irinotecan, 5-FU, leucovorin, and bev Combined with HAI FUDR + dex	None	49	Overall response rate 76% 47% of patients achieved conversion to resection Median OS 38 months Median PFS 13 months Conversion to resection only factor associated with prolonged OS and PFS Bev increased biliary toxicity and stopped
Pak et al. 2018	Patients who received <2 cycles of oxaliplatin with no symptoms received oxaliplatin, irinotecan, and bev Patients who received >2 cycles of oxaliplatin received irinotecan, 5-FU, leucovorin, and bev Combined with HAI FUDR + dex	None	64	54% of patients were converted to resection Median PFS 13 months Median OS 38 months 5-year OS 36% Conversion to resection only factor associated with prolonged OS and PFS

Abbreviations: *HAI* hepatic artery infusion; *FUDR* floxuridine; *PR* partial response; *CR* complete response; *OS* overall survival; *PFS* progression-free survival; *dex* dexamethasone; *5-FU* fluorouracil; *bev* bevacizumab strate an improvement in overall survival (OS) with a 2-year actuarial survival of 22% (intra-arterial) and 15% (intravenous), $p = 0.27$. The poor survival was attributed to patients ultimately succumbing to systemic disease progression despite good locoregional control. A subset analysis revealed that patients with negative portal nodes at the time of surgery demonstrated a significant improvement in survival ($p = 0.03$). Ultimately, the study opened the next phase of trials examining the combined systemic and regional therapy for patients with unresectable CRLM [57].

To address the efficacy of standard systemic therapy compared to regional therapy, a randomized control trial (CALGB 9481) was performed. The protocol randomized 135 patients to either HAI therapy with FUDR or systemic 5-FU and leucovorin. Compared to systemic therapy, patients receiving HAI experienced a longer OS (24.4 vs. 20 months), a greater response rate (47% vs. 24%), a longer time to progression in the liver (9.8 vs. 7.3 months) but a shorter time to extrahepatic progression (7 vs. 14.8) [58]. Unfortunately, the trial was initiated prior to the establishment of modern chemotherapy consisting of oxaliplatin, irinotecan, or targeted therapies. Nonetheless, the use of 5-FU, the backbone of modern systemic chemotherapy, alone provided a greater response rate and time to liver progression thus providing evidence of efficacy of combined systemic and regional therapies utilizing 5-FU. Although the application of this regimen was questioned, it still established the role of systemic therapy in combination with regional therapy.

The ability to resect metastatic liver disease greatly influences survival. The establishment of modern systemic chemotherapy with the incorporation of oxaliplatin and irinotecan has increased interest in interval downstaging of liver metastasis for potential resection. The modern regimen allowed 30–40% of patients with previously unresectable liver disease to become surgical resection candidates [56, 59]. Similarly, HAI therapy was viewed as a bridging therapy to surgical resection. Utilization of HAI therapy with FUDR in conjunction with systemic oxaliplatin and irinotecan resulted in the conversion of 47% of previously unresectable patients. A prolonged and improved prognosis was observed particularly in chemo-naive patients, compared to chemo-conditioned patients (50.8 vs. 35 months respectively, $p = 0.02$) [60]. It is notable that the high-response rates seen in this study may be due to the use of three distinct chemotherapeutic agents. The addition of bevacizumab to systemic therapy had no impact on survival outcomes. More importantly, bevacizumab was identified as a risk factor for high rates of biliary toxicity [61, 62].

A retrospective review of a single institution experience revealed an improved median survival of 59 months in initially unresectable patients converted to either complete resection or ablation compared to 16 months in patients remaining unresectable [63]. Identified factors associated with conversion to resectability included no prior chemotherapy, a clinical risk score less than 3 (risk score including lymph node-positive tumor, disease-free interval < 12 months, >one hepatic metastasis, largest metastasis >5 cm, and preoperative CEA > 200 ng/mL), treatment on a clinical protocol, younger age, and treatment toward the end of the study period [63, 64]. Additional studies confirmed these findings with conversion to resection being the only significant factor associated with prolonged OS and PFS in patients receiving HAI therapy [61, 62]. This finding remains significant independent of method of resection including two-staged hepatectomy or repeated resection for recurrent disease [65, 66]. This principle holds true even for extrahepatic metastasis. The initial trial excluded patients with extrahepatic disease including positive portal lymphadenopathy. A study by Ammori et al. found that in patients with unresectable CRLM treated with HAI and systemic therapy, patients without extrahepatic disease had a median survival of 32 months compared to 18 months in patients with one site of extrahepatic disease or 9 months in patients with multiple sites of extrahepatic disease [67]. Concurrent resection of both hepatic and extrahepatic disease can be associated with long-term survival [68]. Again these are all based on reporting from a single institution study. In a meta-analysis of HAI for unresectable CRLM including 10 studies and 1277 patients, there was no statistical difference in OS between HAI alone and systemic chemotherapy although HAI had a greater tumor response rate [69].

43.5.2 Resectable Colorectal Cancer Liver Metastasis

Complete surgical resection of liver metastasis results in 5-year survival of 37–50% and remains the only curative therapy [64, 70]. At a median follow-up of 30 months, approximately 60% of patients who underwent surgery for curative intent for colorectal liver metastasis presented with recurrences with 43.2% of recurrences being isolated intrahepatic disease [70]. The failure of locoregional control was predicted by R1 resection and the use of loco-ablative therapy [70]. Conversely, the recurrence rates were significantly lower in patients receiving adjuvant systemic chemotherapy [71, 72]. With a goal to examine the efficacy of liver-directed regional therapy for reduction in postoperative recurrence rates, Kemeny et al. examined the role of adjuvant HAI therapy plus intravenous systemic 5-FU compared to systemic chemotherapy alone following complete surgical resection [73] (Table 43.2). Although not statistically significant, a trend toward improved (OS) with a median survival of 72.2 months in the combination therapy group compared to 59.3 months in the control group ($p = 0.11$) was observed. Multivariate analysis validated the combination therapy as a

Table 43.2 Studies of hepatic artery infusion therapy and resected colorectal liver metastasis

Study	Treatment regimen	Control arm	Sample size	Results
Kemeny et al. 1999	HAI FUDR + systemic 5-FU	Systemic 5-FU	156	Median OS combination therapy 72.2 months and monotherapy 59.3 months ($p = 0.21$) Median hepatic PFS combination not reached versus monotherapy 42.7 months ($p < 0.0001$) Median overall PFS combination 37.4 months versus mono therapy 17.2 months ($p = 0.06$)
Kemeny et al. 2005	HAI FUDR + systemic 5-FU	Systemic 5-FU	156	10-year OS combination 41.1 and monotherapy 27.2 months
Kemeny et al. 2002	HAI FUDR + systemic 5-FU	None	109	4-year recurrence-free rate (RFR) combination therapy 45.7% versus no therapy 25.2% ($p = 0.04$) 4-year liver RFR combination therapy 66.9% versus no therapy 43% ($p = 0.03$) 4-year OS combination therapy 61.5% versus no therapy 52.7% ($p = 0.6$)
House et al. 2011	HAI FUDR + systemic modern chemotherapy	Systemic modern chemotherapy	250	5-year disease-specific survival combination therapy 76% versus systemic therapy 55% ($p < 0.01$) 5-year hepatic RFS combination therapy 79% versus systemic therapy 55% ($p < 0.001$) 5-year RFS combination therapy 48% versus systemic therapy 25% ($p < 0.01$)
Goéré et al. 2013	HAI FUDR + systemic 5-FU	Systemic modern chemotherapy	498	5-year OS combination therapy 54% versus systemic therapy 52% ($p = 0.34$) 3-year DFS combination therapy 33% versus systemic therapy 5% ($p < 0.0001$) 3-year hepatic DFS combination therapy 49% versus systemic therapy 21%, ($p = 0.0008$)
Gholami et al. 2019	HAI FUDR	No HAI	674	5-year OS (KRAS WT) HAI 78% vs no HAI 57% ($p < 0.001$) 5-year OS (KRAS mut) HAI 59% versus no HAI 40% ($p < 0.001$)

Abbreviations: *HAI* hepatic artery infusion; *FUDR* floxuridine; *PR* partial response; *CR* complete response; *OS* overall survival; *PFS* progression-free survival; *dex* dexamethasone; *5-FU* fluorouracil; *bev* bevacizumab

predictor of improved OS (HR 2.34 (1.10–4.98), $p = 0.027$). Despite the lack of significant improvement in OS, the study clearly demonstrated improved hepatic progression-free survival (PFS) with hepatic PFS in the monotherapy arm being 42.7 months while hepatic PFS was not reached for the combination therapy ($p < 0.001$) [73]. At the 10-year follow-up, the authors were revealed a persistent and significant PFS in the combination therapy group (31.3 vs. 17.2 months, $p = 0.02$) [74].

Although the previous study introduced the concept of adjuvant HAI therapy, its true effect as an adjuvant therapy was never examined. In 2002, Kemeny et al. examined the role of adjuvant HAI therapy plus intravenous 5-FU following complete surgical resection versus surgical resection alone [75]. The study demonstrated a 4-year recurrence rate of 46% in the HAI group compared to 25% in the control arm ($p = 0.04$). Additionally, the HAI group had a significant improvement in liver-specific recurrence-free rate compared to the control arm (67% vs. 43% respectively, $p = 0.03$) [75]. With the goal of detecting liver-specific recurrence-free rates, the study was unfortunately not powered to detect OS. Besides the study design, the failure to demonstrate improvement in OS was hypothesized to be secondary to the absence of modern systemic therapy or due to extrahepatic disease progression following reduction of systemic therapy dosing. Nonetheless, HAI pump therapy was established as a multimodality adjuvant treatment option to reduce locoregional recurrence rates in high-risk patients. In a pooled analysis of four consecutive protocols examining adjuvant therapy with the combination of HAI and systemic therapy following liver resection, the 5- and 10-year median survival was found to be 66% and 48% respectively in patients treated after 2003 [76].

While modern adjuvant systemic therapy introduced oxaliplatin and irinotecan to the 5-FU backbone, these two particular drugs were not previously evaluated for regional therapy. With FUDR being the key component of HAI therapy, the question of use and efficacy of different HAI-based chemotherapy was posed. Goere et al. sought to examine two key issues in their study. First, they examined the role of HAI pump therapy delivery of oxaliplatin in patients deemed high risk for recurrent disease following liver metastatectomy. [77] Second, they attempted to compare the role of regional delivery versus systemic delivery of modern chemotherapy

[77]. The study demonstrated a significantly prolonged disease-free survival in patients receiving adjuvant HAI plus systemic 5-FU versus modern systemic chemotherapy after resection of CRLM [77]. Similar to the previous two studies, the author faced difficulties in demonstrating significant improvement in OS. However, the benefits of oxaliplatin infusion including a relatively quick infusion of oxaliplatin (2-h infusion every 2 weeks compared to a 14-day infusion every 4 weeks), decreased rates of biliary toxicity, and a potential decrease in the frequency of oxaliplatin-based neuropathy was established. Although the study did not directly examine the role of HAI + modern chemotherapy versus modern chemotherapy, it provided evidence for potential other chemotherapeutic options through HAI pump delivery.

The efficacy of modern systemic chemotherapy with HAI compared to modern systemic chemotherapy was finally examined by House et al. [78] This key study addressed the concerns of efficacy following the addition of HAI therapy to modern systemic chemotherapy. The study demonstrated that the combination therapy compared to systemic therapy was independently associated with significant improvement in 3- and 5-year disease-specific survival (3 year: 86% vs. 76%, 5 year: 75% vs. 55%, $p < 0.01$) and 5-year hepatic recurrence-free survival (79% vs. 55%, $p < 0.001$). Interestingly, 5-year overall recurrence-free survival was also significantly improved (48% vs. 25% $p < 0.01$), resulting in the question of does avoidance of early liver recurrence potentially prevents other site recurrrence [78]. The exact mechanism of this observation remains debatable. However, the authors explained that the regional disease control by HAI therapy may potentially contribute to the improved survival benefit by providing long-term control of liver metastasis. Although this study hinted at the control of the extrahepatic recurrence with regional therapy, the mechanism of this prevention remains unclear. It is hypothesized that the exposure to systemic chemotherapy alone is the key to the suppression of extrahepatic metastasis. The suppression of extrahepatic recurrence was demonstrated even in the setting of dose reduction of systemic chemotherapy in patients receiving HAI compared to standard dosing [79]. As such, it indirectly suggested that the addition of IV systemic chemotherapy may be responsible for the suppression of extrahepatic disease and not the HAI therapy.

Currently, there is starting to be a greater understanding on how tumor biology may influence response to therapy. KRAS mutations are associated with worse survival and higher recurrence risks after resection of CRLM [80, 81]. Examination of 674 patients stratified by KRAS mutational status (418 KRAS wild-type and 256 KRAS mutant), exposure to adjuvant HAI therapy following resection revealed that use of HAI was independently associated with improved survival regardless of KRAS mutational status ($p < 0.002$). In patients with KRAS wild-type tumors, the 5-year OS was 78% for patients receiving HAI therapy versus 57% in patients without HAI ($P < 0.001$). Similarly, in patients with KRAS mutant tumors, the 5-year OS with HAI was 59% versus 40% in patients treated without HAI therapy ($p < 0.001$) [82]. This indicates that although KRAS mutation is an indicator of poor tumor biology and overall worse prognosis long-term, regional therapy with HAI may trump the genetic prognosis and may improve clinical outcomes.

43.6 Conclusion

Additional studies are required to further define the role of HAI in the treatment of CRLM. The implementation of HAI therapy has been shown to be safe but further efficacy still needs to be established [83, 84]. Based on available data, HAI appears to have a greater impact on disease progression and recurrence without a significant OS benefit. Additionally, HAI is generally not an accepted treatment modality outside of a clinical trial in patients with extrahepatic disease [67, 68, 85]. Basic fundamentals of tumor biology still reign: patients who have progressed on systemic chemotherapy prior to HAI display a more aggressive tumor biology and have decreased survival compared to those who have not previously received systemic therapy. In addition, the application of HAI therapy to other primary or secondary liver tumors is being investigated, such as intrahepatic cholangiocarcinoma (ICC). ICC is the second most common primary liver malignancy following hepatocellular carcinoma (HCC). Resection with portal lymphadenectomy remains the only curative therapy. Unfortunately, most patients are initially diagnosed with either locally advanced or metastatic disease [86]. For these patients, standard systemic therapy consisting of gemcitabine and cisplatin offers a median OS of 11.7 months, median PFS of 8 months, and a tumor control rate of 81.4% [87]. These patients ultimately progress and succumb to their disease due to ineffective therapies and limited treatment options.

HAI therapy has gained interest in the treatment of ICC. (Table 43.3). The ideal candidates for this therapy are evolving but initial studies included patients with advanced, multifocal disease confined to the liver or unresectable nature of the liver tumor. The initial phase II clinical trial included 34 patients with unresectable primary liver cancer (26 ICC and 8 hepatocellular carcinoma (HCC)) for treatment with HAI-FUDR. Of the 34 patients, there were 16 patients with a partial response, 14 patients had stable disease, and 3 patients had progressive disease. Patients who responded to therapy had a median disease-specific survival of 29.5 months. The patients with ICC had a greater response rate than HCC (53.8% vs. 25%, respectively) and thus established a potential role of HAI therapy for ICC [88].

Subsequently, a retrospective review analyzed 104 patients with unresectable ICC with disease confined to the liver who were treated with combined HAI and systemic therapy (gemcitabine-based) or systemic therapy alone [89]. The combination therapy demonstrated a response rate of 59% compared to 39% in systemic therapy alone. Unlike the CRLM studies, the HAI/systemic therapy group exhibited improved OS (30.8 months vs. 18.4 months, $p < 0.001$) regardless of nodal status [89].

Systemic gemcitabine and oxaliplatin were then investigated in combination with FUDR via HAI. The study included patients with unresectable ICC and also allowed for resectable regional lymphadenopathy. Eighty-four percentage of patients achieved disease control at 6 months with 58% of patients showing a partial response. The 6-month PFS was 67% for pretreated patients and 89% for chemotherapy-naive patients. The median OS was 25 months. Similar to prior studies, nodal disease did not alter OS or PFS. Four patients experienced a significant enough response to undergo resection of their tumor [90].

Although there is a limitation in number of studies, the use of HAI pump therapy, particularly in ICC is garnering increased interest. Stemming from these studies, a randomized controlled trial is crucial to clearly establish a role of this therapy for ICC. In addition, we can aim to model additional trials to follow suit for CRLM. Lastly, with the application of HAI for ICC and its comparison to resection, HAI therapy becomes the medium to answer the question of the prognosis role of HAI compared to resection.

Open Questions

The questions remain

- Is there role of hepatic artery infusional therapy for other primary or secondary liver cancers?
- Are there other agents including small molecule inhibitors and immunotherapy that can be utilized as a regional therapy?
- Can we establish a role of neoadjuvant regional therapy in patients with resectable disease?
- Is there a difference in the efficacy of hepatic artery infusion therapy based on the sidedness of the primary colon cancer?
- Is there a role for genetics in degerming candidacy and efficacy of the hepatic artery infusional therapy?

Table 43.3 Studies of hepatic artery infusion therapy and intrahepatic cholangiocarcinoma

Study	Treatment regimen	Control arm	Sample size	Results
Jarnagin et al. 2009	HAI FUDR + dex	None	34 patients: 26 ICC and 8 HCC	16 patients with PR 14 patients with stable disease 3 patients with progressive disease Response rate: ICC 53.8% versus HCC 25% 2-year survival 67%
Kemeny et al. 2011	HAI FUDR with dex + bev	None	22 patients 18 ICC and 4 HCC	31.8% PR 68.2% stable disease Median survival: 31.1 months PFS 8.45 months Hepatic PFS 11.3 months Adding bev appeared to increase toxicity without clear improvement in outcomes
Konstantinidis et al. 2016	Systemic gemcitabine-based chemotherapy with HAI FUDR	Gemcitabine-based chemotherapy	104	Response rate: HAI + Chemo—59%, systemic therapy alone 39% ($p = 0.11$) OS: HAI + Chemo—30.8 months, systemic therapy alone—18.4 months ($P < 0.001$) 8 patients responded enough to undergo resection with median OS of 37 months
Cercek et al. 2019	Systemic gemcitabine and oxaliplatin with HAI FUDR	None	38	58% with partial radiologic response 84% achieved disease control at 6 months 4 patients were able to undergo subsequent resection Median PFS 11.8 months Median OS was 25 months 6-month progression-free rate was 67% for pretreated patients and 89% for chemotherapy-naïve patients.

Abbreviations: *HAI* hepatic artery infusion; *ICC* intrahepatic cholangiocarcinoma; *HCC* hepatocellular carcinoma; *FUDR* floxuridine; *PR* partial response; *OS* overall survival; *PFS* progression-free survival; *dex* dexamethasone; *bev* bevacizumab

References

1. Siegel RL, Miller KD, Jemal A. Cancer statistics, 2020. CA Cancer J Clin. 2020;70(1):7–30.
2. Fong Y, Cohen AM, Fortner JG, et al. Liver resection for colorectal metastases. J Clin Oncol. 1997;15(3):938–46.
3. Foster JH. Treatment of metastatic disease of the liver: a skeptic's view. Semin Liver Dis. 1984;4(2):170–9.
4. van Gestel YR, de Hingh IH, van Herk-Sukel MP, et al. Patterns of metachronous metastases after curative treatment of colorectal cancer. Cancer Epidemiol. 2014;38(4):448–54.
5. van der Geest LG, Lam-Boer J, Koopman M, Verhoef C, Elferink MA, de Wilt JH. Nationwide trends in incidence, treatment and survival of colorectal cancer patients with synchronous metastases. Clin Exp Metastasis. 2015;32(5):457–65.
6. Ito K, Govindarajan A, Ito H, Fong Y. Surgical treatment of hepatic colorectal metastasis: evolving role in the setting of improving systemic therapies and ablative treatments in the 21st century. Cancer J. 2010;16(2):103–10.
7. Akgül Ö, Çetinkaya E, Ersöz Ş, Tez M. Role of surgery in colorectal cancer liver metastases. World J Gastroenterol. 2014;20(20):6113–22.
8. Gallagher DJ, Kemeny N. Metastatic colorectal cancer: from improved survival to potential cure. Oncology. 2010;78(3–4):237–48.
9. Adam R, Delvart V, Pascal G, et al. Rescue surgery for unresectable colorectal liver metastases downstaged by chemotherapy: a model to predict long-term survival. Ann Surg. 2004;240(4):644–57; discussion 657–648.
10. Adam R, Vinet E. Regional treatment of metastasis: surgery of colorectal liver metastases. Ann Oncol. 2004;15(Suppl 4):103–6.
11. Chu DZ, Hutchins L, Lang NP. Regional chemotherapy of liver metastases from colorectal carcinoma: hepatic artery or portal vein infusion? Cancer Treat Rev. 1988;15(4):243–56.
12. Breedis C, Young G. The blood supply of neoplasms in the liver. Am J Pathol. 1954;30(5):969–77.
13. Fortner JG, Beattie EJ Jr, Shiu MH, et al. Surgery in liver tumors. Curr Probl Surg. 1972;9(6):1–56.
14. Nilsson LA. Therapeutic hepatic artery ligation in patients with secondary liver tumors. Rev Surg. 1966;23(5):374–6.
15. Mori W, Masuda M, Miyanaga T. Hepatic artery ligation and tumor necrosis in the liver. Surgery. 1966;59(3):359–63.
16. Almersjö O, Bengmark S, Engevik L, Hafström L, Loughridge B, Nilsson L. Serum enzyme changes after hepatic dearterialization in man. Ann Surg. 1968;167(1):9.
17. Plengvanit U. Treatment of primary carcinoma of the liver by hepatic artery ligation: Preliminary report of 409 cases. Paper presented at International Association for the Study of the Liver. Liver Research; translation of the 3^< rd> International Symposium, Tokyo and Kyoto, 1966, Antwerp, Tijdschrift voor Gastroenterologie; 1967.
18. Madding GF, Kennedy PA, Sogemeier E. Hepatic artery ligation for metastatic tumor in the liver. Am J Surg. 1970;120(1):95–6.
19. Sullivan RD, Norcross JW, Watkins E Jr. Chemotherapy of metastatic liver cancer by prolonged hepatic-artery infusion. N Engl J Med. 1964;270(7):321–7.
20. Wirtanen GW, Bernhardt LC, Mackman S, Ramirez G, Curreri AR, Ansfield FJ. Hepatic artery and celiac axis infusion for the treatment of upper abdominal malignant lesions. Ann Surg. 1968;168(1):137–41.
21. Reed ML, Vaitkevicius VK, Al-Sarraf M, et al. The practicality of chronic hepatic artery infusion therapy of primary and metastatic hepatic malignancies: ten-year results of 124 patients in a prospective protocol. Cancer. 1981;47(2):402–9.
22. Buchwald H, Grage TB, Vassilopoulos PP, Rohde TD, Varco RL, Blackshear PJ. Intraarterial infusion chemotherapy for hepatic carcinoma using a totally implantable infusion pump. Cancer. 1980;45(5):866–9.
23. Heinrich S, Petrowsky H, Schwinnen I, et al. Technical complications of continuous intra-arterial chemotherapy with 5-fluorodeoxyuridine and 5-fluorouracil for colorectal liver metastases. Surgery. 2003;133(1):40–8.
24. Bacchetti S, Pasqual E, Crozzolo E, Pellarin A, Cagol PP. Intraarterial hepatic chemotherapy for unresectable colorectal liver metastases: a review of medical devices complications in 3172 patients. Med Devices (Auckland, NZ). 2009;2:31.
25. Blackshear P, Rohde T, Dorman F, Buchwald H. An implantable pump for long-term intravascular drug infusion. Med Instrum. 1981;15(4):226.
26. Burrows J, Talley R, Drake E, San Diego E, Tucker W. Infusion of fluorinated pyrimidines into hepatic artery for treatment of metastatic carcinoma of liver. Cancer. 1967;20(11):1886–92.
27. Clarkson B, Young C, Dierick W, et al. Effects of continuous hepatic artery infusion of antimetabolites on primary and metastatic cancer of the liver. Cancer. 1962;15(3):472–88.
28. Ensminger WD, Rosowsky A, Raso V, et al. A clinical-pharmacological evaluation of hepatic arterial infusions of 5-fluoro-2′-deoxyuridine and 5-fluorouracil. Cancer Res. 1978;38(11 Part 1):3784–92.
29. Hohn DC, Rayner AA, Economou JS, Ignoffo RJ, Lewis BJ, Stagg RJ. Toxicities and complications of implanted pump hepatic arterial and intravenous floxuridine infusion. Cancer. 1986;57(3):465–70.
30. Kelsen D, Hoffman J, Alcock N, et al. Pharmacokinetics of cisplatin regional hepatic infusions. Am J Clin Oncol. 1982;5(2):173–8.
31. Garnick MB, Ensminger WD, Israel M. A clinical-pharmacological evaluation of hepatic arterial infusion of adriamycin. Cancer Res. 1979;39(10):4105–10.
32. Ensminger W, Gyves J. Clinical pharmacology of hepatic arterial chemotherapy. Paper presented at Seminars in oncology; 1983.
33. Kemeny N, Daly J, Oderman P, et al. Hepatic artery pump infusion: toxicity and results in patients with metastatic colorectal carcinoma. J Clin Oncol. 1984;2(6):595–600.
34. Kemeny N, Seiter K, Niedzwiecki D, et al. A randomized trial of intrahepatic infusion of fluorodeoxyuridine with dexamethasone versus fluorodeoxyuridine alone in the treatment of metastatic colorectal cancer. Cancer. 1992;69(2):327–34.
35. Cercek A, D'Angelica M, Power D, et al. Floxuridine hepatic arterial infusion associated biliary toxicity is increased by concurrent administration of systemic bevacizumab. Ann Surg Oncol. 2014;21(2):479–86.
36. Kemeny MM, Battifora H, Blayney DW, et al. Sclerosing cholangitis after continuous hepatic artery infusion of FUDR. Ann Surg. 1985;202(2):176.
37. Daly JM, Kemeny N, Oderman P, Botet J. Long-term hepatic arterial infusion chemotherapy: anatomic considerations, operative technique, and treatment morbidity. Arch Surg. 1984;119(8):936–41.
38. Kemeny MM, Hogan JM, Goldberg DA, et al. Continuous hepatic artery infusion with an implantable pump: problems with hepatic artery anomalies. Surgery. 1986;99(4):501–4.
39. Rayner AA, Kerlan RK, Stagg RJ, Price DC, Hohn DC. Total hepatic arterial perfusion after occlusion of variant lobar vessels: implications for hepatic arterial chemotherapy. Surgery. 1986;99(6):708–15.
40. Allen PJ, Nissan A, Picon AI, et al. Technical complications and durability of hepatic artery infusion pumps for unresectable colorectal liver metastases: an institutional experience of 544 consecutive cases. J Am Coll Surg. 2005;201(1):57–65.
41. Qadan M, D'Angelica MI, Kemeny NE, Cercek A, Kingham TP. Robotic hepatic arterial infusion pump placement. HPB. 2017;19(5):429–35.
42. Leal JN, Kingham TP. Hepatic artery infusion chemotherapy for liver malignancy. Surg Oncol Clin. 2015;24(1):121–48.

43. Arru M, Aldrighetti L, Gremmo F, et al. Arterial devices for regional hepatic chemotherapy: transaxillary versus laparotomic access. J Vasc Access. 2000;1(3):93–9.
44. Franklin ME, Gonzalez JJ. Laparoscopic placement of hepatic artery catheter for regional chemotherapy infusion: technique, benefits, and complications. Surg Laparosc Endosc Percutan Tech. 2002;12(6):398–407.
45. Dhir M, Zenati MS, Padussis JC, et al. Robotic assisted placement of hepatic artery infusion pump is a safe and feasible approach. J Surg Oncol. 2016;114(3):342–7.
46. Dhir M, Magge D, Novak S, Bartlett DL, Zureikat AH. Robotic-assisted placement of an hepatic artery infusion pump and catheter for regional chemotherapy of the liver. Ann Surg Oncol. 2016;23(5):755–6.
47. Campbell KA, Burns RC, Sitzmann JV, Lipsett PA, Grochow LB, Niederhuber JE. Regional chemotherapy devices: effect of experience and anatomy on complications. J Clin Oncol. 1993;11(5):822–6.
48. Scaife CL, Curley SA, Izzo F, et al. Feasibility of adjuvant hepatic arterial infusion of chemotherapy after radiofrequency ablation with or without resection in patients with hepatic metastases from colorectal cancer. Ann Surg Oncol. 2003;10(4):348–54.
49. Curley SA, Chase JL, Roh MS, Hohn DC. Technical considerations and complications associated with the placement of 180 implantable hepatic arterial infusion devices. Surgery. 1993;114(5):928–35.
50. Sofocleous CT, Schubert J, Kemeny N, et al. Arterial embolization for salvage of hepatic artery infusion pumps. J Vasc Interv Radiol. 2006;17(5):801–6.
51. Aloia TA, Vauthey JN, Loyer EM, et al. Solitary colorectal liver metastasis: resection determines outcome. Arch Surg (Chicago, Ill: 1960). 2006;141(5):460–6; discussion 466–467.
52. Hur H, Ko YT, Min BS, et al. Comparative study of resection and radiofrequency ablation in the treatment of solitary colorectal liver metastases. Am J Surg. 2009;197(6):728–36.
53. Lee WS, Yun SH, Chun HK, et al. Clinical outcomes of hepatic resection and radiofrequency ablation in patients with solitary colorectal liver metastasis. J Clin Gastroenterol. 2008;42(8):945–9.
54. Van Cutsem E, Nordlinger B, Adam R, et al. Towards a pan-European consensus on the treatment of patients with colorectal liver metastases. Eur J Cancer (Oxford, England: 1990). 2006;42(14):2212–21.
55. Kemeny N. Management of liver metastases from colorectal cancer. Oncology (Williston Park). 2006;20(10):1161–76, 1179; discussion 1179–1180, 1185–1166.
56. Alberts SR, Horvath WL, Sternfeld WC, et al. Oxaliplatin, fluorouracil, and leucovorin for patients with unresectable liver-only metastases from colorectal cancer: a north central cancer treatment group phase II study. J Clin Oncol. 2005;23(36):9243–9.
57. Chang AE, Schneider PD, Sugarbaker PH, Simpson C, Culnane M, Steinberg SM. A prospective randomized trial of regional versus systemic continuous 5-fluorodeoxyuridine chemotherapy in the treatment of colorectal liver metastases. Ann Surg. 1987;206(6):685–93.
58. Kemeny NE, Niedzwiecki D, Hollis DR, et al. Hepatic arterial infusion versus systemic therapy for hepatic metastases from colorectal cancer: a randomized trial of efficacy, quality of life, and molecular markers (CALGB 9481). J Clin Oncol. 2006;24(9):1395–403.
59. Pozzo C, Basso M, Cassano A, et al. Neoadjuvant treatment of unresectable liver disease with irinotecan and 5-fluorouracil plus folinic acid in colorectal cancer patients. Ann Oncol. 2004;15(6):933–9.
60. Kemeny NE, Melendez FD, Capanu M, et al. Conversion to resectability using hepatic artery infusion plus systemic chemotherapy for the treatment of unresectable liver metastases from colorectal carcinoma. J Clin Oncol. 2009;27(21):3465–71.
61. D'Angelica MI, Correa-Gallego C, Paty PB, et al. Phase II trial of hepatic artery infusional and systemic chemotherapy for patients with unresectable hepatic metastases from colorectal cancer: conversion to resection and long-term outcomes. Ann Surg. 2015;261(2):353–60.
62. Pak LM, Kemeny NE, Capanu M, et al. Prospective phase II trial of combination hepatic artery infusion and systemic chemotherapy for unresectable colorectal liver metastases: long term results and curative potential. J Surg Oncol. 2018;117(4):634–43.
63. Ammori JB, Kemeny NE, Fong Y, et al. Conversion to complete resection and/or ablation using hepatic artery infusional chemotherapy in patients with unresectable liver metastases from colorectal cancer: a decade of experience at a single institution. Ann Surg Oncol. 2013;20(9):2901–7.
64. Fong Y, Fortner J, Sun RL, Brennan MF, Blumgart LH. Clinical score for predicting recurrence after hepatic resection for metastatic colorectal cancer: analysis of 1001 consecutive cases. Ann Surg. 1999;230(3):309–18; discussion 318–321.
65. Brouquet A, Abdalla EK, Kopetz S, et al. High survival rate after two-stage resection of advanced colorectal liver metastases: response-based selection and complete resection define outcome. J Clin Oncol. 2011;29(8):1083–90.
66. Cardona K, Donataccio D, Kingham TP, et al. Treatment of extensive metastatic colorectal cancer to the liver with systemic and hepatic arterial infusion chemotherapy and two-stage hepatic resection: the role of salvage therapy for recurrent disease. Ann Surg Oncol. 2014;21(3):815–21.
67. Ammori JB, D'Angelica MI, Fong Y, et al. Hepatic artery infusional chemotherapy in patients with unresectable colorectal liver metastases and extrahepatic disease. J Surg Oncol. 2012;106(8):953–8.
68. Carpizo DR, Are C, Jarnagin W, et al. Liver resection for metastatic colorectal cancer in patients with concurrent extrahepatic disease: results in 127 patients treated at a single center. Ann Surg Oncol. 2009;16(8):2138–46.
69. Mocellin S, Pilati P, Lise M, Nitti D. Meta-analysis of hepatic arterial infusion for unresectable liver metastases from colorectal cancer: the end of an era? J Clin Oncol. 2007;25(35):5649–54.
70. de Jong MC, Pulitano C, Ribero D, et al. Rates and patterns of recurrence following curative intent surgery for colorectal liver metastasis: an international multi-institutional analysis of 1669 patients. Ann Surg. 2009;250(3):440–8.
71. Nordlinger B, Sorbye H, Glimelius B, et al. Perioperative FOLFOX4 chemotherapy and surgery versus surgery alone for resectable liver metastases from colorectal cancer (EORTC 40983): long-term results of a randomised, controlled, phase 3 trial. Lancet Oncol. 2013;14(12):1208–15.
72. Wang ZM, Chen YY, Chen FF, Wang SY, Xiong B. Peri-operative chemotherapy for patients with resectable colorectal hepatic metastasis: a meta-analysis. Eur J Surg Oncol. 2015;41(9):1197–203.
73. Kemeny N, Huang Y, Cohen AM, et al. Hepatic arterial infusion of chemotherapy after resection of hepatic metastases from colorectal cancer. N Engl J Med. 1999;341(27):2039–48.
74. Kemeny NE, Gonen M. Hepatic arterial infusion after liver resection. N Engl J Med. 2005;352(7):734–5.
75. Kemeny MM, Adak S, Gray B, et al. Combined-modality treatment for resectable metastatic colorectal carcinoma to the liver: surgical resection of hepatic metastases in combination with continuous infusion of chemotherapy—an intergroup study. J Clin Oncol. 2002;20(6):1499–505.
76. Kemeny NE, Chou JF, Boucher TM, et al. Updated long-term survival for patients with metastatic colorectal cancer treated with liver resection followed by hepatic arterial infusion and systemic chemotherapy. J Surg Oncol. 2016;113(5):477–84.
77. Goéré D, Benhaim L, Bonnet S, et al. Adjuvant chemotherapy after resection of colorectal liver metastases in patients at high risk of hepatic recurrence: a comparative study between hepatic arterial infusion of oxaliplatin and modern systemic chemotherapy. Ann Surg. 2013;257(1):114–20.

78. House MG, Kemeny NE, Gönen M, et al. Comparison of adjuvant systemic chemotherapy with or without hepatic arterial infusional chemotherapy after hepatic resection for metastatic colorectal cancer. Ann Surg. 2011;254(6):851–6.
79. Srouji RM, Narayan RR, Boerner T, et al. Extrahepatic recurrence rates in patients receiving adjuvant hepatic artery infusion and systemic chemotherapy after complete resection of colorectal liver metastases. J Surg Oncol. 2020;122(8):1536–42.
80. Karagkounis G, Torbenson MS, Daniel HD, et al. Incidence and prognostic impact of KRAS and BRAF mutation in patients undergoing liver surgery for colorectal metastases. Cancer. 2013;119(23):4137–44.
81. Kemeny NE, Chou JF, Capanu M, et al. KRAS mutation influences recurrence patterns in patients undergoing hepatic resection of colorectal metastases. Cancer. 2014;120(24):3965–71.
82. Gholami S, Kemeny NE, Boucher TM, et al. Adjuvant hepatic artery infusion chemotherapy is associated with improved survival regardless of kras mutation status in patients with resected colorectal liver metastases: a retrospective analysis of 674 patients. Ann Surg. 2020;272(2):352–6.
83. Buisman FE, Grünhagen DJ, Homs MYV, et al. Adjuvant hepatic arterial infusion pump chemotherapy after resection of colorectal liver metastases: results of a safety and feasibility study in the Netherlands. Ann Surg Oncol. 2019;26(13):4599–607.
84. Buisman FE, Homs MYV, Grünhagen DJ, et al. Adjuvant hepatic arterial infusion pump chemotherapy and resection versus resection alone in patients with low-risk resectable colorectal liver metastases—the multicenter randomized controlled PUMP trial. BMC Cancer. 2019;19(1):327.
85. Karanicolas PJ, Metrakos P, Chan K, et al. Hepatic arterial infusion pump chemotherapy in the management of colorectal liver metastases: expert consensus statement. Curr Oncol. 2014;21(1):e129–36.
86. El-Diwany R, Pawlik TM, Ejaz A. Intrahepatic cholangiocarcinoma. Surg Oncol Clin N Am. 2019;28(4):587–99.
87. Valle J, Wasan H, Palmer DH, et al. Cisplatin plus gemcitabine versus gemcitabine for biliary tract cancer. N Engl J Med. 2010;362(14):1273–81.
88. Jarnagin WR, Schwartz LH, Gultekin DH, et al. Regional chemotherapy for unresectable primary liver cancer: results of a phase II clinical trial and assessment of DCE-MRI as a biomarker of survival. Ann Oncol. 2009;20(9):1589–95.
89. Konstantinidis IT, Groot Koerkamp B, Do RK, et al. Unresectable intrahepatic cholangiocarcinoma: systemic plus hepatic arterial infusion chemotherapy is associated with longer survival in comparison with systemic chemotherapy alone. Cancer. 2016;122(5):758–65.
90. Cercek A, Boerner T, Tan BR, et al. Assessment of hepatic arterial infusion of floxuridine in combination with systemic gemcitabine and oxaliplatin in patients with unresectable intrahepatic cholangiocarcinoma: a phase 2 clinical trial. JAMA Oncol. 2019;6(1):60–7.

Part XII

Cancer Metastasis to the Bone

David M. Joyce and Odion Binitie

Bones are the third most common site for metastatic disease following the lung and liver. Some of the most common cancers that spread to bone include cancers of breast, prostate, lung, thyroid, and kidney as well as multiple myeloma. Other cancers likely to spread to bone that also have a propensity for lymphatic spread include head and neck cancer, squamous cell, cervical, uterine carcinomas, and melanoma. As people survive longer with cancer, more will likely develop significant metastatic bone disease burden necessitating the involvement of specialties such as orthopedics, spine surgery, and interventional radiology. Understanding the mechanisms of cancer spread, the effect of metastatic disease on the bone microscopically and macroscopically impacts the surgical, medical, and percutaneous treatment.

During the workup of a patient with suspected metastatic bone disease, orthogonal radiographs of the affected site; a CT scan involving the chest, abdomen, and pelvis with IV contrast; a bone scan; and/or a skeletal survey (should there be concern for multiple myeloma) are obtained. Cross-sectional imaging of the involved bone with the favored test being an MRI with and without contrast to better evaluate the lesion is sometimes obtained. If a MRI is not feasible a CT with contrast can be substituted. A biopsy will often be obtained for diagnosis prior to any surgical intervention.

Traditionally surgeons have used a scoring system to determine if a patient is at risk for fracture. Mirels score is based on several factors including the location of the lesion, amount of pain, and bone destruction in long bones. A sum of 9 or greater was shown to be a bone at risk for pathologic fracture; three-dimensional analysis of lesion size and extent with CT imaging may improve risk assessment. Prevention of a fracture has the benefit of limiting morbidity from the fracture, and some studies have demonstrated improved overall survival. Spinal instability neoplastic score (SINS) was introduced in 2010 to help guide clinicians, scoring patients into groups of stable, potentially unstable, and unstable. However, there is limited evidence on the use of SINS as a predictive tool for the progression of deformity or fracture.

In this section we review the factors surrounding the metastatic spread of cancer to the bone and its management. Frieling and Lynch in their chapter on the molecular basis of cancer metastasis discuss normal bone physiology, highlight how that homeostasis is affected by the "vicious cycle" of metastatic disease, and review the therapeutic interventions that work to disrupt this. The next group of chapters discuss the surgical management of metastatic bone disease based on the anatomic site. Long bones of the upper (Greene et al.) and lower extremities (Mesko and Nystrom) are reviewed with the indications for procedures based on the site of the bone affected. The use of different forms of fixation, bone augmentation, resection, and reconstruction and their associated complications are discussed. The pelvis, given its distinct anatomic considerations, is covered in a separate chapter (Evans et al.), as is the spine (Wang et al.). The utilization of minimal invasive procedures is highlighted in each surgical chapter and is focused on in the chapter on interventional radiology techniques by Parikh and Ahmed.

44 Molecular Mechanisms of Metastasis to the Bone

Jeremy S. Frieling and Conor C. Lynch

Abstract

Bone metastasis is an incurable and lethal aspect of many advanced-stage malignancies that significantly impacts patient quality of life. Although it is often associated with advanced disease, bone metastasis can occur early in disease progression and is dictated, in part, by a combination of genetic and molecular factors from the heterogeneous primary tumor. In order to prepare future metastatic sites, signals from the primary tumor precondition the bone for the arrival of metastatic cells. Bone marrow niches are further co-opted to assist with the recruitment and protection of metastatic cells, while the physiological activities of osteoclasts and osteoblasts, as well as other host bone cell types, can contribute to metastatic outgrowth and survival. The resulting metastases generate lesions of unique osteolytic and/or osteoblastic pathology, depending on the primary cancer type. Understanding the factors driving bone metastasis has led to the application of current standard of care antiresorptive therapeutics, and the continued elucidation of additional mechanisms is aiding the development of novel strategies to tackle specific aspects of these lesions to more effectively inhibit disease progression.

J. S. Frieling · C. C. Lynch (✉)
Department of Tumor Biology, H. Lee Moffitt Cancer Center & Research Institute, Tampa, FL, USA
e-mail: Jeremy.Frieling@moffitt.org; Conor.Lynch@moffitt.org

Learning Objectives

- Review normal bone physiology with respect to cancer-induced bone disease
- Understand the common genetic and molecular determinants of skeletal metastasis
- Identify the contributions of bone niches to metastatic tumor dissemination and dormancy
- Dissect the vicious cycle interactions between tumor, osteoblasts, osteoclasts, and other cells in the tumor–bone microenvironment
- Describe the mechanisms of action for the current standard of care therapies

44.1 Introduction

Skeletal colonization marks a significant turning point in cancer progression for which prognosis is poor and clinical therapies are limited. Many cancers metastasize to the bone, including breast, prostate, lung, melanoma, kidney, and thyroid. Other skeletal malignancies such as multiple myeloma and lymphoma also colonize the systemic skeleton. For reasons that we are only beginning to comprehend, breast and prostate cancer are among the most prevalent with bone metastases occurring in 65–90% of patients [1]. Yet, the resulting skeletal lesions for each cancer type are characterized by unique pathologies: osteoblastic, osteolytic, and mixed lesions. These lesions are largely generated by the activities of bone-building osteoblasts and bone-resorbing osteoclasts in

S. P. Leong et al. (eds.), *Cancer Metastasis Through the Lymphovascular System*,
https://doi.org/10.1007/978-3-030-93084-4_44

response to tumor signals. In addition to increased patient mortality, colonization of the skeleton significantly deteriorates the quality of life with complications such as pain and skeletal-related events (SREs) including pathological fracture, hypercalcemia, and spinal cord and nerve compression.

Like metastasis to other organ sites, metastasis to bone is highly inefficient, and disseminated tumor cells (DTCs) must overcome a sequence of challenges beginning with localized invasion and migration from the primary site and concluding with extravasation and secondary tumor outgrowth in the bone. Modulation of both primary and secondary organ microenvironments aids in the establishment of metastatic lesions, requiring coordination between the tumor and resident osteoblasts, osteoclasts, and osteocytes, in addition to endothelial and immune cell types. Bone metastasis ultimately usurps normal skeletal remodeling and homeostasis to encourage invasion, homing, and eventual outgrowth of occult metastases [2]. While our understanding of these processes remains incomplete, expanding the knowledge of specific underlying mechanisms has led to the development of several standard of care therapies and ongoing research continues to generate novel treatments.

44.2 The Bone

Understanding normal bone physiology and its cellular players offers insights into the local and systemic changes that result in bone metastasis. Bone is a dynamic tissue which not only provides for support and movement but also serves as a reservoir for minerals such as Ca^{2+} and cellular energy generated by osteoblasts. It is largely comprised of a type I collagen-rich extracellular matrix that is mineralized by deposition of hydroxyapatite, a combination of magnesium, calcium, and phosphate ions. This composition produces a stiff, durable structure where the collagen fibers impart flexibility to prevent brittleness or fracture. Two primary types of bone tissue are found within the skeleton: cortical and trabecular. Cortical bone is dense and serves primarily structural and mechanical purposes whereas the lighter, more metabolically active trabecular bone is located with the bone marrow compartment and assists the maintenance of hematopoietic stem cells (HSCs) and mesenchymal stem cells (MSCs). Surrounding trabecular bone is the bone marrow, a site for hematopoiesis, producing myeloid and lymphoid lineages [3]. In the context of cancer, local factors embedded within the bone matrix, together with systemic factors from the primary tumor, not only regulate the specific cell types in bone but can influence the behavior of DTCs.

The skeleton undergoes routine systematic remodeling whereby old bone is resorbed and replaced via the coupled activities of osteoclasts and osteoblasts at a rate of approximately 10% per year. Remodeling occurs in localized compartments consisting of osteoclasts, osteoblasts, and osteocytes that preserve the tightly regulated process [4]. Osteoclasts are large, multinucleated myeloid lineage cells that adhere to the bone surface and are responsible for resorption of existing bone tissue. Mature osteoclasts form by the fusion of myeloid precursors in response to osteoblast-derived (also osteocyte and immune cells) factors receptor activator of nuclear factor kappa-B ligand (RANKL) and macrophage colony-stimulating factor (M-CSF). Under appropriate environmental cues, such as biochemical signals from mechano-sensing osteocytes, osteoclasts polarize to form a ruffled border, which increases their surface area and form a tight actin ring with the bone surface to create a covered space where acid, collagenases, and proteases are secreted to demineralize the bone matrix. In doing so, components of the bone including embedded growth factors are released into the bone microenvironment. Subsequent to osteoclasts, osteoblasts generate bone in remodeled areas. Osteoblasts are derived from the mesenchymal lineage and differentiate in response to signaling cues such as bone morphogenetic proteins (BMPs), transforming growth factor beta (TGF-β), Wnt/β-catenin, and parathyroid hormone-related protein (PTHrP). Mature osteoblasts are recruited to the remodeling compartment where they secrete type I collagen-rich osteoid matrix. After building bone, a fraction of osteoblasts become embedded into the bone matrix and terminally differentiate into mechano-sensing osteocytes which communicate via long cytoplasmic processes that extend through passages in the bone known as canaliculi. Others reside on the bone surface as bone-lining cells. Osteocytes respond to mechanical stimuli and signals from adjacent cells to regulate bone remodeling by releasing sclerostin and/or Dickkopf-1 (Dkk1) [5]. The cellular composition of bone is highly diverse, comprised of both resident and transit cells such as endothelial cells, mesenchymal stem cells, adipocytes, platelets myeloid/myeloid-derived suppressor cells (MDSC), T cells, and nerve cells. Unique roles for many of these cell types and their activities in bone remodeling have been reviewed elsewhere [6].

44.3 Cancer Bone Tropism

Current evidence indicates that organotropism is dictated by a combination of genetic and molecular factors, circulation patterns, and crosstalk between the primary tumor and metastatic and host microenvironment of the bone (Fig. 44.1).

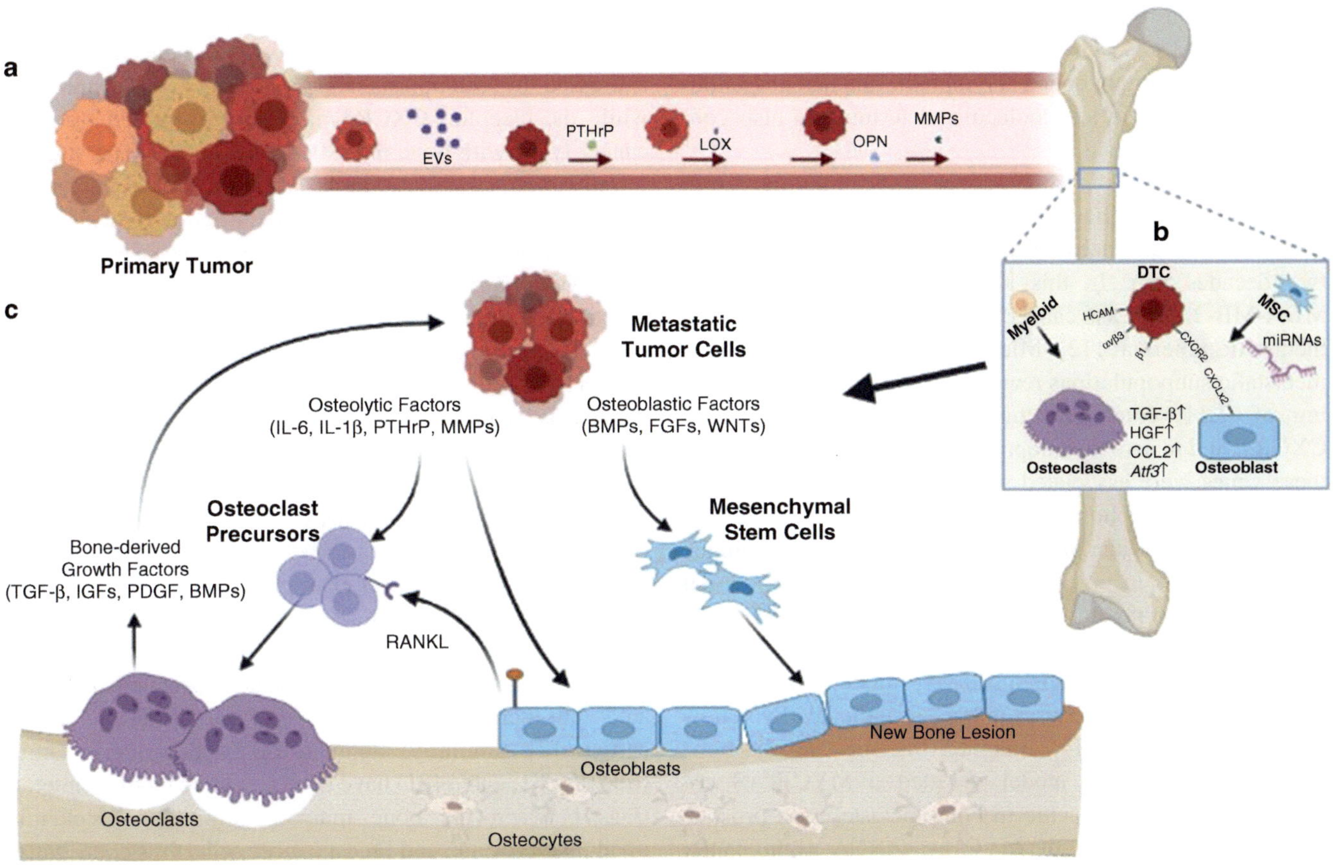

Fig. 44.1 (**a**) Metastatic tumor cells disseminate from primary organ tumor masses via lymphatic and hematogenous vasculature. The primary tumor provides signals in the form of secreted proteins and enzymes as well as cargo-shuttling extracellular vesicles such as exosomes. (**b**) Cues from the primary tumor mediate interactions with stromal components in future distant sites of metastasis to prepare a pre-metastatic niche that offers a suitable environment for the arrival of disseminated tumor cells (DTCs). (**c**) DTCs that successfully colonize the bone microenvironment generate a habitat favorable to growth and survival, where tumor-derived factors contribute to the formation of osteolytic and/or osteoblastic lesions via the "vicious cycle." Illustration created with BioRender

44.3.1 Osteotropism and Gene Signatures/Determinants

The predilection of specific tumor types to metastasize to select organs such as bone ("osteotropism") has long been acknowledged. More than 150 years ago, German pathologist Rudolph Virchow proposed that the locations of metastatic outgrowths were dictated based on where the disseminated cell lodged and arrested within the vasculature, thus explaining why vasculature-rich organs such as the lungs and bones are common sites of metastasis [7]. In the late 1800s, observations of metastatic breast cancer cases led English surgeon Stephen Paget to realize that the metastases occurred in a nonrandom manner. Paget proposed the "seed and soil hypothesis," which suggested that the challenges of metastasis begin before cells leave the primary tumor and that the DTCs (the seed) could only establish and flourish in a secondary site with a conducive microenvironment (the soil) [8]. Paget's hypothesis was not universally accepted, however, and was challenged by others including James Ewing, who posited that metastasis was dictated by the anatomy of the vasculature and lymphatics in which the disseminated cells traveled. The first organ encountered would be the site where most tumor cells arrest due to mechanical limitations of the diameter of the vessels. He speculated that the large number of cells released into circulation from the primary site would allow for a few physically trapped cells to colonize [9].

Modern evidence suggests that these two theories are not mutually exclusive, with bone offering a strong case in point. The red bone marrow has a high rate of blood flow, delivering abundant numbers of DTCs into bone. DTCs invade the sinusoids of the bone marrow cavity and migrate to the bone surface. Bone metastases are most commonly located in bones rich in red marrow such as the vertebrae, ribs, pelvis, and ends of long bones but are rarely found in bones of the hands or feet [10]. The aforementioned sites are also rich in metabolically active trabecular bone, which would present a more conducive soil for growth. Further, tumor cells must express compatible adhesion molecules and produce angiogenic and bone-resorbing factors to remodel the bone microenvironment for successful outgrowth [11]. Therefore, in addition to arriving in bone via lymphatic and hematogenous vasculature, genetic and molecular determinants also contribute to osteotropism.

The importance of genetic cooperation in giving rise to metastatic variants that successfully colonize bone was elegantly demonstrated by Massagué and colleagues almost two decades ago. In this key study, subpopulations of MDA-MB-231 breast cancer cells in vivo with enhanced metastatic potential [12]. Microarray profiling of the bone metastatic subpopulations revealed a gene signature consisting of IL11, connective tissue growth factor (CTGF), CXCR4, and matrix metalloproteinase-1 (MMP-1). Notably, these genes were associated with functions including invasion, bone marrow homing/extravasation, angiogenesis, and osteolysis. A shortage of models that accurately recapitulate bone metastasis in vivo has traditionally hampered similar investigations in prostate cancer. The recent development of NPK^{EYFP} mice (*Nkx3.1*$^{CreERT2/+}$; *Pten*$^{flox/flox}$; *Kras*$^{LSL\text{-}G12D/+}$; *R26R-CAG*-$^{LSL\text{-}EYFP/+}$) offers an in vivo model where DTCs reach and develop metastases in bone as indicated by yellow fluorescent protein (YFP) imaging. Interrogating the bone metastases using this model revealed a MYC/RAS co-activation signature specific to bone metastases in prostate cancer [13]. Silencing of MYC by shRNA significantly reduced bone metastases in the spine, pelvis, femur, tibia, and humerus but did not impact lung metastases. Notably, this signature correlates with human prostate cancer where activation of MYC and RAS is frequently observed in advanced tumors. Further comparative analysis of mouse bone metastasis and human prostate cancer based on these studies has established a human 16-gene signature that is associated with time to metastasis and response to antiandrogen treatment in metastatic disease and could possess clinical utility with continued development. In many cases, patients with advanced disease may already have DTCs in the bone at the time of presentation. The development of predictive gene signatures holds potential for identifying patients at risk of developing bone metastases.

44.3.2 Molecular Pathways/Homing

According to the seed and soil theory, DTCs must be suited to home to and survive in the secondary organ niche, while this niche must reciprocally be a fertile soil that can provide cues and nutrients to DTCs. These cues are often generated via mutual interactions with bone marrow stromal cells through integrins and chemokines. Tumor cells use homing mechanisms similar to those employed by normal cells, such as HSCs, in order to metastasize to the bone. The C–X–C motif chemokine 12 (CXCL12)/C–X–C chemokine receptor type 4 (CXCR4) chemokine axis is one of the best-characterized recruitment mechanisms in bone. CXCL12 is expressed by osteoblasts and bone marrow stromal cells while the receptor, CXCR4, is highly expressed by many cancer types with bone metastatic ability [14]. The study of serially selected MDA-MB-231 human breast cancer cell subpopulations described above identified a gene signature associated with increased capacity to metastasize to bone, and *CXCR4* was among the genes in this signature [12]. In addition to homing, CXCR4/CXCL12 interactions may also promote proliferation of breast and prostate cancer cells in bone [15]. Additional chemokine axes that facilitate bone metastasis include CXCL12/CXCR7, CXCR2/CXCL5, macrophage chemoattract protein-1 (MCP-1)/C–C motif chemokine receptor 2 (CCR2), and CXCL16/CXCR6 [16].

The expression of integrins on the surface of tumor cells can also facilitate interactions between the tumor and extracellular matrix in bone. For example, $\alpha\nu\beta3$ integrin can assist tumor cell adhesion by interacting with osteopontin (OPN), fibronectin, and vitronectin proteins in bone, and expression of $\alpha\nu\beta3$ integrin by breast and prostate cancer cells is associated with increased bone metastasis, osteolysis, and colonization [17]. Similarly, members of the $\beta1$ integrin family ($\alpha5\beta1$, $\alpha2\beta1$, and $\alpha4\beta1$) have been shown to mediate interactions between the bone marrow stroma and leukemia, myeloma, prostate, and breast cancer cells, thus contributing to increased tumor cell colonization and survival in bone. Expression of other cell adhesion factors such as VCAM1, selectins, and galectin 3 has also been implicated [18].

44.3.3 Pre-metastatic Niche

The development of metastasis is often considered a late event in cancer progression, however, biological evidence such as the detection of DTCs in the bones of 4- to 9-week-old breast cancer mouse models (HER2/neu and PyMT transgenic) indicates that the process can occur much earlier [19]. Organotropism arises during the primary stages of malignancy as tumor-derived factors systemically remodel future sites of metastasis in distant organs to prepare a "pre-metastatic niche." The pre-metastatic niche concept was first demonstrated using in vivo models of melanoma where vascular endothelial growth factor receptor 1 (VEGFR1) + bone marrow-derived hematopoietic precursor cells (HPCs) homed to sites of future metastasis in the lung, forming clusters and increasing fibronectin production prior to the arrival of metastatic cancer cells [20]. In bone metastasis, primary tumors condition the bone marrow microenvironment via circulating factors that assist in making the bone more conducive to colonization. For example, PTHrP produced by prostate tumors promotes bone resorption and enhances the production of factors in the bone marrow such as chemokine ligand 2 (CCL2) that in turn can facilitate tumor growth [21]. Primary tumors have also been shown to secrete lysyl oxidase (LOX) which could regulate ECM remodeling and recruiting bone marrow stromal cells to pre-metastatic niches. High LOX expression by primary breast tumors has also been shown to alter NFATc1 expression in osteoclasts, enhancing osteolytic activity in pre-metastatic niches [22]. Tumor-derived OPN is reported to contribute to pre-metastatic niche formation via the induction of growth and chemotactic factors in the bone, namely hepatocyte growth factor (HGF) and TGF-β [23] (Fig. 44.1a).

Emerging data indicate that pre-metastatic niche preparation in bone depends on altering host immune responses [24]. Observations from a novel spontaneous murine model of breast cancer (MMTV-NeuT transgenic) found transcriptional reprogramming of the bone marrow hematopoietic cell compartment prior to metastasis resulted in changes in the cellular composition and stromal architecture of the bone marrow. These changes included increases in Nestin,+ CXCL12-expressing mesenchymal cells and CXCR4-expressing myeloid cells with noted decreases in erythroid and B lymphoid populations. Mechanistically, upregulation of *Atf3* in the bone marrow and epigenetic regulation by circulating miRNAs from the miR-29 and miR-23 families were found to be responsible for the observed changes in this model [24].

Long-distance tumor–stroma communication and pre-metastatic niche formation can be enhanced by the shedding of extracellular vesicles such as exosomes. Exosomes transport a variety of cargo, including functional proteins, mRNA, and miRNA. Pioneering studies in melanoma demonstrated that tumor-derived exosomes home to common secondary sites for melanoma metastasis, and once in the secondary site, horizontal transfer of the cargo could educate bone marrow–derived cells to craft a pre-metastatic niche to support the arrival of DTCs [25]. Prostate cancer-derived exosomes promote pre-metastatic niche formation in the bone by transferring pyruvate kinase M2 (PKM2) to bone marrow stromal cells. This transfer upregulates CXCL12 via HIF1α and enhances prostate cancer seeding and growth in bone [26]. Prostate cancer-derived exosomes also stimulate osteoblast activity, possibly mediating pre-metastatic niche formation and osteoblastic pathology of prostate bone metastases [27]. Collectively, these data point to systemic changes occurring in the bone microenvironment prior to the arrival of cancer cells (Fig. 44.1b).

44.3.4 Bone Niches and Activation/Dormancy

The length of time from diagnosis to development of metastasis is highly variable, ranging from months to decades [28]. This is attributed to DTCs entering states of dormancy in secondary site niches where they are less vulnerable to immune surveillance, radiation, and chemotherapies. A number of dormancy mechanisms have been proposed to date. For example, "tumor cell dormancy" is defined as solitary cells in a quiescent state, harboring decreased expression of proliferation markers. This form of dormancy can be induced by local microenvironment factors, systemic signals from the primary tumor, or in some cases, drug therapies. In contrast, "tumor mass dormancy" arises from DTCs that initially proliferate at the metastatic site but do not develop macrometastases. It is hypothesized that poor angiogenesis and/or active immunosurveillance results in tumor mass dormancy and micrometastases [29]. The steps leading to entry into and exit from dormancy in the bone are not fully characterized due to limited models to study the interactions that take place in bone; however, new technologies and sophisticated models have begun to reveal the key mechanisms involved [30].

DTCs that arrest in bone are protected by specialized niches: hematopoietic, endosteal, and vascular. These niches can contribute to the regulation of dormancy and activation of arrested cells [31]. Dormancy in the hematopoietic and endosteal niches is thought to exploit mechanisms that control the quiescence of HSCs in the bone marrow. For instance, prostate tumor cells have been shown to mimic HSCs and bind to osteoblast-expressed annexin II. This in turn induces the expression of AXL receptor tyrosine kinases. Subsequent engagement of AXL with osteoblast-expressed growth arrest-specific 6 (GAS6) ligand induces dormancy [32]. Heightened expression of AXL and dormancy has also been reported in multiple myeloma cells during in vitro coculture with osteoblasts and these dormant cells could be awakened by pharmacologically targeting AXL [33]. Additional mechanisms exist whereby prostate cancer cell contact with osteoblasts activates Tank-binding kinase 1 (TBK-1) that leads to inhibition of mechanistic target of rapamycin (mTOR). This results in the arrest of protein synthesis and growth, drug resistance, and an increased stemness phenotype [34].

The most studied mechanism of dormancy to date involves the balance between p38 mitogen-activated protein kinase (MAPK) and extracellular signal-regulated kinase (ERK) activity. A high p38/ERK ratio has been shown as a mediator of dormancy in many cancer types, including breast, prostate, melanoma, and fibrosarcoma. However, interactions with microenvironment proteins such as collagen can stimulate a switch to high ERK/p38 ratio and reverse dormancy [35]. Bone marrow–derived TGF-β2 is also implicated in maintaining dormancy via p38 activation, and inhibition of either the TGF-β receptor, TGFβR1, or p38 led to awakening and metastatic outgrowth [36]. Additional bone-derived factors such as Dkk3, BMP1, and growth differentiation factor 10 (GDF10) have also been shown to activate the p38/MAPK pathway and suppress the proliferation of various prostate cancer cell lines both in vitro and in vivo [37]. Sophisticated in vitro techniques are emerging to study dormancy in bone, such as 3D bone-mimicking scaffolds. When cultured in a 3D in vitro endosteal niche model comprised of endothelial, bone marrow stroma, and fetal osteoblasts, parental MDA-MB-231 cells enter a dormant state in a p38 MAPK-dependent manner, strengthening the premise that p38 MAPK is a major mediator of tumor dormancy in the bone microenvironment [38].

In addition to the endosteal niche, the bone marrow vascular niche is also supportive of dormancy. Dormant cells have been identified in perisinusoidal regions rich in E-selectin and CXCL12 proteins, and blocking these proteins mobilized tumor cells [39]. Ghajar and colleagues have provided strong evidence for dormant DTCs in vascular niches of the bone where expression of thrombospondin-1 (TSP-1) by endothelial cells keeps DTCs in a dormant state. However, in regions of vascular sprouting and angiogenesis, TSP-1 expression is lost as periostin and TGF-β expression increase thereby simulating dormancy exit [40]. These findings suggest that active angiogenesis can awaken cells from dormancy, whereas cells that settle near stable microvasculature are likely to be maintained in a quiescent state. Although derived from independent studies, there is likely considerable overlap between hematopoietic, endosteal, and vascular niches and their impact on tumor dormancy. Comprehensive characterization of the spatial, temporal, and molecular niche dynamics remains a high priority to better understand the important and clinically significant problem of dormancy [30].

44.3.5 Invasion and Establishment of the Vicious Cycle

The bone is abundant in growth factors such as TGF-β, insulin growth factors (IGF), BMPs, platelet-derived growth factors (PDGF), fibroblast growth factors (FGF), vascular endothelial growth factors (VEGF), and calcium (Ca2+). Skeletal remodeling by osteoclasts releases many of these factors into the microenvironment, and signals from the primary tumor can initiate the process. Metastatic tumor cells exploit bone remodeling in order to promote survival and growth. The resulting disruption of normal bone physiology and decoupling of bone resorption and apposition has been described as the "vicious cycle" based on the cyclical interactions occurring between metastatic tumor cells and bone stromal cells such as osteoblasts and osteoclasts [2] (Fig. 44.1c).

In the most fundamental version of the vicious cycle, tumor-derived factors stimulate the osteoblast compartment to express RANKL that in turn promotes the fusion of osteoclast precursors into mature, bone-resorbing osteoclasts. Through the process of bone resorption, osteoclasts release sequestered factors such as TGF-β, VEGF, IGFs, and PDGF from the bone matrix, and these factors are potent stimulators of tumor cell proliferation as well as osteoblast activities, either directly or indirectly. The balance of specific factors can dictate osteoblastic or osteolytic lesion pathology. For example, tumor-derived factors such as interleukin 6 (IL-6) and IL-1β act predominantly on osteoclasts to promote an osteolytic vicious cycle. Osteolytic bone disease is a common attribute of breast cancer and multiple myeloma and results from tumor stimulation of osteoclast differentiation and activity rather than direct effect of cancer cells on the skeleton. In addition to IL-6 and IL-1β, the osteolytic vicious cycle is largely mediated by TGF-β, PTHrP, and RANKL. In contrast, osteoblastic lesions tend to arise from the stimulation of osteoblasts mediated by growth factors such as BMPs, FGF, and Wnts. The factors that govern osteoclast and osteoblast biology are well reviewed here: [30, 41]. Osteoclast activity is likely required for bone metastasis even in osteoblastic disease. Increased concentrations of N-terminal telopeptides, a molecular marker of bone turnover, have been detected in osteoblastic disease for instance [42]. Likewise, osteolytic breast cancer metastases may feature pathological bone formation. Overall, however, there is still a substantial net loss of bone in breast cancer skeletal metastasis whereas prostate cancer metastasis results in a gain of poorly formed, woven pathological bone. Bone-derived factors, such as TGF-β released by osteoclasts in the vicious cycle, can also significantly contribute to other complications that impact patient well-being, for example, muscle function. Osteolytic disease often results in significant weakening of skeletal muscle (cachexia), as has been observed in prostate, breast, lung, and multiple myeloma. Bone-derived TGF-β is a primary driver of cancer-induced muscle weakness, upregulating expression of NADPH oxidase 4 (Nox4) which leads to oxidation of the ryanodine receptor (RyR1) and calcium leakage [43].

The vicious cycle continues to evolve as the molecular mechanisms that drive it are revealed. MMPs, for example, regulate the bioavailability of factors such as PTHrP, TGF-β, and RANKL [44–46]. MMPs can be produced by the tumor cells themselves or secreted by other cell types within the stromal compartment. Their processing of growth factors influences a host of other cells. Mesenchymal stem cells (MSCs) comprise 0.001–0.01% of the bone marrow and respond to vicious cycle factors such as TGF-β, BMPs, and Wnts with potential to differentiate into osteoblasts, chondrocytes, and adipocytes [47]. The precise contributions of MSCs in bone metastasis are not fully understood, but reciprocal interactions between tumor cells and MSCs appear to support the proliferation, differentiation, and survival of other stromal cells such as HPCs as well as metastatic tumor cells, depending on disease and context. For example, MSCs recruited to prostate tumors via a CXCR6/CXCL16 axis promote growth and metastasis by secretion of CXCL12 [48].

MSCs can also suppress immune cells by secretion of factors such as prostaglandin E2 (PGE2), IL-6, IL-10, and indoleamine 2,3 deoxygenase (IDO) [47]. Conversely, MSCs have been shown to have anti-tumorigenic effects as well, such as downregulation of Wnt signaling by upregulating Dkk-1 [49]. Further, secretion of BMP-2 by metastatic prostate cancer cells can indirectly induce MSCs to differentiate into osteoblasts and enhance osteoblast activity [50]. This could contribute toward the osteoblastic phenotype characteristic of metastatic prostate cancer. MSC-derived adipocytes have also been shown to influence osteoblast and osteoclast activities in the vicious cycle with new roles emerging [51].

Increasing evidence has revealed vital contributions from immune components in bone remodeling. The bone marrow is involved in systemic immunity and therefore possesses a unique immune environment that provides immune-privileged milieus for tumor cells. Naive immature myeloid cells in the bone marrow can differentiate into macrophages, granulocytes, and neutrophils. In some instances, tumor-derived factors such as VEGF, stem cell factor (SCF), granulocyte-macrophage colony-stimulating factor (GM-CSF), IL-6, cyclooxygenase-2 (COX2), and M-CSF expand a population of MDSCs, which possess the capacity to suppress both innate and adaptive immunity [52]. MDSCs can also contribute to osteolytic lesion formation by serving as osteoclast precursors. For example, MDSCs derived from the bone marrow of breast tumor-bearing mice have been shown to differentiate along an osteoclast lineage [53]. Increased numbers of MDSCs are found in bone metastases compared to both primary breast tumors or lung metastases, and this is regulated in part by interferon regulatory factor (IRF)-7. Silencing IRF7 restricted immunosurveillance and promoted bone metastasis, whereas restoration of IRF7 expression decreased MDSCs while increasing CD8 T cells and NK cells, suggesting that certain patients may benefit from interferon (IFN)-based therapies [54]. MDSCs can also indirectly promote tumor growth by activating tumor angiogenesis and lymphangiogenesis via secretion of VEGF, TNFα, and angiopoietin [55]. Macrophage infiltration is also observed in many cancer types and can contribute to metastasis. Naive macrophages respond to microenvironmental cues to polarize into pro- or anti-inflammatory phenotypes, with the majority acquiring an anti-inflammatory phenotype that is pro-tumorigenic. Tumor-derived factors like CCL2 promote macrophage recruitment to the bone which can increase tumor growth and bone metastasis in vivo [56]. Bone is an abundant reservoir for osteal macrophages, which represent 20% of bone marrow cells. These cells can support tumor growth and aid in immune escape as well as impact the bone microenvironment by stimulating osteoblast and osteoclast activity. For example, macrophages often associate with formation of woven pathological bone in prostate cancer bone metastasis specimens, and ablation of CD169+ macrophages (including osteal macrophages) in an immune-competent mouse model inhibited prostate tumor-induced bone formation [57]. Other myeloid lineage immune cells including neutrophils, megakaryocytes, platelets, and dendritic cells are also involved. For further review, see [58].

The adaptive immune system is tightly linked to bone physiology and consequently is involved in bone metastasis. The percentage of T cells in bone is usually low compared to other cell types. However, active CD4 and CD8 T cells have been found in the bone marrow of breast cancer patients prior to treatment and were demonstrated to provide antitumor effects in bone metastases [58]. Patients with DTCs in bone present with increased numbers of activated NK cells and memory T cells in the bone marrow. The increase was highest in patients with stage II disease and decreased in stage III/IV, suggesting that immune responses may be surmounted as bone metastatic disease progresses [59]. Interestingly, several studies have also reported roles for T cells in stimulating osteoclast activity [6]. Conflicting reports exist for the contributions of regulatory T cells (Tregs) in bone metastasis. Tregs produce RANKL, suggesting that they can promote osteoclast differentiation and as a consequence, bone metastasis [58]. However, under certain conditions, Tregs may inhibit osteoclast activity [60]. Increased numbers of Tregs have been detected in prostate cancer patients with bone metastasis, and the transfer of Tregs in a xenograft mouse model of prostate cancer increased bone mineral density, whereas Treg depletion reduced bone density [61]. Interestingly, patients treated with bisphosphonates that inhibit osteoclast activity have been shown to also undergo an expansion and activation of $\gamma\delta$ T cells. $\gamma\delta$ T cells possess functional characteristics of both innate and adaptive immunity and can exert antitumor effects by releasing pro-inflammatory cytokines. Their role in bone metastasis has not been thoroughly explored, but a European phase I clinical trial treating bone metastatic prostate cancer patients with zoledronate plus interleukin-2 (IL-2) resulted in disease stabilization for six out of nine patients [58].

Crosstalk between immune cells and other cells in the bone microenvironment is multidirectional. Osteoclast secretion of chemokines such as CCL5, CXCL5, and CXCL1 recruits CD8+ T cells in transgenic OT-1 mice. The recruited T cells are stimulated by osteoclasts to proliferate and secrete IL-2, IL-6, and IFN-γ, however, the osteoclast-activated, FoxP3-expressing cells were anergic which suppressed further priming of naïve CD8+ T cells by dendritic cells [62]. Further, osteoclasts can also induce CD4+ T regs or inflammatory CD4+ T cells depending on context, while osteoblast depletion led to an 80% reduction of B cells in the bone specifically [63].

In addition to changes in the tumor–bone microenvironment as discussed previously, bone metastases cause great pain that further contribute to patient morbidity. Many sensory nerve fibers innervate the bone and express acid-sensing ion channels, including transient receptor potential cation channel subfamily V member 1 (TRPV1), acid-sensing ion channel 1 (ASIC1), and ASIC3. These sensors are activated at pH 3–4 and thus respond to osteoclast-induced acidosis [64]. Interestingly, nerve growth factor (NGF) blockade prevented increased sprouting of sensory nerve fibers and significantly reduced early and late-stage pain behaviors in murine models of prostate cancer, however, its beneficial effects did not extend to tumor-induced bone remodeling [64, 65]. Innervation can also contribute to tumor invasion and metastasis via signaling of the autonomic nervous system. Prostate cancer xenografts develop poorly in mice that are pretreated with chemical or surgical sympathectomy of the prostate gland or when stromal β2 and β3 adrenergic receptors were genetically deleted, and tumor invasion and metastasis were reduced following blockade of type 1 muscarinic receptors. The findings of these studies suggest that sympathetic neonerves promote early stages of tumorigenesis whereas parasympathetic nerves contribute to dissemination and metastasis and offer unique opportunities to develop therapeutic approaches that alleviate pain and possibly limit disease progression [66].

44.3.6 Tackling the Vicious Cycle and Cancer-Induced Bone Disease

Understanding the interactions from genetic and molecular perspective can yield therapeutic opportunities. Antiresorptive agents such as bisphosphonates and anti-RANKL antibodies remain a standard of care for patients with bone metastatic disease. Antiresorptives protect skeletal integrity and alleviate pain by disrupting the vicious cycle, however, they have little impact on tumor growth and progression. Bisphosphonates such as zoledronate bind hydroxyapatite crystals in areas of active bone remodeling and are taken up by bone-resorbing osteoclasts which induce their apoptosis [67]. Bisphosphonates increase time to first SRE, however, no significant increases in overall survival have been demonstrated, and the development of new bone metastases and disease progression is seen in 30–50% of patients prescribed bisphosphonates [68]. Although their intended target is osteoclasts, bisphosphonates may also affect other cell types, such as γδ T cells, monocytes, MDSCs, endothelial cells, osteoblasts, and tumor cells [18, 69]. RANKL is a central mediator in bone turnover and metastasis. Denosumab is a fully humanized monoclonal antibody that binds RANKL in order to prevent its binding to RANK on osteoclast precursors, thereby inhibiting osteoclast formation and activity [70]. Denosumab decreases skeletal complications and bone pain, but it did not increase overall survival in bone metastatic breast cancer [71]. However, compared to zoledronic acid, denosumab significantly delayed the onset of SREs in bone metastatic prostate cancer patients [72].

Bone-seeking radiopharmaceuticals have gained traction recently for the treatment of bone metastasis, particularly for osteoblastic lesions due to better uptake and incorporation into newly produced bone. Radionuclides offer better pain alleviation for diffuse pain resulting from multiple metastases and are cleared quickly from nonskeletal sites. Radionuclides emit either α or β radiation. β-emitters primarily provide palliative pain relief, however, α-emitters also provide antitumoral activity with minimal radiation damage to nearby healthy tissue due to their higher energy and soft tissue range of less than 100 μm. In a study of men with bone metastatic prostate cancer previously treated with radiotherapy, the α-emitter Radium-223 delayed time to first SRE and improved overall survival compared to placebo (14.0 months vs. 11.2 months) leading to FDA approval for treatment of castration-resistant bone metastatic prostate cancer without visceral metastases [73, 74]. Based on molecular studies, new agents are being generated and tested for the treatment of bone metastases. For further review, see [75].

44.4 Conclusion

Bone is a vital organ providing structure and support, but the high metabolic and molecular activity combined with well-vascularized nature and signals from the primary tumor appears to prepare a suitable soil for disseminated tumor cells from numerous primary malignancies. Notably, depending on primary origin, the resulting metastatic lesions can have unique osteolytic, osteoblastic, or mixed pathologies, arising from the specific factors at play in a vicious cycle between tumor cells and bone. Recently, the importance of the immune, nervous, and muscular systems has become evident in this process as well, revealing the broad range of cellular players and processes governing bone metastasis. At present bone metastasis remains incurable and significantly impacts patient quality of life with pain, fracture, hypercalcemia, and other skeletal complications. While molecular and genetic understanding has produced several leads for investigation, critical questions and areas remain to be explored.

Open Questions

- How can we define the optimal therapeutic window to tackle dormancy?
- What are the pro- and anti-tumorigenic contributions of an intact immune system in the bone microenvironment during skeletal colonization?
- How does the bone microenvironment contribute to the evolution of resistant disease?
- How do bone metastases regulate cachexia and what strategies can be employed to minimize skeletal muscle loss and improve patient quality of life?
- Are bone metastases the final destination or do bone metastases seed tertiary metastatic sites?

References

1. Ponzetti M, Rucci N. Switching homes: how cancer moves to bone. Int J Mol Sci. 2020;21(11):4124.
2. Mundy GR. Metastasis to bone: causes, consequences and therapeutic opportunities. Nat Rev Cancer. 2002;2(8):584–93.
3. Hvid I, et al. The biology of bone lengthening. J Child Orthop. 2016;10(6):487–92.
4. Andersen TL, et al. A physical mechanism for coupling bone resorption and formation in adult human bone. Am J Pathol. 2009;174(1):239–47.
5. Manolagas SC. Birth and death of bone cells: basic regulatory mechanisms and implications for the pathogenesis and treatment of osteoporosis. Endocr Rev. 2000;21(2):115–37.
6. Fornetti J, Welm AL, Stewart SA. Understanding the bone in cancer metastasis. J Bone Miner Res. 2018;33(12):2099–113.
7. Virchow R. Cellular pathology (1858). Thoemmes Continuum; 2004.
8. Paget G. Remarks on a case of alternate partial anaesthesia. Br Med J. 1889;1(1462):1–3.
9. Neoplastic diseases: a treatise on tumours. By James Ewing, A.M., M.D., Sc.D., Professor of Pathology at Cornell University Medical College, N.Y.; Pathologist to the Memorial Hospital. Third edition. Royal 8vo. Pp. 1127, with 546 illustrations. 1928. Philadelphia and London: W. B. Saunders Co. Ltd. 63s. net. BJS (British Journal of Surgery). 1928;16(61):174–175.
10. Krishnamurthy GT, et al. Distribution pattern of metastatic bone disease: a need for total body skeletal image. JAMA. 1977;237(23):2504–6.
11. Esposito M, Kang Y. Targeting tumor-stromal interactions in bone metastasis. Pharmacol Ther. 2014;141(2):222–33.
12. Kang Y, et al. A multigenic program mediating breast cancer metastasis to bone. Cancer Cell. 2003;3(6):537–49.
13. Arriaga JM, et al. A MYC and RAS co-activation signature in localized prostate cancer drives bone metastasis and castration resistance. Nature Cancer. 2020;1(11):1082–96.
14. Wang J, Loberg R, Taichman RS. The pivotal role of CXCL12 (SDF-1)/CXCR4 axis in bone metastasis. Cancer Metastasis Rev. 2006;25(4):573–87.
15. Lapteva N, et al. CXCR4 knockdown by small interfering RNA abrogates breast tumor growth in vivo. Cancer Gene Ther. 2005;12(1):84–9.
16. Frieling JS, Basanta D, Lynch CC. Current and emerging therapies for bone metastatic castration-resistant prostate cancer. Cancer Control. 2015;22(1):109–20.
17. Schneider JG, Amend SR, Weilbaecher KN. Integrins and bone metastasis: integrating tumor cell and stromal cell interactions. Bone. 2011;48(1):54–65.
18. Weilbaecher KN, Guise TA, McCauley LK. Cancer to bone: a fatal attraction. Nat Rev Cancer. 2011;11(6):411–25.
19. Hüsemann Y, et al. Systemic spread is an early step in breast cancer. Cancer Cell. 2008;13(1):58–68.
20. Kaplan RN, et al. VEGFR1-positive haematopoietic bone marrow progenitors initiate the pre-metastatic niche. Nature. 2005;438(7069):820–7.
21. Li X, et al. A destructive cascade mediated by CCL2 facilitates prostate cancer growth in bone. Cancer Res. 2009;69(4):1685–92.
22. VanHook AM. LOX induces a metastatic niche. Sci Signal. 2015;8(380):ec151.
23. Pang X, et al. Osteopontin as a multifaceted driver of bone metastasis and drug resistance. Pharmacol Res. 2019;144:235–44.
24. Chiodoni C, et al. Transcriptional profiles and stromal changes reveal bone marrow adaptation to early breast cancer in association with deregulated circulating microRNAs. Cancer Res. 2020;80(3):484–98.
25. Peinado H, et al. Melanoma exosomes educate bone marrow progenitor cells toward a pro-metastatic phenotype through MET. Nat Med. 2012;18(6):883–91.
26. Dai J, et al. Primary prostate cancer educates bone stroma through exosomal pyruvate kinase M2 to promote bone metastasis. J Exp Med. 2019;216(12):2883–99.
27. Borel M, et al. Prostate cancer-derived exosomes promote osteoblast differentiation and activity through phospholipase D2. Biochim Biophys Acta (BBA) - Mol Basis Dis. 2020;1866(12):165919.
28. Shiozawa Y, et al. Human prostate cancer metastases target the hematopoietic stem cell niche to establish footholds in mouse bone marrow. J Clin Invest. 2011;121(4):1298–312.
29. Aguirre-Ghiso JA. Models, mechanisms and clinical evidence for cancer dormancy. Nat Rev Cancer. 2007;7(11):834–46.
30. Mayhew V, Omokehinde T, Johnson RW. Tumor dormancy in bone. Cancer Rep. 2020;3(1):e1156.
31. Braun S, et al. Cytokeratin-positive cells in the bone marrow and survival of patients with stage I, II, or III breast cancer. N Engl J Med. 2000;342(8):525–33.
32. Shiozawa Y, et al. GAS6/AXL axis regulates prostate cancer invasion, proliferation, and survival in the bone marrow niche. Neoplasia. 2010;12(2):116–27.
33. Khoo WH, et al. A niche-dependent myeloid transcriptome signature defines dormant myeloma cells. Blood. 2019;134(1):30–43.
34. Kim JK, et al. TBK1 regulates prostate cancer dormancy through mTOR inhibition. Neoplasia. 2013;15(9):1064–74.
35. Aguirre-Ghiso JA, et al. ERK(MAPK) activity as a determinant of tumor growth and dormancy; regulation by p38(SAPK). Cancer Res. 2003;63(7):1684–95.
36. Bragado P, et al. TGF-beta2 dictates disseminated tumour cell fate in target organs through TGF-beta-RIII and p38alpha/beta signalling. Nat Cell Biol. 2013;15(11):1351–61.
37. Yu-Lee L-Y, et al. Bone secreted factors induce cellular quiescence in prostate cancer cells. Sci Rep. 2019;9(1):18635.
38. McGrath J, et al. Identification of genes regulating breast cancer dormancy in 3D bone endosteal niche cultures. Mol Cancer Res. 2019;17(4):860–9.
39. Price TT, et al. Dormant breast cancer micrometastases reside in specific bone marrow niches that regulate their transit to and from bone. Sci Transl Med. 2016;8(340):340ra73.
40. Ghajar CM, et al. The perivascular niche regulates breast tumour dormancy. Nat Cell Biol. 2013;15(7):807–17.
41. Suva LJ, et al. Bone metastasis: mechanisms and therapeutic opportunities. Nat Rev Endocrinol. 2011;7(4):208–18.
42. Coleman RE, et al. Predictive value of bone resorption and formation markers in cancer patients with bone metastases receiving the bisphosphonate zoledronic acid. J Clin Oncol. 2005;23(22):4925–35.
43. Regan JN, et al. The role of TGFβ in bone-muscle crosstalk. Curr Osteoporos Rep. 2017;15(1):18–23.
44. Frieling JS, et al. Matrix metalloproteinase processing of PTHrP yields a selective regulator of osteogenesis, PTHrP1-17. Oncogene. 2017;36:4498–507.
45. Thiolloy S, et al. An osteoblast-derived proteinase controls tumor cell survival via TGF-beta activation in the bone microenvironment. PLoS One. 2012;7(1):e29862.
46. Lynch CC, et al. MMP-7 promotes prostate cancer-induced osteolysis via the solubilization of RANKL. Cancer Cell. 2005;7(5):485–96.
47. Uccelli A, Moretta L, Pistoia V. Mesenchymal stem cells in health and disease. Nat Rev Immunol. 2008;8(9):726–36.
48. Jung Y, et al. Recruitment of mesenchymal stem cells into prostate tumours promotes metastasis. Nat Commun. 2013;4:1795.
49. Qiao L, et al. Dkk-1 secreted by mesenchymal stem cells inhibits growth of breast cancer cells via depression of Wnt signalling. Cancer Lett. 2008;269(1):67–77.

50. Joseph J, et al. Disseminated prostate cancer cells can instruct hematopoietic stem and progenitor cells to regulate bone phenotype. Mol Cancer Res. 2012;10(3):282–92.
51. Cook LM, et al. Integrating new discoveries into the "vicious cycle" paradigm of prostate to bone metastases. Cancer Metastasis Rev. 2014;33(2–3):511–25.
52. Gabrilovich DI, Nagaraj S. Myeloid-derived suppressor cells as regulators of the immune system. Nat Rev Immunol. 2009;9(3):162–74.
53. Sawant A, et al. Myeloid-derived suppressor cells function as novel osteoclast progenitors enhancing bone loss in breast cancer. Cancer Res. 2013;73(2):672–82.
54. Bidwell BN, et al. Silencing of Irf7 pathways in breast cancer cells promotes bone metastasis through immune escape. Nat Med. 2012;18(8)
55. Murdoch C, et al. The role of myeloid cells in the promotion of tumour angiogenesis. Nat Rev Cancer. 2008;8(8):618–31.
56. Mizutani K, et al. The chemokine CCL2 increases prostate tumor growth and bone metastasis through macrophage and osteoclast recruitment. Neoplasia. 2009;11(11):1235–42.
57. Wu AC, et al. CD169(+) macrophages mediate pathological formation of woven bone in skeletal lesions of prostate cancer. J Pathol. 2016;239(2):218–30.
58. Capietto A-H, Faccio R. Immune regulation of bone metastasis. BoneKEy Rep. 2014;3:600.
59. Feuerer M, et al. Enrichment of memory T cells and other profound immunological changes in the bone marrow from untreated breast cancer patients. Int J Cancer. 2001;92(1):96–105.
60. Buchwald ZS, et al. Osteoclast-induced Foxp3+ CD8 T-cells limit bone loss in mice. Bone. 2013;56(1):163–73.
61. Zhao E, et al. Regulatory T cells in the bone marrow microenvironment in patients with prostate cancer. Onco Targets Ther. 2012;1(2):152–61.
62. Kiesel JR, Buchwald ZS, Aurora R. Cross-presentation by osteoclasts induces FoxP3 in CD8+ T cells. J Immunol. 2009;182(9):5477–87.
63. Zhu J, et al. Osteoblasts support B-lymphocyte commitment and differentiation from hematopoietic stem cells. Blood. 2007;109(9):3706–12.
64. Mantyh PW. Mechanisms that drive bone pain across the lifespan. Br J Clin Pharmacol. 2019;85(6):1103–13.
65. Halvorson KG, et al. A blocking antibody to nerve growth factor attenuates skeletal pain induced by prostate tumor cells growing in bone. Cancer Res. 2005;65(20):9426–35.
66. Magnon C, et al. Autonomic nerve development contributes to prostate cancer progression. Science. 2013;341(6142):1236361.
67. Rogers MJ, Watts DJ, Russell RG. Overview of bisphosphonates. Cancer. 1997;80(8 Suppl):1652–60.
68. Saad F, et al. Long-term efficacy of zoledronic acid for the prevention of skeletal complications in patients with metastatic hormone-refractory prostate cancer. J Natl Cancer Inst. 2004;96(11):879–82.
69. Bellido T, Plotkin LI. Novel actions of bisphosphonates in bone: preservation of osteoblast and osteocyte viability. Bone. 2011;49(1):50–5.
70. www.cancer.gov, FDA Approval for Denosumab. National Cancer Institute; 2013.
71. Stopeck AT, et al. Denosumab compared with zoledronic acid for the treatment of bone metastases in patients with advanced breast cancer: a randomized, double-blind study. J Clin Oncol. 2010;28(35):5132–9.
72. Fizazi K, et al. Denosumab versus zoledronic acid for treatment of bone metastases in men with castration-resistant prostate cancer: a randomised, double-blind study. Lancet. 2011;377(9768):813–22.
73. Choi JY. Treatment of bone metastasis with bone-targeting radiopharmaceuticals. Nucl Med Mol Imaging. 2018;52(3):200–7.
74. Parker C, et al. Alpha emitter radium-223 and survival in metastatic prostate cancer. N Engl J Med. 2013;369(3):213–23.
75. Clément-Demange L, Clézardin P. Emerging therapies in bone metastasis. Curr Opin Pharmacol. 2015;22:79–86.

Surgical Management of Metastatic Disease to the Upper Extremity

45

Allison C. Greene, Michael T. Torchia, Daniel C. Austin, John-Erik Bell, and Eric R. Henderson

Abstract

Metastatic disease to bone is common and is best managed by a multidisciplinary team of medical providers. Several interventional modalities exist for metastatic bone disease, including medical therapy, radiation, ablation, embolization, and surgery. The primary goals of these interventions are functional restoration, durable local control, and pain relief. Orthopedic surgeons are integral members of the treatment team. Operative strategies for metastatic bone disease of the upper extremity are unique due to the reduced weight-bearing requirements of the arms compared to the legs. Appropriate selection of an operative strategy requires the careful consideration of several characteristics of the bone and relevant lesion(s) including: (1) is the metastasis isolated; (2) what is the anticipated response to treatment; (3) what is the quality of the available bone; (4) is there a fracture; (5) is the joint surface affected; (6) what are the functional requirements; (7) is embolization warranted? Prior to any intervention, local confirmation of metastatic disease should be obtained unless the patient is failing treatment. Operative strategies include intralesional curettage, extramedullary and intramedullary implants, conventional shoulder arthroplasty, reverse shoulder arthroplasty, and segmental implants. The best strategy will confer the most likely attainment of goals with the least likelihood of morbidity.

A. C. Greene (✉)
Albany Medical Center, Albany, NY, USA
e-mail: greenea2@amc.edu

M. T. Torchia · J.-E. Bell · E. R. Henderson
Dartmouth Hitchcock Medical Center, Lebanon, NH, USA
e-mail: michael.t.torchia@hitchcock.org; john-erik.bell@hitchcock.org; eric.r.henderson@hitchcock.org

D. C. Austin
Dartmouth Hitchcock Medical Center, Lebanon, NH, USA

Learning Objectives

- To understand general strategies for orthopedic surgical management of metastatic bone disease.
- To understand patient- and disease-related considerations that impact surgical planning.
- To understand adjuvant management options that may reduce the morbidity of surgery and/or improve postsurgical outcomes.

45.1 Introduction

Approximately 1.7 M new cases of cancer are diagnosed in the United States each year, of which ~700,000 have a high propensity for skeletal metastasis. Metastatic bone disease (MBD) robs patients of their independence, both by inducing substantial pain and leading to morbid bone destruction and fracture.

Treatment goals for MBD include function restoration and pain relief. When considering the best management of a new bone metastasis, providers must balance the individual patient's global needs and goals. While it is tempting for the surgeon to venture to the operating room to improve a patient's mobility and/or quality of life, surgery will require an interruption in the patient's—usually—ongoing cancer treatment. Decision-making about the surgical management of MBD should, as with most oncological treatment, be made by a multidisciplinary group including specialists from medical oncology, radiation oncology, palliative care, social work, and orthopedic surgery.

Because the upper extremities are not primarily responsible for weightbearing and mobility, the reconstructive options available to surgeons are arguably more diverse than for lower extremity reconstructions, where the ability to bear the body's weight is imperative.

S. P. Leong et al. (eds.), *Cancer Metastasis Through the Lymphovascular System*,
https://doi.org/10.1007/978-3-030-93084-4_45

This chapter is written from the perspective of an orthopedic surgeon who is participating in the care of a patient with biopsy-proven metastatic bone disease, for which appropriate staging and multidisciplinary treatment is underway. For a review of the workup of a new bone lesion or appropriate staging protocols, please review the applicable, preceding chapters.

45.2 General Considerations

When considering operative management of a bone metastasis, we recommend considering the following questions:

- **Is it a metastasis and does it matter?** While the odds that a new, aggressive-appearing bone lesion in an adult patient is a metastasis are ~500:1, a metastatic process should never be assumed unless the discovery of a new primary bone cancer is believed to have no impact on the patient's survival whatsoever, for example, in the setting of a hospice-bound patient with a new pathological fracture. In any other situation, tissue confirmation of the lesion as a metastasis is warranted.
- **Is the metastasis isolated?** The data supporting en bloc resection of isolated bone metastases are not robust, however, this strategy is supported by the National Comprehensive Cancer Network (NCCN) for solitary or oligometastatic disease in the setting of renal cell carcinoma [1]. Before undertaking a substantial en bloc resection, ensure that staging scans are current to avoid unnecessary risk.
- **What is the expected response to antineoplastic therapies?** Newly diagnosed patients who are expected to have an excellent response to treatment may have years or decades of life ahead of them with good potential for bone reconstitution and fracture healing. In this setting, upper extremity lesions are highly amenable to nonoperative management with surveillance and surgical strategies in these patients may rely more on bone reinforcement. Conversely, patients who are failing second-, third-, or last-line treatment have poor healing potential and may benefit from replacement strategies.
- **What is the quality of the remaining bone stock?** This question must be weighed against the expected treatment response—see above—but in general greater degrees of bone loss indicate a heavier reliance on replacement strategies.
- **Is there a fracture?** Pathological fracture healing is always fraught; however, certain primary cancers portend worse healing [2]. As above, anticipated treatment response, and fracture healing, should be considered when choosing a reconstruction strategy.
- **Is a joint surface compromised?** Deformation of the joint surface usually mandates an arthroplasty procedure.
- **What are the patient's functional requirements?** A patient with compromised ability to fully bear weight on the legs may require the use of their upper extremities for an assistive device. Depending on the region of bone involved with metastatic disease, an increased requirement for weightbearing may drive a surgeon toward arthroplasty.
- **Is embolization warranted?** Metastatic kidney and thyroid cancer lesions are often highly vascular. When intervention is taken on axial or proximal appendicular lesions, preoperative embolization should be considered.

45.3 Nonsurgical Management

Because of the lower weight-bearing requirement, nonoperative strategies for MBD are more tenable in the upper extremities than the pelvis or lower extremities. In a treatment-naive patient with a new diagnosis and the expectation of an excellent response, nonoperative management of upper extremity lesions, particularly in the absence of fracture, is reasonable.

45.3.1 Immobilization

In the setting of pathological fracture, casting and splinting produce poor outcomes [3]. Immobilization of impending or existing pathologic fractures is feasible in patients for whom surgery is contraindicated [4]. Additionally, patients without an established diagnosis may require a immobilization prior to definitive intervention [5].

45.3.2 Radiotherapy

Radiotherapy (RT) is employed for most skeletal metastases. RT can decrease pain, maintain and restore function, reduce the risk of pathologic fracture, and improve quality of life [6]. Local tumor control is also possible, although highly dependent on histological origin [7].

45.3.2.1 Photon-Beam Radiation Therapy

Most cancer radiotherapy utilizes high-energy photons (X-ray or gamma-ray). Ionizing radiation produces reactive oxygen species in the targeted tissues, inducing DNA damage in both malignant and healthy tissue. By their nature, cancer cells are less capable of DNA repair and therefore this insult leads to cell death. Cellular damage is dependent on dose; therefore, treatment strategies seek to maximize tumor dose and minimize adjacent tissue exposure. There is no consensus regarding standardization of fractioning [8]; therefore, the trend has been to concentrate dose and reduce fractions, resulting in improved treatment completion.

45.3.2.2 Particle-Beam Radiation Therapy

Heavy ions (protons, carbon ions) are also used for RT with some advantage over photons. Whereas photon dose decreases exponentially upon entering tissue, heavy ions deposit most energy toward the end of their trajectory, termed the Bragg peak [8]. Adjusting initial particle speed and beam intensity allows for precise targeting of a tumor while sparing surrounding tissues, conferring increased relative biological effectiveness (RBE) compared to photons (approximately 3x higher) [8].

45.3.3 Interventional Radiology

In situations where radiotherapy and surgery are contraindicated, percutaneous thermal ablation techniques may be used to manage symptoms and achieve local tumor control. Thermal ablation is a minimally invasive technique in which an applicator is inserted directly into a tumor site under image guidance in order to deliver temperatures (low or high) sufficient to induce tissue necrosis [9, 10].

45.3.3.1 Radiofrequency Ablation

Radiofrequency ablation (RFA) of musculoskeletal lesions was first reported in 1992 and continues to be used widely [11]. RFA uses high-frequency (375–500 kHz) alternating current delivered via 15- to 17-gauge electrodes into a tumor, inducing local ionic agitation, rapid heat formation (>60 °C), resulting in protein denaturation and local hemostasis. Cooled saline is circulated periodically at the site to reduce charring, which can inhibit electrical conduction [10]. Early RFA systems utilize monopolar electrodes with a grounding pad applied to the patient; simultaneous use of three monopolar electrodes has been described [9, 10]. Newer bipolar systems create more precise, larger ablation zones with faster heat production and shorter treatment duration [12].

45.3.3.2 Cryoablation

In contrast to RFA, cryoablation is a thermal technique that relies on temperatures between −20 to −40 °C. A hollow cryoprobe with an uninsulated tip is inserted into the tumor. Pressurized argon and helium gases are delivered at the tip and the resulting gas expansion produces rapid cooling. A single probe may treat a zone 5 cm in length and 3.5 cm in diameter; multiple probes may be used simultaneously to expand the treatment zone. Effective treatment requires >1 min of tissue freezing [10]. Treatment areas can be monitored on CT and MRI [10]. Cryoablation elicits effective MBD-related pain control [10, 13] and has a low adverse event rate [14]. Cryoablation achieved local tumor control in 82% of patients with metastatic renal cell carcinoma [15].

45.3.3.3 Embolization

Metastatic tumors are often hypervascular [16]. Transarterial embolization (TAE) has been investigated as both a preoperative intervention to limit hemorrhage and as a treatment for local tumor control. As a hemorrhage control modality, TAE has greatest utility for highly vascular metastases, spine metastases, and in cases requiring prolonged surgical time [17]. Microcatheters are the preferred conduit for embolic agents, which include proprietary engineered particles, gelfoam, liquid agents, and stainless steel or platinum coils [17]. A recent systematic review reported that 50% of studies (Level III) investigating TAE for metastatic renal cell carcinoma in long bones (primarily femur and humerus) demonstrate a significant reduction in perioperative blood loss [18].

The evidence to support TAE for local tumor control is inconclusive and high-quality randomized controlled trials are necessary to ascertain its effectiveness [18].

45.4 Surgical Management

Surgical management of upper extremity skeletal metastases should restore function and decrease pain [5]. Preoperative planning is paramount to obtaining the desired outcome. Patients with skeletal metastases introduce particular considerations relevant to preoperative planning, specifically osteopenic bone with limited healing potential [2], which therefore places increased loads on the construct. More fundamentally, an accurate diagnosis of the skeletal lesion, which is predicated on a comprehensive workup, is critical to avoiding inadvertent fixation of primary osseous malignancies [19, 20].

Within the upper extremity, the humerus is the most common site of metastatic skeletal lesions [21], therefore, outcomes studies of humeral lesions are represented disproportionally in the literature. As mentioned previously, because it is less responsible for weightbearing, there is a broader range of acceptable operative strategies for MBD of the upper extremity.

45.4.1 Extra-articular, Intralesional Strategies

MBD lesions that do not involve an articular surface and have reasonable bone stock are generally amenable to intralesional operative strategies. These include bone filler-only options, extramedullary—plate-screw—implants, and intramedullary implants (Fig. 45.1).

45.4.1.1 Bone Filler-Only

For limited indications, a bone filler—most commonly polymethylmethacrylate (PMMA)—may be used without a metallic implant. PMMA offers strong resistance to com-

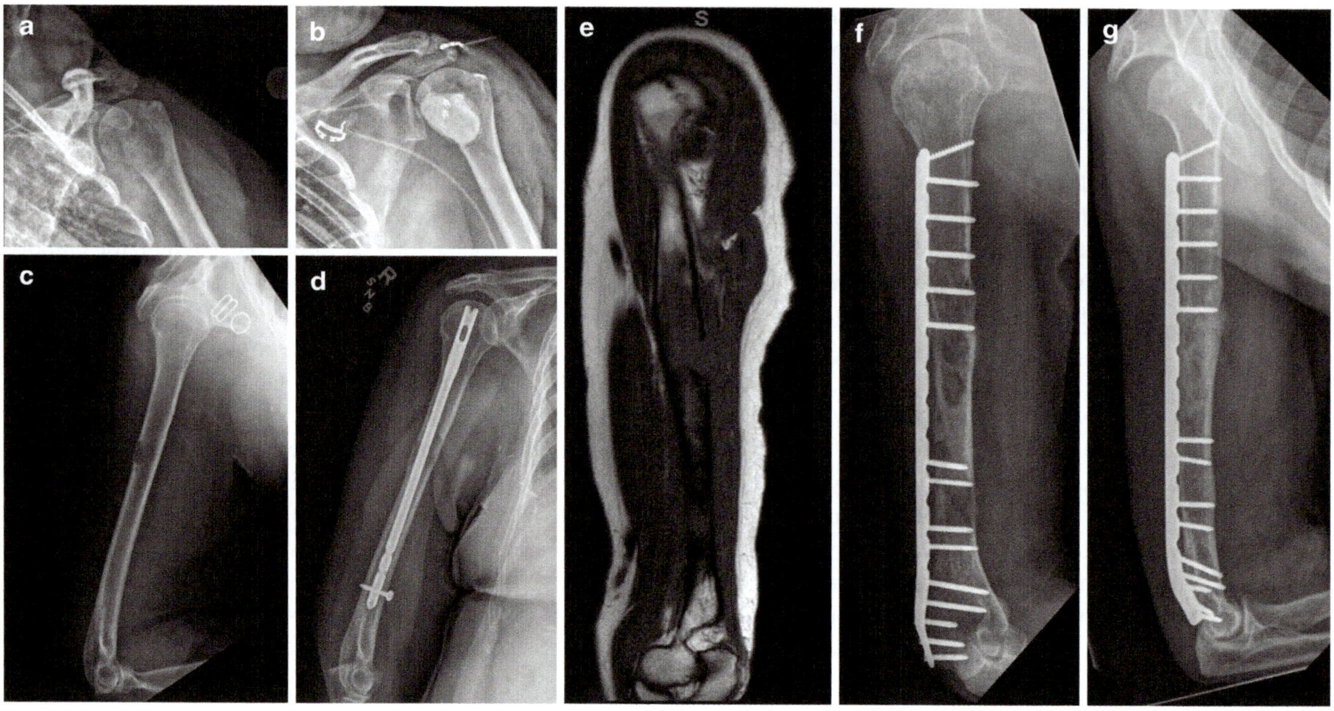

Fig. 45.1 Examples of various strategies of intralesional management of metastatic bone disease: (**a**, **b**) curettage and polymethylmethacrylate cement packing alone without trauma implants; (**c**, **d**) intramedullary fixation; (**e**, **f**) open reduction and internal fixation of a humerus affected by metastatic endometrial cancer, showing bone consolidation following chemotherapy and radiation management (**g**)

pression but does not provide much stability for bending moments. Appropriate indications for use of PMMA only therefore are limited to subchondral or epiphyseal–metaphyseal locations.

45.4.1.2 Extramedullary Instrumentation

The advent of locked plating has improved the stability of plate-screw constructs in osteopenic, and otherwise compromised, bone [22]. Stability can be augmented with the use of PMMA, a strategy that has been used for several decades [23, 24]. PMMA may be used to augment screw pullout strength [25–27] and used as a delivery vehicle for antibiotics and antitumoral drugs [28, 29].

Advantages to plate-screw constructs include generally greater stability than intramedullary options. Furthermore, because the humeral intramedullary space ends several centimeters proximal to the elbow, plate-screw constructs are more appropriate when distal humeral fixation is necessary. Another advantage to open management of MBD is the ability to debulk the lesion through curettage and oversee directly the introduction of PMMA. Plate-screw constructs are the preferred strategy for operative lesions of the forearm.

Disadvantages to plate-screw constructs include more extensive exposure and dissection and higher adverse event rates. A 2009 study prospectively evaluated 187 patients who underwent locked plating of a pathologic proximal humerus fracture [30]. Although clinical outcome scores significantly improved at 1-year follow-up, there were 52 immediate postoperative complications. Similarly, a 2011 review of 63 metastatic bone tumors to the humerus stabilized with plate fixation and cement augmentation found that while all patients reported pain relief and 50/63 (79%) regained independence in activities of daily living, there were 14 complications with seven requiring reoperation [31].

45.4.1.3 Intramedullary Instrumentation

Intramedullary nails (IMNs) are load-sharing devices [32, 33]. IMNs are designed to bear weight initially, with gradual transfer to the healing bone. In the setting of poor bone quality and limited healing potential characteristic of skeletal metastases, this transmission of load sharing may not occur, sometimes resulting in implant failure.

The humerus is the only upper extremity bone readily amenable to IMN. Use of IMNs in the humerus is also limited by lesion location. Lesions suitable for IMNs are limited generally to well-contained proximal metadiaphyseal lesions and mid-diaphyseal lesions [34, 35].

Outcomes following IMN for prophylactic and fracture fixation of humeral MBD are generally good, with reliable and effective pain relief with function restoration and low morbidity [36–38]. A comparison of IMN and IMN with PMMA for pathologic humerus fractures showed reduced pain 6 weeks following surgery in the group receiving PMMA [39].

In the absence of a clear contraindication, RT is generally recommended following intralesional surgical management [40, 41].

45.4.2 Reconstruction: Proximal Humerus Lesions

The proximal one-third of the humerus is the most common site for metastatic disease in the upper extremity [40]. The options for reconstruction of the proximal humerus depend on the volume of diseased humeral bone, the location of the lesion, functional and oncological goals for resection, and the age of the patient. Relatively localized subchondral lesions may be treated with conventional arthroplasty techniques, while more extensive or distal lesions will require more extensive resection and segmental reconstruction. Several implant options are available to reconstruct segmental defects, including primarily metallic endoprostheses (EP), allograft-prosthetic composites (APCs), and osteoarticular allografts [42–44]. The duration following surgery to achieve a good palliative and functional result should be factored into implant selection due to the often short survival of patients with MBD [45].

45.4.3 Arthroplasty: Hemiarthroplasty versus Reverse Total Shoulder Arthroplasty

Conventional hemiarthroplasty has been the traditional strategy for patients with MBD of the humeral head. This procedure is effective for providing patients with pain relief but may result in limited motion [46, 47]. Challenges to postoperative function arise commonly in the setting of MBD due to diseased bone at the footprint of the rotator cuff, creating challenges to RTC repair with a traditional hemiarthroplasty implant [48] and poor outcomes when the rotator cuff is deficient [49]. In a series of 36 patients reconstructed with APC hemiarthroplasty for a variety of tumor-related indications, the postoperative active flexion was only 56° overall [50]. A similar study of 47 patients with endoprosthetic hemiarthroplasties reported average active flexion of 55° [46].

Reverse total shoulder arthroplasty (RTSA) provides an alternative to hemiarthroplasty [49, 51]. This design is effective for cuff-deficient patients, with high relevance to patients with MBD. Outcome reports following RTSA for patients with MBD demonstrate active forward flexion of 163° and an average Constant Shoulder Scores of 76, values similar to those reported for RTSA in the absence of MBD [52]. Another series reports more modest outcomes with a mean Constant score of 52 and flexion of 122° [53].

There are few direct comparisons of outcomes following hemiarthroplasty and RTSA for tumor resection. Grosel et al. reported outcomes for 10 patients who underwent RTSA and 37 with hemiarthroplasty with an average follow-up of 27 months. Postoperative motion was significantly better with RTSA than hemiarthroplasty, with forward flexion averages of 85° and 28°, respectively. Postoperatively, 42% of hemiarthroplasty patients experienced an adverse event compared to 10% of RTSA patients. Instability occurred in 16% of patients with hemiarthroplasty and none of the patients in the RTSA cohort [54].

45.4.4 Addressing Bone Loss

Less conventional implants are required to reestablish anatomical relationships when proceeding with shoulder arthroplasty in the setting of segmental bone insufficiency. Proximal bone loss may be addressed with either an APC or an endoprosthesis, either of which can then be mated to a hemiarthroplasty or RTSA articulation. The endoprosthesis option provides what some consider to be a faster and more straightforward reconstruction [55] where the missing bone is simply replaced with metal. Endoprostheses present challenges including soft-tissue reattachment to the implant to provide stability and function, although creative solutions such as wrapping the implant in a vascular mesh have been proposed [56]. An allograft-prosthetic composite (APC) is a construct in which a portion of the allograft bone is cut to the appropriate length and then mated with either a hemiarthroplasty or RTSA prosthesis embedded within it [57–59]. Advantages of this include the ability to customize the implant to any patient and to provide soft-tissue attachments for repairing the native tendons, increasing stability and functionality [60, 61]. Osteoarticular allografts provide an additional option for reconstruction of large defects and utilize the osseous and articular portions of donor bone to recreate the missing anatomy. This allograft is then mated to the distal host bone with an expectation of osteosynthesis over time [62, 63]. Advantages of this technique include avoiding a metallic implant in younger patients as well as providing allograft sites for reattachment of native tendons. Risks include fracture and infection. While these allografts have been reported to provide excellent or good outcomes in two-thirds of patients [62], they are marked by substantial limitations in movement and lifting at the glenohumeral joint [63]. Despite this, they do provide pain relief and maintain hand functionality [63].

45.4.5 Outcomes and Complications

Shoulder function after tumor resection and reconstruction has traditionally been limited with a goal of achieving a stable, pain-free joint which allows for effective hand function. While the

RTSA has improved range of motion at the shoulder itself, the overall goals of pain relief and stability remain important criteria. When comparing specific bone loss constructs, a large-scale retrospective cohort study and systematic review both indicate that APCs, endoprostheses, and osteoarticular allografts provide similar overall functionality and pain relief [42, 64]. Importantly, these outcomes do not necessarily capture the various complications that a patient may experience. Postoperative complications following tumor resection and reconstruction are common and important to consider [45, 48, 54].

Nonunion at the interface between the host and donor bone can cause construct failure following either an APC or osteoarticular allograft. Rates of nonunion following APC reconstructions vary greatly in the literature from 0 to 26% [58, 59], highlighting the impact of differing surgical techniques or patient populations between studies. Osteoarticular allograft reconstructions can also develop nonunion; a systematic review highlighted that this complication was more common with osteoarticular allografts (11%) than APCs (5.0%). When considering either the APC or osteoarticular allograft techniques that rely on osteosynthesis, it is important to recognize that chemotherapy can be a risk factor for nonunion [60, 65] and may make an endoprosthesis a more attractive option. Postoperative stability of the glenohumeral joint is an important goal of any reconstruction as dislocation is a common reason for reoperation [48, 54]. Theoretical benefits of an APC or osteoarticular allograft include more attachment sites for native soft tissues to help provide stability, however, large-scale studies have not detected meaningful differences between groups [42, 64]. RTSA provides a new, more constrained arthroplasty option, and preliminary studies suggest a lower rate of instability when using RTSA versus hemiarthroplasty following tumor resection [54].

Additional complications to consider include aseptic loosening of an implant and fracture adjacent to, or within, the construct. Varying rates of these complications are presented in heterogeneous small case studies throughout the literature [45, 48, 50, 59] and it is necessary to look to multicenter studies [42] or systematic reviews [64] to observe meaningful trends. A large-scale study demonstrated that endoprosthesis and APC were generally equivalent in terms of the risk of loosening and fracture but that osteoarticular allografts were clearly inferior with a fracture rate of up to 49% [42, 64]. Poor long-term outcomes with osteoarticular allografts in another recent study led the authors to conclude that their use should be avoided [66].

45.4.6 Intercalary Reconstruction: Diaphyseal and Distal Humerus Lesions

In situations where a metastatic humerus lesion is limited to the diaphyseal region of the bone, resection and intercalary reconstruction can be an excellent option if the location is remote from both the proximal and distal joint surfaces. After tumor removal, the resulting void can be bridged with allograft, autograft, or an endoprosthesis. A recent case series of intercalary endoprostheses highlighted the reliability of this technique in the humerus as no non-oncologic failures were observed and patients reported satisfactory functional outcomes scores [67]. Allograft reconstruction presents a biological alternative for filling diaphyseal voids and may be more appropriate for younger patients in whom an endoprosthesis may not be a long-term solution [68]. In a series of 30 patients undergoing humeral reconstruction with allograft, the overall complication rate was 53%, most often due to a nonunion at the host–graft interface [68]. While this technique avoids implantation of a prosthesis and can prove durable if graft incorporation occurs, the overall success can be limited. A final option for a diaphyseal void is a vascularized free-fibula transfer which may provide increased healing potential for the patient. However, the literature highlights that these can also be associated with graft failure and fracture, with a small series demonstrating a 75% reoperation rate for fracture, but ultimately a 100% union rate [69].

45.4.7 Distal Humerus Reconstruction

The need for tumor resection and reconstruction of the distal humerus is rarer than the proximal humerus, but the options for reconstruction are similar and include primary total elbow prostheses, endoprostheses, allograft–prosthetic composites, and osteoarticular allografts, with similar risks and benefits for each approach. Osteoarticular allografts again provide young patients with an alternative to a prosthesis but are marked with high complication rates including fractures and joint instability [68]. When the area of resection is relatively limited, or when combined with an allograft–prosthesis composite construct, primary total elbow replacement components have been associated with good pain relief and function with a 20% revision rate [70]. More extensive reconstruction can be addressed with segmental endoprostheses, which have been found to result in an acceptable 107° arc of motion and reasonable clinical outcomes, but can be associated with aseptic loosening requiring revision [71] (Figs. 45.2 and 45.3).

45.4.8 Amputation

Amputation is rarely performed for malignant lesions of the upper extremity and is particularly uncommon for metastatic disease. Amputation may be utilized for tumors presenting with extensive neurovascular infiltration, ulceration or fungating growth, or infection. Amputation may also be indicated for patients with loss of limb function or unremitting pain. Amputations are characterized by level of limb involvement and include the fol-

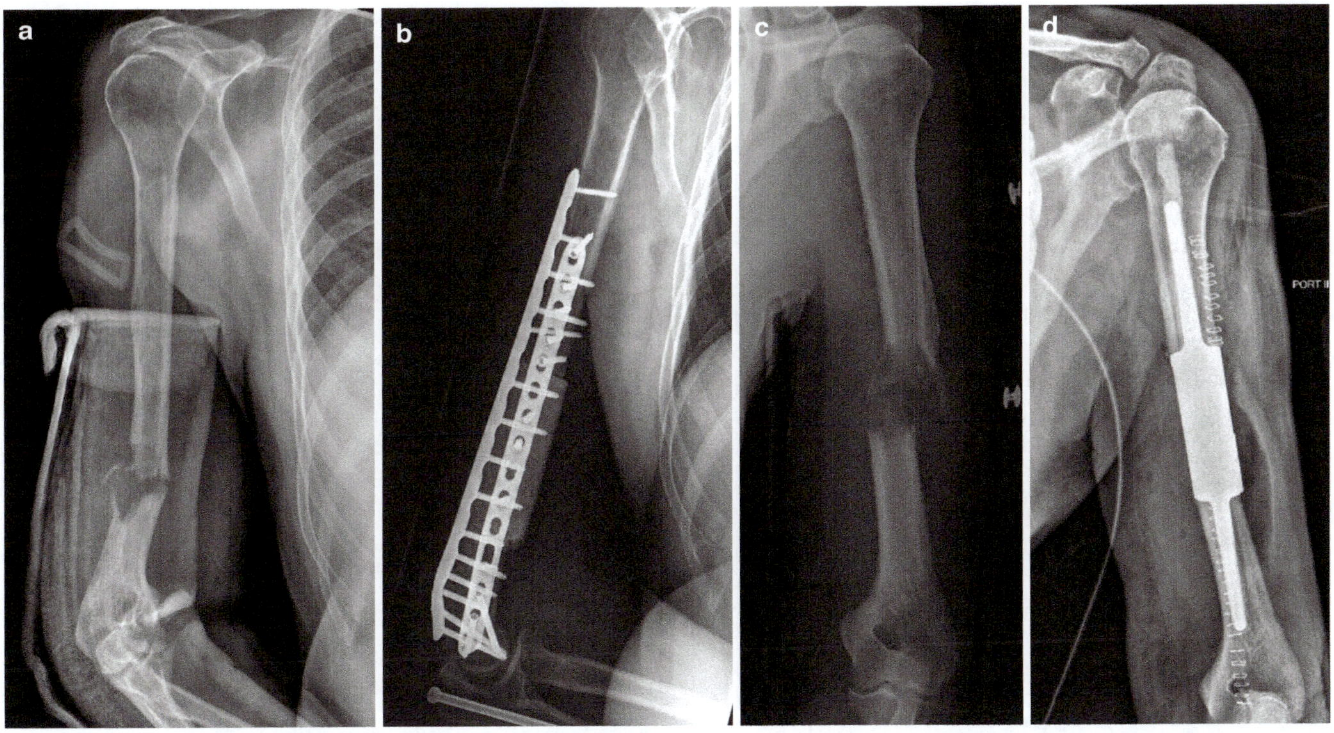

Fig. 45.2 Examples of two strategies of wide-local excision of renal cell metastases of the humerus: (**a**) showing a distal diaphyseal humerus fracture through a renal cell metastasis, (**b**) treated operatively with bicolumnar plating. (**c**) Showing a mid-diaphyseal humerus fracture through a renal cell metastasis, (**d**) treated operatively with a cemented, intercalary implant

lowing: above-shoulder (forequarter), shoulder disarticulation, transhumeral, through-elbow, below-elbow, wrist disarticulation, and at the level of the hand or digits. Resections adjacent to the elbow and wrist are uncommon and are not discussed here.

Forequarter (interscapulothoracic) was first described in the management of major traumatic injuries of the arm and involves surgical resection of the entire upper extremity, shoulder girdle, scapula, and a portion of the clavicle [72]. The technique was first utilized in the management of malignancy in 1836 and has traditionally been used in the treatment of high-grade primary bone sarcomas of the proximal humerus. Forequarter amputation is also indicated for the management of unresectable metastatic lesions arising from the scapula or proximal humerus. Such tumors have failed to respond to adjuvant chemotherapy and/or radiation, are locally aggressive, and lead to debilitating symptoms for patients [73]. Individuals undergoing forequarter amputation have a poor overall prognosis, and time to death following this procedure is short (3–24 months), leading some to advocate against radical amputation [74]. Merimsky et al., however, demonstrate an improvement in performance status and quality of life in patients who underwent palliative forequarter amputation [74]. It is important to bear in mind the potential psychological implications of this procedure [75]. Forequarter amputations are disfiguring, and upper extremity prostheses often do not provide favorable functionality when compared to lower extremity prostheses.

Acrometastases are exceedingly rare, accounting for just 0.1% of bone metastases [76]. These lesions are most commonly seen in patients with primary renal cell disease and typically portend a poor prognosis. Surgical ablative techniques for metastatic disease of the hand and digits have been reported in the literature. Puhaindran et al. report good functional and cosmetic outcomes following single ray resection for malignant tumors of the hand [77]. Not unexpectedly, double ray resection results in worse functional outcomes, though remains a feasible hand-sparing option [78].

45.5 Summary

MBD is common and may result in substantial patient morbidity through the destruction of large portions of the skeleton, causing function and disabling pain. Treatment plans should align with patient goals and be made in the setting of a multidisciplinary team, with attention to the individual patient's global health status and ongoing treatment. Weightbearing and mobility requirements are less of a limitation in the upper extremities, thereby expanding the treatment strategies available to patients with MBD of the upper extremities. In the setting of fracture and/or treatment failure, surgical reconstruction is generally indicated.

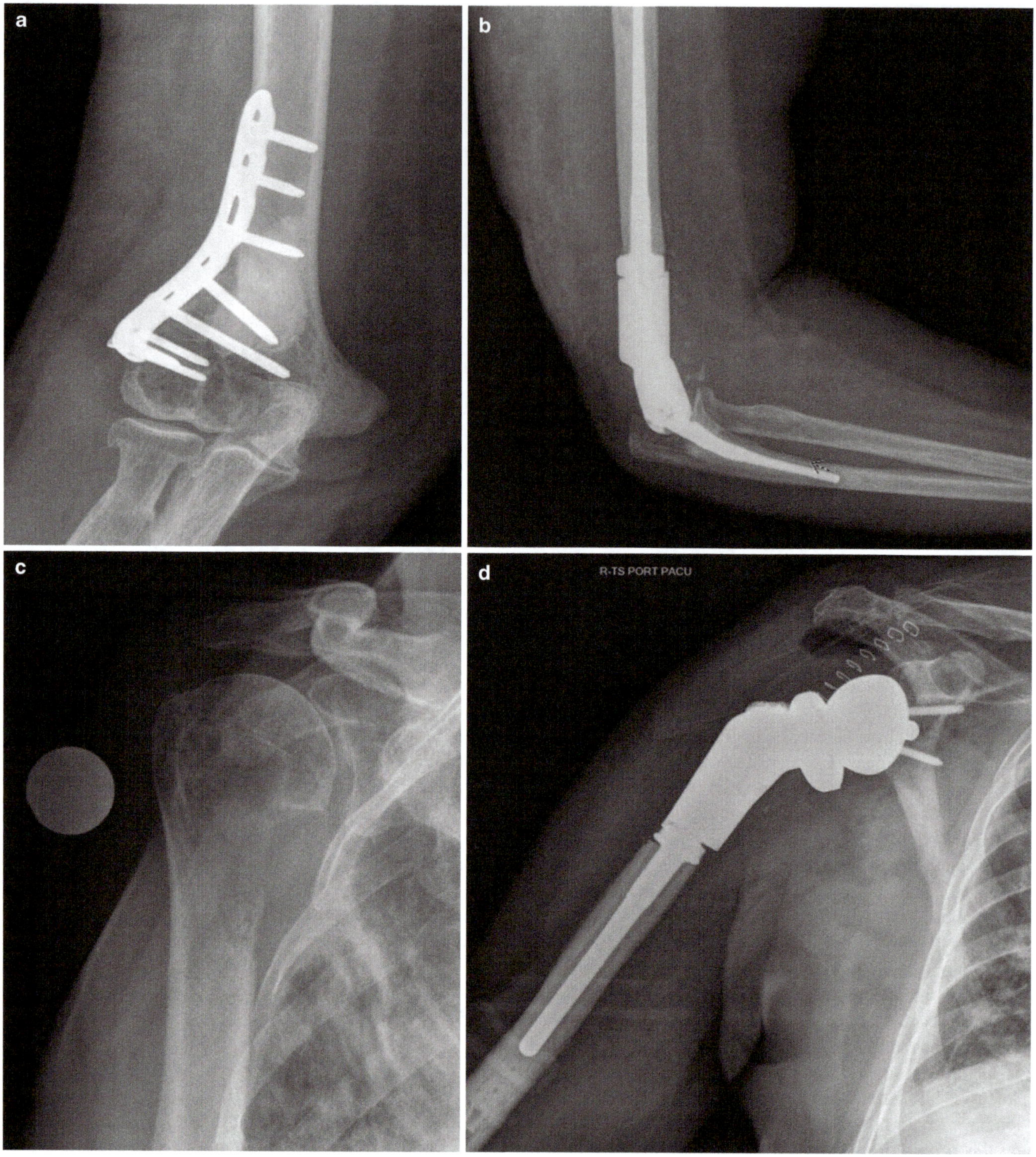

Fig. 45.3 Examples of two strategies of endoprosthesis used in the humerus for metastatic bone disease: (**a**) showing failed intralesional management of a renal cell metastasis, (**b**) later converted to a distal humeral replacement. (**c**) Showing extensive humeral head destruction from metastatic lung cancer, (**d**) treated with a segmental endoprosthesis with reverse total shoulder replacement

Questions

1. What is the most common upper extremity location of skeletal metastatic disease?
   A. Radius
   B. Scaphoid
   C. **Humerus**
   D. Ulna
2. An appropriate indication for using polymethylmethacrylate (PMMA) bone filler would be a metastatic lesion located in the:
   A. Diaphysis
   B. Meta-diaphyseal region
   C. **Subchondral surface**
   D. Articular surface
3. Which location of a metastatic lesion would be most amenable to intramedullary instrumentation?
   A. Distal humerus
   B. Proximal humerus
   C. Diaphysis of the radius
   D. **Diaphysis of the humerus**
4. When the resection of a metastatic lesion from the proximal humerus requires sacrificing the humeral head and tuberosities, which reconstruction option will provide the patient with the best range of motion postoperatively?
   A. Total shoulder arthroplasty
   B. Girdlestone
   C. **Reverse total shoulder arthroplasty**
   D. Hemiarthroplasty
   E. Range of motion is similar between these options
5. When reconstructing large metaphyseal bone defects following metastatic disease resection, multiple reconstruction options are available. Although studies have shown that functional outcomes are similar between options, how do these compare in terms of complication rates?
   A. Endoprostheses, allograft–prosthetic composites, and osteoarticular allografts are all similar
   B. Endoprostheses are marked by higher complication rates
   C. Allograft–prosthetic composites are marked by higher complication rates
   D. **Osteoarticular allografts are marked by higher complication rates**
   E. This question has not been previously studied in the literature

Future Directions

- How will innovations in gene therapy change the surgical management of metastatic disease of bone in the upper extremity?
- How will enhanced understanding of metaphysical wellness practices help patients navigate the physical limitations created by limb reconstruction procedures?

45.6 References

1. National Comprehensive Cancer Network (NCCN). Clinical practice guidelines; n.d.
2. Gainor BJ, Buchert P. Fracture healing in metastatic bone disease. Clin Orthop Relat Res. 1983;178:297–302.
3. Flemming JE, Beals RK. Pathologic fracture of the humerus. Clin Orthop Relat Res. 1986:258–60.
4. Healey JH, Brown HK. Complications of bone metastases. Cancer. 2000;88:2940–51. https://doi.org/10.1002/1097-0142(20000615)88:12+<2940::AID-CNCR10>3.0.CO;2-W.
5. Anract P, Biau D, Boudou-Rouquette P. Metastatic fractures of long limb bones. Orthop Traumatol Surg Res. 2017;103:S41–51. https://doi.org/10.1016/j.otsr.2016.11.001.
6. De Felice F, Piccioli A, Musio D, Tombolini V. The role of radiation therapy in bone metastases management. Oncotarget. 2017;8:25691–9. https://doi.org/10.18632/oncotarget.14823.
7. Agarawal JP, Swangsilpa T, van der Linden Y, et al. The role of external beam radiotherapy in the management of bone metastases. Clin Oncol (R Coll Radiol). 2006;18:747–60. https://doi.org/10.1016/j.clon.2006.09.007.
8. Trikalinos TA, Terasawa T, Ip S, et al. Particle beam radiotherapies for cancer. Agency for Healthcare Research and Quality (US); 2009.
9. Ahmed M, Solbiati L, Brace CL, et al. Image-guided tumor ablation: standardization of terminology and reporting criteria—a 10-year update. Radiology. 2014;273:241–60. https://doi.org/10.1148/radiol.14132958.
10. Moynagh MR, Kurup AN, Callstrom MR. Thermal ablation of bone metastases. Semin Intervent Radiol. 2018;35:299–308. https://doi.org/10.1055/s-0038-1673422.
11. Volkmer D, Sichlau M, Rapp TB. The use of radiofrequency ablation in the treatment of musculoskeletal tumors. J Am Acad Orthop Surg. 2009;17:737–43.
12. Chang W, Lee JM, Lee DH, et al. Comparison of switching bipolar ablation with multiple cooled wet electrodes and switching monopolar ablation with separable clustered electrode in treatment of small hepatocellular carcinoma: a randomized controlled trial. PLoS One. 2018;13:e0192173. https://doi.org/10.1371/journal.pone.0192173.

13. Susa M, Kikuta K, Nakayama R, et al. CT guided cryoablation for locally recurrent or metastatic bone and soft tissue tumor: initial experience. BMC Cancer. 2016;16:798. https://doi.org/10.1186/s12885-016-2852-6.
14. Auloge P, Cazzato RL, Rousseau C, et al. Complications of percutaneous bone tumor cryoablation: a 10-year experience. Radiology. 2019;291:521–8. https://doi.org/10.1148/radiol.2019181262.
15. Gardner CS, Ensor JE, Ahrar K, et al. Cryoablation of bone metastases from renal cell carcinoma for local tumor control. J Bone Joint Surg. 2017;99:1916–26. https://doi.org/10.2106/JBJS.16.01182.
16. Kickuth R, Waldherr C, Hoppe H, et al. Interventional management of hypervascular osseous metastasis: role of embolotherapy before orthopedic tumor resection and bone stabilization. Am J Roentgenol. 2008;191:W240–7. https://doi.org/10.2214/AJR.07.4037.
17. Owen RJT. Embolization of musculoskeletal bone tumors. Semin Interv Radiol. 2010;27:111. https://doi.org/10.1055/s-0030-1253510.
18. Geraets SEW, Bos PK, van der Stok J. Preoperative embolization in surgical treatment of long bone metastasis: a systematic literature review. EFORT Open Rev. 2020;5:17. https://doi.org/10.1302/2058-5241.5.190013.
19. Adams SC, Potter BK, Mahmood Z, et al. Consequences and prevention of inadvertent internal fixation of primary osseous sarcomas. Clin Orthop Relat Res. 2009;467:519–25. https://doi.org/10.1007/s11999-008-0546-3.
20. Spence G, Dunning MT, Cannon SR, Briggs TWR. The hazard of retrograde nailing in pathological fractures. Three cases involving primary musculoskeletal malignancy. Injury. 2002;33(6):533–8. https://doi.org/10.1016/s0020-1383(02)00035-9.
21. Hage W, Aj A, Dm A. Incidence, location, and diagnostic evaluation of metastatic bone disease. The Orthopedic clinics of North America; 2000. https://pubmed.ncbi.nlm.nih.gov/11043092/. Accessed 1 Dec 2020.
22. Tejwani NC, Guerado E. Improving fixation of the osteoporotic fracture: the role of locked plating. J Orthop Trauma. 2011;25:S56. https://doi.org/10.1097/BOT.0b013e31821b8a52.
23. Harrington KD, Johnston JO, Turner RH, Green DL. The use of methylmethacrylate as an adjunct in the internal fixation of malignant neoplastic fractures. J Bone Joint Surg Am. 1972;54:1665–76.
24. Schatzer J, Ha'eri GB, Chapman M. Methylmethacrylate as an adjunct in the internal fixation of intertrochanteric fractures of the femur. J Trauma. 1978;18(10):732–5.
25. Cameron HU, Jacob R, Macnab I, Pilliar RM. Use of polymethylmethacrylate to enhance screw fixation in bone. J Bone Joint Surg. 1975;57:655–6.
26. Motzkin NE, Chao EY, An KN, et al. Pull-out strength of screws from polymethylmethacrylate cement. J Bone Joint Surg Br. 1994;76:320–3.
27. Wittenberg RH, Lee K-S, Shea M, et al. Effect of screw diameter, insertion technique, and bone cement augmentation of pedicular screw fixation strength. Clin Orthop Relat Res. 1993;296:278–87.
28. Wang HM, Galasko CSB, Crank S, et al. Methotrexate loaded acrylic cement in the management of skeletal metastases: biomechanical, biological, and systemic effect. Clin Orthop Relat Res. 1995;312:173–86.
29. Wasserlauf S, Warshawsky A, Arad-Yelin R, et al. The release of cytotoxic drugs from acrylic bone cement. Bull Hosp Jt Dis. 1993;53(1):68–74. https://europepmc.org/article/med/8374496. Accessed 1 Dec 2020.
30. Südkamp N, Bayer J, Hepp P, et al. Open reduction and internal fixation of proximal humeral fractures with use of the locking proximal humerus plate. Results of a prospective, multicenter, observational study. J Bone Joint Surg Am. 2009;91:1320–8. https://doi.org/10.2106/JBJS.H.00006.
31. Weiss KR, Bhumbra R, Biau DJ, et al. Fixation of pathological humeral fractures by the cemented plate technique. J Bone Joint Surg Br. 2011;93:1093–7. https://doi.org/10.1302/0301-620X.93B8.26194.
32. Bong MR, Kummer FJ, Koval KJ, Egol KA. Intramedullary nailing of the lower extremity: biomechanics and biology. J Am Acad Orthop Surg. 2007;15:97–106. https://doi.org/10.5435/00124635-200702000-00004.
33. Cheung G, Zalzal P, Bhandari M, et al. Finite element analysis of a femoral retrograde intramedullary nail subject to gait loading. Med Eng Phys. 2004;26:93–108. https://doi.org/10.1016/j.medengphy.2003.10.006.
34. Redmond BJ, Biermann JS, Blasier RB. Interlocking intramedullary nailing of pathological fractures of the shaft of the humerus. J Bone Joint Surg Am. 1996;78:891–6. https://doi.org/10.2106/00004623-199606000-00012.
35. Scolaro J, Lackman R. Surgical management of metastatic long bone fractures: principles and techniques. J Am Acad Orthop Surg. 2014; https://doi.org/10.5435/JAAOS-22-02-90.
36. Alvi HM, Damron TA. Prophylactic stabilization for bone metastases, myeloma, or lymphoma: do we need to protect the entire bone? Clin Orthop Relat Res. 2013;471:706–14. https://doi.org/10.1007/s11999-012-2656-1.
37. Pretell J, Rodriguez J, Blanco D, et al. Treatment of pathological humeral shaft fractures with intramedullary nailing. A retrospective study. Int Orthop. 2010;34:559–63. https://doi.org/10.1007/s00264-009-0833-y.
38. Spencer SJ, Holt G, Clarke JV, et al. Locked intramedullary nailing of symptomatic metastases in the humerus. J Bone Joint Surg Br. 2010;92:142–5. https://doi.org/10.1302/0301-620X.92B1.22399.
39. Laitinen M, Nieminen J, Pakarinen T-K. Treatment of pathological humerus shaft fractures with intramedullary nails with or without cement fixation. Arch Orthop Trauma Surg. 2011;131:503–8. https://doi.org/10.1007/s00402-010-1172-6.
40. Frassica FJ, Frassica DA. Metastatic bone disease of the humerus. J Am Acad Orthop Surg. 2003;11:282–8. https://doi.org/10.5435/00124635-200307000-00008.
41. Leopold SS. Editor's spotlight/take 5: prophylactic stabilization for bone metastases, myeloma, or lymphoma: do we need to protect the entire bone? (DOI 10.1007/s11999-012-2656-1). Clin Orthop Relat Res. 2013;471:703–5. https://doi.org/10.1007/s11999-012-2714-8.
42. Nota S, Teunis T, Kortlever J, et al. Functional outcomes and complications after oncologic reconstruction of the proximal humerus. J Am Acad Orthop Surg. 2018;26:403–9. https://doi.org/10.5435/JAAOS-D-16-00551.
43. Potter BK, Adams SC, Pitcher DJJ, et al. Proximal humerus reconstructions for tumors. Clin Orthop Relat Res. 2009;467:1035–41. https://doi.org/10.1007/s11999-008-0531-x.
44. Rödl RW, Gosheger G, Gebert C, et al. Reconstruction of the proximal humerus after wide resection of tumours. J Bone Joint Surg Br. 2002;84:1004–8. https://doi.org/10.1302/0301-620x.84b7.12989.
45. Kaa AKS, Jørgensen PH, Søjbjerg JO, Johannsen HV. Reverse shoulder replacement after resection of the proximal humerus for bone tumours. Bone Joint J. 2013;95-B:1551–5. https://doi.org/10.1302/0301-620X.95B11.31545.
46. Kumar D, Grimer RJ, Abudu A, et al. Endoprosthetic replacement of the proximal humerus. Long-term results. J Bone Joint Surg Br. 2003;85:717–22.
47. Ross A, Wilson J, Scales J. Endoprosthetic replacement of the proximal humerus. J Bone Joint Surg Br. 1987;69-B:656–61. https://doi.org/10.1302/0301-620X.69B4.3611177.
48. Raiss P, Kinkel S, Sauter U, et al. Replacement of the proximal humerus with MUTARS tumor endoprostheses. Eur J Surg Oncol (EJSO). 2010;36:371–7. https://doi.org/10.1016/j.ejso.2009.11.001.
49. Boileau P, Watkinson DJ, Hatzidakis AM, Balg F. Grammont reverse prosthesis: design, rationale, and biomechanics. J Shoulder Elb Surg. 2005;14:147S–61S. https://doi.org/10.1016/j.jse.2004.10.006.

50. Abdeen A, Hoang BH, Athanasian EA, et al. Allograft-prosthesis composite reconstruction of the proximal part of the humerus: functional outcome and survivorship. J Bone Joint Surg. 2009;91:2406–15. https://doi.org/10.2106/JBJS.H.00815.
51. Walker M, Brooks J, Willis M, Frankle M. How reverse shoulder arthroplasty works. Clin Orthop Relat Res. 2011;469:2440–51. https://doi.org/10.1007/s11999-011-1892-0.
52. De Wilde L, Boileau P, Van der Bracht H. Does reverse shoulder arthroplasty for tumors of the proximal humerus reduce impairment? Clin Orthop Relat Res. 2011;469:2489–95. https://doi.org/10.1007/s11999-010-1758-x.
53. Bonnevialle N, Mansat P, Lebon J, et al. Reverse shoulder arthroplasty for malignant tumors of proximal humerus. J Shoulder Elb Surg. 2015;24:36–44. https://doi.org/10.1016/j.jse.2014.04.006.
54. Grosel TW, Plummer DR, Everhart JS, et al. Reverse total shoulder arthroplasty provides stability and better function than hemiarthroplasty following resection of proximal humerus tumors. J Shoulder Elb Surg. 2019;28:2147–52. https://doi.org/10.1016/j.jse.2019.02.032.
55. Guven MF, Aslan L, Botanlioglu H, et al. Functional outcome of reverse shoulder tumor prosthesis in the treatment of proximal humerus tumors. J Shoulder Elb Surg. 2016;25:e1–6. https://doi.org/10.1016/j.jse.2015.06.012.
56. Marulanda GA, Henderson E, Cheong D, Letson GD. Proximal and total humerus reconstruction with the use of an aortograft mesh. Clin Orthop Relat Res. 2010;468:2896–903. https://doi.org/10.1007/s11999-010-1418-1.
57. Abdeen A, Healey JH. Allograft-prosthesis composite reconstruction of the proximal part of the humerus: surgical technique. J Bone Joint Surg. 2010;92:188–96. https://doi.org/10.2106/JBJS.J.00167.
58. Chacon A, Virani N, Shannon R, et al. Revision arthroplasty with use of a reverse shoulder prosthesis-allograft composite. J Bone Joint Surg Am. 2009;91:119–27. https://doi.org/10.2106/JBJS.H.00094.
59. Sanchez-Sotelo J, Wagner ER, Houdek MT. Allograft-prosthetic composite reconstruction for massive proximal humeral bone loss in reverse shoulder arthroplasty. JBJS Essent Surg Tech. 2018;8:e3. https://doi.org/10.2106/JBJS.ST.17.00051.
60. Hejna MJ, Gitelis S. Allograft prosthetic composite replacement for bone tumors. Semin Surg Oncol. 1997;13:18–24. https://doi.org/10.1002/(SICI)1098-2388(199701/02)13:1<18::AID-SSU4>3.0.CO;2-7.
61. King JJ, Nystrom LM, Reimer NB, et al. Allograft-prosthetic composite reverse total shoulder arthroplasty for reconstruction of proximal humerus tumor resections. J Shoulder Elb Surg. 2016;25:45–54. https://doi.org/10.1016/j.jse.2015.06.021.
62. Gebhardt MC, Roth YF, Mankin HJ. Osteoarticular allografts for reconstruction in the proximal part of the humerus after excision of a musculoskeletal tumor. J Bone Joint Surg Am. 1990;72:334–45.
63. Getty PJ, Peabody TD. Complications and functional outcomes of reconstruction with an osteoarticular allograft after intra-articular resection of the proximal aspect of the humerus. J Bone Joint Surg Am. 1999;81:1138–46. https://doi.org/10.2106/00004623-199908000-00009.
64. Teunis T, Nota SPFT, Hornicek FJ, et al. Outcome after reconstruction of the proximal humerus for tumor resection: a systematic review. Clin Orthop Relat Res. 2014;472:2245–53. https://doi.org/10.1007/s11999-014-3474-4.
65. Hazan EJ, Hornicek FJ, Tomford W, et al. The effect of adjuvant chemotherapy on osteoarticular allografts. Clin Orthop Relat Res. 2001;385:176–81.
66. Bus MPA, van de Sande MAJ, Taminiau AHM, Dijkstra PDS. Is there still a role for osteoarticular allograft reconstruction in musculoskeletal tumour surgery? a long-term follow-up study of 38 patients and systematic review of the literature. Bone Joint J. 2017;99-B:522–30. https://doi.org/10.1302/0301-620X.99B4.BJJ-2016-0443.R2.
67. Benevenia J, Kirchner R, Patterson F, et al. Outcomes of a modular intercalary endoprosthesis as treatment for segmental defects of the femur, tibia, and humerus. Clin Orthop Relat Res. 2016;474:539–48. https://doi.org/10.1007/s11999-015-4588-z.
68. Ogink PT, Teunissen FR, Massier JR, et al. Allograft reconstruction of the humerus: complications and revision surgery. J Surg Oncol. 2019;119:329–35. https://doi.org/10.1002/jso.25309.
69. Rose PS, Shin AY, Bishop AT, et al. Vascularized free fibula transfer for oncologic reconstruction of the humerus. Clin Orthop Relat Res. 2005;438:80–4. https://doi.org/10.1097/01.blo.0000179586.34727.5b.
70. Athwal GS, Chin PY, Adams RA, Morrey BF. Coonrad-Morrey total elbow arthroplasty for tumours of the distal humerus and elbow. J Bone Joint Surg. 2005;87-B:1369–74. https://doi.org/10.1302/0301-620X.87B10.16569.
71. Weber KL, Lin PP, Yasko AW. Complex Segmental Elbow Reconstruction After Tumor Resection. Clin Orthop Relat Res. 2003;415:31–44. https://doi.org/10.1097/01.blo.0000093894.12372.53.
72. Dimas V, Kargel J, Bauer J, Chang P. Forequarter amputation for malignant tumours of the upper extremity: case report, techniques and indications. Can J Plast Surg. 2007;15(2):83–5. https://doi.org/10.1177/229255030701500204.
73. Wittig JC, Bickels J, Kollender Y, et al. Palliative forequarter amputation for metastatic carcinoma to the shoulder girdle region: indications, preoperative evaluation, surgical technique, and results. J Surg Oncol. 2001;77(2):105–13.
74. Merimsky O, Kollender Y, Inbar M, et al. Is forequarter amputation justified for palliation of intractable cancer symptoms? Oncology. 2001;60:55–9. https://doi.org/10.1159/000055297.
75. Elsner U, Henrichs M, Gosheger G, et al. Forequarter amputation: a safe rescue procedure in a curative and palliative setting in high-grade malignoma of the shoulder girdle. World J Surg Oncol. 2016;14:216. https://doi.org/10.1186/s12957-016-0973-7.
76. Sawada R, Shinoda Y, Niimi A, et al. Multiple acrometastases in a patient with renal pelvic urothelial cancer. Case Reports in Urology; 2017. https://www.hindawi.com/journals/criu/2017/7830207/. Accessed 1 Dec 2020.
77. Puhaindran ME, Healey JH, Athanasian EA. Single ray amputation for tumors of the hand. Clin Orthop Relat Res. 2010;468:1390. https://doi.org/10.1007/s11999-009-1019-z.
78. Puhaindran ME, Athanasian EA. Double ray amputation for tumors of the hand. Clin Orthop Relat Res. 2010;468:2976. https://doi.org/10.1007/s11999-010-1389-2.

46 Surgical Management of Metastatic Disease to the Lower Extremity

Nathan W. Mesko and Lukas M. Nystrom

Abstract

Metastatic skeletal disease of the lower extremity can lead to significant morbidity and quality of life diminishment due to the stresses in the lower extremity during ambulation. In order to maintain mobility and improve pain control, interventions to address lower extremity metastatic lesions are oftentimes considered. These interventions can include radiation therapy, minimally invasive solutions, percutaneous stabilization, open curettage and fixation, or en bloc excision and reconstruction using endoprosthetic joint replacement techniques. Multiple factors should be considered when determining the best solution for each patient situation. Future research will help guide better treatment algorithms as our ability to predict fracture risk and prognosis helps to better balance construct durability and recovery following an intervention.

Learning Objectives

- Understand the indications for different treatment modalities when treating metastatic disease of the lower extremity.
- Recognize the availability of different predictive models in determining the risk for pathologic fracture.
- Recognize the role that anatomic location of skeletal disease plays in determining the type of reconstructive technique utilized.
- Discuss broad categories of complications that can occur when treating metastatic disease of the lower extremities.

N. W. Mesko (✉) · L. M. Nystrom
Cleveland Clinic, Cleveland, OH, USA
e-mail: meskon@ccf.org; nystrol@ccf.org

46.1 Introduction

After the spine, the most common locations for metastatic skeletal involvement are the pelvis, ribs, and proximal femur. Distant metastases in the tibia/foot (distal to the knee) are rare (<10% incidence) and are most commonly represented by lung cancer and renal cell carcinoma [1]. Because of the increased stress loads in the lower extremity, the risk for pathologic fracture is oftentimes high, and surgical intervention is commonly considered. According to a large national registry population analysis quantifying characteristics of surgically treated non-spinal metastases, the most common site of surgically treated skeletal metastases, the proximal femur, humerus, pelvis, and tibia—with the most common histologies involved being that of breast, prostate, kidney, lung, and myeloma [2]. Pragmatically, 70% of patients with metastatic bone disease will experience significant pain [3]. Moreover, up to 30% of patients with metastatic lesions in the pelvis and lower extremity will sustain a pathologic fracture, with 90% requiring surgical intervention [4].

Due to the high incidence of skeletal metastatic involvement in the lower extremity and the important role that the lower extremity anatomy plays in facilitating ambulation, a keen understanding of the balance between the stresses is placed on the human body with varying interventional, the risk for fracture, and overall patient prognosis is mandatory. This chapter will help explore the thought process for determining the type of intervention in these lower extremity skeletal-related events, as well as understand various techniques specific to location in the femur, tibia, and foot/ankle.

46.2 General Treatment Considerations

When considering treatment options, the prioritized goals should drive decisions based on maintaining the highest possible quality of life. The goals should aim to (1) restore function, (2) maintain/restore mobility, and (3) reduce pain—all

S. P. Leong et al. (eds.), *Cancer Metastasis Through the Lymphovascular System*,
https://doi.org/10.1007/978-3-030-93084-4_46

while attempting to minimize recovery time given the oftentimes limited prognosis that can accompany patients displaying advanced skeletal disease burden. Given the importance of the lower extremities with ambulation, maintaining mobility and function directly affects ECOG (Eastern Cooperative Oncology Group) Performance Status measurements—which is a commonly used standard when determining a patient's candidacy for systemic therapy or clinical trial inclusion. In addition, the surgical solution should be a durable construct that takes into account the limited bony healing potential of a pathologic fracture, with the goals to perform a single operation and begin immediate mobilization. Open and percutaneous surgical options, minimally invasive solutions (such as radiofrequency ablation [RFA], cryoablation, or cementoplasty), and nonoperative modalities all have a role in addressing both impending and progressed fractures of the lower extremity. Determining the appropriate course of treatment should draw from a variety of information sources, including surgeon experience, publicly available risk predictor models, patient presentation, and a multidisciplinary discussion surrounding multimodal goals of treatment. When specifically considering surgical versus nonsurgical scenarios, the important patient presentation factors should include patient prognosis, histology, and surgical candidacy based on comorbidities. Surgery should be favored in the lower extremity in the following scenarios:

1. Patients in whom multimodal conservative treatment measures fail (if appropriate)
2. Compromised structural support in the bony anatomy that is unlikely to respond to conservative treatment
3. Metastatic lesions with historically poor response to radiation or chemotherapy (e.g., melanoma, renal cell, lung, etc.)
4. Expected prognosis >4–12 weeks (though this will be surgery and patient situation specific). As a general rule, recovery should not take longer than anticipated prognosis. Surgery should not be discounted as a choice if pain control and OOB transfers can be enhanced with stabilization despite a limited prognosis measured in weeks.

46.3 Nonoperative and Minimally Invasive Options for Treatment

Nonoperative and minimally invasive solutions are appropriate in scenarios with limited prognosis, poor surgical candidacy, or amendable lesions [5–7]. Bisphosphonates and rank-ligand inhibitor drugs have been shown to aid with pain improvement and increase time to a skeletal-related event (SRE) in multiple histologies. Radiation is effective in aiding with pain relief, with single-fraction external beam radiation therapy (EBRT) showing equivalence to multi-fraction regimens [8]. More targeted, focal dosing techniques, such as intensity-modulated radiation therapy [IMRT] or stereotactic body radiation therapy [SBRT] can be delivered in scenarios where either collateral damage to normal tissue should be limited, or in histologies that are historically considered more radioresistant—though the surgeon should realize that radiation techniques rarely diminish the risk for pathologic fracture in isolation [9, 10].

Focal ablation techniques have been shown to be effective and durable techniques for the treatment of appropriately selected lesions [6]. Thermal induced necrosis of the neoplastic tissue, either through heat (RFA) or cold (cryoablation) is meant to gain local control of the tumor, while supplementation with cementoplasty theoretically is meant to help stabilize surrounding microfractures [5]. These techniques described above have limitations, including potential collateral thermal necrosis damage or cement extravasation in periarticular locations or locations adjacent to major neurovascular structures. Furthermore, unique characteristics of the lytic lesion, such as lesser trochanter involvement or >30 mm of axial length cortical destruction have portended to poorer outcomes with minimally invasive techniques [5, 6].

46.4 Surgical Technical Options for Treatment

Surgical techniques may vary, and should be dictated primarily by balancing the need for a durable construct that will outlive expected prognosis and the need for an expedited recovery. Oftentimes, commencement of adjuvant therapies such as immunotherapy, cytotoxic therapy, and radiation therapy is dictated by wound healing and performance status. Intralesional curettage techniques utilizing adjuvant modalities to aid with microscopic tumor burden control (e.g., cryoablation, phenol, electrocautery, concentrated hydrogen peroxide, thermal ablation), typically require larger incisions but allow for improved debulking of gross/microscopic disease. Percutaneous intralesional techniques, such as an intramedullary device, create the advantage of small wounds at the sacrifice of limited tumor debulking ability. Intramedullary vacuum irrigator devices can facilitate some manner of tumor debulking and allow for biopsy material to be obtained. En bloc excisions have a limited role in metastatic disease, leaving larger defects that require reconstruction with endoprosthetic or intercalary implants. Pathologic fracture involving a joint or select scenarios with solitary skeletal metastases can portend to the need for en bloc resection and reconstruction techniques [11–13]. Recent literature would suggest as high as a fourfold longer survival in patients with oligometastatic renal cell carcinoma who undergo en bloc excision (as compared to intralesional surgery) [12, 14].

46.5 Pathologic Fracture Predictive Models

Multiple algorithms have been used historically in determining a patient's overall risk for fracture in the setting of a pathologic lesion. Examples of such models include:

- 1971—Snell and Beals criteria for femoral metastases in the breast cancer population highlighted impending pathologic fracture risk include lytic lesions >2.5 cm and cortical involvement >50% (circumference) [15].
- 1989—The Mirels scoring system predicts risk of pathologic fracture based on (1) lesion size, (2) anatomic location, (3) radiographic appearance, and (4) the presence/severity of pain. This system' reliability and reproducibility have been called into question with multiple suggestions to improve, but still remains a staple for lower extremity risk prediction [16, 17].
- 2005—A randomized trial looking at patients at increased risk for fracture with femoral metastases undergoing radiotherapy used multiple criteria, highlighting pain, lesion size >25 mm, a proximal femoral location, axial cortical involvement >30 mm or circumferential involvement >50% as factors that increase the risk for fracture [18].
- 2015—CT-based rigidity analysis (CTRA) has been more recently demonstrated to be a better predictor of fracture than the Mirels scoring system. This system defines an impending pathologic fracture based on greater than 35% reduction in axial, bending, or torsional rigidities. This system requires CT imaging for every fracture, which is not often necessary in order to plan for interventions. Follow-up studies have also not shown ability to influence surgical decision-making in simulated modeling [19–21].

In more recent times, publicly available, validated predictive nomogram models are available for use at the bedside to assess risk for fracture when taking into account age, gender, lab values, histology, ECOG status, visceral and lymph node metastatic burden, number of skeletal metastases, and the physician's estimate of patient survival [22].

As a simple radiographic predictive determinant, the presence of a lesser trochanter fracture is generally considered to show a strong association with malignant etiology, with the risk for pathologic fracture elevated due to the predicted load-bearing ability being significantly reduced. Because of this, involvement of the lesser trochanter and surrounding posteromedial calcar anatomy serves as a strong indication for operative intervention [23, 24].

46.6 Femur

When considering surgical intervention for femoral metastases it is most helpful to divide the femur into proximal, diaphyseal, and distal segments. This division can aid in surgical decision-making and implant selection. As with all metastatic patients, once the decision to operate has been made, the surgeon should select the procedure that is intended to be definitive in light of the patient's local lesion and ultimate prognosis. As an example, consider two radiographically identical lesions destroying the peritrochanteric femur. One patient with an isolated renal cell metastasis may be treated with a proximal femoral resection and reconstruction with a proximal femur replacement, whereas another patient with progressive widespread osseous and visceral metastases may be treated with an intramedullary nail to prevent pathologic fracture.

46.6.1 Proximal Femur: Head/Neck

The main decision points in determining the appropriate treatment are: (1) Whether there is sufficient bone in the head/neck for internal fixation options and (2) whether fixation options will adequately address the mechanical instability in the bone and thus relieve the patient's pain.

Lesions isolated to the femoral head and neck are reliably treated with arthroplasty. Resection of the head/neck with arthroplasty allows for optimal local surgical management of the disease and an immediately durable reconstruction. There is debate about whether to perform long-stem arthroplasty stabilization [25] versus standard length and also whether to perform hemiarthroplasty versus total hip arthroplasty [26].

Consideration can be given to performing internal fixation for select femoral neck lesions depending on the location. Decision to perform this must take into account size/location of the lesion, blood supply to the femoral head, and durability of fixation.

46.6.2 Proximal Femur: Pertrochanteric

Impending or actual pathologic fractures of the pertrochanteric region become more interesting in terms of the discussion of method of stabilization. The armamentarium includes intramedullary nail, proximal femoral replacement, and long stem arthroplasty. Perhaps extrapolating the orthopedic surgeon's experience with management of intertrochanteric fractures in the non-neoplastic setting, the reflex consideration is to internally fix the majority of these lesions [27]. There is, however, data to suggest that there are better outcomes for patients treated with arthroplasty techniques in this setting as well [28].

As an example, a patient with a pertrochanteric lesion (Fig. 46.1a, lower arrow) may be appropriately treated with an intramedullary device for stabilization. However, if that same patient also has lesions in the femoral head/neck (Fig. 46.1a, upper arrows) and also subsequently develops a pathologic fracture (Fig. 46.1b) that will likely shift the balance in decision-making toward proximal femoral resection and endoprosthetic reconstruction (Fig. 46.1c). While intramedullary fixation could be considered, there would be substantial concerns regarding purchase of fixation in the pathologic femoral head/neck and also ability of the bone to heal given the need for palliative radiotherapy for local control of the disease.

The surgical treatment of the pertrochanteric femoral metastasis should not be prescribed based on the knowledge of the general location in the bone alone. It should be based on the amount of bone destruction, location in relation to the calcar, soft tissue mass, tumor histology, anticipated response to radiotherapy/systemic therapy, the patient's functional status, and overall prognosis.

46.6.3 Diaphyseal Femur

The vast majority of diaphyseal lesions in the femur will be treated with some form of internal fixation, although segmental resection and intercalary prosthetics may be considered in the setting of solitary or oligometastatic disease [29].

The main consideration in the femur will be the mode of fixation. Intramedullary nail fixation is often preferred for purposes of spanning the entire bone, most notably the femoral neck, for prophylactic fixation in the event of future pathologic lesions at a more distant site in the same bone. However, there is evidence that the theoretical benefit of protecting the entire bone is rarely utilized [30, 31]. Given this evidence, it is entirely reasonable to use plate/screw constructs if one already has the entire bone exposed for the purposes of marginal/intralesional management of a metastatic lesion.

The larger question, as it is in many instances of metastatic disease, is whether to perform intralesional management of the tumor simultaneously with the stabilization procedure. This is most often performed in situations where: (1) the primary histology dictates based on history of radioresistance (i.e., renal cell, lung adenocarcinoma) or (2) there is substantial destruction warranting intralesional resection and cement augmentation to further promote skeletal stability.

46.6.4 Distal Femur

As in the proximal femur, the distal femur considerations are large whether there is the ability to salvage the function of the articular structures. If the joint can be salvaged, and predictably maintained based on the natural history of the histologic subtype, then it is the favored approach (Fig. 46.2a). If it cannot, or the underlying condition of the joint (i.e., arthritis and/

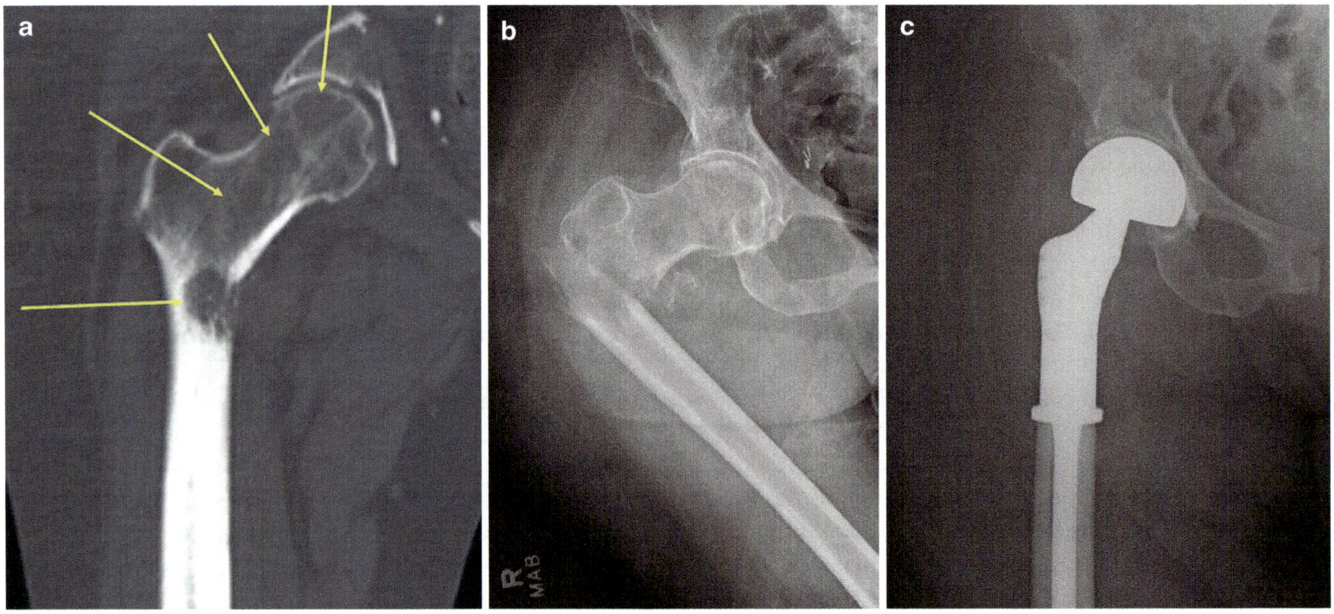

Fig. 46.1 Examples of proximal and diaphyseal femur reconstruction. A large metastatic lytic lesion burden in the calcar and femoral neck (**a**) led to a pathologic subtrochanteric pathologic femur fracture (**b**). This was subsequently treated with a proximal femoral endoprosthetic replacement (**c**)

or intra-articular extension of a pathologic fracture) renders it a major source of disability then an arthroplasty technique is preferred, most often distal femoral replacement (Fig. 46.2b).

46.7 Tibia/Fibula

Though metastatic disease distal to the knee ("acral" metastases) occurs at a <10% incidence, goals of treatment remain similar [1]. The proximal tibial metaphyseal location is the most common location of involvement distal to the knee. As with other locations, treatment categories include an intralesional approach with supplemental stabilization constructs, en bloc excision and reconstruction, or amputation. In the prophylactic setting or minimally displaced fracture with contained subchondral margins, stabilization utilizing a plate/screw construct or intramedullary device is standard, oftentimes coupled with open intralesional curettage, adjuvant therapy treatment (e.g., heat, cold, phenol, etc.) and bone cement supplementation. In the setting of metastatic disease, en bloc excision is a less commonly used technique. Amputation is generally used only in settings where reconstruction techniques do not offer a durable longevity and pain is unable to be effectively addressed with nonoperative measures. Examples of various reconstruction techniques can be found in Fig. 46.2.

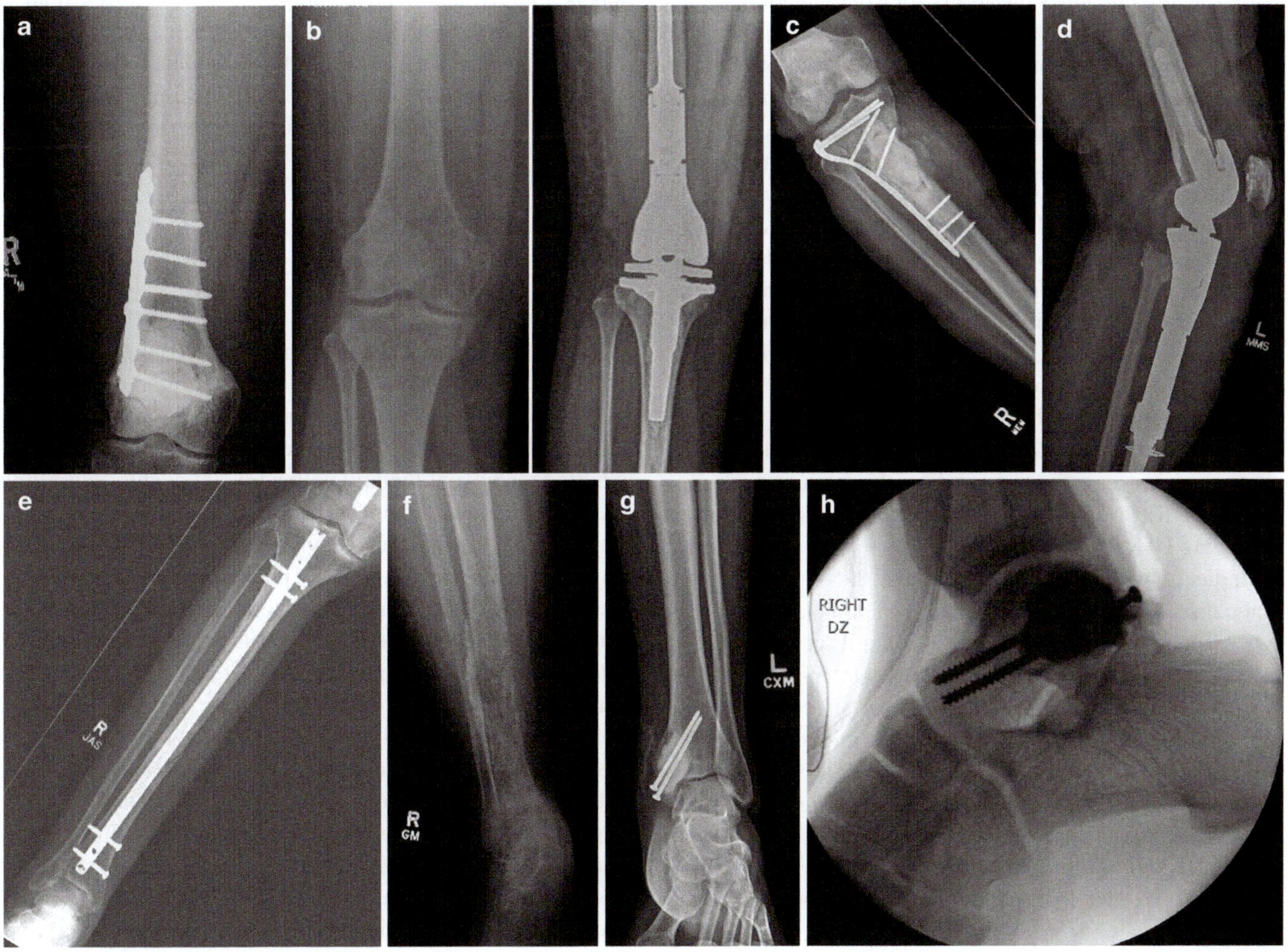

Fig. 46.2 Examples of distal femur, tibia, and foot reconstruction. Reconstructions of the knee and tibia/foot locations should address the tumor burden while the technique should be tailored to anticipate patient prognosis and the required longevity of reconstruction. Curettage with plate/bone cement reconstruction can be used to preserve the knee joint (**a**) or a distal femur replacement may be required when extensive bone replacement by tumor burden is encountered (**b**). Likewise, in the proximal tibia, intralesional procedures reinforced with plate/bone cement techniques (**c**) or replacement with an endoprosthesis (**d**) are options, depending on subchondral bony involvement, then contained defect present, and expected prognosis. Diaphyseal lesions are generally treated with a locked intramedullary device (**e**), though occasionally the burden of tumor leaves such poor bone tock quality that an amputation is the best solution (**f**). Distal tibial lesions and foot lesions are generally treated with intralesional procedures reinforced with plates, screws, and/or bone cement, such as in this metastatic lung cancer medial malleolus lesion (**g**) or this metastatic renal carcinoma talus lesion (**h**)

46.7.1 Proximal

The mainstay "workhorse" of the proximal tibia anatomy is plate/screw reconstruction—utilizing open incisions that facilitate combined techniques of high-speed burr curettage, adjuvant therapy with adjuvant agents, and filling the defect with cement prior to plate supplementation. Intramedullary device solutions are less commonly employed given the lack of ability to gain adequate fixation through the compromised metaphyseal bone, as well as the lack of ability to add reliable support to the subchondral weight-bearing surface.

Arthroplasty solutions can be considered when (1) the subchondral bony anatomy is compromised to a point that would risk cement extravasation into the intra-articular space with plate/bone cement reconstruction, (2) a displaced pathologic fracture precludes the ability for a durable joint salvage reconstruction option, or (3) a peri-implant lesion occurs below a prior knee arthroplasty that can be addressed utilizing a stemmed revision baseplate construct. When considering arthroplasty solutions, if the rim of metaphyseal bone is present to create a "contained" defect after curettage, then conventional tibial revision baseplates with cemented stems can be used to bypass the defect, with potential metaphyseal tibial cone augmentation. With large, uncontained defects or displaced pathologic fractures, endoprosthetic reconstructions with a proximal tibial hinged knee replacement (PTR) can be used—though these come with concerns of high infection rates and soft tissue coverage issues [32–35]. Whenever possible, if the opportunity is available to save the extensor mechanism attachment (i.e., tibial tubercle) without compromising construct durability, this should be prioritized, in order to address the high rate of extensor lag deficits following PTR [36].

Indications for en bloc excision in the setting of metastatic disease are rare. These include oligometastatic disease where potential improvement in overall survival may be afforded by complete excision of certain histologies (i.e., renal cell carcinoma), large lesions that affect the articular weight-bearing surface which does not allow for joint salvage, or peri-articular pathologic fractures with minimal bone stock available for reconstruction. Unlike in the setting of primary bone malignancies, biologic reconstruction with bulk allograft (intercalary allograft, allograft–prosthetic composite) or custom implant solutions are not used as standard reconstructions given the need for limited weight bearing and long recovery requirements. Modular endoprosthesis allowing for PTR with cement fixation is the standard solution that allows for the quickest mobilization.

Fibular lesions are very rare and treated nonoperatively, commonly palliated with radiation and temporary protected weight bearing as a general rule. In contained defects that have failed non-interventional measures, percutaneous cementoplasty or open curettage, and cement augmentation can be used. If en bloc excision is required, reconstruction of the lateral collateral complex is necessary with hinged knee brace protection postoperatively.

46.7.2 Diaphyseal

While intramedullary nail placement generally is used for tibial diaphyseal lesions, plate/screw techniques can play a role. If adjacent implants or hardware are present, such as a diaphyseal tibial lesion below a total knee replacement, plating techniques are more pragmatic to address impending or outright pathologic fracture concerns. Utilizing cement augmentation in large defects can add structural stability to the construct. As with en bloc excisions in the proximal anatomy, this is a rarely employed technique in tibial diaphyseal anatomy. If an en bloc excision is indicated, then a modular intercalary implant can be utilized, utilizing cement fixation for the stems with or without intercalary cross-locking screws.

Fibular diaphyseal lesions failing nonoperative palliative modalities are treated with en bloc excision most commonly.

46.7.3 Distal

As with the proximal anatomy, plate/screw supplementation with bone cement augmentation is the common of the proximal tibia plate/screw reconstruction—utilizing open incisions that facilitate combined techniques of high-speed burr curettage, adjuvant therapy with adjuvant agents, and filling the defect with cement prior to plate supplementation. Smaller contained defects may be addressed with percutaneous techniques, including screw placement with cement supplementation. With limited subcondral destruction in the setting of an uncontained defect, utilizing gel foam over the articular defect can help to contain cement during an intralesional curettage procedure and allow for joint salvage. Arthroplasty is not a reliable solution at the tibiotalar joint—if subchondral anatomic destruction precludes joint salvage, a below-knee amputation (BKA) will precipitate a higher degree of postoperative function in most cases. In rare scenarios, placement of a retrograde tibial-talar-calcaneal (TTC) fusion nail with cement augmentation can be considered—though a BKA may still afford better over functional outcomes given the known reliability of the procedure.

46.7.4 Soft-Tissue Considerations in Tibial Reconstruction

Given the subcutaneous anatomy of the tibia, coverage over a reconstruction construct is vital to prevent again postoperative wound-healing concerns leading to infection. In an intercalary resection and modular implant reconstruction event, a regional soleus or gastrocnemius flap combination can be considered. More commonly, gastrocnemius neurovascular pedicled flaps have become a mainstay to supplement PTR, with the twofold advantage of decreasing infection rates and allowing for improved extensor mechanism augmentation, aiding with postoperative improved extensor mechanism function [34, 36].

If soft-tissue coverage is compromised secondary to adjuvant radiation therapy, and temporization of the soft tissue envelope is needed prior to definitive stabilization and coverage, external fixation is a described potential technique [37].

46.8 Patella

The incidence of metastatic disease in the patella is rare, described in case report literature [38–41]. In small lesions, conservative therapy with antiresorptive agents, radiation, and systemic therapy should be considered as the first line. In larger lesions or lesions that fail conservative therapy, surgery can be performed. If there is a contained that does not communicate with the joint, curettage with cement augmentation can be used. Patellectomy is reserved in large lesions that give poor reconstructive options. This creates weakness with extension given the loss of a native fulcrum, and an adaptive gait will need to be employed to adapt to quadriceps weakness and altered muscle control.

46.9 Foot

Acral metastases to the foot occur rarely, reported at <0.3% incidence [42–44]. The calcaneus and talus are the most commonly involved anatomic locations. Treatment combines a multimodal approach to consist of radiation, temporary immobilization (if there is a chance for healing to occur, such as in myeloma, lymphoma, or breast cancer), orthotics, minimally invasive strategies (in larger bones of the hindfoot), and surgery. Involvement of the distal phalanges portends to the need for amputation, while metatarsal, midfoot, and hindfoot anatomy may benefit from open curettage and cement augmentation +/− plate and screw fixation in contained lesions. As with distal tibial lesions, a BKA is a reproducible operation with reliable functional outcomes in patients where limb salvage surgery will not aid reliably with pain relief or a durable construct.

46.10 Complications

The potential complications that can be experienced in the management of lower extremity metastatic disease are wide-ranging and encompass all of the complications associated with the surgical technique utilized (i.e., hip instability with hip arthroplasty, patellar maltracking with distal femoral replacement, etc.).

While some complications are inherent to the procedure being performed, the more discouraging complications are those that require reoperation and which may have been avoided with optimizing the surgical plan.

46.10.1 Iatrogenic Fracture

When performing prophylactic internal fixation one of the complications that may be encountered is iatrogenic fracture. This can occur for a variety of reasons, not the least of which is the operating environment of pathologic bone. It is important to ensure appropriate nail path to minimize stress on the pathologic bone. Furthermore, if a pathologic fracture is identified it is of the utmost importance that it be appropriately managed. This may mean converting operative approaches to optimize reduction, or potentially even switching implants. A poor reduction of pathologic bone risks nonunion and implant failure (Fig. 46.3a).

46.10.2 Cortical Perforation

By definition, prophylactic internal fixation with an intramedullary device involves placing an intramedullary implant in an unbroken bone. While modern implants have made efforts to normalize the radius of curvature of the implant to most femora, there will be instances where there is a mismatch. Furthermore, if proper technique and starting points are not used there will be a risk of perforating the anterior cortex of the distal femur. In order to minimize the risk of this complication, the starting point on the trochanter should be perfectly at the tip or biased anterior. If cortical perforation does occur the nail should either be redirected and ensured that the defect is spanned or supplemental fixation should be added (Fig. 46.3b).

46.10.3 Disease Progression/Device Failure

The overall risk of device failure is thought to be low, with revision rates reported ranging from 5% to 8.9% [45, 46]. The low rate of revision is largely contributed by the poor overall survival in patients with metastatic cancer. Factors that are felt to contribute to implant failure include renal carcinoma metastases [45], patients with actual fractures (as opposed to impending) [46], a history of radiotherapy prior to the surgical treatment [45, 46].

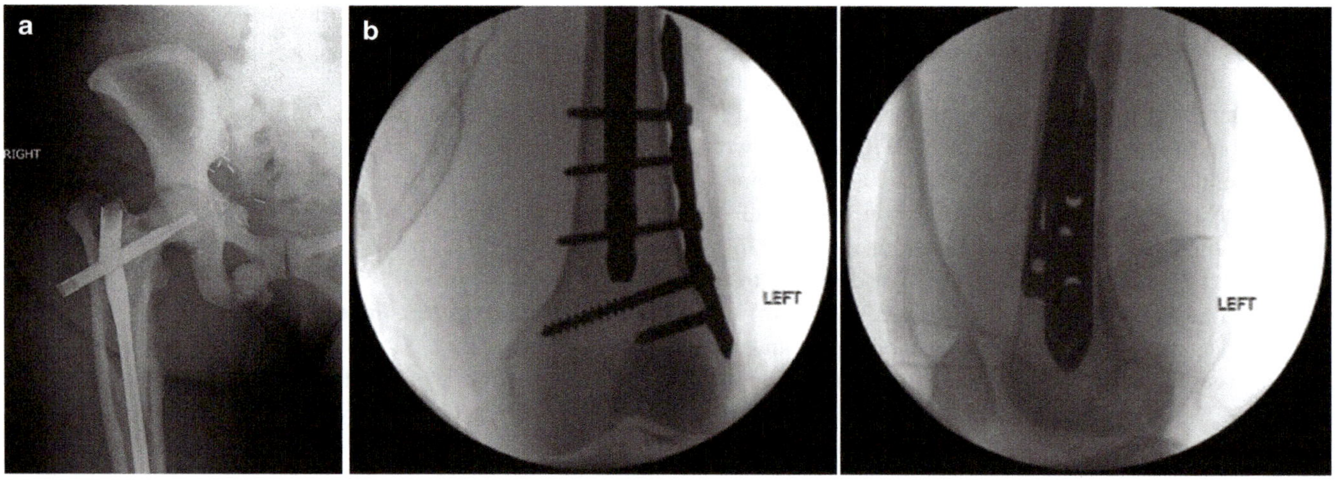

Fig. 46.3 Complications. A poor reduction of a subtrochanteric pathologic femur fracture spanned with an undersized cephalomedullary device culminated in early failure (**a**). A start point place too posterior on the greater trochanter led to distal anterior cortical perforation of the femur, requiring a separate distal incision in order to plate a bypassing plate construct protecting against cortical perforation stress risers

46.11 Conclusion

Lower extremity location for metastatic disease is a common occurrence and patients often present with significant pain-causing ambulatory limitations. Understanding the "complete picture" that encompasses prognosis, tumor characteristics, anatomic support and stress loading, and a multidisciplinary plan for cancer treatment is necessary to tailoring an appropriate algorithm for addressing skeletal-related events in the lower extremity. Predictive models are available and being contemporarily refined to better determine appropriate surgical and nonsurgical solutions for treatment. These models have limitations and do not substitute for multidisciplinary communication between the medical oncology, radiation oncology, radiology, pathology, surgical, palliative medicine, and physical therapy team for coordination of care and goals-directed therapy.

Surgical planning should prioritize a durable construct that outlasts overall prognosis, and allow for immediate weight-bearing and joint motion, utilizing the "least invasive" approach that will accomplish this. Maintaining the quality of life and alleviating pain should be the ultimate goals.

In the future, newer technologies that can allow for more minimally invasive delivery of resection and reconstruction solutions will be considered. These can include robotic technology, virtual/augmented reality, or better predictive models that more reliably stratify and match appropriate treatment techniques with patient presentation.

Open Questions

- When is intralesional treatment of the metastatic lesion warranted and when is it not necessary?
- What are appropriate indications and methods of delivery for using adjuvant treatments to the lesion site in order to offer an advantage in decreasing local recurrence rates? (i.e., topical application, bone cement impregnation with adjuvant, histology-specific chemotherapeutics and/or bone-directed agents, etc.)
- When does wide resection of oligometastatic disease represent a survival advantage?
- Utilizing predictive models, what presentation characteristics (histology, prognosis, age, location, etc.) will accurately guide a surgeon toward choosing between an endoprosthetic device versus an intramedullary nail fixation versus plate + cement technique of a metastatic lesion?

References

1. Leeson MC, Makley JT, Carter JR. Metastatic skeletal disease distal to the elbow and knee. Clin Orthop. 1986;206:94–9.
2. Ratasvuori M, Wedin R, Keller J, et al. Insight opinion to surgically treated metastatic bone disease: Scandinavian Sarcoma Group Skeletal Metastasis Registry report of 1195 operated skeletal metastasis. Surg Oncol. 2013;22(2):132–8. https://doi.org/10.1016/j.suronc.2013.02.008.

3. Varadhachary GR, Abbruzzese JL, Lenzi R. Diagnostic strategies for unknown primary cancer. Cancer. 2004;100(9):1776–85. https://doi.org/10.1002/cncr.20202.
4. Harrington KD. Orthopaedic management of extremity and pelvic lesions. Clin Orthop. 1995;312:136–47.
5. Mesko NW, Lawrenz JM, Lietman SA, Joyce MJ, Winalski CS, Ilaslan H. Minimally invasive techniques for pain palliation in extraspinal bone metastases: a review of cementoplasty and radiofrequency ablation. Curr Orthop Pract. 2016;27(6):686–95. https://doi.org/10.1097/BCO.0000000000000440.
6. Lawrenz J, Ilaslan H, Lietman S, Joyce M, Winalski C, Mesko N. Minimally invasive techniques for pain palliation in extraspinal bone metastases: a review of conventional methods and cryoablation. Curr Orthop Pract. 2016;27(5):547–52. https://doi.org/10.1097/BCO.0000000000000417.
7. Ogbata O, Portnoy D. Nonoperative management of patients with carcinoma metastatic to bone. Curr Orthop Pract. 2014;25:520–4. https://doi.org/10.1097/BCO.0000000000000175.
8. Lutz S, Chow E. Palliative radiotherapy: past, present and future-where do we go from here? Ann Palliat Med. 2014;3(4):286–90. https://doi.org/10.3978/j.issn.2224-5820.2014.10.04.
9. Francolini G, Detti B, Ingrosso G, et al. Stereotactic body radiation therapy (SBRT) on renal cell carcinoma, an overview of technical aspects, biological rationale and current literature. Crit Rev Oncol Hematol. 2018;131:24–9. https://doi.org/10.1016/j.critrevonc.2018.08.010.
10. Rühle A, Andratschke N, Siva S, Guckenberger M. Is there a role for stereotactic radiotherapy in the treatment of renal cell carcinoma? Clin Transl Radiat Oncol. 2019;18:104–12. https://doi.org/10.1016/j.ctro.2019.04.012.
11. Baloch KG, Grimer RJ, Carter SR, Tillman RM. Radical surgery for the solitary bony metastasis from renal-cell carcinoma. J Bone Joint Surg Br. 2000;82-B(1):62–7. https://doi.org/10.1302/0301-620X.82B1.0820062.
12. Ouzaid I, Capitanio U, Staehler M, et al. Surgical metastasectomy in renal cell carcinoma: a systematic review. Eur Urol Oncol. 2019;2(2):141–9. https://doi.org/10.1016/j.euo.2018.08.028.
13. Ljungberg B. The role of metastasectomy in renal cell carcinoma in the era of targeted therapy. Curr Urol Rep. 2013;14(1):19–25. https://doi.org/10.1007/s11934-012-0293-6.
14. Ratasvuori M, Wedin R, Hansen BH, et al. Prognostic role of en-bloc resection and late onset of bone metastasis in patients with bone-seeking carcinomas of the kidney, breast, lung, and prostate: SSG study on 672 operated skeletal metastases. J Surg Oncol. 2014;110(4):360–5. https://doi.org/10.1002/jso.23654.
15. Beals RK, Lawton GD, Snell WE. Prophylactic internal fixation of the femur in metastatic breast cancer. Cancer. 1971;28(5):1350–4. https://doi.org/10.1002/1097-0142(1971)28:5<1350::aid-cncr2820280539>3.0.co;2-6.
16. Mirels H. Metastatic disease in long bones. A proposed scoring system for diagnosing impending pathologic fractures. Clin Orthop. 1989;249:256–64.
17. Howard EL, Shepherd KL, Cribb G, Cool P. The validity of the Mirels score for predicting impending pathological fractures of the lower limb. Bone Jt J. 2018;100-B(8):1100–5. https://doi.org/10.1302/0301-620X.100B8.BJJ-2018-0300.R1.
18. Van der Linden YM, Dijkstra PDS, Kroon HM, et al. Comparative analysis of risk factors for pathological fracture with femoral metastases. J Bone Joint Surg Br. 2004;86(4):566–73.
19. Damron TA, Nazarian A, Entezari V, et al. CT-based structural rigidity analysis is more accurate than mirels scoring for fracture prediction in metastatic femoral lesions. Clin Orthop. 2016;474(3):643–51. https://doi.org/10.1007/s11999-015-4453-0.
20. Nazarian A, Entezari V, Zurakowski D, et al. Treatment planning and fracture prediction in patients with skeletal metastasis with ct-based rigidity analysis. Clin Cancer Res. 2015;21(11):2514–9. https://doi.org/10.1158/1078-0432.CCR-14-2668.
21. Nazarian A, Entezari V, Villa-Camacho JC, et al. Does CT-based rigidity analysis influence clinical decision-making in simulations of metastatic bone disease? Clin Orthop. 2016;474(3):652–9. https://doi.org/10.1007/s11999-015-4371-1.
22. Home. PathFX. https://www.pathfx.org/. Accessed 19 Dec 2020.
23. Shinoda Y, Sawada R, Fujiwara S, Inokuchi H, Karasawa Y, Haga N. Prediction of the pathological fracture risk during stance and fall-loading configurations for metastases in the proximal femur, using a computed tomography-based finite element method. Ann Phys Rehabil Med. 2018;61:e412–3. https://doi.org/10.1016/j.rehab.2018.05.961.
24. Cho HS, Lee Y-K, Yoon B-H, Park J-W, Ha Y-C, Koo K-H. Isolated avulsion fracture of the lesser trochanter in adults. In Vivo. 2020;34(6):3519–26. https://doi.org/10.21873/invivo.12193.
25. Patterson BM, Healey JH, Cornell CN, Sharrock NE. Cardiac arrest during hip arthroplasty with a cemented long-stem component. A report of seven cases. J Bone Joint Surg Am. 1991;73(2):271–7.
26. Reif TJ, Strotman PK, Kliethermes SA, Miller BJ, Nystrom LM. No consensus on implant choice for oligometastatic disease of the femoral head and neck. J Bone Oncol. 2018;12:14–8. https://doi.org/10.1016/j.jbo.2018.02.006.
27. Steensma M, Healey JH. Trends in the surgical treatment of pathologic proximal femur fractures among Musculoskeletal Tumor Society members. Clin Orthop. 2013;471(6):2000–6. https://doi.org/10.1007/s11999-012-2724-6.
28. Steensma M, Boland PJ, Morris CD, Athanasian E, Healey JH. Endoprosthetic treatment is more durable for pathologic proximal femur fractures. Clin Orthop. 2012;470(3):920–6. https://doi.org/10.1007/s11999-011-2047-z.
29. Huang H-C, Hu Y-C, Lun D-X, et al. Outcomes of intercalary prosthetic reconstruction for pathological diaphyseal femoral fractures secondary to metastatic tumors. Orthop Surg. 2017;9(2):221–8. https://doi.org/10.1111/os.12327.
30. Alvi HM, Damron TA. Prophylactic stabilization for bone metastases, myeloma, or lymphoma: do we need to protect the entire bone? Clin Orthop. 2013;471(3):706–14. https://doi.org/10.1007/s11999-012-2656-1.
31. Moon B, Lin P, Satcher R, Bird J, Lewis V. Intramedullary nailing of femoral diaphyseal metastases: is it necessary to protect the femoral neck? Clin Orthop. 2015;473(4):1499–502. https://doi.org/10.1007/s11999-014-4064-1.
32. Jeys LM, Kulkarni A, Grimer RJ, Carter SR, Tillman RM, Abudu A. Endoprosthetic reconstruction for the treatment of musculoskeletal tumors of the appendicular skeleton and pelvis. J Bone Joint Surg. 2008;90(6):1265–71. https://doi.org/10.2106/JBJS.F.01324.
33. Henderson ER, Groundland JS, Pala E, et al. Failure mode classification for tumor endoprostheses: retrospective review of five institutions and a literature review. JBJS. 2011;93(5):418–29. https://doi.org/10.2106/JBJS.J.00834.
34. Myers GJC, Abudu AT, Carter SR, Tillman RM, Grimer RJ. The long-term results of endoprosthetic replacement of the proximal tibia for bone tumours. J Bone Joint Surg Br. 2007;89-B(12):1632–7. https://doi.org/10.1302/0301-620X.89B12.19481.
35. Capanna R, Scoccianti G, Frenos F, Vilardi A, Beltrami G, Campanacci DA. What was the survival of megaprostheses in lower limb reconstructions after tumor resections? Clin Orthop. 2015;473(3):820–30. https://doi.org/10.1007/s11999-014-3736-1.
36. Jentzsch T, Erschbamer M, Seeli F, Fuchs B. Extensor function after medial gastrocnemius flap reconstruction of the proximal tibia. Clin Orthop. 2013;471(7):2333–9. https://doi.org/10.1007/s11999-013-2851-8.

37. Oskouei SV, Walton ZJ. Treatment of metastatic disease to the tibia, ankle, and foot. Curr Orthop Pract. 2014;25(6):554–7. https://doi.org/10.1097/BCO.0000000000000176.
38. Gasagranda B, Leeman K, Heller MT. Knee pain and swelling: an atypical presentation of metastatic colon cancer to the patella. Radiol Case Rep. 2014;9(1):e00031. https://doi.org/10.2484/rcr.v9i1.890.
39. Benedek TG. Lysis of the patella due to metastatic carcinoma. Arthritis Rheum. 1965;8:560–7. https://doi.org/10.1002/art.1780080410.
40. Pauzner R, Istomin V, Segal-Lieberman G, Matetzky S, Farfel Z. Bilateral patellar metastases as the clinical presentation of bronchogenic adenocarcinoma. J Rheumatol. 1996;23(5):939–41.
41. Li G, Shan C, Sun R, et al. Patellar metastasis from primary tumor. Oncol Lett. 2018;15(2):1389–96. https://doi.org/10.3892/ol.2017.7478.
42. Ellington JK, Ellington K, Kneisl JS. Acrometastasis to the foot: three case reports with primary colon cancer. Foot Ankle Spec. 2009;2(3):140–5. https://doi.org/10.1177/1938640009335271.
43. Hattrup SJ, Amadio PC, Sim FH, Lombardi RM. Metastatic tumors of the foot and ankle. Foot Ankle. 1988;8(5):243–7. https://doi.org/10.1177/107110078800800503.
44. Maheshwari AV, Chiappetta G, Kugler CD, Pitcher JD, Temple HT. Metastatic skeletal disease of the foot: case reports and literature review. Foot Ankle Int. 2008;29(7):699–710. https://doi.org/10.3113/FAI.2008.0699.
45. Miller BJ, Soni EEC, Gibbs CP, Scarborough MT. Intramedullary nails for long bone metastases: why do they fail? Orthopedics. 2011;34(4) https://doi.org/10.3928/01477447-20110228-12.
46. Willeumier JJ, Kaynak M, van der Zwaal P, et al. What factors are associated with implant breakage and revision after intramedullary nailing for femoral metastases? Clin Orthop. 2018;476(9):1823–33. https://doi.org/10.1007/s11999.0000000000000201.

Surgical Management of Metastatic Disease to the Pelvis

47

Timothy J. Evans, Odion Binitie, and David M. Joyce

Abstract

This chapter reviews surgical treatment of metastatic disease to the bones of the pelvis from cancers that have a proclivity for lymph node spread. The anatomy of the pelvis is correlated to both nonsurgical and surgical management. The surgical techniques are reviewed as well as their reported outcomes.

Learning Objectives

- Outline common primary cancers that are commonly found to have both lymphovascular spread as well as bony metastatic disease of the pelvis
- Discuss the initial workup of patients found to have bony metastatic disease of the pelvis
- Relate the relevant anatomy of the pelvis to common locations of bony metastatic disease and how this helps direct treatment recommendations
- Discuss common nonoperative and operative interventions used in the management of this disease

47.1 Introduction

The pelvis, possessing large proportions of the body's red bone marrow, is one of the most common sites of metastatic disease of bone [1]. In the following paragraphs, we will outline the common types of cancer that spread through lymphatic pathways and present with metastasis to the pelvis, the particular characteristics of the disease type, and other features which have implications on surgical decision making.

T. J. Evans · O. Binitie · D. M. Joyce (✉)
H. Lee Moffitt Cancer Center, Tampa, FL, USA
e-mail: odion.binitie@moffitt.org; david.joyce@moffitt.org

We will also discuss the variations in management based on those patients presenting with oligometastatic disease (though poorly defined, often <5 metastatic lesions) versus diffuse metastatic disease (>5 metastatic lesions). This concept is especially important in patients with cervical, uterine, melanoma, thyroid, and renal cell cancers whose prognosis and treatment strategy will depend on the extent of metastatic disease at the time of diagnosis [2–4]. The most common metastasis to all bones, including the pelvis are breast, prostate, thyroid, lung, and kidney. Common cancers that specifically spread via lymphatic pathways include lymphoma, head and neck cancers, breast cancer, gynecological cancers, melanoma, kidney cancer, and lung cancer. Renal cell carcinoma (RCC) is one of the most common primary cancers to spread to lymph node (LN) and bone, with a 12% risk of metastases to LN and 34% risk to bone at the time of diagnosis. RCC metastases tend to be highly vascular tumors and often necessitate preoperative embolization in the pelvis to minimize intraoperative blood loss. Lymphoma, melanoma, and lung cancer can also present with large associated soft tissue masses in the pelvis and associated bone destruction. Squamous cell carcinoma of the head and neck develops metastatic bone disease in about 20% of patients. The lesions in these patients were found to be located in the pelvis or femur more than 50% of the time [5, 6].

47.2 Pelvic Metastases and Lytic Pelvic Bone Lesion Workup

Evaluation of the patient with suspected pelvic metastases should begin with a comprehensive physical examination including evaluation of all four extremities with respect to pain, sensorimotor and vascular function as well as range of motion. An assessment of the spine, especially lumbar is critical. Orthogonal radiographs should be obtained of any painful areas. In terms of radiographic evaluation of the pelvis, an anterior–posterior view should be the initial image

S. P. Leong et al. (eds.), *Cancer Metastasis Through the Lymphovascular System*,
https://doi.org/10.1007/978-3-030-93084-4_47

obtained. The iliopectineal and ilioischial lines, as well as the anterior and posterior walls, and the sourcil (acetabular roof) should be evaluated for continuity. Additional imaging depends on the area of concern within the pelvis, with Judet views and inlet/outlet views being the most common. Judet views of the pelvis provide unobstructed views of the anterior and posterior columns, the anterior and posterior walls of the acetabulum. Inlet and outlet views provide orthogonal imaging of the pelvic ring as well as additional views of the sacrum.

Cross-sectional imaging is a valuable tool in the evaluation and treatment of pelvic metastases due to the complex, three-dimensional anatomy of the pelvis as well as the proximity of major neurovascular structures. These scans should be obtained in three planes—coronal, sagittal, and transverse—and the combination of findings on all three planes helps determine the true nature of the metastatic lesions. CT scans provide the best detail of bony anatomy and structural integrity of the pelvis. MRI is used to help delineate the neurovascular structures and soft tissues. MRI can also evaluate for inflammation and infiltrative neoplasms that can explain a patient's symptoms when X-rays are inconclusive. T2 hyperintense on fluid-weighted sequences and contrasted images can delineate tumor extent in the bone and soft tissues. Radiographs, CT, MRI can help evaluate for fracture, tumor extent, and aid in intervention planning.

47.3 Pelvis Structure and Anatomic Considerations

The hemipelvis is made up of three bones, all of which fuse at the triradiate cartilage during growth. Due to the dynamic nature of the forces experienced by the pelvis, much of the stability comes from the ligamentous structures connecting the two hemipelves to each other and the sacrum. The pelvis is unique with respect to metastatic disease due to its complex, three-dimensional anatomy and its proximity to multiple, critical neurovascular structures. The areas of the pelvis that are directly involved in weight bearing include the sacrum, sciatic buttress, and acetabulum while the remaining iliac wing and pubic rami do not contribute to force transmission for weight bearing and serve as muscle origin and insertion locations. Knowledge of this anatomy is vital when evaluating the radiographic characteristics of a lesion or fracture and guides the decision for intervention and surgical techniques used. No definitive classification exists for metastatic disease in the pelvis, however, several classification systems can be used to guide one in the decision of surgical versus nonsurgical treatment.

The sacrum is a wedge-shaped bone that acts as the "keystone" of the pelvis and essentially connects the upper body through the spine to the lower body. It is secured to the hemipelves via strong sacroiliac ligaments and connected to the more caudal aspects of the pelvis via the sacrospinous and sacrotuberous ligaments. Interestingly, the sacral bony architecture is such that the densest bone lies in the trajectory of the sacral pedicles and on either side of this dense bone, there are regions with minimal cancellous bone. This architecture is important to keep in mind when evaluating the sacrum for lytic metastatic lesions as some of this perceived decreased density can be mistaken for lytic lesions. This understanding of the bone quality in the area often can support nonoperative management.

The ilium provides the connection from the axial spine to the legs through the sciatic buttress, which forms the anterior and superior walls of the greater sciatic notch and is responsible for transmitting forces between the acetabulum and hip joint and the axial skeleton. It also serves as the origin of most of the muscles required for mobilization. The sciatic buttress is the thickest and strongest part of the iliac bone and can be approximated by the pelvic brim, between the sacroiliac joint and superomedial acetabulum [7]. If the sciatic buttress is uninvolved by metastasis nonoperative treatment can be utilized to manage metastatic lesions in this area (Fig. 47.1). The ligamentous anatomy of the ilium has been studied extensively in the orthopedic literature as it has significant implications when determining stability after pelvic ring fractures, [8] which in turn can help the orthopedic surgeon determine the need for stabilization. Even with significant metastatic burden of the pelvis, and in large lesions, the ligamentous anatomy can maintain the pelvic ring stability.

The acetabulum is made up of anterior and posterior columns in an inverted-Y pattern in which the convalescence of the columns occurs in the supra-acetabular region. The anterior, posterior, superior, and medial walls of the acetabulum all play important roles in maintaining the congruency and stability of the hip joint. The competency of the medial and superolateral acetabulum, as the weight-bearing dome, helps determine if a lytic lesion can be treated non-operatively or dictate the characteristics of the surgical reconstruction.

The ischium provides the main origin for many of the hamstring and adductor muscles. Similarly, the pubic bone provides the anterior wall of the acetabulum and the insertion site for the abdominal muscles and origin of the adductor musculature. The superior and inferior pubic rami along with the pubic symphysis form the obturator foramen. This part of the pelvis provides for insertions of the abdominal musculature and the origins of the adductor muscle compartment along with some of the rotators of the femur. This part of the pelvis is not weight supporting. It serves only for origins and insertions of muscles. Patients can have pain from these lesions while walking, not because the involved bone is weight supporting, but because the muscles are pulling on weakened bone causing pain.

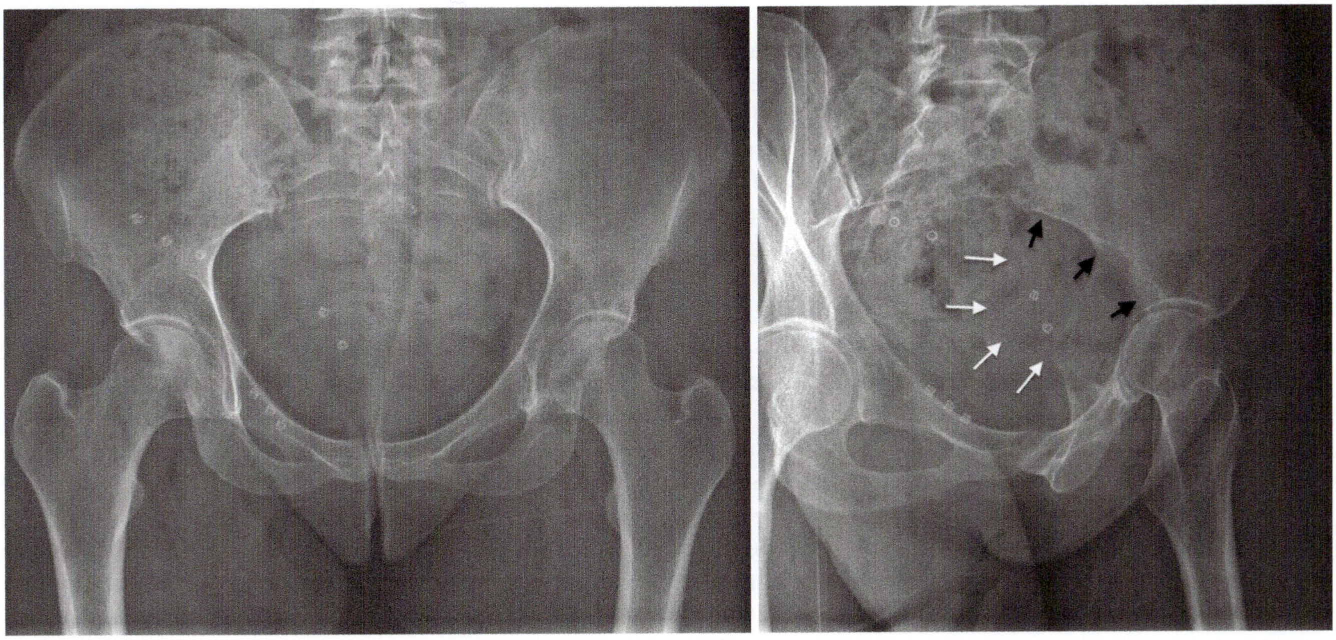

Fig. 47.1 Nonoperative management of pelvic lesions can be recommended when the region of the sciatic buttress and superior acetabulum are left unaffected. The black arrows outline the anatomic location of the sciatic buttress, which is the caudal portion of the iliac bone and is responsible for force transmission from the acetabulum to the sacrum and axial skeleton. Below this region (outlined by white arrows) there is a large lytic lesion within the posterior column of the acetabulum. This patient was treated nonoperatively as the lesion did not compromise the structural integrity of the weight-bearing hemipelvis

An understanding of this anatomy is imperative in the management of patients with metastatic disease of the pelvis, as the first step in their treatment is the determination of whether the location of the metastatic disease compromises the structural integrity of the weight-bearing portions of the pelvis. Succinctly, if the axis of force transmission from the acetabulum into the sacrum is maintained, the patient should still be able to safely bear weight on the extremity without an unacceptable risk of fracture (Fig. 47.1). We can affirm that lesions within the superior and inferior pubic rami, iliac wing (not involving the sciatic buttress) and medial and inferior aspects of the acetabulum do not significantly impact the structural integrity of the hemipelvis. Thus, these regions can be safely considered for nonoperative management from a structural standpoint.

47.4 Noninvasive Treatment Options

When presented with a patient with pelvic metastatic disease of bone it is important to take a multidisciplinary approach involving surgical oncology, radiation oncology, medical oncology, and others to create a personalized, comprehensive plan. Equally important is the need to have a discussion with the patient regarding their goals of care. This is especially important in this patient cohort for several reasons. As will be outlined in the coming pages, there are multiple treatment modalities that can be implemented and the invasiveness and risk associated with these methods vary considerably. It is important the patient understands their diagnosis of metastatic disease and prognosis, especially in patients with advanced cancer [9, 10]. Attempts have been made to develop algorithms to predict survival in patients with metastatic disease, but none have been shown to be consistently reliable and/or widely applicable [9–12]. Thus, this conversation provides valuable information to the surgeon regarding the patient's insight into their own disease and acceptance of their diagnosis. The goal of any intervention in the pelvis is to relieve pain and improve the patient's mobility. Any surgical procedure in the pelvis for diffuse metastatic disease should allow immediate weight bearing.

Radiation therapy may be an option for patients who want to avoid surgery, or for those who are otherwise deemed to unhealthy to undergo surgery, as long as the patient is without a displaced fracture and is able to mobilize safely. Radiation therapy is effective in palliating pain and is used often in metastatic disease of the pelvis. External beam radiation therapy (EBRT) is the typical option for nonoperative treatment for painful osseous metastases and is effective in 60–70% of patients. There has been much interest in the optimal radiation regimen for these patients and, according to the American Society for Radiation Oncology, all these regimens are equally efficacious [13]. A single fraction of 8 Gy to a bony metastasis has been shown to have the same pain relief as higher doses split up into several fractions [13, 14]. This offers a more convenient option for many patients; however, it is important to inform patients interested in a single fraction therapy that they do run a higher risk of having recurrent pain in the treatment field which would require repeat radiation therapy [13]. Patients with recurrence of their pain after single fraction treatment have been found to develop recurrent pain around 3–4 months after treatment [15, 16]. Interest in stereotactic body radiotherapy (SBRT) is increasing given evidence that it has similar local control and pain relief rates when compared to EBRT [17].

47.5 Minimally Invasive Treatment Options

Embolization can be used as either the primary treatment modality or as a neoadjuvant treatment for certain primary cancer types in preparation for future surgical management. This treatment is associated with a 95–97% clinical response rate and there is often healing of the bone lesions in greater than 50% of the cases [18, 19]. It should be noted that, if performed as a neoadjuvant treatment, surgery should be performed within 48 h of embolization to ensure there is no revascularization of the lesion prior to surgery.

Percutaneous techniques exist that can be used in locations of the pelvis that are not amenable to surgery or radiation. Radiofrequency ablation (RFA) and cryoablation are forms of image-guided tumor ablation that can provide palliation to patients with painful pelvic metastases. Both techniques are similar in the sense that they use variations in temperature to cause cell death within the metastatic lesion. These techniques have shown efficacy in the treatment of patients with oligometastatic disease and have led to acceptable pain relief as well as progression-free survival [20]. With this technique, it should be noted that the location and size of the lesion are of utmost importance, as the thermal ablation occurs in a centrifugal pattern and can cause damage to nearby neurovascular structures. The percutaneous techniques chapter covers this in much detail.

Cementoplasty was based on the principles used for kyphoplasty in the spinal column. These procedures offer several benefits over open techniques. The percutaneous approach reduces the blood loss and operative time significantly without sacrificing considerable pain relief postoperatively [21, 22]. Second, the risks of wound complications or infection are reduced due to the size of the incisions and the minimal soft tissue dissection performed throughout the procedure. This, in theory, could facilitate quicker transitions back to medical and radiation adjuvants postoperatively. Limitations in using this treatment involve the treatment of uncontained defects, which could allow extravasation of cement into the surrounding soft tissues or, potentially, into the joint, neural foramen, and pelvic space. One study quoted a 22–28% risk of extra-osseous extravasation of cement in their cohort [23]. This extravasation can result in radiculopathy from impingement on the sciatic nerve and/or pain with ambulation if cement spreads within the hip joint itself. The location of the lesions can influence the ease of this procedure. Sacral lesions are traditionally very difficult to treat in this way given the difficulty with obtained adequate imaging and the three-dimensional anatomy in this region. This method has been taken one step further when percutaneous screws are inserted, as previously described by Starr et al. [24], prior to cement insertion in periacetabular defects (Fig. 47.2). The percutaneous screws are thought to

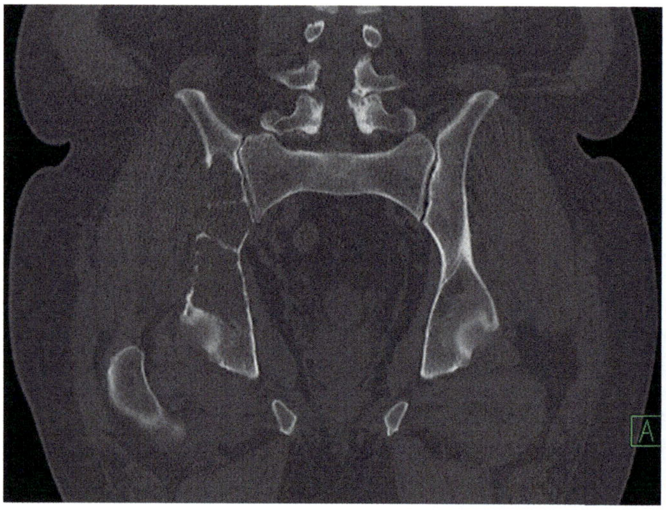

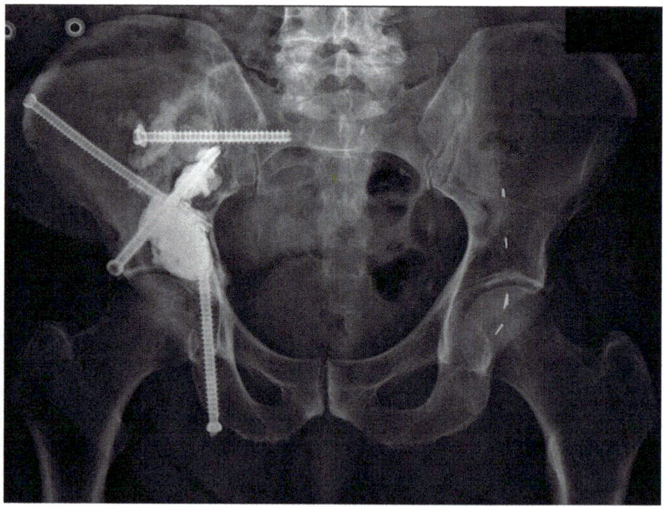

Fig. 47.2 This patient illustrates the indications for percutaneous techniques as shown in the postoperative image. In the preoperative CT scan, there is extension of the lesion cranially into the sciatic buttress and medial ilium; however, due to the extent of the disease this patient was not eligible for acetabular reconstruction with a cage construct. Percutaneous cannulated 6.5 mm screws and cementoplasty were utilized to provide pain relief and allow early ambulation for this patient

provide a scaffold for cementation of the defects, adding stability to the construct [22].

47.6 Invasive Treatment Options

The Harrington classification system was first proposed in 1981, not only as a means of compartmentalizing and guiding surgical management of periacetabular lesions but also as a way of proposing novel treatment strategies for these complex tumors [25]. The driving force behind this classification system is the presence and quality of bone within the periacetabular region of the pelvis and how this affects the load transmission from the acetabulum to the axial skeleton. The classification system is based on the integrity of the medial, lateral, and/or superior walls. This system is still used often to describe lesions and help guide treatment within this region of the pelvis. However, the treatment strategies have continued to evolve as our understanding of the disease process improves.

Total hip arthroplasty in metastatic disease of the periacetabular region is a challenging procedure for several reasons. First, there is often significant bone loss which leads to lack of adequate support for conventional total hip arthroplasty components. In addition, many of these patients have undergone radiation and/or chemotherapeutic treatments which have had an impact on the local bone's ability to in grow into the implanted prostheses, increasing the risk of aseptic loosening postoperatively. Considering this, it is necessary to create a construct that reestablishes (or reduces the need for) column support and provides articular stability. The treatment for most acetabular lesions with significant metastatic defects is an acetabular reconstruction involving a cage. The reconstruction cage is adjusted to fit the patient's anatomy and to provide as many fixation points as possible. These cages prevent medial or superior migration of the hip joint and fill the weakened, lytic areas of the metastatic lesions.

Reconstruction cages come in many varieties; however, the general concept is the same throughout all options (Fig. 47.3). The cage serves as a foundation on which to reconstruct the hip joint and the cage-screw construct act as a sort of rebar to strengthen the cementation of the acetabulum. Reconstruction cages are designed to have at least one point of fixation on either side of the acetabulum—both cranial and caudal to the joint [26–28]. The goal of this cage construct is to provide fixation into healthier bone, distributing the joint forces to the axial skeleton. The cranial aspect often consists of a cluster of screws placed through one or several flanges extending up the iliac wing. The caudal portion consists of an ischial hook, blade, or screw-hole cluster, which are all meant to gain purchase in either the ischium or around the inferior aspect of the cotyloid fossa. Rapidly drying cement is then placed into the cage to anchor the construct to the remaining pelvis. If needed, a small corticotomy can be made in the outer table of the ilium to allow for cement filling of defects in the iliac bone. A cemented femoral stem is typically used.

Although the Harrington reconstruction techniques are still used by many surgeons to this day, there has been continued interest in finding a solution to the loosening seen with this type of reconstruction. As such, techniques used to manage massive acetabular bone loss in revision hip arthroplasty have been adapted to oncologic scenarios. Porous tantalum shells are advantageous given their high coefficient of friction which allows for an immediate scratch fit of the implant in the remaining acetabular bone, facilitating early weight bearing for these patients. In addition, these augments are designed to accept multiple screws that can be inserted to increase the fixation to the bone and improve the compression of the implant to native bone, theoretically enhancing bone ingrowth. In fact, these tantalum implants have shown adequate amounts of bony

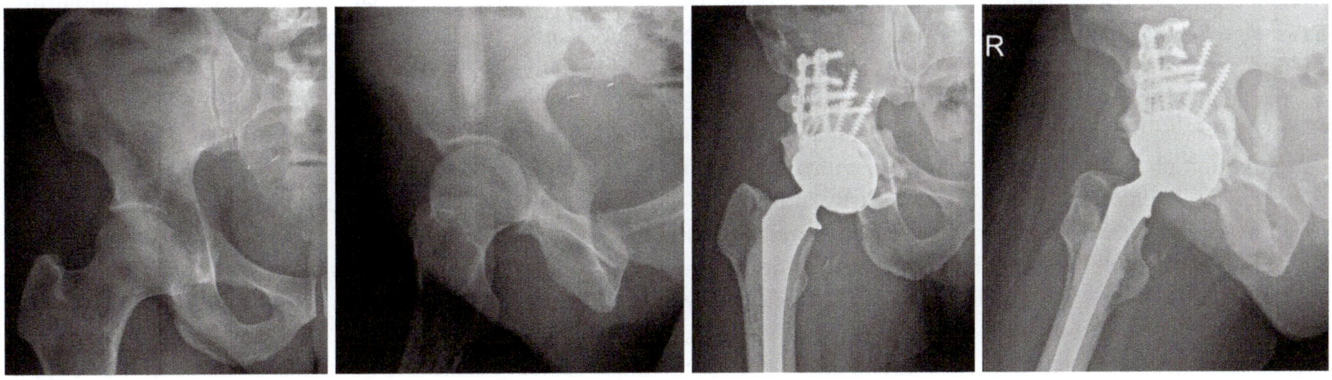

Fig. 47.3 Operative treatment of pelvic lesions can be considered when the morphology of the lesion is such that it impacts the anatomy responsible for force transmission and increases the risk of fracture due to these high amounts of force. This patient underwent acetabular reconstruction with a reconstruction cage, total hip arthroplasty with a dual mobility liner, and a cemented stem. This allows the transmission of the weight-bearing forces from the acetabulum to the healthier, stronger portions of the iliac bone

ingrowth even in the setting of previously irradiated bone. This technique is often used after radiation for fracture deformity or arthritis with defects present with no gross disease present. When compared to the Harrington reconstructions, these uncemented reconstructions achieved similar outcome scores while having a lower rate of loosening—0% rate in one study [29].

There are several options with regards to the type of constraint used in these periacetabular reconstructions. Constrained polyethylene liners are used in a patient population who is at high risk of instability of the hip postoperatively. However, due to the increased restraint, the polyethylene–cement interface experiences high amounts of shear stress and, thus, component loosening may be seen mid-to-long term. As dual mobility technology has improved there has been increased interest in these patients. The dual mobility design provides less constraint compared to the tripolar constrained liners; however, it is more constrained than primary total hip components, in a way possessing some of the desirable aspects of both options. Similarly, the risk of instability after implantation of dual mobility components appears to be similar to constrained liners [28].

There are unique scenarios and disease types in which a single metastatic lesion can be resected in the pelvis to render the patient a status of no evidence of disease (NED). In periacetabular lesions in which a resection of the acetabulum is required for treatment there are several reconstructive options available. The pedestal cup (also known as the "ice cream cone" prosthesis) is a modular design composed of a stem, which is inserted into medial ilium, and a metal cup that is locked into the stem with the desired orientation. The medial ilium along the sciatic buttress is ideal for insertion as it has the straightest geometry of the iliac bone, thus providing the most predictable and reproducible insertion [30]. The stem can be inserted in a cemented or uncemented fashion. The design of this implant is such that it recreates parts of the resected sciatic buttress, facilitating native force transfer from the hip joint into the medial pelvis and sacrum. An advantage lies in the fact that the cup can be adjusted to the desired version at the time of the surgery, helping the surgeon take measures to reduce the likelihood of postoperative instability. Having said that, instability remains the most common complication of these implants, with dislocation rates ranging from 15 to 39% with the majority of these dislocations occurring within the first year of the initial surgery [30, 31]. However, with the advent of dual mobility hip components, the dislocation rates have decreased significantly [31, 32]. Saddle prostheses were designed to provide a quick reconstruction option that allows for tensioning of the hip musculature and restoration of limb length. One disadvantage of this prosthesis is that these patients have strict activity and weight-bearing precautions postoperatively for the first several months. These restrictions are meant to maintain the position of the prosthesis while the soft tissues heal and scar in to provide added stability prior to resumption of normal activity. Additionally, these patients commonly have neuropraxias in the perioperative period secondary to change in their limb length [33, 34]. Proximal migration of the prosthesis through the ilium is commonly seen. Unfortunately, this particular type of prosthesis has not yielded the desired results, with most patients failing to return to ambulation with minimal or no assistance; as such has fallen out of favor.

Due to the poor functional outcomes related to resection arthroplasty for pelvic sarcomas there has always been an interest in endoprosthetic reconstruction. However, early attempts at developing these megaprostheses were unsuccessful mostly due to the aseptic loosening seen secondary to the high shear forces experienced at the prosthesis-native bone interface as well as the limited fixation options. With the advent of 3D printing there is renewed interest in endoprosthetic reconstruction as these implants can now be custom-made to accommodate the patient's unique anatomy. Another downstream effect of 3D printing is the improvement in ingrowth surfaces on these implants. In certain scenarios, these prostheses can be used for specific disease types that have demonstrated disease stability over time.

Resection alone for palliation can be used if other options are not possible. This could be due to the lack of residual bone to allow for reconstruction. Although reconstructing the acetabulum is often the goal of surgery to provide a stable weight-bearing construct these reconstructive procedures are technically demanding, time-consuming and associated with a high rate of complications. Thus, resection arthroplasty, internal and external hemipelvectomy remain viable options for pain control in the setting of single-site metastatic disease. The proximal femur is stabilized to the remaining hemipelvis to create a pseudarthrosis, as a "flail limb." This type of reconstruction does require enough remaining iliac bone in order to provide a stable platform to which the proximal femur will articulate. It also requires preoperative education of the patient regarding limitations of the extremity as well as an expected limb length discrepancy postoperatively.

In a systematic review performed by Brown et al. [35], they compared many of the reconstructive options. Their study showed that there was a 50% risk of complications overall, with deep prosthetic infection and joint instability being most common in all techniques. In addition, the improvement in the MSTS score varied from 62 to 72%; however, all techniques did show a sustained improvement in pain and function at final follow-up.

47.7 Conclusion

- When there is a concern for metastatic disease of the pelvis, the appropriate prognosis, imaging, and goals of care should be determined prior to providing possible treatment recommendations.
- Location of the metastatic lesions within the pelvis, and how these lesions compromise the structural integrity of the acetabulum and pelvic ring during weight bearing, will direct which treatment modalities are possible.
- There are nonoperative and operative treatment options for each patient and the outcomes of each of these options should be provided to help with patient education and decision-making.

Open Questions

Future research in the area of surgical management of metastatic bone disease of the pelvis should address the following questions:What is the quantitative limit of nonoperative management of metastatic disease of the pelvis?

- What are the anatomic characteristics that necessitate the need for operative intervention?
- Likewise, what anatomic characteristics correlate with the best postoperative outcomes of each of the above-mentioned surgical management options?
- What is the role of antiresorptive medication in the management of patients with metastatic disease of the pelvis?
- In the event of nonoperative management of these patients, what is the appropriate surveillance timeline?

References

1. Onken JS, et al. Metastatic dissemination patterns of different primary tumors to the spine and other bones. Clin Exp Metastasis. 2019;36(6):493–8. Print.
2. Lievens Y, et al. Defining oligometastatic disease from a radiation oncology perspective: an estro-astro consensus document. Radiother Oncol. 2020;148:157–66. Print.
3. Lu X, et al. Oligometastatic state predicts a favorable outcome for renal cell carcinoma patients with bone metastasis under the treatment of sunitinib. Oncotarget. 2016;7(18):26879–87. Print.
4. Yin Z, et al. Impact of sites versus number of metastases on survival of patients with organ metastasis from newly diagnosed cervical cancer. Cancer Manag Res. 2019;11:7759–66. Print.
5. Basu D, et al. Detection of occult bone metastases from head and neck squamous cell carcinoma: impact of positron emission tomography computed tomography with fluorodeoxyglucose F 18. Arch Otolaryngol Head Neck Surg. 2007;133(8):801–5. Print.
6. Bhandari V. Incidence of bone metastasis in carcinoma buccal mucosa. Indian J Med Paediatr Oncol. 2016;37(2):70–3. Print.
7. Ebraheim NA, et al. Anatomic consideration in the anterior approach to the sacro-iliac joint. Spine (Phila Pa 1976). 1994;19(6):721–5. Print.
8. Browner BD. Skeletal trauma: basic science, management, and reconstruction. 6th ed. St. Louis, MO: Elsevier; 2019. Print.
9. Kirkinis MN, et al. Metastatic bone disease: a review of survival, prognostic factors and outcomes following surgical treatment of the appendicular skeleton. Eur J Surg Oncol. 2016;42(12):1787–97. Print.
10. Kirkinis MN, et al. Metastatic bone disease of the pelvis and extremities: rationalizing orthopaedic treatment. ANZ J Surg. 2017;87(11):940–4. Print.
11. Chow E, et al. Validation of a predictive model for survival in patients with advanced cancer: secondary analysis of rtog 9714. World J Oncol. 2011;2(4):181–90. Print.
12. Paulino Pereira NR, et al. Development of a prognostic survival algorithm for patients with metastatic spine disease. J Bone Joint Surg Am. 2016;98(21):1767–76. Print.
13. Lutz S, et al. Palliative radiation therapy for bone metastases: update of an astro evidence-based guideline. Pract Radiat Oncol. 2017;7(1):4–12. Print.
14. Rich SE, et al. Update of the systematic review of palliative radiation therapy fractionation for bone metastases. Radiother Oncol. 2018;126(3):547–57. Print.
15. van der Linden YM, et al. Single fraction radiotherapy is efficacious: a further analysis of the dutch bone metastasis study controlling for the influence of retreatment. Int J Radiat Oncol Biol Phys. 2004;59(2):528–37. Print.
16. van der Velden JM, et al. Evaluation of effectiveness of palliative radiotherapy for bone metastases: a prospective cohort study. J Radiat Oncol. 2018;7(4):325–33. Print.
17. Spencer KL, et al. Systematic review of the role of stereotactic radiotherapy for bone metastases. J Natl Cancer Inst. 2019;111(10):1023–32. Print.
18. Rossi G, et al. Embolisation of bone metastases from renal cancer. Radiol Med. 2013;118(2):291–302. Print.
19. Rossi G, et al. Selective arterial embolisation for bone tumours: experience of 454 cases. Radiol Med. 2011;116(5):793–808. Print.
20. Cazzato RL, et al. Complications following percutaneous image-guided radiofrequency ablation of bone tumors: a 10-year dual-center experience. Radiology. 2020;296(1):227–35. Print.
21. Durfee RA, et al. Percutaneous acetabuloplasty for metastatic lesions to the pelvis. Orthopedics. 2017;40(1):e170–e75. Print.
22. Yang R, et al. A novel tripod percutaneous reconstruction technique in periacetabular lesions caused by metastatic cancer. J Bone Joint Surg Am. 2020;102(7):592–9. Print.
23. Weill A, Kobaiter H, Chiras J. Acetabulum malignancies: technique and impact on pain of percutaneous injection of acrylic surgical cement. Eur Radiol. 1998;8(1):123–9. Print.
24. Starr AJ, Reinert CM, Jones AL. Percutaneous fixation of the columns of the acetabulum: a new technique. J Orthop Trauma. 1998;12(1):51–8. Print.
25. Harrington KD. The management of acetabular insufficiency secondary to metastatic malignant disease. J Bone Joint Surg Am. 1981;63(4):653–64. Print.
26. Assi C, et al. The cross technique for the positioning of kerboull plate in acetabular reconstruction surgery. SICOT J. 2018;4:20. Print.
27. Rowell P, et al. Is an acetabular cage and cement fixation sufficiently durable for the treatment of destructive acetabular metastases? Clin Orthop Relat Res. 2019;477(6):1459–65. Print.
28. Wegrzyn J, et al. Total hip arthroplasty for periacetabular metastatic disease. an original technique of reconstruction according to the harrington classification. J Arthroplast. 2018;33(8):2546–55. Print.

29. Houdek MT, et al. Comparison of porous tantalum acetabular implants and harrington reconstruction for metastatic disease of the acetabulum. J Bone Joint Surg Am. 2020;102(14):1239–47. Print.
30. Bus MP, et al. Clinical outcome of pedestal cup endoprosthetic reconstruction after resection of a peri-acetabular tumour. Bone Joint J. 2014;96-B(12):1706–12. Print.
31. Bus MP, et al. Lumic((R)) endoprosthetic reconstruction after peri-acetabular tumor resection: short-term results. Clin Orthop Relat Res. 2017;475(3):686–95. Print.
32. Issa SP, et al. Pelvic reconstructions following peri-acetabular bone tumour resections using a cementless ice-cream cone prosthesis with dual mobility cup. Int Orthop. 2018;42(8):1987–97. Print.
33. Aljassir F, et al. Outcome after pelvic sarcoma resection reconstructed with saddle prosthesis. Clin Orthop Relat Res. 2005;438:36–41. Print.
34. Renard AJ, et al. The saddle prosthesis in pelvic primary and secondary musculoskeletal tumors: functional results at several postoperative intervals. Arch Orthop Trauma Surg. 2000;120(3–4):188–94. Print.
35. Brown TS, et al. Reconstruction of the hip after resection of peri-acetabular oncological lesions: a systematic review. Bone Joint J. 2018;100-B(1 Supple A):22–30. Print.

Surgical Management of Metastatic Disease to the Spine

48

Christopher P. Wang, Amanda Brisco, and James K. C. Liu

Abstract

The surgical management of spinal metastases has unique challenges not encountered in other aspects of spine surgery due to the needs of the oncologic patient. Among the key steps in the surgical management is the decision-making process, which must take into account the systemic and adjuvant therapies that are also necessary for systemic and local control of the disease. Decision-making frameworks, such as the NOMS system, have been designed to take into account the multiple factors that can affect the prognosis of the oncologic patient and reach an optimal surgical plan that incorporates available adjuvant treatment, such as radiation therapy. Surgical intervention for spinal metastases has evolved from large reconstructive procedures to less invasive techniques, which still allow for the preservation of neural function and restoration of spinal stability, but with the strategy of selecting the optimal intervention which can best synergize with available adjuvant therapies for effective local disease control. Advancement of surgical strategies such as "separation surgery" and minimally invasive percutaneous stabilization techniques can achieve surgical goals with decreased morbidity and shortened recovery times, which allows for minimal disruption of adjuvant or systemic therapies. The development of conformal radiation techniques such as radiosurgery has greatly assisted in achieving these goals.

C. P. Wang · A. Brisco · J. K. C. Liu (✉)
H. Lee Moffitt Cancer Center & Research Institute, Tampa, FL, USA
e-mail: cpwang@usf.edu; abrisco@usf.edu; james.liu@moffitt.org

Learning Objectives

- Readers will understand the indications and goals of surgical treatment for spinal metastases, including strategies to assist in optimal surgical intervention.
- Readers will be able to understand the different surgical approaches that can be employed for the surgical management of metastatic disease to the spine.
- Readers will develop an understanding of the application of radiation as an important adjunct to the surgical management of spine metastases.

48.1 Introduction

Metastatic spine tumors are the most common type of spinal column tumors [1]. The malignancies that most commonly develop metastases to the spine are cancers of the lung, breast, and kidney. Hematological malignancies can spread to the spine and are often considered under the category of spinal metastases and these are predominantly composed of multiple myeloma and lymphoma [2]. The surgical treatment of spine metastases requires an understanding of the unique needs of the oncologic patient, which must be taken into account when determining the optimal surgical treatment and approach. There are a variety of surgical techniques that can be tailored to provide the most effective treatment strategy with minimal disruption to the overall goal of continuation of systemic therapy, including minimally invasive strategies as well as percutaneous stabilization, ablative, and radiation techniques. Here we will outline the surgical approach to the metastatic spine patient, from individualizing decision-making to various surgical strategies and how each may be optimally used to treat metastatic spine disease.

S. P. Leong et al. (eds.), *Cancer Metastasis Through the Lymphovascular System*,
https://doi.org/10.1007/978-3-030-93084-4_48

48.2 Surgical Decision-Making

Among the unique challenges that are encountered in the surgical management of spinal metastases is selecting the intervention that best incorporates into the treatment plan for the oncologic patient. Preserving neurologic function through decompression of neural elements and achieving stability remains the primary goal for surgical management. The type of surgical intervention that is employed to achieve those goals must take into account factors that affect the oncologic patient, including the state of systemic disease, prognosis, ongoing therapies, and options of adjuvant therapy that may assist in the treatment of the metastatic disease. Several clinical decision-making tools have been established to provide guidelines in regard to surgical management of metastatic spine disease. Early scoring systems provided a framework to suggest the degree of surgical intervention based on the aggressiveness of underlying malignancy. The modified Bauer scoring system provided a framework for surgical decision-making based on the presence of visceral metastasis [3]. The Tomita scoring system considered the type of primary malignancy, as well as the presence of visceral and bony metastases [4]. The revised Tokuhashi scoring system was similar to the Tomita system, but added number of vertebral body metastases, performance status, and neurological deficits as factors to consider [5]. Based on the extent of the disease, these scoring systems would suggest palliative treatments that may not include surgery, to invasive excisional surgery that may involve extensive tumor resection. The evolution of these scoring systems underscores the importance of understanding the natural course of the primary malignancy and its potential response to systemic therapy, as well as the extent of metastasis present to determine whether a complex surgical intervention is indicated and beneficial to the patient.

48.2.1 NOMS

Previous treatment decision tools were based on assessing how surgery alone can play a role in local management of metastatic spine disease. The emergence of adjuvant therapies, particularly stereotactic radiation therapy, has created additional options for local control of metastatic spine disease and therefore shifted the approach for surgical management.

Laufer et al. established the NOMS system to provide a clinical treatment framework for treating spinal metastases based on the four categories: Neurologic, Oncologic, Mechanical, and Systemic (NOMS) [6] (Fig. 48.1). "Neurologic," refers to the degree of neurological compromise present, as measured radiographically by the amount of spinal cord compression by the tumor, as well as neurological deficits present on physical examination. To help characterize the degree of spinal cord compression, the Spinal Oncology Study Group (SOSG) devised the Epidural Spinal Cord Compression (ESCC) grading scale, which provides a 6-point grading system that describes the extent of spinal cord compression [7]. This scale quantifies the degree of epidural extension of tumor from isolation only within the vertebral body, to compression on the spinal cord (Fig. 48.1). Grade 0 indicates disease confined to the bone. Grade 1 indicates the presence of disease in the epidural space and is subcategorized in 1a–1c. 1a indicates epidural disease without impingement on the thecal sac, 1b indicates the presence of compression of the thecal sac without compression of the spinal cord, and 1c indicates thecal sac compression and abutment of the spinal cord without compression. Grade 2 indicates compression of the spinal cord with CSF visible around the spinal cord and Grade 3 indicates compression of the spinal cord without evidence of surrounding CSF. The ESCC provides a universal scale for assessing the degree of neurological compromise present, which may indicate present or potential neurologic deficits, but also dictates radiation options. Clinical evidence of neural compromise also contributes to surgical decision-making. This includes spinal cord compression resulting in myelopathy, manifesting as hyperreflexia and clonus, to loss of sensation, bowel or bladder dysfunction, and motor weakness. Compression of nerve roots may manifest as radiculopathy along a given dermatomal distribution, but also may be present only upon axial loading and is consistent with mechanical radiculopathy, indicating potential instability of the spine. Together, the radiographic and clinical evidence of the "neurologic" status of the patient determines the potential need and urgency for decompression of the neural elements.

"Oncologic," is defined by the tumor histology of the primary malignancy. This provides insight into the overall prognosis of the patient as well as the responsiveness to radiation. Radiation therapy has become an important adjuvant therapy in the local treatment of spinal metastases but not all pathologies respond equally [8]. Tumor pathologies can be divided in their response to radiation therapy by radiosensitive or radioresistant tumors. Tumors that are considered radiosensitive, including lymphoma, myeloma, seminoma, breast cancer, and prostate cancer, demonstrate good overall control to conventional external beam radiation therapy (cEBRT), which consists of non-conformal radiation applied to a general area with low risk of toxicity to adjacent neural structures [9]. Many solid malignancies such as renal, gastrointestinal, non-small cell lung, thyroid, and sarcomas, demonstrate poor response to cEBRT [8]. These pathologies require conformal radiation in the forms of hypo-fractionated radiation therapy or stereotactic radiosurgery [10]. An understanding of the tumor histology and its responsiveness to radiation therapy will help to optimize surgical intervention so that it can work with the radiation therapy available.

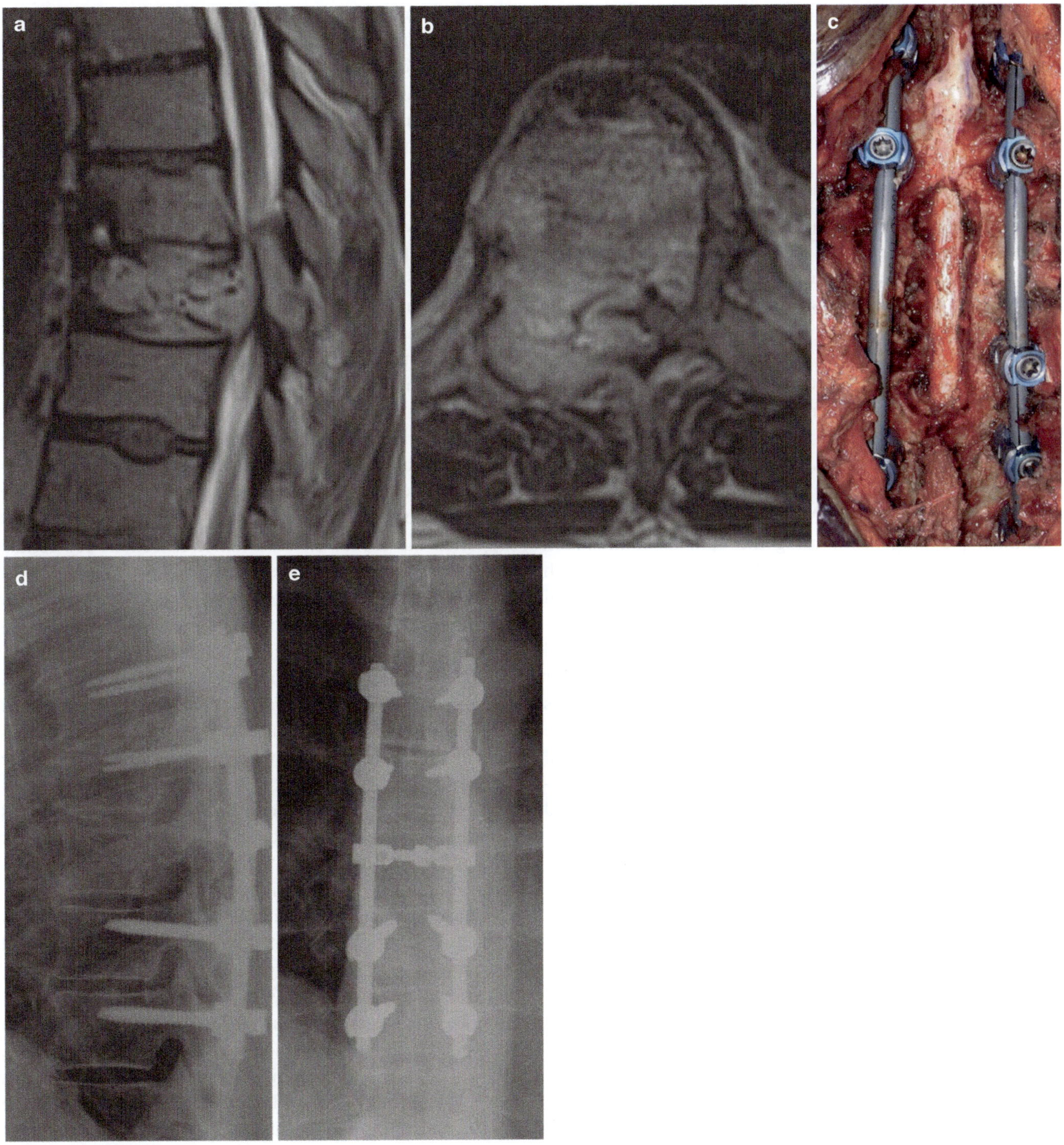

Fig. 48.1 (**a**, **b**) T2-weighted MRI demonstrating grade 3 epidural spinal cord compression. (**c**) Intraoperative view of separation surgery, with the removal of the left facet joint and pedicle, demonstrating circumferential decompression of the thecal sac and posterolateral instrumentation two levels above and below the level of metastasis. (**d**, **e**) Postoperative X-rays demonstrating posterolateral instrumented fixation followed decompression via separation surgery

"Mechanical" refers to the stability of the spine at the level of disease and serves as an indication for surgical intervention. It is assessed based on the clinical assessment of mechanical and postural pain. Restoration of stability is one of the main goals in spine surgery and is a primary indication for intervention. To determine the level of instability and indication for intervention, the Spinal Instability Neoplastic Score (SINS) classification system was derived to provide a framework as to the degree of instability secondary to tumor involvement in the spine. It is composed of six elements:

location of the lesion, mechanical/postural pain, type of bone lesion, radiographic spinal alignment, vertebral body collapse, and posterolateral involvement of the spinal elements. Each element is assigned a score, then tallied to provide a total SINS score, which categorizes the spine as stable, unstable, or of intermediate instability [11]. In general, the presence of mechanical pain represents spinal instability that likely requires stabilization. These can manifest as back pain on axial loading and movement and is relieved with rest. In addition to mechanical back pain, tumor involvement in the posterior elements may manifest as pain when lying flat due to loading of the facet joints, which may also be an indication for stabilization.

"Systemic" characterizes the disease burden and related medical comorbidities and serves to assess a patient's prognosis and fitness for treatment [12]. Disease burden is determined by tumor staging, and the prognosis is estimated using a patient's functional status, extent of disease, and options for systemic treatment.

Taken together, the NOMS framework helps to direct treatment planning based using the optimal combination of therapies. For example, for radiosensitive tumors, such as lymphoma, myeloma, and prostate cancer, conventional radiation therapy alone may be the ideal treatment regardless of neurologic compromise or ESCC, given the rapid response of the tumor to conventional radiation therapy. For radioresistant tumors, high-dose stereotactic radiation is needed to achieve local tumor control. Therefore, the type of surgical intervention will depend on the need for neural decompression, based on the degree of ESCC or neurological symptoms, as well as the stability of the spine, measured by the SINS score. In grade 0 or 1 ESCC, surgery is necessary only if there is evidence of instability. For grade 2 or 3 ESCC, surgical decompression is necessary to provide adequate clearance of tumor from the spinal cord so that the appropriate radiation dosage can be applied without risk of toxicity to the spinal cord. The NOMS framework emphasizes the importance of the histology of the tumor and its responsiveness to radiation to determine whether surgery is necessary for local control.

48.3 Open Surgical Approaches

Surgery for spinal metastases in its infancy consisted primarily of posterior decompression. This was prior to the availability of spinal instrumentation, which allowed for circumferential decompression and reconstruction of the spine. Several studies examined laminectomy for epidural spinal cord compression and found that decompressive laminectomy in addition to conventional external beam radiation (cEBRT) did not demonstrate an advantage compared to radiotherapy alone [13, 14]. These results were unsurprising given that the majority of metastatic disease arises from the vertebral body resulting in anterior neural compression, therefore posterior decompression alone may not provide adequate decompression of the neural elements and would provide little in overall disease control. These studies resulted in a shift of the treatment paradigm away from surgical decompression to radiation alone. This approach remained the standard until Patchell et al. demonstrated in 2005 that direct surgical decompression in addition to radiation therapy provided improvement in ambulatory status, steroid use, pain score, and overall survival compared to radiation alone [15]. The benefit of surgery in addition to radiation noted in this study was likely due to the use of circumferential decompression, which was made possible due to the emergence of spinal instrumentation that allowed for the reconstruction of the spine, which started in the mid-1980s [16].

With the ability to reconstruct the spinal column, more aggressive surgical approaches were taken to provide long-term local control for metastatic spine disease. Prognostic scoring systems such as the Tomita and Tokuhashi systems supported the use of en bloc tumor resections for certain types of oligometastatic disease to achieve optimal local control [4, 5]. The challenges of en bloc resection are that it is difficult to achieve in the spine given the extent of disease and proximity to neural structures, and that it is also technically demanding and can involve significant morbidity, which may not be well tolerated by oncologic patients [17]. Along with the risk for distance recurrence in metastatic disease, en bloc resection may not prove to be worthwhile in the long term [18].

An alternative to en bloc or marginal tumor resections requiring circumferential reconstruction of the spine for metastatic spine disease is "separation surgery," which is a surgical approach with the intent of neurological preservation, the establishment of spinal stability, and preparation for adjunctive therapy. Instead of attempting to achieve marginal resections, separation surgery involves decompression of the spinal cord to provide a 2–3 mm clearance of tumor away from the spinal cord so that an adequate radiation dosage can be applied to the tumor without a risk of neural toxicity [10, 19]. This is achieved primarily through a posterior approach to the spine, followed by a laminectomy to decompress the posterior aspect of the thecal sac. Further decompression of the lateral and anterior aspect of the spinal cord is performed through unilateral or bilateral facetectomies, resection of the pedicles, and sectioning of the anterior longitudinal ligament to resect tumor adjacent to the spinal cord. Removal of the facet joints and pedicles allows access to the anterior vertebral body through a posterolateral approach, avoiding the need for a direct lateral or anterior approach, which involves much greater morbidity, while reducing surgical time and blood loss. Removal of the facet joints requires instrumented stabilization, often through pedicle screw fixation that extends two vertebral body levels above and below the level of decompression (Fig. 48.1). In cases of significant loss of bony integrity in the anterior column secondary to erosion by tumor involvement, the anterior column support can be augmented with placement of polymethylmethacrylate from a posterior approach or placement of an expandable cage from through a transpedicular or costotransversectomy approach, depending on the level and degree of bony destruction [20].

48.4 Minimally Invasive Approaches

The development of tubular retractor systems and percutaneous screw placement techniques has led to an exponential increase in the adoption of minimally invasive techniques for the treatment of spinal pathologies. Smaller incisions with reduced degree of paraspinal muscle destruction have allowed placement of instrumentation into the spine with decreased length of hospital stays and faster recovery following surgery compared to traditional open approaches [21]. By taking advantage of the effectiveness of radiation therapy, minimally invasive approaches can effectively achieve surgical goals of care with minimal interruption of adjuvant and systemic therapies.

48.4.1 Percutaneous Pedicle Screw Fixation

Percutaneous pedicle screw fixation can serve as an effective alternative to open fixation techniques for unstable fractures secondary to metastatic disease without the need for neural decompression. Percutaneous fixation has been associated with lower postoperative pain scores and faster recovery time compared to open approaches [22, 23]. They allow for stabilization of pathology affecting both the anterior and posterior columns, without disruption of the paraspinal musculature necessary in an open posterior approach, and therefore preserving the posterior tension brand. This allows for placement of shorter segmental constructs achieving stabilization through fixation one level above and below the diseased level compared to conventional fixation strategies which necessitate longer constructs [24]. An adjunct to percutaneous fixation is the use of fenestrated pedicle screws which allow for cement augmentation of the screw tips within the vertebral bodies [25]. Patients with pathological fractures secondary to metastatic spine disease often have multiple levels of vertebral body disease involvement preventing optimal screw purchase, necessitating longer constructs to obtain more points of fixation to decrease the risk of hardware pullout. The development of fenestrated screw systems which allow for cement augmentation allows improved bony purchase and short segment constructs even in patients with poor bone quality. A significant advantage of percutaneous fixation is the lack of a long midline incision with disruption of the paraspinal musculature, which allows for rapid restoration of systemic therapy or application of local radiation therapy with little risk for wound complications [26] (Fig. 48.2).

A hybrid approach consisting of percutaneous fixation combined with "mini-open" laminectomies or unilateral facetectomies can also be utilized when focused neural decompression is also indicated [27]. A small midline or paramedian incision can be utilized with or without tubular retractors to allow for focused neural decompression, which still prevents significant paraspinal injury compared to traditional open techniques.

48.4.2 Percutaneous Cement Augmentation

Metastatic spinal disease contained primarily within the anterior vertebral body resulting in mechanical pain may be treated with cement augmentation. Cement augmentation was originally applied to the spine for the treatment of painful hemangiomas and then became widely used primarily for osteoporotic compression fractures [28, 29]. Cement augmentation can also be effectively applied for the treatment of pathologic vertebral compression fractures, or loss of bone quality secondary to lytic metastatic disease. The Cancer Patient Fracture Evaluation (CAFE) study was a randomized controlled trial that involved 22 international sites, which compared balloon kyphoplasty to nonsurgical fracture management of pathologic compression fractures [30]. The study demonstrated definitively that kyphoplasty provided rapid pain reduction and improvement in patient function with low rates of adverse events. Subsequent studies have demonstrated the value of cement augmentation for malignant compression fractures resulting in improvement in functional status with minimal risk [31].

The indication for cement augmentation is localized pain secondary to metastatic disease secondary involving the anterior vertebral body without significant involvement of the posterior column. In addition to proper localization to isolate the correct level as the pain generator, the morphology of the vertebral body dictates whether successful cement augmentation can be performed. Relative contraindications include disruptions in the cortical walls of the vertebral body, particularly the posterior wall, which may lead to the extravasation of cement into the spinal canal [32]. Augmentation can be performed in the presence of cortical wall breaches in experienced hands by adjusting needle placement to minimize extravasation. Although primarily performed in the thoracic and lumbar levels, through a posterior transpedicular approach, augmentation of the cervical levels through an open anterior approach can also be performed [33]. The most common form of cement used is polymethylmethac-

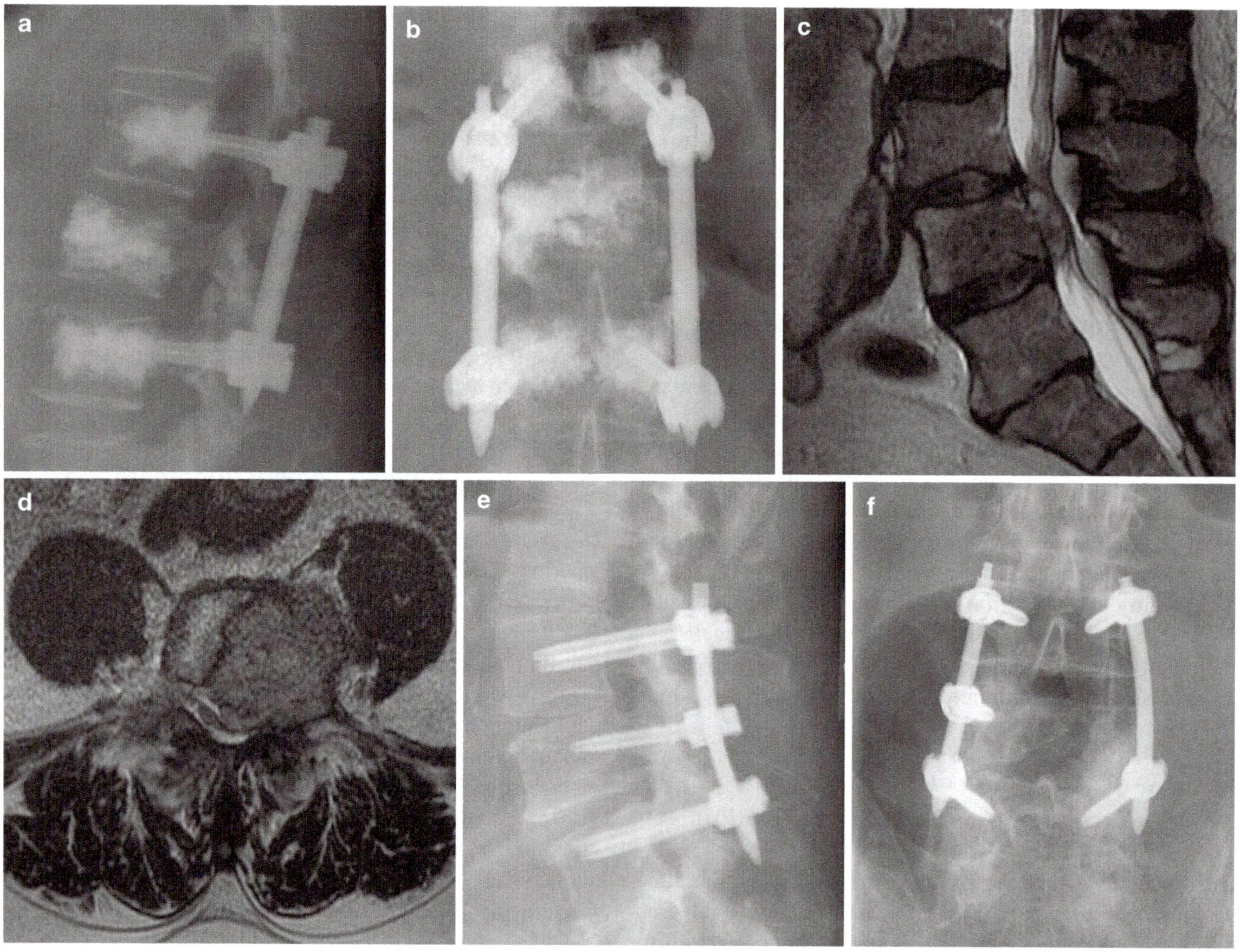

Fig. 48.2 (**a**, **b**) Postoperative X-rays demonstrating percutaneously placed pedicle screw fixation with cement augmentation of the pedicle screws and kyphoplasty at the level of metastatic disease. (**c**, **d**) T2-weighted MRI demonstrated epidural tumor involves the left lateral recess of the thecal sac and involvement of the left pedicle causing instability and nerve root compression. (**e**, **f**) Postoperative X-rays following percutaneously placed pedicle screws and a mini-open laminectomy and a left facetectomy to decompress the exiting nerve root

rylate (PMMA), an acrylic bone cement. Polymerization of PMMA results in an exothermic reaction within the vertebral body which may also contribute to pain relief secondary to necrosis of tumor tissue [34].

Percutaneous cement augmentation can be performed in the form of vertebroplasty or kyphoplasty. In vertebroplasty, cement is directly injected into the vertebral body under fluoroscopic visualization. In kyphoplasty, a balloon tamp is inserted into the vertebral body and inflated to create a cavity, followed by injection of cement. The intention of the cavity creation is to allow for an area of low-pressure cement filling and for potential expansion of the vertebral body endplates for restoration of height [35]. Studies have demonstrated a slight advantage in kyphoplasty in terms of reduced rates of cement extravasation and improvement in the restoration of vertebral body height, although there was no significant difference noted in pain or disability score improvements [36]. An alternative to cavity creation with a balloon tamp is the use of a directional osteotome, which creates tracts within the vertebral body allowing cement to fill across the vertebral body through a unilateral approach [37] (Fig. 48.3).

Despite being a percutaneous procedure, there is the potential for significant complications. Cement extravasation can occur in the presence of a cortical breach. Cement extravasation rates are higher for pathologic fractures compared to osteoporotic fractures, likely due to the lytic nature of the metastatic disease. Leakage rates can be as high as 37.9% and 13.6% in vertebroplasty and kyphoplasty for metastatic spine disease, respectively, although symptomatic leaks are typically around 1% [31]. Cement extravasation into the vasculature is a rare but potentially devastating complication of cement augmentation. Most vascular extravasations are asymptomatic and remain in the distal vasculature

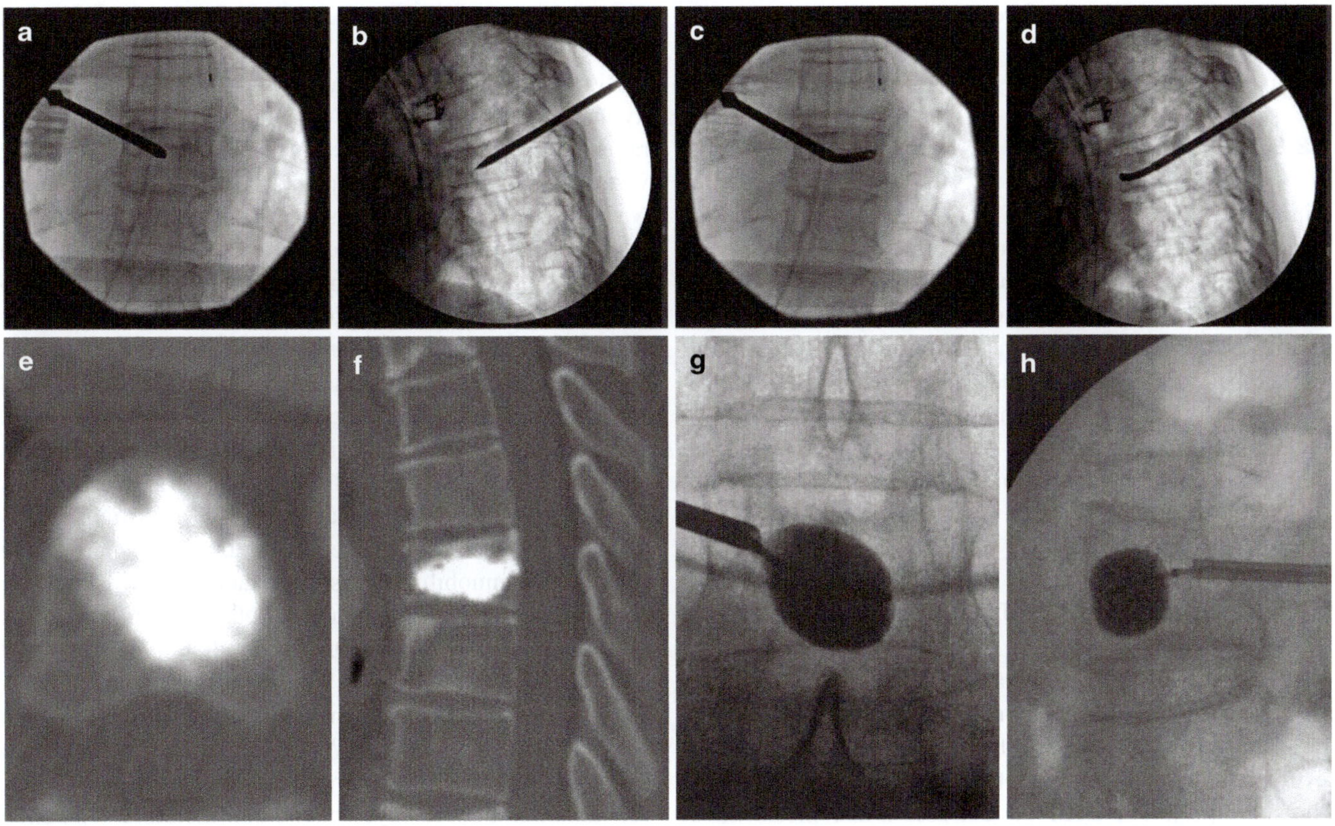

Fig. 48.3 (**a–d**) Intraoperative fluoroscopy imaging demonstrating cannulation of the vertebral body with a Jamishidi needle (**a**, **b**), followed by extention of a directional osteotome to create a tract to the contralateral vertebral body (**d**, **e**). (**e**, **f**) Postoperative CT scan demonstrating PMMA cement after kyphoplasty. (**g**, **h**) Intraoperative fluoroscopy imaging demonstrating expansion of a directional balloon tamp in the vertebral body for midline cavity creation prior to cement placement

near the vertebral bodies, but pulmonary embolism and cardiac thrombus have been reported [38–40].

48.4.3 Selection of Percutaneous Techniques

The introduction of minimally invasive techniques has allowed spinal stabilization of metastatic spine disease and symptomatic relief through percutaneous strategies. Among the challenges is determining the optimal percutaneous strategy and whether cement augmentation alone will provide adequate stability and pain relief, or whether the addition of percutaneous pedicle screw fixation is needed. Although one of the components of the SINS score is the presence of unilateral or bilateral involvement of the posterior elements, it does not specifically address what comprises of tumor involvement. The presence of tumor involvement visualized as signal changes in the posterior elements on MRI may not necessarily correlate with loss of bony integrity resulting in instability. CT imaging is therefore critical in properly assessing the degree of bony destruction of the posterior elements, particularly within the pedicles and facet joints. The presence of lytic lesions in these areas is a strong indication that instrumentation is required for stabilization of the posterior column [41]. Another indication for instrumented fixation is the presence of mechanical radiculopathy, which is radiculopathy present upon weight bearing and relieved with rest. This occurs due to nerve root compression secondary to instability in the posterior elements as axial loading is applied and can often be alleviated with stabilization.

48.5 Percutaneous Ablative Procedures

48.5.1 Radiofrequency Ablation

Radiofrequency ablation (RFA) is a percutaneous technique that uses high-frequency alternative current to generate heat to damage tumor cells at a temperature of 60–100 °C [42]. This can be used along or in conjunction with cement augmentation, prior to cement delivery [43]. RFA is primarily used for the treatment of osteolytic lesions as sclerotic lesions possess higher levels of impedance which prevents reaching cytotoxic temperatures [44]. RFA combined with cement augmentation has demonstrated an improvement in pain as well as disability score, with local tumor control of up to 70% [43, 45].

48.5.2 Cryoablation

Similar to RFA, cryoblation is a percutaneous technique which results in cytotoxic tumor effect although through cold temperatures. Using liquid argon, a cryoprobe that is placed into the tumor undergoes a rapid freeze, thaw, and freeze cycle, which results in the formation of an enlarging ice ball. This results in the creation of an osmotic gradient which leads to cell membrane injury and death [42]. Cryoablation is preferred for metastatic tumors with large soft tissue components and preferred over RFA for osteoblastic lesions.

48.5.3 Laser Interstitial Thermal Therapy

Laser interstitial thermal therapy (LITT) is a technique of percutaneous thermal ablation initially developed for the treatment of malignant brain tumors [46]. This therapy has evolved to include metastatic tumors of the brain as well as the spine [47]. The LITT system uses a nanometer wavelength diode that emits heat circumferentially or directionally, inducing DNA damage and tissue necrosis in the applied area. Thermal ablation is applied, with multiple cycles of heat application under direct visualization with MRI thermal imaging to track temperature changes within the target region. LITT has been used as an alternative to SRS in patients with epidural spinal cord compression. A recent review from a single cancer center demonstrated an 8-month local tumor control rate of 81.7% with LITT for varying grades of epidural spinal cord compression in a cohort of 120 patients [48].

48.6 Radiation Therapy

Although not a form of surgical intervention, radiation therapy has nonetheless become an important interventional adjunct and has significantly shaped the surgical strategies employed against metastatic spine disease and therefore deserves mention here. As noted previously, the understanding of sensitivity to radiation therapy of different tumor histologies and the emergence of conformal radiation in the form of stereotactic radiosurgery, has allowed the evolution of surgical interventions to be tailored toward neural decompression and spinal stabilization through more minimally invasive approaches which are better suited for oncologic patients.

Traditional radiation therapy involves conventional external beam radiation therapy (cEBRT), which delivers wide-field radiation in a fractionated dose, commonly 30 Gray in 10 fractions. Fractionation allows tissue to repair after each session, allowing for recovery from post-procedural toxicity and a greater cumulative dose delivery. cEBRT delivers radiation to a large nonspecific field, often encompassing vertebral body levels above and below the intended target, as well as surrounding normal tissue. The presence of the spinal cord and other adjacent organs at risk limits the radiation dosage. Studies have demonstrated that responsiveness to cEBRT depends on the histology of the tumor being treated, with lymphoma, seminoma, myeloma, and prostate demonstrating the greatest radiosensitivity, and the majority of solid tumors demonstrating varying degrees of radioresistance [8].

The need to overcome this clinical challenge led to the emergence of stereotactic radiosurgery (SRS) , which allows for the delivery of high dosage radiation in a tightly conformal three-dimensional area in the brain or spine. In the spine, rigid immobilization is important to allow for precise dose delivery and often requires the placement of the patient in an immobilizing frame. Once the radiation delivery path has been simulated, the machine can deliver ionizing photon radiation in a three-dimensional plane to within 1–2 mm of the calculated margin. While extremely accurate, the resultant small overspill margin puts the spinal cord a risk for radiation-induced myelopathy, and therefore separation surgery is needed to provide adequate clearance of critical neural structures [49]. SRS has been shown to have much better local control rates compared to EBRT for metastatic epidural spinal cord compression. Single high-dose SRS administration of 24 Gy has reported 1-year local control rates of approximately 95% regardless of tumor histology [50]. Lower dose rates, such as 14–18 Gy, have a reported local control rate of 90% [51]. The 1-year local control rate with EBRT has ranged from 61 to 86%, but can be as low as 46% for radioresistant solid tumors [52]. Similar to SRS is fractionated stereotactic radiotherapy (FSRT), which utilizes the volumetric specificity of SRS to deliver contoured doses over multiple fractions (24–30 Gy in 3–5 fractions). The benefit of FSRT is the high dosage delivery with a decreased risk of injuring adjacent structures. High-dose FSRT is associated with a 4% local progression rate after 1 year, compared with 22% in low-dose hypo-fractionated SRS across all tumor histologies [10].

48.6.1 Intraoperative Radiation Therapy

An alternative or supplement to traditional external radiation therapy is the use of intraoperative radiation therapy or brachytherapy to the tumor itself or to peri-tumoral structures. These procedures are typically performed in the operating room in conjunction with open procedures and allow for more direct application of radiation to minimize toxicity to nearby critical structures. One type is plaque brachytherapy, which has been utilized with ^{192}Ir- and ^{90}Y-based sources with a polycarbonate backing [53]. Folkert et al. used a more flexible silicone-coated plaque chemically bound with ^{32}P [54]. The plaques were applied to the dural surface following separation surgery to reduce local recurrence. The authors noted that ^{32}P plaques could directly deliver radiation without risk to nearby critical structures and could serve as an effective adjunct to radiation therapy in patient who has already received failed previous radiation.

More recently percutaneous intraoperative radiation therapy combined with kyphoplasty has been tested for localized delivery of radiation therapy along with cement augmentation (Kypho-IORT). A phase I/II trial demonstrated improvement in pain score along with a >90% local progression-free survival [55].

48.7 Conclusion and Perspective

- Surgical management for metastasis to the spine requires an understanding of the unique requirements of the oncologic patient.
- The NOMS framework considers the neurologic, oncologic, mechanical, and systemic status of the patient to determine the optimal treatment for spinal metastases.
- Separation surgery allows for neural decompression without significant reconstruction of the spine, combined with conformal radiation therapy to achieve effective local tumor control.
- The application of minimally invasive techniques in spinal oncology has provided strategies that achieve surgical goals of care with decreased morbidity and rapid resumption of oncologic care.

Open Questions

- Can molecular profiling of tumor and targeted therapies be incorporated into decision-making for surgical treatment of spinal metastasis to minimize intervention?
- Can endoscopic approaches be incorporated into decompression for metastatic disease?
- Can intraoperative radiation techniques provide adequate long-term control to obviate the need for external radiation therapy?

References

1. Bollen L, et al. Prognostic factors associated with survival in patients with symptomatic spinal bone metastases: a retrospective cohort study of 1 043 patients. Neuro-Oncology. 2014;16(7):991–8.
2. Groenen KHJ, et al. The Dutch national guideline on metastases and hematological malignancies localized within the spine; a multidisciplinary collaboration towards timely and proactive management. Cancer Treat Rev. 2018;69:29–38.
3. Leithner A, et al. Predictive value of seven preoperative prognostic scoring systems for spinal metastases. Eur Spine J. 2008;17(11):1488–95.
4. Tomita K, et al. Surgical strategy for spinal metastases. Spine (Phila Pa 1976). 2001;26(3):298–306.
5. Tokuhashi Y, et al. A revised scoring system for preoperative evaluation of metastatic spine tumor prognosis. Spine (Phila Pa 1976). 2005;30(19):2186–91.
6. Laufer I, et al. The NOMS framework: approach to the treatment of spinal metastatic tumors. Oncologist. 2013;18(6):744–51.
7. Bilsky MH, et al. Reliability analysis of the epidural spinal cord compression scale. J Neurosurg Spine. 2010;13(3):324–8.
8. Gerszten PC, Mendel E, Yamada Y. Radiotherapy and radiosurgery for metastatic spine disease: what are the options, indications, and outcomes? Spine (Phila Pa 1976). 2009;34(22 Suppl):S78–92.
9. Lovelock DM, et al. Correlation of local failure with measures of dose insufficiency in the high-dose single-fraction treatment of bony metastases. Int J Radiat Oncol Biol Phys. 2010;77(4):1282–7.
10. Laufer I, et al. Local disease control for spinal metastases following "separation surgery" and adjuvant hypofractionated or high-dose single-fraction stereotactic radiosurgery: outcome analysis in 186 patients. J Neurosurg Spine. 2013;18(3):207–14.
11. Fisher CG, et al. A novel classification system for spinal instability in neoplastic disease: an evidence-based approach and expert consensus from the Spine Oncology Study Group. Spine (Phila Pa 1976). 2010;35(22):E1221–9.
12. Newman WC, Laufer I, Bilsky MH. Neurologic, oncologic, mechanical, and systemic and other decision frameworks for spinal disease. Neurosurg Clin N Am. 2020;31(2):151–66.
13. Young RF, Post EM, King GA. Treatment of spinal epidural metastases. Randomized prospective comparison of laminectomy and radiotherapy. J Neurosurg. 1980;53(6):741–8.
14. Gilbert RW, Kim JH, Posner JB. Epidural spinal cord compression from metastatic tumor: diagnosis and treatment. Ann Neurol. 1978;3(1):40–51.
15. Patchell RA, et al. Direct decompressive surgical resection in the treatment of spinal cord compression caused by metastatic cancer: a randomised trial. Lancet. 2005;366(9486):643–8.
16. Steffee AD, Biscup RS, Sitkowski DJ. Segmental spine plates with pedicle screw fixation. A new internal fixation device for disorders of the lumbar and thoracolumbar spine. Clin Orthop Relat Res. 1986;203:45–53.
17. Demura S, et al. Total en bloc spondylectomy for spinal metastases in thyroid carcinoma. J Neurosurg Spine. 2011;14(2):172–6.
18. Cloyd JM, et al. En bloc resection for primary and metastatic tumors of the spine: a systematic review of the literature. Neurosurgery. 2010;67(2):435–44; discussion 444–5.
19. Al-Omair A, et al. Surgical resection of epidural disease improves local control following postoperative spine stereotactic body radiotherapy. Neuro-Oncology. 2013;15(10):1413–9.
20. Bilsky MH, et al. Single-stage posterolateral transpedicle approach for spondylectomy, epidural decompression, and cir-

cumferential fusion of spinal metastases. Spine (Phila Pa 1976). 2000;25(17):2240–9; discussion 250.
21. Pennington Z, et al. Minimally invasive versus conventional spine surgery for vertebral metastases: a systematic review of the evidence. Ann Transl Med. 2018;6(6):103.
22. Than KD, et al. Complication rates associated with open versus percutaneous pedicle screw instrumentation among patients undergoing minimally invasive interbody fusion for adult spinal deformity. Neurosurg Focus. 2017;43(6):E7.
23. Wang H, et al. Comparison of open versus percutaneous pedicle screw fixation using the sextant system in the treatment of traumatic thoracolumbar fractures. Clin Spine Surg. 2017;30(3):E239–e246.
24. Moussazadeh N, et al. Short-segment percutaneous pedicle screw fixation with cement augmentation for tumor-induced spinal instability. Spine J. 2015;15(7):1609–17.
25. Barzilai O, et al. Utility of cement augmentation via percutaneous fenestrated pedicle screws for stabilization of cancer-related spinal instability. Oper Neurosurg (Hagerstown). 2019;16(5):593–9.
26. Versteeg AL, et al. Complications after percutaneous pedicle screw fixation for the treatment of unstable spinal metastases. Ann Surg Oncol. 2016;23(7):2343–9.
27. Barzilai O, Bilsky MH, Laufer I. The role of minimal access surgery in the treatment of spinal metastatic tumors. Global Spine J. 2020;10(2 Suppl):79s–87s.
28. Dousset V, et al. Asymptomatic cervical haemangioma treated by percutaneous vertebroplasty. Neuroradiology. 1996;38:392–4.
29. Amar AP, et al. Percutaneous transpedicular polymethylmethacrylate vertebroplasty for the treatment of spinal compression fractures. Neurosurgery. 2001;49(5):1105–14; discussion 1114–5.
30. Berenson J, et al. Balloon kyphoplasty versus non-surgical fracture management for treatment of painful vertebral body compression fractures in patients with cancer: a multicentre, randomised controlled trial. Lancet Oncol. 2011;12(3):225–35.
31. Sorensen ST, et al. Vertebroplasty or kyphoplasty as palliative treatment for cancer-related vertebral compression fractures: a systematic review. Spine J. 2019;19(6):1067–75.
32. Lee MJ, et al. Percutaneous treatment of vertebral compression fractures: a meta-analysis of complications. Spine (Phila Pa 1976). 2009;34(11):1228–32.
33. Stangenberg M, et al. Cervical vertebroplasty for osteolytic metastases as a minimally invasive therapeutic option in oncological surgery: outcome in 14 cases. Neurosurg Focus. 2017;43(2):E3.
34. Lieberman IH, Togawa D, Kayanja MM. Vertebroplasty and kyphoplasty: filler materials. Spine J. 2005;5(6 Suppl):305S–16S.
35. Lieberman IH, et al. Initial outcome and efficacy of "kyphoplasty" in the treatment of painful osteoporotic vertebral compression fractures. Spine (Phila Pa 1976). 2001;26(14):1631–8.
36. Chang X, et al. Vertebroplasty versus kyphoplasty in osteoporotic vertebral compression fracture: a meta-analysis of prospective comparative studies. Int Orthop. 2015;39(3):491–500.
37. Dalton BE, et al. Radiofrequency-targeted vertebral augmentation versus traditional balloon kyphoplasty: radiographic and morphologic outcomes of an ex vivo biomechanical pilot study. Clin Interv Aging. 2012;7:525–31.
38. Krueger A, et al. Management of pulmonary cement embolism after percutaneous vertebroplasty and kyphoplasty: a systematic review of the literature. Eur Spine J. 2009;18(9):1257–65.
39. Wang LJ, et al. Pulmonary cement embolism associated with percutaneous vertebroplasty or kyphoplasty: a systematic review. Orthop Surg. 2012;4(3):182–9.
40. Gosev I, et al. Right ventricular perforation and pulmonary embolism with polymethylmethacrylate cement after percutaneous kyphoplasty. Circulation. 2013;127(11):1251–3.
41. Barzilai O, et al. Minimal access surgery for spinal metastases: prospective evaluation of a treatment algorithm using patient-reported outcomes. World Neurosurg. 2018;120:e889–901.
42. Tomasian A, et al. Percutaneous thermal ablation of spinal metastases: recent advances and review. AJR Am J Roentgenol. 2018;210(1):142–52.
43. Wallace AN, et al. Radiographic local control of spinal metastases with percutaneous radiofrequency ablation and vertebral augmentation. AJNR Am J Neuroradiol. 2016;37(4):759–65.
44. Singh S, Saha S. Electrical properties of bone. A review. Clin Orthop Relat Res. 1984;186:249–71.
45. Reyes M, et al. Multicenter clinical and imaging evaluation of targeted radiofrequency ablation (t-RFA) and cement augmentation of neoplastic vertebral lesions. J Neurointerv Surg. 2018;10(2):176–82.
46. Sloan AE, et al. Results of the NeuroBlate System first-in-humans Phase I clinical trial for recurrent glioblastoma: clinical article. J Neurosurg. 2013;118(6):1202–19.
47. Tatsui CE, et al. Spinal laser interstitial thermal therapy: a novel alternative to surgery for metastatic epidural spinal cord compression. Neurosurgery. 2016;79(Suppl 1):S73–82.
48. Bastos DCA, et al. Spinal laser interstitial thermal therapy: single-center experience and outcomes in the first 120 cases. J Neurosurg Spine. 2020:1–10.
49. Yamada Y, et al. The impact of histology and delivered dose on local control of spinal metastases treated with stereotactic radiosurgery. Neurosurg Focus. 2017;42(1):E6.
50. Moulding HD, et al. Local disease control after decompressive surgery and adjuvant high-dose single-fraction radiosurgery for spine metastases. J Neurosurg Spine. 2010;13(1):87–93.
51. Yamada Y, et al. High-dose, single-fraction image-guided intensity-modulated radiotherapy for metastatic spinal lesions. Int J Radiat Oncol Biol Phys. 2008;71(2):484–90.
52. Tseng CL, et al. Spine stereotactic body radiotherapy: indications, outcomes, and points of caution. Global Spine J. 2017;7(2):179–97.
53. DeLaney TF, et al. Intraoperative dural irradiation by customized 192iridium and 90yttrium brachytherapy plaques. Int J Radiat Oncol Biol Phys. 2003;57(1):239–45.
54. Folkert MR, et al. Intraoperative 32P high-dose rate brachytherapy of the dura for recurrent primary and metastatic intracranial and spinal tumors. Neurosurgery. 2012;71(5):1003–10; discussion 1010–1.
55. Bludau F, et al. Phase I/II trial of combined kyphoplasty and intraoperative radiotherapy in spinal metastases. Spine J. 2018;18(5):776–81.

Interventional Radiology Techniques for Management of Skeletal Metastases

49

Nainesh Parikh and Altan Ahmed

Abstract

Osseous events in patients with metastatic disease can contribute to significant impairments in quality of life. Advances in minimally invasive image-guided interventions have led to increasing utilization as adjunctive, palliative, and occasionally definitive therapies for osseous metastatic disease. In this chapter, we discuss the clinical applications, technique, and outcomes of percutaneous ablation, transarterial embolization, cementoplasty, and percutaneous osteosynthesis.

Learning Objectives

- Understand the role of interventional radiology in the management of skeletal metastases
- Learn indications for image-guided interventions for skeletal metastases
- Describe technical considerations and outcomes of percutaneous ablation, transarterial embolization, cementoplasty, and percutaneous osteosynthesis

49.1 Introduction

Although the field of Interventional Radiology is relatively young, rapid advancements of minimally invasive techniques have strengthened the armamentarium of therapies available to patients [1]. Imaging modalities such as ultrasound (US), computed tomography (CT), magnetic resonance (MR), and fluoroscopy may be used separately or in combination for performing image-guided procedures. In addition to selecting an imaging modality, several techniques are available for the image-guided management of skeletal metastases. The technique or techniques utilized is dependent on multidisciplinary discussion, location and type of lesion (i.e., lytic, blastic, or mixed), and treatment intent. While transarterial embolization of a hypervascular skeletal metastasis may be utilized to reduce operative blood loss, it can also be used to palliate pain even if surgery is not an option. Similarly, percutaneous ablation might be used for disease control/definitive treatment in addition to palliation of pain, depending on the clinical scenario. Finally, nonoperative percutaneous consolidation (e.g., cementoplasty and/or osteosynthesis) might be utilized in patients in whom surgery is not feasible to improve quality of life [2, 3].

Ultimately the goals of intervention, the ability to perform the intervention safely and with image guidance, the possible requirement of general anesthesia, and the suitability of an image-guided intervention compared to other types of nonoperative intervention (such as systemic or radiotherapy) indicate which type of nonoperative management is best for the patient. This chapter reviews the types of image-guided therapies, indications, and outcomes of these therapies in the management of skeletal metastases.

49.2 Percutaneous Ablation

Percutaneous ablation is performed by using a single or combination guidance modality such as US, CT, or MRI to place a needle or probe into a bony structure to deliver a cytotoxic substance or thermal energy and cause cellular destruction. Various substances and technologies are available including ethanol, radiofrequency (RF), microwave (MW), and cryotherapy. Thermal ablation refers to the direct application of either extreme heat (>50 °C) or extreme cold (between −20 and −40 °C) to the target tissue. The choice of substance/

N. Parikh (✉) · A. Ahmed
Department of Diagnostic Imaging and Interventional Radiology, H. Lee Moffitt Cancer Center and Research Institute, Tampa, FL, USA
e-mail: Nainesh.Parikh@moffitt.org; Altan.Ahmed@moffitt.org

S. P. Leong et al. (eds.), *Cancer Metastasis Through the Lymphovascular System*,
https://doi.org/10.1007/978-3-030-93084-4_49

technology depends on the underlying histology, morphology, and the location of adjacent critical structures. RF and cryotherapy are the most frequently utilized forms of ablation for skeletal metastases. RF ablation works by passing electrical current in the range of radiofrequency waves between the electrodes. This current produces heat around and between the electrodes, resulting in tissue destruction. Cryoablation works by cycling inert gas down the shaft of a probe, rapidly producing ice and locally freezing the tissue. By cycling through freezing and thawing phases, the cells are disrupted to such an extent that they are destroyed (Fig. 49.1) [2–4].

Ablation procedures are typically performed under CT guidance and with either general anesthesia or moderate sedation (fentanyl and midazolam) depending on the patient performance status and associated comorbidities, as well as the planned procedure. A detailed neurologic examination and pain survey are paramount prior to and following the procedure. If the target ablation site is close to the skin, warm or cold saline compresses can be applied for protection. Air, fluid, or catheter-mounted balloons may also be utilized to displace critical structures in the vicinity of the planned ablation.

Because the destructive properties of ablation are limited to the tissue adjacent to and/or in between probes, it becomes challenging to provide definitive tumor control for large tumors. Small tumors, typically less than 5 cm, are ideally treated with ablative technology with the goals of local control and/or palliation of symptoms. If "cancer pain" is felt to be secondary to lytic metastasis, ablative technologies are helpful for palliation. Most ablative procedures can be performed on an outpatient basis, with the option to monitor patients overnight for pain control.

While skeletal metastases often imply advanced stage malignancy, oligometastatic disease (up to five sites) may represent an opportunity to achieve tumor control. While the data are challenging to interpret due to patient heterogeneity and confounding effect of other therapies, one study reports median overall survival of 47 months and median disease-free survival of 7 months after osseous ablation [5]. Prognostic factors that influence the outcome of percutaneous ablation include the size and number of metastases, length of disease-free survival before the onset of metastases, adequacy of treatment of the primary tumor, and the presence of multiple metastatic sites [6, 7].

All ablation modalities are thought to relieve pain via similar multifactorial processes: the destruction of sensory fibers that supply the periosteum, decompression of tumor volume, eradication of cytokine-producing tumor cells, and inhibition of osteoclast activity [8]. Reported outcomes vary; however, most describe significant improvement in bone-related pain refractory to standard therapy [9, 10].

49.3 Transarterial Embolization

Many tumors demonstrate neoangiogenesis with varying degrees of vascularity. Metastases from melanoma, renal cell, thyroid, breast, and hepatocellular carcinoma are considered among the most vascular. TAE aims to devascularize tumors by blocking arterial blood flow to the capillary bed. The most common application of TAE is as a surgical adjunct to reduce perioperative blood loss. Tumor histology, type of surgery planned, and surgeon preference are factors that should be considered in deciding whether adjunctive TAE should be performed. TAE may also be applied as a palliative intervention for symptomatic metastases refractory or not amenable to other treatments.

A thorough understanding of regional vascular anatomy is essential to safe and efficacious TAE. Pre-procedural cross-sectional imaging is useful to delineate tumor anatomy and potential sources of vascularization. In some cases, there

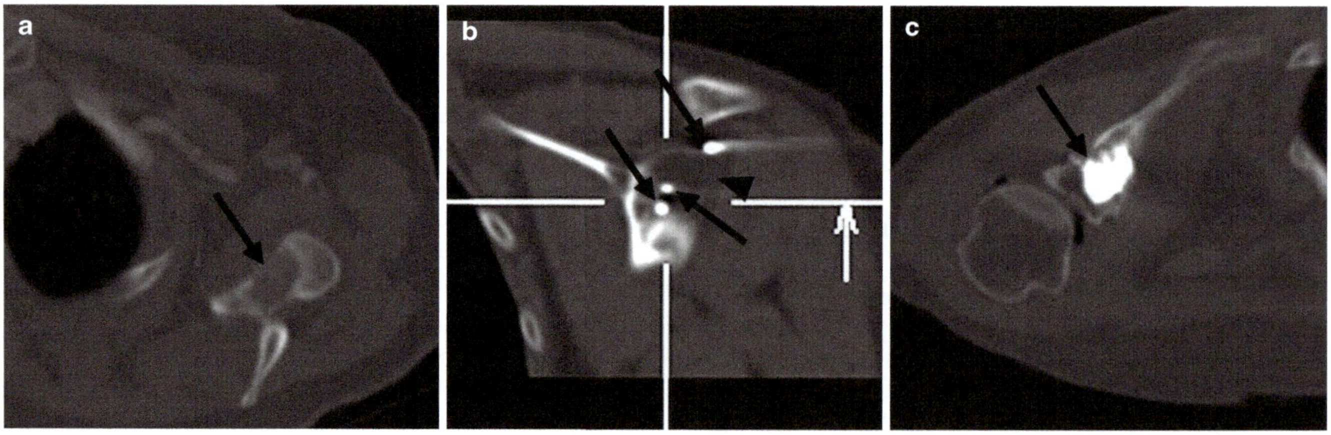

Fig. 49.1 CT demonstrates a lytic renal cell carcinoma left scapular metastasis (arrow) (**a**). CT-guided cryoablation with three probes (arrows) was performed. The arrowhead marks the hypodense ablation zone (**b**). Cementoplasty was subsequently performed due to suboptimal pain response to ablation alone (arrow) (**c**)

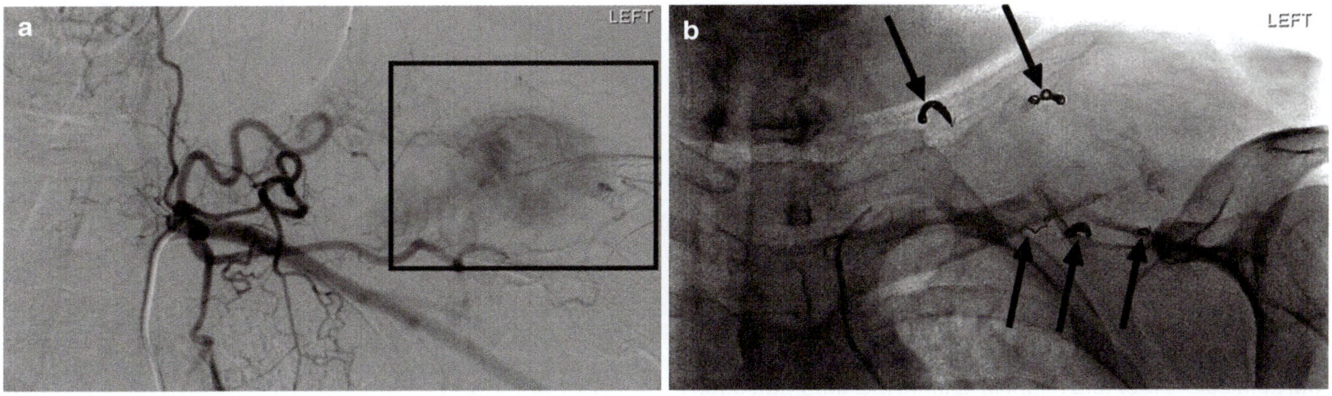

Fig. 49.2 Digital subtraction angiogram demonstrates a vascular renal cell carcinoma metastasis involving the lateral left clavicle (rectangle) with arterial supply arising from branches of the thyrocervical trunk (**a**). The vascular supply was coiled (arrows) without significant residual tumor enhancement (**b**)

may discordance between vascularity identified on cross-sectional imaging versus that seen angiographically [11]. Contraindications to angiography and embolization include uncorrectable coagulopathy, underlying infection, renal failure, or inability to isolate tumoral blood supply from that of critical tissue.

TAE is typically performed under conscious sedation. Arterial access is typically obtained via the femoral or radial artery. After placing a vascular sheath, a 4–5f diagnostic catheter is maneuvered over a guidewire to the area of interest. An initial non-selective angiogram is performed to delineate vascular anatomy. It is critical to identify blood supply to both tumor and normal tissue as embolization of nontarget vessels can result in devastating clinical outcomes such as sensorimotor deficits and tissue or limb loss. Once the target vessel or vessels are identified, a coaxial microcatheter is typically advanced as distal as possible from the parent vessel to reduce the risk of nontarget embolization. After confirming catheter position, embolic material is administered in small aliquots under fluoroscopic observation (Fig. 49.2). The angiographic end point of embolization in this setting has been described as ≥70% reduction in tumor staining [12, 13]. Post-embolization syndrome characterized by fever, fatigue, pain, and/or nausea/vomiting is common after TAE and is usually self-limited.

Embolic materials are categorized as either temporary or permanent agents. The primary temporary agent is gelatin sponge. Permanent embolic materials include particulates (e.g., polyvinyl alcohol (PVA) and microspheres), coils, and liquid embolics (e.g., N-butyl cyanoacrylate and ethylene vinyl alcohol). Overall, liquid materials afford greater penetration into the target vascular bed but require greater operator experience and may carry a higher risk of nontarget embolization. A detailed discussion of each embolic agent is beyond the scope of this text. The choice of embolic material is largely dependent on vascular anatomy, presence of arteriovenous shunting, risk of nontarget embolization, and operator experience. There is insufficient data to support the use of one embolic material over another.

Multiple studies have demonstrated the safety of transarterial embolization of skeletal metastases [11, 13–16]. In addition, data suggest adjunctive TAE is associated with lower operative blood loss [13, 14, 17–21]. There are no consistent data demonstrating significant differences in operative time between patients who undergo adjunctive embolization and those who do not [19]. The optimal interval between TAE and surgery has been reported as <3 days to reduce the formation of collateral circulation [17, 18, 22]. Further investigation is needed to refine selection criteria for patients who may benefit from preoperative TAE of skeletal metastases [23, 24].

In palliative settings, the mean duration of pain improvement has been reported in the range of 1–15 months [15, 17, 25, 26]. Combination therapy with external beam therapy may provide more durable pain relief [26].

49.4 Percutaneous Cementoplasty and Osteosynthesis

Pathologic fractures, pain, and decreased mobility are major morbidities of osseous metastatic disease [27]. Cementoplasty (also referred to as vertebroplasty or kyphoplasty in the spine) involves the injection of polymethyl methacrylate (PMMA) into bone to palliate painful metastases and/or prevent pathologic fracture (Fig. 49.1). In areas where torsional forces are prominent (e.g., long bones), the consolidative effect of cementoplasty alone is limited despite pain reduction and may be supplemented by percutaneous osteosynthesis when appropriate [28]. The analgesic effect of cementoplasty is predominantly thought to arise from fracture consolidation. The exothermic polymerization reaction of PMMA may also contribute to analgesic effect through thermal injury to local nociceptive fibers [29]. Patients with limited osseous metastatic disease may benefit from a combined approach to treat the tumor with thermal ablation, radiation, or systemic therapy followed by cementoplasty to stabilize the bone [30].

Understanding whether compressive or torsional forces predominate in the area of concern is a key component of pre-procedural assessment in addition to quantifying baseline pain and analgesic use. Cross-sectional imaging is essential prior to intervention to establish the safest access and delineate tumor and regional anatomy. Contraindications include uncorrectable coagulopathy, infection, allergy to cement components, neurologic compromise, sagittal spinal instability. Cementoplasty may be performed under conscious sedation if a limited level vertebroplasty is planned. However, general anesthesia is often necessary if patient status dictates and if or multiple sites or types of intervention are planned. In the spine, the most common access routes are transpedicular for the lumbar and lower thoracic, intercostopedicular for mid and upper thoracic, and anterior or transoral for the cervical vertebrae. Unilateral or bilateral access may be obtained. Outside the spine, access routes vary considerably depending on the target bone. A bone access needle is advanced into the bone along the chosen tract under fluoroscopic and/or CT guidance. Once appropriate position is confirmed, PMMA is injected under observation to ensure appropriate filling and to monitor for extravasation or intravasation [31–35].

Percutaneous osteosynthesis utilizes similar principles of bone access. A coaxial needle is placed across a planned trajectory followed by the advancement of a Kirschner wire. Alternatively, a Kirschner wire may be placed directly. After measuring the required length, the appropriate cannulated screw is advanced over the wire which is subsequently removed. If percutaneous osteosynthesis is to be performed at the same time as cementoplasty, it can be done prior to the cement setting using coaxial technique through the existing bone access or prior to cementoplasty using separate access [28].

Cementoplasty in the spine consistently demonstrates a durable pain response in the setting of spinal metastatic disease [32, 33]. In subset analysis of a prospective series, 644 patients with vertebral metastatic disease reported a significant pain reduction from 8.3 to 2.4 on an 11-point visual analog scale (VAS) [36]. There is growing evidence of similar pain response in extra-spinal cementoplasty [37–39]. In a series of 27 patients with painful osseous metastases or impending pathologic fracture who underwent percutaneous osteosynthesis, Pusceddu et al. reported a mean decrease in VAS score from 8 pre-treatment to 1.6 and 1.4 one and six months posttreatment, respectively [40]. Similar findings have been reported in other studies [37, 41–44].

49.5 Conclusion

- Percutaneous image-guided therapies comprise a number of minimally invasive and efficacious interventions for osseous metastatic disease
- Regardless of technique or techniques selected, a multidisciplinary approach to patient selection is essential for successful outcomes

Open Questions

Future controlled studies are needed to determine:
- Which patients would benefit most from minimally invasive management of bone metastases?
- Do differences exist between embolization materials or ablation modalities?
- What is the role of percutaneous techniques for stabilization of pathologic fractures?

References

1. Murphy TP, Soares GM. The evolution of interventional radiology. Semin Intervent Radiol. 2005;22:6–9. https://doi.org/10.1055/s-2005-869570.
2. Barile A, Arrigoni F, Zugaro L, et al. Minimally invasive treatments of painful bone lesions: state of the art. Med Oncol. 2017;34:53. https://doi.org/10.1007/s12032-017-0909-2.
3. Munk PL, Legiehn GM. Musculoskeletal interventional radiology: applications to oncology. Semin Roentgenol. 2007;42:164–74. https://doi.org/10.1053/j.ro.2007.04.004.
4. Errani C, Bazzocchi A, Spinnato P, et al. What's new in management of bone metastases? Eur J Orthop Surg Traumatol. 2019;29:1367–75. https://doi.org/10.1007/s00590-019-02446-y.
5. McMenomy BP, Kurup AN, Johnson GB, et al. Percutaneous cryoablation of musculoskeletal oligometastatic disease for complete remission. J Vasc Interv Radiol. 2013;24:207–13. https://doi.org/10.1016/j.jvir.2012.10.019.
6. Nicholas Kurup A, Callstrom MR. Ablation of musculoskeletal metastases: pain palliation, fracture risk reduction, and oligometastatic disease. Tech Vasc Interv Radiol. 2013;16:253–61. https://doi.org/10.1053/j.tvir.2013.08.007.
7. Weichselbaum RR, Hellman S. Oligometastases revisited. Nat Rev Clin Oncol. 2011;8:378–82. https://doi.org/10.1038/nrclinonc.2011.44.
8. Callstrom MR, Charboneau JW, Goetz MP, et al. Painful metastases involving bone: feasibility of percutaneous CT- and US-guided radio-frequency ablation. Radiology. 2002;224:87–97. https://doi.org/10.1148/radiol.2241011613.
9. Dupuy DE, Liu D, Hartfeil D, et al. Percutaneous radiofrequency ablation of painful osseous metastases: a multicenter American College of Radiology Imaging Network trial. Cancer. 2010;116:989–97. https://doi.org/10.1002/cncr.24837.
10. Goetz MP, Callstrom MR, Charboneau JW, et al. Percutaneous image-guided radiofrequency ablation of painful metastases involving bone: a multicenter study. J Clin Oncol. 2004;22:300–6. https://doi.org/10.1200/JCO.2004.03.097.

11. Prabhu VC, Bilsky MH, Jambhekar K, et al. Results of preoperative embolization for metastatic spinal neoplasms. J Neurosurg. 2003;98:156–64. https://doi.org/10.3171/spi.2003.98.2.0156.
12. Kwon JH, Shin JH, Kim JH, et al. Preoperative transcatheter arterial embolization of hypervascular metastatic tumors of long bones. Acta Radiol. 2010;51:396–401. https://doi.org/10.3109/02841851003660081.
13. Sun S, Lang EV. Bone metastases from renal cell carcinoma: preoperative embolization. J Vasc Interv Radiol. 1998;9:263–9. https://doi.org/10.1016/S1051-0443(98)70267-2.
14. Pazionis TJC, Papanastassiou ID, Maybody M, Healey JH. Embolization of hypervascular bone metastases reduces intraoperative blood loss: a case-control study. Clin Orthop Relat Res. 2014;472:3179–87. https://doi.org/10.1007/s11999-014-3734-3.
15. Radeleff B, Eiers M, Lopez-Benitez R, et al. Transarterial embolization of primary and secondary tumors of the skeletal system. Eur J Radiol. 2006;58:68–75. https://doi.org/10.1016/j.ejrad.2005.12.008.
16. Rossi G, Mavrogenis AF, Rimondi E, et al. Selective embolization with N-butyl cyanoacrylate for metastatic bone disease. J Vasc Interv Radiol. 2011;22:462–70. https://doi.org/10.1016/j.jvir.2010.12.023.
17. Barton PP, Waneck RE, Karnel FJ, et al. Embolization of bone metastases. J Vasc Interv Radiol. 1996;7:81–8. https://doi.org/10.1016/S1051-0443(96)70738-8.
18. Gellad FE, Sadato N, Numaguchi Y, Levine AM. Vascular metastatic lesions of the spine: preoperative embolization. Radiology. 1990;176:683–6. https://doi.org/10.1148/radiology.176.3.2389026.
19. Geraets SEW, Bos PK, van der Stok J. Preoperative embolization in surgical treatment of long bone metastasis: a systematic literature review. EFORT Open Rev. 2020;5:17–25. https://doi.org/10.1302/2058-5241.5.190013.
20. Ma J, Tullius T, Van Ha TG. Update on preoperative embolization of bone metastases. Semin Intervent Radiol. 2019;36:241–8. https://doi.org/10.1055/s-0039-1693120.
21. Olerud C, Jónsson H, Löfberg AM, et al. Embolization of spinal metastases reduces peroperative blood loss: 21 patients operated on for renal cell carcinoma. Acta Orthop. 1993;64:9–12. https://doi.org/10.3109/17453679308994517.
22. Gottfried ON, Schloesser PE, Schmidt MH, Stevens EA. Embolization of metastatic spinal tumors. Neurosurg Clin N Am. 2004;15:391–9. https://doi.org/10.1016/j.nec.2004.04.003.
23. Jernigan EW, Tennant JN, Esther RJ. Not all patients undergoing stabilization of impending pathologic fractures for renal cell carcinoma metastases to the femur need preoperative embolization. Clin Orthop Relat Res. 2018;476:529–34. https://doi.org/10.1007/s11999.0000000000000058.
24. Kickuth R, Waldherr C, Hoppe H, et al. Interventional management of hypervascular osseous metastasis: role of embolotherapy before orthopedic tumor resection and bone stabilization. Am J Roentgenol. 2008;191:240–7. https://doi.org/10.2214/AJR.07.4037.
25. Forauer AR, Kent E, Cwikiel W, et al. Selective palliative transcatheter embolization of bony metastases from renal cell carcinoma. Acta Oncol (Madr). 2007;46:1012–8. https://doi.org/10.1080/02841860701280725.
26. Uemura A, Fujimoto H, Yasuda S, et al. Transcatheter arterial embolization for bone metastases from hepatocellular carcinoma. Eur Radiol. 2001;11:1457–62. https://doi.org/10.1007/s003300000792.
27. Filippiadis DK, Tselikas L, Bazzocchi A, et al. Percutaneous management of cancer pain. Curr Oncol Rep. 2020;22:43. https://doi.org/10.1007/s11912-020-00906-x.
28. Deschamps F, Yevich S, Gravel G, et al. Percutaneous fixation by internal cemented screw for the treatment of unstable osseous disease in cancer patients. Semin Intervent Radiol. 2018;35:238–47. https://doi.org/10.1055/s-0038-1673359.
29. Deramond H, Wright NT, Belkoff SM. Temperature elevation caused by bone cement polymerization during vertebroplasty. Bone. 1999;25:17S–21S. https://doi.org/10.1016/S8756-3282(99)00127-1.
30. Cazzato RL, Buy X, Grasso RF, et al. Interventional radiologist's perspective on the management of bone metastatic disease. Eur J Surg Oncol. 2015;41:967–74. https://doi.org/10.1016/j.ejso.2015.05.010.
31. Anselmetti GC. Musculoskeletal interventions osteoplasty percutaneous bone cement injection beyond the spine. Semin Intervent Radiol. 2010;27:199–208.
32. Buy X, Catena V, Roubaud G, et al. Image-guided bone consolidation in oncology. Semin Intervent Radiol. 2018;35:221–8. https://doi.org/10.1055/s-0038-1669468.
33. Katsanos K, Sabharwal T, Adam A. Percutaneous cementoplasty. Semin Intervent Radiol. 2010;27:137–47. https://doi.org/10.1055/s-0030-1253512.
34. Mastier C, Gjorgjievska A, Thivolet A, et al. Musculoskeletal metastases management: the interventional radiologist's toolbox. Semin Intervent Radiol. 2018;35:281–9. https://doi.org/10.1055/s-0038-1673420.
35. Yevich S, Tselikas L, Gravel G, et al. Percutaneous cement injection for the palliative treatment of osseous metastases: a technical review. Semin Intervent Radiol. 2018;35:268–80. https://doi.org/10.1055/s-0038-1673418.
36. Anselmetti GC, Marcia S, Saba L, et al. Percutaneous vertebroplasty: multi-centric results from EVEREST experience in large cohort of patients. Eur J Radiol. 2012;81:4083–6. https://doi.org/10.1016/j.ejrad.2012.07.005.
37. Deschamps F, de Baere T, Hakime A, et al. Percutaneous osteosynthesis in the pelvis in cancer patients. Eur Radiol. 2016;26:1631–9. https://doi.org/10.1007/s00330-015-3971-1.
38. Deschamps F, Farouil G, Hakime A, et al. Cementoplasty of metastases of the proximal femur: is it a safe palliative option? J Vasc Interv Radiol. 2012a;23:1311–6. https://doi.org/10.1016/j.jvir.2012.06.027.
39. Wang Z, Zhen Y, Wu C, et al. CT fluoroscopy-guided percutaneous osteoplasty for the treatment of osteolytic lung cancer bone metastases to the spine and pelvis. J Vasc Interv Radiol. 2012;23:1135–42. https://doi.org/10.1016/j.jvir.2012.06.007.
40. Pusceddu C, Fancellu A, Ballicu N, et al. CT-guided percutaneous screw fixation plus cementoplasty in the treatment of painful bone metastases with fractures or a high risk of pathological fracture. Skelet Radiol. 2017;46:539–45. https://doi.org/10.1007/s00256-017-2584-y.
41. Amoretti N, Huwart L, Hauger O, et al. Percutaneous screw fixation of acetabular roof fractures by radiologists under CT and fluoroscopy guidance. Am J Roentgenol. 2013;200:447–50. https://doi.org/10.2214/AJR.11.7968.
42. Deschamps F, Farouil G, Hakime A, et al. Percutaneous stabilization of impending pathological fracture of the proximal femur. Cardiovasc Intervent Radiol. 2012b;35:1428–32. https://doi.org/10.1007/s00270-011-0330-8.
43. Fischer S, Vogl TJ, Marzi I, et al. Percutaneous cannulated screw fixation of sacral fractures and sacroiliac joint disruptions with CT-controlled guidewires performed by interventionalists: single center experience in treating posterior pelvic instability. Eur J Radiol. 2015;84:290–4. https://doi.org/10.1016/j.ejrad.2014.11.017.
44. Strobl FF, Haeussler SM, Paprottka PM, et al. Technical and clinical outcome of percutaneous CT fluoroscopy-guided screw placement in unstable injuries of the posterior pelvic ring. Skelet Radiol. 2014;43:1093–100. https://doi.org/10.1007/s00256-014-1890-x.

Part XIII

Cancer Metastasis to the Brain and the Nervous System

Ian Lee

Central nervous system (CNS) metastases are a common occurrence in the course of disease for cancer patients, reported to affect between 9% and 17% of all cancer patients, although this likely underestimates the true prevalence of disease [1]. Although combinatorial treatments can effectively manage disease in many patients, CNS metastases still represent a source of significant decrease in the quality of life, due to the effects of metastases as well as their treatment. In addition, a sizable portion of cancer patients will succumb due to the CNS manifestations of their cancer. Some types of cancer, like metastatic melanoma, have a particularly poor prognosis [2]. As patients are living longer with metastatic cancer due to improved treatments, it is also likely that the incidence of CNS metastases is increasing over time [3]. Taken together, CNS metastases represent a large burden of disease in cancer patients, although not necessarily a terminal condition with the advent of new treatment modalities.

CNS metastases most typically occur in the brain parenchyma. Other less common sites include the meninges, pituitary or pineal gland, and choroid plexus. Leptomeningeal dissemination of cancer through the cerebrospinal fluid carries an especially poor prognosis for most patients, as current treatments still have limited efficacy [4]. The most common primary sites of cancer are lung, breast, and melanoma. Other types of cancer such as pancreas and prostate rarely cause brain metastases. The overall prevalence of lung and breast cancer likely contributes to the high prevalence of brain metastases from those sites. In comparison, the relatively high incidence of CNS metastatic disease from certain histologies (despite their rarity) suggests a CNS tropism depending on the cell type. For instance, in melanoma patients up to 50% of patients with metastatic disease will develop CNS involvement [5]. Another example is the inverse relationship of testicular cancer, which has a high incidence of brain metastases, compared to prostate, which uncommonly spreads to the CNS [6, 7].

One of the hallmarks of CNS metastases is the simultaneous discovery of multiple CNS lesions; however, a significant minority will present with a solitary lesion [8]. Some types of cancer, like lung and melanoma, tend to present as multiple tumors. In contrast, other types such as breast, renal, and colorectal are associated with solitary brain metastases [8, 9]. Most patients will present with neurological symptoms specific to disease location (i.e., numbness, weakness, or aphasia) or with increased intracranial pressure manifesting as headache or decreased level of consciousness [10]. In addition, certain histologies like breast, lung, thyroid, renal, and melanoma tend to present as hemorrhagic lesions in the brain, presenting both diagnostic and treatment challenges. In fact, in a patient who presents with multiple intracerebral hemorrhages, metastatic cancer should be highly suspected [11].

With no treatment, prognosis is thought to be approximately 3 months. For patients with oligometastatic disease, the combination of radiation and surgery can typically control disease. However, overall prognosis remains poor, in the range of 4–6 months, likely as a result of widespread disease outside the CNS [12]. In a subset of patients with single CNS metastases and limited extracranial disease, life expectancy may be more favorable, exceeding a year [13]. Nonetheless, the development of CNS metastases is a harbinger of poor survival.

The following section will focus on the pathophysiology of CNS metastases, including the molecular mechanisms of CNS tropism. We will review the molecular cascades underpinning the spread of cancer to the CNS, which have been an active area of research. Furthermore, these mechanisms may present targets for treatment. Next, the current mainstays of treatment, radiation, and/or surgery will be discussed. The combination of radiation and surgery can typically control CNS disease, especially in cases of oligometastases. However, they do have attendant risks due to their invasive nature as well as an expected decrement in quality of life. As such, newer treatments such as stereotactic radiosurgery represent a paradigm shift in the treatment of CNS metastatic disease in that it is efficacious and (relatively) spares the surrounding normal parenchyma, thus partially preventing the cognitive effects of whole-brain radiation therapy. Surgery is typically reserved for larger tumors or those that have progressed despite radiation treatment. However, with the advent of less invasive treatments like laser interstitial thermal ablation, the surgical indications for CNS metastases have expanded.

Classically, CNS metastases are thought to be unresponsive to systemic chemotherapy due to the blood-brain barrier preventing adequate penetration of chemotherapeutics. However, this viewpoint is gradually shifting with advances in the development of chemotherapeutic agents and especially with the introduction of immunotherapy. Finally, the pathophysiology and treatment of leptomeningeal carcinomatosis, the most dreaded form of CNS metastatic disease, will also be reviewed.

References

1. Nayak L, Lee EQ, Wen PY. Epidemiology of brain metastases. Curr Oncol Rep. 2012;14:48–54. https://doi.org/10.1007/s11912-011-0203-y
2. McWilliams RR, Brown PD, Buckner JC, et al. Treatment of brain metastases from melanoma. Mayo Clin Proc. 2003;78:1529–36. https://doi.org/10.4065/78.12.1529
3. Bachmann C, Schmidt S, Staebler A, et al. CNS metastases in breast cancer patients: prognostic implications of tumor subtype. Med Oncol. 2015;32:400. https://doi.org/10.1007/s12032-014-0400-2
4. Morris PG, Reiner AS, Szenberg OR, et al. Leptomeningeal metastasis from non-small cell lung cancer: survival and the impact of whole brain radiotherapy. J Thorac Oncol. 2012;7:382–5. https://doi.org/10.1097/JTO.0b013e3182398e4f
5. Spagnolo F, Picasso V, Lambertini M, et al. Survival of patients with metastatic melanoma and brain metastases in the era of MAP-kinase inhibitors and immunologic checkpoint blockade antibodies: a systematic review. Cancer Treat Rev. 2016;45:38–45. https://doi.org/10.1016/j.ctrv.2016.03.003
6. Gandaglia G, Abdollah F, Schiffmann J, et al. Distribution of metastatic sites in patients with prostate cancer: a population-based analysis. Prostate. 2014;74:210–6. https://doi.org/10.1002/pros.22742
7. Bredael JJ, Vugrin D, Whitmore WF. Autopsy findings in 154 patients with germ cell tumors of the testis. Cancer. 1982;50:548–51. https://doi.org/10.1002/1097-0142(19820801)50:3<548::aid-cncr2820500327>3.0.co;2-2
8. Nussbaum ES, Djalilian HR, Cho KH, Hall WA. Brain metastases. Histology, multiplicity, surgery, and survival. Cancer. 1996;78:1781–8.
9. Delattre JY, Krol G, Thaler HT, Posner JB. Distribution of brain metastases. Arch Neurol. 1988;45:741–4. https://doi.org/10.1001/archneur.1988.00520310047016
10. Boire A. Metastasis to the central nervous system. CONTINUUM: Lifelong Learning in Neurology. 2020;26:1584. https://doi.org/10.1212/CON.0000000000000939
11. Boissonneau S, Graillon T, Rolland A, et al. Management of patients suffering from hemorrhagic intracranial metastases: propositions to help the neurosurgeon in emergencies situations based on a literature review. Neurochirurgie. 2020; https://doi.org/10.1016/j.neuchi.2020.11.005.
12. Johnson JD, Young B. Demographics of brain metastasis. Neurosurg Clin N Am. 1996;7:337–44.
13. Stelzer KJ. Epidemiology and prognosis of brain metastases. Surg Neurol Int. 2013;4:S192–202. https://doi.org/10.4103/2152-7806.111296

50 Brain Metastases: Overview and Molecular Mechanisms

S. Haider, J. Snyder, and I. Lee

Abstract

Both the incidence and prevalence of brain metastases (BM) have increased, partly due to improved radiographic detection of lesions and prolonged survival from primary cancers. Overall prognosis of BM remains poor despite emerging therapies. Metastasis to the central nervous system (CNS) requires that a cancer cell harbor the unique molecular machinery necessary to overcome protective CNS mechanisms ordinarily designed to impede pathogen invasion and to facilitate tissue repair. A primary tumor's ability to successfully invade and thrive in the central nervous system is dependent on subclonal evolution of cancer cell populations, with genetic changes conferring selective advantages to overcome each step in the multistep model of tumor metastasis. Organotropism refers to the predilection of circulating tumor cells to target specific secondary sites, often involving a bidirectional interaction where the primary tumor influences changes at a distant site to create an environment conducive to metastatic proliferation. Once in the brain, cancer cells hijack glial and CNS-stromal cellular machinery to help create a tumor microenvironment where cancer cells continue to thrive. A better understanding of the molecular mechanisms of BM will help drive future targeted therapy as well as identifying those at greater risk for BM.

S. Haider · J. Snyder · I. Lee (✉)
Henry Ford Health System, Detroit, USA
e-mail: jsnyder5@hfhs.org; ilee1@hfhs.org

Learning Objectives

- Appreciate the multistep model of tumor metastasis
- Intratumoral genetic heterogeneity and clonal evolution promote driver mutations favoring neurotropism
- Cancer cells establish a favorable tumor microenvironment by hijacking glial and stromal molecular machinery of the brain

50.1 Introduction

Brain metastases portend a bleak prognosis in cancer patients and negatively contribute to morbidity and mortality. Extant systemic chemotherapy options offer limited central nervous system (CNS) penetration owing to their intrinsic hydrophobicity; thus surgery or radiation often serve as the mainstay of treatment for appropriately selected patients [1] [Klotz 2020]. With 10–30% of cancer patients developing brain metastases and a median overall survival of under 1 year from the time of diagnosis, *there exists an acute need to understand underlying mechanisms of brain metastasis to develop novel therapeutic approaches*.

We outline herein a brief overview of the multi-step model of tumor spread from a primary site to the brain as well as idiosyncrasies of the central nervous system microenvironment.

S. P. Leong et al. (eds.), *Cancer Metastasis Through the Lymphovascular System*,
https://doi.org/10.1007/978-3-030-93084-4_50

50.2 The Multistep Model of Tumor Metastasis

The pathophysiology of tumor proliferation from a primary to distant site is best described as a series of sequential steps, each exerting a selective pressure among a heterogeneous population of cancer cells. The CNS microenvironment is an intrinsically privileged milieu by virtue of a tightly regulated blood–brain barrier (BBB) that resists the entry of most circulating immune and tumor cells. Thus, infiltration of the BBB plays a key role in the pathogenesis of metastatic spread. Once breached, proliferation in the CNS microenvironment still brings forth its own challenges for tumor cells. Cancer cells that have deposited in the brain employ vascular co-option and other changes to modify the BBB to a *blood–tumor barrier* (BTB) that fosters neoangiogenesis and may modify the tumor local microenvironment to a partially open state. This BTB is likely a critical factor in drug delivery and/or tumor behavior and an evolving area of research [2].

The multistep process of tumor metastasis, first described by Dr. Fidler [3], outlines eight key steps beginning with (1) tumor growth at the primary site, (2) neovascular recruitment (angiogenesis), (3) acquired motility of tumor cells, (4) lymphovascular infiltration, (5) intravasation into circulation, (6) survival in circulation, (7) extravasation out of circulation, and ultimately (8) proliferation at a distant site with secondary tumor growth [3, 4] [Fidler 1975, Hart and Fidler 1980]. The focus of subsequent discussions will be on extravasation out of circulation and proliferation within the central nervous system.

50.3 Clonal Evolution and Intermetastatic Genomic Heterogeneity

The genomic composition of brain metastases harbors key differences from cancer cells at the primary site of tumor. For instance, breast cancer with synchronous bone and brain metastases show genomic differences that lend a selective advantage for proliferation at that particular target organ [1, 5]. One of the hallmarks of cancer is genomic instability leading to dysregulation of replication and repair. The downstream result of these replicative aberrations creates genetic heterogeneity, a minority of which will contribute to mutations that offer a selective growth advantage in comparison to healthy tissue.

Many solid tumors and brain metastases exhibit a growth pattern consistent with branched evolution; a process by which subclonal populations of cancer cells diverge from a common ancestral population. The resultant intratumoral heterogeneity results in genetic variation over time (temporal heterogeneity) as well as an unbalanced metastatic distribution across distant sites (spatial heterogeneity) [5, 6]. Despite improvements in extracranial disease control with cytotoxic therapy and next-generation targeted therapy [7–9], many patients go on to develop CNS metastases, recurrence, or otherwise disproportionate response for intracranial disease burden. While the BBB is partially responsible, next-generation therapies with excellent CNS penetration still have considerable intracranial progression despite extracranial efficacy [10, 11]. Studies that have investigated the molecular makeup of cancer throughout a treatment course have further demonstrated that systemic therapy itself may lead to selective pressures that favor subclonal expansion of treatment-resistant tumors [5, 12–14]. Genomic heterogeneity is therefore a key mechanism by which tumor cells ultimately invade and proliferate in the CNS environment.

50.4 Tropism for the Central Nervous System

The population of circulating tumor cells (CTC) in the lymphovascular system also has subtle genomic and phenotypic differences compared to tumor cells of the primary site. Those differences are partly responsible for overcoming the initial barriers to tumor spread. Yet still, not all CTC are destined for CNS deposition. A growing body of evidence suggests that an additional subset of all CTC harbor the molecular machinery necessary to overcome the BBB and subsequently adapt and thrive in the brain microenvironment. Vascular co-option in the CNS is postulated to be an early initiating factor in BM when CTCs migrate to the abluminal vessel wall where they may travel in a dormant state along vasculature or establish a focus of brain invasion [15, 16]. As this BM focus grows, neoangiogenesis may ensue resulting in tortuous vessels, a hypoxic shift in the tumor microenvironment (TME), and other changes outlined below that result in the heterogeneous blood–tumor barrier that is a hallmark of BM disease [16].

50.5 Tumor Microenvironment of the Brain

The unique microenvironment of the brain is characterized by a host of nonneuronal cell types that maintain physiologic homeostasis to support neurons. The glial cells of the brain, in order of decreasing frequency, include oligodendrocytes, astrocytes, microglia, endothelial cells, pericytes, ependymal cells, and resident/recruited immune cells [17]. Under normal circumstances, perceived threats to the CNS, such as pathogens or tissue damage, initiate an inflammatory response to sequester and thwart the pathogen or aid in tissue repair. Local astrocytes and microglia represent the first line of defense and are extremely proficient at sequestering an

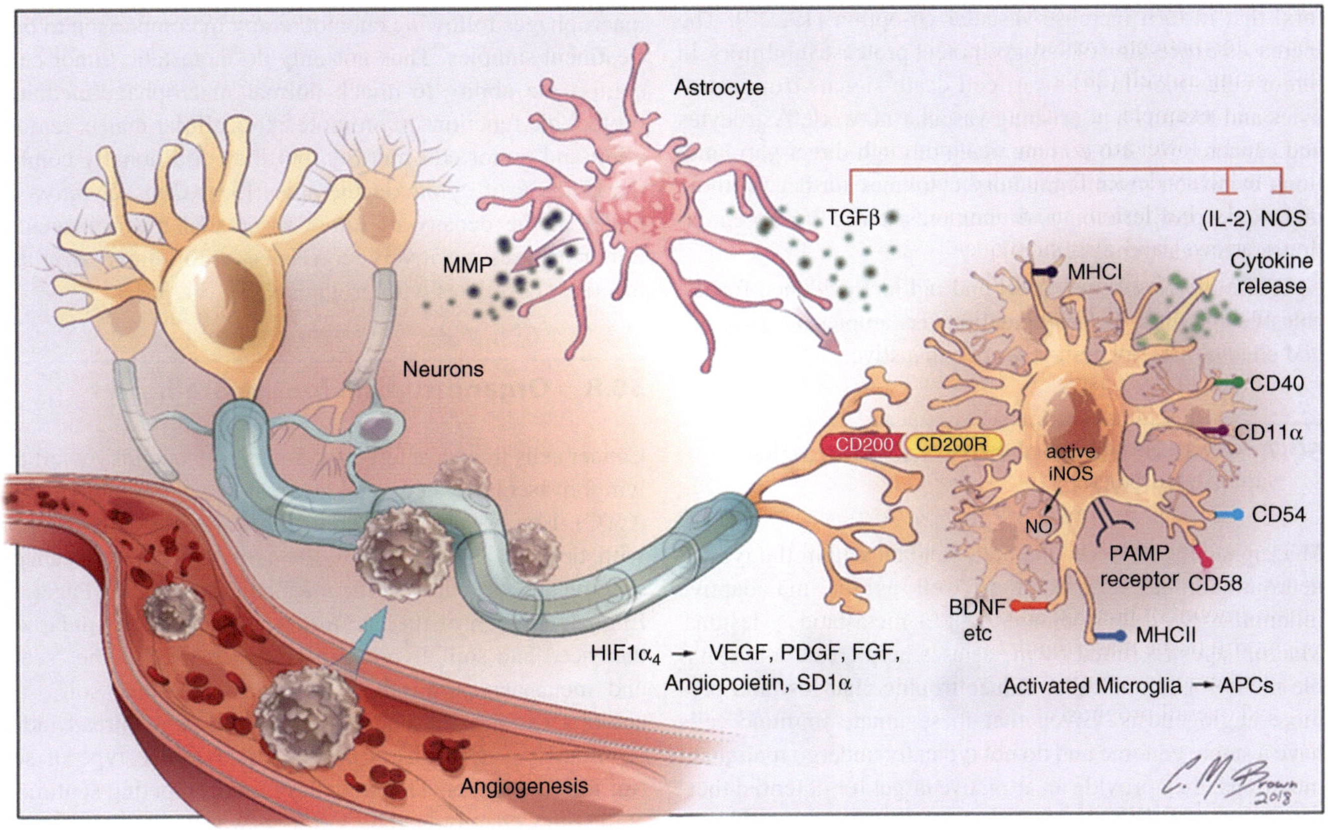

Fig. 50.1 Adapted from: *The role of the immune system in brain metastasis* [20] from Curr Neurobiol and used under a CC-BY 3.0 license. A depiction of tumor cells in circulation extravasating and interacting with glia and neuropil

insult (Fig. 50.1). When unsuccessful at stopping a pathogen or for large areas of tissue damage, the inflammatory cascade intensifies with elevated proinflammatory cytokines and chemokines. The downstream recruitment of immune cells in addition to persistent activation of astrocytes and microglia promote increased blood vessel permeability, cytotoxicity, and later astrogliosis to protect normal brain [18]. Extensive preclinical studies, mostly on murine models, have demonstrated that metastatic cancer cells hijack this physiologic inflammatory response to create a "metastatic niche" in which cancer cells may thrive [19].

50.6 Role of Astrocytes in the TME

The initiation of the neuroinflammatory cascade involves activation of astrocytes and upregulation of proinflammatory genes [21]. Morphological changes in reactive astrocytes promote scarring, known as gliosis, whereby elongated cell processes form a functional demarcation around an area of injury. Brain metastases similarly demonstrate astrogliosis at the interface between normal parenchyma and tumor; however, key astrocytic functions are parasitized to the advantage of the tumor.

Bidirectional signaling between tumor cells and reactive astrocytes creates an environment that both promotes tumor survival and enhances resistance to chemotherapy [22, 23]. While the functional interactions of specific cancer subtypes, cytokines, and chemokines are incompletely understood, there is a significant body of direct and indirect evidence from preclinical models. In the example of lung cancer, proinflammatory cytokines TNF-a, IL-1beta, and IL-6 enhance in vitro tumor proliferation [24]. Interleukin-1beta has been implicated in breast cancer metastasis to the brain whereby astrocytes activated by IL-1beta in turn enhanced cancer stem-cell (CSC) self-renewal both in vitro and in vivo while IL-1beta knockdown attenuated metastatic growth [25]. In a clinical context, breast cancer patients with lesions harboring lower expression of IL-1beta also had longer brain metastasis-free survival compared to those with lesions expressing higher levels of IL-1beta [25]. Breast cancer cells have additionally been shown to release prostaglandin COX2 to induce astrocytic release of CCL7, another potent promotor of CSC renewal [26].

While the early astrocytic response to brain tumor invasion is considered tumoricidal by inducing apoptosis via plasmin activator (PA) secretion, both breast and lung cancer metastasis have been shown to secrete serpins (PA inhibitors) that in turn increase vascular co-option [19, 27]. The increased expression of serpins, potent protease inhibitors, in tumor cells, therefore, thwart cell death signals from astrocytes and recruits a nourishing vascular network. Astrocytes and cancer cells also communicate through direct gap junctions by which proinflammatory cytokines further promote metastatic proliferation and chemoresistance by upregulation of survival genes [28–30].

While offering some direct and indirect evidence for the role of astrocytes, the aforementioned examples of cytokines and chemokines are certainly not exhaustive.

50.7 Role of Macrophages and Microglia in the TME

Macrophages and microglia are abundant within the typical neuroinflammatory cascade as well as the maladaptive inflammatory environment of metastatic lesions. Macrophages recruited by metastases help create a hospitable niche for tumor cells, induce trophic changes, and promote angiogenesis. Given that these innate immune cells have a stable genome and do not typically undergo malignant mutations, they provide an attractive target for potential therapeutic interventions in the intracranial compartment [31].

One particular protein, Cathepsin S, has been shown to be expressed by both tumor cells and macrophages and exerts a strong influence in promoting BBB transmigration as well as proliferation in the brain metastatic microenvironment. The early stages of metastatic invasion show high expression of cathepsin S initially within tumor cells which subsequently declines in late-stage lesions where Cathepsin S expression shifts to stromal macrophages. In a clinical context, high levels of cathepsin S at the primary tumor site are associated with reduced brain metastasis-free survival. While macrophages are the dominant source of cathepsin S, knockdown models of Cathepsin S show a reduction in experimental brain metastasis only occurs when both tumor- and macrophage-derived cathepsin S are depleted [32]. The proteolytic network provided by tumor-associated macrophages additionally promotes remodeling of the stromal extracellular matrix to create an environment suitable for tumor expansion [33].

In the instance of breast cancer metastasis, tumor-associated macrophages also exert a protective effect against cytotoxic chemotherapies such as paclitaxel, etoposide, and doxorubicin through both cathepsin-dependent and cathepsin-independent mechanisms. Further evidence is supported by the observation that cathepsin inhibition consequently improved the efficacy of paclitaxel and etoposide in vivo. In an orthotopic murine model using breast cancer samples before and after neoadjuvant chemotherapy, investigators found a significant accumulation of tumor-associated macrophages following chemotherapy in comparison to pretreatment samples. Thus not only do metastatic tumor cells exhibit the ability to hijack normal macrophage-mediated proteolytic functions to promote extracellular matrix remodeling and tumor cell motility, but they additionally combat the efficacy of cytotoxic therapies [33]. Consequently, we find that the density of tumor-associated macrophages in tumor masses positively correlates with tumor size and inversely correlates with prognosis [31, 34, 35].

50.8 Organotropism for the Brain

Cancer cells that have left the primary site but not yet left the lymphovascular system are known as circulating tumor cells (CTC). Extravasation and transmigration out of circulation with deposition at a distant site are mediated via cancer-specific and organic-specific mechanisms. Stephen Paget, an English surgeon of the late nineteenth century, popularized the "seed and soil" hypothesis in which CTC is the "seed" and metastasis-permissive end organs are the "soil"; the analogy extends the concept that seeds may spread indiscriminately yet require a particular, favorable type of soil (milieu) to prosper. The prevailing and competing sentiment at the time, championed by Virchow, was that metastasis was entirely explained by hematogenous spread alone. Paget, however, in examining over 700 autopsies of breast cancer patients found significant discrepancies between metastases to liver and bone with relative sparing of the spleen—a finding incompatible with Virchow's theory alone [36].

More recent literature provides evidence that a complex, bidirectional interplay exists between tumor cells and destination organs. The notion of a premetastatic niche (PMN) by which soluble factors derived from a primary tumor site modify the microenvironment of distant sites to make them more hospitable to CTCs has gained traction for multiple cancer and target organ subtypes. An example of the premetastatic niche is provided by the observation that breast cancer cells at the primary site secrete extracellular vesicles containing noncoding DNA that modulate distant brain endothelial cells to downregulate phosphoinositide-dependent kinase 1 (PDPK1), thereby weakening the BBB and facilitating transmigration of CTCs [37, 38].

The most common primary tumors to metastasize to brain include lung, melanoma, breast, renal, and colorectal cancer with significantly lower rates of skin and soft tissue metastasis relative to population incidence [21]. Initiation of metastatic colonization necessitates the expression of molecular machinery that (a) overcomes the BBB and (b) thrives in conditions of relative hypoxia and hypoglycemia [21, 39–41]. Transcriptome analysis of brain-metastatic tumor cells indeed reveals driver mutations unappreciated in matched primary tumors or extracranial metastatic cells [6]. Additional epigenetic alterations at distinct tumor-specific loci are also found in similar comparisons [21].

While our understanding of why a particular tumor may metastasize to the brain more frequently than other sites is still incomplete, studies have identified aberrant genetic and epigenetic phenotypes geared toward homing CTCs specifically to the CNS and promoting their subsequent proliferation. Primary tumor subtypes have also been implicated in differential organotropism for the brain; the epidermal growth factor receptor (EGFR) mutation and anaplastic lymphoma kinase (ALK) rearrangement show increased neurotropism in non-small cell lung cancer [42, 43], BRAF mutations driving the MAPK pathway in melanoma metastases [42], and both triple-negative and Her-2 overexpression in breast cancer [44, 45]. The aforementioned cancer subtypes harbor actionable mutations-specific targeted therapies currently available and under investigation can be found in subsequent chapters within this section.

50.9 Conclusion

The prognosis for patients with brain metastases remains poor. A primary tumor's ability to successfully invade and thrive in the central nervous system is dependent on genetic and epigenetic alterations conferring selective advantages to overcome each step in the multistep model of tumor metastasis. A better understanding of the mechanisms for neurotropism as well as key interactions between tumor and glia within the CNS tumor microenvironment holds promising avenues as the basis for developing novel therapeutics as well as identifying those at increased risk for brain metastasis.

Open Questions

- How can circulating tumor cells be screened to determine the risk of brain metastasis?
- What factors promote the survival of circulating tumor cells?
- How may resident immune cells of the central nervous system be targeted to treat BM?
- What is the role of the immune system with respect to the pre-metastatic niche?
- How may vascular co-option by tumor cells be targeted to treat BM?

References

1. Klotz R, Thomas A, Teng T, et al. Circulating tumor cells exhibit metastatic tropism and reveal brain metastasis drivers. Cancer Discov. 2020;10(1):86–103. https://doi.org/10.1158/2159-8290.CD-19-0384.
2. Lockman PR, Mittapalli RK, Taskar KS, et al. Heterogeneous blood-tumor barrier permeability determines drug efficacy in experimental brain metastases of breast cancer. Clin Cancer Res. 2010;16(23):5664–78. https://doi.org/10.1158/1078-0432.CCR-10-1564.
3. Fidler IJ. Mechanisms of cancer invasion and metastasis. In: Becker FF, editor. Biology of tumors: surfaces, immunology, and comparative pathology. Springer US; 1975. p. 101–31. https://doi.org/10.1007/978-1-4684-2736-3_3.
4. Hart IR, Fidler IJ. Cancer invasion and metastasis. Q Rev Biol. 1980;55(2):121–42. https://doi.org/10.1086/411730.
5. Dagogo-Jack I, Carter SL, Brastianos PK. Brain metastasis: clinical implications of branched evolution. Trends in Cancer. 2016;2(7):332–7. https://doi.org/10.1016/j.trecan.2016.06.005.
6. Brastianos PK, Carter SL, Santagata S, et al. Genomic characterization of brain metastases reveals branched evolution and potential therapeutic targets. Cancer Discov. 2015;5(11):1164–77. https://doi.org/10.1158/2159-8290.CD-15-0369.
7. Kang J, Zhang C, Zhong W-Z. Neoadjuvant immunotherapy for non-small cell lung cancer: state of the art. Cancer Commun (Lond). 2021;41(4):287–302. https://doi.org/10.1002/cac2.12153.
8. Keilson JM, Knochelmann HM, Paulos CM, Kudchadkar RR, Lowe MC. The evolving landscape of immunotherapy in solid tumors. J Surg Oncol. 2021;123(3):798–806. https://doi.org/10.1002/jso.26416.
9. Park YH, Park MJ, Ji SH, et al. Trastuzumab treatment improves brain metastasis outcomes through control and durable prolongation of systemic extracranial disease in HER2-overexpressing breast cancer patients. Br J Cancer. 2009;100(6):894–900. https://doi.org/10.1038/sj.bjc.6604941.
10. Ou S-HI, Ahn JS, De Petris L, et al. Alectinib in crizotinib-refractory ALK-rearranged non-small-cell lung cancer: a phase II global study. J Clin Oncol. 2016;34(7):661–8. https://doi.org/10.1200/jco.2015.63.9443.
11. Niu J, Gelbspan D, Weitz D, Markman M, Quan W. HER2-positive, trastuzumab-resistant metastatic esophageal cancer presenting with brain metastasis after durable response to dual HER2 blockade: a case report. J Gastrointest Oncol. 2014;5(6):E103–8. https://doi.org/10.3978/j.issn.2078-6891.2014.045.
12. Bhang HC, Ruddy DA, Krishnamurthy Radhakrishna V, et al. Studying clonal dynamics in response to cancer therapy using high-complexity barcoding. Nat Med. 2015;21(5):440–8. https://doi.org/10.1038/nm.3841.
13. Sequist LV, Waltman BA, Dias-Santagata D, et al. Genotypic and histological evolution of lung cancers acquiring resistance to EGFR inhibitors. Sci Transl Med. 2011;3(75):75ra26. https://doi.org/10.1126/scitranslmed.3002003.
14. Yates LR, Gerstung M, Knappskog S, et al. Subclonal diversification of primary breast cancer revealed by multiregion sequencing. Nat Med. 2015;21(7):751–9. https://doi.org/10.1038/nm.3886.
15. Kienast Y, von Baumgarten L, Fuhrmann M, et al. Real-time imaging reveals the single steps of brain metastasis formation. Nat Med. 2010;16(1):116–22. https://doi.org/10.1038/nm.2072.
16. Arvanitis CD, Ferraro GB, Jain RK. The blood–brain barrier and blood–tumour barrier in brain tumours and metastases.

Nat Rev Cancer. 2020;20(1):26–41. https://doi.org/10.1038/s41568-019-0205-x.
17. DiSabato DJ, Quan N, Godbout JP. Neuroinflammation: the devil is in the details. J Neurochem. 2016;139(Suppl 2):136–53. https://doi.org/10.1111/jnc.13607.
18. Schwartz M, Raposo C. Protective autoimmunity: a unifying model for the immune network involved in CNS repair. Neuroscientist. 2014;20(4):343–58. https://doi.org/10.1177/1073858413516799.
19. Doron H, Pukrop T, Erez N. A blazing landscape: neuroinflammation shapes brain metastasis. Cancer Res. 2019;79(3):423–36. https://doi.org/10.1158/0008-5472.CAN-18-1805.
20. Leibold AT, Monaco GN, Dey M. The role of the immune system in brain metastasis. Curr Neurobiol. 2019;10(2):33–48.
21. Yuzhalin AE, Yu D. Brain metastasis organotropism. Cold Spring Harb Perspect Med. 2020;10(5):a037242. https://doi.org/10.1101/cshperspect.a037242.
22. Langley RR, Fan D, Guo L, et al. Generation of an immortalized astrocyte cell line from H-2Kb-tsA58 mice to study the role of astrocytes in brain metastasis. Int J Oncol. 2009;35(4):665–72. https://doi.org/10.3892/ijo_00000378.
23. Lin Q, Balasubramanian K, Fan D, et al. Reactive astrocytes protect melanoma cells from chemotherapy by sequestering intracellular calcium through gap junction communication channels. Neoplasia. 2010;12(9):748–54. https://doi.org/10.1593/neo.10602.
24. Seike T, Fujita K, Yamakawa Y, et al. Interaction between lung cancer cells and astrocytes via specific inflammatory cytokines in the microenvironment of brain metastasis. Clin Exp Metastasis. 2011;28(1):13–25. https://doi.org/10.1007/s10585-010-9354-8.
25. Xing F, Kobayashi A, Okuda H, et al. Reactive astrocytes promote the metastatic growth of breast cancer stem-like cells by activating Notch signalling in brain. EMBO Mol Med. 2013;5(3):384–96. https://doi.org/10.1002/emmm.201201623.
26. Wu K, Fukuda K, Xing F, et al. Roles of the cyclooxygenase 2 matrix metalloproteinase 1 pathway in brain metastasis of breast cancer. J Biol Chem. 2015;290(15):9842–54. https://doi.org/10.1074/jbc.M114.602185.
27. Valiente M, Obenauf AC, Jin X, et al. Serpins promote cancer cell survival and vascular co-option in brain metastasis. Cell. 2014;156(5):1002–16. https://doi.org/10.1016/j.cell.2014.01.040.
28. Dai W, Zhu H, Chen G, et al. Orchestration of the crosstalk between astrocytes and cancer cells affects the treatment and prognosis of lung cancer sufferers with brain metastasis. J Thorac Dis. 2016;8(11):E1450–4. https://doi.org/10.21037/jtd.2016.11.11.
29. Kim S-J, Kim J-S, Park ES, et al. Astrocytes upregulate survival genes in tumor cells and induce protection from chemotherapy. Neoplasia. 2011;13(3):286–98. https://doi.org/10.1593/neo.11112.
30. Hanibuchi M, Kim S-J, Fidler I, Nishioka Y. The molecular biology of lung cancer brain metastasis: an overview of current comprehensions and future perspectives. J Med Investig. 2014;61(3.4):241–53. https://doi.org/10.2152/jmi.61.241.
31. Condeelis J, Pollard JW. Macrophages: obligate partners for tumor cell migration, invasion, and metastasis. Cell. 2006;124(2):263–6. https://doi.org/10.1016/j.cell.2006.01.007.
32. Sevenich L, Bowman RL, Mason SD, et al. Analysis of tumour- and stroma-supplied proteolytic networks reveals a brain-metastasis-promoting role for cathepsin S. Nat Cell Biol. 2014;16(9):876–88. https://doi.org/10.1038/ncb3011.
33. Shree T, Olson OC, Elie BT, et al. Macrophages and cathepsin proteases blunt chemotherapeutic response in breast cancer. Genes Dev. 2011;25(23):2465–79. https://doi.org/10.1101/gad.180331.111.
34. Medrek C, Pontén F, Jirström K, Leandersson K. The presence of tumor associated macrophages in tumor stroma as a prognostic marker for breast cancer patients. BMC Cancer. 2012;12:306. https://doi.org/10.1186/1471-2407-12-306.
35. Morrison C. Immuno-oncologists eye up macrophage targets. Nat Rev Drug Discov. 2016;15(6):373–4. https://doi.org/10.1038/nrd.2016.111.
36. Fidler IJ, Poste G. The "seed and soil" hypothesis revisited. Lancet Oncol. 2008;9(8):808. https://doi.org/10.1016/S1470-2045(08)70201-8.
37. Tominaga N, Kosaka N, Ono M, et al. Brain metastatic cancer cells release microRNA-181c-containing extracellular vesicles capable of destructing blood-brain barrier. Nat Commun. 2015;6:6716. https://doi.org/10.1038/ncomms7716.
38. Zhou W, Fong MY, Min Y, et al. Cancer-secreted miR-105 destroys vascular endothelial barriers to promote metastasis. Cancer Cell. 2014;25(4):501–15. https://doi.org/10.1016/j.ccr.2014.03.007.
39. Peters A, Schweiger U, Pellerin L, et al. The selfish brain: competition for energy resources. Neurosci Biobehav Rev. 2004;28(2):143–80. https://doi.org/10.1016/j.neubiorev.2004.03.002.
40. Winkler F. The brain metastatic niche. J Mol Med (Berl). 2015;93(11):1213–20. https://doi.org/10.1007/s00109-015-1357-0.
41. Lowery FJ, Yu D. Brain metastasis: unique challenges and open opportunities. Biochim Biophys Acta Rev Cancer. 2017;1867(1):49–57. https://doi.org/10.1016/j.bbcan.2016.12.001.
42. Ni W, Chen W, Lu Y. Emerging findings into molecular mechanism of brain metastasis. Cancer Med. 2018;7(8):3820–33. https://doi.org/10.1002/cam4.1667.
43. Rangachari D, Yamaguchi N, VanderLaan PA, et al. Brain metastases in patients with EGFR-mutated or ALK-rearranged non-small-cell lung cancers. Lung Cancer. 2015;88(1):108–11. https://doi.org/10.1016/j.lungcan.2015.01.020.
44. Shen Q, Sahin AA, Hess KR, et al. Breast cancer with brain metastases: clinicopathologic features, survival, and paired biomarker analysis. Oncologist. 2015;20(5):466–73. https://doi.org/10.1634/theoncologist.2014-0107.
45. Jin L, Han B, Siegel E, Cui Y, Giuliano A, Cui X. Breast cancer lung metastasis: molecular biology and therapeutic implications. Cancer Biol Ther. 2018;19(10):858–68. https://doi.org/10.1080/15384047.2018.1456599.

Indications and Techniques for Surgical Intervention in Patients with Metastatic Brain Tumors

51

Jacob A. Pawloski, Omar Awan, Mateo Ziu, and Adam M. Robin

Abstract

Brain metastases occur in up to 30% of cancer patients and may be the presenting problem in as many as 20%. Advances in our understanding of cancer biology and precision therapeutics have resulted in improved systemic disease control for many cancers, making the treatment of metastatic disease to the central nervous system (CNS) more critical than ever. Surgery and adjuvant radiation have long been the mainstay of treatment. Surgery can provide a diagnosis, symptom relief, and ideally locally curative benefits. Surgical innovations and techniques such as single or multiple craniotomies for microsurgical excision, intraoperative brain mapping, minimally invasive parafascicular surgery (MIPS), intraoperative image-guided surgical resection, and laser interstitial thermal therapy (LITT) allow for the optimization of tumor removal and patient safety as never before. In this chapter, the reader will become familiar with the rationale for the surgical management of the patient with a brain metastasis and many of the techniques driving better surgical outcomes for our patients.

J. A. Pawloski · A. M. Robin (✉)
Department of Neurosurgery, Hermelin Brain Tumor Center, Henry Ford Hospital, Detroit, MI, USA
e-mail: JPAWLOS1@hfhs.org; arobin1@hfhs.org

O. Awan · M. Ziu
Inova Neurosciences and Spine Institute, University of Virginia Medical School, Falls Church, VA, USA
e-mail: Omar.Awan@inova.org; Mateo.ziu@inova.org

Learning Objectives

- Describe the indications for surgical intervention in the patient with a brain metastasis
- Gain familiarity with surgical techniques and their applications for treatment for the patient with a brain metastasis
- Understand the rationale for surgical adjuncts that help surgeons optimize tumor removal and safety

51.1 Introduction

Brain metastases occur in up to 30% of cancer patients and account for over 50% of intracranial tumors [1, 2]. Surgery can provide a diagnosis, symptom relief, and ideally locally curative benefits. Furthermore, the ability to obtain tissue for histopathological and sequencing analysis is an important consideration. While patients with known brain metastatic disease are often found to have lesions which are more easily accessible for tissue diagnosis, there is evidence to suggest that brain metastases can contain therapeutically relevant molecular alterations not found in the primary lesion [3]. Patients with known systemic disease may have other non-metastatic intracranial lesions (e.g., primary brain tumor or abscess) for which diagnosis would prompt a shift in treatment. Surgical resection is the cornerstone of the management of brain metastases and provides unmatched direct cytoreduction of local disease. Craniotomy for resection can

S. P. Leong et al. (eds.), *Cancer Metastasis Through the Lymphovascular System*,
https://doi.org/10.1007/978-3-030-93084-4_51

be a valuable tool for relief of symptomatic mass effect, malignant edema and intracranial pressure, or obstruction of cerebrospinal fluid (CSF) flow, which can all contribute to significant mortality and morbidity. In addition to tumor resection, surgery plays a critical role in a number of other ways. In the case of leptomeningeal disease, CSF diversion in the form of ventriculoperitoneal shunting (VPS) or endoscopic third ventriculostomy (ETV) can be employed to provide palliation for symptoms of hydrocephalus. Furthermore, surgical intervention can be used to implant intraventricular devices for infusion of intrathecal therapeutics or following resection for placement of material imbued with chemotherapeutics or radioactive isotopes for delivery of brachytherapy, allowing for higher drug concentrations in the CSF or peritumoral region than can be achieved in the plasma or for local radiotherapy in the setting of previously radiated lesions.

Over the last 20 years, innovations in the field of brain tumor surgery have provided the neurosurgeon with an expansive armamentarium to treat brain metastases, allowing for the design of surgical plans that are tailored to the patient and their disease. Herein, we will review the role of surgical intervention in patients with brain metastases with a special focus on techniques that are increasingly utilized to optimize tumor removal while minimizing perioperative morbidity.

51.2 Surgical Indications

51.2.1 Single or Dominant Brain Metastasis

Routine use of craniotomy for resection of a solitary metastatic lesion was first shown to be beneficial in the landmark randomized controlled trial performed by Patchell and colleagues [4]. Surgery followed by whole-brain radiation therapy (WBRT) improved overall survival, functional independence, and progression-free survival compared to WBRT alone [5]. However, studies have also shown that surgical resection alone is insufficient, with local recurrence rates ranging from 40 to 70% [6, 7]. Thus, surgery should be considered as a cornerstone of combination therapy with adjuvant radiation.

Indications favoring surgical resection include tumor size >3 cm, location in the posterior fossa associated with or at risk for development of obstructive hydrocephalus, and lesions causing significant symptoms due to mass effect. Individuals with good functional status in whom the systemic disease is well-controlled are the best candidates for surgical intervention, while those with limited life expectancy due to their advanced oncologic disease or significant medical comorbidities may benefit more from less invasive treatment. Current evidence suggests that stereotactic radiosurgery (SRS) alone is as effective as surgery plus WBRT, and thus for tumors smaller than 3 cm, surgery is often reserved as salvage therapy for those that recur following SRS [8].

The histology and location of the metastatic lesions are both important considerations in determining treatment options for single or multiple metastases. Certain histologies such as small cell lung cancers or CNS lymphomas may be effectively treated with radiation and/or chemotherapy alone, such that surgery is not indicated as a first-line therapy even for large tumors while melanoma, renal cell carcinoma, and sarcoma are relatively radioresistant and are more amenable to surgical resection. With regard to tumor location, the functional anatomy of the surrounding brain tissue must be considered. Epidemiological studies have shown that 80–85% of metastatic lesions are supratentorial and 10–15% affect the cerebellum or brainstem [9]. Involvement of eloquent brain areas may result in surgical morbidity that could delay or even preclude adjuvant therapies and offset any potential survival benefit conferred by surgery [10]. As will be discussed later, neurosurgical advances now allow for access to lesions affecting almost any area of the brain safely but the decision to operate should always be made within the appropriate clinical context including the patient's tumor characteristics, prognosis from their systemic disease, likelihood of neurologic improvement following surgery, and other comorbid medical conditions that may impact the risks of surgical intervention.

Discussion to this point has centered around surgical intervention involving craniotomy and removal of lesions with the objective of decreasing mass effect and obtaining local control via direct cytoreduction. Alternatively, stereotactic biopsy is a minimally invasive procedure used to obtain tissue for diagnosis only. Biopsy is indicated in instances where tumors are inaccessible surgically, there is an unknown primary cancer, or craniotomy is contraindicated due to poor prognosis or overall health. Epidemiologically, brain metastasis is more than twice as common as primary brain tumors yet in as many as 15% of cases a primary original of brain metastasis cannot be identified [11, 12]. Further, almost 20% of patients can present with brain metastasis as the first symptom of metastatic cancer [13]. Due to the relatively low extracranial tumor burden, patients with brain metastasis associated with an unknown primary present with a therapeutic window where early biopsy and tissue diagnosis may be associated with improved outcomes when targeted therapy is available [13]. This point is underscored when treating lesions in which a diagnosis of a chemo or radiation-sensitive histology is suspected, such as lymphoma or small-cell lung cancer.

In summary, surgical resection remains the foundation of management for metastatic disease to the central nervous system and has been shown to increase survival and decrease rates of local recurrence, especially when used in combination with adjuvant radiation therapy. Unless patient factors preclude open surgery, resective surgery can improve functional status and halt or improve progressive neurologic deficit in as many as 90% of cases [14, 15].

51.2.2 Multiple Brain Metastases

As many as half of patients with brain metastasis present with multiple lesions [16]. For the most part, studies evaluating the role of surgical resection for patients with multiple brain metastases have been limited. These studies have shown that patients who had resection of all accessible lesions had survival rates similar to patients with single lesion who underwent resection, although patients in these studies typically had resection of three or fewer lesions [14, 17]. Patients with expected overall survival of less than 3 months due to systemic disease should not be considered for surgical intervention. Instead, patients with limited or controlled systemic disease and accessible tumors should be considered for surgical resection—even when not all tumors can be resected, considering that complications from surgery are as low as 7–9% [14, 17]. In general, the same selection criteria for surgical candidacy for single lesion resection may also be applied to patients with multiple metastasis (namely, age, KPS < 70, and extracranial disease burden) with similar outcomes even if all lesions are not amendable to resection. Studies suggest that patients with more than three or four lesions, especially with poor preoperative performance status, are unlikely to benefit from surgical resection and should be considered for radiation therapy instead.

51.2.3 Progressive or Recurrent Brain Metastasis

As individuals with brain metastases continue to live longer after their initial treatment, cases of tumor recurrence or progression are becoming more common and are reported in 10–40% of individuals with brain metastases [18, 19]. Local control is impacted by a variety of factors including histology, tumor size, adjuvant radiotherapy to the resection cavity, as well as method of surgical removal (e.g., en bloc vs. piecemeal resection) [19]. There is some scarce evidence that reoperation can lead to long-term survival in select patients: younger patients, good functional status, and a longer time from initial surgery to recurrence [15, 17, 20].. Aggressive surgical resection of recurrent metastatic lesions is a reasonable option in patients who have exhausted noninvasive treatment options and retain good functional status with controlled systemic disease.

Brain biopsy of a suspected recurrent metastasis is useful in patients who have previously received radiotherapy, as tumor recurrence and radiation necrosis (RN) can both manifest with nearly identical clinical and radiographic presentations. Radiation necrosis can occur at almost any time from a few months to years following radiation therapy and can appear radiographically identical to disease recurrence with increasing contrast enhancement, worsening edema, and mass effect. While both tumor progression and RN can present similarly, distinguishing one from the other may have profound implications on care. Moreover, with increasing recognition of the heterogeneity between brain metastases molecular profiling may inform selection of precision therapeutics. When encountering suspected local tumor recurrence after radiotherapy, surgical management is dependent on the size, location, and symptoms attributable to the suspected recurrence. If the lesion recurs with significant edema and mass effect in a patient who is a good surgical candidate, craniotomy for resection of the recurrent tumor or RN is reasonable. Less impressive progressive enhancement and associated edema on serial MRIs following radiation therapy should be considered for stereotactic biopsy to differentiate recurrence from RN. In these cases, the addition of LITT has become a valuable surgical tool as it has the capacity to treat both tumor recurrence and RN in a minimally invasive fashion with little surgical footprint beyond that which is required for stereotactic needle biopsy alone.

51.3 Surgical Techniques

51.3.1 Decision-Making in the Surgical Treatment of Brain Metastases

When operating on cancer patients with brain metastases, the risks of resective surgery must be balanced with the patient's overall prognosis from their systemic disease, their likelihood of neurologic improvement or decline with surgery, and the alternative treatment options available, among other factors. Maximal resection and adjuvant radiotherapy can provide durable local control and improved quality of life. Overall survival following resection of brain metastasis averages from 10 to 14 months, but can exceed 5 years in select patient populations [1, 10, 17, 21]. Likewise, neurologic complications from surgery can have significant negative consequences, with one study finding overall survival was reduced from 12.6 to 2.4 months in individuals with new postoperative neurologic deficits [10]. This underscores the importance of using tailored surgical approaches and surgical adjuncts to minimize operative morbidity (Fig. 51.1).

51.3.2 Craniotomy for Solitary or Single Dominant Metastases in Eloquent Locations

Brain metastases occurring in speech or primary somatosensory regions inherently carry a higher risk of neurologic dysfunction. The risk of neurologic deficit without treatment or during treatment, particularly surgery can be quite high. SRS is a reasonable option for tumors in eloquent regions, but due

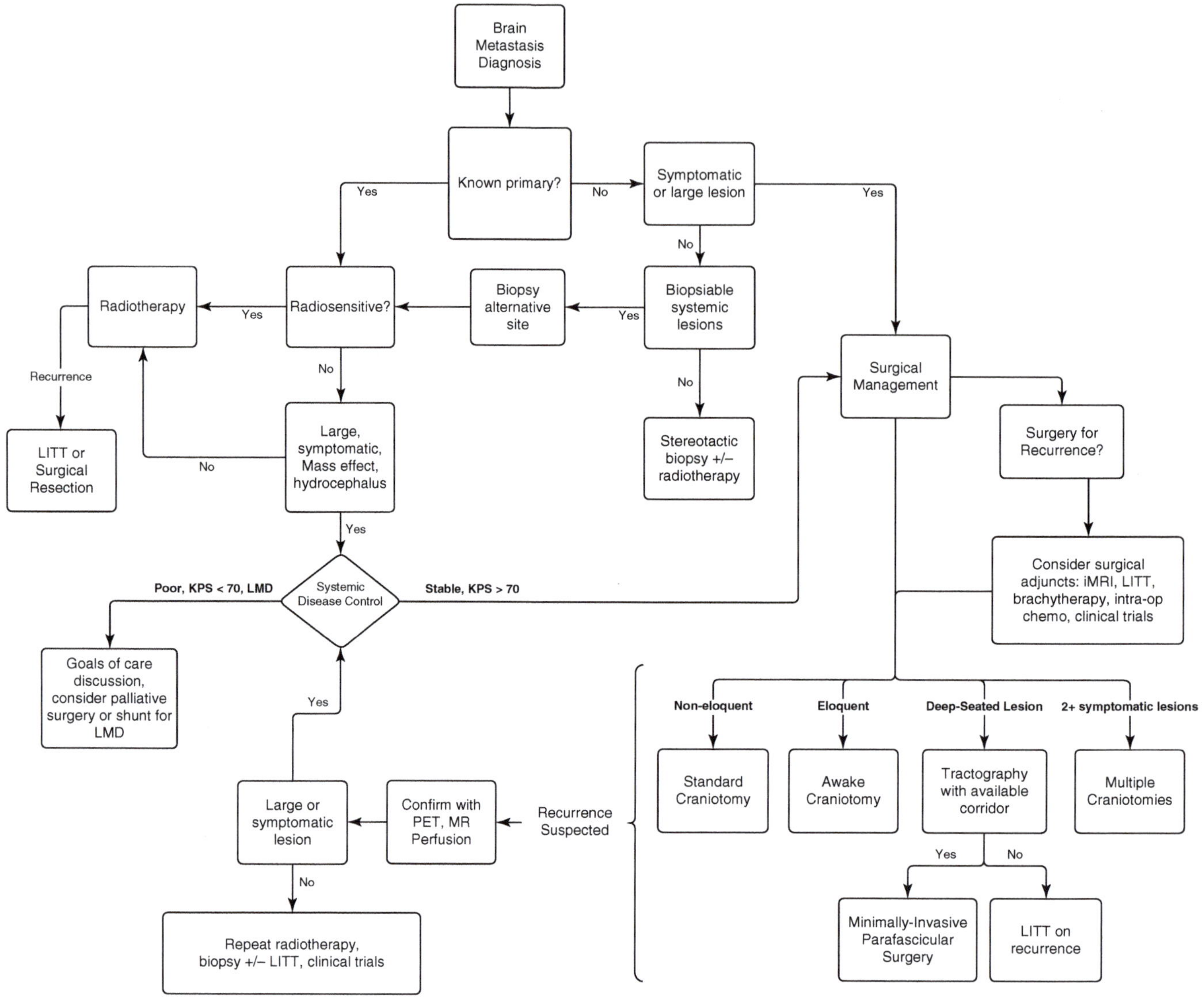

Fig. 51.1 Recommended Treatment Paradigm for the patient presenting with a brain metastasis

to their location, these tumors often manifest with debilitating symptoms that are exacerbated initially with radiation therapy. Moreover, radiation necrosis in eloquent regions can have catastrophic effects on neurologic function and quality of life. Increasingly, metastases in the primary sensorimotor cortex are considered for surgical intervention. Successfully dealing with these tumors requires careful planning and patient evaluation. Patient improvement with preoperative steroid administration is a good prognostic indicator. Stereotactic image guidance, intraoperative ultrasound, and intraoperative neurophysiologic monitoring with or without awake craniotomy are surgical adjuncts that help to localize the tumor augment and tumor removal. The size of the craniotomy should be planned to provide adequate access for resection of the lesion as well as enough exposure to adequately map the adjacent motor or speech centers. Motor mapping can be completed using awake or asleep techniques with intraoperative electrophysiologic monitoring and direct cortical or subcortical stimulation, while speech mapping requires awake craniotomy and the support of a speech and language pathologist. Various paradigms have been described to ensure high fidelity results while mapping function in surgery. A combination of direct visualization, stereotactic image-guided navigation, and ultrasound guides the approach to the lesion (Fig. 51.2). In the case of a central sulcus metastasis, the authors prefer arachnoid dissection of the central vein allowing for a transsulcal approach to the tumor. Using a transsulcal approach obviates the need for dissection through eloquent sensorimotor cortex. Care is taken to preserve critical vascular structures within the sulcus. Use of intraoperative mapping can reliably reduce the associated morbidity of metastatic tumor resection in these regions [22, 23].

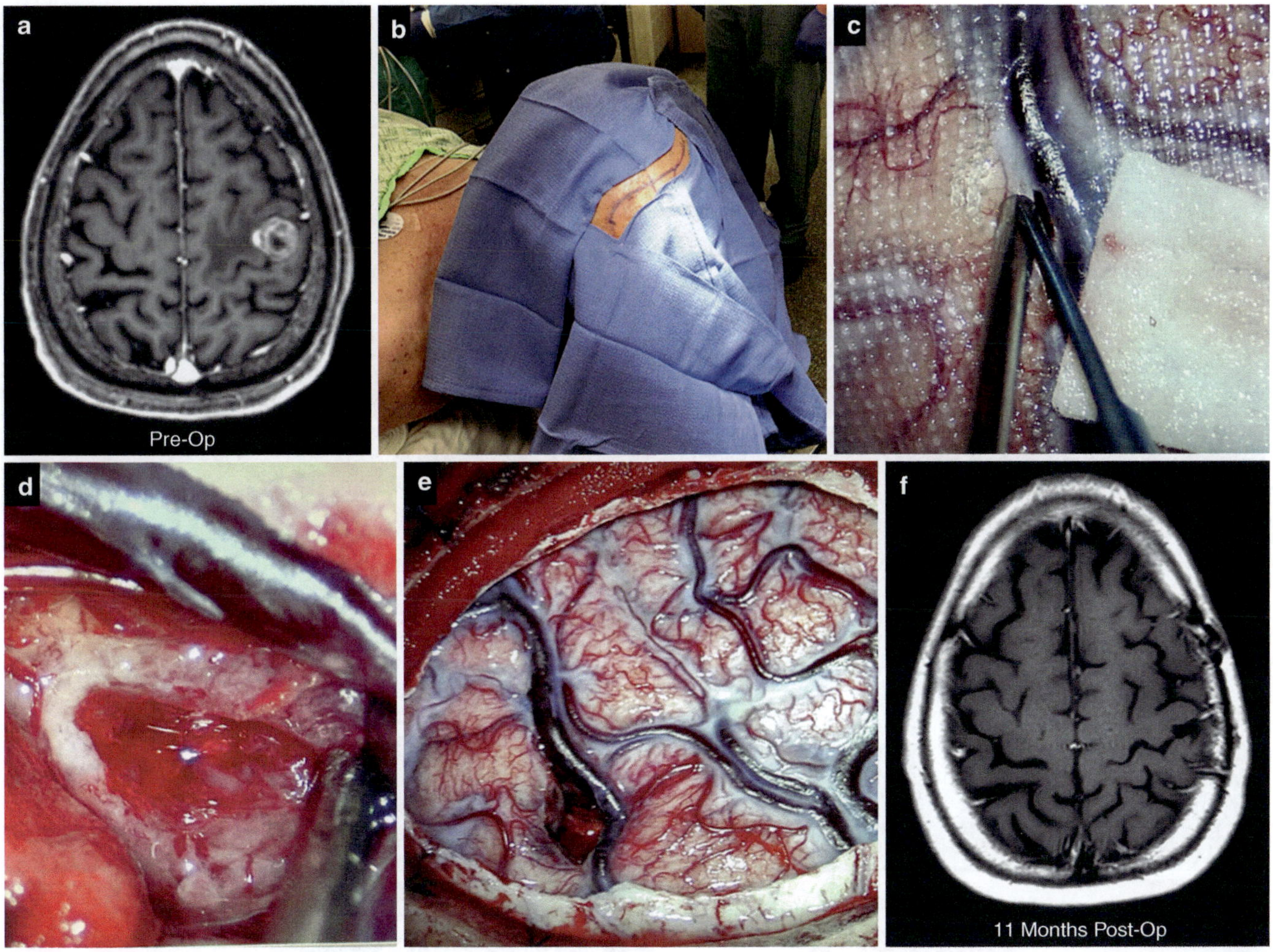

Fig. 51.2 Panel (**a**) T1 axial post-contrast MRI demonstrating a left-sided brain metastasis of unknown origin interposed between sensory and motor cortices. Case example of a 48-year-old man with a central sulcus tumor and no known primary disease who underwent an awake craniotomy for sensorimotor mapping and trans-central sulcus approach with complete resection. Panel (**b**) demonstrates the "lazy S" incision used for exposure of the sensorimotor region. Note the absence of a ventilator circuit and endotracheal tube. The patient's head is fixed in skull pins following a complete scalp block. Panel (**c**) demonstrates dissection of the pia adjacent to the central sulcal vein. For reference, the cotton patty pictured is 12 mm in size. Panel (**d**) shows the dissected central sulcus, splaying of the sensory and motor cortex, and mobilization of the central vein and arterioles after opening the tumor capsule. Panel (**e**) shows the post-operative appearance after mobilization of the vein and sensory motor cortex to allow for resection of the brain metastasis

51.3.3 Craniotomy for Multiple Metastases

The safety of surgical resection has improved, and in practice, individuals undergo craniotomy for multiple brain metastases even if not all lesions can be removed. Iwadate et al. reviewed 138 patients with single or multiple metastasis treated with surgery and WBRT and concluded that there was no difference in overall survival in the single or multiple metastases subgroups [24]. Of note, patients with multiple lesions who underwent surgery but not all lesions were resected had shorter overall survival, however, treating the remaining lesions with SRS is equally effective [25]. Based on these and other evidence, the most recent guidelines from the Congress of Neurological Surgeons recommend surgical resection for patients with good systemic disease control and multiple metastases that are accessible via surgery [26].

Patient selection is crucial and should focus on patients who have multiple lesions that are not amenable to SRS and are likely to tolerate one or more craniotomies with a good chance of stable or improved neurologic function after surgery. The authors do not recommend routinely performing multiple operations in eloquent regions at the same time as the multiple sites required for craniotomy can interfere with the feasibility of brain mapping. Additionally, care should be taken to accurately register stereotactic neuronavigation for incision and craniotomy planning. Incisions must not further compromise blood flow and wound healing, particularly in patients with brain metastases who

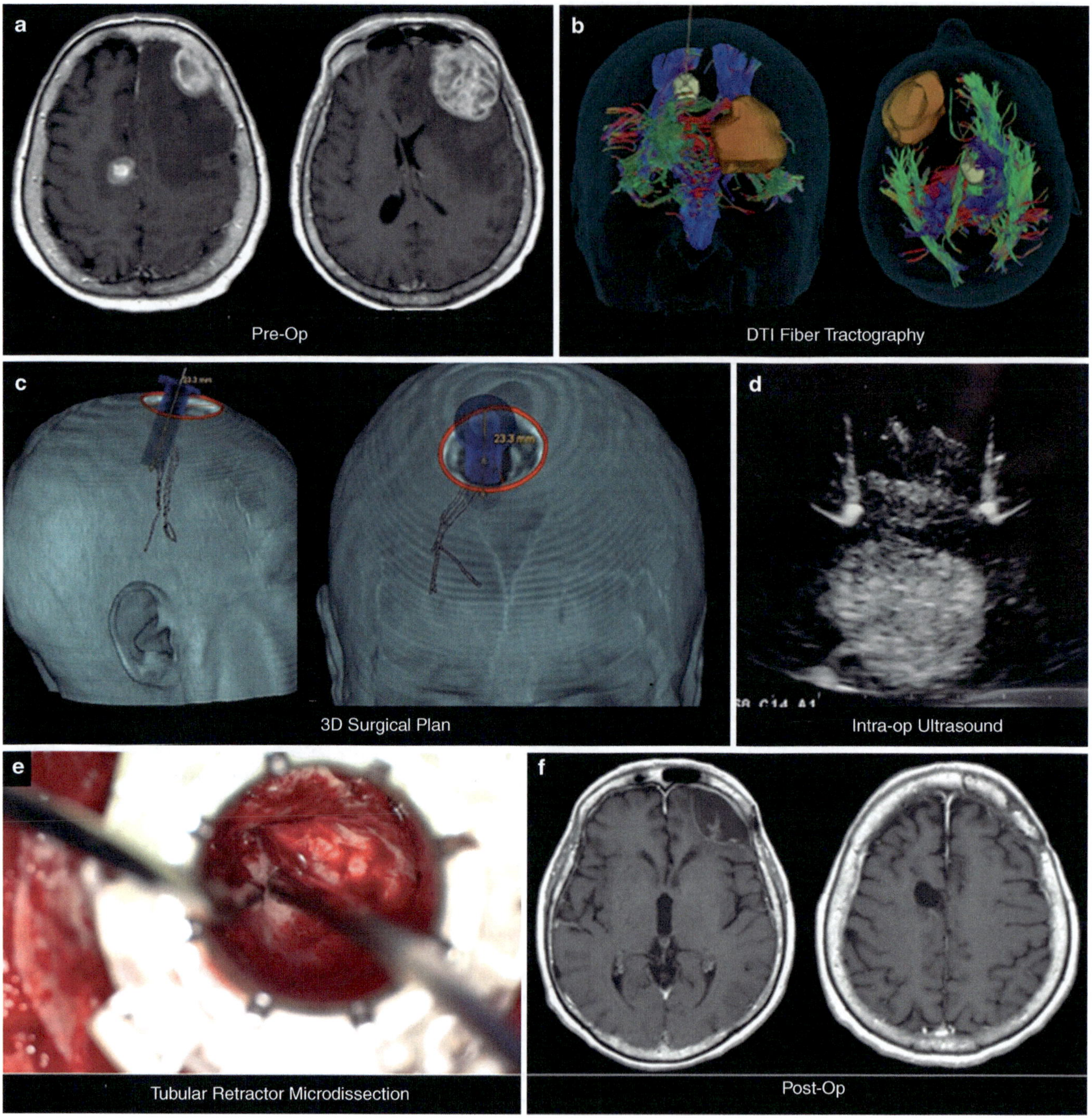

Fig. 51.3 Case example of a 57-year-old woman with bifrontal breast metastases treated with multiple craniotomies in a single session utilizing a traditional craniotomy and a minimally invasive parafascicular surgical (MIPS) approach for complete resection of both lesions. 3D tractography/surgical planning was utilized before and during surgery as was intraoperative ultrasound to optimize safety and tumor removal

likely will undergo multiple radiotherapy treatments and when possible, the authors utilize a single incision. As discussed at length in later sections, minimally invasive options such as laser ablation and minimally invasive tubular retractor systems can be especially useful in treating multiple lesions with minimal manipulation of adjacent structures (Fig. 51.3).

51.3.4 Minimally Invasive Approaches to Metastatic Tumors

Metastatic tumors in deep-white matter areas pose a challenge to the surgeon due to the high risk of injuring traversing fiber tracts while accessing and resecting the tumor. Radiation has been utilized in circumstances where tumors are not easily accessible via standard craniotomy. Alternatively, laser ablation may be used to provide pathologic diagnosis as well as cytoreduction of deep-seated lesions. Minimally invasive parafascicular microsurgery (MIPS) has also been employed to retain the advantages and flexibility of open craniotomy while reducing the amount of white matter injury and retraction on adjacent or superficial structures. Preoperative planning in these cases is important and advances in MRI diffusion tensor imaging (DTI) and fiber tractography allow for planning surgical approaches with the intent of avoiding crucial white matter tracts.

51.3.4.1 Minimally Invasive Parafascicular Surgery (MIPS)

Tubular retractor systems, also known as port-based or channel-based systems, are cylindrical or ovoid retractors, which allow for brain cannulation using a sulcus or small corticotomy to provide access to deep-seated subcortical lesions. Intensive preplanning utilizing high-resolution anatomic studies and DTI white matter tractography are critical for this technique. Algorithms that enable 3D visualization of important fiber bundles aid in the identification of safe surgical corridors along white matter tracts. Accessing these corridors allows for safer access to lesions previously associated with prohibitive morbidity using more traditional surgical approaches. This can be achieved with minimal disruption and manipulation of the surrounding brain (Fig. 51.3). A key benefit of the cylindrical retractors with an atraumatic cannulating tip is that they are minimally traumatic during brain cannulation and distribute the retraction force circumferentially, preventing the sheering of white matter and pressure-related ischemia that can be associated with bladed retractors used in a standard open craniotomy, often with very favorable patient outcomes [27, 28]. A larger series of 50 patients with metastatic and primary CNS lesions with median depth of 5.4 cm below the cortex underwent port-based resection with 94% of patients having a stable or improved KPS postoperatively [29]. Compared to laser ablation techniques, tubular retractors provide the added benefit of debulking, which makes them more suitable for lesions that present with significant mass effect. Despite their diminutive size, some systems as small as 11 mm in diameter, there has been a proliferation of bayonetted, single-shaft instruments some of which are lighted, allowing for two-handed microsurgery. Overall, tubular retractor-based approaches are a promising new addition to surgical management of metastatic disease by offering a safer alternative for resection of deep-seated lesions as compared to traditional craniotomy.

51.3.4.2 Laser Ablation for Brain Metastasis

Laser interstitial thermal therapy (LITT), also known as stereotactic laser ablation (SLA) is a minimally invasive surgical technique, increasingly utilized for patients with brain metastases. The goal of tumor cytoreduction remains, only it is achieved via laser ablation rather than excision, and the tumor is accessed with stereotactic placement of a laser fiber via stab incision and twist drill hole in the skull in lieu of a scalp flap and standard craniotomy (Fig. 51.4). Preoperative planning is paramount and focuses on utilizing high-resolution anatomic imaging, tractography, three-dimensional volumetric analysis with build out of volumes for tumor and other nearby eloquent structures to help ensure that the laser fibers are placed to maximize lesion ablation. Much of the planning is based on surgeon experience whereby heat sinks such as the ventricular system, CSF cisterns, and large vessels are accounted for in order to optimize heat dispersion out from the laser in a controlled fashion to achieve cytoreduction and ablation of tumor or radiation necrosis. Using multiple circumferentially oriented or directionally oriented laser fibers and withdrawing the fibers during the ablation process, the surgeon can define an ablation zone that correlates well with lesions up to 10–20 cc and beyond. Briefly, the technique involves placement of a catheter using stereotactic guidance, through which a laser fiber is used to induce thermal injury to target tissue by controlled heating. A critical step involves intraoperative MRI for confirmation of adequate laser fiber position and for precise temperature control using MR thermometry. The placement of the fiber is nearly identical to that of placing a biopsy needle and ablative therapy can be performed during the same procedure following tissue sampling.

The majority of literature pertaining to the use of LITT for metastatic disease is in the context of recurrence or neoadjuvant ablation following SRS. In treatment of recurrence following SRS, a situation complicated by relatively few treatment options, one systematic review found median overall survival following complete ablation ranging from 6 to 20 months [30]. Significantly, in this study no patient in whom complete ablation occurred suffered from disease progression during the follow-up period. A prospective study of patients undergoing LITT for metastatic tumor recurrence found an overall survival of 64.5% at 26 weeks [31].

Several studies have shown the most significant limitation regarding the success of LITT to be tumor size, and accordingly the proportion of tumor which is within the ablated area. While multiple staged procedures may be undertaken to address larger tumors, this is less favorable as it can increase the risk of procedural complications and delay initiation of adjuvant therapies. Although there are no specific guidelines regarding tumor size, studies have shown that tumors as large as 5 cm may be associated with benefit in terms of progression-free survival [32]. Further, stereotactic trajectories may be limited due to the patient's anatomy. Larger and more vascular lesions are considered to be poor targets for thermal ablation and while generally considered as safe or

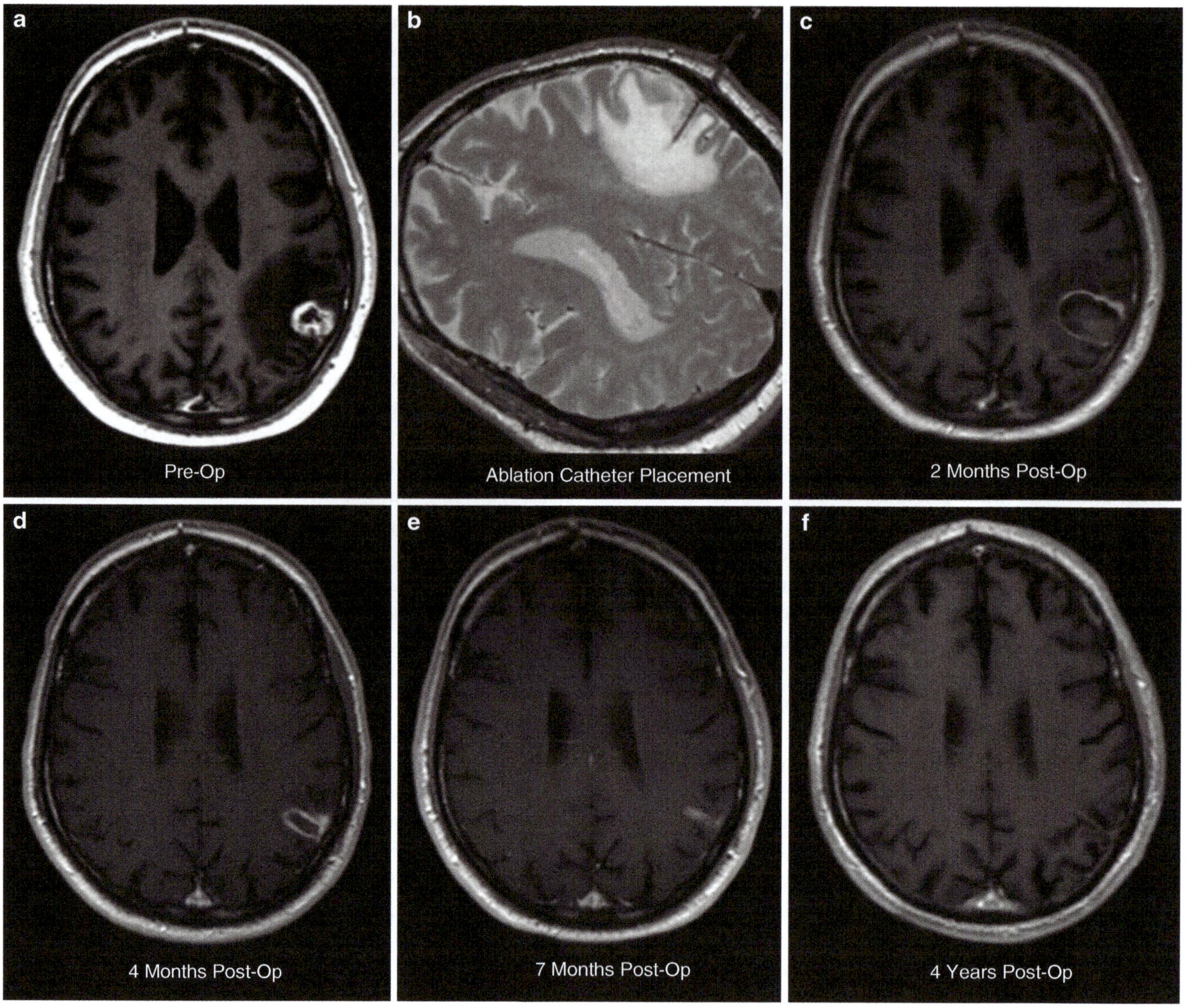

Fig. 51.4 Case example of a 48-year-old man with 4-year radiographic follow-up in a patient with a melanoma metastasis after treatment with stereotactic radiosurgery (SRS) and then laser interstitial thermal therapy (LITT). Biopsy at the time of LITT treatment was consistent with radiation necrosis

safer than standard craniotomy, LITT can be associated with post-procedural hemorrhage or swelling, and is not as suitable for treatment of lesions causing significant mass effect.

In one institutional series, there was a 4% risk of hematoma formation at the ablation site requiring evacuation [33]. Lesions with volume >10 cc are associated with an increased risk of perioperative complications such as seizures or neurologic deficit and a lower rate of gross total ablation [34, 35]. In these cases, there is evidence that adjuvant SRS to the residual tumor within a month following surgery can decrease the rate of tumor progression [35].

Overall, LITT is a novel therapy which has been shown to be particularly useful in treating patients suffering from recurrent metastatic disease, or disease that is refractory to standard treatment, for deeply seated, surgically inaccessible lesions, and patients too frail to tolerate surgical stress. As it offers significant local control and survival benefits in patients who have already failed radiation therapy and may not be candidates for craniotomy, LITT may be a valuable tool for salvage therapy. While promising for other indications and clinical scenarios, further study is yet necessary.

51.3.5 Intraoperative MRI and Optical Guidance Technology

There is substantial evidence demonstrating that gross total resection of brain metastasis improves overall survival and decreases risk of local progression as compared to subtotal resection [8]. Considerable advances have been made in intraoperative imaging technologies to increase the likelihood of identifying and resecting all malignant tissue in the operative field. iMRI has been heavily utilized in glioma surgery to increase extent of resection, leading to improved outcomes in glioma patients [36, 37]. iMRI has not been utilized as heavily for metastatic disease, likely because these lesions are generally less infiltrative than intrinsic tumors and are therefore more readily delineated. However, in cases of recurrent metastases or lesions in eloquent locations, iMRI can be useful to increase the likelihood of gross total resection. Intraoperative MRI has also been utilized with good results as an adjunct to MIPS to increase extent of resection in deep-seated lesions which may be difficult to visualize intraoperatively [38].

In addition to iMRI, there have been advancements in optical technologies that allow for greater visualization of tumor such as large-field florescence with 5-Aminolevulinic acid (5-ALA) and fluorescein sodium, confocal microscopy, and Raman spectroscopy. While a detailed discussion of this diverse group of optical imaging modalities is outside the scope of this discussion these technologies are potentially transformative for their ability to provide instant tissue characterization in the operating room via in vivo or ex vivo evaluation [39].

51.3.6 Methods of Surgical Resection: En Bloc Versus Piecemeal Removal

Resection of intra-axial lesions is typically done via an en bloc resection where the entire lesion is separated from the surrounding parenchyma and removed as a contiguous mass or by a piecemeal resection in which the lesion is entered and then taken out in pieces until all malignant tissue has been removed. Largely speaking, en bloc resection is more favorable when it is possible due to lower complication rates and less risk of local recurrence and leptomeningeal spread [8]. The location and size of the mass may make en bloc resection impossible without causing significant retraction injury or disruption of eloquent structures, and in those instances piecemeal removal may be the only option. One large series looking at complication rates between the two approaches found that resection via the piecemeal approach increased the overall 30-day complication rate by 6% across all tumor types and sizes. Notably, in that study there was a significant difference in tumor size as well as tumor functional grade, such that larger and more eloquent lesions were more likely to be removed via the piecemeal approach, suggesting that the en bloc subgroup consisted of tumors with lower risk characteristics [40].

Leptomeningeal spread as a result of surgery is also a significant concern that is more associated with piecemeal resection strategies, particularly in the posterior fossa. In a large series of patients with a single metastasis the patients who underwent piecemeal resection were 2.7 times more likely to develop leptomeningeal disease (LMD) than those with lesions that were removed en bloc [41]. Melanoma in particular was nearly nine times more likely to progress to LMD when resected via piecemeal versus en bloc technique.

51.3.7 Surgical Adjuncts and Palliative Procedures

51.3.7.1 Surgical Treatment of Leptomeningeal Disease

Leptomeningeal disease (LMD), or carcinomatous meningitis, is a dreaded complication for cancer patients due to the poor prognosis it affords, with overall survival of 3–6 months after diagnosis of LMD. Current management of LMD focuses primarily on high-dose intrathecal and systemic chemotherapy, with or without radiotherapy. Surgical interventions for LMD have focused largely on palliation of pain related to hydrocephalus or to provide a route for the safe administration of intrathecal chemotherapy. Ommaya reservoir (OR) placement simplifies intrathecal chemotherapy administration and allows for more standard and uniform dosing as opposed to serial lumbar punctures. CSF diversion via VPS can also be beneficial in patients with an obstructive CSF flow pattern or symptomatic communicating hydrocephalus. Patient selection for VPS placement is important as headaches are the most common presenting symptom for LMD, and 40% of patients have an opening pressure >20 cmH20 on initial lumbar puncture [42, 43]. While symptoms of headache or nausea may be due to symptomatic hydrocephalus, direct dural and parenchymal invasion by malignant cells can also account for symptoms of LMD [44]. As with any surgical procedure, the decision to operate should be based on a holistic evaluation that includes the patient's functional status, systemic disease control, overall prognosis, and need/capacity for palliation. One series of patients undergoing placement of a reservoir-shunt system for CSF diversion and intermittent chemotherapy administration reported functional improvement in 80% of patients that was sustained for 6 months postoperatively [45]. A trial lumbar puncture, or "tap test" may be useful in differentiating patients with potential to benefit from CSF diversion from those that have headaches or cranial nerve palsies from direct leptomeningeal invasion (unpublished data).

51.3.7.2 Brachytherapy

Brachytherapy consists of internalized radiotherapy seeds deposited at the site of surgical resection, and relies on a constant, low dose of radiation directly to the tumor bed to improve local control after surgery. I-125 and Cs-131 are two of the most common forms of brachytherapy implant, however, I-125 is less widespread due to higher rates of radiation necrosis reported up to 15% [46]. Cs-131 seeds have similar efficacy with regard to local control and overall survival with lower rates of radiation necrosis than I-125 or SRS [47]. Although there is a paucity of large studies with long-term follow-up utilizing brachytherapy, the available evidence suggests there is utility in preventing local recurrence and minimizing the toxic effects of traditional radiation. Brachytherapy may be useful as a salvage therapy, with one series of nine patients with recurrent metastases who underwent repeat operation and Cs-131 seed placement demonstrating no cases of local recurrence at a median follow-up of 10.3 months [48]. The role of brachytherapy in management of recurrent brain metastasis is particularly appealing in patients who have already received traditional radiation, as a means to re-irradiate the tumor bed with less risk of radiation necrosis [49]. Limitations of brachytherapy include decreased propensity to affect rates of distant recurrence as much as traditional field radiation and availability and access of the radioactive materials make broad adoption in clinical practice challenging.

51.4 Conclusion

As cancer patients achieve better control of their systemic disease, effective treatment of intracranial metastases is crucial to improving quality of life and overall survival. Neurosurgical advances are paving the way for the next generation of neuro-oncologic care in which the treatment plan and surgical approach are tailored to the patient's pathology, treatment goals, and tumor characteristics. Minimally invasive and neurophysiology-based approaches now allow for resection of previously inaccessible lesions, serial biopsy for tissue characterization and advanced tumor visualization technologies are helping to increase the effectiveness and safety of tumor resection. Selected patients with recurrent and multiple metastases can still hope to benefit from surgical intervention to improve their overall survival and neurologic function. Further advances in laser-based treatment, optically-based intraoperative tumor visualization technologies, and other intraoperative adjuncts such as brachytherapy have remarkable potential to shape the next decade of neurosurgical treatment of brain metastasis.

Future Perspectives

- To what degree will increased use of minimally-invasive surgical techniques change how practitioners view surgically accessible versus inaccessible tumors?
- Which of the available intra-operative surgical adjuncts (brachytherapy, 5-ALA, etc.) will have the greatest impact when utilized as salvage therapy for recurrent metastases?
- What factors can assist in differentiating patients with leptomeningeal spread that will benefit from CSF diversion via shunting?
- How much can biopsy of a metastatic brain tumor of known pathology for additional molecular characterization inform systemic chemotherapy and radiation treatment regimens?
- Are there certain tumor types in which laser ablation may be beneficial as first-line therapy instead of radiotherapy for appropriately selected small brain metastases?

Patient Information and Guidelines

American Cancer Society https://www.cancer.org

Clinical Trials in Metastatic Brain Tumors https://www.cancer.gov/about-cancer/treatment/clinical-trials/adult-metastatic-brain-tumors

National Brain Tumor Foundation https://braintumor.org

Family Caregiver Alliance National Center on Caregiving https://www.caregiver.org/brain-tumor

References

1. Bindal RK, Sawaya R, Leavens ME, Hess KR, Taylor SH. Reoperation for recurrent metastatic brain tumors. J Neurosurg. 1995;83(4):600–4. https://doi.org/10.3171/jns.1995.83.4.0600.
2. Barnholtz-Sloan JS, Sloan AE, Davis FG, Vigneau FD, Lai P, Sawaya RE. Incidence proportions of brain metastases in patients diagnosed (1973 to 2001) in the Metropolitan Detroit Cancer Surveillance System. J Clin Oncol. 2004;22(14):2865–72. https://doi.org/10.1200/JCO.2004.12.149.
3. Brastianos PK, Carter SL, Santagata S, et al. Genomic characterization of brain metastases reveals branched evolution and potential therapeutic targets. Cancer Discov. 2015;5(11):1164–77. https://doi.org/10.1158/2159-8290.CD-15-0369.
4. Patchell RA, Tibbs PA, Walsh JW, Dempsey RJ, Maruyama Y, Kryscio RJ, Markesbery WR, Macdonald JS, Young B. A randomized trial of surgery in the treatment of single metastases to the brain. N Engl J Med. 1990;322(8):494–500. https://doi.org/10.1056/NEJM199002223220802.
5. Patchell RA, Tibbs PA, Walsh JW, et al. A randomized trial of surgery in the treatment of single metastases to the brain. N

Engl J Med. 1990;322(8):494–500. https://doi.org/10.1056/NEJM199002223220802.
6. Patchell RA, Tibbs PA, Regine WF, et al. Postoperative radiotherapy in the treatment of single metastases to the brain: A randomized trial. J Am Med Assoc. 1998;280(17):1485–9. https://doi.org/10.1001/jama.280.17.1485.
7. Mahajan A, Ahmed S, McAleer MF, et al. Post-operative stereotactic radiosurgery versus observation for completely resected brain metastases: a single-centre, randomised, controlled, phase 3 trial. Lancet Oncol. 2017;18(8):1040–8. https://doi.org/10.1016/S1470-2045(17)30414-X.
8. Nahed BV, Alvarez-Breckenridge C, Brastianos PK, et al. Congress of neurological surgeons systematic review and evidence-based guidelines on the role of surgery in the management of adults with metastatic brain tumors. Neurosurgery. 2019;84(3):E152–5. https://doi.org/10.1093/neuros/nyy542.
9. Delattre JY, Krol G, Thaler HT, Posner JB. Distribution of brain metastases. Arch Neurol. 1988;45(7):741–4. https://doi.org/10.1001/archneur.1988.00520310047016.
10. Medikonda R, Jackson CM, Feghali J, Lim M. The effects of postoperative neurological deficits on survival in patients with single brain metastasis. [Review]. Oper Neurosurg. 2020;19(6):628–34. https://doi.org/10.1093/ons/opaa224.
11. Balestrino R, Rudà R, Soffietti R. Brain metastasis from unknown primary tumour: Moving from old retrospective studies to clinical trials on targeted agents. Cancers. 2020;12(11):1–18. https://doi.org/10.3390/cancers12113350.
12. Schroeder T, Bittrich P, Kuhne JF, et al. Mapping distribution of brain metastases: does the primary tumor matter? J Neuro-Oncol. 2020;147(1):229–35. https://doi.org/10.1007/s11060-020-03419-6.
13. Füreder LM, Widhalm G, Gatterbauer B, et al. Brain metastases as first manifestation of advanced cancer: exploratory analysis of 459 patients at a tertiary care center. Clin Exp Metastasis. 2018;35(8):727–38. https://doi.org/10.1007/s10585-018-9947-1.
14. Schackert G, Steinmetz A, Meier U, Sobottka SB. Surgical management of single and multiple brain metastases: Results of a retrospective study. Onkologie. 2001;24(3):246–55. https://doi.org/10.1159/000055087.
15. Stark AM, Tscheslog H, Buhl R, Held-Feindt J, Mehdorn HM. Surgical treatment for brain metastases: prognostic factors and survival in 177 patients. Neurosurg Rev. 2005;28(2):115–9. https://doi.org/10.1007/s10143-004-0364-3.
16. Schackert G, Lindner C, Petschke S, Leimert M, Kirsch M. Retrospective study of 127 surgically treated patients with multiple brain metastases: Indication, prognostic factors, and outcome. Acta Neurochir. 2013;155(3):379–87. https://doi.org/10.1007/s00701-012-1606-8.
17. Bindal RK, Sawaya R, Leavens ME, Lee JJ. Surgical treatment of multiple brain metastases. J Neurosurg. 1993;79(2):210–6. https://doi.org/10.3171/jns.1993.79.2.0210.
18. Ogiwara H, Kalakota K, Rakhra SS, et al. Intracranial relapse rates and patterns, and survival trends following post-resection cavity radiosurgery for patients with single intracranial metastases. J Neuro-Oncol. 2012;108(1):141–6. https://doi.org/10.1007/s11060-012-0808-5.
19. Patel AJ, Suki D, Hatiboglu MA, et al. Factors influencing the risk of local recurrence after resection of a single brain metastasis. J Neurosurg. 2010;113(2):181–9. https://doi.org/10.3171/2009.11.JNS09659.
20. Schackert G, Schmiedel K, Lindner C, Leimert M, Kirsch M. Surgery of recurrent brain metastases: retrospective analysis of 67 patients. Acta Neurochir. 2013;155(10):1823–32. https://doi.org/10.1007/s00701-013-1821-y.
21. Weil RJ, Lonser RR. Selective excision of metastatic brain tumors originating in the motor cortex with preservation of function. J Clin Oncol. 2005;23(6):1209–17. https://doi.org/10.1200/JCO.2005.04.124.
22. Walter J, Kuhn SA, Waschke A, Kalff R, Ewald C. Operative treatment of subcortical metastatic tumours in the central region. J Neuro-Oncol. 2011;103(3):567–73. https://doi.org/10.1007/s11060-010-0420-5.
23. Kellogg RG, Munoz LF. Selective excision of cerebral metastases from the precentral gyrus. Surg Neurol Int. 2013;4:66. https://doi.org/10.4103/2152-7806.112189.
24. Iwadate Y, Namba H, Yamaura A. Significance of surgical resection for the treatment of multiple brain metastases. Anticancer Res. 2000;20(1B):573–7.
25. Pollock BE, Brown PD, Foote RL, Stafford SL, Schomberg PJ. Properly selected patients with multiple brain metastases may benefit from aggressive treatment of their intracranial disease. J Neuro-Oncol. 2003;61(1):73–80. https://doi.org/10.1023/a:1021262218151.
26. Ammirati M, Nahed BV, Andrews D, Chen CC, Olson JJ. Congress of neurological surgeons systematic review and evidence-based guidelines on treatment options for adults with multiple metastatic brain tumors. Neurosurgery. 2019;84(3):E180–2. https://doi.org/10.1093/neuros/nyy548.
27. Marenco-Hillembrand L, Alvarado-Estrada K, Chaichana KL. Contemporary surgical management of deep-seated metastatic brain tumors using minimally invasive approaches. Front Oncol. 2018;8:558. https://doi.org/10.3389/fonc.2018.00558.
28. Bakhsheshian J, Strickland BA, Jackson C, et al. Multicenter investigation of channel-based subcortical trans-sulcal exoscopic resection of metastatic brain tumors: a retrospective case series. Oper Neurosurg (Hagerstown). 2019;16(2):159–66. https://doi.org/10.1093/ons/opy079.
29. Gassie K, Wijesekera O, Chaichana KL. Minimally invasive tubular retractor-assisted biopsy and resection of subcortical intra-axial gliomas and other neoplasms. J Neurosurg Sci. 2018;62(6):682–9. https://doi.org/10.23736/S0390-5616.18.04466-1.
30. Alattar AA, Bartek J, Chiang VL, et al. Stereotactic laser ablation as treatment of brain metastases recurring after stereotactic radiosurgery: a systematic literature review. World Neurosurg. 2019;128:134–42. https://doi.org/10.1016/j.wneu.2019.04.200.
31. Ahluwalia M, Barnett GH, Deng D, et al. Laser ablation after stereotactic radiosurgery: a multicenter prospective study in patients with metastatic brain tumors and radiation necrosis. J Neurosurg. 2019;130(3):804–11. https://doi.org/10.3171/2017.11.JNS171273.
32. Holste KG, Orringer DA. Laser interstitial thermal therapy. Neuro-Oncology Adv. 2020;2(1):vdz035. https://doi.org/10.1093/noajnl/vdz035.
33. Ahluwalia M, Barnett GH, Deng D, et al. Laser ablation after stereotactic radiosurgery: a multicenter prospective study in patients with metastatic brain tumors and radiation necrosis. J Neurosurg. 2018;130(3):804–11. https://doi.org/10.3171/2017.11.JNS171273.
34. Salehi A, Kamath AA, Leuthardt EC, Kim AH. Management of intracranial metastatic disease with laser interstitial thermal therapy. Front Oncol. 2018;8:499. https://doi.org/10.3389/fonc.2018.00499.
35. Ali MA, Carroll KT, Rennert RC, et al. Stereotactic laser ablation as treatment for brain metastases that recur after stereotactic radiosurgery: a multiinstitutional experience. Neurosurg Focus. 2016;41(4):E11. https://doi.org/10.3171/2016.7.FOCUS16227.
36. Kubben PL, ter Meulen KJ, Schijns OEMG, ter Laak-Poort MP, van Overbeeke JJ, van Santbrink H. Intraoperative MRI-guided resection of glioblastoma multiforme: a systematic review. Lancet Oncol. 2011;12(11):1062–70. https://doi.org/10.1016/S1470-2045(11)70130-9.
37. Senft C, Franz K, Blasel S, et al. Influence of iMRI-guidance on the extent of resection and survival of patients with glioblastoma multiforme. Technol Cancer Res Treat. 2010;9(4):339–46.
38. Akbari SHA, Sylvester PT, Kulwin C, et al. Initial experience using intraoperative magnetic resonance imaging during a trans-sulcal tubular retractor approach for the resection of deep-seated

brain tumors: a case series. Oper Neurosurg (Hagerstown). 2019;16(3):292–301. https://doi.org/10.1093/ons/opy108.
39. Valdés PA, Jacobs V, Harris BT, et al. Quantitative fluorescence using 5-aminolevulinic acid–induced protoporphyrin IX biomarker as a surgical adjunct in low-grade glioma surgery. J Neurosurg. 2015;123(3):771–80. https://doi.org/10.3171/2014.12.JNS14391.
40. Patel AJ, Suki D, Hatiboglu MA, Rao VY, Fox BD, Sawaya R. Impact of surgical methodology on the complication rate and functional outcome of patients with a single brain metastasis. J Neurosurg. 2015;122(5):1132–43. https://doi.org/10.3171/2014.9.JNS13939.
41. Suki D, Hatiboglu MA, Patel AJ, et al. Comparative risk of leptomeningeal dissemination of cancer after surgery or stereotactic radiosurgery for a single supratentorial solid tumor metastasis. Neurosurgery. 2009;64(4):664–74; discussion 674–676. https://doi.org/10.1227/01.NEU.0000341535.53720.3E.
42. Hyun J-W, Jeong IH, Joung A, Cho HJ, Kim S-H, Kim HJ. Leptomeningeal metastasis: clinical experience of 519 cases. Eur J Cancer. 2016;56:107–14. https://doi.org/10.1016/j.ejca.2015.12.021.
43. Lee SH, Kong DS, Seol HJ, Nam D-H, Lee J-I. Ventriculoperitoneal shunt for hydrocephalus caused by central nervous system metastasis. J Neuro-Oncol. 2011;104(2):545–51. https://doi.org/10.1007/s11060-010-0512-2.
44. Le Rhun E, Taillibert S, Chamberlain MC. Carcinomatous meningitis: Leptomeningeal metastases in solid tumors. Surg Neurol Int. 2013;4(Suppl 4):S265–88. https://doi.org/10.4103/2152-7806.111304.
45. Lin N, Dunn IF, Glantz M, et al. Benefit of ventriculoperitoneal cerebrospinal fluid shunting and intrathecal chemotherapy in neoplastic meningitis: a retrospective, case-controlled study. J Neurosurg. 2011;115(4):730–6. https://doi.org/10.3171/2011.5.JNS101768.
46. Raleigh DR, Seymour ZA, Tomlin B, et al. Resection and brain brachytherapy with permanent iodine-125 sources for brain metastasis. J Neurosurg. 2017;126(6):1749–55. https://doi.org/10.3171/2016.4.JNS152530.
47. Chitti B, Goyal S, Sherman JH, et al. The role of brachytherapy in the management of brain metastases: a systematic review. J Contemp Brachytherapy. 2020;12(1):67–83. https://doi.org/10.5114/jcb.2020.93543.
48. Xia Y, Mashouf LA, Baker BR, et al. Outcomes of metastatic brain lesions treated with radioactive Cs-131 seeds after surgery: experience from one institution. Cureus. 2018;10(7):e3075. https://doi.org/10.7759/cureus.3075.
49. Chidambaram S, Pannullo SC, Schwartz TH, Wernicke AG. Reirradiation of recurrent brain metastases: where do we stand? World Neurosurg. 2019;125:156–63. https://doi.org/10.1016/j.wneu.2019.01.182.

Brain Metastases: Current and Future Pharmacological Treatment

52

Erika Santos Horta and Tobias Walbert

Abstract

Solid tumor metastasis to the brain is a frequent event that increases morbidity and mortality of patients suffering from an already life-threatening disease. Pharmacological treatment efficacy is challenged by the difficulties of delivering drugs to the central nervous system. Traditional chemotherapy has limited success. Fortunately, immunotherapy and target-based therapies are broadening treatment options with promising results in clinical trials and in real-world settings.

Learning Objectives

- To understand the challenges of pharmacotherapy when treating brain metastases.
- To be able to evaluate the best pharmacotherapy options for different types of brain metastases.
- Be aware of the side effects of pharmacotherapy for brain metastases.

52.1 Introduction

Due to seeding of systemic cancers, 20–40% of patients will eventually present with brain metastases, and the incidence is increasing due to better systemic control of the disease and expansion of brain imaging [1, 2]. As a consequence, metastases of solid tumors to the brain represent the most common intracranial malignancy [3, 4]. Brain metastases increase patient morbidity and mortality [4–6]. The solid tumors that most commonly metastasize to the brain are melanoma, lung cancer, breast cancer, renal cancer, and colorectal cancer [2, 7, 8].

Molecular pathology findings are revealing that some mutations can stratify a patient's risk for developing brain metastases and offer a molecular target for specific therapies. For example, mutations of the epidermal growth factor receptor (EGFR) or anaplastic lymphoma kinase (ALK) in patients with non-small-cell lung cancer (NSCLC), KRAS/RAS mutation in colon cancer, and human epidermal growth factor receptor 2 (HER2) in breast cancer are risk factors for brain metastasis [2, 8–10].

There are many challenges to the treatment of brain metastases, including the blood-brain and the blood-CSF barriers, different tumor clones, and tumor microenvironment, among others [6]. The blood–brain barrier (BBB) is a functional structure formed by endothelial cells connected by tight junctions with astrocytes and pericytes. The BBB regulates the transport of molecules from the blood to the brain parenchyma through the tight junctions and efflux pumps in the endothelial cells. Doing so protects the brain from toxins, as well as foreign substances, maintaining a stable microenvironment [3, 6, 11]. The blood-CSF barrier is similar to the BBB but is formed by choroid plexus epithelial cells and capillaries. Those barriers can alter the delivery, distribution, and accumulation of chemotherapies, targeted therapies, and antibodies [6]. Other challenges are tumor cell-astrocyte gap cell junctions that directly connect the cytoplasm of two cells as well as alterations in microglia that can alter the tumor microenvironment in the central nervous system (CNS). Another obstacle to treat brain metastases is that they can be formed by different clones than the primary tumor and therefore respond differently to treatment than the primary tumor [3, 6]. For all of these reasons, the treatment of brain metastasis has focused on surgery and radiation therapy, while pharmacotherapy is mostly used in the

E. S. Horta (✉)
University of Arkansas for Medical Sciences, Little Rock, AR, USA
e-mail: esantoshorta@uams.edu

T. Walbert
Hermelin Brain Tumor Center, Henry Ford Health System, Detroit, MI, USA
e-mail: twalber1@hfhs.org

S. P. Leong et al. (eds.), *Cancer Metastasis Through the Lymphovascular System*,
https://doi.org/10.1007/978-3-030-93084-4_52

adjuvant setting [6]. On the other hand, target-based therapies have shown promising results when brain metastases demonstrate selective mutations [6].

Traditional chemotherapy aims to achieve a cure or decrease the size of a tumor by interfering in the cell cycle and therefore killing predominantly cancer cells. A common mechanism of action of those substances is to add errors to the DNA of cells. Currently, cancer treatment has expanded with the use of immunotherapy and target-based therapy. Immunotherapy sensitizes the patient's own immune system to recognize and fight cancer through the presentation of cancer antigens. Although virus therapy can also be used to enhance the immune system to fight cancers, immunotherapy is mainly represented by antibodies. Target-based therapy tries to dismantle some of the intracellular pathways that cancerous cells use by aiming at a particular mutation. Tyrosine kinase inhibitors (TKIs) are a good example of target-based therapies as they block enzymes that work in intracellular pathways. Another advantage of TKIs and other targeted therapies is that the molecules are smaller than the traditional chemotherapy substances and therefore can cross the BBB [8, 11].

In this chapter, we will discuss pharmacotherapy (traditional chemotherapy, target-based chemotherapies, and immunotherapy represented by antibodies) for the most common brain metastases.

52.2 Lung Cancer

Lung cancer is divided into small-cell lung carcinoma (SCLC) and non-small-cell lung cancer (NSCLC) which is further subdivided into adenocarcinoma, squamous cell carcinoma, and large-cell carcinoma. Although SCLC metastases have more tropism to the brain than NSCLC, NSCLC metastases are more common due to the higher prevalence of NSCLC [7]. Among the different subtypes of NSCLC, adenocarcinoma and large-cell carcinoma have more tropism to the brain. SCLC metastasizes to the brain earlier than NSCLC and is more sensitive to standard chemotherapeutic agents. In most institutions, patients with SCLC are treated with prophylactic whole-brain radiation therapy, but the increased sensitivity of SCLC can be further targeted when combining chemotherapy with radiotherapy [6, 12]. NSCLC is frequently associated with mutations in EGFR and anaplastic lymphoma kinase (ALK) which can be targeted with systemic therapy and targeted agents [9]. Those mutations are also associated with better survival when compared with wild-type tumors [9].

52.2.1 Chemotherapy

Response rates vary from 30–80% in SCLC to 10–30% in NSCLC. Chemotherapy is a good choice for multifocal disease, when radiation therapy and surgery are not an option, asymptomatic metastases, and SCLC [12]. Traditional chemotherapy protocols include platinum compounds (alkylating agents) in monotherapy or in combination with pemetrexed, vinorelbine, or etoposide as first-line therapy [4, 12].

52.2.2 Target-Based Therapy

Tyrosine kinase inhibitors (TKIs) targeting ALK inhibitor and EGFR are now considered to be first-line therapy options for patients with lung cancer bearing those mutations [9].

ALK inhibitors have been FDA approved for NSCLC therapy since 2011 and have been shown to be efficacious in patients with brain metastases. Among ALK inhibitors, alectinib, brigatinib, ceritinib, and lorlatinib seem to be more efficacious than crizotinib for CNS disease, due to better ability to cross the BBB. In clinical trials, patients with NSCLC receiving treatment with ALK inhibitors for systemic disease presented with CNS metastases later than controls [4, 11, 13].

TKIs that act to reversibly inhibit EGFR, like erlotinib and osimertinib, have shown better response rates than traditional chemotherapy and better overall survival (OS) in patients with EGFR-mutant NSCLC brain metastases [1, 4]. In the phase 3 clinical trial AURA3, the osimertinib group showed a longer CNS progression-free survival (11.7 months) than the platinum-pemetrexed group (5.6 months) [14]. The percentage of patients alive in the 6 and 12 months (75% and 43%, respectively) in the osimertinib group was also improved when compared with the chemotherapy group (45% and 17%, respectively). The group with osimertinib reached CNS response rate of 70% versus 31% in the chemotherapy group. Osimertinib is now one of the first-line treatments for EGFR-positive metastatic NSCLC, and it showed activity in patients with leptomeningeal disease [14–16].

Clinical trials have evaluated if adding radiation therapy to TKI therapy in patients with brain metastases increase OS or progression-free survival (PFS) in patients with brain metastases [2]. A recent meta-analysis concluded that there is no significant increase of OS or PFS, but added toxicty [9].

52.2.3 Immunotherapy

Checkpoint inhibitors (anti-PD1, anti-PDL1, and anti-CTL4) were studied in patients with NSCLC and brain metastasis and showed sustained activity in patients with PDL1 expression over 1%. Intracranial response of 33–36% in chemotherapy-naive patients and in pre-treated patients is seen in NSCLC and in SCLC [1, 17, 18].

The potential benefits of using the anti-VEGF antibody bevacizumab remain unclear. A recent meta-analysis in patients with NSCLC concluded that bevacizumab might protect patients from developing new brain metastases and it might control existing metastases when compared with conventional chemotherapy [19]. The medium progression-free survival of patients with brain metastases being treated with bevacizumab was 7.1 months, and medium overall survival was 13.5 months. A pooled analysis showed better OS and PFS hazard ratio in patients with bevacizumab when compared with other traditional chemotherapies [19].

52.3 Breast Cancer

Breast cancer is the second most common cause of brain metastases [8]. It is often classified by the type of hormone receptors (estrogen [ER], progesterone [PR], human epidermal growth factor 2 [HER2]) because it guides pharmacotherapy choices and predicts the specific tendency to metastasize to the brain. For example, triple-negative breast cancers are usually treated with traditional chemotherapy, as radiosensitivity is low and there is no role for hormone therapy.

52.3.1 Chemotherapy

The great challenge for breast cancer treatment is that the majority of chemotherapy agents used for systemic therapy do not cross the BBB [2]. This is considered one of the reasons the incidence of breast cancer metastasis to the brain is increasing, while therapy for systemic disease, especially in hormone receptor-positive patients, has resulted in improved outcomes [20]. Options for chemotherapy include capecitabine, cyclophosphamide, vincristine, temozolomide, and many others, with response rates varying from 30 to 60%, when chemotherapy was added to radiation therapy [2, 12, 20]. Unfortunately, median OS for breast cancer patients is less than one year in phase 2 and 3 trials [2, 20].

For ER-positive tumors, tamoxifen and aromatase inhibitors should be offered for brain metastasis patients as first-line therapy, as these agents reach high concentrations in the brain parenchyma and increase patient survival [20, 21]. The increased survival was also seen in some patients with leptomeningeal disease [21].

52.3.2 Target-Based Therapy

Dual HER2/EGFR TKIs block tyrosine phosphorylation and intracellular pathways with good brain parenchyma penetrance. Lapatinib, a HER2/EGFR TKIs, is FDA approved for treatment of breast cancer since 2007. Those TKIs have been used in combination with capecitabine and temozolomide with a response rate of 20–60%, decrease in 6–10% in the volume of intracranial lesions, and better OS than monotherapy with trastuzumab [2, 20, 22]. Therefore, for HER2-positive patients with brain metastases, HER2/EGFR TKIs are currently considered an effective therapeutic option [1, 8].

TKI have been associated with increased risks for radiation necrosis when added to radiation therapy [23]. Tucatinib might be associated with less adverse events than the other dual TKIs [8].

PARP inhibitors have been tested in phase 1 trials and have been well tolerated, but they need to be further investigated in patients with brain metastases. Those medications regulate BRCA1-promoter methylation which is thought to play an important role in breast cancer pathogenesis in the setting of BRCA1/2 mutation carriers [22, 24].

CDK4/6 inhibitors (palbociclib, ribociclib, abemaciclib) are FDA-approved medications for metastatic breast cancers that are HER2-negative and estrogen/progesterone-positive. Current clinical trials show that these medications are also efficient in patients with brain metastases [20, 25].

52.3.3 Immunotherapy

Trastuzumab and pertuzumab are antibodies against the HER2 cell receptor and are FDA approved for HER2-positive breast cancer therapy since 1998 and 2012, respectively. Trastuzumab in addition to chemotherapy increases OS and PFS and, when conjugated with a cytotoxic microtubule molecule, has shown to decrease the size of brain metastases [8, 20]. The CLEOPATRA clinical trial showed that pertuzumab added to trastuzumab and docetaxel delays the time to onset of brain metastasis and increases the OS in patients with brain metastasis from breast cancer [26]. Based on these findings, the combination of pertuzumab and trastuzumab should be offered to HER2 breast cancer patients [12].

52.4 Melanoma

Melanomas exhibit high levels of tropism to the brain. This is reflected by the fact that one third of all melanoma patients present with brain metastases [2]. Before the advent of monoclonal antibodies and target-based therapies, the outcome for melanoma patients with brain metastases was dismal with an OS of only weeks [2]. BRAF and checkpoint inhibitors are changing this trajectory with median overall survival of one year in phase 2 clinical trials [2, 6, 27].

52.4.1 Chemotherapy

Cytotoxic chemotherapy has failed to show efficacy and therefore has no significant role in the treatment of patients with melanoma brain metastases, even when combined with radiation therapy [2, 12].

52.4.2 Target-Based Therapy

Melanoma frequently presents with BRAF V600 mutations, and in patients with brain metastases, clinical trials using BRAF inhibitors in monotherapy or associated with a MEK inhibitor showed intracranial response near 40–60% in treatment-naïve patients and over 30% in previously treated patients [1, 6, 27]. Unfortunately, resistance to treatment has been seen after one year [6, 27]. Combination therapies of BRAF plus MEK inhibitors have shown high response rates, decrease in side effects, and without increase in OS [6, 27].

After published case series and case reports summing up over 20 cases of patients with leptomeningeal disease with response to TKIs, the 2019 Consensus of Melanoma Research Foundation suggests that BRAF monotherapy or the combination of BRAF and MEK inhibitors can be considered in the treatment of patients with leptomeningeal disease, but data is still too sparse to make stronger recommendations [28].

52.4.3 Immunotherapy

Anti-CTL4 immunotherapies were FDA approved for metastatic melanoma in 2011. In 2012, an open-label, phase 2 trial showed that ipilimumab could be an option for patients with brain metastasis [29]. Although ipilimumab cannot cross the BBB due to its particle size, it is thought to stimulate the immune system against melanoma, and the stimulated CD8 T cells can then infiltrate and attack metastatic cancer cells in the CNS [30].

Checkpoint inhibitors have been successful in monotherapy for brain metastases and in combination with traditional chemotherapy or with other immunotherapy [28, 30, 31]. Intracranial response rates have been reported to reach 44–56% [1, 32]. The response to therapies that incorporate these inhibitors seems to be more durable than target-based therapies alone [28, 30, 31]. However, checkpoint inhibitors are less effective when steroids are used. The combination of checkpoint inhibitors with radiation therapy, especially stereotactic radiosurgery, is thought to increase efficacy but also the risk of radiation necrosis [28, 30, 31]. The increase of survival with checkpoint inhibitor and SRS might be higher for anti-PDL1 than for anti-CTL4 [29–31].

There are three ongoing clinical trials with checkpoint inhibitors that are now allowing patients with leptomeningeal disease to participate in, but there are case reports describing response and even complete response of leptomeningeal disease. The 2019 Consensus of Melanoma Research Foundation suggests that trials with checkpoint inhibitors are reasonable [28].

52.5 Renal Cancer

Renal cell carcinoma has a 5-year cumulative incidence of brain metastases of 10% with a mean overall survival of 4–11 months after resection, independently if WBRT or traditional chemotherapy is used [2]. Currently, it is considered a good practice to offer TKIs to patients with renal cell carcinoma with brain metastases [12].

52.5.1 Chemotherapy

Renal cell carcinoma is a chemotherapy- and radiation-resistant tumor as neither whole-brain radiation nor systemic traditional chemotherapies have extended the overall survival or progression-free survival [2]. Systemic chemotherapy can be used in an attempt to control extra-CNS disease.

52.5.2 Target-Based Therapy

VEGFR-TKIs, like sunitinib and pazopanib, have brain penetrance and have shown the ability to stabilize brain metastases and increase overall survival in clinical studies, achieving a response rate of 4–13% and stable disease in 33–60% of patients [33–35]. Efficacy seems to increase when combined with stereotactic radiosurgery (SRS), at the cost of radiation necrosis [12, 36]. Retrospective studies and expanded-access trials indicated that patients with renal cancer developed brain metastases later and in fewer numbers when using TKIs [33–37]. While we are waiting for the final results of clinical trials that are evaluating VEGFR-TKIs in patients with metastatic renal cancer, it is considered a good practice to offer VEGFR-TKI to patients with renal cancer and brain metastases [35, 38].

While mTOR TKIs like everolimus are approved for advanced renal cell carcinoma and are considered safe in patients with brain metastases, they have not yet shown to be efficacious [35].

52.5.3 Immunotherapy

Immunotherapy for brain metastases in renal cell carcinoma has had disappointing results. Although case reports describe successful use of interleukin-2, this treatment was evaluated in a cohort of patients with metastatic renal cancer or melanoma [39]. Objective response was low, 5.6 %, with grade 3 or higher neurological toxicity seen in 7–16 % of patients.

Although nivolumab is indicated for renal cell carcinoma, checkpoint inhibitors in monotherapy for brain metastases have not shown any efficacy [1, 22].

52.6 Colon Cancer

Brain metastases are considered a late event of colon cancer and typically present when the patient has metastases to other organs like the liver and lungs resulting in only 30% of patients with brain metastasis surviving more than one year [40, 41]. Risk factors for brain metastases are age <60 years, RAS mutation, and adenocarcinoma [40, 41]. Due to the lack of clinical trials, there is little knowledge on the best pharmaceutical treatment of brain metastasis in colorectal patients; therefore, protocols are based on surgery and radiation oncology.

52.6.1 Chemotherapy

Patients likely had already received chemotherapy before having brain metastases; therefore, options are limited, and protocols are usually a combination of two or three medications [40]. Options include capecitabine, oxaliplatin, fluorouracil, and irinotecan.

52.6.2 Target-Based Therapy

BRAF mutations are found in 5–15% of patients with colorectal cancer. TKIs are being used more and more frequently in patients with colon cancer, but their safety and efficacy are still being studied in the context of brain metastases [40].

52.6.3 Immunotherapy

A retrospective study of patients with brain metastases showed that bevacizumab, an antiangiogenic monoclonal antibody against VEGF, improved survival from 59 to 151 days [41]. In another retrospective study from a single institution, the time to develop brain metastases in patients with colorectal cancer treated with bevacizumab was longer than in patients that did not received them [42].

52.7 Symptomatic Treatment

Symptomatic treatment of patients with brain metastases focuses on the control of peritumoral edema, increased intracranial pressure, and the treatment of seizures.

Corticosteroids, such as dexamethasone, are the first-line treatment for increased intracranial pressure or symptomatic peritumoral edema, and treatment and dosing should be guided by clinical presentation and neurological deficits rather than by MRI findings [12, 38]. A fast down titration is recommended after patients improve to control side effects like diabetes, osteoporosis, insomnia, increased blood pressure, osteonecrosis, gastrointestinal hemorrhage, and immune-suppression [12, 38]. As alternatives, *Boswellia* and osmolar agents can be considered, and a randomized trial using a short course of bevacizumab has been shown to decrease radiation necrosis-induced edema [12, 38].

While brain metastases can cause seizures, anti-epileptic medications are not recommended prophylactically [12, 38]. Non-enzyme-inducer medications like levetiracetam and lamotrigine are preferred over phenytoin, phenobarbital, and carbamazepine due to a lack of interactions with other medications and better side effect profile [38].

52.8 Treatment-Related Toxicity

When treating a patient, not only efficacy must be analyzed but also the potential for treatment-related toxicity. Traditional chemotherapies have many side effects, which include chemotherapy-induced nausea, hair loss, cognitive dysfunction, cardiomyopathy, neuropathy, and many others [2, 5].

Immunotherapy is also known to cause adverse events such as colitis, pneumonitis, and other endocrinological and neurological symptoms [1, 43–45]. Some of those adverse events differ depending on the type of immunotherapy. For example, myasthenia gravis and hypophysitis are seen with anti-CTL4, anti-PD1, and anti-PDL1 [1, 4, 43].Five other adverse events like encephalopathy, headache, and neuropathy are observed in many classes of immunotherapy [1, 43, 45]. Vitiligo secondary to immunotherapy seems to be more prevalent in patients with melanoma. When combined with radiation therapy, immunotherapy can increase the likelihood of radiation necrosis, encephalopathy, and seizures [9, 12, 24, 28, 43]. Target-based therapies are known to cause photosensibility, paronychia, diarrhea, cardiotoxicity, and neurological adverse events [45].

The treatment of immunotherapy-induced toxicity includes temporary or permanent discontinuation of therapy and, when needed, associated corticosteroids in the form of prednisone or dexamethasone [44, 45]. After the patient improves, resuming therapy is possible and should be considered on a case-by-case basis [43–45].

52.9 Conclusions and Future Directions

In the last ten years, basic research and clinical trials have greatly increased the understanding of brain cancer metastases. Pharmacotherapy for the treatment of brain metastases has increased and is now offering new options that have shown meaningful increases of progression-free and overall survival. Traditional chemotherapy has been replaced by immunotherapy and targeted therapies that are better tolerated and have resulted in promising results. However, challenges specific to brain metastases such as limited BBB permeability remain, and newer drug delivery methods like liposomes, nanoparticles, and pegylation or using mannitol to increase therapeutic response need to be explored. Ultrasound for blood BBB disruption and implantable biomaterial devices are also being studied to increase delivery of drugs to the brain parenchyma. Other lines of research are focused on the actual prevention of these deadly malignancies.

Questions to be answered after reading this chapter are:

- How can we overcome some of the challenges of treating brain metastases with systemic pharmacotherapies?
- How do pharmacotherapy options vary per type of tumor?
- Can we integrate pharmacotherapy with other types of treatment?
- My patient has brain metastases. Should I start an anti-seizure medication?
- What are the potential side effects of pharmacotherapy for brain metastases?

References

1. Lauko A, Thapa B, Venur V, Ahluwalia M. Management of brain metastases in the new era of checkpoint inhibition. Curr Neurol Neurosci. 2018;18:70.
2. Lombardi G, Di Stefano AL, Farina P, Zagonel V, Tabouret E. Systemic treatments for brain metastases from breast cancer, non-small cell lung cancer, melanoma and renal cell carcinoma: an overview of the literature. Cancer Treat Rev. 2014; https://doi.org/10.1016/j.ctrv.2014.05.007.
3. Achrol AS, et al. Brain metastases. Nat Rev Dis Primers. 2019; https://doi.org/10.1038/s41572-018-0055-y.
4. Aizer AA, Lee EQ. Brain metastases. Neurol Clin. 2018; https://doi.org/10.1016/j.ncl.2018.04.010.
5. Walbert T, Horta E. Health-related quality of life related to toxicity treatments in central nervous system metastases. In: Ahluwalia M, Metellus PSR, editors. *Central nervous system metastasis*. SpringerLink; 2020. p. 373–82.
6. Lowery FJ, Yu D. Brain metastasis: unique challenges and open opportunities. Biochim Biophys Acta Rev Cancer. 2017; https://doi.org/10.1016/j.bbcan.2016.12.001.
7. Cagney DN, et al. Incidence and prognosis of patients with brain metastases at diagnosis of systemic malignancy: a population-based study. Neuro-Oncology. 2017; https://doi.org/10.1093/neuonc/nox077.
8. Duchnowska R, Loibl S, Jassem J. Tyrosine kinase inhibitors for brain metastases in HER2-positive breast cancer. Cancer Treat Rev. 2018;67:71–7.
9. Singh R, et al. Brain metastases from non-small cell lung cancer with EGFR or ALK mutations: a systematic review and meta-analysis of multidisciplinary approaches. Radiother Oncol. 2020;144:165–79.
10. Christensen TD, Spindler KLG, Palshof JA, Nielsen DL. Systematic review: brain metastases from colorectal cancer-incidence and patient characteristics. BMC Cancer. 2016; https://doi.org/10.1186/s12885-016-2290-5.
11. Dodson C, Richards TJ, Smith DA, Ramaiya NH. Tyrosine kinase inhibitor therapy for brain metastases in non-small-cell lung cancer: a primer for radiologists. Am J Neuroradiol. 2020; https://doi.org/10.3174/AJNR.A6477.
12. Soffietti R, Abacioglu U, Baumert B. Diagnosis and treatment of brain metastases from solid tumors: guidelines from the European association of neuro-oncology (EANO). Neuro-Oncology. 2017; https://doi.org/10.1093/neuonc/now241.
13. Peters S, et al. Alectinib versus Crizotinib in untreated ALK -positive non–small-cell lung cancer. N Engl J Med. 2017; https://doi.org/10.1056/nejmoa1704795.
14. Wu YL, et al. CNS efficacy of osimertinib in patients with T-790M-positive advanced nonsmall-cell lung cancer: data from a randomized phase III trial (AURA 3). J Clin Oncol. 2018;10:2702–9.
15. Reungwetwattana T, et al. CNS response to osimertinib versus standard epidermal growth factor receptor tyrosine kinase inhibitors in patients with untreated EGFR-mutated advanced non-small-cell lung cancer. J Clin Oncol. 2018; https://doi.org/10.1200/JCO.2018.78.3118.
16. Planchard D, et al. Osimertinib vs standard of care (SoC) EGFR-TKI as first-line therapy in patients (pts) with untreated EGFRm advanced NSCLC: FLAURA post-progression outcomes. Ann Oncol. 2018; https://doi.org/10.1093/annonc/mdy425.
17. Wang S, Hu C, Xie F, Liu Y. Use of programmed death receptor-1 and/or programmed death ligand 1 inhibitors for the treatment of brain metastasis of lung cancer. Onco Targets Ther. 2020;13:667–83.
18. Goldman JW, et al. Nivolumab (nivo) in patients (pts) with advanced (adv) NSCLC and central nervous system (CNS) metastases (mets). J Clin Oncol. 2016; https://doi.org/10.1200/jco.2016.34.15_suppl.9038.
19. Liang P, et al. Bevacizumab for non-small cell lung cancer patients with brain metastasis: a meta-analysis. Open Med. 2020;15:589–97.
20. Shah N, et al. Investigational chemotherapy and novel pharmacokinetic mechanisms for the treatment of breast cancer brain metastases. Pharmacol Res. 2018; https://doi.org/10.1016/j.phrs.2018.03.021.
21. Bergen ES, et al. Continued endocrine therapy is associated with improved survival in patients with breast cancer brain metastases. Clin Cancer Res. 2019; https://doi.org/10.1158/1078-0432.CCR-18-1968.
22. Costa R, et al. Developmental therapeutics for patients with breast cancer and central nervous system metastasis: current landscape and future perspectives. Ann Oncol. 2017; https://doi.org/10.1093/annonc/mdw532.
23. Lin NU, et al. Phase I study of lapatinib (L) in combination with whole-brain radiation therapy (WBRT) in patients (pts) with brain metastases from HER2-positive breast cancer. J Clin Oncol. 2010; https://doi.org/10.1200/jco.2010.28.15_suppl.1154.

24. Mehta MP, et al. Veliparib in combination with whole brain radiation therapy in patients with brain metastases: results of a phase 1 study. J Neuro-Oncol. 2015; https://doi.org/10.1007/s11060-015-1733-1.
25. Abraham J, et al. Use of cyclin-dependent kinase (CDK) 4/6 inhibitors for hormone receptor-positive, human epidermal growth factor receptor 2-negative, metastatic breast cancer: a roundtable discussion by The Breast Cancer Therapy Expert Group (BCTEG). Breast Cancer Res Treat. 2018;171:11–20.
26. Baselga J, et al. Pertuzumab plus Trastuzumab plus Docetaxel for metastatic breast cancer. N Engl J Med. 2012; https://doi.org/10.1056/nejmoa1113216.
27. Davies MA, et al. Dabrafenib plus trametinib in patients with BRAFV600-mutant melanoma brain metastases (COMBI-MB): a multicentre, multicohort, open-label, phase 2 trial. Lancet Oncol. 2017; https://doi.org/10.1016/S1470-2045(17)30429-1.
28. Glitza IC, et al. Leptomeningeal disese in melanoma patients: an update to treatment, challenges, and future directions. Pigment Cell Melanoma Res. 2020;33(4):527–41. https://doi.org/10.1111/pcmr.12861.
29. Margolin K, et al. Ipilimumab in patients with melanoma and brain metastases: an open-label, phase 2 trial. Lancet Oncol. 2012;13:459–65.
30. Fonkem E, et al. Melanoma brain metastasis: overview of current management and emerging targeted therapies. Expert Rev Neurother. 2012; https://doi.org/10.1586/ern.12.111.
31. van Opijnen M, Dirven L, Coremans I, Taphoorn M, Kapiteijn E. The impact of current treatment modalities on the outcomes of patients with melanoma brain metastases: a systematic review. Int J Cancer. 2020;146:1479–89.
32. Kluger HM, et al. Long-term survival of patients with melanoma with active brain metastases treated with pembrolizumab on a phase II trial. J Clin Oncol. 2019; https://doi.org/10.1200/JCO.18.00204.
33. Gore ME, et al. Final results from the large sunitinib global expanded-access trial in metastatic renal cell carcinoma. Br J Cancer. 2015; https://doi.org/10.1038/bjc.2015.196.
34. Matrana MR, et al. Outcomes of patients with metastatic clear-cell renal cell carcinoma treated with pazopanib after disease progression with other targeted therapies. Eur J Cancer. 2013; https://doi.org/10.1016/j.ejca.2013.06.003.
35. Matsui Y. Current multimodality treatments against brain metastases from renal cell carcinoma. Cancer. 2020; https://doi.org/10.3390/cancers12102875.
36. Juloori A, et al. Overall survival and response to radiation and targeted therapies among patients with renal cell carcinoma brain metastases. J Neurosurg. 2020; https://doi.org/10.3171/2018.8.JNS182100.
37. Verma J, et al. Impact of tyrosine kinase inhibitors on the incidence of brain metastasis in metastatic renal cell carcinoma. Cancer. 2011;117:4958–65. https://doi.org/10.1002/cncr.26138.
38. Roth P, et al. Neurological and vascular complications of primary and secondary brain tumours: EANO-ESMO Clinical Practice Guidelines for prophylaxis, diagnosis, treatment and follow-up. Ann Oncol. 2021;32:171–82.
39. Guirguis L, et al. Safety and efficacy of high-dose interleukin-2 therapy in patients with brain metastases. J Immunother. 2002;25:82–7.
40. Mege D, Sans A, Ouaissi M, Iannelli A, Sielezneff I. Brain metastases from colorectal cancer: characteristics and management. ANZ J Surg. 2018; https://doi.org/10.1111/ans.14107.
41. Chahine G, et al. Colorectal cancer and brain metastases: an aggressive disease with a different response to treatment. Tumori. 2019; https://doi.org/10.1177/0300891618765541.
42. Fountzilas C, et al. Clinical characteristics and treatment outcomes of patients with colorectal cancer who develop brain metastasis: a single institution experience. J Gastrointest Oncol. 2017;8:55–63.
43. Horta ES, et al. Neural autoantibody clusters aid diagnosis of cancer. Clin Cancer Res. 2014;20:3862–9.
44. Choi J, Lee S. Clinical characteristics and treatment of immune-related adverse events of immune checkpoint inhibitors. Immune Netw. 2020;20:e9.
45. Baldo B, Pham N. Targeted drugs for cancer therapy: small molecules and monoclonal antibodies. Drug Allergy. 2020:595–644. https://doi.org/10.1007/978-3-030-51740-3_14.

Radiotherapy for CNS Metastases

53

Alexander N. Slade, Mark Ashamalla, and Samuel Ryu

Abstract

Brain metastases are common among cancer patients, estimated to impact nearly one in five patients throughout the course of their disease. Radiation therapy, along with surgery, comprises the primary modalities used to manage brain metastases. This chapter considers contemporary prognostic approaches to patients with brain metastases. We then discuss the evolution of radiation therapy in the management of brain metastases from whole brain radiation therapy to stereotactic radiosurgery, in which it is now possible to deliver highly targeted radiation therapy to brain metastases (or the surgery cavities thereof) while sparing normal structures. We also review recent advances in limiting the neurocognitive toxicities in patients receiving whole brain radiation, including hippocampal avoidance strategies. Finally, we discuss a framework whereby various management strategies, ranging from conservative management to surgical resection and radiosurgery, may be considered.

Learning Objectives

- To understand commonly used prognostic classification systems in patients with brain metastases.
- To understand the role of CNS-directed radiation therapy, including whole brain radiation therapy and stereotactic radiosurgery, both in the intact and postoperative setting, in reducing the risk of local recurrence.
- To understand potential management paradigms for patients with brain metastases which integrate radiation therapy.
- To understand emerging paradigms in the use of radiation therapy for brain metastases, including pre-operative stereotactic radiosurgery and hippocampal avoidance whole brain radiation therapy.

53.1 Introduction

Brain metastasis represents the most common type of intracranial malignancy and is estimated to affect approximately one in five cancer patients [1]. While historically the presence of brain metastasis implied an extremely limited prognosis, the advent of an increasing number of systemic and targeted therapies, along with improved stratification for patients diagnosed with metastatic disease to the brain, has led to improvements in the survival of patients diagnosed with brain metastases over time. However, there exist large variations depending on cancer type [2–4].

Whole brain radiation therapy (WBRT) has historically represented the standard of care for individuals with multiple brain metastases. However, with technical progress of focused radiation delivery and clinical evidence of cognitive decline after WBRT, practice patterns have shifted to favor more focused radiation therapy such as stereotactic radiosurgery (SRS), both in the intact and postoperative setting, for patients with a limited number of metastases. In fact, the American Society for Therapeutic Radiology and Oncology (ASTRO) Choosing Wisely guidelines recommend the avoidance of routine use of WBRT in patients receiving SRS for a limited number of metastases [5].

Since brain metastasis constitutes a large spectrum of patients with significant oncologic and clinical heterogeneity, the possible management strategies for brain metastases

A. N. Slade · M. Ashamalla · S. Ryu (✉)
Department of Radiation Oncology, Stony Brook University, Stony Brook, NY, USA
e-mail: Alexander.Slade@stonybrookmedicine.edu; Mark.Ashamalla@stonybrookmedicine.edu; samuel.ryu@stonybrookmedicine.edu

S. P. Leong et al. (eds.), *Cancer Metastasis Through the Lymphovascular System*,
https://doi.org/10.1007/978-3-030-93084-4_53

can range from conservative management (e.g., steroids alone) up to surgical management and/or radiosurgery.

This chapter is organized as follows: we first review prognostic approaches in patients with brain metastases and then focus on the use of radiosurgery for brain metastasis, both in the intact and postoperative setting. We also review some of the emerging data supporting the addition of hippocampal avoidance and memantine to reduce the risk of cognitive decline among patients receiving whole brain radiation therapy.

53.2 Prognostic Approaches in Patients with Brain Metastasis

Given how common brain metastases are in routine clinical practice, efforts have been made to stratify patients into prognostic groups. One of the original approaches was the Recursive Partitioning Analysis (RPA) by Gaspar et al. [6] who used three variables to create three classes of patients. The variables included Karnofsky Performance Status (KPS), presence of extracranial disease and well-controlled primary disease, and age to develop three classes (i.e., partitions) of patients with brain metastasis. Patients meeting the criteria of Class I, consisting of the most favorable characteristics (KPS > = 70, age < 65, and well-controlled primary and no extracranial disease), had a median overall survival of ≈7 months. Alternatively, patients meeting Class III criteria (KPS < 70) had a median overall survival of ≈2 months. All other patients (Class II) had a median overall survival of ≈4 months. Since then, many other groupings have been proposed.

Sperduto et al. [7] further optimized stratification of these patients by developing the Graded Prognostic Assessment (GPA), which incorporated age, KPS, systemic disease, number of lesions, and volume of disease. Across all malignancies in this classification system, KPS is the strongest and most robust predictor for survival. As both prognostic and predictive mutations emerged for specific cancers, cancer-specific Graded Prognostic Assessment [2] was defined to reflect factors such as disease site and histology, number of lesions, tumor gene status (e.g., EGFR and ALK gene alterations in non-small-cell lung cancer [8] and BRAF mutation in melanoma [9]), and other clinical characteristics (e.g., hemoglobin level in renal cell carcinoma). Under the updated GPA classification, for instance, a patient with excellent performance status and a more favorable histologic profile (e.g., breast cancer) may have a survival in excess of 2 years.

53.3 Postsurgical Management of Brain Metastases: Evolution from Whole Brain Radiation Therapy to Radiosurgery

Surgery, along with radiation therapy, remains the mainstay of treatment in patients with most solitary brain metastases or large lesions causing neurological symptoms. While radiation therapy alone is an option for patients even with a single brain metastasis, there are several clinical scenarios which may make a surgical approach attractive. First, it can clarify pathologic diagnosis, especially if a primary diagnosis does not exist or is not consistent with the development of brain metastasis (e.g., a patient with prostate cancer after a prostatectomy with no evidence of disease recurrence). It may also provide molecular or genetic information toward the use of targeting agents when the information is otherwise not available. Finally, surgical management can relieve mass effect from symptomatic metastases much faster than nonsurgical approaches.

Despite the benefits of surgical management of brain metastases, the risk of local failure without additional metastasis-directed radiotherapy remains high in most series; approximately 50% of the resected brain metastasis ultimately develops local recurrence without adjuvant radiotherapy [10]. Indeed, infiltrative growth beyond the metastasis boundary was detected in up to 63% of cases in one pathological series of resected brain metastases [11]. This may partially explain the high recurrence rate after surgery alone and necessity of adjuvant treatment.

Radiotherapy to brain metastases, either in the intact or postoperative setting, is typically delivered in one of two ways. WBRT treats both the surgical resection site and the remaining brain which has been thought to harbor subclinical metastatic or leptomeningeal disease and thus lowers the risk of development of additional metastasis in the brain. Alternatively, stereotactic radiosurgery delivers a very high dose of radiation directed only to the metastasis (or surgical cavity), without treating the uninvolved tissue. Indeed, randomized data has supported the role of postoperative WBRT in reducing the risk of new distant in-brain recurrences but without any improvement of overall survival [12]. However, improved distant disease control is a potential benefit to WBRT over SRS, as both leptomeningeal disease and distant brain failure can be common after postoperative radiosurgery [13]. However, it comes at the cost of several acute and long-term side effects, most notably neurocognitive decline [14]. By using postoperative radiosurgery (often with a single dose of 14–16 Gy to the surgical cavity), most institutional series reported that tumor control of the resected site improved to 80–90% [15,

16], and randomized data confirms similarly excellent levels of local control with postoperative SRS as well [17]. When there is gross residual disease after surgical resection (which often manifests as contrast-enhancing lesions), the gross disease typically needs to be treated to higher doses. A common approach to postoperative radiosurgical planning has been presented by Robbins et al. [18] and summarized in Fig. 53.1, panel A.

Increasing evidence has been put forth to support the routine use of radiosurgery (where possible) over whole brain radiation therapy. A recent randomized clinical trial compared patients with ≤ 3 brain metastases to postoperative SRS (12–20 Gy) vs. whole brain radiation therapy (30–37.5 Gy in 10–15) fractions [19]. Overall, this study found a similar overall survival of ≈1 year for both groups of patients, but a lower rate of cognitive decline among patients receiving SRS compared to whole brain radiation.

In addition to decompressing the mass effect of a large metastasis and relieving neurological symptoms, surgery may render the cavity smaller compared to the pre-surgical tumor volume. An early single-institutional series found that there was an ≈29% reduction in the size of the surgical cavity (compared to pre-operatively) [20]. This was most pronounced in larger tumor sizes, and it was found that most of the surgical cavity shrinkage was seen within 2 weeks postoperatively. As such, this will typically limit the volume of radiation delivered to both the cavity and normal brain parenchyma (Fig. 53.1, panel B). However, especially for small lesions, it is possible that the cavity may actually have a larger volume compared to pre-operatively [21]. In terms of timing of radiosurgery, we recommend commencing treatment relatively expediently postoperatively soon after the patient recovered from surgery, usually within 2–4 weeks following surgery.

An emerging strategy in the management of brain metastases involves treating the intact tumor before surgical resection. This approach allows for relatively straightforward target delineation and also has the theoretical advantages of limiting leptomeningeal spread secondary to surgical intervention and sterilizing the surgical field. Indeed, institutional series have reported promising results with similar rates of local control in pre-operative compared to postoperative SRS [22], with lower rates of symptomatic radionecrosis and leptomeningeal disease [23]. A common approach involves utilizing single-fraction radiosurgery with a reduction of the treatment dose by 10–20%, with short interval from delivery SRS to surgery (median of 2 days) [24].

Finally, an area of continued interest involves the role of radiosurgery in potentiating an abscopal effect along with

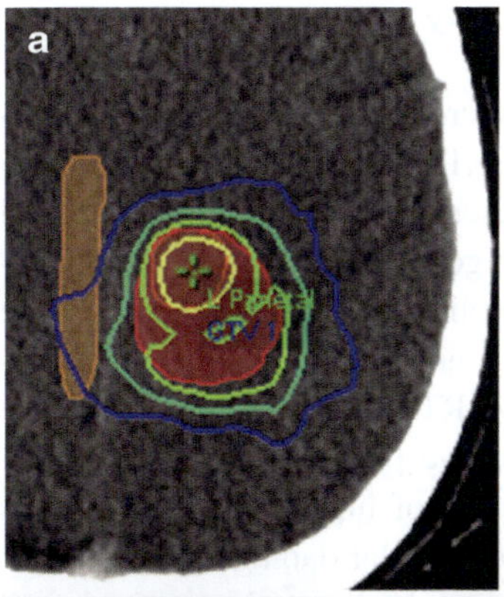

Radiosurgery Target	Dose
Surgical cavity + 2-3 mm margin or clinical decision	Median dose 16 Gy (12-20 Gy), prescribed to 90% isodose line
Gross residual C+ lesion	≥ 18 Gy

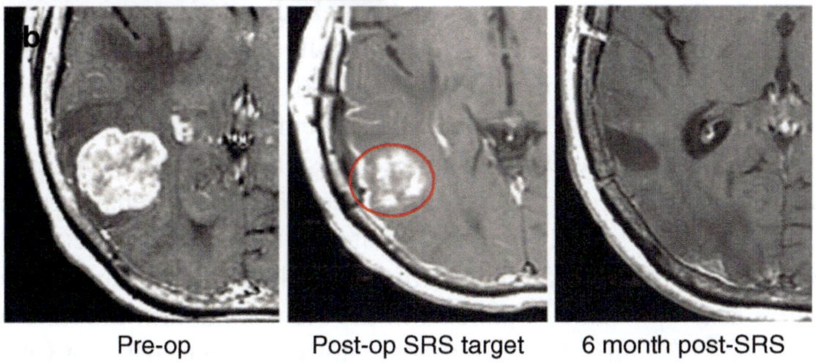

Fig. 53.1 Postoperative treatment with stereotactic radiosurgery. Example of the use of stereotactic radiosurgery in the treatment of a postoperative brain metastasis. Panel **A** depicts an example of a treatment plan in a postoperative cavity (left) and common target volumes and doses (right). In general, the surgical cavity is contoured, and an additional margin of 2–3 mm is added. This volume is prescribed to a dose ranging from 12 to 20 Gy (median 16 Gy) in the single fraction setting. Gross residual disease (typically contrast enhancing) may require higher doses for local control. Panel **B** depicts serial MRI scans showing a brain metastasis pre-operatively (left), the smaller postoperative cavity (middle), and a 6-month follow-up scan after postoperative radiosurgery

systemic immunotherapy, in which an irradiated tumor may induce a regression in non-irradiated sites [25]. It is thought that irradiation induces improved antigen expression and presentation, which combined with increased T cell function can result in an abscopal response [26]. There has been historic belief that extracranial abscopal effects from brain irradiation (either SRS or WBRT) would be uncommon due to a unique immune environment and the blood brain barrier. However, several reports have been noted of this phenomenon, especially related to metastases from lung cancer [27, 28] and melanoma [29]. It remains to be an area of active investigation.

53.4 Radiosurgery for Patients with Intact Brain Metastases

As with radiosurgery delivered postoperatively for resected brain metastases, use of radiosurgery for intact metastases (treatment-naïve or unresected) confers excellent local control, often ranging from 70 to 90% at 1 year in institutional series [30, 31]. The early dose-escalation experience of the RTOG 90–05 trial provided a framework by which SRS doses (in the single-fraction setting) are typically used with acceptable rates of toxicity [32]. Based on the experience of this trial, recommended doses are 24 Gy for tumors 2 cm or smaller and 15 Gy for tumors between 3 and 4 cm. Early experiences explored the value of SRS as a "boost" to whole brain radiation therapy. RTOG 9508 considered patients with a limited number of brain metastases (<3 metastases) who were randomized between WBRT alone and WBRT alone + SRS boost, which found a stable or improved performance status at 6 months follow-up among patients with an SRS compared to those without (43% vs. 27%; $p = 0.03$) [33]. While survival among the entire cohort was similar, patients with favorable histologic status or RPA class 1 had improved overall survival.

More recently, efforts have been made to curtail the use of WBRT in patients who are otherwise good candidates for radiosurgery, such as patients with a good performance status and a limited number of metastases. As such, studies have investigated the role of SRS with or without WBRT. A randomized multi-institutional clinical trial from Japan was performed comparing WBRT+ radiosurgery versus radiosurgery alone in patients with one to four brain metastases each less than 3 cm [34]. This trial found similar tumor control with SRS alone although distant intracranial relapse was more common; however, there was no difference in overall survival. A more recent EORTC 22952–22,601 study randomized patients with one to three brain metastases treated with either surgery or SRS to adjuvant WBRT vs. observation. As with the experience from Japan, the addition of WBRT decreased the rate of intracranial relapse, but did not improve the duration of neurological independence or overall survival [35]. As such, among patients with limited number of metastases that can be treated with radiosurgery alone, the omission of WBRT is often preferable given its morbidity and lack of survival benefit.

While much of the existing literature focused on patients with limited numbers of brain metastases (e.g., four and below), there is increasing credence to using radiosurgical approaches in patients with a higher number and volume of lesions. This is increasingly important with the advent of sophisticated high-resolution MRI scans that can detect very small lesions. As treatment planning techniques have evolved to allow very limited doses to surrounding brain parenchyma and surrounding other critical structures, it is possible to safely treat larger number of lesions. Indeed, institutional series considering treatment of >10 lesions with radiosurgery generally have found outcomes similar to series considering more limited number of metastases [36–38]. Indeed, in the JLGK0901 which enrolled patients with 2–10 newly diagnosed brain metastases treated with SRS alone, the median survival of patients with two to four metastases was not significantly different compared to those with 5–10 metastases (median overall survival was 10.8 months in both groups).

53.5 Approaches to Limit Cognitive Decline in Patients Receiving Whole Brain Radiotherapy

While radiosurgery has been increasingly used in the management of brain metastases, WBRT still has utility, especially in patients with multiple metastases, and offers increased control of subclinical and leptomeningeal disease. However, as discussed above, a common complication of WBRT is cognitive decline, particularly with the memory recall [39]. The precise mechanisms by which WBRT leads to delayed recall are uncertain. However, there are two predominant theories that have been proposed: destruction of hippocampal stem cells [40, 41] and radiation-induced vascular damage [42].

Two trials attempted to address both putative systemic and technical interventions to potentially lower the risk of cognitive decline in WBRT. RTOG 0614 randomized patients receiving WBRT to placebo vs. memantine, an n-methyl-d-aspartate (NMDA) antagonist (classically used for dementia), on the trajectory of cognitive decline after patients receive WBRT [43]. Though the results for this endpoint were not statistically different between the placebo and memantine arms ($p = 0.059$), patients in the memantine arm did have a significantly slower time to cognitive decline. Alternatively, RTOG 0933 considered the role of radiation planning to avoid the hippocampus on cognitive decline, which found the absolute value of cognitive decline significantly lower than historical controls [5, 44]. Hippocampal avoidance may be technically achieved by delineating the hippocampus via image fusion of a T1-weighted MRI sequence to the planning

CT scan. Through intensity-modulated radiation therapy (IMRT) planning, the dose to the hippocampus can be limited. A common dose constraint (used in the NRG CC001 trial discussed below) involves limiting 100% of the hippocampal volume to 9 Gy, with a maximum dose of 16 Gy.

Despite the promising results of these phase II trials, overall uptake of these modalities remained relatively limited across the United States, with a perceived need for randomized, phase III results [45]. The recently completed NRG CC001 trial incorporated both the use of memantine and hippocampal avoidance into the study design [46]. In this trial, patients were randomized to memantine with either hippocampal-sparing WBRT or standard WBRT. The trial found that at a median follow-up of 7.8 months, the risk of cognitive failure was lower after hippocampal-sparing WBRT compared to conventional WBRT.

Finally, in patients with poor performance status and very limited prognoses (e.g., KPS < 70), the omission of metastasis-directed treatment entirely, and using supportive care (including steroids) alone, is a reasonable option. The Quality of Life after Treatment for Brain Metastases (QUARTZ) trial randomized patients deemed ineligible for radiosurgery or surgical resection (many with poor KPS) to optimal supportive care with or without WBRT. The trial found no significant difference in overall survival or quality of life [47].

53.6 Conclusions

With improved systemic therapy and patient stratification, the management of brain metastases is not a "one size fits all" approach; indeed, it represents a spectrum of diseases. Patients with widespread brain metastases may be managed with whole brain radiotherapy, with or without hippocampal sparing, whereas patients with fewer brain metastases and improved performance status may be managed with stereotactic radiosurgery. Still others with few metastases may be managed surgically, especially if these metastases are symptomatic, followed by postoperative radiation therapy. The spectrum of treatment options is displayed in Fig. 53.2.

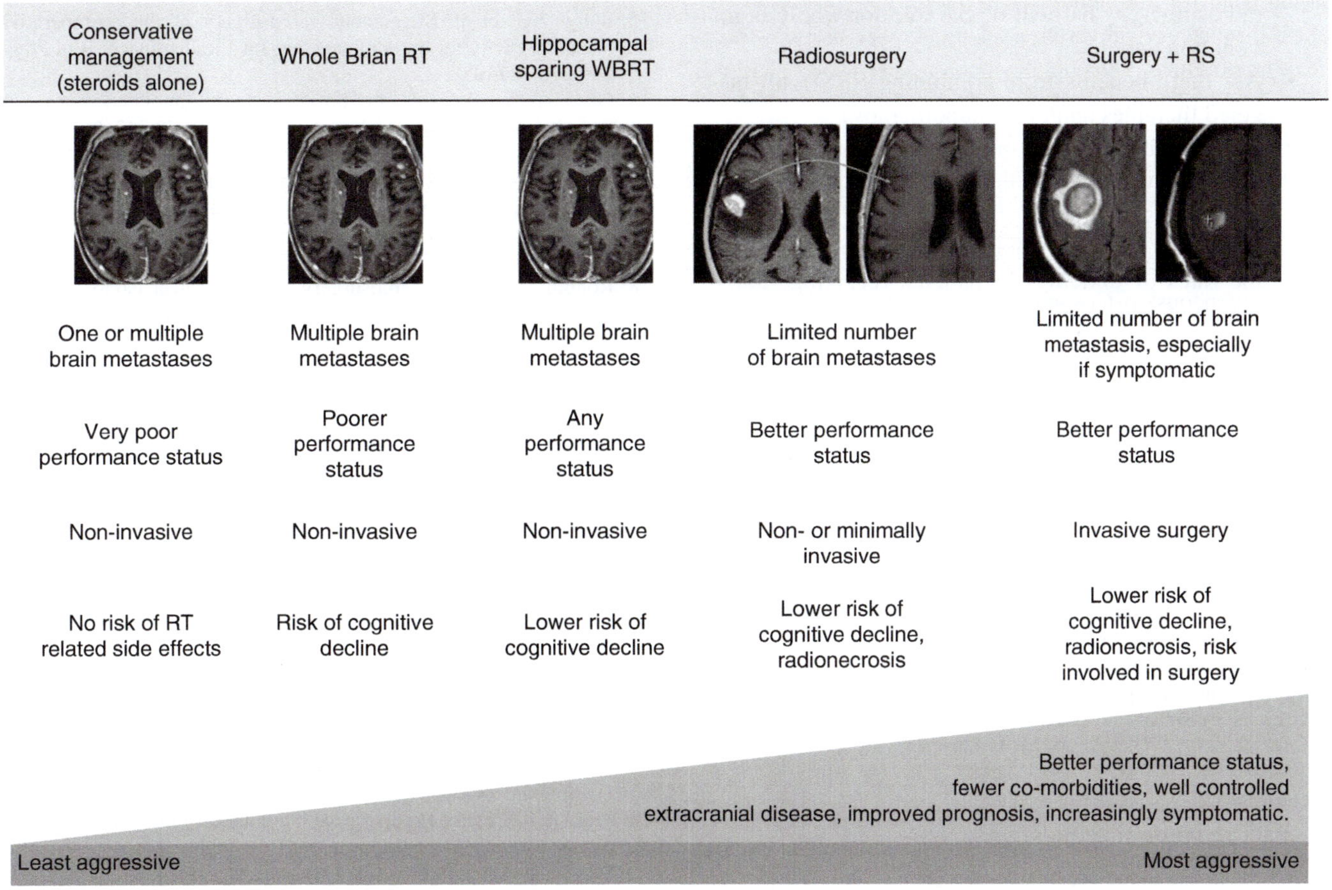

Fig. 53.2 Contemporary management strategies for patients with brain metastases. The spectrum of management options for patients with brain metastases range from no metastasis-directed therapy (i.e., conservative management) to invasive surgical options followed by postoperative stereotactic radiation therapy. Patients with large or symptomatic lesions may benefit from surgical management, which may provide rapid relief of symptoms and provide a smaller cavity volume for irradiation postoperatively. In general, less aggressive options are preferred in patients with a poor prognosis, increased comorbidities, and poorer performance status

For some patients, clear demonstration of reducing the risk of developing new metastasis justifies the option of WBRT. This may be motivated by the fear of symptomatic recurrence or the desire to also reduce the need for future treatment. For others, fear of irreversible toxicity such as neurocognitive decline given the lack of survival benefit drives toward SRS provided all the known tumors are treated.

The clinical and pathologic heterogeneity of patients with brain metastases, combined with diverse treatment options available for its management, suggests a nuanced approach that requires shared decision-making between the patient and his/her surgical, medical, and radiation oncologists to design the optimal treatment plan.

Future Directions

- How will prognostic classifications evolve in the future to more optimally stratify patients with brain metastases?
- As technological advances improve, allowing for faster and more widespread delivery of stereotactic radiosurgery, what will be the role for whole brain radiation therapy?
- Are there benefits from combining SRS with targeted therapies and/or immunotherapy?
- What will be the optimal role for pre-operative SRS, and what will be the optimal timing and dose parameters?
- What are potential future interventions to reduce the rates of toxicities in patients receiving radiotherapy for brain metastases?

References

1. Sacks P, Rahman M. Epidemiology of brain metastases. Neurosurg Clin N Am. 2020;31:481–8.
2. Sperduto PW, et al. Survival in patients with brain metastases: summary report on the updated diagnosis-specific graded prognostic assessment and definition of the eligibility quotient. J Clin Oncol. 2020;38:3773–84.
3. American Society for Radiation Oncology. Choosing wisely: five things physician and patients should question. American Society for Radiation Oncology; 2014. https://www.choosingwisely.org. Accessed 3/1/2022.
4. Kavouridis VK, et al. Survival and prognostic factors in surgically treated brain metastases. J Neuro-Oncol. 2019;143:359–67.
5. Gondi V, et al. Preservation of memory with conformal avoidance of the hippocampal neural stem-cell compartment during whole-brain radiotherapy for brain metastases (RTOG 0933): a phase II multi-institutional trial. J Clin Oncol. 2014;32:3810–6.
6. Gaspar L, et al. Recursive partitioning analysis (RPA) of prognostic factors in three radiation therapy oncology group (RTOG) brain metastases trials. Int J Radiat Oncol Biol Phys. 1997;37:745–51.
7. Sperduto PW, et al. Summary report on the graded prognostic assessment: an accurate and facile diagnosis-specific tool to estimate survival for patients with brain metastases. J Clin Oncol Off J Am Soc Clin Oncol. 2012;30:419–25.
8. Sperduto PW, et al. Estimating survival in patients with lung cancer and brain metastases: an update of the graded prognostic assessment for lung cancer using molecular markers (lung-molGPA). JAMA Oncol. 2017;3:827–31.
9. Sperduto PW, et al. Estimating survival in melanoma patients with brain metastases: an update of the graded prognostic assessment for melanoma using molecular markers (melanoma-molGPA). Int J Radiat Oncol Biol Phys. 2017;99:812–6.
10. Patel AJ, et al. Factors influencing the risk of local recurrence after resection of a single brain metastasis. J Neurosurg. 2010;113:181.
11. Baumert BG, et al. A pathology-based substrate for target definition in radiosurgery of brain metastases. Int J Radiat Oncol Biol Phys. 2006;66:187–94.
12. Patchell RA, et al. Postoperative radiotherapy in the treatment of single metastases to the brain: a randomized trial. JAMA. 1998;280:1485–9.
13. Marcrom SR, et al. Focal management of large brain metastases and risk of leptomeningeal disease. Adv Radiat Oncol. 2020;5:34–42.
14. Brown PD, et al. Effect of radiosurgery alone vs radiosurgery with whole brain radiation therapy on cognitive function in patients with 1 to 3 brain metastases: a randomized clinical trial. JAMA. 2016;316:1–409.
15. Soltys SG, et al. Stereotactic radiosurgery of the postoperative resection cavity for brain metastases. Int J Radiat Oncol Biol Phys. 2008;70:187–93.
16. Gans JH, et al. The role of radiosurgery to the tumor bed after resection of brain metastases. Neurosurgery. 2013;72:317–25. discussion 325-6
17. Mahajan A, et al. Post-operative stereotactic radiosurgery versus observation for completely resected brain metastases: a single-centre, randomised, controlled, phase 3 trial. Lancet Oncol. 2017;18:1040–8.
18. Robbins JR, et al. Radiosurgery to the surgical cavity as adjuvant therapy for resected brain metastasis. Neurosurgery. 2012;71:937–43.
19. Brown PD, et al. Postoperative stereotactic radiosurgery compared with whole brain radiotherapy for resected metastatic brain disease (NCCTG N107C/CEC·3): a multicentre, randomised, controlled, phase 3 trial. Lancet Oncol. 2017;18:1049–60.
20. Atalar B, et al. Cavity volume dynamics after resection of brain metastases and timing of postresection cavity stereotactic radiosurgery. Neurosurgery. 2013;72:180–5. discussion 185
21. Jarvis LA, et al. Tumor bed dynamics after surgical resection of brain metastases: implications for postoperative radiosurgery. Int J Radiat Oncol Biol Phys. 2012;84:943–8.
22. Asher AL, et al. A new treatment paradigm: neoadjuvant radiosurgery before surgical resection of brain metastases with analysis of local tumor recurrence. Int J Radiat Oncol Biol Phys. 2014;88:899–906.
23. Patel KR, et al. Comparing preoperative with postoperative stereotactic radiosurgery for resectable brain metastases: a multi-institutional analysis. Neurosurgery. 2015;79:279–85.
24. Prabhu RS, et al. Preoperative stereotactic radiosurgery before planned resection of brain metastases: updated analysis of efficacy and toxicity of a novel treatment paradigm. J Neurosurg. 2018:1–8.
25. Siva S, et al. Abscopal effects of radiation therapy: a clinical review for the radiobiologist. Cancer Lett. 2015;356:82–90.
26. Reynders K, et al. The abscopal effect of local radiotherapy: using immunotherapy to make a rare event clinically relevant. Cancer Treat Rev. 2015;41:503–10.

27. Lin X, et al. Extracranial abscopal effect induced by combining immunotherapy with brain radiotherapy in a patient with lung adenocarcinoma: a case report and literature review. Thoracic Cancer. 2019;10:1272–5.
28. D'Andrea MA, Reddy GK. Extracranial abscopal effects induced by brain radiation in advanced lung cancer. Am J Clin Oncol. 2019;42:951–7.
29. D'Andrea MA, Reddy GK. Brain radiation induced extracranial abscopal effects in metastatic melanoma. Am J Clin Oncol. 2020;43:836–45.
30. Minniti G, et al. Single-fraction versus multifraction (3Gy) stereotactic radiosurgery for large brain metastases: a comparative analysis of local control and risk of radiation-induced brain necrosis. Int J Radiat Oncol Biol Phys. 2016;95:1142–8.
31. Hasegawa T, et al. Brain metastases treated with radiosurgery alone: an alternative to whole brain radiotherapy? Neurosurgery. 2003;52:1318–26.
32. Shaw E, et al. Single dose radiosurgical treatment of recurrent previously irradiated primary brain tumors and brain metastases: final report of RTOG protocol 90-05. Int J Radiat Oncol Biol Phys. 2000;47:291–8.
33. Andrews DW, et al. Whole brain radiation therapy with or without stereotactic radiosurgery boost for patients with one to three brain metastases: phase III results of the RTOG 9508 randomised trial. Lancet. 2004;363:1665–72.
34. Aoyama H, et al. Stereotactic radiosurgery plus whole-brain radiation therapy vs stereotactic radiosurgery alone for treatment of brain MetastasesA randomized controlled trial. JAMA. 2006;295:2483–91.
35. Kocher M, et al. Adjuvant whole-brain radiotherapy versus observation after radiosurgery or surgical resection of one to three cerebral metastases: results of the EORTC 22952-26001 study. J Clin Oncol. 2010;29:134–41.
36. Grandhi R, et al. Stereotactic radiosurgery using the Leksell gamma knife Perfexion unit in the management of patients with 10 or more brain metastases. J Neurosurg. 2012;117:237–45.
37. Susko MS, et al. Stereotactic radiosurgery to more than 10 brain metastases: evidence to support the role of radiosurgery for ideal hippocampal sparing in the treatment of multiple brain metastases. World Neurosurg. 2020;135:e174–80.
38. Rava P, et al. Survival among patients with 10 or more brain metastases treated with stereotactic radiosurgery. J Neurosurg. 2013;119:457.
39. Chang EL, et al. Neurocognition in patients with brain metastases treated with radiosurgery or radiosurgery plus whole-brain irradiation: a randomised controlled trial. Lancet Oncol. 2009;10:1037–44.
40. Monje ML, et al. Irradiation induces neural precursor-cell dysfunction. Nat Med. 2002;8:955–62.
41. Gondi V, et al. Hippocampal dosimetry predicts neurocognitive function impairment after fractionated stereotactic radiotherapy for benign or low-grade adult brain tumors. Int J Radiat Oncol Biol Phys. 2012;83:e487–93.
42. Brown WR, et al. Vascular damage after fractionated whole-brain irradiation in rats. Radiat Res. 2005;164:662–8.
43. Brown P, et al. Memantine for the prevention of cognitive dysfunction in patients receiving whole-brain radiotherapy: a randomized, double-blind, placebo-controlled trial. Neuro-Oncology. 2013;15:1429–37.
44. Gondi V, et al. Memory preservation with conformal avoidance of the hippocampus during whole-brain radiation therapy for patients with brain metastases: primary endpoint results of RTOG 0933. Int J Radiat Oncol Biol Phys. 2013;87:1186.
45. Slade AN, Stanic S. The impact of RTOG 0614 and RTOG 0933 trials in routine clinical practice: the US survey of utilization of memantine and IMRT planning for hippocampus sparing in patients receiving whole brain radiotherapy for brain metastases. Contemp Clin Trials. 2016;47:74–7.
46. Brown PD, et al. Hippocampal avoidance during whole-brain radiotherapy plus memantine for patients with brain metastases: phase III trial NRG oncology CC001. J Clin Oncol. 2020;38:1019–29.
47. Mulvenna P, et al. Dexamethasone and supportive care with or without whole brain radiotherapy in treating patients with non-small cell lung cancer with brain metastases unsuitable for resection or stereotactic radiotherapy (QUARTZ): results from a phase 3, non-inferiority, randomised trial. Lancet. 2016;388:2004–14.

Leptomeningeal Carcinomatosis

54

Sunny R. K. Singh, Sindhu J. Malapati,
and Ahmad Mattour

Abstract

The prevalence of leptomeningeal carcinomatosis (LC) has increased due to improved survival of cancer patients; however, it continues to be associated with poor outcomes despite multiple advances in cancer-directed therapy. Clinical presentation is highly variable, and its evaluation continues to be largely guided by clinical signs and symptoms, findings on craniospinal magnetic resonance imaging (MRI), and cerebrospinal fluid (CSF) cytology. Recently, the ability to detect circulating cancer cells and tumor DNA in the CSF has improved the accuracy of diagnosing LC in those with false-negative cytology. It has also uncovered the potential of utilizing CSF for molecular analysis using next-generation sequencing. This not only furthers our understanding of this clinical entity but also allows us make treatment decisions based on the presence of certain resistance mutations. To date, there is no standard approach to the treatment of LC. Historically, treatment choices are limited to systemic or intra-CSF cytotoxic chemotherapy and/or radiation. With the advent of novel agents, we now have several potential therapeutic agents being studied prospectively. Also, the inclusion of patients with LC in recent clinical trials marks the beginning of long overdue advances that would lead to improved outcomes in this cohort of patients.

S. R. K. Singh
Department of Hematology/Oncology, Henry Ford Cancer Institute, Detroit, MI, USA

S. J. Malapati
Department of Hematology/Oncology, Ascension St. John Hospital, Detroit, MI, USA

A. Mattour (✉)
Department of Hematology/Oncology, Henry Ford Cancer Institute, Detroit, MI, USA
e-mail: amattou1@hfhs.org

Learning Objectives
- Discuss the epidemiology, pathophysiology, clinical features, and outcomes of leptomeningeal carcinomatosis.
- Describe the workup of leptomeningeal carcinomatosis and its limitations.
- Analyze the role and benefit of currently approved treatment options for leptomeningeal carcinomatosis.
- Outline the recent advances both in terms of evaluation and possible future treatment options for leptomeningeal carcinomatosis.

54.1 Introduction

Leptomeningeal carcinomatosis (LC), also known as neoplastic meningitis, is the spread of malignancy, within the pia mater and/or arachnoid components of the central nervous system (CNS), aka leptomeninges. It is usually a late complication of advanced malignancy with devastating consequences on neurological outcomes, quality of life, and survival. Without treatment, median overall survival has been reported to be around 6–8 weeks. Even with adoption of multiple newer multimodal treatments, survival benefits remain poor (3–9 months with intrathecal chemotherapy) [1–3]. Standard of care for the management of LC is still not well established, and the goal of treatment is largely palliative. Treatment outcomes to be considered while deciding therapy include treatment-associated toxicity, time to neurologic progression (TTNP), impact on quality of life, and overall survival.

S. P. Leong et al. (eds.), *Cancer Metastasis Through the Lymphovascular System*,
https://doi.org/10.1007/978-3-030-93084-4_54

54.2 Epidemiology

The estimated frequency of leptomeningeal carcinomatosis is around 5–8% in solid malignancies and 5–15% in hematological malignancies [4], but this does not account for the significant underdiagnosis of this condition. The combination of newer, life-prolonging systemic therapies and improved access to imaging modalities has resulted in what is described as long-term cancer patients and, inadvertently, higher incidences of late metastasis, including LC [5]. Solid malignancies most commonly associated with LC are breast, lung, melanoma, and gastrointestinal cancers and hematological malignancies.

54.3 Pathogenesis

Several mechanisms are implicated in the development of LC, including direct extension from surrounding structures (bone, brain, dura), hematologic spread through venous plexus, and perineural extension [6]. It must be pointed out that leptomeninges include the arachnoid and pia mater, but not the dura mater (aka pachymeninges). Dura is not protected by the blood-brain barrier and therefore can be accessed by systemic agents, while the leptomeninges are limited in this regard. In a recent preclinical study, complement component 3 (C3) was found to be upregulated and necessary for cancer growth in the leptomeningeal space. C3 derived from cancer cells activates the C3a receptor in choroid plexus epithelium and disrupts the BBB, allowing mitogens to enter the CSF from plasma and promote cancer cell proliferation [7]. Involvement is very common in dependent portions of the CNS like the cauda equina, base of the skull, posterior fossa, and also basal cisterns. Despite extensive study, the exact mechanism of development of LC is not known and is still being evaluated. Malignancy and intrathecal therapies can cause local inflammation resulting in impaired resorption of CSF and obstruction of flow, and this may manifest with symptoms of elevated intracranial pressure [8].

54.4 Diagnostic Evaluation

Diagnosis of LC is based on clinical features, findings on craniospinal MRI, and cerebrospinal fluid (CSF) analysis. While typical clinical and imaging signs are contributory to making the diagnosis, the detection of tumor cells in the CSF is confirmatory in a patient with known malignancy. Among those with known cancer and without CSF confirmation, diagnosis of LC is probable if both typical clinical and imaging signs are *present* and *possible* if only one of them is present [9].

54.4.1 Differential Diagnosis

This condition is frequently misdiagnosed due to vague presenting symptoms and non-specific findings on neurological imaging. Oftentimes, LC is diagnosed at the same time as parenchymal or dural involvement [8]. Differentials include meningitis (infectious or chemical), parenchymal brain metastases at multiple locations, encephalopathy (toxic or metabolic), neuropathy secondary to chemotherapy, cord compression, steroid myopathy, and paraneoplastic syndromes (limbic encephalitis, encephalomyelitis, paraneoplastic cerebellar degeneration).

54.4.2 Clinical Presentation

LC presentation can be highly variable. In cases with high metastatic tumor burden, classical LC symptoms may be more clearly defined. Occasionally LC could even be the initial manifestation of malignancy. The spectrum of symptoms remains broadly vague in nature, and a high index of suspicion remains important to make a timely diagnosis [10]. In a study of 103 patients, more than 80% had a neurological sign or symptom at diagnosis of leptomeningeal disease [11]. Classically, the symptoms of LC have been categorized according to the site of neural involvement [3]:

1. Cerebral involvement usually presents with a variety of symptoms such as headaches, dizziness, nausea, seizures, and encephalopathy. Another manifestation could be a hydrocephalus.
2. Symptoms due to involvement of the cranial nerves by LC manifest with findings of diplopia, vertigo, hearing loss, and facial numbness.
3. Involvement of the spinal cord and nerve roots can present with symptoms of myelopathy such as limb weakness, loss of sensation, pain, and paresthesia.

Proposed mechanisms leading to these symptoms include parenchymal compression or invasion, ischemia secondary to vascular involvement, blood-brain barrier disruption, and metabolic strain [12].

54.4.3 Radiological Evaluation

About 70–80% patient with leptomeningeal disease are noted to have abnormalities on imaging [13]. The test of choice is a T1-weighted MRI of the brain and spine with gadolinium contrast enhancement [14]. Cranial computed tomography can be used when there are contraindications for the use of MRI. Typical findings of neoplastic meningitis on a contrast-enhanced MRI include enhancement or obliteration of sulci, linear ependymal enhancement, cranial nerve root enhancement, and leptomeningeal nodule enhancement, notably of the cauda equina [15] (Fig. 54.1). Four types of LC have been defined based on MRI findings: linear leptomeningeal contrast enhancement only (type A), leptomeningeal nodules only (type B), combination of both (type C), and with no neuroimaging evidence of LC, except for

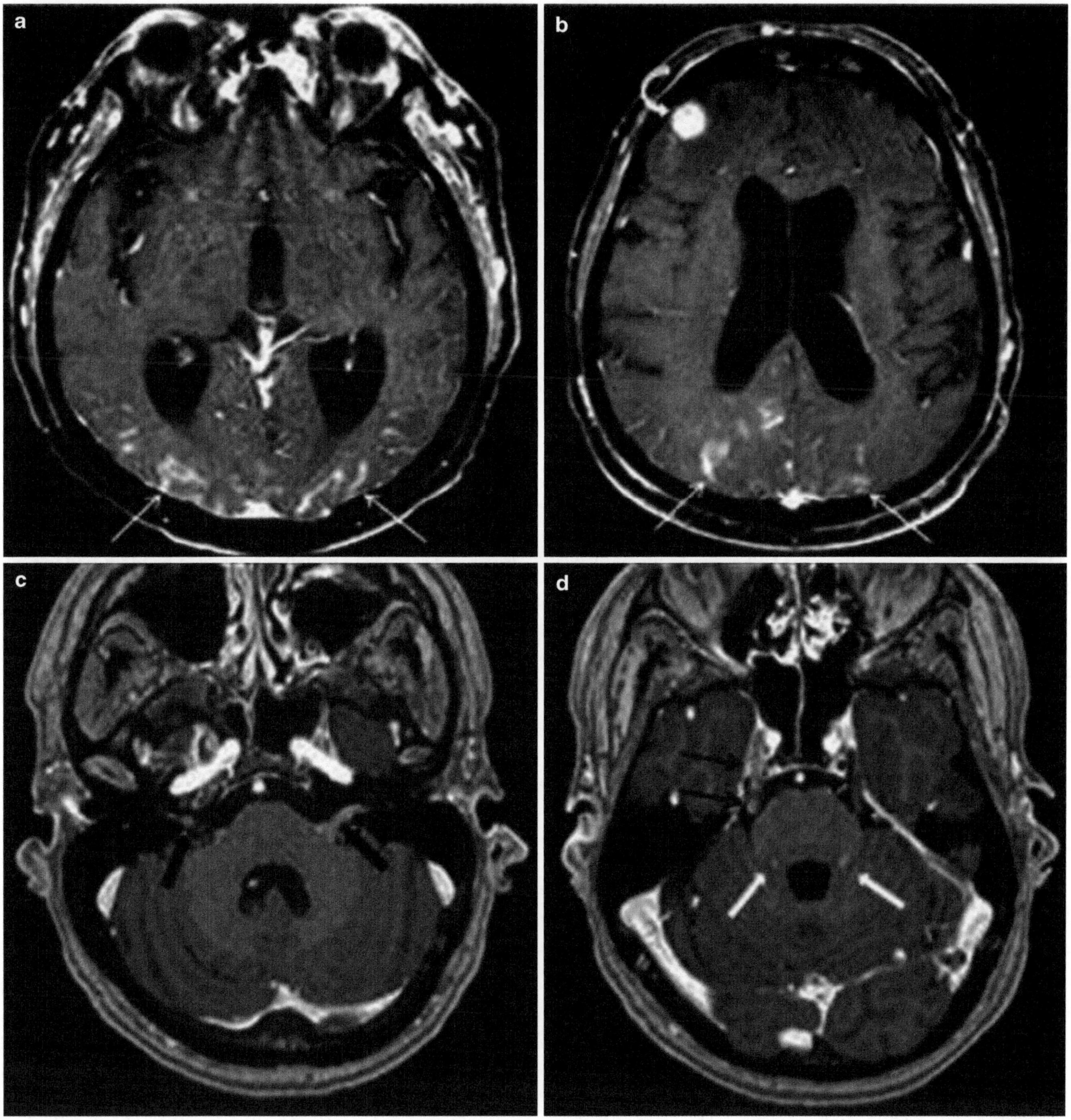

Fig. 54.1 Contrast-enhanced axial imaging of the supratentorial brain (**a, b**) demonstrates curvilinear enhancement of the pial surface of the brain from leptomeningeal disease (LMD). Extensive pial enhancement involves the occipital and parietal cortex bilaterally (white arrows) as well as a focal nodular deposit involving the right frontal cortex (white curved arrow). Contrast-enhanced axial imaging of the posterior fossa (**c, d**) demonstrates nodular enhancing LMD involving cranial nerves VII and VIII within the internal auditory canal and cerebellopontine angle (black block arrows), right cranial nerve V in lateral pontine cistern and Meckel's cave (black arrows), as well as the folia of the cerebellum (white block arrows). Republished with permission of Wiley for Glitza, IC, et al. *Pigment Cell Melanoma Res*. 2020; 33: 527–541; permission conveyed through Copyright Clearance Center

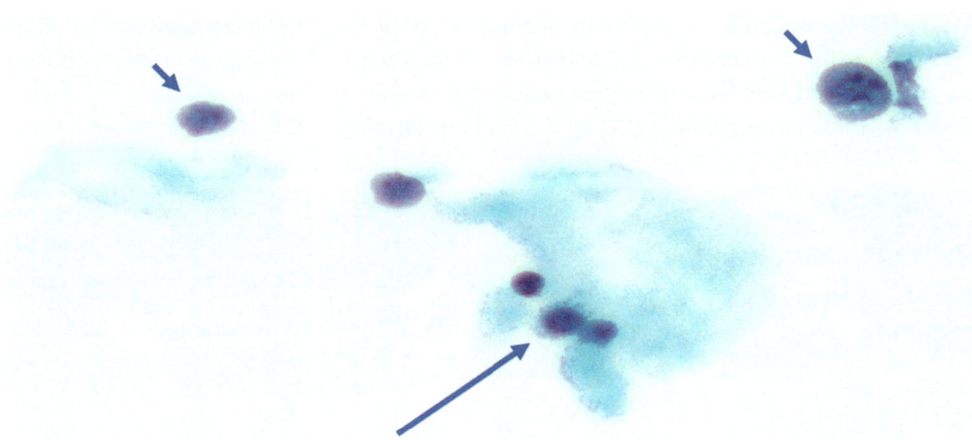

Fig. 54.2 CSF cytological examination (Short arrows: MALIGNANT lymphocytes, Long arrow: Normal lymphocytes)

possibly hydrocephalus (type D) [9]. In order to avoid false positives, cerebrospinal MRI should be obtained before lumbar puncture or ventricular shunt placement because these procedures could lead to contrast enhancement of the meninges. FDG-PET is not considered a standard diagnostic tool in LC, but there are case reports describing its utility in this situation [16, 17].

54.4.4 Cerebrospinal Fluid (CSF) Analysis

Demonstrating tumor cells in the CSF remains the gold standard to diagnose LC (Fig. 54.2). Identifying cancer cells using Papanicolaou staining has a yield of 50–60% in the first lumbar puncture (LP). Sensitivity can be increased to 80% with a second LP and by a further 2–5% in the third attempt. At least 5–10 ml of CSF should be drawn, hemorrhagic contamination should be avoided, and the sample should be processed within 30 min after collection [18, 19]. Flow cytometry to detect cancer cells is used primarily in the setting of hematological malignancies and was found to have severalfold higher sensitivity to detect LC when compared to conventional cytology. Also, it was found to be especially valuable in cases where cytology was equivocal [20]. Even in the absence of pathologic cells, CSF alterations can be observed in more than 90% of the patients with LC. These include increased opening pressure (> 200 mm H_2O) in 21–42%, increased CSF leukocyte count (> 4 mm^{-3}) in 48–77%, elevated protein (> 50 mg/dl) in 56–91%, and decreased glucose (< 60 mg/dl) in 22–63% [13]. Meningeal biopsy is an option when there is a high pretest probability with symptoms and/or imaging concerning for LC but multiple negative CSF cytology testing. This is usually done after ruling out an infectious or inflammatory etiology and documented negative cytology on repeated CSF exams. When biopsy is performed, it should be from a nodular area of the leptomeninges and, if feasible, should be further analyzed with DNA sequencing and/or other molecular techniques [21]. Fluorescent in situ hybridization (FISH) is used in peripheral blood for the enrichment and identification of circulating tumor cells (CTCs) and isolation of tumor DNA, irrespective of cell size variation and specific surface marker expression [22]. This technique has recently been applied to CSF. Circulating tumor DNA (ctDNA) has been shown to be more abundant in CSF than in plasma, and this could translate to a better diagnostic yield. With increasing utilization of driver mutation targeted therapy, it is very helpful that driver and resistance mutations can also be identified in CSF ctDNA in some patients with CNS metastases. This approach had a sensitivity of 95.2% to detect LC, which was higher than that of thin prep cytology (57.1%), MRI (47.6%), and MRI plus cytology (90.5%), respectively [23–25]. Next-generation sequencing of DNA from CSF tumor cells showed a high level of concordance with the primary tumor in terms of genetic profile [23]. Given the ease of access, CSF could be a potential source of a “liquid biopsy” with the option to investigate resistance mutations and determine optimal treatment, especially when there is CNS progression despite stable extra CNS lesions.

54.4.5 Prognostic Factors

While prognosis of LC in general is poor, it is influenced by tumor-specific factors and treatment administered. The INDEX score was recently proposed for patients with LC and a breast primary malignancy. This score correlated overall survival with older age, performance status, luminal subtype, and treatment intensity. Patients classified as having bad prognosis presented a median overall survival of 1.5 months compared to patients with good prognosis who experienced a median overall survival of 9.6 months [26].

Among those with LC in the setting of melanoma, factors associated with poorer survival included ECOG performance status greater than 0, neurological symptoms, elevated LDH, and lack of concurrent systemic disease. Both treatment with targeted agents and intrathecal agents was associated with improved outcomes, while radiation was found to have no impact on survival [27]. In lung cancer patients with LC, based on a retrospective review, good PS, use of EGFR-TKIs, and a nonsmoking status were associated with better outcomes [28].

54.5 Treatment

A generally accepted standard of care for treatment of LC remains an elusive goal. The current management of LC consists of single or multimodality treatment and is individualized on a case-by-case basis. The modalities available include intrathecal chemotherapy, systemic therapy, radiotherapy, or best-supportive care.

54.5.1 Symptom Management

There is no data supporting the use of primary seizure prophylaxis in patients with LC [29]. For signs of increased ICP such as headache, nausea, or encephalopathy, high-volume lumbar puncture can be considered [8]. Ventriculoperitoneal shunts also have a role to relieve symptomatic hydrocephalus [30]. Opioids, steroids, and palliative focal radiation can be used to treat symptoms of meningeal irritation or radicular pain [8, 9].

54.5.2 Radiation Therapy

Treatment of LC needs to be directed towards the entire neuroaxis to be effective, but this modality would be highly toxic. Whole-brain RT is therefore not an appropriate modality and has not been shown to improve survival, except when there is co-existing brain metastasis [31]. However, it has a role in treating nodular/bulky disease to improve CSF flow and in improving penetration of intra-CSF chemotherapy. It is also indicated as a palliative measure for symptomatic sites and cauda equina or skull-base syndromes despite absence of radiographic signs [32].

54.5.3 Chemotherapy

Pharmacokinetic studies involving systemic chemotherapy during treatment of CNS tumors did not demonstrate evidence of penetration of the BBB with effective, therapeutic concentrations [33]. Most chemotherapeutic agents are too large to pass across the BBB and enter the CNS. In addition to this, BBB is lined with drug efflux pumps, such as Pgp, which actively transports toxic substances, including drugs, out of the CNS. These highly functioning pumps ultimately cause drug resistance in the CNS [34–36]. Animal models, for example, have shown that only 3–5% of systemically administered methotrexate penetrates the blood-brain barrier (BBB) at a clearance halftime of 6 hours and requires days of continuous high-dose IV methotrexate infusions to achieve a therapeutic concentration (>1 μM) in CSF [37] [38]. Intrathecal chemotherapy on the other hand aims to bypass the BBB, maximize drug exposure in the CSF, and limit systemic toxicity and remains the cornerstone of treatment of LC. It achieves this by ensuring a higher drug concentration despite a smaller dose due to the lower volume of distribution of the CSF compared to plasma (140 vs 3500 ml) [39]. For example, therapeutic concentration of methotrexate in the CSF with minimal systemic absorption (<0.1 μM) can be achieved for 48–72 hours with a single intraventricular injection (6.25 mg/m^2) [40]. Despite this being a high need area for research, only six randomized clinical trials evaluating treatment strategies for neoplastic meningitis have been completed. These have explored different regimens of intra-CSF chemotherapy or the role of adding intra-CSF chemotherapy to systemic treatment in patients with LC. Contrary to the observations in retrospective studies, OS benefit was not seen in either prospective studies with the addition of IT chemotherapy; however, PFS improvement was noted in one [41–46].

Contrast enhancement in leptomeningeal lesion signal towards the absence of an intact BBB and, therefore, systemic pharmacotherapy, if active, should be able to target these areas. Systemic agents should be selected according to the primary tumor, clinical status, molecular characteristics, prior therapy, and how readily it crosses the BBB.

Chemotherapy can be administered into the CSF through either repeated lumbar punctures or ventricular devices. Ventricular devices offer the benefit of comfort for the patient, ability to achieve an even distribution of the drug, and an extra layer of safety by avoiding inadvertent delivery of the drug into the epidural or subdural space. The most commonly placed device is the Ommaya reservoir, a subcutaneous reservoir attached to an intraventricular catheter. In a 616-patient study reporting an infection rate of 5.5% following placement of Ommaya reservoir, infection was thought to be related to placement of the device in one third and due to subsequent accessing of the reservoir in others [47]. Revision rates of ventricular devices were found to be below 7.4% in different cohorts of patients, and this highlights an overall favorable safety profile of these devices [48]. Before each intra-CSF injection, the same amount of volume as that of the chemotherapy should be removed and sent for analysis. One of the relative contraindications of intra-CSF pharmacotherapy includes patients who have obstructive gross disease or symptomatic hydrocephalus. A ventriculoperitoneal shunt could be considered in select such cases prior to therapy [49]. The most commonly used drugs on a routine basis for intrathecal administration include methotrexate, cytarabine, liposomal cytarabine, and thiotepa. Historically, other agents have been tried, but none have been shown to be better than the abovementioned standard agents [12]. One of the side effects of intra-CSF administration of chemotherapy includes aseptic chemical meningitis, and this is transient. The risk is higher with agents characterized by long half-life, for example, liposomal cytarabine, and can be reduced by the simultaneous administration of dexamethasone [32]. Intra-CSF use of novel agents and targeted therapies in patients with LC (e.g., trastuzumab in breast cancer, interleukin-2 in melanoma, and panitumumab in NSCLC) has been reported; however, long-term results from prospective studies are awaited [50–52].

54.5.4 Assessment of Response to Treatment

The difficulties with assessing treatment response in patients with LC include inconsistencies associated with CSF cytology, lack of quantitative measurement of the disease, and subjectivity associated with LC-related symptoms ([3, 44, 53, 54]. Some of this has been attempted to be addressed by the response assessment in Neuro-Oncology (RANO) working group, which has drafted a consensus proposal for evaluating response in patients with LC. The goal is to standardize response definitions in forthcoming LC trials and permit cross trial comparisons. The tools included in response assessment include a standardized neurological exam, CSF cytology or flow cytometry, and radiographic evaluation. Based on the above, progressive disease is defined by either deteriorating findings on imaging or neurological exam. A persistently positive CSF cytology or flow cytometry is insufficient in itself to define progressive disease, especially when it occurs in the presence of stable or improved clinical or radiographic findings. It is recommended to get an MRI every 2 months for 6 months and then every 3 months if the disease shows stability. Treatment response based on CSF cytology is defined as the conversion of cytology from positive to negative, with negativity confirmed again after 4 weeks. Also, the proposal in its current format does not incorporate patient-reported outcomes and has yet to be validated [14]. Recently however, CSF tumor cell detection using FISH has provided us the ability to measure CSF tumor cells in a dynamic manner. The number of CSF tumor cells was shown to be concordant with headache alleviation and dynamic changes of intracranial pressure [22]. Also, changes in CSF tumor cell counts predicted treatment response [18]. Changes in CSF tumor cell counts can be used to monitor disease progression and treatment response since this parameter correlates well with the leptomeningeal disease state.

54.5.5 Recent Developments and Ongoing Prospective Studies

There are currently many ongoing and recently completed trials that include novel therapies in the management of leptomeningeal disease (Table 54.1; Source: www.clinicaltrials.gov, last accessed on 11/29/2020). Full results and completion of these trials are eagerly awaited in this population which currently has limited options with a minimal to moderate efficacy at best.

54.6 Conclusion

- The frequency of leptomeningeal carcinomatosis is estimated to be at least 5–8% in those with solid malignancies and 5–15% in those with hematological malignancies.
- A majority present with some neurological signs or symptoms.
- Sensitivity of CSF circulating tumor DNA in the diagnosis of LC is much higher than that of CSF cytology.
- DNA sequencing of the CSF circulating tumor DNA offers an attractive avenue for liquid biopsy and DNA sequencing.
- Treatment is based on the clinical context and is primarily through systemic therapy directed at the primary or intra-CSF chemotherapy (methotrexate, cytarabine, thiotepa).

Many promising novel agents are currently under study and will hopefully change the treatment landscape in the near future and hopefully improve outcomes for these patients.

Table 54.1 Recently completed and ongoing prospective studies for patients with Leptomeningeal carcinomatosis (IT Chemo: Intrathecal chemotherapy, IFRT: Involved field radiotherapy, TKI: Tyrosine kinase inhibitor, ICI: Immune checkpoint inhibitor, mAbs: Monoclonal antibodies, IV Chemo: Intravenous chemotherapy)

Phase	Trial identifier	Treatment modality	Study title	Status
I	NCT03101579	IT chemo	Intrathecal Pemetrexed for recurrent Leptomeningeal metastases from non-small cell lung cancer	Completed
II	NCT03082144	RT + IT chemo	Concurrent involved-field radiotherapy and Intrathecal Chemotherapy for Leptomeningeal metastases from solid tumors	Completed
II	NCT02616393	TKI	Study of Tesevatinib in subjects with non-small cell lung cancer, EGFR activating mutation, prior treatment with a tyrosine kinase inhibitor, and brain metastases or Leptomeningeal metastases	Completed
III	NCT01645839	IT chemo	Interest of Intrathecal Chemotherapy with liposomal Cytarabine (DepoCyte®) in meningeal metastasis of breast cancer (DEPOSEIN)	Completed
I	NCT03661424	Cellular therapy	Anti-CD3 x anti-Her2/Neu (Her2Bi) armed activated T cells (ATC) in patients with breast cancer Leptomeningeal metastases	Recruiting
I	NCT03719768	ICI + RT	Avelumab with radiotherapy in patients with Leptomeningeal disease	Recruiting
I	NCT03025256	IV + IT ICI	Intravenous and Intrathecal Nivolumab in treating patients with Leptomeningeal disease	Recruiting
II	NCT02886585	ICI	Pembrolizumab in central nervous system metastases	Recruiting
II	NCT04233021	TKI	Study of Osimertinib in patients with a lung cancer with brain or Leptomeningeal metastases with EGFR mutation (ORBITAL)	Recruiting
II	NCT04425681	TKI + mAbs	Osimertinib with Bevacizumab for Leptomeningeal metastasis from EGFR-mutation non-small cell lung cancer (OWBLM)	Recruiting
II	NCT04343573	RT	Proton Craniospinal radiation therapy vs. partial photon radiation therapy for Leptomeningeal metastasis from solid tumors	Recruiting
II	NCT04420598	Antibody drug conjugate	A multicenter, open-label, single-arm, multicohort phase II clinical trial of Trastuzumab Deruxtecan (DS-8201a) in human epidermal growth factor receptor 2 (HER2)- positive advanced breast cancer with brain metastases and/or Leptomeningeal Carcinomatosis - the DEBBRAH study	Recruiting
II	NCT03501979	TKI + mAbs + IV chemo	Tucatinib and Trastuzumab and Capecitabine for treatment of Leptomeningeal metastases in HER2 positive breast cancer	Recruiting
II	NCT02422641	IV chemo	Prospective evaluation of high-dose systemic methotrexate in patients with breast cancer and Leptomeningeal metastasis	Recruiting
II/III	NCT03275402	Radio-immunotherapy	Intracerebroventricular Radioimmunotherapy using 131I-omburtamab for Neuroblastoma central nervous system/Leptomeningeal metastases	Recruiting

Future Direction

- Broaden the understanding of pathophysiology underlying development of LC in cancer patients.
- Further validation of CSF circulating tumor cells and DNA as a diagnostic tool.
- Application of next-generation sequencing to CSF circulating tumor cells and DNA in order to identify therapeutic targets.
- Standardization of response assessment tools and criteria across clinical trials and practice settings.
- Development of phase III clinical trials which include patients with LC, since this continues to be an exclusion criterion in most prospective studies.

References

1. Chamberlain MC. Leptomeningeal metastases: a review of evaluation and treatment. J Neuro-Oncol. 1998;37:271–84. https://doi.org/10.1023/a:1005976926058.
2. Gwak H-S, Joo J, Kim S, et al. Analysis of treatment outcomes of intraventricular chemotherapy in 105 patients for leptomeningeal carcinomatosis from non-small-cell lung cancer. J Thorac Oncol. 2013;8:599–605.
3. Wasserstrom WR, Glass JP, Posner JB. Diagnosis and treatment of leptomeningeal metastases from solid tumors: experience with 90 patients. Cancer. 1982;49:759–72. https://doi.org/10.1002/1097-0142(19820215)49:4<759::aid-cncr2820490427>3.0.co;2-7.
4. Beauchesne P. Intrathecal chemotherapy for treatment of leptomeningeal dissemination of metastatic tumours. Lancet Oncol. 2010;11:871–9. https://doi.org/10.1016/S1470-2045(10)70034-6.
5. Zhang XH-F, Giuliano M, Trivedi MV, et al. Metastasis dormancy in estrogen receptor-positive breast cancer. Clin Cancer Res. 2013;19:6389–97. https://doi.org/10.1158/1078-0432.CCR-13-0838.

6. Kokkoris CP. Leptomeningeal carcinomatosis. how does cancer reach the pia-arachnoid? Cancer. 1983;51:154–60. https://doi.org/10.1002/1097-0142(19830101)51:1<154::aid-cncr2820510130>3.0.co;2-k.
7. Boire A, Zou Y, Shieh J, et al. Complement component 3 adapts the cerebrospinal fluid for Leptomeningeal metastasis. Cell. 2017;168:1101–113.e13. https://doi.org/10.1016/j.cell.2017.02.025.
8. Wang N, Bertalan MS, Brastianos PK. Leptomeningeal metastasis from systemic cancer: review and update on management. Cancer. 2018;124:21–35. https://doi.org/10.1002/cncr.30911.
9. Le Rhun E, Weller M, Brandsma D, et al. EANO-ESMO clinical practice guidelines for diagnosis, treatment and follow-up of patients with leptomeningeal metastasis from solid tumours. Ann Oncol. 2017c;28:iv84–99. https://doi.org/10.1093/annonc/mdx221.
10. Leal T, Chang JE, Mehta M, Robins HI. Leptomeningeal metastasis: challenges in diagnosis and treatment. Curr Cancer Ther Rev. 2011;7:319–27. https://doi.org/10.2174/157339411797642597.
11. Le Rhun E, Taillibert S, Zairi F, et al. A retrospective case series of 103 consecutive patients with leptomeningeal metastasis and breast cancer. J Neuro-Oncol. 2013b;113:83–92. https://doi.org/10.1007/s11060-013-1092-8.
12. Le Rhun E, Taillibert S, Chamberlain MC. Carcinomatous meningitis: leptomeningeal metastases in solid tumors. Surg Neurol Int. 2013a;4:S265–88. https://doi.org/10.4103/2152-7806.111304.
13. Franzoi MA, Hortobagyi GN. Leptomeningeal carcinomatosis in patients with breast cancer. Crit Rev Oncol Hematol. 2019;135:85–94. https://doi.org/10.1016/j.critrevonc.2019.01.020.
14. Chamberlain M, Junck L, Brandsma D, et al. Leptomeningeal metastases: a RANO proposal for response criteria. Neuro-Oncology. 2017;19:484–92. https://doi.org/10.1093/neuonc/now183.
15. Collie DA, Brush JP, Lammie GA, et al. Imaging features of leptomeningeal metastases. Clin Radiol. 1999;54:765–71. https://doi.org/10.1016/S0009-9260(99)91181-9.
16. Short RG, Bal S, German JP, et al. Potential of F-18 PET/CT in the detection of Leptomeningeal metastasis. Neuroradiol J. 2014;27:685–9. https://doi.org/10.15274/NRJ-2014-10098.
17. Ortapamuk H, Demir MK. Intracranial leptomeningeal carcinomatosis in three cases from breast cancer demonstrated on F-18 Fluorodeoxyglucose positron emission tomography/computerized tomography. Indian J Nucl Med. 2017;32:16–8. https://doi.org/10.4103/0972-3919.198454.
18. Lee JS, Melisko ME, Magbanua MJM, et al. Detection of cerebrospinal fluid tumor cells and its clinical relevance in leptomeningeal metastasis of breast cancer. Breast Cancer Res Treat. 2015;154:339–49. https://doi.org/10.1007/s10549-015-3610-1.
19. Glantz MJ, Cole BF, Glantz LK, et al. Cerebrospinal fluid cytology in patients with cancer: minimizing false-negative results. Cancer. 1998;82:733–9. https://doi.org/10.1002/(sici)1097-0142(19980215)82:4<733::aid-cncr17>3.0.co;2-z.
20. Ahluwalia MS, Wallace PK, Peereboom DM. Flow cytometry as a diagnostic tool in lymphomatous or leukemic meningitis: ready for prime time? Cancer. 2012;118:1747–53. https://doi.org/10.1002/cncr.26335.
21. Glitza IC, Smalley KSM, Brastianos PK, et al. Leptomeningeal disease in melanoma patients: an update to treatment, challenges, and future directions. Pigment Cell Melanoma Res. 2020;33:527–41. https://doi.org/10.1111/pcmr.12861.
22. Li X, Zhang Y, Ding J, et al. Clinical significance of detecting CSF-derived tumor cells in breast cancer patients with leptomeningeal metastasis. Oncotarget. 2018;9:2705–14. https://doi.org/10.18632/oncotarget.23597.
23. Jiang B-Y, Li Y-S, Guo W-B, et al. Detection of driver and resistance mutations in Leptomeningeal metastases of NSCLC by next-generation sequencing of cerebrospinal fluid circulating tumor cells. Clin Cancer Res. 2017;23:5480–8. https://doi.org/10.1158/1078-0432.CCR-17-0047.
24. Murtaza M, Dawson S-J, Pogrebniak K, et al. Multifocal clonal evolution characterized using circulating tumour DNA in a case of metastatic breast cancer. Nat Commun. 2015;6:8760. https://doi.org/10.1038/ncomms9760.
25. Zhao Y, He J-Y, Zou Y-L, et al. Evaluating the cerebrospinal fluid ctDNA detection by next-generation sequencing in the diagnosis of meningeal Carcinomatosis. BMC Neurol. 2019;19:331. https://doi.org/10.1186/s12883-019-1554-5.
26. Niwińska A, Pogoda K, Michalski W, et al. Determinants of prolonged survival for breast cancer patient groups with leptomeningeal metastasis (LM). J Neuro-Oncol. 2018;138:191–8. https://doi.org/10.1007/s11060-018-2790-z.
27. Ferguson SD, Bindal S, Bassett RLJ, et al. Predictors of survival in metastatic melanoma patients with leptomeningeal disease (LMD). J Neuro-Oncol. 2019;142:499–509. https://doi.org/10.1007/s11060-019-03121-2.
28. Yan W, Yu J. The clinical characteristic and prognostic factors of leptomeningeal metastasis in patients with non-small-cell lung cancer: a retrospective study from one single cancer institute. J Clin Oncol. 2019;37:e13502. https://doi.org/10.1200/JCO.2019.37.15_suppl.e13502.
29. Tremont-Lukats IW, Ratilal BO, Armstrong T, Gilbert MR. Antiepileptic drugs for preventing seizures in people with brain tumors. Cochrane Database Syst Rev. 2008:CD004424. https://doi.org/10.1002/14651858.CD004424.pub2.
30. Nigim F, Critchlow JF, Kasper EM. Role of ventriculoperitoneal shunting in patients with neoplasms of the central nervous system: an analysis of 59 cases. Mol Clin Oncol. 2015;3:1381–6. https://doi.org/10.3892/mco.2015.627.
31. Morris PG, Reiner AS, Szenberg OR, et al. Leptomeningeal metastasis from non-small cell lung cancer: survival and the impact of whole brain radiotherapy. J Thorac Oncol. 2012;7:382–5. https://doi.org/10.1097/JTO.0b013e3182398e4f.
32. Le Rhun E, Taillibert S, Chamberlain MC. Neoplastic meningitis due to lung, breast, and melanoma metastases. Cancer Control. 2017b;24:22–32. https://doi.org/10.1177/107327481702400104.
33. Jacus MO, Daryani VM, Harstead KE, et al. Pharmacokinetic properties of anticancer agents for the treatment of central nervous system tumors: update of the literature. Clin Pharmacokinet. 2016;55:297–311. https://doi.org/10.1007/s40262-015-0319-6.
34. Bowman KM, Kumthekar P. Medical management of brain metastases and leptomeningeal disease in patients with breast carcinoma. Future Oncol. 2018;14:391–407. https://doi.org/10.2217/fon-2017-0255.
35. Le Rhun E, Rudà R, Devos P, et al. Diagnosis and treatment patterns for patients with leptomeningeal metastasis from solid tumors across Europe. J Neuro-Oncol. 2017a;133:419–27. https://doi.org/10.1007/s11060-017-2452-6.
36. Scott BJ, Oberheim-Bush NA, Kesari S. Leptomeningeal metastasis in breast cancer - a systematic review. Oncotarget. 2016;7:3740–7. https://doi.org/10.18632/oncotarget.5911.
37. Shapiro WR, Young DF, Mehta BM. Methotrexate: distribution in cerebrospinal fluid after intravenous, ventricular and lumbar injections. N Engl J Med. 1975;293:161–6. https://doi.org/10.1056/NEJM197507242930402.
38. Tetef ML, Margolin KA, Doroshow JH, et al. Pharmacokinetics and toxicity of high-dose intravenous methotrexate in the treatment of leptomeningeal carcinomatosis. Cancer Chemother Pharmacol. 2000;46:19–26. https://doi.org/10.1007/s002800000118.
39. Poplack D, Bleyer WA HM (1980) Pharmacology of antineo- plastic agents in cerebrospinal fluid. https://doi.org/10.1007/978-1-4684-1039-6_39.

40. Taillibert S, Laigle-Donadey F, Chodkiewicz C, et al. Leptomeningeal metastases from solid malignancy: a review. J Neuro-Oncol. 2005;75:85–99. https://doi.org/10.1007/s11060-004-8101-x.
41. Boogerd W, Van Den Bent MJ, Koehler PJ, et al. The relevance of intraventricular chemotherapy for leptomeningeal metastasis in breast cancer: a randomised study. Eur J Cancer. 2004;40:2726–33. https://doi.org/10.1016/j.ejca.2004.08.012.
42. Glantz MJ, Jaeckle KA, Chamberlain MC, et al. A randomized controlled trial comparing intrathecal sustained-release cytarabine (DepoCyt) to intrathecal methotrexate in patients with neoplastic meningitis from solid tumors. Clin Cancer Res. 1999;5:3394–402.
43. Grossman SA, Finkelstein DM, Ruckdeschel JC, et al. Randomized prospective comparison of intraventricular methotrexate and thiotepa in patients with previously untreated neoplastic meningitis. J Clin Oncol. 1993;11:561–9. https://doi.org/10.1200/JCO.1993.11.3.561.
44. Hitchins RN, Bell DR, Woods RL, Levi JA. A prospective randomized trial of single-agent versus combination chemotherapy in meningeal carcinomatosis. J Clin Oncol. 1987;5:1655–62. https://doi.org/10.1200/JCO.1987.5.10.1655.
45. Le Rhun E, Mailliez A, Wallet J, et al. CMET-39. Intra-csf liposomal cytarabine plus systemic therapy as initial treatment of breast cancer leptomeningeal metastasis: a randomised, open-label trial. Neuro-Oncology. 2018;20:vi61-2.
46. Shapiro WR, Schmid M, Glantz M, Miller JJ. A randomized phase III/IV study to determine benefit and safety of cytarabine liposome injection for treatment of neoplastic meningitis. J Clin Oncol. 2006;24:1528. https://doi.org/10.1200/jco.2006.24.18_suppl.1528.
47. Mead PA, Safdieh JE, Nizza P, et al. Ommaya reservoir infections: a 16-year retrospective analysis. J Infect. 2014;68:225–30. https://doi.org/10.1016/j.jinf.2013.11.014.
48. Le Rhun E, Preusser M, van den Bent M, et al. How we treat patients with leptomeningeal metastases. ESMO Open. 2019;4:e000507. https://doi.org/10.1136/esmoopen-2019-000507.
49. Lin N, Dunn IF, Glantz M, et al. Benefit of ventriculoperitoneal cerebrospinal fluid shunting and intrathecal chemotherapy in neoplastic meningitis: a retrospective, case-controlled study. J Neurosurg. 2011;115:730–6. https://doi.org/10.3171/2011.5.JNS101768.
50. Glitza IC, Rohlfs M, Guha-Thakurta N, et al. Retrospective review of metastatic melanoma patients with leptomeningeal disease treated with intrathecal interleukin-2. ESMO Open. 2018;3:e000283. https://doi.org/10.1136/esmoopen-2017-000283.
51. Hollebecque A, Levy A, Broutin S, et al. First case report of intrathecal panitumumab for treatment of meningeal carcinomatousis in an EGFR mutant lung adenocarcinoma patient. Lung Cancer. 2013;80:113–4.
52. Zagouri F, Sergentanis TN, Bartsch R, et al. Intrathecal administration of trastuzumab for the treatment of meningeal carcinomatosis in HER2-positive metastatic breast cancer: a systematic review and pooled analysis. Breast Cancer Res Treat. 2013;139:13–22. https://doi.org/10.1007/s10549-013-2525-y.
53. de Visser BW, Somers R, Nooyen WH, et al. Intraventricular methotrexate therapy of leptomeningeal metastasis from breast carcinoma. Neurology. 1983;33:1565–72. https://doi.org/10.1212/wnl.33.12.1565.
54. Freilich RJ, Krol G, DeAngelis LM. Neuroimaging and cerebrospinal fluid cytology in the diagnosis of leptomeningeal metastasis. Ann Neurol. 1995;38:51–7. https://doi.org/10.1002/ana.410380111.

Part XIV

Urologic Cancer and First Patterns of Metastasis

Samuel L. Washington III, Peter R. Carroll
and Sima P. Porten

In this section of Urologic Cancer and First Patterns of Metastasis, we focus the discussion on two of the most common, and distinct, genitourinary malignancies, prostate cancer and bladder cancer. These two cancers impact organs within close anatomic proximity, and thus share similar patterns of lymphatic drainage. For prostate cancer, we discuss the anatomic considerations surrounding lymphatic drainage and prior efforts to map the lymphatic system of the prostate. Next, we describe the correlation between our anatomic understanding and clinical practice. For bladder cancer we examine the lymphatic drainage and clinical staging at the time of diagnosis. Next, we outline the evaluation for nodal disease and its implications for patients diagnosed with bladder cancer. The comparison of these two genitourinary malignancies affords us the opportunity to gain a greater understanding of how the clinical implications differ despite a similar pattern of lymphatic drainage and metastatic spread.

Urologic Cancer and the First Patterns of Metastasis

55

Samuel L. Washington III, Peter R. Carroll, and Sima P. Porten

Abstract

In this chapter we will describe the patterns of metastasis for two of the most common genitourinary malignancies, prostate cancer and bladder cancer. We will also discuss aspects of management for individuals with node-positive disease to provide insight into the clinical impact of metastasis on oncologic outcomes.

Learning Objectives
- To understand the lymphatic drainage of prostate cancer and bladder cancer.
- To appreciate the patterns of metastatic spread for prostate cancer and bladder cancer.
- To review the current discussion surrounding the surgical management of lymph node metastasis for prostate cancer and bladder cancer.

55.1 Introduction

In this chapter we will review the patterns of metastatic spread for two of the most common genitourinary malignancies, prostate cancer and bladder cancer. These cancers represent distinct histologic tumors in close anatomic proximity to one another within the pelvis. As a result, they share similar patterns of lymphatic drainage.

S. L. Washington III (✉) · P. R. Carroll · S. P. Porten
University of California, San Francisco, CA, USA
e-mail: samuel.washington@ucsf.edu

55.2 Prostate Cancer

Prostate cancer will account for an estimated 248,530 new cancer diagnoses in 2021 and remains the second most common cause of cancer deaths in men in the United States [1]. Although the majority will present with low-grade disease, a subset of patients will present with more aggressive disease. The key principle of timely evaluation and diagnosis of prostate cancer is to identify men with early stage disease who harbor aggressive disease and are at risk of metastasis. Here we describe prior efforts to understand the pathways in which aggressive disease travels from the prostate cancer to regional lymph nodes and distant sites of metastasis.

55.3 Anatomic Considerations for Lymphatic Drainage

The lymphatic drainage of the prostate is thought to mirror the vascular drainage system, with a similarly complex system of the main and smaller branches. Numerous studies have gradually improved our understanding of the lymphatic drainage of the prostate and patterns of metastatic spread in men with prostate cancer [2, 3]. As a result, we now understand that the lymphatic drainage largely depends on the location within the prostate and the tissue types present.

One such study examined five fresh prostate specimens from patients undergoing cystoprostatectomy for bladder cancer [2]. Specimens were sectioned and incubated with a monoclonal antibody specific for desmoplakin, which detects the endothelial layer of lymphatic channels. Five fields within the central, transitional, and peripheral zones were randomly selected to calculate the density of lymphatics per region. Within the glandular prostate, no lymphatics were in close proximity to the glandular acini or intra-acinar connective tissue. A consistent pattern of lymphatic in the interacinar connective tissue septum was noted in each of the three prostatic zones. Some lymphatics ran parallel to the

S. P. Leong et al. (eds.), *Cancer Metastasis Through the Lymphovascular System*,
https://doi.org/10.1007/978-3-030-93084-4_55

prostatic capsule while other accompanied arteries or neural ganglia. Lymphatics were present between smooth muscle bundles of the anterior fibromuscular stroma but rarely detected in the fibrous stoma itself. The greatest density of lymphatics ran parallel and in the longitudinal direction of the ejaculatory ducts compared to other areas within the glandular prostate. The prostatic capsule harbored a slightly greater, yet not statistically significant different, density of lymphatics compared to the glandular prostate itself.

Broadly, larger lymphatic vessels within the gland drain into a subcapsular network, which in turn drains a system of collecting vessels often pairing with arteries [3]. These lymphatic vessel networks are concentrated at the base of the gland, where the highest density of vessels is present. The lymphatic channels are thought to also drain the upper aspect of the posterior prostate, extending along the medial aspect of the seminal vessels and extending towards the ureters, then cross the umbilical artery, and travel laterally to the external iliac lymph nodes. The lymphatic channels draining the inferior aspect of the prostate travel to the base of the gland and then to the hypogastric lymph nodes. Some of these channels communicate with those of the rectum, draining into the pararectal lymph nodes. The anterior prostate has lymphatic channels that run along the apex of the gland and follow the internal pudendal artery, eventually joining the internal iliac lymph nodes. Additional anterior and lateral lymphatics drain to the para-vesical lymph nodes, eventually draining into the bladder lymphatic system.

55.4 Mapping the Lymphatic System of the Prostate

Previous efforts have attempted to map the lymphatic system of the prostate using various methodologies. Prior studies have injected contrast into human glands to patterns of lymphatic spread and have outlined the drainage along larger vessels, but these methods may have limited characterization of smaller vessels that were unable to accommodate the contrast molecule and/or were too small to be visualized adequately on imaging. Other studies have identified three main patterns of spread: superior and lateral to the internal-external iliac lymph nodes; from perineal floor to internal pudendal vessels; and via sacral lymphatics (S2–5) [4, 5]. Others have used radiotracers such as technetium to examine lymphatic spread based on the concept of lymphatic spread occurring through key sentinel lymph nodes, similar to lymphatic spread for breast cancer [6]. This study was limited by the properties of the radiotracer, both in terms of time needed for dispersion and size limitations with respect to the lymphatic vessels it could travel through. The end result of such studies was the confirmation that the most common sites of involvement were generally along the external iliac, internal iliac, and obturator arteries, which make up the standard template for lymph node dissection.

Leveraging prior efforts to map the anatomic pathways of lymphatic drainage, the utility of a sentinel lymph dissection remains a topic of ongoing investigation. The concept of a sentinel lymph node rests on the idea of there being a single lymph node in the direct drainage pathway from the organ of origin to the rest of the body [7]. Conceptually, a sentinel lymph node would provide an accurate assessment of the likelihood of distant nodal metastasis and help justify whether a pelvic lymph node dissection should (or should not) be performed. Yet this is not entirely applicable to prostate cancer, as we know the lymphatic drainage varies based on the location within the gland. In addition, the intrapelvic location of the prostate and relevant lymphatic drainage makes it difficult to accurately visualize the potentially affected lymph nodes [8]. Efforts to incorporate various imaging modalities and lymph node dissection strategies continue, but sentinel lymph node dissection remains largely experimental.

55.5 Clinical Correlation with Anatomic Understanding

Historically, surgical resection of the lymph nodes was utilized to detect spread of cancer beyond the organ of origin. Prior work leveraged autopsy as a primary means of identified patterns of spread in men with advanced disease. By noting the landing sites of metastasis, we were able to gain a greater understanding of the spread. Interestingly, autopsy studies noted that nodal yield of the surgical dissection often underrepresented the total lymph node burden of extrapelvic disease noted at autopsy [9]. Lymph node dissection eventually became a feasible method of sampling, balanced with the risk of morbidity, rather than a thorough examination of the potential full extent of disease. The extent of the dissection evolved over time as surgical approach evolved from a perineal to retropubic approach. This change in approach in turn led to greater lymph node yield and nodal positivity. The observed examination of nodal burden increased as lymph node dissections became more extensive, expanding beyond the obturator and external iliac lymph nodes to include areas such as the internal iliac, common iliac, presacral, and presciatic regions.

Autopsy study of 753 men with metastatic disease demonstrated that 63% had a lymph node metastasis [10]. They noted that even in cases of metastasis confined to the lymph nodes, pelvic nodal involvement was not more frequent than paraaortic nodal involvement. From this study, two patterns of metastatic spread were observed: lymphogenous spread to paraaortic lymph node metastasis with pelvic lymph node involvement and hematogenous spread to paraaortic lymph nodes without pelvic lymph node involvement. Cases involving metastasis to a single organ were noted to frequently harbor metastasis to the lumbar spine and para-aortic lymph nodes, largely thought to occur by way of the vertebral vein, as a method of skipping pelvic lymphatic drainage channels [11].

Another autopsy study of 1589 men with prostate cancer similarly noted lymphatic or hematogenous metastases in 39.7% of men [12]. Paraaortic lymph nodes were most frequently noted, followed by pelvic lymph nodes. A hematogenous metastatic pattern was the most common overall (35%), with frequent sites being the bone (spine, 90%), lung (46%), liver (25%), pleura (21%), and adrenals (13%). Hematogenous metastasis was more frequent in those with paraaortic lymph node involvement than in those with pelvic lymph node involvement (88.9% vs. 63.9%, $p < 0.001$). Spine metastasis was significantly associated with para-aortic lymph node involvement, likely through a hematogenous metastasis pattern. In this study, histologic grade was strongly associated with hematogenous metastatic spread, ranging from 7.5% prevalence in low-grade tumors to 60.7% in high-grade tumors. A similar pattern was noted with pathologic stage, with metastases found in 4.2% of men with pT2 cancers, 41.1% of pT3, and 80.3% of pT4 tumors ($p < 0.001$). Lymph node positivity is rarely associated with anteriorly located tumors, and 30–40% of lymph node metastases are found to be contralateral to the dominant tumor [13].

Using a surgical approach to map nodal metastasis in high-risk prostate cancer, 19 men underwent an extended lymph node dissection and retroperitoneal lymph node dissection at the time of radical prostatectomy [14]. In this study the most common sites of nodal involvement were the obturator (88%), external iliac (83.3%), common iliac (77%), hypogastric (44.4%), and presacral (33.3%). All cases with common iliac lymph node involvement had positive pelvic lymph nodes, and all cases with retroperitoneal disease had positive common iliac lymph nodes. Positive retroperitoneal lymph nodes were found only in cases of positive common iliac lymph nodes.

Given the under-appreciation of the total nodal metastasis burden using a surgical approach, the role and extent of a pelvic lymph node dissection at time of radical prostatectomy continues to be debated. Discussions are commonly centered around appropriate patient selection, the extent of dissection, the morbidity of the procedure, and the diagnostic versus therapeutic value. Utilization of pelvic node dissection has varied across the United States, with rates ranging from 58.9% to 72.1% over time and with greater use in those of intermediate- and high-risk disease compared to those with low-risk features [15]. The diagnostic and therapeutic value of a lymph node dissection is also intrinsically related to the prevalence of nodal disease in the population. For men with nodal disease at time of primary radical prostatectomy, up to 29% may have an undetectable prostate-specific antigen level with minimal differences in survival among those that received adjuvant therapy or not. Those with a detectable prostate-specific antigen level, irrespective of nodal status, were noted to have worse recurrence-free survival [16]. An observational study of 1249 men with high-risk prostate cancer demonstrated that a greater lymph node burden at time of surgery was associated with worse cancer-specific survival [17]. In this setting, a pelvic lymph node dissection provides diagnostic value by providing an estimate of the nodal burden at time of treatment. A systemic review examined the risk-benefit ratio of different extents of lymph node dissection, finding that the extent of lymph node dissection impacted the rates of biochemical recurrence and clinical recurrence with no significant impact on cancer survival outcomes [18]. In addition, more extensive dissections were associated with greater perioperative risks of blood loss, extended hospital stay, and postoperative complications. The findings of this systematic review question the therapeutic value of the lymph node dissection while highlighting the diagnostic benefit.

Most recently, a report from a prospective single-center phase III randomized control trial sought to determine the impact of an extended lymph node dissection on oncologic outcomes for men with intermediate- or high-risk clinically localized prostate cancer [19]. A total of 300 men were randomized to a limited or extended pelvic lymph node dissection. The median biochemical recurrence-free survival was 6.14 months in men who underwent a limited node dissection and had not yet been reached in those with an extended approach. Yet median metastasis-free survival was not reached in either group nor was cancer-specific survival data available because no patient had died due to prostate cancer by the cutoff date. Ultimately, the authors concluded that based on early oncologic outcomes reporting, a more extensive lymph node dissection offers better pathologic staging without demonstrating a clear benefit in oncologic outcomes.

55.6 Bladder Cancer

Bladder cancer is the fourth most common cancer in the United States in men (less common in women) accounting for about 83,730 new cases (64,280 in men and 19,450 in women) and results in 17,200 deaths (12,260 in men and 4940 in women) in 2020 [1]. The primary tumor stage is vital to patient outcomes, and indeed 70% of patients present with non-muscle invasive bladder cancer (Tis, Ta, and T1) [1]. However, 20–40% of patients either present with or progress to muscle-invasive cancer (T2–T4), which in turn has an increased risk of development of metastatic disease [20]. The primary pattern of metastasis in bladder cancer is through lymphovascular spread to locoregional lymph nodes before developing disease to distant sites (most commonly bone and lung) [21]. Clinically, these patients have a markedly worse prognosis with 5-year overall and disease-free survival rates of 35% and 31% compared to 89% and 69% for patients without lymph node involvement [22]. The presence of nodal metastasis (whether occult micro-metastasis or overt disease) is one of the most significant independent prognostic factors of mortality [23].

55.7 Lymphatic Drainage of the Bladder and Clinical Staging

A recent study by Roth et al. prospectively mapped the primary landing sites (first area) for lymphatic drainage in patients with muscle-invasive bladder cancer with 80% of patients having regional lymph node drainage to the obturator and internal iliac areas and external iliac region (bilaterally) [24]. Interestingly, some patients had primary landing sites (first area of drainage) to more distant lymph nodes such as the common iliac (15%) or paraaortic (4%) areas, and the primary location of the tumor can impact lymphatic spread [24, 25].

Similar to other cancers, clinical staging of nodal metastases is based on TNM guidelines set by the American Joint Committee on Cancer and reflects anatomical lymphatic spread [26]. Nodal involvement is denoted as N0-M1a disease with N0 disease defined as no lymph node involvement. N1 disease is defined as one regional pelvic lymph node identified in the obturator, internal iliac, external iliac, perivesical, or presacral chain, and N2 disease is one or more of regional lymph nodes involved. N3 disease is a positive lymph node at either or both common iliac chains. Nodal metastases outside the true pelvis, which are located above the common iliacs, are defined as true metastatic disease (M1a). Identification of nodal disease is based upon histopathology (biopsy or excision) or unequivocal findings on cross-sectional imaging, and the size of nodal involvement is not a part of the current clinical staging system.

55.8 Detection of Nodal Disease and Clinical Implications

At presentation with muscle-invasive disease, patients will undergo cross-sectional imaging to determine if distant metastasis is present, which in turn guides primary treatment and estimates of prognosis. CT (computed tomography) is used most commonly and is performed with and without intravenous contrast. Although distant metastasis can be detected robustly, characterizing lymph node-positive disease on imaging is more elusive. A lymph node in the pelvis is considered involved if it is ≥8 mm in short axis, and other features such roundness and necrosis can be considered in determining lymph node positivity. Unfortunately, the sensitivity for detecting lymph node involvement in bladder cancer is limited at 30–75%, with an overall accuracy of 61% [27, 28]. CT is limited by an inability to detect micro-metastatic disease in normal-sized lymph nodes. Magnetic resonance imaging (MRI) is a potential alternative to CT; however, it has similar limitations in detecting the presence of disease in a normal-sized lymph node. Despite the utilization of multiparametric techniques, which have been revolutionary in characterizing prostate cancer (T1/T2 signal intensity, diffusion characteristics, and timing of enhancement of gadolinium-based contrast), pooled sensitivity was only between 56 and 86% [29, 30].

Attempts to improve sensitivity and overall accuracy for imaging have not proved as fruitful as other advances in imaging such as PSMA-PET for prostate cancer. Although metastatic lymph nodes have increased metabolic activity and radiotracer uptake on PET/CT, spatial resolution limits detection in normal-sized lymph nodes and has the added downside of false-positive findings in the setting of inflammation [31]. Meta-analyses report low and varying sensitivities of 33–70% [32, 33]. Alternative radiotracers such as C11-choline have also been assessed with similarly low sensitivity of 66% in identifying nodal disease [34]. For MRI, the use of intravenous lymphotropic ultrasmall superparamagnetic particles of iron oxide (USPIO) to identify benign lymph nodes (where there is uptake), and thus uncover normal-sized lymph nodes harboring disease, improves sensitivity to 75% in a small prospective trial [35]. Currently, USPIO is not approved for clinical use in many countries and often requires multiple exams.

Unlike other malignancies, sentinel lymph node biopsy has not proven a useful tool in staging for patients with bladder cancer nor for treatment decision-making due to a high false-negative rate. A meta-analysis including 336 patients was performed, and pooled sensitivity was 79% (95% CI 0.69–0.86%) [36]. Recently, Salehi et al. published a consecutive series of 41 patients with sentinel nodes detected in 85% [35] at time of cystectomy, regardless of prior systemic treatment [37]. However, although sentinel lymph node biopsy is feasible, to make decision about proceeding with standard lymphadenectomy, pathologic status for sentinel node had to be performed for both sides of the hemipelvis as contralateral involvement was common. Hence, to forgo a potentially curative lymphadenectomy based on sentinel lymph node biopsy is not aligned with current standard of care.

The current cornerstone of treatment for muscle-invasive bladder cancer is radical cystectomy (complete bladder removal with the prostate in men and potentially uterus/cervix/ovaries in women) with pelvic node dissection. Lymph node removal is thought to be both diagnostic and therapeutic (for some). Although no single template is "standard," extrapolation (based on the abovementioned anatomic studies and staging principles) suggests that boundaries of an adequate lymph node dissection should be the genitofemoral nerve laterally, the bladder wall medially, the inguinal ligament distally/node of Cloquet, and the bifurcation of the common iliac artery proximally [38]. Multiple retrospective studies from institutional and national registries have reported that lymph node status (pathologic staging) along with other quantitative measurements including lymph node yield (number of lymph nodes taken), lymph node involvement (number of positive nodes present), and lymph node density (number of nodes positive per number taken) can affect patient outcomes. Hence, the rationale is that if you remove a larger number of nodes (even if no nodes are positive), and the number of nodes positive is significantly less than the number of nodes taken, patient outcomes are improved significantly by eradicating micro-metastatic disease [39–42]. Extended lymph node dissection (includes presacral lymph nodes and those up to the aortic bifurcation) has been advocated with improvements in survival up to 25% [43, 44]. Recently, a large multicenter randomized phase III study evaluating the extent of pelvic lymph node dissection demonstrated that extended lymphadenectomy was not superior to a standard template based on recurrence-free survival (hazard ratio [HR], 0.84; $P = 0.36$), cancer-specific survival (HR, 0.70; $P = 0.10$), or overall survival (HR, 0.78; $P = 0.12$). However, patients who had an extended node dissection had a higher complication rate, primarily with lymphoceles requiring intervention (3.4% vs. 8.6%; $P = 0.04$) [45]. Another randomized phase III Southwest Oncology Group (SWOG S1011) trial comparing standard to extended node dissection in patients with cystectomy was recently completed and will provide more robust data on the role of extent of lymph node dissection. It remains to be seen if a surgical cure of metastatic disease at the primary landing site (i.e., pelvic lymph nodes) is an attainable goal in patients with invasive bladder cancer. Improvements in imaging may help to identify patients where this could be possible, and a deeper understanding of the biology of metastasis in these patients could result in improved biomarkers of aggressive disease.

55.9 Conclusion

At the conclusion of this chapter, you should have a greater understanding of the anatomy of the lymphatic system, patterns of metastatic spread, and current discussions surrounding the clinical management of metastatic disease for both prostate and bladder cancer.

Open Questions

- How can we improve diagnostic accuracy of imaging for patients with bladder cancer?
- Are there subsets of patients with prostate cancer for whom a pelvic lymph node dissection might confer a therapeutic benefit?
- How can we advance imaging modalities to become accurate noninvasive assessments of nodal disease to replace pelvic lymph node dissection on patients with prostate cancer?
- Are there histologic factors that harbor a greater risk of early distant metastasis in patient with bladder cancer?
- Will we develop biomarkers to accurately assess aggressive disease in prostate and bladder cancer?
- Should sentinel lymph node biopsy be performed for prostate and bladder cancer?

References

1. Siegel RL, Miller KD, Fuchs HE, Jemal A. Cancer statistics, 2021. CA: Cancer J Clin. 2021;71(4):359. First Published online: April 19, 2021. https://doi.org/10.3322/caac.21654.
2. Fukuda H, Yamada T, Kamata S, Saitoh H. Anatomic distribution of intraprostatic lymphatics: implications for the lymphatic spread of prostate cancer – a preliminary study. Prostate. 2000;44:322–7. https://doi.org/10.1002/1097-0045(20000901)44:4<322::aid-pros9>3.0.co;2-8.
3. Rouviere H. Anatomy of the human lymphatic system. Ann Arbor. 1939;27:194–5. https://doi.org/10.1002/bjs.18002710531.
4. Raghavaiah NV, Jordan WP. Prostatic lymphography. J Urol. 1979;121:178–81. https://doi.org/10.1016/s0022-5347(17)56712-9.
5. Brossner C, Ringhofer H, Hernady T, Kuber W, Madersbacher S, Pycha A. Lymphatic drainage of prostatic transition and peripheral zones visualized on a three-dimensional workstation. Urology. 2001;57:389–93. https://doi.org/10.1016/s0090-4295(00)00921-3.
6. Paganelli G, De Cicco C, Cremonesi M, et al. Optimized sentinel node scintigraphy in breast cancer. Q J Nucl Med. 1998;42:49–53.
7. Nieweg OE, Tanis PJ, Kroon BB. The definition of a sentinel node. Ann Surg Oncol. 2001;8:538–41. https://doi.org/10.1007/s10434-001-0538-y.
8. Narayanan R, Wilson TG. Sentinel node evaluation in prostate cancer. Clin Exp Metastasis. 2018;35:471–85. https://doi.org/10.1007/s10585-018-9936-4.
9. Swanson GP, Hubbard JK. A better understanding of lymphatic drainage of the prostate with modern imaging and surgical techniques. Clin Genitourin Cancer. 2013;11:431–40. https://doi.org/10.1016/j.clgc.2013.04.031.
10. Saitoh H, Yoshida K-I, Uchijima Y, Kobayashi N, Suwata J, Kamata S. Two different lymph node metastatic patterns of a prostatic cancer. Cancer. 1990;65:1843–6. https://doi.org/10.1002/1097-0142(19900415)65:8<1843::aid-cncr2820650830>3.0.co;2-4.
11. Batson OV. The function of the vertebral veins and their role in the spread of metastases. Ann Surg. 1940;112:138–49. https://doi.org/10.1097/00000658-194007000-00016.
12. Bubendorf L, Schöpfer A, Wagner U, et al. Metastatic patterns of prostate cancer: an autopsy study of 1,589 patients. Hum Pathol. 2000;31:578–83. https://doi.org/10.1053/hp.2000.6698.

13. Tokuda Y, Carlino LJ, Gopalan A, et al. Prostate cancer topography and patterns of lymph node metastasis. Am J Surg Pathol. 2010;34:1862–7. https://doi.org/10.1097/PAS.0b013e3181fc679e.
14. Briganti A, Suardi N, Capogrosso P, et al. Lymphatic spread of nodal metastases in high-risk prostate cancer: the ascending pathway from the pelvis to the retroperitoneum. Prostate. 2012;72:186–92. https://doi.org/10.1002/pros.21420.
15. Nocera L, Sood A, Dalela D, et al. Rate and extent of pelvic lymph node dissection in the US prostate cancer patients treated with radical prostatectomy. Clin Genitourin Cancer. 2018;16:e451–67. https://doi.org/10.1016/j.clgc.2017.10.003.
16. Washington SL, Cowan JE, Herlemann A, et al. Influence of pelvic lymph node dissection and node-positive disease on biochemical recurrence, secondary treatment, and survival after radical prostatectomy in men with prostate cancer. Prostate. 2021;81:102–8. https://doi.org/10.1002/pros.24085.
17. Moris L, Van den Broeck T, Tosco L, et al. Impact of lymph node burden on survival of high-risk prostate cancer patients following radical prostatectomy and pelvic lymph node dissection. Front Surg. 2016;3:416–9. https://doi.org/10.3389/fsurg.2016.00065.
18. Fossati N, Willemse P-PM, Van den Broeck T, et al. The benefits and harms of different extents of lymph node dissection during radical prostatectomy for prostate cancer: a systematic review. Eur Urol. 2017;72:84–109. https://doi.org/10.1016/j.eururo.2016.12.003.
19. Lestingi JFP, Guglielmetti GB, Trinh QD, et al. Extended versus limited pelvic lymph node dissection during radical prostatectomy for intermediate- and high-risk prostate cancer: early oncological outcomes from a randomized phase 3 trial. Eur Urol. 2021;79:595–604. https://doi.org/10.1016/j.eururo.2020.11.040.
20. Cheng L, Weaver AL, Leibovich BC, et al. Predicting the survival of bladder carcinoma patients treated with radical cystectomy. Cancer. 2000;88:2326–32. https://doi.org/10.1002/(sici)1097-0142(20000515)88:10<2326::aid-cncr17>3.0.co;2-t.
21. Shinagare AB, Ramaiya NH, Jagannathan JP, Fennessy FM, Taplin ME, Van den Abbeele AD. Metastatic pattern of bladder cancer: correlation with the characteristics of the primary tumor. AJR Am J Roentgenol. 2011;196:117–22. https://doi.org/10.2214/AJR.10.5036.
22. Stein JP, Lieskovsky G, Cote R, Groshen S, Feng AC, Boyd S, Skinner E, Bochner B, Thangathurai D, Mikhail M, Raghavan D, Skinner DG. Radical cystectomy in the treatment of invasive bladder cancer: long-term results in 1,054 patients. J Clin Oncol. 2001;19(3):666–75. https://doi.org/10.1200/JCO.2001.19.3.666. PMID: 11157016.
23. Shariat SF, Karakiewicz PI, Palapattu GS, et al. Outcomes of radical cystectomy for transitional cell carcinoma of the bladder: a contemporary series from the bladder cancer research consortium. J Urol. 2006;176:2414–22–discussion2422. https://doi.org/10.1016/j.juro.2006.08.004.
24. Roth B, Wissmeyer MP, Zehnder P, et al. A new multimodality technique accurately maps the primary lymphatic landing sites of the bladder. Eur Urol. 2010;57:205–11. https://doi.org/10.1016/j.eururo.2009.10.026.
25. Paño B, Sebastià C, Buñesch L, et al. Pathways of lymphatic spread in male urogenital pelvic malignancies. Radiographics. 2011;31:135–60. https://doi.org/10.1148/rg.311105072.
26. Paner GP, Stadler WM, Hansel DE, Montironi R, Lin DW, Amin MB. Updates in the eighth edition of the tumor-node-metastasis staging classification for urologic cancers. Eur Urol. 2018;73:560–9. https://doi.org/10.1016/j.eururo.2017.12.018.
27. Horn T, Zahel T, Adt N, et al. Evaluation of computed tomography for lymph node staging in bladder cancer prior to radical cystectomy. Urol Int. 2016;96:51–6. https://doi.org/10.1159/000440889.
28. Ficarra V, Dalpiaz O, Alrabi N, Novara G, Galfano A, Artibani W. Correlation between clinical and pathological staging in a series of radical cystectomies for bladder carcinoma. BJU Int. 2005;95:786–90. https://doi.org/10.1111/j.1464-410X.2005.05401.x.
29. Salminen AP, Jambor I, Syvanen KT, Boström PJ. Update on novel imaging techniques for the detection of lymph node metastases in bladder cancer. Itali J Urol Nephrol. 2016;68:138–49.
30. Woo S, Suh CH, Kim SY, Cho JY, Kim SH. The diagnostic performance of MRI for detection of lymph node metastasis in bladder and prostate cancer: an updated systematic review and diagnostic meta-analysis. AJR Am J Roentgenol. 2018;210:W95–W109. https://doi.org/10.2214/AJR.17.18481.
31. Goodfellow H, Viney Z, Hughes P, et al. Role of fluorodeoxyglucose positron emission tomography (FDG PET)-computed tomography (CT) in the staging of bladder cancer. BJU Int. 2014;114:389–95.
32. Soubra A, Hayward D, Dahm P, et al. The diagnostic accuracy of 18F-fluorodeoxyglucose positron emission tomography and computed tomography in staging bladder cancer: a single-institution study and a systematic review with meta-analysis. World J Urol. 2016;34:1229–37. https://doi.org/10.1007/s00345-016-1772-z.
33. Jeong IG, Hong S, You D, Hong JH, Ahn H, Kim C-S. FDG PET-CT for lymph node staging of bladder cancer: a prospective study of patients with extended pelvic lymphadenectomy. Ann Surg Oncol. 2015;22:3150–6. https://doi.org/10.1245/s10434-015-4369-7.
34. Kim S-J, Koo PJ, Pak K, Kim I-J, Kim K. Diagnostic accuracy of C-11 choline and C-11 acetate for lymph node staging in patients with bladder cancer: a systematic review and meta-analysis. World J Urol. 2018;36:331–40. https://doi.org/10.1007/s00345-017-2168-4.
35. Birkhäuser FD, Studer UE, Froehlich JM, et al. Combined ultrasmall superparamagnetic particles of iron oxide-enhanced and diffusion-weighted magnetic resonance imaging facilitates detection of metastases in normal-sized pelvic lymph nodes of patients with bladder and prostate cancer. Eur Urol. 2013;64:953–60. https://doi.org/10.1016/j.eururo.2013.07.032.
36. Zarifmahmoudi L, Ghorbani H, Sadri K, et al. Sentinel node biopsy in Urothelial carcinoma of the bladder: systematic review and meta-analysis. Urol Int. 2019;103:373–82. https://doi.org/10.1159/000497310.
37. Zarifmahmoudi L, Ghorbani H, Sadeghi R, et al. Sentinel lymph node biopsy in muscle-invasive bladder cancer: single-center experience. Ann Nucl Med. 2020;34:718–24. https://doi.org/10.1007/s12149-020-01502-6.
38. Watts KL, Ristau BT, Yamase HT, Taylor JA. Prognostic implications of lymph node involvement in bladder cancer: are we understaging using current methods? BJU Int. 2011;108:484–92. https://doi.org/10.1111/j.1464-410X.2011.10330.x.
39. Herr HW, Bochner BH, Dalbagni G, Donat SM, Reuter VE, Bajorin DF. Impact of the number of lymph nodes retrieved on outcome in patients with muscle invasive bladder cancer. J Urol. 2002;167:1295–8.
40. Konety BR, Joslyn SA, O'Donnell MA. Extent of pelvic lymphadenectomy and its impact on outcome in patients diagnosed with bladder cancer: analysis of data from the surveillance, epidemiology and end results program data base. J Urol. 2003;169:946–50. https://doi.org/10.1097/01.ju.0000052721.61645.a3.
41. Stein JP, Cai J, Groshen S, Skinner DG. Risk factors for patients with pelvic lymph node metastases following radical cystectomy with en bloc pelvic lymphadenectomy: concept of lymph node density. J Urol. 2003;170:35–41. https://doi.org/10.1097/01.ju.0000072422.69286.0e.
42. Kassouf W, Agarwal PK, Herr HW, et al. Lymph node density is superior to TNM nodal status in predicting disease-specific survival after radical cystectomy for bladder cancer: analysis of pooled data from MDACC and MSKCC. J Clin Oncol. 2008;26:121–6. https://doi.org/10.1200/JCO.2007.12.9247.
43. Dhar NB, Klein EA, Reuther AM, Thalmann GN, Madersbacher S, Studer UE. Outcome after radical cystectomy with limited or extended pelvic lymph node dissection. J Urol. 2008;179:873–8–discussion878. https://doi.org/10.1016/j.juro.2007.10.076.
44. Leissner J, Ghoneim MA, Abol-Enein H, et al. Extended radical lymphadenectomy in patients with urothelial bladder cancer: results of a prospective multicenter study. J Urol. 2004;171:139–44. https://doi.org/10.1097/01.ju.0000102302.26806.fb.
45. Gschwend JE, Heck MM, Lehmann J, et al. Extended versus limited lymph node dissection in bladder cancer patients undergoing radical cystectomy: survival results from a prospective randomized trial. Eur Urol. 2019;75:604–11. https://doi.org/10.1016/j.eururo.2018.09.047.

Part XV

Biology and Clinical Aspects of Sarcoma Progression

William W. Tseng

"Sarcoma" derives from the Greek word meaning "fleshy mass." Sarcoma encompasses a group of rare malignancies that originate from the mesenchymal or connective tissues. These tumors develop from the structural support within the body including the bones and soft tissues. This is in contrast to the more common cancers (e.g., breast, colon carcinoma) which arise from the epithelium that lines the surfaces of hollow organs and structures. Sarcoma is also different on many levels from melanoma, which arises from the pigmented cells of the skin.

With this background it is important to emphasize that sarcoma is therefore (1) *not* organ specific and (2) includes a diversity of histologic subtypes based on the precise tissue of origin. As an example, liposarcoma originates from fat, can occur anywhere in the body (extremities, trunk, or abdomen), and can arguably be regarded as a separate malignancy from another sarcoma subtype (e.g., leiomyosarcoma—originates from smooth muscle). Although recognition of the tissue of origin is very important, ironically the answer is not entirely clear for some subtypes (e.g., desmoid tumor, undifferentiated pleomorphic sarcoma).

While there are well-known exceptions, in general, bone and soft-tissue sarcomas are aggressive malignancies with a high mortality rate. Overall, clinical management of sarcoma can be very challenging. Patients with sarcoma are best managed by a specialist in this disease and ideally in the setting of multidisciplinary team discussion. Recognition of the histologic subtype, as mentioned, is important as a start. Surgery remains the mainstay of treatment for localized disease; however, recurrence (both local and distant) remains a significant issue in most patients. As such, other modalities for disease control such as radiation and systemic therapy are important. Further research is needed to provide better treatments and novel approaches to treatment that are desperately needed in patients with sarcoma.

This section is entirely dedicated to sarcoma. This includes chapters that provide overviews of bone and separately soft-tissue sarcoma, followed by more focused chapters that explore select subtypes. All of the chapters are written by sarcoma specialists and fundamentally grounded with clinical experience: "What is important to the patient with sarcoma?" However, with each chapter, whenever possible, the authors try to discuss disease behavior and clinical observations that deserve further scientific investigation and specific data from translational studies that may help to elucidate mechanisms for disease development and progression. Finally, one specific chapter is dedicated to a relatively newer and promising "frontier" in sarcoma research, the immunologic aspects of the tumor microenvironment. The hope is that the content is relevant to both the practicing clinician and the research scientist and can encourage dialogue between these two audiences.

As the reader will quickly see, different subtypes of sarcoma have unique and biologically fascinating behavior. Liposarcoma, which includes at least three distinct subtypes, can be regarded as a spectrum of aggressiveness and metastatic potential. Desmoid tumors have no metastatic potential and current clinical management includes observation, as a subpopulation of patients will experience spontaneous regression. While the lungs are overall the predominant site of sarcoma metastasis, some subtypes can "ignore" this pattern of dissemination with clear predilection for specific organs (e.g., bone or brain) or even tissues (e.g., fat). With rare exceptions, the vast majority of sarcoma subtypes do *not* spread to the lymph nodes, in stark contrast to paradigms seen in carcinoma and melanoma. Interestingly, some of the subtypes which do demonstrate lymphatic spread either are "epithelioid" or have melanoma markers (e.g., clear cell sarcoma)!

Thus, with this section, the reader is invited to immerse himself/herself in learning about sarcoma. While doing this, the reader should note similarities and differences between sarcoma and other non-sarcoma cancers. Based on this, the reader is also encouraged to formulate scientific hypotheses about this group of rare malignancies that may ultimately—perhaps through interdisciplinary collaboration—help us to better understand fundamental disease mechanisms across all cancers.

Soft Tissue Sarcoma

56

Scott Kizy and Ricardo J. Gonzalez

Abstract

Soft tissue sarcomas are rare mesenchymal malignancies made up of several heterogeneous subtypes with distinct clinicopathologic features. Lymphovascular metastases are rare in sarcoma. However, certain subtypes of sarcoma can develop lymphatic metastases—including rhabdomyosarcoma, synovial sarcoma, epithelioid sarcoma, clear cell sarcoma, and angiosarcoma. Hematogenous spread is more common for most subtypes of sarcoma, with common sites of spread being the lung and liver. Metastases in soft tissue sarcoma portend a poor prognosis and require multimodal therapy for management. The underlying mechanisms for lymphatic and hematogenous spread are complex and understudied but likely include vascular endothelial growth factor (VEGF) and hypoxia inducible factor (HIF) 1α pathways. Further study is necessary to identify potential therapeutic targets in the management of patients with soft tissue sarcoma.

Learning Objectives

- Soft tissue sarcomas are rare malignancies arising from cells of mesenchymal origin with several subtypes with distinct clinicopathologic features.
- Soft tissue sarcomas have a propensity for local recurrence and hematogenous spread (especially to the lungs and liver) but rarely spread to the lymph nodes.
- A combination of surgical resection, radiation therapy, and chemotherapy has been used to treat soft tissue sarcomas.
- Vascular endothelial growth factor (VEGF)- and hypoxia inducible factor (HIF) 1α-dependent pathways have been implicated in sarcoma metastases, but more study is necessary.

56.1 Introduction

Sarcomas are rare malignancies of mesenchymal origin. More than 100 subtypes of sarcomas exist. 80% of sarcomas arise from the soft tissue—i.e., fat, muscle, nerve sheath, and blood vessels. The other 20% arise from the bone. Only about 1% of new cancer diagnoses are sarcomas. The American Cancer Society estimates that 13,130 new soft tissue sarcomas will be diagnosed in 2020 in US, and 5350 people are expected to die of soft tissue sarcoma [1]. According to the Surveillance, Epidemiology, and End Results (SEER) Program of the National Cancer Institute, the incidence of soft tissue sarcoma was 1.3 per 100,000 in 2017. The median age at diagnosis is 61 with a slight male preponderance [2].

As sarcomas arise from cells of mesenchymal origin, they can occur at all anatomic body sites. The most common site of presentation is the extremities, with the lower extremity accounting for 46% of tumors and the upper extremity accounting for 13%. The torso and retroperitoneum account for 18% and 13%, respectively. The head and neck account for 9% of soft tissue sarcomas [3].

In adults, the most common histologic subtype of soft tissue sarcoma is undifferentiated pleomorphic sarcoma, followed by liposarcoma (20%), and leiomyosarcoma (14%). Other common subtypes of sarcoma include gastrointestinal stromal tumors (9%), synovial sarcoma (5%), myxofibrosarcoma (5%), fibrosarcoma (3%), and malignant peripheral nerve sheath tumors (2%). Each subtype has unique clinical pathologic characteristics which affect treatment and prognosis [4].

The cornerstone of treatment of soft tissue sarcoma remains surgical resection with adequate margins. Based on variations in locoregional and distant recurrence among histologic subtypes and disease sites, multimodal therapy (including chemotherapy and radiation) is used to supplement treatment for patients with soft tissue sarcomas. Given the rarity of soft tissue sarcoma and the

S. Kizy (✉) · R. J. Gonzalez
Moffitt Cancer Center, Tampa, FL, USA
e-mail: Scott.Kizy@moffitt.org

S. P. Leong et al. (eds.), *Cancer Metastasis Through the Lymphovascular System*,
https://doi.org/10.1007/978-3-030-93084-4_56

heterogeneity of histologic subtypes, robust clinical data from well-designed randomized prospective clinical trials are difficult to complete.

An important factor when staging and treating patients with a new or recurrent sarcoma is identifying the risk for distant metastases. Most soft tissue sarcomas carry a significant risk for local recurrence, which can be mitigated by adjuvant or neoadjuvant radiation therapy to the primary site. Distant metastases are primarily spread hematogenously, with the lungs being the most likely site for distant disease [5]. Hepatic metastases can also occur in patients with retroperitoneal or visceral sarcomas. Lymphatic metastases are quite rare, although they can occur with certain histologic subtypes—i.e., rhabdomyosarcoma, synovial sarcoma, epithelioid sarcoma, clear cell sarcoma, and vascular sarcoma [6]. In this chapter we present relevant basic science and translational studies on mechanisms of lymphatic metastasis and disease progression in soft tissue sarcomas. Liposarcoma will be covered separately in the upcoming chapter.

56.2 Clinical Implications

Lymphatic metastases from soft tissue sarcomas are rare. Given this rarity, relatively little data exist on the clinical implications and management of lymphatic metastases from soft tissue sarcoma. A prospectively gathered database of Soft Tissue Sarcoma at Memorial Sloan Kettering Cancer Center found that about 2.6% of patients develop lymph node metastases. In this study, tumor subtypes with the highest prevalence of lymph node metastases were angiosarcoma (13.5%), rhabdomyosarcoma (13.6%), and epithelioid sarcoma (16.7%) [6]. A recently published large institutional-based database from the University of Toronto identified 4.5% of patients with extremity sarcoma had developed lymph node metastases. The most common histologic subtypes with lymph node metastases in their dataset were clear cell sarcoma (27.6%), epithelioid sarcoma (21.9%), rhabdomyosarcoma (17.3%), angiosarcoma (14%), and extraskeletal myxoid chondrosarcoma (9.3%). Synovial sarcomas have been identified as having a high rate of lymphatic metastases in other small institutional series, but that was not found in the University of Toronto dataset (2.7%) [7, 8]. Large database studies, including the SEER database, have found that lymphatic metastases in soft tissue sarcoma carry a poor prognosis and signify aggressive tumor biology. An analysis of the SEER database found that 0.9% of patients with soft tissue sarcoma (without distant metastases) had lymphatic metastases [9]. An analysis of the National Cancer Database (NCDB) found that 10.8% of patients with resectable, non-metastatic soft tissue sarcoma underwent LN evaluation, and 1.5% of patients in that cohort had lymph node metastases. In patients with high-risk subtypes (clear cell sarcoma, epithelioid sarcoma, rhabdomyosarcoma, and angiosarcoma), 21.3% of patients underwent lymphadenectomy, and 5.9% were identified to have lymphatic metastases. Lymphatic metastases are associated with a hazard ratio of mortality of 5.1 compared to similar patients without metastases. Lymphatic spread was associated with younger age, tumor grade, tumor size, and subtype [10]. Similar rates of lymph node involvement have been identified in the European studies [11].

In the 2017 revision of the American Joint Committee on Cancer (AJCC), lymph node metastases in sarcomas originating in the trunk and extremities are classified as stage IV disease. For sarcomas of the retroperitoneum or viscera, N1 disease is classified as stage III, given the finding that the prognosis of patients with nodal disease is slightly more favorable compared to patients with systemic metastases [12], although these studies may be biased due to the small incidence of nodal disease in retroperitoneal sarcomas compared to truncal or extremity sarcomas.

For patients with lymph node metastases or suspected lymphatic metastases, management decisions are complex. Given its rarity, no prospective data exists on the management of lymph node metastases in sarcoma. Based on retrospective database studies, omission of lymphatic evaluation in these high-risk cohorts (clear cell sarcoma, epithelioid sarcoma, rhabdomyosarcoma, and angiosarcoma) was found to have negative effects on survival [10, 13, 14]. Based on these retrospective studies, imaging assessment of the regional lymph node basin is recommended, and surgical evaluation can be considered for patients with high-risk subtypes.

The role of imaging to identify lymphatic metastases in sarcoma is unclear. The NCCN guidelines propose that for high-risk histologies, imaging assessment of the regional lymph node basin may be appropriate in the work-up and surveillance of patients with soft tissue sarcoma. The appropriate imaging modality is yet to be identified. CT scan is the most common imaging modality for staging or surveillance of patients with retroperitoneal or extremity sarcoma and can identify lymph nodes that are worrisome for metastatic disease (Fig. 56.1). Its sensitivity and specificity in this capacity are unknown. FDG-PET imaging has been proposed as a possible noninvasive method of lymphatic staging in such subtypes (Fig. 56.2). However, the sensitivity and specificity of PET imaging for the identification of lymphatic metastases in soft tissue sarcoma may actually be quite low, with one study quoting a sensitivity of 57% and a specificity of 52% [15].

Sentinel lymph node biopsy has been proposed as an option for the evaluation of the lymph node basin in patients with high-risk sarcoma histology [16]. Several small series and a

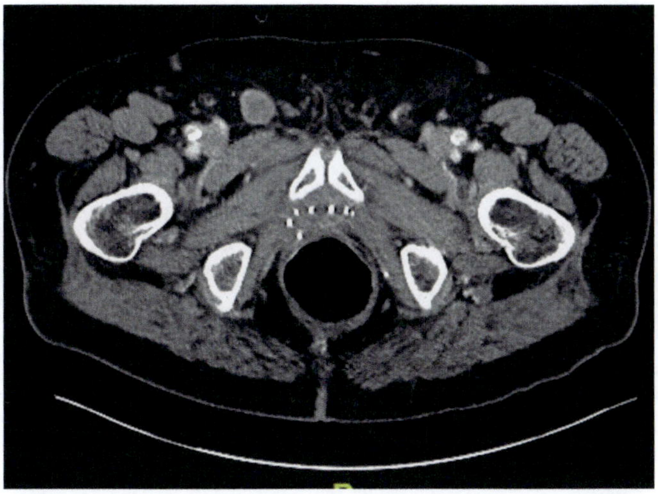

Fig. 56.1 An enlarged right inguinal lymph node identified on CT scan imaging in a patient with a rare prostate sarcoma

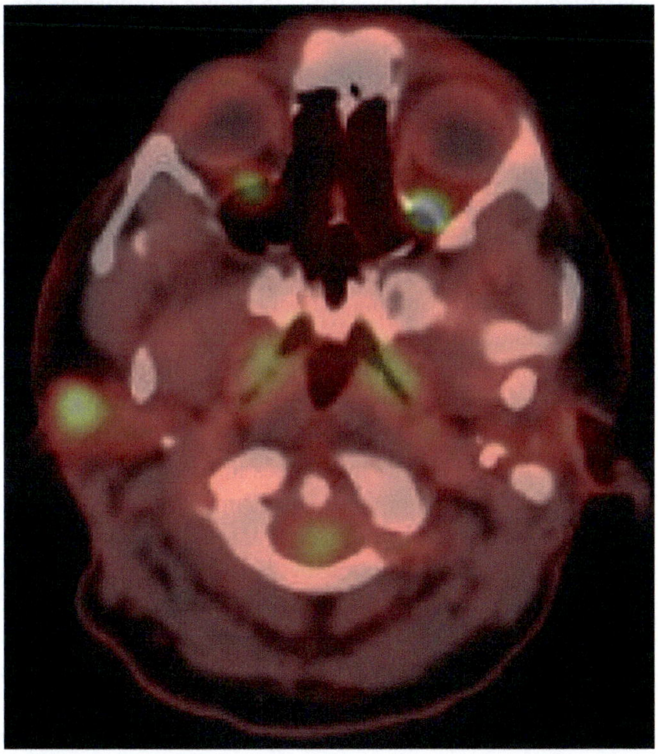

Fig. 56.2 FDG-PET imaging identifying a metabolically active parotid lymph node in a patient with angiosarcoma of the scalp. This was identified to be metastatic angiosarcoma after biopsy

meta-analysis found that patients with clear cell sarcoma may benefit from sentinel lymph node biopsy given a high number of occult metastases, but other subtypes had little benefit [16, 17] (Table 56.1). At our center, if a high-risk lesion is amenable to sentinel node biopsy (i.e., small enough to plan an injection), we will perform such a technique.

For patients with soft tissue sarcoma with lymph node involvement, radical lymphadenectomy may provide survival benefit. The benefit is largely derived from historical single-institution studies, including the MSKCC database study. In this study, patients with lymph node involvement had a median survival of 4.3 months if they did not undergo radical lymphadenectomy compared to 16.3 months in patients who underwent lymph node dissection [6]. In the more recent University of Toronto database study, a similar improvement in survival was seen in patients treated with radical lymphadenectomy [7]. Given the retrospective nature of these studies, there is significant selection bias; however, based on these historical series, the National Comprehensive Cancer Network (NCCN) clinical practice guidelines recommend consideration of regional lymph node dissection at the time of primary surgery for patients with lymph node involvement. As lymph node involvement is considered stage IV disease in extremity sarcoma, and stage III disease in retroperitoneal sarcoma, multimodal therapy, including regional nodal radiation therapy, is also recommended by the NCCN[18]. With patients without clinically evident lymphatic metastases found to have a positive sentinel lymph node, there is currently very little data to guide management. Unlike melanoma, where well done prospective trials have been evaluated, it is unknown if nodal observation can be pursued. Given the risk of further lymphatic metastases, completion lymphadenectomy is generally performed although this is based on the above historical, retrospective data. At this time, we would advocate completion node dissection in this population.

At our institution if a patient with a soft tissue sarcoma is found to have lymphatic metastases, either via sentinel lymph node biopsy or preoperative clinical exam/imaging, most patients are offered a radical lymphadenectomy after discussion at a multidisciplinary tumor board. Treatment would include consideration for neoadjuvant systemic therapy, followed by adjuvant radiation in patients with good performance status. Stage IV disease is a driver for systemic therapy prior to surgery.

Table 56.1 Sentinel lymph node positivity rate in patients with soft tissue sarcoma

Study	Design	*N*	SLN positivity	False-negative rate
Andreou et al. 2013 [17]	Prospective	62	Synovial: 2/42 Clear cell: 6/12 Epithelioid: 0/4 Rhabdomyosarcoma: 0/4	NR
Wright et al. 2012 [29]	Meta-analysis	114	Synovial: 2/34 Clear cell: 6/17 Epithelioid: 0/17 Rhabdomyosarcoma: 5/22	Synovial: 33% (1/3) Clear cell: 25% (2/8) Epithelioid: 100% (1/1) Rhabdomyosarcoma: 43% (3/7)
Wagner et al. 2020 [15]	Prospective	28	Epithelioid: 1/2 Rhabdomyosarcoma 4/8 Other: 2/8	NR
Parida et al. 2012 [30]	Retrospective	23	Total: 1/23 (rhabdomyosarcoma)	NR
Maduekwe et al. 2009 [31]	Retrospective	29	Total: 3/29	NR

56.3 Mechanisms of Lymphatic Metastases

The molecular mechanisms underlying lymphatic metastases in soft tissue sarcoma are not understood. Proposed mechanisms of lymphatic metastases include lymphatic vessel development, tumor cell detachment, and survival within the lymphatic microenvironment. Whether the lack of lymph node metastases in soft tissue sarcoma is due to lower lymphangiogenesis, an inability of tumor cells to detach and move through the lymphatic system, or an inhospitable lymphatic microenvironment is unknown. Considering its rarity, understanding the reason for the resistance of lymph node metastases in soft tissue sarcoma may prove informative in the management of patients with metastatic disease.

The density of intratumoral and peritumoral lymphatics likely plays a role in the development of lymphatic spread in most human carcinomas. In soft tissue sarcoma, while intratumoral lymphatic vessel density has been observed to be similar to that of carcinomas, the correlation with lymphatic spread has not been identified. On the other hand, peritumoral lymphatic density has been identified to be lower in soft tissue sarcoma than that of breast cancer. Moreover, patients with sarcoma lymph node metastases have been identified to have a higher density of peritumoral lymphatics. The decreased peritumoral lymphatic density in soft tissue sarcoma has been proposed as a potential cause for the lack of clinical lymph node metastases [19, 20].

The ability of tumor cells to detach and invade nearby lymphatics is also an important component of lymph node metastases in human malignancy. Lymphovascular invasion, or the identification of tumor cells within lymphatic vessels on pathologic examination, is commonly found in most malignancies with a propensity for lymphatic spread. Although inadequately studied, lymphovascular invasion in soft tissue sarcoma is rare [21]. The reason for this is unclear, and more study is necessary.

A variety of cytokines have been identified to play a role in lymphangiogenesis and lymphovascular invasion. Vascular endothelial growth factor (VEGF) signaling has been identified as an important factor in developing lymph node metastases. In particular VEGF-C and VEGF-D have been identified in multiple tumor types to induce lymphangiogenesis, promote lymphatic vessel permeability, and correlate with lymph node metastases. Certain sarcoma cell lines have been shown to have reduced VEGF-C expression. Overexpression of VEGF-C in a murine fibrosarcoma and human osteosarcoma cell line increased the ability of these cells to metastasize in the lymph nodes. Patients with sarcoma lymph node metastases have also been shown to have a high VEGF-D expression [22]. Unfortunately, this finding has not been confirmed in other sarcoma human cell lines or tissues. Furthermore, many tumor subtypes, including those with rare lymphatic metastases (liposarcoma and leiomyosarcoma), have been shown to express high levels of VEGF-C[20]. These findings have decreased the interest in VEGF-C as a modulator of lymph node metastases in soft tissue sarcoma.

The role of the tumor microenvironment in sarcoma lymphatic metastases is also unknown and understudied. However, the complex interaction between soft tissue sarcoma tumors and the immune system is beginning to be elucidated. Generally, sarcomas are considered immune quiescent as they commonly have low mutational burden. However, the complexity of the immune systems interactions with soft tissue sarcoma has been underscored by advancements in immunotherapy. More recently, soft tissue sarcomas have been subclassified based on immune cell complement within the tumor microenvironment. Immune-type subclass has identified several subgroups of sarcomas with diverse tumor histologies that exhibit different prognoses, virulence, and response to immunotherapy. Tumor subgroups with a higher B-cell lineage have improved overall survival and improved responsiveness to pembrolizumab, a PD-1 inhibitor [23]. Another avenue of interest is the use of tumor-infiltrating lymphocytes (TIL) in the treatment of certain sarcoma subtypes. TIL harvest and culture has been shown to be feasible and demonstrated to have tumor-specific reactivity in recent studies [24]. As a promising new modality of treatment, clinical trials are ongoing evaluating the safety and efficacy of TIL in the treatment of metastatic sarcoma. These findings underscore the heterogeneity of soft tissue sarcoma and complexity of management and may elucidate several mechanisms involved in both lymphatic and systemic metastases of soft tissue sarcomas.

56.4 Systemic Metastasis

Hematogenous spread of soft tissue sarcoma is far more common than lymphatic metastases. Several mechanisms have been implicated in the ability of sarcoma tumors to metastasize. Owing to the heterogeneity of different subtypes of sarcoma, several pathways have been identified to be important in various histologies.

Hypoxia has been identified as an important factor in the induction of metastatic pathways in multiple sarcoma subtypes. Low intratumoral oxygen level and high levels of hypoxia inducible factor 1α (HIF1α) expression have been identified as a key predictor of metastases in soft tissue sarcoma [25]. This is likely dependent on procollagen-lysine,2-oxoglutarate 5-dioxygenase 2 (PLOD2)-regulated deposition of collagen fibers. PLOD2 seems to modify collagen in a way to allow sarcoma cell migration and motility. In cell line and murine models, minoxidil, a PLOD t2 inhibitor, has been found to decrease pulmonary metastases [26]. Underscoring the importance of the extracellular matrix and collagen deposition, matrix metalloproteinases have been identified to play an important role in the metastatic potential of leiomyosarcomas of uterine origin [27].

Gp78, an ubiquitin ligase localized on the endoplasmic reticulum, has also been shown to be implicated in sarcoma metastases. This is likely due to its interaction (ubiquitination and downregulation) with KA11, a transmembrane protein and known metastasis suppressor. This mechanism was identified in a highly metastatic sarcoma tumor cell line but also identified in microarray studies of human sarcoma tissues. KA11 is thought to induce apoptosis of metastatic cells, and its downregulation by Gp78 may allow for survival of tumor cells in its metastatic milieu [28].

VEGF signaling has also been implicated in the ability for sarcomas to metastasize. VEGFA is a regulator of vascularization of tumors and contributes to tumor growth and metastasis. VEGFA isoform expression increases fibrosarcoma metastases likely through direct tumor vascularization and changes in the primary tumor microenvironment. Interest in utilizing a targeted VEGF inhibitor, bevacizumab, in the treatment of patients with metastatic sarcoma has increased. The NCCN currently recommends bevacizumab in certain epithelioid tumors and malignant solitary fibrous tumors [18]. Unfortunately, most studies to date have failed to show improvement in outcomes in other sarcoma subtypes.

56.5 Conclusions

Soft tissue sarcomas are rare tumors of mesenchymal origin. Multiple histologic subtypes of soft tissue sarcomas have been identified with varying clinical/pathologic characteristics and treatment strategies. Lymphatic metastases are rare among soft tissue sarcomas; however, certain histologic subtypes—clear cell sarcoma, epithelioid sarcoma, rhabdomyosarcoma, and angiosarcoma—have a propensity for lymphatic spread. Given its rarity, the underlying mechanisms of lymphatic spread in soft tissue sarcoma remain understudied. Certain molecular pathways including VEGF signaling, HIF1α, and GP78 have been implicated in lymphatic and systemic metastases in soft tissue sarcoma. Further study is necessary to provide potential targeted therapeutic options for patients with soft tissue sarcoma.

Open Questions

- What is the role of sentinel lymph node biopsy for histologic subtypes of soft tissue sarcoma at high risk for lymph node metastasis?
- What signaling pathways underlie lymphatic and systemic metastases in soft tissue sarcoma, and can these pathways be utilized to develop therapeutic interventions?
- Is there a role for immunotherapy and immunomodulation in the treatment and management of patients with metastatic soft tissue sarcoma?

References

1. American Cancer Society. Key statistics for soft tissue sarcomas. 2020. https://www.cancer.org/cancer/soft-tissue-sarcoma/about/key-statistics.html.
2. Surveillance, Epidemiology, and End Results. Cancer Statiscs and Facts: Soft Tissue including heart cancer.
3. Lawrence W, Donegan WL, Natarajan N, Mettlin C, Beart R, Winchester D. Adult soft tissue sarcomas: a pattern of care survey of the American College of Surgeons. Ann Surg. 1987;205:349–59. https://doi.org/10.1097/00000658-198704000-00003.
4. Brennan MF, Antonescu K, Maki A, Maki R. Management of soft tissue sarcoma. 2nd ed. Switzerland: Springer; 2016.
5. Potter DA, Glenn J, Kinsella T, et al. Patterns of recurrence in patients with high-grade soft-tissue sarcomas. J Clin Oncol. 1985;3:353–66. https://doi.org/10.1200/JCO.1985.3.3.353.
6. Fong Y, Coit DG, Woodruff JM, Brennan MF. Lymph node metastasis from soft tissue sarcoma in adults: analysis of data from a prospective database of 1772 sarcoma patients. Ann

Surg. 1993;217:72–7. https://doi.org/10.1097/00000658-199301000-00012.
7. Basile G, Mattei J-C, Alshaygy I, et al. Curability of patients with lymph node metastases from extremity soft-tissue sarcoma. Cancer. 2020;126:5098–108. https://doi.org/10.1002/cncr.33189.
8. Blazer DG, Lazar AJ, Xing Y, et al. Clinical outcomes of molecularly confirmed clear cell sarcoma from a single institution and in comparison with data from the surveillance, epidemiology, and end results registry. Cancer. 2009;115:2971–9. https://doi.org/10.1002/cncr.24322.
9. Johannesmeyer D, Smith V, Cole DJ, Esnaola NF, Camp ER. The impact of lymph node disease in extremity soft-tissue sarcomas: a population-based analysis. Am J Surg. 2013;206:289–95. https://doi.org/10.1016/j.amjsurg.2012.10.043.
10. Ecker BL, Peters MG, McMillan MT, et al. Implications of lymph node evaluation in the management of resectable soft tissue sarcoma. Ann Surg Oncol. 2017;24:425–33. https://doi.org/10.1245/s10434-016-5641-1.
11. Behranwala KA, A'Hern R, Omar AM, Thomas JM. Prognosis of lymph node metastasis in soft tissue sarcoma. Ann Surg Oncol. 2004;11:714–9. https://doi.org/10.1245/ASO.2004.04.027.
12. AJCC. Cancer staging manual. 8th ed; 2017.
13. Sherman KL, Kinnier CV, Farina DA, et al. Examination of national lymph node evaluation practices for adult extremity soft tissue sarcoma. J Surg Oncol. 2014;110:682–8. https://doi.org/10.1002/jso.23687.
14. Nelen SD, Vogelaar FJ, Gilissen F, Van der Linden JC, Bosscha K. Lymph node metastasis after a soft tissue sarcoma of the leg: a case report and a review of the literature. Case Rep Surg. 2013:930361. https://doi.org/10.1155/2013/930361.
15. Wagner LM, Kremer N, Gelfand MJ, et al. Detection of lymph node metastases in pediatric and adolescent/young adult sarcoma: sentinel lymph node biopsy versus fludeoxyglucose positron emission tomography imaging—a prospective trial. Cancer. 2017;123:155–60. https://doi.org/10.1002/cncr.30282.
16. Blazer DG 3rd, Sabel MS, Sondak VK. Is there a role for sentinel lymph node biopsy in the management of sarcoma? Surg Oncol. 2003;12:201–6. https://doi.org/10.1016/s0960-7404(03)00030-6.
17. Andreou D, Boldt H, Werner M, Hamann C, Pink D, Tunn P-U. Sentinel node biopsy in soft tissue sarcoma subtypes with a high propensity for regional lymphatic spread 2014;results of a large prospective trial. Ann Oncol. 2013;24:1400–5. https://doi.org/10.1093/annonc/mds650.
18. National Comprehensive Cancer Network. Soft Tissue Sarcoma. *NCCN*. 2020.
19. Friedrichs N, Hahne JC, Pepper MS, et al. Immunohistochemical quantification of lymph vessels, VEGF-C and VEGF receptor 3 in human sarcomas. Histopathology. 2006;49:87–8. https://doi.org/10.1111/j.1365-2559.2006.02438.x.
20. Lahat G, Lazar A, Wang X, et al. Increased vascular endothelial growth factor-C expression is insufficient to induce lymphatic metastasis in human soft-tissue sarcomas. Clin Cancer Res. 2009;15:2637–46. https://doi.org/10.1158/1078-0432.CCR-08-2442.
21. Ethun CG, Lopez-Aguiar AG, Switchenko JM, et al. The prognostic value of Lymphovascular invasion in Truncal and extremity soft tissue sarcomas: an analysis from the National Cancer Database. Ann Surg Oncol. 2019;26:4723–9. https://doi.org/10.1245/s10434-019-07805-x.
22. Yanagawa T, Shinozaki T, Watanabe H, Saito K, Raz A, Takagishi K. Vascular endothelial growth factor-D is a key molecule that enhances lymphatic metastasis of soft tissue sarcomas. Exp Cell Res. 2012;318:800–8. https://doi.org/10.1016/j.yexcr.2012.01.024.
23. Petitprez F, de Reyniès A, Keung EZ, et al. B cells are associated with survival and immunotherapy response in sarcoma. Nature. 2020;577:556–60. https://doi.org/10.1038/s41586-019-1906-8.
24. Mullinax JE, Hall M, Beatty M, et al. Expanded tumor-infiltrating lymphocytes from soft tissue sarcoma have tumor-specific function. J Immunother. 2021;44:63–70. https://doi.org/10.1097/CJI.0000000000000355.
25. El-Naggar AM, Veinotte CJ, Cheng H, et al. Translational activation of HIF1α by YB-1 promotes sarcoma metastasis. Cancer Cell. 2015;27:682–97. https://doi.org/10.1016/j.ccell.2015.04.003.
26. Eisinger-Mathason TSK, Zhang M, Qiu Q, et al. Hypoxia-dependent modification of collagen networks promotes sarcoma metastasis. Cancer Discov. 2013;3:1190–205. https://doi.org/10.1158/2159-8290.CD-13-0118.
27. Bodner-Adler B, Bodner K, Kimberger O, Czerwenka K, Leodolter S, Mayerhofer K. MMP-1 and MMP-2 expression in uterine leiomyosarcoma and correlation with different clinicopathologic parameters. J Soc Gynecol Investig. 2003;10:443–6. https://doi.org/10.1016/s1071-5576(03)00120-5.
28. Tsai YC, Mendoza A, Mariano JM, et al. The ubiquitin ligase gp78 promotes sarcoma metastasis by targeting KAI1 for degradation. Nat Med. 2007;13:1504–9. https://doi.org/10.1038/nm1686.
29. Wright S, Armeson K, Hill EG, et al. The role of sentinel lymph node biopsy in select sarcoma patients: a meta-analysis. Am J Surg. 2012;204:428–33. https://doi.org/10.1016/j.amjsurg.2011.12.019.
30. Parida L, Morrisson GT, Shammas A, et al. Role of lymphoscintigraphy and sentinel lymph node biopsy in the management of pediatric melanoma and sarcoma. Pediatr Surg Int. 2012;28:571–8. https://doi.org/10.1007/s00383-012-3066-x.
31. Maduekwe UN, Hornicek FJ, Springfield DS, et al. Role of sentinel lymph node biopsy in the staging of synovial, Epithelioid, and clear cell sarcomas. Ann Surg Oncol. 2009;16:1356. https://doi.org/10.1245/s10434-009-0393-9.

Bone Sarcoma

57

Akash A. Shah, Howard Y. Park, and Francis J. Hornicek

Abstract

While relatively rare, primary bone sarcomas are associated with high morbidity and mortality. A comprehensive history, examination, and imaging work-up should be performed in any suspected bone sarcoma. Percutaneous biopsy is preferred to establish a tissue diagnosis. The MSTS system and AJCC system are employed for staging. Osteosarcoma is the most common bone sarcoma, with peaks in the second decade of life and after 59 years. Standard therapy consists of neoadjuvant chemotherapy, wide resection with limb salvage, and adjuvant chemotherapy. Chondrosarcoma is the most common bone sarcoma of adulthood, affecting patients between 40 and 75 years. Wide resection is the mainstay of treatment, as it is relatively resistant to radiation and systemic therapy. Ewing sarcoma is the second most common bone malignancy of childhood, occurring between 5 and 25 years. It is classically associated with a t(11,22) translocation. It is treated with neoadjuvant chemotherapy, wide resection, and adjuvant chemotherapy. Fibrous and vascular sarcomas of the bone are exceedingly rare and treated with wide resection.

Learning Objectives

- Primary bone sarcomas are associated with high morbidity and mortality.
- Imaging work-up includes plain radiographs, CT, and MRI.
- Percutaneous biopsy is preferred for tissue diagnosis.
- Standard treatment for osteosarcoma is neoadjuvant chemotherapy, wide resection, and adjuvant chemotherapy.
- Standard treatment for chondrosarcoma is wide resection.
- Ewing sarcoma is treated with neoadjuvant chemotherapy, wide resection, and adjuvant chemotherapy.
- Treatment for fibrous and vascular sarcomas is wide resection.

57.1 Introduction

Bone tumors can be classified on a spectrum from benign to malignant disease. The vast majority of bone lesions is benign. Benign lesions grow slowly, have a low risk of local recurrence, and do not have metastatic potential.

Primary malignant bone tumors grow rapidly, carry a high risk of local recurrence, and have metastatic potential. Primary malignant tumors of the bone are typically of mesenchymal origin and represent a wide spectrum of disease including sarcoma, lymphoma, and chordoma. There are an estimated 3600 new cases of primary bone malignancies and 1720 deaths per year in the United States [1]. In patients 15–19 years of age, bone malignancies are responsible for 19% and 18% of deaths from cancer in males and females, respectively [2]. Primary bone sarcomas are rare, accounting for <1% of all diagnosed malignancies per year. Nevertheless, they carry a high morbidity and mortality rate—afflicting children, young adults, and older adults.

A. A. Shah · H. Y. Park
Department of Orthopaedic Surgery, David Geffen School of Medicine at UCLA, Los Angeles, CA, USA

F. J. Hornicek (✉)
Department of Orthopaedics, Sarcoma Multidisciplinary Care Group, Sylvester Comprehensive Cancer Center, University of Miami Miller School of Medicine, Los Angeles, CA, USA

Department Orthopaedics, University Miami Miller School of Medicine, Miami, FL, USA
e-mail: FJH21@miami.edu

S. P. Leong et al. (eds.), *Cancer Metastasis Through the Lymphovascular System*,
https://doi.org/10.1007/978-3-030-93084-4_57

57.2 Diagnosis

57.2.1 History and Physical Examination

Obtaining an accurate history is a critical component for evaluation of the patient with a suspected bone malignancy. The presence, nature, and exacerbating factors of pain should be assessed. The patient should be asked about constitutional symptoms. A recent history of trauma should be determined, since pathologic fracture is often the cause of presentation. In addition to a detailed past medical history, a family history of malignancy should be obtained.

A comprehensive physical examination should be performed. If a mass is present, the clinician should examine the mass for presence of a draining sinus or fungation. Size, warmth, tenderness, and firmness should be documented. Presence of lymphadenopathy should be assessed. A complete skin and abdominal examination should also be performed to assess for the presence of syndromes that may be associated with musculoskeletal pathology.

57.2.2 Imaging

Plain radiographs with at least two orthogonal views must be obtained; it is imperative to image the entire bone. Computed tomography (CT) is useful in the pelvis and/or sacrum to evaluate the degree of bony involvement and for pre-operative planning. CT chest is also needed for staging purposes, as the lung is the most common site of metastasis of bone sarcomas. Magnetic resonance imaging (MRI) is a necessary follow-up study if primary bone malignancy is suspected. MRI allows for evaluation of bone marrow, non-calcific soft tissue, and lymph node involvement. MRI of the entire involved bone should be obtained to assess for skip metastases. A whole-body bone scan may be performed to assess for widespread bony metastases.

57.2.3 Biopsy

Biopsy is performed after other staging studies. Percutaneous biopsies are generally favored over open biopsies. There is a higher association of biopsy tract seeding and subsequent local recurrence with open biopsy compared to percutaneous [3]. Among percutaneous biopsy techniques, core biopsy is preferred since it allows for assessment of lesion architecture compared to fine-needle aspiration [4]. Incisional biopsy is not recommended for evaluation of a suspected malignancy. Excisional biopsy may be appropriate for a lesion <3 cm or in a location where wide resection is associated with significant morbidity.

57.3 Staging

There are two commonly used staging systems for primary bone sarcomas: the Musculoskeletal Tumor Society (MSTS) system and the American Joint Committee on Cancer (AJCC) system. Largely unchanged from the Enneking system proposed in 1980, the MSTS system assesses grade, site, and presence of metastases [5]. Low-grade tumors are stage I, and high-grade tumors are stage II; A or B classifiers are assigned if the tumor is confined within the cortex of the bone (intracompartmental) or extends beyond the cortex (extracompartmental), respectively. If metastases are present, the mass is stage III. While the factors employed in MSTS staging are intuitive to surgeons, it rarely impacts the extent of surgery since most tumors are classified as stage IIB [6].

The AJCC staging system is more consistent with the TNM convention employed for other tumor types [7]. Furthermore, it undergoes regular evidence-based updates and includes factors outside the Enneking system. The AJCC system provides different tumor staging criteria for the appendicular skeleton, spine, and pelvis (Table 57.1). Notably, distant metastasis is divided into lung and non-lung sites, acknowledging literature supporting that patients with isolated lung metastasis carry a better prognosis than those with bony metastasis [6].

57.4 Osteosarcoma

Osteosarcoma is the most common bone sarcoma. It is characterized by the production of osteoid and bone by malignant spindle cells. The peak incidence is in the second decade of life; a second peak occurs after 59 years of age. The most common presenting symptom is pain. Intramedullary osteosarcomas develop in the medullary cavity; juxtacortical osteosarcomas arise on the surface of long bones and do not involve the medullary canal (Table 57.2).

On radiographs, a metaphyseal lesion with blastic or lytic features and ill-defined borders is classically encountered [8]. An MRI evaluates soft tissue involvement, bone marrow involvement, and presence of skip metastases. A summary of osteosarcoma subtypes is provided in Table 57.3.

57.4.1 Genetics

Most cases of osteosarcoma are sporadic. Increased expression of vascular endothelial growth factor is correlated with decreased overall survival [9]. Cyclin-dependent kinase 11 is essential for osteosarcoma cell growth and survival in vitro [10], while rare, genetic predisposition to osteosarcoma involves mutations in tumor suppressor genes or DNA helicases [8].

Table 57.1 Definition of primary tumor in the American Joint Committee on Cancer (AJCC) staging system for bone sarcomas

Category	Appendicular skeleton	Spine	Pelvis
TX	Primary tumor cannot be assessed	Primary tumor cannot be assessed	Primary tumor cannot be assessed
T0	No evidence of primary tumor	No evidence of primary tumor	No evidence of primary tumor
T1	Tumor ≤8 cm in greatest dimension	Tumor confined to 1–2 adjacent segments	Tumor confined to 1 segment without extraosseous extension
T1a	–	–	Tumor ≤8 cm in greatest dimension
T1b	–	–	Tumor >8 cm in greatest dimension
T2	Tumor >8 cm in greatest dimension	Tumor confined to 3 adjacent segments	Tumor confined to 1 segment with extraosseous extension or 2 segments without extraosseous extension
T2a	–	–	Tumor ≤8 cm in greatest dimension
T2b	–	–	Tumor >8 cm in greatest dimension
T3	Presence of skip metastases	Tumor confined to ≥4 adjacent segments or any nonadjacent segments	Tumor spanning 2 segments with extraosseous extension
T3a	–	–	Tumor ≤8 cm in greatest dimension
T3b	–	–	Tumor >8 cm in greatest dimension
T4	–	Extension into spinal canal or great vessels	Tumor spanning 3 segments or crossing sacroiliac joint
T4a	–	Extension into spinal canal	Involves sacroiliac joint and extends medial to sacral neuroforamina
T4b	–	Extension into great vessels or presence of tumor thrombus	Tumor encasement of external iliac vessels or presence of tumor thrombus

Table 57.2 Characteristics of common osteosarcoma subtypes

	Conventional	Telangiectatic	Small cell	Low-grade intramedullary	Parosteal	Periosteal	High-grade surface
Proportion of cases	80%	<4%	1.5%	1–2%	1–6%	1–2%	<1%
Age Range	1st and 2nd decades	1st and 2nd decades	1st and 2nd decades	3rd and 4th decades	3rd decade	3rd decade	3rd decade
Common Locations	Distal femur, proximal tibia, proximal humerus	Distal femur, proximal tibia	Distal femur	Distal femur, proximal tibia	Posterior distal femur	Distal femur, proximal tibia	Femur, tibia
Location on bone	Intramedullary	Intramedullary	Intramedullary	Intramedullary	Juxtacortical	Juxtacortical	Juxtacortical
Frequent histologic grade	High grade	High grade	High grade	Low grade	Low grade	Intermediate grade	High grade

Table 57.3 Characteristics of chondrosarcoma subtypes

	Conventional Chondrosarcoma	Mesenchymal chondrosarcoma	Clear cell Chondrosarcoma	Dedifferentiated Chondrosarcoma
Precursor lesion	Osteochondroma, enchondroma	None	None	Conventional Chondrosarcoma
Associated syndromes	Ollier's disease; Maffucci syndrome; MHE	None	None	None
Age range	40s–60s	50–60s	Any age, peak 30s–50s	Any age, peak 30s–50s
Common locations	Pelvis, proximal femur	Axial skeleton, extra-skeletal (meninges)	Femoral and humeral epiphyses	Pelvis, femur
Frequent histologic grade	Low grade	High grade	Low grade	High grade
Prognosis	Good if low grade	Poor	Good	Poor
Chemosensitivity	Low	Sensitive	Low	Low
Radiosensitive	Low	Sensitive	Low	Low

57.4.2 Intramedullary Osteosarcoma Subtypes

57.4.2.1 Conventional Osteosarcoma

Conventional osteosarcoma represents 80% of all osteosarcomas. It affects patients in the first or second decade of life and has a male predominance. The most commonly affected locations are the distal femur, proximal tibia, and proximal humerus. Conventional osteosarcoma tends to be high-grade and originates in the metaphysis; diaphyseal or axial sites are less commonly observed [8].

On radiographs, lytic destruction with cortical disruption is noted. A Codman triangle may be observed. As the tumor extends into the soft tissue, a sunburst pattern of malignant bone may be seen. MRI of the entire bone must be evaluated for the presence of skip metastases.

The defining histopathologic feature of osteosarcoma is the production of osteoid. Malignant mesenchymal cells with pleomorphic nuclei and occasional mitotic figures are typical for conventional osteosarcoma [11]. Conventional osteosarcoma is further subcategorized by the extracellular matrix produced (e.g., osteoblastic, chondroblastic, fibroblastic) [8].

57.4.2.2 Telangiectatic Osteosarcoma

Typically encountered in the second decade of life, telangiectatic osteosarcoma comprises <4% of osteosarcoma cases. A quarter of patients present with pathologic fracture [8]. MRI should be obtained when telangiectatic osteosarcoma is suspected. Care must be taken when interpreting imaging, as findings are similar to benign aneurysmal bone cysts: a radiolucent lesion with fluid-fluid levels. A lesion with an aggressive pattern of bone destruction with partial fluid-fluid levels should prompt core needle biopsy, demonstrating multiple hemorrhagic cavities with osteoid and malignant stromal cells [11–13].

57.4.2.3 Low-Grade Intramedullary Osteosarcoma

Low-grade intramedullary osteosarcoma affects adults in the third or fourth decade of life, most commonly in the distal femur and proximal tibia. On plain radiographs, an unaggressive-appearing lytic and/or blastic lesion with septal sclerosis is encountered [14]. On histology, well-differentiated cells with woven microtrabeculae and fibrous stroma are observed [11].

57.4.2.4 Small Cell Osteosarcoma

Small cell osteosarcoma comprises 1.5% of cases. It has a similar age distribution as conventional osteosarcoma. The distal femur is most commonly involved. On radiographs, a lytic lesion with variable sclerosis is observed [15]. Small cell osteosarcoma is characterized by small, round malignant cells within an osteoid matrix. It is differentiated from Ewing sarcoma by the production of osteoid [11].

57.4.3 Juxtacortical Osteosarcoma

57.4.3.1 Parosteal Osteosarcoma

Parosteal osteosarcoma is the most common juxtacortical osteosarcoma and makes up 1–6% of all osteosarcomas. Peak incidence is in the third decade of life and has a female predominance [8]. Radiographs typically reveal a densely ossified mass on the posterior aspect of the distal femur that has a "stuck-on" appearance [16]. On histology, a low-grade lesion with well-differentiated fibrous stroma is observed [11].

57.4.3.2 Periosteal Osteosarcoma

Periosteal osteosarcoma comprises 1–2% of all osteosarcomas [8]. It is a typically radiolucent mass that most commonly involves the proximal tibia or distal femur [16]. An aggressive periosteal reaction (e.g., sunburst appearance, Codman triangle) may be present. Periosteal osteosarcoma has mostly cartilaginous matrix on histology with small amounts of calcification and osteoid [11].

57.4.3.3 High-Grade Surface Osteosarcoma

High-grade surface osteosarcoma accounts for <1% of all osteosarcomas [11]. It is a surface lesion usually involving the femur or tibia with extension into the surrounding soft tissue with disruption of the underlying cortex [16]. Similar to conventional osteosarcoma, high-grade spindle cells with varying amounts of osteoid are observed on histology [11].

57.4.4 Secondary Osteosarcoma

The second peak in osteosarcoma incidence in older patients is due to malignant transformation in Paget's disease. Pagetic osteosarcomas tend to be high-grade intramedullary lesions. The pelvis, femur, and humerus are most commonly affected. Their histologic appearance reflects the chaotic remodeling of Paget's disease, with multiple atypical osteoblasts and the presence of osteoclast giant cells [17].

Osteosarcoma may arise through a second hit mutation in the setting of an existing genetic predisposition such as Li-Fraumeni syndrome, inherited retinoblastoma, and Rothmund-Thomson syndrome [8]. Radiation-induced osteosarcomas are rare; the mean latency period between radiation and sarcoma occurrence is approximately 15 years. These lesions are almost universally high-grade and carry a dismal prognosis [18].

57.4.5 Treatment and Prognosis

Conventional osteosarcoma requires neoadjuvant chemotherapy, wide resection, and adjuvant chemotherapy. In addition to treating detected or presumed osteosarcoma, chemotherapy may decrease the mass size—allowing for improved tumor demarcation and a less morbid resection. MAP therapy is most commonly used. Chemotherapy response is measured by percent tumor necrosis. Patients with less than 90% necrosis at the time of surgery are at increased risk of local recurrence and decreased overall survival [19]. There is no significant role for radiation therapy; palliative radiation may be employed in the treatment of the painful metastases.

Wide resection with limb salvage is typically performed; amputation is only considered in cases where a functional limb cannot remain or in cases of recurrence. It is imperative that negative margins be obtained. Positive margins or < 2 mm are associated with increased risk of local recurrence and decreased survival [19, 20]. If limb salvage is feasible, several reconstructive options exist including endoprosthetic reconstruction. Low-grade lesions such as parosteal and low-grade intramedullary subtypes can be treated with wide resection alone. Higher-grade osteosarcomas such as periosteal, high-grade surface, and secondary subtypes are treated like conventional osteosarcoma.

Long-term survival rates of patients with localized conventional osteosarcoma range from 60% to 80%; this falls to 20–30% in patients with metastasis [8]. While high-grade surface osteosarcoma has a similar prognosis to conventional osteosarcoma, the other juxtacortical subtypes have an excellent prognosis. Poor prognostic factors include metastatic disease, poor chemotherapy response, axial location, and skip metastases [8, 19, 21].

57.5 Chondrosarcoma

Chondrosarcoma is the most common bone sarcoma of adulthood, with an incidence of 1 in 200,000 per year [22]. It has a slight male predominance and affects patients between 40 and 75 years. Chondrosarcoma is most commonly found in the appendicular skeleton; less than 10% occur in the axial skeleton. The most common presenting complaint is pain due to weakening of the cortex; pathologic fracture may be encountered.

Radiographic signs of aggressive biology include endosteal scalloping >50% of cortical width, cortical thinning, and lucency adjacent to an enchondroma [23]. CT demonstrates a "rings and arcs" pattern of peripheral calcification. MRI assesses intramedullary and soft tissue involvement.

57.5.1 Genetics

Most chondrosarcomas are sporadic. Hedgehog and Rb pathway alterations, matrix metalloproteinases, and hypoxia-inducible factor 1 have been shown to be associated with chondrosarcoma progression [24, 25]. Syndromes associated with secondary chondrosarcoma such as multiple hereditary exostoses (MHE), Ollier's disease, and Maffucci syndrome are due to mutations in EXT1/EXT2, IDH1, and IDH2, respectively.

57.5.2 Primary Chondrosarcoma

57.5.2.1 Conventional Chondrosarcoma

Conventional chondrosarcoma accounts for 85% of all cases, occurring most commonly in the proximal femur and pelvis. Conventional central chondrosarcoma arise de novo within the bone, while peripheral lesions arise from malignant transformation within a pre-existing benign cartilage cap [26].

Conventional chondrosarcoma can be assigned to three grades. Locally aggressive with low metastatic potential, grade I chondrosarcoma contains abundant hyaline cartilage matrix with low cellularity and no pleomorphism. Grade II chondrosarcoma has some pleomorphism and increased cellularity. Grade III lesions are extremely cellular with high pleomorphism; metastasis is observed in up to 70% of cases [27].

57.5.2.2 Clear Cell Chondrosarcoma

Clear cell chondrosarcoma is a low-grade variant, observed in patients between 20 and 50 years. Clear cell chondrosarcoma occurs in the epiphysis with metaphyseal extension, most commonly in the femoral head. The typical radiographic appearance is a well-marginated lucency with sclerotic borders [28]. Histologically, clear cell chondrosarcoma is characterized the presence of clear cells with round nuclei [29].

57.5.2.3 Dedifferentiated Chondrosarcoma

Dedifferentiated chondrosarcoma is the most malignant variant, accounting for approximately 10% of cases. It tends to occur in patients between 50 and 70 years. It is found in a similar anatomic distribution to conventional lesions and is associated with a high risk of pathologic fracture. It is histologically characterized by two distinct neighboring areas of low-grade conventional chondrosarcoma and high-grade heterogeneous sarcoma [30].

57.5.2.4 Mesenchymal Chondrosarcoma

Mesenchymal chondrosarcoma is a rare high-grade variant most commonly occurring in the jaw, spine, and ribs. It has a biphasic histology with subtypes of small round cell components with islands of well-differentiated chondroid tissue [31].

57.5.3 Secondary Chondrosarcoma

The majority of secondary chondrosarcomas arise as osteochondromas in patients with MHE. Malignant transformation occurs in 5–10% of cases; a cartilage cap thicker than 1–2 centimeters should raise suspicion. Ollier's disease carries a 10–40% risk of malignant transformation, and small studies have indicated that nearly all patients with Maffucci syndrome develop chondrosarcoma [32].

57.5.4 Treatment and Prognosis

Usually low-grade, chondrosarcoma is largely unresponsive to radiation and chemotherapy. Radiation does not offer local control and is only employed for unresectable tumors or palliation. Chondrosarcomas have poor vascularity and low associated pH, impeding the delivery of systemic therapy [24]. Furthermore, chondrosarcoma cells express high levels of P-glycoprotein—an efflux pump that removes chemotherapeutic agents from cells [33].

The consensus treatment is wide resection for all grades of conventional chondrosarcoma [34]. Wide resection is indicated for dedifferentiated chondrosarcoma; young patients may benefit from adjuvant chemotherapy [35]. Clear cell chondrosarcoma carries a high risk of local recurrence and requires wide resection. Contrary to other subtypes, mesenchymal chondrosarcoma is radiosensitive and chemosensitive. Treatment includes wide resection with adjuvant chemoradiation [36].

Appendicular chondrosarcomas have improved survival and local recurrence rates compared to axial lesions. Conventional, clear cell, and mesenchymal chondrosarcomas have five-year survival rates of 70%, 100%, and 50% [37]. Most patients with dedifferentiated chondrosarcoma succumb to metastatic disease within 1 year [22].

57.6 Ewing Sarcoma

Ewing sarcoma is the second most common bone malignancy in children after osteosarcoma. It is typically seen between 5 and 25 years with a slight male predominance [38]. The pelvis and femur are most commonly affected. Pain and swelling are classic presenting symptoms. Unlike other bone sarcomas, constitutional symptoms such as fever, weight loss, and fatigue are often present [38, 39]. Elevated serum lactate dehydrogenase (LDH) may be present. A lytic permeative lesion is often seen, often accompanied by a periosteal reaction with an "onion-skin" appearance.

57.6.1 Genetics and Histology

Ewing sarcoma is classically associated with translocations between the EWS and ETS families of genes. In 80–95% of cases, the translocation is t(11;22) and produces the EWS-FLI1 fusion protein [38]. EWS-FLI1 stimulates proliferation, differentiation, and survival of neural crest-derived cells [40]. Ewing sarcoma is a high-grade tumor histologically characterized by sheets of small round blue cells with a high nuclear-to-cytoplasm ratio; pseudo-rosettes may be present [38]. CD99 is expressed almost universally.

57.6.2 Treatment and Prognosis

Ewing sarcoma is treated with neoadjuvant chemotherapy followed by resection and adjuvant chemotherapy. Wide resection with negative margins allows for local control with a low recurrence rate and also avoids the risk of a radiation-induced malignancy. In addition to superior local control and survival, surgical management provides the advantage of assessing tumor response to chemotherapy [38]. Adjuvant radiation may be appropriate for unresectable disease or those with poor chemotherapy response. Radiation of lung metastases improves survival.

Five-year survival is 65–80% and 25–40% in patients with localized and metastatic disease at presentation, respectively [38]. Factors associated with favorable prognosis include peripheral site, normal LDH level at presentation, tumor volume <100 ml, and localized disease [39]. The presence of metastatic disease—especially bony metastasis—is the most significant risk factor for poor prognosis. Recurrent disease is associated with <10% 5-year survival [38].

57.7 Other Bone Sarcomas

57.7.1 Fibrosarcoma of the Bone

Fibrosarcoma incidence peaks in the fourth and sixth decades of life. It most commonly occurs in the metaphyseal distal femur. An eccentric lytic lesion with a permeative appearance is observed on plain radiographs. Fibrosarcoma is characterized by a malignant spindle cell population arranged in a "herringbone" pattern. The most significant prognostic factor is histologic grade. Wide resection is the preferred treatment [41].

57.7.2 Undifferentiated High-Grade Pleomorphic Sarcoma

Historically called malignant fibrous histiocytoma, undifferentiated pleomorphic sarcoma is most commonly seen after the fifth decade of life. On radiographs, a diffusely lytic tumor with cortical destruction is seen in metaphyseal-diaphyseal bone. A high-grade tumor, it is histologically characterized by a mixed population of spindle cells, malignant histiocytes, and multinucleated giant cells. Undifferentiated pleomorphic sarcoma has high rates of local recurrence and metastasis. Treatment is wide resection and adjuvant chemotherapy [42].

57.7.3 Angiosarcoma

Angiosarcoma is a rare malignancy of endothelial origin most commonly involving the proximal femur. Approximately a third of cases are multifocal, usually in the same bone or limb. Angiosarcoma is composed of sheets of atypical epithelioid cells with eosinophilic cytoplasm where vacuoles with erythrocyte fragments are present. Angiosarcoma is often intermediate- or high-grade at presentation and carries a 30% five-year survival. Wide resection offers the best chance of local control. Multifocal disease may benefit from adjuvant chemoradiation [43].

57.8 Advances in Translational Bone Sarcoma Research

In recent years, new investigations into the tumor microenvironment and molecular progression of primary bone sarcoma have led to advances in our understanding of potential therapeutic avenues. The tumor microenvironment plays an important role in malignant transformation as well as disease progression [44, 45]. Tumor-associated macrophages—the primary immune cells present in the tumor microenvironment—are recruited by secretion of chemokines from osteosarcoma cells [46, 47]. M2 macrophages in the tumor microenvironment promote osteosarcoma metastasis through secretion of matrix metalloproteinase 12, which subsequently degrades the extracellular matrix [48]. Tumor-associated macrophages have also been implicated in chemotherapy resistance [49]. Targeting macrophage activity for osteosarcoma immunotherapy is an active area of research. The immune microenvironment associated with Ewing sarcoma also plays a critical role in disease progression. Immunosuppressive fibrocytes present in the Ewing sarcoma cell microenvironment induce angiogenesis and mediate immunosuppression [50]. Multiple surface and intracellular tumor-specific antigens have been identified for Ewing sarcoma; these have been used as targets for monoclonal antibodies, chimeric antigen receptor T cell-mediated approaches, and vaccine-based approaches [51].

57.9 Conclusion

- Primary bone sarcoma is associated with high morbidity and mortality.
- A thorough history and physical examination is critical.
- Radiographs, MRI, and CT are necessary to determine the extent of disease and for staging.
- Percutaneous core biopsy is preferred for tissue diagnosis.
- Bone sarcomas can be staged using the MSTS or AJCC staging systems.
- Osteosarcoma is the most common bone sarcoma, typically affecting patients in the second decade of life. The treatment is neoadjuvant chemotherapy, wide resection, and adjuvant chemotherapy.
- Chondrosarcoma is the most common bone sarcoma of adulthood, typically affecting patients 40–75 years. Wide resection is the mainstay of treatment for chondrosarcoma.
- Ewing sarcoma primarily occurs in patients between 5 and 25 years. It is characterized by the t(11;22) translocation with production of the EWS-FLI1 fusion protein. Standard therapy consists of neoadjuvant chemotherapy, wide resection, and adjuvant chemotherapy.
- Fibrous and vascular sarcomas are treated with wide resection.
- Advances in understanding of the tumor microenvironment and molecular drivers of bone sarcoma progression provide potential targets for therapy.

Open Questions

- Can the genetic drivers of bone sarcoma be better elucidated to allow for targeted therapy?
- What genetic characteristics of mesenchymal chondrosarcoma allow it to be more radiosensitive and chemosensitive compared to other subtypes?
- What soft-tissue coverage options will allow for decreased complications after wide resection for primary bone sarcomas?
- Can advanced machine learning methods be used to more accurately predict survival or outcomes after treatment for primary bone sarcomas?
- Can advances in radiation modalities allow for more targeted therapy for historically radioresistant bone sarcomas?

References

1. Siegel R, Miller K, Jemal A. Cancer statistics, 2020. CA Cancer J Clin. 2020;70(1):7–30.
2. Miller K, Fidler-Benaoudia M, Keegan T, Hipp H, Jemal A, Siegel R. Cancer statistics for adolescents and young adults, 2020. CA Cancer J Clin. 2020;70(6):443–59.
3. Barrientos-Ruiz I, Ortiz-Cruz E, Serrano-Montilla J, Bernabeu-Taboada D, Pozo-Kreilinger J. Are biopsy tracts a concern for seeding and local recurrence in sarcomas? Clin Orthop Relat Res. 2017;475(2):511–8.
4. Mitsuyoshi G, Naito N, Kawai A, et al. Accurate diagnosis of musculoskeletal lesions by core biopsy. J Surg Oncol. 2006;94(1):21–7.
5. Enneking W, Spanier S, Goodman M. A system for the surgical staging of musculoskeletal sarcoma. Clin Orthop Relat Res. 1980;153:106–20. https://doi.org/10.1007/978-1-4471-5451-8_127.
6. Steffner R, Jang E. Staging of bone and soft-tissue sarcomas. J Am Acad Orthop Surg. 2018;26:e269–78.
7. Kneisl J, Rosenberg A, Anderson P, et al. Bone. In: Amin M, Edge S, Greene F, et al., editors. AJCC cancer staging manual. 8th ed. Springer International: Switzerland; 2017. p. 471–86.
8. Messerschmitt P, Garcia R, Abdul-Karim F, Greenfield E, Getty P. Osteosarcoma. J Am Acad Orthop Surg. 2009;17:515–27.
9. Wu H, Zhang J, Dai R, Xu J, Feng H. Transferrin receptor-1 and VEGF are prognostic factors for osteosarcoma. J Orthop Surg Res. 2019;14(1):296.
10. Feng Y, Sassi S, Shen J, et al. Targeting CDK11 in osteosarcoma cells using the CRISPR-Cas9 system. J Orthop Res. 2015;33(2):199–207.
11. Klein M, Siegal G. Osteosarcoma: anatomic and histologic variants. Am J Clin Pathol. 2006;125(4):555–81.
12. Zishan U, Pressney I, Khoo M, Saifuddin A. The differentiation between aneurysmal bone cyst and telangiectatic osteosarcoma: a clinical, radiographic and MRI study. Skelet Radiol. 2020;49(9):1375–86.
13. Hegde V, Burke Z, Park H, et al. Is core needle biopsy reliable in differentiating between aggressive benign and malignant radiolucent bone tumors? Clin Orthop Relat Res. 2018;476(3):568–77.
14. Andresen K, Sundaram M, Unni K, Sim F. Imaging features of low-grade central osteosarcoma of the long bones and pelvis. Skelet Radiol. 2004;33(7):373–9.
15. Nakajima H, Sim F, Bond J, Unni K. Small cell osteosarcoma of bone: review of 72 cases. Cancer. 1997;79(11):2095–106.
16. Harper K, Sathiadoss P, Saifuddin A, Sheikh A. A review of imaging of surface sarcomas of bone. Skelet Radiol. 2021;50:9–28.
17. Hansen M, Seton M, Merchant A. Osteosarcoma in Paget's disease of bone. J Bone Miner Res. 2007;21:P58–63.
18. Mavrogenis A, Pala E, Guerra G, Ruggieri P. Post-radiation sarcomas: clinical outcome of 52 patients. J Surg Oncol. 2012;105(6):570–6.
19. Jeys L, Thorne C, Parry M, Gaston C, Sumathi V, Grimer J. A novel system for the surgical staging of primary high-grade osteosarcoma: the Birmingham classification. Clin Orthop Relat Res. 2017;476(3):842–50.
20. Bertrand T, Cruz A, Binitie O, Cheong D, Letson G. Do surgical margins affect local recurrence and survival in extremity, nonmetastatic, high-grade osteosarcoma? Clin Orthop Relat Res. 2016;474(3):677–83.
21. Sajadi K, Heck R, Neel M, et al. The incidence and prognosis of osteosarcoma skip metastases. Clin Orthop Relat Res. 2004;426:92–6.
22. Giuffrida A, Burguneo J, Koniaris L, Gutierrez J, Duncan R, Scully S. Chondrosarcoma in the United States (1973 to 2003): an analysis of 2890 cases from the SEER database. J Bone Joint Surg Am. 2009;91:1063–72.
23. Marco R, Gitelis S, Brebach G, Healey J. Cartilage tumors: evaluation and treatment. J Am Acad Orthop Surg. 2000;8(5):292–304.
24. Bovee J, Hogendoorn P, Wunder J, Alman B. Cartilage tumors and bone development: molecular pathology and possible therapeutic targets. Nat Rev Cancer. 2010;10(7):481–8.
25. Schrage Y, Lam S, Jochemsen A, et al. Central chondrosarcoma progression is associated with pRb pathway alterations: CDK4 down-regulation and p16 overexpression inhibit cell growth in vitro. J Cell Mol Med. 2009;13(9A):2843–52.
26. Organization WH. Cartilage tumors. In: Fletcher C, Unni K, Mertens F, editors. World Health Organization classification of tumors. Pathology and genetics. Tumors of soft tissue and bone. Lyon, France: IARC Press; 2002. p. 234–57.
27. Angelini A, Guerra G, Mavorgenis A, Pala E, Picci P, Ruggieri P. Clinical outcome of central conventional chondrosarcoma. J Surg Oncol. 2012;106(8):929–37.
28. Collins M, Koyama T, Swee R, Inwards C. Clear cell chondrosarcoma: radiographic, computed tomographic, and magnetic resonance findings in 34 patients with pathologic correlation. Skelet Radiol. 2003;32(12):687–94.
29. Itala A, Leerapun T, Inwards C, Collins M, Scully S. An institutional review of clear cell chondrosarcoma. Clin Orthop Relat Res. 2005;440:209–12.
30. Dornauer K, Soder S, Inwards C, Boveee J, Aigner T. Matrix biochemistry and cell biology of dedifferentiated chondrosarcomas. Pathol Int. 2010;60(5):365–72.
31. Ghafoor S, Hameed M, Tap W, Hwang S. Mesenchymal chondrosarcoma: imaging features and clinical findings. Skelet Radiol. 2021;50(2):333–41.
32. Lin P, Moussallem C, Deavers M. Secondary chondrosarcoma. J Am Acad Orthop Surg. 2010;18(10):608–15.
33. Trieb K, Sulzbacher I, Kubista B. Recurrence rate and progression of chondrosarcoma is correlated with heat shock protein expression. Oncol Lett. 2016;11(1):521–4.
34. Zoccali C, Baldi J, Attala D, et al. Intralesional versus extralesional procedures for low-grade central chondrosarcoma: a systematic review of the literature. Arch Orthop Trauma Surg. 2018;138(7):929–37.
35. Nooij M, Whelan J, Bramwell V, et al. Doxorubicin and cisplatin chemotherapy in high-grade spindle cell sarcomas of the bone, other than osteosarcoma or malignant fibrous histiocytoma: a European osteosarcoma intergroup study. Eur J Cancer. 2005;41(2):225–30.
36. De Amorim BK, DeLaney T. Chordomas and chondrosarcomas - the role of radiation therapy. J Surg Oncol. 2016;114(5):564–9.
37. Schneiderman B, Kliethermes S, Nystrom L. Survival in mesenchymal chondrosarcoma varies based on age and tumor location: a survival analysis of the SEER database. Clin Orthop Relat Res. 2017;475:799–805.
38. Maheshwari A, Cheng E. Ewing sarcoma family of tumors. J Am Acad Orthop Surg. 2010;18:94–107.
39. Biermann J, Chow W, Reed D, et al. NCCN guidelines insights: bone cancer, version 2.2017. J Natl Compr Cancer Netw. 2017;15(2):155–67.
40. Grunewald T, Bernard V, Gilardi-Hebenstreit P, et al. Chimeric EWSR1-FLI1 regulates the Ewing sarcoma susceptibility gene EGR2 via a GGAA microsatellite. Nat Genet. 2015;47(9):1073–8.
41. Papagelopoulos P, Galanis E, Trantafllidis P, Bascainos P, Sim F, Unni K. Clinicopathologic feature, diagnosis, and treatment of fibrosarcoma of bone. Am J Orthop. 2002;31(5):253–7.
42. Kalil R. Malignant fibrous histiocytoma of bone. In: Santini-Araujo E, Kalil R, Bertoni F, Park Y, editors. Tumors and tumor-like lesions of bone. London: Springer; 2015.
43. Kalil R. Angiosarcoma. In: Santini-Araujo E, Kalil R, Bertoni F, Park Y, editors. Tumors and tumor-like lesions of bone. London: Springer; 2015.

44. Binnewies M, Roberts E, Kersten K, Chan V, Fearon D, Merad M. Understanding the tumor immuune microenvironment (TIME) for effective therapy. Nat Med. 2018;24(5):541–50.
45. Quail D, Joyce J. Micorenvironmental regulation of tumor progression and metastasis. Nat Med. 2013;19(11):1423–37.
46. Segaliny A, Mohamadi A, Dizier B, Lokajczyk A, Brion R, Lanel R. Interleukin-34 promotes tumor progression and metastatic process in osteosarcoma through induction of angiogenesis and macrophage recruitment. Int J Cancer. 2015;137(1):73–85.
47. Cassetta L, Pollard J. Targeting macrophages: therapeutic approaches in cancer. Nat Rev Drug Discov. 2018;17(12):887–904.
48. Zhou Q, Xian M, Xiang S, Xiang D, Shao X, Wang J. All-trans retinoic acid prevents osteosarcoma metastasis by inhibiting M2 polarization of tumor-associated macrophages. Cancer Immunol Res. 2017;5(7):547–59.
49. Liang X, Guo W, Ren T, Huang Y, Sun K, Zhang H. Macrophages reduce the sensitivity of osteosarcoma to neoadjuvant chemotherapy drugs by secreting Interleukin-1 beta. Cancer Lett. 2020;480:4–14.
50. Zhang H, Maric I, DiPrima M, et al. Fibrocytes represent a novel MDSC subset circulating in patients with metastatic cancer. Blood. 2013;122(7):1105–13.
51. Morales E, Olson M, Iglesias F, Dahiya S, Luetkens T, Atanackovic D. Role of immunotherapy in Ewing sarcoma. J Immunother Cancer. 2020;8:e000653.

Liposarcoma Metastasis

58

Luke V. Selby, Raphael Pollock, and Valerie Grignol

Abstract

Soft tissue sarcomas are rare tumors originating from mesodermal tissues (fat, muscle connective tissue, and vessels). In-depth molecular analysis has identified at least 50 subtypes, which are often grouped together into larger classifications. Liposarcoma, the most common subtype, accounts for at least 20% of all sarcoma and occurs with a peak incidence of 50–65 years of age without gender predilection. Liposarcoma itself has four subtypes, which occur on a spectrum of biological activity and metastatic potential, though the predominant cause of sarcoma-specific death is locoregional recurrence. Well-differentiated liposarcoma (WDLPS) is locally aggressive but incapable of metastasis. Dedifferentiated liposarcoma, WDLPS's more aggressive counterpart, is locally aggressive and also capable of distant metastasis, primarily to the lung. Myxoid-round cell liposarcoma is a single histology whose metastatic potential (to the spine, axillary fat pads, and the retroperitoneal fat) occurs on a spectrum. Pleomorphic liposarcoma is the most aggressive liposarcoma subtype. Pleomorphic liposarcoma metastasizes early and has a very poor prognosis. Ongoing research suggests metastasis of well-differentiated and dedifferentiated liposarcoma may be mediated by extracellular vesicles, which contain active genetic material and condition the local tumor microenvironment and the distant premetastatic niche to support tumor growth. Further work is needed to further define the role of these extracellular vesicles and develop therapeutic strategies to interrupt this cascade.

L. V. Selby · R. Pollock · V. Grignol (✉)
Division of Surgical Oncology, Department of Surgery, The Ohio State University Wexner Medical Center and James Cancer Hospital, Columbus, OH, USA
e-mail: Valerie.Grignol@osumc.edu

Learning Objectives

- Liposarcoma is the most common subtype of soft tissue sarcomas.
- Liposarcoma exists along a spectrum of biological activity related to subtype and histological grade.
- Dedifferentiated liposarcoma is a locally aggressive histology with metastatic potential.
- Well-differentiated liposarcoma and dedifferentiated liposarcoma secrete biologically active extracellular vesicles that influence the tumor microenvironment and the premetastatic niche to make both more hospitable to tumor growth.

58.1 Introduction

Soft tissue sarcomas (STSs) are a rare group of locally aggressive and potentially metastatic tumors. STSs are derived from tissues of mesodermal origin (fat, muscle, connective tissue, vessels) and depending on histology are either locally aggressive, metastatic, or both. Recent molecular classifications have identified at least 50 different histological subtypes (WHO Classification of Tumours, 5th Edition, Volume 3. Soft Tissue and Bone Tumours [1]), based primarily on the underlying genetic mutations. Given their rarity, varied anatomic locations, and the difficulty in making pathologic diagnoses, surgical pathologists without significant experience making these diagnoses should consult with experienced centers as necessary [2, 3]. Benign tumors of these same tissue types, such as lipomas, are at least 100× more common than their malignant counterparts [4]. Malignant STSs are treated with a combination of surgery, radiation therapy, and systemic therapy, based on the tumor's anatomic location, its histological subtype, and its biological behavior. With the development of effective systemic chemotherapy and immunotherapy, the role of systemic therapy in these diseases is increasing [5–7]. Unfortunately,

S. P. Leong et al. (eds.), *Cancer Metastasis Through the Lymphovascular System*,
https://doi.org/10.1007/978-3-030-93084-4_58

approximately 40% of patients with STS die from complications associated with local recurrence or distant metastatic disease.

Broadly, STSs are characterized by their size, location, and histological grade. Size is broken down into increments of 5 cm, with tumors <5 cm considered small, 5–10 cm considered intermediate, and >10 cm considered large. Though histological subtype influences biological activity, STSs are broadly characterized into low-grade, intermediate-grade, and high-grade tumors. Grade is a histological determination that tracks a tumor's biological activity and aggressiveness. Low-grade tumors are slow growing, and while they can be locally aggressive and recurrent, they have very low risk of metastatic disease. High-grade tumors, on the other hand, are faster growing and frequently metastasize. In addition to the risks of local recurrence, large high-grade tumors can have rates of metastasis as high as 50%. Unfortunately, liposarcoma in general has a poor response to systemic chemotherapy [6, 8].

With the exception of epithelioid sarcoma, rhabdomyosarcoma, clear cell sarcoma, angiosarcoma, and some less frequent histologies, STS generally doesn't metastasize to the regional lymph nodes [9]. While metastatic location is dependent on the particular tumor location and histological type, sarcoma most commonly metastasizes to the lung, though myxoid-round cell liposarcoma also metastasizes to the fat pads and bones.

58.2 Liposarcoma

Liposarcoma, the most common STS subtype, accounts for at least 20% of all STSs. Four distinct liposarcoma subtypes, based on tumor biology and driver mutations, have been described. Well-differentiated (WDLPS) liposarcoma and dedifferentiated (DDLPS) liposarcoma exist along a spectrum of biological activity. WDLPS and DDLPS uniformly have an amplification of chromosome 12q13-15, which involves MDM2, CDK4, and HMGA2, with other amplifications identified less commonly [10, 11]. Overexpression of MDM2 DNA from this amplification leads to excess MDM2 protein expression, which then overrides the patient's wild-type p53 tumor suppression function [12]. WDLPS is a locally aggressive but non-metastatic tumor that occurs within the proximal extremities and the retroperitoneum. WDLPS that occurs within the extremity is cytologically and histologically identical to the retroperitoneal WDLPS but due to much different biological behavior is often referred to as an atypical lipomatous tumor (ALT) (WHO Classification of Tumours, 5th Edition, Volume 3. Soft Tissue and Bone Tumours). Patients with WDLPS or ALT present with a painless slow-growing mass of the extremity or slowly increasing abdominal girth in the case of a retroperitoneal WDLPS (Fig. 58.1). Retrospectively identified gastrointestinal symptoms in patients with retroperitoneal tumors were often sub-

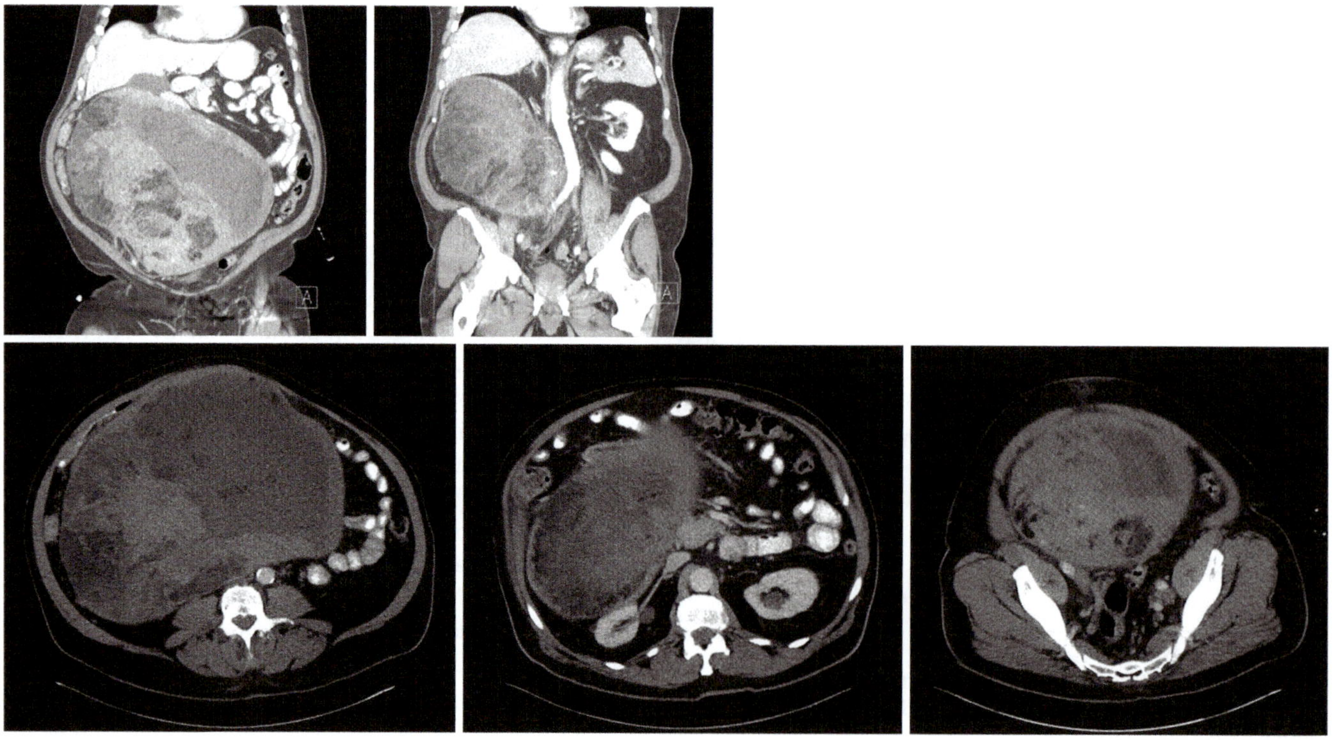

Fig. 58.1 Selected axial and coronal images from a CT scan showing a large retroperitoneal dedifferentiated liposarcoma. The patient underwent resection of the mass without resection of adjacent organs

tle and non-diagnostic. WDLPS of the retroperitoneum can be incidentally diagnosed on cross-sectional imaging or during the workup for an hernia if the mass has protruded through a defect in the abdominal wall [13].

WDLPS and DDLPS occur on a continuum of differentiation and in total account for approximately 60% of all liposarcomas. In combination, WDLPS/DDLPS is the post common liposarcoma histology. DDLPS is identified when a tumor contains a region of non-lipogenic sarcoma of at least several millimeters in diameter. The transition from WDLPS to DDLPS is marked by invasion into, instead of displacement of, adjacent structures and the ability to develop distant metastasis. Radiologically, DDLPS shows both fatty and non-fatty components, while WDLPS appears as a mass of entirely fatty components [14]. DDLPS is more locally aggressive than WDLPS and is frequently diagnosed on re-resection of a locally recurrent WDLPS. Though it has less metastatic potential than other high-grade pleomorphic sarcomas, DDLPS has a high risk of distant metastasis, with series reporting distant recurrent rates of 20–30% [15].

Historically, myxoid liposarcoma and round-cell liposarcoma were thought of as two distinct tumor types [16]; they are now recognized as a single entity with differing clinical behaviors [17] and account for approximately 30–40% of all liposarcomas. They most commonly occur deep in the extremities, with at least 66% occurring in the thigh. More than 90% of all myxoid-round cell tumors have a characteristic t(12;16) (q13-14;p11) translocation [18]. Tumors with high histological grade were previously classified as round-cell liposarcomas; they are defined as having a >5% round cell component and have a much worse prognosis than the pure myxoid subtype, which has 0–5% round cell component [19]. High-grade tumors have a 50% 5-year survival, compared to the >90% 5-year survival of low-grade tumors. Like all liposarcomas, myxoid-round cell tumors metastasize to the lungs. However, they also metastasize to the spine, the axillary fat pads, and the retroperitoneal fat, even in the absence of pulmonary metastasis [20–22]. Consequently, their metastatic workup is more extensive and included CT scan of the primary site, chest, abdomen, and pelvis and finally MRI of the spine. Myxoid-round cell liposarcoma is exquisitely sensitive to radiation therapy, ifosfamide, and trabectedin [23–25].

Pleomorphic liposarcomas are high grade, highly malignant, and account for <5% of all liposarcomas [26]. They have high mitotic activity and due to their aggressiveness often have internal hemorrhage and necrosis. These aggressive tumors metastasize to the lungs early [27], and despite sensitivity to gemcitabine and ifosfamide-based chemotherapy regimens, pleomorphic liposarcomas have high mortality rates.

Though synovial sarcoma, rhabdomyosarcoma, clear cell sarcoma, epithelioid sarcoma, and angiosarcoma are known to have lymphatic metastasis, it has traditionally been thought that liposarcoma achieved distant metastasis via hematogenous, not lymphatic, routes. Recent work from the NCDB, however, suggests that the rate of lymph node metastasis in liposarcoma, regardless of anatomic site of origin, is approximately 2.5% [28]. Given the complexity of making a diagnosis in liposarcoma [2], and the inherent heterogeneity of large databases like the NCDB, it is hard to know if the patients with a lymph node metastasis were diagnosed with the correct histology. It was initially hypothesized that local recurrence following resection of retroperitoneal liposarcoma was the result of incomplete tumor resection and subsequent tumor regrowth. More recently, multiple institutional series have suggested that retroperitoneal sarcoma often occurs in the setting of a field defect within the retroperitoneal fat, placing the entire retroperitoneal compartment at risk of tumor development [29, 30]. However, emerging evidence suggests that tumor produced extracellular vesicles that contribute to tumor regrowth by conditioning the tumor microenvironment.

58.3 Extracellular Vesicles

Extracellular vesicles (EVs) are membrane-bound vesicles that contain cellular cytosol and genetic material and are enclosed in a lipid bilayer [31]. Vesicles between 30 and 150 nm in size are classified as exosomes; those >150 nm in size are called microvesicles. Exosomes are produced within specific compartments of the cell for release and are marked by Tsg101, Alix, CD9, CD63, and CD81 markers, while microvesicles are formed by cell budding and CD59, CD47, and CD55 [31–33]. Initially thought to contain only cell waste, EVs are known to contain mRNA and microRNA in addition to other bioactive components including protein, lipids, nucleic acids, and various metabolites [34, 35].

58.3.1 EVs and the Tumor Microenvironment

Since the demonstration that EVs can transfer their contents to recipient cells, their role in cancer development has garnered great attention (Fig. 58.2). Non-neoplastic cells near a solid tumor make up the tumor microenvironment (TME), and the interaction between these stromal cells and the primary tumor itself is essential to tumor survival and metastasis. Tumor-stromal communication occurs via direct cell-cell interaction as well as indirectly via paracrine signaling involving growth factors, chemokines, and proteases and also via EVs [36]. EVs, containing active RNA, can influence the tumor microenvironment to make the environment supportive of tumor growth and survival. EVs are known to shift fibroblasts and mesenchymal stem cells into a myofi-

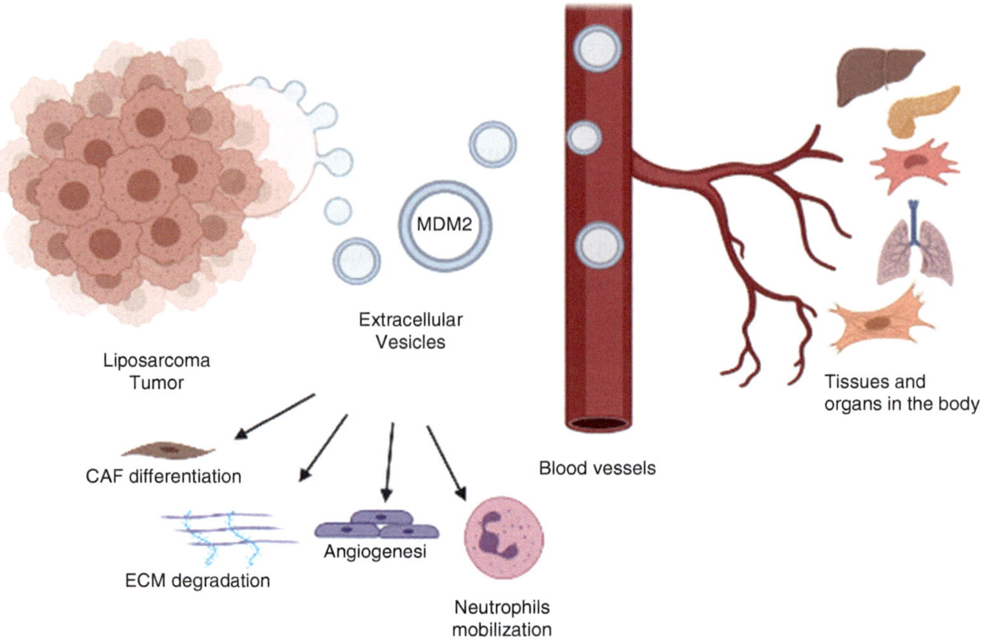

Fig. 58.2 Local and distant effects of liposarcoma-derived extracellular vesicles

broblast phenotype, which facilitates angiogenesis and metastasis [37]. Tumor angiogenesis is supported when EVs transfer specific signals and increase VEGF signaling [38, 39]. As cells in the TME are transformed by EVs, these cells themselves then contribute to the continued development of the TME. EVs from regulatory T lymphocytes (Tregs) have been shown to transfer miRNA containing exosomes to other immune cells, suppressing the anticancer Th1 phenotype by changing cell proliferation and cytokine secretion [40, 41].

DDLPS EVs contain DNA and RNA important in the development of malignancy and metastasis. miR-25-3p and miR-92a-3p regulate their target mRNAs during both transcription and translation, increasing NF-kB-mediated IL-6 secretion by TME macrophages. Cultured DDLPS cells exposed to macrophage condition medium enriched in IL-6 show increased cell proliferation and invasion [42–44]. DDLPS-derived EVs are known to contain amplified copies of MDM2. MDM2 amplification is a core component of WDLPS and DDLPS and results in functional inactivation of the p53 tumor suppressor protein. Transfer of MDM2 DNA to preadipocytes in the TME leads to overexpression of MDM2 in cells without genomic alterations in MDM2 copy number. These cells, genotypically normal preadipocytes from the tumor microenvironment, are also known to produce excess levels of metalloprotease-2 (MMP2), an extracellular collagenase that is vital to metastasis formation [43]. Taken together, this suggests that EVs released by DDLPS increase IL-6-mediated cell proliferation and invasion via miRNA-specific pathways, but they also inactivate wt p53 and increase MMP2 production (which in turns also facilitates cell migration and invasion) via an MDM2-specific pathway.

The liposarcoma TME contains abundant preadipocytes, adipocytes, macrophages, inflammatory cells, endothelial cells, and fibroblasts [45]. Importantly, trabectedin, a medication with activity against myxoid-round cell liposarcoma, acts against the tumor itself and also against the TME. Trabectedin forms covalent bonds with the minor grove of the DNA double helix [46, 47]. As a result, trabectedin modulates cytokine and chemokine production and displaces oncogenic transcription factors from their targeted promoters in tumor cells and cells in the TME. Trabectedin's anti-TME activity seems to come from its activity against tumor-associated macrophages (TAMs) which as quiescent cell are not targeted by traditional cytotoxic chemotherapies [48]. TAMs are monocytes recruited from the circulating blood and activated into a TAM phenotype by the tumor and TME cells through the various mechanisms discussed above [49]. Importantly, trabectedin has activity against slow-growing tumors, but to this point it has shown activity mostly for myxoid-round cell liposarcoma.

58.3.2 EVs and the Premetastatic Niche

Just as EVs can condition the tumor microenvironment to support local tumor growth, so can they condition distant organ sites to transform into a premetastatic nice (PMN). Across tumor types, EVs have been shown to condition the PMN to allow metastasis survival. Breast carcinoma-derived EVs contribute to immunosuppression by pushing myeloid cells towards a myeloid-derived suppressor cell phenotype [50], while melanoma-derived EVs impair dendritic cell maturation in the PMN and shift them towards an immunosuppressive phenotype [51]. Exosomes have been shown to stimulate MMP2 [43] and MMP9 [52] production in the PMN, facilitating metastasis invasion and angiogenesis, which is also aided by EV upregulation of VEGF [53]. In total, just as tumor-derived EVs exert their influence on the TME to allow tumor survival, through the same mechanisms they affect the distant PMN and push it towards supporting metastasis [54].

58.4 Surgery and Postoperative Surveillance

Surgery, the gold standard of treatment for liposarcoma, is often combined with either systemic therapy or radiation therapy in disease-specific settings [55]. The goal of curative-intent surgery is function-preserving complete gross resection. In the retroperitoneum, patients who undergo R2 resection have worse outcome than those undergoing either R0 or R1 resection. It is unclear if there is differential survival between those who undergo R0 or R1 resection [56]. Significant controversy exists surrounding the importance of extended compartmental resection to achieve a gross negative margin of at least 1 cm in the retroperitoneum. Patients undergoing compartmental resection have lower rates of local recurrence, and may have improved overall survival, but higher rates of postoperative complications [56–58]. Additionally, because large tumors lie as close to "unresectable" organs (the central vessels, liver, or duodenum) as they lie to the "resectable" organs (a single kidney, small intestine, large intestine), compartmental resections are generally used selectively when gross negative margins can be achieved with an organ-sparing approach at the initial operation [57, 59]. Rapid local recurrence and invasion into, instead of abutment of, adjacent structures may reflect tumor biology, obviating the benefit of aggressive local resection [60]. Additionally, even in the metastatic setting, most DDLPS death (and all WDLPS death) is the result of locoregional recurrence, not distant metastatic disease.

The principles of resection of extremity myxoid-round cell liposarcomas are identical to those of retroperitoneal tumors: function-sparing complete resection with gross negative margins. Rates of local recurrence for extremity sarcoma are much lower than for retroperitoneal sarcoma [61, 62]. High-grade myxoid-round cell liposarcoma is extremely sensitive to radiation therapy. The decision to administer radiation therapy in the neoadjuvant or adjuvant setting or upon local recurrence is based on the location within the thigh and weighing the risks and benefits of the sequencing of these therapies.

Postoperative surveillance in retroperitoneal liposarcoma consists of physical exam and CT scans of the abdomen and pelvis, with the chest added for grade II and III liposarcomas (DDLPS). In addition to this screening, patients with myxoid-round cells liposarcoma should undergo CT scans of the primary tumor site, chest, abdomen, and pelvis, as well as an MRI of the spine to screen for metastasis to the axillary fat pads and retroperitoneum. MRI is exquisitely sensitive at detecting extrapulmonary metastasis from myxoid-round cell liposarcoma [63]. In the future, it is possible that peripherally detected EVs will be used as biomarkers of disease in WDLPS and DDLPS. Changes in EV levels, either decreased EV levels in response to systemic therapy or rising EV levels during surveillance, may allow patients to be more closely monitored in between cross-sectional imaging.

58.5 Conclusion

Liposarcoma, the most common subtype of soft tissue sarcoma, is made up of four distinct subtypes with unique tumor biology. Liposarcoma metastasis is hematologic, not lymphatic, and is a poorly understood phenomenon. Liposarcoma predominantly metastasizes to the lungs, while myxoid-round cell liposarcoma also possesses the ability to metastasize to the retroperitoneal and axillary fat pads. WDLPS local recurrence and DDLPS metastasis appear to be mediated by tumor-derived extracellular vesicles that influence the stroma of the TME and PMN to undergo a series of pre-metastasis changes that improve tumor survival. Further work is required to more fully define the metastatic steps, and therapeutic interventions to halt metastasis, in these rare tumors.

Perspective

- Is liposarcoma metastasis hematologic or lymphatic?
- Which liposarcoma histologies are prone to metastasis?
- What role do liposarcoma-derived extracellular vesicles play in local recurrence and metastasis?
- What medications can be used to affect the sarcoma tumor microenvironment?
- What are the current first-line systemic therapies for liposarcoma?

References

1. WHO Classification of Tumours, 5th Edition, Volume 3. Soft Tissue and Bone Tumours; (n.d.).
2. Riddle ND, Gonzalez RJ, Bridge JA, Antonia S, Bui MM. A CD117 and CD34 immunoreactive sarcoma masquerading as a gastrointestinal stromal tumor: diagnostic pitfalls of ancillary studies in sarcoma. Cancer Control. 2011;18(3):152–9. https://doi.org/10.1177/107327481101800302.
3. Thway K, Wang J, Mubako T, Fisher C. Histopathological diagnostic discrepancies in soft tissue tumours referred to a specialist centre: reassessment in the era of ancillary molecular diagnosis. Sarcoma. 2014;2014:686902. https://doi.org/10.1155/2014/686902.
4. Rydholm A, Berg NO. Size, site and clinical incidence of lipoma. Factors in the differential diagnosis of lipoma and sarcoma. Acta Orthop Scand. 1983;54(6):929–34. https://doi.org/10.3109/17453678308992936.
5. Bill KL, Casadei L, Prudner BC, Iwenofu H, Strohecker AM, Pollock RE. Liposarcoma: molecular targets and therapeutic implications. Cell Mol Life Sci. 2016;73(19):3711–8. https://doi.org/10.1007/s00018-016-2266-2.
6. Liebner DA. The indications and efficacy of conventional chemotherapy in primary and recurrent sarcoma. J Surg Oncol. 2015;111(5):622–31. https://doi.org/10.1002/jso.23866.
7. McGovern Y, Zhou CD, Jones RL. Systemic therapy in metastatic or unresectable well-differentiated/dedifferentiated liposarcoma. Front Oncol. 2017;7(292) https://doi.org/10.3389/fonc.2017.00292.
8. Anaya DA, Lahat G, Wang X, Xiao L, Tuvin D, Pisters PW, Lev DC, Pollock RE. Establishing prognosis in retroperitoneal sarcoma: a new histology-based paradigm. Ann Surg Oncol. 2009;16(3):667–75. https://doi.org/10.1245/s10434-008-0250-2.
9. Fong Y, Coit DG, Woodruff JM, Brennan MF. Lymph node metastasis from soft tissue sarcoma in adults. Analysis of data from a prospective database of 1772 sarcoma patients. Ann Surg. 1993;217(1):72–7. https://doi.org/10.1097/00000658-199301000-00012.
10. Gisselsson D, Höglund M, Mertens F, Johansson B, Dal Cin P, Van den Berghe H, Earnshaw WC, Mitelman F, Mandahl N. The structure and dynamics of ring chromosomes in human neoplastic and non-neoplastic cells. Hum Genet. 1999;104(4):315–25. https://doi.org/10.1007/s004390050960.
11. Szymanska J, Tarkkanen M, Wiklund T, Virolainen M, Blomqvist C, Asko-Seljavaara S, Tukiainen E, Elomaa I, Knuutila S. Gains and losses of DNA sequences in liposarcomas evaluated by comparative genomic hybridization. Genes Chromosomes Cancer. 1996;15(2):89–94. https://doi.org/10.1002/(sici)1098-2264(199602)15:2<89::Aid-gcc2>3.0.Co;2-#.
12. Sirvent N, Coindre JM, Maire G, Hostein I, Keslair F, Guillou L, Ranchere-Vince D, Terrier P, Pedeutour F. Detection of MDM2-CDK4 amplification by fluorescence in situ hybridization in 200 paraffin-embedded tumor samples: utility in diagnosing adipocytic lesions and comparison with immunohistochemistry and real-time PCR. Am J Surg Pathol. 2007;31(10):1476–89. https://doi.org/10.1097/PAS.0b013e3180581fff.
13. Montgomery E, Fisher C. Paratesticular liposarcoma: a clinicopathologic study. Am J Surg Pathol. 2003;27(1):40–7. https://doi.org/10.1097/00000478-200301000-00005.
14. Messiou C, Morosi C. Imaging in retroperitoneal soft tissue sarcoma. J Surg Oncol. 2018;117(1):25–32. https://doi.org/10.1002/jso.24891.
15. Tseng WW, Chopra S, Engleman EG, Pollock RE. Hypothesis: The intratumoral immune response against a cancer progenitor cell impacts the development of well-differentiated versus dedifferentiated disease in liposarcoma. Front Oncol. 2016;6:134. https://doi.org/10.3389/fonc.2016.00134.
16. Orvieto E, Furlanetto A, Laurino L, Dei Tos AP. Myxoid and round cell liposarcoma: a spectrum of myxoid adipocytic neoplasia. Semin Diagn Pathol. 2001;18(4):267–73.
17. Antonescu CR, Elahi A, Humphrey M, Lui MY, Healey JH, Brennan MF, Woodruff JM, Jhanwar SC, Ladanyi M. Specificity of TLS-CHOP rearrangement for classic myxoid/round cell liposarcoma: absence in predominantly myxoid well-differentiated liposarcomas. J Mol Diagn. 2000;2(3):132–8. https://doi.org/10.1016/s1525-1578(10)60628-9.
18. Knight JC, Renwick PJ, Dal Cin P, Van den Berghe H, Fletcher CD. Translocation t(12;16)(q13;p11) in myxoid liposarcoma and round cell liposarcoma: molecular and cytogenetic analysis. Cancer Res. 1995;55(1):24–7.
19. Fiore M, Grosso F, Lo Vullo S, Pennacchioli E, Stacchiotti S, Ferrari A, Collini P, Lozza L, Mariani L, Casali PG, Gronchi A. Myxoid/round cell and pleomorphic liposarcomas: prognostic factors and survival in a series of patients treated at a single institution. Cancer. 2007;109(12):2522–31. https://doi.org/10.1002/cncr.22720.
20. Dürr HR, Rauh J, Baur-Melnyk A, Knösel T, Lindner L, Roeder F, Jansson V, Klein A. Myxoid liposarcoma: local relapse and metastatic pattern in 43 patients. BMC Cancer. 2018;18(1):304. https://doi.org/10.1186/s12885-018-4226-8.
21. Fuglø HM, Maretty-Nielsen K, Hovgaard D, Keller J, Safwat AA, Petersen MM. Metastatic pattern, local relapse, and survival of patients with myxoid liposarcoma: a retrospective study of 45 patients. Sarcoma. 2013;2013:548628. https://doi.org/10.1155/2013/548628.
22. Spillane AJ, Fisher C, Thomas JM. Myxoid liposarcoma – the frequency and the natural history of nonpulmonary soft tissue metastases. Ann Surg Oncol. 1999;6(4):389–94. https://doi.org/10.1007/s10434-999-0389-5.
23. Chung PW, Deheshi BM, Ferguson PC, Wunder JS, Griffin AM, Catton CN, Bell RS, White LM, Kandel RA, O'Sullivan B. Radiosensitivity translates into excellent local control in extremity myxoid liposarcoma: a comparison with other soft tissue sarcomas. Cancer. 2009;115(14):3254–61. https://doi.org/10.1002/cncr.24375.
24. Crago AM, Dickson MA. Liposarcoma: multimodality management and future targeted therapies. Surg Oncol Clin N Am. 2016;25(4):761–73. https://doi.org/10.1016/j.soc.2016.05.007.
25. Suarez-Kelly LP, Baldi GG, Gronchi A. Pharmacotherapy for liposarcoma: current state of the art and emerging systemic treatments. Expert Opin Pharmacother. 2019;20(12):1503–15. https://doi.org/10.1080/14656566.2019.1618271.
26. Gebhard S, Coindre JM, Michels JJ, Terrier P, Bertrand G, Trassard M, Taylor S, Château MC, Marquès B, Picot V, Guillou L. Pleomorphic liposarcoma: clinicopathologic, immunohistochemical, and follow-up analysis of 63 cases: a study from the French Federation of Cancer Centers Sarcoma Group. Am J Surg Pathol. 2002;26(5):601–16. https://doi.org/10.1097/00000478-200205000-00006.
27. Downes KA, Goldblum JR, Montgomery EA, Fisher C. Pleomorphic liposarcoma: a clinicopathologic analysis of 19 cases. Mod Pathol. 2001;14(3):179–84. https://doi.org/10.1038/modpathol.3880280.
28. Keung EZ, Chiang YJ, Voss RK, Cormier JN, Torres KE, Hunt KK, Feig BW, Roland CL. Defining the incidence and clinical significance of lymph node metastasis in soft tissue sarcoma. Eur J Surg Oncol. 2018;44(1):170–7. https://doi.org/10.1016/j.ejso.2017.11.014.
29. Neuhaus SJ, Barry P, Clark MA, Hayes AJ, Fisher C, Thomas JM. Surgical management of primary and recurrent retroperitoneal liposarcoma. Br J Surg. 2005;92(2):246–52. https://doi.org/10.1002/bjs.4802.
30. Tseng WW, Madewell JE, Wei W, Somaiah N, Lazar AJ, Ghadimi MP, Hoffman A, Pisters PW, Lev DC, Pollock RE. Locoregional

disease patterns in well-differentiated and dedifferentiated retroperitoneal liposarcoma: implications for the extent of resection? Ann Surg Oncol. 2014;21(7):2136–43. https://doi.org/10.1245/s10434-014-3643-4.
31. Colombo M, Raposo G, Théry C. Biogenesis, secretion, and intercellular interactions of exosomes and other extracellular vesicles. Annu Rev Cell Dev Biol. 2014;30:255–89. https://doi.org/10.1146/annurev-cellbio-101512-122326.
32. Gould SJ, Raposo G. As we wait: coping with an imperfect nomenclature for extracellular vesicles. J Extracell Vesicles. 2013;2 https://doi.org/10.3402/jev.v2i0.20389.
33. Simons M, Raposo G. Exosomes – vesicular carriers for intercellular communication. Curr Opin Cell Biol. 2009;21(4):575–81. https://doi.org/10.1016/j.ceb.2009.03.007.
34. Valadi H, Ekström K, Bossios A, Sjöstrand M, Lee JJ, Lötvall JO. Exosome-mediated transfer of mRNAs and microRNAs is a novel mechanism of genetic exchange between cells. Nat Cell Biol. 2007;9(6):654–9. https://doi.org/10.1038/ncb1596.
35. Waldenström A, Gennebäck N, Hellman U, Ronquist G. Cardiomyocyte microvesicles contain DNA/RNA and convey biological messages to target cells. PLoS One. 2012;7(4):e34653. https://doi.org/10.1371/journal.pone.0034653.
36. Roma-Rodrigues C, Mendes R, Baptista PV, Fernandes AR. Targeting tumor microenvironment for cancer therapy. Int J Mol Sci. 2019;20(4) https://doi.org/10.3390/ijms20040840.
37. Zhang X, Yuan X, Shi H, Wu L, Qian H, Xu W. Exosomes in cancer: small particle, big player. J Hematol Oncol. 2015;8:83. https://doi.org/10.1186/s13045-015-0181-x.
38. Al-Nedawi K, Meehan B, Micallef J, Lhotak V, May L, Guha A, Rak J. Intercellular transfer of the oncogenic receptor EGFRvIII by microvesicles derived from tumour cells. Nat Cell Biol. 2008;10(5):619–24. https://doi.org/10.1038/ncb1725.
39. Al-Nedawi K, Meehan B, Rak J. Microvesicles: messengers and mediators of tumor progression. Cell Cycle. 2009;8(13):2014–8. https://doi.org/10.4161/cc.8.13.8988.
40. Okoye IS, Coomes SM, Pelly VS, Czieso S, Papayannopoulos V, Tolmachova T, Seabra MC, Wilson MS. MicroRNA-containing T-regulatory-cell-derived exosomes suppress pathogenic T helper 1 cells. Immunity. 2014;41(3):503. https://doi.org/10.1016/j.immuni.2014.08.008.
41. Smyth LA, Ratnasothy K, Tsang JY, Boardman D, Warley A, Lechler R, Lombardi G. CD73 expression on extracellular vesicles derived from CD4+ CD25+ Foxp3+ T cells contributes to their regulatory function. Eur J Immunol. 2013;43(9):2430–40. https://doi.org/10.1002/eji.201242909.
42. Bartel DP. MicroRNAs: genomics, biogenesis, mechanism, and function. Cell. 2004;116(2):281–97. https://doi.org/10.1016/s0092-8674(04)00045-5.
43. Casadei L, Calore F, Braggio DA, Zewdu A, Deshmukh AA, Fadda P, Lopez G, Wabitsch M, Song C, Leight JL, Grignol VP, Lev D, Croce CM, Pollock RE. MDM2 derived from dedifferentiated liposarcoma extracellular vesicles induces MMP2 production from preadipocytes. Cancer Res. 2019;79(19):4911–22. https://doi.org/10.1158/0008-5472.Can-19-0203.
44. Denli AM, Tops BB, Plasterk RH, Ketting RF, Hannon GJ. Processing of primary microRNAs by the Microprocessor complex. Nature. 2004;432(7014):231–5. https://doi.org/10.1038/nature03049.
45. Casadei L, Calore F, Creighton CJ, Guescini M, Batte K, Iwenofu OH, Zewdu A, Braggio DA, Bill KL, Fadda P, Lovat F, Lopez G, Gasparini P, Chen JL, Kladney RD, Leone G, Lev D, Croce CM, Pollock RE. Exosome-derived miR-25-3p and miR-92a-3p stimulate liposarcoma progression. Cancer Res. 2017;77(14):3846–56. https://doi.org/10.1158/0008-5472.Can-16-2984.
46. D'Incalci M, Badri N, Galmarini CM, Allavena P. Trabectedin, a drug acting on both cancer cells and the tumour microenvironment. Br J Cancer. 2014;111(4):646–50. https://doi.org/10.1038/bjc.2014.149.
47. D'Incalci M, Galmarini CM. A review of trabectedin (ET-743): a unique mechanism of action. Mol Cancer Ther. 2010;9(8):2157–63. https://doi.org/10.1158/1535-7163.Mct-10-0263.
48. Galmarini CM, D'Incalci M, Allavena P. Trabectedin and plitidepsin: drugs from the sea that strike the tumor microenvironment. Mar Drugs. 2014;12(2):719–33. https://doi.org/10.3390/md12020719.
49. Mantovani A, Sozzani S, Locati M, Allavena P, Sica A. Macrophage polarization: tumor-associated macrophages as a paradigm for polarized M2 mononuclear phagocytes. Trends Immunol. 2002;23(11):549–55. https://doi.org/10.1016/s1471-4906(02)02302-5.
50. Xiang X, Poliakov A, Liu C, Liu Y, Deng ZB, Wang J, Cheng Z, Shah SV, Wang GJ, Zhang L, Grizzle WE, Mobley J, Zhang HG. Induction of myeloid-derived suppressor cells by tumor exosomes. Int J Cancer. 2009;124(11):2621–33. https://doi.org/10.1002/ijc.24249.
51. Valenti R, Huber V, Filipazzi P, Pilla L, Sovena G, Villa A, Corbelli A, Fais S, Parmiani G, Rivoltini L. Human tumor-released microvesicles promote the differentiation of myeloid cells with transforming growth factor-beta-mediated suppressive activity on T lymphocytes. Cancer Res. 2006;66(18):9290–8. https://doi.org/10.1158/0008-5472.Can-06-1819.
52. Hiratsuka S, Nakamura K, Iwai S, Murakami M, Itoh T, Kijima H, Shipley JM, Senior RM, Shibuya M. MMP9 induction by vascular endothelial growth factor receptor-1 is involved in lung-specific metastasis. Cancer Cell. 2002;2(4):289–300. https://doi.org/10.1016/s1535-6108(02)00153-8.
53. Grange C, Tapparo M, Collino F, Vitillo L, Damasco C, Deregibus MC, Tetta C, Bussolati B, Camussi G. Microvesicles released from human renal cancer stem cells stimulate angiogenesis and formation of lung premetastatic niche. Cancer Res. 2011;71(15):5346–56. https://doi.org/10.1158/0008-5472.Can-11-0241.
54. Lobb RJ, Lima LG, Möller A. Exosomes: Key mediators of metastasis and pre-metastatic niche formation. Semin Cell Dev Biol. 2017;67:3–10. https://doi.org/10.1016/j.semcdb.2017.01.004.
55. Mansfield SA, Pollock RE, Grignol VP. Surgery for abdominal well-differentiated liposarcoma. Curr Treat Options in Oncol. 2018;19(1):1. https://doi.org/10.1007/s11864-018-0520-6.
56. Bonvalot S, Rivoire M, Castaing M, Stoeckle E, Le Cesne A, Blay JY, Laplanche A. Primary retroperitoneal sarcomas: a multivariate analysis of surgical factors associated with local control. J Clin Oncol. 2009;27(1):31–7. https://doi.org/10.1200/jco.2008.18.0802.
57. Bonvalot S, Miceli R, Berselli M, Causeret S, Colombo C, Mariani L, Bouzaiene H, Le Péchoux C, Casali PG, Le Cesne A, Fiore M, Gronchi A. Aggressive surgery in retroperitoneal soft tissue sarcoma carried out at high-volume centers is safe and is associated with improved local control. Ann Surg Oncol. 2010;17(6):1507–14. https://doi.org/10.1245/s10434-010-1057-5.
58. Gronchi A, Lo Vullo S, Fiore M, Mussi C, Stacchiotti S, Collini P, Lozza L, Pennacchioli E, Mariani L, Casali PG. Aggressive surgical policies in a retrospectively reviewed single-institution case series of retroperitoneal soft tissue sarcoma patients. J Clin Oncol. 2009;27(1):24–30. https://doi.org/10.1200/jco.2008.17.8871.
59. Gronchi A, Pollock R. Surgery in retroperitoneal soft tissue sarcoma: a call for a consensus between Europe and North America. Ann Surg Oncol. 2011;18(8):2107–10. https://doi.org/10.1245/s10434-011-1746-8.

60. Singer S, Antonescu CR, Riedel E, Brennan MF. Histologic subtype and margin of resection predict pattern of recurrence and survival for retroperitoneal liposarcoma. Ann Surg. 2003;238(3):358–70.; discussion 370–351. https://doi.org/10.1097/01.sla.0000086542.11899.38.
61. Cahlon O, Brennan MF, Jia X, Qin LX, Singer S, Alektiar KM. A postoperative nomogram for local recurrence risk in extremity soft tissue sarcomas after limb-sparing surgery without adjuvant radiation. Ann Surg. 2012;255(2):343–7. https://doi.org/10.1097/SLA.0b013e3182367aa7.
62. Hoffman A, Ghadimi MP, Demicco EG, Creighton CJ, Torres K, Colombo C, Peng T, Lusby K, Ingram D, Hornick JL, Wang WL, Ravi V, Lazar AJ, Lev D, Pollock RE. Localized and metastatic myxoid/round cell liposarcoma: clinical and molecular observations. Cancer. 2013;119(10):1868–77. https://doi.org/10.1002/cncr.27847.
63. Gorelik N, Reddy SMV, Turcotte RE, Goulding K, Jung S, Alcindor T, Powell TI. Early detection of metastases using whole-body MRI for initial staging and routine follow-up of myxoid liposarcoma. Skelet Radiol. 2018;47(3):369–79. https://doi.org/10.1007/s00256-017-2845-9.

Desmoid Tumors

59

Gaya Spolverato and Alessandro Gronchi

Abstract

Desmoid tumors (DT) are a rare disease with no metastatic potential, characterized by a local aggressive monoclonal fibroblastic proliferation. DT are characterized by a mutually exclusive CTNNB1 or APC mutation: CTNNB1 mutations are somatic and occur only in the sporadic form, while APC mutations are generally germline and associated with the familiar variant (Gardner syndrome). Surgery has been the standard of care for many years; however, recently we observed a paradigm shift toward a more conservative management. We present herein the available evidences on the management of sporadic and familiar DT, with a particular focus on active surveillance, available treatments, and assessment of response.

Learning Objectives

- Epidemiology and presentation of desmoid tumors.
- Molecular signatures of desmoid tumors.
- Indication for active surveillance of sporadic and familiar desmoids.
- Indication for active treatment and treatment algorithm.
- Assessment of response to treatment and assessment of tumor growth during observation.

G. Spolverato
Department of Surgical, Oncological and Gastroenterological Sciences, University of Padova, Padova, Italy

A. Gronchi (✉)
Department of Surgery, Fondazione IRCCS Istituto Nazionale dei Tumori, Milan, Italy
e-mail: alessandro.gronchi@istitutotumori.mi.it

59.1 Introduction

Desmoid tumors (DT) are a rare disease with no metastatic potential, characterized by a local aggressive monoclonal fibroblastic proliferation. The incidence of DT ranges between 5 and 6 cases per million inhabitants per annum. DT have a female predominance and are more common in the third and fourth life decades. Although DT do not spread distantly, they can be locally aggressive, causing chronic pain, functional deficit, and psychological problems, affecting the overall quality of life. Intestinal obstruction and perforation can also be a complication of intra-abdominal desmoid tumors and life-threatening conditions, mostly in the case of multifocal disease and/or familial adenomatous polyposis (FAP)-associated DT.

Desmoids occur in three locations:

1. Extra-abdominal sites: most commonly extremities, limb girdles (mostly periscapular), thoracic wall, breast, head and neck.
2. Abdominal wall: mostly in women, especially during or after pregnancy.
3. Intra-abdominal sites, such as the bowel wall and mesentery: often associated with FAP.

Sporadic extra-abdominal desmoids can occasionally be multifocal, but mostly confined to one limb. On the contrary FAP-associated desmoids are more often multifocal, and while they predominantly affect the mesentery and abdominal wall, they can be located in multiple extra-abdominal sites (Fig. 59.1).

Unfortunately, the indolent behavior of desmoids is not supported by specific molecular signature, gene mutations, or microenvironment characteristics. No previous studies investigated the metastatic behavior of desmoids and the biological basis of the absent metastatic potential.

S. P. Leong et al. (eds.), *Cancer Metastasis Through the Lymphovascular System*,
https://doi.org/10.1007/978-3-030-93084-4_59

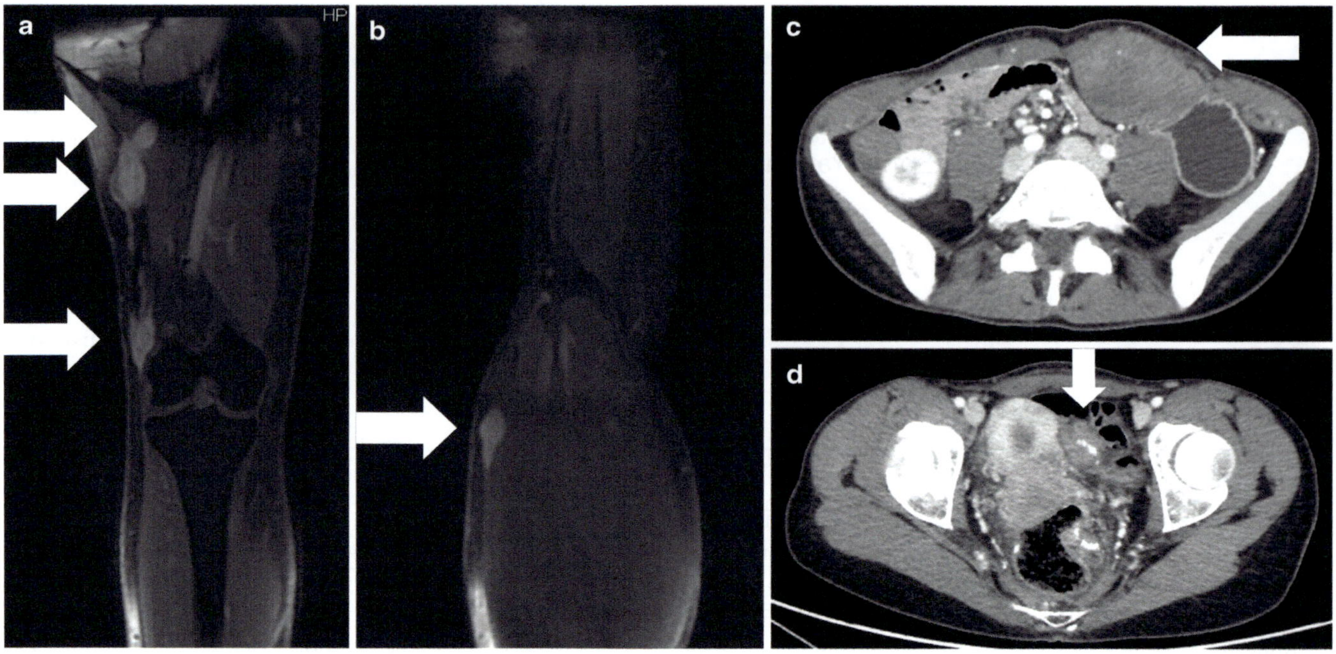

Fig. 59.1 Panel **a**, **b**: contrast-enhanced T1-weighted MRI of the right lower limb, coronal view, showing multiple nodules (white arrows) in the thigh (panel **a**) and leg (panel **b**) in a 26-year-old female with no associated syndromes. Panel **c**, **d**: contrast-enhanced CT scan of the abdomen, axial view, showing masses (white arrows) in the abdominal wall (panel **c**) and mesentery (panel d) in a 31-year-old female with Gardner syndrome

59.2 Molecular Genetics

DT are characterized by a mutually exclusive CTNNB1 or APC mutation. CTNNB1 mutations are somatic and occur only in the sporadic form. Occasionally also APC mutations can be somatic and can occur in the sporadic form. However, they are generally germline and associated with the familiar variant (Gardner syndrome). Hence, if a somatic CTNNB1 is reported, a syndromic condition can be excluded, while in case of CTNNB1 wild-type status, FAP should be investigated with extensive clinical workup (including a colonoscopy or a germline test). CTNNB1 mutation determines a disruption of Wnt/beta-catenin signaling, involved in cell-cell adhesion [1]. Several activating mutations of CTNNB1 are known, all of them in exon 3, causing the following amino acid changes: T41A, S45F, and S45P. In a minority of patients, DT result from a germline or sporadic loss of APC, which is a negative regulator of beta-catenin stability. A germline mutation of the adenomatous polyposis coli gene falls in the context of FAP—Gardner syndrome. The mutation in the APC gene on chromosome 5 is known for the development of hundreds of colonic polyps. DT, manifesting mainly with intra-abdominal multifocal tumors, occur in about 20% of FAP patients [2, 3].

The effect of CTNNB1 mutation on long-term oncological outcome and in particular on recurrence after surgery has been extensively evaluated in sporadic DT [4]. Some studies report that S45F-mutated DT tend to have a higher recurrence rate after surgery, while others report conflicting results [5–10]. A recent meta-analysis evaluating the impact of CTNNB1 mutation type on recurrence and recurrence-free survival in adult patients undergoing surgery for DT [4] found that S45F mutation has a higher risk of recurrence compared to T41A, S45P, and wild-type DT, but the effect modification between tumor size and mutation type suggests that the association between mutation and prognosis is mediated by tumor size.

The prognostic impact of APC mutations in FAP patients is unclear, while the effects of APC mutation on the risk of developing desmoids have been extensively investigated. Mutations between codons 1444 and 1578 were found to correlate with DT, but this observation was not compared to a reference population. In a following systematic review of the genotype-phenotype associations of DT in patients with FAP, APC mutations between codons 543–713 and 1310–2011 were found to confer an increased risk for desmoid tumors in individuals with compared to a reference population of FAP patients with APC gene mutations [11].

Among the epigenetic factors associated with the risk of developing DT, the surgical injury of open colectomy was found to be an independent predictor of DT, and patients receiving laparotomic prophylactic colectomy were found to have a 5-year cumulative risk of developing DT threefold higher than those operated with a laparoscopic approach [12].

In the era of active surveillance for sporadic and FAP-related DT, it is even more important to define the impact of mutation type on risk of progression in order to anticipate the tumor behavior and modulate the treatment plan. The effect of APC or CTNNB1 mutation on tumor aggressiveness or response to medical treatment is still unknown. The correlation between the efficacy of meloxicam treatment and status of CTNNB1 mutations was analyzed in a cohort of 33 patients with extraperitoneal sporadic DT, and S45F mutation was found to be associated with a poor response to treatment [13]. Similarly, the same mutation was found to correlate with response to imatinib, further supporting the idea of a potential predictive impact of the mutation status on DF treatment decision-making [14].

59.3 Active Surveillance

59.3.1 Sporadic Desmoid

Surgery has historically been the primary treatment for patients with resectable DT. Surgical recommendations have traditionally been based on retrospective studies in which surgery was offered when feasible; thus, the conservative approach was neglected for a long time. A cutting-edge study was published by Brennan et al. in 1999, comparing major amputation to observation among patients with recurrent extremity desmoids. In a group of 22 patients who had disease requiring amputation, 7 underwent amputation, and 15 were managed with a conservative approach. Among them, four patients received systemic treatment, and six were just observed. Within a follow-up of 25–92 months, none or insignificant tumor progression was observed, and three patients in the observation group had spontaneous tumor regression. DT were found to be enigmatic: the reason why some tumors continue to grow and others could regress was unclear. For the first time, an accent was put on the need to account for function and structure preservation, while avoiding aggressive attempts to achieve negative resection margins, resulting in unnecessary complications. In parallel with this observation, it became clear over those years how microscopic surgical margins were not associated to a higher risk of local recurrence, shedding some light on a different angle about the presence of a relatively indolent subgroup of diseases. One of the first large studies analyzing the impact of margins on long-term oncological outcome dates back in 2003. Gronchi et al., reporting on a series of 203 patients undergoing surgery for primary or recurrent DT, showed that microscopically positive margins did not affect disease-free survival in patients with extra-abdominal DT [15]. These findings were confirmed by the following studies. In particular, Lev et al., comparing a large series of desmoid patients treated at the University of Texas M.D. Anderson Cancer Center to a previously published series from the same institution, showed that, even if the rate of macroscopic residual disease was higher and the rate of positive microscopic margins was equivalent overtime, 5-year local recurrence rate improved from 30% to 20% [16]. Similarly, Crago et al., reporting on a large series of DT patient treated at Memorial Sloan Kettering Cancer Center, confirmed that margin status was not associated with local recurrence-free survival and excluded it from the prognostic nomogram [17].

Active surveillance was investigated by different authors in retrospective studies reporting on DT in several locations. The first report dates back in 2007, when Bonvalot et al. reported on two groups of patients: those who underwent surgery with or without radiotherapy and those who were managed with nonsurgical strategies, such as systemic treatment alone or active surveillance. Patients with microscopically negative margins resection had similar outcome compared to patients who did not undergo surgery. Growth arrest occurred in two thirds of non-operated patients, and strategy (surgery versus no-surgery) was not correlated with the oncological outcome [18]. Similarly, Fiore et al., reporting on a series of patients who were initially treated with a frontline deliberately conservative policy, showed that half of patients achieved medium-term stable disease. Perhaps more importantly patients managed with active surveillance had a progression-free survival similar to patients managed with medical treatment (mainly hormonal therapy and chemotherapy) [19].

The criteria for selecting patients for active surveillance have been largely discussed. There is consensus that abdominal wall tumors tend to be more indolent. In these patients, surgery can allow for local control [20–25]; however, the post-surgical morbidity associated with abdominal reconstruction needs to be considered. An important study on primary abdominal wall desmoids by Bonvalot et al. [26] reported on 147 patients undergoing either frontline surgery or initial observation/medical treatment. Within 36 months of follow-up, approximately one third of patients that did not undergo surgery had stable disease, and one third experienced spontaneous regression (Fig. 59.2a, b). The authors recommended an initial nonsurgical approach to abdominal wall DT, followed by surgery in case of tumor growth. The safety of a nonsurgical approach was also confirmed in extra-abdominal wall DT. In a study by Colombo et al., analyzing the natural history of patients with extra-abdominal wall DT, among the 122 patients treated with a non-operative management, 5% required surgery, 51% other treatments, while the 20% experienced spontaneous regression (Fig. 59.2c, d) [27]. Tumor site was found to be a major prognostic factor for event-free survival in a following study [28]. The survival of patients with DT in favorable locations, such as the

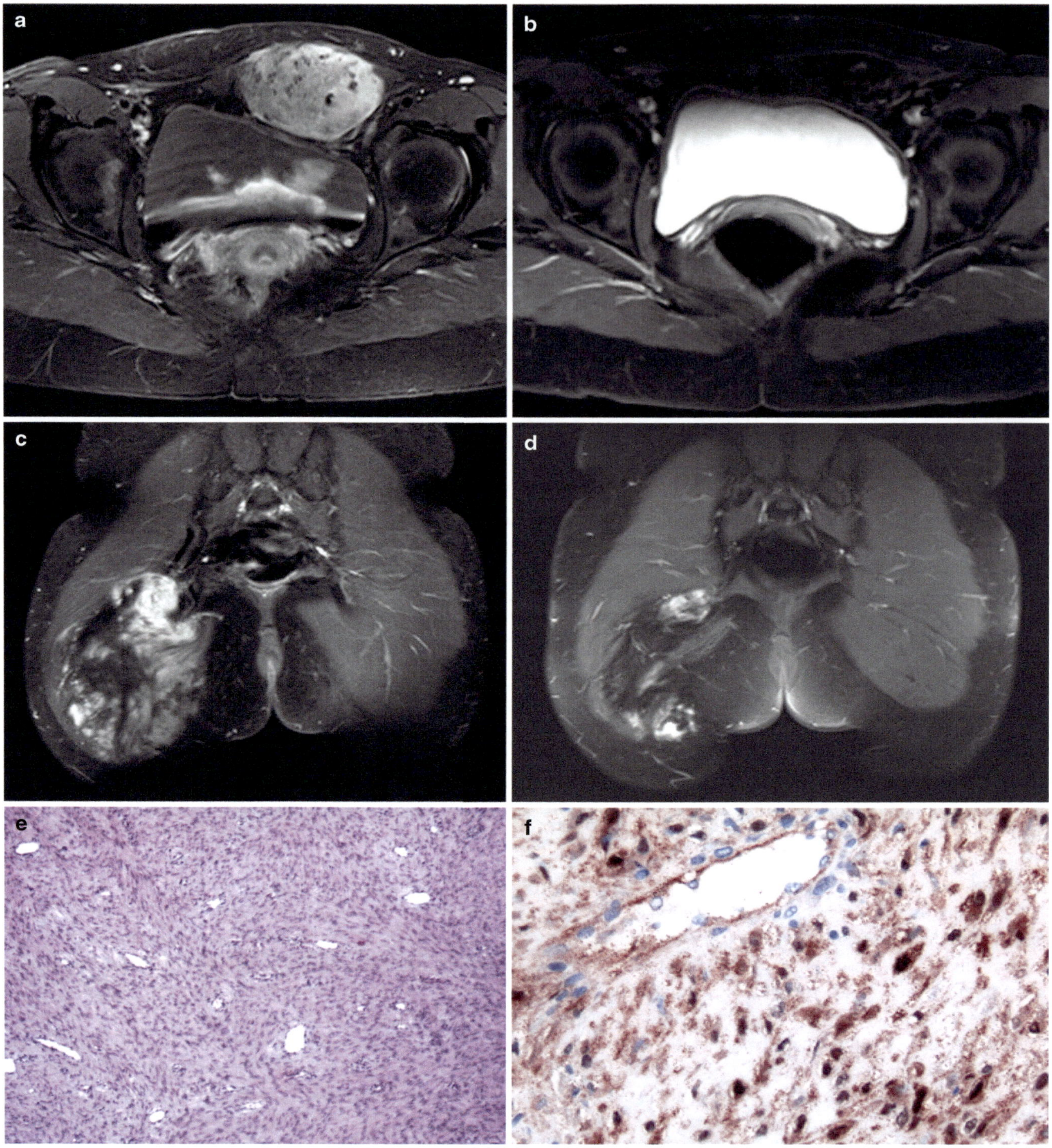

Fig. 59.2 Panel **a**, **b**: contrast-enhanced T1-weighted MRI, axial view, of a left abdominal wall desmoid at presentation (panel **a**) and 6 years after active surveillance (panel **b**), showing spontaneous regression. Panel **c**, **d**: contrast-enhanced T1-weighted MRI, coronal view, of a right buttock desmoid at presentation (panel **c**) and 7 years after active surveillance (panel **d**), showing spontaneous regression; histology image of the desmoid tumor (panel **e**, H&E; panel **f**, IHC)

abdominal wall; intra-abdominal, breast, and digestive viscera; and lower limb, was independent from the treatment strategy (operative versus non-operative). Interestingly, the outcome of patients with unfavorable locations was better if the patients were initially managed with active surveillance [28]. Finally, a randomized clinical trial comparing sorafenib to placebo in patients with progressing DT showed that tumor stabilization or regression occurred in two thirds of patients in the placebo group, further supporting a non-operative first-line approach [29].

The natural history of DT during active observation is difficult to describe, since in most of the studies patients are offered to drop out from active observation early in the management. In a study from Briand et al., DT were found to be able to stabilize spontaneously after 1 year of evolution and within 3 years for the vast majority of patients, while the cumulative probability of failure during active treatment was approximately 10% at 10 years, further confirming active surveillance as a valid first-line management for patients with primary or recurrent extra-abdominal DT [30]. Of note, regressions have been noted also after initial progressions in prospective observational studies on active surveillance [31]. The final results along with the correlation between mutational status and progression are awaited.

Based on these findings, the Desmoid Tumor Working Group reached a consensus about key aspects of the management of DT [32]. According to the most recent guidelines by the Desmoid Tumor Working Group, active surveillance, defined as the continuous monitoring of DT patients with magnetic resonance imaging (MRI), initially performed within 1–2 months and then at 3–6 months interval, does not influence the efficacy of subsequent treatments. Since up to 50–60% of DT do not grow after diagnosis and 20–30% may shrink after initial progression [29], initial management of DT should consider active surveillance while sparing unnecessary surgery. The presence of symptoms should not modify the tumor-directed treatment approach, since pain control can be achieved with a conservative approach. Perhaps more importantly, an extended surgical treatment risks to negatively affect pain control and quality of life.

Among the abandoned indication for upfront surgery, we recognize pain, as previously stated, resection of positive margins, and pregnancy/postpartum. In particular, during pregnancy, due to the secretion of growth factors and hormones, DT can grow or firstly appear. However, the tumor tends to decrease after pregnancy; thus, surgery should not be offered.

The biological basis beyond DT behavior has been investigated, and in particular the characterization of the tumor immune microenvironment and immune checkpoints has been recently described by Siozopoulou et al. [33] Although the authors did not define a molecular and immune pattern characterizing patients experiencing tumor regression under observation, they were the first to show that DT have a strong immune infiltration at the tumor margins, characterized by lymphoid aggregates, and to demonstrate that there is no PD-L1-driven immune suppression present in the tumor cells.

59.3.2 FAP-Related Desmoid

While most of the available evidence is based on sporadic DT, active surveillance is even more appropriate in case of FAP-related DT, which tend to be more aggressive and interest patients who underwent major colorectal surgery. However, the switch to an active treatment may need to be considered sooner, given the higher risk of complications related to the multifocality, anatomic site, and biological behavior.

59.4 Selection for an Active Treatment

The type of intervention mostly depends on tumor site (Fig. 59.3). Abdominal wall is the only site where surgery is the first option in patients failing observation. Surgical morbidity is generally limited, and local control is close to 95% [26]. If surgery is the treatment of choice, the primary goal is to achieve R0 margins [26].

For all other locations, surgery is reserved in case of failure of non-operative management, such as hormonal therapies, chemotherapy, and molecular target therapies. Local modalities such as isolated limb perfusion, radiotherapy, cryotherapy, and HIFU can also be considered depending on location, age, and symptoms.

Several medical treatments are available, but the lack of comparative studies does not allow for a well-defined sequence of systemic options (Table 59.1). Among the systemic therapies, we count: hormonal therapies, nonsteroidal anti-inflammatory drugs (NSAIDs), tyrosine kinase inhibitors (TKIs), low-dose chemotherapy with iv methotrexate + vinblastine/vinorelbine or oral vinorelbine alone, and conventional chemotherapy including liposomal doxorubicin [29, 32, 34–41]. It is intuitive to employ less toxic therapy followed by more toxic ones in a stepwise fashion. Several factors need to be considered when defining the treatment plan, such as (1) level of evidence, (2) overall response rate, (3) progression-free survival rate, (4) the ease of administration, and (5) the expected toxicity, and have been summarized in a five-dimensional model proposed by the Desmoid Tumor Working Group. After several retrospective studies testing tamoxifen or toremifene alone or in combination with NSAIDs and one phase II prospective study showing limited activity of this regimen, there is no evidence to consider antihormonal therapies and NSAIDs in patients with DT [42–45]. Differently, clear clinical benefit was found for TKIs, in particular for sorafenib and pazopanib compared to imatinib. This effect is achievable even with low dosages, limiting the potential side effects (such as fatigue, rash, hypertension, gastrointestinal symptoms) [29, 36, 37, 46–48]. Of note, now that sorafenib and pazopanib are largely offered to DT patients, the long-term risk of hypertension and/or hypothyroidism should be taken into account. These side effects are not negligible since DT patients are often young. Chemotherapy options can include low-dose methotrexate plus vinblastine or vinorelbine

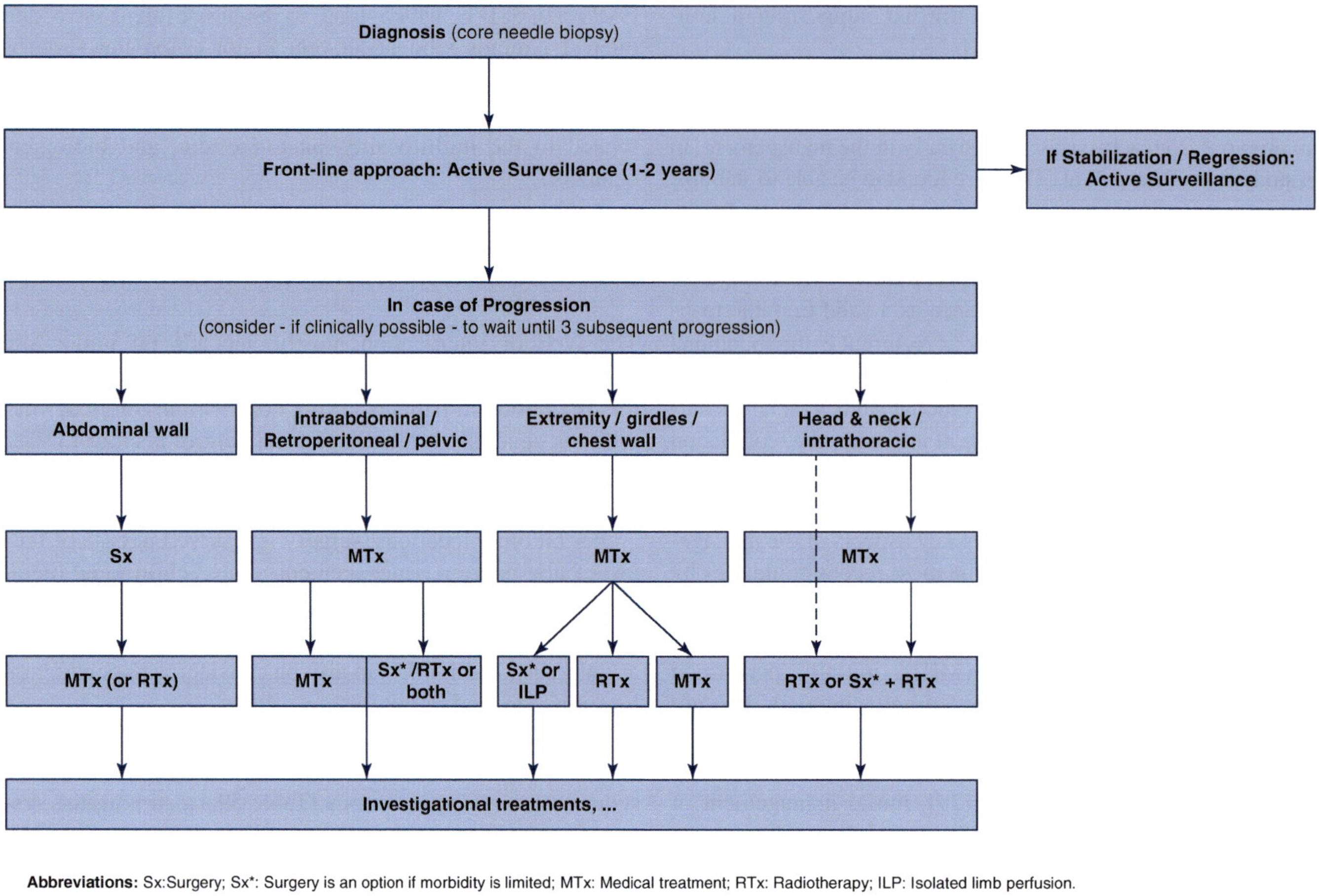

Fig. 59.3 Treatment algorithm of DT. Published with permission from *European Journal of Cancer127 (2020) 96–107*

Table 59.1 Available evidences for the existing tyrosine kinase, γ-secretase inhibitors, and cytotoxic agents for desmoid tumor treatment

Article	Pt	Inclusion criteria	Drug	Duration	Response rate	12 mo PFS
Heinrich 2006	19	Heavily pretreated patients	Imatinib (800 mg)	325 days	16	36
Penel 2011	35	Radiological evidence for progression	Imatinib (400 mg)	1 year	11	67
Chugh 2010	49	Locally advanced disease	Imatinib (200–600 mg)	Until progression	6	66
Kasper 2017	38	RECIST progression	Imatinib (800 mg)	2 years	19	59
Gounder 2018	50	Progressive or symptomatic	Sorafenib (400 mg)	Until progression	33	89
Toulmonde 2019	48	RECIST progression	Pazopanib (800 mg)	1 year	37	86
Kummar 2017	17	Progressive or symptomatic	Nirogacestat (300 mg)	Until progression	29	100
Toulmonde 2019	24	RECIST progression	MTX + VBL	1 year	25	79
Palassini 2017	75	Progressive disease	MTX + VBL MTX + VIN	14 months	48	60

(including oral vinorelbine alone) or conventional anthracycline-based regimens or liposomal doxorubicin. The response to low-dose chemotherapy has proved to occur several months after the start of therapy, to be durable even after treatment withdrawal, and to occur in up to 50–70% of patients with acceptable toxicity [29, 36, 37, 46–48]. Two regimens including doxorubicin plus dacarbazine and low-dose methotrexate plus vinca alkaloids showed to be effective in FAP-associated DT.

A new treatment option for DT patients is nirogacestat, a safe and active drug with effect on the Notch signaling. Nirogacestat (PF-03084014) is an orally available, reversible γ-secretase inhibitor, which proved to be highly active and safe. γ-Secretase cleaves intracellular Notch and results in pathway activation [34]. In a phase I trial, nirogacestat allowed to achieve partial response in five of seven patients [49]. This result was confirmed by a following phase II trial in which the 29% of patients achieved partial response and the 71% stable disease [50]. Finally, tegavivint is currently being tested in the first in-human trial as a new therapeutic option, since it inhibits the Wnt and beta-catenin pathway.

TNF-α and melphalan-based isolated limb perfusion (TM-ILP), considered the standard of care in Europe for patients with limb-threatening sarcomas, are also offered to patients with locally advanced aggressive fibromatosis to avoid limb amputation. The evidences come from a study conducted in three European centers, where 16/25 patients treated with TM-ILP achieved complete or partial response, 22/25 had their limb spared, and the large majority had good functional outcomes [51, 52].

RT is another option to achieve local control, mostly after failure of all available medical treatments, in the elderly, in patients with several comorbidities, or in case of rapidly growing disease threatening vital organs. Definitive RT at moderate doses (i.e., 50 Gy) allows obtaining a long-term progression arrest in up to 70% of the cases. Long-term side effects such as fibrosis, fractures, and radiation-induced tumors should be considered when offering RT to young patients with DT [52].

Among the other local options, cryotherapy and HIFU are gaining popularity in the treatment of DT. However, few case series with a short follow-up and without a matched control group are available to support their indication. In particular cryotherapy has been recently proved to be safe, well tolerated, and effective in a series of 14 patients [53]. Cryoablation demonstrated feasibility in progressive DT patients enrolled in a recent prospective, open-label, non-randomized, non-comparative, multicenter study (CRYODESMO-01) [54].

Similarly, magnetic resonance-guided focused ultrasound has been used to treat extra-abdominal DT due to its noninvasiveness, lack of ionizing radiation, low morbidity, and good safety profile [55]. However, since their promising effects are still under investigation, they should be offered in the context of clinical trials.

59.5 Assessment of Response

Assessment of response to treatment and assessment of tumor growth during observation are still matters of concern. RECIST criteria often fail to answer clinical questions, because tissue response to treatment and in particular cellularity are difficult to define. MRI can be applied as a surrogate for cellular content, since T2 sequences are able to discriminate between neoplastic tissue, fibrous tissue, and extracellular components [56]. Response to systemic therapies has shown to be associated with decrease in T2 signal intensity, and recently hyperintense T2 was also found to correlate with progression-free survival [57]. In the study by Cassidy et al., reporting on 37 patients with extra-abdominal DT managed with observation, 1-year PFS of T2 signal in ≥90% was 55%, compared to 94% of the T2 <90% group. These results need to be confirmed in a larger series of DT, possibly of different sites, to confirm T2 signaling as a valid tool in DT surveillance [57].

59.6 Open Questions

There are still some unresolved issues on DT that require further investigations. First, although a paradigm shift in the management of DT occurred in the last decade, it is necessary to clearly define when to consider upfront surgery. Possible complications, such as obstruction, perforation, and mesenteric ischemia can require upfront surgery; however, if the tumor is not safely resectable with negative margins, one could well consider to manage the complication and start a medical treatment, which—in case of disease control/response—may avoid an R2 resection. Second, the prognostic role of CTNNB1 mutational status among patients undergoing surveillance needs to be clarified in order to integrate the available prognostic tools with reliable biological biomarkers. Similarly, the impact of APC mutations on the development of DT in FAP patients and on their prognosis needs to be further evaluated. Perhaps more importantly, none of the available studies proved a correlation between specific APC mutations and the biological aggressiveness of the disease and in particular the lack of metastatic potential.

Third, although the armamentarium of treatment options for DT is growing, in the absence of comparative studies, the treatment sequence in patients who fail active surveillance has to be defined. Finally, the management of recurrent DT is still debated. To which extent the management of primary DT can be applied to recurrent disease is unclear. Data on recurrent disease are needed to further understand the behavior and the clinical management of these diverse entities.

59.7 Conclusions

In conclusion, desmoids are rare and locally aggressive tumors, with low metastatic potential and often unpredictable clinical course. Surgery has been the standard of care for many years; however, recently we observed a paradigm shift toward a more conservative management. New treatment modalities are awaited for patients who fail active surveillance and for recurrent disease. Health-related quality of life should be adopted as a metric to compare different treatment approaches; validation of a specific tool is in progress. Prognostic and predictive biomarkers could serve as valid tools to stratify patients for progression and response to treatment.

Open Questions

- Although a paradigm shift in the management of DT occurred in the last decade, when is it necessary to consider upfront surgery?
- What is the prognostic role of CTNNB1 mutational status among patients undergoing surveillance?
- What is the impact of APC mutations on the development of DT in FAP patients and on their prognosis?
- What is the treatment sequence in patients who fail active surveillance?
- To which extent the management of primary DT can be applied to recurrent disease?
- What is the biological and mutational basis under the lack of metastatic potential of DT?

References

1. Crago AM, Chmielecki J, Rosenberg M, et al. Near universal detection of alterations in CTNNB1 and Wnt pathway regulators in desmoid-type fibromatosis by whole-exome sequencing and genomic analysis. Genes Chromosomes Cancer. 2015;54(10):606–15.
2. Heiskanen I, Jarvinen HJ. Occurrence of desmoid tumours in familial adenomatous polyposis and results of treatment. Int J Color Dis. 1996;11(4):157–62.
3. Bertario L, Russo A, Sala P, et al. Genotype and phenotype factors as determinants of desmoid tumors in patients with familial adenomatous polyposis. Int J Cancer. 2001;95(2):102–7.
4. Timbergen MJM, Colombo C, Renckens M, et al. The prognostic role of beta-catenin mutations in desmoid-type fibromatosis undergoing resection only: a meta-analysis of individual patient data. Ann Surg. 2021;273(6):1094–01; https://doi.org/10.1097/SLA.0000000000003698.
5. Lazar AJ, Tuvin D, Hajibashi S, et al. Specific mutations in the beta-catenin gene (CTNNB1) correlate with local recurrence in sporadic desmoid tumors. Am J Pathol. 2008;173(5):1518–27.
6. Mullen JT, DeLaney TF, Rosenberg AE, et al. beta-Catenin mutation status and outcomes in sporadic desmoid tumors. Oncologist. 2013;18(9):1043–9.
7. van Broekhoven DL, Verhoef C, Grunhagen DJ, et al. Prognostic value of CTNNB1 gene mutation in primary sporadic aggressive fibromatosis. Ann Surg Oncol. 2015;22(5):1464–70.
8. Colombo C, Miceli R, Lazar AJ, et al. CTNNB1 45F mutation is a molecular prognosticator of increased postoperative primary desmoid tumor recurrence: an independent, multicenter validation study. Cancer. 2013;119(20):3696–702.
9. Garvey PB, Booth JH, Baumann DP, et al. Complex reconstruction of desmoid tumor resections does not increase desmoid tumor recurrence. J Am Coll Surg. 2013;217(3):472–80.
10. Domont J, Salas S, Lacroix L, et al. High frequency of beta-catenin heterozygous mutations in extra-abdominal fibromatosis: a potential molecular tool for disease management. Br J Cancer. 2010;102(6):1032–6.
11. Slowik V, Attard T, Dai H, Shah R, Septer S. Desmoid tumors complicating familial adenomatous polyposis: a meta-analysis mutation spectrum of affected individuals. BMC Gastroenterol. 2015;15:84.
12. Vitellaro M, Sala P, Signoroni S, et al. Risk of desmoid tumours after open and laparoscopic colectomy in patients with familial adenomatous polyposis. Br J Surg. 2014;101(5):558–65.
13. Hamada S, Futamura N, Ikuta K, et al. CTNNB1 S45F mutation predicts poor efficacy of meloxicam treatment for desmoid tumors: a pilot study. PLoS One. 2014;9(5):e96391.
14. Kasper B, Gruenwald V, Reichardt P, Bauer S, Hohenberger P, Haller F. Correlation of CTNNB1 mutation status with progression arrest rate in RECIST Progressive desmoid-type fibromatosis treated with imatinib: translational research results from a phase 2 study of the German Interdisciplinary Sarcoma Group (GISG-01). Ann Surg Oncol. 2016;23(6):1924–7.
15. Gronchi A, Casali PG, Mariani L, et al. Quality of surgery and outcome in extra-abdominal aggressive fibromatosis: a series of patients surgically treated at a single institution. J Clin Oncol. 2003;21(7):1390–7.
16. Lev D, Kotilingam D, Wei C, et al. Optimizing treatment of desmoid tumors. J Clin Oncol. 2007;25(13):1785–91.
17. Crago AM, Denton B, Salas S, et al. A prognostic nomogram for prediction of recurrence in desmoid fibromatosis. Ann Surg. 2013;258(2):347–53.
18. Bonvalot S, Eldweny H, Haddad V, et al. Extra-abdominal primary fibromatosis: aggressive management could be avoided in a subgroup of patients. Eur J Surg Oncol. 2008;34(4):462–8.
19. Fiore M, Rimareix F, Mariani L, et al. Desmoid-type fibromatosis: a front-line conservative approach to select patients for surgical treatment. Ann Surg Oncol. 2009;16(9):2587–93.
20. Peng PD, Hyder O, Mavros MN, et al. Management and recurrence patterns of desmoids tumors: a multi-institutional analysis of 211 patients. Ann Surg Oncol. 2012;19(13):4036–42.
21. Stoeckle E, Coindre JM, Longy M, et al. A critical analysis of treatment strategies in desmoid tumours: a review of a series of 106 cases. Eur J Surg Oncol. 2009;35(2):129–34.
22. Bonvalot S, Desai A, Coppola S, et al. The treatment of desmoid tumors: a stepwise clinical approach. Ann Oncol. 2012;23(Suppl 10):x158–66.
23. Bertani E, Chiappa A, Testori A, et al. Desmoid tumors of the anterior abdominal wall: results from a monocentric surgical experience and review of the literature. Ann Surg Oncol. 2009;16(6):1642–9.
24. Catania G, Ruggeri L, Iuppa G, Di Stefano C, Cardi F, Iuppa A. Abdominal wall reconstruction with intraperitoneal prosthesis in desmoid tumors surgery. Updat Surg. 2012;64(1):43–8.
25. Sutton RJ, Thomas JM. Desmoid tumours of the anterior abdominal wall. Eur J Surg Oncol. 1999;25(4):398–400.
26. Bonvalot S, Ternes N, Fiore M, et al. Spontaneous regression of primary abdominal wall desmoid tumors: more common than previously thought. Ann Surg Oncol. 2013;20(13):4096–102.
27. Colombo C, Miceli R, Le Pechoux C, et al. Sporadic extra abdominal wall desmoid-type fibromatosis: surgical resection can be safely limited to a minority of patients. Eur J Cancer. 2015;51(2):186–92.
28. Penel N, Le Cesne A, Bonvalot S, et al. Surgical versus non-surgical approach in primary desmoid-type fibromatosis patients: a nationwide prospective cohort from the French Sarcoma Group. Eur J Cancer. 2017;83:125–31.
29. Gounder MM, Mahoney MR, Van Tine BA, et al. Sorafenib for advanced and refractory desmoid tumors. N Engl J Med. 2018;379(25):2417–28.
30. Briand S, Barbier O, Biau D, et al. Wait-and-see policy as a first-line management for extra-abdominal desmoid tumors. J Bone Joint Surg Am. 2014;96(8):631–8.
31. Colombo C LVS, Fiore M, Grignani G, Palesandro E, Boccone B, D'Ambrosio L, Bianco A, Collini P, Palassini E, Stacchiotti S, Paolo Dei Tos A, Casali P, Perrone F, Mariani L, Gronchi A. Can wait and see be the standard of care for initial approach to primary sporadic desmoid tumors? Preliminary data from an Italian sarcoma group

prospective study. Connective Tissue Oncology Society Annual Meeting; 2018; Rome.
32. Desmoid Tumor Working G. The management of desmoid tumours: A joint global consensus-based guideline approach for adult and paediatric patients. Eur J Cancer. 2020;127:96–107.
33. Siozopoulou V, Marcq E, Jacobs J, et al. Desmoid tumors display a strong immune infiltration at the tumor margins and no PD-L1-driven immune suppression. Cancer Immunol Immunother. 2019;68(10):1573–83.
34. Kasper B, Raut CP, Gronchi A. Desmoid tumors: to treat or not to treat, that is the question. Cancer. 2020;126(24):5213–21.
35. Heinrich MC, McArthur GA, Demetri GD, et al. Clinical and molecular studies of the effect of imatinib on advanced aggressive fibromatosis (desmoid tumor). J Clin Oncol. 2006;24(7):1195–203.
36. Toulmonde M, Pulido M, Ray-Coquard I, et al. Pazopanib or methotrexate-vinblastine combination chemotherapy in adult patients with progressive desmoid tumours (DESMOPAZ): a non-comparative, randomised, open-label, multicentre, phase 2 study. Lancet Oncol. 2019;20(9):1263–72.
37. Chugh R, Wathen JK, Patel SR, et al. Efficacy of imatinib in aggressive fibromatosis: Results of a phase II multicenter Sarcoma Alliance for Research through Collaboration (SARC) trial. Clin Cancer Res. 2010;16(19):4884–91.
38. Penel N, Le Cesne A, Bui BN, et al. Imatinib for progressive and recurrent aggressive fibromatosis (desmoid tumors): an FNCLCC/French Sarcoma Group phase II trial with a long-term follow-up. Ann Oncol. 2011;22(2):452–7.
39. Palassini E, Frezza AM, Mariani L, et al. Long-term efficacy of methotrexate plus vinblastine/vinorelbine in a large series of patients affected by desmoid-type fibromatosis. Cancer J. 2017;23(2):86–91.
40. Azzarelli A, Gronchi A, Bertulli R, et al. Low-dose chemotherapy with methotrexate and vinblastine for patients with advanced aggressive fibromatosis. Cancer. 2001;92(5):1259–64.
41. Gega M, Yanagi H, Yoshikawa R, et al. Successful chemotherapeutic modality of doxorubicin plus dacarbazine for the treatment of desmoid tumors in association with familial adenomatous polyposis. J Clin Oncol. 2006;24(1):102–5.
42. Fiore M, Colombo C, Radaelli S, et al. Hormonal manipulation with toremifene in sporadic desmoid-type fibromatosis. Eur J Cancer. 2015;51(18):2800–7.
43. Hansmann A, Adolph C, Vogel T, Unger A, Moeslein G. High-dose tamoxifen and sulindac as first-line treatment for desmoid tumors. Cancer. 2004;100(3):612–20.
44. Quast DR, Schneider R, Burdzik E, Hoppe S, Moslein G. Long-term outcome of sporadic and FAP-associated desmoid tumors treated with high-dose selective estrogen receptor modulators and sulindac: a single-center long-term observational study in 134 patients. Familial Cancer. 2016;15(1):31–40.
45. Skapek SX, Anderson JR, Hill DA, et al. Safety and efficacy of high-dose tamoxifen and sulindac for desmoid tumor in children: results of a Children's Oncology Group (COG) phase II study. Pediatr Blood Cancer. 2013;60(7):1108–12.
46. Kasper B, Gruenwald V, Reichardt P, et al. Imatinib induces sustained progression arrest in RECIST progressive desmoid tumours: Final results of a phase II study of the German Interdisciplinary Sarcoma Group (GISG). Eur J Cancer. 2017;76:60–7.
47. Gounder MM, Lefkowitz RA, Keohan ML, et al. Activity of Sorafenib against desmoid tumor/deep fibromatosis. Clin Cancer Res. 2011;17(12):4082–90.
48. Szucs Z, Messiou C, Wong HH, et al. Pazopanib, a promising option for the treatment of aggressive fibromatosis. Anti-Cancer Drugs. 2017;28(4):421–6.
49. Messersmith WA, Shapiro GI, Cleary JM, et al. A Phase I, dose-finding study in patients with advanced solid malignancies of the oral gamma-secretase inhibitor PF-03084014. Clin Cancer Res. 2015;21(1):60–7.
50. Gounder MM. Notch inhibition in desmoids: "Sure it works in practice, but does it work in theory?". Cancer. 2015;121(22):3933–7.
51. van Broekhoven DL, Deroose JP, Bonvalot S, et al. Isolated limb perfusion using tumour necrosis factor alpha and melphalan in patients with advanced aggressive fibromatosis. Br J Surg. 2014;101(13):1674–80.
52. Keus RB, Nout RA, Blay JY, et al. Results of a phase II pilot study of moderate dose radiotherapy for inoperable desmoid-type fibromatosis – an EORTC STBSG and ROG study (EORTC 62991-22998). Ann Oncol. 2013;24(10):2672–6.
53. Saltiel S, Bize PE, Goetti P, et al. Cryoablation of extra-abdominal desmoid tumors: a single-center experience with literature review. Diagnostics (Basel). 2020;10(8):556.
54. Kurtz JE, Buy X, Deschamps F, et al. CRYODESMO-O1: A prospective, open phase II study of cryoablation in desmoid tumour patients progressing after medical treatment. Eur J Cancer. 2020;143:78–87.
55. Griffin MO, Kulkarni NM, O'Connor SD, Sudakoff GS, Lea WB, Tutton SM. Magnetic resonance-guided focused ultrasound: a brief review with emphasis on the treatment of extra-abdominal desmoid tumors. Ultrasound Q. 2019;35(4):346–54.
56. Sundaram M, McGuire MH, Herbold DR, Beshany SE, Fletcher JW. High signal intensity soft tissue masses on T1 weighted pulsing sequences. Skelet Radiol. 1987;16(1):30–6.
57. Cassidy MR, Lefkowitz RA, Long N, et al. Association of MRI T2 signal intensity with desmoid tumor progression during active observation: a retrospective cohort study. Ann Surg. 2020;271(4):748–55.

60 Atypical Patterns of Metastases: How Do Sarcomas Metastasize?

Pia van der Laan, Fabio Tirotta, Valeriya Pankova, Samuel Ford, Paul Huang, and Winan J. van Houdt

Abstract

Soft tissue sarcoma is a rare and heterogeneous group of cancers and comprises around 1% of all malignancies. This group of tumors consists of more than 60 different subtypes, each varying in biology and clinical behavior. In general, between 25% and 50% of the patients develop local or distant recurrences. However, the propensity of sarcoma to metastasize depends on many different factors, such as histological subtype, anatomical site, size, and grade. Most of the STS subtypes have a tendency to spread hematogenously with a predilection for pulmonary seeding, while lymph node metastases are rarely seen. However, there are several subtypes with deviating metastatic patterns, and in this chapter, the different atypical metastatic patterns are being described.

Learning Objectives

- Most soft tissue sarcoma subtypes have a predilection for pulmonary metastases.
- Specific sarcoma subtypes show a spectrum of atypical metastatic patterns.
- Lymph node metastasis occurs only in a small subset of sarcomas, according to the CARES acronym.
- Hypoxia or specific oncogenes might play a role in driving sarcoma metastasis.

60.1 Introduction

Soft tissue sarcoma (STS) is a heterogeneous group of rare solid tumors that arise in mesenchymal tissues and accounts for 1% of all malignancies in adults [1]. STS consists of a broad spectrum of tumors in terms of location, histologic subtype, treatment strategy, age at diagnosis, and prognosis. Nowadays, over 60 subtypes have been described, with undifferentiated pleiomorphic sarcoma (28%), Gastrointestinal stromal tumor (18%), liposarcoma (15%), leiomyosarcoma (12%), and synovial sarcoma (10%) being some of the most frequently diagnosed subtypes [2, 3]. Examples of more rare subtypes are clear cell sarcoma, epithelioid sarcoma, and alveolar soft part sarcoma. Soft tissue sarcoma can occur at all ages and at any anatomic location; however, the majority of STS arise in the extremities, more frequently in the lower compared to the upper extremity [4]. Surgical resection, often with (neo-)adjuvant radiotherapy, is the mainstay of treatment with curative intent or primary STS, while for specific subtypes and for some high-risk sarcoma, also neo-adjuvant chemotherapy is given [5]. However, between 25% and 50% of patients develop distant metastases, and the dogma is that they usually arise as pulmonary metastases without any lymph node metastases. Although a dogma becomes a dogma for a reason, it has become more clear over the years that metastatic patterns vary widely between different sarcoma subtypes and that for some sarcoma subtypes there are atypical metastatic patterns. This is important to realize for surveillance purposes and also for proper staging before starting any treatment. In this chapter, the aim is to clarify these atypical patterns to pro-

P. van der Laan · W. J. van Houdt (✉)
Department of Surgical Oncology, Sarcoma and Melanoma Unit, The Netherlands Cancer Institute – Antoni van Leeuwenhoek, Amsterdam, The Netherlands
e-mail: w.v.houdt@nki.nl

F. Tirotta · S. Ford
Department of Sarcoma Surgery, Queen Elizabeth Hospital, Birmingham, UK

V. Pankova · P. Huang
Division of Molecular Pathology, The Institute of Cancer Research, London, UK

S. P. Leong et al. (eds.), *Cancer Metastasis Through the Lymphovascular System*,
https://doi.org/10.1007/978-3-030-93084-4_60

vide more insight in tumor biology of the different soft tissue sarcoma subtypes.

60.2 Metastases in Sarcoma: General Patterns, Risk Factors, and Prognosis

Despite often adequate locoregional treatment for primary tumors, STS frequently metastasizes, which is the main cause of death in many STS patients. Around 10% of patients have detectable metastases at the time of diagnosis of the primary tumor, whereas as many as 25–40% will develop metastases at a later time point reflecting the high metastatic potential of STS [4]. The majority of these metastases occurs within 2 years after treatment of the primary tumor [6]. In general, most sarcoma subtypes have a tendency to metastasize to the lungs as the only or at least the first site of distant disease. Therefore, in most treatment and follow-up guidelines and local protocols, pulmonary surveillance by chest CT or chest X-ray is recommend after treatment of the primary tumor.

However, in the last decades our knowledge of and insight in the biological behavior of the different sarcoma subtypes have been increased. Also, better prediction tools for the risk of metastases have been developed. Several patient and tumor characteristics have been identified as risk factors for distant metastases. The risk of metastases varies widely per subtype. The location of the primary tumor has also shown to be an important factor in the risk of distant recurrence, since patients with extremity sarcoma have a lower risk of recurrence compared to patients with primary tumors elsewhere. Tumors with a higher histological grade, a larger tumor size, and deeply seated tumors are also more likely to metastasize [7]. Recently, two prediction models have been developed and made available for everyone with an app, the Sarculator and the Persarc apps, which are both available in the App Store or Google Play Store. These models are based on a large number of patients and take into account many patient- and tumor-specific characteristics to better personalize calculations for the risk of metastases [8, 9].

Since the efficiency of systemic treatment options for metastatic STS is limited, survival remains poor. Median survival after detection of metastases is around 1 year [10]. However, there is a wide variety in metastatic presentation, and prognosis is usually better after a long disease-free interval between the primary tumor and development of metastases. Also, oligometastatic disease carries a better prognosis compared to multiple metastatic lesions. Treatment of metastatic disease varies widely between per patients and per subtype, but the mainstay of treatment is usually systemic treatment with older chemotherapy regimens, newer tyrosine kinase inhibitors, or other more modern drugs. The choice of first line and further lines of systemic therapy varies widely per subtype and per country/hospital, but the most common chemotherapeutic therapies for sarcoma are doxorubicin-based regimens. Also, for patients with oligometastatic disease with longer disease-free intervals, local treatment can be beneficial which traditionally used to be resection of mostly pulmonary metastases [10]. For pulmonary metastases, even repeated metastasectomy has been associated with prolonged overall survival [11]. More recently, also other local treatment options such as stereotactic radiotherapy, as well as ablative techniques, have been shown to be effective in this setting.

60.3 Subtypes with Alternative Metastasizing Patterns

Although many sarcoma subtypes have a preference for pulmonary metastases, there are several sarcoma types with deviating metastasizing patterns. To illustrate this, the most important atypical metastatic patterns of STS subtypes are being described below, without claiming to be complete since there are several ultra-rare sarcomas with atypical patterns as well. It is currently poorly understood why many subtypes have the predominance for lung metastases, whereas others show some completely different metastatic patterns.

60.3.1 Leiomyosarcoma

Soft tissue leiomyosarcoma (LMS) originates from cells showing smooth muscle differentiation and predominantly arise in the extremities, abdomen, or retroperitoneum, from which a major part originates from large blood vessels [4]. LMS is a heterogeneous disease, ranging from low-grade tumors located in the dermis that behave relatively benign to very aggressive lesions with a highly aggressive metastatic potential [12, 13]. Grade and size have shown to be the most important factors affecting distant recurrence and survival [14].

Metastatic patterns of LMS are quite distinct from other subtypes. Although the lung is among the most common sites of metastasis, metastases in the liver and in soft tissue like subcutis or muscles are frequently seen, with or without concomitant pulmonary metastases [4, 12]. A study of Wile et al. describe lung and liver metastases as most frequent in LMS each accounting for 30% of metastases [15]. Leiomyosarcoma is also the most common source of sarcoma metastases to the skin [16]. Typical examples are cutaneous or subcutaneous metastases, anywhere on the body but even on the scalp, sometimes clinically mimicking a sebaceous cyst. Treatment of systemic disease will often include chemotherapy regimens such as doxorubicin-DTIC or newer agents like pazopanib. Metastasectomy of pulmonary and hepatic metastases has shown to increase long-term survival in selected cases [17–19].

Follow-up imaging after treatment of the primary tumor should not only consist of chest imaging, but at least a CT chest/abdomen/pelvis is advisable given the aberrant metastatic patterns.

60.3.2 Myxoid Liposarcoma

Myxoid liposarcoma (MLS) accounts for 30–50% of liposarcomas, and most cases are characterized by a t(12,16)(q13;p11) translocation resulting in the FUS-CHOP product [20]. MLS is in general classified as a low-grade sarcoma with a relatively good clinical outcome. However, a subset of MLS tumors is considered as a high-grade version of MLS, containing "hypercellular" or "round cell" areas, showing to have an increased potential to metastasize [4, 21]. Spillane et al. demonstrated that the presence of a round cell component in MLS tumors was predictive of a 5-year metastasis rate of 58%, whereas those without a round cell component had a 5-year metastasis rate of 16% [22].

MLS tumors tend to metastasize in a non-pulmonary fashion. In a study of Shinoda et al., involving 48 metastatic MLS patients, 45 patients (93.8%) having extrapulmonary lesions were identified including lesions in the bone, abdomen, retroperitoneum, and other soft tissues. From this group of patients, the majority showed metastases without the lungs being involved (81.3%). Combining literature and their data, they suggest lung and liver metastases as potential prognostic factors of shorter survival [23]. Because of the distinct pattern of metastasis in MLS, patients should undergo imaging studies of the chest, abdomen, spine, and pelvis during staging or follow-up [20]. Sheah et al. propose MRI for the early detection of metastases, since MRI has shown to be the most sensitive modality to detect soft tissue and bone lesions [24]. A total body MRI could be considered [25]. Metastatic patients can be treated with many different multidisciplinary strategies, varying from radiotherapy for this radiosensitive tumor or resections, and for oligometastatic or oligoregressive lesions, to systemic therapy such as doxorubicin-ifosfamide or trabectedin.

60.3.3 Gastrointestinal Stroma Tumor

Gastrointestinal stromal tumor (GIST) represents 20% of all STS, making it the largest group of single-type STS. GIST tumors are usually caused by activating mutations in KIT and PDGFRA genes, but some other rare molecular aberrations have also been described (so-called wild-type GISTs) [26]. Primary GIST develops throughout the gastrointestinal tract, most commonly in the stomach or small intestine. Up to 20% of the patients are metastasized at initial diagnosis [27]. Metastases occur predominantly intra-abdominally (most frequently involving the liver), while lymph node metastases or metastases outside the abdomen such as in the lungs, bone, or brain are rare, except for some rare GIST subtypes such as SDH-deficient GISTs [27, 28]. Therefore, imaging of the abdomen and pelvis will be sufficient in most cases.

The tyrosine kinase inhibitor (TKI) imatinib is the standard first-line treatment for advanced GIST. The role of surgery in the management of metastatic GIST is controversial: resection of limited disease after good response to imatinib treatment can be considered, whereas surgery for GIST progressing at several sites is not recommended [26]. Resection of oligo-progressive metastases after progression on TKI treatment could be considered; however, the median progression-free survival is fairly short (8–9 months), and the 2-year progression-free survival is only 9.7% [29].

60.3.4 Alveolar Soft-Part Sarcoma

Alveolar soft-part sarcoma (ASPS) is a rare form of STS accounting for less than 1% of the STS. It often affects young adults and in a majority of patients originates from the lower extremities. Although it is known for its relative indolent behavior, ASPS has the greatest metastatic potential compared to other types of STS [30]. Metastases sometimes occur decades after resection of the primary tumor often without evidence of local recurrence [31].

The most common metastatic sites are the lung, bone, and brain. ASPS metastasizes to the brain more commonly compared to other types of STS [32]. Lieberman et al. published follow-up data of 91 ASPS patients and showed a median overall survival of 11 years in patients without metastasis at diagnosis and 3 years in patients with metastatic disease [33]. ASPS is one of the few sarcoma subtypes that has shown some promising responses to anti-PD1 immunotherapy [34].

60.3.5 Low-Grade Fibromyxoid Sarcoma

Low-grade fibromyxoid sarcomas (LGFMS) are tumors histologically consisting of alternating collagenous and myxoid areas. These tumors are genetically characterized by either a t(7,16) or t(11,16) translocation resulting in a FUS-CREB3L2 or FUS-CREB3L1 fusion gene [35]. It commonly arises in the proximal extremities and trunk and typically occurs in young adults with a slight male predilection [4, 36].

Although LGFMS appears histologically indolent and shows low metastatic rates in the first years after resection, the metastatic pattern of this subtype is atypical since this tumor is known for its extreme late metastases with intervals up to

45 years [37]. Prolonged follow-up is therefore recommended in these often relatively young patients. Overall metastatic rate varies between 6% and 26%, and the main sites are the lung, pleura, chest wall, and other soft tissues [35, 36]. Surgical excision is usually performed for locally recurrent lesions and also for late oligometastatic disease [37]. For advanced LGFMS, efficacy of chemotherapy is limited [38].

60.4 Lymph Node Metastases in Sarcoma

Lymph node metastases (LNM) from STS are uncommon, with an overall estimated incidence of approximately 5% [39–46]. Their pathologic evaluation is inconsistent making their real incidence difficult to determine [44], although it is accepted that specific histologies, such as angiosarcoma, clear cell sarcoma, epithelioid sarcoma, and small cell sarcoma, have a higher tendency to give LNM [41–44]. Hence, studies reporting higher incidence rates of LNM are those which considered only high-risk subtype for LNM [42, 43, 46]. The low incidence of LNM in sarcoma is usually explained by the difference in cell of origin, where most malignancies have an epithelial origin, while sarcoma is a mesenchymal tumor which is probably less likely to spread to lymphatic vessel lying underneath epithelial layers.

LNM from STS can affect any anatomical site. Although previously reported series at a single institution described an incidence of LNM in the trunk/extremities up to 70% of cases at the time of presentation [39, 41], two large retrospective studies – one using the Surveillance, Epidemiology, and End Results database and the other using the National Cancer Data Base – found opposite results, reporting an incidence of approximately 2% [44, 47]. Indeed, in the absence of synchronous distant metastases, LNM in extremities were described in just 1% of cases [47]. Moreover, the same authors identified rates of LNM for head and neck, intrathoracic, and intra-abdominal/retroperitoneal of 4.5%, 3.1%, and 4.5%, respectively. This wide difference in distribution across sites may be related to an uneven distribution of the histologies and to a different number of lymph nodes removed during surgery depending on the location, as nondeliberate removal of lymph nodes may occur more often in other sites than in the extremities. In addition, a different staging/treatment related to the anatomical location may also be responsible for such a wide difference.

The clinical impact of LNM in STS is unclear. Several studies have reported worse survival outcomes in patients with LNM, suggesting that it can represent a preliminary manifestation of systemic disease [48–50]. According to the eighth edition of the AJCC staging system, LNM in the extremities/trunk are considered prognostically comparable to distant metastases [51]. This newest edition has been largely criticized for this reason, and different reports have highlighted that a better prognosis is achieved in patients with isolated regional LNM [45, 51–53] and questioned whether a more reliable classification would be achieved by separating LNM from the other sites. To support this data, other studies [41, 44, 47] demonstrated a limited impact on prognosis in patients with isolated LNM when treated properly with resection, compared to those with pulmonary or other systemic metastases. Keung et al. showed that patients with LNM have a poorer prognosis in the absence of distant metastases (N1M0) compared to patients with primary disease only (N0M0) and that LNM affect histologies differently. Interestingly, patients who had favorable histologies and LNM were those showing worse outcomes [44].

An important and unanswered question is how and when to screen lymph nodes in STS. Clinical examination, CT or MRI scan, and sentinel lymph node biopsy (SNLB) can be used. The former seems to be particularly effective for clear cell sarcoma but less so for other histologies [54]. Ecker et al. demonstrated a possible higher risk of death when lymph node evaluations were omitted in patients with epithelioid and clear cell sarcoma [43]. Although there is no clear evidence that routine evaluation of lymph nodes improves survival outcomes in all subtypes or if this should be limited to high-risk histologies only, it does seem logical to at least evaluate for LNM in histologies known to be at higher risk of developing LNM.

Where LNM are present without further metastatic disease (N1M0), the question of whether to perform a locoregional lymphadenectomy still needs to be addressed. In a recent report, Basile et al. demonstrated improved survival outcomes for patients who underwent excision of isolated LNM in extremity STS [45]. This study, along with others [39, 41], recommended treating isolated LNM with en bloc surgical resection. In contrast, Sawamura reported possible benefits for lymphadenectomy at 1.5 years but no difference in survival at 5 years [48]. The role of lymphadenectomy therefore warrants further large, multicenter, and prospective evaluation, but in the meantime locoregional lymph node dissections for LNM in the absence of further distant metastases are considered standard of care and are also incorporated in the NCCN guidelines [55].

60.4.1 Angiosarcoma

Angiosarcoma represents 1% of all soft tissue sarcomas. In half of all cases, it arises from the scalp, originating from deep soft tissues in only 10% of cases [56, 57]. The overall incidence of LNM for angiosarcoma ranges from 6% to 13.5% [39, 44] although a recent study found that regional LNM from the scalp, which is the most frequent site, can reach 52.5% [58]. Distant metastases were less frequent in the absence of lymph nodal spread, and LNM seemed to be able to predict systemic spread within 3 and 6 months. Nevertheless, overall survival did not differ in the presence of LNM. Other anatomical sites such as the breast can be involved, where the number of cases of radiation-induced angiosarcoma is rising. Noticeably, lymphadenectomy is not recommended as lymph nodes are rarely involved [59].

The subtypes more likely to develop lymph node metastases are more epitheiloid like compared to other sarcoma subtypes or are developing in the skin (cutaneous angiosarcoma). This might partly explain why these subtypes more resemble the lymphogenic metastatic patterns of malignancies with an epithelial origin.

60.4.2 Epithelioid Sarcoma

Epithelioid sarcoma represents less than 1% of all STS and have a high tendency to metastasize to lymph nodes [60] with an incidence of LNM of approximately 10% [43, 44]. In contrast to angiosarcoma, lymph nodal evaluation seems to affect survival outcomes, which may be related to a direct benefit from the lymphadenectomy or to better disease staging [43]. The role of lymphadenectomy requires further evaluation. In fact, some studies [43, 47] do not show clear evidence that extended lymphadenectomy improves OS in patients with extremity STS. However, other studies [39] demonstrated opposing results, showing that long-term survival was achieved in those patients who underwent radical resection of involved LNM. Nevertheless, lymph node evaluation should routinely be performed in these patients [43].

60.4.3 Clear Cell Sarcoma

Clear cell sarcoma is a rare type of sarcoma, previously described as "melanoma of soft parts" for its histological similarity and the propensity for lymph nodal spread, which tends to occur in 10–15% of cases [44, 45]. Genetically, clear cell sarcoma is marked by the translocation t(12,22) (q13;q12), found in 70–90% of the cases [4]. In common with other high-risk histologies, the prognostic impact of LNM and the potential benefits of nodal staging and treatment appear to be restricted to the subset of patients with isolated N1 disease in the absence of M0 disease [44, 45]. In a study involving only extremities STS, Bianchi et al. reported a 10-year specific survival rate of 41% for all patients, 53% for localized, and 0% for patients with metastasis, regardless of lymph node involvement [61]. This was consistent with what was reported in larger series [44, 45].

60.4.4 Myxofibrosarcoma

Myxofibrosarcoma is one of the most common STS, especially in the elderly [62]. One of its peculiarities is the high risk of local recurrence despite the fact that it has a better prognosis compared to other histologies, even when a high-grade component is present [62]. LNM from myxofibrosarcoma are usually rare, <10% [44, 45]. Sanfilippo et al. reported that three out of four patients with regional node disease developed distant metastases and one out of five patients had positive LNM before their metastatic spread, suggesting a potential role in predicting metastatic spread in this specific subtype [62].

60.4.5 Rhabdomyosarcoma

LNM in pediatric cases and young adults is particularly relevant in patients with rhabdomyosarcoma, which constitute half of all STS during this age [63]. Different subtypes of rhabdomyosarcoma have been described, including embryonal, alveolar, and pleomorphic rhabdomyosarcomas. Particularly alveolar rhabdomyosarcoma has a distinct genetic background including two patterns of translocations: t(2,13) (q35;q14) creating the PAX3-FOXO1 fusion gene and t(1,13) (p36;q14) with the PAX7-FOXO1 fusion gene [64]. The presence of LNM in rhabdomyosarcoma is approximately 20% and associated with worse prognosis [65, 66]. In this specific case, sentinel node biopsy is safe and effective [63] and can be used in order to potentially offer extended lymphadenectomy to minimize the risk of recurrence and/or metastatic disease. However, the role of SNB for soft tissue sarcoma in general, but also for the subtypes more likely to develop LNM, is not established or proven to be beneficial.

60.4.6 Others

Historically, the abovementioned subtypes were combined with synovial sarcoma and incorporated in the acronym SCARE (synovial sarcoma, clear cell sarcoma, angiosarcoma, rhabdomyosarcoma, and epithelioid sarcoma), also called CARES. These subtypes were always considered to be at high risk for LNM. However, more recent and larger studies demonstrated that the real incidence of LNM in synovial sarcoma is low, representing approximately <5% [43, 45, 67], and this calls into question the accuracy of previously reported series that classified this histology as high risk for LNM.

60.5 Sarcoma Metastases: Biological Mechanisms

60.5.1 Overview of the Metastatic Cascade and Organotropism

The development of metastasis is a multistep process referred to as the metastatic cascade, which enables dissociation of cancer cells from the primary site and secondary outgrowth at distant organs. The metastatic cascade follows sequential steps of local invasion of the primary tissue, intravasation of cancer cells into vessels, survival of circulating cancer cells, extravasation into the secondary sites, and colonization and growth at distant sites [68]. All of these steps must be successfully completed for a detectable lesion to develop; however, in reality metastasis is a highly inefficient process, with <0.01% of tumor cells in the systemic circulation giving rise to macro metastatic lesions [69]. Tumors consist of clones with variable metastatic potential, and it has been argued that there are pre-existing clones present in the primary sarcoma tumor that possess traits necessary to complete all steps in the metastatic cascade resulting in distant metastases which are clonal [70].

Most cancers, including sarcomas, preferentially spread to certain organs in a process known as metastatic organotropism [71]. The reasons for these recurrent patterns of metastatic spread have long been a matter of debate. In 1889 Steven Paget proposed the "seed and soil" theory of metastatic dissemination suggesting that the spread of metastatic cells relies on the compatibility of cancer cells ("seeds") and distant organ or tissue ("soil") [72]. Paget's view was challenged by James Ewing who proposed that metastatic spread is governed purely by mechanical factors such as anatomical structure of vasculature and patterns of blood flow. This "mechanical" spread allows initial delivery and arrest of cancer cells at organs first encountered based on blood-flow patterns from the primary tumor. Growing evidence suggests that these two concepts are not mutually exclusive, and both may play complementary roles in successful metastatic spread and development [73].

60.5.2 Molecular Drivers of Soft Tissue Sarcoma Metastasis

While the biological mechanisms of metastasis have been extensively studied in bone sarcomas, less is known about the molecular drivers of STS metastasis. To date, studies have been largely undertaken in undifferentiated pleomorphic sarcoma (UPS), myxoid liposarcoma (MLS), and synovial sarcoma (SS). Here we summarize our current understanding of the extrinsic and intrinsic molecular factors that play a role in driving metastasis in these subtypes.

60.5.3 Undifferentiated Pleomorphic Sarcoma

Hypoxia occurs when a tumor outgrows its blood supply resulting in insufficient tumor oxygenation that has been linked to distant metastasis development in STS [74, 75]. Moreover, a gene expression signature consisting of upregulated hypoxia-related genes (e.g., HIF1A, PLOD2) predicted risk of metastasis in undifferentiated pleomorphic sarcoma (UPS) [76]. Under hypoxic conditions, the transcription factor hypoxia-inducible factor 1-alpha (HIF1A) induces expression of multiple genes to increase angiogenesis and enable cellular metabolism in oxygen-deprived conditions [77]. In carcinomas, HIF1A is known to promote migration and cancer cell invasion. Procollagen-lysine,2-oxoglutarate 5-dioxygenase 2 (PLOD2) is a downstream target of HIF1A and is an enzyme that catalyzes the hydroxylation of lysine residues, which is a type of posttranslational modification essential for maturation of collagen [78]. Eisinger-Mathason et al. found that HIF1A and PLOD2 were highly expressed in primary tumors from UPS patients who went on to develop metastasis versus those who had low expression levels [79]. Knockdown of either HIF1A or PLOD2 expression reduced sarcoma cell migration *in vitro* and lung metastasis development in a murine model of UPS. Moreover, PLOD2 expression in these models was associated with dense and disorganized collagen network, which could potentially form migration tracks for metastasizing sarcoma cells. It has been further shown by Lewis et al. that in UPS, hypoxia modifies the mechanical properties of collagen through increased PLOD2 expression, and there is a feedback loop between mechanical properties and PLOD2 expression generated through TGFβ-SMAD2 signaling that enhances sarcoma cell migration and lung metastasis [80]. These data demonstrate that hypoxia and the mechanical properties of collagen may be important in promoting UPS metastatic spread, but more investigation is required in order to establish if hypoxia plays a role in determining preferential spread to specific organs in UPS.

60.5.4 Myxoid Liposarcoma

In contrast to the majority of STS subtypes that metastasize to the lung, myxoid liposarcomas (MLS) have atypical patterns of metastasis, with preference for the retroperitoneum, limbs, and bones [81–83]. A genetic hallmark of MLS is a translocation which fuses FUS (FUsed in Sarcoma) to CHOP (C/EBP HOmologous Protein), and the FUS-CHOP oncogene has been implicated in driving metastasis in this disease [84]. Overexpression of FUS-CHOP increased the migration and invasion of liposarcoma and fibrosarcoma cells *in vitro* as well as promoted the formation of liver and lung metastasis in the chicken embryo metastasis (CAM) assay *in vivo* [85]. In another study by Tornin et al., the authors found that in MLS cells, FUS-CHOP activated the proto-oncogene tyrosine-protein kinase Src (SRC) and increased the levels of focal adhesion kinase (FAK), which enhanced MLS cell invasion and extravasation [86]. In MLS cells, inhibition of SRC or FAK reduced 3D collagen spheroid invasion and extravasation in the CAM assay. These studies have shed light on the role of the fusion protein and its downstream effectors in driving metastasis; however, it remains to be determined if the FUS-CHOP-induced pathways govern the atypical metastatic patterns commonly seen in MLS.

60.5.5 Synovial Sarcoma

A few studies suggest the involvement of epithelial-to-mesenchymal transition (EMT) in the synovial sarcoma (SS) metastatic process [87–89]. EMT has been extensively described in carcinomas which increases the propensity to disseminate once cancer cells acquire the mesenchymal phenotype [90]. As sarcomas are mesenchymal tumors, there remains a debate as to whether EMT exists in these tumors. SS exhibit some epithelial differentiation, particularly biphasic SS which have glandular structures, and expression of epithelial markers such as E-cadherin and cytokeratin is seen in SS tumors [91, 92]. In particular, high E-cadherin expression associates with better outcome in SS [93, 94]. Peng et al. showed that in SS cells, the SHC SH2-domain binding protein 1 (SHCBP1) protein enhances adhesion, migration, and invasion through promotion of EMT (decreased levels of E-cadherin and increased levels of vimentin) [87]. In another study by Xu et al., E-cadherin was shown to be targeted and negatively regulated by the microRNA mir-9 in SS [88]. Overexpression of mir-9 enhanced lung metastasis *in vivo* by downregulating E-cadherin. EMT is a complex process with several layers of regulation, and further investigation of the role of this process in driving metastasis in SS and other STS exhibiting some degree of epithelial differentiation is necessary.

60.6 Conclusion

Unfortunately, metastasis is a common phenomenon in STS patients given the high metastatic rate. Although most metastatic STS patients will have pulmonary metastases, several STS subtypes show atypical patterns of metastases leading to different strategies for staging, surveillance, and treatment. In this chapter, we have highlighted different metastatic patterns in terms of anatomical location and timing of metastases in the course of the disease. We also showed the role of lymph node staging and treatment of LNM in a subset of STS.

Despite the relatively poor survival of STS patients compared to many other cancers, metastatic mechanisms are an underexplored area in STS biology. Therefore, metastatic patterns of sarcoma are still poorly understood. Determining the biological mechanisms underpinning STS dissemination and metastatic organotropism might pave the way for more effective treatment strategies for STS patients to either prevent metastases or more effectively treat metastatic disease.

Open Questions

Future directions in research in this field should focus on the many open questions that remain, such as:

- Why is there a preference for pulmonary metastases in most STS subtypes, independent of anatomical location of the primary tumor?
- What explains the variety in metastatic patterns between STS subtypes?
- Why does a minority of specific STS subtypes develop lymphogenic metastases, in contrast to most other subtypes?
- Can we identify biological mechanisms driving STS metastasis?
- How do we personalize treatment and surveillance strategies based on metastatic pattern differences?

Patient Information and Guidelines

ESMO Clinical Practical Guidelines - Soft Tissue and Visceral Sarcomas

https://www.esmo.org/guidelines/sarcoma-and-gist/soft-tissue-and-visceral-sarcomas

References

1. Siegel RL, Miller KD, Jemal A. Cancer statistics, 2019. CA Cancer J Clin. 2019;69(1):7–34.
2. Coindre J-M, et al. Predictive value of grade for metastasis development in the main histologic types of adult soft tissue sarcomas: a study of 1240 patients from the French Federation of Cancer Centers Sarcoma Group. Cancer. 2001;91(10):1914–26.
3. Ducimetière F, et al. Incidence of sarcoma histotypes and molecular subtypes in a prospective epidemiological study with central pathology review and molecular testing. PLoS One. 2011;6(8):e20294. https://doi.org/10.1371/journal.pone.0020294. PMID: 21826194; PMCID: PMC3149593.
4. WHO Classification of Tumours Editorial Board. WHO classification of tumours of soft tissue and bone. 5th ed. Lyon, France: IARC Press; 2020.
5. Roland CL, van Houdt W, Gronchi A. The landmark series: multimodality treatment of extremity sarcoma. Ann Surg Oncol. 2020;27(10):3672–82.
6. Gadd MA, et al. Development and treatment of pulmonary metastases in adult patients with extremity soft tissue sarcoma. Ann Surg. 1993;218(6):705–12.
7. Pisters PWT, et al. Analysis of prognostic factors in 1,041 patients with localized soft tissue sarcomas of the extremities. J Clin Oncol. 1996;14(5):1679–89.
8. van Praag VM, et al. A prediction model for treatment decisions in high-grade extremity soft-tissue sarcomas: Personalised sarcoma care (PERSARC). Eur J Cancer. 2017;83:313–23.

9. Callegaro D, et al. Development and external validation of a dynamic prognostic nomogram for primary extremity soft tissue sarcoma survivors. EClinicalMedicine. 2019;17:100215.
10. Billingsley KG, et al. Multifactorial analysis of the survival of patients with distant metastasis arising from primary extremity sarcoma. Cancer. 1999;85(2):389–95.
11. Chudgar NP, et al. Is repeat pulmonary metastasectomy indicated for soft tissue sarcoma? Ann Thorac Surg. 2017;104(6):1837–45.
12. Gladdy RA, et al. Predictors of survival and recurrence in primary leiomyosarcoma. Ann Surg Oncol. 2013;20(6):1851–7.
13. Pijpe J, et al. The relation between histological, tumor-biological and clinical parameters in deep and superficial leiomyosarcoma and leiomyoma. Sarcoma. 2002;6(3):105–10. https://doi.org/10.1080/1357714021000065404. PMID: 18521340; PMCID: PMC2395483.
14. Miyajima K, et al. Clinicopathological prognostic factors in soft tissue leiomyosarcoma: a multivariate analysis. Histopathology. 2002;40:353–9.
15. Wile AG, Evans HL, Romsdahl MM. Leiomyosarcoma of soft tissue: a clinicopathologic study. Cancer. 1981;48(4):1022–32.
16. Wang WL, et al. Sarcoma metastases to the skin: a clinicopathologic study of 65 patients. Cancer. 2012;118(11):2900–4.
17. Lang H, et al. Hepatic metastases from leiomyosarcoma: a single-center experience with 34 liver resections during a 15-year period. Ann Surg. 2000;231(4):500–5.
18. DeMatteo RP, et al. Results of hepatic resection for sarcoma metastatic to liver. Ann Surg. 2001;234(4):540–7.
19. Burt BM, et al. Repeated and aggressive pulmonary resections for leiomyosarcoma metastases extends survival. Ann Thorac Surg. 2011;92(4):1202–7.
20. Durr HR, et al. Myxoid liposarcoma: local relapse and metastatic pattern in 43 patients. BMC Cancer. 2018;18(1):304.
21. Smith TA, Easley KA, Goldblum JR. Myxoid/round cell liposarcoma of the extremities. A clinicopathologic study of 29 cases with particular attention to extent of round cell liposarcoma. Am J Surg Pathol. 1996;20(2):171–80.
22. Spillane AJ, Fisher C, Meirion Thomas J. Myxoid liposarcoma—the frequency and the natural history of nonpulmonary soft tissue metastases. Ann Surg Oncol. 1999;6(4):389–94.
23. Shinoda Y, et al. Prognostic factors of metastatic myxoid liposarcoma. BMC Cancer. 2020;20(1):883.
24. Sheah K, et al. Metastatic myxoid liposarcomas: imaging and histopathologic findings. Skelet Radiol. 2008;37(3):251–8.
25. Visgauss JD, et al. Staging and surveillance of myxoid liposarcoma: follow-up assessment and the metastatic pattern of 169 patients suggests inadequacy of current practice standards. Ann Surg Oncol. 2021;28(12):7903–11. https://doi.org/10.1245/s10434-021-10091-1. PMID: 33961173.
26. Joensuu H, Hohenberger P, Corless CL. Gastrointestinal stromal tumour. Lancet. 2013;382(9896):973–83.
27. von Mehren M, Joensuu H. Gastrointestinal stromal tumors. J Clin Oncol. 2018;36(2):136–43.
28. DeMatteo RP, et al. Two hundred gastrointestinal stromal tumors: recurrence patterns and prognostic factors for survival. Ann Surg. 2000;231(1):51–8.
29. Mussi C, et al. Post-imatinib surgery in advanced/metastatic GIST: is it worthwhile in all patients? Ann Oncol. 2010;21(2):403–8.
30. Paoluzzi L, Maki RG. Diagnosis, prognosis, and treatment of alveolar soft-part sarcoma: a review. JAMA Oncol. 2019;5(2):254–60.
31. Pennacchioli E, et al. Alveolar soft part sarcoma: clinical presentation, treatment, and outcome in a series of 33 patients at a single institution. Ann Surg Oncol. 2010;17(12):3229–33.
32. Portera CA, et al. Alveolar soft part sarcoma: clinical course and patterns of metastasis in 70 patients treated at a single institution. Cancer. 2001;91(3):585–91.
33. Lieberman PH, et al. Alveolar soft-part sarcoma. A clinico-pathologic study of half a century. Cancer. 1989;63(1):1–13.
34. Italiano A, Bellera C, D'Angelo S. PD1/PD-L1 targeting in advanced soft-tissue sarcomas: a pooled analysis of phase II trials. J Hematol Oncol. 2020;13(1):55. https://doi.org/10.1186/s13045-020-00891-5. PMID: 32430039; PMCID: PMC7236113.
35. Guillou L, et al. Translocation-positive low-grade fibromyxoid sarcoma: clinicopathologic and molecular analysis of a series expanding the morphologic spectrum and suggesting potential relationship to sclerosing epithelioid fibrosarcoma: a study from the French Sarcoma Group. Am J Surg Pathol. 2007;31(9):1387–402.
36. Folpe AL, et al. Low-grade fibromyxoid sarcoma and hyalinizing spindle cell tumor with giant rosettes a clinicopathologic study of 73 cases supporting their identity and assessing the impact of high-grade area. Am J Surg Pathol. 2000;24(10):1353–60.
37. Evans HL. Low-grade fibromyxoid sarcoma: a clinicopathologic study of 33 cases with long-term follow-up. Am J Surg Pathol. 2011;35(10):1450–62.
38. Chamberlain F, et al. Low-grade fibromyxoid sarcoma: treatment outcomes and efficacy of chemotherapy. In Vivo. 2020;34(1):239–45.
39. Fong Y, et al. Lymph node metastasis from soft tissue sarcoma in adults. Analysis of data from a prospective database of 1772 sarcoma patients. Ann Surg. 1993;217(1):72–7.
40. Daigeler A, et al. Lymph node metastases in soft tissue sarcomas - a single center analysis of 1,597 patients. Langenbeck's Arch Surg. 2009;394(2):321–9.
41. Behranwala KA, et al. Prognosis of lymph node metastasis in soft tissue sarcoma. Ann Surg Oncol. 2004;11(7):714–9.
42. Mazeron JJ, Suit HD. Lymph nodes as sites of metastases from sarcomas of soft tissue. Cancer. 1987;60(8):1800–8.
43. Ecker BL, et al. Implications of lymph node evaluation in the management of resectable soft tissue sarcoma. Ann Surg Oncol. 2017;24(2):425–33.
44. Keung EZ, et al. Defining the incidence and clinical significance of lymph node metastasis in soft tissue sarcoma. Eur J Surg Oncol. 2018;44(1):170–7.
45. Basile G, et al. Curability of patients with lymph node metastases from extremity soft-tissue sarcoma. Cancer. 2020;126(23):5098–108.
46. Sherman KL, et al. Examination of national lymph node evaluation practices for adult extremity soft tissue sarcoma. J Surg Oncol. 2014;110(6):682–8.
47. Johannesmeyer D, et al. The impact of lymph node disease in extremity soft-tissue sarcomas: a population-based analysis. Am J Surg. 2013;206(3):289–95.
48. Sawamura C, et al. Lymphadenectomy and histologic subtype affect overall survival of soft tissue sarcoma patients with nodal metastases. Clin Orthop Relat Res. 2013;471(3):926–31.
49. Blazer DG, Sabel MS, Sondak VK. Is there a role for sentinel lymph node biopsy in the management of sarcoma? Surg Oncol. 2003;12(3):201–6.
50. Skinner KA, Eilber FR. Soft tissue sarcoma nodal metastases: biologic significance and therapeutic considerations. Surg Oncol Clin N Am. 1996;5(1):121–7.
51. Cates JMM. The AJCC 8th edition staging system for soft tissue sarcoma of the extremities or trunk: a cohort study of the SEER database. J Natl Compr Canc Netw. 2018;16(2):144–52.
52. Fisher SB, et al. Comparative performance of the 7th and 8th editions of the american joint committee on cancer staging systems for soft tissue sarcoma of the trunk and extremities. Ann Surg Oncol. 2018;25(5):1126–32.
53. Ashamalla M, et al. Challenging AJCC 8 staging for soft tissue sarcoma using the NCDB. Int J Radiat Oncol Biol Phys. 2019;105(2):338–45.

54. Andreou D, et al. Sentinel node biopsy in soft tissue sarcoma subtypes with a high pro- pensity for regional lymphatic spread-results of a large prospective trial. Ann Oncol. 2013;24(5):1400–5.
55. Network, N.C.C. Soft Tissue Sarcoma. 2016; Available from: https://www.nccn.org/professionals/physician_gls/pdf/sarcoma.pdf.
56. Gaballah AH, et al. Angiosarcoma: clinical and imaging features from head to toe. Br J Radiol. 2017;90(1075):20170039.
57. Young RJ, et al. Angiosarcoma. Lancet Oncol. 2010;11(10):983–91.
58. Kang Y, et al. Regional lymph node metastasis of scalp angiosarcoma: a detailed clinical observation study of 40 cases. Ann Surg Oncol. 2020;27(8):3018–27.
59. Torres KE, et al. Long-term outcomes in patients with radiation-associated angiosarcomas of the breast following surgery and radiotherapy for breast cancer. Ann Surg Oncol. 2013;20(4):1267–74.
60. Touati N, et al. European organisation for research and treatment of cancer soft tissue and bone sarcoma group experience with advanced/metastatic epithelioid sarcoma patients treated in prospective trials: clinical profile and response to systemic therapy. Clin Oncol. 2018;30(7):448–54.
61. Bianchi G, et al. Clear cell sarcoma of soft tissue: a retrospective review and analysis of 31 cases treated at Istituto Ortopedico Rizzoli. Eur J Surg Oncol. 2014;40(5):505–10. https://doi.org/10.1016/j.ejso.2014.01.016.
62. Sanfilippo R, et al. Myxofibrosarcoma: prognostic factors and survival in a series of patients treated at a single institution. Ann Surg Oncol. 2011;18(3):720–5.
63. Wagner LM, et al. Detection of lymph node metastases in pediatric and adolescent/young adult sarcoma: Sentinel lymph node biopsy versus fludeoxyglucose positron emission tomography imaging-A prospective trial. Cancer. 2017;123(1):155–60.
64. Nakano K, Takahashi S. Translocation-related sarcomas. Int J Mol Sci. 2018;19(12):3784.
65. Rodeberg DA, et al. Prognostic significance and tumor biology of regional lymph node disease in patients with rhabdomyosarcoma: a report from the Children's Oncology Group. J Clin Oncol. 2011;29(10):1304–11.
66. Meza JL, et al. Analysis of prognostic factors in patients with nonmetastatic rhabdomyosarcoma treated on Intergroup Rhabdomyosarcoma Studies III and IV: the Children's Oncology Group. J Clin Oncol. 2006;24(24):3844–51.
67. Ashamalla M, et al. Clinical presentation and patterns of care in SCARE Soft Tissue Sarcoma. Int J Radiat Oncol Biol Phys. 2017;99(2S):E749.
68. Lambert AW, Pattabiraman DR, Weinberg RA. Emerging biological principles of metastasis. Cell. 2017;168(4):670–91.
69. Fidler IJ. Metastasis: quantitative analysis of distribution and fate of tumor emboli labeled with 125 I-5-iodo-2′-deoxyuridine. J Natl Cancer Inst. 1970;45(4):773–82.
70. Tang YJ, et al. Tracing tumor evolution in sarcoma reveals clonal origin of advanced metastasis. Cell Rep. 2019;28(1):2837–50.
71. Gao Y, et al. Metastasis organotropism: redefining the congenial soil. Dev Cell. 2019;49(3):375–91.
72. Paget S. The distribution of secondary growths in cancer of the breast. Lancet. 1889;133(3421):571–3.
73. Weiss L. Comments on hematogenous metastatic patterns in humans as revealed by autopsy. Clin Exp Metastasis. 1992;10(3):191–9.
74. Brizel DM, et al. Tumor oxygenation predicts for the likelihood of distant metastases in human soft tissue sarcoma. Cancer Res. 1996;56(5):941–3.
75. Lewis DM, et al. Intratumoral oxygen gradients mediate sarcoma cell invasion. Proc Natl Acad Sci U S A. 2016;133(33):9292–7.
76. Francis P, et al. Diagnostic and prognostic gene expression signatures in 177 soft tissue sarcomas: hypoxia-induced transcription profile signifies metastatic potential. BMC Genomics. 2007;8(1):73.
77. Masoud GN, Li W. HIF-1α pathway: role, regulation and intervention for cancer therapy. Acta Pharm Sin B. 2015;5(5):378–89.
78. Du H, et al. PLOD2 in cancer research. Biomed Pharmacother. 2017;90:670–6.
79. Eisinger-Mathason TSK, et al. Hypoxia-dependent modification of collagen networks promotes sarcoma metastasis. Cancer Discov. 2013;3(10):1190–205.
80. Lewis DM, et al. A feedback loop between hypoxia and matrix stress relaxation increases oxygen-axis migration and metastasis in sarcoma. Cancer Res. 2019;79(8):1981–95.
81. Estourgie SH, Nielsen GP, Ott MJ. Metastatic patterns of extremity myxoid liposarcoma and their outcome. J Surg Oncol. 2002;80(2):89–93.
82. Asano N, et al. Metastatic patterns of myxoid/round cell liposarcoma: a review of a 25-year experience. Sarcoma. 2012;2020:1–6.
83. Lee ATJ, et al. Clinical and molecular spectrum of liposarcoma. J Clin Oncol. 2018;36(2):151–9.
84. Crozat A, et al. Fusion of CHOP to a novel RNA-binding protein in human myxoid liposarcoma. Nature. 1993;363(3430):640–4.
85. Patil N, et al. A mechanistic study on the metastasis inducing function of FUS-CHOP fusion protein in liposarcoma. Int J Cancer. 2014;134(12):2808–19.
86. Tornin J, et al. FUS-CHOP promotes invasion in myxoid liposarcoma through a SRC/FAK/RHO/ROCK-dependent pathway. Neoplasia. 2018;20(1):44–56.
87. Peng C, et al. SHCBP1 promotes synovial sarcoma cell metastasis via targeting TGF-β1/Smad signaling pathway and is associated with poor prognosis. J Exp Clin Cancer Res. 2017;36(1):141.
88. Xu X-Z, et al. MiR-9 promotes synovial sarcoma cell migration and invasion by directly targeting CDH1. Int J Biochem Cell Biol. 2019;112:61–71.
89. Qi Y, et al. Transforming growth factor-β1 signaling promotes epithelial-mesenchymal transition-like phenomena, cell motility, and cell invasion in synovial sarcoma cells. PLoS One. 2017;12(8):e0182680.
90. Dongre A, Weinberg RA. New insights into the mechanisms of epithelial-mesenchymal transition and implications for cancer. Nat Rev Mol Cell Biol. 2019;20(2):69–84.
91. Olsen SH, Thomas DG, Lucas DR. Cluster analysis of immunohistochemical profiles in synovial sarcoma, malignant peripheral nerve sheath tumor, and Ewing sarcoma. Mod Pathol. 2006;19(5):659–68.
92. Sato H, et al. Expression of cadherins and their undercoat proteins (alpha-, beta-, and gamma-catenins and p120) and accumulation of beta-catenin with no gene mutations in synovial sarcoma. Virchows Arch. 2001;438(1):23–30.
93. Qi Y, et al. The correlation between morphology and the expression of TGF-β signaling pathway proteins and epithelial-mesenchymal transition-related proteins in synovial sarcomas. Int J Clin Exp Pathol. 2013;6(12):2787–99.
94. Saito T, et al. Prognostic value of the preserved expression of the E-cadherin and catenin families of adhesion molecules and of beta-catenin mutations in synovial sarcoma. J Pathol. 2000;192(3):342–50.

Tumor Immune Microenvironment of Soft Tissue Sarcoma

61

Tom Wei-Wu Chen, Sheng-Fang Su, and William W. Tseng

Abstract

Soft tissue sarcoma (STS) is a group of cancers transformed from mesenchymal cells. The treatment outcome of immunotherapy has not been as successful in STS, and our understanding of the tumor immune microenvironment (TIME) may help us in the search of better treatment. This chapter aims to describe the tumor immune microenvironment of soft tissue sarcoma in detail in terms of specific immune cell types as well as specific histologies. The topics included B cells and tertiary lymphoid structures, T cells, macrophages, and the effects of anticancer treatments on the TIME. In the latter part of this chapter, we will also discuss about the TIME of leiomyosarcoma, liposarcoma, and epigenetic regulation-associated STS and ways to improve the treatment.

Learning Objectives

- The presence of B cells and tertiary lymphoid structures in the tumor immune microenvironment of soft tissue sarcoma are associated with improved prognosis and is considered as the engine of anti-tumor immune response.
- Tumor-associated macrophages are the most abundant immune cell types in the tumor immune microenvironment of various soft tissue sarcoma histologies. Anti-macrophage targets such as CD47 and CSF-1R are under intensive investigations.
- Understanding the alternative effect of anticancer treatments on the tumor immune microenvironment modulation and dynamics provides a broader perspective for future combination treatments with immunotherapy.
- Different histologies in soft tissue sarcoma, such as leiomyosarcoma, liposarcoma, and epigenetic change-related sarcomas, may need different mechanisms to overcome the immunosuppressive tumor immune microenvironment.

61.1 Introduction

Soft tissue sarcomas (STS) are tumors that arise from the mesenchymal cells. It comprised about 1–2% of all solid tumors but could occur in any body parts. Furthermore, STS is not a single disease entity but is composed of over more than 70 different histologies. The advancement in molecular biology has provided useful tools in understanding the underlying causes of various STS histologies. In terms of genomic alterations, STS could be parsed into those with simple pathognomonic genetic rearrangements that could define a specific histology, for instance, the presence of c-kit

T. W.-W. Chen (✉)
Graduate Institute of Oncology, College of Medicine, National Taiwan University, Taipei, Taiwan

Department of Oncology, National Taiwan University Hospital, Taipei, Taiwan
e-mail: weiwuchen@ntu.edu.tw

S.-F. Su
Graduate Institute of Oncology, College of Medicine, National Taiwan University, Taipei, Taiwan

YongLin Institute of Health, National Taiwan University (National Taiwan University YongLin Scholar), Taipei, Taiwan
e-mail: shengfang@stat.sinica.edu.tw

W. W. Tseng
Department of Surgery, Division of Breast, Endocrine and Soft Tissue Surgery, Keck School of Medicine of University of Southern California, Los Angeles, CA, USA
e-mail: William.Tseng@med.usc.edu

S. P. Leong et al. (eds.), *Cancer Metastasis Through the Lymphovascular System*,
https://doi.org/10.1007/978-3-030-93084-4_61

and platelet-derived growth factor receptor (PDGFR) in gastrointestinal stromal tumors (GIST), SS18-SSX gene rearrangement in synovial sarcoma, or MDM2 gene amplification in well-differentiated and dedifferentiated liposarcoma (LPS) or those with a complex genomic alteration or aneuploidy such as leiomyosarcoma and angiosarcoma. However, tumor is not made of only cancer cells; we now understand that the tumor microenvironment is also vital to tumorigenesis as well as treatment resistance. Compared to our understanding of the molecular mechanisms that drive STS cancers, what we comprehend about the tumor microenvironment, or more specifically, the tumor immune microenvironment (TIME), is still very rudimentary. This is reflected in the fact that, as compared to other cancer types, immunotherapy, or mostly immune checkpoint inhibitor (ICI), treatments in STS have not been as successful. To improve the outcome of immunotherapy in STS, deeper understanding of the TIME cannot be overemphasized as it has been proven to be correlated with finding a better solution for ICIs in many other cancer types [1]. This chapter aims to provide the current understanding of the TIME of STS in terms of specific immune cell types, tertiary lymphoid structures, and histologies.

61.2 Common Methods Used for Tumor Immune Microenvironment Investigation

It is essential to first understand the methodologies used for the investigation of TIME, as every methodology has its strengths and pitfalls. The most common and easily approached method to assess the TIME is by immunohistochemistry (IHC). Two methods are commonly used: a full tissue section examination or tissue microarray. The full tissue section is a more conventional approach that resembles clinical pathological examination but is limited by resolution for multiple chromogens and the lack of standards in interpretation. Tissue microarray uses a tiny proportion of the tissue but permits high-throughput analysis that could involve tens or hundreds of specimens in arraying them into paraffin blocks. Tissue microarray saves precious tumor material but may be biased by intratumoral heterogeneity in determination of the true outcome [2]. Recently, multispectral IHC or immunofluorescence staining with up to seven colors on a single slide has gained popularity and in use for TIME analysis [3, 4]. Although flow cytometry is also an important method to determine immune cell type in the lab, the requirement of fresh tissue and laborious effort to filter the tumor-infiltrating immune cells have limited its clinical or translational applicability.

The advent and reduced cost of next-generation sequencing have made transcriptomic-based (RNA-Seq) analysis popular for TIME studies; the openness and public availability of the transcriptomic material also made inter-study comparisons achievable. Although many algorithms for transcriptomic-based TIME analysis exist, these algorithms could be categorized into two conceptually distinct approaches: the marker-gene-based or deconvolution-based approach [5]. Marker-gene-based approach utilizes a prespecified set of genes that were validated as characteristics of the representative immune cells. Nevertheless, marker-gene-based algorithms are mostly semiquantitative, allowing inter-samples comparisons of a specific immune cell type but not intra-sample comparison of different immune cell types [5]. Deconvolution methods use a system of equations to decipher the sample gene expression profile as the weighted sum of expression profiles of different immune cell types. In principle, deconvolution method allows the generation of absolute scores that could be interpreted as different immune cell population fractions, providing detail intratumor immune cell type proportions that could be compared between samples as fractions but not absolute values [6]. Although the analytical results of different algorithms have shown satisfactory correlation on CD8+ T cells, B cells, and NK cells, the correlation of other immune cell types such as regulatory and non-regulatory CD4+ T cells and dendritic cells remains to be improved [5]. Overall, it is essential to understand the computational algorithms used for the analysis of TIME before we can fully interpret and compare the different results.

61.3 Overview of Tumor Immune Microenvironment in Soft Tissue Sarcoma

Much heterogeneity is noted in the TIME of any cancer type, but STS are more likely to have a "cold" TIME that lacks immune cell infiltrations. One of the reasons is that STS have in average a lower tumor mutation burden (TMB) as compared to other carcinoma cancer types [7]. A higher TMB is associated with an increased chance of presenting neoantigens and hence inducing adaptive immunity [8]. The mean somatic mutation in The Cancer Genome Atlas (TCGA) SARC cohort was 1.06 per MB of DNA, while in contrast the TMB of melanoma on average was 14 per Mb of DNA [9]. In addition, STS are more likely to have unbalanced copy number alterations (CNA), which have been associated with a lower cytotoxic immune cell infiltration in the TIME [10]. The genomic feature of STS histology is also associated with different types of TIME. STS with simple karyotypes or genomic alterations such as synovial sarcoma or alveolar rhabdomyosarcoma commonly have a "cold" TIME with fewer immune cell infiltrations [11, 12]. On the contrary, STS with a complex karyotype or genomic alterations such as undifferentiated pleomorphic sarcoma (UPS) or angiosarcoma (AS) are more likely to feature a TIME with higher infiltrating lymphocytes [13, 14].

A recent joint collaborative study by sarcoma researchers from France, Taiwan, and the United States that amassed 936 STS patients' transcriptomic or IHC data provided a broader view of the TIME of STS [15]. Based on the level of different immune cell types and endothelial cell marker expressions in the TIME as computed by the Microenvironment Cell Populations-counter (MCP-counter) [16], STS could be categorized into five distinct Sarcoma Immune Classes (SIC). SICs A, C, and E are more homogeneous and stand for "immune-desert," "vascularized," and "immune and tertiary lymphoid structure (TLS) high." The overall survival (OS) was significantly better for SIC E group as compared to SIC A group. Around 70% of STS patients (including SICs A, B, and C) were considered to have a "cold" TIME with few immune cell infiltrates. Immune checkpoint molecules were also reflective of the level of inflammation in each SIC, with only the SIC E having concomitantly a high expression of multiple immune checkpoint markers including CTLA-4, PD-1, LAG3, and TIM3. Intriguingly, the number of non-synonymous mutations in the STS patients were not correlated with the SIC groups, suggesting that TMB may not be the only factor to determine the TIME of STS.

In the following sections, we discuss in more detail the roles of different immune cell types and specific immune modulating molecules in the TIME and our understandings in specific STS histologies.

61.4 The Role of Specific Immune Cell Types in the Tumor Microenvironment

61.4.1 B Cells/Tertiary Lymphoid Structures

It may be unorthodox to start with a discussion about B cells in TIME rather than the T cells. After all, the cytotoxic T cells are considered the top-tier immune cells responsible for tumor killing. B cells are mostly responsible for the secretion of antibody as an adaptive immunity mechanism. The antibody binds to the surface antigens and induces downstream complement system activation or antibody-dependent cellular cytotoxicity to kill the targeted cells. However, B cells also play an important role as an antigen-presenting cell that could facilitate the maturation of adaptive cytotoxic CD8 T cells. The maturation process mostly occurs in the germinal center (GC) in the secondary lymphoid organs such as the lymph node, tonsils, or spleen. When these GC arise in organs of chronic inflammation or tumorigenesis, these lymphoid follicles are referred to as tertiary lymphoid structures (TLS) [17]. It is postulated that TLS is responsible for B cell productions and, along with dendritic cells and T follicular helper (Tfh) cells, to facilitate the maturation of effector T cells [18].

A positive effect of B cells on STS outcomes was first reported by Sorbye et al. They used tissue microarray to examine the TIME of 249 primary STS. In this study, CD45, CD3, CD4, CD8, and CD20 were used as markers for leukocytes, T cells, CD4+ T cell, cytotoxic T cells, and B cells, respectively. In the multivariate model, CD20 positivity was the only significant positive prognostic marker associated with better disease-specific survival for STS patients who had a wide resection margin (HR 5.5 95% CI 1.6–18.6, $p = 0.006$) [19]. Another study from Karolinska Institute also supported the favorable role of B cells in STS. Primary STS patients with higher CD20+ B cells were associated with better metastasis-free survival (61 vs 29 months, $p = 0.009$) and overall survival (OS; 60 vs 34, months, $p = 0.022$). However, in the validation TCGA cohort, the favorable prognostic role of B cells was only observed in STS tumor with low M2 (protumoral) macrophages, represented by low IL-10, low PTGS2, or low CD163 expression [20].

The concept that in STS the presence of active B cells is critical to support the cytotoxic T cell function is further strengthened in the study by Petitprez et al. [15]. Using the transcriptomic data from 496 STS patients, we confirmed that the B lineage signature (including genes BANK1, CD19, CD22, CD79A, CR2, FCRL2, IGKC, MS4A1, and PAX5) plays a more prominent role than the T cell signature (including genes CD28, CD3D, CD3G, CD5, CD6, CHRM3-AS2, CTLA4, FLT3LG, ICOS, MAL, PBX4, SIRPG, THEMIS, TNFRSF25, and TRAT1) or cytotoxic lymphocyte signature (including genes CD8A, EOMES, FGFBP2, GNLY, KLRC3, KLRC4, KLRD1) in determining the OS outcome. Patients with high cytotoxic lymphocytes/low B lineage or high T cell/low B lineage cells had a worse OS than patients with high cytotoxic lymphocytes/high B lineage ($p = 0.015$) or high T cells/high B lineage ($p = 0.02$), respectively [15].

A plausible explanation for the positive impact of B cells in the TIME of STS is its association with tertiary lymphoid structure (TLS). TLS are observed in chronic inflammation sites and malignant cancers. The formation of TLS requires both lymphoid tissue organizer (LTo) cells and lymphoid tissue inducer (LTi) cells. Where stromal tissues in chronic inflamed sites act as LTo, tissue-resident monocytes, B cells, and Th17 cells could function as LTi in the same microenvironment. The interaction between LTo and LTi stimulates the secretion of chemokines that function as lymphocyte attractant (CCL21 and CXCL12) and also molecules that govern the structure formation of TLS (CXCL13, CCL19, adhesion molecules) [18]. Within TLS microenvironment, mature dendritic cells (DC) present major histocompatibility complex (MHC) class II peptides to CD4+ T cells and facilitate their differentiation into effector T cells and Tfh cells. Mature

Tfh cells express CXCL13 receptor CXCR5, facilitating its migration into the CXCL13-rich B cell zone. The contact of Tfh cells and B cells induces the activation and differentiation of B cells into plasmablasts and the formation of GC in which the B cells undergo class switch recombination [18].

The prognostic impact of TLS has been demonstrated in various cancer types including STS [21]. In a cohort of 93 STS samples including dedifferentiated liposarcoma, leiomyosarcoma, and UPS, 12% (11/93) had at least 1 TLS as defined by the presence of CD20+ B cell follicles juxtaposed to a CD3+ T cell aggregates containing at least 1 DC-LAMP+ mature dendritic cell as determined by IHC [15]. It was confirmed that all but 1 STS patient with TLS was in the SIC E group and 9 out of 11 patients in the SIC E group had TLS, suggesting the potential of TLS as a surrogate biomarker for the identification of SIC E in STS. Indeed, in the same validation cohort, the association of TLS and the presence of anti-tumor immunity were confirmed: STS with TLS had a significantly higher CD3+ cell density ($p = 4.0 \times 10^{-5}$), CD8+ cell density ($p = 1.8 \times 10^{-4}$), and CD20+ cell density ($p = 1.5 \times 10^{-7}$) as compared to those without TLS [15]. The favorable prognostic impact of TLS was also confirmed in GIST [22] and retroperitoneal sarcomas [23].

Different stages of TLS maturation may also signal different levels of anti-tumor immunity [18]. Based on the expression of two different markers, CD21 on follicular dendritic cells and CD23 on both B cells and dendritic cells, the maturation stages of TLS have been proposed as early lymphoid aggregates (E-TLS) that lack both CD21 and CD23 expression, primary follicle-like TLS (PFS-TLS) with CD21-positive but CD23-negative expression, and secondary follicle-like TLS (SFL-TLS) containing both matured CD21-positive dendritic cells and CD23-positive B cells in the GC [18]. Although it is not common practice to differentiate TLS in terms of its maturity stage, the differential impact of the maturity stages of TLS has been shown in non-small-cell lung cancer and colon cancer [24], as well as in GIST [22]. In the study by Petitprez et al., E-TLS, PFL-FLS, and SFL-TLS represented 60.5%, 21.1%, and 18.3% of TLS examined, and the most common histology with SFL-TLS was UPS ($p = 7.76 \times 10^{-5}$) [15] .

In summary, B cells and TLS in the TIME of STS affect the prognosis. What remains to be investigated is how to induce the formation of TLS in different STS histologies and prospective studies to determine the surrogacy of TLS as a biomarker for ICI treatment success.

61.4.2 T Cells

Although T cells remain an integral part in assessing the anti-tumor response in the field of oncology, distinct and partial overlapping assessment methods make it a difficult task to have a consensus interpretation [25]. For instance, the prognostic impact of CD4+ or CD8+ T cells has remained divergent and unsettled in STS [19, 26, 27]. However, it is generally accepted that STS, as compared to other immunogenic cancers such as melanoma or lung cancer, has fewer CD8+ T cells [25, 28]. In addition, CD8+ T cells are more likely to be higher in genomic-complex STS than simple fusion gene-associated STS [12]. Memory effector T cells are matured T cells that could rapidly respond to antigen stimulation and has been shown to be associated with a better prognosis in a variety of cancer types [29, 30]. In STS, pleomorphic sarcoma and myxofibrosarcoma had the highest percentage of memory effector T cells among CD8+ T cells, while GIST had the lowest memory effector T cells [31]. A comprehensive discussion of all the subtypes of T cells is out of the scope of this chapter, but it is safe to say we need to identify the most important T cell subtype that is associated with better prognosis in each STS histology and understand the methods to enhance its infiltration or trafficking into the TIME.

61.4.3 Macrophages

The most common myeloid immune cell type found in the TIME of STS is macrophage [32]. Under normal physiologically, macrophages are responsible for pathogen killing. However, in the TIME, many macrophages differentiate into tumor-associated macrophages (TAMs), retaining features of M2 macrophages such as promoting angiogenesis, breaking down of matrix proteins to facilitate tumor metastasis, and inducing an immunosuppressive TIME by the secretion of IL10 and TGF-β [32, 33].

Studies using either IHC or tissue microarrays have shown that TAMs are the most abundant immune cell type in the TIME of many STS histologies [11, 34]. Although the typical representative marker remains debatable, CD68 and CD163 are the two most common markers to identify TAMs. While CD68 is a marker for pan-macrophage identification, its specificity is low as stromal cells and sarcoma cells could also express CD68 [32]. CD163 is a membranous marker associated with M2 macrophage which is the common form of TAM. In addition, functional studies have demonstrated that mice with CD163-deficient TAM had significantly decreased sarcoma tumor growth as compared with mice with CD163 wild-type TAM. In this model, the production of IL6 from CD163-positive macrophage is the determinant chemokine in facilitating tumor growth [35]. Based on these evidences, CD163 is likely a more specific marker to determine M2-like TAM in the TIME of STS.

In the TIME of STS tumors, the quantity of TAM varies. UPS patients are most likely to have the highest number of TAMs per tumor area or the highest ratio of CD163+ to

CD68+ macrophages, but the variation is also wide [11, 34]. Other STS histologies with higher TAMs included angiosarcoma, dedifferentiated liposarcoma, and leiomyosarcoma [34]. Generally speaking, genomic complex STS have a higher number of TAMs than translocation-associated STS in the TIME. Another possible resistant mechanism for tumor to escape the phagocytic activity of macrophages in the TIME is inducing the expression of CD47. CD47 is an ubiquitous marker on erythrocytes to prevent macrophage phagocytosis. Tumor cell could upregulate CD47 and, when interacts with signal-regulatory protein alpha (SIRPα) on macrophages, suppresses the anti-tumor activity of macrophages [36]. The distribution of CD47 on tumor cells is bimodal, with most cases showing either 0% or higher than 90% of tumor cells positive [34]. Sarcomas with a high percentage of CD47 tumor cell staining included chordoma, angiosarcoma, and pleomorphic liposarcoma [34].

Because of the abundance of TAMs in the TIME, drugs targeting the specific marker are under intense investigation. The leading target is colony-stimulating factor-1 receptor (CSF-1R), which is associated with M2 macrophage polarization [33]. The CSF-1R inhibitor pexidartinib has been approved by the US FDA for the treatment of inoperable tenosynovial giant cell tumor, a rare neoplasm that is associated with high levels of CSF1 and a large amount of macrophage infiltration [37]. However, single-agent CSF-1R inhibitors have not been successful in treating other cancer types. Combination with other systemic treatments such as chemotherapy or ICI is currently under investigations [33].

61.4.4 The Effect of Anticancer Treatments on the Tumor Immune Microenvironment of Soft Tissue Sarcoma

Anticancer treatments, including conventional chemotherapy, radiotherapy, and targeted therapies, not only target cancer cells but may also assert their influence on the elements of TIME in boosting their treatment efficacy. Radiotherapy is an integral part of the treatment for localized STS, and its positive impact on the TIME has been reported in various cancer types [38]. In STS studies, preoperative radiotherapy was associated with increased infiltration of tumor-infiltrating lymphocytes, CD4+ and CD8+ T cells, and FOXP3+ cells [12]. However, some studies suggested radiotherapy did not increase the T cell fraction or clonality as measured by T cell repertoire [14]. In a mouse immunocompetent sarcoma model, radiotherapy was able to ablate the myeloid immune-suppressive cells in the tumor microenvironment of transplantable tumor, but not autochthonous primary tumor. Despite the same histology and similar TMB of the transplantable and autochthonous tumors, the differential effects of radiotherapy suggested that the immune-modulatory effect of radiotherapy also depends on the original TIME status [39]. Currently, the SARC032 (NCT03092323) study is investigating the role of neoadjuvant radiotherapy plus pembrolizumab in localized high-grade STS. The outcome of this study will provide further evidence of the immunomodulatory effect of radiotherapy and combination with ICI in localized STS.

Although chemotherapy-induced immunogenic cell death has been proposed for a long period of time [40], the effect of cytotoxic chemotherapy such as doxorubicin and ifosfamide on transforming an immunosuppressive TIME remains uncertain. Studies examining post-chemotherapy treatment STS TIME have not found either specific change or even an increase in FoxP3+ TILs [12, 14]. One exception is the cytotoxic drug trabectedin. Trabectedin is a cytotoxic agent that binds to the minor groove of DNA which could inhibit transcription factor binding and forces DNA damage [41]. In addition, it has been shown in sarcoma mouse models that trabectedin could apply the cytotoxic effect to TAMs and subsequently decrease chemokine CCL2 secretion and angiogenic processes [42]. This finding was validated in STS patients treated with trabectedin. After trabectedin treatment, reduction of the blood-circulating monocytes decreased, and resected tumors also have decreased CD163+ macrophages in the TIME [42, 43]. The second-generation lurbinectedin also has similar effects limiting the production of inflammatory factors that could subsequently decrease monocyte adhesion and migration [44]. Targeted agents that focus on anti-angiogenic molecules such as vascular endothelial growth factor receptors (VEGFRs) could also facilitate the trafficking of immune cells or maturation of dendritic cells. Combination of anti-angiogenic agents with ICI has shown promising outcomes in advanced STS [45, 46], and biomarkers to elaborate the underlying mechanisms are eagerly awaited.

Lastly, immunotherapy also significantly influences the TIME dynamics. From the biomarker study of SARC 028, in which sarcoma patients were treated with anti-PD-1 antibody pembrolizumab, Keung et al. found that most of the immune cell type densities remained stable after pembrolizumab treatment, except for effector cytotoxic T cells (CD3+ CD8+ CD45RO+) and regulatory T cells [47]. A phase 0 study investigating the impact of systemic administration of interferon-γ on the generally cold TIME of synovial sarcoma and myxoid liposarcoma showed increased MHC class I expression and increased T cell infiltrations after treatment [48]. With limited efficacy of single-agent ICI in STS, reshaping the TIME through other treatments or combination is a promising choice to go forward to improve the efficacy.

61.5 The Tumor Immune Microenvironment in Specific Sarcoma Histologies

61.5.1 Leiomyosarcoma

Leiomyosarcoma (LMS) is one of the most common histologies of STS that could occur in extremities and truncal areas as well as uterine or retroperitoneal areas. Nevertheless, studies have suggested that LMS from different organ sites have different transcriptomic profiles and survival outcomes [49]. A high percentage of LMS has cold TIME [15] and is less likely to respond to ICI treatment [50].

A primary focus of the TIME of LMS has been TAMs and was pioneered by the Stanford groups led by Dr. van de Rijn. They first identified the presence of a high level of TAMS in LMS by gene expression profiling and confirmed by IHC. The criteria for IHC identification of a macrophage were either CD68 or CD163 at the membranous, but not nucleus, staining, and it needed to be a cell with dendritic process. The first study showed that higher-dense infiltrates of either CD68+ or CD168+ macrophages were associated with a worse prognosis in non-uterine, but not uterine, LMS [51]. However, if a composite score (CD163, FCGR3a, and CTSL1) was used to count the macrophages, then the presence of macrophages was associated with a poor prognosis in both uterine and non-uterine LMS [52]. Furthermore, TAMs also play a significant role in angiogenesis within the tumor microenvironment of LMS. While the micro-vessel density (MVD) was associated with a poor prognosis, it was the presence of TAMs, but not vascular endothelial growth factor (VEGF), that was correlated more closely with MVD. In addition, the parallelism between TAM and MVD persisted from primary to metastatic tumors of LMS [53].

Treatments that aim to take advantage of the abundant TAMs in LMS have been explored. One method is to educate the TAMs to perform anti-tumor functions by targeting the CD47-SIRP signaling pathway. LMS have been found to have high levels of CD47, and upon treatment of anti-CD47 antibody in the in vivo mouse model, the primary tumor and its metastatic tumors shrank dramatically as compared with the control treatment [54].

61.5.2 Liposarcoma

Liposarcoma encompasses three subtypes (well differentiated/dedifferentiated or WD/DD, myxoid round cell, and pleomorphic—very rare) each with distinct genetic aberrations and likely unique TIMEs.

WD/DD liposarcoma is characterized genetically by amplification of chromosomal region 12q13-15, which includes important genes such as MDM2 and CDK4. Among the three subtypes, the TIME has been studied the most in WD/DD liposarcoma. Some of the earlier descriptive studies established the presence of B and T cell-rich TLS and also the expression of PD-1 in the tumor-infiltrating T cells in WD/DD [55]. The latter observation would have predicted favorable response to ICI, and in fact, the initial results of the SARC028 trial of anti-PD1 therapy showed a 20% objective response rate (ORR) for the DD cohort [50]. These data were limited by the small number of patients enrolled with this subtype ($n = 10$), and a subsequent expansion cohort study noted a disappointingly lower ORR of 10% in DD (4 of 39 patients) [56]. These results may be explained in part by the low mutational burden seen in WD/DD liposarcoma [57]. In support of low tumor immunogenicity, comprehensive gene expression analysis in WD/DD liposarcoma shows that both TCR clonality and PD-L1 expression are relatively low, despite an inflammatory TIME [14]. Moreover, the TIME in WD/DD liposarcoma is arguably more complex given that these tumors are naturally massive in size, that biologically distinct WD (low grade) and DD (high grade) components may be juxtaposed within the same tumor, and that treatment can significantly alter the TIME—all of which contribute to significant intratumoral and patient-to-patient heterogeneity [23, 58].

Myxoid round cell liposarcoma is defined by a FUS-DDIT3 (or in a minority of cases, FUS-ESWR1) translocation. In contrast to the WD/DD subtype, relatively little is known about the TIME in myxoid liposarcoma. The bulk of immunologic studies in this subtype have focused on the very high tumor expression of NY-ESO-1, a cancer-testis antigen [59] . As a result, ongoing vaccine trials (e.g., CMB305) targeting this antigen have included a modest number of patients specifically with myxoid liposarcoma [60]. Recent preliminary results are promising with in fact the longest progression-free survival (median 5.1 months) in the myxoid liposarcoma cohort.

Importantly, the TIME is likely very different in one liposarcoma subtype versus another, necessitating different immunotherapeutic approaches. As an example of this, recent data have started to highlight a potential key role for macrophages in WD/DD liposarcoma pathogenesis and disease progression [61]. An independent study of the TIME verified that among various sarcoma subtypes, DD liposarcoma has one of the highest infiltrations of TAMs [34]. In contrast, in the same study, TAMs were markedly sparse in myxoid liposarcoma.

61.5.3 Epigenetic Regulation-Associated Soft Tissue Sarcoma

Epigenetic mechanisms have been appreciated as a driver model of tumorigenesis in mesenchymal tumors [62], given the low mutation burden nature of STS [7]. Nacev et al.

addressed multiple levels of epigenetic dysregulation developed in different types of sarcomas through mechanisms of altered epigenetic writers, readers, and chromatin remodelers [62]. Dysfunction of enzymes involved in DNA methylation and histone modifications, mutations in histone genes and remodeling and modifying complexes, and alterations of chromatin pathway drive sarcomas predominantly epigenetic diseases.

One of the effects on chromatin writers is loss-of-function mutations in the polycomb repressive complex 2 (PRC2) subunit suppressor of zeste 12 (*SUZ12*) or embryonic ectoderm development (*EED*) that have been identified in approximately 80% of MPNST [63]. SUZ12 inactivation causes a loss of repressive histone mark H3K27me3 and may induce the recruitment of bromodomain proteins (BRD) and associated transcription factors via an increase in H3K27 acetylation, a transcription-activating signal [64]. The linkage between loss of PRC2-associated H3K27me3 and TIME regulation remains elusive. But studies that had examined the TIME differences between neurofibroma and MPNST have shown increased CD8+ T cells infiltration and tumor cell PD-L1 expression in MPNST but not neurofibroma, which is in parallel of the loss of expression of H3K27me3 [65], providing indirect signal that chromatin writer mutations could affect the TIME components through epigenetic regulation.

In contrast to MPNST, some STS histologies have hyperactivated PRC2 function. However, the hyperactivation of PRC2 is not the result of the gaining of activation mutation but is because of the loss of function of PRC2 suppressors – the SWItch/Sucrose Non-Fermentable (SWI/SNF) complexes [66]. SMARCB1 (INI-1), a core subunit of SWI/SNF, for instance, is frequently inactivated in malignant rhabdoid tumors (MRT) and epithelioid sarcoma (ES). The loss of INI-1 implies the utility of inhibitor of enhancer of zeste homolog 2 (EZH2), the catalytic subunit of PRC2, for epigenetic therapy. In 2020, an open-label, phase 2 basket study confirmed the clinical activity and safety of tazemetostat, a first-in-class EZH2 inhibitor, in advanced ES with loss of INI1/SMARCB1 [67]. Furthermore, recent work showed a correlation of *SMARCB1* loss with global hypomethylation and cytotoxic T cell infiltration in the immune microenvironment, followed by the expression of immune checkpoint regulators [68]. Targeting EZH2 has shown to enhance antigen presentation and anti-tumor immunity in some preclinical models of cancers resistant to immunotherapy [69, 70]. A recent case study of dedifferentiated chordoma with INI-1 loss treated with tazemetostat showed increased T cell infiltration into the TIME after treatment [71], providing evidence that inhibition of the activated EZH2 indeed could reverse the immunosuppressive TIME.

While targeting therapy including kinase inhibitors to selective genetic alteration fails in most STS, targeting epigenetic dysregulation provides another scientific-based therapeutic intervention. The understanding of DNA methylation, histone, and chromatin modifications on genes involved in epigenetic pathways, as well as the potential effects on the immune status in the tumor microenvironment, may help the treatment of STS.

Open Questions

- What are the modalities or methods currently used for the investigation of the tumor immune microenvironment? What are the pros and cons of each different method?
- What is the most important immune cell type or structure that could contribute to a better anti-tumor immunity in soft tissue sarcoma? Are there methods that could enhance the formation of these anti-tumor structures?
- What is the most abundant immune cell type in soft tissue sarcoma immune microenvironment? How could we transform the immunosuppressive tumor microenvironment by targeting these immunosuppressive immune cells?
- How does the histology or their respective genomic feature affect the tumor immune microenvironment?
- What are the anticancer treatments in soft tissue sarcoma that could affect the dynamics of the immune microenvironment? How should they be combined with other immunotherapies and under which clinical scenario would fit the best?

References

1. Binnewies M, Roberts EW, Kersten K, et al. Understanding the tumor immune microenvironment (TIME) for effective therapy. Nat Med. 2018;24:541–50.
2. Lee ATJ, Chew W, Wilding CP, et al. The adequacy of tissue microarrays in the assessment of inter- and intra-tumoural heterogeneity of infiltrating lymphocyte burden in leiomyosarcoma. Sci Rep. 2019;9:14602.
3. Stack EC, Wang C, Roman KA, Hoyt CC. Multiplexed immunohistochemistry, imaging, and quantitation: a review, with an assessment of Tyramide signal amplification, multispectral imaging and multiplex analysis. Methods. 2014;70:46–58.
4. Edin S, Kaprio T, Hagstrom J, et al. The prognostic importance of CD20(+) B lymphocytes in colorectal cancer and the relation to other immune cell subsets. Sci Rep. 2019;9:19997.
5. Sturm G, Finotello F, Petitprez F, et al. Comprehensive evaluation of transcriptome-based cell-type quantification methods for immuno-oncology. Bioinformatics. 2019;35:i436–45.
6. Finotello F, Rieder D, Hackl H, Trajanoski Z. Next-generation computational tools for interrogating cancer immunity. Nat Rev Genet. 2019;20:724–46.
7. Cancer Genome Atlas Research Network. Electronic address edsc, Cancer Genome Atlas Research N. Comprehensive and integrated genomic characterization of adult soft tissue sarcomas. Cell. 2017;171:950–965.e928.

8. Rizvi NA, Hellmann MD, Snyder A, et al. Cancer immunology. Mutational landscape determines sensitivity to PD-1 blockade in non-small cell lung cancer. Science. 2015;348:124–8.
9. Iglesia MD, Parker JS, Hoadley KA, et al. Genomic analysis of immune cell infiltrates across 11 tumor types. J Natl Cancer Inst. 2016;108:djw144.
10. Davoli T, Uno H, Wooten EC, Elledge SJ. Tumor aneuploidy correlates with markers of immune evasion and with reduced response to immunotherapy. Science. 2017;355:eaaf8399.
11. Chen L, Oke T, Siegel N, et al. The immunosuppressive niche of soft-tissue sarcomas is sustained by tumor-associated macrophages and characterized by intratumoral tertiary lymphoid structures. Clin Cancer Res. 2020;26:4018–30.
12. Dancsok AR, Setsu N, Gao D, et al. Expression of lymphocyte immunoregulatory biomarkers in bone and soft-tissue sarcomas. Mod Pathol. 2019;32:1772–85.
13. Fujii H, Arakawa A, Utsumi D, et al. CD8(+) tumor-infiltrating lymphocytes at primary sites as a possible prognostic factor of cutaneous angiosarcoma. Int J Cancer. 2014;134:2393–402.
14. Pollack SM, He Q, Yearley JH, et al. T-cell infiltration and clonality correlate with programmed cell death protein 1 and programmed death-ligand 1 expression in patients with soft tissue sarcomas. Cancer. 2017;123:3291–304.
15. Petitprez F, de Reynies A, Keung EZ, et al. B cells are associated with survival and immunotherapy response in sarcoma. Nature. 2020;577:556–60.
16. Becht E, Giraldo NA, Lacroix L, et al. Estimating the population abundance of tissue-infiltrating immune and stromal cell populations using gene expression. Genome Biol. 2016;17:218.
17. Buckley CD, Barone F, Nayar S, et al. Stromal cells in chronic inflammation and tertiary lymphoid organ formation. Annu Rev Immunol. 2015;33:715–45.
18. Sautes-Fridman C, Verneau J, Sun CM, et al. Tertiary lymphoid structures and B cells: clinical impact and therapeutic modulation in cancer. Semin Immunol. 2020;48:101406.
19. Sorbye SW, Kilvaer T, Valkov A, et al. Prognostic impact of lymphocytes in soft tissue sarcomas. PLoS One. 2011;6:e14611.
20. Tsagozis P, Augsten M, Zhang Y, et al. An immunosuppressive macrophage profile attenuates the prognostic impact of CD20-positive B cells in human soft tissue sarcoma. Cancer Immunol Immunother. 2019;68:927–36.
21. Sautes-Fridman C, Petitprez F, Calderaro J, Fridman WH. Tertiary lymphoid structures in the era of cancer immunotherapy. Nat Rev Cancer. 2019;19:307–25.
22. Lin Q, Tao P, Wang J, et al. Tumor-associated tertiary lymphoid structure predicts postoperative outcomes in patients with primary gastrointestinal stromal tumors. Oncoimmunology. 2020;9:1747339.
23. Yan L, Wang Z, Cui C, et al. Comprehensive immune characterization and T-cell receptor repertoire heterogeneity of retroperitoneal liposarcoma. Cancer Sci. 2019;110(10):3038–48.
24. Posch F, Silina K, Leibl S, et al. Maturation of tertiary lymphoid structures and recurrence of stage II and III colorectal cancer. Oncoimmunology. 2018;7:e1378844.
25. Martin-Broto J, Moura DS, Van Tine BA. Facts and hopes in immunotherapy of soft-tissue sarcomas. Clin Cancer Res. 2020;26:5801–8.
26. D'Angelo SP, Shoushtari AN, Agaram NP, et al. Prevalence of tumor-infiltrating lymphocytes and PD-L1 expression in the soft tissue sarcoma microenvironment. Hum Pathol. 2015;46:357–65.
27. van Erp AEM, Versleijen-Jonkers YMH, Hillebrandt-Roeffen MHS, et al. Expression and clinical association of programmed cell death-1, programmed death-ligand-1 and CD8(+) lymphocytes in primary sarcomas is subtype dependent. Oncotarget. 2017;8:71371–84.
28. Toulmonde M, Penel N, Adam J, et al. Use of PD-1 targeting, macrophage infiltration, and IDO pathway activation in sarcomas: a phase 2 clinical trial. JAMA Oncol. 2018;4:93–7.
29. Pages F, Berger A, Camus M, et al. Effector memory T cells, early metastasis, and survival in colorectal cancer. N Engl J Med. 2005;353:2654–66.
30. Samji T, Khanna KM. Understanding memory CD8(+) T cells. Immunol Lett. 2017;185:32–9.
31. Klaver Y, Rijnders M, Oostvogels A, et al. Differential quantities of immune checkpoint-expressing CD8 T cells in soft tissue sarcoma subtypes. J Immunother Cancer. 2020;8:e000271.
32. Mantovani A, Marchesi F, Malesci A, et al. Tumour-associated macrophages as treatment targets in oncology. Nat Rev Clin Oncol. 2017;14:399–416.
33. Cassetta L, Pollard JW. Targeting macrophages: therapeutic approaches in cancer. Nat Rev Drug Discov. 2018;17:887–904.
34. Dancsok AR, Gao D, Lee AF, et al. Tumor-associated macrophages and macrophage-related immune checkpoint expression in sarcomas. Oncoimmunology. 2020;9:1747340.
35. Shiraishi D, Fujiwara Y, Horlad H, et al. CD163 is required for protumoral activation of macrophages in human and murine sarcoma. Cancer Res. 2018;78:3255–66.
36. Feng M, Jiang W, Kim BYS, et al. Phagocytosis checkpoints as new targets for cancer immunotherapy. Nat Rev Cancer. 2019;19:568–86.
37. Tap WD, Gelderblom H, Palmerini E, et al. Pexidartinib versus placebo for advanced tenosynovial giant cell tumour (ENLIVEN): a randomised phase 3 trial. Lancet. 2019;394:478–87.
38. Weichselbaum RR, Liang H, Deng L, Fu YX. Radiotherapy and immunotherapy: a beneficial liaison? Nat Rev Clin Oncol. 2017;14:365–79.
39. Wisdom AJ, Mowery YM, Hong CS, et al. Single cell analysis reveals distinct immune landscapes in transplant and primary sarcomas that determine response or resistance to immunotherapy. Nat Commun. 2020;11:6410.
40. Galluzzi L, Humeau J, Buque A, et al. Immunostimulation with chemotherapy in the era of immune checkpoint inhibitors. Nat Rev Clin Oncol. 2020;17:725–41.
41. Ueda T, Kakunaga S, Ando M, et al. Phase I and pharmacokinetic study of trabectedin, a DNA minor groove binder, administered as a 24-h continuous infusion in Japanese patients with soft tissue sarcoma. Investig New Drugs. 2014;32:691–9.
42. Allavena P, Signorelli M, Chieppa M, et al. Anti-inflammatory properties of the novel antitumor agent yondelis (trabectedin): inhibition of macrophage differentiation and cytokine production. Cancer Res. 2005;65:2964–71.
43. Germano G, Frapolli R, Belgiovine C, et al. Role of macrophage targeting in the antitumor activity of trabectedin. Cancer Cell. 2013;23:249–62.
44. Belgiovine C, Bello E, Liguori M, et al. Lurbinectedin reduces tumour-associated macrophages and the inflammatory tumour microenvironment in preclinical models. Br J Cancer. 2017;117:628–38.
45. Wilky BA, Trucco MM, Subhawong TK, et al. Axitinib plus pembrolizumab in patients with advanced sarcomas including alveolar soft-part sarcoma: a single-centre, single-arm, phase 2 trial. Lancet Oncol. 2019;20:837–48.
46. Martin-Broto J, Hindi N, Grignani G, et al. Nivolumab and sunitinib combination in advanced soft tissue sarcomas: a multicenter, single-arm, phase Ib/II trial. J Immunother Cancer. 2020;8:e001561.
47. Keung EZ, Burgess M, Salazar R, et al. Correlative analyses of the SARC028 trial reveal an association between sarcoma-associated immune infiltrate and response to pembrolizumab. Clin Cancer Res. 2020;26:1258–66.
48. Zhang S, Kohli K, Black RG, et al. Systemic interferon-gamma increases MHC class I expression and T-cell infiltration in cold tumors: results of a phase 0 clinical trial. Cancer Immunol Res. 2019; https://doi.org/10.1158/2326-6066.CIR-18-0940.
49. Hemming ML, Fan C, Raut CP, et al. Oncogenic gene-expression programs in leiomyosarcoma and characterization of conven-

tional, inflammatory, and uterogenic subtypes. Mol Cancer Res. 2020;18:1302–14.

50. Tawbi HA, Burgess M, Bolejack V, et al. Pembrolizumab in advanced soft-tissue sarcoma and bone sarcoma (SARC028): a multicentre, two-cohort, single-arm, open-label, phase 2 trial. Lancet Oncol. 2017;18:1493–501.
51. Lee CH, Espinosa I, Vrijaldenhoven S, et al. Prognostic significance of macrophage infiltration in leiomyosarcomas. Clin Cancer Res. 2008;14:1423–30.
52. Espinosa I, Beck AH, Lee CH, et al. Coordinate expression of colony-stimulating factor-1 and colony-stimulating factor-1-related proteins is associated with poor prognosis in gynecological and non-gynecological leiomyosarcoma. Am J Pathol. 2009;174:2347–56.
53. Espinosa I, Edris B, Lee CH, et al. CSF1 expression in nongynecological leiomyosarcoma is associated with increased tumor angiogenesis. Am J Pathol. 2011;179:2100–7.
54. Edris B, Weiskopf K, Volkmer AK, et al. Antibody therapy targeting the CD47 protein is effective in a model of aggressive metastatic leiomyosarcoma. Proc Natl Acad Sci U S A. 2012;109:6656–61.
55. Tseng WW, Malu S, Zhang M, et al. Analysis of the intratumoral adaptive immune response in well differentiated and dedifferentiated retroperitoneal liposarcoma. Sarcoma. 2015;2015:547460.
56. Burgess MA, Bolejack V, Schuetze S, et al. Clinical activity of pembrolizumab (P) in undifferentiated pleomorphic sarcoma (UPS) and dedifferentiated/pleomorphic liposarcoma (LPS): final results of SARC028 expansion cohorts. J Clin Oncol. 2019;37:11015.
57. Beird HC, Wu CC, Ingram DR, et al. Genomic profiling of dedifferentiated liposarcoma compared to matched well-differentiated liposarcoma reveals higher genomic complexity and a common origin. Cold Spring Harb Mol Case Stud. 2018;4:a002386.
58. Snow H, Mitchell C, Hendry S, et al. Characterising the immune microenvironment in liposarcoma, its impact on prognosis and the impact of radiotherapy. J Surg Oncol. 2021;123:117–26.
59. Pollack SM, Jungbluth AA, Hoch BL, et al. NY-ESO-1 is a ubiquitous immunotherapeutic target antigen for patients with myxoid/round cell liposarcoma. Cancer. 2012;118:4564–70.
60. Somaiah N, Chawla SP, Block MS, et al. A phase 1b study evaluating the safety, tolerability, and immunogenicity of CMB305, a lentiviral-based prime-boost vaccine regimen, in patients with locally advanced, relapsed, or metastatic cancer expressing NY-ESO-1. Oncoimmunology. 2020;9:1847846.
61. Casadei L, Calore F, Braggio DA, et al. MDM2 derived from dedifferentiated liposarcoma extracellular vesicles induces MMP2 production from preadipocytes. Cancer Res. 2019;79:4911–22.
62. Nacev BA, Jones KB, Intlekofer AM, et al. The epigenomics of sarcoma. Nat Rev Cancer. 2020;20(10):608–23.
63. Lee W, Teckie S, Wiesner T, et al. PRC2 is recurrently inactivated through EED or SUZ12 loss in malignant peripheral nerve sheath tumors. Nat Genet. 2014;46:1227–32.
64. Filippakopoulos P, Picaud S, Mangos M, et al. Histone recognition and large-scale structural analysis of the human bromodomain family. Cell. 2012;149:214–31.
65. Shurell E, Singh AS, Crompton JG, et al. Characterizing the immune microenvironment of malignant peripheral nerve sheath tumor by PD-L1 expression and presence of CD8+ tumor infiltrating lymphocytes. Oncotarget. 2016;7:64300–8.
66. Wilson BG, Wang X, Shen X, et al. Epigenetic antagonism between polycomb and SWI/SNF complexes during oncogenic transformation. Cancer Cell. 2010;18:316–28.
67. Gounder M, Schoffski P, Jones RL, et al. Tazemetostat in advanced epithelioid sarcoma with loss of INI1/SMARCB1: an international, open-label, phase 2 basket study. Lancet Oncol. 2020;21:1423–32.
68. Chun HE, Johann PD, Milne K, et al. Identification and analyses of extra-cranial and cranial rhabdoid tumor molecular subgroups reveal tumors with cytotoxic T cell infiltration. Cell Rep. 2019;29:2338–2354.e2337.
69. Burr ML, Sparbier CE, Chan KL, et al. An evolutionarily conserved function of polycomb silences the MHC class I antigen presentation pathway and enables immune evasion in cancer. Cancer Cell. 2019;36:385–401.e388.
70. Zhou L, Mudianto T, Ma X, et al. Targeting EZH2 enhances antigen presentation, antitumor immunity, and circumvents Anti-PD-1 resistance in head and neck cancer. Clin Cancer Res. 2020;26:290–300.
71. Gounder MM, Zhu G, Roshal L, et al. Immunologic correlates of the abscopal effect in a SMARCB1/INI1-negative poorly differentiated chordoma after EZH2 inhibition and radiotherapy. Clin Cancer Res. 2019;25:2064–71.

Part XVI

The Role of Radiation in the Treatment of Primary and Metastatic Cancer

Baldassarre Stea

Keywords Radiation sensitivity;Stereotactic radiosurgery;Stereotactic body radiation therapy;Image-guided radiation;MR-Linac;Immune checkpoint inhibitors

Along with surgery and chemotherapy, radiation therapy is a widely utilized treatment modality in the field of oncology, with an estimated 50–60% of all cancer patients usually benefitting from its application during the course of their illness [1]. The feasibility of radiation treatments for cancer can be dated back to the discovery of X-rays by Wilhelm Röntgen in 1885, although at that time the potential of X-rays was limited to diagnostic imaging due to their poor penetration through tissues [2]. Today, radiation therapy is used for both palliation of cancer symptoms, such as bleeding, spinal cord compression, or painful bony metastases, and cure of many disease sites such as breast, prostate, lung, and head and neck. As such, radiation therapy has become an integral part of combined multimodality treatment approaches to many cancers.

Part XVI is devoted to multiple aspects of radiotherapy, including a chapter on basic principles of radiobiology, one on technical factors of radiotherapy, another on the definitive treatment of cancer, and three different chapters addressing the role of radiation for metastatic disease. The chapter on radiobiology provides the biological basis for the modality. In this chapter, Harryman and Cress review the basic principles underpinning the generation of radiation damage. The generation and volume of base damage, single-strand DNA breaks (SSBs), and double-strand breaks (DSBs) are dose dependent upon ionizing radiation and the kinetics of repair and cellular outcomes are distinct if the damage is not repaired. The unit of radiation dose is Gray (Gy). One Gray of radiation results in approximately 105 ionization events per cell, which yield approximately 1000 SSBs and 40 DSBs. Radiation-induced DSBs are the lethal events most relevant to clinical radiotherapy. The authors review the five "Rs" of radiotherapy, which are responsible for the differential radiosensitivity of tumor tissue and normal tissue and include Reassortment (cell cycle position), Reoxygenation (of hypoxic cells), Repopulation (of tumor cells after each dose of radiation), Repair of DNA strand breaks, and intrinsic Radiosensitivity of the tissue irradiated. They finally review the tumor microenvironment, long recognized as a dose-limiting factor in the effectiveness of radiotherapy, and how the microenvironment can influence radiation sensitivity [3]. Finally, they review the role of hypoxia in the response to radiotherapy, which not only is a dose-limiting factor in the response but will also select for clones of aggressive disease phenotype, a phenomenon that can be detected in human tumors by multi-parametric MRI [4].

The delivery of radiation treatments relies on the use of targeted high-energy X-rays or charged particles (protons, carbon ions) produced by linear accelerators (Linacs), or radioactive substances contained in receptacles or catheters (brachytherapy) to induce death of malignant cells. When used with curative intent, the goal of radiation is to maximize the radiation dose delivery to the tumor while minimizing the exposure of adjacent normal tissues. It is, thus, imperative to be able to accurately target radiation delivery; this is accomplished by utilizing a variety of imaging technologies during the planning process, including CT, MRI, and PET scans. It is, thus, not surprising that the success of modern radiotherapy has paralleled that of medical imaging. The technology involved in modern treatment planning and delivery and the successful application of radiation are covered in great details in the following chapters of this section by Russell Hamilton. However, two major technical advances are worthy of mention here: one involving treatment planning and the other image guidance. These technical innovations that evolved over the past two decades in areas such as disease localization (CT-MRI-PET fusion), target motion management (4D-CT), treatment planning software, and dose delivery methodologies (stereotactic and intensity-modulated radiation) have indeed revolutionized the field of radiation therapy resulting in better tumor control with fewer complications, thus providing improved patient care. The treatment planning process as well as the multiple imaging modalities available for accurate targeting of the tumor are elegantly reviewed in more detail in the chapter by Russell Hamilton, which also includes a thorough presentation of all major forms of radiation delivery methods (protons, heavy ions, and Flash radiotherapy) and the latest technical innovations that have led to improved patient outcomes.

The second major advance in technology occurred with the introduction of image-guided radiation (IGRT), whereby an image of the target (tumor) is obtained immediately before the delivery of the dose, to ensure correct localization of the disease. In order to avoid a geographic miss, and to account for daily internal motion of the target (variation in tumor position), the target volume (gross tumor volume or GTV) is usually expanded by a variable margin of normal tissue to create a planning target volume (PTV). This "safety" margin around a tumor differs with the degree of technology used and varies between 2 and 10 mm. Modern IGRT technology allows reduction in the margins used around a tumor, thus decreasing the chance of damaging surrounding normal tissues [5]. As the treatment margins became smaller and tighter around the GTV, it became imperative to ascertain the location of the target prior to treatment by taking a snapshot of the tumor just prior to the delivery of radiation. Most modern Linacs are now equipped with onboard imaging systems capable of obtaining a cone beam CT scan prior to each treatment; other forms of image guidance are orthogonal X-ray pairs, optical surface monitoring systems, and more recently MR guidance. This is also reviewed in much greater detail in one of the chapters in this section.

The constant need to continually reduce the PTV margins has led to the latest innovation in radiation oncology, the integration of MR imaging with a linear accelerator (MR-Linacs). MR imaging has superior soft-tissue contrast capabilities when compared to CT imaging. Therefore, MR-Linacs excel in targeting lesions within fleshy organs such as liver metastases, or cancers in the abdominal cavity (e.g., pancreatic cancer) which often blend with surrounding intestinal loops and other abdominal organs [6]. MR-Linacs are equipped with automated beam gating for precise and accurate dosing; they can track a tumor motion in real time and when the tumor moves (due to respiratory motion or patient positioning), the beam stops; this gives radiation oncologists the confidence to shrink the PTV margins down to a few millimeters while simultaneously escalating the dose. In addition to superior soft-tissue visualization, one of the most critical innovations afforded by MR-Linacs is their ability for integrated adaptive treatments with daily on-table re-optimization of the treatment plan. An adaptive treatment may be triggered by a change in tumor size or by the repositioning of an organ at risk close to the tumor, necessitating a new plan to avoid treating the normal tissue unnecessarily. By leveraging their superior imaging features and fast dose calculation algorithms, it is possible to deliver more powerful and effective doses of radiation in fewer sessions (stereotactic body radiation therapy or SBRT), a strategy which in turn paves the way to improved tumor control and patient outcomes [7].

Typical SBRT treatments involve the delivery of 1–5 fractions of radiation, with dose fractions of 6–30 Gy per fraction [8]. The high-precision targeting allows for a higher ablative dose, which improves the likelihood of achieving local control while also allowing for smaller margins for treatment setup, decreasing the amount of normal tissue irradiated. This technique has been increasingly used to treat patients with oligometastatic disease, as described in detail in the chapter by Ho and Hsu. In this setting, radiation therapy has been used to extend survival in patients with metastatic disease. In the recently published phase

2 SABR-COMET study by Palma et al., patients with a controlled primary malignancy and 1–5 extracranial metastatic lesions were randomized to either SBRT to all sites of metastatic disease or palliative standard of care, including palliative doses of radiation at symptomatic sites. Remarkably, not only the median PFS was improved in the SBRT arm compared to the control arm (11.6 vs. 5.4 months, p=0.001), but the overall survival was also improved in the SBRT arm, with median survival of 50 months, compared to 28 months in the standard-of-care arm (p = 0.006); furthermore, the 5-year OS was increased (42.3% vs. 17.7%, respectively). This application of radiation in the metastatic setting can, therefore, influence not only the quality but also the length of life of patients with limited number of metastases. For a more detailed reference on treating metastatic cancer, the reader is referred to a more comprehensive review [9].

One of the most common and painful presentations of metastatic cancer is in the form of spinal metastases which are often treated with low-dose palliative radiation (8–20 Gy). Dr. Ryu and colleagues elegantly review the evolution of spinal radiotherapy over the past three decades, presenting evidence of the efficacy of large single fractions of stereotactic radiation (SRS) or a few large fractions (SBRT) in the treatment of spinal metastases, providing evidence of improved pain control which is often the immediate goal of treatment of spinal metastases. The authors review the radiobiological principles behind the use of single large dose of radiation and critically report on two major clinical trials performed in the past decade, the NRG/RTOG 0631 comparing two different doses of radiation delivered by either standard external beam therapy (EBRT) or SRS and also the CCTG//TROG trial comparing a multi-fraction external beam with 2 fraction SBRT. Although the NRG/RTOG trial showed equivalency of either technique in achieving high degree of pain control (61.3% with SRS versus 43.7% with conventional EBRT at 12 months posttreatment, $p = 0.20$), the second trial reported superior results in pain control when the SBRT technique (24 Gy in 2 fractions) was utilized over the conventional EBRT technique (20 Gy in 5 fractions, $p < 0.001$). These results suggest that the more biologically effective dose of 24 Gy in 2 fractions may be the preferred and more efficient way to treat limited spinal metastases, although long-term results are still pending [10].

Recently, one of the most exciting developments in combined modality therapy of cancer has been seen in the treatment of melanoma brain metastases. Stereotactic radiosurgery is a common technique used for the treatment of small targets in the brain. On the other hand, immune checkpoint inhibitors have been shown to be very effective in the treatment of systemic metastatic melanoma. In a recent retrospective review [11], patients with melanoma metastatic to brain treated with SRS and ipilimumab had a favorable median survival of almost 36 months (compared to historical controls of about 12 months). As for the timing and sequencing of the two modalities, patients receiving immunotherapy concurrently with and within 30 days of SRS had significantly improved local control and regional intracranial progression-free survival compared with patients receiving sequential therapy with more than 30-day separation. This benefit appears to be limited to high-dose radiation, however, suggesting synergy between immune checkpoint inhibitors and radiosurgery.

Finally, one common use of radiation is in the postoperative setting where it is used as a prophylactic strategy to sterilize the operative bed and regional nodes and prevent locoregional failures; this, in turn, increases the cure rate of patients with high-risk features. This application of RT is widely diffused, being recommended from resected brain tumors to head and neck cancers, breast cancer, prostate cancer, sarcomas, and even pediatric cancers (such as Wilms' tumor). One example of the ability of radiotherapy to prevent metastatic spread of cancer and thus increase survival comes from a meta-analysis of 10,801 women in 17 randomized trials of radiotherapy versus no radiotherapy after breast-conserving surgery (BCS). Overall, RT reduced the risk of any recurrence by nearly half; specifically, it reduced the 10-year risk of any (i.e., locoregional or distant) first recurrence from 35.0% to 19.3% (absolute reduction 15.7%, $p < 0.00001$) and reduced the 15-year risk of breast cancer death from 25.2% to 21.4% (absolute reduction 3.8%, $p = 0.00005$). Therefore, with postoperative RT after BCS, one death was avoided at 15 years for every four locoregional recurrences prevented at 10 years [12].

In summary, although the landscape of oncologic treatments continues to evolve with highly promising discoveries in targeted small-molecule inhibitors and new immunotherapy drugs, the role of radiation remains prominent in the management of numerous malignancies in a number of ways including definitive radiotherapy alone, concurrent chemoradiation for locally advanced cancers, adjuvant radiotherapy following surgical resection of primary tumors, and finally in combination with immune checkpoint inhibitors for brain metastases. Following a succinct review of the basic biophysical principles involved in the administration of radiation therapy, this section has separate chapters covering the current approach in the modern treatment of the major malignancies afflicting mankind (breast, prostate, lung, cervical/endometrial cancer) and the palliative role of radiation for metastatic cancer.

References

1. Delaney G, Jacob S, Featherstone C, Barton M. The role of radiotherapy in cancer treatment: estimating optimal utilization from a review of evidence-based clinical guidelines. Cancer. 2005;104(6):1129–37. https://doi.org/10.1002/cncr.21324.
2. Gianfaldoni S, Gianfaldoni R, Wollina U, Lotti J, Tchernev G, Lotti T. An overview on radiotherapy: from its history to its current applications in dermatology. Open Access Maced J Med Sci. 2017;5(4):521–5. https://doi.org/10.3889/oamjms.2017.122.
3. Krisnawan VE, Stanley JA, Schwarz JK, DeNardo DG. Tumor microenvironment as a regulator of radiation therapy: new insights into stromal-mediated Radioresistance. Cancers. 2020;12(10) https://doi.org/10.3390/cancers12102916.
4. Gaustad JV, Hauge A, Wegner CS, Simonsen TG, Lund KV, Hansem LMK, Rofstad EK. DCE-MRI of tumor hypoxia and hypoxia-associated aggressiveness. Cancers. 2020;12(7) https://doi.org/10.3390/cancers12071979.
5. De Los SJ, Popple R, Agazaryan N, Bayouth JE, Bissonnette JP, Bucci MK, Dieterich S, Dong L, Forster KM, Indelicato D, Langen K, Lehmann J, Mayr N, Parsai I, Salter W, Tomblyn M, Yuh WT, Chetty IJ. Image guided radiation therapy (IGRT) technologies for radiation therapy localization and delivery. Int J Radiat Oncol Biol Phys. 2013;87(1):33–45. https://doi.org/10.1016/j.ijrobp.2013.02.021.
6. Witt JS, Rosenberg SA, Bassetti MF. MRI-guided adaptive radiotherapy for liver tumours: visualising the future. Lancet Oncol. 2020;21(2):e74–82. https://doi.org/10.1016/S1470-2045(20)30034-6.
7. Boldrini L, Cusumano D, Cellini F, Azario L, Mattiucci GC, Valentini V. Online adaptive magnetic resonance guided radiotherapy for pancreatic cancer: state of the art, pearls and pitfalls Radiat Oncol;14(1):71. 2019; https://doi.org/10.1186/s13014-019-1275-3.
8. Palma DA, Olson R, Harrow S, Gaede S, Louie AV, Haasbeek C, Mulroy L, Lock M, Rodrigues GB, Yaremko BP, Schellenberg D, Ahmad B, Griffioen G, Senthi S, Swaminath A, Kopek N, Liu M, Moore K, Currie S, Bauman GS, Warner A, Senan S. Stereotactic ablative radiotherapy versus standard of care palliative treatment in patients with oligometastatic cancers (SABR-COMET): a randomised, phase 2, open-label trial. Lancet. 2019;393(10185):2051–8. https://doi.org/10.1016/S0140-6736(18)32487-5.
9. Stea B, Skrepnik T, Hsu CC, Abendroth R. The role of radiation therapy in the treatment of metastatic cancer. Clin Exp Metastasis 2018 Aug;35(5-6):535-546. doi: https://doi.org/10.1007/s10585-018-9926-6.
10. Sahgal A, Myrehaug SD, Siva S, Masucci L, Foote MC, Brundage M, Butler J, Chow E, Fehlings MG, Gabos Z, Greenspoon J, Kerba M, Lee YK, Liu MC, Maralani P, Thibault I, Wong R, Hum M, Ding K, Parulekar W. CCTG SC.24/TROG 17.06: a randomized phase II/III study comparing 24Gy in 2 stereotactic body radiotherapy (SBRT) fractions versus 20Gy in 5 conventional palliative radiotherapy (CRT) fractions for patients with painful spinal metastases. Int J Radiat Oncol Biol Phys. 2020;108(5):1397–8. https://doi.org/10.1016/j.ijrobp.2020.09.019.
11. Skrepnik T, Sundararajan S, Cui H, Stea B. Improved time to disease progression in the brain in patients with melanoma brain metastases treated with concurrent delivery of radiosurgery and ipilimumab. Onco Targets Ther. 2017;19;6(3):e1283461 https://doi.org/10.1080/2162402X.2017.1283461.
12. Early Breast Cancer Trialists' Collaborative Group (EBCTCG), Darby S, McGale P, Correa C, Taylor C, Arriagada R, Clarke M, Cutter D, Davies C, Ewertz M, Godwin J, Gray R, Pierce L, Whelan T, Wang Y, Peto R. Effect of radiotherapy after breast-conserving surgery on 10-year recurrence and 15-year breast cancer death: meta-analysis of individual patient data for 10,801 women in 17 randomised trials. Lancet. 2011;378(9804):1707–16. https://doi.org/10.1016/S0140-6736(11)61629-2.

Basic Principles of Radiobiology and Cancer Metastasis Prevention

62

William L. Harryman and Anne E. Cress

Abstract

Basic biological principles of tumor cells and tissue responses to therapeutic ionizing radiation are presented in the context of metastasis prevention. Focuses include a description of cell survival, reassortment (cell cycle position), reoxygenation, repopulation, repair, radiosensitivity, and the influence of the hypoxic tumor microenvironment. We emphasize understanding the confounding factors of the tumor microenvironment and tumor interaction with ever-changing microenvironments during tumor progression, including examples of how the tumor microenvironment influences tumor sensitivity to ionizing radiation. Microenvironments change in response to tumor invasion, and irradiation influences the microenvironment, suggesting that integrated responses to radiation occur between the tumor and the microenvironment. Future research will require increased understanding of both tumor subtypes and the spatial and temporal interactions of the microenvironment as tumors locally invade and metastasize. Prostate cancer acts as an example in which extraprostatic extension of disease into the periprostatic adipose tissue has unknown influences on radiation therapy responses but may account, in part, for the reported limited therapeutic responses in obese patients. Understanding the integrated responses of tumor with the changing microenvironment during the early stages of metastasis holds potential for identifying new targets to improve tumor eradication by ionizing radiation and preventing tumor metastasis or recurrent disease.

W. L. Harryman
The University of Arizona Cancer Center, Tucson, AZ, USA
e-mail: wharryman@email.arizona.edu

A. E. Cress (✉)
Department of Cellular and Molecular Medicine, The University of Arizona Cancer Center, Tucson, AZ, USA
e-mail: cress@arizona.edu

Learning Objectives

- Fundamental understanding of the biological effects of therapeutic radiation on major cell biological functions.
- Understand five biological factors that influence the radiation response in cells and tissues.
- Understand the dominant DNA damage repair pathways induced by ionizing radiation.
- Familiarity with the potential consequences of unrepaired DNA damage.
- Understanding intrinsic and acquired radiation resistance.
- Define the hypoxic tumor environment, and understand why it influences radiation therapy response.
- Appreciate the complexity and dynamic nature of the cellular and tumor microenvironment.
- Familiarity with new research areas that will address the integrative nature of tumor response to therapy.

62.1 Introduction

The biological effects of radiation principally involve damage to deoxyribonucleic acid (DNA), which is the critical target, damaged via the hydroxyl radical as described in *Radiobiology for the Radiologist* [1]. For the clinical use of ionizing radiation, a major limitation is the consideration of normal tissue tolerance and understanding the unique balance between tumor cell kill and surrounding normal tissue toxicity. The understanding of radiation therapy tolerance limits and the estimation of individual tolerance is a major research area to avoid the severity of late connective tissue damage after therapy [2]. In this section, we will focus our attention on the tumor-specific biological responses to IR

S. P. Leong et al. (eds.), *Cancer Metastasis Through the Lymphovascular System*,
https://doi.org/10.1007/978-3-030-93084-4_62

therapy, including cell survival, DNA damage repair, cell cycle arrest, and tumor biology responses.

62.2 Tumor Cell Survival to Radiation

Tumor cell survival to a radiation dose is determined using a colony-forming assay that measures the clonogenic capacity of individual cells for up to 14 days after a radiation exposure. This measure is chosen since metabolic assays of proliferation will underestimate cell killing since cells can die immediately (called interphase death) or die after several cell divisions (called proliferative death) following exposure to radiation. The colony-forming assay will estimate total cell death from radiation exposure independent of metabolism or the cell death pathway used. In effect, the ability to produce progeny determines cell survival. Typical cell survival curves are shown in Fig. 62.1.

At low doses of radiation, the shape of the survival curve (black line, Fig. 62.1) contains a "shoulder" region, interpreted to mean that within the tumor cell population, effective DNA repair has occurred within 14 days post-irradiation, also referred to as sublethal damage. In contrast, at higher radiation doses, the shape of the survival curve now becomes linear on the logarithmic scale, indicating more lethal damage has occurred and an accumulation of unrepaired DNA damage is expected. The linear portion of the survival curve is used to calculate the slope (called D_0 or mean lethal dose) to provide a comparison between different tumor types. Tumor cells will die from radiation treatment by a variety of mechanisms including mitotic catastrophe, apoptosis (in hematopoietic derived cells), necroptosis, and proliferative death (epithelial derived cells).

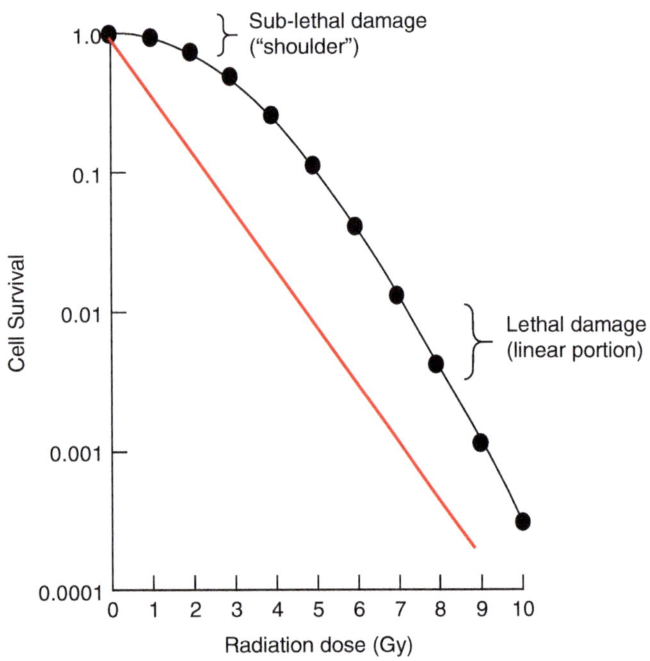

Fig. 62.1 Idealized tumor cell survival data. Cell survival is measured by determining the survival of cells (on a logarithmic scale) 14 days after exposure to increasing doses of radiation (Gy). Typical sublethal damage occurs with low doses of radiation (shoulder), while lethal damage is observed with higher doses (linear portion). In some radiation-sensitive cell types, sublethal damage is not observed (red survival data)

62.3 Several Biological Factors Influence Tumor Cell Survival to Radiation Exposure

These are referred to as the "5 Rs" and include reassortment (cell cycle position), reoxygenation, repopulation, repair, and radiosensitivity. We will take each of these in turn in the following sections.

62.3.1 Reassortment

In response to radiation, a tumor cell population will respond with up to three dominant cell cycle checkpoints. Arrest in G1 (via cyclin D and Cdk4/6) occurs to allow repair of DNA damage before DNA replication can occur. An arrest in S phase (via cyclin E/CdK2) occurs to repair DNA damage early in the DNA replication phase. Arrest in G2 occurs (via cyclin B/CdK1) to repair DNA damage before mitosis. Tumor cells are most sensitive to radiation therapy during mitosis, G1, and G2 phases of the cell cycle and most resistant during S phase. Within the same tumor cells, those cells irradiated in late S phase will result in a shoulder region, whereas cells irradiated within mitosis will display a linear survival curve without a shoulder region. Since the duration of mitosis and S phase are very similar between cell types, tumor cells that have longer cell cycle times most likely have longer G1 intervals. Longer G1 intervals also can result from adverse microenvironments such as low pH, hypoxia, and inflammation. We note that as tumors invade and metastasize to distant sites, the tumor microenvironment will change [3]. The successful radiation treatment of oligometastatic disease [4, 5] will likely require an understanding of the resulting cell cycle distribution or change in radiation responsiveness due to a new and changing microenvironment.

62.3.2 Reoxygenation

Since most DNA damage by radiation depends upon the hydroxyl radical, a dramatic effect of oxygen on radiation sensitivity occurs. Most transplantable tumors in animals have been shown to contain hypoxic cells that limit curability by single doses of X-rays. Hypoxic fractions vary from 0% to 50%, with a tendency to average about 15%. After single doses of radiation, the initial tumor cell kill will be followed by a reoxygenation of the residual tumor that was within the hypoxic regions. The oxygen enhancement of radiation sensitivity can vary up to a threefold increase depending on the cell type. Although it is usually assumed that the oxygen tension of most normal tissues is similar to that of venous blood or lymph (20–40 mmHg), in fact, oxygen probe measurements indicate that the oxygen tension may vary between different tissues over a wide range from 1 to 100 mmHg. There is strong evidence that human tumors contain hypoxic cells that would therefore be radiation resistant. This evidence includes histologic appearance, oxygen probe measurements, the binding of nitroimidazoles, PET and SPECT studies, and pretreatment hemoglobin levels. Many tissues are therefore borderline hypoxic and contain a small proportion of cells that are radio-biologically hypoxic. This is particularly true of, for example, normal liver and muscle or cancer tissue such as human prostate adenocarcinoma. Hypoxic gradients within tumor populations can affect treatment outcomes, and a major need is to identify new markers of these gradients for treatment planning.

62.3.3 Repopulation

Following single doses of radiation, there is a repopulation of the residual tumor, and this recapitulates the survival kinetics of the original tumor. Fractionated radiation schemas have been developed to increase the efficacy of the treatment, in effect, reducing the shoulder region of the survival curve. In model systems, the tumor growth delay induced by the radiation is the gold standard for the measurement of radiation sensitization of an intervention [1]. The use of tumor growth delay is needed since it can reveal a pre-clinical sensitization of the tumor without the confounding factors of tumor necrosis or cell loss. In addition, it has proven to be a reliable estimate of sensitization that is reflective in the clinical situation.

62.3.4 Repair

We consider here the major types of DNA damage induced by ionizing radiation generated by gamma or X-rays for therapeutic intent and the corresponding major types of DNA repair. DNA damage occurs primarily by hydroxyl radical damage to the phosphate-sugar backbone of DNA resulting in base damage, single-stranded breaks (SSB), or double-stranded breaks (DSB). Damaged non-helix-distorting bases are repaired by base excision repair (BER), or bulky helix-distorting lesions are removed by nucleotide excision repair (NER) operating throughout the cell cycle. BER is important for removing damaged bases that could otherwise cause mutations by mispairing or lead to breaks in DNA during replication. The generation and abundance of base damage, SSBs, and DSBs are dose dependent upon ionizing radiation, and the kinetics of repair and cellular outcomes are distinct if the damage is not repaired (Table 62.1). One gray (Gy) of radiation results in approximately 10^5 ionizations per cell, which yields 1000 SSBs and 40 DSBs. Radiation-induced DSBs are the lethal events relevant to clinical radiotherapy.

The two dominant pathways for DSB repair are non-homologous end joining (NHEJ) and homologous recombination (HR). NHEJ is characterized by direct re-joining of the two ends, which frequently generates small mutations at the repair site [6]. This form of DNA repair is active throughout the cell cycle and repairs approximately 85% of the radiation-induced DSBs. NHEJ uses a protein complex of the Ku70/Ku80 proteins as a scaffold structure upon which other enzymatic proteins (DNA PKCs, Artemis endonuclease, ligase IV/XRCC4 complex) are recruited to accomplish repair. NHEJ is also used in normal processes such as the VDJ recombination of IgG or TCR genes in B and T cell maturation.

HR initiates with resection to form single-stranded DNA (ssDNA), which invades "donor" homologous templates for DNA synthesis and repair [7]. HR repairs DNA in the S and G2 phases of the cell cycle before the cell enters mitosis (M phase) [8]. HR uses a protein complex of sensor and effector proteins (RPA, Rad51, BRCA2, and others) to accomplish largely error-free repair.

The cell cycle appears to largely determine whether HR (in S and G2) or NHEJ (all phases, including G_0) is used to repair double-stranded breaks, and since there is a potential for overlap, the determinants of pathway choice are an unresolved question. Given the potential for errors with the NHEJ pathway, one of the remaining questions is how pathway choices are made by targeted cells [9]. If the damage is not repaired or repaired by an error-prone process, the result is chromosome fragmentation, loss of DNA fragments (deletions), chromothripsis, insertion of DNA fragments in a wrong location, or chromosome aberrations, which at sub-

Table 62.1 Comparison of dominant DNA repair pathways in response to ionizing radiation

Repair pathway	DNA damage	Incidence/cell (1 Gy X-ray)	Repair mechanism	Cell death or senescence
BER/NER	Base/nucleotide	1000	Protein complex; easy/fast	NO
HR/NHEJ	SSB	1000	Protein complex; easy/fast	NO
HR/NHEJ	DSB	40	Protein complex; difficult/slow	YES

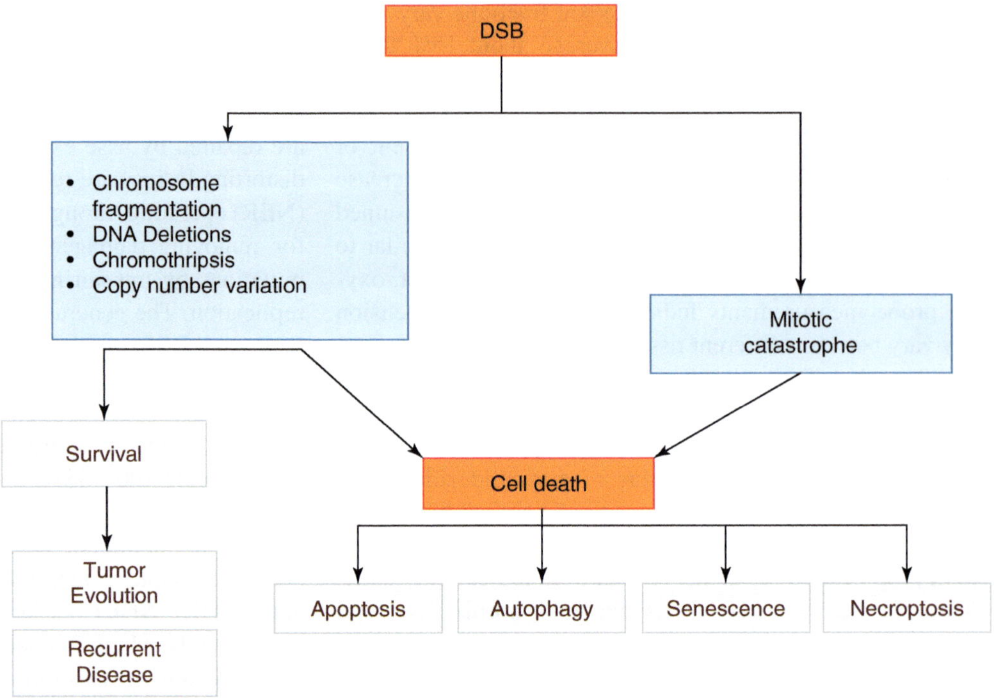

Fig. 62.2 Potential biological consequences of unrepaired or error-prone repair of DSB damage. Different types of chromosome damage have been observed that can persist in surviving tumor cell populations or result in cell death by at least four different pathways. Surviving cell populations with chromosome damage may account for the emergence of aggressive tumors and recurrent disease

lethal radiation doses may be a factor in tumor evolution or recurrent disease. Figure 62.2 illustrates the potential consequences of DSB on tumor biology. Chromatin and nuclear architecture are other emerging considerations in optimization of DNA repair [10] and may reveal additional targeting candidates as adjuvants to radiation-induced tumor cell killing. This is especially important since the basal cell population of epithelial-derived tumors contains heterochromatin, and how this form of chromatin is repaired is largely unknown [11].

Markers of DNA damage repair include the phosphorylated form of histone H2A, called γH2AX, which leads to chromatin decondensation after DNA double-stranded breaks to facilitate repair. The histone variant H2AX constitutes about 10% of the H2A histones in human chromatin and can be detected as soon as 20s after the DSB is induced by ionizing radiation with half of the maximum production of γH2AX occurring in 1 min [12]. γH2AX is required for the accumulation of many sensor/effector proteins at the DSB site [13].

The clinical assessment of DNA repair capacity and determining the most robust assay to complement the ongoing clinical trials targeting specific DNA repair proteins for radiation sensitization is a major unmet need [14]. Patient-level DNA damage and repair pathway profiles have been reported to provide a strong prognostic performance with long-term outcomes and may be useful for risk stratification of prostate cancer patients, a population where radiation therapy is a treatment modality [15]. In another example, loss of NHEJ will sensitize human prostate tumors to radiation [16]. Current research efforts to utilize the emerging technology of single-cell analysis from biopsy specimens and the power of positional digital profiling now used to map the tumor microenvironment [17] will likely assist radiation sensitization efforts. In this strategy, it would be important to apply an in situ analysis of DNA repair complex assembly in the biopsy analysis workflow since single markers are not sufficient; markers such as γH2AX or RPA (replication protein A) will bind DNA independent of the DNA damage response [13, 18].

62.3.5 Radiosensitivity

Radiosensitivity of different cell types is striking when considering that the hematopoietic stem cells are particularly radiosensitive as compared to epithelial cells. Their survival curves have little or no shoulder regions and have a mean lethal dose (D_0) of slightly less than 1 Gy as compared to epithelial cells of 2Gy or higher. Hematopoietic tissues are located primarily in the bone marrow, with 60% located in the pelvis and vertebrae and the remainder in the ribs, skull, sternum, scapula, and proximal sections of the femur and humerus. Lymphocytes are very radiosensitive, largely because of apoptosis, and B cells are more radiosensitive than T cells. Total lymphoid irradiation to a dose of 30–40 Gy is used for the treatment of lymphomas and leads to a long-lasting T cell lymphopenia.

Different cell types within the same organ can also display differences in radiosensitivity. For example, in the normal colon, the differentiated villus cells are radiation resistant as compared to the stem cells in the crypt. This explains the prodromal effects of radiation in which the loss of fluid and massive infections post-irradiation are due to loss of the villus layer. Epithelial tumor cell populations of the lung are another example in which squamous cell carcinoma and adenocarcinoma are more radiation resistant as compared to small-cell carcinoma, in part since one of the required NHEJ DNA repair components, Ku80, is increased [19]. The use of integrated omics profiling of treatment-naïve tumor is likely to reveal more patterns of molecular subtyping that can be exploited to individualize treatments, as shown recently in prostate cancer where the basal cell subtype appears to be more responsive to radiation therapy [20].

62.4 Acquired Radiation Resistance

An enhancement of tumor survival to radiation damage occurs in vivo and in vitro when tumors are maintained under non-growth conditions after treatment. The repair of potentially lethal damage was first uncovered as a variable that was unaccounted for in standard assays of cytotoxic agents [21]. Today this underscores the potential of dormant tumor cells or tumor stem cells within the heterogeneous tumor cell population to be a source of recurrent disease following therapy. The acquired or adaptive response of the tumor population and the ability to modify it are important aspects of fractionated radiotherapy [22].

Previously recognized as a dose-limiting factor for radiation therapy and a component of normal tissue tolerance, it is now recognized that the tumor microenvironment and its response will influence the radiation sensitivity of recurrent disease [23]. Detailed knowledge of these interactions will likely be another biological avenue to exploit for increasing the efficacy of radiation therapy beyond radiation dose fractionation toward improving outcomes for the individual patient [24].

Cell adhesion-mediated radioresistance (CAM-RR) and cell adhesion-mediated drug resistance (CAM-DR) are both driven by integrin- and cytokeratin-dependent mechanisms [25–27]. The blocking of integrin function can enhance radiotherapy efficacy in pre-clinical models of breast cancer [28] and rare incurable cancers such as chordoma [29]. The radiation sensitization of tumor clusters dependent on integrin function and adhesion complexes is an actionable step that avoids the problem of single cell heterogeneity within a tumor cluster. The use of synthetic ligand mimetics or integrin blocking function antibodies can sensitize tumor clusters to ionizing radiation or increase reactive oxygen species resulting in apoptosis and necrotic cell death [29–31]. Recent results suggest that the linkages between the tumor cell surface and the coordination of nuclear events, like HR-dependent DSB repair, may account for the ability of β1 integrin (also known as CD29) antagonists to sensitize tumor cells to the lethal effects of radiation treatment [32]. Alternatively, the repair of heterochromatin in response to radiation treatment directly involves the cytoskeletal system, which has known linkages from the cell surface to the inner nuclear membrane, where heterochromatin resides [11].

62.5 The Hypoxic Tumor Microenvironment

Tumors contain hypoxia gradients which can be observed [33, 34]. Hypoxia not only decreases the tumor response to radiation therapy but also selects for aggressive disease as detected by mpMRI studies in patients [35]. The investigation of 2000 patients with prostate cancer revealed that there is an apparent selection for PTEN-minus aggressive disease by hypoxia [36]. Hypoxic suppression of HR is proposed as a "contextual-lethal" vulnerability in prostate tumors [37]. There is evidence that patients with prostate cancer will have areas of both chronic and acute hypoxia, both detectable by pimonidazole, a compound that is reductively activated in hypoxic cells. An antibody reagent MAb1 binds to the resulting adducts allowing their detection by immunochemical means. The advent of the oral formulation may allow novel clinical trials that intensify cancer therapy to prostate cancer patients whose tumors harbor aggressive genetic changes and hypoxic subregions. Understanding and mapping the hypoxic regions may help to stratify patients into risk categories, aid with appropriate radiation treatment planning, and result in the rational design of combination therapies to prevent metastasis.

62.6 The Complex Cellular and Tumor Microenvironment

The tumor microenvironment contains many different cell types including immune effector cells, normal cell types associated with vessels or nerves, fibroblasts, and smooth muscle cells. For example, the multi-spatial mapping of prostate cancer reveals a wide variety of myeloid and macrophages as detected by multiplex tissue staining in close proximity to the tumor [38]. Thus, it would be expected that irradiation of these cells along with the tumor may impact tumor response to the radiation treatment. As stated earlier, while it is well known that irradiation can increase interstitial collagen deposition, it is now appreciated that in addition to the extracellular structural elements of the tissue, growth factors such as TGF-beta are radiation induced [39] which reorganize the extracellular matrix and induce tumors to alter their phenotypes. Since many different cell types are present in the tumor microenvironment, it remains to be determined how these various cell types in different combinations may affect tumor radiation responses as measured by overall patient survival.

In the case of prostate cancer, when the tumor leaves the gland and extends past the capsule (known as extraprostatic extension, EPE), patient outcomes drop from an overall survival of 100% at 5 years to approximately 30%. EPE is most readily detected by entry of the tumor into the periprostatic adipose tissue (PPAT), since adipose tissue is not present within the prostate gland itself. Thus, EPE of the prostate cancer changes the tumor microenvironment from that of a tensile smooth muscle to adipose tissue [3]. Both the biophysical changes and the biochemical change of the environment will likely be important since adipocytes secrete an inflammatory cytokine called extracellular nicotinamide phosphoribosyltransferase (eNAMPT), and this is associated with prostate cancer progression [40]. The influence of this microenvironment change on prostate tumor radiation responses is unknown, but we note with interest that higher levels of body fat and higher visceral fat are related to poorer outcomes in radiation therapy [41–43]. Future research to understand the influences of the varied tumor microenvironments in combination with molecular subtyping of the tumor will be needed to fully realize the long-term goal of individualizing radiation therapy for a maximum therapeutic response.

Educational Scope of the Chapter

- How can sublethal damage and lethal damage by radiation be determined with data from a survival curve?
- Name and define five biological factors that influence tumor cell survival to radiation exposure.
- What are the three most dominant DNA repair responses to ionizing radiation damage? List at least three factors that distinguish them from each other.
- If tumors do not repair double-stranded DNA breaks and survive it, name at least two potential consequences that would impact radiation therapy.
- Give two examples of intrinsic radiation sensitivity and two examples of acquired radiation resistance.
- Define tumor hypoxia, and provide an example of how knowing this information will alter radiation treatment planning and potentially prevent metastatic disease.
- Explain why molecular subtyping of the tumor and spatial phenotyping of the microenvironment will assist in personalizing radiation therapy in the future.

Future Questions

- How can hypoxic tissues be sensitized to radiotherapy?
- How can DSB repair be blocked to enhance radiation sensitivity of tumor cells?
- What categories of cell types and functions in the varied tumor microenvironments would aid in treatment selection?
- How does determining the radiosensitivity of the variety of cell types found in the tumor microenvironment aid in designing better radiotherapy treatment regimens?
- How might a multi-omics approach to radiotherapy increase its effectiveness?

References

1. Hall E, Giaccia A. Physics and chemistry of radiation absorption. In: Radiobiology for the radiologist. 7th ed. Philadelphia, PA: Lippincott Williams & Wilkins; 2012. p. 3–11.
2. PetersLJ.Radiationtherapytolerancelimits.Foroneorforall?–Janeway Lecture. Cancer. 1996;77(11):2379–85. https://doi.org/10.1002/(sici)1097-0142(19960601)77:11<2379::Aid-cncr29>3.0.Co;2-t.
3. Harryman WL, Warfel NA, Nagle RB, Cress AE. The tumor microenvironments of lethal prostate cancer. Adv Exp Med Biol. 2019;1210:149–70. https://doi.org/10.1007/978-3-030-32656-2_8.
4. Hellman S, Weichselbaum RR. Oligometastases. J Clin Oncol. 1995;13(1):8–10. https://doi.org/10.1200/jco.1995.13.1.8.
5. Rao A, Vapiwala N, Schaeffer EM, Ryan CJ. Oligometastatic prostate cancer: a shrinking subset or an opportunity for cure? American Society of Clinical Oncology educational book American Society of Clinical Oncology Annual Meeting; 2019;39:309–320. https://doi.org/10.1200/edbk_239041.
6. Chang HHY, Pannunzio NR, Adachi N, Lieber MR. Non-homologous DNA end joining and alternative pathways to double-strand break repair. Nat Rev Mol Cell Biol. 2017;18(8):495–506. https://doi.org/10.1038/nrm.2017.48.
7. Kowalczykowski SC. An overview of the molecular mechanisms of recombinational DNA repair. Cold Spring Harb Perspect Biol. 2015;7(11) https://doi.org/10.1101/cshperspect.a016410.
8. Alberts B, Johnson A, Lewis J, Raff M, Roberts K, Walter P. Molecular biology of the cell. 5th ed. New York: Garland Science; 2008.
9. Carballar R, Martínez-Láinez JM, Samper B, Bru S, Bállega E, Mirallas O, Ricco N, Clotet J, Jiménez J. CDK-mediated Yku80 phosphorylation regulates the balance between non-homologous end joining (NHEJ) and homologous directed recombination (HDR). J Mol Biol. 2020;432(24):166715. https://doi.org/10.1016/j.jmb.2020.11.014.
10. Mackenroth B, Alani E. Collaborations between chromatin and nuclear architecture to optimize DNA repair fidelity. DNA Repair. 2020;97:103018. https://doi.org/10.1016/j.dnarep.2020.103018.
11. Rawal CC, Caridi CP, Chiolo I. Actin' between phase separated domains for heterochromatin repair. DNA Repair. 2019;81:102646. https://doi.org/10.1016/j.dnarep.2019.102646.

12. Rogakou EP, Pilch DR, Orr AH, Ivanova VS, Bonner WM. DNA double-stranded breaks induce histone H2AX phosphorylation on serine 139. J Biol Chem. 1998;273(10):5858–68. https://doi.org/10.1074/jbc.273.10.5858.
13. Yuan J, Adamski R, Chen J. Focus on histone variant H2AX: to be or not to be. FEBS Lett. 2010;584(17):3717–24. https://doi.org/10.1016/j.febslet.2010.05.021.
14. Jalal S, Earley JN, Turchi JJ. DNA repair: from genome maintenance to biomarker and therapeutic target. Clin Cancer Res. 2011;17(22):6973–84. https://doi.org/10.1158/1078-0432.Ccr-11-0761.
15. Evans JR, Zhao SG, Chang SL, Tomlins SA, Erho N, Sboner A, Schiewer MJ, Spratt DE, Kothari V, Klein EA, Den RB, Dicker AP, Karnes RJ, Yu X, Nguyen PL, Rubin MA, de Bono J, Knudsen KE, Davicioni E, Feng FY. Patient-level DNA damage and repair pathway profiles and prognosis after prostatectomy for high-risk prostate cancer. JAMA Oncol. 2016;2(4):471–80. https://doi.org/10.1001/jamaoncol.2015.4955.
16. Tarish FL, Schultz N, Tanoglidi A, Hamberg H, Letocha H, Karaszi K, Hamdy FC, Granfors T, Helleday T. Castration radiosensitizes prostate cancer tissue by impairing DNA double-strand break repair. Sci Transl Med. 2015;7(312):312re311. https://doi.org/10.1126/scitranslmed.aac5671.
17. de Vries NL, Mahfouz A, Koning F, de Miranda N. Unraveling the complexity of the cancer microenvironment with multidimensional genomic and cytometric technologies. Front Oncol. 2020;10:1254. https://doi.org/10.3389/fonc.2020.01254.
18. Cerrato A, Morra F, Celetti A. Use of poly ADP-ribose polymerase [PARP] inhibitors in cancer cells bearing DDR defects: the rationale for their inclusion in the clinic. J Exp Clin Cancer Res. 2016;35(1):179. https://doi.org/10.1186/s13046-016-0456-2.
19. Ye J, Ren Z, Gu Q, Wang L, Wang J. Ku80 is differentially expressed in human lung carcinomas and upregulated in response to irradiation in mice. DNA Cell Biol. 2011;30(12):987–94. https://doi.org/10.1089/dna.2010.1196.
20. Ben-Salem S, Hu Q, Liu Y, Alshalalfa M, Zhao X, Wang I, Venkadakrishnan VB, Senapati D, Kumari S, Liu D, Sboner A, Barbieri CE, Feng F, Billaud JN, Davicioni E, Liu S, Heemers HV. Diversity in androgen receptor action among treatment-naïve prostate cancers is reflected in treatment response predictions and molecular subtypes. Eur Urol Open Sci. 2020;22:34–44. https://doi.org/10.1016/j.euros.2020.10.002.
21. Richie JP, Weichselbaum RR, Little JB. Importance of repair of potentially lethal damage in assays of cytotoxic agents. Surgery. 1982;92(2):380–4.
22. Raaphorst GP, Malone S, Szanto J, Gray R. Severe normal tissue complication correlates with increased in vitro fibroblast radiosensitivity in radical prostate radiotherapy: a case report. Int J Cancer. 2000;90(6):336–42.
23. Krisnawan VE, Stanley JA, Schwarz JK, DeNardo DG. Tumor microenvironment as a regulator of radiation therapy: new insights into stromal-mediated radioresistance. Cancers. 2020;12(10) https://doi.org/10.3390/cancers12102916.
24. Schaue D, McBride WH. Opportunities and challenges of radiotherapy for treating cancer. Nat Rev Clin Oncol. 2015;12(9):527–40. https://doi.org/10.1038/nrclinonc.2015.120.
25. Bauman PA, Dalton WS, Anderson JM, Cress AE. Expression of cytokeratin confers multiple drug resistance. Proc Natl Acad Sci U S A. 1994;91(12):5311–4. https://doi.org/10.1073/pnas.91.12.5311.
26. Damiano JS, Cress AE, Hazlehurst LA, Shtil AA, Dalton WS. Cell adhesion mediated drug resistance (CAM-DR): role of integrins and resistance to apoptosis in human myeloma cell lines. Blood. 1999;93(5):1658–67.
27. Deville SS, Cordes N. The extracellular, cellular, and nuclear stiffness, a trinity in the cancer resistome - a review. Front Oncol. 2019;9:1376. https://doi.org/10.3389/fonc.2019.01376.
28. Park CC, Zhang HJ, Yao ES, Park CJ, Bissell MJ. Beta1 integrin inhibition dramatically enhances radiotherapy efficacy in human breast cancer xenografts. Cancer Res. 2008;68(11):4398–405. https://doi.org/10.1158/0008-5472.Can-07-6390.
29. Harryman WL, Gard JMC, Pond KW, Simpson SJ, Heppner LH, Hernandez-Cortes D, Little AS, Eschbacher JM, Cress AE. Targeting the cohesive cluster phenotype in chordoma via β1 integrin increases ionizing radiation efficacy. Neoplasia. 2017;19(11):919–27. https://doi.org/10.1016/j.neo.2017.08.005.
30. Nair RR, Emmons MF, Cress AE, Argilagos RF, Lam K, Kerr WT, Wang HG, Dalton WS, Hazlehurst LA. HYD1-induced increase in reactive oxygen species leads to autophagy and necrotic cell death in multiple myeloma cells. Mol Cancer Ther. 2009;8(8):2441–51. https://doi.org/10.1158/1535-7163.Mct-09-0113.
31. Pawar SC, Dougherty S, Pennington ME, Demetriou MC, Stea BD, Dorr RT, Cress AE. alpha6 integrin cleavage: sensitizing human prostate cancer to ionizing radiation. Int J Radiat Biol. 2007;83(11–12):761–7. https://doi.org/10.1080/09553000701633135.
32. Ahmed KM, Pandita RK, Singh DK, Hunt CR, Pandita TK. β1-integrin impacts Rad51 stability and DNA double-strand break repair by homologous recombination. Mol Cell Biol. 2018;38(9) https://doi.org/10.1128/mcb.00672-17.
33. Horsman MR, Mortensen LS, Petersen JB, Busk M, Overgaard J. Imaging hypoxia to improve radiotherapy outcome. Nat Rev Clin Oncol. 2012;9(12):674–87. https://doi.org/10.1038/nrclinonc.2012.171.
34. Horsman MR, Overgaard J. The impact of hypoxia and its modification of the outcome of radiotherapy. J Radiat Res. 2016;57 Suppl 1(Suppl 1):i90–8. https://doi.org/10.1093/jrr/rrw007.
35. Gaustad JV, Hauge A, Wegner CS, Simonsen TG, Lund KV, Hansem LMK, Rofstad EK. DCE-MRI of tumor hypoxia and hypoxia-associated aggressiveness. Cancers. 2020;12(7) https://doi.org/10.3390/cancers12071979.
36. Bhandari V, Hoey C, Liu LY, Lalonde E, Ray J, Livingstone J, Lesurf R, Shiah YJ, Vujcic T, Huang X, Espiritu SMG, Heisler LE, Yousif F, Huang V, Yamaguchi TN, Yao CQ, Sabelnykova VY, Fraser M, Chua MLK, van der Kwast T, Liu SK, Boutros PC, Bristow RG. Molecular landmarks of tumor hypoxia across cancer types. Nat Genet. 2019;51(2):308–18. https://doi.org/10.1038/s41588-018-0318-2.
37. Ashton J, Bristow R. Bad neighbours: hypoxia and genomic instability in prostate cancer. Br J Radiol. 2020;93(1115):20200087. https://doi.org/10.1259/bjr.20200087.
38. Saylor J, Ma Z, Goodridge HS, Huang F, Cress AE, Pandol SJ, Shiao SL, Vidal AC, Wu L, Nickols NG, Gertych A, Knudsen BS. Spatial mapping of myeloid cells and macrophages by multiplexed tissue staining. Front Immunol. 2018;9:2925. https://doi.org/10.3389/fimmu.2018.02925.
39. Barcellos-Hoff MH. Radiation-induced transforming growth factor beta and subsequent extracellular matrix reorganization in murine mammary gland. Cancer Res. 1993;53(17):3880–6.
40. Sun BL, Sun X, Casanova N, Garcia AN, Oita R, Algotar AM, Camp SM, Hernon VR, Gregory T, Cress AE, Garcia JGN. Role of secreted extracellular nicotinamide phosphoribosyltransferase (eNAMPT) in prostate cancer progression: Novel biomarker and therapeutic target. EBioMedicine. 2020;61:103059. https://doi.org/10.1016/j.ebiom.2020.103059.
41. Di Bella CM, Howard LE, Oyekunle T, De Hoedt AM, Salama JK, Song H, Freedland SJ, Allott EH. Abdominal and pelvic adipose tissue distribution and risk of prostate cancer recurrence after

radiation therapy. Prostate. 2020;80(14):1244–52. https://doi.org/10.1002/pros.24054.

42. Strom SS, Kamat AM, Gruschkus SK, Gu Y, Wen S, Cheung MR, Pisters LL, Lee AK, Rosser CJ, Kuban DA. Influence of obesity on biochemical and clinical failure after external-beam radiotherapy for localized prostate cancer. Cancer. 2006;107(3):631–9. https://doi.org/10.1002/cncr.22025.
43. Wang LS, Murphy CT, Ruth K, Zaorsky NG, Smaldone MC, Sobczak ML, Kutikov A, Viterbo R, Horwitz EM. Impact of obesity on outcomes after definitive dose-escalated intensity-modulated radiotherapy for localized prostate cancer. Cancer. 2015;121(17):3010–7. https://doi.org/10.1002/cncr.29472.

Technical Innovations in the Delivery of Radiation Therapy

63

Russell J. Hamilton

Abstract

Radiation oncology is a technology-driven field quickly adopting technical innovations to improve treatment outcomes. A few months after Röntgen's discovery of X-rays, therapeutic X-ray use was proposed and applied to treat cancer. Since knowledge of the biological effects of radiation increases with better capabilities to control radiation delivery, therapeutic techniques continually evolve as radiotherapy incorporates new technologies. This chapter describes innovations implemented in the current generation of radiotherapy machines. Significant improvements for imaging patients in the treatment room using computed tomography, magnetic resonance imaging, and optical or infrared surface detection combined with more precise radiation delivery are altering practice patterns. The current delivery systems enable treatment regimens having higher doses per fraction and fewer fractions with similar or better outcomes than previous generation machines. Patients' internal and external anatomies are monitored during radiation delivery. Any changes in patient's anatomy or tumor position in excess of pre-established parameters will trigger an automatic stoppage of the radiation delivery system by computer control through analysis of imaging data. Emerging technologies showing promise in preclinical and early clinical studies, including automated real-time treatment delivery to biologically functional regions identified by positron-emission tomography, combining radiotherapy with immunotherapy to treat local and metastatic disease, and ultrafast delivery of radiation at dose rates orders of magnitude greater than currently practiced, are described.

R. J. Hamilton (✉)
University of Arizona, Tucson, AZ, USA
e-mail: rjh@arizona.edu

Learning Objectives

- Know the precision and accuracy of radiation dose delivery achievable by the current generation of medical linear accelerators.
- Identify imaging systems employed during radiotherapy delivery and their purposes.
- Know techniques used for automated control of radiation delivery.
- Understand the principles of biology-guided radiation therapy.
- Know there is evidence for the role of radiation therapy in metastatic disease through the abscopal effect enhanced by immunotherapy.
- Know there is evidence for a radiobiological advantage of FLASH radiotherapy over present conventional radiotherapy.

63.1 Introduction

Radiation therapy is the delivery of radiation dose to a targeted volume in order to achieve a therapeutic outcome. Both malignant and benign diseases are treated with radiation. The dose, number of treatments, and frequency of treatment depend on the clinical situation. Ionizing radiation may penetrate several centimeters into a patient to deliver a lethal dose to a tumor. However, normal tissues such as the skin, muscle, organs, and structures lying in the entrance and exit of the radiation path receive radiation dose too. This could lead to fibrosis, organ failure, or secondary cancers several years after treatment. Thus, the technical goal of radiation therapy is to deliver a therapeutic dose to the targeted volume while minimizing the deleterious effects. This is expressed by the therapeutic ratio which is the ratio of the probability

S. P. Leong et al. (eds.), *Cancer Metastasis Through the Lymphovascular System*,
https://doi.org/10.1007/978-3-030-93084-4_63

of treatment success (e.g., tumor control probability) to the likelihood of normal tissue damage. The higher the therapeutic ratio, the greater the probability of a successful treatment without morbidity.

Shortly after the discovery of X-rays, pioneers of radiation therapy used the novel X-ray tube technology for treating cancers [1]. The X-ray energies were sufficient to pass through the body to produce an image, but much less than needed to treat deep-seated tumors. Treatments had limited efficacy because a toxic skin dose was reached before sufficient dose could be delivered to the targeted tissue within the body and thus a low therapeutic ratio. Microwave technology tremendously advanced medical linear accelerator (linac) design enabling production of X-ray energies >4 MV in a short distance [2]. Entrance doses for these X-ray energies are less than the dose at depth resulting in "skin sparing," greater penetration, and hence a higher therapeutic ratio. Now, linacs typically provide X-ray energies between 6 and 18MV with entrance doses much less than 50% of the maximum dose (D_{max}) which occurs at depths of 1.5–3 cm [3]. For tumors at depths greater than D_{max}, the dose at D_{max} is larger than at the tumor, if radiation is delivered with a single X-ray beam. Therefore, radiation beams from multiple directions are used so that their superimposed doses result in a high tumor dose and lower doses to normal tissues. Even lower entrance dose is achievable with heavy charged particle beams, such as produced by a medical proton accelerator. Heavy charged particles abruptly stop and deposit a significant amount of their energy at the end of their range (Bragg peak) and therefore have D_{max} at the end of their range and deposit most of their dose at the depth of the tumor. Thus, heavy charged particles have minimal exit dose, whereas X-ray dose decreases by only 2–5% per cm with significant exit dose [4].

Accurate radiation dosing of targets within a patient requires precise aiming of an ionizing radiation beam at the target. The radiation beam passes through the patient and necessarily irradiates normal tissues and organs along the entrance and exit of its path. The therapeutic ratio is increased by conforming the high-dose region as closely as possible to the target and maximizing the dose gradient which is the rate at which the dose decreases away from the target. A straightforward geometric approach to improving dose conforming is accomplished by shaping the aperture of the radiation beam to coincide with that of the target. Using beams from several different directions around the patient permits treatment beams that avoid passing through critical normal tissues and organs as well as increases the dose gradient.

Reducing a beam aperture size to as small as possible without missing the target improves the therapeutic ratio. However, radiation delivery systems are comprised of mechanical systems with inherent positional uncertainties that must be accounted for in the beam aperture in order to ensure that the target is irradiated. Precise treatments require accurate identification and geometric localization of targets prior to and during treatment delivery. Gross patient motions or internal physiological processes may cause targets and normal structures to move during radiation therapy delivery. Anything that causes the target to move must be accounted for in the beam aperture. A number of technologies that have been developed to enable accounting for these effects are discussed below.

In this chapter, we describe the first section is devoted to current delivery technologies. Firstly, the state-of-the-art radiation therapy delivery system available, the medical linear accelerator (linac), also the most widely used machine around the globe. This includes the entire process starting with pretreatment determination of the targets and radiation therapy treatment planning. The capabilities of linacs and their ancillary devices are discussed in detail. Examples of radiation treatments which highlight the technical sophistication of radiation therapy delivery are presented. Next, a machine which combines real-time magnetic resonance imaging with a linac, the MR-Linac, is described. A brief description of proton therapy machines concludes the section. The second section describes emerging technologies. First, a machine that combines positron-emission tomography with a linac is described. Next, FLASH radiotherapy, consisting of radiation dose delivery at rates that are orders of magnitude higher than conventional therapies, is introduced. Then a brief mention of carbon ion therapy is given. The promising treatment combination of immunotherapy with radiotherapy ends the section. A summary of the impact of the new technologies on radiotherapy concludes the chapter.

63.2 Radiation Therapy Delivery Technologies

63.2.1 Treatment Planning

The treatment process begins by identifying a target, for example, a tumor. All medical imaging technologies are utilized in the treatment planning process, including computed tomography (CT), magnetic resonance imaging (MR), positron-emission tomography (PET), single photon emission computed tomography (SPECT), and ultrasound. The particular imaging studies used for a given case depend on the disease site and stage. In addition to diagnostic exams, a radiation therapy patient undergoes a treatment planning imaging scan in a dedicated unit having ancillary equipment used in radiotherapy. This is typically performed using a CT simulator, but MR simulators are gaining in popularity. Each patient is fitted with an immobilization device for stabilizing and securing them during the treatment planning CT and each radiation therapy treatment. The CT simulation may include a single scan or a series of scans, for example, several CT series showing the respiratory cycle. The CT simulation scans and diagnostic scans of interest are transferred to a treatment planning computer.

Treatment planning computers utilize sophisticated software for designing radiation therapy treatments. The result is a series of radiation beams used to treat a patient. Firstly, the separate imaging data is registered (aka fused) with the treatment planning CT so that the internal anatomy from each imaging set may be directly compared. A significant technological development in the past decade was deformable registration, permitting registration of data sets accounting for deviations in a patient's position and/or internal displacements, for example, caused by edema or organ filling [5].

Next, a model of the patient is constructed by outlining the internal organs, for example, the lungs, heart, kidneys, spinal cord, etc. The tumor and any clinical or subclinical disease, for example, nodes, must also be outlined (aka segmented). This process may be carried out in the treatment planning software manually at a computer workstation using drawing tools. However, commercially available planning systems offer auto-segmentation of anatomical structures based on customizable atlases [6]. An atlas is comprised of specific anatomical structures defined as ground truth on a series of similar cases by an expert. Treatment planning software is able to automatically segment structures of interest based on an atlas selected by the radiation oncologist. Artificial intelligence in medical imaging might soon reach the point that a deep learning algorithm could perform the segmentation for treatment planning [7].

Next, a radiation oncologist specifies therapeutic doses for the targeted volumes and places dose constraints on the normal tissues with the intent that meeting these dose values will yield a high probability of successful treatment with a low probability of morbidity. The goal is to find a treatment plan, which is a set of radiation beams deliverable by the treatment machine, meeting the prescribed doses for targets and normal tissues. Treatment planning systems are not yet fully autonomous. Human planners must select the number of radiation beams, directions, and prioritize the dose constraints. These choices are usually based on class solutions for common disease sites but for unique or unusual situations requires creativity based on experience. Once these parameters are set, the planning software performs an optimization process to find a solution. It is not known a priori if a plan can be found meeting all constraints. Conversely, if a plan that meets all constraints is found, it is not known if a better plan exists. Knowledge-based treatment planning was developed to address these uncertainties. Knowledge-based treatment planning predicts the expected results, providing mean and standard deviation, for the target and normal tissue doses based on an atlas of similar cases [8]. This approach was shown to be superior to human planning [9, 10]. A goal of future treatment planning systems is to employ artificial intelligence software for treatment plan construction.

63.2.2 Medical Linear Accelerator

A state-of-the-art medical linear accelerator (linac), the TrueBeam (Varian Medical Systems, Palo Alto), is pictured in Fig. 63.1. A multileaf collimator (MLC) which has a minimum leaf width of 2.5 mm and submillimeter positional accuracy shapes the radiation beam. During radiation delivery, the MLC leaves move in and out of the radiation beam modulating the intensity providing intensity-modulated radiation therapy (IMRT). The gantry can rotate around the patient enabling radiation beams to strike targets from all angles (0–360°) with a mechanical precision better than 0.5 mm [11]. Volumetric modulated arc therapy (VMAT) is treatment delivery using IMRT simultaneously with gantry rotation. Another degree of freedom is provided by the patient couch which can rotate ±90° permitting treatment beams to enter in planes that are not coaxial. The couch has three translational degrees of freedom as well as three rotational degrees of freedom (pitch, roll, and yaw), enabling precise patient alignment for treatment delivery.

Treatment planning systems provide dose calculations accurate to better than a few percent, depending on the algorithm, and are accurate to 1% using a Monte Carlo algorithm [12]. In order to realize the therapeutic ratio of the prescribed treatment plan, patients must be positioned in the treatment room such that their radiation beams are directed as specified by the treatment plan. There are several options available to achieve this goal, including imaging systems mounted on the linac, in-room X-ray systems, and 3D cameras with computer vision.

Two imaging systems are mounted on the gantry of the linac shown in Fig. 63.1. The imaging panel in the radiation beam line opposite (180°) the collimator is a megavoltage imaging panel known as an electronic portal imaging device (EPID). It enables capturing images of the patient taken with the radiation beam (energies >4 MV). The radiation beam apertures may be viewed directly on the patient radiographs and compared with those expected from the treatment planning system (so-called digitally reconstructed radiographs). With the EPID, it is possible to reconstruct the dose that the patient actually received during a treatment session by acquiring a sequence of portal images during radiation therapy delivery [13].

The other gantry mounted imaging system is a kilovoltage X-ray system with an imaging panel mounted opposed to an X-ray tube (Fig. 63.1). This system is comparable to a diagnostic X-ray system with X-ray energies ≤150 kV. In addition to planar X-ray radiographs, this system may be used for

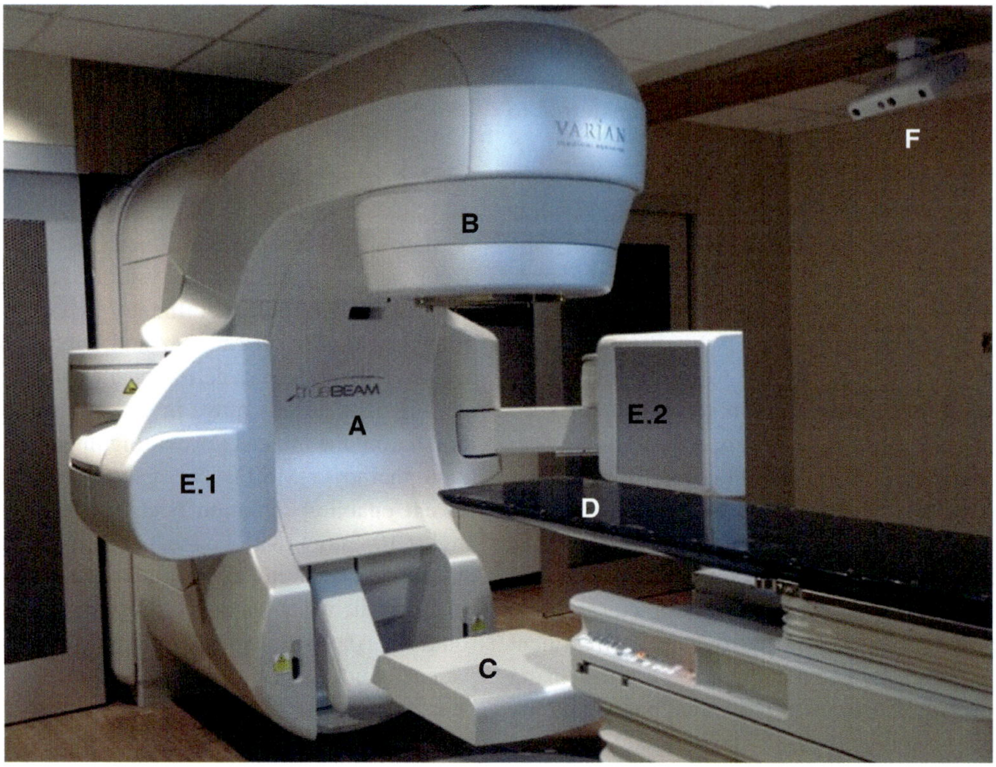

Fig. 63.1 Medial linear accelerator (TrueBeam, Varian Medical Systems, Palo Alto, CA, USA). The C-arm gantry (**a**) support structure rotates around the patient couch (**d**). The therapeutic radiation beam exits the gantry by passing through the collimator (**b**) where the beam aperture is shaped by a multileaf collimator which can also modulate the beam intensity. The electronic portal imaging device (EPID) (**c**) captures megavoltage X-rays that pass through the patient to form digital radiographs. A kilovoltage imaging system comprised of an X-ray tube (**e1**) and an X-ray detector (**e2**) is capable of acquiring radiographs, fluoroscopic images, and cone beam computed tomography (CBCT). One of the three in-room 3D cameras which uses light and computer vision to locate the surface of patients (**f**)

fluoroscopy, and most importantly it can acquire a cone beam computed tomography (CBCT) scan. It is used to precisely position the patient for treatment delivery. The CBCT scan may be obtained by either a full or half gantry rotation, providing three-dimensional anatomical data comparable to a diagnostic CT scan. The CBCT scan is volumetrically registered with the treatment planning CT to determine if there are any translational or rotational differences between a patient's actual position on the treatment couch and the patient's position modelled in the treatment planning system. Differences are rectified by moving the patient by the determined translations and rotations with the six degrees of freedom motion of the couch. The imaging systems permit patient positional accuracy of less than 0.4 mm [14].

An imaging system based on visible light, AlignRT (Vision RT, London, UK), is also part of the treatment room setup (Fig. 63.1). It utilizes 3D cameras and computer vision algorithms to precisely locate the surface of a patient with submillimeter accuracy [15]. The system may be used to position the patient based on the treatment planning CT. It is also used to monitor the patient during radiation delivery. It interfaces with the linac and will turn off the radiation beam if the patient moves out of position. This system is most useful for monitoring treatment sites where the targeted volume can be visualized externally or is closely represented by the skin surface such as in breast irradiation or the treatment of superficial lesions. For monitoring and adjusting a patient based on viewing the external anatomy, this system may be used for any anatomic site. This imaging technology is called surface guided radiation therapy (SGRT) [15–17].

Although the mechanical and geometric accuracies of a linac are submillimeter under static conditions, high-precision radiation therapy delivery requires accounting for the internal motion of targets and organs. Most easily identified is lung tumor motion with respiration. Protocols are available to account for respiratory motion by utilizing four-dimensional computed tomography (4DCT) for the treatment planning simulation scan [18, 19]. During 4DCT acquisition, a method to monitor respiration is employed. The acquired CT imaging detector data are tagged with the current respiratory phase of the patient. This permits reconstruction of separate 3D CT imaging sets covering the range of respiratory motion observed during a 4DCT. Radiation oncologists must then decide how to treat a given patient [20–22]. One method is to create an internal target volume (ITV) consisting of the entire range of tumor motion within all phases of the 4DCT. The greater the distance traveled by the tumor during respiration, the larger the ITV is and the worse the therapeutic ratio is. Another method is to only deliver radiation during certain phases of the respiratory cycle, the gating window. A stable portion of the breathing phase where there is little tumor motion, typically near end of expiration, is used for the gating window. For this method, the ITV is significantly reduced compared to free breathing since it is only defined in the gating window. Thus, much less normal tissue volume is irradiated for the same therapeutic tumor dose.

The treatment delivery system must also account for internal patient respiratory motion during patient setup and treatment delivery. For the linac of Fig. 63.1, this is accomplished with either the respiratory management (RPM) (Varian Medical Systems, Palo Alto, CA USA) or the AlignRT system (Vision RT, London, UK). Either system interfaces with the linac permitting 4D CBCT acquisition [23] as well as imaging triggered at a specific respiratory phase [24]. RPM utilizes a fixture containing infrared reflectors which are monitored by a camera system. The fixture is placed on the patient, typically in the chest or abdomen, and the cameras measure the displacement of the fixture and hence the patient. AlignRT uses light to detect the surface of the patient with a system of 3D cameras employing computer vision technology (SGRT). The process of using precise image guidance during CT simulation, treatment planning, and treatment delivery is called image-guided radiation therapy (IGRT) [25].

Improvements in IGRT have led to a paradigm shift in radiation therapy from treatment courses consisting of daily treatments for a period of several weeks to short radiation courses of 3–5 days of highly conformal large dose per fraction regimens referred to as stereotactic radiation body therapy (SBRT) or stereotactic ablative radiation therapy (SABR) [26, 27]. These techniques employ treatment plans having high doses per fraction and steep dose gradients. Therefore, treatment accuracy is critical. SBRT treatment protocols require the presence of a radiation oncologist and medical physicist at the treatment machine to review imaging for anatomical alignment prior to radiation delivery [28]. Thus, radiation oncologists are becoming more like surgeons with the linac room being the analog of the operating theater.

SBRT has seen use for many disease sites, including to treat primary tumors in the lung, liver, prostate [29], pancreas [30, 31], metastatic tumors in the spine, liver [26], as well as to re-irradiate tumors in the head and neck and pelvis [29]. SBRT offers a therapeutic option for inoperable patients and has become the standard of care for inoperable lung primary tumors.

Both RPM and SGRT technologies permit radiation treatment delivery using the deep inspiration breath hold (DIBH) technique. In this technique, patients take a deep breath and hold it during radiation delivery. For these patients, DIBH is used for the treatment planning CT. During breast radiotherapy, significant reduction of lung and heart dose is achieved [32]. These systems are also used for the expiration breath hold (ExBH) technique where patients exhale and hold during treatment planning CT and treatment delivery. ExBH is typically used in the treatment of lung and abdominal tumors to minimize motion of the target volume, thus reducing the amount of normal tissue receiving a high radiation dose [33].

A particularly powerful treatment technique for treating patients having multiple brain metastases has emerged from the new linac technology. Radiation dose delivery precision of VMAT coupled with the accuracy of IGRT allows radiation oncologists to treat several metastatic brain lesions simultaneously using a single isocenter [34]. Stereotactic radiosurgery (SRS) is a well-known established modality for treating brain lesions in one large highly conformal dose. Historically, a SRS frame was rigidly attached to a patient by screws penetrating into the skull. The emergence of high-precision IGRT led to frameless SRS eliminating the skull-fixed head frame. However, patients with multiple lesions would still endure long treatment sessions with each lesion treated separately. The new single isocenter multiple lesion SRS technique completes treatment of multiple lesions in the same time taken to treat a single lesion.

63.2.3 MR-Linac

Although the imaging systems on a linac have the submillimeter precision of an imaging pixel, they are incapable of real-time three-dimensional imaging. Therefore, verification of patient positioning and anatomical geometry are based on snapshots in time, careful monitoring, and reliance on reproducibility. The ideal situation is having real-time 3D imaging during radiation therapy delivery. This is now possible using an MR-Linac which provides real-time 3D magnetic resonance imaging coupled with radiation therapy delivery.

There are now two commercial MR-Linac systems available: MRIdian (ViewRay, Oakwood Village, Ohio, USA) [35] and Unity (Elekta, Stockholm, Sweden) [36]. The real-time imaging capabilities of the MR-Linac enable reduced radiation beam apertures and an increase in the therapeutic ratio. These systems are capable of monitoring an anatomical structure, for example, a tumor, and turning on or off the radiation beam when the object moves into or out of position. The determination of the location of the structure must be made quickly; therefore, this process operates under computer control. Clinical results indicate that the improved precision of an MR-Linac offers advantages in the treatment of a variety of disease sites [37].

Upon acquiring images to position a patient for conventionally fractionated radiation therapy or for SBRT treatment, it is possible to observe significant anatomical changes from the planning CT that cannot be corrected by repositioning the patient. These could arise, for example, from patient weight gain or loss, more or less edema present, and tumor growth or shrinkage. The original plan therefore no longer applies, and a new plan based on the patient's current anatomy is needed. Either the imaging acquired at the machine or a new planning CT is obtained for the new treatment plan. This adaptive planning process creates a delay in the treatment course. The commercial MR-Linac systems permit online adaptive planning in a matter of minutes with the patient remaining in the treatment room.

The real-time imaging and online adaptive planning further alter the treatment role for radiation oncologists whose presence is essential during MR-Linac treatment delivery.

63.3 Proton Therapy

Proton therapy is external beam radiation therapy using a proton beam instead of photons (X-rays). Because of its mass and charge, interactions of protons with patient tissues are significantly different than photons providing several advantages. Proton beams have a lower entrance (skin) dose and less scatter dose, deposit most of their dose in a narrow range of depths and have minimal exit dose. In fact, proton therapy delivers significantly less (50–60%) total body radiation dose than photons for the same tumor dose [38]. A disadvantage of proton therapy is that the range of a proton beam is strongly dependent on the tissue density in its path. Thus, it is more susceptible to patient anatomical variations than photons.

Protons were first suggested for therapy use in 1946 [39]. It was a major technological achievement over the course of decades to create a practical medical proton machine from the realm of high-energy particle physics experiments. Proton therapy machines are significantly more complex, require a large footprint, and are more than an order of magnitude more expensive than a photon linac.

The most precise method of proton therapy is intensity-modulated proton therapy (IMPT). It is achieved using pencil beam scanning where a narrow proton beam is swept across the target using magnetic control without the need for beam aperture shaping collimators or compensators. Beams from several directions are used to treat a target. This method permits close conforming of the dose to the target while limiting doses to normal tissues.

Not all cancer patients can be treated with protons, because there are a limited number of proton centers. Proton therapy is most useful when normal tissue doses cannot be adequately reduced with photon therapy while still benefiting the patient. Proton therapy is also desirable when sparing normal tissues or organs is of primary importance. Proton therapy is particularly advantageous for pediatric patients, since following successful treatment, these patients are at increased risk for secondary cancers which are significantly less likely following proton therapy because of the overall lower total body dose for the same therapeutic dose. For the treatment of medulloblastoma, requiring treatment of the craniospinal axis, protons have superior outcomes compared to photon therapy [40–42].

63.4 Radiation Therapy Emerging Technologies

63.4.1 Biology-Guided Radiotherapy

Positron-emission tomography (PET) imaging generates volumetric images of functional and molecular biological processes using radiotracers designed to target specific processes. The tracer 18F-FDG is used in clinical oncology for detecting cancer metastasis. This PET scan is commonly used in radiation therapy treatment planning to identify targets. The PET and treatment planning scans are registered in the software, and a treatment plan is generated to treat the tumor(s). For patient delivery, the radiation oncologist relies on the treatment planning CT to guide patient setup.

A new technology has recently been developed to combine PET and CT imaging with a 6MV medical linear accelerator (RefleXion Medical, Hayward, CA, USA) and has been called biology-guided radiation therapy (BgRT) [43]. The machine resembles a PET/CT scanner. A treatment planning process, similar to traditional radiotherapy planning, is employed. A diagnostic PET is registered with a treatment planning CT to designate the targets and subsequently construct a treatment plan. The departure from the conventional radiotherapy is that PET is used to guide the radiation treatment delivery. Prior to treatment, a patient is injected with a PET radiotracer. The patient is placed on the couch which is advanced into the bore of the machine. A CT and PET scan are acquired and compared with the treatment planning scans to confirm patient anatomy. During treatment delivery, the gantry, containing PET detectors spanning symmetrically in opposed 90°arcs, a CT system, and a linear accelerator, rotates at 60 RPM around the patient. PET emissions are detected, and limited time-sampled PET images are continuously reconstructed at 10 frames/s. The linear accelerator delivers radiation beams from 100 firing positions around the circle. The couch remains stationary at a so-called beam station until the desired dose is delivered. It then moves 2.1 mm to another beam station.

BgRT is a promising technology. It removes difficulties associated with tumor tracking since the radiation is delivered to the PET avid volumes, so that the tumors serve as their own fiducial markers. Additionally, PET tracers designed for a specific cancer could be utilized, for example, PSMA-targeted tracers for prostate cancer or CA9-targeted tracers for renal cancers. Possible variations in the PET signal due to inflammation, cancer cell death, or radiotracer administration over the course of treatment have to be accounted for and appear to be manageable [43].

63.4.2 FLASH Radiotherapy

FLASH radiotherapy (FLASH-RT) is the delivery of radiation treatment at ultra-high dose rates (~100 Gy/s) that are several orders of magnitude greater than used in current practice (~0.1 Gy/s). Using this new technology, preliminary pilot and animal studies have shown that normal tissues are protected when exposed to FLASH-RT compared to conventional dose rate RT [44]. This so-called FLASH effect is robust since it is seen in several animal types and several organs and by numerous investigators over a 40-year period. Furthermore, FLASH-RT does not spare tumor cells as the anti-tumor efficacy of FLASH-RT has been found to be equal to conventional RT. Another advantage of FLASH-RT is that the dose is delivered so rapidly, and organ and tumor motion are not important during delivery.

FLASH is believed to cause a rapid consumption of local oxygen at a rate much faster than tissue reoxygenation kinetics. This rapid oxygen depletion creates an induced hypoxia. The depletion rate is optimized at dose rates greater than 100 Gy/s. Additionally, FLASH deposits significantly more energy and ionizes more electrons than conventional RT. Therefore, the redox reactions following FLASH propagate in tissues through a series of biochemical and biophysical reactions that are kinetically different than following conventional RT. The result is that normal tissues have lower pro-oxidant burdens during steady-state metabolism than tumors and eliminate FLASH-mediated free radicals much more effectively, maximizing differences in redox metabolism between tumors and normal tissues [44].

FLASH-RT might be the next technological advancement to significantly improve the therapeutic ratio. The first case report of human use of FLASH-RT has been published [45]. Future randomized clinical trials are needed to prove its efficacy.

63.4.3 Carbon Ion Therapy

Radiation therapy using carbon ions or other heavy nuclei instead of protons is offered in only 13 centers worldwide [46]. Most notably are centers in Germany [47] and Japan [48]. These ions offer reduced doses to normal tissues and increased radiobiological effect on tumors. The technological requirements for these particle accelerators are significantly more challenging than for protons due to the larger mass and charge. The enormous costs of building a heavy particle facility have thus far prevented constructing one in the United States. However, Mayo Clinic announced plans to build one in Florida [49]. Malouff [46] provides an extensive review of the current state of the technology of this modality.

63.4.4 Immunotherapy Boost of the Radiation Abscopal Effect

Although radiation is deposited locally during radiotherapy delivery, there are numerous reports of radiotherapy treatment of a local site of disease leading to regression of metastatic tumors not irradiated [50–52]. This phenomenon, known as the abscopal effect, is believed to be mediated by lymphocytes and enhanced by the checkpoint blockade [53]. Patients receiving radiation therapy to control tumors progressing during checkpoint blockade immunotherapy have shown systemic responses indicating that radiation has an immunostimulatory effect. There are several additional mechanisms proposed to explain the abscopal effect [52, 54]. There is a growing consensus that radiotherapy combined with immunotherapy has a role in treating metastatic disease and that SBRT may provide an advantage over conventional radiotherapy. A comprehensive review of clinical trials investigating this combination therapy concluded that it has the potential to transform cancer treatment [55].

Clinical trials combining immunotherapy with SBRT or the emerging technologies of biology-guided radiotherapy or FLASH-RT might prove valuable in the treatment of metastatic or oligometastatic disease.

63.5 Conclusion

Current technology provides high-precision radiation treatment delivery to static and moving targets, enabling radiation oncologists to treat patients effectively with treatment regimens having only a few high dose fractions (SBRT), producing a paradigm shift in their practice which also necessitates their presence at the treatment console for each treatment fraction. Additional advantages of SBRT over conventional radiotherapy may also be found when used in combination with immunotherapy. Once technology makes online adaptive planning, available today with MR-Linacs, expeditious, SBRT treatment precision will greatly improve.

There are several technologies on the verge of impacting radiotherapy practice. Incorporating artificial intelligence into treatment planning software should improve plan quality over current knowledge-based planning. Biology-guided radiotherapy may improve treatment efficacy of primary and metastatic tumors, especially if novel functional radiotracers are utilized. FLASH-RT-capable treatment delivery machines will offer near-instantaneous radiation delivery as well as biological advantages greatly improving the therapeutic ratio. It is reasonable to expect that one or several of these technologies will be combined yielding even better patient outcomes.

Open Questions

- Will the clinical practice of radiation oncologists morph into that of a surgeon with radiation oncologists working at the treatment machine supervising computer-controlled radiation delivery in real time?
- Will treatment planning systems implement AI software that will take over anatomy contouring and treatment plan construction?
- If FLASH-RT is proven, will it be scalable to clinical machines, significantly reducing treatment delivery times?
- Will biology-guided radiation therapy be widely adopted?
- Will carbon ion therapy prove beneficial enough to justify its cost?

Patient Information and Guidelines

American Society for Radiation Oncology: https://www.astro.org/

American Associate of Physicist in Medicine: https://www.aapm.org

References

1. Grubbe EH. X-rays in the treatment of cancer and other malignant diseases. Med Rec. 1902;62(18):692–5.
2. Thwaites DI, Tuohy JB. Back to the future: the history and development of the clinical linear accelerator. Phys Med Bio. 2006;51:R343–62.
3. Olch AJ, Gerig L, Li H, Mihaylov I, Morgan A. Dosimetric effects caused by couch tops and immobilization devices: Report of AAPM Task Group 176. Med Phys. 2014;41:061501. https://doi.org/10.1118/1.4876299.
4. Khan FM, Gibbons JP. Khan's the physics of radiation therapy. Philadelphia, PA: Lippincott Williams & Wilkins; 2014.
5. Rigaud B, Simon A, Castelli J, Lafond C, Acosta O, Haigron P, Cazoulat G, de Crevoisier R. Deformable image registration for radiation therapy: principle, methods, applications and evaluation. Acta Oncol. 2019;58(9):1225–37. https://doi.org/10.1080/0284186X.2019.1620331. Epub 2019 Jun 3. PMID: 31155990
6. Lee H, Lee E, Kim N, Kim JH, Park K, Lee H, Chun J, Shin JI, Chang JS, Kim JS. Clinical evaluation of commercial atlas-based auto-segmentation in the head and neck region. Front Oncol. 2019;9:239. https://doi.org/10.3389/fonc.2019.00239. PMID: 31024843; PMCID: PMC6465886
7. Sahiner B, Pezeshk A, Hadjiiski LM, Wang X, Drukker K, Cha KH, Summers RM, Giger ML. Deep learning in medical imaging and radiation therapy. Med Phys. 2019;46(1):e1–e36. https://doi.org/10.1002/mp.13264. Epub 2018 Nov 20. PMID: 30367497
8. Ge Y, Wu QJ. Knowledge-based planning for intensity-modulated radiation therapy: a review of data-driven approaches. Med Phys. 2019;46(6):2760–75. https://doi.org/10.1002/mp.13526. Epub 2019 Apr 24. PMID: 30963580; PMCID: PMC6561807
9. Chatterjee A, Serban M, Abdulkarim B, Panet-Raymond V, Souhami L, Shenouda G, Sabri S, Jean-Claude B, Seuntjens J. Performance of knowledge-based radiation therapy planning for the glioblastoma disease site. Int J Radiat Oncol Biol Phys. 2017;99(4):1021–8. https://doi.org/10.1016/j.ijrobp.2017.07.012. Epub 2017 Jul 14. PMID: 28870791
10. Foy JJ, Marsh R, Ten Haken RK, Younge KC, Schipper M, Sun Y, Owen D, Matuszak MM. An analysis of knowledge-based planning for stereotactic body radiation therapy of the spine. Pract Radiat Oncol. 2017;7(5):e355–60. https://doi.org/10.1016/j.prro.2017.02.007. Epub 2017 Mar 2. PMID: 28433524
11. Glide-Hurst C, Bellon M, Foster R, Altunbas C, Speiser M, Altman M, Westerly D, Wen N, Zhao B, Miften M, Chetty IJ, Solberg T. Commissioning of the Varian TrueBeam linear accelerator: a multi-institutional study. Med Phys. 2013;40(3):031719. https://doi.org/10.1118/1.4790563. PMID: 23464314
12. Papanikolaou N, Battista JJ, Boyer AL, Kappas C, Klein E, Mackie TR, Sharpe M, Van Dyk J. AAPM Report No. 85 Tissue Inhomogeneity Corrections for Megavoltage Photon Beams Report of Task Group No. 65 of the Radiation Therapy Committee of the American Association of Physicists in Medicine. Medical Physics Publishing. Madison, WI; 2004. ISBN: 978-1-888340-47-1.
13. van Elmpt W, McDermott L, Nijsten S, Wendling M, Lambin P, Mijnheer B. A literature review of electronic portal imaging for radiotherapy dosimetry. Radiother Oncol. 2008;88(3):289–309. https://doi.org/10.1016/j.radonc.2008.07.008. Epub 2008 Aug 14. PMID: 18706727
14. Graulieres E, Kubler S, Martin E, Ferrand R. Positioning accuracy of a single-isocenter multiple targets SRS treatment: a comparison between Varian TrueBeam CBCT and Brainlab ExacTrac. Phys Med. 2020;80:267–73. https://doi.org/10.1016/j.ejmp.2020.10.022. Epub ahead of print. PMID: 33221708
15. Hoisak JDP, Pawlicki T. The role of optical surface imaging systems in radiation therapy. Semin Radiat Oncol. 2018;28(3):185–93. https://doi.org/10.1016/j.semradonc.2018.02.003. PMID: 29933878
16. Batista V, Meyer J, Kügele M, Al-Hallaq H. Clinical paradigms and challenges in surface guided radiation therapy: where do we go from here? Radiother Oncol. 2020;26:S0167-8140(20)30816-1. https://doi.org/10.1016/j.radonc.2020.09.041. Epub ahead of print. PMID: 32987044.
17. Freislederer P, Kügele M, Öllers M, Swinnen A, Sauer TO, Bert C, Giantsoudi D, Corradini S, Batista V. Recent advanced in Surface Guided Radiation Therapy. Radiat Oncol. 2020;15(1):187. https://doi.org/10.1186/s13014-020-01629-w. Erratum in: Radiat Oncol. 2020 Oct 24;15(1):244. PMID: 32736570; PMCID: PMC7393906.
18. Aznar MC, Warren S, Hoogeman M, Josipovic M. The impact of technology on the changing practice of lung SBRT. Phys Med. 2018;47:129–38.
19. Kwong Y, Mel AO, Wheeler G, Troupis JM. Four-dimensional computed tomography (4DCT): a review of the current status and applications. J Med Imaging Radiat Oncol. 2015;59(5):545–54.
20. Keall PJ, Mageras GS, Balter JM, Emery RS, Forster KM, Jiang SB, Kapatoes JM, Low DA, Murphy MJ, Murray BR, Ramsey CR,

Van Herk MB, Vedam SS, Wong JW, Yorke E. The management of respiratory motion in radiation oncology report of AAPM Task Group 76a. Medical Physics. 2006;33(10):3874–900.
21. Koksal C, Donmez Kesen N, Okutan M, Karaman S, Dagoglu Sakin N, Bilge H. Investigation of approaches for internal target volume definition using 4-dimensional computed tomography in stereotactic body radiotherapy of lung cancer. Med Dosimy. 2021;46(2):136–42.
22. Korreman S, Persson G, Nygaard D, Brink C, Juhler-Nøttrup T. Respiration-correlated image guidance is the most important radiotherapy motion management strategy for most lung cancer patients. Int J Radiat Oncol Biol Phys. 2012;83(4):1338–43.
23. Kadoya N, Nemoto H, Kajikawa T, Nakajima Y, Kanai T, Ieko Y, Ikeda R, Sato K, Dobashi S, Takeda K, Jingu K. Evaluation of four-dimensional cone beam computed tomography ventilation images acquired with two different linear accelerators at various gantry speeds using a deformable lung phantom. Phys Med. 2020;77:75–83. https://doi.org/10.1016/j.ejmp.2020.07.030. Epub 2020 Aug 11. PMID: 32795891
24. King RB, Agnew CE, O'Connell BF, Prise KM, Hounsell AR, McGarry CK. Time-resolved dosimetric verification of respiratory-gated radiotherapy exposures using a high-resolution 2D ionisation chamber array. Phys Med Biol. 2016;61(15):5529–46. https://doi.org/10.1088/0031-9155/61/15/5529. Epub 2016 Jul 6. PMID: 27384459
25. Michalski J, Purdy JA, Gaspar L, Souhami L, Ballow M, Bradley J, Chao CKS, Crane C, Eisbruch A, Fallowil D, Forster K, Fowler J, Gillin MT, Graham ML, Harms WB, Saiful Huq M, Kline RW, Langer M, Mackie TR, Mukherji S, Podgorsak EB, Roach M, Ryu J, Sandler H, Schultz CJ, Schell M, Verhey LJ, Vicini F, Winter KA. Image-guided radiation therapy committee. Int J Radiat Oncol Biol Phys. 2001;51(3):60–5.
26. Folkert MR, Timmerman RD. Stereotactic ablative body radiosurgery (SABR) or Stereotactic body radiation therapy (SBRT). Adv Drug Deliv Rev. 2017;109:3–14. https://doi.org/10.1016/j.addr.2016.11.005. Epub 2016 Dec 5. PMID: 27932046
27. Zindler JD, Thomas CR, Hahn SM, Hoffmann AL, Troost EGC, Lambin P. Increasing the therapeutic ratio of stereotactic ablative radiotherapy by individualized isotoxic dose prescription. J Natl Cancer Inst. 2016;108(2):Djv305. https://doi.org/10.1093/jnci/djv305.
28. Halvorsen PH, Cirino E, Das IJ, Garrett JA, Yang J, Yin FF, Fairobent LA. AAPM-RSS Medical Physics Practice Guideline 9.a. for SRS-SBRT. J Appl Clin Med Phys. 2017;18(5):10–21. https://doi.org/10.1002/acm2.12146. Epub 2017 Aug 8. PMID: 28786239; PMCID: PMC5874865
29. Rubio C, Morera R, Hernando O, Leroy T, Lartigau SE. Extracranial stereotactic body radiotherapy. Review of main SBRT features and indications in primary tumors. Rep Pract Oncol Radiother. 2013;18(6):387–96.
30. De Geus SWL, Eskander MF, Kasumova GG, Ng SC, Kent TS, Mancias JD, Callery MP, Mahadevan A, Tseng JF. Stereotactic body radiotherapy for unresected pancreatic cancer: a nationwide review. Cancer. 2017;123(21):4158–67.
31. Gkika E, Adebahr S, Schimek-Jasch T, Wiehle R, Zirlik K, Wittel U, Nestle U, Grosu A, Brunner TB. Stereotactic body radiotherapy (SBRT) in recurrent or oligometastatic pancreatic cancer: simultaneus intergrated protection (SIP) versus conventional SBRT a toxicity review of two different treatment approaches. J Clin Oncol. 2016;34(15 suppl):E15692.
32. Nissen HD, Appelt AL. Improved heart, lung and target dose with deep inspiration breath hold in a large clinical series of breast cancer patients. Radiother Oncol. 2013;106(1):28–32.
33. Farrugia B, Khor R, Foroudi F, Chao M, Knight K, Wright C. Protocol of a study investigating breath-hold techniques for upper-abdominal radiation therapy (BURDIE): addressing the challenge of a moving target. Radiat Oncol. 2020;15(1):1–7.
34. Clark GM, Popple RA, Young PE, Fiveash JB. Feasibility of single-isocenter volumetric modulated arc radiosurgery for treatment of multiple brain metastases. Int J Radiat Oncol Biol Phys. 2010;76(1):296–302. https://doi.org/10.1016/j.ijrobp.2009.05.029.
35. Mittauer KE, Hill PM, Bassetti MF, Bayouth JE. Validation of an MR-guided online adaptive radiotherapy (MRgoART) program: deformation accuracy in a heterogeneous, deformable, anthropomorphic phantom. Radiother Oncol. 2020;146:97–109. Web.
36. Snyder JE, St-Aubin J, Yaddanapudi S, Boczkowski A, Dunkerley DAP, Graves SA, Hyer DE. Commissioning of a 1.5T Elekta Unity MR-linac: a single institution experience. J Appl Clin Med Phys. 2020;21(7):160–72.
37. Fischer-Valuck BW, Henke L, Green O, Kashani R, Acharya S, Bradley JD, Robinson CG, Thomas M, Zoberi I, Thorstad W, Gay H, Huang J, Roach M, Rodriguez V, Santanam L, Li H, Li H, Contreras J, Mazur T, Hallahan D, Olsen JR, Parikh P, Mutic S, Michalski J. Two-and-a-half-year clinical experience with the world's first magnetic resonance image guided radiation therapy system. Adv Radiat Oncol. 2017;2(3):485–93.
38. Shravan K. Spot-scanning beam proton therapy vs intensity-modulated radiation therapy for ipsilateral head and neck malignancies: a treatment planning comparison. Med Dosim. 2013;38(4):390–4. https://doi.org/10.1016/j.meddos.2013.05.001. PMID 23916884
39. Wilson RR. Radiological use of fast protons. Radiology. 1946;47:487–91.
40. Eaton BR. Endocrine outcomes with proton and photon radiotherapy for standard risk medulloblastoma. Neuro Oncol. 2016;18(6):881–7. https://doi.org/10.1093/neuonc/nov302. PMC 4864263. PMID 26688075
41. Hill-Kayser CE. Outcomes after proton therapy for treatment of pediatric high-risk neuroblastoma. Int J Radiat Oncol Biol Phys. 2019;104(2):401–8. https://doi.org/10.1016/j.ijrobp.2019.01.095. PMID 30738983
42. Kahalley LS. Superior intellectual outcomes after proton radiotherapy compared with photon radiotherapy for pediatric medulloblastoma. J Clin Oncol. 2019;38(5):454–61. https://doi.org/10.1200/JCO.19.01706. PMC 7007288. PMID 31774710
43. Shirvani SM, Huntzinger CJ, Melcher T, Olcott PD, Voronenko Y, Bartlett-Roberto J, et al. Biology-guided radiotherapy: redefining the role of radiotherapy in metastatic cancer. Br J Radiol. 2020;93:20200873.
44. Vozenin MC, Hendry JH, Limoli CL. Biological benefits of ultra-high dose rate FLASH radiotherapy: sleeping beauty awoken. Clin Oncol (R Coll Radiol). 2019;31(7):407–15. https://doi.org/10.1016/j.clon.2019.04.001. Epub 2019 Apr 19. PMID: 31010708; PMCID: PMC6850216
45. Bourhis J, Sozzi WJ, Jorge PG, Gaide O, Bailat C, Duclos F, Patin D, Ozsahin M, Bochud F, Germond JF, Moeckli R, Vozenin MC. Treatment of a first patient with FLASH-radiotherapy. Radiother Oncol. 2019;139:18–22. https://doi.org/10.1016/j.radonc.2019.06.019. Epub 2019 Jul 11. PMID: 31303340
46. Malouff TD, Mahajan A, Krishnan S, Beltran C, Seneviratne DS, Trifiletti DM. Carbon ion therapy: a modern review of an emerging technology. Front Oncol. 2020;10:82. https://doi.org/10.3389/fonc.2020.00082.
47. Jäkel O, Krämer M, Schulz-Ertner D, Heeg P, Karger CP, Didinger B, Nikoghosyan A, Debus J. Treatment planning for carbon ion radiotherapy in Germany: review of clinical trials and treatment planning studies. Radiother Oncol. 2004;73(2):S86–91.
48. Tsujii H, Mizoe J, Kamada T, Baba M, Tsuji H, Kato H, Kato S, et al. Clinical results of carbon ion radiotherapy at NIRS. J Radiat Res. 2007;48(Suppl A):A1–A13.

49. Mayo Clinic plans first carbon ion therapy facility in North America. Forefront Mayo Clinic Cancer Center's Online Magazine; 2020 March;9(1).
50. Ngwa W, Irabor OC, Schoenfeld JD, Hesser J, Demaria S, Formenti SC. Using immunotherapy to boost the abscopal effect. Nat Rev Cancer. 2018;18(5):313–22.
51. Rodríguez-Ruiz ME, Vanpouille-Box C, Melero I, Formenti SC, Demaria S. Immunological mechanisms responsible for radiation-induced abscopal effect. Trends Immunol. 2018;39(8):644–55.
52. Siva S, MacManus MP, Martin RF, Martin OA. Abscopal effects of radiation therapy: a clinical review for the radiobiologist. Cancer Lett. 2015;3569(1):82–90.
53. Buchwald ZS, Wynne J, Nasti TH, Zhu S, Mourad WF, Yan W, Gupta S, Khleif SN, Khan MK. Radiation, immune checkpoint blockade and the abscopal effect: a critical review on timing, dose and fractionation. Front Oncol. 2018;8:612.
54. Kang J, Demaria S, Formenti S. Current clinical trials testing the combination of immunotherapy with radiotherapy. J Immunother of Cancer. 2016;4(1):51.
55. Lubas MJ, Kumar SS. The combined use of SBRT and immunotherapy-a literature review. Curr Oncol Rep. 2020;22(12):117.

64 Radiation Therapy in the Definitive Treatment of Cancer

Alexander N. Garcia and Baldassarre Stea

Abstract

Radiation therapy is an important component of cancer treatment for a variety of malignancies both in the definitive and palliative setting. Radiation can be administered by external means using linear accelerators or internally by brachytherapy; core to the therapeutic effects of radiation is the induction of ionization events within biological molecules which in turn results in irreparable DNA damage and ultimately cell death. Historically, the treatment effectiveness of radiotherapy was limited in part by toxicity to normal tissues; however, recent advances in treatment planning and imaging have allowed for highly conformal delivery of higher doses of radiation that subsequently have improved cancer-related outcomes. Radiation therapy is widely used as a definitive modality, concurrently with chemotherapy for unresectable cancers, in the adjuvant setting after surgery or for consolidation after chemotherapy. With the advent of increasingly specific small-molecule inhibitors and immunotherapy drugs, the oncologic landscape continues to evolve rapidly, creating new opportunities for utilization of radiation therapy to improve cancer-related outcomes.

Learning Objectives

- Understand the biology and physics underlying radiation therapy.
- Understand the treatment planning process and various techniques for radiation therapy administration.
- Understand radiation therapy indications in the management of common malignancies.

64.1 Introduction

Radiation therapy is a widely utilized modality in the treatment of cancer, with an estimated 50% of all patients potentially benefitting from its utilization during the course of their illness [1]. The successful application of radiation treatments relies on the precise delivery of high-energy X-rays, particles (protons or carbon ions), or radioactive substances to induce death of malignant cells. The feasibility of radiation treatments can be dated back to the discovery of X-rays by Wilhelm Röntgen in 1885, although at that time the potential of X-rays was limited to very superficial tumors [1]. It is, thus, not surprising that the evolution of radiation therapy as a cancer treatment modality has followed the same path as that of medical imaging. The field of radiation therapy has constantly evolved over the past century, aiming for more accuracy and fewer side effects with each evolution. Indeed, precise targeting of a tumor while sparing nearby normal tissues has increased steadily with the availability of image guidance before and during the delivery of radiation. Today, most linear accelerators (linacs) come equipped with the ability of obtaining a CT scan prior to treatment to ascertain precision of targeting. A new emergent way to increase precision is by using MR-guided radiation; the MRI-linac combines in one instrument a high-strength

A. N. Garcia (✉)
University of Radiation Oncology, Tucson, USA
e-mail: alexngarcia@arizona.edu

B. Stea (✉)
University of Arizona, Tucson, USA
e-mail: bstea@arizona.edu

S. P. Leong et al. (eds.), *Cancer Metastasis Through the Lymphovascular System*,
https://doi.org/10.1007/978-3-030-93084-4_64

magnetic resonance imaging with a linear accelerator to provide heightened precision, especially when targeting soft tissue tumors such as liver tumors. These new-generation MRI-linacs are capable of monitoring the movement of a tumor in real time and continually adjust the radiation beam to compensate for any shift in its position. Aided by artificial intelligence, it is also possible to alter the daily treatment plan based on tumor volume and/or anatomical changes occurring in real time; this new technology gives a radiation oncologist the ability to adapt the dose of radiation to a changing tumor during the course of radiation (adaptive radiotherapy). Radiation therapy is commonly used to treat primary cancers, such as prostate cancer with the goal of not only controlling the primary but also preventing the insurgence of metastatic disease (vide infra). In addition, radiation therapy plays an important role in controlling symptoms from metastatic disease (such as brain metastases, painful metastases, bleeding, spinal cord compression, airway obstruction) and indeed about fifty percent of patients treated with radiation are treated with palliative intent; this topic is covered in detail in another chapter. Finally, over the past three decades, we have witnessed the benefit of combining chemotherapy with radiation in the treatment of major malignancies (e.g., lung, rectal, head and neck cancers). The past decade has also witnessed the emergence of immune checkpoint inhibitors (ICI) in the treatment of many malignancies. There is now emergent evidence of synergistic interaction between high-dose per fraction radiation and ICI in the treatment of certain malignancies such as melanoma. Finally, as the landscape of oncologic treatments continues to evolve with highly promising discoveries in targeted small-molecule inhibitors and immunotherapy, the role of radiation remains prominent in the management of numerous malignancies in a number of ways including definitive radiotherapy alone, concurrent chemoradiation, and adjuvant radiotherapy following surgical resection of primary tumors. This chapter provides the reader with an overview of the radiobiological and physics aspect of modern radiotherapy in addition to highlighting the successful application of radiation in the treatment of major cancers.

64.2 Physics and Biology of Radiation Therapy

The physics and the biological effects of radiation are complex fields which are covered more extensively in other chapters; however, here, we review the basic principles to provide the reader with the knowledge to better understand the practicality of radiation treatments as well as potential future developments. Core to the generation of radiation-induced damage is the formation of ionization events within atoms which are then able to induce unrepairable DNA damage and subsequent cell death. This can take place following direct ionization events in which photons and/or charged particles (electrons, protons, alpha particles, heavy ions) directly damage DNA to induce single- and double-stranded breaks. More commonly, however, indirect ionization occurs during which high-energy photons react with other molecules (such as water) to generate free radicals which in turn cause DNA damage. The unit of radiation dose is gray (Gy), defined as the absorption of 1 J of radiation per kilogram of matter. Numerous potential molecular and cellular outcomes occur after exposure to ionizing radiation, including chromosomal aberrations, genetic mutations, disruption of cell cycle progression, and apoptosis. Theoretically, a sufficient increase in the total radiation dose should improve the probability of cell kill and tumor control to the point where any tumor could be eradicated; however, in clinical practice this must be balanced with the probability of normal tissue complications. This concept is referred to as the therapeutic ratio and is one of the most fundamental aspects of radiation therapy in clinical practice [2]. Thus, maintenance of the therapeutic ratio within a narrow range is the primary dose-limiting factor in the delivery of radiation treatments. This limitation necessitated fractionated radiation treatments, a practice of administering a smaller daily dose of radiation over a period of weeks to allow normal tissue to repair damage between treatments. In modern practice, advances in treatment planning algorithms, dose conformality, and imaging before or during radiation treatments have dramatically improved our ability to spare normal tissues, thus ensuring targeted delivery of high-dose radiation to the tumor. These technological advances have led to a change in practice patterns whereby most tumors are now being treated with hypofractionated radiation (fewer fractions of larger doses).

64.3 Radiation Treatment Techniques

External beam radiotherapy (EBRT) from linear accelerators (linacs) remains the most utilized technique for radiation treatments worldwide. With this modality a patient lies on a flat surface (couch), while an external treatment machine directs a beam of radiation to an area of interest. Early in the history of radiation therapy, these units utilized radioisotope cobalt-60 as the source of megavoltage radiation, a practice that remains prevalent in developing countries. Modern-day treatment machines, linear accelerators, instead produce high-energy X-rays as well as electrons that deliver radiation to a treatment volume conformal to the tumor (Fig. 64.1).

Over the past 50 years, the evolution of EBRT has paralleled advances in imaging modalities. The development of computed tomography allowed for three-dimensional conformal radiotherapy (3D CRT) in which the radiation

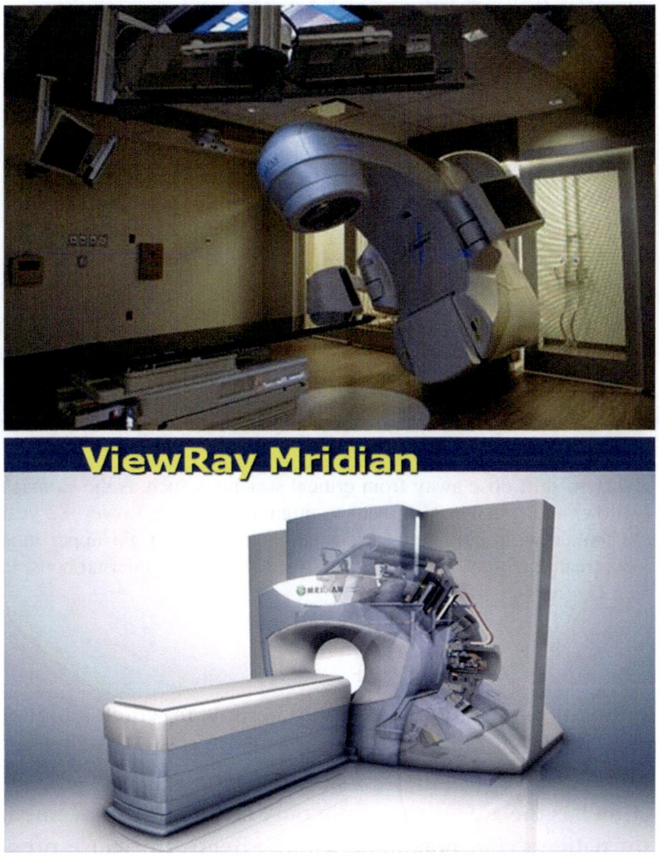

Fig. 64.1 Modern-day linear accelerators. Varian TrueBeam Linear Accelerator and ViewRay MRIdian with MRI guidance

field conforms to a volume of interest in three dimensions [3]. This allows for more precise delivery of radiation that better spares normal tissues using either physical blocks or more commonly multileaf collimators (MLCs) within the treatment machine. MLCs are mechanical lead-leaves that under computer control block the radiation beam to create a conformal field of radiation. More recently, advances in treatment planning software and mechanical engineering have facilitated sophisticated delivery techniques, such as intensity-modulated radiotherapy (IMRT) in which the MLCs, under computer control, are able to variably modulate the intensity of the radiation beam to allow better conformality to the treatment volume [4]. This technology has resulted in profound advances in treatment delivery, particularly in head and neck cancers [5], in which higher dose of radiation can feasibly and more safely be administered to tumors while better sparing critical structures, resulting in reductions of toxicity and increased tumor control (improved therapeutic ratio).

Stereotactic body radiation therapy (SBRT) and stereotactic radiosurgery (SRS) are two other increasingly utilized radiation techniques in which escalated doses of radiation are administered over the course of one to five treatments. Advances in diagnostic imaging, patient immobilization, image guidance, and treatment delivery have greatly facilitated its use [6]. In addition to providing improved patient convenience by avoiding a protracted course of treatment, these techniques provide a greater radiobiological dose of radiation which leads to ablative effects on tumors. The radiobiology underpinning this technology takes advantage of the ability of high-dose radiation to cause small vessel obliteration. This has translated into improved tumor control rates not previously seen with fully fractionated radiotherapy in a number of sites including the lung, liver, pancreas, kidney, prostate, brain metastases, and spine [7] (Fig. 64.2).

Brachytherapy, in contrast to external beam radiotherapy, is an interstitial or intracavitary form of treatment in which a radioactive source is placed into or adjacent to the tumor of interest. Brachytherapy requires the precise placement of the radioactive source within the target tissue, thus allowing delivery of extremely high doses of conformal radiation from within. Furthermore, with brachytherapy there is a rapid dose falloff with increasing distance from the radioactive source (inverse square law) which allows for excellent sparing of adjacent normal tissues to an extent that is not possible with external beam radiotherapy. There are broad applications of this technique, including intracavitary brachytherapy in which the source is placed inside a body cavity (e.g., cervical cancer), interstitial brachytherapy in which sources are surgically inserted into the body tissue (e.g., surgical beds for soft tissue sarcoma), or intraluminal brachytherapy (radioactive plaques placed in the subtenon space for treatment of choroidal melanoma). Furthermore, depending on the disease site and radioisotope used, treatment may further be differentiated as low dose rate (LDR, <2 Gy/h) or high dose rate (HDR, >12 Gy/h). Treatment with LDR is generally accomplished over days to months, such as in the case of permanent radioactive seed implant for early prostate cancer. HDR treatments generally last minutes, as in the case of cervical cancer. Brachytherapy is frequently utilized either alone or in conjunction with external beam radiotherapy in the management of many malignancies, notably cervical, endometrial, prostate, choroidal melanoma, recurrent brain tumors, and breast cancer.

64.4 Modern-Day Applications of Radiotherapy for the Most Common Cancers

64.4.1 Prostate Cancer

Prostate cancer is the most common non-cutaneous malignancy among men worldwide. Management of early-stage disease can involve active surveillance, prostatectomy,

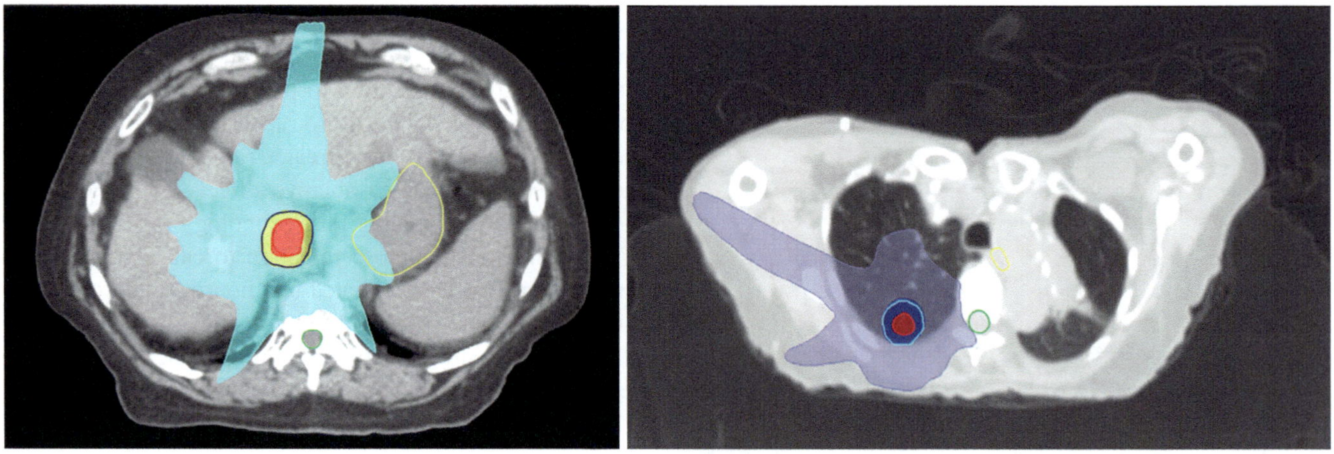

Fig. 64.2 Examples of Stereotactic Body Radiation Treatment. Upper: A 67 year old man presented with hepatocellular carcinoma within the caudate lobe of the liver. Following CT simulation a SBRT plan was generated with a prescription of 50 Gy (Yellow) in 5 daily fractions, and an internal boost to gross tumor to 60 Gy (red). Note the conformality of the prescription isodose lines (Yellow: 50 Gy; Red: 60 Gy) and ability to shift dose away from critical structures such as the stomach (yellow) and spinal cord (green) to minimize toxicity. Lower: 82 year old female with early stage adenocarcinoma of the right upper lung lobe, treated with 50 Gy (blue) in 5 fractions with an internal boost to 57.5 Gy (red)

and radiotherapy, while treatment options for high-risk disease are limited to either radical prostatectomy or radiotherapy, with or without androgen deprivation therapy (ADT). Recently, the UK ProtecT trial randomized patients with localized prostate cancer to active surveillance, radical prostatectomy, or external beam radiotherapy with short-term ADT. At 10-year follow-up, approximately 50% of patients on active surveillance necessitated definitive treatment. No significant difference in prostate cancer-specific survival was noted between the three groups, although a separate analysis of patient-reported outcomes reported significant difference in toxicity between the treatment arms. Radiotherapy was generally associated with worse acute GI toxicity in comparison to radical prostatectomy, which instead was associated with slightly worse urinary and sexual function. Interestingly, despite equivalent survival outcomes patients who did not receive upfront management with radiotherapy or radical prostatectomy were at significantly increased risk of development of metastatic disease (Metastatic disease per 1000 person-years: 6.3 in active surveillance, 2.4 following radical prostatectomy, and 3.0 following definitive radiotherapy). Based on these findings, the authors concluded that 33 patients would require definitive management with radiotherapy to prevent 1 patient developing metastatic disease.

Thus, low-risk and favorable intermediate-risk prostate cancer can be treated with definitive radiotherapy alone. By far the most common treatment technique is EBRT to the prostate gland and some portion or all the seminal vesicles. Dose escalation studies have identified the superiority of regimens treating to 78–80 Gy, administered in daily fractions of 1.8–2 Gy, over 7–8 weeks. More recently, however, studies have evaluated the utility and safety of hypofractionated radiotherapy, in which a higher dose per fraction is given over a smaller number of fractions or days (usually 4 weeks). This mild hypofractionation has the advantage of not only being more convenient for patients, but potentially more effective because prostate cancer is a slow-growing tumor which may inherently be more resistant to smaller fraction size. While these studies have shorter follow-up compared to historic studies, the early 5-year results show non-inferiority to more protracted fractionated regimens. Additionally, low-risk and favorable intermediate-risk prostate cancer may be treated by brachytherapy alone. This is an attractive option for some patients who prefer a single or a few invasive procedures over daily external beam treatments that could potentially last up to 8 weeks. The most common and best studied brachytherapy technique is prostate seed implant (PSI), in which numerous radioactive seeds are inserted in the gland transperineally under ultrasound guidance. This is a form of low-dose rate (LDR) brachytherapy in which the radioactive sources (commonly iodine-125 or palladium-103) are permanently placed within the prostate (Fig. 64.3). High-dose rate (HDR) brachytherapy is another modern technique which takes advantage of modern imaging and treatment planning software. Transperineal catheter insertion in the prostate is performed under spinal anesthesia followed by application of a high-activity radioisotope, iridium-192, over a period of minutes. Literature evaluating prostate HDR brachytherapy continues to mature, although to date biochemical control rates are comparable to those of other radiation techniques. Lastly, as an extrapolation to the theoretical benefits of moderately hypofractionated radiotherapy,

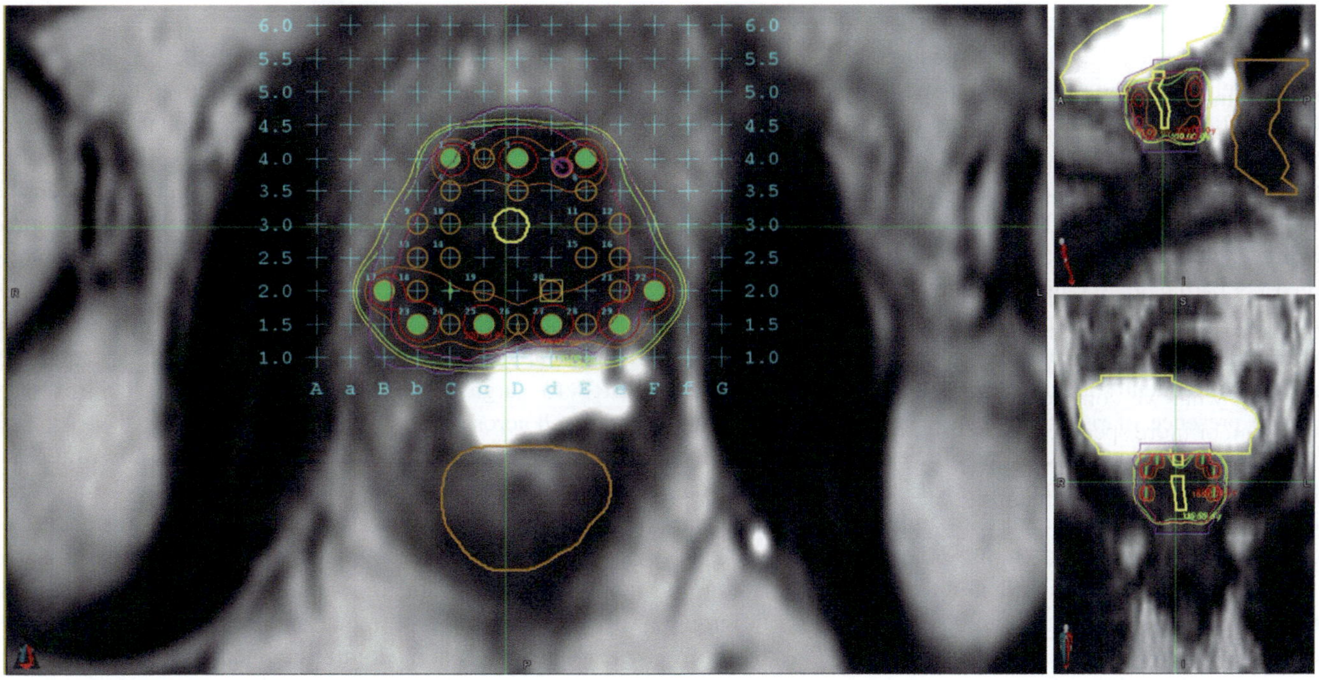

Fig. 64.3 Prostate seed implantation of a patient with high-risk prostate cancer following external beam radiation to the prostate and pelvic lymphatics. A grid is superimposed over the patients imaging. The orange circles correspond to positions of needle insertion for seed placement. Areas adjacent to the urethra (yellow circle) are generally avoided to minimize toxicity. The green isodose line corresponds to prescription dose (110 Gy). Orange line corresponds to 150% isodose line (165 Gy), and red line corresponds to 200% isodose line (220 Gy). The bladder and rectum are contoured (yellow and orange respectively) as avoidance structures to limit dose exposure. Radio-opaque substance in between prostate and rectum represents hydrogel spacer which serves to physically displace the rectum from the prostate to minimize dose exposure

SBRT over 5 fractions is increasingly being utilized in the management of early-stage prostate cancer. Given the proximity of bladder and rectal tissues, extreme care must be taken in treatment delivery, notably accurate daily treatment imaging; however, SBRT is associated with several promising endpoints such as tumor control and patient-reported quality of life.

For patients with unfavorable intermediate- or high-risk prostate cancer, active surveillance is generally not recommended as nearly all patients may potentially benefit from treatment. These patients are treated with some combination of external beam radiotherapy, brachytherapy, ADT, or in select instances of all the above. The most frequently used agents for androgen deprivation are gonadotropin-releasing hormone agonists or oral antiandrogens. For unfavorable intermediate-risk prostate cancer, a 4–6 month course of ADT is used in conjunction with hypofractionated external beam radiotherapy to the prostate and seminal vesicles alone. In contrast, patients with high-risk prostate cancer have been shown to benefit from ADT for upwards of 2 years. This is combined with external beam radiotherapy to both the prostate and regional lymphatics, although the benefits of nodal irradiation are a matter of controversy which is currently being evaluated by ongoing clinical trials. External beam radiotherapy can be administered to a definitive dose of approximately 78–80 Gy; an alternative is treatment of the pelvis with 46 Gy of EBRT followed by a brachytherapy boost which may result in improved cancer-related outcomes [8].

64.4.2 Breast Cancer

Breast cancer remains one of the most common malignancies, as well as the most common cause of cancer death in women. Treatment largely revolves around surgical resection of the tumor, either with breast-conserving therapy or mastectomy, followed by adjuvant radiation therapy in most cases [9]. Numerous large-scale trials have shown the combination of breast-conserving therapy and adjuvant radiotherapy to be comparable in terms of local recurrence and overall survival. These trials have shown adjuvant radiotherapy to reduce local recurrence rates in breast cancer patients by nearly two-thirds, and a meta-analysis performed by the Early Breast Cancer Trialists' Collaborative Group (EBCTG) also reported a survival benefit for postoperative breast radiotherapy; one breast cancer death was prevented at 15 years for every four local recurrence avoided [10]. Breast radiotherapy was administered over a protracted course of 5–6 weeks to a dose of 50–60 Gy, with treatment volumes including the whole breast with or without regional lymph nodes depending on risk factors. More recently, based on radiobiologic analysis of breast cancer cells in relation to normal tissue, it was theorized that hypofractionated courses of radiation over 3–4 weeks may yield a more favorable therapeutic ratio. This has since been proven in numerous clinical trials [11] and remains the current standard of care today. The hypofractionation paradigm is being explored even further with 5-day courses of whole breast or partial breast radiotherapy showing promising results in select patient populations [12]. Regional lymphatics including axillary, supraclavicular, and internal mammary lymph nodes are frequently included in the treatment volume based on the presence of risk factors such as node positivity and hormone-negative breast cancers. Studies such as MA.20 and EORTC 22922 have shown that inclusion of regional nodal irradiation in these patients yields improved local regional control with a trend towards improved overall survival [13].

Based on the observation that most of the in-breast tumor recurrences (IBTR) occur within one centimeter of the surgical bed, accelerated partial breast irradiation (APBI) has been proposed, which dramatically reduces the overall treatment volume and time. To date, there have been multiple trials using different radiation techniques (external beam radiation, interstitial brachytherapy, and balloon brachytherapy) to administer larger dose to a small volume which encompasses the lumpectomy cavity plus a margin. While there is limited long-term follow-up for patients treated with APBI in comparison to conventional whole breast irradiation, early results show comparable rates of IBTR with excellent cosmetic outcomes [14]. Similarly, intraoperative radiotherapy (IORT) has been studied on at least two clinical trials [15] in which a single high dose of radiotherapy is administered to the lumpectomy cavity at the time of surgery. While this offers improved patient convenience and highly localized radiation dose, these clinical trials have shown higher rates of IBTR which has largely precluded its incorporation into treatment guidelines.

Patients with advanced stage disease or high-risk genetics frequently undergo modified radical mastectomy. This procedure removes the entire ipsilateral breast and level one and two axillary lymphatics, and thus adjuvant radiotherapy may not be routinely recommended. However, patients with extensive nodal burden, defined as four or more positive lymph nodes, have been shown to benefit from postmastectomy radiation based on a meta-analysis performed by the Early Breast Cancer Trialists' Collaborative Group. Postmastectomy radiation in this patient population has resulted in a nearly 10% absolute risk reduction in breast cancer mortality [16], although its benefits in patients with less than four positive lymph nodes or those that achieve a pathologic complete response to neoadjuvant chemotherapy remain areas of controversy that warrant further evaluation [17].

64.4.3 Lung Cancer

Lung cancer remains the most common cause of cancer death in the United States [18]. The standard of care for early-stage non-small-cell carcinomas (NSCLC) remains surgery; however, patients frequently present with extensive cardiopulmonary comorbidities which preclude surgical resection. The standard of care for non-surgical patients is definitive radiotherapy, specifically SBRT which has been shown to be superior to conventionally fractionated radiotherapy [19]. There is limited data directly comparing SBRT to surgical resection; however, local control rates compared across trials are generally similar.

Patients with locally advanced disease are often not surgical candidates due to invasive primary tumors or extensive mediastinal nodal disease. The current standard of care for these patients remains conventionally fractionated radiotherapy with concurrent platinum-based chemotherapy. The standard dose of radiation remains 60 Gy/30 daily fractions as a recent randomized clinical trial evaluating dose escalation yielded inferior survival outcomes [20]. Of note this trial was performed prior to the widespread adoption of IMRT, and thus dose escalation may be more feasible and beneficial in the modern era. Patients that undergo surgical resection do not routinely undergo adjuvant radiation therapy; however, a subset of patients with N2 disease or positive margins may benefit from its use [21].

Patients with small-cell lung cancer (SCLC) frequently present with unresectable disease, and definitive chemoradiation remains the standard of care for patients with limited stage disease. In contrast to NSCLC, small-cell lung cancer is routinely treated with accelerated twice daily radiation due the rapid cell division and inherently radiosensitive nature of the disease. This practice pattern is based on the landmark trial by Turrisi, which showed superiority of 45 Gy/30 fractions administered b.i.d compared to 45 Gy in once daily fractions [22]. Criticisms of this trial focused primarily on insufficient dose in the once daily treatment arm, and indeed, further evaluation using escalated doses of 66 Gy showed comparable outcomes, although there was a trend towards superiority in the twice daily arm [23]. Given the high propensity for CNS metastases in patients with SCLC, multiple trials have evaluated the use of prophylactic cranial radiation (PCI). A meta-analysis of these trials showed that patients with limited-stage disease who have a good response to chemoradiation receive an approximate 5% absolute improvement in 3-year overall survival from PCI [24]. However, due to the frequent occurrences of neurocognitive deficits following PCI, a current clinical trial is exploring the use of observation versus SRS in managing these patients.

64.4.4 Gynecologic Cancers

Early-stage cervical cancers are managed with surgery alone, although patients with high-risk pathologic features including lymph-vascular space invasion, deep stromal invasion, and large tumors benefit from the addition of adjuvant radiotherapy [25]. Additionally, concurrent chemotherapy is included for patients with positive margins, parametrial involvement, and positive lymph nodes [26]. For patients with advanced stage disease or contraindications to surgery, definitive radiation is recommended with concurrent platinum-based chemotherapy. Radiotherapy is generally administered in two phases; an initial external beam radiation is administered to the primary tumor and regional lymphatics, with or without a boost to the parametrium, depending on disease stage. This is followed by an intracavitary HDR brachytherapy boost to the primary cervical tumor. Local control rates and survival rates are promising with definitive chemoradiation; however, prolonged treatment times (i.e., beyond 56 days) are associated with significantly reduced survival outcomes as accelerated repopulation of tumors occurs during protracted courses of radiation [27].

In contrast to cervical cancer, endometrial cancer is primarily managed with surgery; however, the presence of high-risk features may necessitate adjuvant radiation, chemotherapy, or both. High-risk features include deep myometrial invasion, high grade, or aggressive histology, such as clear cell and LVSI. Approximately 75% of recurrences after primary surgery occur within the vaginal vault [28]. Studies comparing vaginal vault intracavitary brachytherapy with pelvic external beam pelvic radiation showed no difference in vaginal recurrences or overall survival; as such, given the more favorable toxicity profile of vaginal brachytherapy, this method has become the standard of care to prevent a vaginal apex recurrence. However, patients with numerous high-risk features may still benefit from more aggressive management. The GOG 249 study evaluated patients with high intermediate-risk or early-stage high-risk endometrial cancer; in this trial, patients were randomized to pelvic radiotherapy or vaginal cuff brachytherapy plus chemotherapy. No differences were detected in terms of survival outcomes; however, patients treated with vaginal cuff plus chemotherapy demonstrated greater acute toxicities suggesting that pelvic radiotherapy may be the preferred approach in this patient population [29]. Conversely, the PORTEC 3 trial randomized patients with high-risk endometrial cancer to adjuvant pelvic radiation versus chemoradiation, demonstrating an overall survival benefit in patients treated with chemotherapy and radiation.

64.4.5 Head and Neck Cancer

Head and neck squamous cell carcinomas represent an aggressive heterogeneous group of malignancies. They are generally associated with heavy tobacco, alcohol use, and HPV exposure. Management varies by subsite within the head and neck; however, early-stage disease can be managed with either surgical resection or definitive radiotherapy with or without chemotherapy [30]. The exception to this rule is malignancies located within the oral cavity which are best managed by upfront surgical resection [31, 32], although adjuvant radiotherapy is frequently utilized for high-risk features including advanced stage disease, close or positive margins, deep invasion, or extracapsular extension (ECE). Similarly, concurrent chemotherapy is frequently utilized in instances of positive margins or ECE for any type of head and neck cancer. Concurrent chemotherapy is generally platinum-based which in addition to eradication of potential distant disease serves as a radiosensitizer. Cetuximab has additionally shown some benefit [33], albeit inferior to that seen with cisplatin for patients with HPV-positive oropharyngeal carcinoma [34]. Outcomes with definitive radiation or chemoradiation vary largely based on smoking status, nodal burden, and HPV status. RTOG 0129 [35] demonstrated the significance of these risk factors as low-risk patients (HPV+, <10 pack years smoking, N0-2a) and high-risk patients (HPV-, >10 pack years smoking) resulting in 3-year overall survival rates of approximately 85% and 55%, respectively. Based on these findings, current investigation is underway into potential radiation dose de-escalation in patients with HPV+ head and neck cancer, results of which are expected to be reported in the coming years.

64.5 Conclusion

In addition to the aforementioned sites of disease, radiation therapy has the potential to be impactful to some degree in nearly every malignancy including bladder, gastrointestinal, CNS, skin, lymphoma, and sarcomas. In addition to treatment in the definitive setting, radiotherapy additionally has a role in the management of metastatic disease where it is commonly utilized for the palliation of painful or symptomatic sites in locations such as bone, lung, and brain. Although modest radiation doses with single or multifraction regimens have shown promising pain control, increased survival of patients with metastatic disease has highlighted the potential benefit of aggressive SBRT regimens to maximize local control, particularly in the oligometastatic setting. The role of radiation therapy in the oligometastatic setting is a recent area of interest, with preliminary studies showing promising survival benefits in patients with minimal metastatic burden, although this will be discussed in further detail elsewhere. Common to nearly every application of radiotherapy is an associated local-regional control benefit, although the ability for this to translate into a survival benefit varies based on tumor location and histology. As systemic therapy continues to improve rapidly for many cancers, it is likely that the role of radiation therapy will become even more crucial to patients in the future as a form of consolidation after achieving maximal response to chemotherapy or immunotherapy.

Open Questions

- Given ongoing advances in systemic therapy, will indications for radiotherapy continue to evolve as well? What is the clinical advantage of integrating immunotherapy with stereotactic radiosurgery?
- What will be the effect of more sophisticated imaging modalities on the delivery of radiation treatments?
- In addition to treatments with photons and electrons, will protons and heavy ions see increased use in cancer treatments? Why?
- Despite being correlated with improved patient outcomes, the use of brachytherapy has decreased in recent decades. What are some explanations for this trend?

References

1. Gianfaldoni S, et al. An overview on radiotherapy: from its history to its current applications in dermatology. Open Access Maced J Med Sci. 2017;5(4):521–5. https://doi.org/10.3889/oamjms.2017.122.
2. Hall EJ, Astor M, Bedford J, Borek C, Curtis SB, Fry M, Geard C, Hei T, Mitchell J, Oleinick N, et al. Basic radiobiology. Am J Clin Oncol. 1988;11(3):220–52. https://doi.org/10.1097/00000421-198806000-00003.
3. Tubiana M, Eschwège F. Conformal radiotherapy and intensity-modulated radiotherapy – clinical data. Acta Oncol. 2000;39(5):555–67. https://doi.org/10.1080/028418600750013249.
4. Taylor A, Powell ME. Intensity-modulated radiotherapy – what is it? Cancer Imaging. 2004;4(2):68–73. https://doi.org/10.1102/1470-7330.2004.0003.
5. Ghosh G, Tallari R, Malviya A. Toxicity profile of IMRT Vs. 3D-CRT in head and neck cancer: a retrospective study. J Clin Diagn Res. 2016;10(9):Xc01–xc03. https://doi.org/10.7860/jcdr/2016/21457.8583.
6. Alongi F, Arcangeli S, Filippi AR, Ricardi U, Scorsetti M. Review and uses of stereotactic body radiation therapy for oligometastases. Oncologist. 2012;17(8):1100–7. https://doi.org/10.1634/theoncologist.2012-0092.
7. Chang BK, Timmerman RD. Stereotactic body radiation therapy: a comprehensive review. Am J Clin Oncol. 2007;30(6):637–44. https://doi.org/10.1097/COC.0b013e3180ca7cb1.
8. Morris WJ, Tyldesley S, Rodda S, Halperin R, Pai H, McKenzie M, Duncan G, Morton G, Hamm J, Murray N. Androgen suppression combined with elective nodal and dose escalated radiation therapy (the ASCENDE-RT Trial): an analysis of survival endpoints for a randomized trial comparing a low-dose-rate brachytherapy boost to a dose-escalated external beam boost for high- and intermediate-risk prostate cancer. Int J Radiat Oncol Biol Phys. 2017;98(2):275–85. https://doi.org/10.1016/j.ijrobp.2016.11.026.
9. Hughes KS, Schnaper LA, Bellon JR, Cirrincione CT, Berry DA, McCormick B, Muss HB, Smith BL, Hudis CA, Winer EP, Wood WC. Lumpectomy plus tamoxifen with or without irradiation in women age 70 years or older with early breast cancer: long-term follow-up of CALGB 9343. J Clin Oncol. 2013;31(19):2382–7. https://doi.org/10.1200/jco.2012.45.2615.
10. Early Breast Cancer Trialists' Collaborative Group (EBCTCG). Effects of chemotherapy and hormonal therapy for early breast cancer on recurrence and 15-year survival: an overview of the randomised trials. Lancet. 2005;365(9472):1687–717. https://doi.org/10.1016/s0140-6736(05)66544-0.
11. Whelan TJ, Pignol J-P, Levine MN, Julian JA, MacKenzie R, Parpia S, Shelley W, Grimard L, Bowen J, Lukka H, Perera F, Fyles A, Schneider K, Gulavita S, Freeman C. Long-term results of hypofractionated radiation therapy for breast cancer. N Engl J Med. 2010;362(6):513–20. https://doi.org/10.1056/NEJMoa0906260.
12. Murray Brunt A, Haviland JS, Wheatley DA, Sydenham MA, Alhasso A, Bloomfield DJ, Chan C, Churn M, Cleator S, Coles CE, Goodman A, Harnett A, Hopwood P, Kirby AM, Kirwan CC, Morris C, Nabi Z, Sawyer E, Somaiah N, Stones L, Syndikus I, Bliss JM, Yarnold JR. Hypofractionated breast radiotherapy for 1 week versus 3 weeks (FAST-Forward): 5-year efficacy and late normal tissue effects results from a multicentre, non-inferiority, randomised, phase 3 trial. Lancet. 2020;395(10237):1613–26. https://doi.org/10.1016/s0140-6736(20)30932-6.
13. Whelan TJ, Olivotto IA, Parulekar WR, Ackerman I, Chua BH, Nabid A, Vallis KA, White JR, Rousseau P, Fortin A, Pierce LJ, Manchul L, Chafe S, Nolan MC, Craighead P, Bowen J, McCready DR, Pritchard KI, Gelmon K, Murray Y, Chapman JA, Chen BE, Levine MN. Regional nodal irradiation in early-stage breast cancer. N Engl J Med. 2015;373(4):307–16. https://doi.org/10.1056/NEJMoa1415340.
14. Livi L, Meattini I, Marrazzo L, Simontacchi G, Pallotta S, Saieva C, Paiar F, Scotti V, De Luca CC, Bastiani P, Orzalesi L, Casella D, Sanchez L, Nori J, Fambrini M, Bianchi S. Accelerated partial breast irradiation using intensity-modulated radiotherapy versus whole breast irradiation: 5-year survival analysis of a phase 3 randomised controlled trial. Eur J Cancer. 2015;51(4):451–63. https://doi.org/10.1016/j.ejca.2014.12.013.
15. Silverstein MJ, Fastner G, Maluta S, Reitsamer R, Goer DA, Vicini F, Wazer D. Intraoperative radiation therapy: a critical analysis of the ELIOT and TARGIT trials. Part 2 – TARGIT. Ann Surg Oncol. 2014;21(12):3793–9. https://doi.org/10.1245/s10434-014-3999-5.
16. McGale P, Taylor C, Correa C, Cutter D, Duane F, Ewertz M, Gray R, Mannu G, Peto R, Whelan T, Wang Y, Wang Z, Darby S. Effect of radiotherapy after mastectomy and axillary surgery on 10-year recurrence and 20-year breast cancer mortality: meta-analysis of individual patient data for 8135 women in 22 randomised trials. Lancet. 2014;383(9935):2127–35. https://doi.org/10.1016/s0140-6736(14)60488-8.
17. Remick J, Amin NP. Postmastectomy breast cancer radiation therapy. Treasure Island, FL: StatPearls; 2020.
18. Dela Cruz CS, Tanoue LT, Matthay RA. Lung cancer: epidemiology, etiology, and prevention. Clin Chest Med. 2011;32(4):605–44. https://doi.org/10.1016/j.ccm.2011.09.001.
19. Nyman J, Hallqvist A, Lund J, Brustugun OT, Bergman B, Bergström P, Friesland S, Lewensohn R, Holmberg E, Lax I. SPACE - A randomized study of SBRT vs conventional fractionated radiotherapy in medically inoperable stage I NSCLC. Radiother Oncol. 2016;121(1):1–8. https://doi.org/10.1016/j.radonc.2016.08.015.
20. Bradley JD, Hu C, Komaki RR, Masters GA, Blumenschein GR, Schild SE, Bogart JA, Forster KM, Magliocco AM, Kavadi VS, Narayan S, Iyengar P, Robinson CG, Wynn RB, Koprowski CD, Olson MR, Meng J, Paulus R, Curran WJ Jr, Choy H. Long-term results of NRG oncology RTOG 0617: standard- versus high-dose chemoradiotherapy with or without cetuximab for unresectable

stage III non-small-cell lung cancer. J Clin Oncol. 2020;38(7):706–14. https://doi.org/10.1200/jco.19.01162.
21. Burdett S, Stewart L. Postoperative radiotherapy in non-small-cell lung cancer: update of an individual patient data meta-analysis. Lung Cancer. 2005;47(1):81–3. https://doi.org/10.1016/j.lungcan.2004.09.010.
22. Turrisi AT 3rd, Kim K, Blum R, Sause WT, Livingston RB, Komaki R, Wagner H, Aisner S, Johnson DH. Twice-daily compared with once-daily thoracic radiotherapy in limited small-cell lung cancer treated concurrently with cisplatin and etoposide. N Engl J Med. 1999;340(4):265–71. https://doi.org/10.1056/nejm199901283400403.
23. Faivre-Finn C, Snee M, Ashcroft L, Appel W, Barlesi F, Bhatnagar A, Bezjak A, Cardenal F, Fournel P, Harden S, Le Pechoux C, McMenemin R, Mohammed N, O'Brien M, Pantarotto J, Surmont V, Van Meerbeeck JP, Woll PJ, Lorigan P, Blackhall F. Concurrent once-daily versus twice-daily chemoradiotherapy in patients with limited-stage small-cell lung cancer (CONVERT): an open-label, phase 3, randomised, superiority trial. Lancet Oncol. 2017;18(8):1116–25. https://doi.org/10.1016/s1470-2045(17)30318-2.
24. Aupérin A, Arriagada R, Pignon JP, Le Péchoux C, Gregor A, Stephens RJ, Kristjansen PE, Johnson BE, Ueoka H, Wagner H, Aisner J. Prophylactic cranial irradiation for patients with small-cell lung cancer in complete remission. Prophylactic Cranial Irradiation Overview Collaborative Group. N Engl J Med. 1999;341(7):476–84. https://doi.org/10.1056/nejm199908123410703.
25. Rotman M, Sedlis A, Piedmonte MR, Bundy B, Lentz SS, Muderspach LI, Zaino RJ. A phase III randomized trial of postoperative pelvic irradiation in Stage IB cervical carcinoma with poor prognostic features: follow-up of a gynecologic oncology group study. Int J Radiat Oncol Biol Phys. 2006;65(1):169–76. https://doi.org/10.1016/j.ijrobp.2005.10.019.
26. Monk BJ, Wang J, Im S, Stock RJ, Peters WA 3rd, Liu PY, Barrett RJ 2nd, Berek JS, Souhami L, Grigsby PW, Gordon W Jr, Alberts DS. Rethinking the use of radiation and chemotherapy after radical hysterectomy: a clinical-pathologic analysis of a Gynecologic Oncology Group/Southwest Oncology Group/Radiation Therapy Oncology Group trial. Gynecol Oncol. 2005;96(3):721–8. https://doi.org/10.1016/j.ygyno.2004.11.007.
27. Vitzthum L, Yuan J, Jones D, Boldt A, Dusenbery K. Reducing prolonged chemoradiation treatment times for cervical cancer. BMJ Open Qual. 2019;8(3):e000516. https://doi.org/10.1136/bmjoq-2018-000516.
28. Creutzberg CL, Nout RA, Lybeert ML, Wárlám-Rodenhuis CC, Jobsen JJ, Mens JW, Lutgens LC, Pras E, van de Poll-Franse LV, van Putten WL. Fifteen-year radiotherapy outcomes of the randomized PORTEC-1 trial for endometrial carcinoma. Int J Radiat Oncol Biol Phys. 2011;81(4):e631–8. https://doi.org/10.1016/j.ijrobp.2011.04.013.
29. Randall ME, Filiaci V, McMeekin DS, von Gruenigen V, Huang H, Yashar CM, Mannel RS, Kim JW, Salani R, DiSilvestro PA, Burke JJ, Rutherford T, Spirtos NM, Terada K, Anderson PR, Brewster WR, Small W, Aghajanian CA, Miller DS. Phase III trial: adjuvant pelvic radiation therapy versus vaginal brachytherapy plus paclitaxel/carboplatin in high-intermediate and high-risk early stage endometrial cancer. J Clin Oncol. 2019;37(21):1810–8. https://doi.org/10.1200/jco.18.01575.
30. Kramer S, Gelber RD, Snow JB, Marcial VA, Lowry LD, Davis LW, Chandler R. Combined radiation therapy and surgery in the management of advanced head and neck cancer: final report of study 73-03 of the Radiation Therapy Oncology Group. Head Neck Surg. 1987;10(1):19–30. https://doi.org/10.1002/hed.2890100105.
31. Iyer NG, Tan DS, Tan VK, Wang W, Hwang J, Tan NC, Sivanandan R, Tan HK, Lim WT, Ang MK, Wee J, Soo KC, Tan EH. Randomized trial comparing surgery and adjuvant radiotherapy versus concurrent chemoradiotherapy in patients with advanced, nonmetastatic squamous cell carcinoma of the head and neck: 10-year update and subset analysis. Cancer. 2015;121(10):1599–607. https://doi.org/10.1002/cncr.29251.
32. Robertson AG, Soutar DS, Paul J, Webster M, Leonard AG, Moore KP, McManners J, Yosef HM, Canney P, Errington RD, Hammersley N, Singh R, Vaughan D. Early closure of a randomized trial: surgery and postoperative radiotherapy versus radiotherapy in the management of intra-oral tumours. Clin Oncol (R Coll Radiol). 1998;10(3):155–60. https://doi.org/10.1016/s0936-6555(98)80055-1.
33. Bonner JA, Harari PM, Giralt J, Cohen RB, Jones CU, Sur RK, Raben D, Baselga J, Spencer SA, Zhu J, Youssoufian H, Rowinsky EK, Ang KK. Radiotherapy plus cetuximab for locoregionally advanced head and neck cancer: 5-year survival data from a phase 3 randomised trial, and relation between cetuximab-induced rash and survival. Lancet Oncol. 2010;11(1):21–8. https://doi.org/10.1016/s1470-2045(09)70311-0.
34. Gillison ML, Trotti AM, Harris J, Eisbruch A, Harari PM, Adelstein DJ, Jordan RCK, Zhao W, Sturgis EM, Burtness B, Ridge JA, Ringash J, Galvin J, Yao M, Koyfman SA, Blakaj DM, Razaq MA, Colevas AD, Beitler JJ, Jones CU, Dunlap NE, Seaward SA, Spencer S, Galloway TJ, Phan J, Dignam JJ, Le QT. Radiotherapy plus cetuximab or cisplatin in human papillomavirus-positive oropharyngeal cancer (NRG Oncology RTOG 1016): a randomised, multicentre, non-inferiority trial. Lancet. 2019;393(10166):40–50. https://doi.org/10.1016/s0140-6736(18)32779-x.
35. Ang KK, Harris J, Wheeler R, Weber R, Rosenthal DI, Nguyen-Tân PF, Westra WH, Chung CH, Jordan RC, Lu C, Kim H, Axelrod R, Silverman CC, Redmond KP, Gillison ML. Human papillomavirus and survival of patients with oropharyngeal cancer. N Engl J Med. 2010;363(1):24–35. https://doi.org/10.1056/NEJMoa0912217.

65 Radiation Therapy for Extracranial Oligometastatic Disease

Q. A. Ho and C. C. Hsu

Abstract

Oligometastatic cancer is an intermediary state in which a tumor has spread to a limited number (usually ≤ 5) of metastatic sites rather than widespread (or polymetastatic). The concept first gained traction in 1995 and has served as a rationale for aggressive treatment to improve survival within this select metastatic population. Although not readily distinguishable on presentation, these oligometastatic cancers tend to progress much more slowly than an aggressive cancer in its early polymetastatic stage. The definition has now expanded to include four distinct categories: oligometastatic, oligorecurrent, oligoprogressive, and oligoresidual. Although imaging and blood biomarkers are helpful, time and response to systemic therapy have been keys to identifying those patients with oligometastatic disease who may benefit from aggressive local therapy. New studies on molecular profiling, particularly on specific micro-RNAs, have shown promise in identifying such cases. Although surgical resection, or metastectomy, was the primary treatment of oligometastatic disease initially, stereotactic body radiation therapy (SBRT) has shown to be a safe and effective alternative. Several phase 2 and 3 randomized trials have shown that radiation to the metastatic and/or the primary lesion in patients with limited metastatic burden can improve progression-free survival and even overall survival. As both radiation therapy and systemic therapy improve, so will their synergy and indications for their continued use.

Q. A. Ho · C. C. Hsu (✉)
Department of Radiation Oncology, The University of Arizona, Tucson, AZ, USA
e-mail: cchsu@email.arizona.edu

Learning Objectives

- To define terminology and the characteristics of the oligometastatic state.
- To address the microscopic and biologic origins of oligometastases and their clinical implications.
- To discuss the radiotherapeutic treatment options for oligometastases and future directions.

65.1 Introduction

Oligometastatic cancer, the intermediary between localized and widely disseminated cancer, is a relatively new concept in oncology. Historically, the metastatic state of cancer has been viewed as an inevitably progressive track that begins with a single detectable metastasis but in the presence of widespread subclinical dissemination [1]. Subsequently, the role of local therapy remained purely for palliation, regardless of disease burden, whether encompassing a single site of metastasis or widely disseminated disease. The concept of cancer metastasis consisting of a widely heterogeneous group has only recently been hypothesized. This has been supported by increasingly robust evidence that aggressive local treatment, such as radiation therapy, of patients with limited metastatic burden provides a benefit to both progression-free and overall survival.

65.2 Background History

For an entire century, the Halsted hypothesis of cancer spread was the predominant belief among oncologists [1]. The concept called for a continuous spread of cancer, such that there was a specific order of spread that began from the primary tumor, followed by the lymphatics and lymph nodes, and finally to distant sites. Even among distant sites, there was a belief that "closer" distant sites were more likely to become

S. P. Leong et al. (eds.), *Cancer Metastasis Through the Lymphovascular System*,
https://doi.org/10.1007/978-3-030-93084-4_65

involved. This hypothesis drove radical en bloc surgeries that, although were effective at local and regional control, involved significant morbidity. The systemic (Fisher) hypothesis of cancer behavior later became prominent as an alternative explanation of cancer spread. It stated that cancer begins with systemic dissemination of microscopic metastases (micrometastases), and that treatment should be focused on systemic control of micrometastases before they become macroscopic [1]. Figures 65.1 and 65.2 illustrate these two concepts of metastasis.

In 1995, Hellman and Weichselbaum recognized that because cancers begin and can mutate into a wide variety of different clonogens that behave differently, metastatic cancer represents a spectrum of disease [1]. Thus, there may be a select population among those with metastases whose disease course may warrant more aggressive therapy. The oligometastatic disease state has been classified as having a limited number of visible metastases; this number differs between studies but is generally considered to be 5 or less [2, 3]. The oligometastatic state differs from patients with widely disseminated metastatic disease, who are classified as having polymetastatic disease. In some studies, bone metastases are considered less severe and therefore more characteristic of oligometastatic than visceral metastases [4]. This fits in with the theory of oligometastases as some organ sites have a greater likelihood of receiving metastases based on the location of the primary tumor and histology as well as the distant site's microenvironment. Often, this limited metastatic state is due to a more indolent biology, and there is a therapeutic window for localized therapy before widespread dissemination [1]. For this reason, treatment of oligometastases is becoming more common and supported by literature.

Not all cases of limited metastases can truly be classified as oligometastatic. As imaging represents only a single frame in time, the visibility of a few metastatic lesions could be in

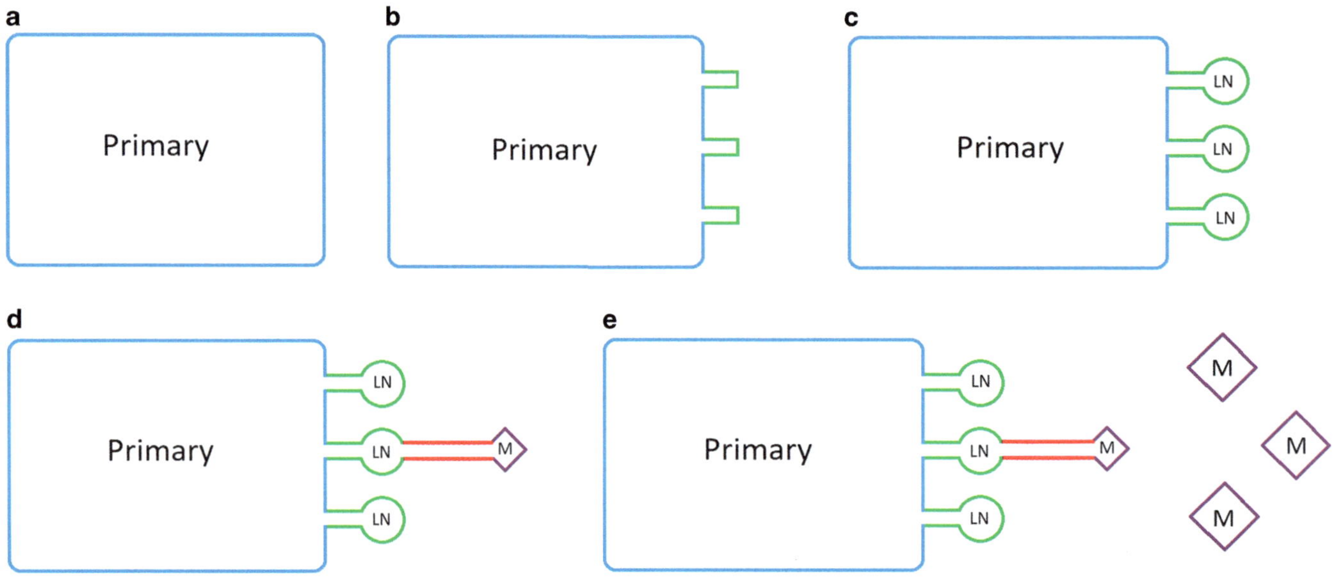

Fig. 65.1 This diagram represents the Halsted hypothesis of cancer progression. (**a**) Cancer begins in a primary location (**b**) before spreading through the lymphatics (**c**) to the lymph nodes. (**d**) Cancer cells further spread hematogenously to distant (metastatic) sites that are nearby the local tumor, (**e**) before widespread dissemination throughout the body

Fig. 65.2 This diagram represents the Fisher hypothesis, also known as the systemic hypothesis of cancer. It states that cancer is a process that begins with systemic dissemination. This hypothesis emphasizes the importance of systemic therapy to target all the tumors, as local therapy would not change the disease course. Although the "primary tumor" is the first one to be visible, (**a**) there is already a background of tumor deposits that are not yet visible. (**b**) Metastatic lesions begin to become visible, (**c**) until there is clear widespread dissemination

the background of numerous microscopic lesions that may inevitably present themselves despite systemic therapy. These cases are not true oligometastases but rather an early wide dissemination that carries a poorer prognosis with an unlikely meaningful response to local therapy. Currently, identifying the oligometastatic state incorporates imaging and blood biomarkers, as well as follow-up and serial imaging to determine systemic therapy response. This allows for better identification of those with a less aggressively metastatic course who may benefit from local therapy. However, the advent of molecular profiling has shown potential to expedite the identification of a true oligometastatic state, one that will benefit tremendously from local consolidative therapy (LCT) [5].

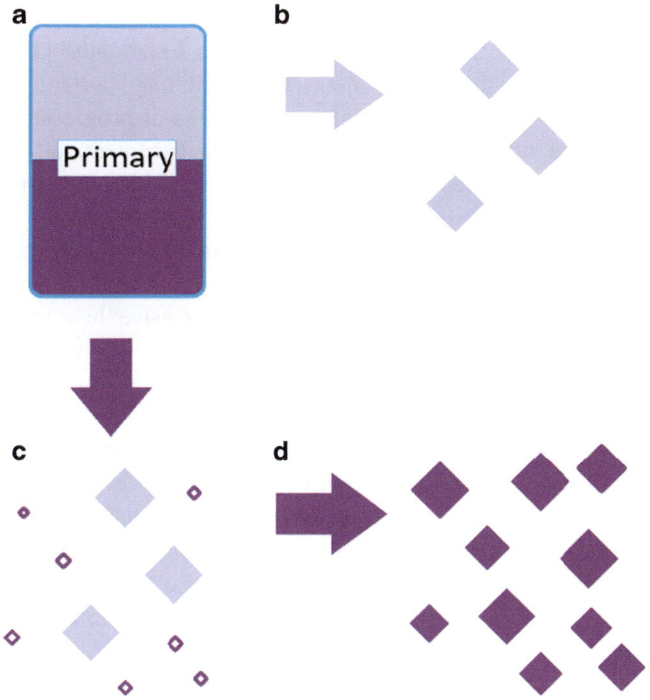

Fig. 65.3 This figure represents the more contemporary view of metastasis, including oligometastasis. (**a**) Tumors may mutate such that they have multiple clonogens that will behave differently, as represented by purple and lavender. (**b**) Some clonogens tend to be more indolent, and the first evidence of metastasis does not precede widespread dissemination of true oligometastatic disease. However, this is indistinguishable from (**c**) limited metastasis that is in a background of microscopic metastases, which will inevitably become (**d**) widespread. In true oligometastases, (**b**) clonogens tend to be limited and more indolent in nature

65.2.1 Definitions and Characteristics of Oligometastatic Disease

As limited metastases can exist without representing a "true" oligometastatic state, it is plausible that some cancers will bypass the oligometastatic intermediary. Two plausible explanations for the oligometastatic state exist. The first explanation states that the progression of metastases is sequential, and that the oligometastatic state is simply a stepping stone to polymetastatic disease. The other explanation states that oligometastatic and polymetastatic diseases are distinct processes originating from distinct genetic phenotypes depending on the particular clonal populations involved (see Fig. 65.3) [6].

It is likely that true metastatic progression is also a spectrum of presentations within the two general explanations. One could argue that any form of oligometastatic disease will become widely disseminated if given a time. This could be due to the inherent indolence of that particular clonogen itself, due to further mutations that occur if the lesion is left untreated, or selective pressures inducing new mutations due to systemic therapy. As both primary and metastatic lesions can further mutate and disseminate additional clonogens, a metastatic patient may have multiple lesions that will behave differently from one another [7, 8]. It is important to define the characteristics of a true oligometastatic state and also the variations that could take place. Better classification of the oligometastatic state would help with identification of specific cases where aggressive local therapy to the primary and/or metastatic lesion (s) would offer not just palliation but also benefits for progression-free and overall survival [1].

The first consideration is quantity, defined as the number of discrete lesions of metastatic disease. Regardless of size of the lesion, it is presumed that each metastatic lesion began as a single deposit of tumor cells. A greater number of distinct discrete lesions increase the likelihood of existing background widespread dissemination, therefore decreasing the potential usefulness of localized treatment. For enrollment purposes, many major studies have specific cutoffs for what is considered oligometastatic, with cutoffs varying between studies. However, five or less metastatic lesions are the common cutoff used by most existing major trials, with many showing an overall survival benefit for treating these limited metastases with focal therapy [2, 9, 10].

It is also important to consider the location of the metastatic disease. As cancer can spread through lymphatics or vasculature, some organs more readily receive metastatic deposits than others. Others, such as malignant pleural effusions or leptomeningeal disease, may not warrant radiation dose escalation as ablation may not be feasible. For instance, prostate metastases are more likely to spread to pelvic lymph nodes and bones than visceral organs and very unlikely to spread to the central nervous system [11]. As it would require a biology that is more aggressive and has higher predilection to spread, a single metastasis in a less common location is more likely to represent underlying dissemination than a single metastasis in a common or "first-echelon" site. These rare metastases are therefore less likely to benefit from localized treatment and pertain a poorer survival. For instance, brain metastases occur in less than 2% of prostate cancer patients and carry a prognosis of about 2.8 months [12]. Furthermore, the site of metastases itself may portend to prognosis based on potential complications of progression, tolerance of treatment, and response to treatment, all of which may be dependent or independent of histology and origin.

The volume of tumor burden is also an important characteristic of oligometastatic disease. Larger lesions (at the time of detection) may represent a more aggressive histology. However, this could be limited to a locally aggressive biology and may not necessarily represent propensity to spread distantly. An isolated large tumor deposit could also represent a true oligometastatic state, for which this single tumor has had a long period of time to grow without evidence of other tumors detected radiographically. Size of lesion is also important to consider for treatment tolerability and response. Larger tumors generally have more areas of hypoxia centrally due to a poor blood supply [13], are therefore more resistant to radiation, and may not be reached by or respond to chemotherapy. Furthermore, it may be more difficult to resect or irradiate large tumors due to associated toxicities of treatment. Rusthoven et al. found that metastatic liver lesions smaller than 3 cm had much better 2-year control than lesions larger than 3 cm of 100% versus 77% ($p = 0.015$) [14]. This study demonstrated not only the importance of size for control but also consideration of toxicity as more healthy liver parenchyma would receive spillover radiation when a larger tumor is treated. In that study, however, treatment with radiation was well tolerated, with only 1 of 47 patients having a grade 3 toxicity to the chest wall requiring debridement and hyperbaric oxygen and none with grade 4 or 5 toxicity [14].

Timing of presentation of metastatic disease can clinically give us clues as to the likelihood of a patient being oligometastatic. Metastatic lesions may present as synchronous lesions along with the primary, or as metachronous lesions, at least 3 months after presentation of the primary tumor [15]. Metachronous lesions are more likely to represent a true oligometastatic state, as it would suggest a slower, step-wise progression of metastasis. This becomes even more likely if the latency between primary and metastatic lesions becomes longer. Synchronous presentations, however, are difficult to determine as it will remain unknown how quickly the existing lesions appeared relative to each other. Studies have shown that metachronous metastatic disease portends a better prognosis than synchronous metastatic disease. Donskov et al. showed that in renal cell carcinoma, synchronous metastases carried an overall survival of 16.7 months, whereas metachronous metastases carried an OS between 23.8 and 41.7 months after the second lesion presents, depending on the duration between presentations [15].

The oligometastatic state is not necessarily a primary or progressing presentation but can also be induced. Generally, there are four distinct cohorts of oligometastatic patients. The three better-known groups include (1) patients who present as oligometastatic, (2) patients who present as nonmetastatic but progress to oligometastatic despite locoregional treatment (oligorecurrence) after definitive therapy and at least 6 month PFS, and (3) oligoprogressive occurring during adjuvant therapy or <6 months after definitive therapy. The fourth group, oligoresidual, represents an induced oligometastatic state when there is at least a partial response in the disease at presentation with induction systemic therapy. The first two scenarios represent an indolent progression that is more classically associated with an oligometastatic biology. Oligoprogressive is seemingly more aggressive based on a shorter PFS period but can still carry a reasonable therapeutic window. Oligoresidual, however, occurs when a limited number of metastases that are resistant to a particular systemic therapy remain after the majority of metastases are effectively treated with that systemic therapy. Although these lesions are less likely to represent an indolent biology, their limited number allows them to be addressed similar to truly indolent oligometastatic lesions [16]. Several studies have included oligometastatic patients at presentation [3], while others have allowed for oligorecurrent or oligoresidual patients also [10, 17]. Table 65.1 and Fig. 65.4 define and summarize the different forms of oligometastatic disease.

65.2.2 Biological Basis of Oligometastases

The fundamental paradigm of tumor growth, invasion, and metastases has evolved to incorporate numerous molecular processes that each contribute to the malignant capability of tumors. Upstream to these molecular processes are genotypes, resulting in various phenotypes. Each individual tumor cell has its own potential to grow, invade, and metastasize at its own rate and also to react differently to radiation and systemic therapy. The metastatic potential of a tumor is determined not only by the tumor itself but also by the microenvironment that will potentially host a metastatic lesion.

Pancreatic adenocarcinoma and its poor prognosis have been the subject of numerous studies at both the clinical and molecular levels. It has been established that these tumors contain geographically separate subclones that can present years before metastasis. Each subclone will differ by possessing different mutations in different intermediate steps in metastases [7, 8]. Different studies in the clear cell variant of renal carcinoma and also breast ductal carcinoma have suggested that metastatic lesions are derived from different, and often unique, clones of the primary site [18, 19]. In short, it

Table 65.1 Summary of the four types of oligometastatic disease, their defining features, and an example clinical scenario of their presentation. The quantity of lesions in an "oligometastatic" state varies with some studies defining the number as ≤3 but will be defined here as up to five metastatic lesions. Similarly, oligorecurrent versus oligoprogressive differs in the time course of presentation. Many definitions use "6 months or more" of progression-free survival to become oligorecurrent as opposed to oligoprogressive. Finally, in some instances, the location of a metastatic lesion may be considered a higher risk for wide dissemination (e.g., peritoneal carcinomatosis or malignant pleural effusions) and will no longer carry an oligometastatic label despite a low number of lesions

Type of oligometastatic disease	Defining features and time course	Clinical presentation example
Oligometastatic	Up to five metastatic lesions at presentation	67 M presents with prostate adenocarcinoma and three vertebral body metastases
Oligorecurrent	Presents as non-metastatic. After definitive treatment and a period of progression-free survival (e.g., ≥6 months), there is progression to having up to five metastatic lesions	68F with breast ductal carcinoma who underwent definitive lumpectomy and radiation, found to have a calvarial metastasis 2 years later
Oligoprogressive	Presents as non-metastatic. Up to five metastatic lesions appear after definitive treatment during adjuvant treatment or within a short period (e.g., ≤6 months) of progression-free survival	75 M found to have an adrenal metastases 2 months after completing chemoradiation and immunotherapy for stage III NSCLC
Oligoresidual	Induced oligometastatic state. Presents as widely metastatic, but disease burden is cytoreduced with systemic therapy to having no more than five metastatic lesions	62 M presents with innumerable bone and visceral metastases from NSCLC, which is reduced to three bony lesions after six cycles of systemic therapy

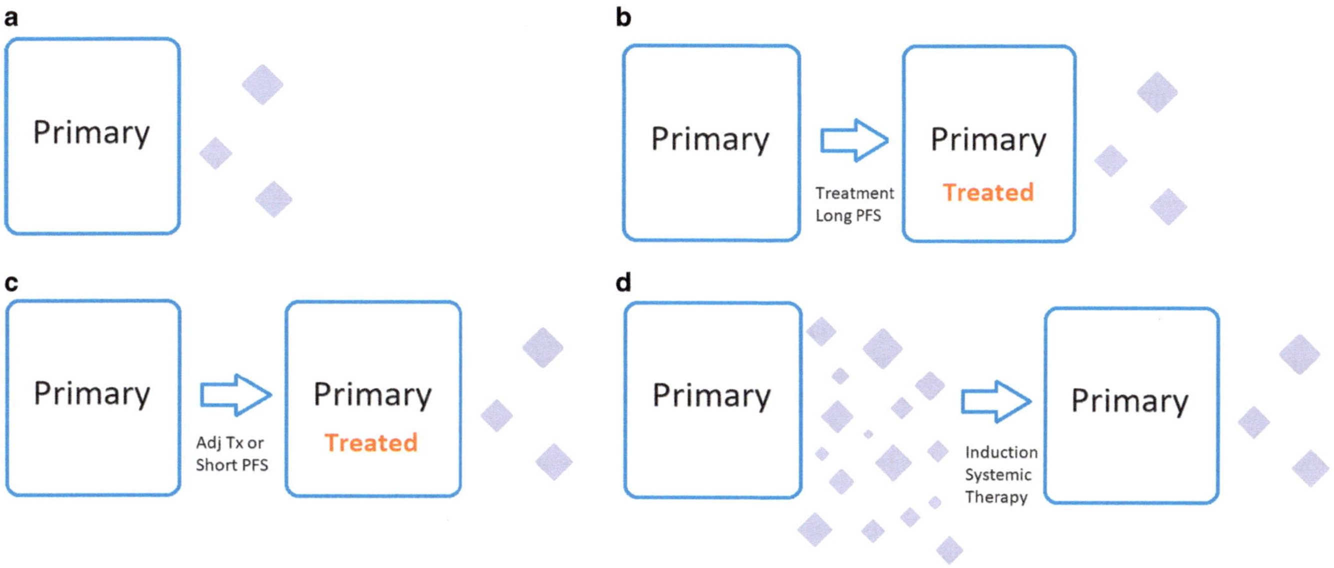

Fig. 65.4 A visual representation of the different subtypes of oligometastatic disease. (**a**) A case can present as oligometastatic with primary tumor and three sites of metastases (diamonds). (**b**) An initially non-metastatic cancer may progress to oligometastasis despite definitive treatment to the primary, and after a longer period of progression-free survival (PFS) in the case of oligorecurrence, (**c**) a non-metastatic case may progress during adjuvant treatment or after a short PFS period, in the case of oligoprogressive disease. (**d**) The oligometastatic state may be derived from a widely metastatic state after induction systemic therapy with oligoresidual disease

is the mutation of an original tumor that leads to different subclones within one primary site, with each subclone behaving differently and having various levels of metastatic potential. Each new lesion also has the same potential to develop subclones, and "re-seed" other sites, resulting in further metastatic dissemination.

The cascade of metastatic development can be traced to the molecular level, with any mutation at each step potentially leading to a bifurcation in behavior and, eventually, metastatic potential. Individually, these steps assist tumor cells in invasion, escape, enter circulation, adhesion to secondary sites, extravasation, micrometastases, and colonization. Micro-RNAs are small RNAs that dictate regulation of gene expression [6]. These micro-RNAs have been shown to be regulated by p53 and myc, leading to either the up- or downregulation of tumor survival and proliferation, respectively [20]. The identification of specific micro-RNAs allowing for the classification of true oligometastatic disease that will benefit from aggressive local therapy is ideal, thus eliminating those with more aggressive phenotypes with shorter time to widespread dissemination that would not benefit from aggressive local therapy.

As the original tumor increases its metastatic potential, it may also create factors that will favor colonization at distant sites. This process is known as the pre-metastatic niche, and upregulation of fibronectin and clustering of bone marrow-derived cellular infiltrates have been implicated in creating a pre-metastatic niche, through the development of vascular endothelial growth factor receptor 1 (VEGFR1) [21].

A study by Lussier et al. evaluated the micro-RNAs of 63 patients with up to five lung metastases. It found 39 micro-RNAs that were expressed at different levels when comparing the patients who had high rates of progression (HRP) and low rates of progression (LRP). Three of these micro-RNAs were downregulated in the LRP group compared to the HRP group, whereas the remainders were upregulated. These suspected micro-RNAs were validated using a separate dataset. Patients in this dataset who were classified as LRP based on the presence of the appropriate micro-RNAs were found to have an improved survival compared to the HRP group [22]. An unsupervised clustering analysis of micro-RNA expression found a divergence of HRP and LRP patients [22]. Similarly, a study by Wong et al. found that the expression levels of three micro-RNAs (miR-23b, miR-449a, and miR-449b) predicted survival for the patients with such data available [23].

65.2.3 Treatment of Oligometastatic Cancer

Systemic therapy remains the mainstay and usually the most important predictor of outcomes for metastatic patient, regardless of disease burden. In cases of true oligometastatic disease, systemic therapy, such as chemotherapy, immunotherapy, or targeted agents, is often the first treatment. This serves two key purposes. The first is for cytoreduction to reduce the overall quantity of metastatic disease. As cases of seemingly limited metastatic burden may truly be in a background of widely disseminated disease, the most effective initial treatment would need to be systemic. This would allow for those patients who would not benefit from systemic therapy and would progress to "declare" their more aggressive disease phenotype. A true oligometastatic state would often also benefit from systemic therapy as these visible tumors would respond to systemic therapy and either remain stable or possibly shrink. However, there are instances where oligoprogressive patients may benefit from aggressive local therapy, such as patients with non-small cell lung cancer (NSCLC) epidermal growth factor receptor (EGFR) mutations with minimally progressive disease; in these instances, the intent of local therapy is to eradicate resistant clonogens and allow for continuation of the same systemic therapy. In these scenarios, the response to systemic therapy could be used as a method to isolate truly oligometastatic disease, either primary oligometastatic, oligoresidual through cytoreduction, oligoprogressive, or oligorecurrent. Those patients with a more aggressive cancer biology will declare themselves as polymetastatic and may not have an overall survival or disease-free survival benefit from aggressive local therapy, thus leading to potential unwarranted treatment-related toxicity from LCT.

After identifying a case of oligometastatic disease, the assumption is that the patient is systemically controlled except for the limited number of metastatic lesions. Although systemic therapy will continue, either sequentially or concurrently with local treatment, there is a window of time for which LCT can be utilized on the known disease sites. Early on, metastectomy, or the surgical resection of a metastatic lesion, was the primary "curative" approach for limited metastatic disease [1, 24–28]. Other local procedures such as radiofrequency ablation, cryotherapy, and chemoembolization have also shown some benefit, particularly in the unresectable scenario.

Local radiotherapy in this setting was initially limited to palliative doses that were much lower than doses used in the curative or definitive setting. As technology improved, higher, ablative doses of radiation that provided local control comparable to resection began to surface [14, 29]. Stereotactic body radiation therapy (SBRT), also known as stereotactic ablative radiotherapy (SABR), was developed following stereotactic radiosurgery (SRS) to treat extracranial sites of disease [29]. Stereotactic radiation's purpose, in general, is for higher-dose radiation to cause tumor death directly, causing DNA double-strand breaks, and indirectly through damage to the tumor vasculature as well as potential activation of the immune cytokine cascade [30–32]. Extensive tumor cell injury and death induce a massive release of tumor-specific antigens and pro-inflammatory and pro-oxidant cytokines. Furthermore, radiation promotes antigen presentation through increasing effector T-cell traffic to tumor, major histocompatibility complex expression, adhesion molecules, costimulatory molecules, heat shock proteins, and death receptors, as well as promoting antigen presentation through activation and maturation of dendritic cells. These multiple pathways ultimately lead to a downstream effect of a tumor-specific heightened immune response, which may be of particular benefit for tumor control when used with immune checkpoint inhibitors [33–35]. These indirect mechanisms may enhance the efficacy of hypofractionated stereotactic treatment beyond its direct effects on irreparable DNA damage to the tumor.

SBRT specifically localizes an extracranial tumor using a stereotactic coordinate system. Higher precision is obtained through three-dimensional planning, as well as image guidance verification, such as a cone-beam CT, prior to delivery of radiation. Typical SBRT doses involve the delivery of one to five fractions of radiation, with dose fractions of 6–30 Gy per fraction [32]. The high precision targeting allows for a higher ablative dose that improves the likelihood of control while also allowing for smaller margins for treatment setup, decreasing the amount of normal tissue irradiated, and thus diminishing the chance of serious side effects. Subsequently, as SBRT is much less invasive than other forms of LCT, it serves to achieve high control with low toxicity. More recently, NRG-BR001, a phase 1 study [36] looking at SBRT for two to four metastatic lesions from the breast, lung, or prostate, found that pre-specified doses for seven anatomic sites could be safely delivered as there were no grade 3 or higher toxicities. This protocol established the modern safety for SBRT of metastatic lesions and is widely used by many institutions. Although SBRT is often preferred due to its high rate of local control and short treatment duration, more fractionated regimens (less radiation per day through a longer treatment duration) have remained useful in some settings, particularly when SBRT does not meet toxicity constraints. Conventional radiation is defined as 1.8–2 Gy per day in the definitive setting and is common in non-metastatic settings but is less common in oligometastatic settings given the long-expected duration of treatment. Hypofractionated radiation (above 2 Gy per day but less than standard SBRT doses) is often considered for larger primary site tumors in the oligometastatic setting, as it is considered safer than SBRT in certain scenarios, but with a shorter duration than conventional radiation, usually 8–15 fractions instead of 25–35 fractions.

Recently, there is increasingly robust data that confirms both progression-free and overall survival benefits from local therapy for oligometastatic cancer. These trials have looked at both treatment of the limited metastatic lesions and treatment of the primary and regional sites. Some trials look at oligometastases from a specific primary site or histology, whereas others combine several groups together. Table 65.2 highlights the key findings from these studies.

65.2.4 Adrenal Metastases

Adrenal metastases are a common and well-studied site of metastatic spread, occurring in a large portion of melanomas and breast, lung, renal, and gastrointestinal cancers [24, 37]. Adrenalectomy was the original modality of treatment, and a meta-analysis of non-small cell lung cancer (NSCLC) patients with isolated adrenal metastases treated with adrenalectomy showed a 5-year OS of 25% for metachronous tumors and 26% for synchronous tumors; however, there were associated complications of adrenalectomy, including infections, cardiovascular events, ileus, and treatment-related mortality [25, 26]. Subsequently, several studies showed that SBRT to adrenal metastases is well tolerated, with two studies showing no grade 3 or higher adverse effects [27, 28] and another showing no grade 2 or higher adverse effects [24]. The long-term data from one of these studies showed a median survival for oligometastatic disease of 33 months [28]. For oligometastatic disease, particularly NSCLC, with a primary site amendable to definitive therapy, the National Comprehensive Cancer Network (NCCN) guidelines allow for either surgical resection or SBRT [38], although SBRT is often preferred due to its noninvasive and less morbid approach.

65.2.5 Hepatic Metastases

Treatment of liver metastases from colorectal cancer has been widely studied for decades, given the unique venous drainage of the gastrointestinal organs. Although surgical resection has remained a consistent option for oligometastatic liver lesions from a colorectal primary, the use of a noninvasive option is safe and efficacious in this setting. A phase 2 trial by Scorsetti et al. [39] found that patients with inoperable liver metastases receiving 75 Gy in three consecutive fractions had a 2-year local control rate of 91%, without grade 3 or higher toxicity. As molecular analyses have matured, studies have identified molecular subtypes that may predict the better prognoses that might benefit from metastasis-direct therapy. Pitroda et al. noted that subtypes that demonstrate VEGFA amplification along with stromal, mesenchymal, and angiogenic signatures or exclusive NOTCH1 and PIK3C2B mutations with E2F/MYC activation had a particularly poor survival. Their conclusion saw that high-risk molecular profiles had a 19% 10-year OS, whereas low-risk had a 94% survival [5].

65.2.6 Oligometastatic Lung Cancer

Patients with oligometastatic NSCLC tend to progress at sites of known disease as opposed to new sites [40]. Subsequently, studies of LCT on oligometastatic NSCLC have shown the most convincing evidence of this phenomenon. An earlier single-arm phase 2 trial by Iyengar et al. examined 24 patients with ≤6 extracranial sites of NSCLC disease who failed at least first-line systemic therapy and underwent concurrent erlotinib and SBRT to all sites of disease. In total 52 sites were treated, with improved median PFS (14.7 months) and median OS (20.4 months) compared to historically reported figures for second-line treatment of median PFS of 2–4 months and median OS of 9 months [17]. None of the tested patients were positive for EGFR mutations; the majority of patients progressed in new distant sites, and treatment was well tolerated with only 2 of 24 patients having grade 3 toxicity [17]. This single-arm study suggested the benefit of local therapy in potentially changing a patient's disease course. A subsequent phase 2 randomized

Table 65.2 Summary of prospective trials of oligometastatic cancer (unless otherwise stated) treated with LCT and their respective outcomes. All trials had a majority of stage IV patients. Three trials were single-arm. Toxicity data (if available) listed at the end of each outcome section, as represented by grading

Author of prospective trial	Site of primary	Study design	Sample size	Enrollment criteria	Treatment	Control arm	Local consolidative therapy arm
Scorsetti [39]	Colorectal	Single-arm phase 2 of liver RT	42 patients; 52 lesions	Up to three liver lesions. No extra-hepatic progressive or untreated gross disease	75 Gy/3 fx to liver lesion	N/A	1-/2-/3-year LC: 95/91/85% Median PFS: 12 months Median OS: 29 months Acute G2/G3 toxicity: 78%/0%
Iyengar [17]	NSCLC	Single-arm phase 2 of SBRT and erlotinib concurrently	24 patients; 52 lesions	Up to six sites of extracranial disease, early systemic chemotherapy failure	14–24 Gy/1 fx, 22–33 Gy/3 fx, 30–40 Gy/5 fx	N/A	Median PFS: 14.7 months Median OS: 20.4 months G3 toxicity: 8%
Iyengar [2]	NSCLC non-EGFR/ALK-targetable	Randomized phase 2 of maintenance chemotherapy +/− SBRT	29 patients; 31 lesions in 14 patients of treatment arm	Up to six extracranial sites including primary, no more than 3 in the liver or lung, did not progress on systemic therapy	16–27 Gy/1 fx, 24.5–33 Gy/3 fx, 28–37.5 Gy/5 fx; 45 Gy/15 fx (if SBRT constraints not met)	Median PFS: 3.5 months Median OS (with crossover): 17 months G3/4/5 toxicity: 2/1/0	Median PFS: 9.7 months ($P = 0.01$) Median OS: Not reached (p-value not reported) G3/4/5 toxicity: 4/0/0
Gomez [3]	NSCLC	Randomized phase 2 of consolidative RT for NSCLC	49 patients	≤3 extra- and intracranial metastatic sites after systemic therapy, did not progress on first-line systemic therapy	Resection or radiation (SBRT, 15 fractions, or conventional allowed) to metastatic sites; chemoradiation allowed for primary and regional site(s)	PFS: 4.4 months OS: 17 months Survival after progression: 9.4 months Median time to new lesion: 6 months G3/4/5 toxicity: 2/0/0	PFS: 14.2 months ($P = 0.022$) OS: 41.2 months ($P = 0.017$) Survival after progression: 37.6 months Median time to new lesion: 14.2 months G3/4/5 toxicity: 4/0/0
Wang [45]	EGFR-mutated stage III–IV NSCLC	Single-arm study	26 patients (21 with stage IV)	0–3 extra- and intracranial metastatic sites	Conformal RT, IMRT, or gamma ray SBRT to thoracic disease concurrently with TKI	N/A	Median TTP: 6.3 months Median PFS: 10.2 months Median OS: 21.8 months G3/4/5 toxicity: 12/3/0
RTOG 0937 [46]	ES-SCLC	Randomized phase 2 of PCI +/− consolidative extracranial RT	86 patients	1–4 extracranial lesions, had at least partial response to four to six cycles of platinum-based chemotherapy to one site, with no progression at other sites	PCI (25 Gy/10 fx) with randomization to consolidative RT to the thorax and metastases (45 Gy/15 fx) or no further treatment	1-year OS: 60.1% Median OS: 15.8 months Median PFS: 2.9 months G3+ toxicity: 23.8%	1-year OS: 50.8% ($P = 0.21$) Median OS: 13.8 months Median PFS: 4.9 months ($P = 0.0148$) G3+ toxicity: 36.4% (one person had G5 pneumonitis)

Table 65.2 (continued)

Author of prospective trial	Site of primary	Study design	Sample size	Enrollment criteria	Treatment	Control arm	Local consolidative therapy arm
HORRAD [49]	Prostate	Randomized study of ADT +/− prostate RT	432 patients	PSA >20 ng/ml with no restrictions on quantity of metastasis	Androgen deprivation therapy randomized to local radiotherapy to the prostate	Median OS: 43 months median time to PSA progression: 15 months	Median OS: 45 months ($P = 0.4$) Median time to PSA progression: 12 months ($P = 0.02$)
Parker – STAMPEDE [4]	Prostate	Randomized phase 3 of systemic therapy +/− prostate RT	2061 patients	Low and high disease burden per CHAARTED definition	Androgen deprivation therapy and docetaxel randomized to local radiotherapy to the prostate 55 Gy/20 fx or 36 Gy/6 weekly fx	PFS (all): 32.4 months FFS (all): 21.4 months OS (all): 41.6 months PFS (low): 39.4 months FFS (low): 27.4 months OS (low): 45.4 months G3+ toxicity 6/12/24 months: 21/12/15%	PFS (all): 33.1 months ($p = 0.468$) FFS (all): 26.2 months ($p < 0.0001$) OS (all): 42.5 months ($p = 0.266$) PFS (low): 42.9 months ($p = 0.033$) FFS (low): 36.1 months ($p < 0.0001$) OS (low): 49.1 months ($p = 0.007$) G3+ toxicity 6/12/24 month: 22/13/13%
Ost [50]	Prostate	Phase 2 randomized trial of surveillance versus MDT	62 patients	Three or fewer extracranial metastases, testosterone over 50 ng/mL	Not on systemic therapy, randomized to metastasis-directed therapy (SBRT 30 Gy/3 fx or surgery)	Median ADT-free survival: 13 months Median PSA progression: 6 months	Median ADT-free survival: 21 months ($P = 0.11$) Median PSA progression: 10 months ($P = 0.03$) G1/2+ toxicity: 17%/0%
Phillips – ORIOLE [51]	Prostate	Phase 2 randomized trial of observation versus MDT SBRT	54 patients	Recurrent hormone-sensitive, 1–3 metastatic sites, no ADT in last 6 months and less than 3 years total	19.5–48 Gy/3–5 fx SBRT	6-month progression: 61% 6-month new lesions: 63% G3+ toxicity: 0	6-month progression: 19% ($P = 0.005$) 6-month new lesions: 16% ($P = 0.006$) G3+ toxicity: 0
Trovo [9]	Breast	Single-arm phase 2 trial of MDT with primary controlled	54 patients with 92 lesions	Up to five extracranial lesions	SBRT 30–45 Gy/3 fx or IMRT 60 Gy/25 fx; 89% received concomitant systemic therapy	N/A	1–/2-year PFS: 75/53% 2-year LC: 97% 2-year OS: 95% G2/3+ toxicity: 2/0

(continued)

Table 65.2 (continued)

Author of prospective trial	Site of primary	Study design	Sample size	Enrollment criteria	Treatment	Control arm	Local consolidative therapy arm
Palma – SABR-COMET [10, 32]	Breast, colorectal, lung, prostate, others	Phase 2 randomized trial of SBRT MDT	99 patients	Up to five lesions, excluding one to three intracranial-only metastatic patients, maximum three lesions per organ	Control: 8 Gy/1 fx, 20 Gy/5 fx, 30 Gy/10 fx SBRT: 16–60 Gy/1–8 fx depending on site as listed in protocol (see citation supplement)	Progression events: 85% Median PFS: 5.4 months Median OS: 28 months G2+ toxicity: 9% Treatment-related deaths: 0	Progression events: 59% Median PFS: 11.6 months ($P = 0.001$) Median OS: 50 months ($P = 0.006$) G2+ toxicity: 29% Treatment-related deaths: 4.5%

PFS progression-free survival, *OS* overall survival, *LC* local control, *FFS* failure-free survival, *TTP* time to progression

control trial (RCT) by Iyengar et al. randomized NSCLC patients who did not possess EGFR- or ALK-targetable mutations with five or fewer metastatic sites, and who did not progress on systemic therapy, to SBRT to all sites of gross disease followed by maintenance chemotherapy or maintenance chemotherapy alone. Although only 29 patients were enrolled, this was due to the accrual terminating early due to a significant improvement in outcomes during the interim analysis which found PFS improving from 3.5 months to 9.7 months when adding SBRT ($p = 0.01$) [2]. A different phase 2 RCT by Gomez et al. randomized patients with three or fewer sites of metastatic disease, without progression after at least 3 months of first-line systemic therapy, to LCT versus maintenance systemic therapy or observation. LCT was defined as ablation of all residual disease, with either surgery or radiation, with concurrent chemoradiation also allowed for primary disease. Patients treated with LCT were permitted to continue maintenance therapy or observed. Patients receiving LCT had both a significant progression-free survival benefit of median 14.2 months compared to 4.4 months ($P = 0.022$) and an overall survival benefit of median 41.2 months versus 17 months ($P = 0.017$) [3]. NRG-LU002 is currently recruiting patients in a phase 2/3, national, multi-institutional setting to determine if LCT does provide survival benefit beyond maintenance therapy alone for oligometastatic/oligoresidual patients with up to three extracranial metastatic sites, regardless of disease burden prior to systemic therapy [41].

The above studies typically exclude EGFR-mutated NSCLC as it represents a different biology and subsequently a better prognosis. The effectiveness of tyrosine kinase inhibitors (TKIs) has imparted a long natural history for these patients, and the majority of patients on erlotinib tend to fail at the initial sites of disease (primary and original metastases) as opposed to new sites of metastases [42]. In a retrospective review by Hu et al., patients with oligometastatic EGFR-mutated NSCLC receiving TKI followed by LCT were found to have a significantly longer PFS of 15 months, compared to 10 months of TKI alone ($P < 0.001$). OS was also longer in the combination group, at 34 months, compared to 21 months ($P = 0.001$) [43]. A different propensity-matched study by Yen et al. looked at EGFR-mutated NSCLC for stages III–IV who responded to TKI and found that patients who received thoracic RT had an adjusted hazard ratio for mortality of 0.72 for all patients ($P = 0.0005$), 0.76 for stage IIIB ($P = 0.042$), and 0.70 for stage IV ($P = 0.005$) [44]. A single-arm prospective study was reported by Wang et al., for which TKI and individualized RT in the stage III–IV setting were well-tolerated, with no mortalities and less than 10% grade 4 toxicity [45]. These data strongly suggest that LCT/SBRT to oligometastatic disease in the setting of NSCLC patients with driver mutations after response to targeted agents may result in improvements in survival without significant toxicity.

However, data in the small-cell lung cancer (SCLC) has not quite developed to the same extent as its NSCLC counterpart. RTOG 0937 was a phase 2 RCT of patients with extensive-stage (ES) SCLC and one to four extracranial metastases, with evidence of response to minimum of one site, without evidence of progression to any site, comparing prophylactic cranial radiation (PCI) alone versus PCI with consolidative radiation to residual intrathoracic and extracranial metastatic sites. There was a 3- and 12- month progression-free survival favoring PCI with LCT of 85.5% and 25% versus 46.7% and 20.4% ($P = 0.01$), respectively. However, the 1-year OS was not different, with 60.1% for PCI and 50.8% for PCI + LCT ($P = 0.21$) [46]. The lack of an OS benefit could be due to the limited options of systemic therapy. The very recent addition of atezolizumab to the standard platinum-based doublet therapy has shown promising results, including an OS benefit [47]. This has prompted the NRG-LU007 RAPTOR study, which seeks to test the addition of LCT to atezolizumab and chemotherapy in ES-SCLC with ten or fewer sites of metastatic disease after induction therapy [48].

65.2.7 Oligometastatic Prostate Cancer

There is strong evidence in favor of treating the primary and distant sites in oligometastatic prostate cancer. However, careful patient selection remains the first step. Studies examining treatment of the primary site alone with RT include the HORRAD Trial and STAMPEDE. The HORRAD Trial examined men with a prostate-specific antigen (PSA) of over 20 ng/ml and primary bone metastases and randomized between androgen deprivation therapy (ADT) alone and ADT with primary treatment of the prostate with external beam radiation (70 Gy/35 fx daily or 57.76 Gy/19 fx three times weekly). It found that there was a median time to PSA progression benefit of 15 versus 12 months ($P = 0.02$), but no survival benefit in the radiotherapy arm. However, 67% of patients had at least five metastatic lesions, and the median PSA level was 142 ng/ml. Furthermore, on unplanned subanalysis, in men with less than five metastatic lesions, there was a trend toward a significant improvement in OS (HR, 0.68; 95% CI, 0.42–1.10; $P < 0.10$) [49]. STAMPEDE, a larger multi-arm, multistage trial comparing EBRT (55 Gy/20 fx daily or 36 Gy/6 fx weekly SBRT) to the prostate with ADT versus ADT alone, also found a failure-free survival benefit (17 vs. 13 months, $P = 0.0001$) but not an overall survival benefit for all comers (48 vs. 46 months, $P = 0.266$). However, the trial stratified patients based on metastatic burden for a preplanned analysis by using the CHAARTED definition of high metastatic burden of four or more bone metastases with one or more outside the pelvis or vertebral body, visceral metastases, or both. Patients who did not meet these criteria were deemed as "low metastatic burden" and had both a 3-year failure-free survival (HR, 0.59; 95% CI, 0.49–0.72; $P < 0.0001$) and a 3-year OS benefit (HR, 0.68; 95% CI, 0.52–0.90; $P = 0.007$). The study also pointed out the safety of local therapy, with only 1% more grade 3+ toxicity at 6 months and 2% less at 2 years compared to systemic therapy alone, with an overall grade 3–4 toxicity rate of 4–5% [4]. These data suggest that treatment of the primary in the setting of metastatic prostate cancer with low metastatic burden may improve both PFS and OS.

Regarding metastasis-directed therapy (MDT) for prostate cancer, there are also a few trials to show benefit. A phase 2 RCT by Ost et al. found that patients previously treated for localized disease who then present with three or fewer extracranial metastases benefited from MDT with an ADT-free survival of 21 months versus 13 months in patients without MDT ($P = 0.11$) [50]. Although not statistically significant, the sample size was only 62. The phase 2 ORIOLE trial looked specifically at patients with castration-sensitive prostate cancer with one to three asymptomatic metastases up to 5 cm in diameter or 250 cm^2. Patients were randomized 2:1 to SBRT or observation to the sites of metastases. Patients had previous definitive surgery or radiation to the prostate. There was a benefit to PFS at 6 months (19% vs. 61%, $P = 0.005$), decrease risk of new lesion at 6 months, no grade 3 or greater toxic effects, and an increase in clonotypic T-cell expansion [51]. These data suggest that MDT for oligoprogressive prostate cancer after prior definitive primary therapy may improve PFS, and further studies are needed to determine potential impacts on OS.

65.2.8 Oligometastatic Breast Cancer

Breast cancer is another malignancy for which an oligometastatic state can carry a favorable prognosis. A phase 2 study by Trovo et al. that enrolled patients with up to five extracranial metastases (the primary tumor controlled) found that treating metastatic lesions with SBRT conferred a 1- and 2-year PFS of 75% and 53%, respectively. The 2-year OS was 95% [9]. Following the initial safety trial BR-001, the NRG has been enrolling patients for BR-002, a randomized phase 2/3 trial for women with up to four sites of metastatic breast cancer, who received up to 12 months of first-line systemic therapy without progression; these patients are randomized to SBRT and/or surgical resection versus no LCT [52]. There is also evidence for surgical resection of the primary breast cancer even in the metastatic setting. An NCDB study by Khan et al. found that margin-negative surgical resection with partial or total mastectomy had a superior prognosis, with HR of 0.61 (95% CI, 0.58–0.65, $P < 0.05$), compared to patients not surgically resected in the metastatic setting [53]. A meta-analysis by Ly et al. found that patients with metastatic breast cancer who received local-regional therapy (surgery with or without radiation) had an overall median survival of 25–42 months compared to 12.6–28.3 months in those without surgery [54].

Surgery has remained the key oncologic treatment of the primary site for breast cancer, and radiation has mainly served in an adjuvant role. However, with the success of SBRT, further technological advances have allowed for the development of GammaPod, which uses stereotactic image guidance, along with vacuum-assisted breast immobilization to deliver high-intensity radiation using Cobalt-60 sources [55]. There is a potential for GammaPod radiotherapy to help with treatment to the primary site, which could prove useful in the metastatic setting, though there is minimal published data.

65.2.9 Oligometastatic Cancers with Controlled Primary

The phase 2 SABR-COMET study by Palma et al. looked at patients with a controlled primary malignancy and one to five extracranial metastatic lesions. The majority of primary tumors were breast, colorectal, lung, and prostate. Patients were randomized to either SBRT to all sites of metastatic disease or palliative standard of care, including palliative doses of radiation at symptomatic sites. The SBRT dosing was lesion- and site-specific per protocol and ranged from 16 Gy/1 fx for vertebral bodies to 60 Gy/8 fx for adrenal, liver, or central lung lesions. Systemic therapy was at the discretion of the medical oncologist but had to be outside of SABR by 4 weeks, unless it was hormonal therapy. Median PFS was improved in the SBRT arm, at 11.6 months, compared to 5.4 months in the control arm ($P = 0.001$). Overall survival was improved in the SBRT arm, with median survival of 50 months, compared to 28 months in the standard-of-care arm ($P = 0.006$) with 5-year OS of 42.3% versus 17.7%, respectively. Grade 2 or higher adverse effects were 29% in the SBRT arm, compared to 9% in the control arm. Furthermore, 4.5% of the deaths in the SBRT group were "possibly, probably, or definitely" related to treatment [10, 32]. Based on these phase 2 results, there are separate phase 3 trials examining these questions among patients with extracranial metastatic sites numbering 1–3 (SABR-COMET-3) and those with four to ten sites (SABR-COMET-10) [56, 57].

65.3 Synergy with Systemic Therapy and Future Directions

Although mechanisms remain unclear, there is a potential synergy between systemic therapy and radiation, as well as the potential abscopal effect of radiation. Radiation therapy can create an abscopal effect, for which not only does the radiation-targeted lesion regress but so do lesions at distant sites. It is hypothesized as a systemic immune-mediated response for which high-dose radiation creates an immunological response that is further upregulated by systemic therapy through antigen processing, as well as generation and trafficking of effector cells [58, 59]. Though enticing, the exact mechanisms of the abscopal effect are not well-characterized, and reproducibility of this response has remained elusive. A greater understanding of these mechanisms could enhance our delivery of treatment to further improve outcomes.

As there is mounting phase 2 evidence of benefit for LCT of up to five sites of extracranial metastatic disease, more mature phase 2/3 studies are still needed, including NRG-BR002 for oligometastatic breast cancer as well as NRG-LU002. Additionally, there is a push to see how many sites of extracranial metastases should be treated for a PFS and OS benefit. In the intracranial metastases setting, there is evidence that treating up to ten intracranial lesions ablatively with stereotactic radiosurgery carries a benefit similar to treating two to four lesions with SRS [60]. Ongoing trials such as SABR-COMET-10 are also looking at focally treating up to ten extracranial lesions, to see if there is a benefit similar to treating up to five extracranial lesions [56].

When compared to surgical metastasectomy, the toxicity profile of radiation therapy is low; however, it is not without risk. As discussed, serious side effects, including treatment-related mortality, can and have occurred, and careful planning at an experienced facility is always recommended. Newer technology brings a promise of safer delivery. Adaptive planning, through the use of fast algorithms and using magnetic resonance guidance, is particularly helpful in areas where the tumor shrinks or nearby normal organs at risk may not be stationary. This new form of MR-guided radiation has been shown to be deliverable and safe [61]. Although the data for heavy particle beam radiotherapy has had mixed result, there remains optimism that less "exit dose", which is the dose of radiation past the tumor that exits the body, could lead to an improved side effect profile and potential dose escalation [62].

As systemic therapies improve, so do the number of cases of oligometastatic/oligoresidual cancer, increasing the role of radiation therapy to extend the usefulness of the systemic therapy, and subsequently improve OS. The advent of molecular profiling will help improve our understanding of the biology of cancer and its response to radiation therapy, and lead to the creation of more targeted agents. LCT, especially radiation therapy, will continue to improve in efficacy and safety, furthering the possibility of long-term survival (and ideally cure in a select few) for oligometastatic cancer.

65.4 Conclusion and Perspective

- Oligometastatic cancer represents an intermediary between non-metastatic and widely metastatic cancer and may have improved progression-free and overall survival with aggressive metastasis-directed therapy.
- Molecular analysis has given us clues on how cancer progresses to the oligometastatic state and may help us determine which patients benefit from metastasis-directed therapy.
- Modern randomized clinical trials have largely shown a benefit to primary/metastasis-directed therapy for oligometastatic cancer, and as both LCT and systemic therapies improve, so will their synergy to benefit survival.

Open Questions

- How do we identify and appropriately counsel patients who are at higher risk for SBRT toxicity, including treatment-related mortality?
- Is there a combined threshold of tumor burden, quantity, and/or location for which SBRT for oligometastatic disease would unlikely improve survival (or even be detrimental)?
- What would be the most appropriate treatment for polymetastatic disease with limited progressive disease?
- What would be the optimal sequencing of systemic therapy and local consolidative therapy based on the type of oligometastatic disease, burden, and response?
- How can we use biomarkers to predict the quantity of unidentified microscopic systemic disease and/or assess response to treatment of oligometastatic disease?

References

1. Hellman S, Weichselbaum RR. Oligometastases. J Clin Oncol. 1995;13:8–10.
2. Iyengar P, Wardak Z, Gerber DE, Tumati V, Ahn C, Hughes RS, Dowell JE, Cheedella N, Nedzi L, Westover KD, Pulipparacharuvil S. Consolidative radiotherapy for limited metastatic non-small-cell lung cancer: a phase 2 randomized clinical trial. JAMA Oncol. 2018 Jan 1;4:e173501.
3. Gomez DR, Tang C, Zhang J, Blumenschein GR Jr, Hernandez M, Lee JJ, Ye R, Palma DA, Louie AV, Camidge DR, Doebele RC. Local consolidative therapy vs. maintenance therapy or observation for patients with oligometastatic non–small-cell lung cancer: long-term results of a multi-institutional, phase II, randomized study. J Clin Oncol. 2019;37:1558.
4. Parker CC, James ND, Brawley CD, Clarke NW, Hoyle AP, Ali A, Ritchie AW, Attard G, Chowdhury S, Cross W, Dearnaley DP. Radiotherapy to the primary tumour for newly diagnosed, metastatic prostate cancer (STAMPEDE): a randomised controlled phase 3 trial. Lancet. 2018;392:2353–66.
5. Pitroda SP, Khodarev NN, Huang L, Uppal A, Wightman SC, Ganai S, Joseph N, Pitt J, Brown M, Forde M, Mangold K. Integrated molecular subtyping defines a curable oligometastatic state in colorectal liver metastasis. Nat Commun. 2018;9:1–8.
6. Uppal A, Ferguson MK, Posner MC, Hellman S, Khodarev NN, Weichselbaum RR. Towards a molecular basis of oligometastatic disease: potential role of micro-RNAs. Clin Exp Metastasis. 2014;31:735–48.
7. Campbell PJ, Yachida S, Mudie LJ, Stephens PJ, Pleasance ED, Stebbings LA, Morsberger LA, Latimer C, McLaren S, Lin ML, McBride DJ. The patterns and dynamics of genomic instability in metastatic pancreatic cancer. Nature. 2010;467:1109–13.
8. Yachida S, Jones S, Bozic I, Antal T, Leary R, Fu B, Kamiyama M, Hruban RH, Eshleman JR, Nowak MA, Velculescu VE. Distant metastasis occurs late during the genetic evolution of pancreatic cancer. Nature. 2010;467:1114–7.
9. Trovo M, Furlan C, Polesel J, Fiorica F, Arcangeli S, Giaj-Levra N, Alongi F, Del Conte A, Militello L, Muraro E, Martorelli D. Radical radiation therapy for oligometastatic breast cancer: results of a prospective phase II trial. Radiother Oncol. 2018;126:177–80.
10. Palma DA, Olson R, Harrow S, Gaede S, Louie AV, Haasbeek C, Mulroy L, Lock M, Rodrigues GB, Yaremko BP, Schellenberg D. Stereotactic ablative radiotherapy versus standard of care palliative treatment in patients with oligometastatic cancers (SABR-COMET): a randomised, phase 2, open-label trial. Lancet. 2019;393:2051–8.
11. Zhao F, Wang J, Chen M, Chen D, Ye S, Li X, Chen X, Ren G, Yan S. Sites of synchronous distant metastases and prognosis in prostate cancer patients with bone metastases at initial diagnosis: a population-based study of 16,643 patients. Clin Transl Med. 2019;8:30.
12. Hatzoglou V, Patel GV, Morris MJ, Curtis K, Zhang Z, Shi W, Huse J, Rosenblum M, Holodny AI, Young RJ. Brain metastases from prostate cancer: an 11-year analysis in the MRI era with emphasis on imaging characteristics, incidence, and prognosis. J Neuroimaging. 2014;24:161–6.
13. Wouters BG, Koritzinsky M, Brown JM, Van der Kogel AJ. The tumour microenvironment and cellular hypoxia responses. In: Joiner MC, Van der Kogel A, editors. Basic clinical radiobiology. 5th ed. Boca Raton, FL: CRC Press; 2009. p. 206–16.
14. Rusthoven KE, Kavanagh BD, Cardenes H, Stieber VW, Burri SH, Feigenberg SJ, Chidel MA, Pugh TJ, Franklin W, Kane M, Gaspar LE. Multi-institutional phase I/II trial of stereotactic body radiation therapy for liver metastases. J Clin Oncol. 2009;27:1572–8.
15. Donskov F, Xie W, Overby A, Wells JC, Fraccon AP, Sacco CS, Porta C, Stukalin I, Lee JL, Koutsoukos K, Yuasa T. Synchronous Versus Metachronous Metastatic Disease: Impact of Time to Metastasis on Patient Outcome—Results From the International Metastatic Renal Cell Carcinoma Database Consortium. European urology oncology. 2020;3(4):530–9.
16. Niibe Y, Hayakawa K. Oligometastases and oligo-recurrence: the new era of cancer therapy. Jpn J Clin Oncol. 2010;40:107–11.
17. Iyengar P, Kavanagh BD, Wardak Z, Smith I, Ahn C, Gerber DE, Dowell J, Hughes R, Abdulrahman R, Camidge DR, Gaspar LE. Phase II trial of stereotactic body radiation therapy combined with erlotinib for patients with limited but progressive metastatic non–small-cell lung cancer. J Clin Oncol. 2014;32:3824–30.
18. Gerlinger M, Rowan AJ, Horswell S, Larkin J, Endesfelder D, Gronroos E, Martinez P, Matthews N, Stewart A, Tarpey P, Varela I. Intratumor heterogeneity and branched evolution revealed by multiregion sequencing. N Engl J Med. 2012;366:883–92.
19. Klein CA. Selection and adaptation during metastatic cancer progression. Nature. 2013;501:365–72.
20. Lujambio A, Lowe SW. The microcosmos of cancer. Nature. 2012;482:347–55.
21. Kaplan RN, Rafii S, Lyden D. Preparing the "soil": the premetastatic niche. Cancer Res. 2006;66:11089–93.
22. Lussier YA, Xing HR, Salama JK, Khodarev NN, Huang Y, Zhang Q, Khan SA, Yang X, Hasselle MD, Darga TE, Malik R. MicroRNA expression characterizes oligometastasis (es). PLoS One. 2011;6:e28650.
23. Wong AC, Watson SP, Pitroda SP, Son CH, Das LC, Stack ME, Uppal A, Oshima G, Khodarev NN, Salama JK, Weichselbaum RR. Clinical and molecular markers of long-term survival after oligometastasis-directed stereotactic body radiotherapy (SBRT). Cancer. 2016;122:2242–50.
24. Chawla S, Chen Y, Katz AW, Muhs AG, Philip A, Okunieff P, Milano MT. Stereotactic body radiotherapy for treatment of adrenal metastases. Int J Radiat Oncol Biol Phys. 2009;75:71–5.
25. Lo CY, Van Heerden JA, Soreide JA, Grant CS, Thompson GB, Lloyd RV, Harmsen WS. Adrenalectomy for metastatic disease to the adrenal glands. Br J Surg. 1996;83:528–31.
26. Tanvetyanon T, Robinson LA, Schell MJ, Strong VE, Kapoor R, Coit DG, Bepler G. Outcomes of adrenalectomy for isolated syn-

chronous versus metachronous adrenal metastases in non-small-cell lung cancer: a systematic review and pooled analysis. In: Database of abstracts of reviews of effects (DARE): quality-assessed reviews. England: Centre for Reviews and Dissemination; 2008.
27. Rudra S, Malik R, Ranck MC, Farrey K, Golden DW, Hasselle MD, Weichselbaum RR, Salama JK. Stereotactic body radiation therapy for curative treatment of adrenal metastases. Technol Cancer Res Treat. 2013 Jun;12:217–24.
28. Buergy D, Rabe L, Siebenlist K, Stieler F, Fleckenstein J, Giordano FA, Wenz F, Boda-Heggemann J. Treatment of adrenal metastases with conventional or hypofractionated image-guided radiation therapy–patterns and outcomes. Anticancer Res. 2018;38:4789–96.
29. Onishi H, Araki T, Shirato H, Nagata Y, Hiraoka M, Gomi K, Yamashita T, Niibe Y, Karasawa K, Hayakawa K, Takai Y. Stereotactic hypofractionated high-dose irradiation for stage I nonsmall cell lung carcinoma: clinical outcomes in 245 subjects in a Japanese multiinstitutional study. Cancer. 2004;101:1623–31.
30. Churilla TM, Chowdhury IH, Handorf E, Collette L, Collette S, Dong Y, Alexander BM, Kocher M, Soffietti R, Claus EB, Weiss SE. Comparison of local control of brain metastases with stereotactic radiosurgery vs surgical resection: a secondary analysis of a randomized clinical trial. JAMA Oncol. 2019;5:243–7.
31. Kim MS, Kim W, Park IH, Kim HJ, Lee E, Jung JH, Cho LC, Song CW. Radiobiological mechanisms of stereotactic body radiation therapy and stereotactic radiation surgery. Radiat Oncol J. 2015;33:265.
32. Palma DA, Olson R, Harrow S, Gaede S, Louie AV, Haasbeek C, Mulroy L, Lock M, Rodrigues GB, Yaremko BP, Schellenberg D. Stereotactic ablative radiotherapy for the comprehensive treatment of oligometastatic cancers: long-term results of the SABR-COMET phase II randomized trial. J Clin Oncol. 2020:1;38(25):2830.
33. Finkelstein SE, Timmerman R, McBride WH, Schaue D, Hoffe SE, Mantz CA, Wilson GD. The confluence of stereotactic ablative radiotherapy and tumor immunology. Clin Dev Immunol. 2011;2011:7. Article ID 439752. https://doi.org/10.1155/2011/439752.
34. Blank C, Gajewski TF, Mackensen A. Interaction of PD-L1 on tumor cells with PD-1 on tumor-specific T cells as a mechanism of immune evasion: implications for tumor immunotherapy. Cancer Immunol Immunother. 2005;54:307–14.
35. Seung SK, Curti BD, Crittenden M, Walker E, Coffey T, Siebert JC, Miller W, Payne R, Glenn L, Bageac A, Urba WJ. Phase 1 study of stereotactic body radiotherapy and interleukin-2—tumor and immunological responses. Sci Transl Med. 2012;4:137ra74.
36. Chmura SJ, Winter K, Salama JK, Robinson CG, Pisansky TM, Borges V, Al-Hallaq HA, Matuszak MM, Park SS, Gonzalez VJ, Hasan Y. Phase I trial of stereotactic body radiation therapy (SBRT) to multiple metastatic sites: a NRG oncology study. Int J Radiat Oncol Biol Phys. 2018;102:S68–9.
37. Lam KY, Lo CY. Metastatic tumours of the adrenal glands: a 30-year experience in a teaching hospital. Clin Endocrinol. 2002;56:95–101.
38. Ettinger DS, Wood DE, Aggarwal C, Aisner DL, Akerley W, Bauman JR, Bharat A, Bruno DS, Chang JY, Chirieac LR, D'Amico TA. NCCN guidelines insights: non–small cell lung cancer, version 1.2020: featured updates to the NCCN guidelines. J Natl Compr Cancer Netw. 2019;17:1464–72.
39. Scorsetti M, Comito T, Tozzi A, Navarria P, Fogliata A, Clerici E, Mancosu P, Reggiori G, Rimassa L, Torzilli G, Tomatis S. Final results of a phase II trial for stereotactic body radiation therapy for patients with inoperable liver metastases from colorectal cancer. J Cancer Res Clin Oncol. 2015;141:543–53.
40. Rusthoven KE, Hammerman SF, Kavanagh BD, Birtwhistle MJ, Stares M, Camidge DR. Is there a role for consolidative stereotactic body radiation therapy following first-line systemic therapy for metastatic lung cancer? A patterns-of-failure analysis. Acta Oncol. 2009;48:578–83.
41. https://clinicaltrials.gov/ct2/show/NCT03137771
42. Patel SH, Rimner A, Foster A, Zhang Z, Woo KM, Helena AY, Riely GJ, Wu AJ. Patterns of initial and intracranial failure in metastatic EGFR-mutant non-small cell lung cancer treated with erlotinib. Lung Cancer. 2017;108:109–14.
43. Hu F, Xu J, Zhang B, Li C, Nie W, Gu P, Hu P, Wang H, Zhang Y, Shen Y, Wang S. Efficacy of local consolidative therapy for oligometastatic lung adenocarcinoma patients harboring epidermal growth factor receptor mutations. Clin Lung Cancer. 2019;20:e81–90.
44. Yen YC, Hsu HL, Chang JH, Lin WC, Chang YC, Chang CL, Chow JM, Yuan KS, Wu AT, Wu SY. Efficacy of thoracic radiotherapy in patients with stage IIIB–IV epidermal growth factor receptor-mutant lung adenocarcinomas who received and responded to tyrosine kinase inhibitor treatment. Radiother Oncol. 2018;129:52–60.
45. Wang J, Xia TY, Wang YJ, Li HQ, Li P, Wang JD, Chang DS, Liu LY, Di YP, Wang X, Wu WZ. Prospective study of epidermal growth factor receptor tyrosine kinase inhibitors concurrent with individualized radiotherapy for patients with locally advanced or metastatic non–small-cell lung cancer. Int J Radiat Oncol Biol Phys. 2011;81:e59–65.
46. Gore EM, Hu C, Sun AY, Grimm DF, Ramalingam SS, Dunlap NE, Higgins KA, Werner-Wasik M, Allen AM, Iyengar P, Videtic GM. Randomized phase II study comparing prophylactic cranial irradiation alone to prophylactic cranial irradiation and consolidative extracranial irradiation for extensive-disease small cell lung cancer (ED SCLC): NRG oncology RTOG 0937. J Thorac Oncol. 2017;12:1561–70.
47. Horn L, Mansfield AS, Szczęsna A, Havel L, Krzakowski M, Hochmair MJ, Huemer F, Losonczy G, Johnson ML, Nishio M, Reck M. First-line atezolizumab plus chemotherapy in extensive-stage small-cell lung cancer. N Engl J Med. 2018;379:2220–9.
48. https://clinicaltrials.gov/ct2/show/NCT04402788
49. Boevé LM, Hulshof MC, Vis AN, Zwinderman AH, Twisk JW, Witjes WP, Delaere KP, van Moorselaar RJ, Verhagen PC, van Andel G. Effect on survival of androgen deprivation therapy alone compared to androgen deprivation therapy combined with concurrent radiation therapy to the prostate in patients with primary bone metastatic prostate cancer in a prospective randomised clinical trial: data from the HORRAD trial. Eur Urol. 2019;75:410–8.
50. Ost P, Reynders D, Decaestecker K, Fonteyne V, Lumen N, DeBruycker A, Lambert B, Delrue L, Bultijnck R, Claeys T, Goetghebeur E. Surveillance or metastasis-directed therapy for oligometastatic prostate cancer recurrence: a prospective, randomized, multicenter phase II trial. J Clin Oncol. 2018;36(5):446–453. https://doi.org/10.1200/JCO.2017.75.4853.
51. Phillips R, Shi WY, Deek M, Radwan N, Lim SJ, Antonarakis ES, Rowe SP, Ross AE, Gorin MA, Deville C, Greco SC. Outcomes of observation vs stereotactic ablative radiation for oligometastatic prostate cancer: the ORIOLE phase 2 randomized clinical trial. JAMA Oncol. 2020;6:650–9.
52. https://clinicaltrials.gov/ct2/show/NCT02364557
53. Khan SA, Stewart AK, Morrow M. Does aggressive local therapy improve survival in metastatic breast cancer? Surgery. 2002;132:620–7.
54. Ly BH, Nguyen NP, Vinh-Hung V, Rapiti E, Vlastos G. Locoregional treatment in metastatic breast cancer patients: is there a survival benefit? Breast Cancer Res Treat. 2010;119:537–45.
55. Yu CX, Shao X, Zhang J, Regine W, Zheng M, Yu YS, Deng J, Duan Z. GammaPod—a new device dedicated for stereotactic radiotherapy of breast cancer. Med Phys. 2013;40:051703.

56. Palma DA, Olson R, Harrow S, Correa RJ, Schneiders F, Haasbeek CJ, Rodrigues GB, Lock M, Yaremko BP, Bauman GS, Ahmad B. Stereotactic ablative radiotherapy for the comprehensive treatment of 4–10 oligometastatic tumors (SABR-COMET-10): study protocol for a randomized phase III trial. BMC Cancer. 2019;19:1–5.
57. Olson R, Mathews L, Liu M, Schellenberg D, Mou B, Berrang T, Harrow S, Correa RJ, Bhat V, Pai H, Mohamed I. Stereotactic ablative radiotherapy for the comprehensive treatment of 1–3 Oligometastatic tumors (SABR-COMET-3): study protocol for a randomized phase III trial. BMC Cancer. 2020;20:1–2.
58. Postow MA, Callahan MK, Barker CA, Yamada Y, Yuan J, Kitano S, Mu Z, Rasalan T, Adamow M, Ritter E, Sedrak C. Immunologic correlates of the abscopal effect in a patient with melanoma. N Engl J Med. 2012;366:925–31.
59. Okwan-Duodu D, Pollack BP, Lawson D, Khan MK. Role of radiation therapy as immune activator in the era of modern immunotherapy for metastatic malignant melanoma. Am J Clin Oncol. 2015;38:119–25.
60. Yamamoto M, Serizawa T, Shuto T, Akabane A, Higuchi Y, Kawagishi J, Yamanaka K, Sato Y, Jokura H, Yomo S, Nagano O. Stereotactic radiosurgery for patients with multiple brain metastases (JLGK0901): a multi-institutional prospective observational study. Lancet Oncol. 2014;15:387–95.
61. Henke L, Kashani R, Robinson C, Curcuru A, DeWees T, Bradley J, Green O, Michalski J, Mutic S, Parikh P, Olsen J. Phase I trial of stereotactic MR-guided online adaptive radiation therapy (SMART) for the treatment of oligometastatic or unresectable primary malignancies of the abdomen. Radiother Oncol. 2018;126:519–26.
62. Sulaiman NS, Fujii O, Demizu Y, Terashima K, Niwa Y, Akagi T, Daimon T, Murakami M, Sasaki R, Fuwa N. Particle beam radiation therapy using carbon ions and protons for oligometastatic lung tumors. Radiat Oncol. 2014;9:183.

Spine Radiotherapy

66

Mark Ashamalla, Alexander N. Slade, Kartik Mani,
and Samuel Ryu

Abstract

A large portion of cancer patients will develop spinal metastases during the natural course of their disease. While **spine metastases** may be asymptomatic in some, they commonly lead to severe pain affecting one's quality of life and may progress to cord compression. The treatment paradigm for spine metastases has continued to evolve over time with overall advancements in imaging which has allowed increasing the use of stereotactic body radiotherapy (SBRT) and stereotactic **radiosurgery** (SRS). These are highly conformal therapies delivering high radiation doses in a single or two to five fraction regimens by using multiple beams or arcs to target the metastasis all while sparing the spinal cord. This chapter reviews the evolution of spinal stereotaxis, the radiobiological basis behind radiosurgery, and several modern-day uses of radiosurgery including de novo spine metastases, reirradiation of spine metastases, radiosurgery in the postoperative setting, and the role of SRS in cases of neurologic compromise.

Learning Objectives

- To understand the historical management of spine metastases and review the evolution of treatment paradigms with the advent of radiosurgery.
- To understand the radiobiological basis behind radiosurgery and its ability to improve the therapeutic ratio.
- To understand the evidence supporting the use of radiosurgery in palliation and local disease control by reviewing both historical and contemporary clinical trials.
- To understand the role of SRS/SBRT in the treatment of spine metastases in multiple clinical scenarios, including de novo spine metastases, reirradiation, and postoperatively in cases of cord compression.

M. Ashamalla (✉) · A. N. Slade · K. Mani · S. Ryu
Department of Radiation Oncology, Stony Brook University Hospital, Stony Brook, NY, USA
e-mail: Mark.Ashamalla@stonybrookmedicine.edu

66.1 Introduction

Spinal metastases are a complicated problem affecting a large portion of adults in the United States across multiple cancer subtypes. These lesions can be asymptomatic, cause bothersome to debilitating pain impacting performance status and quality of life, and in the worst case, progress to metastatic epidural spinal cord compression (MESCC). The latter occurs in 10% of these patients, with more than 20,000 new cases reported every year in the United States [1].

Spinal metastases are ideally managed in a multidisciplinary fashion with pain control usually being the immediate goal of treatment. External beam radiation therapy (EBRT) is widely used with doses of 30 Gy in ten fractions directed to the involved spine including one to two vertebral bodies superior and inferior to the affected spinal level. With the recent technical progress of radiation delivery and immobilization, spine radiosurgery (**SRS**)/stereotactic body radiation therapy (**SBRT**) has been used to deliver large doses in a single fraction, or over the course of a few fractions, directed to the involved spine only. When there is mechanical spine instability, a surgical approach with stabilization is preferred, with kyphoplasty used for pain control of compression fractures. Many patients ultimately receive some combination of the above. The degree and type of intervention depend on several critical factors: performance status, the extent of disease both systemically and at the involved

S. P. Leong et al. (eds.), *Cancer Metastasis Through the Lymphovascular System*,
https://doi.org/10.1007/978-3-030-93084-4_66

spine level, the stability of the underlying spine, and the degree of symptoms the patient is experiencing.

Metastatic epidural spinal cord compression (MESCC) with canal compromise is treated with decompressive surgery followed by EBRT [2]. It is estimated that only 15% of these patients are favorable surgical candidates [1]; however, such a comparison has not been repeated in the era of radiation dose escalation with SRS/SBRT. For other patients showing radiographic compression without acute neurological deficits, or those with non-compressive lesions deemed high risk for potential neurologic compromise, there is no clear consensus on the optimal treatment strategy. Some algorithms, such as the mechanical stability, neurological risk, oncological parameters, and patient wishes (MNOP), have been developed to guide practitioners to preferred options based on many of the factors above [3]. These treatment options also incorporate more advanced radiation techniques. SRS and SBRT provide highly conformal, ablative radiation to spinal tumors and have been shown to be safe and effective [4–6], even in cases of epidural extension or cord involvement. It should be noted that SRS/SBRT can be delivered either as a primary treatment or in the adjuvant setting following decompressive surgery and stabilization. This chapter will focus on radiotherapeutic uses of SRS and SBRT for spine metastasis and MESCC.

66.2 Radiobiology of SRS/SBRT

Oncological **radiobiology** has been mostly studied with regard to cell survival and the dynamics of tissue or tumor mass after daily fractionated doses of 1.8–2 Gy. To compare various different radiation doses or fractionations, we calculate the dose delivered to different tissues, using what is called the biological effective dose (**BED**). BED incorporates not only the total dose given but also the dose delivered per fraction [7]. The calculated BED then predicts the lethality of the delivered radiation, using the linear quadratic cell survival formula, which contains surrogates for radiosensitivity of tumor versus normal tissue to the doses used in different fractionation regimens. Putting all these radiobiological principles together leads us to infer that by delivering the same total dose, but in fewer fractions (i.e., SRS), we will deliver a greater biologically effective dose to the tumor. However, if the goal is to increase the therapeutic ratio, one may be best served by greater total dose fractionation.

There is additional direct experimental evidence that a large dose in a single fraction results in significantly increased tumor cell kill in a transplanted fibrosarcoma tumor model. This increased tumor control was due to microvascular endothelial apoptosis by activation of the ceramide pathway by acid sphingomyelinase [8]. The enhanced radiation response was seen more commonly with radiation doses higher than 8–10 Gy in a single dose as opposed to fractionated regimens [9]. Together with the experimental data and experientially driven linear quadratic BED predictability, large fraction regimens such as SRS and SBRT were brought into clinical use. In fact, such large single doses have been used extensively in radiosurgery of various brain lesions, and the clinical effectiveness of controlling brain metastases is well established. However, such large doses could not be given to the spine where the spinal cord lies within and thus came the need to develop another new method of radiation delivery under accurate stereotactic localization to avoid potential damage to the spinal cord.

66.3 History of Spinal Stereotaxis

The advent of spinal **stereotaxis** dates back to the late nineteenth century when frames were first created by Dittmar and Woroschiloff in an effort to study spinal cord lesions in laboratory animals, which was then followed by Clarke in 1920 with the creation of the first stereotactic device for experimental use in the spine, requiring the fixation of clamps to exposed lamina [10]. Mullan and investigators tried clinical applications of spinal stereotaxis for percutaneous cordotomy, which subsequently lead to the development of the stereotactic frame by Rand in 1965, which was later used by Crue and Hitchcock in various procedures involving trigeminal and spinothalamic tractotomy [11]. More recently, Hamilton et al. developed an anchoring device to the vertebral bone by an open procedure in order to perform stereotactic delivery of high radiation doses using a linear accelerator [12]. In the same time period, a noninvasive method of table-attached or floor-mounted frame was developed for spine stereotaxis by Lax et al. [13]. This became the basis of frameless stereotaxis of spine radiosurgery under image guidance. Since then, the field of spine radiosurgery has welcomed a host of new linear accelerator-based radiation delivery systems.

66.4 Delivery Methods of High Radiation Doses

While radiosurgery was first developed for intracranial lesions due to its precision, the technology of modern linear accelerators (**LINAC**) has allowed this treatment to be utilized in the management of other body sites, particularly the spine. In general, LINAC-based SRS uses multiple arcs delivering photons to a specified target with intensity modulated radiation. Currently, dynamic LINAC systems are mostly used, which involve gantry rotation around the patient, while micromultileaf collimators are simultaneously in motion, three-dimensionally conforming to the tumor, all while the treatment couch is being robotically controlled. There are several currently commercially available LINAC radiosurgery systems.

Another method of treatment delivery is charged particle radiosurgery which can use either the Bragg peak method, where charged particles come to stop within the target volume, or the plateau-beam method, where the target is covered by energy deposition of the charged particles spread across it. Since it also has the potential to be used for delivery of high doses, it may be developed for a greater role in SRS/SBRT.

66.5 SRS for Spine Metastases

EBRT has been used for palliation of pain, commonly utilizing a radiation regimen delivering 30 Gy in ten fractions. Conventional EBRT as monotherapy was used with palliative intent, assuming generally that the spine metastasis was in a patient at a terminal stage of cancer. Thus, EBRT is usually accomplished by a rather simple technique using two or more opposed beams depositing a high dose across the entire width of the patient including the spinal cord and other internal organs within the radiation beam paths [3]. EBRT is given to the involved spine including one to two uninvolved vertebral bodies superior and inferior to the affected vertebral level(s). This can be delivered as monotherapy, adjuvantly following surgery, or salvage in a reirradiation setting [3].

The treatment paradigm for spine metastases has continued to evolve over time with overall advancements in imaging, most notably with MRI. Through the increasingly routine use of spine MRI, we have improved our ability to detect and diagnose early and more localized spine metastases prior to the development of any symptoms or neurological compromise. However, the true incidence of clinically occult spine metastases remains unknown. Advanced imaging also helps visualize anatomical details of the spinal column and the spinal cord when designing SBRT and SRS spine plans. These are highly conformal therapies delivering high radiation doses in a single or two to five fraction regimens by using multiple beams or arcs to target the metastasis all while sparing the spinal cord. In contrast to EBRT, SRS and SBRT target only the involved spine, and thus they are used for only limited or localized spine metastases. In terms of defining the target volume, a consensus target delineation method has been reported considering the extent of involvement and anatomical boundaries of the vertebral body [14]. Early studies as well as a more recent study in the setting of oligometastases by Palma et al. have suggested improvement of local spine tumor control and possible survival benefit [15, 16]. Therefore, the primary goal when treating spine metastases is not only pain palliation but also achieving long-term tumor control of the involved spine and preventing potential neurological complications.

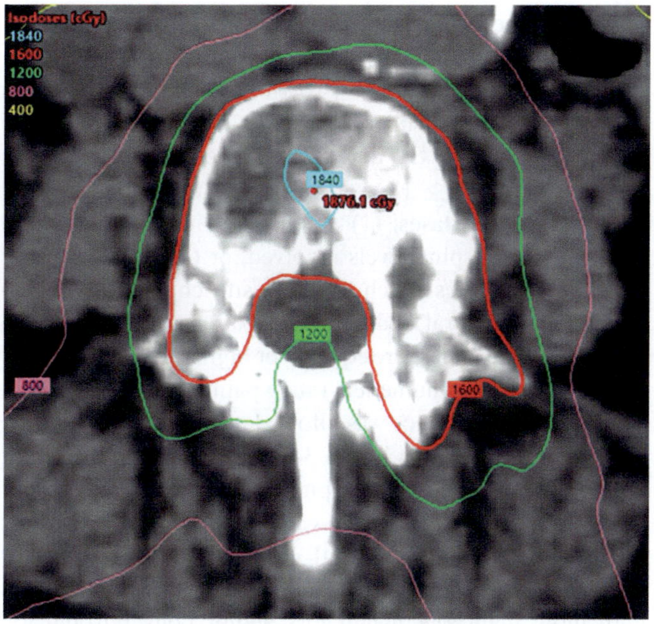

Fig. 66.1 SRS spine treatment plan for metastasis. A lung cancer patient presenting with a painful L4 metastasis causing radiculopathy is depicted. The patient was treated with 16 Gy to the 100% isodose line (red color) surrounding the target volume. Thin green line is 12 Gy and pink line is 8 Gy isodose. At 2 months after radiosurgery, the patient had complete resolution of the presenting pain symptoms and after 1 year, still had a durable tumor response on follow-up imaging. The patient remained fully ambulatory and active

A typical treatment with SRS is shown in Fig. 66.1. Common radiosurgical dose regimens are single-fraction SRS delivering 15–20 Gy, ultra-high single-dose radiotherapy of 24 Gy, or **hypofractionated** regimens such as 3 × 9 Gy. There is some evidence that high single doses may improve local control of **oligometastatic** lesions over more hypofractionated regimens (3 × 9 Gy SBRT) [17]; however this remains to be confirmed.

Since the initial reported clinical experience of spine SRS, there has been an increasing use of SRS/SBRT for spine metastases. Many clinical trials as well as institutions have presented their experiences with SRS/SBRT for spinal metastases reporting rapid and durable pain relief in the range of 85–95%, in addition to demonstrating long-term tumor control in the treated spine region with a median duration of 13 months. Quality of life was also improved secondary to pain control. Based on the spine SRS experiences of many single institutions, multiple randomized clinical trials have been conducted [4, 6, 17–20].

The first and earliest phase II/III randomized study was conducted by the **NRG RTOG 0631** [21]. This study compared a single dose of 16 Gy or 18 Gy with SRS to 8 Gy with EBRT, in 352 enrolled patients. SRS was delivered to the involved spine segment only, while EBRT was given to the involved spine with one additional spine segment above and below the treated spine. For SRS, the spinal cord dose constraints were either 10 Gy to V10% of the spinal cord volume (defined from 6 mm superior to 6 mm inferior to the treated target spine) or 10 Gy to V0.35cc. The eligibility criteria were patients (age ≥ 18 years) with one to three treatment-naïve spine metastases: (1) a solitary spine metastasis, (2) two contiguous spine levels involved, or (3) a maximum of three separate sites. Each separate site could involve two contiguous vertebral bodies. Every patient was required to have a baseline pain score equal to or higher than 5 at presentation using a 0–10 numerical rating pain scale. Pain medication including narcotics was allowed and recorded. Minimal epidural lesions with at least a 3-mm gap between the tumor and the spinal cord and paraspinal masses (< 5 cm) directly attached to the vertebral body were eligible. **Radioresistant** tumors (melanoma, renal cell carcinoma, soft tissue sarcoma) were also eligible. Vertebral compression fractures with more than 50% height loss and/or bony retropulsion were ineligible.

The results demonstrated a surprisingly lower pain response rate in the SRS arm, unexpected from previously reported outcomes. The primary endpoint of patient-reported short-term pain response of the index spine was 41.3% in the SRS arm compared to 60.5% in the cEBRT arm at 3 months ($p = 0.99$). However, long-term pain response after 12 months post-treatment showed an improvement of pain response of 61.3% by SRS compared to 43.7% by cEBRT ($p = 0.20$). Of note, complete pain relief at the index spine was reported as 41.9% in the SRS arm compared to 31.3% in the cEBRT arm at 24 months. Multivariate analysis showed that the number of spine metastases (1 versus 2–3) was the most significant predictor of pain response over time ($p = 0.023$). Pre-treatment pain score did not impact pain response. Radioresistant tumors responded similarly to treatment compared to other histologies.

Another randomized phase II/III study conducted by the Canadian Cancer Trials Group (CCTG) in collaboration with the Trans Tasman Radiation Oncology Group (TROG) recently compared 24 Gy in two SBRT fractions versus 20 Gy in five conventional **palliative** radiotherapy (CRT) fractions [22]. Preliminary results showed that patients in both treatment groups experienced reductions in pain from spinal metastases. At 3 months, 35% of patients in the SBRT arm of the trial, compared to 14% of those in the CRT arm, reported a complete response with no remaining pain from their lesions ($p < 0.001$). After 3 months, 92% of patients in the SBRT arm and 86% of those in the CRT arm were cancer-free at the treated site ($p = 0.4$); the rates at 6 months were 75% and 69% ($p = 0.4$). Long-term results are still pending at this time.

A smaller German randomized phase II trial compared pain response from 24 Gy single-fraction SRS/SBRT to 30 Gy in ten fractions using 3DCRT in a sample of 60 patients, using a study design similar to that used in RTOG 0631. The investigators reported a faster pain response within 3 months and a significant pain reduction at 6 months following SRS over 3DCRT with statistical significance [23].

Taken together, SRS appears to be the preferred treatment for localized spine metastases to achieve short- and long-term pain response and potential tumor cure of the treated spine. Localized limited spine metastases may also be a component of an oligometastatic state where consolidative radiation with SRS/SBRT for metastatic disease has been shown to provide significant survival benefit [16, 24]. It is important to consider the current oncological landscape where clinically occult and asymptomatic solitary or limited spine metastases are more frequently detected as part of oligometastases or longitudinally, during the course of systemic therapy, by a wide use of PET and/or MRI scans. SRS can achieve long-term **tumor control** that can also prevent neurological progression. Choice of treatment should be individualized considering the systemic extent of metastases, tumor burden within the involved spine, presenting symptoms, palliative care measures, and general plan of oncological care.

66.6 SRS for Reirradiation

By some estimates, nearly half of patients who receive conventionally fractionated radiotherapy will be diagnosed with persistent or progressive disease requiring further radiation [25]. This presents a unique and difficult scenario for clinicians because **reirradiation** with further conventionally fractionated radiotherapy, many times just months following the initial course, will require a dose lower than that given during the first course. For this reason, these subsequent courses often provide inferior local control, with overall response rates of approximately 50% and complete response rates reported as less than 15% [26].

Given the inferior efficacy with conventionally fractionated radiotherapy, clinicians turned to radiosurgery instead in cases requiring reirradiation. By most estimates, reirradiating with radiosurgery provides significantly greater local control (>75%) than that observed with conventionally fractionated radiotherapy, all while demonstrating a reasonable toxicity profile, which by most accounts is comparable to de novo SBRT [27]. Thibault et al. evaluated the role of radiosurgery in the **salvage** setting for patients with in-field failure of radiotherapy and reported an 81% local control rate at 1 year with no observed radiation-induced vertebral compression fractures or myelopathies observed [28]. As we continue to improve outcomes in cancer, a greater proportion of patients may require reirradiation in many instances. Radiosurgery has been demonstrated as, and will certainly continue to be, explored as a viable and preferred treatment choice for recurrent tumors requiring multiple courses of radiation.

66.7 SRS Following Surgery

Prospective randomized trials in the early 1990s demonstrated that surgery (decompressive corpectomy) followed by conventional **EBRT** (30 Gy in ten fractions) provided significant improvement in overall ambulatory rates compared to EBRT alone. The results of this era's trials served as proof of concept for future attempts to further improve on these results with SRS/SBRT, which came into practical use in the 2000s. Initial experiences [29], while vitally important for the progression of the field, were limited by the difficulties defining the tumor, surgical bed, and spinal cord when adjacent to neurosurgical hardware. Myelograms were initially used as a substitute for the poor quality of CT and MRI images. Later, consensus guidelines were developed to aid providers and better standardize postoperative target volumes when using radiosurgery [30]. These guidelines include recommendations on the extent of preoperative disease sites to be covered, coverage of postoperative gross and subclinical disease, and different planning target volume (PTV) expansions according to the institution's immobilization systems.

The use of **separation surgery** then grew from a need to protect the cord while simultaneously avoiding potential "underdosing" of the target volume. The intent was to reduce the tumor burden and to provide a necessary gap of a few millimeters between the cord and tumor, thereby allowing a sufficient radiation dose to be given to the tumor. This approach was studied by Memorial Sloan Kettering, with patients receiving 24 Gy in one fraction or 24–30 Gy in three fractions, and demonstrated a low local tumor progression ranging from 4 to 9% at 1 year post-treatment [31].

66.8 Radiation in MESCC or Canal Compromise and Cases of Neurologic Compromise

Not all epidural tumors causing canal compromise cause neurologic deficits; however if left untreated, they most certainly can progress to the point of symptomatic and overt spinal cord compression. These are complicated cases requiring multidisciplinary care among neurosurgery, radiation oncology, and medical oncology. Together with pertinent neurological examination, appropriate MR imaging is required for accurate evaluation. We recommend the patient to undergo complete spine imaging to rule out additional levels of compression and aid in treatment decision-making. When MESCC or canal compromise is present, the usual first course of action is immediate administration of **glucocorticoids**, generally with dexamethasone 10 mg IV bolus followed by 16 mg per day (4 mg every 6 h), tapered over 2 weeks [32], but judicious use of lower doses is recommended to avoid steroid complications.

The treatment goals for patients with spinal cord compression (or compromise) are decompression of the spinal cord, control of the epidural disease, and preservation or improvement of neurological function. Surgical resection or radiation is the mainstay of treatment and offers the best option for epidural tumor control and prevention of cord compression with subsequent neurologic deficits. With surgical resection and postoperative EBRT, the resulting superiority of ambulation over EBRT alone was tested using an endpoint of four steps with walking aid. However, this often requires corpectomy using an anterior and posterior approach with extensive instrumentation. Many oncological patients are not suitable candidates for such involved surgery, but still require a **spinal cord decompression**, ideally using a noninvasive approach, to improve their neurological function. With the progress of spine SRS, one such landmark phase II trial in the 2000s studied SRS (to a dose of 14–20 Gy in a single fraction) in the management of epidural spinal cord compression [33]. Ryu et al. demonstrated a 65% mean reduction in epidural tumor volume seen on 2-month MRI scan with an 80% overall epidural tumor response rate. Of note, this study also showed that the majority (94%) of patients who were intact prior to radiosurgery remained intact following treatment, with an improvement of neurologic deficits in 63% of patients presenting with such deficits. Figure 66.2 shows an example of MR images before and after radiosurgery in a patient with renal cell carcinoma, commonly regarded as a radioresistant tumor, causing MESCC.

While not possible to compare two distinct trials mentioned above in their entireties, one may be able to compare the overall outcomes of preservation or improvement of neurologic function as it relates to each treatment modality. A comparison between subset results of these two trials [2, 33] is seen in Table 66.1, with 94% of patients remaining ambulatory or **neurologically intact** after either therapy (SRS or surgery) and 62–63% of patients presenting as non-ambulatory or with neurological deficits improving to ambulatory or intact following surgery or SRS.

Contrary to current practice, these studies suggest that SRS can be used as the primary treatment for MESCC for patients who do not have neurological deficits, as SRS can take effect gradually to reduce the **epidural** lesion, while upfront surgical resection is better indicated for patients with acute neurological signs as it provides more immediate spinal cord decompression.

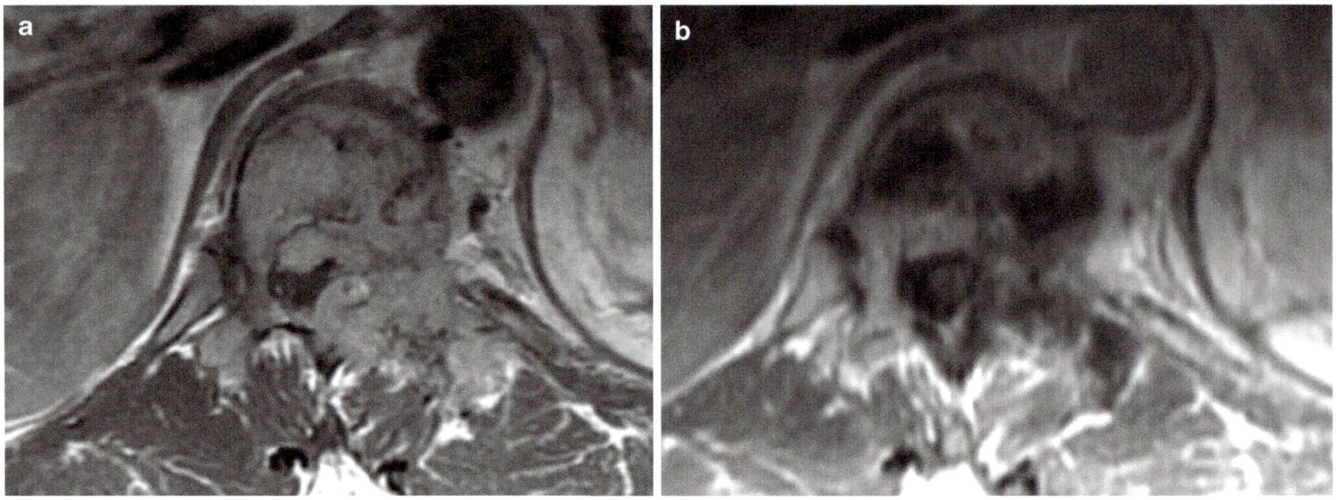

Fig. 66.2 SRS spine treatment plan for MESCC. A renal cell cancer patient with a T12 spinal cord compression causing radiculopathy is depicted. Panel **A** shows significant epidural spinal cord compression with minimal thecal sac patency at presentation. The patient was treated with 18 Gy to the 90% isodose line surrounding the target volume of the epidural and paraspinal tumor and vertebral body. Panel **B** shows that at 3 months after radiosurgery, epidural decompression was achieved and thecal sac patency improved to 95%

Table 66.1 Comparison of clinical trial results of surgery versus radiosurgery cEBRT, conventionally fractionated external beam radiation therapy; SRS, stereotactic radiosurgery

Patchell's phase III trial [2]	cEBRT alone	Surgery + cEBRT	Ryu's phase II trial [33]	SRS alone
Ambulatory patients remain ambulatory	74% (26/35)	94% (32/34)	Intact patients remain intact	94% (33/35)
Overall ambulatory rate	57% (29/51)	84% (42/50)	Overall intact rate	81% (50/62)
Non-ambulatory patients improve to ambulatory	19% (3/16)	62% (10/16)	Deficit patients improve to intact	63% (17/27)

66.9 Adverse Effects of Radiation

When considering the potential adverse effects of radiotherapy for spinal metastases, one is best served by separating effects in the acute setting from those more likely to be present long term.

66.10 Acute Complications

One of the most well-reported adverse effects of radiosurgery in the acute setting is what has been termed **pain flare**, an acute exacerbation of pain usually occurring within 1–5 days following treatment. Pain flares following SBRT have been reported with an incidence in the range of 20–60% and may present anytime starting as early as 24 h following treatment to over 3 weeks later [34]. Compared to SBRT, single-fraction SRS has been reported by some groups as being associated with a higher incidence of pain flares [35]. Pain flares are transient and responsive to short courses of low-dose glucocorticoids (e.g., dexamethasone 4 mg daily for up to 5 days). However, prophylactic use of steroids is not usually advised. The etiology of pain flares is not yet completely understood at this time, but practitioners of spine radiosurgery are generally advised to limit the radiation dose to the spinal nerve root to 14 Gy.

Other side effects from radiation have been reported on and are related to varying doses to neighboring tissues, including the larynx, pharynx, esophagus, bowel, lung, kidneys, etc., which are therefore contoured when planning a radiosurgical treatment. Patients will typically present with symptoms of dysphagia and odynophagia depending on the location of the spinal target. Generally, these side effects are self-limited and will resolve within a few weeks following treatment, but there have been reports of long-term development of tracheoesophageal fistulas [36]; for this reason, we limit the esophageal dose to a total dose of 10–12 Gy in a **single fraction** [37].

66.11 Long-Term Complications

It is of utmost importance to avoid any and all radiation-induced damage to the spinal cord as even minor injury may severely and adversely affect a patient's quality of life. Several analyses and consensus of normal tissue dose constraints for stereotactic radiotherapy have been performed and published by Task Group 101 [38] and the UK groups [39], respectively. The NRG RTOG 0631 study used a partial spinal cord volume with a **tolerance dose** of 10 Gy to the 10% partial volume of the spinal cord defined as including 6 mm above and below the target volume (calculated to be equivalent to 0.35 cc) [15], with the tolerance dose noted to vary according to the dose fractionation being used.

Vertebral compression fractures have been reported in up to 15% of patients after spine SRS [40]. In order to better asses which patients may be at highest risk for such an event, analyses of predictive factors have been performed demonstrating that the patients at highest risk of developing **vertebral compression fractures** are those with prior kyphotic/scoliotic deformities, lytic lesions, and receiving higher doses (≥ 20 Gy).

66.12 Conclusions

Stereotactic radiosurgery is the culmination of years of extensive research into the application of radiobiology, radiation physics, and advanced imaging and technology in radiotherapy. This synergy has enabled radiation oncologists to ablate metastatic lesions in the spine (whether in the initial or reirradiation setting). There are several practical advantages of using spine SRS. For instance, since only the directly involved spine is treated, more functional bone marrow can also be preserved. This may help facilitate the continuation of systemic therapy, which is often required in these patients. A single session of SRS or a few sessions of SBRT will not interfere with an ongoing chemotherapy schedule as opposed to the multiple visits for conventional fractionated radiotherapy. Single or shorter fractionation is indeed more convenient for both the patient and their caregivers all while causing no increased toxicity to the treated spine compared to conventional radiotherapy.

Open Questions

- What is the role of charged particle therapy in spine radiosurgery?
- Is there a synergistic effect between SRS and immunotherapy?
- What is the appropriate sequence of SRS when combined with separation surgery?
- What are the precise criteria to define which treatment (surgery/SRS) a patient with spinal metastases should receive?
- Will advancements in neuroimaging and radiomics/radiogenomics incorporating PET or MR spectroscopy help improve the therapeutic ratio and post-treatment surveillance?

References

1. Byrne TN, Borges LF, Loeffler JS. Metastatic epidural spinal cord compression: update on management. Semin Oncol. 2006;33:307–11.
2. Patchell RA, et al. Direct decompressive surgical resection in the treatment of spinal cord compression caused by metastatic cancer: a randomised trial. Lancet. 2005;366:643–8.
3. Spratt DE, et al. An integrated multidisciplinary algorithm for the management of spinal metastases: an international spine oncology consortium report. Lancet Oncol. 2017;18:e720–30.
4. Ryu S, et al. Pain control by image-guided radiosurgery for solitary spinal metastasis. J Pain Symptom Manag. 2008;35:292–8.
5. Gerszten PC, et al. Single-fraction radiosurgery for the treatment of spinal breast metastases. Cancer. 2005;104:2244–54.
6. Chang EL, et al. Phase I/II study of stereotactic body radiotherapy for spinal metastasis and its pattern of failure. J Neurosurg Spine. 2007;7:151–60.
7. Fowler JF. 21 years of biologically effective dose. Br J Radiol. 2010;83:554–68.
8. Garcia-Barros M, Paris F, Cordon-Cardo C, Lyden D, Rafii S, Haimovitz-Friedman A, Fuks Z, Kolesnick R. Tumor response to radiotherapy regulated by endothelial cell apoptosis. Science 2003;300:1155–1159. doi: https://doi.org/10.1126/science.1082504.: PMID: 12750523.
9. Fuks Z, Kolesnick R. Engaging the vascular component of the tumor response. Cancer Cell. 2005;8:89–91. https://doi.org/10.1016/j.ccr.2005.07.014. PMID: 16098459
10. Gabriel EM, Nashold BS Jr. History of spinal cord stereotaxy. J Neurosurg. 1996;85:725–31.
11. Hitchcock E. Stereotactic cervical myelotomy. J Neurol Neurosurg Psychiatry. 1970;33:224–30.
12. Hamilton AJ, Lulu BA. A prototype device for linear accelerator-based extracranial radiosurgery. Acta Neurochir Suppl. 1995;63:40–3.
13. Lax I, Blomgren H, Naslund I, Svanstrom R. Stereotactic radiotherapy of malignancies in the abdomen. Methodological aspects. Acta Oncol. 1994;33:677–83.
14. Cox BW, Spratt DE, Lovelock M, Bilsky MH, Lis E, Ryu S, Sheehan J, Gerszten PC, Chang E, Gibbs I, Soltys S, Sahgal A, Deasy J, Flickinger J, Quader M, Mindea S, Yamada Y. International Spine Radiosurgery Consortium consensus guidelines for target volume definition in spinal stereotactic radiosurgery. Int J Radiat Oncol Biol Phys. 2012;83:e597–605.
15. Ryu S, Jin JY, Jin R, et al. Partial volume tolerance of the spinal cord and complications of single-dose radiosurgery. Cancer. 2007;109:628–36.
16. Palma DA, Olson R, Harrow S, Gaede S, Louie AV, Haasbeek C, Mulroy L, Lock M, Rodrigues GB, Yaremko BP, Schellenberg D, Ahmad B, Senthi S, Swaminath A, Kopek N, Liu M, Moore K, Currie S, Schlijper R, Bauman GS, Laba J, Qu XM, Warner A, Senan S. Stereotactic ablative radiotherapy for the comprehensive treatment of oligometastatic cancers: long-term results of the SABR-COMET phase II randomized trial. J Clin Oncol. 2020;38:2830–8.
17. Zelefsky MJ, Yamada Y, Greco C, Lis E, Schöder H, Lobaugh S, Zhang Z, Braunstein S, Bilsky MH, Powell SN, Kolesnick R, Fuks Z, Phase III. Multi-center, prospective, randomized trial comparing single dose 24 Gy radiotherapy to a 3-fraction SBRT regimen in the treatment of Oligometastatic cancer. Inte J Radiat Oncol Biol Phys. 2021; https://doi.org/10.1016/j.ijrobp.2021.01.004)
18. Ryu S, Fang Yin F, Rock J, Zhu J, Chu A, Kagan E, Rogers L, Ajlouni M, Rosenblum M, Kim JH. Image-guided and intensity-modulated radiosurgery for patients with spinal metastasis. Cancer. 2003 Apr 15;97:2013–8. https://doi.org/10.1002/cncr.11296. PMID: 12673732

19. Yamada Y, Bilsky MH, Lovelock DM, et al. High-dose, single-fraction image-guided intensity-modulated radiotherapy for metastatic spinal lesions. Int J Radiat Oncol Biol Phys. 2008;71: 484–90.
20. Gerszten PC, Burton SA, Ozhasoglu C, et al. Radiosurgery for spinal metastases: clinical experience in 500 cases from a single institution. Spine (Phila Pa 1976). 2007;32:193–9.
21. Ryu S, Deshmukh S, Timmerman RD, Movsas B, Gerszten PC, Yin FF, et al. Radiosurgery compared to external beam radiotherapy for localized spine metastasis: phase III results of NRG oncology/RTOG 0631. Int J Radiat Oncol Biol Phys. 2019;105: S2–3.
22. Sahgal A, Myrehaug SD, Siva S, Masucci L, Foote MC, Brundage M, et al. CCTG SC. 24/TROG 17.06: a randomized Phase II/III study comparing 24Gy in 2 stereotactic body radiotherapy (SBRT) fractions versus 20Gy in 5 conventional palliative radiotherapy (CRT) fractions for patients with painful spinal metastases. Int J Radiat Oncol Biol Phys. 2020;108:1397–8.
23. Sprave T, Verma V, Förster R, Schlampp I, Bruckner T, Bostel T, Welte SE, Tonndorf-Martini E, Nicolay NH, Debus J, Rief H. Randomized phase II trial evaluating pain response in patients with spinal metastases following stereotactic body radiotherapy versus three-dimensional conformal radiotherapy. Radiother Oncol. 2018 Aug;128:274–82. https://doi.org/10.1016/j.radonc.2018.04.030. Epub 2018 May 26. PMID: 29843899
24. Mitchell KG, Farooqi A, Ludmir EB, Corsini EM, Zhang J, Sepesi B, Vaporciyan AA, Swisher SG, Heymach JV, Zhang J, Gomez DR, Antonoff MB. Improved overall survival with comprehensive local consolidative therapy in synchronous oligometastatic non-small-cell lung cancer. Clin Lung Cancer. 2020;21:37–46.e7. https://doi.org/10.1016/j.cllc.2019.07.007. Epub 2019 Aug 1. PMID: 31447303
25. Hartsell WF, Scott CB, Bruner DW, et al. Randomized trial of short- versus long-course radiotherapy for palliation of painful bone metastases. J Natl Cancer Inst. 2005;97:798–804. https://doi.org/10.1093/jnci/dji139).
26. Chow E, van der Linden YM, Roos D, et al. Single versus multiple fractions of repeat radiation for painful bone metastases: a randomised, controlled, non-inferiority trial. Lancet Oncol. 2014;15:164–71. https://doi.org/10.1016/S1470-2045(13)70556-4.
27. Myrehaug S, Sahgal A, Hayashi M, et al. Reirradiation spine stereotactic body radiation therapy for spinal metastases: systematic review. J Neurosurg Spine. 2017;27:428–35. https://doi.org/10.3171/2017.2.SPINE16976).
28. Thibault I, Campbell M, Tseng C-L, et al. Salvage stereotactic body radiotherapy (SBRT) following in-field failure of initial SBRT for spinal metastases. Int J Radiat Oncol. 2015;93:353–60. https://doi.org/10.1016/j.ijrobp.2015.03.029).
29. Rock JP, Ryu S, Shukairy MS, et al. Postoperative radiosurgery for malignant spinal tumors. Neurosurgery. 2006;58:891–8. https://doi.org/10.1227/01.NEU.0000209913.72761.4F.
30. Redmond KJ, Robertson S, Lo SS, et al. Consensus contouring guidelines for postoperative stereotactic body radiation therapy for metastatic solid tumor malignancies to the spine. Int J Radiat Oncol. 2017;97:64–74. https://doi.org/10.1016/j.ijrobp.2016.09.014.
31. Moussazadeh N, Laufer I, Yamada Y, Bilsky MH. Separation surgery for spinal metastases: effect of spinal radiosurgery on surgical treatment goals. 2014. http://journals.sagepub.com/doi/pdf/10.1177/107327481402100210;21. Accessed October 2, 2018
32. Sørensen S, Helweg-Larsen S, Mouridsen H, Hansen HH. Effect of high-dose dexamethasone in carcinomatous metastatic spinal cord compression treated with radiotherapy: a randomised trial. Eur J Cancer. 1994;30A:22–7. http://www.ncbi.nlm.nih.gov/pubmed/8142159. Accessed September 27, 2018
33. Ryu S, Rock J, Jain R, et al. Radiosurgical decompression of metastatic epidural compression. Cancer. 2010; https://doi.org/10.1002/cncr.24993.
34. Chiang A, Zeng L, Zhang L, et al. Pain flare is a common adverse event in steroid-naïve patients after spine stereotactic body radiation therapy: a prospective clinical trial. Int J Radiat Oncol Biol Phys. 2013;86:638–42. https://doi.org/10.1016/j.ijrobp.2013.03.022).
35. Pan HY, Allen PK, Wang XS, et al. Incidence and predictive factors of pain flare after spine stereotactic body radiation therapy: secondary analysis of Phase 1/2 trials. Int J Radiat Oncol. 2014;90:870–6. https://doi.org/10.1016/J.IJROBP.2014.07.037.
36. Jin R, Rock J, Jin J-Y, et al. Single fraction spine radiosurgery for myeloma epidural spinal cord compression. J Exp Ther Oncol. 2009;8:35–41. http://www.ncbi.nlm.nih.gov/pubmed/19827269. Accessed December 5, 2018
37. Schipani S, Wen W, Jin J-Y, Kim JK, Ryu S. Spine radiosurgery: a dosimetric analysis in 124 patients who received 18 Gy. Int J Radiat Oncol Biol Phys. 2012;84:e571–6. https://doi.org/10.1016/j.ijrobp.2012.06.049].
38. Hanna GG, Murray L, Patel R, et al. UK consensus on Normal tissue dose constraints for stereotactic radiotherapy. Clin Oncol (R Coll Radiol). 2018;30:5–14. https://doi.org/10.1016/j.clon.2017.09.007).
39. Benedict SH, Yenice KM, Followill D, et al. Stereotactic body radiation therapy: the report of AAPM task group 101 [published correction appears in med Phys. 2012 Jan;39(1):563. Dosage error in article text]. Med Phys. 2010;37:4078–101. https://doi.org/10.1118/1.3438081.
40. Boyce-Fappiano D, Elibe E, Schultz L, et al. Analysis of the factors contributing to vertebral compression fractures after spine stereotactic radiosurgery. Int J Radiat Oncol. 2017;97:236–45. https://doi.org/10.1016/j.ijrobp.2016.09.007.

Part XVII

Immunotherapy of Solid Malignancies

Steven O'Day

The human immune system is characterized by power, specificity, and memory. Immune surveillance is a primary defense against cancer. Immunosuppression induced by disease states, exogenous medications, or natural aging increases the incidence of cancer in humans. Extraordinary scientific progress in our understanding of the immune system in the last two decades has led to transformative therapies in oncology. This has resulted in a rapidly increasing proportion of cancer patients experiencing long-term survival.

At the turn of the twentieth century, Dr. William Coley treated cancer patients with a mixture of heat-killed bacteria, later called "Coley's toxins," and reported dramatic tumor regressions in select patients [1]. His rationale for this approach was based on the observation that rare spontaneous cancer regressions occurred after severe bacterial infections. Later, additional nonspecific immune approaches were used to treat bladder cancer and melanoma with the tuberculosis vaccine, Bacillus Calmette-Guérin (BCG) [2]. Toward the end of the century, a greater appreciation for the role of T cells in cancer immunity emerged, and T-cell growth factors (immune cytokines) were discovered [3]. This discovery led to the use of high-dose interleukin-2 (IL-2) and interferon alpha (INF-a) in the treatment of melanoma and renal cell cancer. The twentieth century closed with seminal work from the NCI and Dr. Steven Rosenberg's lab where adoptive autologous tumor-infiltrating lymphocytes were used in conjunction with high-dose IL-2 as a therapeutic strategy against advanced cancer [4]. Across all these therapeutic efforts, only a small percentage of patients were observed to achieve tumor eradication or prolonged tumor control; predictive biomarkers of response and survival were elusive. The significant toxicity of high-dose cytokines and the labor-intensive preparation of adoptive lymphocytes were formidable obstacles to broad clinical benefit and rigorous randomized trials.

The current revolution of immuno-oncology (I-O) began in the first decade of the twenty-first century when the T-cell checkpoint, CTLA-4, was identified as a critical inhibitory T-cell checkpoint which dampens the adaptive immune response, preventing prolonged activation and, ultimately, autoimmunity [9]. Blocking CTLA-4 resulted in more potent T-cell priming, activation, and memory. The seminal discovery of this checkpoint inhibitor was recently recognized with the Nobel Prize in Medicine [5]. Ipilimumab was the first therapeutic CLTA-4-blocking antibody to enter human studies in advanced melanoma, approximately 20 years ago. Early studies with the single agent ipilimumab demonstrated remarkable tumor responses, or long-term disease control, in approximately 25% of patients with widespread disease. In 2010, a large randomized trial with ipilimumab demonstrated improved median and long-term survival in patients with advanced melanoma. This was followed by FDA approval of ipilimumab in 2011 [6]. The revolution of checkpoint inhibitors had begun.

The success of ipilimumab represented a dramatic shift in oncology's treatment paradigm. Focus moved away from cancer cell-directed cytotoxic chemotherapy drugs, whose hallmarks were rapid kinetics of responses, short durations of response, and significant off-target toxicity.

The CTLA-4 treatment paradigm shift was notable for slower kinetics of response, more durable responses, and immune-related toxicity that often correlated with clinical benefit. Immune-mediated toxicity management was optimized; short courses of immunosuppressive medications were used and did not negatively impact ongoing antitumor responses.

Shortly after the discovery of CTLA-4, rapid identification of additional T-cell checkpoints ensued. This began in the last decade with the discovery of a second pivotal immune checkpoint, PD-1, and its ligand PDL-1. Antibodies designed to block PD-1 and PDL-1 have transformed cancer therapy. PD-1 upregulation on T cells indicates an exhausted T-cell phenotype, a much later stage of the adaptive T-cell response than CTLA-4 upregulation. By blocking PD1 on the surface of T cells, the T cell is reactivated and cytolytic capability is restored [7]. The first two FDA-approved anti-PD-1 antibodies were nivolumab and pembrolizumab. These agents demonstrated significant therapeutic benefits including durable responses and long-term survival for patients across a broad spectrum of solid tumors [8]. PD-1 antibodies activate T cells predominately in the tumor microenvironment resulting in a more rapid kinetics of response and less immune-related toxicity than CTLA-4 antibodies because they differ in T-cell biology. On the other hand, CTLA-4's biology may drive more durable responses and shorter treatment courses, due to its enhanced T-cell priming and memory inducement. In the last 5 years, combinations of CTLA-4- and PD-1-blocking antibodies have resulted in further advances in cancer I-O therapies. These combinations have improved overall response, disease control, and durability of response across many solid tumor types [8]. Lower doses of CTLA-4 inhibitors in combination with PD-1 inhibitors appear to maintain efficacy and reduce immune-related toxicity. Long-term survival is now a reality for distinct subsets of patients suffering from advanced cancer.

The clinical validation of CTLA-4 and PD-1/PDL-1 as immune targets led to an explosion of combination immunotherapy treatment approaches leading to further advances in the clinic. These strategies include both I-O/I-O and non-I-O/I-O combinations. Further efforts to optimize T effector function, such as additional agonistic and antagonist T-cell antibodies, as well as genetically engineered adoptive CAR-T cells are ongoing. In addition, many new clinical trials focus on strategies to activate optimally and repolarize both the non-T-cell immune compartment (NK cells, APCs, and suppressive myeloid cells) and tumor stromal cells. Finally, optimal I-O treatments combined with cancer cell-directed cytotoxic therapies (chemotherapy, antibody drug conjugates, molecular targeted agents) are generating significant successes in the clinic and are an area of active clinical research [9].

The following chapters summarize the impact of the immuno-oncology revolution on lung, genitourinary, and skin malignancies. The validated checkpoints, CTLA-4 and PD-1/PDL-1, have positively impacted the treatment of each of these disease areas. However, the development of combination approaches varies between and among each disease category. Melanoma has led to the checkpoint revolution and remains at the forefront of its progress. IO/non-IO combinations in melanoma are focused on both immune checkpoints and molecularly targeted agents (BRAF/MEK). Lung cancer has been at the forefront of the combination of immune checkpoints with chemotherapy. Genitourinary cancers are the leaders of combinations of immune checkpoints with tyrosine kinase and VEGF inhibitors. Overall, significant strides are being made in the development of innovative immunotherapy approaches for patients with all three of these cancer types. Sharing insights across these diseases will contribute tremendously to the progress of treatment for years to come. Our continued drive toward improvement will lead to long-term disease control and may help us eradicate these cancers in the coming decades.

References

1. McCarthy EF. The toxins of William B. Coley and the treatment of bone and soft-tissue sarcomas. Iowa Orthop J. 2006; https://www.ncbi.nlm.nih.gov/pmc/articles/PMC1888599/
2. Zheng YQ, Naguib YW, Dong Y, Shi YC, Bou S, Cui Z. Applications of bacillus Calmette-Guerin and recombinant bacillus Calmette-Guerin in vaccine development and tumor immunotherapy. Expert Rev Vaccines. 2015; https://www.ncbi.nlm.nih.gov/pmc/articles/PMC4920355/
3. Berraondo P, Sanmamed MF, Ochoa MC, et al. Cytokines in clinical cancer immunotherapy. Br J Cancer. 2019; https://www.nature.com/articles/s41416-018-0328-y

4. Mayor P, Starbuck K, Zsiros E. Adoptive cell transfer using autologous tumor infiltrating lymphocytes in gynecologic malignancies. Gynecol Oncol. 2018; https://www.ncbi.nlm.nih.gov/pmc/articles/PMC6167746/
5. Press release: The Nobel Prize in Physiology or Medicine. 2018. https://www.nobelprize.org/prizes/medicine/2018/press-release/
6. Wolchok JD, Hodi FS, Weber JS, Allison JP, Urba WJ, Robert C, O'Day SJ, Hoos A, Humphrey R, Berman DM, Lonberg N, Korman AJ. Development of ipilimumab: a novel immunotherapeutic approach for the treatment of advanced melanoma. Ann N Y Acad Sci. 2013 Jul;1291(1):1–13. https://www.ncbi.nlm.nih.gov/pmc/articles/PMC3910157/
7. Zuazo M, Gato-Canas M.. https://atm.amegroups.com/article/view/15376/html
8. Gong, et al. Development of PD-1 and PD-L1 inhibitors as a form of cancer immunotherapy: a comprehensive review of registration trials and future considerations. J Immunother Cancer. 2018; https://jitc.bmj.com/content/jitc/6/1/8.full.pdf
9. Kamta J, Chaar M, Ande A, et al. Advancing cancer therapy with present and emerging immuno-oncology approaches. Front Oncol. 2017; https://www.frontiersin.org/articles/10.3389/fonc.2017.00064/full

Immunotherapy in Melanoma and Merkel Cell Cancer

67

Melissa Chow, Elizabeth A. Sangalang, Christine Chow, and Adil I. Daud

Abstract

Malignant melanoma is one of the most aggressive and lethal forms of skin cancer, with incidence rates continuing to rise due to increased sun exposure. While primary cutaneous melanoma is often managed through surgery, this alone is not curative for advanced metastatic stages. Within the past few decades, breakthrough discoveries in immunotherapy have led to successful treatment and a better quality of life for patients living with metastatic melanoma. In this chapter, we will explore the history, molecular biomarkers, and revolutionary studies that led to the new era of immunotherapy and targeted therapy for melanoma. Furthermore, melanoma treatment in the adjuvant and neoadjuvant setting is reviewed. Lastly, we briefly examine therapy options for patients with Merkel cell carcinoma.

Learning Objectives

By the end of this chapter, learners will be able to:

- Discuss the history of immunotherapy and studies leading to its use.
- Define the underlying mechanisms that drive immunotherapy.
- Identify key biomarkers that indicate viable treatment options and tumor responsiveness.
- Differentiate between the pros and cons of various therapy options.
- Appraise the capability of treatment in the adjuvant and neoadjuvant setting.

M. Chow · E. A. Sangalang · C. Chow · A. I. Daud (✉)
Department of Medicine, Division of Hematology Oncology, University of California, San Francisco, San Francisco, CA, USA
e-mail: Adil.Daud@ucsf.edu

67.1 Introduction

Melanoma is a cancer characterized by the abnormal proliferation of melanocytes, usually within the skin but also in the mucous membranes and in the pigmented tissues of the eye. Unlike many other more common skin cancers that remain local or regional, melanoma has a propensity to spread regionally and to distant organs. Although only accounting for 1% of skin cancer diagnoses, melanoma is the most lethal form of skin cancer, killing an estimated 6850 Americans in 2020 [1]. When detected early, melanoma is highly curable with surgery and has favorable outcomes. However, with more advanced stages of the disease, surgery alone is not curative in many cases, and additional therapy is needed. While in the past, therapy for melanoma was limited to chemotherapy and cytokine immunotherapy with only limited success, recent advances in targeted therapy and immunotherapy have revolutionized treatments for melanoma, and we will review these advances in the following chapter.

67.2 Principles of Immunotherapy

Modern immunotherapy owes much to William B. Coley, a surgeon, who observed that a patient with lymphoid sarcoma developed a remission following a postoperative wound infection. In 1893, he treated his first cancer patient with a cocktail of heat-killed bacterial toxins called "Coley's toxins" [2]. He subsequently went on to treat scores of cancer cases and published his experience with this approach. In subsequent decades, advances in chemotherapy and targeted therapy drove significant advances in the treatment of cancer, but in recent years, advances in immunotherapy have driven gains in treatment.

In the 1970s, researchers recognized the role of T cells, a type of white blood cell which helps the body target its immune response to specific pathogens, in combating cancer. An important cytokine driving the growth of T cells was

S. P. Leong et al. (eds.), *Cancer Metastasis Through the Lymphovascular System*,
https://doi.org/10.1007/978-3-030-93084-4_67

interleukin-2 (IL-2) which was cloned in 1982. Between 1985 and 1993, a series of clinical trials were completed that assessed the efficacy of this newly discovered protein as a way to fight melanoma [3]. A total of 270 patients were entered into eight clinical trials, and patients received high-dose bolus IL-2. The combined analysis of all patients treated showed the overall response rate (ORR) in patients with melanoma to be 16% (95% confidence interval [CI], 12–21%) where 6% of patients obtained a complete response (CR) and 10% of patients had a partial response (PR). The median progression-free survival (PFS) for all patients with a response was more than 1 year. High-dose IL-2 was associated with severe adverse events including capillary leak syndrome and the effects of secondary cytokine release [4]. Adverse effects (AEs) seen in patients who received IL-2 included hypotension, nausea and vomiting, diarrhea, confusion, shortness of breath, pulmonary edema, abnormal liver function tests, renal failure, pancytopenia, rash, fever, chills, malaise, and infections [5]. Nonetheless, the durable CRs and PRs seen in patients on IL-2 therapy lead to its FDA approval as the first immunotherapy treatment for metastatic melanoma in 1998.

Another major advance in melanoma treatment was interferon-alpha (IFN-α) which is a member of a class of proteins called cytokines which trigger the activation of the innate and adaptive immune system [6]. Its role in immune activation led to its utilization in various diseases including viral infections and in cancer [7]. IFN-α therapy has been tested in a variety of regimens including low dose, intermediate dose, and high dose as an adjuvant treatment for patients with stage II–III high-risk melanoma. Based on the Eastern Cooperative Oncology Group and intergroup trials E1684 and E1694, a significant positive impact on relapse-free survival and overall survival (OS) was seen in patients taking the IFN-α high-dose regimen [8]. However, the E1690 trial which compared low-dose and high-dose interferon to placebo was negative for OS although some improvement in relapse-free survival was seen. Based on meta-data analysis pooled from 12 randomized clinical trials, significantly longer relapse-free survival was recorded in patients with IFN-α compared with the observation group (hazard ratio [HR], 0.83; 95% CI, 0.77–0.90; p = 0.000003); however, OS was not improved (HR, 0.93; 95% CI, 0.85–1.02; $p = 0.1$) [8]. Another meta-data analysis which included 13 trials also demonstrated an IFN-α effect on recurrence (HR, 0.87; 95% CI, 0.81–0.93; $p < 0.0001$) [9]. A third meta-analysis included 14 randomized clinical trials that showed similar results where IFN-α patients had significant improvement in disease-free survival rates (HR, 0.82; 95% CI, 0.77–0.87; $p < 0.001$) [10]. The toxicity profile of high-dose IFN-α includes flu-like symptoms, chronic fatigue, myelosuppression, elevated transaminases, and neurologic symptoms [11, 12]. Ultimately, interferon fell out of favor as the effects on survival were not conclusive and its toxicity was considerable.

67.3 Tumor Immunity

Immune checkpoint inhibitors (ICIs) have revolutionized the treatment of advanced melanoma. These agents block negative regulators of T-cell priming or T-cell activity in the tumor microenvironment [13]. Currently, it is thought that in many cases, the presence of cancer in the body triggers the immune system. Tumors release pathogen-associated molecular patterns (PAMPs) and danger-associated molecular patterns (DAMPs), which are sensed by and stimulate dendritic cells (DCs) which then present tumor antigens to T cells usually within the draining lymph node with co-stimulation. T cells encounter DCs and form an immunological synapse which leads to activation and traffic into the tumor to release cytokines and granzymes and kill tumor cells. Tumor cells in some cases can produce programmed death ligand 1 (PD-L1) which can bind to programmed death 1 (PD-1) on the T cell and inhibit T-cell function, thus evading the body's immune system.

The tumor microenvironment (TME) is used to designate the local environment surrounding a tumor which includes stromal cells, blood vessels, tumor cells, and a variety of infiltrating and resident inflammatory cells. Immune cells found within the TME include tumor-infiltrating lymphocytes (TILs), myeloid-derived suppressor cells, tumor-associated macrophages, T cells, B cells, and dendritic cells. Tumors manipulate the microenvironment by promoting growth and immune tolerance, ultimately inducing tumor progression [14–16]. Generally, tumor microenvironments can be classified into "hot" or immune-infiltrated TME or "cold" immune desert TME [17]. More sophisticated classifications have been proposed such a three-category system: infiltrated-excluded TME, infiltrated inflamed TME, and infiltrated tertiary lymphoid structures [18]. The TME helps elucidate how mutational burden, presence of TILs, host factors such as obesity and gender, presence of liver metastasis, and other factors affect patients' immune responsiveness.

Tumor-infiltrating lymphocytes (TILs) are T cells that invade the tumor environment and actively engage in tumor destruction. TILs can directly migrate, interact, and proliferate within the tumor. As a result, TILs have been explored as a biomarker to predict immunotherapy response [19]. The presence of checkpoint-positive TIL appears to correlate with clinical response to PD-1 blockade. Treatment with anti-PD-1 resulted in increased activation of the exhausted T-cell population [20]. Their data demonstrates a correlation between the subsets of patients who have an abundance of partially exhausted tumor-infiltrating CD8+ T cells to ICI therapy response. This suggests that patients with a lower population of exhausted T cells are less likely to respond to anti-PD-1 monotherapy, and the risk-benefit profile may point to other treatment options.

Tumor mutation burden (TMB) is a measure of the total quantity of mutations within a tumor [21]. A high TMB in tumors can result in neoantigens that are not subject to self-tolerance and enhance immune cell recognition. Snyder et al. initially showed that high mutational burden correlates with a sustained clinical benefit from cytotoxic T-lymphocyte-associated antigen 4 (CTLA-4) blockade in patients with metastatic melanoma [22]. Rizvi et al. demonstrated that a higher mutation burden was associated with ORR, durable clinical benefit, and prolonged PFS in patients with non-small cell lung cancer patients treated with anti-PD-1 [23]. The International Cancer Genome Consortium identified malignant melanoma as the tumor with the highest mutation prevalence among 27 histotypes, indicating the genotoxic effects of UV-A exposure are comparable to smoking signatures [24]. Melanoma patients not coincidentally also had one of the highest response rates to single-agent PD-1/PD-L1 inhibition [25, 26]. Nevertheless, additional large-sized trials are warranted to confirm TMB assessment as a novel biomarker to aid in patient selection.

The intestinal commensal microbiota plays an important role in helping digestion and benefitting the immune system. The microbiota is composed of microorganisms, bacteria, fungi, viruses, protozoa, and their collective genetic material present in the gastrointestinal tract. Recently, the intestinal commensal microbiota has also been shown to influence the response in patients undergoing anti-PD-1 immunotherapy [15, 27–31]. Wargo and colleagues studied the diversity of oral and gut microbiomes in melanoma patients and showed a correlation between diversity of gut flora and response to checkpoint blockade ($n = 112$) [27]. Differences were observed in the patient gut microbiome of responders (R) versus non-responders (NR); analysis of the fecal microbiome samples ($n = 43$, 30R, 13NR) revealed significantly higher alpha diversity ($P < 0.01$) and an abundance of Ruminococcaceae ($p < 0.01$) in responding patients. Similar findings have been reported by other groups; however, this remains an area of active investigation.

PD-L1 expressions are a good surrogate for immune infiltration and have been used as a biomarker for PD-1 response [32–35]. While PD-L1 expression is highly correlated with PD-1 inhibitor response in patients with melanoma, it has been observed that a lack of PD-L1 expression does not ensure a lack of response and vice versa high-level expression of PD-L1 does not ensure response [32]. When T cells attack tumors, interferon-γ (IFN-γ) is also produced and secreted into the TME and can be used as a marker for immune response [36].

67.4 Immune Checkpoint Therapy in Melanoma

67.4.1 CTLA-4 Blockade

Ipilimumab is a humanized monoclonal antibody to a T-cell surface checkpoint known as cytotoxic T-lymphocyte-associated antigen 4 (CTLA-4). The discovery of ipilimumab by James Allison in 1982 and subsequent clinical trials with ipilimumab started the current era of immunotherapy [37]. CTLA-4 functions during the T-cell priming phase by binding B-7 at a higher affinity than CD-28, a T-cell co-stimulatory protein, and this displacement of CD-28 by CTLA-4 reduces T-cell activation [38]. This mechanism reduces T-cell priming and limits response to antigens presented on antigen-presenting cells. When CTLA-4 protein is blocked by ipilimumab from interacting with its ligands, B7–1 and B7–2, on antigen-presenting cells, the interruption of the inhibitory signal leads to the activation of T cells [24].

Phase I and II clinical trials of ipilimumab and tremelimumab (another CTLA-4 antibody) showed a low ORR but durable response rates in patients with responses [39]. In a pivotal phase III study comparing ipilimumab or ipilimumab plus gp100 to gp100 vaccine alone, ipilimumab was found to improve survival in patients with melanoma [40]. The ORR was 10.9% in patients who received ipilimumab alone, compared to 5.7% in the ipilimumab plus gp100 regimen and 1.5% in the gp100 alone group. However, ipilimumab was associated with a high incidence of adverse effects where 12.9% of patients had severe or fatal AEs. The most common immune-related AEs seen in patients taking ipilimumab are colitis, hepatitis, dermatitis, neuropathy, and endocrinopathy. Based on these results, in 2011, the FDA approved ipilimumab for the treatment of metastatic melanoma.

67.4.2 PD-1 Blockade

The programmed death 1 (PD-1) protein was initially identified in 1992 by Honjo and colleagues at the University of Kyoto [41]. In 2000, Honjo and colleagues identified the ligands that interacted with PD-1, programmed death ligand 1 (PD-L1). In 2002, Freeman et al. determined that blocking the PD-1 interaction with its ligand stimulated the immune system and thus created a powerful anti-tumor response. Based on these groundbreaking findings, a fully human anti-PD-1 antibody was developed in 2005 and studied in patients. PD-1 has emerged as a key immune regulator acting as a checkpoint restraining T-cell-mediated autoimmunity [42]. When the PD-L1 ligand on tumor or infiltrating immune cells engages with the PD-1 protein on activated T cells, apoptosis in regulatory T cells is reduced and thus, continues to suppress immune response. Tumor cells are able to evade T-cell destruction through exploitation of this pathway.

Immune checkpoint inhibitors block the interaction between cancer cells and the immune system with the use of PD-1 antibodies (Fig. 67.1). Other pathways targeted for similar immune checkpoint inhibition include cytotoxic T-lymphocyte-associated protein 4, lymphocyte-activation gene 3, and killer-cell immunoglobulin-like receptor [43].

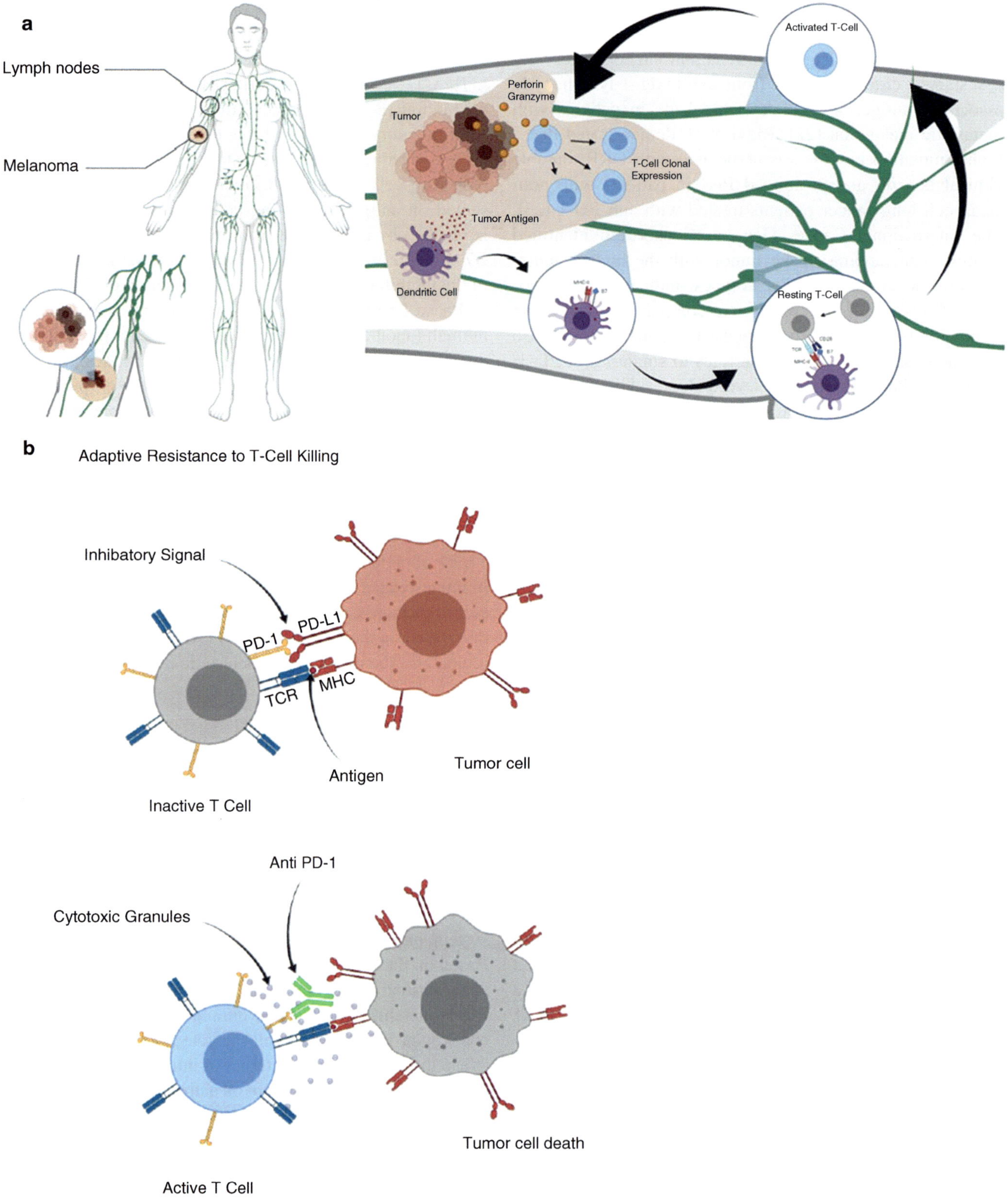

Fig. 67.1 (**a**) Evasion of PD-1 immune checkpoint through upregulated expression of PD-L1 receptor by tumor cell versus activation of T-cell by immune checkpoint inhibitor PD-L1 antibody. (**b**) Activation mechanism of T-cells by antigen-presenting cell ligand CD-28, a T-cell co-stimulatory protein (Source: BioRender.com)

67.4.3 Pembrolizumab

In 2015, pembrolizumab, an anti-PD-1 blocking antibody, gained FDA approval for the treatment of patients with metastatic melanoma [44]. This approval was based on two studies comparing pembrolizumab with ipilimumab or investigator's choice chemotherapy. In the KEYNOTE-006 study, 834 ipilimumab-naive metastatic melanoma patients were equally randomized to pembrolizumab 10 mg/kg intravenously every 2 or 3 weeks or up to four doses of ipilimumab 3 mg/kg intravenously every 3 weeks. In the KN002 study, 540 metastatic melanoma patients previously treated with ipilimumab were randomized to pembrolizumab 2 or 10 mg/kg intravenously every 3 weeks or investigator's choice of chemotherapy: dacarbazine, temozolomide, carboplatin plus paclitaxel, paclitaxel, or carboplatin. In KN006, patients receiving pembrolizumab every 2 weeks demonstrated an ORR of 34% (HR, 0.63; 95% CI, 0.47–0.83; $P < 0.001$) and 33% (HR, 0.69; 95% CI, 0.52–0.90; $P = 0.004$) in the every 3-week dosage schedule compared to 12% in the ipilimumab arm. Similarly improved ORR for pembrolizumab was seen in the KN002 study where the overall response was 21% in the pembrolizumab 2 mg/kg arm and 25% in the pembrolizumab 10 mg/kg arm versus only 5% in the chemotherapy arm. The improved overall response and statistically significant PFS seen in patients taking pembrolizumab led to its rapid FDA approval. Pembrolizumab was relatively tolerable, with common adverse effects including fatigue, cough, nausea, rash, itching, decreased appetite, shortness of breath, constipation, joint pain, and diarrhea. These AEs typically resolved when patients were treated with systemic corticosteroids.

67.4.4 Nivolumab

Nivolumab is a human IgG4 monoclonal antibody that blocks PD-1 binding to PD-L1. It was active in phase I clinical trial in a 2-weekly IV dose [45]. In a pivotal trial in treatment-naïve melanoma, 418 patients were randomized to nivolumab or dacarbazine; overall survival, PFS, and ORR were all statistically superior for nivolumab. In the CheckMate 037 trial, the ORR was 32% (95% CI, 23.5–40.8) in nivolumab-treated patients compared to the ORR of 10.6% in the chemotherapy arm. The most common immune-related AEs were pneumonitis, colitis, hepatitis, hypothyroidism, hyperthyroidism, nephritis, rash, pruritus, dermatitis, and vitiligo. Based on this data, the FDA granted accelerated approval for nivolumab for advanced melanoma.

67.4.5 Ipilimumab/Nivolumab

The combination of ipilimumab and nivolumab has been explored in several trials for melanoma. CTLA-4 functions during the priming phase of immune response, while in contrast, PD-1 works to maintain peripheral tolerance [46]. Targeting both immune system pathways using anti-CTLA-4 and anti-PD-1 antibodies to stimulate a more powerful immune response was the rationale for combination therapy.

In CheckMate 067, a phase III study, 945 treatment-naive patients were randomized (1:1:1) to receive nivolumab 3 mg/kg intravenously every 2 weeks; nivolumab 1 mg/kg intravenously every 3 weeks plus ipilimumab 3 mg/kg intravenously every 3 weeks for four doses, followed by nivolumab 3 mg/kg every 2 weeks; or ipilimumab 3 mg/kg every 3 weeks for four doses [47]. ORRs were higher in the ipilimumab plus nivolumab arm at 57.6% with 11.5% as a CR compared to a 19% ORR (2.2% CR) in the ipilimumab arm and a 43.7% ORR (8.9% CR) in the nivolumab arm. Additionally, at the 60-month follow-up time point, median OS was >60.0 months for the ipilimumab plus nivolumab arm versus 36.9 months and 19.9 months in the nivolumab alone and ipilimumab alone arms, respectively. PFS was also significantly improved in the ipilimumab plus nivolumab arm at 11.5 months compared to 2.9 months (99.5% CI, 0.31–0.57; $p < 0.001$) in the ipilimumab arm and 6.9 months (95% CI, 0.60–0.92) in the nivolumab arm. Unfortunately, more grade 3+ treatment-related AEs, 55.0%, were seen in the ipilimumab plus nivolumab arm compared to 27.3% in the ipilimumab arm and 16.3% in the nivolumab arm. In 2015, the promising PFS and ORR data led this to become the first ever combination immunotherapy treatment approved by the FDA for metastatic melanoma.

67.5 Immunotherapy and Targeted Therapy Combinations

The concept of immunotherapy and targeted therapy combination has preclinical and some clinical support. This has been explored in three recent trials.

KEYNOTE-022, a randomized phase II study, examined pembrolizumab and BRAF/MEK-targeted therapies dabrafenib and trametinib in patients with BRAF-mutant melanoma [48]. One hundred twenty patients were randomly assigned to either the pembrolizumab 2 mg/kg arm or placebo intravenously every 3 weeks combined with dabrafenib 150 mg orally twice a day and trametinib 2 mg orally once a day. At 36.6 months of long-term follow-up, the median PFS was improved in the triple combination therapy arm at

16.9 months (95% CI, 11.3–27.9) compared to 10.7 months (95% CI, 7.2–16.8) in patients that received placebo, dabrafenib, and trametinib. The OS data at 24 months was 63.0% (95% CI, 49.4–73.9%) for triple combination compared to 51.7% (95% CI, 38.4–63.4%) in patients taking dabrafenib and trametinib alone. The ORR was 63% (95% CI, 24.8–8.3) in the triple combination arm, with over 20% achieving a CR versus 72% (95% CI, 24.8–8.3) in the dabrafenib and trametinib arm with 15% reaching a CR. The triple combination arm recorded significantly more grade 3–5 treatment-related AEs in a total of 35 patients (58%) compared to 15 patients (25%) in the dabrafenib and trametinib arm. Despite the trends toward improved PFS, the trial did not meet its primary endpoint.

IMspire150, a multicenter phase III study, randomized 514 treatment-naive patients with metastatic melanoma to PD-L1 inhibitor atezolizumab, BRAF inhibitor vemurafenib, and MEK inhibitor cobimetinib (triple regimen) or vemurafenib and cobimetinib (control group) [49]. Patients in the control arm received vemurafenib 960 mg orally twice a day with cobimetinib 60 mg/d orally on days 1 to 21 of a 4-week cycle. The triple regimen arm received atezolizumab 840 mg intravenously on days 1 and 15 of 4-week cycles plus the same dose of vemurafenib and cobimetinib for cycle 1; during cycle 2, the dose of vemurafenib was lowered to 720 mg. Longer median PFS was found in the triple combination arm at 15.1 months compared to 10.6 months in the control arm ($p = 0.025$). The OS was slightly higher (28.8 months) in the triple combination arm versus 25.1 months in the control. Common treatment-related AEs in the triple combination arm versus the control were blood creatinine phosphokinase increase (51.3% vs 44.8%, respectively), diarrhea (42.2% vs. 46.6%), rash (40.9% in both arms), arthralgia (39.1% vs. 28.1%), pyrexia (38.7% vs 26.0%), alanine aminotransferase increase (33.9% vs. 22.8%), and lipase (32.2% vs. 27.4%). Serious treatment-related AEs were comparable for both groups, 33.5% in the triple combination arm versus 28.8% in the control arm.

A third trial was also reported. In COMBI-i, phase III study, 532 untreated stage IIIC or IV patients with melanoma received PD-1 inhibitor, spartalizumab 400 mg intravenously every 4 weeks in combination with dabrafenib 150 mg orally twice a day and trametinib 2 mg orally once a day or placebo in combination with dabrafenib and trametinib [50]. Calculated PFS for patients placed in the spartalizumab, dabrafenib, and trametinib arm was not significantly improved compared to the control arm. At 27.2 months of follow-up, PFS was 16.2 months in the triple combination arm and 12.0 months on the control arm (HR = 0.82; $P = 0.042$). Objective response rates were similar in both study arms, 68.5% responded in the triple combination arm with 19.9% CRs versus 64.2% in the control arm with 17.7% CRs. Additionally, the triple combination arm had more grade 3+ treatment-related AEs with 54.7% of patients experiencing AEs compared to 33.3% in the control arm.

Considering all three trials, the results of combination immunotherapy and targeted therapy in melanoma are disappointing, and outside of clinical trials or certain specific scenarios, currently there is not a role for these combinations in melanoma.

67.6 Adjuvant Therapy for Melanoma

67.6.1 Ipilimumab

Given the success of CTLA-4 blockade in advanced melanoma, there have been extensive efforts to treat patients with surgically resectable melanoma in the post-surgical (adjuvant) setting or in the pre-surgical (neoadjuvant) setting. Given the effectiveness of checkpoint inhibitors in later-stage melanoma, these agents have been tested extensively in these earlier stages as well. An important adjuvant trial with ipilimumab was EORTC 18071, a phase III trial that randomized 951 patients with stage III melanoma to receive ipilimumab 10 mg/kg intravenously (n-475) or placebo ($n = 476$), [51]. The median RFS was significantly prolonged in the adjuvant ipilimumab group at 26.1 months (95% CI, 19.3–39.3) compared with 17.1 months (95% CI, 13.4–21.6) in the placebo group (HR, 0.75; $p = 0.0013$). Additional follow-up on this trial corroborated the benefit of ipilimumab [52]. In the ipilimumab group, the most common grade 3–4 immune-related AEs were gastrointestinal (16% vs. <1% in the placebo group), hepatic (11% vs. <1%), and endocrine (8% vs. none). AEs led to discontinuation of treatment in 245 (52%) of 471 patients who started ipilimumab (182 patients discontinued during the initial four doses). Five patients (1%) died due to drug-related AEs in the ipilimumab group: three from colitis, one from myocarditis, and one from multiorgan failure with Guillain-Barré syndrome. Based on the improved RFS, DMFS, and OS seen in this study, adjuvant ipilimumab for high-risk patients with stage III cutaneous melanoma after having undergone a complete regional lymph node dissection was FDA approved in 2015.

67.6.2 Pembrolizumab

A recent phase III trial, KEYNOTE-054, randomized patients with complete lymph node dissection stage III cutaneous melanoma to receive pembrolizumab (514 patients) or placebo (505 patients) [53]. At the 3-year median follow-up, significantly prolonged RFS was seen in the pembrolizumab group (190 RFS events) compared with placebo (283 RFS events) (HR, 0.56; 95% CI, 0.47 to 0.68; $P < 0.001$) [54]. Treatment-related AEs occurred in 398 (78.8%) patients in the pembrolizumab group and in 333 (66.3%) in the placebo group. Grade 3–5 treatment-related AEs were observed in 74 (14.5%) patients in the pembrolizumab group and 17 (3.4%) in the placebo group. These data lead to FDA approval in 2019 [55].

67.6.3 Nivolumab

In CheckMate 238, 906 patients with completely resected metastatic melanoma were randomized to receive nivolumab 3 mg/kg intravenously every 2 weeks or ipilimumab 10 mg/kg intravenously every 3 weeks for four doses and then every 12 weeks starting at week 24 [56]. Most notable, at 4 years, the patients receiving nivolumab had a RFS rate of 51.7% compared to 41.2% in the ipilimumab arm (HR, 0.71; 95% CI, 0.60–0.86; $p = 0.0003$). Treatment-related grade 3–4 AEs were reported in 14.4% of patients receiving nivolumab and 45.9% of those receiving ipilimumab [57]. The results from this trial led to the rapid FDA approval for nivolumab for patients with metastatic melanoma in 2017.

67.7 Neoadjuvant Therapy

Neoadjuvant therapy is now being explored as a promising treatment option for high-risk resectable melanoma [58]. It is possible that T-cell checkpoint blockade given neoadjuvantly may induce a stronger and broader tumor-specific T-cell response as compared to adjuvant therapy due to the presence of tumor [59]. Additionally, neoadjuvant therapy may assist in surgical resectability [60]. We summarize some recent neoadjuvant trials of ICI.

Huang et al. conducted a single-arm trial of neoadjuvant pembrolizumab in melanoma and showed that ≈30% of patients experienced a complete or major pathologic response [61]. Comparing neoadjuvant nivolumab to ipilimumab/nivolumab, Amaria et al. showed that the combination yielded high response rates (ORR, 73%; pCR, 45%) but substantial toxicity (73% grade 3 treatment-related adverse events (trAEs)), whereas nivolumab monotherapy yielded modest responses (ORR, 25%; pCR, 25%) and low toxicity (8% grade 3 trAEs) [62]. In the OpACIN-neo phase II trial, stage III melanoma patients with resectable lymph nodes were randomly assigned to one of three neoadjuvant treatment arms (group A, ipilimumab plus nivolumab; group B, two cycles of ipilimumab 1 mg/kg plus nivolumab 3 mg/kg; and group C, two cycles of ipilimumab 3 mg/kg once followed by two cycles of nivolumab 3 mg/kg) [63]. Grade 3–4 immune-related AEs were observed in 12 (40%) of 30 patients in group A, six (20%) of 30 in group B, and 13 (50%) of 26 in group C. None of the grade 3–4 AEs were observed in more than one patient in the group B treatment arm, which was identified as a tolerable neoadjuvant dosing regimen. Results showed that 63% (95% CI, 44–80) of patients in group A, 57% (95% CI, 37–75) of patients in group B, and 35% (95% CI, 17–56) of patients in group C achieved radiological objective response. Based on response and the toxicity profile, group B was determined to be the preferred regimen for neoadjuvant ipilimumab/nivolumab. The PRADO study is an extension cohort of the OpACIN-neo trial and confirmed the response rate and safety of neoadjuvant ipilimumab 1 mg/kg plus nivolumab 3 mg/kg (group B in the OpACIN study) [59]. For the total cohort of patients, 70 (71%) achieved a pathologic response in the index lymph node (95% CI, 0.61–0.79): a CR in 49 patients (50%) of those patients and near pathologic CR in 11 (11%). At week 12 of follow-up, there was a 22% rate of grade 3 to 4 immune-related adverse events, commonly liver enzyme elevation and colitis. Surgery-related adverse events of any grade were observed in 54% patients.

67.8 Merkel Cell Carcinoma

Merkel cell carcinoma (MCC) is a rare, aggressive cutaneous neuroendocrine neoplasm with a high risk of recurrence and death. MCC can be caused by integration of the Merkel cell polyomavirus (MCPyV) [64]. Recently, PD-1 blockade has been shown to be extremely effective in MCC [65]. Infection by MCPyV can lead to MCC in immunosuppressed individuals.

Prior to the advent of PD-1 blockade, cytotoxic chemotherapy regimens were used, but these rarely lead to durable responses [66]. Lipson et al. studied 67 MCC specimens and analyzed MCC PD-L1 expression and its association with TILs and showed that PD-L1 expression was observed in 49% of patients [67]. Significant associations were also observed between the presence of MCPyV DNA, a brisk inflammatory response, and tumor cell PD-L1 expression.

67.8.1 Avelumab

Avelumab is a PD-L1 antibody that was initially explored in MCC. In a single-group, phase II trial, 88 patients with stage IV chemotherapy refractory MCC were enrolled to receive avelumab 10 mg/kg intravenously every 2 weeks [68]. The estimated proportion of responders with ≥1 year duration of response was 74% (95% CI, 53–87). Estimated 1-year PFS rate was 30% (95% CI, 21–41), and 1 year OS rate was 52% (95% CI, 41–62). Median OS was 12.9 months (95% CI, 7.5–not estimable). There were no grade 4 or 5 treatment-related AEs. MCC patients taking avelumab were seen to have durable responses beyond 1 year, and the treatment was well tolerated. Based on these results, avelumab became the first FDA-approved treatment for MCC in 2017.

67.8.2 Pembrolizumab

In the single-arm phase II KEYNOTE-017 study, 50 patients with recurrent locally advanced metastatic MCC received pembrolizumab 2 mg/kg intravenously every 3 weeks [69]. The ORR was 56% (95% CI, 41–70) with 24% CRs. The median response duration was not reached; however, among the 28 patients with responses, 96% had response durations of greater than 6 months. The efficacy and safety information provided in this study led the FDA to grant accelerated approval to pembrolizumab for advanced MCC in 2018.

Based on the data above, PD-1 and PD-L1 blockade is the most commonly employed therapy for MCC. Adjuvant trials in MCC are ongoing.

67.9 Conclusion

Over the past few decades, the emergence of melanoma and Merkel cell carcinoma treatments has been fruitful in achieving remission and quality of life for cancer patients. Although many of these approaches spark optimism into the field as a whole, much remains undiscovered. Continuation of the development in melanoma and Merkel cell carcinoma treatments offers a potentially rich area of future research.

Perspective
- When should immunotherapy be discontinued in a patient who is having a partial or complete response?
- Is neoadjuvant immunotherapy followed by surgery more advantageous than surgery followed by adjuvant therapy?
- How should targeted therapy versus immunotherapy be decided for each individual patient?
- In what cases should triple combination therapy be considered over combination immunotherapy in patients with adverse prognostic factors?
- How might personalized immunotherapy change the direction of melanoma treatments?

References

1. Cancer.org. Melanoma skin cancer statistics. https://www.cancer.org/cancer/melanoma-skin-cancer/about/key-statistics.html. In: . Accessed 29 Jan 2021.
2. Carlson RD, Flickinger JC, Snook AE. Talkin' toxins: from Coley's to modern cancer immunotherapy. Toxins. 2020;12:241.
3. Atkins MB, Lotze MT, Dutcher JP, et al. High-dose recombinant interleukin 2 therapy for patients with metastatic melanoma: analysis of 270 patients treated between 1985 and 1993. J Clin Oncol. 1999;17:2105.
4. Curti B, Daniels GA, McDermott DF, et al. Improved survival and tumor control with Interleukin-2 is associated with the development of immune-related adverse events: data from the PROCLAIMSM registry. J Immunother Cancer. 2017;5:102.
5. Pachella LA, Madsen LT, Dains JE. The toxicity and benefit of various dosing strategies for Interleukin-2 in metastatic melanoma and renal cell carcinoma. J Adv Pract Oncol. 2015;6:212–21.
6. Ivashkiv LB, Donlin LT. Regulation of type I interferon responses. Nat Rev Immunol. 2014;14:36–49.
7. Minn AJ, Wherry EJ. Combination cancer therapies with immune checkpoint blockade: convergence on interferon signaling. Cell. 2016;165:272–5.
8. Tarhini AA, Gogas H, Kirkwood JM. IFN-α in the treatment of melanoma. J Immunol. 2012;189:3789–93.
9. Wheatley K, Ives N, Eggermont A, et al. Interferon-α as adjuvant therapy for melanoma: an individual patient data meta-analysis of randomised trials. J Clin Oncol. 2007;25:8526.
10. Mocellin S, Pasquali S, Rossi CR, Nitti D. Interferon alpha adjuvant therapy in patients with high-risk melanoma: a systematic review and meta-analysis. J Natl Cancer Inst. 2010;102:493–501.
11. Sleijfer S, Bannink M, Van Gool AR, et al. Side effects of interferon-α therapy. Pharm World Sci. 2005;27:423.
12. Kirkwood JM, Bender C, Agarwala S, et al. Mechanisms and management of toxicities associated with high-dose interferon alfa-2b therapy. J Clin Oncol. 2002;20:3703–18.
13. Ribas A, Wolchok JD. Cancer immunotherapy using checkpoint blockade. Science. 2018;359:1350–5.
14. Gajewski TF, Corrales L, Williams J, Horton B, Sivan A, Spranger S. Cancer immunotherapy targets based on understanding the T cell-inflamed versus non-T cell-inflamed tumor microenvironment. Adv Exp Med Biol. 2017;1036:19–31.
15. Spranger S, Sivan A, Corrales L, Gajewski TF. Tumor and host factors controlling antitumor immunity and efficacy of cancer immunotherapy. Adv Immunol. 2016;130:75–93.
16. Strausberg RL. Tumor microenvironments, the immune system and cancer survival. Genome Biol. 2005;6:211.
17. Mujal AM, Krummel MF. Immunity as a continuum of archetypes. Science. 2019;364:28–9.
18. Binnewies M, Roberts EW, Kersten K, et al. Understanding the tumor immune microenvironment (TIME) for effective therapy. Nat Med. 2018;24:541–50.
19. Tumeh PC, Harview CL, Yearley JH, et al. PD-1 blockade induces responses by inhibiting adaptive immune resistance. Nature. 2014;515:568–71.
20. Daud AI, Loo K, Pauli ML, et al. Tumor immune profiling predicts response to anti-PD-1 therapy in human melanoma. J Clin Invest. 2016;126:3447–52.
21. Havel JJ, Chowell D, Chan TA. The evolving landscape of biomarkers for checkpoint inhibitor immunotherapy. Nat Rev Cancer. 2019;19:133–50.
22. Snyder A, Makarov V, Merghoub T, et al. Genetic basis for clinical response to CTLA-4 blockade in melanoma. N Engl J Med. 2014;371:2189–99.
23. Rizvi NA, Hellmann MD, Snyder A, et al. Mutational landscape determines sensitivity to PD-1 blockade in non–small cell lung cancer. Science. 2015;348:124–8.
24. Alexandrov LB, Nik-Zainal S, Wedge DC, et al. Signatures of mutational processes in human cancer. Nature. 2013;500:415–21.
25. Weber JS, D'Angelo SP, Minor D, et al. Nivolumab versus chemotherapy in patients with advanced melanoma who progressed after anti-CTLA-4 treatment (CheckMate 037): a randomised, controlled, open-label, phase 3 trial. Lancet Oncol. 2015;16:75–384.
26. Robert C, Long GV, Brady B, et al. Nivolumab in previously untreated melanoma without BRAF mutation. N Engl J Med. 2015;372:320–30.

27. Gopalakrishnan V, Spencer CN, Nezi L, et al. Gut microbiome modulates response to anti–PD-1 immunotherapy in melanoma patients. Science. 2018;359:97–103.
28. Sivan A, Corrales L, Hubert N, Williams JB, Aquino-Michaels K, Earley ZM, Benyamin FW, Lei YM, Jabri B, Alegre M-L, Chang EB, Gajewski TF. Commensal Bifidobacterium promotes antitumor immunity and facilitates anti-PD-L1 efficacy. Science. 2015;350:1084–9.
29. Vétizou M, Pitt JM, Daillère R, Lepage P, Waldschmitt N, Flament C, Rusakiewicz S, Routy B, Roberti MP, Duong CPM, Poirier-Colame V, Roux A, Becharef S, Formenti S, Golden E, Cording S, Eberl G, Schlitzer A, Ginhoux F, Mani S, Yamazaki T, Jacquelot N, Enot DP, Bérard M, Nigou J, Opolon P, Eggermont A, Woerther P-L, Chachaty E, Chaput N, Robert C, Mateus C, Kroemer G, Raoult D, Boneca IG, Carbonnel F, Chamaillard M, Zitvogel L. Anticancer immunotherapy by CTLA-4 blockade relies on the gut microbiota. Science. 2015;350:1079–84.
30. Wargo JA, Reddy SM, Reuben A, Sharma P. Monitoring immune responses in the tumor microenvironment. Curr Opin Immunol. 2016;41:23–31.
31. Zitvogel L, Ma Y, Raoult D, Kroemer G, Gajewski TF. The microbiome in cancer immunotherapy: diagnostic tools and therapeutic strategies. Science. 2018;359:1366–70.
32. Daud AI, Wolchok JD, Robert C, Hwu W-J, Weber JS, Ribas A, Hodi FS, Joshua AM, Kefford R, Hersey P, Joseph R, Gangadhar TC, Dronca R, Patnaik A, Zarour H, Roach C, Toland G, Lunceford JK, Li XN, Emancipator K, Dolled-Filhart M, Kang SP, Ebbinghaus S, Hamid O. Programmed death-ligand 1 expression and response to the anti-programmed death 1 antibody Pembrolizumab in melanoma. J Clin Oncol. 2016b;34:4102–9.
33. Lyford-Pike S, Peng S, Young GD, Taube JM, Westra WH, Akpeng B, Bruno TC, Richmon JD, Wang H, Bishop JA, Chen L, Drake CG, Topalian SL, Pardoll DM, Pai SI. Evidence for a role of the PD-1:PD-L1 pathway in immune resistance of HPV-associated head and neck squamous cell carcinoma. Cancer Res. 2013;73:1733–41.
34. Taube JM, Anders RA, Young GD, Xu H, Sharma R, McMiller TL, Chen S, Klein AP, Pardoll DM, Topalian SL, Chen L. Colocalization of inflammatory response with B7-h1 expression in human melanocytic lesions supports an adaptive resistance mechanism of immune escape. Sci Transl Med. 2012;4:127ra37.
35. Taube JM, Klein A, Brahmer JR, Xu H, Pan X, Kim JH, Chen L, Pardoll DM, Topalian SL, Anders RA. Association of PD-1, PD-1 ligands, and other features of the tumor immune microenvironment with response to anti-PD-1 therapy. Clin Cancer Res. 2014;20:5064–74.
36. Ayers M, Lunceford J, Nebozhyn M, Murphy E, Loboda A, Kaufman DR, Albright A, Cheng JD, Kang SP, Shankaran V, Piha-Paul SA, Yearley J, Seiwert TY, Ribas A, McClanahan TK. IFN-γ-related mRNA profile predicts clinical response to PD-1 blockade. J Clin Invest. 2017;127(8):2930–40.
37. Abhyankar D. Ipilimumab: a tale of twists and turns. South Asian J Cancer. 2018;7:193–4.
38. Krummel MF, Allison JP. CD28 and CTLA-4 have opposing effects on the response of T cells to stimulation. J Exp Med. 1995;182:459–65.
39. Wolchok JD, Hodi FS, Weber JS, Allison JP, Urba WJ, Robert C, O'Day SJ, Hoos A, Humphrey R, Berman DM, Lonberg N, Korman AJ. Development of ipilimumab: a novel immunotherapeutic approach for the treatment of advanced melanoma. Ann N Y Acad Sci. 2013;1291:1–13.
40. Fellner C. Ipilimumab (Yervoy) prolongs survival in advanced melanoma. P T. 2012;37:503–30.
41. Kudo M. Scientific rationale for combination immunotherapy of hepatocellular carcinoma with anti-PD-1/PD-L1 and anti-CTLA-4 antibodies. Liver Cancer. 2019;8:413–26.
42. Okazaki T, Chikuma S, Iwai Y, Fagarasan S, Honjo T. A rheostat for immune responses: the unique properties of PD-1 and their advantages for clinical application. Nat Immunol. 2013;14:1212–8.
43. O'Malley DP, Yang Y, Boisot S, et al. Immunohistochemical detection of PD-L1 among diverse human neoplasms in a reference laboratory: observations based upon 62,896 cases. Mod Pathol. 2019;32:929–42.
44. Barone A, Hazarika M, Theoret MR, et al. FDA approval summary: pembrolizumab for the treatment of patients with Unresectable or metastatic melanoma. Clin Cancer Res. 2017;23:5661–5.
45. Hazarika M, Chuk MK, Theoret MR, et al. U.S. FDA approval summary: Nivolumab for treatment of Unresectable or metastatic melanoma following progression on Ipilimumab. Clin Cancer Res. 2017;23:3484–8.
46. Qin S, Xu L, Yi M, et al. Novel immune checkpoint targets: moving beyond PD-1 and CTLA-4. Mol Cancer. 2019;18:155.
47. Hodi FS, Chiarion-Sileni V, Gonzalez R, et al. Nivolumab plus ipilimumab or nivolumab alone versus ipilimumab alone in advanced melanoma (CheckMate 067): 4-year outcomes of a multicentre, randomised, phase 3 trial. Lancet Oncol. 2018;19:1480–92.
48. Ferrucci PF, Giacomo AMD, Vecchio MD, et al. KEYNOTE-022 part 3: a randomized, double-blind, phase 2 study of pembrolizumab, dabrafenib, and trametinib in BRAF-mutant melanoma. J Immunother Cancer. 2020;8:1789–801.
49. Goodman A. Triplet regimen improves progression-free survival in advanced melanoma. In: The ASCO Post; 2020. https://ascopost.com/issues/may-25-2020/triplet-regimen-improves-progression-free-survival-in-advanced-melanoma/ Accessed 30 Jan 2021.
50. Helwick C. COMBI-i trial: no statistical benefit to adding Spartalizumab to Dabrafenib/Trametinib in melanoma. In: The ASCO Post; 2020. https://ascopost.com/issues/october-25-2020/combi-i-trial-no-statistical-benefit-to-adding-spartalizumab-to-dabrafenibtrametinib-in-melanoma/ Accessed 30 Jan 2021.
51. Eggermont AMM, Chiarion-Sileni V, Grob J-J, et al. Adjuvant ipilimumab versus placebo after complete resection of high-risk stage III melanoma (EORTC 18071): a randomised, double-blind, phase 3 trial. Lancet Oncol. 2015;16:522–30.
52. Eggermont AMM, Chiarion-Sileni V, Grob JJ, et al. Ipilimumab versus placebo after complete resection of stage III melanoma: Long-term follow-up results the EORTC 18071 double-blind phase 3 randomized trial. JCO. 2019;37:2512.
53. Eggermont AMM, Blank CU, Mandala M, et al. Adjuvant Pembrolizumab versus placebo in resected stage III melanoma. N Engl J Med. 2018;378:1789–801.
54. Eggermont AMM, Blank CU, Mandala M, et al. Longer follow-up confirms recurrence-free survival benefit of adjuvant Pembrolizumab in high-risk stage III melanoma: updated results from the EORTC 1325-MG/KEYNOTE-054 trial. J Clin Oncol. 2020;38:3925–36.
55. FDA (2019) Approves pembrolizumab for adjuvant treatment of melanoma. https://www.fda.gov/drugs/drug-approvals-and-databases/fda-approves-pembrolizumab-adjuvant-treatment-melanoma Accessed 30 Jan 2021.
56. Ascierto PA, Vecchio MD, Mandalá M, et al. Adjuvant nivolumab versus ipilimumab in resected stage IIIB–C and stage IV melanoma (CheckMate 238): 4-year results from a multicentre, double-blind, randomised, controlled, phase 3 trial. Lancet Oncol. 2020;21:1465–77.
57. FDA (2017) Grants regular approval to nivolumab for adjuvant treatment of melanoma. https://www.fda.gov/drugs/resources-information-approved-drugs/fda-grants-regular-approval-nivolumab-adjuvant-treatment-melanoma Accessed 2 Feb 2021.
58. Kelly ZR, Gorantla VC, Davar D. The role of Neoadjuvant therapy in melanoma. Curr Oncol Rep. 2020;22:80.

59. practiceupdate.com. ASCO 2020: neoadjuvant ipilimumab and nivolumab may spare therapeutic lymph node dissection in stage iii melanoma. In: Practice update; 2020. https://www.practiceupdate.com/content/asco-2020-neoadjuvant-ipilimumab-and-nivolumab-may-spare-therapeutic-lymph-node-dissection-in-stage-iii-melanoma/101960 Accessed 30 Jan 2021.
60. Khunger A, Buchwald ZS, Lowe M, et al. Neoadjuvant therapy of locally/regionally advanced melanoma. Ther Adv Med Oncol. 2019;11:1–19.
61. Huang AC, Orlowski RJ, Xu X, Mick R, George SM, Yan PK, Manne S, Kraya AA, Wubbenhorst B, Dorfman L, D'Andrea K, Wenz BM, Liu S, Chilukuri L, Kozlov A, Carberry M, Giles L, Kier MW, Quagliarello F, McGettigan S, Kreider K, Annamalai L, Zhao Q, Mogg R, Xu W, Blumenschein WM, Yearley JH, Linette GP, Amaravadi RK, Schuchter LM, Herati RS, Bengsch B, Nathanson KL, Farwell MD, Karakousis GC, Wherry EJ, Mitchell TC. A single dose of neoadjuvant PD-1 blockade predicts clinical outcomes in resectable melanoma. Nat Med. 2019;25:454–61. https://doi.org/10.1038/s41591-019-0357-y.
62. Amaria RN, Reddy SM, Tawbi HA, Davies MA, Ross MI, Glitza IC, Cormier JN, Lewis C, Hwu W-J, Hanna E, Diab A, Wong MK, Royal R, Gross N, Weber R, Lai SY, Ehlers R, Blando J, Milton DR, Woodman S, Kageyama R, Wells DK, Hwu P, Patel SP, Lucci A, Hessel A, Lee JE, Gershenwald J, Simpson L, Burton EM, Posada L, Haydu L, Wang L, Zhang S, Lazar AJ, Hudgens CW, Gopalakrishnan V, Reuben A, Andrews MC, Spencer CN, Prieto V, Sharma P, Allison J, Tetzlaff MT, Wargo JA. Neoadjuvant immune checkpoint blockade in high-risk resectable melanoma. Nat Med. 2018;24:1649–54. https://doi.org/10.1038/s41591-018-0197-1.
63. Rozeman EA, Menzies AM, van Akkooi ACJ, et al. Identification of the optimal combination dosing schedule of neoadjuvant ipilimumab plus nivolumab in macroscopic stage III melanoma (OpACIN-neo): a multicentre, phase 2, randomised, controlled trial. Lancet Oncol. 2019;20:948–60.
64. Feng H, Shuda M, Chang Y, Moore PS. Clonal integration of a Polyomavirus in human Merkel cell carcinoma. Science. 2008;319:1096–100.
65. Van Akkooi, Alexander Christopher, Paolo Antonio Ascierto, Paul D. Nathan, Paul Nghiem, Philip Reimann, Frank Hermann, and Jürgen C. Becker. "A Phase II, Open-Label Study to Investigate the Efficacy and Safety of Domatinostat in Combination with Avelumab in Patients with Advanced Unresectable/Metastatic Merkel Cell Carcinoma Progressing on Anti-PD-(L)1 Antibody Therapy: The Merklin 2 Study." J Clin Oncol. 2021;31:S767–S768.
66. Kaufman Howard L., et al. "Abstract CT079: Durable Responses to Avelumab (Anti-PD-L1) in Patients with Merkel Cell Carcinoma Progressed after Chemotherapy: 1-Year Efficacy Update." Clinical Trials, 2017;77:13.
67. Lipson EJ, Vincent JG, Loyo M, et al. PD-L1 expression in the Merkel cell carcinoma microenvironment: association with inflammation, Merkel cell polyomavirus and overall survival. Cancer Immunol Res. 2013;1:54–63.
68. cancer.gov (2017) Avelumab first approved drug for merkel cell carcinoma National Cancer Institute. https://www.cancer.gov/news-events/cancer-currents-blog/2017/avelumab-fda-merkel-cell. Accessed 30 Jan 2021.
69. Bradford D, Demko S, Jin S, et al. FDA accelerated approval of Pembrolizumab for recurrent locally advanced or metastatic Merkel cell carcinoma. Oncologist. 2020;25:e1077–82.

Landscape of Immunotherapy in Lung Cancer

68

Nirali Sanghavi, Umme Farwa, Faisal Khurshid, and Hatim Husain

Abstract

Lung cancer is the leading cause of cancer-related mortality not only in the USA but also worldwide. There have been many advances and research in treatment protocols for non-small cell lung cancer in the last few years. Screening with low-dose CT scan for smokers aged ≥65 years has offered a mortality benefit of 20% and a 6.7% decrease in all-cause mortality. The changing landscape of treatment from platinum compounds to tyrosine kinase inhibitors to immune checkpoint inhibitors has dramatically helped in increasing the survival of those high-risk individuals. Different trials that have been carried out to study the efficacy and effectiveness of different drugs give us a different perspective in determining the most appropriate treatment for an individual patient. The checkpoint inhibitors have become the first-line treatment option for patients with NSCLC as monotherapy or in combination with chemotherapeutic drugs. The role of biomarkers such as PD-L1 in tailoring the treatment for NSCLC has been very crucial. However, there is still a need to study different biomarkers and the correct combination and sequence of immune checkpoint inhibitors to be administered.

Learning Objectives

- Trials related to immune checkpoint inhibitors have given promising results when compared to chemotherapy drugs and a favorable toxicity profile.
- The mechanism of action of different immune checkpoint inhibitors.
- The importance of biomarkers and different biomarkers that have been studied so far in relation to NSCLC.
- Strategies used when prescribing immunotherapeutic drugs as monotherapy or in combination therapy.
- Recent advances with combinations of immunotherapies.

68.1 Introduction

Lung cancer according to the World Health Organization (WHO) is the leading cause of cancer-related mortality worldwide with 1.5 million deaths each year [1]. Industrialized countries such as Europe and the USA have the highest mortality rates. The most common type of lung cancer is non-small cell lung cancer that accounts for 85% of all cases. Non-small cell lung cancer (NSCLC) is the primary cause of cancer death worldwide. Among men, lung cancer is the leading cause of cancer death for those aged 40 years or older. While cigarette smoking is the leading cause of lung cancer, other risk factors include secondhand or passive smoking, alcohol, crystalline silica, chrysotile asbestos, and uranium. It is suggested that primary causes of recurrence and death in patients with lung cancer are invasion and metastasis of tumors [2]. Previous approaches to use nonspecific cytotoxic chemotherapy for lung cancer may be associated with severe adverse effects and failure to resolve metastatic or subclinical disease [3]. Almost 75% of all the patients with NSCLC are diagnosed with the disease at an advanced stage [4]. Historically these patients typically have 4–5 months of median survival time after diagnosis with only 10% surviving for 1 year [5].

New approvals for immunotherapy have improved outcomes for patients with metastatic lung cancer. The checkpoints

N. Sanghavi · U. Farwa · F. Khurshid · H. Husain (✉)
University of California San Diego, La Jolla, CA, USA

S. P. Leong et al. (eds.), *Cancer Metastasis Through the Lymphovascular System*,
https://doi.org/10.1007/978-3-030-93084-4_68

of PD-1, PD-L1, and CTLA-4 have been particularly important with several others being currently explored. Pembrolizumab, atezolizumab, nivolumab, ipilimumab, and tremelimumab are all inhibitory immune checkpoint inhibitors that are capable of binding and inactivating the effects of the above-mentioned proteins. This class of drugs has demonstrated positive results in the treatment of many cancers and has been studied for use in NSCLC. This chapter will focus on the role and evidence of immune response in NSCLC pathology, the mechanisms of immune checkpoint inhibitors, several clinical trials, and their results with respect to efficacy and safety.

68.2 Chemotherapy as a Historical Standard

In 1978, cisplatin was approved by the US Food and Drug Administration (FDA) as the first platinum compound for use. Since then several platinum drugs have entered clinical trials, with carboplatin and cisplatin approved for chemotherapy for lung cancer worldwide. Platinum drugs are used to treat advanced non-small cell lung cancer (NSCLC) without mutations that were sensitive to targeted therapy for the past two decades as the standard of care and as first-line treatment [6].

Platinum drugs work by targeting nuclear DNA, resulting in a broad spectrum of both anti-cancer and adverse effects. These agents form adducts preferentially with the N7 atom on guanine and adenosine bases. This binding stops DNA replication and transcription, which then initiates cellular apoptosis [7]. Platinum compounds attack indiscriminately on all rapidly dividing cells leading to severe side effects and inducing the ability of cancers to develop drug resistance. Common side effects of cisplatin and carboplatin include nausea and vomiting, myelosuppression, neuropathy, ototoxicity, hepatotoxicity, and nephrotoxicity. Cisplatin is a highly emetogenic agent, with considerable variability between individuals.

Another treatment option is the combination of the taxane docetaxel and ramucirumab, a recombinant human immunoglobulin G1 monoclonal antibody inhibiting vascular endothelial growth factor receptor-2 in patients with stage IV NSCLC as per the results of REVEL trial [8]. This has been approved after first-line platinum-based chemotherapy and clinical trials with checkpoint inhibitors have compared with docetaxel in the second line.

68.3 Mechanism of Immune Checkpoint Inhibitors

Antigen recognition by the T-cell receptor (TCR) is regulated by a balance between co-stimulatory and inhibitory signals which are called immune checkpoints. The two immune checkpoint receptors that have been most actively studied in the context of clinical cancer immunotherapy are cytotoxic T-lymphocyte-associated antigen 4 (CTLA-4, also known as CD152) and programmed cell death protein 1 (PD-1, also known as CD279). They are both inhibitory receptors and regulate immune responses at different levels and by different mechanisms. By altering this equilibrium, malignancies can escape immune surveillance and thrive.

68.3.1 PD-1/PD-L1 Pathway

PD-1 is expressed by activated T cells and is a key immune checkpoint receptor that mediates immunosuppression. It attenuates T-cell receptor signaling by binding to its ligands PD-L1 (B7-H1) and PD-L2 (B7-DC). PD-L1 expression on tumor and tumor-infiltrating immune cells is thought to be primarily induced by Th1 cytokines (in particular, interferon-γ) released by infiltrating tumor immune cells. Clinical trials have produced promising results evaluating antagonist antibodies that target the coinhibitory immune checkpoint, PD-1 expressed on activated antitumor T cells, and its primary ligand, PD-L1, expressed on tumor cells and immune cells within the tumor microenvironment (Fig. 68.1). Targeting this pathway can be categorized as directed against PD-1 such as nivolumab and pembrolizumab or targeted against PD-L1, such as atezolizumab and durvalumab [9–11].

68.3.2 Anti-CTLA-4 Pathway

CTLA-4 was the first immune checkpoint receptor to be clinically targeted and is expressed exclusively on T cells where it primarily regulates the amplitude of the early stages of T-cell activation. Primarily, CTLA-4 counteracts the activity of the T-cell co-stimulatory receptor, CD28. CTLA-4 expression on the surface of T cells dampens the activation of T cells by outcompeting CD28 in binding CD80 and CD86, as well as actively delivering inhibitory signals to the T cell [12]. CTLA-4 has also been shown to regulate tumor immunity through regulatory T cells (Tregs), which express high levels of surface CTLA-4, and therefore suppress activation and expansion of effector cells. Tregs are upregulated on tumor cells, and blocking Treg function through anti-CTLA-4 antibodies may be an effective mechanism to enhance antitumor activity. Ipilimumab is a fully human immunoglobulin G1 anti-CTLA-4 monoclonal antibody, now approved for melanoma, RCC, HCC, NSCLC, CRC, and mesothelioma. Tremelimumab is a fully human IgG2 anti-CTLA-4 antibody also under investigation (Fig. 68.2).

68.3.3 TIGIT Pathway

TIGIT is an immune checkpoint that inhibits the activation of both T cells and NK cells. The TIGIT receptor has

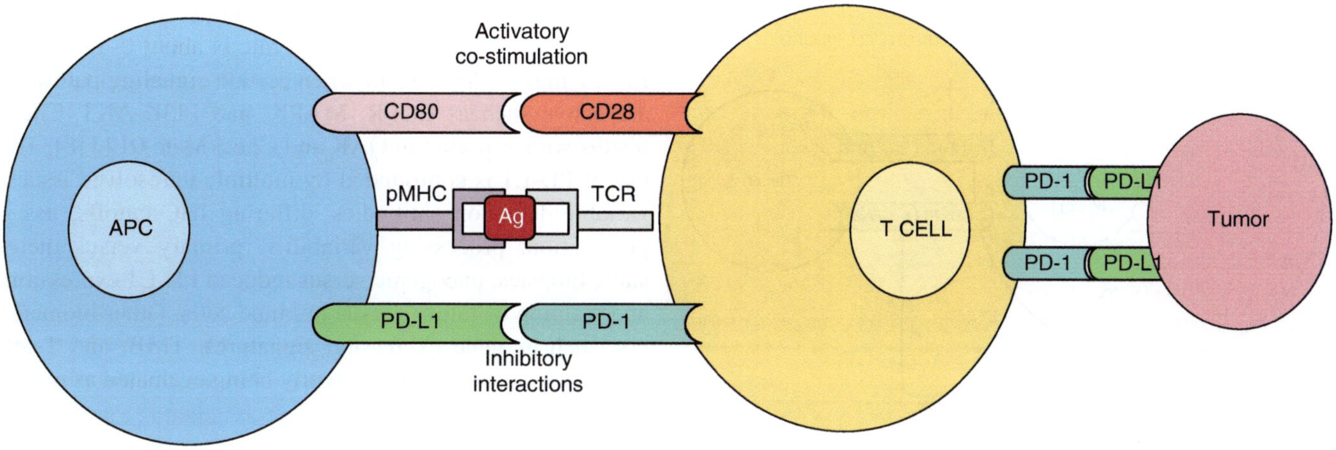

Fig. 68.1 The PD-1/PD-L1pathway. PD-1 expressed on activated T cells interacts with its ligands PD-L1 and PD-L2 when an immune response is triggered. The above figure depicts that the PD-L1 ligand found on antigen-presenting cells (dendritic cells) inhibits the T-cell function by binding to PD-1. MHC, major histocompatibility complex; TCR, T-cell receptor

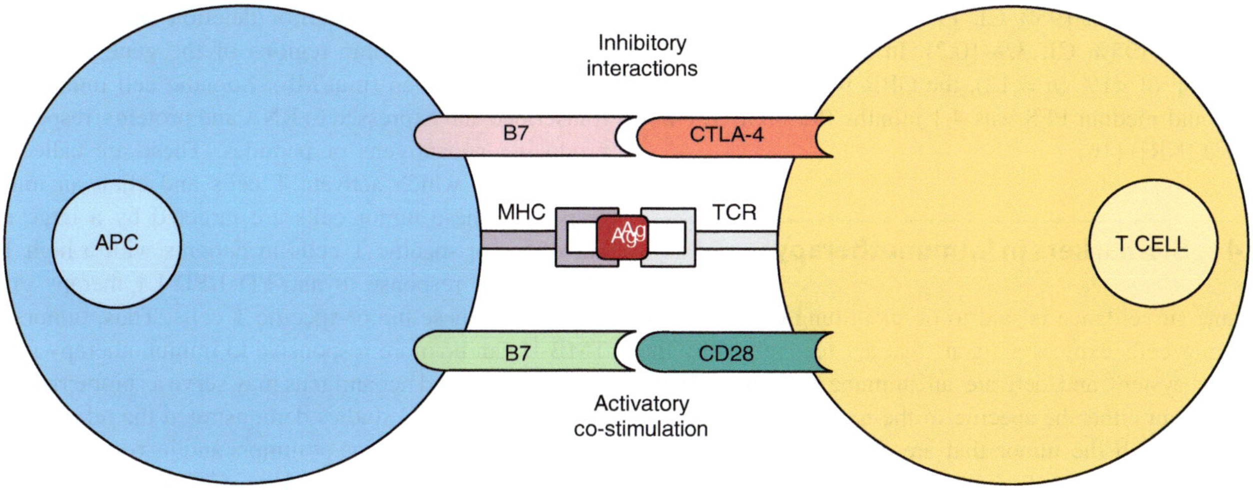

Fig. 68.2 The cytotoxic T-lymphocyte-associated antigen 4 (CTLA-4) pathway. T cell is activated when there is co-stimulatory signal from interaction between B7 on antigen presentation cells and CD28 on a T cell. CTLA-4 translocates to the plasma membrane of the T cells which subsequently inhibits activation of the T cells. MHC, major histocompatibility complex; TCR, T-cell receptor

a transmembrane IgV domain. It is upregulated in tumor antigen-specific CD8+ T cells and CD8+ TILs involved in adaptive immune tumor surveillance. TIGIT shares its two ligands, CD155 (PVR) and CD112 (PVRL2, nectin-2), with the co-stimulatory molecule CD226, and together they comprise a pathway that strongly resembles the well-known B7/CD28/CTLA-4 pathway [13]. Similar to CTLA-4, TIGIT can inhibit immune responses by delivering inhibitory signals to effector T and NK cells, by inducing tolerogenic dendritic cells, and by enhancing the suppressive capacity of Tregs. Both TIGIT and PD-1 are also upregulated in a variety of different cancers. Anti-TIGIT antibodies have demonstrated synergy with anti-PD-1/PD-L1 antibodies in preclinical models. Currently, there are multiple first-in-man phase 1 trials hoping to exploit this new pathway and improve response rates with existing immunotherapies [14]. In a phase 2 CITYSCAPE trial, tiragolumab, anti-TIGIT antibody, plus atezolizumab was compared to placebo plus atezolizumab as first-line treatment in NSCLC. The combination of tiragolumab and atezolizumab had better ORR (31.3% vs 16.2%) and PFS (5.4 months vs 3.6 months) as compared to placebo with atezolizumab [15]. These data have supported the initiation of a phase 3 trial in PD-L1 > 50% 1 L NSCLC (SKYSCRAPER-01).

In a phase 1 trial, another anti-TIGIT Ab, vibostolimab, in combination with pembrolizumab was used for metastatic NSCLC who had not previously received anti-PD-1/PD-L1 therapy, but the majority of whom had received ≥1 prior lines of therapy (Fig. 68.3). This trial showed that in patients whose tumors expressed PD-L1 ≥ 1% (n = 13),

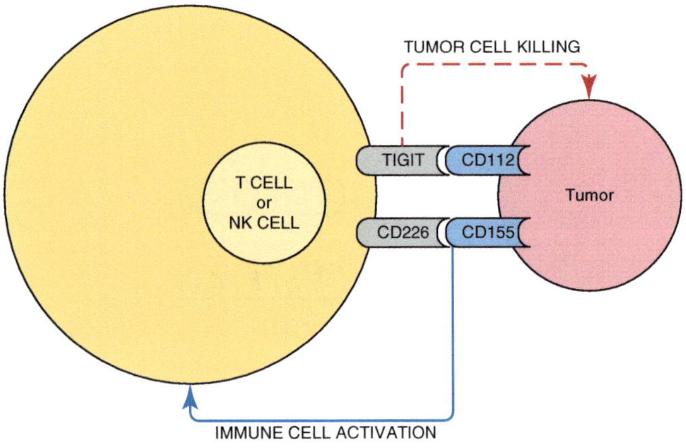

Fig. 68.3 Tiragolumab binds to TIGIT, leaving the CD226 receptor free to bind to CD155 and CD112 without competition. This blocks activation of TIGIT's immunosuppressive function and activates antitumor immune cells

the ORR was 46% (95% CI, 19–75), and median PFS was 8.4 months (95% CI, 3.9–10.2). In patients with PD-L1 expression of <1% (n = 12), the ORR was 25% (95% CI, 6–57), and median PFS was 4.1 months (95% CI, 1.9–not reached [NR]) [16].

68.4 Biomarkers in Immunotherapy

Immune surveillance is said to occur in lung cancer where the tumor cells express antigens that are recognized by the immune system and activate an immune response. These antigens can either be specific to the neoplasm or antigens associated with the tumor that are also expressed on normal cells but have an aberrant or deregulated expression on tumors [17]. PD-L1 and tumor mutational burden (TMB) are currently predictors of response for this class of immunotherapy in lung cancer.

68.4.1 Expression of PD-L1

PD-L1 expression on tumor cells has been a selection criteria for monotherapy [18]. KEYNOTE-024 proved that patients with advanced NSCLC with high PD-L1 expression (Tumor Proportion Score by 22C3 ≥ 50%) had a superior overall survival (OS) with pembrolizumab compared to chemotherapy [19]. However, the results of KEYNOTE-042 suggested that the efficacy of immunotherapy was similar to chemotherapy when PD-L1 expression was 1–49%. This indicates that if there is a higher expression of PD-L1, the effect of immunotherapy is also better [20]. Similarly, it has been shown that patients with PD-L1 over-expressing NSCLC have a 67–100% response rate in contrast to PD-L1-negative NSCLC, where the response rate is about 0–15% [21]. PD-L1 may be less useful when certain signaling pathways are active such as EGFR, MAPK, and PI3K-AKT. These results were reported in OAK and CheckMate 017 [18]. The use of PD-L1 is confounded by multiple unresolved issues: variable detection antibodies, differing IHC cutoffs, tissue preparation, processing variability, primary versus metastatic biopsies, oncogenic versus induced PD-L1 expression, and staining of tumor versus immune cells. Other biomarkers, such as gene expression signatures, TMB, and T-cell receptor sequencing, are currently being evaluated as potential markers of response.

68.4.2 TMB and Neoantigen Burden

Tumor mutation burden is defined as the total number of mutations, including insertion/deletion and replacement, per megabyte of exonic regions of the gene examined in the tumor specimen (mut/Mb). Somatic cell mutations are transcribed and expressed to RNA and proteins, respectively, producing neoantigens or peptides. These are called non-autoantigens which activate T cells and elicit an immune response. These tumor cells are targeted by a large number of tumor-specific T cells in patients with a high TMB (TMB). The response to anti-PD-1/PD-L1 therapy greatly depends on these tumor-specific T cells. Thus, tumors with TMB-H can be more responsive to immunotherapy indicating that TMB and neoantigens may serve as biomarkers [22]. In addition, several studies demonstrated the relationship of the mutational landscape of tumors and its responsiveness to PD-1 blockade in NSCLC and anti-CTLA-4 therapy in melanoma. These two malignancies often have a high TMB, the former due to increased carcinogen exposure via cigarette smoke and the latter due to exposure to DNA-damaging UV light [23]. CheckMate 026 revealed that nivolumab resulted in significantly better ORR (47 vs 28%) and PFS (9.7 vs 5.8 months) in the TMB-H arm (10 mut/MB cutoff) compared with that of platinum-based chemotherapy [6]. The results of CheckMate 227 showed that the PFS of patients with TMB-H (≥10 mut/Mb) treated with nivolumab and ipilimumab was superior to the chemotherapy. Surprisingly, in the same trial, PFS was improved regardless of PD-L1 levels. Similar results were found in CheckMate 012 and CheckMate 026, where no notable correlation was found between TMB and PD-L1 expression. This indicates that TMB may be an independent biomarker, but together with PD-L1, it may predict the efficacy of immunotherapy [24]. Pembrolizumab has recently received FDA approval for a TMB high indication across tumor types.

68.4.3 Tumor-Infiltrating Lymphocytes (TILs)

Apart from TMB, TILs can also be used to predict the efficacy of anti-PD-1/PD-L1 therapy. The PD-1/PD-L1 inhibitors require lymphocytes near the tumor, i.e., TILs, particularly infiltration by CD8 + T cells for their efficacy and favorable outcome [25]. The proliferation of CD8 + T cells has been directly associated with the shrinkage of tumors on imaging after immune checkpoint inhibitor treatment [26]. This is currently under investigation (Fig. 68.4).

68.4.3.1 Immunotherapy as First-Line Treatment

Immunotherapy Monotherapy

Pembrolizumab was established as the standard of care after the results of the phase 3 KEYNOTE-024 trial. The first-line trial had compared the investigator's choice of platinum-based chemotherapy with pembrolizumab in patients with NSCLC who had PD-L1 expression of >50% by 22C3 TPS. Pembrolizumab had shown increase in RR (45% vs 28%), PFS (10.3 vs 6 months), HR (0.5 0; 95% CI, 0.37–0.68; $P < 0.001$), and OS (30 vs 14.2 months) [27]. However, CheckMate 026, a phase 3 trial, studied nivolumab in patients with NSCLC with PD-L1 expression of >1% by 28–8. The trial compared the investigator's choice of platinum doublet against nivolumab. The primary efficacy analysis showed there was no advantage of treatment with nivolumab in RR (26% vs 33%), PFS (4.2 vs 5.9 months), HR (1.15; 95% CI, 0.91–1.45; $p = 0.25$) [6].

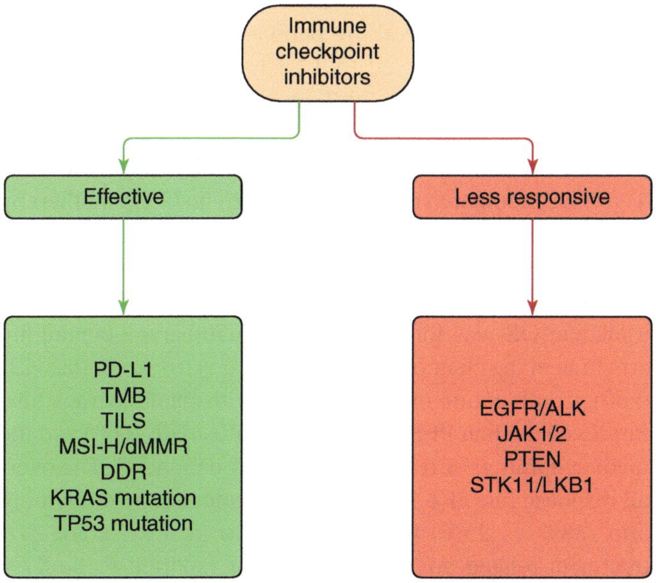

Fig. 68.4 Overview of biomarkers for immune checkpoint inhibition in non-small cell lung cancer. ICIs, immune checkpoint inhibitors; PD-L1, programmed death ligand 1; TMB, tumor mutational burden; TILs, tumor-infiltrating lymphocytes; MSI-H, microsatellite instability, high; dMMR, DNA mismatch repair; DDR, DNA damage repair; KRAS, Kirsten rat sarcoma viral oncogene homolog; TP53, tumor protein p53; EGFR/ALK, epidermal growth factor receptor/ALK tyrosine kinase receptor; JAK1/JAK2, Janus kinase 1/ Janus kinase 2; PTEN, phosphate and tension homology deleted on chromosome ten; STK11, serine/threonine kinase 11; LKB1, Lkb1 kinase

Another phase 3 trial CheckMate 057 compared nivolumab against docetaxel in patients with advanced non-squamous NSCLC. Nivolumab showed a clinically meaningful survival benefit, with an improved safety profile compared to docetaxel (27% lower risk of death at a minimum follow-up of 13.2 months), which persisted with extended follow-up (28% lower risk of death at a minimum follow-up of 17.2 months). The frequencies of all-causality, any-grade adverse events were similar between nivolumab and docetaxel, but grade 3–4 adverse events were fewer with nivolumab than docetaxel [3].

Durvalumab is a selective high-affinity human IgG1 monoclonal antibody that blocks PD-L1 binding to PD-1 and CD80 which allows T cells to recognize and kill tumor cells. The PACIFIC trial reported a PFS benefit in patients treated with durvalumab versus the placebo arm (16.8 vs 5.6 months; $P < 0.001$) and a higher response rate (RR) (28.4% vs 16%; $P < 0.001$). In addition, an OS benefit was observed (hazard ratio [HR] of death, 0.68; 95% CI, 0.54–0.86). The tolerability of durvalumab was similar to that in previous studies. These findings led to the Food and Drug Administration's (FDA) approval of durvalumab for unresectable stage III NSCLC [28].

On the basis of the results of IMpower110, atezolizumab was approved by FDA as a monotherapy in patients with PD-L1 expression of ≥50% on TPS by SP142 [29]. The study showed a statistically significant improvement in OS for patients with high PD-L1 tumor expression receiving atezolizumab compared to those treated with platinum-based chemotherapy. The median OS was 20.2 with atezolizumab versus 13.1 months with chemotherapy (HR, 0.59; 95% CI, 0.40–0.89; $P = 0.0106$) at a median follow-up of 15.7 months [30].

Most patients with advanced NSCLC do not respond to PD-1 inhibitor monotherapy, but in a phase 1 study, cemiplimab plus radiotherapy (RT) demonstrated antitumor activity in pretreated pts. with NSCLC. Overall response rate (ORR; complete response [CR] + partial response [PR]) was 18.2% (0 CR and 6 PRs) with a median duration of response of 14.9 months (95% CI, 5.5–14.9). Disease control rate (ORR + stable disease [SD]) was 72.7% (6 PRs + 18 SDs) [31].

Also, the EMPOWER-Lung 1, a phase 3 trial, compared cemiplimab monotherapy against platinum-doublet chemotherapy in patients with NSCLC with PD-1 ≥ 50%. The results showed that with median follow-up of 10.8 months, median OS was not reached with cemiplimab (95% CI, 17.9–not evaluable [NE]) versus 14.2 months (95% CI, 11.2–17.5) with chemotherapy (HR, 0.57; 95% CI, 0.42–0.77; $p = 0.0002$). Median PFS was 8.2 months (95% CI, 6.1–8.8) with cemiplimab versus 5.7 months (95% CI, 4.5–6.2) with chemotherapy (HR, 0.54; 95% CI, 0.43–0.68; $P < 0.0001$) [32].

Combination Chemoimmunotherapy

The KEYNOTE-189 for metastatic non-squamous NSCLC suggested that the combination of pembrolizumab with platinum doublet is more effective than chemotherapy alone [19, 33]. The ORR (47.6% vs 18.9%) and PFS (8.8 months vs 4.9 months) were significantly improved with the addition of pembrolizumab. Similarly, KEYNOTE-407 for metastatic squamous NSCLC proved that the combination of pembrolizumab and chemotherapy with carboplatin plus paclitaxel or nab-paclitaxel resulted in significantly longer overall survival (15.9 months vs 11.3 months) and progression-free survival (6.4 months vs 4.8 months) than chemotherapy alone [34] with benefit observed across PDL1 expression levels.

In a phase 3 trial, IMpower150, atezolizumab, anti-PD-L1 antibody, was studied in patients with non-squamous NSCLC. The patients were randomized to carboplatin, paclitaxel, and bevacizumab or carboplatin, paclitaxel, and atezolizumab or the same chemoimmunotherapy plus bevacizumab. The results showed improvement in the OS, PFS, and ORR when compared to chemotherapy alone [35]. Atezolizumab was approved by the FDA as the first-line agent in combination with carboplatin, paclitaxel, and bevacizumab in patients with metastatic non-squamous NSCLC with no EGFR or ALK driver mutations.

For squamous NSCLC, the IMpower131 trial showed that atezolizumab plus chemotherapy achieved a median overall survival of 14.2 months versus 13.5 months seen in patients who received chemotherapy alone [36].

Combination Immunotherapy

The phase 3 MYSTIC trial compared results of treatment with durvalumab against durvalumab and tremelimumab combination against chemotherapy alone in patients with metastatic NSCLC. It did not meet its primary endpoints of improved OS with durvalumab versus chemotherapy or improved OS or PFS with durvalumab plus tremelimumab versus chemotherapy in patients with 25% of tumor cells expressing PD-L1. Exploratory analyses identified a TMB threshold of 20 mutations per megabase for optimal OS benefit with durvalumab plus tremelimumab [37].

In phase 1 of CheckMate 012, chemotherapy-naïve patients were treated with different schedules and doses of nivolumab and anti-CTLA4 antibody ipilimumab. The ORR was 21% in patients with less than 1% tumoral PD-L1 expression, 57% in PD-L1 at least 1%, and 92% in patients with 50% PD-L1. PFS also showed improvement in patients that were treated with combination rather than with nivolumab alone [38].

In CheckMate 227 two treatment groups were formed depending on the expression of PD-L1 [39]. In patients with PD-L1-negative tumors (PD-L1 < 1%), a combination of nivolumab and ipilimumab was compared with the combination of nivolumab with chemotherapy and chemotherapy alone. Overall survival rates at 2 years were 40.4% for nivolumab in combination with ipilimumab and 23% with chemotherapy. The results showed that the progression-free interval was significantly higher with the combination of nivolumab with ipilimumab than chemotherapy alone. HR for death was 0.62 (95% CI, 0.48 to 0.78). The median duration of survival was significantly better in nivolumab with ipilimumab group (17.2 months) versus chemotherapy alone (12.2 months).

In patients with ≥1% PD-L1 expression, the combination of nivolumab with ipilimumab was compared with nivolumab alone and chemotherapy alone. The trial increased overall survival to 17.1 months in nivolumab with ipilimumab as compared to 14.9 months in the chemotherapy group alone. The overall survival rates with nivolumab and ipilimumab at 1 year 62.6% and 40% at 2 years, while with chemotherapy at 1 year 56.2% and 32.8% at 2 years. The objective response rate was greater with immunotherapy combination (35.9%), while with chemotherapy it was 30.0%. The median duration of response was 23.2 months with nivolumab and ipilimumab and 6.2 months with chemotherapy.

In addition, the CheckMate 227 also proved that the median overall survival in patients with baseline brain metastasis was improved to 17.4 months with nivolumab and ipilimumab versus 13.7 months with chemotherapy alone [40].

The CheckMate 9LA, a phase 3 trial, in patients with stage 4 NSCLC, compared nivolumab+ipilimumab+chemotherapy of two cycles against four cycles of chemotherapy alone. The combination of chemotherapy included four drugs: carboplatin, cisplatin, paclitaxel, and pemetrexed. The results showed a median OS was longer in the immunotherapy-containing arm than in the chemotherapy arm (15.6 vs 10.9 months; HR, 0.66). Combination immunotherapy with chemotherapy also improved median PFS (6.7 vs 5.0 months; HR, 0.68) and the median duration of response (11.3 vs 5.6 months). The overall response rate also favored the immunotherapy-containing arm (38% vs 25%) [41]. In regard to toxicity, grade 3–4 treatment-related adverse events were reported to be 47% in nivolumab+ipilimumab+chemotherapy versus 38% in the chemotherapy alone arm [15].

68.5 Cellular Therapies

Another advancement in this field is adoptive cellular therapy which relies on the ex vivo generation of tumor-reactive immune cells and administration of the cells in large numbers in the autologous host. Different forms of natural tumor-reactive immune cells have been generated for NSCLC in clinical trials such as TILs, cytokine-induced killer (CIK), CD+ T cells, natural killer (NK) cells, and chimeric antigen receptors (CAR) [42, 43].

68.5.1 The TIL Therapy

Naturally occurring T cells that have already infiltrated patients' tumors are first harvested and subsequently activated and expanded. This is followed by re-infusing the large number of previously activated cells into patients, where they then seek out and destroy the tumor. The ongoing phase 2 trial is studying the efficacy, safety, and toxicity of TIL therapy in patients with NSCLC [41].

68.5.2 The CAR Therapy

It has a key advantage compared to TIL therapy because the latter targets tumor cells only when their antigens are bound by the major histocompatibility complex, while CAR therapy has the ability to bind to cancer cells even if their antigens are not presented on the surface via MHC [44]. The preclinical studies have shown that PD-L1 CAR-T cells can be activated and have significant killing effects on PD-L1-positive tumor cells in vitro (Table 68.1). They inhibited tumor growth with

Table 68.1 Results of different trials comparing various chemotherapy and immunotherapy agents

Trial	Subset	n	OS	PFS	ORR
KEYNOTE trial 189	Carboplatin + pemetrexed + pembrolizumab)	410	Not reached	8.8 months	47.6%
	Placebo + pemetrexed-cisplatin/carboplatin	206	11.3 months	4.9 months	18.9%
			HR, 0.49 (95% CI, 0.38–0.64) $P < 0.001$	HR, 0.52 (95% CI, 0.43–0.64) $P < 0.001$	$P < 0.001$
KEYNOTE-021	Pembrolizumab + pemetrexed-carboplatin		Not reached	24.0 months	56.7%
	Pemetrexed + carboplatin		21.1 months	9.3 months	30.2%
			HR, 0.56 (95% CI, 0.32–0.95) $P = 0.0151$	HR, 0.53 (95% CI, 0.33–0.86) $P = 0.0049$	
KEYNOTE-407	Pembrolizumab + carboplatin + paclitaxel/nab-paclitaxel	278	15.9 months	6.4 months	57.9%
	Carboplatin + paclitaxel/nab –paclitaxel	281	11.3 months	4.8 months	38.4%
			HR, 0.64 (95% CI, 0.49–0.85) $P < 0.001$	HR, 0.56 (95% CI, 0.45–0.70) $P < 0.001$	
CheckMate 9LA	Nivolumab + ipilimumab + Carboplatin, paclitaxel, pemetrexed, Cisplatin	361	14.13 months	6.83 months	37.7%
	Carboplatin, paclitaxel, pemetrexed, Cisplatin	358	10.74 months	4.96 months	25.1%
			HR, 0.69 (95% CI, 0.56–0.86) $P < 0.001$	HR, 0.70 (95% CI, 0.58–0.84) $P < 0.001$	
CheckMate 227	Nivolumab + ipilimumab PD-L1 >/+ 1%	396	17.1 months		35.9%
	Chemotherapy alone PD-L1 >/= 1%	397	14.9 months		30%
	Nivolumab + ipilimumab PD-L1 < 1%	187	17.2 months		40.4%
	Chemotherapy alone PD-L1 < 1%	186	12.2 months		23%
CheckMate 057	Nivolumab	292	12.2 months (95% confidence interval [CI], 9.7 to 15.0	19 months (95% CI, 14–23)	19% (95% CI, 15 to 24)
	Docetaxel	290	9.4 months (95% CI, 8.1 to 10.7) with docetaxel	8 months (95% CI, 5–12)	12% (95% CI, 9 to 17)
IMpower130	Atezolizumab + carboplatin + nab-paclitaxel	483	18.6 months	7 months	49%
	Carboplatin + nab-paclitaxel	240	13.9 months	5.5 months	32%
			HR, 0.79 (95% CI, 0.64–0.98) $P = 0.033$	HR, 0.64 (95% CI,0.54–0.77) $P < 0.0001$	

(continued)

Table 68.1 (continued)

Trial	Subset	n	OS	PFS	ORR
IMpower131	Atezolizumab + carboplatin + nab-paclitaxel	343	14.2 months	6.5 months	
		338	12.6 months	5.6 months	
	Carboplatin+ nab-paclitaxel	340	13.5 months	5.6 months	
			HR, 0.96 (95% CI, 0.78–1.18) $P = 0.6931$	HR, 0.60 (95% CI, 0.60–0.85) $P = 0.0001$	
IMpower132	Atezolizumab + pemetrexed + cisplatin/carboplatin	292	18.1 months	7.6 months	47%
	Pemetrexed + cisplatin/carboplatin	286	13.6 months	5.2 months	32%
			HR, 0.81 (95% CI, 0.64–1.03) $P = 0.0797$	HR, 0.60 (95% CI, 0.49–0.72) $P < 0.0001$	
IMpower133	Atezolizumab +carboplatin-etoposide	201	12.3 months	5.2 months	60.2%
	Placebo + carboplatin-etoposide	202	10.3 months	4.3 months	64.4%
			HR, 0.70 (95% CI, 0.54–0.91) $P = 0.007$	HR, 0.77 (95% CI, 0.62–0.96) $P = 0.02$	
IMpower150	Atezolizumab + carboplatin + paclitaxel	402	19.0 (15.7–21.5) months	8.3 months	49.3%
	Atezolizumab + bevacizumab + paclitaxel + carboplatin	400	19.2 (17.0–23.8) months	8.5 (7.7–9.7) months	63.5%
	Bevacizumab-carboplatin-paclitaxel	400	14.7 (12.9–17.1) months	6.8 (5.8–7.3) months	48.0%
			HR, 0.78 (95% CI, 0.64–0.96) $P = 0.02$	HR, 0.62 (95% CI, 0.52–0.74) $P < 0.001$	

OS overall survival, *PFS* progression-free survival, *ORR* objective response rate, *HR* hazard ratio, *CI* confidence interval

no obvious observed toxicity in mouse xenograft models. In the ongoing phase 1 clinical trial, investigators are exploring the safety and efficacy of the new CAR-T cells in the treatment of advanced NSCLC patients [45].

68.6 Conclusion

Considering all the aspects discussed in this chapter, it is evident that the evolution of immunotherapy has established a new era for the treatment of NSCLC. Recent trials and studies of checkpoint inhibitors have shown promising preliminary results in attaining consequential and durable treatment responses with manageable toxicity. Identifying patient populations that can derive the greatest benefit to treatment with immunotherapeutic agents and combinations of therapy remains one of the most important challenging questions to answer. Exploration of novel checkpoints and cellular-based therapy has recently yielded encouraging responses that will need to be further tested in clinical trials. Antigen-specific vaccines in combination with immunomodulatory agents may have synergistic effects in augmenting the antitumor immune response.

Open Questions

- What are the right predictive biomarkers to guide sequence and combination of these novel immune checkpoint inhibitors?
- What are the evolving trials of targeted molecular therapies in combinations with immunotherapies?
- What are strategies to anticipate and treat toxicities with these agents?
- How the use of immunotherapy affects the rate of recurrence of early-stage NSCLC?

References

1. Torre LA, Bray F, Siegel RL, Ferlay J, Lortet-Tieulent J, Jemal A. Global cancer statistics, 2012. CA Cancer J Clin. 2015;65(2):87–108.
2. Ada G. The coming of age of tumour immunotherapy. Immunol Cell Biol. 1999;77(2):180–5.
3. Borghaei H, Paz-Ares L, Horn L, Spigel DR, Steins M, Ready NE, et al. Nivolumab versus docetaxel in advanced nonsquamous non-small-cell lung cancer. N Engl J Med. 2015;373(17):1627–39.

4. JM TR, Kmieciak M, Manjili MH, Knutson KL. Tumor immunoediting and immunosculpting pathways to cancer progression. Sem Cancer Biol. 2009;17(4):275–87.
5. Page DB, Bourla AB, Daniyan A, Naidoo J, Smith E, Smith M, et al. Tumor immunology and cancer immunotherapy: summary of the 2014 SITC primer. J Immunother Cancer. 2015;3(1):1–10.
6. Carbone DP, Reck M, Paz-Ares L, Creelan B, Horn L, Steins M, et al. First-line nivolumab in stage IV or recurrent non-small-cell lung cancer. N Engl J Med. 2017;376(25):2415–26.
7. Rossi A, Di Maio M. Platinum-based chemotherapy in advanced non-small-cell lung cancer: optimal number of treatment cycles. Expert Rev Anticancer Ther. 2016;16(6):653–60.
8. Garon EB, Cao D, Alexandris E, John WJ, Yurasov S, Perol M. A randomized, double-blind, phase III study of docetaxel and ramucirumab versus docetaxel and placebo in the treatment of stage IV non-small-cell lung cancer after disease progression after 1 previous platinum-based therapy (REVEL): treatment rationale an. Clin Lung Cancer. 2012;13(6):505–9.
9. Miyauchi E, Inoue A. Immune checkpoint therapy for non-small-cell lung cancer. Japanese J Cancer Chemother. 2016;43(6): 666–71.
10. Postow MA, Callahan MK, Wolchok JD. Immune checkpoint blockade in cancer therapy. J Clin Oncol. 2015;33(17):1974–82.
11. Xia B, Herbst R. Immune checkpoint therapy for non-small-cell lung cancer: an update. Immunotherapy. 2016;8(3):279–98.
12. Pardoll DM. The blockade of immune checkpoints in cancer immunotherapy. Nat Rev Cancer. 2012;12(4):252–64.
13. Dixon KO, Schorer M, Nevin J, Etminan Y, Amoozgar Z, Kondo T, et al. Functional anti-TIGIT antibodies regulate development of autoimmunity and antitumor immunity. J Immunol. 2018;200(8):3000–7.
14. Solomon BL, Garrido-Laguna I. TIGIT: a novel immunotherapy target moving from bench to bedside. Cancer Immunol Immunother. 2018;67(11):1659–67.
15. ascopubs.org. https://ascopubs.org/doi/abs/10.1200/JCO.2020.38.15_suppl.9503
16. merck.com. https://www.merck.com/news/merck-presents-promising-new-data-for-three-investigational-medicines-from-diverse-and-expansive-oncology-pipeline-at-esmo-virtual-congress-2020/
17. Duma N, Santana-Davila R, Molina JR. Non–small cell lung cancer: epidemiology, screening, diagnosis, and treatment. Mayo Clin Proc. 2019;94(8):1623–40.
18. Wang L, Yue HU, Wang S, Shen J, Wang X. Biomarkers of immunotherapy in non-small cell lung cancer (review). Oncol Lett. 2020;20(5):1–14.
19. Gandhi L, Rodríguez-Abreu D, Gadgeel S, Esteban E, Felip E, De Angelis F, et al. Pembrolizumab plus chemotherapy in metastatic non-small-cell lung cancer. N Engl J Med. 2018;378(22):2078–92.
20. Nosaki K, Saka H, Hosomi Y, Baas P, de Castro G, Reck M, et al. Safety and efficacy of pembrolizumab monotherapy in elderly patients with PD-L1–positive advanced non–small-cell lung cancer: pooled analysis from the KEYNOTE-010, KEYNOTE-024, and KEYNOTE-042 studies. Lung Cancer. 2019;135(July):188–95. Available from: https://doi.org/10.1016/j.lungcan.2019.07.004
21. Patel SP, Kurzrock R. PD-L1 expression as a predictive biomarker in cancer immunotherapy. Mol Cancer Ther. 2015;14(4):847–56.
22. Johnson DB, Frampton GM, Rioth MJ, Yusko E, Xu Y, Guo X, et al. Targeted next generation sequencing identifies markers of response to PD-1 blockade. Cancer Immunol Res. 2016;4(11):959–67.
23. Tray N, Weber JS, Adams S. Predictive biomarkers for checkpoint immunotherapy: current status and challenges for clinical application. Cancer Immunol Res. 2018;6(10):1122–8.
24. Rizvi H, Sanchez-Vega F, La K, Chatila W, Jonsson P, Halpenny D, et al. Molecular determinants of response to anti-programmed cell death (PD)-1 and anti-programmed death-ligand 1 (PD-L1) blockade in patients with non-small-cell lung cancer profiled with targeted next-generation sequencing. J Clin Oncol. 2018;36(7):633–41.
25. Geng Y, Shao Y, He W, Hu W, Xu Y, Chen J, et al. Prognostic role of tumor-infiltrating lymphocytes in lung cancer: a meta-analysis. Cell Physiol Biochem. 2015;37(4):1560–71.
26. Tumeh PC, Harview CL, Yearley JH, Shintaku IP, Emma J, Taylor M, et al. Immune Resistance. Nature. 2015;515(7528):568–71.
27. Reck M, Rodriguez-Abreu D, Robinson AG, Hui R, Csöszi T, Fülöp A, et al. Pembrolizumab versus chemotherapy for PD-L1-positive non-small-cell lung cancer. N Engl J Med. 2016;375(19):1823–33.
28. Antonia SJ, Villegas A, Daniel D, Vicente D, Murakami S, Hui R, et al. Durvalumab after chemoradiotherapy in stage III non–small-cell lung cancer. N Engl J Med. 2017;377(20):1919–29.
29. fda.gov. https://www.fda.gov/drugs/resources-information-approved-drugs/fda-approves-atezolizumab-first-line-treatment-metastatic-nsclc-high-pd-l1-expression
30. Spigel D, de Marinis F, Giaccone G, Reinmuth N, Vergnenegre A, Barrios CH, et al. IMpower110: interim overall survival (OS) analysis of a phase III study of atezolizumab (atezo) vs platinum-based chemotherapy (chemo) as first-line (1L) treatment (tx) in PD-L1–selected NSCLC. Ann Oncol. 2019;30(October):v915.
31. Moreno V, Gil-Martin M, Johnson ML, Aljumaily R, Lopez Criado P, Northfelt DW, et al. Cemiplimab, a human monoclonal anti-PD-1, plus radiotherapy (RT) in advanced non-small cell lung cancer (NSCLC): results from a phase I expansion cohort (EC 2). Ann Oncol. 2018;29(December):x26.
32. Sezer A, Kilickap S, Gümüş M, Bondarenko I, Özgüroğlu M, Gogishvili M, et al. LBA52 EMPOWER-lung 1: phase III first-line (1L) cemiplimab monotherapy vs platinum-doublet chemotherapy (chemo) in advanced non-small cell lung cancer (NSCLC) with programmed cell death-ligand 1 (PD-L1) ≥50%. Ann Oncol. 2020;31:S1182–3.
33. Patnaik A, Powell SF, Gentzler RD, Martins RG, Stevenson JP, Jalal SI, et al. study. 2019;17(11):1497–1508.
34. Paz-Ares L, Luft A, Vicente D, Tafreshi A, Gümüş M, Mazières J, et al. Pembrolizumab plus chemotherapy for squamous non-small-cell lung cancer. N Engl J Med. 2018;379(21):2040–51.
35. Socinski MA, Jotte RM, Cappuzzo F, Orlandi F, Stroyakovskiy D, Nogami N, et al. Atezolizumab for first-line treatment of metastatic nonsquamous NSCLC. N Engl J Med. 2018;378(24):2288–301.
36. Stinchcombe TE. IMpower 131: the exception to the rule. J Thorac Oncol. 2020;15(8):1258–60.
37. Rizvi NA, Cho BC, Reinmuth N, Lee KH, Luft A, Ahn MJ, et al. Durvalumab with or without Tremelimumab vs standard chemotherapy in first-line treatment of metastatic non-small cell lung cancer: the MYSTIC phase 3 randomized clinical trial. JAMA Oncol. 2020;6(5):661–74.
38. Med CC. Nivolumab plus ipilimumab as first-line treatment for advanced non-small-cell lung cancer (CheckMate 012): results of an open- label, phase 1, multicohort study. 2015;40(4):1291–1296.
39. Hellmann MD, Paz Ares L, Bernabe Caro R, Zurawski B, Kim SW, Carcereny Costa E, et al. Nivolumab plus ipilimumab in advanced non-small-cell lung cancer. N Engl J Med. 2019;381(21):2020–31.
40. ascopost.com. https://ascopost.com/issues/july-25-2020/nivolumabipilimumab-shows-benefit-in-patients-with-non-small-cell-lung-cancer-and-brain-metastases/
41. clinicaltrials.gov. https://clinicaltrials.gov/ct2/show/results/NCT03215706
42. Rohaan MW, Wilgenhof S, Haanen JBAG. Adoptive cellular therapies: the current landscape. Virchows Arch. 2019;474(4):449–61.
43. Wang J, Wang X. Beyond chemotherapy and targeted therapy: adoptive cellular therapy in non-small cell lung cancer. Mol Biol Rep. 2014;41(9):6317–23.
44. cancerresearch.org. https://www.cancerresearch.org/immunotherapy/treatment-types/adoptive-cell-therapy
45. https://clinicaltrials.gov/ct2/show/NCT03330834

69 Immuno-Oncologic Treatment of Genitourinary Malignancies

Przemyslaw Twardowski

Abstract

Tumors of the genitourinary tract represent almost 30% of malignancies in the United States. Prostate cancer accounts for majority of cases followed by urothelial, kidney, and testicular cancers. Despite their anatomical proximity, the biology of these tumors is quite heterogeneous. In the last 10 years, immunotherapy became one of the main cancer treatment modalities, and the field of genitourinary malignancies benefitted tremendously from that development although to a variable degree depending on the specific tumor type. Management of advanced kidney cancer already relies on checkpoint inhibitors as the main component of treatment, and urothelial cancers to a large extent are still dependent on platinum chemotherapies; however checkpoint blockade plays an increasingly critical role in their therapy. Significant progress remains to be accomplished to realize the promise of immunotherapy in the area of prostate cancer, the most prevalent cancer type of the group.

Learning Objectives

- Understand the role of immunotherapy in the management of genitourinary malignancies based on the review of pivotal clinical trial results that established current standard of care.
- Review the distinct differences in the application of various immunotherapeutic agents in the management of kidney, bladder, and prostate cancers.
- Understand the directions of ongoing clinical research aiming at future expansion of immunotherapy in the treatment of genitourinary cancers.

P. Twardowski (✉)
Medical Oncology and Urologic Oncology, Department of Urologic Oncology, John Wayne Cancer Institute, Santa Monica, CA, USA
e-mail: Przemyslaw.Twardowski@providence.org

69.1 Introduction

Genitourinary malignancies encompass a diverse group of neoplasms originating from the urinary and reproductive tracts. Prostate, urothelial (bladder, renal pelvis, ureter, urethra), kidney, and testicular cancers constitute the majority of genitourinary tumors; penile and adrenal cancers while included in that category represent a small minority in the United States. Despite anatomical and functional connections between genitourinary tissues, the biological characteristics of their neoplastic counterparts are extremely heterogeneous and defy unified therapeutic strategies. Historically kidney cancer represented one of the most promising targets for immunotherapy and continues to exemplify this today. Indeed, current frontline standard of care for advanced kidney cancer is based on immunotherapeutic drugs. Urothelial cancers are also responsive to immunotherapeutic strategies, but currently immunotherapy still remains complimentary to chemotherapy rather than the preeminent treatment option. While significant advances were made in the treatment of metastatic prostate cancer, they are based primarily on androgen hormone manipulation, and immunotherapy is relegated to niche applications. Testicular cancer is exquisitely sensitive to and curable with surgery and/or chemotherapy and radiation, and no role for immuno-oncology drugs has emerged as a potential treatment option thus far. In this chapter, we will review the current state of immuno-oncology drugs in the treatment of genitourinary malignancies.

69.2 Kidney Cancer

Kidney cancer represents a tumor that historically was considered susceptible to recognition and elimination by the immune system [1]. Kidney cancer is very resistant to standard chemotherapeutic agents, and rare cases of spontaneous regression of metastatic renal cell carcinoma (RCC) raised

S. P. Leong et al. (eds.), *Cancer Metastasis Through the Lymphovascular System*,
https://doi.org/10.1007/978-3-030-93084-4_69

the intriguing possibility of immune-mediated anticancer effect and instigated interest in pharmacological modulation of the immune system in the therapy of RCC [2]. Immune cytokines were investigated and provided a therapeutic option since the 1980s. Interferon alpha (IFNα) resulted in very modest response rates and improvement in overall survival when compared to chemotherapy and progesterone [3]. In 1992, another cytokine interleukin-2 (IL-2) gained FDA approval for the treatment of RCC [4]. This agent when given intravenously at high doses produced an overall response rate of about 20% with 8–10% complete responses. What was remarkable about these responses was that they were frequently durable and resulted in a curative effect in a small subset of patients with metastatic RCC. Unfortunately, high-dose IL-2 is associated with significant side effects requiring unique expertise and access to intensive care resources and is limited to patients with excellent performance status and intact cardiopulmonary function. After 2 decades since the approval of IL-2 without major breakthroughs, the modern era of immunotherapy began with the development of checkpoint inhibitors (CPIs) such as anti-programmed death receptor 1 (anti-PD-1), anti-programmed death receptor ligand 1 (anti-PD-L1), and anti-CTLA-4 antibodies. Early CPI clinical investigation efforts in kidney cancer focused on pretreated patients who progressed on standard vascular endothelial growth factor (VEGF) targeting therapy. The first Food and Drug Administration (FDA) approval was granted in 2015 for PD-1 inhibitor nivolumab which demonstrated a survival advantage over mTOR inhibitor everolimus (standard of care at that time) in patients with metastatic clear cell RCC who progressed on one to two anti-angiogenic regimens. The median overall survival (OS) was 25.0 months (95% confidence interval [CI] 21.8–not estimable) with nivolumab and 19.6 months (95% CI 17.6–23.1) with everolimus. The hazard ratio for death with nivolumab versus everolimus was 0.73 (98.5% CI 0.57–0.93; $p = 0.002$). The objective response rate (ORR) was greater with nivolumab than with everolimus (25% vs. 5%; odds ratio, 5.98 [95% CI 3.68–9.72]; $p < 0.001$) [5]. Once the efficacy of CPI was established in the second-line setting and beyond, they were rapidly tested and incorporated into the first-line regimens. Combination of nivolumab and CTLA-4 inhibitor ipilimumab demonstrated broad therapeutic advantage over the previous standard-of-care VEGF receptor tyrosine kinase inhibitor (VEGFR TKI) sunitinib in a CheckMate 214 trial [6]. Data was particularly strong in patients with intermediate- and poor-risk cancer based on the International Metastatic Renal Cell Carcinoma Database Consortium (IMDC) criteria which are used to assess the prognosis of patients with metastatic RCC utilizing simple clinical and laboratory criteria. One thousand ninety-six (of which 847 fell into the intermediate- or high-risk group) patients with clear cell RCC were randomized to receive either nivolumab (3 mg per kilogram) plus ipilimumab (1 mg per kilogram) intravenously every 3 weeks for four doses, followed by nivolumab (3 mg per kilogram) every 2 weeks, or sunitinib (50 mg) orally once daily for 4 weeks (6-week cycle). The co-primary endpoints were OS (alpha level, 0.04), ORR (alpha level, 0.001), and progression-free survival (PFS) (alpha level, 0.009) among patients with intermediate or poor prognostic risk. The median OS was not reached with nivolumab plus ipilimumab versus 26.0 months with sunitinib (hazard ratio for death, 0.63; $p < 0.001$). The ORR was 42% versus 27% ($p < 0.001$), and the complete response (CR) rate was 9% with CPI combination versus 1% with sunitinib among intermediate- and poor-risk patients with previously untreated advanced renal cell carcinoma. The results of this clinical trial established a combination of nivolumab and ipilimumab as a standard frontline regimen in patients with intermediate- and poor-risk RCC.

VEGF-targeted agents became incorporated into the frontline management of kidney cancer since the approval of sorafenib in 2005 [7]. Multiple agents in that category (sunitinib, pazopanib, axitinib, cabozantinib, lenvatinib) received regulatory approval in various lines of treatment of kidney cancer. These compounds work primarily by inhibition of the VEGF axis which is central in the pathophysiology of RCC, but they differ in their effect on other pathways affecting redundancies of angiogenic and cancer cell signaling [8]. In the context of the pivotal role assumed by checkpoint inhibitors in the treatment of RCC, simultaneous targeting of angiogenic pathways including VEGF, fibroblast growth factor (FGF), and platelet-derived growth factor (PDGF) has emerged as a strategy to modulate and enhance the immune response. It turns out that in addition to stimulation of angiogenesis, VEGF signaling results in the increase of myeloid-derived suppressor cells (MDSCs) and regulatory T cells (Tregs) which negatively impacts immune response [9]. VEGF also inhibits the maturation of dendritic cells into mature antigen-presenting cells (APCs) and reduces lymphocyte infiltration [9]. Direct and indirect evidence suggests that VEGF inhibitors and multikinase inhibitors that include VEGF targeting can modulate the tumor microenvironment and reverse VEGF-mediated immune-inhibitory mechanisms [10]. That hypothesis was validated in multiple randomized clinical trials comparing this combination strategy to sunitinib control (Table 69.1).

All three immuno-oncology (IO)/VEGF targeting combinations substantially outperformed sunitinib with respect to key endpoints of OS, PFS, and ORR. Even though these combinations have not been directly compared to single-agent PD-1 or programmed death ligand 1 (PD-L1) inhibitors, phase II data with single CPIs yields responses of < 40% [13], suggesting that combination strategies are more effective.

There is now a broad consensus that combination therapy incorporating checkpoint inhibitors represents a modern

Table 69.1 First-line combination IO/VEGF targeting therapy (phase III trials in RCC)

	Pembrolizumab/axitinib KN-426 [11]	Avelumab/axitinib Javelin 101 [12]	Nivolumab/cabozantinib 9ER (Choueiri T, et al. ESMO 2020, Abstract 6960)
OS	HR 0.53 (95% CI 0.38–0.74; $p < 0.0001$)	Immature	HR 0.6 (95% CI 0.4–0.89; $p = 0.001$)
PFS	HR 0.69 (95% CI 0.57–0.84; $p < 0.001$)	HR 0.69 (95% CI 0.574–0.825; $p < 0.0001$)	0.51 (95% CI 0.41–0.64; $p < 0.0001$)
ORR	59.3%	51.4%	55.7%
CR	5.8%	3.4%	8%

standard of care for frontline therapy of advanced RCC. Whether pure immunotherapy combo like nivolumab/ipilimumab or CPI/VEGF inhibitor combination should be used is a subject of debate and depends on specific circumstances and nuanced patient selection. In general, IO/IO combination is a strong contender in patients with intermediate- or poor-risk metastatic RCC by the International Metastatic Renal Cell Carcinoma Database Consortium (IMDC); however its superiority over a single VEGF inhibitor (sunitinib) in the favorable-risk subgroup is unclear [6, 14]. IO/VEGF inhibitor combinations maintain their effectiveness across a whole spectrum of IMDC risk groups (including favorable-risk group), but VEGF targeting component results in a different adverse profile than IO/IO combination, and that may influence the choice of therapy in an individual patient depending on underlying comorbidities and patient and physician preferences. The ongoing clinical trial COSMIC-313 will attempt to investigate whether the triple combination of nivolumab with ipilimumab and cabozantinib (IO/IO/VEGF inhibitor) is superior to nivolumab with ipilimumab (IO/IO). As of now, there is no accepted biomarker predicting responsiveness to immunotherapy in RCC. Tumor mutational burden (TMB) is relatively low as compared to other immune-responsive tumors: T-cell infiltration and PD-L1 expression also fall short in their ability to select patients for IO therapy in RCC. There is an ongoing effort to address these shortcomings and to better understand the genomic and immune landscape affecting response to immunotherapy in kidney cancer [15]. Another area of significant interest focuses on the role of immunotherapy in the management of less common variants of kidney cancer (non-clear cell RCC) which constitute approximately 25% of kidney tumors and were less intensely studied as compared to the dominant clear cell variant with respect to the role of immunotherapy.

69.3 Urothelial Cancer

From the historical perspective, application of the Bacillus Calmette–Guerin (BCG) vaccine in the therapy of superficial bladder cancer represents the earliest approved indication for the use of immunotherapy in genitourinary malignancies [16]. BCG was originally developed as a vaccine against tuberculosis (TB) and is derived from attenuated *Mycobacterium bovis*. Early autopsy observations of the potential decrease in cancer incidence in tuberculosis patients ultimately led to the investigation of BCG for the treatment of cancer including non-muscle-invasive bladder cancer (NMIBC) [17]. Intravesical administration of BCG in the setting of superficial bladder cancer maximizes direct exposure of cancer cells to therapy and minimizes, although not completely prevents, toxic systemic effects. NMIBC represents around 70% of the 70,000 new bladder cancer cases diagnosed in the United States each year [18]. The development of BCG as an immunotherapy for NMIBC and its proven effects of reducing recurrence and progression have transformed the treatment of this malignancy, and its continuous importance is demonstrated by a significant negative impact of recent shortages of BCG on delivery of appropriate care for patients with NMIBC [19, 20]. The effects of BCG on the immune function are very complex and result in activation of innate and adaptive immune systems and involvement of multiple cytokines including IL-2, IL-6, IL-8, IL-10, IL-12, IFN, and tumor necrosis factor-α (TNF-α). Prior to the development of BCG immunotherapy, 50% of patients with carcinoma in situ (CIS) developed the muscle-invasive disease. A large, randomized controlled trial by the Southwest Oncology Group (SWOG) comparing induction plus a 3-week maintenance course versus induction alone using intravesical BCG in CIS showed that the addition of a 3-week maintenance course beginning at 3 months produced a complete response rate of 84%, and, with ongoing 3-monthly installations, over 70% of those who responded remained disease-free for over 5 years [21]. There are some data supporting the need to continue maintenance for a period of at least 3 years, as evidenced by a study comparing a 1-year with a 3-year maintenance regimen for high-risk disease only [22].

Frontline therapy for muscle-invasive and advanced/metastatic urothelial cancer (UC) has not changed for decades and incorporates platinum (preferable cisplatin) chemotherapy combinations either with gemcitabine or alternatively a combination of dose-dense methotrexate, vinblastine, adriamycin, and cisplatin (ddMVAC). This strategy results in response rates of 50–70% and achieves occasional durable remissions [23]. In many ways, UC represents an attractive target for the development of IO drugs based on the biology of neoantigen-rich tumors (likely related to high mutational load from tobacco exposure) and a desperate need for improved treatments especially beyond first-line combinations where historically no effective therapies existed [24]. Initial clinical development focused exactly on that scenario: patients who progressed after prior platinum-based chemotherapy or who were not medically fit for chemotherapy. Over a short period of time, five checkpoint inhibitors (atezolizumab, pembrolizumab, nivolumab, durvalumab, and avelumab) were approved in that setting [25]. The approval was based on a response rate (15–21%) that was considered favorable and durable as compared to historical controls of taxane chemotherapy. Subsequently, pembrolizumab demonstrated overall survival (OS) benefit in a randomized phase III study (KEYNOTE-045) comparing it to standard second-line taxane or vinflunine chemotherapy [26]. Median OS was 10.3 months (95% CI 8.0–11.8) in the pembrolizumab group, as compared to 7.4 months (95% CI 6.1–8.3) in the chemotherapy group [hazard ratio (HR) for death, 0.73; 95% CI 0.59–0.91; $p = 0.002$]. A survival benefit was demonstrated regardless of PD-L1 expression [combined positive score (CPS) of < 10% or ≥ 10%). No difference in PFS was observed, and toxicity profile favored pembrolizumab. Atezolizumab was also evaluated in a randomized phase III study (IMvigor211) comparing it to taxane or vinflunine chemotherapy in the second-line setting [27]. The hierarchical design of the study had a primary endpoint that focused on OS in a subset of patients with high PD-L1 expression, and it was negative making additional statistical analyses only exploratory. It is unclear whether a positive OS outcome with pembrolizumab versus less apparent benefit of atezolizumab in these randomized phase III studies represents a true difference in drug activity or is a result of dissimilar statistical design of respected trials. Despite high hopes for quick expansion of CPI therapy into first-line and adjuvant space, the emerging results of clinical trials are mixed and have not resulted in the replacement of platin-based chemotherapy in the frontline setting. IMvigor130 study randomized patients with metastatic and locally advanced UC to three arms: (A) atezolizumab + platinum/gemcitabine, (B) atezolizumab monotherapy, and (C) placebo + platinum/gemcitabine combination [28]. Interim analysis of that study demonstrated that patients with low PD-L1 expression treated on atezolizumab monotherapy arm (B) had inferior OS as compared to chemotherapy arms and resulted in restriction in accrual to that arm to high PD-L1 expressors only. It also resulted in an adjustment of FDA label for frontline use of atezolizumab to cisplatin-ineligible patients whose tumors exhibit high expression of PD-L1 [PD-L1-stained tumor-infiltrating immune cells (ICs) covering ≥ 5% of the tumor area by companion diagnostic VENTANA PD-L1 SP142 assay] or patients who are ineligible for any platinum chemotherapy. Ultimately Imvigor130 did result in improvement in PFS in the chemo-immunotherapy arm (A) versus chemotherapy alone (arm C) (hazard ratio 0.82; 95% CI 0.7–0.96), but the OS data has not reached the level of significance at the time of latest analysis [28]. In addition, KEYNOTE-361 study with a very similar design to Imvigor130 but utilizing pembrolizumab has failed to meet the primary study endpoint of OS (Alva A, et al. Pembrolizumab combined with chemotherapy versus chemotherapy alone as first-line therapy for advanced urothelial carcinoma: KEYNOTE-361. ESMO 2020, Abstract LBA23). At a time of earlier interim analysis of KEYNOTE-361, the similar observation to Imvigor130 was made indicating inferior OS in patients with low PD-L1 expression treated with single-arm pembrolizumab and resulted in label update of pembrolizumab restricting its use in first-line setting to any platinum-ineligible patients or cisplatin-ineligible patients whose tumors exhibit high expression of PD-L1 (based on combined positive score (CPS ≥ 10) by Dako PD-L1 IHC 22C3 assay). While the results of these two large randomized clinical trials evaluating frontline application of concurrent chemo-immunotherapy failed to unequivocally change the standard of care, a different strategy of CPI maintenance after platinum chemotherapy induction did just that. JAVELIN Bladder 100 Phase III switch-maintenance study randomized patients with metastatic UC to standard four to six cycles of platin-gemcitabine combination or the same chemotherapy followed by maintenance of PD-L1 inhibitor avelumab in those patients who did not progress after four to six cycles of chemotherapy. Eighty-five percent of patients did not progress after initial chemotherapy, and the study demonstrated significant improvement in OS (median 21.4 months in CPI maintenance arm versus 14.3 months in standard chemotherapy alone arm; the hazard ratio for death, 0.69 (95% CI 0.56–0.86; $p = 0.001$) favoring the experimental arm [29]. The results were positive in the entire intention to treat the study population, but the high PD-L1 expressors appeared to experience the most robust benefit. On June 30, 2020, avelumab received FDA approval as a maintenance treatment for patients with locally advanced or metastatic urothelial carcinoma that has not progressed with first-line platinum-containing chemotherapy. Therefore, while platin-based combination chemotherapy remains a mainstay of frontline treatment of metastatic UC, subsequent switch-maintenance avelumab represents a new modification of standard of care in that circumstance. Multiple clinical trials are evaluating the role of CPI therapy in the perioperative setting in patients with muscle-invasive disease following cystectomy or nephroureterectomy. IMvigor010 randomized patients to 1 year of adjuvant atezolizumab or observation and was the first reported trial of that group. Unfortunately, primary endpoint of disease-free survival

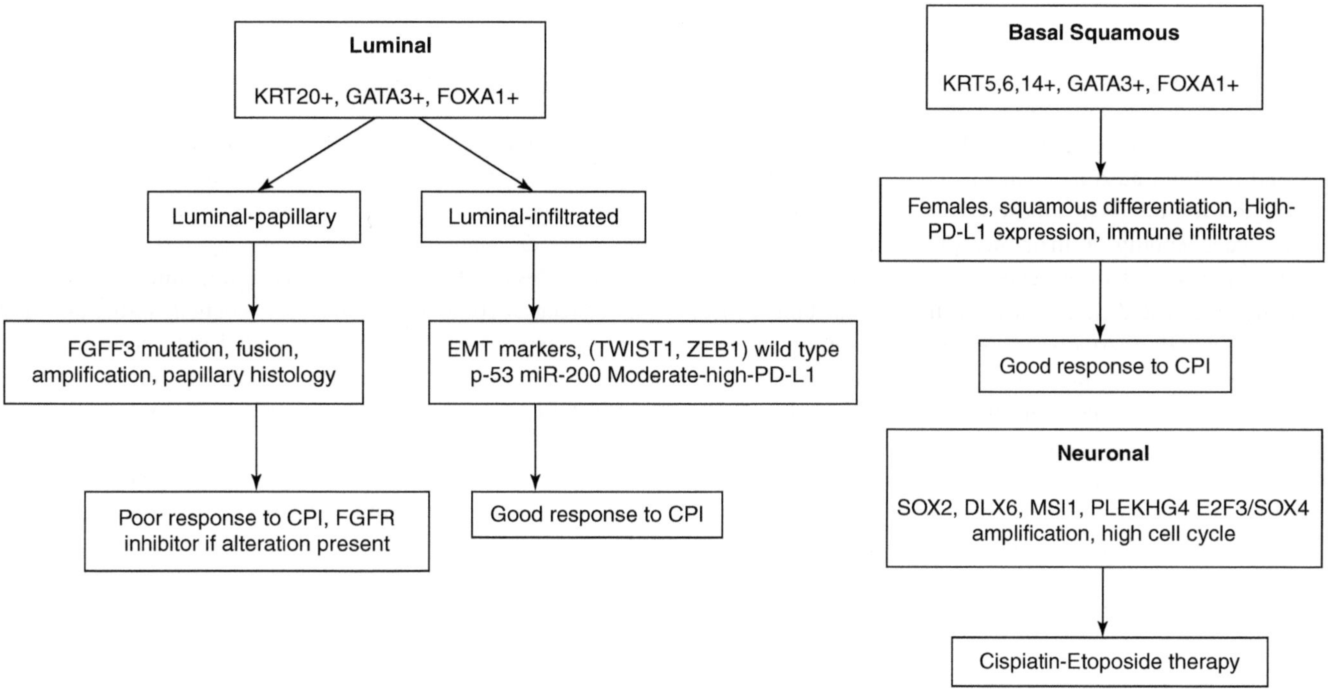

Fig. 69.1 Proposed molecular subtyping of urothelial carcinoma into luminal, basal squamous, and neuronal

(DFS) did not differ between the arms, and OS data is immature (Hussain M, et al. IMvigor010: Primary analysis from a phase III randomized study of adjuvant atezolizumab versus observation in high-risk muscle-invasive urothelial carcinoma. ASCO 2020, Abstract 5000). On the other hand, a press release from September 24, 2020, indicated that CheckMate 274, a pivotal phase III trial evaluating nivolumab after surgery in patients with high-risk, muscle-invasive urothelial carcinoma, met its primary endpoints of improving disease-free survival (DFS) versus placebo in both all randomized patients and in patients whose tumor cells express PD-L1 $\geq$ 1%. Details of the data have not been presented as of writing of this manuscript. Future directions of immunotherapeutics in urothelial cancer involve novel combinations and refinement of identification of appropriate patients for immunotherapy. There are multiple new combinations in development including IO/IO, IO/VEGF TKIs, and IO with antibody-drug conjugates (ADCs). Of particular interest is very impressive phase II data of enfortumab vedotin (ADC targeting nectin-4 and already FDA approved as a single drug in third-line setting) combined with pembrolizumab in the first-line UC demonstrating 73% response rate (Rosenberg JE, et al. GU Ca Symp 2020. Abstract 441). This combination is currently being tested in a pivotal randomized phase III study. Biomarker-driven selection for treatment with immunotherapeutics (particularly CPIs) has been met with mixed success. While higher expression of PD-L1 on tumor cells and/or infiltrating immune cells (ICs) has been associated with a trend of stronger clinical benefit (particularly in frontline setting), it has not achieved discriminatory significance for patient selection in the majority of clinical scenarios. Classification of urothelial cancer by molecular markers into luminal (divided into luminal-papillary, luminal-infiltrated), basal/squamous, and neuronal subtypes [30]. Figure 69.1 represents one of the latest efforts of identifying unique subsets of urothelial carcinoma with important therapeutic implications. Luminal infiltrated (accounting for about 19% of cases) and basal squamous (accounting for 35% of cases) subtypes appear to respond particularly well to CPI therapy, but these observations require prospective validation.

69.4 Prostate Cancer

Prostate cancer has largely missed the recent checkpoint inhibitor revolution likely as a result of being a low mutational burden, immunologically "cold" tumor, thus not possessing immune infiltrate substrate upon which the CPI therapy can perform its function [31]. An exception is a small subset (approximately 2%) of prostate cancers associated with mismatch repair alterations in which pembrolizumab demonstrated significant activity in a study combining multiple tumor types with underlying mismatch repair genotype [32]. This garnered FDA approval which represented a unique circumstance where tissue agnostic but molecularly defined malignancies were combined for drug approval [33]. Outside of this scenario studies of checkpoint inhibitors have failed to achieve results that would justify their approval in advanced prostate cancer [34]. Despite these setbacks, there are ongoing research efforts to attempt to convert the "cold" prostate cancer microenvironment to one that is immunologically more active and receptive to CPI therapy. One of the examples that has advanced to phase III study is a combination of PD-L1 inhibitor atezolizumab and a broad-spectrum VEGFR TKI cabozantinib. While these agents have limited anti-prostate cancer activity on their own [35, 36], a combination of both demonstrated a very respectable response rate of 32% in heavily pretreated metastatic castration-resistant prostate cancer (mCRPC) (Agarwal N, et al. Cabozantinib in combination with atezolizumab in patients with metastatic castration-resistant prostate cancer: results of cohort 6 of COSMIC-021 study. ASCO, 2020 Abstract 5564). Multiple other immunotherapeutic strategies in the early development of prostate cancer include bispecific antibodies and chimeric antigen receptor therapy (CART) platforms [31]. Historically a different type of immunotherapeutic approach has been successful: as one of the first cancer therapeutic vaccines sipuleucel-T was approved for the treatment of advanced mCRPC in 2010. Sipuleucel-T is an active cellular immunotherapy consisting of autologous peripheral blood mononuclear cells (PBMCs), including antigen-presenting cells (APCs), that have been activated in vivo with a recombinant fusion protein (PA2024) [37]. PA2024 consists of a prostate antigen, prostate acid phosphatase (PAP), that is fused to granulocyte-macrophage colony-stimulating factor, an immune cell activator [37]. In a phase III trial evaluating patients with asymptomatic or minimally symptomatic mCRPC, sipuleucel-T demonstrated 22% reduction in the risk of death as compared with placebo group (hazard ratio 0.78; 95% CI 0.61–0.98, $p = 0.03$) [38]. This reduction represented 4.1 months of improvement in median OS (25.8 months for sipuleucel-T versus 21.7 months in the placebo group). No significant improvement in PFS and objective or PSA responses was observed. Sipuleucel-T is a viable treatment option especially in patients with low-volume, asymptomatic mCRPC; however, lack of meaningful responses makes it less attractive for many patients with advanced, clinically symptomatic prostate cancer [39]. Sipuleucel-T is currently studied in a phase III randomized trial in patients with low-risk and favorable intermediate-risk prostate cancer who are otherwise managed with active surveillance.

69.5 Testicular Cancer/Germ Cell Tumors

Germ cell tumors (GCT) represent very unique malignancies in which the majority of patients achieve a curative outcome with cisplatin-based combination chemotherapy even in cases of extensive, metastatic disease [40]. Unfortunately, a fraction of patients who fail first- and especially second-line salvage therapy have a very poor prognosis and lack of effective therapeutic options. Several studies investigated the potential role of CPIs in that setting although theoretically testicular cancer may not represent an ideal target for CPI treatment [41]. Testicular cancer has a low tumor mutational burden, and mammalian testis represents an immunologically privileged site with the immunological microenvironment that protects germ cells from autoimmune attack; it is associated with the deficient immune response against antigens [42]. Nevertheless, PD-L1 expression is significantly higher in testis cancer as compared to normal testicular tissue, and high PD-L1 expression has been associated with poor prognosis [43]. Even though case reports of isolated durable response with CPI were published unfortunately, pembrolizumab, nivolumab, duralumin +/− tremelimumab, and avelumab did not demonstrate activity in phase II studies that would warrant their further development in this disease [44].

69.6 Conclusion

Genitourinary malignancies represent anatomically connected but biologically diverse group of tumors with very different molecular and immunological landscapes. Historically renal cell cancer was subject of particular interest for the immunotherapeutic approaches, and it remains so today with CPIs providing a cornerstone of effective treatment for advanced disease. Urothelial cancer also witnessed the significant application of CPIs in many clinical scenarios although immunotherapy has not displaced chemotherapy as a frontline treatment strategy and arguably the benefits of CPIs in urothelial cancers are less generalized than in kidney cancers. Immunotherapy plays only a niche role in the management of prostate cancer with the availability of autologous vaccine sipuleucel-T for low-volume castration-resistant prostate cancer, but no definitive role for CPIs has emerged so far in prostate cancer outside of rare cases associated with mismatch repair mutations. No application for any immunotherapy has materialized yet for the management of germ cell tumors.

Ongoing and future research efforts aiming at enhancing the benefits of immunotherapy in the management of genitourinary malignancies will focus on the following areas:

- Discovery and validation of biomarkers aiming at more precise selection of patients for immunotherapy.
- Refinements of currently available treatments with respect to dosing and duration of immunotherapy.
- Strategies to minimize toxicities of immunotherapy.
- New combinations of immunotherapy drugs and immunotherapy-enhancing compounds aiming at converting immunologically "cold" tumors to more immunologically responsive ones.
- Better understanding of mechanism of intrinsic and acquired resistance to immunotherapy.

References

1. Hammers H. Immunotherapy in kidney cancer: the past, present, and future. Curr Opin Urol. 2016;26(6):543–7.
2. Muraoka K, Nishikawa Y, Isoyama T. Spontaneous regression of metastatic renal cell carcinoma after cytoreductive nephrectomy: a case report. Urol Case Rep. 2020;31:101179.
3. Naitoh J, Belldegrun AS. Interferon and kidney cancer: a promise unfulfilled? Cancer J Sci Am. 1998;4(3):155–6.
4. Dutcher JP. Interleukin-2 based therapy for kidney cancer. Cancer Treat Res. 2003;116:155–2.
5. Motzer RJ, et al. Nivolumab versus everolimus in advanced renal-cell carcinoma. N Engl J Med. 2015;373(19):1803–13.
6. Motzer RJ, et al. Nivolumab plus ipilimumab versus sunitinib in advanced renal-cell carcinoma. N Engl J Med. 2018;378(14):1277–90.
7. Escudier B, et al. Sorafenib in advanced clear-cell renal-cell carcinoma. N Engl J Med. 2007;356(2):125–4.
8. Vachhani P, George S. VEGF inhibitors in renal cell carcinoma. Clin Adv Hematol Oncol. 2016;14(12):1016–28.
9. Bergerot P, et al. Cabozantinib in combination with immunotherapy for advanced renal cell carcinoma and urothelial carcinoma: rationale and clinical evidence. Mol Cancer Ther. 2019;18(12):2185–93.
10. Incio J, et al. PlGF/VEGFR-1 signaling promotes macrophage polarization and accelerated tumor progression in obesity. Clin Cancer Res. 2016;22(12):2993–04.
11. Rini BI, et al. Pembrolizumab plus axitinib versus sunitinib for advanced renal-cell carcinoma. N Engl J Med. 2019;380(12):1116–27.
12. Choueiri TK, et al. Updated efficacy results from the JAVELIN Renal 101 trial: first-line avelumab plus axitinib versus sunitinib in patients with advanced renal cell carcinoma. Ann Oncol. 2020;31(8):1030–9.
13. Xu W, Atkins MB, McDermott DF. Checkpoint inhibitor immunotherapy in kidney cancer. Nat Rev Urol. 2020;17(3):137–50.
14. Heng DY, et al. External validation and comparison with other models of the International Metastatic Renal-Cell Carcinoma Database Consortium prognostic model: a population-based study. Lancet Oncol. 2013;14(2):141–8.
15. Lopez-Beltran A, et al. The identification of immunological biomarkers in kidney cancers. Front Oncol. 2018;8:456.
16. Fuge O, et al. Immunotherapy for bladder cancer. Res Rep Urol. 2015;7:65–79.
17. Morales A. Adjuvant immunotherapy in superficial bladder cancer. Natl Cancer Inst Monogr. 1978;49:315–9.
18. Siegel RL, Miller KD, Jemal A. Cancer statistics, 2020. CA Cancer J Clin. 2020;70(1):7–30.
19. Sylvester RJ, van der Meijden AP, Lamm DL. Intravesical Bacillus Calmette–Guerin reduces the risk of progression in patients with superficial bladder cancer: a meta-analysis of the published results of randomized clinical trials. J Urol. 2002;168(5):1964–70.
20. Nikas CV, Smith AB. Goldilocks and the BCG: Bacillus Calmette–Guerin dose reduction in the age of shortage. Eur Urol. 2020;78(5):699–700.
21. Lamm DL, et al. Maintenance Bacillus Calmette–Guerin immunotherapy for recurrent TA, T1 and carcinoma in situ transitional cell carcinoma of the bladder: a randomized Southwest Oncology Group Study. J Urol. 2000;163(4):1124–9.
22. Brausi M, et al. Side effects of Bacillus Calmette–Guerin (BCG) in the treatment of intermediate- and high-risk Ta, T1 papillary carcinoma of the bladder: results of the EORTC genito-urinary cancers group randomised phase 3 study comparing one-third dose with full dose and 1 year with 3 years of maintenance BCG. Eur Urol. 2014;65(1):69–76.
23. Bellmunt J, et al. Optimizing therapeutic strategies in advanced bladder cancer: update on chemotherapy and the role of targeted agents. Crit Rev Oncol Hematol. 2009;69(3):211–22.
24. Farina MS, Lundgren KT, Bellmunt J. Immunotherapy in urothelial cancer: recent results and future perspectives. Drugs. 2017;77(10):1077–89.
25. Alhalabi O, et al. Current and future landscape of immune checkpoint inhibitors in urothelial cancer. Oncology (Williston Park). 2019;33(1):11–8.
26. Bellmunt J, et al. Pembrolizumab as second-line therapy for advanced urothelial carcinoma. N Engl J Med. 2017;376(11):1015–26.
27. Powles T, et al. Atezolizumab versus chemotherapy in patients with platinum-treated locally advanced or metastatic urothelial carcinoma (IMvigor211): a multicentre, open-label, phase 3 randomised controlled trial. Lancet. 2018;391(10122):748–57.
28. Galsky MD, et al. Atezolizumab with or without chemotherapy in metastatic urothelial cancer (IMvigor130): a multicentre, randomised, placebo-controlled phase 3 trial. Lancet. 2020;395(10236):1547–57.
29. Powles T, et al. Avelumab maintenance therapy for advanced or metastatic urothelial carcinoma. N Engl J Med. 2020;383(13):1218–30.
30. Robertson AG, et al. Comprehensive molecular characterization of muscle-invasive bladder cancer. Cell. 2018;174(4):540–56.
31. Rizzo A, et al. Is there a role for immunotherapy in prostate cancer? Cells. 2020;9(9):2051.
32. Marabelle A, et al. Efficacy of pembrolizumab in patients with noncolorectal high microsatellite instability/mismatch repair-deficient cancer: results from the phase II KEYNOTE-158 study. J Clin Oncol. 2020;38(1):1–10.
33. Prasad V, Kaestner V, Mailankody S. Cancer drugs approved based on biomarkers and not tumor type—FDA approval of pembrolizumab for mismatch repair-deficient solid cancers. JAMA Oncol. 2018;4(2):157–8.
34. Claps M, et al. Immune-checkpoint inhibitors and metastatic prostate cancer therapy: learning by making mistakes. Cancer Treat Rev. 2020;88:102057.
35. Smith MR, et al. Cabozantinib in chemotherapy-pretreated metastatic castration-resistant prostate cancer: results of a phase II nonrandomized expansion study. J Clin Oncol. 2014;32(30):3391–9.
36. Heidegger I, et al. A systematic review of the emerging role of immune checkpoint inhibitors in metastatic castration-resistant prostate cancer: will combination strategies improve efficacy? Eur Urol Oncol. 2021;4(5):745–54.
37. Harzstark AL, Small EJ. Sipuleucel-T for the treatment of prostate cancer. Drugs Today (Barc). 2008;44(4):271–8.
38. Kantoff PW, et al. Sipuleucel-T immunotherapy for castration-resistant prostate cancer. N Engl J Med. 2010;363(5):411–22.
39. Graff JN, Chamberlain ED. Sipuleucel-T in the treatment of prostate cancer: an evidence-based review of its place in therapy. Core Evid. 2015;10:1–10.
40. Hanna N, Einhorn LH. Testicular cancer: a reflection on 50 years of discovery. J Clin Oncol. 2014;32(28):3085–92.
41. Semaan A, et al. Immunotherapy: last bullet in platinum refractory germ cell testicular cancer. Future Oncol. 2019;15(5):533–41.
42. Zhao S, et al. Testicular defense systems: immune privilege and innate immunity. Cell Mol Immunol. 2014;11(5):428–37.
43. Cierna Z, et al. Prognostic value of programmed-death-1 receptor (PD-1) and its ligand 1 (PD-L1) in testicular germ cell tumors. Ann Oncol. 2016;27(2):300–5.
44. Kalavska K, et al. Immunotherapy in testicular germ cell tumors. Front Oncol. 2020;10:573977.

Part XVIII

Current Topics

Stanley P. Leong

In this section of Current Topics, each topic has been uniquely chosen to highlight the future directions of cancer metastasis. With promising biomarkers discovered in cancer, a newly emerging field of personalized cancer therapy is taking hold in treating cancer patients and is addressed by Drs. Kim and Kashani-Sabet. Cancer stem cells have been found to be associated with cancer recurrence and metastasis. Drs. Ghadially, Kim, and Charruyer detail the origin of stem cells and discuss their role in cancer metastasis. Certain cancers may be caused by viruses such as hepatitis viruses for hepatocellular carcinoma of the liver and human papillomavirus (HPV) causing anogenital and oropharyngeal cancer in ways distinctly different from other spontaneous-arising cancers in most of the cases. Drs. Dickey, Binning, and Rathwell describe the epidemiology and mechanisms of cancer induction by HPV, thus, illustrating the unique virus-induced causes of cancer. Drs. White, Wargo, and McQuade summarize the evidence regarding the significant association between microbiome and cancer progression. Artificial intelligence (AI), a recent development in computer science consisting of a number of supervised and unsupervised computational techniques, is addressed by Dr. Benjamin Franc to illustrate AI's potential application in cancer research. These important novel topics may become important frontiers in the study of cancer metastasis.

Personalized Systemic Cancer Therapy

70

Kevin B. Kim and Mohammed Kashani-Sabet

Abstract

Over the past several decades, the landscape of systemic cancer therapy has changed drastically owing to advances in our understanding of cancer biology, immunology, and genetics. As a result, an increasing number of patients with advanced stages of cancer now live longer than ever. The ultimate goal of personalized cancer therapy is to optimize clinical outcomes by selecting the "right" drug therapies for each patient on the basis of his or her tumor genetics and/or epigenetics and by avoiding unnecessary toxicity of ineffective therapy. Advances in genetic sequencing and molecular cytogenetic techniques have identified oncogenic driver mutations and epigenetic abnormalities, which in turn have enabled the development of effective molecular targeted drugs. In this chapter, the successful development of molecular targeted therapies in malignant melanoma is described to exemplify the concept of personalized cancer therapy.

Learning Objectives
- To understand the principles of molecular targeted cancer therapy.
- To understand mechanisms of resistance to targeted therapy.

70.1 Introduction

Cancer therapy has evolved drastically over the past 2 decades as our knowledge of cancer genetics and immunology has advanced. Although cytotoxic chemotherapy is still a mainstay of treatment for a majority of cancer types in the metastatic setting, molecular targeted therapies and checkpoint inhibitors have been revolutionizing cancer therapy, with improved tumor control and longer survival in an increasing number of advanced malignancies. Targeted drug therapy utilizes drugs that inhibit molecules and proteins prominently involved in growth and survival of tumor cells and thus lead to cell death and clinical regression of tumor and prolongation of patients' survival. Early examples of molecular targeted drugs are trastuzumab for advanced breast cancer and imatinib for chronic myelogenous leukemia (CML), which obtained Food and Drug Administration (FDA) approval in 1998 and 2002, respectively, in the United States. For targeted therapy to be effective, it is essential that tumor cells contain relevant oncogenic molecules. For example, HER2 is a receptor tyrosine kinase, and HER2 gene amplification, which occurs in ~ 15–20% of breast cancers, results in overexpression of a functional HER2 receptor [1]. The activated HER2 signaling pathway, in turn, leads to enhanced cell proliferation and invasion and suppression of apoptosis [2]. Accordingly, trastuzumab, a monoclonal antibody that binds to the extracellular domain of the HER2 receptor, is effective preferentially in cancer cells with HER2 overexpression and/or amplification and is not beneficial to patients without HER2-positive breast cancer. In addition, a hallmark of chronic myelogenous leukemia (CML) is the translocation involving the BCR and ABL genes, creating a constitutively active tyrosine kinase that stimulates proliferative and survival signaling pathways [3]. Imatinib, which is an inhibitor of BCR-ABL, as well as KIT and PDGF receptor-alpha, is very effective in treating this disease [4]. Similarly, vascular endothelial growth factor- (VEGF) or VEGF receptor-targeting drugs are indicated in advanced renal cell carcinoma (RCC) because of prevalent deficient von Hippel-Lindau (VHL) gene functions, resulting in activation of the VEGFR pathway in a vast majority of RCCs [5]. Recently, non-small cell lung cancer, especially non-squamous cell carcinoma of the lung, has received well-deserved attention in regard to the emergence of successful molecularly targeting therapies.

K. B. Kim (✉) · M. Kashani-Sabet
California Pacific Medical Center Research Institute, San Francisco, CA, USA
e-mail: kimkb@sutterhealth.org

S. P. Leong et al. (eds.), *Cancer Metastasis Through the Lymphovascular System*,
https://doi.org/10.1007/978-3-030-93084-4_70

KRAS mutations are the most common driver genetic alterations, accounting for 20–30% among patients with non-squamous cell lung cancers [6]. *EGFR* mutations (12–15%), *ALK* gene rearrangement (2–8%), *BRAF* mutation (1–5%), *MET* alterations (4–5%), *ROS1* gene alterations (1–2%), and *RET* gene arrangements (1–2%) also comprise a substantial proportion of driver oncogenic alterations in this population [6]. Interestingly, the prevalence of these genetic alterations vary depending on patients' ethnicity and demographic or pathological characteristics. For instance, *EGFR* mutations are more frequent among never-smokers, East Asian women, and those with adenocarcinoma histology. Fortunately, there are an increasing number of molecularly targeted drugs for these alterations, which have shown a significant clinical benefit. Examples of successful targeted therapies for non-small cell lung cancers include erlotinib, gefitinib, afatinib, dacomitinib, and osimertinib for *EGFR* mutations; crizotinib, ceritinib, alectinib, brigatinib, lorlatinib, and entrectinib for *ALK* gene rearrangements; dabrafenib/trametinib for *BRAF* mutations; crizotinib, capmatinib, and tepotinib for *MET* alterations (including exon 14 skipping mutations); and crizotinib, ceritinib, lorlatinib, entrectinib, repotrectinib, and taletrectinib for *ROS1* gene rearrangements [6]. Long considered non-targetable and non-actionable *KRAS* mutations are now considered a likely target for novel tyrosine kinase inhibitors. Several small molecule inhibitors have shown promising activity in patients with a *G12C KRAS* mutation in early-phase trials and are currently in registration trials with a hope for FDA approval in the near future [7].

In this chapter, we will review advances in molecular targeted therapies in malignant melanoma in detail to exemplify the concept of personalized cancer therapy.

70.2 Melanoma

Skin cancer is the most common type of malignancy in the United States, and melanoma is the deadliest among the skin cancers. The incidence of melanoma has been increasing significantly and steadily over the past decades [8]. In 2020, approximately 100,350 patients are estimated to be diagnosed with melanoma, and 6859 patients are expected to die from the disease in the United States [9]. Fortunately, the estimated number of deaths has decreased since 2016, in which more than 10,000 patients were expected to die. This is owing in a large part to the development of novel therapies, such as checkpoint inhibitors and molecular targeted drugs. A number of large randomized clinical trials demonstrated that overall survival (OS) of patients with advanced, metastatic melanoma has improved with these novel drugs compared to conventional cytotoxic chemotherapy, and significantly higher proportion of patients are alive at 5 years in the metastatic setting. For example, the 5-year survival rate for patients with advanced, metastatic melanoma was approximately 10% in 2009 [10], whereas the 5-year survival rates with the checkpoint inhibitor and/or inhibitors of BRAF and MEK kinases range between 34 and 52% [11–13].

In terms of the genetics of melanoma, the most common mutations in cutaneous melanoma occur within the mitogen-activated protein (MAP) kinase signal transduction pathway that leads to activation of the extracellular signal-regulated kinase (ERK), leading to proliferation, survival, invasion, and metastasis of melanoma cells. Approximately 50–60% of cutaneous melanomas harbor a single-base substitution mutation in the *BRAF* gene, and about 20% harbor a mutation in the *NRAS* gene, which are largely mutually exclusive [14–16]. Interestingly, non-cutaneous melanomas have different genetic profiles. In acral lentiginous and mucosal melanoma, *BRAF* and *NRAS* mutations occur much less frequently; multiple reports showed that 3–8% and 13–21% of mucosal and acral lentiginous melanomas harbor *BRAF* mutations, respectively, and 5–15% and 15–30% of mucosal and acral melanomas have *NRAS* mutations [17–21]. Interestingly, oncogenic mutations in the *KIT* gene are more clinically relevant, as these mutations occur in 11–21% of cases [21]. In uveal melanoma, mutations occurring in *GNAQ* or *GNA11* gene are found in approximately 80% of the cases, and *BRAF* and *NRAS* gene mutations are rare [22, 23]. These genetic alterations in melanoma are kinase-activating mutations, with a potential for developing molecularly targeted therapeutic drugs. For *BRAF* mutations in patients with cutaneous melanoma and *KIT* mutations in those with acral or mucosal melanoma, effective targeted drugs are now available.

70.2.1 *BRAF*-Mutant Melanoma

70.2.1.1 BRAF Inhibitors

BRAF is a serine–threonine kinase protein substrate in the MAP kinase pathway, with a critical function in melanoma cells. Oncogenic mutations in *BRAF* occur in nearly 50% of melanomas, and a vast majority of the mutations occur at codon 600, among which *V600E* and *V600K* mutations account for 80 and 15%, respectively [16]. The *V600 BRAF* mutations result in constitutive activation of the MEK and ERK protein in the MAP kinase pathway, which in turn leads to proliferation, invasion, and survival of melanoma cells (Fig. 70.1a). Accordingly, inhibition of *V600 BRAF* kinase blocks the activation of MEK and ERK proteins and suppresses the MAP kinase pathway, causing cell growth arrest [16] (Fig. 70.1b). An initial attempt to inhibit the BRAF kinase in melanoma included a trial of sorafenib, an inhibitor of RAF kinases. However, in a phase II study of sorafenib, there was no clinical response among 37 patients

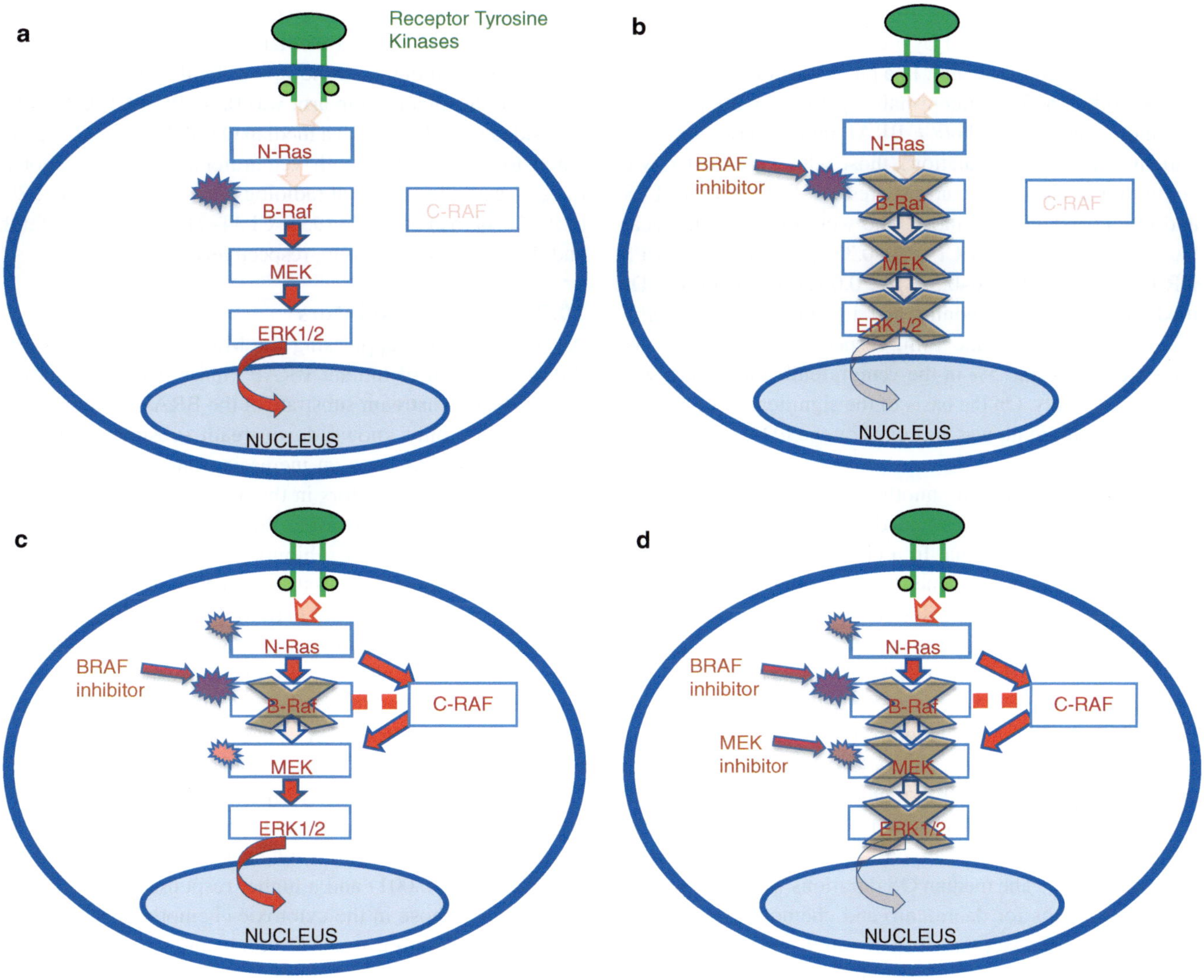

Fig. 70.1 **a** *V600 BRAF* mutation constitutively activates the MAP kinase pathway, leading to cell proliferation and survival. **b** Selective BRAF inhibitors suppress the MAP kinase pathway, leading to cell growth arrest and apoptosis. **c** Over time, cells acquire resistance to BRAF inhibitors, through multiple mechanisms, many of which lead to reactivation of the MAP kinase pathway. **d** Addition of MEK inhibitors suppresses the MAP kinase pathway more completely, resulting in a delay of resistance to BRAF inhibitor therapy

with metastatic melanoma [24]. Likewise, a combination of sorafenib with carboplatin and paclitaxel in a large phase III randomized study failed to show improvement in progression-free survival (PFS) and OS compared to chemotherapy alone [25]. These results indicated that sorafenib cannot sufficiently inhibit the activation of the mutated *BRAF*-driven MAP kinase pathway because it is more potent in inhibiting wild-type BRAF and CRAF protein than *V600 BRAF*, causing significant toxicity before it inhibits the MAPK pathway in patients with *V600 BRAF*-mutant melanoma.

With the failure of sorafenib as a treatment strategy for melanoma, a search for more specific inhibitors of the mutant BRAF kinase wad intensified. Vemurafenib was the first selective inhibitor of *V600 BRAF*-mutant melanoma to be investigated clinically. Vemurafenib was designed to bind and inhibit *V600E BRAF* more specifically compared to wild-type BRAF, using a scaffold-based crystallographic drug design approach, and as a consequence, it has greater potency for *V600 BRAF* than for wild-type BRAF (IC_{50} 31 nM/L versus 100 nM/L) [26, 27]. However, despite promising preclinical efficacy, the initial formulation of vemurafenib (crystalline formulation) was not effective in eliciting clinical responses because it was not able to achieve the adequate serum concentration necessary for tumor cell killing. Only after the drug was reformulated as a micro-precipitated bulk powder with a higher bioavailability, clinical activity was observed. In a phase I study, 32 patients with advanced melanoma harboring a *V600E BRAF* mutation were treated with 960 mg of vemurafenib orally twice a day, and 2 and 24 patients had a complete and partial response, respectively, with an overall response rate of 81%. In many patients, metabolic responses assessed by PET/CT scan were observed within 2 weeks of treatment. In a phase III randomized study that enrolled 675 treatment-naive patients with advanced melanoma harboring a *V600E* BRAF mutation (including 19 patients with *V600K* mutation), those who were treated with vemurafenib had significant progression-free survival (PFS) and OS advantage over those who were treated with dacarbazine (HR 0.26 [95% CI 0.20–0.33, $p < 0.001$] for PFS; HR 0.37 [95% CI 0.26–0.55, $p < 0.001$) for OS) [28]. The median PFS was 5.3 months and 1.6 months in the vemurafenib and dacarbazine arm, respectively. The response rates were 48% and 5% in the vemurafenib and dacarbazine arm, respectively. On the basis of the significant PFS and OS benefit, vemurafenib was approved by the FDA in the United States in 2011.

Likewise, dabrafenib, another selective oral inhibitor of *V600 BRAF*, was investigated in patients with unresectable metastatic melanoma. In a phase III randomized study comparing the clinical benefit of dabrafenib with that of dacarbazine that enrolled 250 treatment-naïve patients, those who received dabrafenib 150 mg twice daily treatment had a superior PFS (HR 0.30 [95% CI 0.18–0.51; $p < 0.0001$]), the primary endpoint of the study, with median PFS durations of 5.1 months and 2.7 months in the dabrafenib arm and the dacarbazine arm, respectively [29]. The response rate of dabrafenib was higher (50%) than that of dacarbazine (6%). At a median follow-up duration of 16.9 months, OS was also longer with dabrafenib treatment, with HR of 0.77 (95% CI 0.52–1.13) [30]. The median OS durations were 20.0 months and 15.6 months for dabrafenib and chemotherapy, respectively, despite the fact that 59% of patients treated with chemotherapy crossed over to dabrafenib treatment. In 2013, the FDA approved dabrafenib for treatment of advanced *V600E/K BRAF*-mutant melanoma.

Encorafenib is another selective oral BRAF inhibitor with a significantly longer dissociation half-life from *V600E*-mutant *BRAF* than the other two BRAF inhibitors (30 h for encorafenib versus 2 h for dabrafenib and 0.5 h for vemurafenib). This is the only BRAF inhibitor that has been compared directly with another BRAF inhibitor, vemurafenib, in a clinical trial. In the COLUMBUS trial, which randomized 577 patients with *V600E/K BRAF*-mutant melanoma at 1:1:1 ratio to receive vemurafenib, encorafenib, or a combination of encorafenib and binimetinib (a selective MEK) inhibitor, the clinical benefit of encorafenib was compared to that of vemurafenib, separately from the comparison with the combination treatment. The updated analysis, with a median follow-up of 36.8 months, revealed statistically significantly longer PFS and OS in the encorafenib arm (dose of 300 mg once daily) over the vemurafenib arm [31]. The HR for PFS in the encorafenib arm was 0.68 (95% CI 0.52–0.88; two-sided $p = 0.0038$) with a median PFS of 9.6 months, compared with 7.3 months for the vemurafenib arm [32]. The HR for OS in the encorafenib arm was 0.76 (95% CI 0.58–0.98; two-sided $p = 0.033$) with a median OS of 23.5 months, compared to 16.9 months with vemurafenib. Overall response rates assessed by a central radiology review panel were 52% (95% CI 44–59) and 41% (95% CI 34–41) in the encorafenib and the vemurafenib group, respectively.

70.2.1.2 MEK Inhibitors

An alternative to suppressing the MAP kinase pathway by direct inhibition of mutant BRAF kinase is through inhibition of the downstream substrate of the BRAF protein. MEK kinase is the only known downstream substrate of BRAF, and thus there has been an increasing interest in the investigation of MEK inhibitors in the treatment of *V600 BRAF*-mutant melanoma. However, the clinical development of earlier versions of MEK inhibitors, including, CI-1040 and PD0325901, was not successful because of intolerable toxicity and/or a lack of clinical activity [33] until trametinib, a specific inhibitor of MEK1 and MEK2 proteins, was shown to have promising clinical activity in phase I and II studies [34, 35]. In a randomized phase III study, 322 patients with metastatic *V600E* or *V600K BRAF*-mutant melanoma were assigned to receive either trametinib treatment or cytotoxic chemotherapy [36], with the primary and the secondary endpoint of PFS and OS, respectively. Patients in the trametinib arm had a longer PFS (HR 0.45 [95% CI 0.33–0.63] $p < 0.001$) and a higher response rate (22% versus 8%) than those in the cytotoxic chemotherapy arm. The median PFS duration was 4.8 months in the trametinib arm and 1.5 months in the chemotherapy arm. Trametinib was also associated with superior OS over chemotherapy, with a HR of 0.54 (95% CI 0.32–0.92; $p = 0.01$). The OS rates at 1 and 2 years were 60.9% and 32.0%, respectively, in the trametinib arm, whereas they were 49.6% and 29.4% in the chemotherapy arm [37]. On the basis of the superior clinical benefit over the then-standard chemotherapy, trametinib was approved by the FDA for treatment of *V600E or K BRAF*-mutant advanced melanoma in 2013. It is important to note that trametinib has low clinical activity in patients with *V600 BRAF*-mutant melanoma whose disease had progressed on a previous BRAF inhibitor treatment. In a phase II study of trametinib, none of 40 patients who were previously treated with a selective BRAF inhibitor responded to trametinib treatment [35]. These data strongly suggest that for *V600 BRAF*-mutant melanoma, selective BRAF inhibitors are favored over trametinib as initial targeted therapy despite a lack of head-to-head comparison of these two classes of drugs.

70.2.1.3 Combination of BRAF Inhibitor and MEK Inhibitor

Although the selective BRAF inhibitors significantly improve clinical responses and overall survival, most patients who are treated with BRAF inhibitors ultimately have disease progression. There are multiple mechanisms by which melanoma cells escape BRAF inhibitor-induced cytotoxicity over time. These include development of new kinase-activating mutations in the *NRAS* gene, alternate splicing of *BRAF*, amplification of *BRAF* and/or *CRAF*, acquisition of new *MEK* mutations, or overexpression of *COT*. Importantly, all of these molecular alterations can bypass the inhibition of V600 BRAF kinase by BRAF inhibitors [38–42] (Fig. 70.1c). Although the resistance to BRAF inhibitors can be induced by the activation of receptor tyrosine kinases, such as the MET receptor, and *NRAS* mutation, which also activates the alternate survival pathway, the PI3K/AKT/mTOR pathway [43], approximately 70% of the resistance mechanisms involve the reactivation of the MAP kinase pathway despite the continued use of BRAF inhibitors. To date, no new mutations in the *BRAF* gene have been reported in the setting of treatment resistance. These observations were corroborated by preclinical studies which showed that addition of a MEK inhibitor to a BRAF inhibitor delayed resistance to treatment in *V600E BRAF*-mutant melanoma cells [38, 40–42] (Fig. 70.1d). To date, three different combination regimens of MEK inhibitor and BRAF inhibitor have been investigated clinically in patients with *V600 BRAF*-mutant advanced melanoma: dabrafenib (BRAF inhibitor) and trametinib (MEK inhibitor), vemurafenib (BRAF inhibitor) and cobimetinib (MEK inhibitor), and encorafenib (BRAF inhibitor) and binimetinib (MEK inhibitor). In separate large randomized phase III trials, these three regimens were compared to vemurafenib alone with primary endpoints of OS [44–46]. Each of these combination regimens was associated with higher response rates, as well as prolonged PFS and OS, than vemurafenib alone (Table 70.1). The HRs for PFS were 0.51–0.61 with the combination regimens, with median PFS

Table 70.1 Pivotal phase III studies of targeted therapies for advanced *BRAF*-mutant melanoma

Study	Treatment	Primary endpoint	No. pts	OS	PFS	ORR
BRIM3 study [28]	Vemurafenib vs dacarbazine	OS and PFS	675	HR 0.70 [95% CI 0.57–0.87]; $p = 0.0008$ Median 13.6 mo (Vem) vs 9.7 mo (dacarbazine)	HR 0.38 [95% CI 0.32–0.46]; $p < 0.0001$ Median 6.9 mo (Vem) vs 1.6 mo (dacarbazine)	57% (Vem) vs 9% (dacarbazine)
BREAK3 study [29, 30]	Dabrafenib vs dacarbazine	PFS	250	HR 0.82 [95% CI 0.57–1.18]; *p*, n/a Median 20.0 mo (Dab) vs 15.6 mo (dacarbazine)	HR 0.30 [95% CI 0.18–0.51]; $p < 0.0001$ Median 5.1 mo (Dab) vs 2.7 mo (dacarbazine)	50% (Dab) vs 6% (dacarbazine)
COMBI-D study [48]	Dabrafenib + trametinib vs dabrafenib	PFS	423	HR 0.71 [95% CI 0.55–0.92]; $p = 0.0107$ Median 25.1 mo (Dab + Tram) vs 18.7 mo (Dab)	HR 0.67 [95% CI 0.53–0.84]; $p = 0.0004$ Median 11 mo (Dab + Tram) vs 8.8 mo (Dab)	69% (Dab + Tram) vs 53% (Dab)
COMBI-V study [44]	Dabrafenib + trametinib vs vemurafenib	OS	704	HR 0.69 [95% CI 0.53–0.89]; $p = 0.005$ Median 26 mo (Dab + Tram) vs 16.8 mo (Vem)	HR 0.56 [95% CI 0.46–0.69]; $p < 0.001$ Median 11.4 mo (Dab + Tram) vs 7.4 mo (Vem)	64% (Dab + Tram) vs 51% (Vem)
Co-BRIM study [45]	Vemurafenib + cobimetinib vs vemurafenib	PFS	495	HR 0.70 [95% CI 0.55–0.90]; $p = 0.005$ Median 22.3 mo (Vem + Cobi) vs 17.4 mo (Vem)	HR 0.58 [95% CI 0.46–0.72]; $p < 0.0001$ Median 12.3 mo (Vem + Cobi) vs 7.2 mo (Vem)	69.6% (Vem + Cobi) vs 50.0% (Vem)
COLUMBUS study [31, 46]	Encorafenib + binimetinib vs encorafenib vs vemurafenib	PFS	577	For Enco + Bini vs Vem: HR 0.61 [95% CI 0.47–0.79]; $p < 0.0001$ Median 33.6 mo (Enco + Bini) vs 16.9 mo (Vem)	For Enco + Bini vs Vem: HR 0.51 [95% CI 0.39–0.67]; $p < 0.0001$ Median 14.9 mo (Enco + Bini) vs 7.3 mo (Vem)	64% (Enco + Bini) vs 41% (Enco) vs 40% (Vem)

No. Pts number of patients, *OS* overall survival, *PFS* progression-free survival, *ORR* objective response rate, *mo* months, *HR* hazard ratio, *CI* confidence interval, *p* p-value, *Vem* vemurafenib, *Dab* dabrafenib, *Tram* trametinib, *Cobi* cobimetinib, *Enco* encorafenib, *Bini* binimetinib, *n/a* not available durations of 11–14.9 months, compared to 7.2–7.3 months with vemurafenib treatment. The HRs for OS were 0.61–0.70 with the combination regimens, with median OS durations of 22.3–33.6 months, compared to 16.9–18.0 months with vemurafenib treatment [31]. Although the combination of encorafenib and binimetinib had numerically the longest PFS and OS durations among the three combination regimens, it is difficult to compare the overall clinical benefit of the three regimens as they have not been compared head to head within the same trial.

In general, the combination regimens are well tolerated. Grades 3 or 4 adverse events occurred in 41–62% of patients, and most of these events were reversible upon treatment discontinuation [44–46]. There were no appreciable differences in frequency of severe adverse events between the combination regimens and single-agent BRAF inhibitor. However, each of the combination regimens has somewhat different safety profiles. For example, pyrexia and chills were more the common adverse events for the combination of dabrafenib and trametinib, whereas diarrhea, nausea, and skin rash were the most common for the vemurafenib and cobimetinib regimen [45]. For the encorafenib and binimetinib regimen, diarrhea, nausea/vomiting, and fatigue were the most common toxicities [46].

Interestingly, certain adverse events occur less frequently with the combination approach than with the BRAF inhibitors alone. Development of keratoacanthomas and/or cutaneous squamous cell carcinomas occurred in 9–26% of patients during treatment with single-agent vemurafenib or dabrafenib [47, 48], whereas they occurred only in 1–3% of patients when they were treated with a BRAF inhibitor in combination with a MEK inhibitor [31, 44, 45, 48]. This observation is consistent with the hypothesis that BRAF inhibitors that are more specific for *V600 BRAF* kinase lead to a paradoxical activation of the MAP kinase signal transduction pathway through a dimerization with other RAF kinases (such as CRAF) in keratinocytes that do not harbor mutant BRAF [49]. Accordingly, when a MEK inhibitor is added to a BRAF inhibitor, the MAP kinase pathway is inhibited at the level downstream of the RAF kinases, and this inhibition, in turn, suppresses the development of keratoacanthoma or cutaneous squamous cell carcinoma. Fortunately, occurrence of other serious secondary malignancies is rare with either single-agent BRAF treatment or the combination regimens.

On the basis of the significant PFS and OS benefits, each of the three BRAF/MEK inhibitor combination regimens was approved by the FDA for treatment of advanced *V600 BRAF*-mutant melanoma: dabrafenib and trametinib in 2013, vemurafenib and cobimetinib in 2015, and encorafenib and binimetinib in 2018.

In addition to the significant clinical benefit including OS benefits in patients with metastatic disease, a co-inhibition of BRAF and MEK kinases improved relapse-free survival in the adjuvant setting in patients with surgically resected *V600E/K BRAF*-mutant stage III melanoma [50]. In a large phase III study with 870 patients who had resected stage III melanoma, the combination of dabrafenib and trametinib was compared to matching placebo treatments; at 5 years, 52% (95% CI 48–58) and 36% (95% CI 32–41) of patients were alive without melanoma relapse in the BRAF/MEK inhibitor treatment group and placebo group, respectively. As of April 2018, FDA granted approval to dabrafenib and trametinib in combination for adjuvant treatment of this patient population.

70.2.2 *KIT*-Mutant Melanoma

Although BRAF mutations are the most frequent genetic abnormalities found in cutaneous melanomas, different genetic mutations are more relevant in other melanoma subtypes. In the case of acral lentiginous or mucosal melanomas, mutations in the KIT receptor kinase are more common than *BRAF* mutations: *KIT* mutations are found in 15–21% and 11% of mucosal and acral lentiginous melanoma, respectively [21, 51]. In comparison, *V600 BRAF* mutation occurs only in approximately 3% of mucosal melanomas, with a higher percentage (13%) of non-*V600 BRAF* mutations [51]. In the early 2000s, there was a great interest in investigating KIT inhibitors in patients with advanced melanoma because of a high frequency of KIT overexpression in melanoma [52]. There were three phase II studies of imatinib, a potent oral KIT, BCR-ABL, and PDGFR inhibitor, conducted in the United States and in Europe, enrolling patients with unresectable, metastatic melanoma. Among 62 patients treated with imatinib in the three studies, there was only one patient with a clinical response, concluding that imatinib was deemed ineffective in melanoma [53–55]. However, these studies enrolled patients with melanoma regardless of *KIT* mutation status, and it is likely that the vast majority of patients in these studies did not have a *KIT*-mutant melanoma.

Since the reporting of the frequency of *KIT* mutations in acral and mucosal melanoma [21], enthusiasm for studying imatinib in patients with *KIT*-mutant advanced melanoma was revived. In three initial single-arm phase II trials of a potent inhibitor, imatinib, response rates were 16–29%, with a median PFS duration of 3–3.7 months among 118 patients with advanced melanoma harboring a KIT mutation [56–58]. Among the mutations occurring in the *KIT* kinase, clinical responses to imatinib treatment were observed mostly in patients with mutations occurring in exon 9 or 11. Because of the rather small numbers of patients in each of these single-arm trials, imatinib has not been filed for FDA review for treatment of *KIT*-mutant melanoma. However, imatinib has been widely used by many oncologists for patients with a *KIT*-mutant melanoma.

Nilotinib is a second-generation tyrosine kinase inhibitor of BCR-ABL, KIT, and PDGFR-alpha which was approved for treatment of imatinib-resistant CLL. In a phase II study of nilotinib, 42 patients with advanced melanoma harboring a *KIT* mutation were treated. While a response rate of 26.2% was observed, with a median PFS of 4.2 months, nilotinib did not appear to be more effective than imatinib [59].

70.2.3 Future Direction

The development of selective BRAF and MEK inhibitors epitomizes the success of molecular targeted therapy for *V600 BRAF*-mutant melanoma and, at the same time, poses new challenges. Despite the high response rates and prolonged tumor control and survival of patients treated with this combination approach, a majority of patients ultimately experience disease progression and succumb to the disease. Likewise, the KIT inhibitors render a meaningful clinical benefit in patients with *KIT*-mutant melanoma, but the degree of clinical benefit is suboptimal. A key to making further advances in targeted therapy for melanoma lies in a better understanding of the molecular biology of melanoma that would enable the development of more effective and less toxic drugs. Currently, many clinical trials incorporate tumor and/or blood collections for various biomarker analyses to discover novel mechanisms of drug resistance, and these efforts will likely lead to more prolonged disease control and longer survival in melanoma patients. In addition, identification of new actionable molecular targets through genetic and epigenetic analyses will result in development of novel classes of effective targeted therapy in patients whose diseases do not harbor a *BRAF* or *KIT* mutation or become resistant to the currently available targeted therapy regimens. Furthermore, it will be essential to understand the effect of the molecular targeted drugs on activation of immune cells and to develop optimal sequences of targeted therapy and checkpoint immunotherapy in order to further improve outcomes associated with melanoma therapy.

Conclusion and Perspective

- This chapter has covered the general principle of personalized cancer therapy, including the identification of *BRAF* and *KIT* mutations in melanoma and the development of molecular targeted drugs for melanomas harboring these mutations. In addition, the development of various mechanisms of drug resistance and strategies to overcome acquired drug resistance were discussed. The reader should understand the concept of molecular targeted drugs and how these drugs can improve the clinical outcome of patients with advanced cancer.

References

1. King CR, Kraus MH, Aaronson SA. Amplification of a novel v-erbB-related gene in a human mammary carcinoma. Science. 1985;229(4717):974–6.
2. Karunagaran D, Tzahar E, Beerli RR, Chen X, Graus-Porta D, Ratzkin BJ, et al. ErbB-2 is a common auxiliary subunit of NDF and EGF receptors: implications for breast cancer. EMBO J. 1996;15(2):254–64.
3. Sawyers CL. Signal transduction pathways involved in BCR-ABL transformation. Baillieres Clin Haematol. 1997;10(2):223–31.
4. O'Brien SG, Guilhot F, Larson RA, Gathmann I, Baccarani M, Cervantes F, et al. Imatinib compared with interferon and low-dose cytarabine for newly diagnosed chronic-phase chronic myeloid leukemia. N Engl J Med. 2003;348(11):994–1004.
5. Nickerson ML, Jaeger E, Shi Y, Durocher JA, Mahurkar S, Zaridze D, et al. Improved identification of von Hippel-Lindau gene alterations in clear cell renal tumors. Clin Cancer Res. 2008;14(15): 4726–34.
6. Konig D, Savic Prince S, Rothschild SI. Targeted therapy in advanced and metastatic non-small cell lung cancer. An update on treatment of the most important actionable oncogenic driver alterations. Cancers (Basel). 2021;13(4):804.
7. Uprety D, Adjei AA. KRAS: From undruggable to a druggable cancer target. Cancer Treat Rev. 2020;89:102070.
8. Guy GP Jr, Thomas CC, Thompson T, Watson M, Massetti GM, Richardson LC, et al. Vital signs: melanoma incidence and mortality trends and projections—United States, 1982–2030. MMWR Morb Mortal Wkly Rep. 2015;64(21):591–6.
9. Siegel RL, Miller KD, Jemal A. Cancer statistics, 2020. CA Cancer J Clin. 2020;70(1):7–30.
10. Balch CM, Gershenwald JE, Soong SJ, Thompson JF, Atkins MB, Byrd DR, et al. Final version of 2009 AJCC melanoma staging and classification. J Clin Oncol. 2009;27(36):6199–206.
11. Robert C, Ribas A, Schachter J, Arance A, Grob JJ, Mortier L, et al. Pembrolizumab versus ipilimumab in advanced melanoma (KEYNOTE-006): post-hoc 5-year results from an open-label, multicentre, randomised, controlled, phase 3 study. Lancet Oncol. 2019;20(9):1239–51.
12. Larkin J, Chiarion-Sileni V, Gonzalez R, Grob JJ, Rutkowski P, Lao CD, et al. Five-year survival with combined nivolumab and ipilimumab in advanced melanoma. N Engl J Med. 2019;381(16):1535–46.
13. Robert C, Grob JJ, Stroyakovskiy D, Karaszewska B, Hauschild A, Levchenko E, et al. Five-year outcomes with dabrafenib plus trametinib in metastatic melanoma. N Engl J Med. 2019;381(7):626–36.
14. Sensi M, Nicolini G, Petti C, Bersani I, Lozupone F, Molla A, et al. Mutually exclusive NRASQ61R and BRAFV600E mutations at the single-cell level in the same human melanoma. Oncogene. 2006;25(24):3357–64.
15. Reddy BY, Miller DM, Tsao H. Somatic driver mutations in melanoma. Cancer. 2017;123(S11):2104–17.
16. Davies H, Bignell GR, Cox C, Stephens P, Edkins S, Clegg S, et al. Mutations of the BRAF gene in human cancer. Nature. 2002;417(6892):949–54.
17. Tacastacas JD, Bray J, Cohen YK, Arbesman J, Kim J, Koon HB, et al. Update on primary mucosal melanoma. J Am Acad Dermatol. 2014;71(2):366–75.
18. Cosgarea I, Ugurel S, Sucker A, Livingstone E, Zimmer L, Ziemer M, et al. Targeted next generation sequencing of mucosal melanomas identifies frequent NF1 and RAS mutations. Oncotarget. 2017;8(25):40683–92.
19. Dika E, Veronesi G, Altimari A, Riefolo M, Ravaioli GM, Piraccini BM, et al. BRAF, KIT, and NRAS mutations of acral melanoma in white patients. Am J Clin Pathol. 2020;153(5):664–71.

20. Zebary A, Omholt K, Vassilaki I, Hoiom V, Linden D, Viberg L, et al. KIT, NRAS, BRAF and PTEN mutations in a sample of Swedish patients with acral lentiginous melanoma. J Dermatol Sci. 2013;72(3):284–9.
21. Curtin JA, Busam K, Pinkel D, Bastian BC. Somatic activation of KIT in distinct subtypes of melanoma. J Clin Oncol. 2006;24(26):4340–6.
22. Van Raamsdonk CD, Bezrookove V, Green G, Bauer J, Gaugler L, O'Brien JM, et al. Frequent somatic mutations of GNAQ in uveal melanoma and blue naevi. Nature. 2009;457(7229):599–602.
23. Van Raamsdonk CD, Griewank KG, Crosby MB, Garrido MC, Vemula S, Wiesner T, et al. Mutations in GNA11 in uveal melanoma. N Engl J Med. 2010;363(23):2191–9.
24. Eisen T, Ahmad T, Flaherty KT, Gore M, Kaye S, Marais R, et al. Sorafenib in advanced melanoma: a phase II randomised discontinuation trial analysis. Br J Cancer. 2006;95(5):581–6.
25. Flaherty KT, Lee SJ, Zhao F, Schuchter LM, Flaherty L, Kefford R, et al. Phase III trial of carboplatin and paclitaxel with or without sorafenib in metastatic melanoma. J Clin Oncol. 2013;31(3):373–9.
26. Tsai J, Lee JT, Wang W, Zhang J, Cho H, Mamo S, et al. Discovery of a selective inhibitor of oncogenic B-Raf kinase with potent antimelanoma activity. Proc Natl Acad Sci USA. 2008;105(8):3041–6.
27. Bollag G, Hirth P, Tsai J, Zhang J, Ibrahim PN, Cho H, et al. Clinical efficacy of a RAF inhibitor needs broad target blockade in BRAF-mutant melanoma. Nature. 2010;467(7315):596–9.
28. Chapman PB, Hauschild A, Robert C, Haanen JB, Ascierto P, Larkin J, et al. Improved survival with vemurafenib in melanoma with BRAF V600E mutation. N Engl J Med. 2011;364(26):2507–16.
29. Hauschild A, Grob JJ, Demidov LV, Jouary T, Gutzmer R, Millward M, et al. Dabrafenib in BRAF-mutated metastatic melanoma: a multicentre, open-label, phase 3 randomised controlled trial. Lancet. 2012;380(9839):358–65.
30. Hauschild A, Grobb J, Demidov L, Jouary T, Gutzmer R, Millward M, et al. An update on overall survival (OS) and follow-on therapies in BREAK-3, a phase III, randomized trial: dabrafenib (D) vs. dacarbazine (DTIC) in patients (pts) with BRAF V600E mutation-positive metastatic melanoma (MM). Ann Oncol. 2014;25(Supplement 4):4374–493.
31. Dummer R, Ascierto PA, Gogas HJ, Arance A, Mandala M, Liszkay G, et al. Overall survival in patients with BRAF-mutant melanoma receiving encorafenib plus binimetinib versus vemurafenib or encorafenib (COLUMBUS): a multicentre, open-label, randomised, phase 3 trial. Lancet Oncol. 2018;19(10):1315–27.
32. Dummer R, Ascierto PA, Gogas H, Arance AM, Mandalà M, Liszkay G, et al. Overall survival in COLUMBUS: a phase 3 trial of encorafenib (ENCO) plus binimetinib (BINI) vs vemurafenib (VEM) or enco in BRAF-mutant melanoma. J Clin Oncol. 2018;36(15_suppl):9504.
33. Salama AK, Kim KB. MEK inhibition in the treatment of advanced melanoma. Curr Oncol Rep. 2013;15(5):473–82.
34. Falchook GS, Lewis KD, Infante JR, Gordon MS, Vogelzang NJ, DeMarini DJ, et al. Activity of the oral MEK inhibitor trametinib in patients with advanced melanoma: a phase 1 dose-escalation trial. Lancet Oncol. 2012;13(8):782–9.
35. Kim KB, Kefford R, Pavlick AC, Infante JR, Ribas A, Sosman JA, et al. Phase II study of the MEK1/MEK2 inhibitor trametinib in patients with metastatic BRAF-mutant cutaneous melanoma previously treated with or without a BRAF inhibitor. J Clin Oncol. 2013;31(4):482–9.
36. Flaherty KT, Robert C, Hersey P, Nathan P, Garbe C, Milhem M, et al. Improved survival with MEK inhibition in BRAF-mutated melanoma. N Engl J Med. 2012;367(2):107–14.
37. Robert C, Flaherty K, Nathan P, Hersey P, Garbe C, Milhem M, et al. Five-year outcomes from a phase 3 METRIC study in patients with BRAF V600 E/K-mutant advanced or metastatic melanoma. Eur J Cancer. 2019;109:61–9.
38. Nazarian R, Shi H, Wang Q, Kong X, Koya RC, Lee H, et al. Melanomas acquire resistance to B-RAF(V600E) inhibition by RTK or N-RAS upregulation. Nature. 2010;468(7326):973–7.
39. Emery CM, Vijayendran KG, Zipser MC, Sawyer AM, Niu L, Kim JJ, et al. MEK1 mutations confer resistance to MEK and B-RAF inhibition. Proc Natl Acad Sci USA. 2009;106(48):20411–6.
40. Johannessen CM, Boehm JS, Kim SY, Thomas SR, Wardwell L, Johnson LA, et al. COT drives resistance to RAF inhibition through MAP kinase pathway reactivation. Nature. 2010;468(7326):968–72.
41. Poulikakos PI, Persaud Y, Janakiraman M, Kong X, Ng C, Moriceau G, et al. RAF inhibitor resistance is mediated by dimerization of aberrantly spliced BRAF(V600E). Nature. 2011;480(7377):387–90.
42. Trunzer K, Pavlick AC, Schuchter L, Gonzalez R, McArthur GA, Hutson TE, et al. Pharmacodynamic effects and mechanisms of resistance to vemurafenib in patients with metastatic melanoma. J Clin Oncol. 2013;31(14):1767–74.
43. Straussman R, Morikawa T, Shee K, Barzily-Rokni M, Qian ZR, Du J, et al. Tumour micro-environment elicits innate resistance to RAF inhibitors through HGF secretion. Nature. 2012;487(7408):500–4.
44. Robert C, Karaszewska B, Schachter J, Rutkowski P, Mackiewicz A, Stroiakovski D, et al. Improved overall survival in melanoma with combined dabrafenib and trametinib. N Engl J Med. 2015;372(1):30–9.
45. Larkin J, Ascierto PA, Dreno B, Atkinson V, Liszkay G, Maio M, et al. Combined vemurafenib and cobimetinib in BRAF-mutated melanoma. N Engl J Med. 2014;371(20):1867–76.
46. Dummer R, Ascierto PA, Gogas HJ, Arance A, Mandala M, Liszkay G, et al. Encorafenib plus binimetinib versus vemurafenib or encorafenib in patients with BRAF-mutant melanoma (COLUMBUS): a multicentre, open-label, randomised phase 3 trial. Lancet Oncol. 2018;19(5):603–15.
47. Lacouture ME, Duvic M, Hauschild A, Prieto VG, Robert C, Schadendorf D, et al. Analysis of dermatologic events in vemurafenib-treated patients with melanoma. Oncologist. 2013;18(3):314–22.
48. Long GV, Stroyakovskiy D, Gogas H, Levchenko E, de Braud F, Larkin J, et al. Combined BRAF and MEK inhibition versus BRAF inhibition alone in melanoma. N Engl J Med. 2014;371(20):1877–88.
49. Su F, Viros A, Milagre C, Trunzer K, Bollag G, Spleiss O, et al. RAS mutations in cutaneous squamous-cell carcinomas in patients treated with BRAF inhibitors. N Engl J Med. 2012;366(3):207–15.
50. Dummer R, Hauschild A, Santinami M, Atkinson V, Mandala M, Kirkwood JM, et al. Five-year analysis of adjuvant dabrafenib plus trametinib in stage III melanoma. N Engl J Med. 2020;383(12):1139–48.
51. Garrido MC, Bastian BC. KIT as a therapeutic target in melanoma. J Investig Dermatol. 2010;130(1):20–7.
52. Shen SS, Zhang PS, Eton O, Prieto VG. Analysis of protein tyrosine kinase expression in melanocytic lesions by tissue array. J Cutan Pathol. 2003;30(9):539–47.
53. Ugurel S, Hildenbrand R, Zimpfer A, La Rosee P, Paschka P, Sucker A, et al. Lack of clinical efficacy of imatinib in metastatic melanoma. Br J Cancer. 2005;92(8):1398–405.
54. Wyman K, Atkins MB, Prieto V, Eton O, McDermott DF, Hubbard F, et al. Multicenter phase II trial of high-dose imatinib mesylate in metastatic melanoma: significant toxicity with no clinical efficacy. Cancer. 2006;106(9):2005–11.
55. Kim KB, Eton O, Davis DW, Frazier ML, McConkey DJ, Diwan AH, et al. Phase II trial of imatinib mesylate in patients with metastatic melanoma. Br J Cancer. 2008;99(5):734–40.

56. Carvajal RD, Antonescu CR, Wolchok JD, Chapman PB, Roman RA, Teitcher J, et al. KIT as a therapeutic target in metastatic melanoma. JAMA. 2011;305(22):2327–34.
57. Guo J, Si L, Kong Y, Flaherty KT, Xu X, Zhu Y, et al. Phase II, open-label, single-arm trial of imatinib mesylate in patients with metastatic melanoma harboring c-Kit mutation or amplification. J Clin Oncol. 2011;29(21):2904–9.
58. Hodi FS, Corless CL, Giobbie-Hurder A, Fletcher JA, Zhu M, Marino-Enriquez A, et al. Imatinib for melanomas harboring mutationally activated or amplified KIT arising on mucosal, acral, and chronically sun-damaged skin. J Clin Oncol. 2013;31(26):3182–90.
59. Guo J, Carvajal RD, Dummer R, Hauschild A, Daud A, Bastian BC, et al. Efficacy and safety of nilotinib in patients with KIT-mutated metastatic or inoperable melanoma: final results from the global, single-arm, phase II TEAM trial. Ann Oncol. 2017;28(6):1380–7.

Cancer Stem Cells and Their Role in Metastasis

71

Ruby Ghadially, Richard W. Kim, and Alexandra Charruyer-Reinwald

Abstract

Cancer stem cells have been a major focus of study in recent years. However, the more we learn about them, the more challenges we seem to face. Cancer stem cells, cells responsible for ongoing tumor propagation, have been isolated from many tumors. In many tumors the cancer stem cell or a subset of cancer stem cells is the cell responsible for metastasis. Findings make it clear that the cancer stem cell is a key therapeutic target, but given clonal evolution and phenotype instability of the cancer stem cell, the non-cancer stem cells must be targeted simultaneously. Targeting cancer stem cells can be difficult due to their rarity and our inability to propagate them in culture. However, targeting the pathways that maintain stemness or targeting microenvironmental controls of the stem cell state can circumvent this problem.

Abbreviations

SC Stem cell
CSC Cancer stem cell

R. Ghadially (✉) · A. Charruyer-Reinwald
Department of Dermatology, Eli and Edythe Broad Center of Regeneration Medicine and Stem Cell Research, University of California San Francisco, San Francisco, CA, USA
e-mail: ruby.ghadially@ucsf.edu

R. W. Kim
School of Medicine, University of California San Francisco, San Francisco, CA, USA

Learning Objectives
- Understand the concept of a cancer stem cell.
- Understand how markers can be used to identify cancer stem cells.
- Understand the signaling pathways used by cancer stem cells to produce metastases.
- Understand the role of cancer stem cells in cancer progression and metastasis.

71.1 Introduction

Not all cells in a cancer are equally proliferative and invasive. In fact, most are not capable of producing a further tumor if separated from the parent tumor. Those that are capable of initiating further tumors in xenograft models have been termed tumor-initiating cells or cancer stem cells (CSCs). These cells usually constitute a small proportion of the total cancer cell population.

It was initially believed that CSCs arise only from normal stem cells (SCs). However, it was soon shown that more differentiated progenitor cells could acquire mutations that resulted in the abilities to proliferate indefinitely and to generate tumors when transplanted into a new host, qualifying them as CSCs. The CSC state can be regulated by intrinsic and extrinsic factors. Intrinsic regulators include genetic and epigenetic factors. Extrinsic factors include the niche and the immune system [42].

71.2 Markers of Cancer Stem Cells

Different types of markers are used to identify and/or isolate CSCs and include cell surface molecules, functional markers, transcription factors, and signaling pathway molecules.

S. P. Leong et al. (eds.), *Cancer Metastasis Through the Lymphovascular System*,
https://doi.org/10.1007/978-3-030-93084-4_71

Functional markers include ALDH1, ABCG2, and side population cells. ALDH1 catalyzes the oxidation of retinal (retinaldehyde), the retinol metabolite, to retinoic acid (RA), and ALDH1 plays a vital role as a marker of SCs and CSCs [75]. CSCs from melanoma [8, 50, 83], breast cancer [21], pancreatic adenocarcinoma [63], prostate cancer [31], colorectal cancer [33], lung cancer [38, 76], and endometrioid adenocarcinoma [62] express high levels of ALDH1.

ABCG2 is expressed in both normal SCs and CSCs and is a membrane transporter. ABCG2 is a CSC marker in breast cancer [70, 74], hepatocellular carcinoma [85], and ovarian cancer [16]. ABCG2 has been shown to reduce the intracellular concentrations of multiple anticancer chemotherapeutic agents in vitro [3].

Side population cells are cells that are able to exclude Hoechst 33342 dye reflecting their high drug efflux capacity. Side population cells constituted CSCs in neuroblastomas, breast cancer, lung cancer, and glioblastoma cell lines [28]. Side population cells also expressed high levels of ABCG2.

Commonly used cell surface markers include CD44, CD133, CD24, CD34, CD90, and CD38 [42]. Studies indicate a role for CD133 in determining cellular fate and maintaining the SC-like state [6, 7, 51]. CD133 identifies CSC populations in multiple solid tumor types [22]. CD133 was found to be a CSC marker in colon cancer [66], brain cancers [71, 72], hepatocellular carcinoma [73, 81], lung cancer [17], and pancreatic cancer cell lines [59].

CD44 appears to have a significant regulatory role in most cancer types, although the signaling pathways in tumor cells remain unresolved. CD24, being an epithelial marker, may not be expressed in all cancers but seems to have a significant role in those tumors expressing CD24 [36]. Breast [2] and prostate [34] CSCs are CD44+ CD24−, while pancreatic [44], ovarian [20, 86], and colorectal [80] CSCs are CD44+ CD24+.

Transcription factors, including Oct4, Sox2, Nanog, c-Myc, and Klf4, are part of the core network that regulates pluripotency of embryonic SCs [24]. Research has demonstrated the role of these pluripotency factors in CSC phenotypes and provided new insights into tumorigenesis and metastasis.

Oct4 is essential for maintenance of embryonic SC properties [57], and studies indicate that Oct4 plays important roles in tumorigenesis and tumor metastasis. In murine lung carcinoma cells and human breast cancer MCF7 cells, ablation of Oct4 expression leads to apoptosis of CSCs (through Oct4/Tcl1/Akt1) and inhibition of tumor growth. Also, expression of Oct4 in prostate CSCs maintained CSC properties [64]. These studies demonstrate a role for Oct4 in CSC survival.

Sox2 is thought to be critical for the maintenance of Oct4 expression and hence SC properties [52]. In melanoma CSCs, Sox2 is expressed at high levels [69], and Sox2 is implicated in the promotion of cell migration and invasion in ovarian cancer [49]. Furthermore, knockdown of Sox2 decreases the growth of prostate CSCs in anchorage-independent cultures [67].

Nanog is a transcription factor that helps maintain embryonic SC properties. The level of Nanog expression is higher in CSCs than in non-stem cancer cells in many cancer types, including renal carcinoma [9], breast cancer [55], colorectal cancer [35], prostate cancer [37], gliomas [58, 84], and endometrial adenocarcinoma [88].

Klf4 belongs to the Kruppel-like transcription factor family and is important in cell cycle regulation and pluripotency [24]. Klf4 is expressed at high levels in colorectal CSCs, and knockdown of Klf4 leads to a decrease in sphere formation, migration, invasion, and epithelial–mesenchymal transition, demonstrating a role for Klr4 in maintaining colorectal CSCs [43]. Klf4 is also required for maintenance of breast CSCs. Altogether, studies of Klf4 indicate that Klf4 has an important role not only for self-renewal of CSCs but for cell invasion and metastasis [24].

C-Myc is a member of the Myc family of genes and is a regulator of cell proliferation and key for maintaining SC pluripotency and proliferation [41]. Myc is elevated in a large number of poorly differentiated aggressive malignancies [82]. c-Myc was expressed at high levels in gliomas [79] and in small-cell lung cancer [68]. c-Myc induced self-renewal capacity in hepatocarcinoma cells [1].

In summary, pluripotency factors Oct4, Nanog, Sox2, Klf4, and c-Myc are markers of CSCs that appear to play crucial roles in cancer growth and progression. Further investigation is needed to understand the roles of these transcription factors in the CSC phenotype, tumorigenesis, and tumor progression/metastasis.

Micro-RNAs (miRNAs), including let-7, miR-22, miR-34a, miR-128, the miR-200 family, and miR-451, are expressed in CSCs and are known to regulate the self-renewal, differentiation, epithelial–mesenchymal transition, tumorigenicity, migration, and invasion of CSCs, contributing to metastatic capability. Thus, through these miRNAs, tumor progression and metastasis can be promoted or inhibited [24].

71.3 Stem Cell Divisions and Cancer

Understanding the patterns of SC divisions has been an area of extensive research in the past years. In many human organs, tissue maintenance appears to involve a balance of SC self-renewal and differentiation (Fig. 71.1). Asymmetric SC divisions produce one SC and one more differentiated committed progenitor. Symmetric SC divisions lead to two SCs or to two differentiated committed progenitor cells. Asymmetric SC divisions are identified by asymmetric

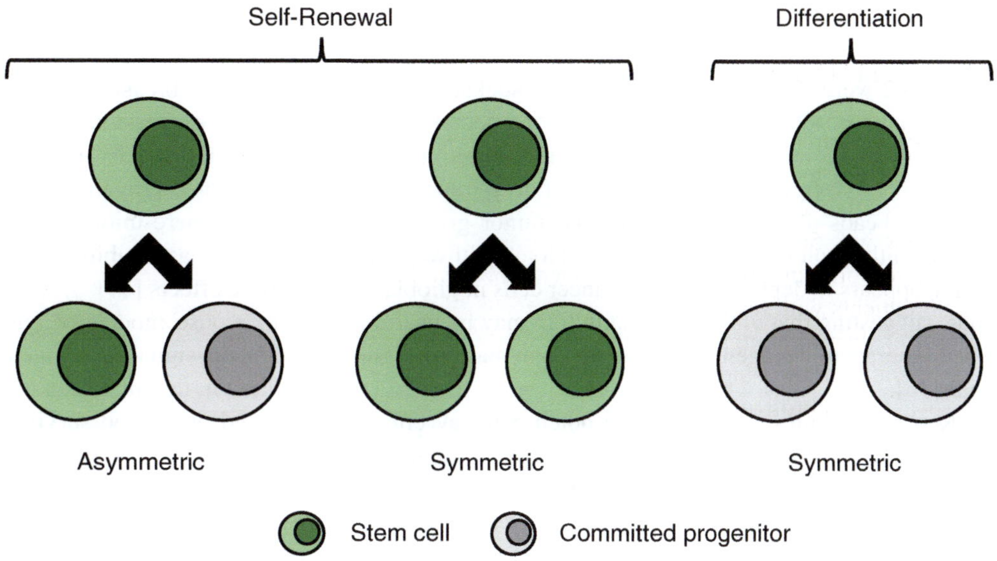

Fig. 71.1 Schematic of stem cell self-renewal and differentiation divisions

segregation of fate determinant proteins such as Numb. The existence of both symmetric and asymmetric SC divisions allows for multiple functions. For example, SCs can be lost by symmetric differentiation, replaced by asymmetric self-renewal division [39], or expanded by symmetric self-renewal division. In response to injury, symmetric divisions could be "switched on" in response to a sudden SC loss. Meanwhile asymmetric division is operational during normal tissue homeostasis.

A dysregulation in either symmetric or asymmetric divisions may lead to the loss of homeostatic control and unchecked cell growth. Alteration in the control of proliferation/differentiation divisions can lead to abnormal SC expansion [65]. Similarly, disruption of asymmetric divisions can result in excess SC self-renewal divisions and subsequent increases in SC number and can be responsible for cancerous growth of undifferentiated cells [10].

71.4 Signaling Pathways of Cancer Stem Cells

Self-renewal is required for maintenance of malignant clones. Multiple signal transduction pathways are involved in the self-renewal of CSCs. These pathways include, but are not limited to, Notch, Hedgehog, Wnt, FGF, and TGF-β [24]. Interruption of these pathways provides a promise for therapeutics that can successfully kill the CSC [60].

Notch Signaling and Cancer Stem Cells: The biological consequences of Notch activation in SCs range from SC maintenance to SC expansion and to promotion of SC differentiation [45]. Initiation of Notch signaling occurs when ligands (DLL/JAG) and receptors (Notch1–Notch4) interact on adjacent cells. This interaction leads to proteolytic cleavage of Notch and release of the Notch intracellular domain (NICD) from the cell membrane. NICD translocates to the nucleus to join a complex that regulates transcription of its target genes [26].

The role of Notch regulation in CSCs and its importance for tumor growth has been shown in multiple cancers, including colon cancer, ovarian cancer, breast cancer, glioblastoma, and chronic myeloid leukemia (CML).

Inhibition of the expression of Notch target genes in human colon cancer xenografts, with the use of anti-human neutralizing DLL4 antibodies, was associated with reduced CSC frequency, as shown by flow cytometric and in vivo tumorigenicity studies. Furthermore, in breast cancer xenografts, Taxol increased CSCs threefold, while DLL4 antibodies, either alone or in combination with Taxol, decreased CSCs [29].

In ovarian cancer Notch3 overexpression resulted in expansion of CSCs and increased tumor resistance to platinum. In contrast, both γ-secretase inhibitor (Notch pathway inhibitor) and Notch3 siRNA knockdown depleted CSCs, as shown by decreased side population cells and gene transcription of SC markers. This was accompanied by increased tumor sensitivity to platinum [53]. These studies indicate a role for Notch in maintaining CSCs and in tumor resistance to platinum.

In breast cancer, Notch4 signaling activity was eightfold higher in CSC-enriched cell populations compared with differentiated tumor cells. Pharmacologic (γ-secretase inhibitor) or genetic inhibition (siRNA) of Notch4 reduced ESA+/CD44+/CD24low CSCs (flow cytometry), decreased mammosphere forming units, and reduced xenograft tumors in vivo in mice [25].

Notch signaling regulates normal SCs in the brain, and glioblastomas contain stem-like cells with high Notch activity. Notch blockade by γ-secretase inhibitors reduced neurosphere growth and clonogenicity in vitro, whereas overexpression of an active form of Notch2 increased tumor growth. The putative CSC markers CD133, Nestin, Bmi1, and Olig2 were reduced following Notch blockade in glioblastoma neurosphere cultures. Injection of γ-secretase subcutaneously in mice inhibited tumor formation. In vivo delivery of γ-secretase by implantation of drug-impregnated polymer beads also effectively blocked tumor growth and significantly prolonged survival. Thus, Notch pathway inhibition appears to deplete stem-like cancer cells in glioblastomas, suggesting that γ-secretase inhibitors may be useful as chemotherapeutic reagents to target CSCs in malignant gliomas [18].

Retroviral expression of Hes1 (a notch pathway effector protein) in common myeloid progenitors and granulocyte-macrophage progenitors in the presence of interleukin-3 immortalizes these cells in vitro. The protein coded for by the abnormal BCR-ABL1 fusion gene (Philadelphia chromosome) is responsible for the uncontrolled growth of leukemic cells. The combination of Hes1 and BCR-ABL in common myeloid progenitors and granulocyte-macrophage progenitors caused acute leukemia resembling the blast crisis of chronic myelogenous leukemia and resulted in rapid death of the recipient mice. BCR-ABL alone caused chronic myeloid leukemia-like disease when expressed in (c-Kit-positive, Sca-1-positive, lineage-negative) hematopoietic SCs, but not in committed progenitors (common myeloid progenitors or granulocyte-macrophage progenitors). Leukemic cells derived from Hes1 and BCR-ABL-expressing common myeloid progenitors and granulocyte-macrophage progenitors were more immature than those derived from BCR-ABL-expressing hematopoietic SCs. The authors thus believe Hes1 may be a key molecule in blast crisis transition [56].

Hedgehog Signaling and Cancer Stem Cells: The hedgehog signaling pathway influences the transcription of many target genes, both during development and in adults [32]. Hedgehog has a major role in maintaining the self-renewing capacity of adult somatic SCs and has been widely implicated in CSC function and maintenance [11]. Activation of hedgehog occurs in multiple cancers, and hedgehog signaling has been shown to regulate CSCs in multiple tumors including pancreatic adenocarcinoma, glioblastoma, breast cancer, multiple myeloma, and chronic myeloid leukemia (CML) [54].

In a pancreatic adenocarcinoma xenograft model, cyclopamine-induced blockade of hedgehog signaling preferentially reduced "ALDH-high" cells threefold. Metastatic spread was profoundly inhibited, with only one of seven cyclopamine-treated mice developing pulmonary micrometastases versus seven of seven control mice developing macrometastases [19].

Shh ligand was expressed in some primary glioblastomas and in glioblastoma-derived neurospheres. Cyclopropylamine completely inhibited glioblastoma-derived neurosphere formation and glioblastoma cells injected intracranially following hedgehog blockade were unable to form tumors in mice, indicating a role for hedgehog in CSC growth [5].

Human breast CSCs ($CD44^{+}CD24^{-/low}Lin^{-}$) showed increased expression of hedgehog pathway members, including PTCH1, Gli1, and Gli2 compared with the non-CSC population. Activation of hedgehog signaling increased mammosphere-initiating cell numbers and mammosphere size, whereas inhibition of hedgehog resulted in a reduction of these effects [47].

In mouse models of chronic myeloid leukemia, loss of hedgehog signaling by genetically disrupting Smoothened resulted in the inhibition of BCR-ABL expressing leukemic SCs and prolonged survival. Loss of Smoothened (component of the hedgehog pathway) in mice impaired hematopoietic SC self-renewal and decreased induction of chronic myelogenous leukemia (CML) by the BCR-ABL1 oncoprotein. Loss of Smoothened caused depletion of chronic myeloid leukemia CSCs, whereas control hematopoietic cells transduced with BCR-ABL1 developed chronic myeloid leukemia in 94% of recipients within 3 months, similarly transduced Smoothened −/− cells developed chronic myeloid leukemia in only 47%, and showed increased tumor latency. These data suggest that hedgehog is required for the initiation and propagation of CML. Further, constitutively active Smoothened augments chronic myelogenous leukemia CSC number and accelerates disease. As a possible mechanism for Smoothened-induced effects, the cell fate determinant Numb, which depletes chronic myelogenous leukemia CSCs, is increased in the absence of Smoothened activity [87].

The subset of multiple myeloma cells with hedgehog pathway activity is located within the CSC compartment. Hedgehog ligand promotes expansion of multiple myeloma CSCs without differentiation, whereas hedgehog pathway blockade inhibits clonal expansion and is accompanied by terminal differentiation of purified multiple myeloma CSCs. Thus, hedgehog pathway activation is heterogeneous across the spectrum of multiple myeloma CSCs and their more differentiated progeny [61].

The above studies validate hedgehog pathway inhibition as a therapeutic strategy for cancers and their metastatic spread. By targeting specific cellular subpopulations likely involved in tumor initiation at metastatic sites, hedgehog inhibitors provide a new paradigm for therapy of disseminated malignancies, particularly when used in combination with conventional antimetabolites that reduce "bulk" tumor size [19].

WNT/B-Catenin Signaling and CSCs: More than 30 extracellular Wnt ligands bind to the receptor complex of Frizzled and LRP5/6 (member of the LDL receptor family) [48]. Wnt signaling regulates the activity of somatic SCs, including those in the skin, blood, brain, and the mammary gland, whereas aberrant signaling results in neoplasia [30].

The Wnt pathway is activated in CD133(+) colon CSCs [13, 14]. Also, in colon adenocarcinomas, high activity of the Wnt pathway is observed preferentially in tumor cells located close to stromal myofibroblasts, indicating that Wnt activity and cancer stemness may be regulated by extrinsic cues. In agreement with this notion, myofibroblast-secreted factors, specifically hepatocyte growth factor, activate beta-catenin-dependent transcription and subsequently CSC clonogenicity. More significantly, myofibroblast-secreted factors also restore the CSC phenotype in more differentiated tumor cells both in vitro and in vivo. Thus, microenvironmental Wnt signaling also has a role in CSC behavior and maintenance [77].

71.5 The Role of CSCs in Cancer Progression and Metastasis

CSCs can, by multiple paradigms, account for tumor heterogeneity. A cancer may arise from a phenotypically uniform single CSC population. However, in some cases, more than one CSC population may exist within a tumor. In other cases, CSCs may stay dormant for months or years, often after surviving cancer therapies that kill the non-CSCs and become activated to form tumors at a later date. Also, CSC populations, distinct from the parent population, may develop as a tumor progresses, due to genetic or epigenetic modifications and often associated with greater tumor aggression. Finally, the CSC phenotype can be unstable, resulting in phenotypic reversion and switching of the CSC phenotype. Thus, CSCs are dynamic in nature, and it is possible for clonal evolution and CSCs to exist and act together [78].

Metastasis is a multifaceted event requiring loss of cell–cell adhesion, intravasation into blood vessels or lymphatic system, extravasation, as well as a receptive tissue environment. Not all cells in a tumor are able to move out of the primary tumor and metastasize, and only a small subpopulation of those that enter the circulation initiate a metastatic lesion [4]. This subpopulation is thought to be of the CSC phenotype, also known as metastasis-initiating cells (MICs). Using bioluminescence imaging, Liu et al. showed that breast cancer CD44+ CSCs are directly involved in metastasis [46].

CSC subsets within primary tumors may harbor CSC subsets with tumor-propagating and/or metastatic capacity. For example, in pancreatic cancer, both CD133+CXCR4+ cells and CD133+CXCR4− cells were tumor-initiating cells, but only CD133+CXCR4+ cells produced metastases. Inhibition of CXCR4 signaling reduced the metastatic potential of the tumors [27]. Likewise, in human colon cancer, murine xenografts showed that tumor initiation, self-renewal, and metastasis formation are limited to particular subpopulations of CSCs [15]. Thus, substantial evidence shows that subpopulations of CSCs within the tumor have specific capabilities in regard to tumor maintenance, growth, and metastasis.

71.6 Targeting CSCs for Cancer Therapy

Many cancer treatments kill most cancer cells, but CSCs exhibit resistance to therapy, and this may lead to tumor relapse and metastasis. Multiple mechanisms exist whereby CSCs can resist therapeutics.

Induction of epithelial–mesenchymal transition can lead to the acquisition of resistance to both chemotherapy and radiotherapy, as shown in ovarian cancer [40] and in breast cancer [23]. How does epithelial–mesenchymal transition lead to chemo- and radio-resistance? Resistance through epithelial–mesenchymal transition may be attributable to the lower proliferative rate that results from the acquisition of mesenchymal properties. Also resistance to chemo in normal SCs may be due to high levels of anti-apoptotic proteins as well as the ABC transporters [60].

It is challenging to perform chemical screens for therapy of CSCs specifically, first because they are a rare population of cancer cells and second because if they are successfully isolated by flow cytometry a pure population is not able to be maintained in culture.

Since CSCs are so hard to target, targeting the tumor microenvironment is an attractive approach. In colon cancer, high Wnt pathway activity is observed preferentially in tumor cells located close to stromal myofibroblasts, and myofibroblast-secreted factors activate β-catenin-dependent transcription and subsequently CSC clonogenicity. Furthermore, myofibroblast-secreted factors also restore the CSC phenotype in more differentiated tumor cells both in vitro and in vivo [77]. Thus, stroma cells appear to have a role in both inducing and maintaining a SC-like state in the CSC.

As CSCs and normal SCs share a number of properties, signaling pathways such as Wnt, TGF-b, Notch, and hedgehog pathways are being explored as targets to eliminate CSCs. Several inhibitors of the Wnt pathway have been developed. They may be effective in preventing secretion of active Wnt molecules by stromal cells, blocking paracrine signaling that induces and maintains CSCs [60]. Several molecules targeting the hedgehog and Notch pathways have also been developed. Sonidegib, a small molecule that inhibits Smoothened (hedgehog pathway activator), was approved for adults with locally advanced basal cell carcinoma in 2015. Also, a small molecule gamma secretase inhibitor (prevents Notch cleavage), MK0752, has been developed, and phase 1 studies were published in 2018) [12].

Another therapeutic approach is to cause CSCs to differentiate into mature non-CSCs. These agents rely on the premise that a defect in terminal differentiation is characteristic of cancer. Finally, there is the idea of using the CSC-specific antigens for immunotherapy, activating adaptive immune cells against specific antigens [60].

Conclusion and Perspective

- Not all cells in a cancer are equally proliferative and invasive, and most are not capable of producing a further tumor if separated from the parent tumor.
- A small proportion of the total cancer cell population is capable of initiating further tumors in xenograft models, and these cells have been termed tumor-initiating cells or cancer stem cells (CSCs).
- Many cancer treatments kill most cancer cells, but CSCs exhibit resistance to therapy, and this may lead to tumor relapse and metastasis.
- Since CSCs are so hard to target, targeting the tumor microenvironment is an attractive approach, as is targeting signaling pathways such as Wnt, TGF-b, Notch, and hedgehog pathways that are key to CSC survival and growth.
- Can we improve our ability to isolate CSCs in order to better study this small population of cells?
- Can we target CSC signaling to decrease metastatic ability?
- Can we combine CSC-targeted treatments with other treatment modalities to improve cancer survival?

References

1. Akita H, et al. MYC activates stem-like cell potential in hepatocarcinoma by a p53-dependent mechanism. Cancer Res. 2014;74(20):5903–13. https://doi.org/10.1158/0008-5472.CAN-14-0527.
2. Al-Hajj M, et al. Prospective identification of tumorigenic breast cancer cells. Proc Natl Acad Sci USA. 2003;100(7):3983–8. https://doi.org/10.1073/pnas.0530291100.
3. An Y, Ongkeko WM. ABCG2: the key to chemoresistance in cancer stem cells? Expert Opin Drug Metab Toxicol. 2009;5(12):1529–42. https://doi.org/10.1517/17425250903228834.
4. Baccelli I, et al. Identification of a population of blood circulating tumor cells from breast cancer patients that initiates metastasis in a xenograft assay. Nat Biotechnol. 2013;31(6):539–44. https://doi.org/10.1038/nbt.2576.
5. Bar EE, et al. Cyclopamine-mediated hedgehog pathway inhibition depletes stem-like cancer cells in glioblastoma. Stem Cells. 2007;25(10):2524–33. https://doi.org/10.1634/stemcells.2007-0166.
6. Bauer N, et al. New insights into the cell biology of hematopoietic progenitors by studying prominin-1 (CD133). Cells Tissues Organs. 2008;188(1–2):127–38. https://doi.org/10.1159/000112847.
7. Bauer N, et al. Haematopoietic stem cell differentiation promotes the release of prominin-1/CD133-containing membrane vesicles—a role of the endocytic–exocytic pathway. EMBO Mol Med. 2011;3(7):398–09. https://doi.org/10.1002/emmm.201100147.
8. Boonyaratanakornkit JB, et al. Selection of tumorigenic melanoma cells using ALDH. J Investig Dermatol. 2010;130(12):2799–08. https://doi.org/10.1038/jid.2010.237.
9. Bussolati B, et al. Identification of a tumor-initiating stem cell population in human renal carcinomas. FASEB J. 2008;22(10):3696–705. https://doi.org/10.1096/fj.08-102590.
10. Clarke RB, et al. Regulation of human breast epithelial stem cells. Cell Prolif. 2003;36(Suppl 1):45–8. https://doi.org/10.1046/j.1365-2184.36.s.1.5.x.
11. Cochrane CR, et al. Hedgehog signaling in the maintenance of cancer stem cells. Cancers. 2015;7(3):1554–85. https://doi.org/10.3390/cancers7030851.
12. Cook N, et al. A phase I trial of the γ-secretase inhibitor MK-0752 in combination with gemcitabine in patients with pancreatic ductal adenocarcinoma. Br J Cancer. 2018;118(6):793–01. https://doi.org/10.1038/bjc.2017.495.
13. Corbo C, et al. Protein cross-talk in CD133+ colon cancer cells indicates activation of the Wnt pathway and upregulation of SRp20 that is potentially involved in tumorigenicity. Proteomics. 2012;12(12):2045–59. https://doi.org/10.1002/pmic.201100370.
14. Deng Y-H, et al. 5-Fluorouracil upregulates the activity of Wnt signaling pathway in CD133-positive colon cancer stem-like cells. Chin J Cancer. 2010;29(9):810–5. https://doi.org/10.5732/cjc.010.10134.
15. Dieter SM, et al. Distinct types of tumor-initiating cells form human colon cancer tumors and metastases. Cell Stem Cell. 2011;9(4):357–65. https://doi.org/10.1016/j.stem.2011.08.010.
16. Dou J, et al. Using ABCG2-molecule-expressing side population cells to identify cancer stem-like cells in a human ovarian cell line. Cell Biol Int. 2011;35(3):227–34. https://doi.org/10.1042/CBI20100347.
17. Eramo A, et al. Identification and expansion of the tumorigenic lung cancer stem cell population. Cell Death Differ. 2008;15(3):504–14. https://doi.org/10.1038/sj.cdd.4402283.
18. Fan X, et al. NOTCH pathway blockade depletes CD133-positive glioblastoma cells and inhibits growth of tumor neurospheres and xenografts. Stem Cells (Dayton, Ohio). 2010;28(1):5–16. https://doi.org/10.1002/stem.254.
19. Feldmann G, et al. Blockade of hedgehog signaling inhibits pancreatic cancer invasion and metastases: a new paradigm for combination therapy in solid cancers. Cancer Res. 2007;67(5):2187–96. https://doi.org/10.1158/0008-5472.CAN-06-3281.
20. Gao M-Q, et al. CD24+ cells from hierarchically organized ovarian cancer are enriched in cancer stem cells. Oncogene. 2010;29(18):2672–80. https://doi.org/10.1038/onc.2010.35.
21. Ginestier C, et al. ALDH1 is a marker of normal and malignant human mammary stem cells and a predictor of poor clinical outcome. Cell Stem Cell. 2007;1(5):555–67. https://doi.org/10.1016/j.stem.2007.08.014.
22. Glumac PM, LeBeau AM. The role of CD133 in cancer: a concise review. Clin Transl Med. 2018;7(1):18. https://doi.org/10.1186/s40169-018-0198-1.
23. Gupta PB, et al. Identification of selective inhibitors of cancer stem cells by high-throughput screening. Cell. 2009;138(4):645–59. https://doi.org/10.1016/j.cell.2009.06.034.
24. Hadjimichael C, et al. Common stemness regulators of embryonic and cancer stem cells. World J Stem Cells. 2015;7(9):1150–84. https://doi.org/10.4252/wjsc.v7.i9.1150.
25. Harrison H, et al. Regulation of breast cancer stem cell activity by signaling through the Notch4 receptor. Cancer Res. 2010;70(2):709–18. https://doi.org/10.1158/0008-5472.CAN-09-1681.
26. Henrique D, Schweisguth F. Mechanisms of Notch signaling: a simple logic deployed in time and space. Development. 2019;146(3):172148. https://doi.org/10.1242/dev.172148.
27. Hermann PC, et al. Distinct populations of cancer stem cells determine tumor growth and metastatic activity in human pancreatic cancer. Cell Stem Cell. 2007;1(3):313–23. https://doi.org/10.1016/j.stem.2007.06.002.

28. Hirschmann-Jax C, et al. A distinct "side population" of cells with high drug efflux capacity in human tumor cells. Proc Natl Acad Sci USA. 2004;101(39):14228–33. https://doi.org/10.1073/pnas.0400067101.
29. Hoey T, et al. DLL4 blockade inhibits tumor growth and reduces tumor-initiating cell frequency. Cell Stem Cell. 2009;5(2):168–77. https://doi.org/10.1016/j.stem.2009.05.019.
30. Holland JD, et al. Wnt signaling in stem and cancer stem cells. Curr Opin Cell Biol. 2013;25(2):254–64. https://doi.org/10.1016/j.ceb.2013.01.004.
31. van den Hoogen C, et al. High aldehyde dehydrogenase activity identifies tumor-initiating and metastasis-initiating cells in human prostate cancer. Cancer Res. 2010;70(12):5163–73. https://doi.org/10.1158/0008-5472.CAN-09-3806.
32. Hooper JE, Scott MP. Communicating with Hedgehogs. Nat Rev Mol Cell Biol. 2005;6(4):306–17. https://doi.org/10.1038/nrm1622.
33. Huang EH, et al. Aldehyde dehydrogenase 1 is a marker for normal and malignant human colonic stem cells (SC) and tracks SC overpopulation during colon tumorigenesis. Cancer Res. 2009;69(8):3382–9. https://doi.org/10.1158/0008-5472.CAN-08-4418.
34. Hurt EM, et al. CD44 + CD24 – prostate cells are early cancer progenitor/stem cells that provide a model for patients with poor prognosis. Br J Cancer. 2008;98(4):756–65. https://doi.org/10.1038/sj.bjc.6604242.
35. Ibrahim EE, et al. Embryonic NANOG activity defines colorectal cancer stem cells and modulates through AP1- and TCF-dependent mechanisms. Stem Cells (Dayton, Ohio). 2012;30(10):2076–87. https://doi.org/10.1002/stem.1182.
36. Jaggupilli A, Elkord E. Significance of CD44 and CD24 as cancer stem cell markers: an enduring ambiguity. Clin Dev Immunol. 2012;2012:708036. https://doi.org/10.1155/2012/708036.
37. Jeter CR, et al. NANOG promotes cancer stem cell characteristics and prostate cancer resistance to androgen deprivation. Oncogene. 2011;30(36):3833–45. https://doi.org/10.1038/onc.2011.114.
38. Jiang F, et al. Aldehyde dehydrogenase 1 is a tumor stem cell-associated marker in lung cancer. Mol Cancer Res. 2009;7(3):330–8. https://doi.org/10.1158/1541-7786.MCR-08-0393.
39. Klein AM, Simons BD. Universal patterns of stem cell fate in cycling adult tissues. Development. 2011;138(15):3103–11. https://doi.org/10.1242/dev.060103.
40. Kurrey NK, et al. Snail and slug mediate radioresistance and chemoresistance by antagonizing p53-mediated apoptosis and acquiring a stem-like phenotype in ovarian cancer cells. Stem Cells (Dayton, Ohio). 2009;27(9):2059–68. https://doi.org/10.1002/stem.154.
41. Laurenti E, Wilson A, Trumpp A. Myc's other life: stem cells and beyond. Curr Opin Cell Biol. 2009;21(6):844–54. https://doi.org/10.1016/j.ceb.2009.09.006.
42. Lee SY, et al. Induction of metastasis, cancer stem cell phenotype, and oncogenic metabolism in cancer cells by ionizing radiation. Mol Cancer. 2017;16(1):10. https://doi.org/10.1186/s12943-016-0577-4.
43. Leng Z, et al. Krüppel-like factor 4 acts as an oncogene in colon cancer stem cell-enriched spheroid cells. PLoS One. 2013;8(2):e56082. https://doi.org/10.1371/journal.pone.0056082.
44. Li C, et al. Identification of pancreatic cancer stem cells. Cancer Res. 2007;67(3):1030–7. https://doi.org/10.1158/0008-5472.CAN-06-2030.
45. Liu H, et al. Cancer stem cells from human breast tumors are involved in spontaneous metastases in orthotopic mouse models. Proc Natl Acad Sci. 2010;107(42):18115–20. https://doi.org/10.1073/pnas.1006732107.
46. Liu J, et al. Notch signaling in the regulation of stem cell self-renewal and differentiation. Curr Top Dev Biol. 2010;92:367–409. https://doi.org/10.1016/S0070-2153(10)92012-7.
47. Liu S, et al. Hedgehog signaling and Bmi-1 regulate self-renewal of normal and malignant human mammary stem cells. Cancer Res. 2006;66(12):6063–71. https://doi.org/10.1158/0008-5472.CAN-06-0054.
48. Logan CY, Nusse R. The Wnt signaling pathway in development and disease. Annu Rev Cell Dev Biol. 2004;20:781–10. https://doi.org/10.1146/annurev.cellbio.20.010403.113126.
49. Lou X, et al. SOX2 targets fibronectin 1 to promote cell migration and invasion in ovarian cancer: new molecular leads for therapeutic intervention. Omics J Integr Biol. 2013;17(10):510–8. https://doi.org/10.1089/omi.2013.0058.
50. Luo Y, et al. ALDH1A isozymes are markers of human melanoma stem cells and potential therapeutic targets. Stem Cells (Dayton, Ohio). 2012;30(10):2100–13. https://doi.org/10.1002/stem.1193.
51. Marzesco A-M, et al. Release of extracellular membrane particles carrying the stem cell marker prominin-1 (CD133) from neural progenitors and other epithelial cells. J Cell Sci. 2005;118(Pt 13):2849–58. https://doi.org/10.1242/jcs.02439.
52. Masui S, et al. Pluripotency governed by Sox2 via regulation of Oct3/4 expression in mouse embryonic stem cells. Nat Cell Biol. 2007;9(6):625–35. https://doi.org/10.1038/ncb1589.
53. McAuliffe SM, et al. Targeting Notch, a key pathway for ovarian cancer stem cells, sensitizes tumors to platinum therapy. Proc Natl Acad Sci USA. 2012;109(43):E2939–48. https://doi.org/10.1073/pnas.1206400109.
54. Merchant AA, Matsui W. Targeting Hedgehog—a cancer stem cell pathway. Clin Cancer Res. 2010;16(12):3130–40. https://doi.org/10.1158/1078-0432.CCR-09-2846.
55. Nagata T, et al. Prognostic significance of NANOG and KLF4 for breast cancer. Breast Cancer (Tokyo, Japan). 2014;21(1):96–101. https://doi.org/10.1007/s12282-012-0357-y.
56. Nakahara F, et al. Hes1 immortalizes committed progenitors and plays a role in blast crisis transition in chronic myelogenous leukemia. Blood. 2010;115(14):2872–81. https://doi.org/10.1182/blood-2009-05-222836.
57. Nichols J, Evans EP, Smith AG. Establishment of germ-line-competent embryonic stem (ES) cells using differentiation inhibiting activity. Development. 1990;110(4):1341–8.
58. Niu C-S, et al. Expression of NANOG in human gliomas and its relationship with undifferentiated glioma cells. Oncol Rep. 2011;26(3):593–601. https://doi.org/10.3892/or.2011.1308.
59. Olempska M, et al. Detection of tumor stem cell markers in pancreatic carcinoma cell lines. Hepatobiliary Pancreat Dis Int. 2007;6(1):92–7.
60. Pattabiraman DR, Weinberg RA. Tackling the cancer stem cells—what challenges do they pose? Nat Rev Drug Discov. 2014;13(7):497–512. https://doi.org/10.1038/nrd4253.
61. Peacock CD, et al. Hedgehog signaling maintains a tumor stem cell compartment in multiple myeloma. Proc Natl Acad Sci USA. 2007;104(10):4048–53. https://doi.org/10.1073/pnas.0611682104.
62. Rahadiani N, et al. Expression of aldehyde dehydrogenase 1 (ALDH1) in endometrioid adenocarcinoma and its clinical implications. Cancer Sci. 2011;102(4):903–8. https://doi.org/10.1111/j.1349-7006.2011.01864.x.
63. Rasheed ZA, et al. Prognostic significance of tumorigenic cells with mesenchymal features in pancreatic adenocarcinoma. J Natl Cancer Inst. 2010;102(5):340–51. https://doi.org/10.1093/jnci/djp535.
64. Rentala S, Mangamoori LN. Oct-4 expression maintained stem cell properties in prostate cancer-derived CD133+MDR1+ cells. Trop J Pharm Res. 2009;8(1):3–9. https://doi.org/10.4314/tjpr.v8i1.14706.
65. Reya T, Clevers H. Wnt signalling in stem cells and cancer. Nature. 2005;434(7035):843–50. https://doi.org/10.1038/nature03319.

66. Ricci-Vitiani L, et al. Identification and expansion of human colon-cancer-initiating cells. Nature. 2007;445(7123):111–5. https://doi.org/10.1038/nature05384.
67. Rybak AP, Tang D. SOX2 plays a critical role in EGFR-mediated self-renewal of human prostate cancer stem-like cells. Cell Signal. 2013;25(12):2734–42. https://doi.org/10.1016/j.cellsig.2013.08.041.
68. Salcido CD, et al. Molecular characterisation of side population cells with cancer stem cell-like characteristics in small-cell lung cancer. Br J Cancer. 2010;102(11):1636–44. https://doi.org/10.1038/sj.bjc.6605668.
69. Santini R, et al. SOX2 regulates self-renewal and tumorigenicity of human melanoma-initiating cells. Oncogene. 2014;33(38):4697–708. https://doi.org/10.1038/onc.2014.71.
70. Sicchieri RD, et al. ABCG2 is a potential marker of tumor-initiating cells in breast cancer. Tumour Biol. 2015;36(12):9233–43. https://doi.org/10.1007/s13277-015-3647-0.
71. Singh SK, et al. Identification of a cancer stem cell in human brain tumors. Cancer Res. 2003;63(18):5821–8.
72. Singh SK, et al. Identification of human brain tumour initiating cells. Nature. 2004;432(7015):396–401. https://doi.org/10.1038/nature03128.
73. Suetsugu A, et al. Characterization of CD133+ hepatocellular carcinoma cells as cancer stem/progenitor cells. Biochem Biophys Res Commun. 2006;351(4):820–4. https://doi.org/10.1016/j.bbrc.2006.10.128.
74. Tiezzi DG, et al. ABCG2 as a potential cancer stem cell marker in breast cancer. J Clin Oncol. 2013;31(15_suppl):e12007. https://doi.org/10.1200/jco.2013.31.15_suppl.e12007.
75. Tomita H, et al. Aldehyde dehydrogenase 1A1 in stem cells and cancer. Oncotarget. 2016;7(10):11018–32. https://doi.org/10.18632/oncotarget.6920.
76. Ucar D, et al. Aldehyde dehydrogenase activity as a functional marker for lung cancer. Chem Biol Interact. 2009;178(1):48–55. https://doi.org/10.1016/j.cbi.2008.09.029.
77. Vermeulen L, et al. Wnt activity defines colon cancer stem cells and is regulated by the microenvironment. Nat Cell Biol. 2010;12(5):468–76. https://doi.org/10.1038/ncb2048.
78. Visvader JE, Lindeman GJ. Cancer stem cells: current status and evolving complexities. Cell Stem Cell. 2012;10(6):717–28. https://doi.org/10.1016/j.stem.2012.05.007.
79. Wang J, et al. c-Myc is required for maintenance of glioma cancer stem cells. PLoS One. 2008;3(11):e3769. https://doi.org/10.1371/journal.pone.0003769.
80. Yeung TM, et al. Cancer stem cells from colorectal cancer-derived cell lines. Proc Natl Acad Sci USA. 2010;107(8):3722–7. https://doi.org/10.1073/pnas.0915135107.
81. Yin S, et al. CD133 positive hepatocellular carcinoma cells possess high capacity for tumorigenicity. Int J Cancer. 2007;120(7):1444–50. https://doi.org/10.1002/ijc.22476.
82. Yin XY, et al. Inverse regulation of cyclin B1 by c-Myc and p53 and induction of tetraploidy by cyclin B1 overexpression. Cancer Res. 2001;61(17):6487–93.
83. Yue L, et al. Targeting ALDH1 to decrease tumorigenicity, growth and metastasis of human melanoma. Melanoma Res. 2015;25(2):138–48. https://doi.org/10.1097/CMR.0000000000000144.
84. Zbinden M, et al. NANOG regulates glioma stem cells and is essential in vivo acting in a cross-functional network with GLI1 and p53. EMBO J. 2010;29(15):2659–74. https://doi.org/10.1038/emboj.2010.137.
85. Zhang G, et al. Expression of potential cancer stem cell marker ABCG2 is associated with malignant behaviors of hepatocellular carcinoma. Gastroenterol Res Pract. 2013;2013:782581. https://doi.org/10.1155/2013/782581.
86. Zhang S, et al. Identification and characterization of ovarian cancer-initiating cells from primary human tumors. Cancer Res. 2008;68(11):4311–20. https://doi.org/10.1158/0008-5472.CAN-08-0364.
87. Zhao C, et al. Hedgehog signalling is essential for maintenance of cancer stem cells in myeloid leukaemia. Nature. 2009;458(7239):776–9. https://doi.org/10.1038/nature07737.
88. Zhou X, et al. Expression of the stem cell marker Nanog in human endometrial adenocarcinoma. Int J Gynecol Pathol. 2011;30(3):262–70. https://doi.org/10.1097/PGP.0b013e3182055a1f.

HPV-Induced Cancers

72

Brittney L. Dickey, Jennifer M. Binning, and Julie Rathwell

Abstract

Highly prevalent human papillomavirus (HPV) can lead to cancer in the anogenital and oropharyngeal regions. Oncogenic, or high-risk, HPV types are attributed to 690,000 cancers worldwide. HPV infects basal epithelial cells and is propagated solely through host cellular differentiation. As viral amplification occurs, virions are shed from infected cells and continues the transmission process. While most HPV infections go unnoticed and clear within 2 years, persistent infections can lead to precancerous and cancerous lesions. Malignant transformation is driven by viral oncoproteins E5, E6, and E7, and these oncoproteins remain important for tumor maintenance. Cervical cancer has been shown to metastasize via the lymphovascular system due to close proximity of the lymphatic spaces, though direct local extension is also common. Oropharyngeal cancer (OPC), though, has more diverse etiology with some cases attributed to HPV, while others are due to tobacco and alcohol use. HPV-associated OPC cases are known to have better survival, and early studies indicate differences in rate or site of metastases by HPV status may exist, though much is still unknown. Compared to other oncoviruses, there are proven tools including vaccine and screening that can not only prevent HPV-associated cancers but potentially eliminate them.

B. L. Dickey (✉) · J. M. Binning · J. Rathwell
Moffitt Cancer Center, Tampa, USA
e-mail: Brittney.Dickey@Moffitt.org;
Jennifer.Binning@Moffitt.org; Julie.Rathwell@Moffitt.org

Learning Objectives

- Understand the process by which HPV infects cells to cause cancer.
- Describe the epidemiology of HPV-associated cancers.
- Explore the patterns of metastases in the presence of an HPV infection.
- Discuss how HPV-related cancers can be eliminated.

72.1 Introduction

Human papillomaviruses (HPV) are epitheliotropic, double-stranded DNA viruses and are highly prevalent. The majority of HPV infections are benign; however, infection with oncogenic types that persist can lead to cancer at six anatomic sites: cervix, vagina, vulva, penis, anus, and oropharynx. In this chapter, we will describe the types of HPV that can cause cancer, how the virus is transmitted, the cellular changes caused by an HPV infection that leads to cancer, metastasis of HPV-related cancers, and the effort to eliminate these cancers.

72.2 HPV Infection

72.2.1 HPV Types

Approximately 200 papillomavirus genotypes have been identified and are grouped into five genera: alpha, beta, gamma, mu, and nu [1]. Among these, alpha and beta types

S. P. Leong et al. (eds.), *Cancer Metastasis Through the Lymphovascular System*,
https://doi.org/10.1007/978-3-030-93084-4_72

Table 72.1 HPV types associated with disease

Risk	HPV type	Genus	Associated disease
High-risk HPV or oncogenic	16, 18, 31, 33, 35, 39, 45, 51, 52, 56, 58, 59	Alpha	Cervical, anal, vaginal, vulvar, penile, and oropharyngeal cancers and associated precursor lesions
Probable carcinogen	68	Alpha	Cervical cancer
Possible carcinogen	5, 8	Beta	Squamous cell carcinoma of the skin in persons with epidermodysplasia verruciformis
	26, 30, 34, 53, 66, 67, 69, 70, 73, 82, 85, 97	Alpha	Uncertain
Low risk	6, 11	Alpha	Genital warts, recurrent respiratory papillomatosis

HPV types are listed with their associated oncogenic risk, genus type, and disease. Oncogenic risk is defined according to the criteria from the International Agency for Research on Cancer Evaluation [3, 4]

have been the most extensively studied [2]. Alpha-HPVs include the genotypes associated with cancer, while infections with all other types are asymptomatic. Twelve types (HPV 16, 18, 31, 33, 35, 39, 45, 51, 52, 56, 58, and 59) that exclusively infect the inner tissue lining are considered high-risk HPVs and group 1 carcinogens by the *International Agency for Research on Cancer* (IARC), and other HPV types are considered probably oncogenic (i.e., HPV68 for cervical cancer) or uncertain. In contrast, low-risk types (HPV6 and 11) cause genital warts and are considered to not be a cause of cancer [3] (Table 72.1).

72.2.2 Viral Transmission

HPV infection occurs when virions infect basal epithelial cells at cutaneous or mucosal sites via direct access to the basement membrane of the epithelium (reviewed in [4]). Mucosal infections are mostly restricted to the anogenital and oral cavities as well as laryngeal epithelium, whereas all cutaneous tissues are susceptible to HPV infections. Natural transmission of mucosal infections primarily involves direct genital mucosa contact [5] and is believed to occur following the mechanical wounding that arises during vaginal and anal sex [6]. The unique nature of the tonsillar crypt epithelium, with a disrupted basal cell layer and basement membrane, likely allows viral infection of the basal cells without the prerequisite wounding [6]. Natural transmission of cutaneous infections involves direct contact of an infected keratinocyte, such as a wart, with uninfected skin cells possibly via microabrasions in the skin.

Once the virus infects those cells most susceptible to infection, usually keratinocytes, the viral life cycle for HPV is strictly dependent upon cellular differentiation. As the virus does not possess its own replication machinery, it relies on cellular division and stratification of the epithelium to successfully propagate. This occurs in the basal layers and progresses toward the suprabasal layers, to maintain cell proliferation in highly differentiated suprabasal regions. This vertical maturation of infected keratinocytes completes the viral life cycle and allows viral genome amplification to occur [4]. Viruses produced through this process are shed from infected cells and can begin the transmission process anew (Fig. 72.1).

72.2.3 Epidemiology of HPV Infection

Anogenital HPV infection is the most common sexually transmitted infection in the USA, with a prevalence of approximately 70 million cases [7]. Due to limited reporting of genital infection, HPV incidence rates and prevalence are likely underestimated, especially globally. However, it is expected that nearly all sexually active individuals will acquire HPV at some time [8]. The principal risk factor is sexual behavior, particularly early age at sexual exposure and multiple partners [7]. Due to such transmission patterns, anogenital HPV infections are very common. Among young women, prevalence peaks at approximately 20–25 years. Variations in peak prevalence globally are slight and are mostly attributed to a difference in average age at sexual initiation. However, a second peak after menopause is observed in some countries, though the reasons for this peak are not well understood [3]. Interestingly, anogenital HPV infection prevalence among men does not vary as much by age, with relatively flat age-specific prevalence curves, and varies by country, specific anatomical site, and risk group (i.e., high-risk groups, such as HIV-infected men, compared to low-risk groups) [9].

Unlike anogenital infections, oral HPV infections are rare. In the USA, the prevalence of any oral HPV infection was estimated to be 11.5% among men and 3.3% among women, whereas the prevalence of genital HPV was 45.2% among men and 39.9% among women [10]. Oral sex may lead to an HPV infection of the oral cavity, with men who perform oral sex on women being more susceptible. The key factor for an oral HPV infection is sexual behavior, though it appears to be more closely associated with the number of lifetime oral sex partners, with an overall odds of infection of 5.7 for six or more lifetime partners [7].

Most individuals with an HPV infection are unaware of the infection and do not incur noticeable symptoms. Approximately half of all HPV infections clear within 6 months, and more than 90% clear within a few years [3, 11]. Infections may present as asymptomatic infections, small epithelial lesions, large cutaneous lesions, or cancerous lesions. The factors that determine whether an HPV infection will be cleared are poorly defined, yet the interplay

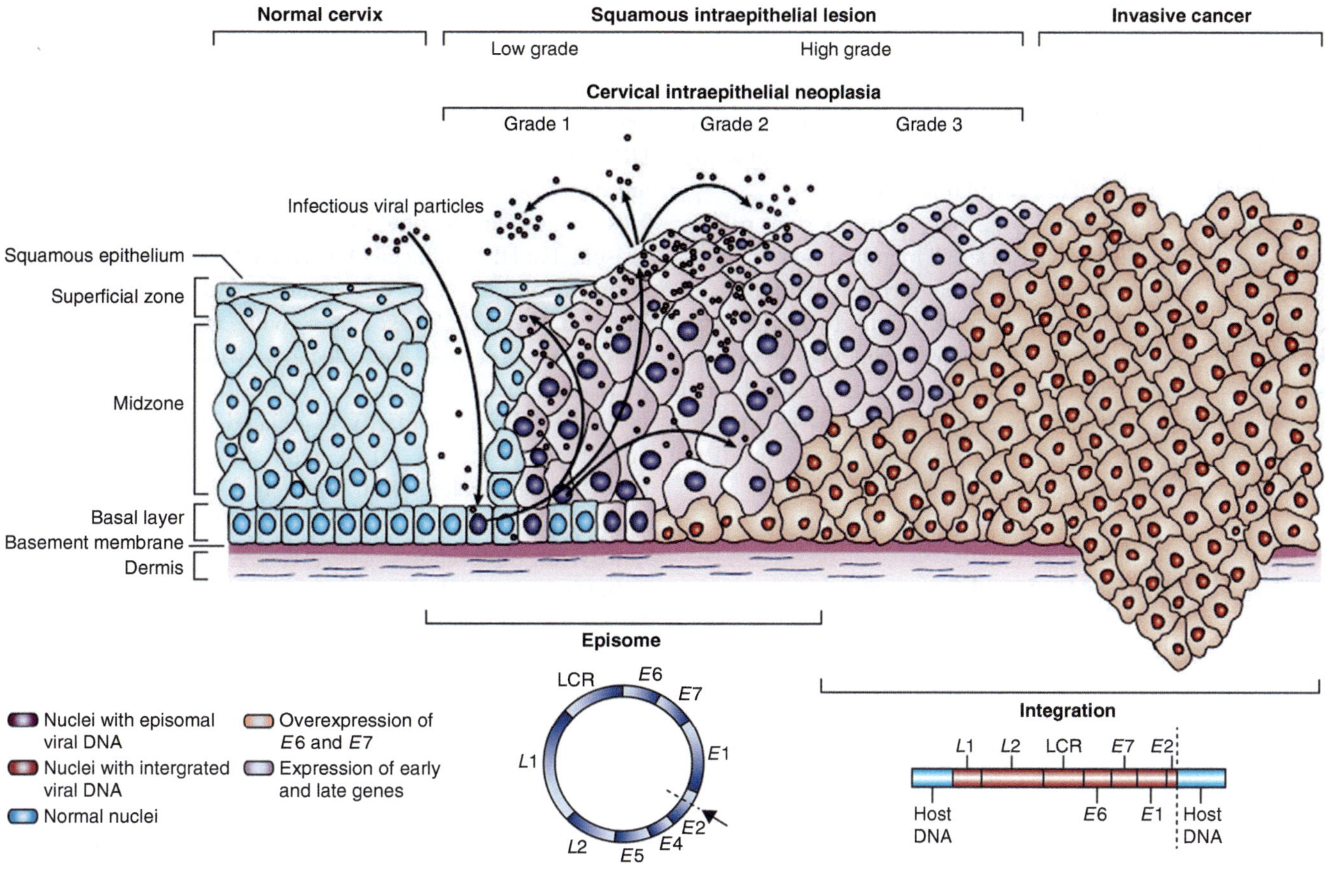

Fig. 72.1 HPV viral integration and transformation to cervical cancer. The viral integration and cellular changes to cancer process are similar for all sites of HPV-induced cancers. Through abrasions in the cervical epithelium, HPV is able to access and infect basal cells. After infection, expression of HPV genes E1, E2, E4, E5, E6, and E7 allows viral DNA to replicate from episomal DNA. Replication continues through the upper layers of the epithelium as late genes, L1 and L2, are expressed and encapsulate the viral genome in order to reproduce in the nucleus until the virus is shed and continues this replication process in nearby cells. Low-grade intraepithelial lesions support productive viral replication and are considered low-risk cervical lesions. Infection with high-risk HPV types will progress to high-grade intraepithelial lesions or precancer. Untreated lesions will progress to microinvasion and invasive cancer and is associated with integration of HPV into host chromosomes (red nuclei), with loss or disruption of E2, and upregulation of E6 and E7 oncogene expression. Reproduced from Crosbie et al. [17]

between HPV defense mechanisms, tissue tropism, and the host immune system undoubtedly plays an important role in this process.

72.3 HPV and Cancer

72.3.1 Transformation from Infection to Cancer

The process of HPV-related carcinogenesis starts with infection and is followed by persistent infection, development of a high-grade precursor or "precancerous" lesion, and invasion (Fig. 72.1). The most important viral factor for progression to precancerous lesion or invasion is the HPV type, specifically high-risk types. In total, this process from HPV infection to cancer can take many years (~ 10–20) years to develop [11].

To better understand carcinogenesis following HPV infection, it is essential to also understand several key HPV proteins that drive this process. HPV encodes for eight proteins, of which HPVs E5, E6, and E7 are known viral oncoproteins. E6 and E7 are major oncogenic drivers and together are sufficient to immortalize primary human keratinocytes [12, 13]. E6 primarily functions by interfering with cell survival pathways, and E7 primarily functions by promoting cellular proliferation. Arguably the most important cellular interactions allow E6 and E7 to inhibit two well-established tumor suppressors, p53 and retinoblastoma protein (pRB), respectively. Specifically, E6 simultaneously binds to p53 and E6-associated protein (E6AP), a ubiquitin ligase. Thus, p53 is targeted for degradation via the ubiquitin-proteasome pathway, thereby preventing it from inducing apoptosis [14]. Similarly, E7 also targets pRb for degradation via the ubiquitin-proteasome pathway. This activity of E7 is important for driving cell cycle progression in a differentiating epithelium, which, as mentioned above, is essential for HPV-DNA replication. In addition to the pRb interaction, E7 also binds to the other pocket proteins (i.e., p107 and p130) using its LXCXE motif. Collectively, E7 is able to regulate cellular proliferation, differentiation, and apoptosis by interacting with and deregulating these cellular proteins.

In addition to E6 and E7, E5 plays an important role in early carcinogenesis [15]. E5, a small membrane-bound protein, itself cannot immortalize human primary cells but can enhance the immortalization of keratinocytes by E6/E7 [16]. E5 promotes tumorigenesis in part by enhancing the expression of epidermal growth factor receptor-1 (EGFR-1) and increasing phosphorylation of protein kinase B (AKT). This in turn activates the MEK/ERK1/2 pathway to induce HPV-associated carcinogenesis [16].

72.3.2 HPV and Associated Cancers

In 2018 approximately 690,000 cancer cases were attributed to infection with high-risk HPV types (age-standardized incidence rate of 8.0 per 100,000) [16]. It is second only to *Helicobacter pylori* among cancer-causing infections. HPV is associated most frequently with cervical cancer (570,000 attributable cases in 2018), oropharyngeal cancer (OPC) (42,000 attributable cases), and anal squamous cell carcinoma (29,000 cases). Other sites associated with HPV infection include the oral cavity, larynx, penis, vagina, and vulva [18].

Cervical cancer represents 80% of all cases attributed to HPV [18]. It is the fourth most common cancer and the fourth leading cause of cancer death among women worldwide. Due to limited access to screening and treatment of precancerous lesions, cervical cancer disproportionately affects women in low- and middle-income countries (LMICs), with age-standardized rates as high as 43 per 100,000 in regions of Africa, but nearly 6 per 100,000 in North America, Australia, and portions of Europe. In high-resource countries, maximum incidence occurs at age 40, compared to a global average age at diagnosis of 53 years [19]. Nearly all cervical cancer is caused by HPV and specifically high-risk HPV types 16 and 18. Together they cause an estimated 72% of all cervical cancer cases, while other high-risk types 31, 33, 45, 52, and 58 cause another 17% of cases [18]. Current treatment for cervical cancer depends on the stage at diagnosis. Early-stage treatment usually involves surgical techniques and may include modified radical hysterectomy with pelvic lymphadenectomy; however, HPV-related cervical cancers are radiosensitive, and radiation therapy may be considered as well. Persistent, recurrent, or metastatic cervical cancer is treated with chemotherapy and vascular endothelial growth factor (VEGF) inhibitor. Options for immunotherapy, including therapeutic vaccines, are currently emerging and under investigation [7].

Oropharyngeal cancer (OPC), a subset of head and neck squamous cell cancers (HNSCCs), includes cancers arising at the base of the tongue, tonsils, posterior pharyngeal wall, and soft palate and represents the second most common HPV-induced cancer site. While nearly all cervical cancers are caused by HPV infection, OPC has two distinct etiologies: tobacco and alcohol use or HPV infection. Both may be present, but HPV-induced OPC (HPV-OPC) tends to occur in younger males with less tobacco and alcohol use [7]. Interestingly, OPC associated with tobacco/alcohol has decreased over the last decade, while HPV-OPC has increased, particularly among men. Similar to cervical cancer, HPV-OPC is primarily associated with infection from HPV types 16 and 18 (90%), while the other high-risk HPV types are associated with the remaining cases of HPV-OPC. Unlike cervical cancer, HPV-OPC is increasing primarily in high-income countries, with rates in LMICs remaining low [20]. When compared to HPV-negative OPC, HPV-OPC is highly responsive to treatment and exhibits superior treatment outcomes and survival. Because of this, the American Joint Committee on Cancer recently revised guidance indicating that OPCs should be staged, with different staging for HPV-positive and HPV-negative OPC [21]. However, treatment for OPC is the same regardless of HPV status. Ongoing studies are examining options for de-escalation among HPV-OPC with many omitting chemotherapies in favor of RT alone when HPV-positive patients present with low risk of metastatic disease [7].

Immunosuppression is a key factor in the progression from HPV infection to precancer or cancer, and, thus, people living with HIV (PLWH) represent an important, high-risk group. Due to similar risk and transmission factors, many PLWH are also infected with HPV, and women living with HIV (WHIV) are more likely to be diagnosed with high-grade precancerous lesions or invasive cervical cancer compared to HIV-uninfected women. In fact, cervical cancer is up to five times more likely to develop among WHIV. This is due to a multitude of factors including increased risk of infection with HPV, reduced HPV clearance, simultaneous infection with multiple types of HPV, and compromised immune function [7]. Because HIV infection increases the risk of progression to cancer, cervical cancer is considered an acquired immunodeficiency syndrome (AIDS)-defining condition [7].

The relationship between HIV status and HPV-related OPC is less clear. HNSCC, regardless of HPV status, is threefold higher in PLWH in the USA. When stratified by HPV status, the incidence of HPV-related HNSCCs increased, while HPV-unrelated HNSCC decreased [22]. The risk of developing OPC is higher for PLWH, though it is not yet understood whether the distribution of HPV types in OPCs differs between PLWH and the general population [23]. In the USA, recent estimates show a 1.6-fold increase in HPV-related OPCs among PLWH during the period of 1996–2012, though HPV-unrelated OPCs exhibited a similar increase (2.2-fold) [24]. Worth mentioning, though relatively rare in the general population, anal cancer associated with HPV infection is much higher among PLWH and is the most common non-AIDS-defining cancer among PLWH. As with cervical cancer, immunosuppression facilitates the progression of HPV infection to cancer at the anal canal. In the USA anal cancer was 19 times more likely to occur in PLWH compared to the general population [24].

72.4 HPV-Related Cancer Metastasis

HPV-related cancers are particularly responsive to radiotherapy and, thus, show favorable survival outcomes. However, as with all cancers, a proportion of HPV-related cancers recur and metastasize, and diagnoses at later stages make this outcome more likely. As previously mentioned, the HPV E6 and E7 proteins directly contribute to malignant transformation via their ability to impair p53 and pRB. In both oral and cervical epithelial cells, E6 and E7 from high-risk HPVs lead to upregulation of proteins associated with the epithelial-to-mesenchymal transition (EMT) which modulates the progression of associated carcinomas [8]. Additionally, HPV type may influence the outcome. Data from The Cancer Genome Atlas suggest improved survival for HPV-OPC may be limited to cases caused by HPV16 [25]. Studies of cervical cancer have found HPV 18 more closely associated with cervical adenocarcinoma than squamous cell carcinoma, and adenocarcinoma can be more aggressive [4]. This is an area of ongoing study.

72.4.1 Cervical Cancer Metastasis

Nearly all cervical cancer is caused by HPV. In these cancers, spread primarily occurs via direct local extension and lymphatic embolization. Approximately 5% of cervical cancer cases will have hematogenous spread via veins and lymphatic spaces that lie close to the basement membrane with the most common sites of spread including the lungs, liver, and bone. However, hematogenous spread usually occurs only in the advanced stage, poorly differentiated tumors, and aggressive cell types [26].

Lymphatic spread is one of the major routes of cervical cancer metastases due to the close proximity of lymphatic spaces to the basement membrane (1–2 mm) and, as such, is even more prevalent with large-volume tumors. The most common site of the first draining lymph node is the external iliac, followed by the obturator, parametrial, and common iliac nodes [27]. Pelvic lymph node metastasis (LNM) occurs in approximately 15–25% of cervical cancer cases with stage IB–IIA disease and is an important prognostic factor, reducing 5-year survival rates from 80 to 53% when present.

In addition to lymphatic metastases, cervical cancer can also metastasize via vascular pathways. Several factors facilitate lymphovascular metastases from primary cervical cancer. Lymphatic vessel density (LVD), for example, is positively associated with lymph node metastases in cases of cervical cancer. Specifically, increasing peritumoral LVD has been linked with growth factor expression, lymphovascular space invasion, LNMs, and overall survival. Vascular endothelial growth factor (VEGF), a key protein in cervical cancer metastases, especially through the lymphovascular system, plays a role in the blood and lymph vessel formation. Many studies have shown that VEGF is a promoter of cervical cancer progression through many oncogenic signaling pathways. VEGF ligand C (VEGF-C) secreted by the primary tumor is known to induce expansion of the lymphatic network within the primary sentinel lymph node. Tumors expressing VEGF-C, such as cervical tumors, are more likely to metastasize from the tumor-involved lymph node. Two oncogenic signaling pathways associated with VEGF-C, for example, include cyclooxygenase-2 (COX-2) and semaphorin 4D (Sema4D). VEGF-C is correlated with COX-2 expression, lymphangiogenesis, and lymph node metastasis, and VEGF-C/D is associated with angiogenesis inducing SEMA4D, which together are positively correlated with lymphatic invasion, lymph node metastases, and FIGO stage [27].

The relative simplicity of HPV compared to other oncogenic viruses, such as *Epstein-Barr virus*, and its unique ability to elicit a host immune response limits its direct role in cancer progression. However, HPV E5, E6, and E7 proteins are associated with many roles necessary to cancer progression including regulation of cell cycle, cellular senescence, cell motility, and invasion and compromised cellular genomic stability. The cellular changes from these proteins during HPV integration are associated with the malignant behavior of HPV-induced cancers. Metastasis is often a late event in HPV-induced cancers that mostly relies on genomic instability and chromosome aberrations inversely associated with the transforming capability of the various high-risk strains [8].

72.4.2 Metastases in OPC

As previously described, patients with HPV-OPC have better overall and disease-specific survival compared to HPV-negative OPC [25]. However, HPV-OPC has not been studied as extensively as cervical cancer, and many questions remain to be answered, particularly regarding the differences and similarities between HPV-driven OPC and OPC caused by tobacco and alcohol use.

One area of study is whether HPV-positive and HPV-negative OPCs have different rates and sites of distant metastases. Patients with HPV-OPC are more likely to present with an enlarged lymph node as the primary symptom. Approximately half of all HPV-OPC cases have an enlarged lymph node, compared to only 18% of HPV-negative OPC cases, and HPV-OPC is known to metastasize to lymph nodes earlier in the disease stage (with smaller primary tumors) than HPV-negative OPC [7, 28].

Regardless of HPV status, studies have found no significant difference in the site of first progression (recurrence), whether locoregional or distant, and distant metastases tend to occur within 3 years of the primary OPC. Some retrospective analyses have suggested HPV-OPC is more likely to have atypical sites for distant metastases, though this may be due to longer survival among HPV-OPC patients, allowing more time to develop metastases at sites not frequently observed among HPV-negative patients [25]. Common sites for distant metastases of OPC, regardless of etiology, are the lungs, liver, and bone [29, 30], although a retrospective review of ~ 1500 HNSCC cases from a single institution showed a higher rate of metastases to the lung in HPV-positive HNSCC (5.5%) compared to HPV-negative HNSCC (1.8%) [29]. Recent data have also found that 36% of HPV-positive tumors metastasized to multiple sites, compared to only 2% of HPV-negative tumors. In studies investigating timing and location of distant progression, HPV-OPC had a longer time to distant progression than HPV-negative OPC, though not statistically different, and cases of HPV-OPC were significantly more likely to progress to nonregional lymph nodes than HPV-negative OPC. Overall, the end-organs of metastases were significantly different between HPV-positive and HPV-negative OPCs, but when compared by specific organs, only nonregional lymph nodes were significantly different. This suggests that HPV-OPCs may metastasize through the lymphatic system, whereas HPV-negative OPCs may not to the same degree [30].

The molecular events governing this process are not completely understood, but growing evidence supports the role of HPV viral proteins in promoting cellular invasion and migration potential. For instance, HPV E6 and E7 regulate EMT through direct (i.e., downregulating E-cadherin) and indirect mechanisms (through Jagged1 and the PI3-K pathway). Furthermore, HPV E7 leads to the upregulation of p16 via the loss of pRB with reduced negative feedback [31, 32]. Overexpression of p16 has been shown to regulate vascular invasiveness and stimulate nodal spread. This occurs by promoting growth of tumor cells via inhibition of angiogenesis and enhancement of lymphangiogenesis [33]. In addition to E6 and E7, a recent HPV interactome study discovered that HPV late capsid protein (L2) interacts with RNF20 and RNF40, which form a heterodimeric ubiquitin ligase complex that monoubiquitinates histone H2B at lysine 120. Decreased activity of the RNF20/RNF40 complex is associated with increased invasiveness and tumor progression, and expression of L2 upregulates several genes in the EMT and tumor invasiveness pathways [34]. Though L2 has not been shown to contribute to metastasis in vivo, these data support a potential role for this protein in viral-mediated EMT that needs to be explored. In addition to viral proteins, numerous studies have implicated miRNA in all stages of cancer progression [35]. HPV oncoproteins can dysregulate the expression of several cellular miRNAs and in doing so contribute to the development of HPV-positive malignancies.

72.5 HPV-Related Cancers: From Control to Elimination

As the understanding of the mechanisms by which HPV causes cancer has improved, tools to control and prevent those cancers have been developed, making it possible to eliminate HPV-related cancers.

72.5.1 Vaccine

Primary prevention of HPV-related cancer was made possible with the advent of HPV vaccines. Three licensed, prophylactic HPV vaccines have been developed, all providing protection against HPV types 16 and 18, the types responsible for ~ 70% of cervical cancers and ~ 90% of oropharyngeal cancers. The bivalent vaccine only provides protection against HPV types 16 and 18, while the quadrivalent vaccine provides additional protection against HPV6 and 11 which cause genital warts. Finally, the 9-valent vaccine includes protection against five additional high-risk HPV types: 31, 33, 45, 52, and 58. Long-term follow-up has found that all three vaccines provide sustained immunogenicity and effectiveness against HPV infection [36–38].

The HPV vaccines have also been shown to protect against other HPV-related diseases, including vulvar, vaginal, and anal cancers and genital warts [39]. Based on evidence from ancillary studies to vaccine trials, in 2020 the US Food and Drug Administration granted provisional licensure of the 9-valent vaccine against oropharyngeal and other head and neck cancers caused by the HPV types targeted by the vaccine [40–42]. At the time of this writing, a definitive, phase III international, multi-center randomized, double-blind, placebo-controlled clinical trial to determine vaccine effectiveness against persistent oral HPV infections (NCT04199689) is underway and will be completed December 31, 2025.

72.5.2 Screening and Treatment for Preneoplastic Lesions

Currently, cervical cancer is the only HPV-related cancer for which there exist evidence-based screening methods. Cervical intraepithelial neoplasia (CIN) is a precancerous lesion to cervical cancer that may be detected via screening and treated to prevent progression to cancer. Several methods exist to detect CIN. Cytology-based screening (Pap testing) was implemented in high-income countries as early as the 1950s and has resulted in dramatic declines in cervical cancer incidence due to its ability to detect morphologic changes in the cervical epithelium and early HPV infections. There is now a shift from cytology-based screening to testing for high-risk HPV as it is more effective, particularly in populations with high rates of HPV vaccine uptake [7, 43].

The use of high-risk HPV testing will also benefit women living in low- and middle-income countries (LMICs), where a lack of national screening programs has contributed to high rates of cervical cancer and associated morbidity and mortality. In low-resource settings, the *American Society of Clinical Oncology* (ASCO) recommends two-lifetime screenings with HPV testing [44]. Researchers are also studying the efficacy of HPV testing from self-collected swabs for cervical cancer screening. This method could not only improve screening among women in low-resource countries, it can help reduce the screening disparities that have persisted in high-resource countries.

Screening is only effective for secondary prevention if women receive treatment for precancerous and cancerous lesions. Effective ablative treatments exist, including methods that are relatively affordable in LMICS, and can extirpate and therefore cure 75–85% of high-grade cervical lesions before ever progressing to cancer [45]. However, disparities in treatment persist, and efforts are needed to ensure all women with cervical lesions are connected to care.

72.5.3 Elimination of HPV-Related Cancers

As described above, metastatic and recurrent cancers associated with HPV result in significant morbidity and mortality worldwide. Without further intervention, modelers predict 44.4 million cases of cervical cancer will be diagnosed globally between 2020 and 2069 [46]. In May 2018, the World Health Organization (WHO), led by Director-General Tedros Adhanom Ghebreyesus, issued a Call to Action to eliminate cervical cancer with adoption by the member states in August 2020 [47]. The WHO strategy for eliminating cervical cancer consists of 90% of girls vaccinated with the HPV vaccine by the age of 15; 70% of women screened using a high-performance test by the age of 35 and again by the age of 45; and 90% of women identified with cervical disease treated.

Elimination of a disease is defined as a reduction to zero, or near zero, incidence in a defined geographic area [48]. Control is the reduction of disease incidence, prevalence, morbidity, and mortality to acceptable levels. In high-resource countries such as the USA, Australia, and the UK, cervical cancer screening programs have been established for decades, and cervical cancer is considered controlled.

The critical tools for eliminating HPV-related cancers are vaccine, screening, and treatment for precancers and cancers. The vaccine will reduce incidence and transmission of HPV infection, which will in turn reduce HPV-related diseases. Mathematical modeling has shown that we can eliminate cervical cancer in some countries by 2030 with other countries to follow; however, the fastest declines in cervical cancer incidence will only be seen if there is a rapid scale-up of cervical cancer screening. HPV 16, 18, 6, and 11 can be eliminated with gender-neutral vaccination of $\geq 80\%$ by the quadrivalent or 9-valent vaccine, which would prevent the related diseases that affect many men and women worldwide [46, 49, 50].

72.6 Conclusion

Cervical and oropharyngeal cancer represent a unique circumstance in cancer prevention and care. Diagnoses of either often accompanies severe mental and physical challenges, with late-stage diagnoses comes the potential loss of fertility (cervical cancer) and extreme physical transformations of the head and neck from treatment (OPC). In spite of these challenges, most cases of cervical cancer, half of OPC, and many other anogenital cancers can be prevented with the HPV vaccine, potentially preventing the devastating consequences of these cancers. However, we have not yet reached universal vaccination against HPV, and while treatable precancerous lesions exist for cervical and anal cancer, many at-risk persons still do not participate in screening. Further, there are no FDA-approved screening options for OPC or cancers at other anogenital sites. While we continue to make advancements in HPV-related cancers, there is still much to be understood about the associated progression and management of disease.

Open Questions

- How do we improve vaccination and screening rates for cervical cancer?
- Can we develop a screening test for OPC to improve early detection of disease?
- Determine differences in metastases of OPC based on HPV status?
- What molecular differences drive changes in metastasis for HPV-positive cancers?
- Will de-escalation reduce undesirable physical challenges associated with OPC treatment while maintaining treatment efficacy?

References

1. International Human Papillomavirus (HPV) Reference Center. 2018. https://www.hpvcenter.se/. Accessed 10 Dec 2020.
2. Egawa N, Doorbar J. The low-risk papillomaviruses. Virus Res. 2017;231:119–27.
3. de Sanjosé S, Brotons M, Pavón MA. The natural history of human papillomavirus infection. Best Pract Res Clin Obstet Gynaecol. 2018;47:2–13.
4. Doorbar J, Quint W, Banks L, Bravo IG, Stoler M, Broker TR, et al. The biology and life-cycle of human papillomaviruses. Vaccine. 2012;30:F55–70.
5. Bosch FX, Castellsagué X, Muñoz N, De Sanjosé S, Ghaffari AM, González LC, et al. Male sexual behavior and human papillomavirus DNA: key risk factors for cervical cancer in Spain. J Natl Cancer Inst. 1996;88(15):1060–7.
6. Howard JD, Chung CH. Biology of human papillomavirus-related oropharyngeal cancer. Semin Radiat Oncol. 2012;22:187–93.
7. Berman TA, Schiller JT. Human papillomavirus in cervical cancer and oropharyngeal cancer: one cause, two diseases. Cancer. 2017;123:2219–29.
8. Müller-Coan BG, Caetano BFR, Pagano JS, de Elgui OD. Cancer progression goes viral: the role of oncoviruses in aggressiveness of malignancies. Trends Cancer. 2018;4:485–98.
9. Smith JS, Gilbert PA, Melendy A, Rana RK, Pimenta JM. Age-specific prevalence of human papillomavirus infection in males: a global review. J Adolesc Health. 2011;48:540–52.
10. Mcquillan G, Kruszon-Moran D, Markowitz LE, Unger ER, Paulose-Ram R. United States, 2011–2014 National Health and Nutrition Examination Survey (NHANES) 2011. https://www.cdc.gov/nchs/data/databriefs/db280_table.pdf#1. Accessed 2020 Nov 24.
11. Schiffman M, Castle PE, Jeronimo J, Rodriguez AC, Wacholder S. Human papillomavirus and cervical cancer. Lancet. 2007;370:890–907.
12. Münger K, Phelps WC, Bubb V, Howley PM, Schlegel R. The E6 and E7 genes of the human papillomavirus type 16 together are necessary and sufficient for transformation of primary human keratinocytes. J Virol. 1989;63(10):4417–21.
13. Hawley-Nelson P, Vousden KH, Hubbert NL, Lowy DR, Schiller JT. HPV16 E6 and E7 proteins cooperate to immortalize human foreskin keratinocytes. EMBO J. 1989;8(12):3905–10.
14. Pal A, Kundu R. Human papillomavirus E6 and E7: the cervical cancer hallmarks and targets for therapy. Front Microbiol. 2020;10:3116.
15. Estêvão D, Costa NR, da Gil Costa RM, Medeiros R. Hallmarks of HPV carcinogenesis: the role of E6, E7 and E5 oncoproteins in cellular malignancy. Biochim Biophys Acta Gene Regul Mech. 2019;1862:153–62.
16. Zhang L, Wu J, Ling MT, Zhao L, Zhao KN. The role of the PI3K/Akt/mTOR signalling pathway in human cancers induced by infection with human papillomaviruses. Mol Cancer. 2015;14:87.
17. Crosbie EJ, Einstein MH, Franceschi S, Kitchener HC. Human papillomavirus and cervical cancer. Lancet. 2013;382:889–99.
18. de Martel C, Georges D, Bray F, Ferlay J, Clifford GM. Global burden of cancer attributable to infections in 2018: a worldwide incidence analysis. Lancet Glob Health. 2020;8(2):e180–90.
19. Arbyn M, Weiderpass E, Bruni L, de Sanjosé S, Saraiya M, Ferlay J, et al. Estimates of incidence and mortality of cervical cancer in 2018: a worldwide analysis. Lancet Glob Health. 2020;8(2):e191–203.
20. Gillison ML, Chaturvedi AK, Anderson WF, Fakhry C. Epidemiology of human papillomavirus-positive head and neck squamous cell carcinoma. J Clin Oncol. 2015;33:3235–42.
21. Lydiatt WM, Patel SG, O'Sullivan B, Brandwein MS, Ridge JA, Migliacci JC, et al. Head and neck cancers-major changes in the American Joint Committee on cancer eighth edition cancer staging manual. CA Cancer J Clin. 2017;67(2):122–37.
22. Beachler DC, Abraham AG, Silverberg MJ, Jing Y, Fakhry C, Gill MJ, et al. Incidence and risk factors of HPV-related and HPV-unrelated head and neck squamous cell carcinoma in HIV-infected individuals. Oral Oncol. 2014;50(12):1169–76.
23. Wang CC, Palefsky JM. Human papillomavirus-related oropharyngeal cancer in the HIV-infected population. Oral Dis. 2016;22:98–106.
24. Hernández-Ramírez RU, Shiels MS, Dubrow R, Engels EA. Cancer risk in HIV-infected people in the USA from 1996 to 2012: a population-based, registry-linkage study. Lancet HIV. 2017;4(11):e495–504.
25. Faraji F, Eisele DW, Fakhry C. Emerging insights into recurrent and metastatic human papillomavirus-related oropharyngeal squamous cell carcinoma. Laryngosc Investig Otolaryngol. 2017;2:10–8.
26. Gallup D. The spread and staging of cervical cancer. London: The Global Library of Women's Medicine's; 2009.
27. Ahmad A. Introduction to cancer metastasis. Cambridge: Academic Press; 2016. p. 1–403.
28. Paver EC, Currie AM, Gupta R, Dahlstrom JE. Human papilloma virus related squamous cell carcinomas of the head and neck: diagnosis, clinical implications and detection of HPV. Pathology. 2020;52:179–91.
29. Sacks R, Law JY, Zhu H, Beg MS, Gerber DE, Sumer BD, et al. Unique patterns of distant metastases in HPV-positive head and neck cancer. Oncology. 2020;98(3):179–85.
30. Tiedemann D, Jakobsen KK, von Buchwald C, Grønhøj C. Systematic review on location and timing of distant progression in human papillomavirus-positive and human papillomavirus-negative oropharyngeal squamous cell carcinomas. Head Neck. 2019;41:793–8.
31. Lawrence MS, Sougnez C, Lichtenstein L, Cibulskis K, Lander E, Gabriel SB, et al. Comprehensive genomic characterization of head and neck squamous cell carcinomas. Nature. 2015;517(7536):576–82.
32. Chung CH, Gillison ML. Human papillomavirus in head and neck cancer: its role in pathogenesis and clinical implications. Clin Cancer Res. 2009;15:6758–62.
33. Dok R, Glorieux M, Holacka K, Bamps M, Nuyts S. Dual role for p16 in the metastasis process of HPV positive head and neck cancers. Mol Cancer. 2017;16:113.
34. Eckhardt M, Zhang W, Gross AM, Von Dollen J, Johnson JR, Franks-Skiba KE, et al. Multiple routes to oncogenesis are promoted by the human papillomavirus-host protein network. Cancer Discov. 2018;8(11):1474–89.
35. Santos JMO, da Silva SP, Costa NR, da Gil Costa RM, Medeiros R. The role of microRNAs in the metastatic process of high-risk HPV-induced cancers. Cancers. 2018;10:493.
36. Olsson SE, Restrepo JA, Reina JC, Pitisuttithum P, Ulied A, Varman M, et al. Long-term immunogenicity, effectiveness, and safety of nine-valent human papillomavirus vaccine in girls and boys 9 to 15 years of age: Interim analysis after 8 years of follow-up. Papillomavirus Res. 2020;10:100203.
37. Kjaer SK, Nygård M, Dillner J, Marshall JB, Radley D, Li M, et al. A 12-year follow-up on the long-term effectiveness of the quadrivalent human papillomavirus vaccine in 4 Nordic countries. Clin Infect Dis. 2018;66(3):339–45.
38. Schwarz TF, Galaj A, Spaczynski M, Wysocki J, Kaufmann AM, Poncelet S, et al. Ten-year immune persistence and safety of the HPV-16/18 AS04-adjuvanted vaccine in females vaccinated at 15–55 years of age. Cancer Med. 2017;6(11):2723–31.
39. Luostarinen T, Apter D, Dillner J, Eriksson T, Harjula K, Natunen K, et al. Vaccination protects against invasive HPV-associated cancers. Int J Cancer. 2018;142:2186–7.

40. Lehtinen M, Apter D, Eriksson T, Harjula K, Hokkanen M, Lehtinen T, et al. Effectiveness of the AS04-adjuvanted HPV-16/18 vaccine in reducing oropharyngeal HPV infections in young females—Results from a community-randomized trial. Int J Cancer. 2020;147(1):170–4.
41. Mehanna H, Bryant TS, Babrah J, Louie K, Bryant JL, Spruce RJ, et al. Human papillomavirus (HPV) vaccine effectiveness and potential herd immunity for reducing oncogenic oropharyngeal HPV-16 prevalence in the United Kingdom: a cross-sectional study. Clin Infect Dis. 2019;69(8):1296–302.
42. Chaturvedi AK, Graubard BI, Broutian T, Pickard RKL, Tong ZY, Xiao W, et al. Effect of prophylactic human papillomavirus (HPV) vaccination on oral HPV infections among young adults in the United States. J Clin Oncol. 2018;36(3):262–7.
43. Wentzensen N, Arbyn M. HPV-based cervical cancer screening-facts, fiction, and misperceptions. Prev Med (Baltim). 2017;98:33–5.
44. Ogilvie G, Nakisige C, Huh WK, Mehrotra R, Franco EL, Jeronimo J. Optimizing secondary prevention of cervical cancer: recent advances and future challenges. Int J Gynecol Obstet. 2017;138:15–9.
45. Castle PE, Murokora D, Perez C, Alvarez M, Quek SC, Campbell C. Treatment of cervical intraepithelial lesions. Int J Gynecol Obstet. 2017;138:20–5.
46. Simms KT, Steinberg J, Caruana M, Smith MA, Lew JB, Soerjomataram I, et al. Impact of scaled up human papillomavirus vaccination and cervical screening and the potential for global elimination of cervical cancer in 181 countries, 2020–99: a modelling study. Lancet Oncol. 2019;20(3):394–407.
47. World Health Organization. 73rd World Health Assembly Decisions. Newsroom. 2020. https://www.who.int/news-room/feature-stories/detail/73rd-world-health-assembly-decisions. Accessed 23 Nov 2020.
48. Dowdle WR. The principles of disease elimination and eradication. Bull World Health Organ. 1998;76(Suppl. 2):22–5.
49. Brisson M, Bénard É, Drolet M, Bogaards JA, Baussano I, Vänskä S, et al. Population-level impact, herd immunity, and elimination after human papillomavirus vaccination: a systematic review and meta-analysis of predictions from transmission-dynamic models. Lancet Public Health. 2016;1(1):e8–17.
50. Hall MT, Simms KT, B L, Smith MA, Brotherton JM, Saville M, et al. The projected timeframe until cervical cancer elimination in Australia: a modelling study. Lancet Public Health. 2019;4(1):e19–27.

Microbiome and Cancer Metastasis

73

Michael G. White, Jennifer A. Wargo, and Jennifer L. McQuade

Abstract

The commensal microbiome is a unique set of bacteria, viruses, and fungi that exist on and in a person's body. As these organisms are increasingly understood as a necessary and symbiotic part of human health, they have also been associated with the development and progression of a variety of cancers. Additionally, these organisms have been noted to aid malignancies in their ability to evade host immune surveillance functions as well as inactivate cytotoxic chemotherapies. As the role of the tumor microenvironment and immune system is increasingly recognized as vital to the treatment of malignant disease, the gut, tumoral, and circulating microbiome must be included in this equation as well.

Learning Objectives

- Understand the role of the gut, circulating, and tumoral microbiome in the development of malignancy, response to treatment, and evasion of therapeutics.
- Describe the role of preclinical models of the gut microbiome in understanding links between correlative associations and causative effects.
- Articulate future directions of study including fecal microbial transplant, consortia microbiome, and metabolomics.

M. G. White · J. A. Wargo
Department of Surgical Oncology, The University of Texas MD Anderson Cancer Center, Houston, TX, USA
e-mail: mgwhite@mdanderson.org; jwargo@mdanderson.org

J. L. McQuade (✉)
Department of Melanoma Medical Oncology, The University of Texas MD Anderson Cancer Center, Houston, TX, USA
e-mail: jmcquade@mdanderson.org

73.1 Introduction

The commensal microbiome is composed of communities of bacterial and viral flora constituting a complex ecosystem that exists in various configurations at various sites throughout the body. It is unique to each individual, varies over time, and responds to physiologic stress as well as changes in a person's environment. This system is multifaceted, particularly when considering its interactions with the host immune system in maintaining and modulating homeostasis [1, 2]. As this symbiotic relationship between the gut microbiome and host immune systems continues to be defined, the effect of both the gut and tumor microbiome on cancer and cancer therapies can be categorized as follows: (1) *the direct effect of the gut microbiome on cancer initiation and progression*, (2) *the effect of the tumor microbiome on the progression of disease*, and (3) *the effect of both the gut and tumor microbiome on the response to treatment* (Fig. 73.1).

73.1.1 The Microbiome and Carcinogenesis

The gut microbiome is a community of trillions of bacteria with increasing complexity as one travels distally through the gastrointestinal tract. These organisms play an essential role in digestion, through fermentation, of various substrates and in enzymatic metabolism. These processes not only aid in nutrient absorption but also produce various metabolites known to have local and systemic effects [3]. These metabolites have been shown to act as systemic mediators of overall health and may also have direct local effects on cancer initiation and progression and immune modulation. Similarly, gut bacteria have been associated with local inflammation and have been directly linked to carcinogenesis through a variety of mechanisms.

Currently, at least 20% of cancers are linked to microbes, and many more associations are being discovered given new tools for analysis [4]. One of the most notable associations to

S. P. Leong et al. (eds.), *Cancer Metastasis Through the Lymphovascular System*,
https://doi.org/10.1007/978-3-030-93084-4_73

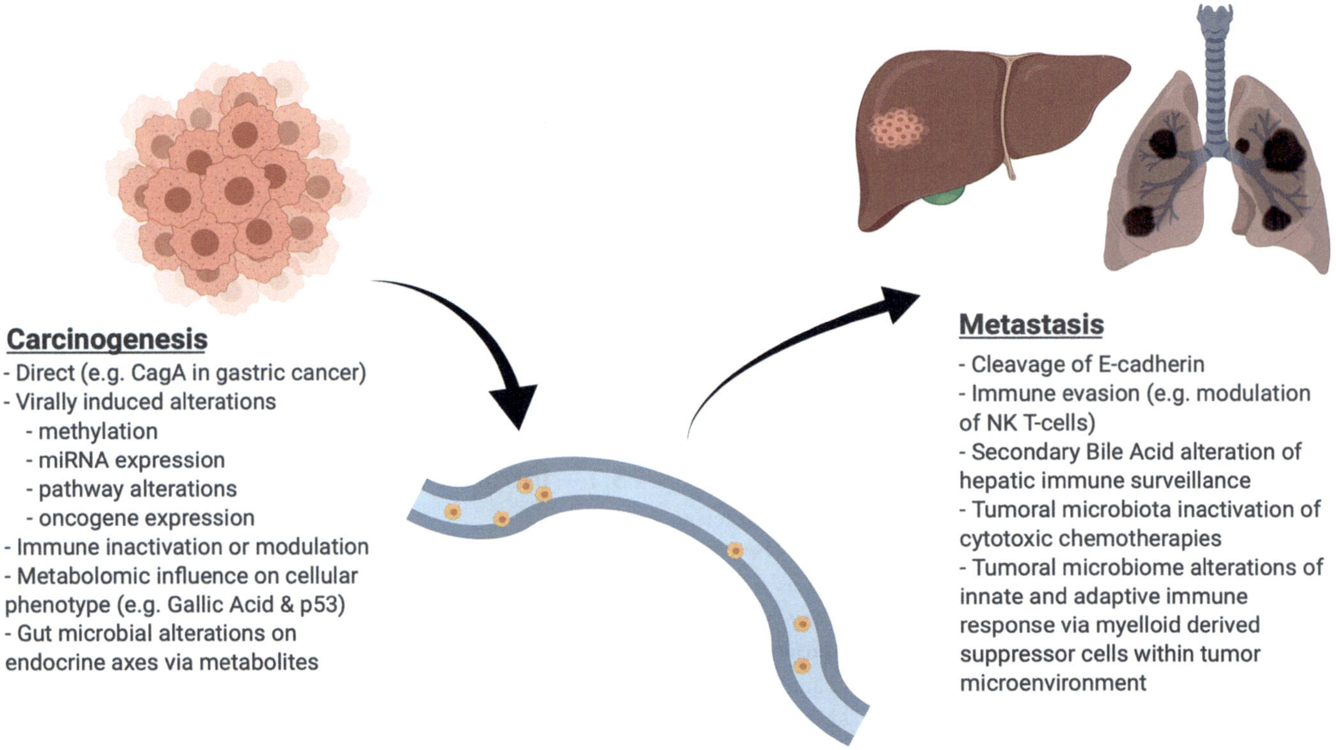

Fig. 73.1 Mechanisms through which commensal microbial bacteria lead to carcinogenesis and metastasis. Created by BioRender.com

date includes the implication of *Helicobacter pylori* in gastric cancer carcinogenesis—for which Barry Marshall and Robin Warren won the Nobel Prize in Medicine and Physiology in 2005. As a result of the dysregulated interface between an overactive host immune system, triggered by a pathogenic bacterium, this example illustrates how host microbial interactions can contribute to malignancy.

Initial correlative work studying the effect alterations in the gut microbiome have on initiation and/or progression of malignancies demonstrated associations between a preponderance or paucity of various bacterial taxa and malignancy. Ongoing work is elucidating causative associations, although these associations are largely preliminary and true causality in the clinical setting remains elusive. Causality is perhaps best demonstrated in the association between *H. pylori* and the development of gastric cancer. This interaction involves the virulence factors of specific strains of *H. pylori*, germline risk, and environmental factors including diet and the microbiome inducing a local inflammatory response and carcinogenesis in a subset of individuals with *H. pylori* colonization [5]. The role of germline risk has been well demonstrated by Kodaman and colleagues. Their work demonstrated that ancestral background correlated with gastric cancer risk. Moreover, this risk is predicated on the strain of *H. pylori* with which a patient is colonized [6]. While African *H. pylori* ancestry was relatively benign in those patients with African ancestry, it conferred substantial risk for those patients with Amerindian ancestry.

Along with the germline risk, *H. pylori* can directly induce carcinogenesis through production of the CagA oncogene. CagA induces cell proliferation through activation of PI3K-AKT, MEK-ERK, and β-catenin-WNT pathways leading to proliferation and potential malignant degeneration [7–9]. Similarly, CagA interferes with p53 and RUNX3, decreasing apoptosis and further increasing the risk of malignancy [10, 11]. Alterations of the host immune system through induction of increased local IL8 by CagA have also been suggested as a factor leading to the development of gastric cancer in these patients [12]. Finally, on a somatic mutational level, colonization by *H. pylori* has been shown to induce genetic instability in cell line, murine, and clinical specimens [5, 13]. The multifactorial nature of this oncogenic process is indicative of the complex interplay between host and commensal flora. In the instance of *H. pylori*, carcinogenesis involves local immune modulation, increased proliferation, and induction of somatic mutations leading to the de-differentiation of terminally differentiated gastric cells, subsequent hyperplasia, and carcinogenesis.

Beyond bacterial associations, the relationship between pathogenic viral colonization and carcinogenesis has been clearly defined in a number of malignancies. Infection with hepatitis B or C viruses has been shown to induce a cascade of changes resulting in alteration of DNA methylation, miRNA expression, and activation of a number of signal transduction pathways and can lead to hepatocellular carcinoma (HCC). Additionally, immune-mediated changes and effects of oxidation on the tumor immune microenvironment induced by hepatitis B and C viruses allow these tumors to evade the host immune system [14].

Anal squamous cell carcinoma, a variety of genitourinary carcinomas, and head and neck squamous cell carcinoma have been linked to human papillomavirus (HPV). Mechanistically, this has been shown to be largely due to the over-expression of oncogenes E6 and E7 leading to the alteration of various signaling pathways including p53, PI3K, Akt, and Wnt/Notch. Meanwhile oncogene E5 has been implicated in evasion of the host immune system [15]. Unlike their bacterial counterparts, these pathogenic viruses can be prevented by vaccination leading to a decreased disease incidence, as has been demonstrated with HPV-related cervical cancer [16, 17]. Unfortunately, eradication of hepatitis C virus with antivirals does not appear to significantly alter the risk of development of hepatocellular carcinoma after an initial infection [18].

In delineating mechanisms of carcinogenesis, the direct effect of the gut microbiome in the development of mucosal tumors, including colorectal adenomas and colorectal cancers, has been of significant interest given their inherent proximity. While initial studies failed to demonstrate a difference in commensal flora between cases and controls [19] gut dysbiosis as measured by either a preponderance of specific taxa or lack of diversity has since been noted to be associated with the generation of adenomas and cancers in further studies [20–23].

Although the vast majority of the microbiome exists within the gastrointestinal tract, mucosal surfaces throughout the body harbor bacterial and viral flora. In the case of lung cancer, the diversity and specific taxa within lower bronchial commensal flora have been associated with carcinogenesis and higher-stage disease. Moreover, these findings were able to be recapitulated within a murine model and demonstrated to be mediated by IL17's effects on the tumor's immune infiltrate [24].

The commensal microbiome has been shown to play a role in the development of malignancies, once developed tumors themselves host an array of viruses and bacteria associated with and living within tumors themselves. These organisms are unique in their ability to thrive in an often-hostile and hypoxic environment. Their role in the development of malignancy is an area of active study. Despite the knowledge that bacteria exist within tumors, little work was done in studying the potential role of these bacteria until the recent surge of interest in microbiome research.

The relationship between tumors and their microbiome has been most robustly delineated in the case of *Fusobacterium* within colorectal cancers in observational and mechanistic studies [25]. Higher levels of *Fusobacterium* have been found in colorectal tumors as compared to normally matched colonic mucosa. Interestingly, intra-tumoral *Fusobacterium* is more commonly seen in right-sided cancers and in BRAF-mutated tumors. Additionally, taxonomic associations were noted by canonical molecular subtype and micro-satellite instability [26]. Suggested mechanisms of carcinogenesis are multifactorial. *Fusobacterium* has been suggested as a chemoattractant for myeloid cells, has been shown to increase β-catenin signaling, inactivates NK cells, acts as an immunomodulator, and alters the production of miRNAs [27]. The causality of these effects is emphasized by the ability to negate these risks in *Fusobacterium*-derived colorectal cancer murine model with treatment with antibiotics [28].

Pancreatic cancer also demonstrates a unique tumoral microbial signature that effects the surrounding tumor microenvironment. As with the previously described malignancies, these tumoral bacteria play a number of roles. This includes potentially beneficial alterations that assist in altering both the innate and adaptive immune responses by reducing myeloid-derived suppressor cells, increasing M1 macrophage differentiation, and enhancing differentiation of CD4 and CD8 T cells [29]. Detrimental effects of tumoral pancreatic bacteria and their inactivation of cytotoxic therapies are discussed later in the chapter. While associations with tumoral bacteria are numerous and exist across histologies, the question of correlation versus causation persists and requires systematic investigation [30, 31].

73.1.2 The Role of the Microbiome in Metastatic Disease

The tumor microbiome may also play a role in metastatic disease. Interestingly, resected liver metastases from patients with *Fusobacterium*-positive primary CRCs have been noted to have *Fusobacterium* within the distant metastases as well as their primary tumors [28]. These *Fusobacterium*-associated colorectal cancers were found to behave more aggressively with worse overall survival.

Mechanistically there are variable ways through which *Fusobacterium* can aid or induce metastates [26, 28, 32]. While progression from a localized to metastatic lesion is complex and multifactorial, requiring detachment from a primary lesion, proliferation, and entry into a metastatic site, a number of bacteria have recently been demonstrated to potentially aid in this process [33, 34]. Both *Fusobacterium* and *Bacteroides* have been shown to bind E-cadherin and disrupt epithelial integrity leading to an increased ability of a tumor to detach from a primary malignancy [35, 36]. In the case of *Fusobacterium*, this has been shown to be mediated by a virulence factor, FadA, and disruption of the *Wnt* pathway as well as upregulating CARD3 and promoting metastasis [34, 37]. While *Bacteroides* works via toxin secretion that has been shown to cleave E-cadherin [38]. Once liberated from a primary tumor, *E. coli* and *E. faecium* have been shown to promote organization of metastatic cells through cross talk of quorum-sensing peptides that promote invasion and angiogenesis [39]. Finally, the gut microbiome has been tied to immune evasion through the manipulation of the hepatic immune microenvironment modulated by NK T cells [40]. While certainly not the sole driver of metastases, this evidence increasingly suggests these microbes are much more than bystanders.

Microbial signatures for cancer histologies beyond GI malignancies exist as well. In the case of breast cancer, distinct tumor microbial signatures exist based on the subtype of breast cancer [41]. Moreover, although not in physical contact, the gut microbiome has a significant effect on the progression of murine-modeled breast cancer that exemplifies the global effect the gut microbiome has on not only carcinogenesis but also disease progression. In a murine model of hormone-responsive breast cancer, gut dysbiosis induced by antibiotic administration was shown to increase peritumoral inflammation as well as metastases to both lymph nodes and the lung [42]. Findings such as these call into question the potential sinister effects antibiotic administration during cancer treatment may have.

Beyond individual bacterial taxa, the gut microbiome has long had a clearly defined role in mediating the levels of circulating metabolites, with variability induced by a patient's diet or antibiotics [43–45]. As part of its symbiotic relationship with a host, the gut microbiome breaks down gut contents to aid in nutrient bioavailability. By-products of these reactions have local effects and are also released into the portal and then systemic circulation where they can have distant effects. These metabolites have been shown to affect a host subject's metabolism, body weight, and a variety of health factors [45]. In the case of oncologic health, changes in the gut microbiome can have global effects such as altering the metabolization of estrogens, thereby increasing breast cancer risk [46].

Recent work has suggested metabolites may also play a crucial role in the determination of the function of molecules such as *p53* as alternatively a tumor suppressor or oncogene. This has significant implications in the development of colorectal adenomas and adenocarcinoma as induced dysbiosis in murine and organoid models demonstrated a switch of *p53* from a tumor suppressor to an oncogene. This modeled effect was shown to be driven by a single metabolite, gallic acid, recapitulating the effect of a complex gut microbiome switching the effect of *p53* from a tumor suppressor to an oncogene [47]. Although these effects are likely much more complex in a clinical setting, the ability of a single metabolite to alter *p53* in this reductive model is impressive.

Direct effects have also been noted in the case of the metabolization of bile acids by the gut microbiome increasing the likelihood of hepatocellular carcinoma development. These bile acids have been shown to affect the recruitment of natural killer T cells to cancerous liver cells [40]. Alterations in the gut microbiome and subsequent alterations in the immune cells within the liver have the potential to allow for the development of primary liver tumors as well as metastases through impaired immunologic anticancer surveillance.

Recently, distinct microbial signatures by cancer histology have also been noted in patients' circulating plasma [48]. While circulating bacterial DNA is an intriguing finding on its own, it was also shown that these signals were prognostic of cancer-specific outcomes across multiple histologies [48]. The effects these bacteria have on local and distant tissues continue to be an area of study as they certainly play critical roles in the tumoral immune microenvironment and in the evasion of therapies.

73.1.3 Gut Microbiome: Response to and Evasion of Treatments

An additional layer of complexity is added to the host-microbiome interaction with the addition of treatment, whether immunotherapy or cytotoxic chemotherapy. To date, significant work has begun to delineate these multifaceted interfaces, leading to the initiation of a number of clinical trials studying the effect of therapeutic alterations to the microbiome in order to enhance existing therapies [49]. At the level of direct effect, tumoral microbial flora has been shown to induce resistance to cytotoxic chemotherapy. In the case of pancreatic adenocarcinoma, gemcitabine levels were found to be lower in the tumor than the surrounding tissues. This decrease was then shown to be mediated by tumoral *Gammaproteobacteria* that was able to enzymatically degrade gemcitabine, thereby decreasing its efficacy [50]. Similarly, it has been observed that irinotecan dose variability correlates with the variability of the gut microbiome between patients. Investigators have therefore been able to define a "metabolizer" gut microbe phenotype [51]. While in the case of *Fusobacterium*-associated colorectal cancer, miRNA-activating autophagy can lead to resistance to chemotherapy [52]. These early clinical observations are strengthened by associations in murine models demonstrating an association with the gut microbiome having an impact on the efficacy of cyclophosphamide, a chemotherapeutic that also is an immunomodulator. Further complicating this association is the fact that cyclophosphamide also effects the gut microbiome itself [53]. These associations are demonstrative of the many degrees of freedom that must be accounted for in delineating these interactions.

Variations in the gut microbiome were first associated with response to immunotherapy in murine models that were later translated to clinical associations demonstrating the gut microbiome-modulated response to anti-PD-1 therapy [54]. The gut flora has since been shown to shape response to checkpoint inhibitors across multiple tumor types: epithelial cancers, pancreatic adenocarcinoma, and melanoma [54–57]. This is of particular clinical importance as the use of antibiotics at time of checkpoint inhibitor initiation has been associated with worse outcomes in multiple studies, again suggesting dysbiosis-mediated effects on treatment response [58]. As antibiotic administration is not uncommon among patients receiving systemic therapy for their malignancy, its effect on induced gut dysbiosis and downstream outcomes should be considered. In elucidating the mechanism of this effect, murine models of ICB have been invaluable in demonstrating improved results with fecal microbiota transplantation (FMT) in immunotherapy-treated mice [59]. These findings have been recently translated to a phase 1 clinical trial by *Baruch* and colleagues demonstrating FMT is able to induce a response to ICB in otherwise immunotherapy-refractory melanoma patients. This work was able to find that in a group of ten heavily pretreated patients with metastatic melanoma, FMT resulted in a more diverse microbiome, improved immune infiltrate into tumors, and three clinical responses to treatment including one complete response [60].

There are a number of potential mechanisms actively being investigated to delineate the effects the gut microbiome has on immunotherapies. These mechanisms include both immunostimulatory and immunosuppressive effects. In the case of immunostimulatory effects, proposed mechanisms involve altered antigenicity through T-cell activation. This has been proposed to occur through antigenic, adjuvantic, or bystander effects via unactivated or conditioned dendritic cells and cytotoxic T lymphocytes. Similarly, immunosuppressive effects are thought to be mediated through local suppressor cells or through locally produced bacterial metabolites. As previously mentioned, cytokines or microbial-derived metabolites can also have significant immunostimulatory or immunosuppressive effects with downstream effects on pharmacologically invoked immunomodulation [61].

73.2 Conclusion

The diverse interactions between a patient and their microbiome clearly play an important role as a determinant of overall and oncologic health. The complexity of a system that varies between patients, within patients, and over time, however, makes its study difficult. Nevertheless, the functional basis of these interactions and their effect on carcinogenesis and treatment response is beginning to be elucidated. In order to optimally deliver care to their patients, it is increasingly important for medical and surgical oncologists to remain abreast of these developments.

While induction of gut dysbiosis and downstream detrimental effects can readily be obtained and modeled, the largest hurdle the investigators are facing has been the definition and induction of the "ideal" microbiome with appropriate diversity and specific taxa to optimize both general and oncologic health. The question of what is the optimal microbiome remains to be answered. In fact, there likely is no "ideal" microbiome, but one can optimize an outcome of interest at the expense of other outcomes. To date, interventions to improve a patient's microbiome have largely treated benign disease such as *Clostridium difficile* infections or irritable bowel disease. However, microbiome modulation via fecal transplant is now being tested in clinical trials, while promising results have been noted in optimizing oncologic care in a variety of murine models. Moving forward, the effect diet may have on optimization of immunomodulatory cancer therapies has the potential to improve outcomes and possibly affect the incidence of malignancies at a population level. More directed alterations of the microbiome through the development of fecal consortia, designer probiotics, or individual bacterial metabolites may also lead to further advancements in the field and will likely be adapted to the optimization of oncology care to optimize cancer-specific outcomes as well as minimize adverse effects of treatments.

Significant questions exist studying the link between the commensal microbiome and malignancy. The investigation of these questions and their answers will direct the future of this field:

- What is the optimal composition of the gut microbiome to optimize oncologic outcomes?
- How can the commensal gut microbial composition and the associated metabolome be influenced to optimize outcomes whether through dietary changes, bacterial consortia, or FMT?
- Delineation of microbial mechanisms behind carcinogenesis, progression of disease, evasion of host immune surveillance, and cytotoxic therapies.
- How can the circulating or stool microbial signatures be leveraged to augment or supplant existing screening or surveillance strategies for various malignancies?
- Can the tumoral virome offer insight correlations and potential further causative associations with malignancy?

References

1. Sender R, Fuchs S, Milo R. Revised estimates for the number of human and bacteria cells in the body. PLoS Biol. 2016;14(8):e1002533.
2. Nejman D, Livyatan I, Fuks G, et al. The human tumor microbiome is composed of tumor type-specific intracellular bacteria. Science. 2020;368(6494):973–80.
3. Valdes AM, Walter J, Segal E, Spector TD. Role of the gut microbiota in nutrition and health. BMJ. 2018;361:k 2179.
4. de Martel C, Ferlay J, Franceschi S, et al. Global burden of cancers attributable to infections in 2008: a review and synthetic analysis. Lancet Oncol. 2012;13(6):607–15.
5. Amieva M, Peek RM Jr. Pathobiology of helicobacter pylori-induced gastric cancer. Gastroenterology. 2016;150(1):64–78.
6. Kodaman N, Pazos A, Schneider BG, et al. Human and helicobacter pylori coevolution shapes the risk of gastric disease. Proc Natl Acad Sci. 2014;111(4):1455–60.
7. Nagy TA, Frey MR, Yan F, Israel DA, Polk DB, Peek RM Jr. Helicobacter pylori regulates cellular migration and apoptosis by activation of phosphatidylinositol 3-kinase signaling. J Infect Dis. 2009;199(5):641–51.
8. Suzuki M, Mimuro H, Kiga K, et al. Helicobacter pylori Cag A phosphorylation-independent function in epithelial proliferation and inflammation. Cell Host Microbe. 2009;5(1):23–34.
9. Higashi H, Tsutsumi R, Fujita A, et al. Biological activity of the helicobacter pylori virulence factor Cag A is determined by variation in the tyrosine phosphorylation sites. Proc Natl Acad Sci. 2002;99(22):14428–33.
10. Tsang YH, Lamb A, Romero-Gallo J, et al. Helicobacter pylori Cag A targets gastric tumor suppressor RUNX3 for proteasome-mediated degradation. Oncogene. 2010;29(41):5643–50.
11. Buti L, Spooner E, Van der Veen AG, Rappuoli R, Covacci A, Ploegh HL. Helicobacter pylori cytotoxin-associated gene A (Cag A) subverts the apoptosis-stimulating protein of p 53 (ASPP2) tumor suppressor pathway of the host. Proc Natl Acad Sci. 2011;108(22):9238–43.
12. Barrozo RM, Cooke CL, Hansen LM, et al. Functional plasticity in the type IV secretion system of Helicobacter pylori. PLoS Pathog. 2013;9(2):e1003189.
13. Machado AM, Figueiredo C, Touati E, et al. Helicobacter pylori infection induces genetic instability of nuclear and mitochondrial DNA in gastric cells. Clin Cancer Res. 2009;15(9):2995–3002.
14. Arzumanyan A, Reis HM, Feitelson MA. Pathogenic mechanisms in HBV- and HCV-associated hepatocellular carcinoma. Nat Rev Cancer. 2013;13(2):123–35.
15. Gupta S, Kumar P, Das BC. HPV: molecular pathways and targets. Curr Probl Cancer. 2018;42(2):161–74.
16. Guo F, Cofie LE, Berenson AB. Cervical cancer Incidence in young U.S. females after human papillomavirus vaccine introduction. Am J Prev Med. 2018;55(2):197–04.
17. Lei J, Ploner A, Elfstrom KM, et al. HPV vaccination and the risk of invasive cervical cancer. N Engl J Med. 2020;383(14):1340–8.
18. Calvaruso V, Cabibbo G, Cacciola I, et al. Incidence of hepatocellular carcinoma in patients with HCV-associated cirrhosis treated with direct-acting antiviral agents. Gastroenterology. 2018;155(2):411–421 e414.
19. Goedert JJ, Gong Y, Hua X, et al. Fecal microbiota characteristics of patients with colorectal adenoma detected by screening: a population-based study. EBioMedicine. 2015;2(6):597–03.
20. Shen XJ, Rawls JF, Randall T, et al. Molecular characterization of mucosal adherent bacteria and associations with colorectal adenomas. Gut Microbes. 2010;1(3):138–47.
21. Mira-Pascual L, Cabrera-Rubio R, Ocon S, et al. Microbial mucosal colonic shifts associated with the development of colorectal cancer reveal the presence of different bacterial and archaeal biomarkers. J Gastroenterol. 2015;50(2):167–79.
22. Zackular JP, Rogers MA, Ruffin MTt, Schloss PD. The human gut microbiome as a screening tool for colorectal cancer. Cancer Prev Res (Phila). 2014;7(11):1112–21.
23. Flemer B, Lynch DB, Brown JM, et al. Tumour-associated and non-tumour-associated microbiota in colorectal cancer. Gut. 2017;66(4):633–43.
24. Tsay JJ, Wu BG, Sulaiman I, et al. Lower airway dysbiosis affects lung cancer progression. Cancer Discov. 2021;11(2):293–307.
25. Kostic AD, Gevers D, Pedamallu CS, et al. Genomic analysis identifies association of fusobacterium with colorectal carcinoma. Genome Res. 2012;22(2):292–8.
26. Hamada T, Zhang X, Mima K, et al. Fusobacterium nucleatum in colorectal cancer relates to immune response differentially by tumor microsatellite instability Status. Cancer Immunol Res. 2018;6(11):1327–36.
27. Hashemi Goradel N, Heidarzadeh S, Jahangiri S, et al. Fusobacterium nucleatum and colorectal cancer: a mechanistic overview. J Cell Physiol. 2019;234(3):2337–44.
28. Bullman S, Pedamallu CS, Sicinska E, et al. Analysis of fusobacterium persistence and antibiotic response in colorectal cancer. Science. 2017;358(6369):1443–8.
29. Pushalkar S, Hundeyin M, Daley D, et al. The pancreatic cancer microbiome promotes oncogenesis by induction of innate and adaptive immune suppression. Cancer Discov. 2018;8(4):403–16.
30. Xavier JB, Young VB, Skufca J, et al. The cancer microbiome: distinguishing direct and indirect effects requires a systemic view. Trends Cancer. 2020;6(3):192–04.
31. Riquelme E, Zhang Y, Zhang L, et al. Tumor microbiome diversity and composition influence pancreatic cancer outcomes. Cell. 2019;178(4):795–806 e712.
32. Mima K, Nishihara R, Qian ZR, et al. Fusobacterium nucleatum in colorectal carcinoma tissue and patient prognosis. Gut. 2016;65(12):1973–80.
33. Tahara T, Yamamoto E, Suzuki H, et al. Fusobacterium in colonic flora and molecular features of colorectal carcinoma. Cancer Res. 2014;74(5):1311–8.
34. Rubinstein MR, Wang X, Liu W, Hao Y, Cai G, Han YW. Fusobacterium nucleatum promotes colorectal carcinogenesis by modulating E-cadherin/beta-catenin signaling via its Fad A adhesin. Cell Host Microbe. 2013;14(2):195–06.
35. Purcell RV, Pearson J, Aitchison A, Dixon L, Frizelle FA, Keenan JI. Colonization with enterotoxigenic Bacteroides fragilis is associated with early-stage colorectal neoplasia. PLoS One. 2017;12(2):e0171602.
36. Kostic AD, Chun E, Robertson L, et al. Fusobacterium nucleatum potentiates intestinal tumorigenesis and modulates the tumor-immune microenvironment. Cell Host Microbe. 2013;14(2):207–15.
37. Chen Y, Chen Y, Zhang J, et al. Fusobacterium nucleatum promotes metastasis in colorectal cancer by activating autophagy signaling via the upregulation of CARD3 expression. Theranostics. 2020;10(1):323–39.
38. Wu S, Lim KC, Huang J, Saidi RF, Sears CL. Bacteroides fragilis enterotoxin cleaves the Zonula Adherens protein, E-cadherin. Proc Natl Acad Sci. 1998;95(25):14979–84.
39. Wynendaele E, Verbeke F, D'Hondt M, et al. Crosstalk between the microbiome and cancer cells by quorum sensing peptides. Peptides. 2015;64:40–8.
40. Ma C, Han M, Heinrich B, et al. Gut microbiome-mediated bile acid metabolism regulates liver cancer via NKT cells. Science. 2018;360(6391).
41. Banerjee S, Tian T, Wei Z, et al. Distinct microbial signatures associated with different breast cancer types. Front Microbiol. 2018;9:951.

42. Buchta Rosean C, Bostic RR, Ferey JCM, et al. Preexisting commensal dysbiosis is a host-intrinsic regulator of tissue inflammation and tumor cell dissemination in hormone receptor-positive breast cancer. Cancer Res. 2019;79(14):3662–75.
43. Arends J. Metabolism in cancer patients. Anticancer Res. 2010;30(5):1863–8.
44. Fujisaka S, Avila-Pacheco J, Soto M, et al. Diet, genetics, and the gut microbiome drive dynamic changes in plasma metabolites. Cell Rep. 2018;22(11):3072–86.
45. Schulfer AF, Schluter J, Zhang Y, et al. The impact of early-life sub-therapeutic antibiotic treatment (STAT) on excessive weight is robust despite transfer of intestinal microbes. ISME J. 2019;13(5):1280–92.
46. Kwa M, Plottel CS, Blaser MJ, Adams S. The intestinal microbiome and estrogen receptor-positive female breast cancer. J Natl Cancer Inst. 2016;108(8):djw02.
47. Kadosh E, Snir-Alkalay I, Venkatachalam A, et al. The gut microbiome switches mutant p 53 from tumour-suppressive to oncogenic. Nature. 2020;586(7827):133–138.
48. Poore GD, Kopylova E, Zhu Q, et al. Microbiome analyses of blood and tissues suggest cancer diagnostic approach. Nature. 2020;579(7800):567–74.
49. Gopalakrishnan V, Helmink BA, Spencer CN, Reuben A, Wargo JA. The influence of the gut microbiome on cancer, immunity, and cancer immunotherapy. Cancer Cell. 2018;33(4):570–80.
50. Geller LT, Barzily-Rokni M, Danino T, et al. Potential role of intratumor bacteria in mediating tumor resistance to the chemotherapeutic drug gemcitabine. Science. 2017;357(6356):1156–60.
51. Guthrie L, Gupta S, Daily J, Kelly L. Human microbiome signatures of differential colorectal cancer drug metabolism. NPJ Biofilms Microbiomes. 2017;3:27.
52. Yu T, Guo F, Yu Y, et al. Fusobacterium nucleatum promotes chemoresistance to colorectal cancer by modulating autophagy. Cell. 2017;170(3):548–563 e516.
53. Viaud S, Saccheri F, Mignot G, et al. The intestinal microbiota modulates the anticancer immune effects of cyclophosphamide. Science. 2013;342(6161):971–6.
54. Gopalakrishnan V, Spencer CN, Nezi L, et al. Gut microbiome modulates response to anti-PD-1 immunotherapy in melanoma patients. Science. 2018;359(6371):97–103.
55. Matson V, Fessler J, Bao R, et al. The commensal microbiome is associated with anti-PD-1 efficacy in metastatic melanoma patients. Science. 2018;359(6371):104–8.
56. Temraz S, Nassar F, Nasr R, Charafeddine M, Mukherji D, Shamseddine A. Gut microbiome: a promising biomarker for immunotherapy in colorectal cancer. Int J Mol Sci. 2019;20(17):4155.
57. Routy B, Le Chatelier E, Derosa L, et al. Gut microbiome influences efficacy of PD-1-based immunotherapy against epithelial tumors. Science. 2018;359(6371):91–7.
58. Pinato DJ, Howlett S, Ottaviani D, et al. Association of prior antibiotic treatment with survival and response to immune checkpoint inhibitor therapy in patients with cancer. JAMA Oncol. 2019;5(12):1774–78
59. Vetizou M, Pitt JM, Daillere R, et al. Anticancer immunotherapy by CTLA-4 blockade relies on the gut microbiota. Science. 2015;350(6264):1079–84.
60. Baruch EN, Youngster I, Ben-Betzalel G, et al. Fecal microbiota transplant promotes response in immunotherapy-refractory melanoma patients. Science. 2021;371(6529):602–609.
61. Fessler J, Matson V, Gajewski TF. Exploring the emerging role of the microbiome in cancer immunotherapy. J Immunother Cancer. 2019;7(1):108.

Application of Artificial Intelligence in Research on Cancer and Its Metastasis

74

Benjamin Franc

Abstract

Artificial intelligence (AI) is composed of a number of supervised and unsupervised computational learning techniques. At present, cancer research utilizes some of these techniques to analyze and cluster large datasets, such as those resulting from genomic or proteomic analyses. AI is also beginning to enable automation of research tasks, increasing efficiency and reproducibility in classifications, and extending capabilities in basic science and clinical research beyond what is readily achievable by human investigators alone. The complex and nonlinear interactions between cancer and its environment can be modeled with AI techniques, allowing new discoveries in the molecular pathways and signaling networks leading to cancer and its metastasis to local and distant sites. Meanwhile, in cancer drug discovery and clinical trials, AI models predict the chemical, biologic, and pharmaceutical properties of small-molecule candidates and new applications of existing drugs and identify characteristics of patients in whom higher drug efficacy might be expected in clinical trials. Through facilitating analysis of sources of big data and increasing the rate of discovery in cancer research, AI promises to drive the field of cancer research forward at an exponential rate over the upcoming decade.

B. Franc (✉)
Department of Radiology, Division of Nuclear Medicine and Molecular Imaging, Stanford University, Stanford, CA, USA
e-mail: benster@stanford.edu

Learning Objectives

- To understand the basic tools of artificial intelligence and how their selection relates to the available data sources.
- To identify key areas where artificial intelligence techniques are already used in the analysis and clustering of data in cancer research.
- To demonstrate areas in cancer research where artificial intelligence techniques can contribute greater speed and reproducibility.
- To discuss the role of artificial intelligence in cancer drug discovery and drug selection.
- To highlight artificial intelligence research as an emerging field within the broader field of cancer research.

74.1 Introduction

Artificial intelligence (AI) includes a wide array of computational techniques, some of which have become embedded in cancer research already. Others are only just beginning to be explored as tools to better understand the complex and interdependent signals and events that contribute to cancer genesis and growth. An improvement over traditional computational modeling techniques, AI can address the systems biology inherent in studying cancer as well as cancer's continuing evolutionary path. Moreover, AI can accommodate the complex and nonlinear interactions between cancer and its environment that underlie metastasis [1]. Following a brief overview of a sampling of AI tools, this chapter will explore a range of areas where AI can uniquely contribute to research on cancer and its metastases.

S. P. Leong et al. (eds.), *Cancer Metastasis Through the Lymphovascular System*,
https://doi.org/10.1007/978-3-030-93084-4_74

74.1.1 The Tools of AI

AI denotes a process of observational learning of patterns by a computer. The distinction between modern AI and earlier model-driven computational algorithms is the current lack of a requirement for a priori knowledge of an underlying process. Machine learning (ML), a subfield of AI, focuses on developing algorithms that learn to perform tasks or make predictions from data.

ML can be subdivided into supervised and unsupervised learning techniques (Fig. 74.1). Supervised techniques are applied in learning tasks when data is organized as input examples (features) with corresponding outputs (labels), and learning consists of building a specific statistical inference model. Supervised learning techniques include regression and classification algorithms such as linear and logistic regression, discriminant analysis, decision tree, random forest, naïve Bayes, support vector machines (SVM), and neural networks. These types of ML models have been evolving for decades and continue to be the most commonly use in healthcare [3]. Preparing datasets with features and labels can be time-consuming; however once accomplished, supervised ML algorithms can be trained and tested rapidly. Unsupervised techniques find patterns in data without the need for data annotation and produce classes within a dataset through structural learning. Clustering analysis and principal component analysis (PCA) are techniques typically used for these tasks.

The availability of large digital datasets and modern computational speed, a subset of ML techniques, classified as deep learning (DL), have rapidly evolved and matured. DL techniques are often supervised learning that uses artificial neural networks (ANNs). ANNs use multiple layers of nonlinear modules that accept features as inputs, progressively extract higher-level features, and ultimately predict labeled outputs. Examples of ANN include convoluted neural networks (CNNs), deep belief networks (DBN), and recurrent neural networks (RNN). DL may also be used in unsupervised or semi-supervised learning tasks. DL techniques are capable of discovering complex nonlinear relationships between features and labels and have, for example, shown problem-solving capabilities similar to or exceeding those of humans [3–5].

74.2 AI as a Basic Tool to Increase the Capabilities and Efficiency of Cancer Research

74.2.1 Driving Data Analysis and Clustering in Cancer Research

Genomics, proteomics, and many other "-omics" generate extremely large datasets that AI routinely helps researchers in parsing. For example, in the field of proteomics, ML is used to predict and quantify proteins present within mass spectrometry data and identify biomarkers of disease as well as in parameter estimation and sequence modeling in protein's multiple sequence alignment [6]. In analyzing microarray data, many ML-based methods have been developed to assist in the selection of reliable gene predictors for disease and the final complete set of genes for the classification task [7].

AI has been key to the proliferation of the field of cancer radiomics, a field focused on identifying imaging phenotypes that can be used in diagnosis, characterization, and treatment management. These imaging phenotypes can be defined by pre-defined statistical imaging features, of which over 5000 have been described. Such radiomic features can identify key components of tumor phenotype for one or multiple lesions at a single or multiple time points over the course of treatment [8], potentially allowing enhanced patient stratification, prognostication, and treatment monitoring for targeted therapies. Radiogenomics, the translation of intratumoral phenotypic features to genotypes, requires the development of new methodologies to summarize phenotypes of large heterogeneous populations of cells within a single tumor and look for underlying genotypic similarities [9].

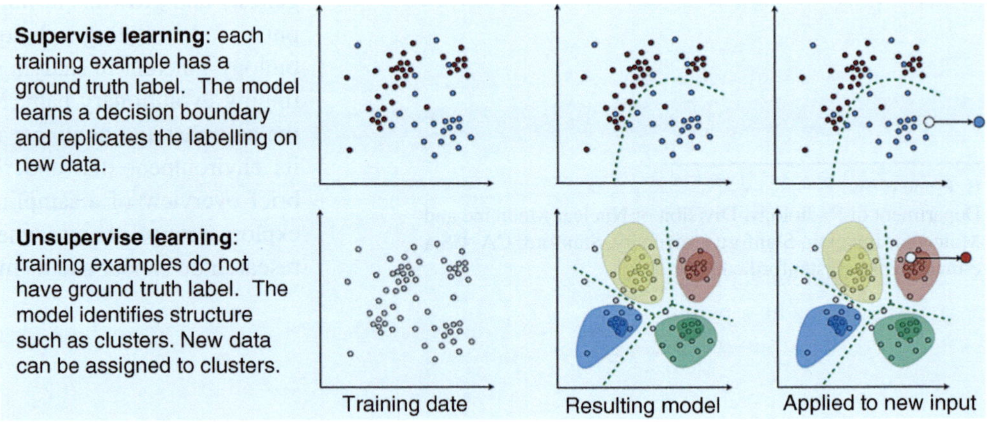

Fig. 74.1 Machine learning techniques. Machine learning techniques include two major classes: supervised and unsupervised (From https://link.springer.com/article/10.1007%2Fs00117-018-0407-3) [2]

By enabling automated extraction of phenotype information from electronic medical records, natural language processing (NLP), a type of machine learning, has evolved to become an important tool in research seeking to find associations between genomic data and the manifestations and timeline of various cancer subtypes. Data discovery can be on primary data or even data from the literature, for example, using NLP to identify every study in the literature about a specified gene-cancer association and using that data to optimize calculation of accurate risk estimates of developing cancer for a cancer susceptibility gene of interest [10]. Many NLP tools are open source [11].

74.2.2 Automation of Tasks in Cancer Research

AI-driven automated identification of cancer on imaging is useful in both basic and clinical cancer research settings. The ability to use AI techniques to rapidly and objectively identify and measure cancer and metastatic disease in animal models on whole-body tomographic imaging, such as computed tomography (CT) or positron emission tomography (PET) [12, 13], or detect and grade cancers on histologic sections opens the doors to higher research efficiency and reproducibility [14]. The power of AI over the experience of any single researcher, imaging physician, or pathologist is the ability to cross-reference imaging or other data from individual tumors to databases of comparisons [15]. As with many ML applications, finding sources of labeled data on which to train models has posed a challenge. Creative approaches have been applied to overcome the challenge of a relatively small number of well-annotated slides, such as an intermediate step where a separate DL model predicts and segments the area of the slide where a cancer would most likely be found, thus allowing the histologic classifier model greater predictive accuracy [16].

AI solutions for pathology have been shown to make diagnoses over tenfold faster than pathologists [17]. In a study comparing the accuracy of various AI algorithms and experienced pathologists in diagnosing metastatic disease on whole-slide images (WSIs) of lymph nodes from women with breast cancer, the accuracies of some AI algorithms in classifying WSIs as having metastatic disease or not were comparable to the accuracies of an experienced pathologist, and the mean area under the receiver operating characteristic curve (AUC) of the five most accurate AI algorithms was comparable to that of a pathologist interpreting the slides without time constraints [18]. Some considerations must be made when contemplating the use of a pretrained model to identify cancer on histopathological images including differences in age of the sample, method of slide preparation, and specifications of the imaging device and associated post-processing software, challenges to which various solutions have been proposed [19].

By making the entire slide available for analysis, whole-slide imaging enables a computer to select the field to analyze rather than leaving this task to a researcher or clinical pathologist. For example, many DL convolutional neural network (CNN) approaches have been applied to the histological classifications of gastric cancer (adenocarcinoma, adenoma, and non-neoplastic) from whole-slide images (WSIs), some of which have focused on local details, while others have analyzed the morphological features from the whole image [16]. One of the most common ML model-based approaches to segmentation is nuclei segmentation to extract features for nuclear morphometrics, an area to which deep learning techniques have recently been applied with significant success [19]. When applying segmentation models utilized to automatically select a region of interest for subsequent classification of the nucleus by the diagnostic AI model, such models may not be able to be generalized to sets of WSIs when the training and test images belong to different organs. This challenge may be addressed by training nuclei segmentation algorithms on images taken from different organs [19].

In developing a CNN capable of identifying breast cancer metastases in lymph nodes on WSIs, Litjens and colleagues used small image regions around areas annotated by a pathologist to train the algorithm to provide per-pixel predictions of the presence of metastases. The authors found that, using the trained algorithm, all micro- and macro-metastases of breast cancer could be identified automatically. The authors also reported that, of those slides containing benign or normal tissue, 30–40% were excluded without any additional immunohistochemical markers or human intervention [20]. This points out the issue of false positives in the task of classifying regions in lymph nodes, often secondary to the presence of reactive lymphoid follicles. In order to address this complexity, a two-step process has been developed wherein a WSI is analyzed by a deep learning algorithm to exclude regions containing lymphoid follicles, followed by a second algorithm that detects malignant cells, resulting in an error reduction of 89% in slides with reactive lymphoid follicles [21].

In addition to the task of differentiating malignant from non-malignant tissue on histopathology, ML methods have been utilized to detect tubular formation, nuclear pleomorphism, and other tasks of tumor grading as well as biomarker assessment in immunohistochemically stained images [22]. An example of DL applied to grading prostate cancer tissue microarray spot images is provided in Fig. 74.2. Automated AI-based cytologic analyses have also been developed for various cancer and specimen types [24].

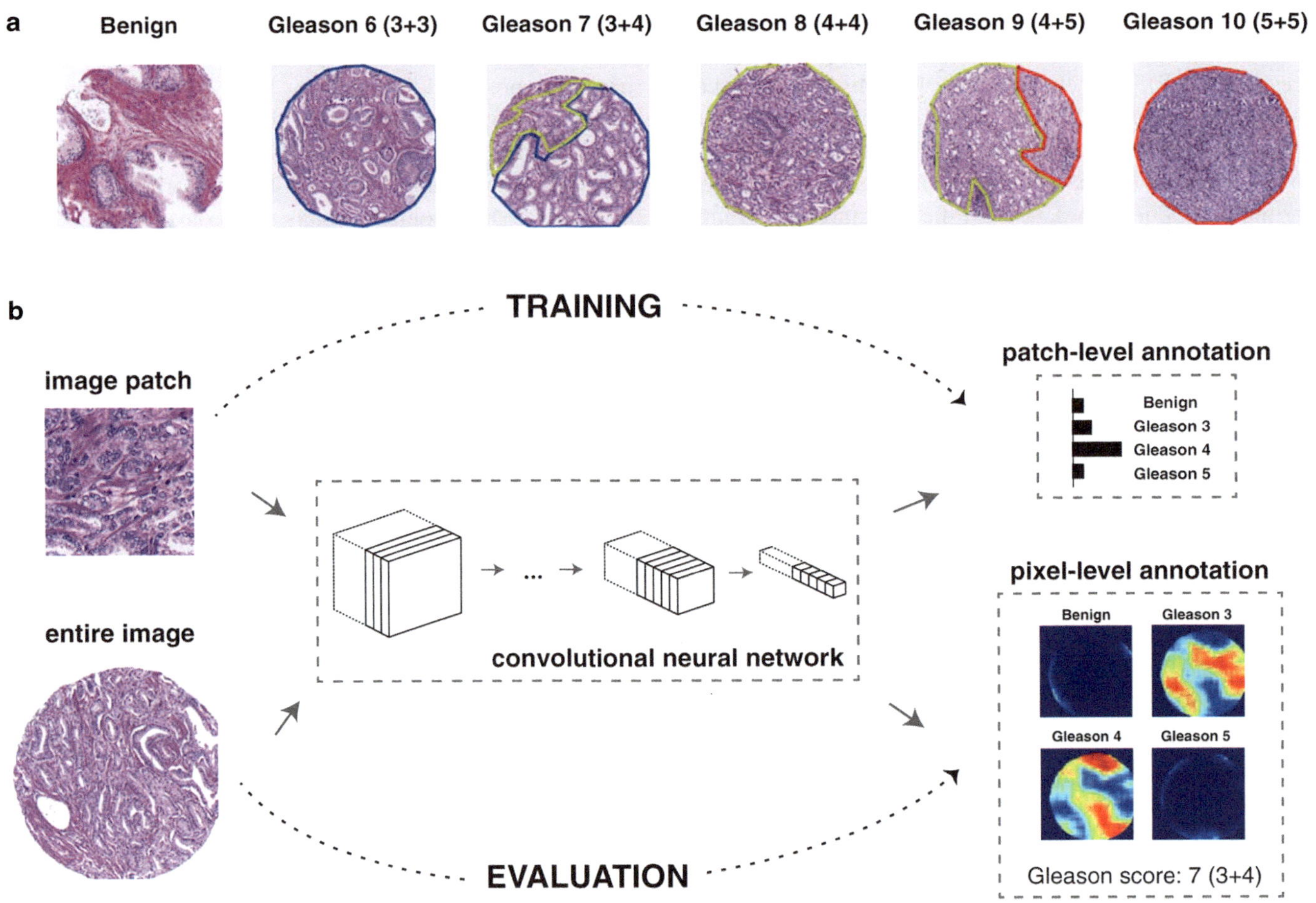

Fig. 74.2 Schematic of deep learning model training used by Arvaniti and colleagues to enable automated Gleason grading of prostate cancer tissue microarrays. (**a**) Pathologists annotate grades of prostate cancer on tissue microarray (TMA) spot images (blue, Gleason pattern 3 region; yellow, Gleason pattern 4 region; red, Gleason pattern 5 region). (**b**) (Top row) A deep neural network trains as a patch-level classifier using an architecture based on convolution blocks to extract increasingly complex features from the input image, a subsequent global average pooling layer that summarizes the local patterns across the entire image, and a final output layer that produces the final classification decision for each input image patch by computing a probability distribution over the four Gleason classes. (Bottom row) The trained patch-level convolutional neural network is then applied to the entire TMA spot image, generating pixel-level probability maps for each of the four Gleason classes. Finally, a TMA spot is assigned a Gleason score (the sum of the primary and secondary Gleason patterns detected in the pixel-level maps) (From https://doi.org/10.1101/280024) [23]

74.2.3 Enabling New Technologies for Cancer Research

In addition to increasing efficiency and reproducibility in classifications that can also be made by humans, AI can improve on and extend capabilities by performing tasks that humans simply can't. For example, Oei and colleagues have recently applied CNNs to classify the malignancy of cells based on intracellular actin patterns on actin-labeled fluorescence microscopy images [25]. Zheng and colleagues have applied DL to identify the origins of unknown primary cancers from their metastases and identify cancer types of circulating cancer cells based on DNA methylation [26]. DL has also been employed to train an algorithm that can identify the cancer subtype (as well as healthy tissue) based upon mRNA transcription profiles [27]. Finally, some initial work suggests the ability of AI methods to differentiate between cancer cells and their radioresistant clones [28].

Flow cytometry, a critical tool that uses anisotropic elastic light scattering of individual cells and emission of fluorescent labels to enable large-scale cell analysis in the laboratory, has the limitation of providing only a single value per detection channel and potential adverse effects of conjugated fluorophores on cellular viability and cell signaling. Label-free cell assay techniques aim to remedy this, but sample sizes are limited, and the assays often cannot sufficiently differentiate cells because they rely on a single feature. Using a high-throughput system that acquired quantitative optical phase and intensity images of individual cells and extracted from them multiple biophysical measurements, Chen and colleagues integrated feature extraction and supervised learning to perform accurate classification of cells. As the authors themselves indicate, this type of label-free high-throughput cell analysis will be critical to the future of personalized genomics, cancer diagnostics, and pharmaceutical development [29]. Similarly, DL techniques have advanced single-cell analysis techniques based on morphological behavior on functionalized substrates [30] and 2D cell tracking and whole-cell segmentation technology in cell cultures, enabling efficient studies of cellular dynamics characterization and drug efficacy [31].

A very exciting set of new diagnostic tools may result from AI-driven correlation of digital pathology and radiology, also referred to as radiology-pathology correlation. Registration of WSIs from adjacent tissue slices allows three-dimensional pathology datasets containing both tumors and their environments [19]. Registration between tissue slices enables colocalization of pathology data with radiographic imaging data. For example, in prostate cancer, ML techniques have enabled complex histopathologic spatial analysis with radiologic imaging, revealing that multiparametric magnetic resonance imaging (MRI) signal characteristics are associated with tissue composition density of stroma, epithelium, epithelial nuclei, and lumen, and these densities can be used to delineate cancerous regions of varying grades. In a similar analysis, epithelium and lumen density can be used to distinguish cancerous regions of varying grade [32]. Such technologies have the potential to enable in vivo monitoring and characterization of tissues at the histologic level via medical imaging techniques [19].

74.2.4 Creating Efficiencies in Clinical Research

AI is beginning to be employed in prospective clinical trial recruitment enriching – strategies to help define candidates most likely to benefit from an intervention. For example, ML models can elucidate the most predictive features associated with recurrence, such as in studies seeking population-specific predictors using datasets of a large number of breast cancer patients, and may then be used to define a set of subjects who may benefit from a novel adjuvant therapy [33]. NLP is being looked to as a means of identifying a more diverse set of clinical trial candidates and doing it more rapidly by using standard clinical trial recruitment processes [34].

In the analysis of data collected from clinical trials, indicators of response of cancer to a therapy are not represented by low-dimensional data that can be easily analyzed; instead, response is now measured through changes on imaging, dose-response curves, or other high-dimensional data for which analysis is much more complex [35]. Trained ML algorithms can determine tumor volume and its change over time and tracking of multiple lesions [15]. Automated PET segmentation of nodules based on neural networks trained in the spatial and wavelet domains has been shown to be reproducible and volumetrically accurate and demonstrate lower absolute relative error when compared to other automated techniques [36]. Other ML approaches have been useful in dealing with segmentation of larger and more complicated tumors of the head and neck, particularly in the setting of heterogeneous radiopharmaceutical uptake, in segmenting brain tumors and classifying brain scans [37–39]. Given the immense amount of data available in the medical record, NLP makes it possible to parse out data about the response of cancer to various treatments directly from patient records, allowing insight into drug efficacy in its "real-world" application, and enables pharmacovigilance and drug-safety surveillance studies after regulatory approval [40].

74.3 Elucidating the Origins and Molecular Signaling Pathways of Cancer and Its Metastasis Using AI

Understanding the molecular pathways that lead to the genesis and growth of cancers is key to uncovering biomarkers useful in diagnosis and targets for new therapies. For example, ANNs have been used to predict subcellular localization of protein sequences using the amino acid sequence and features predicted from it, such as the N-terminal-sorting signal [41]. In the study of the loss of differentiation associated with cancers, ML techniques have uncovered novel dedifferentiation mechanisms and novel associations of immune checkpoint expression and infiltrating immune cells with the level of cancer cell dedifferentiation [42].

Clues to molecular mechanisms of a cancer can be found in its genome, but the relationship between the presence of specific mutations and carcinogenesis is rarely linear in nature, making it particularly attractive to utilize current AI techniques to help unravel these relationships. In the task of differentiating driver mutations (those that directly contribute to a cancer's development or growth) from passenger mutations (those that do not directly contribute), several ML approaches have been applied. For example, the cancer-specific high-throughput annotation of somatic mutation (CHASM) approach trained an RF model using known driver mutations from a curated database or literature indicating a biologic oncogenic role and known passenger mutations defined by background base substitution rates. Models were trained based on the features of amino acid substitution properties, alignment-based estimates of conservation at the mutated position, predicted local structure at the mutated position, and others [43].

In addition, while traditionally cancer was viewed from the perspective of its organ of origin, it is now appreciated that cancer of a single organ may consist of numerous subtypes, each expressing different biologic behavior based on its genetic profile. The problem of identifying these molecular subtypes requires analysis and clustering of gene expression profiles, a task that has also benefited from application of ML approaches. For example, instead of addressing high-dimensional gene expression data head-on using traditional clustering methods, an ANN model with a single layer of hidden nodes can be trained, and the observations at the hidden nodes can then be used for the clustering analysis [44]. Likewise, AI has helped elucidate the role of the tumor microenvironment surrounding cancers [45].

Beyond its contributions to better understand carcinogenesis, AI has been used to elucidate the phenomenon of metastasis, a process involving both the tumor itself and its surroundings, including its immune microenvironment [46]. In oral squamous cell carcinoma, ML techniques combined with discovery proteomics on neoplastic islands and stroma from the invasive tumor front and targeted proteomic analysis of proteins in saliva yielded a peptide-based signature capable of differentiating between patients who will have lymph node metastases [47].

74.4 Cancer Drug Discovery, Development, and Selection with AI

In drug development, AI models predict the chemical, biologic, and pharmaceutical properties of small-molecule candidates and new applications of existing drugs and identify characteristics of patients in whom higher drug efficacy might be expected in clinical trials [48–51].

One of the greatest strengths of AI is being able to draw upon multiple large datasets – such as databases of cancer drugs, drug resistance, next-generation sequencing, genetic variants, and structural biology – at once to solve a problem (Fig. 74.3). AI in combination with sources of a large amount of data in the biosciences has been used to build in silico models of cancer, to enable computer-aided design and testing of potential therapeutic compounds [53]. For example, an RF-based model has been trained to screen compounds for their ability to serve as substrates or inhibitors to proteins conferring multi-drug resistance [54].

One of the most rapidly advancing classes of cancer therapies includes drugs modulating the immune response to cancer as well as antitumor vaccines. One key to design

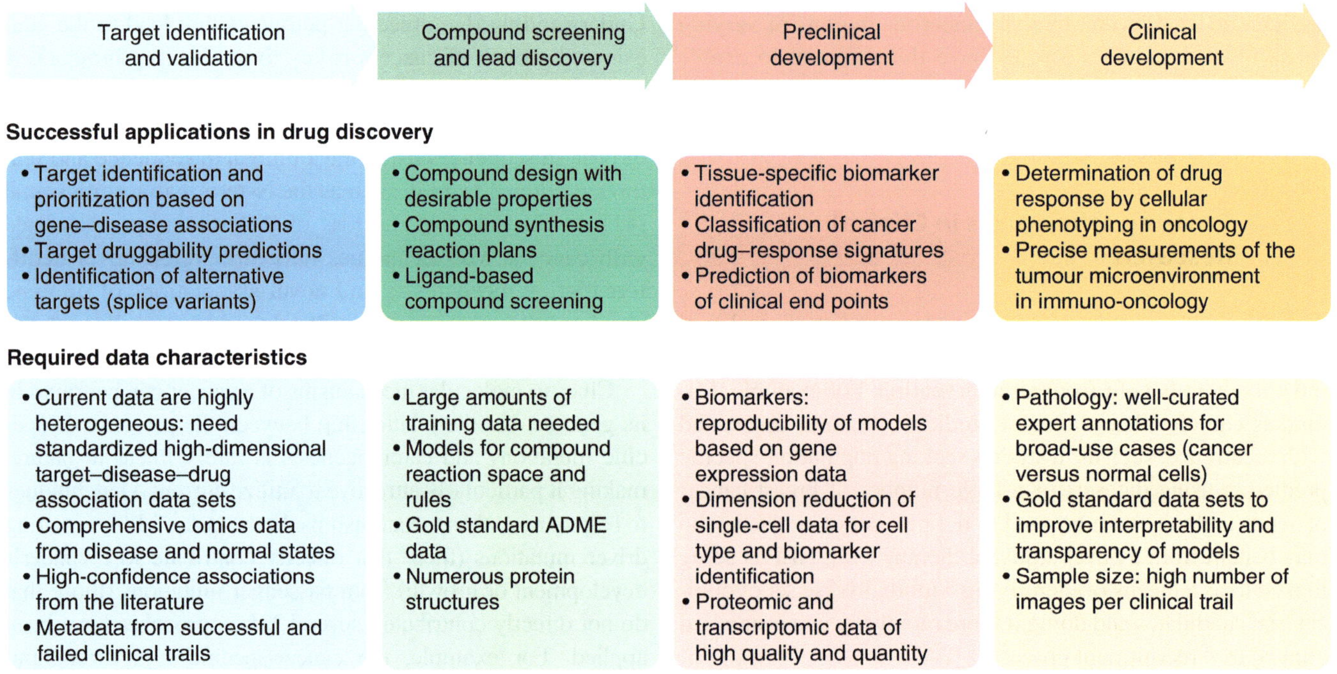

Fig. 74.3 AI applications in cancer drug discovery and development. Applications of machine learning in various steps of the drug development process and associated data requirements. (ADME = absorption, distribution, metabolism, and excretion) (From https://www-nature-com.stanford.idm.oclc.org/articles/s41573-019-0024-5) [52]

immune-mediated strategies involves identifying tumor-associated antigens and understanding their role in eliciting an immune response. Large immunoinformatics databases, such as the publicly available Human Potential Tumor Associated Antigen.

(HPtaa) database (http://www.hptaa.org), have been assembled. ML can be helpful in the task of identifying the small number of antigens expressed by specific cancer cells that can be targeted to trigger an immune response to them. In addition to techniques such as quantitative structure-activity relationship (QSAR) analysis, homology modeling, protein threading, and docking methodologies, machine learning also is used to predict epitopes of B and T cells as well as vaccine candidates. Relatively accurate ML-based prediction models have been trained for both continuous epitopes (those based on linear sequences of amino acids) and discontinuous ones. To advance T-cell epitope prediction, several ANNs have been trained to predict the binding affinity of novel peptide sequences to specific major histocompatibility complex (MHC) class I and II molecules [55].

AI has also enabled an interesting approach to appropriate drug selection that takes into account features of the drug and the cell line. Using a combination of ML techniques, the response of specific cancer cell lines to a therapy has been predicted based on both the genomic features of the cell lines and the chemical properties of the drugs, making possible in silico efficacy studies and patient-specific drug selection based on genomic characteristics [56]. In the testing of lead drug candidates on actual cancer cell lines, AI assists in decoding the large amount of data generated from high-throughput cancer-on-a-chip studies [57]. AI models have also been trained to identify cancer patients likely to benefit from adjuvant chemotherapy based on features of the cancer cells' nuclei and perinuclear regions on pathology slides [58].

In the era of precision medicine where therapies selected specifically for a single patient based upon tumor genetics, patient comorbidities, and many other clinical factors are envisaged, AI is being looked to in order to identify drugs associated with decreased rates of toxicity and increased efficacy for any given patient in a matter of hours rather than months. Specifically, ANNs have been employed to identify genetic variants and mutations from genomic data, to perform virtual screening of drug candidates based on ligand-receptor binding or on overall structure, and to increase the efficiency of molecular dynamics simulations, for example, by making it possible to model thousands of atoms simultaneously, model solvation, classify chemical trajectory data, model molecular properties based on excited state electrons, rapidly screen for novel materials and catalysts, estimate thermochemical properties of active molecules, and model free energy changes during chemical reactions [59].

74.5 AI as an Emerging Field in Cancer Research

74.5.1 Cancer Diagnosis

Given its particularly complicated origins including genetic and environmental factors interacting with the immune system and other normal tissues in the body, cancer diagnosis is an area of AI application, using data from general health screening and diagnostic tests, including blood testing, imaging, and pathology [60].

74.5.2 Cancer Treatment

AI has become a pillar of personalized healthcare whereby ML algorithms identify patterns of disease within a host of data sources to inform therapeutic choices [61, 62]. For example, using an ML algorithm trained on computed tomography (CT) radiomic features and dosimetric parameters, the response of patients with esophageal cancer (EC) undergoing concurrent chemoradiation (CRT) can be predicted [63].

74.5.3 Development of Models to Predict Cancer Outcomes and Prognosis

Outcomes such as overall survival, disease-free survival, cancer recurrence, or cancer metastasis can be predicted after training ML models, sometimes more accurately than when using more traditional methods, such as the tumor-node-metastasis staging system defined by the American Joint Committee on Cancer [16]. Examples of ML-based outcome prediction include the prediction of lung cancer histopathology and metastases based on radiomic features from CT imaging [64], cervical nodal metastases in patients with squamous cell carcinoma of the head and neck using RF models trained on texture analysis of multi-energy CT images [65], invasiveness of pulmonary adenocarcinomas on CT using 3D DL [66], metastasis of rectal cancer to the liver based on the primary tumor's MRI radiomic features [67], and radioresistance of primary nasopharyngeal cancers prior to IMRT using radiomic features from CT, MR, and PET [68].

The features found by ML to be most predictive of outcomes are often not intuitive and are independent of classical cancer predictors. For example, multimodality DL, combining clinical, pathological, and imaging information, increased the predictive power of clinical outcomes in glioblastoma multiforme, where survival is poor and ranges from 1 to

2 years in most patients [69]. In another study, tissue microarray samples are used to classify breast cancer patients as low-risk versus high-risk yielded groups with significant survival differences independent of multiple grades, tumor sizes, hormone receptor status, and the presence/absence of lymph nodes [70]. Pan-cancer prognostication may also be possible using clinical data, mRNA expression data, microRNA expression data, and whole-slide histopathology images [71].

74.5.4 Understanding and Predicting Lymphatic-Based Metastasis

The direct application of ML to identify clinical and pathologic features associated with metastasis through the lymphatic system and to predict nodal involvement in cancer has been explored across many cancer types. Nodal involvement in gastric cancer patients, one of the key prognostic indicators in that population, has been predicted using an ANN-based algorithm using input features of depth of infiltration, size, location, and shape (polypoid, fungating, ulcerating, or infiltrating) [72]. With the aim of identifying breast cancer patients who are at low risk of metastasis at presentation and who may not require a sentinel node biopsy as part of their surgery, Dihge and colleagues have recently developed machine learning–based prediction models of axillary nodal metastasis with predictors from RNA sequencing and/or clinicopathologic features [73]. Zheng and colleagues have recently used DL to predict axillary nodal positivity in breast cancer patients preoperatively by training on conventional ultrasound, shear wave elastography of the breast, and clinicopathologic features [74]. Finally, Rice and colleagues have used ML (RF) to identify esophageal cancer features predictive of nodal disease extent with the aim of separating patients who require limited versus extensive lymphadenectomy [75].

74.6 Conclusion and Future Directions

AI is already an integral aspect of many facets of research on cancer and its metastasis, including tasks of classification, regression, and clustering (Table 74.1). However, when employed using large-scale computing facilities, AI techniques such as DL are on the cusp of transforming cancer research. Industry and academia are actively developing powerful DL frameworks that target a range of tasks, and the US Department of Energy (DOE) Exascale Computing Project's Cancer Distributed Learning Environment (CANDLE) is building software infrastructure to support scalable deep learning on DOE supercomputing resources. The CANDLE initiative has begun to address problems such as understanding protein interactions in the RAS/RAF pathway, predict response of cancer to drug candidates in a patient-specific manner, and analyze millions of medical records from cancer patients with varying lifestyles, environmental exposures, cancer types, and healthcare providers to identify optimal cancer treatment strategies [76]. AI applied at these scales promises to completely transform our understanding of cancer and its metastasis.

Table 74.1 Overview of ML tools for supervised and unsupervised learning tasks and examples of their application in research on cancer and its metastases

	Supervised learning		Unsupervised learning
General task	Classification	Regression	Clustering
Techniques	Support vector machines (SVM) Discriminant analysis Naïve Bayes Nearest neighbor Random forest Neural networks[a]	Linear regression Generalized linear model (GLM) Ground-penetrating radar (GPR) Support vector regression (SVR) Ensemble methods Decision trees Random forest Neural networks[a]	K-means K-medoids Fuzzy C means Hierarchical Gaussian mixture Hidden Markov model Neural networks[a]
Example applications in cancer research	Cell classification Tumor/metastasis detection and segmentation on imaging Biomarker signatures predicting lymph node metastases Prediction of tumor-associated antigens to target for immunotherapy Screening therapeutic candidates for tumor target binding and/or predicted efficacy	Elucidating molecular pathways of cancer genesis, growth, and dedifferentiation Identification of driver mutations Models for prediction of cancer survival, recurrence, and progression	Development of tumor profiles: genomics, transcriptomics, metabolomics Discovery of cancer subtypes Prediction of second primary cancers

[a] Includes artificial neural networks for deep learning, such as convoluted neural networks (CNNs), deep belief networks (DBN), and recurrent neural networks (RNN)

Questions Addressing the Future of the Field of Artificial Intelligence in Cancer Research

- Using observational data extracted from the medical records and images of a large number of cancer patients, can artificial intelligence techniques identify heretofore unknown cancer phenotypes that could benefit from alterations in therapeutic approach?
- Will outcome models of cancer and metastases generated by artificial intelligence techniques lead to changes in our approach to cancer treatment?
- Can artificial intelligence techniques improve the equity and future applicability of randomized clinical trials?
- Will artificial intelligence techniques elucidate mechanisms of interaction between the immune system and cancer that can lead to new approaches to immune-based therapeutics?
- How can artificial intelligence integrate clinical outcomes data, molecular structural data, pathology, and medical imaging to drive drug discovery?

References

1. La Porta CAM, Zapperi S. Explaining the dynamics of tumor aggressiveness: at the crossroads between biology, artificial intelligence and complex systems. Semin Cancer Biol. 2018;53:42–7.
2. Langs G, Rohrich S, Hofmanninger J, et al. Machine learning: from radiomics to discovery and routine. Radiologe. 2018;58:1–6.
3. Jiang F, Jiang Y, Zhi H, et al. Artificial intelligence in healthcare: past, present and future. Stroke Vasc Neurol. 2017;2:230–43.
4. Esteva A, Kuprel B, Novoa RA, et al. Dermatologist-level classification of skin cancer with deep neural networks. Nature. 2017;542:115–8.
5. Gulshan V, Peng L, Coram M, et al. Development and validation of a deep learning algorithm for detection of diabetic retinopathy in retinal fundus photographs. JAMA. 2016;316:2402–10.
6. Do CB, Katoh K. Protein multiple sequence alignment. Methods Mol Biol. 2008;484:379–413.
7. Liu Q, Sung AH, Chen Z, et al. Gene selection and classification for cancer microarray data based on machine learning and similarity measures. BMC Genomics. 2011;12(Suppl 5):S1.
8. Huang SY, Franc BL, Harnish RJ, et al. Exploration of PET and MRI radiomic features for decoding breast cancer phenotypes and prognosis. NPJ Breast Cancer. 2018;4:24.
9. Chidester B, Do MN, Ma J. Discriminative bag-of-cells for imaging-genomics. Pac Symp Biocomput. 2018;23:319–30.
10. Hughes KS, Zhou J, Bao Y, Singh P, Wang J, Yin K. Natural language processing to facilitate breast cancer research and management. Breast J. 2020;26:92–9.
11. Savova GK, Tseytlin E, Finan S, et al. DeepPhe: a natural language processing system for extracting cancer phenotypes from clinical records. Cancer Res. 2017;77:e115–e8.
12. Schwyzer M, Martini K, Benz DC, et al. Artificial intelligence for detecting small FDG-positive lung nodules in digital PET/CT: impact of image reconstructions on diagnostic performance. Eur Radiol. 2020;30:2031–40.
13. Schwyzer M, Ferraro DA, Muehlematter UJ, et al. Automated detection of lung cancer at ultralow dose PET/CT by deep neural networks - initial results. Lung Cancer. 2018;126:170–3.
14. Sikpa D, Fouquet JP, Lebel R, Diamandis P, Richer M, Lepage M. Automated detection and quantification of breast cancer brain metastases in an animal model using democratized machine learning tools. Sci Rep. 2019;9:17333.
15. Bi WL, Hosny A, Schabath MB, et al. Artificial intelligence in cancer imaging: clinical challenges and applications. CA Cancer J Clin. 2019;69:127–57.
16. Niu PH, Zhao LL, Wu HL, Zhao DB, Chen YT. Artificial intelligence in gastric cancer: application and future perspectives. World J Gastroenterol. 2020;26:5408–19.
17. Pokkalla H, Pethia K, Glass B, et al. Machine learning models accurately interpret liver histology in patients with nonalcoholic steatohepatitis (NASH). Hepatology. 2019;70:187.
18. Ehteshami Bejnordi B, Veta M, Johannes van Diest P, et al. Diagnostic assessment of deep learning algorithms for detection of lymph node metastases in women with breast cancer. JAMA. 2017;318:2199–210.
19. Niazi MKK, Parwani AV, Gurcan MN. Digital pathology and artificial intelligence. Lancet Oncol. 2019;20:e253–e61.
20. Litjens G, Sanchez CI, Timofeeva N, et al. Deep learning as a tool for increased accuracy and efficiency of histopathological diagnosis. Sci Rep. 2016;6:26286.
21. Pham HHN, Futakuchi M, Bychkov A, Furukawa T, Kuroda K, Fukuoka J. Detection of lung cancer lymph node metastases from whole-slide histopathologic images using a two-step deep learning approach. Am J Pathol. 2019;189:2428–39.
22. Robertson S, Azizpour H, Smith K, Hartman J. Digital image analysis in breast pathology-from image processing techniques to artificial intelligence. Transl Res. 2018;194:19–35.
23. Arvaniti E, Fricker KS, Moret M, et al. Automated Gleason grading of prostate cancer tissue microarrays via deep learning. Sci Rep. 2018;8:12054.
24. Landau MS, Pantanowitz L. Artificial intelligence in cytopathology: a review of the literature and overview of commercial landscape. J Am Soc Cytopathol. 2019;8:230–41.
25. Oei RW, Hou G, Liu F, et al. Convolutional neural network for cell classification using microscope images of intracellular actin networks. PLoS One. 2019;14:e0213626.
26. Zheng C, Xu R. Predicting cancer origins with a DNA methylation-based deep neural network model. PLoS One. 2020;15:e0226461.
27. Azarkhalili B, Saberi A, Chitsaz H, Sharifi-Zarchi A. DeePathology: deep multi-task learning for inferring molecular pathology from cancer transcriptome. Sci Rep. 2019;9:16526.
28. Toratani M, Konno M, Asai A, et al. A convolutional neural network uses microscopic images to differentiate between mouse and human cell lines and their radioresistant clones. Cancer Res. 2018;78:6703–7.
29. Chen CL, Mahjoubfar A, Tai LC, et al. Deep learning in label-free cell classification. Sci Rep. 2016;6:21471.
30. Hasan MR, Hassan N, Khan R, Kim YT, Iqbal SM. Classification of cancer cells using computational analysis of dynamic morphology. Comput Methods Prog Biomed. 2018;156:105–12.
31. Al-Kofahi Y, Zaltsman A, Graves R, Marshall W, Rusu M. A deep learning-based algorithm for 2-D cell segmentation in microscopy images. BMC Bioinform. 2018;19:365.
32. Harmon SA, Tuncer S, Sanford T, Choyke PL, Turkbey B. Artificial intelligence at the intersection of pathology and radiology in prostate cancer. Diagn Interv Radiol. 2019;25:183–8.

33. Mosayebi A, Mojaradi B, Bonyadi Naeini A, Khodadad Hosseini SH. Modeling and comparing data mining algorithms for prediction of recurrence of breast cancer. PLoS One. 2020;15:e0237658.
34. Woo M. An AI boost for clinical trials. Nature. 2019;573:S100–S2.
35. Schneider P, Walters WP, Plowright AT, et al. Rethinking drug design in the artificial intelligence era. Nat Rev Drug Discov. 2020;19:353–64.
36. Sharif MS, Abbod M, Amira A, Zaidi H. Artificial neural etwork-based system for PET volume segmentation. Int J Biomed Imaging. 2010;2010. https://doi.org/10.1155/2010/105610.
37. Belhassen S, Zaidi H. A novel fuzzy C-means algorithm for unsupervised heterogeneous tumor quantification in PET. Med Phys. 2010;37:1309–24.
38. Blanc-Durand P, Van Der Gucht A, Schaefer N, Itti E, Prior JO. Automatic lesion detection and segmentation of 18F-FET PET in gliomas: a full 3D U-Net convolutional neural network study. PLoS One. 2018;13:e0195798.
39. Nobashi T, Zacharias C, Ellis JK, et al. Performance comparison of individual and ensemble CNN models for the classification of brain 18F-FDG-PET scans. J Digit Imaging. 2020;33:447–55.
40. Savova GK, Danciu I, Alamudun F, et al. Use of natural language processing to extract clinical cancer phenotypes from electronic medical records. Cancer Res. 2019;79:5463–70.
41. Nair R, Rost B. Protein subcellular localization prediction using artificial intelligence technology. Methods Mol Biol. 2008;484:435–63.
42. Malta TM, Sokolov A, Gentles AJ, et al. Machine learning identifies stemness features associated with oncogenic dedifferentiation. Cell. 2018;173(338–54):e15.
43. Zhang J, Liu J, Sun J, Chen C, Foltz G, Lin B. Identifying driver mutations from sequencing data of heterogeneous tumors in the era of personalized genome sequencing. Brief Bioinform. 2014;15:244–55.
44. Wang K, Duan X, Gao F, Wang W, Liu L, Wang X. Dissecting cancer heterogeneity based on dimension reduction of transcriptomic profiles using extreme learning machines. PLoS One. 2018;13:e0203824.
45. Wang S, Rong R, Yang DM, et al. Computational staining of pathology images to study the tumor microenvironment in lung cancer. Cancer Res. 2020;80:2056–66.
46. Van den Eynde M, Mlecnik B, Bindea G, et al. The link between the multiverse of immune microenvironments in metastases and the survival of colorectal cancer patients. Cancer Cell. 2018;34(1012–26):e3.
47. Carnielli CM, Macedo CCS, De Rossi T, et al. Combining discovery and targeted proteomics reveals a prognostic signature in oral cancer. Nat Commun. 2018;9:3598.
48. Zhavoronkov A, Ivanenkov YA, Aliper A, et al. Deep learning enables rapid identification of potent DDR1 kinase inhibitors. Nat Biotechnol. 2019;37:1038–40.
49. Mamoshina P, Volosnikova M, Ozerov IV, et al. Machine learning on human muscle transcriptomic data for biomarker discovery and tissue-specific drug target identification. Front Genet. 2018;9:242.
50. Aliper A, Plis S, Artemov A, Ulloa A, Mamoshina P, Zhavoronkov A. Deep learning applications for predicting pharmacological properties of drugs and drug repurposing using transcriptomic data. Mol Pharm. 2016;13:2524–30.
51. Fleming N. How artificial intelligence is changing drug discovery. Nature. 2018;557:S55–S7.
52. Vamathevan J, Clark D, Czodrowski P, et al. Applications of machine learning in drug discovery and development. Nat Rev Drug Discov. 2019;18:463–77.
53. Bhattacharya T, Brettin T, Doroshow JH, et al. AI meets exascale computing: advancing cancer research with large-scale high performance computing. Front Oncol. 2019;9:984.
54. Kadioglu O, Efferth T. A machine learning-based prediction platform for P-glycoprotein modulators and its validation by molecular docking. Cells. 2019;8(10):1286. https://doi.org/10.3390/cells8101286.
55. Tomar N, De RK. Immunoinformatics: a brief review. Methods Mol Biol. 2014;1184:23–55.
56. Menden MP, Iorio F, Garnett M, et al. Machine learning prediction of cancer cell sensitivity to drugs based on genomic and chemical properties. PLoS One. 2013;8:e61318.
57. Fetah KL, DiPardo BJ, Kongadzem EM, et al. Cancer modeling-on-a-chip with future artificial intelligence integration. Small. 2019;15:e1901985.
58. Bera K, Schalper KA, Rimm DL, Velcheti V, Madabhushi A. Artificial intelligence in digital pathology - new tools for diagnosis and precision oncology. Nat Rev Clin Oncol. 2019;16:703–15.
59. Nagarajan N, Yapp EKY, Le NQK, Kamaraj B, Al-Subaie AM, Yeh HY. Application of computational biology and artificial intelligence technologies in cancer precision drug discovery. Biomed Res Int. 2019;2019:8427042.
60. Putcha G. Blood-based detection of early-stage colorectal cancer using multiomics and machine learning. Proceedings of the American Society of Clinical Oncology Gastrointestinal Cancers Symposium; 2020.
61. Mamoshina P, Kochetov K, Cortese F, et al. Blood biochemistry analysis to detect smoking status and quantify accelerated aging in smokers. Sci Rep. 2019;9:142.
62. Ahmad MA, Eckert C, McKelvey G, Zolfagar K, Zahid A, Teredesai A. Death vs. data science: predicting end of life. Proceedings of the Association for the Advancement of Artificial Intelligence Conference on Artificial Intelligence; 2018.
63. Jin X, Zheng X, Chen D, et al. Prediction of response after chemoradiation for esophageal cancer using a combination of dosimetry and CT radiomics. Eur Radiol. 2019;29:6080–8.
64. Ferreira Junior JR, Koenigkam-Santos M, Cipriano FEG, Fabro AT, Azevedo-Marques PM. Radiomics-based features for pattern recognition of lung cancer histopathology and metastases. Comput Methods Prog Biomed. 2018;159:23–30.
65. Forghani R, Chatterjee A, Reinhold C, et al. Head and neck squamous cell carcinoma: prediction of cervical lymph node metastasis by dual-energy CT texture analysis with machine learning. Eur Radiol. 2019;29:6172–81.
66. Zhao W, Yang J, Sun Y, et al. 3D deep learning from CT scans predicts tumor invasiveness of subcentimeter pulmonary adenocarcinomas. Cancer Res. 2018;78:6881–9.
67. Shu Z, Fang S, Ding Z, et al. MRI-based radiomics nomogram to detect primary rectal cancer with synchronous liver metastases. Sci Rep. 2019;9:3374.
68. Li S, Wang K, Hou Z, et al. Use of radiomics combined with machine learning method in the recurrence patterns after intensity-modulated radiotherapy for nasopharyngeal carcinoma: a preliminary study. Front Oncol. 2018;8:648.
69. Peeken JC, Goldberg T, Pyka T, et al. Combining multimodal imaging and treatment features improves machine learning-based prognostic assessment in patients with glioblastoma multiforme. Cancer Med. 2019;8:128–36.
70. Turkki R, Byckhov D, Lundin M, et al. Breast cancer outcome prediction with tumour tissue images and machine learning. Breast Cancer Res Treat. 2019;177:41–52.
71. Cheerla A, Gevaert O. Deep learning with multimodal representation for pancancer prognosis prediction. Bioinformatics. 2019;35:i446–i54.

72. Bollschweiler EH, Monig SP, Hensler K, Baldus SE, Maruyama K, Holscher AH. Artificial neural network for prediction of lymph node metastases in gastric cancer: a phase II diagnostic study. Ann Surg Oncol. 2004;11:506–11.
73. Dihge L, Vallon-Christersson J, Hegardt C, et al. Prediction of lymph node metastasis in breast cancer by gene expression and clinicopathological models: development and validation within a population-based cohort. Clin Cancer Res. 2019;25:6368–81.
74. Zheng X, Yao Z, Huang Y, et al. Deep learning radiomics can predict axillary lymph node status in early-stage breast cancer. Nat Commun. 2020;11:1236.
75. Rice TW, Ishwaran H, Hofstetter WL, et al. Esophageal cancer: associations with (pN+) lymph node metastases. Ann Surg. 2017;265:122–9.
76. Wozniak JM, Jain R, Balaprakash P, et al. CANDLE/Supervisor: a workflow framework for machine learning applied to cancer research. BMC Bioinform. 2018;19:491.

Glossary

Abscopal effect Radiotherapy treatment of a local site of disease leading to regression of metastatic tumors not irradiated.

Acoustic impedance A physical property of tissue which describes how much resistance an ultrasound beam encounters as it passes through the tissue. It is dependent on the density of the tissue (d, in kg/m^3) and the speed of the sound wave (c, in m/s), with acoustic impedance $(Z) = d \times c$.

Acral metastases Metastatic disease burden that travels distal to the elbows or the knees, involving forearm/hand or leg/ankle/foot anatomy.

Androgen deprivation therapy (ADT) Therapy to reduce the levels of androgen hormones with drugs or surgery to treat prostate cancer.

AJCC staging Established in 1959, the American Joint Committee on Cancer (AJCC) is an organization that formulates cancer classification systems, including cancer staging (TNM staging). Through collaborations with 20 member organizations that are involved in cancer epidemiology, patient care, cancer control, cancer registration, professional education, research, or biostatistics, AJCC facilitates an evidence-based process to support classifications that are biologically and prognostically relevant: https://cancerstaging.org/Pages/default.aspx.

AKT Is a serine/threonine protein kinase also known as protein kinase B (PKB).

Allograft-prosthetic composite This is a segmental bone-replacing hybrid implant that includes human allograft bone that is coupled with a metallic implant.

Anaplastic lymphoma kinase (ALK) The ALK receptor has function on cellular communication, CNS development, and neuromuscular junction synapse development. ALK mutations were also identified in lymphomas and lung cancer.

Androgen Male sex hormone (e.g., testosterone).

Angiotropism In cancer literature, angiotropism defines tumor cells closely apposed to the external surfaces of the endothelial cells of blood vessels in a pericyte-like location without evidence of intravasation. Angiotropism is linked to pericyte mimicry and extravascular migratory metastasis.

Antibody drug conjugates (ADC) A class of highly potent biological drugs built by attaching a small-molecule anticancer drug or another therapeutic agent to an antibody, with either a permanent or a labile linker. The antibody targets a specific antigen only found on target cells.

Anti-resorptives Therapeutics that target bone resorption in order to preserve skeletal health and improve patient quality of life.

APC Adenomatous polyposis coli gene.

Androgen receptor (AR) Receptor protein that binds androgen and regulates gene expression in the nucleus.

Artificial intelligence Development of computer systems capable of performing classification or other tasks that would otherwise require human intelligence.

ATM Serine/threonine kinase involved in DNA repair.

Autosomal dominant A pattern of inheritance characteristic of some genetic diseases. "Autosomal" means that the gene in question is located on one of the numbered, or nonsex, chromosomes. "Dominant" means that a single copy of the disease-associated mutation is enough to cause the disease.

Autosomal recessive A genetic condition that appears only in individuals who have received two copies of an autosomal gene, one copy from each parent. The gene is on an autosome, a nonsex chromosome. The parents are carriers who have only one copy of the gene and do not exhibit the trait because the gene is recessive to its normal counterpart gene.

Axillary lymph node dissection Procedure in which the lymph nodes in levels I and II and sometimes level III nodes of the axilla are removed.

BCBM Breast cancer brain metastasis (BCBM) is the type of brain metastasis occurring in patients diagnosed with breast cancer.

Binding affinity It is widely used as one of the parameters for quantitative evaluation of action and indicates the strength of interaction between a biomolecule and its ligand/binding partner.

S. P. Leong et al. (eds.), *Cancer Metastasis Through the Lymphovascular System*,
https://doi.org/10.1007/978-3-030-93084-4

Blood-tumor barrier The disrupted semipermeable blood-brain barrier in the setting of tumor development. Also referred to as the blood-brain-tumor barrier.

Bone metastases The spread of cancer cells to bone from a different site.

Bone sarcoma Mesenchymal tumors of the skeleton.

Bone scan Indirect imaging for bone metastases using bone-seeking radiopharmaceuticals such as ^{99m}Tc-MDP and ^{18}F-NaF PET/CT.

Brachytherapy Utilization of internalized radiotherapy seeds deposited at the site of surgical resection to provide a constant, low dose of radiation directly to the tumor bed to improve local control after surgery.

BRAF A member of the Raf kinase family that is downstream of RAS and activates its substrate, the MEK protein, in the MAP kinase pathway within tumor cells. Some mutations in BRAF gene can cause birth defects and other mutations, and cancers like lymphomas, colorectal cancer, melanoma, lung cancer, and brain tumors. Approximately, 50% of melanomas harbor a mutation in exon 15, codon 600, which constitutively activates the MAP kinase pathway.

Brain metastasis Brain metastases are growths that spread to the brain from a cancer occurring in another part of the body. They are distinct from primary brain tumors which initiate in the brain. Brain metastases are much more common than primary brain tumors: an estimated 200,000 people are diagnosed with brain metastasis each year (US alone), compared to about 17,000 diagnoses for primary brain tumors. Furthermore, the number of brain metastasis diagnoses has dramatically risen in recent years worldwide. Many oncologists believe that this is due to better early detection of brain metastasis, as well as better treatments for primary cancers. As more patients live longer with a primary cancer, the disease has more time to spread to the brain. This has led to the renewed and important focus on treating brain metastasis.

BRCA1 Breast cancer gene 1, which is involved in DNA repair, involved in hereditary breast, ovarian, and other malignancies.

BRCA2 Breast cancer gene 2, which is involved in DNA repair, involved in hereditary breast, ovarian, and other malignancies.

Breslow The term associated with the measured thickness of a melanoma. It is distance between the epidermal granular layer and point at which the lowermost melanoma cell is seen.

Cancer-associated fibroblasts (CAFs) These are cells with an elongated morphology and are negative for epithelial, endothelial, and leukocyte markers and lack the mutations found within cancer cells.

Cell survival curve Cell survival curves are used to describe the relationship between radiation dose and proportion of cells that survive.

Checkpoint inhibitors Drugs that are manufactured as antibodies against CTLA4, or PD1 or PDL1 cell receptors, which block checkpoint proteins expressed by certain immune system cells, such as T cells, and also cancer cells. These checkpoints help keep immune responses from being too strong and sometimes can keep T cells from killing cancer cells. When these checkpoints are blocked, T cells can attack cancer cells more effectively. Examples of checkpoint proteins found on T cells or cancer cells include PD-1/PD-L1 and CTLA-4/B7-1/B7-2.

CDK 4/6 Cyclin-dependent kinase 4 and cyclin-dependent kinase 6 are proteins that regulate cell cycle. Many drugs known as CDK 4/6 inhibitors can block those proteins, affecting multiple pathways, and halt cell cycle in G1 phase.

Cell-free DNA Circulating DNA, typically as nucleosome-associated fragments or exosomal, secreted, or released by dying cells into plasma and other body fluids.

Cell-free tumor DNA It refers to free DNA that has been spilled from the tumor and circulates in the bloodstream.

Cell signaling It is the mechanism of information transfer that governs cellular functions and activities, self, and cell-to-cell communication. Some components include, but are not limited to, growth factors, receptors, transfer of signals through a cascade of enzymatic activities, and proteins from an activated state to an inactivated state, controlled through an array of positive and negative signaling loops.

Cement augmentation Filling of the vertebral body with cement, such as polymethylmethacrylate, through a percutaneous technique aimed at alleviating pain from pathologic compression fractures or lytic lesions.

Cement osteoplasty Percutaneous injection of acrylic cement (polymethylmethacrylate or PMMA) in a vertebral body with the goal of treating pain and preventing further collapse in the setting of a vertebral body fracture.

Chemical biliary sclerosis Inflammation and scarring of the biliary system secondary to chemical insults to the liver including chemotherapy.

Chimeric antigen receptor T cell Genetically engineered T cells to produce an artificial T-cell receptor for use in immunotherapy.

Circulating tumor cells (CTCs) These are viable and intact tumor cells that have shed into the vasculature after detaching from primary and/or metastatic tumors, and circulate in blood. CTCs can extravasate and become *"seeds"* for the subsequent growth of additional tumors in distant organs (metastases), a mechanism that is responsible for the vast majority of cancer-related deaths.

Circulating tumor DNA (ctDNA) DNA fragments released by tumor cells but not associated with cancer cells and detectable in the bloodstream of cancer patients. Testing allows for personalized surveillance based on a cancer patient's unique set of gene mutations.

CTL4 Cytotoxic T-lymphocyte-associated protein or CD152. This receptor member of the immunoglobulin superfamily is expressed in T cells and participates in the regulation of T cells, inhibiting their function. Drugs have been manufactured as antagonist antibodies (anti-CTL4) to target this receptor, thus blocking its activity and stimulating T-cell response.

Codman triangle An aggressive periosteal reaction due to rapid growth of a mass that pushes the periosteum off faster than new periosteum can lay down new bone.

Color Doppler ultrasound A method of visually detecting and depicting motion, usually blood flow, using a color map that is incorporated into a standard B-mode image. Doppler shifts of returning ultrasound waves within a defined region of interest are color-coded based on the average velocity and direction.

Commensal Nonpathogenic, normally occurring microbial flora.

Complete decongestive therapy (CDT) Gold standard for conservative lymphedema management used to decrease the limb volume (in phase I) and to maintain the obtained effect (in phase II). The components of CDT are as follows: skin care, manual lymphatic drainage, compression therapy, and exercise.

Contrast resolution The ability to distinguish between differences in image intensity. The contrast resolution of MRI is superior to CT.

Convolutional neural network A class of deep neural networks, the connectivity between which resembles the organization of the visual cortex.

CTNNB1 Catenin beta-1.

Cystoprostatectomy Surgical removal of the prostate and urinary bladder.

D2 lymphadenectomy Surgical removal of the N2 nodal basin; nodes of the right and left gastric cardia, greater and lesser curvature, supra- and infrapyloric areas, left gastric, common hepatic, celiac, and splenic arteries. Left paracardial nodes omitted for distal gastrectomy.

DNA damage response (DDR) Denotes a network of cellular pathways that sense, signal, and repair DNA lesions.

Dedifferentiation Cells grow from a terminally differentiated cell to a less differentiated, less specialized state.

Deep learning Supervised learning that uses multiple layers of nonlinear modules that accept features as inputs, progressively extracting higher level features, and ultimately predicting labeled outputs.

Dermatolymphangioadenitis Acute inflammation of skin, subcutaneous tissue, lymphatic vessels, and lymph nodes.

Diapedesis Passage of cells across an intact endothelium.

Differentiation Simply the process of becoming different. In embryology a single-celled zygote differentiates into a multicellular embryo that develops into a more complex multisystem of distinct cell types. Differentiated lymphatic endothelial cells (LECs) retain their identity even if removed from the body and maintained in culture.

Diffusion-weighted imaging A method of signal contrast generation based on the differences in Brownian motion.

Diagnostic leukapheresis (DLA) A short form of leukapheresis that collects white blood cells from peripheral blood via density centrifugation and co-collects CTCs.

DNA damage repair DNA repair pathways that are active throughout different stages of the cell cycle, allowing the cells to repair DNA damage.

Doppler shift (Doppler effect) The change in frequency of a sound wave due to a reflector moving towards or away from an object, which in ultrasound is the transducer. Named after the Austrian physicist, Christian Andreas Doppler (1803–1853).

Downstaging The process by which a patient's risk of developing recurrent disease is deemed to be lower than initially thought, due to either additional information becoming available or a review of preexisting information resulting in a change of parameters.

Dysplasia A condition in which the arrangement and morphology of cells in a tissue are disrupted. It is considered a preliminary stage of cancer.

Early metastatic A limited number of metastases that may be indolent (true oligometastasis), or may precede widespread dissemination.

Edema Excess fluid collection in body parts and cavities (e.g., ascites in abdominal cavity) resulting from high or low flow imbalance between lymph formation and absorption.

EGFR Epidermal growth factor receptor. It is a surface protein that stimulates cells to divide when epidermal growth factor and epidermal growth factor-like peptides bind to it. EGFR mutations and amplifications are found in multiple cancers such as lung cancers and glioblastomas.

Endocrine therapy Treatment that blocks the effect of or inhibits the production of estrogen and/or progesterone.

Endoprosthesis A modular metallic implant—composed usually of cobalt, chromium, and molybdenum—that may be assembled on-demand in the operating room to fill large long-bone defects of any size.

Endosteal niche Site within bone that is rich in cells of hematopoietic stem cell and mesenchymal stem cell origin, including osteoblasts and osteoclasts. The endosteal niche participates in recruitment, dormancy, and metastatic outgrowth.

Epigenetics A field that studies the heritable changes in the expression of genes that do not involve alterations in the DNA sequence.

Epithelial-mesenchymal transition (EMT) An EMT is a biologic process that allows an epithelial cell to undergo multiple biochemical, morphologic, and functional changes that enable it to assume a mesenchymal cell phenotype conferring invasiveness, dissemination, and chemo/immunotherapy resistance. This change allows cancer cells to escape from the primary site and invade surrounding tissue.

Epithelioid sarcoma Epithelioid mesenchymal tumors that commonly arise in the extremities or truncal area. The pathognomonic feature is the loss of INI-1 (SMARCB1) protein expression.

Exosome Exosomes are extracellular vesicles that are shed by the parent cells into ECM with double-layered lipid membrane containing molecules of proteins, lipids, and nucleic acids.

External beam radiation therapy Form of radiation treatment in which a patient lies flat on a treatment table as an external treatment machine directs a beam of radiation to an area of interest within the patient.

Extracellular vesicle Membranous vesicles produced by cells that contain cellular cytosol and cellular genetic material.

Extravascular migratory metastasis Extravascular migratory metastasis or EVMM is a non-hematogenous, non-intravascular mode of tumor metastasis via angiotropism and pericyte mimicry. When angiotropic cancer cells are engaged in pericyte mimicry and EVMM, they migrate away from the advancing front of the tumor to form regional and distant metastases, along blood vessels as a scaffolding to guide dissemination.

Extrinsic regulators Factors external to the cell, including the niche and systemic factors, that can influence stem cell behavior.

FAP Familial adenomatous polyposis.

FDA United States Federal Drug Administration. This agency regulates drug approval and oversees the safety and efficacy of chemotherapies, immunotherapies, and target-based drugs.

Fisher systemic theory of cancer Dr. Bernard Fisher articulated in a lecture in 1980 his theory that cancer is a systemic disease, and that systemic dissemination of cancer cells has frequently already occurred at the time of diagnosis of localized disease.

Fluoroscopy Imaging technique that utilizes real-time X-ray imaging for both diagnostic and interventional procedures. Fluoroscopy systems usually contain an X-ray tube and an image intensifier or a flat panel detector coupled with a monitor.

FLAIR A T2-weighted sequence (on MRI) where the signal of CSF is subtracted, to highlight brain edema.

Future liver remnant (FLR) The volume of functional liver parenchyma remaining after resection, which is associated with both the risk of postoperative liver failure and morbidity and mortality. Using standard radiographic software, the proposed line of resection is drawn, and the remaining liver volume is calculated as a percentage of the whole liver volume, with exclusion of tumors, cysts, and ablation cavities.

Gantry Ring-shaped structure that contains the X-ray tube, collimators, and detectors in CT equipment. Patients are positioned on a table that moves through the central opening of the gantry during image acquisition.

Gastric cancer lymph node stations Numerical classification of the regional lymph nodes for gastric cancer, as defined by the Japanese Gastric Cancer Association.

Germline mutations A gene change in a body's reproductive cell (egg or sperm) that becomes incorporated into the DNA of every cell in the body of the offspring.

Glycolysis The pathway leading from intracellular glucose directly to pyruvate, resulting in the generation of two moles of pyruvate, ATP and NADH, from one mole of glucose.

Graded prognostic assessment A contemporary set of prognostic classification systems to predict outcomes of patients with brain metastases.

Hallmarks of cancer Common traits, characteristics, and capabilities that normal cells acquire to become autonomous through stepwise gain or loss of functions of genes conferring survival advantage and dissemination to distant sites.

Halsted model of cancer Based on his understanding of breast cancer, Dr. William Stewart Halsted postulated that cancer is an orderly disease that spreads from the primary site, by direct extension through the lymphatics to the lymph nodes, and then to distant metastatic sites.

Haptotactic gradient Solid-phase chemokine gradient immobilized in extracellular matrix.

HCC Hepatocellular carcinoma.

Hepatic artery infusion (HAI) pump A surgically implanted pump into the gastroduodenal artery, which allows for the delivery of chemotherapy to liver tumors via the hepatic artery. As the hepatic artery accounts for the majority of the blood supply of tumors, HAI chemotherapy allows for higher drug concentrations with less systemic effects.

Hepatic artery infusion pump therapy Cancer regional therapy consisting of a surgically placed subcutaneous pump with vascular access to the common hepatic artery designed to deliver chemotherapy.

HIFU High-intensity focused ultrasound.

Histopathology Microscopic evaluation of tissue, ranging in scale from the overall tissue architecture to the qualitative features of chromatin in cell nuclei.

HPB Hepatopancreaticobiliary.

HER2 Receptor tyrosine-protein kinase erbB-2. It is also known as HER2/neu. Human epidermal growth factor receptor 2 (HER2) is a receptor of the EGFR family. Its mutation has been seen in breast, ovarian, uterine, gastrointestinal, and other cancers.

HU Hounsfield units are named after Sir Godfrey Hounsfield, the Nobel Prize winner and inventor of CT. HU is a dimensionless relative quantitative measuring unit of radio density unit used in CT scanning to express CT numbers in a standardized and convenient form.

Human papillomavirus (HPV) A double-stranded DNA virus that can infect epithelium. There are hundreds of types of HPVs, some of which can cause cancer.

Iliopectineal line An anatomic and radiographic line that approximates the pelvic brim; is used to evaluate the integrity of the anterior column of the acetabulum based upon its continuity.

Ilioischial line A radiographic line that is used to evaluate the integrity of the posterior column of the acetabulum based upon its integrity.

Image-guided radiation therapy (IGRT) The process of using precise image guidance during CT simulation, treatment planning, and treatment delivery.

Image segmentation Technique of dividing an image into parts, often for the purpose of image compression or object recognition.

IMRT Intensity-modulated radiation therapy. Technique of external beam radiation in which metal leaves within the treatment machine variably modulate the intensity of the radiation beam to allow better conformality of dose to a treatment volume.

Immediate reconstruction The placement of either a prosthetic or an autologous tissue for breast reconstruction at the time of oncologic surgery. This can be done as the first stage of a multistage reconstruction, or as a definitive single-stage reconstruction.

Immune checkpoints Key regulators of the immune system that when stimulated can dampen the immune response to an immunological stimulus.

Immune checkpoint inhibitors A class of drugs that blocks the proteins that regulate and keeps the immune system in check.

Immune-mediated therapy A type of targeted therapy which optimizes the host immune system to recognize and eradicate invading tumor cells.

Immunocytochemistry Immunocytochemistry (ICC) (also known as immunohistochemistry—IHC) is a technique for the detection and visualization of proteins, or other antigens, in cells using antibodies specifically recognizing the target of interest.

Immunogenicity The ability to elicit an immune response in the human body to an antigen that enters the body.

Immunomodulation Defined for this chapter as three major immune cells, i.e., TAMs, MDSCs, and regulatory T (Treg) cells, in TME that help regulate host immune response and are "friendly" to tumor.

Immunoscore A score assigned based on the quantification of T-cell infiltration within the tumor center and tumor invasive margin, with some studies showing a correlation between increasing scores and improved patient survival.

Immunotherapy A type of treatment that activates or suppresses a person's own immune system with substances that stimulate the immune response to fight diseases such as cancer.

International Metastatic Renal Cell Carcinoma Database Consortium (IMDC) International Metastatic Renal Cell Carcinoma Database Consortium (IMDC) criteria are used to assess the prognosis of patients with metastatic RCC utilizing simple clinical and laboratory criteria: (1) KPS (<80%), (2) <1 year from time of diagnosis, (3) hemoglobin < LLN, (4) corrected calcium > ULN, (5) neutrophils > ULN, and (6) platelets > ULN. Presence of 0 of these factors defines favorable risk group, 1–2 factors intermediate risk group, and ≥3 factors poor risk group.

In-transit melanoma Melanoma involvement of the dermis, subcutaneous tissue, and/or lymphatics that occurs between the site of the primary tumor and regional lymph node basin.

Intrathecal Introduced into or occurring in the space under the arachnoid membrane of the brain or spinal cord.

Intratumor heterogeneity A phenomenon in which multiple subclones with different genomes exist in a single tumor.

Intrinsic regulators Stem cell autonomous factors, including genetic factors, that are intrinsic to the cell and not influenced by extrinsic factors.

Ionizing radiation Radiation, traveling as a particle or electromagnetic wave, that carries sufficient energy to detach electrons from atoms or molecules, thereby ionizing an atom or a molecule.

Karnofsky performance status (KPS) A scale from 0 to 100 that assesses a patient's functional ability, including for self-care.

KIT It is a receptor tyrosine kinase protein found on the cell membrane of different cell types. Mutations in *KIT* gene, especially those occurring in exons 9 and 11, drive cell growth and survival in a number of cancer cells, including melanoma and gastrointestinal stromal tumor.

KRAS/RAS It is a gene that encodes K-RAS, a protein part of the RAS/MAPK pathway. KRAS mutations and amplifications are found in colorectal and lung cancers.

Laser interstitial thermal therapy (LITT) Minimally invasive approach to brain tumor surgery featuring stereotactic placement of a laser fiber and thermal ablation of a brain lesion via twist-drill burr hole.

Lauren classification General classification of gastric adenocarcinoma into two histologic subtypes, intestinal and diffuse type, according to the tendency for the formation of recognizable gland structures.

Leiomyosarcoma Soft-tissue sarcoma transformed from mesenchymal cells with smooth muscle features. Commonly occurs in extremities and truncal areas, retroperitoneum, and uterus.

Leptomeningeal disease A late complication of metastatic cancer in which tumor cells spread via the CSF to the arachnoid and pia of the brain and spinal cord.

Limited metastases A limited number of metastatic lesions.

Liposarcoma The most common subtype of soft-tissue sarcoma with varying degrees of adipocytic differentiation. Based on the histology and molecular differences, liposarcoma could be classified as well-differentiated/dedifferentiated liposarcoma (with MDM2 gene amplification), myxoid round-cell liposarcoma (with FUS-DDIT3 gene rearrangement), and pleomorphic liposarcoma (commonly with TP53 mutation). Different types of liposarcoma have different clinical behavior, prognosis, and tumor immune microenvironment compositions.

Liquid biopsy The compendium of tests performed on a sample of blood to look for cancer cells that are circulating in the blood (CTCs), or for fragments of DNA from tumor cells that are in the blood (ctDNA), or any biological fluid containing tumoral material. A liquid biopsy may be used to help find cancer at an early stage. It may also be used to help plan treatment or to understand out how well treatment is working and/or if cancer has relapsed. Being able to take multiple samples of blood over time may also help oncologists understand what type of molecular changes are taking place in a tumor.

Local consolidative therapy Focal treatment (usually surgery or radiation) to the primary and/or metastatic lesions.

Local recurrence Recurrence of any cancer in the primary site after prior surgical resection. In breast cancer this refers to a regrowth in the ipsilateral breast. Also referred to as in-breast tumor recurrence, or IBTR.

Locus (plural loci) The specific physical location of a gene or other DNA sequence on a chromosome, like a genetic street address.

Loss of heterozygosity If there is one normal and one abnormal allele at a particular locus, as might be seen in an inherited autosomal dominant cancer susceptibility disorder, loss of the normal allele produces a locus with no normal function. When the loss of heterozygosity involves the normal allele, it creates a cell that is more likely to show malignant growth if the altered gene is a tumor-suppressor gene. Also called LOH.

Lung cancer Malignancy that occurs in the parenchyma of the lung.

Lymphadenectomy A surgical procedure which involves the complete removal of the lymph nodes from a defined surgical area. This could be *elective (ELND)* where the lymph nodes are removed where there is suspicion of involved nodes, but nothing is palpable; *therapeutic (TLND)*, where lymph nodes have been palpated clinically or detected on imaging; and *completion (CLND)*, whereby, following the removal of an involved sentinel node, the remaining nodes from that lymph node basin are removed, on the suspicion that other nodes may be harboring micrometastatic disease.

Lymphadenopathy An enlarged lymph node.

Lymphangion Lymphatic functional contractile unit between two valves.

Lymphangiogenesis Growth of new lymphatic vessels by budding and sprouting from existing vessels.

Lymphatic-venous bypass Known also as lymphaticovenular anastomosis—surgical connection of lymph vessel with vein to restore the flow of lymph back to the venous system.

Lymphedema Abnormal swelling that generally occurs in the arms or legs.

Lymphovascular invasion The identification of tumor cells within lymphatic vessels.

Machine learning A subfield of AI, which focuses on developing algorithms that learn to perform tasks or make predictions from data.

Malignancy of unknown primary A tumor presenting as a metastasis without a known site of origin.

Mammosphere-forming cells In culture, many cell types, benign and malignant, spontaneously grow in spherical formations, a stem cell or CSC being at the origin. The stem cell or sphere-forming cell in mammary cells is known as the mammosphere-forming cell.

MAP therapy Methotrexate, doxorubicin, cisplatin.

Margin The outside edge of a surgical specimen. A positive margin is one that is involved with cancer cells.

Margin width The distance between the tumor cells and the edge of the surgical specimen.

Maturation Process by which the immature lymphatic plexus becomes hierarchically organized into lymphatic capillaries that collect lymph and collecting lymphatic vessels that transport the fluid.

Metastatic castration-resistant prostate cancer (mCRPC) Metastatic prostate cancer with rising serum levels of PSA or progression of disease despite androgen deprivation therapy (ADT).

MD Anderson Bone Response Criteria A set of rules that are designed to standardize the interpretation of the imaging of bone metastases following therapy.

Medical linear accelerator A particle accelerator that accelerates charged particles to megavoltage energies using a series of oscillating electric fields along a linear beamline adopted which is adapted for medical use.

MEK An essential kinase protein involved in the MAP kinase signaling pathway. MEK is typically activated by the upstream RAF kinases, and in turn activates the downstream ERK protein, resulting in increased cell growth.

Mesenchymal tissue Tissue originating from the mesoderm, which can develop into multiple tissue types including bone, cartilage, muscle, fat cells, and connective tissue.

MET Process by which motile cells acquire cell junctions to associate with each other and form stable clusters.

MR-Linac A radiation therapy delivery machine that combines real-time 3D magnetic resonance imaging with radiation treatment delivery.

Metastasectomy Surgical resection of a metastatic focus of a malignant tumor.

Metastasis The migration and spread of cancer cells from a primary site to regional lymph nodes and/or distant organs in the body, resulting in the development of secondary tumors and a major contributor to the deaths of cancer patients.

Metastatic organotropism Predilection of tumors to metastasize to specific rather than random organs.

Microbiome The collection of microbes that exist in, on, and throughout the human body.

Micrometastatic The presence of submillimeter metastatic disease.

Microsatellite instability A phenomenon in which a sequence shows a different number of repetitions in tumor tissues than in normal tissues due to an impaired ability to repair base sequence errors that occur during DNA replication.

Minimally invasive parafascicular surgery (MIPS) Utilization of tubular cylindrical or ovoid retractors, also known as port-based or channel-based systems, to allow for brain cannulation using a sulcus or small corticotomy to provide access to deep-seated subcortical brain tumors with minimal disruption of the adjacent cortex.

Minimally invasive solutions Interventional techniques for treating metastatic skeletal or soft-tissue disease that shortens recovery time and improves both functionality and pain control. Such techniques are generally delivered through percutaneous on noninvasive routes, and can include cryoablation, radiofrequency ablation, cementoplasty, or MRI-guided high-frequency ultrasound.

MGMT 0(6)-methylguanine-DNA-methyltransferase is a repair protein. Methylation of MGMT has been shown to affect the efficacy of alkylating agents in brain tumors, and the survival of patients with lung cancer.

mTOR Mammalian target of rapamycin is a protein kinase that regulates multiple cell functions including cell growth and proliferation. Its activity is dysfunctional in multiple cancers including breast, lung, melanoma, and kidney cancers.

Multicentric disease Disease in more than one quadrant of the breast.

Multikinase inhibitors Drugs that work by inhibiting multiple intracellular and cell surface enzymes called kinases, some of which are implicated in tumor growth and metastatic progression of cancer, thus decreasing tumor growth and replication. Examples include sunitinib, sorafenib, cabozantinib, and lenvatinib.

Natural language processing A subset of machine learning techniques that help computers analyze, interpret, and manipulate human language.

Necrotic or apoptotic cells Accidental cell death or programmed cell death.

Neoadjuvant therapy Treatment given as a first step to shrink a tumor before the main treatment, which is usually surgery, is given.

Neoantigen A cancer antigen that has newly emerged due to genetic mutations that occur in cancer cells.

Next-generation sequencing Next-generation sequencing (NGS) is a massively parallel sequencing technology that offers ultrahigh throughput, scalability, and speed. The technology is used to determine the order of nucleotides in entire genomes or targeted regions of DNA or RNA.

NOMS A decision-making framework which takes into account the neurologic, oncologic, mechanical, and systemic status of a patient to guide surgical and/or radiation treatment of metastasis disease of the spine.

Non-small cell lung cancer The most common type of lung cancer.

NSAIDs Nonsteroidal anti-inflammatory drugs.

Oligometastatic (at presentation) An indolent early metastatic state with a limited number of (often up to 5) metastatic lesions.

Oligometastatic disease Distant disease that is limited in number of metastases, usually between 1 and 5 different lesions.

Oligometastatic theory of cancer The theory that intermediate steps exist between initial tumor cell dispersion and fulminant metastatic disease, and that patients who develop a small number of confined metastatic lesions may experience improved survival through local management of these metastases.

Oligoprogressive A nonmetastatic case that quickly progresses to an oligometastatic state after definitive therapy.

Oligorecurrent A nonmetastatic case that slowly progresses to an oligometastatic state after a longer (usually at least 6 months) period of progression-free survival.

Oligoresidual A state of oligometastatic disease formed after a more widely metastatic state is cytoreduced with systemic therapy.

Ommaya Reservoir Intraventricular catheter system that can be used for the aspiration of cerebrospinal fluid or for the delivery of drugs into the cerebrospinal fluid.

Oncovirus A virus that can cause cancer.

Organoid Three-dimensional cell culture to grow similar to human tissues, organs, or tumor.

Organotropism The predilection of circulating tumor cells to target specific secondary sites, often involving a bidirectional interaction where the primary tumor influences changes at a distant site to create an environment conducive to metastatic proliferation.

Oropharyngeal cancer A type of head and neck cancer, occurring in the oropharynx (base of the tongue, tonsils, posterior pharyngeal wall, and soft palate).

Osteoarticular allograft This is a portion of a bone that is harvested from a human cadaver and includes the cartilage of the adjacent joint surface.

Osteotropism The predilection of disseminated tumor cells to metastasize to the skeleton compared to other organs.

Parenchyma The functional tissue of an organ as distinguished from the connective and supporting tissue. When used in the context of a sentinel node a tumor deposit is within the central anatomical area of the node.

Poly (ADP-ribose) polymerase (PARP) A family of proteins mainly involved in DNA repair, but also other cellular processes such as programmed cell death.

Pathologic fracture predictive models An assortment of models developed through retrospective or prospective collection of pathologic fracture events, image-based predictive algorithms, or nomogram predictive models that allows a cancer practitioner to predict the risk for fracture. Newer models also help a practitioner balance the overall prognosis with fracture risk in order to better guide treatment solutions.

PD1/PDL1 Programmed cell death protein 1/programmed cell death 1 ligand. The protein regulates apoptosis of T cells in lymph nodes and regulatory T cells. The ligand is highly expressed in multiple cancers and is thought to be one of the mechanisms behind cancer evasion from the immune system. Anti-PD1/PDL1 are drugs that try to stimulate the immune system, in particular CD8 T cells, to target cancer antigens.

Pelvic lymph node dissection Surgical removal of the pelvic lymph nodes, which may include regions such as obturator, external iliac, internal iliac, common iliac, presacral, and periaortic.

Percutaneous ablation Percutaneous chemical or thermal mediated tissue destruction.

Percutaneous cementoplasty Procedure by which bone is consolidated by percutaneous injection of cement.

Percutaneous osteosynthesis Percutaneous fixation of a fracture or area in which there is an impending fracture.

Pericyte mimicry See "angiotropism."

Permeative process Characterized by multiple endosteal lucencies with cortical sparing ("moth-eaten").

Pharmacodynamic endpoint An outcome measure to capture the potential clinical benefit of a therapeutic intervention.

PI3K Phosphoinositide 3-kinase.

PI3K/AKT pathway Signal transduction pathway, which promotes cellular processes such as proliferation, cell survival, and growth.

Polyclonal seeding Multiple cancer cells from different subclones of a tumor tissue (primary or metastatic) disseminate to secondary sites and proliferate in parallel.

Polymetastatic A metastatic state with a diffuse number (usually >5–10) of metastatic lesions.

Pre-analytical phase It is the phase between sampling a biological specimen (such as a tissue biopsy or a vein puncture) in a patient and processing the sample in the laboratory before using a method analysis.

Precision medicine Personalized medicine refers to the practice of providing optimal treatment based on individual patient data.

Pre-metastatic niche The microenvironment of a secondary organ site is modified by tumor-secreted factors with systemic effect, making that organ more conducive to the proliferation of metastatic tumor cells at that site.

Prostate A gland of the male reproductive system that secretes seminal fluid.

PSA Prostate-specific antigen is a protein produced by normal and cancer cells of the prostate gland.

PSMA Prostate-specific membrane antigen. A protein expressed in all forms of prostate tissue, including carcinoma.

PTEN Phosphatase and tensin homolog. A protein that acts as a tumor suppressor.

Quality of life In the context of metastatic skeletal disease, measures of treatment that enhance the body's functional condition, such as maintaining mobility and improving pain.

Radical mastectomy Surgical removal of the breast together with the overlying skin, underlying pectoralis muscle, and axillary lymph nodes.

Radical prostatectomy Surgical removal of the prostate gland.

Radiobiology Radiobiology is a field of clinical and basic medical sciences that involves the study of the action of ionizing radiation on living things, especially health effects of radiation.

Radiomics A field focused on identifying imaging phenotypes based on mathematical features within images.

Radiosensitive A lesion which is likely to respond to radiotherapy treatment.

RECIST criteria Response Evaluation Criteria in Solid Tumors.

Recursive partitioning analysis (RPA) A prognostic classification for patients with brain metastasis consisting of three groups based on patients' performance status, extracranial disease, and age.

Regional recurrence Recurrence of breast cancer in the regional lymph node basin of the index cancer.

Reverse total shoulder arthroplasty An alternative design to conventional shoulder arthroplasty implants that inverts the normal arthroplasty design, employing a convex glenoid component and concave humeral component.

Sarcoma A malignant tumor of the cells of mesenchymal origin.

SBRT Stereotactic body radiation therapy. Form of external beam radiation therapy in which relatively high doses of radiation are accurately delivered to a treatment volume over 1–5 daily fractions.

Sciatic buttress Caudal part of the iliac bone which connects the acetabulum to the sacrum/spine and is responsible for transmitting force from the hip joint to the axial skeleton.

Secretomes Denotes the collection of proteins released by a cell into the extracellular space as defined by TME.

Separation surgery A strategy of open surgical decompression that clears tumor away from the spinal cord and thecal sac, typically through a posterior approach, to provide 2–3 mm margin of clearance, with the intent of allowing adjunctive radiation therapy to be safely applied to the tumor without the risk of radiation toxicity to the neural structures.

Sentinel lymph node The first node or nodes draining the region of a primary tumor, most likely to harbor metastasis.

Sentinel lymph node biopsy Procedure to identify and remove the first node in which cancer cells are likely to spread.

Sentinel lymph node dissection Surgical procedure in which a tumor drain is removed from the first lymph node and histopathologically evaluated to check if cancer has spread beyond a primary site of tumor.

Shannon index A statistical diversity index.

Shear stress Frictional force exerted on endothelial cells due to the movement of fluids. Shear stress regulates various aspects of lymphatic vascular development.

Siewert classification Classification of gastroesophageal junction (GEJ) adenocarcinomas according to the epicenter of the tumor. Type I: 1–5 cm above GEJ, type II: from 1 cm above to 2 cm below GEJ, and type III: 2–5 cm below GEJ.

Small cell lung cancer Highly malignant and rapidly dividing type of lung cancer characterized by early metastasis.

Soft-tissue sarcoma (STS) Mesenchymal tumors developing in the soft tissues.

Somatic mutations Mutations in individual body cells, but not in the germ cells. The mutations only have effects on the body cells concerned and are therefore not transmitted to progeny.

Spatial resolution The ability to distinguish between objects or structures that differ in density. A high spatial resolution is important to discriminate between structures that are located within a proximity to each other. The spatial resolution of CT is superior to MRI.

Specification The labile phase of lymphatic endothelial cell (LEC) development during which PROX1 is expressed in a subset of blood vessel endothelial cells (BECs). These LEC progenitors could revert back to BECs in the absence of signals such as VEGF-C.

Spectrum theory of cancer First described in 1994 by Dr. Samuel Hellman, who theorized that cancer biology is a heterogenous spectrum, with the growth and progression of different tumors being a function of that specific cancer's biology.

SPECT Single-photon emission computed tomography.

Sporadic lymphatic anomaly Non-inherited diseases caused by errors in the development of the lymphatic system.

Stage How extensive the amount of cancer is within the body, which often is defined clinically and/or pathologically by the size of the primary tumor in the organ of origin, the presence of regional lymph node metastasis, and the documentation of systemic metastases.

Staged colorectal cancer metastasectomy For patients with synchronous colorectal cancer and hepatic metastases, the primary tumor and hepatic metastases are resected in separate operations. This may be done by resection of the primary tumor first, followed by resection of the hepatic metastases at a later date, or vice versa via a liver-first approach.

Stage migration Improvements in stage-specific survival for a type of cancer due to reclassification of staging groups and more accurate staging, but without demonstration of improvements in survival from the cancer as a whole.

Stereotactic body radiotherapy A noninvasive treatment modality in which high-dose targeted radiation is delivered to a tumor with narrow margins in a hypofractionation schedule while minimizing radiation to adjacent normal tissue.

Stereotactic body radiation therapy (SBRT)/stereotactic ablative radiotherapy (SABR) Highly focused and precise radiation treatment delivered using stereotactic targeting and image guidance, typically delivered over 1–5 treatments, with an ablative/near-ablative dose to tumor, while limiting dose to adjacent organs at risk.

Stereotactic radiosurgery (SRS) A technique that delivers radiation in a highly conformal manner with high spatial precision typically delivered over a single dose, but can also be delivered in a hypofractionated schedule of 3–5 doses.

STIR An example of a fat suppression technique with MRI classically used to differentiate fat from edema on T2 sequences.

Subcapsular Used in the context of a deposit of recurrent disease located directly under the outer lining capsule of a sentinel node.

Supervised learning Machine learning techniques applied in learning tasks when data is organized as input examples (features) with corresponding outputs (labels).

Synchronous colorectal cancer metastasectomy For patients with synchronous colorectal cancer and hepatic metastases, synchronous resection involves removal of the primary tumor and hepatic metastases in the same operation.

T1-weighted sequence A MRI sequence timed to show fat, proteinaceous fluid, and contrast as hyperintense.

T2-weighted sequence A MRI sequence timed to show water and fat as hyperintense. Commonly used with fat suppression techniques to highlight edema.

Targeted drug therapy A type of drug therapy that intends to inhibit abnormal molecules in cancer cells that activate the signaling transduction pathway to drive cell growth and survival.

Targeted molecular therapy A form of molecular medicine which functions to impede the growth and progression of cancer cells through inhibition of specific targeted molecules necessary for cellular pathways driving tumor growth.

Tertiary lymphoid structure Tertiary lymphoid organs that contain B cell and T cell zones and, when matured, contain germinal centers. Antigen-presenting cells in tertiary lymphoid structure are responsible for instructing T and B cells towards tumor-associated antigens.

Therapeutic ratio The ratio of the probability of treatment success (e.g., tumor control probability) to the probability of normal tissue damage.

Thermal ablation A minimally invasive approach to the treatment of tumors that involves the use of high-intensity hyperthermic procedures to apply energy to tissue and cause necrosis. Commonly used thermal ablation techniques include radiofrequency ablation, microwave ablation, and high-frequency ultrasound ablation.

Thoracic duct Principal central lymphatic trunk collecting lymph from most of the body and emptying into the central venous system in the neck.

TKI Tyrosine kinase inhibitors are medications that inhibit enzymes called tyrosine kinases that are involved in multiple intracellular mechanisms.

TM-ILP Isolated limb perfusion with TNF-alpha and melphalan.

Tolerogenic dendritic cells Heterogenous pool of dendritic cell with immunosuppressive properties, priming the immune system into tolerogenic state against different antigens through regulation of T cells such as inducing T cell anergy, T cell apoptosis, and induction of Tregs.

TP53 A tumor-suppressor gene, which is the most frequently mutated gene in human cancer. Induces growth arrest or apoptosis in response to cellular stress.

Transarterial embolization Occlusion of arterial blood supply, typically that of a tumor.

Transendothelial migration Migration of cancer cells from circulation through endothelium of blood vessels in order to access distant organs and form metastatic colonies.

Triple-negative breast cancer Biologic subtype of breast cancer which does not express estrogen receptor, progesterone receptor, or HER2.

Tumor-associated macrophages Macrophages that are commonly observed in the tumor microenvironment. Most tumor-associated macrophages are M2 macrophages that are associated with tumor angiogenesis, matrix protein degradation, and induction of an immunosuppressive tumor microenvironment.

Tumor biomarker A biological molecule found in body fluids or tissues indicating the possibility of a cancer somewhere in the body.

Tumor immune microenvironment The compositions of the immune cells and their associated chemokines or cytokines in the tumor microenvironment.

Tumor microenvironment (TME) Denotes the compartment that provides a space rich in lympho-vasculature, host cells, stroma, extracellular matrix and metabolic collaborator of proliferating tumor cells, and a signaling network for cellular and molecular cross talks, which is pivotal for tumor establishment, evasion of host immune surveillance, growth, invasion, and metastasis.

Tumor mutational burden (TMB) Tumor mutational burden, defined as the total number of somatic mutations per coding area of a tumor genome, is an emerging clinical biomarker associated with the response to immune checkpoint inhibitor (ICI) therapy. TMB has been shown to vary markedly among tumor types as well as among patients within tumor types.

Two-stage hepatectomy A treatment strategy for bilobar disease with curative intent in which tumors of one hemi-liver are resected during the first stage, followed by a period of convalescence that allows for liver hypertrophy to occur, and subsequent resection of the tumors of the contralateral hemi-liver.

Unsupervised learning Machine learning techniques that find patterns in data without the need for data annotation and produce classes within a dataset through structural learning.

Variant This refers to the individual differences that are recognized when the structure of the DNA of each person is examined in detail.

Vascular co-option Cells (tumor, cancer, circulating tumor cells) interacting physically, cell-to-cell, with pre-existing vessels in the absence of neoangiogenesis. May precede neoangiogenesis.

Vascular niche Sites within bone that are rich in vasculature and endothelial cells. In the context of bone metastasis, the vascular niche can regulate dormancy and metastatic outgrowth via provision of oxygen, nutrients, and growth factors.

Vasogenic edema Edema secondary to venous obstruction, in turn secondary to mass effect. To be distinguished from cytotoxic edema, more commonly seen following infarct in strokes.

VEGFR Vascular endothelial growth factor receptors are tyrosine kinases that bind the protein vascular endothelial growth factor (VEGF) to stimulate the formation of new blood vessels (angiogenesis and vasculogenesis).

Warburg effect Enhanced conversion of glucose into lactate in cancer cells even in the presence of oxygen, a condition also called aerobic glycolysis and now recognized to be a common phenotype of many cancer cells.

WGS Whole-genome sequencing by means of next-generation sequencing (NGS).

Whole-brain radiation therapy (WBRT) A technique that delivers radiation to the entire brain in a uniform fashion for the treatment of brain metastases.

Wide resection Resection of the tumor, pseudocapsule, and a surrounding cuff of normal tissue.

Index

S. P. Leong et al. (eds.), *Cancer Metastasis Through the Lymphovascular System*, https://doi.org/10.1007/978-3-030-93084-4

D

E

GPSR Compliance

The European Union's (EU) General Product Safety Regulation (GPSR) is a set of rules that requires consumer products to be safe and our obligations to ensure this.

If you have any concerns about our products, you can contact us on ProductSafety@springernature.com

In case Publisher is established outside the EU, the EU authorized representative is:

Springer Nature Customer Service Center GmbH
Europaplatz 3
69115 Heidelberg, Germany

Batch number: 08998701

Printed by Printforce, the Netherlands